CALCULUS

FOR BUSINESS, ECONOMICS, LIFE SCIENCES, AND SOCIAL SCIENCES

Thirteenth Edition

RAYMOND A. BARNETT Merritt College

MICHAEL R. ZIEGLER Marquette University

KARL E. BYLEEN Marquette University

Boston   Columbus   Indianapolis   New York   San Francisco   Upper Saddle River
Amsterdam   Cape Town   Dubai   London   Madrid   Milan   Munich   Paris   Montréal   Toronto
Delhi   Mexico City   São Paulo   Sydney   Hong Kong   Seoul   Singapore   Taipei   Tokyo

Editor in Chief: Deirdre Lynch
Executive Editor: Jennifer Crum
Project Manager: Kerri Consalvo
Editorial Assistant: Joanne Wendelken
Senior Managing Editor: Karen Wernholm
Senior Production Supervisor: Ron Hampton
Associate Design Director: Andrea Nix
Interior and Cover Design: Beth Paquin
Executive Manager, Course Production: Peter Silvia
Associate Media Producer: Christina Maestri
Digital Assets Manager: Marianne Groth
Executive Marketing Manager: Jeff Weidenaar
Marketing Assistant: Brooke Smith
Rights and Permissions Advisor: Joseph Croscup
Senior Manufacturing Buyer: Carol Melville
Production Coordination and Composition: Integra

Cover photo: Leigh Prather/Shutterstock; Dmitriy Raykin/Shutterstock;
          Image Source/Getty Images
Photo credits: Page 2, iStockphoto/Thinkstock; Page 94, Purestock/Thinkstock; Page 180,
          Vario Images/Alamy; Page 237, P. Amedzro/Alamy; Page 319, Anonymous Donor/
          Alamy; Page 381, Shime/Fotolia; Page 424, Aurora Photos/Alamy; Page 494, Gary
          Whitton/Fotolia

Many of the designations used by manufacturers and sellers to distinguish their products are
claimed as trademarks. Where those designations appear in this book, and Pearson was aware
of a trademark claim, the designations have been printed in initial caps or all caps.

Library of Congress Cataloging-in-Publication Data

Calculus for business, economics, life sciences, and social sciences /
Raymond A. Barnett ... [et al.].—13th ed.
     p. cm.
   Includes index.
   ISBN-13: 978-0-321-86983-8
   ISBN-10: 0-321-86983-4
1. Calculus—Textbooks   I. Ziegler, Michael R.   II. Byleen, Karl E.   III. Title
   QA303.2.B285   2015
   515—dc23   2013023206

5 6 7 8 9 10—V011—18 17 16 15

www.pearsonhighered.com

ISBN-10: 0-321-86983-4
ISBN-13: 978-0-321-86983-8

CONTENTS

iii

Available separately: Calculus Topics to Accompany Calculus, 13e, and College Mathematics, 13e

PREFACE

The thirteenth edition of *Calculus for Business, Economics, Life Sciences, and Social Sciences* is designed for a one-term course in Calculus for students who have had one to two years of high school algebra or the equivalent. The book's overall approach, refined by the authors' experience with large sections of college freshmen, addresses the challenges of teaching and learning when prerequisite knowledge varies greatly from student to student.

The authors had three main goals when writing this text:

▶ To write a text that students can easily comprehend

▶ To make connections between what students are learning and how they may apply that knowledge

▶ To give flexibility to instructors to tailor a course to the needs of their students.

Many elements play a role in determining a book's effectiveness for students. Not only is it critical that the text be accurate and readable, but also, in order for a book to be effective, aspects such as the page design, the interactive nature of the presentation, and the ability to support and challenge all students have an incredible impact on how easily students comprehend the material. Here are some of the ways this text addresses the needs of students at all levels:

▶ Page layout is clean and free of potentially distracting elements.

▶ *Matched Problems* that accompany each of the completely worked examples help students gain solid knowledge of the basic topics and assess their own level of understanding before moving on.

▶ Review material (Appendix A and Chapter 1) can be used judiciously to help remedy gaps in prerequisite knowledge.

▶ A *Diagnostic Prerequisite Test* prior to Chapter 1 helps students assess their skills, while the *Basic Algebra Review* in Appendix A provides students with the content they need to remediate those skills.

▶ *Explore and Discuss* problems lead the discussion into new concepts or build upon a current topic. They help students of all levels gain better insight into the mathematical concepts through thought-provoking questions that are effective in both small and large classroom settings.

▶ Instructors are able to easily craft homework assignments that best meet the needs of their students by taking advantage of the variety of types and difficulty levels of the exercises. Exercise sets at the end of each section consist of a ***Skills Warm-up*** (four to eight problems that review prerequisite knowledge specific to that section) followed by problems divided into categories A, B, and C by level of difficulty, with level-C exercises being the most challenging.

▶ The MyMathLab course for this text is designed to help students help themselves and provide instructors with actionable information about their progress. The immediate feedback students receive when doing homework and practice in MyMathLab is invaluable, and the easily accessible e-book enhances student learning in a way that the printed page sometimes cannot.

Most important, all students get substantial experience in modeling and solving real-world problems through application examples and exercises chosen from business and economics, life sciences, and social sciences. Great care has been taken to write a book that is mathematically correct, with its emphasis on computational skills, ideas, and problem solving rather than mathematical theory.

Finally, the choice and independence of topics make the text readily adaptable to a variety of courses (see the chapter dependencies chart on page xi). This text is one of three books in the authors' college mathematics series. The others are *Finite Mathematics for Business, Economics, Life Sciences, and Social Sciences*, and *College Mathematics for Business, Economics, Life Sciences, and Social Sciences*; the latter contains selected content from the other two books. *Additional Calculus Topics*, a supplement written to accompany the Barnett/Ziegler/Byleen series, can be used in conjunction with any of these books.

New to This Edition

Fundamental to a book's effectiveness is classroom use and feedback. Now in its thirteenth edition, *Calculus for Business, Economics, Life Sciences, and Social Sciences* has had the benefit of a substantial amount of both. Improvements in this edition evolved out of the generous response from a large number of users of the last and previous editions as well as survey results from instructors, mathematics departments, course outlines, and college catalogs. In this edition,

▶ The Diagnostic Prerequisite Test has been revised to identify the specific deficiencies in prerequisite knowledge that cause students the most difficulty with calculus.

▶ Chapters 1 and 2 of the previous edition have been revised and combined to create a single introductory chapter (Chapter 1) on functions and graphs.

▶ Most exercise sets now begin with a *Skills Warm-up*—four to eight problems that review prerequisite knowledge specific to that section in a just-in-time approach. References to review material are given in the answer section of the text for the benefit of students who struggle with the warm-up problems and need a refresher.

▶ Section 6.4 has been rewritten to cover the trapezoidal rule and Simpson's rule.

▶ Examples and exercises have been given up-to-date contexts and data.

▶ Exposition has been simplified and clarified throughout the book.

▶ An Annotated Instructor's Edition is now available, providing answers to exercises directly on the page (whenever possible). *Teaching Tips* provide less-experienced instructors with insight on common student pitfalls, suggestions for how to approach a topic, or reminders of which prerequisite skills students will need. Lastly, the difficulty level of exercises is indicated only in the instructor's edition so as not to discourage students from attempting the most challenging "C" level exercises.

▶ *MyMathLab* for this text has been enhanced greatly in this revision. Most notably, a "Getting Ready for Chapter X" has been added to each chapter as an optional resource for instructors and students as a way to address the prerequisite skills that students need, and are often missing, for each chapter. Many more improvements have been made. See the detailed description on pages xiv and xv for more information.

Trusted Features

Emphasis and Style

As was stated earlier, this text is written for student comprehension. To that end, the focus has been on making the book both mathematically correct and accessible to students. Most derivations and proofs are omitted, except where their inclusion adds significant insight into a particular concept as the emphasis is on computational skills, ideas, and problem solving rather than mathematical theory. General concepts and results are typically presented only after particular cases have been discussed.

Design

One of the hallmark features of this text is the **clean, straightforward design** of its pages. Navigation is made simple with an obvious hierarchy of key topics and a judicious use of call-outs and pedagogical features. We made the decision to maintain a two-color design to help students stay focused on the mathematics and applications. Whether students start in the chapter opener or in the exercise sets, they can easily reference the content, examples, and *Conceptual Insights* they need to understand the topic at hand. Finally, a functional use of color improves the clarity of many illustrations, graphs, and explanations, and guides students through critical steps (see pages 22, 75, and 306).

Examples and Matched Problems

More than 300 completely worked examples are used to introduce concepts and to demonstrate problem-solving techniques. Many examples have multiple parts, significantly increasing the total number of worked examples. The examples are annotated using blue text to the right of each step, and the problem-solving steps are clearly identified. **To give students extra help** in working through examples, dashed boxes are used to enclose steps that are usually performed mentally and rarely mentioned in other books (see Example 4 on page 9). Though some students may not need these additional steps, many will appreciate the fact that the authors do not assume too much in the way of prior knowledge.

EXAMPLE 2   Tangent Lines   Let $f(x) = (2x - 9)(x^2 + 6)$.

(A) Find the equation of the line tangent to the graph of $f(x)$ at $x = 3$.

(B) Find the value(s) of x where the tangent line is horizontal.

SOLUTION

(A) First, find $f'(x)$:

$$f'(x) = (2x - 9)(x^2 + 6)' + (x^2 + 6)(2x - 9)'$$
$$= (2x - 9)(2x) + (x^2 + 6)(2)$$

Then, find $f(3)$ and $f'(3)$:

$$f(3) = [2(3) - 9](3^2 + 6) = (-3)(15) = -45$$
$$f'(3) = [2(3) - 9]2(3) + (3^2 + 6)(2) = -18 + 30 = 12$$

Now, find the equation of the tangent line at $x = 3$:

$$y - y_1 = m(x - x_1) \quad y_1 = f(x_1) = f(3) = -45$$
$$y - (-45) = 12(x - 3) \quad m = f'(x_1) = f'(3) = 12$$
$$y = 12x - 81 \quad \text{Tangent line at } x = 3$$

(B) The tangent line is horizontal at any value of x such that $f'(x) = 0$, so

$$f'(x) = (2x - 9)2x + (x^2 + 6)2 = 0$$
$$6x^2 - 18x + 12 = 0$$
$$x^2 - 3x + 2 = 0$$
$$(x - 1)(x - 2) = 0$$
$$x = 1, 2$$

The tangent line is horizontal at $x = 1$ and at $x = 2$.

Matched Problem 2   Repeat Example 2 for $f(x) = (2x + 9)(x^2 - 12)$.

Each example is followed by a similar *Matched Problem* for the student to work while reading the material. This actively involves the student in the learning process. The answers to these matched problems are included at the end of each section for easy reference.

Explore and Discuss

Most every section contains *Explore and Discuss* problems at appropriate places to encourage students to think about a relationship or process before a result is stated or to investigate additional consequences of a development in the text. This serves to foster critical thinking and communication skills. The Explore and Discuss material can be used for in-class discussions or out-of-class group activities and is effective in both small and large class settings.

New to this edition, annotations in the instructor's edition provide tips for less-experienced instructors on how to engage students in these Explore and Discuss activities, expand on the topic, or simply guide student responses.

Explore and Discuss 1 Let $F(x) = x^2$, $S(x) = x^3$, and $f(x) = F(x)S(x) = x^5$. Which of the following is $f'(x)$?

(A) $F'(x)S'(x)$ (B) $F(x)S'(x)$

(C) $F'(x)S(x)$ (D) $F(x)S'(x) + F'(x)S(x)$

Exercise Sets

The book contains over 4,500 carefully selected and graded exercises. Many problems have multiple parts, significantly increasing the total number of exercises. Exercises are paired so that consecutive odd- and even-numbered exercises are of the same type and difficulty level. Each exercise set is designed to allow instructors to craft just the right assignment for students. Exercise sets are categorized as Skills Warm-up (review of prerequisite knowledge), and within the Annotated Instructor's Edition only, as A (routine, easy mechanics), B (more difficult mechanics), and C (difficult mechanics and some theory) to make it easy for instructors to create assignments that are appropriate for their classes. The *writing exercises*, indicated by the icon ✎, provide students with an opportunity to express their understanding of the topic in writing. Answers to all odd-numbered problems are in the back of the book.

Applications

A major objective of this book is to give the student substantial experience in modeling and solving real-world problems. Enough applications are included to convince even the most skeptical student that mathematics is really useful (see the Index of Applications at the back of the book). Almost every exercise set contains application problems, including applications from business and economics, life sciences, and social sciences. An instructor with students from all three disciplines can let them choose applications from their own field of interest; if most students are from one of the three areas, then special emphasis can be placed there. Most of the applications are simplified versions of actual real-world problems inspired by professional journals and books. No specialized experience is required to solve any of the application problems.

Additional Pedagogical Features

The following features, while helpful to any student, are particularly helpful to students enrolled in a large classroom setting where access to the instructor is more challenging or just less frequent. These features provide much-needed guidance for students as they tackle difficult concepts.

▶ **Call-out boxes** highlight important definitions, results, and step-by-step processes (see pages 57, 63–64).

▶ **Caution statements** appear throughout the text where student errors often occur (see pages 82, 274, and 332).

⚠ **CAUTION** The expression $0/0$ does not represent a real number and should never be used as the value of a limit. If a limit is a $0/0$ indeterminate form, further investigation is always required to determine whether the limit exists and to find its value if it does exist.   ▲

▶ **Conceptual Insights**, appearing in nearly every section, often make explicit connections to previous knowledge, but sometimes encourage students to think beyond the particular skill they are working on and attain a more enlightened view of the concepts at hand (see pages 56–57, 116, and 306).

CONCEPTUAL INSIGHT

Rather than list the points where a function is discontinuous, sometimes it is useful to state the intervals on which the function is continuous. Using the set operation **union**, denoted by $\cup$, we can express the set of points where the function in Example 1 is continuous as follows:

$$(-\infty, -4) \cup (-4, -2) \cup (-2, 1) \cup (1, 3) \cup (3, \infty)$$

▶ The newly revised **Diagnostic Prerequisite Test**, located at the front of the book, provides students with a tool to assess their prerequisite skills prior to taking the course. The **Basic Algebra Review**, in Appendix A, provides students with seven sections of content to help them remediate in specific areas of need. Answers to the Diagnostic Prerequisite Test are at the back of the book and reference specific sections in the Basic Algebra Review or Chapter 1 for students to use for remediation.

Graphing Calculator and Spreadsheet Technology

Although access to a graphing calculator or spreadsheets is not assumed, it is likely that many students will want to make use of this technology. To assist these students, optional graphing calculator and spreadsheet activities are included in appropriate places. These include brief discussions in the text, examples or portions of examples solved on a graphing calculator or spreadsheet, and exercises for the student to solve. For example, linear and quadratic regression are introduced in Section 1.3, and regression techniques on a graphing calculator are used at appropriate points to illustrate mathematical modeling with real data. All the optional graphing calculator material is clearly identified with the icon 📈 and can be omitted without loss of continuity, if desired. Optional spreadsheet material is identified with the icon ▦. Graphing calculator screens displayed in the text are actual output from the TI-84 Plus graphing calculator.

Chapter Reviews

Often it is during the preparation for a chapter exam that concepts gel for students, making the chapter review material particularly important. The chapter review sections in this text include a comprehensive summary of important terms, symbols, and concepts, keyed to completely worked examples, followed by a comprehensive set of Review Exercises. Answers to Review Exercises are included at the back of the book; *each answer contains a reference to the section in which that type of problem is discussed* so students can remediate any deficiencies in their skills on their own.

Chapter Dependencies

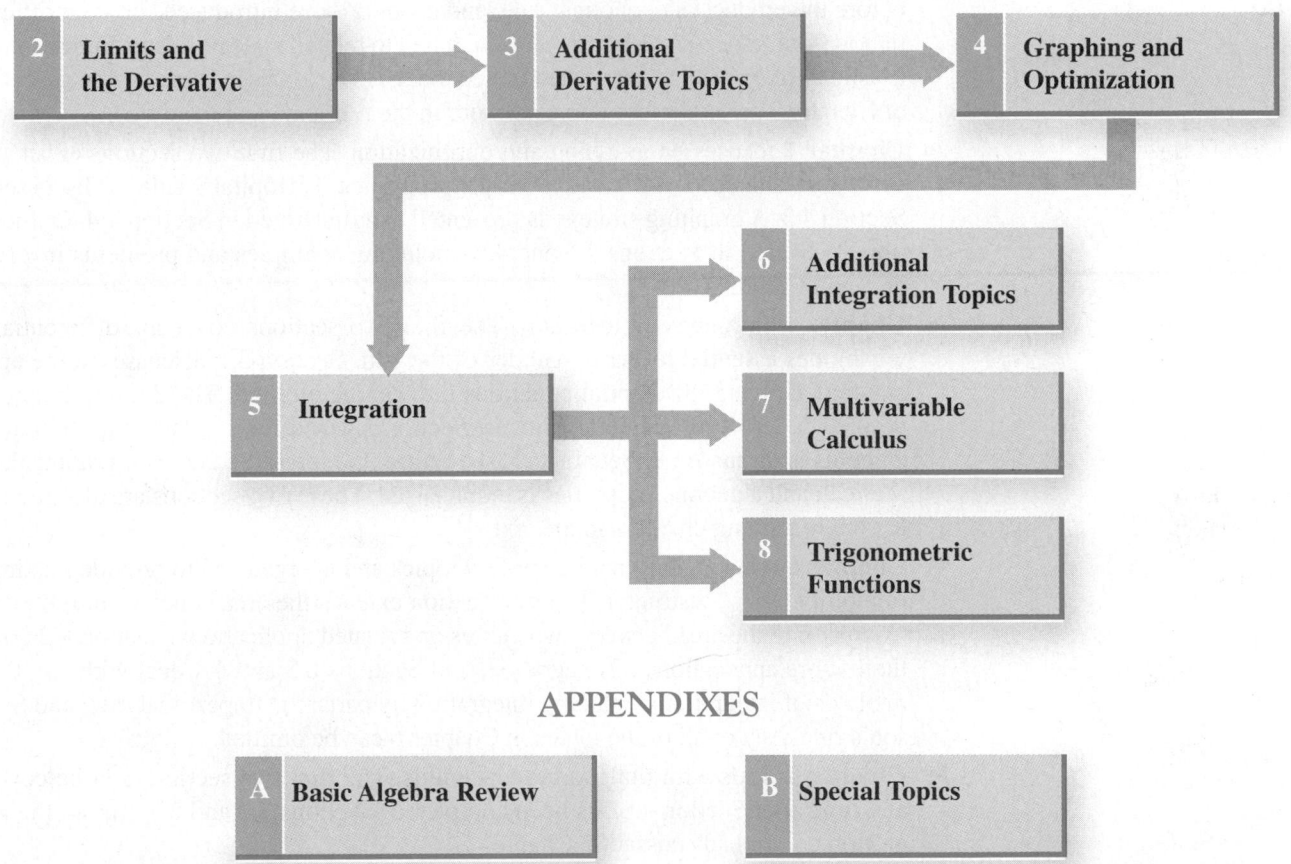

Diagnostic
Prerequisite Test

PART ONE: A LIBRARY OF ELEMENTARY FUNCTIONS*

1   Functions and Graphs

PART TWO: CALCULUS

2   Limits and
     the Derivative

3   Additional
     Derivative Topics

4   Graphing and
     Optimization

6   Additional
     Integration Topics

5   Integration

7   Multivariable
     Calculus

8   Trigonometric
     Functions

APPENDIXES

A   Basic Algebra Review

B   Special Topics

*Selected topics from Part One may be referred to as needed in
 Part Two or reviewed systematically before starting Part Two.

Content

The text begins with the development of a library of elementary functions in **Chapter 1**, including their properties and applications. Many students will be familiar with most, if not all, of the material in this introductory chapter. Depending on students' preparation and the course syllabus, an instructor has several options for using the first chapter, including the following:

(i) Skip Chapter 1 and refer to it only as necessary later in the course;

(ii) Cover Section 1.3 quickly in the first week of the course, emphasizing price–demand equations, price–supply equations, and linear regression, but skip the rest of Chapter 1;

(iii) Cover Chapter 1 systematically before moving on to other chapters.

The material in Part Two (Calculus) consists of differential calculus (Chapters 2–4), integral calculus (Chapters 5 and 6), multivariable calculus (Chapter 7), and a brief discussion of differentiation and integration of trigonometric functions (Chapter 8). In general, Chapters 2–5 must be covered in sequence; however, certain sections can be omitted or given brief treatments, as pointed out in the discussion that follows (see the Chapter Dependencies chart on page xi).

▶ **Chapter 2** introduces the derivative. The first three sections cover limits (including infinite limits and limits at infinity), continuity, and the limit properties that are essential to understanding the definition of the derivative in Section 2.4. The remaining sections of the chapter cover basic rules of differentiation, differentials, and applications of derivatives in business and economics. The interplay between graphical, numerical, and algebraic concepts is emphasized here and throughout the text.

▶ In **Chapter 3** the derivatives of exponential and logarithmic functions are obtained before the product rule, quotient rule, and chain rule are introduced. Implicit differentiation is introduced in Section 3.5 and applied to related rates problems in Section 3.6. Elasticity of demand is introduced in Section 3.7. The topics in these last three sections of Chapter 3 are not referred to elsewhere in the text and can be omitted.

▶ **Chapter 4** focuses on graphing and optimization. The first two sections cover first-derivative and second-derivative graph properties. L'Hôpital's rule is discussed in Section 4.3. A graphing strategy is presented and illustrated in Section 4.4. Optimization is covered in Sections 4.5 and 4.6, including examples and problems involving end-point solutions.

▶ **Chapter 5** introduces integration. The first two sections cover antidifferentiation techniques essential to the remainder of the text. Section 5.3 discusses some applications involving differential equations that can be omitted. The definite integral is defined in terms of Riemann sums in Section 5.4 and the fundamental theorem of calculus is discussed in Section 5.5. As before, the interplay between graphical, numerical, and algebraic properties is emphasized. These two sections are also required for the remaining chapters in the text.

▶ **Chapter 6** covers additional integration topics and is organized to provide maximum flexibility for the instructor. The first section extends the area concepts introduced in Chapter 6 to the area between two curves and related applications. Section 6.2 covers three more applications of integration, and Sections 6.3 and 6.4 deal with additional methods of integration, including integration by parts, the trapezoidal rule, and Simpson's rule. Any or all of the topics in Chapter 6 can be omitted.

▶ **Chapter 7** deals with multivariable calculus. The first five sections can be covered any time after Section 4.6 has been completed. Sections 7.6 and 7.7 require the integration concepts discussed in Chapter 5.

▶ **Chapter 8** provides brief coverage of trigonometric functions that can be incorporated into the course, if desired. Section 8.1 provides a review of basic trigonometric concepts. Section 8.2 can be covered any time after Section 4.3 has been completed. Section 8.3 requires the material in Chapter 5.

▶ **Appendix A** contains a concise review of basic algebra that may be covered as part of the course or referenced as needed. As mentioned previously, **Appendix B** contains additional topics that can be covered in conjunction with certain sections in the text, if desired.

Accuracy Check

Because of the careful checking and proofing by a number of mathematics instructors (acting independently), the authors and publisher believe this book to be substantially error free. If an error should be found, the authors would be grateful if notification were sent to Karl E. Byleen, 9322 W. Garden Court, Hales Corners, WI 53130; or by e-mail to kbyleen@wi.rr.com.

Student Supplements

Student's Solutions Manual

- ▶ By Garret J. Etgen, University of Houston
- ▶ This manual contains detailed, carefully worked-out solutions to all odd-numbered section exercises and all Chapter Review exercises. Each section begins with Things to Remember, a list of key material for review.
- ▶ ISBN-13: 978-0-321-93173-3

Additional Calculus Topics to Accompany Calculus, 13e, and College Mathematics, 13e

- ▶ This separate book contains three unique chapters: Differential Equations, Taylor Polynomials and Infinite Series, and Probability and Calculus.
- ▶ ISBN 13: 978-0-321-93169-6; ISBN 10: 0-321-931696

Graphing Calculator Manual for Applied Math

- ▶ By Victoria Baker, Nicholls State University
- ▶ This manual contains detailed instructions for using the TI-83/TI-83 Plus/TI-84 Plus C calculators with this textbook. Instructions are organized by mathematical topics.
- ▶ Available in MyMathLab.

Excel Spreadsheet Manual for Applied Math

- ▶ By Stela Pudar-Hozo, Indiana University–Northwest
- ▶ This manual includes detailed instructions for using Excel spreadsheets with this textbook. Instructions are organized by mathematical topics.
- ▶ Available in MyMathLab.

Guided Lecture Notes

- ▶ By Salvatore Sciandra, Niagara County Community College
- ▶ These worksheets for students contain unique examples to enforce what is taught in the lecture and/or material covered in the text. Instructor worksheets are also available and include answers.
- ▶ Available in MyMathLab or through Pearson Custom Publishing.

Videos with Optional Captioning

- ▶ The video lectures with optional captioning for this text make it easy and convenient for students to watch videos from a computer at home or on campus. The complete set is ideal for distance learning or supplemental instruction.
- ▶ Every example in the text is represented by a video.
- ▶ Available in MyMathLab.

Instructor Supplements

New! Annotated Instructor's Edition

- ▶ This book contains answers to all exercises in the text on the same page as the exercises whenever possible. In addition, Teaching Tips are provided for less-experienced instructors. Exercises are coded by level of difficulty only in the AIE so students are not dissuaded from trying more challenging exercises.
- ▶ ISBN-13: 978-0-321-92416-2

Online Instructor's Solutions Manual (downloadable)

- ▶ By Garret J. Etgen, University of Houston
- ▶ This manual contains detailed solutions to all even-numbered section problems.
- ▶ Available in MyMathLab or through http://www.pearsonhighered.com/educator.

Mini Lectures (downloadable)

- ▶ By Salvatore Sciandra, Niagara County Community College
- ▶ Mini Lectures are provided for the teaching assistant, adjunct, part-time or even full-time instructor for lecture preparation by providing learning objectives, examples (and answers) not found in the text, and teaching notes.
- ▶ Available in MyMathLab or through http://www.pearsonhighered.com/educator.

PowerPoint® Lecture Slides

- ▶ These slides present key concepts and definitions from the text. They are available in MyMathLab or at http://www.pearsonhighered.com/educator.

Technology Resources

MyMathLab® Online Course
(access code required)

MyMathLab delivers **proven results** in helping individual students succeed.

▶ MyMathLab has a consistently positive impact on the quality of learning in higher education math instruction. MyMathLab can be successfully implemented in any environment—lab based, hybrid, fully online, traditional—and demonstrates the quantifiable difference that integrated usage has on student retention, subsequent success, and overall achievement.

▶ MyMathLab's comprehensive online gradebook automatically tracks your students' results on tests, quizzes, homework, and in the study plan. You can use the gradebook to quickly intervene if your students have trouble or to provide positive feedback on a job well done. The data within MyMathLab is easily exported to a variety of spreadsheet programs, such as Microsoft Excel. You can determine which points of data you want to export and then analyze the results to determine success.

MyMathLab provides **engaging experiences** that personalize, stimulate, and measure learning for each student.

▶ **Personalized Learning:** MyMathLab offers two important features that support adaptive learning—personalized homework and the adaptive study plan. These features allow your students to work on what they need to learn when it makes the most sense, maximizing their potential for understanding and success.

▶ **Exercises:** The homework and practice exercises in MyMathLab are correlated to the exercises in the textbook, and they regenerate algorithmically to give students unlimited opportunity for practice and mastery. The software offers immediate, helpful feedback when students enter incorrect answers.

▶ **Chapter-Level, Just-in-Time Remediation:** The MyMathLab course for these texts includes a short diagnostic, called Getting Ready, prior to each chapter to assess students' prerequisite knowledge. This diagnostic can then be tied to personalized homework so that each student receives a homework assignment specific to his or her prerequisite skill needs.

▶ **Multimedia Learning Aids:** Exercises include guided solutions, sample problems, animations, videos, and eText access for extra help at the point of use.

And, MyMathLab comes from an **experienced partner** with educational expertise and an eye on the future.

▶ Knowing that you are using a Pearson product means that you are using quality content. That means that our eTexts are accurate and our assessment tools work. It means we are committed to making MyMathLab as accessible as possible. MyMathLab is compatible with the JAWS 12/13 screen reader, and enables multiple-choice and free-response problem types to be read and interacted with via keyboard controls and math notation input. More information on this functionality is available at http://mymathlab.com/accessibility.

▶ Whether you are just getting started with MyMathLab or you have a question along the way, we're here to help you learn about our technologies and how to incorporate them into your course.

▶ To learn more about how MyMathLab combines proven learning applications with powerful assessment and continuously adaptive capabilities, visit www.mymathlab.com or contact your Pearson representative.

MyMathLab® Ready-to-Go Course
(access code required)

These new Ready-to-Go courses provide students with all the same great MyMathLab features but make it easier for instructors to get started. Each course includes preassigned homework and quizzes to make creating a course even simpler. In addition, these prebuilt courses include a course-level Getting Ready diagnostic that helps pinpoint student weaknesses in prerequisite skills. Ask your Pearson representative about the details for this particular course or to see a copy of this course.

MyLabsPlus®

MyLabsPlus combines proven results and engaging experiences from MyMathLab® and MyStatLab™ with convenient management tools and a dedicated services team. Designed to support growing math and statistics programs, it includes additional features such as

▶ **Batch Enrollment:** Your school can create the login name and password for every student and instructor, so everyone can be ready to start class on the first day. Automation of this process is also possible through integration with your school's Student Information System.

▶ **Login from your campus portal:** You and your students can link directly from your campus portal into your MyLabsPlus courses. A Pearson service team works with your institution to create a single sign-on experience for instructors and students.

▶ **Advanced Reporting:** MyLabsPlus advanced reporting allows instructors to review and analyze students' strengths and weaknesses by tracking their performance on tests, assignments, and tutorials. Administrators can review grades and assignments across all courses on your MyLabsPlus campus for a broad overview of program performance.

▶ **24/7 Support:** Students and instructors receive 24/7 support, 365 days a year, by email or online chat.

MyLabsPlus is available to qualified adopters. For more information, visit our website at **www.mylabsplus.com** or contact your Pearson representative.

MathXL® Online Course
(access code required)

MathXL is the homework and assessment engine that runs MyMathLab. (MyMathLab is MathXL plus a learning-management system.)

With MathXL, instructors can

▶ Create, edit, and assign online homework and tests using algorithmically generated exercises correlated at the objective level to the textbook.

▶ Create and assign their own online exercises and import TestGen tests for added flexibility.

▶ Maintain records of all student work tracked in MathXL's online gradebook.

With MathXL, students can

▶ Take chapter tests in MathXL and receive personalized study plans and/or personalized homework assignments based on their test results.

▶ Use the study plan and/or the homework to link directly to tutorial exercises for the objectives they need to study.

▶ Access supplemental animations and video clips directly from selected exercises.

MathXL is available to qualified adopters. For more information, visit our website at **www.mathxl.com** or contact your Pearson representative.

TestGen®

TestGen (**www.pearsoned.com/testgen**) enables instructors to build, edit, print, and administer tests using a computerized bank of questions developed to cover all the objectives of the text. TestGen is algorithmically based, allowing instructors to create multiple, but equivalent, versions of the same question or test with the click of a button. Instructors can also modify test bank questions or add new questions. The software and test bank are available for download from Pearson Education's online catalog.

Acknowledgments

In addition to the authors, many others are involved in the successful publication of a book. We wish to thank the following reviewers:

Mark Barsamian, *Ohio University*
Kathleen Coskey, *Boise State University*
Tim Doyle, *DePaul University*
J. Robson Eby, *Blinn College–Bryan Campus*
Irina Franke, *Bowling Green State University*
Andrew J. Hetzel, *Tennessee Tech University*
Timothy Kohl, *Boston University*
Dan Krulewich, *University of Missouri, Kansas City*
Scott Lewis, *Utah Valley University*
Saliha Shah, *Ventura College*
Jerimi Ann Walker, *Moraine Valley Community College*

We also express our thanks to

Mark Barsamian, Theresa Schille, J. Robson Eby, John Samons, and Gary Williams for providing a careful and thorough accuracy check of the text, problems, and answers.

Garret Etgen, Salvatore Sciandra, Victoria Baker, and Stela Pudar-Hozo for developing the supplemental materials so important to the success of a text.

All the people at Pearson Education who contributed their efforts to the production of this book.

Diagnostic Prerequisite Test

Work all of the problems in this self-test without using a calculator. Then check your work by consulting the answers in the back of the book. Where weaknesses show up, use the reference that follows each answer to find the section in the text that provides the necessary review.

1. Replace each question mark with an appropriate expression that will illustrate the use of the indicated real number property:

   (A) Commutative $(\cdot): x(y + z) = ?$

   (B) Associative $(+): 2 + (x + y) = ?$

   (C) Distributive: $(2 + 3)x = ?$

Problems 2–6 refer to the following polynomials:

   (A) $3x - 4$ (B) $x + 2$

   (C) $2 - 3x^2$ (D) $x^3 + 8$

2. Add all four.

3. Subtract the sum of (A) and (C) from the sum of (B) and (D).

4. Multiply (C) and (D).

5. What is the degree of each polynomial?

6. What is the leading coefficient of each polynomial?

In Problems 7 and 8, perform the indicated operations and simplify.

7. $5x^2 - 3x[4 - 3(x - 2)]$

8. $(2x + y)(3x - 4y)$

In Problems 9 and 10, factor completely.

9. $x^2 + 7x + 10$

10. $x^3 - 2x^2 - 15x$

11. Write 0.35 as a fraction reduced to lowest terms.

12. Write $\dfrac{7}{8}$ in decimal form.

13. Write in scientific notation:

   (A) 4,065,000,000,000   (B) 0.0073

14. Write in standard decimal form:

   (A) 2.55×10^8 (B) 4.06×10^{-4}

15. Indicate true (T) or false (F):

   (A) A natural number is a rational number.

   (B) A number with a repeating decimal expansion is an irrational number.

16. Give an example of an integer that is not a natural number.

In Problems 17–24, simplify and write answers using positive exponents only. All variables represent positive real numbers.

17. $6(xy^3)^5$

18. $\dfrac{9u^8v^6}{3u^4v^8}$

19. $(2 \times 10^5)(3 \times 10^{-3})$

20. $(x^{-3}y^2)^{-2}$

21. $u^{5/3}u^{2/3}$

22. $(9a^4b^{-2})^{1/2}$

23. $\dfrac{5^0}{3^2} + \dfrac{3^{-2}}{2^{-2}}$

24. $(x^{1/2} + y^{1/2})^2$

In Problems 25–30, perform the indicated operation and write the answer as a simple fraction reduced to lowest terms. All variables represent positive real numbers.

25. $\dfrac{a}{b} + \dfrac{b}{a}$

26. $\dfrac{a}{bc} - \dfrac{c}{ab}$

27. $\dfrac{x^2}{y} \cdot \dfrac{y^6}{x^3}$

28. $\dfrac{x}{y^3} \div \dfrac{x^2}{y}$

29. $\dfrac{\dfrac{1}{7 + h} - \dfrac{1}{7}}{h}$

30. $\dfrac{x^{-1} + y^{-1}}{x^{-2} - y^{-2}}$

31. Each statement illustrates the use of one of the following real number properties or definitions. Indicate which one.

Commutative $(+, \cdot)$	Associative $(+, \cdot)$	Distributive
Identity $(+, \cdot)$	Inverse $(+, \cdot)$	Subtraction
Division	Negatives	Zero

   (A) $(-7) - (-5) = (-7) + [-(-5)]$

   (B) $5u + (3v + 2) = (3v + 2) + 5u$

   (C) $(5m - 2)(2m + 3) = (5m - 2)2m + (5m - 2)3$

   (D) $9 \cdot (4y) = (9 \cdot 4)y$

   (E) $\dfrac{u}{-(v - w)} = \dfrac{u}{w - v}$

   (F) $(x - y) + 0 = (x - y)$

32. Round to the nearest integer:

   (A) $\dfrac{17}{3}$ (B) $-\dfrac{5}{19}$

33. Multiplying a number x by 4 gives the same result as subtracting 4 from x. Express as an equation, and solve for x.

34. Find the slope of the line that contains the points $(3, -5)$ and $(-4, 10)$.

35. Find the x and y coordinates of the point at which the graph of $y = 7x - 4$ intersects the x axis.

36. Find the x and y coordinates of the point at which the graph of $y = 7x - 4$ intersects the y axis.

In Problems 37 and 38, factor completely.

37. $x^2 - 3xy - 10y^2$

38. $6x^2 - 17xy + 5y^2$

In Problems 39–42, write in the form $ax^p + by^q$ where a, b, p, and q are rational numbers.

39. $\dfrac{3}{x} + 4\sqrt{y}$

40. $\dfrac{8}{x^2} - \dfrac{5}{y^4}$

41. $\dfrac{2}{5x^{3/4}} - \dfrac{7}{6y^{2/3}}$

42. $\dfrac{1}{3\sqrt{x}} + \dfrac{9}{\sqrt[3]{y}}$

In Problems 43 and 44, write in the form $a + b\sqrt{c}$ where a, b, and c are rational numbers.

43. $\dfrac{1}{4 - \sqrt{2}}$

44. $\dfrac{5 - \sqrt{3}}{5 + \sqrt{3}}$

In Problems 45–50, solve for x.

45. $x^2 = 5x$

46. $3x^2 - 21 = 0$

47. $x^2 - x - 20 = 0$

48. $-6x^2 + 7x - 1 = 0$

49. $x^2 + 2x - 1 = 0$

50. $x^4 - 6x^2 + 5 = 0$

PART 1

A LIBRARY OF ELEMENTARY FUNCTIONS

1

Functions and Graphs

Introduction

The function concept is one of the most important ideas in mathematics. The study of mathematics beyond the elementary level requires a firm understanding of a basic list of elementary functions, their properties, and their graphs. See the inside back cover of this book for a list of the functions that form our library of elementary functions. Most functions in the list will be introduced to you by the end of Chapter 1. For example, in Section 1.3 you will learn how to apply quadratic functions to model the effect of tire pressure on mileage (see Problems 73 and 75 on page 48).

1.1 Functions

After a brief review of the Cartesian (rectangular) coordinate system in the plane and graphs of equations, we discuss the concept of function, one of the most important ideas in mathematics.

Cartesian Coordinate System

Recall that to form a **Cartesian** or **rectangular coordinate system**, we select two real number lines—one horizontal and one vertical—and let them cross through their origins as indicated in the figure below. Up and to the right are the usual choices for the positive directions. These two number lines are called the **horizontal axis** and the **vertical axis**, or, together, the **coordinate axes**. The horizontal axis is usually referred to as the *x* **axis** and the vertical axis as the *y* **axis**, and each is labeled accordingly. The coordinate axes divide the plane into four parts called **quadrants**, which are numbered counterclockwise from I to IV (see the figure).

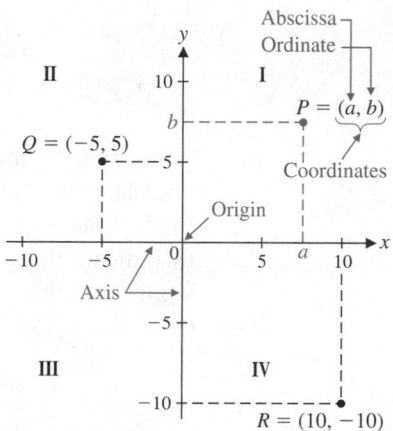

The Cartesian (rectangular) coordinate system

Now we want to assign *coordinates* to each point in the plane. Given an arbitrary point P in the plane, pass horizontal and vertical lines through the point (see figure). The vertical line will intersect the horizontal axis at a point with coordinate a, and the horizontal line will intersect the vertical axis at a point with coordinate b. These two numbers, written as the **ordered pair** (a, b) form the **coordinates** of the point P. The first coordinate, a, is called the **abscissa** of P; the second coordinate, b, is called the **ordinate** of P. The abscissa of Q in the figure is -5, and the ordinate of Q is 5. The coordinates of a point can also be referenced in terms of the axis labels. The *x* **coordinate** of R in the figure is 10, and the *y* **coordinate** of R is -10. The point with coordinates $(0, 0)$ is called the **origin**.

The procedure we have just described assigns to each point P in the plane a unique pair of real numbers (a, b). Conversely, if we are given an ordered pair of real numbers (a, b), then, reversing this procedure, we can determine a unique point P in the plane. Thus,

> There is a one-to-one correspondence between the points in a plane and the elements in the set of all ordered pairs of real numbers.

This is often referred to as the **fundamental theorem of analytic geometry**.

Graphs of Equations

A solution to an equation in one variable is a number. For example, the equation $4x - 13 = 7$ has the solution $x = 5$; when 5 is substituted for x, the left side of the equation is equal to the right side.

A solution to an equation in two variables is an ordered pair of numbers. For example, the equation $y = 9 - x^2$ has the solution $(4, -7)$; when 4 is substituted for x and -7 is substituted for y, the left side of the equation is equal to the right side. The solution $(4, -7)$ is one of infinitely many solutions to the equation $y = 9 - x^2$. The set of all solutions of an equation is called the **solution set**. Each solution forms the coordinates of a point in a rectangular coordinate system. To **sketch the graph** of an equation in two variables, we plot sufficiently many of those points so that the shape of the graph is apparent, and then we connect those points with a smooth curve. This process is called **point-by-point plotting**.

EXAMPLE 1   Point-by-Point Plotting   Sketch the graph of each equation.

(A) $y = 9 - x^2$ (B) $x^2 = y^4$

SOLUTIONS

(A) Make up a table of solutions—that is, ordered pairs of real numbers that satisfy the given equation. For easy mental calculation, choose integer values for x.

x	-4	-3	-2	-1	0	1	2	3	4
y	-7	0	5	8	9	8	5	0	-7

After plotting these solutions, if there are any portions of the graph that are unclear, plot additional points until the shape of the graph is apparent. Then join all the plotted points with a smooth curve (Fig. 1). Arrowheads are used to indicate that the graph continues beyond the portion shown here with no significant changes in shape.

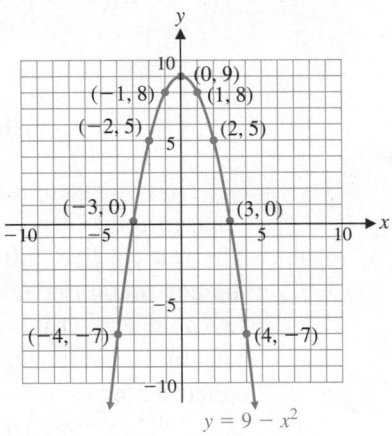

Figure 1 $y = 9 - x^2$

(B) Again we make a table of solutions—here it may be easier to choose integer values for y and calculate values for x. Note, for example, that if $y = 2$, then $x = \pm 4$; that is, the ordered pairs $(4, 2)$ and $(-4, 2)$ are both in the solution set.

x	± 9	± 4	± 1	0	± 1	± 4	± 9
y	-3	-2	-1	0	1	2	3

We plot these points and join them with a smooth curve (Fig. 2).

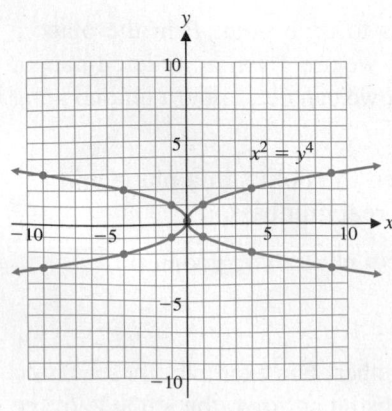

Figure 2 $x^2 = y^4$

Matched Problem 1   Sketch the graph of each equation.

(A) $y = x^2 - 4$ (B) $y^2 = \dfrac{100}{x^2 + 1}$

Explore and Discuss 1 To graph the equation $y = -x^3 + 3x$, we use point-by-point plotting to obtain

x	y
-1	-2
0	0
1	2

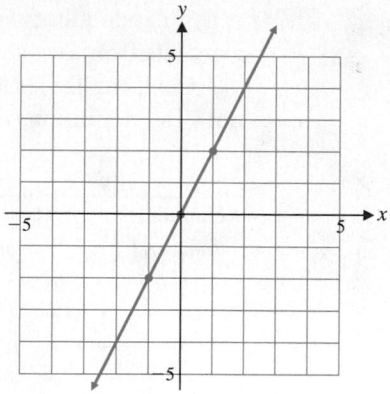

(A) Do you think this is the correct graph of the equation? Why or why not?

(B) Add points on the graph for $x = -2, -1.5, -0.5, 0.5, 1.5,$ and 2.

(C) Now, what do you think the graph looks like? Sketch your version of the graph, adding more points as necessary.

(D) Graph this equation on a graphing calculator and compare it with your graph from part (C).

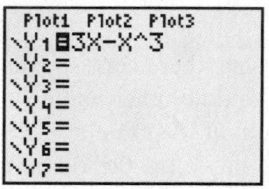

(A)

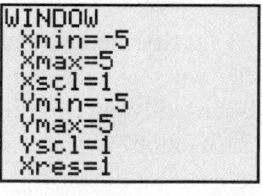

(B)

Figure 3

The icon in the margin is used throughout this book to identify optional graphing calculator activities that are intended to give you additional insight into the concepts under discussion. You may have to consult the manual for your graphing calculator for the details necessary to carry out these activities. For example, to graph the equation in Explore and Discuss 1 on most graphing calculators, you must enter the equation (Fig. 3A) and the window variables (Fig. 3B).

As Explore and Discuss 1 illustrates, the shape of a graph may not be apparent from your first choice of points. Using point-by-point plotting, it may be difficult to find points in the solution set of the equation, and it may be difficult to determine when you have found enough points to understand the shape of the graph. We will supplement the technique of point-by-point plotting with a detailed analysis of several basic equations, giving you the ability to sketch graphs with accuracy and confidence.

Definition of a Function

Central to the concept of function is correspondence. You are familiar with correspondences in daily life. For example,

To each person, there corresponds an annual income.

To each item in a supermarket, there corresponds a price.

To each student, there corresponds a grade-point average.

To each day, there corresponds a maximum temperature.

For the manufacture of x items, there corresponds a cost.

For the sale of x items, there corresponds a revenue.

To each square, there corresponds an area.

To each number, there corresponds its cube.

One of the most important aspects of any science is the establishment of correspondences among various types of phenomena. Once a correspondence is known, predictions can be made. A cost analyst would like to predict costs for various levels of output in a manufacturing process; a medical researcher would like to know the correspondence between heart disease and obesity; a psychologist would like to predict the level of performance after a subject has repeated a task a given number of times; and so on.

What do all of these examples have in common? Each describes the matching of elements from one set with the elements in a second set.

Consider Tables 1–3. Tables 1 and 2 specify functions, but Table 3 does not. Why not? The definition of the term *function* will explain.

Table 1

Domain	Range
Number	*Cube*
−2	−8
−1	−1
0	0
1	1
2	8

Table 2

Domain	Range
Number	*Square*
−2	4
−1	1
0	0
1	
2	

Table 3

Domain	Range
Number	*Square root*
0	0
	1
1	−1
	2
4	−2
	3
9	−3

> **DEFINITION   Function**
> A **function** is a correspondence between two sets of elements such that to each element in the first set, there corresponds one and only one element in the second set.
>   The first set is called the **domain**, and the set of corresponding elements in the second set is called the **range**.

Tables 1 and 2 specify functions since to each domain value, there corresponds exactly one range value (for example, the cube of −2 is −8 and no other number). On the other hand, Table 3 does not specify a function since to at least one domain value, there corresponds more than one range value (for example, to the domain value 9, there corresponds −3 and 3, both square roots of 9).

Explore and Discuss 2   Consider the set of students enrolled in a college and the set of faculty members at that college. Suppose we define a correspondence between the two sets by saying that a student corresponds to a faculty member if the student is currently enrolled in a course taught by that faculty member. Is this correspondence a function? Discuss.

Functions Specified by Equations

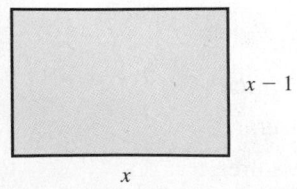

Figure 4

Most of the functions in this book will have domains and ranges that are (infinite) sets of real numbers. The **graph** of such a function is the set of all points (x, y) in the Cartesian plane such that x is an element of the domain and y is the corresponding element in the range. The correspondence between domain and range elements is often specified by an equation in two variables. Consider, for example, the equation for the area of a rectangle with width 1 inch less than its length (Fig. 4). If x is the length, then the area y is given by

$$y = x(x - 1) \qquad x \geq 1$$

For each **input** x (length), we obtain an **output** y (area). For example,

If $x = 5$,   then $y = 5(5 - 1) = 5 \cdot 4 = 20.$

If $x = 1$,   then $y = 1(1 - 1) = 1 \cdot 0 = 0.$

If $x = \sqrt{5}$,   then $y = \sqrt{5}(\sqrt{5} - 1) = 5 - \sqrt{5}$

$$\approx 2.7639.$$

The input values are domain values, and the output values are range values. The equation assigns each domain value x a range value y. The variable x is called an *independent variable* (since values can be "independently" assigned to x from the domain), and y is called a *dependent variable* (since the value of y "depends" on the value assigned to x). In general, any variable used as a placeholder for domain values is called an **independent variable**; any variable that is used as a placeholder for range values is called a **dependent variable**.

When does an equation specify a function?

DEFINITION Functions Specified by Equations

If in an equation in two variables, we get exactly one output (value for the dependent variable) for each input (value for the independent variable), then the equation specifies a function. The graph of such a function is just the graph of the specifying equation.

If we get more than one output for a given input, the equation does not specify a function.

EXAMPLE 2 Functions and Equations Determine which of the following equations specify functions with independent variable x.

(A) $4y - 3x = 8$, x a real number   (B) $y^2 - x^2 = 9$, x a real number

SOLUTION

(A) Solving for the dependent variable y, we have

$$4y - 3x = 8$$
$$4y = 8 + 3x \tag{1}$$
$$y = 2 + \frac{3}{4}x$$

Since each input value x corresponds to exactly one output value $(y = 2 + \frac{3}{4}x)$, we see that equation (1) specifies a function.

(B) Solving for the dependent variable y, we have

$$y^2 - x^2 = 9$$
$$y^2 = 9 + x^2 \tag{2}$$
$$y = \pm\sqrt{9 + x^2}$$

Since $9 + x^2$ is always a positive real number for any real number x, and since each positive real number has two square roots,* then to each input value x there corresponds two output values $(y = -\sqrt{9 + x^2}$ and $y = \sqrt{9 + x^2})$. For example, if $x = 4$, then equation (2) is satisfied for $y = 5$ and for $y = -5$. So equation (2) does not specify a function.

Matched Problem 2 Determine which of the following equations specify functions with independent variable x.

(A) $y^2 - x^4 = 9$, x a real number   (B) $3y - 2x = 3$, x a real number

Since the graph of an equation is the graph of all the ordered pairs that satisfy the equation, it is very easy to determine whether an equation specifies a function by examining its graph. The graphs of the two equations we considered in Example 2 are shown in Figure 5.

In Figure 5A, notice that any vertical line will intersect the graph of the equation $4y - 3x = 8$ in exactly one point. This shows that to each x value, there corresponds

---

*Recall that each positive real number N has two square roots: $\sqrt{N}$, the principal square root; and $-\sqrt{N}$, the negative of the principal square root (see Appendix A, Section A.6).

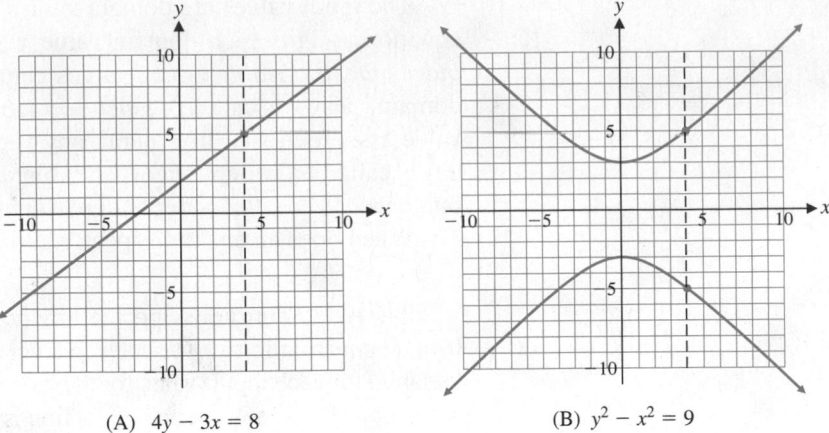

(A) $4y - 3x = 8$ (B) $y^2 - x^2 = 9$

Figure 5

exactly one y value, confirming our conclusion that this equation specifies a function. On the other hand, Figure 5B shows that there exist vertical lines that intersect the graph of $y^2 - x^2 = 9$ in two points. This indicates that there exist x values to which there correspond two different y values and verifies our conclusion that this equation does not specify a function. These observations are generalized in Theorem 1.

> ### THEOREM 1   Vertical-Line Test for a Function
>
> An equation specifies a function if each vertical line in the coordinate system passes through, at most, one point on the graph of the equation.
>
> If any vertical line passes through two or more points on the graph of an equation, then the equation does not specify a function.

The function graphed in Figure 5A is an example of a *linear function*. The vertical-line test implies that equations of the form $y = mx + b$, where $m \neq 0$, specify functions; they are called **linear functions**. Similarly, equations of the form $y = b$ specify functions; they are called **constant functions**, and their graphs are horizontal lines. The vertical-line test implies that equations of the form $x = a$ do not specify functions; note that the graph of $x = a$ is a vertical line.

In Example 2, the domains were explicitly stated along with the given equations. In many cases, this will not be done. Unless stated to the contrary, we shall adhere to the following convention regarding domains and ranges for functions specified by equations:

> If a function is specified by an equation and the domain is not indicated, then we assume that the domain is the set of all real-number replacements of the independent variable (inputs) that produce real values for the dependent variable (outputs). The range is the set of all outputs corresponding to input values.

EXAMPLE 3   Finding a Domain   Find the domain of the function specified by the equation $y = \sqrt{4 - x}$, assuming that x is the independent variable.

SOLUTION   For y to be real, $4 - x$ must be greater than or equal to 0; that is,

$$4 - x \geq 0$$
$$-x \geq -4$$
$$x \leq 4 \qquad \text{Sense of inequality reverses when both sides are divided by } -1.$$

Domain: $x \leq 4$ (inequality notation) or $(-\infty, 4]$ (interval notation)

Matched Problem 3   Find the domain of the function specified by the equation $y = \sqrt{x - 2}$, assuming x is the independent variable.

Function Notation

We have seen that a function involves two sets, a domain and a range, and a correspondence that assigns to each element in the domain exactly one element in the range. Just as we use letters as names for numbers, now we will use letters as names for functions. For example, f and g may be used to name the functions specified by the equations $y = 2x + 1$ and $y = x^2 + 2x - 3$:

$$f: \quad y = 2x + 1$$
$$g: \quad y = x^2 + 2x - 3 \tag{3}$$

If x represents an element in the domain of a function f, then we frequently use the symbol

$$f(x)$$

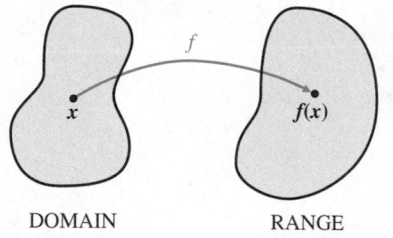

DOMAIN   RANGE

Figure 6

in place of y to designate the number in the range of the function f to which x is paired (Fig. 6). This symbol does *not* represent the product of f and x. The symbol $f(x)$ is read as "f of x," "f at x," or "the value of f at x." Whenever we write $y = f(x)$, we assume that the variable x is an independent variable and that both y and $f(x)$ are dependent variables.

Using function notation, we can now write functions f and g in equation (3) as

$$f(x) = 2x + 1 \quad \text{and} \quad g(x) = x^2 + 2x - 3$$

Let us find $f(3)$ and $g(-5)$. To find $f(3)$, we replace x with 3 wherever x occurs in $f(x) = 2x + 1$ and evaluate the right side:

$$f(x) = 2x + 1$$
$$f(3) = 2 \cdot 3 + 1$$
$$= 6 + 1 = 7 \quad \text{For input 3, the output is 7.}$$

Therefore,

$$f(3) = 7 \quad \text{The function } f \text{ assigns the range value 7 to the domain value 3.}$$

To find $g(-5)$, we replace each x by -5 in $g(x) = x^2 + 2x - 3$ and evaluate the right side:

$$g(x) = x^2 + 2x - 3$$
$$g(-5) = (-5)^2 + 2(-5) - 3$$
$$= 25 - 10 - 3 = 12 \quad \text{For input } -5, \text{ the output is 12.}$$

Therefore,

$$g(-5) = 12 \quad \text{The function } g \text{ assigns the range value 12 to the domain value } -5.$$

It is very important to understand and remember the definition of $f(x)$:

For any element x in the domain of the function f, the symbol $f(x)$ represents the element in the range of f corresponding to x in the domain of f. If x is an input value, then $f(x)$ is the corresponding output value. If x is an element that is not in the domain of f, then f is *not defined at* x and $f(x)$ *does not exist.*

EXAMPLE 4 Function Evaluation For $f(x) = 12/(x - 2), g(x) = 1 - x^2$, and $h(x) = \sqrt{x - 1}$, evaluate:

(A) $f(6)$ (B) $g(-2)$ (C) $h(-2)$ (D) $f(0) + g(1) - h(10)$

SOLUTION

(A) $f(6) \; = \dfrac{12}{6 - 2}^{\,*} = \dfrac{12}{4} = 3$

(B) $g(-2) = 1 - (-2)^2 = 1 - 4 = -3$

*Dashed boxes are used throughout the book to represent steps that are usually performed mentally.

(C) $h(-2) = \sqrt{-2 - 1} = \sqrt{-3}$

But $\sqrt{-3}$ is not a real number. Since we have agreed to restrict the domain of a function to values of x that produce real values for the function, -2 is not in the domain of h, and $h(-2)$ does not exist.

(D) $f(0) + g(1) - h(10) = \dfrac{12}{0 - 2} + (1 - 1^2) - \sqrt{10 - 1}$

$$= \dfrac{12}{-2} + 0 - \sqrt{9}$$

$$= -6 - 3 = -9$$

Matched Problem 4 Use the functions in Example 4 to find

(A) $f(-2)$ (B) $g(-1)$ (C) $h(-8)$ (D) $\dfrac{f(3)}{h(5)}$

EXAMPLE 5 Finding Domains   Find the domains of functions f, g, and h:

$$f(x) = \dfrac{12}{x - 2} \qquad g(x) = 1 - x^2 \qquad h(x) = \sqrt{x - 1}$$

SOLUTION *Domain of f:* $12/(x - 2)$ represents a real number for all replacements of x by real numbers except for $x = 2$ (division by 0 is not defined). Thus, $f(2)$ does not exist, and the domain of f is the set of all real numbers except 2. We often indicate this by writing

$$f(x) = \dfrac{12}{x - 2} \qquad x \neq 2$$

Domain of g: The domain is R, the set of all real numbers, since $1 - x^2$ represents a real number for all replacements of x by real numbers.

Domain of h: The domain is the set of all real numbers x such that $\sqrt{x - 1}$ is a real number, so

$$x - 1 \geq 0$$

$$x \geq 1 \quad \text{or, in interval notation,} \quad [1, \infty)$$

Matched Problem 5 Find the domains of functions F, G, and H:

$$F(x) = x^2 - 3x + 1 \qquad G(x) = \dfrac{5}{x + 3} \qquad H(x) = \sqrt{2 - x}$$

In addition to evaluating functions at specific numbers, it is important to be able to evaluate functions at expressions that involve one or more variables. For example, the **difference quotient**

$$\dfrac{f(x + h) - f(x)}{h} \qquad \text{x and $x + h$ in the domain of f, $h \neq 0$}$$

is studied extensively in calculus.

CONCEPTUAL INSIGHT

In algebra, you learned to use parentheses for grouping variables. For example,

$$2(x + h) = 2x + 2h$$

Now we are using parentheses in the function symbol $f(x)$. For example, if $f(x) = x^2$, then

$$f(x + h) = (x + h)^2 = x^2 + 2xh + h^2$$

Note that $f(x) + f(h) = x^2 + h^2 \neq f(x + h)$. That is, the function name f does not distribute across the grouped variables $(x + h)$, as the "2" does in $2(x + h)$. (see Appendix A, Section A.2).

EXAMPLE 6 Using Function Notation For $f(x) = x^2 - 2x + 7$, find

(A) $f(a)$

(B) $f(a + h)$

(C) $f(a + h) - f(a)$

(D) $\dfrac{f(a + h) - f(a)}{h}$, $h \neq 0$

SOLUTION

(A) $f(a) = a^2 - 2a + 7$

(B) $f(a + h) = (a + h)^2 - 2(a + h) + 7 = a^2 + 2ah + h^2 - 2a - 2h + 7$

(C) $f(a + h) - f(a) = (a^2 + 2ah + h^2 - 2a - 2h + 7) - (a^2 - 2a + 7)$

$$= 2ah + h^2 - 2h$$

(D) $\dfrac{f(a + h) - f(a)}{h} = \dfrac{2ah + h^2 - 2h}{h} = \dfrac{h(2a + h - 2)}{h}$ Because $h \neq 0$, $\dfrac{h}{h} = 1$.

$$= 2a + h - 2$$

Matched Problem 6 Repeat Example 6 for $f(x) = x^2 - 4x + 9$.

Applications

We now turn to the important concepts of **break-even** and **profit–loss** analysis, which we will return to a number of times in this book. Any manufacturing company has **costs**, C, and **revenues**, R. The company will have a **loss** if $R < C$, will **break even** if $R = C$, and will have a **profit** if $R > C$. Costs include **fixed costs** such as plant overhead, product design, setup, and promotion; and **variable costs**, which are dependent on the number of items produced at a certain cost per item. In addition, **price–demand** functions, usually established by financial departments using historical data or sampling techniques, play an important part in profit–loss analysis. We will let x, the number of units manufactured and sold, represent the independent variable. Cost functions, revenue functions, profit functions, and price–demand functions are often stated in the following forms, where a, b, m, and n are constants determined from the context of a particular problem:

Cost Function

$$C = (\text{fixed costs}) + (\text{variable costs})$$
$$= a + bx$$

Price–Demand Function

$$p = m - nx \quad x \text{ is the number of items that can be sold at } \$p \text{ per item.}$$

Revenue Function

$$R = (\text{number of items sold}) \times (\text{price per item})$$
$$= xp = x(m - nx)$$

Profit Function

$$P = R - C$$
$$= x(m - nx) - (a + bx)$$

Example 7 and Matched Problem 7 explore the relationships among the algebraic definition of a function, the numerical values of the function, and the graphical representation of the function. The interplay among algebraic, numeric, and graphic viewpoints is an important aspect of our treatment of functions and their use. In Example 7, we will see how a function can be used to describe data from the real world, a process that is often referred to as *mathematical modeling*. Note that the domain of such a function is determined by practical considerations within the problem.

EXAMPLE 7 Price–Demand and Revenue Modeling   A manufacturer of a popular digital camera wholesales the camera to retail outlets throughout the United States. Using statistical methods, the financial department in the company produced the price–demand data in Table 4, where p is the wholesale price per camera at which x million cameras are sold. Notice that as the price goes down, the number sold goes up.

Table 4 **Price–Demand**

x (millions)	p($)
2	87
5	68
8	53
12	37

Using special analytical techniques (regression analysis), an analyst obtained the following price–demand function to model the Table 4 data:

$$p(x) = 94.8 - 5x \qquad 1 \le x \le 15 \qquad (4)$$

(A) Plot the data in Table 4. Then sketch a graph of the price–demand function in the same coordinate system.

(B) What is the company's revenue function for this camera, and what is its domain?

(C) Complete Table 5, computing revenues to the nearest million dollars.

(D) Plot the data in Table 5. Then sketch a graph of the revenue function using these points.

(E) Plot the revenue function on a graphing calculator.

SOLUTION

(A)

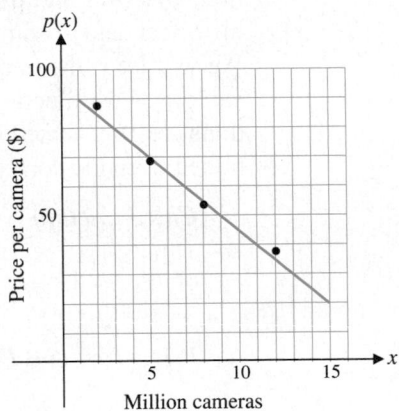

Figure 7 **Price–demand**

In Figure 7, notice that the model approximates the actual data in Table 4, and it is assumed that it gives realistic and useful results for all other values of x between 1 million and 15 million.

(B) $R(x) = xp(x) = x(94.8 - 5x)$ million dollars
Domain: $1 \le x \le 15$
[Same domain as the price–demand function, equation (4).]

(C)

Table 5 **Revenue**

x (millions)	R(x) (million $)
1	90
3	239
6	389
9	448
12	418
15	297

Table 5 **Revenue**

x (millions)	R(x) (million $)
1	90
3	
6	
9	
12	
15	

(D)

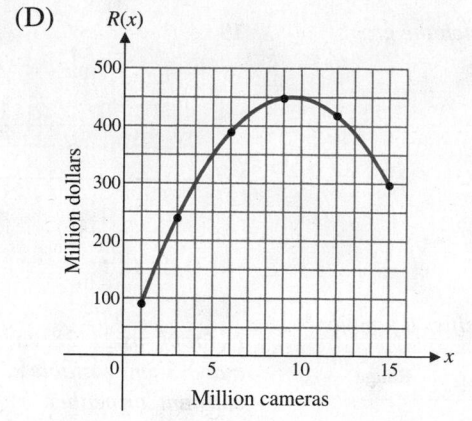

(E)

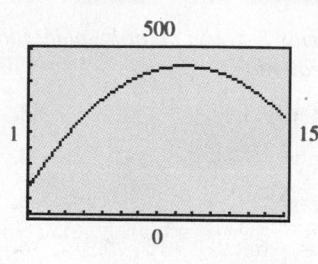

Matched Problem 7 The financial department in Example 7, using statistical techniques, produced the data in Table 6, where $C(x)$ is the cost in millions of dollars for manufacturing and selling x million cameras.

Table 6 **Cost Data**

x (millions)	$C(x)$ (million $)
1	175
5	260
8	305
12	395

Using special analytical techniques (regression analysis), an analyst produced the following cost function to model the Table 6 data:

$$C(x) = 156 + 19.7x \qquad 1 \leq x \leq 15 \tag{5}$$

(A) Plot the data in Table 6. Then sketch a graph of equation (5) in the same coordinate system.

(B) Using the revenue function from Example 7(B), what is the company's profit function for this camera, and what is its domain?

(C) Complete Table 7, computing profits to the nearest million dollars.

Table 7 **Profit**

x (millions)	$P(x)$ (million $)
1	−86
3	
6	
9	
12	
15	

(D) Plot the data in Table 7. Then sketch a graph of the profit function using these points.

(E) Plot the profit function on a graphing calculator.

In Problems 1–8, use point-by-point plotting to sketch the graph of each equation.

1. $y = x + 1$

2. $x = y + 1$

3. $x = y^2$

4. $y = x^2$

5. $y = x^3$

6. $x = y^3$

7. $xy = -6$

8. $xy = 12$

Indicate whether each table in Problems 9–14 specifies a function.

9.
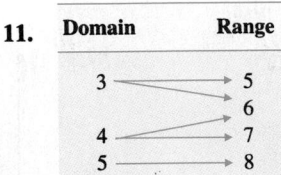

10.

11.
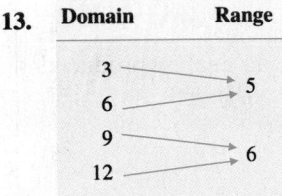

12.

13.

14.

Indicate whether each graph in Problems 15–20 specifies a function.

15.
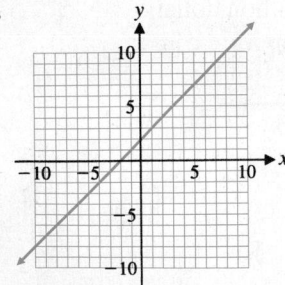

16.
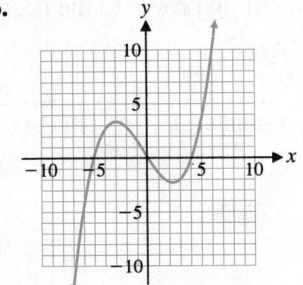

17.
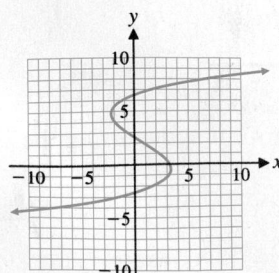

18.
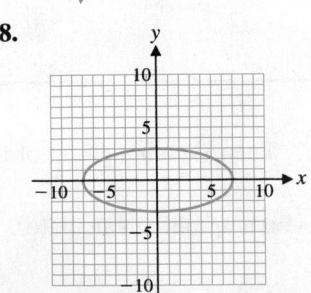

19.
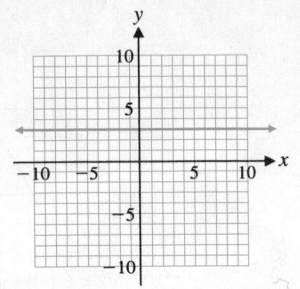

20.
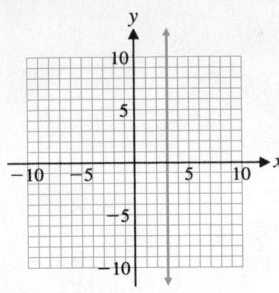

In Problems 21–28, each equation specifies a function with independent variable x. Determine whether the function is linear, constant, or neither.

21. $y - 2x = 7$

22. $y = 10 - 3x$

23. $xy - 4 = 0$

24. $x^2 - y = 8$

25. $y = 5x + \dfrac{1}{2}(7 - 10x)$

26. $y = \dfrac{2 + x}{3} + \dfrac{2 - x}{3}$

27. $3x + 4y = 5$

28. $9x - 2y + 6 = 0$

In Problems 29–36, use point-by-point plotting to sketch the graph of each function.

29. $f(x) = 1 - x$

30. $f(x) = \dfrac{x}{2} - 3$

31. $f(x) = x^2 - 1$

32. $f(x) = 3 - x^2$

33. $f(x) = 4 - x^3$

34. $f(x) = x^3 - 2$

35. $f(x) = \dfrac{8}{x}$

36. $f(x) = \dfrac{-6}{x}$

In Problems 37 and 38, the three points in the table are on the graph of the indicated function f. Do these three points provide sufficient information for you to sketch the graph of $y = f(x)$? Add more points to the table until you are satisfied that your sketch is a good representation of the graph of $y = f(x)$ for $-5 \le x \le 5$.

37.

x	-1	0	1	
$f(x)$	-1	0	1	$f(x) = \dfrac{2x}{x^2 + 1}$

38.

x	0	1	2	
$f(x)$	0	1	2	$f(x) = \dfrac{3x^2}{x^2 + 2}$

In Problems 39–46, use the following graph of a function f to determine x or y to the nearest integer, as indicated. Some problems may have more than one answer.

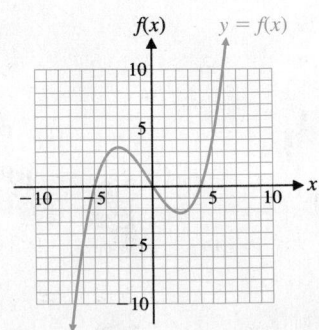

39. $y = f(-5)$ **40.** $y = f(4)$

41. $y = f(5)$ **42.** $y = f(-2)$

43. $0 = f(x)$ **44.** $3 = f(x), x < 0$

45. $-4 = f(x)$ **46.** $4 = f(x)$

In Problems 47–52, find the domain of each function.

47. $F(x) = 2x^3 - x^2 + 3$ **48.** $H(x) = 7 - 2x^2 - x^4$

49. $f(x) = \dfrac{x - 2}{x + 4}$ **50.** $g(x) = \dfrac{x + 1}{x - 2}$

51. $g(x) = \sqrt{7 - x}$ **52.** $F(x) = \dfrac{1}{\sqrt{5 + x}}$

In Problems 53–60, does the equation specify a function with independent variable x? If so, find the domain of the function. If not, find a value of x to which there corresponds more than one value of y.

53. $2x + 5y = 10$ **54.** $6x - 7y = 21$

55. $y(x + y) = 4$ **56.** $x(x + y) = 4$

57. $x^{-3} + y^3 = 27$ **58.** $x^2 + y^2 = 9$

59. $x^3 - y^2 = 0$ **60.** $\sqrt{x} - y^3 = 0$

In Problems 61–72, find and simplify the expression if $f(x) = x^2 - 4$.

61. $f(4)$ **62.** $f(-5)$

63. $f(x + 1)$ **64.** $f(x - 2)$

65. $f(-6x)$ **66.** $f(10x)$

67. $f(x^3)$ **68.** $f(\sqrt{x})$

69. $f(2) + f(h)$ **70.** $f(-3) + f(h)$

71. $f(2 + h)$ **72.** $f(-3 + h)$

73. $f(2 + h) - f(2)$ **74.** $f(-3 + h) - f(-3)$

In Problems 75–80, find and simplify each of the following, assuming $h \neq 0$ in (C).

(A) $f(x + h)$

(B) $f(x + h) - f(x)$

(C) $\dfrac{f(x + h) - f(x)}{h}$

75. $f(x) = 4x - 3$ **76.** $f(x) = -3x + 9$

77. $f(x) = 4x^2 - 7x + 6$ **78.** $f(x) = 3x^2 + 5x - 8$

79. $f(x) = x(20 - x)$ **80.** $f(x) = x(x + 40)$

Problems 81–84 refer to the area A and perimeter P of a rectangle with length l and width w (see the figure).

$$A = lw$$
$$P = 2l + 2w$$

w

l

81. The area of a rectangle is 25 sq in. Express the perimeter $P(w)$ as a function of the width w, and state the domain of this function.

82. The area of a rectangle is 81 sq in. Express the perimeter $P(l)$ as a function of the length l, and state the domain of this function.

83. The perimeter of a rectangle is 100 m. Express the area $A(l)$ as a function of the length l, and state the domain of this function.

84. The perimeter of a rectangle is 160 m. Express the area $A(w)$ as a function of the width w, and state the domain of this function.

Applications

85. Price–demand. A company manufactures memory chips for microcomputers. Its marketing research department, using statistical techniques, collected the data shown in Table 8, where p is the wholesale price per chip at which x million chips can be sold. Using special analytical techniques (regression analysis), an analyst produced the following price–demand function to model the data:

$$p(x) = 75 - 3x \qquad 1 \leq x \leq 20$$

Table 8 **Price–Demand**

x (millions)	p($)
1	72
4	63
9	48
14	33
20	15

(A) Plot the data points in Table 8, and sketch a graph of the price–demand function in the same coordinate system.

(B) What would be the estimated price per chip for a demand of 7 million chips? For a demand of 11 million chips?

86. Price–demand. A company manufactures notebook computers. Its marketing research department, using statistical techniques, collected the data shown in Table 9, where p is the wholesale price per computer at which x thousand computers can be sold. Using special analytical techniques (regression analysis), an analyst produced the following price–demand function to model the data:

$$p(x) = 2,000 - 60x \qquad 1 \leq x \leq 25$$

Table 9 **Price–Demand**

x (thousands)	p($)
1	1,940
8	1,520
16	1,040
21	740
25	500

(A) Plot the data points in Table 9, and sketch a graph of the price–demand function in the same coordinate system.

(B) What would be the estimated price per computer for a demand of 11,000 computers? For a demand of 18,000 computers?

87. Revenue.

(A) Using the price–demand function

$$p(x) = 75 - 3x \qquad 1 \leq x \leq 20$$

from Problem 85, write the company's revenue function and indicate its domain.

(B) Complete Table 10, computing revenues to the nearest million dollars.

Table 10 **Revenue**

x (millions)	R(x) (million $)
1	72
4	
8	
12	
16	
20	

(C) Plot the points from part (B) and sketch a graph of the revenue function using these points. Choose millions for the units on the horizontal and vertical axes.

88. Revenue.

(A) Using the price–demand function

$$p(x) = 2{,}000 - 60x \qquad 1 \leq x \leq 25$$

from Problem 86, write the company's revenue function and indicate its domain.

(B) Complete Table 11, computing revenues to the nearest thousand dollars.

Table 11 **Revenue**

x (thousands)	R(x) (thousand $)
1	1,940
5	
10	
15	
20	
25	

(C) Plot the points from part (B) and sketch a graph of the revenue function using these points. Choose thousands for the units on the horizontal and vertical axes.

89. Profit. The financial department for the company in Problems 85 and 87 established the following cost function for producing and selling x million memory chips:

$$C(x) = 125 + 16x \text{ million dollars}$$

(A) Write a profit function for producing and selling x million memory chips and indicate its domain.

(B) Complete Table 12, computing profits to the nearest million dollars.

Table 12 **Profit**

x (millions)	P(x) (million $)
1	−69
4	
8	
12	
16	
20	

(C) Plot the points in part (B) and sketch a graph of the profit function using these points.

90. Profit. The financial department for the company in Problems 86 and 88 established the following cost function for producing and selling x thousand notebook computers:

$$C(x) = 4{,}000 + 500x \text{ thousand dollars}$$

(A) Write a profit function for producing and selling x thousand notebook computers and indicate its domain.

(B) Complete Table 13, computing profits to the nearest thousand dollars.

Table 13 **Profit**

x (thousands)	P(x) (thousand $)
1	−2,560
5	
10	
15	
20	
25	

(C) Plot the points in part (B) and sketch a graph of the profit function using these points.

91. Packaging. A candy box will be made out of a piece of cardboard that measures 8 by 12 in. Equal-sized squares x inches on a side will be cut out of each corner, and then the ends and sides will be folded up to form a rectangular box.

(A) Express the volume of the box $V(x)$ in terms of x.

(B) What is the domain of the function V (determined by the physical restrictions)?

(C) Complete Table 14.

Table 14 **Volume**

x	V(x)
1	
2	
3	

(D) Plot the points in part (C) and sketch a graph of the volume function using these points.

92. Packaging. Refer to Problem 91.

(A) Table 15 shows the volume of the box for some values of x between 1 and 2. Use these values to estimate to one decimal place the value of x between 1 and 2 that would produce a box with a volume of 65 cu in.

Table 15 **Volume**

x	$V(x)$
1.1	62.524
1.2	64.512
1.3	65.988
1.4	66.976
1.5	67.5
1.6	67.584
1.7	67.252

(B) Describe how you could refine this table to estimate x to two decimal places.

(C) Carry out the refinement you described in part (B) and approximate x to two decimal places.

93. Packaging. Refer to Problems 91 and 92.

(A) Examine the graph of $V(x)$ from Problem 91D and discuss the possible locations of other values of x that would produce a box with a volume of 65 cu in.

(B) Construct a table like Table 15 to estimate any such value to one decimal place.

(C) Refine the table you constructed in part (B) to provide an approximation to two decimal places.

94. Packaging. A parcel delivery service will only deliver packages with length plus girth (distance around) not exceeding 108 in. A rectangular shipping box with square ends x inches on a side is to be used.

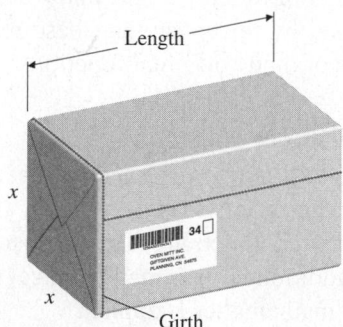

(A) If the full 108 in. is to be used, express the volume of the box $V(x)$ in terms of x.

(B) What is the domain of the function V (determined by the physical restrictions)?

(C) Complete Table 16.

Table 16 **Volume**

x	$V(x)$
5	
10	
15	
20	
25	

(D) Plot the points in part (C) and sketch a graph of the volume function using these points.

95. Muscle contraction. In a study of the speed of muscle contraction in frogs under various loads, British biophysicist A. W. Hill determined that the weight w (in grams) placed on the muscle and the speed of contraction v (in centimeters per second) are approximately related by an equation of the form

$$(w + a)(v + b) = c$$

where a, b, and c are constants. Suppose that for a certain muscle, $a = 15$, $b = 1$, and $c = 90$. Express v as a function of w. Find the speed of contraction if a weight of 16 g is placed on the muscle.

96. Politics. The percentage s of seats in the House of Representatives won by Democrats and the percentage v of votes cast for Democrats (when expressed as decimal fractions) are related by the equation

$$5v - 2s = 1.4 \qquad 0 < s < 1, \quad 0.28 < v < 0.68$$

(A) Express v as a function of s and find the percentage of votes required for the Democrats to win 51% of the seats.

(B) Express s as a function of v and find the percentage of seats won if Democrats receive 51% of the votes.

Answers to Matched Problems

1. (A)   (B)

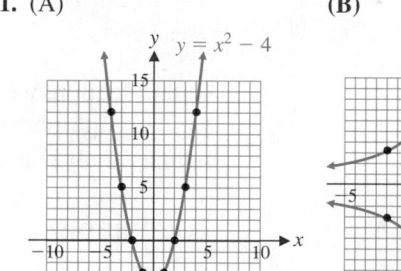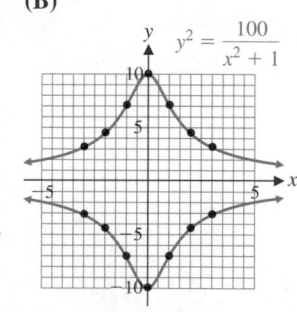

2. (A) Does not specify a function

   (B) Specifies a function

3. $x \geq 2$ (inequality notation) or $[2, \infty)$ (interval notation)

4. (A) -3 (B) 0 (C) Does not exist   (D) 6

5. Domain of F: R; domain of G: all real numbers except -3; domain of H: $x \leq 2$ (inequality notation) or $(-\infty, 2]$ (interval notation)

6. (A) $a^2 - 4a + 9$ (B) $a^2 + 2ah + h^2 - 4a - 4h + 9$

   (C) $2ah + h^2 - 4h$ (D) $2a + h - 4$

7. (A)

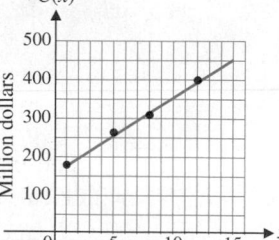

(B) $P(x) = R(x) - C(x)$
$= x(94.8 - 5x) - (156 + 19.7x)$;

domain: $1 \le x \le 15$

(C) Table 7 **Profit**

x (millions)	$P(x)$ (million $)
1	−86
3	24
6	115
9	115
12	25
15	−155

(D)

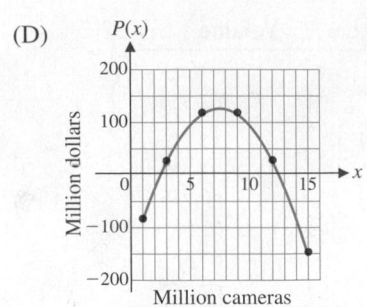

(E)

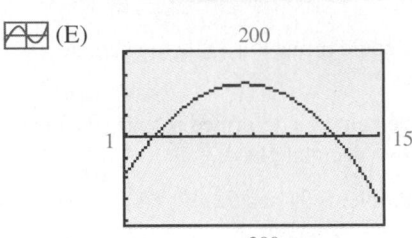

1.2   Elementary Functions: Graphs and Transformations

- A Beginning Library of Elementary Functions
- Vertical and Horizontal Shifts
- Reflections, Stretches, and Shrinks
- Piecewise-Defined Functions

Each of the functions

$$g(x) = x^2 - 4 \qquad h(x) = (x - 4)^2 \qquad k(x) = -4x^2$$

can be expressed in terms of the function $f(x) = x^2$:

$$g(x) = f(x) - 4 \qquad h(x) = f(x - 4) \qquad k(x) = -4f(x)$$

In this section, we will see that the graphs of functions g, h, and k are closely related to the graph of function f. Insight gained by understanding these relationships will help us analyze and interpret the graphs of many different functions.

A Beginning Library of Elementary Functions

As you progress through this book, you will repeatedly encounter a small number of elementary functions. We will identify these functions, study their basic properties, and include them in a library of elementary functions (see the inside front cover). This library will become an important addition to your mathematical toolbox and can be used in any course or activity where mathematics is applied.

We begin by placing six basic functions in our library.

> **DEFINITION**   Basic Elementary Functions
>
> $$f(x) = x \qquad \text{Identity function}$$
> $$h(x) = x^2 \qquad \text{Square function}$$
> $$m(x) = x^3 \qquad \text{Cube function}$$
> $$n(x) = \sqrt{x} \qquad \text{Square root function}$$
> $$p(x) = \sqrt[3]{x} \qquad \text{Cube root function}$$
> $$g(x) = |x| \qquad \text{Absolute value function}$$

These elementary functions can be evaluated by hand for certain values of x and with a calculator for all values of x for which they are defined.

EXAMPLE 1   Evaluating Basic Elementary Functions   Evaluate each basic elementary function at

(A) $x = 64$ (B) $x = -12.75$

Round any approximate values to four decimal places.

SOLUTION

(A) $f(64) = 64$

$h(64) = 64^2 = 4{,}096$ Use a calculator.

$m(64) = 64^3 = 262{,}144$ Use a calculator.

$n(64) = \sqrt{64} = 8$

$p(64) = \sqrt[3]{64} = 4$

$g(64) = |64| = 64$

(B) $f(-12.75) = -12.75$

$h(-12.75) = (-12.75)^2 = 162.5625$ Use a calculator.

$m(-12.75) = (-12.75)^3 \approx -2{,}072.6719$ Use a calculator.

$n(-12.75) = \sqrt{-12.75}$ Not a real number.

$p(-12.75) = \sqrt[3]{-12.75} \approx -2.3362$ Use a calculator.

$g(-12.75) = |-12.75| = 12.75$

Matched Problem 1   Evaluate each basic elementary function at

(A) $x = 729$ (B) $x = -5.25$

Round any approximate values to four decimal places.

Remark—Most computers and graphing calculators use ABS(x) to represent the absolute value function. The following representation can also be useful:

$$|x| = \sqrt{x^2}$$

Figure 1 shows the graph, range, and domain of each of the basic elementary functions.

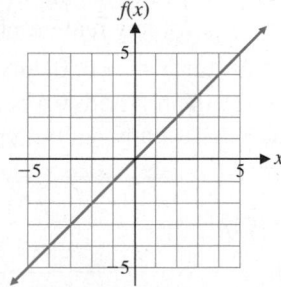

(A) **Identity function**
$f(x) = x$
Domain: R
Range: R

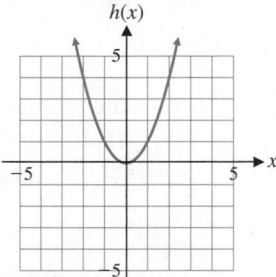

(B) **Square function**
$h(x) = x^2$
Domain: R
Range: $[0, \infty)$

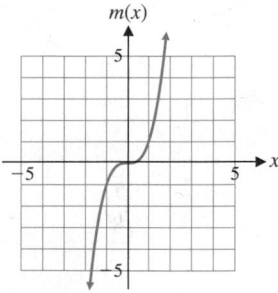

(C) **Cube function**
$m(x) = x^3$
Domain: R
Range: R

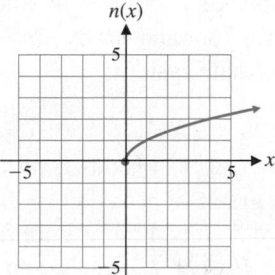

(D) **Square root function**
$n(x) = \sqrt{x}$
Domain: $[0, \infty)$
Range: $[0, \infty)$

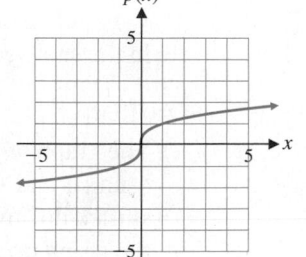

(E) **Cube root function**
$p(x) = \sqrt[3]{x}$
Domain: R
Range: R

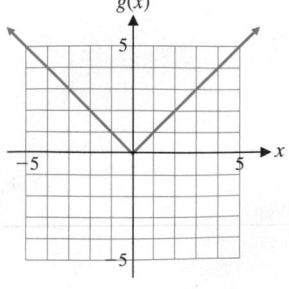

(F) **Absolute value function**
$g(x) = |x|$
Domain: R
Range: $[0, \infty)$

Figure 1 **Some basic functions and their graphs***

Note: Letters used to designate these functions may vary from context to context; R is the set of all real numbers.

> ## CONCEPTUAL INSIGHT
>
> **Absolute Value** In beginning algebra, absolute value is often interpreted as distance from the origin on a real number line (see Appendix A, Section A.1).
>
>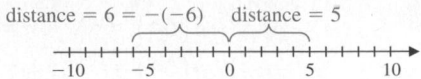
>
> If $x < 0$, then $-x$ is the *positive* distance from the origin to x, and if $x > 0$, then x is the positive distance from the origin to x. Thus,
>
> $$|x| = \begin{cases} -x & \text{if } x < 0 \\ x & \text{if } x \geq 0 \end{cases}$$

Vertical and Horizontal Shifts

If a new function is formed by performing an operation on a given function, then the graph of the new function is called a **transformation** of the graph of the original function. For example, graphs of $y = f(x) + k$ and $y = f(x + h)$ are transformations of the graph of $y = f(x)$.

Explore and Discuss 1   Let $f(x) = x^2$.

(A) Graph $y = f(x) + k$ for $k = -4, 0$, and 2 simultaneously in the same coordinate system. Describe the relationship between the graph of $y = f(x)$ and the graph of $y = f(x) + k$ for any real number k.

(B) Graph $y = f(x + h)$ for $h = -4, 0$, and 2 simultaneously in the same coordinate system. Describe the relationship between the graph of $y = f(x)$ and the graph of $y = f(x + h)$ for any real number h.

EXAMPLE 2   Vertical and Horizontal Shifts

(A) How are the graphs of $y = |x| + 4$ and $y = |x| - 5$ related to the graph of $y = |x|$? Confirm your answer by graphing all three functions simultaneously in the same coordinate system.

(B) How are the graphs of $y = |x + 4|$ and $y = |x - 5|$ related to the graph of $y = |x|$? Confirm your answer by graphing all three functions simultaneously in the same coordinate system.

SOLUTION

(A) The graph of $y = |x| + 4$ is the same as the graph of $y = |x|$ shifted upward 4 units, and the graph of $y = |x| - 5$ is the same as the graph of $y = |x|$ shifted downward 5 units. Figure 2 confirms these conclusions. [It appears that the graph of $y = f(x) + k$ is the graph of $y = f(x)$ shifted up if k is positive and down if k is negative.]

(B) The graph of $y = |x + 4|$ is the same as the graph of $y = |x|$ shifted to the left 4 units, and the graph of $y = |x - 5|$ is the same as the graph of $y = |x|$ shifted to the right 5 units. Figure 3 confirms these conclusions. [It appears that the graph of $y = f(x + h)$ is the graph of $y = f(x)$ shifted right if h is negative and left if h is positive—the opposite of what you might expect.]

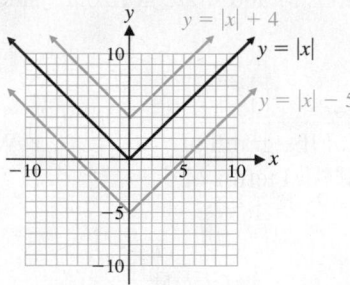

Figure 2 **Vertical shifts**

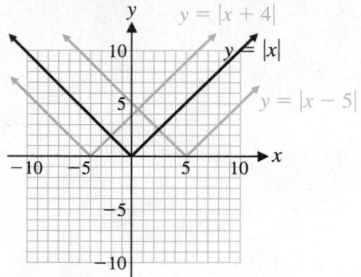

Figure 3 **Horizontal shifts**

Matched Problem 2

(A) How are the graphs of $y = \sqrt{x} + 5$ and $y = \sqrt{x} - 4$ related to the graph of $y = \sqrt{x}$? Confirm your answer by graphing all three functions simultaneously in the same coordinate system.

(B) How are the graphs of $y = \sqrt{x + 5}$ and $y = \sqrt{x - 4}$ related to the graph of $y = \sqrt{x}$? Confirm your answer by graphing all three functions simultaneously in the same coordinate system.

Comparing the graphs of $y = f(x) + k$ with the graph of $y = f(x)$, we see that the graph of $y = f(x) + k$ can be obtained from the graph of $y = f(x)$ by **vertically translating** (shifting) the graph of the latter upward k units if k is positive and downward $|k|$ units if k is negative. Comparing the graphs of $y = f(x + h)$ with the graph of $y = f(x)$, we see that the graph of $y = f(x + h)$ can be obtained from the graph of $y = f(x)$ by **horizontally translating** (shifting) the graph of the latter h units to the left if h is positive and $|h|$ units to the right if h is negative.

EXAMPLE 3   Vertical and Horizontal Translations (Shifts)   The graphs in Figure 4 are either horizontal or vertical shifts of the graph of $f(x) = x^2$. Write appropriate equations for functions H, G, M, and N in terms of f.

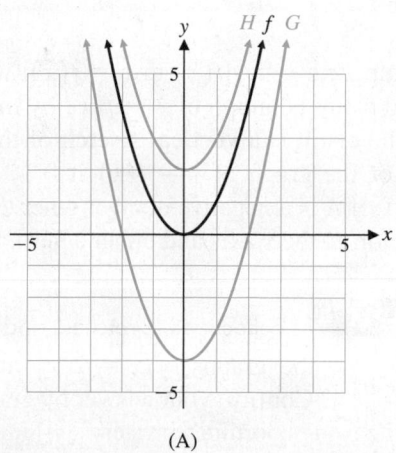

(A)

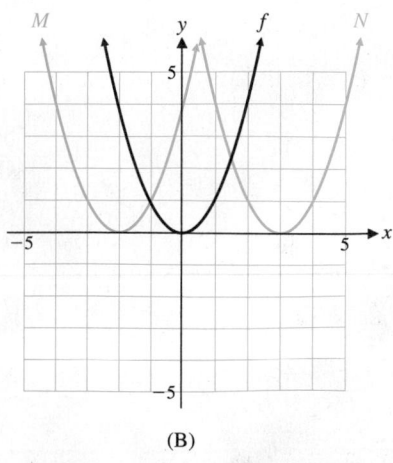

(B)

Figure 4 **Vertical and horizontal shifts**

SOLUTION Functions H and G are vertical shifts given by

$$H(x) = x^2 + 2 \qquad G(x) = x^2 - 4$$

Functions M and N are horizontal shifts given by

$$M(x) = (x + 2)^2 \qquad N(x) = (x - 3)^2$$

<u>Matched Problem 3</u> The graphs in Figure 5 are either horizontal or vertical shifts of the graph of $f(x) = \sqrt[3]{x}$. Write appropriate equations for functions H, G, M, and N in terms of f.

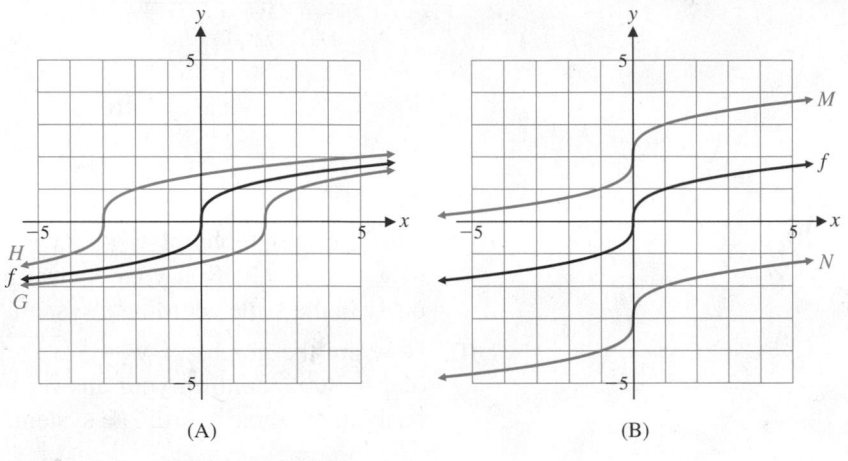

(A)   (B)

Figure 5 **Vertical and horizontal shifts**

Reflections, Stretches, and Shrinks

We now investigate how the graph of $y = Af(x)$ is related to the graph of $y = f(x)$ for different real numbers A.

Explore and Discuss 2 (A) Graph $y = Ax^2$ for $A = 1, 4$, and $\frac{1}{4}$ simultaneously in the same coordinate system.

(B) Graph $y = Ax^2$ for $A = -1, -4$, and $-\frac{1}{4}$ simultaneously in the same coordinate system.

(C) Describe the relationship between the graph of $h(x) = x^2$ and the graph of $G(x) = Ax^2$ for any real number A.

Comparing $y = Af(x)$ to $y = f(x)$, we see that the graph of $y = Af(x)$ can be obtained from the graph of $y = f(x)$ by multiplying each ordinate value of the latter by A. The result is a **vertical stretch** of the graph of $y = f(x)$ if $A > 1$, a **vertical shrink** of the graph of $y = f(x)$ if $0 < A < 1$, and a **reflection in the x axis** if $A = -1$. If A is a negative number other than -1, then the result is a combination of a reflection in the x axis and either a vertical stretch or a vertical shrink.

EXAMPLE 4   Reflections, Stretches, and Shrinks

(A) How are the graphs of $y = 2|x|$ and $y = 0.5|x|$ related to the graph of $y = |x|$? Confirm your answer by graphing all three functions simultaneously in the same coordinate system.

(B) How is the graph of $y = -2|x|$ related to the graph of $y = |x|$? Confirm your answer by graphing both functions simultaneously in the same coordinate system.

SOLUTION

(A) The graph of $y = 2|x|$ is a vertical stretch of the graph of $y = |x|$ by a factor of 2, and the graph of $y = 0.5|x|$ is a vertical shrink of the graph of $y = |x|$ by a factor of 0.5. Figure 6 confirms this conclusion.

(B) The graph of $y = -2|x|$ is a reflection in the x axis and a vertical stretch of the graph of $y = |x|$. Figure 7 confirms this conclusion.

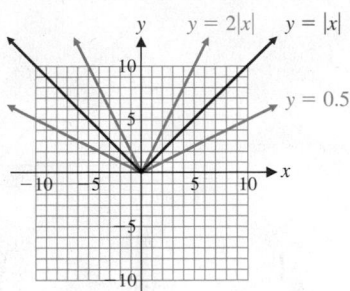

Figure 6 **Vertical stretch and shrink**

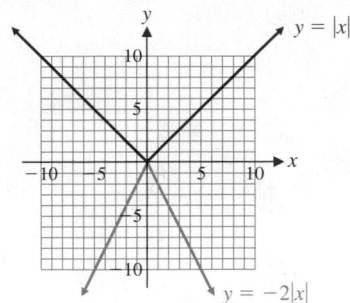

Figure 7 **Reflection and vertical stretch**

Matched Problem 4

(A) How are the graphs of $y = 2x$ and $y = 0.5x$ related to the graph of $y = x$? Confirm your answer by graphing all three functions simultaneously in the same coordinate system.

(B) How is the graph of $y = -0.5x$ related to the graph of $y = x$? Confirm your answer by graphing both functions in the same coordinate system.

The various transformations considered above are summarized in the following box for easy reference:

SUMMARY Graph Transformations

Vertical Translation:

$$y = f(x) + k \quad \begin{cases} k > 0 & \text{Shift graph of } y = f(x) \text{ up } k \text{ units.} \\ k < 0 & \text{Shift graph of } y = f(x) \text{ down } |k| \text{ units.} \end{cases}$$

Horizontal Translation:

$$y = f(x + h) \quad \begin{cases} h > 0 & \text{Shift graph of } y = f(x) \text{ left } h \text{ units.} \\ h < 0 & \text{Shift graph of } y = f(x) \text{ right } |h| \text{ units.} \end{cases}$$

Reflection:

$$y = -f(x) \quad \text{Reflect the graph of } y = f(x) \text{ in the } x \text{ axis.}$$

Vertical Stretch and Shrink:

$$y = Af(x) \quad \begin{cases} A > 1 & \text{Stretch graph of } y = f(x) \text{ vertically} \\ & \text{by multiplying each ordinate value by } A. \\ 0 < A < 1 & \text{Shrink graph of } y = f(x) \text{ vertically} \\ & \text{by multiplying each ordinate value by } A. \end{cases}$$

Explore and Discuss 3 Explain why applying any of the graph transformations in the summary box to a linear function produces another linear function.

EXAMPLE 5 Combining Graph Transformations Discuss the relationship between the graphs of $y = -|x - 3| + 1$ and $y = |x|$. Confirm your answer by graphing both functions simultaneously in the same coordinate system.

SOLUTION The graph of $y = -|x - 3| + 1$ is a reflection of the graph of $y = |x|$ in the x axis, followed by a horizontal translation of 3 units to the right and a vertical translation of 1 unit upward. Figure 8 confirms this description.

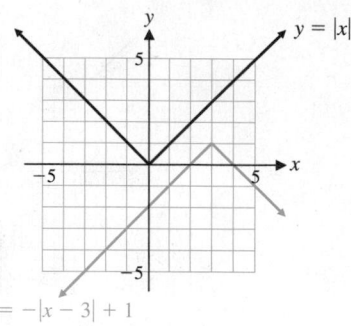

Figure 8 **Combined transformations**

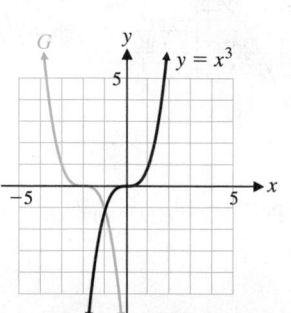

Figure 9 **Combined transformations**

Matched Problem 5 The graph of $y = G(x)$ in Figure 9 involves a reflection and a translation of the graph of $y = x^3$. Describe how the graph of function G is related to the graph of $y = x^3$ and find an equation of the function G.

Piecewise-Defined Functions

Earlier we noted that the absolute value of a real number x can be defined as

$$|x| = \begin{cases} -x & \text{if } x < 0 \\ x & \text{if } x \geq 0 \end{cases}$$

Notice that this function is defined by different rules for different parts of its domain. Functions whose definitions involve more than one rule are called **piecewise-defined functions**. Graphing one of these functions involves graphing each rule over the appropriate portion of the domain (Fig. 10). In Figure 10C, notice that an open dot is used to show that the point $(0, -2)$ is not part of the graph and a solid dot is used to show that $(0, 2)$ is part of the graph.

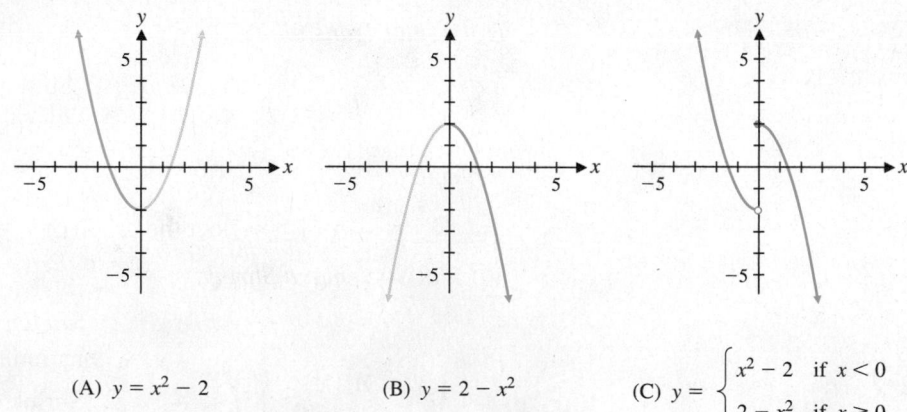

(A) $y = x^2 - 2$ (B) $y = 2 - x^2$ (C) $y = \begin{cases} x^2 - 2 & \text{if } x < 0 \\ 2 - x^2 & \text{if } x \geq 0 \end{cases}$

Figure 10 **Graphing a piecewise-defined function**

EXAMPLE 6 Graphing Piecewise-Defined Functions Graph the piecewise-defined function

$$g(x) = \begin{cases} x + 1 & \text{if } 0 \leq x < 2 \\ 0.5x & \text{if } x \geq 2 \end{cases}$$

SOLUTION If $0 \le x < 2$, then the first rule applies and the graph of g lies on the line $y = x + 1$ (a vertical shift of the identity function $y = x$). If $x = 0$, then $(0, 1)$ lies on $y = x + 1$; we plot $(0, 1)$ with a solid dot (Fig. 11) because $g(0) = 1$. If $x = 2$, then $(2, 3)$ lies on $y = x + 1$; we plot $(2, 3)$ with an open dot because $g(2) \ne 3$. The line segment from $(0, 1)$ to $(2, 3)$ is the graph of g for $0 \le x < 2$. If $x \ge 2$, then the second rule applies and the graph of g lies on the line $y = 0.5x$ (a vertical shrink of the identity function $y = x$). If $x = 2$, then $(2, 1)$ lies on the line $y = 0.5x$; we plot $(2, 1)$ with a solid dot because $g(2) = 1$. The portion of $y = 0.5x$ that starts at $(2, 1)$ and extends to the right is the graph of g for $x \ge 2$.

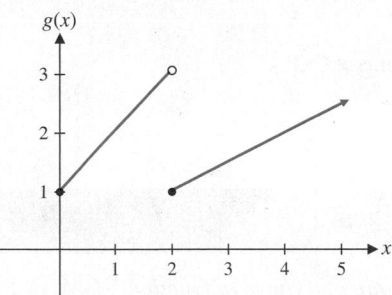

Figure 11

Matched Problem 6 Graph the piecewise-defined function

$$h(x) = \begin{cases} -2x + 4 & \text{if } 0 \le x \le 2 \\ x - 1 & \text{if } x > 2 \end{cases}$$

As the next example illustrates, piecewise-defined functions occur naturally in many applications.

EXAMPLE 7 Natural Gas Rates Easton Utilities uses the rates shown in Table 1 to compute the monthly cost of natural gas for each customer. Write a piecewise definition for the cost of consuming x CCF (cubic hundred feet) of natural gas and graph the function.

Table 1 **Charges per Month**

$0.7866 per CCF for the first 5 CCF
$0.4601 per CCF for the next 35 CCF
$0.2508 per CCF for all over 40 CCF

SOLUTION If $C(x)$ is the cost, in dollars, of using x CCF of natural gas in one month, then the first line of Table 1 implies that

$$C(x) = 0.7866x \quad \text{if } 0 \le x \le 5$$

Note that $C(5) = 3.933$ is the cost of 5 CCF. If $5 < x \le 40$, then $x - 5$ represents the amount of gas that cost $0.4601 per CCF, $0.4601(x - 5)$ represents the cost of this gas, and the total cost is

$$C(x) = 3.933 + 0.4601(x - 5)$$

If $x > 40$, then

$$C(x) = 20.0365 + 0.2508(x - 40)$$

where $20.0365 = C(40)$, the cost of the first 40 CCF. Combining all these equations, we have the following piecewise definition for $C(x)$:

$$C(x) = \begin{cases} 0.7866x & \text{if } 0 \le x \le 5 \\ 3.933 + 0.4601(x - 5) & \text{if } 5 < x \le 40 \\ 20.0365 + 0.2508(x - 40) & \text{if } x > 40 \end{cases}$$

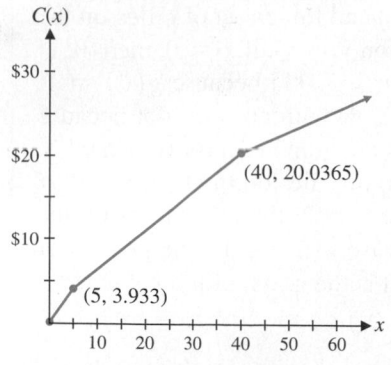

Figure 12 **Cost of purchasing x CCF of natural gas**

To graph C, first note that each rule in the definition of C represents a transformation of the identity function $f(x) = x$. Graphing each transformation over the indicated interval produces the graph of C shown in Figure 12.

Matched Problem 7) Trussville Utilities uses the rates shown in Table 2 to compute the monthly cost of natural gas for residential customers. Write a piecewise definition for the cost of consuming x CCF of natural gas and graph the function.

Table 2 **Charges per Month**

$0.7675 per CCF for the first 50 CCF
$0.6400 per CCF for the next 150 CCF
$0.6130 per CCF for all over 200 CCF

Exercises 1.2

In Problems 1–8, find the domain and range of function.

1. $f(x) = 5x - 10$ **2.** $f(x) = -4x + 12$

3. $f(x) = 15 - \sqrt{x}$ **4.** $f(x) = 3 + \sqrt{x}$

5. $f(x) = 2|x| + 7$ **6.** $f(x) = -5|x| + 2$

7. $f(x) = \sqrt[3]{x} + 100$ **8.** $f(x) = 20 - 10\sqrt[3]{x}$

In Problems 9–24, graph each of the functions using the graphs of functions f and g below.

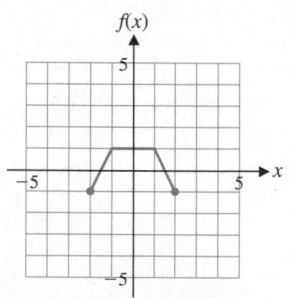

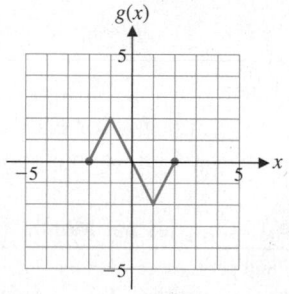

9. $y = f(x) + 2$ **10.** $y = g(x) - 1$

11. $y = f(x + 2)$ **12.** $y = g(x - 1)$

13. $y = g(x - 3)$ **14.** $y = f(x + 3)$

15. $y = g(x) - 3$ **16.** $y = f(x) + 3$

17. $y = -f(x)$ **18.** $y = -g(x)$

19. $y = 0.5g(x)$ **20.** $y = 2f(x)$

21. $y = 2f(x) + 1$ **22.** $y = -0.5g(x) + 3$

23. $y = 2(f(x) + 1)$ **24.** $y = -(0.5g(x) + 3)$

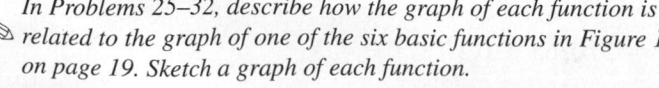 *In Problems 25–32, describe how the graph of each function is related to the graph of one of the six basic functions in Figure 1 on page 19. Sketch a graph of each function.*

25. $g(x) = -|x + 3|$ **26.** $h(x) = -|x - 5|$

27. $f(x) = (x - 4)^2 - 3$

28. $m(x) = (x + 3)^2 + 4$

29. $f(x) = 7 - \sqrt{x}$ **30.** $g(x) = -6 + \sqrt[3]{x}$

31. $h(x) = -3|x|$ **32.** $m(x) = -0.4x^2$

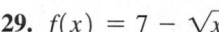 *Each graph in Problems 33–40 is the result of applying a sequence of transformations to the graph of one of the six basic functions in Figure 1 on page 19. Identify the basic function and describe the transformation verbally. Write an equation for the given graph.*

33.

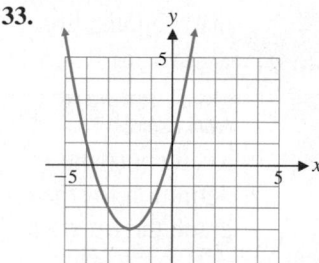

34.

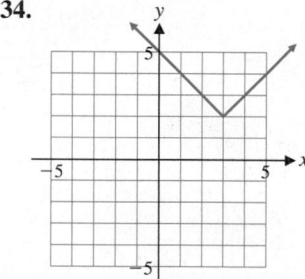

35.

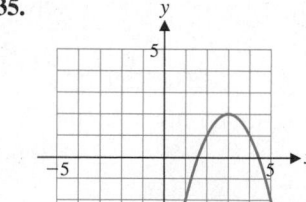

36.

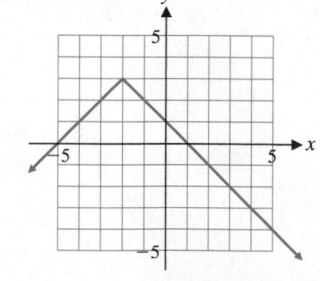

37.

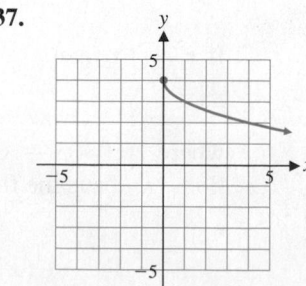

38.

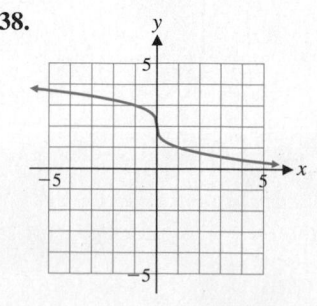

39.

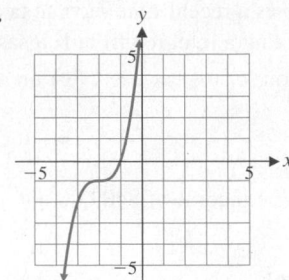

40.

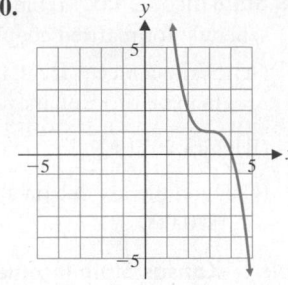

In Problems 41–46, the graph of the function g is formed by applying the indicated sequence of transformations to the given function f. Find an equation for the function g and graph g using $-5 \leq x \leq 5$ *and* $-5 \leq y \leq 5$.

41. The graph of $f(x) = \sqrt{x}$ is shifted 2 units to the right and 3 units down.

42. The graph of $f(x) = \sqrt[3]{x}$ is shifted 3 units to the left and 2 units up.

43. The graph of $f(x) = |x|$ is reflected in the x axis and shifted to the left 3 units.

44. The graph of $f(x) = |x|$ is reflected in the x axis and shifted to the right 1 unit.

45. The graph of $f(x) = x^3$ is reflected in the x axis and shifted 2 units to the right and down 1 unit.

46. The graph of $f(x) = x^2$ is reflected in the x axis and shifted to the left 2 units and up 4 units.

Graph each function in Problems 47–52.

47. $f(x) = \begin{cases} 2 - 2x & \text{if } x < 2 \\ x - 2 & \text{if } x \geq 2 \end{cases}$

48. $g(x) = \begin{cases} x + 1 & \text{if } x < -1 \\ 2 + 2x & \text{if } x \geq -1 \end{cases}$

49. $h(x) = \begin{cases} 5 + 0.5x & \text{if } 0 \leq x \leq 10 \\ -10 + 2x & \text{if } x > 10 \end{cases}$

50. $h(x) = \begin{cases} 10 + 2x & \text{if } 0 \leq x \leq 20 \\ 40 + 0.5x & \text{if } x > 20 \end{cases}$

51. $h(x) = \begin{cases} 2x & \text{if } 0 \leq x \leq 20 \\ x + 20 & \text{if } 20 < x \leq 40 \\ 0.5x + 40 & \text{if } x > 40 \end{cases}$

52. $h(x) = \begin{cases} 4x + 20 & \text{if } 0 \leq x \leq 20 \\ 2x + 60 & \text{if } 20 < x \leq 100 \\ -x + 360 & \text{if } x > 100 \end{cases}$

Each of the graphs in Problems 53–58 involves a reflection in the x axis and/or a vertical stretch or shrink of one of the basic functions in Figure 1 on page 19. Identify the basic function, and describe the transformation verbally. Write an equation for the given graph.

53.

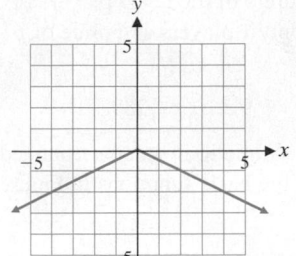

54.

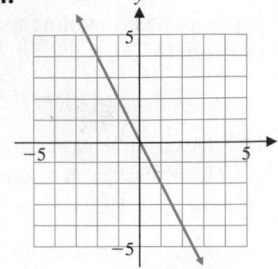

55.

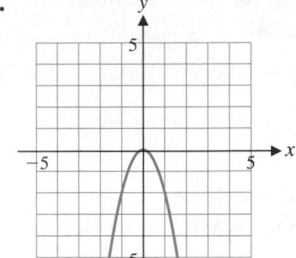

56.

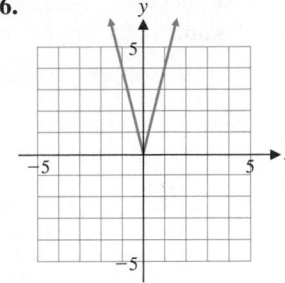

57.

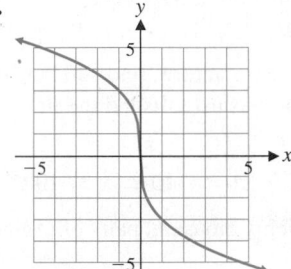

58.

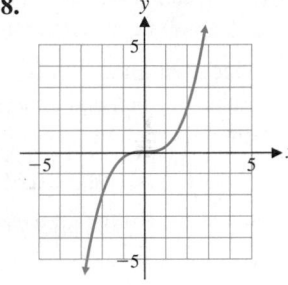

Changing the order in a sequence of transformations may change the final result. Investigate each pair of transformations in Problems 59–64 to determine if reversing their order can produce a different result. Support your conclusions with specific examples and/or mathematical arguments. (The graph of $y = f(-x)$ is the reflection of $y = f(x)$ in the y axis.)

59. Vertical shift; horizontal shift

60. Vertical shift; reflection in y axis

61. Vertical shift; reflection in x axis

62. Vertical shift; vertical stretch

63. Horizontal shift; reflection in y axis

64. Horizontal shift; vertical shrink

Applications

65. Price–demand. A retail chain sells DVD players. The retail price $p(x)$ (in dollars) and the weekly demand x for a particular model are related by

$$p(x) = 115 - 4\sqrt{x} \qquad 9 \leq x \leq 289$$

(A) Describe how the graph of function p can be obtained from the graph of one of the basic functions in Figure 1 on page 19.

(B) Sketch a graph of function p using part (A) as an aid.

66. Price–supply. The manufacturers of the DVD players in Problem 65 are willing to supply x players at a price of $p(x)$ as given by the equation

$$p(x) = 4\sqrt{x} \qquad 9 \le x \le 289$$

(A) Describe how the graph of function p can be obtained from the graph of one of the basic functions in Figure 1 on page 19.

(B) Sketch a graph of function p using part (A) as an aid.

67. Hospital costs. Using statistical methods, the financial department of a hospital arrived at the cost equation

$$C(x) = 0.00048(x - 500)^3 + 60,000 \quad 100 \le x \le 1,000$$

where $C(x)$ is the cost in dollars for handling x cases per month.

(A) Describe how the graph of function C can be obtained from the graph of one of the basic functions in Figure 1 on page 19.

(B) Sketch a graph of function C using part (A) and a graphing calculator as aids.

68. Price–demand. A company manufactures and sells in-line skates. Its financial department has established the price–demand function

$$p(x) = 190 - 0.013(x - 10)^2 \quad 10 \le x \le 100$$

where $p(x)$ is the price at which x thousand pairs of in-line skates can be sold.

(A) Describe how the graph of function p can be obtained from the graph of one of the basic functions in Figure 1 on page 19.

(B) Sketch a graph of function p using part (A) and a graphing calculator as aids.

69. Electricity rates. Table 3 shows the electricity rates charged by Monroe Utilities in the summer months. The base is a fixed monthly charge, independent of the kWh (kilowatt-hours) used during the month.

(A) Write a piecewise definition of the monthly charge $S(x)$ for a customer who uses x kWh in a summer month.

(B) Graph $S(x)$.

Table 3 **Summer (July–October)**

Base charge, $8.50
First 700 kWh or less at 0.0650/kWh
Over 700 kWh at 0.0900/kWh

70. Electricity rates. Table 4 shows the electricity rates charged by Monroe Utilities in the winter months.

(A) Write a piecewise definition of the monthly charge $W(x)$ for a customer who uses x kWh in a winter month.

Table 4 **Winter (November–June)**

Base charge, $8.50
First 700 kWh or less at 0.0650/kWh
Over 700 kWh at 0.0530/kWh

(B) Graph $W(x)$.

71. State income tax. Table 5 shows a recent state income tax schedule for married couples filing a joint return in Kansas.

(A) Write a piecewise definition for the tax due $T(x)$ on an income of x dollars.

(B) Graph $T(x)$.

(C) Find the tax due on a taxable income of $40,000. Of $70,000.

Table 5 **Kansas State Income Tax**

SCHEDULE I—MARRIED FILING JOINT		
If taxable income is		
Over	But Not Over	Tax Due Is
$0	$30,000	3.50% of taxable income
$30,000	$60,000	$1,050 plus 6.25% of excess over $30,000
$60,000		$2,925 plus 6.45% of excess over $60,000

72. State income tax. Table 6 shows a recent state income tax schedule for individuals filing a return in Kansas.

Table 6 **Kansas State Income Tax**

SCHEDULE II—SINGLE, HEAD OF HOUSEHOLD, OR MARRIED FILING SEPARATE		
If taxable income is		
Over	But Not Over	Tax Due Is
$0	$15,000	3.50% of taxable income
$15,000	$30,000	$525 plus 6.25% of excess over $15,000
$30,000		$1,462.50 plus 6.45% of excess over $30,000

(A) Write a piecewise definition for the tax due $T(x)$ on an income of x dollars.

(B) Graph $T(x)$.

(C) Find the tax due on a taxable income of $20,000. Of $35,000.

(D) Would it be better for a married couple in Kansas with two equal incomes to file jointly or separately? Discuss.

73. Human weight. A good approximation of the normal weight of a person 60 inches or taller but not taller than 80 inches is given by $w(x) = 5.5x - 220$, where x is height in inches and $w(x)$ is weight in pounds.

(A) Describe how the graph of function w can be obtained from the graph of one of the basic functions in Figure 1, page 19.

(B) Sketch a graph of function w using part (A) as an aid.

74. Herpetology. The average weight of a particular species of snake is given by $w(x) = 463x^3, 0.2 \le x \le 0.8$, where x is length in meters and $w(x)$ is weight in grams.

(A) Describe how the graph of function w can be obtained from the graph of one of the basic functions in Figure 1, page 19.

(B) Sketch a graph of function w using part (A) as an aid.

75. Safety research. Under ideal conditions, if a person driving a vehicle slams on the brakes and skids to a stop, the speed of the vehicle $v(x)$ (in miles per hour) is given approximately by $v(x) = C\sqrt{x}$, where x is the length of skid marks (in feet) and C is a constant that depends on the road conditions and the weight of the vehicle. For a particular vehicle, $v(x) = 7.08\sqrt{x}$ and $4 \le x \le 144$.

(A) Describe how the graph of function v can be obtained from the graph of one of the basic functions in Figure 1, page 19.

(B) Sketch a graph of function v using part (A) as an aid.

76. Learning. A production analyst has found that on average it takes a new person $T(x)$ minutes to perform a particular assembly operation after x performances of the operation, where $T(x) = 10 - \sqrt[3]{x}, 0 \le x \le 125$.

(A) Describe how the graph of function T can be obtained from the graph of one of the basic functions in Figure 1, page 19.

(B) Sketch a graph of function T using part (A) as an aid.

Answers to Matched Problems

1. (A) $f(729) = 729, h(729) = 531,441,$
$m(729) = 387,420,489, n(729) = 27, p(729) = 9,$
$g(729) = 729$

(B) $f(-5.25) = -5.25, \quad h(-5.25) = 27.5625,$
$m(-5.25) = -144.7031, n(-5.25)$ is not a real number,
$p(-5.25) = -1.7380, g(-5.25) = 5.25$

2. (A) The graph of $y = \sqrt{x} + 5$ is the same as the graph of $y = \sqrt{x}$ shifted upward 5 units, and the graph of $y = \sqrt{x} - 4$ is the same as the graph of $y = \sqrt{x}$ shifted downward 4 units. The figure confirms these conclusions.

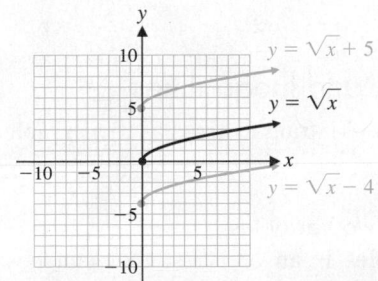

(B) The graph of $y = \sqrt{x + 5}$ is the same as the graph of $y = \sqrt{x}$ shifted to the left 5 units, and the graph of $y = \sqrt{x - 4}$ is the same as the graph of $y = \sqrt{x}$ shifted to the right 4 units. The figure confirms these conclusions.

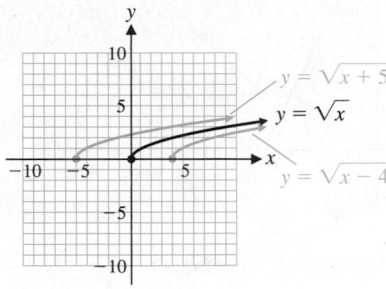

3. $H(x) = \sqrt[3]{x} + 3, G(x) = \sqrt[3]{x} - 2, M(x) = \sqrt[3]{x} + 2,$
$N(x) = \sqrt[3]{x} - 3$

4. (A) The graph of $y = 2x$ is a vertical stretch of the graph of $y = x$, and the graph of $y = 0.5x$ is a vertical shrink of the graph of $y = x$. The figure confirms these conclusions.

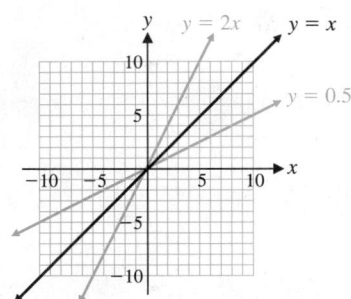

(B) The graph of $y = -0.5x$ is a vertical shrink and a reflection in the x axis of the graph of $y = x$. The figure confirms this conclusion.

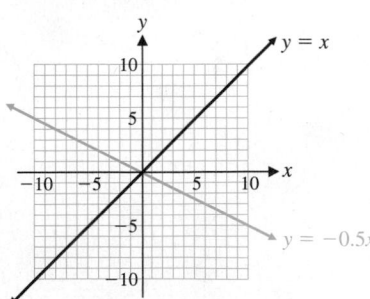

5. The graph of function G is a reflection in the x axis and a horizontal translation of 2 units to the left of the graph of $y = x^3$. An equation for G is $G(x) = -(x + 2)^3$.

6.

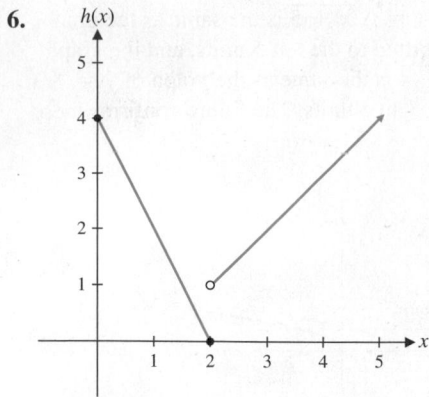

7. $C(x) = \begin{cases} 0.7675x & \text{if } 0 \le x \le 50 \\ 38.375 + 0.64\,(x - 50) & \text{if } 50 < x \le 200 \\ 134.375 + 0.613\,(x - 200) & \text{if } 200 < x \end{cases}$

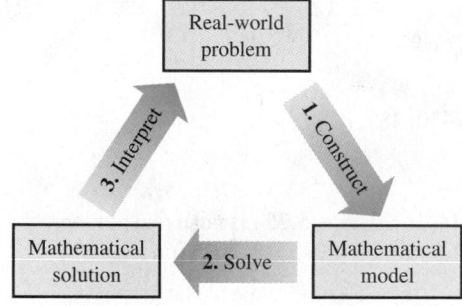

1.3   Linear and Quadratic Functions

- Linear Functions, Equations, and Inequalities

- Quadratic Functions, Equations, and Inequalities

- Properties of Quadratic Functions and Their Graphs

- Applications

- Linear and Quadratic Regression

Mathematical modeling is the process of using mathematics to solve real-world problems. This process can be broken down into three steps (Fig. 1):

Step 1   *Construct* **mathematical model** (that is, a mathematics problem that, when solved, will provide information about the real-world problem).

Step 2   *Solve* the mathematical model.

Step 3   *Interpret* the solution to the mathematical model in terms of the original real-world problem.

```
                  ┌──────────────┐
                  │  Real-world  │
                  │   problem    │
                  └──────────────┘
            3. Interpret        1. Construct
┌──────────────┐              ┌──────────────┐
│ Mathematical │  2. Solve    │ Mathematical │
│   solution   │ ◄─────────── │    model     │
└──────────────┘              └──────────────┘
```

Figure 1

In more complex problems, this cycle may have to be repeated several times to obtain the required information about the real-world problem. In this section, we will show how linear functions and quadratic functions can be used to construct mathematical models of real-world problems.

Linear Functions, Equations, and Inequalities

Linear equations in two variables have (straight) lines as their graphs.

> **DEFINITION**   Linear Equations in Two Variables
> A **linear equation in two variables** is an equation that can be written in the **standard form**
>
> $$Ax + By = C$$
>
> where A, B, and C are constants (A and B not both 0), and x and y are variables.

> **THEOREM 1   Graph of a Linear Equation in Two Variables**
>
> The graph of any equation of the form
>
> $$Ax + By = C \quad \text{(A and B not both O)} \tag{1}$$
>
> is a line, and any line in a Cartesian coordinate system is the graph of an equation of this form.

If $B = 0$ and $A \neq 0$, then equation (1) can be written as $x = \frac{C}{A}$ and its graph is a vertical line. If $B \neq 0$, then $-A/B$ is the *slope* of the line (see Problems 57–58 in Exercises 1.3).

> **DEFINITION   Slope of a Line**
>
> If (x_1, y_1) and (x_2, y_2) are two points on a line with $x_1 \neq x_2$, then the **slope** of the line is
>
> $$m = \frac{y_2 - y_1}{x_2 - x_1}$$

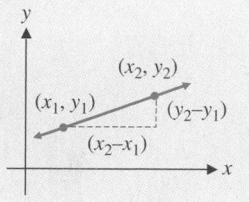

The slope measures the steepness of a line. The vertical change $y_2 - y_1$ is often called the **rise**, and the horizontal change is often called the **run**. The slope may be positive, negative, zero, or undefined (see Table 1).

Table 1 **Geometric Interpretation of Slope**

Line	Rising as x moves from left to right	Falling as x moves from left to right	Horizontal	Vertical
Slope	Positive	Negative	0	Not defined
Example				

If $B \neq 0$ in the standard form of the equation of a line, then solving for y gives the **slope-intercept form** $y = mx + b$, where m is the slope and b is the y intercept.* If a line has slope m and passes through the point (x_1, y_1), then $y - y_1 = m(x - x_1)$ is the **point-slope** form of the equation of the line. The various forms of the equation of a line are summarized in Table 2.

Table 2 **Equations of a Line**

Standard form	$Ax + By = C$	A and B not both 0
Slope-intercept form	$y = mx + b$	Slope: m; y intercept: b
Point-slope form	$y - y_1 = m(x - x_1)$	Slope: m; point: (x_1, y_1)
Horizontal line	$y = b$	Slope: 0
Vertical line	$x = a$	Slope: undefined

---

*If a line passes through the points $(a, 0)$ and $(0, b)$, then a is called the **x intercept** and b is called the **y intercept**. It is common practice to refer to either a or $(a, 0)$ as the x intercept, and either b or $(0, b)$ as the y intercept.

EXAMPLE 1   Equations of lines   A line has slope 4 and passes through the point $(3, 8)$. Find an equation of the line in point-slope form, slope-intercept form, and standard form.

SOLUTION   Let $m = 4$ and $(x_1, y_1) = (3, 8)$. Substitute into the point-slope form $y - y_1 = m(x - x_1)$:

$$\text{Point-slope form:} \quad y - 8 = 4(x - 3) \qquad \text{Add 8 to both sides.}$$
$$y = 4(x - 3) + 8 \qquad \text{Simplify.}$$
$$\text{Slope-intercept form:} \quad y = 4x - 4 \qquad \text{Subtract } 4x \text{ from both sides.}$$
$$\text{Standard form:} \quad -4x + y = -4$$

Matched Problem 1   A line has slope -3 and passes through the point $(-2, 10)$. Find an equation of the line in point-slope form, slope-intercept form, and standard form.

DEFINITION   Linear Function
If m and b are real numbers with $m \neq 0$, then the function

$$f(x) = mx + b$$

is a **linear function**.

So the linear function $f(x) = mx + b$ is the function that is specified by the linear equation $y = mx + b$.

EXAMPLE 2   Graphing a Linear Function

(A) Use intercepts to graph the equation $3x - 4y = 12$.

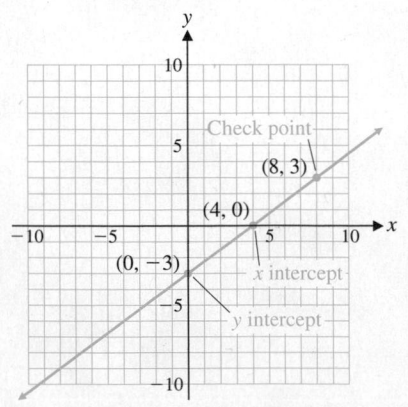* (B) Use a graphing calculator to graph the function $f(x)$ that is specified by $3x - 4y = 12$, and to find its x and y intercepts.

(C) Solve the inequality $f(x) \geq 0$.

SOLUTION

(A) To find the y intercept, let $x = 0$ and solve for y. To find the x intercept, let $y = 0$ and solve for x. It is a good idea to find a third point on the line as a check point.

x	y	
0	-3	y intercept
4	0	x intercept
8	3	Check point

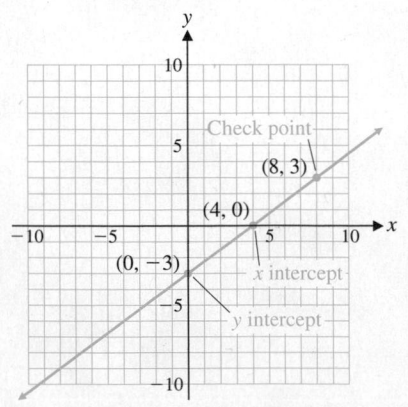

---

*The icon in the margin is used throughout the book to identify optional graphing calculator activities that are intended to give additional insight. We used a Texas Instruments graphing calculator from the TI-83/84 family to produce the graphing calculator screens in this book. You may need to consult the manual for your calculator for the details necessary to carry out these activities.

(B) To find $f(x)$, we solve $3x - 4y = 12$ for y.

$$3x - 4y = 12 \qquad \text{Add } -3x \text{ to both sides.}$$

$$-4y = -3x + 12 \qquad \text{Divide both sides by } -4.$$

$$y = \frac{-3x + 12}{-4} \qquad \text{Simplify.}$$

$$y = f(x) = \frac{3}{4}x - 3 \qquad\qquad (2)$$

Now we enter the right side of equation (2) in a calculator (Fig. 2A), enter values for the window variables (Fig 2B), and graph the line (Fig. 2C). (The numerals to the left and right of the screen in Figure 2C are Xmin and Xmax, respectively. Similarly, the numerals below and above the screen are Ymin and Ymax.)

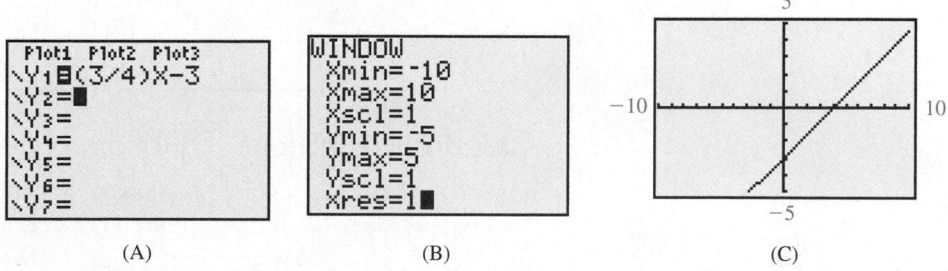

(A)   (B)   (C)

Figure 2 **Graphing a line on a graphing calculator**

Next we use two calculator commands to find the intercepts: TRACE (Fig. 3A) and ZERO (Fig. 3B).

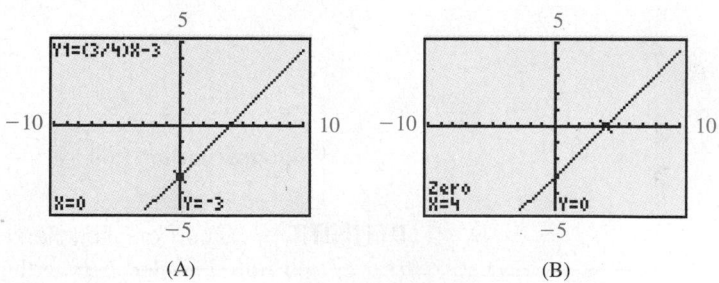

(A)   (B)

Figure 3 **Using TRACE and ZERO on a graphing calculator**

(C) $f(x) \geq 0$ if the graph of $f(x) = \frac{3}{4}x - 3$ in the figures in parts (A) or (B) is on or above the x axis. This occurs if $x \geq 4$, so the solution of the inequality, in interval notation,* is $[4, \infty)$.

Matched Problem 2

(A) Use intercepts to graph the equation $4x - 3y = 12$.

(B) Use a graphing calculator to graph the function $f(x)$ that is specified by $4x - 3y = 12$, and to find its x and y intercepts.

(C) Solve the inequality $f(x) \geq 0$.

---

*We use standard **interval notation** (see Table 3). The numbers a and b in Table 3 are called the **endpoints** of the interval. An interval is **closed** if it contains its endpoints and **open** if it does not contain any of its endpoints. The symbol ∞ (read "infinity") is not a number, and is not considered to be an endpoint. The notation $[b, \infty)$ simply denotes the interval that starts at b and continues indefinitely to the right. We never write $[b, \infty]$. The interval $(-\infty, \infty)$ is the entire real line.

Table 3 Interval Notation

Interval Notation	Inequality Notation	Line Graph
$[a, b]$	$a \leq x \leq b$	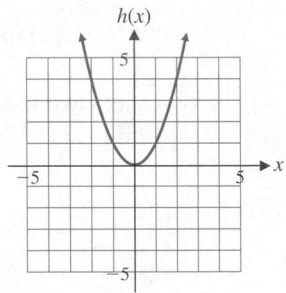
$[a, b)$	$a \leq x < b$	
$(a, b]$	$a < x \leq b$	
(a, b)	$a < x < b$	
$(-\infty, a]$	$x \leq a$	
$(-\infty, a)$	$x < a$	
$[b, \infty)$	$x \geq b$	
(b, ∞)	$x > b$	

Quadratic Functions, Equations, and Inequalities

One of the basic elementary functions of Section 1.2 is the square function $h(x) = x^2$. Its graph is the parabola shown in Figure 4. It is an example of a *quadratic function*.

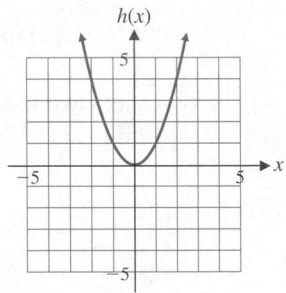

Figure 4 **Square function** $h(x) = x^2$

DEFINITION Quadratic Functions

If a, b, and c are real numbers with $a \neq 0$, then the function

$$f(x) = ax^2 + bx + c \quad \text{Standard form}$$

is a **quadratic function** and its graph is a **parabola**.

The domain of any quadratic function is the set of all real numbers. We will discuss methods for determining the range of a quadratic function later in this section. Typical graphs of quadratic functions are shown in Figure 5.

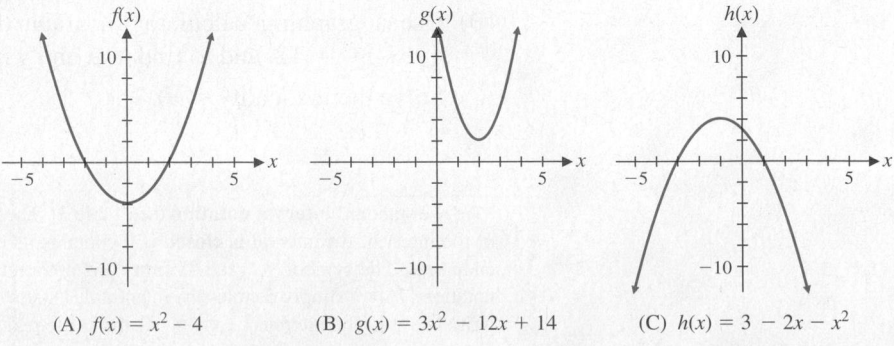

(A) $f(x) = x^2 - 4$ (B) $g(x) = 3x^2 - 12x + 14$ (C) $h(x) = 3 - 2x - x^2$

Figure 5 **Graphs of quadratic functions**

CONCEPTUAL INSIGHT

An x intercept of a function is also called a **zero** of the function. The x intercepts of a quadratic function can be found by solving the quadratic equation $y = ax^2 + bx + c = 0$ for x, $a \neq 0$. Several methods for solving quadratic equations are discussed in Appendix A, Section A.7. The most popular of these is the **quadratic formula**.

If $ax^2 + bx + c = 0$, $a \neq 0$, then

$$x = \frac{-b \pm \sqrt{b^2 - 4ac}}{2a}, \text{ provided } b^2 - 4ac \geq 0$$

EXAMPLE 3   Intercepts, Equations, and Inequalities

(A) Sketch a graph of $f(x) = -x^2 + 5x + 3$ in a rectangular coordinate system.

(B) Find x and y intercepts algebraically to four decimal places.

(C) Graph $f(x) = -x^2 + 5x + 3$ in a standard viewing window.

(D) Find the x and y intercepts to four decimal places using TRACE and ZERO on your graphing calculator.

(E) Solve the quadratic inequality $-x^2 + 5x + 3 \geq 0$ graphically to four decimal places using the results of parts (A) and (B) or (C) and (D).

(F) Solve the equation $-x^2 + 5x + 3 = 4$ graphically to four decimal places using INTERSECT on your graphing calculator.

SOLUTION

(A) Hand-sketching a graph of f:

x	y
-1	-3
0	3
1	7
2	9
3	9
4	7
5	3
6	-3

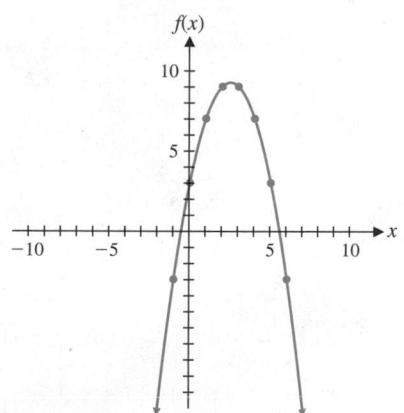

(B) Finding intercepts algebraically:

$$y \text{ intercept: } f(0) = -(0)^2 + 5(0) + 3 = 3$$
$$x \text{ intercepts: } f(x) = 0$$
$$-x^2 + 5x + 3 = 0 \qquad \text{Quadratic equation}$$
$$x = \frac{-b \pm \sqrt{b^2 - 4ac}}{2a} \qquad \text{Quadratic formula (see Appendix A.7)}$$
$$x = \frac{-(5) \pm \sqrt{5^2 - 4(-1)(3)}}{2(-1)}$$
$$= \frac{-5 \pm \sqrt{37}}{-2} = -0.5414 \quad \text{or} \quad 5.5414$$

(C) Graphing in a graphing calculator:

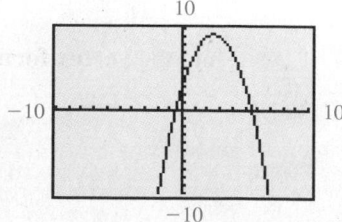

(D) Finding intercepts graphically using a graphing calculator:

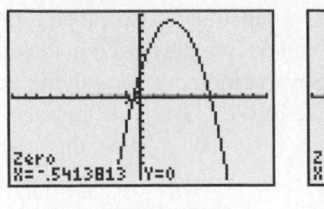 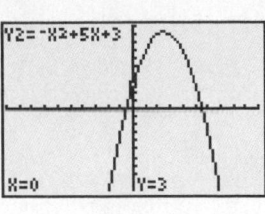

x intercept: −0.5414   x intercept: 0.5414   y intercept: 3

(E) Solving $-x^2 + 5x + 3 \geq 0$ graphically: The quadratic inequality

$$-x^2 + 5x + 3 \geq 0$$

holds for those values of x for which the graph of $f(x) = -x^2 + 5x + 3$ in the figures in parts (A) and (C) is at or above the x axis. This happens for x between the two x intercepts [found in part (B) or (D)], including the two x intercepts. The solution set for the quadratic inequality is $-0.5414 \leq x \leq 5.5414$ or, in interval notation, $[-0.5414, 5.5414]$.

(F) Solving the equation $-x^2 + 5x + 3 = 4$ using a graphing calculator:

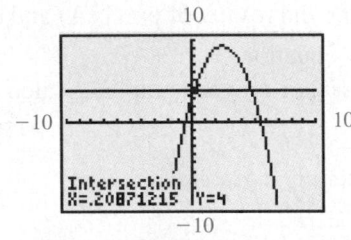 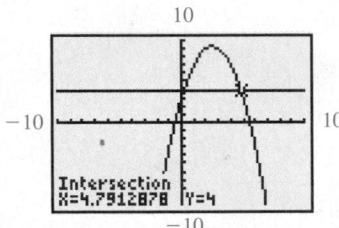

$-x^2 + 5x + 3 = 4$ at $x = 0.2087$ $-x^2 + 5x + 3 = 4$ at $x = 4.7913$

Matched Problem 3

(A) Sketch a graph of $g(x) = 2x^2 - 5x - 5$ in a rectangular coordinate system.

(B) Find x and y intercepts algebraically to four decimal places.

(C) Graph $g(x) = 2x^2 - 5x - 5$ in a standard viewing window.

(D) Find the x and y intercepts to four decimal places using TRACE and the ZERO command on your graphing calculator.

(E) Solve $2x^2 - 5x - 5 \geq 0$ graphically to four decimal places using the results of parts (A) and (B) or (C) and (D).

(F) Solve the equation $2x^2 - 5x - 5 = -3$ graphically to four decimal places using INTERSECT on your graphing calculator.

Explore and Discuss 1   How many x intercepts can the graph of a quadratic function have? How many y intercepts? Explain your reasoning.

Properties of Quadratic Functions and Their Graphs

Many useful properties of the quadratic function can be uncovered by transforming

$$f(x) = ax^2 + bx + c \qquad a \neq 0$$

into the **vertex form**

$$f(x) = a(x - h)^2 + k$$

The process of *completing the square* (see Appendix A.7) is central to the transformation. We illustrate the process through a specific example and then generalize the results.

Consider the quadratic function given by

$$f(x) = -2x^2 + 16x - 24 \tag{3}$$

We use completing the square to transform this function into vertex form:

$$f(x) = -2x^2 + 16x - 24$$
$$= -2(x^2 - 8x) - 24$$
$$= -2(x^2 - 8x + \ ?) - 24$$

Factor the coefficient of x^2 out of the first two terms.

$$= -2(x^2 - 8x + \mathbf{16}) - 24 + \mathbf{32}$$

Add 16 to complete the square inside the parentheses. Because of the -2 outside the parentheses, we have actually added -32, so we must add 32 to the outside.

$$= -2(x - 4)^2 + 8$$

The transformation is complete and can be checked by multiplying out.

Therefore,

$$f(x) = -2(x - 4)^2 + 8 \tag{4}$$

If $x = 4$, then $-2(x - 4)^2 = 0$ and $f(4) = 8$. For any other value of x, the negative number $-2(x - 4)^2$ is added to 8, making it smaller. Therefore,

$$f(4) = 8$$

is the *maximum value* of $f(x)$ for all x. Furthermore, if we choose any two x values that are the same distance from 4, we will obtain the same function value. For example, $x = 3$ and $x = 5$ are each one unit from $x = 4$ and their function values are

$$f(3) = -2(3 - 4)^2 + 8 = 6$$
$$f(5) = -2(5 - 4)^2 + 8 = 6$$

Therefore, the vertical line $x = 4$ is a line of symmetry. That is, if the graph of equation (3) is drawn on a piece of paper and the paper is folded along the line $x = 4$, then the two sides of the parabola will match exactly. All these results are illustrated by graphing equations (3) and (4) and the line $x = 4$ simultaneously in the same coordinate system (Fig. 6).

From the preceding discussion, we see that as x moves from left to right, $f(x)$ is increasing on $(-\infty, 4]$, and decreasing on $[4, \infty)$, and that $f(x)$ can assume no value greater than 8. Thus,

$$\text{Range of } f: \quad y \le 8 \quad \text{or} \quad (-\infty, 8]$$

In general, the graph of a quadratic function is a parabola with line of symmetry parallel to the vertical axis. The lowest or highest point on the parabola, whichever exists, is called the **vertex**. The maximum or minimum value of a quadratic function always occurs at the vertex of the parabola. The line of symmetry through the vertex is called the **axis** of the parabola. In the example above, $x = 4$ is the axis of the parabola and $(4, 8)$ is its vertex.

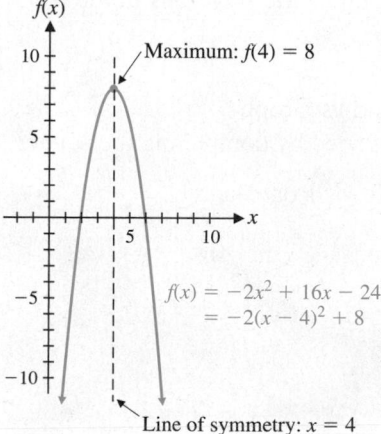

Figure 6 **Graph of a quadratic function**

CONCEPTUAL INSIGHT

Applying the graph transformation properties discussed in Section 1.2 to the transformed equation,

$$f(x) = -2x^2 + 16x - 24$$
$$= -2(x - 4)^2 + 8$$

we see that the graph of $f(x) = -2x^2 + 16x - 24$ is the graph of $h(x) = x^2$ vertically stretched by a factor of 2, reflected in the x axis, and shifted to the right 4 units and up 8 units, as shown in Figure 7.

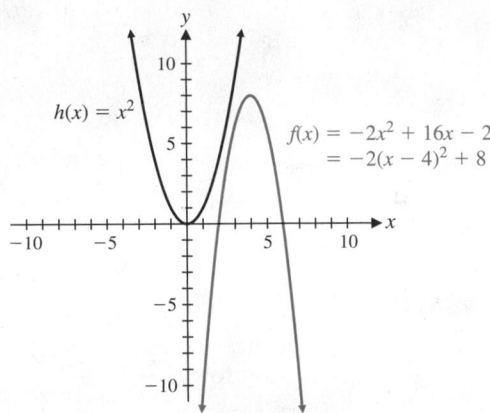

Figure 7 **Graph of *f* is the graph of *h* transformed**

Note the important results we have obtained from the vertex form of the quadratic function *f*:

- The vertex of the parabola
- The axis of the parabola
- The maximum value of $f(x)$
- The range of the function *f*
- The relationship between the graph of $h(x) = x^2$ and the graph of $f(x) = -2x^2 + 16x - 24$

The preceding discussion is generalized to all quadratic functions in the following summary:

SUMMARY Properties of a Quadratic Function and Its Graph
Given a quadratic function and the vertex form obtained by completing the square

$$f(x) = ax^2 + bx + c \qquad a \neq 0 \qquad \text{Standard form}$$
$$= a(x - h)^2 + k \qquad\qquad \text{Vertex form}$$

we summarize general properties as follows:

1. The graph of *f* is a parabola:

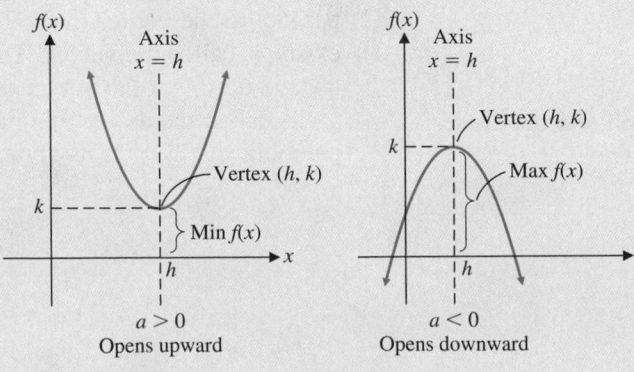

2. Vertex: (h, k) (parabola increases on one side of the vertex and decreases on the other)
3. Axis (of symmetry): $x = h$ (parallel to y axis)
4. $f(h) = k$ is the minimum if $a > 0$ and the maximum if $a < 0$
5. Domain: All real numbers. Range: $(-\infty, k]$ if $a < 0$ or $[k, \infty)$ if $a > 0$
6. The graph of f is the graph of $g(x) = ax^2$ translated horizontally h units and vertically k units.

EXAMPLE 4   Analyzing a Quadratic Function   Given the quadratic function

$$f(x) = 0.5x^2 - 6x + 21$$

(A) Find the vertex form for f.

(B) Find the vertex and the maximum or minimum. State the range of f.

(C) Describe how the graph of function f can be obtained from the graph of $h(x) = x^2$ using transformations.

(D) Sketch a graph of function f in a rectangular coordinate system.

(E) Graph function f using a suitable viewing window.

(F) Find the vertex and the maximum or minimum using the appropriate graphing calculator command.

SOLUTION

(A) Complete the square to find the vertex form:

$$\begin{aligned} f(x) &= 0.5x^2 - 6x + 21 \\ &= 0.5(x^2 - 12x + ?) + 21 \\ &= 0.5(x^2 - 12x + 36) + 21 - 18 \\ &= 0.5(x - 6)^2 + 3 \end{aligned}$$

(B) From the vertex form, we see that $h = 6$ and $k = 3$. Thus, vertex: $(6, 3)$; minimum: $f(6) = 3$; range: $y \geq 3$ or $[3, \infty)$.

(C) The graph of $f(x) = 0.5(x - 6)^2 + 3$ is the same as the graph of $h(x) = x^2$ vertically shrunk by a factor of 0.5, and shifted to the right 6 units and up 3 units.

(D) Graph in a rectangular coordinate system:

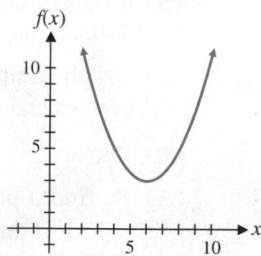

(E) Graph in a graphing calculator:

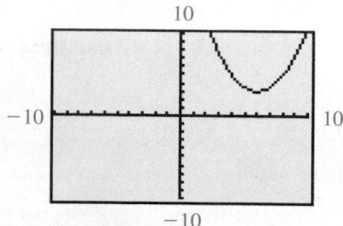

(F) Find the vertex and minimum using the *minimum* command:

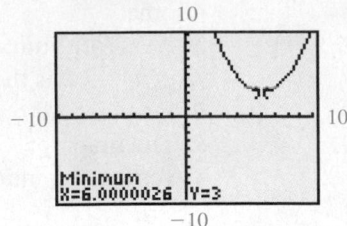

Vertex: $(6, 3)$; minimum: $f(6) = 3$

Matched Problem 4 Given the quadratic function $f(x) = -0.25x^2 - 2x + 2$

(A) Find the vertex form for f.

(B) Find the vertex and the maximum or minimum. State the range of f.

(C) Describe how the graph of function f can be obtained from the graph of $h(x) = x^2$ using transformations.

(D) Sketch a graph of function f in a rectangular coordinate system.

(E) Graph function f using a suitable viewing window.

(F) Find the vertex and the maximum or minimum using the appropriate graphing calculator command.

Applications

In a free competitive market, the price of a product is determined by the relationship between supply and demand. If there is a surplus—that is, the supply is greater than the demand—the price tends to come down. If there is a shortage—that is, the demand is greater than the supply—the price tends to go up. The price tends to move toward an equilibrium price at which the supply and demand are equal. Example 5 introduces the basic concepts.

EXAMPLE 5   Supply and Demand At a price of $9.00 per box of oranges, the supply is 320,000 boxes and the demand is 200,000 boxes. At a price of $8.50 per box, the supply is 270,000 boxes and the demand is 300,000 boxes.

(A) Find a price–supply equation of the form $p = mx + b$, where p is the price in dollars and x is the corresponding supply in thousands of boxes.

(B) Find a price–demand equation of the form $p = mx + b$, where p is the price in dollars and x is the corresponding demand in thousands of boxes.

(C) Graph the price–supply and price–demand equations in the same coordinate system and find their point of intersection.

SOLUTION

(A) To find a price–supply equation of the form $p = mx + b$, we must find two points of the form (x, p) that are on the supply line. From the given supply data, $(320, 9)$ and $(270, 8.5)$ are two such points. First, find the slope of the line:

$$m = \frac{9 - 8.5}{320 - 270} = \frac{0.5}{50} = 0.01$$

Now use the point-slope form to find the equation of the line:

$$p - p_1 = m(x - x_1) \qquad (x_1, p_1) = (320, 9)$$
$$p - 9 = 0.01(x - 320)$$
$$p - 9 = 0.01x - 3.2$$
$$p = 0.01x + 5.8 \qquad \text{Price–supply equation}$$

(B) From the given demand data, (200, 9) and (300, 8.5) are two points on the demand line.

$$m = \frac{8.5 - 9}{300 - 200} = \frac{-0.5}{100} = -0.005$$

$$p - p_1 = m(x - x_1)$$
$$p - 9 = -0.005(x - 200) \quad (x_1, p_1) = (200, 9)$$
$$p - 9 = -0.005x + 1$$
$$p = -0.005x + 10 \qquad \text{Price–demand equation}$$

(C) From part (A), we plot the points (320, 9) and (270, 8.5) and then draw line through them. We do the same with the points (200, 9) and (300, 8.5) from part (B) (Fig. 8). (Note that we restricted the axes to intervals that contain these data points.) To find the intersection point of the two lines, we equate the right-hand sides of the price–supply and price–demand equations and solve for x:

$$\begin{array}{cc} \textbf{Price–supply} & \textbf{Price–demand} \end{array}$$
$$0.01x + 5.8 = -0.005x + 10$$
$$0.015x = 4.2$$
$$x = 280$$

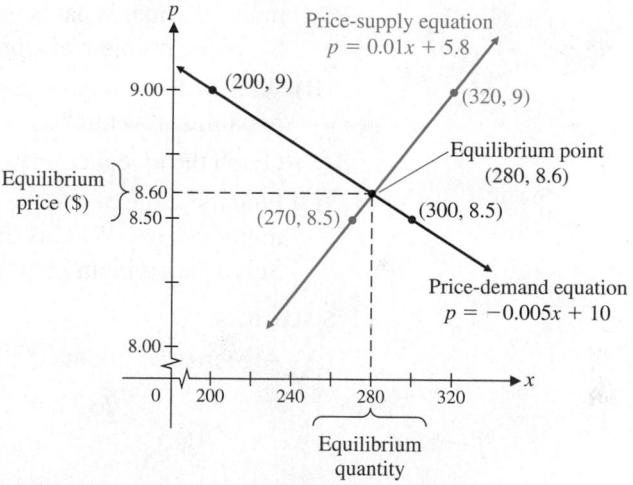

Figure 8 **Graphs of price–supply and price–demand equations**

Now use the price–supply equation to find p when $x = 280$:

$$p = 0.01x + 5.8$$
$$p = 0.01(280) + 5.8 = 8.6$$

As a check, we use the price–demand equation to find p when $x = 280$:

$$p = -0.005x + 10$$
$$p = -0.005(280) + 10 = 8.6$$

The lines intersect at (280, 8.6). The intersection point of the price–supply and price–demand equations is called the **equilibrium point**, and its coordinates are the **equilibrium quantity** (280) and the **equilibrium price** ($8.60). These terms are illustrated in Figure 8. The intersection point can also be found by using the **intersect** command on a graphing calculator (Fig. 9). To summarize, the price of a box of oranges tends toward the equilibrium price of $8.60, at which the supply and demand are both equal to 280,000 boxes.

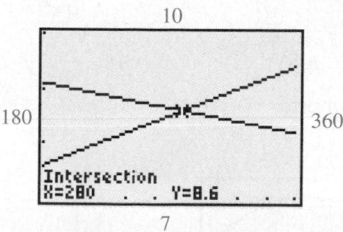

Figure 9 **Finding an intersection point**

Matched Problem 5) At a price of $12.59 per box of grapefruit, the supply is 595,000 boxes and the demand is 650,000 boxes. At a price of $13.19 per box, the supply is 695,000 boxes and the demand is 590,000 boxes. Assume that the relationship between price and supply is linear and that the relationship between price and demand is linear.

(A) Find a price–supply equation of the form $p = mx + b$.

(B) Find a price–demand equation of the form $p = mx + b$.

(C) Find the equilibrium point.

EXAMPLE 6 Maximum Revenue This is a continuation of Example 7 in Section 1.1. Recall that the financial department in the company that produces a digital camera arrived at the following price–demand function and the corresponding revenue function:

$$p(x) = 94.8 - 5x \qquad \text{Price–demand function}$$
$$R(x) = xp(x) = x(94.8 - 5x) \qquad \text{Revenue function}$$

where $p(x)$ is the wholesale price per camera at which x million cameras can be sold and $R(x)$ is the corresponding revenue (in millions of dollars). Both functions have domain $1 \le x \le 15$.

(A) Find the value of x to the nearest thousand cameras that will generate the maximum revenue. What is the maximum revenue to the nearest thousand dollars? Solve the problem algebraically by completing the square.

(B) What is the wholesale price per camera (to the nearest dollar) that generates the maximum revenue?

(C) Graph the revenue function using an appropriate viewing window.

(D) Find the value of x to the nearest thousand cameras that will generate the maximum revenue. What is the maximum revenue to the nearest thousand dollars? Solve the problem graphically using the **maximum** command.

SOLUTION

(A) Algebraic solution:

$$R(x) = x(94.8 - 5x)$$
$$= -5x^2 + 94.8x$$
$$= -5(x^2 - 18.96x + ?)$$
$$= -5(x^2 - 18.96x + 89.8704) + 449.352$$
$$= -5(x - 9.48)^2 + 449.352$$

The maximum revenue of 449.352 million dollars ($449,352,000) occurs when $x = 9.480$ million cameras (9,480,000 cameras).

(B) Finding the wholesale price per camera: Use the price–demand function for an output of 9.480 million cameras:

$$p(x) = 94.8 - 5x$$
$$p(9.480) = 94.8 - 5(9.480)$$
$$= \$47 \text{ per camera}$$

(C) Graph on a graphing calculator:

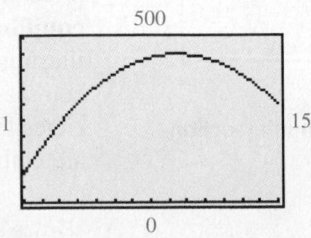

(D) Graphical solution using a graphing calculator:

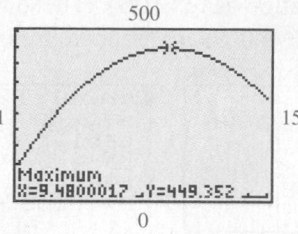

The manufacture and sale of 9.480 million cameras (9,480,000 cameras) will generate a maximum revenue of 449.352 million dollars ($449,352,000).

Matched Problem 6 The financial department in Example 6, using statistical and analytical techniques (see Matched Problem 7 in Section 1.1), arrived at the cost function

$$C(x) = 156 + 19.7x \quad \text{Cost function}$$

where $C(x)$ is the cost (in millions of dollars) for manufacturing and selling x million cameras.

(A) Using the revenue function from Example 6 and the preceding cost function, write an equation for the profit function.

(B) Find the value of x to the nearest thousand cameras that will generate the maximum profit. What is the maximum profit to the nearest thousand dollars? Solve the problem algebraically by completing the square.

(C) What is the wholesale price per camera (to the nearest dollar) that generates the maximum profit?

(D) Graph the profit function using an appropriate viewing window.

(E) Find the output to the nearest thousand cameras that will generate the maximum profit. What is the maximum profit to the nearest thousand dollars? Solve the problem graphically using the **maximum** command.

Linear and Quadratic Regression

Price–demand and price–supply equations (see Example 5), or cost functions (see Example 6), can be obtained from data using **regression analysis**. **Linear regression** produces the linear function (line) that is the **best fit*** for a data set; **quadratic regression** produces the quadratic function (parabola) that is the best fit for a data set; and so on. Examples 7 and 8 illustrate how a graphing calculator can be used to produce a **scatter plot**, that is, a graph of the points in a data set, and to find a linear or quadratic regression model.

EXAMPLE 7 Diamond Prices Table 4 gives diamond prices for round-shaped diamonds. Use linear regression to find the best linear model of the form $y = ax + b$ for the price y (in dollars) of a diamond as a function of its weight x (in carats).

Table 4 **Round-Shaped Diamond Prices**

Weight (carats)	Price
0.5	$2,790
0.6	$3,191
0.7	$3,694
0.8	$4,154
0.9	$5,018
1.0	$5,898

Source: www.tradeshop.com

---

*The definition of "best fit" is given in Section 7.5, where calculus is used to justify regression methods.

SOLUTION Enter the data in a graphing calculator (Fig. 10A) and find the linear regression equation (Fig. 10B). The data set and the model, that is, the line $y = 6{,}137.4x - 478.9$, are graphed in Figure 10C.

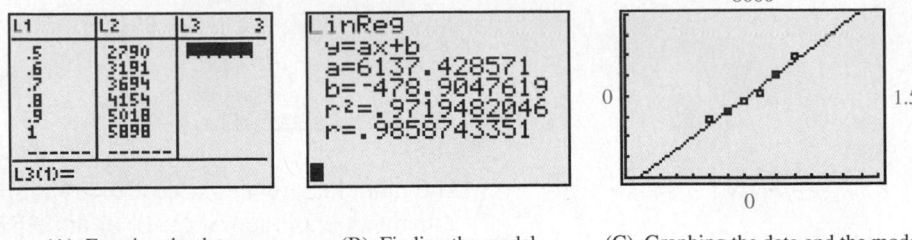

(A) Entering the data   (B) Finding the model   (C) Graphing the data and the model

Figure 10   **Linear regression on a graphing calculator**

Table 5 Emerald-Shaped Diamond Prices

Weight (carats)	Price
0.5	$1,677
0.6	$2,353
0.7	$2,718
0.8	$3,218
0.9	$3,982
1.0	$4,510

Source: www.tradeshop.com

Matched Problem 7 Prices for emerald-shaped diamonds are given in Table 5. Repeat Example 7 for this data set.

A visual inspection of the plot of a data set might indicate that a parabola would be a better model of the data than a straight line. In that case, we would use quadratic, rather than linear, regression.

EXAMPLE 8   Outboard Motors Table 6 gives performance data for a boat powered by an Evinrude outboard motor. Use quadratic regression to find the best model of the form $y = ax^2 + bx + c$ for fuel consumption y (in miles per gallon) as a function of speed x (in miles per hour). Estimate the fuel consumption (to one decimal place) at a speed of 12 miles per hour.

Table 6

rpm	mph	mpg
2,500	10.3	4.1
3,000	18.3	5.6
3,500	24.6	6.6
4,000	29.1	6.4
4,500	33.0	6.1
5,000	36.0	5.4
5,400	38.9	4.9

SOLUTION Enter the data in a graphing calculator (Fig. 11A) and find the quadratic regression equation (Fig. 11B). The data set and the regression equation are graphed in Figure 11C. Using TRACE, we see that the estimated fuel consumption at a speed of 12 mph is 4.5 mpg.

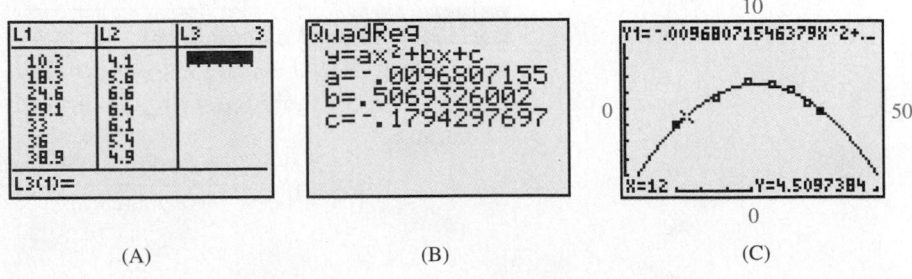

(A)   (B)   (C)

Figure 11

Matched Problem 8 Refer to Table 6. Use quadratic regression to find the best model of the form $y = ax^2 + bx + c$ for boat speed y (in miles per hour) as a function of engine speed x (in revolutions per minute). Estimate the boat speed (in miles per hour, to one decimal place) at an engine speed of 3,400 rpm.

Exercises 1.3

In Problems 1–4, sketch a graph of each equation in a rectangular coordinate system.

1. $y = 2x - 3$ **2.** $y = \dfrac{x}{2} + 1$

3. $2x + 3y = 12$ **4.** $8x - 3y = 24$

In Problems 5–8, find the slope and y intercept of the graph of each equation.

5. $y = 5x - 7$ **6.** $y = 3x + 2$

7. $y = -\dfrac{5}{2}x - 9$ **8.** $y = -\dfrac{10}{3}x + 4$

In Problems 9–12, write an equation of the line with the indicated slope and y intercept.

9. Slope $= 2$
y intercept $= 1$

10. Slope $= 1$
y intercept $= 5$

11. Slope $= -\dfrac{1}{3}$
y intercept $= 6$

12. Slope $= \dfrac{6}{7}$
y intercept $= -\dfrac{9}{2}$

In Problems 13–16, use the graph of each line to find the x intercept, y intercept, and slope. Write the slope-intercept form of the equation of line.

13.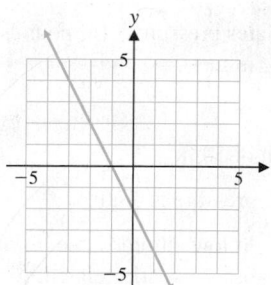

14.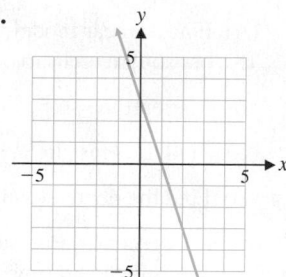

15.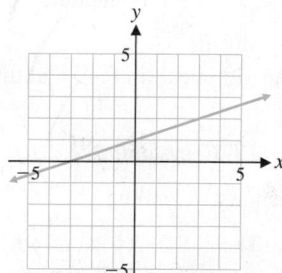

16.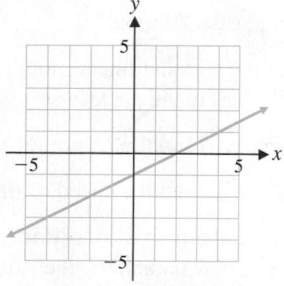

17. Match each equation with a graph of one of the functions f, g, m, or n in the figure.

(A) $y = -(x + 2)^2 + 1$ (B) $y = (x - 2)^2 - 1$

(C) $y = (x + 2)^2 - 1$ (D) $y = -(x - 2)^2 + 1$

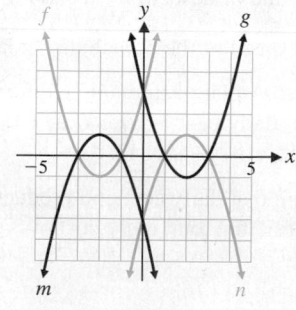

18. Match each equation with a graph of one of the functions f, g, m, or n in the figure.

(A) $y = (x - 3)^2 - 4$ (B) $y = -(x + 3)^2 + 4$

(C) $y = -(x - 3)^2 + 4$ (D) $y = (x + 3)^2 - 4$

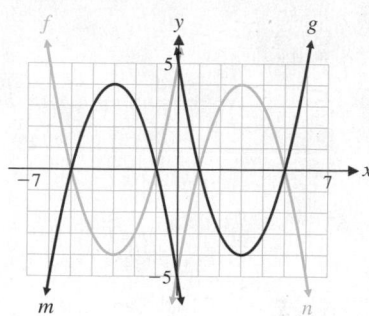

For the functions indicated in Problems 19–22, find each of the following to the nearest integer by referring to the graphs for Problems 17 and 18.

(A) Intercepts   (B) Vertex

(C) Maximum or minimum   (D) Range

19. Function n in the figure for Problem 17

20. Function m in the figure for Problem 18

21. Function f in the figure for Problem 17

22. Function g in the figure for Problem 18

In Problems 23–26, find each of the following:

(A) Intercepts   (B) Vertex

(C) Maximum or minimum   (D) Range

23. $f(x) = -(x - 3)^2 + 2$

24. $g(x) = -(x + 2)^2 + 3$

25. $m(x) = (x + 1)^2 - 2$

26. $n(x) = (x - 4)^2 - 3$

In Problems 27–34,

(A) Find the slope of the line that passes through the given points.

(B) Find the point-slope form of the equation of the line.

(C) Find the slope-intercept form of the equation of the line.

(D) Find the standard form of the equation of the line.

27. $(2, 5)$ and $(5, 7)$ **28.** $(1, 2)$ and $(3, 5)$

29. $(-2, -1)$ and $(2, -6)$ **30.** $(2, 3)$ and $(-3, 7)$

31. $(5, 3)$ and $(5, -3)$ **32.** $(1, 4)$ and $(0, 4)$

33. $(-2, 5)$ and $(3, 5)$ **34.** $(2, 0)$ and $(2, -3)$

In Problems 35–40, find the vertex form for each quadratic function. Then find each of the following:

(A) Intercepts   (B) Vertex

(C) Maximum or minimum   (D) Range

35. $f(x) = x^2 - 8x + 12$

36. $g(x) = x^2 - 6x + 5$

37. $r(x) = -4x^2 + 16x - 15$

38. $s(x) = -4x^2 - 8x - 3$

39. $u(x) = 0.5x^2 - 2x + 5$

40. $v(x) = 0.5x^2 + 4x + 10$

In Problems 41–48, use interval notation to write the solution set of the inequality.

41. $2x + 7 < 0$

42. $6x - 13 > 0$

43. $-5x + 3 \geq 6$

44. $-4x - 21 \leq 1$

45. $x^2 - 10x > 0$

46. $x^2 + 16x \leq 0$

47. $x^2 + 20x + 50 \leq 0$

48. $x^2 - 4x + 5 > 0$

49. Graph $y = 25x + 200, x \geq 0$.

50. Graph $y = 40x + 160, x \geq 0$.

51. (A) Graph $y = 1.2x - 4.2$ in a rectangular coordinate system.

  (B) Find the x and y intercepts algebraically to one decimal place.

  (C) Graph $y = 1.2x - 4.2$ in a graphing calculator.

  (D) Find the x and y intercepts to one decimal place using TRACE and the zero command.

52. (A) Graph $y = -0.8x + 5.2$ in a rectangular coordinate system.

  (B) Find the x and y intercepts algebraically to one decimal place.

  (C) Graph $y = -0.8x + 5.2$ in a graphing calculator.

  (D) Find the x and y intercepts to one decimal place using TRACE and the zero command.

  (E) Using the results of parts (A) and (B), or (C) and (D), find the solution set for the linear inequality

$$-0.8x + 5.2 < 0$$

53. Let $f(x) = 0.3x^2 - x - 8$. Solve each equation graphically to two decimal places.

  (A) $f(x) = 4$ (B) $f(x) = -1$ (C) $f(x) = -9$

54. Let $g(x) = -0.6x^2 + 3x + 4$. Solve each equation graphically to two decimal places.

  (A) $g(x) = -2$ (B) $g(x) = 5$ (C) $g(x) = 8$

55. Let $f(x) = 125x - 6x^2$. Find the maximum value of f to four decimal places graphically.

56. Let $f(x) = 100x - 7x^2 - 10$. Find the maximum value of f to four decimal places graphically.

57. If $B \neq 0$, $Ax_1 + By_1 = C$, and $Ax_2 + By_2 = C$, show that the slope of the line through (x_1, y_1) and (x_2, y_2) is equal to $-\dfrac{A}{B}$.

58. If $B \neq 0$ and (x_1, y_1), (x_2, y_2), and (x_3, y_3) are all solutions of $Ax + By = C$, show that the slope of the line through (x_1, y_1) and (x_2, y_2) is equal to the slope of the line through (x_1, y_1) and (x_3, y_3).

Applications

59. Underwater pressure. At sea level, the weight of the atmosphere exerts a pressure of 14.7 pounds per square inch, commonly referred to as 1 **atmosphere of pressure**. As an object descends in water, pressure P and depth d are linearly related. In salt water, the pressure at a depth of 33 ft is 2 atms, or 29.4 pounds per square inch.

  (A) Find a linear model that relates pressure P (in pounds per square inch) to depth d (in feet).

  (B) Interpret the slope of the model.

  (C) Find the pressure at a depth of 50 ft.

  (D) Find the depth at which the pressure is 4 atms.

60. Underwater pressure. Refer to Problem 59. In fresh water, the pressure at a depth of 34 ft is 2 atms, or 29.4 pounds per square inch.

  (A) Find a linear model that relates pressure P (in pounds per square inch) to depth d (in feet).

  (B) Interpret the slope of the model.

  (C) Find the pressure at a depth of 50 ft.

  (D) Find the depth at which the pressure is 4 atms.

61. Rate of descent—Parachutes. At low altitudes, the altitude of a parachutist and time in the air are linearly related. A jump at 2,880 ft using the U.S. Army's T-10 parachute system lasts 120 secs.

  (A) Find a linear model relating altitude a (in feet) and time in the air t (in seconds).

  (B) Find the rate of descent for a T-10 system.

  (C) Find the speed of the parachutist at landing.

62. Rate of descent—Parachutes. The U.S Army is considering a new parachute, the Advanced Tactical Parachute System (ATPS). A jump at 2,880 ft using the ATPS system lasts 180 secs.

  (A) Find a linear model relating altitude a (in feet) and time in the air t (in seconds).

  (B) Find the rate of descent for an ATPS system parachute.

  (C) Find the speed of the parachutist at landing.

63. Cost analysis. A plant can manufacture 80 golf clubs per day for a total daily cost of $7,647 and 100 golf clubs per day for a total daily cost of $9,147.

  (A) Assuming that daily cost and production are linearly related, find the total daily cost of producing x golf clubs.

(B) Graph the total daily cost for $0 \leq x \leq 200$.

(C) Interpret the slope and y intercept of this cost equation.

64. **Cost analysis.** A plant can manufacture 50 tennis rackets per day for a total daily cost of $3,855 and 60 tennis rackets per day for a total daily cost of $4,245.

(A) Assuming that daily cost and production are linearly related, find the total daily cost of producing x tennis rackets.

(B) Graph the total daily cost for $0 \leq x \leq 100$.

(C) Interpret the slope and y intercept of this cost equation.

65. **Business—Depreciation.** A farmer buys a new tractor for $157,000 and assumes that it will have a trade-in value of $82,000 after 10 years. The farmer uses a constant rate of depreciation (commonly called **straight-line depreciation**—one of several methods permitted by the IRS) to determine the annual value of the tractor.

(A) Find a linear model for the depreciated value V of the tractor t years after it was purchased.

(B) What is the depreciated value of the tractor after 6 years?

(C) When will the depreciated value fall below $70,000?

(D) Graph V for $0 \leq t \leq 20$ and illustrate the answers from parts (B) and (C) on the graph.

66. **Business—Depreciation.** A charter fishing company buys a new boat for $224,000 and assumes that it will have a trade-in value of $115,200 after 16 years.

(A) Find a linear model for the depreciated value V of the boat t years after it was purchased.

(B) What is the depreciated value of the boat after 10 years?

(C) When will the depreciated value fall below $100,000?

(D) Graph V for $0 \leq t \leq 30$ and illustrate the answers from (B) and (C) on the graph.

67. **Flight conditions.** In stable air, the air temperature drops about 3.6°F for each 1,000-foot rise in altitude. (*Source:* Federal Aviation Administration)

(A) If the temperature at sea level is 70°F, write a linear equation that expresses temperature T in terms of altitude A in thousands of feet.

(B) At what altitude is the temperature 34°F?

68. **Flight navigation.** The airspeed indicator on some aircraft is affected by the changes in atmospheric pressure at different altitudes. A pilot can estimate the true airspeed by observing the indicated airspeed and adding to it about 1.6% for every 1,000 feet of altitude. (*Source:* Megginson Technologies Ltd.)

(A) A pilot maintains a constant reading of 200 miles per hour on the airspeed indicator as the aircraft climbs from sea level to an altitude of 10,000 feet. Write a linear equation that expresses true airspeed T (in miles per hour) in terms of altitude A (in thousands of feet).

(B) What would be the true airspeed of the aircraft at 6,500 feet?

69. **Supply and demand.** At a price of $2.28 per bushel, the supply of barley is 7,500 million bushels and the demand is 7,900 million bushels. At a price of $2.37 per bushel, the supply is 7,900 million bushels and the demand is 7,800 million bushels.

(A) Find a price–supply equation of the form $p = mx + b$.

(B) Find a price–demand equation of the form $p = mx + b$.

(C) Find the equilibrium point.

(D) Graph the price–supply equation, price–demand equation, and equilibrium point in the same coordinate system.

70. **Supply and demand.** At a price of $1.94 per bushel, the supply of corn is 9,800 million bushels and the demand is 9,300 million bushels. At a price of $1.82 per bushel, the supply is 9,400 million bushels and the demand is 9,500 million bushels.

(A) Find a price–supply equation of the form $p = mx + b$.

(B) Find a price–demand equation of the form $p = mx + b$.

(C) Find the equilibrium point.

(D) Graph the price–supply equation, price–demand equation, and equilibrium point in the same coordinate system.

71. **Licensed drivers.** The table contains the state population and the number of licensed drivers in the state (both in millions) for the states with population under 1 million in 2010. The regression model for this data is

$$y = 0.75x$$

where x is the state population and y is the number of licensed drivers in the state.

Licensed Drivers in 2010

State	Population	Licensed Drivers
Alaska	0.71	0.52
Delaware	0.90	0.70
Montana	0.99	0.74
North Dakota	0.67	0.48
South Dakota	0.81	0.60
Vermont	0.63	0.51
Wyoming	0.56	0.42

Source: Bureau of Transportation Statistics

(A) Draw a scatter plot of the data and a graph of the model on the same axes.

(B) If the population of Idaho in 2010 was about 1.6 million, use the model to estimate the number of licensed drivers in Idaho in 2010 to the nearest thousand.

(C) If the number of licensed drivers in Rhode Island in 2006 was about 0.75 million, use the model to estimate the population of Rhode Island in 2010 to the nearest thousand.

72. **Licensed drivers.** The table contains the state population and the number of licensed drivers in the state (both in millions) for the states with population over 10 million in 2010. The regression model for this data is

$$y = 0.63x + 0.31$$

where x is the state population and y is the number of licensed drivers in the state.

Licensed Drivers in 2010

State	Population	Licensed Drivers
California	37	24
Florida	19	14
Illinois	13	8
New York	19	11
Ohio	12	8
Pennsylvania	13	9
Texas	25	15

Source: Bureau of Transportation Statistics

(A) Draw a scatter plot of the data and a graph of the model on the same axes.

(B) If the population of Minnesota in 2010 was about 5.3 million, use the model to estimate the number of licensed drivers in Minnesota in 2010 to the nearest thousand.

(C) If the number of licensed drivers in Wisconsin in 2010 was about 4.1 million, use the model to estimate the population of Wisconsin in 2010 to the nearest thousand.

73. **Tire mileage.** An automobile tire manufacturer collected the data in the table relating tire pressure x (in pounds per square inch) and mileage (in thousands of miles):

x	Mileage
28	45
30	52
32	55
34	51
36	47

A mathematical model for the data is given by

$$f(x) = -0.518x^2 + 33.3x - 481$$

(A) Complete the following table. Round values of $f(x)$ to one decimal place.

x	Mileage	$f(x)$
28	45	
30	52	
32	55	
34	51	
36	47	

(B) Sketch the graph of f and the mileage data in the same coordinate system.

(C) Use values of the modeling function rounded to two decimal places to estimate the mileage for a tire pressure of 31 lb/sq in. and for 35 lb/sq in.

✎ (D) Write a brief description of the relationship between tire pressure and mileage.

74. **Automobile production.** The table shows the retail market share of passenger cars from Ford Motor Company as a percentage of the U.S. market.

Year	Market Share
1980	17.2%
1985	18.8%
1990	20.0%
1995	20.7%
2000	20.2%
2005	17.4%
2010	16.4%

A mathematical model for this data is given by

$$f(x) = -0.0169x^2 + 0.47x + 17.1$$

where $x = 0$ corresponds to 1980.

(A) Complete the following table. Round values of $f(x)$ to one decimal place.

x	Market Share	$f(x)$
0	17.2	
5	18.8	
10	20.0	
15	20.7	
20	20.2	
25	17.4	
30	16.4	

(B) Sketch the graph of f and the market share data in the same coordinate system.

(C) Use values of the modeling function f to estimate Ford's market share in 2020 and in 2025.

✎ (D) Write a brief description of Ford's market share from 1980 to 2010.

75. **Tire mileage.** Using quadratic regression on a graphing calculator, show that the quadratic function that best fits the data on tire mileage in Problem 73 is

$$f(x) = -0.518x^2 + 33.3x - 481$$

 **76. Automobile production.** Using quadratic regression on a graphing calculator, show that the quadratic function that best fits the data on market share in Problem 74 is

$$f(x) = -0.0169x^2 + 0.47x + 17.1$$

77. Revenue. The marketing research department for a company that manufactures and sells memory chips for microcomputers established the following price–demand and revenue functions:

$$p(x) = 75 - 3x \qquad \text{Price–demand function}$$
$$R(x) = xp(x) = x(75 - 3x) \quad \text{Revenue function}$$

where $p(x)$ is the wholesale price in dollars at which x million chips can be sold, and $R(x)$ is in millions of dollars. Both functions have domain $1 \le x \le 20$.

(A) Sketch a graph of the revenue function in a rectangular coordinate system.

(B) Find the value of x that will produce the maximum revenue. What is the maximum revenue?

(C) What is the wholesale price per chip that produces the maximum revenue?

78. Revenue. The marketing research department for a company that manufactures and sells notebook computers established the following price–demand and revenue functions:

$$p(x) = 2{,}000 - 60x \qquad \text{Price–demand function}$$
$$R(x) = xp(x) \qquad\qquad \text{Revenue function}$$
$$ = x(2{,}000 - 60x)$$

where $p(x)$ is the wholesale price in dollars at which x thousand computers can be sold, and $R(x)$ is in thousands of dollars. Both functions have domain $1 \le x \le 25$.

(A) Sketch a graph of the revenue function in a rectangular coordinate system.

(B) Find the value of x that will produce the maximum revenue. What is the maximum revenue to the nearest thousand dollars?

(C) What is the wholesale price per computer (to the nearest dollar) that produces the maximum revenue?

79. Forestry. The figure contains a scatter plot of 100 data points for black spruce trees and the linear regression model for this data.

(A) Interpret the slope of the model.

(B) What is the effect of a 1-in. increase in the diameter at breast height (Dbh)?

(C) Estimate the height of a black spruce with a Dbh of 15 in. Round your answer to the nearest foot.

(D) Estimate the Dbh of a black spruce that is 25 ft tall. Round your answer to the nearest inch.

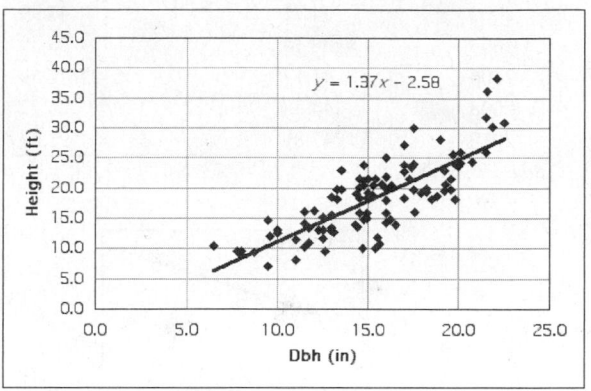

black spruce
Source: Lakehead University

80. Forestry. The figure contains a scatter plot of 100 data points for black walnut trees and the linear regression model for this data.

(A) Interpret the slope of the model.

(B) What is the effect of a 1-in. increase in Dbh?

(C) Estimate the height of a black walnut with a Dbh of 12 in. Round your answer to the nearest foot.

(D) Estimate the Dbh of a black walnut that is 25 ft tall. Round your answer to the nearest inch.

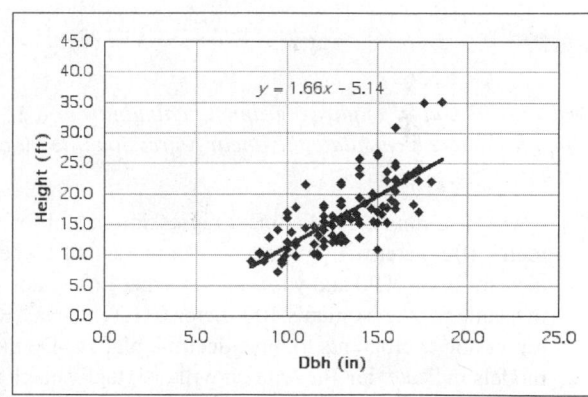

black walnut
Source: Kagen Research

81. Cable television. The table shows the increase in both price and revenue for cable television in the United States. The figure shows a scatter plot and a linear regression model for the average monthly price data in the table.

(A) Interpret the slope of the model.

(B) Use the model to predict the average monthly price (to the nearest dollar) in 2024.

Table for 81 and 82 Cable Television Price and Revenue

Year	Average Monthly Price (dollars)	Annual Total Revenue (billions of dollars)
2000	30.37	36.43
2002	34.71	47.99
2004	38.14	58.59
2006	41.17	71.89
2008	44.28	85.23
2010	47.89	93.37

Source: SNL Kagan

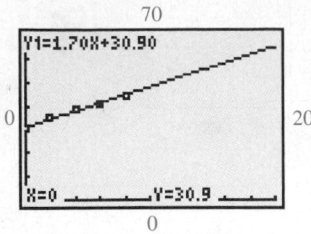

82. Cable television. The figure shows a scatter plot and a linear regression model for the annual revenue data in the table.

(A) Interpret the slope of the model.

(B) Use the model to predict the annual revenue (to the nearest billion dollars) in 2024.

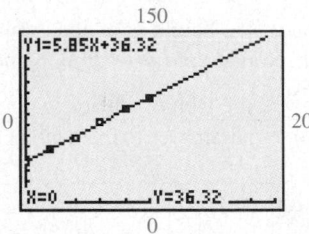

Problems 83 and 84 require a graphing calculator or a computer that can calculate the linear regression line for a given data set.

83. Olympic Games. Find a linear regression model for the men's 100-meter freestyle data given in the table, where x is years since 1990 and y is winning time (in seconds). Do the same for the women's 100-meter freestyle data. (Round regression coefficients to three decimal places.) Do these models indicate that the women will eventually catch up with the men?

Table for 83 and 84 Winning Times in Olympic Swimming Events

	100-Meter Freestyle		200-Meter Backstroke	
	Men	Women	Men	Women
1992	49.02	54.65	1:58.47	2:07.06
1996	48.74	54.50	1:58.54	2:07.83
2000	48.30	53.83	1:56.76	2:08.16
2004	48.17	53.84	1:54.76	2:09.16
2008	47.21	53.12	1:53.94	2:05.24
2012	47.52	53.00	1:53.41	2:04.06

Source: www.infoplease.com

84. Olympic Games. Find a linear regression model for the men's 200-meter backstroke data given in the table, where x is years since 1990 and y is winning time (in seconds). Do the same for the women's 200-meter backstroke data. (Round regression coefficients to three decimal places.) Do these models indicate that the women will eventually catch up with the men?

85. Outboard motors. The table gives performance data for a boat powered by an Evinrude outboard motor. Find a quadratic regression model ($y = ax^2 + bx + c$) for boat speed y (in miles per hour) as a function of engine speed (in revolutions per minute). Estimate the boat speed at an engine speed of 3,100 revolutions per minute.

Table for 85 and 86

rpm	mph	mpg
1,500	4.5	8.2
2,000	5.7	6.9
2,500	7.8	4.8
3,000	9.6	4.1
3,500	13.4	3.7

86. Outboard motors. The table gives performance data for a boat powered by an Evinrude outboard motor. Find a quadratic regression model ($y = ax^2 + bx + c$) for fuel consumption y (in miles per gallon) as a function of engine speed (in revolutions per minute). Estimate the fuel consumption at an engine speed of 2,300 revolutions per minute.

Answers to Matched Problems

1. $y - 10 = -3(x + 2); y = -3x + 4; 3x + y = 4$

2. (A)

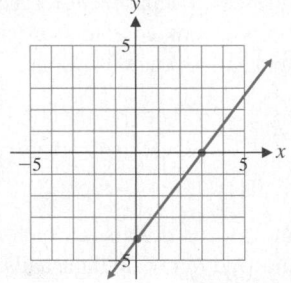

(B) y intercept $= -4$, x intercept $= 3$

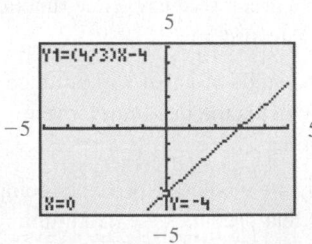

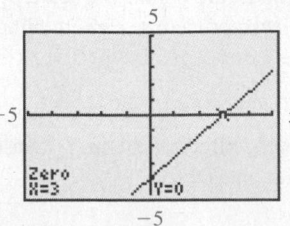

(C) $[3, \infty)$

3. (A)

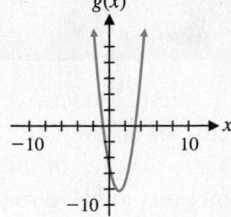

(B) x intercepts: $-0.7656, 3.2656$; y intercept: -5

(C)

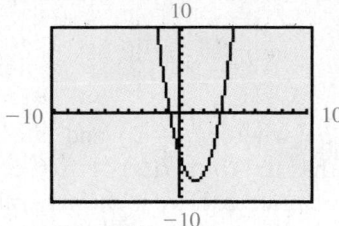

(D) x intercepts: $-0.7656, 3.2656$; y intercept: -5

(E) $x \leq -0.7656$ or $x \geq 3.2656$; or $(-\infty, -0.7656]$ or $[3.2656, \infty)$

(F) $x = -0.3508, 2.8508$

4. (A) $f(x) = -0.25(x + 4)^2 + 6$.

(B) Vertex: $(-4, 6)$; maximum: $f(-4) = 6$; range: $y \leq 6$ or $(-\infty, 6]$

(C) The graph of $f(x) = -0.25(x + 4)^2 + 6$ is the same as the graph of $h(x) = x^2$ vertically shrunk by a factor of 0.25, reflected in the x axis, and shifted 4 units to the left and 6 units up.

(D)

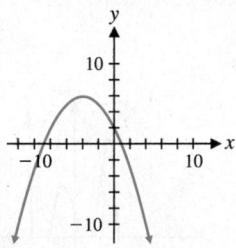

(E)

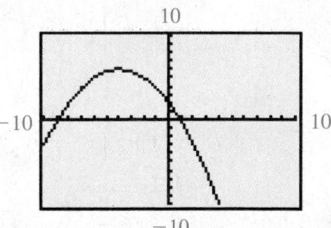

(F) Vertex: $(-4, 6)$; maximum: $f(-4) = 6$

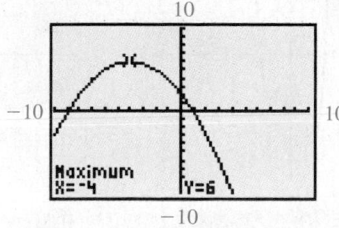

5. (A) $p = 0.006x + 9.02$

(B) $p = -0.01x + 19.09$

(C) $(629, 12.80)$

6. (A) $P(x) = R(x) - C(x) = -5x^2 + 75.1x - 156$

(B) $P(x) = R(x) - C(x) = -5(x - 7.51)^2 + 126.0005$; the manufacture and sale of 7,510,000 million cameras will produce a maximum profit of \$126,001,000.

(C) $p(7.510) = \$57$

(D)

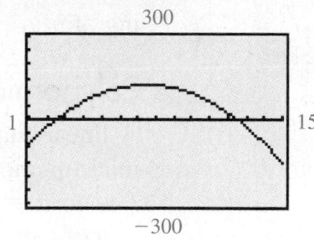

(E)

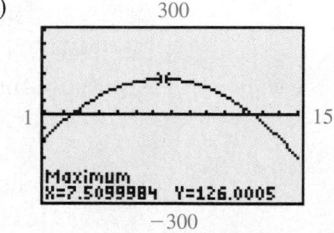

The manufacture and sale of 7,510,000 million cameras will produce a maximum profit of \$126,001,000. (Notice that maximum profit does not occur at the same value of x where maximum revenue occurs.)

7. $y = 5,586x - 1,113$

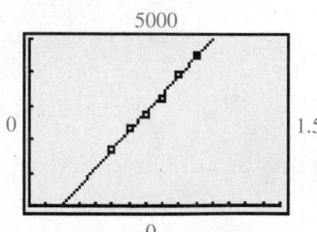

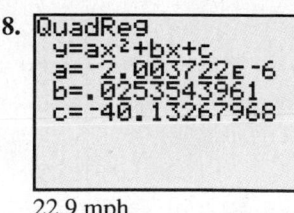

8.

22.9 mph

1.4 Polynomial and Rational Functions

- Polynomial Functions
- Regression Polynomials
- Rational Functions
- Applications

Linear and quadratic functions are special cases of the more general class of *polynomial functions*. Polynomial functions are a special case of an even larger class of functions, the *rational functions*. We will describe the basic features of the graphs of polynomial and rational functions. We will use these functions to solve real-world problems where linear or quadratic models are inadequate; for example, to determine the relationship between length and weight of a species of fish, or to model the training of new employees.

Polynomial Functions

A linear function has the form $f(x) = mx + b$ (where $m \neq 0$) and is a polynomial function of degree 1. A quadratic function has the form $f(x) = ax^2 + bx + c$ (where $a \neq 0$) and is a polynomial function of degree 2. Here is the general definition of a polynomial function.

> **DEFINITION** Polynomial Function
> A **polynomial function** is a function that can be written in the form
> $$f(x) = a_n x^n + a_{n-1} x^{n-1} + \cdots + a_1 x + a_0$$
> for n a nonnegative integer, called the **degree** of the polynomial. The coefficients $a_0, a_1, \ldots, a_n$ are real numbers with $a_n \neq 0$. The **domain** of a polynomial function is the set of all real numbers.

Figure 1 shows graphs of representative polynomial functions of degrees 1 through 6. The figure, which also appears on the inside back cover, suggests some general properties of graphs of polynomial functions.

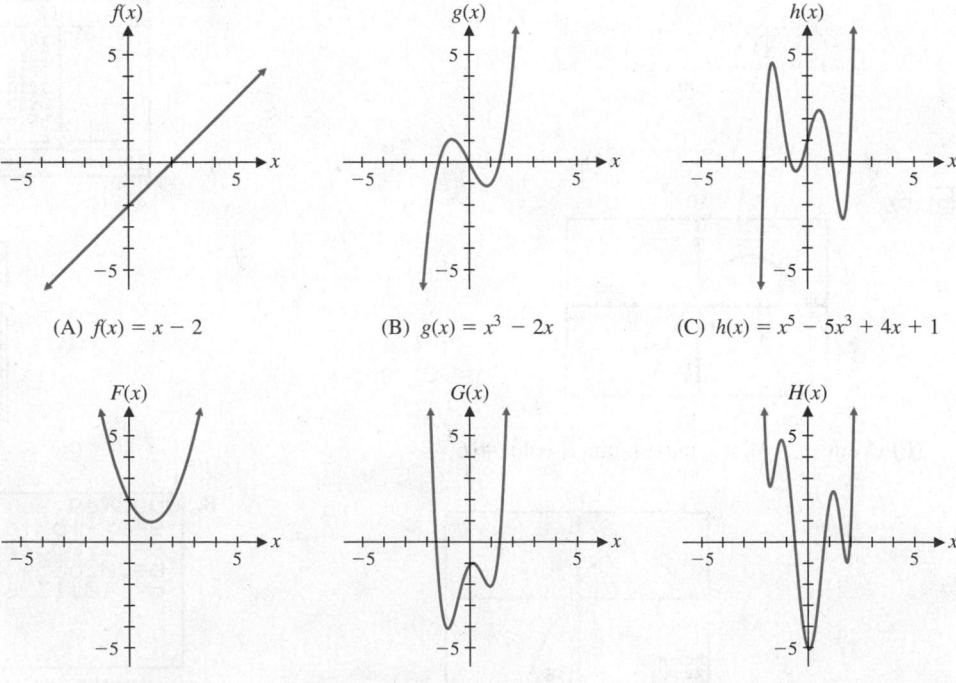

(A) $f(x) = x - 2$ (B) $g(x) = x^3 - 2x$ (C) $h(x) = x^5 - 5x^3 + 4x + 1$

(D) $F(x) = x^2 - 2x + 2$ (E) $G(x) = 2x^4 - 4x^2 + x - 1$ (F) $H(x) = x^6 - 7x^4 + 14x^2 - x - 5$

Figure 1 **Graphs of polynomial functions**

Notice that the odd-degree polynomial graphs start negative, end positive, and cross the x axis at least once. The even-degree polynomial graphs start positive, end positive, and may not cross the x axis at all. In all cases in Figure 1, the **leading coefficient**—that is, the coefficient of the highest-degree term—was chosen positive. If any leading coefficient had been chosen negative, then we would have a similar graph but reflected in the x axis.

A polynomial of degree n can have, at most, n linear factors. Therefore, the graph of a polynomial function of positive degree n can intersect the x axis at most n times. Note from Figure 1 that a polynomial of degree n may intersect the x axis fewer than n times. An x intercept of a function is also called a **zero*** or **root** of the function.

The graph of a polynomial function is **continuous**, with no holes or breaks. That is, the graph can be drawn without removing a pen from the paper. Also, the graph of a polynomial has no sharp corners. Figure 2 shows the graphs of two functions— one that is not continuous, and the other that is continuous but with a sharp corner. Neither function is a polynomial.

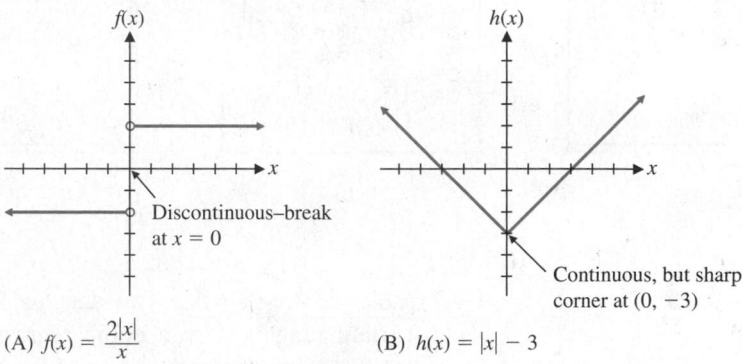

$$(A) \ f(x) = \frac{2|x|}{x} \qquad (B) \ h(x) = |x| - 3$$

Figure 2 **Discontinuous and sharp-corner functions**

 ## Regression Polynomials

In Section 1.3, we saw that regression techniques can be used to fit a straight line to a set of data. Linear functions are not the only ones that can be applied in this manner. Most graphing calculators have the ability to fit a variety of curves to a given set of data. We will discuss polynomial regression models in this section and other types of regression models in later sections.

EXAMPLE 1 Estimating the Weight of a Fish Using the length of a fish to estimate its weight is of interest to both scientists and sport anglers. The data in Table 1 give the average weights of lake trout for certain lengths. Use the data and regression techniques to find a polynomial model that can be used to estimate the weight of a lake trout for any length. Estimate (to the nearest ounce) the weights of lake trout of lengths 39, 40, 41, 42, and 43 inches, respectively.

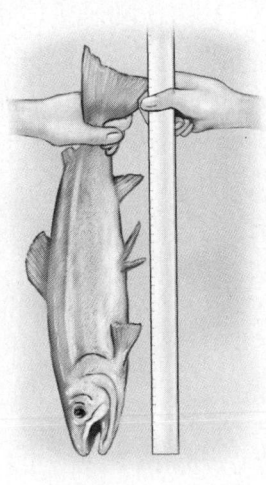

Table 1 **Lake Trout**

Length (in.)	Weight (oz)	Length (in.)	Weight (oz)
x	y	x	y
10	5	30	152
14	12	34	226
18	26	38	326
22	56	44	536
26	96		

*Only real numbers can be x intercepts. Functions may have complex zeros that are not real numbers, but such zeros, which are not x intercepts, will not be discussed in this book.

SOLUTION   The graph of the data in Table 1 (Fig. 3A) indicates that a linear regression model would not be appropriate in this case. And, in fact, we would not expect a linear relationship between length and weight. Instead, it is more likely that the weight would be related to the cube of the length. We use a cubic regression polynomial to model the data (Fig. 3B). (Consult your manual for the details of calculating regression polynomials on your graphing utility.) Figure 3C adds the graph of the polynomial model to the graph of the data. The graph in Figure 3C shows that this cubic polynomial does provide a good fit for the data. (We will have more to say about the choice of functions and the accuracy of the fit provided by regression analysis later in the book.) Figure 3D shows the estimated weights for the lengths requested.

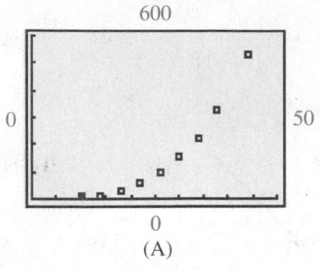

(A)

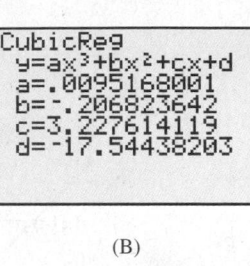

(B)

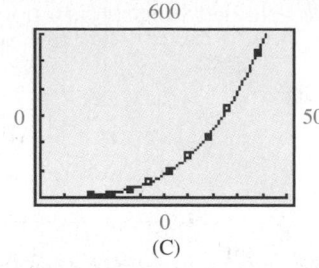

(C)

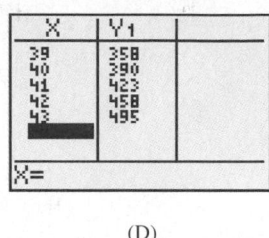

(D)

Figure 3

Matched Problem 1   The data in Table 2 give the average weights of pike for certain lengths. Use a cubic regression polynomial to model the data. Estimate (to the nearest ounce) the weights of pike of lengths 39, 40, 41, 42, and 43 inches, respectively.

Table 2 **Pike**

Length (in.)	Weight (oz)	Length (in.)	Weight (oz)
x	y	x	y
10	5	30	108
14	12	34	154
18	26	38	210
22	44	44	326
26	72	52	522

Rational Functions

Just as rational numbers are defined in terms of quotients of integers, *rational functions* are defined in terms of quotients of polynomials. The following equations specify rational functions:

$$f(x) = \frac{1}{x} \quad g(x) = \frac{x-2}{x^2-x-6} \quad h(x) = \frac{x^3-8}{x}$$

$$p(x) = 3x^2 - 5x \quad q(x) = 7 \quad r(x) = 0$$

DEFINITION   Rational Function

A **rational function** is any function that can be written in the form

$$f(x) = \frac{n(x)}{d(x)} \qquad d(x) \neq 0$$

where $n(x)$ and $d(x)$ are polynomials. The **domain** is the set of all real numbers such that $d(x) \neq 0$.

Figure 4 shows the graphs of representative rational functions. Note, for example, that in Figure 4A the line $x = 2$ is a *vertical asymptote* for the function. The graph of f gets closer to this line as x gets closer to 2. The line $y = 1$ in Figure 4A is a *horizontal asymptote* for the function. The graph of f gets closer to this line as x increases or decreases without bound.

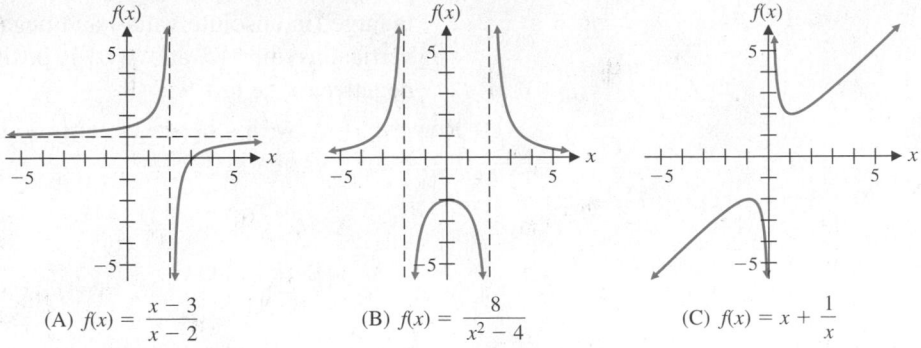

(A) $f(x) = \dfrac{x - 3}{x - 2}$ (B) $f(x) = \dfrac{8}{x^2 - 4}$ (C) $f(x) = x + \dfrac{1}{x}$

Figure 4 **Graphs of rational functions**

The number of vertical asymptotes of a rational function $f(x) = n(x)/d(x)$ is at most equal to the degree of $d(x)$. A rational function has at most one horizontal asymptote (note that the graph in Fig. 4C does not have a horizontal asymptote). Moreover, the graph of a rational function approaches the horizontal asymptote (when one exists) both as x increases and decreases without bound.

EXAMPLE 2   Graphing Rational Functions Given the rational function

$$f(x) = \frac{3x}{x^2 - 4}$$

(A) Find the domain.

(B) Find the x and y intercepts.

(C) Find the equations of all vertical asymptotes.

(D) If there is a horizontal asymptote, find its equation.

(E) Using the information from (A)–(D) and additional points as necessary, sketch a graph of f.

SOLUTION

(A) $x^2 - 4 = (x - 2)(x + 2)$, so the denominator is 0 if $x = -2$ or $x = 2$. Therefore the domain is the set of all real numbers except -2 and 2.

(B) *x intercepts:* $f(x) = 0$ only if $3x = 0$, or $x = 0$. So the only x intercept is 0.

   y intercept:

$$f(0) = \frac{3 \cdot 0}{0^2 - 4} = \frac{0}{-4} = 0$$

So the y intercept is 0.

(C) Consider individually the values of x for which the denominator is 0, namely, 2 and -2, found in part (A).

   (i) If $x = 2$, the numerator is 6, and the denominator is 0, so $f(2)$ is undefined. But for numbers just to the right of 2 (like 2.1, 2.01, 2.001), the numerator is close to 6, and the denominator is a positive number close to 0, so the fraction $f(x)$ is large and positive. For numbers just to the left of 2 (like 1.9, 1.99, 1.999), the numerator is close to 6, and the denominator is a negative number close to 0, so the fraction $f(x)$ is large (in absolute value) and negative. Therefore, the line $x = 2$ is a vertical asymptote, and $f(x)$ is positive to the right of the asymptote, and negative to the left.

(ii) If $x = -2$, the numerator is -6, and the denominator is 0, so $f(2)$ is undefined. But for numbers just to the right of -2 (like $-1.9, -1.99, -1.999$), the numerator is close to -6, and the denominator is a negative number close to 0, so the fraction $f(x)$ is large and positive. For numbers just to the left of -2 (like $-2.1, -2.01, -2.001$), the numerator is close to -6, and the denominator is a positive number close to 0, so the fraction $f(x)$ is large (in absolute value) and negative. Therefore, the line $x = -2$ is a vertical asymptote, and $f(x)$ is positive to the right of the asymptote and negative to the left.

(D) Rewrite $f(x)$ by dividing each term in the numerator and denominator by the highest power of x in $f(x)$.

$$f(x) = \frac{3x}{x^2 - 4} = \frac{\dfrac{3x}{x^2}}{\dfrac{x^2}{x^2} - \dfrac{4}{x^2}} = \frac{\dfrac{3}{x}}{1 - \dfrac{4}{x^2}}$$

As x increases or decreases without bound, the numerator tends to 0 and the denominator tends to 1; so, $f(x)$ tends to 0. The line $y = 0$ is a horizontal asymptote.

(E) Use the information from parts (A)–(D) and plot additional points as necessary to complete the graph, as shown in Figure 5.

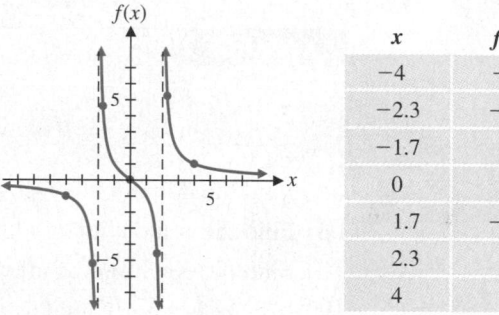

x	$f(x)$
-4	-1
-2.3	-5.3
-1.7	4.6
0	0
1.7	-4.6
2.3	5.3
4	1

Figure 5

Matched Problem 2 | Given the rational function $g(x) = \dfrac{3x + 3}{x^2 - 9}$

(A) Find the domain.

(B) Find the x and y intercepts.

(C) Find the equations of all vertical asymptotes.

(D) If there is a horizontal asymptote, find its equation.

(E) Using the information from parts (A)–(D) and additional points as necessary, sketch a graph of g.

CONCEPTUAL INSIGHT

Consider the rational function

$$g(x) = \frac{3x^2 - 12x}{x^3 - 4x^2 - 4x + 16} = \frac{3x(x - 4)}{(x^2 - 4)(x - 4)}$$

The numerator and denominator of g have a common zero, $x = 4$. If $x \neq 4$, then we can cancel the factor $x - 4$ from the numerator and denominator, leaving the function $f(x)$ of Example 2. So the graph of g (Fig. 6) is identical to the graph of f (Fig. 5), except that the graph of g has an open dot at $(4, 1)$, indicating that 4 is not in the domain of g. In particular, f and g have the same asymptotes. Note that the line $x = 4$ is *not* a vertical asymptote of g, even though 4 is a zero of its denominator.

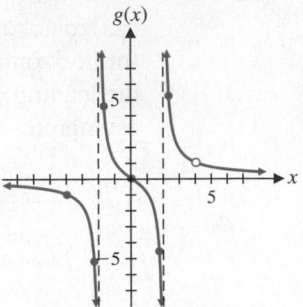

Figure 6

Graphing rational functions is aided by locating vertical and horizontal asymptotes first, if they exist. The following general procedure is suggested by Example 2 and the Conceptual Insight above.

PROCEDURE   Vertical and Horizontal Asymptotes of Rational Functions

Consider the rational function

$$f(x) = \frac{n(x)}{d(x)}$$

where $n(x)$ and $d(x)$ are polynomials.

Vertical asymptotes:

Case 1. Suppose $n(x)$ and $d(x)$ have no real zero in common. If c is a real number such that $d(c) = 0$, then the line $x = c$ is a vertical asymptote of the graph of f.

Case 2. If $n(x)$ and $d(x)$ have one or more real zeros in common, cancel common linear factors, and apply Case 1 to the reduced function. (The reduced function has the same asymptotes as f.)

Horizontal asymptote:

Case 1. If degree $n(x) <$ degree $d(x)$, then $y = 0$ is the horizontal asymptote.

Case 2. If degree $n(x) =$ degree $d(x)$, then $y = a/b$ is the horizontal asymptote, where a is the leading coefficient of $n(x)$, and b is the leading coefficient of $d(x)$.

Case 3. If degree $n(x) >$ degree $d(x)$, there is no horizontal asymptote.

Example 2 illustrates Case 1 of the procedure for horizontal asymptotes. Cases 2 and 3 are illustrated in Example 3 and Matched Problem 3.

EXAMPLE 3   Finding Asymptotes Find the vertical and horizontal asymptotes of the rational function

$$f(x) = \frac{3x^2 + 3x - 6}{2x^2 - 2}$$

SOLUTION   Vertical asymptotes   We factor the numerator $n(x)$ and the denominator $d(x)$:

$$n(x) = 3(x^2 + x - 2) = 3(x - 1)(x + 2)$$
$$d(x) = 2(x^2 - 1) = 2(x - 1)(x + 1)$$

The reduced function is

$$\frac{3(x + 2)}{2(x + 1)}$$

which, by the procedure, has the vertical asymptote $x = -1$. Therefore, $x = -1$ is the only vertical asymptote of f.

Horizontal asymptote   Both $n(x)$ and $d(x)$ have degree 2 (Case 2 of the procedure for horizontal asymptotes). The leading coefficient of the numerator $n(x)$ is 3, and the leading coefficient of the denominator $d(x)$ is 2. So $y = 3/2$ is the horizontal asymptote.

Matched Problem 3   Find the vertical and horizontal asymptotes of the rational function

$$f(x) = \frac{x^3 - 4x}{x^2 + 5x}$$

Explore and Discuss 1   A function f is **bounded** if the entire graph of f lies between two horizontal lines. The only polynomials that are bounded are the constant functions, but there are many rational functions that are bounded. Give an example of a bounded rational function, with domain the set of all real numbers, that is not a constant function.

Applications

Rational functions occur naturally in many types of applications.

EXAMPLE 4   Employee Training   A company that manufactures computers has established that, on the average, a new employee can assemble $N(t)$ components per day after t days of on-the-job training, as given by

$$N(t) = \frac{50t}{t + 4} \quad t \geq 0$$

Sketch a graph of N, $0 \leq t \leq 100$, including any vertical or horizontal asymptotes. What does $N(t)$ approach as t increases without bound?

SOLUTION   Vertical asymptotes   None for $t \geq 0$

Horizontal asymptote

$$N(t) = \frac{50t}{t + 4} = \frac{50}{1 + \dfrac{4}{t}}$$

$N(t)$ approaches 50 (the leading coefficient of $50t$ divided by the leading coefficient of $t + 4$) as t increases without bound. So $y = 50$ is a horizontal asymptote.

Sketch of graph   The graph is shown in the margin.

$N(t)$ approaches 50 as t increases without bound. It appears that 50 components per day would be the upper limit that an employee would be expected to assemble.

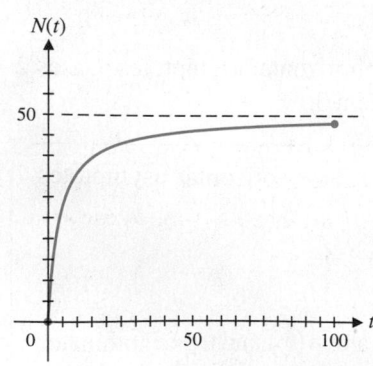

Matched Problem 4   Repeat Example 4 for $N(t) = \dfrac{25t + 5}{t + 5} \quad t \geq 0$.

Exercises 1.4

In Problems 1–10, for each polynomial function find the following:

(A) Degree of the polynomial

(B) All x intercepts

(C) The y intercept

1. $f(x) = 50 - 5x$

2. $f(x) = 72 + 12x$

3. $f(x) = x^4(x - 1)$

4. $f(x) = x^3(x + 5)$

5. $f(x) = x^2 + 3x + 2$

6. $f(x) = x^2 - 4x - 5$

7. $f(x) = (x^2 - 1)(x^2 - 9)$

8. $f(x) = (x^2 - 4)(x^3 + 27)$

9. $f(x) = (2x + 3)^4(x - 5)^5$

10. $f(x) = (x + 3)^2(8x - 4)^6$

Each graph in Problems 11–18 is the graph of a polynomial function. Answer the following questions for each graph:

(A) What is the minimum degree of a polynomial function that could have the graph?

(B) Is the leading coefficient of the polynomial negative or positive?

11.
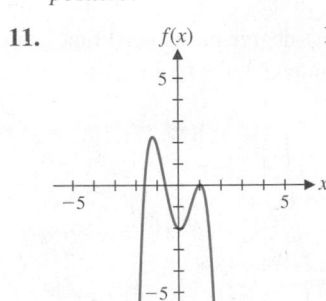

12.
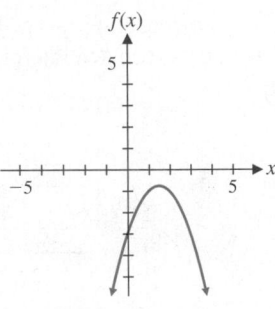

13.
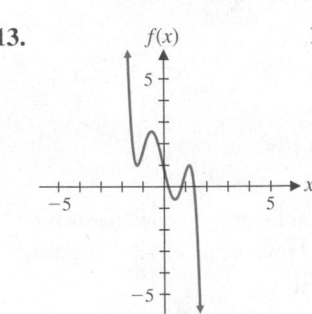

14.
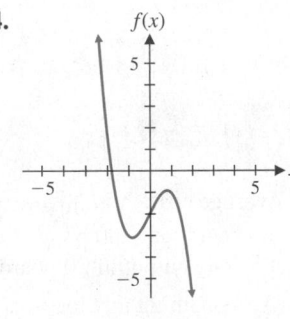

15. & **16.**
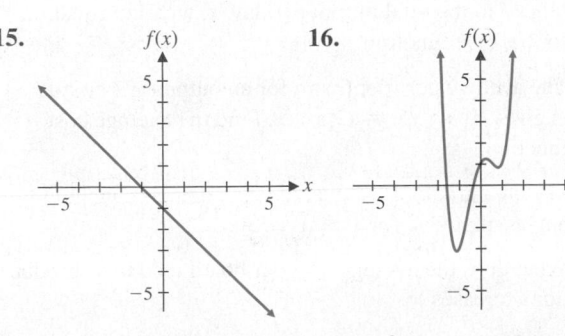

17.
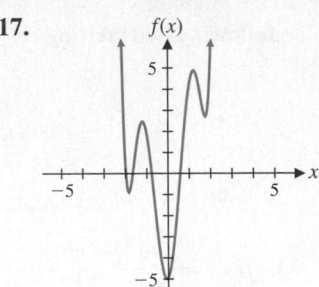

18.
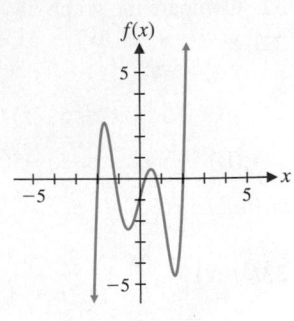

19. What is the maximum number of x intercepts that a polynomial of degree 10 can have?

20. What is the maximum number of x intercepts that a polynomial of degree 7 can have?

21. What is the minimum number of x intercepts that a polynomial of degree 9 can have? Explain.

22. What is the minimum number of x intercepts that a polynomial of degree 6 can have? Explain.

For each rational function in Problems 23–28,

(A) Find the intercepts for the graph.

(B) Determine the domain.

(C) Find any vertical or horizontal asymptotes for the graph.

(D) Sketch any asymptotes as dashed lines. Then sketch a graph of $y = f(x)$.

(E) Graph $y = f(x)$ in a standard viewing window using a graphing calculator.

23. $f(x) = \dfrac{x + 2}{x - 2}$

24. $f(x) = \dfrac{x - 3}{x + 3}$

25. $f(x) = \dfrac{3x}{x + 2}$

26. $f(x) = \dfrac{2x}{x - 3}$

27. $f(x) = \dfrac{4 - 2x}{x - 4}$

28. $f(x) = \dfrac{3 - 3x}{x - 2}$

29. Compare the graph of $y = 2x^4$ to the graph of $y = 2x^4 - 5x^2 + x + 2$ in the following two viewing windows:

   (A) $-5 \le x \le 5, -5 \le y \le 5$

   (B) $-5 \le x \le 5, -500 \le y \le 500$

30. Compare the graph of $y = x^3$ to the graph of $y = x^3 - 2x + 2$ in the following two viewing windows:

   (A) $-5 \le x \le 5, -5 \le y \le 5$

   (B) $-5 \le x \le 5, -500 \le y \le 500$

31. Compare the graph of $y = -x^5$ to the graph of $y = -x^5 + 4x^3 - 4x + 1$ in the following two viewing windows:

   (A) $-5 \le x \le 5, -5 \le y \le 5$

   (B) $-5 \le x \le 5, -500 \le y \le 500$

32. Compare the graph of $y = -x^5$ to the graph of 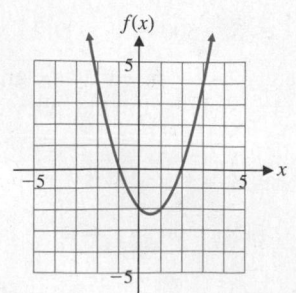 $y = -x^5 + 5x^3 - 5x + 2$ in the following two viewing windows:

(A) $-5 \le x \le 5, -5 \le y \le 5$

(B) $-5 \le x \le 5, -500 \le y \le 500$

In Problems 33–40, find the equation of any horizontal asymptote.

33. $f(x) = \dfrac{5x^3 + 2x - 3}{6x^3 - 7x + 1}$

34. $f(x) = \dfrac{6x^4 - x^3 + 2}{4x^4 + 10x + 5}$

35. $f(x) = \dfrac{1 - 5x + x^2}{2 + 3x + 4x^2}$

36. $f(x) = \dfrac{8 - x^3}{1 + 2x^3}$

37. $f(x) = \dfrac{x^4 + 2x^2 + 1}{1 - x^5}$

38. $f(x) = \dfrac{3 + 5x}{x^2 + x + 3}$

39. $f(x) = \dfrac{x^2 + 6x + 1}{x - 5}$

40. $f(x) = \dfrac{x^2 + x^4 + 1}{x^3 + 2x - 4}$

In Problems 41–46, find the equations of any vertical asymptotes.

41. $f(x) = \dfrac{x^2 + 1}{(x^2 - 1)(x^2 - 9)}$

42. $f(x) = \dfrac{2x + 5}{(x^2 - 4)(x^2 - 16)}$

43. $f(x) = \dfrac{x^2 - x - 6}{x^2 - 3x - 10}$

44. $f(x) = \dfrac{x^2 - 8x + 7}{x^2 + 7x - 8}$

45. $f(x) = \dfrac{x^2 + 3x}{x^3 - 36x}$

46. $f(x) = \dfrac{x^2 + x - 2}{x^3 - 3x^2 + 2x}$

For each rational function in Problems 47–52,

(A) *Find any intercepts for the graph.*

(B) *Find any vertical and horizontal asymptotes for the graph.*

(C) *Sketch any asymptotes as dashed lines. Then sketch a graph of f.*

(D) *Graph the function in a standard viewing window using a graphing calculator.*

47. $f(x) = \dfrac{2x^2}{x^2 - x - 6}$

48. $f(x) = \dfrac{3x^2}{x^2 + x - 6}$

49. $f(x) = \dfrac{6 - 2x^2}{x^2 - 9}$

50. $f(x) = \dfrac{3 - 3x^2}{x^2 - 4}$

51. $f(x) = \dfrac{-4x + 24}{x^2 + x - 6}$

52. $f(x) = \dfrac{5x - 10}{x^2 + x - 12}$

53. Write an equation for the lowest-degree polynomial function with the graph and intercepts shown in the figure.

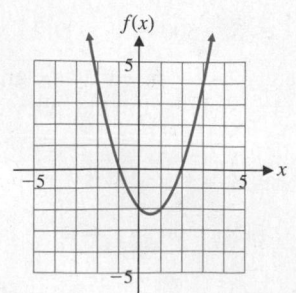

54. Write an equation for the lowest-degree polynomial function with the graph and intercepts shown in the figure.

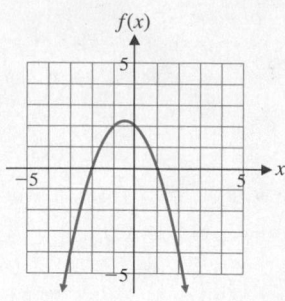

55. Write an equation for the lowest-degree polynomial function with the graph and intercepts shown in the figure.

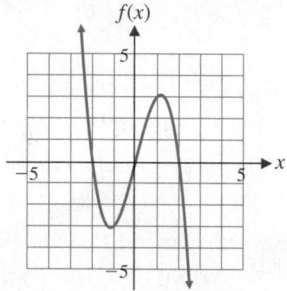

56. Write an equation for the lowest-degree polynomial function with the graph and intercepts shown in the figure.

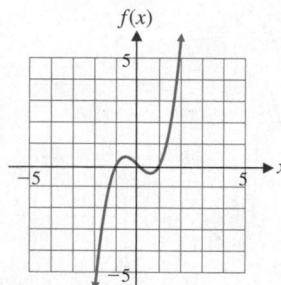

Applications

57. **Average cost.** A company manufacturing snowboards has fixed costs of $200 per day and total costs of $3,800 per day at a daily output of 20 boards.

(A) Assuming that the total cost per day, $C(x)$, is linearly related to the total output per day, x, write an equation for the cost function.

(B) The average cost per board for an output of x boards is given by $\overline{C}(x) = C(x)/x$. Find the average cost function.

(C) Sketch a graph of the average cost function, including any asymptotes, for $1 \le x \le 30$.

(D) What does the average cost per board tend to as production increases?

58. **Average cost.** A company manufacturing surfboards has fixed costs of $300 per day and total costs of $5,100 per day at a daily output of 20 boards.

(A) Assuming that the total cost per day, $C(x)$, is linearly related to the total output per day, x, write an equation for the cost function.

(B) The average cost per board for an output of x boards is given by $\overline{C}(x) = C(x)/x$. Find the average cost function.

(C) Sketch a graph of the average cost function, including any asymptotes, for $1 \le x \le 30$.

(D) What does the average cost per board tend to as production increases?

59. Replacement time. An office copier has an initial price of $2,500. A service contract costs $200 for the first year and increases $50 per year thereafter. It can be shown that the total cost of the copier after n years is given by

$$C(n) = 2,500 + 175n + 25n^2$$

The average cost per year for n years is given by $\overline{C}(n) = C(n)/n$.

(A) Find the rational function $\overline{C}$.

(B) Sketch a graph of $\overline{C}$ for $2 \le n \le 20$.

(C) When is the average cost per year at a minimum, and what is the minimum average annual cost? *[Hint: Refer to the sketch in part (B) and evaluate $\overline{C}(n)$ at appropriate integer values until a minimum value is found.]* The time when the average cost is minimum is frequently referred to as the **replacement time** for the piece of equipment.

(D) Graph the average cost function $\overline{C}$ on a graphing calculator and use an appropriate command to find when the average annual cost is at a minimum.

60. Minimum average cost. Financial analysts in a company that manufactures DVD players arrived at the following daily cost equation for manufacturing x DVD players per day:

$$C(x) = x^2 + 2x + 2,000$$

The average cost per unit at a production level of x players per day is $\overline{C}(x) = C(x)/x$.

(A) Find the rational function $\overline{C}$.

(B) Sketch a graph of $\overline{C}$ for $5 \le x \le 150$.

(C) For what daily production level (to the nearest integer) is the average cost per unit at a minimum, and what is the minimum average cost per player (to the nearest cent)? *[Hint: Refer to the sketch in part (B) and evaluate $\overline{C}(x)$ at appropriate integer values until a minimum value is found.]*

(D) Graph the average cost function $\overline{C}$ on a graphing calculator and use an appropriate command to find the daily production level (to the nearest integer) at which the average cost per player is at a minimum. What is the minimum average cost to the nearest cent?

61. Minimum average cost. A consulting firm, using statistical methods, provided a veterinary clinic with the cost equation

$$C(x) = 0.00048(x - 500)^3 + 60,000$$

$$100 \le x \le 1,000$$

where $C(x)$ is the cost in dollars for handling x cases per month. The average cost per case is given by $\overline{C}(x) = C(x)/x$.

(A) Write the equation for the average cost function $\overline{C}$.

(B) Graph $\overline{C}$ on a graphing calculator.

(C) Use an appropriate command to find the monthly caseload for the minimum average cost per case. What is the minimum average cost per case?

62. Minimum average cost. The financial department of a hospital, using statistical methods, arrived at the cost equation

$$C(x) = 20x^3 - 360x^2 + 2,300x - 1,000$$

$$1 \le x \le 12$$

where $C(x)$ is the cost in thousands of dollars for handling x thousand cases per month. The average cost per case is given by $\overline{C}(x) = C(x)/x$.

(A) Write the equation for the average cost function $\overline{C}$.

(B) Graph $\overline{C}$ on a graphing calculator.

(C) Use an appropriate command to find the monthly caseload for the minimum average cost per case. What is the minimum average cost per case to the nearest dollar?

63. Diet. Table 3 shows the per capita consumption of ice cream and eggs in the United States for selected years since 1980.

(A) Let x represent the number of years since 1980 and find a cubic regression polynomial for the per capita consumption of ice cream.

(B) Use the polynomial model from part (A) to estimate (to the nearest tenth of a pound) the per capita consumption of ice cream in 2025.

Table 3 **Per Capita Consumption of Ice Cream and Eggs**

Year	Ice Cream (pounds)	Eggs (number)
1980	17.5	266
1985	18.1	251
1990	15.8	231
1995	15.5	229
2000	16.5	247
2005	14.4	252
2010	13.3	242

Source: U.S. Department of Agriculture

64. Diet. Refer to Table 3.

(A) Let x represent the number of years since 1980 and find a cubic regression polynomial for the per capita consumption of eggs.

(B) Use the polynomial model from part (A) to estimate (to the nearest integer) the per capita consumption of eggs in 2022.

65. Physiology. In a study on the speed of muscle contraction in frogs under various loads, researchers W. O. Fems and J. Marsh found that the speed of contraction decreases with increasing loads. In particular, they found that the relationship between speed of contraction v (in centimeters per second) and load x (in grams) is given approximately by

$$v(x) = \frac{26 + 0.06x}{x} \qquad x \ge 5$$

(A) What does $v(x)$ approach as x increases?

(B) Sketch a graph of function v.

66. Learning theory. In 1917, L. L. Thurstone, a pioneer in quantitative learning theory, proposed the rational function

$$f(x) = \frac{a(x + c)}{(x + c) + b}$$

to model the number of successful acts per unit time that a person could accomplish after x practice sessions. Suppose that for a particular person enrolled in a typing class,

$$f(x) = \frac{55(x + 1)}{(x + 8)} \quad x \geq 0$$

where $f(x)$ is the number of words per minute the person is able to type after x weeks of lessons.

(A) What does $f(x)$ approach as x increases?

(B) Sketch a graph of function f, including any vertical or horizontal asymptotes.

67. Marriage. Table 4 shows the marriage and divorce rates per 1,000 population for selected years since 1960.

(A) Let x represent the number of years since 1960 and find a cubic regression polynomial for the marriage rate.

(B) Use the polynomial model from part (A) to estimate the marriage rate (to one decimal place) for 2025.

Table 4 **Marriages and Divorces (per 1,000 population)**

Date	Marriages	Divorces
1960	8.5	2.2
1970	10.6	3.5
1980	10.6	5.2
1990	9.8	4.7
2000	8.5	4.1
2010	6.8	3.6

Source: National Center for Health Statistics

68. Divorce. Refer to Table 4.

(A) Let x represent the number of years since 1950 and find a cubic regression polynomial for the divorce rate.

(B) Use the polynomial model from part (A) to estimate the divorce rate (to one decimal place) for 2025.

Answers to Matched Problems

1.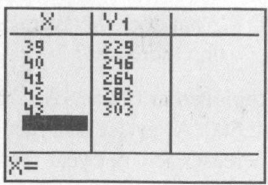

2. (A) Domain: all real numbers except -3 and 3

   (B) x intercept: -1; y intercept: $-\dfrac{1}{3}$

   (C) Vertical asymptotes: $x = -3$ and $x = 3$;

   (D) Horizontal asymptote: $y = 0$

   (E)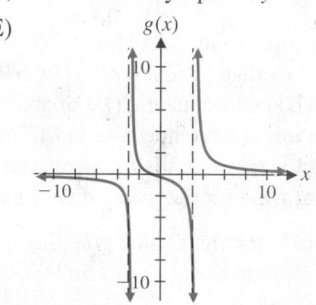

3. Vertical asymptote: $x = -5$

   Horizontal asymptote: none

4. No vertical asymptotes for $t \geq 0$; $y = 25$ is a horizontal asymptote. $N(t)$ approaches 25 as t increases without bound. It appears that 25 components per day would be the upper limit that an employee would be expected to assemble.

   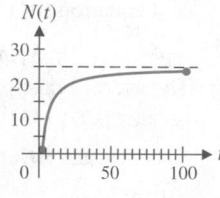

1.5 Exponential Functions

- Exponential Functions
- Base e Exponential Functions
- Growth and Decay Applications
- Compound Interest

This section introduces an important class of functions called *exponential functions*. These functions are used extensively in modeling and solving a wide variety of real-world problems, including growth of money at compound interest, growth of populations, radioactive decay and learning associated with the mastery of such devices as a new computer or an assembly process in a manufacturing plant.

Exponential Functions

We start by noting that

$$f(x) = 2^x \quad \text{and} \quad g(x) = x^2$$

are not the same function. Whether a variable appears as an exponent with a constant base or as a base with a constant exponent makes a big difference. The function g is a quadratic function, which we have already discussed. The function f is a new type of function called an *exponential function*. In general,

> **DEFINITION** Exponential Function
>
> The equation
>
> $$f(x) = b^x \quad b > 0, b \neq 1$$
>
> defines an **exponential function** for each different constant b, called the **base**. The **domain** of f is the set of all real numbers, and the **range** of f is the set of all positive real numbers.

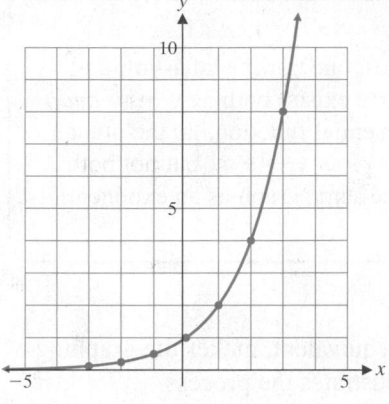

Figure 1 $y = 2^x$

We require the base b to be positive to avoid imaginary numbers such as $(-2)^{1/2} = \sqrt{-2} = i\sqrt{2}$. We exclude $b = 1$ as a base, since $f(x) = 1^x = 1$ is a constant function, which we have already considered.

When asked to hand-sketch graphs of equations such as $y = 2^x$ or $y = 2^{-x}$, many students do not hesitate. [*Note:* $2^{-x} = 1/2^x = (1/2)^x$.] They make tables by assigning integers to x, plot the resulting points, and then join these points with a smooth curve as in Figure 1. The only catch is that we have not defined 2^x for all real numbers. From Appendix A, Section A.6, we know what 2^5, 2^{-3}, $2^{2/3}$, $2^{-3/5}$, $2^{1.4}$, and $2^{-3.14}$ mean (that is, 2^p, where p is a rational number), but what does

$$2^{\sqrt{2}}$$

mean? The question is not easy to answer at this time. In fact, a precise definition of $2^{\sqrt{2}}$ must wait for more advanced courses, where it is shown that

$$2^x$$

names a positive real number for x any real number, and that the graph of $y = 2^x$ is as indicated in Figure 1.

It is useful to compare the graphs of $y = 2^x$ and $y = 2^{-x}$ by plotting both on the same set of coordinate axes, as shown in Figure 2A. The graph of

$$f(x) = b^x \quad b > 1 \text{ (Fig. 2B)}$$

looks very much like the graph of $y = 2^x$, and the graph of

$$f(x) = b^x \quad 0 < b < 1 \text{ (Fig. 2B)}$$

looks very much like the graph of $y = 2^{-x}$. Note that in both cases the x axis is a horizontal asymptote for the graphs.

The graphs in Figure 2 suggest the following general properties of exponential functions, which we state without proof:

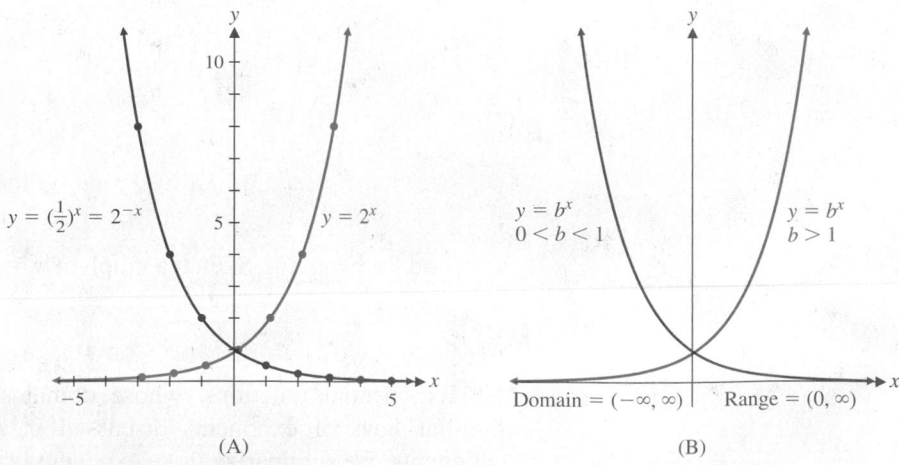

(A)   (B)

Figure 2 **Exponential functions**

> **THEOREM 1**   Basic Properties of the Graph of $f(x) = b^x$, $b > 0$, $b \neq 1$
>
> 1. All graphs will pass through the point $(0,1)$. $b^0 = 1$ for any permissible base b.
> 2. All graphs are continuous curves, with no holes or jumps.
> 3. The x axis is a horizontal asymptote.
> 4. If $b > 1$, then b^x increases as x increases.
> 5. If $0 < b < 1$, then b^x decreases as x increases.

CONCEPTUAL INSIGHT

Recall that the graph of a rational function has at most one horizontal asymptote and that it approaches the horizontal asymptote (if one exists) both as $x \to \infty$ *and* as $x \to -\infty$ (see Section 1.4). The graph of an exponential function, on the other hand, approaches its horizontal asymptote as $x \to \infty$ *or* as $x \to -\infty$, but not both. In particular, there is no rational function that has the same graph as an exponential function.

The use of a calculator with the key $\boxed{y^x}$, or its equivalent, makes the graphing of exponential functions almost routine. Example 1 illustrates the process.

EXAMPLE 1   Graphing Exponential Functions   Sketch a graph of $y = \left(\frac{1}{2}\right)4^x$, $-2 \leq x \leq 2$.

SOLUTION   Use a calculator to create the table of values shown. Plot these points, and then join them with a smooth curve as in Figure 3.

x	y
-2	0.031
-1	0.125
0	0.50
1	2.00
2	8.00

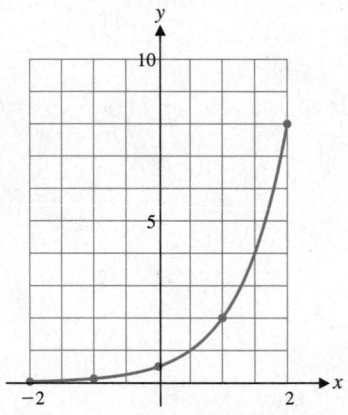

Figure 3   **Graph of $y = \left(\frac{1}{2}\right)4^x$**

Matched Problem 1   Sketch a graph of $y = \left(\frac{1}{2}\right)4^{-x}$, $-2 \leq x \leq 2$.

Exponential functions, whose domains include irrational numbers, obey the familiar laws of exponents discussed in Appendix A, Section A.6 for rational exponents. We summarize these exponent laws here and add two other important and useful properties.

THEOREM 2 Properties of Exponential Functions

For a and b positive, $a \neq 1$, $b \neq 1$, and x and y real,

1. Exponent laws:

$$a^x a^y = a^{x+y} \qquad \frac{a^x}{a^y} = a^{x-y} \qquad \qquad \frac{4^{2y}}{4^{5y}} = 4^{2y-5y} = 4^{-3y}$$

$$(a^x)^y = a^{xy} \qquad (ab)^x = a^x b^x \qquad \left(\frac{a}{b}\right)^x = \frac{a^x}{b^x}$$

2. $a^x = a^y$ if and only if $x = y$

If $7^{5t+1} = 7^{3t-3}$, then
$5t + 1 = 3t - 3$, and $t = -2$.

3. For $x \neq 0$,
$a^x = b^x$ if and only if $a = b$

If $a^5 = 2^5$, then $a = 2$.

Base e Exponential Functions

Of all the possible bases b we can use for the exponential function $y = b^x$, which ones are the most useful? If you look at the keys on a calculator, you will probably see $\boxed{10^x}$ and $\boxed{e^x}$. It is clear why base 10 would be important, because our number system is a base 10 system. But what is e, and why is it included as a base? It turns out that base e is used more frequently than all other bases combined. The reason for this is that certain formulas and the results of certain processes found in calculus and more advanced mathematics take on their simplest form if this base is used. This is why you will see e used extensively in expressions and formulas that model real-world phenomena. In fact, its use is so prevalent that you will often hear people refer to $y = e^x$ as *the* exponential function.

The base e is an irrational number and, like π, it cannot be represented exactly by any finite decimal or fraction. However, e can be approximated as closely as we like by evaluating the expression

$$\left(1 + \frac{1}{x}\right)^x \tag{1}$$

for sufficiently large values of x. What happens to the value of expression (1) as x increases without bound? Think about this for a moment before proceeding. Maybe you guessed that the value approaches 1, because

$$1 + \frac{1}{x}$$

approaches 1, and 1 raised to any power is 1. Let us see if this reasoning is correct by actually calculating the value of the expression for larger and larger values of x. Table 1 summarizes the results.

Table 1

x	$\left(1 + \dfrac{1}{x}\right)^x$
1	2
10	$2.593\,74\ldots$
100	$2.704\,81\ldots$
1,000	$2.716\,92\ldots$
10,000	$2.718\,14\ldots$
100,000	$2.718\,27\ldots$
1,000,000	$2.718\,28\ldots$

Interestingly, the value of expression (1) is never close to 1 but seems to be approaching a number close to 2.7183. In fact, as x increases without bound, the value of expression (1) approaches an irrational number that we call e. The irrational number e to 12 decimal places is

$$e = 2.718\ 281\ 828\ 459$$

Compare this value of e with the value of e^1 from a calculator.

DEFINITION   Exponential Function with Base e

Exponential function with base e and base $1/e$, respectively, are defined by

$$y = e^x \quad \text{and} \quad y = e^{-x}$$

Domain: $(-\infty, \infty)$

Range: $(0, \infty)$

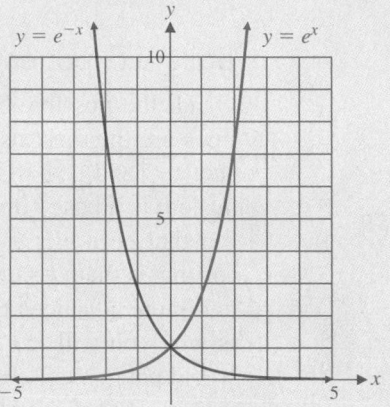

Explore and Discuss 1   Graph the functions $f(x) = e^x$, $g(x) = 2^x$, and $h(x) = 3^x$ on the same set of coordinate axes. At which values of x do the graphs intersect? For positive values of x, which of the three graphs lies above the other two? Below the other two? How does your answer change for negative values of x?

Growth and Decay Applications

Functions of the form $y = ce^{kt}$, where c and k are constants and the independent variable t represents time, are often used to model population growth and radioactive decay. Note that if $t = 0$, then $y = c$. So the constant c represents the initial population (or initial amount). The constant k is called the **relative growth rate** and has the following interpretation: Suppose that $y = ce^{kt}$ models the population of a country, where y is the number of persons and t is time in years. If the relative growth rate is $k = 0.02$, then at any time t, the population is growing at a rate of $0.02\,y$ persons (that is, 2% of the population) per year.

We say that **population is growing continuously at relative growth rate k** to mean that the population y is given by the model $y = ce^{kt}$.

EXAMPLE 2   Exponential Growth   Cholera, an intestinal disease, is caused by a cholera bacterium that multiplies exponentially. The number of bacteria grows continuously at relative growth rate 1.386, that is,

$$N = N_0 e^{1.386t}$$

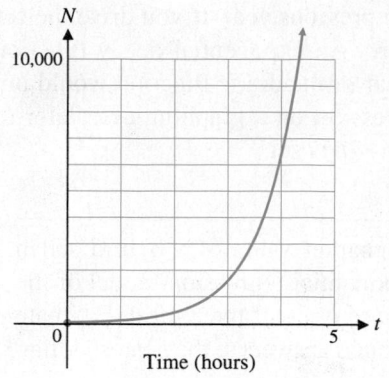

Figure 4

where N is the number of bacteria present after t hours and N_0 is the number of bacteria present at the start $(t = 0)$. If we start with 25 bacteria, how many bacteria (to the nearest unit) will be present

(A) In 0.6 hour?   (B) In 3.5 hours?

SOLUTION Substituting $N_0 = 25$ into the preceding equation, we obtain

$$N = 25e^{1.386t} \quad \text{The graph is shown in Figure 4.}$$

(A) Solve for N when $t = 0.6$:

$$N = 25e^{1.386(0.6)} \quad \text{Use a calculator.}$$
$$= 57 \text{ bacteria}$$

(B) Solve for N when $t = 3.5$:

$$N = 25e^{1.386(3.5)} \quad \text{Use a calculator.}$$
$$= 3{,}197 \text{ bacteria}$$

Matched Problem 2 Refer to the exponential growth model for cholera in Example 2. If we start with 55 bacteria, how many bacteria (to the nearest unit) will be present

(A) In 0.85 hour?   (B) In 7.25 hours?

EXAMPLE 3 Exponential Decay Cosmic-ray bombardment of the atmosphere produces neutrons, which in turn react with nitrogen to produce radioactive carbon-14 (^{14}C). Radioactive ^{14}C enters all living tissues through carbon dioxide, which is first absorbed by plants. As long as a plant or animal is alive, ^{14}C is maintained in the living organism at a constant level. Once the organism dies, however, ^{14}C decays according to the equation

$$A = A_0 e^{-0.000124t}$$

where A is the amount present after t years and A_0 is the amount present at time $t = 0$.

(A) If 500 milligrams of ^{14}C is present in a sample from a skull at the time of death, how many milligrams will be present in the sample in 15,000 years? Compute the answer to two decimal places.

(B) The **half-life** of ^{14}C is the time t at which the amount present is one-half the amount at time $t = 0$. Use Figure 5 to estimate the half-life of ^{14}C.

SOLUTION Substituting $A_0 = 500$ in the decay equation, we have

$$A = 500e^{-0.000124t} \quad \text{See the graph in Figure 5.}$$

(A) Solve for A when $t = 15{,}000$:

$$A = 500e^{-0.000124(15{,}000)} \quad \text{Use a calculator.}$$
$$= 77.84 \text{ milligrams}$$

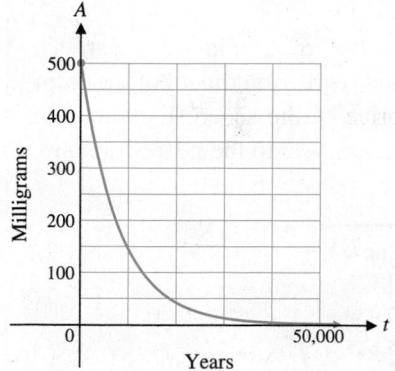

Figure 5

(B) Refer to Figure 5, and estimate the time t at which the amount A has fallen to 250 milligrams: $t \approx 6{,}000$ years. (Finding the intersection of $y_1 = 500e^{-0.000124x}$ and $y_2 = 250$ on a graphing calculator gives a better estimate: $t \approx 5{,}590$ years.)

Matched Problem 3 Refer to the exponential decay model in Example 3. How many milligrams of ^{14}C would have to be present at the beginning in order to have 25 milligrams present after 18,000 years? Compute the answer to the nearest milligram.

If you buy a new car, it is likely to depreciate in value by several thousand dollars during the first year you own it. You would expect the value of the car to decrease in each subsequent year, but not by as much as in the previous year. If you drive the car long enough, its resale value will get close to zero. An exponential decay function will often be a good model of depreciation; a linear or quadratic function would not be suitable (why?). We can use **exponential regression** on a graphing calculator to find the function of the form $y = ab^x$ that best fits a data set.

EXAMPLE 4   Depreciation   Table 2 gives the market value of a hybrid sedan (in dollars) x years after its purchase. Find an exponential regression model of the form $y = ab^x$ for this data set. Estimate the purchase price of the hybrid. Estimate the value of the hybrid 10 years after its purchase. Round answers to the nearest dollar.

Table 2

x	Value ($)
1	12,575
2	9,455
3	8,115
4	6,845
5	5,225
6	4,485

SOLUTION   Enter the data into a graphing calculator (Fig. 6A) and find the exponential regression equation (Fig. 6B). The estimated purchase price is $y_1(0) = \$14,910$. The data set and the regression equation are graphed in Figure 6C. Using TRACE, we see that the estimated value after 10 years is $1,959.

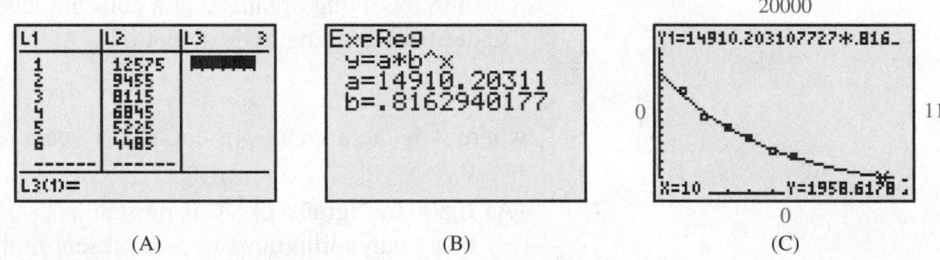

(A)   (B)   (C)

Figure 6

Matched Problem 4   Table 3 gives the market value of a midsize sedan (in dollars) x years after its purchase. Find an exponential regression model of the form $y = ab^x$ for this data set. Estimate the purchase price of the sedan. Estimate the value of the sedan 10 years after its purchase. Round answers to the nearest dollar.

Table 3

x	Value ($)
1	23,125
2	19,050
3	15,625
4	11,875
5	9,450
6	7,125

Compound Interest

The fee paid to use another's money is called **interest**. It is usually computed as a percent (called **interest rate**) of the principal over a given period of time. If, at the end of a payment period, the interest due is reinvested at the same rate, then the interest earned as well as the principal will earn interest during the next payment period. Interest paid on interest reinvested is called **compound interest** and may be calculated using the following compound interest formula:

If a **principal** *P* (**present value**) is invested at an annual **rate** *r* (expressed as a decimal) compounded *m* times a year, then the **amount** *A* (**future value**) in the account at the end of *t* years is given by

$$A = P\left(1 + \frac{r}{m}\right)^{mt} \quad \text{Compound interest formula}$$

For given *r* and *m*, the amount *A* is equal to the principal *P* multiplied by the exponential function b^t, where $b = (1 + r/m)^m$.

EXAMPLE 5 Compound Growth If $1,000 is invested in an account paying 10% compounded monthly, how much will be in the account at the end of 10 years? Compute the answer to the nearest cent.

SOLUTION We use the compound interest formula as follows:

$$A = P\left(1 + \frac{r}{m}\right)^{mt}$$

$$= 1,000\left(1 + \frac{0.10}{12}\right)^{(12)(10)} \quad \text{Use a calculator.}$$

$$= \$2,707.04$$

The graph of

$$A = 1,000\left(1 + \frac{0.10}{12}\right)^{12t}$$

for $0 \le t \le 20$ is shown in Figure 7.

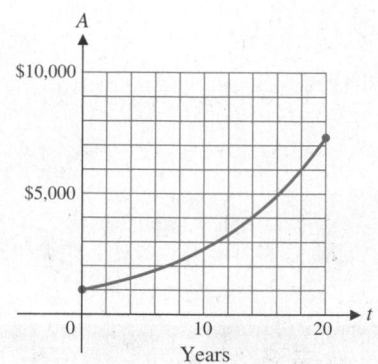

A

$10,000

$5,000

0

10

20

t

Years

Figure 7

Matched Problem 5 If you deposit $5,000 in an account paying 9% compounded daily, how much will you have in the account in 5 years? Compute the answer to the nearest cent.

Explore and Discuss 2 Suppose that $1,000 is deposited in a savings account at an annual rate of 5%. Guess the amount in the account at the end of 1 year if interest is compounded (1) quarterly, (2) monthly, (3) daily, (4) hourly. Use the compound interest formula to compute the amounts at the end of 1 year to the nearest cent. Discuss the accuracy of your initial guesses.

Explore and Discuss 2 suggests that if $1,000 were deposited in a savings account at an annual interest rate of 5%, then the amount at the end of 1 year would be less than $1,051.28, even if interest were compounded every minute or every second. The limiting value, approximately $1,051.271 096, is said to be the amount in the account if interest were compounded continuously.

If a principal, *P*, is invested at an annual rate, *r*, and compounded continuously, then the amount in the account at the end of *t* years is given by

$$A = Pe^{rt} \quad \text{Continuous compound interest formula}$$

where the constant $e \approx 2.718\ 28$ is the base of the exponential function.

EXAMPLE 6 Continuous Compound Interest If $1,000 is invested in an account paying 10% compounded continuously, how much will be in the account at the end of 10 years? Compute the answer to the nearest cent.

SOLUTION We use the continuous compound interest formula:

$$A = Pe^{rt} = 1000e^{0.10(10)} = 1000e = \$2{,}718.28$$

Compare with the answer to Example 5.

Matched Problem 6 ⎤ If you deposit \$5,000 in an account paying 9% compounded continuously, how much will you have in the account in 5 years? Compute the answer to the nearest cent.

The formulas for compound interest and continuous compound interest are summarized below for convenient reference.

SUMMARY

Compound Interest: $A = P\left(1 + \dfrac{r}{m}\right)^{mt}$

Continuous Compound Interest: $A = Pe^{rt}$

where $A = $ amount (future value) at the end of t years

$P = $ principal (present value)

$r = $ annual rate (expressed as a decimal)

$m = $ number of compounding periods per year

$t = $ time in years

Exercises 1.5

1. Match each equation with the graph of f, g, h, or k in the figure.

(A) $y = 2^x$ (B) $y = (0.2)^x$

(C) $y = 4^x$ (D) $y = \left(\frac{1}{3}\right)^x$

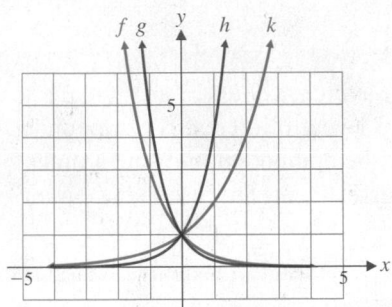

2. Match each equation with the graph of f, g, h, or k in the figure.

(A) $y = \left(\frac{1}{4}\right)^x$ (B) $y = (0.5)^x$

(C) $y = 5^x$ (D) $y = 3^x$

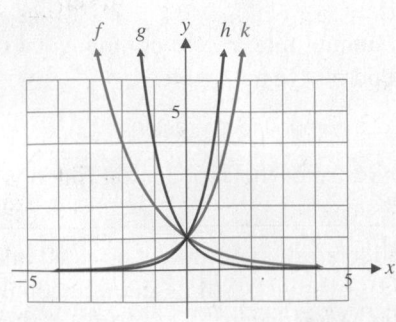

Graph each function in Problems 3–10 over the indicated interval.

3. $y = 5^x$; $[-2, 2]$ **4.** $y = 3^x$; $[-3, 3]$

5. $y = \left(\frac{1}{5}\right)^x = 5^{-x}$; $[-2, 2]$ **6.** $y = \left(\frac{1}{3}\right)^x = 3^{-x}$; $[-3, 3]$

7. $f(x) = -5^x$; $[-2, 2]$ **8.** $g(x) = -3^{-x}$; $[-3, 3]$

9. $y = -e^{-x}$; $[-3, 3]$ **10.** $y = -e^x$; $[-3, 3]$

In Problems 11–18, describe verbally the transformations that can be used to obtain the graph of g from the graph of f (see Section 1.2).

11. $g(x) = -2^x$; $f(x) = 2^x$

12. $g(x) = 2^{x-2}$; $f(x) = 2^x$

13. $g(x) = 3^{x+1}$; $f(x) = 3^x$

14. $g(x) = -3^x$; $f(x) = 3^x$

15. $g(x) = e^x + 1$; $f(x) = e^x$

16. $g(x) = e^x - 2$; $f(x) = e^x$

17. $g(x) = 2e^{-(x+2)}$; $f(x) = e^{-x}$

18. $g(x) = 0.5e^{-(x-1)}$; $f(x) = e^{-x}$

19. Use the graph of f shown in the figure to sketch the graph of each of the following.

(A) $y = f(x) - 1$ (B) $y = f(x + 2)$

(C) $y = 3f(x) - 2$ (D) $y = 2 - f(x - 3)$

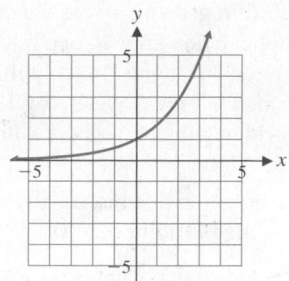

Figure for 19 and 20

20. Use the graph of f shown in the figure to sketch the graph of each of the following.

(A) $y = f(x) + 2$ (B) $y = f(x - 3)$

(C) $y = 2f(x) - 4$ (D) $y = 4 - f(x + 2)$

In Problems 21–26, graph each function over the indicated interval.

21. $f(t) = 2^{t/10}; [-30, 30]$ **22.** $G(t) = 3^{t/100}; [-200, 200]$

23. $y = -3 + e^{1+x}; [-4, 2]$ **24.** $y = 2 + e^{x-2}; [-1, 5]$

25. $y = e^{|x|}; [-3, 3]$ **26.** $y = e^{-|x|}; [-3, 3]$

27. Find all real numbers a such that $a^2 = a^{-2}$. Explain why this does not violate the second exponential function property in Theorem 2 on page 65.

28. Find real numbers a and b such that $a \neq b$ but $a^4 = b^4$. Explain why this does not violate the third exponential function property in Theorem 2 on page 65.

Solve each equation in Problems 29–34 for x.

29. $10^{2-3x} = 10^{5x-6}$ **30.** $5^{3x} = 5^{4x-2}$

31. $4^{5x-x^2} = 4^{-6}$ **32.** $7^{x^2} = 7^{2x+3}$

33. $5^3 = (x + 2)^3$ **34.** $(1 - x)^5 = (2x - 1)^5$

In Problems 35–42, solve each equation for x. (Remember: $e^x \neq 0$ and $e^{-x} \neq 0$ for all values of x).

35. $xe^{-x} + 7e^{-x} = 0$ **36.** $10xe^x - 5e^x = 0$

37. $2x^2e^x - 8e^x = 0$ **38.** $x^2e^{-x} - 9e^{-x} = 0$

39. $e^{4x} - e = 0$ **40.** $e^{4x} + e = 0$

41. $e^{3x-1} + e = 0$ **42.** $e^{3x-1} - e = 0$

Graph each function in Problems 43–46 over the indicated interval.

43. $h(x) = x(2^x); [-5, 0]$ **44.** $m(x) = x(3^{-x}); [0, 3]$

45. $N = \dfrac{100}{1 + e^{-t}}; [0, 5]$ **46.** $N = \dfrac{200}{1 + 3e^{-t}}; [0, 5]$

Applications

In all problems involving days, a 365-day year is assumed.

47. Continuous compound interest. Find the value of an investment of $10,000 in 12 years if it earns an annual rate of 3.95% compounded continuously.

48. Continuous compound interest. Find the value of an investment of $24,000 in 7 years if it earns an annual rate of 4.35% compounded continuously.

49. Compound growth. Suppose that $2,500 is invested at 7% compounded quarterly. How much money will be in the account in

(A) $\frac{3}{4}$ year?   (B) 15 years?

Compute answers to the nearest cent.

50. Compound growth. Suppose that $4,000 is invested at 6% compounded weekly. How much money will be in the account in

(A) $\frac{1}{2}$ year?   (B) 10 years?

Compute answers to the nearest cent.

51. Finance. A person wishes to have $15,000 cash for a new car 5 years from now. How much should be placed in an account now, if the account pays 6.75% compounded weekly? Compute the answer to the nearest dollar.

52. Finance. A couple just had a baby. How much should they invest now at 5.5% compounded daily in order to have $40,000 for the child's education 17 years from now? Compute the answer to the nearest dollar.

53. Money growth. BanxQuote operates a network of websites providing real-time market data from leading financial providers. The following rates for 12-month certificates of deposit were taken from the websites:

(A) Stonebridge Bank, 0.95% compounded monthly

(B) DeepGreen Bank, 0.80% compounded daily

(C) Provident Bank, 0.85% compounded quarterly

Compute the value of $10,000 invested in each account at the end of 1 year.

54. Money growth. Refer to Problem 53. The following rates for 60-month certificates of deposit were also taken from BanxQuote websites:

(A) Oriental Bank & Trust, 1.35% compounded quarterly

(B) BMW Bank of North America, 1.30% compounded monthly

(C) BankFirst Corporation, 1.25% compounded daily

Compute the value of $10,000 invested in each account at the end of 5 years.

55. Advertising. A company is trying to introduce a new product to as many people as possible through television advertising in a large metropolitan area with 2 million possible viewers. A model for the number of people N (in millions) who are aware of the product after t days of advertising was found to be

$$N = 2\left(1 - e^{-0.037t}\right)$$

Graph this function for $0 \leq t \leq 50$. What value does N approach as t increases without bound?

56. Learning curve. People assigned to assemble circuit boards for a computer manufacturing company undergo on-the-job training. From past experience, the learning curve for the average employee is given by

$$N = 40(1 - e^{-0.12t})$$

where N is the number of boards assembled per day after t days of training. Graph this function for $0 \le t \le 30$. What is the maximum number of boards an average employee can be expected to produce in 1 day?

57. Sports salaries. Table 4 shows the average salaries for players in Major League Baseball (MLB) and the National Basketball Association (NBA) in selected years since 1990.

(A) Let x represent the number of years since 1990 and find an exponential regression model ($y = ab^x$) for the average salary in MLB. Use the model to estimate the average salary (to the nearest thousand dollars) in 2022.

(B) The average salary in MLB in 2000 was 1.984 million. How does this compare with the value given by the model of part (A)?

Table 4 **Average Salary (thousand $)**

Year	MLB	NBA
1990	589	750
1993	1,062	1,300
1996	1,101	2,000
1999	1,724	2,400
2002	2,300	4,500
2005	2,633	5,000
2008	3,155	5,585
2011	3,298	4,755

58. Sports salaries. Refer to Table 4.

(A) Let x represent the number of years since 1990 and find an exponential regression model ($y = ab^x$) for the average salary in the NBA. Use the model to estimate the average salary (to the nearest thousand dollars) in 2022.

(B) The average salary in the NBA in 1997 was $2.2 million. How does this compare with the value given by the model of part (A)?

59. Marine biology. Marine life depends on the microscopic plant life that exists in the photic zone, a zone that goes to a depth where only 1% of surface light remains. In some waters with a great deal of sediment, the photic zone may go down only 15 to 20 feet. In some murky harbors, the intensity of light d feet below the surface is given approximately by

$$I = I_0 e^{-0.23d}$$

What percentage of the surface light will reach a depth of

(A) 10 feet?   (B) 20 feet?

60. Marine biology. Refer to Problem 59. Light intensity I relative to depth d (in feet) for one of the clearest bodies of water in the world, the Sargasso Sea, can be approximated by

$$I = I_0 e^{-0.00942d}$$

where I_0 is the intensity of light at the surface. What percentage of the surface light will reach a depth of

(A) 50 feet?   (B) 100 feet?

61. World population growth. From the dawn of humanity to 1830, world population grew to one billion people. In 100 more years (by 1930) it grew to two billion, and 3 billion more were added in only 60 years (by 1990). In 2013, the estimated world population was 7.1 billion with a relative growth rate of 1.1%.

(A) Write an equation that models the world population growth, letting 2013 be year 0.

(B) Based on the model, what is the expected world population (to the nearest hundred million) in 2025? In 2035?

62. Population growth in Ethiopia. In 2012, the estimated population in Ethiopia was 94 million people with a relative growth rate of 3.2%.

(A) Write an equation that models the population growth in Ethiopia, letting 2012 be year 0.

(B) Based on the model, what is the expected population in Ethiopia (to the nearest million) in 2025? In 2035?

63. Internet growth. The number of Internet hosts grew very rapidly from 1994 to 2012 (Table 5).

(A) Let x represent the number of years since 1994. Find an exponential regression model ($y = ab^x$) for this data set and estimate the number of hosts in 2022 (to the nearest million).

(B) Discuss the implications of this model if the number of Internet hosts continues to grow at this rate.

Table 5 **Internet Hosts (millions)**

Year	Hosts
1994	2.4
1997	16.1
2000	72.4
2003	171.6
2006	394.0
2009	625.2
2012	888.2

Source: Internet Software Consortium

64. Life expectancy. Table 6 shows the life expectancy (in years) at birth for residents of the United States from 1970 to 2010. Let x represent years since 1970. Find an exponential regression model for this data and use it to estimate the life expectancy for a person born in 2025.

Table 6

Year of Birth	Life Expectancy
1970	70.8
1975	72.6
1980	73.7
1985	74.7
1990	75.4
1995	75.9
2000	76.9
2005	77.7
2010	78.2

Answers to Matched Problems

1.

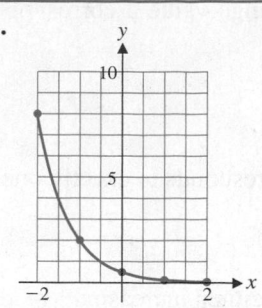

2. (A) 179 bacteria   (B) 1,271,659 bacteria

3. 233 mg

4. Purchase price: $30,363; value after 10 yr: $2,864

5. $7,841.13

6. $7,841.56

1.6 Logarithmic Functions

- Inverse Functions
- Logarithmic Functions
- Properties of Logarithmic Functions
- Calculator Evaluation of Logarithms
- Applications

Find the exponential function keys $\boxed{10^x}$ and $\boxed{e^x}$ on your calculator. Close to these keys you will find the $\boxed{\text{LOG}}$ and $\boxed{\text{LN}}$ keys. The latter two keys represent *logarithmic functions,* and each is closely related to its nearby exponential function. In fact, the exponential function and the corresponding logarithmic function are said to be *inverses* of each other. In this section we will develop the concept of inverse functions and use it to define a logarithmic function as the inverse of an exponential function. We will then investigate basic properties of logarithmic functions, use a calculator to evaluate them for particular values of x, and apply them to real-world problems.

Logarithmic functions are used in modeling and solving many types of problems. For example, the decibel scale is a logarithmic scale used to measure sound intensity, and the Richter scale is a logarithmic scale used to measure the force of an earthquake. An important business application has to do with finding the time it takes money to double if it is invested at a certain rate compounded a given number of times a year or compounded continuously. This requires the solution of an exponential equation, and logarithms play a central role in the process.

Inverse Functions

Look at the graphs of $f(x) = \dfrac{x}{2}$ and $g(x) = \dfrac{|x|}{2}$ in Figure 1:

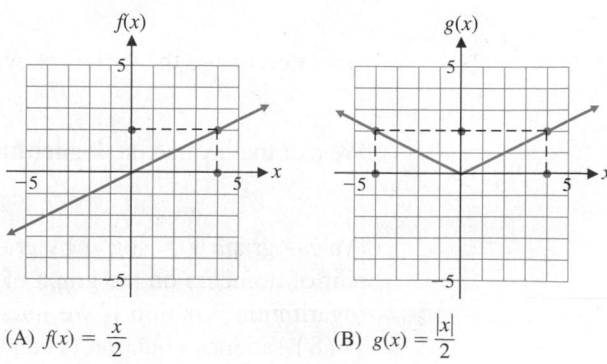

(A) $f(x) = \dfrac{x}{2}$ (B) $g(x) = \dfrac{|x|}{2}$

Figure 1

Because both f and g are functions, each domain value corresponds to exactly one range value. For which function does each range value correspond to exactly one domain value? This is the case only for function *f*. Note that for function *f*, the range value 2 corresponds to the domain value 4. For function *g* the range value 2 corresponds to both −4 and 4. Function *f* is said to be *one-to-one*.

> **DEFINITION**   One-to-One Functions
> A function *f* is said to be **one-to-one** if each range value corresponds to exactly one domain value.

It can be shown that any continuous function that is either increasing* or decreasing for all domain values is one-to-one. If a continuous function increases for some domain values and decreases for others, then it cannot be one-to-one. Figure 1 shows an example of each case.

Explore and Discuss 1   Graph $f(x) = 2^x$ and $g(x) = x^2$. For a range value of 4, what are the corresponding domain values for each function? Which of the two functions is one-to-one? Explain why.

Starting with a one-to-one function *f*, we can obtain a new function called the *inverse* of *f*.

> **DEFINITION**   Inverse of a Function
> If *f* is a one-to-one function, then the **inverse** of *f* is the function formed by interchanging the independent and dependent variables for *f*. Thus, if (a, b) is a point on the graph of *f*, then (b, a) is a point on the graph of the inverse of *f*.
>
> *Note:* If *f* is not one-to-one, then *f* **does not have an inverse**.

In this course, we are interested in the inverses of exponential functions, called *logarithmic functions*.

Logarithmic Functions

If we start with the exponential function *f* defined by

$$y = 2^x \tag{1}$$

and interchange the variables, we obtain the inverse of *f*:

$$x = 2^y \tag{2}$$

We call the inverse the **logarithmic function with base 2**, and write

$$y = \log_2 x \quad \text{if and only if} \quad x = 2^y$$

We can graph $y = \log_2 x$ by graphing $x = 2^y$ since they are equivalent. Any ordered pair of numbers on the graph of the exponential function will be on the graph of the logarithmic function if we interchange the order of the components. For example, $(3, 8)$ satisfies equation (1) and $(8, 3)$ satisfies equation (2). The graphs of $y = 2^x$ and $y = \log_2 x$ are shown in Figure 2. Note that if we fold the paper along the dashed line $y = x$ in Figure 2, the two graphs match exactly. The line $y = x$ is a line of symmetry for the two graphs.

---

*Formally, we say that the function *f* is **increasing** on an interval (a, b) if $f(x_2) > f(x_1)$ whenever $a < x_1 < x_2 < b$; and *f* is **decreasing** on (a, b) if $f(x_2) < f(x_1)$ whenever $a < x_1 < x_2 < b$.

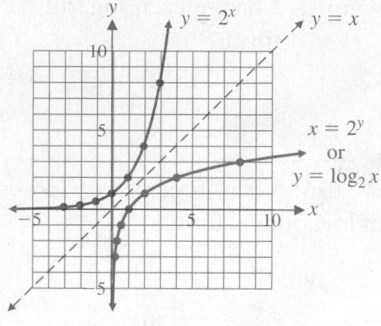

Figure 2

Exponential Function		Logarithmic Function	
x	$y = 2^x$	$x = 2^y$	y
-3	$\frac{1}{8}$	$\frac{1}{8}$	-3
-2	$\frac{1}{4}$	$\frac{1}{4}$	-2
-1	$\frac{1}{2}$	$\frac{1}{2}$	-1
0	1	1	0
1	2	2	1
2	4	4	2
3	8	8	3

$$\begin{bmatrix} \text{Ordered} \\ \text{pairs} \\ \text{reversed} \end{bmatrix}$$

In general, since the graphs of all exponential functions of the form $f(x) = b^x$, $b \neq 1, b > 0$, are either increasing or decreasing, exponential functions have inverses.

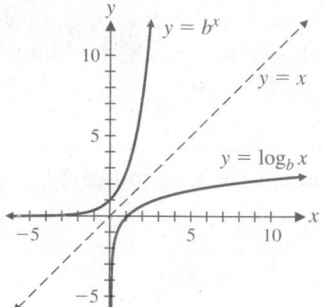

Figure 3

DEFINITION Logarithmic Functions

The inverse of an exponential function is called a **logarithmic function**. For $b > 0$ and $b \neq 1$,

Logarithmic form   Exponential form

$\quad\quad y = \log_b x \quad$ is equivalent to $\quad x = b^y$

The **log to the base b of x** is the exponent to which b must be raised to obtain x. [*Remember:* A logarithm is an exponent.] The **domain** of the logarithmic function is the set of all positive real numbers, which is also the range of the corresponding exponential function; and the **range** of the logarithmic function is the set of all real numbers, which is also the domain of the corresponding exponential function. Typical graphs of an exponential function and its inverse, a logarithmic function, are shown in Figure 3.

CONCEPTUAL **INSIGHT**

Because the domain of a logarithmic function consists of the positive real numbers, the entire graph of a logarithmic function lies to the right of the y axis. In contrast, the graphs of polynomial and exponential functions intersect every vertical line, and the graphs of rational functions intersect all but a finite number of vertical lines.

The following examples involve converting logarithmic forms to equivalent exponential forms, and vice versa.

EXAMPLE 1 Logarithmic–Exponential Conversions Change each logarithmic form to an equivalent exponential form:

(A) $\log_5 25 = 2$ (B) $\log_9 3 = \frac{1}{2}$ (C) $\log_2\left(\frac{1}{4}\right) = -2$

SOLUTION

(A) $\log_5 25 = 2$ is equivalent to $25 = 5^2$

(B) $\log_9 3 = \frac{1}{2}$ is equivalent to $3 = 9^{1/2}$

(C) $\log_2\left(\frac{1}{4}\right) = -2$ is equivalent to $\frac{1}{4} = 2^{-2}$

Matched Problem 1 Change each logarithmic form to an equivalent exponential form:

(A) $\log_3 9 = 2$ (B) $\log_4 2 = \frac{1}{2}$ (C) $\log_3\left(\frac{1}{9}\right) = -2$

EXAMPLE 2 Exponential–Logarithmic Conversions Change each exponential form to an equivalent logarithmic form:

(A) $64 = 4^3$ (B) $6 = \sqrt{36}$ (C) $\frac{1}{8} = 2^{-3}$

SOLUTION

(A) $64 = 4^3$ is equivalent to $\log_4 64 = 3$

(B) $6 = \sqrt{36}$ is equivalent to $\log_{36} 6 = \frac{1}{2}$

(C) $\frac{1}{8} = 2^{-3}$ is equivalent to $\log_2\left(\frac{1}{8}\right) = -3$

Matched Problem 2 Change each exponential form to an equivalent logarithmic form:

(A) $49 = 7^2$ (B) $3 = \sqrt{9}$ (C) $\frac{1}{3} = 3^{-1}$

To gain a deeper understanding of logarithmic functions and their relationship to exponential functions, we consider a few problems where we want to find x, b, or y in $y = \log_b x$, given the other two values. All values are chosen so that the problems can be solved exactly without a calculator.

EXAMPLE 3 Solutions of the Equation $y = \log_b x$ Find y, b, or x, as indicated.

(A) Find y: $y = \log_4 16$ (B) Find x: $\log_2 x = -3$

(C) Find b: $\log_b 100 = 2$

SOLUTION

(A) $y = \log_4 16$ is equivalent to $16 = 4^y$. So,

$$y = 2$$

(B) $\log_2 x = -3$ is equivalent to $x = 2^{-3}$. So,

$$x = \frac{1}{2^3} = \frac{1}{8}$$

(C) $\log_b 100 = 2$ is equivalent to $100 = b^2$. So,

$$b = 10 \quad \text{Recall that } b \text{ cannot be negative.}$$

Matched Problem 3 Find y, b, or x, as indicated.

(A) Find y: $y = \log_9 27$ (B) Find x: $\log_3 x = -1$

(C) Find b: $\log_b 1{,}000 = 3$

Properties of Logarithmic Functions

The properties of exponential functions (Section 1.5) lead to properties of logarithmic functions. For example, consider the exponential property $b^x b^y = b^{x+y}$. Let $M = b^x$, $N = b^y$. Then

$$\log_b MN = \log_b(b^x b^y) = \log_b b^{x+y} = x + y = \log_b M + \log_b N$$

So $\log_b MN = \log_b M + \log_b N$, that is, the logarithm of a product is the sum of the logarithms. Similarly, the logarithm of a quotient is the difference of the logarithms. These properties are among the eight useful properties of logarithms that are listed in Theorem 1.

THEOREM 1 Properties of Logarithmic Functions

If b, M, and N are positive real numbers, $b \neq 1$, and p and x are real numbers, then

1. $\log_b 1 = 0$

2. $\log_b b = 1$

3. $\log_b b^x = x$

4. $b^{\log_b x} = x, \quad x > 0$

5. $\log_b MN = \log_b M + \log_b N$

6. $\log_b \dfrac{M}{N} = \log_b M - \log_b N$

7. $\log_b M^p = p \log_b M$

8. $\log_b M = \log_b N$ if and only if $M = N$

EXAMPLE 4 Using Logarithmic Properties

(A) $\log_b \dfrac{wx}{yz}$ $\quad = \log_b wx - \log_b yz$
$$= \log_b w + \log_b x - (\log_b y + \log_b z)$$
$$= \log_b w + \log_b x - \log_b y - \log_b z$$

(B) $\log_b (wx)^{3/5}$ $\quad = \frac{3}{5} \log_b wx \quad = \frac{3}{5}(\log_b w + \log_b x)$

(C) $e^{x \log_e b} = e^{\log_e b^x} = b^x$

(D) $\dfrac{\log_e x}{\log_e b} = \dfrac{\log_e (b^{\log_b x})}{\log_e b} = \dfrac{(\log_b x)(\log_e b)}{\log_e b} = \log_b x$

Matched Problem 4 Write in simpler forms, as in Example 4.

(A) $\log_b \dfrac{R}{ST}$ (B) $\log_b \left(\dfrac{R}{S}\right)^{2/3}$ (C) $2^{u \log_2 b}$ (D) $\dfrac{\log_2 x}{\log_2 b}$

The following examples and problems will give you additional practice in using basic logarithmic properties.

EXAMPLE 5 Solving Logarithmic Equations Find x so that

$$\tfrac{3}{2}\log_b 4 - \tfrac{2}{3}\log_b 8 + \log_b 2 = \log_b x$$

SOLUTION

$$\tfrac{3}{2}\log_b 4 - \tfrac{2}{3}\log_b 8 + \log_b 2 = \log_b x$$

$$\log_b 4^{3/2} - \log_b 8^{2/3} + \log_b 2 = \log_b x \quad \text{Property 7}$$

$$\log_b 8 - \log_b 4 + \log_b 2 = \log_b x$$

$$\log_b \frac{8 \cdot 2}{4} = \log_b x \quad \text{Properties 5 and 6}$$

$$\log_b 4 = \log_b x$$

$$x = 4 \quad \text{Property 8}$$

Matched Problem 5 Find x so that $3 \log_b 2 + \tfrac{1}{2}\log_b 25 - \log_b 20 = \log_b x$.

EXAMPLE 6 Solving Logarithmic Equations Solve: $\log_{10} x + \log_{10}(x + 1) = \log_{10} 6$.

SOLUTION

$$\log_{10} x + \log_{10}(x + 1) = \log_{10} 6$$

$$\log_{10}[x(x + 1)] = \log_{10} 6 \quad \text{Property 5}$$

$$x(x + 1) = 6 \quad \text{Property 8}$$

$$x^2 + x - 6 = 0 \quad \text{Solve by factoring.}$$

$$(x + 3)(x - 2) = 0$$

$$x = -3, 2$$

We must exclude $x = -3$, since the domain of the function $\log_{10} x$ is $(0, \infty)$; so $x = 2$ is the only solution.

Matched Problem 6   Solve: $\log_3 x + \log_3(x - 3) = \log_3 10$.

Calculator Evaluation of Logarithms

Of all possible logarithmic bases, e and 10 are used almost exclusively. Before we can use logarithms in certain practical problems, we need to be able to approximate the logarithm of any positive number either to base 10 or to base e. And conversely, if we are given the logarithm of a number to base 10 or base e, we need to be able to approximate the number. Historically, tables were used for this purpose, but now calculators make computations faster and far more accurate.

Common logarithms are logarithms with base 10. **Natural logarithms** are logarithms with base e. Most calculators have a key labeled "log" (or "LOG") and a key labeled "ln" (or "LN"). The former represents a common (base 10) logarithm and the latter a natural (base e) logarithm. In fact, "log" and "ln" are both used extensively in mathematical literature, and whenever you see either used in this book without a base indicated, they will be interpreted as follows:

> **Common logarithm:** $\log x$ means $\log_{10} x$
>
> **Natural logarithm:** $\ln x$ means $\log_e x$

Finding the common or natural logarithm using a calculator is very easy. On some calculators, you simply enter a number from the domain of the function and press $\boxed{\text{LOG}}$ or $\boxed{\text{LN}}$. On other calculators, you press either $\boxed{\text{LOG}}$ or $\boxed{\text{LN}}$, enter a number from the domain, and then press $\boxed{\text{ENTER}}$. Check the user's manual for your calculator.

EXAMPLE 7   Calculator Evaluation of Logarithms   Use a calculator to evaluate each to six decimal places:

(A) $\log 3{,}184$ (B) $\ln 0.000\ 349$ (C) $\log(-3.24)$

SOLUTION

(A) $\log 3{,}184 = 3.502\ 973$

(B) $\ln 0.000\ 349 = -7.960\ 439$

(C) $\log(-3.24) = $ Error* -3.24 is not in the domain of the log function.

Matched Problem 7   Use a calculator to evaluate each to six decimal places:

(A) $\log 0.013\ 529$ (B) $\ln 28.693\ 28$ (C) $\ln(-0.438)$

Given the logarithm of a number, how do you find the number? We make direct use of the logarithmic-exponential relationships, which follow from the definition of logarithmic function given at the beginning of this section.

> $\log x = y$ is equivalent to $x = 10^y$
>
> $\ln x = y$ is equivalent to $x = e^y$

EXAMPLE 8   Solving $\log_b x = y$ for x Find x to four decimal places, given the indicated logarithm:

(A) $\log x = -2.315$ (B) $\ln x = 2.386$

SOLUTION

(A) $\log x = -2.315$ Change to equivalent exponential form.

$\quad\quad x = 10^{-2.315}$ Evaluate with a calculator.

$\quad\quad\ \ = 0.0048$

---

*Some calculators use a more advanced definition of logarithms involving complex numbers and will display an ordered pair of real numbers as the value of $\log(-3.24)$. You should interpret such a result as an indication that the number entered is not in the domain of the logarithm function as we have defined it.

(B) $\ln x = 2.386$ Change to equivalent exponential form.

$\qquad x = e^{2.386}$ Evaluate with a calculator.

$\qquad = 10.8699$

Matched Problem 8 Find x to four decimal places, given the indicated logarithm:

(A) $\ln x = -5.062$ (B) $\log x = 2.0821$

We can use logarithms to solve exponential equations.

EXAMPLE 9 Solving Exponential Equations   Solve for x to four decimal places:

(A) $10^x = 2$ (B) $e^x = 3$ (C) $3^x = 4$

SOLUTION

(A) $\qquad 10^x = 2$ Take common logarithms of both sides.

$\qquad \log 10^x = \log 2$ Property 3

$\qquad x = \log 2$ Use a calculator.

$\qquad = 0.3010$ To four decimal places

(B) $\qquad e^x = 3$ Take natural logarithms of both sides.

$\qquad \ln e^x = \ln 3$ Property 3

$\qquad x = \ln 3$ Use a calculator.

$\qquad = 1.0986$ To four decimal places

(C) $\qquad 3^x = 4$ Take either natural or common logarithms of both sides. (We choose common logarithms.)

$\qquad \log 3^x = \log 4$ Property 7

$\qquad x \log 3 = \log 4$ Solve for x.

$\qquad x = \dfrac{\log 4}{\log 3}$ Use a calculator.

$\qquad = 1.2619$ To four decimal places

Matched Problem 9 Solve for x to four decimal places:

(A) $10^x = 7$ (B) $e^x = 6$ (C) $4^x = 5$

Exponential equations can also be solved graphically by graphing both sides of an equation and finding the points of intersection. Figure 4 illustrates this approach for the equations in Example 9.

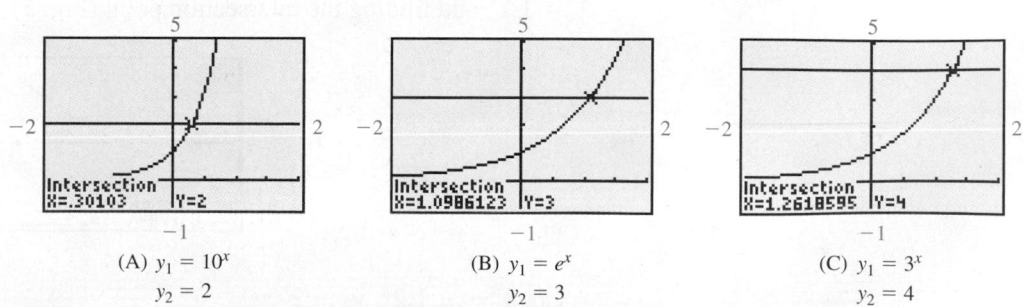

(A) $y_1 = 10^x$
     $y_2 = 2$

(B) $y_1 = e^x$
     $y_2 = 3$

(C) $y_1 = 3^x$
     $y_2 = 4$

Figure 4 **Graphical solution of exponential equations**

Explore and Discuss 2   Discuss how you could find $y = \log_5 38.25$ using either natural or common logarithms on a calculator. [*Hint:* Start by rewriting the equation in exponential form.]

Remark—In the usual notation for natural logarithms, the simplifications of Example 4, parts (C) and (D) on page 77, become

$$e^{x \ln b} = b^x \qquad \text{and} \qquad \frac{\ln x}{\ln b} = \log_b x$$

With these formulas, we can change an exponential function with base b, or a logarithmic function with base b, to expressions involving exponential or logarithmic functions, respectively, to the base e. Such **change-of-base formulas** are useful in calculus.

Applications

A convenient and easily understood way of comparing different investments is to use their **doubling times**—the length of time it takes the value of an investment to double. Logarithm properties, as you will see in Example 10, provide us with just the right tool for solving some doubling-time problems.

EXAMPLE 10   Doubling Time for an Investment   How long (to the next whole year) will it take money to double if it is invested at 20% compounded annually?

SOLUTION   We use the compound interest formula discussed in Section 1.5:

$$A = P\left(1 + \frac{r}{m}\right)^{mt} \qquad \text{Compound interest}$$

The problem is to find t, given $r = 0.20$, $m = 1$, and $A = 2P$; that is,

$$2P = P(1 + 0.2)^t$$
$$2 = 1.2^t$$
$$1.2^t = 2 \qquad \qquad \text{Solve for } t \text{ by taking the natural or}$$
$$\ln 1.2^t = \ln 2 \qquad \qquad \text{common logarithm of both sides (we choose}$$
$$\qquad\qquad\qquad\qquad \text{the natural logarithm).}$$
$$t \ln 1.2 = \ln 2 \qquad \qquad \text{Property 7}$$
$$t = \frac{\ln 2}{\ln 1.2} \qquad \qquad \text{Use a calculator.}$$
$$\approx 3.8 \text{ years} \qquad [\textit{Note:} \ (\ln 2)/(\ln 1.2) \neq \ln 2 - \ln 1.2]$$
$$\approx 4 \text{ years} \qquad \text{To the next whole year}$$

When interest is paid at the end of 3 years, the money will not be doubled; when paid at the end of 4 years, the money will be slightly more than doubled.

Example 10 can also be solved graphically by graphing both sides of the equation $2 = 1.2^t$, and finding the intersection point (Fig. 5).

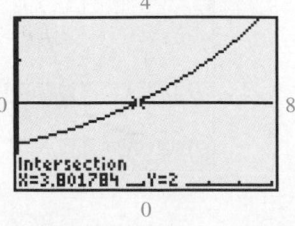

Figure 5 $y_1 = 1.2^x$, $y_2 = 2$

Matched Problem 10)   How long (to the next whole year) will it take money to triple if it is invested at 13% compounded annually?

It is interesting and instructive to graph the doubling times for various rates compounded annually. We proceed as follows:

$$A = P(1 + r)^t$$
$$2P = P(1 + r)^t$$
$$2 = (1 + r)^t$$
$$(1 + r)^t = 2$$
$$\ln(1 + r)^t = \ln 2$$
$$t \ln(1 + r) = \ln 2$$
$$t = \frac{\ln 2}{\ln(1 + r)}$$

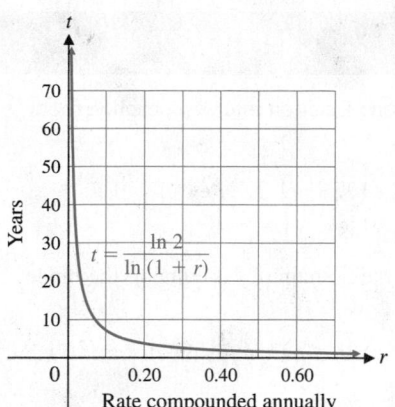

Figure 6

Figure 6 shows the graph of this equation (doubling time in years) for interest rates compounded annually from 1 to 70% (expressed as decimals). Note the dramatic change in doubling time as rates change from 1 to 20% (from 0.01 to 0.20).

Among increasing functions, the logarithmic functions (with bases $b > 1$) increase much more slowly for large values of x than either exponential or polynomial functions. When a visual inspection of the plot of a data set indicates a slowly increasing function, a logarithmic function often provides a good model. We use **logarithmic regression** on a graphing calculator to find the function of the form $y = a + b \ln x$ that best fits the data.

EXAMPLE 11   Home Ownership Rates   The U.S. Census Bureau published the data in Table 1 on home ownership rates. Let x represent time in years with $x = 0$ representing 1900. Use logarithmic regression to find the best model of the form $y = a + b \ln x$ for the home ownership rate y as a function of time x. Use the model to predict the home ownership rate in the United States in 2025 (to the nearest tenth of a percent).

Table 1 **Home Ownership Rates**

Year	Rate (%)
1950	55.0
1960	61.9
1970	62.9
1980	64.4
1990	64.2
2000	67.4
2010	66.9

SOLUTION   Enter the data in a graphing calculator (Fig. 7A) and find the logarithmic regression equation (Fig. 7B). The data set and the regression equation are graphed in Figure 7C. Using TRACE, we predict that the home ownership rate in 2025 would be 69.8%.

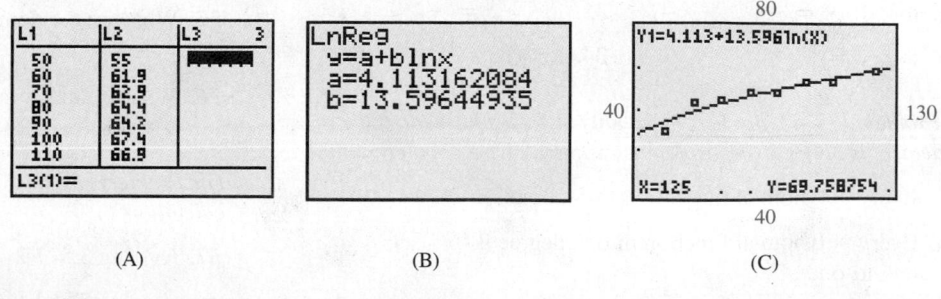

(A)   (B)   (C)

Figure 7

Matched Problem 11)   Refer to Example 11. Use the model to predict the home ownership rate in the United States in 2030 (to the nearest tenth of a percent).

⚠ **CAUTION** Note that in Example 11 we let $x = 0$ represent 1900. If we let $x = 0$ represent 1940, for example, we would obtain a different logarithmic regression equation. We would *not* let $x = 0$ represent 1950 (the first year in Table 1) or any later year, because logarithmic functions are undefined at 0.   ▲

Exercises 1.6

For Problems 1–6, rewrite in equivalent exponential form.

1. $\log_3 27 = 3$

2. $\log_2 32 = 5$

3. $\log_{10} 1 = 0$

4. $\log_e 1 = 0$

5. $\log_4 8 = \frac{3}{2}$

6. $\log_9 27 = \frac{3}{2}$

For Problems 7–12, rewrite in equivalent logarithmic form.

7. $49 = 7^2$

8. $36 = 6^2$

9. $8 = 4^{3/2}$

10. $9 = 27^{2/3}$

11. $A = b^u$

12. $M = b^x$

In Problems 13–20, evaluate the expression without using a calculator.

13. $\log_{10} 100$

14. $\log_{10} 100{,}000$

15. $\log_2 16$

16. $\log_3 \frac{1}{3}$

17. $\log_5 \frac{1}{25}$

18. $\log_4 1$

19. $\ln \frac{1}{e^4}$

20. $\ln e^{-5}$

For Problems 21–26, write in terms simpler forms, as in Example 4.

21. $\log_b \dfrac{P}{Q}$

22. $\log_b FG$

23. $\log_b L^5$

24. $\log_b w^{15}$

25. $3^{p \log_3 q}$

26. $\dfrac{\log_3 P}{\log_3 R}$

For Problems 27–34, find x, y, or b without a calculator.

27. $\log_3 x = 2$

28. $\log_2 x = 2$

29. $\log_7 49 = y$

30. $\log_3 27 = y$

31. $\log_b 10^{-4} = -4$

32. $\log_b e^{-2} = -2$

33. $\log_4 x = \frac{1}{2}$

34. $\log_{25} x = \frac{1}{2}$

✎ *In Problems 35–42, discuss the validity of each statement. If the statement is always true, explain why. If not, give a counterexample.*

35. Every polynomial function is one-to-one.

36. Every polynomial function of odd degree is one-to-one.

37. If g is the inverse of a function f, then g is one-to-one.

38. The graph of a one-to-one function intersects each vertical line exactly once.

39. The inverse of $f(x) = 2x$ is $g(x) = x/2$.

40. The inverse of $f(x) = x^2$ is $g(x) = \sqrt{x}$.

41. If f is one-to-one, then the domain of f is equal to the range of f.

42. If g is the inverse of a function f, then f is the inverse of g.

Find x in Problems 43–50.

43. $\log_b x = \frac{2}{3}\log_b 8 + \frac{1}{2}\log_b 9 - \log_b 6$

44. $\log_b x = \frac{2}{3}\log_b 27 + 2\log_b 2 - \log_b 3$

45. $\log_b x = \frac{3}{2}\log_b 4 - \frac{2}{3}\log_b 8 + 2\log_b 2$

46. $\log_b x = 3\log_b 2 + \frac{1}{2}\log_b 25 - \log_b 20$

47. $\log_b x + \log_b (x - 4) = \log_b 21$

48. $\log_b(x + 2) + \log_b x = \log_b 24$

49. $\log_{10}(x - 1) - \log_{10}(x + 1) = 1$

50. $\log_{10}(x + 6) - \log_{10}(x - 3) = 1$

Graph Problems 51 and 52 by converting to exponential form first.

51. $y = \log_2 (x - 2)$

52. $y = \log_3 (x + 2)$

✎ **53.** Explain how the graph of the equation in Problem 51 can be obtained from the graph of $y = \log_2 x$ using a simple transformation (see Section 1.2).

✎ **54.** Explain how the graph of the equation in Problem 52 can be obtained from the graph of $y = \log_3 x$ using a simple transformation (see Section 1.2).

55. What are the domain and range of the function defined by $y = 1 + \ln(x + 1)$?

56. What are the domain and range of the function defined by $y = \log (x - 1) - 1$?

For Problems 57 and 58, evaluate to five decimal places using a calculator.

57. (A) $\log 3{,}527.2$ (B) $\log 0.006\,913\,2$

(C) $\ln 277.63$ (D) $\ln 0.040\,883$

58. (A) log 72.604 (B) log 0.033 041

(C) ln 40,257 (D) ln 0.005 926 3

For Problems 59 and 60, find x to four decimal places.

59. (A) $\log x = 1.1285$ (B) $\log x = -2.0497$

(C) $\ln x = 2.7763$ (D) $\ln x = -1.8879$

60. (A) $\log x = 2.0832$ (B) $\log x = -1.1577$

(C) $\ln x = 3.1336$ (D) $\ln x = -4.3281$

For Problems 61–66, solve each equation to four decimal places.

61. $10^x = 12$ **62.** $10^x = 153$

63. $e^x = 4.304$ **64.** $e^x = 0.3059$

65. $1.005^{12t} = 3$ **66.** $1.02^{4t} = 2$

Graph Problems 67–74 using a calculator and point-by-point plotting. Indicate increasing and decreasing intervals.

67. $y = \ln x$ **68.** $y = -\ln x$

69. $y = |\ln x|$ **70.** $y = \ln|x|$

71. $y = 2\ln(x + 2)$ **72.** $y = 2\ln x + 2$

73. $y = 4\ln x - 3$ **74.** $y = 4\ln(x - 3)$

75. Explain why the logarithm of 1 for any permissible base is 0.

76. Explain why 1 is not a suitable logarithmic base.

77. Let $p(x) = \ln x$, $q(x) = \sqrt{x}$, and $r(x) = x$. Use a graphing calculator to draw graphs of all three functions in the same viewing window for $1 \le x \le 16$. Discuss what it means for one function to be larger than another on an interval, and then order the three functions from largest to smallest for $1 < x \le 16$.

78. Let $p(x) = \log x$, $q(x) = \sqrt[3]{x}$, and $r(x) = x$. Use a graphing calculator to draw graphs of all three functions in the same viewing window for $1 \le x \le 16$. Discuss what it means for one function to be smaller than another on an interval, and then order the three functions from smallest to largest for $1 < x \le 16$.

Applications

79. Doubling time. In its first 10 years the Gabelli Growth Fund produced an average annual return of 21.36%. Assume that money invested in this fund continues to earn 21.36% compounded annually. How long (to the nearest year) will it take money invested in this fund to double?

80. Doubling time. In its first 10 years the Janus Flexible Income Fund produced an average annual return of 9.58%. Assume that money invested in this fund continues to earn 9.58% compounded annually. How long (to the nearest year) will it take money invested in this fund to double?

81. Investing. How many years (to two decimal places) will it take $1,000 to grow to $1,800 if it is invested at 6% compounded quarterly? Compounded daily?

82. Investing. How many years (to two decimal places) will it take $5,000 to grow to $7,500 if it is invested at 8% compounded semiannually? Compounded monthly?

83. Continuous compound interest. How many years (to two decimal places) will it take an investment of $35,000 to grow to $50,000 if it is invested at 4.75% compounded continuously?

84. Continuous compound interest. How many years (to two decimal places) will it take an investment of $17,000 to grow to $41,000 if it is invested at 2.95% compounded continuously?

85. Supply and demand. A cordless screwdriver is sold through a national chain of discount stores. A marketing company established price–demand and price–supply tables (Tables 2 and 3), where x is the number of screwdrivers people are willing to buy and the store is willing to sell each month at a price of p dollars per screwdriver.

(A) Find a logarithmic regression model $(y = a + b\ln x)$ for the data in Table 2. Estimate the demand (to the nearest unit) at a price level of $50.

Table 2 **Price–Demand**

x	$p = D(x)\,(\$)$
1,000	91
2,000	73
3,000	64
4,000	56
5,000	53

(B) Find a logarithmic regression model $(y = a + b\ln x)$ for the data in Table 3. Estimate the supply (to the nearest unit) at a price level of $50.

Table 3 **Price–Supply**

x	$p = S(x)\,(\$)$
1,000	9
2,000	26
3,000	34
4,000	38
5,000	41

(C) Does a price level of $50 represent a stable condition, or is the price likely to increase or decrease? Explain.

86. Equilibrium point. Use the models constructed in Problem 85 to find the equilibrium point. Write the equilibrium price to the nearest cent and the equilibrium quantity to the nearest unit.

87. Sound intensity: decibels. Because of the extraordinary range of sensitivity of the human ear (a range of over 1,000 million millions to 1), it is helpful to use a logarithmic scale, rather than an absolute scale, to measure sound intensity over this range. The unit of measure is called the *decibel*, after the inventor of the telephone, Alexander Graham Bell. If we let N be the number of decibels, I the power of the sound in question (in watts per square centimeter), and I_0 the power of sound just below the threshold of hearing (approximately 10^{-16} watt per square centimeter), then

$$I = I_0 10^{N/10}$$

Show that this formula can be written in the form

$$N = 10 \log \frac{I}{I_0}$$

88. Sound intensity: decibels. Use the formula in Problem 87 (with $I_0 = 10^{-16}$ W/cm^2) to find the decibel ratings of the following sounds:

(A) Whisper: 10^{-13} W/cm^2

(B) Normal conversation: 3.16×10^{-10} W/cm^2

(C) Heavy traffic: 10^{-8} W/cm^2

(D) Jet plane with afterburner: 10^{-1} W/cm^2

89. Agriculture. Table 4 shows the yield (in bushels per acre) and the total production (in millions of bushels) for corn in the United States for selected years since 1950. Let x

Table 4 **United States Corn Production**

Year	x	Yield (bushels per acre)	Total Production (million bushels)
1950	50	38	2,782
1960	60	56	3,479
1970	70	81	4,802
1980	80	98	6,867
1990	90	116	7,802
2000	100	140	10,192
2010	110	153	12,447

represent years since 1900. Find a logarithmic regression model ($y = a + b \ln x$) for the yield. Estimate (to the nearest bushel per acre) the yield in 2024.

90. Agriculture. Refer to Table 4. Find a logarithmic regression model ($y = a + b \ln x$) for the total production. Estimate (to the nearest million) the production in 2024.

91. World population. If the world population is now 7.1 billion people and if it continues to grow at an annual rate of 1.1% compounded continuously, how long (to the nearest year) would it take before there is only 1 square yard of land per person? (The Earth contains approximately 1.68×10^{14} square yards of land.)

92. Archaeology: carbon-14 dating. The radioactive carbon-14 $\left({}^{14}\text{C} \right)$ in an organism at the time of its death decays according to the equation

$$A = A_0 e^{-0.000124t}$$

where t is time in years and A_0 is the amount of ^{14}C present at time $t = 0$. (See Example 3 in Section 1.5.) Estimate the age of a skull uncovered in an archaeological site if 10% of the original amount of ^{14}C is still present. [*Hint:* Find t such that $A = 0.1A_0$.]

Answers to Matched Problems

1. (A) $9 = 3^2$ (B) $2 = 4^{1/2}$ (C) $\frac{1}{9} = 3^{-2}$
2. (A) $\log_7 49 = 2$ (B) $\log_9 3 = \frac{1}{2}$ (C) $\log_3 \left(\frac{1}{3} \right) = -1$
3. (A) $y = \frac{3}{2}$ (B) $x = \frac{1}{3}$ (C) $b = 10$
4. (A) $\log_b R - \log_b S - \log_b T$ (B) $\frac{2}{3}(\log_b R - \log_b S)$
    (C) b^u (D) $\log_b x$
5. $x = 2$ 6. $x = 5$
7. (A) $-1.868\,734$ (B) $3.356\,663$ (C) Not defined
8. (A) 0.0063 (B) 120.8092
9. (A) 0.8451 (B) 1.7918 (C) 1.1610
10. 9 yr   11. 70.3%

Chapter 1   Summary and Review

Important Terms, Symbols, and Concepts

1.1 Functions

EXAMPLES

- A **Cartesian or rectangular coordinate system** is formed by the intersection of a horizontal real number line, usually called the **x axis**, and a vertical real number line, usually called the **y axis**, at their origins. The axes determine a plane and divide this plane into four **quadrants**. Each point in the plane corresponds to its **coordinates**—an ordered pair (a, b) determined by passing horizontal and vertical lines through the point. The **abscissa** or **x coordinate** a is the coordinate of the intersection of the vertical line and the x axis, and the **ordinate** or **y coordinate** b is the coordinate of the intersection of the horizontal line and the y axis. The point with coordinates $(0, 0)$ is called the **origin**.

- **Point-by-point plotting** may be used to **sketch the graph** of an equation in two variables: Plot enough points from its **solution set** in a rectangular cordinate system so that the total graph is apparent and then connect these points with a smooth curve.

Ex. 1, p. 4

- A **function** is a correspondence between two sets of elements such that to each element in the first set there corresponds one and only one element in the second set. The first set is called the **domain** and the set of corresponding elements in the second set is called the **range**.

- If x is a placeholder for the elements in the domain of a function, then x is called the **independent variable** or the **input**. If y is a placeholder for the elements in the range, then y is called the **dependent variable** or the **output**.

- If in an equation in two variables we get exactly one output for each input, then the equation specifies a function. The graph of such a function is just the graph of the specifying equation. If we get more than one output for a given input, then the equation does not specify a function.

Ex. 2, p. 7

- The **vertical-line test** can be used to determine whether or not an equation in two variables specifies a function (Theorem 1, p. 8).

- The functions specified by equations of the form $y = mx + b$, where $m \neq 0$, are called **linear functions**. Functions specified by equations of the form $y = b$ are called **constant functions**.

- If a function is specified by an equation and the domain is not indicated, we agree to assume that the domain is the set of all inputs that produce outputs that are real numbers.

Ex. 3, p. 8
Ex. 5, p. 10

- The symbol $f(x)$ represents the element in the range of f that corresponds to the element x of the domain.

Ex. 4, p. 9
Ex. 6, p. 11

- **Break-even** and **profit–loss** analysis use a cost function C and a revenue function R to determine when a company will have a loss ($R < C$), will break even ($R = C$), or will have a profit ($R > C$). Typical **cost**, **revenue**, **profit**, and **price–demand functions** are given on page 11.

Ex. 7, p. 12

1.2 Elementary Functions: Graphs and Transformations

- The graphs of **six basic elementary functions** (the identity function, the square and cube functions, the square root and cube root functions, and the absolute value function) are shown on page 19.

Ex. 1, p. 18

- Performing an operation on a function produces a **transformation** of the graph of the function. The basic graph transformations, **vertical and horizontal translations** (shifts), **reflection in the x axis**, and **vertical stretches and shrinks**, are summarized on pages 21 and 23.

Ex. 2, p. 20
Ex. 3, p. 21
Ex. 4, p. 22

- A **piecewise-defined function** is a function whose definition uses different rules for different parts of its domain.

Ex. 5, p. 24
Ex. 6, p. 24
Ex. 7, p. 25

1.3 Linear and Quadratic Functions

- A **mathematical model** is a mathematics problem that, when solved, will provide information about a real-world problem.

- A **linear equation in two variables** is an equation that can be written in the **standard form** $Ax + By = C$, where A, B, and C are constants (A and B are not both zero), and x and y are variables.

- The graph of a linear equation in two variables is a line, and every line in a Cartesian coordinate system is the graph of an equation of the form $Ax + By = C$.

Ex. 2, p. 32

- If (x_1, y_1) and (x_2, y_2) are two points on a line with $x_1 \neq x_2$, then the **slope** of the line is $m = \dfrac{y_2 - y_1}{x_2 - x_1}$.

- The **point-slope form** of the line with slope m that passes through the point (x_1, y_1) is $y - y_1 = m(x - x_1)$.

Ex. 1, p. 32

- The **slope-intercept form** of the line with slope m that has y intercept b is $y = mx + b$.

- The graph of the equation $x = a$ is a **vertical line**, and the graph of $y = b$ is a **horizontal line**.

- A function of the form $f(x) = mx + b$, where $m \neq 0$, is a **linear function**.

- A function of the form $f(x) = ax^2 + bx + c$, where $a \neq 0$, is a **quadratic function** in **standard form**, and its graph is a **parabola**.

Ex. 3, p. 35

1.3 Linear and Quadratic Functions (*Continued*)

- Completing the square in the standard form of a quadratic function produces the **vertex form**
Ex. 4, p. 39

$$f(x) = a(x - h)^2 + k \quad \text{Vertex form}$$
Ex. 6, p. 42

- From the vertex form of a quadratic function, we can read off the vertex, axis of symmetry, maximum or minimum, and range, and can easily sketch the graph (p. 38).

- In a competitive market, the intersection of the supply equation and the demand equation is called the **equilibrium point**, the corresponding price is called the **equilibrium price**, and the common value of supply and demand is called the **equilibrium quantity**.
Ex. 5, p. 40

- A graph of the points in a data set is called a **scatter plot**. **Linear regression** can be used to find the linear function (line) that is the best fit for a data set. **Quadratic regression** can be used to find the quadratic function (parabola) that is the best fit.
Ex. 7, p. 43
Ex. 8, p. 44

1.4 Polynomial and Rational Functions

- A **polynomial function** is a function that can be written in the form

$$f(x) = a_n x^n + a_{n-1} x^{n-1} + \cdots + a_1 x + a_0$$

for n a nonnegative integer called the **degree** of the polynomial. The coefficients $a_0, a_1, \ldots, a_n$ are real numbers with **leading coefficient** $a_n \neq 0$. The **domain** of a polynomial function is the set of all real numbers. Graphs of representative polynomial functions are shown on page 52 and inside the front cover.

- The graph of a polynomial function of degree n can intersect the x axis at most n times. An x intercept is also called a **zero** or **root**.

- The graph of a polynomial function has no sharp corners and is **continuous**, that is, it has no holes or breaks.

- **Polynomial regression** produces a polynomial of specified degree that best fits a data set.
Ex. 1, p. 53

- A **rational function** is any function that can be written in the form

$$f(x) = \frac{n(x)}{d(x)} \qquad d(x) \neq 0$$

where $n(x)$ and $d(x)$ are polynomials. The **domain** is the set of all real numbers such that $d(x) \neq 0$. Graphs of representative rational functions are shown on page 55 and inside the front cover.

- Unlike polynomial functions, a rational function can have vertical asymptotes [but not more than the degree of the denominator $d(x)$] and at most one horizontal asymptote.
Ex. 2, p. 55

- A procedure for finding the vertical and horizontal asymptotes of a rational function is given on page 57.
Ex. 3, p. 57

1.5 Exponential Functions

- An **exponential function** is a function of the form

$$f(x) = b^x$$

where $b \neq 1$ is a positive constant called the **base**. The **domain** of f is the set of all real numbers, and the **range** is the set of positive real numbers.

- The graph of an exponential function is continuous, passes through $(0, 1)$, and has the x axis as a horizontal asymptote. If $b > 1$, then b^x increases as x increases; if $0 < b < 1$, then b^x decreases as x increases (Theorem 1, p. 64).
Ex. 1, p. 64

- Exponential functions obey the familiar laws of exponents and satisfy additional properties (Theorem 2, p. 65).

- The base that is used most frequently in mathematics is the irrational number $e \approx 2.7183$.
Ex. 2, p. 66

- Exponential functions can be used to model population growth and radioactive decay.
Ex. 3, p. 67

- **Exponential regression** on a graphing calculator produces the function of the form $y = ab^x$ that best fits a data set.
Ex. 4, p. 68

- Exponential functions are used in computations of **compound interest** and **continuous compound interest**:

$$A = P\left(1 + \frac{r}{m}\right)^{mt} \quad \text{Compound interest}$$

$$A = Pe^{rt} \quad \text{Continuous compound interest}$$

(see summary on p. 70).

1.6 Logarithmic Functions

- A function is said to be **one-to-one** if each range value corresponds to exactly one domain value.

- The **inverse** of a one-to-one function f is the function formed by interchanging the independent and dependent variables of f. That is, (a, b) is a point on the graph of f if and only if (b, a) is a point on the graph of the inverse of f. A function that is not one-to-one does not have an inverse.

- The inverse of the exponential function with base b is called the **logarithmic function with base b**, denoted $y = \log_b x$. The **domain** of $\log_b x$ is the set of all positive real numbers (which is the range of b^x), and the range of $\log_b x$ is the set of all real numbers (which is the domain of b^x).

- Because $\log_b x$ is the inverse of the function b^x,

Logarithmic form		Exponential form
$y = \log_b x$	is equivalent to	$x = b^y$

- Properties of logarithmic functions can be obtained from corresponding properties of exponential functions (Theorem 1, p. 77).

- Logarithms to the base 10 are called **common logarithms**, often denoted simply by log x. Logarithms to the base e are called **natural logarithms**, often denoted by ln x.

- Logarithms can be used to find an investment's **doubling time**—the length of time it takes for the value of an investment to double.

- **Logarithmic regression** on a graphing calculator produces the function of the form $y = a + b \ln x$ that best fits a data set.

Review Exercises

Work through all the problems in this chapter review and check your answers in the back of the book. Answers to all review problems are there along with section numbers in italics to indicate where each type of problem is discussed. Where weaknesses show up, review appropriate sections in the text.

In Problems 1–3, use point-by-point plotting to sketch the graph of each equation.

1. $y = 5 - x^2$

2. $x^2 = y^2$

3. $y^2 = 4x^2$

4. Indicate whether each graph specifies a function:

(A)

(B)

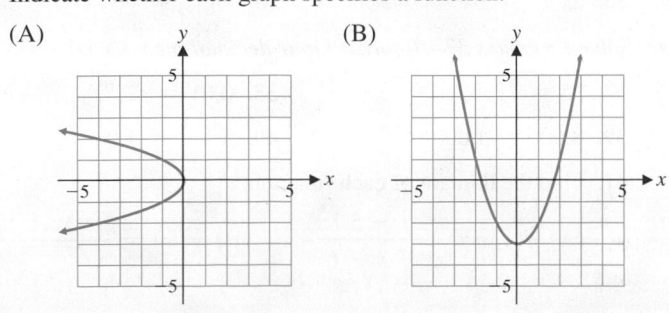

(C)

(D)

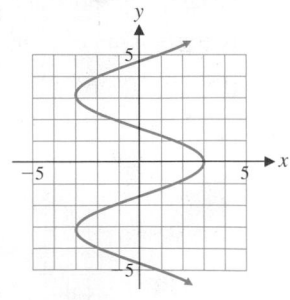

5. For $f(x) = 2x - 1$ and $g(x) = x^2 - 2x$, find:

(A) $f(-2) + g(-1)$ (B) $f(0) \cdot g(4)$

(C) $\dfrac{g(2)}{f(3)}$ (D) $\dfrac{f(3)}{g(2)}$

6. Sketch a graph of $3x + 2y = 9$.

7. Write an equation of a line with x intercept 6 and y intercept 4. Write the final answer in the form $Ax + By = C$.

8. Sketch a graph of $2x - 3y = 18$. What are the intercepts and slope of the line?

9. Write an equation in the form $y = mx + b$ for a line with slope $-\dfrac{2}{3}$ and y intercept 6.

10. Write the equations of the vertical line and the horizontal line that pass through $(-6, 5)$.

11. Write the equation of a line through each indicated point with the indicated slope. Write the final answer in the form $y = mx + b$.

   (A) $m = -\dfrac{2}{3}; (-3, 2)$ (B) $m = 0; (3, 3)$

12. Write the equation of the line through the two indicated points. Write the final answer in the form $Ax + By = C$.

   (A) $(-3, 5), (1, -1)$ (B) $(-1, 5), (4, 5)$

   (C) $(-2, 7), (-2, -2)$

13. Write in logarithmic form using base e: $u = e^v$.

14. Write in logarithmic form using base 10: $x = 10^y$.

15. Write in exponential form using base e: $\ln M = N$.

16. Write in exponential form using base 10: $\log u = v$.

Solve Problems 17–19 for x exactly without using a calculator.

17. $\log_3 x = 2$ 18. $\log_x 36 = 2$

19. $\log_2 16 = x$

Solve problems 20–23 for x to three decimal places.

20. $10^x = 143.7$ 21. $e^x = 503{,}000$

22. $\log x = 3.105$ 23. $\ln x = -1.147$

24. Use the graph of function f in the figure to determine (to the nearest integer) x or y as indicated.

   (A) $y = f(0)$ (B) $4 = f(x)$

   (C) $y = f(3)$ (D) $3 = f(x)$

   (E) $y = f(-6)$ (F) $-1 = f(x)$

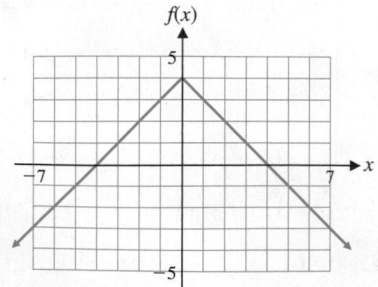

25. Sketch a graph of each of the functions in parts (A)–(D) using the graph of function f in the figure below.

   (A) $y = -f(x)$ (B) $y = f(x) + 4$

   (C) $y = f(x - 2)$ (D) $y = -f(x + 3) - 3$

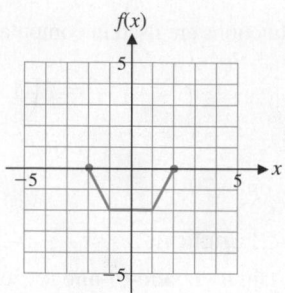

26. Complete the square and find the vertex form for the quadratic function

$$f(x) = -x^2 + 4x$$

   Then write a brief description of the relationship between the graph of f and the graph of $y = x^2$.

27. Match each equation with a graph of one of the functions f, g, m, or n in the figure.

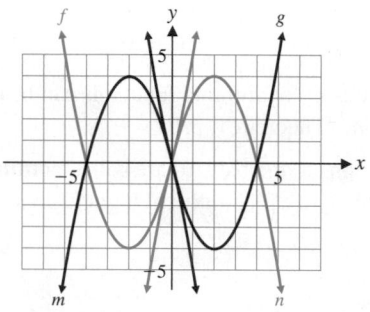

   (A) $y = (x - 2)^2 - 4$ (B) $y = -(x + 2)^2 + 4$

   (C) $y = -(x - 2)^2 + 4$ (D) $y = (x + 2)^2 - 4$

28. Referring to the graph of function f in the figure for Problem 27 and using known properties of quadratic functions, find each of the following to the nearest integer:

   (A) Intercepts   (B) Vertex

   (C) Maximum or minimum   (D) Range

In Problems 29–32, each equation specifies a function. Determine whether the function is linear, quadratic, constant, or none of these.

29. $y = 4 - x + 3x^2$ 30. $y = \dfrac{1 + 5x}{6}$

31. $y = \dfrac{7 - 4x}{2x}$ 32. $y = 8x + 2(10 - 4x)$

Solve Problems 33–36 for x exactly without using a calculator.

33. $\log(x + 5) = \log(2x - 3)$ 34. $2\ln(x - 1) = \ln(x^2 - 5)$

35. $2x^2 e^x = 3xe^x$ 36. $\log_{1/3} 9 = x$

Solve Problems 37–40 for x to four decimal places.

37. $35 = 7(3^x)$ 38. $0.01 = e^{-0.05x}$

39. $8{,}000 = 4{,}000(1.08^x)$ 40. $5^{2x-3} = 7.08$

41. Find the domain of each function:

   (A) $f(x) = \dfrac{2x - 5}{x^2 - x - 6}$ (B) $g(x) = \dfrac{3x}{\sqrt{5 - x}}$

42. Find the vertex form for $f(x) = 4x^2 + 4x - 3$ and then find the intercepts, the vertex, the maximum or minimum, and the range.

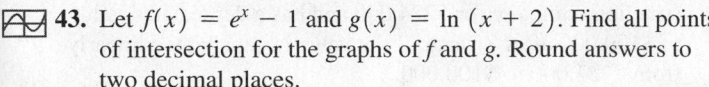 **43.** Let $f(x) = e^x - 1$ and $g(x) = \ln(x + 2)$. Find all points of intersection for the graphs of f and g. Round answers to two decimal places.

In Problems 44 and 45, use point-by-point plotting to sketch the graph of each function.

44. $f(x) = \dfrac{50}{x^2 + 1}$ **45.** $f(x) = \dfrac{-66}{2 + x^2}$

In Problems 46–49, if $f(x) = 5x + 1$, find and simplify.

46. $f(f(0))$ **47.** $f(f(-1))$

48. $f(2x - 1)$ **49.** $f(4 - x)$

50. Let $f(x) = 3 - 2x$. Find

(A) $f(2)$ (B) $f(2 + h)$

(C) $f(2 + h) - f(2)$ (D) $\dfrac{f(2 + h) - f(2)}{h}, h \neq 0$

51. Explain how the graph of $m(x) = -|x - 4|$ is related to the graph of $y = |x|$.

52. Explain how the graph of $g(x) = 0.3x^3 + 3$ is related to the graph of $y = x^3$.

In Problems 53–55, find the equation of any horizontal asymptote.

53. $f(x) = \dfrac{5x + 4}{x^2 - 3x + 1}$ **54.** $f(x) = \dfrac{3x^2 + 2x - 1}{4x^2 - 5x + 3}$

55. $f(x) = \dfrac{x^2 + 4}{100x + 1}$

In Problems 56 and 57, find the equations of any vertical asymptotes.

56. $f(x) = \dfrac{x^2 + 100}{x^2 - 100}$ **57.** $f(x) = \dfrac{x^2 + 3x}{x^2 + 2x}$

In Problems 58–60, discuss the validity of each statement. If the statement is always true, explain why. If not, give a counterexample.

58. Every polynomial function is a rational function.

59. Every rational function is a polynomial function.

60. The graph of every rational function has at least one vertical asymptote.

61. There exists a rational function that has both a vertical and horizontal asymptote.

62. Sketch the graph of f for $x \geq 0$.

$$f(x) = \begin{cases} 9 + 0.3x & \text{if } 0 \leq x \leq 20 \\ 5 + 0.2x & \text{if } x > 20 \end{cases}$$

63. Sketch the graph of g for $x \geq 0$.

$$g(x) = \begin{cases} 0.5x + 5 & \text{if } 0 \leq x \leq 10 \\ 1.2x - 2 & \text{if } 10 < x \leq 30 \\ 2x - 26 & \text{if } x > 30 \end{cases}$$

64. Write an equation for the graph shown in the form $y = a(x - h)^2 + k$, where a is either -1 or $+1$ and h and k are integers.

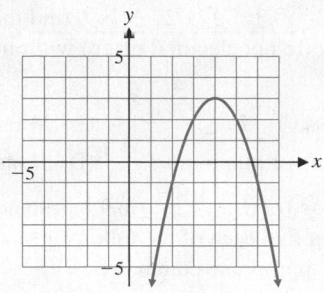

65. Given $f(x) = -0.4x^2 + 3.2x + 1.2$, find the following algebraically (to one decimal place) without referring to a graph:

(A) Intercepts   (B) Vertex

(C) Maximum or minimum   (D) Range

66. Graph $f(x) = -0.4x^2 + 3.2x + 1.2$ in a graphing calculator and find the following (to one decimal place) using TRACE and appropriate commands:

(A) Intercepts   (B) Vertex

(C) Maximum or minimum   (D) Range

67. Noting that $\pi = 3.141\,592\,654\ldots$ and $\sqrt{2} = 1.414\,213\,562\ldots$ explain why the calculator results shown here are obvious. Discuss similar connections between the natural logarithmic function and the exponential function with base e.

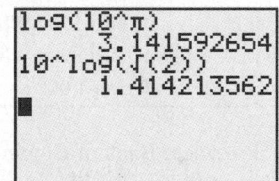

Solve Problems 68–71 exactly without using a calculator.

68. $\log x - \log 3 = \log 4 - \log(x + 4)$

69. $\ln(2x - 2) - \ln(x - 1) = \ln x$

70. $\ln(x + 3) - \ln x = 2 \ln 2$

71. $\log 3x^2 = 2 + \log 9x$

72. Write $\ln y = -5t + \ln c$ in an exponential form free of logarithms. Then solve for y in terms of the remaining variables.

73. Explain why 1 cannot be used as a logarithmic base.

74. The following graph is the result of applying a sequence of transformations to the graph of $y = \sqrt[3]{x}$. Describe the transformations and write an equation for the graph.

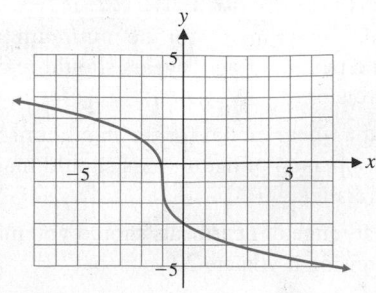

75. Given $G(x) = 0.3x^2 + 1.2x - 6.9$, find the following algebraically (to one decimal place) without the use of a graph:

  (A) Intercepts   (B) Vertex

  (C) Maximum or minimum   (D) Range

76. Graph $G(x) = 0.3x^2 + 1.2x - 6.9$ in a standard viewing window. Then find each of the following (to one decimal place) using appropriate commands.

  (A) Intercepts   (B) Vertex

  (C) Maximum or minimum   (D) Range

Applications

In all problems involving days, a 365-day year is assumed.

77. Electricity rates. The table shows the electricity rates charged by Easton Utilities in the summer months.

  (A) Write a piecewise definition of the monthly charge $S(x)$ (in dollars) for a customer who uses x kWh in a summer month.

  (B) Graph $S(x)$.

Energy Charge (June–September)
\$3.00 for the first 20 kWh or less
5.70¢ per kWh for the next 180 kWh
3.46¢ per kWh for the next 800 kWh
2.17¢ per kWh for all over 1,000 kWh

78. Money growth. Provident Bank of Cincinnati, Ohio, offered a certificate of deposit that paid 1.25% compounded quarterly. If a \$5,000 CD earns this rate for 5 years, how much will it be worth?

79. Money growth. Capital One Bank of Glen Allen, Virginia, offered a certificate of deposit that paid 1.05% compounded daily. If a \$5,000 CD earns this rate for 5 years, how much will it be worth?

80. Money growth. How long will it take for money invested at 6.59% compounded monthly to triple?

81. Money growth. How long will it take for money invested at 7.39% compounded continuously to double?

82. Sports medicine. A simple rule of thumb for determining your maximum safe heart rate (in beats per minute) is to subtract your age from 220. While exercising, you should maintain a heart rate between 60% and 85% of your maximum safe rate.

  (A) Find a linear model for the minimum heart rate m that a person of age x years should maintain while exercising.

  (B) Find a linear model for the maximum heart rate M that a person of age x years should maintain while exercising.

  (C) What range of heartbeats should you maintain while exercising if you are 20 years old?

  (D) What range of heartbeats should you maintain while exercising if you are 50 years old?

83. Linear depreciation. A bulldozer was purchased by a construction company for \$224,000 and has a depreciated value of \$100,000 after 8 years. If the value is depreciated linearly from \$224,000 to \$100,000,

  (A) Find the linear equation that relates value V (in dollars) to time t (in years).

  (B) What would be the depreciated value after 12 years?

84. High school dropout rates. The table gives U.S. high school dropout rates as percentages for selected years since 1980. A linear regression model for the data is

$$r = -0.198t + 14.2$$

where t represents years since 1980 and r is the dropout rate expressed as a percentage.

High School Dropout Rates (%)

1980	1985	1990	1995	2000	2005	2010
14.1	12.6	12.1	12.0	10.9	9.4	7.4

  (A) Interpret the slope of the model.

  (B) Draw a scatter plot of the data and the model in the same coordinate system.

  (C) Use the model to predict the first year for which the dropout rate is less than 5%.

85. Consumer Price Index. The U.S. Consumer Price Index (CPI) in recent years is given in the table. A scatter plot of the data and linear regression line are shown in the figure, where x represents years since 2000.

Consumer Price Index (1982–1984 = 100)

Year	CPI
2000	172.2
2002	179.9
2004	188.9
2006	198.3
2008	211.1
2010	218.1

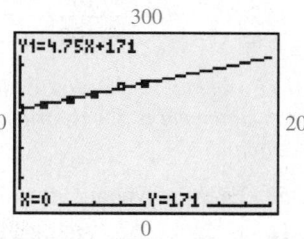

Source: U.S. Bureau of Labor Statistics

  (A) Interpret the slope of the model.

  (B) Predict the CPI in 2024.

86. Construction. A construction company has 840 feet of chain-link fence that is used to enclose storage areas for equipment and materials at construction sites. The supervisor wants to set up two identical rectangular storage areas sharing a common fence (see the figure).

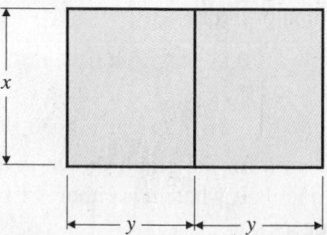

Assuming that all fencing is used,

(A) Express the total area $A(x)$ enclosed by both pens as a function of x.

(B) From physical considerations, what is the domain of the function A?

(C) Graph function A in a rectangular coordinate system.

(D) Use the graph to discuss the number and approximate locations of values of x that would produce storage areas with a combined area of 25,000 square feet.

(E) Approximate graphically (to the nearest foot) the values of x that would produce storage areas with a combined area of 25,000 square feet.

(F) Determine algebraically the dimensions of the storage areas that have the maximum total combined area. What is the maximum area?

87. **Equilibrium point.** A company is planning to introduce a 10-piece set of nonstick cookware. A marketing company established price–demand and price–supply tables for selected prices (Tables 1 and 2), where x is the number of cookware sets people are willing to buy and the company is willing to sell each month at a price of p dollars per set.

Table 1 **Price–Demand**

x	$p = D(x)(\$)$
985	330
2,145	225
2,950	170
4,225	105
5,100	50

Table 2 **Price–Supply**

x	$p = S(x)(\$)$
985	30
2,145	75
2,950	110
4,225	155
5,100	190

(A) Find a quadratic regression model for the data in Table 1. Estimate the demand at a price level of $180.

(B) Find a linear regression model for the data in Table 2. Estimate the supply at a price level of $180.

(C) Does a price level of $180 represent a stable condition, or is the price likely to increase or decrease? Explain.

(D) Use the models in parts (A) and (B) to find the equilibrium point. Write the equilibrium price to the nearest cent and the equilibrium quantity to the nearest unit.

88. **Crime statistics.** According to data published by the FBI, the crime index in the United States has shown a downward trend since the early 1990s (see table).

Crime Index

Year	Crimes per 100,000 Inhabitants
1987	5,550
1992	5,660
1997	4,930
2002	4,125
2007	3,749
2009	3,466

(A) Find a cubic regression model for the crime index if $x = 0$ represents 1987.

(B) Use the cubic regression model to predict the crime index in 2022.

89. **Medicine.** One leukemic cell injected into a healthy mouse will divide into 2 cells in about $\frac{1}{2}$ day. At the end of the day these 2 cells will divide into 4. This doubling continues until 1 billion cells are formed; then the animal dies with leukemic cells in every part of the body.

(A) Write an equation that will give the number N of leukemic cells at the end of t days.

(B) When, to the nearest day, will the mouse die?

90. **Marine biology.** The intensity of light entering water is reduced according to the exponential equation

$$I = I_0 e^{-kd}$$

where I is the intensity d feet below the surface, I_0 is the intensity at the surface, and k is the coefficient of extinction. Measurements in the Sargasso Sea have indicated that half of the surface light reaches a depth of 73.6 feet. Find k (to five decimal places), and find the depth (to the nearest foot) at which 1% of the surface light remains.

91. **Agriculture.** The number of dairy cows on farms in the United States is shown in the table for selected years since 1950. Let 1940 be year 0.

Dairy Cows on Farms in the United States

Year	Dairy Cows (thousands)
1950	23,853
1960	19,527
1970	12,091
1980	10,758
1990	10,015
2000	9,190
2010	9,117

(A) Find a logarithmic regression model $(y = a + b \ln x)$ for the data. Estimate (to the nearest thousand) the number of dairy cows in 2023.

(B) Explain why it is not a good idea to let 1950 be year 0.

92. Population growth. The population of some countries has a relative growth rate of 3% (or more) per year. At this rate, how many years (to the nearest tenth of a year) will it take a population to double?

93. Medicare. The annual expenditures for Medicare (in billions of dollars) by the U.S. government for selected years since 1980 are shown in the table. Let x represent years since 1980.

Medicare Expenditures

Year	Billion $
1980	37
1985	72
1990	111
1995	181
2000	197
2005	299
2010	452

(A) Find an exponential regression model $(y = ab^x)$ for the data. Estimate (to the nearest billion) the annual expenditures in 2022.

(B) When will the annual expenditures reach two trillion dollars?

CALCULUS

2

Limits and the Derivative

Introduction

How do algebra and calculus differ? The two words *static* and *dynamic* probably come as close as any to expressing the difference between the two disciplines. In algebra, we solve equations for a particular value of a variable—a static notion. In calculus, we are interested in how a change in one variable affects another variable—a dynamic notion.

Isaac Newton (1642–1727) of England and Gottfried Wilhelm von Leibniz (1646–1716) of Germany developed calculus independently to solve problems concerning motion. Today calculus is used not just in the physical sciences, but also in business, economics, life sciences, and social sciences—any discipline that seeks to understand dynamic phenomena.

In Chapter 2 we introduce the *derivative*, one of the two key concepts of calculus. The second, the *integral*, is the subject of Chapter 5. Both key concepts depend on the notion of *limit*, which is explained in Sections 2.1 and 2.2. We consider many applications of limits and derivatives. See, for example, Problems 89 and 90 in Section 2.2 on the concentration of a drug in the bloodstream.

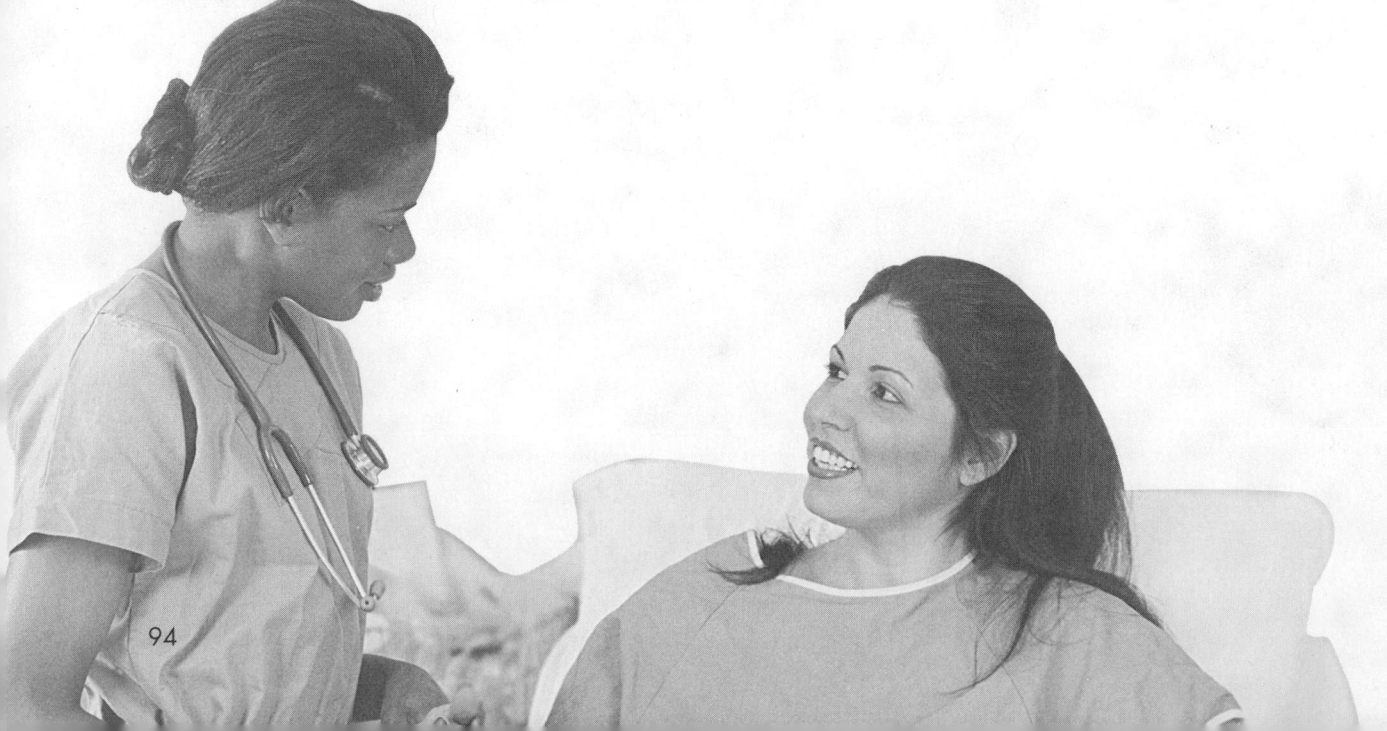

2.1 Introduction to Limits

- Functions and Graphs: Brief Review
- Limits: A Graphical Approach
- Limits: An Algebraic Approach
- Limits of Difference Quotients

Basic to the study of calculus is the concept of a *limit*. This concept helps us to describe, in a precise way, the behavior of $f(x)$ when x is close, but not equal, to a particular value c. In this section, we develop an intuitive and informal approach to evaluating limits.

Functions and Graphs: Brief Review

The graph of the function $y = f(x) = x + 2$ is the graph of the set of all ordered pairs $(x, f(x))$. For example, if $x = 2$, then $f(2) = 4$ and $(2, f(2)) = (2, 4)$ is a point on the graph of f. Figure 1 shows $(-1, f(-1))$, $(1, f(1))$, and $(2, f(2))$ plotted on the graph of f. Notice that the domain values -1, 1, and 2 are associated with the x axis and the range values $f(-1) = 1, f(1) = 3$, and $f(2) = 4$ are associated with the y axis.

Given x, it is sometimes useful to read $f(x)$ directly from the graph of f. Example 1 reviews this process.

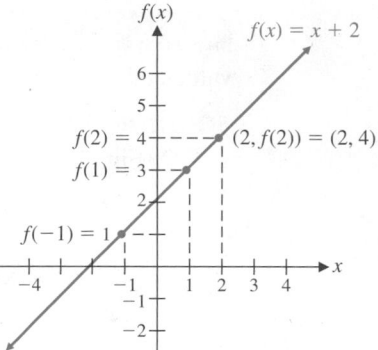

Figure 1

EXAMPLE 1 Finding Values of a Function from Its Graph   Complete the following table, using the given graph of the function g.

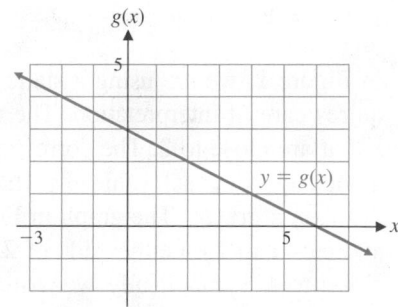

x	$g(x)$
-2	
1	
3	
4	

SOLUTION To determine $g(x)$, proceed vertically from the x value on the x axis to the graph of g and then horizontally to the corresponding y value $g(x)$ on the y axis (as indicated by the dashed lines).

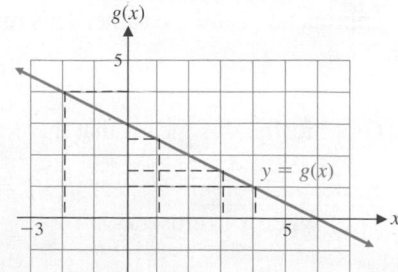

x	$g(x)$
-2	4.0
1	2.5
3	1.5
4	1.0

Matched Problem 1 Complete the following table, using the given graph of the function h.

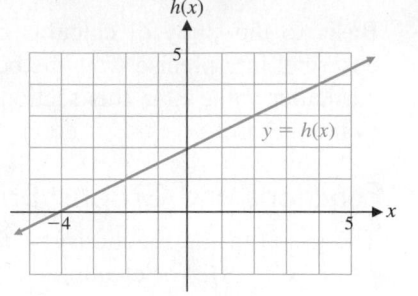

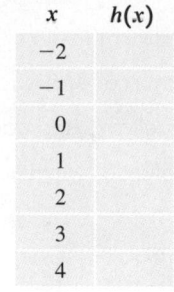

x	$h(x)$
-2	
-1	
0	
1	
2	
3	
4	

Limits: A Graphical Approach

We introduce the important concept of a *limit* through an example, which leads to an intuitive definition of the concept.

EXAMPLE 2 Analyzing a Limit Let $f(x) = x + 2$. Discuss the behavior of the values of $f(x)$ when x is close to 2.

SOLUTION We begin by drawing a graph of f that includes the domain value $x = 2$ (Fig. 2).

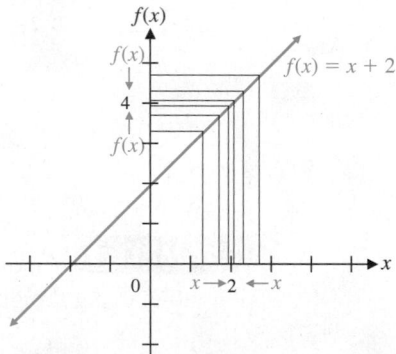

Figure 2

In Figure 2, we are using a static drawing to describe a dynamic process. This requires careful interpretation. The thin vertical lines in Figure 2 represent values of x that are close to 2. The corresponding horizontal lines identify the value of $f(x)$ associated with each value of x. [Example 1 dealt with the relationship between x and $f(x)$ on a graph.] The graph in Figure 2 indicates that as the values of x get closer and closer to 2 on either side of 2, the corresponding values of $f(x)$ get closer and closer to 4. Symbolically, we write

$$\lim_{x \to 2} f(x) = 4$$

This equation is read as "The limit of $f(x)$ as x approaches 2 is 4." Note that $f(2) = 4$. That is, the value of the function at 2 and the limit of the function as x approaches 2 are the same. This relationship can be expressed as

$$\lim_{x \to 2} f(x) = f(2) = 4$$

Graphically, this means that there is no hole or break in the graph of f at $x = 2$.

Matched Problem 2 Let $f(x) = x + 1$. Discuss the behavior of the values of $f(x)$ when x is close to 1.

We now present an informal definition of the important concept of a limit. A precise definition is not needed for our discussion, but one is given in a footnote.*

> **DEFINITION** Limit
> We write
>
> $$\lim_{x \to c} f(x) = L \quad \text{or} \quad f(x) \to L \text{ as } x \to c$$
>
> if the functional value $f(x)$ is close to the single real number L whenever x is close, but not equal, to c (on either side of c).
>
> **Note:** The existence of a limit at c has nothing to do with the value of the function at c. In fact, c may not even be in the domain of f. However, the function must be defined on both sides of c.

The next example involves the **absolute value function**:

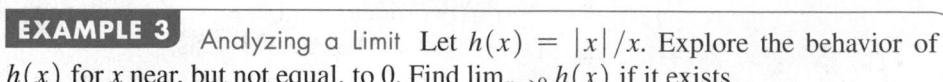

$$f(x) = |x| = \begin{cases} -x & \text{if } x < 0 \\ x & \text{if } x \ge 0 \end{cases} \quad \begin{array}{l} f(-2) = |-2| = -(-2) = 2 \\ f(3) \ = |3| = 3 \end{array}$$

The graph of f is shown in Figure 3.

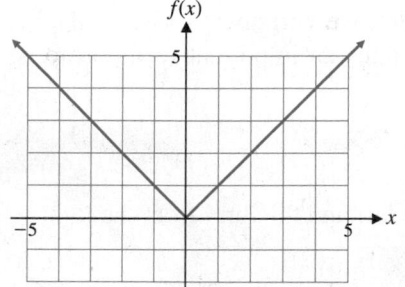

Figure 3 $f(x) = |x|$

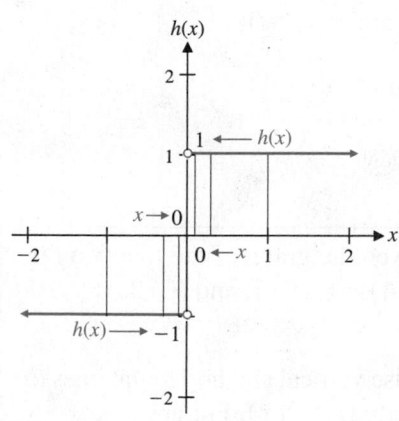

Figure 4

EXAMPLE 3 Analyzing a Limit Let $h(x) = |x|/x$. Explore the behavior of $h(x)$ for x near, but not equal, to 0. Find $\lim_{x \to 0} h(x)$ if it exists.

SOLUTION The function h is defined for all real numbers except 0 [$h(0) = |0|/0$ is undefined]. For example,

$$h(-2) = \frac{|-2|}{-2} = \frac{2}{-2} = -1$$

Note that if x is any negative number, then $h(x) = -1$ (if $x < 0$, then the numerator $|x|$ is positive but the denominator x is negative, so $h(x) = |x|/x = -1$). If x is any positive number, then $h(x) = 1$ (if $x > 0$, then the numerator $|x|$ is equal to the denominator x, so $h(x) = |x|/x = 1$). Figure 4 illustrates the behavior of $h(x)$ for x near 0. Note that the absence of a solid dot on the vertical axis indicates that h is not defined when $x = 0$.

When x is near 0 (on either side of 0), is $h(x)$ near one specific number? The answer is "No," because $h(x)$ is -1 for $x < 0$ and 1 for $x > 0$. Consequently, we say that

$$\lim_{x \to 0} \frac{|x|}{x} \text{ does not exist}$$

Neither $h(x)$ nor the limit of $h(x)$ exists at $x = 0$. However, the limit from the left and the limit from the right both exist at 0, but they are not equal.

Matched Problem 3 Graph

$$h(x) = \frac{x - 2}{|x - 2|}$$

and find $\lim_{x \to 2} h(x)$ if it exists.

In Example 3, we see that the values of the function $h(x)$ approach two different numbers, depending on the direction of approach, and it is natural to refer to these values as "the limit from the left" and "the limit from the right." These experiences suggest that the notion of **one-sided limits** will be very useful in discussing basic limit concepts.

---

*To make the informal definition of *limit* precise, we must make the word *close* more precise. This is done as follows: We write $\lim_{x \to c} f(x) = L$ if, for each $e > 0$, there exists a $d > 0$ such that $|f(x) - L| < e$ whenever $0 < |x - c| < d$. This definition is used to establish particular limits and to prove many useful properties of limits that will be helpful in finding particular limits.

DEFINITION One-Sided Limits

We write

$$\lim_{x \to c^-} f(x) = K \qquad \begin{array}{l} x \to c^- \text{ is read "}x \text{ approaches } c \text{ from} \\ \text{the left" and means } x \to c \text{ and } x < c. \end{array}$$

and call K the **limit from the left** or the **left-hand limit** if $f(x)$ is close to K whenever x is close to, but to the left of, c on the real number line. We write

$$\lim_{x \to c^+} f(x) = L \qquad \begin{array}{l} x \to c^+ \text{ is read "}x \text{ approaches } c \text{ from} \\ \text{the right" and means } x \to c \text{ and } x > c. \end{array}$$

and call L the **limit from the right** or the **right-hand limit** if $f(x)$ is close to L whenever x is close to, but to the right of, c on the real number line.

If no direction is specified in a limit statement, we will always assume that the limit is **two-sided** or **unrestricted**. Theorem 1 states an important relationship between one-sided limits and unrestricted limits.

THEOREM 1 On the Existence of a Limit

For a (two-sided) limit to exist, the limit from the left and the limit from the right must exist and be equal. That is,

$$\lim_{x \to c} f(x) = L \text{ if and only if } \lim_{x \to c^-} f(x) = \lim_{x \to c^+} f(x) = L$$

In Example 3,

$$\lim_{x \to 0^-} \frac{|x|}{x} = -1 \qquad \text{and} \qquad \lim_{x \to 0^+} \frac{|x|}{x} = 1$$

Since the left- and right-hand limits are *not* the same,

$$\lim_{x \to 0} \frac{|x|}{x} \text{ does not exist}$$

EXAMPLE 4 Analyzing Limits Graphically Given the graph of the function f in Figure 5, discuss the behavior of $f(x)$ for x near (A) -1, (B) 1, and (C) 2.

SOLUTION

(A) Since we have only a graph to work with, we use vertical and horizontal lines to relate the values of x and the corresponding values of $f(x)$. For any x near -1 on either side of -1, we see that the corresponding value of $f(x)$, determined by a horizontal line, is close to 1.

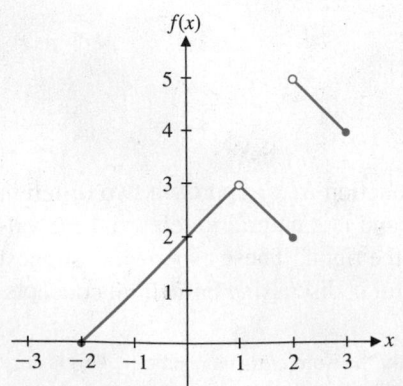

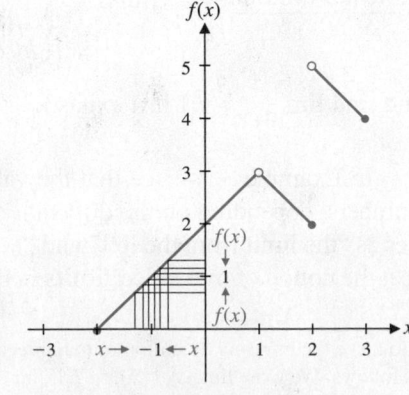

$$\lim_{x \to -1^-} f(x) = 1$$
$$\lim_{x \to -1^+} f(x) = 1$$
$$\lim_{x \to -1} f(x) = 1$$
$$f(-1) = 1$$

Figure 5

(B) Again, for any *x* near, but not equal to, 1, the vertical and horizontal lines indicate that the corresponding value of $f(x)$ is close to 3. The open dot at (1, 3), together with the absence of a solid dot anywhere on the vertical line through $x = 1$, indicates that $f(1)$ is not defined.

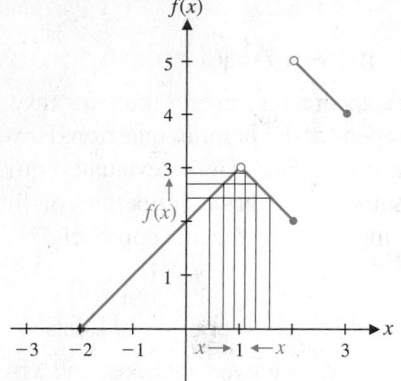

$$\lim_{x \to 1^-} f(x) = 3$$

$$\lim_{x \to 1^+} f(x) = 3$$

$$\lim_{x \to 1} f(x) = 3$$

$f(1)$ not defined

(C) The abrupt break in the graph at $x = 2$ indicates that the behavior of the graph near $x = 2$ is more complicated than in the two preceding cases. If *x* is close to 2 on the left side of 2, the corresponding horizontal line intersects the *y* axis at a point close to 2. If *x* is close to 2 on the right side of 2, the corresponding horizontal line intersects the *y* axis at a point close to 5. This is a case where the one-sided limits are different.

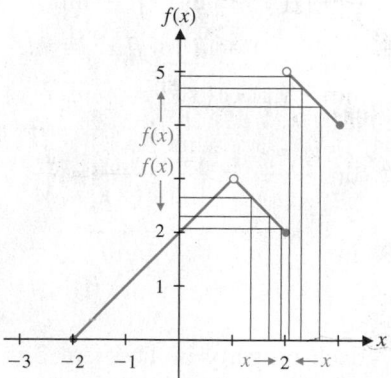

$$\lim_{x \to 2^-} f(x) = 2$$

$$\lim_{x \to 2^+} f(x) = 5$$

$$\lim_{x \to 2} f(x) \text{ does not exist}$$

$$f(2) = 2$$

Matched Problem 4 Given the graph of the function *f* shown in Figure 6, discuss the following, as we did in Example 4:

(A) Behavior of $f(x)$ for *x* near 0

(B) Behavior of $f(x)$ for *x* near 1

(C) Behavior of $f(x)$ for *x* near 3

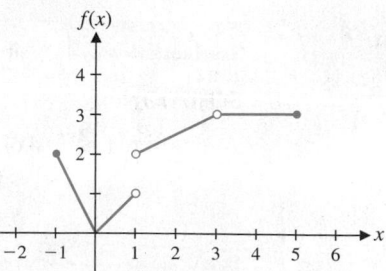

Figure 6

> **CONCEPTUAL INSIGHT**
>
> In Example 4B, note that $\lim_{x \to 1} f(x)$ exists even though f is not defined at $x = 1$ and the graph has a hole at $x = 1$. In general, the value of a function at $x = c$ has no effect on the limit of the function as x approaches c.

Limits: An Algebraic Approach

Graphs are very useful tools for investigating limits, especially if something unusual happens at the point in question. However, many of the limits encountered in calculus are routine and can be evaluated quickly with a little algebraic simplification, some intuition, and basic properties of limits. The following list of properties of limits forms the basis for this approach:

THEOREM 2 Properties of Limits

Let f and g be two functions, and assume that

$$\lim_{x \to c} f(x) = L \qquad \lim_{x \to c} g(x) = M$$

where L and M are real numbers (both limits exist). Then

1. $\lim_{x \to c} k = k$ for any constant k
2. $\lim_{x \to c} x = c$
3. $\lim_{x \to c} [f(x) + g(x)] = \lim_{x \to c} f(x) + \lim_{x \to c} g(x) = L + M$
4. $\lim_{x \to c} [f(x) - g(x)] = \lim_{x \to c} f(x) - \lim_{x \to c} g(x) = L - M$
5. $\lim_{x \to c} kf(x) = k \lim_{x \to c} f(x) = kL$ for any constant k
6. $\lim_{x \to c} [f(x) \cdot g(x)] = [\lim_{x \to c} f(x)][\lim_{x \to c} g(x)] = LM$
7. $\lim_{x \to c} \dfrac{f(x)}{g(x)} = \dfrac{\lim_{x \to c} f(x)}{\lim_{x \to c} g(x)} = \dfrac{L}{M}$ if $M \neq 0$
8. $\lim_{x \to c} \sqrt[n]{f(x)} = \sqrt[n]{\lim_{x \to c} f(x)} = \sqrt[n]{L}$ $L > 0$ for n even

---

Each property in Theorem 2 is also valid if $x \to c$ is replaced everywhere by $x \to c^-$ or replaced everywhere by $x \to c^+$.

Explore and Discuss 1 The properties listed in Theorem 2 can be paraphrased in brief verbal statements. For example, property 3 simply states that *the limit of a sum is equal to the sum of the limits*. Write brief verbal statements for the remaining properties in Theorem 2.

EXAMPLE 5 Using Limit Properties Find $\lim_{x \to 3} (x^2 - 4x)$.

SOLUTION

$$\lim_{x \to 3} (x^2 - 4x) = \lim_{x \to 3} x^2 - \lim_{x \to 3} 4x \qquad \text{Property 4}$$

$$= \left(\lim_{x \to 3} x \right) \cdot \left(\lim_{x \to 3} x \right) - 4 \lim_{x \to 3} x \qquad \text{Properties 5 and 6}$$

$$= \left(\lim_{x \to 3} x \right)^2 - 4 \lim_{x \to 3} x \qquad \text{Definition of exponent}$$

$$= 3^2 - 4 \cdot 3 = -3$$

So, omitting the steps in the dashed boxes,

$$\lim_{x \to 3} (x^2 - 4x) = 3^2 - 4 \cdot 3 = -3$$

Matched Problem 5) Find $\lim_{x \to -2} (x^2 + 5x)$.

If $f(x) = x^2 - 4$ and c is any real number, then, just as in Example 5

$$\lim_{x \to c} f(x) = \lim_{x \to c} (x^2 - 4x) = c^2 - 4c = f(c)$$

So the limit can be found easily by evaluating the function f at c.

This simple method for finding limits is very useful, because there are many functions that satisfy the property

$$\lim_{x \to c} f(x) = f(c) \tag{1}$$

Any polynomial function

$$f(x) = a_n x^n + a_{n-1} x^{n-1} + \cdots + a_0$$

satisfies (1) for any real number c. Also, any rational function

$$r(x) = \frac{n(x)}{d(x)}$$

where $n(x)$ and $d(x)$ are polynomials, satisfies (1) provided c is a real number for which $d(c) \neq 0$.

> **THEOREM 3 Limits of Polynomial and Rational Functions**
>
> 1. $\lim_{x \to c} f(x) = f(c)$ for f any polynomial function.
>
> 2. $\lim_{x \to c} r(x) = r(c)$ for r any rational function with a nonzero denominator at $x = c$.

If Theorem 3 is applicable, the limit is easy to find: *Simply evaluate the function at c.*

EXAMPLE 6 Evaluating Limits Find each limit.

(A) $\lim_{x \to 2} (x^3 - 5x - 1)$ (B) $\lim_{x \to -1} \sqrt{2x^2 + 3}$ (C) $\lim_{x \to 4} \dfrac{2x}{3x + 1}$

SOLUTION

(A) $\lim_{x \to 2} (x^3 - 5x - 1) = 2^3 - 5 \cdot 2 - 1 = -3$ Theorem 3

(B) $\lim_{x \to -1} \sqrt{2x^2 + 3} = \sqrt{\lim_{x \to -1} (2x^2 + 3)}$ Property 8

$$= \sqrt{2(-1)^2 + 3} \qquad \text{Theorem 3}$$

$$= \sqrt{5}$$

(C) $\lim_{x \to 4} \dfrac{2x}{3x + 1} = \dfrac{2 \cdot 4}{3 \cdot 4 + 1}$ Theorem 3

$$= \frac{8}{13}$$

Matched Problem 6) Find each limit.

(A) $\lim_{x \to -1} (x^4 - 2x + 3)$ (B) $\lim_{x \to 2} \sqrt{3x^2 - 6}$ (C) $\lim_{x \to -2} \dfrac{x^2}{x^2 + 1}$

EXAMPLE 7 Evaluating Limits Let

$$f(x) = \begin{cases} x^2 + 1 & \text{if } x < 2 \\ x - 1 & \text{if } x > 2 \end{cases}$$

Find:

(A) $\lim_{x \to 2^-} f(x)$ 　　(B) $\lim_{x \to 2^+} f(x)$ 　　(C) $\lim_{x \to 2} f(x)$ 　　(D) $f(2)$

SOLUTION

(A) $\lim_{x \to 2^-} f(x) = \lim_{x \to 2^-} (x^2 + 1)$ 　If $x < 2, f(x) = x^2 + 1$.
$\qquad\qquad\quad = 2^2 + 1 = 5$

(B) $\lim_{x \to 2^+} f(x) = \lim_{x \to 2^+} (x - 1)$ 　If $x > 2, f(x) = x - 1$.
$\qquad\qquad\quad = 2 - 1 = 1$

(C) Since the one-sided limits are not equal, $\lim_{x \to 2} f(x)$ does not exist.

(D) Because the definition of f does not assign a value to f for $x = 2$, only for $x < 2$ and $x > 2$, $f(2)$ does not exist.

Matched Problem 7 Let

$$f(x) = \begin{cases} 2x + 3 & \text{if } x < 5 \\ -x + 12 & \text{if } x > 5 \end{cases}$$

Find:

(A) $\lim_{x \to 5^-} f(x)$ 　　(B) $\lim_{x \to 5^+} f(x)$ 　　(C) $\lim_{x \to 5} f(x)$ 　　(D) $f(5)$

It is important to note that there are restrictions on some of the limit properties. In particular, if

$$\lim_{x \to c} f(x) = 0 \quad \text{and} \quad \lim_{x \to c} g(x) = 0, \quad \text{then finding} \lim_{x \to c} \frac{f(x)}{g(x)}$$

may present some difficulties, since limit property 7 (the limit of a quotient) does not apply when $\lim_{x \to c} g(x) = 0$. The next example illustrates some techniques that can be useful in this situation.

EXAMPLE 8 Evaluating Limits Find each limit.

(A) $\lim_{x \to 2} \dfrac{x^2 - 4}{x - 2}$ 　　　　　　　　　(B) $\lim_{x \to -1} \dfrac{x|x + 1|}{x + 1}$

SOLUTION

(A) Note that $\lim_{x \to 2} x^2 - 4 = 2^2 - 4 = 0$ and $\lim_{x \to 2} x - 2 = 2 - 2 = 0$. Algebraic simplification is often useful in such a case when the numerator and denominator both have limit 0.

$$\lim_{x \to 2} \frac{x^2 - 4}{x - 2} = \lim_{x \to 2} \frac{(x - 2)(x + 2)}{x - 2} = \lim_{x \to 2} (x + 2) = 4$$

(B) One-sided limits are helpful for limits involving the absolute value function.

$$\lim_{x \to -1^+} \frac{x|x + 1|}{x + 1} = \lim_{x \to -1^+} (x) = -1 \quad \text{If } x > -1, \text{ then } \frac{|x + 1|}{x + 1} = 1.$$

$$\lim_{x \to -1^-} \frac{x|x + 1|}{x + 1} = \lim_{x \to -1^-} (-x) = 1 \quad \text{If } x < -1, \text{ then } \frac{|x + 1|}{x + 1} = -1.$$

Since the limit from the left and the limit from the right are not the same, we conclude that

$$\lim_{x \to -1} \frac{x|x + 1|}{x + 1} \quad \text{does not exist}$$

Matched Problem 8) Find each limit.

(A) $\lim_{x \to -3} \dfrac{x^2 + 4x + 3}{x + 3}$

(B) $\lim_{x \to 4} \dfrac{x^2 - 16}{|x - 4|}$

CONCEPTUAL INSIGHT

In the solution to Example 8A we used the following algebraic identity:

$$\frac{x^2 - 4}{x - 2} = \frac{(x - 2)(x + 2)}{x - 2} = x + 2, \quad x \neq 2$$

The restriction $x \neq 2$ is necessary here because the first two expressions are not defined at $x = 2$. Why didn't we include this restriction in the solution? When x approaches 2 in a limit problem, it is assumed that x is close, but not equal, to 2. It is important that you understand that both of the following statements are valid:

$$\lim_{x \to 2} \frac{x^2 - 4}{x - 2} = \lim_{x \to 2} (x + 2) \quad \text{and} \quad \frac{x^2 - 4}{x - 2} = x + 2, \quad x \neq 2$$

Limits like those in Example 8 occur so frequently in calculus that they are given a special name.

DEFINITION Indeterminate Form

If $\lim_{x \to c} f(x) = 0$ and $\lim_{x \to c} g(x) = 0$, then $\lim_{x \to c} \dfrac{f(x)}{g(x)}$ is said to be **indeterminate**, or, more specifically, a **0/0 indeterminate form**.

The term *indeterminate* is used because the limit of an indeterminate form may or may not exist (see Example 8A and 8B).

⚠ **CAUTION** The expression 0/0 does not represent a real number and should never be used as the value of a limit. If a limit is a 0/0 indeterminate form, further investigation is always required to determine whether the limit exists and to find its value if it does exist. ▲

If the denominator of a quotient approaches 0 and the numerator approaches a nonzero number, then the limit of the quotient is not an indeterminate form. In fact, in this case the limit of the quotient does not exist.

THEOREM 4 Limit of a Quotient

If $\lim_{x \to c} f(x) = L, L \neq 0$, and $\lim_{x \to c} g(x) = 0$,

then

$$\lim_{x \to c} \frac{f(x)}{g(x)} \qquad \text{does not exist}$$

EXAMPLE 9 Indeterminate Forms Is the limit expression a 0/0 indeterminate form? Find the limit or explain why the limit does not exist.

(A) $\lim_{x \to 1} \dfrac{x - 1}{x^2 + 1}$

(B) $\lim_{x \to 1} \dfrac{x - 1}{x^2 - 1}$

(C) $\lim_{x \to 1} \dfrac{x + 1}{x^2 - 1}$

SOLUTION

(A) $\lim\limits_{x \to 1} (x - 1) = 0$ but $\lim\limits_{x \to 1} (x^2 + 1) = 2$. So no, the limit expression is not a $0/0$ indeterminate form. By property 7 of Theorem 2,

$$\lim_{x \to 1} \frac{x - 1}{x^2 + 1} = \frac{0}{2} = 0$$

(B) $\lim\limits_{x \to 1} (x - 1) = 0$ and $\lim\limits_{x \to 1} (x^2 - 1) = 0$. So yes, the limit expression is a $0/0$ indeterminate form. We factor $x^2 - 1$ to simplify the limit expression and find the limit:

$$\lim_{x \to 1} \frac{x - 1}{x^2 - 1} = \lim_{x \to 1} \frac{x - 1}{(x - 1)(x + 1)} = \lim_{x \to 1} \frac{1}{x + 1} = \frac{1}{2}$$

(C) $\lim\limits_{x \to 1} (x + 1) = 2$ and $\lim\limits_{x \to 1} (x^2 - 1) = 0$. So no, the limit expression is not a $0/0$ indeterminate form. By Theorem 4,

$$\lim_{x \to 1} \frac{x + 1}{x^2 - 1} \quad \text{does not exist}$$

Matched Problem 9 | Is the limit expression a $0/0$ indeterminate form? Find the limit or explain why the limit does not exist.

(A) $\lim\limits_{x \to 3} \dfrac{x + 1}{x + 3}$ (B) $\lim\limits_{x \to 3} \dfrac{x - 3}{x^2 + 9}$ (C) $\lim\limits_{x \to 3} \dfrac{x^2 - 9}{x - 3}$

Limits of Difference Quotients

Let the function f be defined in an open interval containing the number a. One of the most important limits in calculus is the limit of the **difference quotient**,

$$\lim_{h \to 0} \frac{f(a + h) - f(a)}{h} \tag{2}$$

If

$$\lim_{h \to 0} [f(a + h) - f(a)] = 0$$

as it often does, then limit (2) is an indeterminate form.

EXAMPLE 10 Limit of a Difference Quotient Find the following limit for $f(x) = 4x - 5$:

$$\lim_{h \to 0} \frac{f(3 + h) - f(3)}{h}$$

SOLUTION

$$\lim_{h \to 0} \frac{f(3 + h) - f(3)}{h} = \lim_{h \to 0} \frac{[4(3 + h) - 5] - [4(3) - 5]}{h}$$

$$= \lim_{h \to 0} \frac{12 + 4h - 5 - 12 + 5}{h}$$

$$= \lim_{h \to 0} \frac{4h}{h} = \lim_{h \to 0} 4 = 4$$

Since this is a 0/0 indeterminate form and property 7 in Theorem 2 does not apply, we proceed with algebraic simplification.

Matched Problem 10 | Find the following limit for $f(x) = 7 - 2x$:

$$\lim_{h \to 0} \frac{f(4 + h) - f(4)}{h}.$$

Explore and Discuss 2   If $f(x) = \dfrac{1}{x}$, explain why $\displaystyle\lim_{h\to 0}\dfrac{f(3+h)-f(3)}{h} = -\dfrac{1}{9}$.

Exercises 2.1

Skills Warm-up Exercises

W   *In Problems 1–8, factor each polynomial into the product of first-degree factors with integer coefficients. (If necessary, review Section A.3).*

1. $x^2 - 81$

2. $x^2 - 64$

3. $x^2 - 4x - 21$

4. $x^2 + 5x - 36$

5. $x^3 - 7x^2 + 12x$

6. $x^3 + 15x^2 + 50x$

7. $6x^2 - x - 1$

8. $20x^2 + 11x - 3$

In Problems 9–16, use the graph of the function f shown to estimate the indicated limits and function values.

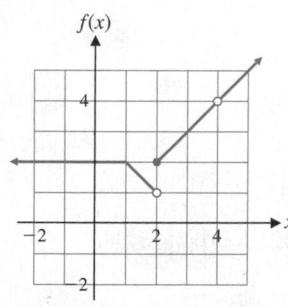

Figure for 9–16

9. $f(-0.5)$

10. $f(-1.5)$

11. $f(1.75)$

12. $f(1.25)$

13. (A) $\displaystyle\lim_{x\to 0^-} f(x)$ (B) $\displaystyle\lim_{x\to 0^+} f(x)$

   (C) $\displaystyle\lim_{x\to 0} f(x)$ (D) $f(0)$

14. (A) $\displaystyle\lim_{x\to 1^-} f(x)$ (B) $\displaystyle\lim_{x\to 1^+} f(x)$

   (C) $\displaystyle\lim_{x\to 1} f(x)$ (D) $f(1)$

15. (A) $\displaystyle\lim_{x\to 2^-} f(x)$ (B) $\displaystyle\lim_{x\to 2^+} f(x)$

   (C) $\displaystyle\lim_{x\to 2} f(x)$ (D) $f(2)$

✎ (E) Is it possible to redefine $f(2)$ so that $\displaystyle\lim_{x\to 2} f(x) = f(2)$? Explain.

16. (A) $\displaystyle\lim_{x\to 4^-} f(x)$ (B) $\displaystyle\lim_{x\to 4^+} f(x)$

   (C) $\displaystyle\lim_{x\to 4} f(x)$ (D) $f(4)$

✎ (E) Is it possible to define $f(4)$ so that $\displaystyle\lim_{x\to 4} f(x) = f(4)$? Explain.

In Problems 17–24, use the graph of the function g shown to estimate the indicated limits and function values.

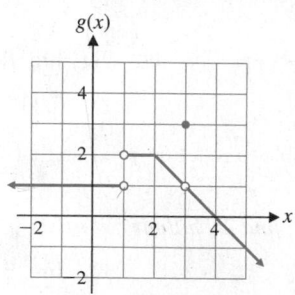

Figure for 17–24

17. $g(1.9)$

18. $g(0.1)$

19. $g(3.5)$

20. $g(2.5)$

21. (A) $\displaystyle\lim_{x\to 1^-} g(x)$ (B) $\displaystyle\lim_{x\to 1^+} g(x)$

   (C) $\displaystyle\lim_{x\to 1} g(x)$ (D) $g(1)$

✎ (E) Is it possible to define $g(1)$ so that $\displaystyle\lim_{x\to 1} g(x) = g(1)$? Explain.

22. (A) $\displaystyle\lim_{x\to 2^-} g(x)$ (B) $\displaystyle\lim_{x\to 2^+} g(x)$

   (C) $\displaystyle\lim_{x\to 2} g(x)$ (D) $g(2)$

23. (A) $\displaystyle\lim_{x\to 3^-} g(x)$ (B) $\displaystyle\lim_{x\to 3^+} g(x)$

   (C) $\displaystyle\lim_{x\to 3} g(x)$ (D) $g(3)$

✎ (E) Is it possible to redefine $g(3)$ so that $\displaystyle\lim_{x\to 3} g(x) = g(3)$? Explain.

24. (A) $\displaystyle\lim_{x\to 4^-} g(x)$ (B) $\displaystyle\lim_{x\to 4^+} g(x)$

   (C) $\displaystyle\lim_{x\to 4} g(x)$ (D) $g(4)$

In Problems 25–28, use the graph of the function f shown to estimate the indicated limits and function values.

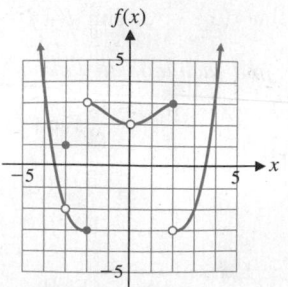

Figure for 25–28

. (A) $\lim\limits_{x \to -3^+} f(x)$ (B) $\lim\limits_{x \to -3^-} f(x)$

(C) $\lim\limits_{x \to -3} f(x)$ (D) $f(-3)$

(E) Is it possible to redefine $f(-3)$ so that $\lim\limits_{x \to -3} f(x) = f(-3)$? Explain.

26. (A) $\lim\limits_{x \to -2^+} f(x)$ (B) $\lim\limits_{x \to -2^-} f(x)$

(C) $\lim\limits_{x \to -2} f(x)$ (D) $f(-2)$

(E) Is it possible to redefine $f(-2)$ so that $\lim\limits_{x \to -2} f(x) = f(-2)$? Explain.

27. (A) $\lim\limits_{x \to 0^+} f(x)$ (B) $\lim\limits_{x \to 0^-} f(x)$

(C) $\lim\limits_{x \to 0} f(x)$ (D) $f(0)$

(E) Is it possible to define $f(0)$ so that $\lim\limits_{x \to 0} f(x) = f(0)$? Explain.

28. (A) $\lim\limits_{x \to 2^+} f(x)$ (B) $\lim\limits_{x \to 2^-} f(x)$

(C) $\lim\limits_{x \to 2} f(x)$ (D) $f(2)$

(E) Is it possible to redefine $f(2)$ so that $\lim\limits_{x \to 2} f(x) = f(2)$? Explain.

In Problems 29–38, find each limit if it exists.

29. $\lim\limits_{x \to 3} 4x$ 30. $\lim\limits_{x \to -2} 3x$

31. $\lim\limits_{x \to -4} (x + 5)$ 32. $\lim\limits_{x \to 5} (x - 3)$

33. $\lim\limits_{x \to 2} x(x - 4)$ 34. $\lim\limits_{x \to -1} x(x + 3)$

35. $\lim\limits_{x \to -3} \dfrac{x}{x + 5}$ 36. $\lim\limits_{x \to 4} \dfrac{x - 2}{x}$

37. $\lim\limits_{x \to 1} \sqrt{5x + 4}$ 38. $\lim\limits_{x \to 0} \sqrt{16 - 7x}$

Given that $\lim\limits_{x \to 1} f(x) = -5$ and $\lim\limits_{x \to 1} g(x) = 4$, find the indicated limits in Problems 39–46.

39. $\lim\limits_{x \to 1} (-3)f(x)$ 40. $\lim\limits_{x \to 1} 2g(x)$

41. $\lim\limits_{x \to 1} [2f(x) + g(x)]$ 42. $\lim\limits_{x \to 1} [g(x) - 3f(x)]$

43. $\lim\limits_{x \to 1} \dfrac{2 - f(x)}{x + g(x)}$ 44. $\lim\limits_{x \to 1} \dfrac{3 - f(x)}{1 - 4g(x)}$

45. $\lim\limits_{x \to 1} \sqrt{g(x) - f(x)}$ 46. $\lim\limits_{x \to 1} \sqrt[3]{2x + 2f(x)}$

In Problems 47–50, sketch a possible graph of a function that satisfies the given conditions.

47. $f(0) = 1; \lim\limits_{x \to 0^-} f(x) = 3; \lim\limits_{x \to 0^+} f(x) = 1$

48. $f(1) = -2; \lim\limits_{x \to 1^-} f(x) = 2; \lim\limits_{x \to 1^+} f(x) = -2$

49. $f(-2) = 2; \lim\limits_{x \to -2^-} f(x) = 1; \lim\limits_{x \to -2^+} f(x) = 1$

50. $f(0) = -1; \lim\limits_{x \to 0^-} f(x) = 2; \lim\limits_{x \to 0^+} f(x) = 2$

In Problems 51–66, find each indicated quantity if it exists.

51. Let $f(x) = \begin{cases} 1 - x^2 & \text{if } x \le 0 \\ 1 + x^2 & \text{if } x > 0 \end{cases}$. Find

(A) $\lim\limits_{x \to 0^+} f(x)$ (B) $\lim\limits_{x \to 0^-} f(x)$

(C) $\lim\limits_{x \to 0} f(x)$ (D) $f(0)$

52. Let $f(x) = \begin{cases} 2 + x & \text{if } x \le 0 \\ 2 - x & \text{if } x > 0 \end{cases}$. Find

(A) $\lim\limits_{x \to 0^+} f(x)$ (B) $\lim\limits_{x \to 0^-} f(x)$

(C) $\lim\limits_{x \to 0} f(x)$ (D) $f(0)$

53. Let $f(x) = \begin{cases} x^2 & \text{if } x < 1 \\ 2x & \text{if } x > 1 \end{cases}$. Find

(A) $\lim\limits_{x \to 1^-} f(x)$ 2   (B) $\lim\limits_{x \to 1^+} f(x)$ 1

(C) $\lim\limits_{x \to 1} f(x)$ DNE   (D) $f(1)$ DNE

54. Let $f(x) = \begin{cases} x + 3 & \text{if } x < -2 \\ \sqrt{x + 2} & \text{if } x > -2 \end{cases}$. Find

(A) $\lim\limits_{x \to -2^+} f(x)$ (B) $\lim\limits_{x \to -2^-} f(x)$

(C) $\lim\limits_{x \to -2} f(x)$ (D) $f(-2)$

55. Let $f(x) = \begin{cases} \dfrac{x^2 - 9}{x + 3} & \text{if } x < 0 \\ \dfrac{x^2 - 9}{x - 3} & \text{if } x > 0 \end{cases}$. Find

(A) $\lim\limits_{x \to -3} f(x)$ (B) $\lim\limits_{x \to 0} f(x)$

(C) $\lim\limits_{x \to 3} f(x)$

56. Let $f(x) = \begin{cases} \dfrac{x}{x + 3} & \text{if } x < 0 \\ \dfrac{x}{x - 3} & \text{if } x > 0 \end{cases}$. Find

(A) $\lim\limits_{x \to -3} f(x)$ (B) $\lim\limits_{x \to 0} f(x)$

(C) $\lim\limits_{x \to 3} f(x)$

57. Let $f(x) = \dfrac{|x - 1|}{x - 1}$. Find

(A) $\lim\limits_{x \to 1^+} f(x)$ (B) $\lim\limits_{x \to 1^-} f(x)$

(C) $\lim\limits_{x \to 1} f(x)$ (D) $f(1)$

58. Let $f(x) = \dfrac{x - 3}{|x - 3|}$. Find

(A) $\lim\limits_{x \to 3^+} f(x)$ (B) $\lim\limits_{x \to 3^-} f(x)$

(C) $\lim\limits_{x \to 3} f(x)$ (D) $f(3)$

59. Let $f(x) = \dfrac{x - 2}{x^2 - 2x}$. Find

(A) $\lim\limits_{x \to 0} f(x)$ DNE   (B) $\lim\limits_{x \to 2} f(x)$ $\frac{1}{2}$

(C) $\lim\limits_{x \to 4} f(x)$ $\frac{1}{4}$

60. Let $f(x) = \dfrac{x + 3}{x^2 + 3x}$. Find

(A) $\lim\limits_{x \to -3} f(x)$ (B) $\lim\limits_{x \to 0} f(x)$

(C) $\lim\limits_{x \to 3} f(x)$

61. Let $f(x) = \dfrac{x^2 - x - 6}{x + 2}$. Find

(A) $\lim\limits_{x \to -2} f(x)$ (B) $\lim\limits_{x \to 0} f(x)$

(C) $\lim\limits_{x \to 3} f(x)$

62. Let $f(x) = \dfrac{x^2 + x - 6}{x + 3}$. Find

(A) $\lim\limits_{x \to -3} f(x)$ (B) $\lim\limits_{x \to 0} f(x)$

(C) $\lim\limits_{x \to 2} f(x)$

63. Let $f(x) = \dfrac{(x+2)^2}{x^2-4}$. Find

   (A) $\lim\limits_{x \to -2} f(x)$ (B) $\lim\limits_{x \to 0} f(x)$

   (C) $\lim\limits_{x \to 2} f(x)$

64. Let $f(x) = \dfrac{x^2-1}{(x+1)^2}$. Find

   (A) $\lim\limits_{x \to -1} f(x)$ (B) $\lim\limits_{x \to 0} f(x)$

   (C) $\lim\limits_{x \to 1} f(x)$

65. Let $f(x) = \dfrac{2x^2-3x-2}{x^2+x-6}$. Find

   (A) $\lim\limits_{x \to 2} f(x)$ (B) $\lim\limits_{x \to 0} f(x)$

   (C) $\lim\limits_{x \to 1} f(x)$

66. Let $f(x) = \dfrac{3x^2+2x-1}{x^2+3x+2}$. Find

   (A) $\lim\limits_{x \to -3} f(x)$ (B) $\lim\limits_{x \to -1} f(x)$

   (C) $\lim\limits_{x \to 2} f(x)$

In Problems 67–72, discuss the validity of each statement. If the statement is always true, explain why. If not, give a counterexample.

67. If $\lim\limits_{x \to 1} f(x) = 0$ and $\lim\limits_{x \to 1} g(x) = 0$, then $\lim\limits_{x \to 1} \dfrac{f(x)}{g(x)} = 0$.

68. If $\lim\limits_{x \to 1} f(x) = 1$ and $\lim\limits_{x \to 1} g(x) = 1$, then $\lim\limits_{x \to 1} \dfrac{f(x)}{g(x)} = 1$.

69. If f is a polynomial, then, as x approaches 0, the right-hand limit exists and is equal to the left-hand limit.

70. If f is a rational function, then, as x approaches 0, the right-hand limit exists and is equal to the left-hand limit.

71. If f is a function such that $\lim\limits_{x \to 0} f(x)$ exists, then $f(0)$ exists.

72. If f is a function such that $f(0)$ exists, then $\lim\limits_{x \to 0} f(x)$ exists.

In Problems 73–80, is the limit expression a 0/0 indeterminate form? Find the limit or explain why the limit does not exist.

73. $\lim\limits_{x \to 7} \dfrac{(x-7)^2}{x^2-4x-21}$ **74.** $\lim\limits_{x \to 2} \dfrac{x-5}{x+2}$

75. $\lim\limits_{x \to 4} \dfrac{x^2+4}{(x+4)^2}$ **76.** $\lim\limits_{x \to 9} \dfrac{x^2-5x-36}{x-9}$

77. $\lim\limits_{x \to -6} \dfrac{x^2+36}{x+6}$ **78.** $\lim\limits_{x \to 10} \dfrac{x^2-15x+50}{(x-10)^2}$

79. $\lim\limits_{x \to 8} \dfrac{x-8}{x^2-64}$ **80.** $\lim\limits_{x \to -3} \dfrac{x+3}{x-3}$

Compute the following limit for each function in Problems 81–88.

$$\lim\limits_{h \to 0} \dfrac{f(2+h)-f(2)}{h}$$

81. $f(x) = 3x+1$ **82.** $f(x) = 5x-1$

83. $f(x) = x^2+1$ **84.** $f(x) = x^2-2$

85. $f(x) = -7x+9$ **86.** $f(x) = -4x+13$

87. $f(x) = |x+1|$ **88.** $f(x) = -3|x|$

89. Let f be defined by

$$f(x) = \begin{cases} 1+mx & \text{if } x \le 1 \\ 4-mx & \text{if } x > 1 \end{cases}$$

where m is a constant.

   (A) Graph f for $m = 1$, and find

$$\lim\limits_{x \to 1^-} f(x) \quad \text{and} \quad \lim\limits_{x \to 1^+} f(x)$$

   (B) Graph f for $m = 2$, and find

$$\lim\limits_{x \to 1^-} f(x) \quad \text{and} \quad \lim\limits_{x \to 1^+} f(x)$$

   (C) Find m so that

$$\lim\limits_{x \to 1^-} f(x) = \lim\limits_{x \to 1^+} f(x)$$

   and graph f for this value of m.

   (D) Write a brief verbal description of each graph. How does the graph in part (C) differ from the graphs in parts (A) and (B)?

90. Let f be defined by

$$f(x) = \begin{cases} -3m+0.5x & \text{if } x \le 2 \\ 3m-x & \text{if } x > 2 \end{cases}$$

where m is a constant.

   (A) Graph f for $m = 0$, and find

$$\lim\limits_{x \to 2^-} f(x) \quad \text{and} \quad \lim\limits_{x \to 2^+} f(x)$$

   (B) Graph f for $m = 1$, and find

$$\lim\limits_{x \to 2^-} f(x) \quad \text{and} \quad \lim\limits_{x \to 2^+} f(x)$$

   (C) Find m so that

$$\lim\limits_{x \to 2^-} f(x) = \lim\limits_{x \to 2^+} f(x)$$

   and graph f for this value of m.

   (D) Write a brief verbal description of each graph. How does the graph in part (C) differ from the graphs in parts (A) and (B)?

Applications

91. Telephone rates. A long-distance telephone service charges $0.99 for the first 20 minutes or less of a call and $0.07 per minute thereafter.

   (A) Write a piecewise definition of the charge $F(x)$ for a long-distance call lasting x minutes.

   (B) Graph $F(x)$ for $0 < x \le 40$.

   (C) Find $\lim\limits_{x \to 20^-} F(x)$, $\lim\limits_{x \to 20^+} F(x)$, and $\lim\limits_{x \to 20} F(x)$, whichever exist.

92. Telephone rates. A second long-distance telephone service charges $0.09 per minute for calls lasting 10 minutes or more and $0.18 per minute for calls lasting less than 10 minutes.

   (A) Write a piecewise definition of the charge $G(x)$ for a long-distance call lasting x minutes.

   (B) Graph $G(x)$ for $0 < x \le 40$.

   (C) Find $\lim\limits_{x \to 10^-} G(x)$, $\lim\limits_{x \to 10^+} G(x)$, and $\lim\limits_{x \to 10} G(x)$, whichever exist.

93. Telephone rates. Refer to Problems 91 and 92. Write a brief verbal comparison of the two services described for calls lasting 20 minutes or less.

94. Telephone rates. Refer to Problems 91 and 92. Write a brief verbal comparison of the two services described for calls lasting more than 20 minutes.

A company sells custom embroidered apparel and promotional products. Table 1 shows the volume discounts offered by the company, where x is the volume of a purchase in dollars. Problems 95 and 96 deal with two different interpretations of this discount method.

Table 1 **Volume Discount (Excluding Tax)**

Volume (x)	Discount Amount
$300 \leq x < \$1,000$	3%
$\$1,000 \leq x < \$3,000$	5%
$\$3,000 \leq x < \$5,000$	7%
$\$5,000 \leq x$	10%

95. Volume discount. Assume that the volume discounts in Table 1 apply to the entire purchase. That is, if the volume x satisfies $\$300 \leq x < \$1,000$, then the entire purchase is discounted 3%. If the volume x satisfies $\$1,000 \leq x < \$3,000$, the entire purchase is discounted 5%, and so on.

(A) If x is the volume of a purchase before the discount is applied, then write a piecewise definition for the discounted price $D(x)$ of this purchase.

(B) Use one-sided limits to investigate the limit of $D(x)$ as x approaches $1,000. As x approaches $3,000.

96. Volume discount. Assume that the volume discounts in Table 1 apply only to that portion of the volume in each interval. That is, the discounted price for a $4,000 purchase would be computed as follows:

$$300 + 0.97(700) + 0.95(2,000) + 0.93(1,000) = 3,809$$

(A) If x is the volume of a purchase before the discount is applied, then write a piecewise definition for the discounted price $P(x)$ of this purchase.

(B) Use one-sided limits to investigate the limit of $P(x)$ as x approaches $1,000. As x approaches $3,000.

(C) Compare this discount method with the one in Problem 95. Does one always produce a lower price than the other? Discuss.

97. Pollution A state charges polluters an annual fee of $20 per ton for each ton of pollutant emitted into the atmosphere, up to a maximum of 4,000 tons. No fees are charged for emissions beyond the 4,000-ton limit. Write a piecewise definition of the fees $F(x)$ charged for the emission of x tons of pollutant in a year. What is the limit of $F(x)$ as x approaches 4,000 tons? As x approaches 8,000 tons?

98. Pollution Refer to Problem 97. The average fee per ton of pollution is given by $A(x) = F(x)/x$. Write a piecewise definition of $A(x)$. What is the limit of $A(x)$ as x approaches 4,000 tons? As x approaches 8,000 tons?

99. Voter turnout. Statisticians often use piecewise-defined functions to predict outcomes of elections. For the following functions f and g, find the limit of each function as x approaches 5 and as x approaches 10.

$$f(x) = \begin{cases} 0 & \text{if } x \leq 5 \\ 0.8 - 0.08x & \text{if } 5 < x < 10 \\ 0 & \text{if } 10 \leq x \end{cases}$$

$$g(x) = \begin{cases} 0 & \text{if } x \leq 5 \\ 0.8x - 0.04x^2 - 3 & \text{if } 5 < x < 10 \\ 1 & \text{if } 10 \leq x \end{cases}$$

Answers to Matched Problems

1.

x	-2	-1	0	1	2	3	4
$h(x)$	1.0	1.5	2.0	2.5	3.0	3.5	4.0

2. $\lim\limits_{x \to 1} f(x) = 2$

3.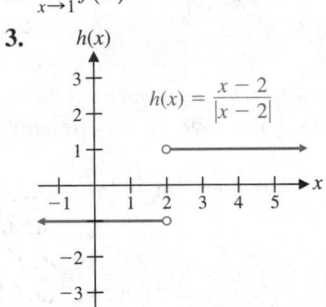

$\lim\limits_{x \to 2} \dfrac{x - 2}{|x - 2|}$ does not exist

4. (A) $\lim\limits_{x \to 0^-} f(x) = 0$ (B) $\lim\limits_{x \to 1^-} f(x) = 1$
 $\lim\limits_{x \to 0^+} f(x) = 0$ $\lim\limits_{x \to 1^+} f(x) = 2$
 $\lim\limits_{x \to 0} f(x) = 0$ $\lim\limits_{x \to 1} f(x)$ does not exist
 $f(0) = 0$ $f(1)$ not defined

 (C) $\lim\limits_{x \to 3^-} f(x) = 3$
 $\lim\limits_{x \to 3^+} f(x) = 3$
 $\lim\limits_{x \to 3} f(x) = 3$ $f(3)$ not defined

5. -6

6. (A) 6
 (B) $\sqrt{6}$
 (C) $\frac{4}{5}$

7. (A) 13
 (B) 7
 (C) Does not exist
 (D) Not defined

8. (A) -2
 (B) Does not exist

9. (A) No; $\dfrac{2}{3}$
 (B) No; 0
 (C) Yes; 6

10. -2

2.2 Infinite Limits and Limits at Infinity

- Infinite Limits
- Locating Vertical Asymptotes
- Limits at Infinity
- Finding Horizontal Asymptotes

In this section, we consider two new types of limits: infinite limits and limits at infinity. Infinite limits and vertical asymptotes are used to describe the behavior of functions that are unbounded near $x = a$. Limits at infinity and horizontal asymptotes are used to describe the behavior of functions as x assumes arbitrarily large positive values or arbitrarily large negative values. Although we will include graphs to illustrate basic concepts, we postpone a discussion of graphing techniques until Chapter 4.

Infinite Limits

The graph of $f(x) = \dfrac{1}{x-1}$ (Fig. 1) indicates that

$$\lim_{x \to 1^+} \frac{1}{x-1}$$

does not exist. There does not exist a real number L that the values of $f(x)$ approach as x approaches 1 from the right. Instead, as x approaches 1 from the right, the values of $f(x)$ are positive and become larger and larger; that is, $f(x)$ increases without bound (Table 1). We express this behavior symbolically as

$$\lim_{x \to 1^+} \frac{1}{x-1} = \infty \quad \text{or} \quad f(x) = \frac{1}{x-1} \to \infty \quad \text{as} \quad x \to 1^+ \qquad (1)$$

Since ∞ is a not a real number, *the limit in (1) does not exist*. We are using the symbol ∞ to describe the manner in which the limit fails to exist, and we call this situation an **infinite limit**. If x approaches 1 from the left, the values of $f(x)$ are negative and become larger and larger in absolute value; that is, $f(x)$ decreases through negative values without bound (Table 2). We express this behavior symbolically as

$$\lim_{x \to 1^-} \frac{1}{x-1} = -\infty \quad \text{or} \quad f(x) = \frac{1}{x-1} \to -\infty \quad \text{as} \quad x \to 1^- \qquad (2)$$

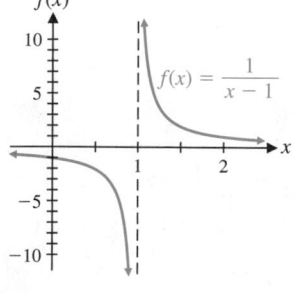

Figure 1

Table 1

x	$f(x) = \dfrac{1}{x-1}$
1.1	10
1.01	100
1.001	1,000
1.0001	10,000
1.00001	100,000
1.000001	1,000,000

Table 2

x	$f(x) = \dfrac{1}{x-1}$
0.9	−10
0.99	−100
0.999	−1,000
0.9999	−10,000
0.99999	−100,000
0.999999	−1,000,000

The one-sided limits in (1) and (2) describe the behavior of the graph as $x \to 1$ (Fig. 1). Does the two-sided limit of $f(x)$ as $x \to 1$ exist? No, because neither of the one-sided limits exists. Also, there is no reasonable way to use the symbol ∞ to describe the behavior of $f(x)$ as $x \to 1$ on both sides of 1. We say that

$$\lim_{x \to 1} \frac{1}{x-1} \text{ does not exist}$$

Explore and Discuss 1

Let $g(x) = \dfrac{1}{(x-1)^2}$.

Construct tables for $g(x)$ as $x \to 1^+$ and as $x \to 1^-$. Use these tables and infinite limits to discuss the behavior of $g(x)$ near $x = 1$.

We used the dashed vertical line $x = 1$ in Figure 1 to illustrate the infinite limits as x approaches 1 from the right and from the left. We call this line a *vertical asymptote*.

DEFINITION   Infinite Limits and Vertical Asymptotes

The vertical line $x = a$ is a **vertical asymptote** for the graph of $y = f(x)$ if

$$f(x) \to \infty \quad \text{or} \quad f(x) \to -\infty \quad \text{as} \quad x \to a^+ \quad \text{or} \quad x \to a^-$$

[That is, if $f(x)$ either increases or decreases without bound as x approaches a from the right or from the left].

Locating Vertical Asymptotes

How do we locate vertical asymptotes? If f is a polynomial function, then $\lim\limits_{x \to a} f(x)$ is equal to the real number $f(a)$ [Theorem 3, Section 2.1]. So *a polynomial function has no vertical asymptotes*. Similarly (again by Theorem 3, Section 2.1), *a vertical asymptote of a rational function can occur only at a zero of its denominator*. Theorem 1 provides a simple procedure for locating the vertical asymptotes of a rational function.

THEOREM 1   Locating Vertical Asymptotes of Rational Functions

If $f(x) = n(x)/d(x)$ is a rational function, $d(c) = 0$ and $n(c) \neq 0$, then the line $x = c$ is a vertical asymptote of the graph of f.

If $f(x) = n(x)/d(x)$ and both $n(c) = 0$ and $d(c) = 0$, then the limit of $f(x)$ as x approaches c involves an indeterminate form and Theorem 1 does not apply:

$$\lim_{x \to c} f(x) = \lim_{x \to c} \frac{n(x)}{d(x)} \quad \frac{0}{0} \text{ indeterminate form}$$

Algebraic simplification is often useful in this situation.

EXAMPLE 1   Locating Vertical Asymptotes   Let $f(x) = \dfrac{x^2 + x - 2}{x^2 - 1}$.

Describe the behavior of f at each zero of the denominator. Use ∞ and $-\infty$ when appropriate. Identify all vertical asymptotes.

SOLUTION Let $n(x) = x^2 + x - 2$ and $d(x) = x^2 - 1$. Factoring the denominator, we see that

$$d(x) = x^2 - 1 = (x - 1)(x + 1)$$

has two zeros: $x = -1$ and $x = 1$.

First, we consider $x = -1$. Since $d(-1) = 0$ and $n(-1) = -2 \neq 0$, Theorem 1 tells us that the line $x = -1$ is a vertical asymptote. So at least one of the one-sided limits at $x = -1$ must be either ∞ or $-\infty$. Examining tables of values of f for x near -1 or a graph on a graphing calculator will show which is the case. From Tables 3 and 4, we see that

$$\lim_{x \to -1^-} \frac{x^2 + x - 2}{x^2 - 1} = -\infty \quad \text{and} \quad \lim_{x \to -1^+} \frac{x^2 + x - 2}{x^2 - 1} = \infty$$

Table 3

x	$f(x) = \dfrac{x^2 + x - 2}{x^2 - 1}$
-1.1	-9
-1.01	-99
-1.001	-999
-1.0001	$-9{,}999$
-1.00001	$-99{,}999$

Table 4

x	$f(x) = \dfrac{x^2 + x - 2}{x^2 - 1}$
-0.9	11
-0.99	101
-0.999	$1{,}001$
-0.9999	$10{,}001$
-0.99999	$100{,}001$

Now we consider the other zero of $d(x)$, $x = 1$. This time $n(1) = 0$ and Theorem 1 does not apply. We use algebraic simplification to investigate the behavior of the function at $x = 1$:

$$\lim_{x \to 1} f(x) = \lim_{x \to 1} \frac{x^2 + x - 2}{x^2 - 1} \qquad \frac{0}{0} \text{ indeterminate form}$$

$$= \lim_{x \to 1} \frac{(x - 1)(x + 2)}{(x - 1)(x + 1)}$$

$$= \lim_{x \to 1} \frac{x + 2}{x + 1} \qquad \text{Reduced to lowest terms (see Appendix A.4)}$$

$$= \frac{3}{2}$$

Since the limit exists as x approaches 1, f does not have a vertical asymptote at $x = 1$. The graph of f (Fig. 2) shows the behavior at the vertical asymptote $x = -1$ and also at $x = 1$.

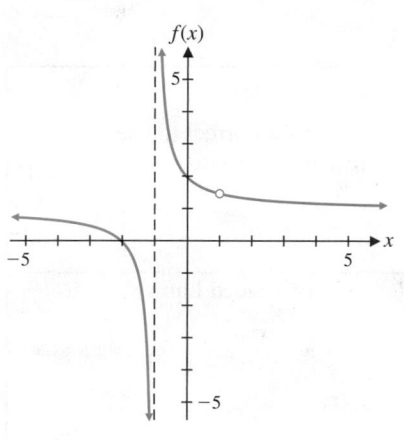

Figure 2 $f(x) = \dfrac{x^2 + x - 2}{x^2 - 1}$

Matched Problem 1 ⌋ Let $f(x) = \dfrac{x - 3}{x^2 - 4x + 3}$.

Describe the behavior of f at each zero of the denominator. Use ∞ and $-\infty$ when appropriate. Identify all vertical asymptotes.

EXAMPLE 2 Locating Vertical Asymptotes   Let $f(x) = \dfrac{x^2 + 20}{5(x - 2)^2}$.

Describe the behavior of f at each zero of the denominator. Use ∞ and $-\infty$ when appropriate. Identify all vertical asymptotes.

SOLUTION Let $n(x) = x^2 + 20$ and $d(x) = 5(x - 2)^2$. The only zero of $d(x)$ is $x = 2$. Since $n(2) = 24 \neq 0$, f has a vertical asymptote at $x = 2$ (Theorem 1). Tables 5 and 6 show that $f(x) \to \infty$ as $x \to 2$ from either side, and we have

$$\lim_{x \to 2^+} \frac{x^2 + 20}{5(x - 2)^2} = \infty \quad \text{and} \quad \lim_{x \to 2^-} \frac{x^2 + 20}{5(x - 2)^2} = \infty$$

Table 5

x	$f(x) = \dfrac{x^2 + 20}{5(x - 2)^2}$
2.1	488.2
2.01	48,080.02
2.001	4,800,800.2

Table 6

x	$f(x) = \dfrac{x^2 + 20}{5(x - 2)^2}$
1.9	472.2
1.99	47,920.02
1.999	4,799,200.2

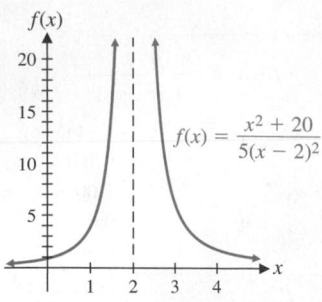

$f(x) = \dfrac{x^2 + 20}{5(x-2)^2}$

Figure 3

The denominator d has no other zeros, so f does not have any other vertical asymptotes. The graph of f (Fig. 3) shows the behavior at the vertical asymptote $x = 2$. Because the left- and right-hand limits are both infinite, we write

$$\lim_{x \to 2} \frac{x^2 + 20}{5(x-2)^2} = \infty$$

Matched Problem 2 Let $f(x) = \dfrac{x-1}{(x+3)^2}$.

Describe the behavior of f at each zero of the denominator. Use ∞ and $-\infty$ when appropriate. Identify all vertical asymptotes.

CONCEPTUAL INSIGHT

When is it correct to say that a limit does not exist, and when is it correct to use $\pm\infty$? It depends on the situation. Table 7 lists the infinite limits that we discussed in Examples 1 and 2.

Table 7

Right-Hand Limit	Left-Hand Limit	Two-Sided Limit
$\lim\limits_{x \to -1^+} \dfrac{x^2 + x - 2}{x^2 - 1} = \infty$	$\lim\limits_{x \to -1^-} \dfrac{x^2 + x - 2}{x^2 - 1} = -\infty$	$\lim\limits_{x \to -1} \dfrac{x^2 + x - 2}{x^2 - 1}$ does not exist
$\lim\limits_{x \to 2^+} \dfrac{x^2 + 20}{5(x-2)^2} = \infty$	$\lim\limits_{x \to 2^-} \dfrac{x^2 + 20}{5(x-2)^2} = \infty$	$\lim\limits_{x \to 2} \dfrac{x^2 + 20}{5(x-2)^2} = \infty$

The instructions in Examples 1 and 2 said that we should use infinite limits to describe the behavior at vertical asymptotes. If we had been asked to *evaluate* the limits, with no mention of ∞ or asymptotes, then the correct answer would be that **all of these limits do not exist**. Remember, ∞ is a symbol used to describe the behavior of functions at vertical asymptotes.

Limits at Infinity

The symbol ∞ can also be used to indicate that an independent variable is increasing or decreasing without bound. We write $x \to \infty$ to indicate that x is increasing without bound through positive values and $x \to -\infty$ to indicate that x is decreasing without bound through negative values. We begin by considering power functions of the form x^p and $1/x^p$ where p is a positive real number.

If p is a positive real number, then x^p increases as x increases. There is no upper bound on the values of x^p. We indicate this behavior by writing

$$\lim_{x \to \infty} x^p = \infty \quad \text{or} \quad x^p \to \infty \quad \text{as} \quad x \to \infty$$

Since the reciprocals of very large numbers are very small numbers, it follows that $1/x^p$ approaches 0 as x increases without bound. We indicate this behavior by writing

$$\lim_{x \to \infty} \frac{1}{x^p} = 0 \quad \text{or} \quad \frac{1}{x^p} \to 0 \quad \text{as} \quad x \to \infty$$

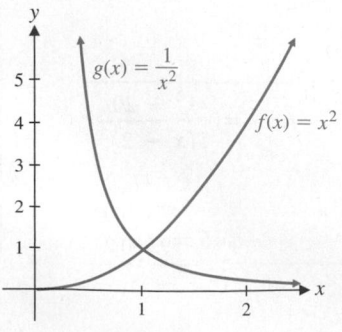

Figure 4

Figure 4 illustrates the preceding behavior for $f(x) = x^2$ and $g(x) = 1/x^2$, and we write

$$\lim_{x \to \infty} f(x) = \infty \quad \text{and} \quad \lim_{x \to \infty} g(x) = 0$$

Limits of power forms as x decreases without bound behave in a similar manner, with two important differences. First, if x is negative, then x^p is not defined for all values of p. For example, $x^{1/2} = \sqrt{x}$ is not defined for negative values of x. Second, if x^p is defined, then it may approach ∞ or $-\infty$, depending on the value of p. For example,

$$\lim_{x \to -\infty} x^2 = \infty \quad \text{but} \quad \lim_{x \to -\infty} x^3 = -\infty$$

For the function g in Figure 4, the line $y = 0$ (the x axis) is called a *horizontal asymptote*. In general, a line $y = b$ is a **horizontal asymptote** of the graph of $y = f(x)$ if $f(x)$ approaches b as either x increases without bound or x decreases without bound. Symbolically, $y = b$ is a horizontal asymptote if either

$$\lim_{x \to -\infty} f(x) = b \quad \text{or} \quad \lim_{x \to \infty} f(x) = b$$

In the first case, the graph of f will be close to the horizontal line $y = b$ for large (in absolute value) negative x. In the second case, the graph will be close to the horizontal line $y = b$ for large positive x. Figure 5 shows the graph of a function with two horizontal asymptotes: $y = 1$ and $y = -1$.

Theorem 2 summarizes the various possibilities for limits of power functions as x increases or decreases without bound.

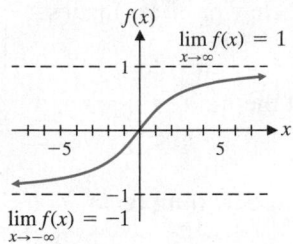

$\lim\limits_{x \to \infty} f(x) = 1$

$\lim\limits_{x \to -\infty} f(x) = -1$

Figure 5

THEOREM 2 Limits of Power Functions at Infinity

If p is a positive real number and k is any real number except 0, then

1. $\displaystyle\lim_{x \to -\infty} \frac{k}{x^p} = 0$ 2. $\displaystyle\lim_{x \to \infty} \frac{k}{x^p} = 0$

3. $\displaystyle\lim_{x \to -\infty} kx^p = \pm\infty$ 4. $\displaystyle\lim_{x \to \infty} kx^p = \pm\infty$

provided that x^p is a real number for negative values of x. The limits in 3 and 4 will be either $-\infty$ or ∞, depending on k and p.

How can we use Theorem 2 to evaluate limits at infinity? It turns out that the limit properties listed in Theorem 2, Section 2.1, are also valid if we replace the statement $x \to c$ with $x \to \infty$ or $x \to -\infty$.

EXAMPLE 3 Limit of a Polynomial Function at Infinity Let $p(x) = 2x^3 - x^2 - 7x + 3$. Find the limit of $p(x)$ as x approaches ∞ and as x approaches $-\infty$.

SOLUTION Since limits of power functions of the form $1/x^p$ approach 0 as x approaches ∞ or $-\infty$, it is convenient to work with these reciprocal forms whenever possible. If we factor out the term involving the highest power of x, then we can write $p(x)$ as

$$p(x) = 2x^3\left(1 - \frac{1}{2x} - \frac{7}{2x^2} + \frac{3}{2x^3}\right)$$

Using Theorem 2 above and Theorem 2 in Section 2.1, we write

$$\lim_{x \to \infty}\left(1 - \frac{1}{2x} - \frac{7}{2x^2} + \frac{3}{2x^3}\right) = 1 - 0 - 0 + 0 = 1$$

For large values of x,

$$\left(1 - \frac{1}{2x} - \frac{7}{2x^2} + \frac{3}{2x^3}\right) \approx 1$$

and

$$p(x) = 2x^3\left(1 - \frac{1}{2x} - \frac{7}{2x^2} + \frac{3}{2x^3}\right) \approx 2x^3$$

Since $2x^3 \to \infty$ as $x \to \infty$, it follows that

$$\lim_{x \to \infty} p(x) = \lim_{x \to \infty} 2x^3 = \infty$$

Similarly, $2x^3 \to -\infty$ as $x \to -\infty$ implies that

$$\lim_{x \to -\infty} p(x) = \lim_{x \to -\infty} 2x^3 = -\infty$$

So the behavior of $p(x)$ for large values is the same as the behavior of the highest-degree term, $2x^3$.

Matched Problem 3 | Let $p(x) = -4x^4 + 2x^3 + 3x$. Find the limit of $p(x)$ as x approaches ∞ and as x approaches $-\infty$.

The term with highest degree in a polynomial is called the **leading term**. In the solution to Example 3, the limits at infinity of $p(x) = 2x^3 - x^2 - 7x + 3$ were the same as the limits of the leading term $2x^3$. Theorem 3 states that this is true for any polynomial of degree greater than or equal to 1.

THEOREM 3   Limits of Polynomial Functions at Infinity

If

$$p(x) = a_n x^n + a_{n-1} x^{n-1} + \cdots + a_1 x + a_0, a_n \neq 0, n \geq 1$$

then

$$\lim_{x \to \infty} p(x) = \lim_{x \to \infty} a_n x^n = \pm \infty$$

and

$$\lim_{x \to -\infty} p(x) = \lim_{x \to -\infty} a_n x^n = \pm \infty$$

Each limit will be either $-\infty$ or ∞, depending on a_n and n.

A polynomial of degree 0 is a constant function $p(x) = a_0$, and its limit as x approaches ∞ or $-\infty$ is the number a_0. For any polynomial of degree 1 or greater, Theorem 3 states that the limit as x approaches ∞ or $-\infty$ cannot be equal to a number. This means that **polynomials of degree 1 or greater never have horizontal asymptotes**.

A pair of limit expressions of the form

$$\lim_{x \to \infty} f(x) = A, \quad \lim_{x \to -\infty} f(x) = B$$

where A and B are ∞, $-\infty$, or real numbers, describes the **end behavior** of the function f. The first of the two limit expressions describes the **right end behavior** and the second describes the **left end behavior**. By Theorem 3, the end behavior of any nonconstant polynomial function is described by a pair of infinite limits.

EXAMPLE 4   End Behavior of a Polynomial   Give a pair of limit expressions that describe the end behavior of each polynomial.

(A) $p(x) = 3x^3 - 500x^2$ (B) $p(x) = 3x^3 - 500x^4$

SOLUTION

(A) By Theorem 3,

$$\lim_{x \to \infty} (3x^3 - 500x^2) = \lim_{x \to \infty} 3x^3 = \infty \qquad \text{Right end behavior}$$

and

$$\lim_{x \to -\infty} (3x^3 - 500x^2) = \lim_{x \to -\infty} 3x^3 = -\infty \qquad \text{Left end behavior}$$

(B) By Theorem 3,

$$\lim_{x \to \infty} (3x^3 - 500x^4) = \lim_{x \to \infty} (-500x^4) = -\infty \qquad \text{Right end behavior}$$

and

$$\lim_{x \to -\infty} (3x^3 - 500x^4) = \lim_{x \to -\infty} (-500x^4) = -\infty \qquad \text{Left end behavior}$$

Matched Problem 4) Give a pair of limit expressions that describe the end behavior of each polynomial.

(A) $p(x) = 300x^2 - 4x^5$ (B) $p(x) = 300x^6 - 4x^5$

Finding Horizontal Asymptotes

Since a rational function is the ratio of two polynomials, it is not surprising that reciprocals of powers of x can be used to analyze limits of rational functions at infinity. For example, consider the rational function

$$f(x) = \frac{3x^2 - 5x + 9}{2x^2 + 7}$$

Factoring the highest-degree term out of the numerator and the highest-degree term out of the denominator, we write

$$f(x) = \frac{3x^2}{2x^2} \cdot \frac{1 - \dfrac{5}{3x} + \dfrac{3}{x^2}}{1 + \dfrac{7}{2x^2}}$$

$$\lim_{x \to \infty} f(x) = \lim_{x \to \infty} \frac{3x^2}{2x^2} \cdot \lim_{x \to \infty} \frac{1 - \dfrac{5}{3x} + \dfrac{3}{x^2}}{1 + \dfrac{7}{2x^2}} = \frac{3}{2} \cdot \frac{1 - 0 + 0}{1 + 0} = \frac{3}{2}$$

The behavior of this rational function as x approaches infinity is determined by the ratio of the highest-degree term in the numerator ($3x^2$) to the highest-degree term in the denominator ($2x^2$). Theorem 2 can be used to generalize this result to any rational function. Theorem 4 lists the three possible outcomes.

THEOREM 4 Limits of Rational Functions at Infinity and Horizontal Asymptotes of Rational Functions

(A) If $f(x) = \dfrac{a_m x^m + a_{m-1} x^{m-1} + \cdots + a_1 x + a_0}{b_n x^n + b_{n-1} x^{n-1} + \cdots + b_1 x + b_0}, \ a_m \neq 0, b_n \neq 0$

then $\displaystyle\lim_{x \to \infty} f(x) = \lim_{x \to \infty} \frac{a_m x^m}{b_n x^n}$ and $\displaystyle\lim_{x \to -\infty} f(x) = \lim_{x \to -\infty} \frac{a_m x^m}{b_n x^n}$

(B) There are three possible cases for these limits:

1. If $m < n$, then $\displaystyle\lim_{x \to \infty} f(x) = \lim_{x \to -\infty} f(x) = 0$, and the line $y = 0$ (the x axis) is a horizontal asymptote of $f(x)$.

2. If $m = n$, then $\displaystyle\lim_{x \to \infty} f(x) = \lim_{x \to -\infty} f(x) = \frac{a_m}{b_n}$, and the line $y = \dfrac{a_m}{b_n}$ is a horizontal asymptote of $f(x)$.

3. If $m > n$, then each limit will be ∞ or $-\infty$, depending on m, n, a_m, and b_n, and $f(x)$ does not have a horizontal asymptote.

Notice that in cases 1 and 2 of Theorem 4, the limit is the same if x approaches ∞ or $-\infty$. So, **a rational function can have at most one horizontal asymptote** (see Fig. 6).

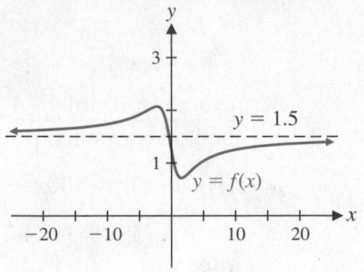

Figure 6 $f(x) = \dfrac{3x^2 - 5x + 9}{2x^2 + 7}$

CONCEPTUAL INSIGHT

The graph of f in Figure 6 dispels the misconception that the graph of a function cannot cross a horizontal asymptote. Horizontal asymptotes give us information about the graph of a function only as $x \to \infty$ and $x \to -\infty$, not at any specific value of x.

EXAMPLE 5   Finding Horizontal Asymptotes   Find all horizontal asymptotes, if any, of each function.

(A) $f(x) = \dfrac{5x^3 - 2x^2 + 1}{4x^3 + 2x - 7}$
(B) $f(x) = \dfrac{3x^4 - x^2 + 1}{8x^6 - 10}$

(C) $f(x) = \dfrac{2x^5 - x^3 - 1}{6x^3 + 2x^2 - 7}$

SOLUTION   We will make use of part A of Theorem 4.

(A) $\displaystyle\lim_{x \to \infty} f(x) = \lim_{x \to \infty} \dfrac{5x^3 - 2x^2 + 1}{4x^3 + 2x - 7} = \lim_{x \to \infty} \dfrac{5x^3}{4x^3} = \dfrac{5}{4}$

The line $y = 5/4$ is a horizontal asymptote of $f(x)$.

(B) $\displaystyle\lim_{x \to \infty} f(x) = \lim_{x \to \infty} \dfrac{3x^4 - x^2 + 1}{8x^6 - 10} = \lim_{x \to \infty} \dfrac{3x^4}{8x^6} = \lim_{x \to \infty} \dfrac{3}{8x^2} = 0$

The line $y = 0$ (the x axis) is a horizontal asymptote of $f(x)$.

(C) $\displaystyle\lim_{x \to \infty} f(x) = \lim_{x \to \infty} \dfrac{2x^5 - x^3 - 1}{6x^3 + 2x^2 - 7} = \lim_{x \to \infty} \dfrac{2x^5}{6x^3} = \lim_{x \to \infty} \dfrac{x^2}{3} = \infty$

The function $f(x)$ has no horizontal asymptotes.

Matched Problem 5   Find all horizontal asymptotes, if any, of each function.

(A) $f(x) = \dfrac{4x^3 - 5x + 8}{2x^4 - 7}$
(B) $f(x) = \dfrac{5x^6 + 3x}{2x^5 - x - 5}$

(C) $f(x) = \dfrac{2x^3 - x + 7}{4x^3 + 3x^2 - 100}$

An accurate sketch of the graph of a rational function requires knowledge of both vertical and horizontal asymptotes. As we mentioned earlier, we are postponing a detailed discussion of graphing techniques until Section 4.4.

EXAMPLE 6 Find all vertical and horizontal asymptotes of the function

$$f(x) = \frac{2x^2 - 5}{x^2 + 5x + 4}$$

SOLUTION Let $n(x) = 2x^2 - 5$ and $d(x) = x^2 + 5x + 4 = (x + 1)(x + 4)$. The denominator $d(x) = 0$ at $x = -1$ and $x = -4$. Since the numerator $n(x)$ is not zero at these values of x [$n(-1) = -3$ and $n(-4) = 27$], by Theorem 1 there are two vertical asymptotes of f: the line $x = -1$ and the line $x = -4$. Since

$$\lim_{x \to \infty} f(x) = \lim_{x \to \infty} \frac{2x^2 - 5}{x^2 + 5x + 4} = \lim_{x \to \infty} \frac{2x^2}{x^2} = 2$$

the horizontal asymptote is the line $y = 2$ (Theorem 3).

Matched Problem 6 Find all vertical and horizontal asymptotes of the function

$$f(x) = \frac{x^2 - 9}{x^2 - 4}.$$

Exercises 2.2

Skills Warm-up Exercises

W *In Problems 1–8, find an equation of the form $Ax + By = C$ for the given line. (If necessary, review Section 1.3).*

1. The horizontal line through $(0, 4)$

2. The vertical line through $(5, 0)$

3. The vertical line through $(-6, 3)$

4. The horizontal line through $(7, 1)$

5. The line through $(-2, 9)$ that has slope 2

6. The line through $(8, -4)$ that has slope -3

7. The line through $(9, 0)$ and $(0, 7)$

8. The line through $(-1, 20)$ and $(1, 30)$

Problems 9–16 refer to the following graph of $y = f(x)$.

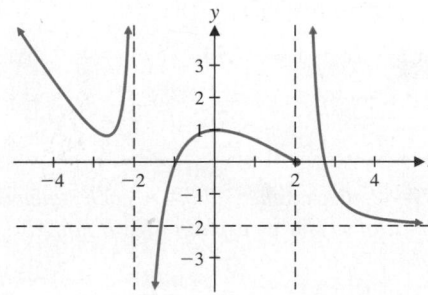

Figure for 9–16

9. $\lim_{x \to \infty} f(x) = ?$

10. $\lim_{x \to -\infty} f(x) = ?$

11. $\lim_{x \to -2^+} f(x) = ?$

12. $\lim_{x \to -2^-} f(x) = ?$

13. $\lim_{x \to -2} f(x) = ?$

14. $\lim_{x \to 2^+} f(x) = ?$

15. $\lim_{x \to 2} f(x) = ?$

16. $\lim_{x \to 2} f(x) = ?$

In Problems 17–24, find each limit. Use $-\infty$ and ∞ when appropriate.

17. $f(x) = \dfrac{x}{x - 5}$

    (A) $\lim_{x \to 5^-} f(x)$ (B) $\lim_{x \to 5^+} f(x)$ (C) $\lim_{x \to 5} f(x)$

18. $f(x) = \dfrac{x^2}{x + 3}$

    (A) $\lim_{x \to -3^-} f(x)$ (B) $\lim_{x \to -3^+} f(x)$ (C) $\lim_{x \to -3} f(x)$

19. $f(x) = \dfrac{2x - 4}{(x - 4)^2}$

    (A) $\lim_{x \to 4^-} f(x)$ (B) $\lim_{x \to 4^+} f(x)$ (C) $\lim_{x \to 4} f(x)$

20. $f(x) = \dfrac{2x + 2}{(x + 2)^2}$

    (A) $\lim_{x \to -2^-} f(x)$ (B) $\lim_{x \to -2^+} f(x)$ (C) $\lim_{x \to -2} f(x)$

21. $f(x) = \dfrac{x^2 + x - 2}{x - 1}$

    (A) $\lim_{x \to 1^-} f(x)$ (B) $\lim_{x \to 1^+} f(x)$ (C) $\lim_{x \to 1} f(x)$

22. $f(x) = \dfrac{x^2 + x + 2}{x - 1}$

    (A) $\lim_{x \to 1^-} f(x)$ (B) $\lim_{x \to 1^+} f(x)$ (C) $\lim_{x \to 1} f(x)$

23. $f(x) = \dfrac{x^2 - 3x + 2}{x + 2}$

    (A) $\lim_{x \to -2^-} f(x)$ (B) $\lim_{x \to -2^+} f(x)$ (C) $\lim_{x \to -2} f(x)$

24. $f(x) = \dfrac{x^2 + x - 2}{x + 2}$

    (A) $\lim_{x \to -2^-} f(x)$ (B) $\lim_{x \to -2^+} f(x)$ (C) $\lim_{x \to -2} f(x)$

In Problems 25–32, find (A) the leading term of the polynomial, (B) the limit as x approaches ∞, and (C) the limit as x approaches −∞.

25. $p(x) = 15 + 3x^2 - 5x^3$

26. $p(x) = 10 - x^6 + 7x^3$

27. $p(x) = 9x^2 - 6x^4 + 7x$

28. $p(x) = -x^5 + 2x^3 + 9x$

29. $p(x) = x^2 + 7x + 12$

30. $p(x) = 5x + x^3 - 8x^2$

31. $p(x) = x^4 + 2x^5 - 11x$

32. $p(x) = 1 + 4x^2 + 4x^4$

In Problems 33–42, use −∞ or ∞ where appropriate to describe the behavior at each zero of the denominator and identify all vertical asymptotes.

33. $f(x) = \dfrac{1}{x + 3}$

34. $g(x) = \dfrac{x}{4 - x}$

35. $h(x) = \dfrac{x^2 + 4}{x^2 - 4}$

36. $k(x) = \dfrac{x^2 - 9}{x^2 + 9}$

37. $F(x) = \dfrac{x^2 - 4}{x^2 + 4}$

38. $G(x) = \dfrac{x^2 + 9}{9 - x^2}$

39. $H(x) = \dfrac{x^2 - 2x - 3}{x^2 - 4x + 3}$

40. $K(x) = \dfrac{x^2 + 2x - 3}{x^2 - 4x + 3}$

41. $T(x) = \dfrac{8x - 16}{x^4 - 8x^3 + 16x^2}$

42. $S(x) = \dfrac{6x + 9}{x^4 + 6x^3 + 9x^2}$

In Problems 43–50, find each function value and limit. Use −∞ or ∞ where appropriate.

43. $f(x) = \dfrac{4x + 7}{5x - 9}$

(A) $f(10)$ (B) $f(100)$ (C) $\lim\limits_{x \to \infty} f(x)$

44. $f(x) = \dfrac{2 - 3x^3}{7 + 4x^3}$

(A) $f(5)$ (B) $f(10)$ (C) $\lim\limits_{x \to \infty} f(x)$

45. $f(x) = \dfrac{5x^2 + 11}{7x - 2}$

(A) $f(20)$ (B) $f(50)$ (C) $\lim\limits_{x \to \infty} f(x)$

46. $f(x) = \dfrac{5x + 11}{7x^3 - 2}$

(A) $f(-8)$ (B) $f(-16)$ (C) $\lim\limits_{x \to -\infty} f(x)$

47. $f(x) = \dfrac{7x^4 - 14x^2}{6x^5 + 3}$

(A) $f(-6)$ (B) $f(-12)$ (C) $\lim\limits_{x \to -\infty} f(x)$

48. $f(x) = \dfrac{4x^7 - 8x}{6x^4 + 9x^2}$

(A) $f(-3)$ (B) $f(-6)$ (C) $\lim\limits_{x \to -\infty} f(x)$

49. $f(x) = \dfrac{10 - 7x^3}{4 + x^3}$

(A) $f(-10)$ (B) $f(-20)$ (C) $\lim\limits_{x \to -\infty} f(x)$

50. $f(x) = \dfrac{3 + x}{5 + 4x}$

(A) $f(-50)$ (B) $f(-100)$ (C) $\lim\limits_{x \to -\infty} f(x)$

In Problems 51–64, find all horizontal and vertical asymptotes.

51. $f(x) = \dfrac{2x}{x + 2}$

52. $f(x) = \dfrac{3x + 2}{x - 4}$

53. $f(x) = \dfrac{x^2 + 1}{x^2 - 1}$

54. $f(x) = \dfrac{x^2 - 1}{x^2 + 2}$

55. $f(x) = \dfrac{x^3}{x^2 + 6}$

56. $f(x) = \dfrac{x}{x^2 - 4}$

57. $f(x) = \dfrac{x}{x^2 + 4}$

58. $f(x) = \dfrac{x^2 + 9}{x}$

59. $f(x) = \dfrac{x^2}{x - 3}$

60. $f(x) = \dfrac{x + 5}{x^2}$

61. $f(x) = \dfrac{2x^2 + 3x - 2}{x^2 - x - 2}$

62. $f(x) = \dfrac{2x^2 + 7x + 12}{2x^2 + 5x - 12}$

63. $f(x) = \dfrac{2x^2 - 5x + 2}{x^2 - x - 2}$

64. $f(x) = \dfrac{x^2 - x - 12}{2x^2 + 5x - 12}$

In Problems 65–68, give a limit expression that describes the right end behavior of the function.

65. $f(x) = \dfrac{x + 3}{x^2 - 5}$

66. $f(x) = \dfrac{3 + 4x + x^2}{5 - x}$

67. $f(x) = \dfrac{x^2 - 5}{x + 3}$

68. $f(x) = \dfrac{4x + 1}{5x - 7}$

In Problems 69–72, give a limit expression that describes the left end behavior of the function.

69. $f(x) = \dfrac{5 - 2x^2}{1 + 8x^2}$

70. $f(x) = \dfrac{2x + 3}{x^2 - 1}$

71. $f(x) = \dfrac{x^2 + 4x}{3x + 2}$

72. $f(x) = \dfrac{6 - x^4}{1 + 2x}$

In Problems 73–76, give a pair of limit expressions that describe the end behavior of the function.

73. $f(x) = x^3 - 3x + 1$

74. $f(x) = 4 - 5x - x^3$

75. $f(x) = \dfrac{2 + 5x}{1 - x}$

76. $f(x) = \dfrac{9x^2 + 6x + 1}{4x^2 + 4x + 1}$

In Problems 77–82, discuss the validity of each statement. If the statement is always true, explain why. If not, give a counterexample.

77. A rational function has at least one vertical asymptote.

78. A rational function has at most one vertical asymptote.

79. A rational function has at least one horizontal asymptote.

80. A rational function has at most one horizontal asymptote.

81. A polynomial function of degree ≥ 1 has neither horizontal nor vertical asymptotes.

82. The graph of a rational function cannot cross a horizontal asymptote.

83. Theorem 3 states that

$$\lim_{x \to \infty} (a_n x^n + a_{n-1} x^{n-1} + \cdots + a_0) = \pm\infty.$$

What conditions must n and a_n satisfy for the limit to be ∞? For the limit to be $-\infty$?

84. Theorem 3 also states that

$$\lim_{x \to -\infty} (a_n x^n + a_{n-1} x^{n-1} + \cdots + a_0) = \pm\infty.$$

What conditions must n and a_n satisfy for the limit to be ∞? For the limit to be $-\infty$?

Applications

85. Average cost. A company manufacturing snowboards has fixed costs of $200 per day and total costs of $3,800 per day for a daily output of 20 boards.

(A) Assuming that the total cost per day $C(x)$ is linearly related to the total output per day x, write an equation for the cost function.

(B) The average cost per board for an output of x boards is given by $\overline{C}(x) = C(x)/x$. Find the average cost function.

(C) Sketch a graph of the average cost function, including any asymptotes, for $1 \le x \le 30$.

(D) What does the average cost per board tend to as production increases?

86. Average cost. A company manufacturing surfboards has fixed costs of $300 per day and total costs of $5,100 per day for a daily output of 20 boards.

(A) Assuming that the total cost per day $C(x)$ is linearly related to the total output per day x, write an equation for the cost function.

(B) The average cost per board for an output of x boards is given by $\overline{C}(x) = C(x)/x$. Find the average cost function.

(C) Sketch a graph of the average cost function, including any asymptotes, for $1 \le x \le 30$.

(D) What does the average cost per board tend to as production increases?

87. Energy costs. Most appliance manufacturers produce conventional and energy-efficient models. The energy-efficient models are more expensive to make but cheaper to operate. The costs of purchasing and operating a 23-cubic-foot refrigerator of each type are given in Table 8. These costs do not include maintenance charges or changes in electricity prices.

Table 8 **23-ft³ Refrigerators**

	Energy-Efficient Model	Conventional Model
Initial cost	$950	$900
Total volume	23 ft³	23 ft³
Annual cost of electricity	$56	$66

(A) Express the total cost $C_e(x)$ and the average cost $\overline{C}_e(x) = C_e(x)/x$ of purchasing and operating an energy-efficient model for x years.

(B) Express the total cost $C_c(x)$ and the average cost $\overline{C}_c(x) = C_c(x)/x$ of purchasing and operating a conventional model for x years.

(C) Are the total costs for an energy-efficient model and for a conventional model ever the same? If so, when?

(D) Are the average costs for an energy-efficient model and for a conventional model ever the same? If so, when?

(E) Find the limit of each average cost function as $x \to \infty$ and discuss the implications of the results.

88. Energy costs. Most appliance manufacturers produce conventional and energy-efficient models. The energy-efficient models are more expensive to make but cheaper to operate. The costs of purchasing and operating a 36,000-Btu central air conditioner of each type are given in Table 9. These costs do not include maintenance charges or changes in electricity prices.

Table 9 **36,000 Btu Central Air Conditioner**

	Energy-Efficient Model	Conventional Model
Initial cost	$4,000	$2,700
Total capacity	36,000 Btu	36,000 Btu
Annual cost of electricity	$932	$1,332

(A) Express the total cost $C_e(x)$ and the average cost $\overline{C}_e(x) = C_e(x)/x$ of purchasing and operating an energy-efficient model for x years.

(B) Express the total cost $C_c(x)$ and the average cost $\overline{C}_c(x) = C_c(x)/x$ of purchasing and operating a conventional model for x years.

(C) Are the total costs for an energy-efficient model and for a conventional model ever the same? If so, when?

(D) Are the average costs for an energy-efficient model and for a conventional model ever the same? If so, when?

(E) Find the limit of each average cost function as $x \to \infty$ and discuss the implications of the results.

89. Drug concentration. A drug is administered to a patient through an injection. The drug concentration (in milligrams/milliliter) in the bloodstream t hours after the injection is given by $C(t) = \dfrac{5t^2(t + 50)}{t^3 + 100}$. Find and interpret $\lim_{t \to \infty} C(t)$.

90. Drug concentration. A drug is administered to a patient through an IV drip. The drug concentration (in milligrams/milliliter) in the bloodstream t hours after the drip was started is given by $C(t) = \dfrac{5t(t + 50)}{t^3 + 100}$. Find and interpret $\lim_{t \to \infty} C(t)$.

91. Pollution. In Silicon Valley, a number of computer-related manufacturing firms were contaminating underground water supplies with toxic chemicals stored in leaking underground containers. A water quality control agency ordered the companies to take immediate corrective action and contribute to a monetary pool for the testing and cleanup of the underground contamination. Suppose that the monetary pool (in millions of dollars) for the testing and cleanup is given by

$$P(x) = \frac{2x}{1-x} \qquad 0 \le x < 1$$

where x is the percentage (expressed as a decimal) of the total contaminant removed.

(A) How much must be in the pool to remove 90% of the contaminant?

(B) How much must be in the pool to remove 95% of the contaminant?

✎ (C) Find $\lim\limits_{x \to 1^{-}} P(x)$ and discuss the implications of this limit.

92. **Employee training.** A company producing computer components has established that, on average, a new employee can assemble $N(t)$ components per day after t days of on-the-job training, as given by

$$N(t) = \frac{100t}{t+9} \qquad t \ge 0$$

(A) How many components per day can a new employee assemble after 6 days of on-the-job training?

(B) How many days of on-the-job training will a new employee need to reach the level of 70 components per day?

✎ (C) Find $\lim\limits_{t \to \infty} N(t)$ and discuss the implications of this limit.

93. **Biochemistry.** In 1913, biochemists Leonor Michaelis and Maude Menten proposed the rational function model (see figure)

$$v(s) = \frac{V_{max}\, s}{K_M + s}$$

for the velocity of the enzymatic reaction v, where s is the substrate concentration. The constants V_{max} and K_M are determined from experimental data.

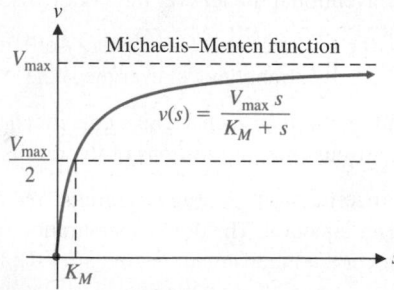

(A) Show that $\lim\limits_{s \to \infty} v(s) = V_{max}$.

(B) Show that $v(K_M) = \dfrac{V_{max}}{2}$.

(C) Table 10* lists data for the substrate saccharose treated with an enzyme.
Plot the points in Table 10 on graph paper and estimate V_{max} to the nearest integer. To estimate K_M, add the horizontal line $v = \dfrac{V_{max}}{2}$ to your graph, connect successive points on the graph with straight-line segments, and

Table 10

s	v
5.2	0.866
10.4	1.466
20.8	2.114
41.6	2.666
83.3	3.236
167	3.636
333	3.636

estimate the value of s (to the nearest multiple of 10) that satisfies $v(s) = \dfrac{V_{max}}{2}$.

(D) Use the constants V_{max} and K_M from part (C) to form a Michaelis–Menten function for the data in Table 10.

(E) Use the function from part (D) to estimate the velocity of the enzyme reaction when the saccharose is 15 and to estimate the saccharose when the velocity is 3.

94. **Biochemistry.** Table 11* lists data for the substrate sucrose treated with the enzyme invertase. We want to model these data with a Michaelis–Menten function.

Table 11

s	v
2.92	18.2
5.84	26.5
8.76	31.1
11.7	33
14.6	34.9
17.5	37.2
23.4	37.1

(A) Plot the points in Table 11 on graph paper and estimate V_{max} to the nearest integer. To estimate K_M, add the horizontal line $v = \dfrac{V_{max}}{2}$ to your graph, connect successive points on the graph with straight-line segments, and estimate the value of s (to the nearest integer) that satisfies $v(s) = \dfrac{V_{max}}{2}$.

(B) Use the constants V_{max} and K_M from part (A) to form a Michaelis–Menten function for the data in Table 11.

(C) Use the function from part (B) to estimate the velocity of the enzyme reaction when the sucrose is 9 and to estimate the sucrose when the velocity is 32.

95. **Physics.** The coefficient of thermal expansion (CTE) is a measure of the expansion of an object subjected to extreme temperatures. To model this coefficient, we use a Michaelis–Menten function of the form

$$C(T) = \frac{C_{max}\, T}{M + T} \qquad \text{(Problem 93)}$$

where C = CTE, T is temperature in K (degrees Kelvin), and C_{max} and M are constants. Table 12[†] lists the coefficients of thermal expansion for nickel and for copper at various temperatures.

---

*Michaelis and Menten (1913) *Biochem. Z.* 49, 333–369.

*Institute of Chemistry, Macedonia.
[†]National Physical Laboratory

Table 12 **Coefficients of Thermal Expansion**

T (K)	Nickel	Copper
100	6.6	10.3
200	11.3	15.2
293	13.4	16.5
500	15.3	18.3
800	16.8	20.3
1,100	17.8	23.7

(A) Plot the points in columns 1 and 2 of Table 12 on graph paper and estimate C_{max} to the nearest integer. To estimate M, add the horizontal line CTE $= \dfrac{C_{max}}{2}$ to your graph, connect successive points on the graph with straight-line segments, and estimate the value of T (to the nearest multiple of fifty) that satisfies $C(T) = \dfrac{C_{max}}{2}$.

(B) Use the constants $\dfrac{C_{max}}{2}$ and M from part (A) to form a Michaelis–Menten function for the CTE of nickel.

(C) Use the function from part (B) to estimate the CTE of nickel at 600 K and to estimate the temperature when the CTE of nickel is 12.

96. Physics. Repeat Problem 95 for the CTE of copper (column 3 of Table 12).

Answers to Matched Problems

1. Vertical asymptote: $x = 1$; $\lim\limits_{x \to 1^+} f(x) = \infty$, $\lim\limits_{x \to 1^-} f(x) = -\infty$
   $\lim\limits_{x \to 3} f(x) = 1/2$ so f does not have a vertical asymptote at $x = 3$

2. Vertical asymptote: $x = -3$; $\lim\limits_{x \to -3^+} f(x) = \lim\limits_{x \to -3^-} f(x) = -\infty$

3. $\lim\limits_{x \to \infty} p(x) = \lim\limits_{x \to -\infty} p(x) = -\infty$

4. (A) $\lim\limits_{x \to \infty} p(x) = -\infty$, $\lim\limits_{x \to -\infty} p(x) = \infty$
   (B) $\lim\limits_{x \to \infty} p(x) = \infty$, $\lim\limits_{x \to -\infty} p(x) = \infty$

5. (A) $y = 0$ (B) No horizontal asymptotes
   (C) $y = 1/2$

6. Vertical asymptotes: $x = -2$, $x = 2$; horizontal asymptote: $y = 1$

2.3 Continuity

- Continuity
- Continuity Properties
- Solving Inequalities Using Continuity Properties

Theorem 3 in Section 2.1 states that if f is a polynomial function or a rational function with a nonzero denominator at $x = c$, then

$$\lim_{x \to c} f(x) = f(c) \tag{1}$$

Functions that satisfy equation (1) are said to be *continuous* at $x = c$. A firm understanding of continuous functions is essential for sketching and analyzing graphs. We will also see that continuity properties provide a simple and efficient method for solving inequalities—a tool that we will use extensively in later sections.

Continuity

Compare the graphs shown in Figure 1. Notice that two of the graphs are broken; that is, they cannot be drawn without lifting a pen off the paper. Informally, a function is *continuous over an interval* if its graph over the interval can be drawn without removing a pen from the paper. A function whose graph is broken (disconnected) at $x = c$ is said to be *discontinuous* at $x = c$. Function f (Fig. 1A) is continuous for all x. Function g (Fig. 1B) is discontinuous at $x = 2$ but is continuous over any interval that does not include 2. Function h (Fig. 1C) is discontinuous at $x = 0$ but is continuous over any interval that does not include 0.

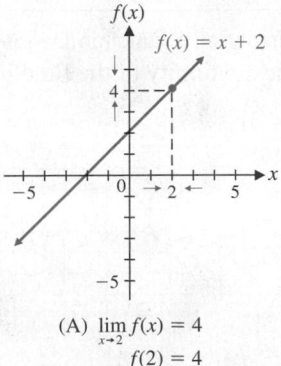

(A) $\lim\limits_{x \to 2} f(x) = 4$
   $f(2) = 4$

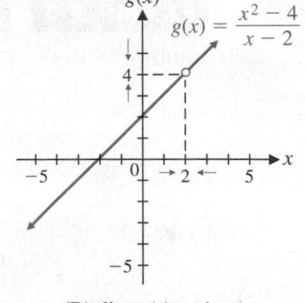
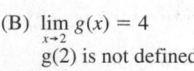

(B) $\lim\limits_{x \to 2} g(x) = 4$
   $g(2)$ is not defined

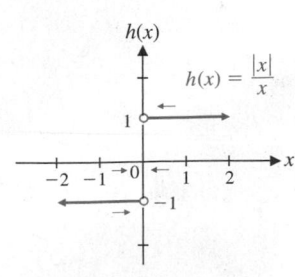

(C) $\lim\limits_{x \to 0} h(x)$ does not exist
   $h(0)$ is not defined

Figure 1

Most graphs of natural phenomena are continuous, whereas many graphs in business and economics applications have discontinuities. Figure 2A illustrates temperature variation over a 24-hour period—a continuous phenomenon. Figure 2B illustrates warehouse inventory over a 1-week period—a discontinuous phenomenon.

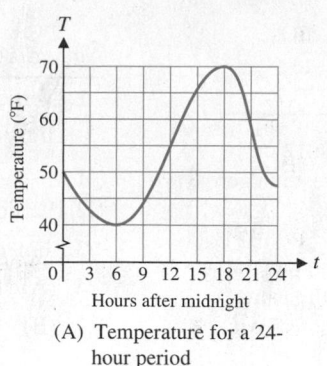

(A) Temperature for a 24-hour period

(B) Inventory in a warehouse during 1 week

Figure 2

Explore and Discuss 1

(A) Write a brief verbal description of the temperature variation illustrated in Figure 2A, including estimates of the high and low temperatures during the period shown and the times at which they occurred.

(B) Write a brief verbal description of the changes in inventory illustrated in Figure 2B, including estimates of the changes in inventory and the times at which those changes occurred.

The preceding discussion leads to the following formal definition of continuity:

DEFINITION   Continuity

A function f is **continuous at the point** $x = c$ **if**

1. $\lim_{x \to c} f(x)$ exists   2. $f(c)$ exists   3. $\lim_{x \to c} f(x) = f(c)$

A function is **continuous on the open interval*** (a, b) if it is continuous at each point on the interval.

*See Table 3 in section 1.3 for a review of interval notation.

If one or more of the three conditions in the definition fails, then the function is **discontinuous** at $x = c$.

EXAMPLE 1   Continuity of a Function Defined by a Graph   Use the definition of continuity to discuss the continuity of the function whose graph is shown in Figure 3.

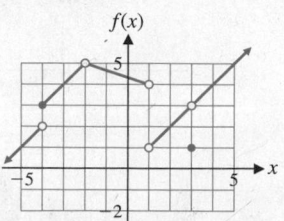

Figure 3

SOLUTION We begin by identifying the points of discontinuity. Examining the graph, we see breaks or holes at $x = -4, -2, 1$, and 3. Now we must determine which conditions in the definition of continuity are not satisfied at each of these points. In each case, we find the value of the function and the limit of the function at the point in question.

Discontinuity at $x = -4$:

$$\lim_{x \to -4^-} f(x) = 2 \qquad \text{Since the one-sided limits are different,}$$
$$\lim_{x \to -4^+} f(x) = 3 \qquad \text{the limit does not exist (Section 2.1).}$$
$$\lim_{x \to -4} f(x) \text{ does not exist}$$
$$f(-4) = 3$$

So, f is not continuous at $x = -4$ because condition 1 is not satisfied.

Discontinuity at $x = -2$:

$$\lim_{x \to -2^-} f(x) = 5 \qquad \text{The hole at } (-2, 5) \text{ indicates that 5 is not the value of } f$$
$$\lim_{x \to -2^+} f(x) = 5 \qquad \text{at } -2. \text{ Since there is no solid dot elsewhere on the}$$
$$\lim_{x \to -2} f(x) = 5 \qquad \text{vertical line } x = -2, f(-2) \text{ is not defined.}$$
$$f(-2) \text{ does not exist}$$

So even though the limit as x approaches -2 exists, f is not continuous at $x = -2$ because condition 2 is not satisfied.

Discontinuity at $x = 1$:

$$\lim_{x \to 1^-} f(x) = 4$$
$$\lim_{x \to 1^+} f(x) = 1$$
$$\lim_{x \to 1} f(x) \text{ does not exist}$$
$$f(1) \text{ does not exist}$$

This time, f is not continuous at $x = 1$ because neither of conditions 1 and 2 is satisfied.

Discontinuity at $x = 3$:

$$\lim_{x \to 3^-} f(x) = 3 \qquad \text{The solid dot at } (3, 1) \text{ indicates that } f(3) = 1.$$
$$\lim_{x \to 3^+} f(x) = 3$$
$$\lim_{x \to 3} f(x) = 3$$
$$f(3) = 1$$

Conditions 1 and 2 are satisfied, but f is not continuous at $x = 3$ because condition 3 is not satisfied.

Having identified and discussed all points of discontinuity, we can now conclude that f is continuous except at $x = -4, -2, 1$, and 3.

CONCEPTUAL INSIGHT

Rather than list the points where a function is discontinuous, sometimes it is useful to state the intervals on which the function is continuous. Using the set operation **union,** denoted by $\cup$, we can express the set of points where the function in Example 1 is continuous as follows:

$$(-\infty, -4) \cup (-4, -2) \cup (-2, 1) \cup (1, 3) \cup (3, \infty)$$

Matched Problem 1 Use the definition of continuity to discuss the continuity of the function whose graph is shown in Figure 4.

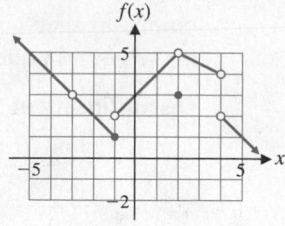

Figure 4

For functions defined by equations, it is important to be able to locate points of discontinuity by examining the equation.

EXAMPLE 2   Continuity of Functions Defined by Equations   Using the definition of continuity, discuss the continuity of each function at the indicated point(s).

(A) $f(x) = x + 2$ at $x = 2$

(B) $g(x) = \dfrac{x^2 - 4}{x - 2}$ at $x = 2$

(C) $h(x) = \dfrac{|x|}{x}$ at $x = 0$ and   at $x = 1$

SOLUTION

(A) f is continuous at $x = 2$, since

$$\lim_{x \to 2} f(x) = 4 = f(2) \quad \text{See Figure 1A.}$$

(B) g is not continuous at $x = 2$, since $g(2) = 0/0$ is not defined (see Fig. 1B).

(C) h is not continuous at $x = 0$, since $h(0) = |0|/0$ is not defined; also, $\lim_{x \to 0} h(x)$ does not exist.

h is continuous at $x = 1$, since

$$\lim_{x \to 1} \frac{|x|}{x} = 1 = h(1) \quad \text{See Figure 1C.}$$

Matched Problem 2 Using the definition of continuity, discuss the continuity of each function at the indicated point(s).

(A) $f(x) = x + 1$ at $x = 1$

(B) $g(x) = \dfrac{x^2 - 1}{x - 1}$ at $x = 1$

(C) $h(x) = \dfrac{x - 2}{|x - 2|}$ at $x = 2$ and at $x = 0$

We can also talk about one-sided continuity, just as we talked about one-sided limits. For example, a function is said to be **continuous on the right** at $x = c$ if $\lim_{x \to c^+} f(x) = f(c)$ and **continuous on the left** at $x = c$ if $\lim_{x \to c^-} f(x) = f(c)$. A function is **continuous on the closed interval [a, b]** if it is continuous on the open interval (a, b) and is continuous both on the right at a and on the left at b.

Figure 5A illustrates a function that is continuous on the closed interval $[-1, 1]$. Figure 5B illustrates a function that is continuous on the half-closed interval $[0, \infty)$.

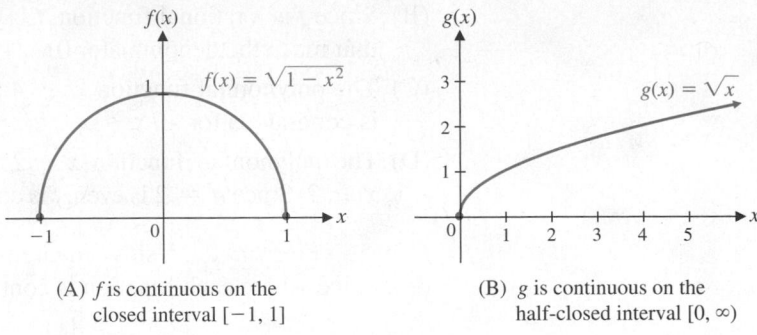

(A) f is continuous on the closed interval $[-1, 1]$

(B) g is continuous on the half-closed interval $[0, \infty)$

Figure 5 **Continuity on closed and half-closed intervals**

Continuity Properties

Functions have some useful **general continuity properties**:

> If two functions are continuous on the same interval, then their sum, difference, product, and quotient are continuous on the same interval except for values of x that make a denominator 0.

These properties, along with Theorem 1, enable us to determine intervals of continuity for some important classes of functions without having to look at their graphs or use the three conditions in the definition.

THEOREM 1   Continuity Properties of Some Specific Functions

(A) A constant function $f(x) = k$, where k is a constant, is continuous for all x.
$f(x) = 7$ is continuous for all x.

(B) For n a positive integer, $f(x) = x^n$ is continuous for all x.
$f(x) = x^5$ is continuous for all x.

(C) A polynomial function is continuous for all x.
$2x^3 - 3x^2 + x - 5$ is continuous for all x.

(D) A rational function is continuous for all x except those values that make a denominator 0.
$\dfrac{x^2 + 1}{x - 1}$ is continuous for all x except $x = 1$, a value that makes the denominator 0.

(E) For n an odd positive integer greater than 1, $\sqrt[n]{f(x)}$ is continuous wherever $f(x)$ is continuous.
$\sqrt[3]{x^2}$ is continuous for all x.

(F) For n an even positive integer, $\sqrt[n]{f(x)}$ is continuous wherever $f(x)$ is continuous and nonnegative.
$\sqrt[4]{x}$ is continuous on the interval $[0, \infty)$.

Parts (C) and (D) of Theorem 1 are the same as Theorem 3 in Section 2.1. They are repeated here to emphasize their importance.

EXAMPLE 3   Using Continuity Properties   Using Theorem 1 and the general properties of continuity, determine where each function is continuous.

(A) $f(x) = x^2 - 2x + 1$ 　　　　(B) $f(x) = \dfrac{x}{(x + 2)(x - 3)}$

(C) $f(x) = \sqrt[3]{x^2 - 4}$ 　　　　(D) $f(x) = \sqrt{x - 2}$

SOLUTION

(A) Since f is a polynomial function, f is continuous for all x.

(B) Since f is a rational function, f is continuous for all x except -2 and 3 (values that make the denominator 0).

(C) The polynomial function $x^2 - 4$ is continuous for all x. Since $n = 3$ is odd, f is continuous for all x.

(D) The polynomial function $x - 2$ is continuous for all x and nonnegative for $x \geq 2$. Since $n = 2$ is even, f is continuous for $x \geq 2$, or on the interval $[2, \infty)$.

__Matched Problem 3__ Using Theorem 1 and the general properties of continuity, determine where each function is continuous.

(A) $f(x) = x^4 + 2x^2 + 1$

(B) $f(x) = \dfrac{x^2}{(x + 1)(x - 4)}$

(C) $f(x) = \sqrt{x - 4}$

(D) $f(x) = \sqrt[3]{x^3 + 1}$

Solving Inequalities Using Continuity Properties

One of the basic tools for analyzing graphs in calculus is a special line graph called a *sign chart*. We will make extensive use of this type of chart in later sections. In the discussion that follows, we use continuity properties to develop a simple and efficient procedure for constructing sign charts.

Suppose that a function f is continuous over the interval $(1, 8)$ and $f(x) \neq 0$ for any x in $(1, 8)$. Suppose also that $f(2) = 5$, a positive number. Is it possible for $f(x)$ to be negative for any x in the interval $(1, 8)$? The answer is "no." If $f(7)$ were -3, for example, as shown in Figure 6, then how would it be possible to join the points $(2, 5)$ and $(7, -3)$ with the graph of a continuous function without crossing the x axis between 1 and 8 at least once? [Crossing the x axis would violate our assumption that $f(x) \neq 0$ for any x in $(1, 8)$.] We conclude that $f(x)$ must be positive for all x in $(1, 8)$. If $f(2)$ were negative, then, using the same type of reasoning, $f(x)$ would have to be negative over the entire interval $(1, 8)$.

In general, **if f is continuous and $f(x) \neq 0$ on the interval (a, b), then $f(x)$ cannot change sign on (a, b).** This is the essence of Theorem 2.

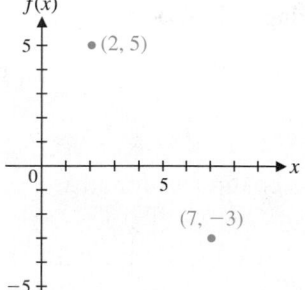

Figure 6

THEOREM 2   Sign Properties on an Interval (a, b)

If f is continuous on (a, b) and $f(x) \neq 0$ for all x in (a, b), then either $f(x) > 0$ for all x in (a, b) or $f(x) < 0$ for all x in (a, b).

Theorem 2 provides the basis for an effective method of solving many types of inequalities. Example 4 illustrates the process.

EXAMPLE 4 Solving an Inequality Solve $\dfrac{x + 1}{x - 2} > 0$.

SOLUTION We start by using the left side of the inequality to form the function f.

$$f(x) = \frac{x + 1}{x - 2}$$

The denominator is equal to 0 if $x = 2$, and the numerator is equal to 0 if $x = -1$. So the rational function f is discontinuous at $x = 2$, and $f(x) = 0$ for $x = -1$ (a fraction is 0 when the numerator is 0 and the denominator is not 0). We plot

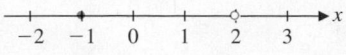

Figure 7

Test Numbers

x	$f(x)$
-2	$\frac{1}{4}$ (+)
0	$-\frac{1}{2}$ (−)
3	4 (+)

$x = 2$ and $x = -1$, which we call *partition numbers,* on a real number line (Fig. 7). (Note that the dot at 2 is open because the function is not defined at $x = 2$.) The partition numbers 2 and -1 determine three open intervals: $(-\infty, -1)$, $(-1, 2)$, and $(2, \infty)$. The function f is continuous and nonzero on each of these intervals. From Theorem 2, we know that $f(x)$ does not change sign on any of these intervals. We can find the sign of $f(x)$ on each of the intervals by selecting a **test number** in each interval and evaluating $f(x)$ at that number. Since any number in each subinterval will do, we choose test numbers that are easy to evaluate: $-2, 0$, and 3. The table in the margin shows the results.

The sign of $f(x)$ at each test number is the same as the sign of $f(x)$ over the interval containing that test number. Using this information, we construct a **sign chart** for $f(x)$ as shown in Figure 8.

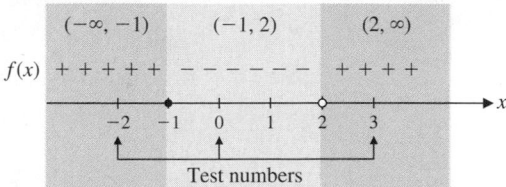

Figure 8

From the sign chart, we can easily write the solution of the given nonlinear inequality:

$$f(x) > 0 \quad \text{for} \quad \begin{array}{ll} x < -1 \quad \text{or} \quad x > 2 & \text{Inequality notation} \\ (-\infty, -1) \cup (2, \infty) & \text{Interval notation} \end{array}$$

Matched Problem 4) Solve $\dfrac{x^2 - 1}{x - 3} < 0$.

Most of the inequalities we encounter will involve strict inequalities ($>$ or $<$). If it is necessary to solve inequalities of the form $\geq$ or $\leq$, we simply include the endpoint x of any interval if f is defined at x and $f(x)$ satisfies the given inequality. For example, from the sign chart in Figure 8, the solution of the inequality

$$\dfrac{x + 1}{x - 2} \geq 0 \quad \text{is} \quad \begin{array}{ll} x \leq -1 \quad \text{or} \quad x > 2 & \text{Inequality notation} \\ (-\infty, -1] \cup (2, \infty) & \text{Interval notation} \end{array}$$

Example 4 illustrates a general procedure for constructing sign charts.

DEFINITION

A real number x is a **partition number** for a function f if f is discontinuous at x or $f(x) = 0$.

Suppose that p_1 and p_2 are consecutive partition numbers for f, that is, there are no partition numbers in the open interval (p_1, p_2). Then f is continuous on (p_1, p_2) [since there are no points of discontinuity in that interval], so f does not change sign on (p_1, p_2) [since $f(x) \neq 0$ for x in that interval]. In other words, **partition numbers determine open intervals on which f does not change sign**. By using a test number from each interval, we can construct a sign chart for f on the real number line. It is then easy to solve the inequality $f(x) < 0$ or the inequality $f(x) > 0$.

We summarize the procedure for constructing sign charts in the following box.

PROCEDURE Constructing Sign Charts

Given a function f,

Step 1 Find all partition numbers of f:

(A) Find all numbers x such that f is discontinuous at x. (Rational functions are discontinuous at values of x that make a denominator 0.)

(B) Find all numbers x such that $f(x) = 0$. (For a rational function, this occurs where the numerator is 0 and the denominator is not 0.)

Step 2 Plot the numbers found in step 1 on a real number line, dividing the number line into intervals.

Step 3 Select a test number in each open interval determined in step 2 and evaluate $f(x)$ at each test number to determine whether $f(x)$ is positive $(+)$ or negative $(-)$ in each interval.

Step 4 Construct a sign chart, using the real number line in step 2. This will show the sign of $f(x)$ on each open interval.

There is an alternative to step 3 in the procedure for constructing sign charts that may save time if the function $f(x)$ is written in factored form. The key is to determine the sign of each factor in the numerator and denominator of $f(x)$. We will illustrate with Example 4. The partition numbers -1 and 2 divide the x axis into three open intervals. If $x > 2$, then both the numerator and denominator are positive, so $f(x) > 0$. If $-1 < x < 2$, then the numerator is positive but the denominator is negative, so $f(x) < 0$. If $x < -1$, then both the numerator and denominator are negative, so $f(x) > 0$. Of course both approaches, the test number approach and the sign of factors approach, give the same sign chart.

Exercises 2.3

Skills Warm-up Exercises

W *In Problems 1–8, use interval notation to specify the given interval. (If necessary, review Table 3 in Section 1.3).*

1. The set of all real numbers from -3 to 5, including -3 and 5

2. The set of all real numbers from -8 to -4, excluding -8 but including -4

3. $\{x \mid -10 < x < 100\}$ **4.** $\{x \mid 0.1 \le x \le 0.3\}$

5. $\{x \mid x^2 > 25\}$ **6.** $\{x \mid x^2 \ge 16\}$

7. $\{x \mid x \le -1 \text{ or } x > 2\}$

8. $\{x \mid x < 6 \text{ or } x \ge 9\}$

In Problems 9–14, sketch a possible graph of a function that satisfies the given conditions at $x = 1$ and discuss the continuity of f at $x = 1$.

9. $f(1) = 2$ and $\lim_{x \to 1} f(x) = 2$

10. $f(1) = -2$ and $\lim_{x \to 1} f(x) = 2$

11. $f(1) = 2$ and $\lim_{x \to 1} f(x) = -2$

12. $f(1) = -2$ and $\lim_{x \to 1} f(x) = -2$

13. $f(1) = -2$, $\lim_{x \to 1^-} f(x) = 2$, and $\lim_{x \to 1^+} f(x) = -2$

14. $f(1) = 2$, $\lim_{x \to 1^-} f(x) = 2$, and $\lim_{x \to 1^+} f(x) = -2$

Problems 15–22 refer to the function f shown in the figure. Use the graph to estimate the indicated function values and limits.

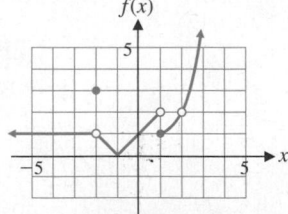

Figure for 15–22

15. $f(0.9)$ **16.** $f(0.1)$

17. $f(-1.9)$ **18.** $f(-0.9)$

19. (A) $\lim_{x \to 1^-} f(x)$ (B) $\lim_{x \to 1^+} f(x)$

(C) $\lim_{x \to 1} f(x)$ (D) $f(1)$

(E) Is f continuous at $x = 1$? Explain.

20. (A) $\lim_{x \to 2^-} f(x)$ (B) $\lim_{x \to 2^+} f(x)$

(C) $\lim_{x \to 2} f(x)$ (D) $f(2)$

(E) Is f continuous at $x = 2$? Explain.

21. (A) $\lim_{x \to -2^-} f(x)$ (B) $\lim_{x \to -2^+} f(x)$

   (C) $\lim_{x \to -2} f(x)$ (D) $f(-2)$

✎ (E) Is f continuous at $x = -2$? Explain.

22. (A) $\lim_{x \to -1^-} f(x)$ (B) $\lim_{x \to -1^+} f(x)$

   (C) $\lim_{x \to -1} f(x)$ (D) $f(-1)$

✎ (E) Is f continuous at $x = -1$? Explain.

Problems 23–30 refer to the function g shown in the figure. Use the graph to estimate the indicated function values and limits.

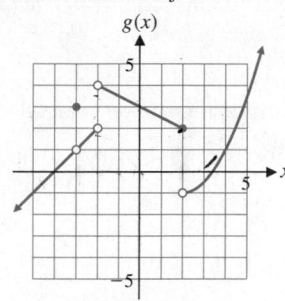

Figure for 23–30

23. $g(-3.1)$ 24. $g(-2.1)$

25. $g(1.9)$ 26. $g(-1.9)$

27. (A) $\lim_{x \to -3^-} g(x)$ (B) $\lim_{x \to -3^+} g(x)$

   (C) $\lim_{x \to -3} g(x)$ (D) $g(-3)$

✎ (E) Is g continuous at $x = -3$? Explain.

28. (A) $\lim_{x \to -2^-} g(x)$ (B) $\lim_{x \to -2^+} g(x)$

   (C) $\lim_{x \to -2} g(x)$ (D) $g(-2)$

✎ (E) Is g continuous at $x = -2$? Explain.

29. (A) $\lim_{x \to 2^-} g(x)$ (B) $\lim_{x \to 2^+} g(x)$

   (C) $\lim_{x \to 2} g(x)$ (D) $g(2)$

✎ (E) Is g continuous at $x = 2$? Explain.

30. (A) $\lim_{x \to 4^-} g(x)$ (B) $\lim_{x \to 4^+} g(x)$

   (C) $\lim_{x \to 4} g(x)$ (D) $g(4)$

✎ (E) Is g continuous at $x = 4$? Explain.

Use Theorem 1 to determine where each function in Problems 31–40 is continuous.

31. $f(x) = 3x - 4$ 32. $h(x) = 4 - 2x$

33. $g(x) = \dfrac{3x}{x + 2}$ 34. $k(x) = \dfrac{2x}{x - 4}$

35. $m(x) = \dfrac{x + 1}{x^2 + 3x - 4}$ 36. $n(x) = \dfrac{x - 2}{x^2 - 2x - 3}$

37. $F(x) = \dfrac{2x}{x^2 + 9}$ 38. $G(x) = \dfrac{1 - x^2}{x^2 + 1}$

39. $M(x) = \dfrac{x - 1}{4x^2 - 9}$ 40. $N(x) = \dfrac{x^2 + 4}{4 - 25x^2}$

In Problems 41–46, find all partition numbers of the function.

41. $f(x) = \dfrac{3x + 8}{x - 4}$ $\dfrac{-8}{3}, 4$ 42. $f(x) = \dfrac{2x + 7}{5x - 1}$

43. $f(x) = \dfrac{1 - x^2}{1 + x^2}$ 44. $f(x) = \dfrac{x^2 + 4}{x^2 - 9}$

45. $f(x) = \dfrac{x^2 + 4x - 45}{x^2 + 6x}$ 46. $f(x) = \dfrac{x^3 + x}{x^2 - x - 42}$

In Problems 47–54, use a sign chart to solve each inequality. Express answers in inequality and interval notation.

47. $x^2 - x - 12 < 0$ 48. $x^2 - 2x - 8 < 0$

49. $x^2 + 21 > 10x$ 50. $x^2 + 7x > -10$

51. $x^3 < 4x$ 52. $x^4 - 9x^2 > 0$

53. $\dfrac{x^2 + 5x}{x - 3} > 0$ 54. $\dfrac{x - 4}{x^2 + 2x} < 0$

55. Use the graph of f to determine where

   (A) $f(x) > 0$ (B) $f(x) < 0$

   Express answers in interval notation.

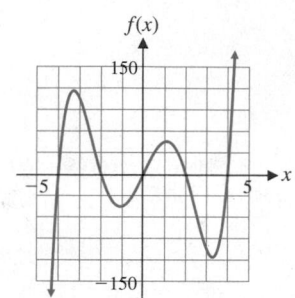

56. Use the graph of g to determine where

   (A) $g(x) > 0$ (B) $g(x) < 0$

   Express answers in interval notation.

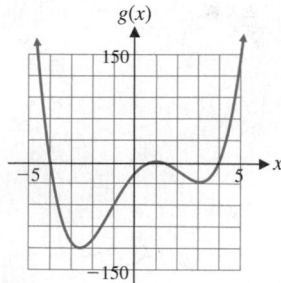

🖩 *In Problems 57–60, use a graphing calculator to approximate the partition numbers of each function f(x) to four decimal places. Then solve the following inequalities:*

   (A) $f(x) > 0$ (B) $f(x) < 0$

Express answers in interval notation.

57. $f(x) = x^4 - 6x^2 + 3x + 5$

58. $f(x) = x^4 - 4x^2 - 2x + 2$

59. $f(x) = \dfrac{3 + 6x - x^3}{x^2 - 1}$ 60. $f(x) = \dfrac{x^3 - 5x + 1}{x^2 - 1}$

Use Theorem 1 to determine where each function in Problems 61–68 is continuous. Express the answer in interval notation.

61. $\sqrt{x - 6}$ 62. $\sqrt{7 - x}$

63. $\sqrt[3]{5 - x}$ 64. $\sqrt[3]{x - 8}$

65. $\sqrt{x^2 - 9}$ **66.** $\sqrt{4 - x^2}$

67. $\sqrt{x^2 + 1}$ **68.** $\sqrt[3]{x^2 + 2}$

In Problems 69–74, graph f, locate all points of discontinuity, and discuss the behavior of f at these points.

69. $f(x) = \begin{cases} 1 + x & \text{if } x < 1 \\ 5 - x & \text{if } x \geq 1 \end{cases}$

70. $f(x) = \begin{cases} x^2 & \text{if } x \leq 1 \\ 2x & \text{if } x > 1 \end{cases}$

71. $f(x) = \begin{cases} 1 + x & \text{if } x \leq 2 \\ 5 - x & \text{if } x > 2 \end{cases}$

72. $f(x) = \begin{cases} x^2 & \text{if } x \leq 2 \\ 2x & \text{if } x > 2 \end{cases}$

73. $f(x) = \begin{cases} -x & \text{if } x < 0 \\ 1 & \text{if } x = 0 \\ x & \text{if } x > 0 \end{cases}$

74. $f(x) = \begin{cases} 1 & \text{if } x < 0 \\ 0 & \text{if } x = 0 \\ 1 + x & \text{if } x > 0 \end{cases}$

*Problems 75 and 76 refer to the **greatest integer function**, which is denoted by $[\![x]\!]$ and is defined as*

$$[\![x]\!] = greatest\ integer \leq x$$

For example,

$$[\![-3.6]\!] = greatest\ integer \leq -3.6 = -4$$

$$[\![2]\!] = greatest\ integer \leq 2 = 2$$

$$[\![2.5]\!] = greatest\ integer \leq 2.5 = 2$$

The graph of $f(x) = [\![x]\!]$ is shown. There, we can see that

$$[\![x]\!] = -2 \quad for \quad -2 \leq x < -1$$

$$[\![x]\!] = -1 \quad for \quad -1 \leq x < 0$$

$$[\![x]\!] = 0 \quad for \quad 0 \leq x < 1$$

$$[\![x]\!] = 1 \quad for \quad 1 \leq x < 2$$

$$[\![x]\!] = 2 \quad for \quad 2 \leq x < 3$$

and so on.

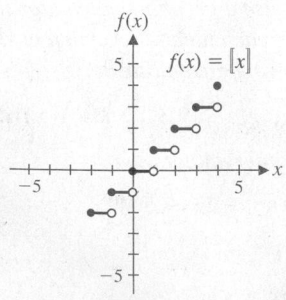

Figure for 75 and 76

75. (A) Is f continuous from the right at $x = 0$?

(B) Is f continuous from the left at $x = 0$?

(C) Is f continuous on the open interval $(0, 1)$?

(D) Is f continuous on the closed interval $[0, 1]$?

(E) Is f continuous on the half-closed interval $[0, 1)$?

76. (A) Is f continuous from the right at $x = 2$?

(B) Is f continuous from the left at $x = 2$?

(C) Is f continuous on the open interval $(1, 2)$?

(D) Is f continuous on the closed interval $[1, 2]$?

(E) Is f continuous on the half-closed interval $[1, 2)$?

In Problems 77–82, discuss the validity of each statement. If the statement is always true, explain why. If not, give a counterexample.

77. A polynomial function is continuous for all real numbers.

78. A rational function is continuous for all but finitely many real numbers.

79. If f is a function that is continuous at $x = 0$ and $x = 2$, then f is continuous at $x = 1$.

80. If f is a function that is continuous on the open interval $(0, 2)$, then f is continuous at $x = 1$.

81. If f is a function that has no partition numbers in the interval (a, b), then f is continuous on (a, b).

82. The greatest integer function (see Problem 75) is a rational function.

In Problems 83–86, sketch a possible graph of a function f that is continuous for all real numbers and satisfies the given conditions. Find the x intercepts of f.

83. $f(x) < 0$ on $(-\infty, -5)$ and $(2, \infty)$; $f(x) > 0$ on $(-5, 2)$

84. $f(x) > 0$ on $(-\infty, -4)$ and $(3, \infty)$; $f(x) < 0$ on $(-4, 3)$

85. $f(x) < 0$ on $(-\infty, -6)$ and $(-1, 4)$; $f(x) > 0$ on $(-6, -1)$ and $(4, \infty)$

86. $f(x) > 0$ on $(-\infty, -3)$ and $(2, 7)$; $f(x) < 0$ on $(-3, 2)$ and $(7, \infty)$

87. The function $f(x) = 2/(1 - x)$ satisfies $f(0) = 2$ and $f(2) = -2$. Is f equal to 0 anywhere on the interval $(-1, 3)$? Does this contradict Theorem 2? Explain.

88. The function $f(x) = 6/(x - 4)$ satisfies $f(2) = -3$ and $f(7) = 2$. Is f equal to 0 anywhere on the interval $(0, 9)$? Does this contradict Theorem 2? Explain.

Applications

89. Postal rates. First-class postage in 2009 was $0.44 for the first ounce (or any fraction thereof) and $0.17 for each additional ounce (or fraction thereof) up to a maximum weight of 3.5 ounces.

(A) Write a piecewise definition of the first-class postage $P(x)$ for a letter weighing x ounces.

(B) Graph $P(x)$ for $0 < x \leq 3.5$.

(C) Is $P(x)$ continuous at $x = 2.5$? At $x = 3$? Explain.

90. Telephone rates. A long-distance telephone service charges $0.07 for the first minute (or any fraction thereof) and $0.05 for each additional minute (or fraction thereof).

(A) Write a piecewise definition of the charge $R(x)$ for a long-distance call lasting x minutes.

(B) Graph $R(x)$ for $0 < x \le 6$.

(C) Is $R(x)$ continuous at $x = 3.5$? At $x = 3$? Explain.

91. Postal rates. Discuss the differences between the function $Q(x) = 0.44 + 0.17[x]$ and the function $P(x)$ defined in Problem 89.

92. Telephone rates. Discuss the differences between the function $S(x) = 0.07 + 0.05[x]$ and the function $R(x)$ defined in Problem 90.

93. Natural-gas rates. Table 1 shows the rates for natural gas charged by the Middle Tennessee Natural Gas Utility District during summer months. The base charge is a fixed monthly charge, independent of the amount of gas used per month.

Table 1 **Summer (May–September)**

Base charge	$5.00
First 50 therms	0.63 per therm
Over 50 therms	0.45 per therm

(A) Write a piecewise definition of the monthly charge $S(x)$ for a customer who uses x therms* in a summer month.

(B) Graph $S(x)$.

(C) Is $S(x)$ continuous at $x = 50$? Explain.

94. Natural-gas rates. Table 2 shows the rates for natural gas charged by the Middle Tennessee Natural Gas Utility District during winter months. The base charge is a fixed monthly charge, independent of the amount of gas used per month.

Table 2 **Winter (October– April)**

Base charge	$5.00
First 5 therms	0.69 per therm
Next 45 therms	0.65 per therm
Over 50 therms	0.63 per therm

*A British thermal unit (Btu) is the amount of heat required to raise the temperature of 1 pound of water 1 degree Fahrenheit, and a therm is 100,000 Btu.

(A) Write a piecewise definition of the monthly charge $S(x)$ for a customer who uses x therms in a winter month.

(B) Graph $S(x)$.

(C) Is $S(x)$ continuous at $x = 5$? At $x = 50$? Explain.

95. Income. A personal-computer salesperson receives a base salary of $1,000 per month and a commission of 5% of all sales over $10,000 during the month. If the monthly sales are $20,000 or more, then the salesperson is given an additional $500 bonus. Let $E(s)$ represent the person's earnings per month as a function of the monthly sales s.

(A) Graph $E(s)$ for $0 \le s \le 30,000$.

(B) Find $\lim_{s \to 10,000} E(s)$ and $E(10,000)$.

(C) Find $\lim_{s \to 20,000} E(s)$ and $E(20,000)$.

(D) Is E continuous at $s = 10,000$? At $s = 20,000$?

96. Equipment rental. An office equipment rental and leasing company rents copiers for $10 per day (and any fraction thereof) or for $50 per 7-day week. Let $C(x)$ be the cost of renting a copier for x days.

(A) Graph $C(x)$ for $0 \le x \le 10$.

(B) Find $\lim_{x \to 4.5} C(x)$ and $C(4.5)$.

(C) Find $\lim_{x \to 8} C(x)$ and $C(8)$.

(D) Is C continuous at $x = 4.5$? At $x = 8$?

97. Animal supply. A medical laboratory raises its own rabbits. The number of rabbits $N(t)$ available at any time t depends on the number of births and deaths. When a birth or death occurs, the function N generally has a discontinuity, as shown in the figure.

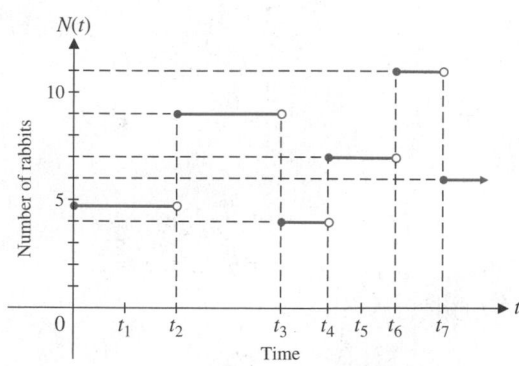

(A) Where is the function N discontinuous?

(B) $\lim_{t \to t_5} N(t) = ?$; $N(t_5) = ?$

(C) $\lim_{t \to t_3} N(t) = ?$; $N(t_3) = ?$

98. Learning. The graph shown represents the history of a person learning the material on limits and continuity in this book. At time t_2, the student's mind goes blank during a quiz. At time t_4, the instructor explains a concept particularly well, then suddenly a big jump in understanding takes place.

(A) Where is the function p discontinuous?

(B) $\lim_{t \to t_1} p(t) = ?$; $p(t_1) = ?$

(C) $\lim_{t \to t_2} p(t) = ?$; $p(t_2) = ?$

(D) $\lim_{t \to t_4} p(t) = ?$; $p(t_4) = ?$

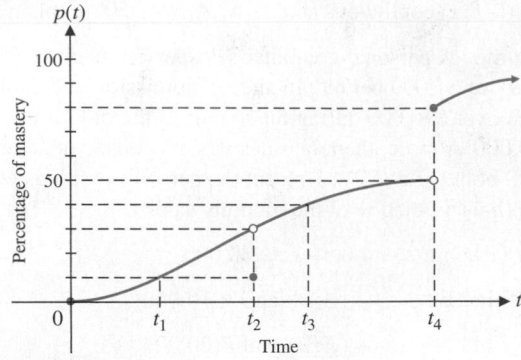

Answers to Matched Problems

1. f is not continuous at $x = -3, -1, 2,$ and 4.

$x = -3$: $\lim_{x \to -3} f(x) = 3$, but $f(-3)$ does not exist

$x = -1$: $f(-1) = 1$, but $\lim_{x \to -1} f(x)$ does not exist

$x = 2$: $\lim_{x \to 2} f(x) = 5$, but $f(2) = 3$

$x = 4$: $\lim_{x \to 4} f(x)$ does not exist, and $f(4)$ does not exist

2. (A) f is continuous at $x = 1$, since $\lim_{x \to 1} f(x) = 2 = f(1)$.

(B) g is not continuous at $x = 1$, since $g(1)$ is not defined.

(C) h is not continuous at $x = 2$ for two reasons: $h(2)$ does not exist and $\lim_{x \to 2} h(x)$ does not exist.

h is continuous at $x = 0$, since $\lim_{x \to 0} h(x) = -1 = h(0)$.

3. (A) Since f is a polynomial function, f is continuous for all x.

(B) Since f is a rational function, f is continuous for all x except -1 and 4 (values that make the denominator 0).

(C) The polynomial function $x - 4$ is continuous for all x and nonnegative for $x \geq 4$. Since $n = 2$ is even, f is continuous for $x \geq 4$, or on the interval $[4, \infty)$.

(D) The polynomial function $x^3 + 1$ is continuous for all x. Since $n = 3$ is odd, f is continuous for all x.

4. $-\infty < x < -1$ or $1 < x < 3$; $(-\infty, -1) \cup (1, 3)$

2.4 The Derivative

- Rate of Change
- Slope of the Tangent Line
- The Derivative
- Nonexistence of the Derivative

We will now make use of the limit concepts developed in Sections 2.1, 2.2, and 2.3 to solve the two important problems illustrated in Figure 1. The solution of each of these apparently unrelated problems involves a common concept called the *derivative*.

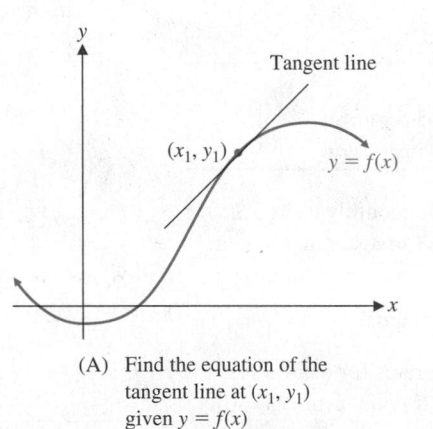

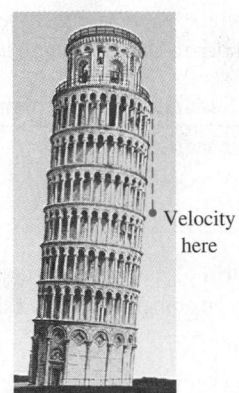

(A) Find the equation of the tangent line at (x_1, y_1) given $y = f(x)$

(B) Find the instantaneous velocity of a falling object

Figure 1 **Two basic problems of calculus**

Rate of Change

If you pass mile marker 120 on the interstate highway at 9 a.m. and mile marker 250 at 11 a.m., then the *average rate of change* of distance with respect to time, also known as *average velocity*, is

$$\frac{250 - 120}{11 - 9} = \frac{130}{2} = 65 \text{ miles per hour}$$

Of course your speedometer reading, that is, the *instantaneous rate of change*, or *instantaneous velocity*, might well have been 75 mph at some moment between 9 a.m. and 11 a.m.

We will define the concepts of average rate of change and instantaneous rate of change more generally, and will apply them in situations that are unrelated to velocity.

DEFINITION Average Rate of Change

For $y = f(x)$, the **average rate of change from $x = a$ to $x = a + h$** is

$$\frac{f(a + h) - f(a)}{(a + h) - a} = \frac{f(a + h) - f(a)}{h} \qquad h \neq 0 \qquad (1)$$

Note that the numerator and denominator in (1) are differences, so (1) is a **difference quotient** (see Section 1.1).

EXAMPLE 1 Revenue Analysis The revenue (in dollars) from the sale of x plastic planter boxes is given by

$$R(x) = 20x - 0.02x^2 \qquad 0 \leq x \leq 1,000$$

and is graphed in Figure 2.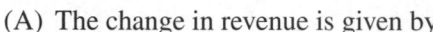

(A) What is the change in revenue if production is changed from 100 planters to 400 planters?

(B) What is the average rate of change in revenue for this change in production?

SOLUTION

(A) The change in revenue is given by

$$R(400) - R(100) = 20(400) - 0.02(400)^2 - [20(100) - 0.02(100)^2]$$
$$= 4,800 - 1,800 = \$3,000$$

Increasing production from 100 planters to 400 planters will increase revenue by \$3,000.

(B) To find the average rate of change in revenue, we divide the change in revenue by the change in production:

$$\frac{R(400) - R(100)}{400 - 100} = \frac{3,000}{300} = \$10$$

The average rate of change in revenue is \$10 per planter when production is increased from 100 to 400 planters.

Matched Problem 1 Refer to the revenue function in Example 1.

(A) What is the change in revenue if production is changed from 600 planters to 800 planters?

(B) What is the average rate of change in revenue for this change in production?

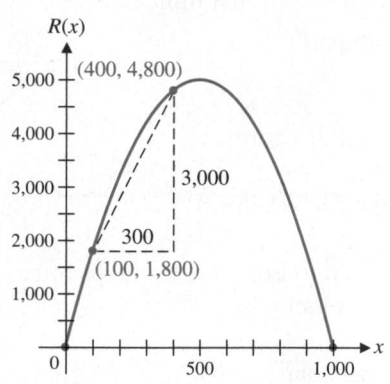

Figure 2 $R(x) = 20x - 0.02x^2$

EXAMPLE 2 Velocity A small steel ball dropped from a tower will fall a distance of y feet in x seconds, as given approximately by the formula

$$y = f(x) = 16x^2$$

Figure 3 shows the position of the ball on a coordinate line (positive direction down) at the end of 0, 1, 2, and 3 seconds.

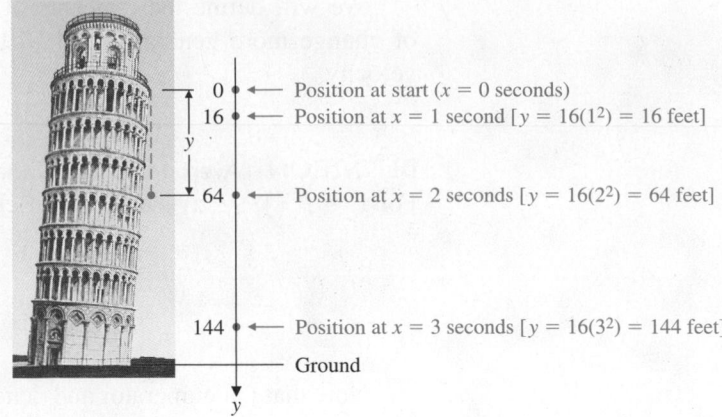

0 •← Position at start ($x = 0$ seconds)

16 •← Position at $x = 1$ second [$y = 16(1^2) = 16$ feet]

64 •← Position at $x = 2$ seconds [$y = 16(2^2) = 64$ feet]

144 •← Position at $x = 3$ seconds [$y = 16(3^2) = 144$ feet]

Ground

Figure 3 **Note: Positive y direction is down.**

(A) Find the average velocity from $x = 2$ seconds to $x = 3$ seconds.

(B) Find and simplify the average velocity from $x = 2$ seconds to $x = 2 + h$ seconds, $h \neq 0$.

(C) Find the limit of the expression from part (B) as $h \to 0$ if that limit exists.

(D) Discuss possible interpretations of the limit from part (C).

SOLUTION

(A) Recall the formula $d = rt$, which can be written in the form

$$r = \frac{d}{t} = \frac{\text{Distance covered}}{\text{Elapsed time}} = \text{Average velocity}$$

For example, if a person drives from San Francisco to Los Angeles (a distance of about 420 miles) in 7 hours, then the average velocity is

$$r = \frac{d}{t} = \frac{420}{7} = 60 \text{ miles per hour}$$

Sometimes the person will be traveling faster and sometimes slower, but the average velocity is 60 miles per hour. In our present problem, the average velocity of the steel ball from $x = 2$ seconds to $x = 3$ seconds is

$$\begin{aligned}
\text{Average velocity} &= \frac{\text{Distance covered}}{\text{Elapsed time}} \\
&= \frac{f(3) - f(2)}{3 - 2} \\
&= \frac{16(3)^2 - 16(2)^2}{1} = 80 \text{ feet per second}
\end{aligned}$$

We see that if $y = f(x)$ is the position of the falling ball, then the average velocity is simply the average rate of change of $f(x)$ with respect to time x.

(B) Proceeding as in part (A), we have

$$\begin{aligned}
\text{Average velocity} &= \frac{\text{Distance covered}}{\text{Elapsed time}} \\
&= \frac{f(2 + h) - f(2)}{h} \qquad \text{Difference quotient} \\
&= \frac{16(2 + h)^2 - 16(2)^2}{h} \qquad \begin{array}{l}\text{Simplify this 0/0} \\ \text{indeterminate form.}\end{array}
\end{aligned}$$

$$= \frac{64 + 64h + 16h^2 - 64}{h}$$

$$= \frac{h(64 + 16h)}{h} = 64 + 16h \qquad h \neq 0$$

Notice that if $h = 1$, the average velocity is 80 feet per second, which is the result in part (A).

(C) The limit of the average velocity expression from part (B) as $h \to 0$ is

$$\lim_{h \to 0} \frac{f(2 + h) - f(2)}{h} = \lim_{h \to 0} (64 + 16h)$$

$$= 64 \text{ feet per second}$$

(D) The average velocity over smaller and smaller time intervals approaches 64 feet per second. This limit can be interpreted as the velocity of the ball at the *instant* that the ball has been falling for exactly 2 seconds. Therefore, 64 feet per second is referred to as the **instantaneous velocity** at $x = 2$ seconds, and we have solved one of the basic problems of calculus (see Fig. 1B).

Matched Problem 2 For the falling steel ball in Example 2, find

(A) The average velocity from $x = 1$ second to $x = 2$ seconds

(B) The average velocity (in simplified form) from $x = 1$ second to $x = 1 + h$ seconds, $h \neq 0$

(C) The instantaneous velocity at $x = 1$ second

The ideas in Example 2 can be applied to the average rate of change of any function.

DEFINITION Instantaneous Rate of Change
For $y = f(x)$, the **instantaneous rate of change at $x = a$** is

$$\lim_{h \to 0} \frac{f(a + h) - f(a)}{h} \qquad (2)$$

if the limit exists.

The adjective *instantaneous* is often omitted with the understanding that the phrase **rate of change** always refers to the instantaneous rate of change and not the average rate of change. Similarly, **velocity** always refers to the instantaneous rate of change of distance with respect to time.

Slope of the Tangent Line

So far, our interpretations of the difference quotient have been numerical in nature. Now we want to consider a geometric interpretation.

In geometry, a line that intersects a circle in two points is called a *secant line*, and a line that intersects a circle in exactly one point is called a *tangent line* (Fig. 4, page 136). If the point Q in Figure 4 is moved closer and closer to the point P, then the angle between the secant line PQ and the tangent line at P gets smaller and smaller. We will generalize the geometric concepts of secant line and tangent line of a circle and will use them to study graphs of functions.

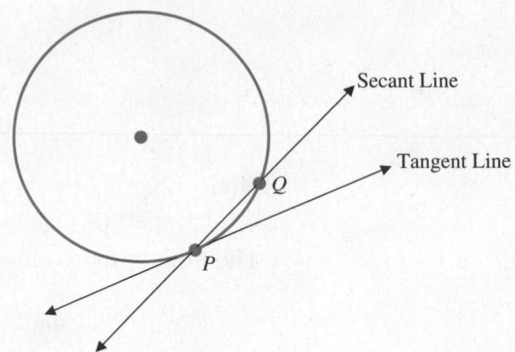

Figure 4 Secant line and tangent line of a circle

A line through two points on the graph of a function is called a **secant line**. If $(a, f(a))$ and $(a + h, f(a + h))$ are two points on the graph of $y = f(x)$, then we can use the slope formula from Section 1.3 to find the slope of the secant line through these points (Fig. 5).

Slope of secant line $= \dfrac{y_2 - y_1}{x_2 - x_1} = \dfrac{f(a + h) - f(a)}{(a + h) - a}$

$\qquad\qquad\qquad = \dfrac{f(a + h) - f(a)}{h}$ \quad Difference quotient

The difference quotient can be interpreted as both the average rate of change and the slope of the secant line.

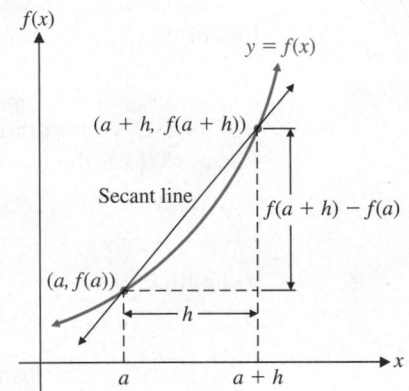

Figure 5 Secant line

EXAMPLE 3 Slope of a Secant Line Given $f(x) = x^2$,

(A) Find the slope of the secant line for $a = 1$ and $h = 2$ and 1, respectively. Graph $y = f(x)$ and the two secant lines.

(B) Find and simplify the slope of the secant line for $a = 1$ and h any nonzero number.

(C) Find the limit of the expression in part (B).

(D) Discuss possible interpretations of the limit in part (C).

SOLUTION

(A) For $a = 1$ and $h = 2$, the secant line goes through $(1, f(1)) = (1, 1)$ and $(3, f(3)) = (3, 9)$, and its slope is

$$\frac{f(1 + 2) - f(1)}{2} = \frac{3^2 - 1^2}{2} = 4$$

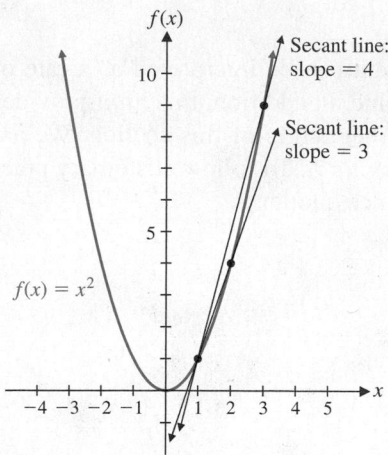

Figure 6 **Secant lines**

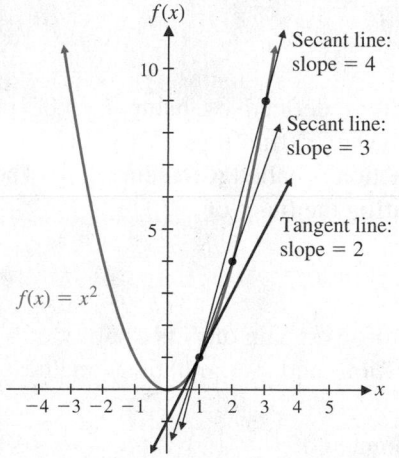

Figure 7 **Tangent line**

For $a = 1$ and $h = 1$, the secant line goes through $(1, f(1)) = (1, 1)$ and $(2, f(2)) = (2, 4)$, and its slope is

$$\frac{f(1 + 1) - f(1)}{1} = \frac{2^2 - 1^2}{1} = 3$$

The graphs of $y = f(x)$ and the two secant lines are shown in Figure 6.

(B) For $a = 1$ and h any nonzero number, the secant line goes through $(1, f(1)) = (1, 1)$ and $(1 + h, f(1 + h)) = (1 + h, (1 + h)^2)$, and its slope is

$$\frac{f(1 + h) - f(1)}{h} = \frac{(1 + h)^2 - 1^2}{h} \qquad \text{Square the binomial.}$$

$$= \frac{1 + 2h + h^2 - 1}{h} \qquad \begin{array}{l}\text{Combine like terms and}\\\text{factor the numerator.}\end{array}$$

$$= \frac{h(2 + h)}{h} \qquad \text{Cancel.}$$

$$= 2 + h \qquad h \neq 0$$

(C) The limit of the secant line slope from part (B) is

$$\lim_{h \to 0} \frac{f(1 + h) - f(1)}{h} = \lim_{h \to 0}(2 + h)$$

$$= 2$$

(D) In part (C), we saw that the limit of the slopes of the secant lines through the point $(1, f(1))$ is 2. If we graph the line through $(1, f(1))$ with slope 2 (Fig. 7), then this line is the limit of the secant lines. The slope obtained from the limit of slopes of secant lines is called the *slope of the graph* at $x = 1$. The line through the point $(1, f(1))$ with this slope is called the *tangent line*. We have solved another basic problem of calculus (see Fig. 1A on page 132).

Matched Problem 3 Given $f(x) = x^2$,

(A) Find the slope of the secant line for $a = 2$ and $h = 2$ and 1, respectively.

(B) Find and simplify the slope of the secant line for $a = 2$ and h any nonzero number.

(C) Find the limit of the expression in part (B).

(D) Find the slope of the graph and the slope of the tangent line at $a = 2$.

The ideas introduced in the preceding example are summarized next:

DEFINITION   Slope of a Graph and Tangent Line
Given $y = f(x)$, the **slope of the graph** at the point $(a, f(a))$ is given by

$$\lim_{h \to 0} \frac{f(a + h) - f(a)}{h} \tag{3}$$

provided the limit exists. In this case, the **tangent line** to the graph is the line through $(a, f(a))$ with slope given by (3).

CONCEPTUAL INSIGHT

If the function f is continuous at a, then

$$\lim_{h \to 0} f(a + h) = f(a)$$

and limit (3) will be a $0/0$ indeterminate form. As we saw in Examples 2 and 3, evaluating this type of limit typically involves algebraic simplification.

The Derivative

We have seen that the limit of a difference quotient can be interpreted as a rate of change, as a velocity, or as the slope of a tangent line. In addition, this limit provides solutions to the two basic problems stated at the beginning of this section. We are now ready to introduce some terms that refer to that limit. To follow customary practice, we use x in place of a and think of the difference quotient

$$\frac{f(x + h) - f(x)}{h}$$

as a function of h, with x held fixed as h tends to 0.

> **DEFINITION**   The Derivative
>
> For $y = f(x)$, we define the **derivative of f at x**, denoted $f'(x)$, by
>
> $$f'(x) = \lim_{h \to 0} \frac{f(x + h) - f(x)}{h} \quad \text{if the limit exists}$$
>
> If $f'(x)$ exists for each x in the open interval (a, b), then f is said to be **differentiable** over (a, b).

(Differentiability from the left or from the right is defined by using $h \to 0^-$ or $h \to 0^+$, respectively, in place of $h \to 0$ in the preceding definition.)

The process of finding the derivative of a function is called **differentiation**. The derivative of a function is obtained by **differentiating** the function.

> **SUMMARY**   Interpretations of the Derivative
>
> The derivative of a function f is a new function f'. The domain of f' is a subset of the domain of f. The derivative has various applications and interpretations, including the following:
>
> 1. *Slope of the tangent line.* For each x in the domain of f', $f'(x)$ is the slope of the line tangent to the graph of f at the point $(x, f(x))$.
> 2. *Instantaneous rate of change.* For each x in the domain of f', $f'(x)$ is the instantaneous rate of change of $y = f(x)$ with respect to x.
> 3. *Velocity.* If $f(x)$ is the position of a moving object at time x, then $v = f'(x)$ is the velocity of the object at that time.

Example 4 illustrates the **four-step process** that we use to find derivatives in this section. In subsequent sections, we develop rules for finding derivatives that do not involve limits. However, it is important that you master the limit process in order to fully comprehend and appreciate the various applications we will consider.

EXAMPLE 4   Finding a Derivative   Find $f'(x)$, the derivative of f at x, for $f(x) = 4x - x^2$.

SOLUTION   To find $f'(x)$, we use a four-step process.

Step 1   Find $f(x + h)$.

$$\begin{aligned}
f(x + h) &= 4(x + h) - (x + h)^2 \\
&= 4x + 4h - x^2 - 2xh - h^2
\end{aligned}$$

Step 2   Find $f(x + h) - f(x)$.

$$\begin{aligned}
f(x + h) - f(x) &= 4x + 4h - x^2 - 2xh - h^2 - (4x - x^2) \\
&= 4h - 2xh - h^2
\end{aligned}$$

Step 3 Find $\dfrac{f(x+h)-f(x)}{h}$.

$$\frac{f(x+h)-f(x)}{h} = \frac{4h-2xh-h^2}{h} = \frac{h(4-2x-h)}{h}$$

$$= 4-2x-h, \quad h \neq 0$$

Step 4 Find $f'(x) = \lim\limits_{h\to 0}\dfrac{f(x+h)-f(x)}{h}$.

$$f'(x) = \lim_{h\to 0}\frac{f(x+h)-f(x)}{h} = \lim_{h\to 0}(4-2x-h) = 4-2x$$

So if $f(x) = 4x - x^2$, then $f'(x) = 4 - 2x$. The function f' is a new function derived from the function f.

Matched Problem 4 Find $f'(x)$, the derivative of f at x, for $f(x) = 8x - 2x^2$.

The four-step process used in Example 4 is summarized as follows for easy reference:

PROCEDURE The four-step process for finding the derivative of a function f:

Step 1 Find $f(x+h)$.

Step 2 Find $f(x+h) - f(x)$.

Step 3 Find $\dfrac{f(x+h)-f(x)}{h}$.

Step 4 Find $\lim\limits_{h\to 0}\dfrac{f(x+h)-f(x)}{h}$.

EXAMPLE 5 Finding Tangent Line Slopes In Example 4, we started with the function $f(x) = 4x - x^2$ and found the derivative of f at x to be $f'(x) = 4 - 2x$. So the slope of a line tangent to the graph of f at any point $(x, f(x))$ on the graph is

$$m = f'(x) = 4 - 2x$$

(A) Find the slope of the graph of f at $x = 0$, $x = 2$, and $x = 3$.

(B) Graph $y = f(x) = 4x - x^2$ and use the slopes found in part (A) to make a rough sketch of the lines tangent to the graph at $x = 0$, $x = 2$, and $x = 3$.

SOLUTION

(A) Using $f'(x) = 4 - 2x$, we have

$$f'(0) = 4 - 2(0) = 4 \qquad \text{Slope at } x = 0$$
$$f'(2) = 4 - 2(2) = 0 \qquad \text{Slope at } x = 2$$
$$f'(3) = 4 - 2(3) = -2 \qquad \text{Slope at } x = 3$$

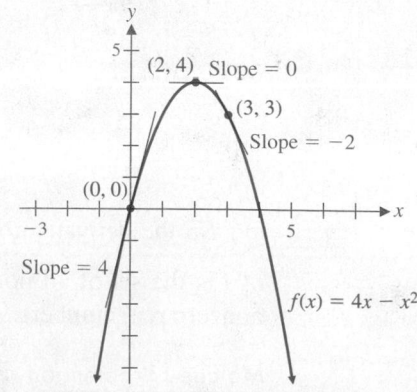

Matched Problem 5] In Matched Problem 4, we started with the function $f(x) = 8x - 2x^2$. Using the derivative found there,

(A) Find the slope of the graph of f at $x = 1$, $x = 2$, and $x = 4$.

(B) Graph $y = f(x) = 8x - 2x^2$, and use the slopes from part (A) to make a rough sketch of the lines tangent to the graph at $x = 1$, $x = 2$, and $x = 4$.

Explore and Discuss 1   In Example 4, we found that the derivative of $f(x) = 4x - x^2$ is $f'(x) = 4 - 2x$. In Example 5, we graphed $f(x)$ and several tangent lines.

(A) Graph f and f' on the same set of axes.

(B) The graph of f' is a straight line. Is it a tangent line for the graph of f? Explain.

(C) Find the x intercept for the graph of f'. What is the slope of the line tangent to the graph of f for this value of x? Write a verbal description of the relationship between the slopes of the tangent lines of a function and the x intercepts of the derivative of the function.

EXAMPLE 6   Finding a Derivative   Find $f'(x)$, the derivative of f at x, for $f(x) = \dfrac{1}{x}$.

SOLUTION

Step 1 Find $f(x + h)$.

$$f(x + h) = \frac{1}{x + h}$$

Step 2 Find $f(x + h) - f(x)$.

$$f(x + h) - f(x) = \frac{1}{x + h} - \frac{1}{x} \qquad \text{Add fractions. (Section A.4)}$$

$$= \frac{x - (x + h)}{x(x + h)} \qquad \text{Simplify.}$$

$$= \frac{-h}{x(x + h)}$$

Step 3 Find $\dfrac{f(x + h) - f(x)}{h}$

$$\frac{f(x + h) - f(x)}{h} = \frac{\dfrac{-h}{x(x + h)}}{h} \qquad \text{Simplify.}$$

$$= \frac{-1}{x(x + h)} \qquad h \neq 0$$

Step 4 Find $\lim\limits_{h \to 0} \dfrac{f(x + h) - f(x)}{h}$.

$$\lim_{h \to 0} \frac{f(x + h) - f(x)}{h} = \lim_{h \to 0} \frac{-1}{x(x + h)}$$

$$= \frac{-1}{x^2} \qquad x \neq 0$$

So the derivative of $f(x) = \dfrac{1}{x}$ is $f'(x) = \dfrac{-1}{x^2}$, a new function. The domain of f is the set of all nonzero real numbers. The domain of f' is also the set of all nonzero real numbers.

Matched Problem 6] Find $f'(x)$ for $f(x) = \dfrac{1}{x + 2}$.

EXAMPLE 7 Finding a Derivative Find $f'(x)$, the derivative of f at x, for $f(x) = \sqrt{x} + 2$.

SOLUTION We use the four-step process to find $f'(x)$.

Step 1 Find $f(x + h)$.

$$f(x + h) = \sqrt{x + h} + 2$$

Step 2 Find $f(x + h) - f(x)$.

$$f(x + h) - f(x) = \sqrt{x + h} + 2 - (\sqrt{x} + 2) \quad \text{Combine like terms.}$$
$$= \sqrt{x + h} - \sqrt{x}$$

Step 3 Find $\dfrac{f(x + h) - f(x)}{h}$.

$$\frac{f(x + h) - f(x)}{h} = \frac{\sqrt{x + h} - \sqrt{x}}{h}$$

We rationalize the numerator (Appendix A, Section A.6) to change the form of this fraction.

$$= \frac{\sqrt{x + h} - \sqrt{x}}{h} \cdot \frac{\sqrt{x + h} + \sqrt{x}}{\sqrt{x + h} + \sqrt{x}}$$

$$= \frac{x + h - x}{h(\sqrt{x + h} + \sqrt{x})}$$

Combine like terms.

$$= \frac{h}{h(\sqrt{x + h} + \sqrt{x})}$$

Cancel.

$$= \frac{1}{\sqrt{x + h} + \sqrt{x}} \quad h \neq 0$$

Step 4 Find $f'(x) = \lim\limits_{h \to 0} \dfrac{f(x + h) - f(x)}{h}$.

$$\lim_{h \to 0} \frac{f(x + h) - f(x)}{h} = \lim_{h \to 0} \frac{1}{\sqrt{x + h} + \sqrt{x}}$$

$$= \frac{1}{\sqrt{x} + \sqrt{x}} = \frac{1}{2\sqrt{x}} \quad x > 0$$

So the derivative of $f(x) = \sqrt{x} + 2$ is $f'(x) = 1/(2\sqrt{x})$, a new function. The domain of f is $[0, \infty)$. Since $f'(0)$ is not defined, the domain of f' is $(0, \infty)$, a subset of the domain of f.

Matched Problem 7 Find $f'(x)$ for $f(x) = \sqrt{x + 4}$.

EXAMPLE 8 Sales Analysis A company's total sales (in millions of dollars) t months from now are given by $S(t) = \sqrt{t} + 2$. Find and interpret $S(25)$ and $S'(25)$. Use these results to estimate the total sales after 26 months and after 27 months.

SOLUTION The total sales function S has the same form as the function f in Example 7. Only the letters used to represent the function and the independent variable have been changed. It follows that S' and f' also have the same form:

$$S(t) = \sqrt{t} + 2 \qquad f(x) = \sqrt{x} + 2$$

$$S'(t) = \frac{1}{2\sqrt{t}} \qquad f'(x) = \frac{1}{2\sqrt{x}}$$

Evaluating S and S' at $t = 25$, we have

$$S(25) = \sqrt{25} + 2 = 7 \qquad S'(25) = \frac{1}{2\sqrt{25}} = 0.1$$

So 25 months from now, the total sales will be $7 million and will be increasing at the rate of $0.1 million ($100,000) per month. If this instantaneous rate of change of sales remained constant, the sales would grow to $7.1 million after 26 months, $7.2 million after 27 months, and so on. Even though $S'(t)$ is not a constant function in this case, these values provide useful estimates of the total sales.

Matched Problem 8) A company's total sales (in millions of dollars) t months from now are given by $S(t) = \sqrt{t} + 4$. Find and interpret $S(12)$ and $S'(12)$. Use these results to estimate the total sales after 13 months and after 14 months. (Use the derivative found in Matched Problem 7.)

In Example 8, we can compare the estimates of total sales by using the derivative with the corresponding exact values of $S(t)$:

Exact values   Estimated values

$$S(26) = \sqrt{26} + 2 = 7.099\ldots \approx 7.1$$
$$S(27) = \sqrt{27} + 2 = 7.196\ldots \approx 7.2$$

For this function, the estimated values provide very good approximations to the exact values of $S(t)$. For other functions, the approximation might not be as accurate.

Using the instantaneous rate of change of a function at a point to estimate values of the function at nearby points is an important application of the derivative.

Nonexistence of the Derivative

The existence of a derivative at $x = a$ depends on the existence of a limit at $x = a$, that is, on the existence of

$$f'(a) = \lim_{h \to 0} \frac{f(a + h) - f(a)}{h} \tag{4}$$

If the limit does not exist at $x = a$, we say that the function f is **nondifferentiable at** $x = a$, or $f'(\mathbf{a})$ **does not exist.**

Explore and Discuss 2 Let $f(x) = |x - 1|$.

(A) Graph f.

(B) Complete the following table:

h	-0.1	-0.01	-0.001	$\to 0 \leftarrow$	0.001	0.01	0.1
$\dfrac{f(1 + h) - f(1)}{h}$	?	?	?	$\to ? \leftarrow$	?	?	?

(C) Find the following limit if it exists:

$$\lim_{h \to 0} \frac{f(1 + h) - f(1)}{h}$$

(D) Use the results of parts (A)–(C) to discuss the existence of $f'(1)$.

(E) Repeat parts (A)–(D) for $\sqrt[3]{x} - 1$.

How can we recognize the points on the graph of f where $f'(a)$ does not exist? It is impossible to describe all the ways that the limit of a difference quotient can fail to exist. However, we can illustrate some common situations where $f'(a)$ fails to exist (see Fig. 8):

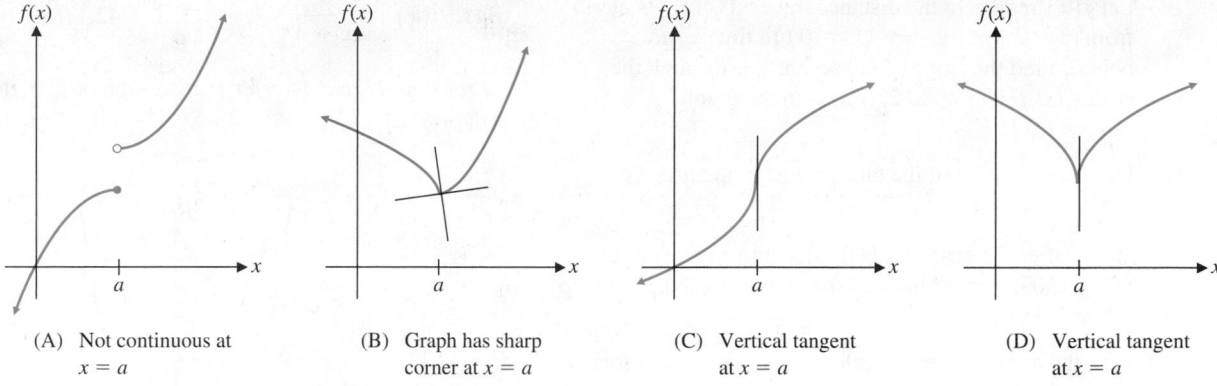

(A) Not continuous at $x = a$

(B) Graph has sharp corner at $x = a$

(C) Vertical tangent at $x = a$

(D) Vertical tangent at $x = a$

Figure 8 **The function f is nondifferentiable at $x = a$.**

1. If the graph of f has a hole or a break at $x = a$, then $f'(a)$ does not exist (Fig. 8A).

2. If the graph of f has a sharp corner at $x = a$, then $f'(a)$ does not exist, and the graph has no tangent line at $x = a$ (Fig. 8B). (In Fig. 8B, the left- and right-hand derivatives exist but are not equal.)

3. If the graph of f has a vertical tangent line at $x = a$, then $f'(a)$ does not exist (Fig. 8C and D).

Exercises 2.4

Skills Warm-up Exercises

W *In Problems 1–4, find the slope of the line through the given points. Write the slope as a reduced fraction, and also give its decimal form. (If necessary, review Section 1.3).*

1. $(2, 7)$ and $(6, 16)$

2. $(-1, 11)$ and $(1, 8)$

3. $(10, 14)$ and $(0, 68)$

4. $(-12, -3)$ and $(4, 3)$

In Problems 5–8, write the expression in the form $a + b\sqrt{n}$ where a and b are reduced fractions and n is an integer. (If necessary, review Section A.6).

5. $\dfrac{1}{\sqrt{3}}$

6. $\dfrac{2}{\sqrt{5}}$

7. $\dfrac{5}{3 + \sqrt{7}}$

8. $\dfrac{1 - \sqrt{2}}{5 + \sqrt{2}}$

In Problems 9 and 10, find the indicated quantity for $y = f(x) = 5 - x^2$ and interpret that quantity in terms of the following graph.

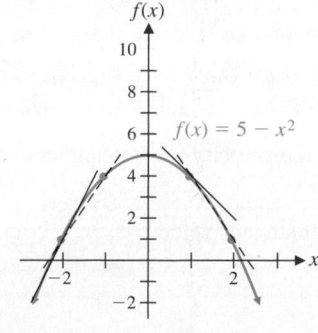

9. (A) $\dfrac{f(2) - f(1)}{2 - 1}$ (B) $\dfrac{f(1 + h) - f(1)}{h}$

(C) $\lim\limits_{h \to 0} \dfrac{f(1 + h) - f(1)}{h}$

10. (A) $\dfrac{f(-1) - f(-2)}{-1 - (-2)}$ (B) $\dfrac{f(-2 + h) - f(-2)}{h}$

(C) $\lim\limits_{h \to 0} \dfrac{f(-2 + h) - f(-2)}{h}$

11. Find the indicated quantities for $f(x) = 3x^2$.

(A) The slope of the secant line through the points $(1, f(1))$ and $(4, f(4))$ on the graph of $y = f(x)$.

(B) The slope of the secant line through the points $(1, f(1))$ and $(1 + h, f(1 + h)), h \neq 0$. Simplify your answer.

(C) The slope of the graph at $(1, f(1))$.

12. Find the indicated quantities for $f(x) = 3x^2$.

(A) The slope of the secant line through the points $(2, f(2))$ and $(5, f(5))$ on the graph of $y = f(x)$.

(B) The slope of the secant line through the points $(2, f(2))$ and $(2 + h, f(2 + h)), h \neq 0$. Simplify your answer.

(C) The slope of the graph at $(2, f(2))$.

13. Two hours after the start of a 100-kilometer bicycle race, a cyclist passes the 80-kilometer mark while riding at a velocity of 45 kilometers per hour.

(A) Find the cyclist's average velocity during the first two hours of the race.

(B) Let $f(x)$ represent the distance traveled (in kilometers) from the start of the race ($x = 0$) to time x (in hours). Find the slope of the secant line through the points $(0, f(0))$ and $(2, f(2))$ on the graph of $y = f(x)$.

(C) Find the equation of the tangent line to the graph of $y = f(x)$ at the point $(2, f(2))$.

14. Four hours after the start of a 600-mile auto race, a driver's velocity is 150 miles per hour as she completes the 352nd lap on a 1.5-mile track.

(A) Find the driver's average velocity during the first four hours of the race.

(B) Let $f(x)$ represent the distance traveled (in miles) from the start of the race ($x = 0$) to time x (in hours). Find the slope of the secant line through the points $(0, f(0))$ and $(4, f(4))$ on the graph of $y = f(x)$.

(C) Find the equation of the tangent line to the graph of $y = f(x)$ at the point $(4, f(4))$.

15. For $f(x) = \frac{1}{1 + x^2}$, the slope of the graph of $y = f(x)$ is known to be $-\frac{1}{2}$ at the point with x coordinate 1. Find the equation of the tangent line at that point.

16. For $f(x) = \frac{1}{1 + x^2}$, the slope of the graph of $y = f(x)$ is known to be -0.16 at the point with x coordinate 2. Find the equation of the tangent line at that point.

17. For $f(x) = x^4$, the instantaneous rate of change is known to be -32 at $x = -2$. Find the equation of the tangent line to the graph of $y = f(x)$ at the point with x coordinate -2.

18. For $f(x) = x^4$, the instantaneous rate of change is known to be -4 at $x = -1$. Find the equation of the tangent line to the graph of $y = f(x)$ at the point with x coordinate -1.

In Problems 19–42, use the four-step process to find $f'(x)$ and then find $f'(1), f'(2)$, and $f'(3)$.

19. $f(x) = -5$

20. $f(x) = 9$

21. $f(x) = 3x - 7$

22. $f(x) = 4 - 6x$

23. $f(x) = 2 - 3x^2$

24. $f(x) = 2x^2 + 8$

25. $f(x) = x^2 + 6x - 10$

26. $f(x) = x^2 + 4x + 7$

27. $f(x) = 2x^2 - 7x + 3$

28. $f(x) = 2x^2 + 5x + 1$

29. $f(x) = -x^2 + 4x - 9$

30. $f(x) = -x^2 + 9x - 2$

31. $f(x) = 2x^3 + 1$

32. $f(x) = -2x^3 + 5$

33. $f(x) = 4 + \frac{4}{x}$

34. $f(x) = \frac{6}{x} - 2$

35. $f(x) = 5 + 3\sqrt{x}$

36. $f(x) = 3 - 7\sqrt{x}$

37. $f(x) = 10\sqrt{x + 5}$

38. $f(x) = 16\sqrt{x + 9}$

39. $f(x) = \dfrac{1}{x - 4}$

40. $f(x) = \dfrac{1}{x + 4}$

41. $f(x) = \dfrac{x}{x + 1}$

42. $f(x) = \dfrac{x}{x + 2}$

Problems 43 and 44 refer to the graph of $y = f(x) = x^2 + x$ shown.

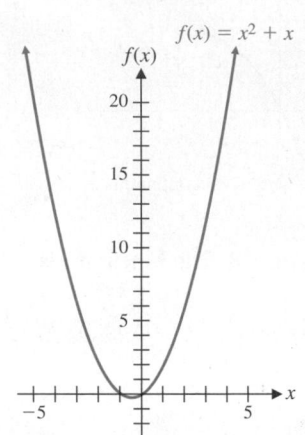

43. (A) Find the slope of the secant line joining $(1, f(1))$ and $(3, f(3))$.

(B) Find the slope of the secant line joining $(1, f(1))$ and $(1 + h, f(1 + h))$.

(C) Find the slope of the tangent line at $(1, f(1))$.

(D) Find the equation of the tangent line at $(1, f(1))$.

44. (A) Find the slope of the secant line joining $(2, f(2))$ and $(4, f(4))$.

(B) Find the slope of the secant line joining $(2, f(2))$ and $(2 + h, f(2 + h))$.

(C) Find the slope of the tangent line at $(2, f(2))$.

(D) Find the equation of the tangent line at $(2, f(2))$.

In Problems 45 and 46, suppose an object moves along the y axis so that its location is $y = f(x) = x^2 + x$ at time x (y is in meters and x is in seconds). Find

45. (A) The average velocity (the average rate of change of y with respect to x) for x changing from 1 to 3 seconds

(B) The average velocity for x changing from 1 to $1 + h$ seconds

(C) The instantaneous velocity at $x = 1$ second

46. (A) The average velocity (the average rate of change of y with respect to x) for x changing from 2 to 4 seconds

(B) The average velocity for x changing from 2 to $2 + h$ seconds

(C) The instantaneous velocity at $x = 2$ seconds

Problems 47–54, refer to the function F in the graph shown. Use the graph to determine whether $F'(x)$ exists at each indicated value of x.

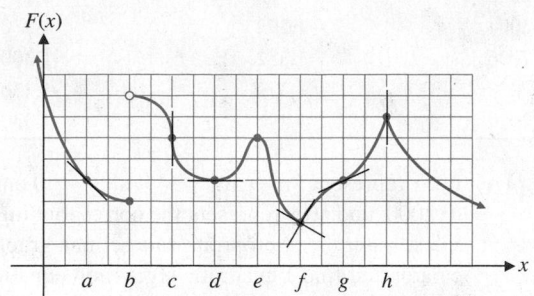

47. $x = a$

48. $x = b$

49. $x = c$

50. $x = d$

51. $x = e$

52. $x = f$

53. $x = g$

54. $x = h$

55. Given $f(x) = x^2 - 4x$,

(A) Find $f'(x)$.

(B) Find the slopes of the lines tangent to the graph of f at $x = 0, 2$, and 4.

(C) Graph f and sketch in the tangent lines at $x = 0, 2$, and 4.

56. Given $f(x) = x^2 + 2x$,

(A) Find $f'(x)$.

(B) Find the slopes of the lines tangent to the graph of f at $x = -2, -1$, and 1.

(C) Graph f and sketch in the tangent lines at $x = -2, -1$, and 1.

57. If an object moves along a line so that it is at $y = f(x) = 4x^2 - 2x$ at time x (in seconds), find the instantaneous velocity function $v = f'(x)$ and find the velocity at times $x = 1, 3$, and 5 seconds (y is measured in feet).

58. Repeat Problem 57 with $f(x) = 8x^2 - 4x$.

59. Let $f(x) = x^2$, $g(x) = x^2 - 1$, and $h(x) = x^2 + 2$.

(A) How are the graphs of these functions related? How would you expect the derivatives of these functions to be related?

(B) Use the four-step process to find the derivative of $m(x) = x^2 + C$, where C is any real constant.

60. Let $f(x) = -x^2$, $g(x) = -x^2 - 1$, and $h(x) = -x^2 + 2$.

(A) How are the graphs of these functions related? How would you expect the derivatives of these functions to be related?

(B) Use the four-step process to find the derivative of $m(x) = -x^2 + C$, where C is any real constant.

In Problems 61–66, discuss the validity of each statement. If the statement is always true, explain why. If not, give a counterexample.

61. If $f(x) = C$ is a constant function, then $f'(x) = 0$.

62. If $f(x) = mx + b$ is a linear function, then $f'(x) = m$.

63. If a function f is continuous on the interval (a, b), then f is differentiable on (a, b).

64. If a function f is differentiable on the interval (a, b), then f is continuous on (a, b).

65. The average rate of change of a function f from $x = a$ to $x = a + h$ is less than the instantaneous rate of change at $x = a + \dfrac{h}{2}$.

66. If the graph of f has a sharp corner at $x = a$, then f is not continuous at $x = a$.

In Problems 67–70, sketch the graph of f and determine where f is nondifferentiable.

67. $f(x) = \begin{cases} 2x & \text{if } x < 1 \\ 2 & \text{if } x \geq 1 \end{cases}$

68. $f(x) = \begin{cases} 2x & \text{if } x < 2 \\ 6 - x & \text{if } x \geq 2 \end{cases}$

69. $f(x) = \begin{cases} x^2 + 1 & \text{if } x < 0 \\ 1 & \text{if } x \geq 0 \end{cases}$

70. $f(x) = \begin{cases} 2 - x^2 & \text{if } x \leq 0 \\ 2 & \text{if } x > 0 \end{cases}$

In Problems 71–76, determine whether f is differentiable at $x = 0$ by considering

$$\lim_{h \to 0} \frac{f(0 + h) - f(0)}{h}$$

71. $f(x) = |x|$

72. $f(x) = 1 - |x|$

73. $f(x) = x^{1/3}$

74. $f(x) = x^{2/3}$

75. $f(x) = \sqrt{1 - x^2}$

76. $f(x) = \sqrt{1 + x^2}$

77. A ball dropped from a balloon falls $y = 16x^2$ feet in x seconds. If the balloon is 576 feet above the ground when the ball is dropped, when does the ball hit the ground? What is the velocity of the ball at the instant it hits the ground?

78. Repeat Problem 77 if the balloon is 1,024 feet above the ground when the ball is dropped.

Applications

79. **Revenue.** The revenue (in dollars) from the sale of x infant car seats is given by

$$R(x) = 60x - 0.025x^2 \qquad 0 \leq x \leq 2,400$$

(A) Find the average change in revenue if production is changed from 1,000 car seats to 1,050 car seats.

(B) Use the four-step process to find $R'(x)$.

(C) Find the revenue and the instantaneous rate of change of revenue at a production level of 1,000 car seats, and write a brief verbal interpretation of these results.

80. Profit. The profit (in dollars) from the sale of x infant car seats is given by

$$P(x) = 45x - 0.025x^2 - 5,000 \qquad 0 \le x \le 2,400$$

(A) Find the average change in profit if production is changed from 800 car seats to 850 car seats.

(B) Use the four-step process to find $P'(x)$.

✎ (C) Find the profit and the instantaneous rate of change of profit at a production level of 800 car seats, and write a brief verbal interpretation of these results.

81. Sales analysis. A company's total sales (in millions of dollars) t months from now are given by

$$S(t) = 2\sqrt{t} + 10$$

(A) Use the four-step process to find $S'(t)$.

✎ (B) Find $S(15)$ and $S'(15)$. Write a brief verbal interpretation of these results.

(C) Use the results in part (B) to estimate the total sales after 16 months and after 17 months.

82. Sales analysis. A company's total sales (in millions of dollars) t months from now are given by

$$S(t) = 2\sqrt{t} + 6$$

(A) Use the four-step process to find $S'(t)$.

✎ (B) Find $S(10)$ and $S'(10)$. Write a brief verbal interpretation of these results.

(C) Use the results in part (B) to estimate the total sales after 11 months and after 12 months.

83. Mineral consumption. The U.S. consumption of tungsten (in metric tons) is given approximately by

$$p(t) = 138t^2 + 1,072t + 14,917$$

where t is time in years and $t = 0$ corresponds to 2010.

(A) Use the four-step process to find $p'(t)$.

✎ (B) Find the annual consumption in 2020 and the instantaneous rate of change of consumption in 2020, and write a brief verbal interpretation of these results.

84. Mineral consumption. The U.S. consumption of refined copper (in thousands of metric tons) is given approximately by

$$p(t) = 48t^2 - 37t + 1,698$$

where t is time in years and $t = 0$ corresponds to 2010.

(A) Use the four-step process to find $p'(t)$.

✎ (B) Find the annual consumption in 2022 and the instantaneous rate of change of consumption in 2022, and write a brief verbal interpretation of these results.

85. Electricity consumption. Table 1 gives the retail sales of electricity (in billions of kilowatt-hours) for the residential and commercial sectors in the United States. (*Source:* Energy Information Administration)

Table 1 **Electricity Sales**

Year	Residential	Commercial
2000	1,192	1,055
2002	1,265	1,104
2004	1,292	1,230
2006	1,352	1,300
2008	1,379	1,336
2010	1,446	1,330

(A) Let x represent time (in years) with $x = 0$ corresponding to 2000, and let y represent the corresponding residential sales. Enter the appropriate data set in a graphing calculator and find a quadratic regression equation for the data.

(B) If $y = R(x)$ denotes the regression equation found in part (A), find $R(20)$ and $R'(20)$, and write a brief verbal interpretation of these results. Round answers to the nearest tenth of a billion.

86. Electricity consumption. Refer to the data in Table 1.

(A) Let x represent time (in years) with $x = 0$ corresponding to 2000, and let y represent the corresponding commercial sales. Enter the appropriate data set in a graphing calculator and find a quadratic regression equation for the data.

✎ (B) If $y = C(x)$ denotes the regression equation found in part (A), find $C(20)$ and $C'(20)$, and write a brief verbal interpretation of these results. Round answers to the nearest tenth of a billion.

87. Air pollution. The ozone level (in parts per billion) on a summer day in a metropolitan area is given by

$$P(t) = 80 + 12t - t^2$$

where t is time in hours and $t = 0$ corresponds to 9 A.M.

(A) Use the four-step process to find $P'(t)$.

✎ (B) Find $P(3)$ and $P'(3)$. Write a brief verbal interpretation of these results.

88. Medicine. The body temperature (in degrees Fahrenheit) of a patient t hours after taking a fever-reducing drug is given by

$$F(t) = 98 + \frac{4}{t + 1}$$

(A) Use the four-step process to find $F'(t)$.

✎ (B) Find $F(3)$ and $F'(3)$. Write a brief verbal interpretation of these results.

Answers to Matched Problems

1. (A) $-\$1,600$ (B) $-\$8$ per planter
2. (A) 48 ft/s   (B) $32 + 16h$
   (C) 32 ft/s
3. (A) 6, 5   (B) $4 + h$
   (C) 4   (D) Both are 4
4. $f'(x) = 8 - 4x$

5. (A) $f'(1) = 4, f'(2) = 0, f'(4) = -8$

(B) $f(x)$

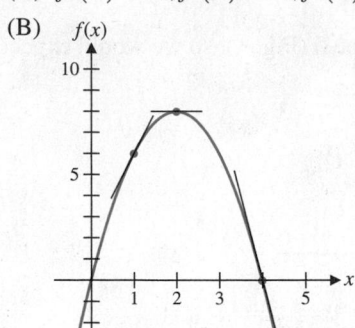

6. $f'(x) = -1/(x + 2)^2, x \neq -2$

7. $f'(x) = 1/(2\sqrt{x + 4}), x \geq -4$

8. $S(12) = 4, S'(12) = 0.125$; 12 months from now, the total sales will be \$4 million and will be increasing at the rate of \$0.125 million (\$125,000) per month. The estimated total sales are \$4.125 million after 13 months and \$4.25 million after 14 months.

2.5 Basic Differentiation Properties

- Constant Function Rule
- Power Rule
- Constant Multiple Property
- Sum and Difference Properties
- Applications

In Section 2.4, we defined the derivative of f at x as

$$f'(x) = \lim_{h \to 0} \frac{f(x + h) - f(x)}{h}$$

if the limit exists, and we used this definition and a four-step process to find the derivatives of several functions. Now we want to develop some rules of differentiation. These rules will enable us to find the derivative of many functions without using the four-step process.

Before exploring these rules, we list some symbols that are often used to represent derivatives.

> **NOTATION** The Derivative
> If $y = f(x)$, then
>
> $$f'(x) \qquad y' \qquad \frac{dy}{dx}$$
>
> all represent the derivative of f at x.

Each of these derivative symbols has its particular advantage in certain situations. All of them will become familiar to you after a little experience.

Constant Function Rule

If $f(x) = C$ is a constant function, then the four-step process can be used to show that $f'(x) = 0$. Therefore,

The derivative of any constant function is 0.

> **THEOREM 1** Constant Function Rule
> If $y = f(x) = C$, then
>
> $$f'(x) = 0$$
>
> Also, $y' = 0$ and $dy/dx = 0$.
>
> **Note:** When we write $C' = 0$ or $\dfrac{d}{dx}C = 0$, we mean that $y' = \dfrac{dy}{dx} = 0$ when $y = C$.

CONCEPTUAL INSIGHT

The graph of $f(x) = C$ is a horizontal line with slope 0 (Fig. 1), so we would expect that $f'(x) = 0$.

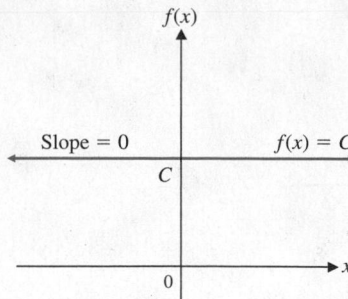

Figure 1

EXAMPLE 1   Differentiating Constant Functions

(A) If $f(x) = 3$, then $f'(x) = 0$.   (B) If $y = -1.4$, then $y' = 0$.

(C) If $y = \pi$, then $\dfrac{dy}{dx} = 0$.   (D) $\dfrac{d}{dx} 23 = 0$

Matched Problem 1 Find

(A) $f'(x)$ for $f(x) = -24$ (B) y' for $y = 12$

(C) $\dfrac{dy}{dx}$ for $y = -\sqrt{7}$ (D) $\dfrac{d}{dx}(-\pi)$

Power Rule

A function of the form $f(x) = x^k$, where k is a real number, is called a **power function**. The following elementary functions are examples of power functions:

$$f(x) = x \qquad h(x) = x^2 \qquad m(x) = x^3 \tag{1}$$
$$n(x) = \sqrt{x} \qquad p(x) = \sqrt[3]{x}$$

Explore and Discuss 1 (A) It is clear that the functions f, h, and m in (1) are power functions. Explain why the functions n and p are also power functions.

(B) The domain of a power function depends on the power. Discuss the domain of each of the following power functions:

$$r(x) = x^4 \qquad s(x) = x^{-4} \qquad t(x) = x^{1/4}$$
$$u(x) = x^{-1/4} \qquad v(x) = x^{1/5} \qquad w(x) = x^{-1/5}$$

The definition of the derivative and the four-step process introduced in Section 2.4 can be used to find the derivatives of many power functions. For example, it can be shown that

$$\begin{array}{llll} \text{If} & f(x) = x^2, & \text{then} & f'(x) = 2x. \\ \text{If} & f(x) = x^3, & \text{then} & f'(x) = 3x^2. \\ \text{If} & f(x) = x^4, & \text{then} & f'(x) = 4x^3. \\ \text{If} & f(x) = x^5, & \text{then} & f'(x) = 5x^4. \end{array}$$

Notice the pattern in these derivatives. In each case, the power in f becomes the coefficient in f' and the power in f' is 1 less than the power in f. In general, for any positive integer n,

$$\text{If} \quad f(x) = x^n, \quad \text{then} \quad f'(x) = nx^{n-1}. \tag{2}$$

In fact, more advanced techniques can be used to show that (2) holds for *any* real number n. We will assume this general result for the remainder of the book.

THEOREM 2 Power Rule

If $y = f(x) = x^n$, where n is a real number, then

$$f'(x) = nx^{n-1}$$

Also, $y' = nx^{n-1}$ and $dy/dx = nx^{n-1}$.

EXAMPLE 2 Differentiating Power Functions

(A) If $f(x) = x^5$, then $f'(x) = 5x^{5-1} = 5x^4$.

(B) If $y = x^{25}$, then $y' = 25x^{25-1} = 25x^{24}$.

(C) If $y = t^{-3}$, then $\dfrac{dy}{dt} = -3t^{-3-1} = -3t^{-4} = -\dfrac{3}{t^4}$.

(D) $\dfrac{d}{dx} x^{5/3} = \dfrac{5}{3} x^{(5/3)-1} = \dfrac{5}{3} x^{2/3}$.

Matched Problem 2 Find

(A) $f'(x)$ for $f(x) = x^6$ (B) y' for $y = x^{30}$

(C) $\dfrac{dy}{dt}$ for $y = t^{-2}$ (D) $\dfrac{d}{dx} x^{3/2}$

In some cases, properties of exponents must be used to rewrite an expression before the power rule is applied.

EXAMPLE 3 Differentiating Power Functions

(A) If $f(x) = 1/x^4$, we can write $f(x) = x^{-4}$ and

$$f'(x) = -4x^{-4-1} = -4x^{-5}, \quad \text{or} \quad \frac{-4}{x^5}$$

(B) If $y = \sqrt{u}$, we can write $y = u^{1/2}$ and

$$y' = \frac{1}{2} u^{(1/2)-1} = \frac{1}{2} u^{-1/2}, \quad \text{or} \quad \frac{1}{2\sqrt{u}}$$

(C) $\dfrac{d}{dx} \dfrac{1}{\sqrt[3]{x}} = \dfrac{d}{dx} x^{-1/3} = -\dfrac{1}{3} x^{(-1/3)-1} = -\dfrac{1}{3} x^{-4/3}$, or $\dfrac{-1}{3\sqrt[3]{x^4}}$

Matched Problem 3 Find

(A) $f'(x)$ for $f(x) = \dfrac{1}{x}$ (B) y' for $y = \sqrt[3]{u^2}$ (C) $\dfrac{d}{dx} \dfrac{1}{\sqrt{x}}$

Constant Multiple Property

Let $f(x) = ku(x)$, where k is a constant and u is differentiable at x. Using the four-step process, we have the following:

Step 1 $f(x + h) = ku(x + h)$

Step 2 $f(x + h) - f(x) = ku(x + h) - ku(x) = k[u(x + h) - u(x)]$

Step 3 $\dfrac{f(x + h) - f(x)}{h} = \dfrac{k[u(x + h) - u(x)]}{h} = k\left[\dfrac{u(x + h) - u(x)}{h}\right]$

Step 4 $f'(x) = \lim\limits_{h \to 0} \dfrac{f(x + h) - f(x)}{h}$

$\qquad\qquad = \lim\limits_{h \to 0} k\left[\dfrac{u(x + h) - u(x)}{h}\right]$ $\lim\limits_{x \to c} kg(x) = k \lim\limits_{x \to c} g(x)$

$\qquad\qquad = k \lim\limits_{h \to 0} \left[\dfrac{u(x + h) - u(x)}{h}\right]$ Definition of $u'(x)$

$\qquad\qquad = ku'(x)$

Therefore,

> The derivative of a constant times a differentiable function is the constant times the derivative of the function.

THEOREM 3   Constant Multiple Property

If $y = f(x) = ku(x)$, then

$$f'(x) = ku'(x)$$

Also,

$$y' = ku' \qquad \dfrac{dy}{dx} = k\dfrac{du}{dx}$$

EXAMPLE 4   Differentiating a Constant Times a Function

(A) If $f(x) = 3x^2$,   then $f'(x) = 3 \cdot 2x^{2-1} = 6x$.

(B) If $y = \dfrac{t^3}{6} = \dfrac{1}{6}t^3$,   then $\dfrac{dy}{dt} = \dfrac{1}{6} \cdot 3t^{3-1} = \dfrac{1}{2}t^2$.

(C) If $y = \dfrac{1}{2x^4} = \dfrac{1}{2}x^{-4}$,   then $y' = \dfrac{1}{2}(-4x^{-4-1}) = -2x^{-5}$, or $\dfrac{-2}{x^5}$.

(D) $\dfrac{d}{dx}\dfrac{0.4}{\sqrt{x^3}} = \dfrac{d}{dx}\dfrac{0.4}{x^{3/2}} = \dfrac{d}{dx} 0.4x^{-3/2} = 0.4\left[-\dfrac{3}{2}x^{(-3/2)-1}\right]$

$\qquad\qquad\qquad\qquad\qquad\qquad = -0.6x^{-5/2},$ or $-\dfrac{0.6}{\sqrt{x^5}}$

Matched Problem 4   Find

(A) $f'(x)$ for $f(x) = 4x^5$

(B) $\dfrac{dy}{dt}$ for $y = \dfrac{t^4}{12}$

(C) y' for $y = \dfrac{1}{3x^3}$

(D) $\dfrac{d}{dx}\dfrac{0.9}{\sqrt[3]{x}}$

Sum and Difference Properties

Let $f(x) = u(x) + v(x)$, where $u'(x)$ and $v'(x)$ exist. Using the four-step process (see Problems 87 and 88 in Exercises 2.5):

$$f'(x) = u'(x) + v'(x)$$

Therefore,

The derivative of the sum of two differentiable functions is the sum of the derivatives of the functions.

Similarly, we can show that

The derivative of the difference of two differentiable functions is the difference of the derivatives of the functions.

Together, we have the **sum and difference property** for differentiation:

THEOREM 4 Sum and Difference Property

If $y = f(x) = u(x) \pm v(x)$, then

$$f'(x) = u'(x) \pm v'(x)$$

Also,

$$y' = u' \pm v' \qquad \frac{dy}{dx} = \frac{du}{dx} \pm \frac{dv}{dx}$$

Note: This rule generalizes to the sum and difference of any given number of functions.

With Theorems 1 through 4, we can compute the derivatives of all polynomials and a variety of other functions.

EXAMPLE 5   Differentiating Sums and Differences

(A) If $f(x) = 3x^2 + 2x$,   then

$$f'(x) = (3x^2)' + (2x)' = 3(2x) + 2(1) = 6x + 2$$

(B) If $y = 4 + 2x^3 - 3x^{-1}$,   then

$$y' = (4)' + (2x^3)' - (3x^{-1})' = 0 + 2(3x^2) - 3(-1)x^{-2} = 6x^2 + 3x^{-2}$$

(C) If $y = \sqrt[3]{w} - 3w$,   then

$$\frac{dy}{dw} = \frac{d}{dw}w^{1/3} - \frac{d}{dw}3w = \frac{1}{3}w^{-2/3} - 3 = \frac{1}{3w^{2/3}} - 3$$

(D) $\dfrac{d}{dx}\left(\dfrac{5}{3x^2} - \dfrac{2}{x^4} + \dfrac{x^3}{9}\right) = \dfrac{d}{dx}\dfrac{5}{3}x^{-2} - \dfrac{d}{dx}2x^{-4} + \dfrac{d}{dx}\dfrac{1}{9}x^3$

$$= \frac{5}{3}(-2)x^{-3} - 2(-4)x^{-5} + \frac{1}{9}\cdot 3x^2$$

$$= -\frac{10}{3x^3} + \frac{8}{x^5} + \frac{1}{3}x^2$$

Matched Problem 5 Find

(A) $f'(x)$ for $f(x) = 3x^4 - 2x^3 + x^2 - 5x + 7$

(B) y' for $y = 3 - 7x^{-2}$

(C) $\dfrac{dy}{dv}$ for $y = 5v^3 - \sqrt[4]{v}$

(D) $\dfrac{d}{dx}\left(-\dfrac{3}{4x} + \dfrac{4}{x^3} - \dfrac{x^4}{8}\right)$

Some algebraic rewriting of a function is sometimes required before we can apply the rules for differentiation.

EXAMPLE 6   Rewrite before Differentiating   Find the derivative of
$$f(x) = \frac{1 + x^2}{x^4}.$$

SOLUTION   It is helpful to rewrite $f(x) = \dfrac{1 + x^2}{x^4}$, expressing $f(x)$ as the sum of terms, each of which can be differentiated by applying the power rule.

$$f(x) = \frac{1 + x^2}{x^4} \qquad \text{Write as a sum of two terms}$$

$$= \frac{1}{x^4} + \frac{x^2}{x^4} \qquad \text{Write each term as a power of } x$$

$$= x^{-4} + x^{-2}$$

Note that we have rewritten $f(x)$, but we have not used any rules of differentiation. Now, however, we can apply those rules to find the derivative:

$$f'(x) = -4x^{-5} - 2x^{-3}$$

Matched Problem 6   Find the derivative of $f(x) = \dfrac{5 - 3x + 4x^2}{x}$.

Applications

EXAMPLE 7   Instantaneous Velocity   An object moves along the y axis (marked in feet) so that its position at time x (in seconds) is
$$f(x) = x^3 - 6x^2 + 9x$$

(A) Find the instantaneous velocity function v.
(B) Find the velocity at $x = 2$ and $x = 5$ seconds.
(C) Find the time(s) when the velocity is 0.

SOLUTION
(A) $v = f'(x) = (x^3)' - (6x^2)' + (9x)' = 3x^2 - 12x + 9$
(B) $f'(2) = 3(2)^2 - 12(2) + 9 = -3$ feet per second
    $f'(5) = 3(5)^2 - 12(5) + 9 = 24$ feet per second
(C) $v = f'(x) = 3x^2 - 12x + 9 = 0$ Factor 3 out of each term.
    $\qquad\quad 3(x^2 - 4x + 3) = 0$ Factor the quadratic term.
    $\qquad\quad 3(x - 1)(x - 3) = 0$ Use the zero property.
    $\qquad\qquad\qquad\qquad x = 1, 3$

So, $v = 0$ at $x = 1$ and $x = 3$ seconds.

Matched Problem 7   Repeat Example 7 for $f(x) = x^3 - 15x^2 + 72x$.

EXAMPLE 8   Tangents   Let $f(x) = x^4 - 6x^2 + 10$.

(A) Find $f'(x)$.
(B) Find the equation of the tangent line at $x = 1$.
(C) Find the values of x where the tangent line is horizontal.

SOLUTION

(A) $f'(x) = (x^4)' - (6x^2)' + (10)'$

$\quad\quad\quad\quad = 4x^3 - 12x$

(B) $y - y_1 = m(x - x_1)$ $\quad y_1 = f(x_1) = f(1) = (1)^4 - 6(1)^2 + 10 = 5$

$\quad\quad y - 5 = -8(x - 1)$ $\quad m = f'(x_1) = f'(1) = 4(1)^3 - 12(1) = -8$

$\quad\quad\quad\quad y = -8x + 13$ $\quad$ Tangent line at $x = 1$

(C) Since a horizontal line has 0 slope, we must solve $f'(x) = 0$ for x:

$\quad\quad\quad\quad f'(x) = 4x^3 - 12x = 0$ $\quad\quad$ Factor $4x$ out of each term.

$\quad\quad\quad\quad 4x(x^2 - 3) = 0$ $\quad\quad$ Factor the difference of two squares.

$\quad\quad 4x(x + \sqrt{3})(x - \sqrt{3}) = 0$ $\quad\quad$ Use the zero property.

$\quad\quad\quad\quad x = 0, -\sqrt{3}, \sqrt{3}$

<u>Matched Problem 8</u> Repeat Example 8 for $f(x) = x^4 - 8x^3 + 7$.

Exercises 2.5

Skills Warm-up Exercises

W *In Problems 1–8, write the expression in the form x^n. (If necessary, review Section A.6).*

1. $\sqrt{x}$ $\quad\quad$ **2.** $\sqrt[3]{x}$ $\quad\quad$ **3.** $\dfrac{1}{x^5}$ $\quad\quad$ **4.** $\dfrac{1}{x}$

5. $(x^4)^3$ $\quad\quad$ **6.** $\dfrac{1}{(x^5)^2}$ $\quad\quad$ **7.** $\dfrac{1}{\sqrt[4]{x}}$ $\quad\quad$ **8.** $\dfrac{1}{\sqrt[5]{x}}$

Find the indicated derivatives in Problems 9–26.

9. $f'(x)$ for $f(x) = 7$ $\quad\quad$ **10.** $\dfrac{d}{dx}3$

11. $\dfrac{dy}{dx}$ for $y = x^9$ $\quad\quad$ **12.** y' for $y = x^6$

13. $\dfrac{d}{dx}x^3$ $\quad\quad$ **14.** $g'(x)$ for $g(x) = x^5$

15. y' for $y = x^{-4}$ $\quad\quad$ **16.** $\dfrac{dy}{dx}$ for $y = x^{-8}$

17. $g'(x)$ for $g(x) = x^{8/3}$ $\quad$ **18.** $f'(x)$ for $f(x) = x^{9/2}$

19. $\dfrac{dy}{dx}$ for $y = \dfrac{1}{x^{10}}$ $\quad\quad$ **20.** y' for $y = \dfrac{1}{x^{12}}$

21. $f'(x)$ for $f(x) = 5x^2$ $\quad$ **22.** $\dfrac{d}{dx}(-2x^3)$

23. y' for $y = 0.4x^7$

24. $f'(x)$ for $f(x) = 0.8x^4$

25. $\dfrac{d}{dx}\left(\dfrac{x^3}{18}\right)$ $\quad\quad$ **26.** $\dfrac{dy}{dx}$ for $y = \dfrac{x^5}{25}$

Problems 27–32 refer to functions f and g that satisfy $f'(2) = 3$ and $g'(2) = -1$. In each problem, find $h'(2)$ for the indicated function h.

27. $h(x) = 4f(x)$ $\quad\quad$ **28.** $h(x) = 5g(x)$

29. $h(x) = f(x) + g(x)$ $\quad\quad$ **30.** $h(x) = g(x) - f(x)$

31. $h(x) = 2f(x) - 3g(x) + 7$

32. $h(x) = -4f(x) + 5g(x) - 9$

Find the indicated derivatives in Problems 33–56.

33. $\dfrac{d}{dx}(2x - 5)$ $\quad\quad$ **34.** $\dfrac{d}{dx}(-4x + 9)$

35. $f'(t)$ if $f(t) = 2t^2 - 3t + 1$

36. $\dfrac{dy}{dt}$ if $y = 2 + 5t - 8t^3$

37. y' for $y = 5x^{-2} + 9x^{-1}$

38. $g'(x)$ if $g(x) = 5x^{-7} - 2x^{-4}$

39. $\dfrac{d}{du}(5u^{0.3} - 4u^{2.2})$

40. $\dfrac{d}{du}(2u^{4.5} - 3.1u + 13.2)$

41. $h'(t)$ if $h(t) = 2.1 + 0.5t - 1.1t^3$

42. $F'(t)$ if $F(t) = 0.2t^3 - 3.1t + 13.2$

43. y' if $y = \dfrac{2}{5x^4}$

44. w' if $w = \dfrac{7}{5u^2}$

45. $\dfrac{d}{dx}\left(\dfrac{3x^2}{2} - \dfrac{7}{5x^2}\right)$

46. $\dfrac{d}{dx}\left(\dfrac{5x^3}{4} - \dfrac{2}{5x^3}\right)$

47. $G'(w)$ if $G(w) = \dfrac{5}{9w^4} + 5\sqrt[3]{w}$

48. $H'(w)$ if $H(w) = \dfrac{5}{w^6} - 2\sqrt{w}$

49. $\dfrac{d}{du}(3u^{2/3} - 5u^{1/3})$

50. $\dfrac{d}{du}(8u^{3/4} + 4u^{-1/4})$

51. $h'(t)$ if $h(t) = \dfrac{3}{t^{3/5}} - \dfrac{6}{t^{1/2}}$

52. $F'(t)$ if $F(t) = \dfrac{5}{t^{1/5}} - \dfrac{8}{t^{3/2}}$

53. y' if $y = \dfrac{1}{\sqrt[3]{x}}$

54. w' if $w = \dfrac{10}{\sqrt[5]{u}}$

55. $\dfrac{d}{dx}\left(\dfrac{1.2}{\sqrt{x}} - 3.2x^{-2} + x\right)$

56. $\dfrac{d}{dx}\left(2.8x^{-3} - \dfrac{0.6}{\sqrt[3]{x^2}} + 7\right)$

For Problems 57–60, find

(A) $f'(x)$

(B) *The slope of the graph of f at $x = 2$ and $x = 4$*

(C) *The equations of the tangent lines at $x = 2$ and $x = 4$*

(D) *The value(s) of x where the tangent line is horizontal*

57. $f(x) = 6x - x^2$ **58.** $f(x) = 2x^2 + 8x$

59. $f(x) = 3x^4 - 6x^2 - 7$ **60.** $f(x) = x^4 - 32x^2 + 10$

If an object moves along the y axis (marked in feet) so that its position at time x (in seconds) is given by the indicated functions in Problems 61–64, find

(A) *The instantaneous velocity function $v = f'(x)$*

(B) *The velocity when $x = 0$ and $x = 3$ seconds*

(C) *The time(s) when $v = 0$*

61. $f(x) = 176x - 16x^2$ **62.** $f(x) = 80x - 10x^2$

63. $f(x) = x^3 - 9x^2 + 15x$ **64.** $f(x) = x^3 - 9x^2 + 24x$

Problems 65–72 require the use of a graphing calculator. For each problem, find $f'(x)$ and approximate (to four decimal places) the value(s) of x where the graph of f has a horizontal tangent line.

65. $f(x) = x^2 - 3x - 4\sqrt{x}$

66. $f(x) = x^2 + x - 10\sqrt{x}$

67. $f(x) = 3\sqrt[3]{x^4} - 1.5x^2 - 3x$

68. $f(x) = 3\sqrt[3]{x^4} - 2x^2 + 4x$

69. $f(x) = 0.05x^4 + 0.1x^3 - 1.5x^2 - 1.6x + 3$

70. $f(x) = 0.02x^4 - 0.06x^3 - 0.78x^2 + 0.94x + 2.2$

71. $f(x) = 0.2x^4 - 3.12x^3 + 16.25x^2 - 28.25x + 7.5$

72. $f(x) = 0.25x^4 - 2.6x^3 + 8.1x^2 - 10x + 9$

73. Let $f(x) = ax^2 + bx + c, a \neq 0$. Recall that the graph of $y = f(x)$ is a parabola. Use the derivative $f'(x)$ to derive a formula for the x coordinate of the vertex of this parabola.

74. Now that you know how to find derivatives, explain why it is no longer necessary for you to memorize the formula for the x coordinate of the vertex of a parabola.

75. Give an example of a cubic polynomial function that has

(A) No horizontal tangents

(B) One horizontal tangent

(C) Two horizontal tangents

76. Can a cubic polynomial function have more than two horizontal tangents? Explain.

Find the indicated derivatives in Problems 77–82.

77. $f'(x)$ if $f(x) = (2x - 1)^2$

78. y' if $y = (2x - 5)^2$

79. $\dfrac{d}{dx}\dfrac{10x + 20}{x}$

80. $\dfrac{dy}{dx}$ if $y = \dfrac{x^2 + 25}{x^2}$

81. $\dfrac{dy}{dx}$ if $y = \dfrac{3x - 4}{12x^2}$

82. $f'(x)$ if $f(x) = \dfrac{2x^5 - 4x^3 + 2x}{x^3}$

In Problems 83–86, discuss the validity of each statement. If the statement is always true, explain why. If not, give a counterexample.

83. The derivative of a product is the product of the derivatives.

84. The derivative of a quotient is the quotient of the derivatives.

85. The derivative of a constant is 0.

86. The derivative of a constant times a function is 0.

87. Let $f(x) = u(x) + v(x)$, where $u'(x)$ and $v'(x)$ exist. Use the four-step process to show that $f'(x) = u'(x) + v'(x)$.

88. Let $f(x) = u(x) - v(x)$, where $u'(x)$ and $v'(x)$ exist. Use the four-step process to show that $f'(x) = u'(x) - v'(x)$.

Applications

89. Sales analysis. A company's total sales (in millions of dollars) t months from now are given by

$$S(t) = 0.03t^3 + 0.5t^2 + 2t + 3$$

(A) Find $S'(t)$.

(B) Find $S(5)$ and $S'(5)$ (to two decimal places). Write a brief verbal interpretation of these results.

(C) Find $S(10)$ and $S'(10)$ (to two decimal places). Write a brief verbal interpretation of these results.

90. **Sales analysis.** A company's total sales (in millions of dollars) t months from now are given by

$$S(t) = 0.015t^4 + 0.4t^3 + 3.4t^2 + 10t - 3$$

(A) Find $S'(t)$.

(B) Find $S(4)$ and $S'(4)$ (to two decimal places). Write a brief verbal interpretation of these results.

(C) Find $S(8)$ and $S'(8)$ (to two decimal places). Write a brief verbal interpretation of these results.

91. **Advertising.** A marine manufacturer will sell $N(x)$ power boats after spending $\$x$ thousand on advertising, as given by

$$N(x) = 1{,}000 - \frac{3{,}780}{x} \qquad 5 \le x \le 30$$

(see figure).

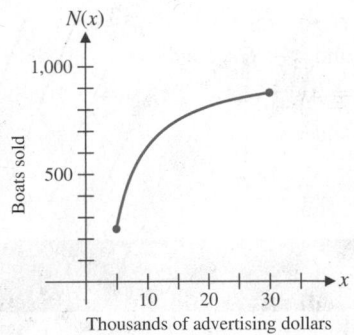

Thousands of advertising dollars

(A) Find $N'(x)$.

(B) Find $N'(10)$ and $N'(20)$. Write a brief verbal interpretation of these results.

92. **Price–demand equation.** Suppose that, in a given gourmet food store, people are willing to buy x pounds of chocolate candy per day at $\$p$ per quarter pound, as given by the price–demand equation

$$x = 10 + \frac{180}{p} \qquad 2 \le p \le 10$$

This function is graphed in the figure. Find the demand and the instantaneous rate of change of demand with respect to price when the price is $\$5$. Write a brief verbal interpretation of these results.

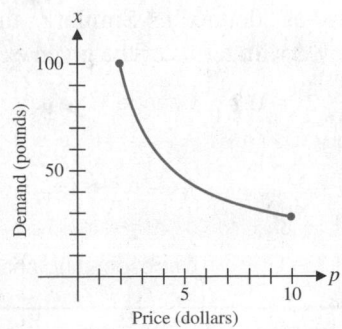

Price (dollars)

93. **College enrollment.** The percentages of male high-school graduates who enrolled in college are given in the second column of Table 1.

Table 1 **College enrollment percentages**

Year	Male	Female
1970	41.0	25.5
1980	33.5	30.3
1990	40.0	38.3
2000	40.8	45.6
2010	45.9	50.5

(A) Let x represent time (in years) since 1970, and let y represent the corresponding percentage of male high-school graduates who enrolled in college. Enter the data in a graphing calculator and find a cubic regression equation for the data.

(B) If $y = M(x)$ denotes the regression equation found in part (A), find $M(50)$ and $M'(50)$ (to the nearest tenth), and write a brief verbal interpretation of these results.

94. **College enrollment.** The percentages of female high-school graduates who enrolled in college are given in the third column of Table 1.

(A) Let x represent time (in years) since 1970, and let y represent the corresponding percentage of female high-school graduates who enrolled in college. Enter the data in a graphing calculator and find a cubic regression equation for the data.

(B) If $y = F(x)$ denotes the regression equation found in part (A), find $F(50)$ and $F'(50)$ (to the nearest tenth), and write a brief verbal interpretation of these results.

95. **Medicine.** A person x inches tall has a pulse rate of y beats per minute, as given approximately by

$$y = 590x^{-1/2} \qquad 30 \le x \le 75$$

What is the instantaneous rate of change of pulse rate at the

(A) 36-inch level?

(B) 64-inch level?

96. **Ecology.** A coal-burning electrical generating plant emits sulfur dioxide into the surrounding air. The concentration $C(x)$, in parts per million, is given approximately by

$$C(x) = \frac{0.1}{x^2}$$

where x is the distance from the plant in miles. Find the instantaneous rate of change of concentration at

(A) $x = 1$ mile

(B) $x = 2$ miles

97. **Learning.** Suppose that a person learns y items in x hours, as given by

$$y = 50\sqrt{x} \qquad 0 \le x \le 9$$

(see figure). Find the rate of learning at the end of

Time (hours)

(A) 1 hour

(B) 9 hours

98. Learning. If a person learns y items in x hours, as given by

$$y = 21\sqrt[3]{x^2} \qquad 0 \le x \le 8$$

find the rate of learning at the end of

(A) 1 hour

(B) 8 hours

1. All are 0.
2. (A) $6x^5$ (B) $30x^{29}$
   (C) $-2t^{-3} = -2/t^3$ (D) $\frac{3}{2}x^{1/2}$
3. (A) $-x^{-2}$, or $-1/x^2$ (B) $\frac{2}{3}u^{-1/3}$, or $2/(3\sqrt[3]{u})$
   (C) $-\frac{1}{2}x^{-3/2}$, or $-1/(2\sqrt{x^3})$
4. (A) $20x^4$ (B) $t^3/3$
   (C) $-x^{-4}$, or $-1/x^4$ (D) $-0.3x^{-4/3}$, or $-0.3/\sqrt[3]{x^4}$
5. (A) $12x^3 - 6x^2 + 2x - 5$ (B) $14x^{-3}$, or $14/x^3$
   (C) $15v^2 - \frac{1}{4}v^{-3/4}$, or $15v^2 - 1/(4v^{3/4})$
   (D) $3/(4x^2) - (12/x^4) - (x^3/2)$
6. $f'(x) = -5x^{-2} + 4$
7. (A) $v = 3x^2 - 30x + 72$
   (B) $f'(2) = 24$ ft/s; $f'(5) = -3$ ft/s
   (C) $x = 4$ and $x = 6$ seconds
8. (A) $f'(x) = 4x^3 - 24x^2$ (B) $y = -20x + 20$
   (C) $x = 0$ and $x = 6$

2.6 Differentials

- Increments
- Differentials
- Approximations Using Differentials

In this section, we introduce increments and differentials. Increments are useful and they provide an alternative notation for defining the derivative. Differentials are often easier to compute than increments and can be used to approximate increments.

Increments

In Section 2.4, we defined the derivative of f at x as the limit of the difference quotient

$$f'(x) = \lim_{h \to 0} \frac{f(x + h) - f(x)}{h}$$

We considered various interpretations of this limit, including slope, velocity, and instantaneous rate of change. Increment notation enables us to interpret the numerator and denominator of the difference quotient separately.

Given $y = f(x) = x^3$, if x changes from 2 to 2.1, then y will change from $y = f(2) = 2^3 = 8$ to $y = f(2.1) = 2.1^3 = 9.261$. The change in x is called the *increment in x* and is denoted by Δx (read as "delta x").* Similarly, the change in y is called the *increment in y* and is denoted by Δy. In terms of the given example, we write

$$\Delta x = 2.1 - 2 = 0.1 \qquad \text{Change in } x$$
$$\Delta y = f(2.1) - f(2) \qquad f(x) = x^3$$
$$= 2.1^3 - 2^3 \qquad \text{Use a calculator.}$$
$$= 9.261 - 8$$
$$= 1.261 \qquad \text{Corresponding change in } y$$

CONCEPTUAL INSIGHT

The symbol Δx does not represent the product of Δ and x but is the symbol for a single quantity: the *change in x*. Likewise, the symbol Δy represents a single quantity: the *change in y*.

*Δ is the uppercase Greek letter delta.

DEFINITION Increments

For $y = f(x)$, $\Delta x = x_2 - x_1$,   so $x_2 = x_1 + \Delta x$, and

$$\Delta y = y_2 - y_1$$
$$= f(x_2) - f(x_1)$$
$$= f(x_1 + \Delta x) - f(x_1)$$

Δy represents the change in y corresponding to a change Δx in x. Δx can be either positive or negative.

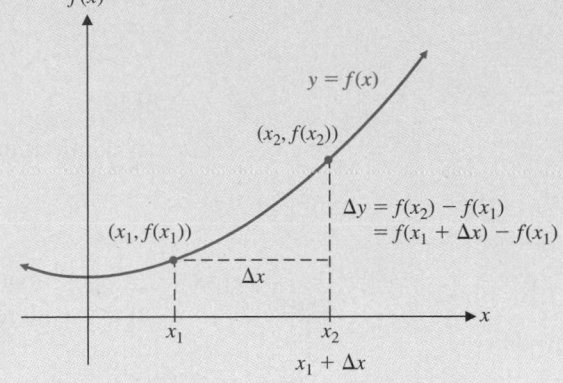

[**Note:** Δy depends on the function f, the input x_1, and the increment Δx.]

EXAMPLE 1 Increments Given the function $y = f(x) = \dfrac{x^2}{2}$,

(A) Find Δx, Δy, and $\Delta y/\Delta x$ for $x_1 = 1$ and $x_2 = 2$.

(B) Find $\dfrac{f(x_1 + \Delta x) - f(x_1)}{\Delta x}$ for $x_1 = 1$ and $\Delta x = 2$.

SOLUTION

(A) $\Delta x = x_2 - x_1 = 2 - 1 = 1$

$\Delta y = f(x_2) - f(x_1)$

$= f(2) - f(1) = \dfrac{4}{2} - \dfrac{1}{2} = \dfrac{3}{2}$

$\dfrac{\Delta y}{\Delta x} = \dfrac{f(x_2) - f(x_1)}{x_2 - x_1} = \dfrac{\frac{3}{2}}{1} = \dfrac{3}{2}$

(B) $\dfrac{f(x_1 + \Delta x) - f(x_1)}{\Delta x} = \dfrac{f(1 + 2) - f(1)}{2}$

$= \dfrac{f(3) - f(1)}{2} = \dfrac{\frac{9}{2} - \frac{1}{2}}{2} = \dfrac{4}{2} = 2$

Matched Problem 1 Given the function $y = f(x) = x^2 + 1$,

(A) Find Δx, Δy, and $\Delta y/\Delta x$ for $x_1 = 2$ and $x_2 = 3$.

(B) Find $\dfrac{f(x_1 + \Delta x) - f(x_1)}{\Delta x}$ for $x_1 = 1$ and $\Delta x = 2$.

In Example 1, we observe another notation for the difference quotient

$$\dfrac{f(x + h) - f(x)}{h} \tag{1}$$

It is common to refer to h, the change in x, as Δx. Then the difference quotient (1) takes on the form

$$\dfrac{f(x + \Delta x) - f(x)}{\Delta x} \qquad \text{or} \qquad \dfrac{\Delta y}{\Delta x} \quad \Delta y = f(x + \Delta x) - f(x)$$

and the derivative is defined by

$$f'(x) = \lim_{\Delta x \to 0} \frac{f(x + \Delta x) - f(x)}{\Delta x}$$

or

$$f'(x) = \lim_{\Delta x \to 0} \frac{\Delta y}{\Delta x} \tag{2}$$

if the limit exists.

Explore and Discuss 1   Suppose that $y = f(x)$ defines a function whose domain is the set of all real numbers. If every increment Δy is equal to 0, then what is the range of f?

Differentials

Assume that the limit in equation (2) exists. Then, for small Δx, the difference quotient $\Delta y / \Delta x$ provides a good approximation for $f'(x)$. Also, $f'(x)$ provides a good approximation for $\Delta y / \Delta x$. We write

$$\frac{\Delta y}{\Delta x} \approx f'(x) \qquad \Delta x \text{ is small, but} \neq 0 \tag{3}$$

Multiplying both sides of (3) by Δx gives us

$$\Delta y \approx f'(x)\,\Delta x \qquad \Delta x \text{ is small, but} \neq 0 \tag{4}$$

From equation (4), we see that $f'(x)\,\Delta x$ provides a good approximation for Δy when Δx is small.

Because of the practical and theoretical importance of $f'(x)\,\Delta x$, we give it the special name **differential** and represent it with the special symbol dy or df:

$$dy = f'(x)\,\Delta x \qquad \text{or} \qquad df = f'(x)\,\Delta x$$

For example,

$$d(2x^3) = (2x^3)'\,\Delta x = 6x^2\,\Delta x$$
$$d(x) = (x)'\,\Delta x = 1\,\Delta x = \Delta x$$

In the second example, we usually drop the parentheses in $d(x)$ and simply write

$$dx = \Delta x$$

In summary, we have the following:

DEFINITION Differentials

If $y = f(x)$ defines a differentiable function, then the **differential dy, or df**, is defined as the product of $f'(x)$ and dx, where $dx = \Delta x$. Symbolically,

$$dy = f'(x)\,dx, \qquad \text{or} \qquad df = f'(x)\,dx$$

where

$$dx = \Delta x$$

Note: The differential dy (or df) is actually a function involving two independent variables, x and dx. A change in either one or both will affect dy (or df).

EXAMPLE 2 Differentials Find dy for $f(x) = x^2 + 3x$. Evaluate dy for

(A) $x = 2$ and $dx = 0.1$

(B) $x = 3$ and $dx = 0.1$

(C) $x = 1$ and $dx = 0.02$

SOLUTION

$dy = f'(x)\, dx$

$\quad = (2x + 3)\, dx$

(A) When $x = 2$ and $dx = 0.1$,
$$dy = \big[2(2) + 3\big]0.1 = 0.7$$

(B) When $x = 3$ and $dx = 0.1$,
$$dy = \big[2(3) + 3\big]0.1 = 0.9$$

(C) When $x = 1$ and $dx = 0.02$,
$$dy = \big[2(1) + 3\big]0.02 = 0.1$$

Matched Problem 2 Find dy for $f(x) = \sqrt{x} + 3$. Evaluate dy for

(A) $x = 4$ and $dx = 0.1$

(B) $x = 9$ and $dx = 0.12$

(C) $x = 1$ and $dx = 0.01$

We now have two interpretations of the symbol dy/dx. Referring to the function $y = f(x) = x^2 + 3x$ in Example 2 with $x = 2$ and $dx = 0.1$, we have

$$\frac{dy}{dx} = f'(2) = 7 \quad \text{Derivative}$$

and

$$\frac{dy}{dx} = \frac{0.7}{0.1} = 7 \quad \text{Ratio of differentials}$$

Approximations Using Differentials

Earlier, we noted that for small Δx,

$$\frac{\Delta y}{\Delta x} \approx f'(x) \quad \text{and} \quad \Delta y \approx f'(x)\Delta x$$

Also, since

$$dy = f'(x)\, dx$$

it follows that

$$\Delta y \approx dy$$

and dy can be used to approximate Δy.

To interpret this result geometrically, we need to recall a basic property of the slope. The vertical change in a line is equal to the product of the slope and the horizontal change, as shown in Figure 1 on page 160.

Now consider the line tangent to the graph of $y = f(x)$, as shown in Figure 2 on page 160. Since $f'(x)$ is the slope of the tangent line and dx is the horizontal change in the tangent line, it follows that the vertical change in the tangent line is given by $dy = f'(x)\, dx$, as indicated in Figure 2.

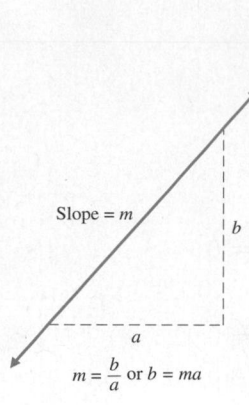

Figure 1

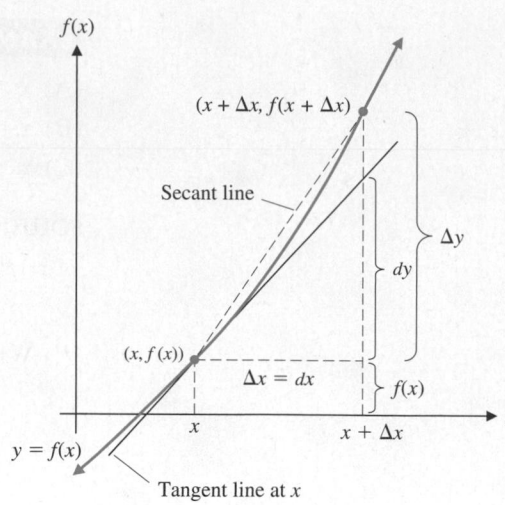

Figure 2

EXAMPLE 3   Comparing Increments and Differentials   Let $y = f(x) = 6x - x^2$.
(A) Find Δy and dy when $x = 2$.
(B) Compare Δy and dy from part (A) for $\Delta x = 0.1, 0.2,$ and 0.3.

SOLUTION

(A) $\Delta y = f(2 + \Delta x) - f(2)$

$\quad\quad = 6(2 + \Delta x) - (2 + \Delta x)^2 - (6 \cdot 2 - 2^2)$ Remove parentheses.

$\quad\quad = 12 + 6\Delta x - 4 - 4\Delta x - \Delta x^2 - 12 + 4$ Collect like terms.

$\quad\quad = 2\Delta x - \Delta x^2$

Since $f'(x) = 6 - 2x, f'(2) = 2,$ and $dx = \Delta x, dy = f'(2)\ dx = 2\Delta x$

(B) Table 1 compares the values of Δx and dy for the indicated values of Δx.

Table 1

Δx	Δy	dy
0.1	0.19	0.2
0.2	0.36	0.4
0.3	0.51	0.6

Matched Problem 3 ⌋   Repeat Example 3 for $x = 4$ and $\Delta x = dx = -0.1, -0.2,$ and -0.3.

EXAMPLE 4   Cost–Revenue   A company manufactures and sells x transistor radios per week. If the weekly cost and revenue equations are

$$C(x) = 5,000 + 2x \quad\quad R(x) = 10x - \frac{x^2}{1,000} \quad\quad 0 \le x \le 8,000$$

then use differentials to approximate the changes in revenue and profit if production is increased from 2,000 to 2,010 units per week.

SOLUTION   We will approximate ΔR and ΔP with dR and dP, respectively, using $x = 2,000$ and $dx = 2,010 - 2,000 = 10$.

$$R(x) = 10x - \frac{x^2}{1,000} \qquad P(x) = R(x) - C(x) = 10x - \frac{x^2}{1,000} - 5,000 - 2x$$

$$dR = R'(x)\, dx \qquad\qquad = 8x - \frac{x^2}{1,000} - 5,000$$

$$= \left(10 - \frac{x}{500}\right) dx \qquad dP = P'(x)\, dx$$

$$= \left(10 - \frac{2,000}{500}\right) 10 \qquad\qquad = \left(8 - \frac{x}{500}\right) dx$$

$$= \$60 \text{ per week} \qquad\qquad = \left(8 - \frac{2,000}{500}\right) 10$$

$$= \$40 \text{ per week}$$

Matched Problem 4 Repeat Example 4 with production increasing from 6,000 to 6,010.

Comparing the results in Example 4 and Matched Problem 4, we see that an increase in production results in a revenue and profit increase at the 2,000 production level but a revenue and profit loss at the 6,000 production level.

Exercises 2.6

Skills Warm-up Exercises

W

In Problems 1–4, let $f(x) = 0.1x + 3$ and find the given values without using a calculator. (If necessary, review Section 1.1.)

1. $f(0); f(0.1)$

2. $f(7); f(7.1)$

3. $f(-2); f(-2.01)$

4. $f(-10); f(-10.01)$

In Problems 5–8, let $g(x) = x^2$ and find the given values without using a calculator.

5. $g(0); g(0.1)$

6. $g(1); g(1.1)$

7. $g(10); g(10.1)$

8. $g(5); g(4.9)$

In Problems 9–14, find the indicated quantities for $y = f(x) = 3x^2$.

9. Δx, Δy, and $\Delta y/\Delta x$; given $x_1 = 1$ and $x_2 = 4$

10. Δx, Δy, and $\Delta y/\Delta x$; given $x_1 = 2$ and $x_2 = 5$

11. $\dfrac{f(x_1 + \Delta x) - f(x_1)}{\Delta x}$; given $x_1 = 1$ and $\Delta x = 2$

12. $\dfrac{f(x_1 + \Delta x) - f(x_1)}{\Delta x}$; given $x_1 = 2$ and $\Delta x = 1$

13. $\Delta y/\Delta x$; given $x_1 = 1$ and $x_2 = 3$

14. $\Delta y/\Delta x$; given $x_1 = 2$ and $x_2 = 3$

In Problems 15–20, find dy for each function.

15. $y = 30 + 12x^2 - x^3$

16. $y = 200x - \dfrac{x^2}{30}$

17. $y = x^2\left(1 - \dfrac{x}{9}\right)$

18. $y = x^3(60 - x)$

19. $y = \dfrac{590}{\sqrt{x}}$

20. $y = 52\sqrt{x}$

In Problems 21 and 22, find the indicated quantities for $y = f(x) = 3x^2$.

21. (A) $\dfrac{f(2 + \Delta x) - f(2)}{\Delta x}$ (simplify)

(B) What does the quantity in part (A) approach as Δx approaches 0?

22. (A) $\dfrac{f(3 + \Delta x) - f(3)}{\Delta x}$ (simplify)

(B) What does the quantity in part (A) approach as Δx approaches 0?

In Problems 23–26, find dy for each function.

23. $y = (2x + 1)^2$

24. $y = (3x + 5)^2$

25. $y = \dfrac{x^2 + 9}{x}$

26. $y = \dfrac{(x - 1)^2}{x^2}$

In Problems 27–30, evaluate dy and Δy for each function for the indicated values.

27. $y = f(x) = x^2 - 3x + 2$; $x = 5$, $dx = \Delta x = 0.2$

28. $y = f(x) = 30 + 12x^2 - x^3$; $x = 2$, $dx = \Delta x = 0.1$

29. $y = f(x) = 75\left(1 - \dfrac{2}{x}\right)$; $x = 5$, $dx = \Delta x = -0.5$

30. $y = f(x) = 100\left(x - \dfrac{4}{x^2}\right)$; $x = 2$, $dx = \Delta x = -0.1$

31. A cube with 10-inch sides is covered with a coat of fiberglass 0.2 inch thick. Use differentials to estimate the volume of the fiberglass shell.

32. A sphere with a radius of 5 centimeters is coated with ice 0.1 centimeter thick. Use differentials to estimate the volume of the ice. $\left[\text{Recall that } V = \frac{4}{3}\pi r^3.\right]$

In Problems 33–36,

(A) Find Δy and dy for the function f at the indicated value of x.

(B) Graph Δy and dy from part (A) as functions of Δx.

(C) Compare the values of Δy and dy from part (A) at the indicated values of Δx.

33. $f(x) = x^2 + 2x + 3$; $x = -0.5$, $\Delta x = dx = 0.1, 0.2, 0.3$

34. $f(x) = x^2 + 2x + 3$; $x = -2$, $\Delta x = dx = -0.1, -0.2, -0.3$

35. $f(x) = x^3 - 2x^2$; $x = 1$, $\Delta x = dx = 0.05, 0.10, 0.15$

36. $f(x) = x^3 - 2x^2$; $x = 2$, $\Delta x = dx = -0.05, -0.10, -0.15$

In Problems 37–40, discuss the validity of each statement. If the statement is always true, explain why. If not, give a counterexample.

37. If the graph of the function $y = f(x)$ is a line, then the functions Δy and dy (of the independent variable $\Delta x = dx$) for $f(x)$ at $x = 3$ are identical.

38. If the graph of the function $y = f(x)$ is a parabola, then the functions Δy and dy (of the independent variable $\Delta x = dx$) for $f(x)$ at $x = 0$ are identical.

39. Suppose that $y = f(x)$ defines a differentiable function whose domain is the set of all real numbers. If every differential dy at $x = 2$ is equal to 0, then $f(x)$ is a constant function.

40. Suppose that $y = f(x)$ defines a function whose domain is the set of all real numbers. If every increment at $x = 2$ is equal to 0, then $f(x)$ is a constant function.

41. Find dy if $y = (1 - 2x)\sqrt[3]{x^2}$.

42. Find dy if $y = (2x^2 - 4)\sqrt{x}$.

43. Find dy and Δy for $y = 52\sqrt{x}$, $x = 4$, and $\Delta x = dx = 0.3$.

44. Find dy and Δy for $y = 590/\sqrt{x}$, $x = 64$, and $\Delta x = dx = 1$.

Applications

Use differential approximations in the following problems.

45. Advertising. A company will sell N units of a product after spending $\$x$ thousand in advertising, as given by

$$N = 60x - x^2 \qquad 5 \le x \le 30$$

Approximately what increase in sales will result by increasing the advertising budget from $\$10,000$ to $\$11,000$? From $\$20,000$ to $\$21,000$?

46. Price–demand. Suppose that the daily demand (in pounds) for chocolate candy at $\$x$ per pound is given by

$$D = 1,000 - 40x^2 \qquad 1 \le x \le 5$$

If the price is increased from $\$3.00$ per pound to $\$3.20$ per pound, what is the approximate change in demand?

47. Average cost. For a company that manufactures tennis rackets, the average cost per racket $\overline{C}$ is

$$\overline{C} = \frac{400}{x} + 5 + \frac{1}{2}x \qquad x \ge 1$$

where x is the number of rackets produced per hour. What will the approximate change in average cost per racket be if production is increased from 20 per hour to 25 per hour? From 40 per hour to 45 per hour?

48. Revenue and profit. A company manufactures and sells x televisions per month. If the cost and revenue equations are

$$C(x) = 72,000 + 60x$$

$$R(x) = 200x - \frac{x^2}{30} \qquad 0 \le x \le 6,000$$

what will the approximate changes in revenue and profit be if production is increased from 1,500 to 1,510? From 4,500 to 4,510?

49. Pulse rate. The average pulse rate y (in beats per minute) of a healthy person x inches tall is given approximately by

$$y = \frac{590}{\sqrt{x}} \qquad 30 \le x \le 75$$

Approximately how will the pulse rate change for a change in height from 36 to 37 inches? From 64 to 65 inches?

50. Measurement. An egg of a particular bird is nearly spherical. If the radius to the inside of the shell is 5 millimeters and the radius to the outside of the shell is 5.3 millimeters, approximately what is the volume of the shell? [Remember that $V = \frac{4}{3}\pi r^3$.]

51. Medicine. A drug is given to a patient to dilate her arteries. If the radius of an artery is increased from 2 to 2.1 millimeters, approximately how much is the cross-sectional area increased? [Assume that the cross section of the artery is circular; that is, $A = \pi r^2$.]

52. Drug sensitivity. One hour after x milligrams of a particular drug are given to a person, the change in body temperature T (in degrees Fahrenheit) is given by

$$T = x^2\left(1 - \frac{x}{9}\right) \qquad 0 \le x \le 6$$

Approximate the changes in body temperature produced by the following changes in drug dosages:

(A) From 2 to 2.1 milligrams

(B) From 3 to 3.1 milligrams

(C) From 4 to 4.1 milligrams

53. Learning. A particular person learning to type has an achievement record given approximately by

$$N = 75\left(1 - \frac{2}{t}\right) \qquad 3 \le t \le 20$$

where N is the number of words per minute typed after t weeks of practice. What is the approximate improvement from 5 to 5.5 weeks of practice?

54. Learning. If a person learns y items in x hours, as given approximately by

$$y = 52\sqrt{x} \qquad 0 \leq x \leq 9$$

what is the approximate increase in the number of items learned when x changes from 1 to 1.1 hours? From 4 to 4.1 hours?

55. Politics. In a new city, the voting population (in thousands) is given by

$$N(t) = 30 + 12t^2 - t^3 \qquad 0 \leq t \leq 8$$

where t is time in years. Find the approximate change in votes for the following changes in time:

(A) From 1 to 1.1 years

(B) From 4 to 4.1 years

(C) From 7 to 7.1 years

Answers to Matched Problems

1. (A) $\Delta x = 1$, $\Delta y = 5$, $\Delta y/\Delta x = 5$ (B) 4

2. $dy = \dfrac{1}{2\sqrt{x}}\,dx$

  (A) 0.025   (B) 0.02   (C) 0.005

3. (A) $\Delta y = -2\Delta x - \Delta x^2$; $dy = -2\Delta x$

  (B)

Δx	Δy	dy
-0.1	0.19	0.2
-0.2	0.36	0.4
-0.3	0.51	0.6

4. $dR = -\$20/\text{wk}$; $dP = -\$40/\text{wk}$

2.7 Marginal Analysis in Business and Economics

- Marginal Cost, Revenue, and Profit
- Application
- Marginal Average Cost, Revenue, and Profit

Marginal Cost, Revenue, and Profit

One important application of calculus to business and economics involves *marginal analysis*. In economics, the word *marginal* refers to a rate of change—that is, to a derivative. Thus, if $C(x)$ is the total cost of producing x items, then $C'(x)$ is called the *marginal cost* and represents the instantaneous rate of change of total cost with respect to the number of items produced. Similarly, the *marginal revenue* is the derivative of the total revenue function, and the *marginal profit* is the derivative of the total profit function.

> **DEFINITION** Marginal Cost, Revenue, and Profit
>
> If x is the number of units of a product produced in some time interval, then
>
> $$\text{total cost} = C(x)$$
> $$\textbf{marginal cost} = C'(x)$$
> $$\text{total revenue} = R(x)$$
> $$\textbf{marginal revenue} = R'(x)$$
> $$\text{total profit} = P(x) = R(x) - C(x)$$
> $$\textbf{marginal profit} = P'(x) = R'(x) - C'(x)$$
> $$= (\text{marginal revenue}) - (\text{marginal cost})$$
>
> Marginal cost (or revenue or profit) is the instantaneous rate of change of cost (or revenue or profit) relative to production at a given production level.

To begin our discussion, we consider a cost function $C(x)$. It is important to remember that $C(x)$ represents the *total* cost of producing x items, not the cost of producing a *single* item. To find the cost of producing a single item, we use the difference of two successive values of $C(x)$:

$$\text{Total cost of producing } x + 1 \text{ items} = C(x + 1)$$
$$\text{Total cost of producing } x \text{ items} = C(x)$$
$$\text{Exact cost of producing the } (x + 1)\text{st item} = C(x + 1) - C(x)$$

EXAMPLE 1   Cost Analysis A company manufactures fuel tanks for cars. The total weekly cost (in dollars) of producing x tanks is given by

$$C(x) = 10,000 + 90x - 0.05x^2$$

(A) Find the marginal cost function.

(B) Find the marginal cost at a production level of 500 tanks per week.

(C) Interpret the results of part (B).

(D) Find the exact cost of producing the 501st item.

SOLUTION

(A) $C'(x) = 90 - 0.1x$

(B) $C'(500) = 90 - 0.1(500) = \$40$ Marginal cost

(C) At a production level of 500 tanks per week, the total production costs are increasing at the rate of $40 per tank.

(D)
$$C(501) = 10,000 + 90(501) - 0.05(501)^2$$
$$= \$42,539.95 \quad \text{Total cost of producing 501 tanks per week}$$
$$C(500) = 10,000 + 90(500) - 0.05(500)^2$$
$$= \$42,500.00 \quad \text{Total cost of producing 500 tanks per week}$$
$$C(501) - C(500) = 42,539.95 - 42,500.00$$
$$= \$39.95 \quad \text{Exact cost of producing the 501st tank}$$

Matched Problem 1 A company manufactures automatic transmissions for cars. The total weekly cost (in dollars) of producing x transmissions is given by

$$C(x) = 50,000 + 600x - 0.75x^2$$

(A) Find the marginal cost function.

(B) Find the marginal cost at a production level of 200 transmissions per week.

(C) Interpret the results of part (B).

(D) Find the exact cost of producing the 201st transmission.

In Example 1, we found that the cost of the 501st tank and the marginal cost at a production level of 500 tanks differ by only a nickel. Increments and differentials will help us understand the relationship between marginal cost and the cost of a single item. If $C(x)$ is any total cost function, then

$$C'(x) \approx \frac{C(x + \Delta x) - C(x)}{\Delta x} \quad \text{See Section 2.6}$$

$$C'(x) \approx C(x + 1) - C(x) \qquad \Delta x = 1$$

We see that the marginal cost $C'(x)$ approximates $C(x + 1) - C(x)$, the exact cost of producing the $(x + 1)$st item. These observations are summarized next and are illustrated in Figure 1.

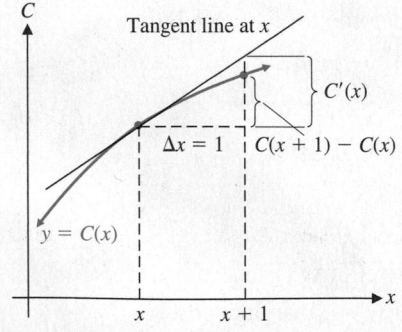

Figure 1 $C'(x) \approx C(x + 1) - C(x)$

THEOREM 1 Marginal Cost and Exact Cost

If $C(x)$ is the total cost of producing x items, then the marginal cost function approximates the exact cost of producing the $(x + 1)$st item:

Marginal cost   Exact cost

$$C'(x) \approx C(x + 1) - C(x)$$

Similar statements can be made for total revenue functions and total profit functions.

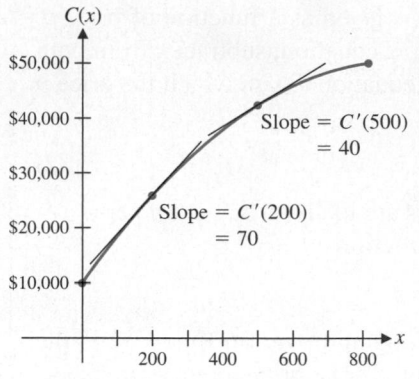

Figure 2
$C(x) = 10,000 + 90x - 0.05x^2$

Theorem 1 states that the marginal cost at a given production level x approximates the cost of producing the $(x + 1)$st, or *next,* item. In practice, the marginal cost is used more frequently than the exact cost. One reason for this is that the marginal cost is easily visualized when one is examining the graph of the total cost function. Figure 2 shows the graph of the cost function discussed in Example 1, with tangent lines added at $x = 200$ and $x = 500$. The graph clearly shows that as production increases, the slope of the tangent line decreases. Thus, the cost of producing the next tank also decreases, a desirable characteristic of a total cost function. We will have much more to say about graphical analysis in Chapter 4.

EXAMPLE 2 Exact Cost and Marginal Cost The total cost of producing x bicycles is given by the cost function

$$C(x) = 10,000 + 150x - 0.2x^2$$

(A) Find the exact cost of producing the 121st bicycle.

(B) Use marginal cost to approximate the cost of producing the 121st bicycle.

SOLUTION

(A) The cost of producing 121 bicycles is

$$C(121) = 10,000 + 150(121) - 0.2(121)^2 = \$25,221.80$$

and the cost of producing 120 bicycles is

$$C(120) = 10,000 + 150(120) - 0.2(120)^2 = \$25,120.00$$

So the exact cost of producing the 121st bicycle is

$$C(121) - C(120) = \$25,221.80 - 25,120.00 = \$101.80$$

(B) By Theorem 1, the marginal cost function $C'(x)$, evaluated at $x = 120$, approximates the cost of producing the 121st bicycle:

$$C'(x) = 150 - 0.4x$$
$$C'(120) = 150 - 0.4(120) = \$102.00$$

Note that the marginal cost, $102.00, at a production level of 120 bicycles, is a good approximation to the exact cost, $101.80, of producing the 121st bicycle.

Matched Problem 2 For the cost function $C(x)$ in Example 2

(A) Find the exact cost of producing the 141st bicycle.

(B) Use marginal cost to approximate the cost of producing the 141st bicycle.

Application

Now we discuss how price, demand, revenue, cost, and profit are tied together in typical applications. Although either price or demand can be used as the independent variable in a price–demand equation, it is common to use demand as the independent variable when marginal revenue, cost, and profit are also involved.

EXAMPLE 3 Production Strategy A company's market research department recommends the manufacture and marketing of a new headphone set for MP3 players. After suitable test marketing, the research department presents the following **price–demand equation:**

$$x = 10,000 - 1,000p \quad \text{x is demand at price p.} \qquad (1)$$

In the price–demand equation (1), the demand x is given as a function of price p. By solving (1) for p (add $1{,}000p$ to both sides of the equation, subtract x from both sides, and divide both sides by 1,000), we obtain equation (2), in which the price p is given as a function of demand x:

$$p = 10 - 0.001x \qquad (2)$$

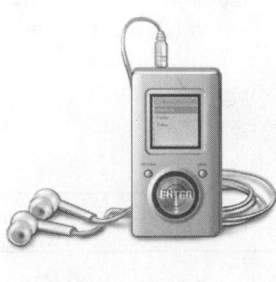

where x is the number of headphones that retailers are likely to buy at \$$p$ per set.

The financial department provides the **cost function**

$$C(x) = 7{,}000 + 2x \qquad (3)$$

where \$7,000 is the estimate of fixed costs (tooling and overhead) and \$2 is the estimate of variable costs per headphone set (materials, labor, marketing, transportation, storage, etc.).

(A) Find the domain of the function defined by the price–demand equation (2).

(B) Find and interpret the marginal cost function $C'(x)$.

(C) Find the revenue function as a function of x and find its domain.

(D) Find the marginal revenue at $x = 2{,}000$, 5,000, and 7,000. Interpret these results.

(E) Graph the cost function and the revenue function in the same coordinate system. Find the intersection points of these two graphs and interpret the results.

(F) Find the profit function and its domain and sketch the graph of the function.

(G) Find the marginal profit at $x = 1{,}000$, 4,000, and 6,000. Interpret these results.

SOLUTION

(A) Since price p and demand x must be nonnegative, we have $x \geq 0$ and

$$p = 10 - 0.001x \geq 0$$
$$10 \geq 0.001x$$
$$10{,}000 \geq x$$

Thus, the permissible values of x are $0 \leq x \leq 10{,}000$.

(B) The marginal cost is $C'(x) = 2$. Since this is a constant, it costs an additional \$2 to produce one more headphone set at any production level.

(C) The **revenue** is the amount of money R received by the company for manufacturing and selling x headphone sets at \$$p$ per set and is given by

$$R = (\text{number of headphone sets sold})(\text{price per headphone set}) = xp$$

In general, the revenue R can be expressed as a function of p using equation (1) or as a function of x using equation (2). As we mentioned earlier, when using marginal functions, we will always use the number of items x as the independent variable. Thus, the **revenue function** is

$$R(x) = xp = x(10 - 0.001x) \quad \text{Using equation (2)} \qquad (4)$$
$$= 10x - 0.001x^2$$

Since equation (2) is defined only for $0 \leq x \leq 10{,}000$, it follows that the domain of the revenue function is $0 \leq x \leq 10{,}000$.

(D) The **marginal revenue** is

$$R'(x) = 10 - 0.002x$$

For production levels of $x = 2{,}000$, 5,000, and 7,000, we have

$$R'(2{,}000) = 6 \qquad R'(5{,}000) = 0 \qquad R'(7{,}000) = -4$$

This means that at production levels of 2,000, 5,000, and 7,000, the respective approximate changes in revenue per unit change in production are \$6, \$0, and $-\$4$. That is, at the 2,000 output level, revenue increases as production increases; at the 5,000 output level, revenue does not change with a "small" change in production; and at the 7,000 output level, revenue decreases with an increase in production.

(E) Graphing $R(x)$ and $C(x)$ in the same coordinate system results in Figure 3 on page 168. The intersection points are called the **break-even points**, because revenue equals cost at these production levels. The company neither makes nor loses money, but just breaks even. The break-even points are obtained as follows:

$$C(x) = R(x)$$
$$7,000 + 2x = 10x - 0.001x^2$$
$$0.001x^2 - 8x + 7,000 = 0 \quad \text{Solve by the quadratic formula}$$
$$x^2 - 8,000x + 7,000,000 = 0 \quad \text{(see Appendix A.7).}$$
$$x = \frac{8,000 \pm \sqrt{8,000^2 - 4(7,000,000)}}{2}$$
$$= \frac{8,000 \pm \sqrt{36,000,000}}{2}$$
$$= \frac{8,000 \pm 6,000}{2}$$
$$= 1,000, \quad 7,000$$
$$R(1,000) = 10(1,000) - 0.001(1,000)^2 = 9,000$$
$$C(1,000) = 7,000 + 2(1,000) = 9,000$$
$$R(7,000) = 10(7,000) - 0.001(7,000)^2 = 21,000$$
$$C(7,000) = 7,000 + 2(7,000) = 21,000$$

The break-even points are $(1,000, 9,000)$ and $(7,000, 21,000)$, as shown in Figure 3. Further examination of the figure shows that cost is greater than revenue for production levels between 0 and 1,000 and also between 7,000 and 10,000. Consequently, the company incurs a loss at these levels. By contrast, for production levels between 1,000 and 7,000, revenue is greater than cost, and the company makes a profit.

(F) The **profit function** is

$$P(x) = R(x) - C(x)$$
$$= (10x - 0.001x^2) - (7,000 + 2x)$$
$$= -0.001x^2 + 8x - 7,000$$

The domain of the cost function is $x \geq 0$, and the domain of the revenue function is $0 \leq x \leq 10,000$. The domain of the profit function is the set of x values for which both functions are defined—that is, $0 \leq x \leq 10,000$. The graph of the profit function is shown in Figure 4 on page 168. Notice that the x coordinates of the break-even points in Figure 3 are the x intercepts of the profit function. Furthermore, the intervals on which cost is greater than revenue and on which revenue is greater than cost correspond, respectively, to the intervals on which profit is negative and on which profit is positive.

(G) The **marginal profit** is

$$P'(x) = -0.002x + 8$$

For production levels of 1,000, 4,000, and 6,000, we have

$$P'(1,000) = 6 \qquad P'(4,000) = 0 \qquad P'(6,000) = -4$$

This means that at production levels of 1,000, 4,000, and 6,000, the respective approximate changes in profit per unit change in production are $6, $0, and -$4. That is, at the 1,000 output level, profit will be increased if production is increased; at the 4,000 output level, profit does not change for "small" changes in production;

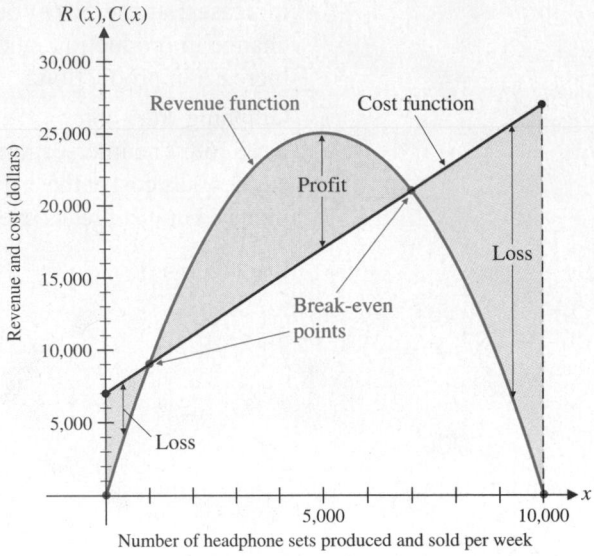

Figure 3

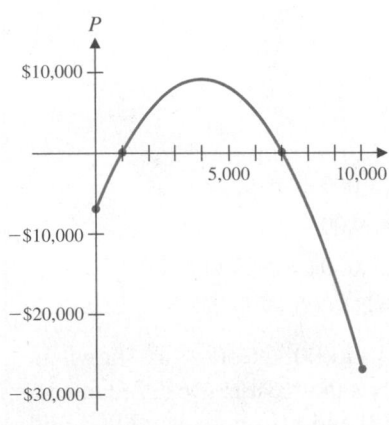

Figure 4

and at the 6,000 output level, profits will decrease if production is increased. It seems that the best production level to produce a maximum profit is 4,000.

Example 3 requires careful study since a number of important ideas in economics and calculus are involved. In the next chapter, we will develop a systematic procedure for finding the production level (and, using the demand equation, the selling price) that will maximize profit.

Matched Problem 3 Refer to the revenue and profit functions in Example 3.

(A) Find $R'(3,000)$ and $R'(6,000)$. Interpret the results.

(B) Find $P'(2,000)$ and $P'(7,000)$. Interpret the results.

Marginal Average Cost, Revenue, and Profit

Sometimes it is desirable to carry out marginal analysis relative to **average cost (cost per unit), average revenue (revenue per unit),** and **average profit (profit per unit).**

DEFINITION Marginal Average Cost, Revenue, and Profit

If x is the number of units of a product produced in some time interval, then

Cost per unit: average cost = $\overline{C}(x) = \dfrac{C(x)}{x}$

marginal average cost = $\overline{C}'(x) = \dfrac{d}{dx}\overline{C}(x)$

Revenue per unit: average revenue = $\overline{R}(x) = \dfrac{R(x)}{x}$

marginal average revenue = $\overline{R}'(x) = \dfrac{d}{dx}\overline{R}(x)$

Profit per unit: average profit = $\overline{P}(x) = \dfrac{P(x)}{x}$

marginal average profit = $\overline{P}'(x) = \dfrac{d}{dx}\overline{P}(x)$

EXAMPLE 4 Cost Analysis A small machine shop manufactures drill bits used in the petroleum industry. The manager estimates that the total daily cost (in dollars) of producing x bits is

$$C(x) = 1,000 + 25x - 0.1x^2$$

(A) Find $\overline{C}(x)$ and $\overline{C}'(x)$.

(B) Find $\overline{C}(10)$ and $\overline{C}'(10)$. Interpret these quantities.

(C) Use the results in part (B) to estimate the average cost per bit at a production level of 11 bits per day.

SOLUTION

(A) $\overline{C}(x) = \dfrac{C(x)}{x} = \dfrac{1,000 + 25x - 0.1x^2}{x}$

$\qquad = \dfrac{1,000}{x} + 25 - 0.1x \qquad\qquad$ Average cost function

$\qquad \overline{C}'(x) = \dfrac{d}{dx}\overline{C}(x) = -\dfrac{1,000}{x^2} - 0.1 \qquad$ Marginal average cost function

(B) $\overline{C}(10) = \dfrac{1,000}{10} + 25 - 0.1(10) = \$124$

$\qquad \overline{C}'(10) = -\dfrac{1,000}{10^2} - 0.1 = -\$10.10$

At a production level of 10 bits per day, the average cost of producing a bit is $124. This cost is decreasing at the rate of $10.10 per bit.

(C) If production is increased by 1 bit, then the average cost per bit will decrease by approximately $10.10. So, the average cost per bit at a production level of 11 bits per day is approximately $124 − $10.10 = $113.90.

Matched Problem 4 Consider the cost function for the production of headphone sets from Example 3: $\qquad C(x) = 7,000 + 2x$

(A) Find $\overline{C}(x)$ and $\overline{C}'(x)$.

(B) Find $\overline{C}(100)$ and $\overline{C}'(100)$. Interpret these quantities.

(C) Use the results in part (B) to estimate the average cost per headphone set at a production level of 101 headphone sets.

Explore and Discuss 1 A student produced the following solution to Matched Problem 4:

$\qquad C(x) = 7,000 + 2x \qquad$ Cost

$\qquad C'(x) = 2 \qquad\qquad\qquad$ Marginal cost

$\qquad \dfrac{C'(x)}{x} = \dfrac{2}{x} \qquad\qquad$ "Average" of the marginal cost

Explain why the last function is not the same as the marginal average cost function.

⚠ **CAUTION**

1. The marginal average cost function is computed by first finding the average cost function and then finding its derivative. As Explore and Discuss 1 illustrates, reversing the order of these two steps produces a different function that does not have any useful economic interpretations.

2. Recall that the marginal cost function has two interpretations: the usual interpretation of any derivative as an instantaneous rate of change and the special interpretation as an approximation to the exact cost of the $(x + 1)$st item. This special interpretation does not apply to the marginal average cost function. Referring to Example 4, we would be incorrect to interpret $\overline{C}'(10) = -\$10.10$ to mean that the average cost of the next bit is approximately $-\$10.10$. In fact, the phrase "average cost of the next bit" does not even make sense. Averaging is a concept applied to a collection of items, not to a single item.

These remarks also apply to revenue and profit functions.   ▲

Exercises 2.7

Skills Warm-up Exercises

W   *In Problems 1–8, let $C(x) = 10,000 + 150x - 0.2x^2$ be the total cost in dollars of producing x bicycles. (If necessary, review Section 1–1).*

1. Find the total cost of producing 99 bicycles.

2. Find the total cost of producing 100 bicycles.

3. Find the cost of producing the 100th bicycle.

4. Find the total cost of producing 199 bicycles.

5. Find the total cost of producing 200 bicycles.

6. Find the cost of producing the 200th bicycle.

7. Find the average cost per bicycle of producing 100 bicycles.

8. Find the average cost per bicycle of producing 200 bicycles.

In Problems 9–12, find the marginal cost function.

9. $C(x) = 175 + 0.8x$ 10. $C(x) = 4,500 + 9.5x$

11. $C(x) = 210 + 4.6x - 0.01x^2$

12. $C(x) = 790 + 13x - 0.2x^2$

In Problems 13–16, find the marginal revenue function.

13. $R(x) = 4x - 0.01x^2$ 14. $R(x) = 36x - 0.03x^2$

15. $R(x) = x(12 - 0.04x)$ 16. $R(x) = x(25 - 0.05x)$

In Problems 17–20, find the marginal profit function if the cost and revenue, respectively, are those in the indicated problems.

17. Problem 9 and Problem 13

18. Problem 10 and Problem 14

19. Problem 11 and Problem 15

20. Problem 12 and Problem 16

In Problems 21–28, find the indicated function if cost and revenue are given by $C(x) = 145 + 1.1x$ and $R(x) = 5x - 0.02x^2$, respectively.

21. Average cost function

22. Average revenue function

23. Marginal average cost function

24. Marginal average revenue function

25. Profit function

26. Marginal profit function

27. Average profit function

28. Marginal average profit function

In Problems 29–32, discuss the validity of each statement. If the statement is always true, explain why. If not, give a counterexample.

29. If a cost function is linear, then the marginal cost is a constant.

30. If a price–demand equation is linear, then the marginal revenue function is linear.

31. Marginal profit is equal to marginal cost minus marginal revenue.

32. Marginal average cost is equal to average marginal cost.

Applications

33. **Cost analysis.** The total cost (in dollars) of producing x food processors is

$$C(x) = 2,000 + 50x - 0.5x^2$$

(A) Find the exact cost of producing the 21st food processor.

(B) Use marginal cost to approximate the cost of producing the 21st food processor.

34. **Cost analysis.** The total cost (in dollars) of producing x electric guitars is

$$C(x) = 1,000 + 100x - 0.25x^2$$

(A) Find the exact cost of producing the 51st guitar.

(B) Use marginal cost to approximate the cost of producing the 51st guitar.

35. **Cost analysis.** The total cost (in dollars) of manufacturing x auto body frames is

$$C(x) = 60,000 + 300x$$

(A) Find the average cost per unit if 500 frames are produced.

(B) Find the marginal average cost at a production level of 500 units and interpret the results.

(C) Use the results from parts (A) and (B) to estimate the average cost per frame if 501 frames are produced.

36. **Cost analysis.** The total cost (in dollars) of printing x dictionaries is

$$C(x) = 20,000 + 10x$$

(A) Find the average cost per unit if 1,000 dictionaries are produced.

(B) Find the marginal average cost at a production level of 1,000 units and interpret the results.

(C) Use the results from parts (A) and (B) to estimate the average cost per dictionary if 1,001 dictionaries are produced.

37. **Profit analysis.** The total profit (in dollars) from the sale of x skateboards is

$$P(x) = 30x - 0.3x^2 - 250 \qquad 0 \le x \le 100$$

(A) Find the exact profit from the sale of the 26th skateboard. 14.70 $P(26) - P(25) = 14.70$

(B) Use marginal profit to approximate the profit from the sale of the 26th skateboard. 15 $30 - 0.6(25) = 15$

38. **Profit analysis.** The total profit (in dollars) from the sale of x calendars is

$$P(x) = 22x - 0.2x^2 - 400 \qquad 0 \le x \le 100$$

(A) Find the exact profit from the sale of the 41st calendar.

(B) Use the marginal profit to approximate the profit from the sale of the 41st calendar.

39. **Profit analysis.** The total profit (in dollars) from the sale of x DVDs is

$$P(x) = 5x - 0.005x^2 - 450 \qquad 0 \le x \le 1,000$$

Evaluate the marginal profit at the given values of x, and interpret the results.

(A) $x = 450$ (B) $x = 750$

40. **Profit analysis.** The total profit (in dollars) from the sale of x cameras is

$$P(x) = 12x - 0.02x^2 - 1,000 \qquad 0 \le x \le 600$$

Evaluate the marginal profit at the given values of x, and interpret the results.

(A) $x = 200$ (B) $x = 350$

41. **Profit analysis.** The total profit (in dollars) from the sale of x lawn mowers is

$$P(x) = 30x - 0.03x^2 - 750 \qquad 0 \le x \le 1,000$$

(A) Find the average profit per mower if 50 mowers are produced.

(B) Find the marginal average profit at a production level of 50 mowers and interpret the results.

(C) Use the results from parts (A) and (B) to estimate the average profit per mower if 51 mowers are produced.

42. **Profit analysis.** The total profit (in dollars) from the sale of x gas grills is

$$P(x) = 20x - 0.02x^2 - 320 \qquad 0 \le x \le 1,000$$

(A) Find the average profit per grill if 40 grills are produced.

(B) Find the marginal average profit at a production level of 40 grills and interpret the results.

(C) Use the results from parts (A) and (B) to estimate the average profit per grill if 41 grills are produced.

43. **Revenue analysis.** The price p (in dollars) and the demand x for a brand of running shoes are related by the equation

$$x = 4,000 - 40p$$

(A) Express the price p in terms of the demand x, and find the domain of this function.

(B) Find the revenue $R(x)$ from the sale of x pairs of running shoes. What is the domain of R?

(C) Find the marginal revenue at a production level of 1,600 pairs and interpret the results.

(D) Find the marginal revenue at a production level of 2,500 pairs, and interpret the results.

44. **Revenue analysis.** The price p (in dollars) and the demand x for a particular steam iron are related by the equation

$$x = 1,000 - 20p$$

(A) Express the price p in terms of the demand x, and find the domain of this function.

(B) Find the revenue $R(x)$ from the sale of x steam irons. What is the domain of R?

(C) Find the marginal revenue at a production level of 400 steam irons and interpret the results.

(D) Find the marginal revenue at a production level of 650 steam irons and interpret the results.

45. **Revenue, cost, and profit.** The price–demand equation and the cost function for the production of table saws are given, respectively, by

$$x = 6,000 - 30p \qquad \text{and} \qquad C(x) = 72,000 + 60x$$

where x is the number of saws that can be sold at a price of $\$p$ per saw and $C(x)$ is the total cost (in dollars) of producing x saws.

(A) Express the price p as a function of the demand x, and find the domain of this function.

(B) Find the marginal cost.

(C) Find the revenue function and state its domain.

(D) Find the marginal revenue.

✎ (E) Find $R'(1,500)$ and $R'(4,500)$ and interpret these quantities.

(F) Graph the cost function and the revenue function on the same coordinate system for $0 \leq x \leq 6,000$. Find the break-even points, and indicate regions of loss and profit.

(G) Find the profit function in terms of x.

(H) Find the marginal profit.

✎ (I) Find $P'(1,500)$ and $P'(3,000)$ and interpret these quantities.

46. Revenue, cost, and profit. The price–demand equation and the cost function for the production of HDTVs are given, respectively, by

$$x = 9,000 - 30p \quad \text{and} \quad C(x) = 150,000 + 30x$$

where x is the number of HDTVs that can be sold at a price of $\$p$ per TV and $C(x)$ is the total cost (in dollars) of producing x TVs.

(A) Express the price p as a function of the demand x, and find the domain of this function.

(B) Find the marginal cost.

(C) Find the revenue function and state its domain.

(D) Find the marginal revenue.

✎ (E) Find $R'(3,000)$ and $R'(6,000)$ and interpret these quantities.

(F) Graph the cost function and the revenue function on the same coordinate system for $0 \leq x \leq 9,000$. Find the break-even points and indicate regions of loss and profit.

(G) Find the profit function in terms of x.

(H) Find the marginal profit.

✎ (I) Find $P'(1,500)$ and $P'(4,500)$ and interpret these quantities.

47. Revenue, cost, and profit. A company is planning to manufacture and market a new two-slice electric toaster. After conducting extensive market surveys, the research department provides the following estimates: a weekly demand of 200 toasters at a price of $\$16$ per toaster and a weekly demand of 300 toasters at a price of $\$14$ per toaster. The financial department estimates that weekly fixed costs will be $\$1,400$ and variable costs (cost per unit) will be $\$4$.

(A) Assume that the relationship between price p and demand x is linear. Use the research department's estimates to express p as a function of x and find the domain of this function.

(B) Find the revenue function in terms of x and state its domain.

(C) Assume that the cost function is linear. Use the financial department's estimates to express the cost function in terms of x.

(D) Graph the cost function and revenue function on the same coordinate system for $0 \leq x \leq 1,000$. Find the break-even points and indicate regions of loss and profit.

(E) Find the profit function in terms of x.

✎ (F) Evaluate the marginal profit at $x = 250$ and $x = 475$ and interpret the results.

48. Revenue, cost, and profit. The company in Problem 47 is also planning to manufacture and market a four-slice toaster. For this toaster, the research department's estimates are a weekly demand of 300 toasters at a price of $\$25$ per toaster and a weekly demand of 400 toasters at a price of $\$20$. The financial department's estimates are fixed weekly costs of $\$5,000$ and variable costs of $\$5$ per toaster.

(A) Assume that the relationship between price p and demand x is linear. Use the research department's estimates to express p as a function of x, and find the domain of this function.

(B) Find the revenue function in terms of x and state its domain.

(C) Assume that the cost function is linear. Use the financial department's estimates to express the cost function in terms of x.

(D) Graph the cost function and revenue function on the same coordinate system for $0 \leq x \leq 800$. Find the break-even points and indicate regions of loss and profit.

(E) Find the profit function in terms of x.

✎ (F) Evaluate the marginal profit at $x = 325$ and $x = 425$ and interpret the results.

49. Revenue, cost, and profit. The total cost and the total revenue (in dollars) for the production and sale of x ski jackets are given, respectively, by

$$C(x) = 24x + 21,900 \quad \text{and} \quad R(x) = 200x - 0.2x^2$$
$$0 \leq x \leq 1,000$$

(A) Find the value of x where the graph of $R(x)$ has a horizontal tangent line.

(B) Find the profit function $P(x)$.

(C) Find the value of x where the graph of $P(x)$ has a horizontal tangent line.

(D) Graph $C(x)$, $R(x)$, and $P(x)$ on the same coordinate system for $0 \leq x \leq 1,000$. Find the break-even points. Find the x intercepts of the graph of $P(x)$.

50. Revenue, cost, and profit. The total cost and the total revenue (in dollars) for the production and sale of x hair dryers are given, respectively, by

$$C(x) = 5x + 2,340 \quad \text{and} \quad R(x) = 40x - 0.1x^2$$
$$0 \leq x \leq 400$$

(A) Find the value of x where the graph of $R(x)$ has a horizontal tangent line.

(B) Find the profit function $P(x)$.

(C) Find the value of x where the graph of $P(x)$ has a horizontal tangent line.

(D) Graph $C(x)$, $R(x)$, and $P(x)$ on the same coordinate system for $0 \le x \le 400$. Find the break-even points. Find the x intercepts of the graph of $P(x)$.

51. Break-even analysis. The price–demand equation and the cost function for the production of garden hoses are given, respectively, by

$$p = 20 - \sqrt{x} \quad \text{and} \quad C(x) = 500 + 2x$$

where x is the number of garden hoses that can be sold at a price of \$$p$ per unit and $C(x)$ is the total cost (in dollars) of producing x garden hoses.

(A) Express the revenue function in terms of x.

(B) Graph the cost function and revenue function in the same viewing window for $0 \le x \le 400$. Use approximation techniques to find the break-even points correct to the nearest unit.

52. Break-even analysis. The price–demand equation and the cost function for the production of handwoven silk scarves are given, respectively, by

$$p = 60 - 2\sqrt{x} \quad \text{and} \quad C(x) = 3{,}000 + 5x$$

where x is the number of scarves that can be sold at a price of \$$p$ per unit and $C(x)$ is the total cost (in dollars) of producing x scarves.

(A) Express the revenue function in terms of x.

(B) Graph the cost function and the revenue function in the same viewing window for $0 \le x \le 900$. Use approximation techniques to find the break-even points correct to the nearest unit.

53. Break-even analysis. Table 1 contains price–demand and total cost data for the production of projectors, where p is the wholesale price (in dollars) of a projector for an annual demand of x projectors and C is the total cost (in dollars) of producing x projectors.

Table 1

x	$p(\$)$	$C(\$)$
3,190	581	1,130,000
4,570	405	1,241,000
5,740	181	1,410,000
7,330	85	1,620,000

(A) Find a quadratic regression equation for the price–demand data, using x as the independent variable.

(B) Find a linear regression equation for the cost data, using x as the independent variable. Use this equation to estimate the fixed costs and variable costs per projector. Round answers to the nearest dollar.

(C) Find the break-even points. Round answers to the nearest integer.

(D) Find the price range for which the company will make a profit. Round answers to the nearest dollar.

54. Break-even analysis. Table 2 contains price–demand and total cost data for the production of treadmills, where p is the wholesale price (in dollars) of a treadmill for an annual demand of x treadmills and C is the total cost (in dollars) of producing x treadmills.

Table 2

x	$p(\$)$	$C(\$)$
2,910	1,435	3,650,000
3,415	1,280	3,870,000
4,645	1,125	4,190,000
5,330	910	4,380,000

(A) Find a linear regression equation for the price–demand data, using x as the independent variable.

(B) Find a linear regression equation for the cost data, using x as the independent variable. Use this equation to estimate the fixed costs and variable costs per treadmill. Round answers to the nearest dollar.

(C) Find the break-even points. Round answers to the nearest integer.

(D) Find the price range for which the company will make a profit. Round answers to the nearest dollar.

Answers to Matched Problems

1. (A) $C'(x) = 600 - 1.5x$
(B) $C'(200) = 300$.
(C) At a production level of 200 transmissions, total costs are increasing at the rate of \$300 per transmission.
(D) $C(201) - C(200) = \$299.25$

2. (A) \$93.80   (B) \$94.00

3. (A) $R'(3{,}000) = 4$. At a production level of 3,000, a unit increase in production will increase revenue by approximately \$4.
$R'(6{,}000) = -2$. At a production level of 6,000, a unit increase in production will decrease revenue by approximately \$2.

(B) $P'(2{,}000) = 4$. At a production level of 2,000, a unit increase in production will increase profit by approximately \$4.
$P'(7{,}000) = -6$. At a production level of 7,000, a unit increase in production will decrease profit by approximately \$6.

4. (A) $\overline{C}(x) = \dfrac{7{,}000}{x} + 2;\ \overline{C}'(x) = -\dfrac{7{,}000}{x^2}$

(B) $\overline{C}(100) = \$72;\ \overline{C}'(100) = -\$0.70$. At a production level of 100 headphone sets, the average cost per headphone set is \$72. This average cost is decreasing at a rate of \$0.70 per headphone set.

(C) Approx. \$71.30.

Chapter 2   Summary and Review

Important Terms, Symbols, and Concepts

2.1 Introduction to Limits

- The graph of the function $y = f(x)$ is the graph of the set of all ordered pairs $(x, f(x))$.
- The limit of the function $y = f(x)$ as x approaches c is L, written as $\lim\limits_{x \to c} f(x) = L$, if the functional value $f(x)$ is close to the single real number L whenever x is close, but not equal, to c (on either side of c).
- The limit of the function $y = f(x)$ as x approaches c from the left is K, written as $\lim\limits_{x \to c^-} f(x) = K$, if $f(x)$ is close to K whenever x is close to, but to the left of, c on the real-number line.
- The limit of the function $y = f(x)$ as x approaches c from the right is L, written as $\lim\limits_{x \to c^+} f(x) = L$, if $f(x)$ is close to L whenever x is close to, but to the right of, c on the real-number line.
- The limit of the difference quotient $[f(a + h) - f(a)]/h$ is often a $0/0$ indeterminate form. Algebraic simplification is often required to evaluate this type of limit.

2.2 Infinite Limits and Limits at Infinity

- If $f(x)$ increases or decreases without bound as x approaches a from either side of a, then the line $x = a$ is a **vertical asymptote** of the graph of $y = f(x)$.
- If $f(x)$ gets close to L as x increases without bound or decreases without bound, then L is called the limit of f at ∞ or $-\infty$.
- The **end behavior** of a function is described by its limits at infinity.
- If $f(x)$ approaches L as $x \to \infty$ or as $x \to -\infty$, then the line $y = L$ is a **horizontal asymptote** of the graph of $y = f(x)$. Polynomial functions never have horizontal asymptotes. A rational function can have at most one.

2.3 Continuity

- Intuitively, the graph of a continuous function can be drawn without lifting a pen off the paper. By definition, a function f is **continuous at c** if

  1. $\lim\limits_{x \to c} f(x)$ exists,   **2.** $f(c)$ exists, and   **3.** $\lim\limits_{x \to c} f(x) = f(c)$
- Continuity properties are useful for determining where a function is continuous and where it is discontinuous.
- Continuity properties are also useful for solving inequalities.

2.4 The Derivative

- Given a function $y = f(x)$, the **average rate of change** is the ratio of the change in y to the change in x.
- The **instantaneous rate of change** is the limit of the average rate of change as the change in x approaches 0.
- The slope of the secant line through two points on the graph of a function $y = f(x)$ is the ratio of the change in y to the change in x. The **slope of the graph** at the point $(a, f(a))$ is the limit of the slope of the secant line through the points $(a, f(a))$ and $(a + h, f(a + h))$ as h approaches 0, provided the limit exists. In this case, the **tangent line** to the graph is the line through $(a, f(a))$ with slope equal to the limit.
- The **derivative of $y = f(x)$ at x**, denoted $f'(x)$, is the limit of the difference quotient $[f(x + h) - f(x)]/h$ as $h \to 0$ (if the limit exists).
- The four-step process is used to find derivatives.
- If the limit of the difference quotient does not exist at $x = a$, then f is **nondifferentiable at a** and $f'(a)$ does not exist.

2.5 Basic Differentiation Properties

- The derivative of a constant function is 0.   Ex. 1, p. 148
- For any real number n, the derivative of $f(x) = x^n$ is nx^{n-1}.   Ex. 2, p. 149
- If f is a differentiable function, then the derivative of $kf(x)$ is $kf'(x)$.   Ex. 3, p. 149
- The derivative of the sum or difference of two differentiable functions is the sum or   Ex. 4, p. 150
  difference of the derivatives of the functions.   Ex. 5, p. 151

2.6 Differentials

- Given the function $y = f(x)$, the change in x is also called the **increment of x** and is   Ex. 1, p. 157
  denoted as Δx. The corresponding change in y is called the **increment of y** and is given
  by $\Delta y = f(x + \Delta x) - f(x)$.   Ex. 2, p. 159
- If $y = f(x)$ is differentiable at x, then the **differential of x** is $dx = \Delta x$ and the **differ-**   Ex. 3, p. 160
  ential of $y = f(x)$ is $dy = f'(x)dx$, or $df = f'(x)dx$. In this context, x and dx are both
  independent variables.

2.7 Marginal Analysis in Business and Economics

- If $y = C(x)$ is the total cost of producing x items, then $y = C'(x)$ is the **marginal**   Ex. 1, p. 164
  cost and $C(x + 1) - C(x)$ is the exact cost of producing item $x + 1$. Furthermore,
  $C'(x) \approx C(x + 1) - C(x)$. Similar statements can be made regarding total revenue and   Ex. 2, p. 165
  total profit functions.   Ex. 3, p. 165
- If $y = C(x)$ is the total cost of producing x items, then the **average cost**, or cost per unit,
  is $\overline{C}(x) = \dfrac{C(x)}{x}$ and the **marginal average cost** is $\overline{C}'(x) = \dfrac{d}{dx}\overline{C}(x)$. Similar statements
  can be made regarding total revenue and total profit functions.

Review Exercises

Work through all the problems in this chapter review, and check your answers in the back of the book. Answers to all review problems are there, along with section numbers in italics to indicate where each type of problem is discussed. Where weaknesses show up, review appropriate sections of the text.

Many of the problems in this exercise set ask you to find a derivative. Most of the answers to these problems contain both an unsimplified form and a simplified form of the derivative. When checking your work, first check that you applied the rules correctly, and then check that you performed the algebraic simplification correctly.

1. Find the indicated quantities for $y = f(x) = 2x^2 + 5$:

(A) The change in y if x changes from 1 to 3

(B) The average rate of change of y with respect to x if x changes from 1 to 3

(C) The slope of the secant line through the points $(1, f(1))$ and $(3, f(3))$ on the graph of $y = f(x)$

(D) The instantaneous rate of change of y with respect to x at $x = 1$

(E) The slope of the line tangent to the graph of $y = f(x)$ at $x = 1$

(F) $f'(1)$

2. Use the four-step process to find $f'(x)$ for $f(x) = -3x + 2$.

3. If $\lim\limits_{x \to 1} f(x) = 2$ and $\lim\limits_{x \to 1} g(x) = 4$, find

(A) $\lim\limits_{x \to 1} (5f(x) + 3g(x))$ (B) $\lim\limits_{x \to 1} [f(x)g(x)]$

(C) $\lim\limits_{x \to 1} \dfrac{g(x)}{f(x)}$ (D) $\lim\limits_{x \to 1} [5 + 2x - 3g(x)]$

In Problems 4–10, use the graph of f to estimate the indicated limits and function values.

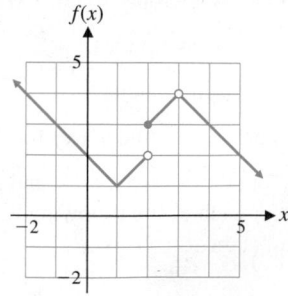

Figure for 4–10

4. $f(1.5)$ **5.** $f(2.5)$ **6.** $f(2.75)$ **7.** $f(3.25)$

8. (A) $\lim\limits_{x \to 1^-} f(x)$ (B) $\lim\limits_{x \to 1^+} f(x)$

   (C) $\lim\limits_{x \to 1} f(x)$ (D) $f(1)$

9. (A) $\lim\limits_{x \to 2^-} f(x)$ (B) $\lim\limits_{x \to 2^+} f(x)$

   (C) $\lim\limits_{x \to 2} f(x)$ (D) $f(2)$

10. (A) $\lim\limits_{x \to 3^-} f(x)$ (B) $\lim\limits_{x \to 3^+} f(x)$

   (C) $\lim\limits_{x \to 3} f(x)$ (D) $f(3)$

In Problems 11–13, use the graph of the function f shown in the figure to answer each question.

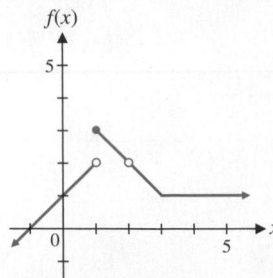

Figure for 11–13

11. (A) $\lim\limits_{x \to 1} f(x) = ?$ (B) $f(1) = ?$

    (C) Is f continuous at $x = 1$?

12. (A) $\lim\limits_{x \to 2} f(x) = ?$ (B) $f(2) = ?$

    (C) Is f continuous at $x = 2$?

13. (A) $\lim\limits_{x \to 3} f(x) = ?$ (B) $f(3) = ?$

    (C) Is f continuous at $x = 3$?

In Problems 14–23, refer to the following graph of $y = f(x)$:

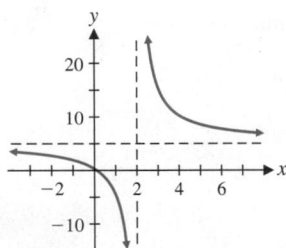

Figure for 14–23

14. $\lim\limits_{x \to \infty} f(x) = ?$ 15. $\lim\limits_{x \to -\infty} f(x) = ?$

16. $\lim\limits_{x \to 2^+} f(x) = ?$ 17. $\lim\limits_{x \to 2^-} f(x) = ?$

18. $\lim\limits_{x \to 0^-} f(x) = ?$ 19. $\lim\limits_{x \to 0^+} f(x) = ?$

20. $\lim\limits_{x \to 0} f(x) = ?$

21. Identify any vertical asymptotes.

22. Identify any horizontal asymptotes.

23. Where is $y = f(x)$ discontinuous?

24. Use the four-step process to find $f'(x)$ for $f(x) = 5x^2$.

25. If $f(5) = 4, f'(5) = -1, g(5) = 2$, and $g'(5) = -3$, then find $h'(5)$ for each of the following functions:

    (A) $h(x) = 3f(x)$

    (B) $h(x) = -2g(x)$

    (C) $h(x) = 2f(x) + 5$

    (D) $h(x) = -g(x) - 1$

    (E) $h(x) = 2f(x) + 3g(x)$

In Problems 26–31, find $f'(x)$ and simplify.

26. $f(x) = \dfrac{1}{3}x^3 - 5x^2 + 1$ 27. $f(x) = 2x^{1/2} - 3x$

28. $f(x) = 5$ 29. $f(x) = \dfrac{3}{2x} + \dfrac{5x^3}{4}$

30. $f(x) = \dfrac{0.5}{x^4} + 0.25x^4$

31. $f(x) = (3x^3 - 2)(x + 1)$ (*Hint:* Multiply and then differentiate.)

In Problems 32–35, find the indicated quantities for $y = f(x) = x^2 + x$.

32. $\Delta x, \Delta y$, and $\Delta y/\Delta x$ for $x_1 = 1$ and $x_2 = 3$.

33. $[f(x_1 + \Delta x) - f(x_1)]/\Delta x$ for $x_1 = 1$ and $\Delta x = 2$.

34. dy for $x_1 = 1$ and $x_2 = 3$.

35. Δy and dy for $x = 1, \Delta x = dx = 0.2$.

Problems 36–38 refer to the function.

$$f(x) = \begin{cases} x^2 & \text{if } 0 \le x < 2 \\ 8 - x & \text{if } x \ge 2 \end{cases}$$

which is graphed in the figure.

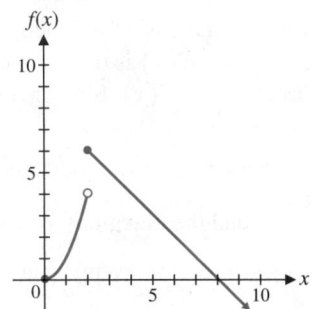

Figure for 36–38

36. (A) $\lim\limits_{x \to 2^-} f(x) = ?$ (B) $\lim\limits_{x \to 2^+} f(x) = ?$

    (C) $\lim\limits_{x \to 2} f(x) = ?$ (D) $f(2) = ?$

    (E) Is f continuous at $x = 2$?

37. (A) $\lim\limits_{x \to 5^-} f(x) = ?$ (B) $\lim\limits_{x \to 5^+} f(x) = ?$

    (C) $\lim\limits_{x \to 5} f(x) = ?$ (D) $f(5) = ?$

    (E) Is f continuous at $x = 5$?

38. Solve each inequality. Express answers in interval notation.

    (A) $f(x) < 0$ (B) $f(x) \ge 0$

In Problems 39–41, solve each inequality. Express the answer in interval notation. Use a graphing calculator in Problem 41 to approximate partition numbers to four decimal places.

39. $x^2 - x < 12$ 40. $\dfrac{x - 5}{x^2 + 3x} > 0$

41. $x^3 + x^2 - 4x - 2 > 0$

42. Let $f(x) = 0.5x^2 - 5$.

    (A) Find the slope of the secant line through $(2, f(2))$ and $(4, f(4))$.

    (B) Find the slope of the secant line through $(2, f(2))$ and $(2 + h, f(2 + h)), h \ne 0$.

    (C) Find the slope of the tangent line at $x = 2$.

In Problems 43–46, find the indicated derivative and simplify.

43. $\dfrac{dy}{dx}$ for $y = \dfrac{1}{3}x^{-3} - 5x^{-2} + 1$

44. y' for $y = \dfrac{3\sqrt{x}}{2} + \dfrac{5}{3\sqrt{x}}$

45. $g'(x)$ for $g(x) = 1.8\sqrt[3]{x} + \dfrac{0.9}{\sqrt[3]{x}}$

46. $\dfrac{dy}{dx}$ for $y = \dfrac{2x^3 - 3}{5x^3}$

47. For $y = f(x) = x^2 + 4$, find

   (A) The slope of the graph at $x = 1$

   (B) The equation of the tangent line at $x = 1$ in the form $y = mx + b$

In Problems 48 and 49, find the value(s) of x where the tangent line is horizontal.

48. $f(x) = 10x - x^2$

49. $f(x) = x^3 + 3x^2 - 45x - 135$

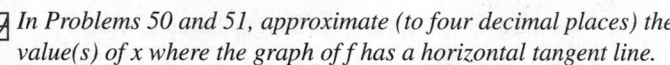 *In Problems 50 and 51, approximate (to four decimal places) the value(s) of x where the graph of f has a horizontal tangent line.*

50. $f(x) = x^4 - 2x^3 - 5x^2 + 7x$

51. $f(x) = x^5 - 10x^3 - 5x + 10$

52. If an object moves along the y axis (scale in feet) so that it is at $y = f(x) = 8x^2 - 4x + 1$ at time x (in seconds), find

   (A) The instantaneous velocity function

   (B) The velocity at time $x = 3$ seconds

53. An object moves along the y axis (scale in feet) so that at time x (in seconds) it is at $y = f(x) = -5x^2 + 16x + 3$. Find

   (A) The instantaneous velocity function

   (B) The time(s) when the velocity is 0

54. Let $f(x) = x^3$, $g(x) = (x - 4)^3$, and $h(x) = (x + 3)^3$.

   (A) How are the graphs of f, g, and h related? Illustrate your conclusion by graphing f, g, and h on the same coordinate axes.

   (B) How would you expect the graphs of the derivatives of these functions to be related? Illustrate your conclusion by graphing f', g', and h' on the same coordinate axes.

In Problems 55–59, determine where f is continuous. Express the answer in interval notation.

55. $f(x) = x^2 - 4$

56. $f(x) = \dfrac{x + 1}{x - 2}$

57. $f(x) = \dfrac{x + 4}{x^2 + 3x - 4}$

58. $f(x) = \sqrt[3]{4 - x^2}$

59. $f(x) = \sqrt{4 - x^2}$

In Problems 60–69, evaluate the indicated limits if they exist.

60. Let $f(x) = \dfrac{2x}{x^2 - 3x}$. Find

   (A) $\lim\limits_{x \to 1} f(x)$ (B) $\lim\limits_{x \to 3} f(x)$ (C) $\lim\limits_{x \to 0} f(x)$

61. Let $f(x) = \dfrac{x + 1}{(3 - x)^2}$. Find

   (A) $\lim\limits_{x \to 1} f(x)$ (B) $\lim\limits_{x \to -1} f(x)$ (C) $\lim\limits_{x \to 3} f(x)$

62. Let $f(x) = \dfrac{|x - 4|}{x - 4}$. Find

   (A) $\lim\limits_{x \to 4^-} f(x)$ (B) $\lim\limits_{x \to 4^+} f(x)$ (C) $\lim\limits_{x \to 4} f(x)$

63. Let $f(x) = \dfrac{x - 3}{9 - x^2}$. Find

   (A) $\lim\limits_{x \to 3} f(x)$ (B) $\lim\limits_{x \to -3} f(x)$ (C) $\lim\limits_{x \to 0} f(x)$

64. Let $f(x) = \dfrac{x^2 - x - 2}{x^2 - 7x + 10}$. Find

   (A) $\lim\limits_{x \to -1} f(x)$ (B) $\lim\limits_{x \to 2} f(x)$ (C) $\lim\limits_{x \to 5} f(x)$

65. Let $f(x) = \dfrac{2x}{3x - 6}$. Find

   (A) $\lim\limits_{x \to \infty} f(x)$ (B) $\lim\limits_{x \to -\infty} f(x)$ (C) $\lim\limits_{x \to 2} f(x)$

66. Let $f(x) = \dfrac{2x^3}{3(x - 2)^2}$. Find

   (A) $\lim\limits_{x \to \infty} f(x)$ (B) $\lim\limits_{x \to -\infty} f(x)$ (C) $\lim\limits_{x \to 2} f(x)$

67. Let $f(x) = \dfrac{2x}{3(x - 2)^3}$. Find

   (A) $\lim\limits_{x \to \infty} f(x)$ (B) $\lim\limits_{x \to -\infty} f(x)$ (C) $\lim\limits_{x \to 2} f(x)$

68. $\lim\limits_{h \to 0} \dfrac{f(2 + h) - f(2)}{h}$ for $f(x) = x^2 + 4$

69. $\lim\limits_{h \to 0} \dfrac{f(x + h) - f(x)}{h}$ for $f(x) = \dfrac{1}{x + 2}$

In Problems 70 and 71, use the definition of the derivative and the four-step process to find $f'(x)$.

70. $f(x) = x^2 - x$ **71.** $f(x) = \sqrt{x} - 3$

Problems 72–77 refer to the function f in the figure. Determine whether f is differentiable at the indicated value of x.

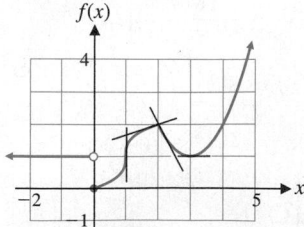

72. $x = -1$ **73.** $x = 0$ **74.** $x = 1$

75. $x = 2$ **76.** $x = 3$ **77.** $x = 4$

In Problems 78–82, find all horizontal and vertical asymptotes.

78. $f(x) = \dfrac{5x}{x - 7}$ **79.** $f(x) = \dfrac{-2x + 5}{(x - 4)^2}$

80. $f(x) = \dfrac{x^2 + 9}{x - 3}$ **81.** $f(x) = \dfrac{x^2 - 9}{x^2 + x - 2}$

82. $f(x) = \dfrac{x^3 - 1}{x^3 - x^2 - x + 1}$

✎ **83.** The domain of the power function $f(x) = x^{1/5}$ is the set of all real numbers. Find the domain of the derivative $f'(x)$. Discuss the nature of the graph of $y = f(x)$ for any x values excluded from the domain of $f'(x)$.

84. Let f be defined by

$$f(x) = \begin{cases} x^2 - m & \text{if } x \le 1 \\ -x^2 + m & \text{if } x > 1 \end{cases}$$

where m is a constant.

(A) Graph f for $m = 0$, and find

$$\lim_{x \to 1^-} f(x) \quad \text{and} \quad \lim_{x \to 1^+} f(x)$$

(B) Graph f for $m = 2$, and find

$$\lim_{x \to 1^-} f(x) \quad \text{and} \quad \lim_{x \to 1^+} f(x)$$

(C) Find m so that

$$\lim_{x \to 1^-} f(x) = \lim_{x \to 1^+} f(x)$$

and graph f for this value of m.

✎ (D) Write a brief verbal description of each graph. How does the graph in part (C) differ from the graphs in parts (A) and (B)?

85. Let $f(x) = 1 - |x - 1|, 0 \le x \le 2$ (see the figure).

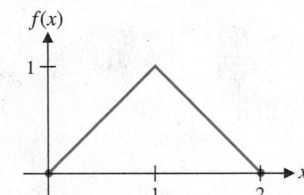

(A) $\displaystyle \lim_{h \to 0^-} \frac{f(1 + h) - f(1)}{h} = ?$

(B) $\displaystyle \lim_{h \to 0^+} \frac{f(1 + h) - f(1)}{h} = ?$

(C) $\displaystyle \lim_{h \to 0} \frac{f(1 + h) - f(1)}{h} = ?$

(D) Does $f'(1)$ exist?

Applications

86. Natural-gas rates. Table 1 shows the winter rates for natural gas charged by the Bay State Gas Company. The base charge is a fixed monthly charge, independent of the amount of gas used per month.

Table 1 **Natural Gas Rates**

Base charge	$7.47
First 90 therms	$0.4000 per therm
All usage over 90 therms	$0.2076 per therm

(A) Write a piecewise definition of the monthly charge $S(x)$ for a customer who uses x therms in a winter month.

(B) Graph $S(x)$.

(C) Is $S(x)$ continuous at $x = 90$? Explain.

87. Cost analysis. The total cost (in dollars) of producing x HDTVs is

$$C(x) = 10,000 + 200x - 0.1x^2$$

(A) Find the exact cost of producing the 101st TV.

(B) Use the marginal cost to approximate the cost of producing the 101st TV.

88. Cost analysis. The total cost (in dollars) of producing x bicycles is

$$C(x) = 5,000 + 40x + 0.05x^2$$

(A) Find the total cost and the marginal cost at a production level of 100 bicycles and interpret the results.

(B) Find the average cost and the marginal average cost at a production level of 100 bicycles and interpret the results.

89. Cost analysis. The total cost (in dollars) of producing x laser printers per week is shown in the figure. Which is greater, the approximate cost of producing the 201st printer or the approximate cost of producing the 601st printer? Does this graph represent a manufacturing process that is becoming more efficient or less efficient as production levels increase? Explain.

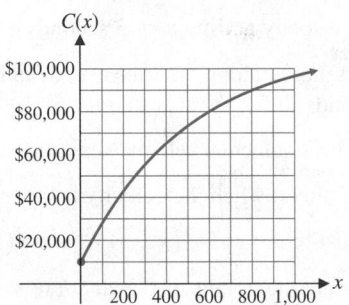

90. Cost analysis. Let

$$p = 25 - 0.01x \quad \text{and} \quad C(x) = 2x + 9,000$$
$$0 \le x \le 2,500$$

be the price–demand equation and cost function, respectively, for the manufacture of umbrellas.

(A) Find the marginal cost, average cost, and marginal average cost functions.

(B) Express the revenue in terms of x, and find the marginal revenue, average revenue, and marginal average revenue functions.

(C) Find the profit, marginal profit, average profit, and marginal average profit functions.

(D) Find the break-even point(s).

✎ (E) Evaluate the marginal profit at $x = 1,000, 1,150$, and $1,400$, and interpret the results.

(F) Graph $R = R(x)$ and $C = C(x)$ on the same coordinate system, and locate regions of profit and loss.

91. Employee training. A company producing computer components has established that, on average, a new employee can assemble $N(t)$ components per day afte r t days of on-the-job training, as given by

$$N(t) = \frac{40t - 80}{t}, t \geq 2$$

(A) Find the average rate of change of $N(t)$ from 2 days to 5 days.

(B) Find the instantaneous rate of change of $N(t)$ at 2 days.

 92. Sales analysis. The total number of swimming pools, N (in thousands), sold during a year is given by

$$N(t) = 2t + \frac{1}{3}t^{3/2}$$

where t is the number of months since the beginning of the year. Find $N(9)$ and $N'(9)$, and interpret these quantities.

 93. Natural-gas consumption. The data in Table 2 give the U.S. consumption of natural gas in trillions of cubic feet.

Table 2

Year	Natural-Gas Consumption
1960	12.0
1970	21.1
1980	19.9
1990	18.7
2000	21.9
2010	24.1

(A) Let x represent time (in years), with $x = 0$ corresponding to 1960, and let y represent the corresponding U.S. consumption of natural gas. Enter the data set in a graphing calculator and find a cubic regression equation for the data.

(B) If $y = N(x)$ denotes the regression equation found in part (A), find $N(60)$ and $N'(60)$, and write a brief verbal interpretation of these results.

94. Break-even analysis. Table 3 contains price–demand and total cost data from a bakery for the production of kringles (a Danish pastry), where p is the price (in dollars) of a kringle for a daily demand of x kringles and C is the total cost (in dollars) of producing x kringles.

Table 3

x	$p(\$)$	$C(\$)$
125	9	740
140	8	785
170	7	850
200	6	900

(A) Find a linear regression equation for the price–demand data, using x as the independent variable.

(B) Find a linear regression equation for the cost data, using x as the independent variable. Use this equation to estimate the fixed costs and variable costs per kringle.

(C) Find the break-even points.

(D) Find the price range for which the bakery will make a profit.

95. Pollution. A sewage treatment plant uses a pipeline that extends 1 mile toward the center of a large lake. The concentration of effluent $C(x)$ in parts per million, x meters from the end of the pipe is given approximately by

$$C(x) = \frac{500}{x^2}, x \geq 1$$

What is the instantaneous rate of change of concentration at 10 meters? At 100 meters?

96. Medicine. The body temperature (in degrees Fahrenheit) of a patient t hours after taking a fever-reducing drug is given by

$$F(t) = 0.16t^2 - 1.6t + 102$$

Find $F(4)$ and $F'(4)$. Write a brief verbal interpretation of these quantities.

97. Learning. If a person learns N items in t hours, as given by

$$N(t) = 20\sqrt{t}$$

find the rate of learning after

(A) 1 hour   (B) 4 hours

98. Physics: The coefficient of thermal expansion (CTE) is a measure of the expansion of an object subjected to extreme temperatures. We want to use a Michaelis–Menten function of the form

$$C(T) = \frac{C_{max}T}{M + T}$$

where $C = $ CTE, T is temperature in K (degrees Kelvin), and C_{max} and M are constants. Table 4 lists the coefficients of thermal expansion for titanium at various temperatures.

Table 4 **Coefficients of Thermal Expansion**

$T(K)$	Titanium
100	4.5
200	7.4
293	8.6
500	9.9
800	11.1
1100	11.7

(A) Plot the points in columns 1 and 2 of Table 4 on graph paper and estimate C_{max} to the nearest integer. To estimate M, add the horizontal line CTE $= \dfrac{C_{max}}{2}$ to your graph, connect successive points on the graph with straight-line segments, and estimate the value of T (to the nearest multiple of fifty) that satisfies

$$C(T) = \frac{C_{max}}{2}.$$

(B) Use the constants $\dfrac{C_{max}}{2}$ and M from part (A) to form a Michaelis–Menten function for the CTE of titanium.

(C) Use the function from part (B) to estimate the CTE of titanium at 600 K and to estimate the temperature when the CTE of titanium is 10.

3 Additional Derivative Topics

Introduction

In this chapter, we develop techniques for finding derivatives of a wide variety of functions, including exponential and logarithmic functions. There are straightforward procedures—the product rule, quotient rule, and chain rule—for writing down the derivative of any function that is the product, quotient, or composite of functions whose derivatives are known. With the ability to calculate derivatives easily, we consider a wealth of applications involving rates of change. For example, we apply the derivative to study population growth, radioactive decay, elasticity of demand, and environmental crises (see Problem 39 in Section 3.6 or Problem 87 in Section 3.7). Before starting this chapter, you may find it helpful to review the basic properties of exponential and logarithmic functions in Sections 1.5 and 1.6.

3.1 The Constant *e* and Continuous Compound Interest

- The Constant *e*
- Continuous Compound Interest

In Chapter 1, both the exponential function with base *e* and continuous compound interest were introduced informally. Now, with an understanding of limit concepts, we can give precise definitions of *e* and continuous compound interest.

The Constant *e*

The irrational number *e* is a particularly suitable base for both exponential and logarithmic functions. The reasons for choosing this number as a base will become clear as we develop differentiation formulas for the exponential function e^x and the natural logarithmic function ln *x*.

In precalculus treatments (Chapter 1), the number *e* is defined informally as the irrational number that can be approximated by the expression $[1 + (1/n)]^n$ for *n* sufficiently large. Now we will use the limit concept to formally define *e* as either of the following two limits. [*Note:* If $s = 1/n$, then as $n \to \infty$, $s \to 0$.]

> **DEFINITION The Number *e***
>
> $$e = \lim_{n \to \infty} \left(1 + \frac{1}{n}\right)^n \qquad \text{or, alternatively,} \qquad e = \lim_{s \to 0}(1 + s)^{1/s}$$

Both limits are equal to $e = 2.718\ 281\ 828\ 459\ldots$

Proof that the indicated limits exist and represent an irrational number between 2 and 3 is not easy and is omitted.

> **CONCEPTUAL INSIGHT**
>
> The two limits used to define *e* are unlike any we have encountered so far. Some people reason (incorrectly) that both limits are 1, since $1 + s \to 1$ as $s \to 0$ and 1 to any power is 1. An ordinary scientific calculator with a y^x key can convince you otherwise. Consider the following table of values for *s* and $f(s) = (1 + s)^{1/s}$ and Figure 1 for *s* close to 0. Compute the table values with a calculator yourself, and try several values of *s* even closer to 0. Note that the function is discontinuous at $s = 0$.
>
>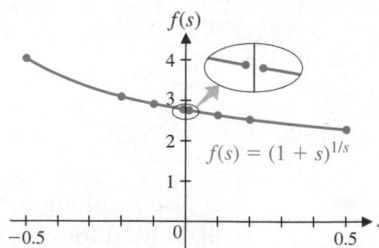
>
> Figure 1
>
> s approaches 0 from the left $\to 0 \leftarrow$ s approaches 0 from the right
>
> | s | -0.5 | -0.2 | -0.1 | $-0.01 \to 0 \leftarrow 0.01$ | 0.1 | 0.2 | 0.5 |
> |---|---|---|---|---|---|---|---|
> | $(1 + s)^{1/s}$ | 4.0000 | 3.0518 | 2.8680 | $2.7320 \to e \leftarrow 2.7048$ | 2.5937 | 2.4883 | 2.2500 |

Continuous Compound Interest

Now we can see how *e* appears quite naturally in the important application of compound interest. Let us start with simple interest, move on to compound interest, and then proceed on to continuous compound interest.

On one hand, if a principal P is borrowed at an annual rate r,* then after t years at simple interest, the borrower will owe the lender an amount A given by

$$A = P + Prt = P(1 + rt) \quad \text{Simple interest} \tag{1}$$

On the other hand, if interest is compounded m times a year, then the borrower will owe the lender an amount A given by

$$A = P\left(1 + \frac{r}{m}\right)^{mt} \quad \text{Compound interest} \tag{2}$$

where r/m is the interest rate per compounding period and mt is the number of compounding periods. Suppose that P, r, and t in equation (2) are held fixed and m is increased. Will the amount A increase without bound, or will it tend to approach some limiting value?

Let us perform a calculator experiment before we attack the general limit problem. If $P = \$100$, $r = 0.06$, and $t = 2$ years, then

$$A = 100\left(1 + \frac{0.06}{m}\right)^{2m}$$

We compute A for several values of m in Table 1. The biggest gain appears in the first step, then the gains slow down as m increases. The amount A appears to approach $\$112.75$ as m gets larger and larger.

Table 1

Compounding Frequency	m	$A = 100\left(1 + \dfrac{0.06}{m}\right)^{2m}$
Annually	1	$112.3600
Semiannually	2	112.5509
Quarterly	4	112.6493
Monthly	12	112.7160
Weekly	52	112.7419
Daily	365	112.7486
Hourly	8,760	112.7496

Keeping P, r, and t fixed in equation (2), we compute the following limit and observe an interesting and useful result:

$$\lim_{m \to \infty} P\left(1 + \frac{r}{m}\right)^{mt} = P \lim_{m \to \infty} \left(1 + \frac{r}{m}\right)^{(m/r)rt}$$

Insert r/r in the exponent and let $s = r/m$. Note that $m \to \infty$ implies $s \to 0$.

$$= P \lim_{s \to 0}[(1 + s)^{1/s}]^{rt}$$

Use a limit property.[†]

$$= P[\lim_{s \to 0}(1 + s)^{1/s}]^{rt}$$

$\lim_{s \to 0}(1 + s)^{1/s} = e$

$$= Pe^{rt}$$

The resulting formula is called the **continuous compound interest formula**, a widely used formula in business and economics.

THEOREM 1   Continuous Compound Interest Formula

If a principal P is invested at an annual rate r (expressed as a decimal) compounded continuously, then the amount A in the account at the end of t years is given by

$$A = Pe^{rt}$$

---

*If r is the interest rate written as a decimal, then $100r\%$ is the rate in percent. For example, if $r = 0.12$, then $100r\% = 100(0.12)\% = 12\%$. The expressions 0.12 and 12% are equivalent. Unless stated otherwise, all formulas in this book use r in decimal form.

[†]The following new limit property is used: If $\lim_{x \to c}f(x)$ exists, then $\lim_{x \to c}[f(x)]^p = [\lim_{x \to c} f(x)]^p$, provided that the last expression names a real number.

EXAMPLE 1 Computing Continuously Compounded Interest If $100 is invested at 6% compounded continuously,* what amount will be in the account after 2 years? How much interest will be earned?

SOLUTION
$$A = Pe^{rt}$$
$$= 100e^{(0.06)(2)} \quad \text{6% is equivalent to } r = 0.06.$$
$$\approx \$112.7497$$

Compare this result with the values calculated in Table 1. The interest earned is $112.7497 − $100 = $12.7497.

Matched Problem 1 ⌋ What amount (to the nearest cent) will be in an account after 5 years if $100 is invested at an annual nominal rate of 8% compounded annually? Semiannually? Continuously?

EXAMPLE 2 Graphing the Growth of an Investment Union Savings Bank offers a 5-year certificate of deposit (CD) that earns 5.75% compounded continuously. If $1,000 is invested in one of these CDs, graph the amount in the account as a function of time for a period of 5 years.

SOLUTION We want to graph
$$A = 1,000e^{0.0575t} \quad 0 \le t \le 5$$

Using a calculator, we construct a table of values (Table 2). Then we graph the points from the table and join the points with a smooth curve (Fig. 2).

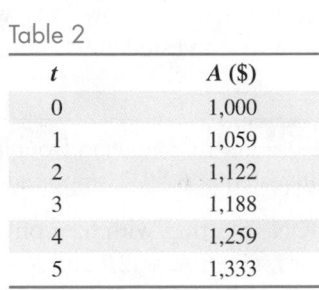

Table 2

t	A ($)
0	1,000
1	1,059
2	1,122
3	1,188
4	1,259
5	1,333

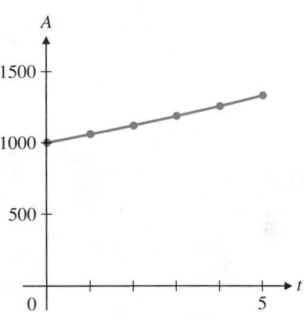

Figure 2

CONCEPTUAL INSIGHT

Depending on the domain, the graph of an exponential function can appear to be linear. Table 2 shows that the graph in Figure 2 is *not* linear. The slope determined by the first two points (for $t = 0$ and $t = 1$) is 59 but the slope determined by the first and third points (for $t = 0$ and $t = 2$) is 61. For a linear graph, the slope determined by any two points is constant.

Matched Problem 2 ⌋ If $5,000 is invested in a Union Savings Bank 4-year CD that earns 5.61% compounded continuously, graph the amount in the account as a function of time for a period of 4 years.

---

*Following common usage, we will often write "at 6% compounded continuously," understanding that this means "at an annual rate of 6% compounded continuously."

EXAMPLE 3   Computing Growth Time   How long will it take an investment of $5,000 to grow to $8,000 if it is invested at 5% compounded continuously?

SOLUTION   Starting with the continous compound interest formula $A = Pe^{rt}$, we must solve for t:

$$A = Pe^{rt}$$
$$8,000 = 5,000e^{0.05t}$$ Divide both sides by 5,000 and
$$e^{0.05t} = 1.6$$ reverse the equation.
$$\ln e^{0.05t} = \ln 1.6$$ Take the natural logarithm of both
$$0.05t = \ln 1.6$$ sides—recall that $\log_b b^x = x$.
$$t = \frac{\ln 1.6}{0.05}$$
$$t \approx 9.4 \text{ years}$$

Figure 3 shows an alternative method for solving Example 3 on a graphing calculator.

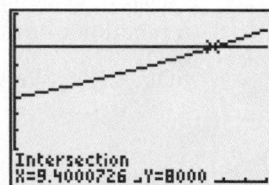

Figure 3

$$y_1 = 5{,}000e^{0.05x}$$
$$y_2 = 8{,}000$$

Matched Problem 3   How long will it take an investment of $10,000 to grow to $15,000 if it is invested at 9% compounded continuously?

EXAMPLE 4   Computing Doubling Time   How long will it take money to double if it is invested at 6.5% compounded continuously?

SOLUTION   Starting with the continuous compound interest formula $A = Pe^{rt}$, we solve for t, given $A = 2P$ and $R = 0.065$:

$$2P = Pe^{0.065t}$$ Divide both sides by P and reverse the equation.
$$e^{0.065t} = 2$$ Take the natural logarithm of both sides.
$$\ln e^{0.065t} = \ln 2$$
$$0.065t = \ln 2$$
$$t = \frac{\ln 2}{0.065}$$
$$t \approx 10.66 \text{ years}$$

Matched Problem 4   How long will it take money to triple if it is invested at 5.5% compounded continuously?

Explore and Discuss 1   You are considering three options for investing $10,000: at 7% compounded annually, at 6% compounded monthly, and at 5% compounded continuously.

(A) Which option would be the best for investing $10,000 for 8 years?

(B) How long would you need to invest your money for the third option to be the best?

Exercises 3.1

Skills Warm-up Exercise

W | *In Problems 1–8, solve for the variable to two decimal places. (If necessary, review Section 1.5).*

1. $A = 1,200e^{0.04(5)}$

2. $A = 3,000e^{0.07(10)}$

3. $9827.30 = Pe^{0.025(3)}$

4. $50,000 = Pe^{0.054(7)}$

5. $6,000 = 5,000e^{0.0325t}$

6. $10,0000 = 7,500e^{0.085t}$

7. $956 = 900e^{1.5r}$

8. $4,840 = 3,750e^{4.25r}$

Use a calculator to evaluate A to the nearest cent in Problems 9 and 10.

9. $A = \$1,000e^{0.1t}$ for $t = 2, 5,$ and 8

10. $A = \$5,000e^{0.08t}$ for $t = 1, 4,$ and 10

11. If $6,000 is invested at 10% compounded continuously, graph the amount in the account as a function of time for a period of 8 years.

12. If $4,000 is invested at 8% compounded continuously, graph the amount in the account as a function of time for a period of 6 years.

In Problems 13–18, solve for t or r to two decimal places.

13. $2 = e^{0.06t}$

14. $2 = e^{0.03t}$

15. $3 = e^{0.1r}$

16. $3 = e^{0.25t}$

17. $2 = e^{5r}$

18. $3 = e^{10r}$

In Problems 19 and 20, use a calculator to complete each table to five decimal places.

19.

n	$[1 + (1/n)]^n$
10	2.593 74
100	
1,000	
10,000	
100,000	
1,000,000	
10,000,000	
↓	↓
∞	$e = 2.718\,281\,828\,459\ldots$

20.

s	$(1 + s)^{1/s}$
0.01	2.704 81
−0.01	
0.001	
−0.001	
0.000 1	
−0.000 1	
0.000 01	
−0.000 01	
↓	↓
0	$e = 2.718\,281\,828\,459\ldots$

21. Use a calculator and a table of values to investigate

$$\lim_{n \to \infty} (1 + n)^{1/n}$$

Do you think this limit exists? If so, what do you think it is?

22. Use a calculator and a table of values to investigate

$$\lim_{s \to 0^+} \left(1 + \frac{1}{s}\right)^s$$

Do you think this limit exists? If so, what do you think it is?

23. It can be shown that the number e satisfies the inequality

$$\left(1 + \frac{1}{n}\right)^n < e < \left(1 + \frac{1}{n}\right)^{n+1} \qquad n \geq 1$$

Illustrate this condition by graphing

$$y_1 = (1 + 1/n)^n$$
$$y_2 = 2.718\,281\,828 \approx e$$
$$y_3 = (1 + 1/n)^{n+1}$$

in the same viewing window, for $1 \leq n \leq 20$.

24. It can be shown that

$$e^s = \lim_{n \to \infty} \left(1 + \frac{s}{n}\right)^n$$

for any real number s. Illustrate this equation graphically for $s = 2$ by graphing

$$y_1 = (1 + 2/n)^n$$
$$y_2 = 7.389\,056\,099 \approx e^2$$

in the same viewing window, for $1 \leq n \leq 50$.

Applications

25. **Continuous compound interest.** Provident Bank offers a 10-year CD that earns 2.15% compounded continuously.

(A) If $10,000 is invested in this CD, how much will it be worth in 10 years?

(B) How long will it take for the account to be worth $18,000?

26. **Continuous compound interest.** Provident Bank also offers a 3-year CD that earns 1.64% compounded continuously.

(A) If $10,000 is invested in this CD, how much will it be worth in 3 years?

(B) How long will it take for the account to be worth $11,000?

27. **Present value.** A note will pay $20,000 at maturity 10 years from now. How much should you be willing to pay for the note now if money is worth 5.2% compounded continuously?

28. Present value. A note will pay $50,000 at maturity 5 years from now. How much should you be willing to pay for the note now if money is worth 6.4% compounded continuously?

29. Continuous compound interest. An investor bought stock for $20,000. Five years later, the stock was sold for $30,000. If interest is compounded continuously, what annual nominal rate of interest did the original $20,000 investment earn?

30. Continuous compound interest. A family paid $99,000 cash for a house. Fifteen years later, the house was sold for $195,000. If interest is compounded continuously, what annual nominal rate of interest did the original $99,000 investment earn?

31. Present value. Solving $A = Pe^{rt}$ for P, we obtain

$$P = Ae^{-rt}$$

which is the present value of the amount A due in t years if money earns interest at an annual nominal rate r compounded continuously.

(A) Graph $P = 10,000e^{-0.08t}, 0 \leq t \leq 50$.

(B) $\lim\limits_{t \to \infty} 10,000e^{-0.08t} = ?$ [Guess, using part (A).]

[*Conclusion:* The longer the time until the amount A is due, the smaller is its present value, as we would expect.]

32. Present value. Referring to Problem 31, in how many years will the $10,000 be due in order for its present value to be $5,000?

33. Doubling time. How long will it take money to double if it is invested at 4% compounded continuously?

34. Doubling time. How long will it take money to double if it is invested at 5% compounded continuously?

35. Doubling rate. At what nominal rate compounded continuously must money be invested to double in 8 years?

36. Doubling rate. At what nominal rate compounded continuously must money be invested to double in 10 years?

37. Growth time. A man with $20,000 to invest decides to diversify his investments by placing $10,000 in an account that earns 7.2% compounded continuously and $10,000 in an account that earns 8.4% compounded annually. Use graphical approximation methods to determine how long it will take for his total investment in the two accounts to grow to $35,000.

38. Growth time. A woman invests $5,000 in an account that earns 8.8% compounded continuously and $7,000 in an account that earns 9.6% compounded annually. Use graphical approximation methods to determine how long it will take for her total investment in the two accounts to grow to $20,000.

39. Doubling times

(A) Show that the doubling time t (in years) at an annual rate r compounded continuously is given by

$$t = \frac{\ln 2}{r}$$

(B) Graph the doubling-time equation from part (A) for $0.02 \leq r \leq 0.30$. Is this restriction on r reasonable? Explain.

(C) Determine the doubling times (in years, to two decimal places) for $r = 5\%, 10\%, 15\%, 20\%, 25\%,$ and 30%.

40. Doubling rates

(A) Show that the rate r that doubles an investment at continuously compounded interest in t years is given by

$$r = \frac{\ln 2}{t}$$

(B) Graph the doubling-rate equation from part (A) for $1 \leq t \leq 20$. Is this restriction on t reasonable? Explain.

(C) Determine the doubling rates for $t = 2, 4, 6, 8, 10,$ and 12 years.

41. Radioactive decay. A mathematical model for the decay of radioactive substances is given by

$$Q = Q_0 e^{rt}$$

where

$$Q_0 = \text{amount of the substance at time } t = 0$$
$$r = \text{continuous compound rate of decay}$$
$$t = \text{time in years}$$
$$Q = \text{amount of the substance at time } t$$

If the continuous compound rate of decay of radium per year is $r = -0.000\ 433\ 2$, how long will it take a certain amount of radium to decay to half the original amount? (This period is the *half-life* of the substance.)

42. Radioactive decay. The continuous compound rate of decay of carbon-14 per year is $r = -0.000\ 123\ 8$. How long will it take a certain amount of carbon-14 to decay to half the original amount? (Use the radioactive decay model in Problem 41.)

43. Radioactive decay. A cesium isotope has a half-life of 30 years. What is the continuous compound rate of decay? (Use the radioactive decay model in Problem 41.)

44. Radioactive decay. A strontium isotope has a half-life of 90 years. What is the continuous compound rate of decay? (Use the radioactive decay model in Problem 41.)

45. World population. A mathematical model for world population growth over short intervals is given by

$$P = P_0 e^{rt}$$

where

$$P_0 = \text{population at time } t = 0$$
$$r = \text{continuous compound rate of growth}$$
$$t = \text{time in years}$$
$$P = \text{population at time } t$$

How long will it take world population to double if it continues to grow at its current continuous compound rate of 1.3% per year?

46. **U.S. population.** How long will it take for the U.S. population to double if it continues to grow at a rate of 0.975% per year?

47. **Population growth.** Some underdeveloped nations have population doubling times of 50 years. At what continuous compound rate is the population growing? (Use the population growth model in Problem 45.)

48. **Population growth.** Some developed nations have population doubling times of 200 years. At what continuous compound rate is the population growing? (Use the population growth model in Problem 45.)

Answers to Matched Problems

1. $146.93; $148.02; $149.18
2. $A = 5{,}000e^{0.0561t}$

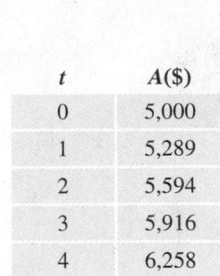

t	$A(\$)$
0	5,000
1	5,289
2	5,594
3	5,916
4	6,258

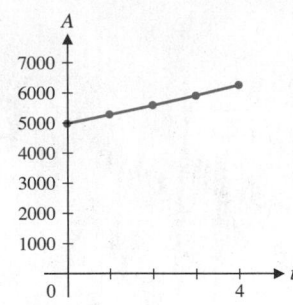

3. 4.51 yr
4. 19.97 yr

3.2 Derivatives of Exponential and Logarithmic Functions

- The Derivative of e^x
- The Derivative of ln x
- Other Logarithmic and Exponential Functions
- Exponential and Logarithmic Models

In this section, we find formulas for the derivatives of logarithmic and exponential functions. A review of Sections 1.5 and 1.6 may prove helpful. In particular, recall that $f(x) = e^x$ is the exponential function with base $e \approx 2.718$, and the inverse of the function e^x is the natural logarithm function ln x. More generally, if b is a positive real number, $b \neq 1$, then the exponential function b^x with base b, and the logarithmic function $\log_b x$ with base b, are inverses of each other.

The Derivative of e^x

In the process of finding the derivative of e^x, we use (without proof) the fact that

$$\lim_{h \to 0} \frac{e^h - 1}{h} = 1 \tag{1}$$

Explore and Discuss 1   Complete Table 1.

Table 1

h	−0.1	−0.01	−0.001	→ 0 ←	0.001	0.01	0.1
$\dfrac{e^h - 1}{h}$							

Do your calculations make it reasonable to conclude that

$$\lim_{h \to 0} \frac{e^h - 1}{h} = 1?$$

Discuss.

We now apply the four-step process (Section 2.4) to the exponential function $f(x) = e^x$.

Step 1 Find $f(x + h)$.

$$f(x + h) = e^{x+h} = e^x e^h \qquad \text{See Section 1.4.}$$

Step 2 Find $f(x + h) - f(x)$.

$$f(x + h) - f(x) = e^x e^h - e^x \qquad \text{Factor out } e^x.$$
$$= e^x(e^h - 1)$$

Step 3 Find $\dfrac{f(x + h) - f(x)}{h}$.

$$\frac{f(x + h) - f(x)}{h} = \frac{e^x(e^h - 1)}{h} = e^x \left(\frac{e^h - 1}{h} \right)$$

Step 4 Find $f'(x) = \lim\limits_{h \to 0} \dfrac{f(x + h) - f(x)}{h}$.

$$f'(x) = \lim_{h \to 0} \frac{f(x + h) - f(x)}{h}$$
$$= \lim_{h \to 0} e^x \left(\frac{e^h - 1}{h} \right)$$
$$= e^x \lim_{h \to 0} \left(\frac{e^h - 1}{h} \right) \qquad \text{Use the limit in (1).}$$
$$= e^x \cdot 1 = e^x$$

Therefore,

$$\frac{d}{dx} e^x = e^x \qquad \begin{array}{l} \text{The derivative of the exponential} \\ \text{function is the exponential function.} \end{array}$$

EXAMPLE 1 Finding Derivatives Find $f'(x)$ for

(A) $f(x) = 5e^x - 3x^4 + 9x + 16$ (B) $f(x) = -7x^e + 2e^x + e^2$

SOLUTIONS

(A) $f'(x) = 5e^x - 12x^3 + 9$ (B) $f'(x) = -7ex^{e-1} + 2e^x$

Remember that e is a real number, so the power rule (Section 2.5) is used to find the derivative of x^e. The derivative of the exponential function e^x, however, is e^x. Note that $e^2 \approx 7.389$ is a constant, so its derivative is 0.

Matched Problem 1 Find $f'(x)$ for

(A) $f(x) = 4e^x + 8x^2 + 7x - 14$ (B) $f(x) = x^7 - x^5 + e^3 - x + e^x$

⚠ **CAUTION**

$$\frac{d}{dx} e^x \neq xe^{x-1} \qquad \frac{d}{dx} e^x = e^x$$

The power rule cannot be used to differentiate the exponential function. The power rule applies to exponential forms x^n, where the exponent is a constant and the base is a variable. In the exponential form e^x, the base is a constant and the exponent is a variable.   ▲

The Derivative of ln x

We summarize some important facts about logarithmic functions from Section 1.6:

SUMMARY

Recall that the inverse of an exponential function is called a **logarithmic function**. For $b > 0$ and $b \neq 1$,

Logarithmic form		Exponential form
$y = \log_b x$	is equivalent to	$x = b^y$
Domain: $(0, \infty)$		Domain: $(-\infty, \infty)$
Range: $(-\infty, \infty)$		Range: $(0, \infty)$

The graphs of $y = \log_b x$ and $y = b^x$ are symmetric with respect to the line $y = x$. (See Figure 1.)

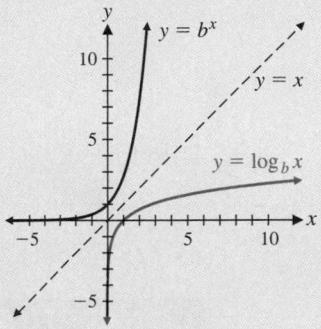

Figure 1

Of all the possible bases for logarithmic functions, the two most widely used are

$$\log x = \log_{10} x \quad \text{Common logarithm (base 10)}$$
$$\ln x = \log_e x \quad \text{Natural logarithm (base } e\text{)}$$

We are now ready to use the definition of the derivative and the four-step process discussed in Section 2.4 to find a formula for the derivative of $\ln x$. Later we will extend this formula to include $\log_b x$ for any base b.

Let $f(x) = \ln x, x > 0$.

Step 1 Find $f(x + h)$.

$$f(x + h) = \ln(x + h) \quad \ln(x + h) \text{ cannot be simplified.}$$

Step 2 Find $f(x + h) - f(x)$.

$$f(x + h) - f(x) = \ln(x + h) - \ln x \quad \text{Use } \ln A - \ln B = \ln \frac{A}{B}.$$
$$= \ln \frac{x + h}{x}$$

Step 3 Find $\dfrac{f(x + h) - f(x)}{h}$.

$$\frac{f(x + h) - f(x)}{h} = \frac{\ln(x + h) - \ln x}{h}$$
$$= \frac{1}{h} \ln \frac{x + h}{x} \quad \text{Multiply by } 1 = x/x \text{ to change form.}$$
$$= \frac{x}{x} \cdot \frac{1}{h} \ln \frac{x + h}{x}$$
$$= \frac{1}{x} \left[\frac{x}{h} \ln \left(1 + \frac{h}{x} \right) \right] \quad \text{Use } p \ln A = \ln A^p.$$
$$= \frac{1}{x} \ln \left(1 + \frac{h}{x} \right)^{x/h}$$

Step 4 Find $f'(x) = \lim\limits_{h \to 0} \dfrac{f(x+h) - f(x)}{h}$.

$$f'(x) = \lim_{h \to 0} \frac{f(x+h) - f(x)}{h}$$

$$= \lim_{h \to 0} \left[\frac{1}{x} \ln\left(1 + \frac{h}{x}\right)^{x/h} \right] \qquad \text{Let } s = h/x. \text{ Note that } h \to 0 \text{ implies } s \to 0.$$

$$= \frac{1}{x} \lim_{s \to 0} \left[\ln(1 + s)^{1/s} \right] \qquad \text{Use a new limit property.}^{*}$$

$$= \frac{1}{x} \ln\left[\lim_{s \to 0} (1 + s)^{1/s} \right] \qquad \text{Use the definition of } e.$$

$$= \frac{1}{x} \ln e \qquad\qquad\qquad \ln e = \log_e e = 1$$

$$= \frac{1}{x}$$

Therefore,

$$\frac{d}{dx} \ln x = \frac{1}{x} \quad (x > 0)$$

CONCEPTUAL INSIGHT

In finding the derivative of ln x, we used the following properties of logarithms:

$$\ln \frac{A}{B} = \ln A - \ln B \qquad \ln A^p = p \ln A$$

We also noted that there is no property that simplifies $\ln(A + B)$. (See Theorem 1 in Section 1.6 for a list of properties of logarithms.)

EXAMPLE 2 Finding Derivatives Find y' for

(A) $y = 3e^x + 5 \ln x$ \qquad\qquad (B) $y = x^4 - \ln x^4$

SOLUTIONS

(A) $y' = 3e^x + \dfrac{5}{x}$

(B) Before taking the derivative, we use a property of logarithms (see Theorem 1, Section 1.6) to rewrite y.

$$y = x^4 - \ln x^4 \qquad \text{Use } \ln M^p = p \ln M.$$
$$y = x^4 - 4 \ln x \qquad \text{Now take the derivative of both sides.}$$
$$y' = 4x^3 - \frac{4}{x}$$

Matched Problem 2 Find y' for

(A) $y = 10x^3 - 100 \ln x$ \qquad\qquad (B) $y = \ln x^5 + e^x - \ln e^2$

Other Logarithmic and Exponential Functions

In most applications involving logarithmic or exponential functions, the number e is the preferred base. However, in some situations it is convenient to use a base other than e. Derivatives of $y = \log_b x$ and $y = b^x$ can be obtained by expressing these functions in terms of the natural logarithmic and exponential functions.

---

*The following new limit property is used: If $\lim_{x \to c} f(x)$ exists and is positive, then $\lim_{x \to c} [\ln f(x)] = \ln[\lim_{x \to c} f(x)]$.

We begin by finding a relationship between $\log_b x$ and $\ln x$ for any base b such that $b > 0$ and $b \ne 1$.

$$y = \log_b x \qquad \text{Change to exponential form.}$$
$$b^y = x \qquad \text{Take the natural logarithm of both sides.}$$
$$\ln b^y = \ln x \qquad \text{Recall that } \ln b^y = y \ln b.$$
$$y \ln b = \ln x \qquad \text{Solve for y.}$$
$$y = \frac{1}{\ln b} \ln x$$

Therefore,

$$\log_b x = \frac{1}{\ln b} \ln x \qquad \text{Change-of-base formula for logarithms*} \qquad (2)$$

Similarly, we can find a relationship between b^x and e^x for any base b such that $b > 0, b \ne 1$.

$$y = b^x \qquad \text{Take the natural logarithm of both sides.}$$
$$\ln y = \ln b^x \qquad \text{Recall that } \ln b^x = x \ln b.$$
$$\ln y = x \ln b \qquad \text{Take the exponential function of both sides.}$$
$$y = e^{x \ln b}$$

Therefore,

$$b^x = e^{x \ln b} \qquad \text{Change-of-base formula for exponential functions} \qquad (3)$$

Differentiating both sides of equation (2) gives

$$\frac{d}{dx} \log_b x = \frac{1}{\ln b} \frac{d}{dx} \ln x = \frac{1}{\ln b} \left(\frac{1}{x} \right) \quad (x > 0)$$

It can be shown that the derivative of the function e^{cx}, where c is a constant, is the function ce^{cx} (see Problems 61–62 in Exercise 3.2 or the more general results of Section 3.4). Therefore, differentiating both sides of equation (3), we have

$$\frac{d}{dx} b^x = e^{x \ln b} \ln b = b^x \ln b$$

For convenience, we list the derivative formulas for exponential and logarithmic functions:

> **Derivatives of Exponential and Logarithmic Functions**
> For $b > 0, b \ne 1$,
>
> $$\frac{d}{dx} e^x = e^x \qquad \frac{d}{dx} b^x = b^x \ln b$$
>
> For $b > 0, b \ne 1$, and $x > 0$,
>
> $$\frac{d}{dx} \ln x = \frac{1}{x} \qquad \frac{d}{dx} \log_b x = \frac{1}{\ln b} \left(\frac{1}{x} \right)$$

EXAMPLE 3 Finding Derivatives Find $g'(x)$ for

(A) $g(x) = 2^x - 3^x$

(B) $g(x) = \log_4 x^5$

---

*Equation (2) is a special case of the **general change-of-base formula** for logarithms (which can be derived in the same way): $\log_b x = (\log_a x)/(\log_a b)$.

SOLUTIONS

(A) $g'(x) = 2^x \ln 2 - 3^x \ln 3$

(B) First, use a property of logarithms to rewrite $g(x)$.

$$g(x) = \log_4 x^5 \qquad \text{Use } \log_b M^p = p \log_b M.$$
$$g(x) = 5 \log_4 x \qquad \text{Take the derivative of both sides.}$$
$$g'(x) = \frac{5}{\ln 4}\left(\frac{1}{x}\right)$$

Matched Problem 3 Find $g'(x)$ for

(A) $g(x) = x^{10} + 10^x$ (B) $g(x) = \log_2 x - 6 \log_5 x$

Explore and Discuss 2 (A) The graphs of $f(x) = \log_2 x$ and $g(x) = \log_4 x$ are shown in Figure 2. Which graph belongs to which function?

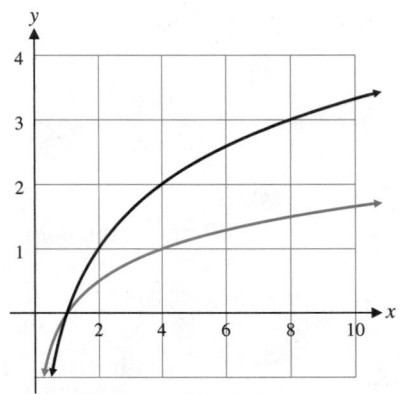

Figure 2

(B) Sketch graphs of $f'(x)$ and $g'(x)$.

(C) The function $f(x)$ is related to $g(x)$ in the same way that $f'(x)$ is related to $g'(x)$. What is that relationship?

Exponential and Logarithmic Models

EXAMPLE 4 Price–Demand Model An Internet store sells Australian wool blankets. If the store sells x blankets at a price of $\$p$ per blanket, then the price–demand equation is $p = 350(0.999)^x$. Find the rate of change of price with respect to demand when the demand is 800 blankets and interpret the result.

SOLUTION

$$\frac{dp}{dx} = 350(0.999)^x \ln 0.999$$

If $x = 800$, then

$$\frac{dp}{dx} = 350(0.999)^{800} \ln 0.999 \approx -0.157, \text{ or } -\$0.16$$

When the demand is 800 blankets, the price is decreasing by $\$0.16$ per blanket.

Matched Problem 4 The store in Example 4 also sells a reversible fleece blanket. If the price–demand equation for reversible fleece blankets is $p = 200(0.998)^x$, find the rate of change of price with respect to demand when the demand is 400 blankets and interpret the result.

EXAMPLE 5 Continuous Compound Interest An investment of $1,000 earns interest at an annual rate of 4% compounded continuously.

(A) Find the instantaneous rate of change of the amount in the account after 2 years.

(B) Find the instantaneous rate of change of the amount in the account at the time the amount is equal to $2,000.

SOLUTION

(A) The amount $A(t)$ at time t (in years) is given by $A(t) = 1,000e^{0.04t}$. Note that $A(t) = 1,000b^t$, where $b = e^{0.04}$. The instantaneous rate of change is the derivative $A'(t)$, which we find by using the formula for the derivative of the exponential function with base b:

$$A'(t) = 1,000b^t \ln b = 1,000e^{0.04t}(0.04) = 40e^{0.04t}$$

After 2 years, $A'(2) = 40e^{0.04(2)} = \$43.33$ per year.

(B) From the calculation of the derivative in part (A), we note that $A'(t) = (0.04)1,000e^{0.04t} = 0.04A(t)$. In other words, the instantaneous rate of change of the amount is always equal to 4% of the amount. So if the amount is $2,000, then the instantaneous rate of change is $(0.04)\$2,000 = \$80$ per year.

Matched Problem 5 An investment of $5,000 earns interest at an annual rate of 6% compounded continuously.

(A) Find the instantaneous rate of change of the amount in the account after 3 years.

(B) Find the instantaneous rate of change of the amount in the account at the time the amount is equal to $8,000.

EXAMPLE 6 Cable TV Subscribers A statistician used data from the U.S. Census Bureau to construct the model

$$S(t) = 11 \ln t + 29$$

where $S(t)$ is the number of cable TV subscribers (in millions) in year ($t = 0$) corresponds to 1980). Use this model to estimate the number of cable TV subscribers in 2020 and the rate of change of the number of subscribers in 2020. Round both to the nearest tenth of a million. Interpret these results.

SOLUTION Since 2020 corresponds to $t = 40$, we must find $S(40)$ and $S'(40)$.

$$S(40) = 11 \ln 40 + 29 = 69.6 \text{ million}$$

$$S'(t) = 11\frac{1}{t} = \frac{11}{t}$$

$$S'(40) = \frac{11}{40} = 0.3 \text{ million}$$

In 2020 there will be approximately 69.6 million subscribers, and this number is growing at the rate of 0.3 million subscribers per year.

Matched Problem 6 A model for a newspaper's circulation is

$$C(t) = 83 - 9 \ln t$$

where $C(t)$ is the circulation (in thousands) in year t ($t = 0$ corresponds to 1980). Use this model to estimate the circulation and the rate of change of circulation in 2020. Round both to the nearest hundred. Interpret these results.

| CONCEPTUAL INSIGHT |

On most graphing calculators, exponential regression produces a function of the form $y = a \cdot b^x$. Formula (3) on page 191 allows you to change the base b (chosen by the graphing calculator) to the more familiar base e:

$$y = a \cdot b^x = a \cdot e^{x \ln b}$$

On most graphing calculators, logarithmic regression produces a function of the form $y = a + b \ln x$. Formula (2) on page 191 allows you to write the function in terms of logarithms to any base d that you may prefer:

$$y = a + b \ln x = a + b(\ln d) \log_d x$$

Exercises 3.2

Skills Warm-up Exercises

In Problems 1–8, solve for the variable without using a calculator. (If necessary, review Section 1.6).

1. $y = \log_2 128$ **2.** $y = \ln e^5$

3. $\log_3 x = 4$ **4.** $\log_5 x = -2$

5. $\log_b 64 = 2$ **6.** $\log_b 5 = \dfrac{1}{3}$

7. $y = \ln \sqrt{e}$ **8.** $y = \ln (\ln e)$

In Problems 9–26, find $f'(x)$.

9. $f(x) = 5e^x + 3x + 1$ **10.** $f(x) = -7e^x - 2x + 5$

11. $f(x) = -2 \ln x + x^2 - 4$ **12.** $f(x) = 6 \ln x - x^3 + 2$

13. $f(x) = x^3 - 6e^x$ **14.** $f(x) = 9e^x + 2x^2$

15. $f(x) = e^x + x - \ln x$ **16.** $f(x) = \ln x + 2e^x - 3x^2$

17. $f(x) = \ln x^3$ **18.** $f(x) = \ln x^8$

19. $f(x) = 5x - \ln x^5$ **20.** $f(x) = 4 + \ln x^9$

21. $f(x) = \ln x^2 + 4e^x$ **22.** $f(x) = \ln x^{10} + 2 \ln x$

23. $f(x) = e^x + x^e$ **24.** $f(x) = 3x^e - 2e^x$

25. $f(x) = xx^e$ **26.** $f(x) = ee^x$

In Problems 27–34, find the equation of the line tangent to the graph of f at the indicated value of x.

27. $f(x) = 3 + \ln x; x = 1$ **28.** $f(x) = 2 \ln x; x = 1$

29. $f(x) = 3e^x; x = 0$ **30.** $f(x) = e^x + 1; x = 0$

31. $f(x) = \ln x^3; x = e$ **32.** $f(x) = 1 + \ln x^4; x = e$

33. $f(x) = 2 + e^x; x = 1$ **34.** $f(x) = 5e^x; x = 1$

35. A student claims that the line tangent to the graph of $f(x) = e^x$ at $x = 3$ passes through the point $(2, 0)$ (see the figure). Is she correct? Will the line tangent at $x = 4$ pass through $(3, 0)$? Explain.

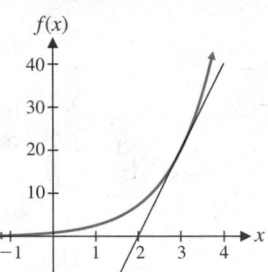

36. Refer to Problem 35. Does the line tangent to the graph of $f(x) = e^x$ at $x = 1$ pass through the origin? Are there any other lines tangent to the graph of f that pass through the origin? Explain.

37. A student claims that the line tangent to the graph of $g(x) = \ln x$ at $x = 3$ passes through the origin (see the figure). Is he correct? Will the line tangent at $x = 4$ pass through the origin? Explain.

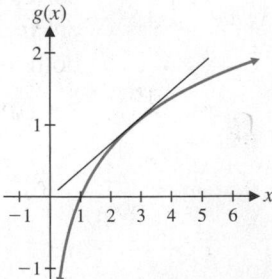

38. Refer to Problem 37. Does the line tangent to the graph of $f(x) = \ln x$ at $x = e$ pass through the origin? Are there any other lines tangent to the graph of f that pass through the origin? Explain.

In Problems 39–42, first use appropriate properties of logarithms to rewrite $f(x)$, and then find $f'(x)$.

39. $f(x) = 10x + \ln 10x$ **40.** $f(x) = 2 + 3 \ln \dfrac{1}{x}$

41. $f(x) = \ln \dfrac{4}{x^3}$ **42.** $f(x) = x + 5 \ln 6x$

In Problems 43–54, find $\dfrac{dy}{dx}$ *for the indicated function y.*

43. $y = \log_2 x$

44. $y = 3 \log_5 x$

45. $y = 3^x$

46. $y = 4^x$

47. $y = 2x - \log x$

48. $y = \log x + 4x^2 + 1$

49. $y = 10 + x + 10^x$

50. $y = x^5 - 5^x$

51. $y = 3 \ln x + 2 \log_3 x$

52. $y = -\log_2 x + 10 \ln x$

53. $y = 2^x + e^2$

54. $y = e^3 - 3^x$

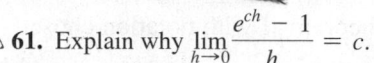 *In Problems 55–60, use graphical approximation methods to find the points of intersection of $f(x)$ and $g(x)$ (to two decimal places).*

55. $f(x) = e^x; g(x) = x^4$

[Note that there are three points of intersection and that e^x is greater than x^4 for large values of x.]

56. $f(x) = e^x; g(x) = x^5$

[Note that there are two points of intersection and that e^x is greater than x^5 for large values of x.]

57. $f(x) = (\ln x)^2; g(x) = x$

58. $f(x) = (\ln x)^3; g(x) = x$

59. $f(x) = \ln x; g(x) = x^{1/5}$

60. $f(x) = \ln x; g(x) = x^{1/4}$

61. Explain why $\lim\limits_{h \to 0} \dfrac{e^{ch} - 1}{h} = c.$

62. Use the result of Problem 61 and the four-step process to show that if $f(x) = e^{cx}$, then $f'(x) = ce^{cx}$.

Applications

63. **Salvage value.** The estimated salvage value S (in dollars) of a company airplane after t years is given by

$$S(t) = 300,000(0.9)^t$$

What is the rate of depreciation (in dollars per year) after 1 year? 5 years? 10 years?

64. **Resale value.** The estimated resale value R (in dollars) of a company car after t years is given by

$$R(t) = 20,000(0.86)^t$$

What is the rate of depreciation (in dollars per year) after 1 year? 2 years? 3 years?

65. **Bacterial growth.** A single cholera bacterium divides every 0.5 hour to produce two complete cholera bacteria. If we start with a colony of 5,000 bacteria, then after t hours, there will be

$$A(t) = 5,000 \cdot 2^{2t} = 5,000 \cdot 4^t$$

bacteria. Find $A'(t), A'(1)$, and $A'(5)$, and interpret the results.

66. **Bacterial growth.** Repeat Problem 65 for a starting colony of 1,000 bacteria such that a single bacterium divides every 0.25 hour.

67. **Blood pressure.** An experiment was set up to find a relationship between weight and systolic blood pressure in children. Using hospital records for 5,000 children, the experimenters found that the systolic blood pressure was given approximately by

$$P(x) = 17.5(1 + \ln x) \qquad 10 \le x \le 100$$

where $P(x)$ is measured in millimeters of mercury and x is measured in pounds. What is the rate of change of blood pressure with respect to weight at the 40-pound weight level? At the 90-pound weight level?

68. **Blood pressure.** Refer to Problem 67. Find the weight (to the nearest pound) at which the rate of change of blood pressure with respect to weight is 0.3 millimeter of mercury per pound.

69. **Psychology: stimulus/response.** In psychology, the Weber–Fechner law for the response to a stimulus is

$$R = k \ln \frac{S}{S_0}$$

where R is the response, S is the stimulus, and S_0 is the lowest level of stimulus that can be detected. Find dR/dS.

70. **Psychology: learning.** A mathematical model for the average of a group of people learning to type is given by

$$N(t) = 10 + 6 \ln t \qquad t \ge 1$$

where $N(t)$ is the number of words per minute typed after t hours of instruction and practice (2 hours per day, 5 days per week). What is the rate of learning after 10 hours of instruction and practice? After 100 hours?

71. **Continuous Compound Interest.** An investment of $10,000 earns interest at an annual rate of 7.5% compounded continuously.

(A) Find the instantaneous rate of change of the amount in the account after 1 year.

(B) Find the instantaneous rate of change of the amount in the account at the time the amount is equal to $12,500.

72. **Continuous Compound Interest.** An investment of $25,000 earns interest at an annual rate of 8.4% compounded continuously.

(A) Find the instantaneous rate of change of the amount in the account after 2 years.

(B) Find the instantaneous rate of change of the amount in the account at the time the amount is equal to $30,000.

Answers to Matched Problems

1. (A) $4e^x + 16x + 7$ (B) $7x^6 - 5x^4 - 1 + e^x$

2. (A) $30x^2 - \dfrac{100}{x}$ (B) $\dfrac{5}{x} + e^x$

3. (A) $10x^9 + 10^x \ln 10$ (B) $\left(\dfrac{1}{\ln 2} - \dfrac{6}{\ln 5} \right)\dfrac{1}{x}$

4. The price is decreasing at the rate of $0.18 per blanket.

5. (A) $359.17 per year   (B) $480 per year

6. The circulation in 2020 is approximately 49,800 and is decreasing at the rate of 200 per year.

3.3   Derivatives of Products and Quotients

- Derivatives of Products
- Derivatives of Quotients

The derivative properties discussed in Section 2.5 add substantially to our ability to compute and apply derivatives to many practical problems. In this and the next two sections, we add a few more properties that will increase this ability even further.

Derivatives of Products

In Section 2.5, we found that the derivative of a sum is the sum of the derivatives. Is the derivative of a product the product of the derivatives?

Explore and Discuss 1   Let $F(x) = x^2$, $S(x) = x^3$, and $f(x) = F(x)S(x) = x^5$. Which of the following is $f'(x)$?

(A) $F'(x)S'(x)$ (B) $F(x)S'(x)$

(C) $F'(x)S(x)$ (D) $F(x)S'(x) + F'(x)S(x)$

Comparing the various expressions computed in Explore and Discuss 1, we see that the derivative of a product is not the product of the derivatives.

Using the definition of the derivative and the four-step process, we can show that

The derivative of the product of two functions is the first times the derivative of the second, plus the second times the derivative of the first.

This **product rule** is expressed more compactly in Theorem 1, with notation chosen to aid memorization (F for "first", S for "second").

THEOREM 1   Product Rule

If

$$y = f(x) = F(x)S(x)$$

and if $F'(x)$ and $S'(x)$ exist, then

$$f'(x) = F(x)S'(x) + S(x)F'(x)$$

Using simplified notation,

$$y' = FS' + SF' \qquad \text{or} \qquad \frac{dy}{dx} = F\frac{dS}{dx} + S\frac{dF}{dx}$$

EXAMPLE 1   Differentiating a Product   Use two different methods to find $f'(x)$ for

$$f(x) = 2x^2(3x^4 - 2).$$

SOLUTION

Method 1. Use the product rule with $F(x) = 2x^2$ and $S(x) = 3x^4 - 2$:

$$\begin{aligned}
f'(x) &= 2x^2(3x^4 - 2)' + (3x^4 - 2)(2x^2)' \quad &\text{First times derivative of} \\
&= 2x^2(12x^3) + (3x^4 - 2)(4x) \quad &\text{second, plus second times} \\
&= 24x^5 + 12x^5 - 8x \quad &\text{derivative of first} \\
&= 36x^5 - 8x
\end{aligned}$$

Method 2. Multiply first; then find the derivative:

$$f(x) = 2x^2(3x^4 - 2) = 6x^6 - 4x^2$$
$$f'(x) = 36x^5 - 8x$$

Matched Problem 1 Use two different methods to find $f'(x)$ for
$$f(x) = 3x^3(2x^2 - 3x + 1).$$

Some products we encounter can be differentiated by either method illustrated in Example 1. In other situations, the product rule *must* be used. Unless instructed otherwise, you should use the product rule to differentiate all products in this section in order to gain experience with this important differentiation rule.

EXAMPLE 2 Tangent Lines Let $f(x) = (2x - 9)(x^2 + 6)$.
(A) Find the equation of the line tangent to the graph of $f(x)$ at $x = 3$.
(B) Find the value(s) of x where the tangent line is horizontal.

SOLUTION
(A) First, find $f'(x)$:

$$f'(x) = (2x - 9)(x^2 + 6)' + (x^2 + 6)(2x - 9)'$$
$$= (2x - 9)(2x) + (x^2 + 6)(2)$$

Then, find $f(3)$ and $f'(3)$:

$$f(3) = [2(3) - 9](3^2 + 6) = (-3)(15) = -45$$
$$f'(3) = [2(3) - 9]2(3) + (3^2 + 6)(2) = -18 + 30 = 12$$

Now, find the equation of the tangent line at $x = 3$:

$$y - y_1 = m(x - x_1) \quad \text{$y_1 = f(x_1) = f(3) = -45$}$$
$$y - (-45) = 12(x - 3) \quad \text{$m = f'(x_1) = f'(3) = 12$}$$
$$y = 12x - 81 \quad \text{Tangent line at $x = 3$}$$

(B) The tangent line is horizontal at any value of x such that $f'(x) = 0$, so

$$f'(x) = (2x - 9)2x + (x^2 + 6)2 = 0$$
$$6x^2 - 18x + 12 = 0$$
$$x^2 - 3x + 2 = 0$$
$$(x - 1)(x - 2) = 0$$
$$x = 1, 2$$

The tangent line is horizontal at $x = 1$ and at $x = 2$.

Matched Problem 2 Repeat Example 2 for $f(x) = (2x + 9)(x^2 - 12)$.

CONCEPTUAL INSIGHT

As Example 2 illustrates, the way we write $f'(x)$ depends on what we want to do. If we are interested only in evaluating $f'(x)$ at specified values of x, then the form in part (A) is sufficient. However, if we want to solve $f'(x) = 0$, we must multiply and collect like terms, as we did in part (B).

EXAMPLE 3 Finding Derivatives Find $f'(x)$ for
(A) $f(x) = 2x^3 e^x$ (B) $f(x) = 6x^4 \ln x$

SOLUTIONS

(A) $f'(x) = 2x^3(e^x)' + e^x(2x^3)'$

$= 2x^3 e^x + e^x(6x^2)$

$= 2x^2 e^x(x + 3)$

(B) $f'(x) = 6x^4(\ln x)' + (\ln x)(6x^4)'$

$= 6x^4 \dfrac{1}{x} + (\ln x)(24x^3)$

$= 6x^3 + 24x^3 \ln x$

$= 6x^3(1 + 4 \ln x)$

Matched Problem 3) Find $f'(x)$ for

(A) $f(x) = 5x^8 e^x$

(B) $f(x) = x^7 \ln x$

Derivatives of Quotients

The derivative of a quotient of two functions is not the quotient of the derivatives of the two functions.

Explore and Discuss 2 Let $T(x) = x^5$, $B(x) = x^2$, and

$$f(x) = \frac{T(x)}{B(x)} = \frac{x^5}{x^2} = x^3$$

Which of the following is $f'(x)$?

(A) $\dfrac{T'(x)}{B'(x)}$

(B) $\dfrac{T'(x)B(x)}{[B(x)]^2}$

(C) $\dfrac{T(x)B'(x)}{[B(x)]^2}$

(D) $\dfrac{T'(x)B(x)}{[B(x)]^2} - \dfrac{T(x)B'(x)}{[B(x)]^2} = \dfrac{B(x)T'(x) - T(x)B'(x)}{[B(x)]^2}$

The expressions in Explore and Discuss 2 suggest that the derivative of a quotient leads to a more complicated quotient than expected.

If $T(x)$ and $B(x)$ are any two differentiable functions and

$$f(x) = \frac{T(x)}{B(x)}$$

then

$$f'(x) = \frac{B(x)T'(x) - T(x)B'(x)}{[B(x)]^2}$$

Therefore,

The derivative of the quotient of two functions is the denominator times the derivative of the numerator, minus the numerator times the derivative of the denominator, divided by the denominator squared.

This **quotient rule** is expressed more compactly in Theorem 2, with notation chosen to aid memorization (T for "top", B for "bottom").

THEOREM 2 Quotient Rule

If

$$y = f(x) = \frac{T(x)}{B(x)}$$

and if $T'(x)$ and $B'(x)$ exist, then

$$f'(x) = \frac{B(x)T'(x) - T(x)B'(x)}{[B(x)]^2}$$

Using simplified notation,

$$y' = \frac{BT' - TB'}{B^2} \quad \text{or} \quad \frac{dy}{dx} = \frac{B\dfrac{dT}{dx} - T\dfrac{dB}{dx}}{B^2}$$

EXAMPLE 4 Differentiating Quotients

(A) If $f(x) = \dfrac{x^2}{2x - 1}$, find $f'(x)$.   (B) If $y = \dfrac{t^2 - t}{t^3 + 1}$, find y'.

(C) Find $\dfrac{d}{dx}\dfrac{x^2 - 3}{x^2}$ by using the quotient rule and also by splitting the fraction into two fractions.

SOLUTION

(A) Use the quotient rule with $T(x) = x^2$ and $B(x) = 2x - 1$;

$$f'(x) = \frac{(2x - 1)(x^2)' - x^2(2x - 1)'}{(2x - 1)^2}$$

The denominator times the derivative of the numerator, minus the numerator times the derivative of the denominator, divided by the square of the denominator

$$= \frac{(2x - 1)(2x) - x^2(2)}{(2x - 1)^2}$$

$$= \frac{4x^2 - 2x - 2x^2}{(2x - 1)^2}$$

$$= \frac{2x^2 - 2x}{(2x - 1)^2}$$

(B) $y' = \dfrac{(t^3 + 1)(t^2 - t)' - (t^2 - t)(t^3 + 1)'}{(t^3 + 1)^2}$

$$= \frac{(t^3 + 1)(2t - 1) - (t^2 - t)(3t^2)}{(t^3 + 1)^2}$$

$$= \frac{2t^4 - t^3 + 2t - 1 - 3t^4 + 3t^3}{(t^3 + 1)^2}$$

$$= \frac{-t^4 + 2t^3 + 2t - 1}{(t^3 + 1)^2}$$

(C) Method 1. Use the quotient rule:

$$\frac{d}{dx}\frac{x^2 - 3}{x^2} = \frac{x^2\dfrac{d}{dx}(x^2 - 3) - (x^2 - 3)\dfrac{d}{dx}x^2}{(x^2)^2}$$

$$= \frac{x^2(2x) - (x^2 - 3)2x}{x^4}$$

$$= \frac{2x^3 - 2x^3 + 6x}{x^4} = \frac{6x}{x^4} = \frac{6}{x^3}$$

Method 2. Split into two fractions:

$$\frac{x^2 - 3}{x^2} = \frac{x^2}{x^2} - \frac{3}{x^2} = 1 - 3x^{-2}$$

$$\frac{d}{dx}(1 - 3x^{-2}) = 0 - 3(-2)x^{-3} = \frac{6}{x^3}$$

Comparing methods 1 and 2, we see that it often pays to change an expression algebraically before choosing a differentiation formula.

Matched Problem 4 Find

(A) $f'(x)$ for $f(x) = \dfrac{2x}{x^2 + 3}$

(B) y' for $y = \dfrac{t^3 - 3t}{t^2 - 4}$

(C) $\dfrac{d}{dx} \dfrac{2 + x^3}{x^3}$ in two ways

EXAMPLE 5   Finding Derivatives   Find $f'(x)$ for

(A) $f(x) = \dfrac{3e^x}{1 + e^x}$

(B) $f(x) = \dfrac{\ln x}{2x + 5}$

SOLUTIONS

(A) $f'(x) = \dfrac{(1 + e^x)(3e^x)' - 3e^x(1 + e^x)'}{(1 + e^x)^2}$

$= \dfrac{(1 + e^x)3e^x - 3e^x e^x}{(1 + e^x)^2}$

$= \dfrac{3e^x}{(1 + e^x)^2}$

(B) $f'(x) = \dfrac{(2x + 5)(\ln x)' - (\ln x)(2x + 5)'}{(2x + 5)^2}$

$= \dfrac{(2x + 5) \cdot \dfrac{1}{x} - (\ln x)(2)}{(2x + 5)^2}$ Multiply by $\dfrac{x}{x}$

$= \dfrac{2x + 5 - 2x \ln x}{x(2x + 5)^2}$

Matched Problem 5 Find $f'(x)$ for

(A) $f(x) = \dfrac{x^3}{e^x + 2}$

(B) $f(x) = \dfrac{4x}{1 + \ln x}$

EXAMPLE 6   Sales Analysis   The total sales S (in thousands of games) of a video game t months after the game is introduced are given by

$$S(t) = \frac{125t^2}{t^2 + 100}$$

(A) Find $S'(t)$.
(B) Find $S(10)$ and $S'(10)$. Write a brief interpretation of these results.
(C) Use the results from part (B) to estimate the total sales after 11 months.

SOLUTION

(A) $S'(t) = \dfrac{(t^2 + 100)(125t^2)' - 125t^2(t^2 + 100)'}{(t^2 + 100)^2}$

$= \dfrac{(t^2 + 100)(250t) - 125t^2(2t)}{(t^2 + 100)^2}$

$= \dfrac{250t^3 + 25{,}000t - 250t^3}{(t^2 + 100)^2}$

$= \dfrac{25{,}000t}{(t^2 + 100)^2}$

(B) $S(10) = \dfrac{125(10)^2}{10^2 + 100} = 62.5$ and $S'(10) = \dfrac{25,000(10)}{(10^2 + 100)^2} = 6.25.$

Total sales after 10 months are 62,500 games, and sales are increasing at the rate of 6,250 games per month.

(C) Total sales will increase by approximately 6,250 games during the next month, so the estimated total sales after 11 months are $62,500 + 6,250 = 68,750$ games.

Matched Problem 6) Refer to Example 6. Suppose that the total sales S (in thousands of games) t months after the game is introduced are given by

$$S(t) = \frac{150t}{t + 3}$$

(A) Find $S'(t)$.

(B) Find $S(12)$ and $S'(12)$. Write a brief interpretation of these results.

(C) Use the results from part (B) to estimate the total sales after 13 months.

Exercises 3.3

Skills Warm-up Exercises

W   *In Problems 1–4, find functions $F(x)$ and $S(x)$, neither a constant function, such that the given function $f(x)$ is equal to the product $F(x)S(x)$. (If necessary, review Section A.2).*

1. $f(x) = 5x^3 - 4x^3 \ln x$

2. $f(x) = 6e^x + 2x^2e^x + 3x^4e^x$

3. $f(x) = x^3e^x + 2x^3 + 3e^x + 6$

4. $f(x) = 20 + 5 \ln x + 4x^2 + x^2 \ln x$

In Problems 5–8, find functions $T(x)$ and $B(x)$, neither a constant function, such that the given function $f(x)$ is equal to the quotient $T(x)/B(x)$. (If necessary, review Section A.2).

5. $f(x) = 9x^2e^{-5x}$ **6.** $f(x) = 8x^{-2} \ln x$

7. $f(x) = \dfrac{3}{x^2} + \dfrac{e^x}{x^4}$ **8.** $f(x) = \dfrac{1}{x} + \dfrac{2}{x^2} + \dfrac{4}{x^3}$

Answers to most of the following problems in this exercise set contain both an unsimplified form and a simplified form of the derivative. When checking your work, first check that you applied the rules correctly and then check that you performed the algebraic simplification correctly. Unless instructed otherwise, when differentiating a product, use the product rule rather than performing the multiplication first.

In Problems 9–34, find $f'(x)$ and simplify.

9. $f(x) = 2x^3(x^2 - 2)$ **10.** $f(x) = 5x^2(x^3 + 2)$

11. $f(x) = (x - 3)(2x - 1)$

12. $f(x) = (3x + 2)(4x - 5)$

13. $f(x) = \dfrac{x}{x - 3}$ **14.** $f(x) = \dfrac{3x}{2x + 1}$

15. $f(x) = \dfrac{2x + 3}{x - 2}$ **16.** $f(x) = \dfrac{3x - 4}{2x + 3}$

17. $f(x) = 3xe^x$ **18.** $f(x) = x^2e^x$

19. $f(x) = x^3 \ln x$ **20.** $f(x) = 5x \ln x$

21. $f(x) = (x^2 + 1)(2x - 3)$

22. $f(x) = (3x + 5)(x^2 - 3)$

23. $f(x) = (0.4x + 2)(0.5x - 5)$

24. $f(x) = (0.5x - 4)(0.2x + 1)$

25. $f(x) = \dfrac{x^2 + 1}{2x - 3}$ **26.** $f(x) = \dfrac{3x + 5}{x^2 - 3}$

27. $f(x) = (x^2 + 2)(x^2 - 3)$

28. $f(x) = (x^2 - 4)(x^2 + 5)$

29. $f(x) = \dfrac{x^2 + 2}{x^2 - 3}$ **30.** $f(x) = \dfrac{x^2 - 4}{x^2 + 5}$

31. $f(x) = \dfrac{e^x}{x^2 + 1}$ **32.** $f(x) = \dfrac{1 - e^x}{1 + e^x}$

33. $f(x) = \dfrac{\ln x}{1 + x}$ **34.** $f(x) = \dfrac{2x}{1 + \ln x}$

In Problems 35–46, find $h'(x)$, where $f(x)$ is an unspecified differentiable function.

35. $h(x) = xf(x)$ **36.** $h(x) = x^2f(x)$

37. $h(x) = x^3f(x)$ **38.** $h(x) = \dfrac{f(x)}{x}$

39. $h(x) = \dfrac{f(x)}{x^2}$ **40.** $h(x) = \dfrac{f(x)}{x^3}$

41. $h(x) = \dfrac{x}{f(x)}$ **42.** $h(x) = \dfrac{x^2}{f(x)}$

43. $h(x) = e^x f(x)$

44. $h(x) = \dfrac{e^x}{f(x)}$

45. $h(x) = \dfrac{\ln x}{f(x)}$

46. $h(x) = \dfrac{f(x)}{\ln x}$

In Problems 47–56, find the indicated derivatives and simplify.

47. $f'(x)$ for $f(x) = (2x + 1)(x^2 - 3x)$

48. y' for $y = (x^3 + 2x^2)(3x - 1)$

49. $\dfrac{dy}{dt}$ for $y = (2.5t - t^2)(4t + 1.4)$

50. $\dfrac{d}{dt}[(3 - 0.4t^3)(0.5t^2 - 2t)]$

51. y' for $y = \dfrac{5x - 3}{x^2 + 2x}$

52. $f'(x)$ for $f(x) = \dfrac{3x^2}{2x - 1}$

53. $\dfrac{d}{dw} \dfrac{w^2 - 3w + 1}{w^2 - 1}$

54. $\dfrac{dy}{dw}$ for $y = \dfrac{w^4 - w^3}{3w - 1}$

55. y' for $y = (1 + x - x^2) e^x$

56. $\dfrac{dy}{dt}$ for $y = (1 + e^t) \ln t$

In Problems 57–60:

(A) *Find $f'(x)$ using the quotient rule, and*

(B) *Explain how $f'(x)$ can be found easily without using the quotient rule.*

57. $f(x) = \dfrac{1}{x}$

58. $f(x) = \dfrac{-1}{x^2}$

59. $f(x) = \dfrac{-3}{x^4}$

60. $f(x) = \dfrac{2}{x^3}$

In Problems 61–66, find $f'(x)$ and find the equation of the line tangent to the graph of f at $x = 2$.

61. $f(x) = (1 + 3x)(5 - 2x)$

62. $f(x) = (7 - 3x)(1 + 2x)$

63. $f(x) = \dfrac{x - 8}{3x - 4}$

64. $f(x) = \dfrac{2x - 5}{2x - 3}$

65. $f(x) = \dfrac{x}{2^x}$

66. $f(x) = (x - 2) \ln x$

In Problems 67–70, find $f'(x)$ and find the value(s) of x where $f'(x) = 0$.

67. $f(x) = (2x - 15)(x^2 + 18)$

68. $f(x) = (2x - 3)(x^2 - 6)$

69. $f(x) = \dfrac{x}{x^2 + 1}$

70. $f(x) = \dfrac{x}{x^2 + 9}$

In Problems 71–74, find $f'(x)$ in two ways: (1) using the product or quotient rule and (2) simplifying first.

71. $f(x) = x^3(x^4 - 1)$

72. $f(x) = x^4(x^3 - 1)$

73. $f(x) = \dfrac{x^3 + 9}{x^3}$

74. $f(x) = \dfrac{x^4 + 4}{x^4}$

In Problems 75–92, find each indicated derivative and simplify.

75. $f(w) = (w + 1)2^w$

76. $g(w) = (w - 5) \log_3 w$

77. $\dfrac{dy}{dx}$ for $y = 9x^{1/3}(x^3 + 5)$

78. $\dfrac{d}{dx}[(4x^{1/2} - 1)(3x^{1/3} + 2)]$

79. y' for $y = \dfrac{\log_2 x}{1 + x^2}$

80. $\dfrac{dy}{dx}$ for $y = \dfrac{10^x}{1 + x^4}$

81. $f'(x)$ for $f(x) = \dfrac{6^3 \sqrt{x}}{x^2 - 3}$

82. y' for $y = \dfrac{2\sqrt{x}}{x^2 - 3x + 1}$

83. $g'(t)$ if $g(t) = \dfrac{0.2t}{3t^2 - 1}$

84. $h'(t)$ if $h(t) = \dfrac{-0.05t^2}{2t + 1}$

85. $\dfrac{d}{dx}[4x \log x^5]$

86. $\dfrac{d}{dt}[10^t \log t]$

87. $\dfrac{d}{dx} \dfrac{x^3 - 2x^2}{\sqrt[3]{x^2}}$

88. $\dfrac{dy}{dx}$ for $y = \dfrac{x^2 - 3x + 1}{\sqrt[4]{x}}$

89. $f'(x)$ for $f(x) = \dfrac{(2x^2 - 1)(x^2 + 3)}{x^2 + 1}$

90. y' for $y = \dfrac{2x - 1}{(x^3 + 2)(x^2 - 3)}$

91. $\dfrac{dy}{dt}$ for $y = \dfrac{t \ln t}{e^t}$

92. $\dfrac{dy}{du}$ for $y = \dfrac{u^2 e^u}{1 + \ln u}$

Applications

93. **Sales analysis.** The total sales S (in thousands of DVDs) of a DVD are given by

$$S(t) = \dfrac{90t^2}{t^2 + 50}$$

where t is the number of months since the release of the DVD.

(A) Find $S'(t)$.

(B) Find $S(10)$ and $S'(10)$. Write a brief interpretation of these results.

(C) Use the results from part (B) to estimate the total sales after 11 months.

94. Sales analysis. A communications company has installed a new cable television system in a city. The total number N (in thousands) of subscribers t months after the installation of the system is given by

$$N(t) = \frac{180t}{t + 4}$$

(A) Find $N'(t)$.

(B) Find $N(16)$ and $N'(16)$. Write a brief interpretation of these results.

(C) Use the results from part (B) to estimate the total number of subscribers after 17 months.

95. Price–demand equation. According to economic theory, the demand x for a quantity in a free market decreases as the price p increases (see the figure). Suppose that the number x of DVD players people are willing to buy per week from a retail chain at a price of p is given by

$$x = \frac{4{,}000}{0.1p + 1} \qquad 10 \le p \le 70$$

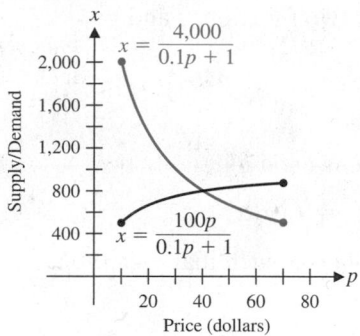

Figure for 95 and 96

(A) Find dx/dp.

(B) Find the demand and the instantaneous rate of change of demand with respect to price when the price is \$40. Write a brief interpretation of these results.

(C) Use the results from part (B) to estimate the demand if the price is increased to \$41.

96. Price–supply equation. According to economic theory, the supply x of a quantity in a free market increases as the price p increases (see the figure). Suppose that the number x of DVD players a retail chain is willing to sell per week at a price of p is given by

$$x = \frac{100p}{0.1p + 1} \qquad 10 \le p \le 70$$

(A) Find dx/dp.

(B) Find the supply and the instantaneous rate of change of supply with respect to price when the price is \$40. Write a brief verbal interpretation of these results.

(C) Use the results from part (B) to estimate the supply if the price is increased to \$41.

97. Medicine. A drug is injected into a patient's bloodstream through her right arm. The drug concentration (in milligrams per cubic centimeter) in the bloodstream of the left arm t hours after the injection is given by

$$C(t) = \frac{0.14t}{t^2 + 1}$$

(A) Find $C'(t)$.

(B) Find $C'(0.5)$ and $C'(3)$, and interpret the results.

98. Drug sensitivity. One hour after a dose of x milligrams of a particular drug is administered to a person, the change in body temperature $T(x)$, in degrees Fahrenheit, is given approximately by

$$T(x) = x^2\left(1 - \frac{x}{9}\right) \qquad 0 \le x \le 7$$

The rate $T'(x)$ at which T changes with respect to the size of the dosage x is called the *sensitivity* of the body to the dosage.

(A) Use the product rule to find $T'(x)$.

(B) Find $T'(1)$, $T'(3)$, and $T'(6)$.

Answers to Matched Problems

1. $30x^4 - 36x^3 + 9x^2$

2. (A) $y = 84x - 297$ (B) $x = -4, x = 1$

3. (A) $5x^8 e^x + e^x(40x^7) = 5x^7(x + 8)e^x$

   (B) $x^7 \cdot \dfrac{1}{x} + \ln x\,(7x^6) = x^6\,(1 + 7\ln x)$

4. (A) $\dfrac{(x^2 + 3)2 - (2x)(2x)}{(x^2 + 3)^2} = \dfrac{6 - 2x^2}{(x^2 + 3)^2}$

   (B) $\dfrac{(t^2 - 4)(3t^2 - 3) - (t^3 - 3t)(2t)}{(t^2 - 4)^2} = \dfrac{t^4 - 9t^2 + 12}{(t^2 - 4)^2}$

   (C) $-\dfrac{6}{x^4}$

5. (A) $\dfrac{(e^x + 2)\,3x^2 - x^3 e^x}{(e^x + 2)^2}$

   (B) $\dfrac{(1 + \ln x)\,4 - 4x\,\dfrac{1}{x}}{(1 + \ln x)^2} = \dfrac{4\ln x}{(1 + \ln x)^2}$

6. (A) $S'(t) = \dfrac{450}{(t + 3)^2}$

   (B) $S(12) = 120; S'(12) = 2$. After 12 months, the total sales are 120,000 games, and sales are increasing at the rate of 2,000 games per month.

   (C) 122,000 games

3.4 The Chain Rule

- Composite Functions
- General Power Rule
- The Chain Rule

The word *chain* in the name "chain rule" comes from the fact that a function formed by composition involves a chain of functions—that is, a function of a function. The *chain rule* enables us to compute the derivative of a composite function in terms of the derivatives of the functions making up the composite. In this section, we review composite functions, introduce the chain rule by means of a special case known as the *general power rule,* and then discuss the chain rule itself.

Composite Functions

The function $m(x) = (x^2 + 4)^3$ is a combination of a quadratic function and a cubic function. To see this more clearly, let

$$y = f(u) = u^3 \quad \text{and} \quad u = g(x) = x^2 + 4$$

We can express y as a function of x:

$$y = f(u) = f[g(x)] = [x^2 + 4]^3 = m(x)$$

The function m is the *composite* of the two functions f and g.

> **DEFINITION** Composite Functions
> A function m is a **composite** of functions f and g if
> $$m(x) = f[g(x)]$$
> The domain of m is the set of all numbers x such that x is in the domain of g, and $g(x)$ is in the domain of f.

The composite m of functions f and g is pictured in Figure 1. The domain of m is the shaded subset of the domain of g (Fig. 1); it consists of all numbers x such that x is in the domain of g and $g(x)$ is in the domain of f. Note that the functions f and g play different roles. The function g, which is on the *inside* or *interior* of the square brackets in $f[g(x)]$, is applied first to x. Then function f, which appears on the *outside* or *exterior* of the square brackets, is applied to $g(x)$, provided $g(x)$ is in the domain of f. Because f and g play different roles, the composite of f and g is usually a different function than the composite of g and f, as illustrated by Example 1.

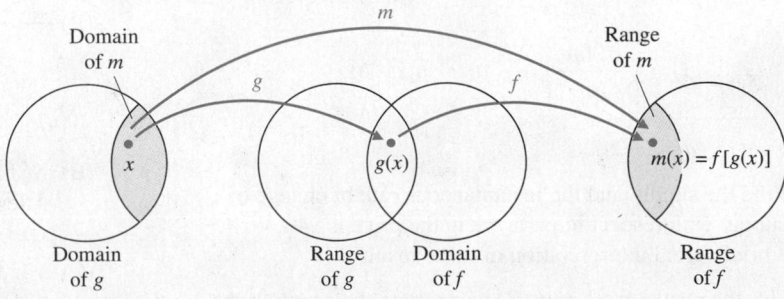

Figure 1　**The composite m of f and g**

EXAMPLE 1 Composite Functions Let $f(u) = e^u$ and $g(x) = -3x$. Find $f[g(x)]$ and $g[f(u)]$.

SOLUTION

$$f[g(x)] = f(-3x) = e^{-3x}$$
$$g[f(u)] = g(e^u) = -3e^u$$

Matched Problem 1 Let $f(u) = 2u$ and $g(x) = e^x$. Find $f[g(x)]$ and $g[f(u)]$.

EXAMPLE 2 Composite Functions Write each function as a composite of two simpler functions.

(A) $y = 100e^{0.04x}$ (B) $y = \sqrt{4 - x^2}$

SOLUTION

(A) Let

$$y = f(u) = 100e^u$$
$$u = g(x) = 0.04x$$

Check: $y = f[g(x)] = f(0.04x) = 100e^{0.04x}$

(B) Let

$$y = f(u) = \sqrt{u}$$
$$u = g(x) = 4 - x^2$$

Check: $y = f[g(x)] = f(4 - x^2) = \sqrt{4 - x^2}$

Matched Problem 2 Write each function as a composite of two simpler functions.

(A) $y = 50e^{-2x}$ (B) $y = \sqrt[3]{1 + x^3}$

CONCEPTUAL INSIGHT

There can be more than one way to express a function as a composite of simpler functions. Choosing $y = f(u) = 100u$ and $u = g(x) = e^{0.04x}$ in Example 2A produces the same result:

$$y = f[g(x)] = 100g(x) = 100e^{0.04x}$$

Since we will be using composition as a means to an end (finding a derivative), usually it will not matter which functions you choose for the composition.

General Power Rule

We have already made extensive use of the power rule,

$$\frac{d}{dx}x^n = nx^{n-1} \tag{1}$$

Can we apply rule (1) to find the derivative of the composite function $m(x) = p[u(x)] = [u(x)]^n$, where p is the power function $p(u) = u^n$ and $u(x)$ is a differentiable function? In other words, is rule (1) valid if x is replaced by $u(x)$?

Explore and Discuss 1 Let $u(x) = 2x^2$ and $m(x) = [u(x)]^3 = 8x^6$. Which of the following is $m'(x)$?

(A) $3[u(x)]^2$ (B) $3[u'(x)]^2$ (C) $3[u(x)]^2 u'(x)$

The calculations in Explore and Discuss 1 show that we cannot find the derivative of $[u(x)]^n$ simply by replacing x with $u(x)$ in equation (1).

How can we find a formula for the derivative of $[u(x)]^n$, where $u(x)$ is an arbitrary differentiable function? Let's begin by considering the derivatives of $[u(x)]^2$ and $[u(x)]^3$ to see if a general pattern emerges. Since $[u(x)]^2 = u(x)u(x)$, we use the product rule to write

$$\frac{d}{dx}[u(x)]^2 = \frac{d}{dx}[u(x)u(x)]$$

$$= u(x)u'(x) + u(x)u'(x)$$

$$= 2u(x)u'(x) \tag{2}$$

Because $[u(x)]^3 = [u(x)]^2 u(x)$, we use the product rule and the result in equation (2) to write

$$\frac{d}{dx}[u(x)]^3 = \frac{d}{dx}\{[u(x)]^2 u(x)\} \qquad \text{Use equation (2) to}$$
$$\text{substitute for}$$

$$= [u(x)]^2 \frac{d}{dx}u(x) + u(x)\frac{d}{dx}[u(x)]^2 \qquad \frac{d}{dx}[u(x)]^2.$$

$$= [u(x)]^2 u'(x) + u(x)[2u(x)u'(x)]$$

$$= 3[u(x)]^2 u'(x)$$

Continuing in this fashion, we can show that

$$\frac{d}{dx}[u(x)]^n = n[u(x)]^{n-1}u'(x) \qquad n \text{ a positive integer} \tag{3}$$

Using more advanced techniques, we can establish formula (3) for all real numbers n, obtaining the **general power rule**.

THEOREM 1 General Power Rule

If $u(x)$ is a differentiable function, n is any real number, and

$$y = f(x) = [u(x)]^n$$

then

$$f'(x) = n[u(x)]^{n-1}u'(x)$$

Using simplified notation,

$$y' = nu^{n-1}u' \qquad \text{or} \qquad \frac{d}{dx}u^n = nu^{n-1}\frac{du}{dx} \qquad \text{where } u = u(x)$$

EXAMPLE 3 Using the General Power Rule Find the indicated derivatives:

(A) $f'(x)$ if $f(x) = (3x + 1)^4$

(B) y' if $y = (x^3 + 4)^7$

(C) $\dfrac{d}{dt}\dfrac{1}{(t^2 + t + 4)^3}$

(D) $\dfrac{dh}{dw}$ if $h(w) = \sqrt{3 - w}$

SOLUTION

(A) $f(x) = (3x + 1)^4$ Let $u = 3x + 1, n = 4$.

$$f'(x) = 4(3x + 1)^3(3x + 1)' \quad\quad nu^{n-1}\frac{du}{dx}$$

$$= 4(3x + 1)^3\, 3 \quad\quad\quad\quad \frac{du}{dx} = 3$$

$$= 12(3x + 1)^3$$

(B) $y = (x^3 + 4)^7$ Let $u = (x^3 + 4), n = 7$.

$$y' = 7(x^3 + 4)^6(x^3 + 4)' \quad\quad nu^{n-1}\frac{du}{dx}$$

$$= 7(x^3 + 4)^6\, 3x^2 \quad\quad\quad \frac{du}{dx} = 3x^2$$

$$= 21x^2(x^3 + 4)^6$$

(C) $\dfrac{d}{dt}\dfrac{1}{(t^2 + t + 4)^3}$

$$= \frac{d}{dt}(t^2 + t + 4)^{-3} \quad\quad\quad\quad \text{Let } u = t^2 + t + 4, n = -3.$$

$$= -3(t^2 + t + 4)^{-4}(t^2 + t + 4)' \quad\quad nu^{n-1}\frac{du}{dt}$$

$$= -3(t^2 + t + 4)^{-4}(2t + 1) \quad\quad\quad \frac{du}{dt} = 2t + 1$$

$$= \frac{-3(2t + 1)}{(t^2 + t + 4)^4}$$

(D) $h(w) = \sqrt{3 - w} = (3 - w)^{1/2}$ Let $u = 3 - w, n = \dfrac{1}{2}$.

$$\frac{dh}{dw} = \frac{1}{2}(3 - w)^{-1/2}(3 - w)' \quad\quad nu^{n-1}\frac{du}{dw}$$

$$= \frac{1}{2}(3 - w)^{-1/2}(-1) \quad\quad\quad\quad \frac{du}{dw} = -1$$

$$= -\frac{1}{2(3 - w)^{1/2}} \quad \text{or} \quad -\frac{1}{2\sqrt{3 - w}}$$

Matched Problem 3   Find the indicated derivatives:

(A) $h'(x)$ if $h(x) = (5x + 2)^3$

(B) y' if $y = (x^4 - 5)^5$

(C) $\dfrac{d}{dt}\dfrac{1}{(t^2 + 4)^2}$

(D) $\dfrac{dg}{dw}$ if $g(w) = \sqrt{4 - w}$

    Notice that we used two steps to differentiate each function in Example 3. First, we applied the general power rule, and then we found du/dx. As you gain experience with the general power rule, you may want to combine these two steps. If you do this, be certain to multiply by du/dx. For example,

$$\frac{d}{dx}(x^5 + 1)^4 = 4(x^5 + 1)^3 5x^4 \quad \text{Correct}$$

$$\frac{d}{dx}(x^5 + 1)^4 \neq 4(x^5 + 1)^3 \quad \text{$du/dx = 5x^4$ is missing}$$

If we let $u(x) = x$, then $du/dx = 1$, and the general power rule reduces to the (ordinary) power rule discussed in Section 2.5. Compare the following:

$$\frac{d}{dx}x^n = nx^{n-1} \qquad \text{Yes—power rule}$$

$$\frac{d}{dx}u^n = nu^{n-1}\frac{du}{dx} \qquad \text{Yes—general power rule}$$

$$\frac{d}{dx}u^n \neq nu^{n-1} \qquad \text{Unless } u(x) = x + k, \text{ so that } du/dx = 1$$

The Chain Rule

We have used the general power rule to find derivatives of composite functions of the form $f[g(x)]$, where $f(u) = u^n$ is a power function. But what if f is not a power function? Then a more general rule, the *chain rule,* enables us to compute the derivatives of many composite functions of the form $f[g(x)]$.

Suppose that

$$y = m(x) = f[g(x)]$$

is a composite of f and g, where

$$y = f(u) \qquad \text{and} \qquad u = g(x)$$

To express the derivative dy/dx in terms of the derivatives of f and g, we use the definition of a derivative (see Section 2.4).

$$m'(x) = \lim_{h \to 0}\frac{m(x+h) - m(x)}{h} \qquad \begin{array}{l}\text{Substitute } m(x+h) = f[g(x+h)] \\ \text{and } m(x) = f[g(x)].\end{array}$$

$$= \lim_{h \to 0}\frac{f[g(x+h)] - f[g(x)]}{h} \qquad \text{Multiply by } 1 = \frac{g(x+h) - g(x)}{g(x+h) - g(x)}.$$

$$= \lim_{h \to 0}\left[\frac{f[g(x+h)] - f[g(x)]}{h} \cdot \frac{g(x+h) - g(x)}{g(x+h) - g(x)}\right]$$

$$= \lim_{h \to 0}\left[\frac{f[g(x+h)] - f[g(x)]}{g(x+h) - g(x)} \cdot \frac{g(x+h) - g(x)}{h}\right] \qquad (4)$$

We recognize the second factor in equation (4) as the difference quotient for $g(x)$. To interpret the first factor as the difference quotient for $f(u)$, we let $k = g(x+h) - g(x)$. Since $u = g(x)$, we write

$$u + k = g(x) + g(x+h) - g(x) = g(x+h)$$

Substituting in equation (4), we have

$$m'(x) = \lim_{h \to 0}\left[\frac{f(u+k) - f(u)}{k} \cdot \frac{g(x+h) - g(x)}{h}\right] \qquad (5)$$

If we assume that $k = [g(x+h) - g(x)] \to 0$ as $h \to 0$, we can find the limit of each difference quotient in equation (5):

$$m'(x) = \left[\lim_{k \to 0}\frac{f(u+k) - f(u)}{k}\right]\left[\lim_{h \to 0}\frac{g(x+h) - g(x)}{h}\right]$$

$$= f'(u)g'(x)$$

$$= f'[g(x)]g'(x)$$

Therefore, referring to f and g in the composite function $f[g(x)]$ as the exterior function and interior function, respectively,

The derivative of the composite of two functions is the derivative of the exterior, evaluated at the interior, times the derivative of the interior.

This **chain rule** is expressed more compactly in Theorem 2, with notation chosen to aid memorization (E for "exterior", I for "interior").

THEOREM 2 Chain Rule

If $m(x) = E[I(x)]$ is a composite function, then

$$m'(x) = E'[I(x)]I'(x)$$

provided that $E'[I(x)]$ and $I'(x)$ exist.
Equivalently, if $y = E(u)$ and $u = I(x)$, then

$$\frac{dy}{dx} = \frac{dy}{du}\frac{du}{dx}$$

provided that $\dfrac{dy}{du}$ and $\dfrac{du}{dx}$ exist.

EXAMPLE 4 Using the Chain Rule Find the derivative $m'(x)$ of the composite function $m(x)$.

(A) $m(x) = (3x^2 + 1)^{3/2}$ (B) $m(x) = e^{2x^3+5}$ (C) $m(x) = \ln(x^2 - 4x + 2)$

SOLUTION

(A) The function m is the composite of $E(u) = u^{3/2}$ and $I(x) = 3x^2 + 1$. Then $E'(u) = \frac{3}{2}u^{1/2}$ and $I'(x) = 6x$, so by the chain rule,
$m'(x) = \frac{3}{2}(3x^2 + 1)^{1/2}(6x) = 9x(3x^2 + 1)^{1/2}$.

(B) The function m is the composite of $E(u) = e^u$ and $I(x) = 2x^3 + 5$. Then $E'(u) = e^u$ and $I'(x) = 6x^2$, so by the chain rule,
$m'(x) = e^{2x^3+5}(6x^2) = 6x^2e^{2x^3+5}$.

(C) The function m is the composite of $E(u) = \ln u$ and $I(x) = x^2 - 4x + 2$. Then $E'(u) = \frac{1}{u}$ and $I'(x) = 2x - 4$, so by the chain rule,
$m'(x) = \frac{1}{x^2 - 4x + 2}(2x - 4) = \frac{2x - 4}{x^2 - 4x + 2}$.

Matched Problem 4) Find the derivative $m'(x)$ of the composite function $m(x)$.
(A) $m(x) = (2x^3 + 4)^{-5}$ (B) $m(x) = e^{3x^4+6}$ (C) $m(x) = \ln(x^2 + 9x + 4)$

Explore and Discuss 2 Let $m(x) = f[g(x)]$. Use the chain rule and Figures 2 and 3 to find

(A) $f(4)$ (B) $g(6)$ (C) $m(6)$
(D) $f'(4)$ (E) $g'(6)$ (F) $m'(6)$

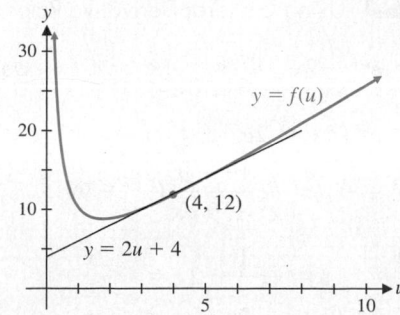

Figure 2

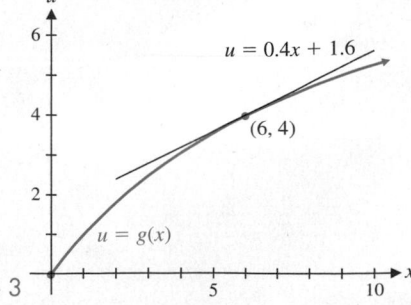

Figure 3

The chain rule can be extended to compositions of three or more functions. For example, if $y = f(w)$, $w = g(u)$, and $u = h(x)$, then

$$\frac{dy}{dx} = \frac{dy}{dw}\frac{dw}{du}\frac{du}{dx}$$

EXAMPLE 5 Using the Chain Rule For $y = h(x) = e^{1+(\ln x)^2}$, find dy/dx.

SOLUTION Note that h is of the form $y = e^w$, where $w = 1 + u^2$ and $u = \ln x$.

$$\frac{dy}{dx} = \frac{dy}{dw}\frac{dw}{du}\frac{du}{dx}$$

$$= e^w(2u)\left(\frac{1}{x}\right)$$

$$= e^{1+u^2}(2u)\left(\frac{1}{x}\right) \qquad \text{Since } w = 1 + u^2$$

$$= e^{1+(\ln x)^2}(2\ln x)\left(\frac{1}{x}\right) \qquad \text{Since } u = \ln x$$

$$= \frac{2}{x}(\ln x)e^{1+(\ln x)^2}$$

Matched Problem 5 For $y = h(x) = [\ln(1 + e^x)]^3$, find dy/dx.

The chain rule generalizes basic derivative rules. We list three general derivative rules here for convenient reference [the first, equation (6), is the general power rule of Theorem 1].

General Derivative Rules

$$\frac{d}{dx}[f(x)]^n = n[f(x)]^{n-1}f'(x) \tag{6}$$

$$\frac{d}{dx}\ln[f(x)] = \frac{1}{f(x)}f'(x) \tag{7}$$

$$\frac{d}{dx}e^{f(x)} = e^{f(x)}f'(x) \tag{8}$$

Unless directed otherwise, you now have a choice between the chain rule and the general derivative rules. However, practicing with the chain rule will help prepare you for concepts that appear later in the text. Examples 4 and 5 illustrate the chain rule method, and the next example illustrates the general derivative rules method.

EXAMPLE 6 Using General Derivative Rules

(A) $\dfrac{d}{dx}e^{2x} = e^{2x}\dfrac{d}{dx}2x$ \qquad Using equation (8)

$\qquad\qquad = e^{2x}(2) = 2e^{2x}$

(B) $\dfrac{d}{dx}\ln(x^2 + 9) = \dfrac{1}{x^2 + 9}\dfrac{d}{dx}(x^2 + 9)$ \qquad Using equation (7)

$\qquad\qquad = \dfrac{1}{x^2 + 9}2x = \dfrac{2x}{x^2 + 9}$

(C) $\dfrac{d}{dx}(1 + e^{x^2})^3 = 3(1 + e^{x^2})^2\dfrac{d}{dx}(1 + e^{x^2})$ Using equation (6)

$\qquad\qquad\qquad = 3(1 + e^{x^2})^2 e^{x^2}\dfrac{d}{dx}x^2$ Using equation (8)

$\qquad\qquad\qquad = 3(1 + e^{x^2})^2 e^{x^2}(2x)$

$\qquad\qquad\qquad = 6xe^{x^2}(1 + e^{x^2})^2$

Matched Problem 6) Find

(A) $\dfrac{d}{dx}\ln(x^3 + 2x)$ (B) $\dfrac{d}{dx}e^{3x^2 + 2}$ (C) $\dfrac{d}{dx}(2 + e^{-x^2})^4$

Exercises 3.4

For many of the problems in this exercise set, the answers in the back of the book include both an unsimplified form and a simplified form. When checking your work, first check that you applied the rules correctly, and then check that you performed the algebraic simplification correctly.

Skills Warm-up Exercises

W   *In Problems 1–4, find f [g(x)]. (If necessary, review Section 1.1).*

1. $f(u) = 3u + 5; g(x) = x^3$

2. $f(u) = u^2 + u; g(x) = 2x - 7$

3. $f(u) = 2u + \ln u; g(x) = x^2 e^x$

4. $f(u) = 4ue^u; g(x) = x^3 + 1$

A *In Problems 5–8, find functions E(u) and I(x) so that y = E[I(x)].*

5. $y = \ln(x^3 - 6x + 10)$ **6.** $y = (2x - 9)^8$

7. $y = \sqrt{x^2 + 4}$ **8.** $y = e^{2x} + 3e^x - 10$

In Problems 9–16, replace ? with an expression that will make the indicated equation valid.

9. $\dfrac{d}{dx}(3x + 4)^4 = 4(3x + 4)^3$ ___?___

10. $\dfrac{d}{dx}(5 - 2x)^6 = 6(5 - 2x)^5$ ___?___

11. $\dfrac{d}{dx}(4 - 2x^2)^3 = 3(4 - 2x^2)^2$ ___?___

12. $\dfrac{d}{dx}(3x^2 + 7)^5 = 5(3x^2 + 7)^4$ ___?___

13. $\dfrac{d}{dx}e^{x^2 + 1} = e^{x^2 + 1}$ ___?___

14. $\dfrac{d}{dx}e^{4x - 2} = e^{4x - 2}$ ___?___

15. $\dfrac{d}{dx}\ln(x^4 + 1) = \dfrac{1}{x^4 + 1}$ ___?___

16. $\dfrac{d}{dx}\ln(x - x^3) = \dfrac{1}{x - x^3}$ ___?___

In Problems 17–38, find f'(x) and simplify.

17. $f(x) = (5 - 2x)^4$ **18.** $f(x) = (9 - 5x)^2$

19. $f(x) = (4 + 0.2x)^5$ **20.** $f(x) = (6 - 0.5x)^4$

21. $f(x) = (3x^2 + 5)^5$ **22.** $f(x) = (5x^2 - 3)^6$

23. $f(x) = 5e^x$ **24.** $f(x) = 10 - 4e^x$

25. $f(x) = e^{5x}$ **26.** $f(x) = 6e^{-2x}$

27. $f(x) = 3e^{-6x}$ **28.** $f(x) = e^{x^2 + 3x + 1}$

29. $f(x) = (2x - 5)^{1/2}$ **30.** $f(x) = (4x + 3)^{1/2}$

31. $f(x) = (x^4 + 1)^{-2}$ **32.** $f(x) = (x^5 + 2)^{-3}$

33. $f(x) = 4 - 2\ln x$ **34.** $f(x) = 8\ln x$

35. $f(x) = 3\ln(1 + x^2)$

36. $f(x) = 2\ln(x^2 - 3x + 4)$

37. $f(x) = (1 + \ln x)^3$

38. $f(x) = (x - 2\ln x)^4$

In Problems 39–44, find f'(x) and the equation of the line tangent to the graph of f at the indicated value of x. Find the value(s) of x where the tangent line is horizontal.

39. $f(x) = (2x - 1)^3$; $x = 1$

40. $f(x) = (3x - 1)^4$; $x = 1$

41. $f(x) = (4x - 3)^{1/2}$; $x = 3$

42. $f(x) = (2x + 8)^{1/2}$; $x = 4$

43. $f(x) = 5e^{x^2 - 4x + 1}$; $x = 0$

44. $f(x) = \ln(1 - x^2 + 2x^4)$; $x = 1$

In Problems 45–60, find the indicated derivative and simplify.

45. y' if $y = 3(x^2 - 2)^4$

46. y' if $y = 2(x^3 + 6)^5$

47. $\dfrac{d}{dt}2(t^2 + 3t)^{-3}$

48. $\dfrac{d}{dt} 3(t^3 + t^2)^{-2}$

49. $\dfrac{dh}{dw}$ if $h(w) = \sqrt{w^2 + 8}$

50. $\dfrac{dg}{dw}$ if $g(w) = \sqrt[3]{3w - 7}$

51. $g'(x)$ if $g(x) = 4xe^{3x}$

52. $h'(x)$ if $h(x) = \dfrac{e^{2x}}{x^2 + 9}$

53. $\dfrac{d}{dx} \dfrac{\ln(1 + x)}{x^3}$

54. $\dfrac{d}{dx}[x^4 \ln(1 + x^4)]$

55. $F'(t)$ if $F(t) = (e^{t^2+1})^3$

56. $G'(t)$ if $G(t) = (1 - e^{2t})^2$

57. y' if $y = \ln(x^2 + 3)^{3/2}$

58. y' if $y = [\ln(x^2 + 3)]^{3/2}$

59. $\dfrac{d}{dw} \dfrac{1}{(w^3 + 4)^5}$

60. $\dfrac{d}{dw} \dfrac{1}{(w^2 - 2)^6}$

In Problems 61–66, find f'(x) and find the equation of the line tangent to the graph of f at the indicated value of x.

61. $f(x) = x(4 - x)^3$; $x = 2$

62. $f(x) = x^2(1 - x)^4$; $x = 2$

63. $f(x) = \dfrac{x}{(2x - 5)^3}$; $x = 3$

64. $f(x) = \dfrac{x^4}{(3x - 8)^2}$; $x = 4$

65. $f(x) = \sqrt{\ln x}$; $x = e$

66. $f(x) = e^{\sqrt{x}}$; $x = 1$

In Problems 67–72, find f'(x) and find the value(s) of x where the tangent line is horizontal.

67. $f(x) = x^2(x - 5)^3$

68. $f(x) = x^3(x - 7)^4$

69. $f(x) = \dfrac{x}{(2x + 5)^2}$

70. $f(x) = \dfrac{x - 1}{(x - 3)^3}$

71. $f(x) = \sqrt{x^2 - 8x + 20}$

72. $f(x) = \sqrt{x^2 + 4x + 5}$

73. A student reasons that the functions $f(x) = \ln[5(x^2 + 3)^4]$ and $g(x) = 4\ln(x^2 + 3)$ must have the same derivative since he has entered $f(x)$, $g(x)$, $f'(x)$, and $g'(x)$ into a graphing calculator, but only three graphs appear (see the figure). Is his reasoning correct? Are $f'(x)$ and $g'(x)$ the same function? Explain.

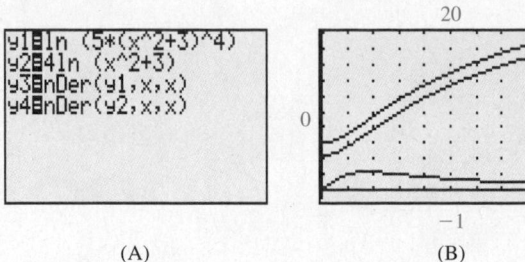

(A)   (B)

Figure for 73

74. A student reasons that the functions $f(x) = (x + 1)\ln(x + 1) - x$ and $g(x) = (x + 1)^{1/3}$ must have the same derivative since she has entered $f(x)$, $g(x)$, $f'(x)$, and $g'(x)$ into a graphing calculator, but only three graphs appear (see the figure). Is her reasoning correct? Are $f'(x)$ and $g'(x)$ the same function? Explain.

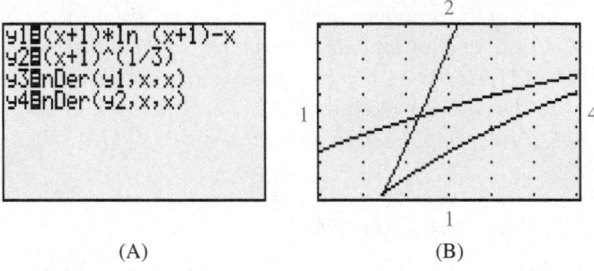

(A)   (B)

In Problems 75–78, give the domain of f, the domain of g, and the domain of m, where m(x) = f[g(x)].

75. $f(u) = \ln u$; $g(x) = 4 - x^2$

76. $f(u) = \ln u$; $g(x) = 2x + 10$

77. $f(u) = \dfrac{1}{u^2 - 1}$; $g(x) = \ln x$

78. $f(u) = \dfrac{1}{u}$; $g(x) = x^2 - 9$

In Problems 79–90, find each derivative and simplify.

79. $\dfrac{d}{dx}[3x(x^2 + 1)^3]$

80. $\dfrac{d}{dx}[2x^2(x^3 - 3)^4]$

81. $\dfrac{d}{dx} \dfrac{(x^3 - 7)^4}{2x^3}$

82. $\dfrac{d}{dx} \dfrac{3x^2}{(x^2 + 5)^3}$

83. $\dfrac{d}{dx} \log_2(3x^2 - 1)$

84. $\dfrac{d}{dx} \log(x^3 - 1)$

85. $\dfrac{d}{dx} 10^{x^2 + x}$

86. $\dfrac{d}{dx} 8^{1 - 2x^2}$

87. $\dfrac{d}{dx} \log_3(4x^3 + 5x + 7)$

88. $\dfrac{d}{dx} \log_5(5^{x^2 - 1})$

89. $\dfrac{d}{dx} 2^{x^3 - x^2 + 4x + 1}$

90. $\dfrac{d}{dx} 10^{\ln x}$

Applications

91. Cost function. The total cost (in hundreds of dollars) of producing x cell phones per day is

$$C(x) = 10 + \sqrt{2x + 16} \qquad 0 \le x \le 50$$

(see the figure).

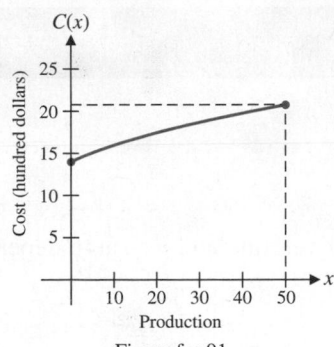

Figure for 91

(A) Find $C'(x)$.

✎ (B) Find $C'(24)$ and $C'(42)$. Interpret the results.   12.50

92. Cost function. The total cost (in hundreds of dollars) of producing x cameras per week is

$$C(x) = 6 + \sqrt{4x + 4} \qquad 0 \le x \le 30$$

(A) Find $C'(x)$.

✎ (B) Find $C'(15)$ and $C'(24)$. Interpret the results.

93. Price–supply equation. The number x of bicycle helmets a retail chain is willing to sell per week at a price of $\$p$ is given by

$$x = 80\sqrt{p + 25} - 400 \qquad 20 \le p \le 100$$

(see the figure).

(A) Find dx/dp.

✎ (B) Find the supply and the instantaneous rate of change of supply with respect to price when the price is $75. Write a brief interpretation of these results.

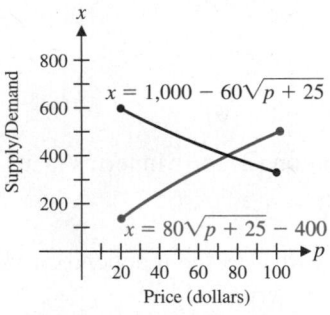

Figure for 93 and 94

94. Price–demand equation. The number x of bicycle helmets people are willing to buy per week from a retail chain at a price of $\$p$ is given by

$$x = 1,000 - 60\sqrt{p + 25} \qquad 20 \le p \le 100$$

(see the figure).

(A) Find dx/dp.

✎ (B) Find the demand and the instantaneous rate of change of demand with respect to price when the price is $75. Write a brief interpretation of these results.

95. Drug concentration. The drug concentration in the bloodstream t hours after injection is given approximately by

$$C(t) = 4.35e^{-t} \qquad 0 \le t \le 5$$

where $C(t)$ is concentration in milligrams per milliliter.

(A) What is the rate of change of concentration after 1 hour? After 4 hours?

(B) Graph C.

96. Water pollution. The use of iodine crystals is a popular way of making small quantities of water safe to drink. Crystals placed in a 1-ounce bottle of water will dissolve until the solution is saturated. After saturation, half of the solution is poured into a quart container of water, and after about an hour, the water is usually safe to drink. The half-empty 1-ounce bottle is then refilled, to be used again in the same way. Suppose that the concentration of iodine in the 1-ounce bottle t minutes after the crystals are introduced can be approximated by

$$C(t) = 250(1 - e^{-t}) \qquad t \ge 0$$

where $C(t)$ is the concentration of iodine in micrograms per milliliter.

(A) What is the rate of change of the concentration after 1 minute? After 4 minutes?

(B) Graph C for $0 \le t \le 5$.

97. Blood pressure and age. A research group using hospital records developed the following mathematical model relating systolic blood pressure and age:

$$P(x) = 40 + 25 \ln(x + 1) \qquad 0 \le x \le 65$$

$P(x)$ is pressure, measured in millimeters of mercury, and x is age in years. What is the rate of change of pressure at the end of 10 years? At the end of 30 years? At the end of 60 years?

98. Biology. A yeast culture at room temperature (68°F) is placed in a refrigerator set at a constant temperature of 38°F. After t hours, the temperature T of the culture is given approximately by

$$T = 30e^{-0.58t} + 38 \qquad t \ge 0$$

What is the rate of change of temperature of the culture at the end of 1 hour? At the end of 4 hours?

Answers to Matched Problems

1. $f[g(x)] = 2e^x, \quad g[f(u)] = e^{2u}$

2. (A) $f(u) = 50e^u, \quad u = -2x$

   (B) $f(u) = \sqrt[3]{u}, \quad u = 1 + x^3$

   [*Note:* There are other correct answers.]

3. (A) $15(5x + 2)^2$ (B) $20x^3(x^4 - 5)^4$

   (C) $-4t/(t^2 + 4)^3$ (D) $-1/(2\sqrt{4 - w})$

4. (A) $m'(x) = -30x^2(2x^3 + 4)^{-6}$

   (B) $m'(x) = 12x^3e^{3x^4+6}$ (C) $m'(x) = \dfrac{2x + 9}{x^2 + 9x + 4}$

5. $\dfrac{3e^x[\ln(1 + e^x)]^2}{1 + e^x}$

6. (A) $\dfrac{3x^2 + 2}{x^3 + 2x}$ (B) $6xe^{3x^2+2}$

   (C) $-8xe^{-x^2}(2 + e^{-x^2})^3$

3.5 Implicit Differentiation

Special Function Notation

The equation

$$y = 2 - 3x^2 \tag{1}$$

defines a function f with y as a dependent variable and x as an independent variable. Using function notation, we would write

$$y = f(x) \qquad \text{or} \qquad f(x) = 2 - 3x^2$$

In order to minimize the number of symbols, we will often write equation (1) in the form

$$y = 2 - 3x^2 = y(x)$$

where y is *both* a dependent variable and a function symbol. This is a convenient notation, and no harm is done as long as one is aware of the double role of y. Other examples are

$$x = 2t^2 - 3t + 1 = x(t)$$
$$z = \sqrt{u^2 - 3u} = z(u)$$
$$r = \frac{1}{(s^2 - 3s)^{2/3}} = r(s)$$

Until now, we have considered functions involving only one independent variable. There is no reason to stop there: The concept can be generalized to functions involving two or more independent variables, and this will be done in detail in Chapter 7. For now, we will "borrow" the notation for a function involving two independent variables. For example,

$$F(x, y) = x^2 - 2xy + 3y^2 - 5$$

specifies a function F involving two independent variables.

Implicit Differentiation

Consider the equation

$$3x^2 + y - 2 = 0 \tag{2}$$

and the equation obtained by solving equation (2) for y in terms of x,

$$y = 2 - 3x^2 \tag{3}$$

Both equations define the same function with x as the independent variable and y as the dependent variable. For equation (3), we write

$$y = f(x)$$

where

$$f(x) = 2 - 3x^2 \tag{4}$$

and we have an **explicit** (directly stated) rule that enables us to determine y for each value of x. On the other hand, the y in equation (2) is the same y as in equation (3), and equation (2) **implicitly** gives (implies, though does not directly express) y as a function of x. We say that equations (3) and (4) define the function f explicitly and equation (2) defines f implicitly.

Using an equation that defines a function implicitly to find the derivative of the function is called **implicit differentiation**. Let's differentiate equation (2) implicitly and equation (3) directly, and compare results.

Starting with

$$3x^2 + y - 2 = 0$$

we think of y as a function of x and write

$$3x^2 + y(x) - 2 = 0$$

Then we differentiate both sides with respect to x:

$$\frac{d}{dx}[(3x^2 + y(x) - 2)] = \frac{d}{dx}0 \qquad \text{Since } y \text{ is a function of } x, \text{ but is not explicitly given, we simply write}$$

$$\frac{d}{dx}3x^2 + \frac{d}{dx}y(x) - \frac{d}{dx}2 = 0 \qquad \frac{d}{dx}y(x) = y' \text{ to indicate its derivative.}$$

$$6x + y' - 0 = 0$$

Now we solve for y':

$$y' = -6x$$

Note that we get the same result if we start with equation (3) and differentiate directly:

$$y = 2 - 3x^2$$
$$y' = -6x$$

Why are we interested in implicit differentiation? Why not solve for y in terms of x and differentiate directly? The answer is that there are many equations of the form

$$F(x, y) = 0 \qquad (5)$$

that are either difficult or impossible to solve for y explicitly in terms of x (try it for $x^2y^5 - 3xy + 5 = 0$ or for $e^y - y = 3x$, for example). But it can be shown that, under fairly general conditions on F, equation (5) will define one or more functions in which y is a dependent variable and x is an independent variable. To find y' under these conditions, we differentiate equation (5) implicitly.

Explore and Discuss 1   (A) How many tangent lines are there to the graph in Figure 1 when $x = 0$? When $x = 1$? When $x = 2$? When $x = 4$? When $x = 6$?

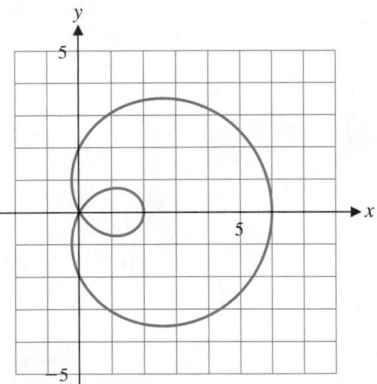

Figure 1

(B) Sketch the tangent lines referred to in part (A), and estimate each of their slopes.

(C) Explain why the graph in Figure 1 is not the graph of a function.

EXAMPLE 1   Differentiating Implicitly   Given

$$F(x, y) = x^2 + y^2 - 25 = 0 \qquad (6)$$

find y' and the slope of the graph at $x = 3$.

SOLUTION We start with the graph of $x^2 + y^2 - 25 = 0$ (a circle, as shown in Fig. 2) so that we can interpret our results geometrically. From the graph, it is clear that equation (6) does not define a function. But with a suitable restriction on the variables, equation (6) can define two or more functions. For example, the upper half and the lower half of the circle each define a function. On each half-circle, a point that corresponds to $x = 3$ is found by substituting $x = 3$ into equation (6) and solving for y:

$$x^2 + y^2 - 25 = 0$$
$$(3)^2 + y^2 = 25$$
$$y^2 = 16$$
$$y = \pm 4$$

The point $(3, 4)$ is on the upper half-circle, and the point $(3, -4)$ is on the lower half-circle. We will use these results in a moment. We now differentiate equation (6) implicitly, treating y as a function of x [i.e., $y = y(x)$]:

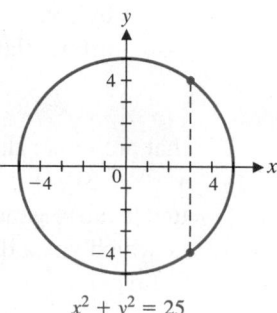

Figure 2 $x^2 + y^2 = 25$

$$x^2 + y^2 - 25 = 0$$
$$x^2 + [y(x)]^2 - 25 = 0$$
$$\frac{d}{dx}\{x^2 + [y(x)]^2 - 25\} = \frac{d}{dx}0$$
$$\frac{d}{dx}x^2 + \frac{d}{dx}[y(x)]^2 - \frac{d}{dx}25 = 0 \qquad \text{Use the chain rule.}$$
$$2x + 2[y(x)]^{2-1}y'(x) - 0 = 0$$
$$2x + 2yy' = 0 \qquad \text{Solve for } y' \text{ in terms of } x \text{ and } y.$$
$$y' = -\frac{2x}{2y}$$
$$y' = -\frac{x}{y} \qquad \text{Leave the answer in terms of } x \text{ and } y.$$

We have found y' without first solving $x^2 + y^2 - 25 = 0$ for y in terms of x. And by leaving y' in terms of x and y, we can use $y' = -x/y$ to find y' for *any* point on the graph of $x^2 + y^2 - 25 = 0$ (except where $y = 0$). In particular, for $x = 3$, we found that $(3, 4)$ and $(3, -4)$ are on the graph. The slope of the graph at $(3, 4)$ is

$$y'|_{(3, 4)} = -\tfrac{3}{4} \qquad \text{The slope of the graph at } (3, 4)$$

and the slope at $(3, -4)$ is

$$y'|_{(3, -4)} = -\tfrac{3}{-4} = \tfrac{3}{4} \qquad \text{The slope of the graph at } (3, -4)$$

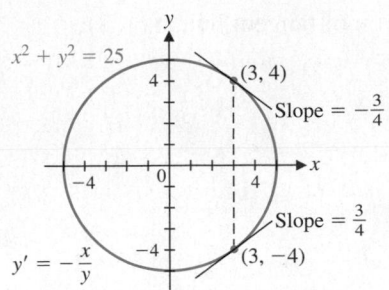

$x^2 + y^2 = 25$

Slope $= -\frac{3}{4}$

Slope $= \frac{3}{4}$

$(3, 4)$

$(3, -4)$

$y' = -\dfrac{x}{y}$

Figure 3

The symbol

$$y'|_{(a, b)}$$

is used to indicate that we are evaluating y' at $x = a$ and $y = b$.

The results are interpreted geometrically in Figure 3 on the original graph.

Matched Problem 1 ⏐ Graph $x^2 + y^2 - 169 = 0$, find y' by implicit differentiation, and find the slope of the graph when $x = 5$.

CONCEPTUAL INSIGHT

When differentiating implicitly, the derivative of y^2 is $2yy'$, not just $2y$. This is because y represents a function of x, so the chain rule applies. Suppose, for example, that y represents the function $y = 5x + 4$. Then

$$(y^2)' = [(5x + 4)^2]' = 2(5x + 4) \cdot 5 = 2yy'$$

So, when differenting implicitly, the derivative of y is y', the derivative of y^2 is $2yy'$, the derivative of y^3 is $3y^2y'$, and so on.

EXAMPLE 2   Differentiating Implicitly   Find the equation(s) of the tangent line(s) to the graph of

$$y - xy^2 + x^2 + 1 = 0 \qquad (7)$$

at the point(s) where $x = 1$.

SOLUTION   We first find y when $x = 1$:

$$y - xy^2 + x^2 + 1 = 0$$
$$y - (1)y^2 + (1)^2 + 1 = 0$$
$$y - y^2 + 2 = 0$$
$$y^2 - y - 2 = 0$$
$$(y - 2)(y + 1) = 0$$
$$y = -1 \quad \text{or} \quad 2$$

So there are two points on the graph of (7) where $x = 1$, namely, $(1, -1)$ and $(1, 2)$. We next find the slope of the graph at these two points by differentiating equation (7) implicitly:

$$y - xy^2 + x^2 + 1 = 0$$

$$\frac{d}{dx}y - \frac{d}{dx}xy^2 + \frac{d}{dx}x^2 + \frac{d}{dx}1 = \frac{d}{dx}0$$

Use the product rule and the chain rule for $\dfrac{d}{dx}xy^2$.

$$y' - (x \cdot 2yy' + y^2) + 2x = 0$$
$$y' - 2xyy' - y^2 + 2x = 0$$
$$y' - 2xyy' = y^2 - 2x$$

Solve for y' by getting all terms involving y' on one side. Factor out y'.

$$(1 - 2xy)y' = y^2 - 2x$$
$$y' = \frac{y^2 - 2x}{1 - 2xy}$$

Now find the slope at each point:

$$y'|_{(1, -1)} = \frac{(-1)^2 - 2(1)}{1 - 2(1)(-1)} = \frac{1 - 2}{1 + 2} = \frac{-1}{3} = -\frac{1}{3}$$

$$y'|_{(1, 2)} = \frac{(2)^2 - 2(1)}{1 - 2(1)(2)} = \frac{4 - 2}{1 - 4} = \frac{2}{-3} = -\frac{2}{3}$$

Equation of tangent line at $(1, -1)$:   Equation of tangent line at $(1, 2)$:

$$y - y_1 = m(x - x_1)$$ $$y - y_1 = m(x - x_1)$$
$$y + 1 = -\tfrac{1}{3}(x - 1)$$ $$y - 2 = -\tfrac{2}{3}(x - 1)$$
$$y + 1 = -\tfrac{1}{3}x + \tfrac{1}{3}$$ $$y - 2 = -\tfrac{2}{3}x + \tfrac{2}{3}$$
$$y = -\tfrac{1}{3}x - \tfrac{2}{3}$$ $$y = -\tfrac{2}{3}x + \tfrac{8}{3}$$

Matched Problem 2 | Repeat Example 2 for $x^2 + y^2 - xy - 7 = 0$ at $x = 1$.

EXAMPLE 3 Differentiating Implicitly Find x' for $x = x(t)$ defined implicitly by

$$t \ln x = xe^t - 1$$

and evaluate x' at $(t, x) = (0, 1)$.

SOLUTION It is important to remember that x is the dependent variable and t is the independent variable. Therefore, we differentiate both sides of the equation with respect to t (using product and chain rules where appropriate) and then solve for x':

$$t \ln x = xe^t - 1 \qquad \text{Differentiate implicitly with respect to } t.$$

$$\frac{d}{dt}(t \ln x) = \frac{d}{dt}(xe^t) - \frac{d}{dt}1 \qquad \text{Use the product rule twice.}$$

$$t\frac{x'}{x} + \ln x = xe^t + e^t x' \qquad \text{Clear fractions.}$$

$$x \cdot t\frac{x'}{x} + x \cdot \ln x = x \cdot xe^t + x \cdot e^t x' \qquad x \neq 0$$

$$tx' + x \ln x = x^2 e^t + xe^t x' \qquad \text{Solve for } x'.$$

$$tx' - xe^t x' = x^2 e^t - x \ln x \qquad \text{Factor out } x'.$$

$$(t - xe^t)x' = x^2 e^t - x \ln x$$

$$x' = \frac{x^2 e^t - x \ln x}{t - xe^t}$$

Now we evaluate x' at $(t, x) = (0, 1)$, as requested:

$$x'\big|_{(0, 1)} = \frac{(1)^2 e^0 - 1 \ln 1}{0 - 1e^0}$$

$$= \frac{1}{-1} = -1$$

Matched Problem 3 | Find x' for $x = x(t)$ defined implicitly by

$$1 + x \ln t = te^x$$

and evaluate x' at $(t, x) = (1, 0)$.

Exercises 3.5

Skills Warm-up Exercises

W | *In Problems 1–8, if it is possible to solve for y in terms of x, do so. If not, write "Impossible". (If necessary, review Section 1.1).*

1. $3x + 2y - 20 = 0$

2. $-4x^2 + 3y + 12 = 0$

3. $\dfrac{x^2}{9} + \dfrac{y^2}{16} = 1$

4. $4y^2 - x^2 = 36$

5. $x^2 + xy + y^2 = 1$

6. $2 \ln y + y \ln x = 3x$

7. $5x + 3y = e^y$

8. $y^2 + e^x y + x^3 = 0$

In Problems 9–12, find y' in two ways:

(A) *Differentiate the given equation implicitly and then solve for y'.*

(B) *Solve the given equation for y and then differentiate directly.*

9. $3x + 5y + 9 = 0$

10. $-2x + 6y - 4 = 0$

11. $3x^2 - 4y - 18 = 0$ **12.** $2x^3 + 5y - 2 = 0$

In Problems 13–30, use implicit differentiation to find y' and evaluate y' at the indicated point.

13. $y - 5x^2 + 3 = 0;\ (1, 2)$

14. $5x^3 - y - 1 = 0;\ (1, 4)$

15. $x^2 - y^3 - 3 = 0;\ (2, 1)$

16. $y^2 + x^3 + 4 = 0;\ (-2, 2)$

17. $y^2 + 2y + 3x = 0;\ (-1, 1)$

18. $y^2 - y - 4x = 0;\ (0, 1)$

19. $xy - 6 = 0;\ (2, 3)$

20. $3xy - 2x - 2 = 0;\ (2, 1)$

21. $2xy + y + 2 = 0;\ (-1, 2)$

22. $2y + xy - 1 = 0;\ (-1, 1)$

23. $x^2 y - 3x^2 - 4 = 0;\ (2, 4)$

24. $2x^3 y - x^3 + 5 = 0;\ (-1, 3)$

25. $e^y = x^2 + y^2;\ (1, 0)$ **26.** $x^2 - y = 4e^y;\ (2, 0)$

27. $x^3 - y = \ln y;\ (1, 1)$ **28.** $\ln y = 2y^2 - x;\ (2, 1)$

29. $x \ln y + 2y = 2x^3;\ (1, 1)$ **30.** $xe^y - y = x^2 - 2;\ (2, 0)$

In Problems 31 and 32, find x' for $x = x(t)$ defined implicitly by the given equation. Evaluate x' at the indicated point.

31. $x^2 - t^2 x + t^3 + 11 = 0;\ (-2, 1)$

32. $x^3 - tx^2 - 4 = 0;\ (-3, -2)$

Problems 33 and 34 refer to the equation and graph shown in the figure.

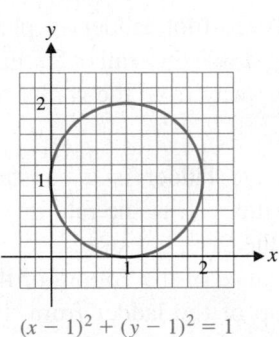

$$(x - 1)^2 + (y - 1)^2 = 1$$

Figure for 33 and 34

33. Use implicit differentiation to find the slopes of the tangent lines at the points on the graph where $x = 1.6$. Check your answers by visually estimating the slopes on the graph in the figure.

34. Find the slopes of the tangent lines at the points on the graph where $x = 0.2$. Check your answers by visually estimating the slopes on the graph in the figure.

In Problems 35–38, find the equation(s) of the tangent line(s) to the graphs of the indicated equations at the point(s) with the given value of x.

35. $xy - x - 4 = 0;\ x = 2$ **36.** $3x + xy + 1 = 0;\ x = -1$

37. $y^2 - xy - 6 = 0;\ x = 1$ **38.** $xy^2 - y - 2 = 0;\ x = 1$

39. If $xe^y = 1$, find y' in two ways, first by differentiating implicitly and then by solving for y explicitly in terms of x. Which method do you prefer? Explain.

40. Explain the difficulty that arises in solving $x^3 + y + xe^y = 1$ for y as an explicit function of x. Find the slope of the tangent line to the graph of the equation at the point $(0, 1)$.

In Problems 41–48, find y' and the slope of the tangent line to the graph of each equation at the indicated point.

41. $(1 + y)^3 + y = x + 7;\ (2, 1)$

42. $(y - 3)^4 - x = y;\ (-3, 4)$

43. $(x - 2y)^3 = 2y^2 - 3;\ (1, 1)$

44. $(2x - y)^4 - y^3 = 8;\ (-1, -2)$

45. $\sqrt{7 + y^2} - x^3 + 4 = 0;\ (2, 3)$

46. $6\sqrt{y^3 + 1} - 2x^{3/2} - 2 = 0;\ (4, 2)$

47. $\ln(xy) = y^2 - 1;\ (1, 1)$

48. $e^{xy} - 2x = y + 1;\ (0, 0)$

49. Find the equation(s) of the tangent line(s) at the point(s) on the graph of the equation

$$y^3 - xy - x^3 = 2$$

where $x = 1$. Round all approximate values to two decimal places.

50. Refer to the equation in Problem 49. Find the equation(s) of the tangent line(s) at the point(s) on the graph where $y = -1$. Round all approximate values to two decimal places.

Applications

For the demand equations in Problems 51–54, find the rate of change of p with respect to x by differentiating implicitly (x is the number of items that can be sold at a price of $p).

51. $x = p^2 - 2p + 1,000$ **52.** $x = p^3 - 3p^2 + 200$

53. $x = \sqrt{10,000 - p^2}$ **54.** $x = \sqrt[3]{1,500 - p^3}$

55. **Biophysics.** In biophysics, the equation

$$(L + m)(V + n) = k$$

is called the *fundamental equation of muscle contraction*, where m, n, and k are constants and V is the velocity of the shortening of muscle fibers for a muscle subjected to a load L. Find dL/dV by implicit differentiation.

56. **Biophysics.** In Problem 55, find dV/dL by implicit differentiation.

57. **Speed of sound.** The speed of sound in air is given by the formula

$$v = k\sqrt{T}$$

where v is the velocity of sound, T is the temperature of the air, and k is a constant. Use implicit differentiation to find $\dfrac{dT}{dv}$.

58. Gravity. The equation

$$F = G\frac{m_1 m_2}{r^2}$$

is Newton's law of universal gravitation. G is a constant and F is the gravitational force between two objects having masses m_1 and m_2 that are a distance r from each other. Use implicit differentiation to find $\dfrac{dr}{dF}$. Assume that m_1 and m_2 are constant.

59. Speed of sound. Refer to Problem 57. Find $\dfrac{dv}{dT}$ and discuss the connection between $\dfrac{dv}{dT}$ and $\dfrac{dT}{dv}$.

60. Gravity. Refer to Problem 58. Find $\dfrac{dF}{dr}$ and discuss the connection between $\dfrac{dF}{dr}$ and $\dfrac{dr}{dF}$.

Answers to Matched Problems

1. $y' = -x/y$. When $x = 5$, $y = \pm 12$; thus, $y'|_{(5,12)} = -\frac{5}{12}$ and $y'|_{(5,-12)} = \frac{5}{12}$

2. $y' = \dfrac{y - 2x}{2y - x}$; $y = \frac{4}{5}x - \frac{14}{5}$, $y = \frac{1}{5}x + \frac{14}{5}$

3. $x' = \dfrac{te^x - x}{t \ln t - t^2 e^x}$; $x'|_{(1,0)} = -1$

3.6   Related Rates

Union workers are concerned that the rate at which wages are increasing is lagging behind the rate of increase in the company's profits. An automobile dealer wants to predict how much an anticipated increase in interest rates will decrease his rate of sales. An investor is studying the connection between the rate of increase in the Dow Jones average and the rate of increase in the gross domestic product over the past 50 years.

In each of these situations, there are two quantities—wages and profits, for example—that are changing with respect to time. We would like to discover the precise relationship between the rates of increase (or decrease) of the two quantities. We begin our discussion of such *related rates* by considering familiar situations in which the two quantities are distances and the two rates are velocities.

EXAMPLE 1   Related Rates and Motion   A 26-foot ladder is placed against a wall (Fig. 1). If the top of the ladder is sliding down the wall at 2 feet per second, at what rate is the bottom of the ladder moving away from the wall when the bottom of the ladder is 10 feet away from the wall?

SOLUTION Many people think that since the ladder is a constant length, the bottom of the ladder will move away from the wall at the rate that the top of the ladder is moving down the wall. This is not the case, however.

At any moment in time, let x be the distance of the bottom of the ladder from the wall and let y be the distance of the top of the ladder from the ground (see Fig. 1). Both x and y are changing with respect to time and can be thought of as functions of time; that is, $x = x(t)$ and $y = y(t)$. Furthermore, x and y are related by the Pythagorean relationship:

$$x^2 + y^2 = 26^2 \tag{1}$$

Differentiating equation (1) implicitly with respect to time t and using the chain rule where appropriate, we obtain

$$2x\frac{dx}{dt} + 2y\frac{dy}{dt} = 0 \tag{2}$$

The rates dx/dt and dy/dt are related by equation (2). This is a **related-rates problem**.

Our problem is to find dx/dt when $x = 10$ feet, given that $dy/dt = -2$ (y is decreasing at a constant rate of 2 feet per second). We have all the quantities we

Figure 1

26 ft

y

x

need in equation (2) to solve for dx/dt, except y. When $x = 10$, y can be found from equation (1):

$$10^2 + y^2 = 26^2$$
$$y = \sqrt{26^2 - 10^2} = 24 \text{ feet}$$

Substitute $dy/dt = -2$, $x = 10$, and $y = 24$ into (2). Then solve for dx/dt:

$$2(10)\frac{dx}{dt} + 2(24)(-2) = 0$$

$$\frac{dx}{dt} = \frac{-2(24)(-2)}{2(10)} = 4.8 \text{ feet per second}$$

The bottom of the ladder is moving away from the wall at a rate of 4.8 feet per second.

CONCEPTUAL INSIGHT

In the solution to Example 1, we used equation (1) in two ways: first, to find an equation relating dy/dt and dx/dt, and second, to find the value of y when $x = 10$. These steps must be done in this order. Substituting $x = 10$ and then differentiating does not produce any useful results:

$$x^2 + y^2 = 26^2 \qquad$$ Substituting 10 for x has the effect of
$$100 + y^2 = 26^2 \qquad$$ stopping the ladder.
$$0 + 2yy' = 0 \qquad$$ The rate of change of a stationary object
$$y' = 0 \qquad$$ is always 0, but that is not the rate of change of the moving ladder.

Matched Problem 1 Again, a 26-foot ladder is placed against a wall (Fig. 1). If the bottom of the ladder is moving away from the wall at 3 feet per second, at what rate is the top moving down when the top of the ladder is 24 feet above ground?

Explore and Discuss 1 (A) For which values of x and y in Example 1 is dx/dt equal to 2 (i.e., the same rate that the ladder is sliding down the wall)?

(B) When is dx/dt greater than 2? Less than 2?

DEFINITION Suggestions for Solving Related-Rates Problems

Step 1 Sketch a figure if helpful.

Step 2 Identify all relevant variables, including those whose rates are given and those whose rates are to be found.

Step 3 Express all given rates and rates to be found as derivatives.

Step 4 Find an equation connecting the variables identified in step 2.

Step 5 Implicitly differentiate the equation found in step 4, using the chain rule where appropriate, and substitute in all given values.

Step 6 Solve for the derivative that will give the unknown rate.

EXAMPLE 2 Related Rates and Motion Suppose that two motorboats leave from the same point at the same time. If one travels north at 15 miles per hour and the other travels east at 20 miles per hour, how fast will the distance between them be changing after 2 hours?

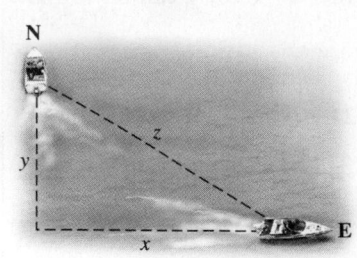

Figure 2

SOLUTION First, draw a picture, as shown in Figure 2.

All variables, x, y, and z, are changing with time. They can be considered as functions of time: $x = x(t)$, $y = y(t)$, and $z = z(t)$, given implicitly. It now makes sense to find derivatives of each variable with respect to time. From the Pythagorean theorem,

$$z^2 = x^2 + y^2 \tag{3}$$

We also know that

$$\frac{dx}{dt} = 20 \text{ miles per hour} \qquad \text{and} \qquad \frac{dy}{dt} = 15 \text{ miles per hour}$$

We want to find dz/dt at the end of 2 hours—that is, when $x = 40$ miles and $y = 30$ miles. To do this, we differentiate both sides of equation (3) with respect to t and solve for dz/dt:

$$2z\frac{dz}{dt} = 2x\frac{dx}{dt} + 2y\frac{dy}{dt} \tag{4}$$

We have everything we need except z. From equation (3), when $x = 40$ and $y = 30$, we find z to be 50. Substituting the known quantities into equation (4), we obtain

$$2(50)\frac{dz}{dt} = 2(40)(20) + 2(30)(15)$$

$$\frac{dz}{dt} = 25 \text{ miles per hour}$$

The boats will be separating at a rate of 25 miles per hour.

Matched Problem 2   Repeat Example 2 for the same situation at the end of 3 hours.

EXAMPLE 3   Related Rates and Motion Suppose that a point is moving along the graph of $x^2 + y^2 = 25$ (Fig. 3). When the point is at $(-3, 4)$, its x coordinate is increasing at the rate of 0.4 unit per second. How fast is the y coordinate changing at that moment?

SOLUTION Since both x and y are changing with respect to time, we can consider each as a function of time, namely,

$$x = x(t) \qquad \text{and} \qquad y = y(t)$$

but restricted so that

$$x^2 + y^2 = 25 \tag{5}$$

We want to find dy/dt, given $x = -3$, $y = 4$, and $dx/dt = 0.4$. Implicitly differentiating both sides of equation (5) with respect to t, we have

$$x^2 + y^2 = 25$$

$$2x\frac{dx}{dt} + 2y\frac{dy}{dt} = 0 \qquad \text{Divide both sides by 2.}$$

$$x\frac{dx}{dt} + y\frac{dy}{dt} = 0 \qquad \begin{array}{l}\text{Substitute } x = -3, y = 4, \text{ and} \\ dx/dt = 0.4, \text{ and solve for } dy/dt.\end{array}$$

$$(-3)(0.4) + 4\frac{dy}{dt} = 0$$

$$\frac{dy}{dt} = 0.3 \text{ unit per second}$$

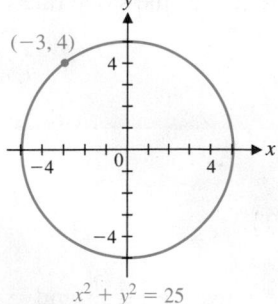

Figure 3

Matched Problem 3   A point is moving on the graph of $y^3 = x^2$. When the point is at $(-8, 4)$, its y coordinate is decreasing by 2 units per second. How fast is the x coordinate changing at that moment?

EXAMPLE 4 Related Rates and Business Suppose that for a company manufacturing flash drives, the cost, revenue, and profit equations are given by

$$C = 5,000 + 2x \qquad \text{Cost equation}$$
$$R = 10x - 0.001x^2 \qquad \text{Revenue equation}$$
$$P = R - C \qquad \text{Profit equation}$$

where the production output in 1 week is x flash drives. If production is increasing at the rate of 500 flash drives per week when production is 2,000 flash drives, find the rate of increase in

(A) Cost   (B) Revenue   (C) Profit

SOLUTION If production x is a function of time (it must be, since it is changing with respect to time), then C, R, and P must also be functions of time. These functions are given implicitly (rather than explicitly). Letting t represent time in weeks, we differentiate both sides of each of the preceding three equations with respect to t and then substitute $x = 2,000$ and $dx/dt = 500$ to find the desired rates.

(A) $C = 5,000 + 2x$ Think: $C = C(t)$ and $x = x(t)$.

$\dfrac{dC}{dt} = \dfrac{d}{dt}(5,000) + \dfrac{d}{dt}(2x)$ Differentiate both sides with respect to t.

$\dfrac{dC}{dt} = 0 + 2\dfrac{dx}{dt} = 2\dfrac{dx}{dt}$

Since $dx/dt = 500$ when $x = 2,000$,

$$\frac{dC}{dt} = 2(500) = \$1,000 \text{ per week}$$

Cost is increasing at a rate of $1,000 per week.

(B) $R = 10x - 0.001x^2$

$\dfrac{dR}{dt} = \dfrac{d}{dt}(10x) - \dfrac{d}{dt}0.001x^2$

$\dfrac{dR}{dt} = 10\dfrac{dx}{dt} - 0.002x\dfrac{dx}{dt}$

$\dfrac{dR}{dt} = (10 - 0.002x)\dfrac{dx}{dt}$

Since $dx/dt = 500$ when $x = 2,000$,

$$\frac{dR}{dt} = [10 - 0.002(2,000)](500) = \$3,000 \text{ per week}$$

Revenue is increasing at a rate of $3,000 per week.

(C) $P = R - C$

$\dfrac{dP}{dt} = \dfrac{dR}{dt} - \dfrac{dC}{dt}$ Results from parts (A) and (B)

$\phantom{\dfrac{dP}{dt}} = \$3,000 - \$1,000$

$\phantom{\dfrac{dP}{dt}} = \$2,000 \text{ per week}$

Profit is increasing at a rate of $2,000 per week.

Matched Problem 4 Repeat Example 4 for a production level of 6,000 flash drives per week.

Exercises 3.6

Skills Warm-up Exercises

W *For Problems 1–8, review the geometric formulas in Appendix C, if necessary.*

1. A circular flower bed has an area of 300 square feet. Find its diameter to the nearest tenth of a foot.

2. A central pivot irrigation system covers a circle of radius 400 meters. Find the area of the circle to the nearest square meter.

3. The hypotenuse of a right triangle has length 50 meters, and another side has length 20 meters. Find the length of the third side to the nearest meter.

4. The legs of a right triangle have lengths 54 feet and 69 feet. Find the length of the hypotenuse to the nearest foot.

5. A person 69 inches tall stands 40 feet from the base of a streetlight. The streetlight casts a shadow of length 96 inches. How far above the ground is the streetlight?

6. The radius of a spherical balloon is 3 meters. Find its volume to the nearest tenth of a cubic meter.

7. A right circular cylinder and a sphere both have radius 12 feet. If the volume of the cylinder is twice the volume of the sphere, find the height of the cylinder.

8. The height of a right circular cylinder is twice its radius. If the volume is 1,000 cubic meters, find the radius and height to the nearest hundredth of a meter.

In Problems 9–14, assume that $x = x(t)$ and $y = y(t)$. Find the indicated rate, given the other information.

9. $y = x^2 + 2$; $dx/dt = 3$ when $x = 5$; find dy/dt

10. $y = x^3 - 3$; $dx/dt = -2$ when $x = 2$; find dy/dt

11. $x^2 + y^2 = 1$; $dy/dt = -4$ when $x = -0.6$ and $y = 0.8$; find dx/dt

12. $x^2 + y^2 = 4$; $dy/dt = 5$ when $x = 1.2$ and $y = -1.6$; find dx/dt

13. $x^2 + 3xy + y^2 = 11$; $dx/dt = 2$ when $x = 1$ and $y = 2$; find dy/dt

14. $x^2 - 2xy - y^2 = 7$; $dy/dt = -1$ when $x = 2$ and $y = -1$; find dx/dt

15. A point is moving on the graph of $xy = 36$. When the point is at $(4, 9)$, its x coordinate is increasing by 4 units per second. How fast is the y coordinate changing at that moment?

16. A point is moving on the graph of $4x^2 + 9y^2 = 36$. When the point is at $(3, 0)$, its y coordinate is decreasing by 2 units per second. How fast is its x coordinate changing at that moment?

17. A boat is being pulled toward a dock as shown in the figure. If the rope is being pulled in at 3 feet per second, how fast is

the distance between the dock and the boat decreasing when it is 30 feet from the dock?

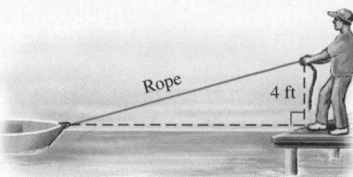

Figure for 17 and 18

18. Refer to Problem 17. Suppose that the distance between the boat and the dock is decreasing by 3.05 feet per second. How fast is the rope being pulled in when the boat is 10 feet from the dock?

19. A rock thrown into a still pond causes a circular ripple. If the radius of the ripple is increasing by 2 feet per second, how fast is the area changing when the radius is 10 feet?

20. Refer to Problem 19. How fast is the circumference of a circular ripple changing when the radius is 10 feet?

21. The radius of a spherical balloon is increasing at the rate of 3 centimeters per minute. How fast is the volume changing when the radius is 10 centimeters?

22. Refer to Problem 21. How fast is the surface area of the sphere increasing when the radius is 10 centimeters?

23. Boyle's law for enclosed gases states that if the volume is kept constant, the pressure P and temperature T are related by the equation

$$\frac{P}{T} = k$$

where k is a constant. If the temperature is increasing at 3 kelvins per hour, what is the rate of change of pressure when the temperature is 250 kelvins and the pressure is 500 pounds per square inch?

24. Boyle's law for enclosed gases states that if the temperature is kept constant, the pressure P and volume V of a gas are related by the equation

$$VP = k$$

where k is a constant. If the volume is decreasing by 5 cubic inches per second, what is the rate of change of pressure when the volume is 1,000 cubic inches and the pressure is 40 pounds per square inch?

25. A 10-foot ladder is placed against a vertical wall. Suppose that the bottom of the ladder slides away from the wall at a constant rate of 3 feet per second. How fast is the top of the ladder sliding down the wall when the bottom is 6 feet from the wall?

26. A weather balloon is rising vertically at the rate of 5 meters per second. An observer is standing on the ground 300 meters from where the balloon was released. At what rate is the distance between the observer and the balloon changing when the balloon is 400 meters high?

27. A streetlight is on top of a 20-foot pole. A person who is 5 feet tall walks away from the pole at the rate of 5 feet per second. At what rate is the tip of the person's shadow moving away from the pole when he is 20 feet from the pole?

28. Refer to Problem 27. At what rate is the person's shadow growing when he is 20 feet from the pole?

29. Helium is pumped into a spherical balloon at a constant rate of 4 cubic feet per second. How fast is the radius increasing after 1 minute? After 2 minutes? Is there any time at which the radius is increasing at a rate of 100 feet per second? Explain.

30. A point is moving along the x axis at a constant rate of 5 units per second. At which point is its distance from $(0, 1)$ increasing at a rate of 2 units per second? At 4 units per second? At 5 units per second? At 10 units per second? Explain.

31. A point is moving on the graph of $y = e^x + x + 1$ in such a way that its x coordinate is always increasing at a rate of 3 units per second. How fast is the y coordinate changing when the point crosses the x axis?

32. A point is moving on the graph of $x^3 + y^2 = 1$ in such a way that its y coordinate is always increasing at a rate of 2 units per second. At which point(s) is the x coordinate increasing at a rate of 1 unit per second?

Applications

33. Cost, revenue, and profit rates. Suppose that for a company manufacturing calculators, the cost, revenue, and profit equations are given by

$$C = 90,000 + 30x \qquad R = 300x - \frac{x^2}{30}$$

$$P = R - C$$

where the production output in 1 week is x calculators. If production is increasing at a rate of 500 calculators per week when production output is 6,000 calculators, find the rate of increase (decrease) in

(A) Cost   (B) Revenue   (C) Profit

34. Cost, revenue, and profit rates. Repeat Problem 33 for

$$C = 72,000 + 60x \qquad R = 200x - \frac{x^2}{30}$$

$$P = R - C$$

where production is increasing at a rate of 500 calculators per week at a production level of 1,500 calculators.

35. Advertising. A retail store estimates that weekly sales s and weekly advertising costs x (both in dollars) are related by

$$s = 60,000 - 40,000e^{-0.0005x}$$

The current weekly advertising costs are $2,000, and these costs are increasing at the rate of $300 per week. Find the current rate of change of sales.

36. Advertising. Repeat Problem 35 for

$$s = 50,000 - 20,000e^{-0.0004x}$$

37. Price–demand. The price p (in dollars) and demand x for a product are related by

$$2x^2 + 5xp + 50p^2 = 80,000$$

(A) If the price is increasing at a rate of $2 per month when the price is $30, find the rate of change of the demand.

(B) If the demand is decreasing at a rate of 6 units per month when the demand is 150 units, find the rate of change of the price.

38. Price–demand. Repeat Problem 37 for

$$x^2 + 2xp + 25p^2 = 74,500$$

39. Pollution. An oil tanker aground on a reef is forming a circular oil slick about 0.1 foot thick (see the figure). To estimate the rate dV/dt (in cubic feet per minute) at which the oil is leaking from the tanker, it was found that the radius of the slick was increasing at 0.32 foot per minute $(dR/dt = 0.32)$ when the radius R was 500 feet. Find dV/dt.

Tanker
R
Oil slick
$A = \pi R^2$
$V = 0.1 A$

40. Learning. A person who is new on an assembly line performs an operation in T minutes after x performances of the operation, as given by

$$T = 6\left(1 + \frac{1}{\sqrt{x}}\right)$$

If $dx/dt = 6$ operations per hours, where t is time in hours, find dT/dt after 36 performances of the operation.

Answers to Matched Problems

1. $dy/dt = -1.25$ ft/sec   **2.** $dz/dt = 25$ mi/hr

3. $dx/dt = 6$ units/sec

4. (A) $dC/dt = \$1,000/$wk   (B) $dR/dt = -\$1,000/$wk

(C) $dP/dt = -\$2,000/$wk

3.7 Elasticity of Demand

- Relative Rate of Change
- Elasticity of Demand

When will a price increase lead to an increase in revenue? To answer this question and study relationships among price, demand, and revenue, economists use the notion of *elasticity of demand*. In this section, we define the concepts of *relative rate of change*, *percentage rate of change*, and *elasticity of demand*.

Relative Rate of Change

Explore and Discuss 1

A broker is trying to sell you two stocks: Biotech and Comstat. The broker estimates that Biotech's price per share will increase $2 per year over the next several years, while Comstat's price per share will increase only $1 per year. Is this sufficient information for you to choose between the two stocks? What other information might you request from the broker to help you decide?

Interpreting rates of change is a fundamental application of calculus. In Explore and Discuss 1, Biotech's price per share is increasing at twice the rate of Comstat's, but that does not automatically make Biotech the better buy. The obvious information that is missing is the current price of each stock. If Biotech costs $100 a share and Comstat costs $25 a share, then which stock is the better buy? To answer this question, we introduce two new concepts: *relative rate of change* and *percentage rate of change*.

> **DEFINITION** Relative and Percentage Rates of Change
>
> The **relative rate of change** of a function $f(x)$ is $\dfrac{f'(x)}{f(x)}$, or equivalently, $\dfrac{d}{dx} \ln f(x)$.
>
> The **percentage rate of change** is $100 \times \dfrac{f'(x)}{f(x)}$, or equivalently, $100 \times \dfrac{d}{dx} \ln f(x)$.

The alternative form for the relative rate of change, $\dfrac{d}{dx} \ln f(x)$, is called the **logarithmic derivative** of $f(x)$.

Note that

$$\frac{d}{dx} \ln f(x) = \frac{f'(x)}{f(x)}$$

by the chain rule. So the relative rate of change of a function $f(x)$ is its logarithmic derivative, and the percentage rate of change is 100 times the logarithmic derivative.

Returning to Explore and Discuss 1, the table shows the relative rate of change and percentage rate of change for Biotech and Comstat. We conclude that Comstat is the better buy.

	Relative rate of change	Percentage rate of change
Biotech	$\dfrac{2}{100} = 0.02$	2%
Comstat	$\dfrac{1}{25} = 0.04$	4%

EXAMPLE 1 Percentage Rate of Change Table 1 lists the GDP (gross domestic product expressed in billions of 2005 dollars) and U.S. population from 2000 to 2012. A model for the GDP is

$$f(t) = 209.5t + 11{,}361$$

where t is years since 2000. Find and graph the percentage rate of change of $f(t)$ for $0 \leq t \leq 12$.

Table 1

Year	Real GDP (billions of 2005 dollars)	Population (in millions)
2000	$11,226	282.2
2004	$12,264	292.9
2008	$13,312	304.1
2012	$13,670	313.9

SOLUTION If $p(t)$ is the percentage rate of change of $f(t)$, then

$$p(t) = 100 \times \frac{d}{dx} \ln{(209.5t + 11{,}361)}$$

$$= \frac{20{,}950}{209.5t + 11{,}361}$$

The graph of $p(t)$ is shown in Figure 1 (graphing details omitted). Notice that $p(t)$ is decreasing, even though the GDP is increasing.

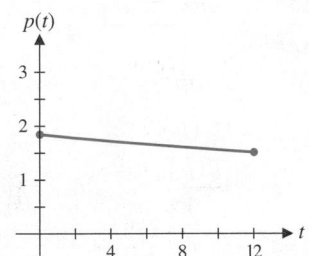

Figure 1

Matched Problem 1 A model for the population data in Table 1 is

$$f(t) = 2.7t + 282$$

where t is years since 2000. Find and graph $p(t)$, the percentage rate of change of $f(t)$ for $0 \leq t \leq 12$.

CONCEPTUAL INSIGHT

If $10,000 is invested at an annual rate of 4.5% compounded continuously, what is the relative rate of change of the amount in the account? The answer is the logarithmic derivative of $A(t) = 10{,}000e^{0.045t}$, namely

$$\frac{d}{dx} \ln{(10{,}000e^{0.045t})} = \frac{10{,}000e^{0.045t}(0.045)}{10{,}000e^{0.045t}} = 0.045$$

So the relative rate of change of $A(t)$ is 0.045, and the percentage rate of change is just the annual interest rate, 4.5%.

Elasticity of Demand

Explore and Discuss 2 In both parts below, assume that increasing the price per unit by $1 will decrease the demand by 500 units. If your objective is to increase revenue, should you increase the price by $1 per unit?

(A) At the current price of $8.00 per baseball cap, there is a demand for 6,000 caps.

(B) At the current price of $12.00 per baseball cap, there is a demand for 4,000 caps.

In Explore and Discuss 2, the rate of change of demand with respect to price was assumed to be −500 units per dollar. But in one case, part (A), you should increase the price, and in the other, part (B), you should not. Economists use the concept of *elasticity of demand* to answer the question "When does an increase in price lead to an increase in revenue?"

DEFINITION   Elasticity of Demand

Let the price p and demand x for a product be related by a price–demand equation of the form $x = f(p)$. Then the **elasticity of demand at price p**, denoted by $E(p)$, is

$$E(p) = -\frac{\text{relative rate of change of demand}}{\text{relative rate of change of price}}$$

Using the definition of relative rate of change, we can find a formula for $E(p)$:

$$E(p) = -\frac{\text{relative rate of change of demand}}{\text{relative rate of change of price}} = -\frac{\dfrac{d}{dp}\ln f(p)}{\dfrac{d}{dp}\ln p}$$

$$= -\frac{\dfrac{f'(p)}{f(p)}}{\dfrac{1}{p}}$$

$$= -\frac{pf'(p)}{f(p)}$$

THEOREM 1   Elasticity of Demand

If price and demand are related by $x = f(p)$, then the elasticity of demand is given by

$$E(p) = -\frac{pf'(p)}{f(p)}$$

CONCEPTUAL INSIGHT

Since p and $f(p)$ are nonnegative and $f'(p)$ is negative (demand is usually a decreasing function of price), $E(p)$ is nonnegative. This is why elasticity of demand is defined as the negative of a ratio.

EXAMPLE 2   Elasticity of Demand   The price p and the demand x for a product are related by the price–demand equation

$$x + 500p = 10{,}000 \tag{1}$$

Find the elasticity of demand, $E(p)$, and interpret each of the following:

(A) $E(4)$ (B) $E(16)$ (C) $E(10)$

SOLUTION   To find $E(p)$, we first express the demand x as a function of the price p by solving (1) for x:

$$x = 10{,}000 - 500p$$
$$= 500(20 - p) \qquad \text{Demand as a function of price}$$

or

$$x = f(p) = 500(20 - p) \qquad 0 \le p \le 20 \tag{2}$$

Since x and p both represent nonnegative quantities, we must restrict p so that $0 \le p \le 20$. Note that the demand is a decreasing function of price. That is, a price increase results in lower demand, and a price decrease results in higher demand (see Figure 2).

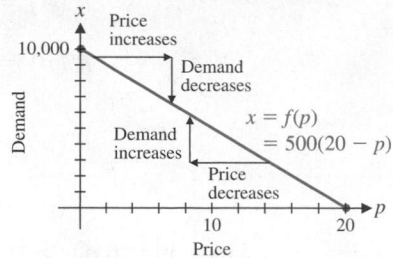

Figure 2

$$E(p) = -\frac{pf'(p)}{f(p)} = -\frac{p(-500)}{500(20 - p)} = \frac{p}{20 - p}$$

In order to interpret values of $E(p)$, we must recall the definition of elasticity:

$$E(p) = -\frac{\text{relative rate of change of demand}}{\text{relative rate of change of price}}$$

or

$$-\left(\begin{array}{c}\text{relative rate of}\\\text{change of demand}\end{array}\right) \approx E(p)\left(\begin{array}{c}\text{relative rate of}\\\text{change of price}\end{array}\right)$$

(A) $E(4) = \frac{4}{16} = 0.25 < 1$. If the $4 price changes by 10%, then the demand will change by approximately $0.25(10\%) = 2.5\%$.

(B) $E(16) = \frac{16}{4} = 4 > 1$. If the $16 price changes by 10%, then the demand will change by approximately $4(10\%) = 40\%$.

(C) $E(10) = \frac{10}{10} = 1$. If the $10 price changes by 10%, then the demand will also change by approximately 10%.

Matched Problem 2 Find $E(p)$ for the price–demand equation

$$x = f(p) = 1{,}000(40 - p)$$

Find and interpret each of the following:

(A) $E(8)$ (B) $E(30)$ (C) $E(20)$

The three cases illustrated in the solution to Example 2 are referred to as **inelastic demand**, **elastic demand**, and **unit elasticity**, as indicated in Table 2.

Table 2

$E(p)$	Demand	Interpretation	Revenue
$0 < E(p) < 1$	Inelastic	Demand is not sensitive to changes in price, that is, percentage change in price produces a smaller percentage change in demand.	A price increase will increase revenue.
$E(p) > 1$	Elastic	Demand is sensitive to changes in price, that is, a percentage change in price produces a larger percentage change in demand.	A price increase will decrease revenue.
$E(p) = 1$	Unit	A percentage change in price produces the same percentage change in demand.	

To justify the connection between elasticity of demand and revenue as given in the fourth column of Table 2, we recall that revenue R is the demand x (number of items sold) multiplied by p (price per item). Assume that the price–demand equation is written in the form $x = f(p)$. Then

$$R(p) = xp = f(p)p \qquad \text{Use the product rule.}$$
$$R'(p) = f(p) \cdot 1 + pf'(p) \qquad \text{Multiply and divide by } f(p).$$
$$R'(p) = f(p) + pf'(p)\frac{f(p)}{f(p)} \qquad \text{Factor out } f(p).$$
$$R'(p) = f(p)\left[1 + \frac{pf'(p)}{f(p)}\right] \qquad \text{Use Theorem 1.}$$
$$R'(p) = f(p)[1 - E(p)]$$

Since $x = f(p) > 0$, it follows that $R'(p)$ and $1 - E(p)$ have the same sign. So if $E(p) < 1$, then $R'(p)$ is positive and revenue is increasing (Fig. 3). Similarly, if $E(p) > 1$, then $R'(p)$ is negative, and revenue is decreasing (Fig. 3).

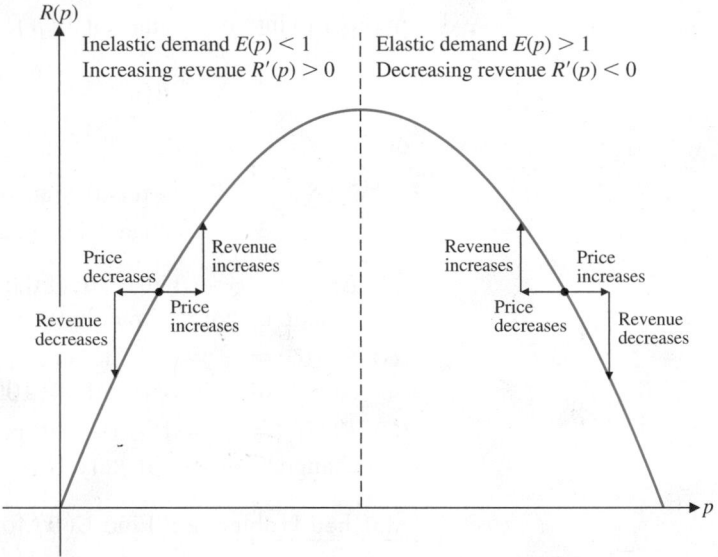

Figure 3 **Revenue and elasticity**

EXAMPLE 3   Elasticity and Revenue   A manufacturer of sunglasses currently sells one type for $15 a pair. The price p and the demand x for these glasses are related by

$$x = f(p) = 9{,}500 - 250p$$

If the current price is increased, will revenue increase or decrease?

SOLUTION
$$E(p) = -\frac{pf'(p)}{f(p)}$$
$$= -\frac{p(-250)}{9{,}500 - 250p}$$
$$= \frac{p}{38 - p}$$
$$E(15) = \frac{15}{23} \approx 0.65$$

At the $15 price level, demand is inelastic and a price increase will increase revenue.

Matched Problem 3   Repeat Example 3 if the current price for sunglasses is $21 a pair.

In summary, if demand is inelastic, then a price increase will increase revenue. But if demand is elastic, then a price increase will decrease revenue.

Exercises 3.7

Skills Warm-up Exercises

W *In Problems 1–8, use the given equation, which expresses price p as a function of demand x, to find a function $f(p)$ that expresses demand x as a function of price p. Give the domain of $f(p)$. (If necessary, review Section 1.6).*

1. $p = 42 - 0.4x, 0 \le x \le 105$

2. $p = 125 - 0.02x, 0 \le x \le 6{,}250$

3. $p = 50 - 0.5x^2, 0 \le x \le 10$

4. $p = 180 - 0.8x^2, 0 \le x \le 15$

5. $p = 25e^{-x/20}, 0 \le x \le 20$

6. $p = 45 - e^{x/4}, 0 \le x \le 12$

7. $p = 80 - 10 \ln x, 1 \le x \le 30$

8. $p = \ln(500 - 5x), 0 \le x \le 90$

In Problems 9–14, find the relative rate of change of $f(x)$.

9. $f(x) = 35x - 0.4x^2$ **10.** $f(x) = 60x - 1.2x^2$

11. $f(x) = 7 + 4e^{-x}$ **12.** $f(x) = 15 - 3e^{-0.5x}$

13. $f(x) = 12 + 5 \ln x$ **14.** $f(x) = 25 - 2 \ln x$

In Problems 15–24, find the relative rate of change of $f(x)$ at the indicated value of x. Round to three decimal places.

15. $f(x) = 45; x = 100$

16. $f(x) = 580; x = 300$

17. $f(x) = 420 - 5x; x = 25$

18. $f(x) = 500 - 6x; x = 40$

19. $f(x) = 420 - 5x; x = 55$

20. $f(x) = 500 - 6x; x = 75$

21. $f(x) = 4x^2 - \ln x; x = 2$

22. $f(x) = 9x - 5 \ln x; x = 3$

23. $f(x) = 4x^2 - \ln x; x = 5$

24. $f(x) = 9x - 5 \ln x; x = 7$

In Problems 25–32, find the percentage rate of change of $f(x)$ at the indicated value of x. Round to the nearest tenth of a percent.

25. $f(x) = 225 + 65x; x = 5$

26. $f(x) = 75 + 110x; x = 4$

27. $f(x) = 225 + 65x; x = 15$

28. $f(x) = 75 + 110x; x = 16$

29. $f(x) = 5{,}100 - 3x^2; x = 35$

30. $f(x) = 3{,}000 - 8x^2; x = 12$

31. $f(x) = 5{,}100 - 3x^2; x = 41$

32. $f(x) = 3{,}000 - 8x^2; x = 18$

In Problems 33–38, use the price–demand equation to find $E(p)$, the elasticity of demand.

33. $x = f(p) = 25{,}000 - 450p$

34. $x = f(p) = 10{,}000 - 190p$

35. $x = f(p) = 4{,}800 - 4p^2$

36. $x = f(p) = 8{,}400 - 7p^2$

37. $x = f(p) = 98 - 0.6e^p$

38. $x = f(p) = 160 - 35 \ln p$

In Problems 39–46, find the logarithmic derivative.

39. $A(t) = 500e^{0.07t}$ **40.** $A(t) = 2{,}000e^{0.052t}$

41. $A(t) = 3{,}500e^{0.15t}$ **42.** $A(t) = 900e^{0.24t}$

43. $f(x) = xe^x$ **44.** $f(x) = x^2e^x$

45. $f(x) = \ln x$ **46.** $f(x) = x \ln x$

In Problems 47–50, use the price–demand equation to determine whether demand is elastic, is inelastic, or has unit elasticity at the indicated values of p.

47. $x = f(p) = 12{,}000 - 10p^2$

(A) $p = 10$ (B) $p = 20$

(C) $p = 30$

48. $x = f(p) = 1{,}875 - p^2$

(A) $p = 15$ (B) $p = 25$

(C) $p = 40$

49. $x = f(p) = 950 - 2p - 0.1p^2$

(A) $p = 30$ (B) $p = 50$

(C) $p = 70$

50. $x = f(p) = 875 - p - 0.05p^2$

(A) $p = 50$ (B) $p = 70$

(C) $p = 100$

51. Given the price–demand equation

$$p + 0.005x = 30$$

(A) Express the demand x as a function of the price p.

(B) Find the elasticity of demand, $E(p)$.

(C) What is the elasticity of demand when $p = \$10$? If this price is increased by 10%, what is the approximate percentage change in demand?

(D) What is the elasticity of demand when $p = \$25$? If this price is increased by 10%, what is the approximate percentage change in demand?

(E) What is the elasticity of demand when $p = \$15$? If this price is increased by 10%, what is the approximate percentage change in demand?

52. Given the price–demand equation

$$p + 0.01x = 50$$

(A) Express the demand x as a function of the price p.

(B) Find the elasticity of demand, $E(p)$.

(C) What is the elasticity of demand when $p = \$10$? If this price is decreased by 5%, what is the approximate change in demand?

(D) What is the elasticity of demand when $p = \$45$? If this price is decreased by 5%, what is the approximate change in demand?

(E) What is the elasticity of demand when $p = \$25$? If this price is decreased by 5%, what is the approximate change in demand?

53. Given the price–demand equation

$$0.02x + p = 60$$

(A) Express the demand x as a function of the price p.

(B) Express the revenue R as a function of the price p.

(C) Find the elasticity of demand, $E(p)$.

(D) For which values of p is demand elastic? Inelastic?

(E) For which values of p is revenue increasing? Decreasing?

(F) If $p = \$10$ and the price is decreased, will revenue increase or decrease?

(G) If $p = \$40$ and the price is decreased, will revenue increase or decrease?

54. Repeat Problem 53 for the price–demand equation

$$0.025x + p = 50$$

In Problems 55–62, use the price–demand equation to find the values of p for which demand is elastic and the values for which demand is inelastic. Assume that price and demand are both positive.

55. $x = f(p) = 210 - 30p$ **56.** $x = f(p) = 480 - 8p$

57. $x = f(p) = 3{,}125 - 5p^2$ **58.** $x = f(p) = 2{,}400 - 6p^2$

59. $x = f(p) = \sqrt{144 - 2p}$ **60.** $x = f(p) = \sqrt{324 - 2p}$

61. $x = f(p) = \sqrt{2{,}500 - 2p^2}$

62. $x = f(p) = \sqrt{3{,}600 - 2p^2}$

In Problems 63–68, use the demand equation to find the revenue function. Sketch the graph of the revenue function, and indicate the regions of inelastic and elastic demand on the graph.

63. $x = f(p) = 20(10 - p)$ **64.** $x = f(p) = 10(16 - p)$

65. $x = f(p) = 40(p - 15)^2$ **66.** $x = f(p) = 10(p - 9)^2$

67. $x = f(p) = 30 - 10\sqrt{p}$ **68.** $x = f(p) = 30 - 5\sqrt{p}$

If a price–demand equation is solved for p, then price is expressed as $p = g(x)$ and x becomes the independent variable. In this case, it can be shown that the elasticity of demand is given by

$$E(x) = -\frac{g(x)}{xg'(x)}$$

In Problems 69–72, use the price–demand equation to find $E(x)$ at the indicated value of x.

69. $p = g(x) = 50 - 0.1x, x = 200$

70. $p = g(x) = 30 - 0.05x, x = 400$

71. $p = g(x) = 50 - 2\sqrt{x}, x = 400$

72. $p = g(x) = 20 - \sqrt{x}, x = 100$

In Problems 73–76, use the price–demand equation to find the values of x for which demand is elastic and for which demand is inelastic.

73. $p = g(x) = 180 - 0.3x$ **74.** $p = g(x) = 640 - 0.4x$

75. $p = g(x) = 90 - 0.1x^2$ **76.** $p = g(x) = 540 - 0.2x^2$

77. Find $E(p)$ for $x = f(p) = Ap^{-k}$, where A and k are positive constants.

78. Find $E(p)$ for $x = f(p) = Ae^{-kp}$, where A and k are positive constants.

Applications

79. Rate of change of cost. A fast-food restaurant can produce a hamburger for $2.50. If the restaurant's daily sales are increasing at the rate of 30 hamburgers per day, how fast is its daily cost for hamburgers increasing?

80. Rate of change of cost. The fast-food restaurant in Problem 79 can produce an order of fries for $0.80. If the restaurant's daily sales are increasing at the rate of 45 orders of fries per day, how fast is its daily cost for fries increasing?

81. Revenue and elasticity. The price–demand equation for hamburgers at a fast-food restaurant is

$$x + 400p = 3{,}000$$

Currently, the price of a hamburger is $3.00. If the price is increased by 10%, will revenue increase or decrease?

82. Revenue and elasticity. Refer to Problem 81. If the current price of a hamburger is $4.00, will a 10% price increase cause revenue to increase or decrease?

83. Revenue and elasticity. The price–demand equation for an order of fries at a fast-food restaurant is

$$x + 1{,}000p = 2{,}500$$

Currently, the price of an order of fries is $0.99. If the price is decreased by 10%, will revenue increase or decrease?

84. Revenue and elasticity. Refer to Problem 83. If the current price of an order of fries is $1.49; will a 10% price decrease cause revenue to increase or decrease?

85. Maximum revenue. Refer to Problem 81. What price will maximize the revenue from selling hamburgers?

86. Maximum revenue. Refer to Problem 83. What price will maximize the revenue from selling fries?

87. Population growth. A model for Canada's population (Table 3) is

$$f(t) = 0.31t + 18.5$$

where t is years since 1960. Find and graph the percentage rate of change of $f(t)$ for $0 \le t \le 50$.

Table 3 **Population**

Year	Canada (millions)	Mexico (millions)
1960	18	39
1970	22	53
1980	25	68
1990	28	85
2000	31	100
2010	34	112

88. Population growth. A model for Mexico's population (Table 3) is

$$f(t) = 1.49t + 38.8$$

where t is years since 1960. Find and graph the percentage rate of change of $f(t)$ for $0 \le t \le 50$.

89. Crime. A model for the number of robberies in the United States (Table 4) is

$$r(t) = 3.3 - 0.7 \ln t$$

where t is years since 1990. Find the relative rate of change for robberies in 2020.

Table 4 **Number of Victimizations per 1,000 Population**

	Robbery	Aggravated Assault
1995	2.21	4.18
2000	1.45	3.24
2005	1.41	2.91
2010	1.19	2.52

90. Crime. A model for the number of assaults in the United States (Table 4) is

$$a(t) = 6.0 - 1.2 \ln t$$

where t is years since 1990. Find the relative rate of change for assaults in 2020.

Answers to Matched Problems

1. $p(t) = \dfrac{270}{2.7t + 282}$

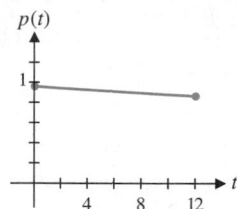

2. $E(p) = \dfrac{p}{40 - p}$

(A) $E(8) = 0.25$; demand is inelastic.

(B) $E(30) = 3$; demand is elastic.

(C) $E(20) = 1$; demand has unit elasticity.

3. $E(21) = \dfrac{21}{17} \approx 1.2$; demand is elastic. Increasing price will decrease revenue.

Chapter 3   Summary and Review

Important Terms, Symbols, and Concepts

3.1 The Constant e and Continuous Compound Interest

EXAMPLES

- The number e is defined as

$$\lim_{x \to \infty} \left(1 + \frac{1}{n}\right)^n = \lim_{x \to 0} (1 + s)^{1/s} = 2.718\ 281\ 828\ 459 \ldots$$

- If a principal P is invested at an annual rate r (expressed as a decimal) compounded continuously, then the amount A in the account at the end of t years is given by the **compound interest formula**

$$A = Pe^{rt}$$

Ex. 1, p. 183
Ex. 2, p. 183
Ex. 3, p. 184
Ex. 4, p. 184

3.2 Derivatives of Exponential and Logarithmic Functions

- For $b > 0, b \ne 1$,

$$\frac{d}{dx} e^x = e^x \qquad \frac{d}{dx} b^x = b^x \ln b$$

For $b > 0, b \ne 1$, and $x > 0$,

$$\frac{d}{dx} \ln x = \frac{1}{x} \qquad \frac{d}{dx} \log_b x = \frac{1}{\ln b} \frac{1}{x}$$

Ex. 1, p. 188
Ex. 2, p. 190
Ex. 3, p. 191
Ex. 4, p. 192
Ex. 5, p. 193

3.2 Derivatives of Exponential and Logarithmic Functions (*Continued*)

- The **change-of-base formulas** allow conversion from base e to any base b, $b > 0$, $b \neq 1$:

$$b^x = e^{x \ln b} \qquad \log_b x = \frac{\ln x}{\ln b}$$

3.3 Derivatives of Products and Quotients

- Product rule. If $y = f(x) = F(x)\, S(x)$, then $f'(x) = F(x)S'(x) + S(x)F'(x)$, provided that both $F'(x)$ and $S'(x)$ exist.

- Quotient rule. If $y = f(x) = \dfrac{T(x)}{B(x)}$, then $f'(x) = \dfrac{B(x)\, T'(x) - T(x)\, B'(x)}{\left[B(x) \right]^2}$ provided that both $T'(x)$ and $B'(x)$ exist.

3.4 The Chain Rule

- A function m is a **composite** of functions f and g if $m(x) = f[g(x)]$.

- The **chain rule** gives a formula for the derivative of the composite function $m(x) = E[I(x)]$:

$$m'(x) = E'[I(x)]I'(x)$$

- A special case of the chain rule is called the **general power rule:**

$$\frac{d}{dx}[f(x)]^n = n[f(x)]^{n-1}f'(x)$$

- Other special cases of the chain rule are the following **general derivative rules:**

$$\frac{d}{dx} \ln [f(x)] = \frac{1}{f(x)}f'(x)$$

$$\frac{d}{dx}\, e^{f(x)} = e^{f(x)}f'(x)$$

3.5 Implicit Differentiation

- If $y = y(x)$ is a function defined implicitly by the equation $F(x, y) = 0$, then we use **implicit differentiation** to find an equation in x, y, and y'.

3.6 Related Rates

- If x and y represent quantities that are changing with respect to time and are related by the equation $F(x, y) = 0$, then implicit differentiation produces an equation that relates x, y, dy/dt, and dx/dt. Problems of this type are called **related-rates problems**.

- Suggestions for solving related-rates problems are given on page 221.

3.7 Elasticity of Demand

- The **relative rate of change**, or the **logarithmic derivative**, of a function $f(x)$ is $f'(x)/f(x)$, and the **percentage rate of change** is $100 \times [f'(x)/f(x)]$.

- If price and demand are related by $x = f(p)$, then the **elasticity of demand** is given by

$$E(p) = -\frac{pf'(p)}{f(p)} = -\frac{\text{relative rate of change of demand}}{\text{relative rate of change of price}}$$

- **Demand is inelastic** if $0 < E(p) < 1$. (Demand is not sensitive to changes in price; a percentage change in price produces a smaller percentage change in demand.) **Demand is elastic** if $E(p) > 1$. (Demand is sensitive to changes in price; a percentage change in price produces a larger percentage change in demand.) **Demand has unit elasticity** if $E(p) = 1$. (A percentage change in price produces the same percentage change in demand.)

- If $R(p) = pf(p)$ is the revenue function, then $R'(p)$ and $[1 - E(p)]$ always have the same sign. If demand is inelastic, then a price increase will increase revenue. If demand is elastic, then a price increase will decrease revenue.

Review Exercises

Work through all the problems in this chapter review, and check your answers in the back of the book. Answers to all review problems are there, along with section numbers in italics to indicate where each type of problem is discussed. Where weaknesses show up, review appropriate sections of the text.

1. Use a calculator to evaluate $A = 2,000e^{0.09t}$ to the nearest cent for $t = 5, 10,$ and 20.

In Problems 2–4, find functions $E(u)$ and $I(x)$ so that $f(x) = E[I(x)]$.

2. $f(x) = (6x + 5)^{3/2}$ 3. $f(x) = \ln(x^2 + 4)$

4. $f(x) = e^{0.02x}$

In Problems 5–8, find the indicated derivative.

5. $\dfrac{d}{dx}(2\ln x + 3e^x)$ 6. $\dfrac{d}{dx}e^{2x-3}$

7. y' for $y = \ln(2x + 7)$

8. $f'(x)$ for $f(x) = \ln(3 + e^x)$

9. Find y' for $y = y(x)$ defined implicity by the equation $2y^2 - 3x^3 - 5 = 0$, and evaluate at $(x, y) = (1, 2)$.

10. For $y = 3x^2 - 5$, where $x = x(t)$ and $y = y(t)$, find dy/dt if $dx/dt = 3$ when $x = 12$.

11. Given the demand equation $25p + x = 1,000$,

   (A) Express the demand x as a function of the price p.

   (B) Find the elasticity of demand, $E(p)$.

   (C) Find $E(15)$ and interpret.

   (D) Express the revenue function as a function of price p.

   (E) If $p = \$25$, what is the effect of a small price cut on revenue?

12. Find the slope of the line tangent to $y = 100e^{-0.1x}$ when $x = 0$.

✎ 13. Use a calculator and a table of values to investigate

$$\lim_{n \to \infty}\left(1 + \frac{2}{n}\right)^n$$

Do you think the limit exists? If so, what do you think it is?

Find the indicated derivatives in Problems 14–19.

14. $\dfrac{d}{dz}[(\ln z)^7 + \ln z^7]$ 15. $\dfrac{d}{dx}(x^6 \ln x)$

16. $\dfrac{d}{dx}\dfrac{e^x}{x^6}$ 17. y' for $y = \ln(2x^3 - 3x)$

18. $f'(x)$ for $f(x) = e^{x^3 - x^2}$ 19. dy/dx for $y = e^{-2x}\ln 5x$

20. Find the equation of the line tangent to the graph of $y = f(x) = 1 + e^{-x}$ at $x = 0$. At $x = -1$.

21. Find y' for $y = y(x)$ defined implicitly by the equation $x^2 - 3xy + 4y^2 = 23$, and find the slope of the graph at $(-1, 2)$.

22. Find x' for $x = x(t)$ defined implicitly by $x^3 - 2t^2x + 8 = 0$, and evaluate at $(t, x) = (-2, 2)$.

23. Find y' for $y = y(x)$ defined implicitly by $x - y^2 = e^y$, and evaluate at $(1, 0)$.

24. Find y' for $y = y(x)$ defined implicitly by $\ln y = x^2 - y^2$, and evaluate at $(1, 1)$.

In Problems 25–27, find the logarithmic derivatives.

25. $A(t) = 400e^{0.049t}$ 26. $f(p) = 100 - 3p$

27. $f(x) = 1 + x^2$

28. A point is moving on the graph of $y^2 - 4x^2 = 12$ so that its x coordinate is decreasing by 2 units per second when $(x, y) = (1, 4)$. Find the rate of change of the y coordinate.

29. A 17-foot ladder is placed against a wall. If the foot of the ladder is pushed toward the wall at 0.5 foot per second, how fast is the top of the ladder rising when the foot is 8 feet from the wall?

30. Water is leaking onto a floor. The resulting circular pool has an area that is increasing at the rate of 24 square inches per minute. How fast is the radius R of the pool increasing when the radius is 12 inches?

31. Find the values of p for which demand is elastic and the values for which demand is inelastic if the price–demand equation is

$$x = f(p) = 20(p - 15)^2 \qquad 0 \le p \le 15$$

32. Graph the revenue function as a function of price p, and indicate the regions of inelastic and elastic demand if the price–demand equation is

$$x = f(p) = 5(20 - p) \qquad 0 \le p \le 20$$

33. Let $y = w^3$, $w = \ln u$, and $u = 4 - e^x$.

   (A) Express y in terms of x.

   (B) Use the chain rule to find dy/dx.

Find the indicated derivatives in Problems 34–36.

34. y' for $y = 5^{x^2 - 1}$ 35. $\dfrac{d}{dx}\log_5(x^2 - x)$

36. $\dfrac{d}{dx}\sqrt{\ln(x^2 + x)}$

37. Find y' for $y = y(x)$ defined implicitly by the equation $e^{xy} = x^2 + y + 1$, and evaluate at $(0, 0)$.

✎ 38. A rock thrown into a still pond causes a circular ripple. The radius is increasing at a constant rate of 3 feet per second. Show that the area does not increase at a constant rate. When is the rate of increase of the area the smallest? The largest? Explain.

✎ 39. A point moves along the graph of $y = x^3$ in such a way that its y coordinate is increasing at a constant rate of 5 units per second. Does the x coordinate ever increase at a faster rate than the y coordinate? Explain.

Applications

40. Doubling time. How long will it take money to double if it is invested at 5% interest compounded

(A) Annually?   (B) Continuously?

41. Continuous compound interest. If $100 is invested at 10% interest compounded continuously, then the amount (in dollars) at the end of t years is given by

$$A = 100e^{0.1t}$$

Find $A'(t), A'(1),$ and $A'(10)$.

42. Continuous compound interest. If $12,000 is invested in an account that earns 3.95% compounded continuously, find the instantaneous rate of change of the amount when the account is worth $25,000.

43. Marginal analysis. The price–demand equation for 14-cubic-foot refrigerators at an appliance store is

$$p(x) = 1,000e^{-0.02x}$$

where x is the monthly demand and p is the price in dollars. Find the marginal revenue equation.

44. Demand equation. Given the demand equation

$$x = \sqrt{5,000 - 2p^3}$$

find the rate of change of p with respect to x by implicit differentiation (x is the number of items that can be sold at a price of p per item).

45. Rate of change of revenue. A company is manufacturing kayaks and can sell all that it manufactures. The revenue (in dollars) is given by

$$R = 750x - \frac{x^2}{30}$$

where the production output in 1 day is x kayaks. If production is increasing at 3 kayaks per day when production is 40 kayaks per day, find the rate of increase in revenue.

✎ **46. Revenue and elasticity.** The price–demand equation for home-delivered large pizzas is

$$p = 38.2 - 0.002x$$

where x is the number of pizzas delivered weekly. The current price of one pizza is $21. In order to generate additional revenue from the sale of large pizzas, would you recommend a price increase or a price decrease? Explain.

47. Average income. A model for the average income per household before taxes are paid is

$$f(t) = 1,700t + 20,500$$

where t is years since 1980. Find the relative rate of change of household income in 2015.

48. Drug concentration. The drug concentration in the bloodstream t hours after injection is given approximately by

$$C(t) = 5e^{-0.3t}$$

where $C(t)$ is concentration in milligrams per milliliter. What is the rate of change of concentration after 1 hour? After 5 hours?

49. Wound healing. A circular wound on an arm is healing at the rate of 45 square millimeters per day (the area of the wound is decreasing at this rate). How fast is the radius R of the wound decreasing when $R = 15$ millimeters?

50. Psychology: learning. In a computer assembly plant, a new employee, on the average, is able to assemble

$$N(t) = 10(1 - e^{-0.4t})$$

units after t days of on-the-job training.

(A) What is the rate of learning after 1 day? After 5 days?

(B) Find the number of days (to the nearest day) after which the rate of learning is less than 0.25 unit per day.

51. Learning. A new worker on the production line performs an operation in T minutes after x performances of the operation, as given by

$$T = 2\left(1 + \frac{1}{x^{3/2}}\right)$$

If, after performing the operation 9 times, the rate of improvement is $dx/dt = 3$ operations per hour, find the rate of improvement in time dT/dt in performing each operation.

4

Graphing and Optimization

Introduction

Since the derivative is associated with the slope of the graph of a function at a point, we might expect that it is also related to other properties of a graph. As we will see in this chapter, the derivative can tell us a great deal about the shape of the graph of a function. In particular, we will study methods for finding absolute maximum and minimum values. Manufacturing companies can use these methods to find production levels that will minimize cost or maximize profit, pharmacologists can use them to find levels of drug dosages that will produce maximum sensitivity, and advertisers can use them to determine the number of ads that will maximize the rate of change of sales (see, for example, Problem 93 in Section 4.2).

4.1   First Derivative and Graphs

- Increasing and Decreasing Functions
- Local Extrema
- First-Derivative Test
- Economics Applications

Increasing and Decreasing Functions

Sign charts will be used throughout this chapter. You may find it helpful to review the terminology and techniques for constructing sign charts in Section 2.3.

Explore and Discuss 1   Figure 1 shows the graph of $y = f(x)$ and a sign chart for $f'(x)$, where

$$f(x) = x^3 - 3x$$

and

$$f'(x) = 3x^2 - 3 = 3(x + 1)(x - 1)$$

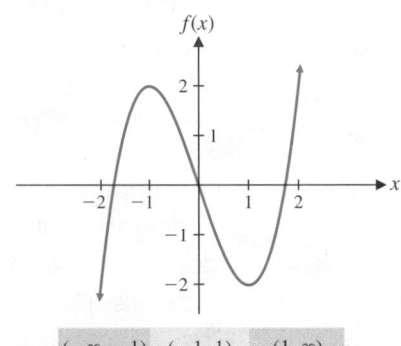

Figure 1

Discuss the relationship between the graph of f and the sign of $f'(x)$ over each interval on which $f'(x)$ has a constant sign. Also, describe the behavior of the graph of f at each partition number for f'.

As they are scanned from left to right, graphs of functions generally have rising and falling sections. If you scan the graph of $f(x) = x^3 - 3x$ in Figure 1 from left to right, you will observe the following:

- On the interval $(-\infty, -1)$, the graph of f is rising, $f(x)$ is increasing,* and tangent lines have positive slope $[f'(x) > 0]$.
- On the interval $(-1, 1)$, the graph of f is falling, $f(x)$ is decreasing, and tangent lines have negative slope $[f'(x) < 0]$.
- On the interval $(1, \infty)$, the graph of f is rising, $f(x)$ is increasing, and tangent lines have positive slope $[f'(x) > 0]$.
- At $x = -1$ and $x = 1$, the slope of the graph is 0 $[f'(x) = 0]$.

If $f'(x) > 0$ (is positive) on the interval (a, b) (Fig. 2), then $f(x)$ increases ($\nearrow$) and the graph of f rises as we move from left to right over the interval. If $f'(x) < 0$ (is negative) on an interval (a, b), then $f(x)$ decreases ($\searrow$) and the graph of f falls as we move from left to right over the interval. We summarize these important results in Theorem 1.

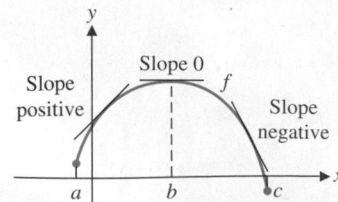

Figure 2

---

*Formally, we say that the function f is **increasing** on an interval (a, b) if $f(x_2) > f(x_1)$ whenever $a < x_1 < x_2 < b$, and f is **decreasing** on (a, b) if $f(x_2) < f(x_1)$ whenever $a < x_1 < x_2 < b$.

THEOREM 1 Increasing and Decreasing Functions

For the interval (a, b), if $f' > 0$, then f is increasing, and if $f' < 0$, then f is decreasing.

$f'(x)$	$f(x)$	Graph of f	Examples
+	Increases ↗	Rises ↗	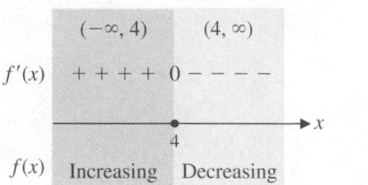
−	Decreases ↘	Falls ↘	

EXAMPLE 1 Finding Intervals on Which a Function Is Increasing or Decreasing
Given the function $f(x) = 8x - x^2$,

(A) Which values of x correspond to horizontal tangent lines?

(B) For which values of x is $f(x)$ increasing? Decreasing?

(C) Sketch a graph of f. Add any horizontal tangent lines.

SOLUTION

(A) $f'(x) = 8 - 2x = 0$

$$x = 4$$

So, a horizontal tangent line exists at $x = 4$ only.

(B) We will construct a sign chart for $f'(x)$ to determine which values of x make $f'(x) > 0$ and which values make $f'(x) < 0$. Recall from Section 2.3 that the partition numbers for a function are the numbers at which the function is 0 or discontinuous. When constructing a sign chart for $f'(x)$, we must locate all points where $f'(x) = 0$ or $f'(x)$ is discontinuous. From part (A), we know that $f'(x) = 8 - 2x = 0$ at $x = 4$. Since $f'(x) = 8 - 2x$ is a polynomial, it is continuous for all x. So, 4 is the only partition number for f'. We construct a sign chart for the intervals $(-\infty, 4)$ and $(4, \infty)$, using test numbers 3 and 5:

$f'(x)$ $(-\infty, 4)$ $(4, \infty)$
          $+ + + + \; 0 \; - - - -$
                    4
$f(x)$ Increasing   Decreasing

Test Numbers	
x	$f'(x)$
3	2 (+)
5	−2 (−)

Therefore, $f(x)$ is increasing on $(-\infty, 4)$ and decreasing on $(4, \infty)$.

(C)

x	$f(x)$
0	0
2	12
4	16
6	12
8	0

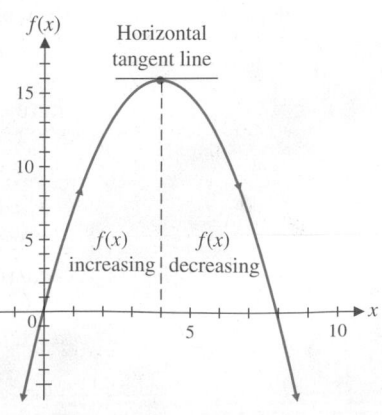

Matched Problem 1 Repeat Example 1 for $f(x) = x^2 - 6x + 10$.

As Example 1 illustrates, the construction of a sign chart will play an important role in using the derivative to analyze and sketch the graph of a function f. The partition numbers for f' are central to the construction of these sign charts and also to the analysis of the graph of $y = f(x)$. The partition numbers for f' that belong to the domain of f are called **critical numbers** of f.*

> **DEFINITION** Critical Numbers
>
> A real number x in the domain of f such that $f'(x) = 0$ or $f'(x)$ does not exist is called a **critical number** of f.

> **CONCEPTUAL INSIGHT**
>
> The critical numbers of f belong to the domain of f and are partition numbers for f'. But f' may have partition numbers that do not belong to the domain of f, so are not critical numbers of f.
>
> If f is a polynomial, then both the partition numbers for f' and the critical numbers of f are the solutions of $f'(x) = 0$.

EXAMPLE 2   Partition Numbers for f' and Critical Numbers of f Find the critical numbers of f, the intervals on which f is increasing, and those on which f is decreasing, for $f(x) = 1 + x^3$.

SOLUTION   Begin by finding the partition numbers for $f'(x)$ [since $f'(x) = 3x^2$ is continuous we just need to solve $f'(x) = 0$]

$$f'(x) = 3x^2 = 0 \quad \text{only if } x = 0$$

The partition number 0 for f' is in the domain of f, so 0 is the only critical number of f.

The sign chart for $f'(x) = 3x^2$ (partition number is 0) is

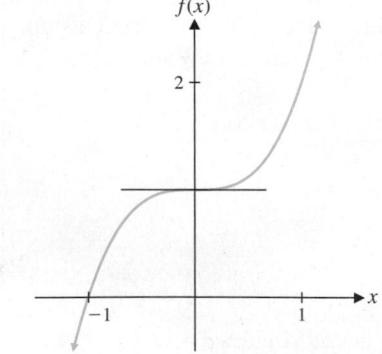

$f(x)$

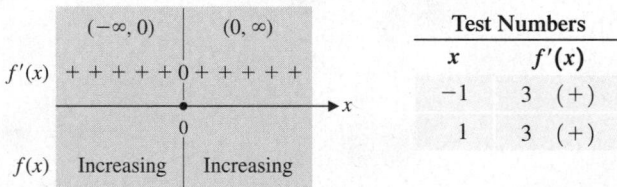

	$(-\infty, 0)$	$(0, \infty)$	**Test Numbers**	
			x	$f'(x)$
$f'(x)$	$+ + + + + 0 + + + + +$		-1	3 (+)
	0		1	3 (+)
$f(x)$	Increasing	Increasing		

The sign chart indicates that $f(x)$ is increasing on $(-\infty, 0)$ and $(0, \infty)$. Since f is continuous at $x = 0$, it follows that $f(x)$ is increasing for all x. The graph of f is shown in Figure 3.

Figure 3

Matched Problem 2   Find the critical numbers of f, the intervals on which f is increasing, and those on which f is decreasing, for $f(x) = 1 - x^3$.

EXAMPLE 3   Partition Numbers for f' and Critical Numbers of f Find the critical numbers of f, the intervals on which f is increasing, and those on which f is decreasing, for $f(x) = (1 - x)^{1/3}$.

SOLUTION $$f'(x) = -\frac{1}{3}(1 - x)^{-2/3} = \frac{-1}{3(1 - x)^{2/3}}$$

*We are assuming that $f'(c)$ does not exist at any point of discontinuity of f'. There do exist functions f such that f' is discontinuous at $x = c$, yet $f'(c)$ exists. However, we do not consider such functions in this book.

To find the partition numbers for f', we note that f' is continuous for all x, except for values of x for which the denominator is 0; that is, $f'(1)$ does not exist and f' is discontinuous at $x = 1$. Since the numerator of f' is the constant -1, $f'(x) \neq 0$ for any value of x. Thus, $x = 1$ is the only partition number for f'. Since 1 is in the domain of f, $x = 1$ is also the only critical number of f. When constructing the sign chart for f' we use the abbreviation ND to note the fact that $f'(x)$ is *not defined* at $x = 1$.

The sign chart for $f'(x) = -1/[3(1 - x)^{2/3}]$ (partition number for f' is 1) is as follows:

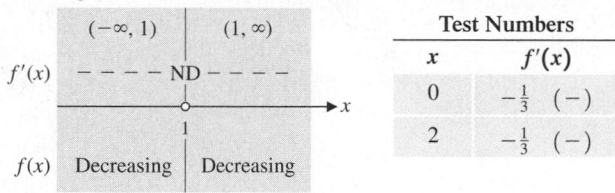

	$(-\infty, 1)$	$(1, \infty)$
$f'(x)$	$- - - -$ ND	$- - - -$
		1
$f(x)$	Decreasing	Decreasing

Test Numbers

x	$f'(x)$
0	$-\frac{1}{3}$ $(-)$
2	$-\frac{1}{3}$ $(-)$

The sign chart indicates that f is decreasing on $(-\infty, 1)$ and $(1, \infty)$. Since f is continuous at $x = 1$, it follows that $f(x)$ is decreasing for all x. **A continuous function can be decreasing (or increasing) on an interval containing values of x where $f'(x)$ does not exist.** The graph of f is shown in Figure 4. Notice that the undefined derivative at $x = 1$ results in a vertical tangent line at $x = 1$. **A vertical tangent will occur at $x = c$ if f is continuous at $x = c$ and if $|f'(x)|$ becomes larger and larger as x approaches c.**

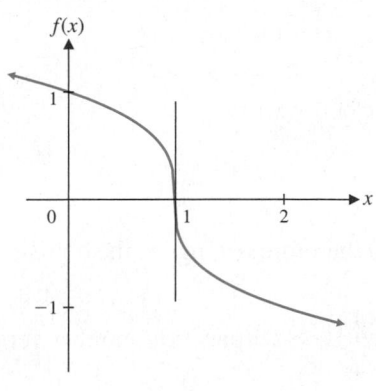

Figure 4

Matched Problem 3 Find the critical numbers of f, the intervals on which f is increasing, and those on which f is decreasing, for $f(x) = (1 + x)^{1/3}$.

EXAMPLE 4 Partition Numbers for f' and Critical Numbers of f Find the critical numbers of f, the intervals on which f is increasing, and those on which f is decreasing, for $f(x) = \dfrac{1}{x - 2}$.

SOLUTION
$$f(x) = \frac{1}{x - 2} = (x - 2)^{-1}$$

$$f'(x) = -(x - 2)^{-2} = \frac{-1}{(x - 2)^2}$$

To find the partition numbers for f', note that $f'(x) \neq 0$ for any x and f' is not defined at $x = 2$. Thus, $x = 2$ is the only partition number for f'. However, $x = 2$ is *not* in the domain of f. Consequently, $x = 2$ is *not* a critical number of f. The function f has no critical numbers.

The sign chart for $f'(x) = -1/(x - 2)^2$ (partition number for f' is 2) is as follows:

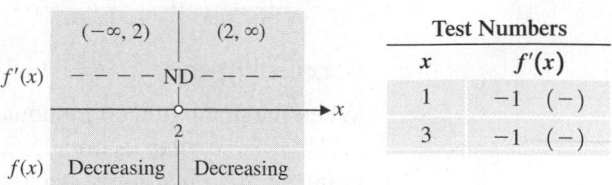

	$(-\infty, 2)$	$(2, \infty)$
$f'(x)$	$- - - -$ ND	$- - - -$
		2
$f(x)$	Decreasing	Decreasing

Test Numbers

x	$f'(x)$
1	-1 $(-)$
3	-1 $(-)$

Therefore, f is decreasing on $(-\infty, 2)$ and $(2, \infty)$. The graph of f is shown in Figure 5.

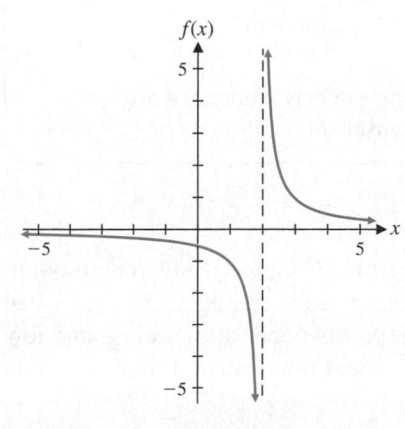

Figure 5

Matched Problem 4 Find the critical numbers of f, the intervals on which f is increasing, and those on which f is decreasing, for $f(x) = \dfrac{1}{x}$.

EXAMPLE 5   Partition Numbers for f' and Critical Numbers of f Find the critical numbers of f, the intervals on which f is increasing, and those on which f is decreasing, for $f(x) = 8 \ln x - x^2$.

SOLUTION   The natural logarithm function $\ln x$ is defined on $(0, \infty)$, or $x > 0$, so $f(x)$ is defined only for $x > 0$.

$$f(x) = 8 \ln x - x^2, x > 0$$

$$f'(x) = \frac{8}{x} - 2x \qquad \text{Find a common denominator.}$$

$$= \frac{8}{x} - \frac{2x^2}{x} \qquad \text{Subtract numerators.}$$

$$= \frac{8 - 2x^2}{x} \qquad \text{Factor numerator.}$$

$$= \frac{2(2 - x)(2 + x)}{x}, \quad x > 0$$

The only partition number for f' that is positive, and therefore belongs to the domain of f, is 2. So 2 is the only critical number of f.

The sign chart for $f'(x) = \dfrac{2(2 - x)(2 + x)}{x}, x > 0$ (partition number for f' is 2), is as follows:

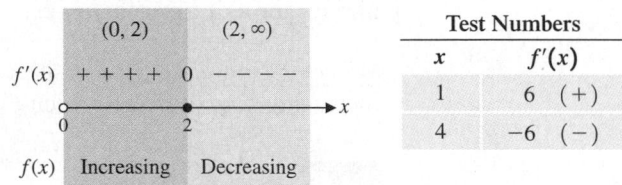

	Test Numbers	
	x	$f'(x)$
	1	6 (+)
	4	−6 (−)

Therefore, f is increasing on $(0, 2)$ and decreasing on $(2, \infty)$. The graph of f is shown in Figure 6.

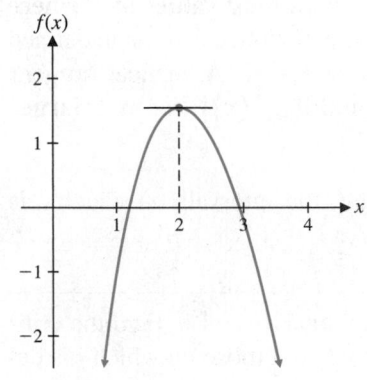

$f(x)$

Figure 6

Matched Problem 5   Find the critical numbers of f, the intervals on which f is increasing, and those on which f is decreasing, for $f(x) = 5 \ln x - x$.

CONCEPTUAL INSIGHT

Examples 4 and 5 illustrate two important ideas:

1. Do not assume that all partition numbers for the derivative f' are critical numbers of the function f. To be a critical number of f, a partition number for f' must also be in the domain of f.

2. The intervals on which a function f is increasing or decreasing must always be expressed in terms of open intervals that are subsets of the domain of f.

Local Extrema

When the graph of a continuous function changes from rising to falling, a high point, or *local maximum,* occurs. When the graph changes from falling to rising, a low point, or *local minimum,* occurs. In Figure 7, high points occur at c_3 and c_6, and low points occur at c_2 and c_4. In general, we call $f(c)$ a **local maximum** if there exists an interval (m, n) containing c such that

$$f(x) \le f(c) \qquad \text{for all } x \text{ in } (m, n)$$

Note that this inequality need hold only for numbers x near c, which is why we use the term *local.* So the y coordinate of the high point $(c_3, f(c_3))$ in Figure 7 is a local maximum, as is the y coordinate of $(c_6, f(c_6))$.

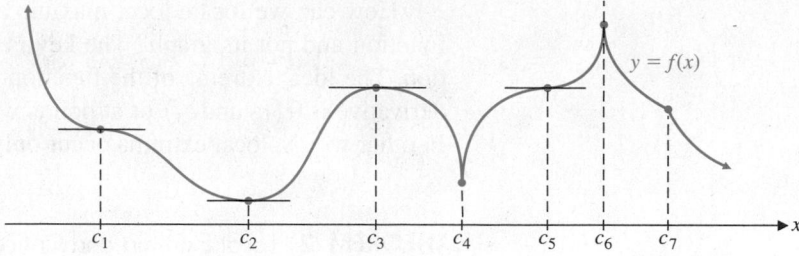

Figure 7

The value $f(c)$ is called a **local minimum** if there exists an interval (m, n) containing c such that

$$f(x) \geq f(c) \qquad \text{for all } x \text{ in } (m, n)$$

The value $f(c)$ is called a **local extremum** if it is either a local maximum or a local minimum. A point on a graph where a local extremum occurs is also called a **turning point**. In Figure 7 we see that local maxima occur at c_3 and c_6, local minima occur at c_2 and c_4, and all four values produce local extrema. Also, the local maximum $f(c_3)$ is not the largest y coordinate of points on the graph in Figure 7. Later in this chapter, we consider the problem of finding *absolute extrema*, the y coordinates of the highest and lowest points on a graph. For now, we are concerned only with locating *local* extrema.

EXAMPLE 6   Analyzing a Graph   Use the graph of f in Figure 8 to find the intervals on which f is increasing, those on which f is decreasing, any local maxima, and any local minima.

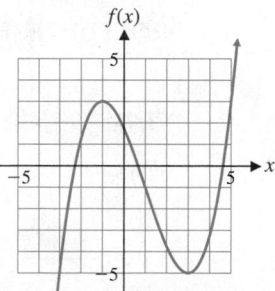

Figure 8

SOLUTION The function f is increasing (the graph is rising) on $(-\infty, -1)$ and on $(3, \infty)$ and is decreasing (the graph is falling) on $(-1, 3)$. Because the graph changes from rising to falling at $x = -1$, $f(-1) = 3$ is a local maximum. Because the graph changes from falling to rising at $x = 3$, $f(3) = -5$ is a local minimum.

Matched Problem 6 Use the graph of g in Figure 9 to find the intervals on which g is increasing, those on which g is decreasing, any local maxima, and any local minima.

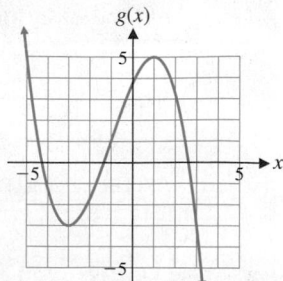

Figure 9

How can we locate local maxima and minima if we are given the equation of a function and not its graph? The key is to examine the critical numbers of the function. The local extrema of the function f in Figure 7 occur either at points where the derivative is 0 (c_2 and c_3) or at points where the derivative does not exist (c_4 and c_6). In other words, local extrema occur only at critical numbers of f.

THEOREM 2   Local Extrema and Critical Numbers

If $f(c)$ is a local extremum of the function f, then c is a critical number of f.

Theorem 2 states that a local extremum can occur only at a critical number, but it does not imply that every critical number produces a local extremum. In Figure 7, c_1 and c_5 are critical numbers (the slope is 0), but the function does not have a local maximum or local minimum at either of these numbers.

Our strategy for finding local extrema is now clear: We find all critical numbers of f and test each one to see if it produces a local maximum, a local minimum, or neither.

First-Derivative Test

If $f'(x)$ exists on both sides of a critical number c, the sign of $f'(x)$ can be used to determine whether the point $(c, f(c))$ is a local maximum, a local minimum, or neither. The various possibilities are summarized in the following box and are illustrated in Figure 10:

PROCEDURE   First-Derivative Test for Local Extrema

Let c be a critical number of f [$f(c)$ is defined and either $f'(c) = 0$ or $f'(c)$ is not defined]. Construct a sign chart for $f'(x)$ close to and on either side of c.

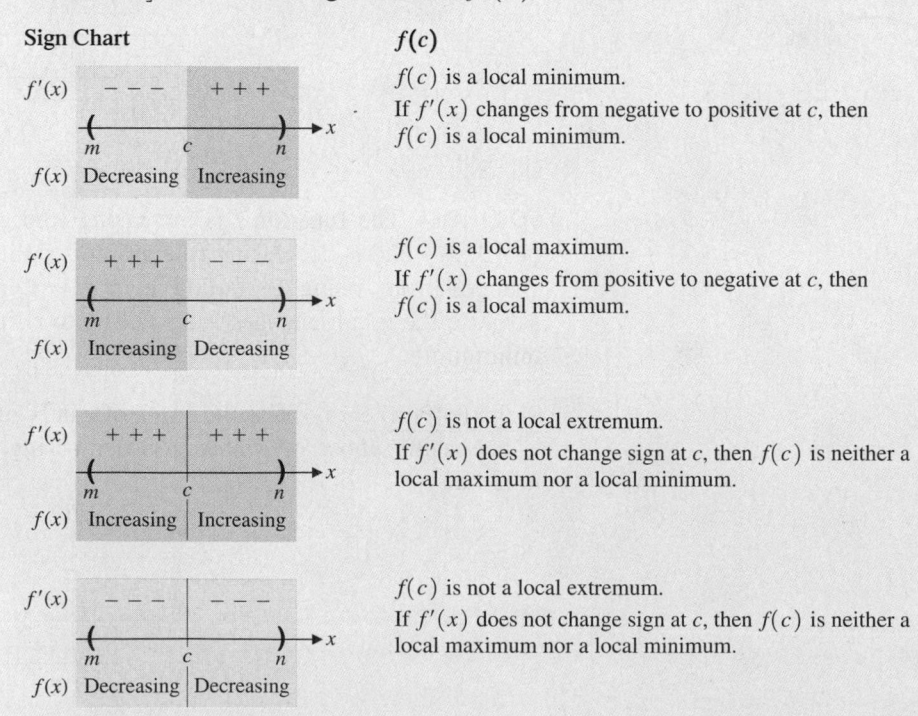

Sign Chart	$f(c)$
$f'(x)$: $---$ (at m to c), $+++$ (c to n); $f(x)$ Decreasing / Increasing	$f(c)$ is a local minimum. If $f'(x)$ changes from negative to positive at c, then $f(c)$ is a local minimum.
$f'(x)$: $+++$ (at m to c), $---$ (c to n); $f(x)$ Increasing / Decreasing	$f(c)$ is a local maximum. If $f'(x)$ changes from positive to negative at c, then $f(c)$ is a local maximum.
$f'(x)$: $+++$ (at m to c), $+++$ (c to n); $f(x)$ Increasing / Increasing	$f(c)$ is not a local extremum. If $f'(x)$ does not change sign at c, then $f(c)$ is neither a local maximum nor a local minimum.
$f'(x)$: $---$ (at m to c), $---$ (c to n); $f(x)$ Decreasing / Decreasing	$f(c)$ is not a local extremum. If $f'(x)$ does not change sign at c, then $f(c)$ is neither a local maximum nor a local minimum.

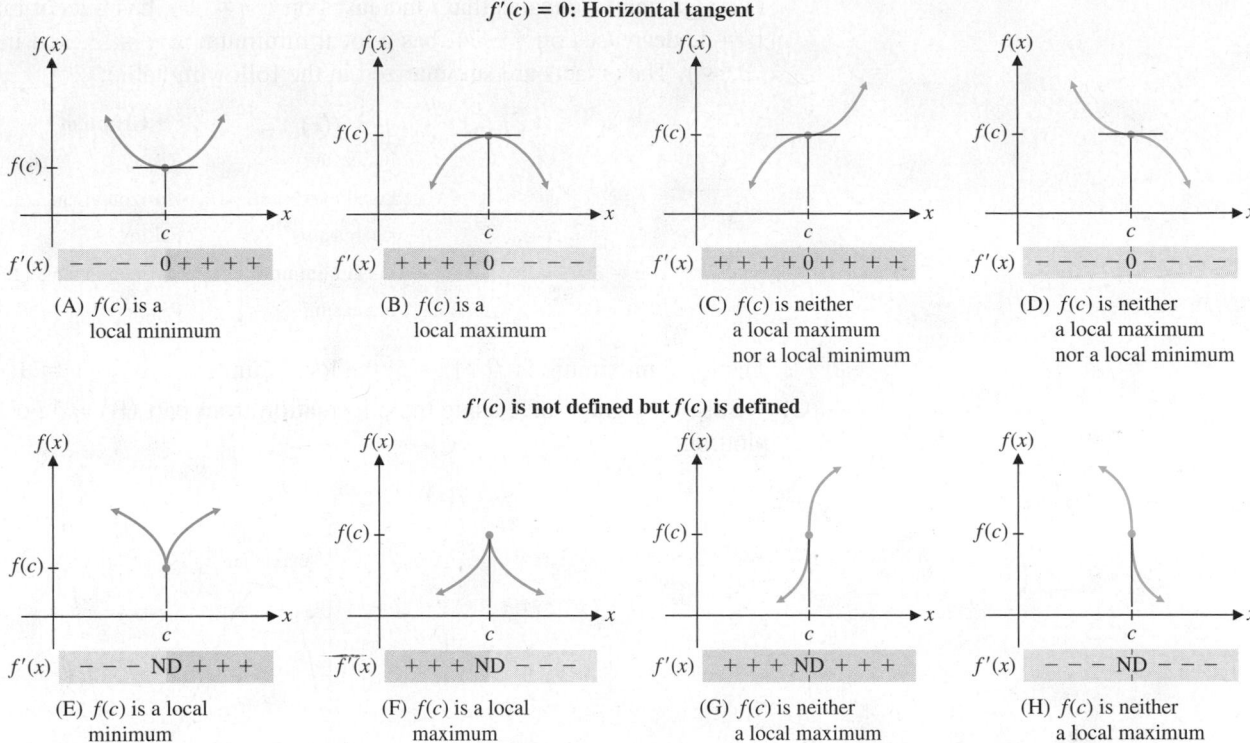

Figure 10 **Local extrema**

EXAMPLE 7 Locating Local Extrema   Given $f(x) = x^3 - 6x^2 + 9x + 1$,

(A) Find the critical numbers of f.

(B) Find the local maxima and local minima of f.

(C) Sketch the graph of f.

SOLUTION

(A) Find all numbers x in the domain of f where $f'(x) = 0$ or $f'(x)$ does not exist.

$$f'(x) = 3x^2 - 12x + 9 = 0$$
$$3(x^2 - 4x + 3) = 0$$
$$3(x - 1)(x - 3) = 0$$
$$x = 1 \quad \text{or} \quad x = 3$$

$f'(x)$ exists for all x; the critical numbers of f are $x = 1$ and $x = 3$.

(B) The easiest way to apply the first-derivative test for local maxima and minima is to construct a sign chart for $f'(x)$ for all x. Partition numbers for $f'(x)$ are $x = 1$ and $x = 3$ (which also happen to be critical numbers of f).

Sign chart for $f'(x) = 3(x - 1)(x - 3)$:

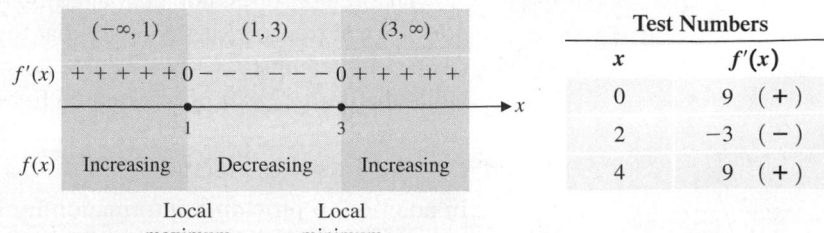

Test Numbers	
x	$f'(x)$
0	9 $(+)$
2	-3 $(-)$
4	9 $(+)$

The sign chart indicates that f increases on $(-\infty, 1)$, has a local maximum at $x = 1$, decreases on $(1, 3)$, has a local minimum at $x = 3$, and increases on $(3, \infty)$. These facts are summarized in the following table:

x	$f'(x)$	$f(x)$	Graph of f
$(-\infty, 1)$	$+$	Increasing	Rising
$x = 1$	0	Local maximum	Horizontal tangent
$(1, 3)$	$-$	Decreasing	Falling
$x = 3$	0	Local minimum	Horizontal tangent
$(3, \infty)$	$+$	Increasing	Rising

The local maximum is $f(1) = 5$; the local minimum is $f(3) = 1$.

(C) We sketch a graph of f, using the information from part (B) and point-by-point plotting.

x	$f(x)$
0	1
1	5
2	3
3	1
4	5

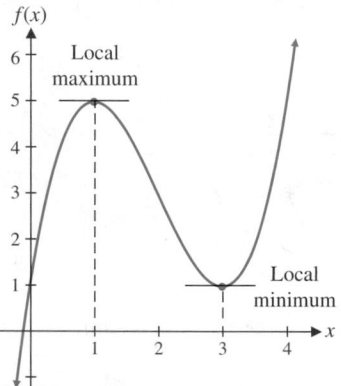

Matched Problem 7 Given $f(x) = x^3 - 9x^2 + 24x - 10$,

(A) Find the critical numbers of f.

(B) Find the local maxima and local minima of f.

(C) Sketch a graph of f.

How can you tell if you have found all the local extrema of a function? In general, this can be a difficult question to answer. However, in the case of a polynomial function, there is an easily determined upper limit on the number of local extrema. Since the local extrema are the x intercepts of the derivative, this limit is a consequence of the number of x intercepts of a polynomial. The relevant information is summarized in the following theorem, which is stated without proof:

> **THEOREM 3   Intercepts and Local Extrema of Polynomial Functions**
> If $f(x) = a_n x^n + a_{n-1}x^{n-1} + \cdots + a_1 x + a_0, a_n \neq 0$, is a polynomial function of degree $n \geq 1$, then f has at most n x intercepts and at most $n - 1$ local extrema.

Theorem 3 does not guarantee that every nth-degree polynomial has exactly $n - 1$ local extrema; it says only that there can never be more than $n - 1$ local extrema. For example, the third-degree polynomial in Example 7 has two local extrema, while the third-degree polynomial in Example 2 does not have any.

Economics Applications

In addition to providing information for hand-sketching graphs, the derivative is an important tool for analyzing graphs and discussing the interplay between a function and its rate of change. The next two examples illustrate this process in the context of economics.

EXAMPLE 8   Agricultural Exports and Imports   Over the past few decades, the United States has exported more agricultural products than it has imported, maintaining a positive balance of trade in this area. However, the trade balance fluctuated considerably during that period. The graph in Figure 11 approximates the rate of change of the balance of trade over a 15-year period, where $B(t)$ is the balance of trade (in billions of dollars) and t is time (in years).

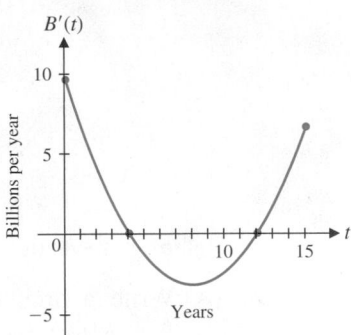

Figure 11 **Rate of change of the balance of trade**

(A) Write a brief description of the graph of $y = B(t)$, including a discussion of any local extrema.

(B) Sketch a possible graph of $y = B(t)$.

SOLUTION

(A) The graph of the derivative $y = B'(t)$ contains the same essential information as a sign chart. That is, we see that $B'(t)$ is positive on $(0, 4)$, 0 at $t = 4$, negative on $(4, 12)$, 0 at $t = 12$, and positive on $(12, 15)$. The trade balance increases for the first 4 years to a local maximum, decreases for the next 8 years to a local minimum, and then increases for the final 3 years.

(B) Without additional information concerning the actual values of $y = B(t)$, we cannot produce an accurate graph. However, we can sketch a possible graph that illustrates the important features, as shown in Figure 12. The absence of a scale on the vertical axis is a consequence of the lack of information about the values of $B(t)$.

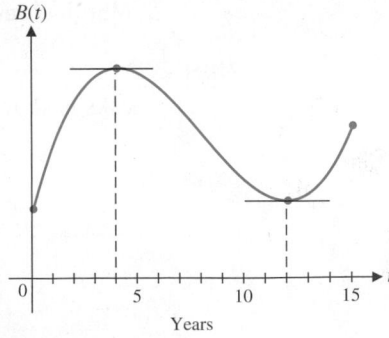

Figure 12 **Balance of trade**

<u>Matched Problem 8</u>   The graph in Figure 13 approximates the rate of change of the U.S. share of the total world production of motor vehicles over a 20-year period, where $S(t)$ is the U.S. share (as a percentage) and t is time (in years).

(A) Write a brief description of the graph of $y = S(t)$, including a discussion of any local extrema.

(B) Sketch a possible graph of $y = S(t)$.

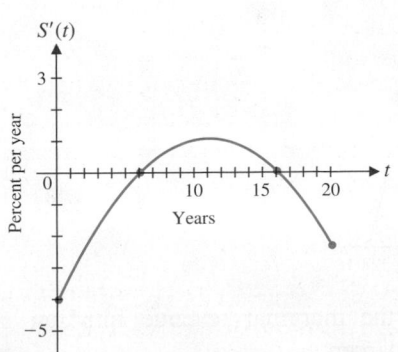

Figure 13

EXAMPLE 9   Revenue Analysis The graph of the total revenue $R(x)$ (in dollars) from the sale of x bookcases is shown in Figure 14.

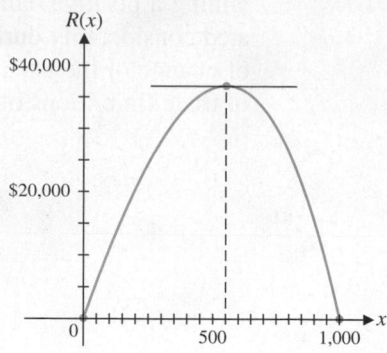

Figure 14 **Revenue**

(A) Write a brief description of the graph of the marginal revenue function $y = R'(x)$, including a discussion of any x intercepts.

(B) Sketch a possible graph of $y = R'(x)$.

SOLUTION

(A) The graph of $y = R(x)$ indicates that $R(x)$ increases on $(0, 550)$, has a local maximum at $x = 550$, and decreases on $(550, 1,000)$. Consequently, the marginal revenue function $R'(x)$ must be positive on $(0, 550)$, 0 at $x = 550$, and negative on $(550, 1,000)$.

(B) A possible graph of $y = R'(x)$ illustrating the information summarized in part (A) is shown in Figure 15.

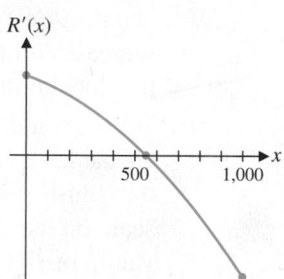

Figure 15 **Marginal revenue**

Matched Problem 9 The graph of the total revenue $R(x)$ (in dollars) from the sale of x desks is shown in Figure 16.

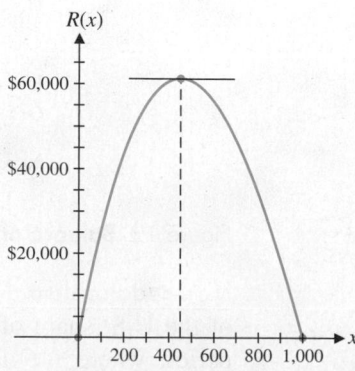

Figure 16

(A) Write a brief description of the graph of the marginal revenue function $y = R'(x)$, including a discussion of any x intercepts.

(B) Sketch a possible graph of $y = R'(x)$.

Comparing Examples 8 and 9, we see that we were able to obtain more information about the function from the graph of its derivative (Example 8) than we were when the process was reversed (Example 9). In the next section, we introduce some ideas that will help us obtain additional information about the derivative from the graph of the function.

Exercises 4.1

Skills Warm-up Exercises

W *In Problems 1–8, inspect the graph of the function to determine whether it is increasing or decreasing on the given interval. (If necessary, review Section 1.2).*

1. $g(x) = |x|$ on $(-\infty, 0)$ **2.** $m(x) = x^3$ on $(0, \infty)$

3. $f(x) = x$ on $(-\infty, \infty)$ **4.** $k(x) = -x^2$ on $(0, \infty)$

5. $p(x) = \sqrt[3]{x}$ on $(-\infty, 0)$

6. $h(x) = x^2$ on $(-\infty, 0)$

7. $r(x) = 4 - \sqrt{x}$ on $(0, \infty)$

8. $g(x) = |x|$ on $(0, \infty)$

Problems 9–16, refer to the following graph of $y = f(x)$:

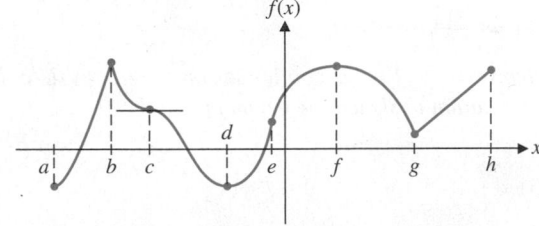

Figure for 9–16

9. Identify the intervals on which $f(x)$ is increasing.

10. Identify the intervals on which $f(x)$ is decreasing.

11. Identify the intervals on which $f'(x) < 0$.

12. Identify the intervals on which $f'(x) > 0$.

13. Identify the x coordinates of the points where $f'(x) = 0$.

14. Identify the x coordinates of the points where $f'(x)$ does not exist.

15. Identify the x coordinates of the points where $f(x)$ has a local maximum.

16. Identify the x coordinates of the points where $f(x)$ has a local minimum.

In Problems 17 and 18, $f(x)$ is continuous on $(-\infty, \infty)$ and has critical numbers at $x = a, b, c,$ and d. Use the sign chart for $f'(x)$ to determine whether f has a local maximum, a local minimum, or neither at each critical number.

17.

 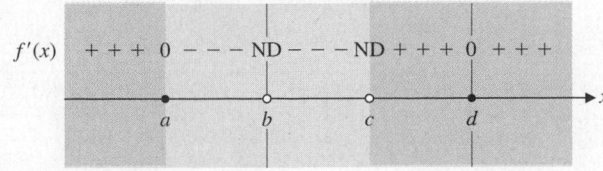

18.

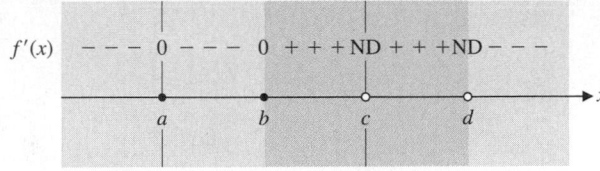

In Problems 19–26, give the local extrema of f and match the graph of f with one of the sign charts a–h in the figure on page 250.

19.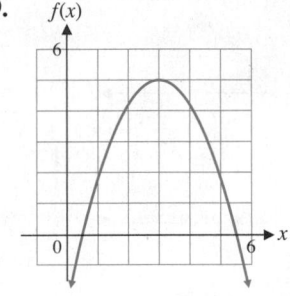

20.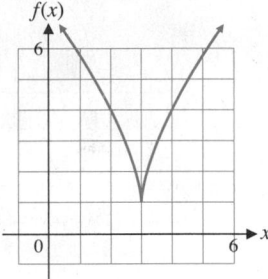

21.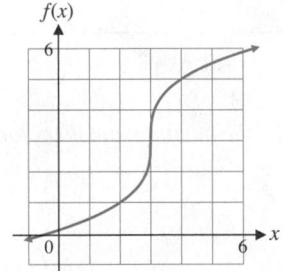

22.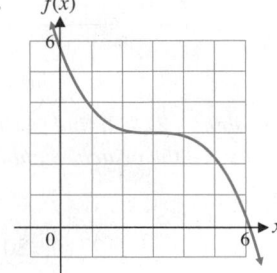

23.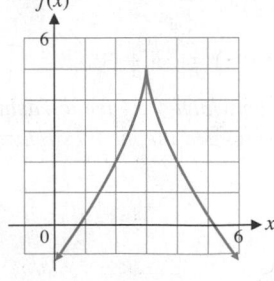

24.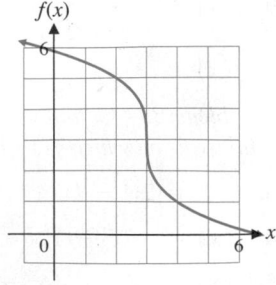

25.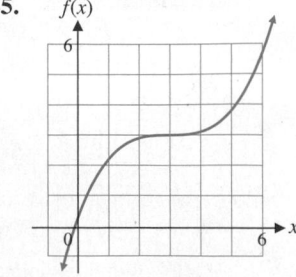

26.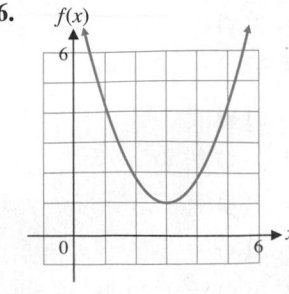

(a)
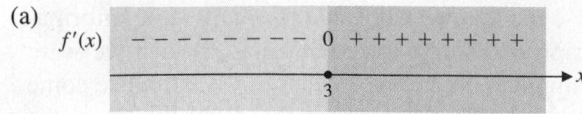

(b)
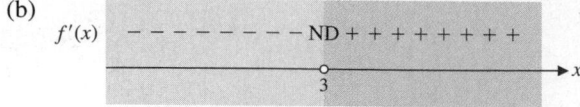

(c)

(d)

(e)
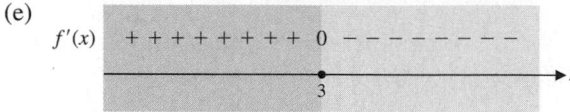

(f)
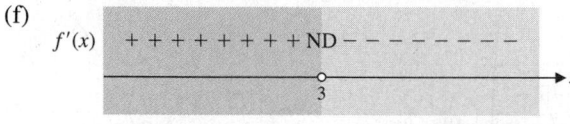

(g)
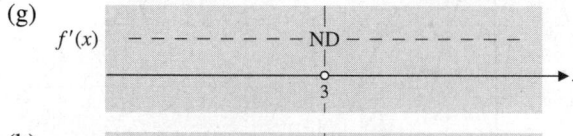

(h)
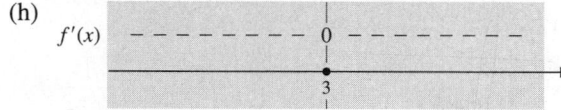

Figure for 19–26

In Problems 27–32, find (A) $f'(x)$, (B) the partition numbers for f', and (C) the critical numbers of f.

27. $f(x) = x^3 - 12x + 8$

28. $f(x) = x^3 - 27x + 30$

29. $f(x) = \dfrac{6}{x + 2}$

30. $f(x) = \dfrac{5}{x - 4}$

31. $f(x) = |x|$

32. $f(x) = |x + 3|$

In Problems 33–46, find the intervals on which $f(x)$ is increasing, the intervals on which $f(x)$ is decreasing, and the local extrema.

33. $f(x) = 2x^2 - 4x$

34. $f(x) = -3x^2 - 12x$

35. $f(x) = -2x^2 - 16x - 25$

36. $f(x) = -3x^2 + 12x - 5$

37. $f(x) = x^3 + 4x - 5$

38. $f(x) = -x^3 - 4x + 8$

39. $f(x) = 2x^3 - 3x^2 - 36x$

40. $f(x) = -2x^3 + 3x^2 + 120x$

41. $f(x) = 3x^4 - 4x^3 + 5$

42. $f(x) = x^4 + 2x^3 + 5$

43. $f(x) = (x - 1)e^{-x}$

44. $f(x) = x \ln x - x$

45. $f(x) = 4x^{1/3} - x^{2/3}$

46. $f(x) = (x^2 - 9)^{2/3}$

In Problems 47–52, use a graphing calculator to approximate the critical numbers of $f(x)$ to two decimal places. Find the intervals on which $f(x)$ is increasing, the intervals on which $f(x)$ is decreasing, and the local extrema.

47. $f(x) = x^4 - 4x^3 + 9x$

48. $f(x) = x^4 + 5x^3 - 15x$

49. $f(x) = x \ln x - (x - 2)^3$

50. $f(x) = e^{-x} - 3x^2$

51. $f(x) = e^x - 2x^2$

52. $f(x) = \dfrac{\ln x}{x} - 5x + x^2$

In Problems 53–60, find the intervals on which $f(x)$ is increasing and the intervals on which $f(x)$ is decreasing. Then sketch the graph. Add horizontal tangent lines.

53. $f(x) = 4 + 8x - x^2$

54. $f(x) = 2x^2 - 8x + 9$

55. $f(x) = x^3 - 3x + 1$

56. $f(x) = x^3 - 12x + 2$

57. $f(x) = 10 - 12x + 6x^2 - x^3$

58. $f(x) = x^3 + 3x^2 + 3x$

59. $f(x) = x^4 - 18x^2$

60. $f(x) = -x^4 + 50x^2$

In Problems 61–68, $f(x)$ is continuous on $(-\infty, \infty)$. Use the given information to sketch the graph of f.

61.
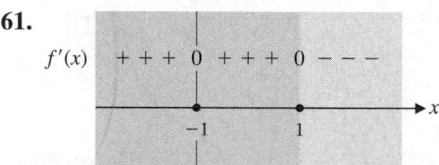

x	-2	-1	0	1	2
$f(x)$	-1	1	2	3	1

62.
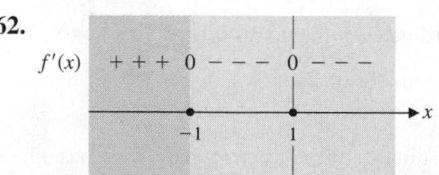

x	-2	-1	0	1	2
$f(x)$	1	3	2	1	-1

63.
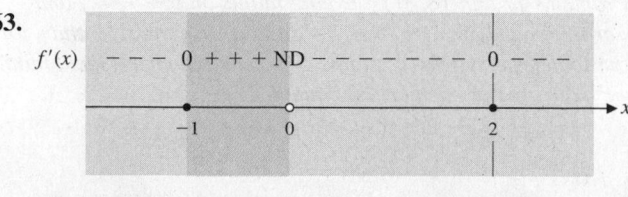

x	-2	-1	0	2	4
$f(x)$	2	1	2	1	0

64.

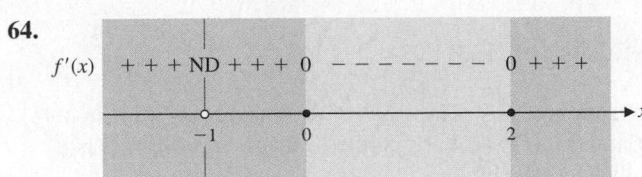

x	-2	-1	0	2	3
f(x)	-3	0	2	-1	0

65. $f(-2) = 4, f(0) = 0, f(2) = -4;$

$f'(-2) = 0, f'(0) = 0, f'(2) = 0;$

$f'(x) > 0$ on $(-\infty, -2)$ and $(2, \infty);$

$f'(x) < 0$ on $(-2, 0)$ and $(0, 2)$

66. $f(-2) = -1, f(0) = 0, f(2) = 1;$

$f'(-2) = 0, f'(2) = 0;$

$f'(x) > 0$ on $(-\infty, -2), (-2, 2),$ and $(2, \infty)$

67. $f(-1) = 2, f(0) = 0, f(1) = -2;$

$f'(-1) = 0, f'(1) = 0, f'(0)$ is not defined;

$f'(x) > 0$ on $(-\infty, -1)$ and $(1, \infty);$

$f'(x) < 0$ on $(-1, 0)$ and $(0, 1)$

68. $f(-1) = 2, f(0) = 0, f(1) = 2;$

$f'(-1) = 0, f'(1) = 0, f'(0)$ is not defined;

$f'(x) > 0$ on $(-\infty, -1)$ and $(0, 1);$

$f'(x) < 0$ on $(-1, 0)$ and $(1, \infty)$

Problems 69–74 involve functions f_1–f_6 and their derivatives, g_1–g_6. Use the graphs shown in figures (A) and (B) to match each function f_i with its derivative g_j.

69. f_1 **70.** f_2 **71.** f_3

72. f_4 **73.** f_5 **74.** f_6

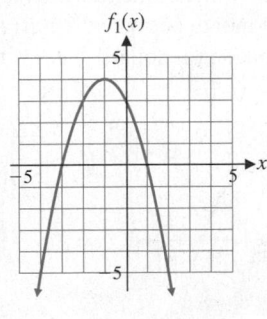
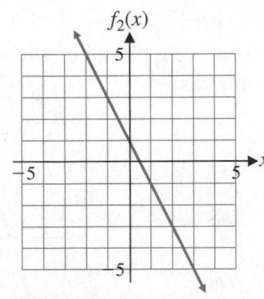
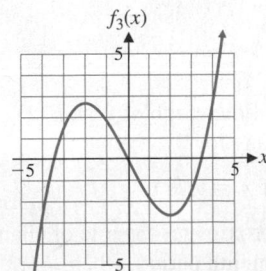
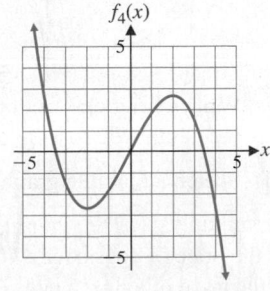

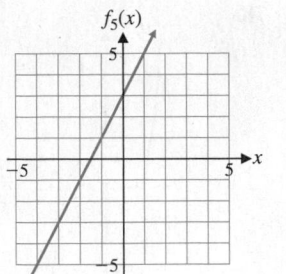
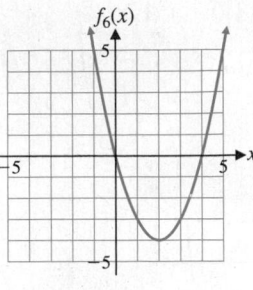

Figure (A) for 69–74

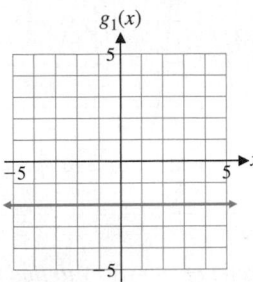
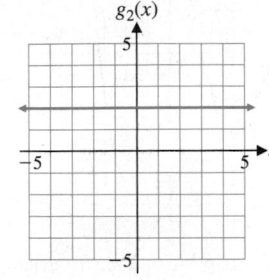
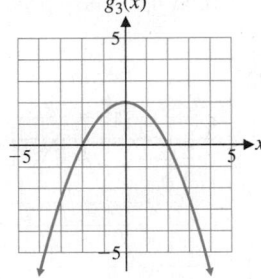
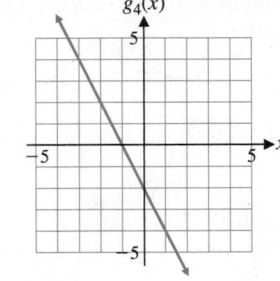
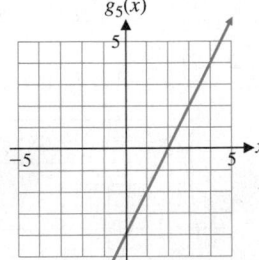
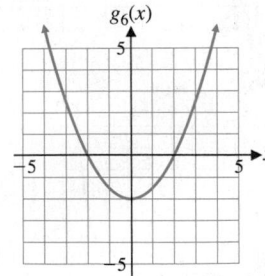

Figure (B) for 69–74

In Problems 75–80, use the given graph of $y = f'(x)$ to find the intervals on which f is increasing, the intervals on which f is decreasing, and the x coordinates of the local extrema of f. Sketch a possible graph of $y = f(x)$.

75.   **76.**

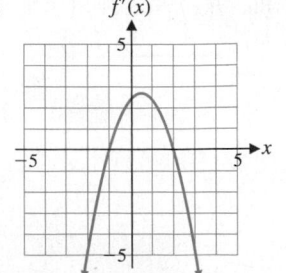
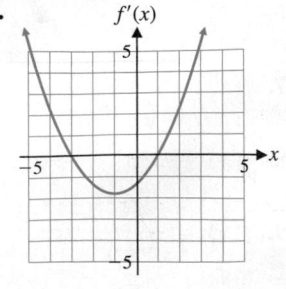

77.

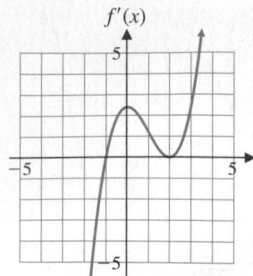

78.

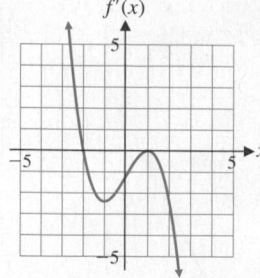

79.

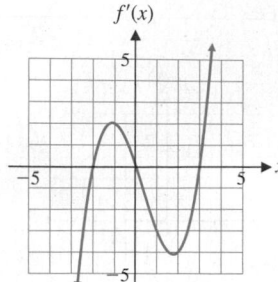

80.

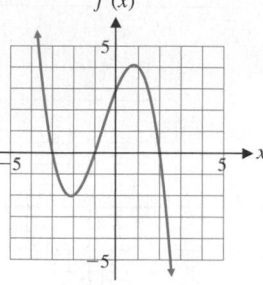

In Problems 81–84, use the given graph of $y = f(x)$ to find the intervals on which $f'(x) > 0$, the intervals on which $f'(x) < 0$, and the values of x for which $f'(x) = 0$. Sketch a possible graph of $y = f'(x)$.

81.

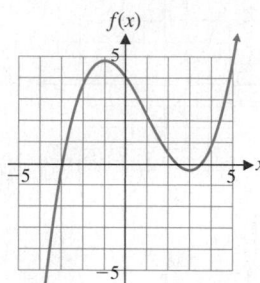

82.

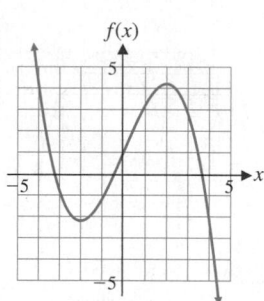

83.

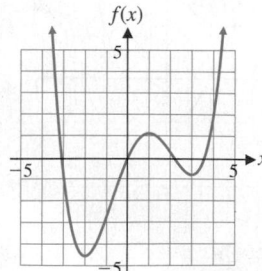

84.

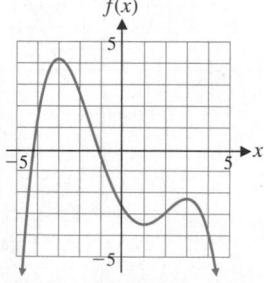

In Problems 85–90, find the critical numbers, the intervals on which $f(x)$ is increasing, the intervals on which $f(x)$ is decreasing, and the local extrema. Do not graph.

85. $f(x) = x + \dfrac{4}{x}$

86. $f(x) = \dfrac{9}{x} + x$

87. $f(x) = 1 + \dfrac{1}{x} + \dfrac{1}{x^2}$

88. $f(x) = 3 - \dfrac{4}{x} - \dfrac{2}{x^2}$

89. $f(x) = \dfrac{x^2}{x - 2}$

90. $f(x) = \dfrac{x^2}{x + 1}$

Applications

91. Profit analysis. The graph of the total profit $P(x)$ (in dollars) from the sale of x cordless electric screwdrivers is shown in the figure.

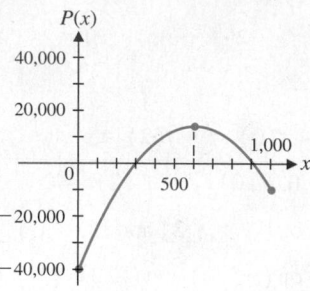

(A) Write a brief description of the graph of the marginal profit function $y = P'(x)$, including a discussion of any x intercepts.

(B) Sketch a possible graph of $y = P'(x)$.

92. Revenue analysis. The graph of the total revenue $R(x)$ (in dollars) from the sale of x cordless electric screwdrivers is shown in the figure.

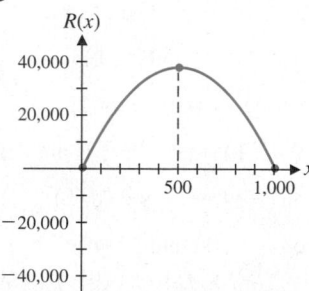

(A) Write a brief description of the graph of the marginal revenue function $y = R'(x)$, including a discussion of any x intercepts.

(B) Sketch a possible graph of $y = R'(x)$.

93. Price analysis. The figure approximates the rate of change of the price of bacon over a 70-month period, where $B(t)$ is the price of a pound of sliced bacon (in dollars) and t is time (in months).

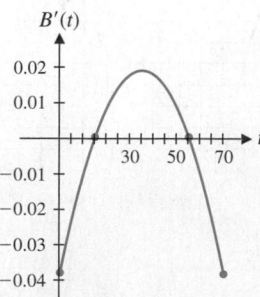

(A) Write a brief description of the graph of $y = B(t)$, including a discussion of any local extrema.

(B) Sketch a possible graph of $y = B(t)$.

94. Price analysis. The figure approximates the rate of change of the price of eggs over a 70-month period, where $E(t)$ is the price of a dozen eggs (in dollars) and t is time (in months).

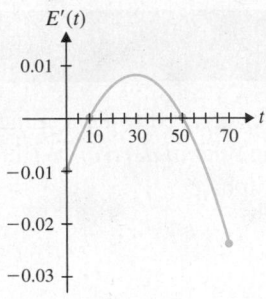

Figure for 94

✎ (A) Write a brief description of the graph of $y = E(t)$, including a discussion of any local extrema.

(B) Sketch a possible graph of $y = E(t)$.

95. Average cost. A manufacturer incurs the following costs in producing x water ski vests in one day, for $0 < x < 150$: fixed costs, $320; unit production cost, $20 per vest; equipment maintenance and repairs, $0.05x^2$ dollars. So, the cost of manufacturing x vests in one day is given by

$$C(x) = 0.05x^2 + 20x + 320 \qquad 0 < x < 150$$

(A) What is the average cost $\overline{C}(x)$ per vest if x vests are produced in one day?

(B) Find the critical numbers of $\overline{C}(x)$, the intervals on which the average cost per vest is decreasing, the intervals on which the average cost per vest is increasing, and the local extrema. Do not graph.

96. Average cost. A manufacturer incurs the following costs in producing x rain jackets in one day for $0 < x < 200$: fixed costs, $450; unit production cost, $30 per jacket; equipment maintenance and repairs, $0.08x^2$ dollars.

(A) What is the average cost $\overline{C}(x)$ per jacket if x jackets are produced in one day?

(B) Find the critical numbers of $\overline{C}(x)$, the intervals on which the average cost per jacket is decreasing, the intervals on which the average cost per jacket is increasing, and the local extrema. Do not graph.

97. Medicine. A drug is injected into the bloodstream of a patient through the right arm. The drug concentration in the bloodstream of the left arm t hours after the injection is approximated by

$$C(t) = \frac{0.28t}{t^2 + 4} \qquad 0 < t < 24$$

Find the critical numbers of $C(t)$, the intervals on which the drug concentration is increasing, the intervals on which the concentration of the drug is decreasing, and the local extrema. Do not graph.

98. Medicine. The concentration $C(t)$, in milligrams per cubic centimeter, of a particular drug in a patient's bloodstream is given by

$$C(t) = \frac{0.3t}{t^2 + 6t + 9} \qquad 0 < t < 12$$

where t is the number of hours after the drug is taken orally. Find the critical numbers of $C(t)$, the intervals on which

the drug concentration is increasing, the intervals on which the drug concentration is decreasing, and the local extrema. Do not graph.

Answers to Matched Problems

1. (A) Horizontal tangent line at $x = 3$.

(B) Decreasing on $(-\infty, 3)$; increasing on $(3, \infty)$

(C)

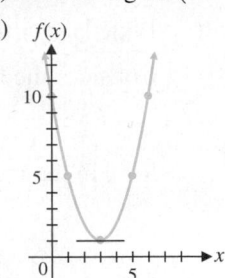

2. Partition number for f': $x = 0$; critical number of f: $x = 0$; decreasing for all x

3. Partition number for f': $x = -1$; critical number of f: $x = -1$; increasing for all x

4. Partition number for f': $x = 0$; no critical number of f; decreasing on $(-\infty, 0)$ and $(0, \infty)$

5. Partition number for f': $x = 5$; critical number of f: $x = 5$; increasing on $(0, 5)$; decreasing on $(5, \infty)$

6. Increasing on $(-3, 1)$; decreasing on $(-\infty, -3)$ and $(1, \infty)$; $f(1) = 5$ is a local maximum; $f(-3) = -3$ is a local minimum

7. (A) Critical numbers of f: $x = 2, x = 4$

(B) $f(2) = 10$ is a local maximum; $f(4) = 6$ is a local minimum

(C)

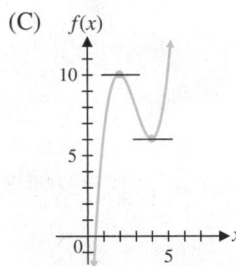

8. (A) The U.S. share of the world market decreases for 6 years to a local minimum, increases for the next 10 years to a local maximum, and then decreases for the final 4 years.

(B)

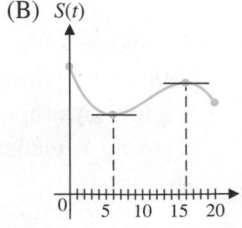

9. (A) The marginal revenue is positive on $(0, 450)$, 0 at $x = 450$, and negative on $(450, 1{,}000)$.

(B)

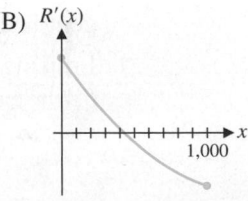

4.2 Second Derivative and Graphs

- Using Concavity as a Graphing Tool
- Finding Inflection Points
- Analyzing Graphs
- Curve Sketching
- Point of Diminishing Returns

In Section 4.1, we saw that the derivative can be used to determine when a graph is rising or falling. Now we want to see what the *second derivative* (the derivative of the derivative) can tell us about the shape of a graph.

Using Concavity as a Graphing Tool

Consider the functions

$$f(x) = x^2 \quad \text{and} \quad g(x) = \sqrt{x}$$

for x in the interval $(0, \infty)$. Since

$$f'(x) = 2x > 0 \quad \text{for } 0 < x < \infty$$

and

$$g'(x) = \frac{1}{2\sqrt{x}} > 0 \quad \text{for } 0 < x < \infty$$

both functions are increasing on $(0, \infty)$.

Explore and Discuss 1 (A) Discuss the difference in the shapes of the graphs of f and g shown in Figure 1.

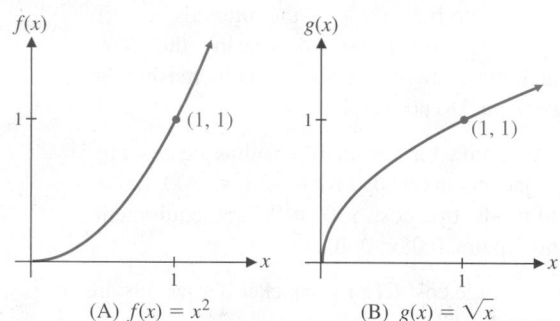

Figure 1   (A) $f(x) = x^2$ (B) $g(x) = \sqrt{x}$

(B) Complete the following table, and discuss the relationship between the values of the derivatives of f and g and the shapes of their graphs:

x	0.25	0.5	0.75	1
$f'(x)$				
$g'(x)$				

We use the term *concave upward* to describe a graph that opens upward and *concave downward* to describe a graph that opens downward. Thus, the graph of f in Figure 1A is concave upward, and the graph of g in Figure 1B is concave downward. Finding a mathematical formulation of concavity will help us sketch and analyze graphs.

We examine the slopes of f and g at various points on their graphs (see Fig. 2) and make two observations about each graph:

1. Looking at the graph of f in Figure 2A, we see that $f'(x)$ (the slope of the tangent line) is *increasing* and that the graph lies *above* each tangent line;

2. Looking at Figure 2B, we see that $g'(x)$ is *decreasing* and that the graph lies *below* each tangent line.

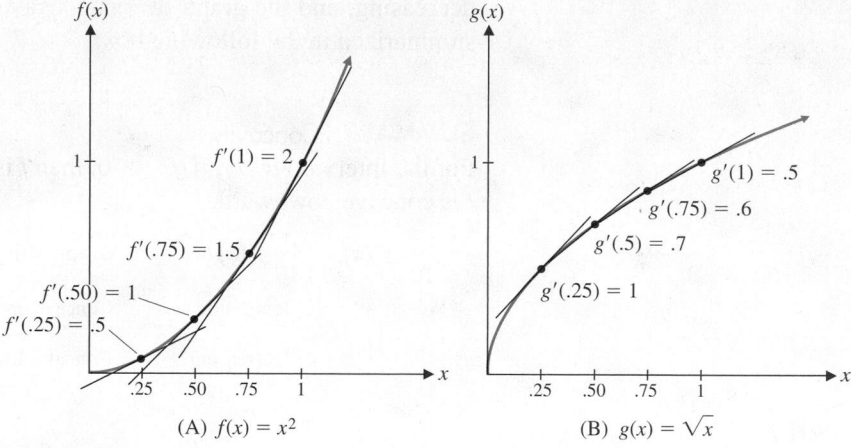

Figure 2

> **DEFINITION** Concavity
>
> The graph of a function f is **concave upward** on the interval (a, b) if $f'(x)$ is *increasing* on (a, b) and is **concave downward** on the interval (a, b) if $f'(x)$ is *decreasing* on (a, b).

Geometrically, the graph is concave upward on (a, b) if it lies above its tangent lines in (a, b) and is concave downward on (a, b) if it lies below its tangent lines in (a, b).

How can we determine when $f'(x)$ is increasing or decreasing? In Section 4.1, we used the derivative to determine when a function is increasing or decreasing. To determine when the function $f'(x)$ is increasing or decreasing, we use the derivative of $f'(x)$. The derivative of the derivative of a function is called the *second derivative* of the function. Various notations for the second derivative are given in the following box:

> **NOTATION** Second Derivative
>
> For $y = f(x)$, the **second derivative** of f, provided that it exists, is
>
> $$f''(x) = \frac{d}{dx} f'(x)$$
>
> Other notations for $f''(x)$ are
>
> $$\frac{d^2y}{dx^2} \quad \text{and} \quad y''$$

Returning to the functions f and g discussed at the beginning of this section, we have

$$f(x) = x^2 \qquad\qquad g(x) = \sqrt{x} = x^{1/2}$$

$$f'(x) = 2x \qquad\qquad g'(x) = \frac{1}{2}x^{-1/2} = \frac{1}{2\sqrt{x}}$$

$$f''(x) = \frac{d}{dx}2x = 2 \qquad g''(x) = \frac{d}{dx}\frac{1}{2}x^{-1/2} = -\frac{1}{4}x^{-3/2} = -\frac{1}{4\sqrt{x^3}}$$

For $x > 0$, we see that $f''(x) > 0$; so, $f'(x)$ is increasing, and the graph of f is concave upward (see Fig. 2A). For $x > 0$, we also see that $g''(x) < 0$; so, $g'(x)$ is decreasing, and the graph of g is concave downward (see Fig. 2B). These ideas are summarized in the following box:

SUMMARY   Concavity

For the interval (a, b), if $f'' > 0$, then f is concave upward, and if $f'' < 0$, then f is concave downward.

$f''(x)$	$f'(x)$	Graph of $y = f(x)$	Examples
$+$	Increasing	Concave upward	
$-$	Decreasing	Concave downward	

CONCEPTUAL INSIGHT

Be careful not to confuse concavity with falling and rising. A graph that is concave upward on an interval may be falling, rising, or both falling and rising on that interval. A similar statement holds for a graph that is concave downward. See Figure 3.

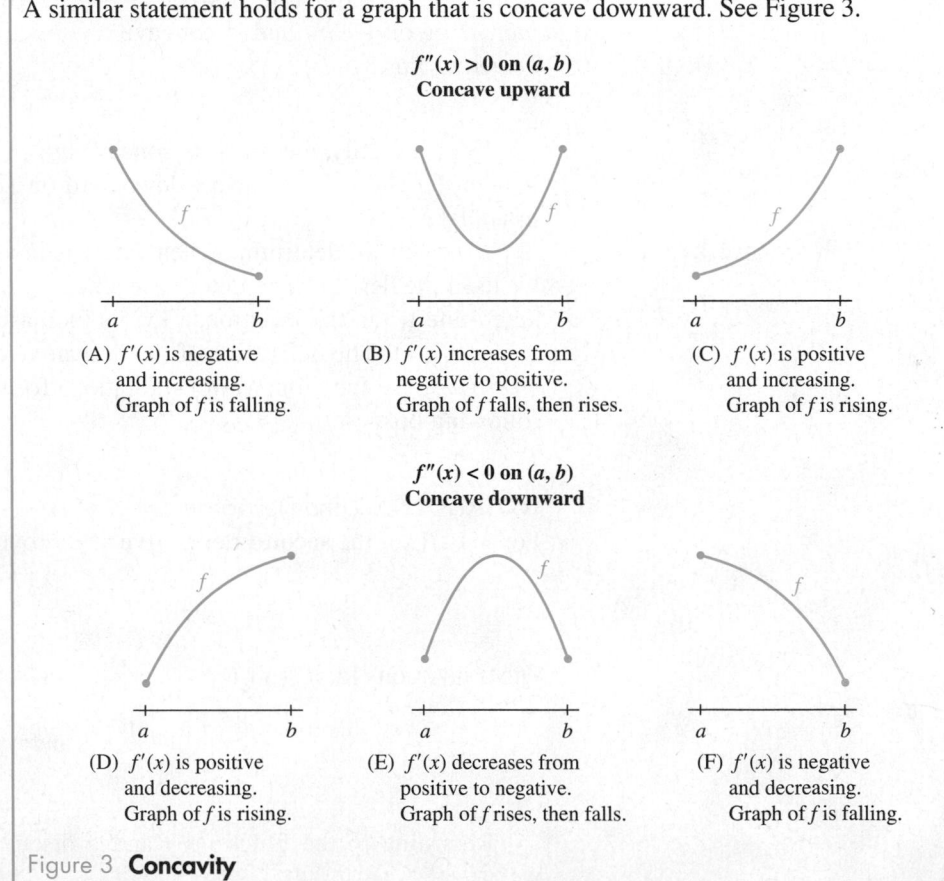

$f''(x) > 0$ on (a, b)
Concave upward

(A) $f'(x)$ is negative and increasing. Graph of f is falling.

(B) $f'(x)$ increases from negative to positive. Graph of f falls, then rises.

(C) $f'(x)$ is positive and increasing. Graph of f is rising.

$f''(x) < 0$ on (a, b)
Concave downward

(D) $f'(x)$ is positive and decreasing. Graph of f is rising.

(E) $f'(x)$ decreases from positive to negative. Graph of f rises, then falls.

(F) $f'(x)$ is negative and decreasing. Graph of f is falling.

Figure 3 **Concavity**

EXAMPLE 1   Concavity of Graphs   Determine the intervals on which the graph of each function is concave upward and the intervals on which it is concave downward. Sketch a graph of each function.

(A) $f(x) = e^x$

(B) $g(x) = \ln x$

(C) $h(x) = x^3$

SOLUTION

(A) $f(x) = e^x$

$f'(x) = e^x$

$f''(x) = e^x$

Since $f''(x) > 0$ on $(-\infty, \infty)$, the graph of $f(x) = e^x$ [Fig. 4(A)] is concave upward on $(-\infty, \infty)$.

(B) $g(x) = \ln x$

$g'(x) = \dfrac{1}{x}$

$g''(x) = -\dfrac{1}{x^2}$

The domain of $g(x) = \ln x$ is $(0, \infty)$ and $g''(x) < 0$ on this interval, so the graph of $g(x) = \ln x$ [Fig. 4(B)] is concave downward on $(0, \infty)$.

(C) $h(x) = x^3$

$h'(x) = 3x^2$

$h''(x) = 6x$

Since $h''(x) < 0$ when $x < 0$ and $h''(x) > 0$ when $x > 0$, the graph of $h(x) = x^3$ [Fig. 4(C)] is concave downward on $(-\infty, 0)$ and concave upward on $(0, \infty)$.

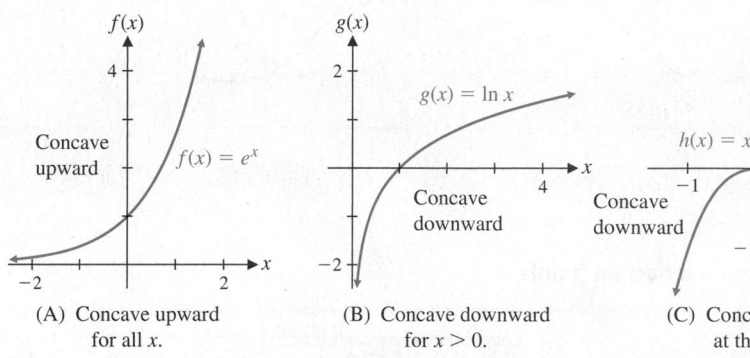

| (A) Concave upward for all x. | (B) Concave downward for $x > 0$. | (C) Concavity changes at the origin. |

Figure 4

Matched Problem 1   Determine the intervals on which the graph of each function is concave upward and the intervals on which it is concave downward. Sketch a graph of each function.

(A) $f(x) = -e^{-x}$ (B) $g(x) = \ln \dfrac{1}{x}$ (C) $h(x) = x^{1/3}$

Refer to Example 1. The graphs of $f(x) = e^x$ and $g(x) = \ln x$ never change concavity. But the graph of $h(x) = x^3$ changes concavity at $(0, 0)$. This point is called an *inflection point*.

Finding Inflection Points

An **inflection point** is a point on the graph of a function where the concavity changes (from upward to downward or from downward to upward). For the concavity to change at a point, $f''(x)$ must change sign at that point. But in Section 2.2, we saw that the partition numbers identify the points where a function can change sign.

THEOREM 1   Inflection Points

If $(c, f(c))$ is an inflection point of f, then c is a partition number for f''.

Our strategy for finding inflection points is clear: We find all partition numbers c for f'' and ask

1. Does $f''(x)$ change sign at c?

2. Is c is in the domain of f?

If both answers are "yes", then $(c, f(c))$ is an inflection point of f. Figure 5 illustrates several typical cases.

If $f'(c)$ exists and $f''(x)$ changes sign at $x = c$, then the tangent line at an inflection point $(c, f(c))$ will always lie below the graph on the side that is concave upward and above the graph on the side that is concave downward (see Fig. 5A, B, and C).

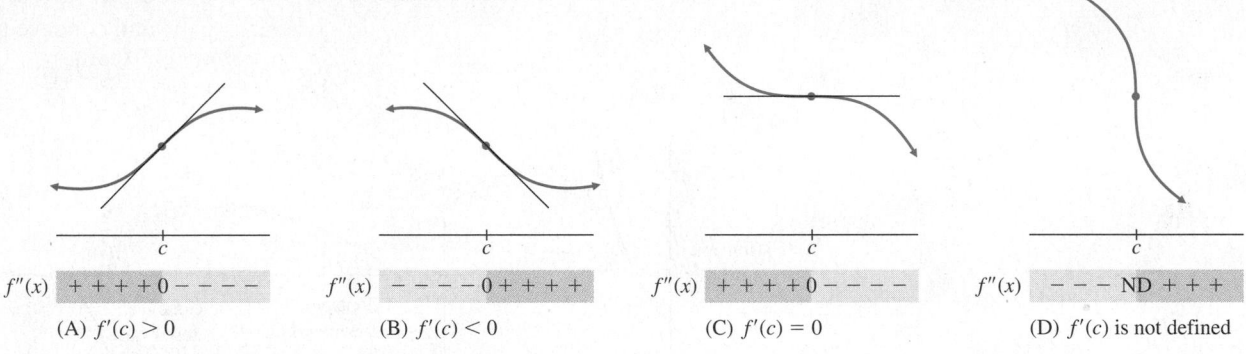

| (A) $f'(c) > 0$ | (B) $f'(c) < 0$ | (C) $f'(c) = 0$ | (D) $f'(c)$ is not defined |

Figure 5   Inflection points

EXAMPLE 2   Locating Inflection Points   Find the inflection point(s) of

$$f(x) = x^3 - 6x^2 + 9x + 1$$

SOLUTION   Since inflection points occur at values of x where $f''(x)$ changes sign, we construct a sign chart for $f''(x)$.

$$f(x) = x^3 - 6x^2 + 9x + 1$$
$$f'(x) = 3x^2 - 12x + 9$$
$$f''(x) = 6x - 12 = 6(x - 2)$$

The sign chart for $f''(x) = 6(x - 2)$ (partition number is 2) is as follows:

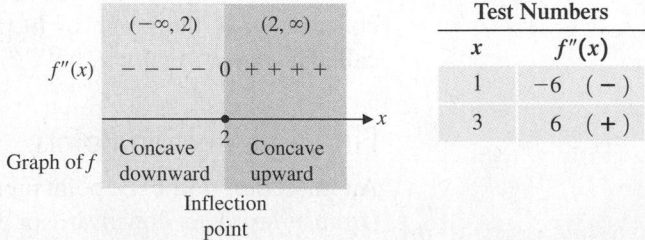

Test Numbers	
x	$f''(x)$
1	-6 $(-)$
3	6 $(+)$

From the sign chart, we see that the graph of f has an inflection point at $x = 2$. That is, the point

$$(2, f(2)) = (2, 3) \quad f(2) = 2^3 - 6 \cdot 2^2 + 9 \cdot 2 + 1 = 3$$

is an inflection point on the graph of f.

Matched Problem 2   Find the inflection point(s) of

$$f(x) = x^3 - 9x^2 + 24x - 10$$

EXAMPLE 3 Locating Inflection Points Find the inflection point(s) of

$$f(x) = \ln(x^2 - 4x + 5)$$

SOLUTION First we find the domain of f. Since $\ln x$ is defined only for $x > 0$, f is defined only for

$$x^2 - 4x + 5 > 0 \quad \text{Complete the square (Section A.7).}$$
$$(x - 2)^2 + 1 > 0 \quad \text{True for all } x.$$

So the domain of f is $(-\infty, \infty)$. Now we find $f''(x)$ and construct a sign chart for it.

$$f(x) = \ln(x^2 - 4x + 5)$$

$$f'(x) = \frac{2x - 4}{x^2 - 4x + 5}$$

$$f''(x) = \frac{(x^2 - 4x + 5)(2x - 4)' - (2x - 4)(x^2 - 4x + 5)'}{(x^2 - 4x + 5)^2}$$

$$= \frac{(x^2 - 4x + 5)2 - (2x - 4)(2x - 4)}{(x^2 - 4x + 5)^2}$$

$$= \frac{2x^2 - 8x + 10 - 4x^2 + 16x - 16}{(x^2 - 4x + 5)^2}$$

$$= \frac{-2x^2 + 8x - 6}{(x^2 - 4x + 5)^2}$$

$$= \frac{-2(x - 1)(x - 3)}{(x^2 - 4x + 5)^2}$$

The partition numbers for $f''(x)$ are $x = 1$ and $x = 3$.
Sign chart for $f''(x)$:

		(−∞, 1)	(1, 3)	(3, ∞)
$f''(x)$		− − − −	0 + + + + 0	− − − −

Concave downward (−∞, 1) | Concave upward (1, 3) | Concave downward (3, ∞)

Inflection point | Inflection point

Test Numbers	
x	$f''(x)$
0	$-\dfrac{6}{25}$ (−)
2	2 (+)
4	$-\dfrac{6}{25}$ (−)

The sign chart shows that the graph of f has inflection points at $x = 1$ and $x = 3$.
Since $f(1) = \ln 2$ and $f(3) = \ln 2$, the inflection points are $(1, \ln 2)$ and $(3, \ln 2)$.

Matched Problem 3 Find the inflection point(s) of

$$f(x) = \ln(x^2 - 2x + 5)$$

CONCEPTUAL INSIGHT

It is important to remember that the partition numbers for f'' are only *candidates* for inflection points. The function f must be defined at $x = c$, and the second derivative must change sign at $x = c$ in order for the graph to have an inflection point at $x = c$. For example, consider

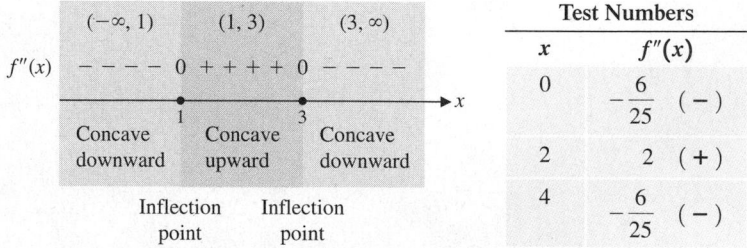

$$f(x) = x^4 \qquad g(x) = \frac{1}{x}$$

$$f'(x) = 4x^3 \qquad g'(x) = -\frac{1}{x^2}$$

$$f''(x) = 12x^2 \qquad g''(x) = \frac{2}{x^3}$$

In each case, $x = 0$ is a partition number for the second derivative, but neither the graph of $f(x)$ nor the graph of $g(x)$ has an inflection point at $x = 0$. Function f does not have an inflection point at $x = 0$ because $f''(x)$ does not change sign at $x = 0$ (see Fig. 6A). Function g does not have an inflection point at $x = 0$ because $g(0)$ is not defined (see Fig. 6B).

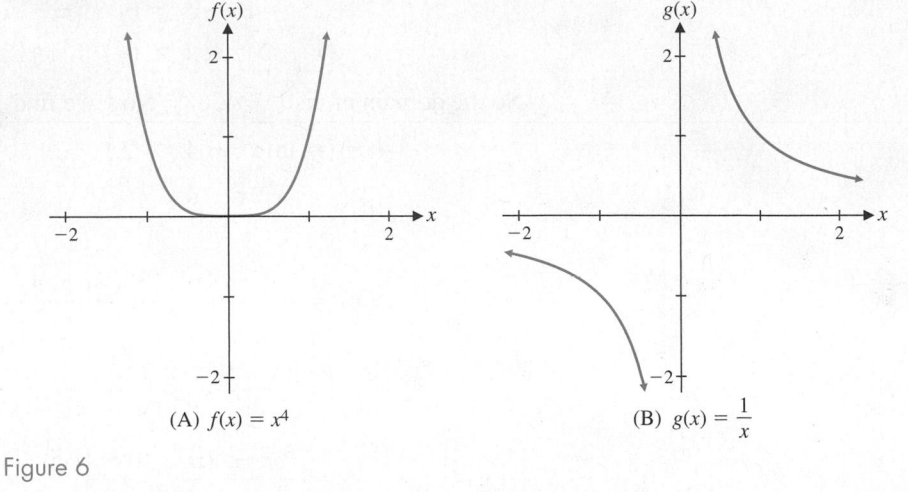

(A) $f(x) = x^4$ (B) $g(x) = \dfrac{1}{x}$

Figure 6

Analyzing Graphs

In the next example, we combine increasing/decreasing properties with concavity properties to analyze the graph of a function.

EXAMPLE 4   Analyzing a Graph   Figure 7 shows the graph of the derivative of a function f. Use this graph to discuss the graph of f. Include a sketch of a possible graph of f.

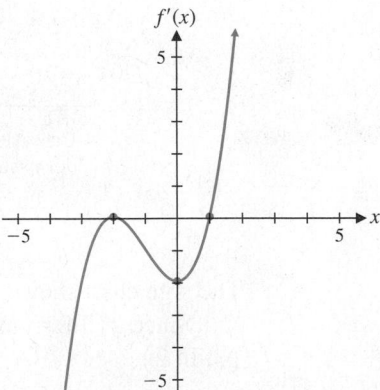

Figure 7

SOLUTION The sign of the derivative determines where the original function is increasing and decreasing, and the increasing/decreasing properties of the derivative determine the concavity of the original function. The relevant information obtained from the graph of f' is summarized in Table 1, and a possible graph of f is shown in Figure 8.

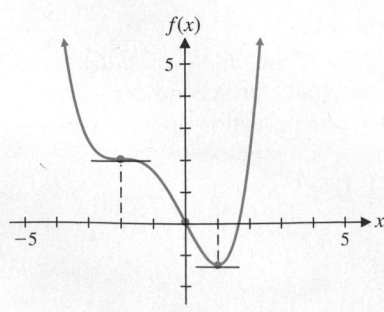

Figure 8

Table 1

x	$f'(x)$ (Fig. 7)	$f(x)$ (Fig. 8)
$-\infty < x < -2$	Negative and increasing	Decreasing and concave upward
$x = -2$	Local maximum	Inflection point
$-2 < x < 0$	Negative and decreasing	Decreasing and concave downward
$x = 0$	Local minimum	Inflection point
$0 < x < 1$	Negative and increasing	Decreasing and concave upward
$x = 1$	x intercept	Local minimum
$1 < x < \infty$	Positive and increasing	Increasing and concave upward

<u>Matched Problem 4</u> Figure 9 shows the graph of the derivative of a function f. Use this graph to discuss the graph of f. Include a sketch of a possible graph of f.

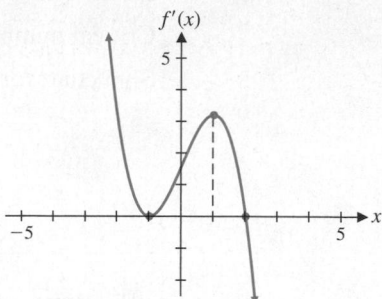

Figure 9

Curve Sketching

Graphing calculators and computers produce the graph of a function by plotting many points. However, key points on a plot many be difficult to identify. Using information gained from the function $f(x)$ and its derivatives, and plotting the key points—intercepts, local extrema, and inflection points—we can sketch by hand a very good representation of the graph of $f(x)$. This graphing process is called **curve sketching**.

PROCEDURE Graphing Strategy (First Version)*

Step 1 *Analyze $f(x)$.* Find the domain and the intercepts. The x intercepts are the solutions of $f(x) = 0$, and the y intercept is $f(0)$.

Step 2 *Analyze $f'(x)$.* Find the partition numbers for f' and the critical numbers of f. Construct a sign chart for $f'(x)$, determine the intervals on which f is increasing and decreasing, and find the local maxima and minima of f.

Step 3 *Analyze $f''(x)$.* Find the partition numbers for $f''(x)$. Construct a sign chart for $f''(x)$, determine the intervals on which the graph of f is concave upward and concave downward, and find the inflection points of f.

Step 4 *Sketch the graph of f.* Locate intercepts, local maxima and minima, and inflection points. Sketch in what you know from steps 1–3. Plot additional points as needed and complete the sketch.

EXAMPLE 5 Using the Graphing Strategy Follow the graphing strategy and analyze the function

$$f(x) = x^4 - 2x^3$$

State all the pertinent information and sketch the graph of f.

SOLUTION

Step 1 *Analyze $f(x)$.* Since f is a polynomial, its domain is $(-\infty, \infty)$.

x intercept: $f(x) = 0$
$$x^4 - 2x^3 = 0$$
$$x^3(x - 2) = 0$$
$$x = 0, 2$$

y intercept: $f(0) = 0$

---

*We will modify this summary in Section 4.4 to include additional information about the graph of f.

Step 2   *Analyze* $f'(x)$. $f'(x) = 4x^3 - 6x^2 = 4x^2(x - \frac{3}{2})$

Partition numbers for $f'(x)$: 0 and $\frac{3}{2}$

Critical numbers of $f(x)$: 0 and $\frac{3}{2}$

Sign chart for $f'(x)$:

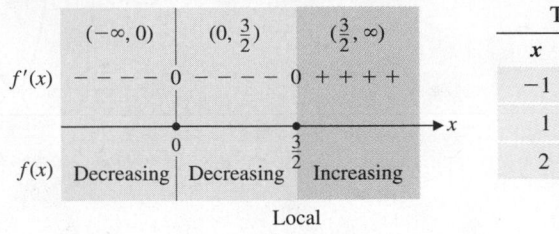

Test Numbers	
x	$f'(x)$
-1	-10 $(-)$
1	-2 $(-)$
2	8 $(+)$

So $f(x)$ is decreasing on $(-\infty, \frac{3}{2})$, is increasing on $(\frac{3}{2}, \infty)$, and has a local minimum at $x = \frac{3}{2}$. The local minimum is $f(\frac{3}{2}) = -\frac{27}{16}$.

Step 3   *Analyze* $f''(x)$. $f''(x) = 12x^2 - 12x = 12x(x - 1)$

Partition numbers for $f''(x)$: 0 and 1

Sign chart for $f''(x)$:

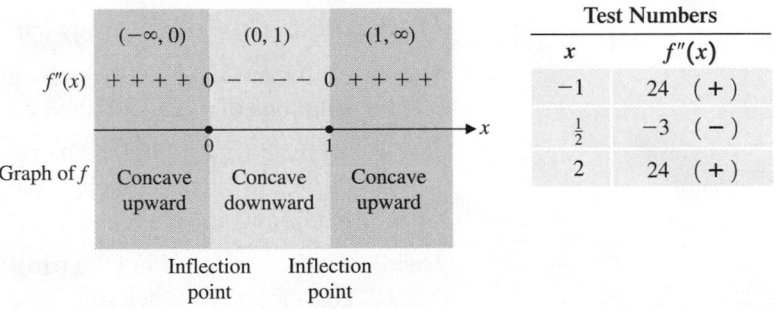

Test Numbers	
x	$f''(x)$
-1	24 $(+)$
$\frac{1}{2}$	-3 $(-)$
2	24 $(+)$

So the graph of f is concave upward on $(-\infty, 0)$ and $(1, \infty)$, is concave downward on $(0, 1)$, and has inflection points at $x = 0$ and $x = 1$. Since $f(0) = 0$ and $f(1) = -1$, the inflection points are $(0, 0)$ and $(1, -1)$.

Step 4   *Sketch the graph of f.*

Key Points	
x	$f(x)$
0	0
1	-1
$\frac{3}{2}$	$-\frac{27}{16}$
2	0

Matched Problem 5   Follow the graphing strategy and analyze the function $f(x) = x^4 + 4x^3$. State all the pertinent information and sketch the graph of f.

CONCEPTUAL INSIGHT

Refer to the solution of Example 5. Combining the sign charts for $f'(x)$ and $f''(x)$ (Fig. 10) partitions the real-number line into intervals on which neither $f'(x)$ nor $f''(x)$ changes sign. On each of these intervals, the graph of $f(x)$ must have one of four basic shapes (see also Fig. 3, parts A, C, D, and F on page 256). This reduces sketching the graph of a function to plotting the points identified in the graphing strategy and connecting them with one of the basic shapes.

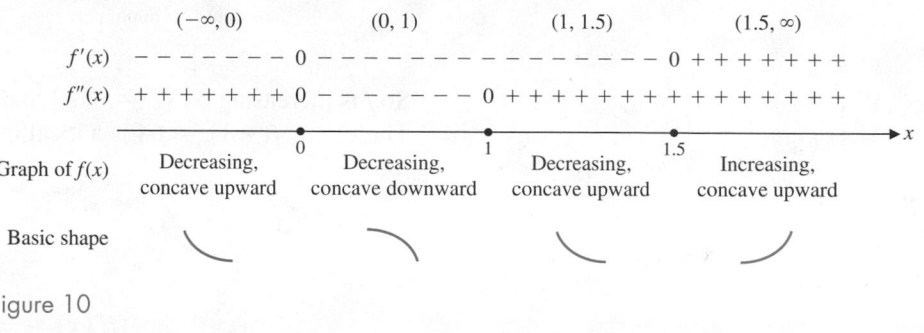

Figure 10

EXAMPLE 6 Using the Graphing Strategy Follow the graphing strategy and analyze the function
$$f(x) = 3x^{5/3} - 20x$$

State all the pertinent information and sketch the graph of f. Round any decimal values to two decimal places.

SOLUTION

Step 1 Analyze $f(x)$. $f(x) = 3x^{5/3} - 20x$

Since x^p is defined for any x and any positive p, the domain of f is $(-\infty, \infty)$.

x intercepts: Solve $f(x) = 0$
$$3x^{5/3} - 20x = 0$$
$$3x\left(x^{2/3} - \frac{20}{3}\right) = 0 \quad (a^2 - b^2) = (a - b)(a + b)$$
$$3x\left(x^{1/3} - \sqrt{\frac{20}{3}}\right)\left(x^{1/3} + \sqrt{\frac{20}{3}}\right) = 0$$

The x intercepts of f are
$$x = 0, \quad x = \left(\sqrt{\frac{20}{3}}\right)^3 \approx 17.21, \quad x = \left(-\sqrt{\frac{20}{3}}\right)^3 \approx -17.21$$

y intercept: $f(0) = 0$.

Step 2 Analyze $f'(x)$.
$$f'(x) = 5x^{2/3} - 20$$
$$= 5(x^{2/3} - 4) \qquad \text{Again, } a^2 - b^2 = (a - b)(a + b)$$
$$= 5(x^{1/3} - 2)(x^{1/3} + 2)$$

Partition numbers for f': $x = 2^3 = 8$ and $x = (-2)^3 = -8$.

Critical numbers of f: $-8, 8$

Sign chart for $f'(x)$:

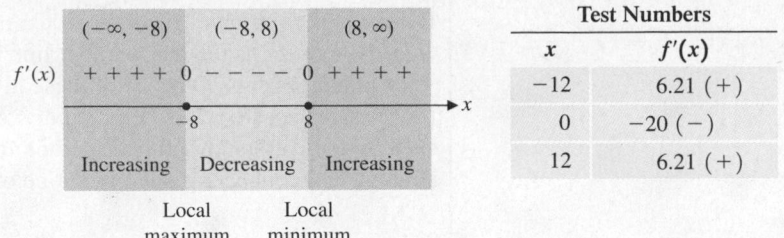

	Test Numbers	
x	$f'(x)$	
-12	6.21 (+)	
0	-20 (−)	
12	6.21 (+)	

So f is increasing on $(-\infty, -8)$ and $(8, \infty)$ and decreasing on $(-8, 8)$. Therefore, $f(-8) = 64$ is a local maximum, and $f(8) = -64$ is a local minimum.

Step 3 *Analyze $f''(x)$.*

$$f'(x) = 5x^{2/3} - 20$$

$$f''(x) = \frac{10}{3}x^{-1/3} = \frac{10}{3x^{1/3}}$$

Partition number for f'': 0

Sign chart for $f''(x)$:

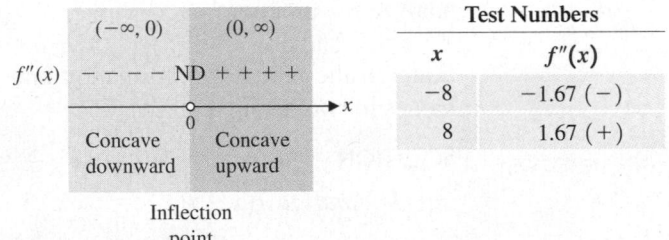

	Test Numbers	
x	$f''(x)$	
-8	-1.67 (−)	
8	1.67 (+)	

So f is concave downward on $(-\infty, 0)$, is concave upward on $(0, \infty)$, and has an inflection point at $x = 0$. Since $f(0) = 0$, the inflection point is $(0, 0)$.

Step 4 *Sketch the graph of f.*

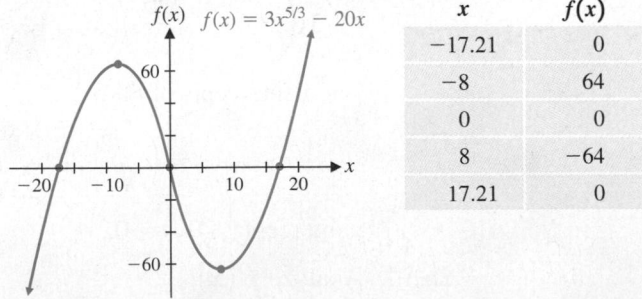

x	$f(x)$
-17.21	0
-8	64
0	0
8	-64
17.21	0

Matched Problem 6 Follow the graphing strategy and analyze the function $f(x) = 3x^{2/3} - x$. State all the pertinent information and sketch the graph of f. Round any decimal values to two decimal places.

Point of Diminishing Returns

If a company decides to increase spending on advertising, it would expect sales to increase. At first, sales will increase at an increasing rate and then increase at a decreasing rate. The dollar amount x at which the rate of change of sales goes from increasing to decreasing is called the **point of diminishing returns**. This is also the amount at which the rate of change has a maximum value. Money spent beyond this amount may increase sales but at a lower rate.

EXAMPLE 7   Maximum Rate of Change   Currently, a discount appliance store is selling 200 large-screen TVs monthly. If the store invests $x thousand in an advertising campaign, the ad company estimates that monthly sales will be given by

$$N(x) = 3x^3 - 0.25x^4 + 200 \qquad 0 \le x \le 9$$

When is the rate of change of sales increasing and when is it decreasing? What is the point of diminishing returns and the maximum rate of change of sales? Graph N and N' on the same coordinate system.

SOLUTION   The rate of change of sales with respect to advertising expenditures is

$$N'(x) = 9x^2 - x^3 = x^2(9 - x)$$

To determine when $N'(x)$ is increasing and decreasing, we find $N''(x)$, the derivative of $N'(x)$:

$$N''(x) = 18x - 3x^2 = 3x(6 - x)$$

The information obtained by analyzing the signs of $N'(x)$ and $N''(x)$ is summarized in Table 2 (sign charts are omitted).

Table 2

x	$N''(x)$	$N'(x)$	$N'(x)$	$N(x)$
$0 < x < 6$	+	+	Increasing	Increasing, concave upward
$x = 6$	0	+	Local maximum	Inflection point
$6 < x < 9$	−	+	Decreasing	Increasing, concave downward

Examining Table 2, we see that $N'(x)$ is increasing on $(0, 6)$ and decreasing on $(6, 9)$. The point of diminishing returns is $x = 6$ and the maximum rate of change is $N'(6) = 108$. Note that $N'(x)$ has a local maximum and $N(x)$ has an inflection point at $x = 6$ [the inflection point of $N(x)$ is $(6, 524)$].

So if the store spends $6,000 on advertising, monthly sales are expected to be 524 TVs, and sales are expected to increase at a rate of 108 TVs per thousand dollars spent on advertising. Money spent beyond the $6,000 would increase sales, but at a lower rate.

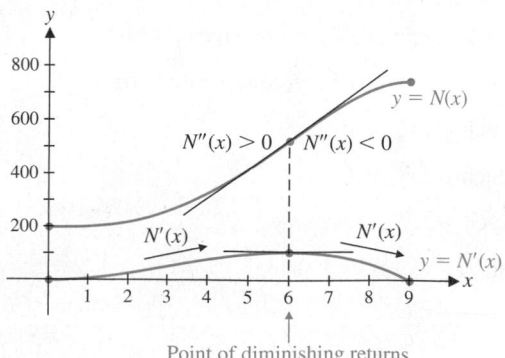

Matched Problem 7   Repeat Example 7 for

$$N(x) = 4x^3 - 0.25x^4 + 500 \qquad 0 \le x \le 12$$

Exercises 4.2

Skills Warm-up Exercises

W *In Problems 1–8, inspect the graph of the function to determine whether it is concave up, concave down, or neither, on the given interval. (If necessary, review Section 1.2).*

1. The square function, $h(x) = x^2$, on $(-\infty, \infty)$

2. The identity function, $f(x) = x$, on $(-\infty, \infty)$

3. The cube function, $m(x) = x^3$, on $(-\infty, 0)$

4. The cube function, $m(x) = x^3$, on $(0, \infty)$

5. The square root function, $n(x) = \sqrt{x}$, on $(0, \infty)$

6. The cube root function, $p(x) = \sqrt[3]{x}$, on $(-\infty, 0)$

7. The absolute value function, $g(x) = |x|$, on $(-\infty, 0)$

8. The cube root function, $p(x) = \sqrt[3]{x}$, on $(0, \infty)$

9. Use the graph of $y = f(x)$, assuming $f''(x) > 0$ if $x = b$ or f, to identify

   (A) Intervals on which the graph of f is concave upward

   (B) Intervals on which the graph of f is concave downward

   (C) Intervals on which $f''(x) < 0$

   (D) Intervals on which $f''(x) > 0$

   (E) Intervals on which $f'(x)$ is increasing

   (F) Intervals on which $f'(x)$ is decreasing

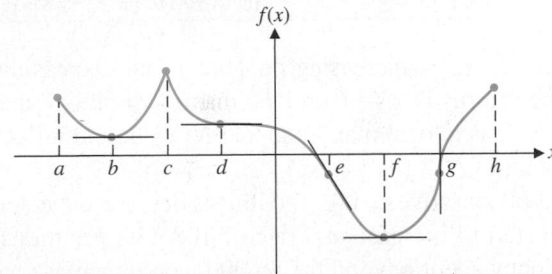

10. Use the graph of $y = g(x)$, assuming $g''(x) > 0$ if $x = c$ or g, to identify

    (A) Intervals on which the graph of g is concave upward

    (B) Intervals on which the graph of g is concave downward

    (C) Intervals on which $g''(x) < 0$

    (D) Intervals on which $g''(x) > 0$

    (E) Intervals on which $g'(x)$ is increasing

    (F) Intervals on which $g'(x)$ is decreasing

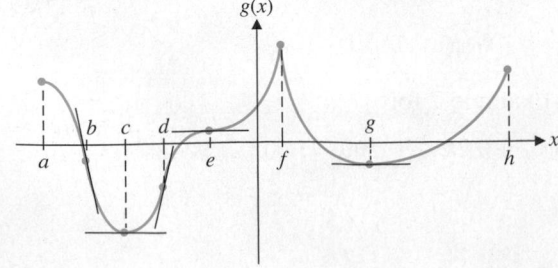

11. Use the graph of $y = f(x)$ to identify

    (A) The local extrema of $f(x)$.

    (B) The inflection points of $f(x)$.

    (C) The numbers u for which $f'(u)$ is a local extremum of $f'(x)$.

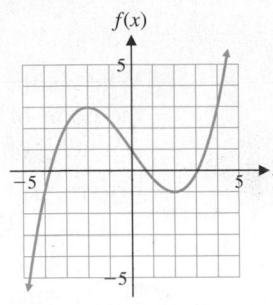

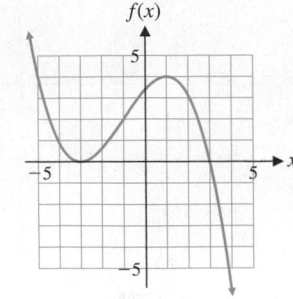

Figure for 11   Figure for 12

12. Use the graph of $y = f(x)$ to identify

    (A) The local extrema of $f(x)$.

    (B) The inflection points of $f(x)$.

    (C) The numbers u for which $f'(u)$ is a local extremum of $f'(x)$.

In Problems 13–16, match the indicated conditions with one of the graphs (A)–(D) shown in the figure.

13. $f'(x) > 0$ and $f''(x) > 0$ on (a, b)

14. $f'(x) > 0$ and $f''(x) < 0$ on (a, b)

15. $f'(x) < 0$ and $f''(x) > 0$ on (a, b)

16. $f'(x) < 0$ and $f''(x) < 0$ on (a, b)

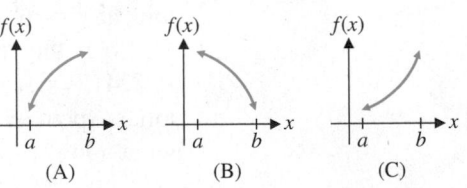

(A)   (B)   (C)   (D)

In Problems 17–24, find the indicated derivative for each function.

17. $f''(x)$ for $f(x) = 2x^3 - 4x^2 + 5x - 6$

18. $g''(x)$ for $g(x) = -x^3 + 2x^2 - 3x + 9$

19. $h''(x)$ for $h(x) = 2x^{-1} - 3x^{-2}$

20. $k''(x)$ for $k(x) = -6x^{-2} + 12x^{-3}$

21. d^2y/dx^2 for $y = x^2 - 18x^{1/2}$

22. d^2y/dx^2 for $y = x^3 - 24x^{1/3}$

23. y'' for $y = (x^2 + 9)^4$

24. y'' for $y = (x^2 - 16)^5$

In Problems 25–30, find the x and y coordinates of all inflection points.

25. $f(x) = x^3 + 30x^2$ 26. $f(x) = x^3 - 24x^2$

27. $f(x) = x^{5/3} + 2$

28. $f(x) = 5 - x^{4/3}$

29. $f(x) = 1 + x + x^{2/5}$

30. $f(x) = x^{3/5} - 6x + 7$

In Problems 31–40, find the intervals on which the graph of f is concave upward, the intervals on which the graph of f is concave downward, and the x coordinates of the inflection points.

31. $f(x) = x^4 + 6x^2$

32. $f(x) = x^4 + 6x$

33. $f(x) = x^3 - 4x^2 + 5x - 2$

34. $f(x) = -x^3 - 5x^2 + 4x - 3$

35. $f(x) = -x^4 + 12x^3 - 12x + 24$

36. $f(x) = x^4 - 2x^3 - 36x + 12$

37. $f(x) = \ln(x^2 - 2x + 10)$

38. $f(x) = \ln(x^2 + 6x + 13)$

39. $f(x) = 8e^x - e^{2x}$

40. $f(x) = e^{3x} - 9e^x$

In Problems 41–48, f(x) is continuous on $(-\infty, \infty)$. Use the given information to sketch the graph of f.

41.

x	−4	−2	−1	0	2	4
f(x)	0	3	1.5	0	−1	−3

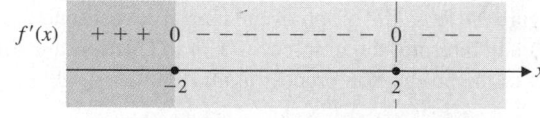

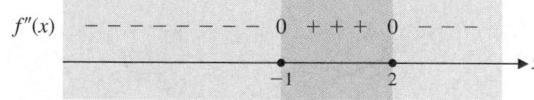

42.

x	−4	−2	−1	0	2	4
f(x)	0	−2	−1	0	1	3

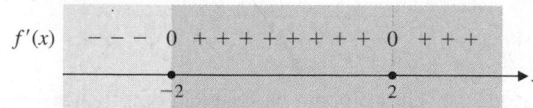

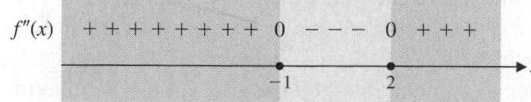

43.

x	−3	0	1	2	4	5
f(x)	−4	0	2	1	−1	0

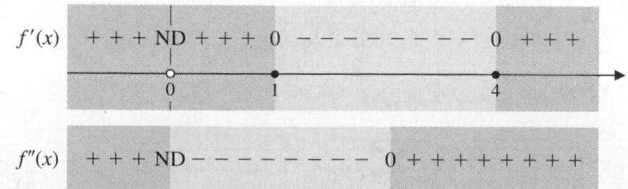

44.

x	−4	−2	0	2	4	6
f(x)	0	3	0	−2	0	3

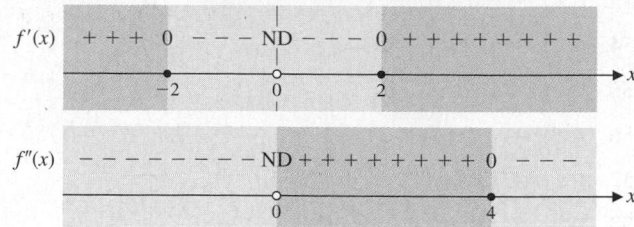

45. $f(0) = 2, f(1) = 0, f(2) = -2;$

$f'(0) = 0, f'(2) = 0;$

$f'(x) > 0$ on $(-\infty, 0)$ and $(2, \infty);$

$f'(x) < 0$ on $(0, 2);$

$f''(1) = 0;$

$f''(x) > 0$ on $(1, \infty);$

$f''(x) < 0$ on $(-\infty, 1)$

46. $f(-2) = -2, f(0) = 1, f(2) = 4;$

$f'(-2) = 0, f'(2) = 0;$

$f'(x) > 0$ on $(-2, 2);$

$f'(x) < 0$ on $(-\infty, -2)$ and $(2, \infty);$

$f''(0) = 0;$

$f''(x) > 0$ on $(-\infty, 0);$

$f''(x) < 0$ on $(0, \infty)$

47. $f(-1) = 0, f(0) = -2, f(1) = 0;$

$f'(0) = 0, f'(-1)$ and $f'(1)$ are not defined;

$f'(x) > 0$ on $(0, 1)$ and $(1, \infty);$

$f'(x) < 0$ on $(-\infty, -1)$ and $(-1, 0);$

$f''(-1)$ and $f''(1)$ are not defined;

$f''(x) > 0$ on $(-1, 1);$

$f''(x) < 0$ on $(-\infty, -1)$ and $(1, \infty)$

48. $f(0) = -2, f(1) = 0, f(2) = 4;$

$f'(0) = 0, f'(2) = 0, f'(1)$ is not defined;

$f'(x) > 0$ on $(0, 1)$ and $(1, 2);$

$f'(x) < 0$ on $(-\infty, 0)$ and $(2, \infty);$

$f''(1)$ is not defined;

$f''(x) > 0$ on $(-\infty, 1);$

$f''(x) < 0$ on $(1, \infty)$

In Problems 49–70, summarize the pertinent information obtained by applying the graphing strategy and sketch the graph of $y = f(x)$.

49. $f(x) = (x - 2)(x^2 - 4x - 8)$

50. $f(x) = (x - 3)(x^2 - 6x - 3)$

51. $f(x) = (x + 1)(x^2 - x + 2)$

52. $f(x) = (1 - x)(x^2 + x + 4)$

53. $f(x) = -0.25x^4 + x^3$

54. $f(x) = 0.25x^4 - 2x^3$

55. $f(x) = 16x(x - 1)^3$

56. $f(x) = -4x(x + 2)^3$

57. $f(x) = (x^2 + 3)(9 - x^2)$

58. $f(x) = (x^2 + 3)(x^2 - 1)$

59. $f(x) = (x^2 - 4)^2$

60. $f(x) = (x^2 - 1)(x^2 - 5)$

61. $f(x) = 2x^6 - 3x^5$

62. $f(x) = 3x^5 - 5x^4$

63. $f(x) = 1 - e^{-x}$

64. $f(x) = 2 - 3e^{-2x}$

65. $f(x) = e^{0.5x} + 4e^{-0.5x}$

66. $f(x) = 2e^{0.5x} + e^{-0.5x}$

67. $f(x) = -4 + 2 \ln x$

68. $f(x) = 5 - 3 \ln x$

69. $f(x) = \ln(x + 4) - 2$

70. $f(x) = 1 - \ln(x - 3)$

In Problems 71–74, use the graph of $y = f'(x)$ to discuss the graph of $y = f(x)$. Organize your conclusions in a table (see Example 4), and sketch a possible graph of $y = f(x)$.

71.

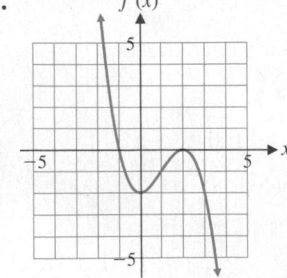

72.

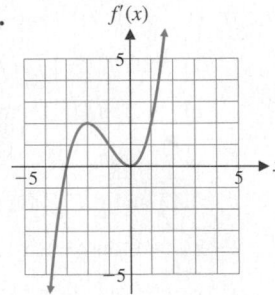

73.

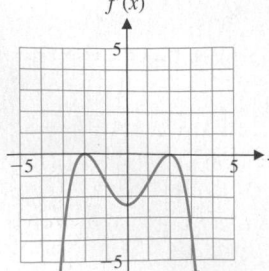

74.

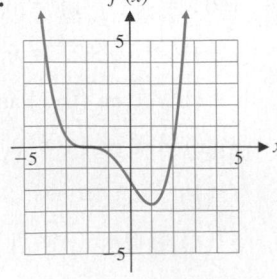

 In Problems 75–82, apply steps 1–3 of the graphing strategy to $f(x)$. Use a graphing calculator to approximate (to two decimal places) x intercepts, critical numbers, and inflection points. Summarize all the pertinent information.

75. $f(x) = x^4 - 5x^3 + 3x^2 + 8x - 5$

76. $f(x) = x^4 + 2x^3 - 5x^2 - 4x + 4$

77. $f(x) = x^4 - 21x^3 + 100x^2 + 20x + 100$

78. $f(x) = x^4 - 12x^3 + 28x^2 + 76x - 50$

79. $f(x) = -x^4 - x^3 + 2x^2 - 2x + 3$

80. $f(x) = -x^4 + x^3 + x^2 + 6$

81. $f(x) = 0.1x^5 + 0.3x^4 - 4x^3 - 5x^2 + 40x + 30$

82. $f(x) = x^5 + 4x^4 - 7x^3 - 20x^2 + 20x - 20$

Applications

83. **Inflation.** One commonly used measure of inflation is the annual rate of change of the Consumer Price Index (CPI). A TV news story says that the annual rate of change of the CPI is increasing. What does this say about the shape of the graph of the CPI?

84. **Inflation.** Another commonly used measure of inflation is the annual rate of change of the Producer Price Index (PPI). A government report states that the annual rate of change of the PPI is decreasing. What does this say about the shape of the graph of the PPI?

85. **Cost analysis.** A company manufactures a variety of camp stoves at different locations. The total cost $C(x)$ (in dollars) of producing x camp stoves per week at plant A is shown in the figure. Discuss the graph of the marginal cost function $C'(x)$ and interpret the graph of $C'(x)$ in terms of the efficiency of the production process at this plant.

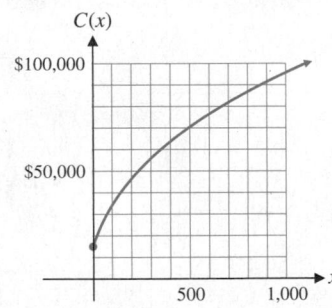

Production costs at plant A

86. **Cost analysis.** The company in Problem 85 produces the same camp stove at another plant. The total cost $C(x)$ (in dollars) of producing x camp stoves per week at plant B is shown in the figure. Discuss the graph of the marginal cost function $C'(x)$ and interpret the graph of $C'(x)$ in terms of the efficiency of the production process at plant B. Compare the production processes at the two plants.

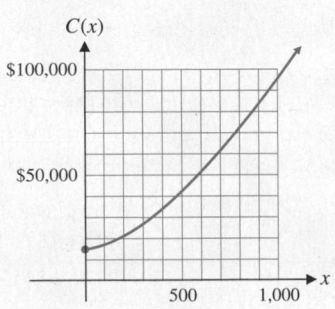

Production costs at plant B

87. Revenue. The marketing research department of a computer company used a large city to test market the firm's new laptop. The department found that the relationship between price p (dollars per unit) and the demand x (units per week) was given approximately by

$$p = 1,296 - 0.12x^2 \qquad 0 < x < 80$$

So, weekly revenue can be approximated by

$$R(x) = xp = 1,296x - 0.12x^3 \qquad 0 < x < 80$$

(A) Find the local extrema for the revenue function.

(B) On which intervals is the graph of the revenue function concave upward? Concave downward?

88. Profit. Suppose that the cost equation for the company in Problem 87 is

$$C(x) = 830 + 396x$$

(A) Find the local extrema for the profit function.

(B) On which intervals is the graph of the profit function concave upward? Concave downward?

89. Revenue. A dairy is planning to introduce and promote a new line of organic ice cream. After test marketing the new line in a large city, the marketing research department found that the demand in that city is given approximately by

$$p = 10e^{-x} \qquad 0 \le x \le 5$$

where x thousand quarts were sold per week at a price of $\$p$ each.

(A) Find the local extrema for the revenue function.

(B) On which intervals is the graph of the revenue function concave upward? Concave downward?

90. Revenue. A national food service runs food concessions for sporting events throughout the country. The company's marketing research department chose a particular football stadium to test market a new jumbo hot dog. It was found that the demand for the new hot dog is given approximately by

$$p = 8 - 2 \ln x \qquad 5 \le x \le 50$$

where x is the number of hot dogs (in thousands) that can be sold during one game at a price of $\$p$.

(A) Find the local extrema for the revenue function.

(B) On which intervals is the graph of the revenue function concave upward? Concave downward?

91. Production: point of diminishing returns. A T-shirt manufacturer is planning to expand its workforce. It estimates that the number of T-shirts produced by hiring x new workers is given by

$$T(x) = -0.25x^4 + 5x^3 \qquad 0 \le x \le 15$$

When is the rate of change of T-shirt production increasing and when is it decreasing? What is the point of diminishing returns and the maximum rate of change of T-shirt production? Graph T and T' on the same coordinate system.

92. Production: point of diminishing returns. A baseball cap manufacturer is planning to expand its workforce. It estimates that the number of baseball caps produced by hiring x new workers is given by

$$T(x) = -0.25x^4 + 6x^3 \qquad 0 \le x \le 18$$

When is the rate of change of baseball cap production increasing and when is it decreasing? What is the point of diminishing returns and the maximum rate of change of baseball cap production? Graph T and T' on the same coordinate system.

93. Advertising: point of diminishing returns. A company estimates that it will sell $N(x)$ units of a product after spending $\$x$ thousand on advertising, as given by

$$N(x) = -0.25x^4 + 23x^3 - 540x^2 + 80,000 \qquad 24 \le x \le 45$$

When is the rate of change of sales increasing and when is it decreasing? What is the point of diminishing returns and the maximum rate of change of sales? Graph N and N' on the same coordinate system.

94. Advertising: point of diminishing returns. A company estimates that it will sell $N(x)$ units of a product after spending $\$x$ thousand on advertising, as given by

$$N(x) = -0.25x^4 + 13x^3 - 180x^2 + 10,000 \qquad 15 \le x \le 24$$

When is the rate of change of sales increasing and when is it decreasing? What is the point of diminishing returns and the maximum rate of change of sales? Graph N and N' on the same coordinate system.

95. Advertising. An automobile dealer uses TV advertising to promote car sales. On the basis of past records, the dealer arrived at the following data, where x is the number of ads placed monthly and y is the number of cars sold that month:

Number of Ads x	Number of Cars y
10	325
12	339
20	417
30	546
35	615
40	682
50	795

(A) Enter the data in a graphing calculator and find a cubic regression equation for the number of cars sold monthly as a function of the number of ads.

(B) How many ads should the dealer place each month to maximize the rate of change of sales with respect to the number of ads, and how many cars can the dealer expect to sell with this number of ads? Round answers to the nearest integer.

96. Advertising. A sporting goods chain places TV ads to promote golf club sales. The marketing director used past records to determine the following data, where x is the number of ads placed monthly and y is the number of golf clubs sold that month.

Number of Ads x	Number of Golf Clubs y
10	345
14	488
20	746
30	1,228
40	1,671
50	1,955

(A) Enter the data in a graphing calculator and find a cubic regression equation for the number of golf clubs sold monthly as a function of the number of ads.

(B) How many ads should the store manager place each month to maximize the rate of change of sales with respect to the number of ads, and how many golf clubs can the manager expect to sell with this number of ads? Round answers to the nearest integer.

97. Population growth: bacteria. A drug that stimulates reproduction is introduced into a colony of bacteria. After t minutes, the number of bacteria is given approximately by

$$N(t) = 1,000 + 30t^2 - t^3 \qquad 0 \le t \le 20$$

(A) When is the rate of growth, $N'(t)$, increasing? Decreasing?

(B) Find the inflection points for the graph of N.

(C) Sketch the graphs of N and N' on the same coordinate system.

(D) What is the maximum rate of growth?

98. Drug sensitivity. One hour after x milligrams of a particular drug are given to a person, the change in body temperature $T(x)$, in degrees Fahrenheit, is given by

$$T(x) = x^2\left(1 - \frac{x}{9}\right) \qquad 0 \le x \le 6$$

The rate $T'(x)$ at which $T(x)$ changes with respect to the size of the dosage x is called the *sensitivity* of the body to the dosage.

(A) When is $T'(x)$ increasing? Decreasing?

(B) Where does the graph of T have inflection points?

(C) Sketch the graphs of T and T' on the same coordinate system.

(D) What is the maximum value of $T'(x)$?

99. Learning. The time T (in minutes) it takes a person to learn a list of length n is

$$T(n) = 0.08n^3 - 1.2n^2 + 6n \qquad n \ge 0$$

(A) When is the rate of change of T with respect to the length of the list increasing? Decreasing?

(B) Where does the graph of T have inflection points?

(C) Graph T and T' on the same coordinate system.

(D) What is the minimum value of $T'(n)$?

Answers to Matched Problems

1. (A) Concave downward on $(-\infty, \infty)$

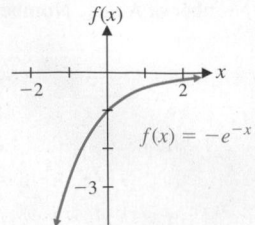

(B) Concave upward on $(0, \infty)$

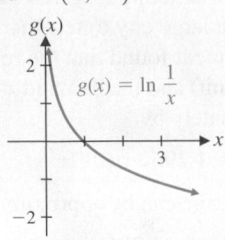

(C) Concave upward on $(-\infty, 0)$ and concave downward on $(0, \infty)$

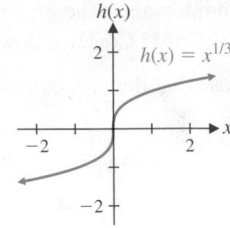

2. The only inflection point is $(3, f(3)) = (3, 8)$.

3. The inflection points are $(-1, f(-1)) = (-1, \ln 8)$ and $(3, f(3)) = (3, \ln 8)$.

4.

x	$f'(x)$	$f(x)$
$-\infty < x < -1$	Positive and decreasing	Increasing and concave downward
$x = -1$	Local minimum	Inflection point
$-1 < x < 1$	Positive and increasing	Increasing and concave upward
$x = 1$	Local maximum	Inflection point
$1 < x < 2$	Positive and decreasing	Increasing and concave downward
$x = 2$	x intercept	Local maximum
$2 < x < \infty$	Negative and decreasing	Decreasing and concave downward

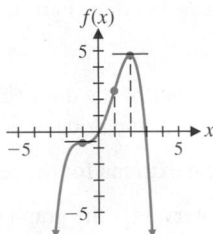

5. Domain: All real numbers

x intercepts: $-4, 0$; y intercept: $f(0) = 0$

Decreasing on $(-\infty, -3)$; increasing on $(-3, \infty)$; local minimum: $f(-3) = -27$

Concave upward on $(-\infty, -2)$ and $(0, \infty)$; concave downward on $(-2, 0)$

Inflection points: $(-2, -16)$, $(0, 0)$

x	$f(x)$
-4	0
-3	-27
-2	-16
0	0

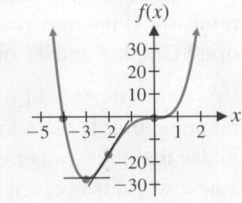

6. Domain: All real numbers

x intercepts: 0, 27; y intercept: $f(0) = 0$

Decreasing on $(-\infty, 0)$ and $(8, \infty)$; increasing on $(0, 8)$; local minimum: $f(0) = 0$; local maximum: $f(8) = 4$

Concave downward on $(-\infty, 0)$ and $(0, \infty)$; no inflection points

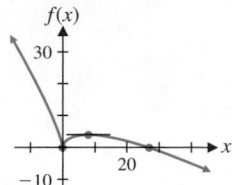

x	f(x)
0	0
8	4
27	0

7. $N'(x)$ is increasing on $(0, 8)$ and decreasing on $(8, 12)$.

The point of diminishing returns is $x = 8$ and the maximum rate of change is $N'(8) = 256$.

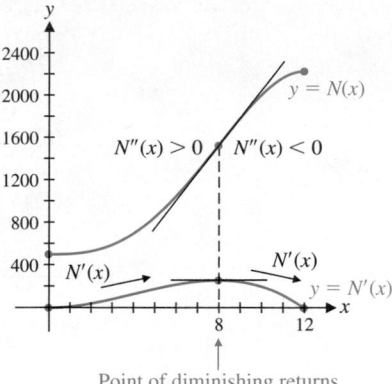

4.3 L'Hôpital's Rule

- Introduction

- L'Hôpital's Rule and the Indeterminate Form 0/0

- One-Sided Limits and Limits at ∞

- L'Hôpital's Rule and the Indeterminate Form ∞/∞

Introduction

The ability to evaluate a wide variety of different types of limits is one of the skills that are necessary to apply the techniques of calculus successfully. Limits play a fundamental role in the development of the derivative and are an important graphing tool. In order to deal effectively with graphs, we need to develop some more methods for evaluating limits.

In this section, we discuss a powerful technique for evaluating limits of quotients called *L'Hôpital's rule*. The rule is named after the French mathematician Marquis de L'Hôpital (1661–1704). To use L'Hôpital's rule, it is necessary to be familiar with the limit properties of some basic functions. Figure 1 reviews some limits involving powers of x that were discussed earlier.

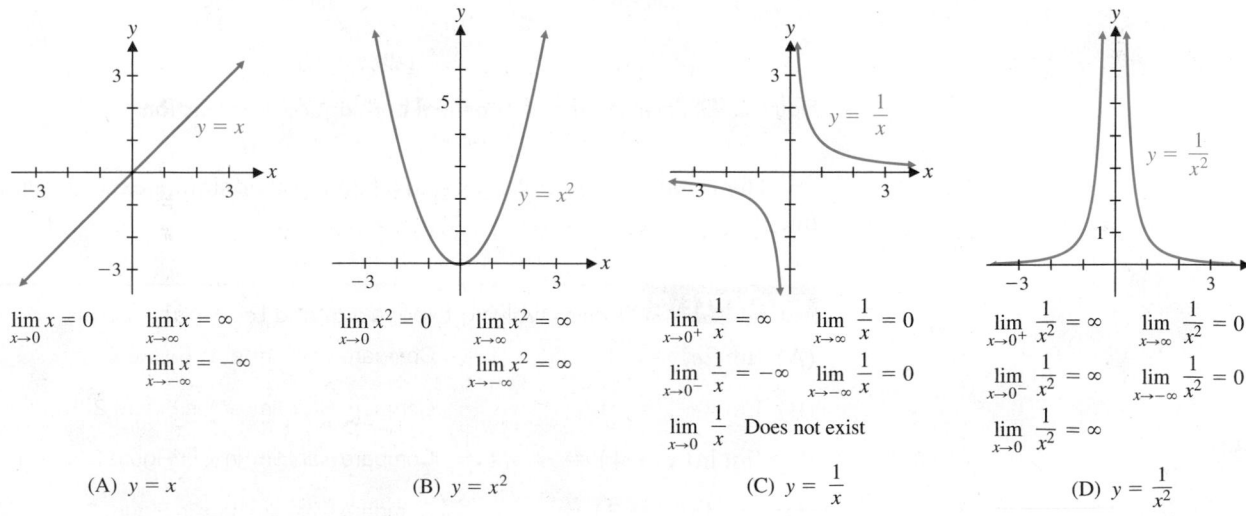

Figure 1 **Limits involving powers of x**

The limits in Figure 1 are easily extended to functions of the form $f(x) = (x - c)^n$ and $g(x) = 1/(x - c)^n$. In general, if n is an odd integer, then limits involving $(x - c)^n$ or $1/(x - c)^n$ as x approaches c (or $\pm\infty$) behave, respectively, like the limits of x and $1/x$ as x approaches 0 (or $\pm\infty$). If n is an even integer, then limits involving these expressions behave, respectively, like the limits of x^2 and $1/x^2$ as x approaches 0 (or $\pm\infty$).

EXAMPLE 1   Limits Involving Powers of $x - c$

(A) $\lim\limits_{x\to 2} \dfrac{5}{(x - 2)^4} = \infty$ Compare with $\lim\limits_{x\to 0} \dfrac{1}{x^2}$ in Figure 1.

(B) $\lim\limits_{x\to -1^-} \dfrac{4}{(x + 1)^3} = -\infty$ Compare with $\lim\limits_{x\to 0^-} \dfrac{1}{x}$ in Figure 1.

(C) $\lim\limits_{x\to \infty} \dfrac{4}{(x - 9)^6} = 0$ Compare with $\lim\limits_{x\to \infty} \dfrac{1}{x^2}$ in Figure 1.

(D) $\lim\limits_{x\to -\infty} 3x^3 = -\infty$ Compare with $\lim\limits_{x\to -\infty} x$ in Figure 1.

Matched Problem 1   Evaluate each limit.

(A) $\lim\limits_{x\to 3^+} \dfrac{7}{(x - 3)^5}$ (B) $\lim\limits_{x\to -4} \dfrac{6}{(x + 4)^6}$

(C) $\lim\limits_{x\to -\infty} \dfrac{3}{(x + 2)^3}$ (D) $\lim\limits_{x\to \infty} 5x^4$

Figure 2 reviews limits of exponential and logarithmic functions.

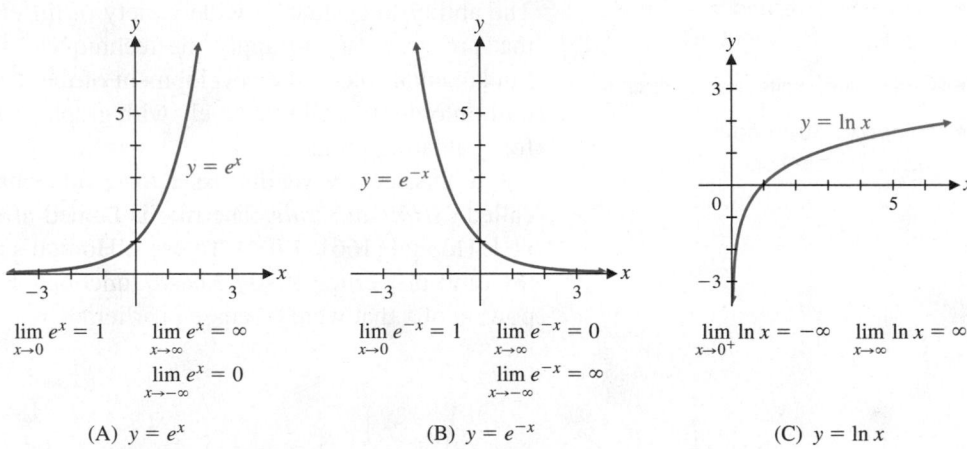

$\lim\limits_{x\to 0} e^x = 1$ $\lim\limits_{x\to \infty} e^x = \infty$

$\lim\limits_{x\to -\infty} e^x = 0$

$\lim\limits_{x\to 0} e^{-x} = 1$ $\lim\limits_{x\to \infty} e^{-x} = 0$

$\lim\limits_{x\to -\infty} e^{-x} = \infty$

$\lim\limits_{x\to 0^+} \ln x = -\infty$ $\lim\limits_{x\to \infty} \ln x = \infty$

(A) $y = e^x$ (B) $y = e^{-x}$ (C) $y = \ln x$

Figure 2 **Limits involving exponential and logarithmic functions**

The limits in Figure 2 also generalize to other simple exponential and logarithmic forms.

EXAMPLE 2   Limits Involving Exponential and Logarithmic Forms

(A) $\lim\limits_{x\to \infty} 2e^{3x} = \infty$ Compare with $\lim\limits_{x\to \infty} e^x$ in Figure 2.

(B) $\lim\limits_{x\to \infty} 4e^{-5x} = 0$ Compare with $\lim\limits_{x\to \infty} e^{-x}$ in Figure 2.

(C) $\lim\limits_{x\to \infty} \ln(x + 4) = \infty$ Compare with $\lim\limits_{x\to \infty} \ln x$ in Figure 2.

(D) $\lim\limits_{x\to 2^+} \ln(x - 2) = -\infty$ Compare with $\lim\limits_{x\to 0^+} \ln x$ in Figure 2.

Matched Problem 2   Evaluate each limit.

(A) $\lim\limits_{x\to -\infty} 2e^{-6x}$ (B) $\lim\limits_{x\to -\infty} 3e^{2x}$

(C) $\lim\limits_{x\to -4^+} \ln(x + 4)$ (D) $\lim\limits_{x\to \infty} \ln(x - 10)$

Now that we have reviewed the limit properties of some basic functions, we are ready to consider the main topic of this section: L'Hôpital's rule.

L'Hôpital's Rule and the Indeterminate Form 0/0

Recall that the limit

$$\lim_{x \to c} \frac{f(x)}{g(x)}$$

is a 0/0 indeterminate form if

$$\lim_{x \to c} f(x) = 0 \quad \text{and} \quad \lim_{x \to c} g(x) = 0$$

The quotient property for limits in Section 2.1 does not apply since $\lim_{x \to c} g(x) = 0$.

If we are dealing with a 0/0 indeterminate form, the limit may or may not exist, and we cannot tell which is true without further investigation.

Each of the following is a 0/0 indeterminate form:

$$\lim_{x \to 2} \frac{x^2 - 4}{x - 2} \quad \text{and} \quad \lim_{x \to 1} \frac{e^x - e}{x - 1} \tag{1}$$

The first limit can be evaluated by performing an algebraic simplification:

$$\lim_{x \to 2} \frac{x^2 - 4}{x - 2} = \lim_{x \to 2} \frac{(x - 2)(x + 2)}{x - 2} = \lim_{x \to 2} (x + 2) = 4$$

The second cannot. Instead, we turn to the powerful **L'Hôpital's rule**, which we state without proof. This rule can be used whenever a limit is a 0/0 indeterminate form, so can be used to evaluate both of the limits in (1).

THEOREM 1   L'Hôpital's Rule for 0/0 Indeterminate Forms: Version 1

For c a real number,
if $\lim_{x \to c} f(x) = 0$ and $\lim_{x \to c} g(x) = 0$, then

$$\lim_{x \to c} \frac{f(x)}{g(x)} = \lim_{x \to c} \frac{f'(x)}{g'(x)}$$

provided that the second limit exists or is ∞ or $-\infty$.

By L'Hôpital's rule,

$$\lim_{x \to 2} \frac{x^2 - 4}{x - 2} = \lim_{x \to 2} \frac{2x}{1} = 4$$

which agrees with the result obtained by algebraic simplification.

EXAMPLE 3   L'Hôpital's Rule   Evaluate $\lim_{x \to 1} \frac{e^x - e}{x - 1}$.

SOLUTION

Step 1   *Check to see if L'Hôpital's rule applies:*

$$\lim_{x \to 1} (e^x - e) = e^1 - e = 0 \quad \text{and} \quad \lim_{x \to 1} (x - 1) = 1 - 1 = 0$$

L'Hôpital's rule does apply.

Step 2 *Apply L'Hôpital's rule:*

0/0 form

$$\lim_{x \to 1} \frac{e^x - e}{x - 1} = \lim_{x \to 1} \frac{\dfrac{d}{dx}(e^x - e)}{\dfrac{d}{dx}(x - 1)} \qquad \text{Apply L'Hôpital's rule.}$$

$$= \lim_{x \to 1} \frac{e^x}{1} \qquad e^x \text{ is continuous at } x = 1.$$

$$= \frac{e^1}{1} = e$$

Matched Problem 3 Evaluate $\displaystyle\lim_{x \to 4} \frac{e^x - e^4}{x - 4}$.

⚠ **CAUTION** In L'Hôpital's rule, the symbol $f'(x)/g'(x)$ represents the derivative of $f(x)$ divided by the derivative of $g(x)$, not the derivative of the quotient $f(x)/g(x)$.

When applying L'Hôpital's rule to a **0/0** indeterminate form, do not use the quotient rule. Instead, evaluate the limit of the derivative of the numerator divided by the derivative of the denominator.   ▲

The functions

$$y_1 = \frac{e^x - e}{x - 1} \quad \text{and} \quad y_2 = \frac{e^x}{1}$$

of Example 3 are different functions (see Fig. 3), but both functions have the same limit e as x approaches 1. Although y_1 is undefined at $x = 1$, the graph of y_1 provides a check of the answer to Example 3.

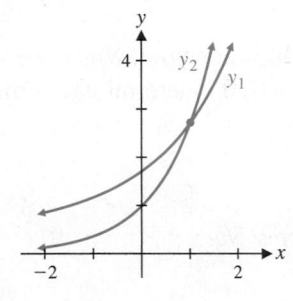

Figure 3

EXAMPLE 4 L'Hôpital's Rule Evaluate $\displaystyle\lim_{x \to 0} \frac{\ln(1 + x^2)}{x^4}$.

SOLUTION

Step 1 *Check to see if L'Hôpital's rule applies:*

$$\lim_{x \to 0} \ln(1 + x^2) = \ln 1 = 0 \qquad \text{and} \qquad \lim_{x \to 0} x^4 = 0$$

L'Hôpital's rule does apply.

Step 2 *Apply L'Hôpital's rule:*

0/0 form

$$\lim_{x \to 0} \frac{\ln(1 + x^2)}{x^4} = \lim_{x \to 0} \frac{\dfrac{d}{dx}\ln(1 + x^2)}{\dfrac{d}{dx}x^4} \qquad \text{Apply L'Hôpital's rule.}$$

$$\lim_{x \to 0} \frac{\ln(1 + x^2)}{x^4} = \lim_{x \to 0} \frac{\dfrac{2x}{1 + x^2}}{4x^3} \qquad \begin{array}{l}\text{Multiply numerator and denominator}\\ \text{by } 1/4x^3.\end{array}$$

$$= \lim_{x \to 0} \frac{\dfrac{2x}{1 + x^2} \dfrac{1}{4x^3}}{4x^3 \dfrac{1}{4x^3}} \qquad \text{Simplify.}$$

$$= \lim_{x \to 0} \frac{1}{2x^2(1 + x^2)}$$

Apply Theorem 1 in Section 2.2 and compare with Fig. 1(D).

$$= \infty$$

Matched Problem 4) Evaluate $\lim_{x \to 1} \dfrac{\ln x}{(x - 1)^3}$.

EXAMPLE 5 L'Hôpital's Rule May Not Be Applicable Evaluate $\lim_{x \to 1} \dfrac{\ln x}{x}$.

SOLUTION

Step 1 *Check to see if L'Hôpital's rule applies:*

$$\lim_{x \to 1} \ln x = \ln 1 = 0, \qquad \text{but} \qquad \lim_{x \to 1} x = 1 \neq 0$$

L'Hôpital's rule does not apply.

Step 2 *Evaluate by another method.* The quotient property for limits from Section 2.1 does apply, and we have

$$\lim_{x \to 1} \frac{\ln x}{x} = \frac{\displaystyle\lim_{x \to 1} \ln x}{\displaystyle\lim_{x \to 1} x} = \frac{\ln 1}{1} = \frac{0}{1} = 0$$

Note that applying L'Hôpital's rule would give us an incorrect result:

$$\lim_{x \to 1} \frac{\ln x}{x} \neq \lim_{x \to 1} \frac{\dfrac{d}{dx} \ln x}{\dfrac{d}{dx} x} = \lim_{x \to 1} \frac{1/x}{1} = 1$$

Matched Problem 5) Evaluate $\lim_{x \to 0} \dfrac{x}{e^x}$.

⚠ **CAUTION** As Example 5 illustrates, some limits involving quotients are not $0/0$ indeterminate forms.

You must always check to see if L'Hôpital's rule applies before you use it. ▲

EXAMPLE 6 Repeated Application of L'Hôpital's Rule Evaluate

$$\lim_{x \to 0} \frac{x^2}{e^x - 1 - x}$$

SOLUTION

Step 1 *Check to see if L'Hôpital's rule applies:*

$$\lim_{x \to 0} x^2 = 0 \qquad \text{and} \qquad \lim_{x \to 0} (e^x - 1 - x) = 0$$

L'Hôpital's rule does apply.

Step 2 Apply L'Hôpital's rule:

$$\underset{0/0 \text{ form}}{\lim_{x \to 0} \frac{x^2}{e^x - 1 - x}} = \lim_{x \to 0} \frac{\dfrac{d}{dx} x^2}{\dfrac{d}{dx}(e^x - 1 - x)} = \lim_{x \to 0} \frac{2x}{e^x - 1}$$

Since $\lim_{x \to 0} 2x = 0$ and $\lim_{x \to 0}(e^x - 1) = 0$, the new limit obtained is also a $0/0$ indeterminate form, and L'Hôpital's rule can be applied again.

Step 3 *Apply L'Hôpital's rule again:*

0/0 form

$$\lim_{x \to 0} \frac{2x}{e^x - 1} = \lim_{x \to 0} \frac{\frac{d}{dx} 2x}{\frac{d}{dx}(e^x - 1)} = \lim_{x \to 0} \frac{2}{e^x} = \frac{2}{e^0} = 2$$

Therefore,

$$\lim_{x \to 0} \frac{x^2}{e^x - 1 - x} = \lim_{x \to 0} \frac{2x}{e^x - 1} = \lim_{x \to 0} \frac{2}{e^x} = 2$$

Matched Problem 6 Evaluate $\displaystyle\lim_{x \to 0} \frac{e^{2x} - 1 - 2x}{x^2}$

One-Sided Limits and Limits at ∞

In addition to examining the limit as x approaches c, we have discussed one-sided limits and limits at ∞ in Chapter 3. L'Hôpital's rule is valid in these cases also.

THEOREM 2 L'Hôpital's Rule for 0/0 Indeterminate Forms: Version 2 (for one-sided limits and limits at infinity)

The first version of L'Hôpital's rule (Theorem 1) remains valid if the symbol $x \to c$ is replaced everywhere it occurs with one of the following symbols:

$$x \to c^+ \qquad x \to c^- \qquad x \to \infty \qquad x \to -\infty$$

For example, if $\displaystyle\lim_{x \to \infty} f(x) = 0$ and $\displaystyle\lim_{x \to \infty} g(x) = 0$, then

$$\lim_{x \to \infty} \frac{f(x)}{g(x)} = \lim_{x \to \infty} \frac{f'(x)}{g'(x)}$$

provided that the second limit exists or is $+\infty$ or $-\infty$. Similar rules can be written for $x \to c^+$, $x \to c^-$, and $x \to -\infty$.

EXAMPLE 7 L'Hôpital's Rule for One-Sided Limits Evaluate $\displaystyle\lim_{x \to 1^+} \frac{\ln x}{(x - 1)^2}$.

SOLUTION

Step 1 *Check to see if L'Hôpital's rule applies:*

$$\lim_{x \to 1^+} \ln x = 0 \qquad \text{and} \qquad \lim_{x \to 1^+} (x - 1)^2 = 0$$

L'Hôpital's rule does apply.

Step 2 *Apply L'Hôpital's rule:*

0/0 form

$$\lim_{x \to 1^+} \frac{\ln x}{(x - 1)^2} = \lim_{x \to 1^+} \frac{\frac{d}{dx}(\ln x)}{\frac{d}{dx}(x - 1)^2} \qquad \text{Apply L'Hôpital's rule.}$$

$$= \lim_{x \to 1^+} \frac{1/x}{2(x - 1)} \qquad \text{Simplify.}$$

$$= \lim_{x \to 1^+} \frac{1}{2x(x - 1)}$$

$$= \infty$$

The limit as $x \to 1^+$ is ∞ because $1/2x(x-1)$ has a vertical asymptote at $x = 1$ (Theorem 1, Section 2.2) and $x(x-1) > 0$ for $x > 1$.

Matched Problem 7) Evaluate $\lim\limits_{x \to 1} \dfrac{\ln x}{(x-1)^2}$.

EXAMPLE 8 L'Hôpital's Rule for Limits at Infinity Evaluate $\lim\limits_{x \to \infty} \dfrac{\ln(1 + e^{-x})}{e^{-x}}$.

SOLUTION

Step 1 *Check to see if L'Hôpital's rule applies:*

$$\lim\limits_{x \to \infty} \ln(1 + e^{-x}) = \ln(1 + 0) = \ln 1 = 0 \text{ and } \lim\limits_{x \to \infty} e^{-x} = 0$$

L'Hôpital's rule does apply.

Step 2 *Apply L'Hôpital's rule:*

$$\lim\limits_{x \to \infty} \overset{\text{0/0 form}}{\dfrac{\ln(1 + e^{-x})}{e^{-x}}} = \lim\limits_{x \to \infty} \dfrac{\dfrac{d}{dx}[\ln(1 + e^{-x})]}{\dfrac{d}{dx}e^{-x}} \quad \text{Apply L'Hôpital's rule.}$$

$$= \lim\limits_{x \to \infty} \dfrac{-e^{-x}/(1 + e^{-x})}{-e^{-x}} \quad \begin{array}{l}\text{Multiply numerator and}\\ \text{denominator by } -e^{x}.\end{array}$$

$$= \lim\limits_{x \to \infty} \dfrac{1}{1 + e^{-x}} \quad \lim\limits_{x \to \infty} e^{-x} = 0$$

$$= \dfrac{1}{1 + 0} = 1$$

Matched Problem 8) Evaluate $\lim\limits_{x \to -\infty} \dfrac{\ln(1 + 2e^x)}{e^x}$.

L'Hôpital's Rule and the Indeterminate Form ∞/∞

In Section 2.2, we discussed techniques for evaluating limits of rational functions such as

$$\lim\limits_{x \to \infty} \dfrac{2x^2}{x^3 + 3} \qquad \lim\limits_{x \to \infty} \dfrac{4x^3}{2x^2 + 5} \qquad \lim\limits_{x \to \infty} \dfrac{3x^3}{5x^3 + 6} \tag{2}$$

Each of these limits is an ∞/∞ *indeterminate form.* In general, if $\lim_{x \to c} f(x) = \pm\infty$ and $\lim_{x \to c} g(x) = \pm\infty$, then

$$\lim\limits_{x \to c} \dfrac{f(x)}{g(x)}$$

is called an **∞/∞ indeterminate form.** Furthermore, $x \to c$ can be replaced in all three limits above with $x \to c^+, x \to c^-, x \to \infty$, or $x \to -\infty$. It can be shown that L'Hôpital's rule also applies to these ∞/∞ indeterminate forms.

THEOREM 3 L'Hôpital's Rule for the Indeterminate Form ∞/∞: Version 3

Versions 1 and 2 of L'Hôpital's rule for the indeterminate form $0/0$ are also valid if the limit of f and the limit of g are both infinite; that is, both $+\infty$ and $-\infty$ are permissible for either limit.

For example, if $\lim_{x \to c^+} f(x) = \infty$ and $\lim_{x \to c^+} g(x) = -\infty$, then L'Hôpital's rule can be applied to $\lim_{x \to c^+} [f(x)/g(x)]$.

Explore and Discuss 1

Evaluate each of the limits in (2) on page 277 in two ways:

1. Use Theorem 4 in Section 2.2.
2. Use L'Hôpital's rule.

Given a choice, which method would you choose? Why?

EXAMPLE 9 L'Hôpital's Rule for the Indeterminate Form ∞/∞ Evaluate $\lim_{x \to \infty} \dfrac{\ln x}{x^2}$.

SOLUTION

Step 1 *Check to see if L'Hôpital's rule applies:*

$$\lim_{x \to \infty} \ln x = \infty \qquad \text{and} \qquad \lim_{x \to \infty} x^2 = \infty$$

L'Hôpital's rule does apply.

Step 2 *Apply L'Hôpital's rule:*

∞/∞ form

$$\lim_{x \to \infty} \frac{\ln x}{x^2} = \lim_{x \to \infty} \frac{\dfrac{d}{dx}(\ln x)}{\dfrac{d}{dx}x^2} \qquad \text{Apply L'Hôpital's rule.}$$

$$= \lim_{x \to \infty} \frac{1/x}{2x} \qquad \text{Simplify.}$$

$$\lim_{x \to \infty} \frac{\ln x}{x^2} = \lim_{x \to \infty} \frac{1}{2x^2} \qquad \text{See Figure 1(D).}$$

$$= 0$$

Matched Problem 9 Evaluate $\lim_{x \to \infty} \dfrac{\ln x}{x}$.

EXAMPLE 10 L'Hôpital's Rule for the Indeterminate Form ∞/∞ Evaluate $\lim_{x \to \infty} \dfrac{e^x}{x^2}$.

SOLUTION

Step 1 *Check to see if L'Hôpital's rule applies:*

$$\lim_{x \to \infty} e^x = \infty \qquad \text{and} \qquad \lim_{x \to \infty} x^2 = \infty$$

L'Hôpital's rule does apply.

Step 2 *Apply L'Hôpital's rule:*

∞/∞ form

$$\lim_{x \to \infty} \frac{e^x}{x^2} = \lim_{x \to \infty} \frac{\dfrac{d}{dx}e^x}{\dfrac{d}{dx}x^2} = \lim_{x \to \infty} \frac{e^x}{2x}$$

Since $\lim_{x \to \infty} e^x = \infty$ and $\lim_{x \to \infty} 2x = \infty$, this limit is an ∞/∞ indeterminate form and L'Hôpital's rule can be applied again.

Step 3 *Apply L'Hôpital's rule again:*

∞/∞ form

$$\lim_{x\to\infty}\frac{e^x}{2x}=\lim_{x\to\infty}\frac{\dfrac{d}{dx}\,e^x}{\dfrac{d}{dx}\,2x}=\lim_{x\to\infty}\frac{e^x}{2}=\infty$$

Therefore,

$$\lim_{x\to\infty}\frac{e^x}{x^2}=\lim_{x\to\infty}\frac{e^x}{2x}=\lim_{x\to\infty}\frac{e^x}{2}=\infty$$

Matched Problem 10 ⌋ Evaluate $\lim_{x\to\infty}\dfrac{e^{2x}}{x^2}$.

CONCEPTUAL INSIGHT

The three versions of L'Hôpital's rule cover a multitude of limits—far too many to remember case by case. Instead, we suggest you use the following pattern, common to all versions, as a memory aid:

1. All versions involve three limits: $\lim\,[f(x)/g(x)]$, $\lim f(x)$, and $\lim g(x)$.

2. The independent variable x must behave the same way in all three limits. The acceptable behaviors are $x\to c$, $x\to c^+$, $x\to c^-$, $x\to\infty$, or $x\to-\infty$.

3. The form of $\lim\,[f(x)/g(x)]$ must be $\frac{0}{0}$ or $\frac{\pm\infty}{\pm\infty}$ and both $\lim f(x)$ and $\lim g(x)$ must approach 0 or both must approach $\pm\infty$.

Exercises 4.3

Skills Warm-up Exercises

W   *In Problems 1–8, round each expression to the nearest integer without using a calculator. (If necessary, review Section A.1).*

1. $\dfrac{5}{0.01}$

2. $\dfrac{8}{0.002}$

3. $\dfrac{3}{1,000}$

4. $\dfrac{2^8}{8}$

5. $\dfrac{1}{2(1.01-1)}$

6. $\dfrac{47}{106}$

7. $\dfrac{\ln 100}{100}$

8. $\dfrac{e^5+5^2}{e^5}$

In Problems 9–16, even though the limit can be found using algebraic simplification as in Section 2.1, use L'Hôpital's rule to find the limit.

9. $\lim_{x\to3}\dfrac{x^2-9}{x-3}$

10. $\lim_{x\to-3}\dfrac{x^2-9}{x+3}$

11. $\lim_{x\to-5}\dfrac{x+5}{x^2-25}$

12. $\lim_{x\to4}\dfrac{x-4}{x^2-16}$

13. $\lim_{x\to1}\dfrac{x^2+5x-6}{x-1}$

14. $\lim_{x\to10}\dfrac{x^2-5x-50}{x-10}$

15. $\lim_{x\to-9}\dfrac{x+9}{x^2+13x+36}$

16. $\lim_{x\to-1}\dfrac{x+1}{x^2-7x-8}$

In Problems 17–24, even though the limit can be found using Theorem 4 of Section 2.2, use L'Hôpital's rule to find the limit.

17. $\lim_{x\to\infty}\dfrac{2x+3}{5x-1}$

18. $\lim_{x\to\infty}\dfrac{6x-7}{7x-6}$

19. $\lim_{x\to\infty}\dfrac{3x^2-1}{x^3+4}$

20. $\lim_{x\to\infty}\dfrac{5x^2+10x+1}{x^4+x^2+1}$

21. $\lim_{x\to-\infty}\dfrac{x^2-9}{x-3}$

22. $\lim_{x\to-\infty}\dfrac{x^4-16}{x^2+4}$

23. $\lim_{x\to\infty}\dfrac{2x^2+3x+1}{3x^2-2x+1}$

24. $\lim_{x\to\infty}\dfrac{5-4x^3}{1+7x^3}$

In Problems 25–32, use L'Hôpital's rule to find the limit. Note that in these problems, neither algebraic simplification nor Theorem 4 of Section 2.2 provides an alternative to L'Hôpital's rule.

25. $\lim_{x\to0}\dfrac{e^x-1}{2x}$

26. $\lim_{x\to0}\dfrac{5x}{e^x-1}$

27. $\lim_{x\to1}\dfrac{x-1}{\ln x}$

28. $\lim_{x\to1}\dfrac{x-1}{\ln x^4}$

29. $\lim_{x\to\infty}\dfrac{x^2}{e^x}$

30. $\lim_{x\to\infty}\dfrac{x^3}{\ln x}$

31. $\lim_{x\to0}\dfrac{e^{4x}-1}{x}$

32. $\lim_{x\to0}\dfrac{\ln(1-3x)}{x}$

✎ *In Problems 33–36, explain why L'Hôpital's rule does not apply. If the limit exists, find it by other means.*

33. $\lim_{x\to1}\dfrac{x^2+5x+4}{x^3+1}$

34. $\lim_{x\to\infty}\dfrac{e^{-x}}{\ln x}$

35. $\lim_{x\to2}\dfrac{x+2}{(x-2)^4}$

36. $\lim_{x\to-3}\dfrac{x^2}{(x+3)^5}$

Find each limit in Problems 37–60. Note that L'Hôpital's rule does not apply to every problem, and some problems will require more than one application of L'Hôpital's rule.

37. $\lim\limits_{x\to 0}\dfrac{e^{4x}-1-4x}{x^2}$

38. $\lim\limits_{x\to 0}\dfrac{3x+1-e^{3x}}{x^2}$

39. $\lim\limits_{x\to 2}\dfrac{\ln(x-1)}{x-1}$

40. $\lim\limits_{x\to -1}\dfrac{\ln(x+2)}{x+2}$

41. $\lim\limits_{x\to 0^+}\dfrac{\ln(1+x^2)}{x^3}$

42. $\lim\limits_{x\to 0^-}\dfrac{\ln(1+2x)}{x^2}$

43. $\lim\limits_{x\to 0^+}\dfrac{\ln(1+\sqrt{x})}{x}$

44. $\lim\limits_{x\to 0^+}\dfrac{\ln(1+x)}{\sqrt{x}}$

45. $\lim\limits_{x\to -2}\dfrac{x^2+2x+1}{x^2+x+1}$

46. $\lim\limits_{x\to 1}\dfrac{2x^3-3x^2+1}{x^3-3x+2}$

47. $\lim\limits_{x\to -1}\dfrac{x^3+x^2-x-1}{x^3+4x^2+5x+2}$

48. $\lim\limits_{x\to 3}\dfrac{x^3+3x^2-x-3}{x^2+6x+9}$

49. $\lim\limits_{x\to 2}\dfrac{x^3-12x+16}{x^3-6x^2+12x-8}$

50. $\lim\limits_{x\to 1^+}\dfrac{x^3+x^2-x+1}{x^3+3x^2+3x-1}$

51. $\lim\limits_{x\to \infty}\dfrac{3x^2+5x}{4x^3+7}$

52. $\lim\limits_{x\to \infty}\dfrac{4x^2+9x}{5x^2+8}$

53. $\lim\limits_{x\to \infty}\dfrac{x^2}{e^{2x}}$

54. $\lim\limits_{x\to \infty}\dfrac{e^{3x}}{x^3}$

55. $\lim\limits_{x\to \infty}\dfrac{1+e^{-x}}{1+x^2}$

56. $\lim\limits_{x\to -\infty}\dfrac{1+e^{-x}}{1+x^2}$

57. $\lim\limits_{x\to \infty}\dfrac{e^{-x}}{\ln(1+4e^{-x})}$

58. $\lim\limits_{x\to \infty}\dfrac{\ln(1+2e^{-x})}{\ln(1+e^{-x})}$

59. $\lim\limits_{x\to 0}\dfrac{e^x-e^{-x}-2x}{x^3}$

60. $\lim\limits_{x\to 0}\dfrac{e^{2x}-1-2x-2x^2}{x^3}$

61. Find $\lim\limits_{x\to 0^+}(x\ln x)$.

[*Hint*: Write $x\ln x=(\ln x)/x^{-1}$.]

62. Find $\lim\limits_{x\to 0^+}(\sqrt{x}\ln x)$.

[*Hint*: Write $\sqrt{x}\ln x=(\ln x)/x^{-1/2}$.]

In Problems 63–66, n is a positive integer. Find each limit.

63. $\lim\limits_{x\to \infty}\dfrac{\ln x}{x^n}$

64. $\lim\limits_{x\to \infty}\dfrac{x^n}{\ln x}$

65. $\lim\limits_{x\to \infty}\dfrac{e^x}{x^n}$

66. $\lim\limits_{x\to \infty}\dfrac{x^n}{e^x}$

In Problems 67–70, show that the repeated application of L'Hôpital's rule does not lead to a solution. Then use algebraic manipulation to evaluate each limit. [Hint: If $x > 0$ and $n > 0$, then $\sqrt[n]{x^n} = x$.]

67. $\lim\limits_{x\to \infty}\dfrac{\sqrt{1+x^2}}{x}$

68. $\lim\limits_{x\to -\infty}\dfrac{x}{\sqrt{4+x^2}}$

69. $\lim\limits_{x\to -\infty}\dfrac{\sqrt[3]{x^3+1}}{x}$

70. $\lim\limits_{x\to \infty}\dfrac{x^2}{\sqrt[3]{(x^3+1)^2}}$

Answers to Matched Problems

1. (A) ∞ (B) ∞ (C) 0   (D) ∞

2. (A) ∞ (B) 0   (C) $-\infty$ (D) ∞

3. e^4 4. ∞ 5. 0   6. 2

7. $-\infty$ 8. 2   9. 0   10. ∞

4.4 Curve-Sketching Techniques

- Modifying the Graphing Strategy
- Using the Graphing Strategy
- Modeling Average Cost

When we summarized the graphing strategy in Section 4.2, we omitted one important topic: asymptotes. Polynomial functions do not have any asymptotes. Asymptotes of rational functions were discussed in Section 2.3, but what about all the other functions, such as logarithmic and exponential functions? Since investigating asymptotes always involves limits, we can now use L'Hôpital's rule (Section 4.3) as a tool for finding asymptotes of many different types of functions.

Modifying the Graphing Strategy

The first version of the graphing strategy in Section 4.2 made no mention of asymptotes. Including information about asymptotes produces the following (and final) version of the graphing strategy.

PROCEDURE Graphing Strategy (Final Version)

Step 1 *Analyze f(x).*

(A) Find the domain of f.

(B) Find the intercepts.

(C) Find asymptotes.

Step 2 *Analyze $f'(x)$.* Find the partition numbers for f' and the critical numbers of f. Construct a sign chart for $f'(x)$, determine the intervals on which f is increasing and decreasing, and find local maxima and minima of f.

Step 3 *Analyze $f''(x)$.* Find the partition numbers for $f''(x)$. Construct a sign chart for $f''(x)$, determine the intervals on which the graph of f is concave upward and concave downward, and find the inflection points of f.

Step 4 *Sketch the graph of f.* Draw asymptotes and locate intercepts, local maxima and minima, and inflection points. Sketch in what you know from steps 1–3. Plot additional points as needed and complete the sketch.

Using the Graphing Strategy

We will illustrate the graphing strategy with several examples. From now on, you should always use the final version of the graphing strategy. If a function does not have any asymptotes, simply state this fact.

EXAMPLE 1   Using the Graphing Strategy Use the graphing strategy to analyze the function $f(x) = (x - 1)/(x - 2)$. State all the pertinent information and sketch the graph of f.

SOLUTION

Step 1 *Analyze $f(x)$.* $f(x) = \dfrac{x - 1}{x - 2}$

(A) Domain: All real x, except $x = 2$

(B) y intercept: $f(0) = \dfrac{0 - 1}{0 - 2} = \dfrac{1}{2}$

x intercepts: Since a fraction is 0 when its numerator is 0 and its denominator is not 0, the x intercept is $x = 1$.

(C) Horizontal asymptote: $\dfrac{a_m x^m}{b_n x^n} = \dfrac{x}{x} = 1$

So the line $y = 1$ is a horizontal asymptote.
Vertical asymptote: The denominator is 0 for $x = 2$, and the numerator is not 0 for this value. Therefore, the line $x = 2$ is a vertical asymptote.

Step 2 *Analyze* $f'(x)$. $f'(x) = \dfrac{(x - 2)(1) - (x - 1)(1)}{(x - 2)^2} = \dfrac{-1}{(x - 2)^2}$

Partition number for $f'(x)$: $x = 2$
Critical numbers of $f(x)$:   None
Sign chart for $f'(x)$:

		Test Numbers	
		x	$f'(x)$
$f'(x)$ $- - - -$ ND $- - - -$		1	-1 $(-)$
$(-\infty, 2)$ $(2, \infty)$		3	-1 $(-)$
$f(x)$ Decreasing Decreasing			

So $f(x)$ is decreasing on $(-\infty, 2)$ and $(2, \infty)$. There are no local extrema.

Step 3 *Analyze $f''(x)$.* $f''(x) = \dfrac{2}{(x - 2)^3}$

Partition number for $f''(x)$: $x = 2$
Sign chart for $f''(x)$:

	$(-\infty, 2)$	$(2, \infty)$
$f''(x)$	$- - - -$ ND	$+ + +$

Test Numbers

x	$f''(x)$
1	-2 $(-)$
3	2 $(+)$

$\xrightarrow{\hspace{3cm}} x$

2

Graph of f	Concave downward	Concave upward

The graph of f is concave downward on $(-\infty, 2)$ and concave upward on $(2, \infty)$. Since $f(2)$ is not defined, there is no inflection point at $x = 2$, even though $f''(x)$ changes sign at $x = 2$.

Step 4   *Sketch the graph of f.* Insert intercepts and asymptotes, and plot a few additional points (for functions with asymptotes, plotting additional points is often helpful). Then sketch the graph.

x	$f(x)$
-2	$\frac{3}{4}$
0	$\frac{1}{2}$
1	0
$\frac{3}{2}$	-1
$\frac{5}{2}$	3
3	2
4	$\frac{3}{2}$

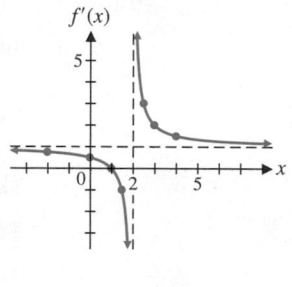

Matched Problem 1   Follow the graphing strategy and analyze the function $f(x) = 2x/(1 - x)$. State all the pertinent information and sketch the graph of f.

EXAMPLE 2   *Using the Graphing Strategy*   Use the graphing strategy to analyze the function

$$g(x) = \frac{2x - 1}{x^2}$$

State all pertinent information and sketch the graph of g.

SOLUTION

Step 1   *Analyze $g(x)$.*

    (A) Domain: All real x, except $x = 0$

    (B) x intercept: $x = \dfrac{1}{2} = 0.5$

          y intercept: Since 0 is not in the domain of g, there is no y intercept.

    (C) Horizontal asymptote: $y = 0$ (the x axis)

          Vertical asymptote: The denominator of $g(x)$ is 0 at $x = 0$ and the numerator is not. So the line $x = 0$ (the y axis) is a vertical asymptote.

Step 2   *Analyze $g'(x)$.*

$$g(x) = \frac{2x - 1}{x^2} = \frac{2}{x} - \frac{1}{x^2} = 2x^{-1} - x^{-2}$$

$$g'(x) = -2x^{-2} + 2x^{-3} = -\frac{2}{x^2} + \frac{2}{x^3} = \frac{-2x + 2}{x^3}$$

$$= \frac{2(1 - x)}{x^3}$$

Partition numbers for $g'(x)$: $x = 0, x = 1$
Critical number of $g(x)$: $x = 1$
Sign chart for $g'(x)$:

	$(-\infty, 0)$	$(0, 1)$	$(1, \infty)$
$g'(x)$	$----$	ND $++++$	0 $----$
		0	1

Function $f(x)$ is decreasing on $(-\infty, 0)$ and $(1, \infty)$, is increasing on $(0, 1)$, and has a local maximum at $x = 1$. The local maximum is $g(1) = 1$.

Step 3   *Analyze $g''(x)$.*

$$g'(x) = -2x^{-2} + 2x^{-3}$$

$$g''(x) = 4x^{-3} - 6x^{-4} = \frac{4}{x^3} - \frac{6}{x^4} = \frac{4x - 6}{x^4} = \frac{2(2x - 3)}{x^4}$$

Partition numbers for $g''(x)$: $x = 0, x = \frac{3}{2} = 1.5$

Sign chart for $g''(x)$:

	$(-\infty, 0)$	$(0, 1.5)$	$(1.5, \infty)$
$g''(x)$	$----$	ND $------$	0 $++++$
		0	1.5

Function $g(x)$ is concave downward on $(-\infty, 0)$ and $(0, 1.5)$, is concave upward on $(1.5, \infty)$, and has an inflection point at $x = 1.5$. Since $g(1.5) = 0.89$, the inflection point is $(1.5, 0.89)$.

Step 4   *Sketch the graph of g.* Plot key points, note that the coordinate axes are asymptotes, and sketch the graph.

x	$g(x)$
-10	-0.21
-1	-3
0.5	0
1	1
1.5	0.89
10	0.19

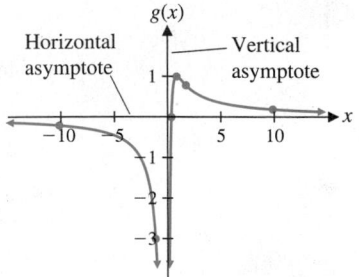

Matched Problem 2   Use the graphing strategy to analyze the function

$$h(x) = \frac{4x + 3}{x^2}$$

State all pertinent information and sketch the graph of h.

EXAMPLE 3   Graphing Strategy   Follow the steps of the graphing strategy and analyze the function $f(x) = xe^x$. State all the pertinent information and sketch the graph of f.

SOLUTION

Step 1   *Analyze* $f(x)$: $f(x) = xe^x$.

(A) Domain: All real numbers

(B) y intercept: $f(0) = 0$

   x intercept: $xe^x = 0$ for $x = 0$ only, since $e^x > 0$ for all x.

(C) Vertical asymptotes: None

   Horizontal asymptotes: We use tables to determine the nature of the graph of f as $x \to \infty$ and $x \to -\infty$:

x	1	5	10	$\to \infty$
$f(x)$	2.72	742.07	220,264.66	$\to \infty$

x	-1	-5	-10	$\to -\infty$
$f(x)$	-0.37	-0.03	$-0.000\,45$	$\to 0$

Step 2   *Analyze* $f'(x)$:

$$f'(x) = x\frac{d}{dx}e^x + e^x\frac{d}{dx}x$$

$$= xe^x + e^x = e^x(x + 1)$$

Partition number for $f'(x)$: -1
Critical number of $f(x)$: -1
Sign chart for $f'(x)$:

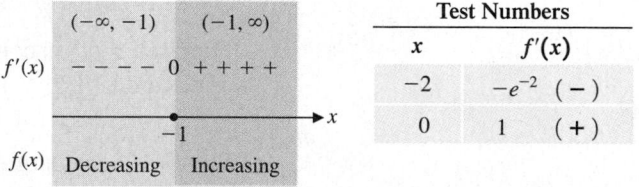

Test Numbers	
x	$f'(x)$
-2	$-e^{-2}$ $(-)$
0	1 $(+)$

So $f(x)$ decreases on $(-\infty, -1)$, has a local minimum at $x = -1$, and increases on $(-1, \infty)$. The local minimum is $f(-1) = -0.37$.

Step 3   *Analyze* $f''(x)$:

$$f''(x) = e^x\frac{d}{dx}(x + 1) + (x + 1)\frac{d}{dx}e^x$$

$$= e^x + (x + 1)e^x = e^x(x + 2)$$

Sign chart for $f''(x)$ (partition number is -2):

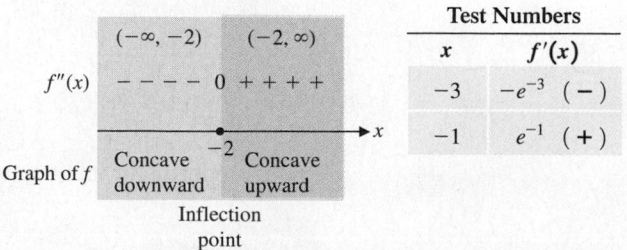

Test Numbers	
x	$f'(x)$
-3	$-e^{-3}$ $(-)$
-1	e^{-1} $(+)$

The graph of f is concave downward on $(-\infty, -2)$, has an inflection point at $x = -2$, and is concave upward on $(-2, \infty)$. Since $f(-2) = -0.27$, the inflection point is $(-2, -0.27)$.

Step 4 *Sketch the graph of f, using the information from steps 1 to 3:*

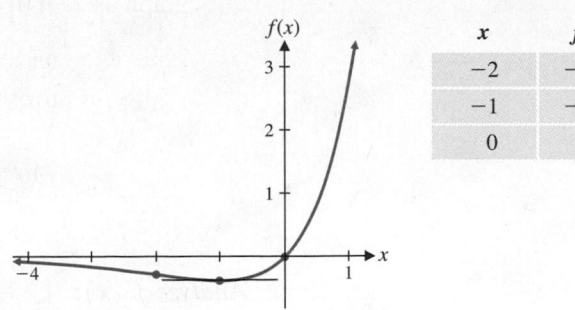

x	f(x)
-2	-0.27
-1	-0.37
0	0

Matched Problem 3 Analyze the function $f(x) = xe^{-0.5x}$. State all the pertinent information and sketch the graph of f.

Explore and Discuss 1 Refer to the discussion of asymptotes in the solution of Example 3. We used tables of values to estimate limits at infinity and determine horizontal asymptotes. In some cases, the functions involved in these limits can be written in a form that allows us to use L'Hôpital's rule.

$$\lim_{x \to -\infty} f(x) = \lim_{x \to -\infty} xe^x \qquad \text{Rewrite as a fraction.}$$

$\qquad\qquad\qquad -\infty \cdot 0$ form

$$= \lim_{x \to -\infty} \frac{x}{e^{-x}} \qquad \text{Apply L'Hôpital's rule.}$$

$\qquad\qquad\qquad -\infty / \infty$ form

$$= \lim_{x \to -\infty} \frac{1}{-e^{-x}} \qquad \text{Simplify.}$$

$$= \lim_{x \to -\infty} (-e^x) \qquad \text{Property of } e^x$$

$$= 0$$

Use algebraic manipulation and L'Hôpital's rule to verify the value of each of the following limits:

(A) $\displaystyle\lim_{x \to \infty} xe^{-0.5x} = 0$

(B) $\displaystyle\lim_{x \to 0^+} x^2(\ln x - 0.5) = 0$

(C) $\displaystyle\lim_{x \to 0^+} x \ln x = 0$

EXAMPLE 4 Graphing Strategy Let $f(x) = x^2 \ln x - 0.5x^2$. Follow the steps in the graphing strategy and analyze this function. State all the pertinent information and sketch the graph of f.

SOLUTION

Step 1 *Analyze f(x):* $f(x) = x^2 \ln x - 0.5x^2 = x^2(\ln x - 0.5)$.

(A) Domain: $(0, \infty)$

(B) y intercept: None [$f(0)$ is not defined.]

  x intercept: Solve $x^2(\ln x - 0.5) = 0$

$\quad \ln x - 0.5 = 0 \qquad$ or $\qquad x^2 = 0 \qquad$ Discard, since 0 is not in the domain of f.

$\qquad\quad \ln x = 0.5 \qquad$ ln $x = a$ if and only if $x = e^a$.

$\qquad\qquad x = e^{0.5} \qquad x$ intercept

(C) Asymptotes: None. The following tables suggest the nature of the graph as $x \to 0^+$ and as $x \to \infty$:

x	0.1	0.01	0.001	$\to 0^+$
$f(x)$	−0.0280	−0.00051	−0.000007	$\to 0$

See Explore & Discuss 1(B).

x	10	100	1,000	$\to \infty$
$f(x)$	180	41,000	6,400,000	$\to \infty$

Step 2   *Analyze $f'(x)$*:

$$f'(x) = x^2 \frac{d}{dx} \ln x + (\ln x) \frac{d}{dx} x^2 - 0.5 \frac{d}{dx} x^2$$

$$= x^2 \frac{1}{x} + (\ln x) 2x - 0.5(2x)$$

$$= x + 2x \ln x - x$$

$$= 2x \ln x$$

Partition number for $f'(x)$: 1
Critical number of $f(x)$: 1
Sign chart for $f'(x)$:

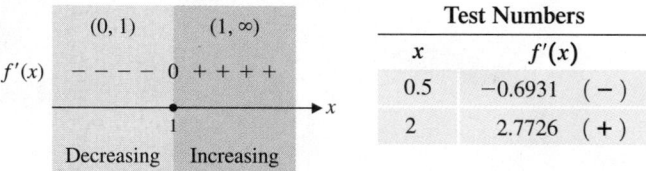

Test Numbers	
x	$f'(x)$
0.5	−0.6931 (−)
2	2.7726 (+)

The function $f(x)$ decreases on $(0, 1)$, has a local minimum at $x = 1$, and increases on $(1, \infty)$. The local minimum is $f(1) = -0.5$.

Step 3   *Analyze $f''(x)$*:

$$f''(x) = 2x \frac{d}{dx} (\ln x) + (\ln x) \frac{d}{dx} (2x)$$

$$= 2x \frac{1}{x} + (\ln x) 2$$

$$= 2 + 2 \ln x = 0$$

$$2 \ln x = -2$$

$$\ln x = -1$$

$$x = e^{-1} \approx 0.3679$$

Sign chart for $f''(x)$ (partition number is e^{-1}):

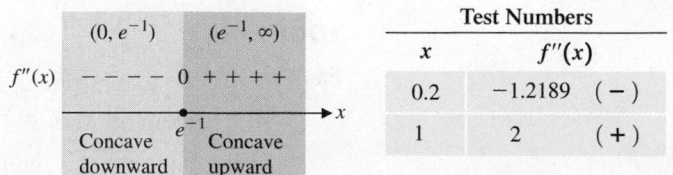

Test Numbers	
x	$f''(x)$
0.2	−1.2189 (−)
1	2 (+)

The graph of $f(x)$ is concave downward on $(0, e^{-1})$, has an inflection point at $x = e^{-1}$, and is concave upward on (e^{-1}, ∞). Since $f(e^{-1}) = -1.5e^{-2} \approx -0.20$, the inflection point is $(0.37, -0.20)$.

Step 4 *Sketch the graph of f, using the information from steps 1 to 3:*

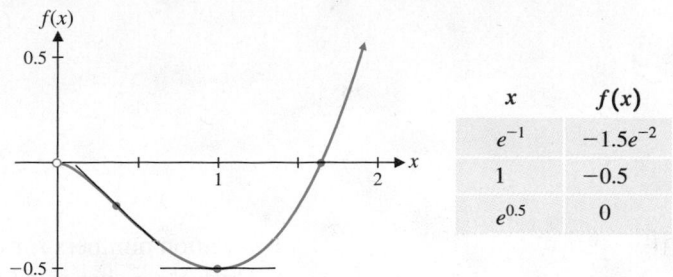

x	$f(x)$
e^{-1}	$-1.5e^{-2}$
1	-0.5
$e^{0.5}$	0

Matched Problem 4   Analyze the function $f(x) = x \ln x$. State all pertinent information and sketch the graph of f.

Modeling Average Cost

EXAMPLE 5 Average Cost Given the cost function $C(x) = 5,000 + 0.5x^2$, where x is the number of items produced, use the graphing strategy to analyze the graph of the average cost function. State all the pertinent information and sketch the graph of the average cost function. Find the marginal cost function and graph it on the same set of coordinate axes.

SOLUTION The average cost function is

$$\overline{C}(x) = \frac{5,000 + 0.5x^2}{x} = \frac{5,000}{x} + 0.5x$$

Step 1 *Analyze $\overline{C}(x)$.*

(A) Domain: Since negative values of x do not make sense and $\overline{C}(0)$ is not defined, the domain is the set of positive real numbers.

(B) Intercepts: None

(C) Horizontal asymptote: $\dfrac{a_m x^m}{b_n x^n} = \dfrac{0.5x^2}{x} = 0.5x$

So there is no horizontal asymptote.

Vertical asymptote: The line $x = 0$ is a vertical asymptote since the denominator is 0 and the numerator is not 0 for $x = 0$.

Oblique asymptotes: If a graph approaches a line that is neither horizontal nor vertical as x approaches ∞ or $-\infty$, then that line is called an **oblique asymptote**. If x is a large positive number, then $5,000/x$ is very small and

$$\overline{C}(x) = \frac{5,000}{x} + 0.5x \approx 0.5x$$

That is,

$$\lim_{x \to \infty} \left[\overline{C}(x) - 0.5x \right] = \lim_{x \to \infty} \frac{5,000}{x} = 0$$

This implies that the graph of $y = \overline{C}(x)$ approaches the line $y = 0.5x$ as x approaches ∞. That line is an oblique asymptote for the graph of $y = \overline{C}(x)$.*

*If $f(x) = n(x)/d(x)$ is a rational function for which the degree of $n(x)$ is 1 more than the degree of $d(x)$, then we can use polynomial long division to write $f(x) = mx + b + r(x)/d(x)$, where the degree of $r(x)$ is less than the degree of $d(x)$. The line $y = mx + b$ is then an oblique asymptote for the graph of $y = f(x)$.

Step 2 *Analyze $\overline{C}'(x)$.*

$$\overline{C}'(x) = -\frac{5,000}{x^2} + 0.5$$

$$= \frac{0.5x^2 - 5,000}{x^2}$$

$$= \frac{0.5(x - 100)(x + 100)}{x^2}$$

Partition numbers for $\overline{C}'(x)$:   0 and 100
Critical number of $\overline{C}(x)$:   100
Sign chart for $\overline{C}'(x)$:

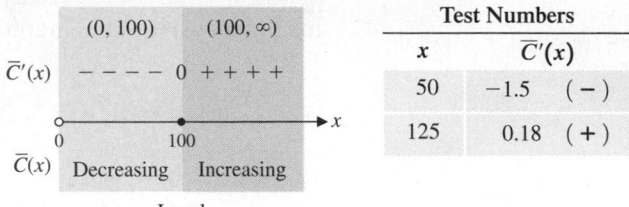

	Test Numbers	
x	$\overline{C}'(x)$	
50	-1.5	$(-)$
125	0.18	$(+)$

So $\overline{C}(x)$ is decreasing on $(0, 100)$, is increasing on $(100, \infty)$, and has a local minimum at $x = 100$. The local minimum is $\overline{C}(100) = 100$.

Step 3 *Analyze $\overline{C}''(x)$:* $\overline{C}''(x) = \dfrac{10,000}{x^3}$.

$\overline{C}''(x)$ is positive for all positive x, so the graph of $y = \overline{C}(x)$ is concave upward on $(0, \infty)$.

Step 4 *Sketch the graph of $\overline{C}$.* The graph of $\overline{C}$ is shown in Figure 1.

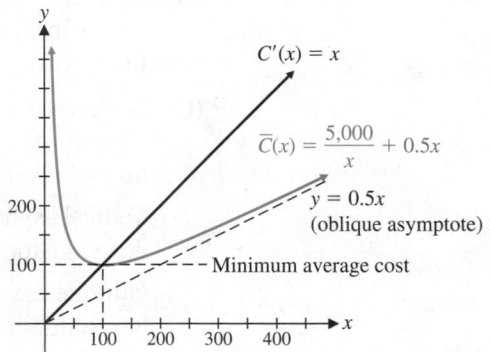

Figure 1

The marginal cost function is $C'(x) = x$. The graph of this linear function is also shown in Figure 1.

Figure 1 illustrates an important principle in economics:

The minimum average cost occurs when the average cost is equal to the marginal cost.

Matched Problem 5] Given the cost function $C(x) = 1,600 + 0.25x^2$, where x is the number of items produced,

(A) Use the graphing strategy to analyze the graph of the average cost function. State all the pertinent information and sketch the graph of the average cost function. Find the marginal cost function and graph it on the same set of coordinate axes. Include any oblique asymptotes.

(B) Find the minimum average cost.

Exercises 4.4

Skills Warm-up Exercises

In Problems 1–8, find the domain of the function and all x or y intercepts. (If necessary, review Section 1.1).

1. $f(x) = 3x + 36$

2. $f(x) = -4x - 28$

3. $f(x) = \sqrt{25 - x}$

4. $f(x) = \sqrt{9 - x^2}$

5. $f(x) = \dfrac{x + 1}{x - 2}$

6. $f(x) = \dfrac{x^2 - 4}{x + 3}$

7. $f(x) = \dfrac{3}{x^2 - 1}$

8. $f(x) = \dfrac{x}{x^2 + 5x + 4}$

9. Use the graph of f in the figure to identify the following (assume that $f''(0) < 0, f''(b) > 0$, and $f''(g) > 0$):

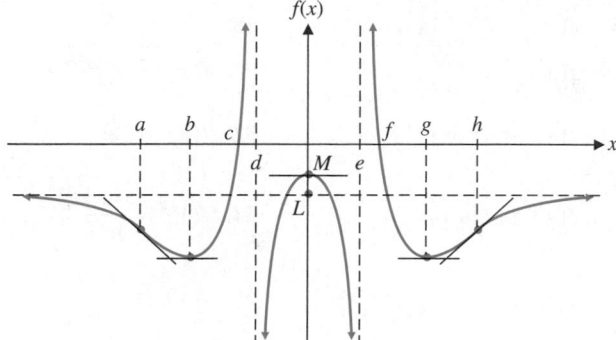

(A) the intervals on which $f'(x) < 0$

(B) the intervals on which $f'(x) > 0$

(C) the intervals on which $f(x)$ is increasing

(D) the intervals on which $f(x)$ is decreasing

(E) the x coordinate(s) of the point(s) where $f(x)$ has a local maximum

(F) the x coordinate(s) of the point(s) where $f(x)$ has a local minimum

(G) the intervals on which $f''(x) < 0$

(H) the intervals on which $f''(x) > 0$

(I) the intervals on which the graph of f is concave upward

(J) the intervals on which the graph of f is concave downward

(K) the x coordinate(s) of the inflection point(s)

(L) the horizontal asymptote(s)

(M) the vertical asymptote(s)

10. Repeat Problem 9 for the following graph of f (assume that $f''(d) < 0$):

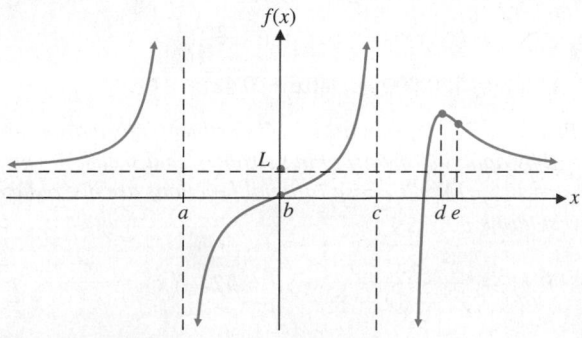

In Problems 11–18, use the given information to sketch the graph of f. Assume that f is continuous on its domain and that all intercepts are included in the table of values.

11. Domain: All real x; $\lim\limits_{x \to \pm\infty} f(x) = 2$

x	−4	−2	0	2	4
$f(x)$	0	−2	0	−2	0

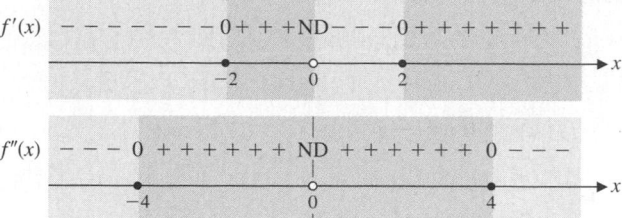

12. Domain: All real x; $\lim\limits_{x \to -\infty} f(x) = -3$; $\lim\limits_{x \to \infty} f(x) = 3$

x	−2	−1	0	1	2
$f(x)$	0	2	0	−2	0

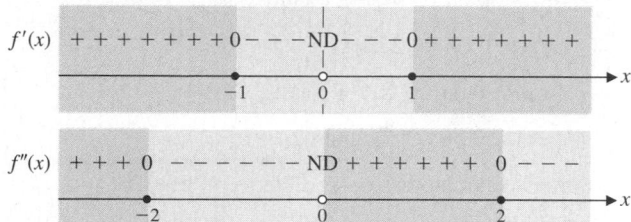

13. Domain: All real x, except $x = -2$;
$\lim\limits_{x \to -2^-} f(x) = \infty$; $\lim\limits_{x \to -2^+} f(x) = -\infty$; $\lim\limits_{x \to \infty} f(x) = 1$

x	−4	0	4	6
$f(x)$	0	0	3	2

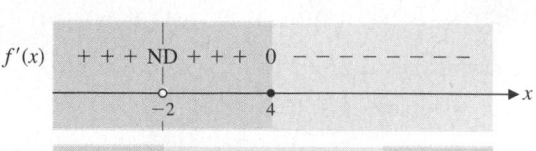

14. Domain: All real x, except $x = 1$;
$\lim\limits_{x \to 1^-} f(x) = \infty$; $\lim\limits_{x \to 1^+} f(x) = \infty$; $\lim\limits_{x \to \infty} f(x) = -2$

x	−4	−2	0	2
$f(x)$	0	−2	0	0

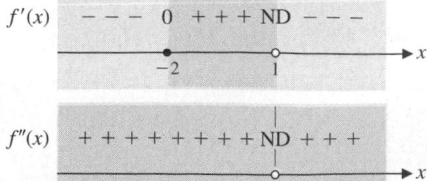

15. Domain: All real x, except $x = -1$;

$f(-3) = 2$, $f(-2) = 3$, $f(0) = -1$, $f(1) = 0$;

$f'(x) > 0$ on $(-\infty, -1)$ and $(-1, \infty)$;

$f''(x) > 0$ on $(-\infty, -1)$; $f''(x) < 0$ on $(-1, \infty)$;

vertical asymptote: $x = -1$;

horizontal asymptote: $y = 1$

16. Domain: All real x, except $x = 1$;

$f(0) = -2$, $f(2) = 0$;

$f'(x) < 0$ on $(-\infty, 1)$ and $(1, \infty)$;

$f''(x) < 0$ on $(-\infty, 1)$;

$f''(x) > 0$ on $(1, \infty)$;

vertical asymptote: $x = 1$;

horizontal asymptote: $y = -1$

17. Domain: All real x, except $x = -2$ and $x = 2$;

$f(-3) = -1$, $f(0) = 0$, $f(3) = 1$;

$f'(x) < 0$ on $(-\infty, -2)$ and $(2, \infty)$;

$f'(x) > 0$ on $(-2, 2)$;

$f''(x) < 0$ on $(-\infty, -2)$ and $(-2, 0)$;

$f''(x) > 0$ on $(0, 2)$ and $(2, \infty)$;

vertical asymptotes: $x = -2$ and $x = 2$;

horizontal asymptote: $y = 0$

18. Domain: All real x, except $x = -1$ and $x = 1$;

$f(-2) = 1$, $f(0) = 0$, $f(2) = 1$;

$f'(x) > 0$ on $(-\infty, -1)$ and $(0, 1)$;

$f'(x) < 0$ on $(-1, 0)$ and $(1, \infty)$;

$f''(x) > 0$ on $(-\infty, -1)$, $(-1, 1)$, and $(1, \infty)$;

vertical asymptotes: $x = -1$ and $x = 1$;

horizontal asymptote: $y = 0$

In Problems 19–58, summarize the pertinent information obtained by applying the graphing strategy and sketch the graph of $y = f(x)$.

19. $f(x) = \dfrac{x + 3}{x - 3}$

20. $f(x) = \dfrac{2x - 4}{x + 2}$

21. $f(x) = \dfrac{x}{x - 2}$

22. $f(x) = \dfrac{2 + x}{3 - x}$

23. $f(x) = 5 + 5e^{-0.1x}$

24. $f(x) = 3 + 7e^{-0.2x}$

25. $f(x) = 5xe^{-0.2x}$

26. $f(x) = 10xe^{-0.1x}$

27. $f(x) = \ln(1 - x)$

28. $f(x) = \ln(2x + 4)$

29. $f(x) = x - \ln x$

30. $f(x) = \ln(x^2 + 4)$

31. $f(x) = \dfrac{x}{x^2 - 4}$

32. $f(x) = \dfrac{1}{x^2 - 4}$

33. $f(x) = \dfrac{1}{1 + x^2}$

34. $f(x) = \dfrac{x^2}{1 + x^2}$

35. $f(x) = \dfrac{2x}{1 - x^2}$

36. $f(x) = \dfrac{2x}{x^2 - 9}$

37. $f(x) = \dfrac{-5x}{(x - 1)^2}$

38. $f(x) = \dfrac{x}{(x - 2)^2}$

39. $f(x) = \dfrac{x^2 + x - 2}{x^2}$

40. $f(x) = \dfrac{x^2 - 5x - 6}{x^2}$

41. $f(x) = \dfrac{x^2}{x - 1}$

42. $f(x) = \dfrac{x^2}{2 + x}$

43. $f(x) = \dfrac{3x^2 + 2}{x^2 - 9}$

44. $f(x) = \dfrac{2x^2 + 5}{4 - x^2}$

45. $f(x) = \dfrac{x^3}{x - 2}$

46. $f(x) = \dfrac{x^3}{4 - x}$

47. $f(x) = (3 - x)e^x$

48. $f(x) = (x - 2)e^x$

49. $f(x) = e^{-0.5x^2}$

50. $f(x) = e^{-2x^2}$

51. $f(x) = x^2 \ln x$

52. $f(x) = \dfrac{\ln x}{x}$

53. $f(x) = (\ln x)^2$

54. $f(x) = \dfrac{x}{\ln x}$

55. $f(x) = \dfrac{1}{x^2 + 2x - 8}$

56. $f(x) = \dfrac{1}{3 - 2x - x^2}$

57. $f(x) = \dfrac{x^3}{3 - x^2}$

58. $f(x) = \dfrac{x^3}{x^2 - 12}$

In Problems 59–66, show that the line $y = x$ is an oblique asymptote for the graph of $y = f(x)$, summarize all pertinent information obtained by applying the graphing strategy, and sketch the graph of $y = f(x)$.

59. $f(x) = x + \dfrac{4}{x}$

60. $f(x) = x - \dfrac{9}{x}$

61. $f(x) = x - \dfrac{4}{x^2}$

62. $f(x) = x + \dfrac{32}{x^2}$

63. $f(x) = x - \dfrac{9}{x^3}$

64. $f(x) = x + \dfrac{27}{x^3}$

65. $f(x) = x + \dfrac{1}{x} + \dfrac{4}{x^3}$

66. $f(x) = x - \dfrac{16}{x^3}$

In Problems 67–70, for the given cost function $C(x)$, find the oblique asymptote of the average cost function $\overline{C}(x)$.

67. $C(x) = 10{,}000 + 90x + 0.02x^2$

68. $C(x) = 7{,}500 + 65x + 0.01x^2$

69. $C(x) = 95{,}000 + 210x + 0.1x^2$

70. $C(x) = 120{,}000 + 340x + 0.4x^2$

In Problems 71–78, summarize all pertinent information obtained by applying the graphing strategy, and sketch the graph of $y = f(x)$. [Note: These rational functions are not reduced to lowest terms.]

71. $f(x) = \dfrac{x^2 + x - 6}{x^2 - 6x + 8}$

72. $f(x) = \dfrac{x^2 + x - 6}{x^2 - x - 12}$

73. $f(x) = \dfrac{2x^2 + x - 15}{x^2 - 9}$ 74. $f(x) = \dfrac{2x^2 + 11x + 14}{x^2 - 4}$

75. $f(x) = \dfrac{x^3 - 5x^2 + 6x}{x^2 - x - 2}$ 76. $f(x) = \dfrac{x^3 - 5x^2 - 6x}{x^2 + 3x + 2}$

77. $f(x) = \dfrac{x^2 + x - 2}{x^2 - 2x + 1}$ 78. $f(x) = \dfrac{x^2 + x - 2}{x^2 + 4x + 4}$

Applications

79. **Revenue.** The marketing research department for a computer company used a large city to test market the firm's new laptop. The department found that the relationship between price p (dollars per unit) and demand x (units sold per week) was given approximately by

$$p = 1,296 - 0.12x^2 \qquad 0 \le x \le 80$$

So, weekly revenue can be approximated by

$$R(x) = xp = 1,296x - 0.12x^3 \qquad 0 \le x \le 80$$

Graph the revenue function R.

80. **Profit.** Suppose that the cost function $C(x)$ (in dollars) for the company in Problem 79 is

$$C(x) = 830 + 396x$$

(A) Write an equation for the profit $P(x)$.

(B) Graph the profit function P.

81. **Pollution.** In Silicon Valley, a number of computer firms were found to be contaminating underground water supplies with toxic chemicals stored in leaking underground containers. A water quality control agency ordered the companies to take immediate corrective action and contribute to a monetary pool for the testing and cleanup of the underground contamination. Suppose that the required monetary pool (in millions of dollars) is given by

$$P(x) = \dfrac{2x}{1 - x} \qquad 0 \le x < 1$$

where x is the percentage (expressed as a decimal fraction) of the total contaminant removed.

(A) Where is $P(x)$ increasing? Decreasing?

(B) Where is the graph of P concave upward? Downward?

(C) Find any horizontal and vertical asymptotes.

(D) Find the x and y intercepts.

(E) Sketch a graph of P.

82. **Employee training.** A company producing dive watches has established that, on average, a new employee can assemble $N(t)$ dive watches per day after t days of on-the-job training, as given by

$$N(t) = \dfrac{100t}{t + 9} \qquad t \ge 0$$

(A) Where is $N(t)$ increasing? Decreasing?

(B) Where is the graph of N concave upward? Downward?

(C) Find any horizontal and vertical asymptotes.

(D) Find the intercepts.

(E) Sketch a graph of N.

83. **Replacement time.** An outboard motor has an initial price of $3,200. A service contract costs $300 for the first year and increases $100 per year thereafter. The total cost of the outboard motor (in dollars) after n years is given by

$$C(n) = 3,200 + 250n + 50n^2 \qquad n \ge 1$$

(A) Write an expression for the average cost per year, $\overline{C}(n)$, for n years.

(B) Graph the average cost function found in part (A).

(C) When is the average cost per year at its minimum? (This time is frequently referred to as the **replacement time** for this piece of equipment.)

84. **Construction costs.** The management of a manufacturing plant wishes to add a fenced-in rectangular storage yard of 20,000 square feet, using a building as one side of the yard (see the figure). If x is the distance (in feet) from the building to the fence, show that the length of the fence required for the yard is given by

$$L(x) = 2x + \dfrac{20,000}{x} \qquad x > 0$$

Storage yard

x

(A) Graph L.

(B) What are the dimensions of the rectangle requiring the least amount of fencing?

85. **Average and marginal costs.** The total daily cost (in dollars) of producing x mountain bikes is given by

$$C(x) = 1,000 + 5x + 0.1x^2$$

(A) Sketch the graphs of the average cost function and the marginal cost function on the same set of coordinate axes. Include any oblique asymptotes.

(B) Find the minimum average cost.

86. **Average and marginal costs.** The total daily cost (in dollars) of producing x city bikes is given by

$$C(x) = 500 + 2x + 0.2x^2$$

(A) Sketch the graphs of the average cost function and the marginal cost function on the same set of coordinate axes. Include any oblique asymptotes.

(B) Find the minimum average cost.

87. Minimizing average costs. The table gives the total daily costs y (in dollars) of producing x pepperoni pizzas at various production levels.

Number of Pizzas	Total Costs
x	y
50	395
100	475
150	640
200	910
250	1,140
300	1,450

(A) Enter the data into a graphing calculator and find a quadratic regression equation for the total costs.

(B) Use the regression equation from part (A) to find the minimum average cost (to the nearest cent) and the corresponding production level (to the nearest integer).

88. Minimizing average costs. The table gives the total daily costs y (in dollars) of producing x deluxe pizzas at various production levels.

Number of Pizzas	Total Costs
x	y
50	595
100	755
150	1,110
200	1,380
250	1,875
300	2,410

(A) Enter the data into a graphing calculator and find a quadratic regression equation for the total costs.

(B) Use the regression equation from part (A) to find the minimum average cost (to the nearest cent) and the corresponding production level (to the nearest integer).

89. Medicine. A drug is injected into the bloodstream of a patient through her right arm. The drug concentration in the bloodstream of the left arm t hours after the injection is given by

$$C(t) = \frac{0.14t}{t^2 + 1}$$

Graph C.

90. Physiology. In a study on the speed of muscle contraction in frogs under various loads, researchers found that the speed of contraction decreases with increasing loads. More precisely, they found that the relationship between speed of contraction, S (in centimeters per second), and load w (in grams) is given approximately by

$$S(w) = \frac{26 + 0.06w}{w} \qquad w \geq 5$$

Graph S.

91. Psychology: retention. Each student in a psychology class is given one day to memorize the same list of 30 special characters. The lists are turned in at the end of the day, and for each succeeding day for 30 days, each student is asked to turn in

a list of as many of the symbols as can be recalled. Averages are taken, and it is found that

$$N(t) = \frac{5t + 20}{t} \qquad t \geq 1$$

provides a good approximation of the average number N(t) of symbols retained after t days. Graph N.

Answers to Matched Problems

1. Domain: All real x, except $x = 1$

y intercept: $f(0) = 0$; x intercept: 0

Horizontal asymptote: $y = -2$;

Vertical asymptote: $x = 1$

Increasing on $(-\infty, 1)$ and $(1, \infty)$

Concave upward on $(-\infty, 1)$;

Concave downward on $(1, \infty)$

x	f(x)
-1	-1
0	0
$\frac{1}{2}$	2
$\frac{3}{2}$	-6
2	-4
5	$-\frac{5}{2}$

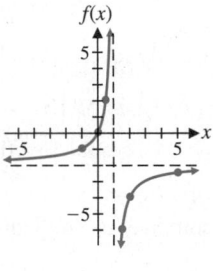

2. Domain: All real x, except $x = 0$

x intercept: $= -\frac{3}{4} = -0.75$

$h(0)$ is not defined

Vertical asymptote: $x = 0$ (the y axis)

Horizontal asymptote: $y = 0$ (the x axis)

Increasing on $(-1.5, 0)$

Decreasing on $(-\infty, -1.5)$ and $(0, \infty)$

Local minimum: $f(-1.5) = -1.33$

Concave upward on $(-2.25, 0)$ and $(0, \infty)$

Concave downward on $(-\infty, -2.25)$

Inflection point: $(-2.25, -1.19)$

x	h(x)
-10	-0.37
-2.25	-1.19
-1.5	-1.33
-0.75	0
2	2.75
10	0.43

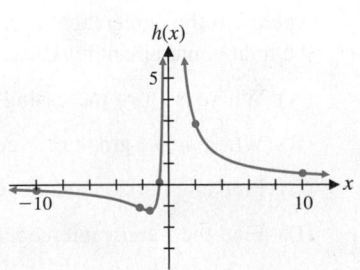

3. Domain: $(-\infty, \infty)$

y intercept: $f(0) = 0$

x intercept: $x = 0$

Horizontal asymptote: $y = 0$ (the x axis)

Increasing on $(-\infty, 2)$

Decreasing on $(2, \infty)$

Local maximum: $f(2) = 2e^{-1} \approx 0.736$

Concave downward on $(-\infty, 4)$

Concave upward on $(4, \infty)$

Inflection point: $(4, 0.541)$

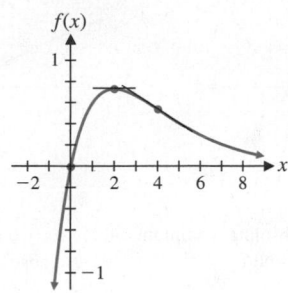

4. Domain: $(0, \infty)$

y intercept: None [$f(0)$ is not defined]

x intercept: $x = 1$

Increasing on (e^{-1}, ∞)

Decreasing on $(0, e^{-1})$

Local minimum: $f(e^{-1}) = -e^{-1} \approx -0.368$

Concave upward on $(0, \infty)$

x	5	10	100	$\rightarrow \infty$
$f(x)$	8.05	23.03	460.52	$\rightarrow \infty$

x	0.1	0.01	0.001	0.000 1	$\rightarrow 0$
$f(x)$	-0.23	-0.046	$-0.006\ 9$	$-0.000\ 92$	$\rightarrow 0$

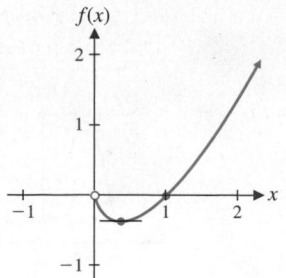

5. (A) Domain: $(0, \infty)$

Intercepts: None

Vertical asymptote: $x = 0$; oblique asymptote: $y = 0.25x$

Decreasing on $(0, 80)$; increasing on $(80, \infty)$;
local minimum at $x = 80$

Concave upward on $(0, \infty)$

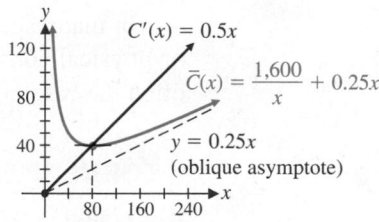

(B) Minimum average cost is 40 at $x = 80$.

4.5 Absolute Maxima and Minima

- Absolute Maxima and Minima
- Second Derivative and Extrema

One of the most important applications of the derivative is to find the absolute maximum or minimum value of a function. An economist may be interested in the price or production level of a commodity that will bring a maximum profit; a doctor may be interested in the time it takes for a drug to reach its maximum concentration in the bloodstream after an injection; and a city planner might be interested in the location of heavy industry in a city in order to produce minimum pollution in residential and business areas. In this section, we develop the procedures needed to find the absolute maximum and absolute minimum values of a function.

Absolute Maxima and Minima

Recall that $f(c)$ is a local maximum if $f(x) \leq f(c)$ for x near c and a local minimum if $f(x) \geq f(c)$ for x near c. Now we are interested in finding the largest and the smallest values of $f(x)$ throughout the domain of f.

> **DEFINITION** Absolute Maxima and Minima
>
> If $f(c) \geq f(x)$ for all x in the domain of f, then $f(c)$ is called the **absolute maximum** of f. If $f(c) \leq f(x)$ for all x in the domain of f, then $f(c)$ is called the **absolute minimum** of f. An absolute maximum or absolute minimum is called an **absolute extremum**.

Figure 1 illustrates some typical examples.

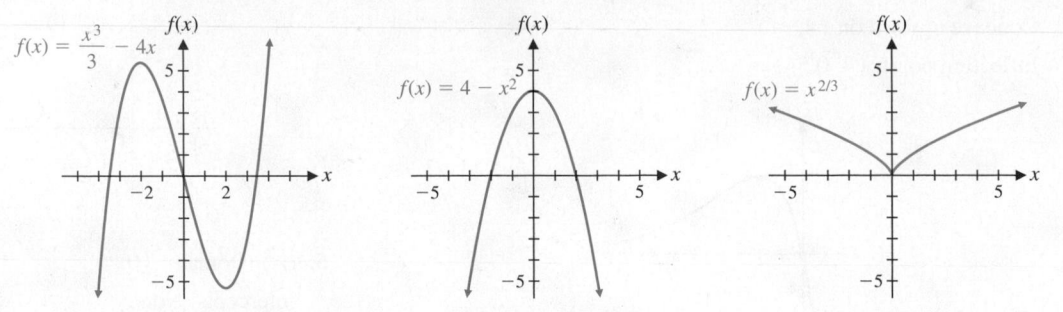

(A) No absolute maximum or minimum

$f(-2) = \frac{16}{3}$ is a local maximum

$f(2) = -\frac{16}{3}$ is a local minimum

(B) $f(0) = 4$ is the absolute maximum
No absolute minimum

(C) $f(0) = 0$ is the absolute minimum
No absolute maximum

Figure 1

In many applications, the domain of a function is restricted because of practical or physical considerations. If the domain is restricted to some closed interval, as is often the case, then Theorem 1 applies.

THEOREM 1   Extreme Value Theorem

A function f that is continuous on a closed interval $[a, b]$ has both an absolute maximum and an absolute minimum on that interval.

It is important to understand that the absolute maximum and absolute minimum depend on both the function f and the interval $[a, b]$. Figure 2 illustrates four cases.

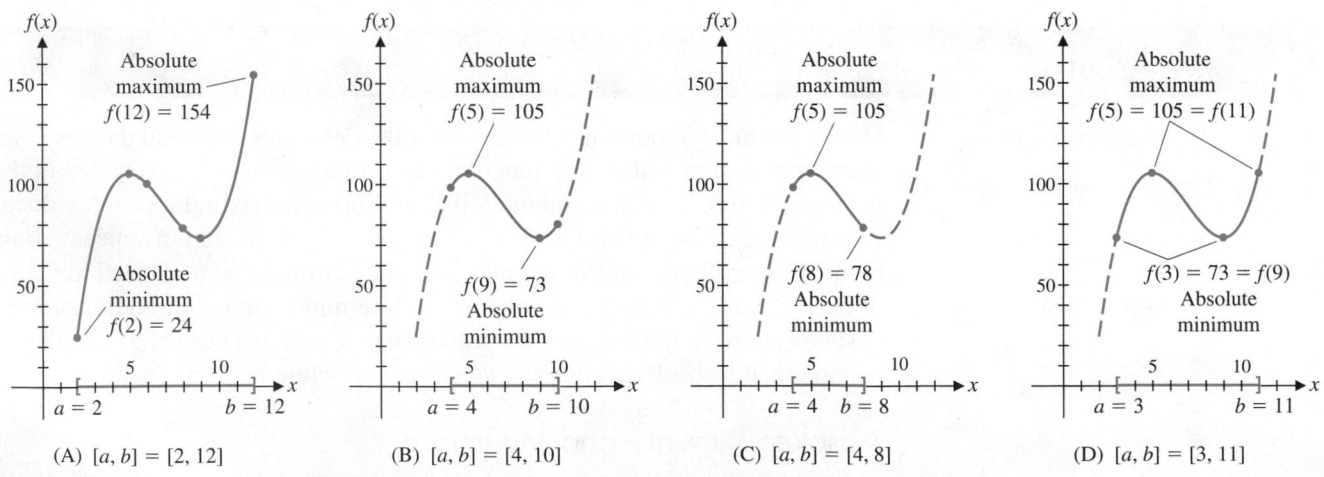

(A) $[a, b] = [2, 12]$

(B) $[a, b] = [4, 10]$

(C) $[a, b] = [4, 8]$

(D) $[a, b] = [3, 11]$

Figure 2   **Absolute extrema for $f(x) = x^3 - 21x^2 + 135x - 170$ on various closed intervals**

In all four cases illustrated in Figure 2, the absolute maximum and absolute minimum occur at a critical number or an endpoint. This property is generalized in Theorem 2. Note that both the absolute maximum and the absolute minimum are unique, but each can occur at more than one point in the interval (Fig. 2D).

THEOREM 2   Locating Absolute Extrema

Absolute extrema (if they exist) must occur at critical numbers or at endpoints.

To find the absolute maximum and minimum of a continuous function on a closed interval, we simply identify the endpoints and critical numbers in the interval, evaluate the function at each, and choose the largest and smallest values.

PROCEDURE Finding Absolute Extrema on a Closed Interval

Step 1 Check to make certain that f is continuous over $[a, b]$.

Step 2 Find the critical numbers in the interval (a, b).

Step 3 Evaluate f at the endpoints a and b and at the critical numbers found in step 2.

Step 4 The absolute maximum of f on $[a, b]$ is the largest value found in step 3.

Step 5 The absolute minimum of f on $[a, b]$ is the smallest value found in step 3.

EXAMPLE 1 Finding Absolute Extrema Find the absolute maximum and absolute minimum of

$$f(x) = x^3 + 3x^2 - 9x - 7$$

on each of the following intervals:

(A) $[-6, 4]$ (B) $[-4, 2]$ (C) $[-2, 2]$

SOLUTION

(A) The function is continuous for all values of x.

$$f'(x) = 3x^2 + 6x - 9 = 3(x - 1)(x + 3)$$

So, $x = -3$ and $x = 1$ are the critical numbers in the interval $(-6, 4)$. Evaluate f at the endpoints and critical numbers $(-6, -3, 1, \text{and } 4)$, and choose the largest and smallest values.

$$f(-6) = -61 \quad \text{Absolute minimum}$$
$$f(-3) = 20$$
$$f(1) = -12$$
$$f(4) = 69 \quad \text{Absolute maximum}$$

The absolute maximum of f on $[-6, 4]$ is 69, and the absolute minimum is -61.

(B) Interval: $[-4, 2]$

x	$f(x)$	
-4	13	
-3	20	Absolute maximum
1	-12	Absolute minimum
2	-5	

The absolute maximum of f on $[-4, 2]$ is 20, and the absolute minimum is -12.

(C) Interval: $[-2, 2]$

x	$f(x)$	
-2	15	Absolute maximum
1	-12	Absolute minimum
2	-5	

Note that the critical number $x = -3$ is not included in the table, because it is not in the interval $[-2, 2]$. The absolute maximum of f on $[-2, 2]$ is 15, and the absolute minimum is -12.

Matched Problem 1 | Find the absolute maximum and absolute minimum of

$$f(x) = x^3 - 12x$$

on each of the following intervals:

(A) $[-5, 5]$ (B) $[-3, 3]$ (C) $[-3, 1]$

Now, suppose that we want to find the absolute maximum or minimum of a function that is continuous on an interval that is not closed. Since Theorem 1 no longer applies, we cannot be certain that the absolute maximum or minimum value exists. Figure 3 illustrates several ways that functions can fail to have absolute extrema.

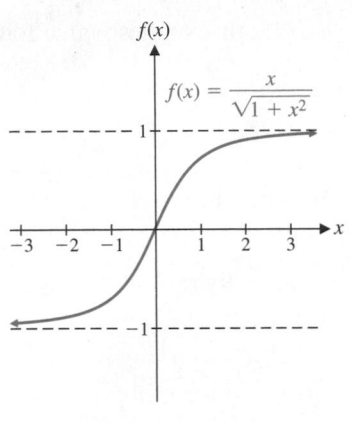

(A) No absolute extrema on $(-\infty, \infty)$:
$-1 < f(x) < 1$ for all x
$[f(x) \neq 1$ or -1 for any $x]$

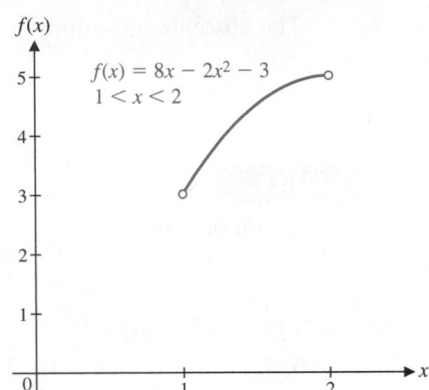

(B) No absolute extrema on $(1, 2)$:
$3 < f(x) < 5$ for $x \in (1, 2)$
$[f(x) \neq 3$ or 5 for any $x \in (1, 2)]$

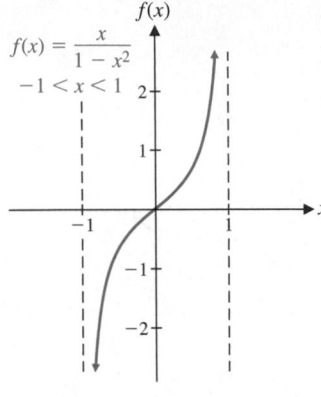

(C) No absolute extrema on $(-1, 1)$:
Graph has vertical asymptotes
at $x = -1$ and $x = 1$

Figure 3 **Functions with no absolute extrema**

In general, the best procedure to follow in searching for absolute extrema on an interval that is not of the form $[a, b]$ is to sketch a graph of the function. However, many applications can be solved with a new tool that does not require any graphing.

Second Derivative and Extrema

The second derivative can be used to classify the local extrema of a function. Suppose that f is a function satisfying $f'(c) = 0$ and $f''(c) > 0$. First, note that if $f''(c) > 0$, then it follows from the properties of limits[*] that $f''(x) > 0$ in some interval (m, n) containing c. Thus, the graph of f must be concave upward in this interval. But this implies that $f'(x)$ is increasing in the interval. Since $f'(c) = 0$, $f'(x)$ must change from negative to positive at $x = c$, and $f(c)$ is a local minimum (see Fig. 4). Reasoning in the same fashion, we conclude that if $f'(c) = 0$ and $f''(c) < 0$, then $f(c)$ is a local maximum. Of course, it is possible that both $f'(c) = 0$ and $f''(c) = 0$. In this case, the second derivative cannot be used to determine the shape of the graph around $x = c$; $f(c)$ may be a local minimum, a local maximum, or neither.

The sign of the second derivative provides a simple test for identifying local maxima and minima. This test is most useful when we do not want to draw the graph of the function. If we are interested in drawing the graph and have already constructed the sign chart for $f'(x)$, then the first-derivative test can be used to identify the local extrema.

---

[*]Actually, we are assuming that $f''(x)$ is continuous in an interval containing c. It is unlikely that we will encounter a function for which $f''(c)$ exists but $f''(x)$ is not continuous in an interval containing c.

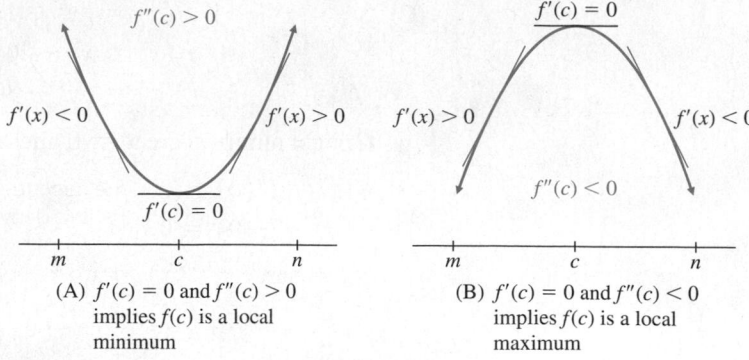

Figure 4 **Second derivative and local extrema**

RESULT Second-Derivative Test for Local Extrema

Let c be a critical number of $f(x)$ such that $f'(c) = 0$. If the second derivative $f''(c) > 0$, then $f(c)$ is a local minimum. If $f''(c) < 0$, then $f(c)$ is a local maximum.

$f'(c)$	$f''(c)$	Graph of f is:	$f(c)$	Example
0	+	Concave upward	Local minimum	$\smile$
0	−	Concave downward	Local maximum	$\frown$
0	0	?	Test does not apply	

EXAMPLE 2 Testing Local Extrema Find the local maxima and minima for each function. Use the second-derivative test for local extrema when it applies.

(A) $f(x) = x^3 - 6x^2 + 9x + 1$

(B) $f(x) = xe^{-0.2x}$

(C) $f(x) = \frac{1}{6}x^6 - 4x^5 + 25x^4$

SOLUTION

(A) Find first and second derivatives and determine critical numbers:

$$f(x) = x^3 - 6x^2 + 9x + 1$$
$$f'(x) = 3x^2 - 12x + 9 = 3(x-1)(x-3)$$
$$f''(x) = 6x - 12 = 6(x-2)$$

Critical numbers are $x = 1$ and $x = 3$.

$$f''(1) = -6 < 0 \quad \text{\scriptsize f has a local maximum at $x = 1$.}$$
$$f''(3) = 6 > 0 \quad \text{\scriptsize f has a local minimum at $x = 3$.}$$

Substituting $x = 1$ in the expression for $f(x)$, we find that $f(1) = 5$ is a local maximum. Similarly, $f(3) = 1$ is a local minimum.

(B)
$$f(x) = xe^{-0.2x}$$
$$f'(x) = e^{-0.2x} + xe^{-0.2x}(-0.2)$$
$$= e^{-0.2x}(1 - 0.2x)$$
$$f''(x) = e^{-0.2x}(-0.2)(1 - 0.2x) + e^{-0.2x}(-0.2)$$
$$= e^{-0.2x}(0.04x - 0.4)$$

Critical number: $x = 1/0.2 = 5$

$$f''(5) = e^{-1}(-0.2) < 0 \quad \text{\scriptsize f has a local maximum at $x = 5$.}$$

So $f(5) = 5e^{-0.2(5)} \approx 1.84$ is a local maximum.

(C)
$$f(x) = \tfrac{1}{6}x^6 - 4x^5 + 25x^4$$
$$f'(x) = x^5 - 20x^4 + 100x^3 = x^3(x - 10)^2$$
$$f''(x) = 5x^4 - 80x^3 + 300x^2$$

Critical numbers are $x = 0$ and $x = 10$.

$$f''(0) = 0 \quad \text{The second-derivative test fails at both critical numbers, so}$$
$$f''(10) = 0 \quad \text{the first-derivative test must be used.}$$

Sign chart for $f'(x) = x^3(x - 10)^2$ (partition numbers for f' are 0 and 10):

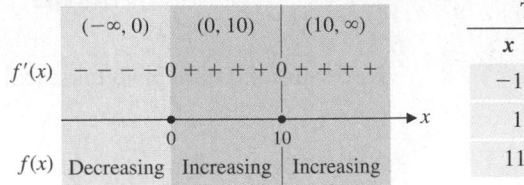

				Test Numbers	
	$(-\infty, 0)$	$(0, 10)$	$(10, \infty)$	x	$f'(x)$
$f'(x)$	$----0++++0+++$			-1	-121 $(-)$
	0	10		1	81 $(+)$
$f(x)$	Decreasing	Increasing	Increasing	11	1,331 $(+)$

From the chart, we see that $f(x)$ has a local minimum at $x = 0$ and does not have a local extremum at $x = 10$. So $f(0) = 0$ is a local minimum.

Matched Problem 2   Find the local maxima and minima for each function. Use the second-derivative test when it applies.

(A) $f(x) = x^3 - 9x^2 + 24x - 10$

(B) $f(x) = e^x - 5x$

(C) $f(x) = 10x^6 - 24x^5 + 15x^4$

CONCEPTUAL INSIGHT

The second-derivative test for local extrema does not apply if $f''(c) = 0$ or if $f''(c)$ is not defined. As Example 2C illustrates, if $f''(c) = 0$, then $f(c)$ may or may not be a local extremum. Some other method, such as the first-derivative test, must be used when $f''(c) = 0$ or $f''(c)$ does not exist.

The solution of many optimization problems involves searching for an absolute extremum. If the function in question has only one critical number, then the second-derivative test for local extrema not only classifies the local extremum but also guarantees that the local extremum is, in fact, the absolute extremum.

THEOREM 3   Second-Derivative Test for Absolute Extrema on an Open Interval

Let f be continuous on an open interval I with only one critical number c in I.

If $f'(c) = 0$ and $f''(c) > 0$, then $f(c)$ is the absolute minimum of f on I.

If $f'(c) = 0$ and $f''(c) < 0$, then $f(c)$ is the absolute maximum of f on I.

Since the second-derivative test for local extrema cannot be applied when $f''(c) = 0$ or $f''(c)$ does not exist, Theorem 3 makes no mention of these cases.

EXAMPLE 3   Finding Absolute Extrema on an Open Interval   Find the absolute extrema of each function on $(0, \infty)$.

(A) $f(x) = x + \dfrac{4}{x}$ 　　　　　　　(B) $f(x) = (\ln x)^2 - 3 \ln x$

SOLUTION

(A) $f(x) = x + \dfrac{4}{x}$

$f'(x) = 1 - \dfrac{4}{x^2} = \dfrac{x^2 - 4}{x^2} = \dfrac{(x-2)(x+2)}{x^2}$ Critical numbers are $x = -2$ and $x = 2$.

$f''(x) = \dfrac{8}{x^3}$

The only critical number in the interval $(0, \infty)$ is $x = 2$. Since $f''(2) = 1 > 0$, $f(2) = 4$ is the absolute minimum of f on $(0, \infty)$.

(B) $f(x) = (\ln x)^2 - 3 \ln x$

$f'(x) = (2 \ln x)\dfrac{1}{x} - \dfrac{3}{x} = \dfrac{2 \ln x - 3}{x}$ Critical number is $x = e^{3/2}$.

$f''(x) = \dfrac{x\dfrac{2}{x} - (2 \ln x - 3)}{x^2} = \dfrac{5 - 2 \ln x}{x^2}$

The only critical number in the interval $(0, \infty)$ is $x = e^{3/2}$. Since $f''(e^{3/2}) = 2/e^3 > 0, f(e^{3/2}) = -2.25$ is the absolute minimum of f on $(0, \infty)$.

Matched Problem 3 Find the absolute extrema of each function on $(0, \infty)$.

(A) $f(x) = 12 - x - \dfrac{5}{x}$ (B) $f(x) = 5 \ln x - x$

Exercises 4.5

Skills Warm-up Exercises

W *In Problems 1–8, by inspecting the graph of the function, find the absolute maximum and absolute minimum on the given interval. (If necessary, review Section 1.2).*

1. $f(x) = x$ on $[-2, 3]$ **2.** $g(x) = |x|$ on $[-1, 4]$

3. $h(x) = x^2$ on $[-5, 3]$ **4.** $m(x) = x^3$ on $[-3, 1]$

5. $n(x) = \sqrt{x}$ on $[3, 4]$ **6.** $p(x) = \sqrt[3]{x}$ on $[-125, 216]$

7. $q(x) = -\sqrt[3]{x}$ on $[27, 64]$ **8.** $r(x) = -x^2$ on $[-10, 11]$

Problems 9–18 refer to the graph of $y = f(x)$ shown here. Find the absolute minimum and the absolute maximum over the indicated interval.

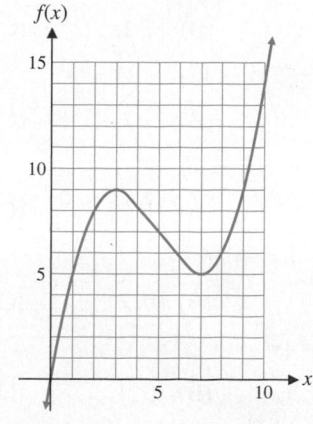

9. $[0, 10]$ **10.** $[2, 8]$ **11.** $[0, 8]$ **12.** $[2, 10]$

13. $[1, 10]$ **14.** $[0, 9]$ **15.** $[1, 9]$ **16.** $[0, 2]$

17. $[2, 5]$ **18.** $[5, 8]$

In Problems 19–22, find the absolute maximum and absolute minimum of each function on the indicated intervals.

19. $f(x) = 2x - 5$

   (A) $[0, 4]$ (B) $[0, 10]$ (C) $[-5, 10]$

20. $f(x) = 8 - x$

   (A) $[0, 1]$ (B) $[-1, 1]$ (C) $[-1, 6]$

21. $f(x) = x^2$

   (A) $[-1, 1]$ (B) $[1, 5]$ (C) $[-5, 5]$

22. $f(x) = 100 - x^2$

   (A) $[-10, 10]$ (B) $[0, 10]$ (C) $[10, 11]$

In Problems 23–26, find the absolute maximum and absolute minimum of each function on the given interval.

23. $f(x) = e^{-x}$ on $[-1, 1]$

24. $f(x) = \ln x$ on $[1, 2]$

25. $f(x) = 9 - x^2$ on $[-4, 4]$

26. $f(x) = x^2 - 6x + 7$ on $[0, 10]$

In Problems 27–42, find the absolute maximum and minimum, if either exists, for each function.

27. $f(x) = x^2 - 2x + 3$

28. $f(x) = x^2 + 4x - 3$

29. $f(x) = -x^2 - 6x + 9$

30. $f(x) = -x^2 + 2x + 4$

31. $f(x) = x^3 + x$

32. $f(x) = -x^3 - 2x$

33. $f(x) = 8x^3 - 2x^4$

34. $f(x) = x^4 - 4x^3$

35. $f(x) = x + \dfrac{16}{x}$

36. $f(x) = x + \dfrac{25}{x}$

37. $f(x) = \dfrac{x^2}{x^2 + 1}$

38. $f(x) = \dfrac{1}{x^2 + 1}$

39. $f(x) = \dfrac{2x}{x^2 + 1}$

40. $f(x) = \dfrac{-8x}{x^2 + 4}$

41. $f(x) = \dfrac{x^2 - 1}{x^2 + 1}$

42. $f(x) = \dfrac{9 - x^2}{x^2 + 4}$

In Problems 43–66, find the indicated extremum of each function on the given interval.

43. Absolute minimum value on $[0, \infty)$ for
$$f(x) = 2x^2 - 8x + 6$$

44. Absolute maximum value on $[0, \infty)$ for
$$f(x) = 6x - x^2 + 4$$

45. Absolute maximum value on $[0, \infty)$ for
$$f(x) = 3x^2 - x^3$$

46. Absolute minimum value on $[0, \infty)$ for
$$f(x) = x^3 - 6x^2$$

47. Absolute minimum value on $[0, \infty)$ for
$$f(x) = (x + 4)(x - 2)^2$$

48. Absolute minimum value on $[0, \infty)$ for
$$f(x) = (2 - x)(x + 1)^2$$

49. Absolute maximum value on $(0, \infty)$ for
$$f(x) = 2x^4 - 8x^3$$

50. Absolute maximum value on $(0, \infty)$ for
$$f(x) = 4x^3 - 8x^4$$

51. Absolute maximum value on $(0, \infty)$ for
$$f(x) = 20 - 3x - \dfrac{12}{x}$$

52. Absolute minimum value on $(0, \infty)$ for
$$f(x) = 4 + x + \dfrac{9}{x}$$

53. Absolute minimum value on $(0, \infty)$ for
$$f(x) = 10 + 2x + \dfrac{64}{x^2}$$

54. Absolute maximum value on $(0, \infty)$ for
$$f(x) = 20 - 4x - \dfrac{250}{x^2}$$

55. Absolute minimum value on $(0, \infty)$ for
$$f(x) = x + \dfrac{1}{x} + \dfrac{30}{x^3}$$

56. Absolute minimum value on $(0, \infty)$ for
$$f(x) = 2x + \dfrac{5}{x} + \dfrac{4}{x^3}$$

57. Absolute minimum value on $(0, \infty)$ for
$$f(x) = \dfrac{e^x}{x^2}$$

58. Absolute maximum value on $(0, \infty)$ for
$$f(x) = \dfrac{x^4}{e^x}$$

59. Absolute maximum value on $(0, \infty)$ for
$$f(x) = \dfrac{x^3}{e^x}$$

60. Absolute minimum value on $(0, \infty)$ for
$$f(x) = \dfrac{e^x}{x}$$

61. Absolute maximum value on $(0, \infty)$ for
$$f(x) = 5x - 2x \ln x$$

62. Absolute minimum value on $(0, \infty)$ for
$$f(x) = 4x \ln x - 7x$$

63. Absolute maximum value on $(0, \infty)$ for
$$f(x) = x^2(3 - \ln x)$$

64. Absolute minimum value on $(0, \infty)$ for
$$f(x) = x^3(\ln x - 2)$$

65. Absolute maximum value on $(0, \infty)$ for
$$f(x) = \ln(xe^{-x})$$

66. Absolute maximum value on $(0, \infty)$ for
$$f(x) = \ln(x^2 e^{-x})$$

In Problems 67–72, find the absolute maximum and minimum, if either exists, for each function on the indicated intervals.

67. $f(x) = x^3 - 6x^2 + 9x - 6$

   (A) $[-1, 5]$ (B) $[-1, 3]$ (C) $[2, 5]$

68. $f(x) = 2x^3 - 3x^2 - 12x + 24$

   (A) $[-3, 4]$ (B) $[-2, 3]$ (C) $[-2, 1]$

69. $f(x) = (x - 1)(x - 5)^3 + 1$

   (A) $[0, 3]$ (B) $[1, 7]$ (C) $[3, 6]$

70. $f(x) = x^4 - 8x^2 + 16$

   (A) $[-1, 3]$ (B) $[0, 2]$ (C) $[-3, 4]$

71. $f(x) = x^4 - 4x^3 + 5$

   (A) $[-1, 2]$ (B) $[0, 4]$ (C) $[-1, 1]$

72. $f(x) = x^4 - 18x^2 + 32$

    (A) $[-4, 4]$ (B) $[-1, 1]$ (C) $[1, 3]$

In Problems 73–80, describe the graph of f at the given point relative to the existence of a local maximum or minimum with one of the following phrases: "Local maximum," "Local minimum," "Neither," or "Unable to determine from the given information." Assume that f(x) is continuous on $(-\infty, \infty)$.

73. $(2, f(2))$ if $f'(2) = 0$ and $f''(2) > 0$

74. $(4, f(4))$ if $f'(4) = 1$ and $f''(4) < 0$

75. $(-3, f(-3))$ if $f'(-3) = 0$ and $f''(-3) = 0$

76. $(-1, f(-1))$ if $f'(-1) = 0$ and $f''(-1) < 0$

77. $(6, f(6))$ if $f'(6) = 1$ and $f''(6)$ does not exist

78. $(5, f(5))$ if $f'(5) = 0$ and $f''(5)$ does not exist

79. $(-2, f(-2))$ if $f'(-2) = 0$ and $f''(-2) < 0$

80. $(1, f(1))$ if $f'(1) = 0$ and $f''(1) > 0$

Answers to Matched Problems

1. (A) Absolute maximum: $f(5) = 65$; absolute minimum: $f(-5) = -65$

    (B) Absolute maximum: $f(-2) = 16$; absolute minimum: $f(2) = -16$

    (C) Absolute maximum: $f(-2) = 16$; absolute minimum: $f(1) = -11$

2. (A) $f(2) = 10$ is a local maximum; $f(4) = 6$ is a local minimum.

    (B) $f(\ln 5) = 5 - 5 \ln 5$ is a local minimum.

    (C) $f(0) = 0$ is a local minimum; there is no local extremum at $x = 1$.

3. (A) $f(\sqrt{5}) = 12 - 2\sqrt{5}$ (B) $f(5) = 5 \ln 5 - 5$

4.6 Optimization

- Area and Perimeter
- Maximizing Revenue and Profit
- Inventory Control

Now we can use calculus to solve **optimization problems**—problems that involve finding the absolute maximum or the absolute minimum of a function. As you work through this section, note that the statement of the problem does not usually include the function to be optimized. Often, it is your responsibility to find the function and then to find the relevant absolute extremum.

Area and Perimeter

The techniques used to solve optimization problems are best illustrated through examples.

EXAMPLE 1 Maximizing Area A homeowner has $320 to spend on building a fence around a rectangular garden. Three sides of the fence will be constructed with wire fencing at a cost of $2 per linear foot. The fourth side will be constructed with wood fencing at a cost of $6 per linear foot. Find the dimensions and the area of the largest garden that can be enclosed with $320 worth of fencing.

SOLUTION To begin, we draw a figure (Fig. 1), introduce variables, and look for relationships among the variables.

Since we don't know the dimensions of the garden, the lengths of fencing are represented by the variables x and y. The costs of the fencing materials are fixed and are represented by constants.

Now we look for relationships among the variables. The area of the garden is

$$A = xy$$

while the cost of the fencing is

$$C = 2y + 2x + 2y + 6x$$
$$= 8x + 4y$$

The problem states that the homeowner has $320 to spend on fencing. We assume that enclosing the largest area will use all the money available for fencing. The problem has now been reduced to

Maximize $\quad A = xy \quad$ subject to $\quad 8x + 4y = 320$

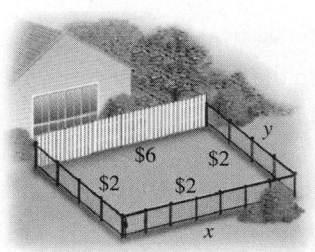

Figure 1

Before we can use calculus to find the maximum area A, we must express A as a function of a single variable. We use the cost equation to eliminate one of the variables in the area expression (we choose to eliminate y—either will work).

$$8x + 4y = 320$$
$$4y = 320 - 8x$$
$$y = 80 - 2x$$
$$A = xy = x(80 - 2x) = 80x - 2x^2$$

Now we consider the permissible values of x. Because x is one of the dimensions of a rectangle, x must satisfy

$$x \geq 0 \quad \text{Length is always nonnegative.}$$

And because $y = 80 - 2x$ is also a dimension of a rectangle, y must satisfy

$$y = 80 - 2x \geq 0 \quad \text{Width is always nonnegative.}$$
$$80 \geq 2x$$
$$40 \geq x \quad \text{or} \quad x \leq 40$$

We summarize the preceding discussion by stating the following model for this optimization problem:

$$\text{Maximize} \quad A(x) = 80x - 2x^2 \quad \text{for } 0 \leq x \leq 40$$

Next, we find any critical numbers of A:

$$A'(x) = 80 - 4x = 0$$
$$80 = 4x$$
$$x = \frac{80}{4} = 20 \quad \text{Critical number}$$

Since $A(x)$ is continuous on $[0, 40]$, the absolute maximum of A, if it exists, must occur at a critical number or an endpoint. Evaluating A at these numbers (Table 1), we see that the maximum area is 800 when

$$x = 20 \quad \text{and} \quad y = 80 - 2(20) = 40$$

Finally, we must answer the questions posed in the problem. The dimensions of the garden with the maximum area of 800 square feet are 20 feet by 40 feet, with one 20-foot side of wood fencing.

Matched Problem 1 ⟩ Repeat Example 1 if the wood fencing costs $8 per linear foot and all other information remains the same.

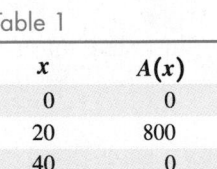

Table 1

x	$A(x)$
0	0
20	800
40	0

We summarize the steps in the solution of Example 1 in the following box:

PROCEDURE Strategy for Solving Optimization Problems

Step 1 Introduce variables, look for relationships among the variables, and construct a mathematical model of the form

$$\text{Maximize (or minimize) } f(x) \text{ on the interval } I$$

Step 2 Find the critical numbers of $f(x)$.

Step 3 Use the procedures developed in Section 4.5 to find the absolute maximum (or minimum) of $f(x)$ on the interval I and the numbers x where this occurs.

Step 4 Use the solution to the mathematical model to answer all the questions asked in the problem.

EXAMPLE 2 Minimizing Perimeter Refer to Example 1. The homeowner judges that an area of 800 square feet for the garden is too small and decides to increase the area to 1,250 square feet. What is the minimum cost of building a fence that will enclose a garden with an area of 1,250 square feet? What are the dimensions of this garden? Assume that the cost of fencing remains unchanged.

SOLUTION Refer to Figure 1 and the solution of Example 1. This time we want to minimize the cost of the fencing that will enclose 1,250 square feet. The problem can be expressed as

$$\text{Minimize} \quad C = 8x + 4y \quad \text{subject to} \quad xy = 1{,}250$$

Since x and y represent distances, we know that $x \geq 0$ and $y \geq 0$. But neither variable can equal 0 because their product must be 1,250.

$$xy = 1{,}250 \qquad \text{Solve the area equation for } y.$$

$$y = \frac{1{,}250}{x}$$

$$C(x) = 8x + 4\frac{1{,}250}{x} \qquad \text{Substitute for } y \text{ in the cost equation.}$$

$$= 8x + \frac{5{,}000}{x} \qquad x > 0$$

The model for this problem is

$$\text{Minimize} \quad C(x) = 8x + \frac{5{,}000}{x} \qquad \text{for } x > 0$$

$$= 8x + 5{,}000x^{-1}$$

$$C'(x) = 8 - 5{,}000x^{-2}$$

$$= 8 - \frac{5{,}000}{x^2} = 0$$

$$8 = \frac{5{,}000}{x^2}$$

$$x^2 = \frac{5{,}000}{8} = 625$$

$$x = \sqrt{625} = 25 \qquad \begin{array}{l}\text{The negative square} \\ \text{root is discarded,} \\ \text{since } x > 0.\end{array}$$

We use the second derivative to determine the behavior at $x = 25$.

$$C'(x) = 8 - 5{,}000x^{-2}$$

$$C''(x) = 0 + 10{,}000x^{-3} = \frac{10{,}000}{x^3}$$

$$C''(25) = \frac{10{,}000}{25^3} = 0.64 > 0$$

The second-derivative test for local extrema shows that $C(x)$ has a local minimum at $x = 25$, and since $x = 25$ is the only critical number of $C(x)$ for $x > 0$, then $C(25)$ must be the absolute minimum for $x > 0$. When $x = 25$, the cost is

$$C(25) = 8(25) + \frac{5{,}000}{25} = 200 + 200 = \$400$$

and

$$y = \frac{1{,}250}{25} = 50$$

The minimum cost for enclosing a 1,250-square-foot garden is $400, and the dimensions are 25 feet by 50 feet, with one 25-foot side of wood fencing.

Matched Problem 2 Repeat Example 2 if the homeowner wants to enclose an 1,800-square-foot garden and all other data remain unchanged.

CONCEPTUAL INSIGHT

The restrictions on the variables in the solutions of Examples 1 and 2 are typical of problems involving areas or perimeters (or the cost of the perimeter):

$$8x + 4y = 320 \quad \text{Cost of fencing (Example 1)}$$
$$xy = 1{,}250 \quad \text{Area of garden (Example 2)}$$

The equation in Example 1 restricts the values of x to

$$0 \leq x \leq 40 \quad \text{or} \quad [0, 40]$$

The endpoints are included in the interval for our convenience (a closed interval is easier to work with than an open one). The area function is defined at each endpoint, so it does no harm to include them.

The equation in Example 2 restricts the values of x to

$$x > 0 \quad \text{or} \quad (0, \infty)$$

Neither endpoint can be included in this interval. We cannot include 0 because the area is not defined when $x = 0$, and we can never include ∞ as an endpoint. Remember, ∞ is not a number; it is a symbol that indicates the interval is unbounded.

Maximizing Revenue and Profit

EXAMPLE 3 Maximizing Revenue An office supply company sells x permanent markers per year at $\$p$ per marker. The price–demand equation for these markers is $p = 10 - 0.001x$. What price should the company charge for the markers to maximize revenue? What is the maximum revenue?

SOLUTION

$$\text{Revenue} = \text{price} \times \text{demand}$$
$$R(x) = (10 - 0.001x)x$$
$$= 10x - 0.001x^2$$

Both price and demand must be nonnegative, so

$$x \geq 0 \quad \text{and} \quad p = 10 - 0.001x \geq 0$$
$$10 \geq 0.001x$$
$$10{,}000 \geq x$$

The mathematical model for this problem is

$$\text{Maximize} \quad R(x) = 10x - 0.001x^2 \quad 0 \leq x \leq 10{,}000$$
$$R'(x) = 10 - 0.002x$$
$$10 - 0.002x = 0$$
$$10 = 0.002x$$
$$x = \frac{10}{0.002} = 5{,}000 \qquad \text{Critical number}$$

Use the second-derivative test for absolute extrema:

$$R''(x) = -0.002 < 0 \quad \text{for all } x$$
$$\text{Max } R(x) = R(5,000) = \$25,000$$

When the demand is $x = 5,000$, the price is

$$10 - 0.001(5,000) = \$5 \quad p = 10 - 0.001x$$

The company will realize a maximum revenue of \$25,000 when the price of a marker is \$5.

Matched Problem 3 ⌋ An office supply company sells x heavy-duty paper shredders per year at \$$p$ per shredder. The price–demand equation for these shredders is

$$p = 300 - \frac{x}{30}$$

What price should the company charge for the shredders to maximize revenue? What is the maximum revenue?

EXAMPLE 4   Maximizing Profit The total annual cost of manufacturing x permanent markers for the office supply company in Example 3 is

$$C(x) = 5,000 + 2x$$

What is the company's maximum profit? What should the company charge for each marker, and how many markers should be produced?

SOLUTION Using the revenue model in Example 3, we have

$$\text{Profit} = \text{Revenue} - \text{Cost}$$
$$P(x) = R(x) - C(x)$$
$$= 10x - 0.001x^2 - 5,000 - 2x$$
$$= 8x - 0.001x^2 - 5,000$$

The mathematical model for profit is

$$\text{Maximize} \quad P(x) = 8x - 0.001x^2 - 5,000 \quad 0 \le x \le 10,000$$

The restrictions on x come from the revenue model in Example 3.

$$P'(x) = 8 - 0.002x = 0$$
$$8 = 0.002x$$
$$x = \frac{8}{0.002} = 4,000 \quad \text{Critical number}$$
$$P''(x) = -0.002 < 0 \quad \text{for all } x$$

Since $x = 4,000$ is the only critical number and $P''(x) < 0$,

$$\text{Max } P(x) = P(4,000) = \$11,000$$

Using the price–demand equation from Example 3 with $x = 4,000$, we find that

$$p = 10 - 0.001(4,000) = \$6 \quad p = 10 - 0.001x$$

A maximum profit of \$11,000 is realized when 4,000 markers are manufactured annually and sold for \$6 each.

The results in Examples 3 and 4 are illustrated in Figure 2.

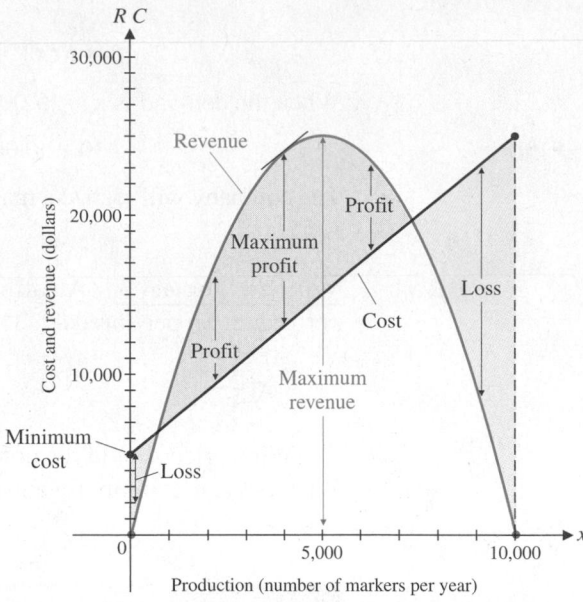

Figure 2

In Figure 2, notice that the maximum revenue and the maximum profit occur at different production levels. The maximum profit occurs when

$$P'(x) = R'(x) - C'(x) = 0$$

that is, when the marginal revenue is equal to the marginal cost. Notice that the slopes
of the revenue function and the cost function are the same at this production level.

Matched Problem 4 The annual cost of manufacturing x paper shredders for the
office supply company in Matched Problem 3 is $C(x) = 90,000 + 30x$. What is
the company's maximum profit? What should it charge for each shredder, and how
many shredders should it produce?

EXAMPLE 5   Maximizing Profit The government decides to tax the company in
Example 4 $2 for each marker produced. Taking into account this additional cost,
how many markers should the company manufacture annually to maximize its
profit? What is the maximum profit? How much should the company charge for the
markers to realize the maximum profit?

SOLUTION The tax of $2 per unit changes the company's cost equation:

$$C(x) = \text{original cost} + \text{tax}$$
$$= 5,000 + 2x + 2x$$
$$= 5,000 + 4x$$

The new profit function is

$$P(x) = R(x) - C(x)$$
$$= 10x - 0.001x^2 - 5,000 - 4x$$
$$= 6x - 0.001x^2 - 5,000$$

So, we must solve the following equation:

$$\text{Maximize} \quad P(x) = 6x - 0.001x^2 - 5,000 \qquad 0 \le x \le 10,000$$
$$P'(x) = 6 - 0.002x$$
$$6 - 0.002x = 0$$
$$x = 3,000 \quad \text{Critical number}$$
$$P''(x) = -0.002 < 0 \quad \text{for all } x$$
$$\text{Max } P(x) = P(3,000) = \$4,000$$

Using the price–demand equation (Example 3) with $x = 3,000$, we find that

$$p = 10 - 0.001(3,000) = \$7 \quad p = 10 - 0.001x$$

The company's maximum profit is $4,000 when 3,000 markers are produced and sold annually at a price of $7.

Even though the tax caused the company's cost to increase by $2 per marker, the price that the company should charge to maximize its profit increases by only $1. The company must absorb the other $1, with a resulting decrease of $7,000 in maximum profit.

Matched Problem 5) The government decides to tax the office supply company in Matched Problem 4 $20 for each shredder produced. Taking into account this additional cost, how many shredders should the company manufacture annually to maximize its profit? What is the maximum profit? How much should the company charge for the shredders to realize the maximum profit?

EXAMPLE 6 Maximizing Revenue When a management training company prices its seminar on management techniques at $400 per person, 1,000 people will attend the seminar. The company estimates that for each $5 reduction in price, an additional 20 people will attend the seminar. How much should the company charge for the seminar in order to maximize its revenue? What is the maximum revenue?

SOLUTION Let x represent the number of $5 price reductions.

$$400 - 5x = \text{price per customer}$$
$$1,000 + 20x = \text{number of customers}$$
$$\text{Revenue} = (\text{price per customer})(\text{number of customers})$$
$$R(x) = (400 - 5x) \times (1,000 + 20x)$$

Since price cannot be negative, we have

$$400 - 5x \ge 0$$
$$400 \ge 5x$$
$$80 \ge x \quad \text{or} \quad x \le 80$$

A negative value of x would result in a price increase. Since the problem is stated in terms of price reductions, we must restrict x so that $x \ge 0$. Putting all this together, we have the following model:

$$\text{Maximize} \quad R(x) = (400 - 5x)(1,000 + 20x) \quad \text{for } 0 \le x \le 80$$
$$R(x) = 400,000 + 3,000x - 100x^2$$
$$R'(x) = 3,000 - 200x = 0$$
$$3,000 = 200x$$
$$x = 15 \quad \text{Critical number}$$

Table 2

x	R(x)
0	400,000
15	422,500
80	0

Since $R(x)$ is continuous on the interval [0, 80], we can determine the behavior of the graph by constructing a table. Table 2 shows that $R(15) = \$422{,}500$ is the absolute maximum revenue. The price of attending the seminar at $x = 15$ is $400 - 5(15) = \$325$. The company should charge \$325 for the seminar in order to receive a maximum revenue of \$422,500.

Matched Problem 6 A walnut grower estimates from past records that if 20 trees are planted per acre, then each tree will average 60 pounds of nuts per year. If, for each additional tree planted per acre, the average yield per tree drops 2 pounds, then how many trees should be planted to maximize the yield per acre? What is the maximum yield?

EXAMPLE 7 Maximizing Revenue After additional analysis, the management training company in Example 6 decides that its estimate of attendance was too high. Its new estimate is that only 10 additional people will attend the seminar for each \$5 decrease in price. All other information remains the same. How much should the company charge for the seminar now in order to maximize revenue? What is the new maximum revenue?

SOLUTION Under the new assumption, the model becomes

$$\text{Maximize} \quad R(x) = (400 - 5x)(1{,}000 + 10x) \qquad 0 \le x \le 80$$
$$= 400{,}000 - 1{,}000x - 50x^2$$
$$R'(x) = -1{,}000 - 100x = 0$$
$$-1{,}000 = 100x$$
$$x = -10 \quad \text{Critical number}$$

Table 3

x	R(x)
0	400,000
80	0

Note that $x = -10$ is not in the interval [0, 80]. Since $R(x)$ is continuous on [0, 80], we can use a table to find the absolute maximum revenue. Table 3 shows that the maximum revenue is $R(0) = \$400{,}000$. The company should leave the price at \$400. Any \$5 decreases in price will lower the revenue.

Matched Problem 7 After further analysis, the walnut grower in Matched Problem 6 determines that each additional tree planted will reduce the average yield by 4 pounds. All other information remains the same. How many additional trees per acre should the grower plant now in order to maximize the yield? What is the new maximum yield?

CONCEPTUAL INSIGHT

The solution in Example 7 is called an **endpoint solution** because the optimal value occurs at the endpoint of an interval rather than at a critical number in the interior of the interval.

Inventory Control

EXAMPLE 8 Inventory Control A multimedia company anticipates that there will be a demand for 20,000 copies of a certain DVD during the next year. It costs the company \$0.50 to store a DVD for one year. Each time it must make additional DVDs, it costs \$200 to set up the equipment. How many DVDs should the company make during each production run to minimize its total storage and setup costs?

SOLUTION This type of problem is called an **inventory control problem**. One of the basic assumptions made in such problems is that the demand is uniform. For example, if there are 250 working days in a year, then the daily demand would be $20{,}000 \div 250 = 80$ DVDs. The company could decide to produce all 20,000 DVDs at the beginning of the year. This would certainly minimize the setup costs but would result in very large storage costs. At the other extreme, the company could produce 80 DVDs each day. This would minimize the storage costs but would result in very large setup costs. Somewhere between these two extremes is the optimal solution that will minimize the total storage and setup costs. Let

$x =$ number of DVDs manufactured during each production run

$y =$ number of production runs

It is easy to see that the total setup cost for the year is $200y$, but what is the total storage cost? If the demand is uniform, then the number of DVDs in storage between production runs will decrease from x to 0, and the average number in storage each day is $x/2$. This result is illustrated in Figure 3.

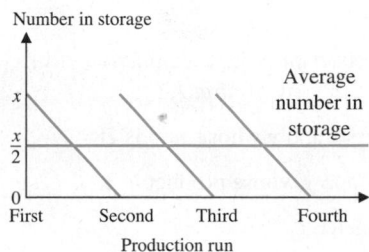

Figure 3

Since it costs $0.50 to store a DVD for one year, the total storage cost is $0.5(x/2) = 0.25x$ and the total cost is

$$\text{total cost} = \text{setup cost} + \text{storage cost}$$
$$C = 200y + 0.25x$$

In order to write the total cost C as a function of one variable, we must find a relationship between x and y. If the company produces x DVDs in each of y production runs, then the total number of DVDs produced is xy.

$$xy = 20{,}000$$
$$y = \frac{20{,}000}{x}$$

Certainly, x must be at least 1 and cannot exceed 20,000. We must solve the following equation:

$$\text{Minimize} \quad C(x) = 200\left(\frac{20{,}000}{x}\right) + 0.25x \qquad 1 \le x \le 20{,}000$$

$$C(x) = \frac{4{,}000{,}000}{x} + 0.25x$$

$$C'(x) = -\frac{4{,}000{,}000}{x^2} + 0.25$$

$$-\frac{4{,}000{,}000}{x^2} + 0.25 = 0$$

$$x^2 = \frac{4{,}000{,}000}{0.25}$$

$$x^2 = 16{,}000{,}000 \qquad \text{−4,000 is not a critical number, since}$$

$$x = 4{,}000 \qquad 1 \le x \le 20{,}000.$$

$$C''(x) = \frac{8{,}000{,}000}{x^3} > 0 \qquad \text{for } x \in (1, 20{,}000)$$

Therefore,

$$\text{Min } C(x) = C(4{,}000) = 2{,}000$$

$$y = \frac{20{,}000}{4{,}000} = 5$$

The company will minimize its total cost by making 4,000 DVDs five times during the year.

Matched Problem 8) Repeat Example 8 if it costs $250 to set up a production run and $0.40 to store a DVD for one year.

Exercises 4.6

Skills Warm-up Exercises

W *In Problems 1–8, express the given quantity as a function f(x) of one variable x. (If necessary, review Section 1.1).*

1. The product of two numbers x and y whose sum is 28

2. The sum of two numbers x and y whose product is 36

3. The area of a circle of diameter x

4. The volume of a sphere of diameter x

5. The volume of a right circular cylinder of radius x and height equal to the radius

6. The volume of a right circular cylinder of diameter x and height equal to twice the diameter

7. The area of a rectangle of length x and width y that has a perimeter of 120 feet

8. The perimeter of a rectangle of length x and width y that has an area of 200 square meters

9. Find two numbers whose sum is 15 and whose product is a maximum.

10. Find two numbers whose sum is 21 and whose product is a maximum.

11. Find two numbers whose difference is 15 and whose product is a minimum.

12. Find two numbers whose difference is 21 and whose product is a minimum.

13. Find two positive numbers whose product is 15 and whose sum is a minimum.

14. Find two positive numbers whose product is 21 and whose sum is a minimum.

15. Find the dimensions of a rectangle with an area of 200 square feet that has the minimum perimeter.

16. Find the dimensions of a rectangle with an area of 108 square feet that has the minimum perimeter.

17. Find the dimensions of a rectangle with a perimeter of 148 feet that has the maximum area.

18. Find the dimensions of a rectangle with a perimeter of 76 feet that has the maximum area.

19. **Maximum revenue and profit.** A company manufactures and sells x smartphones per week. The weekly price–demand and cost equations are, respectively,

$$p = 500 - 0.5x \quad \text{and} \quad C(x) = 20{,}000 + 135x$$

(A) What price should the company charge for the phones, and how many phones should be produced to maximize the weekly revenue? What is the maximum weekly revenue?

(B) What is the maximum weekly profit? How much should the company charge for the phones, and how many phones should be produced to realize the maximum weekly profit?

20. **Maximum revenue and profit.** A company manufactures and sells x digital cameras per week. The weekly price–demand and cost equations are, respectively,

$$p = 400 - 0.4x \quad \text{and} \quad C(x) = 2{,}000 + 160x$$

(A) What price should the company charge for the cameras, and how many cameras should be produced to maximize the weekly revenue? What is the maximum revenue?

(B) What is the maximum weekly profit? How much should the company charge for the cameras, and how many cameras should be produced to realize the maximum weekly profit?

21. **Maximum revenue and profit.** A company manufactures and sells x television sets per month. The monthly cost and price–demand equations are

$$C(x) = 72{,}000 + 60x$$

$$p = 200 - \frac{x}{30} \qquad 0 \le x \le 6{,}000$$

(A) Find the maximum revenue.

(B) Find the maximum profit, the production level that will realize the maximum profit, and the price the company should charge for each television set.

(C) If the government decides to tax the company $5 for each set it produces, how many sets should the company manufacture each month to maximize its profit? What is the maximum profit? What should the company charge for each set?

22. Maximum revenue and profit. Repeat Problem 21 for

$$C(x) = 60,000 + 60x$$

$$p = 200 - \frac{x}{50} \qquad 0 \leq x \leq 10,000$$

23. Maximum profit. The following table contains price–demand and total cost data for the production of extreme-cold sleeping bags, where p is the wholesale price (in dollars) of a sleeping bag for an annual demand of x sleeping bags and C is the total cost (in dollars) of producing x sleeping bags:

(A) Find a quadratic regression equation for the price–demand data, using x as the independent variable.

x	p	C
950	240	130,000
1,200	210	150,000
1,800	160	180,000
2,050	120	190,000

(B) Find a linear regression equation for the cost data, using x as the independent variable.

(C) What is the maximum profit? What is the wholesale price per extreme-cold sleeping bag that should be charged to realize the maximum profit? Round answers to the nearest dollar.

24. Maximum profit. The following table contains price–demand and total cost data for the production of regular sleeping bags, where p is the wholesale price (in dollars) of a sleeping bag for an annual demand of x sleeping bags and C is the total cost (in dollars) of producing x sleeping bags:

x	p	C
2,300	98	145,000
3,300	84	170,000
4,500	67	190,000
5,200	51	210,000

(A) Find a quadratic regression equation for the price–demand data, using x as the independent variable.

(B) Find a linear regression equation for the cost data, using x as the independent variable.

(C) What is the maximum profit? What is the wholesale price per regular sleeping bag that should be charged to realize the maximum profit? Round answers to the nearest dollar.

25. Maximum revenue. A deli sells 640 sandwiches per day at a price of $8 each.

(A) A market survey shows that for every $0.10 reduction in price, 40 more sandwiches will be sold. How much should the deli charge for a sandwich in order to maximize revenue?

(B) A different market survey shows that for every $0.20 reduction in the original $8 price, 15 more sandwiches will be sold. Now how much should the deli charge for a sandwich in order to maximize revenue?

26. Maximum revenue. A university student center sells 1,600 cups of coffee per day at a price of $2.40.

(A) A market survey shows that for every $0.05 reduction in price, 50 more cups of coffee will be sold. How much should the student center charge for a cup of coffee in order to maximize revenue?

(B) A different market survey shows that for every $0.10 reduction in the original $2.40 price, 60 more cups of coffee will be sold. Now how much should the student center charge for a cup of coffee in order to maximize revenue?

27. Car rental. A car rental agency rents 200 cars per day at a rate of $30 per day. For each $1 increase in rate, 5 fewer cars are rented. At what rate should the cars be rented to produce the maximum income? What is the maximum income?

28. Rental income. A 300-room hotel in Las Vegas is filled to capacity every night at $80 a room. For each $1 increase in rent, 3 fewer rooms are rented. If each rented room costs $10 to service per day, how much should the management charge for each room to maximize gross profit? What is the maximum gross profit?

29. Agriculture. A commercial cherry grower estimates from past records that if 30 trees are planted per acre, then each tree will yield an average of 50 pounds of cherries per season. If, for each additional tree planted per acre (up to 20), the average yield per tree is reduced by 1 pound, how many trees should be planted per acre to obtain the maximum yield per acre? What is the maximum yield?

30. Agriculture. A commercial pear grower must decide on the optimum time to have fruit picked and sold. If the pears are picked now, they will bring 30¢ per pound, with each tree yielding an average of 60 pounds of salable pears. If the average yield per tree increases 6 pounds per tree per week for the next 4 weeks, but the price drops 2¢ per pound per week, when should the pears be picked to realize the maximum return per tree? What is the maximum return?

31. Manufacturing. A candy box is to be made out of a piece of cardboard that measures 8 by 12 inches. Squares of equal size will be cut out of each corner, and then the ends and sides will be folded up to form a rectangular box. What size square should be cut from each corner to obtain a maximum volume?

32. Packaging. A parcel delivery service will deliver a package only if the length plus girth (distance around) does not exceed 108 inches.

(A) Find the dimensions of a rectangular box with square ends that satisfies the delivery service's restriction and has maximum volume. What is the maximum volume?

(B) Find the dimensions (radius and height) of a cylindrical container that meets the delivery service's requirement and has maximum volume. What is the maximum volume?

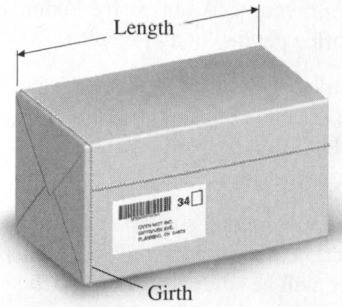

Figure for 32

33. Construction costs. A fence is to be built to enclose a rectangular area of 800 square feet. The fence along three sides is to be made of material that costs $6 per foot. The material for the fourth side costs $18 per foot. Find the dimensions of the rectangle that will allow for the most economical fence to be built.

34. Construction costs. If a builder has only $840 to spend on a fence, but wants to use both $6 and $18 per foot fencing as in Problem 33, what is the maximum area that can be enclosed? What are its dimensions?

35. Construction costs. The owner of a retail lumber store wants to construct a fence to enclose an outdoor storage area adjacent to the store, using all of the store as part of one side of the area (see the figure). Find the dimensions that will enclose the largest area if

(A) 240 feet of fencing material are used.

(B) 400 feet of fencing material are used.

36. Construction costs. If the owner wants to enclose a rectangular area of 12,100 square feet as in Problem 35, what are the dimensions of the area that requires the least fencing? How many feet of fencing are required?

37. Inventory control. A paint manufacturer has a uniform annual demand for 16,000 cans of automobile primer. It costs $4 to store one can of paint for one year and $500 to set up the plant for production of the primer. How many times a year should the company produce this primer in order to minimize the total storage and setup costs?

38. Inventory control. A pharmacy has a uniform annual demand for 200 bottles of a certain antibiotic. It costs $10 to store one bottle for one year and $40 to place an order. How many times during the year should the pharmacy order the antibiotic in order to minimize the total storage and reorder costs?

39. Inventory control. A publishing company sells 50,000 copies of a certain book each year. It costs the company $1 to store a book for one year. Each time that it prints additional copies, it costs the company $1,000 to set up the presses. How many books should the company produce during each printing in order to minimize its total storage and setup costs?

40. Inventory control. A tool company has a uniform annual demand for 9,000 premium chainsaws. It costs $5 to store a chainsaw for a year and $2,500 to set up the plant for manufacture of the premium model. How many chainsaws should be manufactured in each production run in order to minimize the total storage and setup costs?

41. Operational costs. The cost per hour for fuel to run a train is $v^2/4$ dollars, where v is the speed of the train in miles per hour. (Note that the cost goes up as the square of the speed.) Other costs, including labor, are $300 per hour. How fast should the train travel on a 360-mile trip to minimize the total cost for the trip?

42. Operational costs. The cost per hour for fuel to drive a rental truck from Chicago to New York, a distance of 800 miles, is given by

$$f(v) = 0.03v^2 - 2.2v + 72$$

where v is the speed of the truck in miles per hour. Other costs are $40 per hour. How fast should you drive to minimize the total cost?

43. Construction costs. A freshwater pipeline is to be run from a source on the edge of a lake to a small resort community on an island 5 miles offshore, as indicated in the figure.

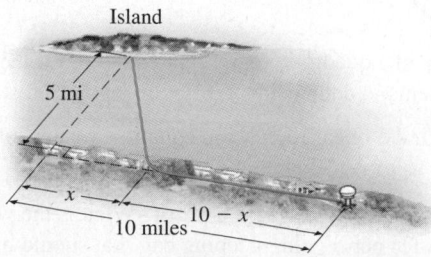

(A) If it costs 1.4 times as much to lay the pipe in the lake as it does on land, what should x be (in miles) to minimize the total cost of the project?

(B) If it costs only 1.1 times as much to lay the pipe in the lake as it does on land, what should x be to minimize the total cost of the project? [*Note:* Compare with Problem 46.]

44. Drug concentration. The concentration $C(t)$, in milligrams per cubic centimeter, of a particular drug in a patient's bloodstream is given by

$$C(t) = \frac{0.16t}{t^2 + 4t + 4}$$

where t is the number of hours after the drug is taken. How many hours after the drug is taken will the concentration be maximum? What is the maximum concentration?

45. Bacteria control. A lake used for recreational swimming is treated periodically to control harmful bacteria growth. Suppose that t days after a treatment, the concentration of bacteria per cubic centimeter is given by

$$C(t) = 30t^2 - 240t + 500 \qquad 0 \le t \le 8$$

How many days after a treatment will the concentration be minimal? What is the minimum concentration?

46. Bird flights. Some birds tend to avoid flights over large bodies of water during daylight hours. Suppose that an adult bird with this tendency is taken from its nesting area on the edge of a large lake to an island 5 miles offshore and is then released (see the figure).

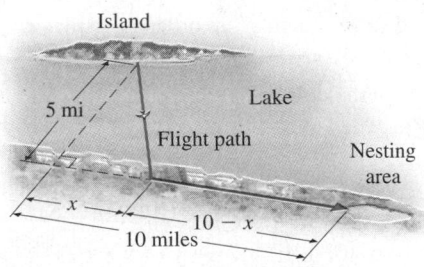

(A) If it takes 1.4 times as much energy to fly over water as land, how far up the shore (x, in miles) should the bird head to minimize the total energy expended in returning to the nesting area?

(B) If it takes only 1.1 times as much energy to fly over water as land, how far up the shore should the bird head to minimize the total energy expended in returning to the nesting area? [*Note:* Compare with Problem 43.]

47. Botany. If it is known from past experiments that the height (in feet) of a certain plant after t months is given approximately by

$$H(t) = 4t^{1/2} - 2t \qquad 0 \le t \le 2$$

then how long, on average, will it take a plant to reach its maximum height? What is the maximum height?

48. Pollution. Two heavily industrial areas are located 10 miles apart, as shown in the figure. If the concentration of particulate matter (in parts per million) decreases as the reciprocal of the square of the distance from the source, and if area A_1 emits eight times the particulate matter as A_2, then the concentration of particulate matter at any point between the two areas is given by

$$C(x) = \frac{8k}{x^2} + \frac{k}{(10-x)^2} \qquad 0.5 \le x \le 9.5, \quad k > 0$$

How far from A_1 will the concentration of particulate matter between the two areas be at a minimum?

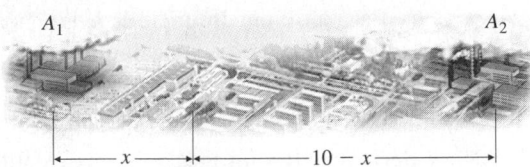

49. Politics. In a newly incorporated city, the voting population (in thousands) is estimated to be

$$N(t) = 30 + 12t^2 - t^3 \qquad 0 \le t \le 8$$

where t is time in years. When will the rate of increase of $N(t)$ be most rapid?

50. Learning. A large grocery chain found that, on average, a checker can recall $P\%$ of a given price list x hours after starting work, as given approximately by

$$P(x) = 96x - 24x^2 \qquad 0 \le x \le 3$$

At what time x does the checker recall a maximum percentage? What is the maximum?

Answers to Matched Problems

1. The dimensions of the garden with the maximum area of 640 square feet are 16 feet by 40 feet, with one 16-foot side with wood fencing.

2. The minimum cost for enclosing a 1,800-square-foot garden is $480, and the dimensions are 30 feet by 60 feet, with one 30-foot side with wood fencing.

3. The company will realize a maximum revenue of $675,000 when the price of a shredder is $150.

4. A maximum profit of $456,750 is realized when 4,050 shredders are manufactured annually and sold for $165 each.

5. A maximum profit of $378,750 is realized when 3,750 shredders are manufactured annually and sold for $175 each.

6. The maximum yield is 1,250 pounds per acre when 5 additional trees are planted on each acre.

7. The maximum yield is 1,200 pounds when no additional trees are planted.

8. The company should produce 5,000 DVDs four times a year.

Important Terms, Symbols, and Concepts

4.1 First Derivative and Graphs

- A function f is **increasing** on an interval (a, b) if $f(x_2) > f(x_1)$ whenever $a < x_1 < x_2 < b$, and f is **decreasing** on (a, b) if $f(x_2) < f(x_1)$ whenever $a < x_1 < x_2 < b$.

- For the interval (a, b), if $f' > 0$, then f is increasing, and if $f' < 0$, then f is decreasing. So a sign chart for f' can be used to tell where f is increasing or decreasing.

- A real number x in the domain of f such that $f'(x) = 0$ or $f'(x)$ does not exist is called a **critical number** of f. So a critical number of f is a partition number for f' that also belongs to the domain of f.

- A value $f(c)$ is a **local maximum** if there is an interval (m, n) containing c such that $f(x) \leq f(c)$ for all x in (m, n). A value $f(c)$ is a **local minimum** if there is an interval (m, n) containing c such that $f(x) \geq f(c)$ for all x in (m, n). A local maximum or local minimum is called a **local extremum**.

- If $f(c)$ is a local extremum, then c is a critical number of f.

- The **first-derivative test for local extrema** identifies local maxima and minima of f by means of a sign chart for f'.

4.2 Second Derivative and Graphs

- The graph of f is **concave upward** on (a, b) if f' is increasing on (a, b), and is **concave downward** on (a, b) if f' is decreasing on (a, b).

- For the interval (a, b), if $f'' > 0$, then f is concave upward, and if $f'' < 0$, then f is concave downward. So a sign chart for f'' can be used to tell where f is concave upward or concave downward.

- An **inflection point** of f is a point $(c, f(c))$ on the graph of f where the concavity changes.

- The graphing strategy on page 261 is used to organize the information obtained from f' and f'' in order to sketch the graph of f.

- If sales $N(x)$ are expressed as a function of the amount x spent on advertising, then the dollar amount at which $N'(x)$, the rate of change of sales, goes from increasing to decreasing is called the **point of diminishing returns**. If d is the point of diminishing returns, then $(d, N(d))$ is an inflection point of $N(x)$.

4.3 L'Hôpital's Rule

- L'Hôpital's rule for $0/0$ indeterminate forms: If $\lim\limits_{x \to c} f(x) = 0$ and $\lim\limits_{x \to c} g(x) = 0$, then

$$\lim_{x \to c} \frac{f(x)}{g(x)} = \lim_{x \to c} \frac{f'(x)}{g'(x)}$$

provided the second limit exists or is ∞ or $-\infty$.

- Always check to make sure that L'Hôpital's rule is applicable before using it.

- L'Hôpital's rule remains valid if the symbol $x \to c$ is replaced everywhere it occurs by one of

$$x \to c^+ \qquad x \to c^- \qquad x \to \infty \qquad x \to -\infty$$

- L'Hôpital's rule is also valid for indeterminate forms $\dfrac{\pm \infty}{\pm \infty}$.

4.4 Curve-Sketching Techniques

- The graphing strategy on pages 280 and 281 incorporates analyses of f, f', and f'' in order to sketch a graph of f, including intercepts and asymptotes.

- If $f(x) = n(x)/d(x)$ is a rational function and the degree of $n(x)$ is 1 more than the degree of $d(x)$, then the graph of $f(x)$ has an **oblique asymptote** of the form $y = mx + b$.

4.5 Absolute Maxima and Minima

- If $f(c) \geq f(x)$ for all x in the domain of f, then $f(c)$ is called the **absolute maximum** of f. If $f(c) \leq f(x)$ for all x in the domain of f, then $f(c)$ is called the **absolute minimum** of f. An absolute maximum or absolute minimum is called an **absolute extremum**.

- A function that is continuous on a closed interval $[a, b]$ has both an absolute maximum and an absolute minimum on that interval.

- Absolute extrema, if they exist, must occur at critical numbers or endpoints.
- To find the absolute maximum and absolute minimum of a continuous function f on a closed interval, identify the endpoints and critical numbers in the interval, evaluate the function f at each of them, and choose the largest and smallest values of f.
- **Second-derivative test for local extrema:** If $f'(c) = 0$ and $f''(c) > 0$, then $f(c)$ is a local minimum. If $f'(c) = 0$ and $f''(c) < 0$, then $f(c)$ is a local maximum. No conclusion can be drawn if $f''(c) = 0$. Ex. 2, p. 297
- The **second-derivative test for absolute extrema on an open interval** is applicable when there is only one critical number c in an open interval I and $f'(c) = 0$ and $f''(c) \neq 0$. Ex. 3, p. 298

4.6 Optimization

- The procedure on page 302 for solving optimization problems involves finding the absolute maximum or absolute minimum of a function $f(x)$ on an interval I. If the absolute maximum or absolute minimum occurs at an endpoint, not at a critical number in the interior of I, the extremum is called an **endpoint solution**. The procedure is effective in solving problems in business, including **inventory control problems**, manufacturing, construction, engineering, and many other fields.

Review Exercises

Work through all the problems in this chapter review, and check your answers in the back of the book. Answers to all review problems are there, along with section numbers in italics to indicate where each type of problem is discussed. Where weaknesses show up, review appropriate sections in the text.

Problems 1–8 refer to the following graph of $y = f(x)$. Identify the points or intervals on the x axis that produce the indicated behavior.

1. $f(x)$ is increasing.

2. $f'(x) < 0$

3. The graph of f is concave downward.

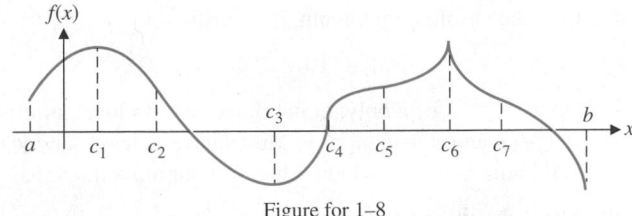

Figure for 1–8

4. Local minima

5. Absolute maxima

6. $f'(x)$ appears to be 0.

7. $f'(x)$ does not exist.

8. Inflection points

In Problems 9 and 10, use the given information to sketch the graph of f. Assume that f is continuous on its domain and that all intercepts are included in the information given.

9. Domain: All real x

x	-3	-2	-1	0	2	3
$f(x)$	0	3	2	0	-3	0

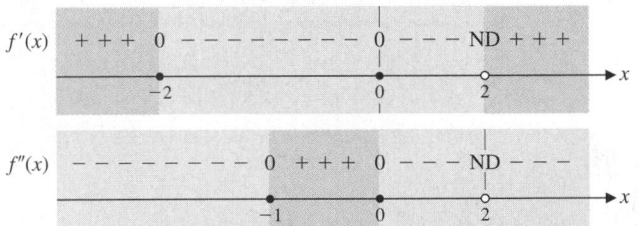

10. Domain: All real x

$$f(-2) = 1, f(0) = 0, f(2) = 1;$$
$$f'(0) = 0; f'(x) < 0 \text{ on } (-\infty, 0);$$
$$f'(x) > 0 \text{ on } (0, \infty);$$
$$f''(-2) = 0, f''(2) = 0;$$
$$f''(x) < 0 \text{ on } (-\infty, -2) \text{ and } (2, \infty);$$
$$f''(x) > 0 \text{ on } (-2, 2);$$
$$\lim_{x \to -\infty} f(x) = 2; \lim_{x \to \infty} f(x) = 2$$

11. Find $f''(x)$ for $f(x) = x^4 + 5x^3$.

12. Find y'' for $y = 3x + \dfrac{4}{x}$.

In Problems 13 and 14, find the domain and intercepts.

13. $f(x) = \dfrac{5 + x}{4 - x}$

14. $f(x) = \ln(x + 2)$

In Problems 15 and 16, find the horizontal and vertical asymptotes.

15. $f(x) = \dfrac{x + 3}{x^2 - 4}$

16. $f(x) = \dfrac{2x - 7}{3x + 10}$

In Problems 17 and 18, find the x and y coordinates of all inflection points.

17. $f(x) = x^4 - 12x^2$ **18.** $f(x) = (2x + 1)^{1/3} - 6$

In Problems 19 and 20, find (A) $f'(x)$, (B) the partition numbers for f', and (C) the critical numbers of f.

19. $f(x) = x^{1/5}$ **20.** $f(x) = x^{-1/5}$

In Problems 21–30, summarize all the pertinent information obtained by applying the final version of the graphing strategy (Section 4–4) to f, and sketch the graph of f.

21. $f(x) = x^3 - 18x^2 + 81x$

22. $f(x) = (x + 4)(x - 2)^2$

23. $f(x) = 8x^3 - 2x^4$ **24.** $f(x) = (x - 1)^3(x + 3)$

25. $f(x) = \dfrac{3x}{x + 2}$ **26.** $f(x) = \dfrac{x^2}{x^2 + 27}$

27. $f(x) = \dfrac{x}{(x + 2)^2}$ **28.** $f(x) = \dfrac{x^3}{x^2 + 3}$

29. $f(x) = 5 - 5e^{-x}$ **30.** $f(x) = x^3 \ln x$

Find each limit in Problems 31–40.

31. $\displaystyle\lim_{x \to 0} \frac{e^{3x} - 1}{x}$ **32.** $\displaystyle\lim_{x \to 2} \frac{x^2 - 5x + 6}{x^2 + x - 6}$

33. $\displaystyle\lim_{x \to 0^-} \frac{\ln(1 + x)}{x^2}$ **34.** $\displaystyle\lim_{x \to 0} \frac{\ln(1 + x)}{1 + x}$

35. $\displaystyle\lim_{x \to \infty} \frac{e^{4x}}{x^2}$ **36.** $\displaystyle\lim_{x \to 0} \frac{e^x + e^{-x} - 2}{x^2}$

37. $\displaystyle\lim_{x \to 0^+} \frac{\sqrt{1 + x} - 1}{\sqrt{x}}$ **38.** $\displaystyle\lim_{x \to \infty} \frac{\ln x}{x^5}$

39. $\displaystyle\lim_{x \to \infty} \frac{\ln(1 + 6x)}{\ln(1 + 3x)}$ **40.** $\displaystyle\lim_{x \to 0} \frac{\ln(1 + 6x)}{\ln(1 + 3x)}$

41. Use the graph of $y = f'(x)$ shown here to discuss the graph of $y = f(x)$. Organize your conclusions in a table (see Example 4, Section 4.2). Sketch a possible graph of $y = f(x)$.

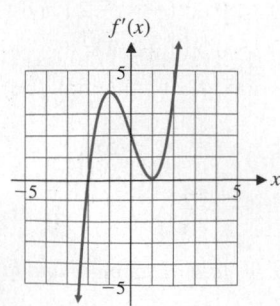

f'(x)

Figure for 41 and 42

42. Refer to the above graph of $y = f'(x)$. Which of the following could be the graph of $y = f''(x)$?

(A)

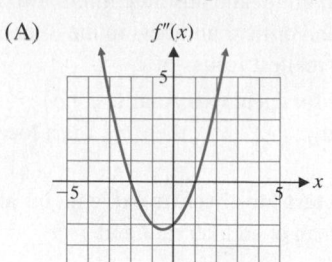

f''(x)

(B)

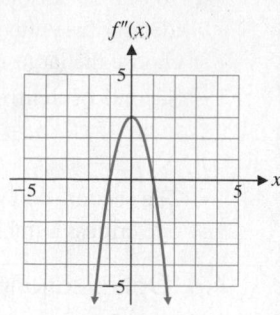

f''(x)

(C)

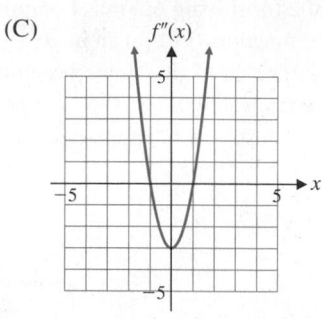

f''(x)

43. Use the second-derivative test to find any local extrema for

$$f(x) = x^3 - 6x^2 - 15x + 12$$

44. Find the absolute maximum and absolute minimum, if either exists, for

$$y = f(x) = x^3 - 12x + 12 \qquad -3 \le x \le 5$$

45. Find the absolute minimum, if it exists, for

$$y = f(x) = x^2 + \frac{16}{x^2} \qquad x > 0$$

46. Find the absolute maximum, if it exists, for

$$f(x) = 11x - 2x \ln x \qquad x > 0$$

47. Find the absolute maximum, if it exists, for

$$f(x) = 10xe^{-2x} \qquad x > 0$$

48. Let $y = f(x)$ be a polynomial function with local minima at $x = a$ and $x = b$, $a < b$. Must f have at least one local maximum between a and b? Justify your answer.

49. The derivative of $f(x) = x^{-1}$ is $f'(x) = -x^{-2}$. Since $f'(x) < 0$ for $x \ne 0$, is it correct to say that $f(x)$ is decreasing for all x except $x = 0$? Explain.

50. Discuss the difference between a partition number for $f'(x)$ and a critical number of $f(x)$, and illustrate with examples.

51. Find the absolute maximum for $f'(x)$ if

$$f(x) = 6x^2 - x^3 + 8$$

Graph f and f' on the same coordinate system for $0 \le x \le 4$.

52. Find two positive numbers whose product is 400 and whose sum is a minimum. What is the minimum sum?

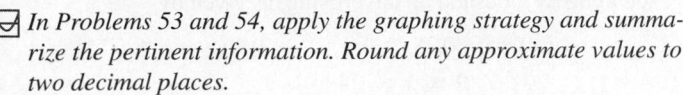 *In Problems 53 and 54, apply the graphing strategy and summarize the pertinent information. Round any approximate values to two decimal places.*

53. $f(x) = x^4 + x^3 - 4x^2 - 3x + 4$

54. $f(x) = 0.25x^4 - 5x^3 + 31x^2 - 70x$

55. Find the absolute maximum, if it exists, for
$$f(x) = 3x - x^2 + e^{-x} \quad x > 0$$

56. Find the absolute maximum, if it exists, for
$$f(x) = \frac{\ln x}{e^x} \quad x > 0$$

Applications

57. Price analysis. The graph in the figure approximates the rate of change of the price of tomatoes over a 60-month period, where $p(t)$ is the price of a pound of tomatoes and t is time (in months).

(A) Write a brief description of the graph of $y = p(t)$, including a discussion of local extrema and inflection points.

(B) Sketch a possible graph of $y = p(t)$.

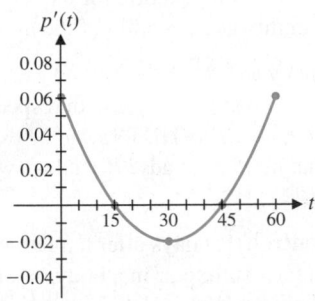

58. Maximum revenue and profit. A company manufactures and sells x e-book readers per month. The monthly cost and price–demand equations are, respectively,

$$C(x) = 350x + 50,000$$
$$p = 500 - 0.025x \quad 0 \le x \le 20,000$$

(A) Find the maximum revenue.

(B) How many readers should the company manufacture each month to maximize its profit? What is the maximum monthly profit? How much should the company charge for each reader?

(C) If the government decides to tax the company $20 for each reader it produces, how many readers should the company manufacture each month to maximize its profit? What is the maximum monthly profit? How much should the company charge for each reader?

59. Construction. A fence is to be built to enclose a rectangular area. The fence along three sides is to be made of material that costs $5 per foot. The material for the fourth side costs $15 per foot.

(A) If the area is 5,000 square feet, find the dimensions of the rectangle that will allow for the most economical fence.

(B) If $3,000 is available for the fencing, find the dimensions of the rectangle that will enclose the most area.

60. Rental income. A 200-room hotel in Reno is filled to capacity every night at a rate of $40 per room. For each $1 increase in the nightly rate, 4 fewer rooms are rented. If each rented room costs $8 a day to service, how much should the management charge per room in order to maximize gross profit? What is the maximum gross profit?

61. Inventory control. A computer store sells 7,200 boxes of storage disks annually. It costs the store $0.20 to store a box of disks for one year. Each time it reorders disks, the store must pay a $5.00 service charge for processing the order. How many times during the year should the store order disks to minimize the total storage and reorder costs?

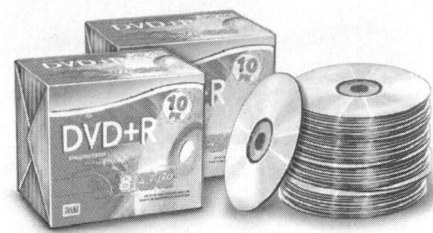

62. Average cost. The total cost of producing x dorm refrigerators per day is given by

$$C(x) = 4,000 + 10x + 0.1x^2$$

Find the minimum average cost. Graph the average cost and the marginal cost functions on the same coordinate system. Include any oblique asymptotes.

63. Average cost. The cost of producing x wheeled picnic coolers is given by

$$C(x) = 200 + 50x - 50 \ln x \quad x \ge 1$$

Find the minimum average cost.

64. Marginal analysis. The price–demand equation for a GPS device is

$$p(x) = 1,000e^{-0.02x}$$

where x is the monthly demand and p is the price in dollars. Find the production level and price per unit that produce the maximum revenue. What is the maximum revenue?

65. Maximum revenue. Graph the revenue function from Problem 64 for $0 \le x \le 100$.

66. Maximum profit. Refer to Problem 64. If the GPS devices cost the store $220 each, find the price (to the nearest cent) that maximizes the profit. What is the maximum profit (to the nearest dollar)?

67. Maximum profit. The data in the table show the daily demand x for cream puffs at a state fair at various price levels p. If it costs \$1 to make a cream puff, use logarithmic regression ($p = a + b \ln x$) to find the price (to the nearest cent) that maximizes profit.

Demand x	Price per Cream Puff($) p
3,125	1.99
3,879	1.89
5,263	1.79
5,792	1.69
6,748	1.59
8,120	1.49

68. Construction costs. The ceiling supports in a new discount department store are 12 feet apart. Lights are to be hung from these supports by chains in the shape of a "Y." If the lights are 10 feet below the ceiling, what is the shortest length of chain that can be used to support these lights?

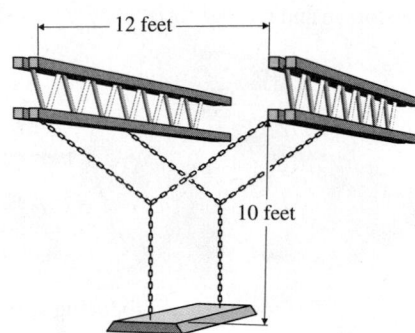

69. Average cost. The table gives the total daily cost y (in dollars) of producing x dozen chocolate chip cookies at various production levels.

Dozens of Cookies x	Total Cost y
50	119
100	187
150	248
200	382
250	505
300	695

(A) Enter the data into a graphing calculator and find a quadratic regression equation for the total cost.

(B) Use the regression equation from part (A) to find the minimum average cost (to the nearest cent) and the corresponding production level (to the nearest integer).

70. Advertising—point of diminishing returns. A company estimates that it will sell $N(x)$ units of a product after spending x thousand on advertising, as given by

$$N(x) = -0.25x^4 + 11x^3 - 108x^2 + 3,000$$
$$9 \le x \le 24$$

When is the rate of change of sales increasing and when is it decreasing? What is the point of diminishing returns and the maximum rate of change of sales? Graph N and N' on the same coordinate system.

71. Advertising. A chain of appliance stores uses TV ads to promote its HDTV sales. Analyzing past records produced the data in the following table, where x is the number of ads placed monthly and y is the number of HDTVs sold that month:

Number of Ads x	Number of HDTVs y
10	271
20	427
25	526
30	629
45	887
48	917

(A) Enter the data into a graphing calculator, set the calculator to display two decimal places, and find a cubic regression equation for the number of HDTVs sold monthly as a function of the number of ads.

(B) How many ads should be placed each month to maximize the rate of change of sales with respect to the number of ads, and how many HDTVs can be expected to be sold with that number of ads? Round answers to the nearest integer.

72. Bacteria control. If t days after a treatment the bacteria count per cubic centimeter in a body of water is given by

$$C(t) = 20t^2 - 120t + 800 \qquad 0 \le t \le 9$$

then in how many days will the count be a minimum?

73. Politics. In a new suburb, the number of registered voters is estimated to be

$$N(t) = 10 + 6t^2 - t^3 \qquad 0 \le t \le 5$$

where t is time in years and N is in thousands. When will the rate of increase of $N(t)$ be at its maximum?

5 Integration

Introduction

In the preceding three chapters, we studied the *derivative* and its applications. In Chapter 5, we introduce the *integral*, the second key concept of calculus. The integral can be used to calculate areas, volumes, the index of income concentration, and consumers' surplus. At first glance, the integral may appear to be unrelated to the derivative. There is, however, a close connection between these two concepts, which is made precise by the *fundamental theorem of calculus* (Section 5.5). We consider many applications of integrals and differential equations in Chapter 5. See, for example, Problem 91 in Section 5.3, which explores how the age of an archaeological site or artifact can be estimated.

5.1   Antiderivatives and Indefinite Integrals

- Antiderivatives
- Indefinite Integrals: Formulas and Properties
- Applications

Many operations in mathematics have reverses—addition and subtraction, multiplication and division, powers and roots. We now know how to find the derivatives of many functions. The reverse operation, *antidifferentiation* (the reconstruction of a function from its derivative), will receive our attention in this and the next two sections.

Antiderivatives

A function F is an **antiderivative** of a function f if $F'(x) = f(x)$.

The function $F(x) = \dfrac{x^3}{3}$ is an antiderivative of the function $f(x) = x^2$ because

$$\frac{d}{dx}\left(\frac{x^3}{3}\right) = x^2$$

However, $F(x)$ is not the only antiderivative of x^2. Note also that

$$\frac{d}{dx}\left(\frac{x^3}{3} + 2\right) = x^2 \qquad \frac{d}{dx}\left(\frac{x^3}{3} - \pi\right) = x^2 \qquad \frac{d}{dx}\left(\frac{x^3}{3} + \sqrt{5}\right) = x^2$$

Therefore,

$$\frac{x^3}{3} + 2 \qquad \frac{x^3}{3} - \pi \qquad \frac{x^3}{3} + \sqrt{5}$$

are also antiderivatives of x^2 because each has x^2 as a derivative. In fact, it appears that

$$\frac{x^3}{3} + C \qquad \text{for any real number } C$$

is an antiderivative of x^2 because

$$\frac{d}{dx}\left(\frac{x^3}{3} + C\right) = x^2$$

Antidifferentiation of a given function does not give a unique function, but an entire family of functions.

Does the expression

$$\frac{x^3}{3} + C \qquad \text{with } C \text{ any real number}$$

include all antiderivatives of x^2? Theorem 1 (stated without proof) indicates that the answer is yes.

> **THEOREM 1   Antiderivatives**
>
> If the derivatives of two functions are equal on an open interval (a, b), then the functions differ by at most a constant. Symbolically, if F and G are differentiable functions on the interval (a, b) and $F'(x) = G'(x)$ for all x in (a, b), then $F(x) = G(x) + k$ for some constant k.

> **CONCEPTUAL   INSIGHT**
>
> Suppose that $F(x)$ is an antiderivative of $f(x)$. If $G(x)$ is any other antiderivative of $f(x)$, then by Theorem 1, the graph of $G(x)$ is a vertical translation of the graph of $F(x)$ (see Section 1.2).

EXAMPLE 1   A Family of Antiderivatives   Note that

$$\frac{d}{dx}\left(\frac{x^2}{2}\right) = x$$

(A) Find all antiderivatives of $f(x) = x$.

(B) Graph the antiderivative of $f(x) = x$ that passes through the point $(0, 0)$; through the point $(0, 1)$; through the point $(0, 2)$.

(C) How are the graphs of the three antiderivatives in part (B) related?

SOLUTION

(A) By Theorem 1, any antiderivative of $f(x)$ has the form

$$F(x) = \frac{x^2}{2} + k$$

where k is a real number.

(B) Because $F(0) = (0^2/2) + k = k$, the functions

$$F_0(x) = \frac{x^2}{2}, \quad F_1(x) = \frac{x^2}{2} + 1, \quad \text{and} \quad F_2(x) = \frac{x^2}{2} + 2$$

pass through the points $(0, 0)$, $(0, 1)$, and $(0, 2)$, respectively (see Fig. 1).

(C) The graphs of the three antiderivatives are vertical translations of each other.

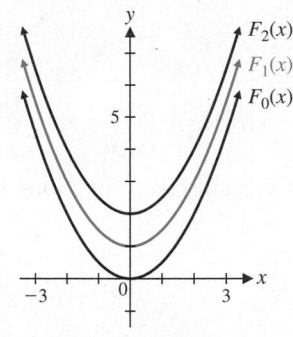

Figure 1

Matched Problem 1   Note that

$$\frac{d}{dx}(x^3) = 3x^2$$

(A) Find all antiderivatives of $f(x) = 3x^2$.

(B) Graph the antiderivative of $f(x) = 3x^2$ that passes through the point $(0, 0)$; through the point $(0, 1)$; through the point $(0, 2)$.

(C) How are the graphs of the three antiderivatives in part (B) related?

Indefinite Integrals: Formulas and Properties

Theorem 1 states that if the derivatives of two functions are equal, then the functions differ by at most a constant. We use the symbol

$$\int f(x)\, dx$$

called the **indefinite integral**, to represent the family of all antiderivatives of $f(x)$, and we write

$$\int f(x)\, dx = F(x) + C \quad \text{if} \quad F'(x) = f(x)$$

The symbol $\int$ is called an **integral sign**, and the function $f(x)$ is called the **integrand**. The symbol dx indicates that the antidifferentiation is performed with respect to the variable x. (We will have more to say about the symbols $\int$ and dx later in the chapter.) The arbitrary constant C is called the **constant of integration**. Referring to the preceding discussion, we can write

$$\int x^2\, dx = \frac{x^3}{3} + C \quad \text{since} \quad \frac{d}{dx}\left(\frac{x^3}{3} + C\right) = x^2$$

Of course, variables other than x can be used in indefinite integrals. For example,

$$\int t^2\, dt = \frac{t^3}{3} + C \quad \text{since} \quad \frac{d}{dt}\left(\frac{t^3}{3} + C\right) = t^2$$

or

$$\int u^2 \, du = \frac{u^3}{3} + C \qquad \text{since} \qquad \frac{d}{du}\left(\frac{u^3}{3} + C\right) = u^2$$

The fact that indefinite integration and differentiation are reverse operations, except for the addition of the constant of integration, can be expressed symbolically as

$$\frac{d}{dx}\left[\int f(x) \, dx\right] = f(x) \qquad \text{The derivative of the indefinite integral of } f(x) \text{ is } f(x).$$

and

$$\int F'(x) \, dx = F(x) + C \qquad \text{The indefinite integral of the derivative of } F(x) \text{ is } F(x) + C.$$

We can develop formulas for the indefinite integrals of certain basic functions from the formulas for derivatives in Chapters 2 and 3.

> **FORMULAS** Indefinite Integrals of Basic Functions
> For C a constant,
>
> **1.** $\displaystyle\int x^n \, dx = \frac{x^{n+1}}{n+1} + C, \qquad n \neq -1$
>
> **2.** $\displaystyle\int e^x \, dx = e^x + C$
>
> **3.** $\displaystyle\int \frac{1}{x} \, dx = \ln|x| + C, \qquad x \neq 0$

Formula 3 involves the natural logarithm of the absolute value of x. Although the natural logarithm function is only defined for $x > 0$, $f(x) = \ln|x|$ is defined for all $x \neq 0$. Its graph is shown in Figure 2A. Note that $f(x)$ is decreasing for $x < 0$ but is increasing for $x > 0$. Therefore the derivative of f, which by formula 3 is $f'(x) = \dfrac{1}{x}$, is negative for $x < 0$ and positive for $x > 0$ (see Fig. 2B).

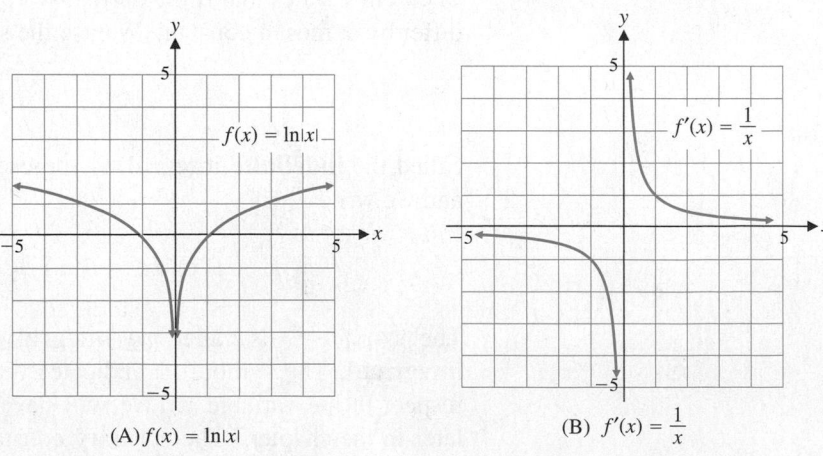

(A) $f(x) = \ln|x|$ (B) $f'(x) = \dfrac{1}{x}$

Figure 2

To justify the three formulas, show that the derivative of the right-hand side is the integrand of the left-hand side (see Problems 75–78 in Exercise 5.1). Note that formula 1 does not give the antiderivative of x^{-1} (because $x^{n+1}/(n+1)$ is undefined when $n = -1$), but formula 3 does.

Explore and Discuss 1   Formulas 1, 2, and 3 do *not* provide a formula for the indefinite integral of the function $\ln x$. Show that if $x > 0$, then

$$\int \ln x \, dx = x \ln x - x + C$$

by differentiating the right-hand side.

We can obtain properties of the indefinite integral from derivative properties that were established in Chapter 2.

PROPERTIES Indefinite Integrals
For k a constant,

4. $\displaystyle\int k f(x) \, dx = k \int f(x) \, dx$

5. $\displaystyle\int [f(x) \pm g(x)] \, dx = \int f(x) \, dx \pm \int g(x) \, dx$

Property 4 states that

The indefinite integral of a constant times a function is the constant times the indefinite integral of the function.

Property 5 states that

The indefinite integral of the sum of two functions is the sum of the indefinite integrals, and the indefinite integral of the difference of two functions is the difference of the indefinite integrals.

To establish property 4, let F be a function such that $F'(x) = f(x)$. Then

$$k \int f(x) \, dx = k \int F'(x) \, dx = k[F(x) + C_1] = kF(x) + kC_1$$

and since $[kF(x)]' = kF'(x) = kf(x)$, we have

$$\int kf(x) \, dx = \int kF'(x) \, dx = kF(x) + C_2$$

But $kF(x) + kC_1$ and $kF(x) + C_2$ describe the same set of functions, because C_1 and C_2 are arbitrary real numbers. Property 4 is established. Property 5 can be established in a similar manner (see Problems 79 and 80 in Exercise 5.1).

⚠ **CAUTION** Property 4 states that **a constant factor can be moved across an integral sign. A variable factor cannot be moved across an integral sign:**

$$\underset{\text{CONSTANT FACTOR}}{\int 5x^{1/2} \, dx = 5 \int x^{1/2} \, dx} \qquad \underset{\text{VARIABLE FACTOR}}{\int xx^{1/2} \, dx \neq x \int x^{1/2} \, dx} \qquad ▲$$

Indefinite integral formulas and properties can be used together to find indefinite integrals for many frequently encountered functions. If $n = 0$, then formula 1 gives

$$\int dx = x + C$$

Therefore, by property 4,

$$\int k \, dx = k(x + C) = kx + kC$$

Because kC is a constant, we replace it with a single symbol that denotes an arbitrary constant (usually C), and write

$$\int k \, dx = kx + C$$

In words,

The indefinite integral of a constant function with value k is $kx + C$.

Similarly, using property 5 and then formulas 2 and 3, we obtain

$$\int \left(e^x + \frac{1}{x} \right) dx = \int e^x \, dx + \int \frac{1}{x} \, dx$$

$$= e^x + C_1 + \ln|x| + C_2$$

Because $C_1 + C_2$ is a constant, we replace it with the symbol C and write

$$\int \left(e^x + \frac{1}{x} \right) dx = e^x + \ln|x| + C$$

EXAMPLE 2   Using Indefinite Integral Properties and Formulas

(A) $\displaystyle\int 5 \, dx = 5x + C$

(B) $\displaystyle\int 9e^x \, dx = 9 \int e^x \, dx = 9e^x + C$

(C) $\displaystyle\int 5t^7 \, dt = 5 \int t^7 \, dt = 5\frac{t^8}{8} + C = \frac{5}{8}t^8 + C$

(D) $\displaystyle\int (4x^3 + 2x - 1) \, dx = \int 4x^3 \, dx + \int 2x \, dx - \int dx$

$$= 4 \int x^3 \, dx + 2 \int x \, dx - \int dx$$

$$= \frac{4x^4}{4} + \frac{2x^2}{2} - x + C$$

$$= x^4 + x^2 - x + C$$

Property 4 can be extended to the sum and difference of an arbitrary number of functions.

(E) $\displaystyle\int \left(2e^x + \frac{3}{x} \right) dx = 2 \int e^x \, dx + 3 \int \frac{1}{x} \, dx$

$$= 2e^x + 3 \ln|x| + C$$

To check any of the results in Example 2, we differentiate the final result to obtain the integrand in the original indefinite integral. When you evaluate an indefinite integral, do not forget to include the arbitrary constant C.

Matched Problem 2   Find each indefinite integral:

(A) $\displaystyle\int 2 \, dx$

(B) $\displaystyle\int 16e^t \, dt$

(C) $\displaystyle\int 3x^4 \, dx$

(D) $\displaystyle\int (2x^5 - 3x^2 + 1) \, dx$

(E) $\displaystyle\int \left(\frac{5}{x} - 4e^x \right) dx$

EXAMPLE 3   Using Indefinite Integral Properties and Formulas

(A) $\displaystyle\int \frac{4}{x^3}\, dx = \int 4x^{-3}\, dx = \frac{4x^{-3+1}}{-3+1} + C = -2x^{-2} + C$

(B) $\displaystyle\int 5\sqrt[3]{u^2}\, du = 5\int u^{2/3}\, du = 5\frac{u^{(2/3)+1}}{\frac{2}{3}+1} + C$

$$= 5\frac{u^{5/3}}{\frac{5}{3}} + C = 3u^{5/3} + C$$

(C) $\displaystyle\int \frac{x^3 - 3}{x^2}\, dx = \int \left(\frac{x^3}{x^2} - \frac{3}{x^2}\right) dx$

$$= \int (x - 3x^{-2})\, dx$$

$$= \int x\, dx - 3\int x^{-2}\, dx$$

$$= \frac{x^{1+1}}{1+1} - 3\frac{x^{-2+1}}{-2+1} + C$$

$$= \tfrac{1}{2}x^2 + 3x^{-1} + C$$

(D) $\displaystyle\int \left(\frac{2}{\sqrt[3]{x}} - 6\sqrt{x}\right) dx = \int (2x^{-1/3} - 6x^{1/2})\, dx$

$$= 2\int x^{-1/3}\, dx - 6\int x^{1/2}\, dx$$

$$= 2\frac{x^{(-1/3)+1}}{-\frac{1}{3}+1} - 6\frac{x^{(1/2)+1}}{\frac{1}{2}+1} + C$$

$$= 2\frac{x^{2/3}}{\frac{2}{3}} - 6\frac{x^{3/2}}{\frac{3}{2}} + C$$

$$= 3x^{2/3} - 4x^{3/2} + C$$

(E) $\displaystyle\int x(x^2 + 2)\, dx = \int (x^3 + 2x)\, dx = \frac{x^4}{4} + x^2 + C$

Matched Problem 3   Find each indefinite integral:

(A) $\displaystyle\int \left(2x^{2/3} - \frac{3}{x^4}\right) dx$

(B) $\displaystyle\int 4\sqrt[5]{w^3}\, dw$

(C) $\displaystyle\int \frac{x^4 - 8x^3}{x^2}\, dx$

(D) $\displaystyle\int \left(8\sqrt[3]{x} - \frac{6}{\sqrt{x}}\right) dx$

(E) $\displaystyle\int (x^2 - 2)(x + 3)\, dx$

⚠ **CAUTION**

1. Note from Example 3E that

$$\int x(x^2 + 2)\, dx \neq \frac{x^2}{2}\left(\frac{x^3}{3} + 2x\right) + C$$

In general, the **indefinite integral of a product is not the product of the indefinite integrals.** (This is expected because the derivative of a product is not the product of the derivatives.)

2.
$$\int e^x \, dx \neq \frac{e^{x+1}}{x+1} + C$$

The power rule applies only to power functions of the form x^n, where the exponent n is a real constant not equal to -1 and the base x is the variable. The function e^x is an exponential function with variable exponent x and constant base e. The correct form is

$$\int e^x \, dx = e^x + C$$

3. Not all elementary functions have elementary antiderivatives. It is impossible, for example, to give a formula for the antiderivative of $f(x) = e^{x^2}$ in terms of elementary functions. Nevertheless, finding such a formula, when it exists, can markedly simplify the solution of certain problems. ▲

Applications

Let's consider some applications of the indefinite integral.

EXAMPLE 4 Curves Find the equation of the curve that passes through $(2, 5)$ if the slope of the curve is given by $dy/dx = 2x$ at any point x.

SOLUTION We want to find a function $y = f(x)$ such that

$$\frac{dy}{dx} = 2x \tag{1}$$

and

$$y = 5 \qquad \text{when} \qquad x = 2 \tag{2}$$

If $dy/dx = 2x$, then

$$y = \int 2x \, dx \tag{3}$$

$$= x^2 + C$$

Since $y = 5$ when $x = 2$, we determine the *particular value of C* so that

$$5 = 2^2 + C$$

So $C = 1$, and

$$y = x^2 + 1$$

is the *particular antiderivative* out of all those possible from equation (3) that satisfies both equations (1) and (2) (see Fig. 3).

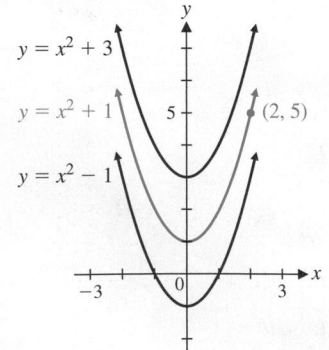

Figure 3 $y = x^2 + C$

Matched Problem 4 Find the equation of the curve that passes through $(2, 6)$ if the slope of the curve is given by $dy/dx = 3x^2$ at any point x.

In certain situations, it is easier to determine the rate at which something happens than to determine how much of it has happened in a given length of time (for example, population growth rates, business growth rates, the rate of healing of a wound, rates of learning or forgetting). If a rate function (derivative) is given and we know the value of the dependent variable for a given value of the independent variable, then we can often find the original function by integration.

EXAMPLE 5   Cost Function   If the marginal cost of producing x units of a commodity is given by

$$C'(x) = 0.3x^2 + 2x$$

and the fixed cost is \$2,000, find the cost function $C(x)$ and the cost of producing 20 units.

SOLUTION   Recall that marginal cost is the derivative of the cost function and that fixed cost is cost at a zero production level. So we want to find $C(x)$, given

$$C'(x) = 0.3x^2 + 2x \qquad C(0) = 2,000$$

We find the indefinite integral of $0.3x^2 + 2x$ and determine the arbitrary integration constant using $C(0) = 2,000$:

$$C'(x) = 0.3x^2 + 2x$$

$$C(x) = \int (0.3x^2 + 2x)\, dx$$

$$= 0.1x^3 + x^2 + K \qquad \text{Since } C \text{ represents the cost, we use}$$
$$\text{K for the constant of integration.}$$

But

$$C(0) = (0.1)0^3 + 0^2 + K = 2,000$$

So $K = 2,000$, and the cost function is

$$C(x) = 0.1x^3 + x^2 + 2,000$$

We now find $C(20)$, the cost of producing 20 units:

$$C(20) = (0.1)20^3 + 20^2 + 2,000$$
$$= \$3,200$$

See Figure 4 for a geometric representation.

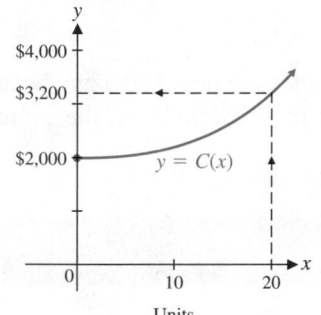

Figure 4

Matched Problem 5   Find the revenue function $R(x)$ when the marginal revenue is

$$R'(x) = 400 - 0.4x$$

and no revenue results at a zero production level. What is the revenue at a production level of 1,000 units?

EXAMPLE 6   Advertising   A satellite radio station is launching an aggressive advertising campaign in order to increase the number of daily listeners. The station currently has 27,000 daily listeners, and management expects the number of daily listeners, $S(t)$, to grow at the rate of

$$S'(t) = 60t^{1/2}$$

listeners per day, where t is the number of days since the campaign began. How long should the campaign last if the station wants the number of daily listeners to grow to 41,000?

SOLUTION   We must solve the equation $S(t) = 41,000$ for t, given that

$$S'(t) = 60t^{1/2} \qquad \text{and} \qquad S(0) = 27,000$$

First, we use integration to find $S(t)$:

$$S(t) = \int 60t^{1/2}\, dt$$

$$= 60\frac{t^{3/2}}{\frac{3}{2}} + C$$

$$= 40t^{3/2} + C$$

Since

$$S(0) = 40(0)^{3/2} + C = 27{,}000$$

we have $C = 27{,}000$ and

$$S(t) = 40t^{3/2} + 27{,}000$$

Now we solve the equation $S(t) = 41{,}000$ for t:

$$40t^{3/2} + 27{,}000 = 41{,}000$$
$$40t^{3/2} = 14{,}000$$
$$t^{3/2} = 350$$
$$t = 350^{2/3} \qquad \text{Use a calculator.}$$
$$= 49.664\,419\ldots$$

The advertising campaign should last approximately 50 days.

Matched Problem 6 There are 64,000 subscribers to an online fashion magazine. Due to competition from a new magazine, the number $C(t)$ of subscribers is expected to decrease at the rate of

$$C'(t) = -600t^{1/3}$$

subscribers per month, where t is the time in months since the new magazine began publication. How long will it take until the number of subscribers to the online fashion magazine drops to 46,000?

Exercises 5.1

Skills Warm-up Exercises

W   *In Problems 1–8, write each function as a sum of terms of the form ax^n, where a is a constant. (If necessary, review Section A.6).*

1. $f(x) = \dfrac{5}{x^4}$

2. $f(x) = -\dfrac{6}{x^9}$

3. $f(x) = \dfrac{3x - 2}{x^5}$

4. $f(x) = \dfrac{x^2 + 5x - 1}{x^3}$

5. $f(x) = \sqrt{x} + \dfrac{5}{\sqrt{x}}$

6. $f(x) = \sqrt[3]{x} - \dfrac{4}{\sqrt[3]{x}}$

7. $f(x) = \sqrt[3]{x}(4 + x - 3x^2)$

8. $f(x) = \sqrt{x}(1 - 5x + x^3)$

In Problems 9–24, find each indefinite integral. Check by differentiating.

9. $\displaystyle\int 7\,dx$

10. $\displaystyle\int 10\,dx$

11. $\displaystyle\int 8x\,dx$

12. $\displaystyle\int 14x\,dx$

13. $\displaystyle\int 9x^2\,dx$

14. $\displaystyle\int 15x^2\,dx$

15. $\displaystyle\int x^5\,dx$

16. $\displaystyle\int x^8\,dx$

17. $\displaystyle\int x^{-3}\,dx$

18. $\displaystyle\int x^{-4}\,dx$

19. $\displaystyle\int 10x^{3/2}\,dx$

20. $\displaystyle\int 8x^{1/3}\,dx$

21. $\displaystyle\int \dfrac{3}{z}\,dz$

22. $\displaystyle\int \dfrac{7}{z}\,dz$

23. $\displaystyle\int 16e^u\,du$

24. $\displaystyle\int 5e^u\,du$

25. Is $F(x) = (x + 1)(x + 2)$ an antiderivative of $f(x) = 2x + 3$? Explain.

26. Is $F(x) = (2x + 5)(x - 6)$ an antiderivative of $f(x) = 4x - 7$? Explain.

27. Is $F(x) = 1 + x \ln x$ an antiderivative of $f(x) = 1 + \ln x$? Explain.

28. Is $F(x) = x \ln x - x + e$ an antiderivative of $f(x) = \ln x$? Explain.

29. Is $F(x) = \dfrac{(2x + 1)^3}{3}$ an antiderivative of $f(x) = (2x + 1)^2$? Explain.

30. Is $F(x) = \dfrac{(3x - 2)^4}{4}$ an antiderivative of $f(x) = (3x - 2)^3$? Explain.

31. Is $F(x) = e^{x^3/3}$ an antiderivative of $f(x) = e^{x^2}$? Explain.

32. Is $F(x) = (e^x - 10)(e^x + 10)$ an antiderivative of $f(x) = 2e^{2x}$? Explain.

In Problems 33–38, discuss the validity of each statement. If the statement is always true, explain why. If not, give a counterexample.

33. The constant function $f(x) = \pi$ is an antiderivative of the constant function $k(x) = 0$.

34. The constant function $k(x) = 0$ is an antiderivative of the constant function $f(x) = \pi$.

35. If n is an integer, then $x^{n+1}/(n+1)$ is an antiderivative of x^n.

36. The constant function $k(x) = 0$ is an antiderivative of itself.

37. The function $h(x) = 5e^x$ is an antiderivative of itself.

38. The constant function $g(x) = 5e^\pi$ is an antiderivative of itself.

In Problems 39–42, could the three graphs in each figure be antiderivatives of the same function? Explain.

39.

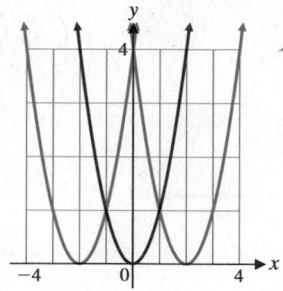

40.

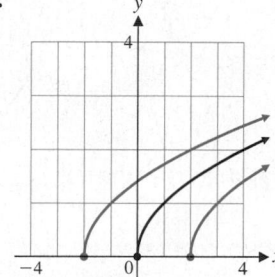

41.

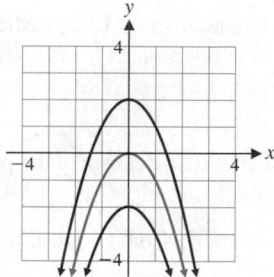

42.

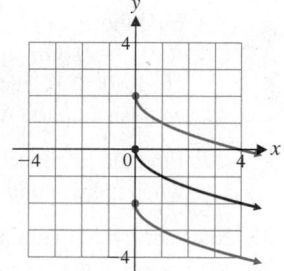

In Problems 43–54, find each indefinite integral. (Check by differentiation.)

43. $\displaystyle \int 5x(1-x)\, dx$

44. $\displaystyle \int x^2(1+x^3)\, dx$

45. $\displaystyle \int \frac{du}{\sqrt{u}}$

46. $\displaystyle \int \frac{dt}{\sqrt[3]{t}}$

47. $\displaystyle \int \frac{dx}{4x^3}$

48. $\displaystyle \int \frac{6\, dm}{m^2}$

49. $\displaystyle \int \frac{4+u}{u}\, du$

50. $\displaystyle \int \frac{1-y^2}{3y}\, dy$

51. $\displaystyle \int (5e^z + 4)\, dz$

52. $\displaystyle \int \frac{e^t - t}{2}\, dt$

53. $\displaystyle \int \left(3x^2 - \frac{2}{x^2} \right) dx$

54. $\displaystyle \int \left(4x^3 + \frac{2}{x^3} \right) dx$

In Problems 55–62, find the particular antiderivative of each derivative that satisfies the given condition.

55. $C'(x) = 6x^2 - 4x;\ C(0) = 3{,}000$

56. $R'(x) = 600 - 0.6x;\ R(0) = 0$

57. $\dfrac{dx}{dt} = \dfrac{20}{\sqrt{t}};\ x(1) = 40$

58. $\dfrac{dR}{dt} = \dfrac{100}{t^2};\ R(1) = 400$

59. $\dfrac{dy}{dx} = 2x^{-2} + 3x^{-1} - 1;\ y(1) = 0$

60. $\dfrac{dy}{dx} = 3x^{-1} + x^{-2};\ y(1) = 1$

61. $\dfrac{dx}{dt} = 4e^t - 2;\ x(0) = 1$

62. $\dfrac{dy}{dt} = 5e^t - 4;\ y(0) = -1$

63. Find the equation of the curve that passes through $(2, 3)$ if its slope is given by

$$\frac{dy}{dx} = 4x - 3$$

for each x.

64. Find the equation of the curve that passes through $(1, 3)$ if its slope is given by

$$\frac{dy}{dx} = 12x^2 - 12x$$

for each x.

In Problems 65–70, find each indefinite integral.

65. $\displaystyle \int \frac{2x^4 - x}{x^3}\, dx$

66. $\displaystyle \int \frac{x^{-1} - x^4}{x^2}\, dx$

67. $\displaystyle \int \frac{x^5 - 2x}{x^4}\, dx$

68. $\displaystyle \int \frac{1 - 3x^4}{x^2}\, dx$

69. $\displaystyle \int \frac{x^2 e^x - 2x}{x^2}\, dx$

70. $\displaystyle \int \frac{1 - xe^x}{x}\, dx$

In Problems 71–74, find the derivative or indefinite integral as indicated.

71. $\dfrac{d}{dx}\left(\displaystyle\int x^3\, dx \right)$

72. $\dfrac{d}{dt}\left(\displaystyle\int \frac{\ln t}{t}\, dt \right)$

73. $\displaystyle \int \frac{d}{dx}(x^4 + 3x^2 + 1)\, dx$

74. $\displaystyle \int \frac{d}{du}\left(e^{u^2} \right) du$

75. Use differentiation to justify the formula

$$\int x^n\, dx = \frac{x^{n+1}}{n+1} + C$$

provided that $n \ne -1$.

76. Use differentiation to justify the formula

$$\int e^x\, dx = e^x + C$$

77. Assuming that $x > 0$, use differentiation to justify the formula

$$\int \frac{1}{x}\, dx = \ln|x| + C$$

78. Assuming that $x < 0$, use differentiation to justify the formula

$$\int \frac{1}{x}\, dx = \ln|x| + C$$

[*Hint:* Use the chain rule after noting that $\ln|x| = \ln(-x)$ for $x < 0$.]

79. Show that the indefinite integral of the sum of two functions is the sum of the indefinite integrals.

[*Hint*: Assume that $\int f(x)\, dx = F(x) + C_1$ and $\int g(x)\, dx = G(x) + C_2$. Using differentiation, show that $F(x) + C_1 + G(x) + C_2$ is the indefinite integral of the function $s(x) = f(x) + g(x)$.]

80. Show that the indefinite integral of the difference of two functions is the difference of the indefinite integrals.

Applications

81. Cost function. The marginal average cost of producing x sports watches is given by

$$\overline{C}'(x) = -\frac{1,000}{x^2} \qquad \overline{C}(100) = 25$$

where $\overline{C}(x)$ is the average cost in dollars. Find the average cost function and the cost function. What are the fixed costs?

82. Renewable energy. In 2012, U.S. consumption of renewable energy was 8.45 quadrillion Btu (or 8.45×10^{15} Btu). Since the 1960s, consumption has been growing at a rate (in quadrillion Btu per year) given by

$$f'(t) = 0.004t + 0.062$$

where t is years after 1960. Find $f(t)$ and estimate U.S. consumption of renewable energy in 2024.

83. Production costs. The graph of the marginal cost function from the production of x thousand bottles of sunscreen per month [where cost $C(x)$ is in thousands of dollars per month] is given in the figure.

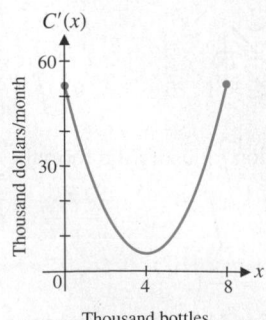

Thousand bottles

(A) Using the graph shown, describe the shape of the graph of the cost function $C(x)$ as x increases from 0 to 8,000 bottles per month.

(B) Given the equation of the marginal cost function,

$$C'(x) = 3x^2 - 24x + 53$$

find the cost function if monthly fixed costs at 0 output are \$30,000. What is the cost of manufacturing 4,000 bottles per month? 8,000 bottles per month?

(C) Graph the cost function for $0 \le x \le 8$. [Check the shape of the graph relative to the analysis in part (A).]

(D) Why do you think that the graph of the cost function is steeper at both ends than in the middle?

84. Revenue. The graph of the marginal revenue function from the sale of x sports watches is given in the figure.

(A) Using the graph shown, describe the shape of the graph of the revenue function $R(x)$ as x increases from 0 to 1,000.

(B) Find the equation of the marginal revenue function (the linear function shown in the figure).

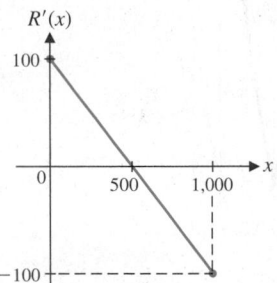

(C) Find the equation of the revenue function that satisfies $R(0) = 0$. Graph the revenue function over the interval $[0, 1,000]$. [Check the shape of the graph relative to the analysis in part (A).]

(D) Find the price–demand equation and determine the price when the demand is 700 units.

85. Sales analysis. Monthly sales of an SUV model are expected to increase at the rate of

$$S'(t) = -24t^{1/3}$$

SUVs per month, where t is time in months and $S(t)$ is the number of SUVs sold each month. The company plans to stop manufacturing this model when monthly sales reach 300 SUVs. If monthly sales now $(t = 0)$ are 1,200 SUVs, find $S(t)$. How long will the company continue to manufacture this model?

86. Sales analysis. The rate of change of the monthly sales of a newly released football game is given by

$$S'(t) = 500t^{1/4} \qquad S(0) = 0$$

where t is the number of months since the game was released and $S(t)$ is the number of games sold each month. Find $S(t)$. When will monthly sales reach 20,000 games?

87. Sales analysis. Repeat Problem 85 if $S'(t) = -24t^{1/3} - 70$ and all other information remains the same. Use a graphing calculator to approximate the solution of the equation $S(t) = 300$ to two decimal places.

88. Sales analysis. Repeat Problem 86 if $S'(t) = 500t^{1/4} + 300$ and all other information remains the same. Use a graphing calculator to approximate the solution of the equation $S(t) = 20,000$ to two decimal places.

89. Labor costs. A defense contractor is starting production on a new missile control system. On the basis of data collected during the assembly of the first 16 control systems, the production manager obtained the following function describing the rate of labor use:

$$L'(x) = 2,400x^{-1/2}$$

For example, after assembly of 16 units, the rate of assembly is 600 labor-hours per unit, and after assembly of 25 units, the rate of assembly is 480 labor-hours per unit. The more units assembled, the more efficient the process. If 19,200 labor-hours are required to assemble of the first 16 units, how many labor-hours $L(x)$ will be required to assemble the first x units? The first 25 units?

90. Labor costs. If the rate of labor use in Problem 89 is

$$L'(x) = 2,000x^{-1/3}$$

and if the first 8 control units require 12,000 labor-hours, how many labor-hours, $L(x)$, will be required for the first x control units? The first 27 control units?

91. Weight–height. For an average person, the rate of change of weight W (in pounds) with respect to height h (in inches) is given approximately by

$$\frac{dW}{dh} = 0.0015h^2$$

Find $W(h)$ if $W(60) = 108$ pounds. Find the weight of an average person who is 5 feet, 10 inches, tall.

92. Wound healing. The area A of a healing wound changes at a rate given approximately by

$$\frac{dA}{dt} = -4t^{-3} \qquad 1 \le t \le 10$$

where t is time in days and $A(1) = 2$ square centimeters. What will the area of the wound be in 10 days?

93. Urban growth. The rate of growth of the population $N(t)$ of a new city t years after its incorporation is estimated to be

$$\frac{dN}{dt} = 400 + 600\sqrt{t} \qquad 0 \le t \le 9$$

If the population was 5,000 at the time of incorporation, find the population 9 years later.

94. Learning. A college language class was chosen for an experiment in learning. Using a list of 50 words, the experiment involved measuring the rate of vocabulary memorization at different times during a continuous 5-hour study session. It was found that the average rate of learning for the entire class was inversely proportional to the time spent studying and was given approximately by

$$V'(t) = \frac{15}{t} \qquad 1 \le t \le 5$$

If the average number of words memorized after 1 hour of study was 15 words, what was the average number of words memorized after t hours of study for $1 \le t \le 5$? After 4 hours of study? Round answer to the nearest whole number.

Answers to Matched Problems

1. (A) $x^3 + C$

   (B)

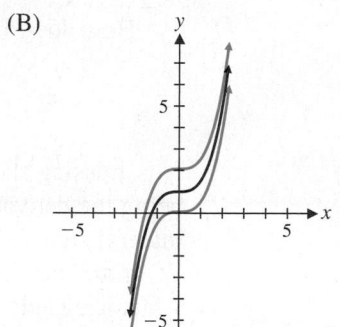

   (C) The graphs are vertical translations of each other.

2. (A) $2x + C$ (B) $16e^t + C$ (C) $\frac{3}{5}x^5 + C$

   (D) $\frac{1}{3}x^6 - x^3 + x + C$ (E) $5 \ln|x| - 4e^x + C$

3. (A) $\frac{6}{5}x^{5/3} + x^{-3} + C$ (B) $\frac{5}{2}w^{8/5} + C$

   (C) $\frac{1}{3}x^3 - 4x^2 + C$ (D) $6x^{4/3} - 12x^{1/2} + C$

   (E) $\frac{1}{4}x^4 + x^3 - x^2 - 6x + C$

4. $y = x^3 - 2$

5. $R(x) = 400x - 0.2x^2; R(1,000) = \$200,000$

6. $t = (40)^{3/4} \approx 16$ mo

5.2 Integration by Substitution

- Reversing the Chain Rule
- Integration by Substitution
- Additional Substitution Techniques
- Application

Many of the indefinite integral formulas introduced in the preceding section are based on corresponding derivative formulas studied earlier. We now consider indefinite integral formulas and procedures based on the chain rule for differentiation.

Reversing the Chain Rule

Recall the chain rule:

$$\frac{d}{dx}f[g(x)] = f'[g(x)]g'(x)$$

The expression on the right is formed from the expression on the left by taking the derivative of the outside function f and multiplying it by the derivative of the inside function g. If we recognize an integrand as a chain-rule form $E'[I(x)]I'(x)$, we can easily find an antiderivative and its indefinite integral:

$$\int E'[I(x)]I'(x)\,dx = E[I(x)] + C \tag{1}$$

We are interested in finding the indefinite integral

$$\int 3x^2 e^{x^3-1}\,dx \tag{2}$$

The integrand appears to be the chain-rule form $e^{g(x)}g'(x)$, which is the derivative of $e^{g(x)}$. Since

$$\frac{d}{dx}e^{x^3-1} = 3x^2 e^{x^3-1}$$

it follows that

$$\int 3x^2 e^{x^3-1}\,dx = e^{x^3-1} + C \tag{3}$$

How does the following indefinite integral differ from integral (2)?

$$\int x^2 e^{x^3-1}\,dx \tag{4}$$

It is missing the constant factor 3. That is, $x^2 e^{x^3-1}$ is within a constant factor of being the derivative of e^{x^3-1}. But because a constant factor can be moved across the integral sign, this causes us little trouble in finding the indefinite integral of $x^2 e^{x^3-1}$. We introduce the constant factor 3 and at the same time multiply by $\frac{1}{3}$ and move the $\frac{1}{3}$ factor outside the integral sign. This is equivalent to multiplying the integrand in integral (4) by 1:

$$\int x^2 e^{x^3-1}\,dx = \int \frac{3}{3}x^2 e^{x^3-1}\,dx \tag{5}$$

$$= \frac{1}{3}\int 3x^2 e^{x^3-1}\,dx = \frac{1}{3}e^{x^3-1} + C$$

The derivative of the rightmost side of equation (5) is the integrand of the indefinite integral (4). Check this.

How does the following indefinite integral differ from integral (2)?

$$\int 3x e^{x^3-1}\,dx \tag{6}$$

It is missing a variable factor x. This is more serious. As tempting as it might be, we *cannot* adjust integral (6) by introducing the variable factor x and moving $1/x$ outside the integral sign, as we did with the constant 3 in equation (5).

⚠ CAUTION   A constant factor can be moved across an integral sign, but a variable factor cannot.   ▲

There is nothing wrong with educated guessing when you are looking for an antiderivative of a given function. You have only to check the result by differentiation. If you are right, you go on your way; if you are wrong, you simply try another approach.

In Section 3.4, we saw that the chain rule extends the derivative formulas for x^n, e^x, and $\ln x$ to derivative formulas for $[f(x)]^n$, $e^{f(x)}$, and $\ln[f(x)]$. The chain rule can also be used to extend the indefinite integral formulas discussed in Section 5.1. Some general formulas are summarized in the following box:

> **FORMULAS** General Indefinite Integral Formulas
>
> **1.** $\displaystyle\int [f(x)]^n f'(x)\,dx = \frac{[f(x)]^{n+1}}{n+1} + C, n \neq -1$
>
> **2.** $\displaystyle\int e^{f(x)} f'(x)\,dx = e^{f(x)} + C$
>
> **3.** $\displaystyle\int \frac{1}{f(x)} f'(x)\,dx = \ln|f(x)| + C$

We can verify each formula by using the chain rule to show that the derivative of the function on the right is the integrand on the left. For example,

$$\frac{d}{dx}\left[e^{f(x)} + C\right] = e^{f(x)} f'(x)$$

verifies formula 2.

EXAMPLE 1   Reversing the Chain Rule

(A) $\displaystyle\int (3x+4)^{10}(3)\,dx = \frac{(3x+4)^{11}}{11} + C$ Formula 1 with $f(x) = 3x + 4$ and $f'(x) = 3$

Check:

$$\frac{d}{dx}\frac{(3x+4)^{11}}{11} = 11\frac{(3x+4)^{10}}{11}\frac{d}{dx}(3x+4) = (3x+4)^{10}(3)$$

(B) $\displaystyle\int e^{x^2}(2x)\,dx = e^{x^2} + C$ Formula 2 with $f(x) = x^2$ and $f'(x) = 2x$

Check:

$$\frac{d}{dx}e^{x^2} = e^{x^2}\frac{d}{dx}x^2 = e^{x^2}(2x)$$

(C) $\displaystyle\int \frac{1}{1+x^3}3x^2\,dx = \ln|1+x^3| + C$ Formula 3 with $f(x) = 1 + x^3$ and $f'(x) = 3x^2$

Check:

$$\frac{d}{dx}\ln|1+x^3| = \frac{1}{1+x^3}\frac{d}{dx}(1+x^3) = \frac{1}{1+x^3}3x^2$$

Matched Problem 1) Find each indefinite integral.

(A) $\displaystyle\int (2x^3 - 3)^{20}(6x^2)\,dx$ (B) $\displaystyle\int e^{5x}(5)\,dx$ (C) $\displaystyle\int \frac{1}{4+x^2}2x\,dx$

Integration by Substitution

The key step in using formulas 1, 2, and 3 is recognizing the form of the integrand. Some people find it difficult to identify $f(x)$ and $f'(x)$ in these formulas and prefer to use a *substitution* to simplify the integrand. The *method of substitution*, which we now discuss, becomes increasingly useful as one progresses in studies of integration.

We start by recalling the definition of the *differential* (see Section 2.6, p. 158). We represent the derivative by the symbol *dy/dx* taken as a whole and now define *dy* and *dx* as two separate quantities with the property that their ratio is still equal to $f'(x)$:

> **DEFINITION** Differentials
>
> If $y = f(x)$ defines a differentiable function, then
>
> 1. The **differential *dx*** of the independent variable x is an arbitrary real number.
> 2. The **differential *dy*** of the dependent variable y is defined as the product of $f'(x)$ and *dx*:
>
> $$dy = f'(x)\, dx$$

Differentials involve mathematical subtleties that are treated carefully in advanced mathematics courses. Here, we are interested in them mainly as a book-keeping device to aid in the process of finding indefinite integrals. We can always check an indefinite integral by differentiating.

EXAMPLE 2   Differentials

(A) If $y = f(x) = x^2$, then

$$dy = f'(x)\, dx = 2x\, dx$$

(B) If $u = g(x) = e^{3x}$, then

$$du = g'(x)\, dx = 3e^{3x}\, dx$$

(C) If $w = h(t) = \ln(4 + 5t)$, then

$$dw = h'(t)\, dt = \frac{5}{4 + 5t}\, dt$$

Matched Problem 2

(A) Find *dy* for $y = f(x) = x^3$.
(B) Find *du* for $u = h(x) = \ln(2 + x^2)$.
(C) Find *dv* for $v = g(t) = e^{-5t}$.

The **method of substitution** is developed through Examples 3–6.

EXAMPLE 3   Using Substitution   Find $\int (x^2 + 2x + 5)^5(2x + 2)\, dx$.

SOLUTION   If

$$u = x^2 + 2x + 5$$

then the differential of u is

$$du = (2x + 2)\, dx$$

Notice that *du* is one of the factors in the integrand. Substitute u for $x^2 + 2x + 5$ and *du* for $(2x + 2)\, dx$ to obtain

$$\int (x^2 + 2x + 5)^5(2x + 2)\, dx = \int u^5\, du$$

$$= \frac{u^6}{6} + C$$

$$= \frac{1}{6}(x^2 + 2x + 5)^6 + C \qquad \text{Since } u = x^2 + 2x + 5$$

Check:

$$\frac{d}{dx}\frac{1}{6}(x^2 + 2x + 5)^6 = \frac{1}{6}(6)(x^2 + 2x + 5)^5\frac{d}{dx}(x^2 + 2x + 5)$$
$$= (x^2 + 2x + 5)^5(2x + 2)$$

Matched Problem 3 Find $\int (x^2 - 3x + 7)^4(2x - 3)\, dx$ by substitution.

The substitution method is also called the **change-of-variable method** since u replaces the variable x in the process. Substituting $u = f(x)$ and $du = f'(x)\, dx$ in formulas 1, 2, and 3 produces the general indefinite integral formulas 4, 5, and 6:

FORMULAS General Indefinite Integral Formulas

4. $\int u^n\, du = \dfrac{u^{n+1}}{n+1} + C, \quad n \neq -1$

5. $\int e^u\, du = e^u + C$

6. $\int \dfrac{1}{u}\, du = \ln|u| + C$

These formulas are valid if u is an independent variable, or if u is a function of another variable and du is the differential of u with respect to that variable.

The substitution method for evaluating certain indefinite integrals is outlined as follows:

PROCEDURE Integration by Substitution

Step 1 Select a substitution that appears to simplify the integrand. In particular, try to select u so that du is a factor in the integrand.

Step 2 Express the integrand entirely in terms of u and du, completely eliminating the original variable and its differential.

Step 3 Evaluate the new integral if possible.

Step 4 Express the antiderivative found in step 3 in terms of the original variable.

EXAMPLE 4 Using Substitution Use a substitution to find each indefinite integral.

(A) $\int (3x + 4)^6(3)\, dx$ (B) $\int e^{t^2}(2t)\, dt$

SOLUTION
(A) If we let $u = 3x + 4$, then $du = 3\, dx$, and

$$\int (3x + 4)^6(3)\, dx = \int u^6\, du \qquad \text{Use formula 4.}$$

$$= \frac{u^7}{7} + C$$

$$= \frac{(3x + 4)^7}{7} + C \qquad \text{Since } u = 3x + 4$$

Check:

$$\frac{d}{dx}\frac{(3x+4)^7}{7} = \frac{7(3x+4)^6}{7}\frac{d}{dx}(3x+4) = (3x+4)^6(3)$$

(B) If we let $u = t^2$, then $du = 2t\,dt$, and

$$\int e^{t^2}(2t)\,dt = \int e^u\,du \qquad \text{Use formula 5.}$$

$$= e^u + C$$
$$= e^{t^2} + C \qquad \text{Since } u = t^2$$

Check:

$$\frac{d}{dt}e^{t^2} = e^{t^2}\frac{d}{dt}t^2 = e^{t^2}(2t)$$

Matched Problem 4   Use a substitution to find each indefinite integral.

(A) $\displaystyle\int (2x^3 - 3)^4(6x^2)\,dx$ (B) $\displaystyle\int e^{5w}(5)\,dw$

Additional Substitution Techniques

In order to use the substitution method, **the integrand must be expressed entirely in terms of u and du.** In some cases, the integrand must be modified before making a substitution and using one of the integration formulas. Example 5 illustrates this process.

EXAMPLE 5   Substitution Techniques   Integrate.

(A) $\displaystyle\int \frac{1}{4x+7}\,dx$ (B) $\displaystyle\int te^{-t^2}\,dt$

(C) $\displaystyle\int 4x^2\sqrt{x^3+5}\,dx$

SOLUTION

(A) If $u = 4x + 7$, then $du = 4\,dx$ and, dividing both sides of the equation $du = 4\,dx$ by 4, we have $dx = \frac{1}{4}\,du$. In the integrand, replace $4x + 7$ by u and replace dx by $\frac{1}{4}\,du$:

$$\int \frac{1}{4x+7}\,dx = \int \frac{1}{u}\left(\frac{1}{4}\,du\right) \qquad \begin{array}{l}\text{Move constant factor across}\\ \text{the integral sign.}\end{array}$$

$$= \frac{1}{4}\int \frac{1}{u}\,du \qquad \text{Use formula 6.}$$

$$= \tfrac{1}{4}\ln|u| + C$$
$$= \tfrac{1}{4}\ln|4x+7| + C \qquad \text{Since } u = 4x+7$$

Check:

$$\frac{d}{dx}\frac{1}{4}\ln|4x+7| = \frac{1}{4}\frac{1}{4x+7}\frac{d}{dx}(4x+7) = \frac{1}{4}\frac{1}{4x+7}4 = \frac{1}{4x+7}$$

(B) If $u = -t^2$, then $du = -2t\,dt$ and, dividing both sides by -2, $-\frac{1}{2}\,du = t\,dt$. In the integrand, replace $-t^2$ by u and replace $t\,dt$ by $-\frac{1}{2}\,du$:

$$\int te^{-t^2}dt = \int e^u\left(-\frac{1}{2}\,du\right) \quad \text{Move constant factor across the integral sign.}$$

$$= -\frac{1}{2}\int e^u\,du \quad \text{Use formula 5.}$$

$$= -\frac{1}{2}e^u + C$$

$$= -\frac{1}{2}e^{-t^2} + C \quad \text{Since } u = -t^2$$

Check:

$$\frac{d}{dt}\left(-\frac{1}{2}e^{-t^2}\right) = -\frac{1}{2}e^{-t^2}\frac{d}{dt}(-t^2) = -\frac{1}{2}e^{-t^2}(-2t) = te^{-t^2}$$

(C) If $u = x^3 + 5$, then $du = 3x^2\,dx$ and, dividing both sides by 3, $\frac{1}{3}\,du = x^2\,dx$. In the integrand, replace $x^3 + 5$ by u and replace $x^2\,dx$ by $\frac{1}{3}\,du$:

$$\int 4x^2\sqrt{x^3 + 5}\,dx = \int 4\sqrt{u}\left(\frac{1}{3}\,du\right) \quad \text{Move constant factors across the integral sign.}$$

$$= \frac{4}{3}\int \sqrt{u}\,du$$

$$= \frac{4}{3}\int u^{1/2}\,du \quad \text{Use formula 4.}$$

$$= \frac{4}{3}\cdot\frac{u^{3/2}}{\frac{3}{2}} + C$$

$$= \frac{8}{9}u^{3/2} + C$$

$$= \frac{8}{9}(x^3 + 5)^{3/2} + C \quad \text{Since } u = x^3 + 5$$

Check:

$$\frac{d}{dx}\left[\frac{8}{9}(x^3 + 5)^{3/2}\right] = \frac{4}{3}(x^3 + 5)^{1/2}\frac{d}{dx}(x^3 + 5)$$

$$= \frac{4}{3}(x^3 + 5)^{1/2}(3x^2) = 4x^2\sqrt{x^3 + 5}$$

Matched Problem 5 | Integrate.

(A) $\displaystyle\int e^{-3x}\,dx$

(B) $\displaystyle\int \frac{x}{x^2 - 9}\,dx$

(C) $\displaystyle\int 5t^2(t^3 + 4)^{-2}\,dt$

Even if it is not possible to find a substitution that makes an integrand match one of the integration formulas exactly, a substitution may simplify the integrand sufficiently so that other techniques can be used.

EXAMPLE 6 Substitution Techniques Find $\displaystyle\int \frac{x}{\sqrt{x+2}}\,dx$.

SOLUTION Proceeding as before, if we let $u = x + 2$, then $du = dx$ and

$$\int \frac{x}{\sqrt{x+2}}\,dx = \int \frac{x}{\sqrt{u}}\,du$$

Notice that this substitution is not complete because we have not expressed the integrand entirely in terms of u and du. As we noted earlier, only a constant factor can be moved across an integral sign, so we cannot move x outside the integral sign. Instead, we must return to the original substitution, solve for x in terms of u, and use the resulting equation to complete the substitution:

$$u = x + 2 \qquad \text{Solve for } x \text{ in terms of } u.$$
$$u - 2 = x \qquad \text{Substitute this expression for } x.$$

Thus,

$$\int \frac{x}{\sqrt{x+2}}\,dx = \int \frac{u-2}{\sqrt{u}}\,du \qquad \text{Simplify the integrand.}$$

$$= \int \frac{u-2}{u^{1/2}}\,du$$

$$= \int (u^{1/2} - 2u^{-1/2})\,du$$

$$\boxed{= \int u^{1/2}\,du - 2\int u^{-1/2}\,du}$$

$$= \frac{u^{3/2}}{\frac{3}{2}} - 2\frac{u^{1/2}}{\frac{1}{2}} + C$$

$$= \tfrac{2}{3}(x+2)^{3/2} - 4(x+2)^{1/2} + C \qquad \text{Since } u = x + 2$$

Check:

$$\frac{d}{dx}\left[\tfrac{2}{3}(x+2)^{3/2} - 4(x+2)^{1/2}\right] = (x+2)^{1/2} - 2(x+2)^{-1/2}$$

$$= \frac{x+2}{(x+2)^{1/2}} - \frac{2}{(x+2)^{1/2}}$$

$$= \frac{x}{(x+2)^{1/2}}$$

Matched Problem 6) Find $\displaystyle\int x\sqrt{x+1}\,dx$.

We can find the indefinite integral of some functions in more than one way. For example, we can use substitution to find

$$\int x(1+x^2)^2\,dx$$

by letting $u = 1 + x^2$. As a second approach, we can expand the integrand, obtaining

$$\int (x + 2x^3 + x^5)\,dx$$

for which we can easily calculate an antiderivative. In such a case, choose the approach that you prefer.

There are also some functions for which substitution is not an effective approach to finding the indefinite integral. For example, substitution is not helpful in finding

$$\int e^{x^2}\, dx \quad \text{or} \quad \int \ln x\, dx$$

Application

EXAMPLE 7 Price–Demand The market research department of a supermarket chain has determined that, for one store, the marginal price $p'(x)$ at x tubes per week for a certain brand of toothpaste is given by

$$p'(x) = -0.015 e^{-0.01x}$$

Find the price–demand equation if the weekly demand is 50 tubes when the price of a tube is \$4.35. Find the weekly demand when the price of a tube is \$3.89.

SOLUTION

$$p(x) = \int -0.015 e^{-0.01x}\, dx$$

$$= -0.015 \int e^{-0.01x}\, dx$$

$$= -0.015 \int e^{-0.01x} \frac{-0.01}{-0.01}\, dx$$

$$= \frac{-0.015}{-0.01} \int e^{-0.01x} (-0.01)\, dx \qquad \text{Substitute } u = -0.01x \text{ and } du = -0.01\ dx.$$

$$= 1.5 \int e^u\, du$$

$$= 1.5 e^u + C$$

$$= 1.5 e^{-0.01x} + C \qquad \text{Since } u = -0.01x$$

We find C by noting that

$$p(50) = 1.5 e^{-0.01(50)} + C = \$4.35$$

$$C = \$4.35 - 1.5 e^{-0.5} \quad \text{Use a calculator.}$$

$$= \$4.35 - 0.91$$

$$= \$3.44$$

So,

$$p(x) = 1.5 e^{-0.01x} + 3.44$$

To find the demand when the price is \$3.89, we solve $p(x) = \$3.89$ for x:

$$1.5 e^{-0.01x} + 3.44 = 3.89$$

$$1.5 e^{-0.01x} = 0.45$$

$$e^{-0.01x} = 0.3$$

$$-0.01x = \ln 0.3$$

$$x = -100 \ln 0.3 \approx 120 \text{ tubes}$$

Matched Problem 7 The marginal price $p'(x)$ at a supply level of x tubes per week for a certain brand of toothpaste is given by

$$p'(x) = 0.001e^{0.01x}$$

Find the price–supply equation if the supplier is willing to supply 100 tubes per week at a price of $3.65 each. How many tubes would the supplier be willing to supply at a price of $3.98 each?

We conclude with two final cautions. The first was stated earlier, but it is worth repeating.

⚠ CAUTION

1. A variable cannot be moved across an integral sign.

2. An integral must be expressed entirely in terms of u and du before applying integration formulas 4, 5, and 6.   ▲

Exercises 5.2

Skills Warm-up Exercises

W In Problems 1–8, use the chain rule to find the derivative of each function. (If necessary, review Section 3.4).

1. $f(x) = (5x + 1)^{10}$
2. $f(x) = (4x - 3)^6$
3. $f(x) = (x^2 + 1)^7$
4. $f(x) = (x^3 - 4)^5$
5. $f(x) = e^{x^2}$
6. $f(x) = 6e^{x^3}$
7. $f(x) = \ln(x^4 - 10)$
8. $f(x) = \ln(x^2 + 5x + 4)$

In Problems 9–44, find each indefinite integral and check the result by differentiating.

9. $\int (3x + 5)^2(3)\,dx$
10. $\int (6x - 1)^3(6)\,dx$

11. $\int (x^2 - 1)^5(2x)\,dx$
12. $\int (x^6 + 1)^4(6x^5)\,dx$

13. $\int (5x^3 + 1)^{-3}(15x^2)\,dx$
14. $\int (4x^2 - 3)^{-6}(8x)\,dx$

15. $\int e^{5x}(5)\,dx$
16. $\int e^{x^3}(3x^2)\,dx$

17. $\int \frac{1}{1 + x^2}(2x)\,dx$
18. $\int \frac{1}{5x - 7}(5)\,dx$

19. $\int \sqrt{1 + x^4}\,(4x^3)\,dx$
20. $\int (x^2 + 9)^{-1/2}(2x)\,dx$

21. $\int (x + 3)^{10}\,dx$
22. $\int (x - 3)^{-4}\,dx$

23. $\int (6t - 7)^{-2}\,dt$
24. $\int (5t + 1)^3\,dt$

25. $\int (t^2 + 1)^5\,t\,dt$
26. $\int (t^3 + 4)^{-2}\,t^2\,dt$

27. $\int xe^{x^2}\,dx$
28. $\int e^{-0.01x}\,dx$

29. $\int \frac{1}{5x + 4}\,dx$
30. $\int \frac{x}{1 + x^2}\,dx$

31. $\int e^{1-t}\,dt$
32. $\int \frac{3}{2 - t}\,dt$

33. $\int \frac{t}{(3t^2 + 1)^4}\,dt$
34. $\int \frac{t^2}{(t^3 - 2)^5}\,dt$

35. $\int x\sqrt{x + 4}\,dx$
36. $\int x\sqrt{x - 9}\,dx$

37. $\int \frac{x}{\sqrt{x - 3}}\,dx$
38. $\int \frac{x}{\sqrt{x + 5}}\,dx$

39. $\int x(x - 4)^9\,dx$
40. $\int x(x + 6)^8\,dx$

41. $\int e^{2x}(1 + e^{2x})^3\,dx$
42. $\int e^{-x}(1 - e^{-x})^4\,dx$

43. $\int \frac{1 + x}{4 + 2x + x^2}\,dx$
44. $\int \frac{x^2 - 1}{x^3 - 3x + 7}\,dx$

In Problems 45–50, the indefinite integral can be found in more than one way. First use the substitution method to find the indefinite integral. Then find it without using substitution. Check that your answers are equivalent.

45. $\int 5(5x + 3)\,dx$
46. $\int -7(4 - 7x)\,dx$

47. $\int 2x(x^2 - 1)\,dx$
48. $\int 3x^2(x^3 + 1)\,dx$

49. $\int 5x^4(x^5)^4\,dx$
50. $\int 8x^7(x^8)^3\,dx$

51. Is $F(x) = x^2 e^x$ an antiderivative of $f(x) = 2xe^x$? Explain.

52. Is $F(x) = \dfrac{1}{x}$ an antiderivative of $f(x) = \ln x$? Explain.

53. Is $F(x) = (x^2 + 4)^6$ an antiderivative of $f(x) = 12x(x^2 + 4)^5$? Explain.

54. Is $F(x) = (x^2 - 1)^{100}$ an antiderivative of $f(x) = 200x(x^2 - 1)^{99}$? Explain.

55. Is $F(x) = e^{2x} + 4$ an antiderivative of $f(x) = e^{2x}$? Explain.

56. Is $F(x) = 1 - 0.2e^{-5x}$ an antiderivative of $f(x) = e^{-5x}$? Explain.

57. Is $F(x) = 0.5(\ln x)^2 + 10$ an antiderivative of $f(x) = \dfrac{\ln x}{x}$? Explain.

58. Is $F(x) = \ln(\ln x)$ an antiderivative of $f(x) = \dfrac{l}{x \ln x}$? Explain.

In Problems 59–70, find each indefinite integral and check the result by differentiating.

59. $\displaystyle\int x\sqrt{3x^2 + 7}\, dx$

60. $\displaystyle\int x^2\sqrt{2x^3 + 1}\, dx$

61. $\displaystyle\int x(x^3 + 2)^2\, dx$

62. $\displaystyle\int x(x^2 + 2)^2\, dx$

63. $\displaystyle\int x^2(x^3 + 2)^2\, dx$

64. $\displaystyle\int (x^2 + 2)^2\, dx$

65. $\displaystyle\int \frac{x^3}{\sqrt{2x^4 + 3}}\, dx$

66. $\displaystyle\int \frac{x^2}{\sqrt{4x^3 - 1}}\, dx$

67. $\displaystyle\int \frac{(\ln x)^3}{x}\, dx$

68. $\displaystyle\int \frac{e^x}{1 + e^x}\, dx$

69. $\displaystyle\int \frac{1}{x^2}e^{-1/x}\, dx$

70. $\displaystyle\int \frac{1}{x \ln x}\, dx$

In Problems 71–76, find the family of all antiderivatives of each derivative.

71. $\dfrac{dx}{dt} = 7t^2(t^3 + 5)^6$

72. $\dfrac{dm}{dn} = 10n(n^2 - 8)^7$

73. $\dfrac{dy}{dt} = \dfrac{3t}{\sqrt{t^2 - 4}}$

74. $\dfrac{dy}{dx} = \dfrac{5x^2}{(x^3 - 7)^4}$

75. $\dfrac{dp}{dx} = \dfrac{e^x + e^{-x}}{(e^x - e^{-x})^2}$

76. $\dfrac{dm}{dt} = \dfrac{\ln(t - 5)}{t - 5}$

Applications

77. Price–demand equation. The marginal price for a weekly demand of x bottles of shampoo in a drugstore is given by

$$p'(x) = \frac{-6,000}{(3x + 50)^2}$$

Find the price–demand equation if the weekly demand is 150 when the price of a bottle of shampoo is $8. What is the weekly demand when the price is $6.50?

78. Price–supply equation. The marginal price at a supply level of x bottles of laundry detergent per week is given by

$$p'(x) = \frac{300}{(3x + 25)^2}$$

Find the price–supply equation if the distributor of the detergent is willing to supply 75 bottles a week at a price of $5.00 per bottle. How many bottles would the supplier be willing to supply at a price of $5.15 per bottle?

79. Cost function. The weekly marginal cost of producing x pairs of tennis shoes is given by

$$C'(x) = 12 + \frac{500}{x + 1}$$

where $C(x)$ is cost in dollars. If the fixed costs are $2,000 per week, find the cost function. What is the average cost per pair of shoes if 1,000 pairs of shoes are produced each week?

80. Revenue function. The weekly marginal revenue from the sale of x pairs of tennis shoes is given by

$$R'(x) = 40 - 0.02x + \frac{200}{x + 1} \qquad R(0) = 0$$

where $R(x)$ is revenue in dollars. Find the revenue function. Find the revenue from the sale of 1,000 pairs of shoes.

81. Marketing. An automobile company is ready to introduce a new line of hybrid cars through a national sales campaign. After test marketing the line in a carefully selected city, the marketing research department estimates that sales (in millions of dollars) will increase at the monthly rate of

$$S'(t) = 10 - 10e^{-0.1t} \qquad 0 \le t \le 24$$

t months after the campaign has started.

(A) What will be the total sales $S(t)$ t months after the beginning of the national campaign if we assume no sales at the beginning of the campaign?

(B) What are the estimated total sales for the first 12 months of the campaign?

(C) When will the estimated total sales reach $100 million? Use a graphing calculator to approximate the answer to two decimal places.

82. Marketing. Repeat Problem 81 if the monthly rate of increase in sales is found to be approximated by

$$S'(t) = 20 - 20e^{-0.05t} \qquad 0 \le t \le 24$$

83. Oil production. Using production and geological data, the management of an oil company estimates that oil will be pumped from a field producing at a rate given by

$$R(t) = \frac{100}{t + 1} + 5 \qquad 0 \le t \le 20$$

where $R(t)$ is the rate of production (in thousands of barrels per year) t years after pumping begins. How many barrels of oil $Q(t)$ will the field produce in the first t years if $Q(0) = 0$? How many barrels will be produced in the first 9 years?

84. Oil production. Assume that the rate in Problem 83 is found to be

$$R(t) = \frac{120t}{t^2 + 1} + 3 \quad 0 \le t \le 20$$

(A) When is the rate of production greatest?

(B) How many barrels of oil $Q(t)$ will the field produce in the first t years if $Q(0) = 0$? How many barrels will be produced in the first 5 years?

(C) How long (to the nearest tenth of a year) will it take to produce a total of a quarter of a million barrels of oil?

85. Biology. A yeast culture is growing at the rate of $w'(t) = 0.2e^{0.1t}$ grams per hour. If the starting culture weighs 2 grams, what will be the weight of the culture $W(t)$ after t hours? After 8 hours?

86. Medicine. The rate of healing for a skin wound (in square centimeters per day) is approximated by $A'(t) = -0.9e^{-0.1t}$. If the initial wound has an area of 9 square centimeters, what will its area $A(t)$ be after t days? After 5 days?

87. Pollution. A contaminated lake is treated with a bactericide. The rate of increase in harmful bacteria t days after the treatment is given by

$$\frac{dN}{dt} = -\frac{2,000t}{1 + t^2} \quad 0 \le t \le 10$$

where $N(t)$ is the number of bacteria per milliliter of water. Since dN/dt is negative, the count of harmful bacteria is decreasing.

(A) Find the minimum value of dN/dt.

(B) If the initial count was 5,000 bacteria per milliliter, find $N(t)$ and then find the bacteria count after 10 days.

(C) When (to two decimal places) is the bacteria count 1,000 bacteria per milliliter?

88. Pollution. An oil tanker aground on a reef is losing oil and producing an oil slick that is radiating outward at a rate given approximately by

$$\frac{dR}{dt} = \frac{60}{\sqrt{t + 9}} \quad t \ge 0$$

where R is the radius (in feet) of the circular slick after t minutes. Find the radius of the slick after 16 minutes if the radius is 0 when $t = 0$.

89. Learning. An average student enrolled in an advanced typing class progressed at a rate of $N'(t) = 6e^{-0.1t}$ words per minute per week t weeks after enrolling in a 15-week course. If, at the beginning of the course, a student could type 40 words per minute, how many words per minute $N(t)$ would the student be expected to type t weeks into the course? After completing the course?

90. Learning. An average student enrolled in a stenotyping class progressed at a rate of $N'(t) = 12e^{-0.06t}$ words per minute per week t weeks after enrolling in a 15-week course. If, at the beginning of the course, a student could stenotype at zero words per minute, how many words per minute $N(t)$ would the student be expected to handle t weeks into the course? After completing the course?

91. College enrollment. The projected rate of increase in enrollment at a new college is estimated by

$$\frac{dE}{dt} = 5,000(t + 1)^{-3/2} \quad t \ge 0$$

where $E(t)$ is the projected enrollment in t years. If enrollment is 2,000 now $(t = 0)$, find the projected enrollment 15 years from now.

Answers to Matched Problems

1. (A) $\frac{1}{21}(2x^3 - 3)^{21} + C$ (B) $e^{5x} + C$
   (C) $\ln|4 + x^2| + C$ or $\ln(4 + x^2) + C$, since $4 + x^2 > 0$
2. (A) $dy = 3x^2\, dx$
   (B) $du = \frac{2x}{2 + x^2}dx$
   (C) $dv = -5e^{-5t}\, dt$
3. $\frac{1}{5}(x^2 - 3x + 7)^5 + C$
4. (A) $\frac{1}{5}(2x^3 - 3)^5 + C$ (B) $e^{5w} + C$
5. (A) $-\frac{1}{3}e^{-3x} + C$ (B) $\frac{1}{2}\ln|x^2 - 9| + C$
   (C) $-\frac{5}{3}(t^3 + 4)^{-1} + C$
6. $\frac{2}{5}(x + 1)^{5/2} - \frac{2}{3}(x + 1)^{3/2} + C$
7. $p(x) = 0.1e^{0.01x} + 3.38$; 179 tubes

5.3 Differential Equations; Growth and Decay

- Differential Equations and Slope Fields
- Continuous Compound Interest Revisited
- Exponential Growth Law
- Population Growth, Radioactive Decay, and Learning
- Comparison of Exponential Growth Phenomena

In the preceding section, we considered equations of the form

$$\frac{dy}{dx} = 6x^2 - 4x \quad y' = -400e^{-0.04x}$$

These are examples of *differential equations*. In general, an equation is a **differential equation** if it involves an unknown function and one or more of its derivatives. Other examples of differential equations are

$$\frac{dy}{dx} = ky \quad y'' - xy' + x^2 = 5 \quad \frac{dy}{dx} = 2xy$$

The first and third equations are called **first-order** equations because each involves a first derivative but no higher derivative. The second equation is called a **second-order** equation because it involves a second derivative but no higher derivative.

A **solution** of a differential equation is a function $f(x)$ which, when substituted for y, satisfies the equation; that is, the left side and right side of the equation are the same function. Finding a solution of a given differential equation may be very difficult. However, it is easy to determine whether or not a given function is a solution of a given differential equation. Just substitute and check whether both sides of the differential equation are equal as functions. For example, even if you have trouble finding a function y that satisfies the differential equation

$$(x - 3)\frac{dy}{dx} = y + 4 \tag{1}$$

it is easy to determine whether or not the function $y = 5x - 19$ is a solution: Since $dy/dx = 5$, the left side of (1) is $(x - 3)5$ and the right side is $(5x - 19) + 4$, so the left and right sides are equal and $y = 5x - 19$ is a solution.

In this section, we emphasize a few special first-order differential equations that have immediate and significant applications. We start by looking at some first-order equations geometrically, in terms of *slope fields*. We then consider continuous compound interest as modeled by a first-order differential equation. From this treatment, we can generalize our approach to a wide variety of other types of growth phenomena.

Differential Equations and Slope Fields

We introduce the concept of *slope field* through an example. Consider the first-order differential equation

$$\frac{dy}{dx} = 0.2y \tag{2}$$

A function f is a solution of equation (2) if $y = f(x)$ satisfies equation (2) for all values of x in the domain of f. Geometrically interpreted, equation (2) gives us the slope of a solution curve that passes through the point (x, y). For example, if $y = f(x)$ is a solution of equation (2) that passes through the point $(0, 2)$, then the slope of f at $(0, 2)$ is given by

$$\frac{dy}{dx} = 0.2(2) = 0.4$$

We indicate this relationship by drawing a short segment of the tangent line at the point $(0, 2)$, as shown in Figure 1A. The procedure is repeated for points $(-3, 1)$ and $(2, 3)$. Assuming that the graph of f passes through all three points, we sketch an approximate graph of f in Figure 1B.

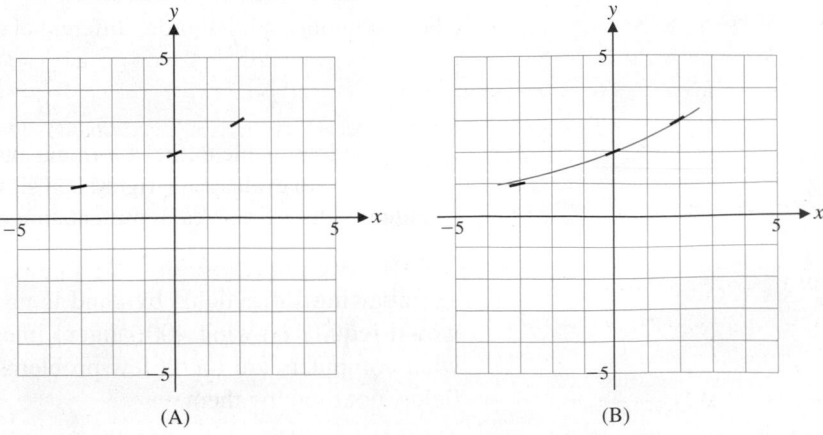

(A)   (B)

Figure 1

If we continue the process of drawing tangent line segments at each point grid in Figure 1—a task easily handled by computers, but not by hand—we obtain a *slope field*. A slope field for differential equation (2), drawn by a computer, is shown in Figure 2. In general, a **slope field** for a first-order differential equation is obtained by drawing tangent line segments determined by the equation at each point in a grid.

Explore and Discuss 1

(A) In Figure 1A (or a copy), draw tangent line segments for a solution curve of differential equation (2) that passes through $(-3, -1), (0, -2)$, and $(2, -3)$.

(B) In Figure 1B (or a copy), sketch an approximate graph of the solution curve that passes through the three points given in part (A). Repeat the tangent line segments first.

(C) What type of function, of all the elementary functions discussed in the first two chapters, appears to be a solution of differential equation (2)?

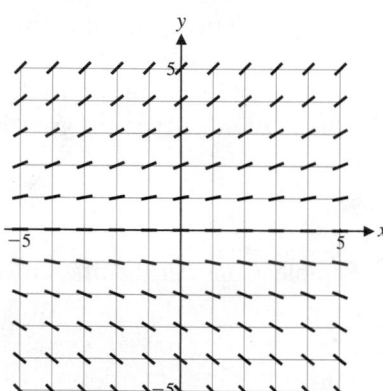

Figure 2

In Explore & Discuss 1, if you guessed that all solutions of equation (2) are exponential functions, you are to be congratulated. We now show that

$$y = Ce^{0.2x} \qquad (3)$$

is a solution of equation (2) for any real number C. We substitute $y = Ce^{0.2x}$ into equation (2) to see if the left side is equal to the right side for all real x:

$$\frac{dy}{dx} = 0.2y$$

Left side: $\dfrac{dy}{dx} = \dfrac{d}{dx}(Ce^{0.2x}) = 0.2Ce^{0.2x}$

Right side: $0.2y = 0.2Ce^{0.2x}$

So equation (3) is a solution of equation (2) for C any real number. Which values of C will produce solution curves that pass through $(0, 2)$ and $(0, -2)$, respectively? Substituting the coordinates of each point into equation (3) and solving for C, we obtain

$$y = 2e^{0.2x} \qquad \text{and} \qquad y = -2e^{0.2x} \qquad (4)$$

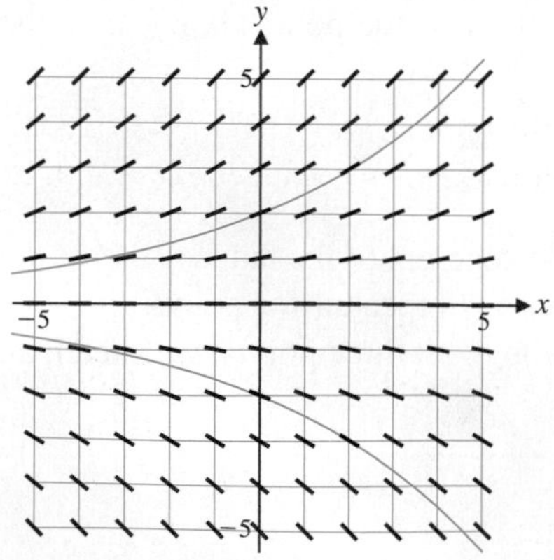

The graphs of equations (4) are shown in Figure 3, and they confirm the results shown in Figure 1B. We say that (3) is the **general solution** of the differential equation (2), and the functions in (4) are the **particular solutions** that satisfy $y(0) = 2$ and $y(0) = -2$, respectively.

Figure 3

CONCEPTUAL INSIGHT

For a complicated first-order differential equation, say,

$$\frac{dy}{dx} = \frac{3 + \sqrt{xy}}{x^2 - 5y^4}$$

it may be impossible to find a formula analogous to (3) for its solutions. Nevertheless, it is routine to evaluate the right-hand side at each point in a grid. The resulting slope field provides a graphical representation of the solutions of the differential equation.

Drawing slope fields by hand is not a task for human beings: A 20-by-20 grid would require drawing 400 tangent line segments! Repetitive tasks of this type are what computers are for. A few problems in Exercises 5.3 involve interpreting slope fields, not drawing them.

Continuous Compound Interest Revisited

Let P be the initial amount of money deposited in an account, and let A be the amount in the account at any time t. Instead of assuming that the money in the account earns a particular rate of interest, suppose we say that the rate of growth of the amount of money in the account at any time t is proportional to the amount present at that time. Since dA/dt is the rate of growth of A with respect to t, we have

$$\frac{dA}{dt} = rA \qquad A(0) = P \qquad A, P > 0 \qquad (5)$$

where r is an appropriate constant. We would like to find a function $A = A(t)$ that satisfies these conditions. Multiplying both sides of equation (5) by $1/A$, we obtain

$$\frac{1}{A}\frac{dA}{dt} = r$$

Now we integrate each side with respect to t:

$$\int \frac{1}{A}\frac{dA}{dt}\,dt = \int r\,dt \qquad \frac{dA}{dt}dt = A'(t)\,dt = dA$$

$$\int \frac{1}{A}\,dA = \int r\,dt$$

$$\ln|A| = rt + C \qquad |A| = A,\ \text{since } A > 0$$

$$\ln A = rt + C$$

We convert this last equation into the equivalent exponential form

$$A = e^{rt+C} \qquad \text{Definition of logarithmic function:}$$
$$\qquad\qquad\qquad y = \ln x \text{ if and only if } x = e^y$$

$$= e^C e^{rt} \qquad \text{Property of exponents: } b^m b^n = b^{m+n}$$

Since $A(0) = P$, we evaluate $A(t) = e^C e^{rt}$ at $t = 0$ and set the result equal to P:

$$A(0) = e^C e^0 = e^C = P$$

Hence, $e^C = P$, and we can rewrite $A = e^C e^{rt}$ in the form

$$A = Pe^{rt}$$

This is the same continuous compound interest formula obtained in Section 4.1, where the principal P is invested at an annual nominal rate r compounded continuously for t years.

Exponential Growth Law

In general, if the rate of change of a quantity Q with respect to time is proportional to the amount of Q present and $Q(0) = Q_0$, then, proceeding in exactly the same way as we just did, we obtain the following theorem:

THEOREM 1 Exponential Growth Law

If $\dfrac{dQ}{dt} = rQ$ and $Q(0) = Q_0$, then $Q = Q_0 e^{rt}$,

where

Q_0 = amount of Q at $t = 0$
r = relative growth rate (expressed as a decimal)
t = time
Q = quantity at time t

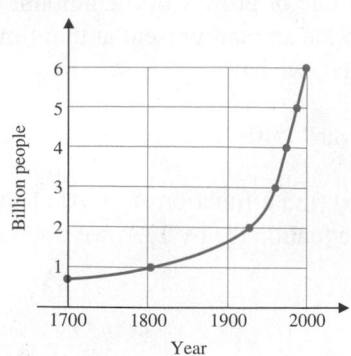

Figure 4 World population growth

The constant r in the exponential growth law is called the **relative growth rate**. If the relative growth rate is $r = 0.02$, then the quantity Q is growing at a rate $dQ/dt = 0.02Q$ (that is, 2% of the quantity Q per unit of time t). Note the distinction between the relative growth rate r and the rate of growth dQ/dt of the quantity Q. If $r < 0$, then $dQ/dt < 0$ and Q is decreasing. This type of growth is called **exponential decay**.

Once we know that the rate of growth is proportional to the amount present, we recognize exponential growth and can use Theorem 1 without solving the differential equation each time. The exponential growth law applies not only to money invested at interest compounded continuously, but also to many other types of problems—population growth, radioactive decay, the depletion of a natural resource, and so on.

Population Growth, Radioactive Decay, and Learning

The world population passed 1 billion in 1804, 2 billion in 1927, 3 billion in 1960, 4 billion in 1974, 5 billion in 1987, and 6 billion in 1999, as illustrated in Figure 4. **Population growth** over certain periods often can be approximated by the exponential growth law of Theorem 1.

EXAMPLE 1 Population Growth India had a population of about 1.2 billion in 2010 ($t = 0$). Let P represent the population (in billions) t years after 2010, and assume a growth rate of 1.5% compounded continuously.

(A) Find an equation that represents India's population growth after 2010, assuming that the 1.5% growth rate continues.

(B) What is the estimated population (to the nearest tenth of a billion) of India in the year 2030?

(C) Graph the equation found in part (A) from 2010 to 2030.

SOLUTION

(A) The exponential growth law applies, and we have

$$\frac{dP}{dt} = 0.015P \qquad P(0) = 1.2$$

Therefore,

$$P = 1.2e^{0.015t} \tag{6}$$

(B) Using equation (6), we can estimate the population in India in 2030 ($t = 20$):

$$P = 1.2e^{0.015(20)} = 1.6 \text{ billion people}$$

(C) The graph is shown in Figure 5.

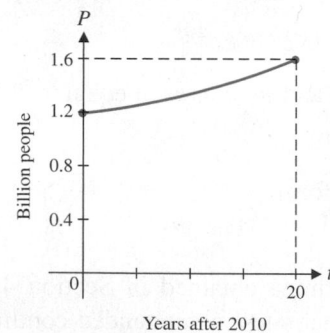

Figure 5 Population of India

Matched Problem 1 Assuming the same continuous compound growth rate as in Example 1, what will India's population be (to the nearest tenth of a billion) in the year 2020?

EXAMPLE 2 Population Growth If the exponential growth law applies to Canada's population growth, at what continuous compound growth rate will the population double over the next 100 years?

SOLUTION We must find r, given that $P = 2P_0$ and $t = 100$:

$$P = P_0e^{rt}$$
$$2P_0 = P_0e^{100r}$$
$$2 = e^{100r} \qquad\qquad \text{Take the natural logarithm}$$
$$100r = \ln 2 \qquad\qquad \text{of both sides and reverse}$$
$$\qquad\qquad\qquad \text{the equation.}$$
$$r = \frac{\ln 2}{100}$$
$$\approx 0.0069 \quad \text{or} \quad 0.69\%$$

Matched Problem 2 | If the exponential growth law applies to population growth in Nigeria, find the doubling time (to the nearest year) of the population if it grows at 2.1% per year compounded continuously.

We now turn to another type of exponential growth: **radioactive decay**. In 1946, Willard Libby (who later received a Nobel Prize in chemistry) found that as long as a plant or animal is alive, radioactive carbon-14 is maintained at a constant level in its tissues. Once the plant or animal is dead, however, the radioactive carbon-14 diminishes by radioactive decay at a rate proportional to the amount present.

$$\frac{dQ}{dt} = rQ \qquad Q(0) = Q_0$$

This is another example of the exponential growth law. The continuous compound rate of decay for radioactive carbon-14 is 0.000 123 8, so $r = -0.000\ 123\ 8$, since decay implies a negative continuous compound growth rate.

EXAMPLE 3   Archaeology   A human bone fragment was found at an archaeological site in Africa. If 10% of the original amount of radioactive carbon-14 was present, estimate the age of the bone (to the nearest 100 years).

SOLUTION   By the exponential growth law for

$$\frac{dQ}{dt} = -0.000\ 123\ 8Q \qquad Q(0) = Q_0$$

we have

$$Q = Q_0 e^{-0.0001238t}$$

We must find t so that $Q = 0.1Q_0$ (since the amount of carbon-14 present now is 10% of the amount Q_0 present at the death of the person).

$$0.1Q_0 = Q_0 e^{-0.0001238t}$$

$$0.1 = e^{-0.0001238t}$$

$$\ln 0.1 = \ln e^{-0.0001238t}$$

$$t = \frac{\ln 0.1}{-0.000\ 123\ 8} \approx 18{,}600 \text{ years}$$

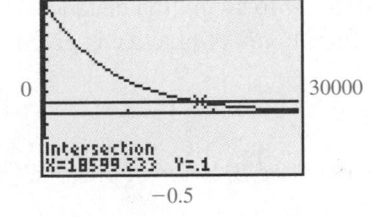

Figure 6 $y_1 = e^{-0.0001238x}$;
$y_2 = 0.1$

See Figure 6 for a graphical solution to Example 3.

Matched Problem 3 | Estimate the age of the bone in Example 3 (to the nearest 100 years) if 50% of the original amount of carbon-14 is present.

In learning certain skills, such as typing and swimming, one often assumes that there is a maximum skill attainable—say, M—and the rate of improvement is proportional to the difference between what has been achieved y and the maximum attainable M. Mathematically,

$$\frac{dy}{dt} = k(M - y) \qquad y(0) = 0$$

We solve this type of problem with the same technique used to obtain the exponential growth law. First, multiply both sides of the first equation by $1/(M - y)$ to get

$$\frac{1}{M - y}\frac{dy}{dt} = k$$

and then integrate each side with respect to t:

$$\int \frac{1}{M-y}\frac{dy}{dt}dt = \int k\,dt$$

$$-\int \frac{1}{M-y}\left(-\frac{dy}{dt}\right)dt = \int k\,dt \qquad \text{Substitute } u = M - y \text{ and}$$

$$-\int \frac{1}{u}du = \int k\,dt \qquad du = -dy = -\frac{dy}{dt}dt.$$

$$-\ln|u| = kt + C \qquad \text{Substitute } M - y, \text{ which is} > 0, \text{ for } u.$$

$$-\ln(M-y) = kt + C \qquad \text{Multiply both sides by } -1.$$

$$\ln(M-y) = -kt - C$$

Change this last equation to an equivalent exponential form:

$$M - y = e^{-kt-C}$$

$$M - y = e^{-C}e^{-kt}$$

$$y = M - e^{-C}e^{-kt}$$

Now, $y(0) = 0$; hence,

$$y(0) = M - e^{-C}e^{0} = 0$$

Solving for e^{-C}, we obtain

$$e^{-C} = M$$

and our final solution is

$$y = M - Me^{-kt} = M(1 - e^{-kt})$$

EXAMPLE 4 Learning For a particular person learning to swim, the distance y (in feet) that the person is able to swim in 1 minute after t hours of practice is given approximately by

$$y = 50(1 - e^{-0.04t})$$

What is the rate of improvement (to two decimal places) after 10 hours of practice?

SOLUTION

$$y = 50 - 50e^{-0.04t}$$

$$y'(t) = 2e^{-0.04t}$$

$$y'(10) = 2e^{-0.04(10)} \approx 1.34 \text{ feet per hour of practice}$$

Matched Problem 4 In Example 4, what is the rate of improvement (to two decimal places) after 50 hours of practice?

Comparison of Exponential Growth Phenomena

Table 1 compares four widely used growth models. Each model (column 2) consists of a first-order differential equation and an **initial condition** that specifies $y(0)$, the value of a solution y when $x = 0$. The differential equation has a family of solutions, but there is only one solution (the particular solution in column 3) that also satisfies the initial condition [just as there is a family, $y = x^2 + k$, of antiderivatives of $g(x) = 2x$, but only one antiderivative (the particular antiderivative $y = x^2 + 5$) that also satisfies the condition $y(0) = 5$]. A graph of the model's solution is shown in column 4 of Table 1, followed by a short (and necessarily incomplete) list of areas in which the model is used.

Table 1 **Exponential Growth**

Description	Model	Solution	Graph	Uses
Unlimited growth: Rate of growth is proportional to the amount present	$\dfrac{dy}{dt} = ky$ $k, t > 0$ $y(0) = c$	$y = ce^{kt}$		• Short-term population growth (people, bacteria, etc.) • Growth of money at continuous compound interest • Price–supply curves
Exponential decay: Rate of growth is proportional to the amount present	$\dfrac{dy}{dt} = -ky$ $k, t > 0$ $y(0) = c$	$y = ce^{-kt}$		• Depletion of natural resources • Radioactive decay • Absorption of light in water • Price–demand curves • Atmospheric pressure (t is altitude)
Limited growth: Rate of growth is proportional to the difference between the amount present and a fixed limit	$\dfrac{dy}{dt} = k(M - y)$ $k, t > 0$ $y(0) = 0$	$y = M(1 - e^{-kt})$		• Sales fads (for example, skateboards) • Depreciation of equipment • Company growth • Learning
Logistic growth: Rate of growth is proportional to the amount present and to the difference between the amount present and a fixed limit	$\dfrac{dy}{dt} = ky(M - y)$ $k, t > 0$ $y(0) = \dfrac{M}{1 + c}$	$y = \dfrac{M}{1 + ce^{-kMt}}$		• Long-term population growth • Epidemics • Sales of new products • Spread of a rumor • Company growth

Exercises 5.3

Skills Warm-up Exercises

W

In Problems 1–8, express the relationship between $f'(x)$ and $f(x)$ in words, and write a differential equation that $f(x)$ satisfies. For example, the derivative of $f(x) = e^{3x}$ is 3 times $f(x)$; $y' = 3y$. (If necessary, review Section 3.4).

1. $f(x) = e^{5x}$

2. $f(x) = e^{-2x}$

3. $f(x) = 10e^{-x}$

4. $f(x) = 25e^{0.04x}$

5. $f(x) = 3.2e^{x^2}$

6. $f(x) = e^{-x^2}$

7. $f(x) = 1 - e^{-x}$

8. $f(x) = 1 - e^{-3x}$

In Problems 9–20, find the general or particular solution, as indicated, for each first-order differential equation.

9. $\dfrac{dy}{dx} = 6x$

10. $\dfrac{dy}{dx} = 3x^{-2}$

11. $\dfrac{dy}{dx} = \dfrac{7}{x}$

12. $\dfrac{dy}{dx} = e^{0.1x}$

13. $\dfrac{dy}{dx} = e^{0.02x}$

14. $\dfrac{dy}{dx} = 8x^{-1}$

15. $\dfrac{dy}{dx} = x^2 - x; y(0) = 0$

16. $\dfrac{dy}{dx} = \sqrt{x}; y(0) = 0$

17. $\dfrac{dy}{dx} = -2xe^{-x^2}; y(0) = 3$

18. $\dfrac{dy}{dx} = e^{x-3}; y(3) = -5$

19. $\dfrac{dy}{dx} = \dfrac{2}{1 + x}; y(0) = 5$

20. $\dfrac{dy}{dx} = \dfrac{1}{4(3 - x)}; y(0) = 1$

In Problems 21–24, give the order (first, second, third, etc.) of each differential equation, where y represents a function of the variable x.

21. $y - 2y' + x^3y'' = 0$

22. $xy' + y^4 = e^x$

23. $y''' - 3y'' + 3y' - y = 0$

24. $y^3 + x^4y'' = \dfrac{5y}{1 + x^2}$

25. Is $y = 5x$ a solution of the differential equation $\dfrac{dy}{dx} = \dfrac{y}{x}$? Explain.

26. Is $y = 8x + 8$ a solution of the differential equation $\dfrac{dy}{dx} = \dfrac{y}{x + 1}$? Explain.

27. Is $y = \sqrt{9 + x^2}$ a solution of the differential equation $y' = \dfrac{x}{y}$? Explain.

28. Is $y = 5e^{x^2/2}$ a solution of the differential equation $y' = xy$? Explain.

29. Is $y = e^{3x}$ a solution of the differential equation $y'' - 4y' + 3y = 0$? Explain.

30. Is $y = -2e^x$ a solution of the differential equation
$y'' - 4y' + 3y = 0$? Explain.

31. Is $y = 100e^{3x}$ a solution of the differential equation
$y'' - 4y' + 3y = 0$? Explain.

32. Is $y = e^{-3x}$ a solution of the differential equation
$y'' - 4y' + 3y = 0$? Explain.

Problems 33–38 refer to the following slope fields:

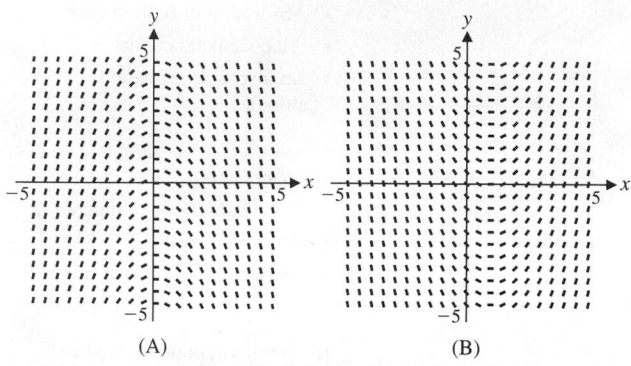

(A)   (B)

Figure for 33–38

33. Which slope field is associated with the differential equation
$dy/dx = x - 1$? Briefly justify your answer.

34. Which slope field is associated with the differential equation
$dy/dx = -x$? Briefly justify your answer.

35. Solve the differential equation $dy/dx = x - 1$ and find the
particular solution that passes through $(0, -2)$.

36. Solve the differential equation $dy/dx = -x$ and find the
particular solution that passes through $(0, 3)$.

37. Graph the particular solution found in Problem 35 in the
appropriate Figure A or B (or a copy).

38. Graph the particular solution found in Problem 36 in the
appropriate Figure A or B (or a copy).

In Problems 39–46, find the general or particular solution, as indicated, for each differential equation.

39. $\dfrac{dy}{dt} = 2y$

40. $\dfrac{dy}{dt} = -3y$

41. $\dfrac{dy}{dx} = -0.5y; y(0) = 100$

42. $\dfrac{dy}{dx} = 0.1y; y(0) = -2.5$

43. $\dfrac{dx}{dt} = -5x$

44. $\dfrac{dx}{dt} = 4t$

45. $\dfrac{dx}{dt} = -5t$

46. $\dfrac{dx}{dt} = 4x$

In Problems 47–50, does the given differential equation model unlimited growth, exponential decay, limited growth, or logistic growth?

47. $y' = 2.5y(300 - y)$

48. $y' = -0.0152y$

49. $y' = 0.43y$

50. $y' = 10,000 - y$

Problems 51–58 refer to the following slope fields:

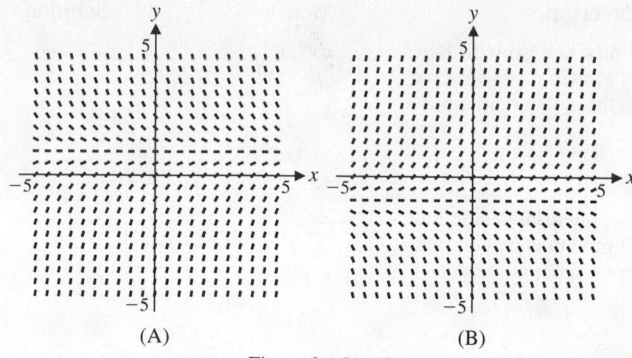

(A)   (B)

Figure for 51–58

51. Which slope field is associated with the differential equation
$dy/dx = 1 - y$? Briefly justify your answer.

52. Which slope field is associated with the differential equation
$dy/dx = y + 1$? Briefly justify your answer.

53. Show that $y = 1 - Ce^{-x}$ is a solution of the differential
equation $dy/dx = 1 - y$ for any real number C. Find the
particular solution that passes through $(0, 0)$.

54. Show that $y = Ce^x - 1$ is a solution of the differential equation $dy/dx = y + 1$ for any real number C. Find the particular solution that passes through $(0, 0)$.

55. Graph the particular solution found in Problem 53 in the
appropriate Figure A or B (or a copy).

56. Graph the particular solution found in Problem 54 in the
appropriate Figure A or B (or a copy).

57. Use a graphing calculator to graph $y = 1 - Ce^{-x}$ for
$C = -2, -1, 1$, and 2, for $-5 \leq x \leq 5, -5 \leq y \leq 5$, all in
the same viewing window. Observe how the solution curves
go with the flow of the tangent line segments in the corresponding slope field shown in Figure A or Figure B.

58. Use a graphing calculator to graph $y = Ce^x - 1$ for
$C = -2, -1, 1$, and 2, for $-5 \leq x \leq 5, -5 \leq y \leq 5$, all in
the same viewing window. Observe how the solution curves
go with the flow of the tangent line segments in the corresponding slope field shown in Figure A or Figure B.

59. Show that $y = \sqrt{C - x^2}$ is a solution of the differential
equation $dy/dx = -x/y$ for any positive real number C. Find
the particular solution that passes through $(3, 4)$.

60. Show that $y = \sqrt{x^2 + C}$ is a solution of the differential
equation $dy/dx = x/y$ for any real number C. Find the particular solution that passes through $(-6, 7)$.

61. Show that $y = Cx$ is a solution of the differential equation
$dy/dx = y/x$ for any real number C. Find the particular
solution that passes through $(-8, 24)$.

62. Show that $y = C/x$ is a solution of the differential equation
$dy/dx = -y/x$ for any real number C. Find the particular
solution that passes through $(2, 5)$.

63. Show that $y = 1/(1 + ce^{-t})$ is a solution of the differential
equation $dy/dt = y(1 - y)$ for any real number c. Find the particular solution that passes through $(0, -1)$.

64. Show that $y = 2/(1 + ce^{-6t})$ is a solution of the differential
equation $dy/dt = 3y(2 - y)$ for any real number c. Find the
particular solution that passes through $(0, 1)$.

In Problems 65–72, use a graphing calculator to graph the given examples of the various cases in Table 1 on page 349.

65. Unlimited growth:

$y = 1,000e^{0.08t}$
$0 \le t \le 15$
$0 \le y \le 3,500$

66. Unlimited growth:

$y = 5,250e^{0.12t}$
$0 \le t \le 10$
$0 \le y \le 20,000$

67. Exponential decay:

$p = 100e^{-0.05x}$
$0 \le x \le 30$
$0 \le p \le 100$

68. Exponential decay:

$p = 1,000e^{-0.08x}$
$0 \le x \le 40$
$0 \le p \le 1,000$

69. Limited growth:

$N = 100(1 - e^{-0.05t})$
$0 \le t \le 100$
$0 \le N \le 100$

70. Limited growth:

$N = 1,000(1 - e^{-0.07t})$
$0 \le t \le 70$
$0 \le N \le 1,000$

71. Logistic growth:

$N = \dfrac{1,000}{1 + 999e^{-0.4t}}$
$0 \le t \le 40$
$0 \le N \le 1,000$

72. Logistic growth:

$N = \dfrac{400}{1 + 99e^{-0.4t}}$
$0 \le t \le 30$
$0 \le N \le 400$

73. Show that the rate of logistic growth, $dy/dt = ky(M - y)$, has its maximum value when $y = M/2$.

74. Find the value of t for which the logistic function

$$y = \frac{M}{1 + ce^{-kMt}}$$

is equal to $M/2$.

75. Let $Q(t)$ denote the population of the world at time t. In 1999, the world population was 6.0 billion and increasing at 1.3% per year; in 2009, it was 6.8 billion and increasing at 1.2% per year. In which year, 1999 or 2009, was dQ/dt (the rate of growth of Q with respect to t) greater? Explain.

76. Refer to Problem 75. Explain why the world population function $Q(t)$ does not satisfy an exponential growth law.

Applications

77. Continuous compound interest. Find the amount A in an account after t years if

$$\frac{dA}{dt} = 0.03A \quad \text{and} \quad A(0) = 1,000$$

78. Continuous compound interest. Find the amount A in an account after t years if

$$\frac{dA}{dt} = 0.02A \quad \text{and} \quad A(0) = 5,250$$

79. Continuous compound interest. Find the amount A in an account after t years if

$$\frac{dA}{dt} = rA \quad A(0) = 8,000 \quad A(2) = 8,260.14$$

80. Continuous compound interest. Find the amount A in an account after t years if

$$\frac{dA}{dt} = rA \quad A(0) = 5,000 \quad A(5) = 5,581.39$$

81. Price–demand. The marginal price dp/dx at x units of demand per week is proportional to the price p. There is no weekly demand at a price of $100 per unit $[p(0) = 100]$, and there is a weekly demand of 5 units at a price of $77.88 per unit $[p(5) = 77.88]$.

(A) Find the price–demand equation.

(B) At a demand of 10 units per week, what is the price?

(C) Graph the price–demand equation for $0 \le x \le 25$.

82. Price–supply. The marginal price dp/dx at x units of supply per day is proportional to the price p. There is no supply at a price of $10 per unit $[p(0) = 10]$, and there is a daily supply of 50 units at a price of $12.84 per unit $[p(50) = 12.84]$.

(A) Find the price–supply equation.

(B) At a supply of 100 units per day, what is the price?

(C) Graph the price–supply equation for $0 \le x \le 250$.

83. Advertising. A company is trying to expose a new product to as many people as possible through TV ads. Suppose that the rate of exposure to new people is proportional to the number of those who have not seen the product out of L possible viewers (limited growth). No one is aware of the product at the start of the campaign, and after 10 days, 40% of L are aware of the product. Mathematically,

$$\frac{dN}{dt} = k(L - N) \quad N(0) = 0 \quad N(10) = 0.4L$$

(A) Solve the differential equation.

(B) What percent of L will have been exposed after 5 days of the campaign?

(C) How many days will it take to expose 80% of L?

(D) Graph the solution found in part (A) for $0 \le t \le 90$.

84. Advertising. Suppose that the differential equation for Problem 83 is

$$\frac{dN}{dt} = k(L - N) \quad N(0) = 0 \quad N(10) = 0.1L$$

(A) Explain what the equation $N(10) = 0.1L$ means.

(B) Solve the differential equation.

(C) How many days will it take to expose 50% of L?

(D) Graph the solution found in part (B) for $0 \le t \le 300$.

85. Biology. For relatively clear bodies of water, the intensity of light is reduced according to

$$\frac{dI}{dx} = -kI \quad I(0) = I_0$$

where I is the intensity of light at x feet below the surface. For the Sargasso Sea off the West Indies, $k = 0.00942$. Find I in terms of x, and find the depth at which the light is reduced to half of that at the surface.

86. Blood pressure. Under certain assumptions, the blood pressure P in the largest artery in the human body (the aorta) changes between beats with respect to time t according to

$$\frac{dP}{dt} = -aP \quad P(0) = P_0$$

where a is a constant. Find $P = P(t)$ that satisfies both conditions.

87. Drug concentration. A single injection of a drug is administered to a patient. The amount Q in the body then decreases at a rate proportional to the amount present. For a particular drug, the rate is 4% per hour. Thus,

$$\frac{dQ}{dt} = -0.04Q \qquad Q(0) = Q_0$$

where t is time in hours.

(A) If the initial injection is 3 milliliters $[Q(0) = 3]$, find $Q = Q(t)$ satisfying both conditions.

(B) How many milliliters (to two decimal places) are in the body after 10 hours?

(C) How many hours (to two decimal places) will it take for only 1 milliliter of the drug to be left in the body?

(D) Graph the solution found in part (A).

88. Simple epidemic. A community of 1,000 people is homogeneously mixed. One person who has just returned from another community has influenza. Assume that the home community has not had influenza shots and all are susceptible. One mathematical model assumes that influenza tends to spread at a rate in direct proportion to the number N who have the disease and to the number $1,000 - N$ who have not yet contracted the disease (logistic growth). Mathematically,

$$\frac{dN}{dt} = kN(1,000 - N) \qquad N(0) = 1$$

where N is the number of people who have contracted influenza after t days. For $k = 0.0004$, $N(t)$ is the logistic growth function

$$N(t) = \frac{1,000}{1 + 999e^{-0.4t}}$$

(A) How many people have contracted influenza after 10 days? After 20 days?

(B) How many days will it take until half the community has contracted influenza?

(C) Find $\lim_{t \to \infty} N(t)$.

(D) Graph $N = N(t)$ for $0 \le t \le 30$.

89. Nuclear accident. One of the dangerous radioactive isotopes detected after the Chernobyl nuclear disaster in 1986 was cesium-137. If 93.3% of the cesium-137 emitted during the disaster was still present 3 years later, find the continuous compound rate of decay of this isotope.

90. Insecticides. Many countries have banned the use of the insecticide DDT because of its long-term adverse effects. Five years after a particular country stopped using DDT, the amount of DDT in the ecosystem had declined to 75% of the amount present at the time of the ban. Find the continuous compound rate of decay of DDT.

91. Archaeology. A skull found in an ancient tomb has 5% of the original amount of radioactive carbon-14 present. Estimate the age of the skull. (See Example 3.)

92. Learning. For a person learning to type, the number N of words per minute that the person could type after t hours of practice was given by the limited growth function

$$N = 100(1 - e^{-0.02t})$$

What is the rate of improvement after 10 hours of practice? After 40 hours of practice?

93. Small-group analysis. In a study on small-group dynamics, sociologists found that when 10 members of a discussion group were ranked according to the number of times each participated, the number $N(k)$ of times that the kth-ranked person participated was given by

$$N(k) = N_1 e^{-0.11(k-1)} \qquad 1 \le k \le 10$$

where N_1 is the number of times that the first-ranked person participated in the discussion. If $N_1 = 180$, in a discussion group of 10 people, estimate how many times the sixth-ranked person participated. How about the 10th-ranked person?

94. Perception. The Weber–Fechner law concerns a person's sensed perception of various strengths of stimulation involving weights, sound, light, shock, taste, and so on. One form of the law states that the rate of change of sensed sensation S with respect to stimulus R is inversely proportional to the strength of the stimulus R. So

$$\frac{dS}{dR} = \frac{k}{R}$$

where k is a constant. If we let R_0 be the threshold level at which the stimulus R can be detected (the least amount of sound, light, weight, and so on, that can be detected), then

$$S(R_0) = 0$$

Find a function S in terms of R that satisfies these conditions.

95. Rumor propagation. Sociologists have found that a rumor tends to spread at a rate in direct proportion to the number x who have heard it and to the number $P - x$ who have not, where P is the total population (logistic growth). If a resident of a 400-student dormitory hears a rumor that there is a case of TB on campus, then $P = 400$ and

$$\frac{dx}{dt} = 0.001x(400 - x) \qquad x(0) = 1$$

where t is time (in minutes). From these conditions, it can be shown that $x(t)$ is the logistic growth function

$$x(t) = \frac{400}{1 + 399e^{-0.4t}}$$

(A) How many people have heard the rumor after 5 minutes? after 20 minutes?

(B) Find $\lim_{t \to \infty} x(t)$.

(C) Graph $x = x(t)$ for $0 \le t \le 30$.

96. Rumor propagation. In Problem 95, how long (to the nearest minute) will it take for half of the group of 400 to have heard the rumor?

Answers to Matched Problems

1. 1.4 billion people   2. 33 yr

3. 5,600 yr   4. 0.27 ft/hr

5.4 The Definite Integral

- Approximating Areas by Left and Right Sums
- The Definite Integral as a Limit of Sums
- Properties of the Definite Integral

The first three sections of this chapter focused on the *indefinite integral*. In this section, we introduce the *definite integral*. The definite integral is used to compute areas, probabilities, average values of functions, future values of continuous income streams, and many other quantities. Initially, the concept of the definite integral may seem unrelated to the notion of the indefinite integral. There is, however, a close connection between the two integrals. The fundamental theorem of calculus, discussed in Section 5.5, makes that connection precise.

Approximating Areas by Left and Right Sums

How do we find the shaded area in Figure 1? That is, how do we find the area bounded by the graph of $f(x) = 0.25x^2 + 1$, the x axis, and the vertical lines $x = 1$ and $x = 5$? [This cumbersome description is usually shortened to "the area under the graph of $f(x) = 0.25x^2 + 1$ from $x = 1$ to $x = 5$."] Our standard geometric area formulas do not apply directly, but the formula for the area of a rectangle can be used indirectly. To see how, we look at a method of approximating the area under the graph by using rectangles. This method will give us any accuracy desired, which is quite different from finding the area exactly. Our first area approximation is made by dividing the interval $[1, 5]$ on the x axis into four equal parts, each of length

$$\Delta x = \frac{5 - 1}{4} = 1*$$

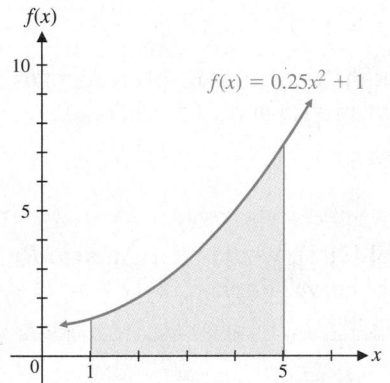

Figure 1 **What is the shaded area?**

We then place a **left rectangle** on each subinterval, that is, a rectangle whose base is the subinterval and whose height is the value of the function at the *left* endpoint of the subinterval (see Fig. 2).

Summing the areas of the left rectangles in Figure 2 results in a **left sum** of four rectangles, denoted by L_4, as follows:

$$L_4 = f(1) \cdot 1 + f(2) \cdot 1 + f(3) \cdot 1 + f(4) \cdot 1$$
$$= 1.25 + 2.00 + 3.25 + 5 = 11.5$$

From Figure 3, since $f(x)$ is increasing, we see that the left sum L_4 underestimates the area, and we can write

$$11.5 = L_4 < \text{Area}$$

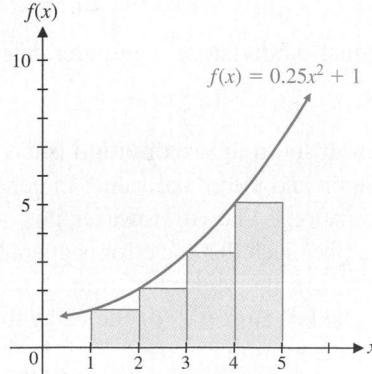

Figure 2 **Left rectangles**

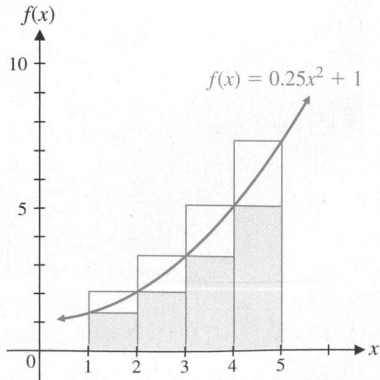

Figure 3 **Left and right rectangles**

*It is customary to denote the length of the subintervals by Δx, which is read "delta x," since Δ is the Greek capital letter delta.

Explore and Discuss 1 If $f(x)$ were decreasing over the interval $[1, 5]$, would the left sum L_4 over- or underestimate the actual area under the curve? Explain.

Similarly, we use the *right* endpoint of each subinterval to find the height of the **right rectangle** placed on the subinterval. Superimposing right rectangles on Figure 2, we get Figure 3 on page 353.

Summing the areas of the right rectangles in Figure 3 results in a **right sum** of four rectangles, denoted by R_4, as follows (compare R_4 with L_4 and note that R_4 can be obtained from L_4 by deleting one rectangular area and adding one more):

$$R_4 = f(2) \cdot 1 + f(3) \cdot 1 + f(4) \cdot 1 + f(5) \cdot 1$$
$$= 2.00 + 3.25 + 5.00 + 7.25 = 17.5$$

From Figure 3, since $f(x)$ is increasing, we see that the right sum R_4 overestimates the area, and we conclude that the actual area is between 11.5 and 17.5. That is,

$$11.5 = L_4 < \text{Area} < R_4 = 17.5$$

Explore and Discuss 2 If $f(x)$ in Figure 3 were decreasing over the interval $[1, 5]$, would the right sum R_4 overestimate or underestimate the actual area under the curve? Explain.

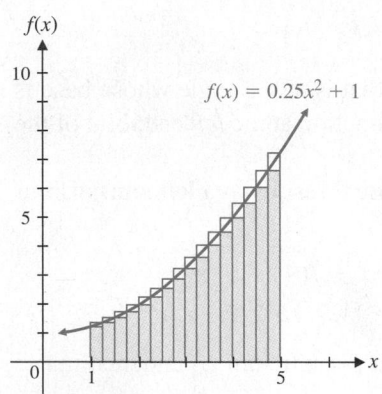

Figure 4

The first approximation of the area under the curve in Figure 1 is fairly coarse, but the method outlined can be continued with increasingly accurate results by dividing the interval $[1, 5]$ into more and more subintervals of equal horizontal length. Of course, this is not a job for hand calculation, but a job that computers are designed to do.* Figure 4 shows left- and right-rectangle approximations for 16 equal subdivisions.

For this case,

$$\Delta x = \frac{5 - 1}{16} = 0.25$$

$$L_{16} = f(1) \cdot \Delta x + f(1.25) \cdot \Delta x + \cdots + f(4.75) \cdot \Delta x$$
$$= 13.59$$

$$R_{16} = f(1.25) \cdot \Delta x + f(1.50) \cdot \Delta x + \cdots + f(5) \cdot \Delta x$$
$$= 15.09$$

Now we know that the area under the curve is between 13.59 and 15.09. That is,

$$13.59 = L_{16} < \text{Area} < R_{16} = 15.09$$

For 100 equal subdivisions, computer calculations give us

$$14.214 = L_{100} < \text{Area} < R_{100} = 14.454$$

The **error in an approximation** is the absolute value of the difference between the approximation and the actual value. In general, neither the actual value nor the error in an approximation is known. However, it is often possible to calculate an **error bound**—a positive number such that the error is guaranteed to be less than or equal to that number.

The error in the approximation of the area under the graph of f from $x = 1$ to $x = 5$ by the left sum L_{16} (or the right sum R_{16}) is less than the sum of the areas of the small rectangles in Figure 4. By stacking those rectangles (see Fig. 5), we see that

$$\text{Error} = |\text{Area} - L_{16}| < |f(5) - f(1)| \cdot \Delta x = 1.5$$

Therefore, 1.5 is an error bound for the approximation of the area under f by L_{16}. We can apply the same stacking argument to any positive function that is increasing on $[a, b]$ or decreasing on $[a, b]$, to obtain the error bound in Theorem 1.

*The computer software that accompanies this book will perform these calculations (see the preface).

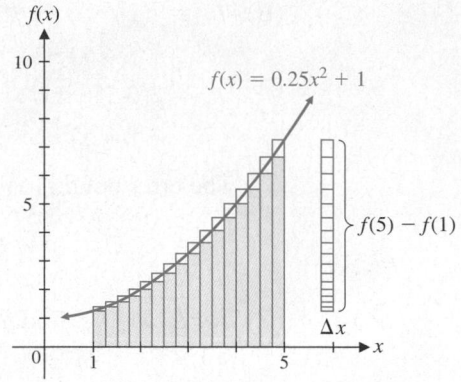

Figure 5

THEOREM 1 Error Bounds for Approximations of Area by Left or Right Sums

If $f(x) > 0$ and is either increasing on $[a, b]$ or decreasing on $[a, b]$, then

$$|f(b) - f(a)| \cdot \frac{b - a}{n}$$

is an error bound for the approximation of the area between the graph of f and the x axis, from $x = a$ to $x = b$, by L_n or R_n.

Because the error bound of Theorem 1 approaches 0 as $n \to \infty$, it can be shown that left and right sums, for certain functions, approach the same limit as $n \to \infty$.

THEOREM 2 Limits of Left and Right Sums

If $f(x) > 0$ and is either increasing on $[a, b]$ or decreasing on $[a, b]$, then its left and right sums approach the same real number as $n \to \infty$.

The number approached as $n \to \infty$ by the left and right sums in Theorem 2 is the area between the graph of f and the x axis from $x = a$ to $x = b$.

EXAMPLE 1 Approximating Areas Given the function $f(x) = 9 - 0.25x^2$, we want to approximate the area under $y = f(x)$ from $x = 2$ to $x = 5$.

(A) Graph the function over the interval $[0, 6]$. Then draw left and right rectangles for the interval $[2, 5]$ with $n = 6$.

(B) Calculate L_6, R_6, and error bounds for each.

(C) How large should n be in order for the approximation of the area by L_n or R_n to be within 0.05 of the true value?

SOLUTION

(A) $\Delta x = 0.5$:

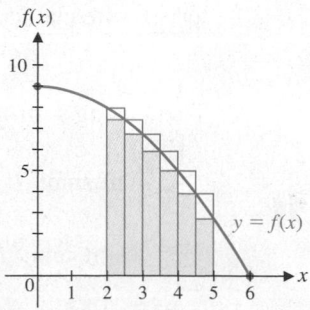

(B) $L_6 = f(2) \cdot \Delta x + f(2.5) \cdot \Delta x + f(3) \cdot \Delta x + f(3.5) \cdot \Delta x + f(4) \cdot \Delta x$
$\qquad + f(4.5) \cdot \Delta x = 18.53$

$R_6 = f(2.5) \cdot \Delta x + f(3) \cdot \Delta x + f(3.5) \cdot \Delta x + f(4) \cdot \Delta x$
$\qquad + f(4.5) \cdot \Delta x + f(5) \cdot \Delta x = 15.91$

The error bound for L_6 and R_6 is

$$\text{error} \leq |f(5) - f(2)|\frac{5 - 2}{6} = |2.75 - 8|(0.5) = 2.625$$

(C) For L_n and R_n, find n such that error ≤ 0.05:

$$|f(b) - f(a)|\frac{b - a}{n} \leq 0.05$$

$$|2.75 - 8|\frac{3}{n} \leq 0.05$$

$$|-5.25|\frac{3}{n} \leq 0.05$$

$$15.75 \leq 0.05n$$

$$n \geq \frac{15.75}{0.05} = 315$$

Matched Problem 1 Given the function $f(x) = 8 - 0.5x^2$, we want to approximate the area under $y = f(x)$ from $x = 1$ to $x = 3$.

(A) Graph the function over the interval $[0, 4]$. Then draw left and right rectangles for the interval $[1, 3]$ with $n = 4$.

(B) Calculate L_4, R_4, and error bounds for each.

(C) How large should n be in order for the approximation of the area by L_n or R_n to be within 0.5 of the true value?

CONCEPTUAL INSIGHT

Note from Example 1C that a relatively large value of n ($n = 315$) is required to approximate the area by L_n or R_n to within 0.05. In other words, 315 rectangles must be used, and 315 terms must be summed, to guarantee that the error does not exceed 0.05. We can obtain a more efficient approximation of the area (fewer terms are summed to achieve a given accuracy) by replacing rectangles with trapezoids. The resulting **trapezoidal rule** and other methods for approximating areas are discussed in Section 6.4.

The Definite Integral as a Limit of Sums

Left and right sums are special cases of more general sums, called *Riemann sums* [named after the German mathematician Georg Riemann (1826–1866)], that are used to approximate areas by means of rectangles.

Let f be a function defined on the interval $[a, b]$. We partition $[a, b]$ into n subintervals of equal length $\Delta x = (b - a)/n$ with endpoints

$$a = x_0 < x_1 < x_2 < \cdots < x_n = b$$

Then, using **summation notation** (see Appendix B.1), we have

Left sum: $L_n = f(x_0)\Delta x + f(x_1)\Delta x + \cdots + f(x_{n-1})\Delta x = \displaystyle\sum_{k=1}^{n} f(x_{k-1})\Delta x$

Right sum: $R_n = f(x_1)\Delta x + f(x_2)\Delta x + \cdots + f(x_n)\Delta x = \displaystyle\sum_{k=1}^{n} f(x_k)\Delta x$

$$\textbf{Riemann sum: } S_n = f(c_1)\Delta x + f(c_2)\Delta x + \cdots + f(c_n)\Delta x = \sum_{k=1}^{n} f(c_k)\Delta x$$

In a **Riemann sum**,* each c_k is required to belong to the subinterval $[x_{k-1}, x_k]$. Left and right sums are the special cases of Riemann sums in which c_k is the left endpoint or right endpoint, respectively, of the subinterval. If $f(x) > 0$, then each term of a Riemann sum S_n represents the area of a rectangle having height $f(c_k)$ and width Δx (see Fig. 6). If $f(x)$ has both positive and negative values, then some terms of S_n represent areas of rectangles, and others represent the negatives of areas of rectangles, depending on the sign of $f(c_k)$ (see Fig. 7).

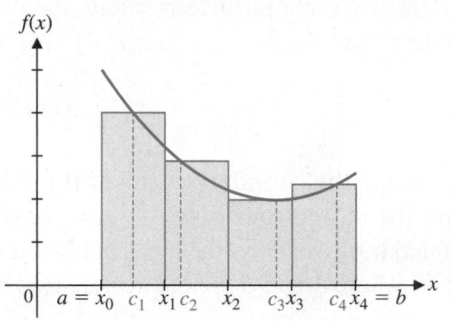

Figure 6

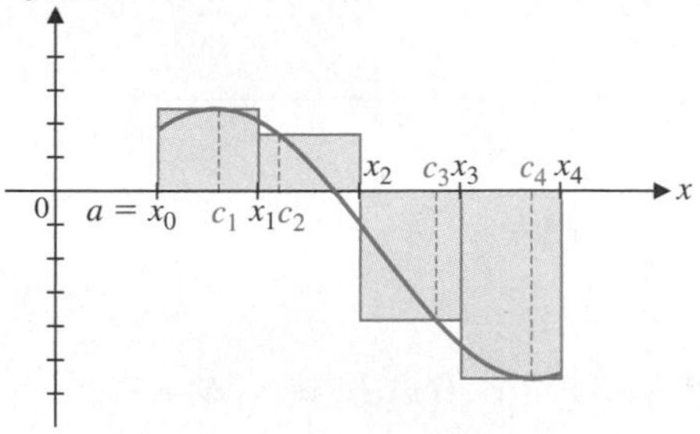

Figure 7

EXAMPLE 2 Riemann Sums Consider the function $f(x) = 15 - x^2$ on $[1, 5]$. Partition the interval $[1, 5]$ into four subintervals of equal length. For each subinterval $[x_{k-1}, x_k]$, let c_k be the midpoint. Calculate the corresponding Riemann sum S_4. (Riemann sums for which the c_k are the midpoints of the subintervals are called **midpoint sums**.)

SOLUTION $\Delta x = \dfrac{5-1}{4} = 1$

$$\begin{aligned} S_4 &= f(c_1) \cdot \Delta x + f(c_2) \cdot \Delta x + f(c_3) \cdot \Delta x + f(c_4) \cdot \Delta x \\ &= f(1.5) \cdot 1 + f(2.5) \cdot 1 + f(3.5) \cdot 1 + f(4.5) \cdot 1 \\ &= 12.75 + 8.75 + 2.75 - 5.25 = 19 \end{aligned}$$

Matched Problem 2 Consider the function $f(x) = x^2 - 2x - 10$ on $[2, 8]$. Partition the interval $[2, 8]$ into three subintervals of equal length. For each subinterval $[x_{k-1}, x_k]$, let c_k be the midpoint. Calculate the corresponding Riemann sum S_3.

By analyzing properties of a continuous function on a closed interval, it can be shown that the conclusion of Theorem 2 is valid if f is continuous. In that case, not just left and right sums, but Riemann sums, have the same limit as $n \to \infty$.

THEOREM 3 Limit of Riemann Sums

If f is a continuous function on $[a, b]$, then the Riemann sums for f on $[a, b]$ approach a real number limit I as $n \to \infty$.†

---

*The term *Riemann sum* is often applied to more general sums in which the subintervals $[x_{k-1}, x_k]$ are not required to have the same length. Such sums are not considered in this book.

†The precise meaning of this limit statement is as follows: For each $e > 0$, there exists some $d > 0$ such that $|S_n - I| < e$ whenever S_n is a Riemann sum for f on $[a, b]$ for which $\Delta x < d$.

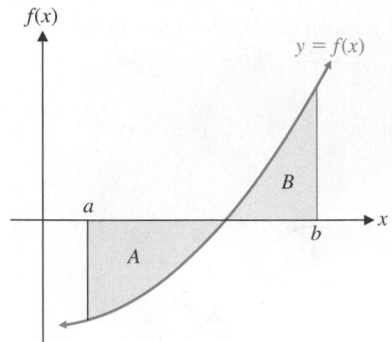

$f(x)$

$y = f(x)$

a B b x

A

Figure 8 $\displaystyle\int_a^b f(x)dx = -A + B$

> **DEFINITION**   Definite Integral
>
> Let f be a continuous function on $[a, b]$. The limit I of Riemann sums for f on $[a, b]$, guaranteed to exist by Theorem 3, is called the **definite integral** of f from a to b and is denoted as
>
> $$\int_a^b f(x)\, dx$$
>
> The **integrand** is $f(x)$, the **lower limit of integration** is a, and the **upper limit of integration** is b.

Because area is a positive quantity, the definite integral has the following geometric interpretation:

$$\int_a^b f(x)\, dx$$

represents the cumulative sum of the signed areas between the graph of f and the x axis from $x = a$ to $x = b$, where the areas above the x axis are counted positively and the areas below the x axis are counted negatively (see Fig. 8, where A and B are the actual areas of the indicated regions).

EXAMPLE 3   Definite Integrals   Calculate the definite integrals by referring to Figure 9.

(A) $\displaystyle\int_a^b f(x)\, dx$

(B) $\displaystyle\int_a^c f(x)\, dx$

(C) $\displaystyle\int_b^c f(x)\, dx$

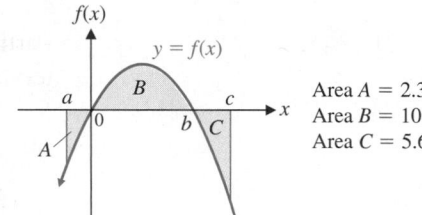

$f(x)$

$y = f(x)$

a B c x

0 b C

A

Area $A = 2.33$
Area $B = 10.67$
Area $C = 5.63$

Figure 9

SOLUTION

(A) $\displaystyle\int_a^b f(x)\, dx = -2.33 + 10.67 = 8.34$

(B) $\displaystyle\int_a^c f(x)\, dx = -2.33 + 10.67 - 5.63 = 2.71$

(C) $\displaystyle\int_b^c f(x)\, dx = -5.63$

Matched Problem 3 ⌋   Referring to the figure for Example 3, calculate the definite integrals.

(A) $\displaystyle\int_a^0 f(x)\, dx$ (B) $\displaystyle\int_0^c f(x)\, dx$ (C) $\displaystyle\int_0^b f(x)\, dx$

Properties of the Definite Integral

Because the definite integral is defined as the limit of Riemann sums, many properties of sums are also properties of the definite integral. Note that Properties 3 and 4 are similar to the indefinite integral properties given in Section 5.1.

Property 5 is illustrated by Figure 9 in Example 3: $2.71 = 8.34 + (-5.63)$. Property 1 follows from the special case of Property 5 in which b and c are both replaced by a. Property 2 follows from the special case of Property 5 in which c is replaced by a.

PROPERTIES   Properties of Definite Integrals

1. $\displaystyle\int_a^a f(x)\,dx = 0$

2. $\displaystyle\int_a^b f(x)\,dx = -\int_b^a f(x)\,dx$

3. $\displaystyle\int_a^b kf(x)\,dx = k\int_a^b f(x)\,dx$, k a constant

4. $\displaystyle\int_a^b [f(x) \pm g(x)]\,dx = \int_a^b f(x)\,dx \pm \int_a^b g(x)\,dx$

5. $\displaystyle\int_a^c f(x)\,dx = \int_a^b f(x)\,dx + \int_b^c f(x)\,dx$

EXAMPLE 4   Using Properties of the Definite Integral   If

$$\int_0^2 x\,dx = 2 \qquad \int_0^2 x^2\,dx = \frac{8}{3} \qquad \int_2^3 x^2\,dx = \frac{19}{3}$$

then

(A) $\displaystyle\int_0^2 12x^2\,dx = 12\int_0^2 x^2\,dx = 12\left(\frac{8}{3}\right) = 32$

(B) $\displaystyle\int_0^2 (2x - 6x^2)\,dx = 2\int_0^2 x\,dx - 6\int_0^2 x^2\,dx = 2(2) - 6\left(\frac{8}{3}\right) = -12$

(C) $\displaystyle\int_3^2 x^2\,dx = -\int_2^3 x^2\,dx = -\frac{19}{3}$

(D) $\displaystyle\int_5^5 3x^2\,dx = 0$

(E) $\displaystyle\int_0^3 3x^2\,dx = 3\int_0^2 x^2\,dx + 3\int_2^3 x^2\,dx = 3\left(\frac{8}{3}\right) + 3\left(\frac{19}{3}\right) = 27$

Matched Problem 4   Using the same integral values given in Example 4, find

(A) $\displaystyle\int_2^3 6x^2\,dx$ (B) $\displaystyle\int_0^2 (9x^2 - 4x)\,dx$

(C) $\displaystyle\int_2^0 3x\,dx$ (D) $\displaystyle\int_{-2}^{-2} 3x\,dx$

(E) $\displaystyle\int_0^3 12x^2\,dx$

Exercises 5.4

Skills Warm-up Exercises

W In Problems 1–6, perform a mental calculation to find the answer and include the correct units. (If necessary, review Appendix C).

1. Find the total area enclosed by 5 non-overlapping rectangles, if each rectangle is 8 inches high and 2 inches wide.

2. Find the total area enclosed by 6 non-overlapping rectangles, if each rectangle is 10 centimeters high and 3 centimeters wide.

3. Find the total area enclosed by 4 non-overlapping rectangles, if each rectangle has width 2 meters and the heights of the rectangles are 3, 4, 5, and 6 meters, respectively.

4. Find the total area enclosed by 5 non-overlapping rectangles, if each rectangle has width 3 feet and the heights of the rectangles are 2, 4, 6, 8, and 10 feet, respectively.

5. A square is inscribed in a circle of radius 1 meter. Is the area inside the circle but outside the square less than 1 square meter?

6. A square is circumscribed around a circle of radius 1 foot. Is the area inside the square but outside the circle less than 1 square foot?

Problems 7–10 refer to the rectangles A, B, C, D, and E in the following figure.

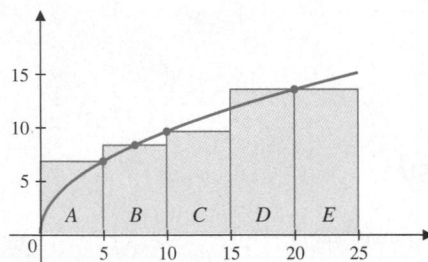

7. Which rectangles are left rectangles?

8. Which rectangles are right rectangles?

9. Which rectangles are neither left nor right rectangles?

10. Which rectangles are both left and right rectangles?

Problems 11–14 refer to the rectangles F, G, H, I, and J in the following figure.

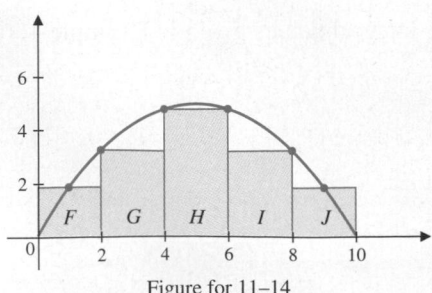

Figure for 11–14

11. Which rectangles are right rectangles?

12. Which rectangles are left rectangles?

13. Which rectangles are both left and right rectangles?

14. Which rectangles are neither left nor right rectangles?

Problems 15–22 involve estimating the area under the curves in Figures A–D from $x = 1$ to $x = 4$. For each figure, divide the interval $[1, 4]$ into three equal subintervals.

15. Draw in left and right rectangles for Figures A and B.

16. Draw in left and right rectangles for Figures C and D.

17. Using the results of Problem 15, compute L_3 and R_3 for Figure A and for Figure B.

18. Using the results of Problem 16, compute L_3 and R_3 for Figure C and for Figure D.

19. Replace the question marks with L_3 and R_3 as appropriate. Explain your choice.

$$? \le \int_1^4 f(x)\, dx \le ? \qquad ? \le \int_1^4 g(x)\, dx \le ?$$

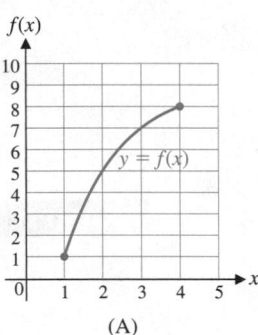

(A)

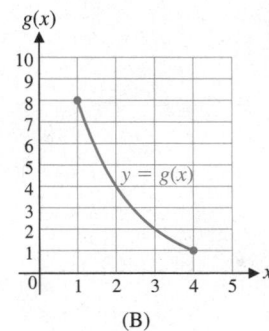

(B)

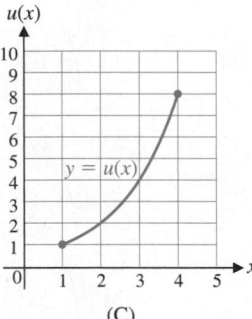

(C)

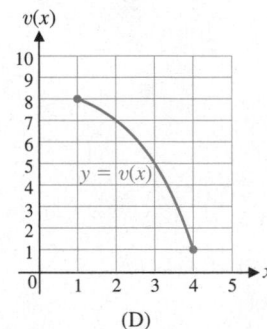

(D)

Figure for 15–22

20. Replace the question marks with L_3 and R_3 as appropriate. Explain your choice.

$$? \le \int_1^4 u(x)\, dx \le ? \qquad ? \le \int_1^4 v(x)\, dx \le ?$$

21. Compute error bounds for L_3 and R_3 found in Problem 17 for both figures.

22. Compute error bounds for L_3 and R_3 found in Problem 18 for both figures.

In Problems 23–26, calculate the indicated Riemann sum S_n for the function $f(x) = 25 - 3x^2$.

23. Partition $[-2, 8]$ into five subintervals of equal length, and for each subinterval $[x_{k-1}, x_k]$, let $c_k = (x_{k-1} + x_k)/2$.

24. Partition $[0, 12]$ into four subintervals of equal length, and for each subinterval $[x_{k-1}, x_k]$, let $c_k = (x_{k-1} + 2x_k)/3$.

25. Partition $[0, 12]$ into four subintervals of equal length, and for each subinterval $[x_{k-1}, x_k]$, let $c_k = (2x_{k-1} + x_k)/3$.

26. Partition $[-5, 5]$ into five subintervals of equal length, and for each subinterval $[x_{k-1}, x_k]$, let $c_k = (x_{k-1} + x_k)/2$.

In Problems 27–30, calculate the indicated Riemann sum S_n for the function $f(x) = x^2 - 5x - 6$.

27. Partition $[0, 3]$ into three subintervals of equal length, and let $c_1 = 0.7, c_2 = 1.8$, and $c_3 = 2.4$.

28. Partition $[0, 3]$ into three subintervals of equal length, and let $c_1 = 0.2, c_2 = 1.5$, and $c_3 = 2.8$.

29. Partition $[1, 7]$ into six subintervals of equal length, and let $c_1 = 1, c_2 = 3, c_3 = 3, c_4 = 5, c_5 = 5$, and $c_6 = 7$.

30. Partition $[1, 7]$ into six subintervals of equal length, and let $c_1 = 2, c_2 = 2, c_3 = 4, c_4 = 4, c_5 = 6$, and $c_6 = 6$.

In Problems 31–42, calculate the definite integral by referring to the figure with the indicated areas.

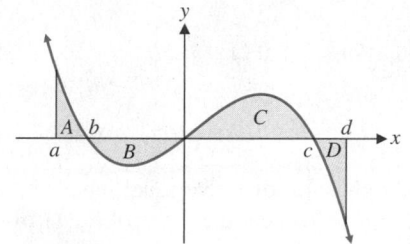

Area $A = 1.408$
Area $B = 2.475$
Area $C = 5.333$
Area $D = 1.792$

Figure for 31–42

31. $\displaystyle\int_b^0 f(x)\, dx$

32. $\displaystyle\int_0^c f(x)\, dx$

33. $\displaystyle\int_a^c f(x)\, dx$

34. $\displaystyle\int_b^d f(x)\, dx$

35. $\displaystyle\int_a^d f(x)\, dx$

36. $\displaystyle\int_0^d f(x)\, dx$

37. $\displaystyle\int_c^0 f(x)\, dx$

38. $\displaystyle\int_d^a f(x)\, dx$

39. $\displaystyle\int_0^a f(x)\, dx$

40. $\displaystyle\int_c^a f(x)\, dx$

41. $\displaystyle\int_d^b f(x)\, dx$

42. $\displaystyle\int_c^b f(x)\, dx$

In Problems 43–54, calculate the definite integral, given that

$$\int_1^4 x\, dx = 7.5 \qquad \int_1^4 x^2\, dx = 21 \qquad \int_4^5 x^2\, dx = \frac{61}{3}$$

43. $\displaystyle\int_1^4 2x\, dx$

44. $\displaystyle\int_1^4 3x^2\, dx$

45. $\displaystyle\int_1^4 (5x + x^2)\, dx$

46. $\displaystyle\int_1^4 (7x - 2x^2)\, dx$

47. $\displaystyle\int_1^4 (x^2 - 10x)\, dx$

48. $\displaystyle\int_1^4 (4x^2 - 9x)\, dx$

49. $\displaystyle\int_1^5 6x^2\, dx$

50. $\displaystyle\int_1^5 -4x^2\, dx$

51. $\displaystyle\int_4^4 (7x - 2)^2\, dx$

52. $\displaystyle\int_5^5 (10 - 7x + x^2)\, dx$

53. $\displaystyle\int_5^4 9x^2\, dx$

54. $\displaystyle\int_4^1 x(1 - x)\, dx$

In Problems 55–60, discuss the validity of each statement. If the statement is always true, explain why. If it is not always true, give a counterexample.

55. If $\int_a^b f(x)\, dx = 0$, then $f(x) = 0$ for all x in $[a, b]$.

56. If $f(x) = 0$ for all x in $[a, b]$, then $\int_a^b f(x)\, dx = 0$.

57. If $f(x) = 2x$ on $[0, 10]$, then there is a positive integer n for which the left sum L_n equals the exact area under the graph of f from $x = 0$ to $x = 10$.

58. If $f(x) = 2x$ on $[0, 10]$ and n is a positive integer, then there is some Riemann sum S_n that equals the exact area under the graph of f from $x = 0$ to $x = 10$.

59. If the area under the graph of f on $[a, b]$ is equal to both the left sum L_n and the right sum R_n for some positive integer n, then f is constant on $[a, b]$.

60. If f is a decreasing function on $[a, b]$, then the area under the graph of f is greater than the left sum L_n and less than the right sum R_n, for any positive integer n.

Problems 61 and 62 refer to the following figure showing two parcels of land along a river:

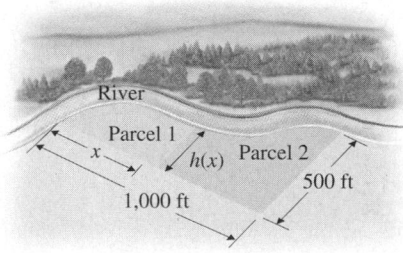

Figure for 61 and 62

61. You want to purchase both parcels of land shown in the figure and make a quick check on their combined area. There is no equation for the river frontage, so you use the average of the left and right sums of rectangles covering the area. The 1,000-foot baseline is divided into 10 equal parts. At the end of each subinterval, a measurement is made from the baseline to the river, and the results are tabulated. Let x be the distance from the left end of the baseline and let $h(x)$ be the distance from the baseline to the river at x. Use L_{10} to estimate the combined area of both parcels, and calculate an error bound for this estimate. How many subdivisions of the baseline would be required so that the error incurred in using L_n would not exceed 2,500 square feet?

x	0	100	200	300	400	500
$h(x)$	0	183	235	245	260	286

x	600	700	800	900	1,000	
$h(x)$	322	388	453	489	500	

62. Refer to Problem 61. Use R_{10} to estimate the combined area of both parcels, and calculate an error bound for this estimate. How many subdivisions of the baseline would be required so that the error incurred in using R_n would not exceed 1,000 square feet?

Problems 63 and 64 refer to the following figure:

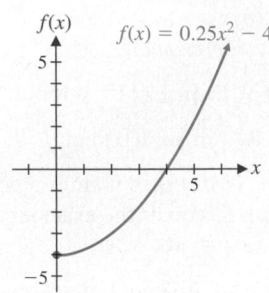

$f(x)$ $f(x) = 0.25x^2 - 4$

Figure for 63 and 64

63. Use L_6 and R_6 to approximate $\int_2^5 (0.25x^2 - 4)\,dx$. Compute error bounds for each. (Round answers to two decimal places.) Describe in geometric terms what the definite integral over the interval $[2, 5]$ represents.

64. Use L_5 and R_5 to approximate $\int_1^6 (0.25x^2 - 4)\,dx$. Compute error bounds for each. (Round answers to two decimal places.) Describe in geometric terms what the definite integral over the interval $[1, 6]$ represents.

For Problems 65–68, use a graphing calculator to determine the intervals on which each function is increasing or decreasing.

65. $f(x) = e^{-x^2}$

66. $f(x) = \dfrac{3}{1 + 2e^{-x}}$

67. $f(x) = x^4 - 2x^2 + 3$

68. $f(x) = e^{x^2}$

In Problems 69–72, the left sum L_n or the right sum R_n is used to approximate the definite integral to the indicated accuracy. How large must n be chosen in each case? (Each function is increasing over the indicated interval.)

69. $\int_1^3 \ln x\,dx = R_n \pm 0.1$

70. $\int_0^{10} \ln(x^2 + 1)\,dx = L_n \pm 0.5$

71. $\int_1^3 x^x\,dx = L_n \pm 0.5$

72. $\int_1^4 x^x\,dx = R_n \pm 0.5$

Applications

73. Employee training. A company producing electric motors has established that, on the average, a new employee can assemble $N(t)$ components per day after t days of on-the-job training, as shown in the following table (a new

employee's productivity increases continuously with time on the job):

t	0	20	40	60	80	100	120
$N(t)$	10	51	68	76	81	84	86

Use left and right sums to estimate the area under the graph of $N(t)$ from $t = 0$ to $t = 60$. Use three subintervals of equal length for each. Calculate an error bound for each estimate.

74. Employee training. For a new employee in Problem 73, use left and right sums to estimate the area under the graph of $N(t)$ from $t = 20$ to $t = 100$. Use four equal subintervals for each. Replace the question marks with the values of L_4 or R_4 as appropriate:

$$? \le \int_{20}^{100} N(t)\,dt \le ?$$

75. Medicine. The rate of healing, $A'(t)$ (in square centimeters per day), for a certain type of skin wound is given approximately by the following table:

t	0	1	2	3	4	5
$A'(t)$	0.90	0.81	0.74	0.67	0.60	0.55
t	6	7	8	9	10	
$A'(t)$	0.49	0.45	0.40	0.36	0.33	

(A) Use left and right sums over five equal subintervals to approximate the area under the graph of $A'(t)$ from $t = 0$ to $t = 5$.

(B) Replace the question marks with values of L_5 and R_5 as appropriate:

$$? \le \int_0^5 A'(t)\,dt \le ?$$

76. Medicine. Refer to Problem 75. Use left and right sums over five equal subintervals to approximate the area under the graph of $A'(t)$ from $t = 5$ to $t = 10$. Calculate an error bound for this estimate.

77. Learning. A psychologist found that, on average, the rate of learning a list of special symbols in a code $N'(x)$ after x days of practice was given approximately by the following table values:

x	0	2	4	6	8	10	12
$N'(x)$	29	26	23	21	19	17	15

Use left and right sums over three equal subintervals to approximate the area under the graph of $N'(x)$ from $x = 6$ to $x = 12$. Calculate an error bound for this estimate.

78. Learning. For the data in Problem 77, use left and right sums over three equal subintervals to approximate the area under the graph of $N'(x)$ from $x = 0$ to $x = 6$. Replace the question marks with values of L_3 and R_3 as appropriate:

$$? \le \int_0^6 N'(x)\,dx \le ?$$

1. (A) $\Delta x = 0.5$:

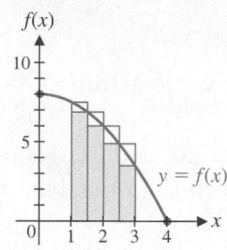

(B) $L_4 = 12.625$, $R_4 = 10.625$; error for L_4 and $R_4 = 2$

(C) $n > 16$ for L_n and R_n

2. $S_3 = 46$

3. (A) -2.33 (B) 5.04 (C) 10.67

4. (A) 38 (B) 16 (C) -6

    (D) 0 (E) 108

5.5 The Fundamental Theorem of Calculus

- Introduction to the Fundamental Theorem

- Evaluating Definite Integrals

- Recognizing a Definite Integral: Average Value

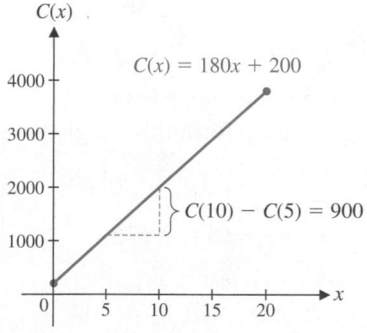

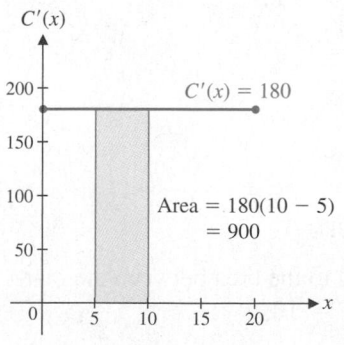

Figure 1

The definite integral of a function f on an interval $[a, b]$ is a number, the area (if $f(x) > 0$) between the graph of f and the x axis from $x = a$ to $x = b$. The indefinite integral of a function is a family of antiderivatives. In this section, we explain the connection between these two integrals, a connection made precise by the fundamental theorem of calculus.

Introduction to the Fundamental Theorem

Suppose that the daily cost function for a small manufacturing firm is given (in dollars) by

$$C(x) = 180x + 200 \qquad 0 \le x \le 20$$

Then the marginal cost function is given (in dollars per unit) by

$$C'(x) = 180$$

What is the change in cost as production is increased from $x = 5$ units to $x = 10$ units? That change is equal to

$$C(10) - C(5) = (180 \cdot 10 + 200) - (180 \cdot 5 + 200)$$
$$= 180(10 - 5)$$
$$= \$900$$

Notice that $180(10 - 5)$ is equal to the area between the graph of $C'(x)$ and the x axis from $x = 5$ to $x = 10$. Therefore,

$$C(10) - C(5) = \int_5^{10} 180 \, dx$$

In other words, the change in cost from $x = 5$ to $x = 10$ is equal to the area between the marginal cost function and the x axis from $x = 5$ to $x = 10$ (see Fig. 1).

CONCEPTUAL INSIGHT

Consider the formula for the slope of a line:

$$m = \frac{y_2 - y_1}{x_2 - x_1}$$

Multiplying both sides of this equation by $x_2 - x_1$ gives

$$y_2 - y_1 = m(x_2 - x_1)$$

The right-hand side, $m(x_2 - x_1)$, is equal to the area of a rectangle of height m and width $x_2 - x_1$. So the change in y coordinates is equal to the area under the constant function with value m from $x = x_1$ to $x = x_2$.

EXAMPLE 1 Change in Cost vs Area under Marginal Cost The daily cost function for a company (in dollars) is given by

$$C(x) = -5x^2 + 210x + 400 \qquad 0 \le x \le 20$$

(A) Graph $C(x)$ for $0 \le x \le 20$, calculate the change in cost from $x = 5$ to $x = 10$, and indicate that change in cost on the graph.

(B) Graph the marginal cost function $C'(x)$ for $0 \le x \le 20$, and use geometric formulas (see Appendix C) to calculate the area between $C'(x)$ and the x axis from $x = 5$ to $x = 10$.

(C) Compare the results of the calculations in parts (A) and (B).

SOLUTION

(A) $C(10) - C(5) = 2{,}000 - 1{,}325 = 675$, and this change in cost is indicated in Figure 2A.

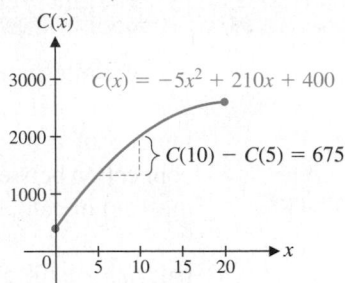

Figure 2A

(B) $C'(x) = -10x + 210$, so the area between $C'(x)$ and the x axis from $x = 5$ to $x = 10$ (see Fig. 2B) is the area of a trapezoid (geometric formulas are given in Appendix C):

$$\text{Area} = \frac{C'(5) + C'(10)}{2}(10 - 5) = \frac{160 + 110}{2}(5) = 675$$

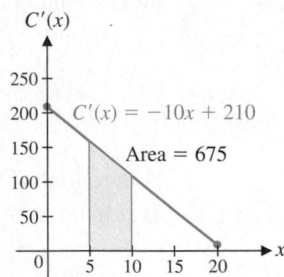

Figure 2B

(C) The change in cost from $x = 5$ to $x = 10$ is equal to the area between the marginal cost function and the x axis from $x = 5$ to $x = 10$.

Matched Problem 1 Repeat Example 1 for the daily cost function

$$C(x) = -7.5x^2 + 305x + 625$$

The connection illustrated in Example 1, between the change in a function from $x = a$ to $x = b$ and the area under the derivative of the function, provides the link between antiderivatives (or indefinite integrals) and the definite integral. This link is known as the fundamental theorem of calculus. (See Problems 67 and 68 in Exercise 5.5 for an outline of its proof.)

THEOREM 1 Fundamental Theorem of Calculus

If f is a continuous function on $[a, b]$, and F is any antiderivative of f, then

$$\int_a^b f(x)\, dx = F(b) - F(a)$$

CONCEPTUAL INSIGHT

Because a definite integral is the limit of Riemann sums, we expect that it would be difficult to calculate definite integrals exactly. The fundamental theorem, however, gives us an easy method for evaluating definite integrals, *provided that we can find an antiderivative $F(x)$ of $f(x)$*: Simply calculate the difference $F(b) - F(a)$. But what if we are unable to find an antiderivative of $f(x)$? In that case, we must resort to left sums, right sums, or other approximation methods to approximate the definite integral. However, it is often useful to remember that such an approximation is also an estimate of the change $F(b) - F(a)$.

Evaluating Definite Integrals

By the fundamental theorem, we can evaluate $\int_a^b f(x)\, dx$ easily and exactly whenever we can find an antiderivative $F(x)$ of $f(x)$. We simply calculate the difference $F(b) - F(a)$. If $G(x)$ is another antiderivative of $f(x)$, then $G(x) = F(x) + C$ for some constant C. So

$$G(b) - G(a) = F(b) + C - [F(a) + C]$$
$$= F(b) - F(a)$$

In other words:

> Any antiderivative of $f(x)$ can be used in the fundamental theorem. One generally chooses the simplest antiderivative by letting $C = 0$, since any other value of C will drop out in computing the difference $F(b) - F(a)$.

Now you know why we studied techniques of indefinite integration before this section—so that we would have methods of finding antiderivatives of large classes of elementary functions for use with the fundamental theorem.

In evaluating definite integrals by the fundamental theorem, it is convenient to use the notation $F(x)\big|_a^b$, which represents the change in $F(x)$ from $x = a$ to $x = b$, as an intermediate step in the calculation. This technique is illustrated in the following examples.

EXAMPLE 2 Evaluating Definite Integrals   Evaluate $\displaystyle\int_1^2 \left(2x + 3e^x - \frac{4}{x}\right) dx.$

SOLUTION $\displaystyle\int_1^2 \left(2x + 3e^x - \frac{4}{x}\right) dx = 2\int_1^2 x\, dx + 3\int_1^2 e^x\, dx - 4\int_1^2 \frac{1}{x}\, dx$

$$= 2\frac{x^2}{2}\bigg|_1^2 + 3e^x\bigg|_1^2 - 4\ln|x|\,\bigg|_1^2$$

$$= (2^2 - 1^2) + (3e^2 - 3e^1) - (4\ln 2 - 4\ln 1)$$

$$= 3 + 3e^2 - 3e - 4\ln 2 \approx 14.24$$

Matched Problem 2 Evaluate $\displaystyle\int_1^3 \left(4x - 2e^x + \frac{5}{x}\right) dx.$

The evaluation of a definite integral is a two-step process: First, find an antiderivative. Then find the change in that antiderivative. If *substitution techniques* are required to find the antiderivative, there are two different ways to proceed. The next example illustrates both methods.

EXAMPLE 3   Definite Integrals and Substitution Techniques   Evaluate

$$\int_0^5 \frac{x}{x^2 + 10}\,dx$$

SOLUTION   We solve this problem using substitution in two different ways.

Method 1.   Use substitution in an indefinite integral to find an antiderivative as a function of x. Then evaluate the definite integral.

$$\int \frac{x}{x^2 + 10}\,dx = \frac{1}{2}\int \frac{1}{x^2 + 10}\,2x\,dx \qquad \text{Substitute } u = x^2 + 10$$
$$\text{and } du = 2x\,dx.$$

$$= \frac{1}{2}\int \frac{1}{u}\,du$$

$$= \tfrac{1}{2}\ln|u| + C$$

$$= \tfrac{1}{2}\ln(x^2 + 10) + C \qquad \text{Since } u = x^2 + 10 > 0$$

We choose $C = 0$ and use the antiderivative $\tfrac{1}{2}\ln(x^2 + 10)$ to evaluate the definite integral.

$$\int_0^5 \frac{x}{x^2 + 10}\,dx = \frac{1}{2}\ln(x^2 + 10)\Big|_0^5$$

$$= \tfrac{1}{2}\ln 35 - \tfrac{1}{2}\ln 10 \approx 0.626$$

Method 2.   Substitute directly into the definite integral, changing both the variable of integration and the limits of integration. In the definite integral

$$\int_0^5 \frac{x}{x^2 + 10}\,dx$$

the upper limit is $x = 5$ and the lower limit is $x = 0$. When we make the substitution $u = x^2 + 10$ in this definite integral, we must change the limits of integration to the corresponding values of u:

$$x = 5 \quad \text{implies} \quad u = 5^2 + 10 = 35 \quad \text{New upper limit}$$
$$x = 0 \quad \text{implies} \quad u = 0^2 + 10 = 10 \quad \text{New lower limit}$$

We have

$$\int_0^5 \frac{x}{x^2 + 10}\,dx = \frac{1}{2}\int_0^5 \frac{1}{x^2 + 10}\,2x\,dx$$

$$= \frac{1}{2}\int_{10}^{35} \frac{1}{u}\,du$$

$$= \frac{1}{2}\left(\ln|u|\,\Big|_{10}^{35}\right)$$

$$= \tfrac{1}{2}(\ln 35 - \ln 10) \approx 0.626$$

Matched Problem 3   Use both methods described in Example 3 to evaluate

$$\int_0^1 \frac{1}{2x + 4}\,dx.$$

EXAMPLE 4 Definite Integrals and Substitution Use method 2 described in Example 3 to evaluate

$$\int_{-4}^{1} \sqrt{5-t}\, dt$$

SOLUTION If $u = 5 - t$, then $du = -dt$, and

$t = 1$	implies	$u = 5 - 1 = 4$	New upper limit
$t = -4$	implies	$u = 5 - (-4) = 9$	New lower limit

Notice that the lower limit for u is larger than the upper limit. Be careful not to reverse these two values when substituting into the definite integral:

$$\int_{-4}^{1} \sqrt{5-t}\, dt = -\int_{-4}^{1} \sqrt{5-t}\,(-dt)$$

$$= -\int_{9}^{4} \sqrt{u}\, du$$

$$= -\int_{9}^{4} u^{1/2}\, du$$

$$= -\left(\frac{u^{3/2}}{\frac{3}{2}}\Big|_{9}^{4}\right)$$

$$= -\left[\tfrac{2}{3}(4)^{3/2} - \tfrac{2}{3}(9)^{3/2}\right]$$

$$= -\left[\tfrac{16}{3} - \tfrac{54}{3}\right] = \tfrac{38}{3} \approx 12.667$$

Matched Problem 4 Use method 2 described in Example 3 to evaluate

$$\int_{2}^{5} \frac{1}{\sqrt{6-t}}\, dt.$$

EXAMPLE 5 Change in Profit A company manufactures x HDTVs per month. The monthly marginal profit (in dollars) is given by

$$P'(x) = 165 - 0.1x \qquad 0 \le x \le 4{,}000$$

The company is currently manufacturing 1,500 HDTVs per month, but is planning to increase production. Find the change in the monthly profit if monthly production is increased to 1,600 HDTVs.

SOLUTION

$$P(1{,}600) - P(1{,}500) = \int_{1{,}500}^{1{,}600} (165 - 0.1x)\, dx$$

$$= \left(165x - 0.05x^2\right)\Big|_{1{,}500}^{1{,}600}$$

$$= \left[165(1{,}600) - 0.05(1{,}600)^2\right]$$

$$\qquad - \left[165(1{,}500) - 0.05(1{,}500)^2\right]$$

$$= 136{,}000 - 135{,}000$$

$$= 1{,}000$$

Increasing monthly production from 1,500 units to 1,600 units will increase the monthly profit by $1,000.

Matched Problem 5 Repeat Example 5 if

$$P'(x) = 300 - 0.2x \qquad 0 \le x \le 3,000$$

and monthly production is increased from 1,400 to 1,500 HDTVs.

EXAMPLE 6 Useful Life   An amusement company maintains records for each video game installed in an arcade. Suppose that $C(t)$ and $R(t)$ represent the total accumulated costs and revenues (in thousands of dollars), respectively, t years after a particular game has been installed. Suppose also that

$$C'(t) = 2 \qquad R'(t) = 9e^{-0.5t}$$

The value of t for which $C'(t) = R'(t)$ is called the **useful life** of the game.
(A) Find the useful life of the game, to the nearest year.
(B) Find the total profit accumulated during the useful life of the game.

SOLUTION

(A) $R'(t) = C'(t)$

$$9e^{-0.5t} = 2$$

$$e^{-0.5t} = \tfrac{2}{9} \qquad\qquad \text{Convert to equivalent logarithmic form.}$$

$$-0.5t = \ln \tfrac{2}{9}$$

$$t = -2 \ln \tfrac{2}{9} \approx 3 \text{ years}$$

Thus, the game has a useful life of 3 years. This is illustrated graphically in Figure 3.

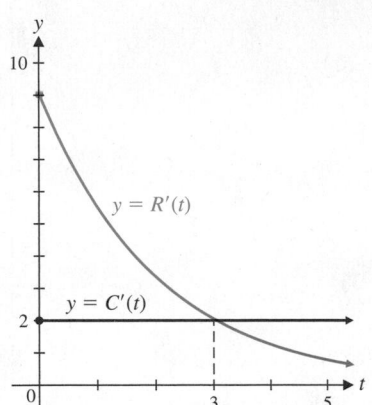

Figure 3 **Useful life**

(B) The total profit accumulated during the useful life of the game is

$$P(3) - P(0) = \int_0^3 P'(t)\, dt$$

$$= \int_0^3 [R'(t) - C'(t)]\, dt$$

$$= \int_0^3 (9e^{-0.5t} - 2)\, dt$$

$$= \left(\frac{9}{-0.5} e^{-0.5t} - 2t \right) \Big|_0^3 \qquad \text{Recall: } \int e^{ax}\, dx = \frac{1}{a} e^{ax} + C$$

$$= (-18e^{-0.5t} - 2t) \big|_0^3$$

$$= (-18e^{-1.5} - 6) - (-18e^0 - 0)$$

$$= 12 - 18e^{-1.5} \approx 7.984 \quad \text{or} \quad \$7,984$$

Matched Problem 6 Repeat Example 6 if $C'(t) = 1$ and $R'(t) = 7.5e^{-0.5t}$.

EXAMPLE 7 Numerical Integration on a Graphing Calculator   Evaluate

$$\int_{-1}^2 e^{-x^2}\, dx \text{ to three decimal places.}$$

SOLUTION The integrand e^{-x^2} does not have an elementary antiderivative, so we are unable to use the fundamental theorem to evaluate the definite integral.

Figure 4

Instead, we use a numerical integration routine that has been preprogrammed into a graphing calculator. (Consult your user's manual for specific details.) Such a routine is an approximation algorithm, more powerful than the left-sum and right-sum methods discussed in Section 5.4. From Figure 4,

$$\int_{-1}^{2} e^{-x^2}\, dx = 1.629$$

Matched Problem 7 Evaluate $\int_{1.5}^{4.3} \dfrac{x}{\ln x}\,dx$ to three decimal places.

Recognizing a Definite Integral: Average Value

Recall that the derivative of a function f was defined in Section 2.4 by

$$f'(x) = \lim_{h \to 0} \frac{f(x + h) - f(x)}{h}$$

This form is generally not easy to compute directly but is easy to recognize in certain practical problems (slope, instantaneous velocity, rates of change, and so on). Once we know that we are dealing with a derivative, we proceed to try to compute the derivative with the use of derivative formulas and rules.

Similarly, evaluating a definite integral with the use of the definition

$$\int_{a}^{b} f(x)\, dx = \lim_{n \to \infty}[f(c_1)\Delta x_1 + f(c_2)\Delta x_2 + \cdots + f(c_n)\Delta x_n] \qquad (1)$$

is generally not easy, but the form on the right occurs naturally in many practical problems. We can use the fundamental theorem to evaluate the definite integral (once it is recognized) if an antiderivative can be found; otherwise, we will approximate it with a rectangle sum. We will now illustrate these points by finding the *average value* of a continuous function.

Suppose that the temperature F (in degrees Fahrenheit) in the middle of a small shallow lake from 8 AM ($t = 0$) to 6 PM ($t = 10$) during the month of May is given approximately by $F(t) = -t^2 + 10t + 50$ as shown in Figure 5.

How can we compute the average temperature from 8 AM to 6 PM? We know that the average of a finite number of values $a_1, a_2, \ldots, a_n$ is given by

$$\text{average} = \frac{a_1 + a_2 + \cdots + a_n}{n}$$

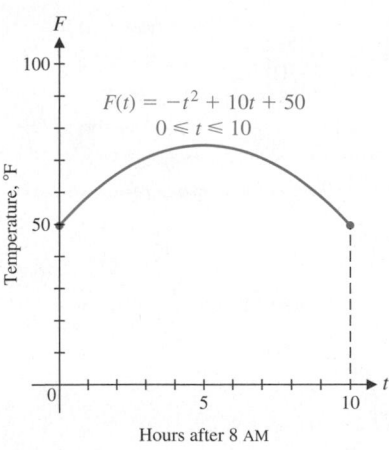

Figure 5

But how can we handle a continuous function with infinitely many values? It would seem reasonable to divide the time interval [0, 10] into n equal subintervals, compute the temperature at a point in each subinterval, and then use the average of the temperatures as an approximation of the average value of the continuous function $F = F(t)$ over [0, 10]. We would expect the approximations to improve as n increases. In fact, we would define the limit of the average of n values as $n \to \infty$ as the *average value of F over* [0, 10] if the limit exists. This is exactly what we will do:

$$\left(\begin{array}{c} \text{average temperature} \\ \text{for } n \text{ values} \end{array} \right) = \frac{1}{n}[F(t_1) + F(t_2) + \cdots + F(t_n)] \qquad (2)$$

Here t_k is a point in the kth subinterval. We will call the limit of equation (2) as $n \to \infty$ the *average temperature over the time interval* [0, 10].

Form (2) resembles form (1), but we are missing the Δt_k. We take care of this by multiplying equation (2) by $(b - a)/(b - a)$, which will change the form of equation (2) without changing its value:

$$\frac{b - a}{b - a} \cdot \frac{1}{n} \left[F(t_1) + F(t_2) + \cdots + F(t_n) \right] = \frac{1}{b - a} \cdot \frac{b - a}{n} \left[F(t_1) + F(t_2) + \cdots + F(t_n) \right]$$

$$= \frac{1}{b - a} \left[F(t_1) \frac{b - a}{n} + F(t_2) \frac{b - a}{n} + \cdots + F(t_n) \frac{b - a}{n} \right]$$

$$= \frac{1}{b - a} \left[F(t_1) \Delta t + F(t_2) \Delta t + \cdots + F(t_n) \Delta t \right]$$

Therefore,

$$\left(\begin{array}{c} \text{average temperature} \\ \text{over } [a, b] = [0, 10] \end{array} \right) = \lim_{n \to \infty} \left\{ \frac{1}{b - a} \left[F(t_1) \Delta t + F(t_2) \Delta t + \cdots + F(t_n) \Delta t \right] \right\}$$

$$= \frac{1}{b - a} \left\{ \lim_{n \to \infty} \left[F(t_1) \Delta t + F(t_2) \Delta t + \cdots + F(t_n) \Delta t \right] \right\}$$

The limit inside the braces is of form (1)—that is, a definite integral. So

$$\left(\begin{array}{c} \text{average temperature} \\ \text{over } [a, b] = [0, 10] \end{array} \right) = \frac{1}{b - a} \int_a^b F(t) \, dt$$

$$= \frac{1}{10 - 0} \int_0^{10} (-t^2 + 10t + 50) \, dt$$

$$= \frac{1}{10} \left(-\frac{t^3}{3} + 5t^2 + 50t \right) \Big|_0^{10}$$

$$= \frac{200}{3} \approx 67°F$$

We now use the fundamental theorem to evaluate the definite integral.

Proceeding as before for an arbitrary continuous function f over an interval $[a, b]$, we obtain the following general formula:

DEFINITION Average Value of a Continuous Function f over $[a, b]$

$$\frac{1}{b - a} \int_a^b f(x) \, dx$$

Explore and Discuss 1   In Figure 6, the rectangle shown has the same area as the area under the graph of $y = f(x)$ from $x = a$ to $x = b$. Explain how the average value of $f(x)$ over the interval $[a, b]$ is related to the height of the rectangle.

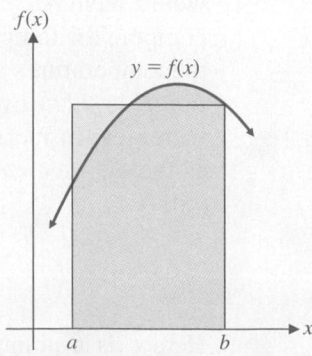

Figure 6

EXAMPLE 8 Average Value of a Function Find the average value of $f(x) = x - 3x^2$ over the interval $[-1, 2]$.

SOLUTION

$$\frac{1}{b-a}\int_a^b f(x)\, dx = \frac{1}{2-(-1)}\int_{-1}^2 (x - 3x^2)\, dx$$

$$= \frac{1}{3}\left(\frac{x^2}{2} - x^3\right)\Big|_{-1}^2 = -\frac{5}{2}$$

Matched Problem 8 Find the average value of $g(t) = 6t^2 - 2t$ over the interval $[-2, 3]$.

EXAMPLE 9 Average Price Given the demand function

$$p = D(x) = 100e^{-0.05x}$$

find the average price (in dollars) over the demand interval $[40, 60]$.

SOLUTION

$$\text{Average price} = \frac{1}{b-a}\int_a^b D(x)\, dx$$

$$= \frac{1}{60-40}\int_{40}^{60} 100e^{-0.05x}\, dx$$

$$= \frac{100}{20}\int_{40}^{60} e^{-0.05x}\, dx \qquad \text{Use } \int e^{ax}\, dx = \frac{1}{a}e^{ax}, a \neq 0.$$

$$= -\frac{5}{0.05}e^{-0.05x}\Big|_{40}^{60}$$

$$= 100(e^{-2} - e^{-3}) \approx \$8.55$$

Matched Problem 9 Given the supply equation

$$p = S(x) = 10e^{0.05x}$$

find the average price (in dollars) over the supply interval $[20, 30]$.

Exercises 5.5

Skills Warm-up Exercises

W In Problems 1–8, use geometric formulas to find the unsigned area between the graph of $y = f(x)$ and the x axis over the indicated interval. (If necessary, review Appendix C.)

1. $f(x) = 100; [1, 6]$

2. $f(x) = -50; [8, 12]$

3. $f(x) = x + 5; [0, 4]$

4. $f(x) = x - 2; [-3, -1]$

5. $f(x) = 3x; [-4, 4]$

6. $f(x) = -10x; [-100, 50]$

7. $f(x) = \sqrt{9 - x^2}; [-3, 3]$

8. $f(x) = -\sqrt{25 - x^2}; [-5, 5]$

In Problems 9–12,

(A) Calculate the change in $F(x)$ from $x = 10$ to $x = 15$.

(B) Graph $F'(x)$ and use geometric formulas (see Appendix C) to calculate the area between the graph of $F'(x)$ and the x axis from $x = 10$ to $x = 15$.

(C) Verify that your answers to (A) and (B) are equal, as is guaranteed by the fundamental theorem of calculus.

9. $F(x) = 3x^2 + 160$

10. $F(x) = 9x + 120$

11. $F(x) = -x^2 + 42x + 240$

12. $F(x) = x^2 + 30x + 210$

Evaluate the integrals in Problems 13–32.

13. $\int_0^{10} 4\,dx$

14. $\int_0^8 9x\,dx$

15. $\int_0^6 x^2\,dx$

16. $\int_0^4 x^3\,dx$

17. $\int_1^4 (5x + 3)\,dx$

18. $\int_2^5 (2x - 1)\,dx$

19. $\int_0^1 e^x\,dx$

20. $\int_0^2 4e^x\,dx$

21. $\int_1^2 \frac{1}{x}\,dx$

22. $\int_1^5 \frac{2}{x}\,dx$

23. $\int_{-2}^2 (x^3 + 7x)\,dx$

24. $\int_0^8 (0.25x - 1)\,dx$

25. $\int_2^5 (2x + 9)\,dx$

26. $\int_1^4 (6x - 5)\,dx$

27. $\int_5^2 (2x + 9)\,dx$

28. $\int_4^1 (6x - 5)\,dx$

29. $\int_2^3 (6 - x^3)\,dx$

30. $\int_6^9 (5 - x^2)\,dx$

31. $\int_6^6 (x^2 - 5x + 1)^{10}\,dx$

32. $\int_{-3}^{-3} (x^2 + 4x + 2)^8\,dx$

Evaluate the integrals in Problems 33–48.

33. $\int_1^2 (2x^{-2} - 3)\,dx$

34. $\int_1^2 (5 - 16x^{-3})\,dx$

35. $\int_1^4 3\sqrt{x}\,dx$

36. $\int_4^{25} \frac{2}{\sqrt{x}}\,dx$

37. $\int_2^3 12(x^2 - 4)^5 x\,dx$

38. $\int_0^1 32(x^2 + 1)^7 x\,dx$

39. $\int_3^9 \frac{1}{x - 1}\,dx$

40. $\int_2^8 \frac{1}{x + 1}\,dx$

41. $\int_{-5}^{10} e^{-0.05x}\,dx$

42. $\int_{-10}^{25} e^{-0.01x}\,dx$

43. $\int_1^e \frac{\ln t}{t}\,dt$

44. $\int_e^{e^2} \frac{(\ln t)^2}{t}\,dt$

45. $\int_0^1 xe^{-x^2}\,dx$

46. $\int_0^1 xe^{x^2}\,dx$

47. $\int_1^1 e^{x^2}\,dx$

48. $\int_{-1}^{-1} e^{-x^2}\,dx$

In Problems 49–56,

(A) *Find the average value of each function over the indicated interval.*

(B) *Use a graphing calculator to graph the function and its average value over the indicated interval in the same viewing window.*

49. $f(x) = 500 - 50x;\ [0, 10]$

50. $g(x) = 2x + 7;\ [0, 5]$

51. $f(t) = 3t^2 - 2t;\ [-1, 2]$

52. $g(t) = 4t - 3t^2;\ [-2, 2]$

53. $f(x) = \sqrt[3]{x};\ [1, 8]$

54. $g(x) = \sqrt{x + 1};\ [3, 8]$

55. $f(x) = 4e^{-0.2x};\ [0, 10]$

56. $f(x) = 64e^{0.08x};\ [0, 10]$

Evaluate the integrals in Problems 57–62.

57. $\int_2^3 x\sqrt{2x^2 - 3}\,dx$

58. $\int_0^1 x\sqrt{3x^2 + 2}\,dx$

59. $\int_0^1 \frac{x - 1}{x^2 - 2x + 3}\,dx$

60. $\int_1^2 \frac{x + 1}{2x^2 + 4x + 4}\,dx$

61. $\int_{-1}^1 \frac{e^{-x} - e^x}{(e^{-x} + e^x)^2}\,dx$

62. $\int_6^7 \frac{\ln(t - 5)}{t - 5}\,dt$

Use a numerical integration routine to evaluate each definite integral in Problems 63–66 (to three decimal places).

63. $\int_{1.7}^{3.5} x\ln x\,dx$

64. $\int_{-1}^1 e^{x^2}\,dx$

65. $\int_{-2}^2 \frac{1}{1 + x^2}\,dx$

66. $\int_0^3 \sqrt{9 - x^2}\,dx$

67. The **mean value theorem** states that if $F(x)$ is a differentiable function on the interval $[a, b]$, then there exists some number c between a and b such that

$$F'(c) = \frac{F(b) - F(a)}{b - a}$$

Explain why the mean value theorem implies that if a car averages 60 miles per hour in some 10-minute interval, then the car's instantaneous velocity is 60 miles per hour at least once in that interval.

68. The fundamental theorem of calculus can be proved by showing that, for every positive integer n, there is a Riemann sum for f on $[a, b]$ that is equal to $F(b) - F(a)$. By the mean value theorem (see Problem 67), within each subinterval $[x_{k-1}, x_k]$ that belongs to a partition of $[a, b]$, there is some c_k such that

$$f(c_k) = F'(c_k) = \frac{F(x_k) - F(x_{k-1})}{x_k - x_{k-1}}$$

Multiplying by the denominator $x_k - x_{k-1}$, we get

$$f(c_k)(x_k - x_{k-1}) = F(x_k) - F(x_{k-1})$$

Show that the Riemann sum

$$S_n = \sum_{k=1}^n f(c_k)(x_k - x_{k-1})$$

is equal to $F(b) - F(a)$.

Applications

69. Cost. A company manufactures mountain bikes. The research department produced the marginal cost function

$$C'(x) = 500 - \frac{x}{3} \qquad 0 \le x \le 900$$

where $C'(x)$ is in dollars and x is the number of bikes produced per month. Compute the increase in cost going from a production level of 300 bikes per month to 900 bikes per month. Set up a definite integral and evaluate it.

70. Cost. Referring to Problem 69, compute the increase in cost going from a production level of 0 bikes per month to 600 bikes per month. Set up a definite integral and evaluate it.

71. Salvage value. A new piece of industrial equipment will depreciate in value, rapidly at first and then less rapidly as time goes on. Suppose that the rate (in dollars per year) at which the book value of a new milling machine changes is given approximately by

$$V'(t) = f(t) = 500(t - 12) \qquad 0 \le t \le 10$$

where $V(t)$ is the value of the machine after t years. What is the total loss in value of the machine in the first 5 years? In the second 5 years? Set up appropriate integrals and solve.

72. Maintenance costs. Maintenance costs for an apartment house generally increase as the building gets older. From past records, the rate of increase in maintenance costs (in dollars per year) for a particular apartment complex is given approximately by

$$M'(x) = f(x) = 90x^2 + 5,000$$

where x is the age of the apartment complex in years and $M(x)$ is the total (accumulated) cost of maintenance for x years. Write a definite integral that will give the total maintenance costs from the end of the second year to the end of the seventh year, and evaluate the integral.

73. Employee training. A company producing computer components has established that, on the average, a new employee can assemble $N(t)$ components per day after t days of on-the-job training, as indicated in the following table (a new employee's productivity usually increases with time on the job, up to a leveling-off point):

t	0	20	40	60	80	100	120
$N(t)$	10	51	68	76	81	84	85

(A) Find a quadratic regression equation for the data, and graph it and the data set in the same viewing window.

(B) Use the regression equation and a numerical integration routine on a graphing calculator to approximate the number of units assembled by a new employee during the first 100 days on the job.

74. Employee training. Refer to Problem 73.

(A) Find a cubic regression equation for the data, and graph it and the data set in the same viewing window.

(B) Use the regression equation and a numerical integration routine on a graphing calculator to approximate the number of units assembled by a new employee during the second 60 days on the job.

75. Useful life. The total accumulated costs $C(t)$ and revenues $R(t)$ (in thousands of dollars), respectively, for a photocopying machine satisfy

$$C'(t) = \tfrac{1}{11}t \qquad \text{and} \qquad R'(t) = 5te^{-t^2}$$

where t is time in years. Find the useful life of the machine, to the nearest year. What is the total profit accumulated during the useful life of the machine?

76. Useful life. The total accumulated costs $C(t)$ and revenues $R(t)$ (in thousands of dollars), respectively, for a coal mine satisfy

$$C'(t) = 3 \qquad \text{and} \qquad R'(t) = 15e^{-0.1t}$$

where t is the number of years that the mine has been in operation. Find the useful life of the mine, to the nearest year. What is the total profit accumulated during the useful life of the mine?

77. Average cost. The total cost (in dollars) of manufacturing x auto body frames is $C(x) = 60,000 + 300x$.

(A) Find the average cost per unit if 500 frames are produced. [*Hint*: Recall that $\overline{C}(x)$ is the average cost per unit.]

(B) Find the average value of the cost function over the interval $[0, 500]$.

(C) Discuss the difference between parts (A) and (B).

78. Average cost. The total cost (in dollars) of printing x dictionaries is $C(x) = 20,000 + 10x$.

(A) Find the average cost per unit if 1,000 dictionaries are produced.

(B) Find the average value of the cost function over the interval $[0, 1,000]$.

(C) Discuss the difference between parts (A) and (B).

79. Cost. The marginal cost at various levels of output per month for a company that manufactures sunglasses is shown in the following table, with the output x given in thousands of units per month and the total cost $C(x)$ given in thousands of dollars per month:

x	0	1	2	3	4	5	6	7	8
$C'(x)$	58	30	18	9	5	7	17	33	51

(A) Find a quadratic regression equation for the data, and graph it and the data set in the same viewing window.

(B) Use the regression equation and a numerical integration routine on a graphing calculator to approximate (to the nearest dollar) the increased cost in going from a production level of 2 thousand sunglasses per month to 8 thousand sunglasses per month.

80. Cost. Refer to Problem 79.

(A) Find a cubic regression equation for the data, and graph it and the data set in the same viewing window.

(B) Use the regression equation and a numerical integration routine on a graphing calculator to approximate (to the nearest dollar) the increased cost in going from a production level of 1 thousand sunglasses per month to 7 thousand sunglasses per month.

81. **Supply function.** Given the supply function

$$p = S(x) = 10(e^{0.02x} - 1)$$

find the average price (in dollars) over the supply interval [20, 30].

82. **Demand function.** Given the demand function

$$p = D(x) = \frac{1,000}{x}$$

find the average price (in dollars) over the demand interval [400, 600].

83. **Labor costs and learning.** A defense contractor is starting production on a new missile control system. On the basis of data collected during assembly of the first 16 control systems, the production manager obtained the following function for the rate of labor use:

$$L'(x) = 2,400x^{-1/2}$$

Approximately how many labor-hours will be required to assemble the 17th through the 25th control units? [*Hint*: Let $a = 16$ and $b = 25$.]

84. **Labor costs and learning.** If the rate of labor use in Problem 83 is

$$L'(x) = 2,000x^{-1/3}$$

then approximately how many labor-hours will be required to assemble the 9th through the 27th control units? [*Hint*: Let $a = 8$ and $b = 27$.]

85. **Inventory.** A store orders 600 units of a product every 3 months. If the product is steadily depleted to 0 by the end of each 3 months, the inventory on hand I at any time t during the year is shown in the following figure:

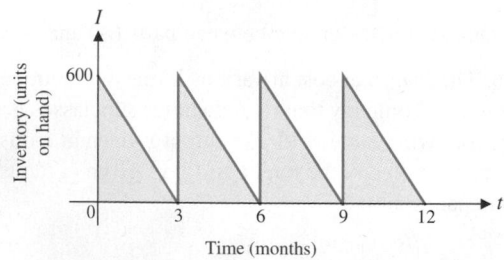

Time (months)

(A) Write an inventory function (assume that it is continuous) for the first 3 months. [The graph is a straight line joining (0, 600) and (3, 0).]

(B) What is the average number of units on hand for a 3-month period?

86. Repeat Problem 85 with an order of 1,200 units every 4 months.

87. **Oil production.** Using production and geological data, the management of an oil company estimates that oil will be pumped from a producing field at a rate given by

$$R(t) = \frac{100}{t+1} + 5 \qquad 0 \le t \le 20$$

where $R(t)$ is the rate of production (in thousands of barrels per year) t years after pumping begins. Approximately how many barrels of oil will the field produce during the first

10 years of production? From the end of the 10th year to the end of the 20th year of production?

88. **Oil production.** In Problem 87, if the rate is found to be

$$R(t) = \frac{120t}{t^2 + 1} + 3 \qquad 0 \le t \le 20$$

then approximately how many barrels of oil will the field produce during the first 5 years of production? The second 5 years of production?

89. **Biology.** A yeast culture weighing 2 grams is expected to grow at the rate of $W'(t) = 0.2e^{0.1t}$ grams per hour at a higher controlled temperature. How much will the weight of the culture increase during the first 8 hours of growth? How much will the weight of the culture increase from the end of the 8th hour to the end of the 16th hour of growth?

90. **Medicine.** The rate at which the area of a skin wound is increasing is given (in square centimeters per day) by $A'(t) = -0.9e^{-0.1t}$. The initial wound has an area of 9 square centimeters. How much will the area change during the first 5 days? The second 5 days?

91. **Temperature.** If the temperature in an aquarium (in degrees Celsius) is given by

$$C(t) = t^3 - 2t + 10 \qquad 0 \le t \le 2$$

over a 2-hour period, what is the average temperature over this period?

92. **Medicine.** A drug is injected into the bloodstream of a patient through her right arm. The drug concentration in the bloodstream of the left arm t hours after the injection is given by

$$C(t) = \frac{0.14t}{t^2 + 1}$$

What is the average drug concentration in the bloodstream of the left arm during the first hour after the injection? During the first 2 hours after the injection?

93. **Politics.** Public awareness of a congressional candidate before and after a successful campaign was approximated by

$$P(t) = \frac{8.4t}{t^2 + 49} + 0.1 \qquad 0 \le t \le 24$$

where t is time in months after the campaign started and $P(t)$ is the fraction of the number of people in the congressional district who could recall the candidate's name. What is the average fraction of the number of people who could recall the candidate's name during the first 7 months of the campaign? During the first 2 years of the campaign?

94. **Population composition.** The number of children in a large city was found to increase and then decrease rather drastically. If the number of children (in millions) over a 6-year period was given by

$$N(t) = -\tfrac{1}{4}t^2 + t + 4 \qquad 0 \le t \le 6$$

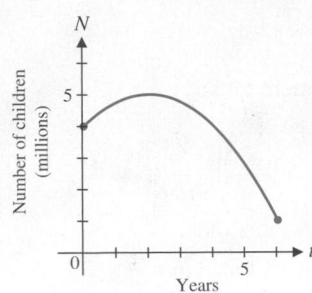

what was the average number of children in the city over the 6-year period? [Assume that $N = N(t)$ is continuous.]

Answers to Matched Problems

1. (A)

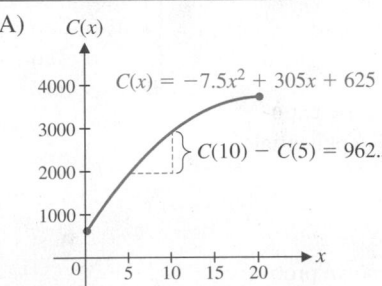

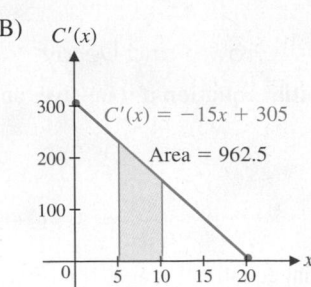

(C) The change in cost from $x = 5$ to $x = 10$ is equal to the area between the marginal cost function and the x axis from $x = 5$ to $x = 10$.

2. $16 + 2e - 2e^3 + 5 \ln 3 \approx -13.241$

3. $\frac{1}{2}(\ln 6 - \ln 4) \approx 0.203$

4. 2

5. $\$1,000$

6. (A) $-2 \ln\frac{2}{15} \approx 4$ yr   **(B)** $11 - 15e^{-2} \approx 8.970$ or $\$8,970$

7. 8.017

8. 13

9. $\$35.27$

Chapter 5   Summary and Review

Important Terms, Symbols, and Concepts

5.1 Antiderivatives and Indefinite Integrals

EXAMPLES

- A function F is an **antiderivative** of a function f if $F'(x) = f(x)$.

Ex. 1, p. 321

- If F and G are both antiderivatives of f, then F and G differ by a constant; that is, $F(x) = G(x) + k$ for some constant k.

- We use the symbol $\int f(x)\, dx$, called an **indefinite integral,** to represent the family of all antiderivatives of f, and we write

$$\int f(x)\, dx = F(x) + C$$

The symbol $\int$ is called an **integral sign,** $f(x)$ is the **integrand,** and C is the **constant of integration.**

Ex. 2, p. 324
Ex. 3, p. 325
Ex. 4, p. 326

- Indefinite integrals of basic functions are given by the formulas on page 322.

- Properties of indefinite integrals are given on page 323; in particular, a constant factor can be moved across an integral sign. However, a variable factor *cannot* be moved across an integral sign.

Ex. 5, p. 327
Ex. 6, p. 327

5.2 Integration by Substitution

- The **method of substitution** (also called the **change-of-variable method**) is a technique for finding indefinite integrals. It is based on the following formula, which is obtained by reversing the chain rule:

Ex. 1, p. 333

$$\int E'[I(x)]I'(x)\, dx = E[I(x)] + C$$

- This formula implies the general indefinite integral formulas on page 333.

- When using the method of substitution, it is helpful to use differentials as a bookkeeping device:

Ex. 2, p. 334
Ex. 3, p. 334

1. The **differential dx** of the independent variable x is an arbitrary real number.

Ex. 4, p. 335

2. The **differential dy** of the dependent variable y is defined by $dy = f'(x)\, dx$.

Ex. 5, p. 336
Ex. 6, p. 338

- Guidelines for using the substitution method are given by the procedure on page 335.

Ex. 7, p. 339

5.3 Differential Equations; Growth and Decay

- An equation is a **differential equation** if it involves an unknown function and one or more of the function's derivatives.

- The equation

$$\frac{dy}{dx} = 3x(1 + xy^2)$$

is a **first-order** differential equation because it involves the first derivative of the unknown function y but no second or higher-order derivative.

- A **slope field** can be constructed for the preceding differential equation by drawing a tangent line segment with slope $3x(1 + xy^2)$ at each point (x, y) of a grid. The slope field gives a graphical representation of the functions that are solutions of the differential equation.

- The differential equation

$$\frac{dQ}{dt} = rQ$$

Ex. 1, p. 346
Ex. 2, p. 346
Ex. 3, p. 347

(in words, the rate at which the unknown function Q increases is proportional to Q) is called the **exponential growth law.** The constant r is called the **relative growth rate.** The solutions of the exponential growth law are the functions

$$Q(t) = Q_0 e^{rt}$$

where Q_0 denotes $Q(0)$, the amount present at time $t = 0$. These functions can be used to solve problems in population growth, continuous compound interest, radioactive decay, blood pressure, and light absorption.

- Table 1 on page 349 gives the solutions of other first-order differential equations that can be used to model the limited or logistic growth of epidemics, sales, and corporations.

Ex. 4, p. 348

5.4 The Definite Integral

- If the function f is positive on $[a, b]$, then the area between the graph of f and the x axis from $x = a$ to $x = b$ can be approximated by partitioning $[a, b]$ into n subintervals $[x_{k-1}, x_k]$ of equal length $\Delta x = (b - a)/n$ and summing the areas of n rectangles. This can be done using **left sums, right sums,** or, more generally, **Riemann sums:**

Ex. 1, p. 355
Ex. 2, p. 357

Left sum: $\quad L_n = \sum_{k=1}^{n} f(x_{k-1}) \Delta x$

Right sum: $\quad R_n = \sum_{k=1}^{n} f(x_k) \Delta x$

Riemann sum: $\quad S_n = \sum_{k=1}^{n} f(c_k) \Delta x$

In a Riemann sum, each c_k is required to belong to the subinterval $[x_{k-1}, x_k]$. Left sums and right sums are the special cases of Riemann sums in which c_k is the left endpoint and right endpoint, respectively, of the subinterval.

- The **error in an approximation** is the absolute value of the difference between the approximation and the actual value. An **error bound** is a positive number such that the error is guaranteed to be less than or equal to that number.

- Theorem 1 on page 355 gives error bounds for the approximation of the area between the graph of a positive function f and the x axis from $x = a$ to $x = b$, by left sums or right sums, if f is either increasing or decreasing.

- If $f(x) > 0$ and is either increasing on $[a, b]$ or decreasing on $[a, b]$, then the left and right sums of $f(x)$ approach the same real number as $n \to \infty$ (Theorem 2, page 355).

- If f is a continuous function on $[a, b]$, then the Riemann sums for f on $[a, b]$ approach a real-number limit I as $n \to \infty$ (Theorem 3, page 357).

- Let f be a continuous function on $[a, b]$. Then the limit I of Riemann sums for f on $[a, b]$, guaranteed to exist by Theorem 3, is called the **definite integral** of f from a to b and is denoted

$$\int_a^b f(x)\, dx$$

  The **integrand** is $f(x)$, the **lower limit of integration** is a, and the **upper limit of integration** is b.   Ex. 3, p. 358

- Geometrically, the definite integral

$$\int_a^b f(x)\, dx$$

  represents the cumulative sum of the signed areas between the graph of f and the x axis from $x = a$ to $x = b$.   Ex. 4, p. 359

- Properties of the definite integral are given on page 359.

5.5 The Fundamental Theorem of Calculus

- If f is a continuous function on $[a, b\}$ and F is any antiderivative of f, then   Ex. 1, p. 364
  Ex. 2, p. 365

$$\int_a^b f(x)\, dx = F(b) - F(a)$$

  This is the fundamental theorem of calculus (see page 365).   Ex. 3, p. 366

- The fundamental theorem gives an easy and exact method for evaluating definite integrals, provided that we can find an antiderivative $F(x)$ of $f(x)$. In practice, we first find an antiderivative $F(x)$ (when possible), using techniques for computing indefinite integrals. Then we calculate the difference $F(b) - F(a)$. If it is impossible to find an antiderivative, we must resort to left or right sums, or other approximation methods, to evaluate the definite integral. Graphing calculators have a built-in numerical approximation routine, more powerful than left- or right-sum methods, for this purpose.   Ex. 4, p. 367
  Ex. 5, p. 367
  Ex. 6, p. 368
  Ex. 7, p. 368

- If f is a continuous function on $[a, b]$, then the **average value** of f over $[a, b]$ is defined to be   Ex. 8, p. 371
  Ex. 9, p. 371

$$\frac{1}{b - a} \int_a^b f(x)\, dx$$

Review Exercises

Work through all the problems in this chapter review and check your answers in the back of the book. Answers to all review problems are there, along with section numbers in italics to indicate where each type of problem is discussed. Where weaknesses show up, review appropriate sections of the text.

Find each integral in Problems 1–6.

1. $\displaystyle\int (6x + 3)\, dx$

2. $\displaystyle\int_{10}^{20} 5\, dx$

3. $\displaystyle\int_0^9 (4 - t^2)\, dt$

4. $\displaystyle\int (1 - t^2)^3 t\, dt$

5. $\displaystyle\int \frac{1 + u^4}{u}\, du$

6. $\displaystyle\int_0^1 xe^{-2x^2}\, dx$

7. Is $F(x) = \ln x^2$ an antiderivative of $f(x) = \ln (2x)$? Explain.

8. Is $F(x) = \ln x^2$ an antiderivative of $f(x) = \dfrac{2}{x}$? Explain.

9. Is $F(x) = (\ln x)^2$ an antiderivative of $f(x) = 2 \ln x$? Explain.

10. Is $F(x) = (\ln x)^2$ an antiderivative of $f(x) = \dfrac{2 \ln x}{x}$? Explain.

11. Is $y = 3x + 17$ a solution of the differential equation $(x + 5)y' = y - 2$? Explain.

12. Is $y = 4x^3 + 7x^2 - 5x + 2$ a solution of the differential equation $(x + 2)y''' - 24x = 48$? Explain.

In Problems 13 and 14, find the derivative or indefinite integral as indicated.

13. $\dfrac{d}{dx}\left(\displaystyle\int e^{-x^2}\, dx \right)$

14. $\displaystyle\int \dfrac{d}{dx}\left(\sqrt{4 + 5x} \right)\, dx$

15. Find a function $y = f(x)$ that satisfies both conditions:

$$\frac{dy}{dx} = 3x^2 - 2 \qquad f(0) = 4$$

16. Find all antiderivatives of

(A) $\dfrac{dy}{dx} = 8x^3 - 4x - 1$ (B) $\dfrac{dx}{dt} = e^t - 4t^{-1}$

17. Approximate $\int_1^5 (x^2 + 1)\, dx$, using a right sum with $n = 2$. Calculate an error bound for this approximation.

18. Evaluate the integral in Problem 17, using the fundamental theorem of calculus, and calculate the actual error $|I - R_2|$ produced by using R_2.

19. Use the following table of values and a left sum with $n = 4$ to approximate $\int_1^{17} f(x)\, dx$:

x	1	5	9	13	17
$f(x)$	1.2	3.4	2.6	0.5	0.1

20. Find the average value of $f(x) = 6x^2 + 2x$ over the interval $[-1, 2]$.

21. Describe a rectangle that has the same area as the area under the graph of $f(x) = 6x^2 + 2x$ from $x = -1$ to $x = 2$ (see Problem 20).

In Problems 22 and 23, calculate the indicated Riemann sum S_n for the function $f(x) = 100 - x^2$.

22. Partition $[3, 11]$ into four subintervals of equal length, and for each subinterval $[x_{i-1}, x_i]$, let $c_i = (x_{i-1} + x_i)/2$.

23. Partition $[-5, 5]$ into five subintervals of equal length and let $c_1 = -4$, $c_2 = -1$, $c_3 = 1$, $c_4 = 2$, and $c_5 = 5$.

Use the graph and actual areas of the indicated regions in the figure to evaluate the integrals in Problems 24–31:

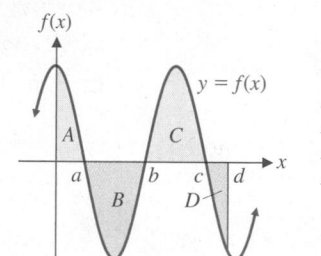

Area $A = 1$
Area $B = 2$
Area $C = 2$
Area $D = 0.6$

Figure for 24–31

24. $\displaystyle\int_a^b 5f(x)\, dx$

25. $\displaystyle\int_b^c \frac{f(x)}{5}\, dx$

26. $\displaystyle\int_b^d f(x)\, dx$

27. $\displaystyle\int_a^c f(x)\, dx$

28. $\displaystyle\int_0^d f(x)\, dx$

29. $\displaystyle\int_b^a f(x)\, dx$

30. $\displaystyle\int_c^b f(x)\, dx$

31. $\displaystyle\int_d^0 f(x)\, dx$

Problems 32–37 refer to the slope field shown in the figure:

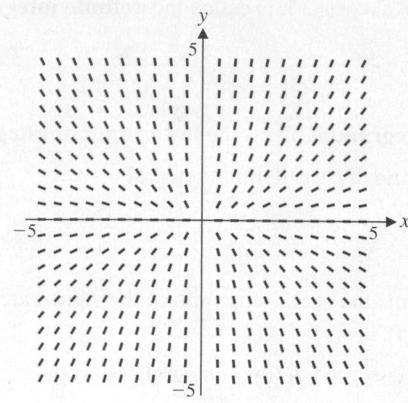

Figure for 32–37

32. (A) For $dy/dx = (2y)/x$, what is the slope of a solution curve at $(2, 1)$? At $(-2, -1)$?

   (B) For $dy/dx = (2x)/y$, what is the slope of a solution curve at $(2, 1)$? At $(-2, -1)$?

33. Is the slope field shown in the figure for $dy/dx = (2x)/y$ or for $dy/dx = (2y)/x$? Explain.

34. Show that $y = Cx^2$ is a solution of $dy/dx = (2y)/x$ for any real number C.

35. Referring to Problem 34, find the particular solution of $dy/dx = (2y)/x$ that passes through $(2, 1)$. Through $(-2, -1)$.

36. Graph the two particular solutions found in Problem 35 in the slope field shown (or a copy).

37. Use a graphing calculator to graph, in the same viewing window, graphs of $y = Cx^2$ for $C = -2, -1, 1$, and 2 for $-5 \le x \le 5$ and $-5 \le y \le 5$.

Find each integral in Problems 38–48.

38. $\displaystyle\int_{-1}^1 \sqrt{1 + x}\, dx$

39. $\displaystyle\int_{-1}^0 x^2(x^3 + 2)^{-2}\, dx$

40. $\displaystyle\int 5e^{-t}\, dt$

41. $\displaystyle\int_1^e \frac{1 + t^2}{t}\, dt$

42. $\displaystyle\int xe^{3x^2}\, dx$

43. $\displaystyle\int_{-3}^1 \frac{1}{\sqrt{2 - x}}\, dx$

44. $\displaystyle\int_0^3 \frac{x}{1 + x^2}\, dx$

45. $\displaystyle\int_0^3 \frac{x}{(1 + x^2)^2}\, dx$

46. $\displaystyle\int x^3(2x^4 + 5)^5\, dx$

47. $\displaystyle\int \frac{e^{-x}}{e^{-x} + 3}\, dx$

48. $\displaystyle\int \frac{e^x}{(e^x + 2)^2}\, dx$

49. Find a function $y = f(x)$ that satisfies both conditions:

$$\frac{dy}{dx} = 3x^{-1} - x^{-2} \qquad f(1) = 5$$

50. Find the equation of the curve that passes through (2, 10) if its slope is given by

$$\frac{dy}{dx} = 6x + 1$$

for each x.

51. (A) Find the average value of $f(x) = 3\sqrt{x}$ over the interval [1, 9].

(B) Graph $f(x) = 3\sqrt{x}$ and its average over the interval [1, 9] in the same coordinate system.

Find each integral in Problems 52–56.

52. $\int \frac{(\ln x)^2}{x}\,dx$

53. $\int x(x^3 - 1)^2\,dx$

54. $\int \frac{x}{\sqrt{6 - x}}\,dx$

55. $\int_0^7 x\sqrt{16 - x}\,dx$

56. $\int_1^1 (x + 1)^9\,dx$

57. Find a function $y = f(x)$ that satisfies both conditions:

$$\frac{dy}{dx} = 9x^2 e^{x^3} \qquad f(0) = 2$$

58. Solve the differential equation

$$\frac{dN}{dt} = 0.06N \qquad N(0) = 800 \qquad N > 0$$

Graph Problems 59–62 on a graphing calculator, and identify each curve as unlimited growth, exponential decay, limited growth, or logistic growth:

59. $N = 50(1 - e^{-0.07t}); 0 \le t \le 80, 0 \le N \le 60$

60. $p = 500e^{-0.03x}; 0 \le x \le 100, 0 \le p \le 500$

61. $A = 200e^{0.08t}; 0 \le t \le 20, 0 \le A \le 1,000$

62. $N = \dfrac{100}{1 + 9e^{-0.3t}}; 0 \le t \le 25, 0 \le N \le 100$

Use a numerical integration routine to evaluate each definite integral in Problems 63–65 (to three decimal places).

63. $\int_{-0.5}^{0.6} \dfrac{1}{\sqrt{1 - x^2}}\,dx$

64. $\int_{-2}^{3} x^2 e^x\,dx$

65. $\int_{0.5}^{2.5} \dfrac{\ln x}{x^2}\,dx$

Applications

66. Cost. A company manufactures downhill skis. The research department produced the marginal cost graph shown in the accompanying figure, where $C'(x)$ is in dollars and x is the number of pairs of skis produced per week. Estimate the increase in cost going from a production level of 200 to 600 pairs of skis per week. Use left and right sums over two equal subintervals. Replace the question marks with the values of L_2 and R_2 as appropriate:

$$? \le \int_{200}^{600} C'(x)\,dx \le ?$$

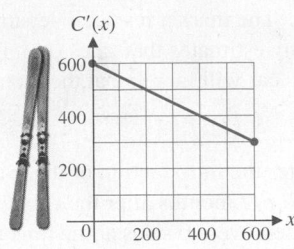

Figure for 66

67. Cost. Assuming that the marginal cost function in Problem 66 is linear, find its equation and write a definite integral that represents the increase in costs going from a production level of 200 to 600 pairs of skis per week. Evaluate the definite integral.

68. Profit and production. The weekly marginal profit for an output of x units is given approximately by

$$P'(x) = 150 - \frac{x}{10} \qquad 0 \le x \le 40$$

What is the total change in profit for a change in production from 10 units per week to 40 units? Set up a definite integral and evaluate it.

69. Profit function. If the marginal profit for producing x units per day is given by

$$P'(x) = 100 - 0.02x \qquad P(0) = 0$$

where $P(x)$ is the profit in dollars, find the profit function P and the profit on 10 units of production per day.

70. Resource depletion. An oil well starts out producing oil at the rate of 60,000 barrels of oil per year, but the production rate is expected to decrease by 4,000 barrels per year. Thus, if $P(t)$ is the total production (in thousands of barrels) in t years, then

$$P'(t) = f(t) = 60 - 4t \qquad 0 \le t \le 15$$

Write a definite integral that will give the total production after 15 years of operation, and evaluate the integral.

71. Inventory. Suppose that the inventory of a certain item t months after the first of the year is given approximately by

$$I(t) = 10 + 36t - 3t^2 \qquad 0 \le t \le 12$$

What is the average inventory for the second quarter of the year?

72. Price–supply. Given the price–supply function

$$p = S(x) = 8(e^{0.05x} - 1)$$

find the average price (in dollars) over the supply interval [40, 50].

73. Useful life. The total accumulated costs $C(t)$ and revenues $R(t)$ (in thousands of dollars), respectively, for a coal mine satisfy

$$C'(t) = 3 \qquad \text{and} \qquad R'(t) = 20e^{-0.1t}$$

where t is the number of years that the mine has been in operation. Find the useful life of the mine, to the nearest year. What is the total profit accumulated during the useful life of the mine?

74. Marketing. The market research department for an automobile company estimates that sales (in millions of dollars) of a new electric car will increase at the monthly rate of

$$S'(t) = 4e^{-0.08t} \qquad 0 \le t \le 24$$

t months after the introduction of the car. What will be the total sales $S(t)$ t months after the car is introduced if we assume that there were 0 sales at the time the car entered the marketplace? What are the estimated total sales during the first 12 months after the introduction of the car? How long will it take for the total sales to reach $40 million?

75. Wound healing. The area of a healing skin wound changes at a rate given approximately by

$$\frac{dA}{dt} = -5t^{-2} \qquad 1 \le t \le 5$$

where t is time in days and $A(1) = 5$ square centimeters. What will be the area of the wound in 5 days?

76. Pollution. An environmental protection agency estimates that the rate of seepage of toxic chemicals from a waste dump (in gallons per year) is given by

$$R(t) = \frac{1,000}{(1 + t)^2}$$

where t is the time in years since the discovery of the seepage. Find the total amount of toxic chemicals that seep from the dump during the first 4 years of its discovery.

77. Population. The population of Mexico was 116 million in 2013 and was growing at a rate of 1.07% per year, compounded continuously.

(A) Assuming that the population continues to grow at this rate, estimate the population of Mexico in the year 2025.

(B) At the growth rate indicated, how long will it take the population of Mexico to double?

78. Archaeology. The continuous compound rate of decay for carbon-14 is $r = -0.000\ 123\ 8$. A piece of animal bone found at an archaeological site contains 4% of the original amount of carbon-14. Estimate the age of the bone.

79. Learning. An average student enrolled in a typing class progressed at a rate of $N'(t) = 7e^{-0.1t}$ words per minute t weeks after enrolling in a 15-week course. If a student could type 25 words per minute at the beginning of the course, how many words per minute $N(t)$ would the student be expected to type t weeks into the course? After completing the course?

6

Additional Integration Topics

Introduction

In Chapter 6 we explore additional applications and techniques of integration. We use the integral to find probabilities and to calculate several quantities that are important in business and economics: the total income and future value produced by a continuous income stream, consumers' and producers' surplus, and the Gini index of income concentration. The Gini index is a single number that measures the equality of a country's income distribution (see Problems 93 and 94, for example, in Section 6.1).

6.1   Area Between Curves

- Area Between Two Curves
- Application: Income Distribution

In Chapter 5, we found that the definite integral $\int_a^b f(x)\,dx$ represents the sum of the signed areas between the graph of $y = f(x)$ and the x axis from $x = a$ to $x = b$, where the areas above the x axis are counted positively and the areas below the x axis are counted negatively (see Fig. 1). In this section, we are interested in using the definite integral to find the actual area between a curve and the x axis or the actual area between two curves. These areas are always nonnegative quantities—**area measure is never negative**.

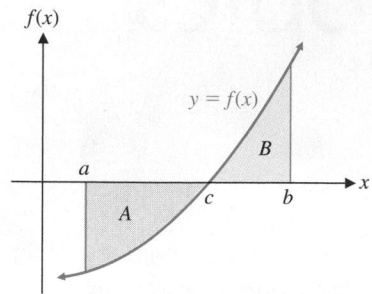

Figure 1 $\int_a^b f(x)\,dx = -A + B$

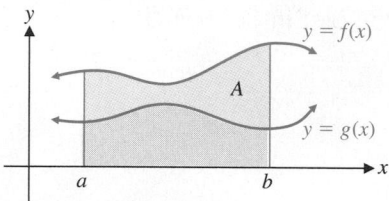

Figure 2

Area Between Two Curves

Consider the area bounded by $y = f(x)$ and $y = g(x)$, where $f(x) \geq g(x) \geq 0$, for $a \leq x \leq b$, as shown in Figure 2.

$$\begin{pmatrix} \text{Area } A \text{ between} \\ f(x) \text{ and } g(x) \end{pmatrix} = \begin{pmatrix} \text{area} \\ \text{under } f(x) \end{pmatrix} - \begin{pmatrix} \text{area} \\ \text{under } g(x) \end{pmatrix}$$ Areas are from $x = a$ to $x = b$ above the x axis.

$$= \int_a^b f(x)\,dx - \int_a^b g(x)\,dx$$ Use definite integral property 4 (Section 5.4).

$$= \int_a^b [f(x) - g(x)]\,dx$$

It can be shown that the preceding result does not require $f(x)$ or $g(x)$ to remain positive over the interval $[a, b]$. A more general result is stated in the following box:

THEOREM 1   Area Between Two Curves

If f and g are continuous and $f(x) \geq g(x)$ over the interval $[a, b]$, then the area bounded by $y = f(x)$ and $y = g(x)$ for $a \leq x \leq b$ is given exactly by

$$A = \int_a^b [f(x) - g(x)]\,dx$$

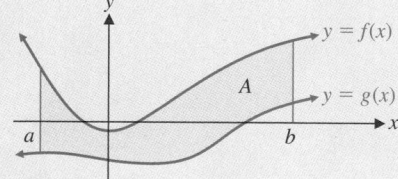

CONCEPTUAL INSIGHT

Theorem 1 requires the graph of f to be *above* (or equal to) the graph of g throughout $[a, b]$, but f and g can be either positive, negative, or 0. In Section 5.4, we considered the special cases of Theorem 1 in which (1) f is positive and g is the zero function on $[a, b]$; and (2) f is the zero function and g is negative on $[a, b]$:

Special case 1. If f is continuous and positive over $[a, b]$, then the area bounded by the graph of f and the x axis for $a \leq x \leq b$ is given exactly by

$$\int_a^b f(x)\,dx$$

Special case 2. If g is continuous and negative over $[a, b]$, then the area bounded by the graph of g and the x axis for $a \leq x \leq b$ is given exactly by

$$\int_a^b [-g(x)]\,dx$$

EXAMPLE 1 Area Between a Curve and the x Axis Find the area bounded by $f(x) = 6x - x^2$ and $y = 0$ for $1 \leq x \leq 4$.

SOLUTION We sketch a graph of the region first (Fig. 3). The solution of every area problem should begin with a sketch. Since $f(x) \geq 0$ on $[1, 4]$,

$$A = \int_1^4 (6x - x^2) \, dx = \left(3x^2 - \frac{x^3}{3} \right) \Big|_1^4$$

$$= \left[3(4)^2 - \frac{(4)^3}{3} \right] - \left[3(1)^2 - \frac{(1)^3}{3} \right]$$

$$= 48 - \frac{64}{3} - 3 + \frac{1}{3}$$

$$= 48 - 21 - 3$$

$$= 24$$

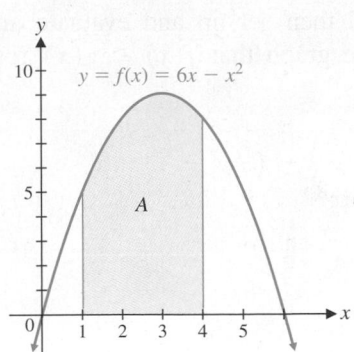

Figure 3

Matched Problem 1 Find the area bounded by $f(x) = x^2 + 1$ and $y = 0$ for $-1 \leq x \leq 3$.

EXAMPLE 2 Area Between a Curve and the x Axis Find the area between the graph of $f(x) = x^2 - 2x$ and the x axis over the indicated intervals:

(A) $[1, 2]$ (B) $[-1, 1]$

SOLUTION We begin by sketching the graph of f, as shown in Figure 4.

(A) From the graph, we see that $f(x) \leq 0$ for $1 \leq x \leq 2$, so we integrate $-f(x)$:

$$A_1 = \int_1^2 [-f(x)] \, dx$$

$$= \int_1^2 (2x - x^2) \, dx$$

$$= \left(x^2 - \frac{x^3}{3} \right) \Big|_1^2$$

$$= \left[(2)^2 - \frac{(2)^3}{3} \right] - \left[(1)^2 - \frac{(1)^3}{3} \right]$$

$$= 4 - \frac{8}{3} - 1 + \frac{1}{3} = \frac{2}{3} \approx 0.667$$

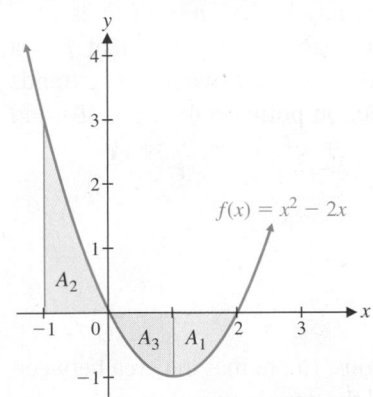

Figure 4

(B) Since the graph shows that $f(x) \geq 0$ on $[-1, 0]$ and $f(x) \leq 0$ on $[0, 1]$, the computation of this area will require two integrals:

$$A = A_2 + A_3$$

$$= \int_{-1}^0 f(x) \, dx + \int_0^1 [-f(x)] \, dx$$

$$= \int_{-1}^0 (x^2 - 2x) \, dx + \int_0^1 (2x - x^2) \, dx$$

$$= \left(\frac{x^3}{3} - x^2 \right) \Big|_{-1}^0 + \left(x^2 - \frac{x^3}{3} \right) \Big|_0^1$$

$$= \frac{4}{3} + \frac{2}{3} = 2$$

Matched Problem 2 Find the area between the graph of $f(x) = x^2 - 9$ and the x axis over the indicated intervals:

(A) $[0, 2]$ (B) $[2, 4]$

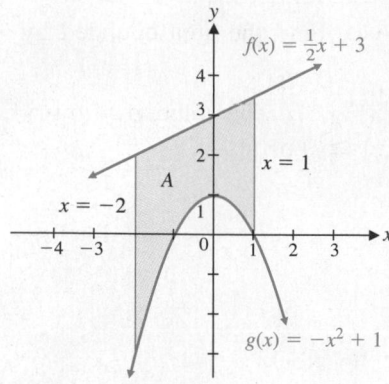

Figure 5

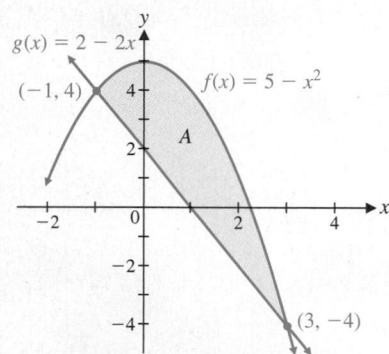

Figure 6

EXAMPLE 3   Area Between Two Curves   Find the area bounded by the graphs of $f(x) = \frac{1}{2}x + 3$, $g(x) = -x^2 + 1$, $x = -2$, and $x = 1$.

SOLUTION   We first sketch the area (Fig. 5) and then set up and evaluate an appropriate definite integral. We observe from the graph that $f(x) \geq g(x)$ for $-2 \leq x \leq 1$, so

$$A = \int_{-2}^{1} [f(x) - g(x)] \, dx = \int_{-2}^{1} \left[\left(\frac{x}{2} + 3 \right) - (-x^2 + 1) \right] dx$$

$$= \int_{-2}^{1} \left(x^2 + \frac{x}{2} + 2 \right) dx$$

$$= \left(\frac{x^3}{3} + \frac{x^2}{4} + 2x \right) \Bigg|_{-2}^{1}$$

$$= \left(\frac{1}{3} + \frac{1}{4} + 2 \right) - \left(\frac{-8}{3} + \frac{4}{4} - 4 \right) = \frac{33}{4} = 8.25$$

Matched Problem 3)   Find the area bounded by $f(x) = x^2 - 1$, $g(x) = -\frac{1}{2}x - 3$, $x = -1$, and $x = 2$.

EXAMPLE 4   Area Between Two Curves   Find the area bounded by $f(x) = 5 - x^2$ and $g(x) = 2 - 2x$.

SOLUTION   First, graph f and g on the same coordinate system, as shown in Figure 6. Since the statement of the problem does not include any limits on the values of x, we must determine the appropriate values from the graph. The graph of f is a parabola and the graph of g is a line. The area bounded by these two graphs extends from the intersection point on the left to the intersection point on the right. To find these intersection points, we solve the equation $f(x) = g(x)$ for x:

$$f(x) = g(x)$$
$$5 - x^2 = 2 - 2x$$
$$x^2 - 2x - 3 = 0$$
$$x = -1, 3$$

You should check these values in the original equations. (Note that the area between the graphs for $x < -1$ is unbounded on the left, and the area between the graphs for $x > 3$ is unbounded on the right.) Figure 6 shows that $f(x) \geq g(x)$ over the interval $[-1, 3]$, so we have

$$A = \int_{-1}^{3} [f(x) - g(x)] \, dx = \int_{-1}^{3} [5 - x^2 - (2 - 2x)] \, dx$$

$$= \int_{-1}^{3} (3 + 2x - x^2) \, dx$$

$$= \left(3x + x^2 - \frac{x^3}{3} \right) \Bigg|_{-1}^{3}$$

$$= \left[3(3) + (3)^2 - \frac{(3)^3}{3} \right] - \left[3(-1) + (-1)^2 - \frac{(-1)^3}{3} \right] = \frac{32}{3} \approx 10.667$$

Matched Problem 4)   Find the area bounded by $f(x) = 6 - x^2$ and $g(x) = x$.

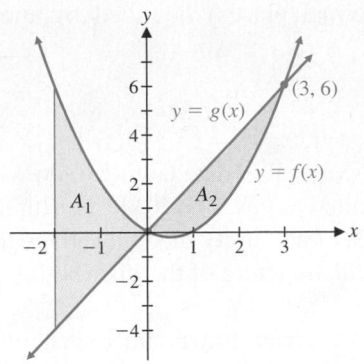

Figure 7

EXAMPLE 5 Area Between Two Curves Find the area bounded by $f(x) = x^2 - x$ and $g(x) = 2x$ for $-2 \leq x \leq 3$.

SOLUTION The graphs of f and g are shown in Figure 7. Examining the graph, we see that $f(x) \geq g(x)$ on the interval $[-2, 0]$, but $g(x) \geq f(x)$ on the interval $[0, 3]$. Thus, two integrals are required to compute this area:

$$A_1 = \int_{-2}^{0} [f(x) - g(x)]\, dx \qquad f(x) \geq g(x) \text{ on } [-2, 0]$$

$$= \int_{-2}^{0} [x^2 - x - 2x]\, dx$$

$$= \int_{-2}^{0} (x^2 - 3x)\, dx$$

$$= \left(\frac{x^3}{3} - \frac{3}{2}x^2 \right) \Big|_{-2}^{0}$$

$$= (0) - \left[\frac{(-2)^3}{3} - \frac{3}{2}(-2)^2 \right] = \frac{26}{3} \approx 8.667$$

$$A_2 = \int_{0}^{3} [g(x) - f(x)]\, dx \qquad g(x) \geq f(x) \text{ on } [0, 3]$$

$$= \int_{0}^{3} [2x - (x^2 - x)]\, dx$$

$$= \int_{0}^{3} (3x - x^2)\, dx$$

$$= \left(\frac{3}{2}x^2 - \frac{x^3}{3} \right) \Big|_{0}^{3}$$

$$= \left[\frac{3}{2}(3)^2 - \frac{(3)^3}{3} \right] - (0) = \frac{9}{2} = 4.5$$

The total area between the two graphs is

$$A = A_1 + A_2 = \frac{26}{3} + \frac{9}{2} = \frac{79}{6} \approx 13.167$$

Matched Problem 5 Find the area bounded by $f(x) = 2x^2$ and $g(x) = 4 - 2x$ for $-2 \leq x \leq 2$.

 EXAMPLE 6 Computing Areas with a Numerical Integration Routine Find the area (to three decimal places) bounded by $f(x) = e^{-x^2}$ and $g(x) = x^2 - 1$.

SOLUTION First, we use a graphing calculator to graph the functions f and g and find their intersection points (see Fig. 8A). We see that the graph of f is bell shaped and the graph of g is a parabola. We note that $f(x) \geq g(x)$ on the interval $[-1.131, 1.131]$ and compute the area A by a numerical integration routine (see Fig. 8B):

$$A = \int_{-1.131}^{1.131} \left[e^{-x^2} - (x^2 - 1) \right] dx = 2.876$$

Figure 8   (A)   (B)

Matched Problem 6) Find the area (to three decimal places) bounded by the graphs of $f(x) = x^2 \ln x$ and $g(x) = 3x - 3$.

Application: Income Distribution

The U.S. Census Bureau compiles and analyzes a great deal of data having to do with the distribution of income among families in the United States. For 2011, the Bureau reported that the lowest 20% of families received 3% of all family income and the top 20% received 51%. Table 1 and Figure 9 give a detailed picture of the distribution of family income in 2011.

The graph of $y = f(x)$ in Figure 9 is called a **Lorenz curve** and is generally found by using *regression analysis*, a technique for fitting a function to a data set over a given interval. The variable x **represents the cumulative percentage of families at or below a given income level,** and y **represents the cumulative percentage of total family income received.** For example, data point (0.40, 0.11) in Table 1 indicates that the bottom 40% of families (those with incomes under $39,000) received 11% of the total income for all families in 2011, data point (0.60, 0.26) indicates that the bottom 60% of families received 26% of the total income for all families that year, and so on.

Table 1 **Family Income Distribution in the United States, 2011**

Income Level	x	y
Under $20,000	0.20	0.03
Under $39,000	0.40	0.11
Under $62,000	0.60	0.26
Under $102,000	0.80	0.49

Source: U.S. Census Bureau

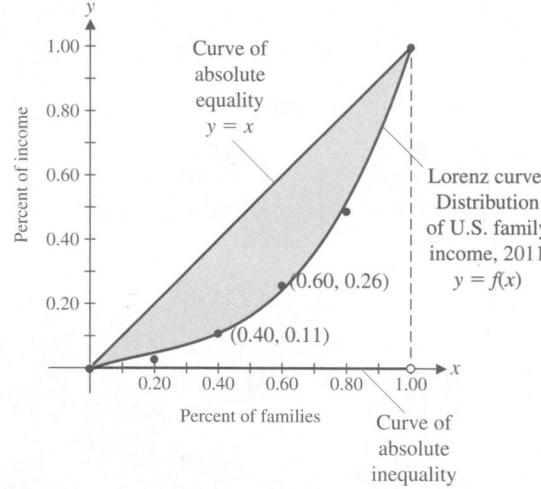

Figure 9 **Lorenz curve**

Absolute equality of income would occur if the area between the Lorenz curve and $y = x$ were 0. In this case, the Lorenz curve would be $y = x$ and all families would receive equal shares of the total income. That is, 5% of the families would receive 5% of the income, 20% of the families would receive 20% of the income, 65% of the families would receive 65% of the income, and so on. The maximum possible area between a Lorenz curve and $y = x$ is $\frac{1}{2}$, the area of the triangle below $y = x$. In this case, we would have **absolute inequality**: All the income would be in the hands of one family and the rest would have none. In actuality, Lorenz curves lie between these two extremes. But as the shaded area increases, the greater is the inequality of income distribution.

We use a single number, the **Gini index** [named after the Italian sociologist Corrado Gini (1884–1965)], to measure income concentration. The Gini index is the ratio of two areas: the area between $y = x$ and the Lorenz curve, and the area between $y = x$ and the x axis, from $x = 0$ to $x = 1$. The first area equals $\int_0^1 [x - f(x)]\, dx$ and the second (triangular) area equals $\frac{1}{2}$, giving the following definition:

DEFINITION Gini Index of Income Concentration

If $y = f(x)$ is the equation of a Lorenz curve, then

$$\textbf{Gini index} = 2 \int_0^1 [x - f(x)]\, dx$$

The Gini index is always a number between 0 and 1:

A Gini index of 0 indicates absolute equality—all people share equally in the income. A Gini index of 1 indicates absolute inequality—one person has all the income and the rest have none.

The closer the index is to 0, the closer the income is to being equally distributed. The closer the index is to 1, the closer the income is to being concentrated in a few hands. The Gini index of income concentration is used to compare income distributions at various points in time, between different groups of people, before and after taxes are paid, between different countries, and so on.

EXAMPLE 7 Distribution of Income The Lorenz curve for the distribution of income in a certain country in 2013 is given by $f(x) = x^{2.6}$. Economists predict that the Lorenz curve for the country in the year 2025 will be given by $g(x) = x^{1.8}$. Find the Gini index of income concentration for each curve, and interpret the results.

SOLUTION The Lorenz curves are shown in Figure 10.

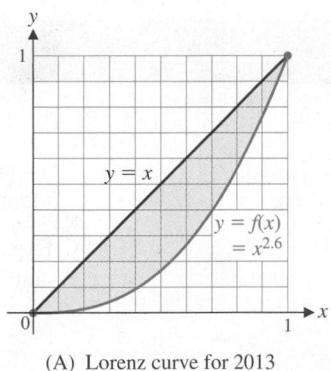

(A) Lorenz curve for 2013

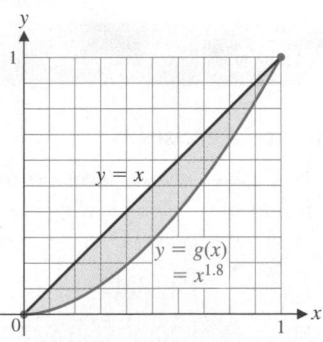

(B) Projected Lorenz curve for 2025

Figure 10

The Gini index in 2013 is (see Fig. 10A)

$$2 \int_0^1 [x - f(x)]\, dx = 2 \int_0^1 [x - x^{2.6}]\, dx = 2\left(\frac{1}{2}x^2 - \frac{1}{3.6}x^{3.6} \right)\Big|_0^1$$

$$= 2\left(\frac{1}{2} - \frac{1}{3.6} \right) \approx 0.444$$

The projected Gini index in 2025 is (see Fig. 10B)

$$2 \int_0^1 [x - g(x)]\, dx = 2 \int_0^1 [x - x^{1.8}]\, dx = 2\left(\frac{1}{2}x^2 - \frac{1}{2.8}x^{2.8} \right)\Big|_0^1$$

$$= 2\left(\frac{1}{2} - \frac{1}{2.8} \right) \approx 0.286$$

If this projection is correct, the Gini index will decrease, and income will be more equally distributed in the year 2025 than in 2013.

Matched Problem 7) Repeat Example 7 if the projected Lorenz curve in the year 2025 is given by $g(x) = x^{3.8}$.

Explore and Discuss 1   Do you agree or disagree with each of the following statements (explain your answers by referring to the data in Table 2):

(A) In countries with a low Gini index, there is little incentive for individuals to strive for success, and therefore productivity is low.

(B) In countries with a high Gini index, it is almost impossible to rise out of poverty, and therefore productivity is low.

Table 2

Country	Gini Index	Per Capita Gross Domestic Product
Brazil	0.52	$12,100
Canada	0.32	43,400
China	0.47	9,300
France	0.33	36,100
Germany	0.27	39,700
India	0.37	3,900
Japan	0.38	36,900
Jordan	0.40	6,100
Mexico	0.48	15,600
Russia	0.42	18,000
Sweden	0.23	41,900
United States	0.45	50,700

Source: The World Factbook, CIA

Exercises 6.1

Skills Warm-up Exercises

W

In Problems 1–8, use geometric formulas to find the area between the graphs of $y = f(x)$ and $y = g(x)$ over the indicated interval. (If necessary, review Appendix C).

1. $f(x) = 60, g(x) = 45; [2, 12]$

2. $f(x) = -30, g(x) = 20; [-3, 6]$

3. $f(x) = 6 + 2x, g(x) = 6 - x; [0, 5]$

4. $f(x) = 0.5x, g(x) = 0.5x - 4; [0, 8]$

5. $f(x) = -3 - x, g(x) = 4 + 2x; [-1, 2]$

6. $f(x) = 100 - 2x, g(x) = 10 + 3x; [5, 10]$

7. $f(x) = x, g(x) = \sqrt{4 - x^2}; [0, \sqrt{2}]$

8. $f(x) = \sqrt{16 - x^2}, g(x) = |x|; [-2\sqrt{2}, 2\sqrt{2}]$

Problems 9–14 refer to Figures A–D. Set up definite integrals in Problems 9–12 that represent the indicated shaded area.

9. Shaded area in Figure B

10. Shaded area in Figure A

11. Shaded area in Figure C

12. Shaded area in Figure D

13. Explain why $\int_a^b h(x)\, dx$ does not represent the area between the graph of $y = h(x)$ and the x axis from $x = a$ to $x = b$ in Figure C.

14. Explain why $\int_a^b [-h(x)]\, dx$ represents the area between the graph of $y = h(x)$ and the x axis from $x = a$ to $x = b$ in Figure C.

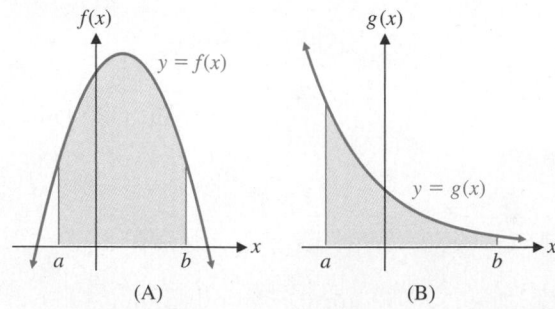

(A)   (B)

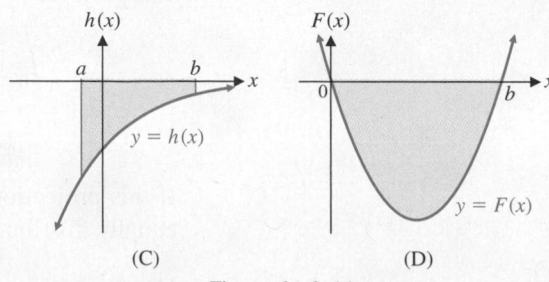

(C)   (D)

Figures for 9–14

In Problems 15–28, find the area bounded by the graphs of the indicated equations over the given interval. Compute answers to three decimal places.

15. $y = x + 4$; $y = 0$; $0 \le x \le 4$

16. $y = -x + 10$; $y = 0$; $-2 \le x \le 2$

17. $y = x^2 - 20$; $y = 0$; $-3 \le x \le 0$

18. $y = x^2 + 2$; $y = 0$; $0 \le x \le 3$

19. $y = -x^2 + 10$; $y = 0$; $-3 \le x \le 3$

20. $y = -2x^2$; $y = 0$; $-6 \le x \le 0$

21. $y = x^3 + 1$; $y = 0$; $0 \le x \le 2$

22. $y = -x^3 + 3$; $y = 0$; $-2 \le x \le 1$

23. $y = x(1 - x)$; $y = 0$; $-1 \le x \le 0$

24. $y = -x(3 - x)$; $y = 0$; $1 \le x \le 2$

25. $y = -e^x$; $y = 0$; $-1 \le x \le 1$

26. $y = e^x$; $y = 0$; $0 \le x \le 1$

27. $y = \dfrac{1}{x}$; $y = 0$; $1 \le x \le e$

28. $y = -\dfrac{1}{x}$; $y = 0$; $-1 \le x \le -\dfrac{1}{e}$

In Problems 29–32, base your answers on the Gini index of income concentration (see Table 2, page 388).

29. In which of Canada, Mexico, or the United States is income most equally distributed? Most unequally distributed?

30. In which of France, Germany, or Sweden is income most equally distributed? Most unequally distributed?

31. In which of Brazil, India, or Jordan is income most equally distributed? Most unequally distributed?

32. In which of China, Japan, or Russia is income most equally distributed? Most unequally distributed?

Problems 33–42 refer to Figures A and B. Set up definite integrals in Problems 33–40 that represent the indicated shaded areas over the given intervals.

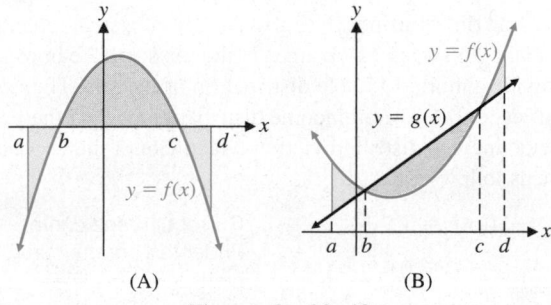

(A)   (B)

Figures for 33–42

33. Over interval $[a, b]$ in Figure A

34. Over interval $[c, d]$ in Figure A

35. Over interval $[b, d]$ in Figure A

36. Over interval $[a, c]$ in Figure A

37. Over interval $[c, d]$ in Figure B

38. Over interval $[a, b]$ in Figure B

39. Over interval $[a, c]$ in Figure B

40. Over interval $[b, d]$ in Figure B

41. Referring to Figure B, explain how you would use definite integrals and the functions f and g to find the area bounded by the two functions from $x = a$ to $x = d$.

42. Referring to Figure A, explain how you would use definite integrals to find the area between the graph of $y = f(x)$ and the x axis from $x = a$ to $x = d$.

In Problems 43–58, find the area bounded by the graphs of the indicated equations over the given intervals (when stated). Compute answers to three decimal places.

43. $y = -x$; $y = 0$; $-2 \le x \le 1$

44. $y = -x + 1$; $y = 0$; $-1 \le x \le 2$

45. $y = x^2 - 4$; $y = 0$; $0 \le x \le 3$

46. $y = 4 - x^2$; $y = 0$; $0 \le x \le 4$

47. $y = x^2 - 3x$; $y = 0$; $-2 \le x \le 2$

48. $y = -x^2 - 2x$; $y = 0$; $-2 \le x \le 1$

49. $y = -2x + 8$; $y = 12$; $-1 \le x \le 2$

50. $y = 2x + 6$; $y = 3$; $-1 \le x \le 2$

51. $y = 3x^2$; $y = 12$

52. $y = x^2$; $y = 9$

53. $y = 4 - x^2$; $y = -5$

54. $y = x^2 - 1$; $y = 3$

55. $y = x^2 + 1$; $y = 2x - 2$; $-1 \le x \le 2$

56. $y = x^2 - 1$; $y = x - 2$; $-2 \le x \le 1$

57. $y = e^{0.5x}$; $y = -\dfrac{1}{x}$; $1 \le x \le 2$

58. $y = \dfrac{1}{x}$; $y = -e^x$; $0.5 \le x \le 1$

In Problems 59–64, set up a definite integral that represents the area bounded by the graphs of the indicated equations over the given interval. Find the areas to three decimal places. [Hint: A circle of radius r, with center at the origin, has equation $x^2 + y^2 = r^2$ and area πr^2].

59. $y = \sqrt{9 - x^2}$; $y = 0$; $-3 \le x \le 3$

60. $y = \sqrt{25 - x^2}$; $y = 0$; $-5 \le x \le 5$

61. $y = -\sqrt{16 - x^2}$; $y = 0$; $0 \le x \le 4$

62. $y = -\sqrt{36 - x^2}$; $y = 0$; $-6 \le x \le 0$

63. $y = -\sqrt{4 - x^2}$; $y = \sqrt{4 - x^2}$; $-2 \le x \le 2$

64. $y = -\sqrt{100 - x^2}$; $y = \sqrt{100 - x^2}$; $-10 \le x \le 10$

In Problems 65–70, find the area bounded by the graphs of the indicated equations over the given interval (when stated). Compute answers to three decimal places.

65. $y = e^x$; $y = e^{-x}$; $0 \le x \le 4$

66. $y = e^x$; $y = -e^{-x}$; $1 \le x \le 2$

67. $y = x^3; y = 4x$

68. $y = x^3 + 1; y = x + 1$

69. $y = x^3 - 3x^2 - 9x + 12; y = x + 12$

70. $y = x^3 - 6x^2 + 9x; y = x$

In Problems 71–76, use a graphing calculator to graph the equations and find relevant intersection points. Then find the area bounded by the curves. Compute answers to three decimal places.

71. $y = x^3 - x^2 + 2; y = -x^3 + 8x - 2$

72. $y = 2x^3 + 2x^2 - x; y = -2x^3 - 2x^2 + 2x$

73. $y = e^{-x}; y = 3 - 2x$

74. $y = 2 - (x + 1)^2; y = e^{x+1}$

75. $y = e^x; y = 5x - x^3$

76. $y = 2 - e^x; y = x^3 + 3x^2$

In Problems 77–80, use a numerical integration routine on a graphing calculator to find the area bounded by the graphs of the indicated equations over the given interval (when stated). Compute answers to three decimal places.

77. $y = e^{-x}; y = \sqrt{\ln x}; 2 \le x \le 5$

78. $y = x^2 + 3x + 1; y = e^{e^x}; -3 \le x \le 0$

79. $y = e^{x^2}; y = x + 2$

80. $y = \ln(\ln x); y = 0.01x$

In Problems 81–84, find the constant c (to 2 decimal places) such that the Lorenz curve $f(x) = x^c$ has the given Gini index of income concentration.

81. 0.52

82. 0.23

83. 0.29

84. 0.65

Applications

In the applications that follow, it is helpful to sketch graphs to get a clearer understanding of each problem and to interpret results. A graphing calculator will prove useful if you have one, but it is not necessary.

85. Oil production. Using production and geological data, the management of an oil company estimates that oil will be pumped from a producing field at a rate given by

$$R(t) = \frac{100}{t + 10} + 10 \qquad 0 \le t \le 15$$

where $R(t)$ is the rate of production (in thousands of barrels per year) t years after pumping begins. Find the area between the graph of R and the t axis over the interval $[5, 10]$ and interpret the results.

86. Oil production. In Problem 85, if the rate is found to be

$$R(t) = \frac{100t}{t^2 + 25} + 4 \qquad 0 \le t \le 25$$

then find the area between the graph of R and the t axis over the interval $[5, 15]$ and interpret the results.

87. Useful life. An amusement company maintains records for each video game it installs in an arcade. Suppose that $C(t)$ and $R(t)$ represent the total accumulated costs and revenues (in thousands of dollars), respectively, t years after a particular game has been installed. If

$$C'(t) = 2 \qquad \text{and} \qquad R'(t) = 9e^{-0.3t}$$

then find the area between the graphs of C' and R' over the interval on the t axis from 0 to the useful life of the game and interpret the results.

88. Useful life. Repeat Problem 87 if

$$C'(t) = 2t \qquad \text{and} \qquad R'(t) = 5te^{-0.1t^2}$$

89. Income distribution. In a study on the effects of World War II on the U.S. economy, an economist used data from the U.S. Census Bureau to produce the following Lorenz curves for the distribution of U.S. income in 1935 and in 1947:

$$f(x) = x^{2.4} \quad \text{Lorenz curve for 1935}$$

$$g(x) = x^{1.6} \quad \text{Lorenz curve for 1947}$$

Find the Gini index of income concentration for each Lorenz curve and interpret the results.

90. Income distribution. Using data from the U.S. Census Bureau, an economist produced the following Lorenz curves for the distribution of U.S. income in 1962 and in 1972:

$$f(x) = \tfrac{3}{10}x + \tfrac{7}{10}x^2 \quad \text{Lorenz curve for 1962}$$

$$g(x) = \tfrac{1}{2}x + \tfrac{1}{2}x^2 \quad \text{Lorenz curve for 1972}$$

Find the Gini index of income concentration for each Lorenz curve and interpret the results.

91. Distribution of wealth. Lorenz curves also can provide a relative measure of the distribution of a country's total assets. Using data in a report by the U.S. Congressional Joint Economic Committee, an economist produced the following Lorenz curves for the distribution of total U.S. assets in 1963 and in 1983:

$$f(x) = x^{10} \quad \text{Lorenz curve for 1963}$$

$$g(x) = x^{12} \quad \text{Lorenz curve for 1983}$$

Find the Gini index of income concentration for each Lorenz curve and interpret the results.

92. Income distribution. The government of a small country is planning sweeping changes in the tax structure in order to provide a more equitable distribution of income. The Lorenz curves for the current income distribution and for the projected income distribution after enactment of the tax changes are as follows:

$$f(x) = x^{2.3} \qquad \text{Current Lorenz curve}$$

$$g(x) = 0.4x + 0.6x^2 \qquad \begin{array}{l}\text{Projected Lorenz curve after}\\ \text{changes in tax laws}\end{array}$$

Find the Gini index of income concentration for each Lorenz curve. Will the proposed changes provide a more equitable income distribution? Explain.

93. Distribution of wealth. The data in the following table describe the distribution of wealth in a country:

x	0	0.20	0.40	0.60	0.80	1
y	0	0.12	0.31	0.54	0.78	1

(A) Use quadratic regression to find the equation of a Lorenz curve for the data.

(B) Use the regression equation and a numerical integration routine to approximate the Gini index of income concentration.

94. Distribution of wealth. Refer to Problem 93.

(A) Use cubic regression to find the equation of a Lorenz curve for the data.

(B) Use the cubic regression equation you found in Part (A) and a numerical integration routine to approximate the Gini index of income concentration.

95. Biology. A yeast culture is growing at a rate of $W'(t) = 0.3e^{0.1t}$ grams per hour. Find the area between the graph of W' and the t axis over the interval $[0, 10]$ and interpret the results.

96. Natural resource depletion. The instantaneous rate of change in demand for U.S. lumber since 1970 ($t = 0$), in billions of cubic feet per year, is given by

$$Q'(t) = 12 + 0.006t^2 \qquad 0 \le t \le 50$$

Find the area between the graph of Q' and the t axis over the interval $[35, 40]$, and interpret the results.

97. Learning. A college language class was chosen for a learning experiment. Using a list of 50 words, the experiment measured the rate of vocabulary memorization at different times during a continuous 5-hour study session. The average rate of learning for the entire class was inversely proportional to the time spent studying and was given approximately by

$$V'(t) = \frac{15}{t} \qquad 1 \le t \le 5$$

Find the area between the graph of V' and the t axis over the interval $[2, 4]$, and interpret the results.

98. Learning. Repeat Problem 97 if $V'(t) = 13/t^{1/2}$ and the interval is changed to $[1, 4]$.

Answers to Matched Problems

1. $A = \int_{-1}^{3} (x^2 + 1)\, dx = \frac{40}{3} \approx 13.333$
2. (A) $A = \int_{0}^{2} (9 - x^2)\, dx = \frac{46}{3} \approx 15.333$
   (B) $A = \int_{2}^{3} (9 - x^2)\, dx + \int_{3}^{4} (x^2 - 9)\, dx = 6$
3. $A = \int_{-1}^{2} \left[(x^2 - 1) - \left(-\frac{x}{2} - 3 \right) \right] dx = \frac{39}{4} = 9.75$
4. $A = \int_{-3}^{2} [(6 - x^2) - x]\, dx = \frac{125}{6} \approx 20.833$
5. $A = \int_{-2}^{1} [(4 - 2x) - 2x^2]\, dx$
   $+ \int_{1}^{2} [2x^2 - (4 - 2x)]\, dx = \frac{38}{3} \approx 12.667$
6. 0.443
7. Gini index of income concentration ≈ 0.583; income will be less equally distributed in 2025.

6.2 Applications in Business and Economics

- Probability Density Functions
- Continuous Income Stream
- Future Value of a Continuous Income Stream
- Consumers' and Producers' Surplus

This section contains important applications of the definite integral to business and economics. Included are three independent topics: probability density functions, continuous income streams, and consumers' and producers' surplus. Any of the three may be covered in any order as time and interests dictate.

Probability Density Functions

We now take a brief, informal look at the use of the definite integral to determine probabilities. A more formal treatment of the subject requires the use of the special "improper" integral form $\int_{-\infty}^{\infty} f(x)\, dx$, which we will not discuss.

Suppose that an experiment is designed in such a way that any real number x on the interval $[c, d]$ is a possible outcome. For example, x may represent an IQ score, the height of a person in inches, or the life of a lightbulb in hours. Technically, we refer to x as a *continuous random variable*.

In certain situations, we can find a function f with x as an independent variable such that the function f can be used to determine Probability $(c \le x \le d)$, that is, the probability that the outcome x of an experiment will be in the interval $[c, d]$. Such a function, called a **probability density function**, must satisfy the following three conditions (see Fig. 1):

1. $f(x) \ge 0$ for all real x.
2. The area under the graph of $f(x)$ over the interval $(-\infty, \infty)$ is exactly 1.
3. If $[c, d]$ is a subinterval of $(-\infty, \infty)$, then

$$\text{Probability } (c \le x \le d) = \int_{c}^{d} f(x)\, dx$$

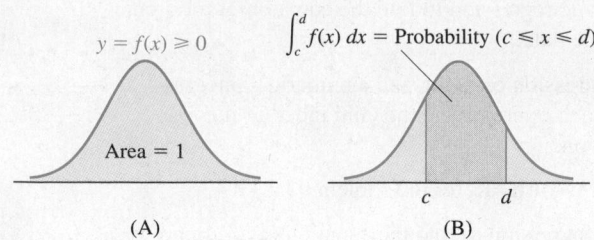

Figure 1 **Probability density function**

EXAMPLE 1   Duration of Telephone Calls Suppose that the length of telephone calls (in minutes) is a continuous random variable with the probability density function shown in Figure 2:

$$f(t) = \begin{cases} \frac{1}{4}e^{-t/4} & \text{if } t \geq 0 \\ 0 & \text{otherwise} \end{cases}$$

(A) Determine the probability that a call selected at random will last between 2 and 3 minutes.

(B) Find b (to two decimal places) so that the probability of a call selected at random lasting between 2 and b minutes is .5.

SOLUTION

(A) Probability $(2 \leq t \leq 3) = \displaystyle\int_2^3 \frac{1}{4}e^{-t/4}\, dt$

$$= \left(-e^{-t/4} \right) \Big|_2^3$$

$$= -e^{-3/4} + e^{-1/2} \approx .13$$

(B) We want to find b such that Probability $(2 \leq t \leq b) = .5$.

$$\int_2^b \frac{1}{4}e^{-t/4}\, dt = .5$$

$$-e^{-b/4} + e^{-1/2} = .5 \qquad \text{Solve for } b.$$

$$e^{-b/4} = e^{-.5} - .5$$

$$-\frac{b}{4} = \ln\left(e^{-.5} - .5\right)$$

$$b = 8.96 \text{ minutes}$$

So the probability of a call selected at random lasting from 2 to 8.96 minutes is .5.

Matched Problem 1

(A) In Example 1, find the probability that a call selected at random will last 4 minutes or less.

(B) Find b (to two decimal places) so that the probability of a call selected at random lasting b minutes or less is .9

CONCEPTUAL INSIGHT

The probability that a phone call in Example 1 lasts exactly 2 minutes (not 1.999 minutes, not 1.999 999 minutes) is given by

$$\text{Probability } (2 \leq t \leq 2) = \int_2^2 \frac{1}{4}e^{-t/4}\, dt \qquad \text{Use Property 1, Section 5.4}$$

$$= 0$$

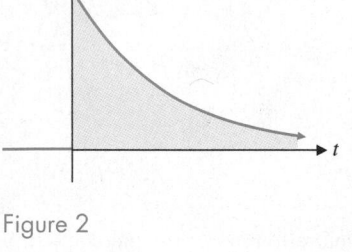

Figure 2

In fact, for any *continuous* random variable x with probability density function $f(x)$, the probability that x is exactly equal to a constant c is equal to 0:

$$\text{Probability } (c \le x \le c) = \int_c^c f(x)\,dx \quad \text{Use Property 1, Section 5.4}$$

$$= 0$$

In this respect, a *continuous* random variable differs from a *discrete* random variable. If x, for example, is the discrete random variable that represents the number of dots that appear on the top face when a fair die is rolled, then

$$\text{Probability } (2 \le x \le 2) = \tfrac{1}{6}$$

One of the most important probability density functions, the **normal probability density function**, is defined as follows and graphed in Figure 3:

$$f(x) = \frac{1}{\sigma\sqrt{2\pi}}e^{-(x-\mu)^2/2\sigma^2} \quad \begin{array}{l}\mu \text{ is the mean.}\\ \sigma \text{ is the standard deviation.}\end{array}$$

It can be shown (but not easily) that the area under the normal curve in Figure 3 over the interval $(-\infty, \infty)$ is exactly 1. Since $\int e^{-x^2}\,dx$ is nonintegrable in terms of elementary functions (that is, the antiderivative cannot be expressed as a finite combination of simple functions), probabilities such as

$$\text{Probability } (c \le x \le d) = \frac{1}{\sigma\sqrt{2\pi}}\int_c^d e^{-(x-\mu)^2/2\sigma^2}\,dx$$

can be determined by making an appropriate substitution in the integrand and then using a table of areas under the standard normal curve (that is, the normal curve with $\mu = 0$ and $\sigma = 1$). As an alternative to a table, calculators and computers can be used to compute areas under normal curves.

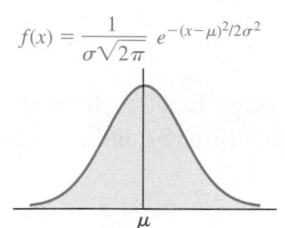

$$f(x) = \frac{1}{\sigma\sqrt{2\pi}}e^{-(x-\mu)^2/2\sigma^2}$$

Figure 3 **Normal curve**

Continuous Income Stream

We start with a simple example having an obvious solution and generalize the concept to examples having less obvious solutions.

Suppose that an aunt has established a trust that pays you $2,000 a year for 10 years. What is the total amount you will receive from the trust by the end of the 10th year? Since there are 10 payments of $2,000 each, you will receive

$$10 \times \$2,000 = \$20,000$$

We now look at the same problem from a different point of view. Let's assume that the income stream is continuous at a rate of $2,000 per year. In Figure 4, the area under the graph of $f(t) = 2,000$ from 0 to t represents the income accumulated t years after the start. For example, for $t = \tfrac{1}{4}$ year, the income would be $\tfrac{1}{4}(2,000) = \$500$; for $t = \tfrac{1}{2}$ year, the income would be $\tfrac{1}{2}(2,000) = \$1,000$; for $t = 1$ year, the income would be $1(2,000) = \$2,000$; for $t = 5.3$ years, the income would be $5.3(2,000) = \$10,600$; and for $t = 10$ years, the income would be $10(2,000) = \$20,000$. The total income over a 10-year period—that is, the area under the graph of $f(t) = 2,000$ from 0 to 10—is also given by the definite integral

$$\int_0^{10} 2,000\,dt = 2,000t\big|_0^{10} = 2,000(10) - 2,000(0) = \$20,000$$

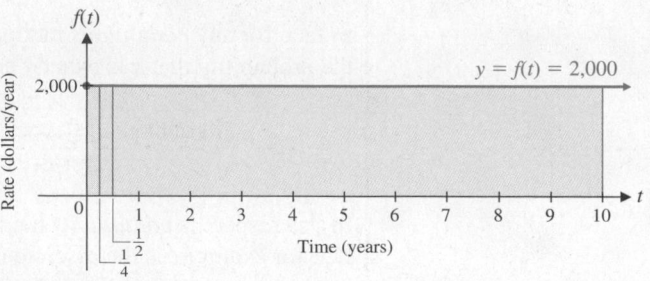

Figure 4 **Continuous income stream**

EXAMPLE 2   Continuous Income Stream   The rate of change of the income produced by a vending machine is given by

$$f(t) = 5{,}000e^{0.04t}$$

where t is time in years since the installation of the machine. Find the total income produced by the machine during the first 5 years of operation.

SOLUTION   The area under the graph of the rate-of-change function from 0 to 5 represents the total change in income over the first 5 years (Fig. 5), and is given by a definite integral:

$$
\begin{aligned}
\text{Total income} &= \int_0^5 5{,}000e^{0.04t}\, dt \\
&= 125{,}000e^{0.04t}\Big|_0^5 \\
&= 125{,}000e^{0.04(5)} - 125{,}000e^{0.04(0)} \\
&= 152{,}675 - 125{,}000 \\
&= \$27{,}675 \qquad \text{Rounded to the nearest dollar}
\end{aligned}
$$

The vending machine produces a total income of \$27,675 during the first 5 years of operation.

Matched Problem 2   Referring to Example 2, find the total income produced (to the nearest dollar) during the second 5 years of operation.

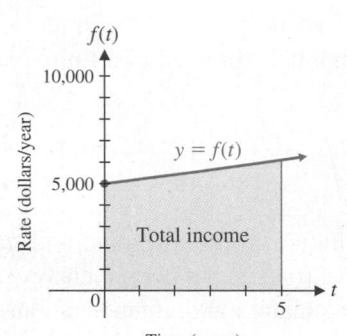

Figure 5 **Continuous income stream**

In Example 2, we assumed that the rate of change of income was given by the continuous function f. The assumption is reasonable because income from a vending machine is often collected daily. In such situations, we assume that income is received in a **continuous stream**; that is, we assume that the rate at which income is received is a continuous function of time. The rate of change is called the **rate of flow** of the continuous income stream.

DEFINITION   Total Income for a Continuous Income Stream
If $f(t)$ is the rate of flow of a continuous income stream, then the **total income** produced during the period from $t = a$ to $t = b$ is

$$\text{Total income} = \int_a^b f(t)\, dt$$

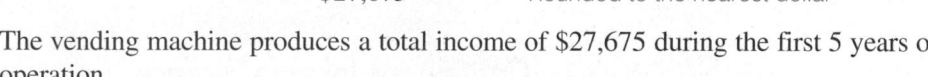

Future Value of a Continuous Income Stream

In Section 3.1, we discussed the continuous compound interest formula

$$A = Pe^{rt}$$

where P is the principal (or present value), A is the amount (or future value), r is the annual rate of continuous compounding (expressed as a decimal), and t is time in years. For example, if money is worth 12% compounded continuously, then the future value of a \$10,000 investment in 5 years is (to the nearest dollar)

$$A = 10{,}000e^{0.12(5)} = \$18{,}221$$

We want to apply the future value concept to the income produced by a continuous income stream. Suppose that $f(t)$ is the rate of flow of a continuous income stream, and the income produced by this continuous income stream is invested as soon as it is received at a rate r, compounded continuously. We already know how to find the total income produced after T years, but how can we find the total of the income produced and the interest earned by this income? Since the income is received in a continuous flow, we cannot just use the formula $A = Pe^{rt}$. This formula is valid only for a single deposit P, not for a continuous flow of income. Instead, we use a Riemann sum approach that will allow us to apply the formula $A = Pe^{rt}$ repeatedly. To begin, we divide the time interval $[0, T]$ into n equal subintervals of length Δt and choose an arbitrary point c_k in each subinterval, as shown in Figure 6.

The total income produced during the period from $t = t_{k-1}$ to $t = t_k$ is equal to the area under the graph of $f(t)$ over this subinterval and is approximately equal to $f(c_k)\,\Delta t$, the area of the shaded rectangle in Figure 6. The income received during this period will earn interest for approximately $T - c_k$ years. So, from the future-value formula $A = Pe^{rt}$ with $P = f(c_k)\,\Delta t$ and $t = T - c_k$, the future value of the income produced during the period from $t = t_{k-1}$ to $t = t_k$ is approximately equal to

$$f(c_k)\,\Delta t\, e^{(T-c_k)r}$$

The total of these approximate future values over n subintervals is then

$$f(c_1)\,\Delta t\, e^{(T-c_1)r} + f(c_2)\,\Delta t\, e^{(T-c_2)r} + \cdots + f(c_n)\,\Delta t\, e^{(T-c_n)r} = \sum_{k=1}^{n} f(c_k)e^{r(T-c_k)}\,\Delta t$$

This equation has the form of a Riemann sum, the limit of which is a definite integral. (See the definition of the definite integral in Section 5.4.) Therefore, the *future value FV* of the income produced by the continuous income stream is given by

$$FV = \int_0^T f(t)e^{r(T-t)}\,dt$$

Since r and T are constants, we also can write

$$FV = \int_0^T f(t)e^{rT}e^{-rt}\,dt = e^{rT}\int_0^T f(t)e^{-rt}\,dt \qquad (1)$$

This last form is preferable, since the integral is usually easier to evaluate than the first form.

> **DEFINITION** Future Value of a Continuous Income Stream
>
> If $f(t)$ is the rate of flow of a continuous income stream, $0 \le t \le T$, and if the income is continuously invested at a rate r, compounded continuously, then the **future value FV** at the end of T years is given by
>
> $$FV = \int_0^T f(t)e^{r(T-t)}\,dt = e^{rT}\int_0^T f(t)e^{-rt}\,dt$$

The future value of a continuous income stream is the total value of all money produced by the continuous income stream (income and interest) at the end of T years.

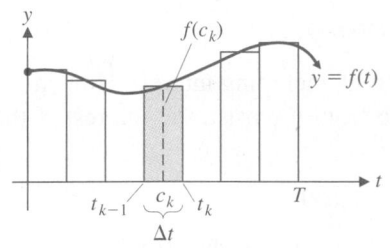

Figure 6

We return to the trust that your aunt set up for you. Suppose that the $2,000 per year you receive from the trust is invested as soon as it is received at 8%, compounded continuously. We consider the trust income to be a continuous income stream with a flow rate of $2,000 per year. What is its future value (to the nearest dollar) by the end of the 10th year? Using the definite integral for future value from the preceding box, we have

$$FV = e^{rT} \int_0^T f(t)e^{-rt}\,dt$$

$$FV = e^{0.08(10)} \int_0^{10} 2{,}000e^{-0.08t}\,dt \qquad r = 0.08,\ T = 10,\ f(t) = 2{,}000$$

$$= 2{,}000e^{0.8} \int_0^{10} e^{-0.08t}\,dt$$

$$= 2{,}000e^{0.8}\left[\frac{e^{-0.08t}}{-0.08}\right]\Bigg|_0^{10}$$

$$= 2{,}000e^{0.8}\left[-12.5e^{-0.8} + 12.5\right] = \$30{,}639$$

At the end of 10 years, you will have received $30,639, including interest. How much is interest? Since you received $20,000 in income from the trust, the interest is the difference between the future value and income. So,

$$\$30{,}639 - \$20{,}000 = \$10{,}639$$

is the interest earned by the income received from the trust over the 10-year period.

EXAMPLE 3   Future Value of a Continuous Income Stream   Using the continuous income rate of flow for the vending machine in Example 2, namely,

$$f(t) = 5{,}000e^{0.04t}$$

find the future value of this income stream at 12%, compounded continuously for 5 years, and find the total interest earned. Compute answers to the nearest dollar.

SOLUTION   Using the formula

$$FV = e^{rT} \int_0^T f(t)e^{-rt}\,dt$$

with $r = 0.12$, $T = 5$, and $f(t) = 5{,}000e^{0.04t}$, we have

$$FV = e^{0.12(5)} \int_0^5 5{,}000e^{0.04t}e^{-0.12t}\,dt$$

$$= 5{,}000e^{0.6} \int_0^5 e^{-0.08t}\,dt$$

$$= 5{,}000e^{0.6}\left(\frac{e^{-0.08t}}{-0.08}\right)\Bigg|_0^5$$

$$= 5{,}000e^{0.6}(-12.5e^{-0.4} + 12.5)$$

$$= \$37{,}545 \qquad \text{Rounded to the nearest dollar}$$

The future value of the income stream at 12% compounded continuously at the end of 5 years is $37,545.

In Example 2, we saw that the total income produced by this vending machine over a 5-year period was $27,675. The difference between future value and income is interest. So,

$$\$37{,}545 - \$27{,}675 = \$9{,}870$$

is the interest earned by the income produced by the vending machine during the 5-year period.

Matched Problem 3   Repeat Example 3 if the interest rate is 9%, compounded continuously.

Consumers' and Producers' Surplus

Let $p = D(x)$ be the price–demand equation for a product, where x is the number of units of the product that consumers will purchase at a price of \$$p$ per unit. Suppose that $\bar{p}$ is the current price and $\bar{x}$ is the number of units that can be sold at that price. Then the price–demand curve in Figure 7 shows that if the price is higher than $\bar{p}$, the demand x is less than $\bar{x}$, but some consumers are still willing to pay the higher price. Consumers who are willing to pay more than $\bar{p}$, but who are still able to buy the product at $\bar{p}$, have saved money. We want to determine the total amount saved by all the consumers who are willing to pay a price higher than $\bar{p}$ for the product.

To do this, consider the interval $[c_k, c_k + \Delta x]$, where $c_k + \Delta x < \bar{x}$. If the price remained constant over that interval, the savings on each unit would be the difference between $D(c_k)$, the price consumers are willing to pay, and $\bar{p}$, the price they actually pay. Since Δx represents the number of units purchased by consumers over the interval, the total savings to consumers over this interval is approximately equal to

$$[D(c_k) - \bar{p}]\, \Delta x \quad \text{(savings per unit)} \times \text{(number of units)}$$

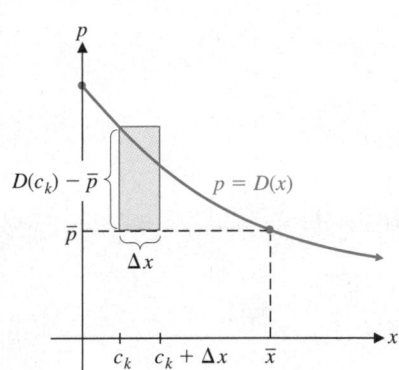

Figure 7

which is the area of the shaded rectangle shown in Figure 7. If we divide the interval $[0, \bar{x}]$ into n equal subintervals, then the total savings to consumers is approximately equal to

$$[D(c_1) - \bar{p}]\, \Delta x + [D(c_2) - \bar{p}]\, \Delta x + \cdots + [D(c_n) - \bar{p}]\, \Delta x = \sum_{k=1}^{n} [D(c_k) - \bar{p}]\, \Delta x$$

which we recognize as a Riemann sum for the integral

$$\int_0^{\bar{x}} [D(x) - \bar{p}]\, dx$$

We define the *consumers' surplus* to be this integral.

Figure 8

> **DEFINITION** Consumers' Surplus
>
> If $(\bar{x}, \bar{p})$ is a point on the graph of the price–demand equation $p = D(x)$ for a particular product, then the **consumers' surplus CS** at a price level of $\bar{p}$ is
>
> $$CS = \int_0^{\bar{x}} [D(x) - \bar{p}]\, dx$$
>
> which is the area between $p = \bar{p}$ and $p = D(x)$ from $x = 0$ to $x = \bar{x}$, as shown in Figure 8.
>
> The consumers' surplus represents the total savings to consumers who are willing to pay more than $\bar{p}$ for the product but are still able to buy the product for $\bar{p}$.

EXAMPLE 4   Consumers' Surplus   Find the consumers' surplus at a price level of \$8 for the price–demand equation

$$p = D(x) = 20 - 0.05x$$

SOLUTION

Step 1 Find $\bar{x}$, the demand when the price is $\bar{p} = 8$:

$$\bar{p} = 20 - 0.05\bar{x}$$
$$8 = 20 - 0.05\bar{x}$$
$$0.05\bar{x} = 12$$
$$\bar{x} = 240$$

Step 2 Sketch a graph, as shown in Figure 9.

Step 3 Find the consumers' surplus (the shaded area in the graph):

$$CS = \int_0^{\bar{x}} [D(x) - \bar{p}]\, dx$$
$$= \int_0^{240} (20 - 0.05x - 8)\, dx$$
$$= \int_0^{240} (12 - 0.05x)\, dx$$
$$= (12x - 0.025x^2)\big|_0^{240}$$
$$= 2{,}880 - 1{,}440 = \$1{,}440$$

The total savings to consumers who are willing to pay a higher price for the product is $1,440.

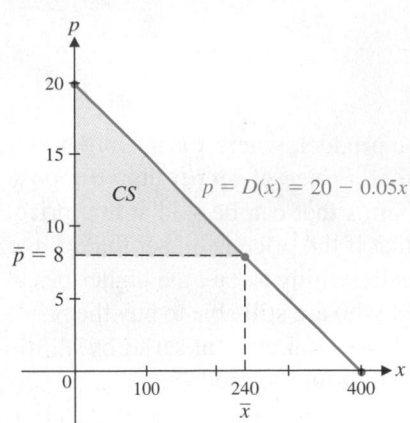

Figure 9

Matched Problem 4 Repeat Example 4 for a price level of $4.

If $p = S(x)$ is the price–supply equation for a product, $\bar{p}$ is the current price, and $\bar{x}$ is the current supply, then some suppliers are still willing to supply some units at a lower price than $\bar{p}$. The additional money that these suppliers gain from the higher price is called the *producers' surplus* and can be expressed in terms of a definite integral (proceeding as we did for the consumers' surplus).

DEFINITION Producers' Surplus

If $(\bar{x}, \bar{p})$ is a point on the graph of the price–supply equation $p = S(x)$, then the **producers' surplus PS** at a price level of $\bar{p}$ is

$$PS = \int_0^{\bar{x}} [\bar{p} - S(x)]\, dx$$

which is the area between $p = \bar{p}$ and $p = S(x)$ from $x = 0$ to $x = \bar{x}$, as shown in Figure 10.

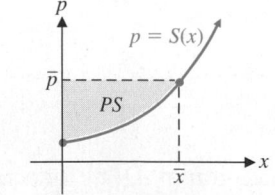

Figure 10

The producers' surplus represents the total gain to producers who are willing to supply units at a lower price than $\bar{p}$ but are still able to supply units at $\bar{p}$.

EXAMPLE 5 Producers' Surplus Find the producers' surplus at a price level of $20 for the price–supply equation

$$p = S(x) = 2 + 0.0002x^2$$

SOLUTION

Step 1 Find $\bar{x}$, the supply when the price is $\bar{p} = 20$:

$$\bar{p} = 2 + 0.0002\bar{x}^2$$
$$20 = 2 + 0.0002\bar{x}^2$$
$$0.0002\bar{x}^2 = 18$$
$$\bar{x}^2 = 90{,}000$$
$$\bar{x} = 300 \qquad \text{There is only one solution, since } \bar{x} \geq 0.$$

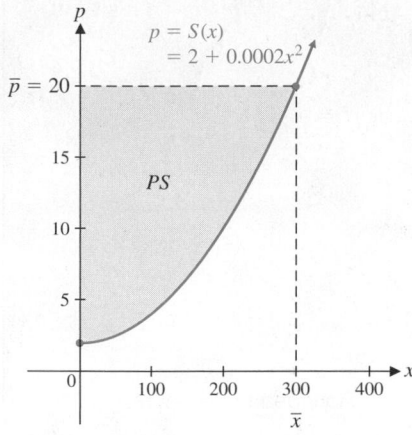

Figure 11

Step 2 Sketch a graph, as shown in Figure 11.

Step 3 Find the producers' surplus (the shaded area in the graph):

$$PS = \int_0^{\bar{x}} [\bar{p} - S(x)]\, dx = \int_0^{300} [20 - (2 + 0.0002x^2)]\, dx$$

$$= \int_0^{300} (18 - 0.0002x^2)\, dx = \left(18x - 0.0002\frac{x^3}{3} \right)\Big|_0^{300}$$

$$= 5,400 - 1,800 = \$3,600$$

The total gain to producers who are willing to supply units at a lower price is $3,600.

Matched Problem 5 Repeat Example 5 for a price level of $4.

In a free competitive market, the price of a product is determined by the relationship between supply and demand. If $p = D(x)$ and $p = S(x)$ are the price–demand and price–supply equations, respectively, for a product and if $(\bar{x}, \bar{p})$ is the point of intersection of these equations, then $\bar{p}$ is called the **equilibrium price** and $\bar{x}$ is called the **equilibrium quantity**. If the price stabilizes at the equilibrium price $\bar{p}$, then this is the price level that will determine both the consumers' surplus and the producers' surplus.

EXAMPLE 6 Equilibrium Price and Consumers' and Producers' Surplus Find the equilibrium price and then find the consumers' surplus and producers' surplus at the equilibrium price level, if

$$p = D(x) = 20 - 0.05x \quad \text{and} \quad p = S(x) = 2 + 0.0002x^2$$

SOLUTION

Step 1 Find the equilibrium quantity. Set $D(x)$ equal to $S(x)$ and solve:

$$D(x) = S(x)$$
$$20 - 0.05x = 2 + 0.0002x^2$$
$$0.0002x^2 + 0.05x - 18 = 0$$
$$x^2 + 250x - 90,000 = 0$$
$$x = 200, -450$$

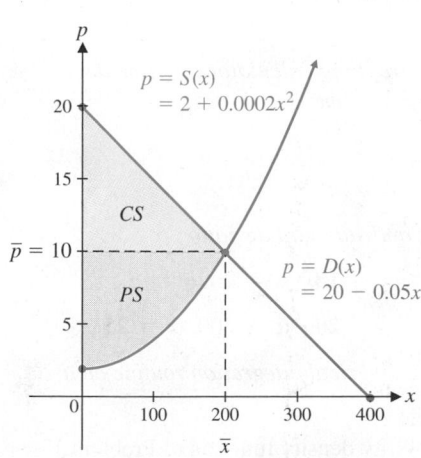

Figure 12

Since x cannot be negative, the only solution is $\bar{x} = 200$. The equilibrium price can be determined by using $D(x)$ or $S(x)$. We will use both to check our work:

$\bar{p} = D(200)$	$\bar{p} = S(200)$
$= 20 - 0.05(200) = 10$	$= 2 + 0.0002(200)^2 = 10$

The equilibrium price is $\bar{p} = 10$, and the equilibrium quantity is $\bar{x} = 200$.

Step 2 Sketch a graph, as shown in Figure 12.

Step 3 Find the consumers' surplus:

$$CS = \int_0^{\bar{x}} [D(x) - \bar{p}]\, dx = \int_0^{200} (20 - 0.05x - 10)\, dx$$

$$= \int_0^{200} (10 - 0.05x)\, dx$$

$$= (10x - 0.025x^2)\Big|_0^{200}$$

$$= 2,000 - 1,000 = \$1,000$$

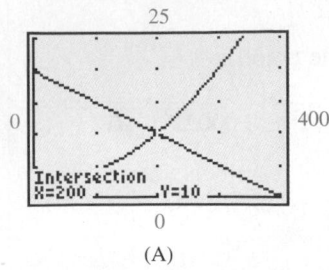

(A)

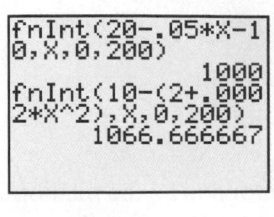

(B)

Figure 13

Step 4 Find the producers' surplus:

$$PS = \int_0^{\bar{x}} [\bar{p} - S(x)] \, dx$$

$$= \int_0^{200} [10 - (2 + 0.0002x^2)] \, dx$$

$$= \int_0^{200} (8 - 0.0002x^2) \, dx$$

$$= \left(8x - 0.0002\frac{x^3}{3} \right) \Big|_0^{200}$$

$$= 1{,}600 - \tfrac{1{,}600}{3} \approx \$1{,}067 \qquad \text{Rounded to the nearest dollar}$$

A graphing calculator offers an alternative approach to finding the equilibrium point for Example 6 (Fig. 13A). A numerical integration command can then be used to find the consumers' and producers' surplus (Fig. 13B).

Matched Problem 6 Repeat Example 6 for

$$p = D(x) = 25 - 0.001x^2 \qquad \text{and} \qquad p = S(x) = 5 + 0.1x$$

Exercises 6.2

Skills Warm-up Exercises

W   *In Problems 1–8, find real numbers b and c such that*
$f(t) = e^b e^{ct}$. (If necessary, review Section 1.5).

1. $f(t) = e^{5(4-t)}$

2. $f(t) = e^{3(15-t)}$

3. $f(t) = e^{0.04(8-t)}$

4. $f(t) = e^{0.02(12-t)}$

5. $f(t) = e^{0.05t}e^{0.08(20-t)}$

6. $f(t) = e^{0.03t}e^{0.09(30-t)}$

7. $f(t) = e^{0.09t}e^{0.07(25-t)}$

8. $f(t) = e^{0.14t}e^{0.11(15-t)}$

In Problems 9–14, evaluate each definite integral to two decimal places.

9. $\int_0^8 e^{0.06(8-t)} \, dt$

10. $\int_1^{10} e^{0.07(10-t)} \, dt$

11. $\int_0^{20} e^{0.08t}e^{0.12(20-t)} \, dt$

12. $\int_0^{15} e^{0.05t}e^{0.06(15-t)} \, dt$

13. $\int_0^{30} 500\, e^{0.02t}e^{0.09(30-t)} \, dt$

14. $\int_0^{25} 900\, e^{0.03t}e^{0.04(25-t)} \, dt$

In Problems 15 and 16, explain which of (A), (B), and (C) are equal before evaluating the expressions. Then evaluate each expression to two decimal places.

15. (A) $\int_0^8 e^{0.07(8-t)} \, dt$ (B) $\int_0^8 (e^{0.56} - e^{0.07t}) \, dt$

   (C) $e^{0.56} \int_0^8 e^{-0.07t} \, dt$

16. (A) $\int_0^{10} 2{,}000 e^{0.05t} e^{0.12(10-t)} \, dt$

   (B) $2{,}000 e^{1.2} \int_0^{10} e^{-0.07t} \, dt$

   (C) $2{,}000 e^{0.05} \int_0^{10} e^{0.12(10-t)} \, dt$

In Problems 17–20, use a graphing calculator to graph the normal probability density function

$$f(x) = \frac{1}{\sigma\sqrt{2\pi}} e^{-(x-\mu)^2/2\sigma^2}$$

that has the given mean μ and standard deviation σ.

17. $\mu = 0, \sigma = 1$

18. $\mu = 20, \sigma = 5$

19. $\mu = 500, \sigma = 100$

20. $\mu = 300, \sigma = 25$

In Problems 21–24, use a numerical integration routine on a graphing calculator.

21. For the normal probability density function of Problem 17, find:

   (A) Probability $(-1 \le x \le 1)$

   (B) Probability $(-2 \le x \le 2)$

   (C) Probability $(-3 \le x \le 3)$

22. For the normal probability density function of Problem 18, find:

   (A) Probability $(15 \le x \le 25)$

   (B) Probability $(10 \le x \le 30)$

   (C) Probability $(5 \le x \le 35)$

23. For the normal probability density function of Problem 19, find:

(A) Probability $(400 \le x \le 600)$

(B) Probability $(300 \le x \le 700)$

(C) Probability $(200 \le x \le 800)$

24. For the normal probability density function of Problem 20, find:

(A) Probability $(275 \le x \le 325)$

(B) Probability $(250 \le x \le 350)$

(C) Probability $(225 \le x \le 375)$

Applications

Unless stated to the contrary, compute all monetary answers to the nearest dollar.

25. The life expectancy (in years) of a microwave oven is a continuous random variable with probability density function

$$f(x) = \begin{cases} 2/(x+2)^2 & \text{if } x \ge 0 \\ 0 & \text{otherwise} \end{cases}$$

(A) Find the probability that a randomly selected microwave oven lasts at most 6 years.

(B) Find the probability that a randomly selected microwave oven lasts from 6 to 12 years.

(C) Graph $y = f(x)$ for [0, 12] and show the shaded region for part (A).

26. The shelf life (in years) of a laser pointer battery is a continuous random variable with probability density function

$$f(x) = \begin{cases} 1/(x+1)^2 & \text{if } x \ge 0 \\ 0 & \text{otherwise} \end{cases}$$

(A) Find the probability that a randomly selected laser pointer battery has a shelf life of 3 years or less.

(B) Find the probability that a randomly selected laser pointer battery has a shelf life of from 3 to 9 years.

(C) Graph $y = f(x)$ for [0, 10] and show the shaded region for part (A).

27. In Problem 25, find d so that the probability of a randomly selected microwave oven lasting d years or less is .8.

28. In Problem 26, find d so that the probability of a randomly selected laser pointer battery lasting d years or less is .5.

29. A manufacturer guarantees a product for 1 year. The time to failure of the product after it is sold is given by the probability density function

$$f(t) = \begin{cases} .01e^{-.01t} & \text{if } t \ge 0 \\ 0 & \text{otherwise} \end{cases}$$

where t is time in months. What is the probability that a buyer chosen at random will have a product failure

(A) During the warranty period?

(B) During the second year after purchase?

30. In a certain city, the daily use of water (in hundreds of gallons) per household is a continuous random variable with probability density function

$$f(x) = \begin{cases} .15e^{-.15x} & \text{if } x \ge 0 \\ 0 & \text{otherwise} \end{cases}$$

Find the probability that a household chosen at random will use

(A) At most 400 gallons of water per day

(B) Between 300 and 600 gallons of water per day

31. In Problem 29, what is the probability that the product will last at least 1 year? [*Hint:* Recall that the total area under the probability density function curve is 1.]

32. In Problem 30, what is the probability that a household will use more than 400 gallons of water per day? [See the hint in Problem 31.]

33. Find the total income produced by a continuous income stream in the first 5 years if the rate of flow is $f(t) = 2,500$.

34. Find the total income produced by a continuous income stream in the first 10 years if the rate of flow is $f(t) = 3,000$.

35. Interpret the results of Problem 33 with both a graph and a description of the graph.

36. Interpret the results of Problem 34 with both a graph and a description of the graph.

37. Find the total income produced by a continuous income stream in the first 3 years if the rate of flow is $f(t) = 400e^{0.05t}$.

38. Find the total income produced by a continuous income stream in the first 2 years if the rate of flow is $f(t) = 600e^{0.06t}$.

39. Interpret the results of Problem 37 with both a graph and a description of the graph.

40. Interpret the results of Problem 38 with both a graph and a description of the graph.

41. Starting at age 25, you deposit $2,000 a year into an IRA account. Treat the yearly deposits into the account as a continuous income stream. If money in the account earns 5%, compounded continuously, how much will be in the account 40 years later, when you retire at age 65? How much of the final amount is interest?

42. Suppose in Problem 41 that you start the IRA deposits at age 30, but the account earns 6%, compounded continuously. Treat the yearly deposits into the account as a continuous income stream. How much will be in the account 35 years later when you retire at age 65? How much of the final amount is interest?

43. Find the future value at 3.25% interest, compounded continuously for 4 years, of the continuous income stream with rate of flow $f(t) = 1,650e^{-0.02t}$.

44. Find the future value, at 2.95% interest, compounded continuously for 6 years, of the continuous income stream with rate of flow $f(t) = 2,000e^{0.06t}$.

45. Compute the interest earned in Problem 43.

46. Compute the interest earned in Problem 44.

47. An investor is presented with a choice of two investments: an established clothing store and a new computer store. Each choice requires the same initial investment and each produces a continuous income stream of 4%, compounded continuously. The rate of flow of income from the clothing store is $f(t) = 12,000$, and the rate of flow of income from the computer store is expected to be $g(t) = 10,000e^{0.05t}$. Compare the future values of these investments to determine which is the better choice over the next 5 years.

48. Refer to Problem 47. Which investment is the better choice over the next 10 years?

49. An investor has $10,000 to invest in either a bond that matures in 5 years or a business that will produce a continuous stream of income over the next 5 years with rate of flow $f(t) = 2,150$. If both the bond and the continuous income stream earn 3.75%, compounded continuously, which is the better investment?

50. Refer to Problem 49. Which is the better investment if the rate of the income from the business is $f(t) = 2,250$?

51. A business is planning to purchase a piece of equipment that will produce a continuous stream of income for 8 years with rate of flow $f(t) = 9,000$. If the continuous income stream earns 6.95%, compounded continuously, what single deposit into an account earning the same interest rate will produce the same future value as the continuous income stream? (This deposit is called the **present value** of the continuous income stream.)

52. Refer to Problem 51. Find the present value of a continuous income stream at 7.65%, compounded continuously for 12 years, if the rate of flow is $f(t) = 1,000e^{0.03t}$.

53. Find the future value at a rate r, compounded continuously for T years, of a continuous income stream with rate of flow $f(t) = k$, where k is a constant.

54. Find the future value at a rate r, compounded continuously for T years, of a continuous income stream with rate of flow $f(t) = ke^{ct}$, where c and k are constants, $c \neq r$.

55. Find the consumers' surplus at a price level of $\bar{p} = \$150$ for the price–demand equation

$$p = D(x) = 400 - 0.05x$$

56. Find the consumers' surplus at a price level of $\bar{p} = \$120$ for the price–demand equation

$$p = D(x) = 200 - 0.02x$$

57. Interpret the results of Problem 55 with both a graph and a description of the graph.

58. Interpret the results of Problem 56 with both a graph and a description of the graph.

59. Find the producers' surplus at a price level of $\bar{p} = \$67$ for the price–supply equation

$$p = S(x) = 10 + 0.1x + 0.0003x^2$$

60. Find the producers' surplus at a price level of $\bar{p} = \$55$ for the price–supply equation

$$p = S(x) = 15 + 0.1x + 0.003x^2$$

61. Interpret the results of Problem 59 with both a graph and a description of the graph.

62. Interpret the results of Problem 60 with both a graph and a description of the graph.

In Problems 63–70, find the consumers' surplus and the producers' surplus at the equilibrium price level for the given price–demand and price–supply equations. Include a graph that identifies the consumers' surplus and the producers' surplus. Round all values to the nearest integer.

63. $p = D(x) = 50 - 0.1x$; $p = S(x) = 11 + 0.05x$

64. $p = D(x) = 25 - 0.004x^2$; $p = S(x) = 5 + 0.004x^2$

65. $p = D(x) = 80e^{-0.001x}$; $p = S(x) = 30e^{0.001x}$

66. $p = D(x) = 185e^{-0.005x}$; $p = S(x) = 25e^{0.005x}$

67. $p = D(x) = 80 - 0.04x$; $p = S(x) = 30e^{0.001x}$

68. $p = D(x) = 190 - 0.2x$; $p = S(x) = 25e^{0.005x}$

69. $p = D(x) = 80e^{-0.001x}$; $p = S(x) = 15 + 0.0001x^2$

70. $p = D(x) = 185e^{-0.005x}$; $p = S(x) = 20 + 0.002x^2$

71. The following tables give price–demand and price–supply data for the sale of soybeans at a grain market, where x is the number of bushels of soybeans (in thousands of bushels) and p is the price per bushel (in dollars):

Tables for 71–72

Price–Demand		Price–Supply	
x	$p = D(x)$	x	$p = S(x)$
0	6.70	0	6.43
10	6.59	10	6.45
20	6.52	20	6.48
30	6.47	30	6.53
40	6.45	40	6.62

Use quadratic regression to model the price–demand data and linear regression to model the price–supply data.

(A) Find the equilibrium quantity (to three decimal places) and equilibrium price (to the nearest cent).

(B) Use a numerical integration routine to find the consumers' surplus and producers' surplus at the equilibrium price level.

72. Repeat Problem 71, using quadratic regression to model both sets of data.

Answers to Matched Problems

1. (A) .63   (B) 9.21 min

2. $33,803   **3.** $FV = \$34,691$; interest = $7,016

4. $2,560   **5.** $133

6. $\bar{p} = 15$; $CS = \$667$; $PS = \$500$

6.3 Integration by Parts

In Section 5.1, we promised to return later to the indefinite integral

$$\int \ln x \, dx$$

since none of the integration techniques considered up to that time could be used to find an antiderivative for ln x. We now develop a very useful technique, called *integration by parts,* that will enable us to find not only the preceding integral, but also many others, including integrals such as

$$\int x \ln x \, dx \quad \text{and} \quad \int x e^x \, dx$$

The method of integration by parts is based on the product formula for derivatives. If f and g are differentiable functions, then

$$\frac{d}{dx}[f(x)g(x)] = f(x)g'(x) + g(x)f'(x)$$

which can be written in the equivalent form

$$f(x)g'(x) = \frac{d}{dx}[f(x)g(x)] - g(x)f'(x)$$

Integrating both sides, we obtain

$$\int f(x)g'(x) \, dx = \int \frac{d}{dx}[f(x)g(x)] \, dx - \int g(x)f'(x) \, dx$$

The first integral to the right of the equal sign is $f(x)g(x) + C$. Why? We will leave out the constant of integration for now, since we can add it after integrating the second integral to the right of the equal sign. So,

$$\int f(x)g'(x) \, dx = f(x)g(x) - \int g(x)f'(x) \, dx$$

This equation can be transformed into a more convenient form by letting $u = f(x)$ and $v = g(x)$; then $du = f'(x) \, dx$ and $dv = g'(x) \, dx$. Making these substitutions, we obtain the **integration-by-parts formula**:

Integration-by-Parts Formula

$$\int u \, dv = uv - \int v \, du$$

This formula can be very useful when the integral on the left is difficult or impossible to integrate with standard formulas. If u and dv are chosen with care—this is the crucial part of the process—then the integral on the right side may be easier to integrate than the one on the left. The formula provides us with another tool that is helpful in many, but not all, cases. We are able to easily check the results by differentiating to get the original integrand, a good habit to develop.

EXAMPLE 1 Integration by Parts Find $\int x e^x \, dx$, using integration by parts, and check the result.

SOLUTION First, write the integration-by-parts formula:

$$\int u \, dv = uv - \int v \, du \qquad (1)$$

Now try to identify u and dv in $\int xe^x \, dx$ so that $\int v \, du$ on the right side of (1) is easier to integrate than $\int u \, dv = \int xe^x \, dx$ on the left side. There are essentially two reasonable choices in selecting u and dv in $\int xe^x \, dx$:

$$\begin{array}{cc} \text{Choice 1} & \text{Choice 2} \\ \overset{u}{\overbrace{}} \; \overset{dv}{\overbrace{}} & \overset{u}{\overbrace{}} \; \overset{dv}{\overbrace{}} \\ \int x \; e^x \, dx & \int e^x \; x \, dx \end{array}$$

We pursue choice 1 and leave choice 2 for you to explore (see Explore and Discuss 1 following this example).

From choice 1, $u = x$ and $dv = e^x \, dx$. Looking at formula (1), we need du and v to complete the right side. Let

$$u = x \qquad dv = e^x \, dx$$

Then,

$$du = dx \qquad \int dv = \int e^x \, dx$$

$$v = e^x$$

Any constant may be added to v, but we will always choose 0 for simplicity. The general arbitrary constant of integration will be added at the end of the process.

Substituting these results into formula (1), we obtain

$$\int u \, dv = uv - \int v \, du$$

$$\int xe^x \, dx = xe^x - \int e^x \, dx \qquad \text{The right integral is easy to integrate.}$$

$$= xe^x - e^x + C \qquad \text{Now add the arbitrary constant } C.$$

Check:

$$\frac{d}{dx}(xe^x - e^x + C) = xe^x + e^x - e^x = xe^x$$

Explore and Discuss 1   Pursue choice 2 in Example 1, using the integration-by-parts formula, and explain why this choice does not work out.

Matched Problem 1⌋ Find $\int xe^{2x} \, dx$.

EXAMPLE 2   Integration by Parts   Find $\int x \ln x \, dx$.

SOLUTION   As before, we have essentially two choices in choosing u and dv:

$$\begin{array}{cc} \text{Choice 1} & \text{Choice 2} \\ \overset{u}{\overbrace{}} \; \overset{dv}{\overbrace{}} & \overset{u}{\overbrace{}} \; \overset{dv}{\overbrace{}} \\ \int x \; \ln x \, dx & \int \ln x \; x \, dx \end{array}$$

Choice 1 is rejected since we do not yet know how to find an antiderivative of $\ln x$. So we move to choice 2 and choose $u = \ln x$ and $dv = x\,dx$. Then we proceed as in Example 1. Let

$$u = \ln x \qquad dv = x\,dx$$

Then,

$$du = \frac{1}{x}dx \qquad \int dv = \int x\,dx$$

$$v = \frac{x^2}{2}$$

Substitute these results into the integration-by-parts formula:

$$\int u\,dv = uv - \int v\,du$$

$$\int x \ln x\,dx = (\ln x)\left(\frac{x^2}{2}\right) - \int \left(\frac{x^2}{2}\right)\left(\frac{1}{x}\right)dx$$

$$= \frac{x^2}{2}\ln x - \int \frac{x}{2}dx \qquad \text{An easy integral to evaluate}$$

$$= \frac{x^2}{2}\ln x - \frac{x^2}{4} + C$$

Check:

$$\frac{d}{dx}\left(\frac{x^2}{2}\ln x - \frac{x^2}{4} + C\right) = x \ln x + \left(\frac{x^2}{2}\cdot\frac{1}{x}\right) - \frac{x}{2} = x \ln x$$

Matched Problem 2   Find $\int x \ln 2x\,dx$.

CONCEPTUAL INSIGHT

As you may have discovered in Explore and Discuss 1, some choices for u and dv will lead to integrals that are more complicated than the original integral. This does not mean that there is an error in either the calculations or the integration-by-parts formula. It simply means that the particular choice of u and dv does not change the problem into one we can solve. When this happens, we must look for a different choice of u and dv. In some problems, it is possible that no choice will work.

Guidelines for selecting u and dv for integration by parts are summarized in the following box:

SUMMARY Integration by Parts: Selection of u and dv

For $\int u\,dv = uv - \int v\,du$,

1. The product $u\,dv$ must equal the original integrand.
2. It must be possible to integrate dv (preferably by using standard formulas or simple substitutions).
3. The new integral $\int v\,du$ should not be more complicated than the original integral $\int u\,dv$.
4. For integrals involving $x^p e^{ax}$, try

$$u = x^p \qquad \text{and} \qquad dv = e^{ax}\,dx$$

5. For integrals involving $x^p(\ln x)^q$, try

$$u = (\ln x)^q \qquad \text{and} \qquad dv = x^p\,dx$$

In some cases, repeated use of the integration-by-parts formula will lead to the evaluation of the original integral. The next example provides an illustration of such a case.

EXAMPLE 3   Repeated Use of Integration by Parts   Find $\int x^2 e^{-x}\, dx$.

SOLUTION   Following suggestion 4 in the box, we choose

$$u = x^2 \qquad dv = e^{-x}\, dx$$

Then,

$$du = 2x\, dx \qquad v = -e^{-x}$$

and

$$\int x^2 e^{-x}\, dx = x^2(-e^{-x}) - \int (-e^{-x})2x\, dx$$

$$= -x^2 e^{-x} + 2\int xe^{-x}\, dx \qquad (2)$$

The new integral is not one we can evaluate by standard formulas, but it is simpler than the original integral. Applying the integration-by-parts formula to it will produce an even simpler integral. For the integral $\int xe^{-x}\, dx$, we choose

$$u = x \qquad dv = e^{-x}\, dx$$

Then,

$$du = dx \qquad v = -e^{-x}$$

and

$$\int xe^{-x}\, dx = x(-e^{-x}) - \int (-e^{-x})\, dx$$

$$= -xe^{-x} + \int e^{-x}\, dx$$

$$= -xe^{-x} - e^{-x} \qquad \text{Choose 0 for the constant.} \qquad (3)$$

Substituting equation (3) into equation (2), we have

$$\int x^2 e^{-x}\, dx = -x^2 e^{-x} + 2(-xe^{-x} - e^{-x}) + C \quad \text{Add an arbitrary constant here.}$$

$$= -x^2 e^{-x} - 2xe^{-x} - 2e^{-x} + C$$

Check:

$$\frac{d}{dx}(-x^2 e^{-x} - 2xe^{-x} - 2e^{-x} + C) = x^2 e^{-x} - 2xe^{-x} + 2xe^{-x} - 2e^{-x} + 2e^{-x}$$

$$= x^2 e^{-x}$$

Matched Problem 3   Find $\int x^2 e^{2x}\, dx$.

EXAMPLE 4   Using Integration by Parts   Find $\int_1^e \ln x\, dx$ and interpret the result geometrically.

SOLUTION   First, we find $\int \ln x\, dx$. Then we return to the definite integral. Following suggestion 5 in the box (with $p = 0$), we choose

$$u = \ln x \qquad dv = dx$$

Then,

$$du = \frac{1}{x}dx \qquad v = x$$

$$\int \ln x \, dx = (\ln x)(x) - \int (x)\frac{1}{x}dx$$

$$= x \ln x - x + C$$

This is the important result we mentioned at the beginning of this section. Now we have

$$\int_1^e \ln x \, dx = (x \ln x - x)\Big|_1^e$$

$$= (e \ln e - e) - (1 \ln 1 - 1)$$

$$= (e - e) - (0 - 1)$$

$$= 1$$

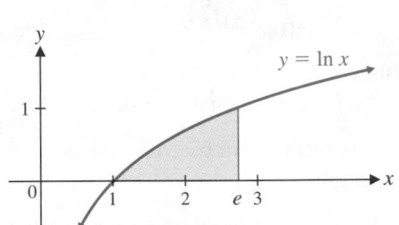

The integral represents the area under the curve $y = \ln x$ from $x = 1$ to $x = e$, as shown in Figure 1.

Figure 1

Matched Problem 4) Find $\int_1^2 \ln 3x \, dx$.

Explore and Discuss 2   Try using the integration-by-parts formula on $\int e^{x^2} \, dx$, and explain why it does not work.

Exercises 6.3

Skills Warm-up Exercises

W   *In Problems 1–8, find the derivative of $f(x)$ and the indefinite integral of $g(x)$. (If necessary, review Sections 3.2 and 5.1).*

1. $f(x) = 5x; g(x) = x^3$ **2.** $f(x) = x^2; g(x) = e^x$

3. $f(x) = x^3; g(x) = 5x$ **4.** $f(x) = e^x; g(x) = x^2$

5. $f(x) = e^{4x}; g(x) = \dfrac{1}{x}$

6. $f(x) = \sqrt{x}; g(x) = e^{-2x}$

7. $f(x) = \dfrac{1}{x}; g(x) = e^{4x}$

8. $f(x) = e^{-2x}; g(x) = \sqrt{x}$

In Problems 9–12, integrate by parts. Assume that $x > 0$ whenever the natural logarithm function is involved.

9. $\displaystyle\int xe^{3x} \, dx$ **10.** $\displaystyle\int xe^{4x} \, dx$

11. $\displaystyle\int x^2 \ln x \, dx$ **12.** $\displaystyle\int x^3 \ln x \, dx$

13. If you want to use integration by parts to find $\int (x + 1)^5 (x + 2) \, dx$, which is the better choice for u: $u = (x + 1)^5$ or $u = x + 2$? Explain your choice and then integrate.

14. If you want to use integration by parts to find $\int (5x - 7)(x - 1)^4 \, dx$, which is the better choice for u: $u = 5x - 7$ or $u = (x - 1)^4$? Explain your choice and then integrate.

Problems 15–28 are mixed—some require integration by parts, and others can be solved with techniques considered earlier. Integrate as indicated, assuming $x > 0$ whenever the natural logarithm function is involved.

15. $\displaystyle\int xe^{-x} \, dx$ **16.** $\displaystyle\int (x - 1)e^{-x} \, dx$

17. $\displaystyle\int xe^{x^2} \, dx$ **18.** $\displaystyle\int xe^{-x^2} \, dx$

19. $\displaystyle\int_0^1 (x - 3)e^x \, dx$ **20.** $\displaystyle\int_0^1 (x + 1)e^x \, dx$

21. $\displaystyle\int_1^3 \ln 2x \, dx$ **22.** $\displaystyle\int_1^2 \ln\left(\frac{x}{2}\right) dx$

23. $\displaystyle\int \frac{2x}{x^2 + 1} \, dx$ **24.** $\displaystyle\int \frac{x^2}{x^3 + 5} \, dx$

25. $\displaystyle\int \frac{\ln x}{x} \, dx$ **26.** $\displaystyle\int \frac{e^x}{e^x + 1} \, dx$

27. $\displaystyle\int \sqrt{x} \ln x \, dx$ **28.** $\displaystyle\int \frac{\ln x}{\sqrt{x}} \, dx$

In Problems 29–32, the integral can be found in more than one way. First use integration by parts, then use a method that does not involve integration by parts. Which method do you prefer?

29. $\int (x - 3)(x + 1)^2 \, dx$ **30.** $\int (x + 2)(x - 1)^2 \, dx$

31. $\int (2x + 1)(x - 2)^2 \, dx$ **32.** $\int (5x - 1)(x + 2)^2 \, dx$

In Problems 33–36, illustrate each integral graphically and describe what the integral represents in terms of areas.

33. Problem 19   **34.** Problem 20

35. Problem 21   **36.** Problem 22

Problems 37–58 are mixed—some may require use of the integration-by-parts formula along with techniques we have considered earlier; others may require repeated use of the integration-by-parts formula. Assume that $g(x) > 0$ whenever $\ln g(x)$ is involved.

37. $\int x^2 e^x \, dx$ **38.** $\int x^3 e^x \, dx$

39. $\int x e^{ax} \, dx, a \neq 0$ **40.** $\int \ln (ax) \, dx, a > 0$

41. $\int_1^e \frac{\ln x}{x^2} \, dx$ **42.** $\int_1^2 x^3 e^{x^2} \, dx$

43. $\int_0^2 \ln (x + 4) \, dx$ **44.** $\int_0^2 \ln (4 - x) \, dx$

45. $\int x e^{x-2} \, dx$ **46.** $\int x e^{x+1} \, dx$

47. $\int x \ln (1 + x^2) \, dx$ **48.** $\int x \ln (1 + x) \, dx$

49. $\int e^x \ln (1 + e^x) \, dx$ **50.** $\int \frac{\ln (1 + \sqrt{x})}{\sqrt{x}} \, dx$

51. $\int (\ln x)^2 \, dx$ **52.** $\int x(\ln x)^2 \, dx$

53. $\int (\ln x)^3 \, dx$ **54.** $\int x(\ln x)^3 \, dx$

55. $\int_1^e \ln (x^2) \, dx$ **56.** $\int_1^e \ln (x^4) \, dx$

57. $\int_0^1 \ln (e^{x^2}) \, dx$ **58.** $\int_1^2 \ln (x e^x) \, dx$

In Problems 59–62, use a graphing calculator to graph each equation over the indicated interval and find the area between the curve and the x axis over that interval. Find answers to two decimal places.

59. $y = x - 2 - \ln x; 1 \le x \le 4$

60. $y = 6 - x^2 - \ln x; 1 \le x \le 4$

61. $y = 5 - x e^x; 0 \le x \le 3$

62. $y = x e^x + x - 6; 0 \le x \le 3$

Applications

63. Profit. If the marginal profit (in millions of dollars per year) is given by

$$P'(t) = 2t - t e^{-t}$$

use an appropriate definite integral to find the total profit (to the nearest million dollars) earned over the first 5 years of operation.

64. Production. An oil field is estimated to produce oil at a rate of $R(t)$ thousand barrels per month t months from now, as given by

$$R(t) = 10t e^{-0.1t}$$

Use an appropriate definite integral to find the total production (to the nearest thousand barrels) in the first year of operation.

65. Profit. Interpret the results of Problem 63 with both a graph and a description of the graph.

66. Production. Interpret the results of Problem 64 with both a graph and a description of the graph.

67. Continuous income stream. Find the future value at 3.95%, compounded continuously, for 5 years of a continuous income stream with a rate of flow of

$$f(t) = 1,000 - 200t$$

68. Continuous income stream. Find the interest earned at 4.15%, compounded continuously, for 4 years for a continuous income stream with a rate of flow of

$$f(t) = 1,000 - 250t$$

69. Income distribution. Find the Gini index of income concentration for the Lorenz curve with equation

$$y = x e^{x-1}$$

70. Income distribution. Find the Gini index of income concentration for the Lorenz curve with equation

$$y = x^2 e^{x-1}$$

71. Income distribution. Interpret the results of Problem 69 with both a graph and a description of the graph.

72. Income distribution. Interpret the results of Problem 70 with both a graph and a description of the graph.

73. Sales analysis. Monthly sales of a particular personal computer are expected to increase at the rate of

$$S'(t) = -4t e^{0.1t}$$

computers per month, where t is time in months and $S(t)$ is the number of computers sold each month. The company plans to stop manufacturing this computer when monthly sales reach 800 computers. If monthly sales now ($t = 0$) are 2,000 computers, find $S(t)$. How long, to the nearest month, will the company continue to manufacture the computer?

74. Sales analysis. The rate of change of the monthly sales of a new basketball game is given by

$$S'(t) = 350 \ln (t + 1) \qquad S(0) = 0$$

where t is the number of months since the game was released and $S(t)$ is the number of games sold each month. Find $S(t)$.

When, to the nearest month, will monthly sales reach 15,000 games?

75. Consumers' surplus. Find the consumers' surplus (to the nearest dollar) at a price level of $\bar{p} = \$2.089$ for the price–demand equation

$$p = D(x) = 9 - \ln(x + 4)$$

Use $\bar{x}$ computed to the nearest higher unit.

76. Producers' surplus. Find the producers' surplus (to the nearest dollar) at a price level of $\bar{p} = \$26$ for the price–supply equation

$$p = S(x) = 5 \ln(x + 1)$$

Use $\bar{x}$ computed to the nearest higher unit.

77. Consumers' surplus. Interpret the results of Problem 75 with both a graph and a description of the graph.

78. Producers' surplus. Interpret the results of Problem 76 with both a graph and a description of the graph.

79. Pollution. The concentration of particulate matter (in parts per million) t hours after a factory ceases operation for the day is given by

$$C(t) = \frac{20 \ln(t + 1)}{(t + 1)^2}$$

Find the average concentration for the period from $t = 0$ to $t = 5$.

80. Medicine. After a person takes a pill, the drug contained in the pill is assimilated into the bloodstream. The rate of assimilation t minutes after taking the pill is

$$R(t) = te^{-0.2t}$$

Find the total amount of the drug that is assimilated into the bloodstream during the first 10 minutes after the pill is taken.

81. Learning. A student enrolled in an advanced typing class progressed at a rate of

$$N'(t) = (t + 6)e^{-0.25t}$$

words per minute per week t weeks after enrolling in a 15-week course. If a student could type 40 words per minute at the beginning of the course, then how many words per minute $N(t)$ would the student be expected to type t weeks into the course? How long, to the nearest week, should it take the student to achieve the 70-word-per-minute level? How many words per minute should the student be able to type by the end of the course?

82. Learning. A student enrolled in a stenotyping class progressed at a rate of

$$N'(t) = (t + 10)e^{-0.1t}$$

words per minute per week t weeks after enrolling in a 15-week course. If a student had no knowledge of stenotyping (that is, if the student could stenotype at 0 words per minute) at the beginning of the course, then how many words per minute $N(t)$ would the student be expected to handle t weeks into the course? How long, to the nearest week, should it take the student to achieve 90 words per minute? How many words per minute should the student be able to handle by the end of the course?

83. Politics. The number of voters (in thousands) in a certain city is given by

$$N(t) = 20 + 4t - 5te^{-0.1t}$$

where t is time in years. Find the average number of voters during the period from $t = 0$ to $t = 5$.

Answers to Matched Problems

1. $\dfrac{x}{2}e^{2x} - \dfrac{1}{4}e^{2x} + C$

2. $\dfrac{x^2}{2}\ln 2x - \dfrac{x^2}{4} + C$

3. $\dfrac{x^2}{2}e^{2x} - \dfrac{x}{2}e^{2x} + \dfrac{1}{4}e^{2x} + C$

4. $2 \ln 6 - \ln 3 - 1 \approx 1.4849$

6.4 Other Integration Methods

- The Trapezoidal Rule
- Simpson's Rule
- Using a Table of Integrals
- Substitution and Integral Tables
- Reduction Formulas
- Application

In Chapter 5 we used left and right sums to approximate the definite integral of a function, and, if an antiderivative could be found, calculated the exact value using the fundamental theorem of calculus. Now we discuss other methods for approximating definite integrals, and a procedure for finding exact values of definite integrals of many standard functions.

Approximation of definite integrals by left sums and right sums is instructive and important, but not efficient. A large number of rectangles must be used, and many terms must be summed, to get good approximations. The *trapezoidal rule* and *Simpson's rule* provide more efficient approximations of definite integrals in the sense that fewer terms must be summed to achieve a given accuracy.

A **table of integrals** can be used to find antiderivatives of many standard functions (see Table II of Appendix C on pages 574–576). Definite integrals of such functions can therefore be found exactly by means of the fundamental theorem of calculus.

The Trapezoidal Rule

The trapezoid in Figure 1 is a more accurate approximation of the area under the graph of f and above the x axis than the left rectangle or the right rectangle. Using Δx to denote $x_1 - x_0$,

$$\text{Area of left rectangle: } f(x_0)\,\Delta x$$
$$\text{Area of right rectangle: } f(x_1)\,\Delta x$$
$$\text{Area of trapezoid: } \frac{f(x_0) + f(x_1)}{2}\Delta x \qquad (1)$$

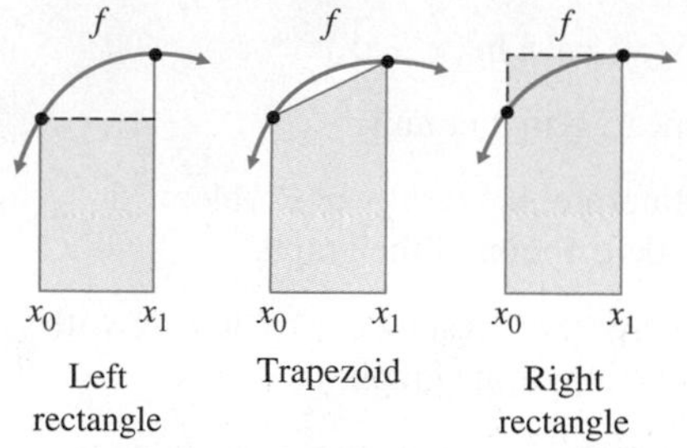

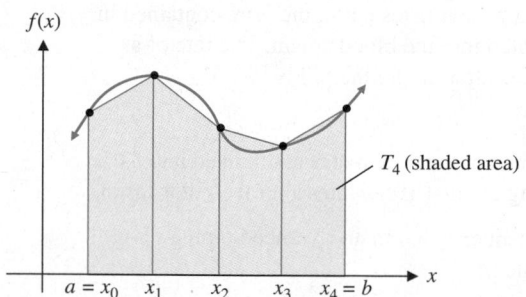

Left rectangle   Trapezoid   Right rectangle

Figure 1

Note that the area of the trapezoid in Figure 1 [also see formula (1)] is the average of the areas of the left and right rectangles. So the average T_4 of the left sum L_4 and the right sum R_4 for a function f on an interval $[a, b]$ is equal to the sum of the areas of four trapezoids (Fig. 2).

$f(x)$

T_4 (shaded area)

$a = x_0$ x_1 x_2 x_3 $x_4 = b$ x

Figure 2

Adding L_4 and R_4 and dividing by 2 gives a formula for T_4:

$$L_4 = [f(x_0) + f(x_1) + f(x_2) + f(x_3)]\Delta x$$
$$R_4 = [f(x_1) + f(x_2) + f(x_3) + f(x_4)]\Delta x$$
$$T_4 = [f(x_0) + 2f(x_1) + 2f(x_2) + 2f(x_3) + f(x_4)]\Delta x/2$$

The **trapezoidal sum** T_4 is the case $n = 4$ of the **trapezoidal rule**.

TRAPEZOIDAL RULE

Let f be a function defined on an interval $[a, b]$. Partition $[a, b]$ into n subintervals of equal length $\Delta x = (b - a)/n$ with endpoints

$$a = x_0 < x_1 < x_2 < \cdots < x_n = b.$$

Then

$$T_n = [f(x_0) + 2f(x_1) + 2f(x_2) + \cdots + 2f(x_{n-1}) + f(x_n)]\Delta x/2$$

is an approximation of $\int_a^b f(x)\,dx$.

EXAMPLE 1 Trapezoidal rule   Use the trapezoidal rule with $n = 5$ to approximate $\int_2^4 \sqrt{100 + x^2}\, dx$. Round function values to 4 decimal places and the final answer to 2 decimal places.

SOLUTION   Partition $[2, 4]$ into 5 equal subintervals of width $(4 - 2)/5 = 0.4$. The endpoints are $x_0 = 2$, $x_1 = 2.4$, $x_2 = 2.8$, $x_3 = 3.2$, $x_4 = 3.6$, and $x_5 = 4$. We calculate the value of the function $f(x) = \sqrt{100 + x^2}$ at each endpoint:

x	$f(x)$
2.0	10.1980
2.4	10.2840
2.8	10.3846
3.2	10.4995
3.6	10.6283
4.0	10.7703

By the trapezoidal rule,

$$\begin{aligned}
T_5 &= [f(2) + 2f(2.4) + 2f(2.8) + 2f(3.2) + 2f(3.6) + f(4)](0.4/2) \\
&= [10.1980 + 2(10.2840) + 2(10.3846) + 2(10.4995) + 2(10.6283) + 10.7703](0.2) \\
&= 20.91
\end{aligned}$$

Matched Problem 1   Use the trapezoidal rule with $n = 5$ to approximate $\int_2^4 \sqrt{81 + x^5}\, dx$ (round function values to 4 decimal places and the final answer to 2 decimal places).

Simpson's Rule

The trapezoidal sum provides a better approximation of the definite integral of a function that is increasing (or decreasing) than either the left or right sum. Similarly, the **midpoint sum,**

$$M_n = \left[f\left(\frac{x_0 + x_1}{2}\right) + f\left(\frac{x_1 + x_2}{2}\right) + \cdots + f\left(\frac{x_{n-1} + x_n}{2}\right) \right] \Delta x$$

(see Example 2, Section 5.4) is a better approximation of the definite integral of a function that is increasing (or decreasing) than either the left or right sum. How do T_n and M_n compare? A midpoint sum rectangle has the same area as the corresponding tangent line trapezoid (the larger trapezoid in Fig. 3). It appears from Figure 3, and can be proved in general, that the trapezoidal sum error is about double the midpoint sum error when the graph of the function is concave up or concave down.

$$T_n \leq \int_a^b f(x)\, dx \leq M_n$$

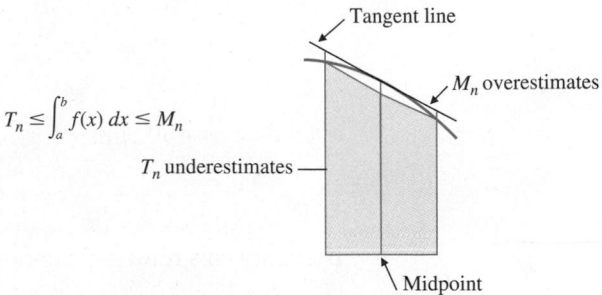

T_n underestimates

Tangent line

M_n overestimates

Midpoint

Figure 3

This suggests that a weighted average of the two estimates, with the midpoint sum being counted double the trapezoidal sum, might be an even better estimate than either separately. This weighted average,

$$S_{2n} = \frac{2M_n + T_n}{3} \tag{2}$$

leads to a formula called *Simpson's rule*. To simplify notation, we agree to divide the interval $[a, b]$ into $2n$ equal subintervals when Simpson's rule is applied. So, if $n = 2$, for example, then $[a, b]$ is divided into $2n = 4$ equal subintervals, of length Δx, with endpoints

$$a = x_0 < x_1 < x_2 < x_3 < x_4 = b.$$

There are two equal subintervals for M_2 and T_2, each of length $2\Delta x$, with endpoints

$$a = x_0 < x_2 < x_4 = b.$$

Therefore,

$$M_2 = [f(x_1) + f(x_3)](2\Delta x)$$
$$T_2 = [f(x_0) + 2f(x_2) + f(x_4)](2\Delta x)/2$$

We use equation (2) to get a formula for S_4:

$$S_4 = [f(x_0) + 4f(x_1) + 2f(x_2) + 4f(x_3) + f(x_4)]\Delta x/3$$

The formula for S_4 is the case $n = 2$ of **Simpson's rule**.

> ### Simpson's Rule
> Let f be a function defined on an interval $[a, b]$. Partition $[a, b]$ into $2n$ subintervals of equal length $\Delta x = (b - a)/n$ with endpoints
>
> $$a = x_0 < x_1 < x_2 < \cdots < x_{2n} = b.$$
>
> Then
>
> $$S_{2n} = [f(x_0) + 4f(x_1) + 2f(x_2) + 4f(x_3) + 2f(x_4) + \cdots + 4f(x_{2n-1}) + f(x_{2n})]\Delta x/3$$
>
> is an approximation of $\int_a^b f(x)\,dx$.

⚠ **CAUTION** Simpson's rule always requires an *even* number of subintervals of $[a, b]$.   ▲

EXAMPLE 2   Simpson's rule   Use Simpson's rule with $n = 2$ to approximate $\int_2^{10} \frac{x^4}{\ln x}\,dx$. Round function values to 4 decimal places and the final answer to 2 decimal places.

SOLUTION   Partition the interval $[2, 10]$ into $2n = 4$ equal subintervals of width $(10 - 2)/4 = 2$. The endpoints are $x_0 = 2, x_1 = 4$, $x_2 = 6$, $x_3 = 8$, and $x_4 = 10$. We calculate the value of the function $f(x) = \frac{x^4}{\ln x}$ at each endpoint:

x	$f(x)$
2	23.0831
4	184.6650
6	723.3114
8	1,969.7596
10	4,342.9448

By Simpson's rule,

$$S_4 = [f(2) + 4f(4) + 2f(6) + 4f(8) + f(10)](2/3)$$
$$= [23.0831 + 4(184.6650) + 2(723.3114) + 4(1,969.7596) + 4,342.9448](2/3)$$
$$= 9,620.23$$

Matched Problem 2   Use Simpson's rule with $n = 2$ to approximate $\int_2^{10} \frac{1}{\ln x}\,dx$ (round function values to 4 decimal places and the final answer to 2 decimal places).

Using a Table of Integrals

The formulas in Table II on pages 574–576 are organized by categories, such as "Integrals Involving $a + bu$," "Integrals Involving $\sqrt{u^2 - a^2}$," and so on. The variable u is the variable of integration. All other symbols represent constants. To use a table to evaluate an integral, you must first find the category that most closely agrees with the form of the integrand and then find a formula in that category that you can make to match the integrand exactly by assigning values to the constants in the formula.

EXAMPLE 3 Integration Using Tables   Use Table II to find

$$\int \frac{x}{(5 + 2x)(4 - 3x)} dx$$

SOLUTION Since the integrand

$$f(x) = \frac{x}{(5 + 2x)(4 - 3x)}$$

is a rational function involving terms of the form $a + bu$ and $c + du$, we examine formulas 15 to 20 in Table II on page 575 to see if any of the integrands in these formulas can be made to match $f(x)$ exactly. Comparing the integrand in formula 16 with $f(x)$, we see that this integrand will match $f(x)$ if we let $a = 5, b = 2, c = 4$, and $d = -3$. Letting $u = x$ and substituting for a, b, c, and d in formula 16, we have

$$\int \frac{u}{(a + bu)(c + du)} du = \frac{1}{ad - bc}\left(\frac{a}{b}\ln|a + bu| - \frac{c}{d}\ln|c + du|\right) \quad \text{Formula 16}$$

$$\int \frac{x}{\underset{a\ \ \ b\ \ \ \ c\ \ \ \ d}{(5 + 2x)(4 - 3x)}} dx = \frac{1}{\underset{a \cdot d - b \cdot c = 5 \cdot (-3) - 2 \cdot 4 = -23}{5 \cdot (-3) - 2 \cdot 4}}\left(\frac{5}{2}\ln|5 + 2x| - \frac{4}{-3}\ln|4 - 3x|\right) + C$$

$$= -\tfrac{5}{46}\ln|5 + 2x| - \tfrac{4}{69}\ln|4 - 3x| + C$$

Notice that the constant of integration, C, is not included in any of the formulas in Table II. However, you must still include C in all antiderivatives.

Matched Problem 3   Use Table II to find $\displaystyle\int \frac{1}{(5 + 3x)^2(1 + x)} dx$.

EXAMPLE 4 Integration Using Tables   Evaluate $\displaystyle\int_3^4 \frac{1}{x\sqrt{25 - x^2}} dx$.

SOLUTION First, we use Table II to find

$$\int \frac{1}{x\sqrt{25 - x^2}} dx$$

Since the integrand involves the expression $\sqrt{25 - x^2}$, we examine formulas 29 to 31 in Table II and select formula 29 with $a^2 = 25$ and $a = 5$:

$$\int \frac{1}{u\sqrt{a^2 - u^2}}\,du = -\frac{1}{a}\ln\left|\frac{a + \sqrt{a^2 - u^2}}{u}\right| \qquad \text{Formula 29}$$

$$\int \frac{1}{x\sqrt{25 - x^2}}\,dx = -\frac{1}{5}\ln\left|\frac{5 + \sqrt{25 - x^2}}{x}\right| + C$$

So

$$\int_3^4 \frac{1}{x\sqrt{25 - x^2}}\,dx = -\frac{1}{5}\ln\left|\frac{5 + \sqrt{25 - x^2}}{x}\right|\Bigg|_3^4$$

$$= -\frac{1}{5}\ln\left|\frac{5 + 3}{4}\right| + \frac{1}{5}\ln\left|\frac{5 + 4}{3}\right|$$

$$= -\tfrac{1}{5}\ln 2 + \tfrac{1}{5}\ln 3 = \tfrac{1}{5}\ln 1.5 \approx 0.0811$$

Matched Problem 4) Evaluate $\displaystyle\int_6^8 \frac{1}{x^2\sqrt{100 - x^2}}\,dx$.

Substitution and Integral Tables

As Examples 3 and 4 illustrate, if the integral we want to evaluate can be made to match one in the table exactly, then evaluating the indefinite integral consists simply of substituting the correct values of the constants into the formula. But what happens if we cannot match an integral with one of the formulas in the table? In many cases, a substitution will change the given integral into one that corresponds to a table entry.

EXAMPLE 5   Integration Using Substitution and Tables   Find $\displaystyle\int \frac{x^2}{\sqrt{16x^2 - 25}}\,dx$.

SOLUTION   In order to relate this integral to one of the formulas involving $\sqrt{u^2 - a^2}$ (formulas 40 to 45 in Table II), we observe that if $u = 4x$, then

$$u^2 = 16x^2 \qquad \text{and} \qquad \sqrt{16x^2 - 25} = \sqrt{u^2 - 25}$$

So, we will use the substitution $u = 4x$ to change this integral into one that appears in the table:

$$\int \frac{x^2}{\sqrt{16x^2 - 25}}\,dx = \frac{1}{4}\int \frac{\frac{1}{16}u^2}{\sqrt{u^2 - 25}}\,du \qquad \begin{array}{l}\text{Substitution:}\\ u = 4x,\ du = 4\,dx,\ x = \tfrac{1}{4}u\end{array}$$

$$= \frac{1}{64}\int \frac{u^2}{\sqrt{u^2 - 25}}\,du$$

This last integral can be evaluated with the aid of formula 44 in Table II with $a = 5$:

$$\int \frac{u^2}{\sqrt{u^2 - a^2}}\,du = \frac{1}{2}\left(u\sqrt{u^2 - a^2} + a^2\ln|u + \sqrt{u^2 - a^2}|\right) \qquad \text{Formula 44}$$

$$\int \frac{x^2}{\sqrt{16x^2 - 25}}\,dx = \frac{1}{64}\int \frac{u^2}{\sqrt{u^2 - 25}}\,du \qquad \text{Use formula 44 with } a = 5.$$

$$= \tfrac{1}{128}\left(u\sqrt{u^2 - 25} + 25\ln|u + \sqrt{u^2 - 25}|\right) + C \qquad \text{Substitute } u = 4x.$$

$$= \tfrac{1}{128}\left(4x\sqrt{16x^2 - 25} + 25\ln|4x + \sqrt{16x^2 - 25}|\right) + C$$

Matched Problem 5) Find $\int \sqrt{9x^2 - 16}\,dx$.

EXAMPLE 6 Integration Using Substitution and Tables Find $\displaystyle\int \frac{x}{\sqrt{x^4 + 1}}\,dx$.

SOLUTION None of the formulas in Table II involve fourth powers; however, if we let $u = x^2$, then

$$\sqrt{x^4 + 1} = \sqrt{u^2 + 1}$$

and this form does appear in formulas 32 to 39. Thus, we substitute $u = x^2$:

$$\int \frac{1}{\sqrt{x^4 + 1}}x\,dx = \frac{1}{2}\int \frac{1}{\sqrt{u^2 + 1}}\,du \qquad \text{Substitution:}\quad u = x^2,\ du = 2x\,dx$$

We recognize the last integral as formula 36 with $a = 1$:

$$\int \frac{1}{\sqrt{u^2 + a^2}}\,du = \ln\left|u + \sqrt{u^2 + a^2}\right| \qquad \text{Formula 36}$$

$$\int \frac{x}{\sqrt{x^4 + 1}}\,dx = \frac{1}{2}\int \frac{1}{\sqrt{u^2 + 1}}\,du \qquad \text{Use formula 36 with } a = 1.$$

$$= \tfrac{1}{2}\ln\left|u + \sqrt{u^2 + 1}\right| + C \qquad \text{Substitute } u = x^2.$$

$$= \tfrac{1}{2}\ln\left|x^2 + \sqrt{x^4 + 1}\right| + C$$

Matched Problem 6 Find $\int x\sqrt{x^4 + 1}\,dx$.

Reduction Formulas

EXAMPLE 7 Using Reduction Formulas Use Table II to find $\int x^2 e^{3x}\,dx$.

SOLUTION Since the integrand involves the function e^{3x}, we examine formulas 46–48 and conclude that formula 47 can be used for this problem. Letting $u = x$, $n = 2$, and $a = 3$ in formula 47, we have

$$\int u^n e^{au}\,du = \frac{u^n e^{au}}{a} - \frac{n}{a}\int u^{n-1} e^{au}\,du \qquad \text{Formula 47}$$

$$\int x^2 e^{3x}\,dx = \frac{x^2 e^{3x}}{3} - \frac{2}{3}\int x e^{3x}\,dx$$

Notice that the expression on the right still contains an integral, but the exponent of x has been reduced by 1. Formulas of this type are called **reduction formulas** and are designed to be applied repeatedly until an integral that can be evaluated is obtained. Applying formula 47 to $\int x e^{3x}\,dx$ with $n = 1$, we have

$$\int x^2 e^{3x}\,dx = \frac{x^2 e^{3x}}{3} - \frac{2}{3}\left(\frac{x e^{3x}}{3} - \frac{1}{3}\int e^{3x}\,dx\right)$$

$$= \frac{x^2 e^{3x}}{3} - \frac{2x e^{3x}}{9} + \frac{2}{9}\int e^{3x}\,dx$$

This last expression contains an integral that is easy to evaluate:

$$\int e^{3x}\,dx = \tfrac{1}{3}e^{3x}$$

After making a final substitution and adding a constant of integration, we have

$$\int x^2 e^{3x}\,dx = \frac{x^2 e^{3x}}{3} - \frac{2x e^{3x}}{9} + \frac{2}{27}e^{3x} + C$$

Matched Problem 7 Use Table II to find $\int (\ln x)^2\,dx$.

Application

EXAMPLE 8   Producers' Surplus   Find the producers' surplus at a price level of $20 for the price–supply equation

$$p = S(x) = \frac{5x}{500 - x}$$

SOLUTION

Step 1   Find $\bar{x}$, the supply when the price is $\bar{p} = 20$:

$$\bar{p} = \frac{5\bar{x}}{500 - \bar{x}}$$

$$20 = \frac{5\bar{x}}{500 - \bar{x}}$$

$$10{,}000 - 20\bar{x} = 5\bar{x}$$

$$10{,}000 = 25\bar{x}$$

$$\bar{x} = 400$$

Step 2   Sketch a graph, as shown in Figure 4.

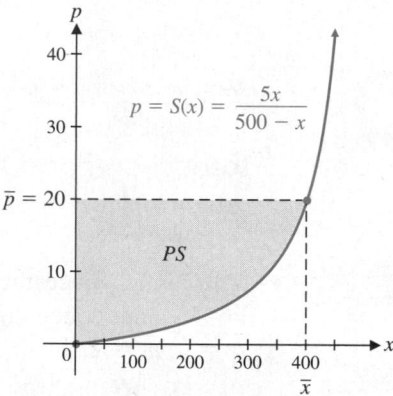

Figure 4

Step 3   Find the producers' surplus (the shaded area of the graph):

$$PS = \int_0^{\bar{x}} [\bar{p} - S(x)]\, dx$$

$$= \int_0^{400} \left(20 - \frac{5x}{500 - x} \right) dx$$

$$= \int_0^{400} \frac{10{,}000 - 25x}{500 - x}\, dx$$

Use formula 20 with $a = 10{,}000$, $b = -25$, $c = 500$, and $d = -1$:

$$\int \frac{a + bu}{c + du}\, du = \frac{bu}{d} + \frac{ad - bc}{d^2} \ln|c + du| \qquad \text{Formula 20}$$

$$PS = (25x + 2{,}500 \ln|500 - x|)\Big|_0^{400}$$

$$= 10{,}000 + 2{,}500 \ln|100| - 2{,}500 \ln|500|$$

$$\approx \$5{,}976$$

Matched Problem 8   Find the consumers' surplus at a price level of $10 for the price–demand equation

$$p = D(x) = \frac{20x - 8{,}000}{x - 500}$$

In Problems 1–8, round function values to 4 decimal places and the final answer to 2 decimal places.

1. Use the trapezoidal rule with $n = 3$ to approximate $\int_0^6 \sqrt{1 + x^4}\, dx$.

2. Use the trapezoidal rule with $n = 2$ to approximate $\int_0^8 \sqrt{1 + x^2}\, dx$.

3. Use the trapezoidal rule with $n = 6$ to approximate $\int_0^6 \sqrt{1 + x^4}\, dx$.

4. Use the trapezoidal rule with $n = 4$ to approximate $\int_0^8 \sqrt{1 + x^2}\, dx$.

5. Use Simpson's rule with $n = 1$ (so there are $2n = 2$ subintervals) to approximate $\int_1^3 \frac{1}{1 + x^2}\, dx$.

6. Use Simpson's rule with $n = 1$ (so there are $2n = 2$ subintervals) to approximate $\int_2^{10} \frac{x^2}{\ln x}\, dx$.

7. Use Simpson's rule with $n = 2$ (so there are $2n = 4$ subintervals) to approximate $\int_1^3 \frac{1}{1 + x^2}\, dx$.

8. Use Simpson's rule with $n = 2$ (so there are $2n = 4$ subintervals) to approximate $\int_2^{10} \frac{x^2}{\ln x}\, dx$.

Use Table II on pages 574–576 to find each indefinite integral in Problems 9–22.

9. $\int \frac{1}{x(1 + x)}\, dx$

10. $\int \frac{1}{x^2(1 + x)}\, dx$

11. $\int \frac{1}{(3 + x)^2(5 + 2x)}\, dx$

12. $\int \frac{x}{(5 + 2x)^2(2 + x)}\, dx$

13. $\int \frac{x}{\sqrt{16 + x}}\, dx$

14. $\int \frac{1}{x\sqrt{16 + x}}\, dx$

15. $\int \frac{1}{x\sqrt{1 - x^2}}\, dx$

16. $\int \frac{\sqrt{9 - x^2}}{x}\, dx$

17. $\int \frac{1}{x\sqrt{x^2 + 4}}\, dx$

18. $\int \frac{1}{x^2\sqrt{x^2 - 16}}\, dx$

19. $\int x^2 \ln x\, dx$

20. $\int x^3 \ln x\, dx$

21. $\int \frac{1}{1 + e^x}\, dx$

22. $\int \frac{1}{5 + 2e^{3x}}\, dx$

Evaluate each definite integral in Problems 23–28. Use Table II on pages 574–576 to find the antiderivative.

23. $\int_1^3 \frac{x^2}{3 + x}\, dx$

24. $\int_2^6 \frac{x}{(6 + x)^2}\, dx$

25. $\int_0^7 \frac{1}{(3 + x)(1 + x)}\, dx$

26. $\int_0^7 \frac{x}{(3 + x)(1 + x)}\, dx$

27. $\int_0^4 \frac{1}{\sqrt{x^2 + 9}}\, dx$

28. $\int_4^5 \sqrt{x^2 - 16}\, dx$

29. Use the trapezoidal rule with $n = 5$ to approximate $\int_3^{13} x^2 dx$ and use the fundamental theorem of calculus to find the exact value of the definite integral.

30. Use the trapezoidal rule with $n = 5$ to approximate $\int_1^{11} x^3 dx$ and use the fundamental theorem of calculus to find the exact value of the definite integral.

31. Use Simpson's rule with $n = 4$ (so there are $2n = 8$ subintervals) to approximate $\int_1^5 \frac{1}{x}\, dx$ and use the fundamental theorem of calculus to find the exact value of the definite integral.

32. Use Simpson's rule with $n = 4$ (so there are $2n = 8$ subintervals) to approximate $\int_1^5 x^4 dx$ and use the fundamental theorem of calculus to find the exact value of the definite integral.

33. Use the trapezoidal rule with $n = 3$ to approximate $\int_5^8 (4x - 3)\, dx$ and use the fundamental theorem of calculus to find the exact value of the definite integral.

34. Use the trapezoidal rule with $n = 3$ to approximate $\int_7^{10} (2 - 9x)\, dx$ and use the fundamental theorem of calculus to find the exact value of the definite integral.

35. Use Simpson's rule with $n = 2$ (so there are $2n = 4$ subintervals) to approximate $\int_5^9 (3x^2 + 5x + 3)\, dx$ and use the fundamental theorem of calculus to find the exact value of the definite integral.

36. Use Simpson's rule with $n = 2$ (so there are $2n = 4$ subintervals) to approximate $\int_{-2}^2 (x^3 - 3x^2 + 2x + 8)\, dx$ and use the fundamental theorem of calculus to find the exact value of the definite integral.

In Problems 37–48, use substitution techniques and Table II to find each indefinite integral.

37. $\int \frac{\sqrt{4x^2 + 1}}{x^2}\, dx$

38. $\int x^2 \sqrt{9x^2 - 1}\, dx$

39. $\int \frac{x}{\sqrt{x^4 - 16}}\, dx$

40. $\int x\sqrt{x^4 - 16}\, dx$

41. $\int x^2 \sqrt{x^6 + 4}\, dx$

42. $\int \frac{x^2}{\sqrt{x^6 + 4}}\, dx$

43. $\int \frac{1}{x^3\sqrt{4 - x^4}}\, dx$

44. $\int \frac{\sqrt{x^4 + 4}}{x}\, dx$

45. $\int \frac{e^x}{(2 + e^x)(3 + 4e^x)}\, dx$

46. $\int \frac{e^x}{(4 + e^x)^2(2 + e^x)}\, dx$

47. $\int \frac{\ln x}{x\sqrt{4 + \ln x}}\, dx$

48. $\int \frac{1}{x \ln x\sqrt{4 + \ln x}}\, dx$

In Problems 49–54, use Table II to find each indefinite integral.

49. $\int x^2 e^{5x}\, dx$

50. $\int x^2 e^{-4x}\, dx$

51. $\int x^3 e^{-x}\, dx$

52. $\int x^3 e^{2x}\, dx$

53. $\int (\ln x)^3\, dx$

54. $\int (\ln x)^4\, dx$

Problems 55–62 are mixed—some require the use of Table II, and others can be solved with techniques considered earlier.

55. $\int_3^5 x\sqrt{x^2 - 9}\, dx$

56. $\int_3^5 x^2\sqrt{x^2 - 9}\, dx$

57. $\int_2^4 \frac{1}{x^2 - 1}\, dx$

58. $\int_2^4 \frac{x}{(x^2 - 1)^2}\, dx$

59. $\int \frac{\ln x}{x^2}\, dx$

60. $\int \frac{(\ln x)^2}{x}\, dx$

61. $\int \frac{x}{\sqrt{x^2 - 1}}\, dx$

62. $\int \frac{x^2}{\sqrt{x^2 - 1}}\, dx$

63. If $f(x) = ax^2 + bx + c$, where a, b, and c are any real numbers, use Simpson's rule with $n = 1$ (so there are $2n = 2$ subintervals) to show that

$$S_2 = \int_{-1}^1 f(x)\, dx.$$

64. If $f(x) = ax^3 + bx^2 + cx + d$, where a, b, c, and d are any real numbers, use Simpson's rule with $n = 1$ (so there are $2n = 2$ subintervals) to show that

$$S_2 = \int_{-1}^1 f(x)\, dx.$$

In Problems 65–68, find the area bounded by the graphs of $y = f(x)$ and $y = g(x)$ to two decimal places. Use a graphing calculator to approximate intersection points to two decimal places.

65. $f(x) = \dfrac{10}{\sqrt{x^2 + 1}};\ g(x) = x^2 + 3x$

66. $f(x) = \sqrt{1 + x^2};\ g(x) = 5x - x^2$

67. $f(x) = x\sqrt{4 + x};\ g(x) = 1 + x$

68. $f(x) = \dfrac{x}{\sqrt{x + 4}};\ g(x) = x - 2$

Applications

Use Table II to evaluate all integrals involved in any solutions of Problems 69–92.

69. Consumers' surplus. Find the consumers' surplus at a price level of $\bar{p} = \$15$ for the price–demand equation

$$p = D(x) = \frac{7{,}500 - 30x}{300 - x}$$

70. Producers' surplus. Find the producers' surplus at a price level of $\bar{p} = \$20$ for the price–supply equation

$$p = S(x) = \frac{10x}{300 - x}$$

71. Consumers' surplus. Graph the price–demand equation and the price-level equation $\bar{p} = 15$ of Problem 69 in the same coordinate system. What region represents the consumers' surplus?

72. Producers' surplus. Graph the price–supply equation and the price-level equation $\bar{p} = 20$ of Problem 70 in the same coordinate system. What region represents the producers' surplus?

73. Cost. A company manufactures downhill skis. It has fixed costs of $25,000 and a marginal cost given by

$$C'(x) = \frac{250 + 10x}{1 + 0.05x}$$

where $C(x)$ is the total cost at an output of x pairs of skis. Find the cost function $C(x)$ and determine the production level (to the nearest unit) that produces a cost of $150,000. What is the cost (to the nearest dollar) for a production level of 850 pairs of skis?

74. Cost. A company manufactures a portable DVD player. It has fixed costs of $11,000 per week and a marginal cost given by

$$C'(x) = \frac{65 + 20x}{1 + 0.4x}$$

where $C(x)$ is the total cost per week at an output of x players per week. Find the cost function $C(x)$ and determine the production level (to the nearest unit) that produces a cost of $52,000 per week. What is the cost (to the nearest dollar) for a production level of 700 players per week?

75. Continuous income stream. Find the future value at 4.4%, compounded continuously, for 10 years for the continuous income stream with rate of flow $f(t) = 50t^2$.

76. Continuous income stream. Find the interest earned at 3.7%, compounded continuously, for 5 years for the continuous income stream with rate of flow $f(t) = 200t$.

77. Income distribution. Find the Gini index of income concentration for the Lorenz curve with equation

$$y = \tfrac{1}{2}x\sqrt{1 + 3x}$$

78. Income distribution. Find the Gini index of income concentration for the Lorenz curve with equation

$$y = \tfrac{1}{2}x^2\sqrt{1 + 3x}$$

79. Income distribution. Graph $y = x$ and the Lorenz curve of Problem 77 over the interval $[0, 1]$. Discuss the effect of the area bounded by $y = x$ and the Lorenz curve getting smaller relative to the equitable distribution of income.

80. Income distribution. Graph $y = x$ and the Lorenz curve of Problem 78 over the interval $[0, 1]$. Discuss the effect of the area bounded by $y = x$ and the Lorenz curve getting larger relative to the equitable distribution of income.

81. Marketing. After test marketing a new high-fiber cereal, the market research department of a major food producer estimates that monthly sales (in millions of dollars) will grow at the monthly rate of

$$S'(t) = \frac{t^2}{(1+t)^2}$$

t months after the cereal is introduced. If we assume 0 sales at the time the cereal is introduced, find $S(t)$, the total sales t months after the cereal is introduced. Find the total sales during the first 2 years that the cereal is on the market.

82. Average price. At a discount department store, the price–demand equation for premium motor oil is given by

$$p = D(x) = \frac{50}{\sqrt{100 + 6x}}$$

where x is the number of cans of oil that can be sold at a price of $\$p$. Find the average price over the demand interval $[50, 250]$.

83. Marketing. For the cereal of Problem 81, show the sales over the first 2 years geometrically, and describe the geometric representation.

84. Price–demand. For the motor oil of Problem 82, graph the price–demand equation and the line representing the average price in the same coordinate system over the interval $[50, 250]$. Describe how the areas under the two curves over the interval $[50, 250]$ are related.

85. Profit. The marginal profit for a small car agency that sells x cars per week is given by

$$P'(x) = x\sqrt{2 + 3x}$$

where $P(x)$ is the profit in dollars. The agency's profit on the sale of only 1 car per week is $-\$2,000$. Find the profit function and the number of cars that must be sold (to the nearest unit) to produce a profit of $\$13,000$ per week. How much weekly profit (to the nearest dollar) will the agency have if 80 cars are sold per week?

86. Revenue. The marginal revenue for a company that manufactures and sells x graphing calculators per week is given by

$$R'(x) = \frac{x}{\sqrt{1 + 2x}} \qquad R(0) = 0$$

where $R(x)$ is the revenue in dollars. Find the revenue function and the number of calculators that must be sold (to the nearest unit) to produce $\$10,000$ in revenue per week. How much weekly revenue (to the nearest dollar) will the company have if 1,000 calculators are sold per week?

87. Pollution. An oil tanker is producing an oil slick that is radiating outward at a rate given approximately by

$$\frac{dR}{dt} = \frac{100}{\sqrt{t^2 + 9}} \qquad t \geq 0$$

where R is the radius (in feet) of the circular slick after t minutes. Find the radius of the slick after 4 minutes if the radius is 0 when $t = 0$.

88. Pollution. The concentration of particulate matter (in parts per million) during a 24-hour period is given approximately by

$$C(t) = t\sqrt{24 - t} \qquad 0 \leq t \leq 24$$

where t is time in hours. Find the average concentration during the period from $t = 0$ to $t = 24$.

89. Learning. A person learns N items at a rate given approximately by

$$N'(t) = \frac{60}{\sqrt{t^2 + 25}} \qquad t \geq 0$$

where t is the number of hours of continuous study. Determine the total number of items learned in the first 12 hours of continuous study.

90. Politics. The number of voters (in thousands) in a metropolitan area is given approximately by

$$f(t) = \frac{500}{2 + 3e^{-t}} \qquad t \geq 0$$

where t is time in years. Find the average number of voters during the period from $t = 0$ to $t = 10$.

91. Learning. Interpret Problem 89 geometrically. Describe the geometric interpretation.

92. Politics. For the voters of Problem 90, graph $y = f(t)$ and the line representing the average number of voters over the interval $[0, 10]$ in the same coordinate system. Describe how the areas under the two curves over the interval $[0, 10]$ are related.

Answers to Matched Problems

1. 38.85   2. 5.20
3. $\dfrac{1}{2}\left(\dfrac{1}{5 + 3x}\right) + \dfrac{1}{4}\ln\left|\dfrac{1 + x}{5 + 3x}\right| + C$
4. $\dfrac{7}{1,200} \approx 0.0058$
5. $\frac{1}{6}\left(3x\sqrt{9x^2 - 16} - 16\ln|3x + \sqrt{9x^2 - 16}|\right) + C$
6. $\frac{1}{4}\left(x^2\sqrt{x^4 + 1} + \ln|x^2 + \sqrt{x^4 + 1}|\right) + C$
7. $x(\ln x)^2 - 2x \ln x + 2x + C$
8. $3,000 + 2,000 \ln 200 - 2,000 \ln 500 \approx \$1,167$

Important Terms, Symbols, and Concepts

6.1 Area Between Curves

EXAMPLES

- If f and g are continuous and $f(x) \geq g(x)$ over the interval $[a, b]$, then the area bounded by $y = f(x)$ and $y = g(x)$ for $a \leq x \leq b$ is given exactly by

Ex. 1, p. 383
Ex. 2, p. 383
Ex. 3, p. 384
Ex. 4, p. 384
Ex. 5, p. 385
Ex. 6, p. 385

$$A = \int_a^b [f(x) - g(x)]\, dx$$

- A graphical representation of the distribution of income among a population can be found by plotting data points (x, y), where **x represents the cumulative percentage of families at or below a given income level** and **y represents the cumulative percentage of total family income received.** Regression analysis can be used to find a particular function $y = f(x)$, called a **Lorenz curve,** that best fits the data.

- A single number, the **Gini index,** measures income concentration:

Ex. 7, p. 387

$$\text{Gini index} = 2 \int_0^1 [x - f(x)]\, dx$$

A Gini index of 0 indicates **absolute equality:** All families share equally in the income. A Gini index of 1 indicates **absolute inequality:** One family has all of the income and the rest have none.

6.2 Applications in Business and Economics

- *Probability Density Functions* If any real number x in an interval is a possible outcome of an experiment, then x is said to be a **continuous random variable.** The probability distribution of a continuous random variable is described by a **probability density function** f that satisfies the following conditions:

Ex. 1, p. 392

1. $f(x) \geq 0$ for all real x.
2. The area under the graph of $f(x)$ over the interval $(-\infty, \infty)$ is exactly 1.
3. If $[c, d]$ is a subinterval of $(-\infty, \infty)$, then

$$\text{Probability } (c \leq x \leq d) = \int_c^d f(x)\, dx$$

- *Continuous Income Stream* If the rate at which income is received—its **rate of flow**—is a continuous function $f(t)$ of time, then the income is said to be a **continuous income stream.** The **total income** produced by a continuous income stream from $t = a$ to $t = b$ is

Ex. 2, p. 394

$$\text{Total income} = \int_a^b f(t)\, dt$$

The **future value** of a continuous income stream that is invested at rate r, compounded continuously, for $0 \leq t \leq T$, is

Ex. 3, p. 396

$$FV = \int_0^T f(t) e^{r(T-t)}\, dt$$

- *Consumers' and Producers' Surplus* If $(\bar{x}, \bar{p})$ is a point on the graph of a price–demand equation $p = D(x)$, then the **consumers' surplus** at a price level of $\bar{p}$ is

Ex. 4, p. 397

$$CS = \int_0^{\bar{x}} [D(x) - \bar{p}]\, dx$$

The consumers' surplus represents the total savings to consumers who are willing to pay more than $\bar{p}$ but are still able to buy the product for $\bar{p}$.

Similarly, for a point $(\bar{x}, \bar{p})$ on the graph of a price–supply equation $p = S(x)$, the **producers' surplus** at a price level of $\bar{p}$ is

$$PS = \int_0^{\bar{x}} [\bar{p} - S(x)]\, dx$$

Ex. 5, p. 398

The producers' surplus represents the total gain to producers who are willing to supply units at a lower price $\bar{p}$, but are still able to supply units at $\bar{p}$.

If $(\bar{x}, \bar{p})$ is the intersection point of a price–demand equation $p = D(x)$ and a price–supply equation $p = S(x)$, then $\bar{p}$ is called the **equilibrium price** and $\bar{x}$ is called the **equilibrium quantity**.

Ex. 6, p. 399

6.3 Integration by Parts

- Some indefinite integrals, but not all, can be found by means of the **integration-by-parts formula**

$$\int u\, dv = uv - \int v\, du$$

Ex. 1, p. 403
Ex. 2, p. 404
Ex. 3, p. 406
Ex. 4, p. 406

- Select u and dv with the help of the guidelines in the summary on page 405.

6.4 Other Integration Methods

- The **trapezoidal rule** and **Simpson's rule** provide approximations of the definite integral that are more efficient than approximations by left or right sums: Fewer terms must be summed to achieve a given accuracy.

- *Trapezoidal Rule* Let f be a function defined on an interval $[a, b]$. Partition $[a, b]$ into n subintervals of equal length $\Delta x = (b - a)/n$ with endpoints

$$a = x_0 < x_1 < x_2 < \cdots < x_n = b.$$

Then

$$T_n = [f(x_0) + 2f(x_1) + 2f(x_2) + \cdots + 2f(x_{n-1}) + f(x_n)]\Delta x/2$$

is an approximation of $\int_a^b f(x)\, dx$.

Ex. 1, p. 411

- *Simpson's Rule* Let f be a function defined on an interval $[a, b]$. Partition $[a, b]$ into $2n$ subintervals of equal length $\Delta x = (b - a)/n$ with endpoints

$$a = x_0 < x_1 < x_2 < \cdots < x_{2n} = b.$$

Then

$$S_{2n} = [f(x_0) + 4f(x_1) + 2f(x_2) + 4f(x_3) + 2f(x_4) + \cdots + 4f(x_{2n-1}) + f(x_{2n})]\Delta x/3$$

is an approximation of $\int_a^b f(x)\, dx$.

Ex. 2, p. 412

- A **table of integrals** is a list of integration formulas that can be used to find indefinite or definite integrals of frequently encountered functions. Such a list appears in Table II of Appendix C on pages 574–576.

Ex. 3, p. 413
Ex. 4, p. 413
Ex. 5, p. 414
Ex. 6, p. 415
Ex. 7, p. 415
Ex. 8, p. 416

Review Exercises

Work through all the problems in this chapter review and check your answers in the back of the book. Answers to all review problems are there, along with section numbers in italics to indicate where each type of problem is discussed. Where weaknesses show up, review appropriate sections of the text.

Compute all numerical answers to three decimal places unless directed otherwise.

In Problems 1–3, set up definite integrals that represent the shaded areas in the figure over the indicated intervals.

1. Interval $[a, b]$ **2.** Interval $[b, c]$

3. Interval $[a, c]$

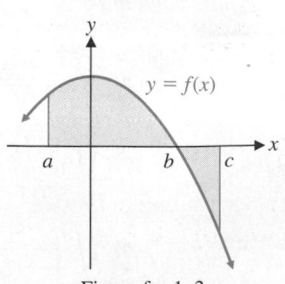

Figure for 1–3

4. Sketch a graph of the area between the graphs of $y = \ln x$ and $y = 0$ over the interval $[0.5, e]$ and find the area.

In Problems 5–10, evaluate each integral.

5. $\displaystyle\int xe^{4x}\, dx$ **6.** $\displaystyle\int x \ln x\, dx$

7. $\displaystyle\int \frac{\ln x}{x}\, dx$ **8.** $\displaystyle\int \frac{x}{1 + x^2}\, dx$

9. $\displaystyle\int \frac{1}{x(1 + x)^2}\, dx$ **10.** $\displaystyle\int \frac{1}{x^2\sqrt{1 + x}}\, dx$

In Problems 11–16, find the area bounded by the graphs of the indicated equations over the given interval.

11. $y = 5 - 2x - 6x^2; y = 0, 1 \le x \le 2$

12. $y = 5x + 7; y = 12, -3 \le x \le 1$

13. $y = -x + 2; y = x^2 + 3, -1 \le x \le 4$

14. $y = \dfrac{1}{x}; y = -e^{-x}, 1 \le x \le 2$

15. $y = x; y = -x^3, -2 \le x \le 2$

16. $y = x^2; y = -x^4; -2 \le x \le 2$

17. The Gini indices of Indonesia and Malaysia are 0.37 and 0.46, respectively. In which country is income more equally distributed?

18. The Gini indices of Thailand and Vietnam are 0.54 and 0.38, respectively. In which country is income more equally distributed?

In Problems 19–22, set up definite integrals that represent the shaded areas in the figure over the indicated intervals.

19. Interval $[a, b]$ **20.** Interval $[b, c]$

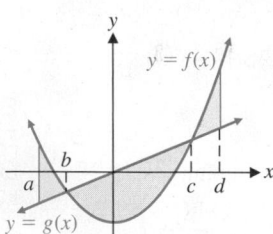

Figure for 19–22

21. Interval $[b, d]$ **22.** Interval $[a, d]$

23. Sketch a graph of the area bounded by the graphs of $y = x^2 - 6x + 9$ and $y = 9 - x$ and find the area.

In Problems 24–29, evaluate each integral.

24. $\displaystyle\int_0^1 xe^x\, dx$ **25.** $\displaystyle\int_0^3 \frac{x^2}{\sqrt{x^2 + 16}}\, dx$

26. $\displaystyle\int \sqrt{9x^2 - 49}\, dx$ **27.** $\displaystyle\int te^{-0.5t}\, dt$

28. $\displaystyle\int x^2 \ln x\, dx$ **29.** $\displaystyle\int \frac{1}{1 + 2e^x}\, dx$

30. Sketch a graph of the area bounded by the indicated graphs, and find the area. In part (B), approximate intersection points and area to two decimal places.

(A) $y = x^3 - 6x^2 + 9x; y = x$

(B) $y = x^3 - 6x^2 + 9x; y = x + 1$

In Problems 31–34, round function values to 4 decimal places and the final answer to 2 decimal places.

31. Use the trapezoidal rule with $n = 3$ to approximate $\int_0^3 e^{x^2} dx$.

32. Use the trapezoidal rule with $n = 5$ to approximate $\int_0^3 e^{x^2} dx$.

33. Use Simpson's rule with $n = 2$ (so there are $2n = 4$ subintervals) to approximate $\int_1^5 (\ln x)^2 dx$.

34. Use Simpson's rule with $n = 4$ (so there are $2n = 8$ subintervals) to approximate $\int_1^5 (\ln x)^2 dx$.

In Problems 35–42, evaluate each integral.

35. $\displaystyle\int \frac{(\ln x)^2}{x}\, dx$ **36.** $\displaystyle\int x(\ln x)^2\, dx$

37. $\displaystyle\int \frac{x}{\sqrt{x^2 - 36}}\, dx$ **38.** $\displaystyle\int \frac{x}{\sqrt{x^4 - 36}}\, dx$

39. $\displaystyle\int_0^4 x \ln(10 - x)\, dx$ **40.** $\displaystyle\int (\ln x)^2\, dx$

41. $\displaystyle\int xe^{-2x^2}\, dx$ **42.** $\displaystyle\int x^2 e^{-2x}\, dx$

43. Use a numerical integration routine on a graphing calculator to find the area in the first quadrant that is below the graph of

$$y = \frac{6}{2 + 5e^{-x}}$$

and above the graph of $y = 0.2x + 1.6$.

Applications

44. Product warranty. A manufacturer warrants a product for parts and labor for 1 year and for parts only for a second year. The time to a failure of the product after it is sold is given by the probability density function

$$f(t) = \begin{cases} 0.21e^{-0.21t} & \text{if } t \ge 0 \\ 0 & \text{otherwise} \end{cases}$$

What is the probability that a buyer chosen at random will have a product failure

(A) During the first year of warranty?

(B) During the second year of warranty?

45. Product warranty. Graph the probability density function for Problem 44 over the interval $[0, 3]$, interpret part (B) of Problem 44 geometrically and describe the geometric representation.

46. Revenue function. The weekly marginal revenue from the sale of x hair dryers is given by

$$R'(x) = 65 - 6\ln(x + 1) \qquad R(0) = 0$$

where $R(x)$ is the revenue in dollars. Find the revenue function and the production level (to the nearest unit) for a revenue of $20,000 per week. What is the weekly revenue (to the nearest dollar) at a production level of 1,000 hair dryers per week?

47. Continuous income stream. The rate of flow (in dollars per year) of a continuous income stream for a 5-year period is given by

$$f(t) = 2,500e^{0.05t} \qquad 0 \le t \le 5$$

(A) Graph $y = f(t)$ over [0, 5] and shade the area that represents the total income received from the end of the first year to the end of the fourth year.

(B) Find the total income received, to the nearest dollar, from the end of the first year to the end of the fourth year.

48. Future value of a continuous income stream. The continuous income stream in Problem 47 is invested at 4%, compounded continuously.

(A) Find the future value (to the nearest dollar) at the end of the 5-year period.

(B) Find the interest earned (to the nearest dollar) during the 5-year period.

49. Income distribution. An economist produced the following Lorenz curves for the current income distribution and the projected income distribution 10 years from now in a certain country:

$$f(x) = 0.1x + 0.9x^2 \qquad \text{Current Lorenz curve}$$

$$g(x) = x^{1.5} \qquad \text{Projected Lorenz curve}$$

(A) Graph $y = x$ and the current Lorenz curve on one set of coordinate axes for [0, 1] and graph $y = x$ and the projected Lorenz curve on another set of coordinate axes over the same interval.

(B) Looking at the areas bounded by the Lorenz curves and $y = x$, can you say that the income will be more or less equitably distributed 10 years from now?

(C) Compute the Gini index of income concentration (to one decimal place) for the current and projected curves. What can you say about the distribution of income 10 years from now? Is it more equitable or less?

50. Consumers' and producers' surplus. Find the consumers' surplus and the producers' surplus at the equilibrium price level for each pair of price–demand and price–supply equations. Include a graph that identifies the consumers' surplus and the producers' surplus. Round all values to the nearest integer.

(A) $p = D(x) = 70 - 0.2x$;
$p = S(x) = 13 + 0.0012x^2$

(B) $p = D(x) = 70 - 0.2x$;
$p = S(x) = 13e^{0.006x}$

51. Producers' surplus. The accompanying table gives price–supply data for the sale of hogs at a livestock market, where x is the number of pounds (in thousands) and p is the price per pound (in cents):

Price–Supply	
x	$p = S(x)$
0	43.50
10	46.74
20	50.05
30	54.72
40	59.18

(A) Using quadratic regression to model the data, find the demand at a price of 52.50 cents per pound.

(B) Use a numerical integration routine to find the producers' surplus (to the nearest dollar) at a price level of 52.50 cents per pound.

52. Drug assimilation. The rate at which the body eliminates a certain drug (in milliliters per hour) is given by

$$R(t) = \frac{60t}{(t + 1)^2(t + 2)}$$

where t is the number of hours since the drug was administered. How much of the drug is eliminated during the first hour after it was administered? During the fourth hour?

53. With the aid of a graphing calculator, illustrate Problem 52 geometrically.

54. Medicine. For a particular doctor, the length of time (in hours) spent with a patient per office visit has the probability density function

$$f(t) = \begin{cases} \dfrac{\frac{4}{3}}{(t + 1)^2} & \text{if } 0 \le t \le 3 \\ 0 & \text{otherwise} \end{cases}$$

(A) What is the probability that this doctor will spend less than 1 hour with a randomly selected patient?

(B) What is the probability that this doctor will spend more than 1 hour with a randomly selected patient?

55. Medicine. Illustrate part (B) in Problem 54 geometrically. Describe the geometric interpretation.

56. Politics. The rate of change of the voting population of a city with respect to time t (in years) is estimated to be

$$N'(t) = \frac{100t}{(1 + t^2)^2}$$

where $N(t)$ is in thousands. If $N(0)$ is the current voting population, how much will this population increase during the next 3 years?

57. Psychology. Rats were trained to go through a maze by rewarding them with a food pellet upon successful completion of the run. After the seventh successful run, the probability density function for length of time (in minutes) until success on the eighth trial was given by

$$f(t) = \begin{cases} .5e^{-.5t} & \text{if } t \ge 0 \\ 0 & \text{otherwise} \end{cases}$$

What is the probability that a rat selected at random after seven successful runs will take 2 or more minutes to complete the eighth run successfully? [Recall that the area under a probability density function curve from $-\infty$ to ∞ is 1.]

7

Multivariable Calculus

Introduction

In previous chapters, we have applied the key concepts of calculus, the derivative and the integral, to functions with one independent variable. The graph of such a function is a curve in the plane. In Chapter 7, we extend the key concepts of calculus to functions with two independent variables. Graphs of such functions are surfaces in a three-dimensional coordinate system. We use functions with two independent variables to study how production depends on both labor and capital; how braking distance depends on both the weight and speed of a car; how resistance in a blood vessel depends on both its length and radius. In Section 7.5, we justify the method of least squares and use the method to construct linear models (see, for example, Problem 37 in Section 7.5 on global warming).

7.1   Functions of Several Variables

- Functions of Two or More Independent Variables
- Examples of Functions of Several Variables
- Three-Dimensional Coordinate Systems

Functions of Two or More Independent Variables

In Section 1.1, we introduced the concept of a function with one independent variable. Now we broaden the concept to include functions with more than one independent variable.

A small manufacturing company produces a standard type of surfboard. If fixed costs are $500 per week and variable costs are $70 per board produced, the weekly cost function is given by

$$C(x) = 500 + 70x \tag{1}$$

where x is the number of boards produced per week. The cost function is a function of a single independent variable x. For each value of x from the domain of C, there exists exactly one value of $C(x)$ in the range of C.

Now, suppose that the company decides to add a high-performance competition board to its line. If the fixed costs for the competition board are $200 per week and the variable costs are $100 per board, then the cost function (1) must be modified to

$$C(x, y) = 700 + 70x + 100y \tag{2}$$

where $C(x, y)$ is the cost for a weekly output of x standard boards and y competition boards. Equation (2) is an example of a function with two independent variables x and y. Of course, as the company expands its product line even further, its weekly cost function must be modified to include more and more independent variables, one for each new product produced.

In general, an equation of the form

$$z = f(x, y)$$

describes a **function of two independent variables** if, for each permissible ordered pair (x, y), there is one and only one value of z determined by $f(x, y)$. The variables x and y are **independent variables**, and the variable z is a **dependent variable**. The set of all ordered pairs of permissible values of x and y is the **domain** of the function, and the set of all corresponding values $f(x, y)$ is the **range** of the function. Unless otherwise stated, we will assume that the domain of a function specified by an equation of the form $z = f(x, y)$ is the set of all ordered pairs of real numbers (x, y) such that $f(x, y)$ is also a real number. It should be noted, however, that certain conditions in practical problems often lead to further restrictions on the domain of a function.

We can similarly define functions of three independent variables, $w = f(x, y, z)$; of four independent variables, $u = f(w, x, y, z)$; and so on. In this chapter, we concern ourselves primarily with functions of two independent variables.

EXAMPLE 1   Evaluating a Function of Two Independent Variables For the cost function $C(x, y) = 700 + 70x + 100y$ described earlier, find $C(10, 5)$.

SOLUTION
$$C(10, 5) = 700 + 70(10) + 100(5)$$
$$= \$1,900$$

Matched Problem 1 ⌋ Find $C(20, 10)$ for the cost function in Example 1.

EXAMPLE 2 Evaluating a Function of Three Independent Variables For the function $f(x, y, z) = 2x^2 - 3xy + 3z + 1$, find $f(3, 0, -1)$.

SOLUTION

$$f(3, 0, -1) = 2(3)^2 - 3(3)(0) + 3(-1) + 1$$
$$= 18 - 0 - 3 + 1 = 16$$

Matched Problem 2 Find $f(-2, 2, 3)$ for f in Example 2.

EXAMPLE 3 Revenue, Cost, and Profit Functions Suppose the surfboard company discussed earlier has determined that the demand equations for its two types of boards are given by

$$p = 210 - 4x + y$$
$$q = 300 + x - 12y$$

where p is the price of the standard board, q is the price of the competition board, x is the weekly demand for standard boards, and y is the weekly demand for competition boards.

(A) Find the weekly revenue function $R(x, y)$, and evaluate $R(20, 10)$.

(B) If the weekly cost function is

$$C(x, y) = 700 + 70x + 100y$$

find the weekly profit function $P(x, y)$ and evaluate $P(20, 10)$.

SOLUTION

(A)

$$\text{Revenue} = \begin{pmatrix} \text{demand for} \\ \text{standard} \\ \text{boards} \end{pmatrix} \times \begin{pmatrix} \text{price of a} \\ \text{standard} \\ \text{board} \end{pmatrix} + \begin{pmatrix} \text{demand for} \\ \text{competition} \\ \text{boards} \end{pmatrix} \times \begin{pmatrix} \text{price of a} \\ \text{competition} \\ \text{board} \end{pmatrix}$$

$$R(x, y) = xp + yq$$
$$= x(210 - 4x + y) + y(300 + x - 12y)$$
$$= 210x + 300y - 4x^2 + 2xy - 12y^2$$
$$R(20, 10) = 210(20) + 300(10) - 4(20)^2 + 2(20)(10) - 12(10)^2$$
$$= \$4{,}800$$

(B)　Profit = revenue − cost

$$P(x, y) = R(x, y) - C(x, y)$$
$$= 210x + 300y - 4x^2 + 2xy - 12y^2 - 700 - 70x - 100y$$
$$= 140x + 200y - 4x^2 + 2xy - 12y^2 - 700$$
$$P(20, 10) = 140(20) + 200(10) - 4(20)^2 + 2(20)(10) - 12(10)^2 - 700$$
$$= \$1{,}700$$

Matched Problem 3 Repeat Example 3 if the demand and cost equations are given by

$$p = 220 - 6x + y$$
$$q = 300 + 3x - 10y$$
$$C(x, y) = 40x + 80y + 1{,}000$$

Examples of Functions of Several Variables

A number of concepts can be considered as functions of two or more variables.

Area of a rectangle	$A(x, y) = xy$	A = area, y, x
Volume of a box	$V(x, y, z) = xyz$	V = volume, z, x, y
Volume of a right circular cylinder	$V(r, h) = \pi r^2 h$	r, h
Simple interest	$A(P, r, t) = P(1 + rt)$	A = amount P = principal r = annual rate t = time in years
Compound interest	$A(P, r, t, n) = P\left(1 + \dfrac{r}{n}\right)^{nt}$	A = amount P = principal r = annual rate t = time in years n = number of compound periods per year
IQ	$Q(M, C) = \dfrac{M}{C}(100)$	Q = IQ = intelligence quotient M = MA = mental age C = CA = chronological age
Resistance for blood flow in a vessel (Poiseuille's law)	$R(L, r) = k\dfrac{L}{r^4}$	R = resistance L = length of vessel r = radius of vessel k = constant

EXAMPLE 4 Package Design A company uses a box with a square base and an open top for a bath assortment (see figure). If x is the length (in inches) of each side of the base and y is the height (in inches), find the total amount of material $M(x, y)$ required to construct one of these boxes, and evaluate $M(5, 10)$.

SOLUTION

$$\text{Area of base} = x^2$$
$$\text{Area of one side} = xy$$
$$\text{Total material} = (\text{area of base}) + 4(\text{area of one side})$$
$$M(x, y) = x^2 + 4xy$$
$$M(5, 10) = (5)^2 + 4(5)(10)$$
$$= 225 \text{ square inches}$$

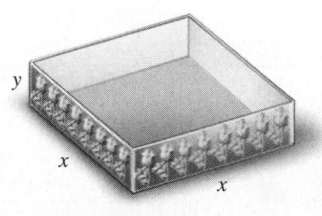

Matched Problem 4 For the box in Example 4, find the volume $V(x, y)$ and evaluate $V(5, 10)$.

The next example concerns the **Cobb–Douglas production function**

$$f(x, y) = kx^m y^n$$

where k, m, and n are positive constants with $m + n = 1$. Economists use this function to describe the number of units $f(x, y)$ produced from the utilization of x units of labor and y units of capital (for equipment such as tools, machinery, buildings, and so on). Cobb–Douglas production functions are also used to describe the productivity of a single industry, of a group of industries producing the same product, or even of an entire country.

EXAMPLE 5   Productivity   The productivity of a steel-manufacturing company is given approximately by the function

$$f(x, y) = 10x^{0.2}y^{0.8}$$

with the utilization of x units of labor and y units of capital. If the company uses 3,000 units of labor and 1,000 units of capital, how many units of steel will be produced?

SOLUTION   The number of units of steel produced is given by

$$f(3,000, 1,000) = 10(3,000)^{0.2}(1,000)^{0.8} \quad \text{Use a calculator.}$$
$$\approx 12,457 \text{ units}$$

Matched Problem 5   Refer to Example 5. Find the steel production if the company uses 1,000 units of labor and 2,000 units of capital.

Three-Dimensional Coordinate Systems

We now take a brief look at graphs of functions of two independent variables. Since functions of the form $z = f(x, y)$ involve two independent variables x and y, and one dependent variable z, we need a *three-dimensional coordinate system* for their graphs. A **three-dimensional coordinate system** is formed by three mutually perpendicular number lines intersecting at their origins (see Fig. 1). In such a system, every ordered **triplet of numbers** (x, y, z) can be associated with a unique point, and conversely.

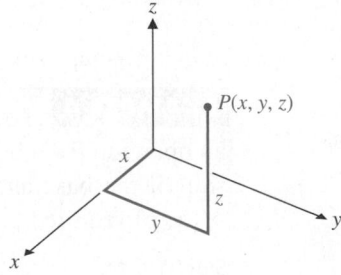

Figure 1 **Rectangular coordinate system**

EXAMPLE 6   Three-Dimensional Coordinates   Locate $(-3, 5, 2)$ in a rectangular coordinate system.

SOLUTION

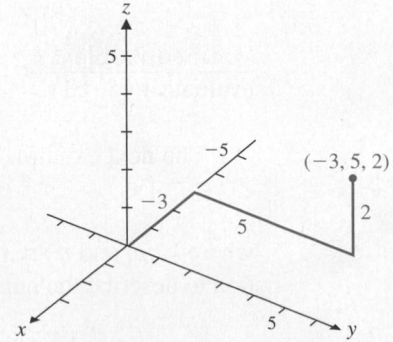

Matched Problem 6 Find the coordinates of the corners *A, C, G,* and *D* of the rectangular box shown in the following figure.

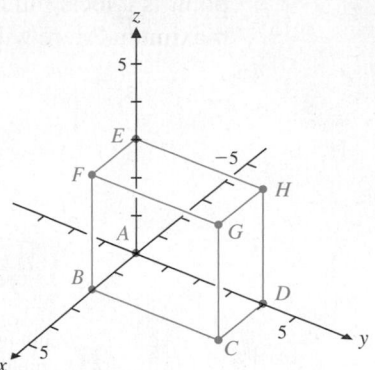

Explore and Discuss 1   Imagine that you are facing the front of a classroom whose rectangular walls meet at right angles. Suppose that the point of intersection of the floor, front wall, and left-side wall is the origin of a three-dimensional coordinate system in which every point in the room has nonnegative coordinates. Then the plane $z = 0$ (or, equivalently, the xy plane) can be described as "the floor," and the plane $z = 2$ can be described as "the plane parallel to, but 2 units above, the floor." Give similar descriptions of the following planes:

(A) $x = 0$ (B) $x = 3$ (C) $y = 0$ (D) $y = 4$ (E) $x = -1$

What does the graph of $z = x^2 + y^2$ look like? If we let $x = 0$ and graph $z = 0^2 + y^2 = y^2$ in the yz plane, we obtain a parabola; if we let $y = 0$ and graph $z = x^2 + 0^2 = x^2$ in the xz plane, we obtain another parabola. The graph of $z = x^2 + y^2$ is either one of these parabolas rotated around the z axis (see Fig. 2). This cup-shaped figure is a *surface* and is called a **paraboloid**.

In general, the graph of any function of the form $z = f(x, y)$ is called a **surface**. The graph of such a function is the graph of all ordered triplets of numbers (x, y, z) that satisfy the equation. Graphing functions of two independent variables is a difficult task, and the general process will not be dealt with in this book. We present only a few simple graphs to suggest extensions of earlier geometric interpretations of the derivative and local maxima and minima to functions of two variables. Note that $z = f(x, y) = x^2 + y^2$ appears (see Fig. 2) to have a local minimum at $(x, y) = (0, 0)$. Figure 3 shows a local maximum at $(x, y) = (0, 0)$.

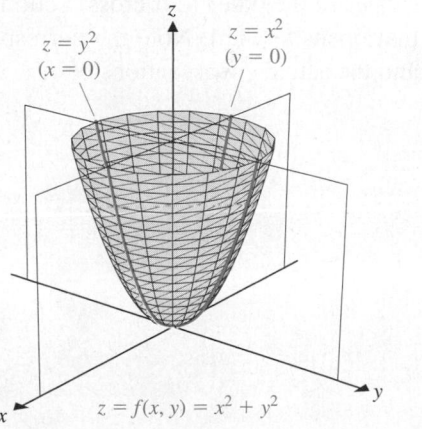

Figure 2 **Paraboloid**

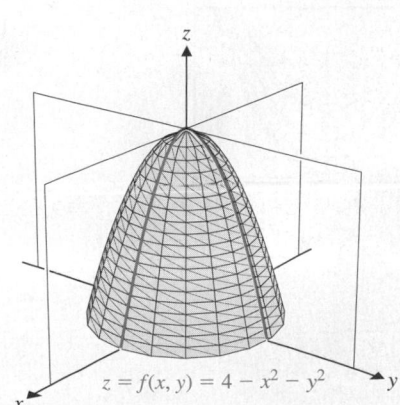

Figure 3 **Local maximum:** $f(0, 0) = 4$

Figure 4 shows a point at $(x, y) = (0, 0)$, called a **saddle point**, that is neither a local minimum nor a local maximum. Note that in the cross section $x = 0$, the saddle point is a local minimum, and in the cross section $y = 0$, the saddle point is a local maximum. More will be said about local maxima and minima in Section 7.3.

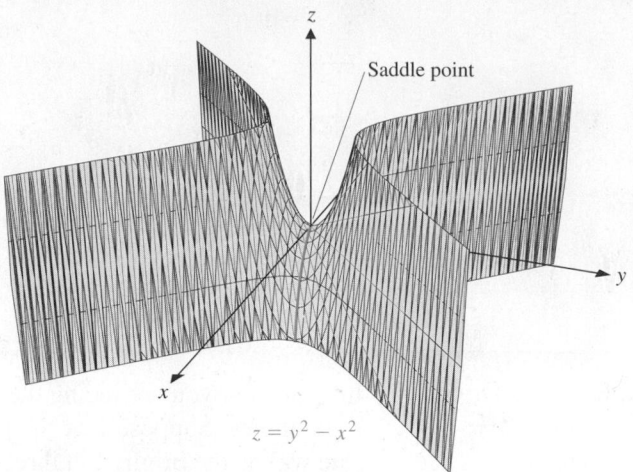

Saddle point

$z = y^2 - x^2$

Figure 4 **Saddle point at (0, 0, 0)**

Some graphing calculators are designed to draw graphs (like those of Figs. 2, 3, and 4) of functions of two independent variables. Others, such as the graphing calculator used for the displays in this book, are designed to draw graphs of functions of one independent variable. When using the latter type of calculator, we can graph cross sections produced by cutting surfaces with planes parallel to the xz plane or yz plane to gain insight into the graph of a function of two independent variables.

EXAMPLE 7   Graphing Cross Sections

(A) Describe the cross sections of $f(x, y) = 2x^2 + y^2$ in the planes $y = 0$, $y = 1, y = 2, y = 3$, and $y = 4$.

(B) Describe the cross sections of $f(x, y) = 2x^2 + y^2$ in the planes $x = 0$, $x = 1, x = 2, x = 3$, and $x = 4$.

SOLUTION

(A) The cross section of $f(x, y) = 2x^2 + y^2$ produced by cutting it with the plane $y = 0$ is the graph of the function $f(x, 0) = 2x^2$ in this plane. We can examine the shape of this cross section by graphing $y_1 = 2x^2$ on a graphing calculator (Fig. 5). Similarly, the graphs of $y_2 = f(x, 1) = 2x^2 + 1$, $y_3 = f(x, 2) = 2x^2 + 4$, $y_4 = f(x, 3) = 2x^2 + 9$, and $y_5 = f(x, 4) = 2x^2 + 16$ show the shapes of the other four cross sections (see Fig. 5). Each of these is a parabola that opens upward. Note the correspondence between the graphs in Figure 5 and the actual cross sections of $f(x, y) = 2x^2 + y^2$ shown in Figure 6.

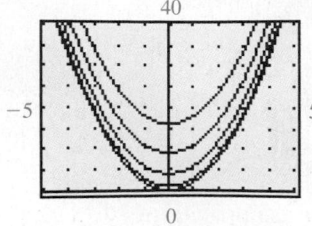

Figure 5

$y_1 = 2x^2$ $y_4 = 2x^2 + 9$

$y_2 = 2x^2 + 1$ $y_5 = 2x^2 + 16$

$y_3 = 2x^2 + 4$

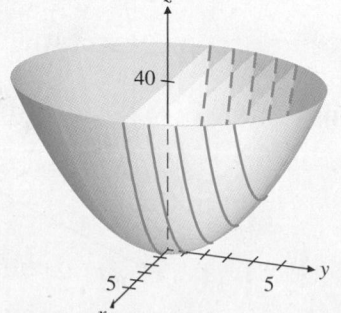

Figure 6

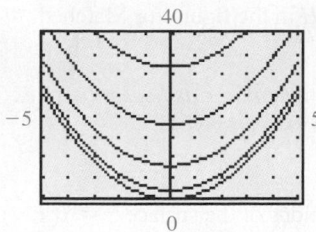

Figure 7

$y_1 = x^2$ $y_4 = 18 + x^2$
$y_2 = 2 + x^2$ $y_5 = 32 + x^2$
$y_3 = 8 + x^2$

(B) The five cross sections are represented by the graphs of the functions $f(0, y) = y^2, f(1, y) = 2 + y^2, f(2, y) = 8 + y^2, f(3, y) = 18 + y^2$, and $f(4, y) = 32 + y^2$. These five functions are graphed in Figure 7. (Note that changing the name of the independent variable from y to x for graphing purposes does not affect the graph displayed.) Each of the five cross sections is a parabola that opens upward.

Matched Problem 7

(A) Describe the cross sections of $g(x, y) = y^2 - x^2$ in the planes $y = 0$, $y = 1, y = 2, y = 3$, and $y = 4$.

(B) Describe the cross sections of $g(x, y) = y^2 - x^2$ in the planes $x = 0$, $x = 1, x = 2, x = 3$, and $x = 4$.

CONCEPTUAL INSIGHT

The graph of the *equation*

$$x^2 + y^2 + z^2 = 4 \tag{3}$$

is the graph of all ordered triplets of numbers (x, y, z) that satisfy the equation. The Pythagorean theorem can be used to show that the distance from the point (x, y, z) to the origin $(0, 0, 0)$ is equal to

$$\sqrt{x^2 + y^2 + z^2}$$

Therefore, the graph of (3) consists of all points that are at a distance 2 from the origin—that is, all points on the sphere of radius 2 and with center at the origin. Recall that a circle in the plane is *not* the graph of a function $y = f(x)$, because it fails the vertical-line test (Section 1.1). Similarly, a sphere is *not* the graph of a *function* $z = f(x, y)$ of two variables.

Exercises 7.1

Skills Warm-up Exercises

W In Problems 1–8, find the indicated value of the function of two or three variables. (If necessary, review Appendix C).

1. The height of a trapezoid is 3 feet and the lengths of its parallel sides are 5 feet and 8 feet. Find the area.

2. The height of a trapezoid is 4 meters and the lengths of its parallel sides are 25 meters and 32 meters. Find the area.

3. The length, width, and height of a rectangular box are 12 inches, 5 inches, and 4 inches, respectively. Find the volume.

4. The length, width, and height of a rectangular box are 30 centimeters, 15 centimeters, and 10 centimeters, respectively. Find the volume.

5. The height of a right circular cylinder is 8 meters and the radius is 2 meters. Find the volume.

6. The height of a right circular cylinder is 6 feet and the diameter is also 6 feet. Find the total surface area.

7. The height of a right circular cone is 48 centimeters and the radius is 20 centimeters. Find the total surface area.

8. The height of a right circular cone is 42 inches and the radius is 7 inches. Find the volume.

In Problems 9–16, find the indicated values of the functions

$$f(x, y) = 2x + 7y - 5 \quad \text{and} \quad g(x, y) = \frac{88}{x^2 + 3y}$$

9. $f(4, -1)$

10. $f(0, 10)$

11. $f(8, 0)$

12. $f(5, 6)$

13. $g(1, 7)$

14. $g(-2, 0)$

15. $g(3, -3)$

16. $g(0, 0)$

In Problems 17–20, find the indicated values of

$$f(x, y, z) = 2x - 3y^2 + 5z^3 - 1$$

17. $f(0, 0, 0)$

18. $f(0, 0, 2)$

19. $f(6, -5, 0)$

20. $f(-10, 4, -3)$

In Problems 21–30, find the indicated value of the given function.

21. $P(13, 5)$ for $P(n, r) = \dfrac{n!}{(n - r)!}$

22. $C(13, 5)$ for $C(n, r) = \dfrac{n!}{r!(n - r)!}$

23. $V(4, 12)$ for $V(R, h) = \pi R^2 h$

24. $T(4, 12)$ for $T(R, h) = 2\pi R(R + h)$

25. $S(3, 10)$ for $S(R, h) = \pi R \sqrt{R^2 + h^2}$

26. $W(3, 10)$ for $W(R, h) = \dfrac{1}{3} \pi R^2 h$

27. $A(100, 0.06, 3)$ for $A(P, r, t) = P + Prt$

28. $A(10, 0.04, 3, 2)$ for $A(P, r, t, n) = P\left(1 + \dfrac{r}{n}\right)^{tn}$

29. $P(0.05, 12)$ for $P(r, T) = \displaystyle\int_0^T 4{,}000e^{-rt}\, dt$

30. $F(0.07, 10)$ for $F(r, T) = \displaystyle\int_0^T 4{,}000e^{r(T-t)}\, dt$

In Problems 31–36, find the indicated function f of a single variable.

31. $f(x) = G(x, 0)$ for $G(x, y) = x^2 + 3xy + y^2 - 7$

32. $f(y) = H(0, y)$ for $H(x, y) = x^2 - 5xy - y^2 + 2$

33. $f(y) = K(4, y)$ for $K(x, y) = 10xy + 3x - 2y + 8$

34. $f(x) = L(x, -2)$ for $L(x, y) = 25 - x + 5y - 6xy$

35. $f(y) = M(y, y)$ for $M(x, y) = x^2y - 3xy^2 + 5$

36. $f(x) = N(x, 2x)$ for $N(x, y) = 3xy + x^2 - y^2 + 1$

37. Let $F(x, y) = 2x + 3y - 6$. Find all values of y such that $F(0, y) = 0$.

38. Let $F(x, y) = 5x - 4y + 12$. Find all values of x such that $F(x, 0) = 0$.

39. Let $F(x, y) = 2xy + 3x - 4y - 1$. Find all values of x such that $F(x, x) = 0$.

40. Let $F(x, y) = xy + 2x^2 + y^2 - 25$. Find all values of y such that $F(y, y) = 0$.

41. Let $F(x, y) = x^2 + e^x y - y^2$. Find all values of x such that $F(x, 2) = 0$.

42. Let $G(a, b, c) = a^3 + b^3 + c^3 - (ab + ac + bc) - 6$. Find all values of b such that $G(2, b, 1) = 0$.

43. For the function $f(x, y) = x^2 + 2y^2$, find

$$\frac{f(x + h, y) - f(x, y)}{h}$$

44. For the function $f(x, y) = x^2 + 2y^2$, find

$$\frac{f(x, y + k) - f(x, y)}{k}$$

45. For the function $f(x, y) = 2xy^2$, find

$$\frac{f(x + h, y) - f(x, y)}{h}$$

46. For the function $f(x, y) = 2xy^2$, find

$$\frac{f(x, y + k) - f(x, y)}{k}$$

47. Find the coordinates of E and F in the figure for Matched Problem 6 on page 429.

48. Find the coordinates of B and H in the figure for Matched Problem 6 on page 429.

In Problems 49–54, use a graphing calculator as necessary to explore the graphs of the indicated cross sections.

49. Let $f(x, y) = x^2$.

   (A) Explain why the cross sections of the surface $z = f(x, y)$ produced by cutting it with planes parallel to $y = 0$ are parabolas.

   (B) Describe the cross sections of the surface in the planes $x = 0$, $x = 1$, and $x = 2$.

   (C) Describe the surface $z = f(x, y)$.

50. Let $f(x, y) = \sqrt{4 - y^2}$.

   (A) Explain why the cross sections of the surface $z = f(x, y)$ produced by cutting it with planes parallel to $x = 0$ are semicircles of radius 2.

   (B) Describe the cross sections of the surface in the planes $y = 0$, $y = 2$, and $y = 3$.

   (C) Describe the surface $z = f(x, y)$.

51. Let $f(x, y) = \sqrt{36 - x^2 - y^2}$.

   (A) Describe the cross sections of the surface $z = f(x, y)$ produced by cutting it with the planes $y = 1$, $y = 2$, $y = 3$, $y = 4$, and $y = 5$.

   (B) Describe the cross sections of the surface in the planes $x = 0$, $x = 1$, $x = 2$, $x = 3$, $x = 4$, and $x = 5$.

   (C) Describe the surface $z = f(x, y)$.

52. Let $f(x, y) = 100 + 10x + 25y - x^2 - 5y^2$.

   (A) Describe the cross sections of the surface $z = f(x, y)$ produced by cutting it with the planes $y = 0$, $y = 1$, $y = 2$, and $y = 3$.

   (B) Describe the cross sections of the surface in the planes $x = 0$, $x = 1$, $x = 2$, and $x = 3$.

   (C) Describe the surface $z = f(x, y)$.

53. Let $f(x, y) = e^{-(x^2 + y^2)}$.

   (A) Explain why $f(a, b) = f(c, d)$ whenever (a, b) and (c, d) are points on the same circle centered at the origin in the xy plane.

   (B) Describe the cross sections of the surface $z = f(x, y)$ produced by cutting it with the planes $x = 0$, $y = 0$, and $x = y$.

   (C) Describe the surface $z = f(x, y)$.

54. Let $f(x, y) = 4 - \sqrt{x^2 + y^2}$.

   (A) Explain why $f(a, b) = f(c, d)$ whenever (a, b) and (c, d) are points on the same circle with center at the origin in the xy plane.

   (B) Describe the cross sections of the surface $z = f(x, y)$ produced by cutting it with the planes $x = 0$, $y = 0$, and $x = y$.

   (C) Describe the surface $z = f(x, y)$.

Applications

55. Cost function. A small manufacturing company produces two models of a surfboard: a standard model and a competition model. If the standard model is produced at a variable cost of $210 each and the competition model at a variable cost of $300 each, and if the total fixed costs per month are $6,000, then the monthly cost function is given by

$$C(x, y) = 6,000 + 210x + 300y$$

where x and y are the numbers of standard and competition models produced per month, respectively. Find $C(20, 10)$, $C(50, 5)$, and $C(30, 30)$.

56. Advertising and sales. A company spends $\$x$ thousand per week on online advertising and $\$y$ thousand per week on TV advertising. Its weekly sales are found to be given by

$$S(x, y) = 5x^2 y^3$$

Find $S(3, 2)$ and $S(2, 3)$.

57. Revenue function. A supermarket sells two brands of coffee: brand A at $\$p$ per pound and brand B at $\$q$ per pound. The daily demand equations for brands A and B are, respectively,

$$x = 200 - 5p + 4q$$
$$y = 300 + 2p - 4q$$

(both in pounds). Find the daily revenue function $R(p, q)$. Evaluate $R(2, 3)$ and $R(3, 2)$.

58. Revenue, cost, and profit functions. A company manufactures 10- and 3-speed bicycles. The weekly demand and cost equations are

$$p = 230 - 9x + y$$
$$q = 130 + x - 4y$$
$$C(x, y) = 200 + 80x + 30y$$

where $\$p$ is the price of a 10-speed bicycle, $\$q$ is the price of a 3-speed bicycle, x is the weekly demand for 10-speed bicycles, y is the weekly demand for 3-speed bicycles, and $C(x, y)$ is the cost function. Find the weekly revenue function $R(x, y)$ and the weekly profit function $P(x, y)$. Evaluate $R(10, 15)$ and $P(10, 15)$.

59. Productivity. The Cobb–Douglas production function for a petroleum company is given by

$$f(x, y) = 20x^{0.4} y^{0.6}$$

where x is the utilization of labor and y is the utilization of capital. If the company uses 1,250 units of labor and 1,700 units of capital, how many units of petroleum will be produced?

60. Productivity. The petroleum company in Problem 59 is taken over by another company that decides to double both the units of labor and the units of capital utilized in the production of petroleum. Use the Cobb–Douglas production function given in Problem 59 to find the amount of petroleum that will be produced by this increased utilization of labor and capital. What is the effect on productivity of doubling both the units of labor and the units of capital?

61. Future value. At the end of each year, $5,000 is invested into an IRA earning 3% compounded annually.

(A) How much will be in the account at the end of 30 years? Use the annuity formula

$$F(P, i, n) = P \frac{(1 + i)^n - 1}{i}$$

where

$$P = \text{periodic payment}$$
$$i = \text{rate per period}$$
$$n = \text{number of payments (periods)}$$
$$F = \text{FV} = \text{future value}$$

(B) Use graphical approximation methods to determine the rate of interest that would produce $300,000 in the account at the end of 30 years.

62. Package design. The packaging department in a company has been asked to design a rectangular box with no top and a partition down the middle (see the figure). Let x, y, and z be the dimensions of the box (in inches). Ignore the thickness of the material from which the box will be made.

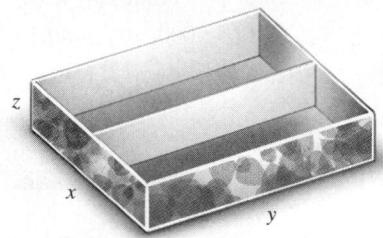

(A) Find the total area of material $M(x, y, z)$ used in constructing one of these boxes, and evaluate $M(10, 12, 6)$.

(B) Suppose that the box will have a square base and a volume of 720 cubic inches. Use graphical approximation methods to determine the dimensions that require the least material.

63. Marine biology. For a diver using scuba-diving gear, a marine biologist estimates the time (duration) of a dive according to the equation

$$T(V, x) = \frac{33V}{x + 33}$$

where

$$T = \text{time of dive in minutes}$$
$$V = \text{volume of air, at sea level pressure,}$$
$$\text{compressed into tanks}$$
$$x = \text{depth of dive in feet}$$

Find $T(70, 47)$ and $T(60, 27)$.

64. Blood flow. Poiseuille's law states that the resistance R for blood flowing in a blood vessel varies directly as the length L of the vessel and inversely as the fourth power of its radius r. Stated as an equation,

$$R(L, r) = k \frac{L}{r^4} \qquad k \text{ a constant}$$

Find $R(8, 1)$ and $R(4, 0.2)$.

65. Physical anthropology. Anthropologists use an index called the *cephalic index*. The cephalic index C varies directly as the width W of the head and inversely as the length L of the head (both viewed from the top). In terms of an equation,

$$C(W, L) = 100\frac{W}{L}$$

where

W = width in inches

L = length in inches

Find $C(6, 8)$ and $C(8.1, 9)$.

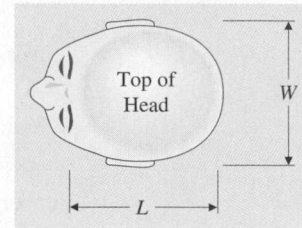

66. Safety research. Under ideal conditions, if a person driving a car slams on the brakes and skids to a stop, the length of the skid marks (in feet) is given by the formula

$$L(w, v) = kwv^2$$

where

k = constant

w = weight of car in pounds

v = speed of car in miles per hour

For $k = 0.000\ 013\ 3$, find $L(2{,}000, 40)$ and $L(3{,}000, 60)$.

67. Psychology. The intelligence quotient (IQ) is defined to be the ratio of mental age (MA), as determined by certain tests, to chronological age (CA), multiplied by 100. Stated as an equation,

$$Q(M, C) = \frac{M}{C} \cdot 100$$

where

$$Q = \text{IQ} \qquad M = \text{MA} \qquad C = \text{CA}$$

Find $Q(12, 10)$ and $Q(10, 12)$.

Answers to Matched Problems

1. $3,100   2. 30
3. (A) $R(x, y) = 220x + 300y - 6x^2 + 4xy - 10y^2$;
        $R(20, 10) = \$4{,}800$
   (B) $P(x, y) = 180x + 220y - 6x^2 + 4xy - 10y^2$
        $- 1{,}000; P(20, 10) = \$2{,}200$
4. $V(x, y) = x^2y; V(5, 10) = 250$ in.^{3}
5. 17,411 units
6. $A(0, 0, 0); C(2, 4, 0); G(2, 4, 3); D(0, 4, 0)$
7. (A) Each cross section is a parabola that opens downward.
   (B) Each cross section is a parabola that opens upward.

7.2 Partial Derivatives

- Partial Derivatives
- Second-Order Partial Derivatives

Partial Derivatives

We know how to differentiate many kinds of functions of one independent variable and how to interpret the derivatives that result. What about functions with two or more independent variables? Let's return to the surfboard example considered on page 426.

For the company producing only the standard board, the cost function was

$$C(x) = 500 + 70x$$

Differentiating with respect to x, we obtain the marginal cost function

$$C'(x) = 70$$

Since the marginal cost is constant, $70 is the change in cost for a 1-unit increase in production at any output level.

For the company producing two types of boards—a standard model and a competition model—the cost function was

$$C(x, y) = 700 + 70x + 100y$$

Now suppose that we differentiate with respect to x, holding y fixed, and denote the resulting function by $C_x(x, y)$; or suppose we differentiate with respect to y, holding x fixed, and denote the resulting function by $C_y(x, y)$. Differentiating in this way, we obtain

$$C_x(x, y) = 70 \qquad C_y(x, y) = 100$$

Each of these functions is called a **partial derivative**, and, in this example, each represents marginal cost. The first is the change in cost due to a 1-unit increase in production of the standard board with the production of the competition model held fixed. The second is the change in cost due to a 1-unit increase in production of the competition board with the production of the standard board held fixed.

In general, if $z = f(x, y)$, then the **partial derivative of f with respect to x**, denoted $\partial z/\partial x, f_x$, or $f_x(x, y)$, is defined by

$$\frac{\partial z}{\partial x} = \lim_{h \to 0} \frac{f(x + h, y) - f(x, y)}{h}$$

provided that the limit exists. We recognize this formula as the ordinary derivative of f with respect to x, holding y constant. We can continue to use all the derivative rules and properties discussed in Chapters 2 to 4 and apply them to partial derivatives.

Similarly, the **partial derivative of f with respect to y**, denoted $\partial z/\partial y, f_y$, or $f_y(x, y)$, is defined by

$$\frac{\partial z}{\partial y} = \lim_{k \to 0} \frac{f(x, y + k) - f(x, y)}{k}$$

which is the ordinary derivative with respect to y, holding x constant.

Parallel definitions and interpretations hold for functions with three or more independent variables.

EXAMPLE 1 Partial Derivatives For $z = f(x, y) = 2x^2 - 3x^2y + 5y + 1$, find

(A) $\partial z/\partial x$ (B) $f_x(2, 3)$

SOLUTION

(A) $z = 2x^2 - 3x^2y + 5y + 1$

Differentiating with respect to x, holding y constant (that is, treating y as a constant), we obtain

$$\frac{\partial z}{\partial x} = 4x - 6xy$$

(B) $f(x, y) = 2x^2 - 3x^2y + 5y + 1$

First, differentiate with respect to x. From part (A), we have

$$f_x(x, y) = 4x - 6xy$$

Then evaluate this equation at $(2, 3)$:

$$f_x(2, 3) = 4(2) - 6(2)(3) = -28$$

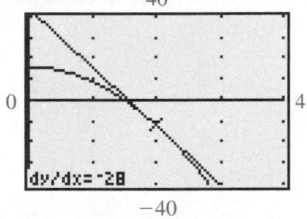

Figure 1 $y_1 = -7x^2 + 16$

In part 1B, an alternative approach would be to substitute $y = 3$ into $f(x, y)$ and graph the function $f(x, 3) = -7x^2 + 16$, which represents the cross section of the surface $z = f(x, y)$ produced by cutting it with the plane $y = 3$. Then determine the slope of the tangent line when $x = 2$. Again, we conclude that $f_x(2, 3) = -28$ (see Fig. 1).

Matched Problem 1 For f in Example 1, find

(A) $\partial z/\partial y$ (B) $f_y(2, 3)$

EXAMPLE 2 Partial Derivatives Using the Chain Rule For $z = f(x, y) = e^{x^2 + y^2}$, find

(A) $\partial z/\partial x$ (B) $f_y(2, 1)$

SOLUTION

(A) Using the chain rule [thinking of $z = e^u, u = u(x); y$ is held constant], we obtain

$$\frac{\partial z}{\partial x} = e^{x^2 + y^2} \frac{\partial(x^2 + y^2)}{\partial x}$$

$$= 2xe^{x^2 + y^2}$$

(B) $f_y(x, y) = e^{x^2+y^2}\dfrac{\partial(x^2 + y^2)}{\partial y} = 2ye^{x^2+y^2}$

$f_y(2, 1) = 2(1)e^{(2)^2 + (1)^2}$

$= 2e^5$

Matched Problem 2) For $z = f(x, y) = (x^2 + 2xy)^5$, find
(A) $\partial z/\partial y$ (B) $f_x(1, 0)$

EXAMPLE 3   Profit   The profit function for the surfboard company in Example 3 of Section 7.1 was

$$P(x, y) = 140x + 200y - 4x^2 + 2xy - 12y^2 - 700$$

Find $P_x(15, 10)$ and $P_x(30, 10)$, and interpret the results.

SOLUTION

$$P_x(x, y) = 140 - 8x + 2y$$
$$P_x(15, 10) = 140 - 8(15) + 2(10) = 40$$
$$P_x(30, 10) = 140 - 8(30) + 2(10) = -80$$

At a production level of 15 standard and 10 competition boards per week, increasing the production of standard boards by 1 unit and holding the production of competition boards fixed at 10 will increase profit by approximately $40. At a production level of 30 standard and 10 competition boards per week, increasing the production of standard boards by 1 unit and holding the production of competition boards fixed at 10 will decrease profit by approximately $80.

Matched Problem 3) For the profit function in Example 3, find $P_y(25, 10)$ and $P_y(25, 15)$, and interpret the results.

EXAMPLE 4   Productivity   The productivity of a major computer manufacturer is given approximately by the Cobb–Douglas production function

$$f(x, y) = 15x^{0.4}y^{0.6}$$

with the utilization of x units of labor and y units of capital. The partial derivative $f_x(x, y)$ represents the rate of change of productivity with respect to labor and is called the **marginal productivity of labor**. The partial derivative $f_y(x, y)$ represents the rate of change of productivity with respect to capital and is called the **marginal productivity of capital**. If the company is currently utilizing 4,000 units of labor and 2,500 units of capital, find the marginal productivity of labor and the marginal productivity of capital. For the greatest increase in productivity, should the management of the company encourage increased use of labor or increased use of capital?

SOLUTION

$$f_x(x, y) = 6x^{-0.6}y^{0.6}$$
$$f_x(4{,}000, 2{,}500) = 6(4{,}000)^{-0.6}(2{,}500)^{0.6}$$
$$\approx 4.53 \qquad \text{Marginal productivity of labor}$$
$$f_y(x, y) = 9x^{0.4}y^{-0.4}$$
$$f_y(4{,}000, 2{,}500) = 9(4{,}000)^{0.4}(2{,}500)^{-0.4}$$
$$\approx 10.86 \qquad \text{Marginal productivity of capital}$$

At the current level of utilization of 4,000 units of labor and 2,500 units of capital, each 1-unit increase in labor utilization (keeping capital utilization fixed at 2,500 units) will increase production by approximately 4.53 units, and each 1-unit increase in capital utilization (keeping labor utilization fixed at 4,000 units) will increase production by approximately 10.86 units. The management of the company should encourage increased use of capital.

Matched Problem 4) The productivity of an airplane-manufacturing company is given approximately by the Cobb–Douglas production function

$$f(x, y) = 40x^{0.3}y^{0.7}$$

(A) Find $f_x(x, y)$ and $f_y(x, y)$.

(B) If the company is currently using 1,500 units of labor and 4,500 units of capital, find the marginal productivity of labor and the marginal productivity of capital.

(C) For the greatest increase in productivity, should the management of the company encourage increased use of labor or increased use of capital?

Partial derivatives have simple geometric interpretations, as shown in Figure 2. If we hold x fixed at $x = a$, then $f_y(a, y)$ is the slope of the curve obtained by intersecting the surface $z = f(x, y)$ with the plane $x = a$. A similar interpretation is given to $f_x(x, b)$.

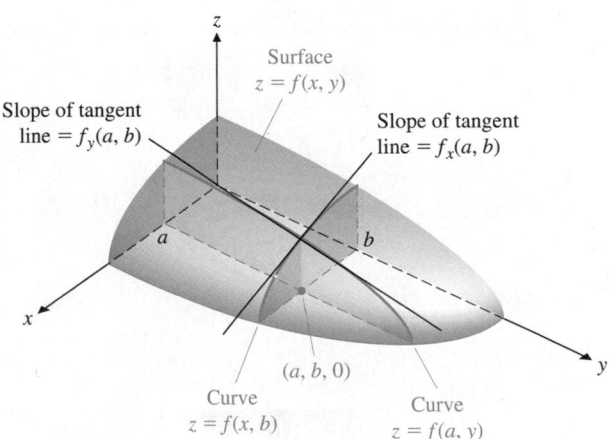

Figure 2

Second-Order Partial Derivatives

The function

$$z = f(x, y) = x^4y^7$$

has two **first-order partial derivatives**:

$$\frac{\partial z}{\partial x} = f_x = f_x(x, y) = 4x^3y^7 \quad \text{and} \quad \frac{\partial z}{\partial y} = f_y = f_y(x, y) = 7x^4y^6$$

Each of these partial derivatives, in turn, has two partial derivatives called **second-order partial derivatives** of $z = f(x, y)$. Generalizing the various notations we have for first-order partial derivatives, we write the four second-order partial derivatives of $z = f(x, y) = x^4y^7$ as

Equivalent notations

$$\overbrace{\qquad\qquad\qquad}$$

$$f_{xx} = f_{xx}(x, y) = \frac{\partial^2 z}{\partial x^2} = \frac{\partial}{\partial x}\left(\frac{\partial z}{\partial x}\right) = \frac{\partial}{\partial x}(4x^3y^7) = 12x^2y^7$$

$$f_{xy} = f_{xy}(x, y) = \frac{\partial^2 z}{\partial y\,\partial x} = \frac{\partial}{\partial y}\left(\frac{\partial z}{\partial x}\right) = \frac{\partial}{\partial y}(4x^3y^7) = 28x^3y^6$$

$$f_{yx} = f_{yx}(x, y) = \frac{\partial^2 z}{\partial x\,\partial y} = \frac{\partial}{\partial x}\left(\frac{\partial z}{\partial y}\right) = \frac{\partial}{\partial x}(7x^4y^6) = 28x^3y^6$$

$$f_{yy} = f_{yy}(x, y) = \frac{\partial^2 z}{\partial y^2} = \frac{\partial}{\partial y}\left(\frac{\partial z}{\partial y}\right) = \frac{\partial}{\partial y}(7x^4y^6) = 42x^4y^5$$

In the mixed partial derivative $\partial^2 z/\partial y\,\partial x = f_{xy}$, we started with $z = f(x, y)$ and first differentiated with respect to x (holding y constant). Then we differentiated with respect to y (holding x constant). In the other mixed partial derivative, $\partial^2 z/\partial x\,\partial y = f_{yx}$, the order of differentiation was reversed; however, the final result was the same—that is, $f_{xy} = f_{yx}$. Although it is possible to find functions for which $f_{xy} \neq f_{yx}$, such functions rarely occur in applications involving partial derivatives. For all the functions in this book, we will assume that $f_{xy} = f_{yx}$.

In general, we have the following definitions:

DEFINITION Second-Order Partial Derivatives

If $z = f(x, y)$, then

$$f_{xx} = f_{xx}(x, y) = \frac{\partial^2 z}{\partial x^2} = \frac{\partial}{\partial x}\left(\frac{\partial z}{\partial x}\right)$$

$$f_{xy} = f_{xy}(x, y) = \frac{\partial^2 z}{\partial y\,\partial x} = \frac{\partial}{\partial y}\left(\frac{\partial z}{\partial x}\right)$$

$$f_{yx} = f_{yx}(x, y) = \frac{\partial^2 z}{\partial x\,\partial y} = \frac{\partial}{\partial x}\left(\frac{\partial z}{\partial y}\right)$$

$$f_{yy} = f_{yy}(x, y) = \frac{\partial^2 z}{\partial y^2} = \frac{\partial}{\partial y}\left(\frac{\partial z}{\partial y}\right)$$

EXAMPLE 5 Second-Order Partial Derivatives For $z = f(x, y) = 3x^2 - 2xy^3 + 1$, find

(A) $\dfrac{\partial^2 z}{\partial x\,\partial y}, \dfrac{\partial^2 z}{\partial y\,\partial x}$ (B) $\dfrac{\partial^2 z}{\partial x^2}$ (C) $f_{yx}(2, 1)$

SOLUTION

(A) First differentiate with respect to y and then with respect to x:

$$\frac{\partial z}{\partial y} = -6xy^2 \qquad \frac{\partial^2 z}{\partial x\,\partial y} = \frac{\partial}{\partial x}\left(\frac{\partial z}{\partial y}\right) = \frac{\partial}{\partial x}(-6xy^2) = -6y^2$$

Now differentiate with respect to x and then with respect to y:

$$\frac{\partial z}{\partial x} = 6x - 2y^3 \qquad \frac{\partial^2 z}{\partial y\,\partial x} = \frac{\partial}{\partial y}\left(\frac{\partial z}{\partial x}\right) = \frac{\partial}{\partial y}(6x - 2y^3) = -6y^2$$

(B) Differentiate with respect to x twice:

$$\frac{\partial z}{\partial x} = 6x - 2y^3 \qquad \frac{\partial^2 z}{\partial x^2} = \frac{\partial}{\partial x}\left(\frac{\partial z}{\partial x}\right) = 6$$

(C) First find $f_{yx}(x, y)$; then evaluate the resulting equation at $(2, 1)$. Again, remember that f_{yx} signifies differentiation first with respect to y and then with respect to x.

$$f_y(x, y) = -6xy^2 \qquad f_{yx}(x, y) = -6y^2$$

and

$$f_{yx}(2, 1) = -6(1)^2 = -6$$

Matched Problem 5 For $z = f(x, y) = x^3y - 2y^4 + 3$, find

(A) $\dfrac{\partial^2 z}{\partial y \, \partial x}$

(B) $\dfrac{\partial^2 z}{\partial y^2}$

(C) $f_{xy}(2, 3)$

(D) $f_{yx}(2, 3)$

CONCEPTUAL INSIGHT

Although the mixed second-order partial derivatives f_{xy} and f_{yx} are equal for all functions considered in this book, it is a good idea to compute both of them, as in Example 5A, as a check on your work. By contrast, the other two second-order partial derivatives, f_{xx} and f_{yy}, are generally not equal to each other. For example, for the function

$$f(x, y) = 3x^2 - 2xy^3 + 1$$

of Example 5,

$$f_{xx} = 6 \qquad \text{and} \qquad f_{yy} = -12xy$$

Exercises 7.2

W **Skills Warm-up Exercises**

In Problems 1–8, find the indicated derivative. (If necessary, review Sections 3.3 and 3.4).

1. $f'(x)$ if $f(x) = \pi x^3 + x\pi^3$

2. $f'(x)$ if $f(x) = (\pi x + 3)^5 - x^4$

3. $f'(x)$ if $f(x) = x^e + e^x$

4. $f'(x)$ if $f(x) = x \ln \pi + \pi \ln x$

5. $\dfrac{dz}{dx}$ if $z = \dfrac{x}{e} + \dfrac{e}{x}$

6. $\dfrac{dz}{dx}$ if $z = x^3 \ln \pi + 4\pi^2 e^x$

7. $\dfrac{dz}{dx}$ if $z = \ln(x^2 + e^2)$

8. $\dfrac{dz}{dx}$ if $z = e^7 - 2ex^7$

In Problems 9–16, find the indicated first-order partial derivative for each function $z = f(x, y)$.

9. $f_x(x, y)$ if $f(x, y) = 4x - 3y + 6$

10. $f_x(x, y)$ if $f(x, y) = 7x + 8y - 2$

11. $f_y(x, y)$ if $f(x, y) = x^2 - 3xy + 2y^2$

12. $f_y(x, y)$ if $f(x, y) = 3x^2 + 2xy - 7y^2$

13. $\dfrac{\partial z}{\partial x}$ if $z = x^3 + 4x^2y + 2y^3$

14. $\dfrac{\partial z}{\partial y}$ if $z = 4x^2y - 5xy^2$

15. $\dfrac{\partial z}{\partial y}$ if $z = (5x + 2y)^{10}$

16. $\dfrac{\partial z}{\partial x}$ if $z = (2x - 3y)^8$

In Problems 17–24, find the indicated value.

17. $f_x(1, 3)$ if $f(x, y) = 5x^3y - 4xy^2$

18. $f_x(4, 1)$ if $f(x, y) = x^2y^2 - 5xy^3$

19. $f_y(1, 0)$ if $f(x, y) = 3xe^y$

20. $f_y(2, 4)$ if $f(x, y) = x^4 \ln y$

21. $f_y(2, 1)$ if $f(x, y) = e^{x^2} - 4y$

22. $f_y(3, 3)$ if $f(x, y) = e^{3x} - y^2$

23. $f_x(1, -1)$ if $f(x, y) = \dfrac{2xy}{1 + x^2y^2}$

24. $f_x(-1, 2)$ if $f(x, y) = \dfrac{x^2 - y^2}{1 + x^2}$

✎ In Problems 25–30, $M(x, y) = 68 + 0.3x - 0.8y$ gives the mileage (in mpg) of a new car as a function of tire pressure x (in psi) and speed (in mph). Find the indicated quantity (include the appropriate units) and explain what it means.

25. $M(32, 40)$

26. $M(22, 40)$

27. $M(32, 50)$

28. $M(22, 50)$

29. $M_x(32, 50)$

30. $M_y(32, 50)$

In Problems 31–42, find the indicated second-order partial derivative for each function $f(x, y)$.

31. $f_{xx}(x, y)$ if $f(x, y) = 6x - 5y + 3$

32. $f_{yx}(x, y)$ if $f(x, y) = -2x + y + 8$

33. $f_{xy}(x, y)$ if $f(x, y) = 4x^2 + 6y^2 - 10$

34. $f_{yy}(x, y)$ if $f(x, y) = x^2 + 9y^2 - 4$

35. $f_{xy}(x, y)$ if $f(x, y) = e^{xy^2}$

36. $f_{yx}(x, y)$ if $f(x, y) = e^{3x+2y}$

37. $f_{yy}(x, y)$ if $f(x, y) = \dfrac{\ln x}{y}$

38. $f_{xx}(x, y)$ if $f(x, y) = \dfrac{3 \ln x}{y^2}$

39. $f_{xx}(x, y)$ if $f(x, y) = (2x + y)^5$

40. $f_{yx}(x, y)$ if $f(x, y) = (3x - 8y)^6$

41. $f_{xy}(x, y)$ if $f(x, y) = (x^2 + y^4)^{10}$

42. $f_{yy}(x, y)$ if $f(x, y) = (1 + 2xy^2)^8$

In Problems 43–52, find the indicated function or value if $C(x, y) = 3x^2 + 10xy - 8y^2 + 4x - 15y - 120$.

43. $C_x(x, y)$

44. $C_y(x, y)$

45. $C_x(3, -2)$

46. $C_y(3, -2)$

47. $C_{xx}(x, y)$

48. $C_{yy}(x, y)$

49. $C_{xy}(x, y)$

50. $C_{yx}(x, y)$

51. $C_{xx}(3, -2)$

52. $C_{yy}(3, -2)$

✎ In Problems 53–58, $S(T, r) = 50(T - 40)(5 - r)$ gives an ice cream shop's daily sales as a function of temperature T (in °F) and rain r (in inches). Find the indicated quantity (include the appropriate units) and explain what it means.

53. $S(60, 2)$

54. $S(80, 0)$

55. $S_r(90, 1)$

56. $S_T(90, 1)$

57. $S_{Tr}(90, 1)$

58. $S_{rT}(90, 1)$

✎ **59.** (A) Let $f(x, y) = y^3 + 4y^2 - 5y + 3$. Show that $\partial f/\partial x = 0$.

   (B) Explain why there are an infinite number of functions $g(x, y)$ such that $\partial g/\partial x = 0$.

✎ **60.** (A) Find an example of a function $f(x, y)$ such that $\partial f/\partial x = 3$ and $\partial f/\partial y = 2$.

   (B) How many such functions are there? Explain.

In Problems 61–66, find $f_{xx}(x, y)$, $f_{xy}(x, y)$, $f_{yx}(x, y)$, and $f_{yy}(x, y)$ for each function f.

61. $f(x, y) = x^2y^2 + x^3 + y$

62. $f(x, y) = x^3y^3 + x + y^2$

63. $f(x, y) = \dfrac{x}{y} - \dfrac{y}{x}$

64. $f(x, y) = \dfrac{x^2}{y} - \dfrac{y^2}{x}$

65. $f(x, y) = xe^{xy}$

66. $f(x, y) = x \ln(xy)$

67. For
$$P(x, y) = -x^2 + 2xy - 2y^2 - 4x + 12y - 5$$
find all values of x and y such that
$$P_x(x, y) = 0 \quad \text{and} \quad P_y(x, y) = 0$$
simultaneously.

68. For
$$C(x, y) = 2x^2 + 2xy + 3y^2 - 16x - 18y + 54$$
find all values of x and y such that
$$C_x(x, y) = 0 \quad \text{and} \quad C_y(x, y) = 0$$
simultaneously.

⊞ **69.** For
$$F(x, y) = x^3 - 2x^2y^2 - 2x - 4y + 10$$
find all values of x and y such that
$$F_x(x, y) = 0 \quad \text{and} \quad F_y(x, y) = 0$$
simultaneously.

⊞ **70.** For
$$G(x, y) = x^2 \ln y - 3x - 2y + 1$$
find all values of x and y such that
$$G_x(x, y) = 0 \quad \text{and} \quad G_y(x, y) = 0$$
simultaneously.

✎ **71.** Let $f(x, y) = 3x^2 + y^2 - 4x - 6y + 2$.

   (A) Find the minimum value of $f(x, y)$ when $y = 1$.

   (B) Explain why the answer to part (A) is not the minimum value of the function $f(x, y)$.

✎ **72.** Let $f(x, y) = 5 - 2x + 4y - 3x^2 - y^2$.

   (A) Find the maximum value of $f(x, y)$ when $x = 2$.

   (B) Explain why the answer to part (A) is not the maximum value of the function $f(x, y)$.

⊞ **73.** Let $f(x, y) = 4 - x^4y + 3xy^2 + y^5$.

   (A) Use graphical approximation methods to find c (to three decimal places) such that $f(c, 2)$ is the maximum value of $f(x, y)$ when $y = 2$.

   (B) Find $f_x(c, 2)$ and $f_y(c, 2)$.

74. Let $f(x, y) = e^x + 2e^y + 3xy^2 + 1$.

(A) Use graphical approximation methods to find d (to three decimal places) such that $f(1, d)$ is the minimum value of $f(x, y)$ when $x = 1$.

(B) Find $f_x(1, d)$ and $f_y(1, d)$.

75. For $f(x, y) = x^2 + 2y^2$, find

(A) $\lim\limits_{h \to 0} \dfrac{f(x + h, y) - f(x, y)}{h}$

(B) $\lim\limits_{k \to 0} \dfrac{f(x, y + k) - f(x, y)}{k}$

76. For $f(x, y) = 2xy^2$, find

(A) $\lim\limits_{h \to 0} \dfrac{f(x + h, y) - f(x, y)}{h}$

(B) $\lim\limits_{k \to 0} \dfrac{f(x, y + k) - f(x, y)}{k}$

Applications

77. Profit function. A firm produces two types of calculators each week, x of type A and y of type B. The weekly revenue and cost functions (in dollars) are

$$R(x, y) = 80x + 90y + 0.04xy - 0.05x^2 - 0.05y^2$$
$$C(x, y) = 8x + 6y + 20{,}000$$

Find $P_x(1{,}200, 1{,}800)$ and $P_y(1{,}200, 1{,}800)$, and interpret the results.

78. Advertising and sales. A company spends $\$x$ per week on online advertising and $\$y$ per week on TV advertising. Its weekly sales were found to be given by

$$S(x, y) = 10x^{0.4}y^{0.8}$$

Find $S_x(3{,}000, 2{,}000)$ and $S_y(3{,}000, 2{,}000)$, and interpret the results.

79. Demand equations. A supermarket sells two brands of coffee: brand A at $\$p$ per pound and brand B at $\$q$ per pound. The daily demands x and y (in pounds) for brands A and B, respectively, are given by

$$x = 200 - 5p + 4q$$
$$y = 300 + 2p - 4q$$

Find $\partial x / \partial p$ and $\partial y / \partial p$, and interpret the results.

80. Revenue and profit functions. A company manufactures 10- and 3-speed bicycles. The weekly demand and cost functions are

$$p = 230 - 9x + y$$
$$q = 130 + x - 4y$$
$$C(x, y) = 200 + 80x + 30y$$

where $\$p$ is the price of a 10-speed bicycle, $\$q$ is the price of a 3-speed bicycle, x is the weekly demand for 10-speed

bicycles, y is the weekly demand for 3-speed bicycles, and $C(x, y)$ is the cost function. Find $R_x(10, 5)$ and $P_x(10, 5)$, and interpret the results.

81. Productivity. The productivity of a certain third-world country is given approximately by the function

$$f(x, y) = 10x^{0.75}y^{0.25}$$

with the utilization of x units of labor and y units of capital.

(A) Find $f_x(x, y)$ and $f_y(x, y)$.

(B) If the country is now using 600 units of labor and 100 units of capital, find the marginal productivity of labor and the marginal productivity of capital.

(C) For the greatest increase in the country's productivity, should the government encourage increased use of labor or increased use of capital?

82. Productivity. The productivity of an automobile-manufacturing company is given approximately by the function

$$f(x, y) = 50\sqrt{xy} = 50x^{0.5}y^{0.5}$$

with the utilization of x units of labor and y units of capital.

(A) Find $f_x(x, y)$ and $f_y(x, y)$.

(B) If the company is now using 250 units of labor and 125 units of capital, find the marginal productivity of labor and the marginal productivity of capital.

(C) For the greatest increase in the company's productivity, should the management encourage increased use of labor or increased use of capital?

Problems 83–86 refer to the following: If a decrease in demand for one product results in an increase in demand for another product, the two products are said to be **competitive,** *or* **substitute, products.** *(Real whipping cream and imitation whipping cream are examples of competitive, or substitute, products.) If a decrease in demand for one product results in a decrease in demand for another product, the two products are said to be* **complementary products.** *(Fishing boats and outboard motors are examples of complementary products.) Partial derivatives can be used to test whether two products are competitive, complementary, or neither. We start with demand functions for two products such that the demand for either depends on the prices for both:*

$$x = f(p, q) \quad \text{Demand function for product } A$$
$$y = g(p, q) \quad \text{Demand function for product } B$$

The variables x and y represent the number of units demanded of products A and B, respectively, at a price p for 1 unit of product A and a price q for 1 unit of product B. Normally, if the price of A increases while the price of B is held constant, then the demand for A will decrease; that is, $f_p(p, q) < 0$. Then, if A and B are competitive products, the demand for B will increase; that is, $g_p(p, q) > 0$. Similarly, if the price of B increases while the price of A is held constant, the demand for B will decrease; that is, $g_q(p, q) < 0$. Then, if A and B are competitive products, the demand for A will increase; that is,

$f_q(p, q) > 0$. *Reasoning similarly for complementary products, we arrive at the following test:*

Test for Competitive and Complementary Products

Partial Derivatives	Products A and B
$f_q(p, q) > 0$ and $g_p(p, q) > 0$	Competitive (substitute)
$f_q(p, q) < 0$ and $g_p(p, q) < 0$	Complementary
$f_q(p, q) \geq 0$ and $g_p(p, q) \leq 0$	Neither
$f_q(p, q) \leq 0$ and $g_p(p, q) \geq 0$	Neither

Use this test in Problems 83–86 to determine whether the indicated products are competitive, complementary, or neither.

83. Product demand. The weekly demand equations for the sale of butter and margarine in a supermarket are

$$x = f(p, q) = 8,000 - 0.09p^2 + 0.08q^2 \quad \text{Butter}$$
$$y = g(p, q) = 15,000 + 0.04p^2 - 0.3q^2 \quad \text{Margarine}$$

84. Product demand. The daily demand equations for the sale of brand A coffee and brand B coffee in a supermarket are

$$x = f(p, q) = 200 - 5p + 4q \quad \text{Brand A coffee}$$
$$y = g(p, q) = 300 + 2p - 4q \quad \text{Brand B coffee}$$

85. Product demand. The monthly demand equations for the sale of skis and ski boots in a sporting goods store are

$$x = f(p, q) = 800 - 0.004p^2 - 0.003q^2 \quad \text{Skis}$$
$$y = g(p, q) = 600 - 0.003p^2 - 0.002q^2 \quad \text{Ski boots}$$

86. Product demand. The monthly demand equations for the sale of tennis rackets and tennis balls in a sporting goods store are

$$x = f(p, q) = 500 - 0.5p - q^2 \quad \text{Tennis rackets}$$
$$y = g(p, q) = 10,000 - 8p - 100q^2 \quad \text{Tennis balls (cans)}$$

87. Medicine. The following empirical formula relates the surface area A (in square inches) of an average human body to its weight w (in pounds) and its height h (in inches):

$$A = f(w, h) = 15.64w^{0.425}h^{0.725}$$

(A) Find $f_w(w, h)$ and $f_h(w, h)$.

(B) For a 65-pound child who is 57 inches tall, find $f_w(65, 57)$ and $f_h(65, 57)$, and interpret the results.

88. Blood flow. Poiseuille's law states that the resistance R for blood flowing in a blood vessel varies directly as the length L of the vessel and inversely as the fourth power of its radius r. Stated as an equation,

$$R(L, r) = k\frac{L}{r^4} \quad k \text{ a constant}$$

Find $R_L(4, 0.2)$ and $R_r(4, 0.2)$, and interpret the results.

89. Physical anthropology. Anthropologists use the cephalic index C, which varies directly as the width W of the head and inversely as the length L of the head (both viewed from the top). In terms of an equation,

$$C(W, L) = 100\frac{W}{L}$$

where

$$W = \text{width in inches}$$
$$L = \text{length in inches}$$

Find $C_W(6, 8)$ and $C_L(6, 8)$, and interpret the results.

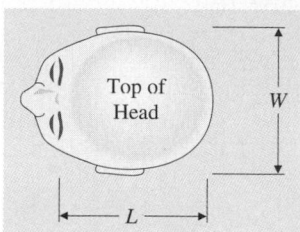

90. Safety research. Under ideal conditions, if a person driving a car slams on the brakes and skids to a stop, the length of the skid marks (in feet) is given by the formula

$$L(w, v) = kwv^2$$

where

$$k = \text{constant}$$
$$w = \text{weight of car in pounds}$$
$$v = \text{speed of car in miles per hour}$$

For $k = 0.000\ 013\ 3$, find $L_w(2,500, 60)$ and $L_v(2,500, 60)$, and interpret the results.

Answers to Matched Problems

1. (A) $\partial z/\partial y = -3x^2 + 5$ (B) $f_y(2, 3) = -7$
2. (A) $10x(x^2 + 2xy)^4$ (B) 10
3. $P_y(25, 10) = 10$: At a production level of $x = 25$ and $y = 10$, increasing y by 1 unit and holding x fixed at 25 will increase profit by approximately \$10; $P_y(25, 15) = -110$: At a production level of $x = 25$ and $y = 15$, increasing y by 1 unit and holding x fixed at 25 will decrease profit by approximately \$110
4. (A) $f_x(x, y) = 12x^{-0.7}y^{0.7}; f_y(x, y) = 28x^{0.3}y^{-0.3}$
   (B) Marginal productivity of labor ≈ 25.89; marginal productivity of capital ≈ 20.14
   (C) Labor
5. (A) $3x^2$ (B) $-24y^2$ (C) 12   (D) 12

7.3 Maxima and Minima

We are now ready to undertake a brief, but useful, analysis of local maxima and minima for functions of the type $z = f(x, y)$. We will extend the second-derivative test developed for functions of a single independent variable. We assume that all second-order partial derivatives exist for the function f in some circular region in the xy plane. This guarantees that the surface $z = f(x, y)$ has no sharp points, breaks, or ruptures. In other words, we are dealing only with smooth surfaces with no edges (like the edge of a box), breaks (like an earthquake fault), or sharp points (like the bottom point of a golf tee). (See Figure 1.)

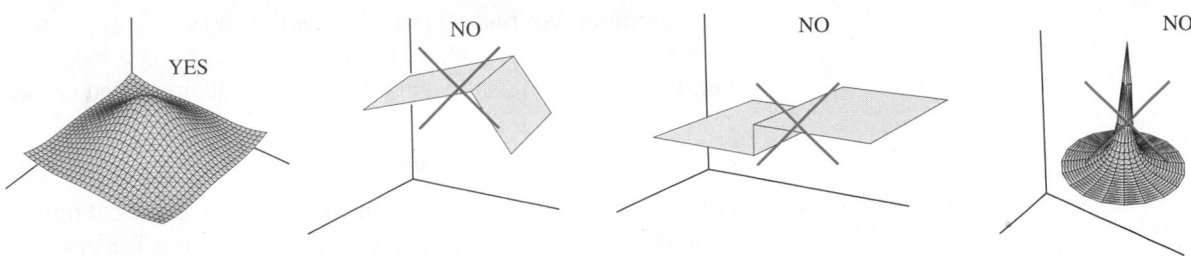

Figure 1

In addition, we will not concern ourselves with boundary points or absolute maxima–minima theory. Despite these restrictions, the procedure we will describe will help us solve a large number of useful problems.

What does it mean for $f(a, b)$ to be a local maximum or a local minimum? We say that $f(a, b)$ **is a local maximum** if there exists a circular region in the domain of f with (a, b) as the center, such that

$$f(a, b) \geq f(x, y)$$

for all (x, y) in the region. Similarly, we say that $f(a, b)$ **is a local minimum** if there exists a circular region in the domain of f with (a, b) as the center, such that

$$f(a, b) \leq f(x, y)$$

for all (x, y) in the region. Figure 2A illustrates a local maximum, Figure 2B a local minimum, and Figure 2C a **saddle point**, which is neither a local maximum nor a local minimum.

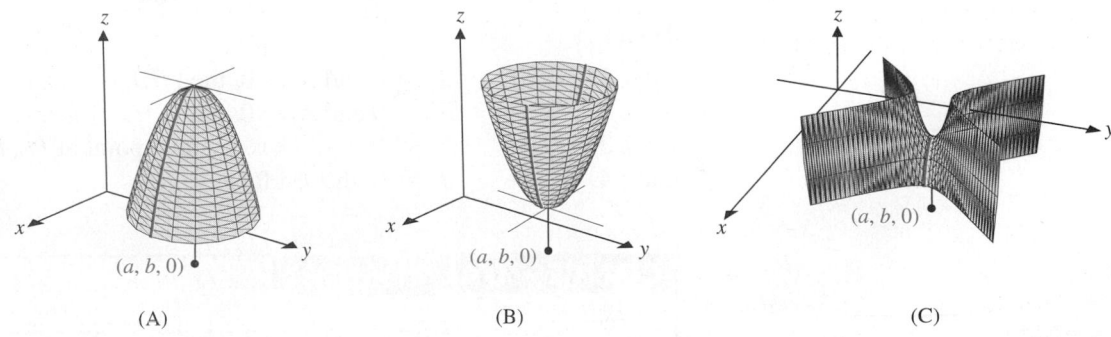

Figure 2

What happens to $f_x(a, b)$ and $f_y(a, b)$ if $f(a, b)$ is a local minimum or a local maximum and the partial derivatives of f exist in a circular region containing (a, b)? Figure 2 suggests that $f_x(a, b) = 0$ and $f_y(a, b) = 0$, since the tangent lines to the given curves are horizontal. Theorem 1 indicates that our intuitive reasoning is correct.

THEOREM 1   Local Extrema and Partial Derivatives

Let $f(a, b)$ be a local extremum (a local maximum or a local minimum) for the function f. If both f_x and f_y exist at (a, b), then

$$f_x(a, b) = 0 \quad \text{and} \quad f_y(a, b) = 0 \tag{1}$$

The converse of this theorem is false. If $f_x(a, b) = 0$ and $f_y(a, b) = 0$, then $f(a, b)$ may or may not be a local extremum; for example, the point $(a, b, f(a, b))$ may be a saddle point (see Fig. 2C).

Theorem 1 gives us *necessary* (but not *sufficient*) conditions for $f(a, b)$ to be a local extremum. We find all points (a, b) such that $f_x(a, b) = 0$ and $f_y(a, b) = 0$ and test these further to determine whether $f(a, b)$ is a local extremum or a saddle point. Points (a, b) such that conditions (1) hold are called **critical points**.

Explore and Discuss 1

(A) Let $f(x, y) = y^2 + 1$. Explain why $f(x, y)$ has a local minimum at every point on the x axis. Verify that every point on the x axis is a critical point. Explain why the graph of $z = f(x, y)$ could be described as a trough.

(B) Let $g(x, y) = x^3$. Show that every point on the y axis is a critical point. Explain why no point on the y axis is a local extremum. Explain why the graph of $z = g(x, y)$ could be described as a slide.

The next theorem, using second-derivative tests, gives us *sufficient* conditions for a critical point to produce a local extremum or a saddle point.

THEOREM 2   Second-Derivative Test for Local Extrema

If

1. $z = f(x, y)$
2. $f_x(a, b) = 0$ and $f_y(a, b) = 0$ [(a, b) is a critical point]
3. All second-order partial derivatives of f exist in some circular region containing (a, b) as center.
4. $A = f_{xx}(a, b), \quad B = f_{xy}(a, b), \quad C = f_{yy}(a, b)$

Then

Case 1.   If $AC - B^2 > 0$ and $A < 0$, then $f(a, b)$ is a local maximum.
Case 2.   If $AC - B^2 > 0$ and $A > 0$, then $f(a, b)$ is a local minimum.
Case 3.   If $AC - B^2 < 0$, then f has a saddle point at (a, b).
Case 4.   If $AC - B^2 = 0$, the test fails.

CONCEPTUAL INSIGHT

The condition $A = f_{xx}(a, b) < 0$ in case 1 of Theorem 2 is analogous to the condition $f''(c) < 0$ in the second-derivative test for local extrema for a function of one variable (Section 4.5), which implies that the function is concave downward and therefore has a local maximum. Similarly, the condition $A = f_{xx}(a, b) > 0$ in case 2 is analogous to the condition $f''(c) > 0$ in the earlier second-derivative test, which implies that the function is concave upward and therefore has a local minimum.

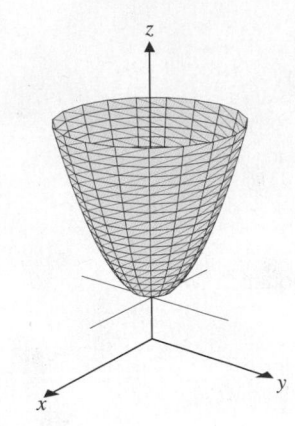

Figure 3

To illustrate the use of Theorem 2, we find the local extremum for a very simple function whose solution is almost obvious: $z = f(x, y) = x^2 + y^2 + 2$. From the function f itself and its graph (Fig. 3), it is clear that a local minimum is found at $(0, 0)$. Let us see how Theorem 2 confirms this observation.

Step 1 Find critical points: Find (x, y) such that $f_x(x, y) = 0$ and $f_y(x, y) = 0$ simultaneously:

$$f_x(x, y) = 2x = 0 \qquad f_y(x, y) = 2y = 0$$
$$x = 0 \qquad\qquad y = 0$$

The only critical point is $(a, b) = (0, 0)$.

Step 2 Compute $A = f_{xx}(0, 0), B = f_{xy}(0, 0)$, and $C = f_{yy}(0, 0)$:

$$f_{xx}(x, y) = 2; \quad \text{so,} \quad A = f_{xx}(0, 0) = 2$$
$$f_{xy}(x, y) = 0; \quad \text{so,} \quad B = f_{xy}(0, 0) = 0$$
$$f_{yy}(x, y) = 2; \quad \text{so,} \quad C = f_{yy}(0, 0) = 2$$

Step 3 Evaluate $AC - B^2$ and try to classify the critical point $(0, 0)$ by using Theorem 2:

$$AC - B^2 = (2)(2) - (0)^2 = 4 > 0 \qquad \text{and} \qquad A = 2 > 0$$

Therefore, case 2 in Theorem 2 holds. That is, $f(0, 0) = 2$ is a local minimum. We will now use Theorem 2 to analyze extrema without the aid of graphs.

EXAMPLE 1 Finding Local Extrema Use Theorem 2 to find local extrema of

$$f(x, y) = -x^2 - y^2 + 6x + 8y - 21$$

SOLUTION

Step 1 Find critical points: Find (x, y) such that $f_x(x, y) = 0$ and $f_y(x, y) = 0$ simultaneously:

$$f_x(x, y) = -2x + 6 = 0 \qquad f_y(x, y) = -2y + 8 = 0$$
$$x = 3 \qquad\qquad y = 4$$

The only critical point is $(a, b) = (3, 4)$.

Step 2 Compute $A = f_{xx}(3, 4), B = f_{xy}(3, 4)$, and $C = f_{yy}(3, 4)$:

$$f_{xx}(x, y) = -2; \quad \text{so,} \quad A = f_{xx}(3, 4) = -2$$
$$f_{xy}(x, y) = 0; \quad \text{so,} \quad B = f_{xy}(3, 4) = 0$$
$$f_{yy}(x, y) = -2; \quad \text{so,} \quad C = f_{yy}(3, 4) = -2$$

Step 3 Evaluate $AC - B^2$ and try to classify the critical point $(3, 4)$ by using Theorem 2:

$$AC - B^2 = (-2)(-2) - (0)^2 = 4 > 0 \qquad \text{and} \qquad A = -2 < 0$$

Therefore, case 1 in Theorem 2 holds, and $f(3, 4) = 4$ is a local maximum.

Matched Problem 1 Use Theorem 2 to find local extrema of

$$f(x, y) = x^2 + y^2 - 10x - 2y + 36$$

EXAMPLE 2 Finding Local Extrema: Multiple Critical Points Use Theorem 2 to find local extrema of

$$f(x, y) = x^3 + y^3 - 6xy$$

SOLUTION

Step 1 Find critical points of $f(x, y) = x^3 + y^3 - 6xy$:

$$f_x(x, y) = 3x^2 - 6y = 0 \qquad \text{Solve for } y.$$
$$6y = 3x^2$$
$$y = \tfrac{1}{2}x^2 \qquad\qquad\qquad (2)$$
$$f_y(x, y) = 3y^2 - 6x = 0$$
$$3y^2 = 6x \qquad \text{Use equation (2) to eliminate } y.$$
$$3\left(\tfrac{1}{2}x^2\right)^2 = 6x$$
$$\tfrac{3}{4}x^4 = 6x \qquad \text{Solve for } x.$$
$$3x^4 - 24x = 0$$
$$3x(x^3 - 8) = 0$$

$$x = 0 \quad \text{or} \quad x = 2$$
$$y = 0 \qquad\qquad y = \tfrac{1}{2}(2)^2 = 2$$

The critical points are $(0, 0)$ and $(2, 2)$.

Since there are two critical points, steps 2 and 3 must be performed twice.

Test (0, 0)

Step 2 Compute $A = f_{xx}(0, 0)$, $B = f_{xy}(0, 0)$, and $C = f_{yy}(0, 0)$:

$$f_{xx}(x, y) = 6x; \qquad \text{so,} \qquad A = f_{xx}(0, 0) = 0$$
$$f_{xy}(x, y) = -6; \qquad \text{so,} \qquad B = f_{xy}(0, 0) = -6$$
$$f_{yy}(x, y) = 6y; \qquad \text{so,} \qquad C = f_{yy}(0, 0) = 0$$

Step 3 Evaluate $AC - B^2$ and try to classify the critical point $(0, 0)$ by using Theorem 2:

$$AC - B^2 = (0)(0) - (-6)^2 = -36 < 0$$

Therefore, case 3 in Theorem 2 applies. That is, f has a saddle point at $(0, 0)$.

Now we will consider the second critical point, $(2, 2)$:

Test (2, 2)

Step 2 Compute $A = f_{xx}(2, 2)$, $B = f_{xy}(2, 2)$, and $C = f_{yy}(2, 2)$:

$$f_{xx}(x, y) = 6x; \qquad \text{so,} \qquad A = f_{xx}(2, 2) = 12$$
$$f_{xy}(x, y) = -6; \qquad \text{so,} \qquad B = f_{xy}(2, 2) = -6$$
$$f_{yy}(x, y) = 6y; \qquad \text{so,} \qquad C = f_{yy}(2, 2) = 12$$

Step 3 Evaluate $AC - B^2$ and try to classify the critical point $(2, 2)$ by using Theorem 2:

$$AC - B^2 = (12)(12) - (-6)^2 = 108 > 0 \qquad \text{and} \qquad A = 12 > 0$$

So, case 2 in Theorem 2 applies, and $f(2, 2) = -8$ is a local minimum.

Our conclusions in Example 2 may be confirmed geometrically by graphing cross sections of the function f. The cross sections of f in the planes $y = 0$, $x = 0$, $y = x$, and $y = -x$ [each of these planes contains $(0, 0)$] are represented by the graphs of the functions $f(x, 0) = x^3$, $f(0, y) = y^3$, $f(x, x) = 2x^3 - 6x^2$, and $f(x, -x) = 6x^2$, respectively, as shown in Figure 4A (note that the first two functions have the same graph). The cross sections of f in the planes $y = 2$, $x = 2$, $y = x$, and $y = 4 - x$ [each of these planes contains $(2, 2)$] are represented by the graphs of $f(x, 2) = x^3 - 12x + 8$, $f(2, y) = y^3 - 12y + 8$, $f(x, x) = 2x^3 - 6x^2$, and $f(x, 4 - x) = x^3 + (4 - x)^3 + 6x^2 - 24x$, respectively, as shown in Figure 4B (the first two functions have the same graph). Figure 4B illustrates the fact that since f has a local minimum at $(2, 2)$, each of the cross sections of f through $(2, 2)$ has a local minimum of -8 at $(2, 2)$. Figure 4A, by contrast, indicates that some cross sections of f through $(0, 0)$ have a local minimum, some a local maximum, and some neither one, at $(0, 0)$.

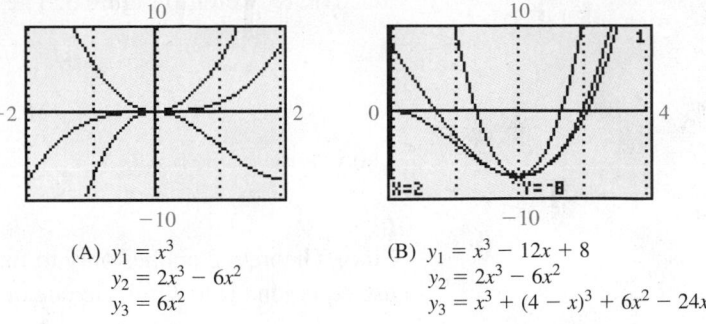

(A) $y_1 = x^3$
$y_2 = 2x^3 - 6x^2$
$y_3 = 6x^2$

(B) $y_1 = x^3 - 12x + 8$
$y_2 = 2x^3 - 6x^2$
$y_3 = x^3 + (4 - x)^3 + 6x^2 - 24x$

Figure 4

Matched Problem 2 Use Theorem 2 to find local extrema for $f(x, y) = x^3 + y^2 - 6xy$.

EXAMPLE 3 Profit Suppose that the surfboard company discussed earlier has developed the yearly profit equation

$$P(x, y) = -22x^2 + 22xy - 11y^2 + 110x - 44y - 23$$

where x is the number (in thousands) of standard surfboards produced per year, y is the number (in thousands) of competition surfboards produced per year, and P is profit (in thousands of dollars). How many of each type of board should be produced per year to realize a maximum profit? What is the maximum profit?

SOLUTION

Step 1 Find critical points:

$$P_x(x, y) = -44x + 22y + 110 = 0$$
$$P_y(x, y) = 22x - 22y - 44 = 0$$

Solving this system, we obtain $(3, 1)$ as the only critical point.

Step 2 Compute $A = P_{xx}(3, 1), B = P_{xy}(3, 1)$, and $C = P_{yy}(3, 1)$:

$$P_{xx}(x, y) = -44; \quad \text{so,} \quad A = P_{xx}(3, 1) = -44$$
$$P_{xy}(x, y) = 22; \quad \text{so,} \quad B = P_{xy}(3, 1) = 22$$
$$P_{yy}(x, y) = -22; \quad \text{so,} \quad C = P_{yy}(3, 1) = -22$$

Step 3 Evaluate $AC - B^2$ and try to classify the critical point $(3, 1)$ by using Theorem 2:

$$AC - B^2 = (-44)(-22) - 22^2 = 484 > 0 \quad \text{and} \quad A = -44 < 0$$

Therefore, case 1 in Theorem 2 applies. That is, $P(3, 1) = 120$ is a local maximum. A maximum profit of $120,000 is obtained by producing and selling 3,000 standard boards and 1,000 competition boards per year.

Matched Problem 3 Repeat Example 3 with

$$P(x, y) = -66x^2 + 132xy - 99y^2 + 132x - 66y - 19$$

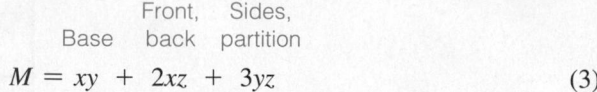

EXAMPLE 4   Package Design   The packaging department in a company is to design a rectangular box with no top and a partition down the middle. The box must have a volume of 48 cubic inches. Find the dimensions that will minimize the area of material used to construct the box.

SOLUTION   Refer to Figure 5. The area of material used in constructing this box is

$$\underset{\substack{\text{Base} \quad \text{back} \quad \text{partition}}}{M = xy \; + \; 2xz \; + \; 3yz}$$

Front, Sides,

$$M = xy \; + \; 2xz \; + \; 3yz \qquad (3)$$

The volume of the box is

$$V = xyz = 48 \qquad (4)$$

Since Theorem 2 applies only to functions with two independent variables, we must use equation (4) to eliminate one of the variables in equation (3):

$$M = xy + 2xz + 3yz \qquad \text{Substitute } z = 48/xy.$$

$$= xy + 2x\left(\frac{48}{xy}\right) + 3y\left(\frac{48}{xy}\right)$$

$$= xy + \frac{96}{y} + \frac{144}{x}$$

So, we must find the minimum value of

$$M(x, y) = xy + \frac{96}{y} + \frac{144}{x} \qquad x > 0 \qquad \text{and} \qquad y > 0$$

Step 1   Find critical points:

$$M_x(x, y) = y - \frac{144}{x^2} = 0$$

$$y = \frac{144}{x^2} \qquad (5)$$

$$M_y(x, y) = x - \frac{96}{y^2} = 0$$

$$x = \frac{96}{y^2} \qquad \text{Solve for } y^2.$$

$$y^2 = \frac{96}{x} \qquad \text{Use equation (5) to eliminate } y \text{ and solve for } x.$$

$$\left(\frac{144}{x^2}\right)^2 = \frac{96}{x}$$

$$\frac{20{,}736}{x^4} = \frac{96}{x} \qquad \text{Multiply both sides by } x^4/96 \text{ (recall that } x > 0).$$

$$x^3 = \frac{20{,}736}{96} = 216$$

$$x = 6 \qquad \text{Use equation (5) to find } y.$$

$$y = \frac{144}{36} = 4$$

Therefore, $(6, 4)$ is the only critical point.

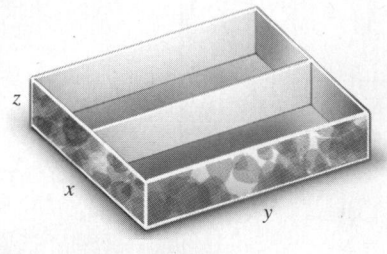

Figure 5

Step 2 Compute $A = M_{xx}(6, 4)$, $B = M_{xy}(6, 4)$, and $C = M_{yy}(6, 4)$:

$$M_{xx}(x, y) = \frac{288}{x^3}; \qquad \text{so,} \qquad A = M_{xx}(6, 4) = \tfrac{288}{216} = \tfrac{4}{3}$$

$$M_{xy}(x, y) = 1; \qquad \text{so,} \qquad B = M_{xy}(6, 4) = 1$$

$$M_{yy}(x, y) = \frac{192}{y^3}; \qquad \text{so,} \qquad C = M_{yy}(6, 4) = \tfrac{192}{64} = 3$$

Step 3 Evaluate $AC - B^2$ and try to classify the critical point $(6, 4)$ by using Theorem 2:

$$AC - B^2 = \left(\tfrac{4}{3}\right)(3) - (1)^2 = 3 > 0 \qquad \text{and} \qquad A = \tfrac{4}{3} > 0$$

Case 2 in Theorem 2 applies, and $M(x, y)$ has a local minimum at $(6, 4)$. If $x = 6$ and $y = 4$, then

$$z = \frac{48}{xy} = \frac{48}{(6)(4)} = 2$$

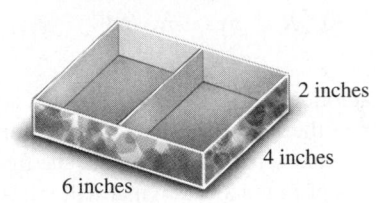

2 inches

4 inches

6 inches

Figure 6

The dimensions that will require the least material are 6 inches by 4 inches by 2 inches (see Fig. 6).

Matched Problem 4 If the box in Example 4 must have a volume of 384 cubic inches, find the dimensions that will require the least material.

Exercises 7.3

Skills Warm-up Exercises

In Problems 1–8, find $f'(0)$, $f''(0)$, and determine whether f has a local minimum, local maximum, or neither at $x = 0$. (If necessary, review the second derivative test for local extrema in Section 4.5).

1. $f(x) = 2x^3 - 9x^2 + 4$

2. $f(x) = 4x^3 + 6x^2 + 100$

3. $f(x) = \dfrac{1}{1 - x^2}$

4. $f(x) = \dfrac{1}{1 + x^2}$

5. $f(x) = e^{-x^2}$

6. $f(x) = e^{x^2}$

7. $f(x) = x^3 - x^2 + x - 1$

8. $f(x) = (3x + 1)^2$

In Problems 9–12, find $f_x(x, y)$ and $f_y(x, y)$, and explain, using Theorem 1, why $f(x, y)$ has no local extrema.

9. $f(x, y) = 4x + 5y - 6$

10. $f(x, y) = 10 - 2x - 3y + x^2$

11. $f(x, y) = 3.7 - 1.2x + 6.8y + 0.2y^3 + x^4$

12. $f(x, y) = x^3 - y^2 + 7x + 3y + 1$

Use Theorem 2 to find local extrema in Problems 13–32.

13. $f(x, y) = 6 - x^2 - 4x - y^2$

14. $f(x, y) = 3 - x^2 - y^2 + 6y$

15. $f(x, y) = x^2 + y^2 + 2x - 6y + 14$

16. $f(x, y) = x^2 + y^2 - 4x + 6y + 23$

17. $f(x, y) = xy + 2x - 3y - 2$

18. $f(x, y) = x^2 - y^2 + 2x + 6y - 4$

19. $f(x, y) = -3x^2 + 2xy - 2y^2 + 14x + 2y + 10$

20. $f(x, y) = -x^2 + xy - 2y^2 + x + 10y - 5$

21. $f(x, y) = 2x^2 - 2xy + 3y^2 - 4x - 8y + 20$

22. $f(x, y) = 2x^2 - xy + y^2 - x - 5y + 8$

23. $f(x, y) = e^{xy}$

24. $f(x, y) = x^2y - xy^2$

25. $f(x, y) = x^3 + y^3 - 3xy$

26. $f(x, y) = 2y^3 - 6xy - x^2$

27. $f(x, y) = 2x^4 + y^2 - 12xy$

28. $f(x, y) = 16xy - x^4 - 2y^2$

29. $f(x, y) = x^3 - 3xy^2 + 6y^2$

30. $f(x, y) = 2x^2 - 2x^2y + 6y^3$

31. $f(x, y) = y^3 + 2x^2y^2 - 3x - 2y + 8$

32. $f(x, y) = x \ln y + x^2 - 4x - 5y + 3$

33. Explain why $f(x, y) = x^2$ has a local extremum at infinitely many points.

34. (A) Find the local extrema of the functions $f(x, y) = x + y$, $g(x, y) = x^2 + y^2$, and $h(x, y) = x^3 + y^3$.

    (B) Discuss the local extrema of the function $k(x, y) = x^n + y^n$, where n is a positive integer.

35. (A) Show that $(0, 0)$ is a critical point of the function $f(x, y) = x^4 e^y + x^2 y^4 + 1$, but that the second-derivative test for local extrema fails.

(B) Use cross sections, as in Example 2, to decide whether f has a local maximum, a local minimum, or a saddle point at $(0, 0)$.

36. (A) Show that $(0, 0)$ is a critical point of the function $g(x, y) = e^{xy^2} + x^2 y^3 + 2$, but that the second-derivative test for local extrema fails.

(B) Use cross sections, as in Example 2, to decide whether g has a local maximum, a local minimum, or a saddle point at $(0, 0)$.

Applications

37. Product mix for maximum profit. A firm produces two types of earphones per year: x thousand of type A and y thousand of type B. If the revenue and cost equations for the year are (in millions of dollars)

$$R(x, y) = 2x + 3y$$
$$C(x, y) = x^2 - 2xy + 2y^2 + 6x - 9y + 5$$

determine how many of each type of earphone should be produced per year to maximize profit. What is the maximum profit?

38. Automation–labor mix for minimum cost. The annual labor and automated equipment cost (in millions of dollars) for a company's production of HDTVs is given by

$$C(x, y) = 2x^2 + 2xy + 3y^2 - 16x - 18y + 54$$

where x is the amount spent per year on labor and y is the amount spent per year on automated equipment (both in millions of dollars). Determine how much should be spent on each per year to minimize this cost. What is the minimum cost?

39. Maximizing profit. A store sells two brands of camping chairs. The store pays $60 for each brand A chair and $80 for each brand B chair. The research department has estimated the weekly demand equations for these two competitive products to be

$x = 260 - 3p + q$ Demand equation for brand A
$y = 180 + p - 2q$ Demand equation for brand B

where p is the selling price for brand A and q is the selling price for brand B.

(A) Determine the demands x and y when $p = \$100$ and $q = \$120$; when $p = \$110$ and $q = \$110$.

(B) How should the store price each chair to maximize weekly profits? What is the maximum weekly profit? [*Hint:* $C = 60x + 80y$, $R = px + qy$, and $P = R - C$.]

40. Maximizing profit. A store sells two brands of laptop sleeves. The store pays $25 for each brand A sleeve and $30 for each brand B sleeve. A consulting firm has estimated

the daily demand equations for these two competitive products to be

$x = 130 - 4p + q$ Demand equation for brand A
$y = 115 + 2p - 3q$ Demand equation for brand B

where p is the selling price for brand A and q is the selling price for brand B.

(A) Determine the demands x and y when $p = \$40$ and $q = \$50$; when $p = \$45$ and $q = \$55$.

(B) How should the store price each brand of sleeve to maximize daily profits? What is the maximum daily profit? [*Hint:* $C = 25x + 30y$, $R = px + qy$, and $P = R - C$.]

41. Minimizing cost. A satellite TV station is to be located at $P(x, y)$ so that the sum of the squares of the distances from P to the three towns A, B, and C is a minimum (see the figure). Find the coordinates of P, the location that will minimize the cost of providing satellite TV for all three towns.

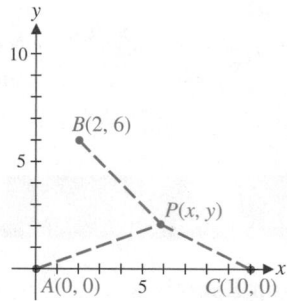

42. Minimizing cost. Repeat Problem 41, replacing the coordinates of B with $B(6, 9)$ and the coordinates of C with $C(9, 0)$.

43. Minimum material. A rectangular box with no top and two parallel partitions (see the figure) must hold a volume of 64 cubic inches. Find the dimensions that will require the least material.

44. Minimum material. A rectangular box with no top and two intersecting partitions (see the figure) must hold a volume of 72 cubic inches. Find the dimensions that will require the least material.

45. Maximum volume. A mailing service states that a rectangular package cannot have the sum of its length and girth exceed 120 inches (see the figure). What are the dimensions of the largest (in volume) mailing carton that can be constructed to meet these restrictions?

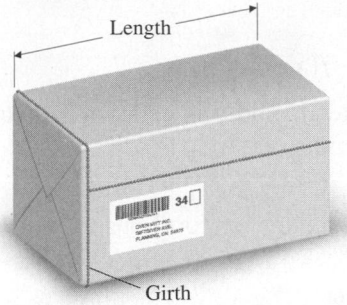

Length

Girth

46. Maximum shipping volume. A shipping box is to be reinforced with steel bands in all three directions, as shown in the figure. A total of 150 inches of steel tape is to be used, with 6 inches of waste because of a 2-inch overlap in each direction.

Find the dimensions of the box with maximum volume that can be taped as described.

Answers to Matched Problems

1. $f(5, 1) = 10$ is a local minimum
2. f has a saddle point at $(0, 0)$; $f(6, 18) = -108$ is a local minimum
3. Local maximum for $x = 2$ and $y = 1$; $P(2, 1) = 80$; a maximum profit of $80,000 is obtained by producing and selling 2,000 standard boards and 1,000 competition boards
4. 12 in. by 8 in. by 4 in.

7.4 Maxima and Minima Using Lagrange Multipliers

- Functions of Two Independent Variables

- Functions of Three Independent Variables

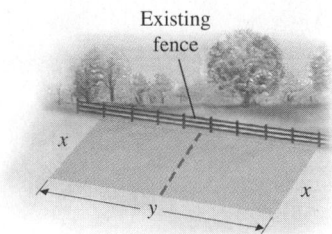

Existing fence

x

y

x

Figure 1

Functions of Two Independent Variables

We now consider a powerful method of solving a certain class of maxima–minima problems. Joseph Louis Lagrange (1736–1813), an eighteenth-century French mathematician, discovered this method, called the **method of Lagrange multipliers**. We introduce the method through an example.

A rancher wants to construct two feeding pens of the same size along an existing fence (see Fig. 1). If the rancher has 720 feet of fencing materials available, how long should x and y be in order to obtain the maximum total area? What is the maximum area?

The total area is given by

$$f(x, y) = xy$$

which can be made as large as we like, provided that there are no restrictions on x and y. But there are restrictions on x and y, since we have only 720 feet of fencing. The variables x and y must be chosen so that

$$3x + y = 720$$

This restriction on x and y, called a **constraint**, leads to the following maxima–minima problem:

$$\text{Maximize} \quad f(x, y) = xy \tag{1}$$
$$\text{subject to} \quad 3x + y = 720, \quad \text{or} \quad 3x + y - 720 = 0 \tag{2}$$

This problem is one of a general class of problems of the form

$$\text{Maximize (or minimize)} \quad z = f(x, y) \tag{3}$$
$$\text{subject to} \quad g(x, y) = 0 \tag{4}$$

Of course, we could try to solve equation (4) for y in terms of x, or for x in terms of y, then substitute the result into equation (3), and use methods developed in Section 4.5 for functions of a single variable. But what if equation (4) is more complicated than equation (2), and solving for one variable in terms of the other is either very difficult or impossible? In the method of Lagrange multipliers, we will work with $g(x, y)$ directly and avoid solving equation (4) for one variable in terms of the other. In addition, the method generalizes to functions of arbitrarily many variables subject to one or more constraints.

Now to the method: We form a new function F, using functions f and g in equations (3) and (4), as follows:

$$F(x, y, \lambda) = f(x, y) + \lambda g(x, y) \tag{5}$$

Here, λ (the Greek lowercase letter lambda) is called a **Lagrange multiplier**. Theorem 1 gives the basis for the method.

THEOREM 1   Method of Lagrange Multipliers for Functions of Two Variables

Any local maxima or minima of the function $z = f(x, y)$ subject to the constraint $g(x, y) = 0$ will be among those points (x_0, y_0) for which (x_0, y_0, λ_0) is a solution of the system

$$F_x(x, y, \lambda) = 0$$
$$F_y(x, y, \lambda) = 0$$
$$F_\lambda(x, y, \lambda) = 0$$

where $F(x, y, \lambda) = f(x, y) + \lambda g(x, y)$, provided that all the partial derivatives exist.

We now use the method of Lagrange multipliers to solve the fence problem.

Step 1   Formulate the problem in the form of equations (3) and (4):

$$\text{Maximize} \quad f(x, y) = xy$$
$$\text{subject to} \quad g(x, y) = 3x + y - 720 = 0$$

Step 2   Form the function F, introducing the Lagrange multiplier λ:

$$F(x, y, \lambda) = f(x, y) + \lambda g(x, y)$$
$$= xy + \lambda(3x + y - 720)$$

Step 3   Solve the system $F_x = 0, F_y = 0, F_\lambda = 0$ (the solutions are **critical points** of F):

$$F_x = y + 3\lambda = 0$$
$$F_y = x + \lambda = 0$$
$$F_\lambda = 3x + y - 720 = 0$$

From the first two equations, we see that

$$y = -3\lambda$$
$$x = -\lambda$$

Substitute these values for x and y into the third equation and solve for λ:

$$-3\lambda - 3\lambda = 720$$
$$-6\lambda = 720$$
$$\lambda = -120$$

So,

$$y = -3(-120) = 360 \text{ feet}$$
$$x = -(-120) = 120 \text{ feet}$$

and $(x_0, y_0, \lambda_0) = (120, 360, -120)$ is the only critical point of F.

Step 4   According to Theorem 1, if the function $f(x, y)$, subject to the constraint $g(x, y) = 0$, has a local maximum or minimum, that maximum or minimum must occur at $x = 120$, $y = 360$. Although it is possible to develop a test similar to Theorem 2 in Section 7.3 to determine the nature of this local extremum, we will not do so. [Note that Theorem 2 cannot be applied to $f(x, y)$ at (120, 360), since this point is not a critical point of the unconstrained function $f(x, y)$.] We simply assume that the maximum value of $f(x, y)$ must occur for $x = 120$, $y = 360$.

$$\text{Max } f(x, y) = f(120, 360)$$
$$= (120)(360) = 43{,}200 \text{ square feet}$$

The key steps in applying the method of Lagrange multipliers are as follows:

PROCEDURE Method of Lagrange Multipliers: Key Steps

Step 1 Write the problem in the form

$$\text{Maximize (or minimize)} \quad z = f(x, y)$$
$$\text{subject to} \quad g(x, y) = 0$$

Step 2 Form the function F:

$$F(x, y, \lambda) = f(x, y) + \lambda g(x, y)$$

Step 3 Find the critical points of F; that is, solve the system

$$F_x(x, y, \lambda) = 0$$
$$F_y(x, y, \lambda) = 0$$
$$F_\lambda(x, y, \lambda) = 0$$

Step 4 If (x_0, y_0, λ_0) is the only critical point of F, we assume that (x_0, y_0) will always produce the solution to the problems we consider. If F has more than one critical point, we evaluate $z = f(x, y)$ at (x_0, y_0) for each critical point (x_0, y_0, λ_0) of F. For the problems we consider, we assume that the largest of these values is the maximum value of $f(x, y)$, subject to the constraint $g(x, y) = 0$, and the smallest is the minimum value of $f(x, y)$, subject to the constraint $g(x, y) = 0$.

EXAMPLE 1 Minimization Subject to a Constraint Minimize $f(x, y) = x^2 + y^2$ subject to $x + y = 10$.

SOLUTION

Step 1
$$\text{Minimize} \quad f(x, y) = x^2 + y^2$$
$$\text{subject to} \quad g(x, y) = x + y - 10 = 0$$

Step 2
$$F(x, y, \lambda) = x^2 + y^2 + \lambda(x + y - 10)$$

Step 3
$$F_x = 2x + \lambda = 0$$
$$F_y = 2y + \lambda = 0$$
$$F_\lambda = x + y - 10 = 0$$

From the first two equations, $x = -\lambda/2$ and $y = -\lambda/2$. Substituting these values into the third equation, we obtain

$$-\frac{\lambda}{2} - \frac{\lambda}{2} = 10$$
$$-\lambda = 10$$
$$\lambda = -10$$

The only critical point is $(x_0, y_0, \lambda_0) = (5, 5, -10)$.

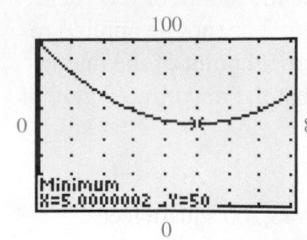

Figure 2 $h(x) = x^2 + (10 - x)^2$

Step 4 Since $(5, 5, -10)$ is the only critical point of F, we conclude that (see step 4 in the box)

$$\text{Min } f(x, y) = f(5, 5) = (5)^2 + (5)^2 = 50$$

Since $g(x, y)$ in Example 1 has a relatively simple form, an alternative to the method of Lagrange multipliers is to solve $g(x, y) = 0$ for y and then substitute into $f(x, y)$ to obtain the function $h(x) = f(x, 10 - x) = x^2 + (10 - x)^2$ in the single variable x. Then we minimize h (see Fig. 2). From Figure 2, we conclude that min $f(x, y) = f(5, 5) = 50$. This technique depends on being able to solve the constraint for one of the two variables and so is not always available as an alternative to the method of Lagrange multipliers.

Matched Problem 1 Maximize $f(x, y) = 25 - x^2 - y^2$ subject to $x + y = 4$.

Figures 3 and 4 illustrate the results obtained in Example 1 and Matched Problem 1, respectively.

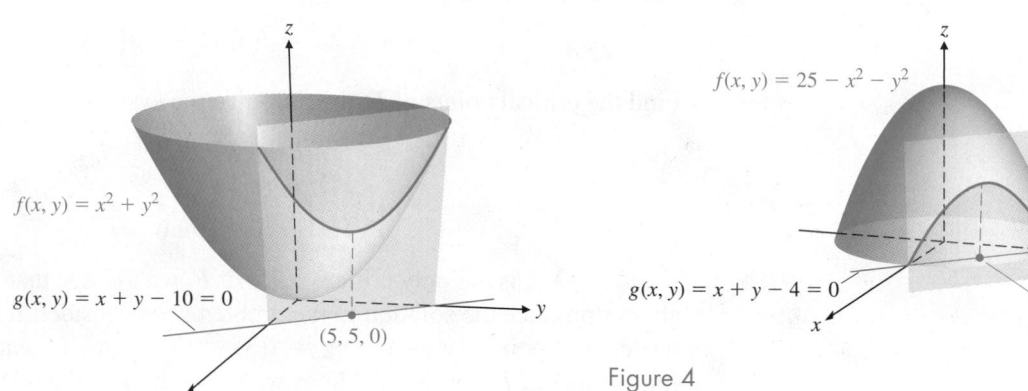

Figure 3

Figure 4

Explore and Discuss 1

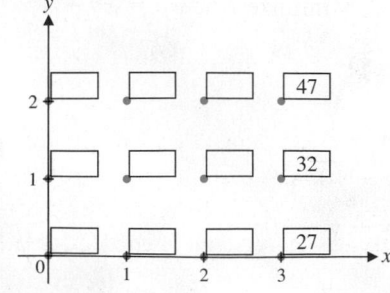

Figure 5

Consider the problem of minimizing $f(x, y) = 3x^2 + 5y^2$ subject to the constraint $g(x, y) = 2x + 3y - 6 = 0$.

(A) Compute the value of $f(x, y)$ when x and y are integers, $0 \le x \le 3, 0 \le y \le 2$. Record your answers in the empty boxes next to the points (x, y) in Figure 5.

(B) Graph the constraint $g(x, y) = 0$.

(C) Estimate the minimum value of f on the basis of your graph and the computations from part (A).

(D) Use the method of Lagrange multipliers to solve the minimization problem.

EXAMPLE 2 Productivity The Cobb–Douglas production function for a new product is given by

$$N(x, y) = 16x^{0.25}y^{0.75}$$

where x is the number of units of labor and y is the number of units of capital required to produce $N(x, y)$ units of the product. Each unit of labor costs \$50 and each unit of capital costs \$100. If \$500,000 has been budgeted for the production of this product, how should that amount be allocated between labor and capital in order to maximize production? What is the maximum number of units that can be produced?

SOLUTION The total cost of using x units of labor and y units of capital is $50x + 100y$. Thus, the constraint imposed by the \$500,000 budget is

$$50x + 100y = 500,000$$

Step 1   Maximize $N(x, y) = 16x^{0.25}y^{0.75}$
subject to $g(x, y) = 50x + 100y - 500,000 = 0$

Step 2 $F(x, y, \lambda) = 16x^{0.25}y^{0.75} + \lambda(50x + 100y - 500,000)$

Step 3 $F_x = 4x^{-0.75}y^{0.75} + 50\lambda = 0$
$F_y = 12x^{0.25}y^{-0.25} + 100\lambda = 0$
$F_\lambda = 50x + 100y - 500,000 = 0$

From the first two equations,

$$\lambda = -\tfrac{2}{25}x^{-0.75}y^{0.75} \qquad \text{and} \qquad \lambda = -\tfrac{3}{25}x^{0.25}y^{-0.25}$$

Therefore,

$$-\tfrac{2}{25}x^{-0.75}y^{0.75} = -\tfrac{3}{25}x^{0.25}y^{-0.25} \qquad \text{Multiply both sides by } x^{0.75}\,y^{0.25}.$$

$$-\tfrac{2}{25}y = -\tfrac{3}{25}x \qquad \text{(We can assume that } x \ne 0 \text{ and } y \ne 0.)$$

$$y = \tfrac{3}{2}x$$

Now substitute for y in the third equation and solve for x:

$$50x + 100\left(\tfrac{3}{2}x\right) - 500,000 = 0$$
$$200x = 500,000$$
$$x = 2,500$$

So,

$$y = \tfrac{3}{2}(2,500) = 3,750$$

and

$$\lambda = -\tfrac{2}{25}(2,500)^{-0.75}(3,750)^{0.75} \approx -0.1084$$

The only critical point of F is $(2,500, 3,750, -0.1084)$.

Step 4   Since F has only one critical point, we conclude that maximum productivity occurs when 2,500 units of labor and 3,750 units of capital are used (see step 4 in the method of Lagrange multipliers).

$$\text{Max } N(x, y) = N(2,500, 3,750)$$
$$= 16(2,500)^{0.25}(3,750)^{0.75}$$
$$\approx 54,216 \text{ units}$$

The negative of the value of the Lagrange multiplier found in step 3 is called the **marginal productivity of money** and gives the approximate increase in production for each additional dollar spent on production. In Example 2, increasing the production budget from \$500,000 to \$600,000 would result in an approximate increase in production of

$$0.1084(100,000) = 10,840 \text{ units}$$

Note that simplifying the constraint equation

$$50x + 100y - 500,000 = 0$$

to

$$x + 2y - 10,000 = 0$$

before forming the function $F(x, y, \lambda)$ would make it difficult to interpret $-\lambda$ correctly. **In marginal productivity problems, the constraint equation should not be simplified.**

Matched Problem 2) The Cobb–Douglas production function for a new product is given by
$$N(x, y) = 20x^{0.5}y^{0.5}$$
where x is the number of units of labor and y is the number of units of capital required to produce $N(x, y)$ units of the product. Each unit of labor costs $40 and each unit of capital costs $120.

(A) If $300,000 has been budgeted for the production of this product, how should that amount be allocated in order to maximize production? What is the maximum production?

(B) Find the marginal productivity of money in this case, and estimate the increase in production if an additional $40,000 is budgeted for production.

Explore and Discuss 2   Consider the problem of maximizing $f(x, y) = 4 - x^2 - y^2$ subject to the constraint $g(x, y) = y - x^2 + 1 = 0$.

(A) Explain why $f(x, y) = 3$ whenever (x, y) is a point on the circle of radius 1 centered at the origin. What is the value of $f(x, y)$ when (x, y) is a point on the circle of radius 2 centered at the origin? On the circle of radius 3 centered at the origin? (See Fig. 6.)

(B) Explain why some points on the parabola $y - x^2 + 1 = 0$ lie inside the circle $x^2 + y^2 = 1$.

(C) In light of part (B), would you guess that the maximum value of $f(x, y)$ subject to the constraint is greater than 3? Explain.

(D) Use Lagrange multipliers to solve the maximization problem.

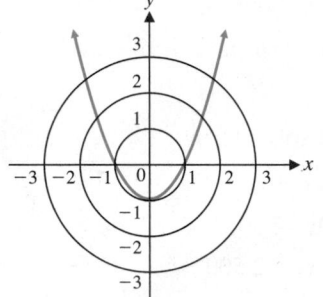

Figure 6

Functions of Three Independent Variables

The method of Lagrange multipliers can be extended to functions with arbitrarily many independent variables with one or more constraints. We now state a theorem for functions with three independent variables and one constraint, and we consider an example that demonstrates the advantage of the method of Lagrange multipliers over the method used in Section 7.3.

THEOREM 2   Method of Lagrange Multipliers for Functions of Three Variables

Any local maxima or minima of the function $w = f(x, y, z)$, subject to the constraint $g(x, y, z) = 0$, will be among the set of points (x_0, y_0, z_0) for which $(x_0, y_0, z_0, \lambda_0)$ is a solution of the system

$$F_x(x, y, z, \lambda) = 0$$
$$F_y(x, y, z, \lambda) = 0$$
$$F_z(x, y, z, \lambda) = 0$$
$$F_\lambda(x, y, z, \lambda) = 0$$

where $F(x, y, z, \lambda) = f(x, y, z) + \lambda g(x, y, z)$, provided that all the partial derivatives exist.

EXAMPLE 3 Package Design A rectangular box with an open top and one partition is to be constructed from 162 square inches of cardboard (Fig. 7). Find the dimensions that result in a box with the largest possible volume.

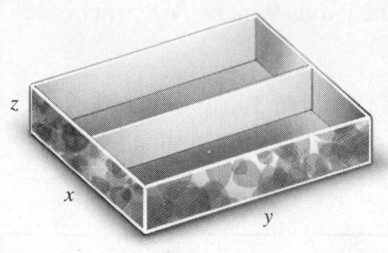

Figure 7

SOLUTION We must maximize

$$V(x, y, z) = xyz$$

subject to the constraint that the amount of material used is 162 square inches. So x, y, and z must satisfy

$$xy + 2xz + 3yz = 162$$

Step 1   Maximize $V(x, y, z) = xyz$

subject to $g(x, y, z) = xy + 2xz + 3yz - 162 = 0$

Step 2 $F(x, y, z, \lambda) = xyz + \lambda(xy + 2xz + 3yz - 162)$

Step 3 $F_x = yz + \lambda(y + 2z) = 0$

$F_y = xz + \lambda(x + 3z) = 0$

$F_z = xy + \lambda(2x + 3y) = 0$

$F_\lambda = xy + 2xz + 3yz - 162 = 0$

From the first two equations, we can write

$$\lambda = \frac{-yz}{y + 2z} \qquad \lambda = \frac{-xz}{x + 3z}$$

Eliminating λ, we have

$$\frac{-yz}{y + 2z} = \frac{-xz}{x + 3z}$$

$$-xyz - 3yz^2 = -xyz - 2xz^2$$

$$3yz^2 = 2xz^2 \qquad \text{We can assume that } z \neq 0.$$

$$3y = 2x$$

$$x = \tfrac{3}{2}y$$

From the second and third equations,

$$\lambda = \frac{-xz}{x + 3z} \qquad \lambda = \frac{-xy}{2x + 3y}$$

Eliminating λ, we have

$$\frac{-xz}{x + 3z} = \frac{-xy}{2x + 3y}$$

$$-2x^2z - 3xyz = -x^2y - 3xyz$$

$$2x^2z = x^2y \qquad \text{We can assume that } x \neq 0.$$

$$2z = y$$

$$z = \tfrac{1}{2}y$$

Substituting $x = \tfrac{3}{2}y$ and $z = \tfrac{1}{2}y$ into the fourth equation, we have

$$\left(\tfrac{3}{2}y\right)y + 2\left(\tfrac{3}{2}y\right)\left(\tfrac{1}{2}y\right) + 3y\left(\tfrac{1}{2}y\right) - 162 = 0$$

$$\tfrac{3}{2}y^2 + \tfrac{3}{2}y^2 + \tfrac{3}{2}y^2 = 162$$

$$y^2 = 36 \qquad \text{We can assume that } y > 0.$$

$$y = 6$$

$$x = \tfrac{3}{2}(6) = 9 \qquad \text{Using } x = \tfrac{3}{2}y$$

$$z = \tfrac{1}{2}(6) = 3 \qquad \text{Using } z = \tfrac{1}{2}y$$

and finally,

$$\lambda = \frac{-(6)(3)}{6 + 2(3)} = -\frac{3}{2} \qquad \text{Using } \lambda = \frac{-yz}{y + 2z}$$

The only critical point of F with x, y, and z all positive is $(9, 6, 3, -\tfrac{3}{2})$.

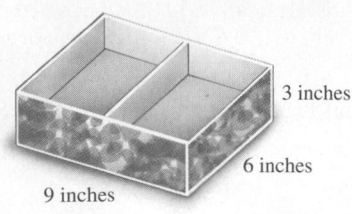

3 inches

6 inches

9 inches

Figure 8

Step 4 The box with the maximum volume has dimensions 9 inches by 6 inches by 3 inches (see Fig. 8).

Matched Problem 3⌋ A box of the same type as in Example 3 is to be constructed from 288 square inches of cardboard. Find the dimensions that result in a box with the largest possible volume.

CONCEPTUAL INSIGHT

An alternative to the method of Lagrange multipliers would be to solve Example 3 by means of Theorem 2 (the second-derivative test for local extrema) in Section 7.3. That approach involves solving the material constraint for one of the variables, say, z:

$$z = \frac{162 - xy}{2x + 3y}$$

Then we would eliminate z in the volume function to obtain a function of two variables:

$$V(x, y) = xy\,\frac{162 - xy}{2x + 3y}$$

The method of Lagrange multipliers allows us to avoid the formidable tasks of calculating the partial derivatives of V and finding the critical points of V in order to apply Theorem 2.

Exercises 7.4

Skills Warm-up Exercises

W *In Problems 1–6, maximize or minimize subject to the constraint without using the method of Lagrange multipliers; instead, solve the constraint for x or y and substitute into f(x, y). (If necessary, review Section 1.3).*

1. Minimize $f(x, y) = x^2 + xy + y^2$
    subject to $y = 4$

2. Maximize $f(x, y) = 64 + x^2 + 3xy - y^2$
    subject to $x = 6$

3. Minimize $f(x, y) = 4xy$
    subject to $x - y = 2$

4. Maximize $f(x, y) = 3xy$
    subject to $x + y = 1$

5. Maximize $f(x, y) = 2x + y$
    subject to $x^2 + y = 1$

6. Minimize $f(x, y) = 10x - y^2$
    subject to $x^2 + y^2 = 25$

Use the method of Lagrange multipliers in Problems 7–10.

7. Maximize $f(x, y) = 2xy$
    subject to $x + y = 6$

8. Minimize $f(x, y) = 6xy$
    subject to $y - x = 6$

9. Minimize $f(x, y) = x^2 + y^2$
    subject to $3x + 4y = 25$

10. Maximize $f(x, y) = 25 - x^2 - y^2$
    subject to $2x + y = 10$

In Problems 11 and 12, use Theorem 1 to explain why no maxima or minima exist.

11. Minimize $f(x, y) = 4y - 3x$
    subject to $2x + 5y = 3$

12. Maximize $f(x, y) = 6x + 5y + 24$
    subject to $3x + 2y = 4$

Use the method of Lagrange multipliers in Problems 13–22.

13. Find the maximum and minimum of $f(x, y) = 2xy$ subject to $x^2 + y^2 = 18$.

14. Find the maximum and minimum of $f(x, y) = x^2 - y^2$ subject to $x^2 + y^2 = 25$.

15. Maximize the product of two numbers if their sum must be 10.

16. Minimize the product of two numbers if their difference must be 10.

17. Minimize $f(x, y, z) = x^2 + y^2 + z^2$
    subject to $2x - y + 3z = -28$

18. Maximize $f(x, y, z) = xyz$
    subject to $2x + y + 2z = 120$

19. Maximize and Minimize $f(x, y, z) = x + y + z$
    subject to $x^2 + y^2 + z^2 = 12$

20. Maximize and Minimize $f(x, y, z) = 2x + 4y + 4z$
    subject to $x^2 + y^2 + z^2 = 9$

21. Maximize $f(x, y) = y + xy^2$

subject to $x + y^2 = 1$

22. Maximize and Minimize $f(x, y) = x + e^y$

subject to $x^2 + y^2 = 1$

In Problems 23 and 24, use Theorem 1 to explain why no maxima or minima exist.

23. Maximize $f(x, y) = e^x + 3e^y$

subject to $x - 2y = 6$

24. Minimize $f(x, y) = x^3 + 2y^3$

subject to $6x - 2y = 1$

25. Consider the problem of maximizing $f(x, y)$ subject to $g(x, y) = 0$, where $g(x, y) = y - 5$. Explain how the maximization problem can be solved without using the method of Lagrange multipliers.

26. Consider the problem of minimizing $f(x, y)$ subject to $g(x, y) = 0$, where $g(x, y) = 4x - y + 3$. Explain how the minimization problem can be solved without using the method of Lagrange multipliers.

27. Consider the problem of maximizing $f(x, y) = e^{-(x^2+y^2)}$ subject to the constraint $g(x, y) = x^2 + y - 1 = 0$.

(A) Solve the constraint equation for y, and then substitute into $f(x, y)$ to obtain a function $h(x)$ of the single variable x. Solve the original maximization problem by maximizing h (round answers to three decimal places).

(B) Confirm your answer by the method of Lagrange multipliers.

28. Consider the problem of minimizing

$$f(x, y) = x^2 + 2y^2$$

subject to the constraint $g(x, y) = ye^{x^2} - 1 = 0$.

(A) Solve the constraint equation for y, and then substitute into $f(x, y)$ to obtain a function $h(x)$ of the single variable x. Solve the original minimization problem by minimizing h (round answers to three decimal places).

(B) Confirm your answer by the method of Lagrange multipliers.

Applications

29. Budgeting for least cost. A manufacturing company produces two models of an HDTV per week, x units of model A and y units of model B at a cost (in dollars) of

$$C(x, y) = 6x^2 + 12y^2$$

If it is necessary (because of shipping considerations) that

$$x + y = 90$$

how many of each type of set should be manufactured per week to minimize cost? What is the minimum cost?

30. Budgeting for maximum production. A manufacturing firm has budgeted $60,000 per month for labor and materials.

If x thousand is spent on labor and y thousand is spent on materials, and if the monthly output (in units) is given by

$$N(x, y) = 4xy - 8x$$

then how should the $60,000 be allocated to labor and materials in order to maximize N? What is the maximum N?

31. Productivity. A consulting firm for a manufacturing company arrived at the following Cobb–Douglas production function for a particular product:

$$N(x, y) = 50x^{0.8}y^{0.2}$$

In this equation, x is the number of units of labor and y is the number of units of capital required to produce $N(x, y)$ units of the product. Each unit of labor costs $40 and each unit of capital costs $80.

(A) If $400,000 is budgeted for production of the product, determine how that amount should be allocated to maximize production, and find the maximum production.

(B) Find the marginal productivity of money in this case, and estimate the increase in production if an additional $50,000 is budgeted for the production of the product.

32. Productivity. The research department of a manufacturing company arrived at the following Cobb–Douglas production function for a particular product:

$$N(x, y) = 10x^{0.6}y^{0.4}$$

In this equation, x is the number of units of labor and y is the number of units of capital required to produce $N(x, y)$ units of the product. Each unit of labor costs $30 and each unit of capital costs $60.

(A) If $300,000 is budgeted for production of the product, determine how that amount should be allocated to maximize production, and find the maximum production.

(B) Find the marginal productivity of money in this case, and estimate the increase in production if an additional $80,000 is budgeted for the production of the product.

33. Maximum volume. A rectangular box with no top and two intersecting partitions is to be constructed from 192 square inches of cardboard (see the figure). Find the dimensions that will maximize the volume.

34. Maximum volume. A mailing service states that a rectangular package shall have the sum of its length and girth not to exceed 120 inches (see the figure). What are the dimensions of the largest (in volume) mailing carton that can be constructed to meet these restrictions?

Figure for 34

35. Agriculture. Three pens of the same size are to be built along an existing fence (see the figure). If 400 feet of fencing is available, what length should x and y be to produce the maximum total area? What is the maximum area?

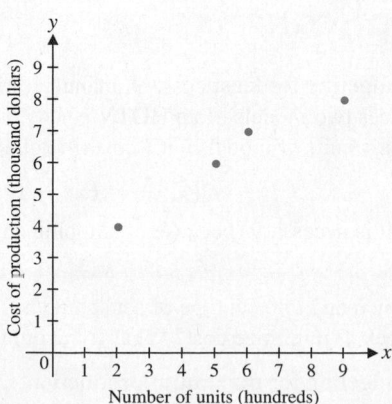

36. Diet and minimum cost. A group of guinea pigs is to receive 25,600 calories per week. Two available foods produce $200xy$ calories for a mixture of x kilograms of type M food and y kilograms of type N food. If type M costs \$1 per kilogram and type N costs \$2 per kilogram, how much of each type of food should be used to minimize weekly food costs? What is the minimum cost?

Note: $x \geq 0, y \geq 0$

Answers to Matched Problems

1. Max $f(x, y) = f(2, 2) = 17$ (see Fig. 4)
2. (A) 3,750 units of labor and 1,250 units of capital;
   Max $N(x, y) = N(3,750, 1,250) \approx 43,301$ units
   (B) Marginal productivity of money ≈ 0.1443; increase in production $\approx 5,774$ units
3. 12 in. by 8 in. by 4 in.

7.5 Method of Least Squares

- Least Squares Approximation
- Applications

Least Squares Approximation

Regression analysis is the process of fitting an elementary function to a set of data points by the **method of least squares**. The mechanics of using regression techniques were introduced in Chapter 1. Now, using the optimization techniques of Section 7.3, we can develop and explain the mathematical foundation of the method of least squares. We begin with **linear regression**, the process of finding the equation of the line that is the "best" approximation to a set of data points.

Suppose that a manufacturer wants to approximate the cost function for a product. The value of the cost function has been determined for certain levels of production, as listed in Table 1. Although these points do not all lie on a line (see Fig. 1), they are very close to being linear. The manufacturer would like to approximate the cost function by a linear function—that is, determine values a and b so that the line

$$y = ax + b$$

is, in some sense, the "best" approximation to the cost function.

Table 1

Number of Units x (hundreds)	Cost y (thousand \$)
2	4
5	6
6	7
9	8

Figure 1

What do we mean by "best"? Since the line $y = ax + b$ will not go through all four points, it is reasonable to examine the differences between the y coordinates of the points listed in the table and the y coordinates of the corresponding points on the line. Each of these differences is called the **residual** at that point (see Fig. 2). For example, at $x = 2$, the point from Table 1 is $(2, 4)$ and the point on the line is $(2, 2a + b)$, so the residual is

$$4 - (2a + b) = 4 - 2a - b$$

All the residuals are listed in Table 2.

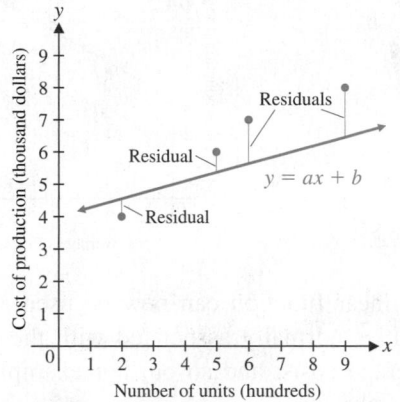

Table 2

x	y	$ax + b$	Residual
2	4	$2a + b$	$4 - 2a - b$
5	6	$5a + b$	$6 - 5a - b$
6	7	$6a + b$	$7 - 6a - b$
9	8	$9a + b$	$8 - 9a - b$

Figure 2

Our criterion for the "best" approximation is the following: Determine the values of a and b that *minimize the sum of the squares* of the residuals. The resulting line is called the **least squares line**, or the **regression line**. To this end, we minimize

$$F(a, b) = (4 - 2a - b)^2 + (6 - 5a - b)^2 + (7 - 6a - b)^2 + (8 - 9a - b)^2$$

Step 1 Find critical points:

$$\begin{aligned}
F_a(a, b) &= 2(4 - 2a - b)(-2) + 2(6 - 5a - b)(-5) \\
&\quad + 2(7 - 6a - b)(-6) + 2(8 - 9a - b)(-9) \\
&= -304 + 292a + 44b = 0 \\
F_b(a, b) &= 2(4 - 2a - b)(-1) + 2(6 - 5a - b)(-1) \\
&\quad + 2(7 - 6a - b)(-1) + 2(8 - 9a - b)(-1) \\
&= -50 + 44a + 8b = 0
\end{aligned}$$

After dividing each equation by 2, we solve the system

$$146a + 22b = 152$$
$$22a + 4b = 25$$

obtaining $(a, b) = (0.58, 3.06)$ as the only critical point.

Step 2 Compute $A = F_{aa}(a, b), B = F_{ab}(a, b)$, and $C = F_{bb}(a, b)$:

$$\begin{aligned}
F_{aa}(a, b) &= 292; &\text{so,}&& A &= F_{aa}(0.58, 3.06) = 292 \\
F_{ab}(a, b) &= 44; &\text{so,}&& B &= F_{ab}(0.58, 3.06) = 44 \\
F_{bb}(a, b) &= 8; &\text{so,}&& C &= F_{bb}(0.58, 3.06) = 8
\end{aligned}$$

Step 3 Evaluate $AC - B^2$ and try to classify the critical point (a, b) by using Theorem 2 in Section 7.3:

$$AC - B^2 = (292)(8) - (44)^2 = 400 > 0 \quad \text{and} \quad A = 292 > 0$$

Therefore, case 2 in Theorem 2 applies, and $F(a, b)$ has a local minimum at the critical point $(0.58, 3.06)$.

So, the least squares line for the given data is

$$y = 0.58x + 3.06 \quad \text{Least squares line}$$

The sum of the squares of the residuals is minimized for this choice of a and b (see Fig. 3).

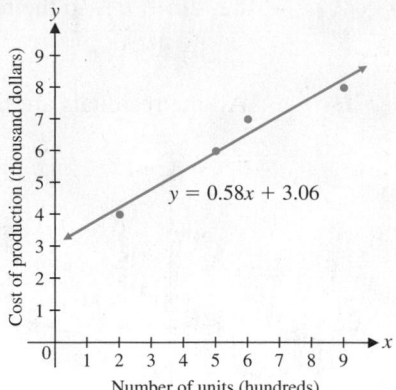

Figure 3

This linear function can now be used by the manufacturer to estimate any of the quantities normally associated with the cost function—such as costs, marginal costs, average costs, and so on. For example, the cost of producing 2,000 units is approximately

$$y = (0.58)(20) + 3.06 = 14.66, \quad \text{or} \quad \$14,660$$

The marginal cost function is

$$\frac{dy}{dx} = 0.58$$

The average cost function is

$$\bar{y} = \frac{0.58x + 3.06}{x}$$

In general, if we are given a set of n points $(x_1, y_1), (x_2, y_2), \ldots, (x_n, y_n)$, we want to determine the line $y = ax + b$ for which the sum of the squares of the residuals is minimized. Using summation notation, we find that the sum of the squares of the residuals is given by

$$F(a, b) = \sum_{k=1}^{n} (y_k - ax_k - b)^2$$

Note that in this expression the variables are a and b, and the x_k and y_k are all known values. To minimize $F(a, b)$, we thus compute the partial derivatives with respect to a and b and set them equal to 0:

$$F_a(a, b) = \sum_{k=1}^{n} 2(y_k - ax_k - b)(-x_k) = 0$$

$$F_b(a, b) = \sum_{k=1}^{n} 2(y_k - ax_k - b)(-1) = 0$$

Dividing each equation by 2 and simplifying, we see that the coefficients a and b of the least squares line $y = ax + b$ must satisfy the following system of *normal equations*:

$$\left(\sum_{k=1}^{n} x_k^2 \right)a + \left(\sum_{k=1}^{n} x_k \right)b = \sum_{k=1}^{n} x_k y_k$$

$$\left(\sum_{k=1}^{n} x_k \right)a + nb = \sum_{k=1}^{n} y_k$$

Solving this system for a and b produces the formulas given in Theorem 1.

THEOREM 1 Least Squares Approximation

For a set of n points $(x_1, y_1), (x_2, y_2), \ldots, (x_n, y_n)$, the coefficients of the least squares line $y = ax + b$ are the solutions of the system of **normal equations**

$$\left(\sum_{k=1}^{n} x_k^2 \right) a + \left(\sum_{k=1}^{n} x_k \right) b = \sum_{k=1}^{n} x_k y_k \tag{1}$$

$$\left(\sum_{k=1}^{n} x_k \right) a + nb = \sum_{k=1}^{n} y_k$$

and are given by the formulas

$$a = \frac{n \left(\sum_{k=1}^{n} x_k y_k \right) - \left(\sum_{k=1}^{n} x_k \right) \left(\sum_{k=1}^{n} y_k \right)}{n \left(\sum_{k=1}^{n} x_k^2 \right) - \left(\sum_{k=1}^{n} x_k \right)^2} \tag{2}$$

$$b = \frac{\sum_{k=1}^{n} y_k - a \left(\sum_{k=1}^{n} x_k \right)}{n} \tag{3}$$

Now we return to the data in Table 1 and tabulate the sums required for the normal equations and their solution in Table 3.

Table 3

	x_k	y_k	$x_k y_k$	x_k^2
	2	4	8	4
	5	6	30	25
	6	7	42	36
	9	8	72	81
Totals	22	25	152	146

The normal equations (1) are then

$$146a + 22b = 152$$
$$22a + 4b = 25$$

The solution of the normal equations given by equations (2) and (3) is

$$a = \frac{4(152) - (22)(25)}{4(146) - (22)^2} = 0.58$$

$$b = \frac{25 - 0.58(22)}{4} = 3.06$$

Compare these results with step 1 on page 461. Note that Table 3 provides a convenient format for the computation of step 1.

Many graphing calculators have a linear regression feature that solves the system of normal equations obtained by setting the partial derivatives of the sum of squares of the residuals equal to 0. Therefore, in practice, we simply enter the given data points and use the linear regression feature to determine the line $y = ax + b$ that best fits the data (see Fig. 4). There is no need to compute partial derivatives or even to tabulate sums (as in Table 3).

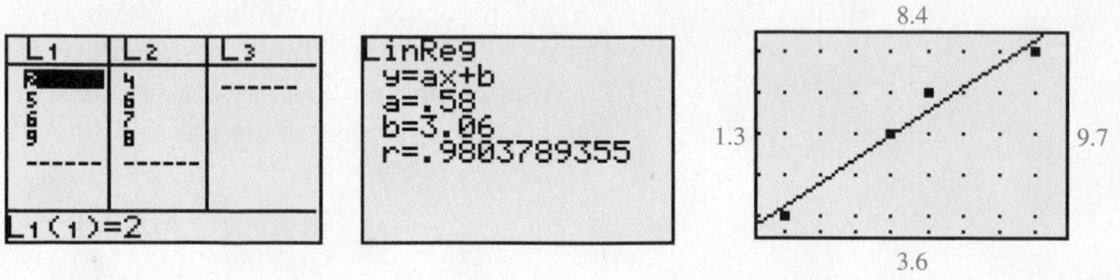

(A)   (B)   (C) $y_1 = 0.58x + 3.06$

Figure 4

Explore and Discuss 1

(A) Plot the four points $(0, 0)$, $(0, 1)$, $(10, 0)$, and $(10, 1)$. Which line would you guess "best" fits these four points? Use formulas (2) and (3) to test your conjecture.

(B) Plot the four points $(0, 0)$, $(0, 10)$, $(1, 0)$ and $(1, 10)$. Which line would you guess "best" fits these four points? Use formulas (2) and (3) to test your conjecture.

(C) If either of your conjectures was wrong, explain how your reasoning was mistaken.

CONCEPTUAL INSIGHT

Formula (2) for a is undefined if the denominator equals 0. When can this happen? Suppose $n = 3$. Then

$$n\left(\sum_{k=1}^{n} x_k^2\right) - \left(\sum_{k=1}^{n} x_k\right)^2$$

$$= 3(x_1^2 + x_2^2 + x_3^2) - (x_1 + x_2 + x_3)^2$$
$$= 3(x_1^2 + x_2^2 + x_3^2) - (x_1^2 + x_2^2 + x_3^2 + 2x_1x_2 + 2x_1x_3 + 2x_2x_3)$$
$$= 2(x_1^2 + x_2^2 + x_3^2) - (2x_1x_2 + 2x_1x_3 + 2x_2x_3)$$
$$= (x_1^2 + x_2^2) + (x_1^2 + x_3^2) + (x_2^2 + x_3^2) - (2x_1x_2 + 2x_1x_3 + 2x_2x_3)$$
$$= (x_1^2 - 2x_1x_2 + x_2^2) + (x_1^2 - 2x_1x_3 + x_3^2) + (x_2^2 - 2x_2x_3 + x_3^2)$$
$$= (x_1 - x_2)^2 + (x_1 - x_3)^2 + (x_2 - x_3)^2$$

and the last expression is equal to 0 if and only if $x_1 = x_2 = x_3$ (i.e., if and only if the three points all lie on the same vertical line). A similar algebraic manipulation works for any integer $n > 1$, showing that, in formula (2) for a, the denominator equals 0 if and only if all n points lie on the same vertical line.

The method of least squares can also be applied to find the quadratic equation $y = ax^2 + bx + c$ that best fits a set of data points. In this case, the sum of the squares of the residuals is a function of three variables:

$$F(a, b, c) = \sum_{k=1}^{n} (y_k - ax_k^2 - bx_k - c)^2$$

There are now three partial derivatives to compute and set equal to 0:

$$F_a(a, b, c) = \sum_{k=1}^{n} 2(y_k - ax_k^2 - bx_k - c)(-x_k^2) = 0$$

$$F_b(a, b, c) = \sum_{k=1}^{n} 2(y_k - ax_k^2 - bx_k - c)(-x_k) = 0$$

$$F_c(a, b, c) = \sum_{k=1}^{n} 2(y_k - ax_k^2 - bx_k - c)(-1) = 0$$

The resulting set of three linear equations in the three variables a, b, and c is called the *set of normal equations for quadratic regression.*

A quadratic regression feature on a calculator is designed to solve such normal equations after the given set of points has been entered. Figure 5 illustrates the computation for the data of Table 1.

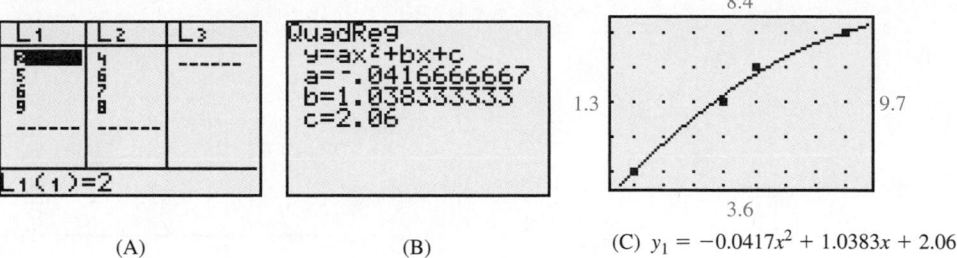

| (A) | (B) | (C) $y_1 = -0.0417x^2 + 1.0383x + 2.06$ |

Figure 5

Explore and Discuss 2 (A) Use the graphs in Figures 4 and 5 to predict which technique, linear regression or quadratic regression, yields the smaller sum of squares of the residuals for the data of Table 1. Explain.

(B) Confirm your prediction by computing the sum of squares of the residuals in each case.

The method of least squares can also be applied to other regression equations—for example, cubic, quartic, logarithmic, exponential, and power regression models. Details are explored in some of the exercises at the end of this section.

Applications

EXAMPLE 1 Exam Scores Table 4 lists the midterm and final examination scores of 10 students in a calculus course.

Table 4

Midterm	Final	Midterm	Final
49	61	78	77
53	47	83	81
67	72	85	79
71	76	91	93
74	68	99	99

(A) Use formulas (1), (2), and (3) to find the normal equations and the least squares line for the data given in Table 4.

(B) Use the linear regression feature on a graphing calculator to find and graph the least squares line.

(C) Use the least squares line to predict the final examination score of a student who scored 95 on the midterm examination.

SOLUTION

(A) Table 5 shows a convenient way to compute all the sums in the formulas for a and b.

Table 5

	x_k	y_k	$x_k y_k$	x_k^2
	49	61	2,989	2,401
	53	47	2,491	2,809
	67	72	4,824	4,489
	71	76	5,396	5,041
	74	68	5,032	5,476
	78	77	6,006	6,084
	83	81	6,723	6,889
	85	79	6,715	7,225
	91	93	8,463	8,281
	99	99	9,801	9,801
Totals	750	753	58,440	58,496

From the last line in Table 5, we have

$$\sum_{k=1}^{10} x_k = 750 \qquad \sum_{k=1}^{10} y_k = 753 \qquad \sum_{k=1}^{10} x_k y_k = 58{,}440 \qquad \sum_{k=1}^{10} x_k^2 = 58{,}496$$

and the normal equations are

$$58{,}496a + 750b = 58{,}440$$
$$750a + 10b = 753$$

Using formulas (2) and (3), we obtain

$$a = \frac{10(58{,}440) - (750)(753)}{10(58{,}496) - (750)^2} = \frac{19{,}650}{22{,}460} \approx 0.875$$

$$b = \frac{753 - 0.875(750)}{10} = 9.675$$

The least squares line is given (approximately) by

$$y = 0.875x + 9.675$$

(B) We enter the data and use the linear regression feature, as shown in Figure 6. [The discrepancy between values of a and b in the preceding calculations and those in Figure 6B is due to rounding in part (A).]

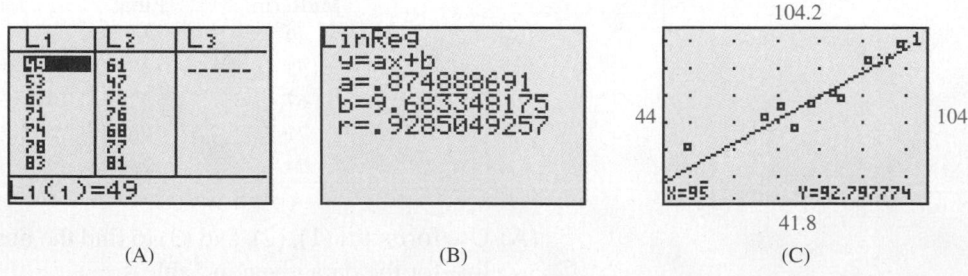

      (A)   (B)   (C)

Figure 6

(C) If $x = 95$, then $y = 0.875(95) + 9.675 \approx 92.8$ is the predicted score on the final exam. This is also indicated in Figure 6C. If we assume that the exam score must be an integer, then we would predict a score of 93.

Matched Problem 1 Repeat Example 1 for the scores listed in Table 6.

Table 6

Midterm	Final	Midterm	Final
54	50	84	80
60	66	88	95
75	80	89	85
76	68	97	94
78	71	99	86

EXAMPLE 2 Energy Consumption The use of fuel oil for home heating in the United States has declined steadily for several decades. Table 7 lists the percentage of occupied housing units in the United States that were heated by fuel oil for various years between 1960 and 2009. Use the data in the table and linear regression to estimate the percentage of occupied housing units in the United States that were heated by fuel oil in the year 1995.

Table 7 **Occupied Housing Units Heated by Fuel Oil**

Year	Percent	Year	Percent
1960	32.4	1989	13.3
1970	26.0	1999	9.8
1979	19.5	2009	7.3

SOLUTION We enter the data, with $x = 0$ representing 1960, $x = 10$ representing 1970, and so on, and use linear regression as shown in Figure 7.

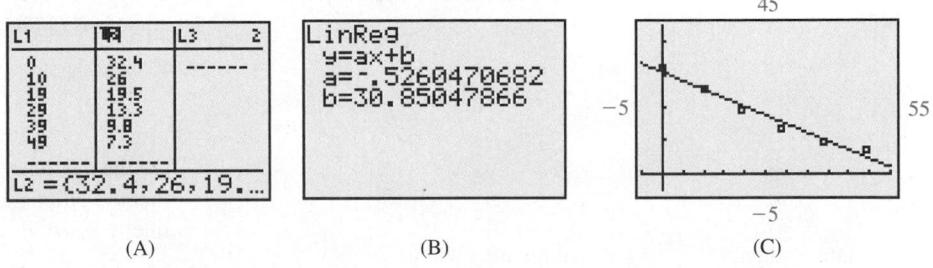

(A)   (B)   (C)

Figure 7

Figure 7 indicates that the least squares line is $y = -0.526x + 30.85$. To estimate the percentage of occupied housing units heated by fuel oil in the year 1995 (corresponding to $x = 35$), we substitute $x = 35$ in the equation of the least squares line: $-0.526(35) + 30.85 = 12.44$. The estimated percentage for 1995 is 12.44%.

Matched Problem 2 In 1950, coal was still a major source of fuel for home energy consumption, and the percentage of occupied housing units heated by fuel oil was only 22.1%. Add the data for 1950 to the data for Example 2, and compute the new least squares line and the new estimate for the percentage of occupied housing units heated by fuel oil in the year 1995. Discuss the discrepancy between the two estimates. (As in Example 2, let $x = 0$ represent 1960.)

Exercises 7.5

Skills Warm-up Exercises

Problems 1–6 refer to the n = 5 data points $(x_1, y_1) = (0, 4)$, $(x_2, y_2) = (1, 5)$, $(x_3, y_3) = (2, 7)$, $(x_4, y_4) = (3, 9)$, and $(x_5, y_5) = (4, 13)$. Calculate the indicated sum or product of sums. (If necessary, review Section B.1).

1. $\displaystyle\sum_{k=1}^{5} x_k$

2. $\displaystyle\sum_{k=1}^{5} y_k$

3. $\displaystyle\sum_{k=1}^{5} x_k y_k$

4. $\displaystyle\sum_{k=1}^{5} x_k^2$

5. $\displaystyle\sum_{k=1}^{5} x_k \sum_{k=1}^{5} y_k$

6. $\left(\displaystyle\sum_{k=1}^{5} x_k\right)^2$

In Problems 7–12, find the least squares line. Graph the data and the least squares line.

7.

x	y
1	1
2	3
3	4
4	3

8.

x	y
1	-2
2	-1
3	3
4	5

9.

x	y
1	8
2	5
3	4
4	0

10.

x	y
1	20
2	14
3	11
4	3

11.

x	y
1	3
2	4
3	5
4	6

12.

x	y
1	2
2	3
3	3
4	2

In Problems 13–20, find the least squares line and use it to estimate y for the indicated value of x. Round answers to two decimal places.

13.

x	y
1	3
2	1
2	2
3	0

Estimate y when x = 2.5.

14.

x	y
1	0
3	1
3	6
3	4

Estimate y when x = 3.

15.

x	y
0	10
5	22
10	31
15	46
20	51

Estimate y when x = 25.

16.

x	y
-5	60
0	50
5	30
10	20
15	15

Estimate y when x = 20.

17.

x	y
-1	14
1	12
3	8
5	6
7	5

Estimate y when x = 2.

18.

x	y
2	-4
6	0
10	8
14	12
18	14

Estimate y when x = 15.

19.

x	y	x	y
0.5	25	9.5	12
2	22	11	11
3.5	21	12.5	8
5	21	14	5
6.5	18	15.5	1

Estimate y when x = 8.

20.

x	y	x	y
0	-15	12	11
2	-9	14	13
4	-7	16	19
6	-7	18	25
8	-1	20	33

Estimate y when x = 10.

21. To find the coefficients of the parabola

$$y = ax^2 + bx + c$$

that is the "best" fit to the points $(1, 2)$, $(2, 1)$, $(3, 1)$, and $(4, 3)$, minimize the sum of the squares of the residuals

$$\begin{aligned}F(a, b, c) = {} & (a + b + c - 2)^2 \\ & + (4a + 2b + c - 1)^2 \\ & + (9a + 3b + c - 1)^2 \\ & + (16a + 4b + c - 3)^2\end{aligned}$$

by solving the system of normal equations

$$F_a(a, b, c) = 0 \qquad F_b(a, b, c) = 0 \qquad F_c(a, b, c) = 0$$

for a, b, and c. Graph the points and the parabola.

22. Repeat Problem 21 for the points $(-1, -2)$, $(0, 1)$, $(1, 2)$, and $(2, 0)$.

Problems 23 and 24 refer to the system of normal equations and the formulas for a and b given on page 463.

23. Verify formulas (2) and (3) by solving the system of normal equations (1) for a and b.

24. If

$$\bar{x} = \frac{1}{n}\sum_{k=1}^{n} x_k \qquad \text{and} \qquad \bar{y} = \frac{1}{n}\sum_{k=1}^{n} y_k$$

are the averages of the x and y coordinates, respectively, show that the point $(\bar{x}, \bar{y})$ satisfies the equation of the least squares line, $y = ax + b$.

25. (A) Suppose that n = 5 and the x coordinates of the data points (x_1, y_1), (x_2, y_2), ..., (x_n, y_n) are −2, −1, 0, 1, 2. Show that system (1) in the text implies that

$$a = \frac{\sum x_k y_k}{\sum x_k^2}$$

and that b is equal to the average of the values of y_k.

(B) Show that the conclusion of part (A) holds whenever the average of the x coordinates of the data points is 0.

26. (A) Give an example of a set of six data points such that half of the points lie above the least squares line and half lie below.

(B) Give an example of a set of six data points such that just one of the points lies above the least squares line and five lie below.

27. (A) Find the linear and quadratic functions that best fit the data points $(0, 1.3)$, $(1, 0.6)$, $(2, 1.5)$, $(3, 3.6)$, and $(4, 7.4)$. Round coefficients to two decimal places.

(B) Which of the two functions best fits the data? Explain.

28. (A) Find the linear, quadratic, and logarithmic functions that best fit the data points $(1, 3.2)$, $(2, 4.2)$, $(3, 4.7)$, $(4, 5.0)$, and $(5, 5.3)$. (Round coefficients to two decimal places.)

  (B) Which of the three functions best fits the data? Explain.

29. Describe the normal equations for cubic regression. How many equations are there? What are the variables? What techniques could be used to solve the equations?

30. Describe the normal equations for quartic regression. How many equations are there? What are the variables? What techniques could be used to solve the equations?

Applications

31. Crime rate. Data on U.S. property crimes (in number of crimes per 100,000 population) are given in the table for the years 2001 through 2011.

U.S. Property Crime Rates

Year	Rate
2001	3,658
2003	3,591
2005	3,431
2007	3,276
2009	3,041
2011	2,908

  (A) Find the least squares line for the data, using $x = 0$ for 2000.

  (B) Use the least squares line to predict the property crime rate in 2024.

32. Cable TV revenue. Data for cable TV revenue are given in the table for the years 2002 through 2010.

Cable TV Revenue

Year	Millions of dollars
2002	47,989
2004	58,586
2006	71,887
2008	85,232
2010	93,368

  (A) Find the least squares line for the data, using $x = 0$ for 2000.

  (B) Use the least squares line to predict cable TV revenue in 2025.

33. Maximizing profit. The market research department for a drugstore chain chose two summer resort areas to test market a new sunscreen lotion packaged in 4-ounce plastic bottles. After a summer of varying the selling price and recording the monthly demand, the research department arrived at the following demand table, where y is the number of bottles purchased per month (in thousands) at x dollars per bottle:

x	y
5.0	2.0
5.5	1.8
6.0	1.4
6.5	1.2
7.0	1.1

  (A) Use the method of least squares to find a demand equation.

  (B) If each bottle of sunscreen costs the drugstore chain $4, how should the sunscreen be priced to achieve a maximum monthly profit? [*Hint:* Use the result of part (A), with $C = 4y$, $R = xy$, and $P = R - C$.]

34. Maximizing profit. A market research consultant for a supermarket chain chose a large city to test market a new brand of mixed nuts packaged in 8-ounce cans. After a year of varying the selling price and recording the monthly demand, the consultant arrived at the following demand table, where y is the number of cans purchased per month (in thousands) at x dollars per can:

x	y
4.0	4.2
4.5	3.5
5.0	2.7
5.5	1.5
6.0	0.7

  (A) Use the method of least squares to find a demand equation.

  (B) If each can of nuts costs the supermarket chain $3, how should the nuts be priced to achieve a maximum monthly profit?

35. Olympic Games. The table gives the winning heights in the pole vault in the Olympic Games from 1980 to 2012.

Olympic Pole Vault Winning Height

Year	Height (ft)
1980	18.96
1984	18.85
1988	19.35
1992	19.02
1996	19.42
2000	19.35
2004	19.52
2008	19.56
2012	19.59

  (A) Use a graphing calculator to find the least squares line for the data, letting $x = 0$ for 1980.

  (B) Estimate the winning height in the pole vault in the Olympic Games of 2024.

36. Biology. In biology, there is an approximate rule, called the *bioclimatic rule for temperate climates*. This rule states that in spring and early summer, periodic phenomena such as the blossoming of flowers, the appearance of insects, and the ripening of fruit usually come about 4 days later for each 500 feet of altitude. Stated as a formula, the rule becomes

$$d = 8h \qquad 0 \le h \le 4$$

where d is the change in days and h is the altitude (in thousands of feet). To test this rule, an experiment was set up to record the difference in blossoming times of the same type of apple tree at different altitudes. A summary of the results is given in the following table:

h	d
0	0
1	7
2	18
3	28
4	33

(A) Use the method of least squares to find a linear equation relating h and d. Does the bioclimatic rule $d = 8h$ appear to be approximately correct?

(B) How much longer will it take this type of apple tree to blossom at 3.5 thousand feet than at sea level? [Use the linear equation found in part (A).]

37. Global warming. Average global temperatures from 1885 to 2005 are given in the table.

Average Global Temperatures

Year	°F	Year	°F
1885	56.65	1955	57.06
1895	56.64	1965	57.05
1905	56.52	1975	57.04
1915	56.57	1985	57.36
1925	56.74	1995	57.64
1935	57.00	2005	58.59
1945	57.13		

(A) Find the least squares line for the data, using $x = 0$ for 1885.

(B) Use the least squares line to estimate the average global temperature in 2085.

38. Organic Farming. The table gives the number of acres of certified organic farmland in the United States from 2000 to 2008.

Certified Organic Farmland in the United States

Year	Acres
2000	1,776,000
2002	1,926,000
2004	3,045,000
2006	2,936,000
2008	4,816,000

(A) Find the least squares line for the data, using $x = 0$ for the year 2000.

(B) Use the least squares line to estimate the number of acres of certified organic farmland in the United States in 2023.

Answers to Matched Problems

1. (A) $y = 0.85x + 9.47$

   (B)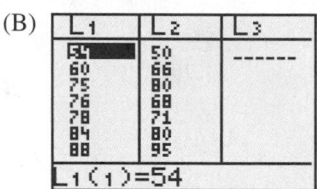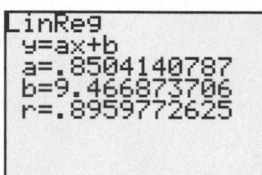

   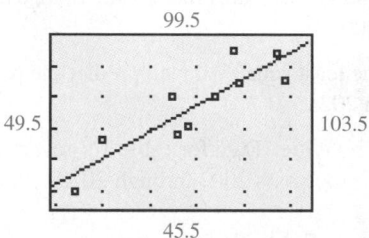

   (C) 90.3

2. $y = -0.37x + 25.86$; 12.83%

7.6 Double Integrals over Rectangular Regions

- Introduction
- Definition of the Double Integral
- Average Value over Rectangular Regions
- Volume and Double Integrals

Introduction

We have generalized the concept of differentiation to functions with two or more independent variables. How can we do the same with integration, and how can we interpret the results? Let's look first at the operation of antidifferentiation. We can antidifferentiate a function of two or more variables with respect to one of the variables by treating all the other variables as though they were constants. Thus, this operation is the reverse operation of partial differentiation, just as ordinary antidifferentiation is the reverse operation of ordinary differentiation. We write $\int f(x, y)\, dx$ to indicate that we are to antidifferentiate $f(x, y)$ with respect to x, holding y fixed; we write $\int f(x, y)\, dy$ to indicate that we are to antidifferentiate $f(x, y)$ with respect to y, holding x fixed.

EXAMPLE 1   Partial Antidifferentiation   Evaluate

(A) $\displaystyle\int (6xy^2 + 3x^2)\, dy$ (B) $\displaystyle\int (6xy^2 + 3x^2)\, dx$

SOLUTION

(A) Treating x as a constant and using the properties of antidifferentiation from Section 5.1, we have

$$\int (6xy^2 + 3x^2)\, dy = \int 6xy^2\, dy + \int 3x^2\, dy$$

> The dy tells us that we are looking for the antiderivative of $6xy^2 + 3x^2$ with respect to y only, holding x constant.

$$= 6x \int y^2\, dy + 3x^2 \int dy$$

$$= 6x \left(\frac{y^3}{3}\right) + 3x^2(y) + C(x)$$

$$= 2xy^3 + 3x^2y + C(x)$$

Note that the constant of integration can be *any function of x alone* since for any such function,

$$\frac{\partial}{\partial y} C(x) = 0$$

Check:

We can verify that our answer is correct by using partial differentiation:

$$\frac{\partial}{\partial y}[2xy^3 + 3x^2y + C(x)] = 6xy^2 + 3x^2 + 0$$

$$= 6xy^2 + 3x^2$$

(B) We treat y as a constant:

$$\int (6xy^2 + 3x^2)\, dx = \int 6xy^2\, dx + \int 3x^2\, dx$$

$$= 6y^2 \int x\, dx + 3 \int x^2\, dx$$

$$= 6y^2 \left(\frac{x^2}{2}\right) + 3\left(\frac{x^3}{3}\right) + E(y)$$

$$= 3x^2y^2 + x^3 + E(y)$$

The antiderivative contains an arbitrary function $E(y)$ of y alone.

Check:

$$\frac{\partial}{\partial x}[3x^2y^2 + x^3 + E(y)] = 6xy^2 + 3x^2 + 0$$

$$= 6xy^2 + 3x^2$$

Matched Problem 1   Evaluate

(A) $\displaystyle\int (4xy + 12x^2y^3)\, dy$ (B) $\displaystyle\int (4xy + 12x^2y^3)\, dx$

Now that we have extended the concept of antidifferentiation to functions with two variables, we also can evaluate definite integrals of the form

$$\int_a^b f(x, y)\, dx \quad \text{or} \quad \int_c^d f(x, y)\, dy$$

EXAMPLE 2 Evaluating a Partial Antiderivative Evaluate, substituting the limits of integration in y if dy is used and in x if dx is used:

(A) $\displaystyle\int_0^2 (6xy^2 + 3x^2)\, dy$

(B) $\displaystyle\int_0^1 (6xy^2 + 3x^2)\, dx$

SOLUTION

(A) From Example 1A, we know that

$$\int (6xy^2 + 3x^2)\, dy = 2xy^3 + 3x^2 y + C(x)$$

According to properties of the definite integral for a function of one variable, we can use any antiderivative to evaluate the definite integral. Thus, choosing $C(x) = 0$, we have

$$\int_0^2 (6xy^2 + 3x^2)\, dy = (2xy^3 + 3x^2 y)\big|_{y=0}^{y=2}$$

$$= [2x(2)^3 + 3x^2(2)] - [2x(0)^3 + 3x^2(0)]$$

$$= 16x + 6x^2$$

(B) From Example 1B, we know that

$$\int (6xy^2 + 3x^2)\, dx = 3x^2 y^2 + x^3 + E(y)$$

Choosing $E(y) = 0$, we have

$$\int_0^1 (6xy^2 + 3x^2)\, dx = (3x^2 y^2 + x^3)\big|_{x=0}^{x=1}$$

$$= [3y^2(1)^2 + (1)^3] - [3y^2(0)^2 + (0)^3]$$

$$= 3y^2 + 1$$

Matched Problem 2 Evaluate

(A) $\displaystyle\int_0^1 (4xy + 12x^2 y^3)\, dy$

(B) $\displaystyle\int_0^3 (4xy + 12x^2 y^3)\, dx$

Integrating and evaluating a definite integral with integrand $f(x, y)$ with respect to y produces a function of x alone (or a constant). Likewise, integrating and evaluating a definite integral with integrand $f(x, y)$ with respect to x produces a function of y alone (or a constant). Each of these results, involving at most one variable, can now be used as an integrand in a second definite integral.

EXAMPLE 3 Evaluating Integrals Evaluate

(A) $\displaystyle\int_0^1 \left[\int_0^2 (6xy^2 + 3x^2)\, dy \right] dx$

(B) $\displaystyle\int_0^2 \left[\int_0^1 (6xy^2 + 3x^2)\, dx \right] dy$

SOLUTION

(A) Example 2A showed that

$$\int_0^2 (6xy^2 + 3x^2)\, dy = 16x + 6x^2$$

Therefore,

$$\int_0^1 \left[\int_0^2 (6xy^2 + 3x^2) \, dy \right] dx = \int_0^1 (16x + 6x^2) \, dx$$

$$= (8x^2 + 2x^3)\big|_{x=0}^{x=1}$$

$$= [8(1)^2 + 2(1)^3] - [8(0)^2 + 2(0)^3] = 10$$

(B) Example 2B showed that

$$\int_0^1 (6xy^2 + 3x^2) \, dx = 3y^2 + 1$$

Therefore,

$$\int_0^2 \left[\int_0^1 (6xy^2 + 3x^2) \, dx \right] dy = \int_0^2 (3y^2 + 1) \, dy$$

$$= (y^3 + y)\big|_{y=0}^{y=2}$$

$$= [(2)^3 + 2] - [(0)^3 + 0] = 10$$

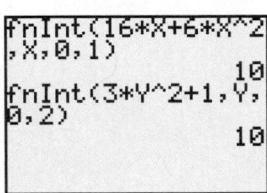

Figure 1

A numerical integration command can be used as an alternative to the fundamental theorem of calculus to evaluate the last integrals in Examples 3A and 3B, $\int_0^1 (16x + 6x^2) \, dx$ and $\int_0^2 (3y^2 + 1) \, dy$, since the integrand in each case is a function of a single variable (see Fig. 1).

Matched Problem 3) Evaluate

(A) $\displaystyle\int_0^3 \left[\int_0^1 (4xy + 12x^2y^3) \, dy \right] dx$ (B) $\displaystyle\int_0^1 \left[\int_0^3 (4xy + 12x^2y^3) \, dx \right] dy$

Definition of the Double Integral

Notice that the answers in Examples 3A and 3B are identical. This is not an accident. In fact, it is this property that enables us to define the *double integral,* as follows:

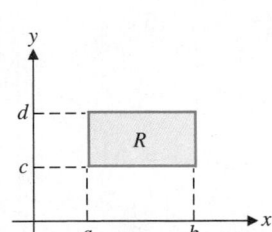

DEFINITION Double Integral
The **double integral** of a function $f(x, y)$ over a rectangle

$$R = \{ (x, y) \,|\, a \leq x \leq b, c \leq y \leq d \}$$

is

$$\iint_R f(x, y) \, dA = \int_a^b \left[\int_c^d f(x, y) \, dy \right] dx$$

$$= \int_c^d \left[\int_a^b f(x, y) \, dx \right] dy$$

In the double integral $\iint_R f(x, y) \, dA, f(x, y)$ is called the **integrand**, and R is called the **region of integration**. The expression dA indicates that this is an integral over a two-dimensional region. The integrals

$$\int_a^b \left[\int_c^d f(x, y) \, dy \right] dx \quad \text{and} \quad \int_c^d \left[\int_a^b f(x, y) \, dx \right] dy$$

are referred to as **iterated integrals** (the brackets are often omitted), and the order in which dx and dy are written indicates the order of integration. This is not the most general definition of the double integral over a rectangular region; however, it is equivalent to the general definition for all the functions we will consider.

EXAMPLE 4   Evaluating a Double Integral   Evaluate

$$\iint\limits_R (x + y)\, dA \qquad \text{over} \qquad R = \{(x, y)\,|\, 1 \le x \le 3, \ -1 \le y \le 2\}$$

SOLUTION   Region R is illustrated in Figure 2. We can choose either order of iteration. As a check, we will evaluate the integral both ways:

$$\iint\limits_R (x + y)\, dA = \int_1^3 \int_{-1}^2 (x + y)\, dy\, dx$$

$$= \int_1^3 \left[\left(xy + \frac{y^2}{2} \right) \Big|_{y=-1}^{y=2} \right] dx$$

$$= \int_1^3 \left[(2x + 2) - \left(-x + \tfrac{1}{2} \right) \right] dx$$

$$= \int_1^3 \left(3x + \tfrac{3}{2} \right) dx$$

$$= \left(\tfrac{3}{2}x^2 + \tfrac{3}{2}x \right) \Big|_{x=1}^{x=3}$$

$$= \left(\tfrac{27}{2} + \tfrac{9}{2} \right) - \left(\tfrac{3}{2} + \tfrac{3}{2} \right) = 18 - 3 = 15$$

$$\iint\limits_R (x + y)\, dA = \int_{-1}^2 \int_1^3 (x + y)\, dx\, dy$$

$$= \int_{-1}^2 \left[\left(\frac{x^2}{2} + xy \right) \Big|_{x=1}^{x=3} \right] dy$$

$$= \int_{-1}^2 \left[\left(\tfrac{9}{2} + 3y \right) - \left(\tfrac{1}{2} + y \right) \right] dy$$

$$= \int_{-1}^2 (4 + 2y)\, dy$$

$$= (4y + y^2) \Big|_{y=-1}^{y=2}$$

$$= (8 + 4) - (-4 + 1) = 12 - (-3) = 15$$

Matched Problem 4   Evaluate

$$\iint\limits_R (2x - y)\, dA \qquad \text{over} \qquad R = \{(x, y)\,|\, -1 \le x \le 5, \ 2 \le y \le 4\}$$

both ways.

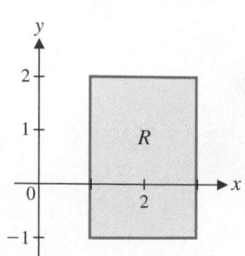

Figure 2

EXAMPLE 5   Double Integral of an Exponential Function   Evaluate

$$\iint\limits_R 2xe^{x^2+y}\, dA \qquad \text{over} \qquad R = \{(x, y)\,|\, 0 \le x \le 1, \ -1 \le y \le 1\}$$

SOLUTION   Region R is illustrated in Figure 3.

$$\iint\limits_R 2xe^{x^2+y}\, dA = \int_{-1}^1 \int_0^1 2xe^{x^2+y}\, dx\, dy$$

$$= \int_{-1}^1 \left[(e^{x^2+y}) \Big|_{x=0}^{x=1} \right] dy$$

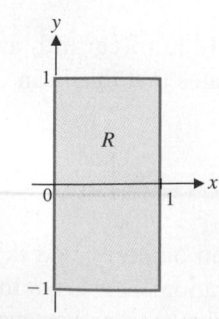

Figure 3

$$= \int_{-1}^{1} (e^{1+y} - e^y) \, dy$$

$$= (e^{1+y} - e^y)\big|_{y=-1}^{y=1}$$

$$= (e^2 - e) - (e^0 - e^{-1})$$

$$= e^2 - e - 1 + e^{-1}$$

Matched Problem 5 Evaluate

$$\iint_R \frac{x}{y^2} e^{x/y} \, dA \qquad \text{over} \qquad R = \{(x, y) \mid 0 \le x \le 1, \ 1 \le y \le 2\}.$$

Average Value over Rectangular Regions

In Section 5.5, the average value of a function $f(x)$ over an interval $[a, b]$ was defined as

$$\frac{1}{b - a} \int_a^b f(x) \, dx$$

This definition is easily extended to functions of two variables over rectangular regions as follows (notice that the denominator $(b - a)(d - c)$ is simply the area of the rectangle R):

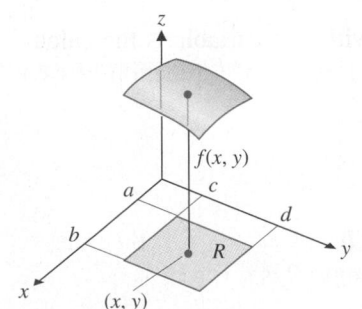

> **DEFINITION** Average Value over Rectangular Regions
> The **average value** of the function $f(x, y)$ over the rectangle
> $$R = \{(x, y) \mid a \le x \le b, \ c \le y \le d\}$$
> is
> $$\frac{1}{(b - a)(d - c)} \iint_R f(x, y) \, dA$$

EXAMPLE 6 Average Value Find the average value of $f(x, y) = 4 - \frac{1}{2}x - \frac{1}{2}y$ over the rectangle $R = \{(x, y) \mid 0 \le x \le 2, \ 0 \le y \le 2\}$.

SOLUTION Region R is illustrated in Figure 4. We have

$$\frac{1}{(b - a)(d - c)} \iint_R f(x, y) \, dA = \frac{1}{(2 - 0)(2 - 0)} \iint_R \left(4 - \frac{1}{2}x - \frac{1}{2}y\right) dA$$

$$= \frac{1}{4} \int_0^2 \int_0^2 \left(4 - \frac{1}{2}x - \frac{1}{2}y\right) dy \, dx$$

$$= \frac{1}{4} \int_0^2 \left[\left(4y - \frac{1}{2}xy - \frac{1}{4}y^2\right)\Big|_{y=0}^{y=2}\right] dx$$

$$= \frac{1}{4} \int_0^2 (7 - x) \, dx$$

$$= \frac{1}{4}\left(7x - \frac{1}{2}x^2\right)\big|_{x=0}^{x=2}$$

$$= \frac{1}{4}(12) = 3$$

Figure 5 illustrates the surface $z = f(x, y)$, and our calculations show that 3 is the average of the z values over the region R.

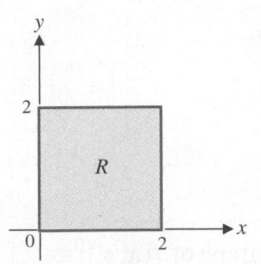

Figure 4

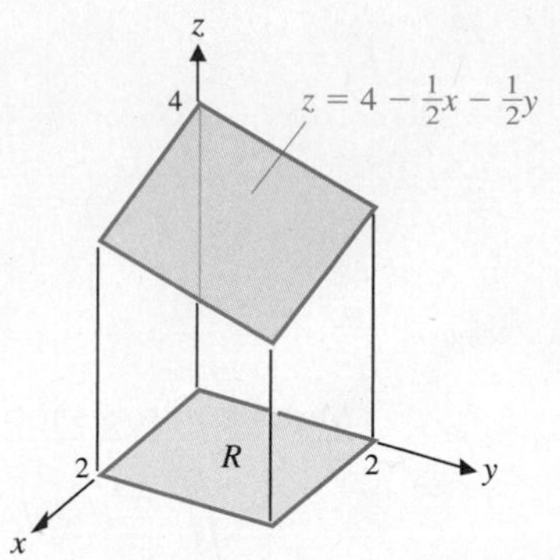

$$z = 4 - \tfrac{1}{2}x - \tfrac{1}{2}y$$

Figure 5

Matched Problem 6 Find the average value of $f(x, y) = x + 2y$ over the rectangle
$R = \{(x, y) \mid 0 \leq x \leq 2, \quad 0 \leq y \leq 1\}$

Explore and Discuss 1

(A) Which of the functions $f(x, y) = 4 - x^2 - y^2$ and $g(x, y) = 4 - x - y$
would you guess has the greater average value over the rectangle
$R = \{(x, y) \mid 0 \leq x \leq 1, \quad 0 \leq y \leq 1\}$? Explain.

(B) Use double integrals to check the correctness of your guess in part (A).

Volume and Double Integrals

One application of the definite integral of a function with one variable is the calculation of areas, so it is not surprising that the definite integral of a function of two variables can be used to calculate volumes of solids.

THEOREM 1 Volume under a Surface

If $f(x, y) \geq 0$ over a rectangle $R = \{(x, y) \mid a \leq x \leq b, \quad c \leq y \leq d\}$, then the volume of the solid formed by graphing f over the rectangle R is given by

$$V = \iint_R f(x, y)\, dA$$

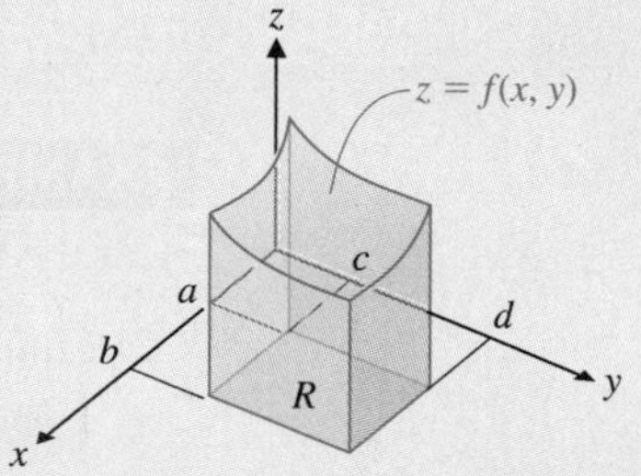

$z = f(x, y)$

EXAMPLE 7 Volume Find the volume of the solid under the graph of $f(x, y) = 1 + x^2 + y^2$ over the rectangle $R = \{(x, y) \mid 0 \leq x \leq 1, \quad 0 \leq y \leq 1\}$.

SOLUTION Figure 6 shows the region R, and Figure 7 illustrates the volume under consideration.

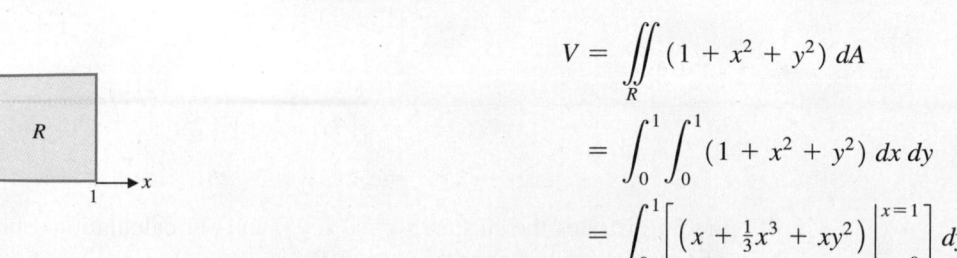

$$V = \iint_R (1 + x^2 + y^2)\, dA$$

$$= \int_0^1 \int_0^1 (1 + x^2 + y^2)\, dx\, dy$$

$$= \int_0^1 \left[\left(x + \tfrac{1}{3}x^3 + xy^2 \right) \Big|_{x=0}^{x=1} \right] dy$$

Figure 6

$$= \int_0^1 \left(\tfrac{4}{3} + y^2\right) dy$$

$$= \left(\tfrac{4}{3}y + \tfrac{1}{3}y^3\right)\big|_{y=0}^{y=1} = \tfrac{5}{3} \text{ cubic units}$$

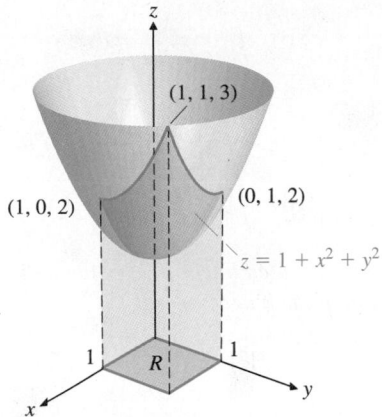

Figure 7

<u>Matched Problem 7</u> Find the volume of the solid under the graph of $f(x, y) = 1 + x + y$ over the rectangle $R = \{(x, y)\,|\,0 \le x \le 1, \quad 0 \le y \le 2\}$.

CONCEPTUAL INSIGHT

Double integrals can be defined over regions that are more general than rectangles. For example, let $R > 0$. Then the function $f(x, y) = \sqrt{R^2 - (x^2 + y^2)}$ can be integrated over the circular region $C = \{(x, y)\,|\,x^2 + y^2 \le R^2\}$. In fact, it can be shown that

$$\iint_C \sqrt{R^2 - (x^2 + y^2)}\, dx\, dy = \frac{2\pi R^3}{3}$$

Because $x^2 + y^2 + z^2 = R^2$ is the equation of a sphere of radius R centered at the origin, the double integral over C represents the volume of the upper hemisphere. Therefore, the volume of a sphere of radius R is given by

$$V = \frac{4\pi R^3}{3} \qquad \text{Volume of sphere of radius } R$$

Double integrals can also be used to obtain volume formulas for other geometric figures (see Table 1, Appendix C).

Exercises 7.6

Skills Warm-up Exercises

W   *In Problems 1–6, find each antiderivative. (If necessary, review Sections 5.1 and 5.2).*

1. $\int (\pi + x)\, dx$

2. $\int (x\pi^2 + \pi x^2)\, dx$

3. $\int \left(1 + \dfrac{\pi}{x}\right) dx$

4. $\int \left(1 + \dfrac{x}{\pi}\right) dx$

5. $\int e^{\pi x} dx$

6. $\int \dfrac{\ln x}{\pi x}\, dx$

In Problems 7–14, find each antiderivative. Then use the antiderivative to evaluate the definite integral.

7. (A) $\int 12x^2 y^3\, dy$ (B) $\int_0^1 12x^2 y^3\, dy$

8. (A) $\int 12x^2 y^3\, dx$ (B) $\int_{-1}^2 12x^2 y^3\, dx$

9. (A) $\int (4x + 6y + 5)\, dx$ (B) $\int_{-2}^3 (4x + 6y + 5)\, dx$

10. (A) $\displaystyle\int (4x + 6y + 5)\, dy$ **(B)** $\displaystyle\int_1^4 (4x + 6y + 5)\, dy$

11. (A) $\displaystyle\int \frac{x}{\sqrt{y + x^2}}\, dx$ **(B)** $\displaystyle\int_0^2 \frac{x}{\sqrt{y + x^2}}\, dx$

12. (A) $\displaystyle\int \frac{x}{\sqrt{y + x^2}}\, dy$ **(B)** $\displaystyle\int_1^5 \frac{x}{\sqrt{y + x^2}}\, dy$

13. (A) $\displaystyle\int \frac{\ln x}{xy}\, dy$ **(B)** $\displaystyle\int_1^{e^2} \frac{\ln x}{xy}\, dy$

14. (A) $\displaystyle\int \frac{\ln x}{xy}\, dx$ **(B)** $\displaystyle\int_1^e \frac{\ln x}{xy}\, dx$

In Problems 15–22, evaluate each iterated integral. (See the indicated problem for the evaluation of the inner integral.)

15. $\displaystyle\int_{-1}^2 \int_0^1 12x^2 y^3\, dy\, dx$

(See Problem 7.)

16. $\displaystyle\int_0^1 \int_{-1}^2 12x^2 y^3\, dx\, dy$

(See Problem 8.)

17. $\displaystyle\int_1^4 \int_{-2}^3 (4x + 6y + 5)\, dx\, dy$

(See Problem 9.)

18. $\displaystyle\int_{-2}^3 \int_1^4 (4x + 6y + 5)\, dy\, dx$

(See Problem 10.)

19. $\displaystyle\int_1^5 \int_0^2 \frac{x}{\sqrt{y + x^2}}\, dx\, dy$

(See Problem 11.)

20. $\displaystyle\int_0^2 \int_1^5 \frac{x}{\sqrt{y + x^2}}\, dy\, dx$

(See Problem 12.)

21. $\displaystyle\int_1^e \int_1^{e^2} \frac{\ln x}{xy}\, dy\, dx$

(See Problem 13.)

22. $\displaystyle\int_1^{e^2} \int_1^e \frac{\ln x}{xy}\, dx\, dy$

(See Problem 14.)

Use both orders of iteration to evaluate each double integral in Problems 23–26.

23. $\displaystyle\iint_R xy\, dA;\ R = \{(x, y)\,|\,0 \le x \le 2,\ \ 0 \le y \le 4\}$

24. $\displaystyle\iint_R \sqrt{xy}\, dA;\ R = \{(x, y)\,|\,1 \le x \le 4,\ \ 1 \le y \le 9\}$

25. $\displaystyle\iint_R (x + y)^5\, dA;\ R = \{(x, y)\,|\,-1 \le x \le 1,\ \ 1 \le y \le 2\}$

26. $\displaystyle\iint_R xe^y\, dA;\ R = \{(x, y)\,|\,-2 \le x \le 3,\ \ 0 \le y \le 2\}$

In Problems 27–30, find the average value of each function over the given rectangle.

27. $f(x, y) = (x + y)^2$;

$R = \{(x, y)\,|\,1 \le x \le 5,\ \ -1 \le y \le 1\}$

28. $f(x, y) = x^2 + y^2$;

$R = \{(x, y)\,|\,-1 \le x \le 2,\ \ 1 \le y \le 4\}$

29. $f(x, y) = x/y;\ R = \{(x, y)\,|\,1 \le x \le 4,\ \ 2 \le y \le 7\}$

30. $f(x, y) = x^2 y^3;\ R = \{(x, y)\,|\,-1 \le x \le 1,\ \ 0 \le y \le 2\}$

In Problems 31–34, find the volume of the solid under the graph of each function over the given rectangle.

31. $f(x, y) = 2 - x^2 - y^2$;

$R = \{(x, y)\,|\,0 \le x \le 1,\ \ 0 \le y \le 1\}$

32. $f(x, y) = 5 - x$;

$R = \{(x, y)\,|\,0 \le x \le 5,\ \ 0 \le y \le 5\}$

33. $f(x, y) = 4 - y^2;\ R = \{(x, y)\,|\,0 \le x \le 2,\ \ 0 \le y \le 2\}$

34. $f(x, y) = e^{-x - y};\ R = \{(x, y)\,|\,0 \le x \le 1,\ \ 0 \le y \le 1\}$

Evaluate each double integral in Problems 35–38. Select the order of integration carefully; each problem is easy to do one way and difficult the other.

35. $\displaystyle\iint_R xe^{xy}\, dA;\ R = \{(x, y)\,|\,0 \le x \le 1,\ \ 1 \le y \le 2\}$

36. $\displaystyle\iint_R xye^{x^2 y}\, dA;\ R = \{(x, y)\,|\,0 \le x \le 1,\ \ 1 \le y \le 2\}$

37. $\displaystyle\iint_R \frac{2y + 3xy^2}{1 + x^2}\, dA$;

$R = \{(x, y)\,|\,0 \le x \le 1,\ \ -1 \le y \le 1\}$

38. $\displaystyle\iint_R \frac{2x + 2y}{1 + 4y + y^2}\, dA$;

$R = \{(x, y)\,|\,1 \le x \le 3,\ \ 0 \le y \le 1\}$

39. Show that $\int_0^2 \int_0^2 (1 - y)\, dx\, dy = 0$. Does the double integral represent the volume of a solid? Explain.

40. (A) Find the average values of the functions $f(x, y) = x + y$, $g(x, y) = x^2 + y^2$, and $h(x, y) = x^3 + y^3$ over the rectangle

$R = \{(x, y)\,|\,0 \le x \le 1,\ \ 0 \le y \le 1\}$

(B) Does the average value of $k(x, y) = x^n + y^n$ over the rectangle

$R_1 = \{(x, y)\,|\,0 \le x \le 1,\ \ 0 \le y \le 1\}$

increase or decrease as n increases? Explain.

(C) Does the average value of $k(x, y) = x^n + y^n$ over the rectangle

$R_2 = \{(x, y)\,|\,0 \le x \le 2,\ \ 0 \le y \le 2\}$

increase or decrease as n increases? Explain.

41. Let $f(x, y) = x^3 + y^2 - e^{-x} - 1$.

 (A) Find the average value of $f(x, y)$ over the rectangle

 $$R = \{(x, y) \mid -2 \le x \le 2, \;\; -2 \le y \le 2\}.$$

 (B) Graph the set of all points (x, y) in R for which
 $f(x, y) = 0$.

 (C) For which points (x, y) in R is $f(x, y)$ greater than 0?
 Less than 0? Explain.

 42. Find the dimensions of the square S centered at the origin
for which the average value of $f(x, y) = x^2 e^y$ over S is equal
to 100.

Applications

43. Multiplier principle. Suppose that Congress enacts a
one-time-only 10% tax rebate that is expected to infuse
$\$y$ billion, $5 \le y \le 7$, into the economy. If every person
and every corporation is expected to spend a proportion
$x, 0.6 \le x \le 0.8$, of each dollar received, then, by the
multiplier principle in economics, the total amount of
spending S (in billions of dollars) generated by this tax
rebate is given by

$$S(x, y) = \frac{y}{1 - x}$$

What is the average total amount of spending for the indicated ranges of the values of x and y? Set up a double integral
and evaluate it.

44. Multiplier principle. Repeat Problem 43 if $6 \le y \le 10$ and
$0.7 \le x \le 0.9$.

45. Cobb–Douglas production function. If an industry invests x thousand labor-hours, $10 \le x \le 20$, and $\$y$ million,
$1 \le y \le 2$, in the production of N thousand units of a certain
item, then N is given by

$$N(x, y) = x^{0.75} y^{0.25}$$

What is the average number of units produced for the
indicated ranges of x and y? Set up a double integral and
evaluate it.

46. Cobb–Douglas production function. Repeat Problem 45 for

$$N(x, y) = x^{0.5} y^{0.5}$$

where $10 \le x \le 30$ and $1 \le y \le 3$.

47. Population distribution. In order to study the population
distribution of a certain species of insect, a biologist has constructed an artificial habitat in the shape of a rectangle 16 feet
long and 12 feet wide. The only food available to the insects
in this habitat is located at its center. The biologist has determined that the concentration C of insects per square foot at
a point d units from the food supply (see the figure) is given
approximately by

$$C = 10 - \tfrac{1}{10} d^2$$

What is the average concentration of insects throughout the
habitat? Express C as a function of x and y, set up a double
integral, and evaluate it.

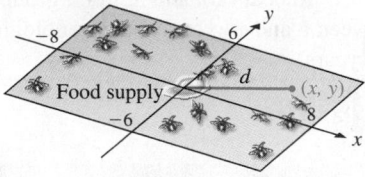

Figure for 47

48. Population distribution. Repeat Problem 47 for a square
habitat that measures 12 feet on each side, where the insect
concentration is given by

$$C = 8 - \tfrac{1}{10} d^2$$

49. Pollution. A heavy industrial plant located in the center of
a small town emits particulate matter into the atmosphere.
Suppose that the concentration of particulate matter (in parts
per million) at a point d miles from the plant (see the figure)
is given by

$$C = 100 - 15 d^2$$

If the boundaries of the town form a rectangle 4 miles long
and 2 miles wide, what is the average concentration of particulate matter throughout the town? Express C as a function
of x and y, set up a double integral, and evaluate it.

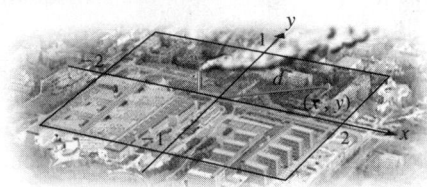

50. Pollution. Repeat Problem 49 if the boundaries of the town
form a rectangle 8 miles long and 4 miles wide and the concentration of particulate matter is given by

$$C = 100 - 3 d^2$$

51. Safety research. Under ideal conditions, if a person driving
a car slams on the brakes and skids to a stop, the length of the
skid marks (in feet) is given by the formula

$$L = 0.000\,013\,3 x y^2$$

where x is the weight of the car (in pounds) and y is the speed
of the car (in miles per hour). What is the average length of
the skid marks for cars weighing between 2,000 and 3,000
pounds and traveling at speeds between 50 and 60 miles per
hour? Set up a double integral and evaluate it.

52. Safety research. Repeat Problem 51 for cars weighing
between 2,000 and 2,500 pounds and traveling at speeds between 40 and 50 miles per hour.

53. Psychology. The intelligence quotient Q for a person with
mental age x and chronological age y is given by

$$Q(x, y) = 100 \frac{x}{y}$$

In a group of sixth-graders, the mental age varies between
8 and 16 years and the chronological age varies between

10 and 12 years. What is the average intelligence quotient for this group? Set up a double integral and evaluate it.

54. **Psychology.** Repeat Problem 53 for a group with mental ages between 6 and 14 years and chronological ages between 8 and 10 years.

7.7 Double Integrals over More General Regions

- Regular Regions
- Double Integrals over Regular Regions
- Reversing the Order of Integration
- Volume and Double Integrals

In this section, we extend the concept of double integration discussed in Section 7.6 to nonrectangular regions. We begin with an example and some new terminology.

Regular Regions

Let R be the region graphed in Figure 1. We can describe R with the following inequalities:

$$R = \{(x, y) \mid x \leq y \leq 6x - x^2, \;\; 0 \leq x \leq 5\}$$

The region R can be viewed as a union of vertical line segments. For each x in the interval $[0, 5]$, the line segment from the point $(x, g(x))$ to the point $(x, f(x))$ lies in the region R. Any region that can be covered by vertical line segments in this manner is called a *regular x region*.

Now consider the region S in Figure 2. It can be described with the following inequalities:

$$S = \{(x, y) \mid y^2 \leq x \leq y + 2, \;\; -1 \leq y \leq 2\}$$

The region S can be viewed as a union of horizontal line segments going from the graph of $h(y) = y^2$ to the graph of $k(y) = y + 2$ on the interval $[-1, 2]$. Regions that can be described in this manner are called *regular y regions*.

In general, *regular regions* are defined as follows:

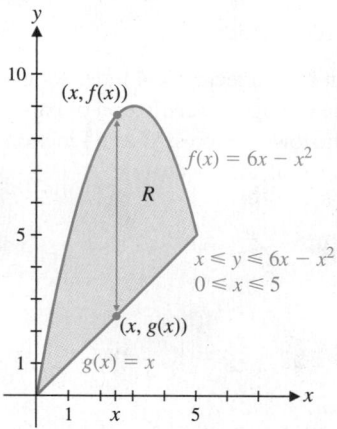

Figure 1

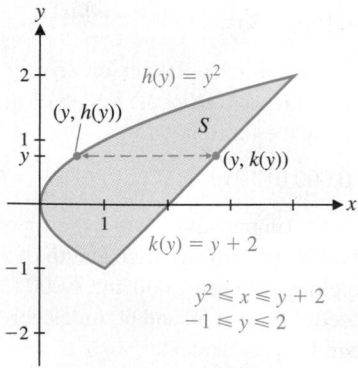

Figure 2

> **DEFINITION Regular Regions**
>
> A region R in the xy plane is a **regular x region** if there exist functions $f(x)$ and $g(x)$ and numbers a and b such that
>
> $$R = \{(x, y) \mid g(x) \leq y \leq f(x), \;\; a \leq x \leq b\}$$
>
> A region R in the xy plane is a **regular y region** if there exist functions $h(y)$ and $k(y)$ and numbers c and d such that
>
> $$R = \{(x, y) \mid h(y) \leq x \leq k(y), \;\; c \leq y \leq d\}$$
>
> See Figure 3 for a geometric interpretation.

> **CONCEPTUAL INSIGHT**
>
> If, for some region R, there is a horizontal line that has a nonempty intersection I with R, and if I is neither a closed interval nor a point, then R is *not* a regular y region. Similarly, if, for some region R, there is a vertical line that has a nonempty intersection I with R, and if I is neither a closed interval nor a point, then R is *not* a regular x region (see Fig. 3).

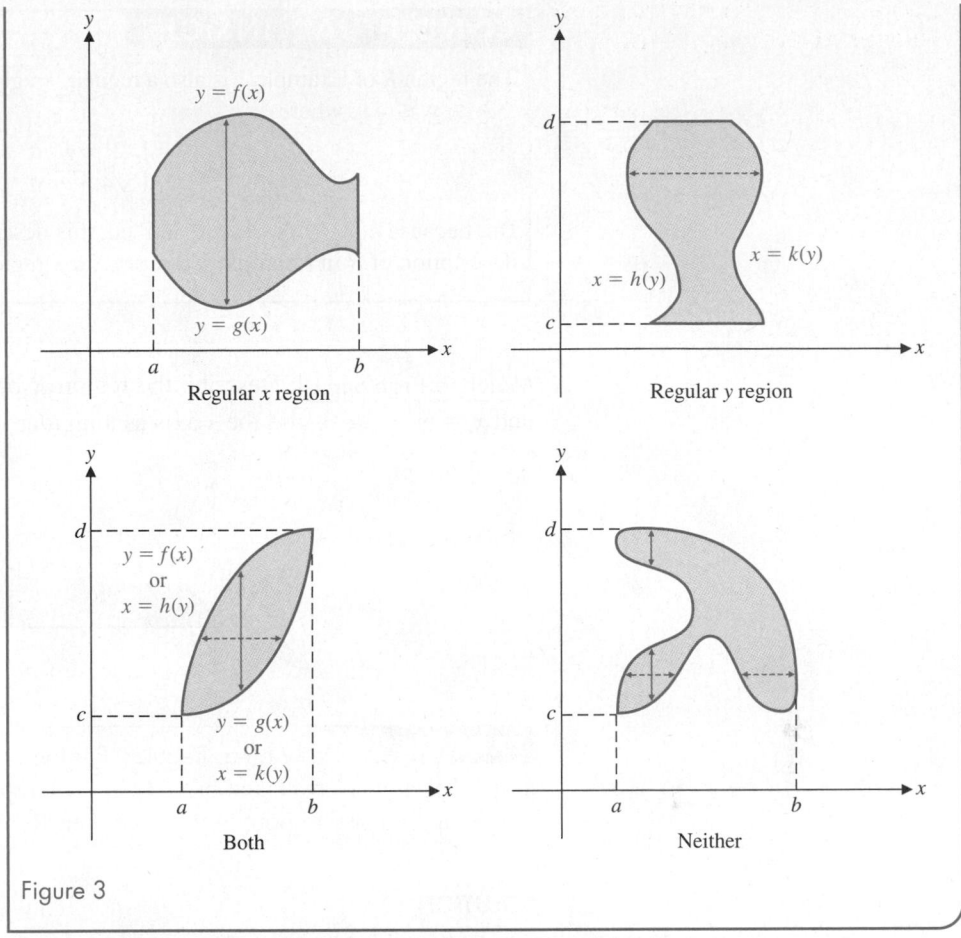

Figure 3

EXAMPLE 1 Describing a Regular x Region The region R is bounded by the graphs of $y = 4 - x^2$ and $y = x - 2$, $x \geq 0$, and the y axis. Graph R and use set notation with double inequalities to describe R as a regular x region.

SOLUTION As the solid line in the following figure indicates, R can be covered by vertical line segments that go from the graph of $y = x - 2$ to the graph of $y = 4 - x^2$. So, R is a regular x region. In terms of set notation with double inequalities, we can write

$$R = \{(x, y) \,|\, x - 2 \leq y \leq 4 - x^2, \ 0 \leq x \leq 2\}$$

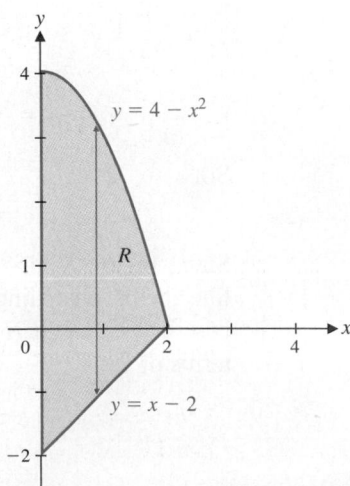

CONCEPTUAL INSIGHT

The region R of Example 1 is also a regular y region, since $R = \{(x, y) \mid 0 \leq x \leq k(y),$ $-2 \leq y \leq 4\}$, where

$$k(y) = \begin{cases} 2 + y & \text{if } -2 \leq y \leq 0 \\ \sqrt{4 - y} & \text{if } 0 \leq y \leq 4 \end{cases}$$

But because $k(y)$ is piecewise defined, this description is more complicated than the description of R in Example 1 as a regular x region.

Matched Problem 1 ⌋ Describe the region R bounded by the graphs of $x = 6 - y$ and $x = y^2$, $y \geq 0$, and the x axis as a regular y region.

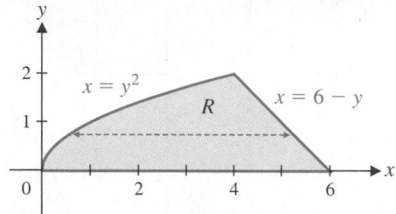

EXAMPLE 2 Describing Regular Regions The region R is bounded by the graphs of $x + y^2 = 9$ and $x + 3y = 9$. Graph R and describe R as a regular x region, a regular y region, both, or neither. Represent R in set notation with double inequalities.

SOLUTION

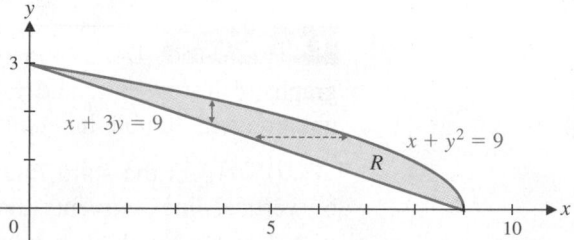

Region R can be covered by vertical line segments that go from the graph of $x + 3y = 9$ to the graph of $x + y^2 = 9$. Thus, R is a regular x region. In order to describe R with inequalities, we must solve each equation for y in terms of x:

$$x + 3y = 9 \qquad\qquad x + y^2 = 9$$
$$3y = 9 - x \qquad\qquad y^2 = 9 - x \qquad \text{We use the positive square}$$
$$y = 3 - \tfrac{1}{3}x \qquad\qquad y = \sqrt{9 - x} \qquad \text{root, since the graph is in the first quadrant.}$$

So,

$$R = \{(x, y) \mid 3 - \tfrac{1}{3}x \leq y \leq \sqrt{9 - x}, \ 0 \leq x \leq 9\}$$

Since region R also can be covered by horizontal line segments (see the dashed line in the preceding figure) that go from the graph of $x + 3y = 9$ to the graph of $x + y^2 = 9$, it is a regular y region. Now we must solve each equation for x in terms of y:

$$x + 3y = 9 \qquad\qquad x + y^2 = 9$$
$$x = 9 - 3y \qquad\qquad x = 9 - y^2$$

Therefore,

$$R = \{(x, y) \mid 9 - 3y \leq x \leq 9 - y^2, \ 0 \leq y \leq 3\}$$

Matched Problem 2 Repeat Example 2 for the region bounded by the graphs of $2y - x = 4$ and $y^2 - x = 4$, as shown in the following figure:

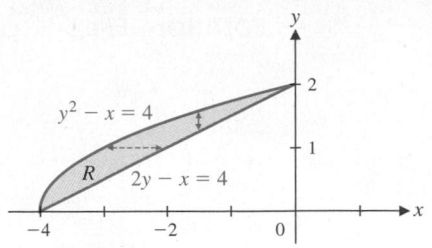

Explore and Discuss 1

A E I O U

Consider the vowels A, E, I, O, U, written in block letters as shown in the margin, to be regions of the plane. One of the vowels is a regular x region, but not a regular y region; one is a regular y region, but not a regular x region; one is both; two are neither. Explain.

Double Integrals over Regular Regions

Now we want to extend the definition of double integration to include regular x regions and regular y regions. The order of integration now depends on the nature of the region R. If R is a regular x region, we integrate with respect to y first, while if R is a regular y region, we integrate with respect to x first.

> Note that the variable limits of integration (when present) are always on the inner integral, and the constant limits of integration are always on the outer integral.

DEFINITION Double Integration over Regular Regions

Regular x Region

If $R = \{(x, y) \,|\, g(x) \leq y \leq f(x), \quad a \leq x \leq b\}$, then

$$\iint\limits_R F(x, y) \, dA = \int_a^b \left[\int_{g(x)}^{f(x)} F(x, y) \, dy \right] dx$$

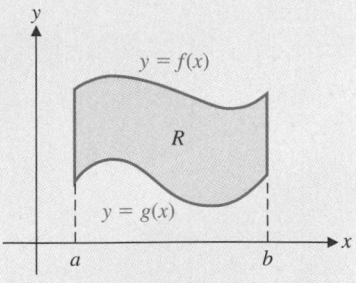

Regular y Region

If $R = \{(x, y) \,|\, h(y) \leq x \leq k(y), \quad c \leq y \leq d\}$, then

$$\iint\limits_R F(x, y) \, dA = \int_c^d \left[\int_{h(y)}^{k(y)} F(x, y) \, dx \right] dy$$

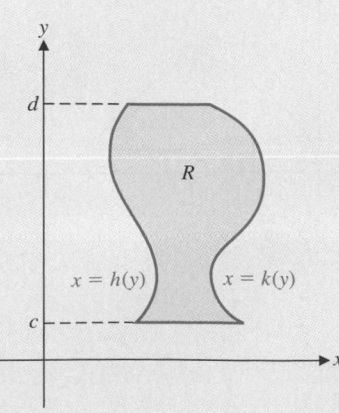

EXAMPLE 3 Evaluating a Double Integral Evaluate $\iint\limits_{R} 2xy \, dA$, where R is the region bounded by the graphs of $y = -x$ and $y = x^2$, $x \geq 0$, and the graph of $x = 1$.

SOLUTION From the graph, we can see that R is a regular x region described by

$$R = \{(x, y) \mid -x \leq y \leq x^2, \ 0 \leq x \leq 1\}$$

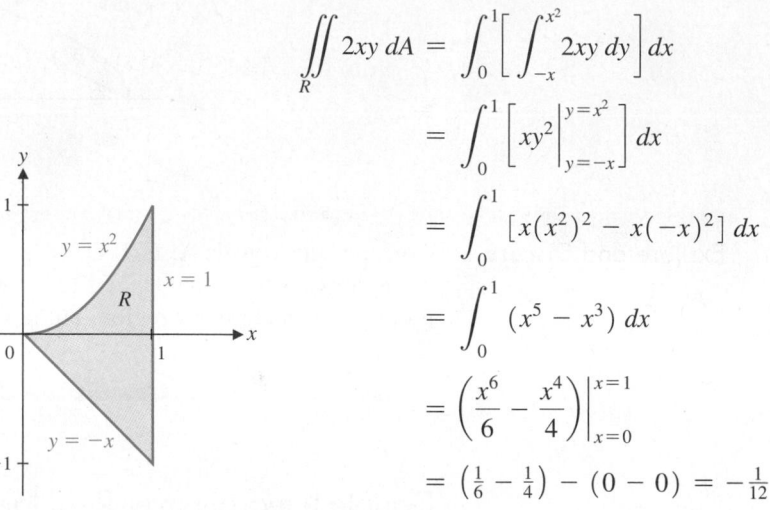

$$\iint\limits_{R} 2xy \, dA = \int_{0}^{1}\left[\int_{-x}^{x^2} 2xy \, dy\right] dx$$

$$= \int_{0}^{1}\left[xy^2\Big|_{y=-x}^{y=x^2}\right] dx$$

$$= \int_{0}^{1}\left[x(x^2)^2 - x(-x)^2\right] dx$$

$$= \int_{0}^{1}(x^5 - x^3) \, dx$$

$$= \left(\frac{x^6}{6} - \frac{x^4}{4}\right)\Big|_{x=0}^{x=1}$$

$$= \left(\tfrac{1}{6} - \tfrac{1}{4}\right) - (0 - 0) = -\tfrac{1}{12}$$

Matched Problem 3 Evaluate $\iint\limits_{R} 3xy^2 \, dA$, where R is the region in Example 3.

EXAMPLE 4 Evaluating a Double Integral Evaluate $\iint\limits_{R}(2x + y) \, dA$, where R is the region bounded by the graphs of $y = \sqrt{x}$, $x + y = 2$, and $y = 0$.

SOLUTION From the graph, we can see that R is a regular y region. After solving each equation for x, we can write

$$R = \{(x, y) \mid y^2 \leq x \leq 2 - y, \ 0 \leq y \leq 1\}$$

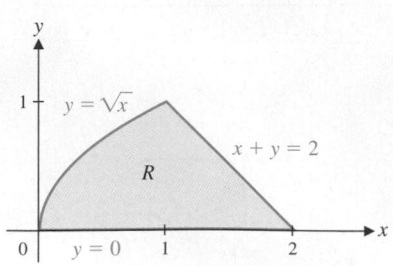

$$\iint\limits_{R}(2x + y) \, dA = \int_{0}^{1}\left[\int_{y^2}^{2-y}(2x + y) \, dx\right] dy$$

$$= \int_{0}^{1}\left[(x^2 + yx)\Big|_{x=y^2}^{x=2-y}\right] dy$$

$$= \int_{0}^{1}\left\{[(2 - y)^2 + y(2 - y)] - [(y^2)^2 + y(y^2)]\right\} dy$$

$$= \int_{0}^{1}(4 - 2y - y^3 - y^4) \, dy$$

$$= \left(4y - y^2 - \tfrac{1}{4}y^4 - \tfrac{1}{5}y^5\right)\Big|_{y=0}^{y=1}$$

$$= \left(4 - 1 - \tfrac{1}{4} - \tfrac{1}{5}\right) - 0 = \tfrac{51}{20}$$

Matched Problem 4 Evaluate $\iint\limits_{R}(y - 4x) \, dA$, where R is the region in Example 4.

EXAMPLE 5 Evaluating a Double Integral The region R is bounded by the graphs of $y = \sqrt{x}$ and $y = \frac{1}{2}x$. Evaluate $\iint\limits_R 4xy^3\,dA$ two different ways.

SOLUTION Region R is both a regular x region and a regular y region:

$$R = \{(x, y) \mid \tfrac{1}{2}x \le y \le \sqrt{x}, \ 0 \le x \le 4\} \quad \text{Regular } x \text{ region}$$
$$R = \{(x, y) \mid y^2 \le x \le 2y, \ 0 \le y \le 2\} \quad \text{Regular } y \text{ region}$$

Using the first representation (a regular x region), we obtain

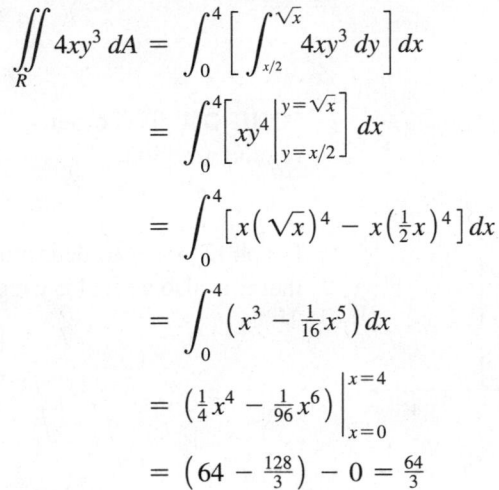

$$\iint\limits_R 4xy^3\,dA = \int_0^4 \left[\int_{x/2}^{\sqrt{x}} 4xy^3\,dy \right] dx$$

$$= \int_0^4 \left[xy^4 \Big|_{y=x/2}^{y=\sqrt{x}} \right] dx$$

$$= \int_0^4 \left[x(\sqrt{x})^4 - x(\tfrac{1}{2}x)^4 \right] dx$$

$$= \int_0^4 \left(x^3 - \tfrac{1}{16}x^5 \right) dx$$

$$= \left(\tfrac{1}{4}x^4 - \tfrac{1}{96}x^6 \right) \Big|_{x=0}^{x=4}$$

$$= \left(64 - \tfrac{128}{3} \right) - 0 = \tfrac{64}{3}$$

Using the second representation (a regular y region), we obtain

$$\iint\limits_R 4xy^3\,dA = \int_0^2 \left[\int_{y^2}^{2y} 4xy^3\,dx \right] dy$$

$$= \int_0^2 \left[2x^2y^3 \Big|_{x=y^2}^{x=2y} \right] dy$$

$$= \int_0^2 \left[2(2y)^2y^3 - 2(y^2)^2y^3 \right] dy$$

$$= \int_0^2 \left(8y^5 - 2y^7 \right) dy$$

$$= \left(\tfrac{4}{3}y^6 - \tfrac{1}{4}y^8 \right) \Big|_{y=0}^{y=2}$$

$$= \left(\tfrac{256}{3} - 64 \right) - 0 = \tfrac{64}{3}$$

Matched Problem 5 The region R is bounded by the graphs of $y = x$ and $y = \frac{1}{2}x^2$. Evaluate $\iint\limits_R 4xy^3\,dA$ two different ways.

Reversing the Order of Integration

Example 5 shows that

$$\iint\limits_R 4xy^3\,dA = \int_0^4 \left[\int_{x/2}^{\sqrt{x}} 4xy^3\,dy \right] dx = \int_0^2 \left[\int_{y^2}^{2y} 4xy^3\,dx \right] dy$$

In general, if R is both a regular x region and a regular y region, then the two iterated integrals are equal. In rectangular regions, reversing the order of integration in an

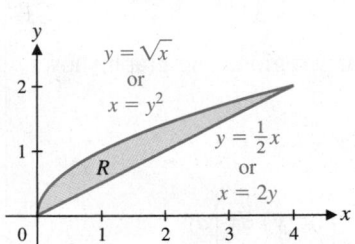

iterated integral was a simple matter. As Example 5 illustrates, the process is more complicated in nonrectangular regions. The next example illustrates how to start with an iterated integral and reverse the order of integration. Since we are interested in the reversal process and not in the value of either integral, the integrand will not be specified.

EXAMPLE 6   Reversing the Order of Integration   Reverse the order of integration in

$$\int_1^3 \left[\int_0^{x-1} f(x, y) \, dy \right] dx$$

SOLUTION   The order of integration indicates that the region of integration is a regular x region:

$$R = \{(x, y) \mid 0 \le y \le x - 1, \quad 1 \le x \le 3\}$$

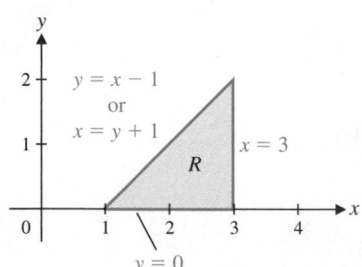

Graph region R to determine whether it is also a regular y region. The graph shows that R is also a regular y region, and we can write

$$R = \{(x, y) \mid y + 1 \le x \le 3, \quad 0 \le y \le 2\}$$

$$\int_1^3 \left[\int_0^{x-1} f(x, y) \, dy \right] dx = \int_0^2 \left[\int_{y+1}^3 f(x, y) \, dx \right] dy$$

Matched Problem 6   Reverse the order of integration in $\int_2^4 \left[\int_0^{4-x} f(x, y) \, dy \right] dx$.

Explore and Discuss 2   Explain the difficulty in evaluating $\int_0^2 \int_{x^2}^4 x e^{y^2} dy \, dx$ and how it can be overcome by reversing the order of integration.

Volume and Double Integrals

In Section 7.6, we used the double integral to calculate the volume of a solid with a rectangular base. In general, if a solid can be described by the graph of a positive function $f(x, y)$ over a regular region R (not necessarily a rectangle), then the double integral of the function f over the region R still represents the volume of the corresponding solid.

EXAMPLE 7   Volume   The region R is bounded by the graphs of $x + y = 1$, $y = 0$, and $x = 0$. Find the volume of the solid under the graph of $z = 1 - x - y$ over the region R.

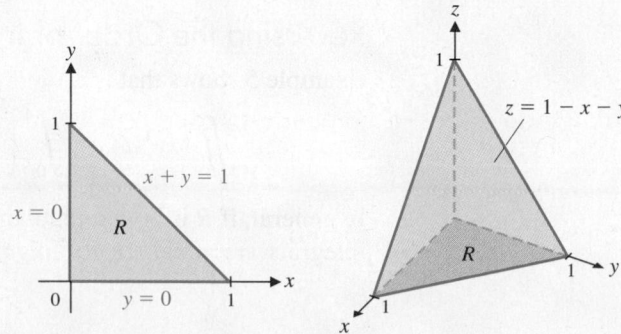

SOLUTION The graph of R shows that R is both a regular x region and a regular y region. We choose to use the regular x region:

$$R = \{(x, y) | 0 \le y \le 1 - x, \quad 0 \le x \le 1\}$$

The volume of the solid is

$$\begin{aligned}
V &= \iint\limits_R (1 - x - y)\, dA = \int_0^1 \left[\int_0^{1-x} (1 - x - y)\, dy \right] dx \\
&= \int_0^1 \left[(y - xy - \tfrac{1}{2}y^2) \Big|_{y=0}^{y=1-x} \right] dx \\
&= \int_0^1 [(1 - x) - x(1 - x) - \tfrac{1}{2}(1 - x)^2]\, dx \\
&= \int_0^1 \left(\tfrac{1}{2} - x + \tfrac{1}{2}x^2 \right) dx \\
&= \left(\tfrac{1}{2}x - \tfrac{1}{2}x^2 + \tfrac{1}{6}x^3 \right) \Big|_{x=0}^{x=1} \\
&= \left(\tfrac{1}{2} - \tfrac{1}{2} + \tfrac{1}{6} \right) - 0 = \tfrac{1}{6}
\end{aligned}$$

Matched Problem 7) The region R is bounded by the graphs of $y + 2x = 2$, $y = 0$, and $x = 0$. Find the volume of the solid under the graph of $z = 2 - 2x - y$ over the region R. [*Hint*: Sketch the region first; the solid does not have to be sketched.]

Exercises 7.7

In Problems 1–6, graph the region R bounded by the graphs of the equations. Use set notation and double inequalities to describe R as a regular x region and a regular y region in Problems 1 and 2, and as a regular x region or a regular y region, whichever is simpler, in Problems 3–6.

1. $y = 4 - x^2, y = 0, 0 \le x \le 2$

2. $y = x^2, y = 9, 0 \le x \le 3$

3. $y = x^3, y = 12 - 2x, x = 0$

4. $y = 5 - x, y = 1 + x, y = 0$

5. $y^2 = 2x, y = x - 4$

6. $y = 4 + 3x - x^2, x + y = 4$

Evaluate each integral in Problems 7–10.

7. $\int_0^1 \int_0^x (x + y)\, dy\, dx$

8. $\int_0^2 \int_0^y xy\, dx\, dy$

9. $\int_0^1 \int_{y^3}^{\sqrt{y}} (2x + y)\, dx\, dy$

10. $\int_1^4 \int_x^{x^2} (x^2 + 2y)\, dy\, dx$

In Problems 11–14, give a verbal description of the region R and determine whether R is a regular x region, a regular y region, both, or neither.

11. $R = \{(x, y) | |x| \le 2, \quad |y| \le 3\}$

12. $R = \{(x, y) | 1 \le x^2 + y^2 \le 4\}$

13. $R = \{(x, y) | x^2 + y^2 \ge 1, \quad |x| \le 2, \quad 0 \le y \le 2\}$

14. $R = \{(x, y) | |x| + |y| \le 1\}$

In Problems 15–20, use the description of the region R to evaluate the indicated integral.

15. $\iint\limits_R (x^2 + y^2)\, dA;$

$R = \{(x, y) | 0 \le y \le 2x, \quad 0 \le x \le 2\}$

16. $\iint\limits_R 2x^2y\, dA;$

$R = \{(x, y) | 0 \le y \le 9 - x^2, \quad -3 \le x \le 3\}$

17. $\iint\limits_R (x + y - 2)^3\, dA;$

$R = \{(x, y) | 0 \le x \le y + 2, \quad 0 \le y \le 1\}$

18. $\iint\limits_R (2x + 3y)\, dA;$

$R = \{(x, y) | y^2 - 4 \le x \le 4 - 2y, \quad 0 \le y \le 2\}$

19. $\iint\limits_R e^{x+y}\, dA;$

$R = \{(x, y) | -x \le y \le x, \quad 0 \le x \le 2\}$

20. $\iint\limits_R \frac{x}{\sqrt{x^2 + y^2}}\, dA;$

$R = \{(x, y) | 0 \le x \le \sqrt{4y - y^2}, \quad 0 \le y \le 2\}$

In Problems 21–26, graph the region R bounded by the graphs of the indicated equations. Describe R in set notation with double inequalities, and evaluate the indicated integral.

21. $y = x + 1, y = 0, x = 0, x = 1;$ $\iint\limits_R \sqrt{1 + x + y}\, dA$

22. $y = x^2,\ y = \sqrt{x}\,;$ $\iint\limits_R 12xy\, dA$

23. $y = 4x - x^2,\ y = 0;$ $\iint\limits_R \sqrt{y + x^2}\, dA$

24. $x = 1 + 3y,\ x = 1 - y,\ y = 1;$ $\iint\limits_R (x + y + 1)^3\, dA$

25. $y = 1 - \sqrt{x}, y = 1 + \sqrt{x}, x = 4;$ $\iint\limits_R x(y - 1)^2\, dA$

26. $y = \frac{1}{2}x, y = 6 - x,\ y = 1;$ $\iint\limits_R \frac{1}{x + y}\, dA$

In Problems 27–32, evaluate each integral. Graph the region of integration, reverse the order of integration, and then evaluate the integral with the order reversed.

27. $\int_0^3 \int_0^{3-x} (x + 2y)\, dy\, dx$

28. $\int_0^2 \int_0^y (y - x)^4\, dx\, dy$

29. $\int_0^1 \int_0^{1-x^2} x\sqrt{y}\, dy\, dx$

30. $\int_0^2 \int_{x^3}^{4x} (1 + 2y)\, dy\, dx$

31. $\int_0^4 \int_{x/4}^{\sqrt{x}/2} x\, dy\, dx$

32. $\int_0^4 \int_{y^2/4}^{2\sqrt{y}} (1 + 2xy)\, dx\, dy$

In Problems 33–36, find the volume of the solid under the graph of $f(x, y)$ over the region R bounded by the graphs of the indicated equations. Sketch the region R; the solid does not have to be sketched.

33. $f(x, y) = 4 - x - y;$ R is the region bounded by the graphs of $x + y = 4, y = 0, x = 0$

34. $f(x, y) = (x - y)^2;$ R is the region bounded by the graphs of $y = x, y = 2, x = 0$

35. $f(x, y) = 4;$ R is the region bounded by the graphs of $y = 1 - x^2$ and $y = 0$ for $0 \le x \le 1$

36. $f(x, y) = 4xy;$ R is the region bounded by the graphs of $y = \sqrt{1 - x^2}$ and $y = 0$ for $0 \le x \le 1$

In Problems 37–40, reverse the order of integration for each integral. Evaluate the integral with the order reversed. Do not attempt to evaluate the integral in the original form.

37. $\int_0^2 \int_{x^2}^4 \frac{4x}{1 + y^2}\, dy\, dx$

38. $\int_0^1 \int_y^1 \sqrt{1 - x^2}\, dx\, dy$

39. $\int_0^1 \int_{y^2}^1 4ye^{x^2}\, dx\, dy$

40. $\int_0^4 \int_{\sqrt{x}}^2 \sqrt{3x + y^2}\, dy\, dx$

In Problems 41–46, use a graphing calculator to graph the region R bounded by the graphs of the indicated equations. Use approximation techniques to find intersection points correct to two decimal places. Describe R in set notation with double inequalities, and evaluate the indicated integral correct to two decimal places.

41. $y = 1 + \sqrt{x},\quad y = x^2,\ x = 0;$ $\iint\limits_R x\, dA$

42. $y = 1 + \sqrt[3]{x}, y = x, x = 0;$ $\iint\limits_R x\, dA$

43. $y = \sqrt[3]{x}, y = 1 - x, y = 0;$ $\iint\limits_R 24xy\, dA$

44. $y = x^3, y = 1 - x, y = 0;$ $\iint\limits_R 48xy\, dA$

45. $y = e^{-x},\ y = 3 - x;$ $\iint\limits_R 4y\, dA$

46. $y = e^x,\ y = 2 + x;$ $\iint\limits_R 8y\, dA$

Answers to Matched Problems

1. $R = \{(x, y)| y^2 \le x \le 6 - y,\ 0 \le y \le 2\}$

2. R is both a regular x region and a regular y region:
$R = \{(x, y)|\frac{1}{2}x + 2 \le y \le \sqrt{x + 4},\ -4 \le x \le 0\}$
$R = \{(x, y)|y^2 - 4 \le x \le 2y - 4,\ 0 \le y \le 2\}$

3. $\frac{13}{40}$ **4.** $-\frac{77}{20}$ **5.** $\frac{64}{15}$

6. $\int_0^2 \int_2^{4-y} f(x, y)\, dx\, dy$

7. $\frac{2}{3}$

Chapter 7   Summary and Review

Important Terms, Symbols, and Concepts

7.1 Functions of Several Variables

EXAMPLES

- An equation of the form $z = f(x, y)$ describes a **function of two independent variables** if, for each permissible ordered pair (x, y), there is one and only one value of z determined by $f(x, y)$. The variables x and y are **independent variables**, and z is a **dependent variable**. The set of all ordered pairs of permissible values of x and y is the **domain** of the function, and the set of all corresponding values $f(x, y)$ is the **range**. Functions of more than two independent variables are defined similarly.

Ex. 1, p. 425
Ex. 2, p. 426
Ex. 3, p. 426
Ex. 4, p. 427
Ex. 5, p. 428
Ex. 6, p. 428
Ex. 7, p. 430

- The graph of $z = f(x, y)$ consists of all triples (x, y, z) in a **three-dimensional coordinate system** that satisfy the equation. The graphs of the functions $z = f(x, y) = x^2 + y^2$ and $z = g(x, y) = x^2 - y^2$, for example, are **surfaces**; the first has a local minimum, and the second has a **saddle point**, at $(0, 0)$.

7.2 Partial Derivatives

- If $z = f(x, y)$, then the **partial derivative of f with respect to x**, denoted as $\partial z / \partial x$, f_x, or $f_x(x, y)$, is

$$\frac{\partial z}{\partial x} = \lim_{h \to 0} \frac{f(x + h, y) - f(x, y)}{h}$$

Ex. 1, p. 435
Ex. 2, p. 435
Ex. 3, p. 436
Ex. 4, p. 436

Similarly, the **partial derivative of f with respect to y**, denoted as $\partial z / \partial y$, f_y, or $f_y(x, y)$, is

$$\frac{\partial z}{\partial y} = \lim_{k \to 0} \frac{f(x, y + k) - f(x, y)}{k}$$

The partial derivatives $\partial z / \partial x$ and $\partial z / \partial y$ are said to be **first-order partial derivatives**.

- There are four **second-order partial derivatives** of $z = f(x, y)$:

Ex. 5, p. 438

$$f_{xx} = f_{xx}(x, y) = \frac{\partial^2 z}{\partial x^2} = \frac{\partial}{\partial x}\left(\frac{\partial z}{\partial x}\right)$$

$$f_{xy} = f_{xy}(x, y) = \frac{\partial^2 z}{\partial y \, \partial x} = \frac{\partial}{\partial y}\left(\frac{\partial z}{\partial x}\right)$$

$$f_{yx} = f_{yx}(x, y) = \frac{\partial^2 z}{\partial x \, \partial y} = \frac{\partial}{\partial x}\left(\frac{\partial z}{\partial y}\right)$$

$$f_{yy} = f_{yy}(x, y) = \frac{\partial^2 z}{\partial y^2} = \frac{\partial}{\partial y}\left(\frac{\partial z}{\partial y}\right)$$

7.3 Maxima and Minima

- If $f(a, b) \geq f(x, y)$ for all (x, y) in a circular region in the domain of f with (a, b) as center, then $f(a, b)$ is a **local maximum**. If $f(a, b) \leq f(x, y)$ for all (x, y) in such a region, then $f(a, b)$ is a **local minimum**.

Ex. 1, p. 445
Ex. 2, p. 445
Ex. 3, p. 447
Ex. 4, p. 448

- If a function $f(x, y)$ has a local maximum or minimum at the point (a, b), and f_x and f_y exist at (a, b), then both first-order partial derivatives equal 0 at (a, b) [Theorem 1, p. 444].

- The second-derivative test for local extrema (Theorem 2, p. 444) gives conditions on the first- and second-order partial derivatives of $f(x, y)$, which guarantee that $f(a, b)$ is a local maximum, local minimum, or saddle point.

7.4 Maxima and Minima Using Lagrange Multipliers

- The **method of Lagrange multipliers** can be used to find local extrema of a function $z = f(x, y)$ subject to the constraint $g(x, y) = 0$. A procedure that lists the key steps in the method is given on page 453.

Ex. 1, p. 453
Ex. 2, p. 454
Ex. 3, p. 456

- The method of Lagrange multipliers can be extended to functions with arbitrarily many independent variables with one or more constraints (see Theorem 1, p. 452, and Theorem 2, p. 456, for the method when there are two and three independent variables, respectively).

7.5 Method of Least Squares

- **Linear regression** is the process of fitting a line $y = ax + b$ to a set of data points $(x_1, y_1), (x_2, y_2), \ldots, (x_n, y_n)$ by using the **method of least squares**.

Ex. 1, p. 465

- We minimize $F(a, b) = \sum_{k=1}^{n} (y_k - ax_k - b)^2$, the **sum of the squares of the residuals**, by computing the first-order partial derivatives of F and setting them equal to 0. Solving for a and b gives the formulas

$$a = \frac{n\left(\sum_{k=1}^{n} x_k y_k\right) - \left(\sum_{k=1}^{n} x_k\right)\left(\sum_{k=1}^{n} y_k\right)}{n\left(\sum_{k=1}^{n} x_k^2\right) - \left(\sum_{k=1}^{n} x_k\right)^2}$$

$$b = \frac{\sum_{k=1}^{n} y_k - a\left(\sum_{k=1}^{n} x_k\right)}{n}$$

- Graphing calculators have built-in routines to calculate linear—as well as quadratic, cubic, quartic, logarithmic, exponential, power, and trigonometric—regression equations.

Ex. 2, p. 467

7.6 Double Integrals over Rectangular Regions

- The **double integral** of a function $f(x, y)$ over a rectangle

$$R = \{(x, y) | a \le x \le b, \quad c \le y \le d\}$$

  is

$$\iint_R f(x, y) \, dA = \int_a^b \left[\int_c^d f(x, y) \, dy \right] dx$$

$$= \int_c^d \left[\int_a^b f(x, y) \, dx \right] dy$$

- In the double integral $\iint_R f(x, y) \, dA$, $f(x, y)$ is called the **integrand** and R is called the **region of integration.** The expression dA indicates that this is an integral over a two-dimensional region. The integrals

$$\int_a^b \left[\int_c^d f(x, y) \, dy \right] dx \qquad \text{and} \qquad \int_c^d \left[\int_a^b f(x, y) \, dx \right] dy$$

  are referred to as **iterated integrals** (the brackets are often omitted), and the order in which dx and dy are written indicates the order of integration.

- The **average value** of the function $f(x, y)$ over the rectangle

$$R = \{(x, y) | a \le x \le b, \quad c \le y \le d\}$$

  is

$$\frac{1}{(b - a)(d - c)} \iint_R f(x, y) \, dA$$

- If $f(x, y) \ge 0$ over a rectangle $R = \{(x, y) | a \le x \le b, c \le y \le d\}$, then the volume of the solid formed by graphing f over the rectangle R is given by

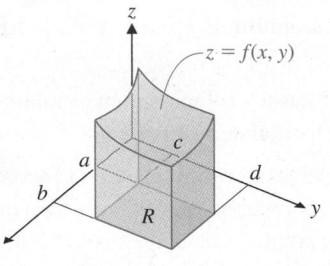

$$V = \iint_R f(x, y) \, dA$$

7.7 Double Integrals over More General Regions

- A region R in the xy plane is a **regular x region** if there exist functions $f(x)$ and $g(x)$ and numbers a and b such that

$$R = \{(x, y) \,|\, g(x) \le y \le f(x), \quad a \le x \le b\}$$

- A region R in the xy plane is a **regular y region** if there exist functions $h(y)$ and $k(y)$ and numbers c and d such that

$$R = \{(x, y) \,|\, h(y) \le x \le k(y), \quad c \le y \le d\}$$

- The double integral of a function $F(x, y)$ over a regular x region $R = \{(x, y) \,|\, g(x) \le y \le f(x),$ $a \le x \le b\}$ is

$$\iint_R F(x, y) \, dA = \int_a^b \left[\int_{g(x)}^{f(x)} F(x, y) \, dy \right] dx$$

- The double integral of a function $F(x, y)$ over a regular y region $R = \{(x, y) \,|\, h(y) \le x \le k(y),$ $c \le y \le d\}$ is

$$\iint_R F(x, y) \, dA = \int_c^d \left[\int_{h(y)}^{k(y)} F(x, y) \, dx \right] dy$$

Review Exercises

Work through all the problems in this chapter review and check your answers in the back of the book. Answers to all review problems are there, along with section numbers in italics to indicate where each type of problem is discussed. Where weaknesses show up, review appropriate sections of the text.

1. For $f(x, y) = 2,000 + 40x + 70y$, find $f(5, 10), f_x(x, y)$, and $f_y(x, y)$.

2. For $z = x^3 y^2$, find $\partial^2 z / \partial x^2$ and $\partial^2 z / \partial x \, \partial y$.

3. Evaluate $\int (6xy^2 + 4y) \, dy$.

4. Evaluate $\int (6xy^2 + 4y) \, dx$.

5. Evaluate $\int_0^1 \int_0^1 4xy \, dy \, dx$.

6. For $f(x, y) = 6 + 5x - 2y + 3x^2 + x^3$, find $f_x(x, y)$, and $f_y(x, y)$, and explain why $f(x, y)$ has no local extrema.

7. For $f(x, y) = 3x^2 - 2xy + y^2 - 2x + 3y - 7$, find $f(2, 3)$ $f_y(x, y)$, and $f_y(2, 3)$.

8. For $f(x, y) = -4x^2 + 4xy - 3y^2 + 4x + 10y + 81$, find $[f_{xx}(2, 3)][f_{yy}(2, 3)] - [f_{xy}(2, 3)]^2$.

9. If $f(x, y) = x + 3y$ and $g(x, y) = x^2 + y^2 - 10$, find the critical points of $F(x, y, \lambda) = f(x, y) + \lambda g(x, y)$.

10. Use the least squares line for the data in the following table to estimate y when $x = 10$.

x	y
2	12
4	10
6	7
8	3

11. For $R = \{(x, y) | -1 \leq x \leq 1, \quad 1 \leq y \leq 2\}$, evaluate the following in two ways:

$$\iint_R (4x + 6y) \, dA$$

12. For $R = \{(x, y) | \sqrt{y} \leq x \leq 1, \quad 0 \leq y \leq 1\}$, evaluate

$$\iint_R (6x + y) \, dA$$

13. For $f(x, y) = e^{x^2 + 2y}$, find f_x, f_y, and f_{xy}.

14. For $f(x, y) = (x^2 + y^2)^5$, find f_x and f_{xy}.

15. Find all critical points and test for extrema for

$$f(x, y) = x^3 - 12x + y^2 - 6y$$

16. Use Lagrange multipliers to maximize $f(x, y) = xy$ subject to $2x + 3y = 24$.

17. Use Lagrange multipliers to minimize $f(x, y, z) = x^2 + y^2 + z^2$ subject to $2x + y + 2z = 9$.

18. Find the least squares line for the data in the following table.

x	y	x	y
10	50	60	80
20	45	70	85
30	50	80	90
40	55	90	90
50	65	100	110

19. Find the average value of $f(x, y) = x^{2/3} y^{1/3}$ over the rectangle

$$R = \{(x, y) | -8 \leq x \leq 8, \quad 0 \leq y \leq 27\}$$

20. Find the volume of the solid under the graph of $z = 3x^2 + 3y^2$ over the rectangle

$$R = \{(x, y) | 0 \leq x \leq 1, \quad -1 \leq y \leq 1\}$$

21. Without doing any computation, predict the average value of $f(x, y) = x + y$ over the rectangle $R = \{(x, y) | -10 \leq x \leq 10, \quad -10 \leq y \leq 10\}$. Then check the correctness of your prediction by evaluating a double integral.

22. (A) Find the dimensions of the square S centered at the origin such that the average value of

$$f(x, y) = \frac{e^x}{y + 10}$$

over S is equal to 5.

    (B) Is there a square centered at the origin over which

$$f(x, y) = \frac{e^x}{y + 10}$$

has average value 0.05? Explain.

23. Explain why the function $f(x, y) = 4x^3 - 5y^3$, subject to the constraint $3x + 2y = 7$, has no maxima or minima.

24. Find the volume of the solid under the graph of $F(x, y) = 60x^2 y$ over the region R bounded by the graph of $x + y = 1$ and the coordinate axes.

Applications

25. **Maximizing profit.** A company produces x units of product A and y units of product B (both in hundreds per month). The monthly profit equation (in thousands of dollars) is given by

$$P(x, y) = -4x^2 + 4xy - 3y^2 + 4x + 10y + 81$$

    (A) Find $P_x(1, 3)$ and interpret the results.

    (B) How many of each product should be produced each month to maximize profit? What is the maximum profit?

26. Minimizing material. A rectangular box with no top and six compartments (see the figure) is to have a volume of 96 cubic inches. Find the dimensions that will require the least amount of material.

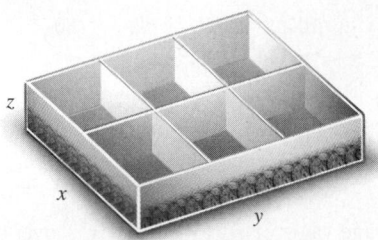

27. Profit. A company's annual profits (in millions of dollars) over a 5-year period are given in the following table. Use the least squares line to estimate the profit for the sixth year.

Year	Profit
1	2
2	2.5
3	3.1
4	4.2
5	4.3

28. Productivity. The Cobb–Douglas production function for a product is

$$N(x, y) = 10x^{0.8}y^{0.2}$$

where x is the number of units of labor and y is the number of units of capital required to produce N units of the product.

(A) Find the marginal productivity of labor and the marginal productivity of capital at $x = 40$ and $y = 50$. For the greatest increase in productivity, should management encourage increased use of labor or increased use of capital?

(B) If each unit of labor costs $100, each unit of capital costs $50, and $10,000 is budgeted for production of this product, use the method of Lagrange multipliers to determine the allocations of labor and capital that will maximize the number of units produced and find the maximum production. Find the marginal productivity of money and approximate the increase in production that would result from an increase of $2,000 in the amount budgeted for production.

(C) If $50 \leq x \leq 100$ and $20 \leq y \leq 40$, find the average number of units produced. Set up a double integral, and evaluate it.

29. Marine biology. When diving using scuba gear, the function used for timing the duration of the dive is

$$T(V, x) = \frac{33V}{x + 33}$$

where T is the time of the dive in minutes, V is the volume of air (in cubic feet, at sea-level pressure) compressed into tanks, and x is the depth of the dive in feet. Find $T_x(70, 17)$ and interpret the results.

30. Pollution. A heavy industrial plant located in the center of a small town emits particulate matter into the atmosphere. Suppose that the concentration of particulate matter (in parts per million) at a point d miles from the plant is given by

$$C = 100 - 24d^2$$

If the boundaries of the town form a square 4 miles long and 4 miles wide, what is the average concentration of particulate matter throughout the town? Express C as a function of x and y, and set up a double integral and evaluate it.

31. Sociology. A sociologist found that the number n of long-distance telephone calls between two cities during a given period varied (approximately) jointly as the populations P_1 and P_2 of the two cities and varied inversely as the distance d between the cities. An equation for a period of 1 week is

$$n(P_1, P_2, d) = 0.001\frac{P_1P_2}{d}$$

Find $n(100{,}000, 50{,}000, 100)$.

32. Education. At the beginning of the semester, students in a foreign language course take a proficiency exam. The same exam is given at the end of the semester. The results for 5 students are shown in the following table. Use the least squares line to estimate the second exam score of a student who scored 40 on the first exam.

First Exam	Second Exam
30	60
50	75
60	80
70	85
90	90

33. Population density. The following table gives the U.S. population per square mile for the years 1960–2010:

U.S. Population Density

Year	Population (per square mile)
1960	50.6
1970	57.5
1980	64.1
1990	70.4
2000	79.7
2010	87.4

(A) Find the least squares line for the data, using $x = 0$ for 1960.

(B) Use the least squares line to estimate the population density in the United States in the year 2025.

(C) Now use quadratic regression and exponential regression to obtain the estimate of part (B).

34. *Life expectancy.* The following table gives life expectancies for males and females in a sample of Central and South American countries:

Life Expectancies for Central and South American Countries

Males	Females	Males	Females
62.30	67.50	70.15	74.10
68.05	75.05	62.93	66.58
72.40	77.04	68.43	74.88
63.39	67.59	66.68	72.80
55.11	59.43		

(A) Find the least squares line for the data.

(B) Use the least squares line to estimate the life expectancy of a female in a Central or South American country in which the life expectancy for males is 60 years.

(C) Now use quadratic regression and logarithmic regression to obtain the estimate of part (B).

8 Trigonometric Functions

Introduction

Business cycles, blood pressure in the aorta, seasonal growth, water waves, and amounts of pollution in the atmosphere are periodic (or cyclical) phenomena. To model such phenomena, we need functions that are periodic. The algebraic, exponential, and logarithmic functions studied in previous chapters do not have this property. But the trigonometric functions do. They are well suited to describe phenomena that repeat in cycles (see, for example, Problem 50 in Section 8.2 on air pollution). In Section 8.1, we provide a brief review of basic topics in trigonometry. In Sections 8.2 and 8.3, we study derivatives and integrals of the trigonometric functions.

8.1 Trigonometric Functions Review

- Angles: Degree–Radian Measure
- Trigonometric Functions
- Graphs of the Sine and Cosine Functions
- Four Other Trigonometric Functions

Angles: Degree—Radian Measure

In a plane, an **angle** is formed by rotating a ray m, called the **initial side** of the angle, around its endpoint until the ray coincides with a ray n, called the **terminal side** of the angle. The common endpoint P of m and n is called the **vertex** (see Fig. 1).

There is no restriction on the amount or direction of rotation. A counterclockwise rotation produces a **positive** angle (Fig. 2A), and a clockwise rotation produces a **negative** angle (Fig. 2B). Two different angles may have the same initial and terminal sides, as shown in Figure 2C. Such angles are said to be **coterminal**.

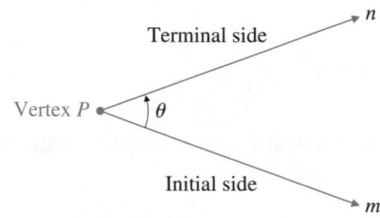

Figure 1 **Angle θ**

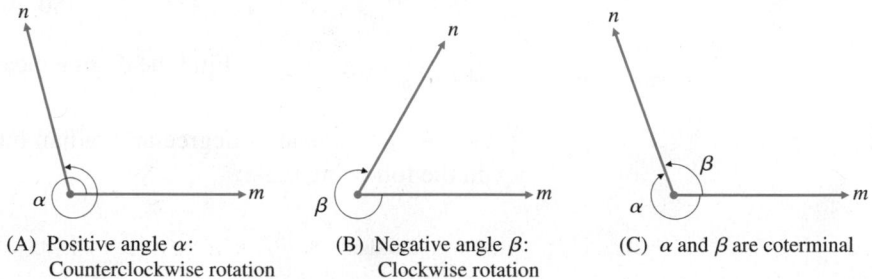

(A) Positive angle α: Counterclockwise rotation

(B) Negative angle β: Clockwise rotation

(C) α and β are coterminal

Figure 2

There are two widely used measures of angles: the *degree* and the *radian*. When a central angle of a circle is subtended by an arc that is $\frac{1}{360}$ the circumference of the circle, the angle is said to have **degree measure 1**, written as **1°** (see Fig. 3A). It follows that a central angle subtended by an arc that is $\frac{1}{4}$ of the circumference has a degree measure of 90, $\frac{1}{2}$ of the circumference has a degree measure of 180, and the whole circumference of a circle has a degree measure of 360.

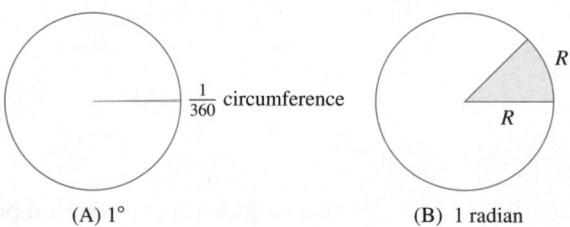

(A) 1°

(B) 1 radian

Figure 3 **Degree and radian measure**

The other measure of angles is radian measure. A central angle subtended by an arc of length equal to the radius (R) of the circle is said to have **radian measure 1**, written as **1 radian** or **1 rad** (see Fig. 3B). In general, a central angle subtended by an arc of length s has radian measure that is determined as follows:

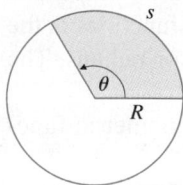

Figure 4

$$\theta_{\text{rad}} = \text{radian measure of } \theta = \frac{\text{arc length}}{\text{radius}} = \frac{s}{R}$$

(See Figure 4.) [*Note*: If $R = 1$, then $\theta_{\text{rad}} = s$.]

What is the radian measure of a 180° angle? A central angle of 180° is subtended by an arc that is $\frac{1}{2}$ the circumference of a circle. Thus,

$$s = \frac{C}{2} = \frac{2\pi R}{2} = \pi R \quad \text{and} \quad \theta_{\text{rad}} = \frac{s}{R} = \frac{\pi R}{R} = \pi \text{ rad}$$

The following proportion can be used to convert degree measure to radian measure and vice versa:

Degree–Radian Conversion

$$\frac{\theta_{\text{deg}}}{180°} = \frac{\theta_{\text{rad}}}{\pi \text{ rad}}$$

EXAMPLE 1   From Degrees to Radians   Find the radian measure of 1°.

SOLUTION

$$\frac{1°}{180°} = \frac{\theta_{\text{rad}}}{\pi \text{ rad}}$$

$$\theta_{\text{rad}} = \frac{\pi}{180} \text{ rad} \approx 0.0175 \text{ rad}$$

Matched Problem 1 Find the degree measure of 1 rad.

A comparison of degree and radian measure for a few important angles is given in the following table:

Radian	0	$\pi/6$	$\pi/4$	$\pi/3$	$\pi/2$	π	2π
Degree	0	30°	45°	60°	90°	180°	360°

An angle in a rectangular coordinate system is said to be in **standard position** if its vertex is at the origin and its initial side is on the positive x axis. Figure 5 shows three angles in standard position.

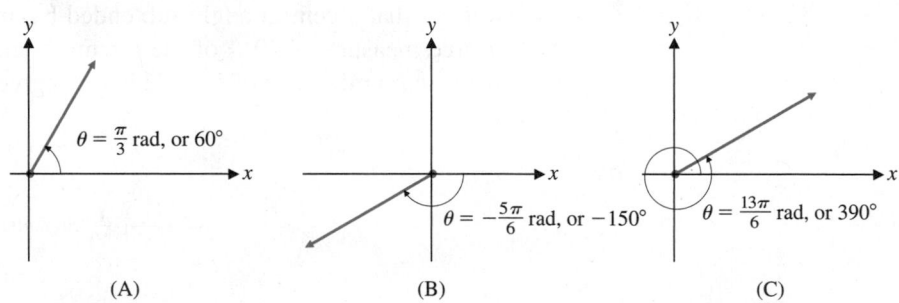

| (A) | (B) | (C) |

$\theta = \frac{\pi}{3}$ rad, or 60° $\theta = -\frac{5\pi}{6}$ rad, or −150° $\theta = \frac{13\pi}{6}$ rad, or 390°

Figure 5 **Angles in standard position**

Trigonometric Functions

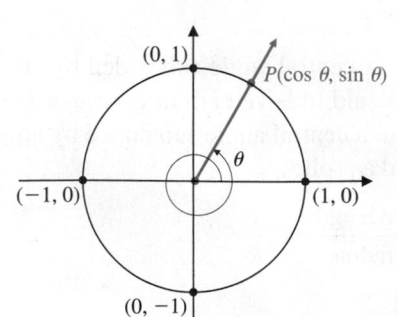

Figure 6

Consider a unit circle (radius 1) in a coordinate system with center at the origin (Fig. 6). The terminal side of any angle in standard position will pass through this circle at some point P. The abscissa of point P is called the **cosine of θ** (abbreviated **cos θ**), and the ordinate of the point is the **sine of θ** (abbreviated **sin θ**). The set of all ordered pairs of the form $(\theta, \cos \theta)$ and the set of all ordered pairs of the form $(\theta, \sin \theta)$ constitute, respectively, the **cosine** and **sine functions**. The **domain** of these two functions is the set of all angles, positive or negative, with measure either in degrees or radians. The **range** is a subset of the set of real numbers.

It is necessary for our work in calculus to define these two trigonometric functions in terms of real-number domains. This is done as follows:

DEFINITION   Sine and Cosine Functions with Real-Number Domains

For any real number x,

$$\sin x = \sin(x \text{ radians}) \quad \text{and} \quad \cos x = \cos(x \text{ radians})$$

EXAMPLE 2   Evaluating Sine and Cosine Functions   Referring to Figure 6, find

(A) $\cos 90°$ (B) $\sin(-\pi/2 \text{ rad})$ (C) $\cos \pi$

SOLUTION

(A) The terminal side of an angle of degree measure 90 passes through $(0, 1)$ on the unit circle. This point has abscissa 0. So

$$\cos 90° = 0$$

(B) The terminal side of an angle of radian measure $-\pi/2$ $(-90°)$ passes through $(0, -1)$ on the unit circle. This point has ordinate -1. So

$$\sin\left(-\frac{\pi}{2} \text{ rad}\right) = -1$$

(C) $\cos \pi = \cos(\pi \text{ rad}) = -1$, since the terminal side of an angle of radian measure $\pi(180°)$ passes through $(-1, 0)$ on the unit circle and this point has abscissa -1.

Matched Problem 2   Referring to Figure 6, find

(A) $\sin 180°$ (B) $\cos(2\pi \text{ rad})$ (C) $\sin(-\pi)$

Finding the value of either the sine or cosine function for any angle or any real number using the definition is not easy. Calculators with sin and cos keys are used. Calculators generally have degree and radian options, so we can use a calculator to evaluate these functions for most of the real numbers in which we might have an interest. The following table includes a few values produced by a calculator in radian mode:

x	1	-7	35.26	-105.9
$\sin x$	0.8415	-0.6570	-0.6461	0.7920
$\cos x$	0.5403	0.7539	-0.7632	0.6105

Explore and Discuss 1   Many errors in trigonometry can be traced to having the calculator in the wrong mode, radian instead of degree, or vice versa, when performing calculations. The calculator screen in Figure 7 gives two different values for cos 30. Experiment with your calculator and explain the discrepancy.

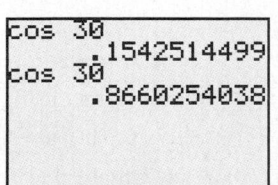

Figure 7

Exact values of the sine and cosine functions can be obtained for multiples of the special angles shown in Figure 8, because these triangles can be used to find the coordinate of the intersection of the terminal side of each angle with the unit circle.

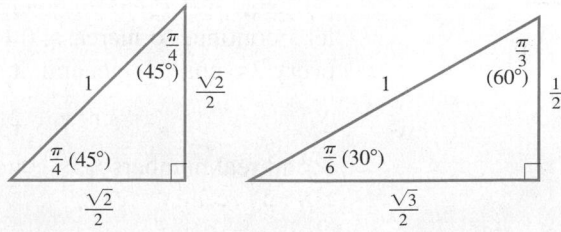

Figure 8

EXAMPLE 3   Finding Exact Values for Special "Angles"   Use Figure 8 to find the exact value of each of the following:

(A) $\cos\dfrac{\pi}{4}$ (B) $\sin\dfrac{\pi}{6}$ (C) $\sin\left(-\dfrac{\pi}{6}\right)$

SOLUTION

(A) $\cos\dfrac{\pi}{4} = \dfrac{\sqrt{2}}{2}$ (B) $\sin\dfrac{\pi}{6} = \dfrac{1}{2}$ (C) $\sin\left(-\dfrac{\pi}{6}\right) = -\dfrac{1}{2}$

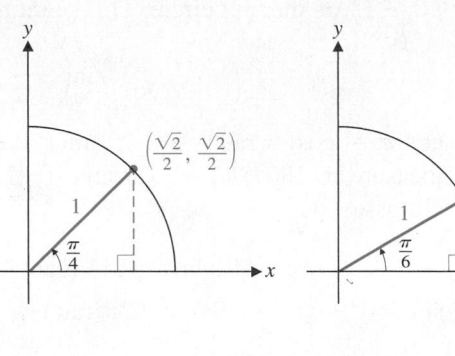 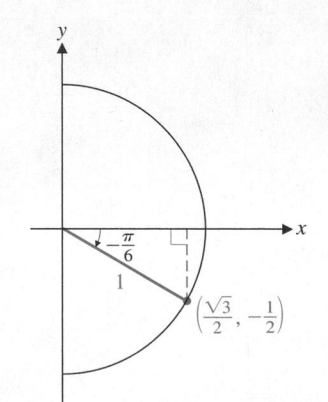

Matched Problem 3   Use Figure 8 to find the exact value of each of the following:

(A) $\sin\dfrac{\pi}{4}$ (B) $\cos\dfrac{\pi}{3}$ (C) $\cos\left(-\dfrac{\pi}{3}\right)$

Graphs of the Sine and Cosine Functions

To graph $y = \sin x$ or $y = \cos x$ for x a real number, we could use a calculator to produce a table and then plot the ordered pairs from the table in a coordinate system. However, we can speed up the process by returning to basic definitions. Referring to Figure 9, since $\cos x$ and $\sin x$ are the coordinates of a point on the unit circle, we see that

$$-1 \le \sin x \le 1 \quad \text{and} \quad -1 \le \cos x \le 1$$

for all real numbers x. Furthermore, as x increases and P moves around the unit circle in a counterclockwise (positive) direction, both $\sin x$ and $\cos x$ behave in uniform ways, as indicated in the following table:

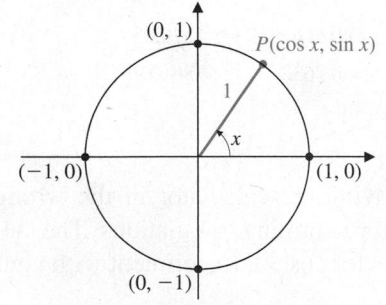

Figure 9

As x Increases from	$y = \sin x$	$y = \cos x$
0 to $\pi/2$	Increases from 0 to 1	Decreases from 1 to 0
$\pi/2$ to π	Decreases from 1 to 0	Decreases from 0 to -1
π to $3\pi/2$	Decreases from 0 to -1	Increases from -1 to 0
$3\pi/2$ to 2π	Increases from -1 to 0	Increases from 0 to 1

Note that P has completed one revolution and is back at its starting place. If we let x continue to increase, the second and third columns in the table will be repeated every 2π units. In general, it can be shown that

$$\sin(x + 2\pi) = \sin x \qquad \cos(x + 2\pi) = \cos x$$

for all real numbers x. Functions such that

$$f(x + p) = f(x)$$

for some positive constant p and all real numbers x for which the functions are defined are said to be **periodic**. The smallest such value of p is called the **period** of the function. Both the sine and cosine functions are periodic with period 2π.

Putting all this information together and adding a few values obtained from a calculator or Figure 8, we obtain the graphs of the sine and cosine functions illustrated in Figure 10. Notice that these curves are continuous. It can be shown that **the sine and cosine functions are continuous for all real numbers.**

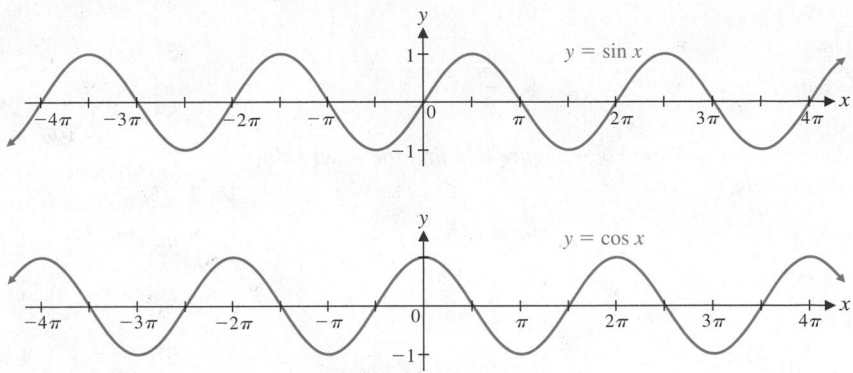

Figure 10

Four Other Trigonometric Functions

The sine and cosine functions are only two of six trigonometric functions. They are, however, the most important of the six for many applications. The other four trigonometric functions are the **tangent, cotangent, secant**, and **cosecant**.

DEFINITION   Four Other Trigonometric Functions

$$\tan x = \frac{\sin x}{\cos x} \quad \cos x \neq 0 \qquad \sec x = \frac{1}{\cos x} \quad \cos x \neq 0$$

$$\cot x = \frac{\cos x}{\sin x} \quad \sin x \neq 0 \qquad \csc x = \frac{1}{\sin x} \quad \sin x \neq 0$$

CONCEPTUAL INSIGHT

The functions $\sin x$ and $\cos x$ are periodic with period 2π, so

$$\tan(x + 2\pi) = \frac{\sin(x + 2\pi)}{\cos(x + 2\pi)} = \frac{\sin x}{\cos x} = \tan x$$

One might guess that $\tan x$ is periodic with period 2π. However, 2π is not the *smallest* positive constant p such that $\tan(x + p) = \tan x$. Because the points $(\cos x, \sin x)$ and $(\cos(x + \pi), \sin(x + \pi))$ are diametrically opposed on the unit circle,

$$\sin(x + \pi) = -\sin x \quad \text{and} \quad \cos(x + \pi) = -\cos x$$

Therefore,

$$\tan(x + \pi) = \frac{\sin(x + \pi)}{\cos(x + \pi)} = \frac{-\sin x}{-\cos x} = \tan x$$

It follows that the functions $\tan x$ and $\cot x$ have period π. The other four trigonometric functions—$\sin x$, $\cos x$, $\sec x$, and $\csc x$—all have period 2π.

Exercises 8.1

Skills Warm-up Exercises

W   *In Problems 1–8, mentally convert each degree measure to radian measure, and each radian measure to degree measure.*

1. 60°

2. 90°

3. 135°

4. −30°

5. $-\dfrac{\pi}{4}$ rad

6. $\dfrac{5\pi}{4}$ rad

7. $\dfrac{3\pi}{2}$ rad

8. $\dfrac{2\pi}{3}$ rad

In Problems 9–24, use Figure 6 or Figure 8 to find the exact value of each expression.

9. sin 60°

10. sin 45°

11. cos 135°

12. cos 120°

13. sin 90°

14. cos 180°

15. cos(−90°)

16. sin(−180°)

17. $\cos\left(\dfrac{5\pi}{4}\right)$

18. $\sin\left(\dfrac{3\pi}{4}\right)$

19. $\sin\left(-\dfrac{\pi}{6}\right)$

20. $\cos\left(-\dfrac{2\pi}{3}\right)$

21. $\sin\left(\dfrac{3\pi}{2}\right)$

22. cos (2π)

23. $\cos\left(-\dfrac{11\pi}{6}\right)$

24. $\sin\left(-\dfrac{5\pi}{6}\right)$

25. Refer to Figure 6 and use the Pythagorean theorem to show that
$$(\sin x)^2 + (\cos x)^2 = 1$$
for all *x*.

26. Use the results of Problem 25 and basic definitions to show that

(A) $(\tan x)^2 + 1 = (\sec x)^2$

(B) $1 + (\cot x)^2 = (\csc x)^2$

In Problems 27–42, use Figure 6 or Figure 8 to find the exact value of each expression.

27. $\tan\left(\dfrac{3\pi}{4}\right)$

28. $\sec\left(\dfrac{\pi}{3}\right)$

29. $\csc\left(\dfrac{2\pi}{3}\right)$

30. $\cot\left(\dfrac{\pi}{4}\right)$

31. sec 90°

32. csc (−30°)

33. cot (−150°)

34. tan (−90°)

35. $\csc\left(\dfrac{7\pi}{6}\right)$

36. $\cot\left(\dfrac{5\pi}{6}\right)$

37. sec (−π)

38. $\csc\left(-\dfrac{3\pi}{2}\right)$

39. tan 120°

40. sec (135°)

41. cot (−45°)

42. tan (−720°)

In Problems 43–54, use a calculator in radian or degree mode, as appropriate, to find the value of each expression to four decimal places.

43. sin 10°

44. tan 141°

45. cos (−52°)

46. sec (−18°)

47. tan 1

48. cot (−2)

49. sec (−1.56)

50. cos 3.13

51. csc 1°

52. csc 182°

53. $\cot\left(\dfrac{\pi}{10}\right)$

54. $\sin\left(\dfrac{4\pi}{5}\right)$

In Problems 55–58, use a graphing calculator set in radian mode to graph each function.

55. $y = 2 \sin \pi x; \ 0 \le x \le 2, -2 \le y \le 2$

56. $y = -0.5 \cos 2x; \ 0 \le x \le 2\pi, -0.5 \le y \le 0.5$

57. $y = 4 - 4 \cos\dfrac{\pi x}{2}; \ 0 \le x \le 8, 0 \le y \le 8$

58. $y = 6 + 6 \sin\dfrac{\pi x}{26}; \ 0 \le x \le 104, 0 \le y \le 12$

59. Find the domain of the tangent function.

60. Find the domain of the cotangent function.

61. Find the domain of the secant function.

62. Find the domain of the cosecant function.

63. Explain why the range of the cosecant function is $(-\infty, -1] \cup [1, \infty)$.

64. Explain why the range of the secant function is $(-\infty, -1] \cup [1, \infty)$.

65. Explain why the range of the cotangent function is $(-\infty, \infty)$.

66. Explain why the range of the tangent function is $(-\infty, \infty)$.

Applications

67. Seasonal business cycle. Suppose that profits on the sale of swimming suits over a 2-year period are given approximately by

$$P(t) = 5 - 5 \cos\dfrac{\pi t}{26} \qquad 0 \le t \le 104$$

where *P* is profit (in hundreds of dollars) for a week of sales *t* weeks after January 1. The graph of the profit function is shown in the figure.

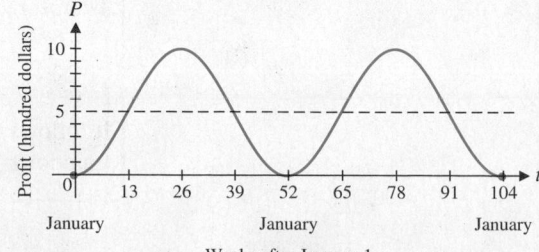

(A) Find the exact values of $P(13), P(26), P(39)$, and $P(52)$ without using a calculator.

✎ (B) Use a calculator to find $P(30)$ and $P(100)$. Interpret the results.

⊞ (C) Use a graphing calculator to confirm the graph shown here for $y = P(t)$.

68. Seasonal business cycle. Revenues from sales of a soft drink over a 2-year period are given approximately by

$$R(t) = 4 - 3\cos\frac{\pi t}{6} \qquad 0 \le t \le 24$$

where $R(t)$ is revenue (in millions of dollars) for a month of sales t months after February 1. The graph of the revenue function is shown in the figure.

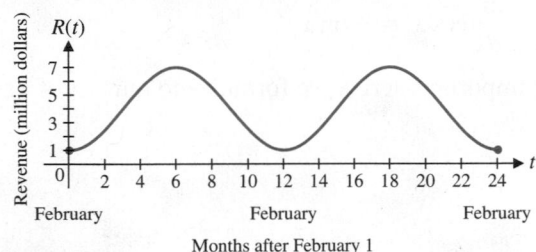

Months after February 1

(A) Find the exact values of $R(0), R(2), R(3)$, and $R(18)$ without using a calculator.

✎ (B) Use a calculator to find $R(5)$ and $R(23)$. Interpret the results.

⊞ (C) Use a graphing calculator to confirm the graph shown here for $y = R(t)$.

69. Physiology. A normal seated adult inhales and exhales about 0.8 liter of air every 4 seconds. The volume $V(t)$ of air in the lungs t seconds after exhaling is given approximately by

$$V(t) = 0.45 - 0.35\cos\frac{\pi t}{2} \qquad 0 \le t \le 8$$

The graph for two complete respirations is shown in the figure.

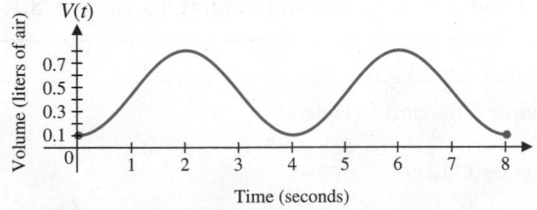

Time (seconds)

(A) Find the exact value of $V(0), V(1), V(2), V(3)$, and $V(7)$ without using a calculator.

✎ (B) Use a calculator to find $V(3.5)$ and $V(5.7)$. Interpret the results.

⊞ (C) Use a graphing calculator to confirm the graph shown here for $y = V(t)$.

70. Pollution. In a large city, the amount of sulfur dioxide pollutant released into the atmosphere due to the burning of coal and oil for heating purposes varies seasonally. Suppose that the number of tons of pollutant released into

the atmosphere during the nth week after January 1 is given approximately by

$$P(n) = 1 + \cos\frac{\pi n}{26} \qquad 0 \le n \le 104$$

The graph of the pollution function is shown in the figure.

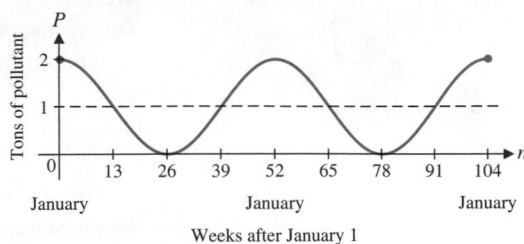

Weeks after January 1

(A) Find the exact values of $P(0), P(39), P(52)$, and $P(65)$ without using a calculator.

✎ (B) Use a calculator to find $P(10)$ and $P(95)$. Interpret the results.

⊞ (C) Use a graphing calculator to confirm the graph shown here for $y = P(n)$.

71. Psychology. Individuals perceive objects differently in different settings. Consider the well-known illusions shown in Figure A. Lines that appear parallel in one setting may appear to be curved in another (the two vertical lines are actually parallel). Lines of the same length may appear to be of different lengths in two different settings (the two horizontal lines are actually the same length). Psychologists Berliner and Berliner reported that when subjects were presented with a large tilted field of parallel lines and were asked to estimate the position of a horizontal line in the field, most of the subjects were consistently off (Figure B). They found that the difference d in degrees between the estimates and the actual horizontal could be approximated by the equation

$$d = a + b\sin 4\theta$$

where a and b are constants associated with a particular person and θ is the angle of tilt of the visual field (in degrees). Suppose that, for a given person, $a = -2.1$ and $b = -4$. Find d if

(A) $\theta = 30°$ (B) $\theta = 10°$

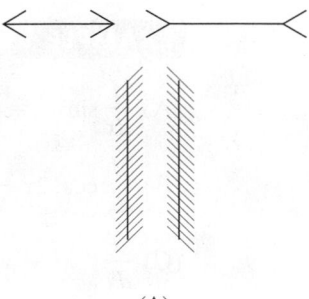

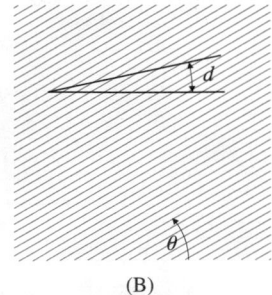

(A)   (B)

Answers to Matched Problems

1. $180/\pi \approx 57.3°$

2. (A) 0   (B) 1   (C) 0

3. (A) $\sqrt{2}/2$ (B) $\frac{1}{2}$ (C) $\frac{1}{2}$

8.2 Derivatives of Trigonometric Functions

- Derivative Formulas
- Application

Derivative Formulas

In this section, we discuss derivative formulas for the sine and cosine functions. Once we have these formulas, we will automatically have integral formulas for the same functions, which we discuss in the next section.

From the definition of the derivative (Section 2.4),

$$\frac{d}{dx}\sin x = \lim_{h \to 0} \frac{\sin(x + h) - \sin x}{h}$$

On the basis of trigonometric identities and some special trigonometric limits, it can be shown that the limit on the right is $\cos x$. Similarly, it can be shown that

$$\frac{d}{dx}\cos x = -\sin x$$

We now add the following important derivative formulas to our list of derivative formulas:

> **Derivative of Sine and Cosine**
> *Basic Form*
>
> $$\frac{d}{dx}\sin x = \cos x \qquad \frac{d}{dx}\cos x = -\sin x$$
>
> *Generalized Form*
> For $u = u(x)$,
>
> $$\frac{d}{dx}\sin u = \cos u \frac{du}{dx} \qquad \frac{d}{dx}\cos u = -\sin u \frac{du}{dx}$$

> **CONCEPTUAL INSIGHT**
>
> The derivative formula for the function $y = \sin x$ implies that each line tangent to the graph of the function has a slope between -1 and 1. Furthermore, the slope of the line tangent to $y = \sin x$ is equal to 1 if and only if $\cos x = 1$—that is, at $x = 0, \pm 2\pi, \pm 4\pi, \ldots$. Similarly, the derivative formula for $y = \cos x$ implies that each line tangent to the graph of the function has a slope between -1 and 1. The slope of the tangent line is equal to 1 if and only if $-\sin x = 1$—that is, at $x = 3\pi/2, (3\pi/2) \pm 2\pi, (3\pi/2) \pm 4\pi, \ldots$. Note that these observations are consistent with the graphs of $y = \sin x$ and $y = \cos x$ shown in Figure 10, Section 8.1.

EXAMPLE 1   Derivatives Involving Sine and Cosine

(A) $\dfrac{d}{dx}\sin x^2 = (\cos x^2)\dfrac{d}{dx}x^2 = (\cos x^2)2x = 2x \cos x^2$

(B) $\dfrac{d}{dx}\cos(2x - 5) = -\sin(2x - 5)\dfrac{d}{dx}(2x - 5) = -2\sin(2x - 5)$

(C) $\dfrac{d}{dx}(3x^2 - x)\cos x = (3x^2 - x)\dfrac{d}{dx}\cos x + (\cos x)\dfrac{d}{dx}(3x^2 - x)$

$\qquad = -(3x^2 - x)\sin x + (6x - 1)\cos x$

$\qquad = (x - 3x^2)\sin x + (6x - 1)\cos x$

Matched Problem 1   Find each of the following derivatives:

(A) $\dfrac{d}{dx}\cos x^3$ 　　　　 (B) $\dfrac{d}{dx}\sin(5 - 3x)$ 　　　　 (C) $\dfrac{d}{dx}\dfrac{\sin x}{x}$

EXAMPLE 2 Slope Find the slope of the graph of $f(x) = \sin x$ at $(\pi/2, 1)$, and sketch in the line tangent to the graph at this point.

SOLUTION Slope at $(\pi/2, 1) = f'(\pi/2) = \cos(\pi/2) = 0$.

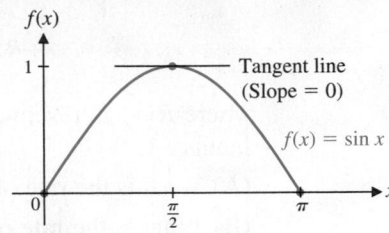

Matched Problem 2 Find the slope of the graph of $f(x) = \cos x$ at $(\pi/6, \sqrt{3}/2)$.

Explore and Discuss 1 From the graph of $y = f'(x)$ shown in Figure 1, describe the shape of the graph of $y = f(x)$ relative to where it is increasing, where it is decreasing, its concavity, and the locations of local maxima and minima. Make a sketch of a possible graph of $y = f(x), 0 \le x \le 2\pi$, given that it has x intercepts at $(0, 0)$, $(\pi, 0)$, and $(2\pi, 0)$. Can you identify $f(x)$ and $f'(x)$ in terms of sine or cosine functions?

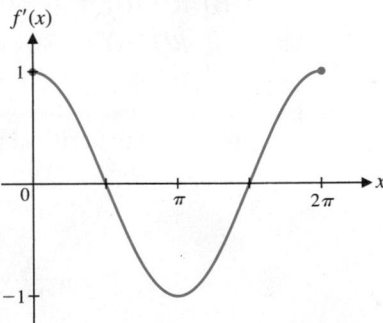

Figure 1

EXAMPLE 3 Derivative of Secant Find $\dfrac{d}{dx} \sec x$.

SOLUTION

$$\frac{d}{dx} \sec x = \frac{d}{dx} \frac{1}{\cos x} \qquad \sec x = \frac{1}{\cos x}$$

$$= \frac{d}{dx} (\cos x)^{-1}$$

$$= -(\cos x)^{-2} \frac{d}{dx} \cos x$$

$$= -(\cos x)^{-2}(-\sin x)$$

$$= \frac{\sin x}{(\cos x)^2}$$

$$= \left(\frac{\sin x}{\cos x}\right)\left(\frac{1}{\cos x}\right) \qquad \tan x = \frac{\sin x}{\cos x}$$

$$= \tan x \sec x$$

Matched Problem 3 Find $\dfrac{d}{dx} \csc x$.

Application

EXAMPLE 4   Revenue   Revenues from the sale of ski jackets are given approximately by

$$R(t) = 1.55 + 1.45 \cos \frac{\pi t}{26} \qquad 0 \le t \le 104$$

where $R(t)$ is revenue (in thousands of dollars) for a week of sales t weeks after January 1.

(A) What is the rate of change of revenue t weeks after the first of the year?

(B) What is the rate of change of revenue 10 weeks after the first of the year? 26 weeks after the first of the year? 40 weeks after the first of the year?

(C) Find all local maxima and minima for $0 < t < 104$.

(D) Find the absolute maximum and minimum for $0 \le t \le 104$.

(E) Illustrate the results from parts (A)–(D) by sketching a graph of $y = R(t)$ with the aid of a graphing calculator.

SOLUTION

(A) $R'(t) = -\dfrac{1.45\pi}{26} \sin \dfrac{\pi t}{26} \qquad 0 \le t \le 104$

(B) $R'(10) \approx -\$0.164$ thousand,   or $-\$164$ per week

$R'(26) = \$0$ per week

$R'(40) \approx \$0.174$ thousand,   or $\$174$ per week

(C) Find the critical points:

$$R'(t) = -\frac{1.45\pi}{26} \sin \frac{\pi t}{26} = 0 \qquad 0 < t < 104$$

$$\sin \frac{\pi t}{26} = 0$$

$$\frac{\pi t}{26} = \pi, 2\pi, 3\pi \quad \text{Note: } 0 < t < 104 \text{ implies that } 0 < \frac{\pi t}{26} < 4\pi.$$

$$t = 26, 52, 78$$

Differentiate $R'(t)$ to get $R''(t)$.

$$R''(t) = -\frac{1.45\pi^2}{26^2} \cos \frac{\pi t}{26}$$

Use the second-derivative test to get the results shown in Table 1.

Table 1

t	$R''(t)$	Graph of R
26	+	Local minimum
52	–	Local maximum
78	+	Local minimum

(D) Evaluate $R(t)$ at endpoints $t = 0$ and $t = 104$ and at the critical points found in part (C), as listed in Table 2. The absolute maximum is 3,000 and it occurs at $t = 0, 52$, and 104; the absolute minimum is 100 at it occurs at $t = 26$ and 78.

Table 2

t	$R(t)$	
0	\$3,000	Absolute maximum
26	\$100	Absolute minimum
52	\$3,000	Absolute maximum
78	\$100	Absolute minimum
104	\$3,000	Absolute maximum

(E) The results from parts (A)–(D) can be visualized as shown in Figure 2.

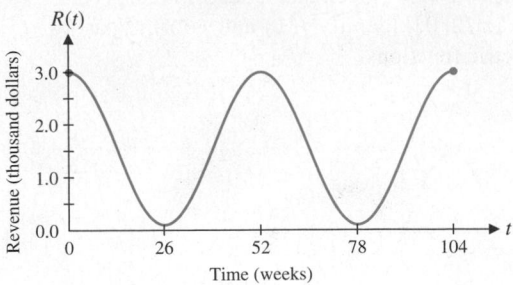

Figure 2

Matched Problem 4 Suppose that revenues from the sale of ski jackets are given approximately by

$$R(t) = 6.2 + 5.8 \cos \frac{\pi t}{6} \qquad 0 \le t \le 24$$

where $R(t)$ is revenue (in thousands of dollars) for a month of sales t months after January 1.

(A) What is the rate of change of revenue t months after the first of the year?

(B) What is the rate of change of revenue 2 months after the first of the year? 12 months after the first of the year? 23 months after the first of the year?

(C) Find all local maxima and minima for $0 < t < 24$.

(D) Find the absolute maximum and minimum for $0 \le t \le 24$.

(E) Illustrate the results from parts (A)–(D) by sketching a graph of $y = R(t)$ with the aid of a graphing calculator.

Exercises 8.2

Skills Warm-up Exercises

W *In Problems 1–4, by inspecting a graph of $y = \sin x$ or $y = \cos x$, determine whether the function is increasing or decreasing on the given interval.*

1. $y = \cos x$ on $(0, \pi)$

2. $y = \sin x$ on $(\pi/2, 3\pi/2)$

3. $y = \sin x$ on $(-\pi/2, \pi/2)$

4. $y = \cos x$ on $(\pi, 2\pi)$

In Problems 5–8, by inspecting a graph of $y = \sin x$ or $y = \cos x$, determine whether the graph is concave up or concave down on the given interval.

5. $y = \sin x$ on $(0, \pi)$

6. $y = \cos x$ on $(\pi/2, 3\pi/2)$

7. $y = \cos x$ on $(-\pi/2, \pi/2)$

8. $y = \sin x$ on $(-\pi, 0)$

Find the indicated derivatives in Problems 9–26.

9. $\dfrac{d}{dx}(5 \cos x)$

10. $\dfrac{d}{dx}(8 \sin x)$

11. $\dfrac{d}{dx} \cos(5x)$

12. $\dfrac{d}{dx} \sin(8x)$

13. $\dfrac{d}{dx} \sin(x^2 + 1)$

14. $\dfrac{d}{dx} \cos(x^3 - 1)$

15. $\dfrac{d}{dw} \sin(w + \pi)$

16. $\dfrac{d}{dt} \cos\left(\dfrac{\pi t}{2}\right)$

17. $\dfrac{d}{dt} t \sin t$

18. $\dfrac{d}{du} u \cos u$

19. $\dfrac{d}{dx} \sin x \cos x$

20. $\dfrac{d}{dx} \dfrac{\sin x}{\cos x}$

21. $\dfrac{d}{dx}(\sin x)^5$

22. $\dfrac{d}{dx}(\cos x)^8$

23. $\dfrac{d}{dx} \sqrt{\sin x}$

24. $\dfrac{d}{dx} \sqrt{\cos x}$

25. $\dfrac{d}{dx} \cos \sqrt{x}$

26. $\dfrac{d}{dx} \sin \sqrt{x}$

27. Find the slope of the graph of $f(x) = \sin x$ at $x = \pi/6$.

28. Find the slope of the graph of $f(x) = \cos x$ at $x = \pi/4$.

29. From the graph of $y = f'(x)$ shown here, describe the shape of the graph of $y = f(x)$ relative to where it is increasing, where it is decreasing, its concavity, and the locations of

local maxima and minima. Make a sketch of a possible graph of $y = f(x)$, $-\pi \le x \le \pi$, given that it has x intercepts at $(-\pi/2, 0)$ and $(\pi/2, 0)$. Identify $f(x)$ and $f'(x)$ as particular trigonometric functions.

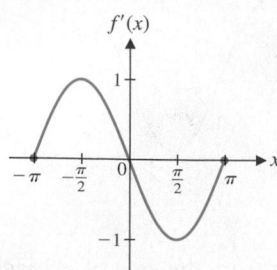

Figure for 29

30. From the graph of $y = f'(x)$ shown here, describe the shape of the graph of $y = f(x)$ relative to where it is increasing, where it is decreasing, its concavity, and the locations of local maxima and minima. Make a sketch of a possible graph of $y = f(x)$, $-\pi \le x \le \pi$, given that it has x intercepts at $(-\pi, 0)$ and $(\pi, 0)$. Identify $f(x)$ and $f'(x)$ as particular trigonometric functions.

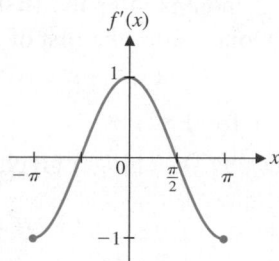

Find the indicated derivatives in Problems 31–38.

31. $\dfrac{d}{dx} \csc (\pi x)$

32. $\dfrac{d}{dx} \sec (x + \pi)$

33. $\dfrac{d}{dx} \cot \left(\dfrac{\pi x}{2} \right)$

34. $\dfrac{d}{dx} \tan (\pi x)$

35. $\dfrac{d}{dx} \cos (xe^x)$

36. $\dfrac{d}{dx} \sin (xe^x)$

37. $\dfrac{d}{dx} \tan (x^2)$

38. $\dfrac{d}{dx} \csc (x^3)$

In Problems 39 and 40, find $f''(x)$.

39. $f(x) = e^x \sin x$

40. $f(x) = e^x \cos x$

In Problems 41–46, graph each function on a graphing calculator.

41. $y = x \sin \pi x$; $0 \le x \le 9$, $-9 \le y \le 9$

42. $y = -x \cos \pi x$; $0 \le x \le 9$, $-9 \le y \le 9$

43. $y = \dfrac{\cos \pi x}{x}$; $0 \le x \le 8$, $-2 \le y \le 3$

44. $y = \dfrac{\sin \pi x}{0.5x}$; $0 \le x \le 8$, $-2 \le y \le 3$

45. $y = e^{-0.3x} \sin \pi x$; $0 \le x \le 10$, $-1 \le y \le 1$

46. $y = e^{-0.2x} \cos \pi x$; $0 \le x \le 10$, $-1 \le y \le 1$

Applications

47. Profit. Suppose that profits on the sale of swimming suits are given approximately by

$$P(t) = 5 - 5 \cos \frac{\pi t}{26} \qquad 0 \le t \le 104$$

where $P(t)$ is profit (in hundreds of dollars) for a week of sales t weeks after January 1.

(A) What is the rate of change of profit t weeks after the first of the year?

(B) What is the rate of change of profit 8 weeks after the first of the year? 26 weeks after the first of the year? 50 weeks after the first of the year?

(C) Find all local maxima and minima for $0 < t < 104$.

(D) Find the absolute maximum and minimum for $0 \le t \le 104$.

(E) Repeat part (C), using a graphing calculator.

48. Revenue. Revenues from sales of a soft drink over a 2-year period are given approximately by

$$R(t) = 4 - 3 \cos \frac{\pi t}{6} \qquad 0 \le t \le 24$$

where $R(t)$ is revenue (in millions of dollars) for a month of sales t months after February 1.

(A) What is the rate of change of revenue t months after February 1?

(B) What is the rate of change of revenue 1 month after February 1? 6 months after February 1? 11 months after February 1?

(C) Find all local maxima and minima for $0 < t < 24$.

(D) Find the absolute maximum and minimum for $0 \le t \le 24$.

(E) Repeat part (C), using a graphing calculator.

49. Physiology. A normal seated adult inhales and exhales about 0.8 liter of air every 4 seconds. The volume of air $V(t)$ in the lungs t seconds after exhaling is given approximately by

$$V(t) = 0.45 - 0.35 \cos \frac{\pi t}{2} \qquad 0 \le t \le 8$$

(A) What is the rate of flow of air t seconds after exhaling?

(B) What is the rate of flow of air 3 seconds after exhaling? 4 seconds after exhaling? 5 seconds after exhaling?

(C) Find all local maxima and minima for $0 < t < 8$.

(D) Find the absolute maximum and minimum for $0 \le t \le 8$.

(E) Repeat part (C), using a graphing calculator.

50. Pollution. In a large city, the amount of sulfur dioxide pollutant released into the atmosphere due to the burning of coal and oil for heating purposes varies seasonally. Suppose that the number of tons of pollutant released into the atmosphere during the *n*th week after January 1 is given approximately by

$$P(n) = 1 + \cos\frac{\pi n}{26} \qquad 0 \le n \le 104$$

(A) What is the rate of change of pollutant *n* weeks after the first of the year?

(B) What is the rate of change of pollutant 13 weeks after the first of the year? 26 weeks after the first of the year? 30 weeks after the first of the year?

(C) Find all local maxima and minima for $0 < t < 104$.

(D) Find the absolute maximum and minimum for $0 \le t \le 104$.

(E) Repeat part (C), using a graphing calculator.

<u>Answers to Matched Problems</u>

1. (A) $-3x^2 \sin x^3$ (B) $-3\cos(5 - 3x)$

   (C) $\dfrac{x \cos x - \sin x}{x^2}$

2. $-\frac{1}{2}$ 3. $-\cot x \csc x$

4. (A) $R'(t) = -\dfrac{5.8\pi}{6}\sin\dfrac{\pi t}{6}, 0 < t < 24$

(B) $R'(2) \approx -\$2.630$ thousand, or $-\$2,630$/month; $R'(12) = \$0$/month; $R'(23) \approx \$1.518$ thousand, or $\$1,518$ month

(C) Local minima at $t = 6$ and $t = 18$; local maximum at $t = 12$

(D)

	t	$R(t)$	
Endpoint	0	\$12,000	Absolute maximum
	6	\$400	Absolute minimum
	12	\$12,000	Absolute maximum
	18	\$400	Absolute minimum
Endpoint	24	\$12,000	Absolute maximum

(E) $R(t)$

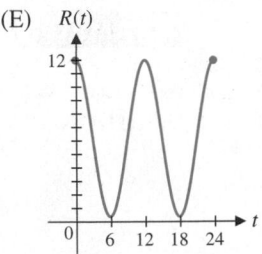

8.3 Integration of Trigonometric Functions

- Integral Formulas
- Application

Integral Formulas

Now that we know the derivative formulas

$$\frac{d}{dx}\sin x = \cos x \quad \text{and} \quad \frac{d}{dx}\cos x = -\sin x$$

we automatically have the two integral formulas from the definition of the indefinite integral of a function (Section 5.1):

$$\int \cos x \, dx = \sin x + C \quad \text{and} \quad \int \sin x \, dx = -\cos x + C$$

EXAMPLE 1 Area Under a Sine Curve   Find the area under the sine curve $y = \sin x$ from 0 to π.

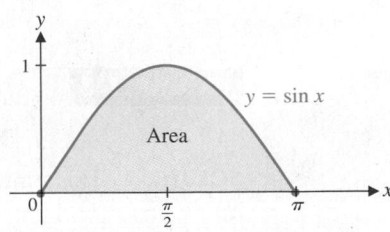

SOLUTION $\text{Area} = \displaystyle\int_0^{\pi} \sin x \, dx = -\cos x \Big|_0^{\pi}$

$$= (-\cos \pi) - (-\cos 0)$$
$$= [-(-1)] - [-(1)] = 2$$

<u>Matched Problem 1</u>   Find the area under the cosine curve $y = \cos x$ from 0 to $\pi/2$.

From the general derivative formulas

$$\frac{d}{dx}\sin u = \cos u \frac{du}{dx} \quad \text{and} \quad \frac{d}{dx}\cos u = -\sin u \frac{du}{dx}$$

we obtain the following general integral formulas:

Indefinite Integrals of Sine and Cosine
For $u = u(x)$,

$$\int \sin u \, du = -\cos u + C \quad \text{and} \quad \int \cos u \, du = \sin u + C$$

EXAMPLE 2   Indefinite Integrals and Trigonometric Functions   Find $\int x \sin x^2 \, dx$.

SOLUTION $\displaystyle \int x \sin x^2 \, dx = \frac{1}{2} \int 2x \sin x^2 \, dx$

$$= \frac{1}{2} \int (\sin x^2) 2x \, dx \quad \text{Let } u = x^2; \text{ then } du = 2x \, dx.$$

$$= \frac{1}{2} \int \sin u \, du$$

$$= -\frac{1}{2} \cos u + C$$

$$= -\frac{1}{2} \cos x^2 + C \quad \text{Since } u = x^2$$

Check:
To check, we differentiate the result to obtain the original integrand:

$$\frac{d}{dx}\left(-\frac{1}{2}\cos x^2\right) = -\frac{1}{2}\frac{d}{dx}\cos x^2$$

$$= -\frac{1}{2}(-\sin x^2)\frac{d}{dx}x^2$$

$$= -\frac{1}{2}(-\sin x^2)(2x)$$

$$= x \sin x^2$$

Matched Problem 2   Find $\int \cos 20\pi t \, dt$.

EXAMPLE 3   Indefinite Integrals and Trigonometric Functions   Find $\int (\sin x)^5 \cos x \, dx$.

SOLUTION   This integrand is of the form $\int u^p \, du$, where $u = \sin x$ and $du = \cos x \, dx$. Thus,

$$\int (\sin x)^5 \cos x \, dx = \frac{(\sin x)^6}{6} + C$$

Matched Problem 3   Find $\int \sqrt{\sin x} \cos x \, dx$.

EXAMPLE 4 Definite Integrals and Trigonometric Functions Evaluate
$\int_2^{3.5} \cos x \, dx.$

SOLUTION
$$\int_2^{3.5} \cos x \, dx = \sin x \Big|_2^{3.5}$$
$$= \sin 3.5 - \sin 2 \qquad \text{Use a calculator in radian mode.}$$
$$= -0.3508 - 0.9093$$
$$= -1.2601$$

Matched Problem 4) Use a calculator to evaluate $\int_1^{1.5} \sin x \, dx.$

CONCEPTUAL INSIGHT

Recall that $y = \sin x$ is a periodic function with period 2π, and let c be any real number. Then

$$\int_c^{c+2\pi} \cos x \, dx = \sin x \Big|_c^{c+2\pi} = \sin(c + 2\pi) - \sin c = 0$$

In other words, over any interval of the form $[c, c + 2\pi]$, the area that is above the x axis, but below the graph of $y = \cos x$, is equal to the area that is below the x axis, but above the graph of $y = \cos x$ (see Figure 10, Section 8.1). Similarly, for any real number c,

$$\int_c^{c+2\pi} \sin x \, dx = 0$$

Application

EXAMPLE 5 Total Revenue In Example 4 of Section 8.2, we were given the following revenue equation from the sale of ski jackets:

$$R(t) = 1.55 + 1.45 \cos \frac{\pi t}{26} \qquad 0 \le t \le 104$$

Here, $R(t)$ is revenue (in thousands of dollars) for a week of sales t weeks after January 1.

(A) Find the total revenue taken in over the 2-year period—that is, from $t = 0$ to $t = 104$.

(B) Find the total revenue taken in from $t = 39$ to $t = 65$.

SOLUTION

(A) The area under the graph of the revenue equation for the 2-year period approximates the total revenue taken in for that period:

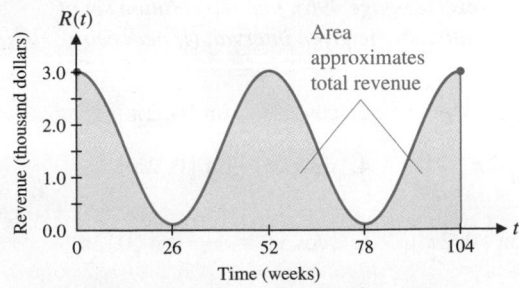

This area (and therefore the total revenue) is given by the following definite integral:

$$\text{Total revenue} \approx \int_0^{104} \left(1.55 + 1.45 \cos \frac{\pi t}{26} \right) dt$$

$$= \left[1.55t + 1.45 \left(\frac{26}{\pi} \right) \sin \frac{\pi t}{26} \right]\Big|_0^{104}$$

$$= \$161.200 \text{ thousand, or } \$161,200$$

(B) The total revenue from $t = 39$ to $t = 65$ is approximated by the area under the curve from $t = 39$ to $t = 65$:

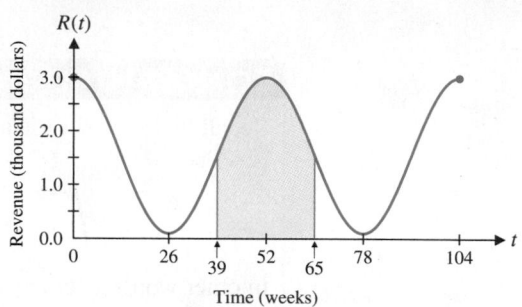

$$\text{Total revenue} \approx \int_{39}^{65} \left(1.55 + 1.45 \cos \frac{\pi t}{26} \right) dt$$

$$= \left[1.55t + 1.45 \left(\frac{26}{\pi} \right) \sin \frac{\pi t}{26} \right]\Big|_{39}^{65}$$

$$= \$64.301 \text{ thousand, or } \$64,301$$

Matched Problem 5 Suppose that revenues from the sale of ski jackets are given approximately by

$$R(t) = 6.2 + 5.8 \cos \frac{\pi t}{6} \quad 0 \le t \le 24$$

where $R(t)$ is revenue (in thousands of dollars) for a month of sales t months after January 1.

(A) Find the total revenue taken in over the 2 year-period—that is, from $t = 0$ to $t = 24$.

(B) Find the total revenue taken in from $t = 4$ to $t = 8$.

Exercises 8.3

Skills Warm-up Exercises

W *In Problems 1–8, by using only the unit circle definitions of the sine and cosine functions (see page 496), find the solution set of each equation or inequality on the given interval. (If necessary, review Section 1.3).*

1. $\sin x \ge 0$ on $[0, 2\pi]$

2. $\cos x \ge 0$ on $[0, 2\pi]$

3. $\cos x > \dfrac{1}{2}$ on $[0, 2\pi]$

4. $\sin x > \dfrac{1}{2}$ on $[0, 2\pi]$

5. $|\sin x| = \dfrac{\sqrt{2}}{2}$ on $[0, 2\pi]$

6. $|\cos x| = \dfrac{\sqrt{2}}{2}$ on $[0, 2\pi]$

7. $\tan x \le 1$ on $(-\pi/2, \pi/2)$

8. $\cot x \ge 1$ on $(0, \pi)$

Find each of the indefinite integrals in Problems 9–18.

9. $\displaystyle\int \sin t \, dt$

10. $\displaystyle\int \cos w \, dw$

11. $\displaystyle\int \cos 3x \, dx$

12. $\displaystyle\int \sin 2x \, dx$

13. $\displaystyle\int (\sin x)^{12} \cos x \, dx$

14. $\displaystyle\int \sin x \cos x \, dx$

15. $\displaystyle\int \sqrt[3]{\cos x} \sin x \, dx$

16. $\displaystyle\int \dfrac{\cos x}{\sqrt{\sin x}} dx$

17. $\displaystyle\int x^2 \cos x^3\, dx$ **18.** $\displaystyle\int (x+1)\sin(x^2+2x)\, dx$

Evaluate each of the definite integrals in Problems 19–22.

19. $\displaystyle\int_0^{\pi/2} \cos x\, dx$ **20.** $\displaystyle\int_0^{\pi/4} \cos x\, dx$

21. $\displaystyle\int_{\pi/2}^{\pi} \sin x\, dx$ **22.** $\displaystyle\int_{\pi/6}^{\pi/3} \sin x\, dx$

23. Find the shaded area under the cosine curve in the figure:

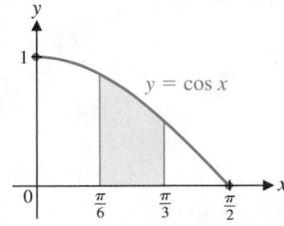

24. Find the shaded area under the sine curve in the figure:

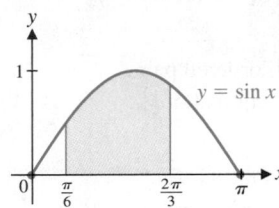

Use a calculator to evaluate the definite integrals in Problems 25–28 after performing the indefinite integration. (Remember that the limits are real numbers, so radian mode must be used on the calculator.)

25. $\displaystyle\int_0^2 \sin x\, dx$ **26.** $\displaystyle\int_0^{0.5} \cos x\, dx$

27. $\displaystyle\int_1^2 \cos x\, dx$ **28.** $\displaystyle\int_1^3 \sin x\, dx$

Find each of the indefinite integrals in Problems 29–34.

29. $\displaystyle\int e^{\sin x} \cos x\, dx$ **30.** $\displaystyle\int e^{\cos x} \sin x\, dx$

31. $\displaystyle\int \frac{\cos x}{\sin x}\, dx$ **32.** $\displaystyle\int \frac{\sin x}{\cos x}\, dx$

33. $\displaystyle\int \tan x\, dx$ **34.** $\displaystyle\int \cot x\, dx$

35. Given the definite integral

$$I = \int_0^3 e^{-x} \sin x\, dx$$

(A) Graph the integrand $f(x) = e^{-x} \sin x$ over $[0, 3]$.

(B) Use the left sum L_6 (see Section 5.5) to approximate I.

36. Given the definite integral

$$I = \int_0^3 e^{-x} \cos x\, dx$$

(A) Graph the integrand $f(x) = e^{-x} \cos x$ over $[0, 3]$.

(B) Use the right sum R_6 (see Section 5.5) to approximate I.

Applications

37. Seasonal business cycle. Suppose that profits on the sale of swimming suits in a department store are given approximately by

$$P(t) = 5 - 5\cos\frac{\pi t}{26} \quad 0 \le t \le 104$$

where $P(t)$ is profit (in hundreds of dollars) for a week of sales t weeks after January 1. Use definite integrals to approximate

(A) The total profit earned during the 2-year period

(B) The total profit earned from $t = 13$ to $t = 26$

(C) Illustrate part (B) graphically with an appropriate shaded region representing the total profit earned.

38. Seasonal business cycle. Revenues from sales of a soft drink over a 2-year period are given approximately by

$$R(t) = 4 - 3\cos\frac{\pi t}{6} \quad 0 \le t \le 24$$

where $R(t)$ is revenue (in millions of dollars) for a month of sales t months after February 1. Use definite integrals to approximate

(A) Total revenues taken in over the 2-year period

(B) Total revenues taken in from $t = 8$ to $t = 14$

(C) Illustrate part (B) graphically with an appropriate shaded region representing the total revenues taken in.

39. Pollution. In a large city, the amount of sulfur dioxide pollutant released into the atmosphere due to the burning of coal and oil for heating purposes is given approximately by

$$P(n) = 1 + \cos\frac{\pi n}{26} \quad 0 \le n \le 104$$

where $P(n)$ is the amount of sulfur dioxide (in tons) released during the nth week after January 1.

(A) How many tons of pollutants were emitted into the atmosphere over the 2-year period?

(B) How many tons of pollutants were emitted into the atmosphere from $n = 13$ to $n = 52$?

(C) Illustrate part (B) graphically with an appropriate shaded region representing the total tons of pollutants emitted into the atmosphere.

Answers to Matched Problems

1. 1   **2.** $\dfrac{1}{20\pi}\sin 20\pi t + C$ **3.** $\dfrac{2}{3}(\sin x)^{3/2} + C$

4. 0.4696

5. (A) \$148.8 thousand, or \$148,800

    (B) \$5.614 thousand, or \$5,614

Important Terms, Symbols, and Concepts

8.1 Trigonometric Functions Review

- In a plane, an **angle** is formed by rotating a ray m, called the **initial side** of the angle, around its endpoint until the ray coincides with a ray n, called the **terminal side** of the angle. The common endpoint of m and n is called the **vertex.**

- A counterclockwise rotation produces a **positive** angle, and a clockwise rotation produces a **negative** angle.

- Two angles with the same initial and terminal sides are said to be **coterminal**.

- An angle of **degree measure 1** is $\frac{1}{360}$ of a complete rotation. In a circle, an angle of **radian measure 1** is the central angle subtended by an arc having the same length as the radius of the circle.

- Degree measure can be converted to radian measure, and vice versa, by the proportion

Ex. 1, p. 496

$$\frac{\theta_{\text{deg}}}{180°} = \frac{\theta_{\text{rad}}}{\pi \text{ rad}}$$

- An angle in a rectangular coordinate system is in **standard position** if its vertex is at the origin and its initial side is on the positive x axis.

- If θ is an angle in standard position, its terminal side intersects the unit circle at a point P. We denote the coordinates of P by $(\cos \theta, \sin \theta)$ [see Fig. 6, p. 496].

- The set of all ordered pairs of the form $(\theta, \sin \theta)$ is the **sine function**, and the set of all ordered pairs of the form $(\theta, \cos \theta)$ is the **cosine function**. For work in calculus, we define these functions for angles measured in radians. (See Fig. 10, p. 499, for the graphs of $y = \sin x$ and $y = \cos x$.)

Ex. 2, p. 497
Ex. 3, p. 498

- A function is **periodic** if $f(x + p) = f(x)$ for some positive constant p and all real numbers x for which $f(x)$ is defined. The smallest such constant p is called the **period**. Both $\sin x$ and $\cos x$ are periodic continuous functions with period 2π.

- Four additional trigonometric functions—the **tangent, cotangent, secant,** and **cosecant** functions—are defined in terms of $\sin x$ and $\cos x$:

$$\tan x = \frac{\sin x}{\cos x} \quad \cos x \neq 0 \qquad \sec x = \frac{1}{\cos x} \quad \cos x \neq 0$$

$$\cot x = \frac{\cos x}{\sin x} \quad \sin x \neq 0 \qquad \csc x = \frac{1}{\sin x} \quad \sin x \neq 0$$

8.2 Derivatives of Trigonometric Functions

- The derivatives of the functions $\sin x$ and $\cos x$ are

Ex. 1, p. 502
Ex. 2, p. 503
Ex. 3, p. 503
Ex. 4, p. 504

$$\frac{d}{dx} \sin x = \cos x \qquad \frac{d}{dx} \cos x = -\sin x$$

For $u = u(x)$,

$$\frac{d}{dx} \sin u = \cos u \frac{du}{dx} \qquad \frac{d}{du} \cos u = -\sin u \frac{du}{dx}$$

8.3 Integration of Trigonometric Functions

- Indefinite integrals of the functions $\sin x$ and $\cos x$ are

Ex. 1, p. 507
Ex. 2, p. 508
Ex. 3, p. 508
Ex. 4, p. 509
Ex. 5, p. 509

$$\int \sin u \, du = -\cos u + C \quad \text{and} \quad \int \cos u \, du = \sin u + C$$

Review Exercises

Work through all the problems in this chapter review and check your answers in the back of the book. Answers to all review problems are there, along with section numbers in italics to indicate where each type of problem is discussed. Where weaknesses show up, review appropriate sections of the text.

1. Convert to radian measure in terms of π:

(A) $30°$ (B) $45°$ (C) $60°$ (D) $90°$

2. Evaluate without using a calculator:

(A) $\cos \pi$ (B) $\sin 0$ (C) $\sin \dfrac{\pi}{2}$

In Problems 3–6, find each derivative or integral.

3. $\dfrac{d}{dm} \cos m$ **4.** $\dfrac{d}{du} \sin u$

5. $\dfrac{d}{dx} \sin (x^2 - 2x + 1)$ **6.** $\displaystyle\int \sin 3t \, dt$

7. Convert to degree measure:

(A) $\pi/6$ (B) $\pi/4$ (C) $\pi/3$ (D) $\pi/2$

8. Evaluate without using a calculator:

(A) $\sin \dfrac{\pi}{6}$ (B) $\cos \dfrac{\pi}{4}$ (C) $\sin \dfrac{\pi}{3}$

9. Evaluate with the use of a calculator:

(A) $\cos 33.7$ (B) $\sin (-118.4)$

In Problems 10–16, find each derivative or integral.

10. $\dfrac{d}{dx}(x^2 - 1) \sin x$ **11.** $\dfrac{d}{dx}(\sin x)^6$

12. $\dfrac{d}{dx} \sqrt[3]{\sin x}$ **13.** $\displaystyle\int t \cos(t^2 - 1) \, dt$

14. $\displaystyle\int_0^\pi \sin u \, du$ **15.** $\displaystyle\int_0^{\pi/3} \cos x \, dx$

16. $\displaystyle\int_1^{2.5} \cos x \, dx$

17. Find the slope of the cosine curve $y = \cos x$ at $x = \pi/4$.

18. Find the area under the sine curve $y = \sin x$ from $x = \pi/4$ to $x = 3\pi/4$.

19. Given the definite integral

$$I = \int_1^5 \frac{\sin x}{x} dx$$

(A) Graph the integrand

$$f(x) = \frac{\sin x}{x}$$

over $[1, 5]$.

(B) Use the right sum R_4 to approximate I.

20. Convert $15°$ to radian measure.

21. Evaluate without using a calculator:

(A) $\sin \dfrac{3\pi}{2}$ (B) $\cos \dfrac{5\pi}{6}$ (C) $\sin \left(\dfrac{-\pi}{6} \right)$

In Problems 22–26, find each derivative or integral.

22. $\dfrac{d}{du} \tan u$ **23.** $\dfrac{d}{dx} e^{\cos x^2}$

24. $\displaystyle\int e^{\sin x} \cos x \, dx$ **25.** $\displaystyle\int \tan x \, dx$

26. $\displaystyle\int_2^5 (5 + 2 \cos 2x) \, dx$

✎ *In Problems 27–29, graph each function on a graphing calculator set in radian mode.*

27. $y = \dfrac{\sin \pi x}{0.2x}; 1 \le x \le 8, -4 \le y \le 4$

28. $y = 0.5x \cos \pi x; 0 \le x \le 8, -5 \le y \le 5$

29. $y = 3 - 2 \cos \pi x; 0 \le x \le 6, 0 \le y \le 5$

Applications

Problems 30–32 refer to the following: Revenues from sweater sales in a sportswear chain are given approximately by

$$R(t) = 3 + 2 \cos \frac{\pi t}{6} \qquad 0 \le t \le 24$$

where $R(t)$ is the revenue (in thousands of dollars) for a month of sales t months after January 1.

30. (A) Find the exact values of $R(0), R(2), R(3)$, and $R(6)$ without using a calculator.

✎ (B) Use a calculator to find $R(1)$ and $R(22)$. Interpret the results.

31. (A) What is the rate of change of revenue t months after January 1?

(B) What is the rate of change of revenue 3 months after January 1? 10 months after January 1? 18 months after January 1?

(C) Find all local maxima and minima for $0 < t < 24$.

(D) Find the absolute maximum and minimum for $0 \le t \le 24$.

(E) Repeat part (C), using a graphing calculator.

32. (A) Find the total revenues taken in over the 2-year period.

(B) Find the total revenues taken in from $t = 5$ to $t = 9$.

(C) Illustrate part (B) graphically with an appropriate shaded region representing the total revenue taken in.

Basic Algebra Review

Appendix A reviews some important basic algebra concepts usually studied in earlier courses. The material may be studied systematically before beginning the rest of the book or reviewed as needed.

A.1 Real Numbers

- Set of Real Numbers
- Real Number Line
- Basic Real Number Properties
- Further Properties
- Fraction Properties

The rules for manipulating and reasoning with symbols in algebra depend, in large measure, on properties of the real numbers. In this section we look at some of the important properties of this number system. To make our discussions here and elsewhere in the book clearer and more precise, we occasionally make use of simple *set* concepts and notation.

Set of Real Numbers

Informally, a **real number** is any number that has a decimal representation. Table 1 describes the set of real numbers and some of its important subsets. Figure 1 illustrates how these sets of numbers are related.

The set of integers contains all the natural numbers and something else—their negatives and 0. The set of rational numbers contains all the integers and something else—noninteger ratios of integers. And the set of real numbers contains all the rational numbers and something else—irrational numbers.

Table 1 **Set of Real Numbers**

Symbol	Name	Description	Examples
N	Natural numbers	Counting numbers (also called positive integers)	$1, 2, 3, \ldots$
Z	Integers	Natural numbers, their negatives, and 0	$\ldots, -2, -1, 0, 1, 2, \ldots$
Q	Rational numbers	Numbers that can be represented as a/b, where a and b are integers and $b \neq 0$; decimal representations are repeating or terminating	$-4, 0, 1, 25, \frac{-3}{5}, \frac{2}{3}, 3.67, -0.33\overline{3}, 5.272\,7\overline{27}$*
I	Irrational numbers	Numbers that can be represented as nonrepeating and nonterminating decimal numbers	$\sqrt{2}, \pi, \sqrt[3]{7}, 1.414\,213\ldots, 2.718\,281\,82\ldots$
R	Real numbers	Rational and irrational numbers	

*The overbar indicates that the number (or block of numbers) repeats indefinitely. The space after every third digit is used to help keep track of the number of decimal places.

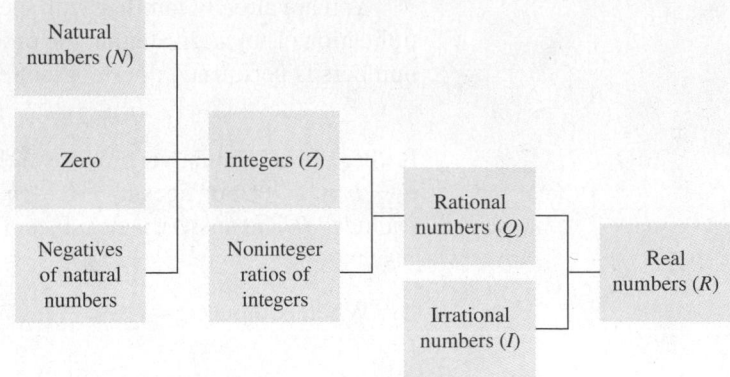

Figure 1 **Real numbers and important subsets**

Real Number Line

A one-to-one correspondence exists between the set of real numbers and the set of points on a line. That is, each real number corresponds to exactly one point, and each point corresponds to exactly one real number. A line with a real number associated with each point, and vice versa, as shown in Figure 2, is called a **real number line**, or simply a **real line**. Each number associated with a point is called the coordinate of the point.

The point with coordinate 0 is called the **origin**. The arrow on the right end of the line indicates a positive direction. The coordinates of all points to the right of the origin are called **positive real numbers**, and those to the left of the origin are called **negative real numbers**. The real number 0 is neither positive nor negative.

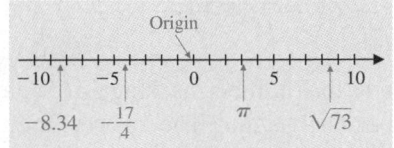

Figure 2 **Real number line**

Basic Real Number Properties

We now take a look at some of the basic properties of the real number system that enable us to convert algebraic expressions into *equivalent forms*.

SUMMARY Basic Properties of the Set of Real Numbers

Let a, b, and c be arbitrary elements in the set of real numbers R.

Addition Properties

Associative: $(a + b) + c = a + (b + c)$

Commutative: $a + b = b + a$

Identity: 0 is the additive identity; that is, $0 + a = a + 0 = a$ for all a in R, and 0 is the only element in R with this property.

Inverse: For each a in R, $-a$, is its unique additive inverse; that is, $a + (-a) = (-a) + a = 0$ and $-a$ is the only element in R relative to a with this property.

Multiplication Properties

Associative: $(ab)c = a(bc)$

Commutative: $ab = ba$

Identity: 1 is the multiplicative identity; that is, $(1)a = a(1) = a$ for all a in R, and 1 is the only element in R with this property.

Inverse: For each a in R, $a \neq 0$, $1/a$ is its unique multiplicative inverse; that is, $a(1/a) = (1/a)a = 1$, and $1/a$ is the only element in R relative to a with this property.

Distributive Properties

$$a(b + c) = ab + ac \quad (a + b)c = ac + bc$$

You are already familiar with the **commutative properties** for addition and multiplication. They indicate that the order in which the addition or multiplication of two numbers is performed does not matter. For example,

$$7 + 2 = 2 + 7 \quad \text{and} \quad 3 \cdot 5 = 5 \cdot 3$$

Is there a commutative property relative to subtraction or division? That is, does $a - b = b - a$ or does $a \div b = b \div a$ for all real numbers a and b (division by 0 excluded)? The answer is no, since, for example,

$$8 - 6 \neq 6 - 8 \quad \text{and} \quad 10 \div 5 \neq 5 \div 10$$

When computing

$$3 + 2 + 6 \quad \text{or} \quad 3 \cdot 2 \cdot 6$$

why don't we need parentheses to indicate which two numbers are to be added or multiplied first? The answer is to be found in the **associative properties**. These properties allow us to write

$$(3 + 2) + 6 = 3 + (2 + 6) \quad \text{and} \quad (3 \cdot 2) \cdot 6 = 3 \cdot (2 \cdot 6)$$

so it does not matter how we group numbers relative to either operation. Is there an associative property for subtraction or division? The answer is no, since, for example,

$$(12 - 6) - 2 \neq 12 - (6 - 2) \quad \text{and} \quad (12 \div 6) \div 2 \neq 12 \div (6 \div 2)$$

Evaluate each side of each equation to see why.

What number added to a given number will give that number back again? What number times a given number will give that number back again? The answers are 0 and 1, respectively. Because of this, 0 and 1 are called the **identity elements** for the real numbers. Hence, for any real numbers a and b,

$$0 + 5 = 5 \quad \text{and} \quad (a + b) + 0 = a + b$$
$$1 \cdot 4 = 4 \quad \text{and} \quad (a + b) \cdot 1 = a + b$$

We now consider **inverses**. For each real number a, there is a unique real number $-a$ such that $a + (-a) = 0$. The number $-a$ is called the **additive inverse** of a, or the **negative** of a. For example, the additive inverse (or negative) of 7 is -7, since $7 + (-7) = 0$. The additive inverse (or negative) of -7 is $-(-7) = 7$, since $-7 + [-(-7)] = 0$.

CONCEPTUAL INSIGHT

Do not confuse negation with the sign of a number. If a is a real number, $-a$ is the negative of a and may be positive or negative. Specifically, if a is negative, then $-a$ is positive and if a is positive, then $-a$ is negative.

For each nonzero real number a, there is a unique real number $1/a$ such that $a(1/a) = 1$. The number $1/a$ is called the **multiplicative inverse** of a, or the **reciprocal** of a. For example, the multiplicative inverse (or reciprocal) of 4 is $\frac{1}{4}$, since $4\left(\frac{1}{4}\right) = 1$. (Also note that 4 is the multiplicative inverse of $\frac{1}{4}$.) The number 0 has no multiplicative inverse.

We now turn to the **distributive properties**, which involve both multiplication and addition. Consider the following two computations:

$$5(3 + 4) = 5 \cdot 7 = 35 \qquad 5 \cdot 3 + 5 \cdot 4 = 15 + 20 = 35$$

Thus,

$$5(3 + 4) = 5 \cdot 3 + 5 \cdot 4$$

and we say that multiplication by 5 *distributes* over the sum $(3 + 4)$. In general, **multiplication distributes over addition** in the real number system. Two more illustrations are

$$9(m + n) = 9m + 9n \qquad (7 + 2)u = 7u + 2u$$

EXAMPLE 1 Real Number Properties State the real number property that justifies the indicated statement.

Statement	Property Illustrated
(A) $x(y + z) = (y + z)x$	Commutative $(\cdot)$
(B) $5(2y) = (5 \cdot 2)y$	Associative $(\cdot)$
(C) $2 + (y + 7) = 2 + (7 + y)$	Commutative $(+)$
(D) $4z + 6z = (4 + 6)z$	Distributive
(E) If $m + n = 0$, then $n = -m$.	Inverse $(+)$

Matched Problem 1 State the real number property that justifies the indicated statement.

(A) $8 + (3 + y) = (8 + 3) + y$

(B) $(x + y) + z = z + (x + y)$

(C) $(a + b)(x + y) = a(x + y) + b(x + y)$

(D) $5xy + 0 = 5xy$

(E) If $xy = 1, x \neq 0$, then $y = 1/x$.

Further Properties

Subtraction and *division* can be defined in terms of addition and multiplication, respectively:

DEFINITION Subtraction and Division

For all real numbers a and b,

Subtraction: $a - b = a + (-b)$ $7 - (-5) = 7 + [-(-5)]$
$$= 7 + 5 = 12$$

Division: $a \div b = a\left(\dfrac{1}{b}\right), b \neq 0$ $9 \div 4 = 9\left(\dfrac{1}{4}\right) = \dfrac{9}{4}$

To subtract b from a, add the negative (the additive inverse) of b to a. To divide a by b, multiply a by the reciprocal (the multiplicative inverse) of b. Note that division by 0 is not defined, since 0 does not have a reciprocal. **0 can never be used as a divisor!**

The following properties of negatives can be proved using the preceding assumed properties and definitions.

THEOREM 1 Negative Properties

For all real numbers a and b,

1. $-(-a) = a$

2. $(-a)b = -(ab)$
$$= a(-b) = -ab$$

3. $(-a)(-b) = ab$

4. $(-1)a = -a$

5. $\dfrac{-a}{b} = -\dfrac{a}{b} = \dfrac{a}{-b}, b \neq 0$

6. $\dfrac{-a}{-b} = -\dfrac{-a}{b} = -\dfrac{a}{-b} = \dfrac{a}{b}, b \neq 0$

We now state two important properties involving 0.

THEOREM 2 Zero Properties

For all real numbers a and b,

1. $a \cdot 0 = 0$ $0 \cdot 0 = 0$ $(-35)(0) = 0$

2. $ab = 0$ if and only if $a = 0$ or $b = 0$
   If $(3x + 2)(x - 7) = 0$, then either $3x + 2 = 0$ or $x - 7 = 0$.

Fraction Properties

Recall that the quotient $a \div b \, (b \neq 0)$ written in the form a/b is called a **fraction**. The quantity a is called the **numerator**, and the quantity b is called the **denominator**.

THEOREM 3   Fraction Properties

For all real numbers a, b, c, d, and k (division by 0 excluded):

1. $\dfrac{a}{b} = \dfrac{c}{d}$ if and only if $ad = bc$ $\dfrac{4}{6} = \dfrac{6}{9}$ since $4 \cdot 9 = 6 \cdot 6$

2. $\dfrac{ka}{kb} = \dfrac{a}{b}$

   $\dfrac{7 \cdot 3}{7 \cdot 5} = \dfrac{3}{5}$

3. $\dfrac{a}{b} \cdot \dfrac{c}{d} = \dfrac{ac}{bd}$

   $\dfrac{3}{5} \cdot \dfrac{7}{8} = \dfrac{3 \cdot 7}{5 \cdot 8}$

4. $\dfrac{a}{b} \div \dfrac{c}{d} = \dfrac{a}{b} \cdot \dfrac{d}{c}$

   $\dfrac{2}{3} \div \dfrac{5}{7} = \dfrac{2}{3} \cdot \dfrac{7}{5}$

5. $\dfrac{a}{b} + \dfrac{c}{b} = \dfrac{a + c}{b}$

   $\dfrac{3}{6} + \dfrac{5}{6} = \dfrac{3 + 5}{6}$

6. $\dfrac{a}{b} - \dfrac{c}{b} = \dfrac{a - c}{b}$

   $\dfrac{7}{8} - \dfrac{3}{8} = \dfrac{7 - 3}{8}$

7. $\dfrac{a}{b} + \dfrac{c}{d} = \dfrac{ad + bc}{bd}$

   $\dfrac{2}{3} + \dfrac{3}{5} = \dfrac{2 \cdot 5 + 3 \cdot 3}{3 \cdot 5}$

A fraction is a quotient, not just a pair of numbers. So if a and b are real numbers with $b \neq 0$, then $\frac{a}{b}$ corresponds to a point on the real number line. For example, $\frac{17}{2}$ corresponds to the point halfway between $\frac{16}{2} = 8$ and $\frac{18}{2} = 9$. Similarly, $-\frac{21}{5}$ corresponds to the point that is $\frac{1}{5}$ unit to the left of -4.

EXAMPLE 2   Estimation   Round $\frac{22}{7} + \frac{18}{19}$ to the nearest integer.

SOLUTION   Note that a calculator is not required: $\frac{22}{7}$ is a little greater than 3, and $\frac{18}{19}$ is a little less than 1. Therefore the sum, rounded to the nearest integer, is 4.

Matched Problem 2   Round $\frac{6}{93}$ to the nearest integer.

Fractions with denominator 100 are called **percentages**. They are used so often that they have their own notation:

$$\frac{3}{100} = 3\% \qquad \frac{7.5}{100} = 7.5\% \qquad \frac{110}{100} = 110\%$$

So 3% is equivalent to 0.03, 7.5% is equivalent to 0.075, and so on.

EXAMPLE 3   State Sales Tax   Find the sales tax that is owed on a purchase of \$947.69 if the tax rate is 6.5%.

SOLUTION $6.5\%(\$947.69) = 0.065(947.69) = \$61.60$

Matched Problem 3   You intend to give a 20% tip, rounded to the nearest dollar, on a restaurant bill of \$78.47. How much is the tip?

Exercises A.1

All variables represent real numbers.

In Problems 1–6, replace each question mark with an appropriate expression that will illustrate the use of the indicated real number property.

1. Commutative property $(\,\cdot\,)$: $uv = ?$

2. Commutative property $(+)$: $x + 7 = ?$

3. Associative property $(+)$: $3 + (7 + y) = ?$

4. Associative property $(\,\cdot\,)$: $x(yz) = ?$

5. Identity property $(\,\cdot\,)$: $1(u + v) = ?$

6. Identity property $(+)$: $0 + 9m = ?$

In Problems 7–26, indicate true (T) or false (F).

7. $5(8m) = (5 \cdot 8)m$

8. $a + cb = a + bc$

9. $5x + 7x = (5 + 7)x$

10. $uv(w + x) = uvw + uvx$

11. $-2(-a)(2x - y) = 2a(-4x + y)$

12. $8 \div (-5) = 8\left(\dfrac{1}{-5}\right)$

13. $(x + 3) + 2x = 2x + (x + 3)$

14. $\dfrac{x}{3y} \div \dfrac{5y}{x} = \dfrac{15y^2}{x^2}$

15. $\dfrac{2x}{-(x + 3)} = -\dfrac{2x}{x + 3}$

16. $-\dfrac{2x}{-(x - 3)} = \dfrac{2x}{x - 3}$

17. $(-3)\left(\dfrac{1}{-3}\right) = 1$

18. $(-0.5) + (0.5) = 0$

19. $-x^2y^2 = (-1)x^2y^2$

20. $[-(x + 2)](-x) = (x + 2)x$

21. $\dfrac{a}{b} + \dfrac{c}{d} = \dfrac{a + c}{b + d}$

22. $\dfrac{k}{k + b} = \dfrac{1}{1 + b}$

23. $(x + 8)(x + 6) = (x + 8)x + (x + 8)6$

24. $u(u - 2v) + v(u - 2v) = (u + v)(u - 2v)$

25. If $(x - 2)(2x + 3) = 0$, then either $x - 2 = 0$ or $2x + 3 = 0$.

26. If either $x - 2 = 0$ or $2x + 3 = 0$, then $(x - 2)(2x + 3) = 0$.

27. If $uv = 1$, does either u or v have to be 1? Explain.

28. If $uv = 0$, does either u or v have to be 0? Explain.

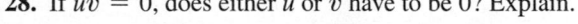

29. Indicate whether the following are true (T) or false (F):

(A) All integers are natural numbers.

(B) All rational numbers are real numbers.

(C) All natural numbers are rational numbers.

30. Indicate whether the following are true (T) or false (F):

(A) All natural numbers are integers.

(B) All real numbers are irrational.

(C) All rational numbers are real numbers.

31. Give an example of a real number that is not a rational number.

32. Give an example of a rational number that is not an integer.

33. Given the sets of numbers N (natural numbers), Z (integers), Q (rational numbers), and R (real numbers), indicate to which set(s) each of the following numbers belongs:

(A) 8   (B) $\sqrt{2}$ (C) -1.414 (D) $\dfrac{-5}{2}$

34. Given the sets of numbers N, Z, Q, and R (see Problem 33), indicate to which set(s) each of the following numbers belongs:

(A) -3 (B) 3.14   (C) π (D) $\dfrac{2}{3}$

35. Indicate true (T) or false (F), and for each false statement find real number replacements for a, b, and c that will provide a counterexample. For all real numbers a, b, and c,

(A) $a(b - c) = ab - c$

(B) $(a - b) - c = a - (b - c)$

(C) $a(bc) = (ab)c$

(D) $(a \div b) \div c = a \div (b \div c)$

36. Indicate true (T) or false (F), and for each false statement find real number replacements for a and b that will provide a counterexample. For all real numbers a and b,

(A) $a + b = b + a$

(B) $a - b = b - a$

(C) $ab = ba$

(D) $a \div b = b \div a$

37. If $c = 0.151515\ldots$, then $100c = 15.1515\ldots$ and

$$100c - c = 15.1515\ldots - 0.151515\ldots$$
$$99c = 15$$
$$c = \frac{15}{99} = \frac{5}{33}$$

Proceeding similarly, convert the repeating decimal $0.090909\ldots$ into a fraction. (All repeating decimals are rational numbers, and all rational numbers have repeating decimal representations.)

38. Repeat Problem 37 for $0.181818\ldots$.

Use a calculator to express each number in Problems 39 and 40 as a decimal to the capacity of your calculator. Observe the repeating decimal representation of the rational numbers and the nonrepeating decimal representation of the irrational numbers.

39. (A) $\dfrac{13}{6}$ (B) $\sqrt{21}$ (C) $\dfrac{7}{16}$ (D) $\dfrac{29}{111}$

40. (A) $\dfrac{8}{9}$ (B) $\dfrac{3}{11}$ (C) $\sqrt{5}$ (D) $\dfrac{11}{8}$

In Problems 41–44, without using a calculator, round to the nearest integer.

41. (A) $\dfrac{43}{13}$ (B) $\dfrac{37}{19}$

42. (A) $\dfrac{9}{17}$ (B) $-\dfrac{12}{25}$

43. (A) $\dfrac{7}{8} + \dfrac{11}{12}$ (B) $\dfrac{55}{9} - \dfrac{7}{55}$

44. (A) $\dfrac{5}{6} - \dfrac{18}{19}$ (B) $\dfrac{13}{5} + \dfrac{44}{21}$

Applications

45. Sales tax. Find the tax owed on a purchase of $182.39 if the state sales tax rate is 9%. (Round to the nearest cent).

46. Sales tax. If you paid $29.86 in tax on a purchase of $533.19, what was the sales tax rate? (Write as a percentage, rounded to one decimal place).

47. Gasoline prices. If the price per gallon of gas jumped from $4.25 to $4.37, what was the percentage increase? (Round to one decimal place).

48. Gasoline prices. The price of gas increased 4% in one week. If the price last week was $4.30 per gallon, what is the price now? (Round to the nearest cent).

Answers to Matched Problems

1. (A) Associative ($+$)   (B) Commutative ($+$)
   (C) Distributive   (D) Identity ($+$)
   (E) Inverse ($\cdot$)

2. 0   **3.** $16

A.2 Operations on Polynomials

- Natural Number Exponents
- Polynomials
- Combining Like Terms
- Addition and Subtraction
- Multiplication
- Combined Operations

This section covers basic operations on *polynomials*. Our discussion starts with a brief review of natural number exponents. Integer and rational exponents and their properties will be discussed in detail in subsequent sections. (Natural numbers, integers, and rational numbers are important parts of the real number system; see Table 1 and Figure 1 in Appendix A.1.)

Natural Number Exponents

We define a **natural number exponent** as follows:

> **DEFINITION**   Natural Number Exponent
> For n a natural number and b any real number,
>
> $$b^n = b \cdot b \cdot \,\cdots\, \cdot b \qquad n \text{ factors of } b$$
> $$3^5 = 3 \cdot 3 \cdot 3 \cdot 3 \cdot 3 \qquad 5 \text{ factors of } 3$$
>
> where n is called the exponent and b is called the **base**.

Along with this definition, we state the **first property of exponents**:

> **THEOREM 1**   First Property of Exponents
> For any natural numbers m and n, and any real number b:
> $$b^m b^n = b^{m+n} \qquad (2t^4)(5t^3) = 2 \cdot 5 t^{4+3} = 10t^7$$

Polynomials

Algebraic expressions are formed by using constants and variables and the algebraic operations of addition, subtraction, multiplication, division, raising to powers, and taking roots. Special types of algebraic expressions are called *polynomials*. A **polynomial in one variable** x is constructed by adding or subtracting constants and terms of the form ax^n, where a is a real number and n is a natural number. A **polynomial in two variables** x and y is constructed by adding and subtracting constants and terms of the form $ax^m y^n$, where a is a real number and m and n are natural numbers. Polynomials in three and more variables are defined in a similar manner.

Polynomials		Not Polynomials	
8	0	$\dfrac{1}{x}$	$\dfrac{x-y}{x^2+y^2}$
$3x^3 - 6x + 7$	$6x + 3$		
$2x^2 - 7xy - 8y^2$	$9y^3 + 4y^2 - y + 4$	$\sqrt{x^3 - 2x}$	$2x^{-2} - 3x^{-1}$
$2x - 3y + 2$	$u^5 - 3u^3v^2 + 2uv^4 - v^4$		

Polynomial forms are encountered frequently in mathematics. For the efficient study of polynomials, it is useful to classify them according to their *degree*. If a term in a polynomial has only one variable as a factor, then the **degree of the term** is the power of the variable. If two or more variables are present in a term as factors, then the **degree of the term** is the sum of the powers of the variables. The **degree of a polynomial** is the degree of the nonzero term with the highest degree in the polynomial. Any nonzero constant is defined to be a **polynomial of degree 0**. The number 0 is also a polynomial but is not assigned a degree.

EXAMPLE 1 Degree

(A) The degree of the first term in $5x^3 + \sqrt{3}x - \frac{1}{2}$ is 3, the degree of the second term is 1, the degree of the third term is 0, and the degree of the whole polynomial is 3 (the same as the degree of the term with the highest degree).

(B) The degree of the first term in $8u^3v^2 - \sqrt{7}uv^2$ is 5, the degree of the second term is 3, and the degree of the whole polynomial is 5.

Matched Problem 1

(A) Given the polynomial $6x^5 + 7x^3 - 2$, what is the degree of the first term? The second term? The third term? The whole polynomial?

(B) Given the polynomial $2u^4v^2 - 5uv^3$, what is the degree of the first term? The second term? The whole polynomial?

In addition to classifying polynomials by degree, we also call a single-term polynomial a **monomial**, a two-term polynomial a **binomial**, and a three-term polynomial a **trinomial**.

Combining Like Terms

The concept of *coefficient* plays a central role in the process of combining *like terms*. A constant in a term of a polynomial, including the sign that precedes it, is called the **numerical coefficient**, or simply, the **coefficient**, of the term. If a constant does not appear, or only a $+$ sign appears, the coefficient is understood to be 1. If only a $-$ sign appears, the coefficient is understood to be -1. Given the polynomial

$$5x^4 - x^3 - 3x^2 + x - 7 \;\boxed{= 5x^4 + (-1)x^3 + (-3)x^2 + 1x + (-7)}$$

the coefficient of the first term is 5, the coefficient of the second term is -1, the coefficient of the third term is -3, the coefficient of the fourth term is 1, and the coefficient of the fifth term is -7.

The following distributive properties are fundamental to the process of combining *like terms*.

THEOREM 2 Distributive Properties of Real Numbers

1. $a(b + c) = (b + c)a = ab + ac$
2. $a(b - c) = (b - c)a = ab - ac$
3. $a(b + c + \cdots + f) = ab + ac + \cdots + af$

Two terms in a polynomial are called **like terms** if they have exactly the same variable factors to the same powers. The numerical coefficients may or may not be the same. Since constant terms involve no variables, all constant terms are like terms. If a polynomial contains two or more like terms, these terms can be combined into a single term by making use of distributive properties. The following example illustrates the reasoning behind the process:

$$
\begin{aligned}
3x^2y - 5xy^2 + x^2y - 2x^2y &= 3x^2y + x^2y - 2x^2y - 5xy^2 \\
&= (3x^2y + 1x^2y - 2x^2y) - 5xy^2 \\
&= (3 + 1 - 2)x^2y - 5xy^2 \\
&= 2x^2y - 5xy^2
\end{aligned}
$$

Note the use of distributive properties.

Free use is made of the real number properties discussed in Appendix A.1.

How can we simplify expressions such as $4(x - 2y) - 3(2x - 7y)$? We clear the expression of parentheses using distributive properties, and combine like terms:

$$
\begin{aligned}
4(x - 2y) - 3(2x - 7y) &= 4x - 8y - 6x + 21y \\
&= -2x + 13y
\end{aligned}
$$

EXAMPLE 2   Removing Parentheses   Remove parentheses and simplify:

(A) $2(3x^2 - 2x + 5) + (x^2 + 3x - 7)$
$$
\begin{aligned}
&= 2(3x^2 - 2x + 5) + 1(x^2 + 3x - 7) \\
&= 6x^2 - 4x + 10 + x^2 + 3x - 7 \\
&= 7x^2 - x + 3
\end{aligned}
$$

(B) $(x^3 - 2x - 6) - (2x^3 - x^2 + 2x - 3)$
$$
\begin{aligned}
&= 1(x^3 - 2x - 6) + (-1)(2x^3 - x^2 + 2x - 3) \\
&= x^3 - 2x - 6 - 2x^3 + x^2 - 2x + 3 \\
&= -x^3 + x^2 - 4x - 3
\end{aligned}
$$

Be careful with the sign here

(C) $[3x^2 - (2x + 1)] - (x^2 - 1) = [3x^2 - 2x - 1] - (x^2 - 1)$
$$
\begin{aligned}
&= 3x^2 - 2x - 1 - x^2 + 1 \\
&= 2x^2 - 2x
\end{aligned}
$$

Remove inner parentheses first.

Matched Problem 2   Remove parentheses and simplify:

(A) $3(u^2 - 2v^2) + (u^2 + 5v^2)$

(B) $(m^3 - 3m^2 + m - 1) - (2m^3 - m + 3)$

(C) $(x^3 - 2) - [2x^3 - (3x + 4)]$

Addition and Subtraction

Addition and subtraction of polynomials can be thought of in terms of removing parentheses and combining like terms, as illustrated in Example 2. Horizontal and vertical arrangements are illustrated in the next two examples. You should be able to work either way, letting the situation dictate your choice.

EXAMPLE 3   Adding Polynomials   Add horizontally and vertically:

$$
x^4 - 3x^3 + x^2, \quad -x^3 - 2x^2 + 3x, \quad \text{and} \quad 3x^2 - 4x - 5
$$

SOLUTION   Add horizontally:

$$
\begin{aligned}
&(x^4 - 3x^3 + x^2) + (-x^3 - 2x^2 + 3x) + (3x^2 - 4x - 5) \\
&= x^4 - 3x^3 + x^2 - x^3 - 2x^2 + 3x + 3x^2 - 4x - 5 \\
&= x^4 - 4x^3 + 2x^2 - x - 5
\end{aligned}
$$

Or vertically, by lining up like terms and adding their coefficients:

$$
\begin{array}{r}
x^4 - 3x^3 + x^2 \qquad\qquad \\
- x^3 - 2x^2 + 3x \quad\;\; \\
3x^2 - 4x - 5 \\
\hline
x^4 - 4x^3 + 2x^2 - x - 5
\end{array}
$$

Matched Problem 3) Add horizontally and vertically:

$$3x^4 - 2x^3 - 4x^2, \quad x^3 - 2x^2 - 5x, \quad \text{and} \quad x^2 + 7x - 2$$

EXAMPLE 4 Subtracting Polynomials Subtract $4x^2 - 3x + 5$ from $x^2 - 8$, both horizontally and vertically.

SOLUTION

$$(x^2 - 8) - (4x^2 - 3x + 5) \qquad \text{or}$$
$$= x^2 - 8 - 4x^2 + 3x - 5$$
$$= -3x^2 + 3x - 13$$

$$\begin{array}{r} x^2 \qquad\quad - 8 \\ \underline{-4x^2 + 3x - 5} \leftarrow \text{Change} \\ -3x^2 + 3x - 13 \end{array}$$
signs and add.

Matched Problem 4) Subtract $2x^2 - 5x + 4$ from $5x^2 - 6$, both horizontally and vertically.

Multiplication

Multiplication of algebraic expressions involves the extensive use of distributive properties for real numbers, as well as other real number properties.

EXAMPLE 5 Multiplying Polynomials Multiply: $(2x - 3)(3x^2 - 2x + 3)$

SOLUTION

$$(2x - 3)(3x^2 - 2x + 3) = 2x(3x^2 - 2x + 3) - 3(3x^2 - 2x + 3)$$
$$= 6x^3 - 4x^2 + 6x - 9x^2 + 6x - 9$$
$$= 6x^3 - 13x^2 + 12x - 9$$

Or, using a vertical arrangement,

$$\begin{array}{r} 3x^2 - \ 2x \ + 3 \\ \underline{2x \ - \ 3} \\ 6x^3 - \ 4x^2 + \ 6x \\ \underline{- \ 9x^2 + \ 6x - 9} \\ 6x^3 - 13x^2 + 12x - 9 \end{array}$$

Matched Problem 5) Multiply: $(2x - 3)(2x^2 + 3x - 2)$

Thus, to multiply two polynomials, multiply each term of one by each term of the other, and combine like terms.

Products of binomial factors occur frequently, so it is useful to develop procedures that will enable us to write down their products by inspection. To find the product $(2x - 1)(3x + 2)$ we proceed as follows:

$$(2x - 1)(3x + 2) = 6x^2 + 4x - 3x - 2$$
$$= 6x^2 + x - 2$$

The inner and outer products are like terms, so combine into a single term.

To speed the process, we do the step in the dashed box mentally.

Products of certain binomial factors occur so frequently that it is useful to learn formulas for their products. The following formulas are easily verified by multiplying the factors on the left.

THEOREM 3 Special Products

1. $(a - b)(a + b) = a^2 - b^2$
2. $(a + b)^2 = a^2 + 2ab + b^2$
3. $(a - b)^2 = a^2 - 2ab + b^2$

EXAMPLE 6 Special Products Multiply mentally, where possible.

(A) $(2x - 3y)(5x + 2y)$ 　　　　　　　(B) $(3a - 2b)(3a + 2b)$

(C) $(5x - 3)^2$ 　　　　　　　　　　　(D) $(m + 2n)^3$

SOLUTION

(A) $(2x - 3y)(5x + 2y)$ $\boxed{= 10x^2 + 4xy - 15xy - 6y^2}$

　　　　　　　　　　　　$= 10x^2 - 11xy - 6y^2$

(B) $(3a - 2b)(3a + 2b)$ $\boxed{= (3a)^2 - (2b)^2}$

　　　　　　　　　　　　$= 9a^2 - 4b^2$

(C) $(5x - 3)^2$ $\boxed{= (5x)^2 - 2(5x)(3) + 3^2}$

　　　　　　$= 25x^2 - 30x + 9$

(D) $(m + 2n)^3 = (m + 2n)^2(m + 2n)$

　　　　　　　　$= (m^2 + 4mn + 4n^2)(m + 2n)$

　　　　　　　　$= m^2(m + 2n) + 4mn(m + 2n) + 4n^2(m + 2n)$

　　　　　　　　$= m^3 + 2m^2n + 4m^2n + 8mn^2 + 4mn^2 + 8n^3$

　　　　　　　　$= m^3 + 6m^2n + 12mn^2 + 8n^3$

Matched Problem 6 ⎪ Multiply mentally, where possible.

(A) $(4u - 3v)(2u + v)$ 　　　　　　　(B) $(2xy + 3)(2xy - 3)$

(C) $(m + 4n)(m - 4n)$ 　　　　　　　(D) $(2u - 3v)^2$

(E) $(2x - y)^3$

Combined Operations

We complete this section by considering several examples that use all the operations just discussed. Note that in simplifying, we usually remove grouping symbols starting from the inside. That is, we remove parentheses ( ) first, then brackets [ ], and finally braces { }, if present. Also,

DEFINITION Order of Operations

Multiplication and division precede addition and subtraction, and taking powers precedes multiplication and division.

$$2 \cdot 3 + 4 = 6 + 4 = 10, \quad \text{not} \quad 2 \cdot 7 = 14$$

$$\frac{10^2}{2} = \frac{100}{2} = 50, \quad \text{not} \quad 5^2 = 25$$

EXAMPLE 7 Combined Operations Perform the indicated operations and simplify:

(A) $3x - \{5 - 3[x - x(3 - x)]\} = 3x - \{5 - 3[x - 3x + x^2]\}$

　　　　　　　　　　　　　　　　　$= 3x - \{5 - 3x + 9x - 3x^2\}$

　　　　　　　　　　　　　　　　　$= 3x - 5 + 3x - 9x + 3x^2$

　　　　　　　　　　　　　　　　　$= 3x^2 - 3x - 5$

(B) $(x - 2y)(2x + 3y) - (2x + y)^2 = 2x^2 - xy - 6y^2 - (4x^2 + 4xy + y^2)$

　　　　　　　　　　　　　　　　　　　$= 2x^2 - xy - 6y^2 - 4x^2 - 4xy - y^2$

　　　　　　　　　　　　　　　　　　　$= -2x^2 - 5xy - 7y^2$

Matched Problem 7 ⎪ Perform the indicated operations and simplify:

(A) $2t - \{7 - 2[t - t(4 + t)]\}$

(B) $(u - 3v)^2 - (2u - v)(2u + v)$

Exercises A.2

Problems 1–8 refer to the following polynomials:

(A) $2x - 3$ (B) $2x^2 - x + 2$ (C) $x^3 + 2x^2 - x + 3$

1. What is the degree of (C)?

2. What is the degree of (A)?

3. Add (B) and (C).

4. Add (A) and (B).

5. Subtract (B) from (C).

6. Subtract (A) from (B).

7. Multiply (B) and (C).

8. Multiply (A) and (C).

In Problems 9–30, perform the indicated operations and simplify.

9. $2(u - 1) - (3u + 2) - 2(2u - 3)$

10. $2(x - 1) + 3(2x - 3) - (4x - 5)$

11. $4a - 2a[5 - 3(a + 2)]$

12. $2y - 3y[4 - 2(y - 1)]$

13. $(a + b)(a - b)$

14. $(m - n)(m + n)$

15. $(3x - 5)(2x + 1)$

16. $(4t - 3)(t - 2)$

17. $(2x - 3y)(x + 2y)$

18. $(3x + 2y)(x - 3y)$

19. $(3y + 2)(3y - 2)$

20. $(2m - 7)(2m + 7)$

21. $-(2x - 3)^2$

22. $-(5 - 3x)^2$

23. $(4m + 3n)(4m - 3n)$

24. $(3x - 2y)(3x + 2y)$

25. $(3u + 4v)^2$

26. $(4x - y)^2$

27. $(a - b)(a^2 + ab + b^2)$

28. $(a + b)(a^2 - ab + b^2)$

29. $[(x - y) + 3z][(x - y) - 3z]$

30. $[a - (2b - c)][a + (2b - c)]$

In Problems 31–44, perform the indicated operations and simplify.

31. $m - \{m - [m - (m - 1)]\}$

32. $2x - 3\{x + 2[x - (x + 5)] + 1\}$

33. $(x^2 - 2xy + y^2)(x^2 + 2xy + y^2)$

34. $(3x - 2y)^2(2x + 5y)$

35. $(5a - 2b)^2 - (2b + 5a)^2$

36. $(2x - 1)^2 - (3x + 2)(3x - 2)$

37. $(m - 2)^2 - (m - 2)(m + 2)$

38. $(x - 3)(x + 3) - (x - 3)^2$

39. $(x - 2y)(2x + y) - (x + 2y)(2x - y)$

40. $(3m + n)(m - 3n) - (m + 3n)(3m - n)$

41. $(u + v)^3$

42. $(x - y)^3$

43. $(x - 2y)^3$

44. $(2m - n)^3$

45. Subtract the sum of the last two polynomials from the sum of the first two: $2x^2 - 4xy + y^2$, $3xy - y^2$, $x^2 - 2xy - y^2$, $-x^2 + 3xy - 2y^2$

46. Subtract the sum of the first two polynomials from the sum of the last two: $3m^2 - 2m + 5$, $4m^2 - m$, $3m^2 - 3m - 2$, $m^3 + m^2 + 2$

In Problems 47–50, perform the indicated operations and simplify.

47. $[(2x - 1)^2 - x(3x + 1)]^2$

48. $[5x(3x + 1) - 5(2x - 1)^2]^2$

49. $2\{(x - 3)(x^2 - 2x + 1) - x[3 - x(x - 2)]\}$

50. $-3x\{x[x - x(2 - x)] - (x + 2)(x^2 - 3)\}$

51. If you are given two polynomials, one of degree m and the other of degree n, where m is greater than n, what is the degree of their product?

52. What is the degree of the sum of the two polynomials in Problem 51?

53. How does the answer to Problem 51 change if the two polynomials can have the same degree?

54. How does the answer to Problem 52 change if the two polynomials can have the same degree?

🖉 55. Show by example that, in general, $(a + b)^2 \neq a^2 + b^2$. Discuss possible conditions on a and b that would make this a valid equation.

🖉 56. Show by example that, in general, $(a - b)^2 \neq a^2 - b^2$. Discuss possible conditions on a and b that would make this a valid equation.

Applications

57. **Investment.** You have $10,000 to invest, part at 9% and the rest at 12%. If x is the amount invested at 9%, write an algebraic expression that represents the total annual income from both investments. Simplify the expression.

58. **Investment.** A person has $100,000 to invest. If $x are invested in a money market account yielding 7% and twice that amount in certificates of deposit yielding 9%, and if the rest is invested in high-grade bonds yielding 11%, write an algebraic expression that represents the total annual income from all three investments. Simplify the expression.

59. Gross receipts. Four thousand tickets are to be sold for a musical show. If x tickets are to be sold for $20 each and three times that number for $30 each, and if the rest are sold for $50 each, write an algebraic expression that represents the gross receipts from ticket sales, assuming all tickets are sold. Simplify the expression.

60. Gross receipts. Six thousand tickets are to be sold for a concert, some for $20 each and the rest for $35 each. If x is the number of $20 tickets sold, write an algebraic expression that represents the gross receipts from ticket sales, assuming all tickets are sold. Simplify the expression.

61. Nutrition. Food mix A contains 2% fat, and food mix B contains 6% fat. A 10-kilogram diet mix of foods A and B is formed. If x kilograms of food A are used, write an algebraic expression that represents the total number of kilograms of fat in the final food mix. Simplify the expression.

62. Nutrition. Each ounce of food M contains 8 units of calcium, and each ounce of food N contains 5 units of calcium. A 160-ounce diet mix is formed using foods M and N. If x is the number of ounces of food M used, write an algebraic expression that represents the total number of units of calcium in the diet mix. Simplify the expression.

Answers to Matched Problems

1. (A) 5, 3, 0, 5   (B) 6, 4, 6
2. (A) $4u^2 - v^2$ (B) $-m^3 - 3m^2 + 2m - 4$
   (C) $-x^3 + 3x + 2$
3. $3x^4 - x^3 - 5x^2 + 2x - 2$
4. $3x^2 + 5x - 10$ 5. $4x^3 - 13x + 6$
6. (A) $8u^2 - 2uv - 3v^2$ (B) $4x^2y^2 - 9$ (C) $m^2 - 16n^2$
   (D) $4u^2 - 12uv + 9v^2$ (E) $8x^3 - 12x^2y + 6xy^2 - y^3$
7. (A) $-2t^2 - 4t - 7$ (B) $-3u^2 - 6uv + 10v^2$

A.3 Factoring Polynomials

- Common Factors
- Factoring by Grouping
- Factoring Second-Degree Polynomials
- Special Factoring Formulas
- Combined Factoring Techniques

A positive integer is **written in factored form** if it is written as the product of two or more positive integers; for example, $120 = 10 \cdot 12$. A positive integer is **factored completely** if each factor is prime; for example, $120 = 2 \cdot 2 \cdot 2 \cdot 3 \cdot 5$. (Recall that an integer $p > 1$ is **prime** if p cannot be factored as the product of two smaller positive integers. So the first ten primes are 2, 3, 5, 7, 11, 13, 17, 19, 23, and 29). A **tree diagram** is a helpful way to visualize a factorization (Fig 1).

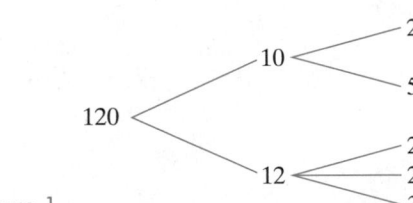

Figure 1

A polynomial is **written in factored form** if it is written as the product of two or more polynomials. The following polynomials are written in factored form:

$$4x^2y - 6xy^2 = 2xy(2x - 3y) \qquad 2x^3 - 8x = 2x(x - 2)(x + 2)$$
$$x^2 - x - 6 = (x - 3)(x + 2) \quad 5m^2 + 20 = 5(m^2 + 4)$$

Unless stated to the contrary, we will limit our discussion of factoring polynomials to polynomials with integer coefficients.

A polynomial with integer coefficients is said to be **factored completely** if each factor cannot be expressed as the product of two or more polynomials with integer coefficients, other than itself or 1. All the polynomials above, as we will see by the conclusion of this section, are factored completely.

Writing polynomials in completely factored form is often a difficult task. But accomplishing it can lead to the simplification of certain algebraic expressions and to the solution of certain types of equations and inequalities. The distributive properties for real numbers are central to the factoring process.

Common Factors

Generally, a first step in any factoring procedure is to factor out all factors common to all terms.

EXAMPLE 1   Common Factors Factor out all factors common to all terms.

(A) $3x^3y - 6x^2y^2 - 3xy^3$

(B) $3y(2y + 5) + 2(2y + 5)$

SOLUTION

(A) $3x^3y - 6x^2y^2 - 3xy^3 = (3xy)x^2 - (3xy)2xy - (3xy)y^2$

$$= 3xy(x^2 - 2xy - y^2)$$

(B) $3y(2y + 5) + 2(2y + 5) = 3y(2y + 5) + 2(2y + 5)$

$$= (3y + 2)(2y + 5)$$

Matched Problem 1 Factor out all factors common to all terms.

(A) $2x^3y - 8x^2y^2 - 6xy^3$ (B) $2x(3x - 2) - 7(3x - 2)$

Factoring by Grouping

Occasionally, polynomials can be factored by grouping terms in such a way that we obtain results that look like Example 1B. We can then complete the factoring following the steps used in that example. This process will prove useful in the next subsection, where an efficient method is developed for factoring a second-degree polynomial as the product of two first-degree polynomials, if such factors exist.

EXAMPLE 2   Factoring by Grouping Factor by grouping.

(A) $3x^2 - 3x - x + 1$

(B) $4x^2 - 2xy - 6xy + 3y^2$

(C) $y^2 + xz + xy + yz$

SOLUTION

(A) $3x^2 - 3x - x + 1$

$$= (3x^2 - 3x) - (x - 1)$$ Group the first two and the last two terms.
$$= 3x(x - 1) - (x - 1)$$ Factor out any common factors from each group. The common factor $(x - 1)$ can be
$$= (x - 1)(3x - 1)$$ taken out, and the factoring is complete.

(B) $4x^2 - 2xy - 6xy + 3y^2 = (4x^2 - 2xy) - (6xy - 3y^2)$

$$= 2x(2x - y) - 3y(2x - y)$$

$$= (2x - y)(2x - 3y)$$

(C) If, as in parts (A) and (B), we group the first two terms and the last two terms of $y^2 + xz + xy + yz$, no common factor can be taken out of each group to complete the factoring. However, if the two middle terms are reversed, we can proceed as before:

$$y^2 + xz + xy + yz = y^2 + xy + xz + yz$$
$$= (y^2 + xy) + (xz + yz)$$
$$= y(y + x) + z(x + y)$$
$$= y(x + y) + z(x + y)$$
$$= (x + y)(y + z)$$

Matched Problem 2 Factor by grouping.

(A) $6x^2 + 2x + 9x + 3$

(B) $2u^2 + 6uv - 3uv - 9v^2$

(C) $ac + bd + bc + ad$

Factoring Second-Degree Polynomials

We now turn our attention to factoring second-degree polynomials of the form

$$2x^2 - 5x - 3 \quad \text{and} \quad 2x^2 + 3xy - 2y^2$$

into the product of two first-degree polynomials with integer coefficients. Since many second-degree polynomials with integer coefficients cannot be factored in this way, it would be useful to know ahead of time that the factors we are seeking actually exist. The factoring approach we use, involving the *ac test*, determines at the beginning whether first-degree factors with integer coefficients do exist. Then, if they exist, the test provides a simple method for finding them.

THEOREM 1   *ac* Test for Factorability

If in polynomials of the form

$$ax^2 + bx + c \quad \text{or} \quad ax^2 + bxy + cy^2 \tag{1}$$

the product ac has two integer factors p and q whose sum is the coefficient b of the middle term; that is, if integers p and q exist so that

$$pq = ac \quad \text{and} \quad p + q = b \tag{2}$$

then the polynomials have first-degree factors with integer coefficients. If no integers p and q exist that satisfy equations (2), then the polynomials in equations (1) will not have first-degree factors with integer coefficients.

If integers p and q exist that satisfy equations (2) in the *ac* test, the factoring always can be completed as follows: Using $b = p + q$, split the middle terms in equations (1) to obtain

$$ax^2 + bx + c = ax^2 + px + qx + c$$
$$ax^2 + bxy + cy^2 = ax^2 + pxy + qxy + cy^2$$

Complete the factoring by grouping the first two terms and the last two terms as in Example 2. This process always works, and it does not matter if the two middle terms on the right are interchanged.

Several examples should make the process clear. After a little practice, you will perform many of the steps mentally and will find the process fast and efficient.

EXAMPLE 3   Factoring Second-Degree Polynomials   Factor, if possible, using integer coefficients.

(A) $4x^2 - 4x - 3$ (B) $2x^2 - 3x - 4$ (C) $6x^2 - 25xy + 4y^2$

SOLUTION

(A) $4x^2 - 4x - 3$

Step 1   Use the *ac* test to test for factorability. Comparing $4x^2 - 4x - 3$ with $ax^2 + bx + c$, we see that $a = 4$, $b = -4$, and $c = -3$. Multiply a and c to obtain

$$ac = (4)(-3) = -12$$

List all pairs of integers whose product is -12, as shown in the margin. These are called **factor pairs** of -12. Then try to find a factor pair that sums to $b = -4$, the coefficient of the middle term in $4x^2 - 4x - 3$. (In practice, this part of Step 1 is often done mentally and can be done rather quickly.) Notice that the factor pair 2 and -6 sums to -4. By the *ac* test, $4x^2 - 4x - 3$ has first-degree factors with integer coefficients.

pq	
$(1)(-12)$	All factor pairs of
$(-1)(12)$	$-12 = ac$
$(2)(-6)$	
$(-2)(6)$	
$(3)(-4)$	
$(-3)(4)$	

Step 2 Split the middle term, using $b = p + q$, and complete the factoring by grouping. Using $-4 = 2 + (-6)$, we split the middle term in $4x^2 - 4x - 3$ and complete the factoring by grouping:

$$
\begin{aligned}
4x^2 - 4x - 3 &= 4x^2 + 2x - 6x - 3 \\
&= (4x^2 + 2x) - (6x + 3) \\
&= 2x(2x + 1) - 3(2x + 1) \\
&= (2x + 1)(2x - 3)
\end{aligned}
$$

The result can be checked by multiplying the two factors to obtain the original polynomial.

(B) $2x^2 - 3x - 4$

Step 1 Use the ac test to test for factorability:

$$ac = (2)(-4) = -8$$

pq

| $(-1)(8)$ |
| $(1)(-8)$ |
| $(-2)(4)$ |
| $(2)(-4)$ |

All factor pairs of $-8 = ac$

Does -8 have a factor pair whose sum is -3? None of the factor pairs listed in the margin sums to $-3 = b$, the coefficient of the middle term in $2x^2 - 3x - 4$. According to the ac test, we can conclude that $2x^2 - 3x - 4$ does not have first-degree factors with integer coefficients, and we say that the polynomial is **not factorable**.

(C) $6x^2 - 25xy + 4y^2$

Step 1 Use the ac test to test for factorability:

$$ac = (6)(4) = 24$$

Mentally checking through the factor pairs of 24, keeping in mind that their sum must be $-25 = b$, we see that if $p = -1$ and $q = -24$, then

$$pq = (-1)(-24) = 24 = ac$$

and

$$p + q = (-1) + (-24) = -25 = b$$

So the polynomial is factorable.

Step 2 Split the middle term, using $b = p + q$, and complete the factoring by grouping. Using $-25 = (-1) + (-24)$, we split the middle term in $6x^2 - 25xy + 4y^2$ and complete the factoring by grouping:

$$
\begin{aligned}
6x^2 - 25xy + 4y^2 &= 6x^2 - xy - 24xy + 4y^2 \\
&= (6x^2 - xy) - (24xy - 4y^2) \\
&= x(6x - y) - 4y(6x - y) \\
&= (6x - y)(x - 4y)
\end{aligned}
$$

The check is left to the reader.

Matched Problem 3 Factor, if possible, using integer coefficients.

(A) $2x^2 + 11x - 6$

(B) $4x^2 + 11x - 6$

(C) $6x^2 + 5xy - 4y^2$

Special Factoring Formulas

The factoring formulas listed in the following box will enable us to factor certain polynomial forms that occur frequently. These formulas can be established by multiplying the factors on the right.

THEOREM 2 Special Factoring Formulas

Perfect square:	1. $u^2 + 2uv + v^2 = (u + v)^2$
Perfect square:	2. $u^2 - 2uv + v^2 = (u - v)^2$
Difference of squares:	3. $u^2 - v^2 = (u - v)(u + v)$
Difference of cubes:	4. $u^3 - v^3 = (u - v)(u^2 + uv + v^2)$
Sum of cubes:	5. $u^3 + v^3 = (u + v)(u^2 - uv + v^2)$

⚠ **CAUTION** Notice that $u^2 + v^2$ is not included in the list of special factoring formulas. In fact,

$$u^2 + v^2 \neq (au + bv)(cu + dv)$$

for any choice of real number coefficients a, b, c, and d. ▲

EXAMPLE 4 Factoring   Factor completely.

(A) $4m^2 - 12mn + 9n^2$ (B) $x^2 - 16y^2$ (C) $z^3 - 1$

(D) $m^3 + n^3$ (E) $a^2 - 4(b + 2)^2$

SOLUTION

(A) $4m^2 - 12mn + 9n^2 = (2m - 3n)^2$

(B) $x^2 - 16y^2 = x^2 - (4y)^2 = (x - 4y)(x + 4y)$

(C) $z^3 - 1 = (z - 1)(z^2 + z + 1)$ Use the *ac* test to verify that $z^2 + z + 1$ cannot be factored.

(D) $m^3 + n^3 = (m + n)(m^2 - mn + n^2)$ Use the *ac* test to verify that $m^2 - mn + n^2$ cannot be factored.

(E) $a^2 - 4(b + 2)^2 = [a - 2(b + 2)][a + 2(b + 2)]$

Matched Problem 4) Factor completely:

(A) $x^2 + 6xy + 9y^2$ (B) $9x^2 - 4y^2$ (C) $8m^3 - 1$

(D) $x^3 + y^3z^3$ (E) $9(m - 3)^2 - 4n^2$

Combined Factoring Techniques

We complete this section by considering several factoring problems that involve combinations of the preceding techniques.

PROCEDURE Factoring Polynomials

Step 1 Take out any factors common to all terms.

Step 2 Use any of the special formulas listed in Theorem 2 that are applicable.

Step 3 Apply the *ac* test to any remaining second-degree polynomial factors.

Note: It may be necessary to perform some of these steps more than once. Furthermore, the order of applying these steps can vary.

EXAMPLE 5 Combined Factoring Techniques   Factor completely.

(A) $3x^3 - 48x$ (B) $3u^4 - 3u^3v - 9u^2v^2$

(C) $3m^2 - 24mn^3$ (D) $3x^4 - 5x^2 + 2$

SOLUTION

(A) $3x^3 - 48x = 3x(x^2 - 16) = 3x(x - 4)(x + 4)$

(B) $3u^4 - 3u^3v - 9u^2v^2 = 3u^2(u^2 - uv - 3v^2)$

(C) $3m^4 - 24mn^3 = 3m(m^3 - 8n^3) = 3m(m - 2n)(m^2 + 2mn + 4n^2)$

(D) $3x^4 - 5x^2 + 2 = (3x^2 - 2)(x^2 - 1) = (3x^2 - 2)(x - 1)(x + 1)$

Matched Problem 5) Factor completely.

(A) $18x^3 - 8x$

(B) $4m^3n - 2m^2n^2 + 2mn^3$

(C) $2t^4 - 16t$

(D) $2y^4 - 5y^2 - 12$

Exercises A.3

In Problems 1–8, factor out all factors common to all terms.

1. $6m^4 - 9m^3 - 3m^2$ **2.** $6x^4 - 8m^3 - 2x^2$

3. $8u^3v - 6u^2v^2 + 4uv^3$ **4.** $10x^3y + 20x^2y^2 - 15xy^3$

5. $7m(2m - 3) + 5(2m - 3)$

6. $5x(x + 1) - 3(x + 1)$

7. $4ab(2c + d) - (2c + d)$

8. $12a(b - 2c) - 15b(b - 2c)$

In Problems 9–18, factor by grouping.

9. $2x^2 - x + 4x - 2$ **10.** $x^2 - 3x + 2x - 6$

11. $3y^2 - 3y + 2y - 2$ **12.** $2x^2 - x + 6x - 3$

13. $2x^2 + 8x - x - 4$ **14.** $6x^2 + 9x - 2x - 3$

15. $wy - wz + xy - xz$ **16.** $ac + ad + bc + bd$

17. $am - 3bm + 2na - 6bn$ **18.** $ab + 6 + 2a + 3b$

In Problems 19–56, factor completely. If a polynomial cannot be factored, say so.

19. $3y^2 - y - 2$ **20.** $2x^2 + 5x - 3$

21. $u^2 - 2uv - 15v^2$ **22.** $x^2 - 4xy - 12y^2$

23. $m^2 - 6m - 3$ **24.** $x^2 + x - 4$

25. $w^2x^2 - y^2$ **26.** $25m^2 - 16n^2$

27. $9m^2 - 6mn + n^2$ **28.** $x^2 + 10xy + 25y^2$

29. $y^2 + 16$ **30.** $u^2 + 81$

31. $4z^2 - 28z + 48$ **32.** $6x^2 + 48x + 72$

33. $2x^4 - 24x^3 + 40x^2$ **34.** $2y^3 - 22y^2 + 48y$

35. $4xy^2 - 12xy + 9x$ **36.** $16x^2y - 8xy + y$

37. $6m^2 - mn - 12n^2$ **38.** $6s^2 + 7st - 3t^2$

39. $4u^3v - uv^3$ **40.** $x^3y - 9xy^3$

41. $2x^3 - 2x^2 + 8x$ **42.** $3m^3 - 6m^2 + 15m$

43. $8x^3 - 27y^3$ **44.** $5x^3 + 40y^3$

45. $x^4y + 8xy$ **46.** $8a^3 - 1$

47. $(x + 2)^2 - 9y^2$ **48.** $(a - b)^2 - 4(c - d)^2$

49. $5u^2 + 4uv - 2v^2$ **50.** $3x^2 - 2xy - 4y^2$

51. $6(x - y)^2 + 23(x - y) - 4$

52. $4(A + B)^2 - 5(A + B) - 6$

53. $y^4 - 3y^2 - 4$

54. $m^4 - n^4$

55. $15y(x - y)^3 + 12x(x - y)^2$

56. $15x^2(3x - 1)^4 + 60x^3(3x - 1)^3$

✎ *In Problems 57–60, discuss the validity of each statement. If the statement is true, explain why. If not, give a counterexample.*

57. If n is a positive integer greater than 1, then $u^n - v^n$ can be factored.

58. If m and n are positive integers and $m \neq n$, then $u^m - v^n$ is not factorable.

59. If n is a positive integer greater than 1, then $u^n + v^n$ can be factored.

60. If k is a positive integer, then $u^{2k+1} + v^{2k+1}$ can be factored.

Answers to Matched Problems

1. (A) $2xy(x^2 - 4xy - 3y^2)$ (B) $(2x - 7)(3x - 2)$

2. (A) $(3x + 1)(2x + 3)$ (B) $(u + 3v)(2u - 3v)$
   (C) $(a + b)(c + d)$

3. (A) $(2x - 1)(x + 6)$ (B) Not factorable
   (C) $(3x + 4y)(2x - y)$

4. (A) $(x + 3y)^2$ (B) $(3x - 2y)(3x + 2y)$
   (C) $(2m - 1)(4m^2 + 2m + 1)$
   (D) $(x + yz)(x^2 - xyz + y^2z^2)$
   (E) $[3(m - 3) - 2n][3(m - 3) + 2n]$

5. (A) $2x(3x - 2)(3x + 2)$ (B) $2mn(2m^2 - mn + n^2)$
   (C) $2t(t - 2)(t^2 + 2t + 4)$
   (D) $(2y^2 + 3)(y - 2)(y + 2)$

A.4   Operations on Rational Expressions

- Reducing to Lowest Terms
- Multiplication and Division
- Addition and Subtraction
- Compound Fractions

We now turn our attention to fractional forms. A quotient of two algebraic expressions (division by 0 excluded) is called a **fractional expression**. If both the numerator and the denominator are polynomials, the fractional expression is called a **rational expression**. Some examples of rational expressions are

$$\frac{1}{x^3 + 2x} \qquad \frac{5}{x} \qquad \frac{x + 7}{3x^2 - 5x + 1} \qquad \frac{x^2 - 2x + 4}{1}$$

In this section, we discuss basic operations on rational expressions. Since variables represent real numbers in the rational expressions we will consider, the properties of real number fractions summarized in Appendix A.1 will play a central role.

> **AGREEMENT**   Variable Restriction
>
> Even though not always explicitly stated, we always assume that variables are restricted so that division by 0 is excluded.

For example, given the rational expression

$$\frac{2x + 5}{x(x + 2)(x - 3)}$$

the variable x is understood to be restricted from being 0, -2, or 3, since these values would cause the denominator to be 0.

Reducing to Lowest Terms

Central to the process of reducing rational expressions to *lowest terms* is the *fundamental property of fractions*, which we restate here for convenient reference:

> **THEOREM 1**   Fundamental Property of Fractions
>
> If a, b, and k are real numbers with $b, k \neq 0$, then
>
> $$\frac{ka}{kb} = \frac{a}{b} \qquad \frac{5 \cdot 2}{5 \cdot 7} = \frac{2}{7} \qquad \frac{x(x + 4)}{2(x + 4)} = \frac{x}{2}, \quad x \neq -4$$

Using this property from left to right to eliminate all common factors from the numerator and the denominator of a given fraction is referred to as **reducing a fraction to lowest terms**. We are actually dividing the numerator and denominator by the same nonzero common factor.

Using the property from right to left—that is, multiplying the numerator and denominator by the same nonzero factor—is referred to as **raising a fraction to higher terms**. We will use the property in both directions in the material that follows.

> **EXAMPLE 1**   Reducing to Lowest Terms   Reduce each fraction to lowest terms.
>
> (A) $\dfrac{1 \cdot 2 \cdot 3 \cdot 4}{1 \cdot 2 \cdot 3 \cdot 4 \cdot 5 \cdot 6} = \dfrac{\cancel{1} \cdot \cancel{2} \cdot \cancel{3} \cdot \cancel{4}}{\cancel{1} \cdot \cancel{2} \cdot \cancel{3} \cdot \cancel{4} \cdot 5 \cdot 6} = \dfrac{1}{5 \cdot 6} = \dfrac{1}{30}$
>
> (B) $\dfrac{2 \cdot 4 \cdot 6 \cdot 8}{1 \cdot 2 \cdot 3 \cdot 4} = \dfrac{\overset{2}{\cancel{2}} \cdot \overset{2}{\cancel{4}} \cdot \overset{2}{\cancel{6}} \cdot \overset{2}{\cancel{8}}}{\cancel{1} \cdot \cancel{2} \cdot \cancel{3} \cdot \cancel{4}} = 2 \cdot 2 \cdot 2 \cdot 2 = 16$

Matched Problem 1   Reduce each fraction to lowest terms.

(A) $\dfrac{1 \cdot 2 \cdot 3 \cdot 4 \cdot 5}{1 \cdot 2 \cdot 1 \cdot 2 \cdot 3}$ (B) $\dfrac{1 \cdot 4 \cdot 9 \cdot 16}{1 \cdot 2 \cdot 3 \cdot 4}$

CONCEPTUAL INSIGHT

Using Theorem 1 to divide the numerator and denominator of a fraction by a common factor is often referred to as **canceling**. This operation can be denoted by drawing a slanted line through each common factor and writing any remaining factors above or below the common factor. Canceling is often incorrectly applied to individual terms in the numerator or denominator, instead of to common factors. For example,

$$\frac{14 - 5}{2} = \frac{9}{2}$$ Theorem 1 does not apply. There are no common factors in the numerator.

$$\frac{14 - 5}{2} \neq \frac{\overset{7}{\cancel{14}} - 5}{\underset{1}{\cancel{2}}} = 2$$ Incorrect use of Theorem 1. To cancel 2 in the denominator, 2 must be a factor of each term in the numerator.

EXAMPLE 2 Reducing to Lowest Terms Reduce each rational expression to lowest terms.

(A) $\dfrac{6x^2 + x - 1}{2x^2 - x - 1} = \dfrac{(2x + 1)(3x - 1)}{(2x + 1)(x - 1)}$ Factor numerator and denominator completely.

$$= \frac{3x - 1}{x - 1}$$ Divide numerator and denominator by the common factor $(2x + 1)$.

(B) $\dfrac{x^4 - 8x}{3x^3 - 2x^2 - 8x} = \dfrac{x(x - 2)(x^2 + 2x + 4)}{x(x - 2)(3x + 4)}$

$$= \frac{x^2 + 2x + 4}{3x + 4}$$

Matched Problem 2 Reduce each rational expression to lowest terms.

(A) $\dfrac{x^2 - 6x + 9}{x^2 - 9}$ (B) $\dfrac{x^3 - 1}{x^2 - 1}$

Multiplication and Division

Since we are restricting variable replacements to real numbers, multiplication and division of rational expressions follow the rules for multiplying and dividing real number fractions summarized in Appendix A.1.

THEOREM 2 Multiplication and Division

If a, b, c, and d are real numbers, then

1. $\dfrac{a}{b} \cdot \dfrac{c}{d} = \dfrac{ac}{bd}$, $b, d \neq 0$ $\dfrac{3}{5} \cdot \dfrac{x}{x + 5} = \dfrac{3x}{5(x + 5)}$

2. $\dfrac{a}{b} \div \dfrac{c}{d} = \dfrac{a}{b} \cdot \dfrac{d}{c}$, $b, c, d \neq 0$ $\dfrac{3}{5} \div \dfrac{x}{x + 5} = \dfrac{3}{5} \cdot \dfrac{x + 5}{x}$

EXAMPLE 3 Multiplication and Division Perform the indicated operations and reduce to lowest terms.

(A) $\dfrac{10x^3 y}{3xy + 9y} \cdot \dfrac{x^2 - 9}{4x^2 - 12x}$ Factor numerators and denominators. Then divide any numerator and any denominator with a like common factor.

$$= \frac{\overset{5x^2}{\cancel{10x^3 y}}}{\underset{3 \cdot 1}{\cancel{3y(x + 3)}}} \cdot \frac{\overset{1 \cdot 1}{\cancel{(x - 3)}\cancel{(x + 3)}}}{\underset{2 \cdot 1}{\cancel{4x}\cancel{(x - 3)}}}$$

$$= \frac{5x^2}{6}$$

(B) $\dfrac{4 - 2x}{4} \div (x - 2) = \dfrac{\overset{1}{2}(2 - x)}{\underset{2}{4}} \cdot \dfrac{1}{x - 2}$ $x - 2 = \dfrac{x - 2}{1}$

$= \dfrac{2 - x}{2(x - 2)} = \dfrac{\overset{-1}{-(x - 2)}}{2\underset{1}{(x - 2)}}$ $b - a = -(a - b)$, a useful change in some problems

$= -\dfrac{1}{2}$

Matched Problem 3 Perform the indicated operations and reduce to lowest terms.

(A) $\dfrac{12x^2y^3}{2xy^2 + 6xy} \cdot \dfrac{y^2 + 6y + 9}{3y^3 + 9y^2}$ (B) $(4 - x) \div \dfrac{x^2 - 16}{5}$

Addition and Subtraction

Again, because we are restricting variable replacements to real numbers, addition and subtraction of rational expressions follow the rules for adding and subtracting real number fractions.

THEOREM 3   Addition and Subtraction

For a, b, and c real numbers,

1. $\dfrac{a}{b} + \dfrac{c}{b} = \dfrac{a + c}{b}$, $b \neq 0$ $\dfrac{x}{x + 5} + \dfrac{8}{x + 5} = \dfrac{x + 8}{x + 5}$

2. $\dfrac{a}{b} - \dfrac{c}{b} = \dfrac{a - c}{b}$, $b \neq 0$ $\dfrac{x}{3x^2y^2} - \dfrac{x + 7}{3x^2y^2} = \dfrac{x - (x + 7)}{3x^2y^2}$

We add rational expressions with the same denominators by adding or subtracting their numerators and placing the result over the common denominator. If the denominators are not the same, we raise the fractions to higher terms, using the fundamental property of fractions to obtain common denominators, and then proceed as described.

Even though any common denominator will do, our work will be simplified if the *least common denominator (LCD)* is used. Often, the LCD is obvious, but if it is not, the steps in the next box describe how to find it.

PROCEDURE   Least Common Denominator

The least common denominator (LCD) of two or more rational expressions is found as follows:

1. Factor each denominator completely, including integer factors.
2. Identify each different factor from all the denominators.
3. Form a product using each different factor to the highest power that occurs in any one denominator. This product is the LCD.

EXAMPLE 4   Addition and Subtraction Combine into a single fraction and reduce to lowest terms.

(A) $\dfrac{3}{10} + \dfrac{5}{6} - \dfrac{11}{45}$ (B) $\dfrac{4}{9x} - \dfrac{5x}{6y^2} + 1$ (C) $\dfrac{1}{x - 1} - \dfrac{1}{x} - \dfrac{2}{x^2 - 1}$

SOLUTION

(A) To find the LCD, factor each denominator completely:

$$\left.\begin{array}{l} 10 = 2 \cdot 5 \\ 6 = 2 \cdot 3 \\ 45 = 3^2 \cdot 5 \end{array}\right\} \quad LCD = 2 \cdot 3^2 \cdot 5 = 90$$

Now use the fundamental property of fractions to make each denominator 90:

$$\frac{3}{10} + \frac{5}{6} - \frac{11}{45} = \frac{9 \cdot 3}{9 \cdot 10} + \frac{15 \cdot 5}{15 \cdot 6} - \frac{2 \cdot 11}{2 \cdot 45}$$

$$= \boxed{\frac{27}{90} + \frac{75}{90} - \frac{22}{90}}$$

$$= \frac{27 + 75 - 22}{90} = \frac{80}{90} = \frac{8}{9}$$

(B) $\left.\begin{array}{l} 9x = 3^2 x \\ 6y^2 = 2 \cdot 3y^2 \end{array}\right\} \quad LCD = 2 \cdot 3^2 xy^2 = 18xy^2$

$$\frac{4}{9x} - \frac{5x}{6y^2} + 1 = \frac{2y^2 \cdot 4}{2y^2 \cdot 9x} - \frac{3x \cdot 5x}{3x \cdot 6y^2} + \frac{18xy^2}{18xy^2}$$

$$= \frac{8y^2 - 15x^2 + 18xy^2}{18xy^2}$$

(C) $\dfrac{1}{x - 1} - \dfrac{1}{x} - \dfrac{2}{x^2 - 1}$

$$= \frac{1}{x - 1} - \frac{1}{x} - \frac{2}{(x - 1)(x + 1)} \qquad LCD = x(x - 1)(x + 1)$$

$$= \frac{x(x + 1) - (x - 1)(x + 1) - 2x}{x(x - 1)(x + 1)}$$

$$= \frac{x^2 + x - x^2 + 1 - 2x}{x(x - 1)(x + 1)}$$

$$= \frac{1 - x}{x(x - 1)(x + 1)}$$

$$= \frac{-\overset{-1}{\cancel{(x - 1)}}}{x\underset{1}{\cancel{(x - 1)}}(x + 1)} = \frac{-1}{x(x + 1)}$$

Matched Problem 4 Combine into a single fraction and reduce to lowest terms.

(A) $\dfrac{5}{28} - \dfrac{1}{10} + \dfrac{6}{35}$ (B) $\dfrac{1}{4x^2} - \dfrac{2x + 1}{3x^3} + \dfrac{3}{12x}$

(C) $\dfrac{2}{x^2 - 4x + 4} + \dfrac{1}{x} - \dfrac{1}{x - 2}$

Compound Fractions

A fractional expression with fractions in its numerator, denominator, or both is called a **compound fraction**. It is often necessary to represent a compound fraction as a **simple fraction**—that is (in all cases we will consider), as the quotient of two polynomials. The process does not involve any new concepts. It is a matter of applying old concepts and processes in the correct sequence.

EXAMPLE 5 Simplifying Compound Fractions Express as a simple fraction reduced to lowest terms:

(A) $\dfrac{\dfrac{1}{5+h} - \dfrac{1}{5}}{h}$

(B) $\dfrac{\dfrac{y}{x^2} - \dfrac{x}{y^2}}{\dfrac{y}{x} - \dfrac{x}{y}}$

SOLUTION We will simplify the expressions in parts (A) and (B) using two different methods—each is suited to the particular type of problem.

(A) We simplify this expression by combining the numerator into a single fraction and using division of rational forms.

$$\frac{\dfrac{1}{5+h} - \dfrac{1}{5}}{h} = \left[\frac{1}{5+h} - \frac{1}{5} \right] \div \frac{h}{1}$$

$$= \frac{5 - 5 - h}{5(5+h)} \cdot \frac{1}{h}$$

$$= \frac{-h}{5(5+h)h} = \frac{-1}{5(5+h)}$$

(B) The method used here makes effective use of the fundamental property of fractions in the form

$$\frac{a}{b} = \frac{ka}{kb} \qquad b, k \neq 0$$

Multiply the numerator and denominator by the LCD of all fractions in the numerator and denominator—in this case, x^2y^2:

$$\frac{x^2y^2\left(\dfrac{y}{x^2} - \dfrac{x}{y^2}\right)}{x^2y^2\left(\dfrac{y}{x} - \dfrac{x}{y}\right)} = \frac{x^2y^2\dfrac{y}{x^2} - x^2y^2\dfrac{x}{y^2}}{x^2y^2\dfrac{y}{x} - x^2y^2\dfrac{x}{y}} = \frac{y^3 - x^3}{xy^3 - x^3y}$$

$$= \frac{(y-x)(y^2 + xy + x^2)}{xy(y-x)(y+x)}$$

$$= \frac{y^2 + xy + x^2}{xy(y+x)} \quad \text{or} \quad \frac{x^2 + xy + y^2}{xy(x+y)}$$

Matched Problem 5 Express as a simple fraction reduced to lowest terms:

(A) $\dfrac{\dfrac{1}{2+h} - \dfrac{1}{2}}{h}$

(B) $\dfrac{\dfrac{a}{b} - \dfrac{b}{a}}{\dfrac{a}{b} + 2 + \dfrac{b}{a}}$

Exercises A.4

In Problems 1–22, perform the indicated operations and reduce answers to lowest terms.

1. $\dfrac{5 \cdot 9 \cdot 13}{3 \cdot 5 \cdot 7}$

2. $\dfrac{10 \cdot 9 \cdot 8}{3 \cdot 2 \cdot 1}$

3. $\dfrac{12 \cdot 11 \cdot 10 \cdot 9}{4 \cdot 3 \cdot 2 \cdot 1}$

4. $\dfrac{15 \cdot 10 \cdot 5}{20 \cdot 15 \cdot 10}$

5. $\dfrac{d^5}{3a} \div \left(\dfrac{d^2}{6a^2} \cdot \dfrac{a}{4d^3} \right)$

6. $\left(\dfrac{d^5}{3a} \div \dfrac{d^2}{6a^2} \right) \cdot \dfrac{a}{4d^3}$

7. $\dfrac{x^2}{12} + \dfrac{x}{18} - \dfrac{1}{30}$

8. $\dfrac{2y}{18} - \dfrac{-1}{28} - \dfrac{y}{42}$

9. $\dfrac{4m-3}{18m^3} + \dfrac{3}{4m} - \dfrac{2m-1}{6m^2}$

10. $\dfrac{3x+8}{4x^2} - \dfrac{2x-1}{x^3} - \dfrac{5}{8x}$

11. $\dfrac{x^2-9}{x^2-3x} \div (x^2 - x - 12)$

12. $\dfrac{2x^2+7x+3}{4x^2-1} \div (x+3)$

13. $\dfrac{2}{x} - \dfrac{1}{x-3}$

14. $\dfrac{5}{m-2} - \dfrac{3}{2m+1}$

15. $\dfrac{2}{(x+1)^2} - \dfrac{5}{x^2-x-2}$

16. $\dfrac{3}{x^2-5x+6} - \dfrac{5}{(x-2)^2}$

17. $\dfrac{x+1}{x-1} - 1$

18. $m - 3 - \dfrac{m-1}{m-2}$

19. $\dfrac{3}{a-1} - \dfrac{2}{1-a}$

20. $\dfrac{5}{x-3} - \dfrac{2}{3-x}$

21. $\dfrac{2x}{x^2-16} - \dfrac{x-4}{x^2+4x}$

22. $\dfrac{m+2}{m^2-2m} - \dfrac{m}{m^2-4}$

In Problems 23–34, perform the indicated operations and reduce answers to lowest terms. Represent any compound fractions as simple fractions reduced to lowest terms.

23. $\dfrac{x^2}{x^2+2x+1} + \dfrac{x-1}{3x+3} - \dfrac{1}{6}$

24. $\dfrac{y}{y^2-y-2} - \dfrac{1}{y^2+5y-14} - \dfrac{2}{y^2+8y+7}$

25. $\dfrac{1-\dfrac{x}{y}}{2-\dfrac{y}{x}}$

26. $\dfrac{2}{5-\dfrac{3}{4x+1}}$

27. $\dfrac{c+2}{5c-5} - \dfrac{c-2}{3c-3} + \dfrac{c}{1-c}$

28. $\dfrac{x+7}{ax-bx} + \dfrac{y+9}{by-ay}$

29. $\dfrac{1+\dfrac{3}{x}}{x-\dfrac{9}{x}}$

30. $\dfrac{1-\dfrac{y^2}{x^2}}{1-\dfrac{y}{x}}$

31. $\dfrac{\dfrac{1}{2(x+h)} - \dfrac{1}{2x}}{h}$

32. $\dfrac{\dfrac{1}{x+h} - \dfrac{1}{x}}{h}$

33. $\dfrac{\dfrac{x}{y} - 2 + \dfrac{y}{x}}{\dfrac{x}{y} - \dfrac{y}{x}}$

34. $\dfrac{1+\dfrac{2}{x}-\dfrac{15}{x^2}}{1+\dfrac{4}{x}-\dfrac{5}{x^2}}$

✎ *In Problems 35–42, imagine that the indicated "solutions" were given to you by a student whom you were tutoring in this class.*

(A) *Is the solution correct? If the solution is incorrect, explain what is wrong and how it can be corrected.*

(B) *Show a correct solution for each incorrect solution.*

35. $\dfrac{x^2+4x+3}{x+3} = \dfrac{x^2+4x}{x} = x + 4$

36. $\dfrac{x^2-3x-4}{x-4} = \dfrac{x^2-3x}{x} = x - 3$

37. $\dfrac{(x+h)^2-x^2}{h} = (x+1)^2 - x^2 = 2x + 1$

38. $\dfrac{(x+h)^3-x^3}{h} = (x+1)^3 - x^3 = 3x^2 + 3x + 1$

39. $\dfrac{x^2-3x}{x^2-2x-3} + x - 3 = \dfrac{x^2-3x+x-3}{x^2-2x-3} = 1$

40. $\dfrac{2}{x-1} - \dfrac{x+3}{x^2-1} = \dfrac{2x+2-x-3}{x^2-1} = \dfrac{1}{x+1}$

41. $\dfrac{2x^2}{x^2-4} - \dfrac{x}{x-2} = \dfrac{2x^2-x^2-2x}{x^2-4} = \dfrac{x}{x+2}$

42. $x + \dfrac{x-2}{x^2-3x+2} = \dfrac{x+x-2}{x^2-3x+2} = \dfrac{2}{x-2}$

Represent the compound fractions in Problems 43–46 as simple fractions reduced to lowest terms.

43. $\dfrac{\dfrac{1}{3(x+h)^2} - \dfrac{1}{3x^2}}{h}$

44. $\dfrac{\dfrac{1}{(x+h)^2} - \dfrac{1}{x^2}}{h}$

45. $x - \dfrac{2}{1-\dfrac{1}{x}}$

46. $2 - \dfrac{1}{1-\dfrac{2}{a+2}}$

Answers to Matched Problems

1. (A) 10 (B) 24

2. (A) $\dfrac{x-3}{x+3}$ (B) $\dfrac{x^2+x+1}{x+1}$

3. (A) $2x$ (B) $\dfrac{-5}{x+4}$

4. (A) $\dfrac{1}{4}$ (B) $\dfrac{3x^2-5x-4}{12x^3}$ (C) $\dfrac{4}{x(x-2)^2}$

5. (A) $\dfrac{-1}{2(2+h)}$ (B) $\dfrac{a-b}{a+b}$

A.5 Integer Exponents and Scientific Notation

- Integer Exponents
- Scientific Notation

We now review basic operations on integer exponents and scientific notation.

Integer Exponents

DEFINITION Integer Exponents

For n an integer and a a real number:

1. For n a positive integer,

$$a^n = a \cdot a \cdot \cdots \cdot a \quad n \text{ factors of } a \qquad 5^4 = 5 \cdot 5 \cdot 5 \cdot 5$$

2. For $n = 0$,

$$a^0 = 1 \quad a \neq 0 \qquad 12^0 = 1$$
$$0^0 \text{ is not defined.}$$

3. For n a negative integer,

$$a^n = \frac{1}{a^{-n}} \quad a \neq 0 \qquad a^{-3} = \frac{1}{a^{-(-3)}} = \frac{1}{a^3}$$

[If n is negative, then $(-n)$ is positive.]

Note: It can be shown that for *all* integers n,

$$a^{-n} = \frac{1}{a^n} \quad \text{and} \quad a^n = \frac{1}{a^{-n}} \quad a \neq 0 \qquad a^5 = \frac{1}{a^{-5}}, \quad a^{-5} = \frac{1}{a^5}$$

The following properties are very useful in working with integer exponents.

THEOREM 1 Exponent Properties

For n and m integers and a and b real numbers,

1. $a^m a^n = a^{m+n}$ $\qquad\qquad a^8 a^{-3} = a^{8+(-3)} = a^5$

2. $(a^n)^m = a^{mn}$ $\qquad\qquad (a^{-2})^3 = a^{3(-2)} = a^{-6}$

3. $(ab)^m = a^m b^m$ $\qquad\qquad (ab)^{-2} = a^{-2} b^{-2}$

4. $\left(\dfrac{a}{b}\right)^m = \dfrac{a^m}{b^m} \quad b \neq 0 \qquad \left(\dfrac{a}{b}\right)^5 = \dfrac{a^5}{b^5}$

5. $\dfrac{a^m}{a^n} = a^{m-n} = \dfrac{1}{a^{n-m}} \quad a \neq 0 \qquad \dfrac{a^{-3}}{a^7} = \dfrac{1}{a^{7-(-3)}} = \dfrac{1}{a^{10}}$

Exponents are frequently encountered in algebraic applications. You should sharpen your skills in using exponents by reviewing the preceding basic definitions and properties and the examples that follow.

EXAMPLE 1 Simplifying Exponent Forms Simplify, and express the answers using positive exponents only.

(A) $(2x^3)(3x^5) = 2 \cdot 3 x^{3+5} = 6x^8$

(B) $x^5 x^{-9} = x^{-4} = \dfrac{1}{x^4}$

(C) $\dfrac{x^5}{x^7} = x^{5-7} = x^{-2} = \dfrac{1}{x^2}$ or $\dfrac{x^5}{x^7} = \dfrac{1}{x^{7-5}} = \dfrac{1}{x^2}$

(D) $\dfrac{x^{-3}}{y^{-4}} = \dfrac{y^4}{x^3}$

(E) $(u^{-3}v^2)^{-2} \boxed{= (u^{-3})^{-2}(v^2)^{-2}} = u^6v^{-4} = \dfrac{u^6}{v^4}$

(F) $\left(\dfrac{y^{-5}}{y^{-2}}\right)^{-2} \boxed{= \dfrac{(y^{-5})^{-2}}{(y^{-2})^{-2}}} = \dfrac{y^{10}}{y^4} = y^6$

(G) $\dfrac{4m^{-3}n^{-5}}{6m^{-4}n^3} \boxed{= \dfrac{2m^{-3-(-4)}}{3n^{3-(-5)}}} = \dfrac{2m}{3n^8}$

__Matched Problem 1__ Simplify, and express the answers using positive exponents only.

(A) $(3y^4)(2y^3)$ (B) m^2m^{-6} (C) $(u^3v^{-2})^{-2}$

(D) $\left(\dfrac{y^{-6}}{y^{-2}}\right)^{-1}$ (E) $\dfrac{8x^{-2}y^{-4}}{6x^{-5}y^2}$

EXAMPLE 2 Converting to a Simple Fraction Write $\dfrac{1-x}{x^{-1}-1}$ as a simple fraction with positive exponents.

SOLUTION First note that

$$\dfrac{1-x}{x^{-1}-1} \neq \dfrac{x(1-x)}{-1} \qquad \text{A common error}$$

The original expression is a compound fraction, and we proceed to simplify it as follows:

$$\dfrac{1-x}{x^{-1}-1} = \dfrac{1-x}{\dfrac{1}{x}-1} \qquad \begin{array}{l}\text{Multiply numerator and denominator}\\ \text{by } x \text{ to clear internal fractions.}\end{array}$$

$$= \dfrac{x(1-x)}{x\left(\dfrac{1}{x}-1\right)}$$

$$= \dfrac{x(1-x)}{1-x} = x$$

__Matched Problem 2__ Write $\dfrac{1+x^{-1}}{1-x^{-2}}$ as a simple fraction with positive exponents.

Scientific Notation

In the real world, one often encounters very large and very small numbers. For example,

- The public debt in the United States in 2013, to the nearest billion dollars, was

$$\$16,739,000,000,000$$

- The world population in the year 2025, to the nearest million, is projected to be

$$7,947,000,000$$

- The sound intensity of a normal conversation is

$$0.000\ 000\ 000\ 316 \text{ watt per square centimeter*}$$

---

*We write 0.000 000 000 316 in place of 0.000000000316, because it is then easier to keep track of the number of decimal places.

It is generally troublesome to write and work with numbers of this type in standard decimal form. The first and last example cannot even be entered into many calculators as they are written. But with exponents defined for all integers, we can now express any finite decimal form as the product of a number between 1 and 10 and an integer power of 10, that is, in the form

$$a \times 10^n \qquad 1 \le a < 10, \quad a \text{ in decimal form}, \quad n \text{ an integer}$$

A number expressed in this form is said to be in **scientific notation**. The following are some examples of numbers in standard decimal notation and in scientific notation:

Decimal and Scientific Notation

$7 = 7 \times 10^0$	$0.5 = 5 \times 10^{-1}$
$67 = 6.7 \times 10$	$0.45 = 4.5 \times 10^{-1}$
$580 = 5.8 \times 10^2$	$0.0032 = 3.2 \times 10^{-3}$
$43,000 = 4.3 \times 10^4$	$0.000\,045 = 4.5 \times 10^{-5}$
$73,400,000 = 7.34 \times 10^7$	$0.000\,000\,391 = 3.91 \times 10^{-7}$

Note that the power of 10 used corresponds to the number of places we move the decimal to form a number between 1 and 10. The power is positive if the decimal is moved to the left and negative if it is moved to the right. Positive exponents are associated with numbers greater than or equal to 10; negative exponents are associated with positive numbers less than 1; and a zero exponent is associated with a number that is 1 or greater, but less than 10.

EXAMPLE 3 Scientific Notation

(A) Write each number in scientific notation:

$$7,320,000 \quad \text{and} \quad 0.000\,000\,54$$

(B) Write each number in standard decimal form:

$$4.32 \times 10^6 \quad \text{and} \quad 4.32 \times 10^{-5}$$

SOLUTION

(A) $7,320,000 = 7.320\,000. \times 10^6 = 7.32 \times 10^6$

  6 places left
  Positive exponent

$0.000\,000\,54 = 0.000\,000\,5.4 \times 10^{-7} = 5.4 \times 10^{-7}$

  7 places right
  Negative exponent

(B) $4.32 \times 10^6 = 4,320,000$ $4.32 \times 10^{-5} = \dfrac{4.32}{10^5} = 0.000\,043\,2$

  6 places right   5 places left
  Positive exponent 6   Negative exponent -5

Matched Problem 3

(A) Write each number in scientific notation: 47,100; 2,443,000,000; 1.45

(B) Write each number in standard decimal form: 3.07×10^8; 5.98×10^{-6}

Exercises A.5

In Problems 1–14, simplify and express answers using positive exponents only. Variables are restricted to avoid division by 0.

1. $2x^{-9}$

2. $3y^{-5}$

3. $\dfrac{3}{2w^{-7}}$

4. $\dfrac{5}{4x^{-9}}$

5. $2x^{-8}x^5$

6. $3c^{-9}c^4$

7. $\dfrac{w^{-8}}{w^{-3}}$

8. $\dfrac{m^{-11}}{m^{-5}}$

9. $(2a^{-3})^2$

10. $7d^{-4}d^4$

11. $(a^{-3})^2$

12. $(5b^{-2})^2$

13. $(2x^4)^{-3}$

14. $(a^{-3}b^4)^{-3}$

In Problems 15–20, write each number in scientific notation.

15. 82,300,000,000

16. 5,380,000

17. 0.783

18. 0.019

19. 0.000 034

20. 0.000 000 007 832

In Problems 21–28, write each number in standard decimal notation.

21. 4×10^4

22. 9×10^6

23. 7×10^{-3}

24. 2×10^{-5}

25. 6.171×10^7

26. 3.044×10^3

27. 8.08×10^{-4}

28. 1.13×10^{-2}

In Problems 29–38, simplify and express answers using positive exponents only. Assume that variables are nonzero.

29. $(22 + 31)^0$

30. $(2x^3y^4)^0$

31. $\dfrac{10^{-3} \cdot 10^4}{10^{-11} \cdot 10^{-2}}$

32. $\dfrac{10^{-17} \cdot 10^{-5}}{10^{-3} \cdot 10^{-14}}$

33. $(5x^2y^{-3})^{-2}$

34. $(2m^{-3}n^2)^{-3}$

35. $\left(\dfrac{-5}{2x^3}\right)^{-2}$

36. $\left(\dfrac{2a}{3b^2}\right)^{-3}$

37. $\dfrac{8x^{-3}y^{-1}}{6x^2y^{-4}}$

38. $\dfrac{9m^{-4}n^3}{12m^{-1}n^{-1}}$

In Problems 39–42, write each expression in the form $ax^p + bx^q$ or $ax^p + bx^q + cx^r$, where a, b, and c are real numbers and p, q, and r are integers. For example,

$$\dfrac{2x^4 - 3x^2 + 1}{2x^3} = \dfrac{2x^4}{2x^3} - \dfrac{3x^2}{2x^3} + \dfrac{1}{2x^3} = x - \dfrac{3}{2}x^{-1} + \dfrac{1}{2}x^{-3}$$

39. $\dfrac{7x^5 - x^2}{4x^5}$

40. $\dfrac{5x^3 - 2}{3x^2}$

41. $\dfrac{5x^4 - 3x^2 + 8}{2x^2}$

42. $\dfrac{2x^3 - 3x^2 + x}{2x^2}$

Write each expression in Problems 43–46 with positive exponents only, and as a single fraction reduced to lowest terms.

43. $\dfrac{3x^2(x-1)^2 - 2x^3(x-1)}{(x-1)^4}$

44. $\dfrac{5x^4(x+3)^2 - 2x^5(x+3)}{(x+3)^4}$

45. $2x^{-2}(x - 1) - 2x^{-3}(x - 1)^2$

46. $2x(x + 3)^{-1} - x^2(x + 3)^{-2}$

In Problems 47–50, convert each number to scientific notation and simplify. Express the answer in both scientific notation and in standard decimal form.

47. $\dfrac{9,600,000,000}{(1,600,000)(0.000\,000\,25)}$

48. $\dfrac{(60,000)(0.000\,003)}{(0.0004)(1,500,000)}$

49. $\dfrac{(1,250,000)(0.000\,38)}{0.0152}$

50. $\dfrac{(0.000\,000\,82)(230,000)}{(625,000)(0.0082)}$

51. What is the result of entering 2^{3^2} on a calculator?

52. Refer to Problem 51. What is the difference between $2^{(3^2)}$ and $(2^3)^2$? Which agrees with the value of 2^{3^2} obtained with a calculator?

53. If $n = 0$, then property 1 in Theorem 1 implies that $a^m a^0 = a^{m+0} = a^m$. Explain how this helps motivate the definition of a^0.

54. If $m = -n$, then property 1 in Theorem 1 implies that $a^{-n}a^n = a^0 = 1$. Explain how this helps motivate the definition of a^{-n}.

Write the fractions in Problems 55–58 as simple fractions reduced to lowest terms.

55. $\dfrac{u + v}{u^{-1} + v^{-1}}$

56. $\dfrac{x^{-2} - y^{-2}}{x^{-1} + y^{-1}}$

57. $\dfrac{b^{-2} - c^{-2}}{b^{-3} - c^{-3}}$

58. $\dfrac{xy^{-2} - yx^{-2}}{y^{-1} - x^{-1}}$

Applications

Problems 59 and 60 refer to Table 1.

Table 1 **U.S. Public Debt, Interest on Debt, and Population**

Year	Public Debt ($)	Interest on Debt ($)	Population
2000	5,674,000,000,000	362,000,000,000	281,000,000
2012	16,066,000,000,000	360,000,000,000	313,000,000

59. Public debt. Carry out the following computations using scientific notation, and write final answers in standard decimal form.

(A) What was the per capita debt in 2012 (to the nearest dollar)?

(B) What was the per capita interest paid on the debt in 2012 (to the nearest dollar)?

(C) What was the percentage interest paid on the debt in 2012 (to two decimal places)?

541

60. Public debt. Carry out the following computations using scientific notation, and write final answers in standard decimal form.

(A) What was the per capita debt in 2000 (to the nearest dollar)?

(B) What was the per capita interest paid on the debt in 2000 (to the nearest dollar)?

(C) What was the percentage interest paid on the debt in 2000 (to two decimal places)?

Air pollution. *Air quality standards establish maximum amounts of pollutants considered acceptable in the air. The amounts are frequently given in parts per million (ppm). A standard of 30 ppm also can be expressed as follows:*

$$30 \text{ ppm} = \frac{30}{1,000,000} = \frac{3 \times 10}{10^6}$$
$$= 3 \times 10^{-5} = 0.00003 = 0.003\%$$

In Problems 61 and 62, express the given standard:

(A) In scientific notation

(B) In standard decimal notation

(C) As a percent

61. 9 ppm, the standard for carbon monoxide, when averaged over a period of 8 hours

62. 0.03 ppm, the standard for sulfur oxides, when averaged over a year

63. Crime. In 2010, the United States had a violent crime rate of 404 per 100,000 people and a population of 309 million people. How many violent crimes occurred that year? Compute the answer using scientific notation and convert the answer to standard decimal form (to the nearest thousand).

64. Population density. The United States had a 2012 population of 313 million people and a land area of 3,539,000 square miles. What was the population density? Compute the answer using scientific notation and convert the answer to standard decimal form (to one decimal place).

Answers to Matched Problems

1. (A) $6y^7$ (B) $\dfrac{1}{m^4}$ (C) $\dfrac{v^4}{u^6}$ (D) y^4 (E) $\dfrac{4x^3}{3y^6}$

2. $\dfrac{x}{x-1}$

3. (A) 4.7×10^4; 2.443×10^9; 1.45×10^0

(B) 307,000,000; 0.000 005 98

A.6 Rational Exponents and Radicals

- *nth* Roots of Real Numbers
- Rational Exponents and Radicals
- Properties of Radicals

Square roots may now be generalized to *nth roots*, and the meaning of exponent may be generalized to include all rational numbers.

nth Roots of Real Numbers

Consider a square of side r with area 36 square inches. We can write

$$r^2 = 36$$

and conclude that side r is a number whose square is 36. We say that r is a **square root** of b if $r^2 = b$. Similarly, we say that r is a **cube root** of b if $r^3 = b$. And, in general,

> **DEFINITION**   *nth* Root
> For any natural number n,
>
> $$r \text{ is an } \textbf{\textit{n}th root} \text{ of } b \text{ if } r^n = b$$

So 4 is a square root of 16, since $4^2 = 16$; -2 is a cube root of -8, since $(-2)^3 = -8$. Since $(-4)^2 = 16$, we see that -4 is also a square root of 16. It can be shown that any positive number has two real square roots, two real 4th roots, and, in general, two real *nth* roots if n is even. Negative numbers have no real square roots, no real 4th roots, and, in general, no real *nth* roots if n is even. The reason is that no real number raised to an even power can be negative. For odd roots, the situation is simpler. Every real number has exactly one real cube root, one real 5th root, and, in general, one real *nth* root if n is odd.

Additional roots can be considered in the *complex number system*. In this book, we restrict our interest to *real roots of real numbers*, and *root* will always be interpreted to mean "real root."

Rational Exponents and Radicals

We now turn to the question of what symbols to use to represent nth roots. For n a natural number greater than 1, we use

$$b^{1/n} \quad \text{or} \quad \sqrt[n]{b}$$

to represent a **real nth root of b.** The exponent form is motivated by the fact that $(b^{1/n})^n = b$ if exponent laws are to continue to hold for rational exponents. The other form is called an ***n*th root radical.** In the expression below, the symbol $\sqrt{}$ is called a **radical,** n is the **index** of the radical, and b is the **radicand:**

Index $\longrightarrow$ $\downarrow$ $\downarrow$ $\longleftarrow$ Radical

$$\sqrt[n]{b}$$

$\uparrow$ $\longleftarrow$ Radicand

When the index is 2, it is usually omitted. That is, when dealing with square roots, we simply use $\sqrt{b}$ rather than $\sqrt[2]{b}$. If there are two real nth roots, both $b^{1/n}$ and $\sqrt[n]{b}$ denote the positive root, called the **principal nth root.**

EXAMPLE 1 Finding nth Roots Evaluate each of the following:

(A) $4^{1/2}$ and $\sqrt{4}$ (B) $-4^{1/2}$ and $-\sqrt{4}$ (C) $(-4)^{1/2}$ and $\sqrt{-4}$

(D) $8^{1/3}$ and $\sqrt[3]{8}$ (E) $(-8)^{1/3}$ and $\sqrt[3]{-8}$ (F) $-8^{1/3}$ and $-\sqrt[3]{8}$

SOLUTION

(A) $4^{1/2} = \sqrt{4} = 2$ $(\sqrt{4} \neq \pm 2)$ (B) $-4^{1/2} = -\sqrt{4} = -2$

(C) $(-4)^{1/2}$ and $\sqrt{-4}$ are not real numbers

(D) $8^{1/3} = \sqrt[3]{8} = 2$ (E) $(-8)^{1/3} = \sqrt[3]{-8} = -2$

(F) $-8^{1/3} = -\sqrt[3]{8} = -2$

Matched Problem 1 Evaluate each of the following:

(A) $16^{1/2}$ (B) $-\sqrt{16}$ (C) $\sqrt[3]{-27}$ (D) $(-9)^{1/2}$ (E) $\left(\sqrt[4]{81}\right)^3$

⚠ **CAUTION** The symbol $\sqrt{4}$ represents the single number 2, not ± 2. Do not confuse $\sqrt{4}$ with the solutions of the equation $x^2 = 4$, which are usually written in the form $x = \pm\sqrt{4} = \pm 2$.   ▲

We now define b^r for any rational number $r = m/n$.

DEFINITION Rational Exponents

If m and n are natural numbers without common prime factors, b is a real number, and b is nonnegative when n is even, then

$$b^{m/n} = \begin{cases} (b^{1/n})^m = (\sqrt[n]{b})^m & 8^{2/3} = (8^{1/3})^2 = (\sqrt[3]{8})^2 = 2^2 = 4 \\ (b^m)^{1/n} = \sqrt[n]{b^m} & 8^{2/3} = (8^2)^{1/3} = \sqrt[3]{8^2} = \sqrt[3]{64} = 4 \end{cases}$$

and

$$b^{-m/n} = \frac{1}{b^{m/n}} \quad b \neq 0 \qquad 8^{-2/3} = \frac{1}{8^{2/3}} = \frac{1}{4}$$

Note that the two definitions of $b^{m/n}$ are equivalent under the indicated restrictions on m, n, and b.

CONCEPTUAL INSIGHT

All the properties for integer exponents listed in Theorem 1 in Section A.5 also hold for rational exponents, provided that b is nonnegative when n is even. This restriction on b is necessary to avoid nonreal results. For example,

$$(-4)^{3/2} = \sqrt{(-4)^3} = \sqrt{-64} \quad \text{Not a real number}$$

To avoid nonreal results, all variables in the remainder of this discussion represent positive real numbers.

EXAMPLE 2   From Rational Exponent Form to Radical Form and Vice Versa   Change rational exponent form to radical form.

(A) $x^{1/7} = \sqrt[7]{x}$

(B) $(3u^2v^3)^{3/5} = \sqrt[5]{(3u^2v^3)^3}$ or $(\sqrt[5]{3u^2v^3})^3$ The first is usually preferred.

(C) $y^{-2/3} = \dfrac{1}{y^{2/3}} = \dfrac{1}{\sqrt[3]{y^2}}$ or $\sqrt[3]{y^{-2}}$ or $\sqrt[3]{\dfrac{1}{y^2}}$

Change radical form to rational exponent form.

(D) $\sqrt[5]{6} = 6^{1/5}$

(E) $-\sqrt[3]{x^2} = -x^{2/3}$

(F) $\sqrt{x^2 + y^2} = (x^2 + y^2)^{1/2}$ Note that $(x^2 + y^2)^{1/2} \ne x + y$. Why?

Matched Problem 2⌋   Convert to radical form.

(A) $u^{1/5}$ (B) $(6x^2y^5)^{2/9}$ (C) $(3xy)^{-3/5}$

Convert to rational exponent form.

(D) $\sqrt[4]{9u}$ (E) $-\sqrt{(2x)^4}$ (F) $\sqrt[3]{x^3 + y^3}$

EXAMPLE 3   Working with Rational Exponents   Simplify each and express answers using positive exponents only. If rational exponents appear in final answers, convert to radical form.

(A) $\left(3x^{1/3}\right)\left(2x^{1/2}\right) = 6x^{1/3+1/2} = 6x^{5/6} = 6\sqrt[6]{x^5}$

(B) $(-8)^{5/3} = [(-8)^{1/3}]^5 = (-2)^5 = -32$

(C) $\left(2x^{1/3}y^{-2/3}\right)^3 = 8xy^{-2} = \dfrac{8x}{y^2}$

(D) $\left(\dfrac{4x^{1/3}}{x^{1/2}}\right)^{1/2} = \dfrac{4^{1/2}x^{1/6}}{x^{1/4}} = \dfrac{2}{x^{1/4-1/6}} = \dfrac{2}{x^{1/12}} = \dfrac{2}{\sqrt[12]{x}}$

Matched Problem 3⌋   Simplify each and express answers using positive exponents only. If rational exponents appear in final answers, convert to radical form.

(A) $9^{3/2}$ (B) $(-27)^{4/3}$ (C) $\left(5y^{1/4}\right)\left(2y^{1/3}\right)$ (D) $\left(2x^{-3/4}y^{1/4}\right)^4$

(E) $\left(\dfrac{8x^{1/2}}{x^{2/3}}\right)^{1/3}$

EXAMPLE 4   Working with Rational Exponents   Multiply, and express answers using positive exponents only.

(A) $3y^{2/3}\left(2y^{1/3} - y^2\right)$ (B) $\left(2u^{1/2} + v^{1/2}\right)\left(u^{1/2} - 3v^{1/2}\right)$

SOLUTION

(A) $3y^{2/3}\left(2y^{1/3} - y^2\right) = 6y^{2/3+1/3} - 3y^{2/3+2}$

$$= 6y - 3y^{8/3}$$

(B) $\left(2u^{1/2} + v^{1/2}\right)\left(u^{1/2} - 3v^{1/2}\right) = 2u - 5u^{1/2}v^{1/2} - 3v$

Matched Problem 4) Multiply, and express answers using positive exponents only.

(A) $2c^{1/4}(5c^3 - c^{3/4})$ (B) $(7x^{1/2} - y^{1/2})(2x^{1/2} + 3y^{1/2})$

EXAMPLE 5 Working with Rational Exponents Write the following expression in the form $ax^p + bx^q$, where a and b are real numbers and p and q are rational numbers:

$$\frac{2\sqrt{x} - 3\sqrt[3]{x^2}}{2\sqrt[3]{x}}$$

SOLUTION $\dfrac{2\sqrt{x} - 3\sqrt[3]{x^2}}{2\sqrt[3]{x}} = \dfrac{2x^{1/2} - 3x^{2/3}}{2x^{1/3}}$ Change to rational exponent form.

$$= \frac{2x^{1/2}}{2x^{1/3}} - \frac{3x^{2/3}}{2x^{1/3}}$$ Separate into two fractions.

$$= x^{1/6} - 1.5x^{1/3}$$

Matched Problem 5) Write the following expression in the form $ax^p + bx^q$, where a and b are real numbers and p and q are rational numbers:

$$\frac{5\sqrt[3]{x} - 4\sqrt{x}}{2\sqrt{x^3}}$$

Properties of Radicals

Changing or simplifying radical expressions is aided by several properties of radicals that follow directly from the properties of exponents considered earlier.

> **THEOREM 1   Properties of Radicals**
>
> If n is a natural number greater than or equal to 2, and if x and y are positive real numbers, then
>
> 1. $\sqrt[n]{x^n} = x$ $\sqrt[3]{x^3} = x$
> 2. $\sqrt[n]{xy} = \sqrt[n]{x}\sqrt[n]{y}$ $\sqrt[5]{xy} = \sqrt[5]{x}\sqrt[5]{y}$
> 3. $\sqrt[n]{\dfrac{x}{y}} = \dfrac{\sqrt[n]{x}}{\sqrt[n]{y}}$ $\sqrt[4]{\dfrac{x}{y}} = \dfrac{\sqrt[4]{x}}{\sqrt[4]{y}}$

EXAMPLE 6 Applying Properties of Radicals Simplify using properties of radicals.

(A) $\sqrt[4]{(3x^4y^3)^4}$ (B) $\sqrt[4]{8}\sqrt[4]{2}$ (C) $\sqrt[3]{\dfrac{xy}{27}}$

SOLUTION

(A) $\sqrt[4]{(3x^4y^3)^4} = 3x^4y^3$ Property 1

(B) $\sqrt[4]{8}\sqrt[4]{2} = \sqrt[4]{16} = \sqrt[4]{2^4} = 2$ Properties 2 and 1

(C) $\sqrt[3]{\dfrac{xy}{27}} = \dfrac{\sqrt[3]{xy}}{\sqrt[3]{27}} = \dfrac{\sqrt[3]{xy}}{3}$ or $\dfrac{1}{3}\sqrt[3]{xy}$ Properties 3 and 1

Matched Problem 6) Simplify using properties of radicals.

(A) $\sqrt[7]{(x^3 + y^3)^7}$ (B) $\sqrt[3]{8y^3}$ (C) $\dfrac{\sqrt[3]{16x^4y}}{\sqrt[3]{2xy}}$

What is the best form for a radical expression? There are many answers, depending on what use we wish to make of the expression. In deriving certain formulas, it is sometimes useful to clear either a denominator or a numerator of radicals.

The process is referred to as **rationalizing** the denominator or numerator. Examples 7 and 8 illustrate the rationalizing process.

EXAMPLE 7   Rationalizing Denominators   Rationalize each denominator.

(A) $\dfrac{6x}{\sqrt{2x}}$ (B) $\dfrac{6}{\sqrt{7} - \sqrt{5}}$ (C) $\dfrac{x - 4}{\sqrt{x} + 2}$

SOLUTION

(A) $\dfrac{6x}{\sqrt{2x}} = \dfrac{6x}{\sqrt{2x}} \cdot \dfrac{\sqrt{2x}}{\sqrt{2x}} = \dfrac{6x\sqrt{2x}}{2x} = 3\sqrt{2x}$

(B) $\dfrac{6}{\sqrt{7} - \sqrt{5}} = \dfrac{6}{\sqrt{7} - \sqrt{5}} \cdot \dfrac{\sqrt{7} + \sqrt{5}}{\sqrt{7} + \sqrt{5}}$

$\qquad = \dfrac{6(\sqrt{7} + \sqrt{5})}{2} = 3(\sqrt{7} + \sqrt{5})$

(C) $\dfrac{x - 4}{\sqrt{x} + 2} = \dfrac{x - 4}{\sqrt{x} + 2} \cdot \dfrac{\sqrt{x} - 2}{\sqrt{x} - 2}$

$\qquad = \dfrac{(x - 4)(\sqrt{x} - 2)}{x - 4} = \sqrt{x} - 2$

Matched Problem 7   Rationalize each denominator.

(A) $\dfrac{12ab^2}{\sqrt{3ab}}$ (B) $\dfrac{9}{\sqrt{6} + \sqrt{3}}$ (C) $\dfrac{x^2 - y^2}{\sqrt{x} - \sqrt{y}}$

EXAMPLE 8   Rationalizing Numerators   Rationalize each numerator.

(A) $\dfrac{\sqrt{2}}{2\sqrt{3}}$ (B) $\dfrac{3 + \sqrt{m}}{9 - m}$ (C) $\dfrac{\sqrt{2 + h} - \sqrt{2}}{h}$

SOLUTION

(A) $\dfrac{\sqrt{2}}{2\sqrt{3}} = \dfrac{\sqrt{2}}{2\sqrt{3}} \cdot \dfrac{\sqrt{2}}{\sqrt{2}} = \dfrac{2}{2\sqrt{6}} = \dfrac{1}{\sqrt{6}}$

(B) $\dfrac{3 + \sqrt{m}}{9 - m} = \dfrac{3 + \sqrt{m}}{9 - m} \cdot \dfrac{3 - \sqrt{m}}{3 - \sqrt{m}} = \dfrac{9 - m}{(9 - m)(3 - \sqrt{m})} = \dfrac{1}{3 - \sqrt{m}}$

(C) $\dfrac{\sqrt{2 + h} - \sqrt{2}}{h} = \dfrac{\sqrt{2 + h} - \sqrt{2}}{h} \cdot \dfrac{\sqrt{2 + h} + \sqrt{2}}{\sqrt{2 + h} + \sqrt{2}}$

$\qquad = \dfrac{h}{h(\sqrt{2 + h} + \sqrt{2})} = \dfrac{1}{\sqrt{2 + h} + \sqrt{2}}$

Matched Problem 8   Rationalize each numerator.

(A) $\dfrac{\sqrt{3}}{3\sqrt{2}}$ (B) $\dfrac{2 - \sqrt{n}}{4 - n}$ (C) $\dfrac{\sqrt{3 + h} - \sqrt{3}}{h}$

Exercises A.6

Change each expression in Problems 1–6 to radical form. Do not simplify.

1. $6x^{3/5}$ **2.** $7y^{2/5}$ **3.** $(32x^2y^3)^{3/5}$

4. $(7x^2y)^{5/7}$ **5.** $(x^2 + y^2)^{1/2}$ **6.** $x^{1/2} + y^{1/2}$

Change each expression in Problems 7–12 to rational exponent form. Do not simplify.

7. $5\sqrt[4]{x^3}$ **8.** $7m\sqrt[5]{n^2}$ **9.** $\sqrt[5]{(2x^2y)^3}$

10. $\sqrt[7]{(8x^4y)^3}$ **11.** $\sqrt[3]{x} + \sqrt[3]{y}$ **12.** $\sqrt[3]{x^2 + y^3}$

In Problems 13–24, find rational number representations for each, if they exist.

13. $25^{1/2}$

14. $64^{1/3}$

15. $16^{3/2}$

16. $16^{3/4}$

17. $-49^{1/2}$

18. $(-49)^{1/2}$

19. $-64^{2/3}$

20. $(-64)^{2/3}$

21. $\left(\dfrac{4}{25}\right)^{3/2}$

22. $\left(\dfrac{8}{27}\right)^{2/3}$

23. $9^{-3/2}$

24. $8^{-2/3}$

In Problems 25–34, simplify each expression and write answers using positive exponents only. All variables represent positive real numbers.

25. $x^{4/5}x^{-2/5}$

26. $y^{-3/7}y^{4/7}$

27. $\dfrac{m^{2/3}}{m^{-1/3}}$

28. $\dfrac{x^{1/4}}{x^{3/4}}$

29. $(8x^3y^{-6})^{1/3}$

30. $(4u^{-2}v^4)^{1/2}$

31. $\left(\dfrac{4x^{-2}}{y^4}\right)^{-1/2}$

32. $\left(\dfrac{w^4}{9x^{-2}}\right)^{-1/2}$

33. $\dfrac{(8x)^{-1/3}}{12x^{1/4}}$

34. $\dfrac{6a^{3/4}}{15a^{-1/3}}$

Simplify each expression in Problems 35–40 using properties of radicals. All variables represent positive real numbers.

35. $\sqrt[5]{(2x+3)^5}$

36. $\sqrt[3]{(7+2y)^3}$

37. $\sqrt{6x}\sqrt{15x^3}\sqrt{30x^7}$

38. $\sqrt[5]{16a^4}\sqrt[5]{4a^2}\sqrt[5]{8a^3}$

39. $\dfrac{\sqrt{6x}\sqrt{10}}{\sqrt{15x}}$

40. $\dfrac{\sqrt{8}\sqrt{12y}}{\sqrt{6y}}$

In Problems 41–48, multiply, and express answers using positive exponents only.

41. $3x^{3/4}(4x^{1/4}-2x^8)$

42. $2m^{1/3}(3m^{2/3}-m^6)$

43. $(3u^{1/2}-v^{1/2})(u^{1/2}-4v^{1/2})$

44. $(a^{1/2}+2b^{1/2})(a^{1/2}-3b^{1/2})$

45. $(6m^{1/2}+n^{-1/2})(6m-n^{-1/2})$

46. $(2x-3y^{1/3})(2x^{1/3}+1)$

47. $(3x^{1/2}-y^{1/2})^2$

48. $(x^{1/2}+2y^{1/2})^2$

Write each expression in Problems 49–54 in the form ax^p+bx^q, where a and b are real numbers and p and q are rational numbers.

49. $\dfrac{\sqrt[3]{x^2}+2}{2\sqrt[3]{x}}$

50. $\dfrac{12\sqrt{x}-3}{4\sqrt{x}}$

51. $\dfrac{2\sqrt[4]{x^3}+\sqrt[3]{x}}{3x}$

52. $\dfrac{3\sqrt[3]{x^2}+\sqrt{x}}{5x}$

53. $\dfrac{2\sqrt[3]{x}-\sqrt{x}}{4\sqrt{x}}$

54. $\dfrac{x^2-4\sqrt{x}}{2\sqrt[3]{x}}$

Rationalize the denominators in Problems 55–60.

55. $\dfrac{12mn^2}{\sqrt{3mn}}$

56. $\dfrac{14x^2}{\sqrt{7x}}$

57. $\dfrac{2(x+3)}{\sqrt{x-2}}$

58. $\dfrac{3(x+1)}{\sqrt{x+4}}$

59. $\dfrac{7(x-y)^2}{\sqrt{x}-\sqrt{y}}$

60. $\dfrac{3a-3b}{\sqrt{a}+\sqrt{b}}$

Rationalize the numerators in Problems 61–66.

61. $\dfrac{\sqrt{5xy}}{5x^2y^2}$

62. $\dfrac{\sqrt{3mn}}{3mn}$

63. $\dfrac{\sqrt{x+h}-\sqrt{x}}{h}$

64. $\dfrac{\sqrt{2(a+h)}-\sqrt{2a}}{h}$

65. $\dfrac{\sqrt{t}-\sqrt{x}}{t^2-x^2}$

66. $\dfrac{\sqrt{x}-\sqrt{y}}{\sqrt{x}+\sqrt{y}}$

Problems 67–70 illustrate common errors involving rational exponents. In each case, find numerical examples that show that the left side is not always equal to the right side.

67. $(x+y)^{1/2}\neq x^{1/2}+y^{1/2}$

68. $(x^3+y^3)^{1/3}\neq x+y$

69. $(x+y)^{1/3}\neq \dfrac{1}{(x+y)^3}$

70. $(x+y)^{-1/2}\neq \dfrac{1}{(x+y)^2}$

In Problems 71–82, discuss the validity of each statement. If the statement is true, explain why. If not, give a counterexample.

71. $\sqrt{x^2}=x$ for all real numbers x

72. $\sqrt{x^2}=|x|$ for all real numbers x

73. $\sqrt[3]{x^3}=|x|$ for all real numbers x

74. $\sqrt[3]{x^3}=x$ for all real numbers x

75. If $r<0$, then r has no cube roots.

76. If $r<0$, then r has no square roots.

77. If $r>0$, then r has two square roots.

78. If $r>0$, then r has three cube roots.

79. The fourth roots of 100 are $\sqrt{10}$ and $-\sqrt{10}$.

80. The square roots of $2\sqrt{6}-5$ are $\sqrt{3}-\sqrt{2}$ and $\sqrt{2}-\sqrt{3}$.

81. $\sqrt{355-60\sqrt{35}}=5\sqrt{7}-6\sqrt{5}$

82. $\sqrt[3]{7-5\sqrt{2}}=1-\sqrt{2}$

In Problems 83–88, simplify by writing each expression as a simple or single fraction reduced to lowest terms and without negative exponents.

83. $-\dfrac{1}{2}(x-2)(x+3)^{-3/2}+(x+3)^{-1/2}$

84. $2(x-2)^{-1/2}-\dfrac{1}{2}(2x+3)(x-2)^{-3/2}$

85. $\dfrac{(x-1)^{1/2}-x(\frac{1}{2})(x-1)^{-1/2}}{x-1}$

86. $\dfrac{(2x-1)^{1/2}-(x+2)(\frac{1}{2})(2x-1)^{-1/2}(2)}{2x-1}$

87. $\dfrac{(x+2)^{2/3}-x(\frac{2}{3})(x+2)^{-1/3}}{(x+2)^{4/3}}$

88. $\dfrac{2(3x-1)^{1/3}-(2x+1)(\frac{1}{3})(3x-1)^{-2/3}(3)}{(3x-1)^{2/3}}$

In Problems 89–94, evaluate using a calculator. (Refer to the instruction book for your calculator to see how exponential forms are evaluated.)

89. $22^{3/2}$

90. $15^{5/4}$

91. $827^{-3/8}$

92. $103^{-3/4}$

93. $37.09^{7/3}$

94. $2.876^{8/5}$

In Problems 95 and 96, evaluate each expression on a calculator and determine which pairs have the same value. Verify these results algebraically.

95. (A) $\sqrt{3} + \sqrt{5}$ (B) $\sqrt{2 + \sqrt{3}} + \sqrt{2 - \sqrt{3}}$

   (C) $1 + \sqrt{3}$ (D) $\sqrt[3]{10 + 6\sqrt{3}}$

   (E) $\sqrt{8 + \sqrt{60}}$ (F) $\sqrt{6}$

96. (A) $2\sqrt[3]{2} + \sqrt{5}$ (B) $\sqrt{8}$

   (C) $\sqrt{3} + \sqrt{7}$ (D) $\sqrt{3 + \sqrt{8}} + \sqrt{3 - \sqrt{8}}$

   (E) $\sqrt{10 + \sqrt{84}}$ (F) $1 + \sqrt{5}$

Answers to Matched Problems

1. (A) 4   (B) -4
   (C) -3 (D) Not a real number   (E) 27

2. (A) $\sqrt[5]{u}$ (B) $\sqrt[9]{(6x^2y^5)^2}$ or $\left(\sqrt[9]{(6x^2y^5)}\right)^2$
   (C) $1/\sqrt[5]{(3xy)^3}$ (D) $(9u)^{1/4}$
   (E) $-(2x)^{4/7}$ (F) $(x^3 + y^3)^{1/3}$ (not $x + y$)

3. (A) 27   (B) 81
   (C) $10y^{7/12} = 10\sqrt[12]{y^7}$ (D) $16y/x^3$
   (E) $2/x^{1/18} = 2/\sqrt[18]{x}$

4. (A) $10c^{13/4} - 2c$ (B) $14x + 19x^{1/2}y^{1/2} - 3y$

5. $2.5x^{-7/6} - 2x^{-1}$

6. (A) $x^3 + y^3$ (B) $2y$ (C) $2x$

7. (A) $4b\sqrt{3ab}$ (B) $3(\sqrt{6} - \sqrt{3})$
   (C) $(x + y)(\sqrt{x} + \sqrt{y})$

8. (A) $\dfrac{1}{\sqrt{6}}$ (B) $\dfrac{1}{2 + \sqrt{n}}$ (C) $\dfrac{1}{\sqrt{3 + h} + \sqrt{3}}$

A.7 Quadratic Equations

- Solution by Square Root
- Solution by Factoring
- Quadratic Formula
- Quadratic Formula and Factoring
- Other Polynomial Equations
- Application: Supply and Demand

In this section we consider equations involving second-degree polynomials.

> **DEFINITION Quadratic Equation**
> A **quadratic equation** in one variable is any equation that can be written in the form
> $$ax^2 + bx + c = 0 \qquad a \neq 0 \qquad \text{Standard form}$$
> where x is a variable and a, b, and c are constants.

The equations

$$5x^2 - 3x + 7 = 0 \qquad \text{and} \qquad 18 = 32t^2 - 12t$$

are both quadratic equations, since they are either in the standard form or can be transformed into this form.

We restrict our review to finding real solutions to quadratic equations.

Solution by Square Root

The easiest type of quadratic equation to solve is the special form where the first-degree term is missing:

$$ax^2 + c = 0 \qquad a \neq 0$$

The method of solution of this special form makes direct use of the square-root property:

> **THEOREM 1 Square-Root Property**
> If $a^2 = b$, then $a = \pm\sqrt{b}$.

EXAMPLE 1 Square-Root Method Use the square-root property to solve each equation.

(A) $x^2 - 7 = 0$ (B) $2x^2 - 10 = 0$
(C) $3x^2 + 27 = 0$ (D) $(x - 8)^2 = 9$

SOLUTION

(A) $x^2 - 7 = 0$

$\qquad x^2 = 7$ What real number squared is 7?

$\qquad x = \pm\sqrt{7}$ Short for $\sqrt{7}$ and $-\sqrt{7}$

(B) $2x^2 - 10 = 0$
$\qquad 2x^2 = 10$
$\qquad x^2 = 5$ What real number squared is 5?
$\qquad x = \pm\sqrt{5}$

(C) $3x^2 + 27 = 0$
$\qquad 3x^2 = -27$
$\qquad x^2 = -9$ What real number squared is -9?

No real solution, since no real number squared is negative.

(D) $(x - 8)^2 = 9$
$\qquad x - 8 = \pm\sqrt{9}$
$\qquad x - 8 = \pm 3$
$\qquad x = 8 \pm 3 = 5 \ \text{ or } \ 11$

Matched Problem 1 Use the square-root property to solve each equation.

(A) $x^2 - 6 = 0$ (B) $3x^2 - 12 = 0$
(C) $x^2 + 4 = 0$ (D) $(x + 5)^2 = 1$

Solution by Factoring

If the left side of a quadratic equation when written in standard form can be factored, the equation can be solved very quickly. The method of solution by factoring rests on a basic property of real numbers, first mentioned in Section A.1.

CONCEPTUAL INSIGHT

Theorem 2 in Section A.1 states that if a and b are real numbers, then $ab = 0$ if and only if $a = 0$ or $b = 0$. To see that this property is useful for solving quadratic equations, consider the following:

$$x^2 - 4x + 3 = 0 \qquad (1)$$
$$(x - 1)(x - 3) = 0$$
$$x - 1 = 0 \quad \text{or} \quad x - 3 = 0$$
$$x = 1 \quad \text{or} \quad x = 3$$

You should check these solutions in equation (1).

If one side of the equation is not 0, then this method cannot be used. For example, consider

$$x^2 - 4x + 3 = 8 \qquad (2)$$
$$(x - 1)(x - 3) = 8$$
$$x - 1 \neq 8 \quad \text{or} \quad x - 3 \neq 8 \qquad \begin{array}{l}ab = 8 \text{ does not imply} \\ \text{that } a = 8 \text{ or } b = 8.\end{array}$$
$$x = 9 \quad \text{or} \quad x = 11$$

Verify that neither $x = 9$ nor $x = 11$ is a solution for equation (2).

EXAMPLE 2   Factoring Method   Solve by factoring using integer coefficients, if possible.

(A) $3x^2 - 6x - 24 = 0$ (B) $3y^2 = 2y$ (C) $x^2 - 2x - 1 = 0$

SOLUTION

(A) $3x^2 - 6x - 24 = 0$ Divide both sides by 3, since 3 is a factor of each coefficient.

$x^2 - 2x - 8 = 0$ Factor the left side, if possible.

$(x - 4)(x + 2) = 0$

$x - 4 = 0$ or $x + 2 = 0$

$x = 4$ or $x = -2$

(B) $3y^2 = 2y$

$3y^2 - 2y = 0$ We lose the solution $y = 0$ if both sides are divided by y

$y(3y - 2) = 0$ ($3y^2 = 2y$ and $3y = 2$ are not equivalent).

$y = 0$ or $3y - 2 = 0$

$3y = 2$

$y = \dfrac{2}{3}$

(C) $x^2 - 2x - 1 = 0$

This equation cannot be factored using integer coefficients. We will solve this type of equation by another method, considered below.

Matched Problem 2 Solve by factoring using integer coefficients, if possible.

(A) $2x^2 + 4x - 30 = 0$ (B) $2x^2 = 3x$ (C) $2x^2 - 8x + 3 = 0$

Note that an equation such as $x^2 = 25$ can be solved by either the square-root or the factoring method, and the results are the same (as they should be). Solve this equation both ways and compare.

Also, note that the factoring method can be extended to higher-degree polynomial equations. Consider the following:

$$x^3 - x = 0$$
$$x(x^2 - 1) = 0$$
$$x(x - 1)(x + 1) = 0$$
$$x = 0 \quad \text{or} \quad x - 1 = 0 \quad \text{or} \quad x + 1 = 0$$
$$\text{Solution: } x = 0, 1, -1$$

Check these solutions in the original equation.

The factoring and square-root methods are fast and easy to use when they apply. However, there are quadratic equations that look simple but cannot be solved by either method. For example, as was noted in Example 2C, the polynomial in

$$x^2 - 2x - 1 = 0$$

cannot be factored using integer coefficients. This brings us to the well-known and widely used *quadratic formula*.

Quadratic Formula

There is a method called *completing the square* that will work for all quadratic equations. After briefly reviewing this method, we will use it to develop the quadratic formula, which can be used to solve any quadratic equation.

The method of **completing the square** is based on the process of transforming a quadratic equation in standard form,

$$ax^2 + bx + c = 0$$

into the form

$$(x + A)^2 = B$$

where A and B are constants. Then, this last equation can be solved easily (if it has a real solution) by the square-root method discussed above.

Consider the equation from Example 2C:

$$x^2 - 2x - 1 = 0 \qquad (3)$$

Since the left side does not factor using integer coefficients, we add 1 to each side to remove the constant term from the left side:

$$x^2 - 2x = 1 \qquad (4)$$

Now we try to find a number that we can add to each side to make the left side a square of a first-degree polynomial. Note the following square of a binomial:

$$(x + m)^2 = x^2 + 2mx + m^2$$

We see that the third term on the right is the square of one-half the coefficient of x in the second term on the right. To complete the square in equation (4), we add the square of one-half the coefficient of x, $\left(-\frac{2}{2}\right)^2 = 1$, to each side. (This rule works only when the coefficient of x^2 is 1, that is, $a = 1$.) Thus,

$$x^2 - 2x + 1 = 1 + 1$$

The left side is the square of $x - 1$, and we write

$$(x - 1)^2 = 2$$

What number squared is 2?

$$x - 1 = \pm\sqrt{2}$$
$$x = 1 \pm \sqrt{2}$$

And equation (3) is solved!

Let us try the method on the general quadratic equation

$$ax^2 + bx + c = 0 \qquad a \neq 0 \qquad (5)$$

and solve it once and for all for x in terms of the coefficients a, b, and c. We start by multiplying both sides of equation (5) by $1/a$ to obtain

$$x^2 + \frac{b}{a}x + \frac{c}{a} = 0$$

Add $-c/a$ to both sides:

$$x^2 + \frac{b}{a}x = -\frac{c}{a}$$

Now we complete the square on the left side by adding the square of one-half the coefficient of x, that is, $(b/2a)^2 = b^2/4a^2$ to each side:

$$x^2 + \frac{b}{a}x + \frac{b^2}{4a^2} = \frac{b^2}{4a^2} - \frac{c}{a}$$

Writing the left side as a square and combining the right side into a single fraction, we obtain

$$\left(x + \frac{b}{2a}\right)^2 = \frac{b^2 - 4ac}{4a^2}$$

Now we solve by the square-root method:

$$x + \frac{b}{2a} = \pm \sqrt{\frac{b^2 - 4ac}{4a^2}}$$

$$x = -\frac{b}{2a} \pm \frac{\sqrt{b^3 - 4ac}}{2a} \qquad \text{Since } \pm\sqrt{4a^2} = \pm 2a \text{ for any real number } a$$

When this is written as a single fraction, it becomes the **quadratic formula**:

Quadratic Formula

If $ax^2 + bx + c = 0, a \neq 0$, then

$$x = \frac{-b \pm \sqrt{b^2 - 4ac}}{2a}$$

This formula is generally used to solve quadratic equations when the square-root or factoring methods do not work. The quantity $b^2 - 4ac$ under the radical is called the **discriminant**, and it gives us the useful information about solutions listed in Table 1.

Table 1

$b^2 - 4ac$	$ax^2 + bx + c = 0$
Positive	Two real solutions
Zero	One real solution
Negative	No real solutions

EXAMPLE 3 Quadratic Formula Method Solve $x^2 - 2x - 1 = 0$ using the quadratic formula.

SOLUTION

$$x^2 - 2x - 1 = 0$$

$$x = \frac{-b \pm \sqrt{b^2 - 4ac}}{2a} \qquad a = 1, b = -2, c = -1$$

$$= \frac{-(-2) \pm \sqrt{(-2)^2 - 4(1)(-1)}}{2(1)}$$

$$= \frac{2 \pm \sqrt{8}}{2} = \frac{2 \pm 2\sqrt{2}}{2} = 1 \pm \sqrt{2} \approx -0.414 \quad \text{or} \quad 2.414$$

CHECK

$$x^2 - 2x - 1 = 0$$
When $x = 1 + \sqrt{2}$,

$$(1 + \sqrt{2})^2 - 2(1 + \sqrt{2}) - 1 = 1 + 2\sqrt{2} + 2 - 2 - 2\sqrt{2} - 1 = 0$$

When $x = 1 - \sqrt{2}$,

$$(1 - \sqrt{2})^2 - 2(1 - \sqrt{2}) - 1 = 1 - 2\sqrt{2} + 2 - 2 + 2\sqrt{2} - 1 = 0$$

Matched Problem 3 Solve $2x^2 - 4x - 3 = 0$ using the quadratic formula.

If we try to solve $x^2 - 6x + 11 = 0$ using the quadratic formula, we obtain

$$x = \frac{6 \pm \sqrt{-8}}{2}$$

which is not a real number. (Why?)

Quadratic Formula and Factoring

As in Section A.3, we restrict our interest in factoring to polynomials with integer coefficients. If a polynomial cannot be factored as a product of lower-degree polynomials with integer coefficients, we say that the polynomial is **not factorable in the integers**.

How can you factor the quadratic polynomial $x^2 - 13x - 2{,}310$? We start by solving the corresponding quadratic equation using the quadratic formula:

$$x^2 - 13x - 2{,}310 = 0$$

$$x = \frac{-(-13) \pm \sqrt{(-13)^3 - 4(1)(-2{,}310)}}{2}$$

$$x = \frac{-(-13) \pm \sqrt{9{,}409}}{2}$$

$$= \frac{13 \pm 97}{2} = 55 \quad \text{or} \quad -42$$

Now we write

$$x^2 - 13x - 2{,}310 = [x - 55][x - (-42)] = (x - 55)(x + 42)$$

Multiplying the two factors on the right produces the second-degree polynomial on the left.

What is behind this procedure? The following two theorems justify and generalize the process:

THEOREM 2 Factorability Theorem

A second-degree polynomial, $ax^2 + bx + c$, with integer coefficients can be expressed as the product of two first-degree polynomials with integer coefficients if and only if $\sqrt{b^2 - 4ac}$ is an integer.

THEOREM 3 Factor Theorem

If r_1 and r_2 are solutions to the second-degree equation $ax^2 + bx + c = 0$, then

$$ax^2 + bx + c = a(x - r_1)(x - r_2)$$

EXAMPLE 4 Factoring with the Aid of the Discriminant Factor, if possible, using integer coefficients.

(A) $4x^2 - 65x + 264$ (B) $2x^2 - 33x - 306$

SOLUTION (A) $4x^2 - 65x + 264$

Step 1   Test for factorability:

$$\sqrt{b^2 - 4ac} = \sqrt{(-65)^2 - 4(4)(264)} = 1$$

Since the result is an integer, the polynomial has first-degree factors with integer coefficients.

Step 2   Factor, using the factor theorem. Find the solutions to the corresponding quadratic equation using the quadratic formula:

$$4x^2 - 65x + 264 = 0 \quad \text{From step 1}$$

$$x = \frac{-(-65) \pm 1}{2 \cdot 4} = \frac{33}{4} \quad \text{or} \quad 8$$

Thus,

$$4x^2 - 65x + 264 = 4\left(x - \frac{33}{4}\right)(x - 8)$$

$$= (4x - 33)(x - 8)$$

(B) $2x^2 - 33x - 306$

Step 1 Test for factorability:

$$\sqrt{b^2 - 4ac} = \sqrt{(-33)^2 - 4(2)(-306)} = \sqrt{3,537}$$

Since $\sqrt{3,537}$ is not an integer, the polynomial is not factorable in the integers.

Matched Problem 4 Factor, if possible, using integer coefficients.

(A) $3x^2 - 28x - 464$ (B) $9x^2 + 320x - 144$

Other Polynomial Equations

There are formulas that are analogous to the quadratic formula, but considerably more complicated, that can be used to solve any cubic (degree 3) or quartic (degree 4) polynomial equation. It can be shown that no such general formula exists for solving quintic (degree 5) or polynomial equations of degree greater than five. Certain polynomial equations, however, can be solved easily by taking roots.

EXAMPLE 5   Solving a Quartic Equation Find all real solutions to $6x^4 - 486 = 0$.

SOLUTION

$$6x^4 - 486 = 0 \quad \text{Add 486 to both sides}$$
$$6x^4 = 486 \quad \text{Divide both sides by 6}$$
$$x^4 = 81 \quad \text{Take the 4th root of both sides}$$
$$x = \pm 3$$

Matched Problem 5 Find all real solutions to $6x^5 + 192 = 0$.

Application: Supply and Demand

Supply-and-demand analysis is a very important part of business and economics. In general, producers are willing to supply more of an item as the price of an item increases and less of an item as the price decreases. Similarly, buyers are willing to buy less of an item as the price increases, and more of an item as the price decreases. We have a dynamic situation where the price, supply, and demand fluctuate until a price is reached at which the supply is equal to the demand. In economic theory, this point is called the **equilibrium point**. If the price increases from this point, the supply will increase and the demand will decrease; if the price decreases from this point, the supply will decrease and the demand will increase.

EXAMPLE 6   Supply and Demand At a large summer beach resort, the weekly supply-and-demand equations for folding beach chairs are

$$p = \frac{x}{140} + \frac{3}{4} \quad \text{Supply equation}$$

$$p = \frac{5,670}{x} \quad \text{Demand equation}$$

The supply equation indicates that the supplier is willing to sell x units at a price of p dollars per unit. The demand equation indicates that consumers are willing to buy x units at a price of p dollars per unit. How many units are required for supply to equal demand? At what price will supply equal demand?

SOLUTION   Set the right side of the supply equation equal to the right side of the demand equation and solve for x.

$$\frac{x}{140} + \frac{3}{4} = \frac{5{,}670}{x} \qquad \text{Multiply by 140x, the LCD.}$$

$$x^2 + 105x = 793{,}800 \qquad \text{Write in standard form.}$$

$$x^2 + 105x - 793{,}800 = 0 \qquad \text{Use the quadratic formula.}$$

$$x = \frac{-105 \pm \sqrt{105^2 - 4(1)(-793{,}800)}}{2}$$

$$x = 840 \text{ units}$$

The negative root is discarded since a negative number of units cannot be produced or sold. Substitute $x = 840$ back into either the supply equation or the demand equation to find the equilibrium price (we use the demand equation).

$$p = \frac{5{,}670}{x} = \frac{5{,}670}{840} = \$6.75$$

At a price of \$6.75 the supplier is willing to supply 840 chairs and consumers are willing to buy 840 chairs during a week.

Matched Problem 6 Repeat Example 6 if near the end of summer, the supply-and-demand equations are

$$p = \frac{x}{80} - \frac{1}{20} \qquad \text{Supply equation}$$

$$p = \frac{1{,}264}{x} \qquad \text{Demand equation}$$

Exercises A.7

Find only real solutions in the problems below. If there are no real solutions, say so.

Solve Problems 1–4 by the square-root method.

1. $2x^2 - 22 = 0$

2. $3m^2 - 21 = 0$

3. $(3x - 1)^2 = 25$

4. $(2x + 1)^2 = 16$

Solve Problems 5–8 by factoring.

5. $2u^2 - 8u - 24 = 0$

6. $3x^2 - 18x + 15 = 0$

7. $x^2 = 2x$

8. $n^2 = 3n$

Solve Problems 9–12 by using the quadratic formula.

9. $x^2 - 6x - 3 = 0$

10. $m^2 + 8m + 3 = 0$

11. $3u^2 + 12u + 6 = 0$

12. $2x^2 - 20x - 6 = 0$

Solve Problems 13–30 by using any method.

13. $\frac{2x^2}{3} = 5x$

14. $x^2 = -\frac{3}{4}x$

15. $4u^2 - 9 = 0$

16. $9y^2 - 25 = 0$

17. $8x^2 + 20x = 12$

18. $9x^2 - 6 = 15x$

19. $x^2 = 1 - x$

20. $m^2 = 1 - 3m$

21. $2x^2 = 6x - 3$

22. $2x^2 = 4x - 1$

23. $y^2 - 4y = -8$

24. $x^2 - 2x = -3$

25. $(2x + 3)^2 = 11$

26. $(5x - 2)^2 = 7$

27. $\frac{3}{p} = p$

28. $x - \frac{7}{x} = 0$

29. $2 - \frac{2}{m^2} = \frac{3}{m}$

30. $2 + \frac{5}{u} = \frac{3}{u^2}$

In Problems 31–38, factor, if possible, as the product of two first-degree polynomials with integer coefficients. Use the quadratic formula and the factor theorem.

31. $x^2 + 40x - 84$

32. $x^2 - 28x - 128$

33. $x^2 - 32x + 144$

34. $x^2 + 52x + 208$

35. $2x^2 + 15x - 108$

36. $3x^2 - 32x - 140$

37. $4x^2 + 241x - 434$

38. $6x^2 - 427x - 360$

39. Solve $A = P(1 + r)^2$ for r in terms of A and P; that is, isolate r on the left side of the equation (with coefficient 1) and end up with an algebraic expression on the right side involving A and P but not r. Write the answer using positive square roots only.

40. Solve $x^2 + 3mx - 3n = 0$ for x in terms of m and n.

✎ **41.** Consider the quadratic equation

$$x^2 + 4x + c = 0$$

where c is a real number. Discuss the relationship between the values of c and the three types of roots listed in Table 1 on page 552.

✎ **42.** Consider the quadratic equation

$$x^2 - 2x + c = 0$$

where c is a real number. Discuss the relationship between the values of c and the three types of roots listed in Table 1 on page 552.

In Problems 43–48, find all real solutions.

43. $x^3 + 8 = 0$

44. $x^3 - 8 = 0$

45. $5x^4 - 500 = 0$

46. $2x^3 + 250 = 0$

47. $x^4 - 8x^2 + 15 = 0$

48. $x^4 - 12x^2 + 32 = 0$

Applications

49. Supply and demand. A company wholesales shampoo in a particular city. Their marketing research department established the following weekly supply-and-demand equations:

$$p = \frac{x}{450} + \frac{1}{2} \quad \text{Supply equation}$$

$$p = \frac{6{,}300}{x} \quad \text{Demand equation}$$

How many units are required for supply to equal demand? At what price per bottle will supply equal demand?

50. Supply and demand. An importer sells an automatic camera to outlets in a large city. During the summer, the weekly supply-and-demand equations are

$$p = \frac{x}{6} + 9 \quad \text{Supply equation}$$

$$p = \frac{24{,}840}{x} \quad \text{Demand equation}$$

How many units are required for supply to equal demand? At what price will supply equal demand?

51. Interest rate. If P dollars are invested at $100r$ percent compounded annually, at the end of 2 years it will grow to $A = P(1 + r)^2$. At what interest rate will \$484 grow to \$625 in 2 years? (*Note:* If $A = 625$ and $P = 484$ find r.)

52. Interest rate. Using the formula in Problem 51, determine the interest rate that will make \$1,000 grow to \$1,210 in 2 years.

53. Ecology. To measure the velocity v (in feet per second) of a stream, we position a hollow L-shaped tube with one end under the water pointing upstream and the other end pointing straight up a couple of feet out of the water. The water will then be pushed up the tube a certain distance h (in feet) above the surface of the stream. Physicists have shown that $v^2 = 64h$. Approximately how fast is a stream flowing if $h = 1$ foot? If $h = 0.5$ foot?

54. Safety research. It is of considerable importance to know the least number of feet d in which a car can be stopped, including reaction time of the driver, at various speeds v (in miles per hour). Safety research has produced the formula $d = 0.044v^2 + 1.1v$. If it took a car 550 feet to stop, estimate the car's speed at the moment the stopping process was started.

Answers to Matched Problems

1. (A) $\pm\sqrt{6}$ (B) ± 2

    (C) No real solution   (D) $-6, -4$

2. (A) $-5, 3$ (B) $0, \frac{3}{2}$

    (C) Cannot be factored using integer coefficients

3. $(2 \pm \sqrt{10})/2$

4. (A) Cannot be factored using integer coefficients

    (B) $(9x - 4)(x + 36)$

5. -2

6. 320 chairs at \$3.95 each

Appendix

B

Special Topics

B.1 Sequences, Series, and Summation Notation

- Sequences
- Series and Summation Notation

If someone asked you to list all natural numbers that are perfect squares, you might begin by writing

$$1, 4, 9, 16, 25, 36$$

But you would soon realize that it is impossible to actually list all the perfect squares, since there are an infinite number of them. However, you could represent this collection of numbers in several different ways. One common method is to write

$$1, 4, 9, \ldots, n^2, \ldots \quad n \in N$$

where N is the set of natural numbers. A list of numbers such as this is generally called a *sequence*.

Sequences

Consider the function f given by

$$f(n) = 2n + 1 \tag{1}$$

where the domain of f is the set of natural numbers N. Note that

$$f(1) = 3, \quad f(2) = 5, \quad f(3) = 7, \quad \ldots$$

The function f is an example of a sequence. In general, a **sequence** is a function with domain a set of successive integers. Instead of the standard function notation used in equation (1), sequences are usually defined in terms of a special notation.

The range value $f(n)$ is usually symbolized more compactly with a symbol such as a_n. Thus, in place of equation (1), we write

$$a_n = 2n + 1$$

and the domain is understood to be the set of natural numbers unless something is said to the contrary or the context indicates otherwise. The elements in the range are

557

called **terms of the sequence**; a_1 is the first term, a_2 is the second term, and a_n is the *n*th **term**, or **general term**.

$$a_1 = 2(1) + 1 = 3 \quad \text{First term}$$
$$a_2 = 2(2) + 1 = 5 \quad \text{Second term}$$
$$a_3 = 2(3) + 1 = 7 \quad \text{Third term}$$
$$\vdots$$
$$a_n = 2n + 1 \quad\quad\quad \text{General term}$$

The ordered list of elements

$$3, 5, 7, \ldots, 2n + 1, \ldots$$

obtained by writing the terms of the sequence in their natural order with respect to the domain values is often informally referred to as a sequence. A sequence also may be represented in the abbreviated form $\{a_n\}$, where a symbol for the *n*th term is written within braces. For example, we could refer to the sequence $3, 5, 7, \ldots, 2n + 1, \ldots$ as the sequence $\{2n + 1\}$.

If the domain of a sequence is a finite set of successive integers, then the sequence is called a **finite sequence**. If the domain is an infinite set of successive integers, then the sequence is called an **infinite sequence**. The sequence $\{2n + 1\}$ discussed above is an infinite sequence.

EXAMPLE 1   Writing the Terms of a Sequence   Write the first four terms of each sequence:

(A) $a_n = 3n - 2$ (B) $\left\{ \dfrac{(-1)^n}{n} \right\}$

SOLUTION

(A) $1, 4, 7, 10$ (B) $-1, \dfrac{1}{2}, \dfrac{-1}{3}, \dfrac{1}{4}$

Matched Problem 1]   Write the first four terms of each sequence:

(A) $a_n = -n + 3$ (B) $\left\{ \dfrac{(-1)^n}{2^n} \right\}$

Now that we have seen how to use the general term to find the first few terms in a sequence, we consider the reverse problem. That is, can a sequence be defined just by listing the first three or four terms of the sequence? And can we then use these initial terms to find a formula for the *n*th term? In general, without other information, the answer to the first question is no. Many different sequences may start off with the same terms. Simply listing the first three terms (or any other finite number of terms) does not specify a particular sequence.

What about the second question? That is, given a few terms, can we find the general formula for at least one sequence whose first few terms agree with the given terms? The answer to this question is a qualified yes. If we can observe a simple pattern in the given terms, we usually can construct a general term that will produce that pattern. The next example illustrates this approach.

EXAMPLE 2   Finding the General Term of a Sequence   Find the general term of a sequence whose first four terms are

(A) $3, 4, 5, 6, \ldots$ (B) $5, -25, 125, -625, \ldots$

SOLUTION

(A) Since these terms are consecutive integers, one solution is $a_n = n, n \geq 3$. If we want the domain of the sequence to be all natural numbers, another solution is $b_n = n + 2$.

(B) Each of these terms can be written as the product of a power of 5 and a power of -1:

$$5 = (-1)^0 5^1 = a_1$$
$$-25 = (-1)^1 5^2 = a_2$$
$$125 = (-1)^2 5^3 = a_3$$
$$-625 = (-1)^3 5^4 = a_4$$

If we choose the domain to be all natural numbers, a solution is

$$a_n = (-1)^{n-1} 5^n$$

Matched Problem 2 Find the general term of a sequence whose first four terms are

(A) $3, 6, 9, 12, \ldots$ (B) $1, -2, 4, -8, \ldots$

In general, there is usually more than one way of representing the nth term of a given sequence (see the solution of Example 2A). However, unless something is stated to the contrary, we assume that the domain of the sequence is the set of natural numbers N.

Series and Summation Notation

If $a_1, a_2, a_3, \ldots, a_n, \ldots$ is a sequence, the expression

$$a_1 + a_2 + a_3 + \cdots + a_n + \cdots$$

is called a **series**. If the sequence is finite, the corresponding series is a **finite series**. If the sequence is infinite, the corresponding series is an **infinite series**. We consider only finite series in this section. For example,

$$1, 3, 5, 7, 9 \qquad \text{Finite sequence}$$
$$1 + 3 + 5 + 7 + 9 \quad \text{Finite series}$$

Notice that we can easily evaluate this series by adding the five terms:

$$1 + 3 + 5 + 7 + 9 = 25$$

Series are often represented in a compact form called **summation notation**. Consider the following examples:

$$\sum_{k=3}^{6} k^2 = 3^2 + 4^2 + 5^2 + 6^2$$

$$= 9 + 16 + 25 + 36 = 86$$

$$\sum_{k=0}^{2} (4k + 1) = (4 \cdot 0 + 1) + (4 \cdot 1 + 1) + (4 \cdot 2 + 1)$$

$$= 1 + 5 + 9 = 15$$

In each case, the terms of the series on the right are obtained from the expression on the left by successively replacing the **summing index k** with integers, starting with the number indicated below the **summation sign Σ** and ending with the number that appears above Σ. The summing index may be represented by letters other than k and may start at any integer and end at any integer greater than or equal to the starting integer. If we are given the finite sequence

$$\frac{1}{2}, \frac{1}{4}, \frac{1}{8}, \ldots, \frac{1}{2^n}$$

the corresponding series is

$$\frac{1}{2} + \frac{1}{4} + \frac{1}{8} + \cdots + \frac{1}{2^n} = \sum_{j=1}^{n} \frac{1}{2^j}$$

where we have used j for the summing index.

EXAMPLE 3   Summation Notation   Write

$$\sum_{k=1}^{5} \frac{k}{k^2 + 1}$$

without summation notation. Do not evaluate the sum.

SOLUTION

$$\sum_{k=1}^{5} \frac{k}{k^2 + 1} = \frac{1}{1^2 + 1} + \frac{2}{2^2 + 1} + \frac{3}{3^2 + 1} + \frac{4}{4^2 + 1} + \frac{5}{5^2 + 1}$$

$$= \frac{1}{2} + \frac{2}{5} + \frac{3}{10} + \frac{4}{17} + \frac{5}{26}$$

Matched Problem 3   Write

$$\sum_{k=1}^{5} \frac{k + 1}{k}$$

without summation notation. Do not evaluate the sum.

If the terms of a series are alternately positive and negative, we call the series an **alternating series**. The next example deals with the representation of such a series.

EXAMPLE 4   Summation Notation   Write the alternating series

$$\frac{1}{2} - \frac{1}{4} + \frac{1}{6} - \frac{1}{8} + \frac{1}{10} - \frac{1}{12}$$

using summation notation with
(A) The summing index k starting at 1
(B) The summing index j starting at 0

SOLUTION

(A) $(-1)^{k+1}$ provides the alternation of sign, and $1/(2k)$ provides the other part of each term. So, we can write

$$\frac{1}{2} - \frac{1}{4} + \frac{1}{6} - \frac{1}{8} + \frac{1}{10} - \frac{1}{12} = \sum_{k=1}^{6} \frac{(-1)^{k+1}}{2k}$$

(B) $(-1)^j$ provides the alternation of sign, and $1/[2(j + 1)]$ provides the other part of each term. So, we can write

$$\frac{1}{2} - \frac{1}{4} + \frac{1}{6} - \frac{1}{8} + \frac{1}{10} - \frac{1}{12} = \sum_{j=0}^{5} \frac{(-1)^j}{2(j + 1)}$$

Matched Problem 4   Write the alternating series

$$1 - \frac{1}{3} + \frac{1}{9} - \frac{1}{27} + \frac{1}{81}$$

using summation notation with
(A) The summing index k starting at 1
(B) The summing index j starting at 0

Summation notation provides a compact notation for the sum of any list of numbers, even if the numbers are not generated by a formula. For example, suppose that the results of an examination taken by a class of 10 students are given in the following list:

$$87, 77, 95, 83, 86, 73, 95, 68, 75, 86$$

If we let $a_1, a_2, a_3, \ldots, a_{10}$ represent these 10 scores, then the average test score is given by

$$\frac{1}{10}\sum_{k=1}^{10} a_k = \frac{1}{10}(87 + 77 + 95 + 83 + 86 + 73 + 95 + 68 + 75 + 86)$$

$$= \frac{1}{10}(825) = 82.5$$

More generally, in statistics, the **arithmetic mean** $\bar{a}$ of a list of n numbers $a_1, a_2, \ldots, a_n$ is defined as

$$\bar{a} = \frac{1}{n}\sum_{k=1}^{n} a_k$$

EXAMPLE 5   Arithmetic Mean   Find the arithmetic mean of 3, 5, 4, 7, 4, 2, 3, and 6.

SOLUTION

$$\bar{a} = \frac{1}{8}\sum_{k=1}^{8} a_k = \frac{1}{8}(3 + 5 + 4 + 7 + 4 + 2 + 3 + 6) = \frac{1}{8}(34) = 4.25$$

Matched Problem 5   Find the arithmetic mean of 9, 3, 8, 4, 3, and 6.

Exercises B.1

Write the first four terms for each sequence in Problems 1–6.

1. $a_n = 2n + 3$

2. $a_n = 4n - 3$

3. $a_n = \dfrac{n + 2}{n + 1}$

4. $a_n = \dfrac{2n + 1}{2n}$

5. $a_n = (-3)^{n+1}$

6. $a_n = \left(-\frac{1}{4}\right)^{n-1}$

7. Write the 10th term of the sequence in Problem 1.

8. Write the 15th term of the sequence in Problem 2.

9. Write the 99th term of the sequence in Problem 3.

10. Write the 200th term of the sequence in Problem 4.

In Problems 11–16, write each series in expanded form without summation notation, and evaluate.

11. $\displaystyle\sum_{k=1}^{6} k$

12. $\displaystyle\sum_{k=1}^{5} k^2$

13. $\displaystyle\sum_{k=4}^{7} (2k - 3)$

14. $\displaystyle\sum_{k=0}^{4} (-2)^k$

15. $\displaystyle\sum_{k=0}^{3} \frac{1}{10^k}$

16. $\displaystyle\sum_{k=1}^{4} \frac{1}{2^k}$

Find the arithmetic mean of each list of numbers in Problems 17–20.

17. 5, 4, 2, 1, and 6

18. 7, 9, 9, 2, and 4

19. 96, 65, 82, 74, 91, 88, 87, 91, 77, and 74

20. 100, 62, 95, 91, 82, 87, 70, 75, 87, and 82

Write the first five terms of each sequence in Problems 21–26.

21. $a_n = \dfrac{(-1)^{n+1}}{2^n}$

22. $a_n = (-1)^n(n - 1)^2$

23. $a_n = n[1 + (-1)^n]$

24. $a_n = \dfrac{1 - (-1)^n}{n}$

25. $a_n = \left(-\dfrac{3}{2}\right)^{n-1}$

26. $a_n = \left(-\dfrac{1}{2}\right)^{n+1}$

In Problems 27–42, find the general term of a sequence whose first four terms agree with the given terms.

27. $-2, -1, 0, 1, \ldots$

28. $4, 5, 6, 7, \ldots$

29. $4, 8, 12, 16, \ldots$

30. $-3, -6, -9, -12, \ldots$

31. $\frac{1}{2}, \frac{3}{4}, \frac{5}{6}, \frac{7}{8}, \ldots$

32. $\frac{1}{2}, \frac{2}{3}, \frac{3}{4}, \frac{4}{5}, \ldots$

33. $1, -2, 3, -4, \ldots$

34. $-2, 4, -8, 16, \ldots$

35. $1, -3, 5, -7, \ldots$

36. $3, -6, 9, -12, \ldots$

37. $1, \frac{2}{5}, \frac{4}{25}, \frac{8}{125}, \ldots$

38. $\frac{4}{3}, \frac{16}{9}, \frac{64}{27}, \frac{256}{81}, \ldots$

39. $x, x^2, x^3, x^4, \ldots$

40. $1, 2x, 3x^2, 4x^3, \ldots$

41. $x, -x^3, x^5, -x^7, \ldots$

42. $x, \dfrac{x^2}{2}, \dfrac{x^3}{3}, \dfrac{x^4}{4}, \ldots$

Write each series in Problems 43–50 in expanded form without summation notation. Do not evaluate.

43. $\displaystyle\sum_{k=1}^{5} (-1)^{k+1}(2k-1)^2$

44. $\displaystyle\sum_{k=1}^{4} \dfrac{(-2)^{k+1}}{2k+1}$

45. $\displaystyle\sum_{k=2}^{5} \dfrac{2^k}{2k+3}$

46. $\displaystyle\sum_{k=3}^{7} \dfrac{(-1)^k}{k^2-k}$

47. $\displaystyle\sum_{k=1}^{5} x^{k-1}$

48. $\displaystyle\sum_{k=1}^{3} \dfrac{1}{k}x^{k+1}$

49. $\displaystyle\sum_{k=0}^{4} \dfrac{(-1)^k x^{2k+1}}{2k+1}$

50. $\displaystyle\sum_{k=0}^{4} \dfrac{(-1)^k x^{2k}}{2k+2}$

Write each series in Problems 51–54 using summation notation with

(A) The summing index k starting at $k = 1$

(B) The summing index j starting at $j = 0$

51. $2 + 3 + 4 + 5 + 6$

52. $1^2 + 2^2 + 3^2 + 4^2$

53. $1 - \frac{1}{2} + \frac{1}{3} - \frac{1}{4}$

54. $1 - \frac{1}{3} + \frac{1}{5} - \frac{1}{7} + \frac{1}{9}$

Write each series in Problems 55–58 using summation notation with the summing index k starting at $k = 1$.

55. $2 + \dfrac{3}{2} + \dfrac{4}{3} + \cdots + \dfrac{n+1}{n}$

56. $1 + \dfrac{1}{2^2} + \dfrac{1}{3^2} + \cdots + \dfrac{1}{n^2}$

57. $\dfrac{1}{2} - \dfrac{1}{4} + \dfrac{1}{8} - \cdots + \dfrac{(-1)^{n+1}}{2^n}$

58. $1 - 4 + 9 - \cdots + (-1)^{n+1}n^2$

In Problems 59–62, discuss the validity of each statement. If the statement is true, explain why. If not, give a counterexample.

59. For each positive integer n, the sum of the series
$1 + \dfrac{1}{2} + \dfrac{1}{3} + \cdots + \dfrac{1}{n}$ is less than 4.

60. For each positive integer n, the sum of the series
$\dfrac{1}{2} + \dfrac{1}{4} + \dfrac{1}{8} + \cdots + \dfrac{1}{2^n}$ is less than 1.

61. For each positive integer n, the sum of the series
$\dfrac{1}{2} - \dfrac{1}{4} + \dfrac{1}{8} - \cdots + \dfrac{(-1)^{n+1}}{2^n}$ is greater than or
equal to $\dfrac{1}{4}$.

62. For each positive integer n, the sum of the series
$1 - \dfrac{1}{2} + \dfrac{1}{3} - \dfrac{1}{4} + \cdots + \dfrac{(-1)^{n+1}}{n}$ is greater than or
equal to $\dfrac{1}{2}$.

*Some sequences are defined by a **recursion formula** —that is, a formula that defines each term of the sequence in terms of one or more of the preceding terms. For example, if $\{a_n\}$ is defined by*

$$a_1 = 1 \quad and \quad a_n = 2a_{n-1} + 1 \quad for \quad n \geq 2$$

then

$$a_2 = 2a_1 + 1 = 2 \cdot 1 + 1 = 3$$
$$a_3 = 2a_2 + 1 = 2 \cdot 3 + 1 = 7$$
$$a_4 = 2a_3 + 1 = 2 \cdot 7 + 1 = 15$$

and so on. In Problems 63–66, write the first five terms of each sequence.

63. $a_1 = 2$ and $a_n = 3a_{n-1} + 2$ for $n \geq 2$

64. $a_1 = 3$ and $a_n = 2a_{n-1} - 2$ for $n \geq 2$

65. $a_1 = 1$ and $a_n = 2a_{n-1}$ for $n \geq 2$

66. $a_1 = 1$ and $a_n = -\frac{1}{3}a_{n-1}$ for $n \geq 2$

If A is a positive real number, the terms of the sequence defined by

$$a_1 = \dfrac{A}{2} \quad and \quad a_n = \dfrac{1}{2}\left(a_{n-1} + \dfrac{A}{a_{n-1}}\right) \quad for\ n \geq 2$$

can be used to approximate $\sqrt{A}$ to any decimal place accuracy desired. In Problems 67 and 68, compute the first four terms of this sequence for the indicated value of A, and compare the fourth term with the value of $\sqrt{A}$ obtained from a calculator.

67. $A = 2$

68. $A = 6$

69. The sequence defined recursively by $a_1 = 1$, $a_2 = 1$, $a_n = a_{n-1} + a_{n-2}$ for $n \geq 3$ is called the *Fibonacci sequence*. Find the first ten terms of the Fibonacci sequence.

70. The sequence defined by $b_n = \dfrac{\sqrt{5}}{5}\left(\dfrac{1+\sqrt{5}}{2}\right)^n$ is related to the Fibonacci sequence. Find the first ten terms (to three decimal places) of the sequence $\{b_n\}$ and describe the relationship.

Answers to Matched Problems

1. (A) $2, 1, 0, -1$ (B) $\frac{-1}{2}, \frac{1}{4}, \frac{-1}{8}, \frac{1}{16}$

2. (A) $a_n = 3n$ (B) $a_n = (-2)^{n-1}$

3. $2 + \frac{3}{2} + \frac{4}{3} + \frac{5}{4} + \frac{6}{5}$

4. (A) $\displaystyle\sum_{k=1}^{5} \dfrac{(-1)^{k-1}}{3^{k-1}}$ (B) $\displaystyle\sum_{j=0}^{4} \dfrac{(-1)^j}{3^j}$

5. 5.5

B.2   Arithmetic and Geometric Sequences

- Arithmetic and Geometric Sequences
- nth-Term Formulas
- Sum Formulas for Finite Arithmetic Series
- Sum Formulas for Finite Geometric Series
- Sum Formula for Infinite Geometric Series
- Applications

For most sequences, it is difficult to sum an arbitrary number of terms of the sequence without adding term by term. But particular types of sequences—*arithmetic sequences* and *geometric sequences*—have certain properties that lead to convenient and useful formulas for the sums of the corresponding *arithmetic series* and *geometric series*.

Arithmetic and Geometric Sequences

The sequence $5, 7, 9, 11, 13, \ldots, 5 + 2(n - 1), \ldots$, where each term after the first is obtained by adding 2 to the preceding term, is an example of an arithmetic sequence. The sequence $5, 10, 20, 40, 80, \ldots, 5(2)^{n-1}, \ldots$, where each term after the first is obtained by multiplying the preceding term by 2, is an example of a geometric sequence.

> **DEFINITION   Arithmetic Sequence**
> A sequence of numbers
> $$a_1, a_2, a_3, \ldots, a_n, \ldots$$
> is called an **arithmetic sequence** if there is a constant d, called the **common difference**, such that
> $$a_n - a_{n-1} = d$$
> That is,
> $$a_n = a_{n-1} + d \quad \text{for every } n > 1$$

> **DEFINITION   Geometric Sequence**
> A sequence of numbers
> $$a_1, a_2, a_3, \ldots, a_n, \ldots$$
> is called a **geometric sequence** if there exists a nonzero constant r, called a **common ratio**, such that
> $$\frac{a_n}{a_{n-1}} = r$$
> That is,
> $$a_n = ra_{n-1} \quad \text{for every } n > 1$$

EXAMPLE 1   Recognizing Arithmetic and Geometric Sequences   Which of the following can be the first four terms of an arithmetic sequence? Of a geometric sequence?

(A) $1, 2, 3, 5, \ldots$ (B) $-1, 3, -9, 27, \ldots$

(C) $3, 3, 3, 3, \ldots$ (D) $10, 8.5, 7, 5.5, \ldots$

SOLUTION

(A) Since $2 - 1 \neq 5 - 3$, there is no common difference, so the sequence is not an arithmetic sequence. Since $2/1 \neq 3/2$, there is no common ratio, so the sequence is not geometric either.

(B) The sequence is geometric with common ratio -3. It is not arithmetic.

(C) The sequence is arithmetic with common difference 0, and is also geometric with common ratio 1.

(D) The sequence is arithmetic with common difference -1.5. It is not geometric.

Matched Problem 1 Which of the following can be the first four terms of an arithmetic sequence? Of a geometric sequence?

(A) $8, 2, 0.5, 0.125, \ldots$ (B) $-7, -2, 3, 8, \ldots$ (C) $1, 5, 25, 100, \ldots$

nth-Term Formulas

If $\{a_n\}$ is an arithmetic sequence with common difference d, then

$$a_2 = a_1 + d$$
$$a_3 = a_2 + d = a_1 + 2d$$
$$a_4 = a_3 + d = a_1 + 3d$$

This suggests that

THEOREM 1 nth Term of an Arithmetic Sequence

$$a_n = a_1 + (n-1)d \quad \text{for all } n > 1 \tag{1}$$

Similarly, if $\{a_n\}$ is a geometric sequence with common ratio r, then

$$a_2 = a_1 r$$
$$a_3 = a_2 r = a_1 r^2$$
$$a_4 = a_3 r = a_1 r^3$$

This suggests that

THEOREM 2 nth Term of a Geometric Sequence

$$a_n = a_1 r^{n-1} \quad \text{for all } n > 1 \tag{2}$$

EXAMPLE 2 Finding Terms in Arithmetic and Geometric Sequences

(A) If the 1st and 10th terms of an arithmetic sequence are 3 and 30, respectively, find the 40th term of the sequence.

(B) If the 1st and 10th terms of a geometric sequence are 3 and 30, find the 40th term to three decimal places.

SOLUTION

(A) First use formula (1) with $a_1 = 3$ and $a_{10} = 30$ to find d:

$$a_n = a_1 + (n-1)d$$
$$a_{10} = a_1 + (10-1)d$$
$$30 = 3 + 9d$$
$$d = 3$$

Now find a_{40}:

$$a_{40} = 3 + 39 \cdot 3 = 120$$

(B) First use formula (2) with $a_1 = 3$ and $a_{10} = 30$ to find r:

$$a_n = a_1 r^{n-1}$$
$$a_{10} = a_1 r^{10-1}$$
$$30 = 3r^9$$
$$r^9 = 10$$
$$r = 10^{1/9}$$

Now find a_{40}:

$$a_{40} = 3(10^{1/9})^{39} = 3(10^{39/9}) = 64{,}633.041$$

Matched Problem 2

(A) If the 1st and 15th terms of an arithmetic sequence are -5 and 23, respectively, find the 73rd term of the sequence.

(B) Find the 8th term of the geometric sequence

$$\frac{1}{64}, \frac{-1}{32}, \frac{1}{16}, \ldots$$

Sum Formulas for Finite Arithmetic Series

If $a_1, a_2, a_3, \ldots, a_n$ is a finite arithmetic sequence, then the corresponding series $a_1 + a_2 + a_3 + \cdots + a_n$ is called a *finite arithmetic series*. We will derive two simple and very useful formulas for the sum of a finite arithmetic series. Let d be the common difference of the arithmetic sequence $a_1, a_2, a_3, \ldots, a_n$ and let S_n denote the sum of the series $a_1 + a_2 + a_3 + \cdots + a_n$. Then

$$S_n = a_1 + (a_1 + d) + \cdots + [a_1 + (n - 2)d] + [a_1 + (n - 1)d]$$

Reversing the order of the sum, we obtain

$$S_n = [a_1 + (n - 1)d] + [a_1 + (n - 2)d] + \cdots + (a_1 + d) + a_1$$

Something interesting happens if we combine these last two equations by addition (adding corresponding terms on the right sides):

$$2S_n = [2a_1 + (n - 1)d] + [2a_1 + (n - 1)d] + \cdots + [2a_1 + (n - 1)d] + [2a_1 + (n - 1)d]$$

All the terms on the right side are the same, and there are n of them. Thus,

$$2S_n = n[2a_1 + (n - 1)d]$$

and we have the following general formula:

THEOREM 3 Sum of a Finite Arithmetic Series: First Form

$$S_n = \frac{n}{2}[2a_1 + (n - 1)d] \tag{3}$$

Replacing

$$[a_1 + (n - 1)d] \quad \text{in} \quad \frac{n}{2}[a_1 + a_1 + (n - 1)d]$$

by a_n from equation (1), we obtain a second useful formula for the sum:

THEOREM 4 Sum of a Finite Arithmetic Series: Second Form

$$S_n = \frac{n}{2}(a_1 + a_n) \tag{4}$$

EXAMPLE 3 Finding a Sum Find the sum of the first 30 terms in the arithmetic sequence:

$$3, 8, 13, 18, \ldots$$

SOLUTION Use formula (3) with $n = 30$, $a_1 = 3$, and $d = 5$:

$$S_{30} = \frac{30}{2}[2 \cdot 3 + (30 - 1)5] = 2{,}265$$

Matched Problem 3 Find the sum of the first 40 terms in the arithmetic sequence:

$$15, 13, 11, 9, \ldots$$

EXAMPLE 4 Finding a Sum Find the sum of all the even numbers between 31 and 87.

SOLUTION First, find n using equation (1):

$$a_n = a_1 + (n - 1)d$$
$$86 = 32 + (n - 1)2$$
$$n = 28$$

Now find S_{28} using formula (4):

$$S_n = \frac{n}{2}(a_1 + a_n)$$

$$S_{28} = \frac{28}{2}(32 + 86) = 1{,}652$$

Matched Problem 4 Find the sum of all the odd numbers between 24 and 208.

Sum Formulas for Finite Geometric Series

If $a_1, a_2, a_3, \ldots, a_n$ is a finite geometric sequence, then the corresponding series $a_1 + a_2 + a_3 + \cdots + a_n$ is called a _finite geometric series_. As with arithmetic series, we can derive two simple and very useful formulas for the sum of a finite geometric series. Let r be the common ratio of the geometric sequence $a_1, a_2, a_3, \ldots, a_n$ and let S_n denote the sum of the series $a_1 + a_2 + a_3 + \cdots + a_n$. Then

$$S_n = a_1 + a_1r + a_1r^2 + \cdots + a_1r^{n-2} + a_1r^{n-1}$$

If we multiply both sides by r, we obtain

$$rS_n = a_1r + a_1r^2 + a_1r^3 + \cdots + a_1r^{n-1} + a_1r^n$$

Now combine these last two equations by subtraction to obtain

$$rS_n - S_n = (a_1r + a_1r^2 + a_1r^3 + \cdots + a_1r^{n-1} + a_1r^n) - (a_1 + a_1r + a_1r^2 + \cdots + a_1r^{n-2} + a_1r^{n-1})$$
$$(r - 1)S_n = a_1r^n - a_1$$

Notice how many terms drop out on the right side. Solving for S_n, we have

THEOREM 5 Sum of a Finite Geometric Series: First Form

$$S_n = \frac{a_1(r^n - 1)}{r - 1} \quad r \neq 1 \tag{5}$$

Since $a_n = a_1r^{n-1}$, or $ra_n = a_1r^n$, formula (5) also can be written in the form

THEOREM 6 Sum of a Finite Geometric Series: Second Form

$$S_n = \frac{ra_n - a_1}{r - 1} \quad r \neq 1 \tag{6}$$

EXAMPLE 5 Finding a Sum Find the sum (to 2 decimal places) of the first ten terms of the geometric sequence:

$$1, 1.05, 1.05^2, \ldots$$

SOLUTION Use formula (5) with $a_1 = 1$, $r = 1.05$, and $n = 10$:

$$S_n = \frac{a_1(r^n - 1)}{r - 1}$$

$$S_{10} = \frac{1(1.05^{10} - 1)}{1.05 - 1}$$

$$\approx \frac{0.6289}{0.05} \approx 12.58$$

<u>Matched Problem 5</u> Find the sum of the first eight terms of the geometric sequence:

$$100, 100(1.08), 100(1.08)^2, \ldots$$

Sum Formula for Infinite Geometric Series

Given a geometric series, what happens to the sum S_n of the first n terms as n increases without stopping? To answer this question, let us write formula (5) in the form

$$S_n = \frac{a_1 r^n}{r - 1} - \frac{a_1}{r - 1}$$

It is possible to show that if $-1 < r < 1$, then r^n will approach 0 as n increases. The first term above will approach 0 and S_n can be made as close as we please to the second term, $-a_1/(r - 1)$ [which can be written as $a_1/(1 - r)$], by taking n sufficiently large. So, if the common ratio r is between -1 and 1, we conclude that the sum of an infinite geometric series is

> **THEOREM 7   Sum of an Infinite Geometric Series**
>
> $$S_\infty = \frac{a_1}{1 - r} \qquad -1 < r < 1 \qquad (7)$$

If $r \leq -1$ or $r \geq 1$, then an infinite geometric series has no sum.

Applications

EXAMPLE 6 Loan Repayment A person borrows \$3,600 and agrees to repay the loan in monthly installments over 3 years. The agreement is to pay 1% of the unpaid balance each month for using the money and \$100 each month to reduce the loan. What is the total cost of the loan over the 3 years?

SOLUTION Let us look at the problem relative to a time line:

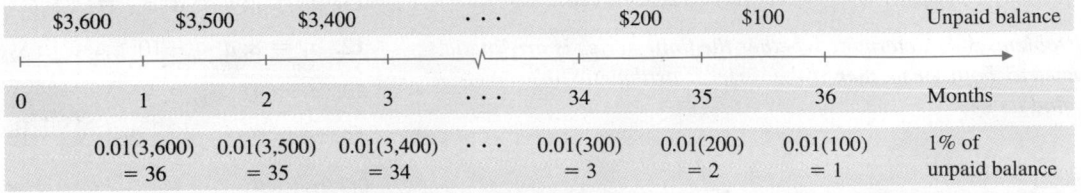

The total cost of the loan is

$$1 + 2 + \cdots + 34 + 35 + 36$$

The terms form a finite arithmetic series with $n = 36$, $a_1 = 1$, and $a_{36} = 36$, so we can use formula (4):

$$S_n = \frac{n}{2}(a_1 + a_n)$$

$$S_{36} = \frac{36}{2}(1 + 36) = \$666$$

We conclude that the total cost of the loan over 3 years is $666.

Matched Problem 6 Repeat Example 6 with a loan of $6,000 over 5 years.

EXAMPLE 7 Economy Stimulation The government has decided on a tax rebate program to stimulate the economy. Suppose that you receive $1,200 and you spend 80% of this, and each of the people who receive what you spend also spend 80% of what they receive, and this process continues without end. According to the **multiplier principle** in economics, the effect of your $1,200 tax rebate on the economy is multiplied many times. What is the total amount spent if the process continues as indicated?

SOLUTION We need to find the sum of an infinite geometric series with the first amount spent being $a_1 = (0.8)(\$1,200) = \$960$ and $r = 0.8$. Using formula (7), we obtain

$$S_\infty = \frac{a_1}{1 - r}$$

$$= \frac{\$960}{1 - 0.8} = \$4,800$$

Assuming the process continues as indicated, we would expect the $1,200 tax rebate to result in about $4,800 of spending.

Matched Problem 7 Repeat Example 7 with a tax rebate of $2,000.

Exercises B.2

In Problems 1 and 2, determine whether the indicated sequence can be the first three terms of an arithmetic or geometric sequence, and, if so, find the common difference or common ratio and the next two terms of the sequence.

1. (A) $-11, -16, -21, \ldots$ (B) $2, -4, 8, \ldots$

   (C) $1, 4, 9, \ldots$ (D) $\frac{1}{2}, \frac{1}{6}, \frac{1}{18}, \ldots$

2. (A) $5, 20, 100, \ldots$ (B) $-5, -5, -5, \ldots$

   (C) $7, 6.5, 6, \ldots$ (D) $512, 256, 128, \ldots$

In Problems 3–8, determine whether the finite series is arithmetic, geometric, both, or neither. If the series is arithmetic or geometric, find its sum.

3. $\displaystyle\sum_{k=1}^{101}(-1)^{k+1}$ **4.** $\displaystyle\sum_{k=1}^{200}3$

5. $1 + \dfrac{1}{2} + \dfrac{1}{3} + \cdots + \dfrac{1}{50}$

6. $3 - 9 + 27 - \cdots - 3^{20}$

7. $5 + 4.9 + 4.8 + \cdots + 0.1$

8. $1 - \dfrac{1}{4} + \dfrac{1}{9} - \cdots - \dfrac{1}{100^2}$

Let $a_1, a_2, a_3, \ldots, a_n, \ldots$ be an arithmetic sequence. In Problems 9–14, find the indicated quantities.

9. $a_1 = 7$; $d = 4$; $a_2 = ?$; $a_3 = ?$

10. $a_1 = -2$; $d = -3$; $a_2 = ?$; $a_3 = ?$

11. $a_1 = 2$; $d = 4$; $a_{21} = ?$; $S_{31} = ?$

12. $a_1 = 8$; $d = -10$; $a_{15} = ?$; $S_{23} = ?$

13. $a_1 = 18$; $a_{20} = 75$; $S_{20} = ?$

14. $a_1 = 203$; $a_{30} = 261$; $S_{30} = ?$

Let $a_1, a_2, a_3, \ldots, a_n, \ldots$ be a geometric sequence. In Problems 15–24, find the indicated quantities.

15. $a_1 = 3$; $r = -2$; $a_2 = ?$; $a_3 = ?$; $a_4 = ?$

16. $a_1 = 32$; $r = -\frac{1}{2}$; $a_2 = ?$; $a_3 = ?$; $a_4 = ?$

17. $a_1 = 1; a_7 = 729; r = -3; S_7 = ?$

18. $a_1 = 3; a_7 = 2,187; r = 3; S_7 = ?$

19. $a_1 = 100; r = 1.08; a_{10} = ?$

20. $a_1 = 240; r = 1.06; a_{12} = ?$

21. $a_1 = 100; a_9 = 200; r = ?$

22. $a_1 = 100; a_{10} = 300; r = ?$

23. $a_1 = 500; r = 0.6; S_{10} = ?; S_\infty = ?$

24. $a_1 = 8,000; r = 0.4; S_{10} = ?; S_\infty = ?$

25. $S_{41} = \sum_{k=1}^{41} (3k + 3) = ?$ **26.** $S_{50} = \sum_{k=1}^{50} (2k - 3) = ?$

27. $S_8 = \sum_{k=1}^{8} (-2)^{k-1} = ?$ **28.** $S_8 = \sum_{k=1}^{8} 2^k = ?$

29. Find the sum of all the odd integers between 12 and 68.

30. Find the sum of all the even integers between 23 and 97.

31. Find the sum of each infinite geometric sequence (if it exists).

(A) $2, 4, 8, \ldots$ (B) $2, -\frac{1}{2}, \frac{1}{8}, \ldots$

32. Repeat Problem 31 for:

(A) $16, 4, 1, \ldots$ (B) $1, -3, 9, \ldots$

33. Find $f(1) + f(2) + f(3) + \cdots + f(50)$ if $f(x) = 2x - 3$.

34. Find $g(1) + g(2) + g(3) + \cdots + g(100)$ if $g(t) = 18 - 3t$.

35. Find $f(1) + f(2) + \cdots + f(10)$ if $f(x) = \left(\frac{1}{2}\right)^x$.

36. Find $g(1) + g(2) + \cdots + g(10)$ if $g(x) = 2^x$.

37. Show that the sum of the first n odd positive integers is n^2, using appropriate formulas from this section.

38. Show that the sum of the first n even positive integers is $n + n^2$, using formulas in this section.

39. If $r = 1$, neither the first form nor the second form for the sum of a finite geometric series is valid. Find a formula for the sum of a finite geometric series if $r = 1$.

40. If all of the terms of an infinite geometric series are less than 1, could the sum be greater than 1,000? Explain.

41. Does there exist a finite arithmetic series with $a_1 = 1$ and $a_n = 1.1$ that has sum equal to 100? Explain.

42. Does there exist a finite arithmetic series with $a_1 = 1$ and $a_n = 1.1$ that has sum equal to 105? Explain.

43. Does there exist an infinite geometric series with $a_1 = 10$ that has sum equal to 6? Explain.

44. Does there exist an infinite geometric series with $a_1 = 10$ that has sum equal to 5? Explain.

Applications

45. Loan repayment. If you borrow $4,800 and repay the loan by paying $200 per month to reduce the loan and 1% of the unpaid balance each month for the use of the money, what is the total cost of the loan over 24 months?

46. Loan repayment. If you borrow $5,400 and repay the loan by paying $300 per month to reduce the loan and 1.5% of the unpaid balance each month for the use of the money, what is the total cost of the loan over 18 months?

47. Economy stimulation. The government, through a subsidy program, distributes $5,000,000. If we assume that each person or agency spends 70% of what is received, and 70% of this is spent, and so on, how much total increase in spending results from this government action? (Let $a_1 = \$3,500,000$.)

48. Economy stimulation. Due to reduced taxes, a person has an extra $1,200 in spendable income. If we assume that the person spends 65% of this on consumer goods, and the producers of these goods in turn spend 65% on consumer goods, and that this process continues indefinitely, what is the total amount spent (to the nearest dollar) on consumer goods?

49. Compound interest. If $1,000 is invested at 5% compounded annually, the amount A present after n years forms a geometric sequence with common ratio $1 + 0.05 = 1.05$. Use a geometric sequence formula to find the amount A in the account (to the nearest cent) after 10 years. After 20 years. (*Hint*: Use a time line.)

50. Compound interest. If P is invested at $100r\%$ compounded annually, the amount A present after n years forms a geometric sequence with common ratio $1 + r$. Write a formula for the amount present after n years. (*Hint*: Use a time line.)

Answers to Matched Problems

1. (A) The sequence is geometric with $r = \frac{1}{4}$. It is not arithmetic.

(B) The sequence is arithmetic with $d = 5$. It is not geometric.

(C) The sequence is neither arithmetic nor geometric.

2. (A) 139   (B) -2

3. -960 **4.** 10,672   **5.** 1,063.66   **6.** $1,830   **7.** $8,000

B.3 Binomial Theorem

- Factorial
- Development of the Binomial Theorem

The binomial form

$$(a + b)^n$$

where n is a natural number, appears more frequently than you might expect. The coefficients in the expansion play an important role in probability studies. The *binomial formula*, which we will derive informally, enables us to expand $(a + b)^n$ directly for *n* any natural number. Since the formula involves *factorials*, we digress for a moment here to introduce this important concept.

Factorial

For *n* a natural number, **n factorial**, denoted by *n*!, is the product of the first *n* natural numbers. **Zero factorial** is defined to be 1. That is,

---

DEFINITION   *n* Factorial

$$n! = n \cdot (n - 1) \cdot \cdots \cdot 2 \cdot 1$$
$$1! = 1$$
$$0! = 1$$

It is also useful to note that *n*! can be defined recursively.

DEFINITION   *n* Factorial—Recursive Definition

$$n! = n \cdot (n - 1)! \quad n \geq 1$$

EXAMPLE 1   Factorial Forms   Evaluate.

(A) $5! = 5 \cdot 4 \cdot 3 \cdot 2 \cdot 1 = 120$

(B) $\dfrac{8!}{7!} = \dfrac{8 \cdot \cancel{7!}}{\cancel{7!}} = 8$

(C) $\dfrac{10!}{7!} = \dfrac{10 \cdot 9 \cdot 8 \cdot \cancel{7!}}{\cancel{7!}} = 720$

Matched Problem 1   Evaluate.

(A) $4!$ (B) $\dfrac{7!}{6!}$ (C) $\dfrac{8!}{5!}$

The following formula involving factorials has applications in many areas of mathematics and statistics. We will use this formula to provide a more concise form for the expressions encountered later in this discussion.

THEOREM 1   For *n* and *r* integers satisfying $0 \leq r \leq n$,

$$_nC_r = \frac{n!}{r!(n - r)!}$$

EXAMPLE 2   Evaluating $_nC_r$

(A) $_9C_2 = \dfrac{9!}{2!(9 - 2)!} = \dfrac{9!}{2!7!} = \dfrac{9 \cdot 8 \cdot \cancel{7!}}{2 \cdot \cancel{7!}} = 36$

(B) $_5C_5 = \dfrac{5!}{5!(5 - 5)!} = \dfrac{5!}{5!0!} = \dfrac{5!}{5!} = 1$

Matched Problem 2   Find

(A) $_5C_2$ (B) $_6C_0$

Development of the Binomial Theorem

Let us expand $(a + b)^n$ for several values of n to see if we can observe a pattern that leads to a general formula for the expansion for any natural number n:

$$(a + b)^1 = a + b$$
$$(a + b)^2 = a^2 + 2ab + b^2$$
$$(a + b)^3 = a^3 + 3a^2b + 3ab^2 + b^3$$
$$(a + b)^4 = a^4 + 4a^3b + 6a^2b^2 + 4ab^3 + b^4$$
$$(a + b)^5 = a^5 + 5a^4b + 10a^3b^2 + 10a^2b^3 + 5ab^4 + b^5$$

CONCEPTUAL INSIGHT

1. The expansion of $(a + b)^n$ has $(n + 1)$ terms.

2. The power of a decreases by 1 for each term as we move from left to right.

3. The power of b increases by 1 for each term as we move from left to right.

4. In each term, the sum of the powers of a and b always equals n.

5. Starting with a given term, we can get the coefficient of the next term by multiplying the coefficient of the given term by the exponent of a and dividing by the number that represents the position of the term in the series of terms. For example, in the expansion of $(a + b)^4$ above, the coefficient of the third term is found from the second term by multiplying 4 and 3, and then dividing by 2 [that is, the coefficient of the third term $= (4 \cdot 3)/2 = 6$].

We now postulate these same properties for the general case:

$$(a + b)^n = a^n + \frac{n}{1}a^{n-1}b + \frac{n(n-1)}{1 \cdot 2}a^{n-2}b^2 + \frac{n(n-1)(n-2)}{1 \cdot 2 \cdot 3}a^{n-3}b^3 + \cdots + b^n$$

$$= \frac{n!}{0!(n-0)!}a^n + \frac{n!}{1!(n-1)!}a^{n-1}b + \frac{n!}{2!(n-2)!}a^{n-2}b^2 + \frac{n!}{3!(n-3)!}a^{n-3}b^3 + \cdots + \frac{n!}{n!(n-n)!}b^n$$

$$= {}_nC_0 a^n + {}_nC_1 a^{n-1}b + {}_nC_2 a^{n-2}b^2 + {}_nC_3 a^{n-3}b^3 + \cdots + {}_nC_n b^n$$

And we are led to the formula in the binomial theorem:

THEOREM 2 Binomial Theorem

For all natural numbers n,

$$(a + b)^n = {}_nC_0 a^n + {}_nC_1 a^{n-1}b + {}_nC_2 a^{n-2}b^2 + {}_nC_3 a^{n-3}b^3 + \cdots + {}_nC_n b^n$$

EXAMPLE 3 Using the Binomial Theorem Use the binomial theorem to expand $(u + v)^6$.

SOLUTION

$$(u + v)^6 = {}_6C_0 u^6 + {}_6C_1 u^5 v + {}_6C_2 u^4 v^2 + {}_6C_3 u^3 v^3 + {}_6C_4 u^2 v^4 + {}_6C_5 u v^5 + {}_6C_6 v^6$$
$$= u^6 + 6u^5 v + 15u^4 v^2 + 20u^3 v^3 + 15u^2 v^4 + 6u v^5 + v^6$$

Matched Problem 3 Use the binomial theorem to expand $(x + 2)^5$.

EXAMPLE 4   Using the Binomial Theorem   Use the binomial theorem to find the sixth term in the expansion of $(x - 1)^{18}$.

SOLUTION   Sixth term $= {}_{18}C_5 x^{13}(-1)^5 = \dfrac{18!}{5!(18 - 5)!}x^{13}(-1)$

$$= -8{,}568x^{13}$$

Matched Problem 4   Use the binomial theorem to find the fourth term in the expansion of $(x - 2)^{20}$.

Exercises B.3

In Problems 1–20, evaluate each expression.

1. $6!$ **2.** $7!$ **3.** $\dfrac{10!}{9!}$

4. $\dfrac{20!}{19!}$ **5.** $\dfrac{12!}{9!}$ **6.** $\dfrac{10!}{6!}$

7. $\dfrac{5!}{2!3!}$ **8.** $\dfrac{7!}{3!4!}$ **9.** $\dfrac{6!}{5!(6 - 5)!}$

10. $\dfrac{7!}{4!(7 - 4)!}$ **11.** $\dfrac{20!}{3!17!}$ **12.** $\dfrac{52!}{50!2!}$

13. ${}_5C_3$ **14.** ${}_7C_3$ **15.** ${}_6C_5$ **16.** ${}_7C_4$

17. ${}_5C_0$ **18.** ${}_5C_5$ **19.** ${}_{18}C_{15}$ **20.** ${}_{18}C_3$

Expand each expression in Problems 21–26 using the binomial theorem.

21. $(a + b)^4$ **22.** $(m + n)^5$

23. $(x - 1)^6$ **24.** $(u - 2)^5$

25. $(2a - b)^5$ **26.** $(x - 2y)^5$

Find the indicated term in each expansion in Problems 27–32.

27. $(x - 1)^{18}$; 5th term   **28.** $(x - 3)^{20}$; 3rd term

29. $(p + q)^{15}$; 7th term   **30.** $(p + q)^{15}$; 13th term

31. $(2x + y)^{12}$; 11th term   **32.** $(2x + y)^{12}$; 3rd term

33. Show that ${}_nC_0 = {}_nC_n$ for $n \geq 0$.

34. Show that ${}_nC_r = {}_nC_{n-r}$ for $n \geq r \geq 0$.

35. The triangle next is called **Pascal's triangle**. Can you guess what the next two rows at the bottom are? Compare these numbers with the coefficients of binomial expansions.

$$
\begin{array}{ccccccccc}
 & & & & 1 & & & & \\
 & & & 1 & & 1 & & & \\
 & & 1 & & 2 & & 1 & & \\
 & 1 & & 3 & & 3 & & 1 & \\
1 & & 4 & & 6 & & 4 & & 1 \\
\end{array}
$$

36. Explain why the sum of the entries in each row of Pascal's triangle is a power of 2. (*Hint:* Let $a = b = 1$ in the binomial theorem.)

37. Explain why the alternating sum of the entries in each row of Pascal's triangle (e.g., $1 - 4 + 6 - 4 + 1$) is equal to 0.

38. Show that ${}_nC_r = \dfrac{n - r + 1}{r}{}_nC_{r-1}$ for $n \geq r \geq 1$.

39. Show that ${}_nC_{r-1} + {}_nC_r = {}_{n+1}C_r$ for $n \geq r \geq 1$.

Answers to Matched Problems

1. (A) 24   (B) 7   (C) 336

2. (A) 10   (B) 1

3. $x^5 + 10x^4 + 40x^3 + 80x^2 + 80x + 32$

4. $-9{,}120x^{17}$

Appendix

C

Tables

Table I Basic Geometric Formulas

1. Similar Triangles

(A) Two triangles are similar if two angles of one triangle have the same measure as two angles of the other.

(B) If two triangles are similar, their corresponding sides are proportional:

$$\frac{a}{a'} = \frac{b}{b'} = \frac{c}{c'}$$

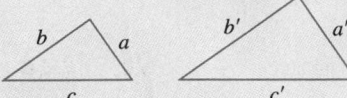

2. Pythagorean Theorem

$$c^2 = a^2 + b^2$$

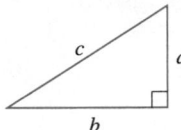

3. Rectangle

$A = ab$ Area

$P = 2a + 2b$ Perimeter

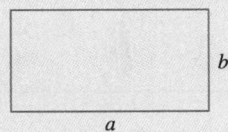

4. Parallelogram

$h = $ height

$A = ah = ab \sin \theta$ Area

$P = 2a + 2b$ Perimeter

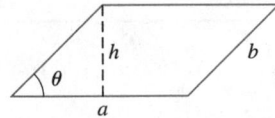

5. Triangle

$h = $ height

$A = \frac{1}{2}hc$ Area

$P = a + b + c$ Perimeter

$s = \frac{1}{2}(a + b + c)$ Semiperimeter

$A = \sqrt{s(s - a)(s - b)(s - c)}$ Area: Heron's formula

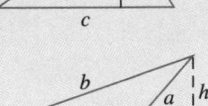

6. Trapezoid

Base a is parallel to base b.

$h = $ height

$A = \frac{1}{2}(a + b)h$ Area

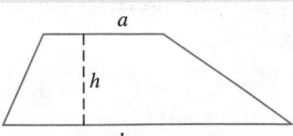

7. Circle

$R = $ radius

$D = $ diameter

$D = 2R$

$A = \pi R^2 = \frac{1}{4}\pi D^2$ Area

$C = 2\pi R = \pi D$ Circumference

$\dfrac{C}{D} = \pi$ For all circles

$\pi \approx 3.14159$

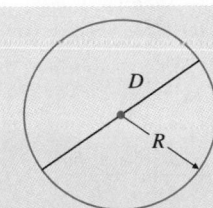

Table I Continued

8. Rectangular Solid

$V = abc$ Volume

$T = 2ab + 2ac + 2bc$ Total surface area

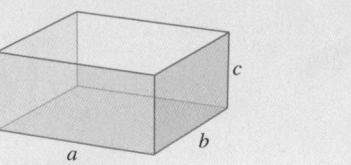

9. Right Circular Cylinder

R = radius of base

h = height

$V = \pi R^2 h$ Volume

$S = 2\pi R h$ Lateral surface area

$T = 2\pi R(R + h)$ Total surface area

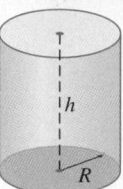

10. Right Circular Cone

R = radius of base

h = height

s = slant height

$V = \frac{1}{3}\pi R^2 h$ Volume

$S = \pi R s = \pi R \sqrt{R^2 + h^2}$ Lateral surface area

$T = \pi R(R + s) = \pi R(R + \sqrt{R^2 + h^2})$ Total surface area

11. Sphere

R = radius

D = diameter

$D = 2R$

$V = \frac{4}{3}\pi R^3 = \frac{1}{6}\pi D^3$ Volume

$S = 4\pi R^2 = \pi D^2$ Surface area

Table II Integration Formulas

Integrals Involving u^n

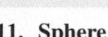

1. $\displaystyle\int u^n \, du = \frac{u^{n+1}}{n + 1}, \quad n \neq -1$

2. $\displaystyle\int u^{-1} \, du = \int \frac{1}{u} \, du = \ln|u|$

Integrals Involving $a + bu, a \neq 0$ and $b \neq 0$

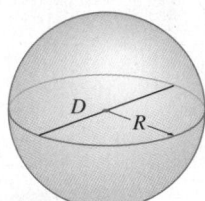

3. $\displaystyle\int \frac{1}{a + bu} \, du = \frac{1}{b} \ln|a + bu|$

4. $\displaystyle\int \frac{u}{a + bu} \, du = \frac{u}{b} - \frac{a}{b^2} \ln|a + bu|$

5. $\displaystyle\int \frac{u^2}{a + bu} \, du = \frac{(a + bu)^2}{2b^3} - \frac{2a(a + bu)}{b^3} + \frac{a^2}{b^3} \ln|a + bu|$

6. $\displaystyle\int \frac{u}{(a + bu)^2} \, du = \frac{1}{b^2}\left(\ln|a + bu| + \frac{a}{a + bu}\right)$

7. $\displaystyle\int \frac{u^2}{(a + bu)^2} \, du = \frac{(a + bu)}{b^3} - \frac{a^2}{b^3(a + bu)} - \frac{2a}{b^3} \ln|a + bu|$

8. $\displaystyle\int u(a + bu)^n \, du = \frac{(a + bu)^{n+2}}{(n + 2)b^2} - \frac{a(a + bu)^{n+1}}{(n + 1)b^2}, \quad n \neq -1, -2$

9. $\displaystyle\int \frac{1}{u(a + bu)} \, du = \frac{1}{a} \ln\left|\frac{u}{a + bu}\right|$

10. $\displaystyle\int \frac{1}{u^2(a + bu)} \, du = -\frac{1}{au} + \frac{b}{a^2} \ln\left|\frac{a + bu}{u}\right|$

11. $\displaystyle\int \frac{1}{u(a + bu)^2} \, du = \frac{1}{a(a + bu)} + \frac{1}{a^2} \ln\left|\frac{u}{a + bu}\right|$

12. $\displaystyle\int \frac{1}{u^2(a + bu)^2} \, du = -\frac{a + 2bu}{a^2 u(a + bu)} + \frac{2b}{a^3} \ln\left|\frac{a + bu}{u}\right|$

Table II **Continued**

Integrals Involving $a^2 - u^2, a > 0$

13. $\displaystyle\int \frac{1}{u^2 - a^2}\, du = \frac{1}{2a} \ln\left|\frac{u - a}{u + a}\right|$

14. $\displaystyle\int \frac{1}{a^2 - u^2}\, du = \frac{1}{2a} \ln\left|\frac{u + a}{u - a}\right|$

Integrals Involving $(a + bu)$ and $(c + du), b \neq 0, d \neq 0$, and $ad - bc \neq 0$

15. $\displaystyle\int \frac{1}{(a + bu)(c + du)}\, du = \frac{1}{ad - bc} \ln\left|\frac{c + du}{a + bu}\right|$

16. $\displaystyle\int \frac{u}{(a + bu)(c + du)}\, du = \frac{1}{ad - bc} \left(\frac{a}{b} \ln|a + bu| - \frac{c}{d} \ln|c + du|\right)$

17. $\displaystyle\int \frac{u^2}{(a + bu)(c + du)}\, du = \frac{1}{bd} u - \frac{1}{ad - bc} \left(\frac{a^2}{b^2} \ln|a + bu| - \frac{c^2}{d^2} \ln|c + du|\right)$

18. $\displaystyle\int \frac{1}{(a + bu)^2(c + du)}\, du = \frac{1}{ad - bc}\frac{1}{a + bu} + \frac{d}{(ad - bc)^2} \ln\left|\frac{c + du}{a + bu}\right|$

19. $\displaystyle\int \frac{u}{(a + bu)^2(c + du)}\, du = -\frac{a}{b(ad - bc)}\frac{1}{a + bu} - \frac{c}{(ad - bc)^2} \ln\left|\frac{c + du}{a + bu}\right|$

20. $\displaystyle\int \frac{a + bu}{c + du}\, du = \frac{bu}{d} + \frac{ad - bc}{d^2} \ln|c + du|$

Integrals Involving $\sqrt{a + bu}, a \neq 0$ and $b \neq 0$

21. $\displaystyle\int \sqrt{a + bu}\, du = \frac{2\sqrt{(a + bu)^3}}{3b}$

22. $\displaystyle\int u\sqrt{a + bu}\, du = \frac{2(3bu - 2a)}{15b^2}\sqrt{(a + bu)^3}$

23. $\displaystyle\int u^2\sqrt{a + bu}\, du = \frac{2(15b^2u^2 - 12abu + 8a^2)}{105b^3}\sqrt{(a + bu)^3}$

24. $\displaystyle\int \frac{1}{\sqrt{a + bu}}\, du = \frac{2\sqrt{a + bu}}{b}$

25. $\displaystyle\int \frac{u}{\sqrt{a + bu}}\, du = \frac{2(bu - 2a)}{3b^2}\sqrt{a + bu}$

26. $\displaystyle\int \frac{u^2}{\sqrt{a + bu}}\, du = \frac{2(3b^2u^2 - 4abu + 8a^2)}{15b^3}\sqrt{a + bu}$

27. $\displaystyle\int \frac{1}{u\sqrt{a + bu}}\, du = \frac{1}{\sqrt{a}} \ln\left|\frac{\sqrt{a + bu} - \sqrt{a}}{\sqrt{a + bu} + \sqrt{a}}\right|, \quad a > 0$

28. $\displaystyle\int \frac{1}{u^2\sqrt{a + bu}}\, du = -\frac{\sqrt{a + bu}}{au} - \frac{b}{2a\sqrt{a}} \ln\left|\frac{\sqrt{a + bu} - \sqrt{a}}{\sqrt{a + bu} + \sqrt{a}}\right|, \quad a > 0$

Integrals Involving $\sqrt{a^2 - u^2}, a > 0$

29. $\displaystyle\int \frac{1}{u\sqrt{a^2 - u^2}}\, du = -\frac{1}{a} \ln\left|\frac{a + \sqrt{a^2 - u^2}}{u}\right|$

30. $\displaystyle\int \frac{1}{u^2\sqrt{a^2 - u^2}}\, du = -\frac{\sqrt{a^2 - u^2}}{a^2 u}$

31. $\displaystyle\int \frac{\sqrt{a^2 - u^2}}{u}\, du = \sqrt{a^2 - u^2} - a \ln\left|\frac{a + \sqrt{a^2 - u^2}}{u}\right|$

Integrals Involving $\sqrt{u^2 + a^2}, a > 0$

32. $\displaystyle\int \sqrt{u^2 + a^2}\, du = \frac{1}{2}\left(u\sqrt{u^2 + a^2} + a^2 \ln|u + \sqrt{u^2 + a^2}|\right)$

33. $\displaystyle\int u^2\sqrt{u^2 + a^2}\, du = \frac{1}{8}\left[u(2u^2 + a^2)\sqrt{u^2 + a^2} - a^4 \ln|u + \sqrt{u^2 + a^2}|\right]$

34. $\displaystyle\int \frac{\sqrt{u^2 + a^2}}{u}\, du = \sqrt{u^2 + a^2} - a \ln\left|\frac{a + \sqrt{u^2 + a^2}}{u}\right|$

35. $\displaystyle\int \frac{\sqrt{u^2 + a^2}}{u^2}\, du = -\frac{\sqrt{u^2 + a^2}}{u} + \ln|u + \sqrt{u^2 + a^2}|$

36. $\displaystyle\int \frac{1}{\sqrt{u^2 + a^2}}\, du = \ln|u + \sqrt{u^2 + a^2}|$

(continued)

Table II   Continued

37. $\displaystyle\int \frac{1}{u\sqrt{u^2 + a^2}}\, du = \frac{1}{a}\ln\left|\frac{u}{a + \sqrt{u^2 + a^2}}\right|$

38. $\displaystyle\int \frac{u^2}{\sqrt{u^2 + a^2}}\, du = \frac{1}{2}\left(u\sqrt{u^2 + a^2} - a^2\ln\left|u + \sqrt{u^2 + a^2}\right|\right)$

39. $\displaystyle\int \frac{1}{u^2\sqrt{u^2 + a^2}}\, du = -\frac{\sqrt{u^2 + a^2}}{a^2 u}$

Integrals Involving $\sqrt{u^2 - a^2}, a > 0$

40. $\displaystyle\int \sqrt{u^2 - a^2}\, du = \frac{1}{2}\left(u\sqrt{u^2 - a^2} - a^2\ln\left|u + \sqrt{u^2 - a^2}\right|\right)$

41. $\displaystyle\int u^2\sqrt{u^2 - a^2}\, du = \frac{1}{8}\left[u(2u^2 - a^2)\sqrt{u^2 - a^2} - a^4\ln\left|u + \sqrt{u^2 - a^2}\right|\right]$

42. $\displaystyle\int \frac{\sqrt{u^2 - a^2}}{u^2}\, du = -\frac{\sqrt{u^2 - a^2}}{u} + \ln\left|u + \sqrt{u^2 - a^2}\right|$

43. $\displaystyle\int \frac{1}{\sqrt{u^2 - a^2}}\, du = \ln\left|u + \sqrt{u^2 - a^2}\right|$

44. $\displaystyle\int \frac{u^2}{\sqrt{u^2 - a^2}}\, du = \frac{1}{2}\left(u\sqrt{u^2 - a^2} + a^2\ln\left|u + \sqrt{u^2 - a^2}\right|\right)$

45. $\displaystyle\int \frac{1}{u^2\sqrt{u^2 - a^2}}\, du = \frac{\sqrt{u^2 - a^2}}{a^2 u}$

Integrals Involving $e^{au}, a \neq 0$

46. $\displaystyle\int e^{au}\, du = \frac{e^{au}}{a}$

47. $\displaystyle\int u^n e^{au}\, du = \frac{u^n e^{au}}{a} - \frac{n}{a}\int u^{n-1} e^{au}\, du$

48. $\displaystyle\int \frac{1}{c + de^{au}}\, du = \frac{u}{c} - \frac{1}{ac}\ln\left|c + de^{au}\right|, \quad c \neq 0$

Integrals Involving $\ln u$

49. $\displaystyle\int \ln u\, du = u\ln u - u$

50. $\displaystyle\int \frac{\ln u}{u}\, du = \frac{1}{2}(\ln u)^2$

51. $\displaystyle\int u^n \ln u\, du = \frac{u^{n+1}}{n+1}\ln u - \frac{u^{n+1}}{(n+1)^2}, \quad n \neq -1$

52. $\displaystyle\int (\ln u)^n\, du = u(\ln u)^n - n\int (\ln u)^{n-1}\, du$

Integrals Involving Trigonometric Functions of $au, a \neq 0$

53. $\displaystyle\int \sin au\, du = -\frac{1}{a}\cos au$

54. $\displaystyle\int \cos au\, du = \frac{1}{a}\sin au$

55. $\displaystyle\int \tan au\, du = -\frac{1}{a}\ln|\cos au|$

56. $\displaystyle\int \cot au\, du = \frac{1}{a}\ln|\sin au|$

57. $\displaystyle\int \sec au\, du = \frac{1}{a}\ln|\sec au + \tan au|$

58. $\displaystyle\int \csc au\, du = \frac{1}{a}\ln|\csc au - \cot au|$

59. $\displaystyle\int (\sin au)^2\, du = \frac{u}{2} - \frac{1}{4a}\sin 2au$

60. $\displaystyle\int (\cos au)^2\, du = \frac{u}{2} + \frac{1}{4a}\sin 2au$

61. $\displaystyle\int (\sin au)^n\, du = -\frac{1}{an}(\sin au)^{n-1}\cos au + \frac{n-1}{n}\int (\sin au)^{n-2}\, du, \quad n \neq 0$

62. $\displaystyle\int (\cos au)^n\, du = \frac{1}{an}\sin au(\cos au)^{n-1} + \frac{n-1}{n}\int (\cos au)^{n-2}\, du, \quad n \neq 0$

[*Note:* **The constant of integration is omitted for each integral, but must be included in any particular application of a formula.** The variable u is the variable of integration; all other symbols represent constants.]

ANSWERS

Diagnostic Prerequisite Test

Section references are provided in parentheses following each answer to guide students to the specific content in the book where they can find help or remediation.

1. (A) $(y + z)x$ (B) $(2 + x) + y$ (C) $2x + 3x$ *(A.1)*   **2.** $x^3 - 3x^2 + 4x + 8$ *(A.2)*   **3.** $x^3 + 3x^2 - 2x + 12$ *(A.2)*   **4.** $-3x^5 + 2x^3 - 24x^2 + 16$ *(A.2)*
5. (A) 1   (B) 1   (C) 2   (D) 3 *(A.2)*   **6.** (A) 3   (B) 1   (C) -3 (D) 1 *(A.2)*   **7.** $14x^2 - 30x$ *(A.2)*   **8.** $6x^2 - 5xy - 4y^2$ *(A.2)*   **9.** $(x + 2)(x + 5)$ *(A.3)*
10. $x(x + 3)(x - 5)$ *(A.3)*   **11.** $7/20$ *(A.1)*   **12.** 0.875 *(A.1)*   **13.** (A) 4.065×10^{12} (B) 7.3×10^{-3} *(A.5)*   **14.** (A) $255{,}000{,}000$ (B) $0{,}000\ 406$ *(A.5)*
15. (A) T   (B) F *(A.1)*   **16.** 0 and -3 are two examples of infinitely many. *(A.1)*   **17.** $6x^5y^{15}$ *(A.5)*   **18.** $3u^4/v^2$ *(A.5)*   **19.** 6×10^2 *(A.5)*   **20.** x^6/y^4 *(A.5)*
21. $u^{7/3}$ *(A.6)*   **22.** $3a^2/b$ *(A.6)*   **23.** $\frac{5}{9}$ *(A.5)*   **24.** $x + 2x^{1/2}y^{1/2} + y$ *(A.6)*   **25.** $\frac{a^2 + b^2}{ab}$ *(A.4)*   **26.** $\frac{a^2 - c^2}{abc}$ *(A.4)*   **27.** $\frac{y^5}{x}$ *(A.4)*   **28.** $\frac{1}{xy^2}$ *(A.4)*
29. $\frac{-1}{7(7 + h)}$ *(A.4)*   **30.** $\frac{xy}{y - x}$ *(A.6)*   **31.** (A) Subtraction   (B) Commutative $(+)$ (C) Distributive   (D) Associative $(\cdot)$ (E) Negatives
(F) Identity $(+)$ *(A.1)*   **32.** (A) 6   (B) 0 *(A.1)*   **33.** $4x = x - 4; x = -4/3$ *(1.1)*   **34.** $-15/7$ *(1.2)*   **35.** $(4/7, 0)$ *(1.2)*   **36.** $(0, -4)$ *(1.2)*
37. $(x - 5y)(x + 2y)$ *(A.3)*   **38.** $(3x - y)(2x - 5y)$ *(A.3)*   **39.** $3x^{-1} + 4y^{1/2}$ *(A.6)*   **40.** $8x^{-2} - 5y^{-4}$ *(A.5)*
41. $\frac{2}{5}x^{-3/4} - \frac{7}{6}y^{-2/3}$ *(A.6)*   **42.** $\frac{1}{3}x^{-1/2} + 9y^{-1/3}$ *(A.6)*   **43.** $\frac{2}{7} + \frac{1}{14}\sqrt{2}$ *(A.6)*   **44.** $\frac{14}{11} - \frac{5}{11}\sqrt{3}$ *(A.6)*   **45.** $x = 0, 5$ *(A.7)*
46. $x = \pm\sqrt{7}$ *(A.7)*   **47.** $x = -4, 5$ *(A.7)*   **48.** $x = 1, \frac{1}{6}$ *(A.7)*   **49.** $x = -1 \pm \sqrt{2}$ *(A.7)*   **50.** $x = \pm 1, \pm\sqrt{5}$ *(A.7)*

Chapter 1

Exercises 1.1

1. 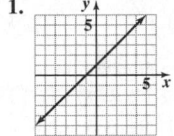   **3.** 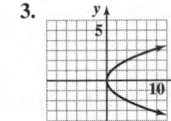   **5.** 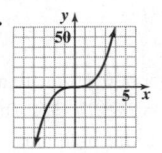   **7.** 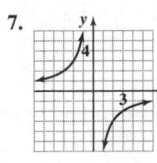   **9.** A function   **11.** Not a function   **13.** A function   **15.** A function
17. Not a function   **19.** A function   **21.** Linear   **23.** Neither
25. Constant   **27.** Linear

29. 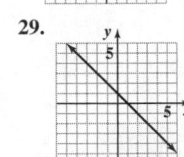   **31.** 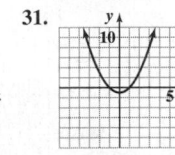   **33.** 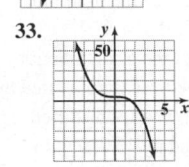   **35.** 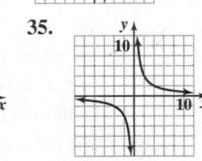   **37.** 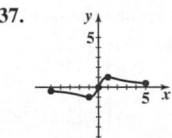   **39.** $y = 0$ **41.** $y = 4$ **43.** $x = -5, 0, 4$
45. $x = -6$ **47.** All real numbers
49. All real numbers except -4
51. $x \leq 7$ **53.** Yes; all real numbers
55. No; for example, when $x = 0$, $y = \pm 2$

57. Yes; all real numbers except 0   **59.** No; when $x = 1$, $y = \pm 1$ **61.** 12   **63.** $x^2 + 2x - 3$ **65.** $36x^2 - 4$ **67.** $x^6 - 4$ **69.** $h^2 - 4$ **71.** $4h + h^2$
73. $4h + h^2$ **75.** (A) $4x + 4h - 3$ (B) $4h$ (C) 4   **77.** (A) $4x^2 + 8xh + 4h^2 - 7x - 7h + 6$ (B) $8xh + 4h^2 - 7h$ (C) $8x + 4h - 7$
79. (A) $20x + 20h - x^2 - 2xh - h^2$ (B) $20h - 2xh - h^2$ (C) $20 - 2x - h$ **81.** $P(w) = 2w + \frac{50}{w}$, $w > 0$ **83.** $A(l) = l(50 - l)$, $0 < l < 50$

85. (A) 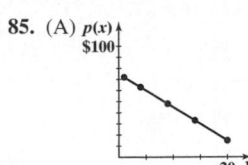   (B) \$54; \$42   **87.** (A) $R(x) = (75 - 3x)x$, $1 \leq x \leq 20$ (B)

x	$R(x)$
1	72
4	252
8	408
12	468
16	432
20	300

(C)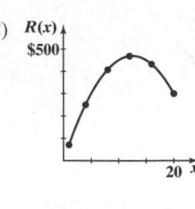

89. (A) $P(x) = 59x - 3x^2 - 125$, $1 \leq x \leq 20$ (B)

x	$P(x)$
1	-69
4	63
8	155
12	151
16	51
20	-145

(C)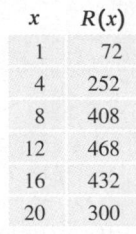

91. (A) $V(x) = x(8 - 2x)(12 - 2x)$ (B) $0 < x < 4$
(C)

x	$V(x)$
1	60
2	64
3	36

(D) $V(x)$

93. (A) The graph indicates that there is a value of x near 2 that will produce a volume of 65.
(B) The table shows $x = 1.9$ to one decimal place:

x	1.8	1.9	2
$V(x)$	66.5	65.4	64

(C) $x = 1.93$ to two decimal places   **95.** $v = \frac{75 - w}{15 + w}$; 1.9032 cm/sec

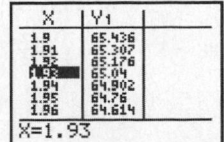

Exercises 1.2   **1.** Domain: all real numbers; range: all real numbers   **3.** Domain: $[0, \infty)$; range: $(-\infty, 15]$

5. Domain: all real numbers; range: $[7, \infty)$ **7.** Domain: all real numbers; range: all real numbers

9. 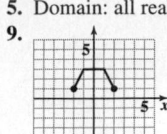   **11.** 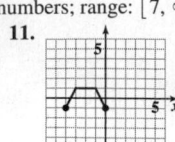   **13.** 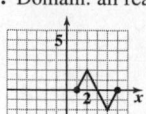   **15.** 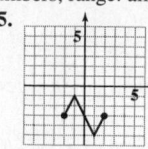   **17.** 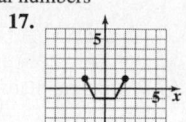   **19.** 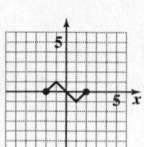   **21.**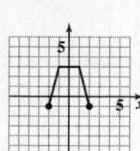

23. 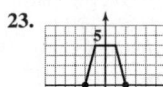   **25.** The graph of $g(x) = -|x + 3|$ is the graph of $y = |x|$ reflected in the x axis and shifted 3 units to the left. 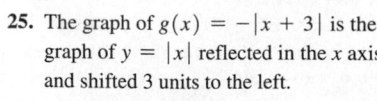   **27.** The graph of $f(x) = (x - 4)^2 - 3$ is the graph of $y = x^2$ shifted 4 units to the right and 3 units down.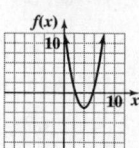

29. The graph of $f(x) = 7 - \sqrt{x}$ is the graph of $y = \sqrt{x}$ reflected in the x axis and shifted 7 units up. 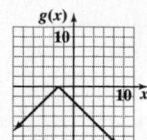   **31.** The graph of $h(x) = -3|x|$ is the graph of $y = |x|$ reflected in the x axis and vertically stretched by a factor of 3. 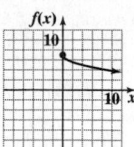   **33.** The graph of the basic function $y = x^2$ is shifted 2 units to the left and 3 units down. Equation: $y = (x + 2)^2 - 3$.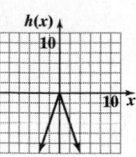

35. The graph of the basic function $y = x^2$ is reflected in the x axis and shifted 3 units to the right and 2 units up. Equation: $y = 2 - (x - 3)^2$.   **37.** The graph of the basic function $y = \sqrt{x}$ is reflected in the x axis and shifted 4 units up. Equation: $y = 4 - \sqrt{x}$.   **39.** The graph of the basic function $y = x^3$ is shifted 2 units to the left and 1 unit down. Equation: $y = (x + 2)^3 - 1$.

41. $g(x) = \sqrt{x - 2} - 3$ **43.** $g(x) = -|x + 3|$ **45.** $g(x) = -(x - 2)^3 - 1$ **47.** 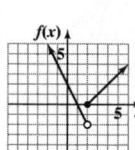   **49.** 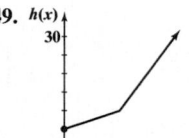   **51.**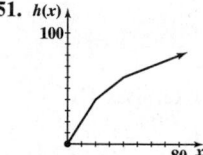

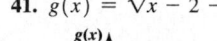 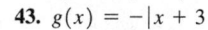

53. The graph of the basic function $y = |x|$ is reflected in the x axis and vertically shrunk by a factor of 0.5. Equation: $y = -0.5|x|$.   **55.** The graph of the basic function $y = x^2$ is reflected in the x axis and vertically stretched by a factor of 2. Equation: $y = -2x^2$.   **57.** The graph of the basic function $y = \sqrt[3]{x}$ is reflected in the x axis and vertically stretched by a factor of 3. Equation: $y = -3\sqrt[3]{x}$.   **59.** Reversing the order does not change the result.   **61.** Reversing the order can change the result.   **63.** Reversing the order can change the result.

65. (A) The graph of the basic function $y = \sqrt{x}$ is reflected in the x axis, vertically expanded by a factor of 4, and shifted up 115 units.   (B) 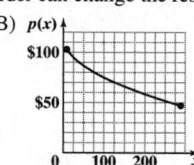   **67.** (A) The graph of the basic function $y = x^3$ is vertically contracted by a factor of 0.000 48 and shifted right 500 units and up 60,000 units.   (B)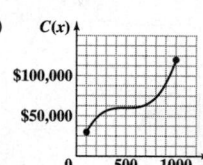

69. (A) $S(x) = \begin{cases} 8.5 + 0.065x & \text{if } 0 \le x \le 700 \\ -9 + 0.09x & \text{if } x > 700 \end{cases}$

(B)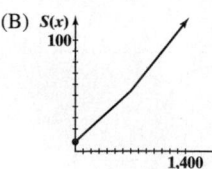

71. (A) $T(x) = \begin{cases} 0.035x & \text{if } 0 \le x \le 30{,}000 \\ 0.0625x - 825 & \text{if } 30{,}000 < x \le 60{,}000 \\ 0.0645x - 945 & \text{if } x > 60{,}000 \end{cases}$

(B)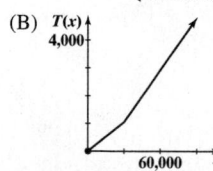

(C) \$1,675; \$3,570

73. (A) The graph of the basic function $y = x$ is vertically stretched by a factor of 5.5 and shifted down 220 units.   (B) 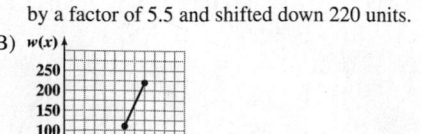   **75.** (A) The graph of the basic function $y = \sqrt{x}$ is vertically stretched by a factor of 7.08.   (B)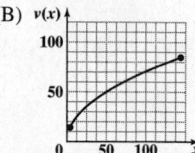

Exercises 1.3   **1.** 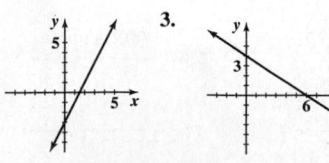   **3.**   **5.** Slope $= 5$; y intercept $= -7$ **7.** Slope $= -\dfrac{5}{2}$; y intercept $= -9$ **9.** $y = 2x + 1$

11. $y = -\dfrac{1}{3}x + 6$ **13.** x int.: -1; y int.: -2; $y = -2x - 2$

15. x int.: -3; y int.: 1; $y = \dfrac{x}{3} + 1$ **17.** (A) m (B) g (C) f (D) n

19. (A) x int.: 1, 3; y int.: -3 (B) Vertex: $(2, 1)$ (C) Max.: 1   (D) Range: $y \le 1$ or $(-\infty, 1]$ **21.** (A) x int.: $-3, -1$; y int.: 3   (B) Vertex: $(-2, -1)$ (C) Min.: -1 (D) Range: $y \ge -1$ or $[-1, \infty)$ **23.** (A) x int.: $3 \pm \sqrt{2}$; y int.: -7 (B) Vertex: $(3, 2)$ (C) Max.: 2   (D) Range: $y \le 2$ or $(-\infty, 2]$

25. (A) x int.: $-1 \pm \sqrt{2}$; y int.: -1 (B) Vertex: $(-1, -2)$ (C) Min.: -2 (D) Range: $y \geq -2$ or $[-2, \infty)$ **27.** (A) $m = \dfrac{2}{3}$ (B) $y - 5 = \dfrac{2}{3}(x - 2)$

(C) $y = \dfrac{2}{3}x + \dfrac{11}{3}$ (D) $-2x + 3y = 11$ **29.** (A) $m = -\dfrac{5}{4}$ (B) $y + 1 = -\dfrac{5}{4}(x + 2)$ (C) $y = -\dfrac{5}{4}x - \dfrac{7}{2}$ (D) $5x + 4y = -14$ **31.** (A) Not defined

(B) None   (C) None   (D) $x = 5$ **33.** (A) $m = 0$ (B) $y - 5 = 0$ (C) $y = 5$ (D) $y = 5$ **35.** Vertex form: $(x - 4)^2 - 4$ (A) x int.: 2, 6; y int.: 12

(B) Vertex: $(4, -4)$ (C) Min.: -4 (D) Range: $y \geq -4$ or $[-4, \infty)$ **37.** Vertex form: $-4(x - 2)^2 + 1$ (A) x int.: 1.5, 2.5; y int.: -15 (B) Vertex: $(2, 1)$

(C) Max.: 1   (D) Range: $y \leq 1$ or $(-\infty, 1]$ **39.** Vertex form: $0.5(x - 2)^2 + 3$ (A) x int.: none; y int.: 5   (B) Vertex: $(2, 3)$ (C) Min.: 3

(D) Range: $y \geq 3$ or $[3, \infty)$ **41.** $\left(-\infty, -\dfrac{7}{2}\right)$ **43.** $\left(-\infty, -\dfrac{3}{5}\right]$ **45.** $(-\infty, 0) \cup (10, \infty)$ **47.** $\left[-10 - 5\sqrt{2}, -10 + 5\sqrt{2}\right]$

49. 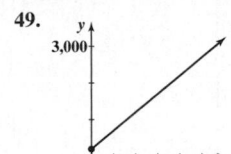   **51.** (A) 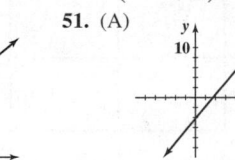   (B) x int.: 3.5; y int.: -4.2 (C) 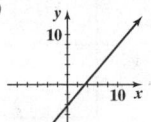 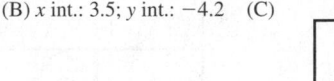   (D) x int.: 3.5; y int.: -4.2

53. (A) $-4.87, 8.21$ (B) $-3.44, 6.78$ (C) No solution   **55.** 651.0417 **59.** (A) $P = 0.44\overline{5}d + 14.7$ (B) The rate of change of pressure with respect to depth is
$0.44\overline{5}$ lb/in^2 per foot.   (C) 37 lb/in^2 (D) 99 ft   **61.** (A) $a = 2,880 - 24t$ (B) -24 ft/sec   (C) 24 ft/sec

63. (A) $C = 75x + 1,647$ **65.** (A) $V = -7,500t + 157,000$ **67.** (A) $T = 70 - 3.6A$ **69.** (A) $p = 0.000\,225x + 0.5925$

(B) 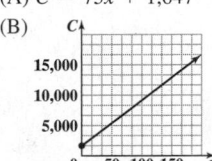   (B) $112,000   (B) 10,000 ft   (B) $p = -0.0009x + 9.39$

(C) During the 12th year   (C) $(7,820, 2.352)$

(C) The y intercept, $1,647, is the   (D) 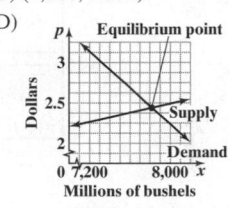   (D)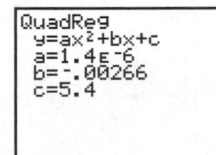
fixed cost and the slope, $75, is
the cost per club.

71. (A) 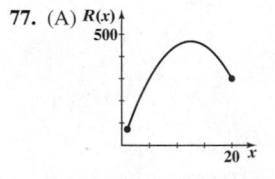   **73.** (A)

x	28	30	32	34	36
Mileage	45	52	55	51	47
$f(x)$	45.3	51.8	54.2	52.4	46.5

(B) 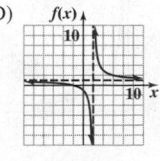   (C) $f(31) = 53.50$ thousand miles;
$f(35) = 49.95$ thousand miles

(B) 1,200,000   (C) 1,000,000

77. (A) 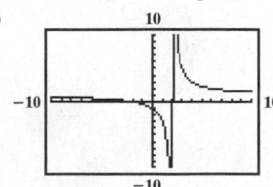   **79.** (A) The rate of change of height with respect to Dbh is 1.37 ft/in.   **85.** 10.6 mph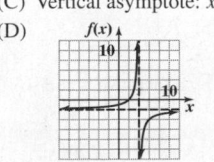
(B) Height increases by approximately 1.37 ft.   (C) 18 ft   (D) 20 in.
81. (A) The monthly price is increasing at a rate of $1.70 per
year.   (B) $71.70   **83.** Men: $y = -0.087x + 49.207$; women:
$y = -0.088x + 54.884$; yes

(B) 12.5 (12,500,000 chips);
$468,750,000   (C) $37.50

Exercises 1.4   **1.** (A) 1   (B) 10   (C) 50   **3.** (A) 5   (B) 0, 1   (C) 0   **5.** (A) 2   (B) $-1, -2$ (C) 2   **7.** (A) 4   (B) $-1, 1, -3, 3$
(C) 9   **9.** (A) 9   (B) $-3/2, 5$ (C) $-253,125$ **11.** (A) 4   (B) Negative   **13.** (A) 5   (B) Negative   **15.** (A) 1   (B) Negative
17. (A) 6   (B) Positive   **19.** 10   **21.** 1   **23.** (A) x int.: -2; y int.: -1 (B) Domain: all real numbers except 2
(C) Vertical asymptote: $x = 2$; horizontal asymptote: $y = 1$ (D) 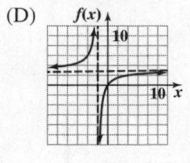   (E)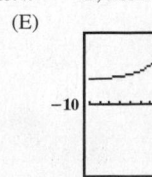

25. (A) x int.: 0; y int.: 0   **27.** (A) x int.: 2; y int.: -1
(B) Domain: all real numbers except -2 (B) Domain: all real numbers except 4
(C) Vertical asymptote: $x = -2$; horizontal asymptote: $y = 3$ (C) Vertical asymptote: $x = 4$; horizontal asymptote: $y = -2$
(D) 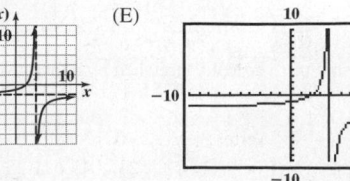   (E)   (D) (E)

29. (A)

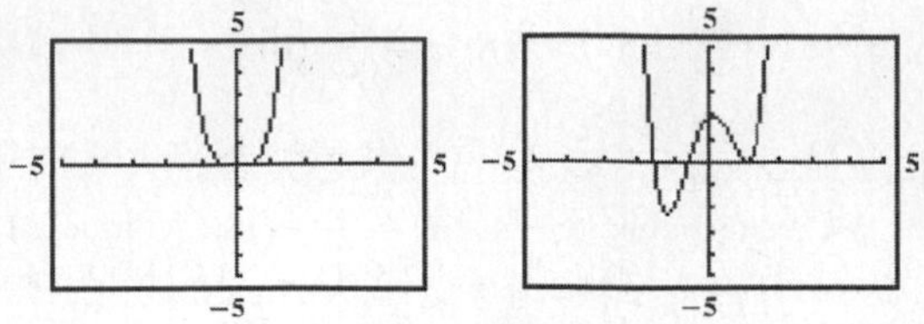

(B)

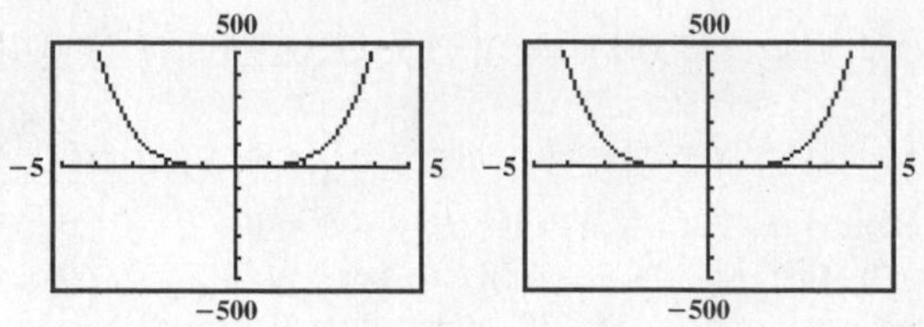

31. (A)

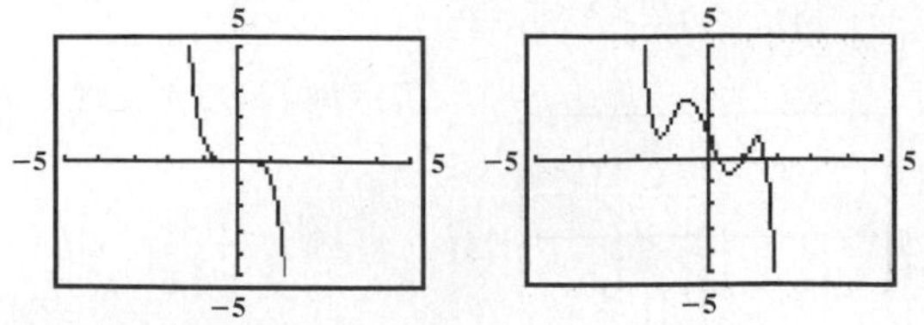

(B)

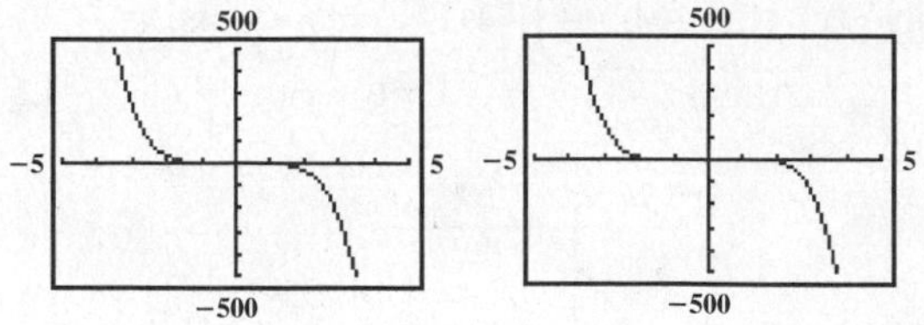

33. $y = \frac{5}{6}$ **35.** $y = \frac{1}{4}$ **37.** $y = 0$ **39.** None **41.** $x = -1, x = 1, x = -3, x = 3$ **43.** $x = 5$ **45.** $x = -6, x = 6$

47. (A) x int.: 0; y int.: 0
(B) Vertical asymptotes: $x = -2, x = 3$; horizontal asymptote: $y = 2$
(C)

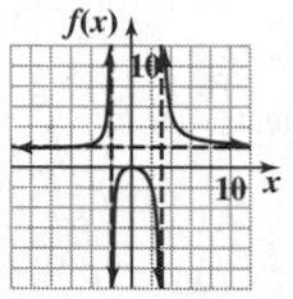

(D)

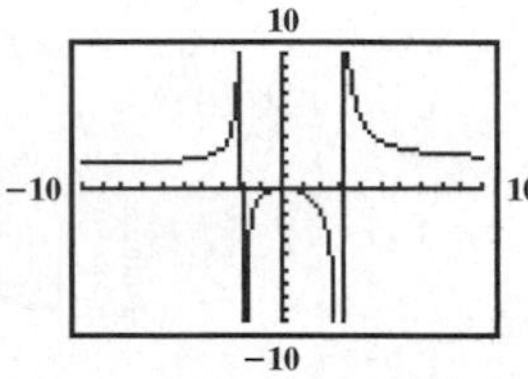

49. (A) x int.: $\pm\sqrt{3}$; y int.: $-\frac{2}{3}$
(B) Vertical asymptotes: $x = -3, x = 3$; horizontal asymptote: $y = -2$
(C)

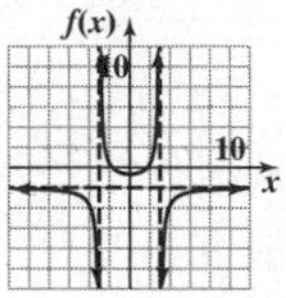

(D)

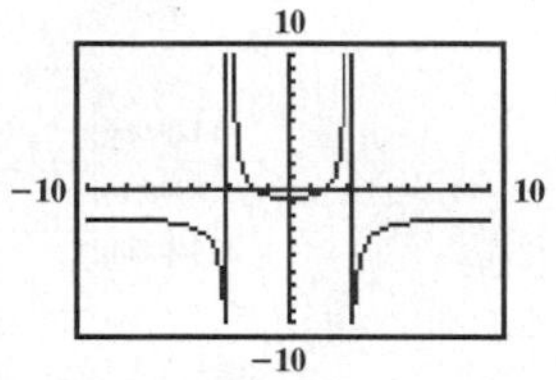

51. (A) x int.: 6; y int.: −4
(B) Vertical asymptotes: $x = -3, x = 2$; horizontal asymptote: $y = 0$
(C)

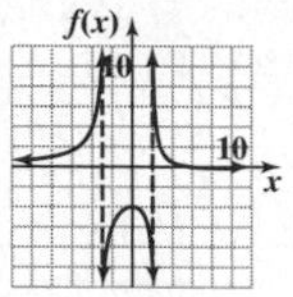

(D)

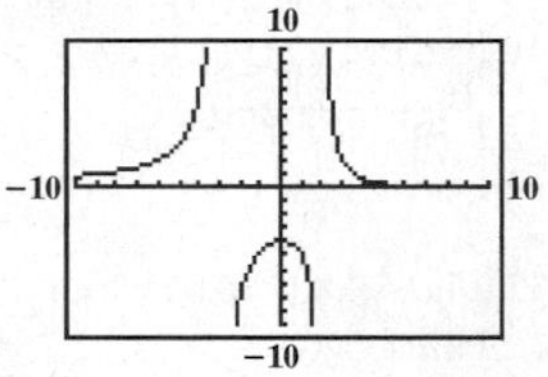

53. $f(x) = x^2 - x - 2$ **55.** $f(x) = 4x - x^3$

57. (A) $C(x) = 180x + 200$ (B) $\overline{C}(x) = \frac{180x + 200}{x}$
(C)

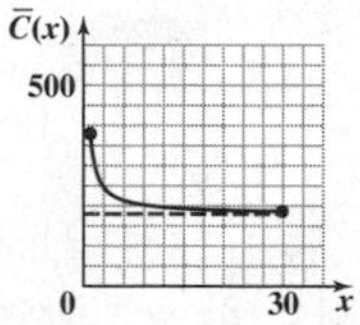

(D) $180 per board

59. (A) $\overline{C}(n) = \frac{2{,}500 + 175n + 25n^2}{n}$ (C) 10 yr; $675.00 per year (D) 10 yr; $675.00 per year
(B)

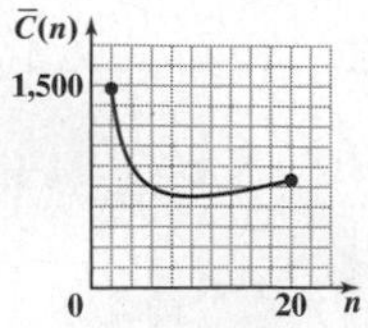

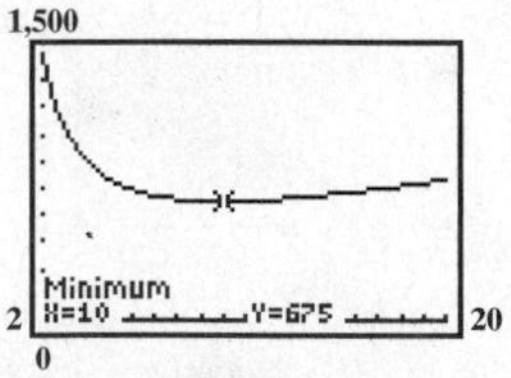

61. (A) $\overline{C}(x) = \frac{0.00048(x - 500)^3 + 60{,}000}{x}$
(B)

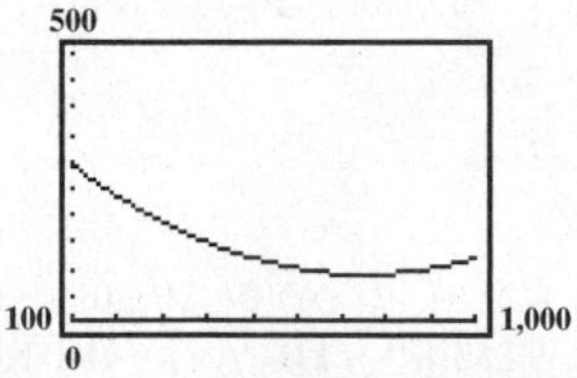

(C) 750 cases per month; $90 per case

63. (A)

```
CubicReg
 y=ax³+bx²+cx+d
 a=-2.666667E-4
 b=.0096666667
 c=-.2011904762
 d=17.84761905
```

(B) 4.1 lb **65.** (A) 0.06 cm/sec (B)

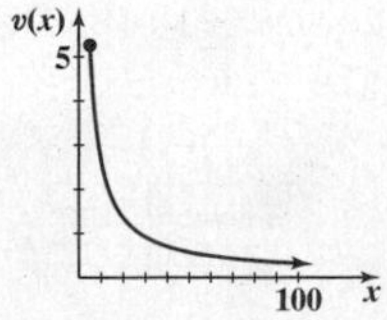

67. (A)

```
CubicReg
 y=ax³+bx²+cx+d
 a=8.7037037E-5
 b=-.0108492063
 c=.2907407407
 d=8.546031746
```

(B) 5.5

Exercises 1.5 **1.** (A) k (B) g (C) h (D) f

3.

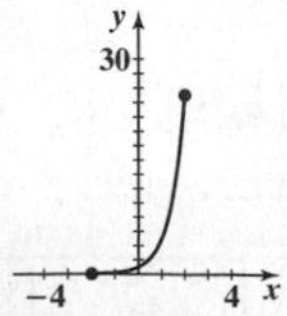

5.

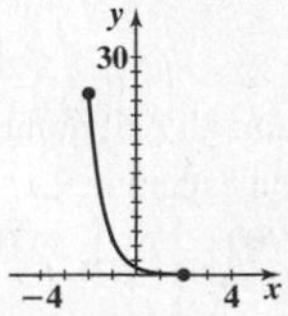

7.

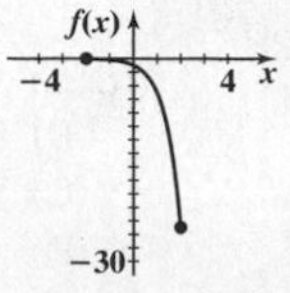

9.

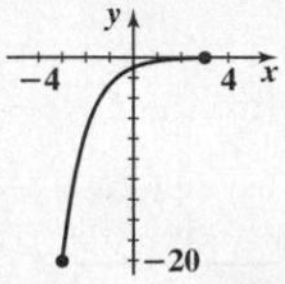

11. The graph of g is the graph of f reflected in the x axis. **13.** The graph of g is the graph of f shifted 1 unit to the left. **15.** The graph of g is the graph of f shifted 1 unit up.

17. The graph of g is the graph of f vertically stretched by a factor of 2 and shifted to the left 2 units.

19. (A) 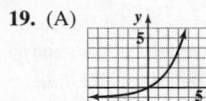 (B) 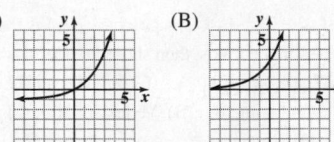 (C) 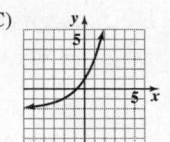 (D) 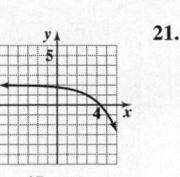   **21.** 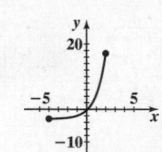 **23.**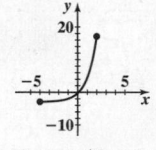

25. 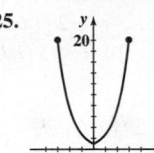   **27.** $a = 1, -1$ **29.** $x = 1$ **43.** 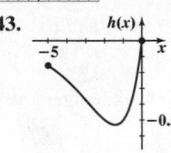   **45.** 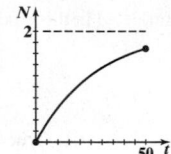   **47.** \$16,064.07   **49.** (A) \$2,633.56
31. $x = -1, 6$ **33.** $x = 3$ (B) \$7,079.54 **51.** \$10,706 **53.** (A) \$10,095.41
35. $x = -7$ **37.** $x = -2, 2$ (B) \$10,080.32 (C) \$10,085.27
39. $x = 1/4$ **41.** No solution

55. N approaches 2 as t increases without bound.   **57.** (A) \$9,781,000   **59.** (A) 10% (B) 1%

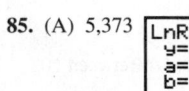   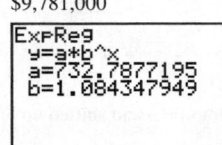   **61.** (A) $P = 7.1e^{0.011t}$ (B) 2025: 8.1 billion; 2035: 9.0 billion

63. (A) 42,772,000,000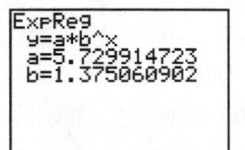

(B) The model gives an annual
salary of \$1,647,000 in 2000.

Exercises 1.6   **1.** $27 = 3^3$ **3.** $10^0 = 1$ **5.** $8 = 4^{3/2}$ **7.** $\log_7 49 = 2$ **9.** $\log_4 8 = \dfrac{3}{2}$ **11.** $\log_b A = u$ **13.** 2   **15.** 4   **17.** -2 **19.** -4

21. $\log_b P - \log_b Q$ **23.** $5 \log_b L$ **25.** q^p **27.** $x = 9$ **29.** $y = 2$ **31.** $b = 10$ **33.** $x = 2$ **35.** False   **37.** True   **39.** True   **41.** False   **43.** $x = 2$
45. $x = 8$ **47.** $x = 7$ **49.** No solution   **51.** 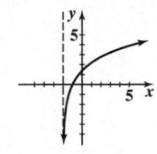   **53.** The graph of $y = \log_2 (x - 2)$ is the graph of $y = \log_2 x$ shifted to the right 2 units.
55. Domain: $(-1, \infty)$; range: all real numbers
57. (A) 3.547 43   (B) $-2.160 32$ (C) 5.626 29   (D) $-3.197 04$
59. (A) 13.4431   (B) 0.0089   (C) 16.0595   (D) 0.1514
61. 1.0792   **63.** 1.4595   **65.** 18.3559

67. Increasing: $(0, \infty)$ **69.** Decreasing: $(0, 1]$ **71.** Increasing: $(-2, \infty)$ **73.** Increasing: $(0, \infty)$ **75.** Because $b^0 = 1$ for any permissible base
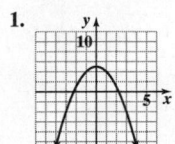   Increasing: $[1, \infty)$ 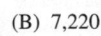      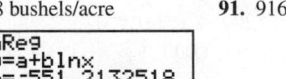 $b(b > 0, b \neq 1)$.   **77.** $x > \sqrt{x} > \ln x$ for $1 < x \leq 16$
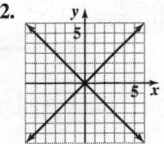   **79.** 4 yr   **81.** 9.87 yr; 9.80 yr   **83.** 7.51 yr

85. (A) 5,373 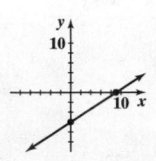   (B) 7,220 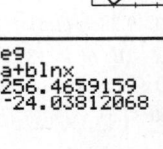   **89.** 168 bushels/acre   **91.** 916 yr

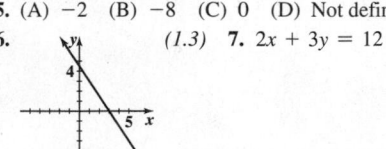

Chapter 1 Review Exercises

1. 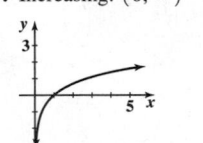 *(1.1)* **2.** 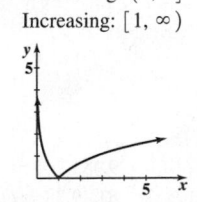 *(1.1)* **3.** 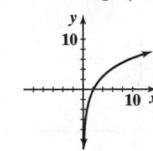 *(1.1)*   **4.** (A) Not a function (B) A function (C) A function (D) Not a function *(1.1)*
5. (A) -2 (B) -8 (C) 0 (D) Not defined *(1.1)*
6.   *(1.3)* **7.** $2x + 3y = 12$ *(1.3)*

8. x intercept $= 9$; y intercept $= -6$; slope $= \dfrac{2}{3}$ *(1.3)*   **9.** $y = -\dfrac{2}{3}x + 6$ *(1.3)* **10.** Vertical line: $x = -6$; horizontal line: $y = 5$ *(1.3)*

11. (A) $y = -\dfrac{2}{3}x$ (B) $y = 3$ *(1.3)*   **12.** (A) $3x + 2y = 1$ (B) $y = 5$ (C) $x = -2$ *(1.3)*
13. $v = \ln u$ *(1.6)*   **14.** $y = \log x$ *(1.6)*   **15.** $M = e^N$ *(1.6)*   **16.** $u = 10^v$ *(1.6)*   **17.** $x = 9$ *(1.6)*
18. $x = 6$ *(1.6)*   **19.** $x = 4$ *(1.6)*   **20.** $x = 2.157$ *(1.6)*   **21.** $x = 13.128$ *(1.6)*   **22.** $x = 1,273.503$ *(1.6)*
23. $x = 0.318$ *(1.6)*   **24.** (A) $y = 4$ (B) $x = 0$ (C) $y = 1$ (D) $x = -1$ or 1
(E) $y = -2$ (F) $x = -5$ or 5 *(1.1)*

25. (A) 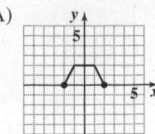 (B) 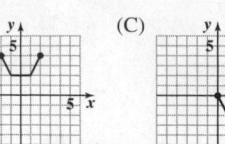 (C) 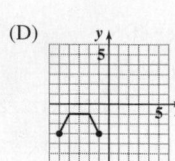 (D) 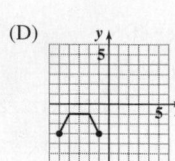 *(1.2)* **26.** $f(x) = -(x - 2)^2 + 4$. The graph of $f(x)$ is the graph of $y = x^2$ reflected in the x axis, then shifted right 2 units and up 4 units. *(1.2)* **27.** (A) g (B) m (C) n (D) f *(1.2, 1.3)*
28. (A) x int.: $-4, 0$; y int.: 0 (B) Vertex: $(-2, -4)$
(C) Min.: -4 (D) Range: $y \geq -4$ or $[-4, \infty)$ *(1.3)*

29. Quadratic *(1.3)* **30.** Linear *(1.1)* **31.** None *(1.1, 1.3)* **32.** Constant *(1.1)* **33.** $x = 8$ *(1.6)* **34.** $x = 3$ *(1.6)* **35.** $x = 0, \frac{3}{2}$ *(1.5)* **36.** $x = -2$ *(1.6)*
37. $x = 1.4650$ *(1.6)* **38.** $x = 92.1034$ *(1.6)* **39.** $x = 9.0065$ *(1.6)* **40.** $x = 2.1081$ *(1.6)* **41.** (A) All real numbers except $x = -2$ and 3 (B) $x < 5$ *(1.1)*

42. Vertex form: $4\left(x + \frac{1}{2}\right)^2 - 4$; x int.: $-\frac{3}{2}$ and $\frac{1}{2}$; y int.: -3; vertex: $(-\frac{1}{2}, -4)$; min.: -4; range: $y \geq -4$ or $[-4, \infty)$ *(1.3)*

43. $(-1.54, -0.79)$; $(0.69, 0.99)$ *(1.5, 1.6)* **44.** 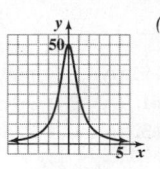 *(1.1)* **45.** 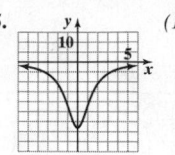 *(1.1)* **46.** 6 *(1.1)* **47.** -19 *(1.1)* **48.** $10x - 4$ *(1.1)*
49. $21 - 5x$ *(1.1)* **50.** (A) -1 (B) $-1 - 2h$
(C) $-2h$ (D) -2 *(1.1)* **51.** The graph of function m is the graph of $y = |x|$ reflected in the x axis and shifted to the right 4 units. *(1.2)*

52. The graph of function g is the graph of $y = x^3$ vertically shrunk by a factor of 0.3 and shifted up 3 units. *(1.2)*   **53.** $y = 0$ *(1.4)*

54. $y = \frac{3}{4}$ *(1.4)* **55.** None *(1.4)* **56.** $x = -10, x = 10$ *(1.4)* **57.** $x = -2$ *(1.4)* **58.** True *(1.3)* **59.** False *(1.3)* **60.** False *(1.3)* **61.** True *(1.3)*

62. 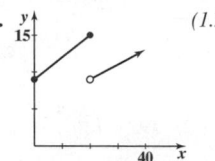 *(1.2)* **63.** 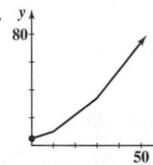 *(1.2)* **64.** $y = -(x - 4)^2 + 3$ *(1.2, 1.3)*
65. $f(x) = -0.4(x - 4)^2 + 7.6$
(A) x int.: $-0.4, 8.4$; y int.: 1.2 (B) Vertex: $(4.0, 7.6)$ (C) Max.: 7.6
(D) Range: $y \leq 7.6$ or $(-\infty, 7.6]$ *(1.3)*

66.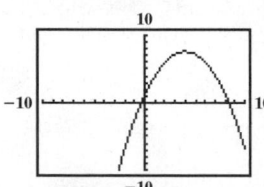
(A) x int.: $-0.4, 8.4$; y int.: 1.2
(B) Vertex: $(4.0, 7.6)$ (C) Max.: 7.6
(D) Range: $y \leq 7.6$ or $(-\infty, 7.6]$ *(1.3)*

67. $\log 10^\pi = \pi$ and $10^{\log \sqrt{2}} = \sqrt{2}$; $\ln e^\pi = \pi$ and $e^{\ln\sqrt{2}} = \sqrt{2}$ *(1.6)* **68.** $x = 2$ *(1.6)* **69.** $x = 2$ *(1.6)*
70. $x = 1$ *(1.6)* **71.** $x = 300$ *(1.6)* **72.** $y = ce^{-5t}$ *(1.6)* **73.** The function $y = 1^x$ is not one-to-one, so has no inverse. *(1.6)* **74.** The graph of $y = \sqrt[3]{x}$ is vertically expanded by a factor of 2, reflected in the x axis, and shifted 1 unit left and 1 unit down. Equation: $y = -2\sqrt[3]{x + 1} - 1$. *(1.2)* **75.** $G(x) = 0.3(x + 2)^2 - 8.1$ (A) x int.: $-7.2, 3.2$; y int.: -6.9 (B) Vertex: $(-2, -8.1)$ (C) Min.: -8.1 (D) Range: $y \geq -8.1$ or $[-8.1, \infty)$ *(1.3)*
76. 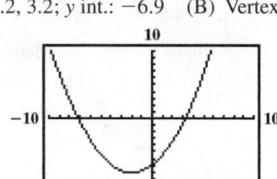 (A) x int.: $-7.2, 3.2$; y int.: -6.9 (B) Vertex: $(-2, -8.1)$
(C) Min.: -8.1 (D) Range: $y \geq -8.1$ or $[-8.1, \infty)$ *(1.3)*

77. (A) $S(x) = \begin{cases} 3 & \text{if } 0 \leq x \leq 20 \\ 0.057x + 1.86 & \text{if } 20 < x \leq 200 \\ 0.0346x + 6.34 & \text{if } 200 < x \leq 1{,}000 \\ 0.0217x + 19.24 & \text{if } x > 1{,}000 \end{cases}$ (B) 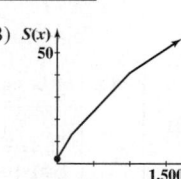 *(1.2)* **78.** \$5,321.95 *(1.5)* **79.** \$5,269.51 *(1.5)* **80.** 201 months (≈ 16.7 years) *(1.5)* **81.** 9.38 yr *(1.5)*
82. (A) $m = 132 - 0.6x$ (B) $M = 187 - 0.85x$
(C) Between 120 and 170 beats per minute (D) Between 102 and 144.5 beats per minute *(1.3)*

83. (A) $V = 224{,}000 - 15{,}500t$ **84.** (A) The dropout rate is decreasing at a rate of 0.198 percentage points per year. (B)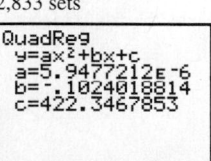
(B) \$38,000 *(1.3)*

(C) 2027 *(1.3)*
85. (A) The CPI is increasing at a rate of 4.75 units per year. (B) 285 *(1.3)*

86. (A) $A(x) = -\frac{3}{2}x^2 + 420x$ (B) Domain: $0 \leq x \leq 280$
(C)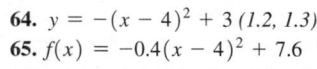

87. (A) 2,833 sets

```
QuadReg
y=ax²+bx+c
a=5.9477212E-6
b=-.1024018814
c=422.3467853
```

(B) 4,836

```
LinReg
y=ax+b
a=.0387421907
b=-7.364689544
```

(D) There are two solutions to the equation $A(x) = 25{,}000$, one near 90 and another near 190.
(E) 86 ft; 194 ft
(F) Maximum combined area is 29,400 ft^2. This occurs for $x = 140$ ft and $y = 105$ ft. *(1.3)*

(C) Price is likely to decrease (D) Equilibrium price: \$131.59; equilibrium quantity: 3,587 cookware sets *(1.3)*

88. (A) 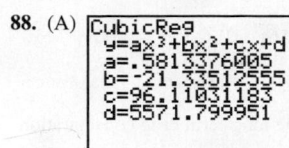   (B) 7,725 *(1.4)*   **89.** (A) $N = 2^{2t}$ or $N = 4^t$ **91.** (A) 6,134,000 *(1.6)*   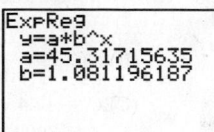   **92.** 23.1 yr *(1.5)*
(B) 15 days *(1.5)*
90. $k = 0.009\,42$; 489 ft *(1.6)*

93. (A) \$1,203 billion   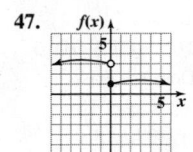   (B) 2028 *(1.5)*

Chapter 2

Exercises 2.1   **1.** $(x - 9)(x + 9)$ **3.** $(x - 7)(x + 3)$ **5.** $x(x - 3)(x - 4)$ **7.** $(2x - 1)(3x + 1)$ **9.** 2   **11.** 1.25   **13.** (A) 2   (B) 2
(C) 2   (D) 2   **15.** (A) 1   (B) 2   (C) Does not exist   (D) 2   (E) No   **17.** 2   **19.** 0.5   **21.** (A) 1   (B) 2   (C) Does not exist   (D) Does not exist   (E) No
23. (A) 1   (B) 1   (C) 1   (D) 3   (E) Yes, define $g(3) = 1$ **25.** (A) -2 (B) -2 (C) -2 (D) 1   (E) Yes, define $f(-3) = -2$ **27.** (A) 2   (B) 2   (C) 2
(D) Does not exist   (E) Yes, define $f(0) = 2$ **29.** 12   **31.** 1   **33.** -4 **35.** -1.5 **37.** 3   **39.** 15   **41.** -6 **43.** $\dfrac{7}{5}$ **45.** 3

47. 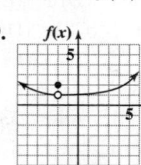   **49.**   **51.** (A) 1   (B) 1   (C) 1   (D) 1   **53.** (A) 2   (B) 1   (C) Does not exist   (D) Does not exist
55. (A) -6 (B) Does not exist   (C) 6   **57.** (A) 1   (B) -1 (C) Does not exist   (D) Does not exist
59. (A) Does not exist   (B) $\dfrac{1}{2}$ (C) $\dfrac{1}{4}$ **61.** (A) -5 (B) -3 (C) 0   **63.** (A) 0   (B) -1 (C) Does not exist
65. (A) 1   (B) $\dfrac{1}{3}$ (C) $\dfrac{3}{4}$ **67.** False   **69.** True   **71.** False   **73.** Yes; 0   **75.** No; 5/16   **77.** No; does not exist

79. Yes; 1/16   **89.** (A) $\begin{array}{l}\lim_{x\to 1^-} f(x) = 2 \\ \lim_{x\to 1^+} f(x) = 3\end{array}$ (B) $\begin{array}{l}\lim_{x\to 1^-} f(x) = 3 \\ \lim_{x\to 1^+} f(x) = 2\end{array}$ (C) $m = 1.5$ (D) The graph in (A) is broken when it jumps from $(1, 2)$ up to
81. 3 $(1, 3)$. The graph in (B) is also broken when it jumps down
83. 4   from $(1, 3)$ to $(1, 2)$. The graph in (C) is one continuous
85. -7 piece, with no breaks or jumps.
87. 1

91. (A) $F(x) = \begin{cases} 0.99 & \text{if } 0 < x \le 20 \\ 0.07x - 0.41 & \text{if } x \ge 20 \end{cases}$ **95.** (A) $D(x) = \begin{cases} x & \text{if } 0 \le x < 300 \\ 0.97x & \text{if } 300 \le x < 1{,}000 \\ 0.95x & \text{if } 1{,}000 \le x < 3{,}000 \\ 0.93x & \text{if } 3{,}000 \le x < 5{,}000 \\ 0.9x & \text{if } x \ge 5{,}000 \end{cases}$ **97.** $F(x) = \begin{cases} 20x & \text{if } 0 < x \le 4{,}000 \\ 80{,}000 & \text{if } x \ge 4{,}000 \end{cases}$
$\lim_{x\to 4{,}000} F(x) = 80{,}000$; $\lim_{x\to 8{,}000} F(x) = 80{,}000$

(B) 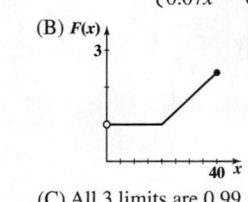   (B) $\lim_{x\to 1{,}000} D(x)$ does not exist because   **99.** $\lim_{x\to 5} f(x)$ does not exit; $\lim_{x\to 10} f(x) = 0$;
$\lim_{x\to 1{,}000^-} D(x) = 970$ and $\lim_{x\to 1{,}000^+} D(x) = 950$; $\lim_{x\to 5} g(x) = 0$; $\lim_{x\to 10} g(x) = 1$
$\lim_{x\to 3{,}000} D(x)$ does not exist because
(C) All 3 limits are 0.99. $\lim_{x\to 3{,}000^-} D(x) = 2{,}850$ and $\lim_{x\to 3{,}000^+} D(x) = 2{,}790$

Exercises 2.2   **1.** $y = 4$ **3.** $x = -6$ **5.** $2x - y = -13$ **7.** $7x + 9y = 63$ **9.** -2 **11.** $-\infty$ **13.** Does not exist   **15.** 0   **17.** (A) $-\infty$
(B) ∞ (C) Does not exist   **19.** (A) ∞ (B) ∞ (C) ∞ **21.** (A) 3   (B) 3   (C) 3   **23.** (A) $-\infty$ (B) ∞ (C) Does not exist   **25.** (A) $-5x^3$
(B) $-\infty$ (C) ∞ **27.** (A) $-6x^4$ (B) $-\infty$ (C) $-\infty$ **29.** (A) x^2 (B) ∞ (C) ∞ **31.** (A) $2x^5$ (B) ∞ (C) $-\infty$

33. $\lim_{x\to -3^-} f(x) = -\infty$; $\lim_{x\to -3^+} f(x) = \infty$; $x = -3$ is a vertical asymptote
35. $\lim_{x\to -2^-} h(x) = \infty$; $\lim_{x\to -2^+} h(x) = -\infty$; $\lim_{x\to 2^-} h(x) = -\infty$; $\lim_{x\to 2^+} h(x) = \infty$; $x = -2$ and $x = 2$ are a vertical asymptote
37. No zeros of denominator; no vertical asymptotes   **39.** $\lim_{x\to 1^-} H(x) = -\infty$; $\lim_{x\to 1^+} H(x) = \infty$; $\lim_{x\to 3} H(x) = 2$; $x = 1$ is a vertical asymptote

41. $\lim_{x\to 0} T(x) = -\infty$; $\lim_{x\to 4} T(x) = \infty$; $x = 0$ and $x = 4$ are vertical asymptotes   **43.** (A) $\dfrac{47}{41} \approx 1.146$ (B) $\dfrac{407}{491} \approx 0.829$ (C) $\dfrac{4}{5} = 0.8$
45. (A) $\dfrac{2{,}011}{138} \approx 14.572$ (B) $\dfrac{12{,}511}{348} \approx 35.951$ (C) ∞ **47.** (A) $-\dfrac{8{,}568}{46{,}653} \approx -0.184$ (B) $-\dfrac{143{,}136}{1{,}492{,}989} \approx -0.096$ (C) 0   **49.** (A) $-\dfrac{7{,}010}{996} \approx -7.038$
(B) $-\dfrac{56{,}010}{7{,}996} \approx -7.005$ (C) -7 **51.** Horizontal asymptote: $y = 2$; vertical asymptote: $x = -2$ **53.** Horizontal asymptote: $y = 1$; vertical asymptotes:
$x = -1$ and $x = 1$ **55.** No horizontal asymptotes; no vertical asymptotes   **57.** Horizontal asymptote: $y = 0$; no vertical asymptotes   **59.** No horizontal
asymptotes; vertical asymptote: $x = 3$ **61.** Horizontal asymptote: $y = 2$; vertical asymptotes: $x = -1$ and $x = 2$ **63.** Horizontal asymptote: $y = 2$; vertical
asymptote: $x = -1$ **65.** $\lim_{x\to\infty} f(x) = 0$ **67.** $\lim_{x\to\infty} f(x) = \infty$ **69.** $\lim_{x\to\infty} f(x) = -\dfrac{1}{4}$ **71.** $\lim_{x\to -\infty} f(x) = -\infty$ **73.** $\lim_{x\to\infty} f(x) = \infty$, $\lim_{x\to -\infty} f(x) = -\infty$
75. $\lim_{x\to\infty} f(x) = -5$, $\lim_{x\to -\infty} f(x) = -5$ **77.** False   **79.** False   **81.** True   **83.** If $n \ge 1$ and $a_n > 0$, then the limit is ∞. If $n \ge 1$ and $a_n < 0$, then the limit is $-\infty$.

85. (A) $C(x) = 180x + 200$ (C) $\overline{C}(x)$ (D) \$180 per board

(B) $\overline{C}(x) = \dfrac{180x + 200}{x}$

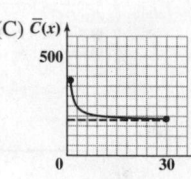

87. (A) $C_e(x) = 950 + 56x$; $\overline{C}_e(x) = \dfrac{950}{x} + 56$ (B) $C_e(x) = 900 + 66x$;

$\overline{C}_e(x) = \dfrac{900}{x} + 66$ (C) At $x = 5$ years (D) At $x = 5$ years

(E) $\lim\limits_{x\to\infty} \overline{C}_e(x) = 56$; $\lim\limits_{x\to\infty} \overline{C}_e(x) = 66$ **89.** The long-term drug concentration

is 5 mg/ml. **91.** (A) \$18 million (B) \$38 million (C) $\lim\limits_{x\to1^-} P(x) = \infty$

93. (A) $V_{max} = 4$, $K_M = 20$ (B) $v(s) = \dfrac{4s}{20+s}$

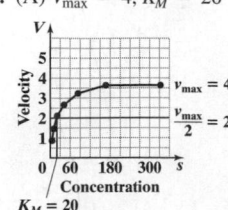

(C) $v = \dfrac{12}{7}$ when $s = 15$; $s = 60$ when $v = 3$

95. (A) $C_{max} = 18$, $M = 150$

(B) $C(T) = \dfrac{18T}{150 + T}$

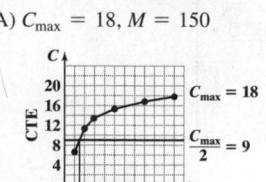

(C) $C = 14.4$ when $T = 600$ K;

$T = 300$ K when $C = 12$

Exercises 2.3

1. $[-3, 5]$ **3.** $(-10, 100)$ **5.** $(-\infty, -5) \cup (5, \infty)$ **7.** $(-\infty, -1] \cup (2, \infty)$

9. f is continuous at $x = 1$, **11.** f is discontinuous at $x = 1$, **13.** f is discontinuous at $x = 1$, **15.** 1.9 **17.** 0.9 **19.** (A) 2 (B) 1

since $\lim_{x\to1} f(x) = f(1)$. since $\lim_{x\to1} f(x) \neq f(1)$. since $\lim_{x\to1} f(x)$ does not exist. (C) Does not exist (D) 1 (E) No

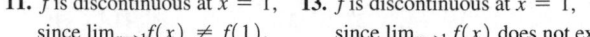
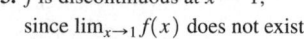
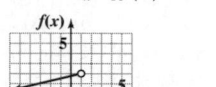

21. (A) 1 (B) 1 (C) 1 (D) 3 (E) No

23. 0.9 **25.** 2.05 **27.** (A) 1 (B) 1 (C) 1 (D) 3

(E) No **29.** (A) 2 (B) -1 (C) Does not exist

(D) 2 (E) No **31.** All x **33.** All x, except $x = -2$

35. All x, except $x = -4$ and $x = 1$ **37.** All x

39. All x, except $x = \pm\dfrac{3}{2}$ **41.** $-\dfrac{8}{3}, 4$ **43.** $-1, 1$ **45.** $-9, -6, 0, 5$ **47.** $-3 < x < 4$; $(-3, 4)$ **49.** $x < 3$ or $x > 7$; $(-\infty, 3) \cup (7, \infty)$

51. $x < -2$ or $0 < x < 2$; $(-\infty, -2) \cup (0, 2)$ **53.** $-5 < x < 0$ or $x > 3$; $(-5, 0) \cup (3, \infty)$ **55.** (A) $(-4, -2) \cup (0, 2) \cup (4, \infty)$

(B) $(-\infty, -4) \cup (-2, 0) \cup (2, 4)$ **57.** (A) $(-\infty, -2.5308) \cup (-0.7198, \infty)$ (B) $(-2.5308, -0.7198)$

59. (A) $(-\infty, -2.1451) \cup (-1, -0.5240) \cup (1, 2.6691)$ (B) $(-2.1451, -1) \cup (-0.5240, 1) \cup (2.6691, \infty)$ **61.** $[6, \infty)$ **63.** $(-\infty, \infty)$

65. $(-\infty, -3] \cup [3, \infty)$ **67.** $(-\infty, \infty)$

69. Since $\lim_{x\to1^-} f(x) = 2$ and **71.** This function is **73.** Since $\lim_{x\to0} f(x) = 0$ and **75.** (A) Yes (B) No (C) Yes

$\lim_{x\to1^+} f(x) = 4$, $\lim_{x\to1} f(x)$ does not continuous for all x. $f(0) = 1$, $\lim_{x\to0} f(x) \neq f(0)$ and f (D) No (E) Yes **77.** True

exist and f is not continuous at $x = 1$. is not continuous at $x = 0$. **79.** False **81.** True

83. x int.: $-5, 2$

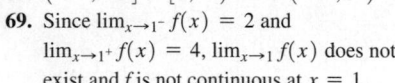

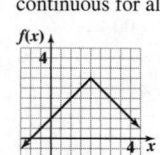

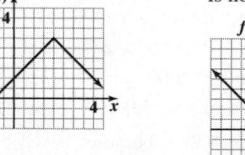

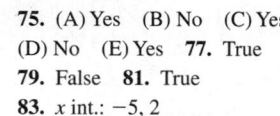

85. x int.: $x = -6, -1, 4$ **87.** No, but this does not contradict Theorem 2, since f is discontinuous at $x = 1$.

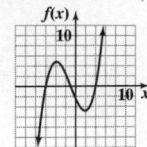

89. (A)
$$P(x) = \begin{cases} 0.44 & \text{if } 0 < x \leq 1 \\ 0.61 & \text{if } 1 < x \leq 2 \\ 0.78 & \text{if } 2 < x \leq 3 \\ 0.95 & \text{if } 3 < x \leq 3.5 \end{cases}$$

(B) $P(x)$ (C) Yes; no

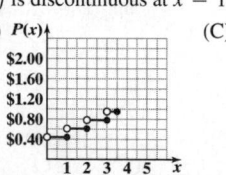

93. (A) $S(x) = \begin{cases} 5 + 0.63x & \text{if } 0 \leq x \leq 50 \\ 14 + 0.45x & \text{if } 50 < x \end{cases}$

(B) $S(x)$ (C) Yes

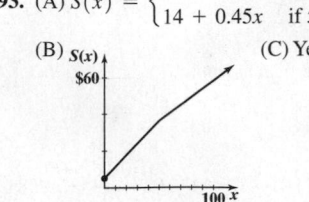

95. (A) $E(s)$

(B) $\lim_{x\to10,000} E(s) = \$1,000$; $E(10,000) = \$1,000$

(C) $\lim_{x\to20,000} E(s)$ does not exist; $E(20,000) = \$2,000$

(D) Yes; no

97. (A) t_2, t_3, t_4, t_6, t_7

(B) $\lim_{t\to t_5} N(t) = 7$; $N(t_5) = 7$

(C) $\lim_{t\to t_3} N(t)$ does not exist;

$N(t_3) = 4$

Exercises 2.4

1. $\dfrac{9}{4} = 2.25$ **3.** $-\dfrac{27}{5} = -5.4$ **5.** $\dfrac{1}{3}\sqrt{3}$ **7.** $\dfrac{15}{2} - \dfrac{5}{2}\sqrt{7}$ **9.** (A) -3; slope of the secant line through $(1, f(1))$ and $(2, f(2))$

(B) $-2 - h$; slope of the secant line through $(1, f(1))$ and $(1 + h, f(1 + h))$ (C) -2; slope of the tangent line at $(1, f(1))$ **11.** (A) 15 (B) $6 + 3h$ (C) 6

13. (A) 40 km/hr (B) 40 (C) $y - 80 = 45(x - 2)$ or $y = 45x - 10$ **15.** $y - \dfrac{1}{2} = -\dfrac{1}{2}(x - 1)$ or $y = -\dfrac{x}{2} + 1$ **17.** $y - 16 = -32(x + 2)$ or

$y = -32x - 48$ **19.** $f'(x) = 0$; $f'(1) = 0$, $f'(2) = 0$, $f'(3) = 0$ **21.** $f'(x) = 3$; $f'(1) = 3$, $f'(2) = 3$, $f'(3) = 3$ **23.** $f'(x) = -6x$; $f'(1) = -6$,

$f'(2) = -12$, $f'(3) = -18$ **25.** $f'(x) = 2x + 6$; $f'(1) = 8$, $f'(2) = 10$, $f'(3) = 12$ **27.** $f'(x) = 4x - 7$; $f'(1) = -3$, $f'(2) = 1$, $f'(3) = 5$

29. $f'(x) = -2x + 4; f'(1) = 2, f'(2) = 0, f'(3) = -2$ **31.** $f'(x) = 6x^2; f'(1) = 6, f'(2) = 24, f'(3) = 54$ **33.** $f'(x) = -\dfrac{4}{x^2}; f'(1) = -4,$

$f'(2) = -1, f'(3) = -\dfrac{4}{9}$ **35.** $f'(x) = \dfrac{3}{2\sqrt{x}}; f'(1) = \dfrac{3}{2}, f'(2) = \dfrac{3}{2\sqrt{2}}$ or $\dfrac{3\sqrt{2}}{4}, f'(3) = \dfrac{3}{2\sqrt{3}}$ or $\dfrac{\sqrt{3}}{2}$ **37.** $f'(x) = \dfrac{5}{\sqrt{x+5}}; f'(1) = \dfrac{5}{\sqrt{6}}$ or $\dfrac{5\sqrt{6}}{6},$

$f'(2) = \dfrac{5}{\sqrt{7}}$ or $\dfrac{5\sqrt{7}}{7}, f'(3) = \dfrac{5}{2\sqrt{2}}$ or $\dfrac{5\sqrt{2}}{4}$ **39.** $f'(x) = -\dfrac{1}{(x-4)^2}; f'(1) = -\dfrac{1}{9}; f'(2) = -\dfrac{1}{4}; f'(3) = -1$ **41.** $f'(x) = \dfrac{1}{(x+1)^2}; f'(1) = \dfrac{1}{4};$

$f'(2) = \dfrac{1}{9}; f'(3) = \dfrac{1}{16}$ **43.** (A) 5   (B) $3 + h$ (C) 3   (D) $y = 3x - 1$ **45.** (A) 5 m/s   (B) $3 + h$ m/s   (C) 3 m/s   **47.** Yes   **49.** No

51. Yes   **55.** (A) $f'(x) = 2x - 4$ **57.** $v = f'(x) = 8x - 2$; 6 ft/s, 22 ft/s, 38 ft/s   **67.** f is nondifferentiable at $x = 1$ **69.** f is differentiable for all

53. Yes   (B) $-4, 0, 4$ **59.** (A) The graphs of g and h are vertical   real numbers

(C) 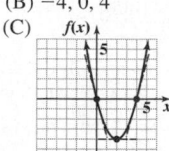   translations of the graph of f. All three functions

should have the same derivative.   (B) $2x$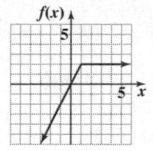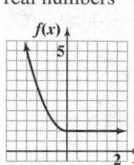

61. True   **63.** False   **65.** False

71. No   **73.** No   **75.** $f'(0) = 0$ **77.** 6 s; 192 ft/s   **79.** (A) $8.75   (B) $R'(x) = 60 - 0.05x$ (C) $R(1,000) = 35,000; R'(1,000) = 10$; At a production

level of 1,000 car seats, the revenue is $35,000 and is increasing at the rate of $10 per seat.   **81.** (A) $S'(t) = 1/\sqrt{t + 10}$ (B) $S(15) = 10; S'(15) = 0.2$;

After 15 months, the total sales are $10 million and are increasing at the rate of $0.2 million, or $200,000, per month.   (C) The estimated total sales are $10.2

million after 16 months and $10.4 million after 17 months.   **83.** (A) $p'(t) = 276t + 1,072$ (B) $p(10) = 39,437, p'(10) = 3,832$; In 2020, 39,437 metric

tons of tungsten are consumed and this quantity is increasing at the rate of 3,832 metric tons per year.

85. (A)
```
QuadReg
y=ax²+bx+c
a=-.1339285714
b=25.225
c=1199.785714
```

(B) $R(20) = 1650.7$ billion kilowatts, $R'(20) = 19.9$ billion kilowatts per year. In 2020, 1650.7 billion kilowatts will be
sold and the amount sold is increasing at the rate of 19.9 billion kilowatts per year.   **87.** (A) $P'(t) = 12 - 2t$
(B) $P(3) = 107; P'(3) = 6$. After 3 hours, the ozone level is 107 ppb and is increasing at the rate of 6 ppb per hour.

Exercises 2.5   **1.** $x^{1/2}$ **3.** x^{-5} **5.** x^{12} **7.** $x^{-1/4}$ **9.** 0   **11.** $9x^8$ **13.** $3x^2$ **15.** $-4x^{-5}$ **17.** $\dfrac{8}{3}x^{5/3}$ **19.** $-\dfrac{10}{x^{11}}$ **21.** $10x$ **23.** $2.8x^6$ **25.** $\dfrac{x^2}{6}$

27. 12   **29.** 2   **31.** 9   **33.** 2   **35.** $4t - 3$ **37.** $-10x^{-3} - 9x^{-2}$ **39.** $1.5u^{-0.7} - 8.8u^{1.2}$ **41.** $0.5 - 3.3t^2$ **43.** $-\dfrac{8}{5}x^{-5}$ **45.** $3x + \dfrac{14}{5}x^{-3}$

47. $-\dfrac{20}{9}w^{-5} + \dfrac{5}{3}w^{-2/3}$ **49.** $2u^{-1/3} - \dfrac{5}{3}u^{-2/3}$ **51.** $-\dfrac{9}{5}t^{-8/5} + 3t^{-3/2}$ **53.** $-\dfrac{1}{3}x^{-4/3}$ **55.** $-0.6x^{-3/2} + 6.4x^{-3} + 1$ **57.** (A) $f'(x) = 6 - 2x$

(B) $f'(2) = 2; f'(4) = -2$ (C) $y = 2x + 4; y = -2x + 16$ (D) $x = 3$ **59.** (A) $f'(x) = 12x^3 - 12x$ (B) $f'(2) = 72; f'(4) = 720$
(C) $y = 72x - 127; y = 720x - 2,215$ (D) $x = -1, 0, 1$ **61.** (A) $v = f'(x) = 176 - 32x$ (B) $f'(0) = 176$ ft/s; $f'(3) = 80$ ft/s   (C) 5.5 s
63. (A) $v = f'(x) = 3x^2 - 18x + 15$ (B) $f'(0) = 15$ ft/s; $f'(3) = -12$ ft/s   (C) $x = 1$ s, $x = 5$ s

65. $f'(x) = 2x - 3 - 2x^{-1/2} = 2x - 3 - \dfrac{2}{x^{1/2}}; x = 2.1777$ **67.** $f'(x) = 4\sqrt[3]{x} - 3x - 3; x = -2.9018$ **69.** $f'(x) = 0.2x^3 + 0.3x^2 - 3x - 1.6;$

$x = -4.4607, -0.5159, 3.4765$ **71.** $f'(x) = 0.8x^3 - 9.36x^2 + 32.5x - 28.25; x = 1.3050$ **77.** $8x - 4$ **79.** $-20x^{-2}$ **81.** $-\dfrac{1}{4}x^{-2} + \dfrac{2}{3}x^{-3}$

83. False   **85.** True   **89.** (A) $S'(t) = 0.09t^2 + t + 2$ (B) $S(5) = 29.25, S'(5) = 9.25$. After 5 months, sales are $29.25 million and are increasing at the
rate of $9.25 million per month.   (C) $S(10) = 103, S'(10) = 21$. After 10 months, sales are $103 million and are increasing at the rate of $21 million
per month.   **91.** (A) $N'(x) = 3,780/x^2$ (B) $N'(10) = 37.8$. At the $10,000 level of advertising, sales are increasing at the rate of 37.8 boats per $1,000 spent
on advertising. $N'(20) = 9.45$. At the $20,000 level of advertising, sales are increasing at the rate of 9.45 boats per $1,000 spent on advertising.

93. (A)
```
CubicReg
y=ax³+bx²+cx+d
a=-8.083333E-4
b=.0624285714
c=-1.081309524
d=40.57571429
```

(B) In 2020, 41.5% of male high-school graduates enroll in   **95.** (A) -1.37 beats/min   (B) -0.58 beats/min
college and the percentage is decreasing at the rate of   **97.** (A) 25 items/hr   (B) 8.33 items/hr
0.9% per year.

Exercises 2.6   **1.** 3; 3.01   **3.** 2.8; 2.799   **5.** 0; 0.01   **7.** 100; 102.01   **9.** $\Delta x = 3; \Delta y = 45; \Delta y/\Delta x = 15$ **11.** 12   **13.** 12   **15.** $dy = (24x - 3x^2)dx$

17. $dy = \left(2x - \dfrac{x^2}{3}\right)dx$ **19.** $dy = -\dfrac{295}{x^{3/2}}dx$ **21.** (A) $12 + 3\Delta x$ (B) 12   **23.** $dy = (8x + 4)dx$ **25.** $dy = (1 - 9x^{-2})dx$ **27.** $dy = 1.4; \Delta y = 1.44$

29. $dy = -3; \Delta y = -\dfrac{10}{3}$ **31.** 120 in.^{3} **33.** (A) $\Delta y = \Delta x + (\Delta x)^2; dy = \Delta x$ (B)   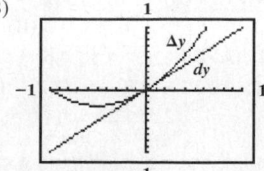   (C)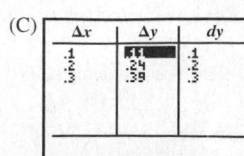

35. (A) $\Delta y = -\Delta x + (\Delta x)^2 + (\Delta x)^3$; $dy = -\Delta x$

(B)

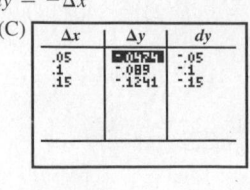

(C)

Δx	Δy	dy
.05	-.0474	-.05
.1	-.089	-.1
.15	-.1241	-.15

37. True **39.** False **41.** $dy = \left(\dfrac{2}{3}x^{-1/3} - \dfrac{10}{3}x^{2/3}\right)dx$ **43.** $dy = 3.9$; $\Delta y = 3.83$

45. 40-unit increase; 20-unit increase **47.** $-\$2.50$; $\$1.25$ **49.** -1.37 beats/min; -0.58 beats/min **51.** $1.26\ \text{mm}^2$ **53.** 3 wpm **55.** (A) 2,100 increase (B) 4,800 increase (C) 2,100 increase

Exercises 2.7 **1.** $\$22,889.80$ **3.** $\$110.20$ **5.** $\$32,000.00$ **7.** $\$230.00$ **9.** $C'(x) = 0.8$ **11.** $C'(x) = 4.6 - 0.02x$ **13.** $R'(x) = 4 - 0.02x$

15. $R'(x) = 12 - 0.08x$ **17.** $P'(x) = 3.2 - 0.02x$ **19.** $P'(x) = 7.4 - 0.06x$ **21.** $\overline{C}(x) = 1.1 + \dfrac{145}{x}$ **23.** $\overline{C}'(x) = -\dfrac{145}{x^2}$

25. $P(x) = 3.9x - 0.02x^2 - 145$ **27.** $\overline{P}(x) = 3.9 - 0.02x - \dfrac{145}{x}$ **29.** True **31.** False **33.** (A) $\$29.50$ (B) $\$30$ **35.** (A) $\$420$ (B) $\overline{C}'(500) = -0.24$. At a production level of 500 frames, average cost is decreasing at the rate of 24¢ per frame. (C) Approximately $\$419.76$ **37.** (A) $\$14.70$ (B) $\$15$

39. (A) $P'(450) = 0.5$. At a production level of 450 DVDs, profit is increasing at the rate of 50¢ per DVD. (B) $P'(750) = -2.5$. At a production level of 750 DVDs, profit is decreasing at the rate of $\$2.50$ per DVD. **41.** (A) $\$13.50$ (B) $\overline{P}'(50) = \$0.27$. At a production level of 50 mowers, the average profit per mower is increasing at the rate of $\$0.27$ per mower. (C) Approximately $\$13.77$ **43.** (A) $p = 100 - 0.025x$, domain: $0 \le x \le 4,000$

(B) $R(x) = 100x - 0.025x^2$, domain: $0 \le x \le 4,000$ (C) $R'(1,600) = 20$. At a production level of 1,600 pairs of running shoes, revenue is increasing at the rate of $\$20$ per pair. (D) $R'(2,500) = -25$. At a production level of 2,500 pairs of running shoes, revenue is decreasing at the rate of $\$25$ per pair.

45. (A) $p = 200 - \dfrac{1}{30}x$, domain: $0 \le x \le 6,000$ (B) $C'(x) = 60$ (C) $R(x) = 200x - (x^2/30)$, domain: $0 \le x \le 6,000$

(D) $R'(x) = 200 - (x/15)$ (E) $R'(1,500) = 100$. At a production level of 1,500 saws, revenue is increasing at the rate of $\$100$ per saw. $R'(4,500) = -100$. At a production level of 4,500 saws, revenue is decreasing at the rate of $\$100$ per saw.

(F) Break-even points: $(600, 108,000)$ and $(3,600, 288,000)$ (G) $P(x) = -(x^2/30) + 140x - 72,000$

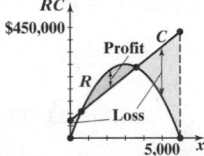

(H) $P'(x) = -(x/15) + 140$

(I) $P'(1,500) = 40$. At a production level of 1,500 saws, profit is increasing at the rate of $\$40$ per saw. $P'(3,000) = -60$. At a production level of 3,000 saws, profit is decreasing at the rate of $\$60$ per saw.

47. (A) $p = 20 - 0.02x$, domain: $0 \le x \le 1,000$

(B) $R(x) = 20x - 0.02x^2$, domain: $0 \le x \le 1,000$ (C) $C(x) = 4x + 1,400$ (D) Break-even points: $(100, 1,800)$ and $(700, 4,200)$

(E) $P(x) = 16x - 0.02x^2 - 1,400$

(F) $P'(250) = 6$. At a production level of 250 toasters, profit is increasing at the rate of $\$6$ per toaster. $P'(475) = -3$. At a production level of 475 toasters, profit is decreasing at the rate of $\$3$ per toaster.

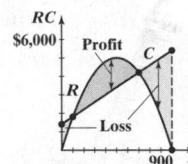

49. (A) $x = 500$

(B) $P(x) = 176x - 0.2x^2 - 21,900$

(C) $x = 440$

(D) Break-even points: $(150, 25,500)$ and $(730, 39,420)$; x intercepts for $P(x)$: $x = 150$ and $x = 730$

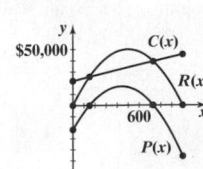

51. (A) $R(x) = 20x - x^{3/2}$

(B) Break-even points: $(44, 588)$, $(258, 1,016)$

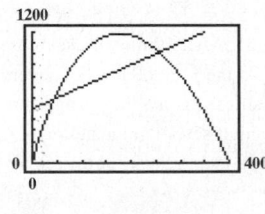

53. (A)

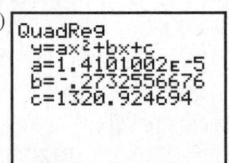

```
QuadReg
y=ax²+bx+c
a=1.4101002ε-5
b=-.2732556676
c=1320.924694
```

(B) Fixed costs $\approx \$721,680$ Variable costs $\approx \$121$

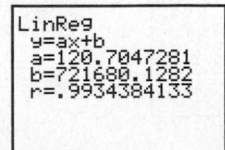

```
LinReg
y=ax+b
a=120.7047281
b=721680.1282
r=.9934384133
```

(C) $(713, 807,703)$, $(5,423, 1,376,227)$

(D) $\$254 \le p \le \$1,133$

Chapter 2 Review Exercises **1.** (A) 16 (B) 8 (C) 8 (D) 4 (E) 4 (F) 4 *(2.2)* **2.** $f'(x) = -3$ *(2.2)* **3.** (A) 22 (B) 8 (C) 2 (D) -5 *(2.1)* **4.** 1.5 *(2.1)* **5.** 3.5 *(2.1)* **6.** 3.75 *(2.1)* **7.** 3.75 *(2.1)* **8.** (A) 1 (B) 1 (C) 1 (D) 1 *(2.1)* **9.** (A) 2 (B) 3 (C) Does not exist (D) 3 *(2.1)* **10.** (A) 4 (B) 4 (C) 4 (D) Does not exist *(2.1)* **11.** (A) Does not exist (B) 3 (C) No *(2.3)* **12.** (A) 2 (B) Not defined (C) No *(2.3)* **13.** (A) 1 (B) 1 (C) Yes *(2.3)* **14.** 5 *(2.2)* **15.** 5 *(2.2)* **16.** ∞ *(2.2)* **17.** $-\infty$ *(2.2)* **18.** 0 *(2.1)* **19.** 0 *(2.1)* **20.** 0 *(2.1)* **21.** Vertical asymptote: $x = 2$ *(2.3)*

22. Horizontal asymptote: $y = 5$ *(2.2)* **23.** $x = 2$ *(2.3)* **24.** $f'(x) = 10x$ *(2.4)* **25.** (A) -3 (B) 6 (C) -2 (D) 3 (E) -11 **26.** $x^2 - 10x$ *(2.5)*

27. $x^{-1/2} - 3 = \dfrac{1}{x^{1/2}} - 3$ *(2.5)* **28.** 0 *(2.5)* **29.** $-\dfrac{3}{2}x^{-2} + \dfrac{15}{4}x^2 = \dfrac{-3}{2x^2} + \dfrac{15x^2}{4}$ *(2.5)* **30.** $-2x^{-5} + x^3 = \dfrac{-2}{x^5} + x^3$ *(2.5)*

31. $f'(x) = 12x^3 + 9x^2 - 2$ *(2.5)* **32.** $\Delta x = 2$, $\Delta y = 10$, $\Delta y/\Delta x = 5$ *(2.6)* **33.** 5 *(2.6)* **34.** 6 *(2.6)* **35.** $\Delta y = 0.64$; $dy = 0.6$ *(2.6)* **36.** (A) 4

(B) 6   (C) Does not exist   (D) 6   (E) No *(2.3)*   **37.** (A) 3   (B) 3   (C) 3   (D) 3   (E) Yes *(2.3)*   **38.** (A) $(8, \infty)$ (B) $[0, 8]$ *(2.3)*   **39.** $(-3, 4)$ *(2.3)*

40. $(-3, 0) \cup (5, \infty)$ *(2.3)*   **41.** $(-2.3429, -0.4707) \cup (1.8136, \infty)$ *(2.3)*   **42.** (A) 3   (B) $2 + 0.5h$ (C) 2 *(2.4)*   **43.** $-x^{-4} + 10x^{-3}$ *(2.4)*

44. $\dfrac{3}{4}x^{-1/2} - \dfrac{5}{6}x^{-3/2} = \dfrac{3}{4\sqrt{x}} - \dfrac{5}{6\sqrt{x^3}}$ *(2.5)*   **45.** $0.6x^{-2/3} - 0.3x^{-4/3} = \dfrac{0.6}{x^{2/3}} - \dfrac{0.3}{x^{4/3}}$ *(2.4)*   **46.** $-\dfrac{3}{5}(-3)x^{-4} = \dfrac{9}{5x^4}$ *(2.5)*   **47.** (A) $m = f'(1) = 2$

(B) $y = 2x + 3$ *(2.4, 2.5)*   **48.** $x = 5$ *(2.4)*   **49.** $x = -5, x = 3$ *(2.5)*   **50.** $x = -1.3401, 0.5771, 2.2630$ *(2.4)*   **51.** ± 2.4824 *(2.5)*

52. (A) $v = f'(x) = 16x - 4$ (B) 44 ft/sec *(2.5)*   **53.** (A) $v = f'(x) = -10x + 16$ (B) $x = 1.6$ sec *(2.5)*

54. (A) The graph of g is the graph of f shifted 4 units to the right, and the graph of h is the graph of f shifted 3 units to the left:

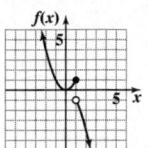

(B) The graph of g' is the graph of f' shifted 4 units to the right, and the graph of h' is the graph of f' shifted 3 units to the left:

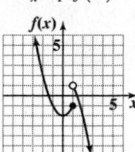

55. $(-\infty, \infty)$ *(2.3)*

56. $(-\infty, 2) \cup (2, \infty)$ *(2.3)*

57. $(-\infty, -4) \cup (-4, 1) \cup (1, \infty)$ *(2.3)*

58. $(-\infty, \infty)$ *(2.3)*   **59.** $[-2, 2]$ *(2.3)*

60. (A) -1 (B) Does not exist   (C) $-\dfrac{2}{3}$ *(2.1)*

61. (A) $\dfrac{1}{2}$ (B) 0   (C) Does not exist *(2.1)*   **62.** (A) -1 (B) 1   (C) Does not exist *(2.1)*   **63.** (A) $-\dfrac{1}{6}$ (B) Does not exist   (C) $-\dfrac{1}{3}$ *(2.1)*   **64.** (A) 0   (B) -1

(C) Does not exist *(2.1)*   **65.** (A) $\dfrac{2}{3}$ (B) $\dfrac{2}{3}$ (C) Does not exist *(2.3)*   **66.** (A) ∞ (B) $-\infty$ (C) ∞ *(2.3)*   **67.** (A) 0   (B) 0   (C) Does not exist *(2.2)*

68. 4 *(2.1)*   **69.** $\dfrac{-1}{(x+2)^2}$ *(2.1)*   **70.** $2x - 1$ *(2.4)*   **71.** $1/(2\sqrt{x})$ *(2.4)*   **72.** Yes *(2.4)*   **73.** No *(2.4)*   **74.** No *(2.4)*   **75.** No *(2.4)*   **76.** Yes *(2.4)*

77. Yes *(2.4)*   **78.** Horizontal asymptote: $y = 5$; vertical asymptote: $x = 7$ *(2.2)*   **79.** Horizontal asymptote: $y = 0$; vertical asymptote: $x = 4$ *(2.2)*

80. No horizontal asymptote; vertical asymptote: $x = 3$ *(2.2)*   **81.** Horizontal asymptote: $y = 1$; vertical asymptotes: $x = -2, x = 1$ *(2.2)*   **82.** Horizontal asymptote: $y = 1$; vertical asymptotes: $x = -1, x = 1$ *(2.2)*   **83.** The domain of $f'(x)$ is all real numbers except $x = 0$. At $x = 0$, the graph of $y = f(x)$ is smooth, but it has a vertical tangent. *(2.4)*

84. (A) $\lim_{x\to 1^-} f(x) = 1$;   (B) $\lim_{x\to 1^-} f(x) = -1$;   (C) $m = 1$

$\lim_{x\to 1^+} f(x) = -1$ $\lim_{x\to 1^+} f(x) = 1$

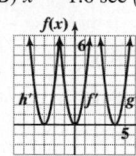

(D) The graphs in (A) and (B) have discontinuities at $x = 1$; the graph in (C) does not. *(2.2)*

85. (A) 1   (B) -1 (C) Does not exist
(D) No *(2.4)*

86. (A) $S(x) = \begin{cases} 7.47 \quad + 0.4x & \text{if } 0 \le x \le 90 \\ 24.786 + 0.2076x & \text{if } 90 < x \end{cases}$

(B) 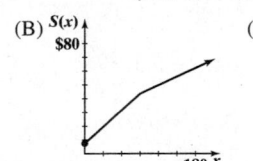   (C) Yes *(2.2)*

87. (A) \$179.90   (B) \$180 *(2.7)*   **88.** (A) $C(100) = 9{,}500$; $C'(100) = 50$. At a production level of 100 bicycles, the total cost is \$9,500, and cost is increasing at the rate of \$50 per bicycle.
(B) $\overline{C}(100) = 95$; $\overline{C}'(100) = -0.45$. At a production level of 100 bicycles, the average cost is \$95, and average cost is decreasing at a rate of \$0.45 per bicycle. *(2.7)*   **89.** The approximate cost of producing the 201st printer is greater than that of the 601st printer. Since these marginal costs are decreasing, the manufacturing process is becoming more efficient. *(2.7)*

90. (A) $C'(x) = 2$; $\overline{C}(x) = 2 + \dfrac{9{,}000}{x}$; $\overline{C}'(x) = \dfrac{-9{,}000}{x^2}$ (B) $R(x) = xp = 25x - 0.01x^2$; $R'(x) = 25 - 0.02x$; $\overline{R}(x) = 25 - 0.01x$; $\overline{R}'(x) = -0.01$

(C) $P(x) = R(x) - C(x) = 23x - 0.01x^2 - 9{,}000$; $P'(x) = 23 - 0.02x$; $\overline{P}(x) = 23 - 0.01x - \dfrac{9{,}000}{x}$; $\overline{P}'(x) = -0.01 + \dfrac{9{,}000}{x^2}$

(D) $(500, 10{,}000)$ and $(1{,}800, 12{,}600)$ (E) $P'(1{,}000) = 3$. Profit is increasing at the rate of \$3 per umbrella.
$P'(1{,}150) = 0$. Profit is flat.
$P'(1{,}400) = -5$. Profit is decreasing at the rate of \$5 per umbrella.

(F)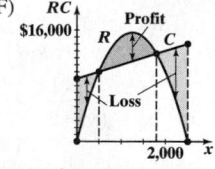

91. (A) 8   (B) 20 *(2.5)*
92. $N(9) = 27$; $N'(t) = 3.5$; After 9 months, 27,000 pools have been sold and the total sales are increasing at the rate of 3,500 pools per month. *(2.5)*

93. (A)
```
CubicReg
y=ax³+bx²+cx+d
a=5.5277778E-4
b=-.0444761905
c=1.084484127
d=12.5452381
```

94. (A)
```
LinReg
y=ax+b
a=-.0384180791
b=13.59887006
r=-.9897782666
```

(B) Fixed costs: \$484.21; variable costs per kringle: \$2.11

```
LinReg
y=ax+b
a=2.107344633
b=484.2090395
r=.9939318704
```

(C) $(51, 591.15), (248, 1{,}007.62)$
(D) $\$4.07 < p < \$11.64$ *(2.7)*

(B) $N(60) = 36.9$; $N'(60) = 1.7$.
In 2020, natural-gas consumption is 36.9 trillion cubic feet and is increasing at the rate of 1.7 trillion cubic feet per year *(2.4)*

95. $C'(10) = -1$; $C'(100) = -0.001$ *(2.5)*
96. $F(4) = 98.16$; $F'(4) = -0.32$; After 4 hours the patient's temperature is 98.16°F and is decreasing at the rate of 0.32°F per hour. *(2.5)*
97. (A) 10 items/h   (B) 5 items/h *(2.5)*

98. (A)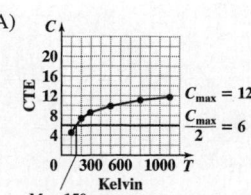

$M = 150$

(B) $C(T) = \dfrac{12T}{150 + T}$

(C) $C = 9.6$ at $T = 600$ K,
$T = 750$ K when $C = 10$ *(2.3)*

Chapter 3

Exercises 3.1   **1.** $A = 1,465.68$ **3.** $P = 9,117.21$ **5.** $t = 5.61$ **7.** $r = 0.04$ **9.** \$1,221.40; \$1,648.72; \$2,225.54

11. 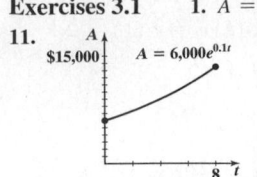   **13.** 11.55   **19.**

15. 10.99

17. 0.14

n	$[1 + (1/n)]^n$
10	2.593 74
100	2.704 81
1,000	2.716 92
10,000	2.718 15
100,000	2.718 27
1,000,000	2.718 28
10,000,000	2.718 28
↓	↓
∞	$e = 2.718\,281\,828\,459\ldots$

21. $\lim_{n \to \infty} (1 + n)^{1/n} = 1$ **23.**

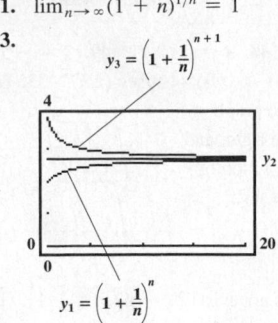

25. (A) \$12,398.62   (B) 27.34 yr   **31.** (A)   (B) $\lim_{t \to \infty} 10,000e^{-0.08t} = 0$ **33.** 17.33 yr   **35.** 8.66%   **37.** 7.3 yr
27. \$11,890.41   **29.** 8.11%

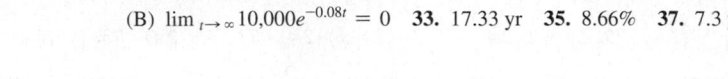

39. (A) $A = Pe^{rt}$ (B)   Although r could be any positive number, the restrictions on   **41.** $t = -(\ln 0.5)/0.000\,433\,2 \approx 1,600$ yr
$2P = Pe^{rt}$ r are reasonable in the sense that most investments would be   **43.** $r = (\ln 0.5)/30 \approx -0.0231$
$2 = e^{rt}$ expected to earn a return of between 2% and 30%.   **45.** 53.3 yr
$rt = \ln 2$ (C) The doubling times (in years) are 13.86, 6.93, 4.62,   **47.** 1.39%
$t = \dfrac{\ln 2}{r}$ 3.47, 2.77, and 2.31, respectively.

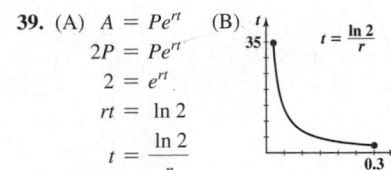

Exercises 3.2   **1.** $y = 7$ **3.** $x = 81$ **5.** $b = 8$ **7.** $y = \dfrac{1}{2}$ **9.** $5e^x + 3$ **11.** $-\dfrac{2}{x} + 2x$ **13.** $3x^2 - 6e^x$ **15.** $e^x + 1 - \dfrac{1}{x}$ **17.** $\dfrac{3}{x}$ **19.** $5 - \dfrac{5}{x}$

21. $\dfrac{2}{x} + 4e^x$ **23.** $f'(x) = e^x + ex^{e-1}$ **25.** $f'(x) = (e + 1)x^e$ **27.** $f'(x) = \dfrac{1}{x}; y = x + 2$ **29.** $f'(x) = 3e^x; y = 3x + 3$ **31.** $f'(x) = \dfrac{3}{x}; y = \dfrac{3x}{e}$

33. $f'(x) = e^x; y = ex + 2$ **35.** Yes; yes   **37.** No; no   **39.** $f(x) = 10x + \ln 10 + \ln x;\ f'(x) = 10 + \dfrac{1}{x}$ **41.** $f(x) = \ln 4 - 3 \ln x; f'(x) = -\dfrac{3}{x}$

43. $\dfrac{1}{x \ln 2}$ **45.** $3^x \ln 3$ **47.** $2 - \dfrac{1}{x \ln 10}$ **49.** $1 + 10^x \ln 10$ **51.** $\dfrac{3}{x} + \dfrac{2}{x \ln 3}$ **53.** $2^x \ln 2$ **55.** $(-0.82, 0.44), (1.43, 4.18), (8.61, 5503.66)$

57. $(0.49, 0.49)$ **59.** $(3.65, 1.30), (332,105.11, 12.71)$ **63.** \$28,447/yr; \$18,664/yr; \$11,021/yr   **65.** $A'(t) = 5,000(\ln 4)4^t$; $A'(1) = 27,726$ bacteria/hr (rate of change at the end of the first hour); $A'(5) = 7,097,827$ bacteria/hr (rate of change at the end of the fifth hour)   **67.** At the 40-lb weight level, blood pressure would increase at the rate of 0.44 mm of mercury per pound of weight gain. At the 90-lb weight level, blood pressure would increase at the rate of 0.19 mm of mercury per pound of weight gain.   **69.** $dR/dS = k/S$ **71.** (A) \$808.41 per year   (B) \$937.50 per year

Exercises 3.3   **1.** One answer is: $F(x) = x^3, S(x) = 5 - 4 \ln x$ **3.** One answer is: $F(x) = x^3 + 3, S(x) = e^x + 2$ **5.** One answer is:
$T(x) = 9x^2, B(x) = e^{5x}$ **7.** One answer is: $T(x) = 3x^2 + e^x, B(x) = x^4$ **9.** $2x^3(2x) + (x^2 - 2)(6x^2) = 10x^4 - 12x^2$

11. $(x - 3)(2) + (2x - 1)(1) = 4x - 7$ **13.** $\dfrac{(x - 3)(1) - x(1)}{(x - 3)^2} = \dfrac{-3}{(x - 3)^2}$ **15.** $\dfrac{(x - 2)(2) - (2x + 3)(1)}{(x - 2)^2} = \dfrac{-7}{(x - 2)^2}$

17. $3xe^x + 3e^x = 3(x + 1)e^x$ **19.** $x^3\left(\dfrac{1}{x}\right) + 3x^2 \ln x = x^2(1 + 3 \ln x)$ **21.** $(x^2 + 1)(2) + (2x - 3)(2x) = 6x^2 - 6x + 2$

23. $(0.4x + 2)(0.5) + (0.5x - 5)(0.4) = 0.4x - 1$ **25.** $\dfrac{(2x - 3)(2x) - (x^2 + 1)(2)}{(2x - 3)^2} = \dfrac{2x^2 - 6x - 2}{(2x - 3)^2}$ **27.** $(x^2 + 2)2x + (x^2 - 3)2x = 4x^3 - 2x$

29. $\dfrac{(x^2 - 3)2x - (x^2 + 2)2x}{(x^2 - 3)^2} = \dfrac{-10x}{(x^2 - 3)^2}$ **31.** $\dfrac{(x^2 + 1)e^x - e^x(2x)}{(x^2 + 1)^2} = \dfrac{(x - 1)^2 e^x}{(x^2 + 1)^2}$ **33.** $\dfrac{(1 + x)\left(\dfrac{1}{x}\right) - \ln x}{(1 + x)^2} = \dfrac{1 + x - x \ln x}{x(1 + x)^2}$ **35.** $xf'(x) + f(x)$

37. $x^3 f'(x) + 3x^2 f(x)$ **39.** $\dfrac{x^2 f'(x) - 2xf(x)}{x^4}$ **41.** $\dfrac{f(x) - xf'(x)}{[f(x)]^2}$ **43.** $e^x f'(x) + f(x)e^x = e^x[f'(x) + f(x)]$

45. $\dfrac{f(x)\left(\dfrac{1}{x}\right) - (\ln x)f'(x)}{f(x)^2} = \dfrac{f(x) - (x \ln x)f'(x)}{xf(x)^2}$ **47.** $(2x + 1)(2x - 3) + (x^2 - 3x)(2) = 6x^2 - 10x - 3$

49. $(2.5t - t^2)(4) + (4t + 1.4)(2.5 - 2t) = -12t^2 + 17.2t + 3.5$ **51.** $\dfrac{(x^2 + 2x)(5) - (5x - 3)(2x + 2)}{(x^2 + 2x)^2} = \dfrac{-5x^2 + 6x + 6}{(x^2 + 2x)^2}$

53. $\dfrac{(w^2 - 1)(2w - 3) - (w^2 - 3w + 1)(2w)}{(w^2 - 1)^2} = \dfrac{3w^2 - 4w + 3}{(w^2 - 1)^2}$ **55.** $(1 + x - x^2)e^x + e^x(1 - 2x) = (2 - x - x^2)e^x$

57. (A) $f'(x) = \dfrac{x \cdot 0 - 1 \cdot 1}{x^2} = -\dfrac{1}{x^2}$ (B) Note that $f(x) = x^{-1}$ and use the power rule: $f'(x) = -x^{-2} = -\dfrac{1}{x^2}$ **59.** (A) $f'(x) = \dfrac{x^4 \cdot 0 - (-3) \cdot 4x^3}{x^8} = \dfrac{12}{x^5}$

(B) Note that $f(x) = -3x^{-4}$ and use the power rule: $f'(x) = 12x^{-5} = \dfrac{12}{x^5}$ **61.** $f'(x) = (1 + 3x)(-2) + (5 - 2x)(3); y = -11x + 29$

63. $f'(x) = \dfrac{(3x - 4)(1) - (x - 8)(3)}{(3x - 4)^2}; y = 5x - 13$ **65.** $f'(x) = \dfrac{2^x - x(2^x \ln 2)}{2^{2x}}; y = \left(\dfrac{1 - 2 \ln 2}{4}\right) x + \ln 2$

67. $f'(x) = (2x - 15)(2x) + (x^2 + 18)(2) = 6(x - 2)(x - 3); x = 2, x = 3$ **69.** $f'(x) = \dfrac{(x^2 + 1)(1) - x(2x)}{(x^2 + 1)^2} = \dfrac{1 - x^2}{(x^2 + 1)^2}; x = -1, x = 1$

71. $7x^6 - 3x^2$ **73.** $-27x^{-4} = -\dfrac{27}{x^4}$ **75.** $(w + 1)2^w \ln 2 + 2^w = [(w + 1) \ln 2 + 1]2^w$ **77.** $9x^{1/3}(3x^2) + (x^3 + 5)(3x^{-2/3}) = \dfrac{30x^3 + 15}{x^{2/3}}$

79. $\dfrac{(1 + x^2)\dfrac{1}{x \ln 2} - 2x \log_2 x}{(1 + x^2)^2} = \dfrac{1 + x^2 - 2x^2 \ln x}{x(1 + x^2)^2 \ln 2}$ **81.** $\dfrac{(x^2 - 3)(2x^{-2/3}) - 6x^{1/3}(2x)}{(x^2 - 3)^2} = \dfrac{-10x^2 - 6}{(x^2 - 3)^2 x^{2/3}}$

83. $g'(t) = \dfrac{(3t^2 - 1)(0.2) - (0.2t)(6t)}{(3t^2 - 1)^2} = \dfrac{-0.6t^2 - 0.2}{(3t^2 - 1)^2}$ **85.** $(20x)\dfrac{1}{x \ln 10} + 20 \log x = \dfrac{20(1 + \ln x)}{\ln 10}$

87. $x^{-2/3}(3x^2 - 4x) + (x^3 - 2x^2)(-\tfrac{2}{3}x^{-5/3}) = -\tfrac{8}{3}x^{1/3} + \tfrac{7}{3}x^{4/3}$

89. $\dfrac{(x^2 + 1)[(2x^2 - 1)(2x) + (x^2 + 3)(4x)] - (2x^2 - 1)(x^2 + 3)(2x)}{(x^2 + 1)^2} = \dfrac{4x^5 + 8x^3 + 16x}{(x^2 + 1)^2}$ **91.** $\dfrac{e^t(1 + \ln t) - (t \ln t)e^t}{e^{2t}} = \dfrac{1 + \ln t - t \ln t}{e^t}$

93. (A) $S'(t) = \dfrac{(t^2 + 50)(180t) - 90t^2(2t)}{(t^2 + 50)^2} = \dfrac{9,000t}{(t^2 + 50)^2}$ (B) $S(10) = 60; S'(10) = 4$. After 10 months, the total sales are 60,000 DVDs and sales are

increasing at the rate of 4,000 DVDs per month. (C) Approximately 64,000 DVDs

95. (A) $\dfrac{dx}{dp} = \dfrac{(0.1p + 1)(0) - 4,000(0.1)}{(0.1p + 1)^2} = \dfrac{-400}{(0.1p + 1)^2}$ (B) $x = 800; dx/dp = -16$. At a price level of \$40, the demand is 800 DVD players per week

and demand is decreasing at the rate of 16 players per dollar. (C) Approximately 784 DVD players

97. (A) $C'(t) = \dfrac{(t^2 + 1)(0.14) - 0.14t(2t)}{(t^2 + 1)^2} = \dfrac{0.14 - 0.14t^2}{(t^2 + 1)^2}$ (B) $C'(0.5) = 0.0672$. After 0.5 hr, concentration is increasing at the rate of

0.0672 mg/cm^3 per hour. $C'(3) = -0.0112$. After 3 hr, concentration is decreasing at the rate of 0.0112 mg/cm^3 per hour.

Exercises 3.4 **1.** $3x^3 + 5$ **3.** $2x^2e^x + \ln x^2e^x$ **5.** $E(u) = \ln u, I(x) = x^3 - 6x + 10$ is one answer. **7.** $E(u) = \sqrt{u}, I(x) = x^2 + 4$ is one answer.

9. 3 **11.** $(-4x)$ **13.** $2x$ **15.** $4x^3$ **17.** $-8(5 - 2x)^3$ **19.** $5(4 + 0.2x)^4(0.2) = (4 + 0.2x)^4$ **21.** $30x(3x^2 + 5)^4$ **23.** $5e^x$ **25.** $5e^{5x}$ **27.** $-18e^{-6x}$

29. $(2x - 5)^{-1/2} = \dfrac{1}{(2x - 5)^{1/2}}$ **31.** $-8x^3(x^4 + 1)^{-3} = \dfrac{-8x^3}{(x^4 + 1)^3}$ **33.** $-\dfrac{2}{x}$ **35.** $\dfrac{6x}{1 + x^2}$ **37.** $\dfrac{3(1 + \ln x)^2}{x}$ **39.** $f'(x) = 6(2x - 1)^2; y = 6x - 5; x = \tfrac{1}{2}$

41. $f'(x) = 2(4x - 3)^{-1/2} = \dfrac{2}{(4x - 3)^{1/2}}; y = \dfrac{2}{3}x + 1$; none **43.** $f'(x) = 10(x - 2)e^{x^2 - 4x + 1}; y = -20ex + 5e; x = 2$

45. $12(x^2 - 2)^3(2x) = 24x(x^2 - 2)^3$ **47.** $-6(t^2 + 3t)^{-4}(2t + 3) = \dfrac{-6(2t + 3)}{(t^2 + 3t)^4}$ **49.** $\dfrac{1}{2}(w^2 + 8)^{-1/2}(2w) = \dfrac{w}{\sqrt{w^2 + 8}}$

51. $12xe^{3x} + 4e^{3x} = 4(3x + 1)e^{3x}$ **53.** $\dfrac{x^3\left(\dfrac{1}{1 + x}\right) - 3x^2 \ln (1 + x)}{x^6} = \dfrac{x - 3(1 + x) \ln (1 + x)}{x^4(1 + x)}$ **55.** $6te^{3(t^2 + 1)}$ **57.** $\dfrac{3x}{x^2 + 3}$

59. $-5(w^3 + 4)^{-6}(3w^2) = \dfrac{-15w^2}{(w^3 + 4)^6}$ **61.** $f'(x) = (4 - x)^3 - 3x(4 - x)^2 = 4(4 - x)^2(1 - x); y = -16x + 48$

63. $f'(x) = \dfrac{(2x - 5)^3 - 6x(2x - 5)^2}{(2x - 5)^6} = \dfrac{-4x - 5}{(2x - 5)^4}; y = -17x + 54$ **65.** $f'(x) = \dfrac{1}{2x\sqrt{\ln x}}; y = \dfrac{x}{2e} + \dfrac{1}{2}$

67. $f'(x) = 2x(x - 5)^3 + 3x^2(x - 5)^2 = 5x(x - 5)^2(x - 2); x = 0, 2, 5$ **69.** $f'(x) = \dfrac{(2x + 5)^2 - 4x(2x + 5)}{(2x + 5)^4} = \dfrac{5 - 2x}{(2x + 5)^3}; x = \tfrac{5}{2}$

71. $f'(x) = (x^2 - 8x + 20)^{-1/2}(x - 4) = \dfrac{x - 4}{(x^2 - 8x + 20)^{1/2}}; x = 4$ **73.** No; yes **75.** Domain of f: $(0, \infty)$; domain of g: $(-\infty, \infty)$; domain of m: $(-2, 2)$

77. Domain of f: all real numbers except ± 1; domain of g: $(0, \infty)$; domain of m: all positive real numbers except e and $\tfrac{1}{e}$

79. $18x^2(x^2 + 1)^2 + 3(x^2 + 1)^3 = 3(x^2 + 1)^2(7x^2 + 1)$ **81.** $\dfrac{24x^5(x^3 - 7)^3 - (x^3 - 7)^4 6x^2}{4x^6} = \dfrac{3(x^3 - 7)^3(3x^3 + 7)}{2x^4}$ **83.** $\dfrac{1}{\ln 2}\left(\dfrac{6x}{3x^2 - 1}\right)$

85. $(2x + 1)(10^{x^2 + x})(\ln 10)$ **87.** $\dfrac{12x^2 + 5}{(4x^3 + 5x + 7) \ln 3}$ **89.** $2^{x^3 - x^2 + 4x + 1}(3x^2 - 2x + 4) \ln 2$ **91.** (A) $C'(x) = (2x + 16)^{-1/2} = \dfrac{1}{(2x + 16)^{1/2}}$

(B) $C'(24) = \tfrac{1}{8}$, or \$12.50. At a production level of 24 cell phones, total cost is increasing at the rate of \$12.50 per cell phone and the cost of producing the 25th

cell phone is approximately \$12.50. $C'(42) = \tfrac{1}{10}$, or \$10.00. At a production level of 42 cell phones, total cost is increasing at the rate of \$10.00 per cell phone

and the cost of producing the 43rd cell phone is approximately \$10.00. **93.** (A) $\dfrac{dx}{dp} = 40(p + 25)^{-1/2} = \dfrac{40}{(p + 25)^{1/2}}$ (B) $x = 400$ and $dx/dp = 4$.

At a price of \$75, the supply is 400 bicycle helmets per week and supply is increasing at

the rate of 4 bicycle helmets per dollar.

95. (A) After 1 hr, the concentration is decreasing at the rate of 1.60 mg/mL per hour; after 4 hr, the concentration is decreasing at the rate of 0.08 mg/mL per hour.   (B) $C(t)$ **97.** 2.27 mm of mercury/yr; 0.81 mm of mercury/yr; 0.41 mm of mercury/yr

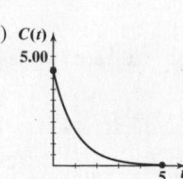

Exercises 3.5   **1.** $y = -\dfrac{3}{2}x + 10$ **3.** $y = \pm\dfrac{4}{3}\sqrt{9 - x^2}$ **5.** $y = \dfrac{-x \pm \sqrt{4 - 3x^2}}{2}$ **7.** Impossible   **9.** $y' = -\dfrac{3}{5}$ **11.** $y' = \dfrac{3x}{2}$ **13.** $y' = 10x; 10$

15. $y' = \dfrac{2x}{3y^2}; \dfrac{4}{3}$ **17.** $y' = -\dfrac{3}{2y + 2}; -\dfrac{3}{4}$ **19.** $y' = -\dfrac{y}{x}; -\dfrac{3}{2}$ **21.** $y' = -\dfrac{2y}{2x + 1}; 4$ **23.** $y' = \dfrac{6 - 2y}{x}; -1$ **25.** $y' = \dfrac{2x}{e^y - 2y}; 2$ **27.** $y' = \dfrac{3x^2y}{y + 1}; \dfrac{3}{2}$

29. $y' = \dfrac{6x^2y - y \ln y}{x + 2y}; 2$ **31.** $x' = \dfrac{2tx - 3t^2}{2x - t^2}; 8$ **33.** $y'|_{(1.6,1.8)} = -\dfrac{3}{4}; y'|_{(1.6,0.2)} = \dfrac{3}{4}$ **35.** $y = -x + 5$ **37.** $y = \dfrac{2}{5}x - \dfrac{12}{5}; y = \dfrac{3}{5}x + \dfrac{12}{5}$

39. $y' = -\dfrac{1}{x}$ **41.** $y' = \dfrac{1}{3(1 + y)^2 + 1}; \dfrac{1}{13}$ **43.** $y' = \dfrac{3(x - 2y)^2}{6(x - 2y)^2 + 4y}; \dfrac{3}{10}$ **45.** $y' = \dfrac{3x^2(7 + y^2)^{1/2}}{y}; 16$ **47.** $y' = \dfrac{y}{2xy^2 - x}; 1$

49. $y = 0.63x + 1.04$ **51.** $p' = -\dfrac{1}{2p - 2}$ **53.** $p' = -\dfrac{x}{p} = -\dfrac{\sqrt{10,000 - p^2}}{p}$ **55.** $\dfrac{dL}{dV} = \dfrac{-(L + m)}{V + n}$ **57.** $\dfrac{dT}{dv} = \dfrac{2}{k}\sqrt{T}$ **59.** $\dfrac{dv}{dT} = \dfrac{k}{2\sqrt{T}}$

Exercises 3.6   **1.** 19.5 ft   **3.** 46 m   **5.** 34.5 ft   **7.** 32 ft   **9.** 30   **11.** $-\dfrac{16}{3}$ **13.** $-\dfrac{16}{7}$ **15.** Decreasing at 9 units/sec   **17.** Approx. -3.03 ft/sec
19. $dA/dt \approx 126$ ft²/sec   **21.** 3,770 cm³/min   **23.** 6 lb/in.²/hr   **25.** $\dfrac{9}{4}$ ft/sec   **27.** $\dfrac{20}{3}$ ft/sec   **29.** 0.0214 ft/sec; 0.0135 ft/sec; yes, at $t = 0.000\ 19$ sec
31. 3.835 units/sec   **33.** (A) $dC/dt = \$15,000/$wk   (B) $dR/dt = -\$50,000/$wk   (C) $dP/dt = -\$65,000/$wk   **35.** $ds/dt = \$2,207/$wk
37. (A) $dx/dt = -12.73$ units/month   (B) $dp/dt = \$1.53/$month   **39.** Approximately 100 ft³/min

Exercises 3.7   **1.** $x = f(p) = 105 - 2.5p, 0 \le p \le 42$ **3.** $x = f(p) = \sqrt{100 - 2p}, 0 \le p \le 50$

5. $x = f(p) = 20(\ln 25 - \ln p), 25/e \approx 9.2 \le p \le 25$ **7.** $x = f(p) = e^{8 - 0.1p}, 80 - 10 \ln 30 \approx 46.0 \le p \le 80$ **9.** $\dfrac{35 - 0.8x}{35x - 0.4x^2}$ **11.** $-\dfrac{4e^{-x}}{7 + 4e^{-x}}$
13. $\dfrac{5}{x(12 + 5 \ln x)}$ **15.** 0   **17.** -0.017 **19.** -0.034 **21.** 1.013   **23.** 0.405   **25.** 11.8%   **27.** 5.4%   **29.** -14.7% **31.** -431.6%

33. $E(p) = \dfrac{450p}{25,000 - 450p}$ **35.** $E(p) = \dfrac{8p^2}{4,800 - 4p^2}$ **37.** $E(p) = \dfrac{0.6pe^p}{98 - 0.6e^p}$ **39.** 0.07   **41.** 0.15   **43.** $\dfrac{x + 1}{x}$ **45.** $\dfrac{1}{x \ln x}$ **47.** (A) Inelastic

(B) Unit elasticity   (C) Elastic   **49.** (A) Inelastic   (B) Unit elasticity   (C) Elastic   **51.** (A) $x = 6,000 - 200p\ \ 0 \le p \le 30$ (B) $E(p) = \dfrac{p}{30 - p}$

(C) $E(10) = 0.5$; 5% decrease   (D) $E(25) = 5$; 50% decrease   (E) $E(15) = 1$; 10% decrease   **53.** (A) $x = 3,000 - 50p\ \ 0 \le p \le 60$

(B) $R(p) = 3,000p - 50p^2$ (C) $E(p) = \dfrac{p}{60 - p}$ (D) Elastic on $(30, 60)$; inelastic on $(0, 30)$ (E) Increasing on $(0, 30)$; decreasing on $(30, 60)$

(F) Decrease   (G) Increase
55. Elastic on $(3.5, 7)$; inelastic on $(0, 3.5)$
57. Elastic on $(25/\sqrt{3}, 25)$; inelastic on $(0, 25/\sqrt{3})$
59. Elastic on $(48, 72)$; inelastic on $(0, 48)$
61. Elastic on $(25, 25\sqrt{2})$; inelastic on $(0, 25)$

63. $R(p) = 20p(10 - p)$ **65.** $R(p) = 40p(p - 15)^2$ **67.** $R(p) = 30p - 10p\sqrt{p}$

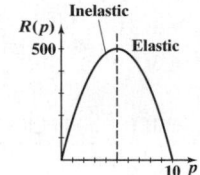
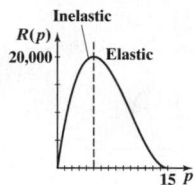
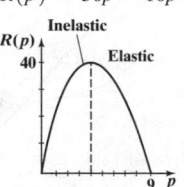

69. $\dfrac{3}{2}$ **71.** $\dfrac{1}{2}$ **73.** Elastic on $(0, 300)$; inelastic on $(300, 600)$ **75.** Elastic on $(0, 10\sqrt{3})$; inelastic on $(10\sqrt{3}, 30)$ **77.** k **79.** \$75 per day

81. Increase   **83.** Decrease   **85.** \$3.75   **87.** $p(t) = \dfrac{31}{0.31t + 18.5}$ **89.** -0.025

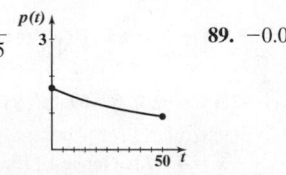

Chapter 3 Review Exercises   **1.** \$3,136.62; \$4,919.21; \$12,099.29 *(3.1)*   **2.** $E(u) = u^{3/2}, I(x) = 6x + 5$ is one answer. *(3.4)*

3. $E(u) = \ln u, I(x) = x^2 + 4$ is one answer. *(3.4)*   **4.** $E(u) = e^u, I(x) = 0.02x$ is one answer. *(3.4)*   **5.** $\dfrac{2}{x} + 3e^x$ *(3.2)*   **6.** $2e^{2x - 3}$ *(3.4)*   **7.** $\dfrac{2}{2x + 7}$ *(3.4)*

8. $\dfrac{e^x}{3 + e^x}$ *(3.4)*   **9.** $y' = \dfrac{9x^2}{4y}; \dfrac{9}{8}$ *(3.5)*   **10.** $dy/dt = 216$ *(3.6)*   **11.** (A) $x = 1,000 - 25p$ (B) $\dfrac{p}{40 - p}$ (C) 0.6; demand is inelastic and insensitive to

small changes in price.   (D) $1,000p - 25p^2$ (E) Revenue increases *(3.7)*   **12.** -10 *(3.2)*   **13.** $\lim\limits_{n \to \infty}\left(1 + \dfrac{2}{n}\right)^n = e^2 \approx 7.389\ 06$ *(3.1)*

14. $\dfrac{7[(\ln z)^6 + 1]}{z}$ *(3.4)*   **15.** $x^5(1 + 6 \ln x)$ *(3.3)*   **16.** $\dfrac{e^x(x - 6)}{x^7}$ *(3.3)*   **17.** $\dfrac{6x^2 - 3}{2x^3 - 3x}$ *(3.4)*   **18.** $(3x^2 - 2x)e^{x^3 - x^2}$ *(3.4)*   **19.** $\dfrac{1 - 2x \ln 5x}{xe^{2x}}$ *(3.4)*

20. $y = -x + 2; y = -ex + 1$ *(3.4)*   **21.** $y' = \dfrac{3y - 2x}{8y - 3x}; \dfrac{8}{19}$ *(3.5)*   **22.** $x' = \dfrac{4tx}{3x^2 - 2t^2}; -4$ *(3.5)*   **23.** $y' = \dfrac{1}{e^y + 2y}; 1$ *(3.5)*

24. $y' = \dfrac{2xy}{1 + 2y^2}; \dfrac{2}{3}$ *(3.5)*   **25.** 0.049 *(3.7)*   **26.** $-\dfrac{3}{100 - }$. 0.27 ft/sec *(3.6)*

30. $dR/dt = 1/\pi \approx 0.318$ in./min *(3.6)*   **31.** Elastic for 5 ·

32. 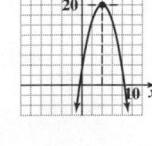   **33.** (A) $y = [\ln(4 - e^x)]^3$ (B   . $\left(\dfrac{1}{\ln 5}\right)\dfrac{2x - 1}{x^2 - x}$ *(3.4)*

36. $\dfrac{2x + 1}{2(x^2 + x)\sqrt{\ln(x^2 + x)}}$ *(3.*   is proportional to the radius R,

so the rate is smallest when $R = 0$ $\bar{}/3$ *(3.6)*

40. (A) 15 yr   (B) 13.9 yr *(3.1)*   .1)   **42.** $987.50/yr *(3.2)*

43. $R'(x) = (1{,}000 - 20x)e^{-0.02x}$ *(3.4)*   **44.** $p' = -\dfrac{x}{3p^2} = \dfrac{\bar{}}{}$ ease price *(3.7)*   **47.** 0.02125 *(3.7)*

48. -1.111 mg/mL per hour; -0.335 mg/mL per hour *(3.4)*   4   asing at the rate of 2.68 units/
day at the end of 1 day of training; increasing at the rate of 0.54 u   ~~...~~   (B) 7 days *(3.4)*   **51.** $dT/dt = -1/27 \approx -0.037$ min/hr *(3.6)*

Chapter 4

Exercises 4.1   **1.** Decreasing   **3.** Increasing   **5.** Increasing   **7.** Decreasing   **9.** (a, b); (d, f); (g, h) **11.** (b, c); (c, d); (f, g) **13.** c, d, f
15. b, f **17.** Local maximum at $x = a$; local minimum at $x = c$; no local extrema at $x = b$ and $x = d$ **19.** $f(3) = 5$ is a local maximum; e
21. No local extremum; d   **23.** $f(3) = 5$ is a local maximum; f   **25.** No local extremum; c   **27.** (A) $f'(x) = 3x^2 - 12$ (B) $-2, 2$ (C) $-2, 2$
29. (A) $f'(x) = -\dfrac{6}{(x + 2)^2}$ (B) -2 (C) None   **31.** (A) $f'(x) = \begin{cases} -1 \text{ if } x < 0 \\ 1 \text{ if } x > 0 \end{cases}$ (B) 0 (C) 0 **33.** Decreasing on $(-\infty, 1)$; increasing on
$(1, \infty)$; $f(1) = -2$ is a local minimum   **35.** Increasing on $(-\infty, -4)$; decreasing on $(-4, \infty)$; $f(-4) = 7$ is a local maximum   **37.** Increasing for all x; no lo-
cal extrema   **39.** Increasing on $(-\infty, -2)$ and $(3, \infty)$; decreasing on $(-2, 3)$; $f(-2) = 44$ is a local maximum, $f(3) = -81$ is a local minimum
41. Decreasing on $(-\infty, 1)$; increasing on $(1, \infty)$; $f(1) = 4$ is a local minimum   **43.** Increasing on $(-\infty, 2)$; decreasing on $(2, \infty)$; $f(2) = e^{-2} \approx 0.135$ is a
local maximum   **45.** Increasing on $(-\infty, 8)$; decreasing on $(8, \infty)$; $f(8) = 4$ is a local maximum   **47.** Critical numbers: $x = -0.77, 1.08, 2.69$;
decreasing on $(-\infty, -0.77)$ and $(1.08, 2.69)$; increasing on $(-0.77, 1.08)$ and $(2.69, \infty)$; $f(-0.77) = -4.75$ and $f(2.69) = -1.29$ are local minima; $f(1.08) = 6.04$ is a local maximum   **49.** Critical numbers: $x = 1.34, 2.82$; decreasing on $(0, 1.34)$ and $(2.82, \infty)$; increasing on $(1.34, 2.82)$;
$f(1.34) = 0.68$ is a local minimum; $f(2.82) = 2.37$ is a local maximum   **51.** Critical numbers: $0.36, 2.15$; increasing on $(-\infty, 0.36)$ and $(2.15, \infty)$;
decreasing on $(0.36, 2.15)$; $f(0.36) = 1.17$ is a local maximum; $f(2.15) = -0.66$ is a local minimum

53. Increasing on $(-\infty, 4)$ **55.** Increasing on $(-\infty, -1)$, $(1, \infty)$ **57.** Decreasing for all x **59.** Decreasing on $(-\infty, -3)$ and $(0, 3)$;
Decreasing on $(4, \infty)$ Decreasing on $(-1, 1)$ Horizontal tangent at $x = 2$ increasing on $(-3, 0)$ and $(3, \infty)$
Horizontal tangent at $x = 4$ Horizontal tangents at $x = -1, 1$ Horizontal tangents at $x = -3, 0, 3$

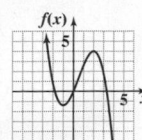   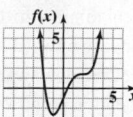   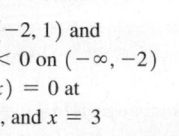   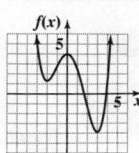

61. 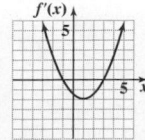   **63.** 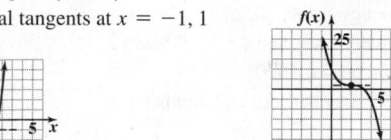   **65.** 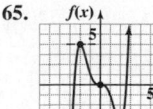   **67.** 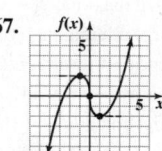   **69.** g_4 **71.** g_6 **73.** g_2

75. Increasing on $(-1, 2)$; decreasing on $(-\infty, -1)$ **77.** Increasing on $(-1, 2)$ and $(2, \infty)$ **79.** Increasing on $(-2, 0)$ and $(3, \infty)$;
and $(2, \infty)$; local minimum at $x = -1$;   decreasing on $(-\infty, -1)$; local   decreasing on $(-\infty, -2)$ and $(0, 3)$;
local maximum at $x = 2$ minimum at $x = -1$ local minima at $x = -2$ and $x = 3$;
                                                                   local maximum at $x = 0$

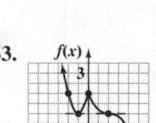   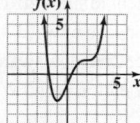   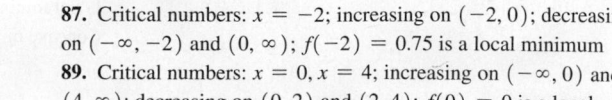

81. $f'(x) > 0$ on $(-\infty, -1)$ and   **83.** $f'(x) > 0$ on $(-2, 1)$ and   **85.** Critical numbers: $x = -2, x = 2$; increasing on $(-\infty, -2)$
$(3, \infty)$; $f'(x) < 0$ on $(-1, 3)$; $f'(x) = 0$ $(3, \infty)$; $f'(x) < 0$ on $(-\infty, -2)$ and $(2, \infty)$; decreasing on $(-2, 0)$ and $(0, 2)$; $f(-2) = -4$ is a
at $x = -1$ and $x = 3$ and $(1, 3)$; $f'(x) = 0$ at   local maximum; $f(2) = 4$ is a local minimum
                             $x = -2, x = 1$, and $x = 3$ **87.** Critical numbers: $x = -2$; increasing on $(-2, 0)$; decreasing
                                                     on $(-\infty, -2)$ and $(0, \infty)$; $f(-2) = 0.75$ is a local minimum
                                                     89. Critical numbers: $x = 0, x = 4$; increasing on $(-\infty, 0)$ and

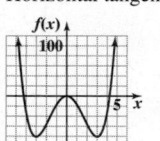

                                                       $(4, \infty)$; decreasing on $(0, 2)$ and $(2, 4)$; $f(0) = 0$ is a local
                                                       maximum; $f(4) = 8$ is a local minimum

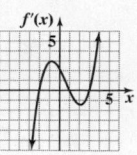

91. (A) The marginal profit is positive on $(0, 600)$, 0 (B) $P'(x)$
at $x = 600$, and negative on $(600, 1,000)$.

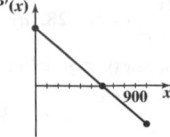

93. (A) The price decreases for the first 15 months to a local minimum, increases for the next 40 months to a local maximum, and then decreases for the remaining 15 months.

(B) $B(t)$

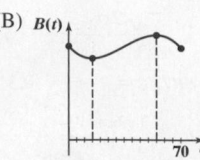

95. (A) $\overline{C}(x) = 0.05x + 20 + \dfrac{320}{x}$ (B) Critical number: $x = 80$; decreasing for $0 < x < 80$; increasing for $80 < x < 150$; $\overline{C}(80) = 28$ is a local minimum

97. Critical number: $t = 2$; increasing on $(0, 2)$; decreasing on $(2, 24)$; $C(2) = 0.07$ is a local maximum.

Exercises 4.2   **1.** Concave up **3.** Concave down **5.** Concave down **7.** Neither **9.** (A) $(a, c), (c, d), (e, g)$ (B) $(d, e), (g, h)$ (C) $(d, e), (g, h)$
(D) $(a, c), (c, d), (e, g)$ (E) $(a, c), (c, d), (e, g)$ (F) $(d, e), (g, h)$ **11.** (A) $f(-2) = 3$ is a local maximum of f; $f(2) = -1$ is a local minimum of f.

(B) $(0, 1)$ (C) 0 **13.** (C) **15.** (D) **17.** $12x - 8$ **19.** $4x^{-3} - 18x^{-4}$ **21.** $2 + \dfrac{9}{2}x^{-3/2}$ **23.** $8(x^2 + 9)^3 + 48x^2(x^2 + 9)^2 = 8(x^2 + 9)^2(7x^2 + 9)$

25. $(-10, 2{,}000)$ **27.** $(0, 2)$ **29.** None **31.** Concave upward for all x; no inflection points **33.** Concave downward on $\left(-\infty, \frac{4}{3}\right)$; concave upward
on $\left(\frac{4}{3}, \infty\right)$; inflection point at $x = \frac{4}{3}$ **35.** Concave downward on $(-\infty, 0)$ and $(6, \infty)$; concave upward on $(0, 6)$; inflection points at $x = 0$ and $x = 6$
37. Concave upward on $(-2, 4)$; concave downward on $(-\infty, -2)$ and $(4, \infty)$; inflection points at $x = -2$ and $x = 4$ **39.** Concave upward on $(-\infty, \ln 2)$;
concave downward on $(\ln 2, \infty)$; inflection point at $x = \ln 2$

41.
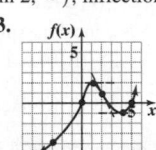

43.
$f(x)$

45.
$f(x)$

47.
$f(x)$

49. Domain: All real numbers
y int.: 16; x int.: $2 - 2\sqrt{3}, 2, 2 + 2\sqrt{3}$
Increasing on $(-\infty, 0)$ and $(4, \infty)$
Decreasing on $(0, 4)$
Local maximum: $f(0) = 16$;
local minimum: $f(4) = -16$
Concave downward on $(-\infty, 2)$
Concave upward on $(2, \infty)$
Inflection point: $(2, 0)$

$f(x)$

51. Domain: All real numbers
y int.: 2; x int.: -1
Increasing on $(-\infty, \infty)$
Concave downward on $(-\infty, 0)$
Concave upward on $(0, \infty)$
Inflection point: $(0, 2)$

$f(x)$

53. Domain: All real numbers
y int.: 0; x int.: 0, 4
Increasing on $(-\infty, 3)$
Decreasing on $(3, \infty)$
Local maximum: $f(3) = 6.75$
Concave upward on $(0, 2)$
Concave downward on $(-\infty, 0)$ and $(2, \infty)$
Inflection points: $(0, 0), (2, 4)$

$f(x)$

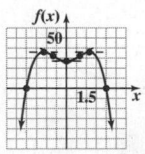

55. Domain: All real numbers
y int.: 0; x int.: 0, 1
Increasing on $(0.25, \infty)$
Decreasing on $(-\infty, 0.25)$
Local minimum: $f(0.25) = -1.6875$
Concave upward on $(-\infty, 0.5)$ and $(1, \infty)$
Concave downward on $(0.5, 1)$
Inflection points: $(0.5, -1), (1, 0)$

$f(x)$

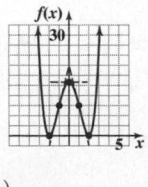

57. Domain: All real numbers
y int.: 27; x int.: $-3, 3$
Increasing on $(-\infty, -\sqrt{3})$ and $(0, \sqrt{3})$
Decreasing on $(-\sqrt{3}, 0)$ and $(\sqrt{3}, \infty)$
Local maxima: $f(-\sqrt{3}) = 36, f(\sqrt{3}) = 36$
Local minimum: $f(0) = 27$
Concave upward on $(-1, 1)$
Concave downward on $(-\infty, -1)$ and $(1, \infty)$
Inflection points: $(-1, 32), (1, 32)$

$f(x)$

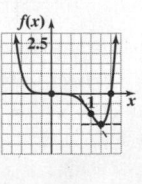

59. Domain: All real numbers
y int.: 16; x int.: $-2, 2$
Decreasing on $(-\infty, -2)$ and $(0, 2)$
Increasing on $(-2, 0)$ and $(2, \infty)$
Local minima: $f(-2) = 0, f(2) = 0$
Local maximum: $f(0) = 16$
Concave upward on $(-\infty, -2\sqrt{3}/3)$ and $(2\sqrt{3}/3, \infty)$
Concave downward on $(-2\sqrt{3}/3, 2\sqrt{3}/3)$
Inflection points: $(-1.15, 7.11), (1.15, 7.11)$

$f(x)$

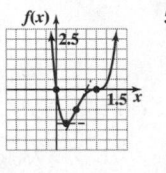

61. Domain: All real numbers
y int.: 0; int.: 0, 1.5
Decreasing on $(-\infty, 0)$ and $(0, 1.25)$
Increasing on $(1.25, \infty)$
Local minimum: $f(1.25) = -1.53$
Concave upward on $(-\infty, 0)$ and $(1, \infty)$
Concave downward on $(0, 1)$
Inflection points: $(0, 0), (1, -1)$

$f(x)$

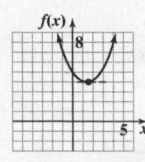

63. Domain: All real numbers
y int.: 0; x int.: 0
Increasing on $(-\infty, \infty)$
Concave downward on $(-\infty, \infty)$

$f(x)$

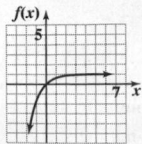

65. Domain: All real numbers
y int.: 5
Decreasing on $(-\infty, \ln 4)$
Increasing on $(\ln 4, \infty)$
Local minimum: $f(\ln 4) = 4$
Concave upward on $(-\infty, \infty)$

$f(x)$

67. Domain: $(0, \infty)$
x int.: e^2
Increasing on $(-\infty, \infty)$
Concave downward on $(-\infty, \infty)$

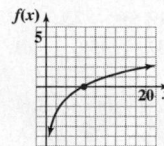

69. Domain: $(-4, \infty)$
y int.: $-2 + \ln 4$; x int.: $e^2 - 4$
Increasing on $(-4, \infty)$
Concave downward on $(-4, \infty)$

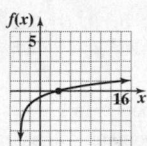

71.

x	$f'(x)$	$f(x)$
$-\infty < x < -1$	Positive and decreasing	Increasing and concave downward
$x = -1$	x intercept	Local maximum
$-1 < x < 0$	Negative and decreasing	Decreasing and concave downward
$x = 0$	Local minimum	Inflection point
$0 < x < 2$	Negative and increasing	Decreasing and concave upward
$x = 2$	Local max., x intercept	Inflection point, horiz. tangent
$2 < x < \infty$	Negative and decreasing	Decreasing and concave downward

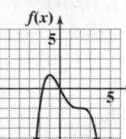

73.

x	$f'(x)$	$f(x)$
$-\infty < x < -2$	Negative and increasing	Decreasing and concave upward
$x = -2$	Local max., x intercept	Inflection point, horiz. tangent
$-2 < x < 0$	Negative and decreasing	Decreasing and concave downward
$x = 0$	Local minimum	Inflection point
$0 < x < 2$	Negative and increasing	Decreasing and concave upward
$x = 2$	Local max., x intercept	Inflection point, horiz. tangent
$2 < x < \infty$	Negative and decreasing	Decreasing and concave downward

75. Domain: All real numbers
x int.: $-1.18, 0.61, 1.87, 3.71$
y int.: -5
Decreasing on $(-\infty, -0.53)$ and $(1.24, 3.04)$
Increasing on $(-0.53, 1.24)$ and $(3.04, \infty)$
Local minima: $f(-0.53) = -7.57$, $f(3.04) = -8.02$
Local maximum: $f(1.24) = 2.36$
Concave upward on $(-\infty, 0.22)$ and $(2.28, \infty)$
Concave downward on $(0.22, 2.28)$
Inflection points: $(0.22, -3.15)$, $(2.28, -3.41)$

77. Domain: All real numbers
y int.: 100; x int.: $8.01, 13.36$
Increasing on $(-0.10, 4.57)$ and $(11.28, \infty)$
Decreasing on $(-\infty, -0.10)$ and $(4.57, 11.28)$
Local minima: $f(-0.10) = 99.02$, $f(11.28) = -901.18$
Local maximum: $f(4.57) = 711.75$
Concave upward on $(-\infty, 1.95)$ and $(8.55, \infty)$
Concave downward on $(1.95, 8.55)$
Inflection points: $(1.95, 377.82)$, $(8.55, -200.66)$

79. Domain: All real numbers
x int.: $-2.40, 1.16$; y int.: 3
Increasing on $(-\infty, -1.58)$
Decreasing on $(-1.58, \infty)$
Local maximum: $f(-1.58) = 8.87$
Concave downward on $(-\infty, -0.88)$ and $(0.38, \infty)$
Concave upward on $(-0.88, 0.38)$
Inflection points: $(-0.88, 6.39)$, $(0.38, 2.45)$

81. Domain: All real numbers
x int.: $-6.68, -3.64, -0.72$; y int.: 30
Decreasing on $(-5.59, -2.27)$ and $(1.65, 3.82)$
Increasing on $(-\infty, -5.59)$, $(-2.27, 1.65)$, and $(3.82, \infty)$
Local minima: $f(-2.27) = -37.84$, $f(3.82) = 32.09$
Local maxima: $f(-5.59) = 95.97$, $f(1.65) = 67.87$
Concave upward on $(-4.31, -0.40)$ and $(2.91, \infty)$
Concave downward on $(-\infty, -4.31)$ and $(-0.40, 2.91)$
Inflection points: $(-4.31, 39.98)$, $(-0.40, 13.54)$, $(2.91, 47.83)$

83. The graph of the CPI is concave upward.   **85.** The graph of $y = C'(x)$ is positive and decreasing. Since marginal costs are decreasing, the production process is becoming more efficient as production increases.   **87.** (A) Local maximum at $x = 60$ (B) Concave downward on the whole interval $(0, 80)$ **89.** (A) Local maximum at $x = 1$ (B) Concave downward on $(-\infty, 2)$; concave upward on $(2, \infty)$

91. Increasing on $(0, 10)$; decreasing on $(10, 15)$; point of diminishing returns is $x = 10$, max $T'(x) = T'(10) = 500$

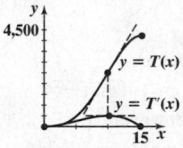

93. Increasing on $(24, 36)$; decreasing on $(36, 45)$; point of diminishing returns is $x = 36$, max $N'(x) = N'(36) = 3,888$

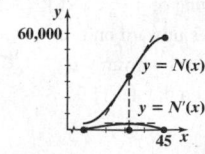

95. (A)

```
CubicReg
y=ax³+bx²+cx+d
a=-.005
b=.485
c=-1.85
d=300
```
(B) 32 ads to sell 574 cars per month

97. (A) Increasing on $(0, 10)$; decreasing on $(10, 20)$
(B) Inflection point: $(10, 3000)$
(C) 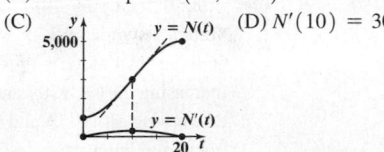   (D) $N'(10) = 300$

99. (A) Increasing on $(5, \infty)$; decreasing on $(0, 5)$
(B) Inflection point: $(5, 10)$
(C) 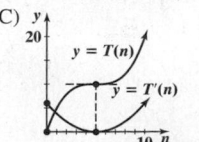   (D) $T'(5) = 0$

Exercises 4.3   **1.** 500   **3.** 0   **5.** 50   **7.** 0   **9.** 6   **11.** $-\frac{1}{10}$ **13.** 7   **15.** $-\frac{1}{5}$ **17.** $\frac{2}{5}$ **19.** 0   **21.** $-\infty$ **23.** $\frac{2}{3}$ **25.** $\frac{1}{2}$ **27.** 1   **29.** 0   **31.** 4
33. 5   **35.** ∞ **37.** 8   **39.** 0   **41.** ∞ **43.** ∞ **45.** $\frac{1}{3}$ **47.** -2 **49.** $-\infty$ **51.** 0   **53.** 0   **55.** 0   **57.** $\frac{1}{4}$ **59.** $\frac{1}{3}$ **61.** 0   **63.** 0   **65.** ∞ **67.** 1   **69.** 1

Exercises 4.4   **1.** Domain: All real numbers; x int.: -12; y int.: 36   **3.** Domain: $(-\infty, 25]$; x int.: 25; y int.: 5   **5.** Domain: All real numbers except 2; x int.: -1; y int.: $-\frac{1}{2}$ **7.** Domain: All real numbers except -1 and 1; No x intercept; y int.: -3 **9.** (A) $(-\infty, b)$, $(0, e)$, (e, g) (B) (b, d), $(d, 0)$, (g, ∞) (C) (b, d), $(d, 0)$, (g, ∞) (D) $(-\infty, b)$, $(0, e)$, (e, g) (E) $x = 0$ (F) $x = b$, $x = g$ (G) $(-\infty, a)$, (d, e), (h, ∞) (H) (a, d), (e, h) (I) (a, d), (e, h) (J) $(-\infty, a)$, (d, e), (h, ∞) (K) $x = a$, $x = h$ (L) $y = L$ (M) $x = d$, $x = e$

11. 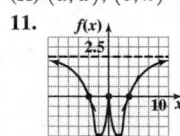   **13.** 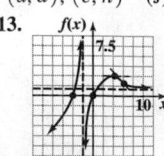   **15.** 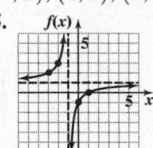   **17.**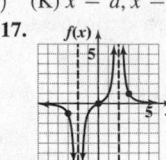

19. Domain: All real numbers, except 3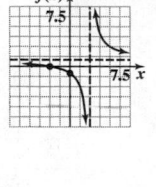
y int.: -1; x int.: -3
Horizontal asymptote: $y = 1$
Vertical asymptote: $x = 3$
Decreasing on $(-\infty, 3)$ and $(3, \infty)$
Concave upward on $(3, \infty)$
Concave downward on $(-\infty, 3)$

21. Domain: All real numbers, except 2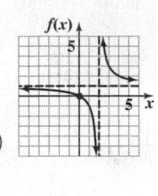
y int.: 0; x int.: 0
Horizontal asymptote: $y = 1$
Vertical asymptote: $x = 2$
Decreasing on $(-\infty, 2)$ and $(2, \infty)$
Concave downward on $(-\infty, 2)$
Concave upward on $(2, \infty)$

23. Domain: $(-\infty, \infty)$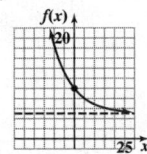
y int.: 10
Horizontal asymptote: $y = 5$
Decreasing on $(-\infty, \infty)$
Concave upward on $(-\infty, \infty)$

25. Domain: $(-\infty, \infty)$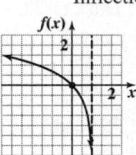
y int.: 0; x int.: 0
Horizontal asymptote: $y = 0$
Increasing on $(-\infty, 5)$
Decreasing on $(5, \infty)$
Local maximum: $f(5) = 9.20$
Concave upward on $(10, \infty)$
Concave downward on $(-\infty, 10)$
Inflection point: $(10, 6.77)$

27. Domain: $(-\infty, 1)$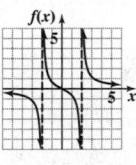
y int.: 0; x int.: 0
Vertical asymptote: $x = 1$
Decreasing on $(-\infty, 1)$
Concave downward on $(-\infty, 1)$

29. Domain: $(0, \infty)$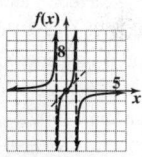
Vertical asymptote: $x = 0$
Increasing on $(1, \infty)$
Decreasing on $(0, 1)$
Local minimum: $f(1) = 1$
Concave upward on $(0, \infty)$

31. Domain: All real numbers, except ± 2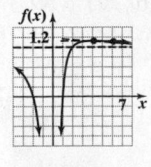
y int.: 0; x int.: 0
Horizontal asymptote: $y = 0$
Vertical asymptotes: $x = -2$, $x = 2$
Decreasing on $(-\infty, -2)$, $(-2, 2)$, and $(2, \infty)$
Concave upward on $(-2, 0)$ and $(2, \infty)$
Concave downward on $(-\infty, -2)$ and $(0, 2)$
Inflection point: $(0, 0)$

33. Domain: All real numbers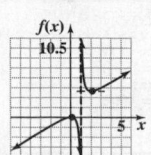
y int.: 1
Horizontal asymptote: $y = 0$
Increasing on $(-\infty, 0)$
Decreasing on $(0, \infty)$
Local maximum: $f(0) = 1$
Concave upward on $(-\infty, -\sqrt{3}/3)$ and $(\sqrt{3}/3, \infty)$
Concave downward on $(-\sqrt{3}/3, \sqrt{3}/3)$
Inflection points: $(-\sqrt{3}/3, 0.75)$, $(\sqrt{3}/3, 0.75)$

35. Domain: All real numbers except -1 and 1
y int.: 0; x int.: 0
Horizontal asymptote: $y = 0$
Vertical asymptote: $x = -1$ and $x = 1$
Increasing on $(-\infty, -1)$, $(-1, 1)$, and $(1, \infty)$
Concave upward on $(-\infty, -1)$ and $(0, 1)$
Concave downward on $(-1, 0)$ and $(1, \infty)$
Inflection point: $(0, 0)$

37. Domain: All real numbers except 1
y int.: 0; x int.: 0
Horizontal asymptote: $y = 0$
Vertical asymptote: $x = 1$
Increasing on $(-\infty, -1)$ and $(1, \infty)$
Decreasing on $(-1, 1)$
Local maximum: $f(-1) = 1.25$
Concave upward on $(-\infty, -2)$
Concave downward on $(-2, 1)$ and $(1, \infty)$
Inflection point: $(-2, 1.11)$

39. Domain: All real numbers except 0
Horizontal asymptote: $y = 1$
Vertical asymptote: $x = 0$
Increasing on $(0, 4)$
Decreasing on $(-\infty, 0)$ and $(4, \infty)$
Local maximum: $f(4) = 1.125$
Concave upward on $(6, \infty)$
Concave downward on $(-\infty, 0)$ and $(0, 6)$
Inflection point: $(6, 1.11)$

41. Domain: All real numbers except 1
y int.: 0; x int.: 0
Vertical asymptote: $x = 1$
Oblique asymptote: $y = x + 1$
Increasing on $(-\infty, 0)$ and $(2, \infty)$
Decreasing on $(0, 1)$ and $(1, 2)$
Local maximum: $f(0) = 0$
Local minimum: $f(2) = 4$
Concave upward on $(1, \infty)$
Concave downward on $(-\infty, 1)$

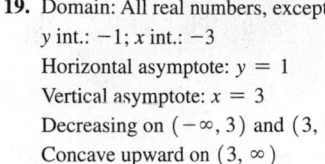
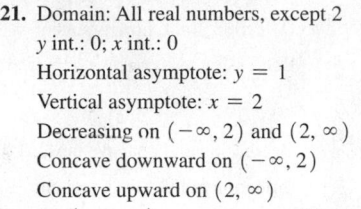
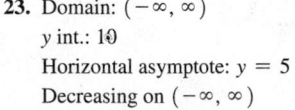
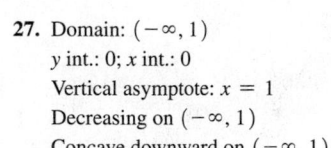
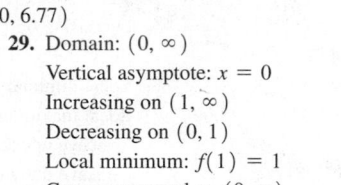
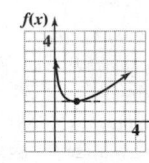
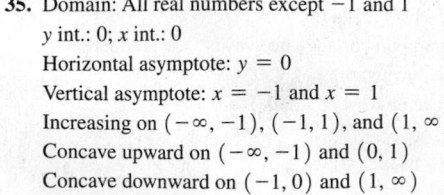
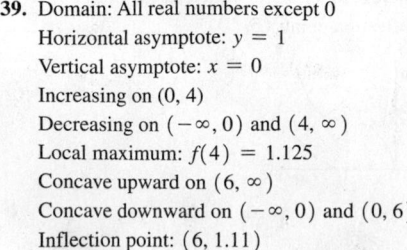

43. Domain: All real numbers except $-3, 3$

y int.: $-\frac{2}{9}$

Horizontal asymptote: $y = 3$

Vertical asymptotes: $x = -3, x = 3$

Increasing on $(-\infty, -3)$ and $(-3, 0)$

Decreasing on $(0, 3)$ and $(3, \infty)$

Local maximum: $f(0) = -0.22$

Concave upward on $(-\infty, -3)$ and $(3, \infty)$

Concave downward on $(-3, 3)$

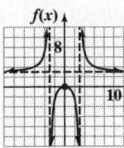

47. Domain: All real numbers

y int.: 3; x int.: 3

Horizontal asymptote: $y = 0$

Increasing on $(-\infty, 2)$

Decreasing on $(2, \infty)$

Local maximum: $f(2) = 7.39$

Concave upward on $(-\infty, 1)$

Concave downward on $(1, \infty)$

Inflection point: $(1, 5.44)$

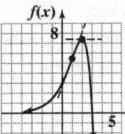

51. Domain: $(0, \infty)$

x int.: 1

Increasing on $(e^{-1/2}, \infty)$

Decreasing on $(0, e^{-1/2})$

Local minimum: $f(e^{-1/2}) = -0.18$

Concave upward on $(e^{-3/2}, \infty)$

Concave downward on $(0, e^{-3/2})$

Inflection point: $(e^{-3/2}, -0.07)$

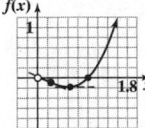

55. Domain: All real numbers except $-4, 2$

y int.: $-\frac{1}{8}$

Horizontal asymptote: $y = 0$

Vertical asymptote: $x = -4, x = 2$

Increasing on $(-\infty, -4)$ and $(-4, -1)$

Decreasing on $(-1, 2)$ and $(2, \infty)$

Local maximum: $f(-1) = -0.11$

Concave upward on $(-\infty, -4)$ and $(2, \infty)$

Concave downward on $(-4, 2)$

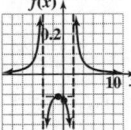

59. Domain: All real numbers except 0

Vertical asymptote: $x = 0$

Oblique asymptote: $y = x$

Increasing on $(-\infty, -2)$ and $(2, \infty)$

Decreasing on $(-2, 0)$ and $(0, 2)$

Local maximum: $f(-2) = -4$

Local minimum: $f(2) = 4$

Concave upward on $(0, \infty)$

Concave downward on $(-\infty, 0)$

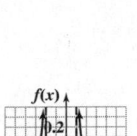

63. Domain: All real numbers except 0

x int.: $-\sqrt{3}, \sqrt{3}$

Vertical asymptote: $x = 0$

Oblique asymptote: $y = x$

Increasing on $(-\infty, 0)$ and $(0, \infty)$

Concave upward on $(-\infty, 0)$

Concave downward on $(0, \infty)$

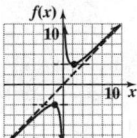

45. Domain: All real numbers except 2

y int.: 0; x int.: 0

Vertical asymptote: $x = 2$

Increasing on $(3, \infty)$

Decreasing on $(-\infty, 2)$ and $(2, 3)$

Local minimum: $f(3) = 27$

Concave upward on $(-\infty, 0)$ and $(2, \infty)$

Concave downward on $(0, 2)$

Inflection point: $(0, 0)$

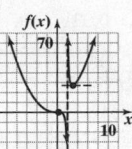

49. Domain: $(-\infty, \infty)$

y int.: 1

Horizontal asymptote: $y = 0$

Increasing on $(-\infty, 0)$

Decreasing on $(0, \infty)$

Local maximum: $f(0) = 1$

Concave upward on $(-\infty, -1)$ and $(1, \infty)$

Concave downward on $(-1, 1)$

Inflection points: $(-1, 0.61), (1, 0.61)$

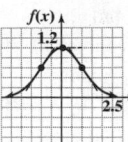

53. Domain: $(0, \infty)$

x int.: 1

Vertical asymptote: $x = 0$

Increasing on $(1, \infty)$

Decreasing on $(0, 1)$

Local minimum: $f(1) = 0$

Concave upward on $(0, e)$

Concave downward on (e, ∞)

Inflection point: $(e, 1)$

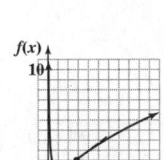

57. Domain: All real numbers except $-\sqrt{3}, \sqrt{3}$

y int.: 0; x int.: 0

Vertical asymptote: $x = -\sqrt{3}, x = \sqrt{3}$

Oblique asymptote: $y = -x$

Increasing on $(-3, -\sqrt{3}), (-\sqrt{3}, \sqrt{3})$, and $(\sqrt{3}, 3)$

Decreasing on $(-\infty, -3)$ and $(3, \infty)$

Local maximum: $f(3) = -4.5$

Local minimum: $f(-3) = 4.5$

Concave upward on $(-\infty, -\sqrt{3})$ and $(0, \sqrt{3})$

Concave downward on $(-\sqrt{3}, 0)$ and $(\sqrt{3}, \infty)$

Inflection point: $(0, 0)$

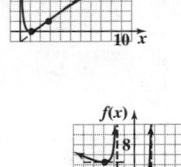

61. Domain: All real numbers except 0

x int.: $\sqrt[3]{4}$

Vertical asymptote: $x = 0$

Oblique asymptote: $y = x$

Increasing on $(-\infty, -2)$ and $(0, \infty)$

Local maximum: $f(-2) = -3$

Decreasing on $(-2, 0)$

Concave downward on $(-\infty, 0)$ and $(0, \infty)$

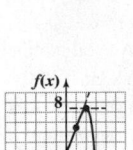

65. Domain: All real numbers except 0

Vertical asymptote: $x = 0$

Oblique asymptote: $y = x$

Increasing on $(-\infty, -2)$ and $(2, \infty)$

Decreasing on $(-2, 0)$ and $(0, 2)$

Local maximum: $f(-2) = -3$

Local minimum: $f(2) = 3$

Concave upward on $(0, \infty)$

Concave downward on $(-\infty, 0)$

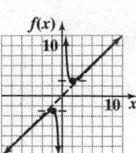

67. $y = 90 + 0.02x$ **69.** $y = 210 + 0.1x$

71. Domain: All real numbers except 2, 4

y int.: $-3/4$; x int.: -3

Vertical asymptote: $x = 4$

Horizontal asymptote: $y = 1$

Decreasing on $(-\infty, 2)$, $(2, 4)$, and $(4, \infty)$

Concave upward on $(4, \infty)$

Concave downward on $(-\infty, 2)$ and $(2, 4)$

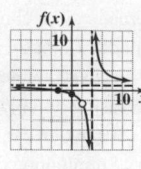

73. Domain: All real numbers except -3, 3

y int.: $5/3$; x int.: 2.5

Vertical asymptote: $x = 3$

Horizontal asymptote: $y = 2$

Decreasing on $(-\infty, -3)$, $(-3, 3)$, and $(3, \infty)$

Concave upward on $(3, \infty)$

Concave downward on $(-\infty, -3)$ and $(-3, 3)$

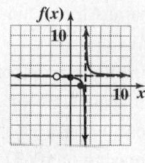

75. Domain: All real numbers except -1, 2

y int.: 0; x int.: 0, 3

Vertical asymptote: $x = -1$

Oblique asymptote: $y = x - 4$

Increasing on $(-\infty, -3)$, $(1, 2)$, and $(2, \infty)$

Decreasing on $(-3, -1)$ and $(-1, 1)$

Local maximum: $f(-3) = -9$

Local minimum: $f(1) = -1$

Concave upward on $(-1, 2)$ and $(2, \infty)$

Concave downward on $(-\infty, -1)$

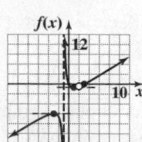

77. Domain: All real numbers except 1

y int.: -2; x int.: -2

Vertical asymptote: $x = 1$

Horizontal asymptote: $y = 1$

Decreasing on $(-\infty, 1)$ and $(1, \infty)$

Concave upward on $(1, \infty)$

Concave downward on $(-\infty, 1)$

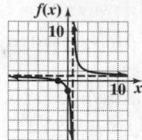

79.

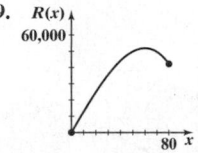

81. (A) Increasing on $(0, 1)$

(B) Concave upward on $(0, 1)$

(C) $x = 1$ is a vertical asymptote

(D) The origin is both an x and a y intercept

(E)

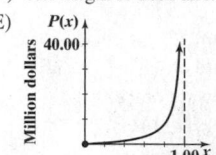

83. (A) $\overline{C}(n) = \dfrac{3{,}200}{n} + 250 + 50n$

(B) $\overline{C}(n)$

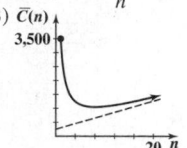

(C) 8 yr

85. (A)

(B) \$25 at $x = 100$

87. (A)

```
QuadReg
y=ax²+bx+c
a=.0100714286
b=.7835714286
c=316
```

(B) Minimum average cost is \$4.35 when 177 pizzas are produced daily.

89.

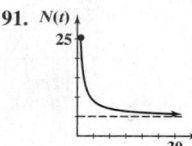

91.

Exercise 4.5   **1.** Max $f(x) = f(3) = 3$; Min $f(x) = f(-2) = -2$ **3.** Max $h(x) = h(-5) = 25$; Min $h(x) = h(0) = 0$

5. Max $n(x) = n(4) = 2$; Min $n(x) = n(3) = \sqrt{3}$ **7.** Max $q(x) = q(27) = -3$; Min $q(x) = q(64) = -4$

9. Min $f(x) = f(0) = 0$; Max $f(x) = f(10) = 14$ **11.** Min $f(x) = f(0) = 0$; Max $f(x) = f(3) = 9$

13. Min $f(x) = f(1) = f(7) = 5$; Max $f(x) = f(10) = 14$ **15.** Min $f(x) = f(1) = f(7) = 5$; Max $f(x) = f(3) = f(9) = 9$

17. Min $f(x) = f(5) = 7$; Max $f(x) = f(3) = 9$ **19.** (A) Max $f(x) = f(4) = 3$; Min $f(x) = f(0) = -5$

(B) Max $f(x) = f(10) = 15$; Min $f(x) = f(0) = -5$ (C) Max $f(x) = f(10) = 15$; Min $f(x) = f(-5) = -15$

21. (A) Max $f(x) = f(-1) = f(1) = 1$; Min $f(x) = f(0) = 0$ (B) Max $f(x) = f(5) = 25$; Min $f(x) = f(1) = 1$

(C) Max $f(x) = f(-5) = f(5) = 25$; Min $f(x) = f(0) = 0$ **23.** Max $f(x) = f(-1) = e \approx 2.718$; Min $f(x) = e^{-1} \approx 0.368$

25. Max $f(x) = f(0) = 9$; Min $f(x) = f(\pm 4) = -7$ **27.** Min $f(x) = f(1) = 2$; no maximum   **29.** Max $f(x) = f(-3) = 18$; no minimum

31. No absolute extrema   **33.** Max $f(x) = f(3) = 54$; no minimum   **35.** No absolute extrema   **37.** Min $f(x) = f(0) = 0$; no maximum

39. Max $f(x) = f(1) = 1$; Min $f(x) = f(-1) = -1$ **41.** Min $f(x) = f(0) = -1$; no maximum   **43.** Min $f(x) = f(2) = -2$

45. Max $f(x) = f(2) = 4$ **47.** Min $f(x) = f(2) = 0$ **49.** No maximum   **51.** Max $f(x) = f(2) = 8$ **53.** Min $f(x) = f(4) = 22$

55. Min $f(x) = f(\sqrt{10}) = 14/\sqrt{10}$ **57.** Min $f(x) = f(2) = \dfrac{e^2}{4} \approx 1.847$ **59.** Max $f(x) = f(3) = \dfrac{27}{e^3} \approx 1.344$

61. Max $f(x) = f(e^{1.5}) = 2e^{1.5} \approx 8.963$ **63.** Max $f(x) = f(e^{2.5}) = \dfrac{e^5}{2} \approx 74.207$ **65.** Max $f(x) = f(1) = -1$

67. (A) Max $f(x) = f(5) = 14$; Min $f(x) = f(-1) = -22$ (B) Max $f(x) = f(1) = -2$; Min $f(x) = f(-1) = -22$

(C) Max $f(x) = f(5) = 14$; Min $f(x) = f(3) = -6$ **69.** (A) Max $f(x) = f(0) = 126$; Min $f(x) = f(2) = -26$

(B) Max $f(x) = f(7) = 49$; Min $f(x) = f(2) = -26$ (C) Max $f(x) = f(6) = 6$; Min $f(x) = f(3) = -15$

71. (A) Max $f(x) = f(-1) = 10$; Min $f(x) = f(2) = -11$ (B) Max $f(x) = f(0) = f(4) = 5$; Min $f(x) = f(3) = -22$

(C) Max $f(x) = f(-1) = 10$; Min $f(x) = f(1) = 2$ **73.** Local minimum   **75.** Unable to determine   **77.** Neither   **79.** Local maximum

Exercises 4.6   **1.** $f(x) = x(28 - x)$ **3.** $f(x) = \pi x^2/4$ **5.** $f(x) = \pi x^3$ **7.** $f(x) = x(60 - x)$ **9.** 7.5 and 7.5   **11.** 7.5 and -7.5

13. $\sqrt{15}$ and $\sqrt{15}$ **15.** $10\sqrt{2}$ ft by $10\sqrt{2}$ ft   **17.** 37 ft by 37 ft   **19.** (A) Maximum revenue is \$125,000 when 500 phones are produced and sold for \$250 each.   (B) Maximum profit is \$46,612.50 when 365 phones are produced and sold for \$317.50 each.   **21.** (A) Max $R(x) = R(3{,}000) = $ \$300,000

(B) Maximum profit is \$75,000 when 2,100 sets are manufactured and sold for \$130 each.   (C) Maximum profit is \$64,687.50 when 2,025 sets are manufactured and sold for \$132.50 each.

23. (A)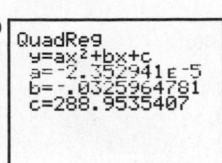

QuadReg
y=ax²+bx+c
a=⁻2.352941ᴇ⁻5
b=⁻.0325964781
c=288.9535407

(B)

LinReg
y=ax+b
a=53.50318471
b=82245.22293

(C) The maximum profit is $118,996 when the price per sleeping bag is $195.
25. (A) $4.80 (B) $8 **27.** $35; $6,125 **29.** 40 trees; 1,600 lb
31. $(10 - 2\sqrt{7})/3 = 1.57$ in. squares **33.** 20 ft by 40 ft (with the expensive side being one of the short sides) **35.** (A) 70 ft by 100 ft (B) 125 ft by 125 ft **37.** 8 production runs per year **39.** 10,000 books in 5 printings **41.** 34.64 mph

43. (A) $x = 5.1$ mi (B) $x = 10$ mi **45.** 4 days; 20 bacteria/cm³ **47.** 1 month; 2 ft **49.** 4 yr from now

Chapter 4 Review Exercises   **1.** $(a, c_1), (c_3, c_6)$ *(4.1, 4.2)* **2.** $(c_1, c_3), (c_6, b)$ *(4.1, 4.2)* **3.** $(a, c_2), (c_4, c_5), (c_7, b)$ *(4.1, 4.2)* **4.** c_3 *(4.1)*
5. c_1, c_6 *(4.1)* **6.** c_1, c_3, c_5 *(4.1)* **7.** c_4, c_6 *(4.1)* **8.** c_2, c_4, c_5, c_7 *(4.2)*
9. *(4.2)* 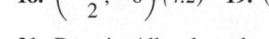 **10.** *(4.2)* 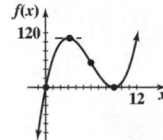 **11.** $f''(x) = 12x^2 + 30x$ *(4.2)* **12.** $y'' = 8/x^3$ *(4.2)* **13.** Domain: All real numbers, except 4
y int.: $\frac{5}{4}$; x int.: -5 *(4.2)* **14.** Domain: $(-2, \infty)$ y int.: ln 2; x int.: -1 *(4.2)*
15. Horizontal asymptote: $y = 0$; Vertical asymptotes: $x = -2, x = 2$ *(4.4)* **16.** Horizontal asymptote:
$y = \frac{2}{3}$; Vertical asymptote: $x = -\frac{10}{3}$ *(4.4)* **17.** $(-\sqrt{2}, -20), (\sqrt{2}, -20)$ *(4.2)*

18. $\left(-\frac{1}{2}, -6\right)$ *(4.2)* **19.** (A) $f'(x) = \frac{1}{5}x^{-4/5}$ (B) 0 (C) 0 *(4.1)* **20.** (A) $f'(x) = -\frac{1}{5}x^{-6/5}$ (B) 0 (C) None *(4.1)*

21. Domain: All real numbers
  y int.: 0; x int.: 0, 9
  Increasing on $(-\infty, 3)$ and $(9, \infty)$
  Decreasing on $(3, 9)$
  Local maximum: $f(3) = 108$
  Local minimum: $f(9) = 0$
  Concave upward on $(6, \infty)$
  Concave downward on $(-\infty, 6)$
  Inflection point: $(6, 54)$ *(4.4)*

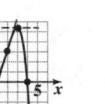

22. Domain: All real numbers
  y int.: 16; x int.: $-4, 2$
  Increasing on $(-\infty, -2)$ and $(2, \infty)$
  Decreasing on $(-2, 2)$
  Local maximum: $f(-2) = 32$
  Local minimum: $f(2) = 0$
  Concave upward on $(0, \infty)$
  Concave downward on $(-\infty, 0)$
  Inflection point: $(0, 16)$ *(4.4)*

23. Domain: All real numbers
  y int.: 0; x int.: 0, 4
  Increasing on $(-\infty, 3)$
  Decreasing on $(3, \infty)$
  Local maximum: $f(3) = 54$
  Concave upward on $(0, 2)$
  Concave downward on $(-\infty, 0)$ and $(2, \infty)$
  Inflection points: $(0, 0), (2, 32)$ *(4.4)*

24. Domain: all real numbers
  y int.: -3; x int.: $-3, 1$
  No vertical or horizontal asymptotes
  Increasing on $(-2, \infty)$
  Decreasing on $(-\infty, -2)$
  Local minimum: $f(-2) = -27$
  Concave upward on $(-\infty, -1)$ and $(1, \infty)$
  Concave downward on $(-1, 1)$
  Inflection points: $(-1, -16), (1, 0)$ *(4.4)*

25. Domain: All real numbers, except -2
  y int.: 0; x int.: 0
  Horizontal asymptote: $y = 3$
  Vertical asymptote: $x = -2$
  Increasing on $(-\infty, -2)$ and $(-2, \infty)$
  Concave upward on $(-\infty, -2)$
  Concave downward on $(-2, \infty)$ *(4.4)*

26. Domain: All real numbers
  y int.: 0; x int.: 0
  Horizontal asymptote: $y = 1$
  Increasing on $(0, \infty)$
  Decreasing on $(-\infty, 0)$
  Local minimum: $f(0) = 0$
  Concave upward on $(-3, 3)$
  Concave downward on $(-\infty, -3)$ and $(3, \infty)$
  Inflection points: $(-3, 0.25), (3, 0.25)$ *(4.4)*

27. Domain: All real numbers except $x = -2$
  y int.: 0; x int.: 0
  Horizontal asymptote: $y = 0$
  Vertical asymptote: $x = -2$
  Increasing on $(-2, 2)$
  Decreasing on $(-\infty, -2)$ and $(2, \infty)$
  Local maximum: $f(2) = 0.125$
  Concave upward on $(4, \infty)$
  Concave downward on $(-\infty, -2)$ and $(-2, 4)$
  Inflection point: $(4, 0.111)$ *(4.4)*

28. Domain: All real numbers
  y int.: 0; x int.: 0
  Oblique asymptote: $y = x$
  Increasing on $(-\infty, \infty)$
  Concave upward on $(-\infty, -3)$ and $(0, 3)$
  Concave downward on $(-3, 0)$ and $(3, \infty)$
  Inflection points: $(-3, -2.25), (0, 0), (3, 2.25)$ *(4.4)*

29. Domain: All real numbers
  y int.: 0; x int.: 0
  Horizontal asymptote: $y = 5$
  Increasing on $(-\infty, \infty)$
  Concave downward on $(-\infty, \infty)$ *(4.4)*

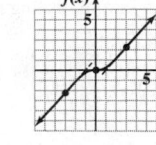

30. Domain: $(0, \infty)$
  x int.: 1
  Increasing on $(e^{-1/3}, \infty)$
  Decreasing on $(0, e^{-1/3})$
  Local minimum: $f(e^{-1/3}) = -0.123$
  Concave upward on $(e^{-5/6}, \infty)$
  Concave downward on $(0, e^{-5/6})$
  Inflection point: $(e^{-5/6}, -0.068)$ *(4.4)*

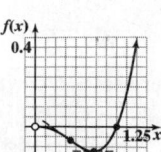

31. 3 *(4.3)* **32.** $-\dfrac{1}{5}$ *(4.3)* **33.** $-\infty$ *(4.3)* **34.** 0 *(4.3)* **35.** ∞ *(4.3)* **36.** 1 *(4.3)* **37.** 0 *(4.3)* **38.** 0 *(4.3)* **39.** 1 *(4.3)* **40.** 2 *(4.3)*

41.

x	$f'(x)$	$f(x)$
$-\infty < x < -2$	Negative and increasing	Decreasing and concave upward
$x = -2$	x intercept	Local minimum
$-2 < x < -1$	Positive and increasing	Increasing and concave upward
$x = -1$	Local maximum	Inflection point
$-1 < x < 1$	Positive and decreasing	Increasing and concave downward
$x = 1$	Local min., x intercept	Inflection point, horiz. tangent
$1 < x < \infty$	Positive and increasing	Increasing and concave upward

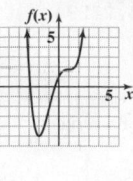 *(4.2)*

42. (C) *(4.2)* **43.** Local maximum: $f(-1) = 20$; local minimum $f(5) = -88$ *(4.5)*
44. Min $f(x) = f(2) = -4$;
Max $f(x) = f(5) = 77$ *(4.5)*
45. Min $f(x) = f(2) = 8$ *(4.5)*
46. Max $f(x) = f(e^{4.5}) = 2e^{4.5} \approx 180.03$ *(4.5)*
47. Max $f(x) = f(0.5) = 5e^{-1} \approx 1.84$ *(4.5)*
48. Yes. Since f is continuous on $[a, b]$, f has an absolute maximum on $[a, b]$. But neither $f(a)$ nor $f(b)$ is an absolute maximum, so the absolute maximum must occur between a and b. *(4.5)*

49. No, increasing/decreasing properties apply to intervals in the domain of f. It is correct to say that $f(x)$ is decreasing on $(-\infty, 0)$ and $(0, \infty)$. *(4.1)*
50. A critical number of $f(x)$ is a partition number for $f'(x)$ that is also in the domain of f. For example, if $f(x) = x^{-1}$, then 0 is a partition number for $f'(x) = -x^{-2}$, but 0 is not a critical number of $f(x)$ since 0 is not in the domain of f. *(4.1)*
51. Max $f'(x) = f'(2) = 12$ *(4.2, 4.5)*

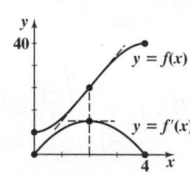

52. Each number is 20; minimum sum is 40 *(4.6)*

53. Domain: All real numbers
x int.: 0.79, 1.64; y int.: 4
Increasing on $(-1.68, -0.35)$ and $(1.28, \infty)$
Decreasing on $(-\infty, -1.68)$ and $(-0.35, 1.28)$
Local minima: $f(-1.68) = 0.97$, $f(1.28) = -1.61$
Local maximum: $f(-0.35) = 4.53$
Concave downward on $(-1.10, 0.60)$
Concave upward on $(-\infty, -1.10)$ and $(0.60, \infty)$
Inflection points: $(-1.10, 2.58)$, $(0.60, 1.08)$

54. Domain: All real numbers
x intercepts: 0, 11.10; y int.: 0
Increasing on $(1.87, 4.19)$ and $(8.94, \infty)$
Decreasing on $(-\infty, 1.87)$ and $(4.19, 8.94)$
Local maximum: $f(4.19) = -39.81$
Local minima: $f(1.87) = -52.14$, $f(8.94) = -123.81$
Concave upward on $(-\infty, 2.92)$ and $(7.08, \infty)$
Concave downward on $(2.92, 7.08)$
Inflection points: $(2.92, -46.41)$, $(7.08, -88.04)$ *(4.4)*

55. Max $f(x) = f(1.373) = 2.487$ *(4.5)* **56.** Max $f(x) = f(1.763) = 0.097$ *(4.5)*

57. (A) For the first 15 months, the graph of the price is increasing and concave downward, with a local maximum at $t = 15$.
For the next 15 months, the graph of the price is decreasing and concave downward, with an inflection point at $t = 30$.
For the next 15 months, the graph of the price is decreasing and concave upward, with a local minimum at $t = 45$.
For the remaining 15 months, the graph of the price is increasing and concave upward.

(B) 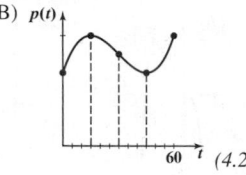 *(4.2)*

58. (A) Max $R(x) = R(10,000) = \$2,500,000$
(B) Maximum profit is \$175,000 when 3,000 readers are manufactured and sold for \$425 each.
(C) Maximum profit is \$119,000 when 2,600 readers are manufactured and sold for \$435 each. *(4.6)*
59. (A) The expensive side is 50 ft; the other side is 100 ft. (B) The expensive side is 75 ft; the other side is 150 ft. *(4.6)* **60.** \$49; \$6,724 *(4.6)*
61. 12 orders/yr *(4.6)*

62. Min $\overline{C}(x) = \overline{C}(200) = \$50$ *(4.4)*

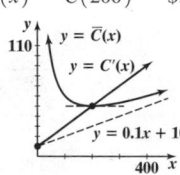

63. Min $\overline{C}(x) = \overline{C}(e^5) \approx \$49.66$ *(4.4)*
64. A maximum revenue of \$18,394 is realized at a production level of 50 units at \$367.88 each. *(4.6)*

65. 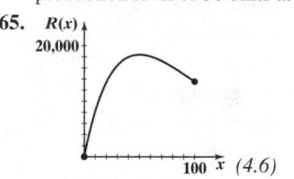 *(4.6)*

66. \$549.15; \$9,864 *(4.6)*
67. \$1.52 *(4.6)*
68. 20.39 feet *(4.6)*
69. (A)

```
QuadReg
y=ax²+bx+c
a=.0061285714
b=.1224285714
c=102.2
```

(B) Min $\overline{C}(x) = \overline{C}(129) = \$1.71$ *(4.4)*

70. Increasing on $(0, 18)$; decreasing on $(18, 24)$; point of diminishing returns is $x = 18$, max $N'(x) = N'(18) = 972$ *(4.2)*

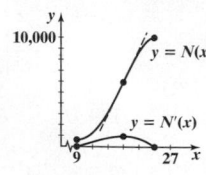

71. (A)
```
CubicReg
y=ax³+bx²+cx+d
a=-.01
b=.83
c=-2.3
d=221
```
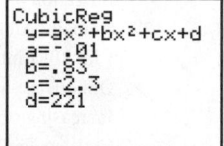

(B) 28 ads to sell 588 refrigerators per month *(4.2)*
72. 3 days *(4.1)*
73. 2 yr from now *(4.1)*

Chapter 5

Exercises 5.1 **1.** $f(x) = 5x^{-4}$ **3.** $f(x) = 3x^{-4} - 2x^{-5}$ **5.** $f(x) = x^{1/2} + 5x^{-1/2}$ **7.** $f(x) = 4x^{1/3} + x^{4/3} - 3x^{7/3}$ **9.** $7x + C$ **11.** $4x^2 + C$
13. $3x^3 + C$ **15.** $(x^6/6) + C$ **17.** $(-x^{-2}/2) + C$ **19.** $4x^{5/2} + C$ **21.** $3 \ln |z| + C$ **23.** $16e^u + C$ **25.** Yes **27.** Yes **29.** No **31.** No
33. True **35.** False **37.** True **39.** No, since one graph cannot be obtained from another by a vertical translation. **41.** Yes, since one graph can be obtained from another by a vertical translation. **43.** $(5x^2/2) - (5x^3/3) + C$ **45.** $2\sqrt{u} + C$ **47.** $-(x^{-2}/8) + C$ **49.** $4 \ln |u| + u + C$ **51.** $5e^z + 4z + C$
53. $x^3 + 2x^{-1} + C$ **55.** $C(x) = 2x^3 - 2x^2 + 3,000$ **57.** $x = 40\sqrt{t}$ **59.** $y = -2x^{-1} + 3 \ln |x| - x + 3$ **61.** $x = 4e^t - 2t - 3$
63. $y = 2x^2 - 3x + 1$ **65.** $x^2 + x^{-1} + C$ **67.** $\frac{1}{2}x^2 + x^{-2} + C$ **69.** $e^x - 2 \ln |x| + C$ **71.** x^3 **73.** $x^4 + 3x^2 + C$

81. $\overline{C}(x) = 15 + \dfrac{1,000}{x}$; $C(x) = 15x + 1,000$; $C(0) = \$1,000$ **83.** (A) The cost function increases from 0 to 8, is concave downward from 0 to 4, and is concave upward from 4 to 8. There is an inflection point at $x = 4$.   (B) $C(x) = x^3 - 12x^2 + 53x + 30$; $C(4) = \$114,000$; $C(8) = \$198,000$

(C)

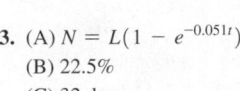

(D) Manufacturing plants are often inefficient at low and high levels of production.
85. $S(t) = 1,200 - 18t^{4/3}$; $50^{3/4} \approx 19$ mo   **87.** $S(t) = 1,200 - 18t^{4/3} - 70t$; $t \approx 8.44$ mo
89. $L(x) = 4,800x^{1/2}$; 24,000 labor-hours   **91.** $W(h) = 0.0005h^3$; 171.5 lb   **93.** 19,400

Exercises 5.2   **1.** $f'(x) = 50(5x + 1)^9$ **3.** $f'(x) = 14x(x^2 + 1)^6$ **5.** $f'(x) = 2xe^{x^2}$ **7.** $f'(x) = \dfrac{4x^3}{x^4 - 10}$ **9.** $\frac{1}{3}(3x + 5)^3 + C$

11. $\frac{1}{6}(x^2 - 1)^6 + C$ **13.** $-\frac{1}{2}(5x^3 + 1)^{-2} + C$ **15.** $e^{5x} + C$ **17.** $\ln|1 + x^2| + C$ **19.** $\frac{2}{3}(1 + x^4)^{3/2} + C$ **21.** $\frac{1}{11}(x + 3)^{11} + C$

23. $-\frac{1}{6}(6t - 7)^{-1} + C$ **25.** $\frac{1}{12}(t^2 + 1)^6 + C$ **27.** $\frac{1}{2}e^{x^2} + C$ **29.** $\frac{1}{5}\ln|5x + 4| + C$ **31.** $-e^{1-t} + C$ **33.** $-\frac{1}{18}(3t^2 + 1)^{-3} + C$

35. $\frac{2}{5}(x + 4)^{5/2} - \frac{8}{3}(x + 4)^{3/2} + C$ **37.** $\frac{2}{3}(x - 3)^{3/2} + 6(x - 3)^{1/2} + C$ **39.** $\frac{1}{11}(x - 4)^{11} + \frac{2}{5}(x - 4)^{10} + C$ **41.** $\frac{1}{8}(1 + e^{2x})^4 + C$

43. $\frac{1}{2}\ln|4 + 2x + x^2| + C$ **45.** $\frac{1}{2}(5x + 3)^2 + C$ **47.** $\frac{1}{2}(x^2 - 1)^2 + C$ **49.** $\frac{1}{5}(x^5)^5 + C$ **51.** No   **53.** Yes   **55.** No   **57.** Yes

59. $\frac{1}{9}(3x^2 + 7)^{3/2} + C$ **61.** $\frac{1}{8}x^8 + \frac{4}{5}x^5 + 2x^2 + C$ **63.** $\frac{1}{9}(x^3 + 2)^3 + C$ **65.** $\frac{1}{4}(2x^4 + 3)^{1/2} + C$ **67.** $\frac{1}{4}(\ln x)^4 + C$ **69.** $e^{-1/x} + C$

71. $x = \frac{1}{3}(t^3 + 5)^7 + C$ **73.** $y = 3(t^2 - 4)^{1/2} + C$ **75.** $p = -(e^x - e^{-x})^{-1} + C$ **77.** $p(x) = \dfrac{2,000}{3x + 50} + 4$; 250 bottles

79. $C(x) = 12x + 500\ln(x + 1) + 2,000$; $\overline{C}(1,000) = \$17.45$ **81.** (A) $S(t) = 10t + 100e^{-0.1t} - 100, 0 \le t \le 24$ (B) \$50 million   (C) 18.41 mo
83. $Q(t) = 100\ln(t + 1) + 5t, 0 \le t \le 20$; 275 thousand barrels   **85.** $W(t) = 2e^{0.1t}$; 4.45 g   **87.** (A) $-1,000$ bacteria/mL per day
(B) $N(t) = 5,000 - 1,000\ln(1 + t^2)$; 385 bacteria/mL   (C) 7.32 days   **89.** $N(t) = 100 - 60e^{-0.1t}, 0 \le t \le 15$; 87 words/min
91. $E(t) = 12,000 - 10,000(t + 1)^{-1/2}$; 9,500 students

Exercises 5.3   **1.** The derivative of $f(x)$ is 5 times $f(x)$; $y' = 5y$ **3.** The derivative of $f(x)$ is -1 times $f(x)$; $y' = -y$
5. The derivative of $f(x)$ is $2x$ times $f(x)$; $y' = 2xy$ **7.** The derivative of $f(x)$ is 1 minus $f(x)$; $y' = 1 - y$ **9.** $y = 3x^2 + C$ **11.** $y = 7\ln|x| + C$

13. $y = 50e^{0.02x} + C$ **15.** $y = \dfrac{x^3}{3} - \dfrac{x^2}{2}$ **17.** $y = e^{-x^2} + 2$ **19.** $y = 2\ln|1 + x| + 5$ **21.** Second-order   **23.** Third-order   **25.** Yes

27. Yes   **29.** Yes   **31.** Yes   **33.** Figure B. When $x = 1$, the slope $dy/dx = 1 - 1 = 0$ for any y. When $x = 0$, the slope $dy/dx = 0 - 1 = -1$
for any y. Both are consistent with the slope field shown in Figure B.   **35.** $y = \dfrac{x^2}{2} - x + C$; $y = \dfrac{x^2}{2} - x - 2$

37.   **39.** $y = Ce^{2t}$ **41.** $y = 100e^{-0.5x}$ **43.** $x = Ce^{-5t}$ **45.** $x = -(5t^2/2) + C$ **47.** Logistic growth   **49.** Unlimited growth
51. Figure A. When $y = 1$, the slope $dy/dx = 1 - 1 = 0$ for any x. When $y = 2$, the slope $dy/dx = 1 - 2 = -1$ for any x. Both are
consistent with the slope field shown in Figure A.   **53.** $y = 1 - e^{-x}$

55.   **57.**   **59.** $y = \sqrt{25 - x^2}$ **65.**   **67.**
61. $y = -3x$
63. $y = 1/(1 - 2e^{-t})$

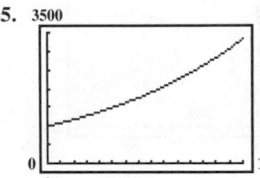

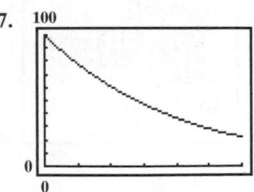

69.   **71.**   **73.** Apply the second-derivative test to $f(y) = ky(M - y)$.   **75.** 2009
77. $A = 1,000e^{0.03t}$ **79.** $A = 8,000e^{0.016t}$ **81.** (A) $p(x) = 100e^{-0.05x}$
(B) \$60.65 per unit   (C)

83. (A) $N = L(1 - e^{-0.051t})$ **85.** $I = I_0e^{-0.00942x}$; $x \approx 74$ ft   **89.** 0.023 117   **91.** Approx. 24,200 yr   **93.** 104 times; 67 times
(B) 22.5%   **87.** (A) $Q = 3e^{-0.04t}$ **95.** (A) 7 people; 353 people   (B) 400   (C)
(C) 32 days   (B) $Q(10) = 2.01$ mL
(D)   (C) 27.47 hr
(D)

Exercises 5.4   **1.** 80 in.^{2} **3.** 36 m^2 **5.** No, $\pi - 2 > 1$ m^2 **7.** C, E   **9.** B   **11.** H, I   **13.** H

15.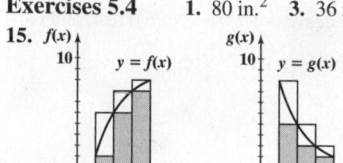

17. Figure A: $L_3 = 13$, $R_3 = 20$; Figure B: $L_3 = 14$, $R_3 = 7$

19. $L_3 \leq \int_1^4 f(x)\,dx \leq R_3$; $R_3 \leq \int_1^4 g(x)\,dx \leq L_3$; since $f(x)$ is increasing, L_3 underestimates the area and R_3 overestimates the area; since $g(x)$ is decreasing, the reverse is true.   **21.** In both figures, the error bound for L_3 and R_3 is 7.   **23.** $S_5 = -260$ **25.** $S_4 = -1{,}194$ **27.** $S_3 = -33.01$ **29.** $S_6 = -38$ **31.** -2.475

33. 4.266   **35.** 2.474   **37.** -5.333 **39.** 1.067   **41.** -1.066 **43.** 15   **45.** 58.5   **47.** -54 **49.** 248   **51.** 0   **53.** -183 **55.** False   **57.** False
59. False   **61.** $L_{10} = 286{,}100$ ft^2; error bound is 50,000 ft^2; $n \geq 200$ **63.** $L_6 = -3.53$, $R_6 = -0.91$; error bound for L_6 and R_6 is 2.63. Geometrically, the definite integral over the interval [2, 5] is the sum of the areas between the curve and the x axis from $x = 2$ to $x = 5$, with the areas below the x axis counted negatively and those above the x axis counted positively.   **65.** Increasing on $(-\infty, 0]$; decreasing on $[0, \infty)$ **67.** Increasing on $[-1, 0]$ and $[1, \infty)$; decreasing on $(-\infty, -1]$ and $[0, 1]$ **69.** $n \geq 22$ **71.** $n \geq 104$ **73.** $L_3 = 2{,}580$, $R_3 = 3{,}900$; error bound for L_3 and R_3 is 1,320

75. (A) $L_5 = 3.72$; $R_5 = 3.37$ (B) $R_5 = 3.37 \leq \int_0^5 A'(t)\,dt \leq 3.72 = L_5$ **77.** $L_3 = 114$, $R_3 = 102$; error bound for L_3 and R_3 is 12

Exercises 5.5   **1.** 500   **3.** 28   **5.** 48   **7.** $4.5\pi \approx 14.14$ **9.** (A) $F(15) - F(10) = 375$ **11.** (A) $F(15) - F(10) = 85$ **13.** 40   **15.** 72   **17.** 46.5

(B) 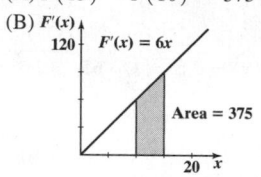   (B)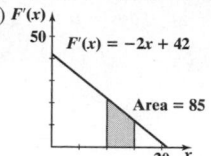

19. $e - 1 \approx 1.718$
21. $\ln 2 \approx 0.693$
23. 0   **25.** 48
27. -48 **29.** -10.25
31. 0   **33.** -2 **35.** 14

37. $5^6 = 15{,}625$ **39.** $\ln 4 \approx 1.386$ **41.** $20(e^{0.25} - e^{-0.5}) \approx 13.550$ **43.** $\frac{1}{2}$ **45.** $\frac{1}{2}(1 - e^{-1}) \approx 0.316$ **47.** 0
49. (A) Average $f(x) = 250$ **51.** (A) Average $f(t) = 2$ **53.** (A) Average $f(x) = \frac{45}{28} \approx 1.61$ **55.** (A) Average $f(x) = 2(1 - e^{-2}) \approx 1.73$
(B) 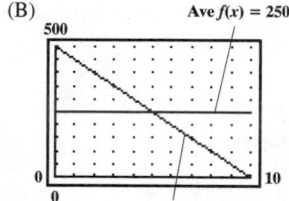   (B) [graph]   (B) [graph]   (B)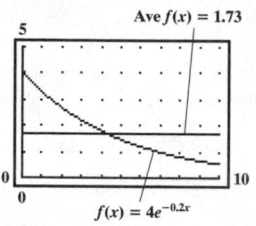

57. $\frac{1}{6}(15^{3/2} - 5^{3/2}) \approx 7.819$ **59.** $\frac{1}{2}(\ln 2 - \ln 3) \approx -0.203$ **61.** 0   **63.** 4.566   **65.** 2.214   **69.** $\int_{300}^{900}\left(500 - \frac{x}{3}\right) dx = \$180{,}000$

71. $\int_0^5 500(t - 12)\,dt = -\$23{,}750$; $\int_5^{10} 500(t - 12)\,dt = -\$11{,}250$

73. (A) 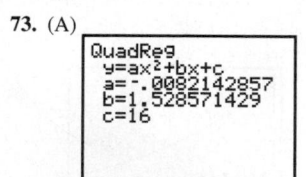   [graph]   **75.** Useful life $= \sqrt{\ln 55} \approx 2$ yr; total profit $= \frac{51}{22} - \frac{5}{2}e^{-4} \approx 2.272$ or \$2,272
77. (A) \$420   (B) \$135,000

(B) 6,505

79. (A) 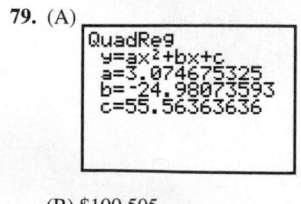   [graph]   **81.** $50e^{0.6} - 50e^{0.4} - 10 \approx \$6.51$ **83.** 4,800 labor-hours   **85.** (A) $I = -200t + 600$
(B) $\frac{1}{3}\int_0^3 (-200t + 600)\,dt = 300$ **87.** $100 \ln 11 + 50 \approx 290$ thousand barrels;
$100 \ln 21 - 100 \ln 11 + 50 \approx 115$ thousand barrels
89. $2e^{0.8} - 2 \approx 2.45$ g; $2e^{1.6} - 2e^{0.8} \approx 5.45$ g
91. 10°C   **93.** $0.6 \ln 2 + 0.1 \approx 0.516$; $(4.2 \ln 625 + 2.4 - 4.2 \ln 49)/24 \approx 0.546$

(B) \$100,505

Chapter 5 Review Exercises   **1.** $3x^2 + 3x + C$ *(5.1)*   **2.** 50 *(5.5)*   **3.** -207 *(5.5)*   **4.** $-\frac{1}{8}(1 - t^2)^4 + C$ *(5.2)*   **5.** $\ln|u| + \frac{1}{4}u^4 + C$ *(5.1)*
6. 0.216 *(5.5)*   **7.** No *(5.1)*   **8.** Yes *(5.1)*   **9.** No *(5.1)*   **10.** Yes *(5.1)*   **11.** Yes *(5.3)*   **12.** Yes *(5.3)*   **13.** e^{-x^2} *(5.1)*   **14.** $\sqrt{4 + 5x} + C$ *(5.1)*
15. $y = f(x) = x^3 - 2x + 4$ *(5.3)*   **16.** (A) $2x^4 - 2x^2 - x + C$ (B) $e^t - 4 \ln|t| + C$ *(5.1)*   **17.** $R_2 = 72$; error bound for R_2 is 48 *(5.4)*
18. $\int_1^5 (x^2 + 1)\,dx = \frac{136}{3} \approx 45.33$; actual error is $\frac{80}{3} \approx 26.67$ *(5.5)*   **19.** $L_4 = 30.8$ *(5.4)*   **20.** 7 *(5.5)*   **21.** Width $= 2 - (-1) = 3$;
height $=$ average $f(x) = 7$ *(5.5)*   **22.** $S_4 = 368$ *(5.4)*   **23.** $S_5 = 906$ *(5.4)*   **24.** -10 *(5.4)*   **25.** 0.4 *(5.4)*   **26.** 1.4 *(5.4)*   **27.** 0 *(5.4)*   **28.** 0.4 *(5.4)*
29. 2 *(5.4)*   **30.** -2 *(5.4)*   **31.** -0.4 *(5.4)*   **32.** (A) 1; 1   (B) 4; 4 *(5.3)*   **33.** $dy/dx = (2y)/x$; at points on the x axis $(y = 0)$ the slopes are 0. *(5.3)*
35. $y = \frac{1}{4}x^2$; $y = -\frac{1}{4}x^2$ *(5.3)*

36. *(5.3)*   **37.** *(5.3)*

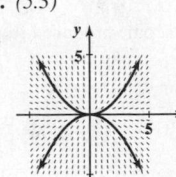

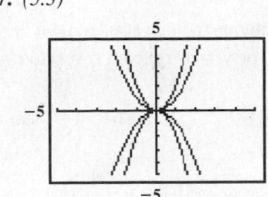

38. $\frac{2}{3}(2)^{3/2} \approx 1.886$ *(5.5)*   **39.** $\frac{1}{6} \approx 0.167$ *(5.5)*   **40.** $-5e^{-t} + C$ *(5.1)*   **41.** $\frac{1}{2}(1 + e^2)$ *(5.1)*

42. $\frac{1}{6}e^{3x^2} + C$ *(5.2)*   **43.** $2(\sqrt{5} - 1) \approx 2.472$ *(5.5)*   **44.** $\frac{1}{2}\ln 10 \approx 1.151$ *(5.5)*

45. 0.45 *(5.5)*   **46.** $\frac{1}{48}(2x^4 + 5)^6 + C$ *(5.2)*   **47.** $-\ln(e^{-x} + 3) + C$ *(5.2)*

48. $-(e^x + 2)^{-1} + C$ *(5.2)*   **49.** $y = f(x) = 3\ln|x| + x^{-1} + 4$ *(5.2, 5.3)*

50. $y = 3x^2 + x - 4$ *(5.3)*

51. (A) Average $f(x) = 6.5$
(B)
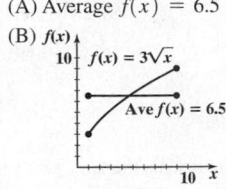

52. $\frac{1}{3}(\ln x)^3 + C$ *(5.2)*   **53.** $\frac{1}{8}x^8 - \frac{2}{5}x^5 + \frac{1}{2}x^2 + C$ *(5.2)*

54. $\frac{2}{3}(6 - x)^{3/2} - 12(6 - x)^{1/2} + C$ *(5.2)*   **55.** $\frac{1,234}{15} \approx 82.267$ *(5.5)*

56. 0 *(5.5)*   **57.** $y = 3e^{x^3} - 1$ *(5.3)*   **58.** $N = 800e^{0.06t}$ *(5.3)*

59. Limited growth

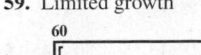

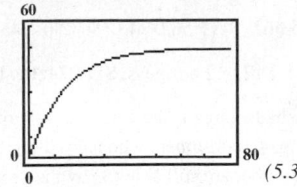

60. Exponential decay   **61.** Unlimited growth   **62.** Logistic growth

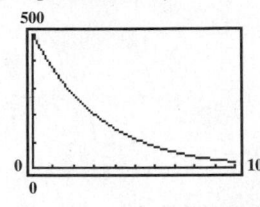   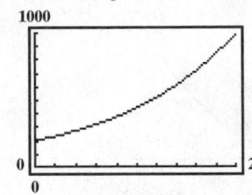   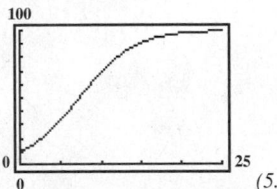

63. 1.167 *(5.5)*   **64.** 99.074 *(5.5)*   **65.** -0.153 *(5.5)*   **66.** $L_2 = \$180,000; R_2 = \$140,000; \$140,000 \le \int_{200}^{600} C'(x)\,dx \le \$180,000$ *(5.4)*

67. $\int_{200}^{600}\left(600 - \frac{x}{2}\right)dx = \$160,000$ *(5.5)*   **68.** $\int_{10}^{40}\left(150 - \frac{x}{10}\right)dx = \$4,425$ *(5.5)*   **69.** $P(x) = 100x - 0.01x^2; P(10) = \$999$ *(5.3)*

70. $\int_0^{15}(60 - 4t)\,dt = 450$ thousand barrels *(5.5)*   **71.** 109 items *(5.5)*   **72.** $16e^{2.5} - 16e^2 - 8 \approx \$68.70$ *(5.5)*   **73.** Useful life $= 10\ln\frac{20}{3} \approx 19$ yr;
total profit $= 143 - 200e^{-1.9} \approx 113.086$ or $\$113,086$ *(5.5)*   **74.** $S(t) = 50 - 50e^{-0.08t}; 50 - 50e^{-0.96} \approx \$31$ million; $-(\ln 0.2)/0.08 \approx 20$ mo *(5.3)*

75. 1 cm^2 *(5.3)*   **76.** 800 gal *(5.5)*   **77.** (A) 132 million   (B) About 65 years *(5.3)*   **78.** $\dfrac{-\ln 0.04}{0.000\,123\,8} \approx 26,000$ yr *(5.3)*

79. $N(t) = 95 - 70e^{-0.1t}; N(15) \approx 79$ words/min *(5.3)*

Chapter 6

Exercises 6.1   **1.** 150   **3.** 37.5   **5.** 25.5   **7.** $\pi/2$ **9.** $\int_a^b g(x)\,dx$ **11.** $\int_a^b[-h(x)]\,dx$ **13.** Since the shaded region in Figure C is below the x axis, $h(x) \le 0$; so, $\int_a^b h(x)\,dx$ represents the negative of the area.   **15.** 24   **17.** 51   **19.** 42   **21.** 6   **23.** 0.833   **25.** 2.350   **27.** 1   **29.** Canada; Mexico

31. India; Brazil   **33.** $\int_a^b[-f(x)]\,dx$ **35.** $\int_b^c f(x)\,dx + \int_c^d[-f(x)]\,dx$ **37.** $\int_c^d[f(x) - g(x)]\,dx$ **39.** $\int_a^b[f(x) - g(x)]\,dx + \int_b^c[g(x) - f(x)]\,dx$

41. Find the intersection points by solving $f(x) = g(x)$ on the interval $[a, d]$ to determine b and c.
Then observe that $f(x) \ge g(x)$ over $[a, b], g(x) \ge f(x)$ over $[b, c]$, and $f(x) \ge g(x)$ over $[c, d]$.
Area $= \int_a^b[f(x) - g(x)]\,dx + \int_b^c[g(x) - f(x)]\,dx + \int_c^d[f(x) - g(x)]\,dx$.

43. 2.5   **45.** 7.667   **47.** 12   **49.** 15   **51.** 32   **53.** 36   **55.** 9   **57.** 2.832   **59.** $\int_{-3}^3\sqrt{9 - x^2}\,dx$; 14.137   **61.** $\int_0^4\sqrt{16 - x^2}\,dx$; 12.566

63. $\int_{-2}^2 2\sqrt{4 - x^2}\,dx$; 12.566   **65.** 52.616   **67.** 8   **69.** 101.75   **71.** 17.979   **73.** 5.113   **75.** 8.290   **77.** 3.166   **79.** 1.385   **81.** 3.17   **83.** 1.82

85. Total production from the end of the fifth year to the end of the 10th year is $50 + 100\ln 20 - 100\ln 15 \approx 79$ thousand barrels.   **87.** Total profit over the 5-yr useful life of the game is $20 - 30e^{-1.5} \approx 13.306$, or $\$13,306$.   **89.** 1935: 0.412; 1947: 0.231; income was more equality distributed in 1947.

91. 1963: 0.818; 1983: 0.846; total assets were less equally distributed in 1983.   **93.** (A) $f(x) = 0.3125x^2 + 0.7175x - 0.015$ (B) 0.104

95. Total weight gain during the first 10 hr is $3e - 3 \approx 5.15$ g.   **97.** Average number of words learned from $t = 2$ hr to $t = 4$ hr is $15\ln 4 - 15\ln 2 \approx 10$.

Exercises 6.2   **1.** $b = 20; c = -5$ **3.** $b = 0.32; c = -0.04$ **5.** $b = 1.6; c = -0.03$ **7.** $b = 1.75; c = 0.02$ **9.** 10.27   **11.** 151.75

13. 93,268.66   **15.** (A) 10.72   (B) 3.28   (C) 10.72

17.

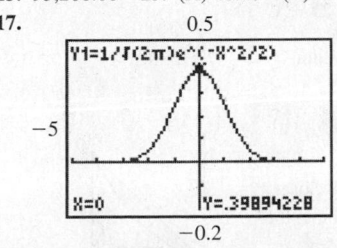

19.

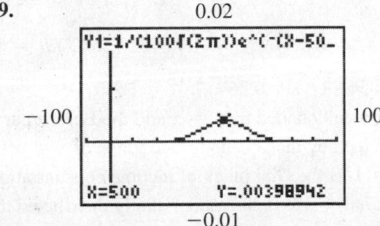

21. (A) 0.6827   (B) 0.9545   (C) 0.9973
23. (A) 0.6827   (B) 0.9545   (C) 0.9973
25. (A) .75   (B) .11   (C)

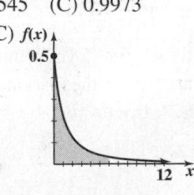

27. 8 yr   **29.** (A) .11   (B) .10   **31.** $P(t \ge 12) = 1 - P(0 \le t \le 12) = .89$ **33.** $\$12,500$

35. If $f(t)$ is the rate of flow of a continuous income stream, then the total income produced from 0 to 5 yr is the area under the graph of $y = f(t)$ from $t = 0$ to $t = 5$.

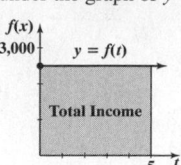

37. $8,000(e^{0.15} - 1) \approx \$1,295$

39. If $f(t)$ is the rate of flow of a continuous income stream, then the total income produced from 0 to 3 yr is the area under the graph of $y = f(t)$ from $t = 0$ to $t = 3$.

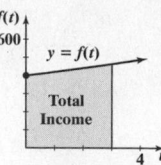

41. $255,562; $175,562   **43.** $6,780   **45.** $437   **47.** Clothing store: $66,421; computer store: $62,623; the clothing store is the better investment.

49. Bond: $12,062 business: $11,824; the bond is the better investment.   **51.** $55,230   **53.** $\frac{k}{r}(e^{rT} - 1)$ **55.** $625,000

57. The shaded area is the consumers' surplus and represents the total savings to consumers who are willing to pay more than $150 for a product but are still able to buy the product for $150.

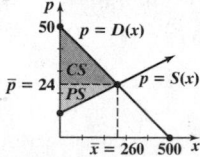

59. $9,900

61. The area of the region PS is the producers' surplus and represents the total gain to producers who are willing to supply units at a lower price than $67 but are still able to supply the product at $67.

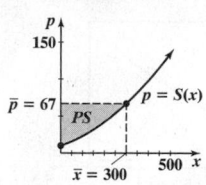

63. $CS = \$3,380; PS = \$1,690$

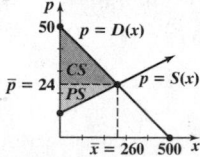

65. $CS = \$6,980; PS = \$5,041$

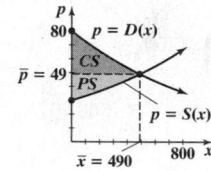

67. $CS = \$7,810; PS = \$8,336$

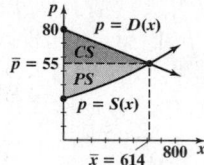

69. $CS = \$8,544; PS = \$11,507$

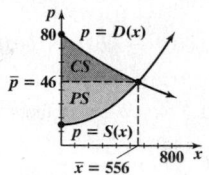

71. (A) $\bar{x} = 21.457; \bar{p} = \$6.51$ (B) $CS = 1.774$ or $1,774; PS = 1.087$ or $1,087$

Exercises 6.3   **1.** $f'(x) = 5; \int g(x)\,dx = \dfrac{x^4}{4} + C$ **3.** $f'(x) = 3x^2; \int g(x)\,dx = \dfrac{5x^2}{2} + C$ **5.** $f'(x) = 4e^{4x}; \int g(x)\,dx = \ln|x| + C$

7. $f'(x) = -x^{-2}; \int g(x)\,dx = \dfrac{1}{4}e^{4x} + C$ **9.** $\frac{1}{3}xe^{3x} - \frac{1}{9}e^{3x} + C$ **11.** $\dfrac{x^3}{3}\ln x - \dfrac{x^3}{9} + C$ **13.** $u = x + 2; \dfrac{(x+2)(x+1)^6}{6} - \dfrac{(x+1)^7}{42} + C$

15. $-xe^{-x} - e^{-x} + C$ **17.** $\dfrac{1}{2}e^{x^2} + C$ **19.** $(xe^x - 4e^x)\big|_0^1 = -3e + 4 \approx -4.1548$ **21.** $(x\ln 2x - x)\big|_1^3 = (3\ln 6 - 3) - (\ln 2 - 1) \approx 2.6821$

23. $\ln(x^2 + 1) + C$ **25.** $(\ln x)^2/2 + C$ **27.** $\frac{2}{3}x^{3/2}\ln x - \frac{4}{9}x^{3/2} + C$ **29.** $\dfrac{(x-3)(x+1)^3}{3} - \dfrac{(x+1)^4}{12} + C$ or $\dfrac{x^4}{4} - \dfrac{x^3}{3} - \dfrac{5x^2}{2} - 3x + C$

31. $\dfrac{(2x+1)(x-2)^3}{3} - \dfrac{(x-2)^4}{6} + C$ or $\dfrac{x^4}{2} - \dfrac{7x^3}{3} + 2x^2 + 4x + C$

33. The integral represents the negative of the area between the graph of $y = (x-3)e^x$ and the x axis from $x = 0$ to $x = 1$.

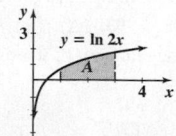

35. The integral represents the area between the graph of $y = \ln 2x$ and the x axis from $x = 1$ to $x = 3$.

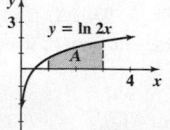

37. $(x^2 - 2x + 2)e^x + C$

39. $\dfrac{xe^{ax}}{a} - \dfrac{e^{ax}}{a^2} + C$

41. $\left(-\dfrac{\ln x}{x} - \dfrac{1}{x}\right)\Big|_1^e = -\dfrac{2}{e} + 1 \approx 0.2642$

43. $6\ln 6 - 4\ln 4 - 2 \approx 3.205$

45. $xe^{x-2} - e^{x-2} + C$

47. $\frac{1}{2}(1 + x^2)\ln(1 + x^2) - \frac{1}{2}(1 + x^2) + C$ **49.** $(1 + e^x)\ln(1 + e^x) - (1 + e^x) + C$ **51.** $x(\ln x)^2 - 2x\ln x + 2x + C$

53. $x(\ln x)^3 - 3x(\ln x)^2 + 6x\ln x - 6x + C$ **55.** 2 **57.** $\dfrac{1}{3}$ **59.** 1.56 **61.** 34.98 **63.** $\int_0^5(2t - te^{-t})\,dt = \$24$ million

65. The total profit for the first 5 yr (in millions of dollars) is the same as the area under the marginal profit function, $P'(t) = 2t - te^{-t}$, from $t = 0$ to $t = 5$.

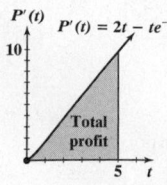

67. $2,854.88   **69.** 0.264

71. The area bounded by $y = x$ and the Lorenz curve $y = xe^{x-1}$, divided by the area under the graph of $y = x$ from $x = 0$ to $x = 1$, is the Gini index of income concentration. The closer this index is to 0, the more equally distributed the income; the closer the index is to 1, the more concentrated the income in a few hands.

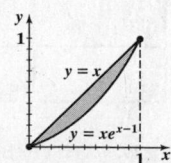

73. $S(t) = 1{,}600 + 400e^{0.1t} - 40te^{0.1t}$; 15 mo   **75.** \$977

77. The area bounded by the price–demand equation, $p = 9 - \ln(x + 4)$, and the price equation, $y = \bar{p} = 2.089$, from $x = 0$ to $x = \bar{x} = 1{,}000$, represents the consumers' surplus. This is the amount consumers who are willing to pay more than \$2.089 save.

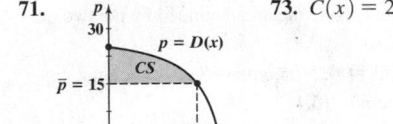

79. 2.1388 ppm   **81.** $N(t) = -4te^{-0.25t} - 40e^{-0.25t} + 80$; 8 wk; 78 words/min

83. 20,980

Exercises 6.4   **1.** 77.32   **3.** 74.15   **5.** 0.47   **7.** 0.46   **9.** $\ln\left|\dfrac{x}{1+x}\right| + C$ **11.** $\dfrac{1}{3+x} + 2\ln\left|\dfrac{5+2x}{3+x}\right| + C$ **13.** $\dfrac{2(x-32)}{3}\sqrt{16+x} + C$

15. $-\ln\left|\dfrac{1+\sqrt{1-x^2}}{x}\right| + C$ **17.** $\dfrac{1}{2}\ln\left|\dfrac{x}{2+\sqrt{x^2+4}}\right| + C$ **19.** $\frac{1}{3}x^3\ln x - \frac{1}{9}x^3 + C$ **21.** $x - \ln|1+e^x| + C$ **23.** $9\ln\frac{3}{2} - 2 \approx 1.6492$

25. $\frac{1}{2}\ln\frac{12}{5} \approx 0.4377$ **27.** $\ln 3 \approx 1.0986$ **29.** $730; 723\frac{1}{3}$ **31.** $1.61; \ln 5 \approx 1.61$ **33.** $69; 69$ **35.** $756; 756$

37. $-\dfrac{\sqrt{4x^2+1}}{x} + 2\ln|2x + \sqrt{4x^2+1}| + C$ **39.** $\frac{1}{2}\ln|x^2 + \sqrt{x^4-16}| + C$ **41.** $\frac{1}{6}(x^3\sqrt{x^6+4} + 4\ln|x^3 + \sqrt{x^6+4}|) + C$

43. $-\dfrac{\sqrt{4-x^4}}{8x^2} + C$ **45.** $\frac{1}{5}\ln\left|\dfrac{3+4e^x}{2+e^x}\right| + C$ **47.** $\frac{2}{3}(\ln x - 8)\sqrt{4 + \ln x} + C$ **49.** $\frac{1}{5}x^2e^{5x} - \frac{2}{25}xe^{5x} + \frac{2}{125}e^{5x} + C$

51. $-x^3e^{-x} - 3x^2e^{-x} - 6xe^{-x} - 6e^{-x} + C$ **53.** $x(\ln x)^3 - 3x(\ln x)^2 + 6x\ln x - 6x + C$ **55.** $\frac{64}{3}$ **57.** $\frac{1}{2}\ln\frac{9}{5} \approx 0.2939$ **59.** $\dfrac{-1-\ln x}{x} + C$

61. $\sqrt{x^2-1} + C$ **65.** 31.38   **67.** 5.48   **69.** $3{,}000 + 1{,}500\ln\frac{1}{3} \approx \$1{,}352$

71.

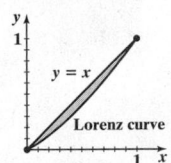

73. $C(x) = 200x + 1{,}000\ln(1 + 0.05x) + 25{,}000$; 608; \$198,773   **75.** \$18,673.95   **77.** 0.1407

79. As the area bounded by the two curves gets smaller, the Lorenz curve approaches $y = x$ and the distribution of income approaches perfect equality—all persons share equally in the income available.

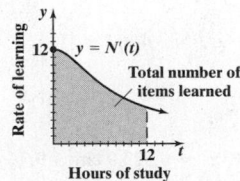

81. $S(t) = 1 + t - \dfrac{1}{1+t} - 2\ln|1+t|$; $24.96 - 2\ln 25 \approx \$18.5$ million

83. The total sales (in millions of dollars) over the first 2 yr (24 mo) is the area under the graph of $y = S'(t)$ from $t = 0$ to $t = 24$.

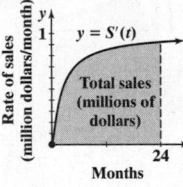

85. $p(x) = \dfrac{2(9x-4)}{135}(2+3x)^{3/2} - 2{,}000.83$; 54; \$37,932   **87.** $100\ln 3 \approx 110$ ft   **89.** $60\ln 5 \approx 97$ items

91. The area under the graph of $y = N'(t)$ from $t = 0$ to $t = 12$ represents the total number of items learned in that time interval.

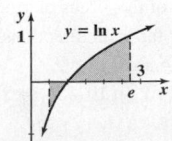

Chapter 6 Review Exercises

1. $\int_a^b f(x)\,dx$ *(6.1)*   **2.** $\int_b^c[-f(x)]\,dx$ *(6.1)*   **3.** $\int_a^b f(x)\,dx + \int_b^c[-f(x)]\,dx$ *(6.1)*

4. Area $= 1.153$ *(6.1)*   **5.** $\frac{1}{4}xe^{4x} - \frac{1}{16}e^{4x} + C$ *(6.3, 6.4)*   **6.** $\frac{1}{2}x^2\ln x - \frac{1}{4}x^2 + C$ *(6.3, 6.4)*   **7.** $\dfrac{(\ln x)^2}{2} + C$ *(5.2)*   **8.** $\dfrac{\ln(1+x^2)}{2} + C$ *(6.2)*

9. $\dfrac{1}{1+x} + \ln\left|\dfrac{x}{1+x}\right| + C$ *(6.4)*   **10.** $-\dfrac{\sqrt{1+x}}{x} - \frac{1}{2}\ln\left|\dfrac{\sqrt{1+x}-1}{\sqrt{1+x}+1}\right| + C$ *(6.4)*   **11.** 12 *(6.1)*   **12.** 40 *(6.1)*   **13.** 34.167 *(6.1)*

14. 0.926 *(6.1)*   **15.** 12 *(6.1)*   **16.** 18.133 *(6.1)*   **17.** Indonesia *(6.1)*   **18.** Vietnam *(6.1)*   **19.** $\int_a^b[f(x) - g(x)]\,dx$ *(6.1)*

20. $\int_b^c[g(x) - f(x)]\,dx$ *(6.1)*   **21.** $\int_b^c[g(x) - f(x)]\,dx + \int_c^d[f(x) - g(x)]\,dx$ *(6.1)*

22. $\int_a^b [f(x) - g(x)]dx + \int_b^c [g(x) - f(x)]dx + \int_c^d [f(x) - g(x)]dx$ *(6.1)*

23. Area $= 20.833$ *(6.1)*   **24.** 1 *(6.3, 6.4)*   **25.** $\frac{15}{2} - 8 \ln 8 + 8 \ln 4 \approx 1.955$ *(6.4)*   **26.** $\frac{1}{6}(3x\sqrt{9x^2 - 49} - 49 \ln|3x + \sqrt{9x^2 - 49}|) + C$ *(6.4)*

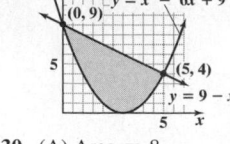

27. $-2te^{-0.5t} - 4e^{-0.5t} + C$ *(6.3, 6.4)*   **28.** $\frac{1}{3}x^3 \ln x - \frac{1}{9}x^3 + C$ *(6.3, 6.4)*   **29.** $x - \ln|1 + 2e^x| + C$ *(6.4)*

30. (A) Area $= 8$ (B) Area $= 8.38$ *(6.1)*

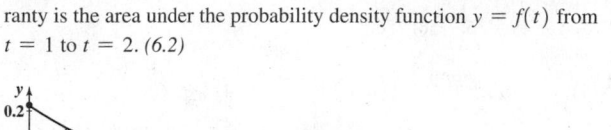

31. $4{,}109.36$ *(6.4)*   **32.** $2{,}640.35$ *(6.4)*   **33.** 4.87 *(6.4)*   **34.** 4.86 *(6.4)*   **35.** $\frac{1}{3}(\ln x)^3 + C$ *(5.2)*

36. $\frac{1}{2}x^2(\ln x)^2 - \frac{1}{2}x^2 \ln x + \frac{1}{4}x^2 + C$ *(6.3, 6.4)*   **37.** $\sqrt{x^2 - 36} + C$ *(5.2)*

38. $\frac{1}{2}\ln|x^2 + \sqrt{x^4 - 36}| + C$ *(6.4)*   **39.** $50 \ln 10 - 42 \ln 6 - 24 \approx 15.875$ *(6.3, 6.4)*

40. $x(\ln x)^2 - 2x \ln x + 2x + C$ *(6.3, 6.4)*   **41.** $-\frac{1}{4}e^{-2x^2} + C$ *(5.2)*

42. $-\frac{1}{2}x^2 e^{-2x} - \frac{1}{2}xe^{-2x} - \frac{1}{4}e^{-2x} + C$ *(6.3, 6.4)*   **43.** 1.703 *(6.1)*   **44.** (A) $.189$ (B) $.154$ *(6.2)*

45. The probability that the product will fail during the second year of warranty is the area under the probability density function $y = f(t)$ from $t = 1$ to $t = 2$. *(6.2)*

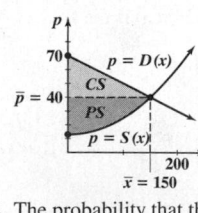

46. $R(x) = 65x - 6[(x + 1)\ln(x + 1) - x]$; 618/wk; \$29,506 *(6.3)*

47. (A) 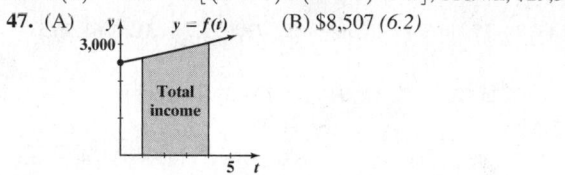 (B) \$8,507 *(6.2)*

48. (A) \$15,656   (B) \$1,454 *(6.2)*   **49.** (A)

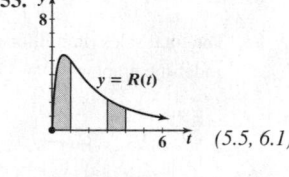

(B) More equitably distributed, since the area bounded by the two curves will have decreased.

(C) Current $= 0.3$; Projected $= 0.2$; income will be more equitably distributed 10 years from now *(6.1)*

50. (A) $CS = \$2{,}250$;   (B) $CS = \$2{,}890$; *(6.2)*   **51.** (A) 25.403 or 25,403 lb   (B) $PS = 121.6$ or \$1,216 *(6.2)*   **52.** 4.522 mL; 1.899 mL *(5.5, 6.4)*

$PS = \$2{,}700$ $PS = \$2{,}278$ **53.** **54.** $.667; .333$ *(6.2)*

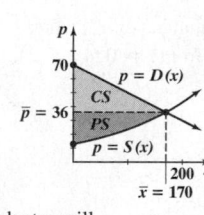

55. The probability that the doctor will spend more than an hour with a randomly selected patient is the area under the probability density function $y = f(t)$ from $t = 1$ to $t = 3$. *(6.2)*

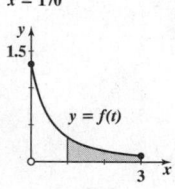

56. 45 thousand *(5.5, 6.1)*   **57.** $.368$ *(6.2)*

Chapter 7

Exercises 7.1   **1.** 19.5 ft^2 **3.** 240 in.^3 **5.** $32\pi \approx 100.5 \text{ m}^3$ **7.** $1{,}440\pi \approx 4{,}523.9 \text{ cm}^2$ **9.** -4 **11.** 11   **13.** 4   **15.** Not defined   **17.** -1
19. -64 **21.** $154{,}440$ **23.** $192\pi \approx 603.2$ **25.** $3\pi\sqrt{109} \approx 98.4$ **27.** 118   **29.** $36{,}095.07$ **31.** $f(x) = x^2 - 7$ **33.** $f(y) = 38y + 20$
35. $f(y) = -2y^3 + 5$ **37.** $y = 2$ **39.** $x = -\frac{1}{2}, 1$ **41.** $-1.926, 0.599$ **43.** $2x + h$ **45.** $2y^2$ **47.** $E(0, 0, 3); F(2, 0, 3)$ **49.** (A) In the plane $y = c$,
c any constant, $z = x^2$.   (B) The y axis; the line parallel to the y axis and passing through the point $(1, 0, 1)$; the line parallel to the y axis and passing through the point $(2, 0, 4)$ (C) A parabolic "trough" lying on top of the y axis   **51.** (A) Upper semicircles whose centers lie on the y axis   (B) Upper semicircles whose centers lie on the x axis   (C) The upper hemisphere of radius 6 with center at the origin   **53.** (A) $a^2 + b^2$ and $c^2 + d^2$ both equal the square of the radius of the circle.   (B) Bell-shaped curves with maximum values of 1 at the origin   (C) A bell, with maximum value 1 at the origin, extending infinitely far in all directions.
55. \$13,200; \$18,000; \$21,300   **57.** $R(p, q) = -5p^2 + 6pq - 4q^2 + 200p + 300q; R(2, 3) = \$1{,}280; R(3, 2) = \$1{,}175$ **59.** 30,065 units
61. (A) \$237,877.08   (B) 4.4%   **63.** $T(70, 47) \approx 29 \text{ min}; T(60, 27) = 33 \text{ min}$ **65.** $C(6, 8) = 75; C(8.1, 9) = 90$ **67.** $Q(12, 10) = 120; Q(10, 12) \approx 83$

Exercises 7.2   **1.** $f'(x) = 3\pi x^2 + \pi^3$ **3.** $f'(x) = exe^{x-1} + e^x$ **5.** $\dfrac{dz}{dx} = \dfrac{1}{e} - \dfrac{e}{x^2}$ **7.** $\dfrac{dz}{dx} = \dfrac{2x}{x^2 + e^2}$ **9.** $f_x(x, y) = 4$ **11.** $f_y(x, y) = -3x + 4y$

13. $\dfrac{\partial z}{\partial x} = 3x^2 + 8xy$ **15.** $\dfrac{\partial z}{\partial y} = 20(5x + 2y)^9$ **17.** 9   **19.** 3   **21.** -4 **23.** 0   **25.** 45.6 mpg; mileage is 45.6 mpg at a tire pressure of 32 psi and a speed of

40 mph   **27.** 37.6 mpg; mileage is 37.6 mpg at a tire pressure of 32 psi and a speed of 50 mph   **29.** 0.3 mpg per psi; mileage increases at a rate of

0.3 mpg per psi of tire pressure   **31.** $f_{xx}(x, y) = 0$ **33.** $f_{xy}(x, y) = 0$ **35.** $f_{xy}(x, y) = y^2 e^{xy^2}(2xy) + e^{xy^2}(2y) = 2y(1 + xy^2)e^{xy^2}$ **37.** $f_{yy}(x, y) = \dfrac{2 \ln x}{y^3}$

39. $f_{xx}(x, y) = 80(2x + y)^3$ **41.** $f_{xy}(x, y) = 720xy^3(x^2 + y^4)^8$ **43.** $C_x(x, y) = 6x + 10y + 4$ **45.** 2   **47.** $C_{xx}(x, y) = 6$ **49.** $C_{xy}(x, y) = 10$

51. 6   **53.** $3,000; daily sales are \$3,000 when the temperature is $60°$ and the rainfall is 2 in.   **55.** $-2,500$ \$/in.; daily sales decrease at a rate of \$2,500 per inch

of rain when the temperature is $90°$ and rainfall is 1 in.   **57.** -50 \$/in. per °F; S_r decreases at a rate of 50 \$/in. per degree of temperature

61. $f_{xx}(x, y) = 2y^2 + 6x; f_{xy}(x, y) = 4xy = f_{yx}(x, y); f_{yy}(x, y) = 2x^2$ **63.** $f_{xx}(x, y) = -2y/x^3; f_{xy}(x, y) = (-1/y^2) + (1/x^2) = f_{yx}(x, y); f_{yy}(x, y) = 2x/y^3$

65. $f_{xx}(x, y) = (2y + xy^2)e^{xy}; f_{xy}(x, y) = (2x + x^2y)e^{xy} = f_{yx}(x, y); f_{yy}(x, y) = x^3 e^{xy}$ **67.** $x = 2$ and $y = 4$ **69.** $x = 1.200$ and $y = -0.695$

71. (A) $-\frac{13}{3}$ (B) The function $f(0, y)$, for example, has values less than $-\frac{3}{13}$.   **73.** (A) $c = 1.145$ (B) $f_x(c, 2) = 0; f_y(c, 2) = 92.021$ **75.** (A) $2x$ (B) $4y$

77. $P_x(1,200, 1,800) = 24$; profit will increase approx. \$24 per unit increase in production of type A calculators at the $(1,200, 1,800)$ output level;

$P_y(1,200, 1,800) = -48$; profit will decrease approx. \$48 per unit increase in production of type B calculators at the $(1,200, 1,800)$ output level

79. $\partial x/\partial p = -5$: a \$1 increase in the price of brand A will decrease the demand for brand A by 5 lb at any price level (p, q); $\partial y/\partial p = 2$: a \$1 increase in the price

of brand A will increase the demand for brand B by 2 lb at any price level (p, q) **81.** (A) $f_x(x, y) = 7.5x^{-0.25}y^{0.25}; f_y(x, y) = 2.5x^{0.75}y^{-0.75}$ (B) Marginal

productivity of labor $= f_x(600, 100) \approx 4.79$; marginal productivity of capital $= f_y(600, 100) \approx 9.58$ (C) Capital   **83.** Competitive   **85.** Complementary

87. (A) $f_w(w, h) = 6.65w^{-0.575}h^{0.725}; f_h(w, h) = 11.34w^{0.425}h^{-0.275}$ (B) $f_w(65, 57) = 11.31$: for a 65-lb child 57 in. tall, the rate of change in surface area

is 11.31 in.^{2} for each pound gained in weight (height is held fixed); $f_h(65, 57) = 21.99$: for a child 57 in. tall, the rate of change in surface area is 21.99 in.^{2}

for each inch gained in height (weight is held fixed)   **89.** $C_W(6, 8) = 12.5$: index increases approx. 12.5 units for a 1-in. increase in width of head (length held

fixed) when $W = 6$ and $L = 8$; $C_L(6, 8) = -9.38$: index decreases approx. 9.38 units for a 1-in. increase in length (width held fixed) when $W = 6$ and $L = 8$.

Exercises 7.3   **1.** $f'(0) = 0; f''(0) = -18$; local maximum   **3.** $f'(0) = 0; f''(0) = 2$; local minimum   **5.** $f'(0) = 0; f''(0) = -2$; local maximum

7. $f'(0) = 1; f''(0) = -2$; neither   **9.** $f_x(x, y) = 4; f_y(x, y) = 5$; the functions $f_x(x, y)$ and $f_y(x, y)$ never have the value 0.   **11.** $f_x(x, y) = -1.2 + 4x^3$;

$f_y(x, y) = 6.8 + 0.6y^2$; the function $f_y(x, y)$ never has the value 0.   **13.** $f(-2, 0) = 10$ is a local maximum.   **15.** $f(-1, 3) = 4$ is a local minimum.

17. f has a saddle point at $(3, -2)$.   **19.** $f(3, 2) = 33$ is a local maximum.   **21.** $f(2, 2) = 8$ is a local minimum.   **23.** f has a saddle point at $(0, 0)$.

25. f has a saddle point at $(0, 0); f(1, 1) = -1$ is a local minimum.   **27.** f has a saddle point at $(0, 0); f(3, 18) = -162$ and $f(-3, -18) = -162$ are

local minima.   **29.** The test fails at $(0, 0); f$ has saddle points at $(2, 2)$ and $(2, -2)$.   **31.** f has a saddle point at $(0.614, -1.105)$.   **33.** $f(x, y)$ is nonnegative

and equals 0 when $x = 0$, so f has the local minimum 0 at each point of the y axis.   **35.** (B) Local minimum   **37.** 2,000 type A and 4,000 type B; max $P =$

$P(2, 4) = \$15$ million   **39.** (A) When $p = \$100$ and $q = \$120$, $x = 80$ and $y = 40$; when $p = \$110$ and $q = \$110$, $x = 40$ and $y = 70$ (B) A maximum

weekly profit of \$4,800 is realized for $p = \$100$ and $q = \$120$.   **41.** $P(x, y) = P(4, 2)$ **43.** 8 in. by 4 in. by 2 in.   **45.** 20 in. by 20 in. by 40 in.

Exercises 7.4   **1.** Min $f(x, y) = f(-2, 4) = 12$ **3.** Min $f(x, y) = f(1, -1) = -4$ **5.** Max $f(x, y) = f(1, 0) = 2$

7. Max $f(x, y) = f(3, 3) = 18$ **9.** Min $f(x, y) = f(3, 4) = 25$ **11.** $F_x = -3 + 2\lambda = 0$ and $F_y = 4 + 5\lambda = 0$ have no simultaneous solution.

13. Max $f(x, y) = f(3, 3) = f(-3, -3) = 18$; min $f(x, y) = f(3, -3) = f(-3, 3) = -18$ **15.** Maximum product is 25 when each number is 5.

17. Min $f(x, y, z) = f(-4, 2, -6) = 56$ **19.** Max $f(x, y, z) = f(2, 2, 2) = 6$; min $f(x, y, z) = f(-2, -2, -2) = -6$

21. Max $f(x, y) = f(0.217, 0.885) = 1.055$ **23.** $F_x = e^x + \lambda = 0$ and $F_y = 3e^y - 2\lambda = 0$ have no simultaneous solution.   **25.** Maximize $f(x, 5)$, a

function of just one independent variable.   **27.** (A) Max $f(x, y) = f(0.707, 0.5) = f(-0.707, 0.5) = 0.47$ **29.** 60 of model A and 30 of model B will yield a

minimum cost of \$32,400 per week.   **31.** (A) 8,000 units of labor and 1,000 units of capital; max $N(x, y) = N(8,000, 1,000) \approx 263,902$ units   (B) Marginal

productivity of money ≈ 0.6598; increase in production $\approx 32,990$ units   **33.** 8 in. by 8 in. by $\frac{8}{3}$ in.   **35.** $x = 50$ ft and $y = 200$ ft; maximum area is 10,000 ft^2

Exercises 7.5   **1.** 10   **3.** 98   **5.** 380   **7.** $y = 0.7x + 1$ **9.** $y = -2.5x + 10.5$ **11.** $y = x + 2$ **13.** $y = -1.5x + 4.5$; $y = 0.75$ when $x = 2.5$

15. $y = 2.12x + 10.8$; $y = 63.8$ when $x = 25$

17. $y = -1.2x + 12.6$; $y = 10.2$ when $x = 2$

19. $y = -1.53x + 26.67$; $y = 14.4$ when $x = 8$

21. $y = 0.75x^2 - 3.45x + 4.75$ **27.** (A) $y = 1.52x - 0.16$; $y = 0.73x^2 - 1.39x + 1.30$ (B) The quadratic function   **29.** The normal equations form a

system of 4 linear equations in the 4 variables a, b, c, and d, which can be solved by Gauss–Jordan elimination.

31. (A) $y = -79.36x + 3793.6$ (B) 1,889 crimes per 100,000 population   **33.** (A) $y = -0.48x + 4.38$

(B) \$6.56 per bottle   **35.** (A) $y = 0.0222x + 18.94$ (B) 19.92 ft   **37.** (A) $y = 0.0121x + 56.35$ (B) 58.77°F

Exercises 7.6   **1.** $\pi x + \dfrac{x^2}{2} + C$ **3.** $x + \pi \ln |x| + C$ **5.** $\dfrac{e^{\pi x}}{\pi} + C$ **7.** (A) $3x^2y^4 + C(x)$ (B) $3x^2$ **9.** (A) $2x^2 + 6xy + 5x + E(y)$ (B) $35 + 30y$

11. (A) $\sqrt{y + x^2} + E(y)$ (B) $\sqrt{y + 4} - \sqrt{y}$ **13.** (A) $\dfrac{\ln x \ln y}{x} + C(x)$ (B) $\dfrac{2 \ln x}{x}$ **15.** 9   **17.** 330   **19.** $(56 - 20\sqrt{5})/3$ **21.** 1   **23.** 16

25. 49   **27.** $\frac{1}{8}\int_1^5 \int_{-1}^1 (x + y)^2 dy\, dx = \frac{32}{3}$ **29.** $\frac{1}{15}\int_1^4 \int_2^7 (x/y)\, dy\, dx = \frac{1}{2}\ln\frac{7}{2} \approx 0.6264$ **31.** $\frac{4}{3}$ cubic units   **33.** $\frac{32}{3}$ cubic units

35. $\int_0^1 \int_1^2 x e^{xy}\, dy\, dx = \frac{1}{2} + \frac{1}{2}e^2 - e$　**37.** $\int_0^1 \int_{-1}^1 \dfrac{2y + 3xy^2}{1 + x^2}\, dy\, dx = \ln 2$　**41.** (A) $\dfrac{1}{3} + \dfrac{1}{4}e^{-2} - \dfrac{1}{4}e^2$　(B)

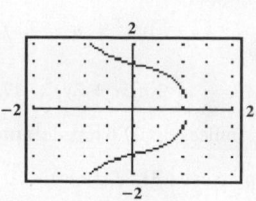

(C) Points to the right of the graph in part (B) are greater than 0; points to the left of the graph are less than 0.

43. $\dfrac{1}{0.4}\int_{0.6}^{0.8}\int_5^7 \dfrac{y}{1 - x}\, dy\, dx = 30 \ln 2 \approx \$20.8$ billion　**45.** $\frac{1}{10}\int_{10}^{20}\int_1^2 x^{0.75}y^{0.25}\, dy\, dx = \frac{8}{175}(2^{1.25} - 1)(20^{1.75} - 10^{1.75}) \approx 8.375$ or $8,375$ units

47. $\dfrac{1}{192}\int_{-8}^8\int_{-6}^6 [10 - \frac{1}{10}(x^2 + y^2)]\, dy\, dx = \frac{20}{3}$ insects/ft^2　**49.** $\frac{1}{8}\int_{-2}^2\int_{-1}^1 [100 - 15(x^2 + y^2)]\, dy\, dx = 75$ ppm

51. $\dfrac{1}{10,000}\int_{2,000}^{3,000}\int_{50}^{60} 0.000\,013\,3xy^2\, dy\, dx \approx 100.86$ ft　**53.** $\frac{1}{16}\int_8^{16}\int_{10}^{12} 100\dfrac{x}{y}\, dy\, dx = 600 \ln 1.2 \approx 109.4$

Exercises 7.7

1. $R = \{(x, y)\,|\,0 \le y \le 4 - x^2, 0 \le x \le 2\}$
　$R = \{(x, y)\,|\,0 \le x \le \sqrt{4 - y}, 0 \le y \le 4\}$

3. R is a regular x region:
　$R = \{(x, y)\,|\,x^3 \le y \le 12 - 2x, 0 \le x \le 2\}$

5. R is a regular y region:
　$R = \{(x, y)\,|\,\frac{1}{2}y^2 \le x \le y + 4, -2 \le y \le 4\}$

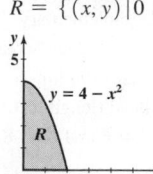

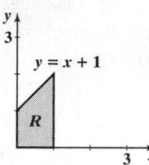

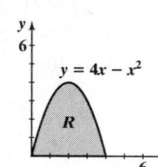

7. $\frac{1}{2}$　**9.** $\frac{39}{70}$　**11.** R consists of the points on or inside the rectangle with corners $(\pm 2, \pm 3)$; both　**13.** R is the arch-shaped region consisting of the points on or inside the rectangle with corners $(\pm 2, 0)$ and $(\pm 2, 2)$ that are not inside the circle of radius 1 centered at the origin; regular x region　**15.** $\frac{56}{3}$　**17.** $-\frac{3}{4}$
19. $\frac{1}{2}e^4 - \frac{5}{2}$　**21.** $R = \{(x, y)\,|\,0 \le y \le x + 1, 0 \le x \le 1\}$　**23.** $R = \{(x, y)\,|\,0 \le y \le 4x - x^2, 0 \le x \le 4\}$

　　$\int_0^1 \int_0^{x+1} \sqrt{1 + x + y}\, dy\, dx = (68 - 24\sqrt{2})/15$　　$\int_0^4 \int_0^{4x - x^2} \sqrt{y + x^2}\, dy\, dx = \frac{128}{5}$

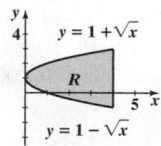

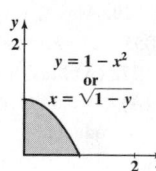

25. $R = \{(x, y)\,|\,1 - \sqrt{x} \le y \le 1 + \sqrt{x}, 0 \le x \le 4\}$　**27.** $\int_0^3\int_0^{3-y}(x + 2y)\, dx\, dy = \frac{27}{2}$　**29.** $\int_0^1\int_0^{\sqrt{1-y}} x\sqrt{y}\, dx\, dy = \frac{2}{15}$　**31.** $\int_0^1\int_{4y^2}^{4y} x\, dx\, dy = \frac{16}{15}$

　　$\int_0^4 \int_{1-\sqrt{x}}^{1+\sqrt{x}} x(y - 1)^2\, dy\, dx = \frac{512}{21}$

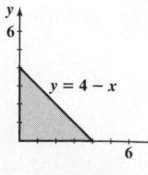

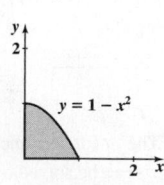

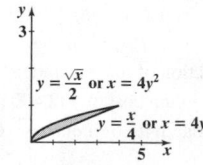

33. $\int_0^4\int_0^{4-x}(4 - x - y)\, dy\, dx = \frac{32}{3}$　**35.** $\int_0^1\int_0^{1-x^2} 4\, dy\, dx = \frac{8}{3}$
37. $\int_0^4 \int_0^{\sqrt{y}} \dfrac{4x}{1 + y^2}\, dx\, dy = \ln 17$　**39.** $\int_0^1\int_0^{\sqrt{x}} 4ye^{x^2}\, dy\, dx = e - 1$

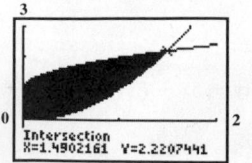

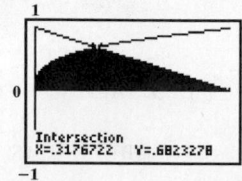

41. $R = \{(x, y)\,|\,x^2 \le y \le 1 + \sqrt{x}, 0 \le x \le 1.49\}$　**43.** $R = \{(x, y)\,|\,y^3 \le x \le 1 - y, 0 \le y \le 0.68\}$

　　$\int_0^{1.49}\int_{x^2}^{1+\sqrt{x}} x\, dy\, dx \approx 0.96$　　$\int_0^{0.68}\int_{y^3}^{1-y} 24xy\, dx\, dy \approx 0.83$

45. $R = \{(x, y) \,|\, e^{-x} \le y \le 3 - x, -1.51 \le x \le 2.95\}$ Regular x region
$R = \{(x, y) \,|\, -\ln y \le x \le 3 - y, 0.05 \le y \le 4.51\}$ Regular y region
$\int_{-1.51}^{2.95} \int_{e^{-x}}^{3-x} 4y \, dy \, dx = \int_{0.05}^{4.51} \int_{-\ln y}^{3-y} 4y \, dx \, dy \approx 40.67$

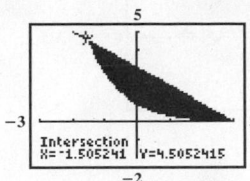

Chapter 7 Review Exercises   **1.** $f(5, 10) = 2,900; f_x(x, y) = 40; f_y(x, y) = 70 \,(7.1, 7.2)$ **2.** $\partial^2 z/\partial x^2 = 6xy^2; \partial^2 z/\partial x \, \partial y = 6x^2 y \,(7.2)$
3. $2xy^3 + 2y^2 + C(x) \,(7.6)$ **4.** $3x^2 y^2 + 4xy + E(y) \,(7.6)$ **5.** $1 \,(7.6)$ **6.** $f_x(x, y) = 5 + 6x + 3x^2; f_y(x, y) = -2;$ the function $f_y(x, y)$ never has the
value 0. (7.3) **7.** $f(2, 3) = 7; f_y(x, y) = -2x + 2y + 3; f_y(2, 3) = 5 \,(7.1, 7.2)$ **8.** $(-8)(-6) - (4)^2 = 32 \,(7.2)$ **9.** $(1, 3, -\frac{1}{2}), (-1, -3, \frac{1}{2}) \,(7.4)$
10. $y = -1.5x + 15.5; y = 0.5$ when $x = 10 \,(7.5)$ **11.** $18 \,(7.6)$ **12.** $\frac{8}{5} \,(7.7)$ **13.** $f_x(x, y) = 2xe^{x^2 + 2y}; f_y(x, y) = 2e^{x^2 + 2y}; f_{xy}(x, y) = 4xe^{x^2 + 2y} \,(7.2)$
14. $f_x(x, y) = 10x(x^2 + y^2)^4; f_{xy}(x, y) = 80xy(x^2 + y^2)^3 \,(7.2)$ **15.** $f(2, 3) = -25$ is a local minimum; f has a saddle point at $(-2, 3) . \,(7.3)$
16. Max $f(x, y) = f(6, 4) = 24 \,(7.4)$ **17.** Min $f(x, y, z) = f(2, 1, 2) = 9 \,(7.4)$ **18.** $y = \frac{116}{165} x + \frac{100}{3} \,(7.5)$ **19.** $\frac{27}{5} \,(7.6)$ **20.** 4 cubic units (7.6)
21. $0 \,(7.6)$ **22.** (A) 12.56   (B) No (7.6) **23.** $F_x = 12x^2 + 3\lambda = 0, F_y = -15y^2 + 2\lambda = 0,$ and $F_\lambda = 3x + 2y - 7 = 0$ have no simultaneous solution. (7.4)
24. $1 \,(7.7)$ **25.** (A) $P_x(1, 3) = 8;$ profit will increase \$8,000 for a 100-unit increase in product A if the production of product B is held fixed at an output level
of $(1, 3)$.   (B) For 200 units of A and 300 units of B, $P(2, 3) = \$100$ thousand is a local maximum. $(7.2, 7.3)$ **26.** $x = 6$ in., $y = 8$ in., $z = 2$ in. (7.3)
27. $y = 0.63x + 1.33;$ profit in sixth year is \$5.11 million (7.4) **28.** (A) Marginal productivity of labor $\approx 8.37;$ marginal productivity of capital $\approx 1.67;$
management should encourage increased use of labor.   (B) 80 units of labor and 40 units of capital; max $N(x, y) = N(80, 40) \approx 696$ units; marginal productiv-
ity of money $\approx 0.0696;$ increase in production ≈ 139 units   (C) $\dfrac{1}{1,000} \int_{50}^{100} \int_{20}^{40} 10x^{0.8} y^{0.2} \, dy \, dx = \dfrac{(40^{1.2} - 20^{1.2})(100^{1.8} - 50^{1.8})}{216} = 621$ items (7.4)

29. $T_x(70, 17) = -0.924$ min/ft increase in depth when $V = 70$ ft^3 and $x = 17$ ft (7.2) **30.** $\frac{1}{16} \int_{-2}^{2} \int_{-2}^{2} [100 - 24(x^2 + y^2)] dy \, dx = 36$ ppm (7.6)
31. $50,000 \,(7.1)$ **32.** $y = \frac{1}{2}x + 48; y = 68$ when $x = 40 \,(7.5)$ **33.** (A) $y = 0.734x + 49.93$ (B) 97.64 people/mi^2 (C) 101.10 people/mi^2;
103.70 people/mi^2 (7.5) **34.** (A) $y = 1.069x + 0.522$ (B) 64.68 yr   (C) 64.78 yr; 64.80 yr (7.5)

Chapter 8

Exercises 8.1   **1.** $\dfrac{\pi}{3}$ rad   **3.** $\dfrac{3\pi}{4}$ rad   **5.** $-45°$ **7.** $270°$ **9.** $\sqrt{3}/2$ **11.** $-1/\sqrt{2}$ **13.** 1   **15.** 0   **17.** $-1/\sqrt{2}$ **19.** $-1/2$ **21.** -1 **23.** $\sqrt{3}/2$
27. -1 **29.** $2/\sqrt{3}$ **31.** Not defined   **33.** $\sqrt{3}$ **35.** -2 **37.** -1 **39.** $-\sqrt{3}$ **41.** -1 **43.** 0.1736   **45.** 0.6157   **47.** 1.5574   **49.** 92.6259   **51.** 57.2987
53. 3.0777   **55.**

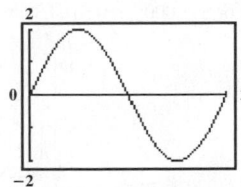

57.

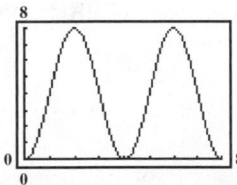

59. $\left\{ x \,|\, x \ne \pm \dfrac{\pi}{2}, \pm \dfrac{3\pi}{2}, \pm \dfrac{5\pi}{2}, \ldots \right\}$ **61.** $\left\{ x \,|\, x \ne \pm \dfrac{\pi}{2}, \pm \dfrac{3\pi}{2}, \pm \dfrac{5\pi}{2}, \ldots \right\}$

67. (A) $P(13) = 5, P(26) = 10, P(39) = 5, P(52) = 0$
(B) $P(30) \approx 9.43, P(100) \approx 0.57;$ 30 weeks after
January 1 the profit on a week's sales of bathing suits
is \$943, and 100 weeks after January 1 the profit on a
week's sales of bathing suits is \$57.

69. (A) $V(0) = 0.10, V(1) = 0.45, V(2) = 0.80, V(3) = 0.45,$
$V(7) = 0.45$
(B) $V(3.5) \approx 0.20, V(5.7) \approx 0.76;$ the volume of air in the lungs of
a normal seated adult 3.5 sec after exhaling is approximately 0.20 L
and 5.7 sec after exhaling is approx. 0.76 L.

71. (A) $-5.6°$
(B) $-4.7°$

(C)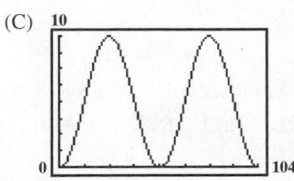

(C)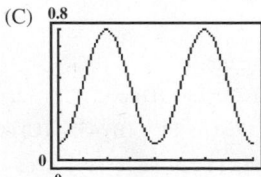

Exercises 8.2   **1.** Decreasing   **3.** Increasing   **5.** Concave down   **7.** Concave down   **9.** $-5 \sin x$ **11.** $-5 \sin(5x)$ **13.** $2x \cos(x^2 + 1)$

15. $\cos(w + \pi)$ **17.** $t \cos t + \sin t$ **19.** $(\cos x)^2 - (\sin x)^2$ **21.** $5(\sin x)^4 \cos x$ **23.** $\dfrac{\cos x}{2\sqrt{\sin x}}$ **25.** $-\dfrac{x^{-1/2}}{2} \sin \sqrt{x} = \dfrac{-\sin \sqrt{x}}{2\sqrt{x}}$

27. $f'\left(\dfrac{\pi}{6}\right) = \cos \dfrac{\pi}{6} = \dfrac{\sqrt{3}}{2}$

29. Increasing on $[-\pi, 0]$; decreasing on $[0, \pi]$; concave upward on $[-\pi, -\pi/2]$ and $[\pi/2, \pi]$; concave downward on $[-\pi/2, \pi/2]$; local maximum at $x = 0$; $f(x) = \cos x$; $f'(x) = -\sin x$

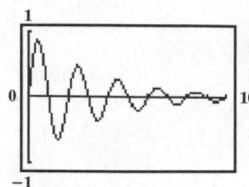

31. $-\pi \csc(\pi x)\cot(\pi x)$ **33.** $-\dfrac{\pi}{2}\csc^2\left(\dfrac{\pi x}{2}\right)$ **35.** $-(x + 1)e^x \sin(xe^x)$ **37.** $2x \sec^2(x^2)$

39. $2e^x \cos x$ **41.**

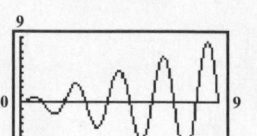

43.

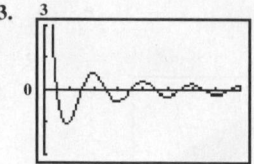

45.

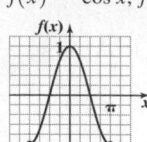

47. (A) $P'(t) = \dfrac{5\pi}{26}\sin\dfrac{\pi t}{26}$, $0 < t < 104$ (B) $P'(8) = \$0.50$ hundred, or $50 per week; $P'(26) = \$0$ per week; $P'(50) = -\$0.14$ hundred, or $-\$14$ per week

(C)

t	$P(t)$	
26	$1,000	Local maximum
52	$0	Local minimum
78	$1,000	Local maximum

(D)

t	$P(t)$	
0	$0	Absolute minimum
26	$1,000	Absolute maximum
52	$0	Absolute minimum
78	$1,000	Absolute maximum
104	$0	Absolute minimum

(E) Same answer as for part (C)

49. (A) $V'(t) = \dfrac{0.35\pi}{2}\sin\dfrac{\pi t}{2}$, $0 \le t \le 8$ (B) $V'(3) = -0.55$ L/sec; $V'(4) = 0.00$ L/sec; $V'(5) = 0.55$ L/sec

(C)

t	$V(t)$	
2	0.80	Local maximum
4	0.10	Local minimum
6	0.80	Local maximum

(D)

t	$V(t)$	
0	0.10	Absolute minimum
2	0.80	Absolute maximum
4	0.10	Absolute minimum
6	0.80	Absolute maximum
8	0.10	Absolute minimum

(E) Same answer as for part (C)

Exercises 8.3   **1.** $[0, \pi]$ **3.** $[0, \pi/3) \cup (5\pi/3, 2\pi]$ **5.** $\{\pi/4, 3\pi/4, 5\pi/4, 7\pi/4\}$ **7.** $(-\pi/2, \pi/4]$ **9.** $-\cos t + C$ **11.** $\frac{1}{3}\sin 3x + C$

13. $\frac{1}{13}(\sin x)^{13} + C$ **15.** $-\frac{3}{4}(\cos x)^{4/3} + C$ **17.** $\frac{1}{3}\sin x^3 + C$ **19.** 1   **21.** 1   **23.** $\sqrt{3}/2 - \frac{1}{2} \approx 0.366$ **25.** 1.4161   **27.** 0.0678   **29.** $e^{\sin x} + C$

31. $\ln|\sin x| + C$ **33.** $-\ln|\cos x| + C$ **35.** (A)

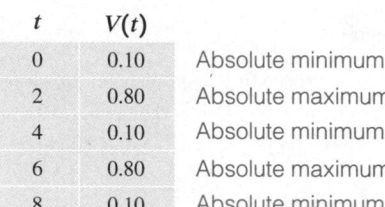

(B) $L_6 \approx 0.498$ **37.** (A) $520 hundred, or $52,000   (B) $106.38 hundred, or $10,638

(C)

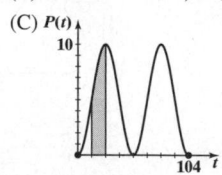

39. (A) 104 tons   (B) 31 tons   (C)

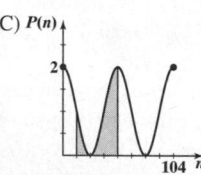

Chapter 8 Review Exercises   **1.** (A) $\pi/6$ (B) $\pi/4$ (C) $\pi/3$ (D) $\pi/2$ *(8.1)*   **2.** (A) -1 (B) 0   (C) 1 *(8.1)*   **3.** $-\sin m$ *(8.2)*   **4.** $\cos u$ *(8.2)*

5. $(2x - 2)\cos(x^2 - 2x + 1)$ *(8.2)*   **6.** $-\frac{1}{3}\cos 3t + C$ *(8.3)*   **7.** (A) $30°$ (B) $45°$ (C) $60°$ (D) $90°$ *(8.1)*   **8.** (A) $\frac{1}{2}$ (B) $\sqrt{2}/2$ (C) $\sqrt{3}/2$ *(8.1)*

9. (A) -0.6543 (B) 0.8308 *(8.1)*   **10.** $(x^2 - 1)\cos x + 2x \sin x$ *(8.2)*   **11.** $6(\sin x)^5 \cos x$ *(8.2)*   **12.** $(\cos x)/[3(\sin x)^{2/3}]$ *(8.2)*

13. $\frac{1}{2}\sin(t^2 - 1) + C$ *(8.3)*   **14.** 2 *(8.3)*   **15.** $\sqrt{3}/2$ *(8.3)*   **16.** -0.243 *(8.3)*   **17.** $-\sqrt{2}/2$ *(8.2)*   **18.** $\sqrt{2}$ *(8.3)*

19. (A)

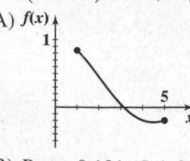

20. $\pi/12$ *(8.1)*   **21.** (A) -1 (B) $-\sqrt{3}/2$ (C) $-\frac{1}{2}$ *(8.1)*

22. $1/(\cos u)^2 = (\sec u)^2$ *(8.2)*   **23.** $-2x(\sin x^2)e^{\cos x^2}$ *(8.2)*

24. $e^{\sin x} + C$ *(8.3)*   **25.** $-\ln|\cos x| + C$ *(8.3)*   **26.** 15.2128 *(8.3)*

(B) $R_4 \approx 0.121$ *(5.4, 8.3)*

27. *(8.2, 8.3)*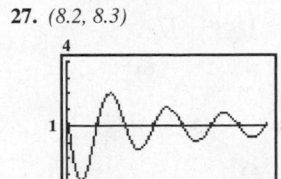

28. *(8.2, 8.3)*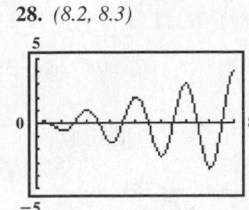

29. *(8.2, 8.3)*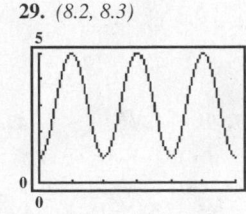

30. (A) $R(0) =$ \$5 thousand; $R(2) =$ \$4 thousand; $R(3) =$ \$3 thousand; $R(6) =$ \$1 thousand
(B) $R(1) =$ \$4.732 thousand; $R(22) =$ \$4 thousand; the revenue is \$4,732 for a month of sweater sales 1 month after January 1, and \$4,000 for a month of sweater sales 22 months after January 1. *(8.1)*

31. (A) $R'(t) = -\dfrac{\pi}{3}\sin\dfrac{\pi t}{6}, 0 \le t \le 24$ (B) $R'(3) = -\$1.047$ thousand, or $-\$1.047/$mo; $R'(10) = \$0.907$ thousand, or \$907/mo; $R'(18) = \$0.000$ thousand

(C)

t	$R(t)$	
6	\$1,000	Local minimum
12	\$5,000	Local maximum
18	\$1,000	Local minimum

(D)

t	$R(t)$	
0	\$5,000	Absolute maximum
6	\$1,000	Absolute minimum
12	\$5,000	Absolute maximum
18	\$1,000	Absolute minimum
24	\$5,000	Absolute maximum

(E) Same answer as for part (C) *(8.2)*

32. (A) \$72 thousand, or \$72,000   (B) \$6.270 thousand, or \$6,270   (C) 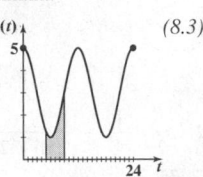   *(8.3)*

Appendix A

Exercises A.1   **1.** vu **3.** $(3 + 7) + y$ **5.** $u + v$ **7.** T   **9.** T   **11.** F   **13.** T   **15.** T   **17.** T   **19.** T   **21.** F   **23.** T   **25.** T   **27.** No
29. (A) F   (B) T   (C) T   **31.** $\sqrt{2}$ and π are two examples of infinitely many.   **33.** (A) N, Z, Q, R (B) R (C) Q, R (D) Q, R
35. (A) F, since, for example, $2(3 - 1) \neq 2 \cdot 3 - 1$ (B) F, since, for example, $(8 - 4) - 2 \neq 8 - (4 - 2)$ (C) T   (D) F, since, for example,
$(8 \div 4) \div 2 \neq 8 \div (4 \div 2)$.   **37.** $\dfrac{1}{11}$ **39.** (A) $2.166\,666\,666\ldots$ (B) $4.582\,575\,69\ldots$ (C) $0.437\,500\,000\ldots$ (D) $0.261\,261\,261\ldots$ **41.** (A) 3   (B) 2
43. (A) 2   (B) 6   **45.** \$16.42   **47.** 2.8%

Exercises A.2   **1.** 3   **3.** $x^3 + 4x^2 - 2x + 5$ **5.** $x^3 + 1$ **7.** $2x^5 + 3x^4 - 2x^3 + 11x^2 - 5x + 6$ **9.** $-5u + 2$ **11.** $6a^2 + 6a$ **13.** $a^2 - b^2$
15. $6x^2 - 7x - 5$ **17.** $2x^2 + xy - 6y^2$ **19.** $9y^2 - 4$ **21.** $-4x^2 + 12x - 9$ **23.** $16m^2 - 9n^2$ **25.** $9u^2 + 24uv + 16v^2$ **27.** $a^3 - b^3$
29. $x^2 - 2xy + y^2 - 9z^2$ **31.** 1   **33.** $x^4 - 2x^2y^2 + y^4$ **35.** $-40ab$ **37.** $-4m + 8$ **39.** $-6xy$ **41.** $u^3 + 3u^2v + 3uv^2 + v^3$
43. $x^3 - 6x^2y + 12xy^2 - 8y^3$ **45.** $2x^2 - 2xy + 3y^2$ **47.** $x^4 - 10x^3 + 27x^2 - 10x + 1$ **49.** $4x^3 - 14x^2 + 8x - 6$ **51.** $m + n$ **53.** No change
55. $(1 + 1)^2 \neq 1^2 + 1^2$; either a or b must be 0   **57.** $0.09x + 0.12(10{,}000 - x) = 1{,}200 - 0.03x$
59. $20x + 30(3x) + 50(4{,}000 - x - 3x) = 200{,}000 - 90x$ **61.** $0.02x + 0.06(10 - x) = 0.6 - 0.04x$

Exercises A.3   **1.** $3m^2(2m^2 - 3m - 1)$ **3.** $2uv(4u^2 - 3uv + 2v^2)$ **5.** $(7m + 5)(2m - 3)$ **7.** $(4ab - 1)(2c + d)$ **9.** $(2x - 1)(x + 2)$
11. $(y - 1)(3y + 2)$ **13.** $(x + 4)(2x - 1)$ **15.** $(w + x)(y - z)$ **17.** $(a - 3b)(m + 2n)$ **19.** $(3y + 2)(y - 1)$ **21.** $(u - 5v)(u + 3v)$
23. Not factorable   **25.** $(wx - y)(wx + y)$ **27.** $(3m - n)^2$ **29.** Not factorable   **31.** $4(z - 3)(z - 4)$ **33.** $2x^2(x - 2)(x - 10)$ **35.** $x(2y - 3)^2$
37. $(2m - 3n)(3m + 4n)$ **39.** $uv(2u - v)(2u + v)$ **41.** $2x(x^2 - x + 4)$ **43.** $(2x - 3y)(4x^2 + 6xy + 9y^2)$ **45.** $xy(x + 2)(x^2 - 2x + 4)$
47. $[(x + 2) - 3y][(x + 2) + 3y]$ **49.** Not factorable   **51.** $(6x - 6y - 1)(x - y + 4)$ **53.** $(y - 2)(y + 2)(y^2 + 1)$
55. $3(x - y)^2(5xy - 5y^2 + 4x)$ **57.** True   **59.** False

Exercises A.4   **1.** 39/7   **3.** 495   **5.** $8d^6$ **7.** $\dfrac{15x^2 + 10x - 6}{180}$ **9.** $\dfrac{15m^2 + 14m - 6}{36m^3}$ **11.** $\dfrac{1}{x(x - 4)}$ **13.** $\dfrac{x - 6}{x(x - 3)}$ **15.** $\dfrac{-3x - 9}{(x - 2)(x + 1)^2}$
17. $\dfrac{2}{x - 1}$ **19.** $\dfrac{5}{a - 1}$ **21.** $\dfrac{x^2 + 8x - 16}{x(x - 4)(x + 4)}$ **23.** $\dfrac{7x^2 - 2x - 3}{6(x + 1)^2}$ **25.** $\dfrac{x(y - x)}{y(2x - y)}$ **27.** $\dfrac{-17c + 16}{15(c - 1)}$ **29.** $\dfrac{1}{x - 3}$ **31.** $\dfrac{-1}{2x(x + h)}$ **33.** $\dfrac{x - y}{x + y}$
35. (A) Incorrect   (B) $x + 1$ **37.** (A) Incorrect   (B) $2x + h$ **39.** (A) Incorrect   (B) $\dfrac{x^2 - x - 3}{x + 1}$ **41.** (A) Correct   **43.** $\dfrac{-2x - h}{3(x + h)^2x^2}$ **45.** $\dfrac{x(x - 3)}{x - 1}$

Exercises A.5   **1.** $2/x^9$ **3.** $3w^7/2$ **5.** $2/x^3$ **7.** $1/w^5$ **9.** $4/a^6$ **11.** $1/a^6$ **13.** $1/8x^{12}$ **15.** 8.23×10^{10} **17.** 7.83×10^{-1} **19.** 3.4×10^{-5}
21. 40,000   **23.** 0.007   **25.** 61,710,000   **27.** 0.000 808   **29.** 1   **31.** 10^{14} **33.** $y^6/25x^4$ **35.** $4x^6/25$ **37.** $4y^3/3x^5$ **39.** $\dfrac{7}{4} - \dfrac{1}{4}x^{-3}$
41. $\dfrac{5}{2}x^2 - \dfrac{3}{2} + 4x^{-2}$ **43.** $\dfrac{x^2(x - 3)}{(x - 1)^3}$ **45.** $\dfrac{2(x - 1)}{x^3}$ **47.** 2.4×10^{10}; 24,000,000,000   **49.** 3.125×10^4; 31,250   **51.** 64   **55.** uv **57.** $\dfrac{bc(c + b)}{c^2 + bc + b^2}$
59. (A) \$51,329   (B) \$1,150   (C) 2.24%   **61.** (A) 9×10^{-6} (B) 0.000 009   (C) 0.0009%   **63.** 1,248,000

Exercises A.6   **1.** $6\sqrt[5]{x^3}$ **3.** $\sqrt[5]{(32x^2y^3)^3}$ **5.** $\sqrt{x^2+y^2}$ (not $x+y$)   **7.** $5x^{3/4}$ **9.** $(2x^2y)^{3/5}$ **11.** $x^{1/3}+y^{1/3}$ **13.** 5   **15.** 64   **17.** -7
19. -16 **21.** $\dfrac{8}{125}$ **23.** $\dfrac{1}{27}$ **25.** $x^{2/5}$ **27.** m **29.** $2x/y^2$ **31.** $xy^2/2$ **33.** $1/(24x^{7/12})$ **35.** $2x+3$ **37.** $30x^5\sqrt{3x}$ **39.** 2   **41.** $12x-6x^{35/4}$
43. $3u-13u^{1/2}v^{1/2}+4v$ **45.** $36m^{3/2}-\dfrac{6m^{1/2}}{n^{1/2}}+\dfrac{6m}{n^{1/2}}-\dfrac{1}{n}$ **47.** $9x-6x^{1/2}y^{1/2}+y$ **49.** $\dfrac{1}{2}x^{1/3}+x^{-1/3}$ **51.** $\dfrac{2}{3}x^{-1/4}+\dfrac{1}{3}x^{-2/3}$ **53.** $\dfrac{1}{2}x^{-1/6}-\dfrac{1}{4}$
55. $4n\sqrt{3mn}$ **57.** $\dfrac{2(x+3)\sqrt{x-2}}{x-2}$ **59.** $7(x-y)(\sqrt{x}+\sqrt{y})$ **61.** $\dfrac{1}{xy\sqrt{5xy}}$ **63.** $\dfrac{1}{\sqrt{x+h}+\sqrt{x}}$ **65.** $\dfrac{1}{(t+x)(\sqrt{t}+\sqrt{x})}$
67. $x=y=1$ is one of many choices.   **69.** $x=y=1$ is one of many choices.   **71.** False   **73.** False   **75.** False   **77.** True   **79.** True   **81.** False
83. $\dfrac{x+8}{2(x+3)^{3/2}}$ **85.** $\dfrac{x-2}{2(x-1)^{3/2}}$ **87.** $\dfrac{x+6}{3(x+2)^{5/3}}$ **89.** 103.2   **91.** 0.0805   **93.** 4,588   **95.** (A) and (E); (B) and (F); (C) and (D)

Exercises A.7   **1.** $\pm\sqrt{11}$ **3.** $-\dfrac{4}{3},2$ **5.** $-2,6$ **7.** 0, 2   **9.** $3\pm2\sqrt{3}$ **11.** $-2\pm\sqrt{2}$ **13.** $0,\dfrac{15}{2}$ **15.** $\pm\dfrac{3}{2}$ **17.** $\dfrac{1}{2},-3$ **19.** $(-1\pm\sqrt{5})/2$

21. $(3\pm\sqrt{3})/2$ **23.** No real solution   **25.** $(-3\pm\sqrt{11})/2$ **27.** $\pm\sqrt{3}$ **29.** $-\dfrac{1}{2},2$ **31.** $(x-2)(x+42)$ **33.** Not factorable in the integers
35. $(2x-9)(x+12)$ **37.** $(4x-7)(x+62)$ **39.** $r=\sqrt{A/P}-1$ **41.** If $c<4$, there are two distinct real roots; if $c=4$, there is one real double root; and if $c>4$, there are no real roots.   **43.** -2 **45.** $\pm\sqrt{10}$ **47.** $\pm\sqrt{3},\pm\sqrt{5}$ **49.** 1,575 bottles at \$4 each   **51.** 13.64%   **53.** 8 ft/sec; $4\sqrt{2}$ or 5.66 ft/sec

Appendix B

Exercises B.1   **1.** 5, 7, 9, 11   **3.** $\dfrac{3}{2},\dfrac{4}{3},\dfrac{5}{4},\dfrac{6}{5}$ **5.** 9, -27, 81, -243 **7.** 23   **9.** $\dfrac{101}{100}$ **11.** $1+2+3+4+5+6=21$ **13.** $5+7+9+11=32$
15. $1+\dfrac{1}{10}+\dfrac{1}{100}+\dfrac{1}{1,000}=\dfrac{1,111}{1,000}$ **17.** 3.6   **19.** 82.5   **21.** $\dfrac{1}{2},-\dfrac{1}{4},\dfrac{1}{8},-\dfrac{1}{16},\dfrac{1}{32}$ **23.** 0, 4, 0, 8, 0   **25.** $1,-\dfrac{3}{2},\dfrac{9}{4},-\dfrac{27}{8},\dfrac{81}{16}$ **27.** $a_n=n-3$
29. $a_n=4n$ **31.** $a_n=(2n-1)/2n$ **33.** $a_n=(-1)^{n+1}n$ **35.** $a_n=(-1)^{n+1}(2n-1)$ **37.** $a_n=\left(\dfrac{2}{5}\right)^{n-1}$ **39.** $a_n=x^n$ **41.** $a_n=(-1)^{n+1}x^{2n-1}$
43. $1-9+25-49+81$ **45.** $\dfrac{4}{7}+\dfrac{8}{9}+\dfrac{16}{11}+\dfrac{32}{13}$ **47.** $1+x+x^2+x^3+x^4$ **49.** $x-\dfrac{x^3}{3}+\dfrac{x^5}{5}-\dfrac{x^7}{7}+\dfrac{x^9}{9}$ **51.** (A) $\displaystyle\sum_{k=1}^{5}(k+1)$ (B) $\displaystyle\sum_{j=0}^{4}(j+2)$
53. (A) $\displaystyle\sum_{k=1}^{4}\dfrac{(-1)^{k+1}}{k}$ (B) $\displaystyle\sum_{j=0}^{3}\dfrac{(-1)^{j}}{j+1}$ **55.** $\displaystyle\sum_{k=1}^{n}\dfrac{k+1}{k}$ **57.** $\displaystyle\sum_{k=1}^{n}\dfrac{(-1)^{k+1}}{2^k}$ **59.** False   **61.** True   **63.** 2, 8, 26, 80, 242   **65.** 1, 2, 4, 8, 16
67. $1,\dfrac{3}{2},\dfrac{17}{12},\dfrac{577}{408};a_4=\dfrac{577}{408}\approx1.414\,216,\sqrt{2}\approx1.414\,214$ **69.** 1, 1, 2, 3, 5, 8, 13, 21, 34, 55

Exercises B.2   **1.** (A) Arithmetic, with $d=-5;-26,-31$ (B) Geometric, with $r=-2;-16,32$ (C) Neither   (D) Geometric, with $r=\dfrac{1}{3};\dfrac{1}{54},\dfrac{1}{162}$
3. Geometric; 1   **5.** Neither   **7.** Arithmetic; 127.5   **9.** $a_2=11,a_3=15$ **11.** $a_{21}=82,S_{31}=1,922$ **13.** $S_{20}=930$ **15.** $a_2=-6,a_3=12,a_4=-24$
17. $S_7=547$ **19.** $a_{10}=199.90$ **21.** $r=1.09$ or -1.09 **23.** $S_{10}=1,242,S_\infty=1,250$ **25.** 2,706   **27.** -85 **29.** 1,120   **31.** (A) Does not exist
(B) $S_\infty=\dfrac{8}{5}=1.6$ **33.** 2,400   **35.** 0.999   **37.** Use $a_1=1$ and $d=2$ in $S_n=(n/2)[2a_1+(n-1)d]$.   **39.** $S_n=na_1$ **41.** No   **43.** Yes
45. $\$48+\$46+\cdots+\$4+\$2=\$600$ **47.** About \$11,670,000   **49.** \$1,628.89; \$2,653.30

Exercises B.3   **1.** 720   **3.** 10   **5.** 1,320   **7.** 10   **9.** 6   **11.** 1,140   **13.** 10   **15.** 6   **17.** 1   **19.** 816
21. ${}_4C_0a^4+{}_4C_1a^3b+{}_4C_2a^2b^2+{}_4C_3ab^3+{}_4C_4b^4=a^4+4a^3b+6a^2b^2+4ab^3+b^4$ **23.** $x^6-6x^5+15x^4-20x^3+15x^2-6x+1$
25. $32a^5-80a^4b+80a^3b^2-40a^2b^3+10ab^4-b^5$ **27.** $3,060x^{14}$ **29.** $5,005p^9q^6$ **31.** $264x^2y^{10}$ **33.** ${}_nC_0=\dfrac{n!}{0!\,n!}=1;{}_nC_n=\dfrac{n!}{n!\,0!}=1$
35. 1 5 10 10 5 1; 1 6 15 20 15 6 1

INDEX

INDEX OF APPLICATIONS

A Library of Elementary Functions

BASIC FUNCTIONS

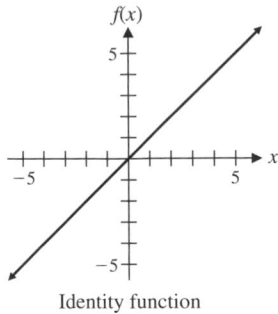

Identity function
$f(x) = x$

Absolute value function
$g(x) = |x|$

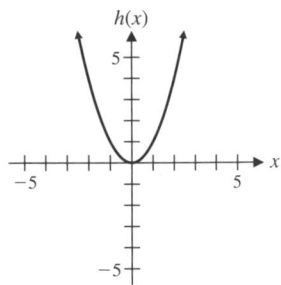

Square function
$h(x) = x^2$

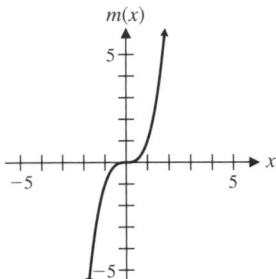

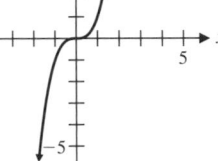

Cube function
$m(x) = x^3$

Square root function
$n(x) = \sqrt{x}$

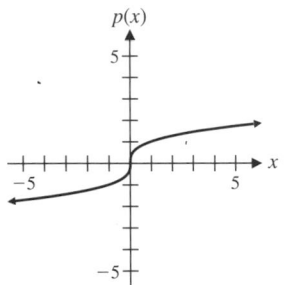

Cube root function
$p(x) = \sqrt[3]{x}$

LINEAR AND CONSTANT FUNCTIONS

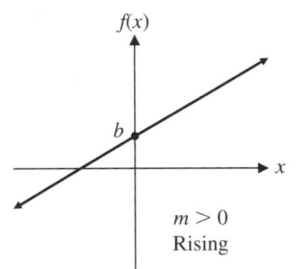

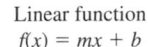

$m > 0$
Rising

Linear function
$f(x) = mx + b$

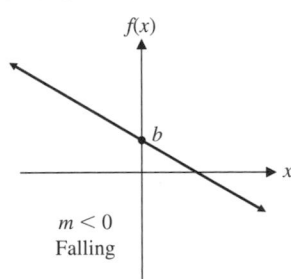

$m < 0$
Falling

Linear function
$f(x) = mx + b$

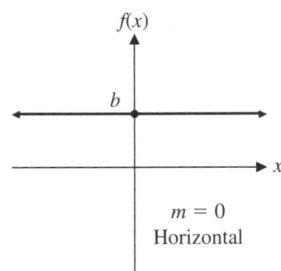

$m = 0$
Horizontal

Constant function
$f(x) = b$

QUADRATIC FUNCTIONS

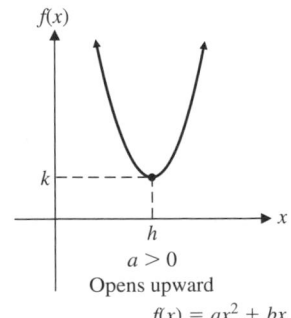

$a > 0$
Opens upward

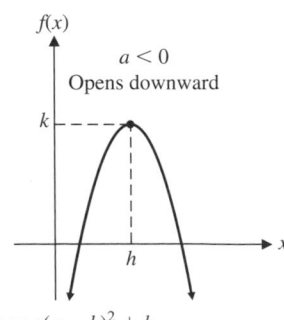

$a < 0$
Opens downward

$$f(x) = ax^2 + bx + c = a(x - h)^2 + k$$

Talcott Parsons
1902–1979
Dean of American sociology during the mid-1900s promoted a conception of society as a social system with subsystems of human action, in which individuals fulfill the systems needs of the societies of which they are members.

Samuel Delbert Clark
1910–2003
Canadian historical sociologist and educator founded the University of Toronto's Department of Sociology (1963).

Robert K. Merton
1910–2003
American sociologist and educator developed middle-range theory, which sought to bridge the gap between high-level theories and low-level observations.

Guy Rocher
b. 1924
Canadian educator and pioneer in the sociology of education, law, and medical ethics has sat on several commissions and boards of inquiry at the provincial and federal levels and wrote a lucid and highly regarded introduction to the discipline, *Introduction à la sociologie* (1968).

Erving Goffman
1922–1982
Canadian-born sociologist advanced microsociology and studied social roles, deviance, stigma, and 'total institutions'.

Michel Foucault
1926–1984
French thinker, famous for historical studies of madness and civilization, imprisonment and sexuality, portrayed science as an arbitrary instrument for control and power, and constructed a theory of power as actions and relations.

Herbert Blumer
1900–1987
American student of Mead, who coined the term 'symbolic interactionism'.

Theodor Adorno
1903–1969
German Frankfurt School philosopher argued that philosophical authoritarianism is inevitably oppressive.

John Porter
1921–1979
Canadian sociologist examined connections between ethnicity and barriers of opportunity in Canadian society, which he characterized as a 'vertical mosaic'.

Dorothy Smith
b. 1926
English-born Canadian sociologist developed standpoint theory, which sought to frame and understand everyday life from a feminist point of view.

Everett C. Hughes
1897–1983
American sociologist studied economic organization, work and occupations, and ethnic relations, including a key study of the 'ethnic division of labour' in Quebec.

Oswald Hall
1908–2007
Canadian educator researched the sociology of work and medicine and served on the Royal Commission on Health Services and the Royal Commission on Bilingualism and Biculturalism.

C. Wright Mills
1916–1962
American critical sociologist studied power structure in the US and coined the term 'sociological imagination'.

Jean Baudrillard
1929–2007
French cultural theorist influenced postmodernism and showed how capitalist consumer society erases distinctions between reality and reference, leading to a loss of meaning.

Margrit Eichler
b. 1942
Canadian sociologist has studied family sociology, feminist research methods, and gender inequality.

1900 — **1940** — **2000**

1969
Doctors and Doctrines: The Ideology of Medical Care in Canada, an examination of Canada's healthcare system in terms of role strains, conflict in values, and relations to the public, by Bernard Blishen (b.1919)

1978
The Double Ghetto: Canadian Women and their Segregated Work, a study of gender inequality in the labour force and the home, by Pat Armstrong (b. 1945) and Hugh Armstrong (b.1943)

1983
Green Gold: The Forest Industry in British Columbia (1983), an early study in the social, political, and economic aspects of a particular staples industry, the BC forest industry, by Patricia Marchak (1936–2010)

2002
The Impact of Feminism on Canadian Sociology, a study of the rise of sociology as a feminist discipline, by Margrit Eichler

1996
The Barbershop Singer: Inside the Social World of a Musical Hobby, a study of leisure and hobbies in society, by Robert Stebbins (b. 1938)

1986
'The "Wets" and the "Drys": Binary Images of Women and Alcohol in Popular Culture', a study of gender inequalities and mass media, by Thelma McCormack (b. 1921)

1968
Introduction to the Mathematics of Population, a landmark contribution to the field of population studies, by Canadian demographer Nathan Keyfitz (1913-2010)

1987
The Everyday World as Problematic: A Feminist Sociology, an argument that sociology has developed without proper insight into women's experiences, by Dorothy Smith

2004
Perspectives de Recherche en Santé des Populations au Moyen de Données Complexes, an analysis of the Quebec healthcare system, by Paul Bernard (1945-2011)

1975
The Rise of a Third Party: A Study in Crisis Politics (1975), a sociological analysis of the growth of nationalist politics in Quebec, by Maurice Pinard (b. 1929)

The Canadian Corporate Elite: An Analysis of Economic Power, a response to *The Vertical Mosaic* examining corporate elites and their impact on class and social stratification, by Wallace Clement

1988
Quebec Society: Tradition, Modernity, and Nationhood, a study of Quebec's rising middle class and the separatist movement, by Hubert Guindon (1929-2002)

1965
Lament for a Nation: The Defeat of Canadian Nationalism, an examination of the dangers of Canadian cultural absorption by the US, by Canadian social philosopher George Grant (1918–1988)

The Vertical Mosaic: An Analysis of Social Class and Power in Canada, a groundbreaking and influential study of Canada's class structure, depicting a complex system of groups organized in hierarchy across lines of ethnicity and class, by John Porter

Families in Canada Today: Recent Changes and Their Policy Implications, a study of how the way we think and talk about gender roles pre-empts useful changes in family policy, by Margrit Eichler

2006
Do Men Mother? Fathering, Care, and Domestic Responsibility, an examination of the changing role of fathers, by Andrea Doucet

1989
The Social Significance of Sport, a study of how individuals take control of and participate in society through voluntary association, by James Curtis (1943–2005)

2008
Canada's Rights Revolution: Social Movements and Social Change, 1937-82, a study of post-war Canadian social movements, by Dominique Clément

Milestones in Canadian Sociology

SOCIOLOGY

SOCIOLOGY

A CANADIAN PERSPECTIVE | THIRD EDITION

Tepperman • Albanese • Curtis

OXFORD

UNIVERSITY PRESS

OXFORD
UNIVERSITY PRESS

Oxford University Press is a department of the University of Oxford.
It furthers the University's objective of excellence in research, scholarship, and education by publishing worldwide.
Oxford is a registered trade mark of Oxford University Press in the UK and in certain other countries.

Published in Canada by
Oxford University Press
8 Sampson Mews, Suite 204,
Don Mills, Ontario M3C 0H5 Canada

www.oupcanada.com

Library and Archives Canada Cataloguing in Publication
Sociology : a Canadian perspective / editors, Lorne Tepperman, Patrizia Albanese & Jim Curtis. — 3rd ed.

Includes bibliographical references and index.
ISBN 978-0-19-544380-6

1. Sociology—Textbooks. 2. Canada—Social conditions—
1991– —Textbooks. I. Tepperman, Lorne, 1943– II. Albanese, Patrizia III. Curtis, James E., 1943–

HM586.S62 2012   301   C2012-900060-4

Photo credits: Page 1: zaragyemo/BigStock.com; page 9: © Corbis Flirt/Alamy; page 10: © Deco/Alamy;
page 32: Norman Hollands/The Bridgeman Art Lilbrary/Getty Images; page 55: Emmanuel Joly/Getty Images;
page 56: © imagebroker/Alamy; page 78: © iStockphoto.com/sefaoncul; page 102: © Gabe Palmer/Corbis;
page 124: © iStockphoto.com/nicole abejon; page 150: © Jim Cornfield/Corbis; page 175: © brianlatino/Alamy;
page 176: © Kamyar Adl/Alamy; page 180: Marco Baass/Getty Images; page 224: © Hugh Smith/Demotix/
Demotix/Corbis; page 248: © iStockphoto.com/Bartosz Hadyniak; page 274: © frans lemmens/Alamy;
page 301: © Agnieszka Pastuszak-Maksin/istock; page 302: Carey Kirkella/Getty Images; page 326: © Marmaduke
St. John/Alamy; page 353: © Darren Modricker/CORBIS; page 378: © Gunter Marx/Alamy ; page 402: © Images
& Stories/Alamy; page 426: © Igor Vidyashev/Alamy; page 450: © Jim West/Alamy; page 473: © Baci/Corbis;
page 474: © iStockphoto.com/Bartosz Hadyniak; page 498: © Andrew Rubtsov/Alamy; page 526: © Megapress/
Alamy; page 548: © Robert Harding World Imagery/Alamy; page 572: © Roderick Chen/Alamy.

Page 95, Figure 4.3: 'Generation M2: Media in the Lives of 8-to-18 Year Olds', (#8010), The Henry J. Kaiser Family
Foundation, January 2010. This information was reprinted with permission from the Henry J. Kaiser Family
Foundation. The Kaiser Family Foundation, a leader in health policy analysis, health journalism and communication,
is dedicated to filling the need for trusted, independent information on the biggest health issues facing our nation and
its people. The Foundation is a non-profit private operation foundation, based in Menlo Park, California.

Cover image: © Ikon Images/Alamy

This book is printed on paper which contains 10% post-consumer waste.

Printed and bound in the United States of America

1 2 3 4 — 15 14 13 12

brief contents

detailed contents

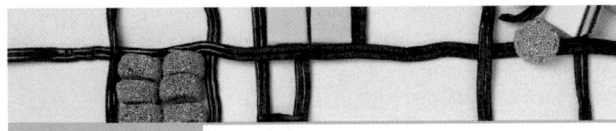

Part II Major Social Processes 55

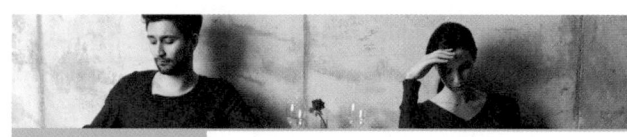

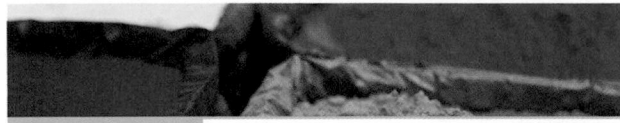

CHAPTER 11: Ethnic and Race Relations 249
Nikolaos I. Liodakis

CHAPTER 12: The Social Aspects of Aging 275
Lynn McDonald

Part IV Social Institutions 301

CHAPTER 13: Families and Personal Life 303
Maureen Baker

CHAPTER 14: Education 327
Terry Wotherspoon

CHAPTER 15: Work and the Economy 353
Pamela Sugiman

CHAPTER 16: Health Issues   379
Juanne Clarke

CHAPTER 17: Religion in Canada   403
Lori G. Beaman

**CHAPTER 18: Politics and Political
Movements   427**
Howard Ramos and Karen Stanbridge

CHAPTER 19: Social Movements   451
John Veugelers and Randle Hart

Part V Canadian Society and the Global Context   473

CHAPTER 20: Challenges of Globalization   475
Pierre Beaudet

tables

figures

boxed features

GLOBAL ISSUES

A Sociological Perspective on Cases from around the World

UNDER the WIRE

Exploration of the Ways in Which Media and Technologies Intersect with Social Behaviours

preface

From the Publisher

While preparing this third edition of *Sociology: A Canadian Perspective*, the general editors, contributing authors, and publisher kept in mind one paramount goal: to produce the most authoritative, comprehensive, yet accessible and interesting introduction to sociology available for Canadian students.

This revision builds on the strengths of the well-received first and second editions and incorporates many new features designed to enhance the book's usefulness for students and instructors alike.

NEW CHAPTER ON AGING

A brand-new chapter dealing with the sociological implications of aging has been added to this edition. In what has become a critically important area of sociological inquiry, Lynn McDonald reflects on the demographic shift and societal changes that an aging population will bring to Western countries.

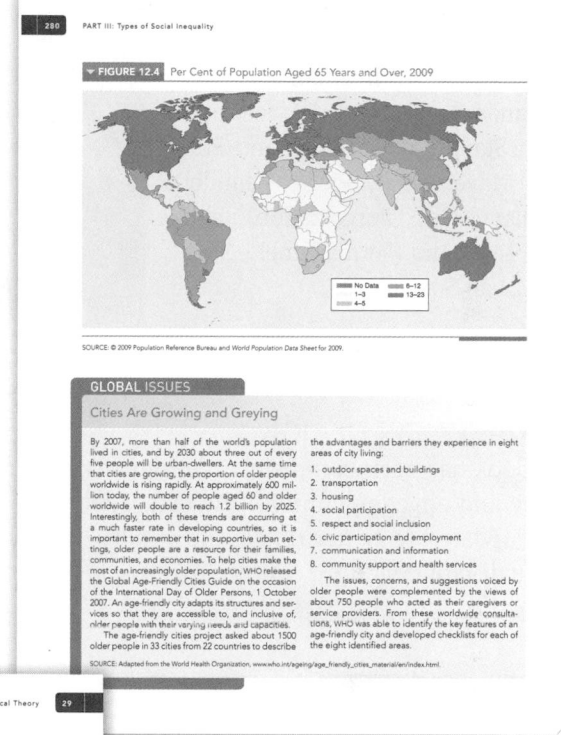

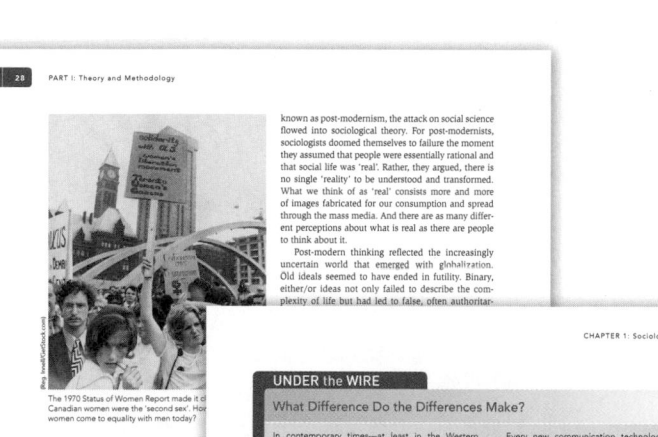

NEW CHAPTER ON THEORY

In response to feedback from reviewers, the chapter on sociological theory was rewritten to better emphasize the linkages between classical theory and contemporary debates. Using clear and accessible language, Anthony Thomson breaks down complex theory and lays the groundwork for successful student engagement with the chapters that follow.

NEW CONTRIBUTING AUTHORS

For the third edition, we welcome aboard several new contributing authors:
Anthony Thomson (Sociological Theory), Barbara A. Mitchell (Socialization), Dorothy Pawluch and William Shaffir (Statuses, Roles, Self, and Identity), Sara J. Cumming and Ann D. Duffy (Class and Status Inequality), Janet Siltanen and Andrea Doucet (Gender Relations), Lynn McDonald (The Social Aspects of Aging), Lori G. Beaman (Religion in Canada), Howard Ramos and Karen Stanbridge (Politics and Political Movements), Pierre Beaudet (Challenges of Globalization), and Louis Guay and Pierre Hamel (Cities and Urban Sociology).

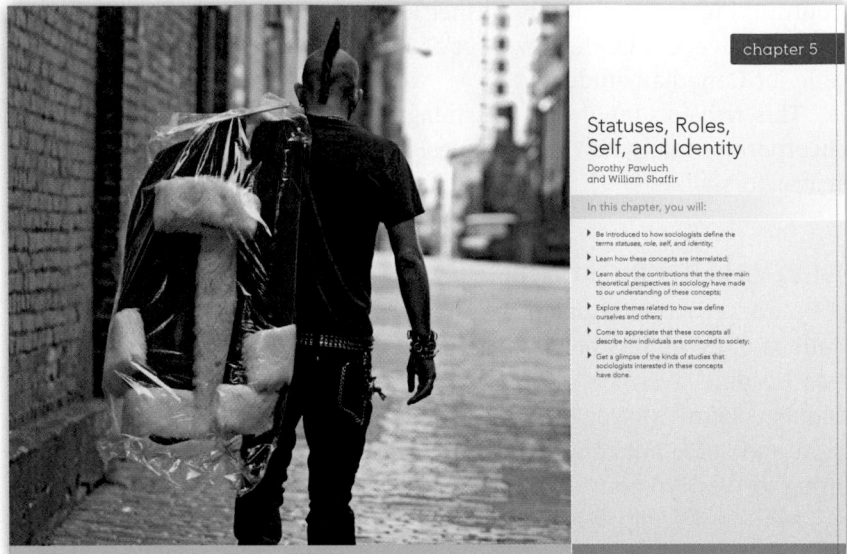

VIBRANT NEW DESIGN

The book has been completely redesigned and modernized to enhance readability and engagement with the text. The vibrant four-colour aesthetic has been updated to better reflect the vitality of Canadian sociology as an academic discipline.

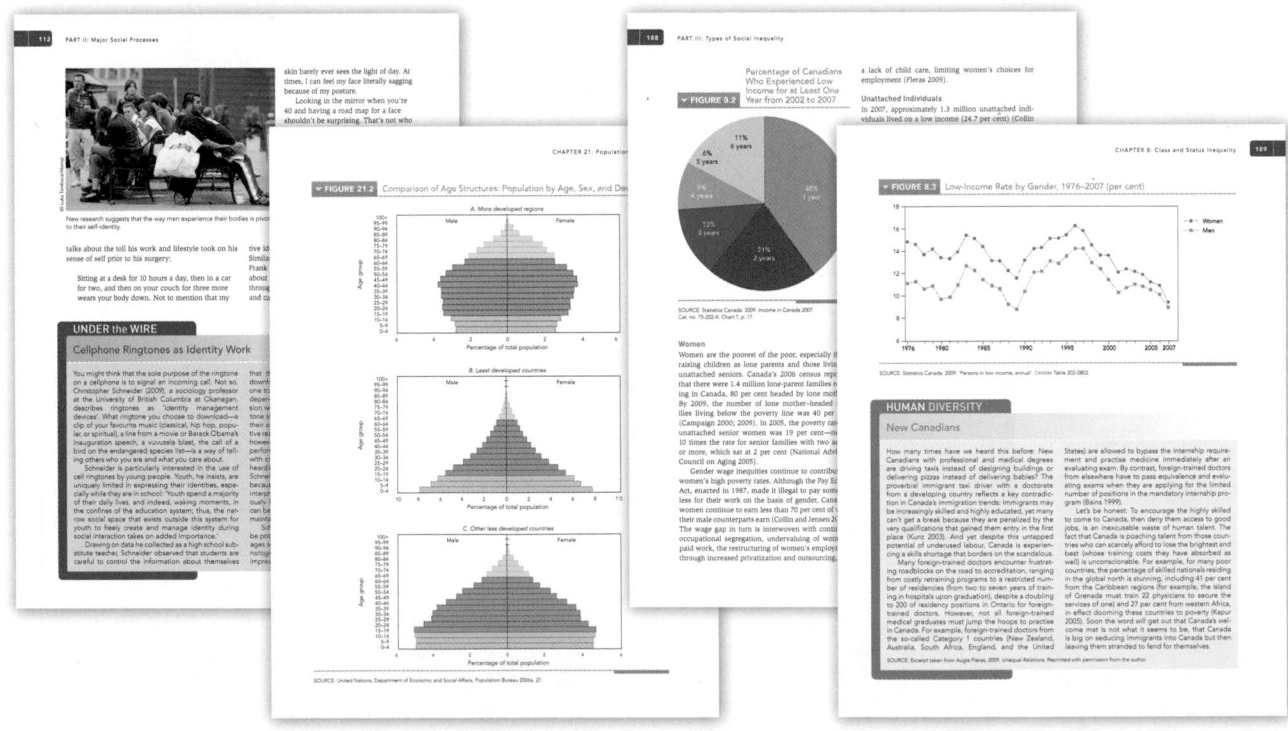

All the features that made the previous editions of *Sociology: A Canadian Perspective* so popular have been retained in this revision:

TOP CANADIAN CONTRIBUTORS

Sociology is a global discipline but one to which sociologists working in Canada have made unique contributions. Not merely an adaptation of a book originally written for American undergraduates, this text was conceived and written from the ground up as a Canadian perspective on this fascinating field. Experts in their particular sub-disciplines not only examine the key concepts and terminology of sociology as an academic discipline but also use those concepts to shed light on the nature of Canadian society and Canada's place in the world.

GLOBAL PERSPECTIVE

Although this is a book written by and for Canadians, the editors and authors never forget that Canada is but one small part of a vast, diverse, and endlessly fascinating social world. Along with Canadian data, examples, and illustrations, a wealth of information about how humans live and interact around the world is presented in every chapter.

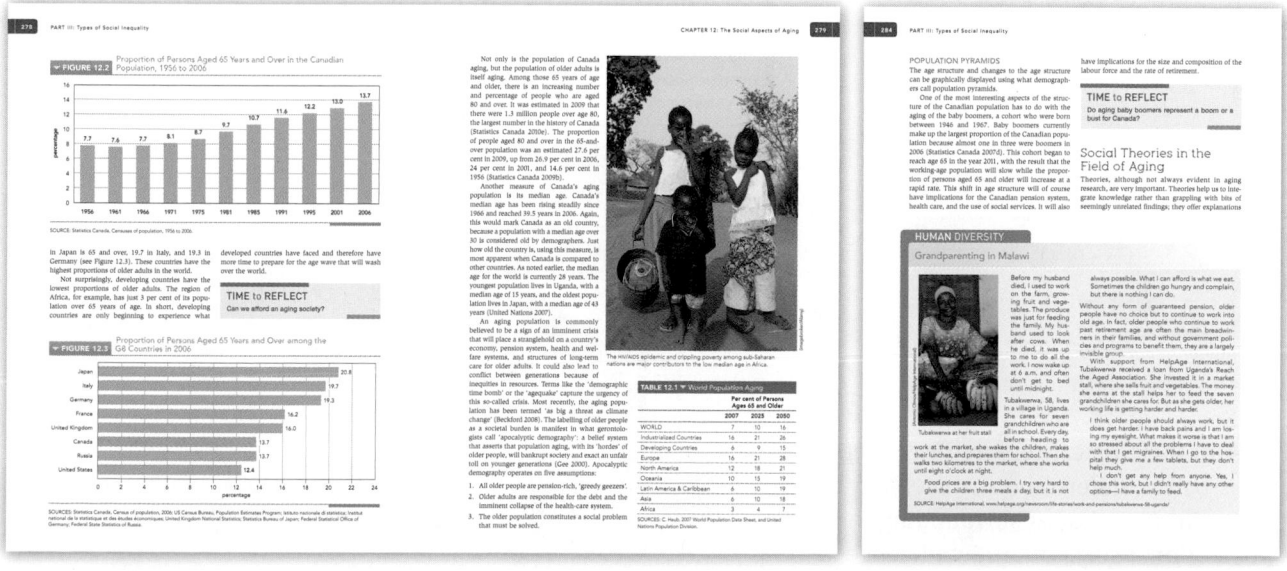

THEORETICAL BALANCE

The overriding goal in *Sociology: A Canadian Perspective* has been not just to make the theories that underpin the discipline comprehensible but to show how they inform an understanding of the data that sociologists gather—and how the choice of which theoretical perspective to employ can yield new and surprising insights. Throughout the text, emerging paradigms are also discussed when they shed new light on long-standing questions.

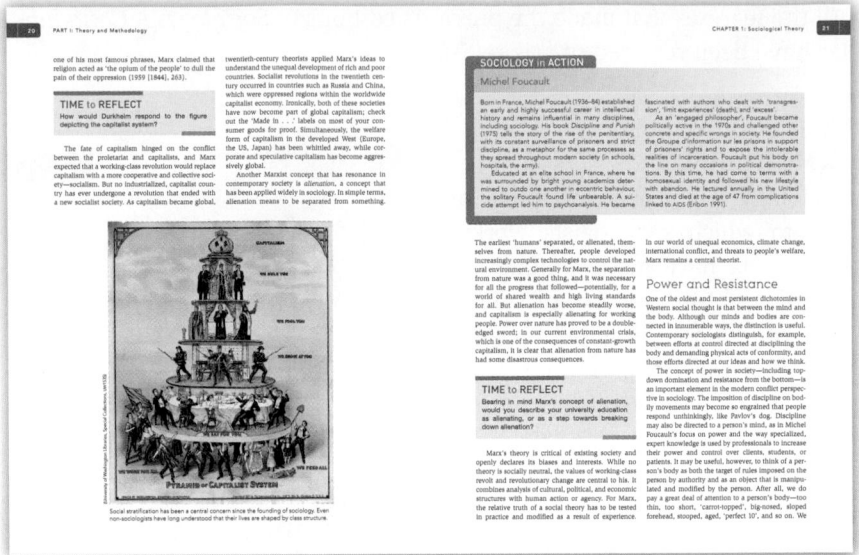

INSIGHTFUL THEME BOXES

'Why study sociology?' is a question frequently asked by students. There are many reasons, of course: sociology provides a unique insight into the nature of the human world; it shows us things about society and ourselves that we might not otherwise know; and the lessons of sociology can be intriguing, touching, tragic, even fun. The dozens of theme boxes scattered throughout the text illustrate all of these dimensions of the discipline.

- **'Open for Discussion' boxes** use contemporary social issues and debates to focus understanding of core sociological concepts. Examples range from sexual assault on campus to medically assisted conception.

- **'Sociology in Action' boxes** show how sociological research can help us better understand the everyday world, from issues surrounding the emotional scars of family violence to neo-liberalism in reality TV.

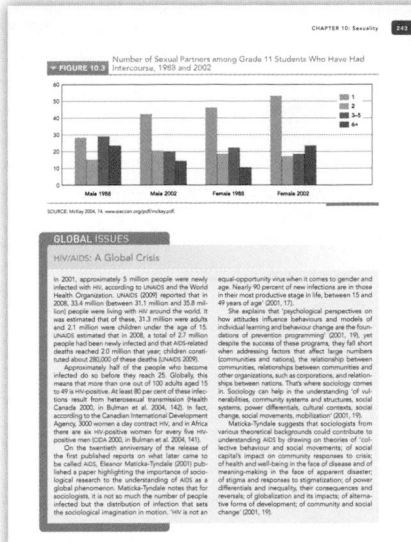

- **'Global Issues' boxes** draw upon examples from around the world to illustrate the effects of globalization and show what sociologists have to say about, for instance, health inequality or the worldwide crisis of HIV/AIDS.

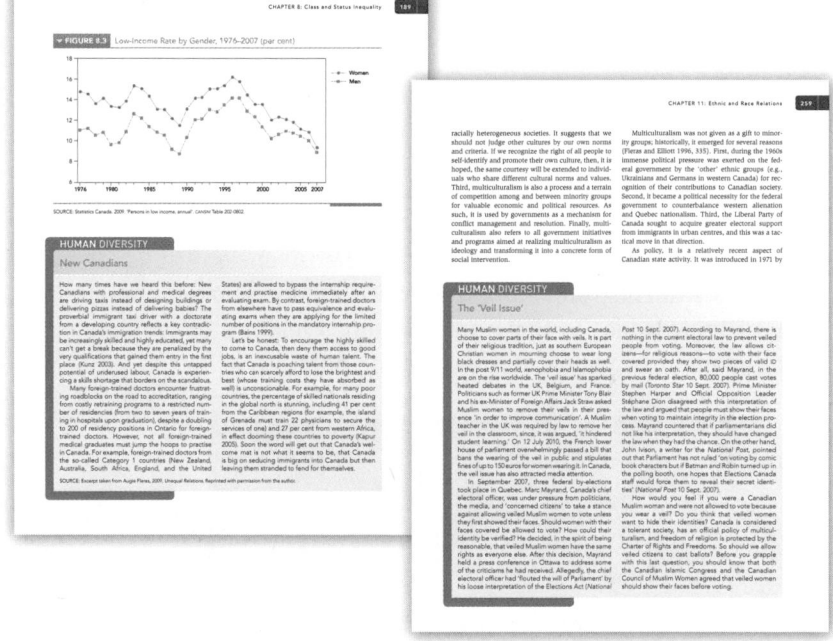

- **'Human Diversity' boxes** recognize the overwhelming and unavoidable fact of human diversity and seek to introduce students to the ways of life and world views of different cultures and social groups, whether they are Aboriginal Canadians or Mexican farm workers.

- **New 'Under the Wire' boxes** analyze the ways in which current media and new technologies influence social patterns and behaviours, such as religious expression on the Internet and environmental cyberactivism.

SOCIOLOGY AS A HUMAN PURSUIT

Sociology: A Canadian Perspective celebrates the fact that while sociology is an academic discipline with a distinguished pedigree, it is also a very human pursuit—a fact that becomes clear in the brief 'In the First Person' narratives included in every chapter. The text's contributors first encountered the discipline at the same age as many of the students now using this text in 'intro soc'. Sociology is, above all, the study of human beings interacting within society in all their wonderful complexity, and 'In the First Person' provides an intriguing glimpse into why this particular group of individuals chose to make sociology a part of their lifework.

AIDS TO STUDENT LEARNING

A textbook must fulfill a double duty: while meeting instructors' expectations for accuracy, currency, and comprehensiveness, it must also speak to today's students, providing them with an accessible introduction to a body of knowledge. To that end, numerous features to promote student learning are incorporated throughout the book.

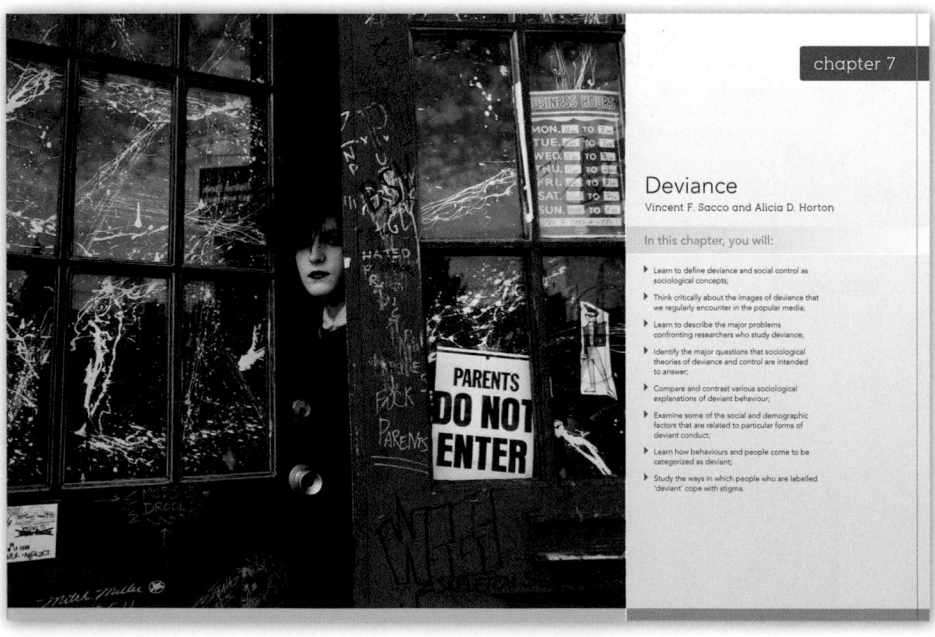

- **Learning objectives** at the start of each chapter provide a concise overview of the key concepts that will be covered.

- **Graphs and tables.** Although qualitative research methods have grown in importance in recent years, one of the characteristics that still distinguishes sociology from other liberal arts and social sciences is its emphasis on using quantitative data and analysis as a crucial tool for understanding society. Colourful graphs and charts make such data clear in a way that sometimes text cannot.

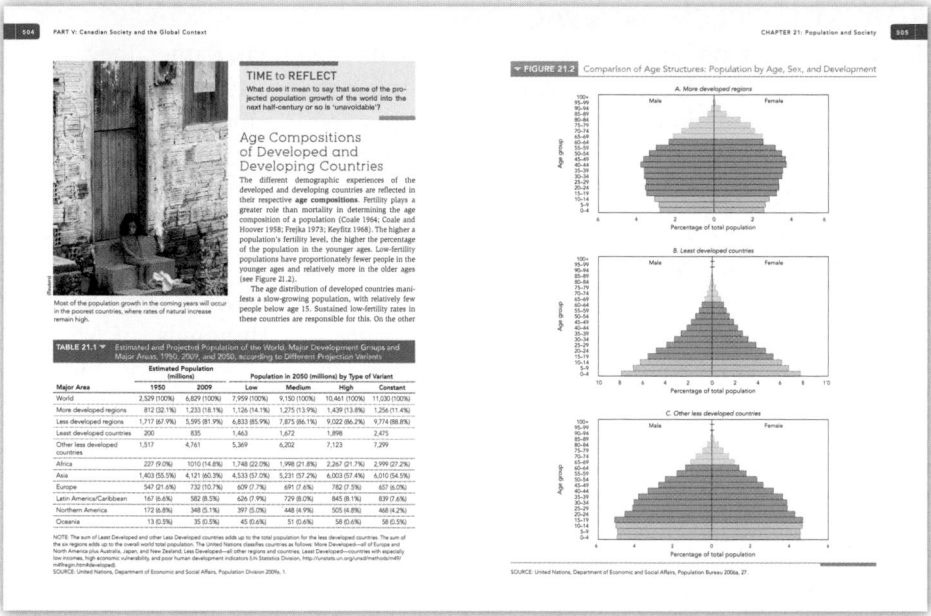

- **'Time to Reflect'** questions placed throughout the text prompt students to analyze the material both in and out of the classroom.

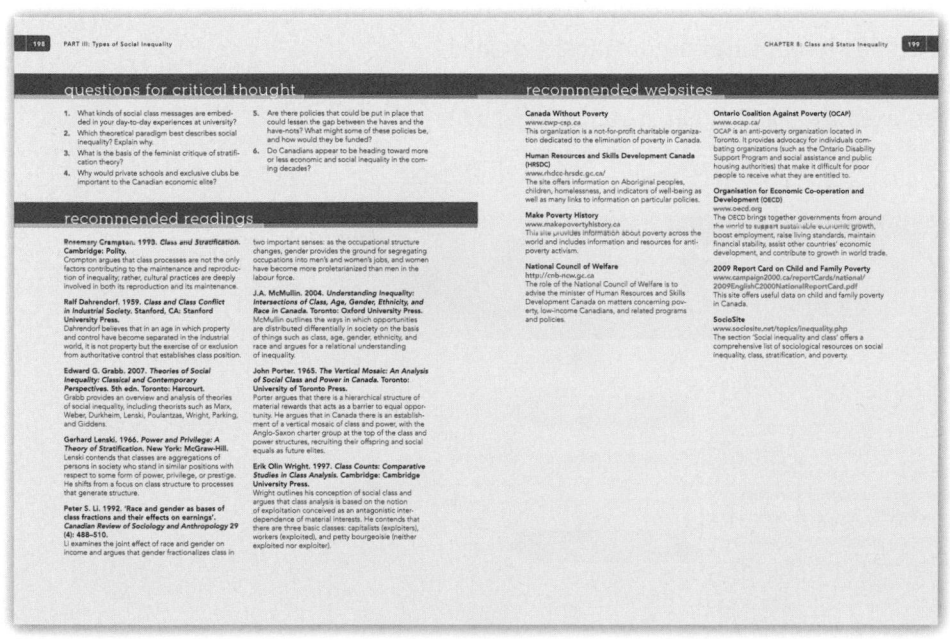

- **Questions for Critical Thought**; **Recommended Readings**; and **Recommended Websites** at the end of each chapter encourage readers to think deeply about key issues and point students toward useful sources for further research.

ROBUST SUPPLEMENTS PACKAGE

Today's texts are no longer volumes that stand on their own. Rather, they are but the central element in a complete learning and teaching package. *Sociology: A Canadian Perspective* is no exception and is supported by an outstanding array of ancillary materials for both students and instructors.

For Instructors

Online Instructor's Manual

This fully revised online resource includes comprehensive outlines of the text's various parts and chapters, additional questions for encouraging class discussion, suggestions on how to use videos to enhance classes, and extra resource material for use in lectures.

Online Test Generator

A comprehensive test generator allows instructors to sort, edit, import, and distribute hundreds of questions in multiple-choice, short-answer, and true/false format.

PowerPoint® Slides

Hundreds of slides for classroom presentation—rewritten for this edition—incorporate graphics and tables from the text, summarize key points from each chapter, and can be edited to suit individual instructor's needs.

DVD

An exciting collection of video clips, provided by the Media Education Foundation and chosen specifically to complement the content in each of the chapters, will enhance students' experiences of critical concepts and issues.

For Students

Companion Website

A comprehensive online study-guide site provides automatically graded study questions, annotated links to other useful Web resources, additional review questions, applied exercises, and other material designed to enhance student learning.

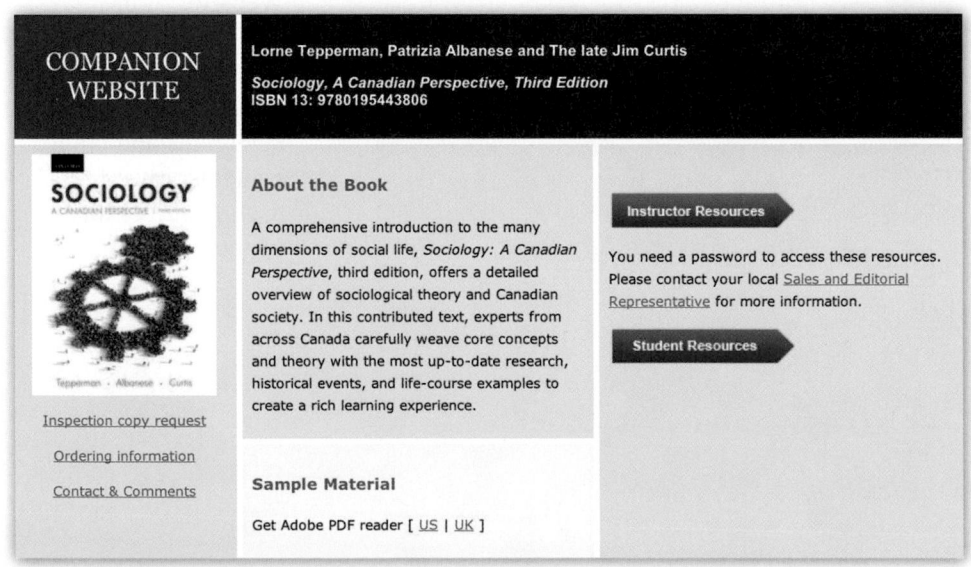

 www.oupcanada.com/sociology3e

From the General Editors

FROM THE GENERAL EDITORS

The two editors of this book—formerly teacher and student, now colleagues—are happy to bring enthusiastic readers a new edition of a text that tries to explain sociology, and Canadian society, to a new generation of Canadian students. We think that in this third edition, *Sociology: A Canadian Perspective* continues to provide an up-to-date picture of Canadian society and Canadian sociology in the early parts of the twenty-first century. As always, we remain committed to providing the best, most reader-friendly presentation of social facts and theories. Our publisher, Oxford University Press Canada, has helped us to do so, and we are grateful for this help. *Sociology: A Canadian Perspective* has received strong support from the topmost levels of Oxford University Press Canada, including David Stover, president of the Canadian branch.

Working with us on this edition of *Sociology: A Canadian Perspective* was editor Patricia Simoes. Patricia has gently and persistently helped us to hone this book so that it is even better than the previous edition. Our main thanks go to the contributing authors, without whom this book simply would not exist. They put up with our (seemingly endless) demands, and somehow we got from A to B without any catastrophes or unseemly emotional displays. It has been a great privilege working with this distinguished group of top Canadian scholars from all over the country. Thank you, authors.

We also remember Jim Curtis, who co-edited the first edition of this book and passed away shortly after its publication. He was a singular collaborator and a remarkable human being.

Finally, in closing, we dedicate this book to sociologist Slobodan Drakulic, Patrizia's husband and Lorne's friend. We will miss his wisdom and gentle humour.

Lorne Tepperman, University of Toronto
Patrizia Albanese, Ryerson University

ACKNOWLEDGEMENTS

We would like to acknowledge the following reviewers, along with those reviewers who chose to remain anonymous, whose insightful comments have helped shape the first, second, and third editions of *Sociology: A Canadian Perspective*:

Kate Bezanson, Brock University

Laurie Forbes, Lakehead University

Ray Foui, University of Manitoba

Jana Grekul, University of Alberta

Helga Hallgrimsdottir, University of Victoria

Robert Hill, Memorial University of Newfoundland

Kate Krug, Cape Breton University

Barry McClinchey, University of Waterloo

Phyllis Rippeyoung, Acadia University

Jesse Seary, Langara College

Tamy Superle, Carleton University

Valerie Zawilski, King's University College at the University of Western Ontario

Li Zong, University of Saskatchewan

contributors

Patrizia Albanese is associate professor and interim chair of the Department of Criminal Justice and Criminology at Ryerson University.

Bruce Arai teaches courses in research methods, statistics, and the sociology of work at Wilfrid Laurier University.

Maureen Baker is professor of sociology at the University of Auckland.

Shyon Baumann is associate professor of sociology at the University of Toronto.

Lori G. Beaman is professor of religious studies at the University of Ottawa.

Pierre Beaudet is replacement professor of sociology at the University of Ottawa.

Juanne Clarke is professor of sociology at Wilfrid Laurier University.

Sara J. Cumming is a PhD candidate in the Department of Sociology and Legal Studies at the University of Waterloo.

Ann D. Duffy is professor of sociology and associate chair of the Department of Sociology at Brock University.

Andrea Doucet is professor of sociology at Carleton University.

Louis Guay is professor of sociology and a member of the Institut en environnement, développement et société at Laval University.

Pierre Hamel is professor of sociology at the University of Montreal and editor of the journal *Sociologie et sociétés*.

Randle Hart is assistant professor in the Department of History and Sociology at Southern Utah University.

Alicia D. Horton is a PhD candidate in the Department of Sociology at Queen's University.

Nikolaos I. Liodakis is assistant professor of sociology at Wilfrid Laurier University.

Lynn MacDonald is professor of social work and director of the Institute for the Life Course and Aging at the University of Toronto.

Barbara A. Mitchell is professor of sociology at Simon Fraser University.

Dorothy Pawluch is associate professor of sociology at McMaster University.

Howard Ramos is a political sociologist in the Department of Sociology and Social Anthropology at Dalhousie University.

Vincent F. Sacco is professor of sociology and teaches courses relating to crime, deviance, and social control at Queen's University.

William Shaffir is professor and associate chair in the Department of Sociology at McMaster University.

Janet Siltanen is professor of sociology at Carleton University and director of the Institute of Political Economy.

Karen Stanbridge is associate professor of sociology at Memorial University of Newfoundland.

Pamela Sugiman is professor of sociology at Ryerson University.

Lorne Tepperman is professor of sociology at the University of Toronto and past president of the Canadian Sociological Association.

Anthony Thomson is professor of sociology at Acadia University.

Frank Trovato is professor of demography and population studies at the University of Alberta.

John Veugelers is associate professor of sociology at the University of Toronto.

G. Keith Warriner is associate professor of sociology at the University of Waterloo.

Terry Wotherspoon is professor and head of sociology at the University of Saskatchewan.

David Young is assistant professor of sociology at McMaster University.

Why Not Become a Sociologist?

Lorne Tepperman
and Patrizia Albanese

Introduction

Why do people become sociologists? There are many answers to this question, and it is likely that everyone at one point or another has been on the brink of becoming a sociologist. We say this because all people experience odd facets of social life that affect their opportunities and they try to understand them. This is where sociology begins for most people. When people go beyond that point, they feel even more motivated to do sociology. What can be more fascinating, more empowering, and more personal than to begin to understand the society that shapes our lives? For these reasons, sociology is an inherently attractive area of study, and many people do study it.

Maybe as a child you noticed that:

- Parents sometimes treat their sons differently from their daughters.
- Teachers often treat pretty little girls better than plain-looking ones.
- Adults treat well-dressed children better than poorly dressed children.
- Movies typically portray people with 'accents' as strange or ridiculous.

If you noticed these things, you may have wondered why they happen. They may even have affected you, as a daughter or son, a plain-looking or attractive person, a poorly dressed or well-dressed person, or a person with or without an 'accent'. You may have felt ashamed, angry, or pleased, depending on whether you identified with the favourably treated or with the unfavourably treated category of people.

Perhaps you grew up in a small town and then moved to a big city, or you grew up in a big city and now live in a small town. You notice that:

- People are not the way the media portray them.
- The ethnic and racial composition of the people around you is not what you are used to.
- The gap between rich and poor is more pronounced.
- People interact with each other differently from the way they do back home.
- They react suspiciously to strangers.
- They talk differently, dress differently, and eat different kinds of foods.

If you noticed these things, you may have wanted to understand them better. These are the kinds of circumstances in which sociological curiosity begins. All sociologists somehow, at some time or another, got hooked on trying to better understand their own lives and the lives of people around them. They came to understand that common sense gave them only incomplete explanations about what happened to people, about people's behaviour, and about the society in which they live. They were not satisfied with the incomplete explanation and wanted to know more.

For many people, and for much of what we do, common-sense understanding is just fine. Still, for anyone who wants to understand how society works, it is not good enough. You may already realize there are many questions common sense cannot answer adequately. For example:

- Why are some people so different from you, and why are some so similar?
- Why do seemingly similar people lead such different lives?
- How is it possible for different people to get along?
- Why do we treat some people as if they are more 'different' than others?
- Why do we often treat 'different' people much worse than others?
- What do people do to escape from being treated badly?
- Why do some parts of society change quickly and others hardly at all?
- What can citizens do to make Canadian society a more equitable place?
- What can young people do to make their elders think differently?
- Can we bring about social change by changing the laws of the country?

Sociologists try to answer these questions by studying societies methodically. In fact, their task is to study people's lives—their own and others'—more carefully than anyone else. Sociologists want to understand how societies change and how people's lives change with them. Social changes, inequalities, and conflicts captivate sociologists because such issues—war and peace, wealth and poverty, environmental destruction and technological innovation, for example—are important for people's lives. Sociologists know that 'personal problems' are similar across many individuals. They also know that many of our personal problems are the private side of public issues. American sociologist C. Wright Mills called this knowledge or ability 'the sociological imagination'. With this ability or approach, we know we need to deal with personal problems collectively and, often, politically—with full awareness that we share these problems and their solutions with others.

However, solving problems entails clear thinking and careful research. So social theorists and social science researchers have developed concepts, theories, and research methods that help them study the social world more effectively. Our goal as sociologists is to be able to explain social life, critique social inequities, and work toward effecting social change. In this book, you will learn how sociologists go about these tasks and some of what sociologists have found out about the social world.

Our starting point here is a formal definition of *sociology*, comparisons of sociology with other related fields of study, and a discussion of sociology's most basic subject matter.

A Definition of Sociology

Scholars have defined sociology in many ways, but most practising sociologists think of their discipline as the systematic study of social behaviour in human societies. Humans are intensely social beings and spend most of their time interacting with other humans. That is why sociologists study the social units people create when they join with others. As we will see in the following chapters, these units range from small groups—comprising as few as two people—to large corporations and even whole societies (see, for example, Chapter 6 on groups, cliques, and bureaucracies). Sociologists are interested in learning about how group membership affects individual behaviour. They are also interested in learning how individuals change the groups of which they are members. In most social life, at least in Canadian society, there is a visible tug-of-war between these two forces: the group and the individual.

In the 1950s, American sociologist C. Wright Mills was a famous rebel with a cause. He pushed the sociological community toward a more historically informed, politically committed engagement with the world.

(© Yaroslava Mills)

However, it is impossible for any sociologist to study all social issues or to become an expert in all the sub-disciplines of sociology. As a result, most sociologists specialize in either macrosociology or microsociology—two related but distinct approaches to studying the social world—and choose problems for study from within these realms.

Macrosociology is the study of large social organizations (for example, the Roman Catholic church, universities, corporations, or government bureaucracies) and large social categories (for example, ethnic minorities, the elderly, or college students). Sociologists who specialize in the macrosociological approach to the social world focus on the complex social patterns that people form over long periods. You can see this in Parts IV and V of this volume, on social institutions and global society, respectively.

On the other hand, **microsociology** focuses on the typical processes and patterns of face-to-face interaction in small groups. A microsociologist might study a marriage, a clique, a business meeting, an argument between friends, or a first date. A microsociologist would study the common, everyday interactions and negotiations that together produce lasting, secure patterns. You can see many examples of this in Part II, Chapters 4 and 5, on socialization and on roles and identities, respectively.

The difference in names—*macro* versus *micro*—refers to the difference in size between the social units of interest. Macrosociologists study large social units—organizations, societies, or even empires—over long periods of time: years, centuries, or millennia. Microsociologists study small social units over short periods of time—for example, what happens during a conversation, a party, a classroom lecture, or a love affair. As in nature, large things move (and change) slowly, and small things move more quickly.

As a result, macrosociologists are likely to stress how slowly things change and how persistent a social pattern is as it plays itself out in one generation after another. An example is the way society tends to be controlled by its elite groups, decade after decade. The connection between business elites and political elites is persistent.

By contrast, microsociologists are likely to stress how quickly things change and how elusive the thing we call 'social life' is. In their eyes, any social unit is constantly being created and reconceived by the

(moodboard/Corbis)

Teams and other social groups, such as bands, gangs, and classrooms, all need communication, co-ordination, leadership, shared identity, and ritual. Many teams have uniforms and other insignia to increase their solidarity.

members of society. An example is the way one's friendship group changes yearly, if not more often, as one moves through the school system or the world of work. Some people remain our close friends over years, but many are close friends for only a short while.

Combining macro and micro approaches improves our understanding of the social world. Consider a common social phenomenon: the domestic division of labour—who does what chores around the home. From the micro perspective, who does what is constantly open to negotiation. It is influenced by personal characteristics, the history of the couple, and many other unique factors. Yet viewed from a macro perspective, different households have similar divisions of labour despite different personal histories. This suggests that the answer lies in a society's history, culture, and economy. It is far from accidental that across millions of households, men enjoy the advantage of a better salary and more social power both in a great many workplaces and at home.

While these approaches are different, they are also connected. They have to be: after all, both macro- and microsociologists are studying the same people in the same society. All of us are leading unique lives within a common social context, facing common problems. The question is, how can sociologists bring these elements together? As noted above, C. Wright Mills (1959) gave the answer when he introduced the notion of the *sociological imagination* as something that enables us to relate personal biographies—the lives of millions of ordinary people like ourselves—to the broad sweep of human history. The sociological imagination is what we need to use to understand how societies control and change their members and, at the same time, are constantly changed by the actions of their members.

All of this is the subject matter of sociology. We may choose to focus on problems of microsociology or macrosociology because of our preference to understand one or the other. However, a full understanding of most problems requires that we consider elements of both, for the two types of processes are closely connected.

How Sociology Differs from Other Academic Fields

Sociology is just one of several fields of study designed to help describe and explain human behaviour; others include journalism, history, philosophy, and psychology. How does sociology differ from these other fields? Canadian sociologist Kenneth Westhues (1982) has compared sociology's approach with those of the others. He stresses that journalism and history describe real events, as does sociology. However, journalism and history only sometimes base their descriptions on a theory or interpretation, and even then it is often an implicit or hidden theory.

Sociology is different. It strives to make its theories clear to test them. Telling a story is important for sociologists, but it is less important than the explanation behind the story. Besides, stories often make the news because they are unusual; sociologists instead are drawn to issues because they are common events or recurring patterns. Sociology may be good preparation for history or journalism, but it differs from these disciplines.

Sociology also differs from philosophy. Both are *analytical*—that is, concerned with testing and refining theory. However, sociology is firmly *empirical*, or concerned with gathering evidence and doing studies, while philosophy is not. Philosophy is more concerned with the internal logic of its arguments. Sociological theories must stand up logically, but they must also stand up to evidence in a way philosophical theories need not. No matter how logical the theory may be, sociologists will not accept a sociological theory if its predictions are not supported by evidence gathered in a sound way.

Finally, sociology differs from psychology, which is also analytical, empirical, and interpretive. The difference here lies in the subject matter. Psychologists study the behaviour of individual humans or, sometimes, animals. Generally, they do so under experimental conditions. Sociology's subject matter is social relationships or groups viewed in society. As you will see, sociologists study the family, schools, workplaces, the media—even the total society. Sociology and psychology come close together in a field called *social psychology*, but this field is defined differently by sociologists and psychologists. Studies by sociologists are more likely to focus on the effects of group living on people's views and behaviours. By contrast, psychologists are more likely to focus on particular individuals and how they respond under certain experimental conditions.

Another way of characterizing sociology and what makes it unique has been put forward by Earl Babbie (1988). He states that sociologists hold some basic or fundamental ideas that set them apart from those in other fields:

1. Society has an existence of its own.

2. Society can be studied scientifically.

3. Society creates itself.

4. Cultures vary over time and place.

5. Individual identity is a product of society.

6. Social structure must satisfy survival requirements.

7. Institutions are inherently conservative.

8. Societies constrain and transform.

9. Multiple paradigms or fundamental models of reality are needed.

As we will see in the chapters that follow, these are many of the most basic ideas or assumptions of sociology. Part I of this collection (Chapters 1 and 2) introduces the theoretical underpinnings and methodologies of sociology. Part II addresses the major social processes: culture (Chapter 3), socialization (Chapter 4), role and identity formation (Chapter 5), group formation (Chapter 6), and deviance (Chapter 7). Part III presents different forms of inequality people experience through chapters on class and status (Chapter 8), gender relations (Chapter 9), sexuality (Chapter 10), ethnic and race relations (Chapter 11), and aging (Chapter 12). In Part IV, you will learn about different social institutions that shape and constrain our lives, including the family (Chapter 13), education (Chapter 14), work and the economy (Chapter 15), health (Chapter 16), religion (Chapter 17), politics (Chapter 18), and social movements (Chapter 19). Increasingly, understanding Canadian society means also understanding global issues. So in Part V of the book you will read about global society (Chapter 20), population (Chapter 21), cities and urbanization (Chapter 22), the mass media (Chapter 23), and the environment (Chapter 24).

What do people do after studying sociology? Obviously, that's a hard question to answer, since every year hundreds of thousands of students take sociology courses in Canada and elsewhere. People who major or specialize in sociology gain valuable skills in critical thinking and research methods. This prepares them for a variety of second-entry college and university programs, including law, social work, teaching, industrial relations, personnel work,

(Volker Kreinacks/iStockphoto)

Sociology is just one of the fields of study designed to help describe and explain human behaviour. It differs from others, such as psychology, in its focus on social relationships or groups observed in society.

opinion polling, public health, public administration, and other fields. People who go on to get an MA or a PhD in sociology often end up teaching in colleges or universities or holding responsible positions as researchers, consultants, and policy planners.

Conclusion

Sociology is a good idea. It pays off by enlightening us, and it has worthy goals. Sociology, as you have heard, is the systematic study of society and the ways patterns of social behaviour change.

It is a broad field of study. This is obvious in the broad theoretical perspectives used to guide most sociological research. Sociology highlights both micro- and macro-level analyses and the complex relationships between the two, as noted in Mills's idea of the sociological imagination. Sociology also covers a broad subject matter—consider the subject matter of the following chapters, ranging across deviance, family, education, religion, politics, the economy, health, and beyond.

Sociology allows people to move beyond a purely common-sense approach to better understanding social life. It gives people more powerful tools to explore the connections between social institutions and processes. In the process, they will learn that much common-sense knowledge is faulty. Sociology will help them to see that things are not always what they seem.

Sociology stresses the relationships among individuals, social structure, and culture. As we will see, social structure and culture both constrain the behaviour of individuals. However, they are both essential for social life. As well, both social structure and culture are created by humans in social interaction. Therefore, they are both subject to future change in the same way. In short, sociology demystifies social life, showing that social arrangements are in our own hands. That said, powerful interest groups play a disproportionate role in controlling the kinds of social and cultural change that take place.

Sociology has obvious personal relevance, since it addresses everyday life issues. And finally, sociology has an important goal overall: to contribute positively to the future of humanity. Our sincere hope is that this text will set you on your way to developing your own sociological imagination.

PART I

Theory and Methodology

chapter 1

Sociological Theory

Anthony Thomson

In this chapter you will:

▶ Learn that you are already a social theorist;

▶ Apply the concept of binary thinking to understand the origins of sociology in the transition from traditional to modern society;

▶ Distinguish between objectivity and subjectivity in social theory;

▶ Identify the key ideas of the major classical theorists: Durkheim, Marx, and Weber;

▶ Relate classical social theory to the modern perspectives in sociology: functionalist, conflict, interactionist, feminist, and post-modernist;

▶ Connect concepts from classical theory to concerns in contemporary sociology.

In the First Person

My earliest experiences with sociological thinking came while I was attending a racially integrated high school in my native Halifax, where I learned about history and racism from a radical Indian immigrant and first heard the music of Bob Dylan from a draft-dodging English teacher from Maine. Later, my wife, Heather Frenette, taught me the principles of women's liberation. When I was an undergraduate student, my interests were shaped by theatre and an art history course I took initially as a 'bird' course. But in the 1960s, only sociologists were talking seriously about the issues of the day—civil rights, the Vietnam War, poverty. Herb Gamberg, a radical American educated at Brandeis and Princeton, drew me into sociology and social theory. After a few years teaching in Newfoundland, I returned to Dalhousie University to study social history, class theory, and international development with Greg Kealey and Tom Bottomore and then worked with Bob Blackburn at Cambridge University. Today, sociology pervades every aspect of my life, even watching films or TV shows. Sociology has the potential to change the way you look at the world, what you do in it, and what you do about it.

—Tony Thomson

Introduction: Why Theory?

The Italian revolutionary Antonio Gramsci believed that everyone is a social theorist. We all have an intellect and use our minds to make sense of the world in which we live. At the same time, even though we may fry an egg for breakfast, we are not all kitchen wizards. In sociology, social theorists are like chefs who concoct ideas that explain the world that people construct and inhabit.

Understanding the world and our places within it is almost as important as eating. The majority of Canadians (though certainly not all) can take eating pretty much for granted. This is a great privilege. In our increasingly connected but greatly unequal globe, a full stomach cannot be assumed. Our privilege is also our responsibility. Not only are we able to use our intellect to understand the world, but it is a crucial ingredient for guiding the actions we take. From a sociological perspective, what we do—our day-to-day actions—is part of the making and remaking of the actual world we live in.

Gramsci's point is not only that anyone can learn to cook up ideas or make judgments about different tastes but that we already use our intellects to explain how society works. Like eating, however, much of what we do and think is simply taken for granted. This is not necessarily a bad thing. We take it for granted that the physical world will pretty much do what we have come to expect; that food will burn if you add too much heat; that the sidewalk won't turn into quicksand; that the couch you sit on won't collapse, fly away, or swallow you whole (unless you're a character in *The Simpsons*).

We also expect the social world to follow rules that we can take for granted: people in Canada will drive on the right-hand side of the road; our mothers will love us; our paid employees will show up for work on time. But the rules of nature—for example, if you are deprived of food long enough, you will die—are not the same as the rules of society—you will not have enough to eat if you don't have the money to pay for it.

Following social rules more or less unthinkingly most of the time is necessary for society to work. If we thought about everything we did, all the time, we would be too paralyzed to do anything. Many social theorists want specifically to understand the taken-for-granted nature of social life: why it is so often unthinkingly orderly, routine, and generally predictable—*naturalistic*, in sociological terms.

Part of the answer is that our thinking itself has become something we accept as natural, not something we question. People live, day-to-day, with simple formulas that help them understand the way things work: the world is the way it is and couldn't be much different; people get what they deserve; each of us is a unique individual; we are endowed with an inner 'self' that truly defines our identity, who we are. We even begin to study sociology with the blasé attitude that sociologist Georg Simmel talks about—the progressive rock group Pink Floyd would call it 'comfortably numb'.

TIME to REFLECT

Would you say that your response to things that happen in the world reflects a blasé, indifferent attitude?

We don't, however, just act blindly or randomly or make up these ways of understanding as we go along. To borrow from the film industry, we follow scripts that are given to us and play the *roles* assigned. We are born into an existing society of things and people and into a world of ideas of what we should and shouldn't think or do. We are told who we are by other people; our identity is not something we make by ourselves, willy-nilly.

Gramsci would say that all these ideas about the world and our place within it reflect theories of society that have been built into our intellect. The process goes on so automatically that, like digestion, it seems to just happen. We begin ingesting ideas about the world along with our mother's milk, as Chapter 4 on socialization demonstrates. But the ideas that we swallow may be open to question. They may not be the best, or the most useful, or the most revealing about how society actually works. They may be popular fictions or myths. They may be part of the problems in the world, not the solutions.

Sociologists argue that although people are physical beings, they are also intellectual beings who can question the rules about such social problems as how food is distributed, to whom, and at what social and environmental cost. In fact, people can even choose to violate what might be seen as the most basic rule of biology: survival. People can and have chosen, in the midst of plenty, to starve themselves to death in support of some political or social cause.

The key to understanding, C. Wright Mills says, is the ability to connect personal problems with larger forces in society. Mills was a critical and caustic sociologist, and his ideas were ignored for some time in mainstream sociology. In *The Sociological Imagination*, Mills said it wasn't enough to focus only on a person's experiences. These experiences must be understood in their social context. Sociology must address social problems by linking an individual's personal troubles—one's own biography—with the way society is organized and structured; sociology should understand an individual's private troubles as rooted in widespread public issues.

A great deal of sociology is concerned with the question of why society operates so imperfectly. As much as the chapters that follow explain how social institutions such as the family, schooling, or religion actually work, they also identify social problems. Sociology is a critical study. Most of the social theorists you will read about in this book have been motivated by a desire not only to understand society but to change it for the better. Sociology challenges existing ways of thinking. This book is as much about helping you to see the world through different lenses as about providing information.

BINARY THINKING

How do theorists go about building a framework for understanding social life? Quite a lot depends on how they approach some basic questions about individuals and society. Are people's actions the result of a choice they have made? Or are people really just the puppets of social forces that work the strings behind their backs, determining what they do or think? Should you only look at the existing facts as they appear, such as how food is unevenly distributed, or should you also ask questions about values, such as how food *should* be distributed? When people do act, do they base their actions on their intellect, or do they respond emotionally or intuitively?

The use of either/or propositions is termed **binary** thinking. One of the most important binary distinctions in contemporary social theory is between 'structure' and 'agency'. On the one hand, we are born into pre-existing social arrangements or structures, including physical objects such as buildings as well as social codes of behaviour and morality. On the other hand, we choose one course of action over others. You didn't choose the degree requirements for the credentials you are seeking, but you chose to enter one program among many that were available to you. And you may toss the whole thing aside tomorrow, drop out, and travel to Bali. People are thinking and acting individuals; in short, they exercise agency.

Typically, thinking of yourself as making choices on your own comes easily. It is part of the taken-for-granted way we have learned to look at ourselves. Yet how much choice have you been able to exercise in your education to date? Whose expectations are you fulfilling? Your parents'? Your teachers'? Your friends'? Simply your own expectation of how the world works—that you need an 'education' to get the job you want? Any decision has major implications for the way you will be able to spend the rest of your life.

TIME to REFLECT

As a theorist yourself, what do you think causes some people to be rich or poor? How would you think sociologically about this question?

Much of our social life, then, is powerfully shaped by existing social institutions, rules, practices, and structures of power and authority. Social theory focuses much of its attention on the working of these pre-existing structures and institutions that set the limits and boundaries of our lives.

Social thinkers have examined *society*—the ways that people organize their lives together—for thousands of years. Every new idea comes into a world that is dominated by old ideas—'new' versus 'old' is one of the most ancient and enduring binaries. Sociology was developed by scholars who were aware that their world was changing rapidly and fundamentally. What was new when sociology was invented about two centuries ago was the idea that society could be studied scientifically. It was a broad and somewhat surprising claim.

The Birth of Sociology in the Age of Revolution

People who are part of the generation now reaching adulthood are used to rapid change, at least in many of the ways they go about their lives—how they communicate, travel, work, and experience diversions and pleasures. There is a sense that the gap between generations is widening, that the old generation can't understand the new one and can't appreciate the ways microchip technology has been inserted into social life. '*I used to be* with it,' Grandpa Simpson says, but now what's 'it' seems pretty weird and scary to him. Sociology thrives in these periods of large-scale social change. It is harder to take things for granted when the ground that had seemed so familiar is shifting under your feet.

GLOBAL ISSUES

Origin of Globalization

Sociology was born in the nineteenth century, when the promise of a better future sometimes seemed overshadowed by the emergence of new, perhaps worse, problems. The original myth of the sorcerer's apprentice concerns a young magician-in-training who can't control the dangerous results of the magic he sets in motion. Many of the social theorists who created sociology saw that their 'new', modern world was running dangerously out of control. It was exhilarating in its rapid change and amazing possibilities but also frightening because of the collapse of so much that had seemed to hold society together. Now theorists talk about the careening '**juggernaut**' of modern globalization—the phrase comes from Anthony Giddens. Globalization is a seemingly neutral word for a complex and troublesome social change through which the globe becomes increasingly integrated. Perhaps it began when our *Homo sapiens* ancestors trudged out of Africa and migrated to the 'corners' of the globe. Perhaps globalization didn't begin in earnest until the late fifteenth century, when a small number of previously insignificant European nations began their scramble for worldwide domination. Now what we call globalization may be thought of, in the rest of the world, as Westernization, Americanization, or imperialism.

Sociology, as an attempt to understand society objectively and scientifically, was a European invention, constructed to enable people to comprehend and, they hoped, control the transformation of traditional society into modernity. The same phenomenon is currently playing out worldwide. The first sociologists had a thoroughly Eurocentric view of the world. While the 'North' (Europe, North America) had progressed to the stage of industrialization, sociologists believed the South (Africa, Latin America, Asia) was mired in a stagnant, traditional, and undeveloped social system. It was the 'white man's burden', poet and novelist Rudyard Kipling said, to 'rescue' the rest of the world from its backwardness. Karl Marx deplored the terrible effects that European colonialism imposed on India, but he believed that the old ways had to be destroyed before modern capitalism could be introduced to Asia as the first step to its full integration into the modern world.

We are a lot closer now to this 'fully integrated' world. In 2010, China became the world's second largest economy after the United States, which is the most important consumer of Chinese products. Many new graduates from Canada will seek their first job opportunities in Japan, Korea, or China. In our megacities such as Toronto and Vancouver, the previously predominant white Europeans will soon become a 'visible minority'. Globalization in the age of classical sociology wasn't—and it isn't now—a gentle, benevolent evolution. The term *globalization* is now likely to evoke images of mass protest and police repression (as in Toronto in 2010), the re-emergence of ethnic violence to the point of genocide, the resurgence of extremism rooted in religious fundamentalism linked to state and individual terrorism, and the widespread environmental destruction caused by endless-growth capitalism. In this world of rapid social and economic change, sociology has never been more relevant.

When European social theorists surveyed their 'new' world and began to apply the word 'modern' to it, they simultaneously created the binary concept, 'traditional', meaning Europe before the modern age. Looking backwards, they believed the key difference separating the traditional from the modern was the way people understood and thought about the world. Traditional society had been a world of magic, mystery, and irrational authority. In contrast, modern society had entered a new world of **Enlightenment**. Through the use of reason or rationality, the human mind could shine light into the darkest caves and discover 'true' knowledge. More than 200 years later, we seem no closer to this elusive and improbable 'universal' understanding.

The French social theorist who invented the term 'sociology', Auguste Comte (1798–1857), intended to create a 'science' of society that would allow us to understand social life the way that biology had enabled us to understand physical life. The word 'science' was used deliberately to give sociology high status among the branches of knowledge, because 'science' ruled. Sociology would be based on facts, evidence, and scientific laws, not just on imagination, abstract philosophy, or fiction. But the 'natural' sciences—astronomy, physics, chemistry—were more advanced in their knowledge at that time than any purported science of social life. There was a lot of catching up to do.

By combining careful, detailed, and systematic observations of real life with logical, systematic thought (theories), scientists had 'discovered' that natural laws made the world orderly and predictable. Knowledge of the natural world could lead to having power over nature. The way forward for sociology, then, was to discover the 'natural' laws that determined social life, just as Sir Isaac Newton had discovered the law of gravity. Once these laws of social life were discovered scientifically, they could be applied to controlling society. And society seemed to need controlling. Political and social unrest was shaking the foundations of traditional institutions, undermining the authority of church and state; the new forces of industrial capitalism were transforming social life rapidly, profoundly, and in many cases disturbingly. The understanding of society—social theory—had to be rooted in science if social life was again to be orderly, peaceful, and secure.

(© Reuters/Corbis)

Juggernauts are gigantic chariots used in Hindu processions that sometimes careen out of control, crushing people in their path. How do the movements of these chariots mimic the effects of globalization on culture?

As Auguste Comte (1974, 19–27) saw it, there was a 'law' according to which social thinking necessarily passed through three stages. In the first stage, it was assumed that the world was run by supernatural powers, by gods. In the Middle Ages, for example, people believed that angels actively pushed the sun as it made its daily circle around the flat, stationary Earth.

With the rise of scientific understanding, religious theory (theology) gave way to philosophy, the second stage. The idea of 'nature' replaced the idea of an active, miracle-working god. The world proceeded on its natural way like clockwork. Finally, in the third, positive stage, people began to apply their scientific knowledge of the laws of nature to change the physical world to suit themselves. Similarly, the knowledge developed by the new social science he termed 'sociology' would give humans power over social change.

Comte's own life, however, didn't quite follow this rational, scientific script. He experienced bouts of insanity, attempted suicide, fell romantically and passionately in love with Clotilde de Vaux, who died of tuberculosis within a year, went through a religious conversion, and tried to establish a new 'religion of humanity' with himself as pope (Andreski 1974, 8–9). Real life was clearly a lot messier than pure scientific reason would suggest, and unreason—the irrational—had to be taken into account. As his social theory evolved backward from science to theology, Comte's own career served to refute his 'law' of three stages.

Throughout most of the nineteenth century, sociological thinking involved the search for law-like certainties that could explain social life. Following Comte, this approach became known as **positivism**. English sociologist Herbert Spencer (1969, 120) believed that society, like nature, was a struggle for existence. The strongest and 'best' individuals inevitably rose to the top of the social pyramid and deserved their privileged status, while the poor and the weak naturally sank to the bottom. Karl Marx also sought to discover the 'laws' of modern, capitalist society (1959 [1844]). In his view, however, these laws would inevitably lead to a working-class revolution, causing the capitalist system to come crashing down: the last would become first. Thus, the search for social 'laws' had led to quite different—indeed opposite—conclusions.

TIME to REFLECT

What does it mean to live in a revolutionary age? How is the word 'revolution' used today?

Classical Sociology

What was a sociologist to do? First, let's consider that many of the questions sociologists are interested in don't have a rational basis. The idea that we frequently act according to habit or custom, unthinkingly, or according to our beliefs and values, suggests that people often do not reason things out and make 'rational' choices. Many of the things we accept as natural, from our experience of time to our views of 'human nature', are actually socially constructed.

Second, many social theorists have sought to understand people's intentions and bring the elements of thinking and choosing into their analysis. German sociologist Max Weber said that people act on the basis of what they intend and what they believe; to understand them, you have to take these **subjective** factors into account. As long as we believe that the laws of the country are basically right and for the good of everyone, we will obey—in Weber's view, we accept authority as legitimate and follow the rules that are given to us (1964 [1920], 325–8). If, instead, we believe that law represents only the interests of the rich and powerful, we might disobey, intentionally breaking store windows and setting police cars on fire.

Giving up the search for the 'laws of society' did not mean giving up the search for (uncertain) prediction and (limited) control. It meant being more modest about what could be understood. We all know someone who 'defied the odds', who 'made it' economically despite being born into poverty. But sociologically speaking, poverty is more likely than not to beget poverty. Even free public schooling for everyone doesn't necessarily create a level playing field; most people settle back into the same social position once they're in the workforce.

Finally, as Comte realized, two apparently contradictory things seem true of society: it basically stays the same over time, and it is constantly changing. The binary of continuity and change has led to two different sociological perspectives. From one point of view, what is most important about society is how it remains, day-to-day, basically the same. Look around your classroom, and think about the things all of your classmates *could be doing together now*. But what they *are* doing in reality allows the lecture to go on, in an orderly and routine way.

An important perspective in sociology seeks to understand this continuity—how society is reproduced. Part of the answer is that society possesses powerful mechanisms that force us to obey, such as police and prisons. But maintaining social order is usually much more subtle than employing direct coercion. Operating more or less behind the scenes are powerful social forces that shape what we do and help determine the consequences of our actions.

Understanding these structural forces is the business of sociological theory. Functionalist theorists, for example, seek to identify the basic **functions** that must be fulfilled in all societies and understand how they are accomplished. From a functionalist perspective, if something exists in society and persists over time—religion, for example, or sports, or even crime—it must perform some necessary function that is important for the reproduction of society. What is necessary is to understand the forces that generate agreement and consensus among people—social *solidarity*.

Reproduction of society over time is one side of the story. But things also change over time, which is the focus of another group of sociologists. Marx famously said that the purpose of social theory is not to understand the world but to change it (although we might argue that you have to understand it *in order* to change it). According to the conflict perspective in sociology, change comes from conflict—between generations, between the government and the people, between rich and poor. Many sociologists adopted the conflict approach, including Max Weber in Germany and, more recently in the United States, George Ritzer.

In short, studying sociological theory entails grasping the perspectives of the early developers of sociology who strove to carve out the new discipline, the new science of understanding social life. We refer to members of the founding generation of sociologists as classical theorists. Two of the most important are Émile Durkheim and Karl Marx; both are cited frequently throughout this text, so at this point it is useful to outline their thinking.

ÉMILE DURKHEIM

Émile Durkheim (1858–1917) would become the most famous French sociologist of the late nineteenth and early twentieth century, but he was originally destined to be a rabbi. He was born in eastern France to a close-knit, strictly moral, and orthodox Jewish family, part of a small and cohesive community. Educated in Paris among the intellectual elite of his generation, Durkheim broke with the Jewish faith and with religious belief generally, but he understood the powerful hold that religion has on people in society (Lukes 1972, 39–46).

As a university lecturer, Durkheim began his lifelong study of the relationship between the individual and society. Above all, he thought, modern society seemed to have lost that solid and shared code of morality that had acted as the glue holding it together—in Durkheim's terms, people no longer shared a *collective conscience* (Durkheim 1964 [1912], 444). Durkheim argued that the simplest societies were held together by such practices as religious celebrations and gift-giving. When wives, or gossip, or stories, or 'gifts' were exchanged among tribal members, relationships were strengthened and given meaning. Even in our culture, gift-giving carries much more symbolic meaning than a merely economic exchange. It stands for friendship or love; it may be a peace offering or a token of respect; it helps to repair or cement our relationships.

A second source of togetherness, Durkheim said, originated in regular sacred gatherings, events when the tribe feasted and celebrated its community. In joyous, often delirious celebration, people experienced a 'collective effervescence' that bound them together and generated a feeling of spirituality—they sensed a power greater than themselves. This feeling, said Durkheim, was the root of religious belief, but it actually originated from the shared social experience itself—that is, not from the supernatural but from society, which is the material foundation of the religious life.

Durkheim hypothesized that over time, the sacred part of life had become overshadowed by the secular. Gift-giving in modern society, for example, is highly ritualized, but it is integrated into the modern rituals of consumer society rather than conventional religion. For Jean Baudrillard (1929–2007), modern individualism is paradoxically expressed by participation in the rites of mass consumption (1998). Through the ostentatious display of goods for sale in shop windows and enormous big-box malls, 'things' take on an almost sacred aura, even when they are merely worshipped through gazing (window-shopping). At every moment in consumer society, 'in the streets . . . on advertising hoardings and neon signs', individuals are trained into a modern collective consciousness of mass consumption.

In modern society, Durkheim felt, the power of religion to uphold a system of moral rules had diminished. People were no longer united by a single code of right and wrong, an uncertainty that Durkheim termed **anomie**. He saw that other traditional institutions, such as marriage and the family, also seemed to be breaking down in modern society. He believed that men had an inborn sexual passion that had to be curbed and regulated by marriage. Otherwise, he thought, men's sexual desires would lead them into the pursuit of novelty and excess. It seemed to him that it was far better for people to live 'courageously' with unhappiness in their marriage than to abandon their duties to their family and society. He might well have understood a line from the 1926 silent film *The Crowd*: 'Marriage isn't a word, it's a sentence.' In his view, marriage was a sentence that had to be endured for the good of society. A typical Victorian man, Durkheim also opposed sex education, which, he said, was like treating a mysterious and private act as if it were no more than digestion (Lukes 1972, 530–4). Women sociologists, on the other hand, soon recognized that the writings of classical male sociologists reflected their culturally induced gender biases.

For Durkheim, people were becoming more individualized as modern society evolved away from the spirit of community. How could millions of increasingly distinct individuals ever create an orderly society? At the same time, Durkheim recognized that the old community spirit could not be brought back to life; it had disappeared forever. Individualism was here to stay, and it had both positive and negative features. It was progressive insofar as it promoted respect for human rights, which Durkheim actively defended. However, individualism created problems because it undermined people's connection to society; it was at the root of modern aimlessness and anomie. The task of sociology, then, was to put an end to anomie and the conflict it engendered.

But how to overcome anomie? Society needed a new moral code, but it could not be based on the old model of everyone thinking and acting in identical ways. That was no longer possible. Instead, people would have to be educated to recognize that each person's individuality should be respected. At the same time, each individual should understand that his or her welfare depends, in a deep and multifaceted way, on everyone else. Just as many organs keep an animal alive, everyone in society plays a part

in maintaining social life, which Durkheim termed *organic solidarity*. The individual, Durkheim said, is 'the organ' of a much larger organism—society—and everyone must learn to perform conscientiously 'one's role as an organ' (Lukes 1972, 102).

Feeling closely connected to others was good for society, but it was also essential for the well-being of an individual, Durkheim felt. His best-known application of sociological methods to understand a social problem was his examination of suicide. Taking one's own life would seem to be an absolutely individual act that could only be understood by examining the psychology of the victim. This was not entirely true, Durkheim reasoned. He used his examination of suicide to demonstrate that sociology could make an important contribution to the study of what was perhaps the most lonely act.

One of Durkheim's closest friends at school, Victor Hommay, committed suicide in 1886. Teaching in isolated provincial towns, Hommay felt himself a stranger in small communities where people observed each other too closely. His life felt 'pale, colourless, monotonous, [and] insipid' (Durkheim 1887). Durkheim

(The Granger Collection)

Moneychanger and Wife, by Marinus Van Reymerwaele. Oil on wood, 1539.

concluded that Hommay's situation might have been typical of cases of suicide in which the individual feels isolated, has few ties to other people, and sees no reason to live (Lukes 1972, 49–51). The key to the sociological explanation for suicide, Durkheim reasoned, was the strength or weakness of the individuals' ties to their community and society (Durkheim 1951 [1897], 208–23). He saw self-destruction as a *social fact* that could be understood scientifically and objectively.

Durkheim's social theory involved examining society as a totality of interconnected parts—an approach that is fundamental to theory. Through his publications, his teaching, and the work of his followers, he stamped his functionalist approach to understanding society on the new discipline he helped to create.

KARL MARX

While he did not describe his work as 'sociological', the theories of Karl Marx (1818–83) have inspired movements of revolution and reform that have had deep and lasting consequences for the study of sociology. Both Émile Durkheim and his German contemporary Max Weber—who is discussed next—developed their ideas in the context of, and often consciously in opposition to, socialist movements that had been inspired by Marx. In the 1960s and 1970s, Marx's theories entered sociology directly as part of a critical and radical reorientation of the discipline that challenged functionalist arguments and emerged as conflict theory. However, the major contemporary theorists, in a way that was similar to that of the classical generation, initially differentiated their ideas from Marx's.

The son of a Jewish lawyer, Marx was born into a respectable middle-class family in Trier, Germany. In 1843 he followed in the footsteps of many middle-class German professionals and married into the minor aristocracy. His wife was Jenny von Westphalen, the daughter of a baron. For the 25-year-old Marx, however, the match was a romance, not a route into the upper class. The groom was not conventionally ambitious but instead was steeped in youthful radicalism. In Germany at that time, open identification with left-wing politics closed the door to any career as a university professor. Marx's writing soon ran afoul of the German authorities, and Marx and Jenny, along with six children, spent the rest of their lives in England, living often from hand to mouth (Berlin 1963, 27–32).

Marx sought to understand the origins of contemporary society and the forces leading to change within it. For Marx, the way to begin the analysis of any society was to examine the way it produced and distributed the basic necessities of life—its economic system. Originally, societies had existed in simple hunting and gathering modes. People in these earliest societies were close to being social equals, didn't live under the thumb of any powerful government, and produced only what they needed to survive.

Over time, however, land and goods, which had been the common property of all, became the private property of a few. Class divisions and conflict became basic features of all societies and the keys to understanding social change. The majority of people, who laboriously tilled the land, were able to produce more goods than they required for their own use. This excess was a **surplus**, and it went to support the elite few who were free from the burden of daily labour. Over time, the few grew into a rich and powerful dominant class, and society became increasingly unequal. Ancient Greece and Rome, for example, were slave societies, divided between slaves and slave-owners. In Europe during the Middle Ages—the traditional society from which modern capitalist society emerged—the majority of the people worked on the land as *serfs* and handed over their surplus production to the various 'noble lords' making up the aristocracy.

Economically, what distinguished traditional from modern society, Marx said, was the rise of capitalism. Capitalists originated as a group of merchants, buyers and sellers—represented in the painting *Moneychanger and Wife*—who occupied space as a 'middle class' between the serfs (lower class) and aristocracy (upper class). Over time, the middle class grew economically wealthy until, through a series of revolutions (such as the French Revolution of 1789), they took political power from the monarchs and aristocracies and established themselves as the new dominant upper class.

Under the control of the capitalists, the economic system was revolutionized. People were drawn from the countryside into the booming, bustling new cities and into the factory system, which became the dominant economic form—now a worldwide phenomenon. They became wage workers or, in Marx's terms, the **proletariat** (Marx and Engels 1959 [1848]). Marx's perspective is illustrated in a figure titled 'Pyramid of Capitalist System'. The labour of the working class toiling at the bottom provides surplus in the form of profit for the livelihood of the opulent capitalist class (shown at dinner), for priests (second from the top), and for the government, which orders soldiers to shoot the workers when they rebel. In this model, religion tells lies to keep the workers pacified. In one of his most famous phrases, Marx claimed that religion acted as 'the opium of the people' to dull the pain of their oppression (1959 [1844], 263).

The fate of capitalism hinged on the conflict between the proletariat and capitalists, and Marx expected that a working-class revolution would replace capitalism with a more cooperative and collective society—socialism. But no industrialized, capitalist country has ever undergone a revolution that ended with a new socialist society. As capitalism became global,

twentieth-century theorists applied Marx's ideas to understand the unequal development of rich and poor countries. Socialist revolutions in the twentieth century occurred in countries such as Russia and China, which were oppressed regions within the worldwide capitalist economy. Ironically, both of these societies have now become part of global capitalism; check out the 'Made in . . .' labels on most of your consumer goods for proof. Simultaneously, the welfare form of capitalism in the developed West (Europe, the US, Japan) has been whittled away, while corporate and speculative capitalism has become aggressively global.

Another Marxist concept that has resonance in contemporary society is *alienation*, a concept that has been applied widely in sociology. In simple terms, alienation means to be separated from something.

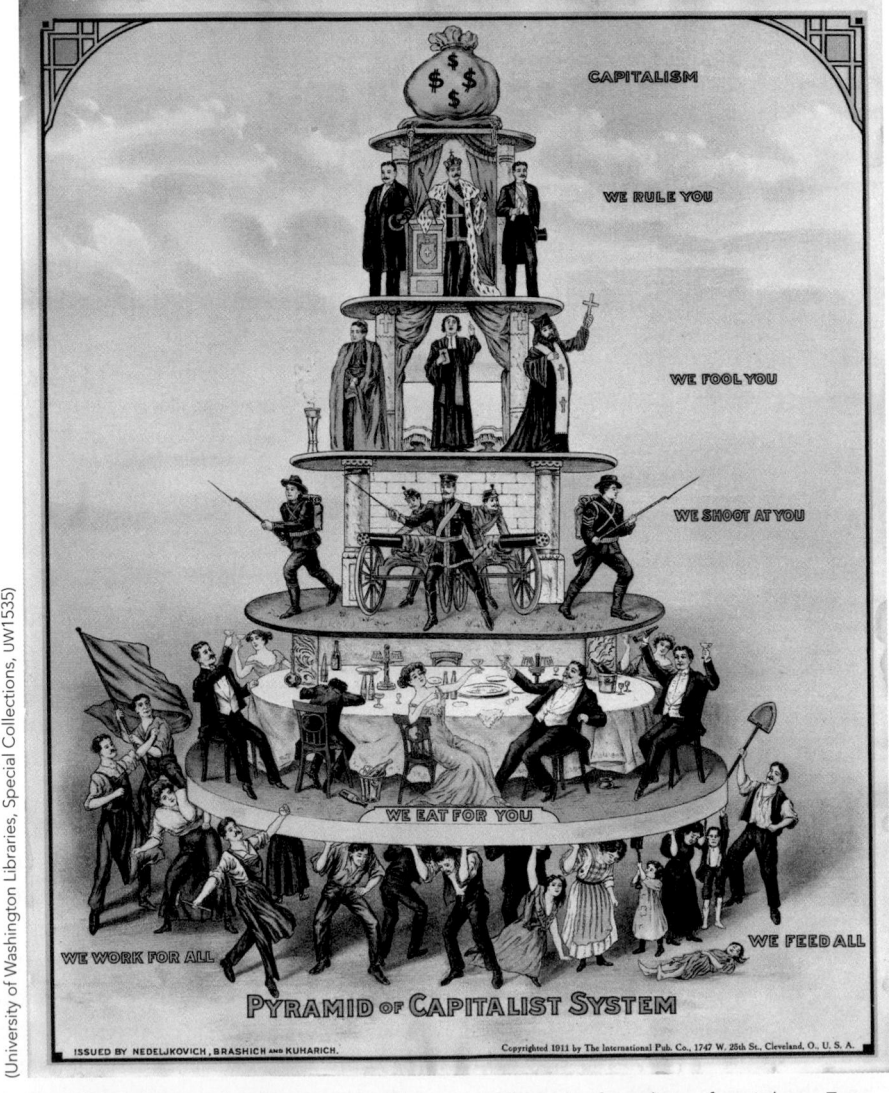

(University of Washington Libraries, Special Collections, UW1535)

Social stratification has been a central concern since the founding of sociology. Even non-sociologists have long understood that their lives are shaped by class structure.

SOCIOLOGY in ACTION

Michel Foucault

Born in France, Michel Foucault (1936–84) established an early and highly successful career in intellectual history and remains influential in many disciplines, including sociology. His book *Discipline and Punish* (1975) tells the story of the rise of the penitentiary, with its constant surveillance of prisoners and strict discipline, as a metaphor for the same processes as they spread throughout modern society (in schools, hospitals, the army).

Educated at an elite school in France, where he was surrounded by bright young academics determined to outdo one another in eccentric behaviour, the solitary Foucault found life unbearable. A suicide attempt led him to psychoanalysis. He became fascinated with authors who dealt with 'transgression', 'limit experiences' (death), and 'excess'.

As an 'engaged philosopher', Foucault became politically active in the 1970s and challenged other concrete and specific wrongs in society. He founded the Groupe d'information sur les prisons in support of prisoners' rights and to expose the intolerable realities of incarceration. Foucault put his body on the line on many occasions in political demonstrations. By this time, he had come to terms with a homosexual identity and followed his new lifestyle with abandon. He lectured annually in the United States and died at the age of 47 from complications linked to AIDS (Eribon 1991).

The earliest 'humans' separated, or alienated, themselves from nature. Thereafter, people developed increasingly complex technologies to control the natural environment. Generally for Marx, the separation from nature was a good thing, and it was necessary for all the progress that followed—potentially, for a world of shared wealth and high living standards for all. But alienation has become steadily worse, and capitalism is especially alienating for working people. Power over nature has proved to be a double-edged sword; in our current environmental crisis, which is one of the consequences of constant-growth capitalism, it is clear that alienation from nature has had some disastrous consequences.

TIME to REFLECT

Bearing in mind Marx's concept of alienation, would you describe your university education as alienating, or as a step towards breaking down alienation?

Marx's theory is critical of existing society and openly declares its biases and interests. While no theory is socially neutral, the values of working-class revolt and revolutionary change are central to his. It combines analysis of cultural, political, and economic structures with human action or agency. For Marx, the relative truth of a social theory has to be tested in practice and modified as a result of experience.

In our world of unequal economics, climate change, international conflict, and threats to people's welfare, Marx remains a central theorist.

Power and Resistance

One of the oldest and most persistent dichotomies in Western social thought is that between the mind and the body. Although our minds and bodies are connected in innumerable ways, the distinction is useful. Contemporary sociologists distinguish, for example, between efforts at control directed at disciplining the body and demanding physical acts of conformity, and those efforts directed at our ideas and how we think.

The concept of power in society—including top-down domination and resistance from the bottom—is an important element in the modern conflict perspective in sociology. The imposition of discipline on bodily movements may become so engrained that people respond unthinkingly, like Pavlov's dog. Discipline may also be directed to a person's mind, as in Michel Foucault's focus on power and the way specialized, expert knowledge is used by professionals to increase their power and control over clients, students, or patients. It may be useful, however, to think of a person's body as both the target of rules imposed on the person by authority and as an object that is manipulated and modified by the person. After all, we do pay a great deal of attention to a person's body—too thin, too short, 'carrot-topped', big-nosed, sloped forehead, stooped, aged, 'perfect 10', and so on. We sculpt our bodies through various means, including exercise and diet. We adorn our bodies with ornaments, tattoos, styled clothing for various occasions, and hairstyles.

The body is the site of control and resistance. Schools, like prisons, subject people to bodily discipline. If kindergarten is referred to in sociology as 'boot camp' for children, it is because teachers impose rules on the way students align their bodies and direct their bodily movements, from lining up at the bell, to sitting straight in specified places (traditionally in rows), to the rule of silence; there are rules about covering parts of the body and controlling such bodily needs as elimination, eating, and drinking.

Rules are external to the individual. They exist in codes of behaviour and dress, in formal requirements of orderly movement, as behavioural imperatives ('silence!'), and as underlying specific demands ('open the text to page 14'). Rules are also internal—in sociological terms, they are internalized and become not merely what 'they' tell us to do but what we feel we 'ought' to do. In Michel Foucault's terms, people learn to regulate themselves.

Nevertheless, no one is powerless. Ask any teacher or parent. People exercise institutional power through the position they occupy, such as a teacher, a judge, or a parole officer. Simultaneously, they provoke 'resistance'—the ways people find to subvert power and achieve their aims against external authority (actions that transgress rules). To escape from the dull routine of the classroom, students often 'need' to go to the bathroom. Teachers understand these mild forms of rebellion and agree to the temporary escape. Students have invented many other strategies to resist school rules. They copy homework, skip classes, keep watch when the teacher is absent, feign sickness, collaborate in passing notes, carve their desks, cheat on tests, plagiarize, pretend to be paying attention (to avoid being singled out)—in short, they learn the informal anti-rules of schooling (Raby 2009, 127).

MAX WEBER

The German sociologist Max Weber (1864–1920) understood that modern society was increasingly individualistic, and he devoted much of his historical research to uncovering the roots of capitalism. In his view, however, neither individualism nor capitalism captured the fundamental way that modern society differed from

TIME to REFLECT

Is the 'colour-line' still a fact of life in Canada, or is racism largely a thing of the past?

HUMAN DIVERSITY

W.E.B. Du Bois and the Veil of Racism

The value orientation of the first prominent black sociologist, W.E.B. Du Bois (1868–1963), was never in doubt. Born in Massachusetts, Du Bois experienced racial discrimination at the hands of his classmates—the 'shadow of racism' crept over him, he wrote. Rejected on racial grounds as an undergraduate by Harvard, Du Bois went below the US 'colour-line' to Fisk, a black university in Tennessee. Once back north, he became the first African American to graduate with a PhD from Harvard. He conducted an extensive sociological study of the black community in Pennsylvania and published *The Philadelphia Negro* in 1899, discussing racism, employment, poverty, and the emergence of a black middle class.

The fundamental problem of the twentieth century in America, he declared, was the colour-line. The laws of segregation (known as 'Jim Crow') were deepening in the United States. Du Bois wrote extensively about racism and the creation of a distinctly racial 'caste' system in America. He was also a social activist and a co-founder of the National Association for the Advancement of Colored People. He sought a broad alliance of progressive people—black as well as white—to bring about legal and social change. Ultimately, he believed, it was up to African Americans to fight for their own liberation. No group had achieved emancipation by simply asking for it—liberation had to be won through struggle.

Since he had studied in Germany, Du Bois's approach to sociology was directly influenced by classical theory in that country. He intended to establish American sociology on a scientific basis, employing concepts such as race, class, and status in the interpretation of grounded empirical research. In practice, Du Bois worked to deepen black pride and independence. He lived through the civil rights movement in the 1950s, but he became increasingly disillusioned by American society and turned his attention to independence movements in Africa. He died in 1963, a self-exiled radical in Ghana.

traditional society. For Weber, the growing importance of rationality was the basic underlying difference.

Weber was born near Weimar, Germany, and grew up under the iron-fisted power of Kaiser Wilhelm. His father was a lawyer who came from a family of industrialists and merchants in the textile business. Content under authoritarian rule, Weber's father was dictatorial, patriarchal, self-centred, and hedonistic (Käsler 1988, 1–3). Weber himself was appointed to his first academic position by the age of 30, but his career was wracked by periodic breakdowns and bouts of depression (Gerth and Mills 1946, 11).

In Germany, Marxism was an influential political force by the time Weber had begun to address sociological issues. As with Marx, Weber's sociology is historical and comparative, two essential elements of good social theory. But Marx was identified with positivism, and Weber used this simplification as a convenient foil for building his own theory of society. Weber sought to uncover the social, cultural, and political factors that shaped modern society independently of economic processes.

For Weber, modern society differed from traditional society in that it was highly rational, and the most direct example was modern science and its use of reason to understand the world. Arising in a traditional society dominated by religion, science tended to erode mystical and supernatural beliefs. Weber referred to this process as the 'disenchantment of the world' (Weber 1946, 350–1). He saw the modern world as characterized by a certain type of rational thinking that he defined as formal. Formal rationality involves calculating the most efficient means to achieve a goal, just as a 'pitchman' figures out the best type of propaganda to sell products such as the Slap Chop or politicians plot how to manipulate public opinion and win votes.

In traditional monarchies, the first-born son of the sovereign automatically becomes the next ruler, but this form of traditional authority does not mean that person is actually fit to rule. In the film *The King's Speech*, the shy, speech-impaired second son of the deceased king learns to rule in place of his older but decadent, pro-Nazi brother. The ability to exercise power over others can also be based on personal charisma—an irrational power Weber termed charismatic authority that compels people to follow a leader such as Jesus or Hitler. Charismatic authority either dies with the leader or is converted into another form of tradition, as demonstrated by the evolution of Christianity following the death of Jesus.

In the modern world, politics has become rationalized. Tradition and charisma have been replaced by legal–rational authority. A set of rational and legal regulations determines how we choose those who govern us and the rules they must follow. All of the institutions of modern life are governed according to a rational set of rules that define the duties of a hierarchy of positions and power—in short, by a bureaucracy. When you get a job as an assistant bank manager, you are at the bottom of a highly organized, complex, and rationally devised career ladder. Your work will be judged according to objective standards of performance, which will determine whether you keep your job and whether you are promoted. Increasingly, traditional types of evaluation and judgment—who you know or to whom you are related—become less important than your qualifications, experience, and work habits.

TIME to REFLECT

Weber says that traditional authority and charismatic authority have been largely superseded by legal–rational authority. Thinking about Canadian elections, would you say this is largely true?

In addition to rational calculations, Weber understood, sociologists have to pay attention to the unintended consequences of people's actions. Things don't always work out the way we intended or rationally expected they would. We may intend to deter criminals from a life of crime by locking them in a maximum-security prison, but prison may be graduate school for criminality. In one of his most controversial historical investigations, Weber concluded that the rise of capitalism in Europe was an unintended consequence of the Protestant Reformation (1958 [1904]).

For Weber, capitalism was not the inevitable outgrowth of European feudalism; rather, capitalism had sprouted in many places in ancient times, but family traditions, religious beliefs, or political values had strangled its development (Weber 1964 [1920]). It took a change in religious values (Protestantism) in Europe for capitalism to emerge as the single, dominant economic system. Once industrialization took off, all aspects of life were transformed.

Furthermore, Weber argued, modern capitalism was busily creating a new middle class of professionals, technicians, and office workers who were employees—like Marx's proletariat—but were paid higher salaries and given more autonomy at work. The 'new middle class' occupied positions of higher **status** than mere factory workers. Weber's attention to class, status, and power as partly independent forms of inequality suggested a multi-variable analysis of modern society.

In contemporary sociology, French social theorist Pierre Bourdieu (1930–2002) redefined Weber's analysis of modern society to examine the various ways in which people can acquire resources of power and control (1984). Money capital is obvious. For Bourdieu, the structure of power in society is also maintained partly through culturally acquired 'tastes' and practices, such as what we eat and how we are entertained—even how we blow our nose—which give individuals a sense of their place in the world. If you are well-educated and have acquired the necessary knowledge and 'taste' to be able to fit seamlessly into higher classes, you have **cultural capital**. If you are well-connected and have an 'in' with important people, you can benefit by using these connections and have **social capital**, says Bourdieu.

TIME to REFLECT

How would you rate your own social and cultural capital? Do they connect closely to your *money* capital?

Weber's analysis of formal rationality—calculating the most efficient means to attain an end such as private profit—is central to globalization. George Ritzer's influential analysis of the spread of rationality in the global economy takes off from Weber. Ritzer calls the process McDonaldization, since the fast-food chain's techniques of production and marketing have become the model for others. McDonaldization is evident in 'education, work, health care, travel, leisure, dieting, politics, the family, and virtually every other aspect of society' (Ritzer 2000, 1–2).

Weberian categories of sociological analysis have many applications, and they recur frequently in the chapters that follow. For Weber, however, the goals that we seek are not necessarily rational ones. We may seek profit, but we may just as likely seek pleasure for its own sake or seek a goal that we value highly on purely emotional grounds, such as justice or religious salvation (Bourdieu 1984, 57). The Holocaust entailed the application of formally rational techniques of systematic persecution, mass execution, and ethnic genocide in pursuit of an essentially irrational belief in racial inferiority.

In fact, as life becomes increasingly dominated by rational calculation, people often try to escape from rationality into other, competing, non-rational realms.

To escape from the routine world of workaday life, which has no deep or ultimate meaning, people strive for self-cultivation through acquiring things that are valued in our culture and express 'good taste'. The value we sometimes put on erotic love, for example, can act as a life-affirming, enchanting goal that gives meaning to our existence. In the modern world, the value of eroticism is accentuated, and erotic love, Weber said, is not only a joyous triumph over the all-too-rational but also appears to be the gateway into the irrational, 'real kernel of life' (1946, 341–7).

The focus on people's values and goals meant that sociology could not, as with Durkheim, concentrate only on 'social facts'. In his discussion of the methods of sociology, Weber emphasized the binary (or dual) nature of facts and values. Facts were open to **objective** analysis or understanding, but values were entirely **subjective**—within the mind of the individual. You could rationally calculate the best means to terminate a pregnancy, but there was no objective way to determine whether this goal was right, or moral, or desirable. Rational arguments are ineffective when they come up against irrationally held, and especially supernaturally based, beliefs. Sociologists, then, had to get inside a person's thinking to understand why they did what they did. Weber used the termed *verstehen* to indicate this need to understand someone subjectively, using the directly untranslatable German word. He did not, however, develop a systematic sociology of subjectivity.

Microsociology

As the challenges of the fact–value distinction and the irrational side of humanity suggest, there are two sides to sociology. It isn't possible to study people the way you study rocks or plants. If you push a rock, the results are quite predictable. You can calculate how far and in what direction it will move. But what happens if you push a human being? Rocks don't push back; they don't run away or turn and fight.

If I predict that you will close this book now in frustration and turn to Twitter instead, even if that is what you actually want to do, you could choose not to, precisely to refute my prediction. Sociologists have to take into account the way people change their behaviour *because* they are being studied. Human beings are not simply objects that are studied and acted upon; they make choices and take actions as a result. They have intentions they want to carry out; they have interpretations of their own actions and those of others; and they have desires and needs they seek to fulfill.

When sociology was brought over to the United States at the end of the nineteenth century, it entered a rapidly industrializing culture. One important problem was assimilating newly arrived immigrants who came from a multitude of European countries and were crowded into pockets of the rapidly expanding northern cities. A group of sociologists centred in Chicago—the Chicago School—undertook firsthand investigations, interviewing people in these neighbourhoods and highlighting the social problems of urban America.

The Chicago School was attuned to the distinctiveness of American society, which was a highly individualistic culture dedicated to upward mobility and material prosperity. From Chicago grew a distinctive sociology that was closely linked to psychology.

Rather than understanding society historically and comparatively and searching for large (macro) social forces that acted on and shaped individuals, the Chicago approach began with the point of view of individuals. Over time, this approach became known as microsociology because it focused attention on the smallest units of society—individual people and their interactions.

In microsociology, society emerges from social interaction—the way people orient their actions to take other people into account. Among the early influences on American social psychology was Charles Cooley, who said that individuals develop a consciousness of their 'self'—answer the question, 'who are you?'—through their interactions with others. For Cooley, social interaction is a complex process. Individuals

OPEN for DISCUSSION

Should/Can Social Theory Be Value-Neutral?

The practice of natural science is said to be value-neutral; the evidence is supposed to speak for itself. Studying physics or chemistry, scientists are expected to leave their values behind, at the laboratory door. Whether smoking causes lung cancer, for a scientist, is a question to be decided by objective evidence. Scientists whose research funding comes from tobacco companies, however, may report only findings that don't show a link between smoking and cancer. Their interest in keeping their research money flowing may influence the results of their work. If you don't believe climate change is caused by carbon emissions, you will search for and believe only the data that confirms your pre-existing belief.

A lot depends on how you look at things. When we look at a typical Canadian map of the Earth, north is up and south is down. Canada sits on top of the United States, and we go 'down' to Florida. Things change if you look at a map drawn from the perspective of 'down under' in Australia. The Australian-drawn globe appears, to Canadians, to be 'upside down'. Canada appears beneath the US, and we have to drive *up* for spring break in Fort Lauderdale. There really is no simple 'up' and 'down'. These binary terms are purely relative to where you are.

The search for generalizations about people's actions and beliefs is not straightforward. Values and beliefs intrude at least as much, and probably more, in sociological inquiry than in natural science. For any social theorist, much depends on the assumptions she or he makes about human nature. Spencer believed that people were naturally and always selfish

and competitive. The dog-eat-dog world of capitalism in which he lived was, for him, a natural result of human nature and couldn't be any different. The world may be cruel and unfair, but it was the way it had to be. Marx, on the other hand, believed that there was no fixed and unchanging 'human nature'. The way humans acted, whether selfishly or compassionately, depended on the circumstances in which they lived. If you change society, you change the circumstances of life and change 'human nature'. For Marx, the world was cruel and unfair, but it could be changed for the better.

Among all the differences between groups, however, there are certain common elements or features that explain a lot—if not everything—about society. The view of the world of the average American may be quite different from that of the average Canadian. But neither may be close to the view of a slum-dweller in Mumbai, India. It is likely that women and men experience and understand the world somewhat differently. The view of how society works from the point of view of the urban underclass is different in understandable ways from the perspective of Conrad Black or members of the corporate elite.

In sociology, 'where you are' is a complicated business. So, too, is what is done about different perspectives and interests. In the residential schools, church and state in Canada attempted to erase the cultural heritage of Canada's indigenous peoples, a practice referred to as 'cultural genocide'. At a minimum, as Max Weber said, in sociology it is necessary to understand your biases and make them clear.

imagine how they are perceived and judged by others and then develop a sense of themselves from these imaginings. In his modern novel *Ulysses*, James Joyce (1968, 602) asks this question about Bloom and Stephen, two of the characters in his novel: 'What . . . were Bloom's thoughts about Stephen's thoughts about Bloom, and Bloom's thoughts about Stephen's thoughts about Bloom's thoughts about Stephen?' In sociology, this complex calculation involves **intersubjectivity**—people orient their action according to what they think (subjectively) others think.

In Chicago, W.I. Thomas recognized that, like Bloom and Stephen, people entering a situation have different points of view of it and their actions result from their interpretation of their own and others' standpoints. For example, in the classroom, what the professor intends to accomplish with a lecture may be quite different from the various motivations that bring the students into the course. Everyone enters a social occasion, Thomas said, with a specific 'definition of the situation', and this definition affects how they act and what they hope to get out of the interaction.

Similarly, G.H. Mead (1863–1931) argued that not only do we learn to see ourselves the way others see us, but we are able to put ourselves in the position of the other person and see what she sees, feel what she feels. For Mead, interaction involves 'taking of the role of the other'. The better socialized we are, the better we are able to stand in others' shoes and see the world from their point of view—in Mead's terms, we are able to think reflexively (1934, 73, 133–4).

It is precisely through reflexive thinking—taking into account what others think—that an individual is able to build up a sense of '**self**'. In large measure, Mead said, we think of ourselves the way we are thought of and are treated by others. The more closely or significantly we are attached to these others, the more their opinion toward us affects how we feel about our self. Your parents were the most significant 'other' people in building your earliest sense of self. Now, perhaps, it's your boy- or girlfriend. It doesn't matter to the Mark Zuckerberg character in *The Social Network* if a neighbour down the hall thinks he's an 'asshole', but it does matter when, in the opening scene, his girlfriend comes to the same conclusion.

TIME to REFLECT

How much do you think your identity is shaped by other people, and how much is under your control?

Mead recognized that we relate differently to different people, being one self to one person and someone different to another: 'We divide ourselves up in all sorts of different selves in relation to our acquaintances,' Mead says. 'A multiple personality is in a certain sense normal' (1934, 140–4). It's complicated, not just in our intimate relationships but in all of our social interactions. *The Presentation of the Self in Everyday Life* is the title of Erving Goffman's investigation of how we attempt to influence other people's views by the way we present ourselves to them (in dress, speech, attitude, and so on). Goffman says that everyday social life is like a theatrical performance. We are all role-players in a continually changing set of dramas, moving from set to set, interacting with a shifting cast of multiple characters, and playing different roles. Sometimes we are in the visible front-stage, where we have to maintain social appearances appropriate for the occasion. Later, we move to the back-stage, where we can 'let our hair down'.

From a micro perspective, we are all highly skilled social performers. The order that exists in society is something that we continually create and recreate through our interactions and shared understandings. Thus, microsociology sees society from the bottom up, as an ensemble of relationships maintained through communication and interpretation. Social institutions only exist because people continually, minute by minute, act in ways that reproduce them.

Feminist Sociology

As more and more women entered sociology and undertook professional studies in the middle of the past century, it became increasingly clear that male theorists not only monopolized the field but also brought their gender biases into their theorizing. Women were absent from their theory, or were apparently subsumed under a male point of view, or were subject to the conservative and patriarchal prejudices of their historical era.

For maternal feminists, women are superior to men by their very nature. Women are nurturing, cooperative, peaceful, and non-competitive—precisely the values that are lacking in the public sphere, dominated by men. It is up to women to enter the public realm, as mothers and sisters, and reform a world that has been ruined by men. Nellie McClung (1972), in Canada, combined the sentiments of maternal feminism (such as temperance and women's rights in and out of marriage) with a feisty campaign for political equality and economic rights. Maternal feminism inspired mostly moral crusades and also—contrary to Durkheim—sought to liberalize divorce laws, especially in situations of domestic abuse.

The logic of socialist feminism emphasizes the economic role of women. Women are oppressed in all spheres, but the key to the liberation and equality of women, Marx believed, is to free them from domestic slavery and integrate them fully with men into productive work. Since capitalism relied on the unpaid household labour of women, however, women would never be equal until society had moved to a socialist stage.

Marx's close collaborator, Friedrich Engels, studied the rise of male dominance (1970). In the earliest society, he said, women held positions of power as mothers. Later, in prehistory, as private property developed, men acquired control over it, which included rights over women. Patriarchy (male dominance) was established as a result of the first social revolution: the historical defeat of women. Since then, male dominance has persisted and shows no sign of disappearing.

In sociology, feminists such as Marianne Weber (1870–1954) have integrated a distinctly female standpoint into sociological research. Weber was a leader in the German women's movement, which may be described as liberal feminism. The earliest feminists' writings, such as Mary Wollstonecraft's *Vindication of the Rights of Women* (1986 [1792]), asserted the equal rationality of women and men, from which they inferred the right of women to be equal in all aspects of society.

To generalize, liberals in the women's movement seek to equalize conditions for women within the existing economic structures (Friedan 1963). There is nothing in the logic of capitalism, for example, that precludes women from achieving equality with men in pay, employment opportunities, educational qualifications, and corporate power. Over time, women in modern society have made great advances, but they have been more successful in acquiring the *rights* to equality than *actual* equality in many spheres of modern life.

In the view of radical feminism, capitalism is a modern form of patriarchal dominance of women maintained through violence, whether by agents of the state (police, soldiers) or by men in the privacy of the household. Many feminists in the 1970s experienced sexism at the hand of the supposedly 'radical' men in the political organizations they had joined that were supposedly working for social change. Radical feminism asserts the need for the independent organization of women for their self-emancipation (Firestone 1970). It campaigns for changes in the everyday relationships between men and women, such as for the equal recognition of alternative forms of families and different sexualities.

Feminist theory in contemporary sociology has understood the differences and divisions among women. The experience of poor, or minority, or sexually differentiated women is likely to be misunderstood when seen through the eyes of middle-class, professional, heterosexual feminists. While there are many kinds of feminism, one of the most pervasive strains within feminist sociology is the need to understand society from women's standpoint and focus on women's subjectivity. In Canada, sociologist Dorothy Smith (b. 1926) developed a perspective that is critical of the top-down model of sociology that dominates the discipline. Typically, (mostly male) sociologists seek to understand social life through a predetermined theoretical lens designed to understand men. This is the reverse of what should happen, Smith argues. Feminist research should begin with the everyday experiences of women and with their understanding of these 'lived experiences'. Feminist sociology begins with the perspectives of women and offers a counterbalance to the values and techniques that have been skewed for so long in favour of men. Radical feminism is engaged, explicit in its goals, and not value-free.

Smith is active both in academics and in the wider women's movement. She practises a kind of 'action research' that addresses the issues raised by the women with whom she works. Positivist sociology involves the study of 'subjects' (individuals) by experts, who seek to acquire objective knowledge about them but not necessarily in the interest of these subjects. Smith's activist research method is consciously interested, not disinterested. She integrates her 'subjects' into all aspects of the research process and produces results that are shared with the participants and are aimed at addressing the issues that sparked the research. Smith's method is applicable to research generally, not only research about women. In her own words, Smith now advocates sociology for 'people'.

Modernism and Post-modernism

Much contemporary feminism is rooted in a wider theoretical attitude known as post-modernism, which rejects the entire theoretical tradition that began with the Enlightenment. Enlightenment thinking, which replaced gods and religions with human rationality, became entrenched in modern politics, economics,

The 1970 Status of Women Report made it clear that Canadian women were the 'second sex'. How close have women come to equality with men today?

law, medicine—and sociology. But as the irrationality of Comte's life suggested, there was another side to humanity. The nineteenth-century world gave birth to theoretical but not identical twins. Growing alongside Enlightenment rationalism was Romanticism, an alternative view, which encompassed the irrational, emotional, and expressive aspects of life. Romanticism found its home, at first, in the ideas of artists, poets, novelists, and philosophers.

While positivist sociologists followed the elusive dream of predicting and controlling human action, Romantics rejected the possibility and desirability of any social science by claiming there were deeper, more profound sources of knowledge than rationality and that people were at heart emotional, feeling, wilful, and desiring beings. The Romantic impulse resurfaced periodically over the next two centuries, among the Decadent artists at the end of the nineteenth century, among the youthful and artistic rebels of the 1950s and 1960s (the beats and hippies), and in much of the contemporary alternative music scene.

From philosophy, the Romantic tradition spread into the analysis of literature and culture (the arts, music, film), and no Great Wall separated theory in the arts from sociology. In the 1980s, in a movement

known as post-modernism, the attack on social science flowed into sociological theory. For post-modernists, sociologists doomed themselves to failure the moment they assumed that people were essentially rational and that social life was 'real'. Rather, they argued, there is no single 'reality' to be understood and transformed. What we think of as 'real' consists more and more of images fabricated for our consumption and spread through the mass media. And there are as many different perceptions about what is real as there are people to think about it.

Post-modern thinking reflected the increasingly uncertain world that emerged with globalization. Old ideals seemed to have ended in futility. Binary, either/or ideas not only failed to describe the complexity of life but had led to false, often authoritarian solutions. New problems had emerged, such as ethnic cleansing, religious conflict, and individual and state-sponsored terrorism. On the one hand, post-modernism led to an extreme relativism—not only did different views exist, but there was no way to validate one view over any other. All claims were equal, and none could claim any special degree of 'truth'; spiritual beliefs, for example, were as valid as scientific ones. In terms of action, all that anyone could do was respond individually, particularly by transgressing (consciously and publicly violating) social regulations about such things as authority, sexuality, and gender roles.

On the other hand, seen as theoretical criticism rather than as a script for effective social change, post-modernism draws our attention to new cultural forces in global society that affect the way people experience, understand, and respond to the world and the problems within it. It denies the existence of an essential and unitary 'self'; it emphasizes diversity, identity

In a show of rejection of contemporary models of international governance and authority, demonstrators damage police cars during the 2010 G20 protests in Toronto.

UNDER the WIRE

What Difference Do the Differences Make?

In contemporary times—at least in the Western world—new technology, international travel, and the Internet have modified the way we experience our world. In the age of the microchip, modern technology is changing the way we live, and the implications our new lifestyle will have for us and future generations are issues of vital public interest.

Looking around your lecture hall, you will see some things that are the same as they were 50 years ago—perhaps a lone lecturer is pontificating at the front—but much about the classroom is different. In particular, the students' use of personal technology in the classroom (laptops, iPhones, iPads, etc.) is certainly different. In the past, students were busy taking notes or drifting pleasantly away in a daydream; now there are many more distractions and much more multi-tasking.

Sociologists are not just interested in what the differences are—they want to know *what difference do the differences make?* Does modern communication technology enable us to be better informed about socially significant events or only about everyday trivia? Do the new media provide platforms for the greater realization of democratic decision-making, or are they tools of Big Brother? Do they create more communities of interest and sharing, or do they more thoroughly individualize us? Do they deepen our actual engagement with the world or merely make even the most horrendous event seem like a spectacle, to be gazed at but not acted upon?

Every new communication technology creates new possibilities for control but also for resistance. Protesters in Egypt and Tunisia in 2011, who drove their dictatorial leaders from power, frequently communicated via social networking. Blurred Vision, a Toronto-based rock group fronted by two Iranian brothers, covered Pink Floyd's transgressive anthem 'Another Brick in the Wall' and posted it on YouTube in 2010. They modified one line, replacing 'teacher' with the defiant, 'Hey, *Ayatollah*, leave those kids alone.' The song was banned by the Iranian authorities, but it created an underground sensation among disaffected youth.

One of the most controversial theorists in contemporary times, Jean Baudrillard, challenged our view of 'the real'. He studied sociology in the hothouse of French student activism in the 1960s. By 1972, he had abandoned an early Marxist influence and began to develop an analysis of modern society that has become increasingly obscure and poetic. For Baudrillard, we perceive society through the veil of mass media so that 'reality' has been overtaken by simulations, such as Disney World, that impose upon us images of what we take to be real. No wonder the Hollywood 'culture industry' is full of questions about what is 'real' versus virtual (*The Matrix, Tron*), sanity versus madness (*Shutter Island, Black Swan*), and what is only a dream (*Dark City, Inception*).

---

politics, risk, and the role of technology in creating the taken-for-granted images that form the spectacles of modern life. An adequate sociology for the present must face the challenges of post-modern thought.

TIME to REFLECT

How large a generation gap exists today? What causes this gap?

Conclusion

Is society best understood as a set of forces external to individuals and constraining them (macro) or as a skilled accomplishment of knowledgeable individuals (micro)? Given this central dilemma in sociology, it is

reasonable that one of the most important developments in contemporary sociology is the attempt to integrate micro and macro approaches into a single perspective. Perhaps more typically, whether you adopt a micro or a macro perspective depends on what you are trying to discover.

The contemporary generation faces a litany of deeply rooted social problems, such as climate change, widening global inequality, nuclear proliferation, and economic crises. We are living in an unprecedented 'new' age that is both vastly different from the past and 'weird and scary' in more ways than one. Sociology, it is apparent, can best approach this new world by both criticizing existing structures and focusing on human agency.

The classical attempt to develop a science of society has been significantly modified over time.

But social science cannot be abandoned. After all, we may have theories of randomness and relativity in physical science, but the technology that humanity has developed—for good or ill—is a testament to the ability to understand and control the most basic forces of nature. Einstein may have said that even as basic an element as time is not constant, but his theory that you can convert small amounts of mass into enormous amounts of energy ($e = mc^2$) was realized in the nuclear bomb. Some form of sociological theory may be equally explosive.

Rather than discovering the 'laws' of society, sociologists make generalizations about social phenomena that are tentative and most likely in the given circumstances. Scientific predictions are actually probabilities, not certainties. Given a certain set of circumstances, a given outcome is *likely* to occur with a certain probability. As with any science, these answers are open to modification or refutation by new evidence. But the questions asked in sociology are of vital importance to our world and our life, as demonstrated by the reasoned critical analyses that follow in this book.

questions for critical thought

1. What do you see as the main social problems caused by globalization, and how do they affect Canada?

2. Thinking about the distinction between 'structure' and 'agency', in what ways do you think of yourself as being a free agent of your own will, and in what ways do you see your actions as affected by society?

3. Are modern communication devices increasing our sense of social collectivity, or do they tend to individualize us more?

4. Would you say that your everyday actions are guided by some kind of moral code? What code of behaviour do you follow, and what is its source?

5. Do the mass media make us reasonably content with our lives, or do they make us critical of our social system?

6. As the world becomes increasingly rational, new forms of irrationality continually arise and compete. What forms of irrationality are important in the contemporary world?

7. What difference does it make to your identity to claim that you have a unique and true 'self' rather than a number of different selves at different times and places?

8. If you are a woman, what type of feminism comes closest to your understanding of the status and potential of women in the world? If you are a man, what difference would it make to you if your intimate partner said, 'I am a feminist'?

recommended readings

Jared Diamond. 1997. *Guns, Germs and Steel*. New York: Norton.
This is a readable and detailed historical account, by a geographer, of the reasons for European dominance over the past 500 years. Technology, including food production, and diseases (smallpox, influenza) helped to establish the foundation for contemporary capitalist globalization.

Barbara Ehrenreich. 2001. *Nickel and Dimed: On (Not) Getting By in America*. New York: Henry Holt.
Ehrenreich is a journalist who went 'undercover' in Florida, working at minimum-wage jobs to understand the lives of women at the bottom of the employment ladder. Her firsthand account is revealing, touching, and heated—and by no means value-neutral.

Harry H. Hiller. 1980. 'Paradigmatic shifts, indigenization, and the development of sociology in Canada'. *Journal of the History of the Behavioral Sciences* 16 (3): 263–74.
Hiller examines the development of sociology in Canada, the importation of sociological paradigms that originate outside the country, and efforts to make sociology a distinctly Canadian enterprise.

Robert D. Putnam. 2000. *Bowling Alone: The Collapse and Revival of American Community*. New York: Simon and Schuster.
Based on extensive research, Putnam argues that community, community feeling, and collective behaviour have declined in American society, to the detriment of grassroots democracy.

Eric Schlosser. 2006. *Fast Food Nation: The Dark Side of the All-American Meal*. New York: Houghton Mifflin.
Schlosser describes the history of the fast-food industry in the United States, covering familiar terrain such as marketing to children and the 'obesity' panic. His most searing indictment is aimed at the meat industry.

Lisa Shannon. 2010. *A Thousand Sisters: My Journey into the Worst Place on Earth to Be a Woman.*
The worst place, Shannon says, is the Congo, where millions of people have been brutally murdered, and women raped and tortured, over 'conflict minerals', including 'blood diamonds'.

Peter Stallybrass and Allon White. 1986. *The Politics and Poetics of Transgression*. Ithaca, NY: Cornell University Press.
The authors argue that hierarchy is a fundamental basis of order and sense-making in European cultures. Transgressions (violations) of the rules of one social hierarchy destabilize all other foundations of social order.

Naomi Wolf. 1991. *The Beauty Myth: How Images of Beauty Are Used against Women.*
Wolf exposes the social pressures behind the manipulation of women's self-image. The image of the 'perfect body' has changed to the point where it is virtually unattainable for most, underscoring many physical and health problems facing girls and women today.

recommended websites

Dead Sociologists Index
http://media.pfeiffer.edu/lridener/DSS
The Dead Sociologists Index provides detailed information on 16 classical theorists, including biographical information, commentary on the theorists' ideas, and links to original works.

Dorothy Smith
http://classiques.uqac.ca/contemporains/smith_dorothy/smith_dorothy_photo/smith_dorothy_photo.html
Dorothy Smith is a famous Canadian sociologist and feminist, and her brief autobiography can be found here. The entry discusses her discovery of sociology, her tribulations as a female academic, her criticism of mainstream sociology, and her creation of a women-centred sociology of the everyday.

Marxist Internet Archive
www.marxists.org
The Marxist Internet Archive is a comprehensive site offering sources and links for Marxism as a worldwide movement. It offers the full text of various writings, as well as information about many theorists within the Marxist perspective and about many other thinkers.

Max Weber
http://homepage.newschool.edu/het//profiles/weber.htm
This site, out of the New School of Social Research in New York, provides a brief overview of Max Weber's life, links to publications and to other online resources, and a 'home page' dedicated to Weber.

Online Dictionary of Sociology
http://bitbucket.icaap.org/
This is a useful site for definitions of sociological terms.

Social Theorists
http://people.brandeis.edu/~teuber/usem.page_3.html
This site offers links to a number of important social theorists, including Sigmund Freud and Canadian-born Erving Goffman.

Sociology Online
www.sociologyonline.co.uk
This site, produced in Britain, is dedicated to providing resources for students of sociology. There is a 'Classics' link to the 'selective tradition' that comprises classical theory. The 'Sociology News' link has many items of interest.

WWW Virtual Library: Sociology
http://socserv.mcmaster.ca/w3virtsoclib/theories.htm
McMaster University in Hamilton is the host for the WWW Virtual Library page on social theorists. It covers a number of classical and contemporary sociologists and includes a link to the Durkheim Pages at the University of Illinois.

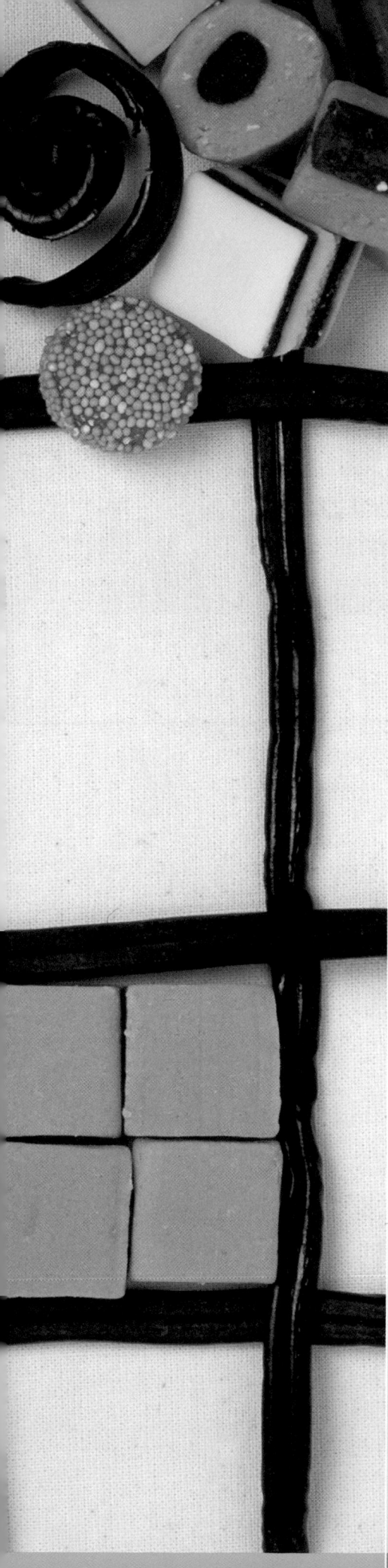

chapter 2

Research Methods

Bruce Arai

In this chapter, you will:

▶ Become familiar with the traditional model of science and with the more sophisticated model that has superseded it;

▶ Differentiate between quantitative and qualitative approaches to sociology;

▶ Differentiate between validity and reliability;

▶ Appreciate the importance of ethics in sociological research;

▶ Learn the different approaches to validity in qualitative and quantitative methods;

▶ Gain a better grasp of social survey design;

▶ Learn about field research, interviewing, and ethnographic research;

▶ Study the different types of existing data and how they are analyzed.

Introduction

In the previous chapter, you read about the major theories that sociologists use to understand social life. In this chapter, we focus on the methods that sociologists use to collect the data that have informed these theories and produced the many findings you will read about throughout this text. Among other things, you will learn about the debate over the scientific status of the discipline of sociology, the major techniques of data collection, and the importance of ethics in research.

Sociology departments in Canada belong to many different faculties. At your university or college, is the sociology department located in the faculty of arts, the faculty of social sciences, or somewhere else? This may seem like an unimportant issue, but it gives you some indication of the different ways in which sociology is perceived. Many sociologists think of sociology as a science or as a social science, others view it as an arts or humanities discipline, and still others choose not to categorize it at all.

How we view sociology is important for this chapter because the methods that we use to produce sociological knowledge are intimately related to whether it is seen as a science, an art, or something else. If sociology is viewed as a science, sociological research is often designed to measure and quantify social life; this is called quantitative sociology. If it is viewed as an arts or humanities discipline, sociological research is often designed to get at the rich meanings that people attribute to their lives; this is called qualitative sociology. This distinction between quantitative and qualitative sociology is not absolute, and many sociologists prefer not to use it. It is used occasionally in this chapter to organize the discussion and also to show how the distinction itself has become irrelevant to some scholars. The chapter begins by exploring a scientific view of sociology, because the founders of the discipline certainly thought of it this way and because it remains one of the most dominant views of the discipline today. The scientific view of sociology is also used as a backdrop for understanding other perspectives on the discipline.

Sociology as a Science

We often imagine that a science has several distinguishing characteristics (see Table 2.1), including:

- Knowledge is based only on facts.
- Facts are part of the real world and can be observed.

"Now, *that's* the sign of a very advanced culture!"

(© Brenda Brown. Reprinted with permission.)

The traditional view of sociology as a science.

- When making scientific observations, we do not let our personal emotions or biases interfere with our observations; this is called 'objectivity'.

- We usually use the scientific method, including experiments, to collect and analyze data and to draw conclusions.

- Science gives us the best understanding of the way the world works. In other words, if we follow the proper procedures of science, we will discover the truth about the world, and this truth will be better than other truths.

We usually base these ideas on what we think of as 'real science'—something like physics, astronomy, chemistry, or perhaps biology. Compared to these fields of study, sociology seems to be a lesser science, perhaps not even a science at all. It is debatable whether any social science can ever live up to the rigorous standards of scientific inquiry set by these other disciplines. So it is not just sociology that fails to meet the standards but also the other social sciences, including psychology, economics, and political science.

However, the perception that the social sciences are lesser sciences than their more hard-nosed cousins is based largely on an outdated, unrealistic understanding of what actually happens in disciplines like physics, astronomy, chemistry, and biology. As Sandra Harding (1986) and others (Beaulieu 2010) have shown, what physicists actually do is only one very particular way of studying science, and it is not clear that most sciences follow this model.

Instead, if we look at what scientists actually do and the methods they use to produce their scientific claims, many of them do not involve experiments or even the collection of data. Some science can be done with only a paper and pencil. Sociologists, anthropologists, and others who have studied the actual behaviour of scientists have shown convincingly that 'doing science' depends heavily on things like informal negotiations between scientists, drama, and rituals (Beaulieu 2010; Latour and Woolgar 1987; Lynch 1985). In other words, it is not just the scientific method and **objective** observations that lead to scientific conclusions; personal opinions, biases, and cultural understandings are at the very heart of most sciences.

In the First Person

The person who really solidified my interest in becoming a sociologist was the late Jim Richardson, my master's supervisor at the University of New Brunswick. Most of us like to think that we make most decisions independently, from the kind of music we listen to, to the kinds of restaurants we prefer, even to our choice of girlfriends, boyfriends, and partners. But the reality is that strong social forces shape and constrain the choices we make. Sociology shows us how our seemingly individual decisions are shaped by these larger social pressures. Jim had a keen eye for this, and one of his extraordinary gifts was to be able to make those processes crystal clear for me and for his many other students.

But the two people who started me thinking sociologically were Tullio Caputo (now at Carleton University) and Lesley Miller at the University of Calgary. Dr Caputo's classes were filled with his energy, humour, and incisiveness, which made sociology both relevant and fun. And the best class I've ever taken at university was my third-year sociological theory course with Dr Miller. As we read works by Marx, Durkheim, Weber, Parsons, Berger and Luckman, and Goffman, she led us to think abstractly and comparatively about their ideas. She was able to show us the power of sociology as a way to understand the world we live in. I feel lucky to have met these three people and to share a passion for sociology; I hope you will have similar experiences.

—Bruce Arai

In addition, completely objective observations of the world are impossible. Nobody can separate himself or herself completely from his or her prior knowledge of the world when making observations. Indeed, it is impossible even to make observations of the world, much less record them and communicate them to others, without knowing a language. And since languages are not unbiased, accurate reflections of the world, the fact that we have to make and express our observations of the world through language makes total objectivity impossible (see Chalmers 1999).

This would seem to be a serious threat to the possibility of science in general. Scientific findings should not be influenced by the emotions, thoughts, prejudices, or language of any individual scientist. Instead, they are supposed to be true, accurate recordings of the real world. Does this make science impossible?

One answer is 'yes': some people argue that if objectivity is impossible, then science is impossible. The reason for this opinion is that without the claim to objective knowledge, science ceases to be useful. We should abandon science or at least alter it radically. This point of view is often associated with **relativism**—the belief that there is no ultimate truth.

8 ITEMS OR LESS

" 9....10....11....12...."

(© Brenda Brown. Reprinted with permission.)

Another view of sociology as a science.

But another answer is 'no': that it is possible for sociology and other disciplines to be scientific. The reason is that the criticisms of objectivity, convincing as they are, do not undermine science, because objectivity is not necessary for something to be called 'scientific'. All that we require is that science be defined as 'guidelines or rules of thumb or conventions that produce claims that are defensible when called upon, though not perfect' (Goldenberg 1992, 19). The 'eight items or less' rule in grocery stores, satirized on page 35, is simply a rule we use to make our lives easier. The rules we use in science are more important than 'eight items or less', but they are not ultimate or final statements.

What this means in practical terms is that sociologists can and do develop rules for what constitutes 'good sociology' and for how it can be differentiated from 'bad sociology'. Physicists, chemists, and psychologists do the same thing, developing rules for 'good' and 'bad' physics, chemistry, and psychology. Many of these rules concern the proper methods used to collect, record, and analyze data. We will cover some of these basic rules or conventions in the next two sections of this chapter, on the connections between theory and research and on the major techniques of sociological research.

Theory and Research

For most sociologists, it is important that their research be closely connected with a theory or set of theories. Briefly, *theories* are abstract ideas about the world. Most sociological research is designed to evaluate a theory, either by testing it or by exploring the applicability of the theory to different situations. As can be seen throughout the many chapters in this text, sociologists investigate substantive problems and try to use their theories to help them understand these problems better. For instance, sociologists may be interested in understanding crime, the family, the environment, or education, and they will almost always use their theories to provide a deeper appreciation of these issues.

Sociologists use theories as models or conceptual maps of how the world works, and they use research methods to gather data that are relevant to these theories. Thus, theories and methods are always intertwined in the research process. There are hundreds of different theories in sociology, but most of them can be grouped into the four main theoretical perspectives that can be found throughout this text: structural functionalism, conflict theory, symbolic interactionism, and feminism. Theories cannot be tested directly, because they are only abstract ideas. Theories must be translated into observable ideas before they can be tested. This process of translation is called **operationalization**.

TIME to REFLECT

To engage in science, or anything else for that matter, we must use language. Imagine that you were asked to determine the temperature at which different liquids reach a boiling point. Presumably, this would tell you something about the properties of these liquids. But to do this experiment and to make your observations, you would need to understand a language. So doing science of any kind is dependent on language. Yet languages are always imperfect and biased, not just reflections of the world around us. This makes 'objective science' impossible. How could you record a boiling point if you didn't know what it means for a liquid to boil or you didn't know what a temperature scale or thermometer was? Does this make you sceptical about science being 'objective'?

OPERATIONALIZATION

Operationalization is the process of translating theories and concepts into hypotheses and variables. Theories are abstract ideas, composed of concepts. **Concepts** are single ideas. Usually, theories explain how two or more concepts are related to each other. For instance, Karl Marx used concepts such as 'alienation', 'exploitation', and 'class' to construct an abstract explanation (theory) of capitalism.

Once we have a theory, we need some way to test it. The problem is, we are not able to test theories directly. We need an observable equivalent of a theory or at least a set of observable statements that are consistent with our theory. These are called hypotheses. In the same way that theories express relationships between concepts, **hypotheses** express relationships between variables. Unlike the typical definition of a hypothesis as simply an 'educated guess', it is important to point out that hypotheses must be observable or testable. This means they must be composed of or express relationships between variables.

Variables are the empirical or observable equivalent of concepts. The two key points about variables are that they must be observable and that they must have a range of different values they can take on. For instance, ethnicity, age, years of schooling, and

HUMAN DIVERSITY

Are Leaders Born or Bred?

Every organization needs leaders. Companies need CEOs who will help them to increase profits or turn around a run of bad years. Churches require strong leaders and universities need visionary presidents to ensure that they remain dynamic and relevant.

But are leaders born or bred? Sociologically, this is an important question, because if they are born (that is, if most of the essential features of a good leader are determined at birth or a young age), then the hundreds of leadership programs available across North America are largely ineffective. If leaders are bred, then a very different story emerges.

This discussion shows how important theory is to the definition of a problem. If leaders are born, we can do little but sit back and hope that the right people end up in the right positions. But if leaders are bred, 'leadership training' as a concept starts to make sense.

The prevailing view at present is that leaders are bred—that leadership can be taught. Perhaps this seems to be the prevailing view because so many organizations are offering leadership training programs, and they all have an interest in this position. But if leaders are bred, then it becomes possible to evaluate the quality or effectiveness of different training programs. A great number of studies attempt to do just that. Sometimes the evidence seems to suggest that these programs work (Mighty and Ashton 2003; Nichols 2002), and at other times it suggests they don't (Allio 2005; Fossey and Shoho 2006).

So despite the fact that leadership is critical to the success of virtually all modern organizations, human diversity makes it almost impossible to predict who will follow a particular leader. And this in turn makes it extremely difficult to evaluate whether or not these programs are successful.

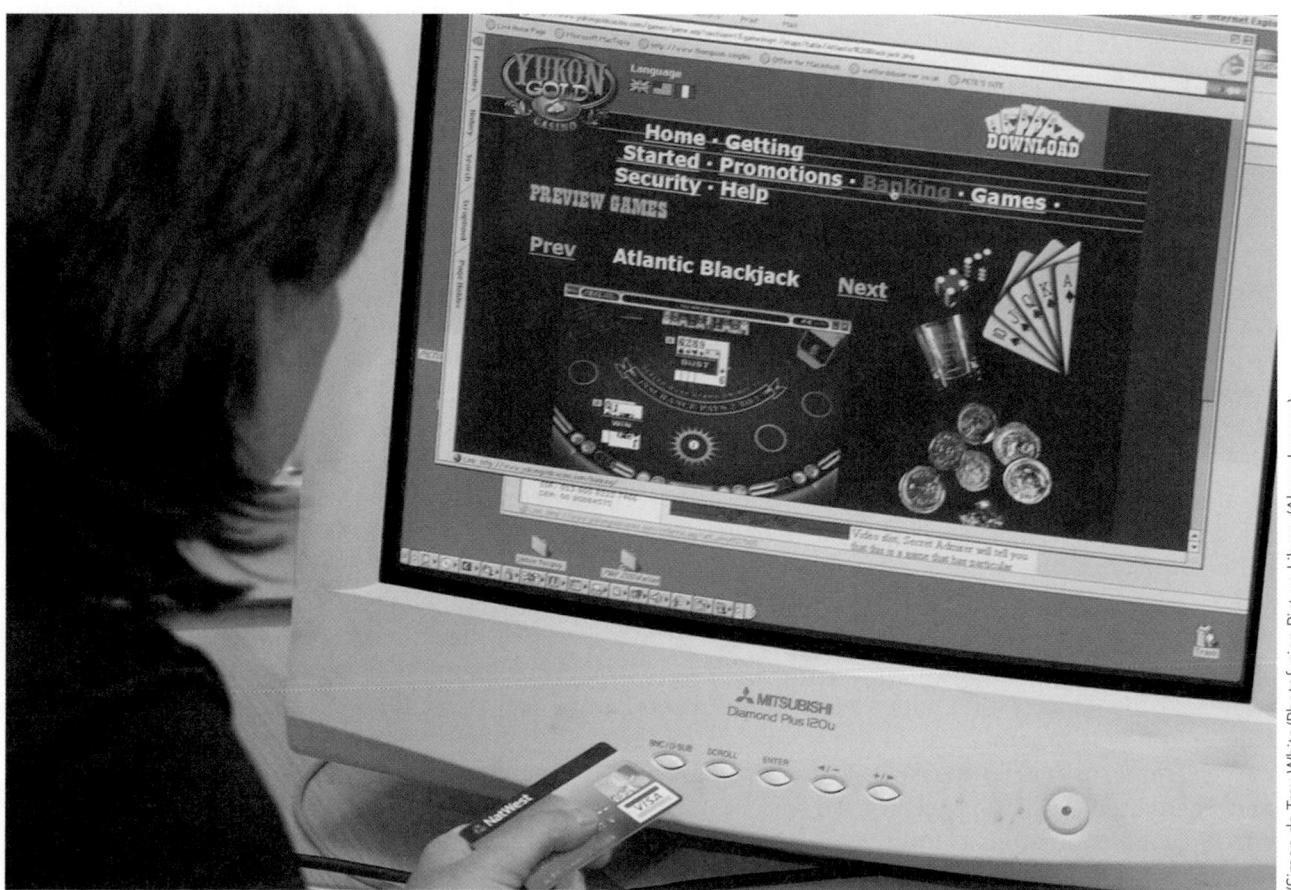

Albert Einstein famously said, 'God does not play dice with the universe', meaning that we can successfully study the world through systemic research. Popular games of chance carry no such guarantee of success.

(Simon de Trey-White/Photofusion Picture Library/Alamy Images)

annual income are variables. We can collect information on all of these items (that is, they are observable), and people can have different ages, ethnicities, and so on. 'French', '45 years old', '12 years of schooling', and '$50,000 per year' are not variables because although they are observable, they do not vary. They are *values* of variables, not variables in themselves, and it is important not to confuse the two.

In most cases, our hypotheses contain a minimum of two types of variables: independent variables and dependent variables. Independent variables are roughly equivalent to causes, and dependent variables are roughly equivalent to effects. Another way to keep them straight is to remember that the value of a dependent variable depends on the value of an independent variable. For instance, if you wanted to investigate differences between the average earnings of men and women, then sex or gender would be the independent variable, and earnings would be the dependent variable. This is because people's earnings may depend on their gender. Indeed, it is easy to keep the independent and dependent variables straight in this example because it makes no sense to say that a person's gender can depend on his or her earnings.

VALIDITY AND RELIABILITY

In the process of operationalizing our theories and concepts into variables and hypotheses, it is important that this process of translation be as clean as possible. That is, we want to ensure that our variables and hypotheses accurately reflect our theories and concepts. In particular, we want to ensure that when we go out and collect our data, we are using measures that are both valid and reliable. The most important step in this process is the construction of *operational definitions* of our concepts. These are definitions that specify what we are going to observe, how we are going to make our observations, and how we are going to differentiate observations from non-observations (or how we will know which possible elements to exclude from our study).

Validity

The **validity** of an empirical indicator is always related to the concept it is supposed to capture. A valid measure is one that adequately represents the concept, and an invalid measure is one that does not. There are no valid or invalid measures as such, and the validity of an indicator is always a matter of degree. A measure can be perfectly valid in relation to one concept but invalid for another. For instance, sociologists often use a person's years of schooling

as an indicator of educational attainment, and most sociologists consider it a valid (if imperfect) indicator of this concept. However, a person's years of schooling are not a valid indicator of ethnic origin.

It is probably fairly obvious to you that a person's years of schooling are a better indicator of her or his educational attainment than of her or his ethnicity. But this is really only one of several validity criteria, one called *face validity*. An indicator has high face validity when it seems to fit nicely with our mental image of the concept it is supposed to measure. There are several other types of validity that are often used in social sciences research, but we will cover only two more of them here.

The *external validity* of a piece of research refers to the extent to which the results from that research can be generalized to a larger population (that is, the generalizability of the findings beyond, or external to, the specific **sample** used in the study). The two main areas of concern in establishing external validity are the proper selection of the sample of people or elements to be studied and the specific techniques used in the research.

Ensuring that a sample is chosen using a properly random procedure usually satisfies concerns about sample selection. It is important, however, to distinguish randomness from haphazardness, because choosing a truly random sample is a much more intentional process than the name suggests. For instance, if you are working for a bottled water company and are asked to find out which of the company's two brands people prefer, you could do this in a number of ways. If you choose to stand in front of a mall and ask people which brand of water they prefer, the external validity of your research would be very low, because there is no obvious connection between the sample of people you talked to and a larger population of bottled water customers. Alternatively, if you had a list of current customers of this company, you could deliberately choose every tenth name on the list, mail them a questionnaire, and analyze the results. In this case, the external validity of your research would be pretty high, because there is a direct and unbiased connection between your sample and the population of current customers.

The most common threat from the procedures themselves is that there may be a reason to suspect that the results are a product of something in the study itself and not completely reflective of real-life situations. In other words, if there is a reason to think that people acted or answered differently during the research than they would in real life, then there is a threat to the external validity of the study. For instance, people may answer questions differently if they know they are being tape-recorded, and people may act differently simply because a researcher is present.

The *internal validity* of a study concerns the degree to which the conclusions from the study are actually supported by the data and methods that were used. Internal validity is threatened when the effects that are attributed to specific variables or processes in the study are actually produced by other factors. For instance, if people perform better on a math test after taking a math class than they did on a test before taking the class, it might be tempting to conclude that the class was a good one because people's grades went up. However, there are many other reasons why math grades might have gone up, including the fact that the students would have a better idea of the types of questions that would be asked on the second test than they did on the first one. If these and other factors are not accounted for in the research, then the internal validity of the study is lower than it could be (Campbell and Stanley 1970).

Reliability

The **reliability** of a measurement process refers to its level of consistency. A reliable measurement process is one that produces the same measurements of the same phenomenon again and again. For instance, you can get a sense of the reliability of a supermarket scale by weighing an apple on it several times in succession. If you get very similar weights each time, the scale is reliable. However, if you get very different weights each time, it is not. In sociology, assessing the reliability of our measurement processes is rarely this straightforward. The details of these assessment procedures are beyond the scope of this chapter, but you can learn more about them in sociology research methods courses.

Another Approach to Validity and Reliability

Another way of assessing the validity and reliability of our measurement processes is an approach most often associated with qualitative research and is based on the depth of understanding that a researcher has of her or his topic. In this approach, researchers usually refer to the validity and reliability of their findings rather than to that of their measurement processes, in part because many qualitative researchers do not see themselves as 'measuring' social phenomena so much as 'recording' or 'understanding' them. Indeed, sociologists often hesitate to speak of *validity* and *reliability*

because these terms are closely associated with the idea of precise, often quantitative, measurement.

One of the main goals of a great deal of qualitative research is to gain a better understanding of the topic or group being studied. But in order to acquire this better understanding, it is important that the results of the research be both valid and reliable. The reason is simply that we cannot have a better understanding of the topic if our results are unreliable and invalid. Wendy Griswold (1987), for instance, has developed one of the only sophisticated models for assessing the validity of qualitative cultural data.

To achieve a level of validity and reliability in qualitative research, we do not usually engage in an extensive process of testing the different dimensions of validity and reliability as we do in quantitative research. Instead, we establish it by making a convincing argument that we have properly understood our topic or group. In other words, we demonstrate to our readers that we have gained enough understanding of our topic to ensure that our results are valid. We do this in several ways, including spending enough time with the people we are studying to fully understand their point of view, interviewing group members until we begin to see the same ideas coming up again and again, or conducting extensive analyses of written records and documents on our topic of interest.

BIAS AND ERROR

Even with solid conceptualization and operationalization, a study can be threatened by the existence of both *biases* and *errors*. Although the words 'bias' and 'error' are often used interchangeably in everyday conversations, they have two distinctly different meanings in social science research.

Error refers to the unintentional, accidental mistakes that inevitably creep into a piece of research. These errors are often referred to as 'random errors', reflecting the understanding that these mistakes are unintentional and unpredictable. Random errors have many sources, including having the wrong people participate in research, respondents making mistakes when they fill out a survey, errors in coding (the process of preparing data for analysis), and errors in analysis. There are ways of checking for some types of random errors, but there are no foolproof checks that will catch all of them. Obviously, our goal is to minimize the number of random errors in our research.

Bias is different from error—it refers to systematic inaccuracies in our data or analysis. Biases are usually unintentional, but they can be more serious than errors because they distort our findings in systematic ways. They also have multiple sources, but most researchers are concerned primarily about *respondent biases*. Two of the most common forms of respondent bias are *acquiescence bias*, whereby respondents simply check off answers to questions without thinking about them, and *social desirability bias*, whereby they try to answer questions the way they think the researcher wants them to instead of answering the way they themselves want to.

Research Techniques

Having covered many of the more general issues in social scientific research, we now turn our attention to some of the specific techniques that sociologists and other social scientists use to conduct research. Again, this chapter will not cover these techniques in great detail, but rather it will try to give you a sense of what they are and when and how they are used. But first, let us return for a moment to the traditional vision of what methods a scientist might use.

A popular image is that the experiment is the primary method scientists use to conduct research. However, even in the natural sciences, experiments often are the exception rather than the rule. A great

deal of biology, astronomy, and other sciences is not done—and in many cases cannot be done—experimentally. Nevertheless, the image persists, and it has traditionally been the standard against which science of all types, natural or social, has been measured.

The main advantages of experiments are that (1) they provide a controlled environment in which it is possible to (2) manipulate specific factors in an attempt to determine their effect on an outcome. Experiments can show the effects of one variable on another variable quite convincingly because of these two features.

However, sociologists do not use experiments very often, for two reasons. First, we cannot manipulate many of the variables we are interested in, for either practical or ethical reasons. Sociologists are often interested in the effects of variables like gender, ethnicity, and social class on other variables like educational outcomes, earnings, or health status. But it is neither ethical nor practical to alter people's ethnicity or gender just so we can observe what happens to their educational outcomes. Nor can we simply move a person from an upper-class home into a lower-class home (or vice versa) just so we can find out what effect this might have on his or her eventual choice of career.

OPEN for DISCUSSION

Max Weber and *Verstehen*

In many of the chapters in this text, you will come across the ideas of Max Weber. One of his most enduring contributions to research methods in sociology is his elaboration of a concept he called *verstehen* (German for 'to understand'). His idea is that in order to properly study the cultures of other peoples, a researcher needs to develop not just knowledge but an 'empathetic understanding' of their lives as well in order to see the world as that group sees it.

Verstehen became a cornerstone of qualitative sociology as researchers tried to understand the lives of others 'from the inside'. In Weber's view, developing *verstehen* was a bit of an art, but in theory anyone who was good at it could understand the world view of any other group. In other words, the views of any group could be understood regardless of the personal characteristics of the researcher.

But this view has been criticized as too simplistic. That is, some researchers have argued that there are limits to *verstehen*, because the personal characteristics of the researcher will affect how the group reacts to her or him. And this will limit the depth

of *verstehen* or understanding that a researcher can achieve. For instance, Margaret Mead's classic anthropological study in Samoa has been criticized because the Samoans later claimed that they were not completely honest with her. Similarly, men will be able to reach only a certain limited level of understanding with women, and because of this, it may not even be appropriate for men to study women, or vice versa. If we relate this to Killingsworth's (2000) study of mom-and-tot groups discussed later in this chapter, it might be the case that as a male, he may not have had the same access to the ongoing discussions around motherhood and child care. So are there factors that would limit the level of *verstehen* that a researcher can achieve? And if so, what are those factors, and how do we identify them? At its extreme, this would mean that a researcher would have to match up with her or his participants on everything from gender, to education level, to hair colour, to fashion sense. So neither extreme position is particularly convincing, but exactly where we draw this line remains open for discussion.

Second, one of the enduring criticisms of experiments is that it is not always clear that what happens under the controlled conditions of an experiment will also happen when we try to apply our findings to the 'real world' (that is, external validity). For instance, many of the experiments in medical research are done first on rats and other animals, and there is always the question of whether or not what happens to rats will also happen to humans. Similarly, in social scientific research it is never clear that what we observe in a controlled social experiment will also happen to people in their daily lives.

Despite the fact that sociologists do not use experiments very often, the logic of the experiment still dominates at least one of the major techniques of sociological research. Surveys almost always collect a great deal of extra information from respondents in an attempt to recreate the controlled environment of the experiment after the fact. Surveys are often referred to as *quasi-experimental* designs, because they are only able to construct a controlled environment after the data have been collected. In other words, in true experiments, the controlled conditions

are set in place, and then the experiment is allowed to run, while in quasi-experiments, observations of 'naturally occurring' phenomena are made, and an attempt is made to remove the effects of confounding variables during the analysis stage.

SURVEYS

Surveys are the most widely used technique in social scientific research. Sociologists, economists, political scientists, psychologists, and others use them regularly (Gray and Guppy 2008). They are an excellent way to gather data on large populations that cannot be studied effectively in a face-to-face manner. The goals of almost all social scientific surveys are to produce detailed data that will allow researchers to describe the characteristics of the group under study, to test theories about that group, and to generalize results beyond the people who responded to the survey.

You have probably participated in a survey or opinion poll before, although perhaps not in one used for social scientific research. Many polls done by political parties or polling organizations to find out about the

The systematic study of public opinion is well developed in sociology and plays an important role in political life and public decision-making.

(Ivy Images)

political preferences of the electorate are very well done. The need to be able to generalize results from a sample of people to the preferences of voters in general is extremely important, and consequently, most of these organizations put a great deal of effort into constructing and administering their polls properly. The one disadvantage of many of these polls is that they are very focused on certain candidates or time-sensitive issues, so they are only useful for a limited range of social scientific research topics. However, studies of voting behaviour and other political processes have profited immensely from public opinion polls.

Many 'surveys' are done for purposes other than social scientific research, and most of them will not produce data that are amenable to social science research. Designing and administering a good survey is much more difficult than it seems, but respondents rarely see all of the work that goes into producing a good survey. This may partly explain why surveys seem easy to create, and it may also contribute to the proliferation of pseudo-surveys in many different forms.

Pseudo-surveys

A great deal of marketing and customer satisfaction research does not meet the standards of a social scientific survey, despite the fact that generalizable results remain an important goal. That is, companies do this research only to find out what their customers and potential customers want and do not want to see in their products and services. But if their results apply only to the specific people who answered the survey and not to their customers and potential customers more generally, then they have very limited value. The

usual problem in these surveys is that not enough attention is paid to the selection of the people to whom the survey will be sent or to ensuring that enough of the surveys are actually returned. Both of these factors jeopardize the generalizability of the results.

Some companies also use surveys as a cover for their sales pitches. These tactics take many forms: people are called and asked to participate in a survey about their buying habits or perhaps about their concerns about their health or the environment. Then a week or two later, they receive calls from companies selling products that fit their buying habits.

To alleviate some of the confusion around surveys and pseudo-surveys, it is useful to distinguish between surveys and questionnaires. A *questionnaire* is any set of questions administered to a group of people. A *survey*, on the other hand, is a properly designed set of questions systematically administered to a randomly chosen sample from a population. In other words, all surveys are questionnaires, but not all questionnaires meet the standards of a survey. What sets a survey apart is the design of the questions, the goal of collecting data rather than manipulating people, the method of administration, and how the sample is chosen.

Constructing Survey Questions

Having summarized what social scientific surveys are not, it is also important to describe some of the key characteristics of a good survey. Surveys consist mainly of questions, and there are many issues to consider in designing good survey questions (see Table 2.1 for a summary of key points to consider when constructing survey questions).

TABLE 2.1 ▼ Guidelines for Designing Good Survey Questions

Focus	Each question should have one specific topic. Questions with more than one topic are difficult to answer, and the answers are often ambiguous.
Brevity	Generally, shorter questions are preferable to longer questions. They are easier for respondents to understand. An important exception to this guideline is when asking about threatening topics, when longer questions are often preferable.
Clarity	Use clear, understandable words, and avoid jargon. This is especially important for general audiences, but if you are surveying a distinct group or population (such as lawyers), then specialized language is often preferable.
Bias	Avoid biased words, phrases, statements, and questions. If one answer to a question is more likely or is more socially acceptable than others, then the question is probably biased and should be reworded. For instance, if you are asking people about their religious preferences, do not use words like 'ungodly', 'heathen', or 'fanatic' in your questions or you will bias your answers.
Relevance	Ensure that the questions you ask of your respondents are relevant to them and to your research. Also, in most surveys, some questions will not be relevant to all respondents; filter questions allow people to skip questions that are not pertinent to them. For instance, if you want to know why some people did not complete high school, you should first filter out high school graduates and ask them not to answer the questions about not completing high school.

SOURCE: Adapted from George Gray and Neil Guppy, *Successful Surveys: Research Methods and Practices*, 4th edn (Toronto: Nelson Thomson, 2008).

At first glance, it might seem that designing good questions for a survey would be easy. The reality is that it is quite difficult—sociologists can spend months trying to figure out what questions they will ask, how they will word them, and the order in which they will ask them. One of the reasons it is so difficult is that each question must be unambiguous for both the respondent and the researcher. A question with several different interpretations is not useful, because respondents may answer it from a different perspective than is intended by the researcher. Similarly, questions that are too complicated for respondents to answer, or that presume a level of knowledge that respondents do not have, will not produce useful data. There are many, many issues to consider in designing good questions and the order they appear on the questionnaire, but we will talk about just three of them in this chapter.

First, sociologists must avoid the use of double-barrelled questions in surveys. *Double-barrelled questions* are those that have two or more referents or subjects and that can therefore be answered honestly in more than one way. For instance, 'Do you think that the government should increase taxes so that it can spend more on environmental protection?' is a double-barrelled question. The problem is that a yes or no answer to this question is impossible to interpret. 'Yes' may mean that a person agrees with the whole statement, or only with the environmental protection component, or only with the increasing taxes component. 'No' may mean that the person disagrees with increased taxes but may nonetheless want more environmental protection or that he or she disagrees with the entire statement. A researcher will not know which of these interpretations is correct for any individual respondent. The way to avoid this confusion is to formulate two separate questions, one about tax increases and one for environmental spending.

Second, in some surveys, it is deemed necessary to ask people about uncomfortable topics—topics that they may perceive as threatening. Surveys of criminal activity, sexual practices, and income imply socially threatening questions. Asking people about their incomes is generally perceived as somewhat threatening, and care must be exercised in how you ask about this.

Strategies for reducing the threat of questions include making the behaviour or attitude you are asking about less threatening by positioning the respondent at some distance—by asking about it in hypothetical terms or asking about it as one of a series of questions. For instance, instead of asking people directly if they have been victims of crime, you could ask people how they feel about their own safety, how much they are exposed to media coverage of criminal activity, and then whether or not they have been victimized. No single strategy will work for every question, topic, or sample.

Finally, the order in which you ask questions can have a significant effect on respondents' answers and even on whether or not they will complete the survey. As a general rule, it is better to ask threatening questions near the end rather than at the beginning of a survey. By asking easier questions upfront, researchers have an opportunity to establish a rapport with the respondent, in phone and interview surveys through conversation and in paper surveys by leading the respondent to identify with the topics and issues on the survey. Second, if threatening questions turn off respondents so much that they refuse to participate further in the research, then at least they have completed part of the survey already. If this happens at the beginning of the survey, no usable data are collected at all.

Random Sampling, Sample Size, and Response Rates

The idea of choosing samples using a random procedure was emphasized in the discussion of validity. Random sampling is so important because it is the only way that we can be confident that our sample is representative of (that is, that it looks like) the population we are interested in. If our sample is representative, then we can be fairly confident that the patterns we find among our sample will also be present in the larger population. If it is not representative, then we have no idea whether what we found in our sample is also present in our population. Using a proper randomization procedure ensures that we do not deliberately bias our sample and guards against any unintentional biases that may creep into our selection process. Further, although randomization does not guarantee that our sample will be representative, by minimizing both intentional and unintentional biases we maximize our chances that the sample will be representative. However, randomization does not solve all problems in sociological research and is not always appropriate or necessary in field research.

TIME to REFLECT
Notice that the word 'bias' is used here instead of 'error'. Why?

There are many procedures for choosing a truly random sample, but all are based on the principle that each person (or element) in a population has an equal (and non-zero) chance of being selected into the sample. The simplest random sampling procedure is known as *simple random sampling*: each person in a population is put on a list, and then a proportion of them are chosen from this list completely at random. The usual way of ensuring that people are chosen at random is to use a table of random numbers either to select all of the people or to select the starting point in the list from which the sample will be chosen.

Actually, the adjective 'simple' in 'simple random sampling' does not refer to the degree of difficulty—simple random samples are quite difficult to construct. Generating truly random samples is not as easy as the word 'simple' suggests. The problem lies not in choosing the actual people or elements but in constructing a complete list of every person or element in the population. For this reason, other sampling techniques, such as stratified sampling and cluster sampling, are frequently used to choose samples, even by large government agencies like Statistics Canada.

It is also important to consider the issue of sample size in designing a proper survey. How big a sample do you need in order to be able to generalize your results? Actually, this is the wrong question to ask—it is not the size of the sample but rather how it is chosen that determines how confident you can be that your results are applicable to the population. That is, even a very large sample, if it is not chosen randomly, offers no guarantee about the generalizability of the results. On the other hand, a small sample, properly chosen, can produce very good results. So never assume that because a sample is large it must be representative. Always make sure you find out about how the sample was chosen before making any judgments about its generalizability.

Another crucial factor in determining the generalizability of survey results is how many people from the original sample actually complete the survey. This percentage is called the *response rate*, and it is an important, although not the only, issue to consider in determining the generalizability of the results of a survey. The reason it is important is that unless a large proportion of the people in the original sample actually complete the survey, it is quite possible that

SOCIOLOGY in ACTION

Asking the Right Questions about Survey Data

Rudner (1999) investigated the relative performance of high school students who graduated from a traditional high school, as compared to those who had been home-schooled, on two standardized tests. Interestingly, his results showed that home-schoolers did as well as or better than their counterparts who went to a traditional school. His results were based on an analysis of more than 20,000 home-schooled students whose test scores were compiled by Bob Jones University Press, the publishing division of a small religious college in South Carolina.

Rudner found that on average, those who were home-schooled scored within the 70th and 80th percentiles on two tests of scholastic ability widely accepted in the US. By definition, students from traditional schools on average score at the 50th percentile. He claimed that these results show that home-schoolers outperform their counterparts who went to regular school.

However, the data in his study cannot be used to support this conclusion. This is because the sample was not representative of the larger population of

US home-schoolers. The sample had low external validity for at least two reasons. First, the test was voluntary for home-schoolers but mandatory for schooled children, so there is good reason to believe that what is being compared here is the top end of home-schooled students against all schooled children. After all, if you are a home-schooler and you know you are a weak student, what incentive do you have to take a voluntary test? Institutionally educated children were not given this choice, so they were being compared to the top end of the home-schoolers and found wanting.

Second, the claim that more than 20,000 home-schooled students were used in the study, thus making the claims generalizable to a larger population, is patently false. Always be wary of studies that emphasize how large the sample is as the basis of their validity. Somebody either doesn't know what he or she is doing in the area of sample selection or is hiding something. Luckily for you, a good grounding in research methods will prevent you from making the same mistakes!

the people who do not respond to it are different from those who do respond.

FIELD RESEARCH

In surveys, the primary aim is to collect quantitative or numerical data that can be generalized to a larger population. In contrast, field researchers are concerned about collecting qualitative or non-numerical data that may or may not be generalized to a larger population. In field research, the aim is to collect rich, nuanced data by going into the 'field' to observe and talk to people directly. Researchers spend time getting to know their subjects in order to be able to capture their world view. Some of the classic sociological field studies, such as William Whyte's *Street Corner Society* (1943) and Elliot Liebow's *Tell Them Who I Am* (1993), are vivid portrayals of what life is like for certain groups of people—in the former case, members of a lower-class urban community and in the latter case, homeless women.

"File these in random order like you usually do."

Randomization.

Several separate techniques fall under the rubric of field research. They include participant observation or ethnography, in-depth interviewing, and documentary analysis. In many studies, more than one of these techniques is used.

Ethnographic or Participant Observation Research

In *ethnographic* or *participant observation* research, the researcher participates in the daily activities of his or her research subjects, usually for an extended period of time. This may include accompanying them on their daily activities (such as following police officers on patrol), interviews and discussions about their lives, and occasionally even living with them. During these activities, researchers take field notes (or make recordings) during and after an episode in the field.

A good example of participant observation research is Ben Killingsworth's study (2006) of how mothers interact with each other in a 'moms and tots' playgroup. He aimed to establish some ideas about what a good mother is, and also about how mothers can reconcile those ideas with the consumption of alcohol. Killingsworth participated in a playgroup of mothers and toddlers in Australia over a period of several months. As is typical of participant observation research, he did not have a rigid research design that he followed strictly over the time he was in the field. Rather, his main interest was in the women's conversations about alcohol and their own personal consumption of it and how they used these conversations to define, alter, and reconstruct ideas about 'good mothers'. He did not direct the women's conversations or ask them to focus their talk on particular issues. Instead, he simply participated in the playgroup and allowed the conversations to occur naturally. He found that the women were able to reconcile their understandings of good motherhood with the consumption of alcohol by recapturing the

(© Brenda Brown. Reprinted with permission.)

importance of alcohol to their previous identities as childless women and using that to build ideals of themselves as women first and good mothers second.

TIME to REFLECT

Now that you understand some of the major points about surveys, perhaps you will be more critical of some of the questionnaires you read about in the newspaper or see on TV. And if you're asked to participate in a survey in a mall or over the phone, would you be willing to ask the interviewer about the sampling procedures and questionnaire design? After all, you are the one giving up your time to answer the questions, so don't you want the results to be meaningful?

Killingworth's research is interesting for several reasons, but one is particularly relevant to his use of participant observation. By focusing on naturally occurring conversations, he was able to show how cultural ideals about things like motherhood and womanhood are embedded in and recreated by seemingly mundane discussions among mothers. In other words, ideals about good mothers do not just appear out of nowhere and exert pressure on people through 'norms' or 'society'. Instead, ideals about good mothers are defined, interpreted, and reconstructed by actual people in actual interactions.

TIME to REFLECT

What role did Killingworth's gender play in his research? Would he have reached different conclusions if he were a woman?

In-Depth Interviews

The in-depth interview is another popular field research technique and may be used in conjunction with participant observation. *In-depth interviews* are extensive interviews that are often tape-recorded and later transcribed into text. In some cases, these interviews are highly structured, and neither the researchers nor the respondents are permitted to deviate from a specific set of questions. At the other extreme, unstructured interviews may seem like ordinary conversations in which researchers and respondents simply explore topics as they arise. In many cases, researchers use semi-structured in-depth interviews that ask all respondents a basic set of questions but that also allow participants to explore other topics

and issues. Striking the right balance between structured and unstructured questions can be difficult for sociologists, as can asking the right questions.

Elizabeth Murphy investigated the connections between health-care conversations (2000), images of childhood (2007), and ideals of good mothers. Obviously, field research techniques can be used to investigate many more issues than motherhood, but Murphy's and Killingsworth's studies provide a nice illustration of how different field methods can be used to study similar topics. In her earlier article, Murphy used theories about how people understand and respond to risks as the basis of her research on breastfeeding and motherhood. In her later article, she focused on how feeding, and breastfeeding in particular, allowed women to negotiate different conceptions of childhood. She interviewed 36 British mothers six times each, from one month before the birth of their babies to two years after birth. Each interview was semi-structured. In Murphy's sample, 31 women breastfed their babies initially, but only six were still breastfeeding four months after birth. This is interesting in light of current medical advice about the importance of exclusively breastfeeding infants for at least four months, and the recommendations of many health practitioners to continue breastfeeding up to two years of age. Did the women in this study who stopped breastfeeding before four months think of themselves as bad mothers? Or were they able to set aside this medical advice and still think of themselves as good mothers? Murphy found that almost all of the women who had stopped breastfeeding recognized formula feeding as inferior but that none of them perceived this as a threat to their status as good mothers. Rather, they were able to justify their decisions because other people were at least partly—and in many cases primarily—responsible for the switch to formula-feeding. Also, breastfeeding and the demands of the children were critical in shaping how mothers viewed their children in relation to 'Appolonian themes of natural goodness and innocence' (Murphy 2007, 122). The interviews revealed that some women encountered health-care workers who were unsupportive of breastfeeding or who did not diagnose medical problems that prevented breastfeeding. Other women had babies who were either uncooperative or could not do it because of 'incompetence' (Murphy 2000, 317).

One of the strengths of Murphy's research is the flexibility of her semi-structured interview technique. By directing the women to discuss their breastfeeding decisions and then following their leads, Murphy was able to gain a much deeper understanding of these choices. Had she not imposed some structure on the interviews, it is possible that the women might not have talked about their breastfeeding choices at all. Instead, her research presents us with a better understanding of how women can reconcile individual decisions to stop breastfeeding with seemingly contradictory ideals about 'good motherhood'.

Documentation

In some field studies, researchers have access not only to people but also to documents. This is more common in the study of formal organizations like police forces or law firms, but it can also be the case with churches, political groups, and even families. These documents (case records, files, posters, diaries, even photos) can be analyzed to provide a more complete picture of the group under study.

Conducting Field Research

The elaborate procedures needed to choose a sample for a survey are not necessary for selecting the research site and the sample in field research. Strictly following a randomization protocol is necessary only if statistical analysis and generalization are the goal of the research. Field research is done to gain greater understanding through the collection of detailed data, not through generalization. Nevertheless, it is important to choose both the research site and the subjects or informants carefully (Bryman, Teevan, and Bell 2009).

The first consideration in choosing a site for field research is the topic of study. A field study of lawyers or police officers likely will take place at the offices and squad rooms of the respective groups. Choosing which offices and squad rooms to study involves many factors, including which ones will be most

(Ivy Images)

Sociologists, like anthropologists, often study subcultures from the inside, using well-established techniques of ethnographic research. At times, however, they run risks in studying deviant subcultures.

useful for the purposes of the research. But a practical element impinges on much field research—the actual choice of research site can come down to which law offices or squad rooms will grant access. This is not a criticism of field research but a recognition of the realities facing scholars doing this kind of research.

Once the site has been chosen, the issues of whom to talk to, what types of data to record, and how long to stay in the field become important. Some things can be planned in advance, but many things are decided during the course of the field research. The selection of key informants—the people who will be most valuable in the course of the study—cannot always be made beforehand. Similarly, figuring out what to write down in field notes, whom to quote, and which observations to record cannot always be determined until after the research has begun.

When to leave the field is almost always determined during the course of the research. Most researchers stay in the field until they get a sense that they are not gaining much new information. In many cases, researchers decide to leave the field when they find that the data coming from new informants merely repeat what they have learned from previous informants. This is often taken as a sign that the researcher has reached a deep enough level of understanding to be confident that he or she will not learn much from additional time in the field (Bryman, Teevan, and Bell 2009).

Time to Reflect

The research by Murphy and Killingsworth reveals that mothers do not simply accept social norms about things like alcohol consumption and childhood. For instance, mothers who drink alcohol are able to talk about how they can drink and still be good mothers despite all of the messages they receive about the harmful effects of alcohol on babies. This is a clear example of why it is wrong to say that 'society tells us . . . .' Society tells us nothing. Rather, people talk about things they hear and believe, and in turn they refine their thoughts. What are some of the messages that you hear that you have modified in your talk with your friends?

This flexibility during the course of the study is one of the great advantages of field research over survey research. Mistakes in research design and the pursuit of new and unexpected opportunities are possible in field research but not usually possible in quantitative survey research. Once a survey has been designed, pre-tested, and administered to a sample, it is impractical to recall the survey to make changes. This is one of the reasons that pre-testing is so important in surveys.

EXISTING DATA

Both in surveys and in field research, sociologists are involved in collecting new, original data. However, a great deal of sociological research is done with data already collected. Of the several different types of existing data, most are amenable to different modes of analysis. Some of the major types of existing data are official statistics and surveys done by other researchers; books, magazines, newspapers, and other media; case files and records; and historical documents.

Secondary Data Analysis

The analysis of official statistics and existing surveys—also known as *secondary data analysis*—has grown immensely with the development of computers and statistical software packages. It has become one of the most common forms of research reported in major sociological journals, such as the *American Journal of Sociology*.

Quantitative data can be presented in tables like Table 2.2. However, tables can be designed in many different ways, and the type of information being presented will determine the types of comparisons that can be made. In Table 2.2, on marital status in Canada, comparisons can be made within or across the values of marital status (for example, how many people are married versus single), by sex, and across five different years.

As an example, we can see that the number of divorced males increased by more than 90,000 between 2003 and 2007 (712,531 − 620,679 = 91,852) and the number of divorced females rose by more than 119,000 (972,183 − 852,277 = 119,906), while the numbers of married men and women increased by 209,161 and 267,727 respectively. However, the table does not tell us anything about why these numbers may have changed, nor can we make any comparisons with the number of married and divorced people in other countries. Also, notice that the numbers are very precise, but the fact is that these are estimates of the true numbers. So while it is possible to discern trends in tables like this, be careful about reading too much into the precision of the numbers.

Personal computers, statistical software packages, and the availability of many national and international datasets have made secondary data analysis possible for almost every social scientist. The advantages of secondary analysis are that the coverage of the data is broad and that the hard work involved in constructing and administering a survey has already been done, usually by an agency with far more expertise and resources than most individual researchers. The disadvantages are that the data collected are often not precise enough to test the specific ideas that interest researchers, and that mastering the techniques to analyze the data properly can be challenging.

Historical Research and Content Analysis

The analysis of historical documents, print and other media, and records and case materials can be done by several methods. The two most common forms of analysis are historical research and content analysis. Historical sociology relies on *historical research* into all kinds of historical documents, from organizational records, old newspapers, and magazines to speeches and sermons, letters and diaries, and even interviews with people who participated in the events of interest. In *content analysis*, documents such as newspapers, magazines, TV shows, and case records are subjected to careful sampling and analysis procedures to reveal patterns. (See Warren and Karner 2009 for a good discussion of historical research and content analysis.)

One of the major issues facing historical sociologists is that someone or some organization created the records used in their analyses, but the potential biases and reasons for recording the information in the documents are not always clear. Further, some documents are lost or destroyed with the passing of time, so the historical sociologist must be aware that the extant documents may not give a complete picture of the events or time period under study. Why have certain documents survived while others have not? Is there any significance to the ordering or cataloguing of the documents? These and other issues must be dealt with continually in historical research.

Content analysis can be done in a number of ways, but it usually involves taking a sample of relevant documents and then carrying out a rigorous procedure of identifying and classifying particular features, words, or images in these documents. For instance, in studying political posters, content analysis could be used to determine whether the posters from particular parties put more emphasis on the positive aspects of their own party or the negative aspects of other parties. These results could then be used to better understand styles of political campaigning in a particular country or time period. In *manifest content analysis*, words, phrases, or images are counted to provide a sense of the importance of different ideas in the documents. In *latent content analysis*, researchers focus

TABLE 2.2 ▼ Population by Marital Status and Sex, Canada, 2003–2007

	2003	2004	2005	2006	2007
Total					
Both sexes	31,676,077	31,995,199	32,312,077	32,649,482	32,976,026
Male	15,688,977	15,846,832	16,003,804	16,170,723	16,332,277
Female	15,987,100	16,148,367	16,308,273	16,478,759	16,643,749
Single					
Both sexes	13,231,209	13,368,674	13,507,149	13,653,059	13,800,997
Male	7,078,089	7,155,622	7,233,428	7,314,611	7,396,835
Female	6,153,120	6,213,052	6,273,721	6,338,448	6,404,162
Married[1]					
Both sexes	15,438,972	15,558,054	15,675,089	15,802,300	15,916,860
Male	7,701,393	7,752,882	7,803,419	7,860,087	7,910,554
Female	7,737,579	7,805,172	7,871,670	7,942,213	8,006,306
Divorced					
Both sexes	1,472,956	1,524,245	1,576,351	1,630,267	1,684,714
Male	620,679	642,882	665,553	688,975	712,531
Female	852,277	881,363	910,798	941,292	972,183

[1]Includes persons legally married, persons legally married and separated, and persons living in common-law unions.
SOURCE: Statistics Canada. CANSIM, Table 051-0010.

less on specific word or phrase counts and more on the themes implicit in the documents.

SELECTING A RESEARCH METHOD

To summarize, all of the methods described here are used by sociologists to collect data on particular research problems, and they use theories to help them understand or solve these problems. Any of these methods can be used to investigate problems from any of the theoretical perspectives encountered in this text, although some methods are almost never used in some perspectives. For instance, symbolic interactionists rarely, if ever, use quantitative surveys, while most conflict theorists prefer surveys to participant observation.

How do you know which method to use with which theory or theoretical perspective? A complete answer to this question is beyond the scope of this chapter, but the rule of thumb in sociology has been that you let the problem determine the method. For example, if you want to find out something about the national divorce rate and how divorced people differ from those who remain married, then you need a method that will give you data from people all over the country, such as a survey. But if you want to find out how nurses manage the many pressures of their jobs, then participant observation is a more appropriate method.

The Context of Sociological Research

We have discussed the scientific status of sociology, the connection between theory and research, and some of the major research techniques used by sociologists, but we have said little about the context in which this research takes place. In this final section, we consider two of the many possible issues that affect the way in which sociologists do their research: the different purposes of sociological research and research ethics.

PURPOSES OF RESEARCH

Sociological research has several different purposes. Some research is *exploratory* or descriptive: the goal is to find out more about a particular group or topic. For instance, research on home-schoolers likely will be exploratory, because not much is known yet about this particular group (Arai 2000). Other studies are designed to be *explanatory*. Usually, these studies test different theories against each other to determine which theory provides the best

explanation for the phenomenon. Explanatory studies may test whether a theory developed from one group or time period applies to another group or time period. Good examples of explanatory studies are Morgan and Sorensen's test of James Coleman's 'social capital' theory, applied to why students in Catholic high schools tend to outperform students from public high schools (Morgan and Sorensen 1999), and Driessen and Smit's (2007) analysis of how social capital explains differences in school performance between immigrants and non-immigrants in the Netherlands.

TIME to REFLECT

If you were to read a piece of research, would you be more inclined to believe the results if they were based on quantitative analysis or on field research? Why?

Other research aims to be able to predict future patterns of behaviour. Many early studies of criminal recidivism (convicted criminals' committing further offences after being released from prison) were designed to enable people to make better predictions about which criminals would be more likely to re-offend once they were released from jail. Note that although they sometimes are related, prediction and explanation are not the same thing. It is fairly easy to predict that night will follow day, but this does not mean that night causes or explains day—the two phenomena have a common cause.

Still other research aims to empower the group being studied. In the past, anthropologists would often become very involved in the concerns of the people they studied and would become advocates for that group. This was generally frowned upon, and anthropologists were accused of 'going native'—becoming one of the people they were studying—instead of objectively studying them. Nowadays, many researchers specifically want to empower the people they study. For instance, *participatory action research*, in which researchers are guided at least as much by the goals and wishes of their respondents as they are by their own theoretical concerns, has become more popular in sociology as well as in such other disciplines as psychology and social work.

The goal of empowering the people being studied is an important component in feminist research. **Feminism** has had an enormous impact on many disciplines, including sociology. This influence has affected the topics that sociologists study, the perspectives they use to study them, and even the methods they use. The topics and perspectives influenced by feminism are covered in other chapters in this book, but it is important here to mention the influence of feminism on research methods in sociology.

RESEARCH ETHICS

Our final topic in this chapter is the ethics of social research. While it is too broad and complicated to cover in any detail here, we can summarize some of the main issues confronting researchers. It is a topic with increasing relevance in the social sciences but also one with a long history.

A turning point in the history of research ethics in the social and natural sciences was the Nuremberg trials of Nazi doctors and concentration camp officials after World War II. Research performed on prisoners in these camps had produced valuable insights into human physiology, but only as a result of the prisoners' having been subjected to horrific experiments (Guillebaud 2002). The moral imperatives on the protection of research subjects during the acquisition of knowledge have guided the development of ethical principles for social and natural scientific research ever since.

In North America, one of the most important principles of research ethics to arise from these trials is the principle of *voluntary participation*. This means that people can be asked to participate in any piece of research, regardless of its potential damage, as long as they voluntarily agree to be a part of that research.

People must not be coerced or tricked into participating, and they must be able to withdraw from the study at any time, without penalty.

The principle of voluntary participation cannot be fully realized unless its complementary principle, *informed consent*, also prevails. Potential participants must give their consent to participate in the research with a full knowledge of the potential costs and benefits to themselves and to the researchers. Indeed, participation cannot be considered voluntary unless a person is fully informed about the research before giving his or her consent.

One of the implications of informed consent is that research should be designed without deceiving the subjects unnecessarily. If deception is used at all, it must outweigh the harms associated with the deception and be absolutely necessary to the project and the benefits of the research. In most cases in sociology, this problem does not arise, but there are topics that require the use of deception. For instance, studying people's discriminatory attitudes is often not possible without deception. One way of doing this that has been particularly successful is to have people evaluate fictitious resumés of job applicants in which all of the details of the resumé are the same except for the gender of the applicant. The idea is to determine whether or not people evaluate the abilities of men and women differently based not on their reported abilities but simply on their gender. Telling people that the gender of the applicants will be switched around beforehand would defeat the purpose of the study (Foschi, Lai, and Sigerson 1994).

GLOBAL ISSUES

Research Ethics in Global Perspective

In the discussion of research ethics, you may have noticed the phrase 'in North America' a couple of times. This may seem unnecessary, for after all, aren't ethical principles by their very nature applicable at all times and in all places? Ironically, in the case of research ethics, this is definitely not true.

Research participants in many parts of the world do not enjoy the protections that have been explained in this chapter. In many countries, particularly those with more repressive political regimes, coercion of participants and violations of confidentiality and of other research protections are commonplace. Additionally, even when rights are respected, some strong cultural factors influence issues of how these rights are established. For instance, establishing that someone has provided informed consent to participate in a study is usually affirmed via signed waivers. But in many cultures, agreements are concluded with a handshake or some other informal mechanism. A signed agreement makes people suspicious, and in many cases people won't sign an agreement even though they're perfectly happy to participate in the study. Research ethics truly is a global matter because of many such differences, and its application, at least, turns out to be very context-oriented and country-specific. As the amount of international research continues to grow, this issue will become increasingly relevant.

The identity of research participants must be protected in North America, either through an assurance of *confidentiality* or through *anonymity*. Sometimes these terms are used interchangeably, but they are quite different. To assure participants of anonymity, a researcher must not collect any identifying information about his or her respondents. This is possible only in mail-in surveys. In most interviews, field research, and telephone surveys, researchers either have met their participants or possess identifying information about them.

It is more common to assure respondents of confidentiality, whereby identifying information is collected but deliberately withheld in the publication of any results. This is possible in all types of research, but note that the protection of a subject's identity is weaker with confidentiality than it is with anonymity. For instance, researchers asked by a court to reveal the identities of their subjects may have to decide between breaking a promise to their subjects or disobeying a court order. If they have not collected identifying information about participants, they will have nothing to reveal.

A final principle, often attributed to the Nuremberg trials, is that research should not involve any unnecessary harm to the participants. The Nazi doctors were completely unconcerned about the well-being of their subjects and did not take any precautions to protect them from harm. Most researchers find this abhorrent, and their research usually only involves some inconvenience to participants. In most sociological research, the greatest inconvenience to participants is the time they have to spend talking to researchers, filling out surveys or diaries, or, to some extent, allowing the researchers into their lives. However, there are instances in which real psychological damage can occur, and researchers have an obligation to be aware of this and guard against it.

Conclusion

This chapter has introduced you to some of the major methods that sociologists use in their research and to some of the issues they face when conducting that research. A simplistic understanding of science is not helpful—for sociology or any other science. But a more sophisticated view of science does underpin how many sociologists view their discipline, although some sociologists reject the idea that sociology is a science at all. Nevertheless, establishing the validity and reliability of their results remains important to almost all sociologists.

The chapter reviewed the three main techniques—surveys, field research, and the analysis of existing data—and ended with a discussion of research ethics. And while it has not been possible to cover all topics in detail, you should have a better appreciation for the difficulties and the joys of finding out about the world sociologically.

questions for critical thought

1. What was your understanding of science before reading this chapter? How has it changed?

2. When you read about a social scientific finding in the newspaper, what kinds of evidence convince you of its veracity? In other words, do you need quantitative, statistical results, are you convinced by detailed accounts of individuals, or are both equally convincing?

3. If you were going to investigate across Canada the effects of a person's ethnicity on his or her educational attainment, what method would be most appropriate? Why?

4. If you wanted to find out more about the motivations of parents who send their children to a particular private religious school, what methods would be most appropriate? Justify your answer.

5. Outline the two different approaches to establishing validity.

6. Summarize the criticisms of science outlined in this chapter. Which one is most convincing, and why?

7. When you read a 'human interest' story in a magazine, what convinces you that the story is true? Does the author need to demonstrate that he or she has a deep understanding of the group? If the story contains numbers or statistics, do you find it more believable? If so, should you rethink this idea, given how hard it is to conduct good quantitative research?

8. It is often difficult to distinguish between a survey and a pseudo-survey, especially if you are a respondent rather than a researcher. Summarize the key points necessary for a proper social scientific survey and how they are violated in each of the types of pseudo-surveys discussed in this chapter.

recommended readings

Earl Babbie. 2009. *The Practice of Social Research.* 12th edn. Belmont, CA: Wadsworth Thomson Learning.
Babbie's books are used in more research methods courses across North America than those of any other author. This one is a comprehensive treatment of research methods.

Bruce Berg. 2008. *Qualitative Research Methods for the Social Sciences.* 6th edn. Boston: Allyn and Bacon.
Berg's book is the current standard for qualitative research methods courses.

Sheldon Goldenberg. 1992. *Thinking Methodologically.* New York: HarperCollins.
This sophisticated research methods text is more difficult to read than the other texts listed here but is well worth the effort.

George Gray and Neil Guppy. 2008. *Successful Surveys: Research Methods and Practices.* 4th edn. Toronto: Nelson Thomson.
Gray and Guppy have written an accessible and comprehensive introduction to survey research methods. The book can be used as a step-by-step guide to conducting a basic survey.

Don G. McTavish and Herman Loether. 2002. *Social Research: An Evolving Process.* 2nd edn. Boston: Allyn and Bacon.
Another popular methods book, this one is more focused on quantitative methods than on qualitative methods. Loether and McTavish have also written statistics texts that have been used in many social statistics courses across North America.

Charles Ragin. 2010. *Constructing Social Research: The Unity and Diversity of Method.* 2nd edn. Thousand Oaks, CA: Sage.
Ragin's book covers the main topics of social research but is shorter and less comprehensive than Babbie, Goldenberg, or McTavish and Loether. Nevertheless, it is an excellent introduction to social research methods.

Shulamit Reinharz. 2010. *Observing the Observer: Understanding Our Selves in Field Research.* New York: Oxford University Press.
A higher-level text that explores in detail the role that researchers play in the qualitative research process.

Stephen Yearley. 2004. *Making Sense of Science: Understanding the Social Study of Science.* Thousand Oaks, CA: Sage.
A very readable introduction to the sociology of science and many of the issues discussed at the beginning of this chapter.

recommended websites

Inter-university Consortium for Political and Social Research
www.icpsr.umich.edu
This is one of the first—and one of the best—social science data archives in the world. As a college or university student, you can use many of the datasets from the ICPSR for class and research purposes at no charge.

LISPOP
www.wlu.ca/lispop
LISPOP is a research centre that focuses on analyzing opinion polling. It is also a great site for following provincial and national elections.

Organisation for Economic Co-operation and Development (OECD)
www.oecd.org
The OECD collects data on economic, political, social, environmental, and industrial conditions in member countries (including Canada) and non-member countries (particularly developing countries). Data, publications (online and print), and special reports are available at this site.

QualPage
www.qualitativeresearch.uga.edu/QualPage
This site is dedicated to qualitative methods. It lists new books, conferences, and many other resources for people interested in qualitative research.

Social Research Methods
www.socialresearchmethods.net
This useful site lists both qualitative and quantitative methods resources, though the focus is on the quantitative. There are also dozens of tutorials on a wide range of methods topics.

Society for the Study of Symbolic Interaction (SSSI)
http://sun.soci.niu.edu/~sssi

This society is open to anyone interested in qualitative social science research—in particular, symbolic interactionist research. The society publishes a journal and holds annual meetings at which members and others present their work. The journal *Symbolic Interaction* publishes high-quality qualitative research.

Statistics Canada
www.statcan.ca

This is one of the most useful sites on the Internet for Canadian sociologists. Many of the surveys listed at this site can be used for class projects and research papers through your university or college, at no charge to you. Ask at your computing centre or library for details on the Data Liberation Initiative. If you go to only one of the sites listed here, make it this one.

US Census Bureau
www.census.gov

Like Statistics Canada, the US Census Bureau provides a wealth of resources and data, some of it free to use. You can also order publications, view them online, and download some US data from this site.

chapter 3

Culture and Culture Change

Shyon Baumann

In this chapter, you will:

▶ See that culture has many meanings and dimensions;

▶ Observe that culture is ubiquitous, thoroughly a part of our lives, and necessary for social life;

▶ Learn that culture is powerful: it integrates members of society but can also cause great conflict;

▶ Learn that culture quintessentially carries meaning and facilitates communication;

▶ See that change in culture is inevitable: it is within the nature of culture to evolve;

▶ Find that sometimes culture change is caused by socio-structural factors (such as political or economic events) and that at other times cultural elements cause changes to one another;

▶ Learn that important social institutions—like the government, the family, and the media, among others—are shaped by culture and, in turn, influence culture.

Why Study Culture?

Why do sociologists care about culture? Briefly, **culture** is an amazingly powerful social force that influences events as diverse as whom we marry and whether we go to war. Marriage and war are interesting examples—while they seem unrelated, they are similar insofar as they both involve the bonds between people, in one case bringing people closer together and in the other pushing them further apart.

Let us consider how culture is implicated in each of these events. How we choose whom to marry is incredibly complicated, but what is clear is that, in general, people like to marry other people with whom they share interests and experiences. Such shared ideas and preferences create a feeling of comfort and familiarity, which are things we enjoy about being with other people. If we like the same music and the same kind of movies, if we share a belief in the importance of family and the role of religion in our lives, if we share a notion of the different roles and responsibilities of men and women, if we support the same political ideals, then we feel more connected to each other. In all these ways, culture influences how we relate to others. Culture includes all these preferences and ideas and notions, and these are the things that allow us in our daily lives to feel connections to other people. Cultural similarities influence our decisions not only about getting married but about all kinds of connections—with whom we become and stay friends, even with whom we work.

Just as we often are brought closer to other people, so too we often experience social divisions, some relatively minor and others quite significant. Like marriage, war is an enormously complicated phenomenon; it can result from a wide array of social, economic, and geopolitical factors. But it is also clear that culture can play a role in creating or worsening the divisions between groups or societies that can lead to war. While a conflict of material interests usually sets the stage for war, culture can play a large role in determining whether war is waged. If we differ in fundamental *beliefs* about such things as democracy and human rights, if we speak different languages and cannot easily communicate, if we cannot understand others' religious concepts and practices or strongly oppose them, if we do not share preferences for what we consider to be the normal and good ways to live our lives, then we feel less connected to each other. In all these ways, culture plays a role in dividing us from others, and it is only in the presence of such divisions, when we feel essentially different and disconnected from others, that we are able to pursue as drastic a course of action as war. In addition, culture plays a role in many more minor social divisions that are not as significant as war, such as the various social cleavages between many **social groups** within the same society.

Culture, then, is important because it is the key to understanding how we relate to each other; specifically, it is behind both what unites us and what divides us. Our cultural differences and similarities are continually coming into play in our daily face-to-face interactions and on a global scale. To truly understand the dynamics of war and peace, love and hate, and more, we need to look at the ways culture facilitates or inhibits the bonds and the rifts between us.

The goals of this chapter are to review the many nuances to the meaning of culture and to explain how culture is implicated in many important social processes. First, we will further specify what culture is through a clear conceptualizing of culture's many dimensions. Second, we will summarize how culture is used in sociological theorizing about society and examine how culture fits into causal explanations of the way society works. In this chapter, we are also interested in a description of those realms of social life that are primarily cultural—the loci of culture. The nature of culture change is a third focus of this chapter, and we will examine the reciprocal relationship between culture change and social change. Finally, we will discuss the insights of this chapter as they pertain to Canadian culture.

TIME to REFLECT

Aside from marriage and war, what other social phenomena are influenced by culture's ability to bring people together and to keep them socially distant?

What Is Culture?

Think of the many ways that you might use the word 'culture' in conversation, from the way that an entire society lives (e.g., 'Thai culture'), to refined aesthetic productions (e.g., symphony concerts), to a phrase such as 'consumer culture', which focuses on a major pattern of people's behaviour and a set of economic institutions in contemporary society. These divergent meanings complicate any attempt to provide a succinct and definitive summary of the sociology of culture. Perhaps the best way to tie together these divergent meanings is to recognize that culture is those elements of social life that have meanings that social actors interpret and can also convey.

To clarify how cultural elements of social life embody meanings, we can create a list of specific things that are always or usually classified as 'culture' in sociology. Languages, symbols, discourses, texts, knowledge, values, attitudes, beliefs, norms, world views, folkways, art, music, ideas, and ideologies are all 'culture', as are the practices through which these things are often performed or put into concrete form. To help clarify why these things are culture, we can consider the distinction between *culture* and *structure*, two terms that have specific meanings within formal sociology. Structural aspects of society are the enduring patterns of social relations and **social institutions** through which society is organized and through which individual and collective actions are carried out. For example, our political system with its elections, taxation, and rule of law is a structural element of society. Our economic system with the role of financial institutions and a free market orientation is another structural element. The patterns of relationships that constitute families, schools of all levels, and health care are further examples of structures in our society.

Recall that cultural elements are those that carry meaning and can be interpreted, and so it is true that the structures named above contain or interact with culture. With regard to the labour market as a structure, the fact that there exists a high degree of occupational segregation by gender, with some jobs (for example, elementary school teachers) primarily done by women and others (for example, elementary school principals) primarily done by men, is not cultural. Rather, this segregation is structural. It is an enduring pattern of social behaviour, existing primarily not at a mental level but at a level of lived experience. The *idea* that it is normal or proper for men to be principals and women to be elementary school teachers is a cultural value. The fact that this pattern exists in our society (although it is changing) is a structural property of our society, but the gender beliefs that underlie this pattern are cultural.

We can find another example of the distinction between culture and structure in the realm of politics. In Canada, the widely held preference for representative democracy and a belief that it is a legitimate and necessary form of self-government represent a deep-rooted aspect of Canadian culture. This political orientation is related in a significant way to many other beliefs about authority, individual rationality,

In the First Person

At the most general level, I chose to study sociology because it helps us to figure out how the world works. More specifically, though, I am interested in understanding the role of culture in shaping social inequality. The reasons why some people have more wealth and status in society are varied and complex. Cultural factors are a part of this story—that is, people can draw on and manipulate cultural resources as a conscious or unconscious strategy for helping them to succeed socially and economically. What is fascinating about the role of culture in social inequality, however, is that culture is not *just* a tool for conflict. People sincerely appreciate and identify with cultural objects and productions—their cultural tastes are genuine. So it is interesting to determine how and when culture performs these different functions.

—Shyon Baumann

and justice, and so it is an element of culture that is enmeshed in a web of other important cultural elements. In contrast, representative democracy is not merely an idea; it is a practice that involves a tremendous amount of material resources and engenders long-standing patterns of social behaviour. Known in the sociological literature as the **state**, our democratic government is a structural dimension of social life. It is related in significant ways to many aspects of citizens' daily existence; it influences, among other things, our work lives, our consumption patterns, our health outcomes, and our educational outcomes, and so it is a material element of social life that is clearly enmeshed in a web of other important structural elements. It does not qualify as culture, because it is not in itself a symbol; it does not exist to be received and understood as having a meaning. However, there is no shortage of politically oriented culture or of political symbols existing in a wide array of forms. Political ideologies of the left, centre, and right, expressed in political discourses, in conversations, newspaper articles, and books, both fiction and non-fiction, are squarely in the realm of culture. The national anthem and the Canadian flag are both explicit political symbols. The neo-Gothic Parliament buildings in Ottawa, while serving as the venue for federal politics, also serve as a political symbol, not only through the 'messages' associated with their stately, traditional,

European style of architecture (they are not pagodas or pyramids) but also because they are sufficiently well known to conjure an association with the federal government and so can represent the country as a whole. While these and other political symbols are culture, the state itself is a key structure in Canadian society.

Why is it sociologically useful to focus on culture, especially with regard to its role in carrying meaning? To answer this question, in the following sections we will first learn about how culture varies over time and between places. By drawing these contrasts, we will then proceed to highlight culture's role in shaping the actions of groups and individuals, as well as culture's role in shaping their lived experiences and identities.

CULTURE IN PLACE AND TIME

In some popular uses, 'culture' can refer to the entire social reality of particular social and geographical groups in comparison to other social groups—Western culture, for example. However, we often think of culture in more specific geographic terms than just Western or Eastern. We frequently think in national terms, with fairly strong ideas of what we mean by, for example, Japanese culture, Italian culture, or Mexican culture. The pervasive use of such terms points to the reality that culture can vary systematically between nations, even in ways we are commonly unaware of. Nation-states have often (although not always) coalesced around a common cultural foundation, or if one was not clearly defined from early on, they have tended to promote such a cultural foundation for the sake of national unity and cohesion.

Although references to national cultures are ubiquitous, on closer inspection we can see that, like the larger generalizations of 'Western culture' or 'Eastern culture', national cultures also entail a great deal of regional and local variation. Obvious examples are the cultural differences between Quebec and Ontario or between Alberta and BC. We can continue to spatially limit our concept of culture by pointing to the general social differences between various cities and even between parts of cities. The culture of downtown Toronto, for example, brings up notions of a lifestyle and a built environment that are business-oriented, cosmopolitan, and culturally rich. Toronto's **urbanism** is often cited for its impersonality and inward-looking character and contrasts with the habits, manners, and interaction styles of, for example, St John's, Newfoundland. It is worth noting that, as is also the case with the differences in regions, such local cultural variations exist in a broader cultural

environment of greater similarities than differences (i.e., they all share a Canadian culture).

Just as we can differentiate cultures with respect to physical space, we can observe that cultures vary according to social space or according to social groupings. Therefore, we can think of the culture of, for example, adolescent males as distinct from that of adolescent females and adult males, whether their geographic location is Vancouver or Halifax. Acknowledging culture's social, not just physical, boundedness provides us with a second dimension for understanding the meaning of culture.

Age and **gender**, the social groupings in the above example, are just two of many social boundaries that can differentiate between cultures. Other social lines along which cultural elements may fall include **race** and **ethnicity**, sexual orientation, religion, and many other ways that people see fit to distinguish themselves. Another social space with important cultural implications is that of social **class**. Stereotypes of distinct working- and upper-class cultures are at least as pervasive as national stereotypes. We have firm ideas about the typical speech, mannerisms, dress, culinary preferences, occupations, and leisure activities of the working class and the upper class.

At this point, it is necessary to point out again that just as the cultures of urban Ontario and rural Alberta share more similarities than differences, the cultures of different social groups within a society likewise share more similarities than differences. By enumerating the ways in which, for example, social classes in Canada differ, we neglect myriad ways in which they are similar: difference in accent is trivial to the overall nature of a language; a similar reliance on automobiles overshadows any consideration of whether those automobiles are foreign or domestic; and a propensity to vote for different political parties cannot diminish the tremendous importance of a shared faith in parliamentary democracy.

In addition, it is necessary to point out that the dimensions of physical and social space are relatively but not entirely independent of each other. In some instances, there is considerable overlap, as when a social grouping exclusively or almost exclusively inhabits a physical space. For example, if we were to study the culture of retirement communities, we would see that these are physical spaces populated mostly by a specific social group defined by age. The social boundedness of culture by age (the culture of an older generation) overlays a physical boundedness of culture by residential location (the culture of a retirement community). Likewise, there is a great deal of overlap between, for example, the physical space of Anglican churches and the social space of Anglicans.

Notice that cases in which the physical and social spatial dimensions of culture intersect to the exclusion of other social groups are fairly narrowly circumscribed. For the most part, our social lives are messier, and different **subcultures** interact with each other all the time. Sometimes the young visit their grandparents in retirement homes; quite frequently, individuals of various social classes occupy the same classrooms, malls, arenas, and workspaces (although with different functions within those workspaces); segregation on the basis of race sometimes occurs residentially, although for the most part the common venues in which daily life is played out are racially integrated.

Adding to the fuzziness of cultural boundaries, borrowing across cultures happens all the time. Often, such borrowing occurs without anyone noticing, but sometimes it can happen in ways that are thought to be illegitimate, leading to charges of cultural appropriation. In those cases, the borrowing of

culture across social boundaries can offend a group's sense of identity and cultural heritage. Other times, culture crosses social and geographic borders through global commerce and communication channels, with the involvement of multinational corporations (see Global Issues box).

Finally, we can recognize that culture varies over time. The temporal dimension is an important qualifier because of the magnitude of differences that accumulate to produce cultures that are vastly different from what came before. In other words, culture evolves.

Leaving aside the precise mechanisms of cultural evolution for now, we can recognize that for the most part, culture is never static. It is always developing new features and characteristics. Therefore, the Western culture of today is remarkably different from that of 500 years ago and is in many ways quite different even from Western culture 10 years ago. The temporal dimension of culture is independent of its physical and social locations—culture changes over time in all countries and regions and for

GLOBAL ISSUES

Global Voices: Skin-Lightening Products Promote Caste System

Madhu's mother scolds her for spending too much money on cosmetics. But the 29-year-old from Rajasthan, India, says she spends much of her modest, factory wages on beauty anyways.

Madhu passes the bottles marked with Dove's logo in the store. The brand's award-winning advertising campaign has made it the leading cleanser in the United States. By purporting real beauty under the slogan, 'Love the skin you're in', Dove has generated millions in profits for its parent company, Unilever.

But, Madhu walks past.

Here in India, Unilever's other advertising campaign is far more powerful. The company's profits have helped pay the salaries of Bollywood celebrity spokespeople. Unlike Dove, Fair & Lovely encourages people to lighten the colour of their skin.

'Beauty means a fair complexion,' says Madhu. 'There is a social prestige to using market products.'

Skin-bleaching is a growing trend in countries throughout Africa, Asia, and the Middle East. Those who can't afford the brand name products often turn to homemade concoctions. As a child, Madhu used talcum powder to turn her face white. When

she can't afford Fair & Lovely, she sometimes uses household bleach.

Critics call the products racist. But, their biggest problem is advertising campaigns that link white skin to getting a better job or finding a mate. Despite the criticism, these campaigns have been successful. Global Industry Analysts reports the market for skin-whitening creams will grow to $10 billion by 2015. That profit is fuelling large-scale, contradictory advertising campaigns.

'We have hundreds of whitening products in the market,' says Sandeep Anirudhan, an Indian man who launched a one-man campaign against skin-whitening advertisements. 'If you sit through an hour of Indian television, you'll have so many advertisements that tell you that you must become white, that I have stopped watching.'

Anirudhan recently targeted Vaseline, another Unilever brand, for a skin-lightening Facebook application. The page, which has almost 9,000 active users, encourages men to upload pictures while a program lightens their complexion because, as their slogan says, 'People see your face first.' . . .

SOURCE: Abridged from Craig and Marc Kielburger, 'Global voices: Skin-lightening products promote caste system', *Toronto Star*, 2 August 2010, www.thestar.com/news/globalvoices/article/841751--global-voices-skin-lightening-products-promote-caste-system.

all social groupings. Norwegian culture today is different from what it was in 1900; French-Canadian culture, irrespective of actual geographic roots, has evolved over the century as well; and the culture of Canadians in their 60s has changed dramatically over time—the leisure and work options and the values and ideals of older Canadians bear little resemblance to what they were in earlier time periods. Many observers of culture argue that cultural changes are occurring more frequently in recent time periods—the rate of cultural change is increasing. When we turn to the specifics of cultural dynamics, we will learn more about the reasons behind this increase in the rate of change.

TIME to REFLECT

Provide examples of things that can serve both cultural and structural roles simultaneously, and explain how these roles can coexist.

The Role of Culture in Social Theory

Now that we have a clear idea of what culture is, we can gain an understanding of how it has figured in the works of some of the major sociological theorists. In this section, we will outline how these thinkers have employed culture in their writings about the fundamental driving forces of society. In doing so, we answer our earlier question of why sociologists study culture.

ORTHODOX MARXIST AND NEO-MARXIST THEORIES

One of the most influential theoretical perspectives in sociology is *Marxism*. In developing his theory of society, Karl Marx was responding directly to previous philosophical arguments about the central role of ideas (squarely cultural) in determining the path of history and the nature of social reality. In such arguments, the general cultural environment worked at the level of ideas to shape people's thoughts and actions and so was in principle the root cause behind events and social change.

In contrast, Marxism argues that the nature of society is determined primarily by the prevailing *mode of economic production*, evolving through history from agrarian societies to slave ownership to feudalism and then to industrial **capitalism**. This perspective is squarely structural, because it argues that all social change is a result of the economic organization of society. In Marxist terminology, the economic mode of production forms the 'base' of society on which the 'superstructure' rests, which includes everything else, including all cultural elements of society.

Neo-Marxist perspectives do not adhere so strictly to the view that culture is entirely dependent on society's mode of production. While they borrow extensively from Marx's insights, they also modify these insights, and in so doing they provide a significantly different view of culture. These perspectives share with Marxism a focus on the role of culture in maintaining and supporting capitalism and inequality, but they differ from Marxism insofar as they view culture as more than simply the reflection of the underlying economic base.

One particularly important neo-Marxist perspective on culture is the argument that our current economic mode of production is accompanied by a **dominant ideology**. This ideology is a system of thoughts, knowledge, and beliefs that serves to legitimate and perpetuate capitalism. Our mental lives and our entire thought modes are shaped to minimize criticism of capitalism and to maximize participation in and support of capitalism.

Where does this dominant ideology come from? Neo-Marxists recognize that culture can be shaped by specific groups and individuals who seek to achieve certain social outcomes. For example, Antonio Gramsci (1992) argued in the 1920s and 1930s that intellectuals within spheres such as politics, religion, the mass media, and education provide knowledge, values, advice, and direction to the general population that serve to perpetuate the status quo and to suppress revolutionary tendencies. To take another example, members of the Frankfurt School, who began writing in the 1920s, identified pro-capitalist functions in much of popular culture, which promotes capitalist ideals and stifles critical, independent thinking. The groups responsible for the creation and promotion of popular culture within the entertainment industry are themselves significant members of the **bourgeoisie**. In the view of the Frankfurt School, the entertainment industry is of great use to the capitalist order through the cultural products it creates.

Growing out of a neo-Marxist perspective, the cultural studies tradition is a field with roots in British literary scholarship and in sociology. The specific insight that cultural studies borrows from neo-Marxists is that culture can be shaped and manipulated by dominant groups and employed to maintain **hegemony**, which is a common-sense understanding that inequality and domination by elites is natural and inevitable. Cultural studies has thus provided a more sophisticated understanding of the ways in which culture can work to reproduce inequality; the meanings that are embedded in cultural works can be hegemonic and can therefore legitimize inequality.

Cultural studies practitioners agree with neo-Marxists that culture can function to maintain social divisions, keeping some groups dominant over others. Where they break from Marxists and early neo-Marxists is in the recognition that class conflict is only one of many sites of ideological dominance. As Philip Smith writes of cultural studies, 'a move has gradually taken place away from Marxism toward an understanding of society as textured with multiple sources of inequality and fragmented local struggles' (2001, 152). Dominant groups can be defined not only by class position but also by race, gender, geography, and sexual orientation.

One of the main figures in this tradition from the Birmingham School is Stuart Hall, who has produced some of the seminal concepts of cultural studies. As Hall (1980) explains, communication of meaning requires both **encoding and decoding**. By this he means that such things as an advertisement or a television show are created in such a way as to convey a particular perspective. The predominant beliefs of the creators are encoded into these cultural productions (or texts) in subtle and sometimes subconscious ways. A fresh, critically informed reading of such texts is required to see how they encode assumptions and messages about such things as gender and social class relations. Another significant insight of Hall's is that meaning does not simply exist as part of cultural creations but instead is constructed by individuals through the process of receiving and interpreting culture. Meaning is created by people while they make sense of the culture they consume or take in. It is important to note that neo-Marxists and those who have further developed their insights make a fundamental advance in their view of culture insofar as they see it as more than simply an artifact of the economic base. Culture, they argue, can also help to determine other facets of social reality—not merely reflective of other things in society, it also helps to shape society. A significant continuity between Marxist and neo-Marxist views of culture is that culture is implicated in the essentially conflictual nature of society. Culture, in a sense, supports dominant groups in their efforts to maintain their dominance.

CULTURAL FUNCTIONALISM

A different approach to understanding culture can be found in work that is based on the theoretical insights of Émile Durkheim. In contrast to the conflictual emphasis of the Marxists and neo-Marxists, the views on culture that are based on Durkheimian sociological insights focus on the integrative ability of culture. Rather than pointing to the ways in which culture can create social fissures, Durkheim (1964 [1912]) identified the ways in which culture can create social stability and solidarity, focusing on how culture unites us rather than on how culture divides us.

Culture, in terms of norms, values, attitudes, and beliefs, is not reflective of the economic mode of production. Instead, these cultural elements are generated according to the needs of society by its form as a more or less complex system. Culture rises out of a particular society's **social structure** to produce a general consensus about the goals and nature of society. As such, our values about, for example, the importance of education evolve in response to the changing needs of a modernizing society in which higher general levels of education allow for a more smoothly functioning society. In this sense, culture serves a necessary function: through common values and beliefs, society is able to remain coherent, and all the different parts of society can effectively carry out their specific purpose.

Durkheim paid special attention to the role of religion as a motivating force in society—one that made possible the affirmation of collective sentiments and ideas and one that could therefore play an important role in strengthening social bonds that then strengthened and reinforced the fabric of society.

SYMBOLIC INTERACTIONIST AND DRAMATURGICAL PERSPECTIVES

A third important perspective treats culture as a product of individuals' interactions. In **symbolic interactionist** thought, culture plays the role of a vehicle for meaning (hence 'symbolic') and is generated by individuals in face-to-face encounters (hence 'interactionist'). Culture is the enacted signals and attitudes that people use to communicate effectively in order to go about their daily lives. Body language and the signals we send through it, however subconsciously, are a clear element of culture in this perspective. The decisions we make and carry out to reveal or to suppress certain pieces of information about ourselves are also culture.

Social interaction can be analyzed to reveal layers of meaning behind routine actions. It becomes evident that there is a communicative element in a great deal of our interactions, although we are not always conscious of its presence or of the nature of the messages we send. The result of our interactions is (usually) the successful management of our relationships with others.

In terms of its view on culture, the symbolic interactionist approach contrasts with Marxist and functionalist approaches insofar as it attributes more responsibility to individuals as the active creators and implementers of culture. Rather than originating from an economic order or indirectly from the general social structure, culture is a product of creative individual agents who use it to manage their everyday tasks and routines.

One of the most influential theorists to write about the interactions of individuals was Erving Goffman. Goffman developed an analytical framework that analogizes social interaction to what goes on in a theatre. For that reason, it is known as a dramaturgical perspective. In a theatre, there are actors with roles to play for an audience. Likewise, when we interact with people, we assume a role for the situation we find ourselves in and perform that role according to a well-known script that defines the boundaries of what is expected and acceptable for the role. We learn these rules of social behaviour through the ordinary process of socialization. We use these rules to create meaningful and effective interaction with others. When we are interacting with others and are in our roles, we are managing impressions and performing in a front-stage area. When we let down our guard and behave informally and in ways that would embarrass us in front of others, we are in the back-stage area.

Culture plays a part in the dramaturgical perspective that is in one sense quite central: social order

(Franklin Carmichael [1890–1945] *Northern Tundra*, 1931. Oil on canvas. 77.4 x 92.5 cm. Gift of Col. R.S. McLaughlin. McMichael Canadian Collection. 1968.7.14)

Sometimes art is employed to achieve ideological ends. This painting by Franklin Carmichael, a member of the Group of Seven, depicts the landscape of the Canadian North. The Group sought to express their nationalistic sentiments through paintings of scenes that were uniquely Canadian.

is constituted by the creation and use of meanings embodied in interaction. The sending and receiving of signals and messages is the key to understanding why society functions at all when there is so much potential for chaos. When you think about it, we are remarkably efficient at maintaining social order most of the time, and this achievement is made possible through the shared meanings in face-to-face interactions.

This view of culture, however, is one that is perhaps less rich than that offered by the cultural functionalist perspective. Rather than culture playing a fundamental role in shaping individuals' very consciousness, as the functionalist perspective would argue, the dramaturgical perspective sees culture as a tool for creative individuals to manipulate strategically. Rather than persons' being fully subject to the influence of culture, culture is subject more to the influence of individuals.

THE 'PRODUCTION OF CULTURE' PERSPECTIVE

The 'production of culture' perspective takes as an object of study those aspects of culture that are created through explicit, intentional, and co-ordinated processes. This approach focuses on material culture, and studies taking this perspective focus on mass media, technology, art, and other material symbol-producing realms such as science and law. The guiding insight of this perspective is that culture is a product of social action in much the same way that non-cultural products are. The implication of this view is that culture is studied best according to the same methods and analysis that are standard in other fields of sociology.

A key figure in the development of the production of culture perspective, Richard A. Peterson (1994), notes that the perspective was developed to account for perceived shortcomings in the prevailing 'mirror' or 'reflection' view, which posited that culture was somehow a manifestation of underlying social-structural needs or realities. This view, held by orthodox Marxists and by functionalists, is quite vague about the specific mechanisms through which culture is created. The metaphor of a mirror is descriptive of the content of culture—it represents the true nature or character of society—but is mute about culture's production.

Such a view would find, for instance, that the contours of Canadian national identity are visible through studying the literary output of Canadian authors. As a body of work, Canadian literature takes

on the characteristics of and reflects the essence of Canadian society. Likewise, Baroque art forms are seen as expressions of society in the Baroque period, and modernist art is explained as an expression of societal sentiments and values in the early decades of the twentieth century.

By contrast, the production of culture perspective is insistent on specifying all the factors involved not only in cultural production per se but also in how culture is 'distributed, evaluated, taught, and preserved' (Peterson 1994, 165). Through a thorough analysis of all these processes, we can better account for the specific content of culture. We need to examine the resources and constraints that specific actors were working with and that influenced the kind of art or other symbols that they created. In this way, the production of culture perspective provides us with the means of explaining the shape of culture.

TIME to REFLECT

Which of the above theoretical traditions do you find most useful for understanding culture's role in social inequality?

CONFLICT, INTEGRATION, AND CULTURE'S ORIGIN

It is useful to compare these various perspectives according to their views on several key features of culture. Marxists and neo-Marxists are clear in their argument that culture is a tool of conflict in society, a view that contrasts with functionalists, who emphasize the integrative function of culture. Functionalists are interested in explaining social order, and they see culture as a key factor in creating social stability.

For symbolic interactionists, culture is primarily the means by which individuals create order out of potentially chaotic and unpredictable social situations, and so they support an integrative view of culture. The production of culture perspective has little to say about characterizing culture as integrative or as implicated in conflict. But while it has the least to say about that dimension of culture, it says the most about another dimension, the origin of culture, because it developed out of dissatisfaction with the views of Marxism and functionalism on the origin of culture. While these older perspectives relied on a 'reflection' metaphor to explain where culture comes from, the production perspective locates cultural origin in 'purposive productive activity' (Peterson 1994,

164). Cultural studies does not provide quite so articulate an account of cultural origin, but neither does it merely rely on vague notions of reflection. Instead, it sees culture as originating in the work of hegemonic leaders who create the texts, symbols, and discourses that embody particular ideologies. Symbolic interactionists also provide an explanation for the origin of culture: it is produced in the meanings that people create through social interaction at the micro level.

TIME to REFLECT

Which of the traditions are most similar, and which are most different? Why?

Cultural Realms

The stage is now set for a discussion of some of the attributes of those realms of social life most commonly located at the core of the sociology of culture. Although we could discuss the cultural dimension of almost any area of society, we will limit our discussion to language and discourse, the mass media, religion, and art. Within each realm, we will highlight the insights that the sociology of culture can bring to bear.

LANGUAGE AND DISCOURSE

Language, a system of words both written and spoken, is but one means of communication. Communication is the sharing of meaning, by which the thoughts of one person are made understandable to another. Communication can occur through a variety of signs and symbols, but we reserve a special place for the study of language because it is the primary means by which our communication takes place.

Languages are complicated systems of many symbols deployed according to a set of rules, and their use gives rise to a number of interesting social phenomena. It is argued, for instance, that the presence of language structures our very thoughts and consciousness, that without a vocabulary with which to label events (as is the case for infants), we cannot remember them. The character of social reality is tied to language insofar as we make sense of all our experiences in terms of the linguistic devices of and the logic made available through the language we speak.

Discourse is a linguistic phenomenon that refers to a set of ideas, concepts, and vocabulary that are regularly used together. We use specific discourses to habitually speak about and understand a topic or issue. Take, for example, the issue of crime. In talking about crime, we might employ an *individualist discourse* that understands crime as the actions of a self-interested individual who is presented with options and makes certain choices. Crime in this discourse is conceived as something that one person does to one or more others, and it occurs in discrete instances. This discourse of crime encourages an understanding of the psychological factors involved in criminal behaviour and leads to solutions that work at the level of the individual. An individualist solution might suggest that if we alter the attractiveness of the option of committing crime by making penalties harsher for those who are caught, the individual will, we hope, no longer choose to commit crime.

In contrast, a *collectivist discourse* of crime also exists. This discourse views crime as a social problem. Crime is conceived as a feature of society that can be more or less prevalent. The focus is on crime rates and on the social conditions that influence the likelihood that crime will be committed in society. This discourse encourages a view more sociological than psychological of the factors contributing to crime, focusing on the social level rather than on the individual level. Just as the problem is conceived at the group level, the ideas and terminology of a collectivist discourse promote a conception of solutions at the group level. For example, an effort to reduce crime might be based on information gained from a comparison of low- and high-crime societies to determine how certain social differences influence crime rates.

As the example of discourses about crime shows, discourses have the potential for great influence. The promotion of certain discourses throughout society, by those with the power to do so, can have the effect of setting the public agenda for certain issues.

Such discourses play a role in the **social construction** of the categories and definitions we use to understand and to analyze social life. They are worth studying sociologically because of the power of these categories and definitions to define how we act. In our daily lives, we constantly refer to these categories and definitions in order to make judgments about good and bad, right and wrong, desirable and undesirable, how to distinguish between 'us' and the 'other', as well as to understand the very nature of things—Is abortion murder? Is killing in warfare murder? Are movies an art form, an educational medium, a propaganda tool, or entertainment? Is eating animals a question of morality? Is 'race' about biological differences? Is crime an individual failing? Is crime a collective failing of the society? For all these questions, our answers will be influenced by the way

OPEN for DISCUSSION

Why Both Social Structure and Culture Matter in a Holistic Analysis of Inner-City Poverty

The vast majority of social scientists agree that as a national cultural frame, racism in its various forms has had harmful effects on African Americans as a group. Indeed, considerable research has been devoted to the effects of racism in American society. However, there is little research on and far less awareness of the impact of emerging cultural traits in the inner city on the social and economic outcomes of poor blacks. . . .

[F]rom a historical perspective it is hard to overstate the cumulative impact of structural impediments on black inner-city neighborhoods. We have to consider, of course, the racialist structural factors such as the enduring effects of slavery, Jim Crow segregation, public school segregation, legalized discrimination, residential segregation, the Federal Housing Administration's redlining of black neighborhoods in the 1940s and 1950s, the construction of public housing projects in poor black urban neighborhoods, employer discrimination, and other racial acts and processes. . . .

Nonetheless, despite the obvious fact that structural changes have adversely affected inner-city neighborhoods, there is a widespread notion in America that the problems plaguing people in the inner city have little to do with racial discrimination or the effects of living in segregated poverty. For many Americans, it is the individual and the family who bear the main responsibility for their low social and economic achievement in society. If unchallenged, this view may suggest that cultural traits are at the root of problems experienced by the ghetto poor, because most Americans tend to focus on the outlooks and modes of behavior shared by many inner-city residents.

Culture provides tools (habits, skills, and styles) and creates constraints (restrictions or limits on outlooks and behavior) in patterns of social interaction. These constraints include cultural frames (shared group constructions of reality) developed over time through the processes of meaning-making (shared views of how the world works) and decision-making (choices that reflect shared definitions of how the world works). For example, in the inner-city ghetto, cultural frames define issues of trust—street smarts and 'acting black' and 'acting white'—that lead to observable group characteristics.

One of the effects of living in a racially segregated, poor neighborhood is the exposure to cultural traits that may not be conducive to facilitating social mobility. For example, some social scientists have discussed the negative effects of a 'cool-pose culture' that has emerged among young black men in the inner city, which includes sexual conquests, hanging out on the street after school, party drugs, and hip-hop music. These patterns of behavior are seen as a hindrance to social mobility in the larger society. . . .

SOURCE: Abridged from William Julius Wilson, 2010, 'Why both social structure and culture matter in a holistic analysis of inner-city poverty', *The Annals of the American Academy of Political and Social Science* 629 (1): 200–19.

predominant discourses shape and frame our notions of the issues central to them.

MASS MEDIA

The mass media are potent social forces that constitute a key realm of cultural production and distribution. The mass media comprise the technologically based methods and institutions that allow a single source to transmit messages to a mass audience. In Canada, the mass media include print (newspapers, magazines, books, and journals), film, radio, television (broadcast, cable, and satellite), and the Internet. The Internet is a special case, because although it can function as a mass medium, it is also much more—a network medium by virtue of its ability to allow multiple message sources. Potentially, every person on the Internet can be a source of mass media content. The Internet is also more than traditional mass media in the sense that it provides features and functions that extend to commerce, education, and politics, among other realms (see Table 3.1 on the ways in which Canadians are using the Internet).

The mass media are a central cultural concern because of the nature of the content that they bring to the vast majority of people. That content can be categorized both as *information* and as *entertainment*.

In addition to information that is delivered as news per se on news programs and in newspapers, the mass media provide us with a wealth of other information about the world that we might never have access to through firsthand experience. Through the mass media, we can read about the modernization

TABLE 3.1 ▼ Internet Use by Individuals at Home,[1] by Type of Activity

	2005	2007	2009
	% of individuals		
Email	91.3	92.0	93.0
Participating in chat groups or using a messenger	37.9	..	..
Using an instant messenger	..	49.9	44.8
Searching for information on Canadian municipal, provincial, or federal governments	52.0	51.4	56.5
Communicating with Canadian municipal, provincial, or federal governments	22.6	25.5	26.9
Searching for medical or health-related information	57.9	58.6	69.9
Education, training, or school work	42.9	49.5	50.3
Obtaining travel information or making travel arrangements	63.1	66.1	66.2
Paying bills	55.0	..	..
Electronic banking	57.8	..	..
Searching for employment	..	32.3	34.9
Researching investments	26.2	25.5	27.1
Playing games	38.7	38.7	42.1
Obtaining or saving music	36.6	44.5	46.5
Obtaining or saving software	31.8	32.5	35.0
Viewing the news or sports	61.7	63.7	67.7
Obtaining information on weather or road conditions	66.6	69.8	74.6
Listening to the radio over the Internet	26.1	28.1	31.8
Downloading or watching television	8.5	15.7	24.7
Downloading or watching a movie	8.3	12.5	19.8
Researching community events	42.3	44.3	50.0
General browsing (surfing)	84.0	76.0	77.7
Researching other matters (family history, parenting)	..	69.5	72.7
Contributing content (blogs, photos, discussion groups)	..	20.3	26.7
Making telephone calls	..	8.7	13.8
Selling goods or services (through auction sites)	..	8.9	13.4
Other Internet activity	10.9	1.5	7.8

.. : Data not available in that year.

NOTE: The target population for the Canadian Internet Use Survey changed from individuals 18 years of age and older in 2005 to individuals 16 years of age and older in 2007.

[1]Internet users at home are individuals who answered that they used the Internet from home during the past 12 months.

SOURCE: Statistics Canada, www40.statcan.gc.ca/l01/cst01/comm29a-eng.htm (from Statistics Canada, CANSIM Table 358-0130 [for fee]).

of industries in China, we can see what the skyline of Buenos Aires looks like, we can hear about the best way to invest money in a sluggish economy, we can find information about our diagnosed disease and likely find a support group willing to share their experiences with the ailment.

As the providers of so much information, the mass media have an enormous amount of influence on people's attitudes and behaviours, which are dependent on the state of our knowledge. For example, some people will alter their eating habits based on information they learn from magazine articles about

the dangers and benefits of certain foods, and some people will form an opinion about strengthening environmental protection regulations based on stories they watch on television news programs. Because the mass media select a limited amount of information to present to audiences from a virtually infinite supply, they serve as informational gatekeepers (White 1950). This gatekeeping function can account for much of their influence. However, just as important as what they present is the question of how they present media content. There is a connection here to the preceding discussion of discourses, because it is

UNDER the WIRE

Sustaining the Luxury Brand on the Internet

If the unique relationship that luxury has with its clients were to be placed in the context of the Internet where the consumer is in total control and expects to be looked up to, it would likely lead to resistance, apprehension, and anxiety from the top (the luxury brand) and confusion, surprise, and disappointment from the bottom (the luxury client). This has been the situation of luxury in the Internet virtual environment in the last decade and explains why several brands were slow in establishing a web presence and why brands like Chanel and Hermès continue to resist integrated e-Commerce, leaving their consumers at the mercy of fake luxury goods traders who are currently rampant online.

A major existing paradox, however, lies in creating and retaining the 'desire' and 'exclusivity' attributes of luxury brands on the mass and classless Internet world and at the same time maintaining and enhancing the equity of the brand. Another contradiction that luxury brands face online is the task of increasing sales and the risk of overexposure while maintaining a fragile perception of limited supply. These factors are inherently peculiar to the Internet, whose central features appear to be the opposite of luxury's core elements. The characteristics of the Internet and e-Retail are a global reach; a pull marketing approach where customers are drawn to information and purchases, rather than a push medium where customers are driven by advertising; a lack of physical contact with the goods and human contact with the sellers; a low switching cost as it takes only one click to switch between websites; fast and convenient; more product variety and access to viewing them; availability and accessibility irrespective of time and location; less powerful sales as it is easy to say no to a computer; a universal appeal; and uniform information. These characteristics indicate that the Internet as a medium of communications and retail is available to a mass consumer base, which is in direct disparity with the niche consumer base that luxury goods have always targeted. The second indication is the lack of physical contact with the goods and their sellers. Luxury goods are regarded as sensory in nature, and this means that the human senses of visuals, smell, touch, and feel are considered imperative in selling luxury goods. These above factors imply that luxury goods are unsuitable to be placed and retailed on the Internet, according to existing research. The reality is that luxury can be successfully positioned online, and as a result, several luxury brands have currently adopted e-Retail and have identified this channel as one of their fastest-growing distribution channels. A disparity and gap between existing literature and current business practice has therefore been identified and remains unexplored.

SOURCE: Abridged from Uché Okonkwo, 2009, 'Sustaining the luxury brand on the Internet', *Journal of Brand Management* 16 (5/6): 302–10.

---

through the mass media that most discourses are disseminated to the general public.

The mass media are also the primary source for popular culture. While high culture (discussed below) is only rarely made available through the mass media, popular culture is everywhere. The popular culture productions brought to us by the mass media are argued to be linked to deep-seated social problems. For example, popular culture productions are seen to have a negative impact on society because of the materialistic values they explicitly and implicitly advocate. By constantly connecting depictions of happiness and success to material wealth, the mass media have been a principal cause of the development of a consumer culture that focuses our attention and energies on gaining and spending money and away from spiritual, moral, ethical, and social issues. The mass media also are blamed for a culture of violence: it is argued that they contribute to high levels of violence in society to the extent that portrayals of violence incite violent acts and desensitize people to the presence of violence. At the same time, it is argued that the mass media contribute to an unhelpful, unrealistic, and shallow understanding of and response to this violence. The list of social problems linked to the ways in which the mass media may distort and negatively influence our culture is long, including such serious issues as body consciousness and eating disorders, racism, and sexism, exacerbated through stereotypical and misleading depictions.

RELIGION

Religion is a sociological subfield of its own, but it merits inclusion in a discussion of culture because it has had such a large impact on the development of values and cultural traditions in most countries, Canada included. The connection between religion and the ways that people find meaning in the world is clear. We often characterize Western countries as belonging to a Judeo-Christian tradition—a tradition that denotes a specific history and related social institutions and dominant values. It is important to realize that it does not require a specifically religious mindset to be influenced by Judeo-Christian values in general. These values permeate our culture and are seen in such things as views on the role of authority; the moral code undergirding much of the legal system; beliefs in the value of punishment and the possibility of rehabilitation; social action on behalf of the underprivileged, sick, disabled, and terminally ill; and attitudes toward work and leisure. As other religious traditions gradually become more prominent, Canadian culture will undoubtedly continue to evolve in response.

Perhaps the best-known thesis in sociology regarding the influence of religion on culture is Max Weber's argument in *The Protestant Ethic and the Spirit of Capitalism* (1958 [1904]). Weber argues that several aspects of Protestant (specifically Calvinist) doctrine specifically encouraged the values and behaviour of economic rationalism, thereby promoting the

(Michael Ochs Archives/Corbis)

Culture carries aesthetic meanings through various signs—clothing, hairstyles, and so on—that go in and out of fashion. Do you recognize this pop cultural icon?

rapid advancement of capitalism in Protestant societies. The accuracy of this argument and its strength as a single explanation of economic development has been questioned, but its importance for an understanding of the cultural role of religion remains.

ART AND AESTHETICS

The realm of art is, above all, an expressive area of social life. Whereas much of our behaviour is oriented toward the practical achievement of a useful goal, art stands out as an activity that is done to communicate through aesthetic means, and the goal may be no more (or less) practical than to induce feelings. The *New Shorter Oxford English Dictionary* (1993) defines **aesthetics** as 'a system of principles for the appreciation of the beautiful', which begs the definition by failing to define beauty! Today, many artists would eschew the word 'beauty' for 'important' (and some might narrow that idea to 'what is important to me'). Art, then, employs a set of rules or principles embodied in many different forms and pertaining to the artist's notions of what is beautiful or important. This makes art—whether visual, musical, or literary—a special form of communication: it is an expression of thoughts and emotions not communicated through the ordinary means of language. Instead, it relies on the much more implicit and intuitive rules that people in general use to assess beauty or truth or, for that matter, societal strengths, values, and shortcomings.

As discussed above, we often distinguish between 'popular culture' and 'high culture'. This distinction points to the existence of a *cultural hierarchy* in which certain forms of culture are granted greater legitimacy and prestige. Oil painting is higher in the hierarchy than filmmaking, which in turn is higher than television. Works by certain artists command higher prices than those of others. It is important to recognize that such status distinctions are themselves cultural productions. That is to say, our categories of 'high' and 'popular' or 'low' are socially constructed. These categorizations represent more than tangible differences in the characteristics of cultural productions. They also reflect differences in the social contexts in which the artifacts of culture are produced, distributed, and received. For example, the formality and opulence of the settings in which opera is produced and appreciated, and the fact that audiences who typically attend the opera are wealthy and highly educated, encourage an understanding of opera as high culture.

This approach to art highlights several of art's sociologically significant features. First, as explained by the production of culture perspective, art does not

SOCIOLOGY in ACTION

Neo-liberalism and the Realities of Reality Television

Informed by the free-market theories of the conservative Chicago School of economics and its acolytes, neo-liberalism represents a strategy of economic growth developed in opposition to the Keynesian approaches that shaped US monetary and fiscal policy during the mid-twentieth century, from the New Deal to the postwar era of economic expansion. Neo-liberal principles are associated with global free trade and the deregulation of industry, the weakening of union labor, a decline in welfare assistance and social service provision, and the privatization of publicly owned resources.

At first glance, neo-liberal dogma and reality television seem worlds apart—that is, until one considers exactly why the entertainment industry developed the genre in the first place.

. . . It bears remembering that TV studios and networks introduced the first generation of reality television shows—notably the law enforcement shows 'COPS' and 'America's Most Wanted'—in response to the 1988 Writers Guild of America strike. Their goal was to create a form of programming that would be largely immune from union tactics from sit-downs to picket lines. Since reality television shows do not rely on traditional scripts, producers avoid the risks and expensive costs associated with hiring unionized writers. By casting amateur participants willing to work for free, rather than professional actors, producers also avoid paying industry-standard union wages to members of the Screen Actors Guild.

. . . Consider also where reality television creators produce their shows. They have increasingly taken advantage of the globalization of markets and flexibility of national borders that neo-liberal policies make possible. It is no accident, for example, that many seasons of 'Survivor' have been shot in Third World countries undergoing rapid economic development, where local authorities regularly relax labor laws, child protections, health codes, and environmental regulations in the interests of remaining 'business friendly'.

. . . While the production of reality television employs neo-liberalism's economic principles, the genre's narrative conventions reflect its morals. Competitive programs celebrate the radical right-wing values championed especially by free-market Republicans. Both 'Survivor' and 'The Apprentice' require 16 or more participants to fiercely compete against one another in winner-take-all contests guaranteed to produce extreme levels of social inequality. Although team members are initially expected to work cooperatively on 'Survivor', they eventually vote their collaborators out of the game in naked displays of individualism and self-interest. . . .

Although the very design of competitive reality programs like 'The Apprentice' or 'Hell's Kitchen' guarantees that nearly all players must lose, such shows inevitably emphasize the moral failings of each contestant just before they are deposed . . . on programs like 'The Biggest Loser' . . . fitness trainers personally criticize the show's overweight (and typically working-class) contestants for their poor health. In such instances, the contributions of neo-liberal federal policy to increased health disparities in the US—notably the continued lack of affordable and universal health care and cutbacks in welfare payments to indigent mothers and their children—are ignored in favor of arguments that blame the victims of poverty for their own misfortune.

SOURCE: Abridged from David Grazian, 2010, 'Neoliberalism and the realities of reality television', *Contexts* 9 (2): 68–71.

---

just spring out of a collective consciousness or even out of an individual's consciousness. Instead, art is a collective activity that requires collaboration between many actors in an art world (Becker 1982). It is this collective activity that helps to determine how artistic genres may be legitimated or become prestigious and how they may be viewed in the wider society.

Second, the socially constructed nature of distinctions between high and low in art also points to the significance of art in helping to determine the contours of social stratification. This link is rooted in the notion of **cultural capital**: the knowledge, preferences, and tastes that people have concerning art and aesthetics. Having abundant cultural capital usually means sharing the knowledge, preferences, and tastes that are common among those of high status in society. The link between cultural capital and stratification is based on the power of high cultural capital to provide access to informal interpersonal connections that can influence our occupational and economic prospects. Sharing similar artistic tastes and consumption patterns with those in economically privileged positions provides us with access to networks and opportunities not open to those who do not have the necessary aesthetic preferences and expertise. In sociological terms, our cultural capital can increase our economic capital. (This is an interesting inversion of Marxist logic in that the cultural realm is seen as determining, or influencing, the economic realm.)

Third, and perhaps even more significant, is the role that artistic consumption plays in creating social groupings in society. The enjoyment of products of aesthetic quality is deeply related to our conceptions of our own identities, of who we really are, and of the kind of people with whom we wish to be associated. In this way, our tastes are profoundly implicated in how our lives are structured. We've already seen how artistic tastes can interact with our class position, but tastes can also be a way of expressing racial, gender, regional, and age-based identities. This last example can be illustrated through reference to 'youth culture', which includes the artistic, leisure, and style preferences and habits of young people, who thereby distinguish themselves from prior generations.

TIME to REFLECT

What characteristics of each of the above four social phenomena—language, mass media, religion, and art—make them particularly cultural rather than structural?

Cultural Dynamics

We have already seen that culture changes over time, but we have yet to fully consider any specific mechanisms of cultural change. There are various perspectives we can take to understand why and how culture changes over time. First, we can view changes in culture as responses to particular social-structural changes; we will focus on the cultural ramifications of economic changes and of technological changes. Second, we can also view changes in culture as responses to other cultural developments—a view that emphasizes the web-like, interconnected nature of culture.

ECONOMIC, TECHNOLOGICAL, AND CULTURAL CHANGE

The discussion of Marxism earlier in this chapter reviewed the case for the economic foundation of culture. In Marxism, culture is a reflection of the underlying economic basis of society. But it is not necessary to adopt Marxists' assumptions of culture merely as a reflection of economics to see that important economic changes are capable of provoking specific changes in culture. An example of such a change is the liberalization of attitudes toward women and work. In the mid to late nineteenth century and in the first half of the twentieth century, there was a strong belief in Western societies that it was most appropriate for women, especially married women, not to work outside the home but rather to find fulfillment in their roles as mothers and housewives. While the reasons for the change that has occurred in this attitude are many, it can be argued that an important cause of the change was economic. Maintenance of a middle-class standard of living increasingly required a second income. Changing attitudes about women and work, in this view, were an adaptive response to a changing economic reality.

Over the same period of time, rising levels of affluence made it possible for teenagers to possess a certain amount of disposable income. The development of youth culture, while deriving from various causes, was facilitated by the economic changes that created consumers out of young people and thereby encouraged cultural producers to target and cater to youth. The continued growth in spending power of teenagers has also allowed them to become the primary demographic target of Hollywood film studios. Because young people see films in theatres more often than do older groups, a great deal of film production is tailored to their tastes and expectations. This dynamic is representative of the more general dependence of the content of cultural industries on economic conditions.

Technological change can also be viewed as the source of a great deal of the change in our culture. Perhaps the clearest and most significant technological influence on culture has been the development of the mass media. The printing press, invented by Johannes Gutenberg in 1452, has been credited with transforming European culture in diverse ways. For example, the printing press—by making reproduction fast—reduced the value of a book and eventually brought the cost down so that many people could personally own Bibles and read and interpret them apart from what priests instructed them to believe and think—a precondition for the Protestant Reformation. The invention of the telegraph, which vastly hastened the speed with which information could travel over great distances, has been cited as changing our attitudes concerning the pace of life and punctuality and even our very definitions of the proportions of space and time.

It would be impossible to enumerate all the ways in which technology has created cultural change. To take an example of a broad cultural pattern, the very idea of 'nightlife' and all its attendant activities is predicated on the existence of electricity and the light bulb. Much more narrowly, the technological innovation of the electrification of musical instruments has influenced tastes in musical styles. Suffice it to say that technological change frequently has the potential to create cultural reverberations, sometimes of limited significance and other times life-transforming.

CHANGE FOR THE SAKE OF CHANGE

Despite the strength of the relationship between culture and social structure, culture also has internal dynamics that can account for cultural change. In this view, cultural change is inevitable, because culture, as representative of individual and collective self-expression, is inherently progressive, evolutionary, volatile, and unstable: it is the nature of culture to evolve.

The validity of this view is perhaps best exemplified by the phenomenon of fashion. *Fashion* is change for the sake of change in the realm of aesthetics. Ongoing change is built into the very idea of fashion. Moreover, fashion is not just the styles of clothes that are popular, although that is one of its most visible manifestations. Rather, elements of fashion can be found in many areas of social life.

Consider, for instance, how vocabulary choices acknowledge that some words are 'in' while others are 'out'. To express approval, one might have heard adjectives in the past such as 'swell', 'groovy', or 'mod'—words that sound dated now despite the fact that the need to express approval has not gone away. New, more fashionable words do the job today. Consider also how changes in furniture and interior design occur gradually but consistently enough to evoke associations with particular decades. Few of these changes are linked to changes in function or technological innovations.

Although aesthetic changes do not serve any practical or functional purposes, they may still be related to a social purpose: they satisfy needs for self-expression. In this sense, the aesthetic dimension of life is symbolic—we communicate to others and articulate (however obliquely) for ourselves certain thoughts, values, identities, and senses of group affiliation.

TIME to REFLECT

Based on the information above, why might the rate of cultural change be increasing?

(Left: Lebrecht Music and Arts Photo Library/Alamy Images; Right: Gonzalo Fuentes/Reuters/Corbis)

Fashions in dress come and go. What social changes can you infer by comparing the dress and look of the '1830 woman' with the '1930 woman'? What might the '2030 woman' look like?

Canadian Culture

The concepts and arguments reviewed in this chapter can help us to understand the current state of Canadian culture, along with some of the more contentious issues facing Canadian society. Because of its unique history, Canadian culture is unlike any other national culture, with a unique set of challenges and a unique set of opportunities.

DISTINCT SOCIETIES

One of the defining features of Canadian culture is its basis in 'two founding peoples', French and English. The term *peoples* refers, of course, not only to the actual members of the French and English colonies but also to their respective ways of life—their cultures. How different or similar are the cultures of French and English Canada? On a global scale, they are quite similar to one another in comparison with, for example, Pakistani or Indonesian culture. However, they differ in important ways still. Most obvious is the linguistic basis for the distinction. As discussed earlier, language is a core component of culture, with significant implications for social life.

(Alamy Images)

The Mounties' iconic red coats were selected when the force was established in 1873 because Native peoples retained respect for the British Army—the 'redcoats'.

The ability to communicate through verbal and written language is a key element in social bonding—without this form of communication, opportunities for social interaction are limited. Differences in other cultural traditions exist as well, ranging from cuisine and leisure activities to political values and views on marriage and family.

The challenge for Canada has been and continues to be the forging of a unified Canadian culture that respects the unique characteristics of both traditions. To this end, we employ a policy of official bilingualism, and we foster cultural events and new traditions that embrace both French and English cultural elements.

MULTICULTURALISM

The conception of two founding peoples can be seen as primarily a legal construct rather than as an accurate historical depiction. In reality, there have always been more than two cultural traditions in Canada. The Aboriginal cultures of **First Nations** and Inuit peoples were, of course, present before the idea of a Canadian society or culture was ever proposed.

More recently, increased immigration from a large number of countries and the formation of an equally large number of ethnic communities in Canada have added to the number of cultural traditions we have to work with. As a society, we have adopted a stance of official **multiculturalism**, although the merits of this position engender a good deal of debate. We should distinguish between multiculturalism as a fact of contemporary Canadian society—there are ethnic subcultures that are thriving—and multiculturalism as a policy—the tolerance and encouragement of the maintenance of the national cultures that immigrants bring with them from their countries of origin.

Proponents of multiculturalism point to its helpfulness in easing the transition of new immigrants into Canadian society. This happens through the fostering of ethnic communities that can provide social support. In addition, proponents argue that multiculturalism is a policy that is properly respectful to all Canadians and that enriches the wider Canadian culture. Detractors, on the other hand, argue that multiculturalism only makes it more difficult to create a unifying Canadian culture. Moreover, they question the wisdom of a policy that encourages, to however small a degree, self-segregation rather than facilitating the full cultural integration of immigrants into Canadian life. Again, just as with the question of two founding peoples, the challenge here is to balance culture's potential for unifying us with our desire to maintain certain cultural partitions.

HUMAN DIVERSITY

China's Youth Look to Seoul for Inspiration

At Korea City, on the top floor of the Xidan Shopping Center, a warren of tiny shops sell hip-hop clothes, movies, music, cosmetics, and other offerings in the South Korean style.

To young Chinese shoppers, it seemed not to matter that some of the products, like New York Yankees caps or Japan's Astro Boy dolls, clearly have little to do with South Korea. Or that most items originated, in fact, in Chinese factories.

'We know that the products at Korea City are made in China,' said Wang Ying, 28, who works for the local branch of an American company. 'But to many young people, "Korea" stands for fashionable or stylish. So they copy the Korean style.'

From clothes to hairstyle, music to television dramas, South Korea has been defining the tastes of many Chinese and other Asians for the past half decade. As part of what the Chinese call the Korean Wave of pop culture, a television drama about a royal cook, 'The Jewel in the Palace', is garnering record ratings throughout Asia, and Rain, a 23-year-old singer from Seoul, drew more than 40,000 fans to a sold-out concert at a sports stadium here in October.

But South Korea's 'soft power' also extends to the material and spiritual spheres. Samsung's cellphones and televisions are symbols of a coveted consumerism for many Chinese. Christianity, in the evangelical form championed by Korean missionaries deployed throughout China, is finding Chinese

converts despite Beijing's efforts to rein in the spread of the religion. South Korea acts as a filter for Western values, experts say, making them more palatable to Chinese and other Asians.

For a country that has been influenced by other cultures, especially China but also Japan and America, South Korea finds itself at a turning point in its new role as exporter.

The transformation began with South Korea's democratization in the late 1980s, which unleashed sweeping domestic changes. As its democracy and economy have matured, its influence on the rest of Asia, negligible until a decade ago, has grown accordingly. Its cultural exports have even caused complaints about cultural invasion in China and Vietnam.

. . . Jin Yaxi, 25, a graduate student at Beijing University, said, 'We like American culture, but we can't accept it directly.'

'And there is no obstacle to our accepting South Korean culture, unlike Japanese culture,' said Ms Jin, who has studied both Korean and Japanese. 'Because of the history between China and Japan, if a young person here likes Japanese culture, the parents will get angry.'

Politics also seems to underlie the Chinese preference for South Korean–filtered American hip-hop culture. Messages about rebelliousness, teenage angst, and freedom appear more palatable to Chinese in their Koreanized versions. . . .

GLOBALIZATION AND AMERICAN CULTURAL IMPERIALISM

Globalization typically refers to the fact that goods, services, information, and people, now more than ever, can easily flow between distant countries. Of particular concern for us is the cultural influence that globalization brings. There are various implications of globalization for Canadian culture. Technological advancements in mass media have made possible easy and abundant access to the sights and sounds of geographically distant locales. Through media representations, we can be made aware of cultural elements from across the globe, and the potential exists to incorporate these elements into Canadian culture. In a sense, one effect of globalization is the internationalization of national cultures as they are increasingly exposed to

one another. The mass media, then, are the key channels of the cultural diffusion occurring through the mutual influence of many national cultures.

Globalization, however, can bring with it many difficult cultural challenges. Chief among these challenges is the need to manage the global export of American popular culture. Popular culture, in the form of films, television shows, music, and websites, is one of the largest American exports, reaching every corner of the globe. The sheer volume of American cultural export has led to the term 'cultural imperialism', describing the scope of the global dominance of American culture.

The reaction to this state of affairs in Canada has been one of alarm, and a concerted effort has been mounted to maintain the integrity of Canadian culture. In order to promote Canadian cultural production, the federal government has for several decades enacted policies that require Canadian broadcasters to make a sizable proportion of their content of Canadian origin. In addition, a variety of programs exist to subsidize Canadian film, television, music, and book production.

This policy of Canadian cultural protectionism has clearly achieved some measured successes. Scores of Canadian artists have achieved a level of success that would have been unlikely if they had been left to compete on the unequal playing field with American artists, who are promoted by vast media conglomerates.

Conclusion

Many different social phenomena can be called 'cultural', but we have learned here that making and conveying meaning is what they have in common. Culture influences people, but people also use culture to shape their actions. Through a sociological analysis of the nature and significance of cultural meanings, we can better understand and explain a wide range of social phenomena. Culture is always evolving and is intimately tied to other social changes and to other cultural changes. Culture is implicated in the social dynamics both of conflict and of people coming together, and for that reason, as well as others, it is an essential subject for sociological analysis.

TIME to REFLECT

What are the arguments for and against multiculturalism in Canada? Which do you find more persuasive?

questions for critical thought

1. What is Canadian culture, and what are its most important or distinctive facets?

2. How much do you know about other cultures, and how did you learn about them? How do you know whether your impressions are accurate?

3. Where does culture come from? How could you begin to research such a question?

4. What role do art and music play in your life? Do art and music bring you closer to some people and differentiate you from others?

5. How do you decide what music and movies are good? Should you listen to experts on these matters, or can you always decide for yourself?

6. How would you go about measuring cultural change? Moreover, how would you try to explain such change?

7. Is cultural change beneficial to society? Is it conceivable to have no changes in our culture?

recommended readings

Victoria D. Alexander. 2003. *Sociology of the Arts: Exploring Fine and Popular Forms.* Oxford: Blackwell.
This book is a clear, engaging, thorough, and sophisticated overview of this area of study.

Peter Berger and Thomas Luckmann. 1966. *The Social Construction of Reality: A Treatise in the Sociology of Knowledge.* Garden City, NY: Doubleday-Anchor.
This is a seminal work in the sociology of culture, laying the groundwork for social constructionist thought.

David Grazian. 2010. *Mix It Up: Popular Culture, Mass Media, and Society.* New York: Norton.
Grazian's book is a highly entertaining and accessible introduction to the sociological significance of the media and cultural production.

Wendy Griswold. 2008. *Cultures and Societies in a Changing World.* 3rd edn. Thousand Oaks, CA: Pine Forge Press.
Griswold's book lays out a systematic analysis of culture's reciprocal influence on other key sociological nodes, including producers, receivers, and distributors.

Eric Klinenberg. 2005. *Cultural Production in a Digital Age: The Annals of the American Academy of Political and Social Science.* **Thousand Oaks, CA: Sage.**
Taking a broad view of what counts as culture, this edited volume investigates how the technological advances of the digital age have influenced the methods and outcomes of cultural production.

Nelson Phillips and Cynthia Hardy. 2002. *Discourse Analysis: Investigating Processes of Social Construction.* **Thousand Oaks, CA: Sage.**
This short book provides a concise and insightful review of the theory and research in sociology and related fields on the role of discourse in social life.

Philip Smith. 2008. *Cultural Theory: An Introduction.* **2nd edn. Oxford: Blackwell.**
This book takes the potentially intimidating realm of cultural theory and explains, in a clear and helpful way, how culture has been conceptualized in different schools of sociological thought.

recommended websites

Canadian Broadcasting Corporation (CBC)
www.cbc.ca
In addition to finding the news, you will also find links to the corporate history of the CBC, the broadcasting entity charged with strengthening Canadian culture and identity.

Canadian Heritage
www.canadianheritage.gc.ca
There are many agencies within this federal government department actively involved in promoting the health of Canadian culture.

Canadian Radio-television and Telecommunications Commission (CRTC)
www.crtc.gc.ca
Here you'll find the public policy behind our broadcasting regimes. The CRTC is the quasi-independent agency responsible for regulating the entire broadcast industry.

Cultural Sociology
http://cus.sagepub.com
This journal was launched in 2007 to provide an outlet for academic work that specializes in taking a cultural approach toward sociological questions.

National Film Board of Canada (NFB)
www.nfb.ca
There are countless interesting links on the website of the NFB, which is especially renowned for its documentary and animated productions.

Poetics
www.sciencedirect.com/science/journal/0304422X
This is the electronic version of the journal *Poetics: Journal of Empirical Research on Culture, Media and the Arts.*

United Nations Educational, Scientific and Cultural Organization (UNESCO)
www.unesco.org
UNESCO deals with, among others, issues of cultural diversity and preservation.

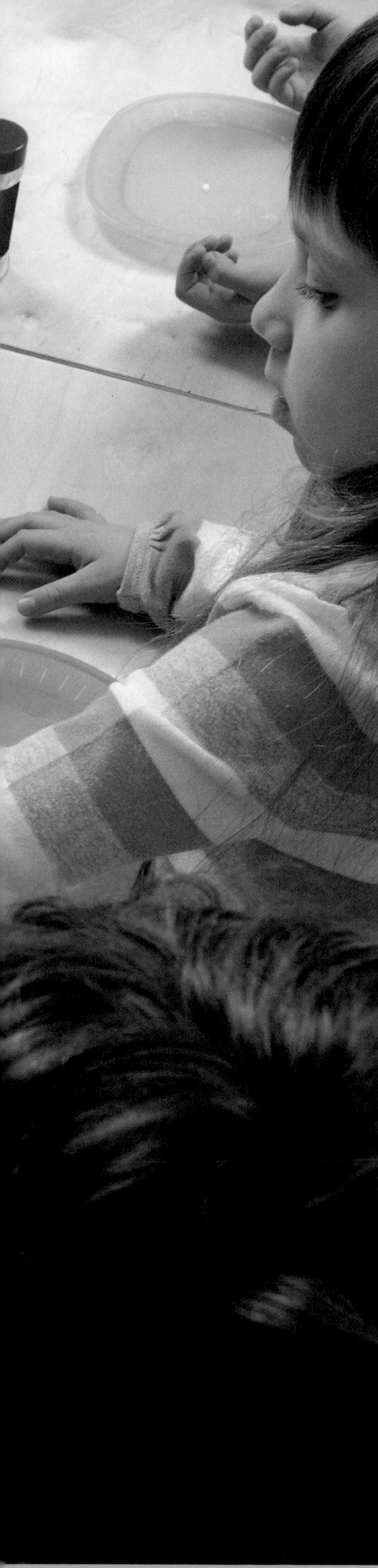

chapter 4

Socialization

Barbara A. Mitchell

In this chapter, you will:

▶ Comprehend basic patterns of socialization and what it means to be human;

▶ Situate the social experience within the nature/nurture debate;

▶ Understand key concepts and consider how they are applied in research and everyday situations;

▶ Learn how different theorists explain processes of socialization;

▶ Examine different agents of socialization, such as the family, the school, the peer group, and the mass media;

▶ Critically evaluate how socialization experiences are socially structured and vary by social class, gender/sexual orientation, ethnicity, generation, and geographical location;

▶ Explore how socialization is a reciprocal, dynamic, and lifelong process that changes over the life course;

▶ Gain appreciation of how early socialization experiences shape later family, work, and health-related outcomes.

Introduction: What Is Socialization?

In the recent Discovery Channel television documentary, 'Into the Universe with Stephen Hawking', the famous British scientist asserts that 'aliens are out there, but it could be too dangerous for humans to interact with extraterrestrial life.' Hawking further speculates that most extraterrestrials would be similar to microbes, or small animals, but that communicating with them could be 'too risky'. But imagine, for a moment, a slightly different scenario. Imagine that you arrive home one day to a neighbourhood that has been newly inhabited by creatures that are biologically identical to adult humans. Intelligent, cooperative, and peaceful (although occasionally combative), they do not recognize our language, or comprehend our customs, norms, or ways of interacting. Instead, they exchange thoughts sonically using suction-like pods, and they make intense eye contact with strangers. After setting up camp in the local park, they quickly begin to learn our language and engage in local community affairs. Over time, they internalize

and copy many of our behaviours, beliefs, and technologies, and most become accepted and productive citizens of our society.

Although this hypothetical situation reads like a scene out of a science fiction movie, it illustrates the process of **socialization**, which can be defined as the acquisition of knowledge, skills, and motivation to participate in social life. In other words, it is the process whereby individuals learn, through interaction with others, that which they must know in order to survive, function, and become members of our society. Moreover, socialization is not confined to babies and children but constitutes a complex, life-long learning process that enables us to develop our selves, roles, and identities.

Yet socialization is not a uniform phenomenon whereby we are all churned out by some kind of giant socialization factory. Although humans share many values and norms, differences are found by geographic region, ethnic/immigration background, gender, religion, and social class. Additionally, each generation experiences socializing effects particular to their birthplace and historical location. Growing up during the

The socialization of each generation is very different due to the social, economic, and technological environment. For instance, growing up during the Great Depression would present challenges unknown to most youths now.

© Everett Collection Inc/Alamy

Great Depression (1929–39) in Canada, for instance, would have been vastly different from what it is today, given the very different social, economic, and technological environments. Further, criminologists point out that socialization does not necessarily mean that what is learned is acceptable to the 'mainstream' or is positive for us. Take, for example, youth gang behaviour whereby a young adult self-identifies with a certain group (e.g., the Crips, the Bloods). Through interaction with other gang members, youths learn and display 'anti-social' norms and behaviours, such as participating in delinquent acts, crime sprees, drug trafficking, and slayings.

TIME to REFLECT

How have your opportunities in life been affected by your historical location (e.g., the generational time and place in which you were born)?

Human Behaviour— Nature or Nurture?

For more than a century, social scientists have argued over the relative contributions of biology and the environment to human development, popularly called 'the nature–nurture debate'. By way of example, consider your own musical abilities. Are you an awesome singer, or can you play a musical instrument such as a guitar or piano with ease and finesse? Do you think that your musical talent (or lack thereof) is the result of your biology (nature) or environmental influences (nurture)? If you lean toward a biological explanation, you might attribute your musical talent to the unfolding of 'hardwired' genetic factors. You might believe, for example, that you inherited strong music genes from your parents or from your Uncle Albert, who might be a professional musician. However, if you lean more toward an environmental explanation, you might focus more on the role of social forces in producing your musical talents. For instance, maybe you had lots of opportunities to take music lessons growing up, and your parents made you diligently practise what you learned.

Alternatively, you might argue that both sides of the debate have some merit. Indeed, recent advances in biology and genetics support the idea that biology and environment interact in dynamic ways to transform us into functioning members of society, each with our own unique sets of skills and abilities. In

In the First Person

Growing up in southwestern Ontario, I had the amazing opportunity to live in both urban and rural environments and to travel throughout the world. I found these diverse settings fascinating, because they opened my eyes to vastly different social worlds. I also learned to appreciate how our daily lives are interwoven by intricate webs of social networks and how norms, values, and family lifestyles can vary not only between communities and cultures but within them as well. As an undergraduate at the University of Waterloo, I naturally gravitated toward the discipline of sociology. For me, sociology offered a fresh, meaningful, and insightful way to make sense of and better the human experience. Consequently, it was an easy decision to pursue an MA degree at Waterloo and then a PhD (McMaster University) and to devote my entire professional career to teaching and 'doing' sociology—a decision I have never regretted.

—Barbara Mitchell

short, we may be predisposed toward certain abilities, but our environment will determine the extent to which these abilities can be realized.

One of the most exciting advances in this area is the emerging science of epigenetics, or the study of how the environment modifies the way that genes are expressed. The *New York Times* recently ran a series of articles on this topic, one of which reviewed Shenk's (2010) best-selling book *The Genius in All of Us: Why Everything You've Been Told about Genetics, Talent, and IQ Is Wrong*. In this article, Paul (2010) discusses Shenk's major thesis that our genes are constantly activated by environmental stimuli, nutrition, and so on. Consequently, there can be 'no guaranteed genetic windfalls or fixed genetic limits . . . instead, there is a continually unfolding interaction between our heredity and our world, a process that may be in some measure under our control.' From this lens, Shenk asserts that epigenetics introduces a 'new paradigm' that 'reveals how bankrupt the phrase "nature versus nurture really is".'

In summary, one useful way to conceptualize socialization is that it provides the link between biology and culture. We are born with the capacity to learn (e.g., music), to use language, and to forge social and emotional bonds, all of which are necessary for normal childhood development. However, our environment may limit (or facilitate) the extent to which innate gifts or propensities are realized. But what happens when our biological potential is not actualized? In order to examine this question more closely, let's consider a case study that illustrates this phenomenon—the effects of social deprivation on human development.

TIME to REFLECT

Many children throughout the world do not have the opportunity to grow up in a safe, secure, and loving environment. For example, they may be living in a war-ravaged country, or they may be forced into child-labour camps. How will these early experiences shape their later adult life?

THE CASE OF GENIE

Throughout history, reports have surfaced of young children who have been isolated from society and who, in one way or another, have lived in a 'wild' state. For example, one study compiled a list of 53 cases of isolated children, beginning with the Hesse Wolf-Child (purportedly raised by wolves) discovered in Germany in 1344 (Newton 2002). More recent examples include the cases of Isabelle, Anna, and Genie. Genie, for instance, was discovered at the age of 13 in her home in Los Angeles, California, on 4 November 1970 (Rymer 1993; Curtiss 1977). Tied to a potty chair during the day, she had been locked in a room alone for more than 10 years. At night she was often straitjacketed into a sleeping bag and placed in an over-sized metal crib.

At first, people were shocked to find out that Genie was 13 years old, since she weighed only 59 pounds and was just 54 inches tall. She could only understand a few words like 'stopit' and 'nomore', and she had a strange bunny-like gait. She was not toilet-trained, and she could not eat solid food. After she was placed in the Children's Hospital in Los Angeles, her mental and physical development started to improve immediately. She began to move more smoothly, and she was always eager to learn new words. A team of scientists (known as the Genie Team) began to work with her and wondered whether Genie would have a normal learning capacity. Notably, could a nurturing, enriched environment compensate for her horrible past? It appeared that this partially happened, since Genie's vocabulary grew significantly. Unfortunately, scientists were never able to fully answer this question. After five years, Genie's mother forbade the team from having contact with Genie because she claimed that their 'tests' constituted cruel treatment. Soon thereafter, Genie was sent to a series of foster homes (because her mother was unable to care for her), and her privacy has been protected ever since.

Although case studies such as this one are obviously extreme, research documents that even milder forms of social isolation can have profound effects for children later in life. Recent studies report that repeated social isolation (measured as a lack of social support and controlling for other factors) leads to poor psychological and physical health, such as an increased risk of cardiovascular disease (e.g., Grant, Hamer, and Steptoe 2009). These health problems are attributed to the cumulative 'wear and tear' caused by weak adaptation to stress. Fortunately, social connections (or social 'capital') are found to have a buffering effect on our ability to handle life's ups and downs, and this can positively influence how we age.

In conclusion, socialization is the essential bridge between the individual and society, and it is the process through which we become human. All of the evidence points to the crucial role of social experience in personality and healthy human development. Children need to be surrounded by people they trust, who care for them, interact with them, and can meet their basic needs. And although humans are resilient creatures, there is a point at which abuse, neglect, or social isolation (especially in infancy) results in irreparable developmental damage.

Theorizing Socialization

In this section, major theoretical approaches to socialization are reviewed. These approaches include: learning/behaviourist theory, Freud's psychoanalytic theory, developmental approaches (Erikson, Piaget, and Kohlberg), the symbolic interactionist view on the development of the self, functionalist and conflict approaches, and feminist theory (especially in relation to gender role socialization). Although there is sometimes overlap in basic assumptions or ideas, these approaches mainly differ in emphasis and conception of what socialization comprises and how learning occurs.

THEORIES ON CHILDHOOD SOCIALIZATION

Learning/Behaviourist Frame of Reference
Learning theory, which has its roots in behaviourism, assumes that the same concepts and principles

that apply to animals apply to humans. Although there are many variations of this theory, socialization as applied to the newborn infant involves changes that result from maturations that include classical or instrumental conditioning. Classical conditioning links a response to a known stimulus. A popular example is Pavlov's dog experiment in which a hungry dog is placed in a soundproof room and hears a tuning fork (a conditioned stimulus) before receiving some meat. After this situation is repeated several times, the dog begins to salivate upon hearing the tuning fork. The same principles are assumed to hold true with an infant upon hearing his or her mother's voice or approaching footsteps.

Operant or instrumental conditioning focuses attention on the response which is not related to any known stimulus. Instead, it functions in an instrumental manner in that one learns to make a certain response on the basis of the outcome that the response produces. According to Skinner (1953), it is the response that correlates with positive reinforcement or a reward. For example, imagine that a baby is picked up after saying 'da-da-da' because the father is convinced that the baby is saying 'Daddy'. Consequently, the baby begins to say 'da-da-da' all day long because there may be lots of rewards. As children grow older, different reinforcements (e.g., praise, candy, allowance) are used as deliberate techniques to teach children approved forms of behaviour.

While there is some usefulness in this theory, there may be limited applicability in generalizing animal behaviour to socialized humans. Humans, unlike animals, have the capacity to share symbolic meanings and symbols in ways that animals cannot.

Psychoanalytic Frame of Reference

Developed by Sigmund Freud (1856–1939) and his followers, psychoanalytic theory stresses the importance of childhood experiences, biological drives and unconscious processes, and cultural influences. Beneath the surface of each individual's consciousness are impulsive, pleasure-seeking, and selfish energies that Freud termed the 'id'. Individuals also have 'egos' and engage in cognitive, conscious thought processes that make each one of us a unique individual. Both the 'id' and the 'ego' are controlled by the individual's gradual internalization of societal restraints (the 'superego'). Parents play a key role in 'impulse taming' by transmitting cultural values and rules that guide the ego and repress the id.

Accordingly, socialization consists of a number of stages of development that occur and are called

the *oral*, *anal*, and *phallic stages*, followed later by a period of *latency* and then a *genital phase*. When individuals pass through all of these stages, it culminates in a healthy, mature personality with a well-developed superego, which channels libidinal forces in appropriate directions. However, Freud theorized that it is also possible for one to remain in one stage for an inordinate length of time (fixation) or return to an earlier stage (regression) and that this can be the source of inappropriate or problematic social behaviour (Freud 1938).

Freudian ideas have received mixed empirical support. Practices such as breastfeeding and bowel and bladder training (which have been so strongly emphasized in the psychoanalytic literature) are found to be negligible in terms of how they affect personality development. Moreover, it should be noted that Freud's work was highly controversial at the time, particularly since he often focused on sexuality during a repressive Victorian era. Indeed, Freud thought that sexuality was a primary motivating force not only for adults but also for children—an idea that sparked public outrage at the time.

Child Development Frames of Reference

Similar to Freud, both Erikson (1963; 1982) and Piaget (1932; 1950) emphasized the early stages of childhood development. Unlike Freud, both extended their stages beyond the early years and focused more attention on social structure and reasoning. Erikson, who was one of Freud's students, viewed socialization as a lifelong process. He developed the 'eight stages of human development', which range from *trust versus mistrust* (first year of infancy) to *integrity versus despair* (old age). As individuals create solutions to developmental concerns, those solutions become institutionalized in our culture. Swiss social psychologist Jean Piaget, who wrote during the 1920s, was also interested in maturational stages. However, his interest was more in cognitive development, characterizing this as the ability to reason abstractly, to think about hypothetical situations logically, and to organize rules into higher-order, complex operations or structures.

Piaget also developed four major cumulative stages of intellectual development, which include the sensorimotor period (birth to two years), preoperational period (two to seven years), concrete operational period (seven to 11 years), and formal operational period (age 11 through adulthood). In Piaget's view, children develop their cognitive abilities through interaction with the world and adaptation to their environment. They adapt by assimilating, which means making new information compatible with their understanding of the world. Children also learn to accommodate by adjusting their cognitive framework to incorporate new experiences as they become socialized into adults.

Kohlberg (1975; 1969) expanded on Piaget's ideas with his stages of moral development. His ideas were based on his research in which children were presented with moral dilemmas that asked what they would do and why they would do it. In one such dilemma, a man's wife was dying, and the druggist (who had invented the only medicine that could save her) was charging 10 times what the medicine cost to produce, a cost that was also considerably higher than what the husband could afford. Subjects were then asked whether the husband had the right to steal the drug and why, the latter of which was of the greatest interest to Kohlberg.

Subject responses clustered into three general levels of moral reasoning, each of which could be subdivided into more specific stages. Kohlberg found that each of the stages generally occurs in a set sequence and that many people do not develop beyond more advanced stages. In the earliest stages, children say 'it's wrong to steal' and 'it's against the law', but they are unable to elaborate any further. By Stage 4 (the most prevalent stage, usually reached as children mature), moral decisions are made from the perspective of society as a whole, since we think from a full-fledged member-of-society perspective. Few people (except for great moral leaders such as Gandhi and Martin Luther King) ever reach Stages 5 or 6. Notably, Stage 6 is the theoretical stage in which people live by principles based on human rights that transcend government and laws and that endorse civil disobedience (Crain 1985).

In summary, so far we have reviewed frames of reference that emphasize overt behaviour (i.e., behaviourism, learning theory), the unconscious role of motives and emotions (i.e., the Freudians), and motor skills, thought, and moral reasoning processes (i.e., child developmentalists). Next we will review perspectives that shine the spotlight on societal influences on socialization.

Symbolic Interactionist Frame of Reference

In sociology, the symbolic internationalist (SI) perspective has had one of the greatest influences on theories of socialization. Central importance is placed on interactions with others and the internalized definitions, meanings, and interpretations of our interactions (Charon 1979; Mitchell 2009). Basic assumptions include:

1. *Humans must be studied on their own level.* Social life involves sharing meanings and communicating symbolically via language, which enables humans alone to deal with events in terms of past, present, or future.

2. *An analysis of society is the most valuable method in understanding society.* Individual behaviour needs to be contextualized within the structure of society. When one is born into a given society, one learns the language, customs, and expectations of that culture. Thus, behaviour that is appropriate in one culture (e.g., spanking a child or kissing in public) may not be appropriate in others.

3. *At birth, the human infant is asocial.* Newborns are born with impulses and needs and with the potential to become a social being. Behaviours and expectations do not begin to take on meaning until babies begin to learn to channel their behaviours in specific directions via training and socialization from their parents.

4. *A socialized being is an actor as well as a reactor.* Humans do not simply react to one another in robotic fashion. Rather, humans are minded beings, actively responding to a symbolic environment that involves responses to interpreted and anticipated stimuli. In this way, they can feel guilt over past behaviours, assess new ways of responding, and dream of future possibilities.

Another key concept is the idea of the development of a social self, which takes place in interaction with others. For example, a young adult may occupy the status of child, student, sister, athlete, and many others. These statuses have expectations (roles) assigned to them and are organized and integrated into the social self. In this way, the social self is never fixed, static, or in a final state. Family members play an important role not only in the development of the social self but also in feelings of self-worth.

Of central importance are the roles of **significant others** and reference groups. Although parents are usually the most significant socializers, other people or groups are important. These individuals can be other family members or even role models presented in the media, such as pop star sensations like Justin Bieber or Lady Gaga. These significant others influence behaviour by what they do and what they say, and every generation is characterized by its own set of role models. Many media sources, for example, report that recently there was a steady parade of boys asking for the 'Bieber bob' at barbershops. Apparently, it was the choice cut for male teens and 'tweens' who wanted to emulate teen idol Justin Bieber's 'side sweep surfer look' in order to attract girls. Reference groups, on the other hand, constitute a source of comparison that operates in a similar fashion. These groups serve as a point of reference and standard for conduct, such as a religious group, a hobby club, peer groups, or a company (e.g., Boy Scouts, Apple).

TIME to REFLECT

Who are your most significant role models? How do they influence what you say and do? How do they contribute to your feelings of self-worth and your self-identity?

Popular SI theorists include George Herbert Mead (1934) and Charles Horton Cooley (1902; 1962). To Mead, social, not biological, forces are the primary source of human behaviour. He maintained that a newborn baby is *tabula rasa*, or a 'blank slate',

OPEN for DISCUSSION

Are Athletes Still Role Models? Should They Be in the Face of Recent Issues?

Tiger Woods was undoubtedly one of the most successful and popular athletes of all time—that is, until November 2009 and the scandal that rocked the sports world.

Woods had crashed his car near his Florida mansion, and what followed were accusations that the famous golfer had been unfaithful in his marriage, to which he eventually confessed and apologized.

Reports of infidelity and bad behavior by our celebrity and sports 'heroes'—such as those surrounding Woods and, more recently, former Miami University and current Pittsburgh Steelers quarterback Ben Roethlisberger, who was suspended after allegations of sexual abuse—are nothing new.

However, talking with your kids about the stories they see on television and read in papers and magazines is important, especially when these figures are sometimes looked up to as role models.

'I think that honesty is really the best policy,' said Lisa Polk, guidance counselor at Marietta Middle School. 'I also think it is best to let [the kids] ask the questions and then ask "what is your opinion" and "what do you think about it."'

While parents undoubtedly have their own opinions about these types of stories, Polk thinks it is probably best to let the kids take the lead and express their feelings.

'My son's attitude was why would someone mess up everything they have and do something like that,' said Lisa Weekley, 43, whose 14-year-old son plays baseball and basketball and plans to study to become a dentist.

'He doesn't really think about [sports figures or celebrities] as role models. He loves LeBron James, but I don't know [if something negative came out about James] if it would affect him.

'My son plays baseball and basketball because he enjoys it, but he is realistic.'

At a press conference in February, Woods apologized for his behavior and admitted that he felt he had earned the right to act inappropriately.

'I thought I could get away with whatever I wanted to,' Woods told reporters. 'I felt that I had worked hard my entire life and deserved to enjoy all the temptations around me . . . thanks to money and fame, I didn't have to go far to find them.'

Unfortunately, Polk believes this sends the wrong message to kids.

'They might feel like they can get away with this behavior. It just spreads this sense of entitlement,' she said.

As far as both Polk and Weekley are concerned, role models can come from all aspects of life.

'I think someone who is genuinely a nice person,' said Polk. 'Princess Di comes to mind because of her humanity. She had problems in her life, but a lot of people looked up to her.'

For Weekley, her role models were a little closer to home.

'I try to teach my kids to be involved in things like band and sports because camaraderie is important,' she said. 'My parents were always involved in everything we did and guess I'm just trying to follow that.'

SOURCE: Erin O'Neil, 2010, 'Are athletes still role models?: Should they be in the face of recent issues?' *Marietta Times* 24 May. www.mariettatimes.com/page/content.detail/id/522107.html

without predisposition to develop any particular type of personality. Mead referred to this spontaneous and unsocialized self as the 'I'. Through interaction, our personalities develop and the socialized ('Me') self emerges. Mead also asserts that although the 'Me' becomes predominant with socialization, the 'I' continues to exist and that it can be the source of unpredictable or 'untamed' social behaviour.

Another central concept is his notion of **generalized other**. According to Mead, children usually pass through three stages in developing a full sense of self-hood: the play stage (whereby the child models significant others); the game stage (whereby children pretend to be other people); and finally, the generalized other stage (learning generalized values and cultural rules). This final stage signifies how individuals become consistent and predictable in their behaviour and how people learn to view themselves from the perspective of others. Thus, behaviour results less from drives and needs, unconscious processes, and biological forces and more from social interaction processes and internalized meanings of self and others.

FUNCTIONALIST AND CONFLICT PERSPECTIVES

While many theories of socialization describe and analyze the *process* of socialization, functionalist and conflict theories place emphasis on understanding the *role* and the *importance* of socialization. A functionalist approach addresses the ways in which conformity helps to preserve and meet the needs of society. It does so by providing knowledge that is passed from generation to generation, which helps society to survive and meet the demands of its environment. Through cultural transmission, values and norms that are widely shared in society (e.g., trust) are critical to solidarity and cooperation, although it is also recognized that society needs to adapt to new conditions and situations. Thus, the fundamental task of any society is to reproduce itself such that the needs of society become the needs of the individual (Newman 2006).

Conflict perspectives (which often have roots in Marxist theories) focus more on issues of power and control and how socialization helps the powerful and wealthy pass on their advantages to the next generation. Socialization does so by supporting ideologies and practices that work to the advantage of dominant groups and by social channelling. It teaches people to accept, rather than to challenge or question, the status quo or ways of society. This creates a 'false consciousness', or a lack of awareness and a distorted perception of class realities (Pines 1993). In this way, children are prepared for future societal roles, and gender, class, and racial inequities are reproduced. Notably, poor children are channelled toward a life of poverty through the educational and employment system. Gender role socialization also prepares women and men for different and unequal roles in society, a topic to which we now turn.

FEMINIST THEORIES AND GENDER ROLE SOCIALIZATION

Although one unified feminist theory does not exist, feminists often critique functionalist views and build upon conflict theorizing by emphasizing how gender is a fundamental organizing feature of social life. Focus is often placed on how social interaction, including discourse (i.e., the usage of language and symbols), is socially constructed and on how gender-role socialization mirrors and perpetuates inequities found throughout society. Gender roles refer to the expectations associated with being masculine or feminine, which may or may not correspond with one's sex, whereas sex roles can be defined as the expectations related to being biologically of one sex or another.

There is little denying that gendered divisions are found in virtually all societies. For many centuries, it was assumed that 'anatomy is destiny' and that these differences were largely innate or inborn. However, many feminists assert that this belief provides a major (functionalistic) ideological justification for a system of stratification that subordinates women and privileges men. Instead, we must also consider socialization processes and the organization and practices of society. Fundamental differences in gender-role socialization and stereotyping continue to exist, and this begins at birth and continues throughout one's life. This is seen in the sexual objectification of girls in society and the media (see Stankiewicz and Rosselli 2008) and the kinds of gender-specific toys and games that continue to be manufactured, marketed, and bought for children.

Indeed, a campaign called 'Pinkstinks' was recently set up in the UK against the marketing of pink toys to girls. The founder of the campaign, Emma Moore, believes that the rows and rows of pink toys, clothes, and other items aimed at girls encourages an obsession with looks and body image, which creates gender inequality. Overall, gender-specific toys are thought to normalize and perpetuate activities directed toward appearance, romance, and the (unpaid) work realm (i.e., home) for girls. Conversely, 'rugged' and aggressive action-oriented activities directed away from the home and toward public life are encouraged for boys (Greenglass 1982).

Although there has been some positive social change in this regard, many feminists observe that progress has been slow. 'Traditionally' (in the 1950s, for example), little girls were often given dolls, sewing machines, and makeup, while little boys were often presented with toy guns, action figures, cars, and games that contain violent content—practices that remain popular. And while the ever-popular Barbie doll (introduced by Mattel in 1959) has evolved to represent the 'career woman' as well as non-white ethnic identities, many feminists continue to bemoan Barbie's unrealistic connotations, cultural ideals that, they claim, are still imposed on little girls by a Western patriarchal society (Mitchell 2009).

Another example of how social institutions are highly gendered is our educational system, which also plays an important role in the formation of gender identities through curriculum and its local culture. Connell (1996) maintains that each school has its own 'gender regime', which contributes to the ongoing negotiation and renegotiation of femininity or masculinity. A school's style of dress can act as a powerful signifier of social acceptability and expression of identity, as well as a signifier of fashion that separates 'the girls from the boys'. For instance, at some schools, 'the look for boys' is to appear somehow connected to sports, athleticism, strength, and power, and this becomes the hegemonic norm (Swain 2004).

Fox (2001) reveals other structural sources of gender differences and how they can resurface beyond childhood (i.e., in young families). Her research uncovered how the transition to parenthood can produce a more conventional division of labour in the home. Gender inequity arises out of the gendered division of paid and unpaid work, and these conditions further shape and constrain options and behaviours. In today's society, a shortage of outside community supports and the privatization of parenthood also mean that women continue to have the ultimate responsibility for their babies' welfare. This creates women's dependence on men or other family members (e.g., on their own mothers), and this further strengthens gendered divisions.

In short, gender socialization does not end in childhood. It continues (and can even deepen) through certain institutional practices and discourses that produce gendered adults and identities. As a result, despite feminist efforts over the past half-century to challenge gender role socialization and conventional gender divisions, many inequities persist. And

while many couples negotiate the changes in their lives, women in a materially strong position (whose bargaining power tends to be relatively good before motherhood) may be better able to resist dynamics that place them in an unequal position within the family (Fox 2001).

TIME to REFLECT

In Naomi Wolf's best-selling book *The Beauty Myth* (first published in 1990), she argues that media images place a great deal of pressure on women to conform to an impossible standard of physical perfection. Do you agree or disagree? Have times changed at all since she wrote the book?

From the proceeding discussion, it is clear that families are a **primary agent of socialization** but that other (secondary) agents, such as the educational system, also contribute to socialization processes. To further illustrate this point, parents may be significant role models for children and can influence whether children will have drug abuse problems. Yet other agents of socialization at the peer group, school, and community/societal level can simultaneously exert risk and protective factors in the process, as depicted in Figure 4.1. Thus, various agents, in conjunction with individual-level factors (e.g., peer resistance skills) can both complement and compete with one another to produce certain behavioural outcomes such as problem drug use.

In the next section, we will examine more closely the role of families in socialization, followed by the **secondary socialization** that takes place in the wider society. We will consider how these secondary agents—school, peer group, and mass media/technology—differentially shape our social experiences by variables such as social class, ethnicity/cultural background, gender/sexual orientation, and age.

The Family

There is little denial that during infancy and childhood, our family constitutes the most significant agent of socialization. Families provide the primary source of our early emotional attachments and learning, although other agents (e.g., daycare, the mass media) also shape children's basic beliefs and values. Families also play a critical role in transmitting culture from one generation to the next. Through

▼ **FIGURE 4.1**   Socializing Agents and Risk/Protective Factors for Substance Abuse

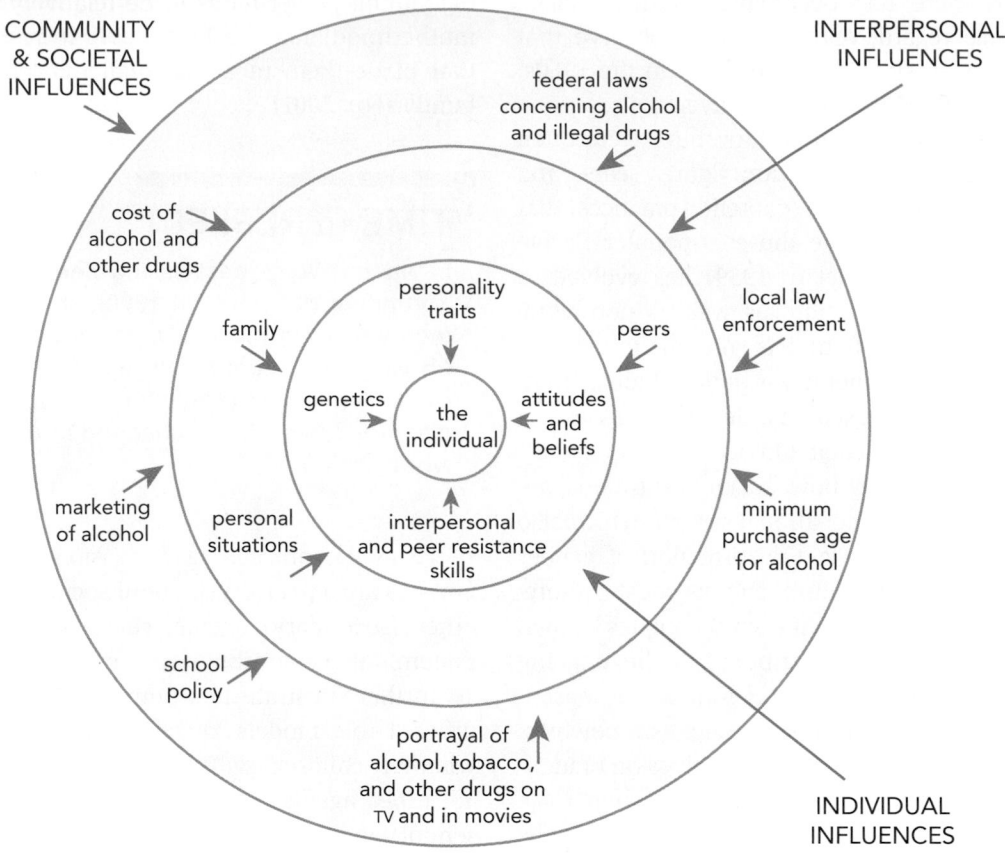

SOURCE: Image taken from 'Educating students about drug use and mental health—Risk and protective factors', retrieved 23 July 2010, from www.camh .net.education/Resources-teachers_school.

countless family rituals (e.g., birthday celebrations, graduation ceremonies) and activities, children are taught to reproduce social patterns and behaviours familiar to adults. Families are also sites of power and control relations prevalent in wider society, in addition to having their own hierarchies based on aspects such as age and gender.

Experiences within contemporary families are very different from what they were in the past. Although family life has never been homogeneous, changes in family structure (e.g., the rise in step- and single-father families), transformations in work and greater gender role equality (e.g., dual-career households), and continuing high rates of immigration contribute to the diversification and experience of 'family'. Despite these changes, much continuity in family life exists. Notably, as shown in Figure 4.2, most Canadian children still grow up in two-parent households. Bibby (2009) also finds that most young people aspire to marry (for life and formalized with some type of religious ceremony) and want to have children.

Regardless of the continuation of many family practices, our experiences are variable, and they occur within unique social/ecological contexts (Wilson 2008). Intra-familial factors, such as the age of the parents when a child is born, the number of other children in the family, and the types of social support received from others, can profoundly influence socialization. Extra-familial factors, such as the neighbourhood in which the family lives, work/employment experiences, social class background, and culture, also play a role in these experiences (Albanese 2009). Growing up in a one-child Jewish family household in Toronto, for example, would be considerably different from growing up in a large Ukrainian farm family in rural Saskatchewan. Indeed, family size can also influence socialization experiences, since siblings can play a part as socializing agents, for example, as role models or as baby-sitters.

Further, social class and ethnic/cultural background can influence the kinds of life chances and values that are being transmitted to children. The classic work conducted by Kohn (1969) concluded that working-class parents tend to stress conformity in their children, whereas middle-class parents model a wider range of behaviour and are more likely to encourage their children to attend university. Parents at a higher socio-economic level are also more likely to provide opportunities and resources for learning, such as through travel and exposure to more 'cultured' lifestyles and activities, which are positively rewarded by the school system. Thus, it is not surprising that the enrolment rates in university are higher among children from higher social classes than among children from lower ones.

Ethnic and immigration background also shape socialization experiences. Many cultural groups value and emphasize high educational and employment achievement. This means that parents inculcate educational expectations and place a great deal of pressure on children to succeed in school and work life. Dimensions of cultural heritage can also produce norms, values, and obligations pertaining to many family roles—for example, with respect to how we treat and care for our senior family members.

Social scientists typically view socialization as a reciprocal or two-way process. Daly, for example (2004, 5), explains that it would be a mistake to think of socialization in terms of a simple linear transmission model whereby culture shapes parents and parents in turn shape children. A more accurate depiction is that of **reciprocal socialization**, or the simultaneity (whereby parents and children are the 'players'

in this case) of parenting and child behaviours and outcomes. For example, Ambert (1997) observes that despite what parents expect their infants to be like, each baby requires different types of care. The type of care consequently influences parents' perceptions and experiences surrounding the 'easiness' of their baby. For instance, 'easy' babies make parents feel happy, adequate, and successful, whereas more 'challenging' babies make parents feel inadequate, stressed, and ineffective. And while all of this is unfolding, parents' own lives and the larger social environment interact to create their attitudes and socialization practices.

The Peer Group

The peer group is commonly regarded as the second most potent socialization agent. The age-grading of our society means that we are often placed in situations with people of similar age and characteristics—for example, we are often segregated by age group in neighbourhoods, schools, and recreational/ leisure settings. However, unlike our families, friends typically do not consciously intend to socialize us. Through our interactions with peers, and because of our need for companionship and approval, there is a mutual learning of information, attitudes, and values. Interactions with friends also allow children to begin to separate themselves from the family's all-encompassing influence. In this way, peers may serve as a source of comparison, and they can influence our preferences for certain kinds of activities.

▼ **FIGURE 4.2** Families with Children, by Family Structure, 1981 and 2006 (%)

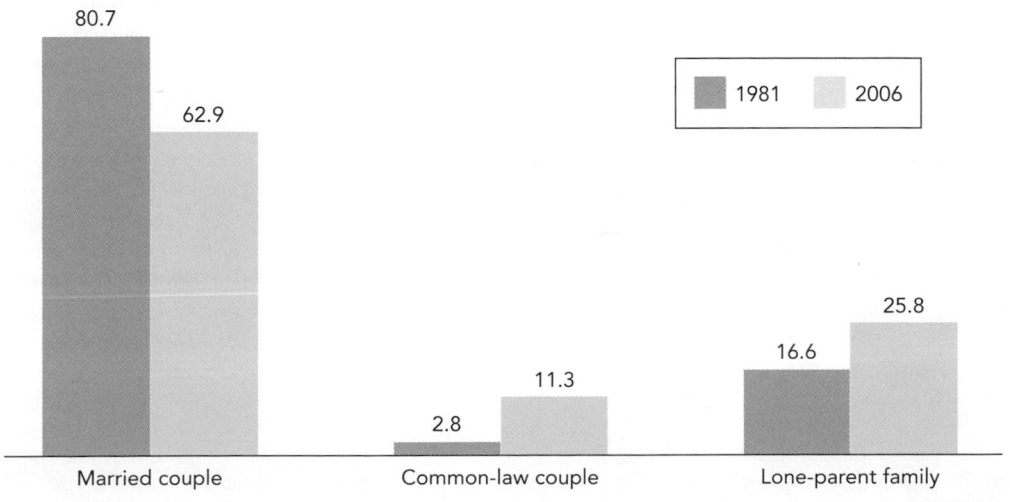

SOURCE: Statistics Canada. 1981, 2006 Census. www4.hrsdc.gc.ca/.3indic.lt.4r@-eng.jsp.

Although friendships are of central importance throughout the life course, adolescence is a time when peer influences are particularly strong. Youth groups and the culture that develops around them form a foundation for creating unique subcultures or cliques. In contrast to the adult-dominated world, this allows youths to establish their own identity, with its own norms, rules, and regulations. Peer groups also provide a setting for engaging in and trying new behaviours (often without adult supervision), which can have both positive and negative socializing aspects. With respect to other positive benefits, friends may serve as important companions and sources of social support.

However, youths may engage in behaviours that parents might not approve of, such as drinking and doing drugs or engaging in certain sexual activities. This is no doubt why many parents have long expressed great worry about who their children's friends are. Peer groups can rival parents in influence, as implied by the familiar phrase 'the generation gap'. This can create some level of intergenerational conflict or tension. Yet research tends to conclude that despite some differences in values and attitudes, peer influences tend to be relatively short-term, and children remain strongly influenced by their families.

Recent studies (see Tirone and Pendlar 2005; Karakayali 2005) on immigrant youths also reveal multiple or 'hybrid' identities of youths and how identities are flexible and subject to multiple influences. A good illustration focuses on the experiences of immigrant youths who have to negotiate their identities amid sometimes conflicting environments. In a Canadian study of the narratives among second-generation youths in their early 20s (whose families originated in India, Pakistan, and Bangladesh), most

SOCIOLOGY in ACTION

Project Teen Canada and the Emerging Millennials

Project Teen Canada has surveyed more than 5000 Canadian teenagers every eight years since 1984. These surveys examine a wide range of topics of relevance to the everyday lives of teens, such as their values, attitudes, and behaviour. Study findings are made available to all participating schools and to the general public, as well as to a wide range of government departments, educators, and youth workers.

In a recent book entitled *The Emerging Millennials: How Canada's Newest Generation Is Responding to Change and Choice* (2009), sociologist and author Reginald Bibby (University of Lethbridge) summarizes Project Teen Canada 2008. These data came from a sample of 5564 teenagers (aged 15 to 19) who were in secondary schools. A special supplemental over-sample of more than 800 students in Aboriginal schools was also included.

A unique feature of the book is that educator Ron Rolheiser offers responses to each chapter, along with Sarah Russell, an RCMP community relations officer and a former Project Teen Canada research associate. The book also examines the impact of the baby-boomer legacy and compares this generation with those born since the 1980s.

Other notable social trends include:

- Top-rated interpersonal values: (1) trust (84 per cent); (2) honesty (81 per cent); (3) humour (75 per cent).
- Sources of influence: Virtually all (92 per cent) of teens perceive that 'the way that they were brought up' has influenced their lives 'a great deal' or 'quite a bit', followed by parents (especially moms), friends, characteristics born with, other adults, music, reading, teachers, television, god/some supernatural force, luck, the Internet, what people in power decide, and advertising.
- Friendship is everything: 72 per cent of teenagers report having four or more close friends; only 1 per cent report having 'no close friends'.
- Keeping in touch (daily): cellphone (54 per cent); text message (44 per cent); Facebook (43 per cent); email (42 per cent); YouTube (27 per cent).
- Let's talk about sex: 56 per cent state that they have never had sex; the highest rates of sexual activity were found in the North and in Quebec.
- Losing their religion: Quebec teens (22 per cent) were the least likely to 'definitely' believe in God, compared to 41 per cent in the rest of Canada, although 54 per cent of Canadian teens acknowledge that they have spiritual needs. Some 85 per cent also say they expect to have a religious wedding ceremony.

SOURCE: Selected material extracted from Project Teen Canada website (www.ptc08.com) and Reginald W. Bibby with Sarah Russell and Ron Rolheiser, 2009, *The Emerging Millennials: How Canada's Newest Generation Is Responding to Change and Choice*, University of Lethbridge: Project Teen Canada Books.

youths had a profound appreciation of their family heritage even as they participated in 'dominant' Canadian youth culture. These findings challenge the prevailing discourse that children of immigrants are a 'problem group' living in 'two worlds'—caught between their 'old' culture and the mainstream (Tyyskä 2009).

In addition to age and ethnic background, peer networks are socially structured by other sociological variables such as gender and social class. Research on girls' friendships uncovers a set of social rules, including 'reliability, reciprocity, commitment, confidentiality, trust, and sharing' (Hey 1997). Tensions are also observed as girls balance the ethics of friendship with both vying for social position and responding to the male gaze. It is further noted that the ethical rules and foci of girls' cliques vary according to their social class. 'Niceness' or getting along with everyone is associated with middle-class femininity, whereas working-class girls tend to present a tougher image and manage their conflicts more openly (Aapola, Gonick, and Harris 2005; Tyyskä 2009).

Schools

As a key socializing agent, schools provide a social environment that is separate from the family, and teachers and schoolmates widen our early learning and experiences. Consequently, a major socializing dimension of schools encompasses its many informal and social elements, including interactions among students and between teachers and students. Moreover, on a more macro level, the school plays an important role in political socialization, since it inculcates children with the basic beliefs and values of their society. The educational system provides an allocation function as it channels students through programs of occupational preparation into various positions in the socio-economic and labour structure of society.

In modern society, the role of the school in the socialization process has become even more pronounced. It is relatively common for children under the age of five to attend some type of daycare or preschool program. According to the Canadian Council

(© Christopher Futcher)

The classroom provides a unique opportunity for socialization separate from the family. Here, interactions between students widen our early learning and experiences.

on Learning, these experiences provide cognitive, language, socio-emotional, and motor learning that can significantly affect future development. Further, the amount of specialized technical and scientific knowledge required to participate in society has expanded well beyond what parents can teach in the home. Youths in Canada and other industrialized countries are more educated than their parents' or grandparents' generations, and a post-secondary education is now an expected part of many young people's lives. And while critics note that this trend has delayed the transition to adulthood and created greater dependency on parents for economic and housing support, benefits are also documented. Living at home with parents, for instance, can help young adults while attending school and create more peer-like intergenerational socialization experiences (Mitchell 2006). Young women also have more options in life as they become increasingly highly educated.

Yet as previously mentioned, not all Canadian children have the opportunity to achieve higher levels of education, since family and cultural background and gender have a significant effect on educational experiences. A recent report on Toronto high school students showed alarmingly high dropout rates among black and Aboriginal youths, as well as those with Portuguese, Hispanic, and Middle Eastern backgrounds (Brown 2010). There is also concern over the sizable inequalities in educational pathways of First Nations youth, which can be traced to a legacy of colonization, marginalization, and discrimination. Overall, as presented in Table 4.1, racialized youths, visible minorities, and immigrants (with the exception of Asian students) are less likely to attend college and university (Thiessen 2009).

It is argued that the continuation of unequal educational patterns and experiences lies in the institutionalization of classism, racism, and sexism. Tyyskä (2009) asserts that this is manifested in a pro–middle-class and Eurocentric male mentality that dominates the educational system. One aspect of this is the 'hidden curriculum', which refers to implicit messages in education that may not be consciously taught or planned. These messages emphasize 'dominant' societal values (such as competition and that our society's way of life is morally good) and social hierarchies based on social class, gender, race, and sexual orientation. The hidden curriculum also perpetuates certain attitudes, values, norms, and practices that create a cultural and social ethos that prevents the full and equal participation of subdominant groups. These practices might include course streaming (e.g., gender tracking), lack of teacher's positive attention, biased testing procedures, and discriminatory practices by guidance counsellors.

In contemporary schools, other pressing issues of concern with respect to student social interactions and how they affect learning and identity include sexual and gender harassment, bullying, and school violence. Bullying, harassment, and social isolation

TABLE 4.1 ▼ Educational Pathways of Population Groups (%)

	No Post-Secondary	Community College	University	N*
Canadian-born				
European	28	35	37	16,342
First Nations	50	29	21	573
African/Latin American	36	41	23	209
East Asian	11	11	78	185
Other Asian	17	26	58	347
Immigrant				
European	28	29	43	406
African/Latin American	31	41	28	121
East Asian	18	14	68	198
Other Asian	34	32	34	331
Total	28	34	38	18,712

*N = unweighted sample size; further methodological details can be found in Statistics Canada, 2006, 'Youth in transition survey (YITS) Cohort B—20–22-year-olds cycle user guide' (Ottawa: Statistics Canada).

SOURCE: Adapted from Victor Thiessen, 2009, 'The pursuit of postsecondary education: A comparison of First Nations, African, Asian, and European Canadian youth', *Canadian Review of Sociology* 46 (1): 5–40 (Table 1, p. 11).

are more commonly experienced by children who differ from the social norm or are perceived as different. 'Gay bashing' and the use of homophobic language, for example, are highly prevalent in many schools. Studies (see Berlan et al. 2010) also reveal that sexual minority students (or teens who self-identify as gay, lesbian, or bisexual) are bullied two to three times more often than heterosexuals. They are also more likely to commit suicide and have other health-related problems than those who self-identify as heterosexual.

TIME to REFLECT

Why do children bully other children? Consider how various socializing agents play a role in this behaviour. What do you think might prevent children from bullying or from being bullied? For example, should we specifically target 'at risk' children, their families, or the school system?

Mass Media

Television, computers, newspapers, radio, magazines, and entertainment such as movies are readily available and constitute another powerful source of socialization. The media environment experienced by children today is vastly different from the one their parents or grandparents faced. As noted by Strasburger, Wilson, and Jordan (2009), terms such as *digital television*, *gangsta rap*, and *Google* did not even exist 20 or 30 years ago. Children can now participate in a much wider range of media-related activities than ever before, including online social networking and shopping, texting, video on demand, and viewing digitally recorded photographs and home movies. Overall, the sheer proliferation of media outlets and technologies and the amount of time people spend exposed to these outlets has risen dramatically.

Although the adoption of social media has been rapid and widespread, rates and types of usage can vary among socio-demographic groups. As shown in Figure 4.3, children aged 8 to 10 are exposed to the media an average of 7.5 hours a day, compared to 11.5 hours for those aged 11 to 14. Other research shows that Internet usage is especially prevalent among younger age groups compared to those aged 65 and over. Further, Canadian adults aged 55+ are more likely to watch television every week compared to those aged 18 to 34, and those with a university

education watch significantly less television per week than those with less education (Ipsos Reid 2010). Moreover, gender and linguistic differences affect the propensity to engage in certain media activities. For instance, females are more likely to use Facebook than males, and French-speaking Canadians are less likely to use social networking sites than English-speaking Canadians (Dewing 2010).

Social scientists observe that media and technology influences have always been controversial and have both positive and negative aspects. On the positive side, the media can be educational, informative, and entertaining and provide new avenues for social interaction. Parents who are physically distant from their children can interact and 'visit' their children via Internet software programs or 'keep tabs' on their whereabouts and safety through cellphones and text messaging. Friendships and social support groups can also be formed and maintained via social network sites on the Internet.

However, there is concern that modern-day media technologies exert too much control over our daily lives and that they contribute to unhealthy behaviours. By way of illustration, a popular video game series called *Grand Theft Auto* encourages the player to take on the role of a criminal in a big city and engage in numerous illegal activities, such as killing police and military personal. Women are typically depicted as prostitutes and men as violent thugs—stereotypes that appear in other media venues (Dill and Thill 2007). Other critics link media exposure to the significant rise in childhood obesity and (at the opposite of the spectrum) to eating-related disorders such as anorexia nervosa and bulimia. It is argued that unhealthy images influence behaviour through advertising (e.g., foods with little nutritive value) and because of celebrity role modelling (e.g., stick-thin actresses and models).

There is also concern that our society is becoming too celebrity-obsessed. Magazines (e.g., *inTouch*), television shows (e.g., *TMZ*), and Internet sites (e.g., Perez Hilton) devoted to the latest gossip on the rich and famous are clearly on the rise. Yet while it is purported that changing media technologies facilitate better access to tabloid stories, the mainstream media have been 'serving up gossip and sexual scandal since ancient Rome' (Hume 2010). Gamson (2001) also takes a critical view by showing how mainstream media seek to frame scandals as more serious philosophical issues in an unconscious attempt to differentiate themselves from the tabloids, bloggers, and Web-based celebrity news sites that

UNDER the WIRE

The Net Generation, Unplugged

Technology and society: Is it really helpful to talk about a new generation of 'digital natives' who have grown up with the Internet?

They are variously known as the Net Generation, Millennials, Generation Y, or Digital Natives. But whatever you call this group of young people—roughly those born between 1980 and 2000—there is widespread consensus among educators, marketers, and policy-makers that digital technologies have given rise to a new generation of students, consumers, and citizens who see the world in a different way. Growing up with the Internet, it is argued, has transformed their approach to education, work, and politics.

'Unlike those of us a shade older, this new generation didn't have to relearn anything to live lives of digital immersion. They learned in digital the first time around,' declare John Palfrey and Urs Gasser of the Berkman Center of Harvard Law School in their 2008 book *Born Digital*, one of the many recent tomes about digital natives. The authors argue that young people like to use new, digital ways to express themselves: shooting a YouTube video where their parents would have written an essay, for instance.

Anecdotes like this are used to back calls for educational systems to be transformed in order to cater to these computer-savvy students, who differ fundamentally from earlier generations of students: professors should move their class discussions to Facebook, for example, where digital natives feel more comfortable. 'Our students have changed radically. Today's students are no longer the people our educational system was designed to teach,' argues Marc Prensky in his book *Digital Natives, Digital Immigrants*, published in 2001. Management gurus, meanwhile, have weighed in to explain how employers should cope with this new generation's preference for collaborative working rather than traditional command-and-control and their need for constant feedback about themselves.

But does it really make sense to generalize about a whole generation in this way? Not everyone thinks it does. 'This is essentially a wrong-headed argument that assumes that our kids have some special path to the witchcraft of "digital awareness" and that they understand something that we, teachers, don't—and we have to catch up with them,' says Siva Vaidhyanathan, who teaches media studies at the University of Virginia. Michael Wesch, who pioneered the use of new media in his cultural anthropology classes at Kansas State University, is also sceptical, saying that many of his incoming students only have a superficial familiarity with the digital tools that they regularly use, especially when it comes to the tools' social and political potential. Only a small fraction of students may count as true digital natives, in other words. The rest are no better or worse at using technology than the rest of the population.

Writing in the *British Journal of Education Technology* in 2008, a group of academics led by Sue Bennett of the University of Wollongong set out to debunk the whole idea of digital natives, arguing that there may be 'as much variation within the digital native generation as between the generations'.

SOURCE: *The Economist*, 6 March 2010, www.lexisnexis.com/us/lnacademic/delivery/PrintDoc.

drive the competitive spiral toward the lowest common denominator. Consequently, Gamson theorizes that mainstream news media reveal not individual but institutional pathologies, not a normative order but institutional decay. From this perspective, news media have a positive socializing function, since they regulate our conduct by ensuring public discussion of these activities as 'transgressions of certain values, norms, or moral codes' (Thompson 1997, 39).

Critics are also concerned about the power of music and lyrics, given that numerous studies have documented potential harmful effects in the areas of violence, smoking and drinking behaviour, and risky sexual activity. Music television (e.g., MTV) is often a target of criticism for its sexual content and for depicting violence, promiscuous sex, and sexism (Strasburger, Wilson, and Jordan 2009). Other critics point out that overall, we live in a society in which sexual images and content (e.g., pornography, the sexualization of children) are predominant because of these new technologies. This makes children vulnerable to a host of new risks and challenges, ranging from health problems (e.g., addictions, depression) to Internet predators.

And as media industries grow, Strasburger et al. (2009) assert that they become increasingly global and commercial in nature. Notably, media corporations target children and youths as a profitable group of consumers. Television networks such as Nickelodeon and the Cartoon Network are designed

▼ FIGURE 4.3   Average Amount of Time Spent with Each Medium in a Typical Day

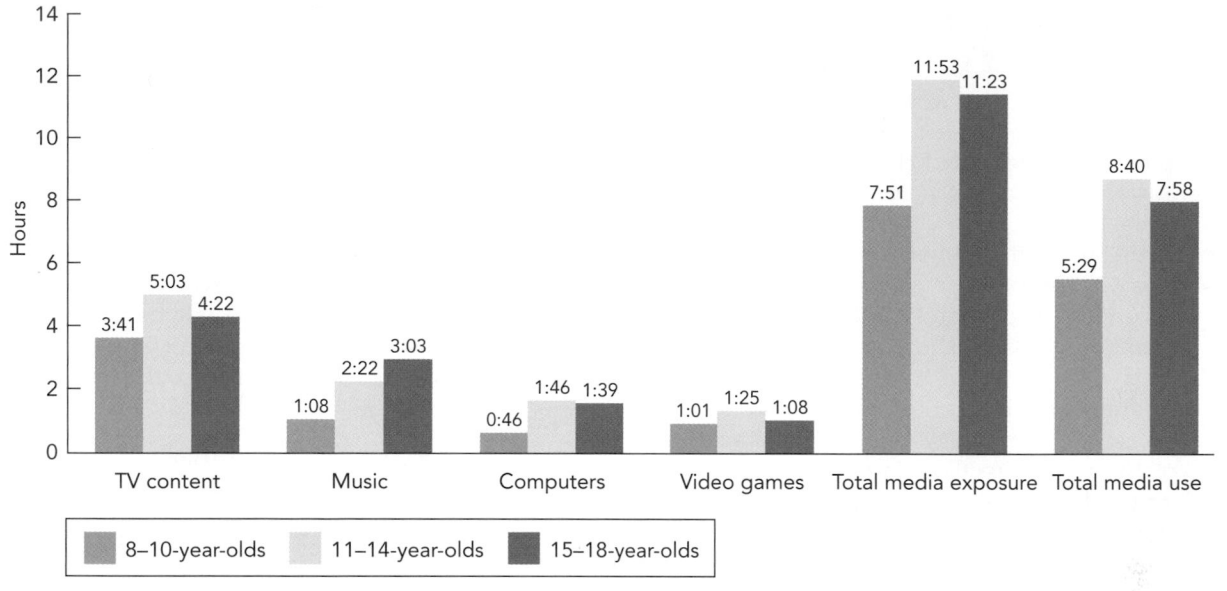

Legend: ■ 8–10-year-olds   ■ 11–14-year-olds   ■ 15–18-year-olds

SOURCE: Victoria J. Rideout, Ulla G. Foehr, and Donald F. Roberts. 2010. *Generation M²: Media in the Lives of 8- to 18-Year-Olds*, p. 5. Menlo Park, CA: Kaiser Family Foundation. www.kff.org/entmedia/upload/8010.pdf.

specifically for young viewers, and teen magazines such as *CosmoGIRL!* and *Sports Illustrated for Kids* are a growing phenomenon. Websites like Nicktropolis (which also contain a lot of advertising) encourage young children to enter an immersive 3-D virtual world where they can create avatars, interact with cartoon characters (e.g., Spongebob Squarepants), and chat with other kids in 'real time'. Consequently, it is deemed that children are increasingly being socialized to become self-indulgent lifelong consumers as well as to form imaginary 'para-social' (one-sided) relationships (Chung, Debuys, and Nam 2007).

TABLE 4.2 ▼ Content Analysis of Video Game Magazine Character Stereotypes[1]

Characterization	Male (%)	Female (%)
Aggressive	82.6	62.1
Sexualized	0.8	59.9
Scantily clad	8.1	38.7
Sex role stereotype*	33.1	62.6
Portrayal of aggression		
Military	4.1	0.0
Fighting	33.2	16.2
Glamorized violence**	31.6	30.6
Wearing armour	41.9	35.6

[1]Video game magazines analyzed were those ranked by Amazon.com as their six top sellers (on sale January 2006).

*Sex role stereotype refers here to the beauty stereotype for the female characters and to the hypermasculine stereotype (i.e., exaggeration of 'macho' characteristics) for the male characters.

**Refers to posing with a weapon.

SOURCE: Data drawn from Karen E. Dill and Kathryn P. Thill. 2007. 'Video game characters and the socialization of gender roles: Young people's perceptions mirror sexist media depictions'. *Sex Roles* 57: 851–64 (Table 1, p. 858).

The debate continues about the extent to which celebrity-watching socializes young people and whether the resulting socialization (e.g., attitudes toward body image, sex, and violence) are benign or malignant.

(© Marco Secchi/Alamy)

GLOBAL ISSUES

Globalization, Children, and Media in Times of War and Conflict

In an increasingly global world, even crises and catastrophes that take place in countries thousands of miles away become part of children's daily lives when there is exposure to the news media. News reports from the conflicts in Afghanistan, Chechnya, Iraq, Palestine, Sudan, the former Yugoslavia, to name but a few, as well as events such as the 11 September 2001 attacks on the United States, the bombings of Madrid and London transportation systems, and suicide bombers in Israel, have all been at the center of world media attention. The sad truth is that war and conflict are an everyday reality for many children all over the globe, either directly or in mediated forms. That violent conflicts have always had detrimental effects on all humans, children included, is self-evident. The physical effects (death, injury, famine, infectious disease, relocation, sexual abuse, etc.), the psychological effects (fear, stress, bereavement, post-traumatic reactions, desensitization to suffering, maladjustment, etc.), or even the distant threat of such have proved to have deep and lasting effects on children, even following incidental encounters (Leavitt and Fox 1993).

In her compassionate reflections on the pain of others, Sontag (2003) discussed the meanings of visual portrayals of the suffering of other people in faraway zones of conflicts viewed by privileged and often safe audiences. In her critique, she offered the following observations:

> Being a spectator of calamities taking place in another country is a quintessential modern experience, the cumulative offering by more than a century and a half's worth of those professional, specialized tourists known as journalists. Wars are now also living room sights and sounds. Information about what is happening elsewhere, called 'news', features conflict and violence—'If it bleeds, it leads' runs the venerable guideline of tabloids and twenty-four-hour headline news shows—to which the response is compassion, or indignation, or titillation, or approval, as each misery heaves into view (p. 18).

What is the meaning of this technological and social development? Children hear about, see, and must cope with these troubling, often frightening events that were once only the preserve of adults. They have to endeavour to assimilate the fragments of information they receive from the media and try to make sense of them. They have to deal emotionally with the suffering of others and with gruesome portrayals of atrocities. Children at various ages, developmental levels, media competencies, and personal life experiences have varying skills and cognitive schemes, as well as interests in and experiences with which to make sense of news reports. Clearly they develop a picture of the events as functions of their personal life history and the media offerings available to them. However, the social-political-cultural environments as well as adult mediation at home and in the educational system also influence them.

SOURCE: Dafna Lemish and May Götz, eds. 2007. *Children and Media in Times of Conflict*, p. 1–2. Cresskill, NJ: Hampton Press.

Finally, socialization also takes place in other institutionalized settings, such as in religious contexts and in the workplace, the latter of which will be discussed in the next section. With respect to religious institutions, Statistics Canada documents that attendance at formal religious services has fallen dramatically over the past several decades, particularly among younger age groups (Lindsay 2008), a trend that has many implications for other socialization processes. Religious norms influence many facets of family life, such as gender roles, parent–child relations, attitudes toward moral issues (e.g., abortion), and how families celebrate rituals and holidays. At the same time, the number of adherents to religions such as Islam, Hinduism, Sikhism, and Buddhism has increased substantially in Canada as the result of changing sources of immigration (Statistics Canada 2005d). Taken together, these trends contribute to more diverse profiles of Canadian families and further support the notion that socialization experiences vary across social groups.

TIME to REFLECT

How have your family and religious/spiritual background influenced your current opinions or attitudes with respect to some controversial social issues (e.g., abortion, same-sex marriage, assisted suicide, the death penalty)?

The Life Course, Aging, and Socialization

Throughout this chapter, it has been emphasized that socialization occurs throughout the life course, although the basic, formative instruction occurs fairly early in life. Young children need to be taught basic values, knowledge, and beliefs. Some of the socialization that takes place during this time is called **anticipatory socialization**: a term used to refer to how individuals acquire the values and orientations they will likely take up in the future. In childhood, this might include doing household chores, a childhood job, sports, dance lessons, and dating—experiences that give youngsters an opportunity to rehearse for the kinds of roles that await them in adulthood (Newman 2006).

As we age, many other kinds of experiences also give us the opportunity to rehearse for the kinds of adult roles that we might eventually adopt. In particular, many educational, training, or occupational settings prepare us for our future work roles. A recent ethnographic study by Chappell and Lanza-Kaduce (2010) on police academy socialization explores the socialization that takes place during training to serve on the police force. The researchers found that despite the philosophical emphasis on 'community policing' and its powerful themes of decentralization and flexibility, the most important lessons learned in police training were those that reinforce the paramilitary structure and culture.

In the study, it was observed that recruits had to adhere to a strict code of conduct, including a dress code (i.e., a uniform). This is viewed as a type of 'symbolic control', since it ensured social distance from 'outsiders'. Many such lessons were taught and internalized. For example, a strong moral code that

HUMAN DIVERSITY

Socialization and Aging in Religious Subcultures

The Druze in Syria and Israel

The Druze, a minority religious sect (Gutmann 1976), live in the highland villages of Syria and Israel. They follow a traditional way of life with an agricultural economy. To coexist with the dominant Muslim world from which they are geographically separated, they raise their sons to be policemen and soldiers for the government of Syria or Israel. Religion is central to their identity and way of life, particularly for men. The basic tenets of the religion are kept secret from the outside world, from all Druze women, and from young Druze males, who are labelled *hajil*, or the 'unknown ones'.

When a Druze man enters late middle age, he is invited to become an *aqil* and receives a copy of the sect's secret religious text. If the invitation is accepted, he gives up alcohol and tobacco and devotes a great deal of time to prayer, and his life becomes almost completely ruled by religious duties. Admittance to the religious sect gives men increasing power as they age because they are thought to serve as a passive interface between their god Allah and the community. As Gutmann (1976, 107) notes, the older Druze 'switches his allegiance from the norms that govern the productive and secular life to those that govern the traditional and moral life'. Religion enables men to continue being active in the community but on a level and for a purpose different from when they were younger.

Old Order Mennonites in Rural Canada and the United States

Many Mennonites who immigrated to North America from Europe settled in rural areas of Canada and the United States. Isolated from mainstream society, the 'Old Order' Mennonites maintain traditional ways of life: clothing is simple but somewhat formal, electricity and farm mechanization is not used, and transportation is by horse and buggy.

Following the traditional teachings and practices of Mennonite law, Old Order Mennonites adhere to codified practices of inheritance and caring for elderly parents that were established in Europe. Parents are respected by their children to the extent that one child will remain single and live in the family home to care for aging parents. Parent–child relations are strong, and most children live close to their parents. Families are large, and religious teachings require that children inherit the family property or an equivalent cash gift. Basic to this process of inheritance is a desire to preserve family stability, to support all members throughout their lives, and to provide security and care for parents in later life. If children are not present, others in the church provide assistance and care, either informally or through church-sponsored nursing homes.

SOURCE: Barry D. McPherson and Andrew Wister (2008). *Aging as a Social Process: Canadian Perspectives*, p. 64. Toronto: Oxford University Press.

categorized and separated 'the good guys' from the 'bad guys' was emphasized, as were values of group solidarity and loyalty. As one recruit stated, 'What happens in the academy stays in the academy' (Chappell and Lanza-Kaduce 2010, 204). Overall, these norms were not only defined formally in the curriculum but also informally through various other means, such as telling stories about chases or drug busts.

Socialization to many new roles continues as we age and face new transitions and responsibilities. Older adults, for example, may have to 'reverse' and learn new family roles as they care for frail and dependent aging parents. Moreover, adjustment to grandparenthood, retirement, and the death of friends and family members, as well as acceptance of the inevitability of one's own death, are part of socialization for aging adults.

It is also possible that during any time of our lives, we may become resocialized because of changing circumstances—a term that reflects the learning of new ways of life. Notably, events (e.g., migration, wars, recessions), social movements (e.g., the women's movement, gay rights), and technological developments can affect our behaviours, relationships, and self-images.

This **resocialization** can cause stress, because old behaviours must be unlearned and new behaviours acquired and the tension between the old and new behaviours may be contradictory or confusing. For instance, a new immigrant to Canada describes learning about this country's food: 'If they gave me a hot dog, I wouldn't eat it! I thought that it was dog meat! You know, if no one tells you. . . .' (Disman 1983, cited in Mackie 1990, 63).

TIME to REFLECT

Think back to your first paid work experience. What kinds of skills and 'lessons' did you learn? How did your interactions with others on the job (e.g., bosses, co-workers, customers) specifically influence your experience and what you learned?

Socialization Processes: Pawns, Puppets, or Free Agents?

In summary, we have learned that socialization is life-long and shapes the individual and society. Societal order and continuity rely on members learning to share norms, values, and language. On an individual level, socialization allows us to realize our potential as human beings. However, because socialization is such a powerful process, you might ask yourself: To what extent are we free, active agents? Or are we just pawns subject to the invisible hands of larger social forces? This latter view has been called the 'oversocialized conception of man', a term coined by American sociologist Dennis Wrong in 1961. It reflects a critique of Parsons's functionalist theory of socialization that assumes we passively accept what is taught to us in order to conform to societal norms.

Fortunately, it is well established that while socialization processes have a significant impact on our lives, we are not necessarily prisoners of this process or non-thinking clones. As Peter Berger points out, 'unlike puppets, we have the possibility of stopping in our movements, looking up and perceiving the machinery by which we have been moved' (1963, 176). Studies also document that some of us are in a better position (e.g., cognitively, materially, or socially) than others to decide whether we want to accept what we learn. This is illustrated by the work of Jamie Keiles, a University of Chicago student who recently decided to 'live by the rules' of *Seventeen*,

(© garysludden)

Socialization to new roles (such as to grandparenthood, retirement, or widowhood) occurs throughout the life course.

a magazine that targets teen girls. Interested in sociology, economics, and gender roles, she conducted her self-imposed 'social experiment' by following every beauty, fashion, diet, activity, and social recommendation for one month, including adorning her walls with 'hot guy' posters supplied by the magazine.

Chronicling her 30-day experiment in a blog called 'The Seventeen Magazine Project', Keiles records many interesting reflections. One major lesson she learned was that just because certain socializing agents (the media, in this case) exert pressure on us to behave in a certain way (for example, the magazine told her to get her eyebrows waxed in preparation for a prom), that doesn't mean that everyone will blindly follow these rules. She states, 'some of the stuff was so ridiculous. I guess the assumption they make is that people are actually going to follow it' (quoted in Proudfoot 2010). She also learned that while she possessed the skills to make critical judgments for herself as to whether she would follow the advice and rules, other young women might not share the same level of self-awareness, cynicism, or resistance. For example, one of her blog's readers wrote: 'hey! i love what you're doing, but i was wondering if you could help me! im looking for the dress on page 58 in the june/july issue the french nautical one. Its from amiclubwear but i cant find the dress! please help if you can!'

Moreover, although we are not completely passive in the socialization process, resocialization by **total institutions** can occur. Developed by Goffman (1961), this term refers to any group or organization that has almost total, continuous control over the individual and that attempts to erase the effects of previous socialization. It often denotes a setting in which people are isolated in some way from the rest of society and manipulated by others (e.g., by administrative staff). This might entail brainwashing, religious conversion, military propaganda, physical brutality, and 'rehabilitation' programs in prisons and mental hospitals designed to change one's personality.

Conclusion

This chapter explored the social experience and what it means to be human. Although there are biological or genetic limits to how socialization shapes us, socialization is an extremely powerful process that makes us functioning, 'civilized' members of society. It is a major link between the individual and society, and social relationships are of fundamental importance to our experiences, health, and well-being. Sociologists theorize and study socialization processes at various levels, ranging from day-to-day interactions between individuals to the organization of society as a whole. On an individual level, socialization moulds our tastes and preferences, attitudes and values, and tendencies to act in particular ways in particular situations. Think back to your last family meal or business meeting, for example, and visualize what could have transpired if there were no customs, rules, or rituals governing everyone's behaviour. Similarly, imagine what our society would be like in the absence of such cultural norms and behavioural guidelines.

Another major theme was that beliefs, values, and norms of society are not transmitted through various agents in a uniform, passive, and unidirectional fashion. Diversity is found not only across cultures but also within cultures according to such aspects as family and ethnic background, gender, and age. Further, in line with popular sociological theory, although socialization performs key functions for the individual and society and can perpetuate inequities in society, individuals are not like the fabled lemmings (small rodents) that blindly or unquestioning follow others. Instead, individuals have some capacity to reflect upon or even resist certain socializing forces, depending upon factors such as social and economic position.

Finally, it was emphasized that socialization is a dynamic, lifelong process and that our social experiences change as we age, along with other societal transformations. Globalization, rapid population aging, and continuing high rates of immigration in Canadian society may provide the potential to learn and experience an even wider array of roles, identities, and behaviours. Technological advancements in communication and medical fields also mean that our social worlds are constantly shifting, although basic beliefs, norms, and values remain relatively stable with the passage of time.

questions for critical thought

1. Identify some unique socialization experiences of various generations within your family. For example, how did the socialization experience of your grandparents differ from that of your parents and from what you have experienced? Also note any gender differences.

2. The representation of social groups (e.g., by age, gender, race/ethnicity, sexual orientation) in the mass media can be an important socializing experience. Consider how certain groups are portrayed in various media settings (e.g., television, the Internet) and how these portrayals shape our social behaviour (e.g., stereotypes and power relations).

3. What kinds of challenges and opportunities do immigrant children face in trying to become socialized into Canadian society? Consider the role of family and ethnic background, cultural traditions, and values.

4. Recall the last time that you were in a toy store or saw an ad for a child's toy or game. What kinds of toys and games appear to specifically target little boys and little girls? How might this shape gender-role socialization and subsequent behaviours and activities?

5. How might social class shape specific tastes and preferences in food, entertainment, and other activities?

6. Make a list of contradictory lessons that different agents of socialization taught you as a child. How have you resolved these contradictions?

recommended readings

Roberta Berns. 2010. *Child, Family, School, Community: Socialization and Support*. 8th edn. Florence, KY: Wadsworth Publishing.
Examining how the school, family, and community shape children's socialization, this text focuses on issues of diversity. Topics related to culture, ethnicity, gender, sexual orientation, and special needs are addressed, drawing from a social ecological approach.

Wendy Griswold. 2008. *Cultures and Societies in a Changing World*. 3rd edn. Thousand Oaks, CA: Sage.
This book focuses on the role that culture plays in shaping our norms, values, beliefs, and practices. Cultural phenomena, including stories, beliefs, media, art, religious practices, fashion, and rituals from a global sociological perspective, are critically examined. The effects of political symbols and rituals are also examined to highlight the interplay of culture and power.

Gerald Handel, Spencer Cahill, and Frederick Elkin. 2007. *Children and Society: The Sociology of Children and Childhood Socialization*. Toronto: Oxford University Press.
From a symbolic interactionist perspective, a central theme of this book is the tension between children's active agency and the socializing forces of family, peer groups, school, and mass media and how socialization is shaped by social class, race and ethnicity, and gender.

Robert Hess and Judith Torney. 2005. *The Development of Political Attitudes in Children*. New Brunswick, NJ: Transaction.
Based on a study of 12,000 elementary school children, this book explores how political attitudes and socialization into citizenship roles are formed in childhood. Attention is paid to social class, teacher attitude, religious membership, and identification with key authority figures (e.g., fathers, police officers). Readers will enjoy the entertaining verbatim interviews with grade-schoolers.

Mizuko Ito, Sonja Baumer, Matteo Bittanti, et al. 2009. *Hanging out, Messing Around, and Geeking Out: Kids Living and Learning with New Media*. Cambridge, MA: MIT Press.
Grounded in a rich three-year ethnographic study, this book investigates how young people are learning with new media in settings such as the home, after-school programs, and online spaces. Integrating 23 different case studies, such as Harry Potter podcasting, video-game playing, music sharing, and online romances, this book offers insight into what it means to grow up in a digital era.

Michael Kimmel. 2007. *The Gendered Society.* 3rd edn. New York: Oxford University Press.
This book investigates contemporary gender relations and shows how gender differences are often exaggerated in society. The author also illustrates that gender is not just an element of individual identity, but also a socially constructed phenomenon. Students will find the chapter on media and image-driven industries and the portrayal of gender interesting.

George Ritzer. 2010. *Enchanting a Disenchanted World: Continuity and Change in the Cathedrals of Consumption.* 3rd edn. Thousand Oaks, CA: Sage.
This book argues that society has created new 'cathedrals' of consumption or places that enchant us so that we stay longer and consume more. The central thesis is that society has undergone significant change because of the way that we consume. Using examples of 'cathedrals' such as Disney, malls, cruise lines, Las Vegas, the Internet, and credit cards, we learn how capitalists profoundly shape our lives.

recommended websites

Child and Family Canada
www.cfc~efc.ca
Child and Family Canada is the website of a public educational organization offering resources (e.g., research reports, articles) from non-profit organizations.

The Goffman Page
http://people.brandeis.edu/~teuber/goffmanbio.html
The Goffman Page outlines the contributions of well-known Canadian-born sociologist Erving Goffman in terms of his insightful analyses of human social interaction.

Invest in Kids Foundation
www.investinkids.ca
Invest in Kids Foundation is a national charitable organization dedicated to the healthy social, emotional, and intellectual development of children from birth to age five by strengthening parenting knowledge, skills, and confidence.

Media Awareness Network
http://media-awareness.ca/english/index.cfm
The network provides information and resources for anyone interested in media and information literacy. It focuses on such topics as media violence, online hate, media stereotyping, and beauty and body image.

Ontario Federation of Teaching Parents
http://ontariohomeschool.org/socialization.shtml
The Ontario Federation of Teaching Parents provides articles and research on the socialization of home-schooled children in addition to Web links related to various socialization issues in schools.

Thrive! The Canadian Centre for Positive Youth Development
www.lionsquest.ca
Thrive! The Canadian Centre for Positive Youth Development was established in 1988 and is dedicated to fostering positive youth development. Tools to empower and unite caring adults are provided, including programs, products, and training and services for youths, parents, educators, and community leaders.

chapter 5

Statuses, Roles, Self, and Identity

Dorothy Pawluch
and William Shaffir

In this chapter, you will:

▶ Be introduced to how sociologists define the
  terms *statuses, role, self,* and *identity*;

▶ Learn how these concepts are interrelated;

▶ Learn about the contributions that the three main
  theoretical perspectives in sociology have made
  to our understanding of these concepts;

▶ Explore themes related to how we define
  ourselves and others;

▶ Come to appreciate that these concepts all
  describe how individuals are connected to society;

▶ Get a glimpse of the kinds of studies that
  sociologists interested in these concepts
  have done.

Introduction

Who am I? Can there be an existential question bigger than this one? Take a moment to think about it. If you were to ask yourself the question seriously, how would you respond? Would you say that you are a woman, man, daughter, son, brother, sister, student, Canadian, atheist, gay? Would you say you are working-class, a Maple Leafs fan, Aboriginal, or immigrant? Would you say that you are honest, caring, outgoing, gifted, shy, athletic, hot, messed up? Now take a moment to think about what difference it makes how you define yourself and how others define you. How do these definitions affect your experience of life—what you do, who you interact with, how they interact with you? Where does the menu of possibilities come from in the first place? Is there any 'you', any essence left after all of these labels are stripped away? This simple exercise begins to capture why the concepts of statuses, roles, self, and identity are of central importance to so many sociologists.

As is true for virtually every topic that sociologists explore, how sociologists think about these concepts depends on how they understand society more generally. As a discipline, sociology has developed several distinct theoretical perspectives. By this point, you will have read about the major perspectives: structural functionalism, symbolic interactionism, and critical theories, including feminist theories. Each has had something to say about statuses, roles, self, and identity, so we divide our discussion according to the contributions that each has made. If the discussions are not perfectly balanced and symbolic interactionism gets more attention than the others, it is largely because these issues have been at the centre of symbolic interactionist thinking. As a result, many of the key concepts that sociologists have used in this area have been generated by symbolic interactionists.

A Functionalist View of Statuses and Roles

A good place to start is by reiterating how structural functionalists understand society. As a macro theoretical perspective, functionalism focuses on large-scale phenomena or even entire societies. Functionalists are impressed with how societies organize themselves and persist over time. They are particularly drawn to the part that large-scale structures or institutions, such as the family, religion, and education, play in ensuring that societies endure. The analogy is often made to society as a living organism in which each part contributes to the survival of the whole by serving its own unique function but also by working with other parts. Just as the heart fulfills the function of pumping blood so that the rest of the organs receive the oxygen they need to operate, so do the 'organs' of society play their part to keep societies healthy. The educational system, the family, and the media, for example, are each essential to a smooth-functioning

In the First Person

As an undergraduate majoring in sociology, I managed to avoid taking even one course in theory or research methods. My ambition, initially, was not to become a sociologist, though I am not at all sorry that this is the career I chose. While still an undergraduate, I was encouraged to begin an ethnography of a popular pool hall in downtown Montreal. I loved the experience. Malcolm Spector, who became my dissertation advisor and to whom I owe an enormous debt, subsequently suggested that I might study the Hasidic community in Montreal. To my father's immense relief, I shelved the pool hall research and wrote my doctoral dissertation on the social organization of the Lubavitch Hasidic sect in the city. I have continued studying this segment of Orthodox Judaism since my days as a graduate student.

—William Shaffir

I knew early on what I wanted to be. From the moment I took my first sociology course in high school and learned what sociologists do, my mind was made up. Maybe it had to do with having a 'marginal identity'. Growing up in Sudbury, the daughter of immigrants, my mother from Italy and my father from Ukraine, I moved in and out of a variety of groups, each with its own way of seeing the world. I think that experience made me a natural people-watcher. Even then there was the perennial question, 'But what will you *do* with a sociology degree?' Undeterred, I followed my heart. Now, decades later, I can say that I have been richly rewarded for making that decision. I too had the very good fortune of working with Malcolm Spector at McGill University. In fact, Billy Shaffir was Malcolm's first PhD student; I was his last. He calls us his bookends.

—Dorothy Pawluch

society for a whole host of reasons, but in terms of the role they play in socializing new members or recruits to a society—another essential component of any society that hopes to be ongoing—these institutions need to work together at least to some degree, reinforcing the messages they send about fundamental values and expectations. Functionalists maintain that if an aspect of social life fails to contribute to a society's stability or survival, it will not be passed on.

How do roles and statuses fit in? Functionalists underscore the patterned ways in which social institutions are integrated to make up and stabilize society. In response to the question of why people behave in ways that contribute to the integration of society, the functionalist's answer focuses on norms—that is, on sets of socially derived expectations about appropriate behaviour in particular settings. The most important of these norms are learned in childhood during socialization. As children, individuals learn what is expected of them from parents, relatives, peers, teachers, and the media. Later in life, people continue to act in ways that are socially approved. As they enter new social institutions, such as universities, corporations, or organizations, they learn new norms.

Norms in turn are organized around statuses and roles. **Status** refers to particular social positions that people hold. All positions occupied by individuals are statuses, whether one is a hockey player, restaurant server, human resource specialist, social worker, or sex trade worker. Attached to every status is one or more roles. **Roles** consist of the responsibilities, behaviours, and privileges that are connected to the position the person occupies. In other words, roles are the action element of status. One can say that a status is something we occupy, while a role is something we play. Or another way to capture the distinction is to say that a status describes what one is, while a role describes what one does. Thus, 'student' is a status, while studying, attending class, and sitting exams are all part of the role. You hold the status of student by virtue of having registered at a particular educational institution, but in bringing that status to life, you are playing the role of student.

STATUSES

Sociologists separate statuses into those that are ascribed and those that are achieved. One way of acquiring a status is to be born into it or having it imposed on you by nature or by chance; this is called an ascribed status. We have little control over whether we are young or old, male or female, black or white. Ascribed statuses are neither earned nor chosen. By contrast, some statuses are more a matter of choice. They can be achieved through hard work and effort. In our society, most individuals can decide whether to become a spouse and can choose whether to pursue a professional degree or some other type of career. In this respect, social worker, electrician, salesperson, and nurse are achieved statuses.

All of us hold many statuses simultaneously. One can be a father, construction worker, jogger, and much more at the same time, though not necessarily in the same situation. The cluster of statuses held by any given individual at one time is called a status set. It is also important to recognize that persons enter into, and exit from, statuses over the course of the lifecycle. As such, status sets are far from fixed—they are regularly reconfigured. A single person becomes a spouse after marriage, a medical student becomes a doctor upon graduation, and a wife becomes a widow upon the death of her spouse.

ROLES

The roles connected with any particular status bring with them both rights and responsibilities. In fact, roles are often organized in such a way that the rights attached to one are linked with the responsibilities attached to another. Sociologists refer to this pattern as the reciprocity of roles. Members of a sports team have the right to expect training opportunities, expertise, and support from their coach, and a coach has the responsibility to provide that expertise and leadership; on the other hand, coaches have the right to expect commitment and effort from their players, and players are obliged to display the required level of dedication to the team. Parents/children, professors/students, rock star/audience, pastor/parishioner, salesperson/customer are all similarly linked.

The fact that people hold different statuses, each of which can have different roles attached to them, has the potential to create situations in which conflict can occur. What happens when the behavioural expectations attached to one role interfere or conflict with one's ability to meet the expectations of another role? Sociologists identify such a challenge as **role conflict**. For example, many women today find themselves juggling motherhood and a career. Their mothering role demands devotion to their children—being present to care for them, guide them, and attend to their every need. If they work outside of the home, their worker role comes with its own set of demands, including going into work early, staying late, or even working on weekends or on days off if the need arises. To play one role well is to feel that they are letting down those who count on them in their other role.

By contrast, the concept of **role strain** describes a situation in which competing demands are built into a single role, causing tension and stress. Going back to the example of a worker, there are multiple expectations built into the role for most workers. Supposing the worker is a lawyer. Lawyers are expected to represent their clients to the best of their ability. At the same time, they are expected to mentor young lawyers, be helpful to their colleagues, generate new clients, attend legal conferences, engage in community work to increase the public profile of their firm, and keep their billing hours up. As in most areas of work, lawyers often feel as though there are not enough hours in the day to properly fulfill the duties connected with their role. Unless some balance or happy medium is found, role strain will result.

THE SICK ROLE

An illustrative example of how functionalists think about roles is offered in Talcott Parsons's (1951) discussion of the **sick role**. According to Parsons, societies are dependent on individuals within them playing their assigned roles. But what happens when sickness strikes? While having individuals unable to fill their roles productively is clearly a problem, it is not in society's interests to force sick people to continue to function as though nothing were amiss. First, people who are sick may be physically incapable of doing what is expected of them. Even if this were not the case, what sort of performance could one expect from someone who is not well? Would you want to undergo surgery at the hands of a doctor who is sick? Then there is the matter of contagion. If the sickness in question is a bad case of the flu, would you want those affected dragging themselves to work only to infect others? Giving individuals who are sick a break from their normal roles and routines so that they can recuperate seems only sensible, but this strategy can present yet another threat to the social order. If it is too easy for people to claim that they cannot do what is expected of them because they are sick, there is the possibility of pervasive use of sickness as a way of evading responsibility.

According to Parsons, the sick role represents a way to deal with the threat of social members being sick and not able to play their roles while at the same time minimizing the disruption that sickness and the temptation to claim illness generates for the social order. The sick role is organized around two rights and two responsibilities. On the rights side, people who are sick are excused from their social responsibilities. If you claim that you are sick and your claim is recognized as legitimate, you are exempted from your usual roles and duties. Students with a doctor's note need not attend class or can have a test or an exam delayed. Sick employees can stay home. Nor are you blamed or penalized for your situation. Instead, you have the right to expect the understanding, sympathy, and support of others. On the responsibility side, however, is the expectation that you will view your situation as undesirable. You must want to get better. Moreover, you are expected to be proactive about it. That is, you must do whatever it takes to get better, including consulting health-care providers and following whatever instructions and treatment advice they give you.

Parsons's ideas about the sick role have been challenged on several fronts. Critics have pointed out that the concept does a better job of explaining acute physical illness rather than long-term, chronic conditions or mental illness. Critics have also pointed out that Parsons's argument does not work in situations where there is no treatment available. Nevertheless, as a model for how functionalists understand roles and the way roles both structure interactions between people and contribute to the 'ongoingness' of society, the sick role is instructive.

In the end, the picture that emerges about statuses and roles from a functionalist perspective is one that emphasizes constraints. Although people have a measure of choice over at least some of the statuses they hold and whether or not to act out institutional roles, they have little choice as to how to act them out. Each role comes with pre-existing scripts, demands, and expectations that shape behaviour. These scripts are more or less agreed upon and do not allow for much flexibility. From a functionalist standpoint, institutions channel behaviour; they 'predefine' for people how to be a mother, a student, a bank employee, or a prison inmate. Institutions precede the individual, and people learn through socialization how to participate in them properly. Thus, the patterns of interaction have a reality independent of the persons who enact them.

Symbolic Interactionism: Roles, Self, and Identity

Just as a functionalist view of roles is shaped by how functionalists understand society, an interactionist view is closely tied to the central ideas of symbolic interactionism. First formulated by George Herbert Mead (1934) and developed further by Herbert Blumer (1969), symbolic interactionism is concerned with how social actors make sense of their worlds. The perspective is captured in three often-cited premises presented by Blumer (see Table 5.1).

From a symbolic interactionist perspective, individuals are constantly and actively involved in assessing and reassessing things around them, defining them, making sense of them, and working out how they are going to act in relation to them. In contrast to the functionalist approach, which portrays individual actors as buffeted by social forces outside them—including role expectations—symbolic interactionism emphasizes how individuals interact to create, sustain, and transform social relationships.

According to symbolic interactionists, human behaviour does not occur in a vacuum but arises out of how social actors define the situations in which they find themselves. The definition of the situation, as Hewitt defines it, is '. . . an organization of perception in which people assemble objects, meanings, and others, and act toward them in a coherent, organized way' (Hewitt 2000, 57). The phrase 'definition of the situation' was coined by W.I. and D.S. Thomas (1928, 572), who maintained that 'If men define situations as real, they are real in their consequences.' Claiming that definitions of the situation are integral to social interaction among human beings underscores the idea that reality is socially constructed and that people respond to the meaning a situation has for them rather than to the objective features of the situation. For example, people do not interpret prayer in the abstract but within contexts in which they expect that things are likely to occur. An individual in dialogue with God while sitting in a restaurant may be regarded as a candidate for a mental facility, yet the same behaviour in a house of worship is entirely appropriate.

The process of defining situations, including people in those situations, is ongoing, and our ability to do so successfully enables interaction to flow smoothly.

TABLE 5.1 ▼	The Three Premises of Symbolic Interactionism
Human beings act toward things (objects, situations, people, themselves) on the basis of the meanings that these things have for them.	
The meaning of things arises out of interaction.	
The meanings of things are handled and modified through a process of interpretation that individuals engage in as they deal with the things they encounter.	

Consider a situation in which one is at a bar and considering the appropriateness of telling a politically insensitive joke to a group of people, not all of whom one knows well. In that situation, our first step normally is to try to size people up. If everyone strikes us as laid-back and mellow, we may decide to tell the joke. If we have some inkling, however, that not everyone in the group shares our sense of humour, we might be inclined to resist the temptation. Indeed, when a definition of the situation is lacking or unclear, we focus on establishing one that is satisfactory and can help guide the ongoing interaction (Shibutani 1961). If you are suddenly stopped behind a lengthy line of cars on the highway, you attempt to define the situation so that the unexpected stoppage makes some sense—to determine, for instance, whether there was an accident or whether the stoppage is due to road construction—in order to plan how to proceed.

The symbolic interactionist approach focuses less on the functioning of the social structure and sets out instead to capture the interactive process among individuals. From a symbolic interactionist standpoint, the conventional approach to roles offers a misleading portrayal of how people actually organize their behaviour and the place of roles in the process. Statuses and roles, for the interactionist, do not determine social interaction, since this is organized and managed by actors themselves. Rather, they provide a fluid and malleable context within which human interaction unfolds. While human beings make use of norms to guide their interactions, they typically focus less on conformity to norms than on an ongoing appraisal of what is going on in the situations in which they find themselves acting. From a symbolic interactionist perspective, a role can be thought of as a resource that persons employ to organize and carry out their activities. Individuals carry around a repertoire of roles and determine which of them to use. Roles are best viewed as a perspective from which behaviour is constructed or a platform from which one acts. While people may act within their roles, they do so in a manner that offers them considerable latitude and flexibility. Far from being locked into particular role configurations or structures, we have the capacity to use roles.

TIME to REFLECT

How do structural functionalists and symbolic interactionists view roles differently?

ROLE-TAKING

Two concepts that further emphasize the symbolic interactionist view are **role-taking** and **role-making**. Role-taking is the process by which we co-ordinate or align our actions with those of others. When we engage in role-taking, we put ourselves in the shoes of others and try to determine how they are defining a given situation. What roles are they projecting, what meanings are they attaching to the situation, what course of action are they likely to follow? Role-taking also entails looking at ourselves from the point of view of others and trying to anticipate the consequences of our own plan of action. This process continues as we initiate a response. We are constantly monitoring how others are reacting and performing their own roles, and as a result we adjust, fine-tune, or modify our role performance. Or we might abandon a particular performance altogether if it does not appear to be working out for us. These ongoing adjustments are what make joint action possible.

ROLE-MAKING

The concept of role-making describes another feature of our interactions. In the way we choose to play our roles, we do not follow rigid, predetermined, or prewritten scripts. The expectations attached to any given role provide us with a rough guideline at best as to how we ought to act. There is room for innovation and creativity. Not everyone who performs the role of student does so in precisely the same way. Indeed, there is remarkable variability in how students choose to play their student role. How you choose to play the role is to some extent up to you. In this sense, the concept of role-making comes closer than does role-playing to describing how we enact our roles.

Role-taking and role-making, although we have defined them separately, are intricately linked. There can be no role-making in the absence of role-taking. The construction of a role is impossible without an ability to view oneself from the vantage point of another. Again, behaviour is not simply a matter of repeating preset lines of a script in which the unfolding action is detailed in advance. In this respect, roles are not merely packages of mandatory behaviour but, as suggested, perspectives from which people organize lines of behaviour that fit the situation.

(© Image Source / Alamy)

Our identity is often shaped by what we imagine others think of us. This reflected self emerges when we consider how we appear to others and how that appearance is judged by society.

THE SELF

Having laid out some of the main ideas in symbolic interactionism, it becomes possible to introduce two additional concepts. These are **self** and **identity**.

We have stressed throughout that symbolic interactionists see human beings as active agents, defining the things to which we respond. Among the things to which we assign meanings are our selves. Our capacity to role-take and step outside of ourselves means that we can treat ourselves as an object to define. 'I am Elena. I have dark hair and eyes. I am into soccer.' Or: 'I am Adam. I am quiet. People say I am smart.' As with all objects, the meaning we attach to our selves is not fixed but constantly changing as we interact with others. One sociologist (Berger 1963, 106) has argued that the self is not 'a solid, given entity that moves from one situation to another. It is rather a process, continuously created and recreated in each social situation that one enters.'

Thus, we acquire our sense of self by imagining how we appear to others. This is what symbolic interactionists mean when they say that individuals and society are in a dialectical relationship. Society cannot exist without individuals, but self-aware individuals cannot exist without others (society). Charles Horton Cooley (1902) captures the interdependence between individuals and society in his concept of 'the looking glass self'. Others (society) are the mirror that reflects back an image of who we are.

According to Mead, the capacity to take the role of others, so critical to being able to define ourselves, is not something that is present at birth. Rather, we develop this capacity in stages.

Once we are able to take the role of the other, we are left with a self that Mead argued is made up of an 'I' and a 'Me'. When we experience an impulse to act, as in, 'It's hot in here; I need a cold drink', the 'Me' kicks in as we consider our social context. 'Yes, a cold drink would be nice, but the store is full of customers, I'm on cash, and my boss is standing right behind me.' The 'Me' takes the significant and generalized other into account, asking itself what norms govern the situation, given one's role, the image one wants to project, others' expectations, and how one wants to be seen. But the 'Me' does not dictate our actions. The 'Me', as Mead (1934, 175) describes it, is the 'response of the organism to the attitudes of others'. The 'Me', in other words, responds to the attitude of the generalized other. The 'I' may be swayed by the 'Me', deciding that to leave for a cold drink at this precise moment would not be a career-enhancing move, or it can choose to ignore the 'Me', arguing, 'Who needs this lousy job anyway?' or 'I'll take my chances. Someone will cover me on cash.'

While most symbolic interactionists stress this fluid, ever-changing feature of the self, there are other interactionists who see the self as more stable. Manford Kuhn, a sociologist at the University of Iowa, used Mead's view of the self as an object to develop the idea of a 'core self', a stable set of meanings that one attaches to oneself. According to Kuhn, while we are responsive to others, we also carry this core self into all of our interactions. Since we typically act in ways consistent with how we view ourselves, this core self shapes our interactions, giving them continuity and predictability. The stability of the core self, Kuhn argued, suggests the possibility of measurement. Kuhn and his student, McPartland, developed the 'Twenty Statements' test as a way of measuring the self. The test asks people to list 20 responses to the question 'Who am I?' (Kuhn and McPartland 1954). Kuhn also believed that it was possible to develop a set of general propositions about the self from which hypotheses could be deduced and tested. Kuhn's quantitative version of symbolic interactionism is known as the 'Iowa school' or 'self-theory'.

IDENTITY

If the self is an object to which we assign meaning, **identity** refers to the names we give ourselves.

TABLE 5.2 ▼ Stages in the Development of Self	
Preparatory Stage	Infants learn to imitate the behaviour of others without understanding the social meaning of the behaviours or that these behaviours may be attached to roles.
Play Stage	Young children learn to take the role of a particular, usually significant, other, as is the case when they pretend to play at being a mom or dad.
Game Stage	Children learn to take the role of several others or the generalized other simultaneously, as is the case when they play a game of baseball and consider the position of everyone else on the team.

Gregory Stone (1962, 93) argues that 'identity' is not a substitute word for self but instead refers to how one casts one's self as a social object, who one tells the self one is, and what one 'announces' to others one is. In 'announcing' ourselves to others, we enact or suppress certain aspects of our self depending on how we want to come across.

A sociologist who has made a significant contribution to understanding how we do this is Erving Goffman. In his classic book, *The Presentation of Self in Everyday Life*, Goffman (1959) analyzes everyday interaction using what he calls a dramaturgical approach. Borrowing in a literal sense from the imagery and language of the theatre, Goffman argues that every encounter between people is an occasion for social actors to present one or more of their social roles to each other, much as in a stage performance. Through a process called **impression management**, actors try to shape how others will define them. University of Manitoba sociologists Daniel and Cheryl Albas offer an example that students should find easy to relate to in identifying the strategies that university students use to create a desired appearance upon receiving their grades on exams. Students are divided into test *bombers* (those who have done poorly or failed) and *aces* (those who have done well). Aces engage in behaviours like 'sitting tall' at their desks, 'broad grins', and 'jaunty walks'. Bombers use these behaviours to figure out who to avoid, since they do not want to come out of the encounter looking (and feeling) lazy, irresponsible, or 'dumb' (Albas and Albas 2003).

Reflecting a symbolic interactionist understanding of roles, Goffman distinguishes between role and role performance. For Goffman, role refers to a person's conduct if he or she did only what the norms attached to the particular position directed. By contrast, role performance reflects the actual behaviour of an individual while acting out the role. For Goffman, each status a person holds can be seen as a part in a play. Using Mead's notion of taking the role of the other, he argues that people try not only to see with others' eyes but also to manage how others define the situation.

Goffman divided the social world into two regions: front-stage and back-stage. In the front-stage region, social actors carefully manage presentations of themselves they project to others in the hope of creating a positive impression so that others respond in a desired manner. Thus, for example, a student who wishes to favourably impress the professor attends classes regularly, participates in class discussions, nods approvingly when the professor emphasizes a particular point, and may seek out the professor during office hours. In the back-stage region, the same student, perhaps in the company of peers, is likely to be more relaxed, venting perhaps about the professor's frustrating lecturing style or admitting to not having a clue what is going on in class. Our back-stage behaviour is generally displayed among those with whom we share close social bonds and whom we trust with often unflattering information about ourselves that we keep from others.

IDENTITY WORK

Sociologists have focused considerable attention on how individuals present themselves and construct others. These processes can be referred to generally as *identity work*. We project our identities using appearance, behaviours, talk, and props of various kinds. An interesting example of this process appears in a recent study of the 'punk' scene. William Force (2009) points out that there is a gendered quality to identity work among punks. Men wear band shirts or pins and baseball caps, while women are more likely to wear mainstream clothes but with a subversive twist, like wearing them extra tightly or tearing out the original seams and re-stitching them with either thread or shoestrings. Hairstyles too vary, but

Dressing in punk clothes and having punk hairstyles illustrate the conscious effort of constructing identity.

(© Janine Wiedel Photolibrary/Alamy)

the messy look that most punks work hard to achieve is described as a 'scene cut' on men and 'JBF' ('just been fucked') hair on women. That those who identify as punk use these aspects of self-presentation consciously to communicate who they are is reflected in the fact that they themselves refer to this look as 'the uniform'. Other ways to communicate membership in the local punk scene include publicizing the ownership of punk goods. Force points out that he repeatedly heard references such as: 'Oh, Mastodon was just here? I have their last album on CD and vinyl, it's really good' or 'Yeah, I know "Murder by Death". I own that record' (Force 2009, 294). Owning the 'right' records serves to establish membership in the local punk scene and allows those who identify with it to 'talk shop'.

TIME to REFLECT

Thinking about a role that is central to how you define yourself, what identity work do you engage in to communicate to others who you are?

Another example of identity work is offered in research on medical students by McMaster University sociologists Jack Haas and William Shaffir (1987). Medical students, they point out, are under great pressure to come across as competent and trustworthy, since the occupational role to which they aspire demands these qualities. After all, as doctors-in-training they have the responsibility of making decisions that affect the well-being of others. Audiences look for cues of personal and/or collective competence, and practitioners, in response, organize a carefully managed presentation of self to create an aura of competence. In communicating competence, the students used props, costumes, and language to demonstrate to their audiences that they possessed the special knowledge and trustworthiness demanded of their role. Just as significantly, this identity work reinforced the students' identification with, and commitment to, medicine. One student, commenting on the dynamic nature of the relationship between the symbols of medicine that he relied on and his self-image, said:

When you wore the jacket, especially in the beginning, people were impressed. After all, it told everyone, including yourself, that you were studying to be a doctor. . . . The other thing about wearing the white jacket is that it does make things more obvious. You know what

you are. . . . You know, it is sort of another way of identifying. There were very few ways that people had to identify with the medical profession, and one of the ways was to begin to look like some of the doctors.

The student makes clear that he recognizes the importance of appearing authoritative in professional situations, but he is also describing how neophyte students are slowly transformed into doctors.

Over the past several decades, the body as part of identity work has attracted increasing attention. Sociologists have studied embodiment, or how we experience our body as part of ourselves. They have also looked at how we use our bodies to communicate to others who we are. For example, McMaster sociologist Leanne Joanisse (2005) studied women who had undergone weight loss surgery for obesity. She found that when the surgery was successful, women reported that they were 'new' people. Removing the layers of fat, they explained, brought out their 'real' self. The group for whom the surgery was unsuccessful also experienced an identity change, but in this case the change was brought about 'by separating body and self into two distinct entities' and constructing 'positive identities independent of their non-normative bodies' (2005, 257). The women adopted a perspective that resisted seeing the fat body as reflective of a flawed self.

On a related theme, University of Toronto sociologist Michael Atkinson has looked at men who undergo plastic surgery as a way of dealing with what they consider to be deficient (aged, overweight, unattractive) bodies. The example is interesting because it is generally assumed that cosmetic surgery is performed almost exclusively on women. Atkinson presents data that show, however, that more than 10,000 Canadian men underwent elective cosmetic surgery between 1998 and 2008, with a 20 per cent jump in the figures between 2003 and 2008 alone (Atkinson 2006, 248). Atkinson links this increase to the 'crisis of masculinity', or uncertainty about what it means to be a man in the face of gender equity movements, ideologies of political correctness, and attitudes of misandry (male-bashing). In an effort to create what Atkinson calls 'a mask of masculinity', men are willing to subject themselves to both invasive (e.g., eyelid surgery, liposuction, hair transplantation) and non-invasive (e.g., chemical peels, hair removal, Botox or collagen injections) procedures. Atkinson's interviews highlight how pivotal men's experiences of their bodies are to their identities. In the following example, Roger

New research suggests that the way men experience their bodies is pivotal to their self-identity.

talks about the toll his work and lifestyle took on his sense of self prior to his surgery:

> Sitting at a desk for 10 hours a day, then in a car for two, and then on your couch for three more wears your body down. Not to mention that my skin barely ever sees the light of day. At times, I can feel my face literally sagging because of my posture.
>
> Looking in the mirror when you're 40 and having a road map for a face shouldn't be surprising. That's not who I am, that's not the image of my inside I want to project (Atkinson 2008, 82).

Also interesting are studies that show that identity work continues in earnest, and perhaps with even greater urgency, among the sick and dying and, after death, on their behalf by those who are left behind. Once people with HIV/AIDS get over painful self-feelings such as grief, guilt, and death anxiety, according to Sandstrom (1990), many embrace and project a new, positive identity as a 'PWA' (person living with HIV/AIDS). Similarly, University of Calgary sociologist Arthur Frank (2002) has written an autobiographical book about the self-transformative experience of suffering through two serious illnesses—a heart attack at age 39 and cancer a year later. In sections of the book, Frank

UNDER the WIRE

Cellphone Ringtones as Identity Work

You might think that the sole purpose of the ringtone on a cellphone is to signal an incoming call. Not so. Christopher Schneider (2009), a sociology professor at the University of British Columbia at Okanagan, describes ringtones as 'identity management devices'. What ringtone you choose to download—a clip of your favourite music (classical, hip hop, popular, or spiritual), a line from a movie or Barack Obama's inauguration speech, a vuvuzela blast, the call of a bird on the endangered species list—is a way of telling others who you are and what you care about.

Schneider is particularly interested in the use of cell ringtones by young people. Youth, he insists, are uniquely limited in expressing their identities, especially while they are in school: 'Youth spend a majority of their daily lives, and indeed, waking moments, in the confines of the education system; thus, the narrow social space that exists outside this system for youth to freely create and manage identity during social interaction takes on added importance.'

Drawing on data he collected as a high school substitute teacher, Schneider observed that students are careful to control the information about themselves that their ringtones communicate. Many students downloaded multiple ringtones and switched from one to another much like we choose different outfits depending on the setting we are in and the impression we want to create. Schneider called this practice *tone shifting*. In certain situations, students would set their cellphones to 'silent' or 'vibrate' to avoid negative reactions. In spite of efforts to control impressions, however, students, like all social actors, are subject to performance blunders—as in a case when ringtones with crude or sexually explicit lyrics are inadvertently heard by teachers or fellow students, who take offence. Schneider calls these situations *identity blitzkriegs* because they involve the threat that the audience will interpret the information as contradicting the previously formed positive impression. Social interaction can be seriously disrupted, and the effort to create or maintain a favourable impression can be undermined.

Schneider concludes that cellphone ringtones can be potent reflections of people's identities. He encourages sociologists to continue to study how new technologies affect the ways in which people manage the impressions they create on others.

literally has his new self talking to his old self about how he has changed and what he has learned about life. Martin (2010) examined the case of murdered children and how family members often compete with police, funeral directors, and other institutional players to construct the identity of the victim. Was he a drug dealer, gangbanger, troubled youth, a reckless or naive kid, or was he basically a good kid in the wrong place at the wrong time, a victim of circumstance? Martin calls this competition an *identity contest*.

SOCIAL VERSUS PERSONAL IDENTITIES

As several examples suggest, the image that people present can be at odds with their self-image. There can be a disjuncture between the roles we play (social identity) and who we understand ourselves to be (personal identity). Where this separation exists, we may try to distance ourselves from the roles we play and that others use to define us. This role distancing work can be contrasted with role embracement—situations in which there is such congruence between roles and self-definition that we feel that we are our roles.

Another concept that builds on the distinction between social and personal identities is the authentic self. We typically become aware of our authentic self when how we view ourselves is at odds with normative guidelines concerning appropriate or desirable behaviours attached to our roles. The sociologist Arlie Hochschild (1983) has written insightfully about the authentic self. She looked at how people align their presentations to conform to employers' expectations, and her research focused on the airline industry's expectations for flight attendants. She found that managers and trainers instruct flight attendants in more than the specific tasks that they are expected to accomplish, such as serving meals and giving safety instructions. They train them to control their own feelings and expressions in order to shape passengers' emotions, keeping them calm and satisfied. Hochschild demonstrates that even when flight attendants are confronted by obnoxious and unruly passengers, they can usually control their inner feelings and keep smiling. Hochschild then asks how such 'deep acting' by flight attendants, what she calls *emotional labour*, affects attendants' internal sentiments. She

OPEN for DISCUSSION

The Making of 'the Blind'

Both role embracement and role distance are demonstrated powerfully in a classic study by Robert Scott (1969) about the blind. Scott's central argument is that there is nothing inherent in blindness that makes those who are blind helpless, docile, and dependent. As with the rest of the human race, there is tremendous variability among the blind in terms of their characteristics, personalities, and reactions. Yet both agencies for the blind and the public tend to have stereotypical views of the blind. Services and workplace opportunities created for the blind are based on these stereotypes. So too are many interactions with the blind. The end result is that those with sight problems learn what society expects from them as a 'blind person' and feel pressured to play the role as others expect them to play it.

But not all blind people necessarily adopt the blind role. Those who embrace society's definition of who they are as blind people—their capacities, limits, preferences, propensities, emotions, and so on—become what Scott calls *true believers*. How society defines them and their sense of who they really are become one. They experience the emotions that those around them think they must feel and genuinely come to see themselves as in need of the assistance of others. There are others, however, who resist a definition of themselves as helpless. Scott calls this group *the expedient blind*. These individuals may in certain situations behave in ways consistent with others' expectations, not because they have internalized the qualities they display, but only to maximize their gains. In relation to agencies for the blind, for example, they know that unless they present themselves as helpless, they may not get the resources for which they have approached the agency in the first place (e.g., information about the latest optical aids, mobility training, or financial assistance). The expedient blind distance themselves from the role of 'blind person', bringing it out only when it is useful but otherwise disposing of the facade. Services and attitudes toward the blind may have changed in the years since Scott wrote his book, but it is worth asking whether the questions that Scott raised about social roles and identity are any less relevant today in relation not only to 'the blind' but also 'the learning disabled', 'autistics', 'the homeless', 'the poor', and other labels we use to construct people. What do you think?

concludes that over time the forced smiles become divorced from their true feelings. Such concerns are far from limited to flight attendants. They are an inherent part of the work performed by employees in a great many jobs in the service sector—for example, car salespersons, teachers, or entertainers. All require emotional labour both to present themselves and to control the responses of the people they serve.

PROTECTIVE IDENTITY WORK

Sociologists have also considered how social actors try to shape the view that others have of them in situations in which their identity or actions may invite a negative interpretation. In these situations, we often offer disclaimers, accounts, excuses, and justifications. We will define each of these concepts, but taken together, these verbal devices fall under the category of vocabularies of motive. The term was first used by C. Wright Mills (1940) to describe the standardized forms employed by people to explain and excuse their behaviour. Here it is essential to distinguish between motivation and motive talk. Motivation refers to whatever internal states, drives, needs, or compulsions create an impulse to act in a particular way. Motive talk is what we *say* about why we have acted in particular ways. Unlike motivation, which is generally inaccessible to an observer, motive talk is completely accessible, since it constitutes the stuff of everyday conversation and interaction. According to Mills, certain stated motives for our actions are more acceptable than others, and how we explain our actions will vary depending on who we are talking to.

On an intuitive level, each of us can relate to this, since we are constantly avowing and imputing motives, trying to explain others' unusual or unpredicted behaviour, and explaining, justifying, or excusing our own. For example, university students occasionally drink to excess. Friends may feel compelled to impute a motive such as 'His girlfriend dumped him; that's why it happened.' Or the student might explain, 'Had I not drunk on an empty stomach, I would have been fine' or 'I was so embarrassed that I apologized to everyone at the table.' The point is that motive talk is a common feature of everyday life. Motive talk arises, as Hewitt observes, 'whenever people are uncertain of the meaning of others' acts or of how others will interpret their own acts' (2000, 154). In each case, the object is to try to negotiate one's identity.

Looking more carefully at different types of motive talk, disclaimers are verbal devices used by persons who want to ward off the potential negative implications of an impending act—something they are about to do or anticipate will be criticized by others. We are all familiar with such remarks as 'I am totally in favour of religious freedom and for Muslim women to dress as they see fit, but . . .' Each phrase introduces a statement that contradicts the premise of the disclaimer. The person espousing religious freedom is explaining his objections to the distinctive garb worn by some Muslim women.

In contrast to disclaimers, which are prospective or future-oriented and used when we are anticipating in advance how others will respond, accounts are retrospective or after-the-fact. They are aimed at what has already happened. According to Scott and Lyman (1968), accounts are verbal statements made by one social actor to another to explain behaviours that were unanticipated. Accounts can take one of two forms: excuses or justifications.

People work especially hard to counter others' views in situations when the labels that may be applied are discrediting and stigmatizing. The damaging consequences of such labels is another area in which Erving Goffman has made a significant contribution. In a book called *Stigma: Notes on the Management of a Spoiled Identity* (1963b), Goffman pointed out that negative evaluations can be based on physical abnormalities or deformities, membership in a discredited group, or behaviour that deviates from a moral standard. The related concept of courtesy stigma describes a situation in which individuals find themselves dealing with a spoiled identity not by virtue of who they are, but because of their connection to an individual who has been labelled as deviant. For example, children of 'alcoholics', as much as their alcoholic parent(s), may find themselves dealing with others' negative reactions.

TABLE 5.3 ▼ Excuses and Justifications	
Excuses	Used when we admit our behaviour is wrong but deny responsibility for it. For example, if you are late picking up a friend, in saying that the traffic was terrible you are offering an excuse.
Justifications	Used when we accept responsibility but suggest that the behaviour should not be seen as wrong. In the same example, you offer a justification if you say that your friend should not complain since he or she always keeps you waiting as well.

A deviant label is consequential because it generally becomes a **master status**. Master status, as defined by Everett Hughes (1945), is a status that overshadows all other statuses in terms of how others see us. Consider someone who suffers from mental illness. Whatever else that person may be—father, business owner, marathon runner—in most situations that person will be judged primarily on the basis of his mental illness. Hughes suggested that statuses such as gender, race, class, and age are master statuses. For better or worse, we generally begin our interactions with assumptions about who people are on the basis of these characteristics. Deviant statuses work in the same way in that they have the power to shape how others see and interact with us.

Accounts offer one way to deal with the stigma attached to deviant labels. In their study of convicted rapists, Scully and Marolla found that some admitted their crime, offering excuses that often had to do with saying that they were under the influence of drugs or alcohol. Other rapists, a group that Scully and Marolla call deniers, offered justifications instead. They argued that they were merely responding to their victims' flirtatious behaviour or that their victims deserved it for dressing or acting in particular ways. Other deniers justified their actions by claiming that rape is only a minor wrongdoing: 'I shouldn't have all of this time [sentence] just for going to bed with a broad,' one rapist insisted (Scully and Marolla 1984, 537). Scully and Marolla end their paper by provocatively suggesting that the excuses and justifications that rapists offer—and their assumption that these will be heard in a way that might make us, their audience, judge them less harshly—reflect societal attitudes that still exist about rape.

Another form that accounts can take are techniques of neutralization. According to Sykes and Matza (1957), techniques of neutralization are ways of thinking that allow social actors to maintain a non-deviant image in the face of social disapproval. Like other accounts, they can be used before or after individuals engage in actions that might invite disapproval. Based on their research on youth offenders, Sykes and Matza identified five such techniques (see Table 5.4).

Techniques of neutralization have been extensively studied by sociologists and criminologists in relation to a variety of groups. Wife batterers use *denial of responsibility* when they claim that as a result of the violence they experienced themselves as children, they do not know how to deal non-violently with conflict in their relationships (Cavanagh et al.

TABLE 5.4 ▼	Sykes and Matza's Techniques of Neutralization
denial of responsibility	'I'm not to blame.'
denial of injury	'No one got hurt.'
denial of victim	'They deserved it.'
condemning the condemners	'Who are you to judge me?'
appealing to a higher loyalty	'I didn't do it for me.'

2001). Pedophiles use *denial of injury* when they argue that their behaviours do not really hurt children (De Young 1988). Shoplifters use *denial of victim* when they say that department stores deserve it for charging such high prices (Cromwell and Thurman 2003). Tax evaders use *condemning the condemners* when they argue that everybody cheats on their taxes, even judges and police officers (Benson 1985). Mothers who put their young children in beauty pageants, where they are often presented in highly sexualized ways, use *appealing to a higher loyalty* when they argue that the money their children earn is significant and may one day pay for their university tuition (Heltsley and Calhoun 2003). The growing number of studies on stigma management capture the endless creativity of social actors in negotiating their identities and protecting a positive sense of self.

TIME to REFLECT

Think of a transgression you have committed—cheating on an exam or partner, buying an essay, lying to a friend, and so on. What technique(s) of neutralization did you use?

IDENTITY CHANGE

Sociologists are also interested in how individuals move through statuses, roles, and identities. Anselm Strauss (1959) has referred to these movements as *status passages* and to the key junctures along the way as *turning points*. According to Strauss (1959, 93), turning points are critical incidents that signal to individuals, 'I am not the same as I was, as I used to be.' How these turning points are experienced depends on whether the changes are voluntary or involuntary, desirable or undesirable, important or insignificant, sudden or gradual, planned or unexpected, reversible or irreversible, and individual or collective (Sandstrom, Martin, and Fine 2006, 77).

SOCIOLOGY in ACTION

Protecting the Self: Stigma Management Strategies

Sociologists who have studied how human beings manage stigmatized identities have shown how endlessly creative we are in developing self-protective strategies:

- Ex-politicians deal with the stigma of defeat by attributing their loss to their political party, the party leader, the party's unpopular decisions, unfair media coverage, or even the timing of the election itself (Shaffir and Kleinknecht 2005).
- Male cheerleaders try to counter the stigma of being men in a 'women's' sport by emphasizing the physical stamina required, displaying aggression, and sexually objectifying their female teammates in comments about the great opportunity the sport affords to meet 'hot' girls and 'grab butts' (Bemiller 2005).
- Funeral directors who suffer the stigma connected with handling the dead in a culture that seems to have issues with death, respond by avoiding telling people what they do for a living in situations when they do not have to, by presenting themselves as upbeat though still professional, and by using symbolic language to redefine what they do so that 'death' becomes 'eternal slumber', 'corpses' become 'loved ones', 'morticians'

become 'funeral directors' or 'grief counsellors', and 'graves' become 'final resting places' (Thompson 1991).
- Pit bull owners who find themselves dealing with courtesy stigma by virtue of the changing attitudes towards pit bulls respond by passing their dogs off as a breed other than pit bull, taking a 'don't blame the dog' approach and faulting owners who do not control their dogs, debunking media accounts of 'vicious pit bulls', emphasizing their dog's personality (cuddly, sweet, docile) over their muscular appearance, and avoiding stereotypical equipment and accessories like studded leather collars and muzzles (Twining, Arluke, and Patronek 2000).
- Students who work as strippers use their socially acceptable identity as 'student' to avoid seeing themselves as they know most others see them—that is, as 'sleazy'. They frame their dancing as a transient occupation, something they do only to make their studies possible. They are particularly eager to make this clear to customers and co-managers. In relation to most others, including family and fellow students, they are discreet and carefully manage information about their stripping (Trautner and Collett 2010).

An interesting example is offered by McMaster University sociologist William Shaffir (1991), who compared the experiences of secular Jews who decide to become Orthodox Jews (*baalei tshuvah* in Hebrew) with the experiences of Jews born into ultra-Orthodox communities who chose to pursue a more secular way of life (*haredim*). There were significant parallels in the two experiences. Both involved a radical transformation in lifestyle, relationships with others, and self-definition. Both experiences, Shaffir pointed out, can be characterized as a conversion in that each 'involves the adoption of a pervasive identity which rests on a change at least in emphasis from one universe of discourse to another' (Travisano 1970). Shaffir also noted a significant difference: the *baalei tshuvah* changed their behaviours and presentation of self before completely redefining themselves. They quickly adopted the appropriate behavioural trappings of Orthodox Judaism, including dress, language, study, and prayer, hoping to eventually acquire the requisite attitudes that corresponded

to their behaviour. While outwardly displaying confidence in their behavioural commitment to Orthodox Judaism, they inwardly engaged in realigning their understanding of themselves and life's meaning to fit the requirements of their new lifestyle. This sequence was reversed for the *haredim*, who left their community and took on a more secular lifestyle only after going through a process of questioning their authenticity as part of the *haredi* community and undergoing attitudinal and emotional transformations.

A related line of inquiry is concerned with the stages through which identities are acquired and shed. An example of the former is provided in a study of the bisexual identity. Weinberg, Williams, and Pryor (1994) described the adoption of a bisexual identity as a gradual process that starts with *initial confusion* about one's identity—confusion heightened by the fact that one has strong sexual feelings toward not opposite-sexed or same-sexed individuals but both. One person recalled: 'When I was young, I didn't know what I was. . . . I thought I had to be either gay or straight' (Weinberg, Williams, and Pryor 1994, 28). The next stage is *discovering and applying the label*. Discovering that the category of bisexuality existed was a turning point for most participants and provided them with a means of making sense of their feelings. The third stage was *settling into the identity* and signalled a more complete transition into self-labelling. Participants also used the term 'self-accepting' to describe this stage. The final stage was *continued uncertainty*, created in part by the intolerance of the heterosexual world but more powerfully by social disapproval from the homosexual world. Segments of the gay community believe that no one is really bisexual and that in not accepting their homosexuality or heterosexuality, bisexuals are clinging to an inauthentic identity. In the words of one participant, 'Their [the gay community's] negation of the concept and the term bisexual has sometimes made

me wonder whether I'm just imagining the whole thing' (Weinberg, Williams, and Pryor 1994, 35).

At the other end of the continuum is how individuals experience leaving roles and identities behind. The process of exiting from a social role and shedding an identity has been studied by Helen Rose Ebaugh (1988). Ebaugh became interested in role exits as a result of her own experience leaving the convent and her life as a Catholic nun. Curious about the extent to which her own experience mirrored what others went through, Ebaugh conducted 185 interviews with various sorts of 'ex's': those who had exited occupational roles (ex-nuns, ex-doctors, alumni, retirees), family roles (divorcees, mothers without custody of their children), and deviant roles (ex-convicts, ex-alcoholics, and transsexuals). Her focus was on voluntary exit from significant roles. On the basis of her interviews, Ebaugh developed a model that divides the exiting process into four stages.

HUMAN DIVERSITY

Constructing a Transsexual Identity

Stories are like containers that hold us together; they give us a sense of coherence and continuity. By telling what happened to us once upon a time, we make sense of who we are today. (Mason-Schrock 1996, 176)

Preoperative transsexuals (transsexuals who have not yet undergone body-changing procedures) provide an intriguing opportunity to study the process of self-construction. As Douglas Mason-Schrock points out, transsexuals typically feel that they are born into the wrong-sexed bodies and want to remedy the problem. But in a culture where the body is taken as an unequivocal sign of gender (sex = gender), it is not easy to claim a 'true self' that is at odds with their body—that is, for those with male bodies to claim that they are truly women and for those with female bodies to claim that they are truly men. Hence, transsexuals must look elsewhere—beyond their natural bodies—for signs of the gendered character of their 'true selves'.

Based on interviews and participant observation with transsexuals in a transgender support group, Mason-Schrock argues that a transsexual identity is achieved by constructing a self-narrative that supports a differently gendered 'true self'. The key building blocks of these self-stories are:

- recollections of childhood events (e.g., real or fantasized cross-dressing) that serve as evidence that the individual has *always* been different;

- accounts that explain away any discrepant biographical data—like male-to-female transsexuals having excelled in sports or female-to-male transsexuals having had an interest in makeup and jewellery—usually by claiming that the individual was in a state of denial at the time.

The more interesting aspect of Mason-Schrock's study, however, is the finding that these self-narratives are not constructed in isolation. Transsexuals did not engage in this identity work alone. Rather, the transgender support group played a critical role. The support group encouraged those within the group who were adept at telling these self-narratives to share them so that newcomers could *model* these tellings. Established members *guided* newcomers through the construction of their own stories by asking specific questions that highlighted certain life events. They *affirmed* the production of the necessary elements of an adequate self-narrative by smiling, sighing, nodding, and so on when certain biographical events were recounted. When the emerging stories were inconsistent, implausible, or filled with 'loose ends', group members engaged in *tactful blindness*, ignoring problematic bits while the tellers smoothed their stories out. In engaging in this work, Mason-Schrock argues, the group members assured the production of a self-narrative that could sustain a transsexual identity.

The initial stage, *first doubts*, is a period through which individuals start to question roles and identities that they have taken for granted. These doubts can be precipitated by disappointments, burnout, organizational changes, or changing relationships with others. The second stage, *seeking and weighing alternatives*, involves considering options, weighing the pros and cons, seeking out new reference groups, and in some cases, rehearsing new roles. The third stage, *turning point*, is reached when the individual decides that there is no turning back. At this point, individuals are ready to 'announce to the world' that they have left an old role behind. The final stage, *creating an ex-role*, involves adjusting to the new self-definition, including how to manage the new presentation of self and dealing with one's own and others' expectations attached to the hangover identity.

TIME to REFLECT

Think of a role that you have exited. Do Ebaugh's four steps describe your experience?

CONSTRUCTING NEW IDENTITY LABELS

We have focused to this point largely on how we define ourselves and others. But where do the labels we use come from? The list of options is not stable or fixed. Previous generations did not have available to them labels like 'gifted child', 'co-dependent', 'ombudsperson', 'queer', or 'drunk driver'. On the other hand, we generally no longer use labels like 'blasphemer', 'apothecary', 'draft dodger', or 'debutante'. Moreover, the meaning of labels can change dramatically over time. What it means to be 'a woman' or 'a man' today is different from what it meant a generation ago. The responsibilities, expectations, behaviours, and privileges associated with these roles have changed. Similarly, the meanings of such labels as 'homosexual' or 'smoker' have changed, the first having become less stigmatizing while the latter has become more stigmatized. How have these changes happened? These are questions that social constructionists have addressed.

An off-shoot of symbolic interactionism and influenced as well by Peter Berger and Thomas Luckmann's book, *The Social Construction of Reality* (1966), social

Aggressive anti-smoking campaigns spearheaded by public health authorities have pushed smoking from a widely accepted behaviour to an increasingly marginalized activity.

© Gianni Muratore/Alamy

constructionism is concerned with how we come to understand the world and the things in it. A basic assumption is that what goes on in the world is not given and self-evident. Reality requires interpretation. As social actors, we have to make sense of what we hear and see around us. In making sense of the world, we are constructing the world in particular ways. For example, we experience childhood as a universally given aspect of life; the idea of childhood, however, did not actually emerge until the eighteenth or nineteenth century. Until then, human beings that we now categorize as 'children' were seen as miniature adults. They were not dressed or treated differently from adults (Aries 1962). If we see children as a separate category with certain identifiable characteristics—dependent, innocent, not responsible, precious—this is because we construct children this way.

Another basic assumption is that when our constructions shift, these shifts do not just happen. In each case, shifts are due to the activities of *claims-makers*—individuals seeking to persuade audiences to take note of aspects of reality they may never have noticed before or to view things differently. If we think differently about what it means to be 'a woman' from the way our grandparents did, the difference is because of the claims-making activity of the women's movement and its claims that traditional views of women—their abilities, proclivities, and desires—were sexist and constraining. If to be called 'a smoker' is a stigmatizing experience today in ways that it was not a few decades ago, we can attribute the change to the successful claims-making activity of the anti-smoking movement. If attitudes toward homosexuality and the rights of homosexuals are different today from what they were in the past, they have changed because of the claims-making activities of the gay rights movement.

Constructing understandings of how the world works and what is going on inevitably involves constructing people in particular ways. In making this point, Loseke (2003, 121) uses the concept of *people categories*. People categories are the slots we create to describe different types of people that exist in the world. We use these categories to define ourselves and others. In our encounters with individuals, defining who any given person is involves figuring out what category to put that person into so that we know how to react. Even in cases when the goal of claims-making activity is to construct particular conditions as problematic, new people categories are created in the process. Take the example of child abuse. Again,

although it is difficult to imagine a time when the beating of children was not seen as a problem, in the past there were alternative understandings of child-beating. At certain points, beating children was seen as a parental prerogative. Since children *belonged* to parents, parents could treat them as they wished. At other points in time, the beating of children was viewed as responsible parenting. If children were not properly disciplined—with discipline understood as a physical act—they would not grow into responsible and contributing adults (Davis 2003). An understanding of the beating of children as a social problem is a relatively recent invention. In fact, the term 'child abuse' itself only appeared in the 1960s.

The child abuse label became available as the result of claims-making on the part of pediatric radiologists who drew attention to the problem of 'beaten children'. Following the lead of pediatric radiologists, other groups mobilized to promote a different understanding of, and societal response to, the beating of children (Pfohl 1977). As a result, child abuse was constructed as a social problem. In the process, however, two new people categories were created: 'the abused child' and 'the child abuser'. When a parallel claims-making process emerged in the 1970s around the wife-battering problem, new categories were created there too: 'the battered wife' and 'the wife batterer' (Loseke and Cahill 1984). Eventually, these parallel developments also gave rise to categories such as 'family violence counsellor', 'shelter worker', and 'child advocate', among others.

TIME to REFLECT

Consider the claims-making efforts of a group like People for the Ethical Treatment of Animals (PETA) or Greenpeace. What new identities or people categories have these claims-making efforts generated?

To observe that child-beating, wife-battering, and other behaviours and conditions are socially constructed is not to challenge these constructions. The point is to recognize the interpretive work that social actors do when they see things in one way as opposed to another and to recognize as well that in creating these understandings, new people categories are created, old people categories disappear, and the meanings of existing people categories change. This has implications, of course, for the menu of possibilities available to us when we, as individuals, define ourselves or others. These people categories are the materials we use to construct identities. And the meaning of any particular people category, including whether it describes a valued identity or not, will have consequences for how we experience ourselves and others.

Conflict Perspectives

As macrosociological approaches, conflict or critical theory perspectives in sociology have paid less attention to roles, self, and identity. The concern among these sociologists has been power—its unequal distribution, forms, sources, and consequences, especially in relation to gender, race, and class. There are exceptions. McMaster University sociologist W. Peter Archibald (1976) used Marx's concept of alienation to understand how inequality is experienced by social actors and how, or even whether, people in groups with differing degrees of power interact with each other. Archibald argued that more privileged groups tend to avoid the less privileged. Many organizations, for example, have separate entrances, washrooms, and eating facilities for those at the top and for those at the bottom of the hierarchy. When interaction does occur, it happens in formal, role-specific ways and tends not to be personal. Similarly, feminist theory, a version of conflict theory that is particularly concerned with gender inequalities, has generated studies of how women experience the forces of patriarchy. For the most part, however, there has been comparatively less interest in looking at micro-level interactions or how social actors understand themselves, except to point out that these actors often do not see the ways in which they are marginalized and oppressed, a phenomenon referred to as false consciousness.

Recent versions of conflict theory (including post-structuralism, cultural studies, feminist theory, and queer theory), however, have shone the spotlight on how identity is implicated in the exercise and experience of power. This growing interest has been part of a change in how many conflict theorists understand power. Rather than viewing power as something that resides in, and is exercised by, certain classes or groups, the new approach is to see power as something that is located in the discourse (i.e., beliefs, ideas, taken-for-granted assumptions about reality, and so on) that people use to understand themselves and the world around them. For example, both women and men are oppressed by the assumptions and expectations surrounding who they are and how they should behave. To the extent that we buy into the notion that certain behaviours are 'naturally' feminine or masculine and take on one identity or the other, our behaviours are likely to fall into line. In this form, power is not coercive but disciplinary. No one has to 'make' us behave in normative ways. We regulate our own behaviour. We do so by defining ourselves in particular ways and then trying to be true to those definitions. The task for those who take this approach is to uncover or deconstruct how these discourses about who we are as individuals of a particular race, class, or gender regulate and control us.

Conflict sociologists who take this approach recognize that social actors can and do push back. In fact, what has sparked the recent interest in questions of self and identity is the organization of those who see themselves as experiencing marginalization and social injustice into groups whose goal it is to challenge dominant, oppressive characterizations of who they are. These efforts to control how certain

GLOBAL ISSUES

The Globalization of Self

How do the forces of globalization shape the self? This is the question raised by Peter Callero (2008). Callero argues that on the one hand, there are top-down forces associated with globalization that serve to *colonize* the self. By this he means that globalization disrupts or erases traditional roles (e.g., steelworker, logger, miner) that are no longer needed when corporations shift production to regions with

lower labour costs, and at the same time creates new roles (e.g., international financier, tour guide) required for global consumption and production. On the other hand, he insists, globalization also creates bottom-up forces. Those hurt by globalization create grassroots organizations and social movements that facilitate resistance. In doing so, they support the emergence of activists and political selves.

individuals and groups will be defined have been referred to as **identity politics**. One of the first such efforts was the civil rights movement in the US, which sought to liberate those who were marginalized on the basis of race. Since then, groups have organized around identities rooted in ethnicity, religion, gender, sexual orientation, socio-economic status, and age, to name just a few.

Conclusion

If sociology is the study of the relationship between individuals and society, there is no question more central than how our statuses, roles, identities, and sense of self connect us to those around us. Our goal in this chapter has been to discuss how sociologists have defined and thought about these concepts. We also sought to introduce readers to the types of sociological studies that these concepts have generated. We started with structural functionalists, explaining the emphasis placed on statuses, roles, and the more or less predetermined role scripts attached to them. Symbolic interactionists, we went on to explain, reject a view of social actors as mere role players, stressing instead the agency that individuals exercise in deciding what roles to take on and how to play those roles, as well as in defining themselves and others. Finally, we described briefly how sociologists inspired by conflict or critical perspectives have focused attention on the ways in which identities work to oppress and discipline individuals and the identity politics that this has given rise to in recent years.

Sociologists continue to debate the nature of selves and identities. Some sociologists have suggested that in a rapidly changing world where technology is dramatically expanding the size of the stage we play on, our relationships are less stable. If who we are is dependent on those with whom we interact, a single, coherent, unified self is less likely. Our sense that there even needs to be a consistent storyline that explains for us and for others who we are, at our core, may have diminished. The self has become fragmented and mutable—that is, constantly changing (Gergen 1991; Giddens 1991; Zurcher 1977).

Other sociologists, while agreeing that the foundations to which our selves are anchored are expanding and changing, insist that the idea of a core or personal self continues to be important. Indeed, with the increasing range of possible identities we can adopt in a rapidly changing world, it may be even more important for us to construct and protect a personal and unitary sense of who we truly are to see us through all of the situations we encounter (Holstein and Gubrium 2000).

Wherever one falls in this debate, and whatever sociological perspective one uses to think about these questions, there is certainty about one thing. For as long as there are individuals who ponder who they are and how they fit in, sociological interest in these issues will continue.

questions for critical thought

1. Thinking about the connection between gender and role performance, can you think how this relationship is shaped by cultural, political, and social contexts both in our society and elsewhere?

2. Can you recall a role you played—son, daughter, boyfriend, girlfriend—that has shifted in how you have played it over time? What prompted the change(s)?

3. Can you think of an instance in which you and/or others felt you had defined a situation inappropriately? What happened? How did you figure out that you had made a mistake? Was the error obvious, or were you able to hide it from others?

4. Motives offered are intended to explain the unexpected and, where necessary, to repair damage to one's identity. How do people know how to respond when asked to explain their actions? How do they know what counts as an acceptable or unacceptable motive?

5. What are typical accounts offered by students for skipping classes, partying to excess, engaging in plagiarism, or handing in assignments late? What kinds of accounts are they—excuses or justifications? How do students learn and perfect these accounts?

recommended readings

Anthony Giddens. 1991. *Modernity and Self-Identity: Self and Society in the Late Modern Age.* **Stanford, CA: Stanford University Press.**
In one of the most clearly argued statements about how the experience of self and identity may be changing, Giddens emphasizes the challenges of finding personal meaningfulness and an authentic self in a society where trust and intimacy have become problematic.

Erving Goffman. 1959. *The Presentation of Self in Everyday Life.* **Garden City, NY: Doubleday.**
This is Goffman's classic statement on the dramaturgical approach. 'Must' reading for anyone interested in identity work.

John P. Hewitt. 2006. *Self and Society: A Symbolic Interactionist Social Psychology.* **10th edn. Boston: Allyn and Bacon.**
A good introduction to the central concepts of symbolic interactionism, filled with useful examples that make the perspective accessible to anyone reading about it for the first time.

Adina Nack. 2008. *Damaged Goods? Women Living with Sexually Transmitted Diseases.* **Philadelphia: Temple University Press.**
An excellent example of how a chronic health condition can affect identity. The book also deals extensively with the stigma attached to women's having a sexually transmitted disease, how women manage this stigma, and what the stigma tells us about the tensions between morality and sexuality in Western cultures.

Sherry Turkle. 1995. *Life on the Screen: Identity in the Age of the Internet.* **New York: Simon and Schuster.**
This is a provocative book, full of interesting case studies of heavy users of the Internet. Turkle explores such issues as the construction of multiple online identities and why people present themselves in ways so radically different from who and what they are in their non-virtual lives.

Dennis Waskul and Phillip Vannini. 2006. *Body/Embodiment: Symbolic Interaction and the Sociology of the Body.* **Hampshire, UK: Ashgate.**
This collection deals with the interrelationship between the body and the self, with papers on body-builders, exotic dancers, female professional football players, opera singers, pregnant/birthing mothers, and the disabled.

recommended websites

Global Sociology
http://globalsociology.com/tag/identity
This is a website maintained by sociologists for those interested in sociological perspectives on globalization, including identity issues.

The Mead Project
www.brocku.ca/MeadProject
This website is a useful inventory of documents by or about the founder of symbolic interactionism, George Herbert Mead, and other Meadians, including Herbert Blumer.

MIT's Initiative on Technology and Self
http://web.mit.edu/sts/people/turkle.html
This is the website of Sherry Turkle, director of MIT's Initiative on Technology and Self, a centre of research on the relationship between artifacts and identity.

Self-Labelling and Identity
www.youtube.com/watch?v=pxbw7dDMX60
Here you will find a clip about how people who have had encounters with the mental health system describe themselves, powerfully demonstrating many of the concepts discussed in this chapter.

Society for the Study of Symbolic Interactionism
www.symbolicinteraction.org
The website of the Society for the Study of Symbolic Interactionism includes news about a range of conferences and other activities related to this theoretical perspective, as well as the contents of the journal *Symbolic Interaction*, which regularly publishes papers on self and identity.

Temple Grandin Website
www.templegrandin.com
This is the official website of Temple Grandin, whose life was depicted in a recently released HBO movie. In making the case that those with autism should be appreciated for their *neurodiversity* by those of us who are *neurotypical*, Grandin represents an intriguing example of someone trying to change the meaning of an existing identity label.

chapter 6

Groups and Organizations

Lorne Tepperman

In this chapter, you will:

▶ Learn about the different sociological 'sets' of people;

▶ Understand how cliques function inside schools and bureaucracies;

▶ Learn the history of the bureaucratic form of organization;

▶ Understand the characteristics of a bureaucracy;

▶ Identify the differences between the ideal bureaucracy and the real-world bureaucracy;

▶ Read about the Hawthorne studies and their impact on sociological thinking;

▶ Distinguish between the main points of organization theory.

Introduction

This chapter is about 'social forms': basic elements of social structure that include groups, networks, communities, and organizations. In fact, with the previous chapter on roles and identities, this chapter maps the most fundamental elements of social structure. Perhaps the chapter should have been titled 'Cooperation', because it is about the benefits of cooperation: in particular, about the ways that cooperation through a division of labour (and specialization) can benefit people. Cooperation in groups and organizations makes people more productive; in this sense, groups and organizations are more than the sum of their parts.

One founder of sociology, Georg Simmel, defined social interactions and social forms—the essential features of groups and organizations—as the basic subject matter of sociology. In his 1908 essay 'Individuality and social forms', Simmel notes that these two elements are only distinguishable analytically. One element is the purpose or motive of an interaction. The other is form, the form of interaction among individuals through which the interaction communicates its meaning. This is important, because as we will see, groups and organizations work similarly despite differences in their size and purpose.

Fashion is a good example of this distinction between content and form, because it speaks to both at the same time. Fashionable dress allows people to display themselves in a way that (they feel) displays their individuality; yet fashion, by nature, is social. The group as a whole decides what is in fashion or out of fashion. Individuals themselves cannot make fashion—they can only select from it and try to personalize it. In other words, fashionable clothing is the social form into which we pour our individual desires for adornment and self-presentation.

Social forms emerge more or less unbidden. One strong proof of this was provided by American sociologist Robert Bales, who in the 1950s studied training groups (known as T-groups) at Harvard University. These groups of undergraduates were recruited to meet once a week over the course of a year and discuss designated topics or solve problems together while being watched through a one-way mirror. Behind the mirror, researchers recorded all the words and actions of the participants, using a specially designed interaction measure.

Bales's study revealed that each of his groups, unexpectedly, produced three social forms: a task leader, an emotional leader, and a joker. The task leader helped the group organize itself to solve the problem that had been posed; he or she helped to set goals and organize the work. The emotional leader helped the group cope with frustration and conflict so that strong feelings did not deflect the group from its task. In effect, the emotional leader was the peacemaker. Finally, the joker was—as the name suggests—the person who helped release tensions in the group by joking and fooling around. This seemingly slack time-waster was as important as all the other group members. Without the periodic tension release of joking, the group might not have performed successfully.

This suggested that groups, to survive, need some of their members to perform special kinds of tasks—as task leader, emotional leader, joker, or otherwise. Eventually, in every group, different people step forward to fill these roles, or the group breaks up. This happens without conscious planning or even a conscious awareness of group needs. If the group survives, the process just happens, and vice versa.

Today, most sociologists would agree that groups have their own logic and follow certain typical patterns. Group members often invest themselves emotionally, socially, and financially in the groups to which they belong, so they willingly take actions to protect the group; in other words, they protect their investment. Second, people with a large stake in the group, who receive particular benefits (for example, authority and esteem) from group membership, will do their best to keep the group alive.

However, despite the benefits that may flow from group membership, some groups are better at surviving than others. Take a simple example: well-functioning families. Family sociologists know that cohesive (or tight-knit) and adaptable (or flexible) families are best able to deal with stressful conditions. Families without a history of cohesion and adaptability are likely to break down under difficult circumstances, leading to divorce, violence, or addiction, for example.

With these ideas about social forms in mind, consider a few general observations. First, all small or primary groups have similar characteristics and patterns, whatever their purpose or goal. For example, they are all based on intense, face-to-face interaction, and the members tend to identify with one another. Often in such groups, the emotional boundaries between members are blurred. People identify closely with the group and with one another and find it hard to leave or betray the group.

Second, in every such group, certain structural changes will significantly affect how the group operates. For example, demographic factors like size and turnover will affect small groups (even more dramatically than they affect entire societies, in fact). Imagine a family of, say, two parents and two children. Now, double the size of that family—two parents and six children—and see what happens. Increase the rate at which parents or children enter and leave the family, and see what happens. Change the mix in the family: for example, substitute a stepfather for the biological father, a stepmother for the biological mother, or stepsiblings for the biological siblings, and see what happens.

Teams, bands, and gangs (I suggest the abbreviation TBG) are three types of larger groups. Unlike families, they do not always command our primary social loyalty. However, many people consider the teams, bands, and gangs they belong to almost like surrogate families. What is interesting is that TBGs are very similar, despite the different goals of their members. A basketball team has different goals from those of a marching band, which in turn has different goals from a motorcycle gang.

TBGs are also different from much smaller groups (e.g., families), larger groups (e.g., college classes), and formal organizations, to be discussed shortly. First, unlike a family situation, people are not born into TBGs; rather, joining them is a matter of choice. Second, they typically join TBGs because they want to be members, not merely as the means to another end (i.e., more than a wish to earn a course credit or paycheque, for example). In every case, people join because they want to identify with, and be identified with, the team, band, or gang.

Each TBG has a clear set of goals and main activities. And given these goals, the TBG has a leadership structure. One or more leaders has the job of setting goals, mobilizing resources to achieve these goals, and motivating members to take part, according to group rules. In these respects, TBGs are like families, with the leaders acting as parents. TBGs also have a simple political structure (with leaders and followers), legal system (with procedures to resolve conflicts), economy (with a treasury and assets), and culture (with a shared memory of great events, heroes, and villains).

Bands, like teams and gangs, work according to certain unwritten rules. For example, band members do not openly criticize one another. If they have concerns about another member's performance, they direct these concerns to the leader, and the leader deals with the problem. Members may also quietly suggest improvements or grumble discreetly to another band member. Often, they use jokes to express irritation or frustration. Even the leader may express criticisms in a gentle or roundabout way, without pointing a finger of blame at the particular offender. A spirit of cooperation and collegiality is hard to preserve if members are criticized harshly, unfairly, or unequally.

THE STUDY OF 'SOCIAL ORGANIZATION'

The study of 'social organization' in its broadest sense, then, touches on two main questions: (1) How do people typically act in social groupings of different size and purpose? (2) How could we organize social groupings to increase the chances that people will achieve their collective goals? We can address these questions to a wide variety of social groupings ranging in size from (two-person) dyads to cliques, small groups, large groups, social networks, communities, and formal organizations. We will start small and

In the First Person

Though I fell into sociology by accident—I didn't want to be a doctor, a dentist, or even a lawyer—three main influences led me in that direction. One was my Aunt Toby. She was what people used to call a 'bohemian'—a political activist as well as an actor, drama teacher, and book author. She always urged me to listen to social critics and take an international viewpoint. My sociology teachers were a second influence—especially Jan Loubser (who pointed me toward Talcott Parsons) and George Homans (who pointed me away). Among other things, they taught me the fun and grandeur of social research. A third influence was music, which I have studied and performed much of my life. Music teaches a love of pure forms. Music and sociology are both about principles of order and variation that underlie complex, seemingly unique events. What Georg Simmel meant by 'social forms' makes intuitive sense to me because of my musical background. Nowhere is this more relevant than in the study of groups, networks, and organizations.

—Lorne Tepperman

build up to bureaucracies, because large and small organizations are more similar than you might think.

First, as we shall see, large groupings (such as bureaucracies) contain networks, groups, and cliques. These small, informal organizations perform much of the work of large, formal bureaucracies. Sometimes they also subvert the plans and efforts of these bureaucracies. Second, many of the same organizational principles that shape small groups, cliques, and communities also shape large, formal organizations.

Large organizations of the kind we see today are still fairly new in human history, and in some ways they are a major human accomplishment. Yet in many respects, they have also become a source of social risk and harm. This chapter discusses this central problem of large organizations: namely, that they are so effective and yet potentially so dangerous.

Sets of People

To better grasp the idea of social forms, imagine five sets of 20 people. Call them *categories, networks, communities, groups,* and *organizations.* Sociologists study these five sets differently because they are organized differently and have different effects on their members.

CATEGORIES

Imagine, first, that these 20 people are a mere collection of people unconnected with one another—say, a random sample of Canadian 19-year-olds—but fall into the same category: in this case, they are the same age.

This sample of teenagers is of interest to sociologists if they represent the attitudes and behaviours of 19-year-olds across the country. Knowing these attitudes and behaviours may help us to predict the future behaviour or explain the current behaviour of 19-year-olds. However, few sociologists will be interested in such samples of people. Since they are unconnected, people in categories have no social structure of interest. It is social structure—the invisible feature of social life that controls and transforms our behaviour—that is mainly of interest to sociologists.

Categories only become sociologically interesting when people dramatize (or socially construct) meanings for the differences between one category and another. No such meaningful boundaries exist for 19-year-olds, compared with 18- and 20-year-olds. However, important cultural boundaries exist between the categories named 'male' and 'female', 'young' and 'old', and, in some societies, 'white' and 'black'. As a result, these categories assume social importance.

TIME to REFLECT

Why do you think male and female categories have more social importance than the 19-year-olds category? Under what conditions do you think the 19-year-olds category would have greater social importance?

NETWORKS

Generally, sociologists are more interested in what they call networks, or **social networks**. Imagine the same 20 people all connected to one another, whether directly or indirectly. By *direct connections*, we mean links of kinship, friendship, and acquaintance among all 20 people, each connected to the other.

Indirect connections are also of interest to sociologists. In fact, some sociologists, such as Mark Granovetter (1974), argue that *weakly tied networks*, based largely on indirect links, may be even more useful than *strongly tied* or *completely connected networks*. Information, social support, and other valuable resources flow through incompletely connected, or weakly tied, networks.

In recent years, Internet-based social networking services such as LinkedIn, Friendster, and Facebook have rapidly increased in popularity. These services collect information from an individual's profile and their list of social contacts to create a display of their personal social network. Such networking services claim that by allowing members to 'get to know one's friends of friends [they can] expand their own social circle' (Adamic and Adar 2005, 188). Increasingly, people are setting up virtual networks of relationships in cyberspace as well as real ones.

As you can see from this glimpse, social networks are important and interesting. However, much of social life is not well understood in terms of networks. That is because networks lack several key characteristics. First, people in networks lack a sense of collective identity such as a community would have. Second, people in networks lack an awareness of their membership and its characteristics, such as a group would have. Third, people in networks lack a collective goal, such as an organization would have.

COMMUNITIES

Sets of people with a common sense of identity are typically called **communities**, and there is a long history of community studies in sociology. Imagine, for the sake of consistency, that we are thinking of a community of only 20 people—say, a community of

PUBLISHED IN 1862 BY CURRIER & IVES

THE CURRIER & IVES FOUNDATION

CENTRAL-PARK, WINTER:
THE SKATING POND

(Courtesy of the Currier and Ives Foundation).

American popular artists Currier and Ives captured nineteenth-century North American conceptions of 'community', in this case, the pleasure of sharing one another's company in a simple holiday pastime—skating on the frozen pond.

like-minded people living together on the land (perhaps a hippie commune in 1960s British Columbia or a utopian farming community in nineteenth-century upstate New York) or in the city (perhaps a community of anarchist or bohemian youth living in a broken-down squat in twenty-first-century Amsterdam).

These are likely to be people drawn together by common sentiments, or they may be people who have grown up together and share strong values uncommon to the rest of society. The nineteenth-century German sociologist Ferdinand Tönnies (1957 [1887]) distinguished community life, which he called *Gemeinschaft*, from non-community life, which he called *Gesellschaft*. Tönnies also associated community life with rural areas and non-community life with urban areas.

Gemeinschaft refers to the typical features of rural and small-town life. They include a stable, homogeneous group of residents with a strong attachment to one particular place. Not only are their lives similar, they are also linked by intimate, enduring relationships of kinship, friendship, neighbouring, and (often) working together. Because rural people share

so much, it is not surprising that they also share similar moral values, and moral guardians such as the church, school, and local upper classes protect these values. The *Gemeinschaft* is marked by dense or highly connected networks, centralized and controlling elites, and multiple social ties.

By contrast, city life is characterized by *Gesellschaft*. This kind of organization brings together a fluid, diverse group of residents with different personal histories and impersonal, brief relationships. They interact around similar interests, not similar characteristics or histories. They share few moral values and few moral guardians to enforce a common moral code. In cities, people's social networks are less connected, less centralized, less cliquish, and less redundant. In short, people who live in a *Gesellschaft* are less cohesive and, largely for this reason, less controlled.

Sociologists since Tönnies have debated whether *Gesellschaft*—especially city life—represents a loss of community or a new kind of community. Most sociologists today believe that people are not as isolated and atomized in large cities as previous sociologists thought. Rather, most city-dwellers form small communities based on friendship, whether they are residentially close or scattered.

Communities, whether urban or rural, real or virtual, are important because people are aware of their membership in them. They want the community to survive and may make large personal sacrifices to see that it does.

GROUPS

What all groups have in common is an awareness of membership. Also, members are all connected with one another (directly or indirectly), and to varying degrees they communicate, interact, and conduct exchanges with one another. To continue our example,

a 20-student classroom is one kind of group. It is more highly connected than a 20-person category, more self-aware than a 20-person network, but shows less solidarity (based on common values) than a 20-person community.

Since Charles Horton Cooley (1962 [1909]) wrote in the early twentieth century, sociologists have distinguished between primary groups and secondary groups. *Primary groups* are small and marked by regular face-to-face interaction; an example is a family household. Cliques, which we will discuss shortly, and work groups also fall into this category of primary groups. *Secondary groups* are larger, and many members may not interact with one another regularly. However, even in secondary groups there

GLOBAL ISSUES

Welcome to Leeds, Home of Poverty, Hatred, and Alienation

Is it about England?

There is a growing feeling among many of the Londoners I know that it was not so much radical Islam that attacked them, or al-Qaeda, or some deranged youth cult. It was Leeds that attacked London. Here was the other England, the impoverished, hateful, culture-devoid England, attacking the country's much more successful, happily pluralist urban pole.

Enough has been written about the brown-skinned Muslims of Leeds, and the social conditions that somehow caused a violent and effective terrorist group, and the beliefs behind such a group, to foment among its youth.

Let us turn for a moment to the white people of Leeds. To understand them, I dropped in on one of their more successful representatives, a 31-year-old retired squash pro named Nick Cass.

Mr Cass, a red-haired giant, is the Yorkshire organizer for the British National Party, the most successful of Britain's far-right political parties.

'We've always said it's not about individual cultures or races or religions. We just think that when you put them all together in the same space, it doesn't work,' he said as his infant son burbled away in the playpen beside him. 'It never has worked, not in any country in the world.'

Osama bin Laden could not have put it better. Infidels out! And Mr Cass has a plan to get them out. Yes, he said, it did complicate things that three of the bombers were not actual immigrants, so his party's back-on-the-boat argument would need some

refining. There would be a government program, a very expensive one, to pay people of Indian and Pakistani descent, along with Jamaicans and Africans (I didn't ask about Welsh and Scots) to return to their motherland.

After all, he explained, this is just how things are in Leeds. This business of racial harmony is strictly poncy London stuff.

'There's never been any integration in Leeds,' he said. 'Except on an individual basis by a small number of people, not as a community. There's been no effort to integrate from either side of the community, white or Muslim, which tells me that people don't want to integrate.'

So what is it about England? What is it that can create, at the same time, the most successful and cooperative multicultural society in the world and one of the most hateful and racially segregated places in Europe?

This is something that the English try to avoid discussing.

The people of Leeds, a post-industrial mill town with little viable economy today, are poor. This is not simply an economic matter but a cultural one.

The poor and the not-poor in Britain are fixed, isolated worlds. If you're well-off, as Londoners tend to be, you get to have the most tolerant and resilient community in the world. If you're not, like both the brown people and the white people of Leeds, you get a permanent isolation cell of mutual, deepening hatred.

SOURCE: Doug Saunders. 2005. 'Welcome to Leeds, home of poverty, hatred and alienation'. *The Globe and Mail*, 16 July. www.theglobeandmail.com/archives/article891400.ece.

is a clear membership. There is also an identifiable normative order and shared sense of collective existence, as in a community.

Secondary groups, though less strongly integrated than primary groups, are no less important. We spend most of our waking hours as members of secondary groups, interacting, communicating, and exchanging resources with other people. Like primary groups, they bind people in fairly stable patterns of social interaction. Formal organizations, which we will discuss at length later, are subtypes of secondary groups. In turn, bureaucracies are subtypes of formal organization. So in the end, almost everything in this chapter—except the discussion of cliques—is about secondary groups. As we have noted, TBGs are hard to classify, because though larger than families, they have many family-like features.

ORGANIZATIONS

As just noted, organizations are secondary groups that have a collective goal or purpose. An organization can be a giant multinational corporation, like General Motors, or a small corner variety store; a political party or a government; a church, a school, a sports club, or a search party. Given the endless variety of organizational forms and the millions of specific examples, what do all organizations have in common?

Every organization comprises a group of people working together, co-ordinated by communication and leadership to achieve a common goal or goals. Within this general definition, however, organizations vary a lot. For example, the organization in question may come together spontaneously or deliberately. The division of labour within that organization may be crude or complex. The communication and leadership may be *informal* or *formal* (these terms will be defined shortly). The organization may have one specific goal or various loosely related goals.

One important distinction to make about organizations is between spontaneous and formal organizations. A **spontaneous organization** arises quickly to meet a single goal and then disbands when the goal is achieved. Perhaps the most commonly cited examples of spontaneous organizations are bucket brigades and search parties. They each have a single goal—keeping a barn from burning down or finding a lost child. Each arises spontaneously, and its leaders emerge informally, without planning. Each has a crude division of labour—for example, filling buckets, passing them along, emptying them on the fire. Each disappears when the job is completed.

Organizations with unstated goals and/or little division of labour are considered **informal organizations**. One familiar example of informal organization is the clique. Cliques seem different from the formal organizations discussed in most of this chapter, yet paradoxically, they share a number of features. Also, as we will discover later, cliques and other informal organizations often nest within formal organizations, doing much of the work.

Cliques

DEFINING THE CLIQUE

Dictionaries variously define *clique* as 'a small exclusive set', a 'faction', a 'gang', or a 'noisy set'. This meaning comes from the French *cliquer*, meaning 'to click', or 'to make a noise'. People in cliques—especially the

TABLE 6.1 ▼ Criteria Used by Enforcement Agencies to Define Youth Gangs

Gang Characteristic	Agencies Selecting As Most Important Criterion	
	Number	Per cent
Commits crimes together	613	50
Has a name	228	19
Hangs out together	119	10
Claims a turf or territory of some sort	104	9
Displays/wears common colours or other insignia	101	8
Has a leader or several leaders	89	7

NOTE: Number of observations = 1,221.

SOURCE: Mike Carlie. 2002. *Into the Abyss: A Personal Journey into the World of Street Gangs.* http://people.missouristate.edu/MichaelCarlie/what_I_learned_about/gangs/Table45.gif.

most popular ones—make a lot of noise, pumping themselves up and ridiculing others.

To come closer to our current sociological meaning, we would define 'clique' as a group of tightly interconnected people—a friendship circle whose members are all connected to one another, and to the outside world, in similar ways. Clique members spend more time with one another than with non-clique members, share their knowledge with one another, and think and behave similarly. They try to ignore or exclude outsiders—people not like themselves and not friends of their friends.

In short, cliques are built on friendship and the exclusion of 'outsiders'. In various respects, cliques are mini-communities, even like mini-states. Like states, they amass power and resources. They receive, censor, and direct information flow. They also produce information, distort it, and send information out as gossip and rumour. Because they produce and control the flow of information effectively, cliques are stable structures (on this, see Carley 1989; 1991). They survive largely through what psychologist Irving Janis (1982) called 'groupthink'.

Though seemingly without goals, cliques have an unstated 'mission' or purpose: to raise the status of clique members at the expense of non-members. Though lacking an organizational chart or stated division of labour, school cliques (for example) have a clear hierarchy of influence and popularity, with the leader on top surrounded by his or her favourites. In this sense, then, a clique is a group of people working together and co-ordinated by communication and leadership to achieve a common goal or goals.

CLIQUES IN SCHOOL SETTINGS

Cliques form in every area of life, even within bureaucracies and other formal organizations. However, cliques are most familiar to us from our childhood school experience. In school settings, cliques typically have a well-defined membership. Clique members are typically similar to one another in background and behaviour (Ennett and Bauman 1996). Cliques also have a leader, who is the most popular member of the clique and who dominates the other members.

Cliques are not only organizations; they are communities and miniature societies. In cliques, children

(Heidi Besner/zefa/Corbis)

Cliques organize the flow of information to ensure that some people are included and other are excluded—even mocked, ridiculed, and mistreated.

first learn the rules and expectations of society outside their family home. Through games and play with clique members, children internalize the beliefs, values, and attitudes of their group. By these means, children also form judgments of themselves. For example, they learn what it means to be 'good-looking', 'sexy', and 'popular', to be chosen or passed over (Crockett, Losoff, and Petersen 1984).

Cliques, though often supportive, can also offer excellent examples of structured cruelty, and they can be found everywhere. It was perhaps unavoidable that the reach of cliques would extend into cyberspace. Online bullying by clique members is a new phenomenon and potentially just as damaging as the bullying that occurs face-to-face. With online or e-bullying, youth can constantly harass their victims over the Internet, through instant text messaging on cellphones and postings on bulletin boards, and on their blogs.

Cliques control their members by defining the behaviours that are proper and acceptable. Leaders are skilled in exercising control. They often do so by building up the clique members and then cutting them down (Adler and Adler 1995). Leaders also take advantage of quarrels to divide and conquer the membership. They degrade and make fun of those who are lower in the hierarchy or outside the group. All of these tactics allow leaders to build up their own power and authority. They also foster clique solidarity by clarifying the norms for acceptance and rejection.

The cohesion of a clique is based on both loyalty to the leader and loyalty to the group. This loyalty, in turn, is based as much on exclusion as it is on inclusion. Clique members use gossip to reinforce their ignorance of outsiders and keep social distance from them. They may even harass outsiders. Doing so instills fear, forcing outsiders to accept their inferior status and discouraging them from rallying together to challenge the power hierarchy. Cliques and the rituals of inclusion and exclusion on which they rely are small-scale models of the way organizations teach and enforce rules. As such, they provide a lesson in social control.

TIME to REFLECT

By what means do cliques control group members? From a clique leader's perspective, what are the advantages and disadvantages of cliques? How would a regular clique member's perspective of cliques differ from the clique leader's?

Bureaucracies

FORMAL ORGANIZATIONS

Organizations are formal if they are deliberately planned and organized. Within formal organizations, formal roles and statuses provide the skeleton for all communication and leadership. Often, formal organizations have multiple goals, and they usually have a long lifespan. The Roman Catholic church is a formal organization that has lasted nearly 2000 years, for example.

We can define a **formal organization** as a deliberately planned social group that co-ordinates people, capital, and tools through formalized roles, statuses, and relationships to gain a specific set of goals. There is a huge literature, containing many lively debates, that addresses the question of why some organizations are more successful and powerful than others. Some explanations cite the degree to which an organization fills a social need (either real or successfully promoted by the organization itself). Other explanations note that successful organizations control more of the necessary resources and technologies. Finally, others argue that successful organizations tailor their goals to match the goals of their members and adapt well to changes in their environment.

The most successful form of organization in the past century or so has been the **bureaucracy**. The term *bureaucracy* has negative connotations for most people. We often hear about how bureaucracies impede business, how people battle against them, and most of all, how frustrating they are. The word calls to mind images of red tape, books of rules and regulations, and inefficient and unwieldy groups moving at a tortoise-like pace. To sociologists, however, bureaucracies are merely formal organizations that thrive in both the public and the private sector and in both capitalist and socialist societies. Whatever their setting, bureaucracies are the main organizational form because they are efficient and effective.

THE EMERGENCE OF THE BUREAUCRATIC FORM OF ORGANIZATION

Consider the basic features of bureaucracy. First, all resources belong to the organization, not to the boss. Office-holding is based on expertise and effectiveness alone, not personal connections. People move through organizational positions, or offices, based on their merit. There are elaborate written rules to govern many (if not all) of the relationships in the organization. Organizational charts define the (ideal) chains of responsibility, authority, and communication between superiors and subordinates.

It was obvious to Max Weber (1978 [1908])—the first sociologist to study bureaucracies—that this form of organization held enormous advantages over earlier organizational forms. Compare them with patrimonialism or clientelism in which clients are tied to their boss or patron by personal loyalty, for example. Because of its formal characteristics, bureaucratic organization holds the potential for rational planning. Bureaucracies can state clear goals, plan team strategies, train the most able people, mobilize the needed resources, evaluate effectiveness, and carry out organizational improvements. How different this makes IBM or the University of British Columbia—both bureaucracies—from the Italian mafia or the court of Louis XIV, which are patron–client organizations!

Bureaucracy in its modern form arose in response to three important historical conditions: European nation-building, capitalism, and industrialization. Nation-building—and by extension imperial conquest and colonization—created the need for effective tax collection and military skill. An honest and effective military was needed to beat down the local aristocrats and fight the armies of other countries.

▼ FIGURE 6.1 Bureacracy

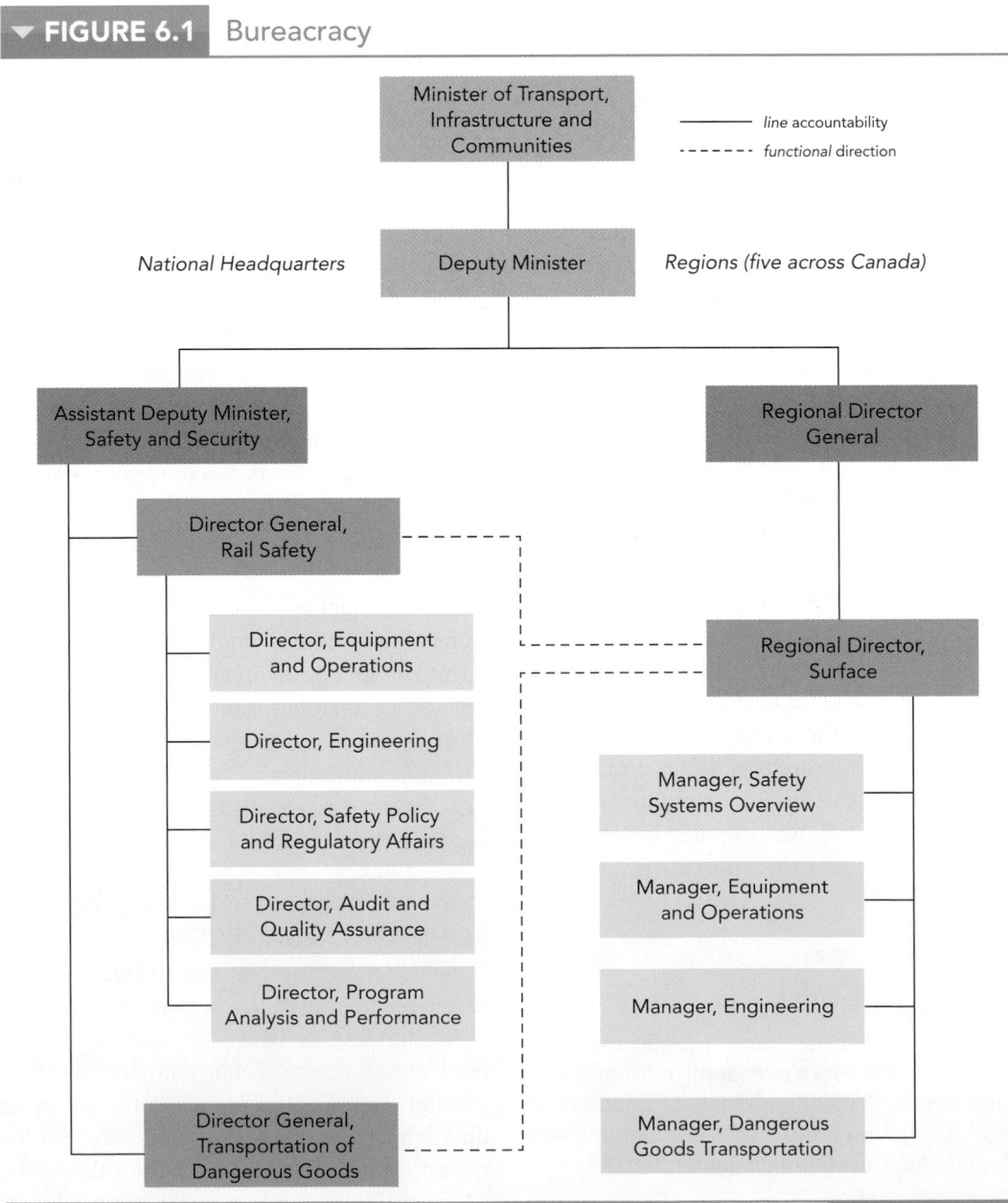

SOURCE: Transport Canada. www.tc.gc.ca/media/images/railsafety/figure3-1.jpg.

In the eighteenth and nineteenth centuries, rulers of Britain, France, and Germany quickly discovered that their armies were not properly organized, provisioned, or led. Command was weak, usually because officer positions were filled through patronage or the purchase of commissions, not level of competence. The results were ineffective leaders and, often, disastrous results, as shown by Britain's lengthy failure to control Napoleon's armies. Eventually, one nation after another recognized the need to reorganize the military and the civil service along bureaucratic lines to achieve their national goals (see Bensman 1987; Gorski 1995; Kiser and Schneider 1995; Spittler 1980; Tyrell 1981).

Capitalism imposed similar demands. Under capitalism—a system devoted to the pursuit of maximum profits—people quickly discover that some forms of social and economic organization yield higher rates of profit on investment than others. Bureaucratic organization, because it is rational, is well suited to the rational pursuit of profits. The legal idea of 'limited liability' also allows a bureaucracy to manage investment and profits impersonally in a way that protects the owners. This impersonality also makes bureaucracy different from clientelist systems.

Finally, industrialization also favoured the rise of bureaucracies. Bureaucracies are good at controlling large workforces—even highly educated and differentiated ones. As organizations get larger, they become more differentiated, with a more complex division of labour. As a result, management and administrative structures have to grow too. Responding to continuing changes in the environment, bureaucracies become adept at adopting new technology and administrative techniques (Kelley and Helper 1997).

As noted, Weber (1978 [1908]) traced the rise of bureaucracies to the rise of capitalism and modern states and also to the secularization of human activities. With the decline of formal religion as a political power, many activities 'rationalized'. Rationalization, in Weber's conception of history, refers to the movement away from mystical and religious interpretations of the world. These practices are replaced by the methodical collection and analysis of evidence. Also associated with rationalization is the rule of law: the rise of impersonal authority based on the universal application of a codified set of rules and laws.

For obvious reasons, the new, rational society prizes efficient, effective administration in business and government. In Weber's view, these values spurred the growth of bureaucracy because bureaucracies organize human activity in a logical, impersonal, and efficient manner. Or so he thought.

THE CHARACTERISTICS OF BUREAUCRACY

Weber (1958 [1922]) was the first to analyze the particular features of the bureaucratic form of organization, and he identified seven essential characteristics of bureaucracy:

- division of labour
- hierarchy of positions
- formal system of rules
- reliance on written documents
- separation of the person from the office
- hiring and promotion based on technical merit
- protection of careers

Division of Labour

In earlier eras, workers handcrafted specific articles from start to finish to produce society's goods. Gradually, this production process gave way to specialization and a detailed division of labour. Adam Smith noted the overwhelming productive superiority of specialization as long ago as 1776 (Smith 1976 [1776]). A specialized division of labour became the foundation of modern industry and bureaucratization. An automotive assembly line is perhaps the typical modern example of such a division of labour.

As on an assembly line, every member of a bureaucracy performs named and identified duties. The bureaucracy itself provides the facilities and resources for carrying out these duties. Workers work with equipment they do not own; in other words, they are separated from the **means of production**. Also, administrators manage what they do not own. Efficiency and productivity both increase when an organization combines specialized labour with centrally controlled resources.

Hierarchy of Positions

We can imagine the structure of an organization as a pyramid, with authority centralized at the top (see, for example, the organizational chart in Figure 6.1). Authority filters down toward the base through a well-defined hierarchy of command. Within this hierarchy, each person is responsible to a specific person one level up the pyramid and for a specific group of people one level down.

The organizational chart of any large corporation is shaped roughly like a Christmas tree. The number of workers increases as you move down toward the bottom of the hierarchy. With the other characteristics of bureaucracy, this feature serves to increase efficiency: all communications flow upward to 'control central' from large numbers of workers 'at ground level'.

TIME to REFLECT

What are the advantages and disadvantages of a bureaucracy? Can you imagine a system that employees would prefer? What about one that bosses would prefer?

Rules

Bureaucracies work according to written rules. The rules allow a bureaucracy to formalize and classify the countless circumstances it routinely confronts. For each situation, decision-makers can find or develop a rule that provides for an objective and impersonal response. The rules, therefore, guarantee impersonal, predictable responses to specific situations. This impersonality and objectivity helps the organization to achieve its objectives.

Separation of the Person from the Office

In a bureaucracy, each person is an office-holder in a hierarchy. The duties, roles, and authority of this office are all clearly defined. For example, the organization spells out the duties of a Level 3 manager in relation to a Level 4 manager (her superior) or Level 2 manager (her subordinate). Thus, the relations between positions in a bureaucracy are impersonal relations between roles, not personal relations between people. This separation of person and office means that people are replaceable functionaries. People come and go, but the organization remains intact. It also means that personal feelings toward other office-holders must be subordinated to the impersonal demands of the office. Equally, relationships are confined to the official duties of office-holders and—ideally—do not invade their private lives.

Hiring and Promotion Based on Technical Merit

An ideal bureaucracy hires employees impartially. They base decisions on the candidates' technical

OPEN for DISCUSSION

It's the Uneconomy, Stupid!

I felt compassion watching US auto CEOs beg for bailouts last week.

So they're incompetent, shortsighted, and venal—nobody's perfect. And at least they produce items you can *use* to get to the mall or the cottage, if a more suitable foreign car is unavailable. You can't say that for the titans of banking and finance, proud devisers of credit instruments, collateralized debt obligations, swaps, etc., literally ad nauseam. (I mean 'literally' literally.) They are the villains here. They gave us the uneconomy.

Yes, you need banking and credit to make an economy work. But their alternate universe shaved to infinitesimal levels the relations between their financial gimmicks and the real world of housing, food, health care, transport—then ground them into tinier bits with even less connection to reality and multiplied the scraps magically into huge profits. It worked on paper, for a while.

Yet where does empathy and aid go? To the uneconomy. In the trillions.

'My focus is on the financial sector,' said US Treasury Secretary Henry Paulson. Auto makers ask for a mere $25-billion and get smacked.

Where did this notion come from, that you don't need to *make* anything people use?

I'll tell you: free trade and globalization. You 'offshore' everything imaginable because desperate foreigners will work for far less. Among the few things you can't offshore are natural resources—so Canada is in relatively better shape—and housing, which leaves it as a rare 'real' source of wealth in the United States.

What's left to do? Reap the profits, for which you need some financial gimmickry that somehow moves the wealth now being created elsewhere into your pockets.

What happens to the mass of American workers in this script? There's not much left to make; it's all been offshored. So their task becomes to consume, not produce. But they make little money at the pathetic, part-time jobs they can still scavenge, so they must consume on credit. It's the uneconomy all over again, but at the bottom end.

The US workforce and its elite weirdly mirror each other. They're both engaged in finance and credit—none of them *produce* anything useful.

SOURCE: Rick Salutin. 2008. 'It's the uneconomy, stupid!' *The Globe and Mail* 28 November. www.theglobeandmail.com/news/opinions/article725090.ece.

(© Jim West/Alamy)

All organizations depend on a division of labour that is usually hierarchical and brings together expertise and technology to complete a specific task.

competence, not on inborn features like gender, race, or ethnicity. They also base promotion on technical competence, or sometimes on seniority. People are neither discriminated against nor favoured because of their personalities or connections to the boss.

Protection of Careers

The final characteristic of bureaucracies is that people's careers are protected within them. People can look forward to long careers in a bureaucracy, because they are not subject to arbitrary dismissal for personal reasons. As long as they follow the rules attached to their office or position, they are secure in their jobs.

MERTON'S BUREAUCRATIC PERSONALITY

What sociologists have found out is that often, bureaucracies do not behave rationally in terms of their long-term interests and survival. This occurs in several ways.

First, it occurs because of the creation of 'bureaucratic personalities'. As sociologist Robert Merton (1957) pointed out, bureaucracy puts a pressure on bureaucrats to act in ways that, in the long run, weaken the organization. In particular, bureaucracies force their members to conform to rigid bureaucratic rules. This pressure, combined with intensive training, overemphasizes members' knowledge of and adherence to the bureaucracy's rules. This, in turn, makes it easy for bureaucrats to act habitually in routine ways. In Merton's words, they follow rules in a methodical, prudent, and disciplined way. Yet inevitably, the routines become similar to blinkers on a horse, keeping bureaucrats from recognizing new situations in which the old rules are inappropriate. Thus, Merton argued, bureaucrats develop a 'trained incapacity' for dealing with new situations.

Also, applying the rules means classifying all situations by objective criteria so they can fit into the correct pigeonhole. The result is that bureaucrats often fail to see their clients as people with unique wants and needs, seeing them only as impersonal categories. This viewpoint is harmful to the organization, because it causes bureaucrats to fail to meet the unique needs of individual clients. In the end, this creates hostility in the public and weakens the authority of the organization.

INFORMAL ORGANIZATIONS IN BUREAUCRACIES: THE HAWTHORNE STUDIES

Bureaucracy is intended to be an impersonal form of organization; however, actual people fill the bureaucratic roles. As human beings, workers resist becoming faceless cogs in the bureaucratic machine (replaceable cogs at that). So they develop complex personal and informal networks that work within the formal organization. Within formal organizations, we find informal organizations, even cliques of the kind discussed earlier in this chapter.

Informal networks among people who interact on the job serve many purposes. First, they humanize the organization. They also provide support and protection to workers at the lower levels of the hierarchy. They serve as active channels of information flow (the grapevine) and mechanisms for exchanging favours. They also direct the flow of information, enforce moral standards, and exclude people they consider inferior. Paradoxically, informal networks within formal organizations—though similar to cliques in many ways—can serve to free people from the limits of formal organization and, occasionally, allow them to protest and subvert their working conditions. They also confer human meaning on otherwise impersonal settings, as we see from the classic Hawthorne studies.

The Hawthorne studies were conducted between 1927 and 1932 at the Western Electric plant in Hawthorne, Illinois, under the direction of Elton Mayo. In a series of experiments, management varied the working conditions to find out how this would affect worker productivity. Mayo held the view that workers were non-rational, emotional beings. The Hawthorne studies first revealed the importance of the informal organization in formal organizations.

Early conclusions drawn from the Hawthorne studies provided the foundation of the human relations school. One of the first conclusions became known as the **Hawthorne effect**. This proposition holds that when people know they are subjects of an important experiment and receive much special attention, they behave the way they think the researchers expect them to. The Hawthorne effect has influenced the design of social-psychological experiments since then as researchers try to control for this distortion.

Other conclusions drawn from the studies dealt with the social aspects of work: the relationships among the members of the informal group, the norms developed by the informal group, and types of supervision. The relationships among the women in Phase II of the research were happy and encouraging—and associated with higher productivity—while those among the men in Phase IV were not. This finding led human relations theorists to infer that happy group relationships may even increase productivity. The Hawthorne studies also found that group norms can limit productivity, especially when formal supervision is lax.

Re-analyses of the Hawthorne data by Brannigan and Zwerman (2001) have raised doubts about the truth and significance of the Hawthorne findings. These authors point out that in the first set of experiments, the productivity of the Western Electric employees increased over time with every variation introduced in their working conditions. Because the workers knew they were being watched, they did their best to show themselves in a positive light. However, in a second set of experiments, the gains in productivity were modest and inconsistent.

This later research has led to the conclusion that informal organizations can either help the formal organization to arrive at its goals or hinder it. Which end it does achieve will depend largely on the quality of the relationship between the workers and their managers. In short, workers are not as non-rational or irrational as managers might have wanted to believe, nor are they entirely focused on the paycheque.

FEMINIST PERSPECTIVES: GENDER MATTERS

For many years, organizational theorists ignored the fact that organizations include women as well as men. A notable exception is Rosabeth Moss Kanter (1977). Kanter took a structural approach in accounting for the disadvantaged positions of women. Women are indeed disadvantaged, considering that they are heavily overrepresented in lower-level clerical and service occupations and under-represented in management (Armstrong and Armstrong 1984).

A new wave of feminist theories has developed to address this organizational issue, using gender as an important focus. Earlier theorists either ignored gender or assumed that organizations were gender-neutral. However, feminist theorists, including Joan Acker (1991), have pointed out that historically, men have dominated organizations, with the result that the organizational image of the manager and the worker is a male image. This bias has several important results.

First, the hierarchical structure of bureaucratic organizations and the accompanying sets of rigid rules and procedures are incompatible with female

SOCIOLOGY in ACTION

Working in Good Company

'You've gotta have friends,' as Bette Midler's song goes. And that goes for the workplace, too. Having a bosom buddy in the office can go a long way toward making that workday more satisfying—and productive. And workplace friends are all the more important in these times, experts say.

'We are social beings, we need those connections. Friends provide social support that helps to buffer employees against stress. Friends help employees weather the stress that comes from the threat of downsizing and job insecurity, higher workloads, and anxious bosses,' said Sandra Robinson, a professor of organizational behaviour at the University of British Columbia's Sauder School of Business.

More than 38 per cent of American workers have colleagues they consider personal friends, 67 per cent believe that having workplace pals makes their job more fun and enjoyable, while 55 per cent say work friends make their job worthwhile and satisfying, according to a new poll of 1017 employees by Ipsos Reid for staffing firm Randstad U.S.

Having a pal at work can boost employees' energy and enthusiasm, provide an ally and fulfill emotional needs so much that they look forward to going to work, Prof. Robinson said. 'And they'll come in early and stay late if they find work more enjoyable.'

It's good for the employer, too: Workplace friendships boost teamwork, morale, communication, motivation, productivity, and commitment to the company and lower turnover, the Randstad survey found.

But as with any relationship, there are pitfalls to befriending your colleagues. If conflicts arise, you still have to work together, said Antoinette Blunt, president of Ironside Consulting Services Inc., a human resources consultancy in Sault Ste Marie, Ont.

Dealing with issues such as favouritism, gossip, conflicts of interest, blurring boundaries, oversocializing, and cliques can make office friendships tricky, Ms Blunt said. 'And a falling out with a friend can have a huge negative impact.'

SOURCE: Jennifer Myers. 2010. 'Working in good company'. *The Globe and Mail* 12 March. www.theglobeandmail.com/report-on-business/managing/weekend-workout/working-in-good-company/article1499013.

---

gender characteristics. Kathy Ferguson (1984) has suggested that women press for more open and democratic organizational systems not only for their own sake but also because they believe such changes would make their organizations more effective. However, women are rarely in positions of enough power and authority to change the organization. Thus, organizations remain bastions of male power.

Second, as Acker (1991) asserts, the dominant male image excludes and marginalizes women. Almost by definition, women cannot achieve the qualities of a 'real' worker because to do so is to become like a man. Also, women's bodies and sexuality are often stigmatized in organizations and used as grounds for control and exclusion.

Third, the male image of the organizational worker causes women's gender roles to be regarded as deviant. Women are seen as being incompatible with organizational life because of their (assumed) ties to marriage and responsibility for children (Cuneo 1990). Cockburn (1990, 92) points out that all women are eventually dismissed as deviant because of the potential role conflicts of some.

TIME to REFLECT

What organizational and political policies would help women to balance their careers with their family lives?

Today, studies continue to probe the effects of gender on hiring, promotion, and career trajectory. We understand now that they are also affected by marital responsibilities—especially child-rearing, housework, and caregiving. So, for example, Maume (2006), using data from the National Study of the Changing Workforce, finds that in the presence of family and work-related controls, women are less likely than men to have unused vacation time. Men's work schedules, supervisory duties, and concerns about job security significantly reduce the duration of their vacations. Even though family factors have no direct impact on women's vacation use, women's concerns about the success of their family lives increase with the number of unused vacation days. That is, they feel guilty about failing to use their available vacation days to provide unpaid family work. These findings suggest the persistence of traditional work–family priorities.

How Bureaucracies Actually Work

Weber's idea of bureaucracy, as we have seen, is a useful model for the study of this complex form of organization. It calls our attention to central features of bureaucracy. However, it is an idealization, like the notion of a perfect vacuum in physics. This section considers a few of the more troubling discrepancies between ideal bureaucracies and actual bureaucracies.

Ideally, every member understands his or her role in a bureaucratic network of reporting relationships. In graphic form, a bureaucracy is a Christmas tree–shaped structure that repeatedly branches out as you go down the hierarchy. Thus, at the bottom of the hierarchy there are a great many people whose job it is to carry out orders from above and report work-related information up the tree to their superiors. At the top of the hierarchy, a few people are responsible for issuing orders, processing information received from below, and preserving links with other organizations. Also at the top, leaders share information between parts of the organization.

In practice, organizations do not work this way, as sociologists since Weber have pointed out. They could not afford to work this way, and human beings aren't built to work this way. Thus, there is the ideal or formal structure, which prescribes how a bureaucracy ought to work. Then there is an actual, informal structure, which is how it really works.

ACTUAL FLOWS OF INFORMATION

In theory, a failure to report information up the hierarchy would never occur. In practice, it occurs all the time, and controlling the flow of information from below is a means of changing the balance of power between superiors and subordinates. And as the French sociologist Michel Crozier (1964) showed, bureaucracies work differently in different societies. This is because people raised in different cultures have different ideas about inequality, deference,

▼ FIGURE 6.2   Computer-Graphed Research Network of Five Important Gambling Researchers

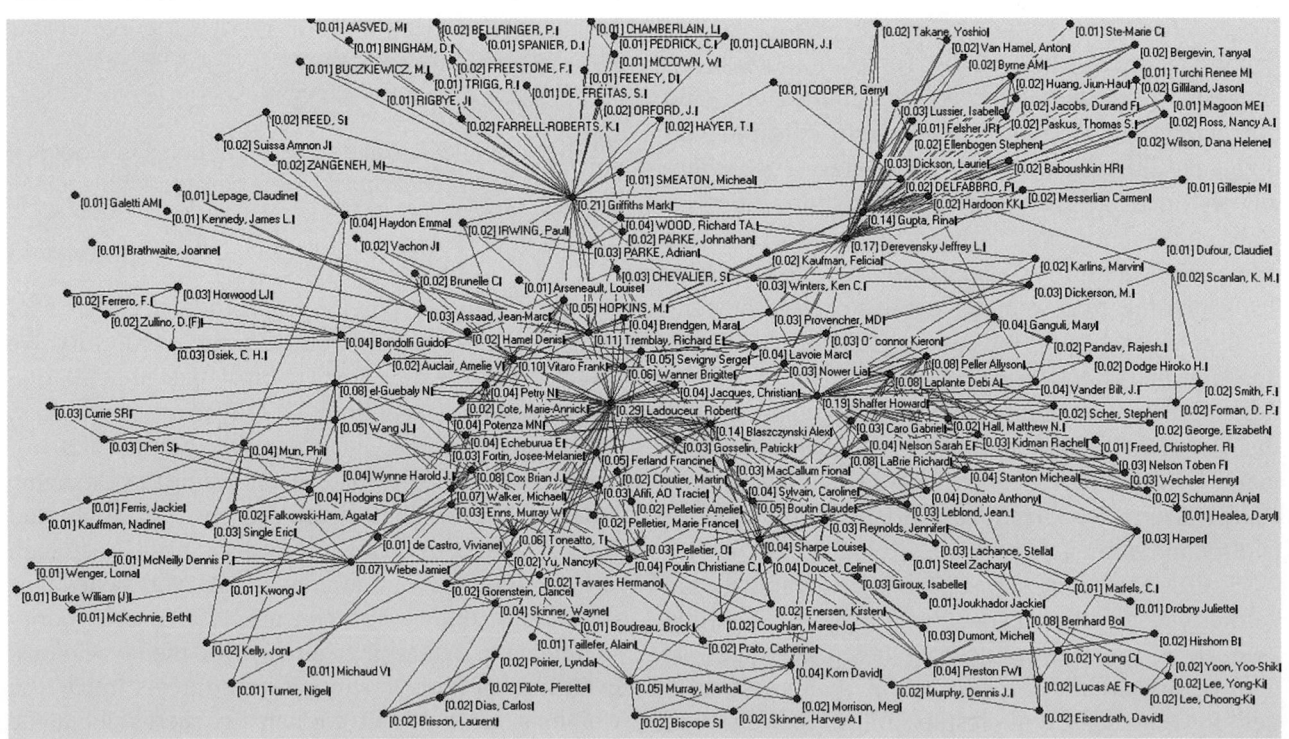

SOURCE: Stark (2009).

openness, and secrecy. For example, people raised in France or Russia are much more alert to the inequality of bureaucratic relations and the power of information control to equalize relations than workers raised in the United States. French and Russian workers behave differently, and as a result, bureaucracies work differently in these countries.

Bureaucracies also appear to work differently for men and women. When playing a managerial role, women often adopt a collaborative approach that draws on qualities learned in family relations. Women's managerial styles stress good employer–employee relations and sharing information and power. By contrast, men stress purely economic considerations (Occhionero 1996).

In practice, workers everywhere make friends and acquaintances. As a result, they casually share work information. Often, workers use information purposely to help one another. Sometimes they may even leak information for personal gain or to subvert their boss or the organization as a whole.

TIME to REFLECT

If bureaucracies function differently in different countries, how might they work when made up of other categories of people? For instance, how might a bureacracy made up of only women function differently from one of only men? Or a bureaucracy of teenagers compared to one of elderly people?

Thus, in organizations where (ideally) strangers relate to other strangers based on written rules, we find workers forming small communities that obey their own rules. Political actors below the top level cannot employ routine channels or resources to negotiate in the idealized manner. The basis of this informal organization is trust, which relies on friendship, acquaintance, and gossip about third parties that strengthens existing ties (Burt and Knez 1996). In the end, the same materials that build cliques build the informal, often hidden infrastructure of bureaucracies.

As in cliques, trust in bureaucracies is built gradually, maintained continuously, and easily destroyed (Lewicki and Bunker 1996). When trust is violated, the result is often revenge or another disruptive response—confrontation, withdrawal, or feuding, for example (Bies and Tripp 1996). Within organizations, managers can oversee and enforce rules of reciprocity.

Across organizations, the rules are harder to enforce. The result is that organizational boundaries tend to limit trust, and this limits the sharing of ideas (Zucker et al. 1996). Within organizations, the flow of information is harder to contain.

To correct this, managers often create temporary team structures to cut across the bureaucratic hierarchy. People in these short-lived groups learn to make do with whatever information is available, and they quickly come to trust one another. Trust develops most quickly when (1) fewer workers are involved; (2) their interaction is based on roles, not personalities; (3) **role expectations** are clear; and (4) dependence on one another is moderate, not high or low (Meyerson, Weick, and Kramer 1996).

New information technology also makes it easier for horizontal groupings to form, since distant employees can easily exchange information through a large organizational computer network (Constant, Sproull, and Kiesler 1996). As well, telecommuting, or teleworking, now occupies an important place in the world of information work, posing new problems (Di Martino 1996). It may reduce costs by externalizing or delocalizing work, but we are far from knowing how it will affect work organization and productivity (Carre and Craipeau 1996). For example, computer-mediated communication (like email) seems to increase user satisfaction in task-oriented organizational cultures. It decreases user satisfaction in person-oriented organizational cultures, where workers prefer face-to-face communication (Kanungo 1998).

Nowhere is the importance of new communication technology more obvious than in the development of huge 'virtual teams' of workers.

TIME to REFLECT

Consider the factors that contribute to creating trust within organizations. How might an increase in 'teleworking' affect development of trust among workers? How could technology be used to increase trust among workers?

ORGANIZATIONAL CULTURES AND FLEXIBILITY

In temporary or other horizontal groupings, a worker reports to more than one superior, which may create conflicts or inconsistent demands. Sometimes it becomes unclear where a worker's main duties lie and, therefore, how that person's work should be evaluated and rewarded.

Organizations need ever-greater flexibility and cooperation from workers. Workers who receive continuing education and training and take part in planning efforts are best able to provide this (de la Torre 1997). Yet worker motivation, recruitment, and training all pose problems for bureaucracies. The motivational problem is greatest in organizations where professional expertise and judgment are most required. These would include organizations like universities, law firms, and technology development firms. There we find the greatest attention given to matters of organizational culture and career development. To keep the most able workers, an organization must give them autonomy and rewards for corporate loyalty. Only in this way can they be induced to join, stay, and carry out their duties in conformity with organizational goals.

Some organizations attempt to correct the flaws of bureaucracy by trying to empower their workers. For example, they preach open management, teamwork, continuous improvement, and a partnership between customers and suppliers, all within a bureaucratic context. In the end, however, these tactics prove to be aimed at obscuring management roles while justifying increased corporate control and a heavier workload.

However, people usually form stronger attachments to other people than they do to 'the organization' as an abstract entity. Thus, patterns of clientelism develop even within bureaucracies. In the end, bureaucracies are organizations where two principles—rule-based rationality and person-based clientelism—compete for dominance, with neither being able to win decisively at the expense of the other.

THE PROBLEM OF RATIONALITY

Bureaucracies are thought to be rational in the ways they make and carry out plans. This is because, over the long term, by making impersonal decisions and rewarding excellence, they are more able to pursue

UNDER the WIRE

Telecommuting Found to Boost Morale, Cut Stress

Tired of traffic jams, late trains, packed buses? Telecommuting can be a big plus for workers and employers because it boosts morale and job satisfaction and cuts stress, researchers said on Monday.

In an analysis of 46 studies on telecommuting, researchers found that working away from the office by using computers, cellphones, or other electronic equipment can have more pluses than negatives for people and the companies that employ them.

'Our results show that telecommuting has an overall beneficial effect because the arrangement provides employees with more control over how they do their work,' said Dr Ravi Gajendran of Pennsylvania State University.

'Telecommuting seems to have some mildly positive effects on employee morale, on work–family balance, and on stress,' he added in an interview.

Gajendran and David Harrison, who reported their findings in the *Journal of Applied Psychology*, studied data on 12,833 telecommuters who spend time working away from the office.

Telecommuting has been a growing trend in the United States since about 2000. Last year an estimated 45 million Americans telecommuted, an increase of 4 million from 2003, according to the magazine *WorldatWork*.

Gajendran believes the numbers will continue to grow as access to broadband increases.

'Over the last couple of years there has been a spike, especially in the number of people who are regularly telecommuting. By regularly I mean people who are telecommuting at least once a month,' he said.

'There has almost been a 60 per cent increase in those numbers.'

Although some companies and workers feared telecommuting could hamper career prospects or lead to a breakdown in relationships with managers and co-workers, the researchers found no evidence to support it.

'Telecommuting by and large does not have any negative relational outcomes as has been commonly believed,' said Gajendran.

There was also no evidence that telecommuting stymied career development.

Telecommuting also has added benefits, according to the researchers, because it cuts commuting costs and relieves congestion on inner-city transport systems, as well as traffic on roads.

'If you could save a long commute, say two days a week or maybe even one, you will see substantial costs saving as well as substantial reductions in terms of pollution,' Gajendran explained.

SOURCE: Patricia Reaney. 2007. 'Telecommuting found to boost morale, cut stress'. *The Globe and Mail* 20 November. www.theglobeandmail.com/news/technology/article798839.ece.

long-term organizational goals with huge amounts of wealth and power.

However, the sheer size of large bureaucracies and their long-term outlook introduces certain types of irrationality that, in the end, may undermine the organization. A concern with the mere survival of the organization may undermine shorter-term concerns with the quality of decisions, products, and services the organization is providing to its customers. Bureaucracies work hard to control subjectivity and individuality, but in the end, this undermines their productivity; by creating boundaries between the institution and its environment, the institution loses touch with its customers—the objects of their efforts (Imershein and Estes 1996).

Bureaucratic organizations rely heavily on official records and procedures in routines that create a separate reality—'papereality'. This world of symbols, separate from other forms of representation, inhibits both forgetting and learning (Dery 1998). Another result is the creation of a *bureaucratic personality*. This type of person is willing to ignore ethical concerns in order to follow corporate rules (Ten Bos 1997). This, and anonymity, make moral indifference almost inevitable. The bureaucratic characteristic of relying on the rules as written can create another problem: as a bureacracy grows and more rules are added, the system becomes increasingly complex. This can lead to a situation in which no one person knows all of the rules and different offices act independently of each other, creating rules that conflict with one another.

Rule by offices undermines personal responsibility for decisions the organization may take. No

HUMAN DIVERSITY

How Hyundai Became the Auto Industry's Pacesetter

While Detroit grew fat and lazy, Hyundai made a science of carmaking. Now the one-time butt of jokes is leveraging its unique culture to outpace the rest of the industry—and to move seriously upmarket.

From the outside, Hyundai Motor Co.'s headquarters near Seoul doesn't look like the home of the world's fastest-growing carmaker. The two 21-storey glass towers, linked by an atrium, wouldn't attract much attention in a suburb of Toronto or Vancouver.

Come here at 5:30 in the morning, however, and you start to see the Hyundai difference. Yes, that's an Olympic-sized swimming pool in the basement and, yes, that's the leadership of the company doing laps or working out in the adjacent gym before they start another marathon day.

And come up to the second floor—if only in your imagination, because outsiders and indeed most employees aren't allowed here—and you'll see another marker of Hyundai's distinct culture: a computerized worldwide command-and-control centre that could pass for a set in a James Bond movie.

And then, if you get to the city of Ulsan, 300 kilometres to the southeast on the Sea of Japan, see how the sheer enormity of Hyundai's highly centralized manufacturing overwhelms you. Here the company operates the world's largest auto plant, spread over 1233 acres, producing up to 5600 cars a day. The place has its own port, capable of docking three 50,000-ton ships at once.

Hyundai is still a relatively young carmaker and essentially remains a family-run company. That's one big reason it has been able to maintain focus and discipline. The company is also one of the key actors in the astonishing economic surge in South Korea since the 1960s. The success story is based on a national business culture that is hard-working, highly educated, ambitious, and export-oriented. South Korea's growth—and no small amount of internal controversy—also stems from central economic planning by authoritarian, pro-US regimes that have worked closely with a tightly knit network of several dozen family-owned conglomerates called *chaebol*.

In 2000, Hyundai adopted the Six Sigma management discipline. The process uses intense statistical analysis to identify flaws in a manufacturing process. Quality specialists rate processes from one sigma (31 per cent of products are flawless) to six sigma (almost 100 per cent). Hyundai Canada CEO Steve Kelleher said a key part of the discipline was 'not just reading your own PR, or your own internal quality reviews, but really looking at third-party assessments of what the customer is saying. One of the great things about management in Korea was the ability to seek out criticism and deal with it.'

That's a striking difference from Detroit, whose downfall was a large element of hubris. And the new attitude was also strikingly different from the Hyundai of old.

SOURCE: John Daly and Mitch Moxley. 2010. 'How Hyundai became the auto industry's pacesetter'. *The Globe and Mail* 29 April. www.theglobeandmail.com/report-on-business/rob-magazine/how-hyundai-became-the-auto-industrys-pacesetter/article1548295.

member of the bureaucracy is asked, or obliged, to take responsibility for collective decisions. As a result, so-called collective decisions—typically taken by the top executives—are liable to be foolish, harmful, or even criminal. Corporate and government bodies are unique in one respect: their deviant behaviour may be a result of systemic patterns in their organization, not merely individual misbehaviour. However, once deviant behaviour has occurred, they are well positioned to evade responsibility.

Organizational Crime

Organizations themselves are rarely penalized for deviant behaviour (Ermann and Lundman 1996). Consider the prosecution of top leaders of Enron for falsifying records and manipulating information for their personal benefit. This came at the expense of thousands of investors and millions of American citizens. Enron's collapse in 2001 wiped out more than 5000 jobs, more than $60 billion in market value, and more than $2 billion in pensions.

Sociologist Edwin Sutherland (1940; 1949) was the first to carry out systematic research on this topic. Unlike amateur crime—for example, shoplifting—business crime victimizes millions of people; it robs businesses of billions of dollars each year, and it undermines the legitimacy of public institutions. Business crime has received ever more attention from investigators, because the number and influence of business organizations has increased dramatically.

A recent and dramatic example of white-collar crime was committed by so-called financier Bernard Madoff. He fraudulently gained an estimated $50 billion (US) from naive investors who were seeking an unusually high rate of return. Investigators have been unable to discover where all this money has gone, much less return it to its original owners. This fraudulent 'Ponzi scheme'—named after a fraudster in the early twentieth century, whose gains were slightly less spectacular—played a part in the general collapse of the American banking and investment system. However, that system was also beleaguered by other forms of fraud and bad judgment. Fraudulent stock issues based on insecure mortgages were at the bottom of the mess.

Fraud, such as insider trading and falsifying account books (e.g., Enron), is the perfect example of business crime. Commonly, frauds misrepresent a product or service. This enables the criminal to sell something of little or no value for a large amount of money and make a huge profit. Fraud relies on manipulating information. The high-tech industry, for example, is a perfect place for fraudulent activity. It enjoys rapid growth, high stakes, and a huge potential for profits.

However, nothing proved as perfect for fraudulent investment as the recent low-interest mortgage meltdown in the US. People without any secure means of payment were induced to take home mortgages at impossibly low rates. These 'toxic' mortgages were then rolled in with other notes and properties to create seductive investment opportunities that were, indeed, 'to good to be true'. When the mortgagees started to default in large numbers, a great many homes were repossessed, the housing market collapsed, and many banks, pension funds, and other investors lost a significant portion of their holdings. Yet some financiers and corporate executives who got out soon enough walked away with millions of dollars in profits. Many businesses failed; many Americans and to a lesser degree Canadians were thrown out of work, some of whom became homeless, and the US government has had to use public funds to secure the banking industry and the mortgage market.

Has the US government prosecuted the people—the so-called financiers—who knowingly created this economic disaster? So far, only Bernie Madoff is in jail.

Since Sutherland's groundbreaking work, business crime has evolved into a visible global problem. Governments

(© vario images GmbH & Co.KG/Alamy)

The disastrous social and economic effects resulting from the collapse of Enron, recent international 'Ponzi schemes', and the real estate collapse of 2008–9 have led governments to increase their efforts to reduce white-collar crime.

around the world are putting more effort into fighting this form of crime, which affects the economy, government, and the public. Many white-collar criminals rely on offshore banking and bank secrecy, as historically practised for Swiss bank accounts, to hide and launder billions of dollars stolen from people throughout the world. US treasury officials believe that 99.9 per cent of foreign criminal and terrorist money sent to the US is placed in secure accounts, making it safe from detection. These shell companies are also known as 'mailbox' companies, international business corporations (IBCs), or personal investment companies (PICs) marketed by banks and accounting firms. They launder money and also hide profits from income taxes.

Some experts calculate that as much as half the world's capital flows are handled in offshore centres. For example, the International Monetary Fund (IMF) estimates that between $600 billion and $1.5 trillion in illicit money is laundered yearly through secret bank accounts. In crimes of this sort, the line is blurred between **corporate crime**, organized crime, and political crime.

Closer to home, a recent example of business crime involved Canadian-born magnate Conrad Black. Black was found to have fraudulently misused millions of dollars from Hollinger Inc. profits to bankroll an extraordinarily lavish lifestyle. Critics said that Black had become unable to distinguish between the company's money and his personal funds. This characteristic, which Max Weber called *patrimonial rule*, was common among kings and other aristocrats before the Industrial Revolution. It has been largely eliminated in companies traded publicly on the stock exchange. There, an expert board of directors is expected to act forcefully to protect the stockholders. At Hollinger, they did not.

Another fraudulent act that came to light recently involved entertainment magnate Garth Drabinsky. On 25 March 2009, Drabinsky and his associate Myron Gottlieb were found guilty on fraud and forgery charges. The CBC reported that 'an Ontario judge said they "systemically manipulated the books" at their now-defunct theatrical production company Livent Inc. . . . Prosecutors had maintained that Drabinsky and Gottlieb directed a scheme to make the ailing company look healthy, bilking investors of roughly $500 million. Judge Benotto said the two men created "spectacular" successes on the stage, productions which brought Toronto, the company's home base, kudos from the worldwide theatre community.' But, the judge said, those successes were built on a platform of lies and manipulation as Gottlieb and Drabinsky artificially inflated the company's profits and depressed its costs to make Livent's financial situation appear rosier than it was.

As these varied examples show, corporate and business crimes are usually carried out inside large organizations, usually by people with more than average social status; most organizations want to keep these crimes secret to protect their public image. As a result, the public is slow to learn about these crimes, and (it would seem) few are prosecuted. This places unusual importance on the role of people within these organizations—so-called whistle-blowers—to step forward and reveal what they know to corporate auditors, the media, or law enforcement personnel.

Why are charges and convictions of white-collar criminals so hard to secure? As a protected 'legal person', the corporation can command many more resources than individuals accused of a crime. This power makes officers in many corporations blind to the implications of their actions. The result may be fraudulent practices, dangerous commercial products, or even, as in Nazi Germany, death camps. Both corporate fraud and racial extermination are facilitated by bureaucratic organization.

TOTAL INSTITUTIONS

As Goffman (1961) pointed out, mental hospitals, convents, prisons, and military installations have a lot in common as organizations. True, they have different institutional goals and provide different services to society; they also employ different kinds of experts and oversee different kinds of 'customers'. However, what they have in common organizationally far outweighs these differences.

First, they have total control over their 'customers'—whether mental patients, nuns, convicts, or soldiers-in-training. Twenty-four hours a day, seven days a week, their staff are able to watch and, if desired, control behaviour within the institution. They can see their customer pool perfectly, but none of them—psychiatrists or nurses, priests or mothers superior, guards or officers—can be watched unknowingly or unwillingly themselves.

Total institutions offer an extreme example of the bureaucratic organization and the bureaucratic society. Indeed, their organizational principles of efficiency and procedural rigidity conflict with democratic values. In particular, they hinder democratic participation by employees and they undermine habits of citizenship (Davis 1996).

What Goffman (1961) tells us about mental institutions and prisons reminds us of what we have heard about life in **totalitarian** societies like Nazi Germany and Soviet Russia. Under both Nazism and communism, people are dominated by uncontrollable rulers through government and party bureaucracies (Maslovski 1996).

The systematic extermination of Jews by European Nazis would not have been possible without large, efficient administrative bureaucracies. They carried out the extermination through steps that included segregation, imprisonment in concentration camps, starvation, and eventual annihilation. Once put into place, the same killing machinery was also used on gypsies, leftists, homosexuals, and Polish prisoners of war. Some managers responsible for this program claimed that they experienced psychological repulsion. However, most justified their behaviour, saying it was part of their duty in the bureaucratic system (Hilberg 1996).

Totalitarian societies are like total institutions. They also make liberal use of total institutions to punish, brainwash, and resocialize uncooperative citizens. Thus, as Weber warned, modern bureaucratic society is an 'iron cage' in which we are all trapped by our ambitions for success, efficiency, and progress (1958 [1904], 181). Bureaucracy has an enormous potential for enslavement, exploitation, and cruelty. It also has an enormous potential for promoting human progress. For example, it can promote economic development and scientific discovery, high-quality mass education, or the delivery of humane social services to the needy. It is to fulfill the second potential that we have risked the first. The jury remains out on whether, in the twentieth century, the gain justified the cost.

In the end, Weber was ambivalent about bureaucracy. Its superiority over organizations based on friendship, kinship, charisma, or tradition impressed him. He realized that a bureaucracy is an extremely powerful tool for whoever controls it. For that reason, he (1978 [1908]) expressed uneasiness over the vast power a bureaucracy can wield in society, citing the domination of the weak German parliament by Otto von Bismarck's bureaucracy before the First World War. In his writings on bureaucracy, Weber also expressed concern about the fate of 'gray-faced bureaucrats' (1978 [1908]; 1958 [1922]). Bureaucracies pose problems because they share many of the shortcomings of cliques, yet they are infinitely more dangerous because they are more likely to achieve their goals.

Conclusion

This chapter has reviewed various 'sets' of people, including categories, networks, communities, groups, cliques, and organizations. Sets of people with a common sense of identity are typically called communities, and there is a long history of community studies in sociology. Communities, whether urban or rural, real or virtual, are important because people are conscious of their membership and make personal investments in remaining members. Formal organizations combine many of the features of networks, groups, cliques, and communities.

The main form of the large, powerful, and long-lived formal organization of the twentieth century is the bureaucracy. The goals of bureaucracy are maximum efficiency and productivity. These goals are to be achieved by dividing the work, developing specialized skills, and stockpiling resources. Largely, bureaucracies achieve their desired goals in the expected ways, but they do so with unwanted side effects.

Finally, this chapter considered total institutions. As Goffman pointed out, mental hospitals, convents, prisons, and military installations have a lot in common as organizations. All of them are organizations that have total control over their 'customers'—whether mental patients, nuns, convicts, or soldiers-in-training. Myths and ideologies are propagated to justify the differences between rulers and ruled. Total institutions offer an extreme example of the bureaucratic organization and the bureaucratized society.

questions for critical thought

1. Considering that weakly tied networks can be more useful than strongly tied ones, how do you think the creation and proliferation of social networking websites, such as Facebook, are affecting the flow of resources and information? Does your experience with social networking bear out your theory?

2. How might Internet communities, such as comment boards or chat rooms, resemble and affect *Gemeinschaft* and *Gesellschaft*?

3. Robert Bales discovered that in discussion groups, three roles regularly emerged: task leader, emotional leader, and joker. He inferred that the groups 'need' these roles to survive. Do you think these particular roles are more likely to appear within primary groups, secondary groups, or equally in both?

4. In what ways are bureaucracies rationally and irrationally designed? What kind of organizational system do you think would work more effectively than a bureaucracy?

5. What are the advantages and disadvantages of clientelism? What factors contributed to the downfall of clientelism and the rise of bureaucracy?

6. What are the characteristics of a bureaucracy? What role do these characteristics play in creating Merton's bureaucratic personality?

recommended readings

Vered Amit, ed. 2002. *Realizing Community: Concepts, Social Relationships and Sentiments.* London and New York: Routledge.
What is 'community'? This book takes a word whose meaning has become murky through overuse and brings anthropological research to bear on its meaning. Using ethnographic accounts and theoretical analyses, Vered Amit explains what community is, how it is created, and what it looks like in the modern world.

Michel Crozier. 1964. *The Bureaucratic Phenomenon.* Chicago: University of Chicago Press.
Crozier is a French sociologist who explored whether bureaucracies can work differently in different societies and cultures despite their formal organization as described by Max Weber. Because societies vary—historically, socially, and politically—bureaucracies vary too.

Paul du Gay. 2005. *The Values of Bureaucracy.* Oxford: Oxford University Press.
This book explores why bureaucracies are such successful, and therefore persistent, organizational structures. The book outlines the characteristics that make bureaucracies efficient in various settings.

Linton C. Freeman. 2004. *The Development of Social Network Analysis: A Study in the Sociology of Science.* Vancouver: Empirical Press.
Humans have an innate need to be social; therefore, social networks and social structure have existed for centuries. This book considers developing sociological research on these topics, which began to gain speed at the beginning of the twentieth century. It also discusses the characteristics of social network composition as vast webs of connections between nodes.

Geert Hofstede. 2010. *Cultures and Organizations: Software for the Mind.* rev. and expanded 3rd edn. New York: McGraw-Hill.
This important study of cultural differences across 70 nations explains how people think as members of groups—and how they sometimes fail to think.

Frank K. Salter. 1995/2007. *Emotions in Command: Biology, Bureaucracy, and Cultural Evolution.* New York: Transaction Publishers.
The author seeks a general theory of organizations that can be applied to all cultures. Salter explores issues of social power, asking why people obey their superiors. His investigation looks at the roots of organizational power, pinpointing particular behavioural details of hierarchical institutions that enforce power relations.

Max Travers. 2007. *The New Bureaucracy: Quality Assurance and Its Critics.* Bristol, UK: The Policy Press.
This book, written from a symbolic interactionist viewpoint, analyzes British bureaucratic organizations. It also reviews the classic works on this topic, including material by Weber, Merton, Durkheim, Marx, Parsons, and Goffman.

recommended websites

Cliques in Organizations
www.eric.ed.gov/ERICWebPortal/search/
detailmini.jsp?_nfpb=true&_&ERICExtSearch_
SearchValue_0=EJ078727&ERICExtSearch_
SearchType_0=no&accno=EJ078727
A number of testable propositions are developed which relate the variables of compliance, mobility, and size to motivation for clique formation and to constraints within which cliques form.

Community-Based Research
http://communityresearchcanada.ca
This website contains resources for people who support and have an interest in community-based research.

Rumor and Gossip Research
www.apa.org/science/about/psa/2005/04/
gossip.aspx
Built by the American Psychological Association, this web page provides useful information about two of the most important means by which people uphold informal social control.

Shame and Psychotherapy
www.columbiapsych.com/shame_miller.html
This study argues that although shame has a central role in people's lives, it has been little studied and is little understood.

Society for Study of Symbolic Interactionism (SSSI)
www.symbolicinteraction.org
This is the website of an organization that promotes communication between people who take a symbolic interactionist approach to studying social organization.

Sociosite: Resources for the Study of Organization
www.sociosite.net/topics/organization.php
This site provides sociological knowledge resources on organizational theory, organizational learning, virtual organization, and intranet.

Theories about Persuasion
http://changingminds.org/explanations/theories/
a_persuading.htm
This website provides various ideas about persuasion, negotiation, and mind-changing strategies.

chapter 7

Deviance

Vincent F. Sacco and Alicia D. Horton

In this chapter, you will:

▶ Learn to define deviance and social control as sociological concepts;

▶ Think critically about the images of deviance that we regularly encounter in the popular media;

▶ Learn to describe the major problems confronting researchers who study deviance;

▶ Identify the major questions that sociological theories of deviance and control are intended to answer;

▶ Compare and contrast various sociological explanations of deviant behaviour;

▶ Examine some of the social and demographic factors that are related to particular forms of deviant conduct;

▶ Learn how behaviours and people come to be categorized as deviant;

▶ Study the ways in which people who are labelled 'deviant' cope with stigma.

Introduction

Consider the following scenarios:

- Two teenagers try to get a fake ID in order to buy some beer for a party.
- The daily newspaper announces a mayor's resignation after it was discovered that she gave untendered contracts to a construction firm owned by someone who made large, regular contributions to her recent political campaign.
- A university student buys a copy of an A+ essay and submits it as his own work in a course he is failing.
- The owners of a clothing company launch a publicity blitz that denies charges that their clothing is made by sweatshop labour instead of considering how they might improve their employees' deplorable working conditions.
- A nurse sneaks outside for a quick cigarette, feeling guilty because she knows her co-workers view smoking as an unhealthy habit of weak-willed people. She hopes a mint will disguise evidence of her 'addiction'.
- A teenager shows off her new tattoos to her family but hides her plan to complete the look with jewelled dermal implants and brandings.

On the surface, it might seem that these situations have very little in common. However, they all raise questions about the nature of disvalued action: why some people engage in it and why others react to it in particular ways. Some important common themes run through these examples that relate to the central concerns of this chapter: the sociological nature of deviance and control.

This chapter has three central objectives. First, it explains the meaning and use of **deviance** and **social control** in sociological discourse. Next, it considers some of the major problems faced by researchers who empirically investigate deviance and social control. Finally, it focuses on the major theoretical questions that occupy the time and attention of sociologists who study deviance.

What Is Deviance?

Any discussion of the sociology of deviance and social control must begin with some consideration of what these terms mean. Formal sociological conceptualizations of deviance can be contrasted with more popular views that define *deviance* by illustration, statistics, and harm.

BY ILLUSTRATION

When asked to define deviance, a typical first response is to list types of people or types of behaviours we think deserve the label. These lists could include criminals, child molesters, drug addicts, alcoholics, cult leaders, chronic liars, and more. Of course, who goes on the list and who does not is very much a function of who is doing the listing and when and where the listing is being done. The major problem with these stand-alone lists is that they are incomplete and tell us nothing about why some types of people and behaviour are (and why other types are not) included. In short, we are left in the dark regarding the definitional criteria being employed.

IN STATISTICAL TERMS

Statistical rarity suggests a more explicit way of thinking about the meaning of *deviance*. At face value, it makes a certain amount of sense to identify deviance by rarity, since many of the kinds of people we think of as deviant are, in a statistical sense, relatively unusual. A major problem with statistical definitions of deviance is illustrated by Figure 7.1. The area between points X_1 and X_2 represents typical performance levels across some characteristic. The shaded area on the far left represents the minority of statistically rare cases that fall well below the average. On an examination, for instance, people who fail very badly would be represented here. We might tend to think of such people as 'deviants' in the conventional sense of 'inferior'.

However, the shaded portion on the far right-hand side also suggests a statistically rare performance—but in the positive direction (Fielding, Hogg,

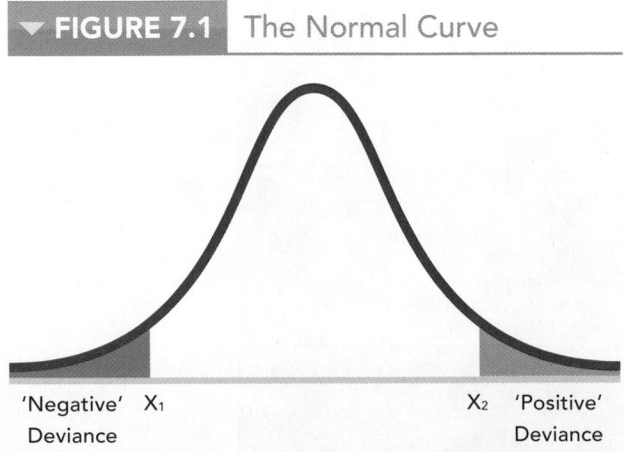

▼ **FIGURE 7.1**   The Normal Curve

'Negative' X_1 X_2 'Positive'
Deviance   Deviance

Statistical definitions of deviance make it difficult to distinguish 'negative' deviance from 'positive' deviance.

and Annandale 2006; Spreitzer and Sonenshein 2004). It could represent people who show superior knowledge on an examination. Statistical definitions obscure distinctions between people who exceed and people who fall short of certain expectations.

AS HARMFUL

Another familiar way of defining *deviance* is by equating deviant action with action that produces destructive outcomes. Again, many of those who would appear on shortlists of deviants seem to be encompassed by this definitional criterion. Murderers, thieves, liars, sexual abusers, and wife-assaulters can all be said to be authors of real and tangible harm. However, like statistical rarity, equations of deviance with harm are fraught with difficulty. While many people consider that deviants do cause harm, many do not. The mentally ill, for example, may be treated as deviant, although it is difficult to ascribe harm to their differentness or document harm that they cause. In contrast, corporate executives and unethical politicians may be able to manage a socially benign image even though their actions might result in considerable damage to life and property. We tend to reserve the label of 'deviant' in our society for other categories of people (Pearce and Snider 1995; Simon 2007).

Considerable disagreement persists in our society about what is harmful and whom we need to fear (Glassner 1999; Siegel 2005). Indeed, historical and anthropological evidence shows that judgments about harm may change over time and from one culture to another (Curra 2011).

Deviance as a Sociological Concept

As sociologists, we are interested in deviance as a product of **social interaction** and group structure; that is, we understand the study of deviance to be the study of people, behaviours, and conditions subject to 'social control'—the myriad ways in which members of **social groups** express their disapproval of people and behaviour. These include name-calling, ridicule, ostracism, incarceration, and even killing. The study of deviance is about ways of acting and ways of being that, within particular social contexts and in particular historical periods, elicit moral condemnation.

To clarify, when the sociologist says, for instance, that LGBT (lesbian, gay, bisexual, transgender) people are an appropriate subject for the scholarly study of deviance, the implication is not that the sociologist thinks of these people as deviant; rather, it is being suggested that members of LGBT groups are the targets of various forms of social control in our society. As sociologists, we are interested in why those with the power to exert social control regard LGBT individuals in this way and the consequences of such actions for these targeted groups (Alden and Parker 2005; Nylund 2004).

Some sociologists think that it is useful to distinguish between the 'ordinary deviance' that most of us engage in and the 'extreme deviance' that only a small number of us commit. The former might include the occasional 'little white lie', the sporadic abuse of alcohol, or a somewhat unusual sexual preference. In contrast, 'extreme deviance' could include believing that one has been kidnapped by extraterrestrials, being a white supremacist, or having and endorsing adult–child sexual contact (Goode and Vail 2008). According to Erich Goode and D. Angus Vail (2008), extreme deviance is behaviour that is so far beyond the norm that it invites an extremely strong negative reaction from almost all sectors of the community.

As the concept of 'extreme deviance' suggests, it is important to distinguish between the *objective* and the *subjective* character of deviance (Loseke 2003). The former refers to particular ways of thinking, acting, and being, the latter to the moral status accorded such thoughts, actions, and characteristics. From a sociological perspective, the 'deviant' character of certain behaviours, world views, or physical features is not implicit in those behaviours, world views, or physical features, but conferred on them by society. To be deemed 'deviant' by a sociologist, a particular behaviour must not only hold the potential for being called deviant (e.g., unusual or rare behaviour) but also must be labelled as 'deviant' by powerful others.

As sociologists, we recognize the need to focus our attention on both sides of the deviance issue. Not everything that could be labelled 'deviant' is necessarily labelled 'deviant' by society. The ability of some in society to use available resources to resist the efforts of others to consider them deviant is also of sociological interest. For example, although corporations engage in activities that undermine health or safety or weaken the economic well-being of many in society, they are able to define themselves as morally respectable by making donations to universities and hospitals or by launching public relations campaigns to promote a positive image.

Why do movies, television, and most other forms of popular culture seem so focused on deviant activities and deviant people?

Researching Deviance

Sociologists who empirically study deviance use the same methodological tools employed in other areas of the discipline. These include experiments, surveys, content analyses, and field research. However, attempting to study the degree to which people might be engaging in behaviours that excite widespread disapproval can create rather formidable problems. The problems discussed in this section represent challenges to all forms of social research, but special difficulties arise when the subject is deviance.

SECRECY

Often, people wish to keep their deviant behaviour secret to protect themselves from social reactions.

How do sociologists undertake valid research in a way that does not intrude excessively into the lives of those under study?

Sometimes researchers attempt to gain the confidence of the subjects by posing as one who shares their deviant status (Whyte 1943). This involves extremely hazardous ethical dilemmas. One much-discussed case in this respect is an early study by sociologist Laud Humphreys. His book *Tearoom Trade* (1970) is a study of impersonal sexual encounters between men in public washrooms ('tearooms'). In order to familiarize himself with the social character of these meetings, Humphreys presented himself to 'tearoom' participants as someone willing to play the role of voyeur/lookout. This deception allowed him to observe the interaction between sexual partners without rousing their suspicion. To compound the ethical problem, Humphreys recorded the licence-plate numbers of 'tearoom' participants and determined their addresses. After disguising his appearance, he went to their homes under the pretence of conducting a public health survey in order to garner more information. Needless to say, when his research techniques

GLOBAL ISSUES

Deviance and Diffusion

Because we can think about deviance as resulting from a claims-making process, it is possible to conceptualize claims about the disreputability of behaviour as separate from the behaviour itself. This means that it is possible to ask questions about the 'performance' of these claims. Such questions have been posed by Joel Best (2008) and other writers who have attempted to determine how claims about deviance and social problems diffuse from one social setting to another. When a way of acting is seen as troublesome in one place, how do these definitions of the behaviour travel within and across national boundaries? The task of the sociologist interested in these issues is to understand the conditions under which such diffusion takes place and why some kinds of claims travel with greater ease or difficulty than others.

An interesting case study of the diffusion process concerns the relationship between Canada and the United States (Sacco and Ismaili, 2001). In this instance, it can be argued that many of the new forms of deviance that are constructed by the American cultural industries work their way northward in rather short order. For example, Canadians found themselves worrying about such problems as school violence,

ritual abuse, stalkers, home invaders, rap music, and Internet predators quite soon after these issues had begun to attract widespread attention south of the border.

There are several reasons why this diffusion occurs as it does. Similarities in the linguistic and political cultures of the two countries facilitate similar styles of discussion and debate on such topics. In addition, the American mass media exert a remarkable influence on Canadian images of what is worrisome in the contemporary world.

However, differences in cultural settings can seriously impede the diffusion of claims about deviance. The moral panic about Satanic crime that originated in the United States in the 1980s travelled easily to Canada, Great Britain, and Australia. Despite a remarkable lack of evidence, many residents of these countries, like the Americans who preceded them, came to believe that Satanists were kidnapping and molesting children, desecrating bodies, influencing the content of youth music and other forms of leisure, and conducting ritual human sacrifices. In France and elsewhere, however, these claims were met with ridicule, and the Satanic panic never took root.

In the First Person

I have been interested in sociology—especially the sociology of crime and deviance—my entire life. Growing up in the border city and tourist destination of Niagara Falls, I became aware early of some of the more peculiar eccentricities and excesses of human behaviour. Like many students, I majored in sociology because I thought the subject matter was, as we used to say, 'pretty cool' and because I did reasonably well in my first-year course. My fascination with the discipline was nurtured by many dedicated and patient teachers and mentors who showed me the real value of careful sociological analysis. However, it was not until I read C. Wright Mills's *The Sociological Imagination* that I began to think seriously about the links between the lives we live and the social structures within which we live them.

—Vince Sacco

Like Dr Sacco's, my interest in the sociology of crime and deviance has been shaped by geography and my personal biography. Growing up outside Vancouver, my ventures into 'the city' brought me into contact with the Downtown East Side (DES)—a stigmatized and devastatingly impoverished area characterized by violence, prostitution, homelessness, and drug use. I was perplexed by the close proximity of the ritzy downtown shopping area to the DES and the unfair hand that people living there seemed to have been dealt. My interest and concern followed me to college and later university, where I met passionate and dedicated scholars who helped me develop a repertoire of conceptual and theoretical tools to think critically about crime, deviance, and social problems such as these. In graduate school, I have been exposed to a number of influential mentors as well as books and ideas that solidified my interest in the way in which deviance is socially created, and the consequences of this for people who experience the stigma of deviant labels.

—Alicia Horton

were discovered, they provoked a firestorm of controversy. Generally, sociologists do not believe that such deception is ever excusable.

A second author undertook research intended to understand the experience of people who engage in extreme forms of body modification. She conducted an ethnography of participants in a radical modification ritual referred to as 'flesh hook pulling'. This practice involves partial nudity, cheek skewering, sewing limes to one's back and piercing the flesh of one's chest with hooks attached to ropes. Participants pull on the ropes against the weight of their bodies until an 'altered state' is achieved. While highly deviant to outsiders, members of this group claim that the practice meets their need for spiritual fulfillment, experimentation with body chemistry, community and connection, and personal validation. The event is organized and carried out in secret at an isolated geographic location where the group's activities are shielded from public view. While the research methods involved participant observation, she avoided deception by being forthright about her identity as a researcher and the goals of the research. She attempted to gain the confidence of participants and strengthen the validity of the research by obtaining the participants' informed consent and respecting the group's need for secrecy by maintaining confidentiality.

DISCOVERY OF REPORTABLE BEHAVIOUR

If research subjects confide in the researcher and reveal information about illegal or harmful circumstances, does the researcher have an obligation to report that wrongdoing to authorities? The problem is brought about by the cross-pressures that the researcher experiences (Bostock 2002). On the one hand, the researcher has a professional obligation to respect the confidentiality of information that research subjects divulge. On the other, one has a social and moral obligation to protect the safety of the public and the research subjects.

The complexities involved in the discovery of reportable behaviour are exemplified in sociologist Sudhir Venkatesh's book *Gang Leader for a Day* (2008). The book concerns the author's research activities while he was a graduate student at the University of Chicago. He describes how he began a rather standard piece of survey research in a high-crime, low-income neighbourhood, where he became acquainted with the members of a large and powerful local gang. Venkatesh developed close personal friendships with many of the gang's members and their families and in so doing became aware of many of the illegal and often violent activities in which the gang engaged. It is difficult to read his book without feeling a strong sense of unease regarding the numerous ethical traps he had to circumvent.

SAFETY

Researchers should take no action that could result in harm to those who participate in the research. While we tend to think only of physical harm in this respect, the injunction is much broader and includes emotional, mental, and economic harm.

In the case of one major survey of female victims of male violence, there was a real concern on the part of researchers that calling women and asking questions about violence in their lives could put them in danger if, for instance, a woman's abuser might be sitting next to her when she received the phone call and started answering questions (Johnson 1996). It was necessary to take several special precautions, such as training interviewers to be sensitive to cues that the respondent might be under some immediate stress.

In a more general sense, it is important to remember that by definition, research into the disvalued nature of people and behaviour involves looking into the lives of the most vulnerable members of society. The sociologist needs to remain aware that research findings can often be used *against* these vulnerable groups, especially when sufficient care is not taken to qualify conclusions or to suggest appropriate interpretations of research evidence.

TIME to REFLECT

Based on your reading, do you think that sufficient ethical safeguards are in place to protect the subjects of deviance research?

The Sociology of Deviant Behaviour

We have defined *deviance* as ways of thinking, acting, and being that are subject to social control—in other words, as kinds of conditions and kinds of people that are viewed by most of the members of a society as wrong, immoral, disreputable, bizarre, or unusual. We recognize that deviance has two distinct yet related dimensions: objective and subjective. *Objective* refers to the behaviour or condition itself; *subjective* refers to the placement of that condition by the members of society in their system of moral stratification. Sociologists do not confuse the physical act of someone smoking marijuana with the designation

of marijuana-smoking as a deviant act. Each suggests a distinct realm of experience, and each is an appropriate object of sociological attention, but why some people smoke marijuana and others consider it deviant are indeed separate questions.

Several theoretical problem areas can be identified. They include questions about (1) causes and forms of deviant behaviour; (2) content and character of moral definitions; and (3) issues that arise over deviant labels.

While sociologists are interested in a broad array of issues, questions about why deviants do what they do have attracted most of the attention. However, the 'why do they do it?' question contains a number of important (if unstated) assumptions. It implies that most of us share a conformist view of the world in which the important thing to understand is why some deviant minority refuses to act the way 'we' act and the moral status of deviant behaviour is never called into question. In a sense, the 'why do they do it?' question proceeds from the assumption that—by and large—society is a stable and orderly place; there is generally widespread agreement about what is right and wrong, and we therefore need to understand what pushes or pulls some off the path of conformity.

Most of the theoretical thought in this respect reflects the influence of functionalist perspectives. Three dominant ways of thinking about 'why they do it' can be identified: strain theory, cultural support theory, and control theory (Cullen, Wright, and Blevins 2007).

STRAIN THEORY

Strain theory derives from the writings of the famous American sociologist Robert Merton, who in 1938 published an influential paper entitled 'Social structure and anomie'. Merton sought to understand why, according to official statistics, so many types of non-conformity such as crime, delinquency, drug addiction, and alcoholism are much more pervasive among members of the lower social classes. As a sociologist, Merton was interested in understanding the structure of society, rather than individual personalities, as the central explanatory mechanism in this issue.

The answer, he argued, resided in the mal-integration of the cultural and social structures of societies. Stated otherwise, the lack of fit between the *cultural goals* people are encouraged to seek and the *means* available to pursue these goals creates a social strain to which deviant behaviour is an adjustment. Merton's logic is elegant and compelling. In a society like the United States, there is

OPEN for DISCUSSION

How Random Is Random Violence?

Mass media coverage of crime tends to focus a great deal of attention on random violence. After all, crimes such as serial murder, carjacking, robbery, home invasion, and child abduction are frightening to most people precisely because they involve elements of random threat. It is the strongly felt sense that such crimes could victimize us as easily as anyone else and that we are at risk whenever and wherever we find ourselves that makes news stories (and movies and television shows) about random violence so powerful and popular.

Despite public beliefs about the random nature of violence, sociologists have known for decades that the risks of becoming a victim of crime are anything but random. Like disease, accidents, and other kinds of negative life events, violence seems more likely to afflict some of us than others.

Sociological researchers have been able to document the 'social structure of violence' through large-scale studies that ask a representative sample of the population about crimes they may have experienced during some defined period of time (e.g., the previous six months or the previous year). These studies allow the researchers to compare the profiles of victims and non-victims and therefore to identify the factors that seem to be associated with the risk of becoming a victim of violence.

The most recent such Canadian study was conducted by Statistics Canada in 2009. Almost 20,000 Canadians over the age of 15 were interviewed about their own victim experiences during the previous 12 months. In addition, the researchers collected a wealth of information about respondents' social, demographic, and lifestyle characteristics and about the circumstances of the crimes. Among the findings relating to the differential risk of violent victimization are the following:

- Younger Canadians were more likely than older Canadians to become victims of violence. People between the ages of 15 and 24 were 15 times more likely to become victims than were people over the age of 65.
- With respect to marital status, single people were most likely to be victims and married people were least likely.
- Risk was also higher for those who self-identified as homosexual, had some form of activity limitation, or participated in evening activities outside the home.
- Among persons who identified as Aboriginal, rates of violent victimization were double those of non-Aboriginal people.
- Rates of violent victimization were lower for people who identified as a visible minority than for non-visible minorities. Rates of victimization were also lower for immigrants than for non-immigrants.
- Violent victimization was associated with higher levels of alcohol consumption.

It should be noted that similar results have been produced by many researchers in many different countries. However, despite the consistency of many of these findings, they raise significant and controversial questions, which interested parties continue to debate. One such question concerns the reasons why such factors are related to an elevated risk of becoming a victim. A second question relates to the potential such research findings might have for blaming the victims for the crimes that befall them. In other words, when we start to focus on the reasons why some people rather than others are likely to be victimized, are we suggesting implicitly that they have done something to 'invite' their victimization?

little recognition of the role that **class** barriers play in social life. As a result, everyone is encouraged to pursue the goal of material success—and everyone is judged a success or a failure based on the ability to become materially successful.

When people steal money or material goods, it can be said that they are attempting to use 'illegitimate means' to achieve the trappings of success. When they take drugs (or become 'societal dropouts'), they can be interpreted as having withdrawn from the competition for stratification outcomes. For Merton, these

problems are most acute in the lower social classes, where people are most likely to experience the disjuncture between the things to which they aspire and the things actually available (see Table 7.1).

Critics note certain problems with Merton's arguments (Downes and Rock 2003; Bernard, Snipes, and Gerould 2009), such as his assumption of the accuracy of official statistics and his failure to account for much middle- and upper-class crime and deviance.

Despite these limitations, Merton's argument has greatly influenced the way sociologists think about

TABLE 7.1 ▼ Robert Merton's Paradigm of Deviant Behaviour

Robert Merton argued that there are five ways of adjusting to a social structure that encourages large numbers of people to seek objectives that are not actually available to them.

	Attitude to Goals	Attitude to Means	Explanation/Example
Conformity	Accept	Accept	Most people accept as legitimate the culturally approved ways of achieving those goals. In Merton's example, most strive for material success by working hard, trying to get a good education, etc.
Innovation	Accept	Reject	The bank robber, drug dealer, or white-collar thief seeks success too but rejects the conventional means for achieving that success.
Ritualism	Reject	Accept	Some people seem to simply be going through the motions of achieving desired social goals. In large organizations, we use the term 'bureaucrat' to describe people who are fixated on procedures at the expense of outcomes.
Retreatism	Reject	Reject	Some people adjust to strain by 'dropping out' of the system. Such dropping out could include losing oneself in a world of alcohol or illegal drugs or adopting some unconventional lifestyle.
Rebellion	Reject/Accept	Reject/Accept	Rebellion includes acts intended to replace the current cultural goals (and means) with new ones. In this category we might include the radical political activist or even the domestic terrorist.

the causes of deviant behaviour (Laufer and Adler 1994). Sociologists Richard Cloward and Lloyd Ohlin (1960) expanded on Merton's ideas to argue that there is a need to explain why different kinds of delinquent behaviour patterns emerge in different types of neighbourhoods. They suggested three types of delinquency adaptations: (1) the *criminal pattern*, characterized by instrumental delinquency activities, particularly delinquency for gain, in which those involved seek to generate illegal profits (such as theft and fencing of stolen goods); (2) the *conflict pattern*, characterized by the presence of 'fighting gangs' who battle over turf and neighbourhood boundaries; and (3) the *retreatist pattern*, organized around the acquisition and use of hard drugs.

More recently, Robert Agnew (1985; 2006) has theorized that in addition to the inability to achieve the things we want in life, a second source of strain involves an inability to avoid or escape some negative condition. For example, youths who cannot avoid abusive parents might use drugs, run away, or become aggressive as a way of coping with this strain. Strain can also result when individuals lose something they value. A child who is forced to move and thus leave important friendships might experience this type of loss strain.

These arguments take as their point of reference the need to explain why some individuals but not others behave in ways that invite social sanction. They share an explanatory logic that focuses on how

the organization of our social relations creates problems that require solutions. In this paradigm, the causes of deviant behaviour are located in patterns of social life that are external to but affect individuals.

CULTURAL SUPPORT THEORY

Cultural support theory focuses on the way patterns of cultural beliefs create and sustain deviant conduct (Cohen 1966). According to cultural arguments, people behave in ways that reflect the cultural values to which they have been exposed and then internalize. In this way, it can be said that you attend university because you value education and come from a cultural setting that values education and learning. If conventional values support conventional behaviour, it should follow that deviant values support deviant behaviour.

Writing in the 1930s, sociologist Edward Sutherland proposed that people become deviant because they have been exposed to learning experiences that make deviance more likely. In short, people end up deviant in the same way that they end up as Catholics, stamp collectors, saxophone players, or French film fans—that is, as a result of exposure to influential learning experiences. According to Sutherland, learning the 'specific direction of drives, motives, attitudes and rationalizations' is important in becoming deviant (1947, 7). In other words, we must learn to think about deviant conduct as acceptable to ourselves. For instance, we most commonly refrain from committing murder not because we don't know how but because we have come to define such action as morally repugnant. For Sutherland, learning to accept or to value criminal or deviant action in a very real sense makes such action possible.

Sutherland's cultural insights help us to understand how people come to value actions the rest of the society might despise. However, we live in a society that simultaneously condemns and supports deviant behaviour. Is it possible, then, both to believe in and to break important social rules? Most of us think that stealing is wrong but have also stolen something at some point. This is possible because we have learned to define certain deviant situations as ones to which the rules do not apply. When we steal a pen from work, we tell ourselves (and others) that we are underpaid and deserve whatever fringe benefits we can get, or that employers expect people to steal and build this cost into their budgets. From this perspective, the broader culture both condemns deviance and makes available cognitive techniques for neutralizing the laws that prohibit deviant action (Fritsche 2005; Matza and Sykes 1957).

Some critics argue that the use of culture to explain deviance is tautological (Maxim and Whitehead 1998).

Cultural theories tell us that deviant beliefs and values are the source of deviant conduct. But how do we know what people's beliefs and values are? If we observe people stealing and infer that they have come to acquire values that are supportive of stealing and that these values explain the stealing, we have reasoned in a circle and explained nothing.

Still, cultural arguments have been very influential in the sociological study of deviant behaviour (Akers and Jensen 2003) and have proven more useful than strain arguments in making sense of so-called 'respectable crimes' such as corporate crimes and 'digital piracy' (Ingram and Hinduja 2008; Morris and Higgins 2009). One may argue that corporate crime, at least to some extent, is rooted in a 'culture of competition' that legitimates organizational wrongdoing (Calavita and Pontell 1991).

CONTROL THEORY

Advocates of **control theory** argue that most types of deviant behaviour do not require a sophisticated form of explanation. People lie, cheat, steal, take drugs, or engage in sexual excess when and if they are free to do so and if these activities can be the

(Chuck Savage/Corbis)

'Deviant' behaviours such as excessive drinking are not always deviant from a statistical standpoint. Among young men, many risky behaviours (such as binge drinking) are celebrated as signs of masculinity and group conformity.

most expeditious ways of getting what we want. The important question, then, is not 'why do some people break rules?' but 'why don't more of us engage in "forbidden" behaviour?' For control theorists, deviant behaviour occurs whenever it is allowed to occur, so we can expect to find deviance when social controls are weak or broken.

In sociology, this is a very venerable idea that can be traced to the writing of Émile Durkheim (1951 [1897]). In his classic study of suicide, Durkheim sought to explain variation in suicide rates among groups and across time. Catholics, he found, have lower suicide rates than Protestants, and married people have lower rates than single people. Suicide rates increase both in times of economic boom and during depressions. What is varying in all of these cases? Durkheim suggested that the crucial variable might be social regulation (or *social control*) that forces people to take others into account and discourages behaviours that are excessively individualistic. In short, suicide is more likely when people are disconnected from social regulation and left to their own resources.

In more recent times, sociologist Travis Hirschi (1969) attempted to use social control logic to explain the conduct of youthful offenders. For Hirschi, the problem of juvenile crime could be understood in reference to the concept of the bond to conventional society. Each of us, to a greater or lesser degree, has a connection to the world of conventional others. For youth, the world of conventional others is represented by parents, teachers, and members of the legitimate adult community. Hirschi reasoned that if youthful bonds to conventional others are strong, youths need to take these others into account when they act; if the bonds are weak, they are free to act in ways that reflect much narrower self-interest. While theories of the bond have been eclipsed by later theoretical developments, the idea that deviance is a product of weak links to conventional society continues to attract attention (Church, Wharton, and Taylor 2009; Ford 2009)

More recently, in collaboration with Michael Gottfredson (Gottfredson and Hirschi 1990), Hirschi has proposed a general theory of crime and deviance (Gottfredson 2006). This theory posits that crimes of all types tend to be committed by people who are impulsive, short-sighted, non-verbal risk-takers. The underlying social-psychological characteristic of such people is low self-control. These people are more likely to commit crime and to engage in a wide range of deviant practices (Baron 2003; Kerley, Xu, and Sirisunyaluck 2008; Nakhaie, Silverman, and LaGrange 2000). For Gottfredson and Hirschi, the problem of low self-control originates in inadequate child-rearing that fails to discourage delinquent outcomes.

Social control theories remain very influential, but they can be criticized for rendering motivation irrelevant to the study of crime and deviance and for inadequately explaining why people with strong bonds to the conventional world also engage in prohibited acts (Deutschmann 2007).

TIME to REFLECT

Do you think that some explanations, such as strain theory, can be faulted for excusing deviant behaviour?

THE TRANSACTIONAL CHARACTER OF DEVIANCE

Despite their sociological character, strain, cultural support, and social control arguments tend to focus attention on the individual. According to these theories, people commit deviant acts because they respond to strain, are exposed to learning environments that support deviance, or are free from social constraints. However, others encourage us to understand deviant behaviour as an interactional, joint, or collective product rather than an individual outcome.

Familiar explanations of murder focus on the murderous acts of the individual (see Figure 7.2 for homicide rates in Canada). As sociologists, we might try to understand how people who commit murder do so in response to social strain (Pratt and Godsey 2003) or as a result of an affiliation with a culture of violence (Chilton 2004; Wolfgang and Ferracuti 1967). Alternatively, we might try to understand how murder results from particular kinds of interactions.

David Luckenbill (1977) has demonstrated that murder can in many cases be understood as a **situated transaction**. Some murder may be seen not as an individual act but an interaction sequence of participants in a common physical territory. Based on a study of 70 homicides in California, Luckenbill identified six common stages of murder:

- *Stage 1*: The transaction starts when the eventual victim does something that the eventual offender could define as an insult or as an offence to 'face'. The victim might call the offender a liar, refuse to share a cigarette, or make a sexually suggestive comment about the offender's partner.

▼ FIGURE 7.2 | Provincial Variations in Rates of Homicide (Number of Homicides per 100,000 Population), 2008

For reasons that are not entirely clear, the rate at which the situated transaction we refer to as homicide occurs varies from province to province.

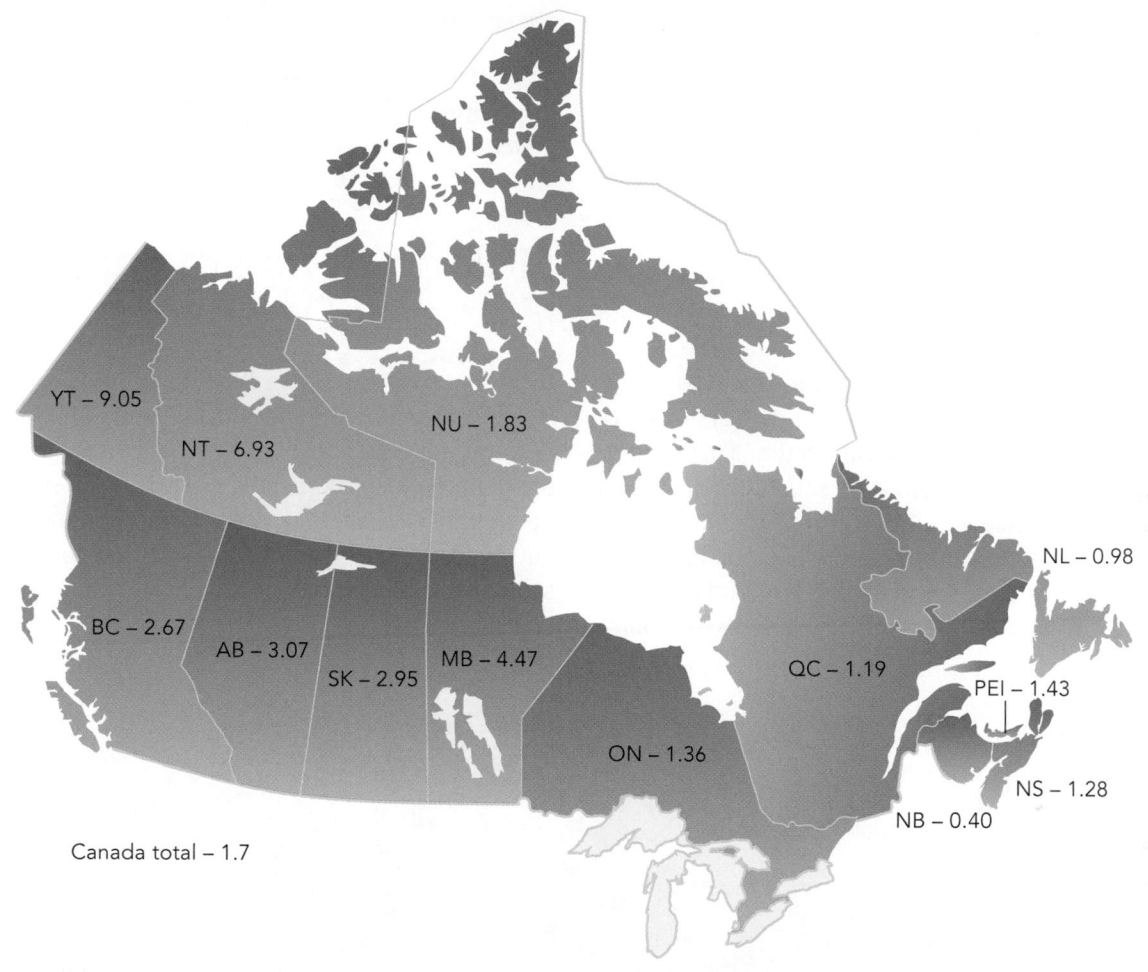

SOURCE: Sara Beattie. 2009. 'Homicide in Canada, 2008'. *Juristat* Catalogue no. 85-002, 29 (2). www.statcan.gc.ca/pub/85-002-x/2009004/article/10929-eng.pdf.

- *Stage 2*: The offender defines what the victim has said or done as threatening or offensive.

- *Stage 3*: The offender makes a countermove intended to respond and save face. This could involve a verbal response or a physical gesture.

- *Stage 4*: The victim responds in an aggressive manner. A working definition of the situation as one that will require a violent resolution seems to be emerging. The problems may be aggravated by onlookers who jeer the participants, hold their jackets, or block convenient exits.

- *Stage 5*: A brief violent exchange occurs. It may involve a fatal blow, stab, or gunshot.

- *Stage 6*: The battle is over; the offender flees or remains at the scene.

Luckenbill's work demonstrates how murder can be understood as a social product. This does not imply an absence of guilt or excuse for killing; rather, it shows that acts of deviance can involve complex and significant interactional dimensions. For this reason, some argue that it is more useful to think of murder and other forms of deviance as 'social events' rather than the acts of individuals in isolation (Sacco and Kennedy 2011). Crime as a social event implies much more than bad behaviour. The outcome of a homicide event is contingent not only on the 'killer's' actions but on such factors as the actions of the bystanders, whether those involved are alone or with friends, and the response time of the police.

Sociologist Randall Collins (2008) has proposed an alternative theoretical approach to the role played by situational factors in the development of troublesome behaviour. Collins argues that a useful situational theory of violence should explain all of its forms by placing interaction in the centre of the analysis. Collins's theory assumes that what matters is the way in which the situation unfolds—not the culture or social backgrounds of the individuals. In violent situations, people are tense and afraid of being hurt and hurting others. For example, researchers consistently find that many soldiers in battle do not fire their weapons but do exhibit signs of extreme trauma. Because the driving force is much more emotional than rational, the eruption of violence depends on a variety of conditions that determine how people negotiate emotional tension.

According to Collins, one of the main pathways around confrontational tension and fear involves identifying and attacking a weak victim. This strategy applies across a wide range of situations, including military raids, violence during arrests, domestic violence, and bullying. Another strategy is making third parties the focus of emotional attention. In these situations, observers play a key role in affecting the course of the violent exchange. For violence to be successful, it must turn emotional tension into emotional energy. This usually involves an expense to one of the parties to the conflict at the hands of the other.

TIME to REFLECT
Are theories of situated transaction guilty of victim-blaming?

Making Sense of the 'Facts' of Deviant Behaviour

Sociologists have repeatedly demonstrated that deviant acts are not randomly distributed in the population. Instead, people with certain social and demographic characteristics are much more likely to be involved in deviant behaviour than others. The task of sociologists of deviance is to explain these levels of differential involvement.

(Michael Newman/PhotoEdit Inc.)

Bullying is a common form of 'deviance' in schools and playgrounds. Typically the work of cliques (discussed in the previous chapter), bullying relies on exclusion, ridicule, and violence to turn ordinary kids into outcasts and victims.

TABLE 7.2 ▼ Apprehension by the Police by Gender, 2005, Rates per 100,000 Population	Males	Females
Violations against the person	1,409	304
Violations against property	1,653	510
Violations against the administration of justice	819	186
Other Criminal Code violations	312	79

SOURCE: Statistics Canada. 2008. 'Female offenders in Canada'. *Juristat* 28 (1). www.statcan.gc.ca/pub/85-002-x/85-002-x2008001-eng.pdf.

GENDER

Gender correlates closely with a wide range of behaviours. With respect to deviance, males and females differ in terms of the amounts and types of disapproved behaviours in which they engage. Males are more likely to be involved in behaviours that most members of Canadian society would say they disapprove of. Males are much more likely to be involved in criminal behaviour (crimes related to prostitution are an important exception in this regard). The differential is greatest in cases of violence but is also significant for other kinds of crime (Gannon et al. 2005). While there has been some narrowing of the gender gap in recent years, crime remains a male-dominated activity (Sacco and Kennedy 2011).

Males are more likely to consume both legal and illegal drugs (Health Canada 2009). They are more likely to commit suicide and are more likely to use guns or explosives to do it (Langlois and Morrison 2002). While overall rates of mental illness do not differ markedly between men and women, women are more likely to be diagnosed with depression and anxiety, while men are more likely to experience problems related to addiction and psychosis (Health Canada 2002).

Several feminist writers have argued that there is a marked tendency in the sociological literature to systematically ignore the deviant behaviour of women (Miller and Mullins 2007). Most of what is written about crime and deviance concerns the behaviour of men, both as deviants and as police and other agents of social control. Many sociologists have assumed that female deviant behaviour could be explained using the same theoretical ideas and models that have been used to make sense of male behaviour—a position with which many feminists do not agree.

The failure to be sufficiently attentive to the gendered nature of criminal and deviant behaviour has been an empirical problem. Historically, sociologists tended to not be terribly interested in acts of crime or deviance without a significant male dimension. Only through the work of feminist social critics did researchers come to focus on problems that affect women more directly. These include forms of deviance that tend to uniquely victimize women, such as intimate violence and sexual harassment (Chasteen 2001; Comack 2008).

AGE

Age is strongly associated with many kinds of deviant behaviour (Tanner 2009). Crime rates are greatest during the late teens and early adulthood and decline very sharply after that (Sacco and Kennedy 2011). This pattern characterizes even violence in the home: young husbands (those under 30) are much more likely than older husbands to treat their wives violently (Mihorean 2005).

However, this pattern does not apply to all kinds of crime. Suicide rates actually tend to be lower among younger Canadians (Langlois and Morrison 2002). While older people are traditionally assumed to be those most likely to experience a variety of forms of mental illness, the onset of most mental illness occurs during adolescence and young adulthood (Health Canada 2002; Kessler et al. 2005).

TABLE 7.3 ▼ Suicide Rates by Sex and Age (Rates per 100,000), 2005	Both Sexes	Males	Females
All ages	11.6	17.9	5.4
10–14	2.0	1.7	2.4
15–19	9.9	13.4	6.3
20–24	13.2	20.1	5.9
25–29	10.4	17.5	3.0
30–34	12.7	19.6	5.7
35–39	16.1	24.9	7.2
40–44	18.0	27.2	8.8
45–49	18.2	28.8	7.5
50–54	17.7	25.7	9.8
55–59	14.6	22.2	7.2
60–64	11.0	17.5	4.7
65–69	11.6	18.3	5.3
70–74	9.5	17.0	2.9
75–79	13.1	22.7	5.5
80–84	10.5	19.4	4.9
85–89	9.6	21.4	3.7
90 and older	7.6	20.1	3.2

SOURCE: Statistics Canada data.

CLASS AND ETHNICITY

A great deal of sociological theorizing about the 'causes' of deviant behaviour has taken as its central issue the need to explain why social and economic precariousness is related to deviant outcomes. Many of the studies indicate that poorer people and people from minority groups are more likely to be involved in many forms of crime and delinquency, use drugs and alcohol, and develop various kinds of mental illness. However, a consensus in the research literature does not exist regarding how concepts such as poverty, economic inequality, ethnicity, or minority-group status should be measured for research purposes (Braithewaite 1979; Wortley 1999). Similarly, while some studies indicate that working-class youth are more likely to be delinquent, others tell us the opposite (Tittle, Villemez, and Smith 1978). While minority-group status seems to be related to higher rates of crime in some cases—for example, with **First Nations** people—it seems to be related to lower rates of crime in other groups such as Asian immigrants in British Columbia (Gordon and Nelson 2000; Brzozowski, Taylor-Butts, and Johnson 2006; Perreault 2009).

Other interpretations of the significance of the relationship between social disadvantage and deviant behaviour point in the direction of a more general fault line that runs through the sociology of deviance. Are poorer or minority people more likely to be deviant, or are they just more likely to get caught and be labelled as 'deviant'? Do our definitions of what constitutes crime and deviance themselves reflect class biases? Poor people, for instance, are less likely to commit many kinds of crimes, such as fraud and embezzlement, and even less likely to manufacture faulty products, engage in false advertising, profit from political corruption, or become involved in stock market swindles.

These observations suggest a need to ask questions about the subjective character of deviance. Why are some ways of thinking, acting, and being more likely than others to excite indignation and disapproval, and why are some people more likely than others to become the objects of social control?

TIME to REFLECT

Are locations and activities that host large numbers of young males more likely to be characterized by high levels of deviance?

The Sociology of Deviant Categories

The sociology of deviance is also the study of moral stratification. To call something or someone 'deviant' is to articulate a judgment that the thing or person is disreputable. An important set of issues in the sociology of deviance relates to the creation of deviant and non-deviant categories into which people and actions are sorted (Loseke 2003; Best 2008).

We tend to treat categorical distinctions as common sense. The deviant qualities of people and acts, we convince ourselves, reside within the people and acts themselves. However, judged from another standpoint, known as **social constructionism** (Miller and Holstein 1993; Spector and Kitsuse 1977), this logic is flawed. Acts and people are not inherently deviant but are defined as such by those with the power to do so. This perspective maintains that there is nothing self-evident or commonsensical about the deviant quality of people and their behaviour. Instead, the deviant quality assigned to people and behaviour is itself problematic and requires investigation.

Further, we need to recognize that the character of social condemnation is fluid and dynamic over time (Curra 2011). For example, only a few years ago being gay might have been considered grounds for social exclusion, but in the contemporary context, it is seen as much less deviant.

Similarly, many ways of acting or being that were once widely tolerated now seem to draw considerable disapproval. One clear example is cigarette-smoking (Bell et al. 2010; Stubera, Galea, and Link 2008; Tilleczek and Hine 2006; Troyer and Markle 1983; Tuggle and Holmes 1997). Only a few decades ago, people smoked in elevators, restaurants, around children—even in sociology classes. Today, smokers are the object of scorn, and their habit is the subject of a variety of forms of legal and extra-legal control (Stubera, Galea, and Link 2008; Blanke and Silva 2004; McNabola and Gill 2009). Other examples of behaviour for which social tolerance has decreased include drinking and driving (Asbridge, Mann, and Flam-Zalcman 2004; Gusfield 1981), wife assault (Johnson 1996), and sexual harassment (Lopez, Hodson, and Roscigno 2009).

DEVIANCE AS A CLAIMS-MAKING PROCESS

Social constructionist writers understand the distinctions that people make between deviant and non-deviant behaviour as part of a **claims-making** process (Best 2008; Spector and Kitsuse 1977). This refers to the process by which groups assert grievances about the troublesome character of 'other' people or behaviours. Claims-making can include many different sorts of activities, such as debating exotic sexual practices on a daytime talk show, marching in protest, or providing expert testimony before a parliamentary committee. Claims-making promotes certain moral visions of social life; in short, it is anything anybody does to propagate a view of who or what is deviant and what needs to be done about it (Loseke 2003).

Claims-making is directed toward three broad types of objectives:

1. *Publicizing the problematic character of the people with the behaviour in question.* People generally need to be convinced that there is some tangible reason to regard others as troublesome. Claims-makers may endeavour to convince us that deviants are dangerous or irresponsible or that their behaviour is contagious (Best 2008; Macek 2006). Claims may be widely accepted as valid statements about the world regardless of objective basis.

2. *Shaping a particular view of the problem.* It matters greatly whether we see people as troubled or as troublesome (Gusfield 1989). Generally, claims-makers want to convince us that certain people are a problem and a problem of a particular type. 'Problem drinking', for instance, can be constructed in many different ways (Holmes and Antell 2001). Interpreted as a sin, it becomes a religious problem. Calling it a crime implies it is a legal problem. We might see it as a sickness, which implies that it is a medical problem. The behaviour in question remains the same, but the kind of deviant the problem drinker is varies. Different constructions have very different implications for what it is we think needs to be done.

3. *Building consensus around new moral categories.* Claims-makers endeavour to build widespread agreement about the correctness of a particular moral vision (Heimer 2002; Macek 2006). This is accomplished by winning the support of the media, officialdom, and the general public (Best 2008; Hilgartner and Bosk 1988). As consensus is built, dissenting views are relegated to the margins of legitimate discourse, and deviant categories take on a common-sense character.

Who Are Claims-Makers?

The movement to 'deviantize' people and behaviour originates in the perception that something is troubling and needs correction. Howard Becker (1963) coined the term *moral entrepreneur* to describe those who 'discover' and attempt to publicize deviant conditions. These are crusading reformers who are

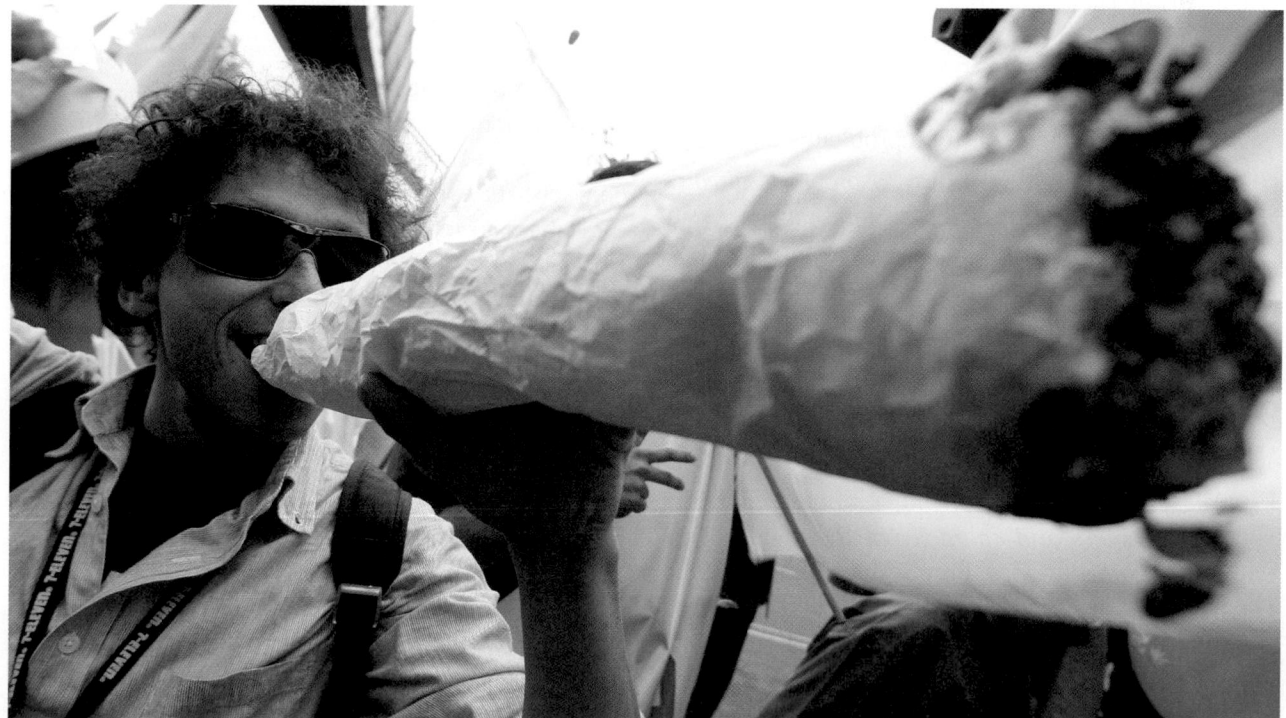

(Reuters/Daniel Aguilar)

Illegal behaviours such as marijuana-smoking are common among young people. Some believe that marijuana use should be decriminalized, since non-enforcement of the law brings all lawmaking into disrepute.

disturbed by some perceived evil and who will not rest until something is done about it.

In the early stages, definitions of deviance are often promoted by those with some direct connection to the problem. For instance, claims-making members of MADD (Mothers Against Drunk Driving) had a powerful emotional stake in initial efforts to criminalize drunk driving (Reinarman 1996). Still, many involved in the construction of deviance have no vested interest or emotional connection to the problem or outcome. Lawmakers, journalists, daytime talk-show hosts, and the producers of television drama frequently play a significant role in the promotion of particular designations of deviance (Hogeveen 2005; Sacco 2005), but their social distance from the issue is often greater than that of victims' groups.

What Are Claims?

When social constructionists speak of *claims*, they are talking about the actual message content that conveys a moral vision of deviance and non-deviance. What do claims-makers say to convey the message that something or someone deserves the appellation of 'deviant'? The study of claims is the study of rhetoric communication, because communication persuades audiences. Successful claims-making rhetoric demonstrates the gravity of a problem in several ways:

- *Using compelling statistics.* Statistics impress upon media consumers the size of a problem and its escalating severity (Best 2004; Gilbert 1997). Statistical estimates of this nature legitimate concern and provide compelling evidence of a problem's urgency.

- *Linking an emergent concern to problems already on the public agenda.* Familiar moral language can be used to provide reference points for emergent problems. For instance, because addiction is widely recognized as a problem in North America, the term 'addiction' has been used in a very liberal way to convey problems of 'pornography addicts', 'gambling addicts', and 'Internet addiction' (Butters and Erickson 1999).

- *Using emotionally compelling examples to typify the seriousness and character of the threat posed by the behaviour* (see Bromley and Shupe 1981; Loseke 2009). For example, the Columbine High School killings are applied in a rhetoric to exemplify the problem of school violence, even though such incidents are extremely rare and most school crime in no way resembles this incident (Fox and Levin 2001).

Deviance Ownership

Claims-making is a matter not only of seeing particular types of people or behaviour as problems but of seeing them as particular *kinds* of problems (Gusfield 1989). What is at stake is problem 'ownership', and how a problem is framed determines who will be responsible for dealing with the problem (Sasson 1995). If problem drinking is understood as a legal problem, we might expect the courts and police to do something about it, but when it is constructed as a medical problem, we turn to doctors and psychiatrists.

Medicalization is a dominant trend in deviance solutions concerns (Adler and Adler 2006; Conrad and Schneider 1980). Many behaviours are constructed as medical disorders requiring treatment rather than punishment (Dworkin 2001). Increasingly, the language of sickness, health, and disease has been applied to conditions as diverse as violence, gambling, obesity, and rampant consumerism. This shift suggests a more benign approach to deviants because it implies that individuals are not entirely to blame for their behaviour, and the stigma associated with deviant conduct is reduced (Appleton 1995). Nonetheless, it can be argued that medicalization encourages us to ignore structural contexts when we consider deviance. Medical models minimize the responsibility of society in general by implying that problems occur because individuals 'get sick' and not because social structural conditions make some kinds of behaviour more likely.

Deviance and Social Conflict

Disagreement exists in society regarding who or what should be seen as disreputable (Hier 2002). These conflicts are evident in the battle over abortion, the movement to legalize marijuana, and efforts to control cigarette-smoking. In other cases, the conflict is less evident only because effective claims-making has resulted in consensus.

Two broad types of **conflict theory** can be distinguished: conservative and radical (Williams and McShane 2009). These theories suggest different ways of understanding the wider social dynamic of the claims-making process. From the perspective of conservative conflict theory, social conflicts regarding the moral meaning of conduct emerge from diverse sources (Turk 1976; Bernard, Snipes, and Gerould 2009). As members of various ethnic, religious, professional, lifestyle, or cultural groups pursue their social interests, they may come into conflict with other groups over scarce resources. In the context of such models, **power** is not concentrated in any one sector of the society (Gusfield 1963). Instead, various **status groups** come into conflict, often over specific issues. From this perspective, the study of moral differentiation is the study of how some groups in society are able to influence systems of social control and more effectively compete in the struggle to achieve their goals. Social control bureaucracies may find that the resources made available to them become more plentiful when they can identify new forms of danger that require control (Becker 1963; Jenkins 1994). Alternatively, new or struggling medical specialties can find their social status enhanced if the members of a society become convinced that they are indispensable to the solution of some pressing social problem (Pfohl 1977).

Often, the struggle to define deviance reflects a much more evident cultural difference regarding moral behaviour. Contemporary debates are, in the end, debates over who gets to call whom 'deviant'. Those whose cultural or religious beliefs lead them to oppose a movement for gay rights may think of gay people as deviants. Similarly, gays and lesbians (and others) may think of those who actively oppose the movement for gay rights as suffering from a psychological malady known as *homophobia*.

In contrast, radical conflict theory draws on the Marxian understanding of society (Spitzer 1975). It views the economic organization of society as the key to understanding moral stratification. The **social construction** of deviance must be understood as reflecting the economic realities of capitalism and the class exploitation it engenders. From this perspective, the internal logic of capitalism gives deviance both its objective and its subjective character. Capitalism requires a large pool of labour that can be exploited by keeping wages low. This means that there will always be more workers than jobs and some people will inevitably be marginalized. These populations have little stake in the system and are at greater risk of criminal labels and involvement.

TIME to REFLECT

Do you think that the concern over 'Internet predators' is another example of exaggerated deviance construction?

THE SOCIOLOGY OF DEVIANT STIGMA

A third key area of study in the sociology of deviance is stigma application and management (see, for example, Table 7.4). This body of research and theory focuses attention on the social interaction between those who exercise social control and those who are thought of as disreputable. In this respect, questions about the application and consequences of deviant stigma tend to be more micro- than macrosociological.

The Process of Labelling

People come to be seen as deviant because of others' perceptions or beliefs. The labels of 'deviant' that are assigned to people are charged with a great deal of emotion. Such labels sort through the thousands of acts in which a person has engaged and indicate that the person's identity is best understood in terms of the act according to which the label is affixed (Erikson 1966).

The assignment of stigma suggests what sociologists refer to as a **master status** or deviant label that overrides all other status considerations (Becker 1963). Being known as a murderer trumps any other status characteristics the person might have (bright, interesting, poor, blonde, left-handed). Sociologists use the term **status degradation ceremony** to refer to the rituals during which the status of 'deviant' is conferred (Garfinkel 1956). Status degradation ceremonies, such as incompetency hearings, psychiatric examinations, and courtroom trials, mark the movement from one social position to another by publicly and officially acknowledging a shift in social **roles** and the emergence of a new, deviant identity. We

---

TABLE 7.4 ▼ Types of Deviant Behaviour

Howard Becker (1963) suggested that once we recognize that deviant stigma is separable from deviant acts, it is possible to recognize at least four types of deviants. These types are created by the contrast between what people between what people actually do (break rules or keep them) and how they are perceived (deviant or not deviant).

		Behaviour	
		Obedient	Rule-Breaking
Perception	Perceived as deviant	Falsely accused	Pure deviant
	Not perceived as deviant	Conforming	Secret deviant

have designed ceremonies to confer the status of 'deviant', but we don't have comparable ceremonies to move deviants back to 'normality'.

Resistance to Labelling

The ability of some in society to confer the status of 'deviant' on others reflects differentials in social power. People with access to power resources are able to more effectively negotiate the status of 'deviant' (Pfuhl and Henry 1993). People might use a range of other strategies to avoid or negotiate a label of deviance. One obvious method involves efforts to undermine social control through *evasion*. 'Successful' deviants learn to engage in prohibited conduct in ways that decrease the likelihood of getting caught (Becker 1963).

Individuals try to avoid or negotiate stigma through what Goffman calls *performance* (Goffman 1959). The dramatic roles we might perform when stopped by a police officer for speeding are part of a performance intended to neutralize the efforts of police to impose a deviant designation (Piliavin and Briar 1964). Similarly, *disclaimer mannerisms* are actions intended to signal to agents of social control that one is not the appropriate target of deviant attribution.

Deviant Careers and Deviant Identities

One potential consequence of the labelling process is *deviancy amplification*: the situation wherein the very attempt to control deviance makes deviance more likely (Lemert 1951; Tannenbaum 1938). Efforts to describe this usually distinguish between primary and secondary deviance (Bernburg, Krohn, and Rivera 2006; Lemert 1951). *Primary deviance* is that in which we all engage that has no real consequence for how we see ourselves or for how other people see us. From time to time all of us might lie, cheat, or engage in some other prohibited behaviour.

Secondary deviance, in contrast, is marked by a life organized around deviance. It suggests emergence in a deviant role rather than ephemeral acts of deviance. While any of us might tell an occasional lie, most of us do not think of ourselves or are thought of by others as liars.

What turns primary deviance into secondary deviance? The answer is societal reaction. It is argued that the ways in which agents of social control respond to initial acts of deviance—through stereotyping, rejection, and the degradation of status—can actually make future deviance more rather than less likely (Markin 2005; Tannenbaum 1938). One of the key intervening mechanisms in this process is the transformation of the **self**. Consistent with social psychological theories, such as the one advanced by Charles Horton Cooley (1902), of how the self emerges and is maintained, labelling theorists have argued that individuals who are consistently stigmatized may come to accept others' definition of their deviant identity. To the extent that individuals increasingly come to see themselves as others see them, they may become much more likely to behave in ways that are consistent with the label of 'deviant'. Individuals become committed to a life of deviance largely because others expected them to—deviance becomes a self-fulfilling prophecy (Tannenbaum 1938).

Managing Stigma

How do people manage 'deviant' labels? Various strategies may be employed to control information about a deviant identity or to alter the meaning of stigma so as to reduce the significance of the deviance in their lives (Durkin 2009; Hathaway 2004; Park 2002).

A useful distinction is made between the *discreditable* and the *discredited* (Mankoff 1971). The former refers to people who might become discredited if their stigma were to become public. Conversely,

HUMAN DIVERSITY

What's in a Name?

Does it matter how we label behaviour? In your view, do the labels in each pair below refer to the same or different kinds of behaviour and people? If they are different, then how are they different?

Terrorist	Freedom fighter
Prostitute	Sex worker
Sex assault victim	Sex assault survivor

Cult leader	Religious leader
Disabled	Differently abled
Pro-life	Anti-choice
Pro-choice	Pro-abortion
Addiction	Bad habit
Alcoholic	Drunk
Modification	Mutilation
Stripper	Exotic dancer

discredited stigma is either evident or assumed to be known.

For the discreditable, stigma management involves a pressing need to control information others have about them. People with a hidden stigma face the constant worry that others they care about may reject them if information about this stigma becomes public (James and Craft 2002). For instance, those suffering from stigmatized diseases might keep this aspect of their life secret because they fear rejection. The discreditable may attempt to 'pass' by fraudulently assuming an identity other than that which is stigmatized. A gay man, for instance, might 'stay in the closet' to allow others to assume that he is 'straight'.

The discredited person's stigma tends to be apparent, so there is no need to keep it secret. Rather, the discredited attempt to restrict stigma relevance to the ways others treat them. They may attempt purification by trying to convince others that they have left a deviant identity behind (Pfuhl and Henry 1993). This redefinition of self is intended to restrict the interactional relevance of the stigma by locating it in the past. One of our contemporary definitions of a hero is someone who has left a deviant stigma behind. Helen Keller and Christopher Reeve, for instance, were thought of as heroic largely because they rose above the stigmatizing character of particular physical conditions.

The discredited may also invoke collective stigma management. Bearers of stigma may join together to form an association intent on changing public perceptions of their disvalued character. Organizations intended to 'undeviantize' behaviour include the National Organization for the Reform of Marijuana Laws (NORML); COYOTE (Call Off Your Old Tired Ethics), which promotes the rights of sex workers; and the National Association to Advance Fat Acceptance (NAAFA), which advocates for the rights of the 'hugely obese' (Gimlin 2008). Collective stigma management may involve attempts to influence media coverage of the group in question or the terms used to describe members of the group (Bullock and Culbert 2002). For instance, groups organized around collective stigma management have advocated that the terms 'disabled', 'retarded', and 'AIDS victim' be replaced in popular and official discourse with 'differently abled', 'developmentally delayed', and 'AIDS survivor', respectively (Titchkosky 2001).

SOCIOLOGY in ACTION

Disclaimer Mannerisms in University Examinations

Sociologists Daniel Albas and Cheryl Albas (1993) undertook a study of how students attempt to distance themselves from charges of academic dishonesty while writing examinations. People writing examinations are at high risk of stigmatization. Invigilators patrol the rooms, and often neither the professor nor graduate assistants have any direct knowledge of the individuals writing the exam. For these reasons, students take steps to ensure that they will not be wrongly accused of cheating.

The authors define two major strategies: actions that students take and those they avoid.

Actions taken include:

- *Picayune over-conformity with regulations.* This involves the demonstration of conformity with even the most minor examination rules. A student who needs to blow his or her nose will be sure to wave the tissue around first so it is clear that it is nothing other than a tissue.
- *The expression of repression of creature releases.* Creature releases are those aspects of the self that steal through the facade of social control, including sneezes and yawns. A student who needs to use the bathroom during the exam might make very exaggerated gestures to impress upon invigilators the urgency of the situation.
- *Shows of innocence.* Because students know that a lack of activity might be read as indicative of a lack of preparation, when they are not writing, they might be underlining or circling words on the exam sheet.

Actions avoided include:

- *Control of eyes.* Students know that they are not supposed to have roving eyes, so they are careful where they look. Strategies include staring at the ceiling or the head of the person in front of them.
- *Control of notes.* Students might frisk themselves before they enter the exam room to ensure that they are not carrying anything that could cause trouble.
- *Morality of place.* Students worry that where they sit can send a message about their trustworthiness. Care is taken not to sit next to very good students or someone they believe is perceived as a potential cheater.

TIME to REFLECT

If deviance is a source of stigma, why are so many kinds of deviants in our society (e.g., gangsta rappers, misbehaving actors and athletes) treated like celebrities?

Deviance and Post-modernism

Post-modernism suggests a more recent theoretical trend that builds upon many earlier insights. Post-modernist theorists of crime and deviance study the way in which language works to marginalize and stigmatize people who come into contact with the criminal justice system or are otherwise labelled deviant. Language is a part of what post-modernists, and other theorists, call 'discourse'—large, specialized units of knowledge (such as the scientific, legal, and medical languages) made up of any number of modes of communication. For post-modern criminologists, that which is 'criminal' or 'deviant' can be understood as resulting from the capacity of powerful groups—such as lawmakers—to 'discursively' control the behaviour of less powerful groups.

Dominant discourses reflect the interests and values of those with the power to have their version of truth 'normalized' or accepted. Normative modes of discourse inform popular notions of what is morally superior or 'true', while minority views go unheard. Post-modernists argue, then, that when the meaning of language is taken for granted, it reflects the ability of

UNDER the WIRE

Deviance in Movies

Crime, delinquency, and deviance permeate movies, television, music videos, the Internet, and other arenas of popular culture. Films about serial killers, gangs, mental illness, or drug trafficking, for example, speak to the nature of disvalued action by presenting images that morally evaluate people and behaviour. Because movies, television, and magazines play a significant role in defining deviance, it is crucial to consider how popular culture constructs certain groups and people as troubled or troublesome (Gusfield 1989) and blameworthy or praiseworthy (Loseke 2003).

For instance, in films, prostitution is almost uniformly depicted as a dangerous, degrading, and devalued line of work that prostitutes ought to 'escape' (Horton 2009). Prostitutes are depicted as routinely raped (*Leaving Las Vegas*), held captive and drugged (*Eastern Promises*), verbally abused (*Hustle & Flow, Deuce Bigalow: Male Gigolo, Pretty Woman*), cut with knives (*Unforgiven*), spat on (*Mysterious Skin*), and killed (*From Hell, The Dead Girl, Very Bad Things, Four Rooms*). In the movies, the problem of violence against prostitutes is often presented as an individual problem (Horton 2009). In other words, the victimization that prostitutes are subjected to is constructed as originating with some personal deficiency that prevents them from leaving the trade. Characters who continue to prostitute end up being jailed, executed, or murdered (*Monster, The Dead Girl*), whereas others avoid repeat victimization by becoming hairdressers, getting married, finding conventional employment, or otherwise leaving prostitution (*Mighty Aphrodite, True Romance, Deuce Bigalow: Male Gigolo*).

Prostitutes' rights groups such as COYOTE argue that violence against prostitutes results from the legal prohibition of acts of prostitution in Canada, which implies that prostitutes are 'bad women'. They claim that the problem of violence against prostitutes is perpetuated by unsafe working conditions and negative cultural attitudes toward sex work. This problem is understood as originating with the legal prohibition of prostitution; thus, part of the solution is to change the law to make safer working conditions for sex workers and to apprehend those who perpetrate violence against them. Popular film, however, constructs this problem in a very different way. It tends to build a consensus around the problem of violence against prostitutes by implicitly claiming it is an individual problem that requires prostitutes to do something about their situation. It excludes explanations of the problem that shift responsibility to social institutions such as the legal system.

As you have learned in this chapter, what we perceive as the solution to a problem depends greatly on what type of problem we understand it to be. To be competent media consumers, we must learn to think critically about the taken-for-granted character of deviance and social problems in popular culture and the often one-dimensional nature of problems and solutions that it presents.

one group to impose its will on others (Derrida 1976; 1978). Accordingly, ways of understanding crime and deviance that fall outside accepted, primary discourses can be understood as disempowered through 'linguistic domination' (Arrigo and Bernard 1997).

For post-modernist criminologists, dominant discourses such as law create categories, or binaries, that are organized in a hierarchal fashion. By 'deconstructing' or breaking down binaries created by language (such as male/female, rational/irrational, deviant/conforming), post-modernists illuminate the way in which categories privilege the truth claims of some groups over others by defining the behaviour of powerless groups as abnormal or criminal (Arrigo 1999). Recently, post-modern theorists of deviance and crime have applied their insights to analyses of housing homeless people (Arrigo 2004), the resistance strategies of female offenders (Geiger 2006), and reality-based television and mental illness (Shon and Arrigo 2006).

Conclusion

Our experience with deviance reflects the influence of the cultural context and the historical period in which we live. As times change, so do the categories of people and behaviour society finds troublesome. While gay and lesbian people were once viewed as deviants, today they are seen as less so. Drunk driving, wife assault, and cigarette-smoking were once regarded as normal; they are now viewed as highly deviant. Deviance is thus a dynamic process, and the future will present further permutations and innovations. By way of example, we need think only about the large number of newly constructed forms of deviance that we already associate with computer use, such as cyberporn, cyberstalking, and Internet addiction.

In the most general terms, the sociology of deviance is concerned with the study of the relationships between people who think, act, or appear in disvalued ways and those who seek to control them (Sacco 1992). It seeks to understand the origins, character, consequences, and broader social contexts of these relationships.

Deviance can be thought of as having two dimensions: the objective and the subjective. The objective aspect is the behaviour, condition, or cognitive style itself. The subjective aspect is the collective understanding of the behaviour, condition, or cognitive style as disreputable. A comprehensive sociology of deviance needs to consider both dimensions. Thus, we want to know why some people rather than others act in ways that society forbids and why some ways of acting rather than others are forbidden.

Correspondingly, it is possible to identify several major types of questions around which theory and research in the sociology of deviance are organized. First, how do we understand the social and cultural factors that make prohibited behaviour possible? Strain theory argues that people engage in deviant behaviour because it is a form of problem-solving; cultural support theories focus on the ways in which people acquire definitions of deviant conduct that are supportive of such behaviour; and control theories maintain that deviance results when the factors that would check or constrain it are absent.

Second, what is and what is not viewed as disreputable is not obvious, and there is a need to explain the prevailing system of moral stratification. Definitions of deviance emerge from a process of claims-making. The establishment of consensus around such definitions gives categories of deviance a taken-for-granted quality.

We also need to ask questions about the application and management of deviant stigma. Being labelled 'deviant' is a complex process that creates numerous problems for people subject to social control attention. It is important, therefore, to understand who gets labelled and how people cope with social control. In particular, we need to be alert to the manner in which the imposition of labels can worsen the very problems that the application of social control is meant to correct.

Finally, we need to consider the role of language itself in the creation of categories of deviance and how these categories create power structures that privilege some ways of understanding crime and deviance over others.

questions for critical thought

1. What images of crime and deviance dominate coverage in the local media in your community? What images do they create of troubled and troublesome people?

2. Why do people cheat on university examinations? How might this question be answered by proponents of strain, cultural support, and control theories?

3. In your view, why are young males so much more likely than other groups to engage in a range of behaviours that many in society find troublesome?

4. What evidence do you see in your own social environment of the disvalued character of cigarette-smoking and smokers?

5. Aside from the examples given in the text, can you suggest behaviours or conditions that have undergone a shift in moral status in the past few years? How would you account for these changes?

6. How might Marxian and more conservative conflict theorists differ in their interpretations of the legal and moral battle regarding the use of 'soft' illegal drugs such as marijuana?

7. How might you explain to an interested layperson the difference between the ways in which sociologists and journalists think about deviance?

8. In your opinion, does it make sense to speak of something called 'positive deviance'? Why or why not?

recommended readings

Joel Best. 2008. *Social Problems*. New York: W.W. Norton.
This is a thorough and accessible introduction to the social constructionist approach to deviance and social problems. The author makes excellent use of practical examples in order to unravel the complex process by which people and behaviour come to be seen as troublesome.

Deborah Brock. 1998. *Making Work, Making Trouble: Prostitution as a Social Problem*. Toronto: University of Toronto Press.
Brock offers a comprehensive treatment of the social problem of prostitution in Canada—the author's analysis illustrates the value of a constructionist approach to the study of specific forms of social deviance.

Francis T. Cullen, John Paul Wright, and Kristie R. Blevins, eds. 2007. *Taking Stock: The Status of Criminological Theory*. Edison, NJ: Transaction.
This collection of essays provides a detailed and comprehensive examination of the major varieties of sociological theories of nonconformity. Each of the theories is explained and assessed with respect to available empirical evidence.

John Curra. 2011. *The Relativity of Deviance*. 2nd edn. Thousand Oaks, CA: Sage.
The author argues that deviance cannot be considered an absolute and that what is subject to social control varies by time and place. The analysis of necessity calls into question many common assumptions about the nature of problematic people.

Frederick J. Desroches. 2005. *The Crime That Pays: Drug Trafficking and Organized Crime in Canada*. Toronto: Canadian Scholars' Press.
This book offers a systematic and readable treatment of the problem of organized drug-trafficking. The author's insightful analysis highlights the social organizational character of this kind of deviant activity.

Erving Goffman. 1963. *Stigma: Notes on the Management of Spoiled Identity*. Englewood Cliffs, NJ: Prentice-Hall.
This is the classic discussion of how people who are defined as 'deviants' manage stigma. The book was formative in the development of the sociology of labels of deviance.

Lorne Tepperman. 2010. *Deviance, Crime, and Control: Beyond the Straight and Narrow*. 2nd edn. Toronto: Oxford University Press.
Tepperman provides a detailed discussion of the contemporary field of the sociology of deviance. The book focuses both on theories that structure the field and their substantive application.

Sudhir A. Venkatesh. 2008. *Gang Leader for a Day: A Rogue Sociologist Takes to the Streets*. New York: Penguin Books.
This is a fascinating first-person narrative in which the author tells of his experiences studying a high-crime neighbourhood on the South Side of Chicago. The book raises a large number of ethical and methodological questions.

recommended websites

Canadian Sociology and Anthropology Association
www.csaa.ca/structure/Code.htm
This website contains the rules, regulations, and principles relating to the ethics of professional sociological research.

Crime Theory
www.crimetheory.com
This site provides a very comprehensive discussion of deviance and crime theory for educational and research purposes.

Sex Professionals of Canada
http://spoc.ca
This site contains resources for professional sex workers and provides some interesting insights into how those involved in professional pursuits that others regard as deviant think about their own lives.

Society for the Study of Social Problems
www.sssp1.org
This is the main page for the Society for the Study of Social Problems (SSSP). The journal of the society, *Social Problems*, has been very influential in the development of the sociology of deviance.

Statistical Literacy
www.statlit.org
This is the website of an organization devoted to the promotion of statistical literacy. It is an invaluable resource for researching the various means by which statistics may be manipulated for political and social purposes.

Statistics Canada
www.statcan.gc.ca
This is the main page for Canada's national statistical agency, Statistics Canada. Many different sorts of reports, tables, and graphs relating to a variety of forms of deviance can be found at this site.

The Surveillance Project
www.queensu.ca/sociology/Surveillance
This page contains a wealth of information relating to the Queen's University Surveillance Project. The project is concerned with the study of the increasingly large number of technologies and social practices employed for the purpose of social control.

SALE
70%
final
clearance
phase-eight.co.uk
SIZE 8 - 18
NOW AVAILABLE

Class and Status Inequality

Sara J. Cumming
and Ann D. Duffy

In this chapter, you will:

▶ Understand the pivotal role that social and class inequality plays in the social construction of reality;

▶ Appreciate the centrality of class and status inequality in the development of sociological thought;

▶ Learn some of the key historical as well as contemporary realities of economic and social inequalities in Canadian society;

▶ Comprehend the behind-the-scenes role of ideology and institutions in maintaining societal patterns of inequality;

▶ Recognize the diverse ways in which social inequalities are dynamic throughout history as a result, for example, of state policies;

▶ Learn more about the complex patterns of global inequalities that structure our lives in the globalized economy;

▶ Understand the contentious future that Canadians face in terms of economic and social inequalities.

Introduction

One of the foundational insights provided by sociology is that our lived realities are constructed socially. We become human through a social process, and our understanding of the world is forever framed by these social experiences. For example, a new friend offers to give us a ride home. When we approach her car in the parking lot, we do not simply register the fact that here is a vehicle with four wheels and an internal combustion engine. Rather, we immediately and unconsciously run through a whole gamut of socially constructed meanings—meanings embedded in patterns of social inequality. The age, make, and upkeep of the car all are instantly noted. That our new friend drives a brand-new, sparkling BMW evokes a whole range of social reactions and connections that are quite different from those we would experience if a rusty, dented Toyota Corolla were sitting there.

Of course, this example is directly related to the centrality of **class and status** inequalities in our day-to-day experiences. When we look at this car, we are generally making assumptions about the relative class and standing of the individual who drives it. The BMW evokes, rightly or wrongly, an impression of wealth and economic well-being, while the rusty Toyota conjures up images of poverty and social marginalization. Clearly, among the most important sorting devices incorporated in our social construction of daily reality are these divisions between those who have and those who do not and, importantly, our relationship to them.

Within the complex diversity of possible responses to class inequalities, there is, however, a typical relationship between high economic class and **power**. Historically, individuals capable of wresting control over a community's assets—land, animals, property—do in the process acquire 'power'. Understood in the simplest terms, the wealthy are in an excellent position to dictate what others do, and in this way they exercise control over the lives of those who are less well-off. This is evident historically in the starkest terms with the rulers in China and Egypt as well as in Central and South America, who were able not only to require others to build enormous monuments in their honour but also to demand that many be killed in order to accompany them as servants into the next life. This is power in its rawest terms.

These relationships between economic advantage and power have become increasingly nuanced and complex over the passage of time. Today, the power associated with social class is likely to be much more indirect and subtle (Lukes 1974). Bill Gates, founder of Microsoft and one of the wealthiest individuals on the planet, is not only an economic force to be reckoned with on the stock market, but he has also created a powerful public persona as a revered philanthropist and global activist. Instead of displaying coercive power, Gates may be said to have successfully mobilized his power indirectly.

The public nature of Gates's status is, however, unusual in contemporary North American societies. The **economic elite** in Canada and internationally today are typically obscured by their corporate connections. The power wielded by a globalized manufacturing conglomerate like General Motors may be widely recognized, but the individuals sitting on its corporate boards are not likely to be well known. When they decide to close a factory and thousands of workers become unemployed, their exercise of power over the lives of others tends to be obscured as the mass media talk about 'GM closings' rather than the actions of individual corporate executives.

Not only are economic power relations often indirect and obscured, they are likely to be muddied by interconnections with other patterns of inequality. Various social factors—gender, ethnicity, race, age, disability, sexual orientation, and immigrant status—play an important mediating role. Certainly, broad categories of individuals—women, ethnic and racial minorities, the disabled, recent immigrants, the elderly—are generally at much greater risk of being both relatively poor and powerless. Even individuals who combine great advantage and disadvantage often find themselves hobbled. For example, a woman who is a member of the upper class may find that her gender limits the opportunities otherwise provided by her class position (Duffy 1986). In short, as sociologists have suggested, there are complex intersections between economic and other forms of inequalities. When examining economic and status inequalities, it is important to keep these **intersectionalities** in mind.

TIME to REFLECT

In what ways do you think age (children, youth, adult, and senior) affects economic inequalities in Canada, and what are the social class implications of the fact that its population is rapidly aging?

Finally, class and status inequalities are further complicated by their ebb and flow through history.

Class position may shift dramatically through the course of an individual's life, and certain historical periods have certainly lent themselves to dramatic change. The recession of 2008 shook up both the RRSP portfolios as well as the economic future of many Canadians, and the long-term implications still remain to be seen. Similarly, the historical trends among the elites suggest change. With corporations going global, along with the expansion of global governance bodies (for example, the World Trade Organization [WTO]), some analysts are suggesting a transition from a Canadian corporate elite to a transnational capitalist class (Carroll 2007; Rothkopf 2008).

In this chapter, we will explore social and economic inequalities in greater detail. Certainly, social class is a pivotal term in much historical and contemporary sociological thought—indeed, many of the 'classics' centre on this concept. Given this centrality, it is not surprising that an understanding of Canadian society requires an appreciation of the patterns of class structure—especially the elites and the poor. As discussed below, these class divisions are sustained not only by various institutions, ranging from elite private schools to political parties, but also by popular ideologies that legitimize and perpetuate specific social class arrangements. Despite these supports, the history of Canadian society reveals dynamic shifts in social class arrangements. Notably, social movements, public agencies, and government policies have addressed and changed the extent of economic inequalities in Canada. Finally, we conclude with a brief examination of future trends. Ominously, it appears that social and economic inequalities are on the brink of increasing in Canada and in other industrialized countries.

Class and Status Inequalities in Sociological Thought

A number of the key concepts in sociology emanate from concern with patterns of social inequality. Indeed, the term **social stratification** is one of the foundational concerns in sociological inquiry. It refers to the hierarchical arrangement of individuals based upon wealth, power, and prestige. Social stratification affects almost every aspect of our lives—from where we live, our material possessions, and our level of education to our health and well-being. Rather than focusing on individual circumstances, it stresses the layering of groups of people according to their relative privilege into social classes. An

individual's position within a social class is referred to as their social status. An individual's social status can be achieved or ascribed.

An ascribed status is typically assigned to a person at birth and is connected to many characteristics other than the income of parents—race, gender, disability/ability, age, and other factors that are not chosen or earned. In almost all instances, the factors that determine your ascribed status cannot be changed, although a few people do change their gender and disabilities may appear or disappear. In contrast, an achieved status is based primarily on earned accomplishments/achievements. If a person goes to university and successfully completes their PhD, that person has earned the title of 'doctor'. If an athlete performs well, he or she can potentially achieve the status of 'professional athlete' and the income that accompanies that status.

Although most people would argue that a **meritocracy**, a system based upon achievement rather than ascribed status, is preferable, understanding individual social status and social class is far more complex than equating success or failures to individual strengths and/or weaknesses. Not all positions within the hierarchy are based solely on merit. For instance, under a meritocracy, entrance into university should be based solely on a student's achieving the grades necessary for acceptance; in fact, however, the best predictor of university entrance is family income. Additionally, students who have university-educated parents are more likely to do well in university themselves. If social status were based primarily on earned achievement, we might expect to see a high degree of social mobility—the movement between classes; however, throughout their lifetime most people remain in the social class into which they were born (Western and Wright 1994).

When we look beyond the borders of Canada, however, we find that compared to some other societies around the world, we appear to have an open stratification system. It is possible in Canada for a person from a poor family to move up in class. A young person from a poor family can win a scholarship or can apply for a student loan to attend college or university. Through hard work and determination, the student may obtain a medical degree and become a practising physician, improving his or her socioeconomic status dramatically.

From a global perspective, we see that Canada offers more opportunity for upward mobility; however, we also need to recognize the degree to which ascribed status limits opportunities for Canadians. As we discuss later, Aboriginal peoples, visible

In the First Person

My brother and I were raised by a single teen mother on welfare during much of our childhood. We lived in a two-bedroom apartment far from any city centre and our school, without a vehicle or accessible public transportation. My mother eventually remarried, and our economic situation changed; however, I moved out on my own a few years later at age 17. I worked a full-time job while finishing high school and then took on another, part-time job when I graduated. It was with a great deal of naiveté that I enrolled in university as the 23-year-old mother of a toddler. After my oldest daughter was born, I decided that I needed an education if I was ever going to escape the impoverished lifestyle I had become accustomed to, and I spent a great deal of time berating myself for our poor economic situation. It wasn't until my first sociology class that I began to understand the complicated institutional web that structured my experience. Sociology taught me that living in poverty is most often *not* the result of some individual deficiency.

—Sara J. Cumming

I grew up in one of the many postwar suburban developments that sprang up as the federal government provided low-cost housing to ex-soldiers. The subdivision was filled with streets of families living in similar economic circumstances. Social class was apparently a non-issue. However, when asked in a first-year sociology course to find out my social class, reality revealed its complexities. From my mother's viewpoint, we were 'working-class', while my father embraced the North American ideal of classlessness—we were 'middle-class'. For my mother, a war bride who had grown up in hard-scrabble working-class London, England, being a member of the working class was a statement of pride—we worked for a living and didn't rely on others. For my father, being a member of the middle class reflected Canada's openness—here everyone was equal and had the same opportunity to advance.

Class was an issue—even within my own family.

—Ann Duffy

minorities, recent immigrants, those with disabilities, and lone parents (especially lone mothers) experience disproportionate poverty. Additionally, it is important to note that social class is not simply a reflection of income but includes other factors such as wealth and prestige. For instance, Heidi Fleiss became a household name in the early 1990s when she was arrested as the madam of an escort service alleged to serve a celebrity clientele. While Fleiss is believed to have been a millionaire at the time of her arrest, she certainly did not have a 'prestigious' career. Similarly, many factory workers earn more money than teachers and nurses, but they are generally not perceived as holding the same social position, because their jobs lack professional prestige. Thus, income and economic assets alone are not clear indicators of social class. Not only are the nature and determinants of social class membership not clear-cut, but

sociologists are divided on the overall societal significance of social stratification.

TIME to REFLECT

Is Canada truly a meritocracy?

SOCIOLOGY AND SOCIAL STRATIFICATION

Conflict Approaches to Social Stratification: Karl Marx and Max Weber

The history of all hitherto existing society is the history of class struggles. . . . society is more and more splitting up into two great hostile camps, into two great classes directly facing each other—bourgeoisie and proletariat (Marx and Engels 1983 [1848], 203–4).

In this famous quotation from the *Communist Manifesto*, Karl Marx (1818–83) outlines the two issues that are central to his work on class: he argues that society is characterized by conflict (class struggles) and a distinguishing feature of capitalism is the division of society into two central classes, the **bourgeoisie** and the **proletariat**. Marx maintained that society is divided into these classes on the basis of their relationship to the **means of production**—their access to the tools, factories, land, and investment capital used to produce wealth (Marx and Engels 1985 [1848]). As the bourgeoisie, who owned the means of production, pursued their self-interest in the form of profit, they necessarily exploited the proletariat, who had little choice other than to sell their labour.

According to Marx, this capitalist mode of organization has several characteristics that distort the structure and meaning of the economic process: private property, expropriation of surplus wealth, division of labour, and alienated labour (Grabb 2007). Marx contended that the drive for private property was primarily responsible for creating the two-class system. Under capitalism, everyone needed to have an income in order to obtain property and ensure survival. Those who existed outside this system—for example, the unemployed who lacked an income—served as a reserve army of labour ready to be called upon if the demand for labour increased or if current workers complained about their exploitation. By paying workers low wages, capitalists were able to expropriate surplus wealth from the labour process. Given the absence of employment opportunities outside the industrial economy, workers had no choice but to exchange their labour for wages that were far below the value of the products they were producing for the owners. While working people were responsible for creating a surplus of wealth, they did not reap the benefits.

Marx maintained that the inequality this system produced was neither desirable nor inevitable (Lindsey and Beach 2003). In fact, he held that class conflict between wage-labourers and the owners of the means of production would be historically inevitable as the inequality between these classes became ever more pronounced. As the proletariat class developed **class consciousness**—an awareness of workers' shared interests and their ability to act in those interests—Marx predicted a socialist revolution, the eradication of capitalist economies, and a new mode of production.

Born almost 50 years later than Marx, Max Weber (1864–1920) had the advantage of seeing a more developed form of industrial capitalism in European society. Like Marx, Weber believed that the ownership of property and economic inequalities were central to the system of social stratification. Weber argued, however, that stratification was based on more than who owned the means of production. He suggested that social stratification could be better understood by looking at economic positions, hierarchies of prestige, and the ability (or inability) to control others. Thus, Weber's theory of stratification introduced three independent factors: class, status, and power.

Class: While Weber agreed with Marx that the ownership of property was important in determining a person's position in society, he believed that it was not necessarily the sole determinant of one's class position. Weber noted that some people had power over the means of production even though they did not own them. Managers and supervisors of large corporations can make decisions (such as awarding themselves bonuses) even though they do not own the means of production.

Status: While Weber called the second component of his theory of stratification 'status', it is often referred to as prestige in contemporary discussions. Status refers to a person's prestige, popularity, or social honour. Status, or prestige, can be the result of property, since people tend to look up to the wealthy and covet the things that they possess. However, status may also be based on other factors. For example, religious leaders, saints, nuns, and even some political leaders can hold influence in society while having few economic possessions. Terry Fox is not only a Canadian icon, he is a striking example of the importance of social honour.

Power: The third component of social class according to Weber is the ability to exert power or control over others despite their objections. While property is a major source of power, so too is prestige. Once retired from politics, Pierre Trudeau continued to wield significant power in Canadian society, not so much because of his property and economic position but because of his social prestige. In moments of historical significance—Meech Lake and the 1995 Quebec referendum—his views were considered so significant to the public debate that media commentators sought out and then publicized his comments. From a Weberian viewpoint, despite his relative lack of economic assets, Trudeau continued to exercise significant power.

In short, since Weber lived in a period marked by the emergence of a more complex and nuanced social and economic order, he conceptualized social stratification as a more complicated phenomenon. After all, he witnessed the dramatic growth of white-collar work, the expansion of trade unions, and the mushrooming of state bureaucracies. From his vantage point, it appeared that social inequalities were based on a variety of factors and were not purely economic. In his formulation, it was possible to exercise power through prestige as much as through economic clout. While Weber shared with Marx a preoccupation with social inequality, he urged attention to non-economic social forces.

Structural Functionalist Approaches to Social Stratification: Durkheim and Davis and Moore

In contrast to Marx and Weber, other sociologists have emphasized the social functions played by social stratification. Émile Durkheim (1858–1917) looked at how social structures within which power and struggle operate are even possible in the first place (Grabb 2007). He argued that early societies were held together by mechanical solidarity—union based on minimal division of labour, similarity of people who have roughly the same life experience, and shared common beliefs. When a society is comprised of similar individuals, the group feels a sense of solidarity based on these common experiences, values, and beliefs. Durkheim suggested that the division of labour was a crucial force in the historical evolution of social structures. Once the division of labour became more extensive, no one could survive without the cooperation of others. This division of labour weakened the old mechanical solidarity and was the key means by which the new form of solidarity emerged: organic solidarity. Under mechanical solidarity, Durkheim argued, there was a moral stability as people recognized each other's obligations to and dependence upon one another. However, when people do not recognize this obligation, anomie may occur. An anomic division of labour—conflict between labour and capital—leads to class polarization. Because this division of labour is forced, it encourages class polarization.

Davis and Moore

Kingsley Davis and Wilbert Moore are best known for their 1945 publication 'Some principles of social stratification', which later became known as the Davis–Moore thesis. Davis and Moore argued that inequalities exist in all societies and thus must be necessary. In order to function properly, a society must somehow distribute its members into various social positions and persuade them to perform the duties of these positions. Not only are some positions more pleasurable than others, but many require special training or talents and are viewed as more important. According to Davis and Moore, the positions that are rewarded with the highest economic gains and highest rank are those that have the greatest importance for society and those that require the greatest training or talent. In order to induce people into undertaking a lengthy, expensive education, such as is required for becoming a medical doctor, the rewards must be high. Davis and Moore contended that this system is based on consensus and shared values, because members of society generally agree that the reward system is fair and just.

While the Davis–Moore thesis sounds like a remarkably reasonable and simple explanation for social stratification, their structural functionalist approach has been severely criticized. For example, while some difference in pay might be reasonable for people who spend long periods of time in school, the functionalist approach does not take into account that it is often people who can afford to attend school who do so, rather than those who are the most talented or gifted. Additionally, there are substantial differences—regardless of educational qualifications—in who gets into the highest-paid jobs, with minorities and women represented disproportionately at the lowest pay tiers. Furthermore, the social inequality between the top and bottom is extreme. Movie stars, bankers, and professional athletes bring home millions of dollars a year, while nurses, teachers, and daycare workers are paid significantly less. There is no doubt that many better-paying occupations require a great deal of training and education; however, this alone does not explain the inequality in our society, which is much greater than simple differences in effort and reward.

TIME to REFLECT

Do you think class inequality is a functional and necessary part of society? Are higher rewards necessary in order to motivate people to fill certain positions?

Symbolic Interactionist Perspectives on Social Stratification: Thorstein Veblen

Symbolic interactionists take a different approach to social stratification. Rather than attempting to explain why stratification exists or how conflict is created because of class inequality, they are interested in how people interpret and represent inequality. As their name suggests, symbolic interactionists consider how meanings and symbols enable people to carry out uniquely human actions and interactions (Ritzer 2000, 357). In reference to social stratification, they pay particular attention to the use of status symbols.

In *Theory of the Leisure Class* (1899), Thorstein Veblen (1857–1929) suggested that there was a distinct difference between the productiveness of the manufacturing industry and the greed of business. Business, he argued, existed only to earn profits for a leisure class. Veblen maintained that the main activity in which the leisure class engaged was **conspicuous consumption**—the purchasing of expensive goods and services primarily for the purpose of putting wealth on display. These purchases were status symbols—various signs that identified a particular social and economic status or position. Diamond tiaras, massive country estates, and large retinues of servants would all be status symbols in Veblen's era. Not surprisingly, he not only drew attention to this conspicuous consumption, but given the dire poverty

that characterized that period of US history, he also harshly criticized the waste and excess that accompanied conspicuous consumption.

Today, status symbols remain apparent everywhere, although they may manifest themselves differently depending on culture and location. Expensive houses, luxury cars and clothing, along with exotic vacations and elite sports, still communicate wealth and social position. Symbolic statements about wealth are also apparent throughout the university, with name brands such as lululemon, Bench, and UGG dotting the hallways even though similar inexpensive apparel is widely available.

Veblen's main contributions, then, were in highlighting the symbolic embodiment of social inequality through the practice of conspicuous consumption and in his contention that most people want to appear as though they live above their actual social location. Today, conspicuous consumption, even at the risk of indebtedness, has become an epidemic, as suggested by the multi-billion-dollar industries dedicated to helping the millions of North Americans live

(Atlantide Phototravel/Corbis)

Despite gains in education and job opportunities, women in traditional female-dominated occupations such as care workers or housekeepers continue to be marginalized members of society.

beyond their means, industries ranging from payday loan companies to the credit card divisions of banks and credit unions.

Feminist Explanations for Social Stratification

Predictably, few theorists in the past included women in class analysis. Perhaps the breadwinner ideology (the assumption that a woman's role in the household was to provide unpaid work for her family while her husband provided the economic resources through his paid labour) excluded women as participants in the social class structures (Nakhaie 2002). Certainly, stratification research has repeatedly been critiqued for being 'malestream'—excluding women from research samples on the basis of their secondary relationship to the labour market, focusing exclusively on class rather than incorporating gender inequality, and assuming that women's economic and social positions are derived from those of their husbands or fathers (Abbott and Sapsford 1987). It has most often been assumed that 'stay-at-home' wives take on their husband's social class, while women who earn a wage are included in what is assumed to be a gender-neutral class system (Acker 1990).

Today, feminist scholars continue to press analysts to recognize that gender intersects in important and complex ways with social class. For example, because of the continued gender segregation of the labour force and women's traditional role in providing unpaid labour in the household, women are at greater risk of poverty than their male counterparts. This **feminization of poverty** is important to an understanding of social class inequalities in Canada and globally. While there have been dramatic changes in women's status over the past 50 years, evidence suggests that women remain disadvantaged both in the world of paid work and in the home. As repeatedly documented in the Canadian census, many continue to undertake the lion's share of domestic labour in the home and must make the necessary adjustments (including less sleep) when they hold paid employment. Within the labour force, despite employment equity legislation, women continue to be ghettoized, not only into specific employment areas but also at the lower rungs of their work milieu and their professions—they tend to be school teachers rather than principals, and so on (Armstrong and Armstrong 1994; Hochschild 1997). In short, feminist analysts continue to underscore the complexities and contradictions embedded in class and status inequalities.

CLASS AND STATUS INEQUALITY IN CANADA

The Wealthy, Elites, and Super Rich

Often when social inequality is discussed, attention immediately turns to those who are deprived and marginalized. The poor, homeless, and low-income earners seem the natural targets of any such analysis. However, the opposite end of the continuum—those who hold disproportionate financial and other assets—are at least as significant in an understanding of patterns of social inequality. Not surprisingly, the elites have long been of interest to sociologists.

One of the most prominent US sociologists of the twentieth century—C. Wright Mills (1916–62)—played a key role in establishing elites as a topic worthy of social research and analysis. Most notably, his popular book *The Power Elite* (1956) challenged the then dominant structural-functional approach to social stratification—a viewpoint that minimized class differences among Americans. Indeed, Mills proposed that the elites in US society were so powerful and so co-ordinated that they jeopardized democratic processes. This book triggered a wide variety of further studies examining the nature and implications of power discrepancies. Most notably in the US, G. William Domhoff (1936–) has produced a steady stream of books profiling the power elite and exploring 'who rules America' (2006).

Canadian sociologists and social analysts have also studied the powerful. English Canada's pre-eminent sociologist of the twentieth century—John Porter (1921–79)—authored *The Vertical Mosaic* (1965), which provided an eye-opening analysis of the concentration of corporate power in the hands of a few—mostly anglophone Canadian males. Wallace Clement (1951–) advanced Porter's work with his book *The Canadian Corporate Elite* (1975). Clement reveals that, in several respects, those holding economic power in Canada have become a more diverse (ethnically) group but many of the traditional structures that bound the elite together—private schools, family relations, cultural and charitable organizations—persist.

More recently, popular commentators have joined with academic researchers in revealing the ways in which wealth and power are socially constructed and perpetuated in Canada. Journalists Peter C. Newman and Diane Francis played particularly important roles in drawing Canadians' attention to the key power-holders in our economy. Francis, for example, wrote *Controlling Interest: Who Owns Canada?* (1986) and

Who Owns Canada Now? (2008). Although she reported in 1986 that 32 families and five conglomerates were in control of 40 per cent of Canadian banking, business, and politics, her more recent work suggests that the elite is not becoming an increasingly closed circle, and she identifies new players who have entered these rarefied spheres. By 2007, '[O]nly 21% of the biggest 500 [companies] were family-controlled, by . . . 75 [families, rather than 32]; 30% of the country's biggest 500 [companies] were foreign owned, 8.5% [were] government owned [down from 22% in 1986]' (2008, 16). Family control, while still centralized, has spread out, foreign ownership has increased, and government control has dropped precipitously in the era of privatization. The economic elite described by Francis is a dynamic and shifting reality—but one that still plays a major role in society.

Canadian social scientists have not, however, left the field to the mass media. Jamie Brownlee's *Ruling Canada: Corporate Cohesion and Democracy* (2005) pulls together much of the more recent information on the 'ruling class'; he explores the various contemporary mechanisms—including conservative think-tanks such as the Fraser Institute and overlapping memberships on various corporate boards—that serve to draw the Canadian corporate elite together into a self-conscious and integrated social force. Most notable among contemporary researchers is sociologist William K. Carroll. His *Corporate Power in a Globalizing World* (2004) uses network analysis to document the social ties (from shared membership in elite clubs to corporate interlocks with universities) that integrate the elite into every facet of Canadian society while also creating the basis for elite solidarity.

In short, elite research has clarified the considerable gap that exists between the top and bottom of our economic hierarchy. Some of this is also communicated through Statistics Canada's income survey (2009a). This analysis divides all Canadian families into quintiles (one-fifths) and determines the average income for those occupying the wealthiest one-fifth all the way down to those at the bottom one-fifth. The latest data (2009) indicate that in 2007 (prior to the 2008 recession), the 'average after tax total' of Canadian families in the richest quintile was $126,700, while those in the lowest quintile averaged $13,900 (2009a, 77). Significantly, despite the economic good times in the early twenty-first century and a reduction in the number of low-income Canadians to 3 million (down by 400,000 in 2006),

the income ratio between the wealthy and the poor was exactly the same in 2000 and in 2007 (2009a, 14).

However, the issue here is not simply wealth inequality but also the presence of a very small number of Canadians who are extremely wealthy, and, often, powerful. The richest 100 Canadians held $172.7 billion in assets in 2009, up from $165.1 billion in 2008 (Taylor 2009, B3). As with many of the global elite, the 2008 recession barely registered on their portfolios, and like most of the wealthy they either continued to make money during the recession or their assets have already rebounded to pre-2008 levels (*Toronto Star* 2010, B4). To put these levels of wealth into perspective, consider the strategy proposed by Linda McQuaig and Neil Brooks (2010). If one of Canada's wealthiest families—the Thomson family—started counting their wealth at $1.00 per second and counted non-stop day and night, they would have it all counted up in approximately 700 years. This is wealth beyond the wildest imaginings of most Canadians. This is wealth that can and is translated into endowments to universities, the creation of cultural institutions, and support for particular political parties. While some might see this pattern as benign, social critics question the desirability of small groups of individuals holding so much sway over the direction of Canadian society.

Further, the very wealthy, as Mills, Domhoff, Porter, Clement, Carroll, and others have pointed out, tend to be bound together by important shared experiences. Many were themselves born into wealthy families. These families tend to live in exclusive neighbourhoods, vacation at elite resorts, belong to the 'best' clubs, send their children to exclusive private schools, and join other wealthy families in participating in specific philanthropic and cultural events. In addition, out in the more public domain, they sit with one another on corporate boards, university governing councils, and political organizations. These shared experiences inevitably lend themselves to friendships, family intermarriage, and a common perspective on social issues. The elite are by no means a homogeneous, in-grown mass, but—to invoke the famous phrase from F. Scott Fitzgerald—'They are different from you and me.'

The bottom line is that there is considerable evidence that quite a small number of Canadians and non-Canadians occupy positions of extreme wealth and privilege, their power often extends beyond our national boundaries, and there is no sign that their economic dominance is waning. These social realities necessarily raise two concerns: what are the implications of this narrow consolidation of wealth and power for our democratic structures, and how much of a gap between haves and have-nots is desirable or acceptable in any society that wants to maintain a range of social mobility and open access? Interestingly, the 2008 recession provoked strong criticism of the financial and corporate elite and calls for an end to the enormous yearly bonuses awarded to corporate executives and leading financiers by fellow executives (*Toronto Star* 2009, A13).

TIME to REFLECT

What are the implications of the fact that the president of the United States earns $400,000 a year (presiding over a $14-trillion economy) while the CEO of the Potash Corporation of Saskatchewan earns $18 million a year? (Olive 2009, B4)

The Poor and Economically Marginalized

Who are the poor and marginalized in Canada? Is there something specific that determines who is poor and who is not? Do you have to be homeless? Jobless? Receiving social assistance? Are there status symbols that visibly mark the poor, such as worn-out clothing that lacks a brand name, a neglect of hygiene, criminal behaviour? Why are people poor? Is it because they are lazy? Drunks and/or drug users? Or is it because they cannot find employment or they have too many children? This next section will attempt to navigate you through some of these much-reported myths about the poor and illuminate how poverty rates are determined in Canada.

The word *poverty* is often used as an all encompassing term to describe situations in which people lack many of the opportunities available to the average citizen (Levitas 1998). There is no shortage of poverty measures. The Canadian federal government has developed five measures, while social planning councils, individual researchers, non-profit organizations, and others have developed their own measures (deGroot-Maggetti 2002). The most common distinctions made between these definitions of poverty are the terms 'absolute' and 'relative'. The former definitions refer to a lack of basic necessities, while the latter emphasize inadequacy compared to average living standards (Mitchell and Shillington 2002; Sarlo 1996). Teasing out the differences between absolute and relative definitions of poverty, Ross and Shillington (1994) suggest that the first approach assumes that we can ascertain an absolute measure of poverty by calculating the cost of goods and services essential for physical survival. The relative approach is grounded in the belief that any definition of poverty should take social and physical well-being into account. This approach argues that someone who has noticeably less than their surrounding community will feel disadvantaged (Ross and Schillington 1994, 4).

Statistics Canada, which collects income data annually from sample surveys of Canadians, relies on three measures to determine which Canadians are poor, living at or below low-income levels: the Low-Income Cut-Offs (LICOs), the Market Basket Measure (MBM), and the low-income measure (LIM) (Hay 2009). These measures are income-based measures that use a formula to calculate an income 'line' below which one is considered to be living in poverty (Hay 2009; Curwood 2009, 3).

The LICO is a relative measure of poverty that defines as low-income those Canadians who spend 20 per cent more of their gross (before tax) income on food, shelter, and clothing than does the average Canadian (Curwood 2009, 3). The LIM, also a relative measure, is equal to half of the median income in a census metropolitan area (CMA) and is adjusted for family size. The MBM is used to reflect the fact that living costs vary in different parts of Canada (HRSDC

During periods of high unemployment, many families come to rely on support programs, such as food banks or soup kitchens, for their meals.

(Richard Levine/Alamy)

2009). The MBM includes five types of expenses for a family of four (two adults and two children)—food, shelter, clothing and footwear, transportation, and household necessities. Any families with a post-tax income lower than the cost of purchasing this basket of goods and services is considered low-income (Curwood 2009, 3).

Applying all three of these statistical measures gives us a fuller picture of low income than can be ascertained from just one measure. In 2006, the percentage of Canadians with incomes under the LICO was 10.5 per cent, under the MBM 11.9 per cent, and under the LIM 21.7 per cent (Curwood 2009, 4). Although various organizations have developed these systems for drawing a dividing line between the haves and the have-nots, it is clear that the actual choice of a poverty line is ultimately rather arbitrary. Further, many analysts argue that poverty is an issue that extends far beyond income. Amartya Sen (2000) has made a significant intellectual contribution to poverty discourse with his assertion that an impoverished life is more than just the lack of money: 'Income may be the most prominent means for a good life without deprivation, but it is not the only influence on the lives we can lead.' Arguing for a relational understanding

of poverty and deprivation, he suggests: 'We must look at impoverished lives, and not just at depleted wallets' (2000, 3). For Sen, poverty is the lack of the capability to live a minimally decent life that, in turn, limits the ability to take part in the life of the community (2000, 4).

Over the past 30 years, poverty rates (as measured by LICOs) have consistently hovered between 15 and 20 per cent of the overall Canadian population (Hay 2009). As of 2007, it was estimated that 3 million Canadians were living in poverty (Statistics Canada 2009a, 14). We do not yet know the impact of the recession of 2008 on poverty rates. However, continued high rates of unemployment along with growing concern about government deficits suggests that when the data for 2008 and beyond does come in, it will not paint a rosier picture. Further, many of the usual victims will likely bear the brunt of this economic downturn.

Groups of Canadians vulnerable to economic marginalization include women (especially those in lone-parent families and seniors), unattached individuals, Aboriginal people, persons with disabilities, recent immigrants and visible minorities, and the working poor (Collin and Jensen 2009; Hay 2009).

▼ FIGURE 8.1   Percentage of Canadians Living on a Low Income, 2007

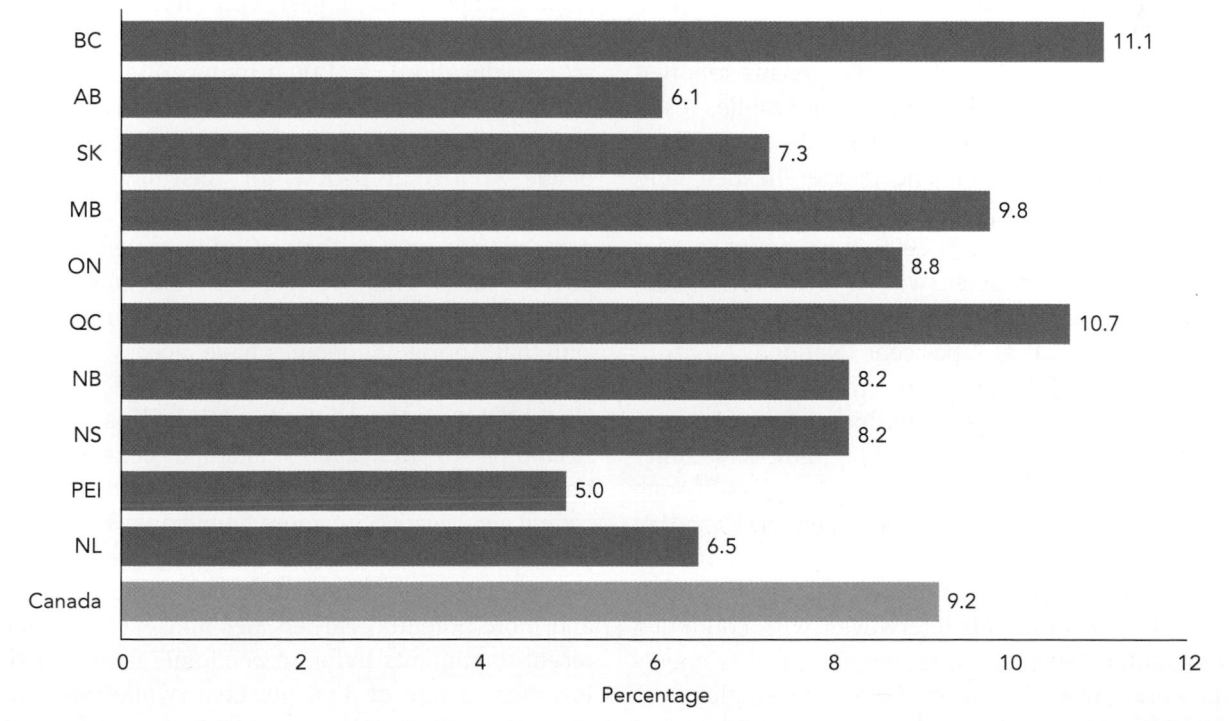

SOURCE: Statistics Canada. 'Table 202-0802—Persons in low income, annual'. CANSIM database.

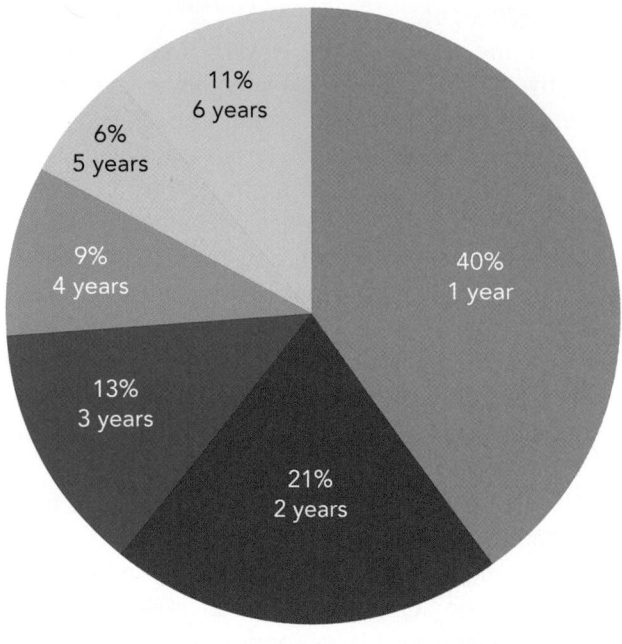

▼ FIGURE 8.2 | Percentage of Canadians Who Experienced Low Income for at Least One Year from 2002 to 2007

- 11% 6 years
- 6% 5 years
- 9% 4 years
- 13% 3 years
- 40% 1 year
- 21% 2 years

SOURCE: Statistics Canada. 2009. *Income in Canada 2007.* Cat. no. 75-202-X. Chart 7, p. 17.

Women

Women are the poorest of the poor, especially those raising children as lone parents and those living as unattached seniors. Canada's 2006 census reported that there were 1.4 million lone-parent families residing in Canada, 80 per cent headed by lone mothers. By 2009, the number of lone mother–headed families living below the poverty line was 40 per cent (Campaign 2000; 2009). In 2005, the poverty rate for unattached senior women was 19 per cent—nearly 10 times the rate for senior families with two adults or more, which sat at 2 per cent (National Advisory Council on Aging 2005).

Gender wage inequities continue to contribute to women's high poverty rates. Although the Pay Equity Act, enacted in 1987, made it illegal to pay someone less for their work on the basis of gender, Canadian women continue to earn less than 70 per cent of what their male counterparts earn (Collin and Jensen 2009). The wage gap in turn is interwoven with continued occupational segregation, undervaluing of women's paid work, the restructuring of women's employment through increased privatization and outsourcing, and a lack of child care, limiting women's choices for employment (Fleras 2009).

Unattached Individuals

In 2007, approximately 1.3 million unattached individuals lived on a low income (24.7 per cent) (Collin and Jensen 2009). Unattached individuals between 45 and 64 years old are particularly vulnerable to poverty (HRSDC 2009). The gap between the low-income rates of unattached females and males has closed. According to Statistics Canada, 35.7 per cent of unattached women had a low income compared to 30 per cent of males in 2000. However, by 2007 the numbers had declined to 24.7 per cent of unattached males compared with 27.5 per cent of unattached females (Collin and Jensen 2009).

Aboriginal People

Most Canadians are well aware that poverty is significantly higher among Aboriginal people (Collin and Jensen 2009). Statistics Canada reports that in 2007, 18.7 per cent of Aboriginal people living in economic families lived in poverty, while 47.8 per cent of unattached Aboriginal people did (compare 27.4 per cent of non-Aboriginal people) (Collin and Jensen 2009). Unemployment partly explains the high rates of poverty; however, a host of other reasons intersect, making Aboriginal people particularly vulnerable to impoverishment (Noel 2009). Many Aboriginal people lack basic education (43.7 per cent of Aboriginal Canadians have less than a secondary-school education), and their living and health conditions, especially on remote northern reserves, remain well below those of the majority (Noel 2009). Many blame Aboriginal people for their own situation, citing rampant alcoholism, substance abuse, and domestic violence as the self-inflicted cause of their poverty rates. This view ignores the long-standing history of colonial domination and cultural oppression that Aboriginal peoples have faced.

New Immigrants and Visible Minorities

According to the 2006 Canadian census, recent immigrants (those who had arrived within the preceding five years) are more vulnerable to living in poverty than other Canadians. They are at greater risk despite having significantly higher education and more potential earners per household. In 2005, recent immigrants living in economic families had a low-income rate of 32.6 per cent, while those who were unattached had a rate of 58.5 per cent (Statistics

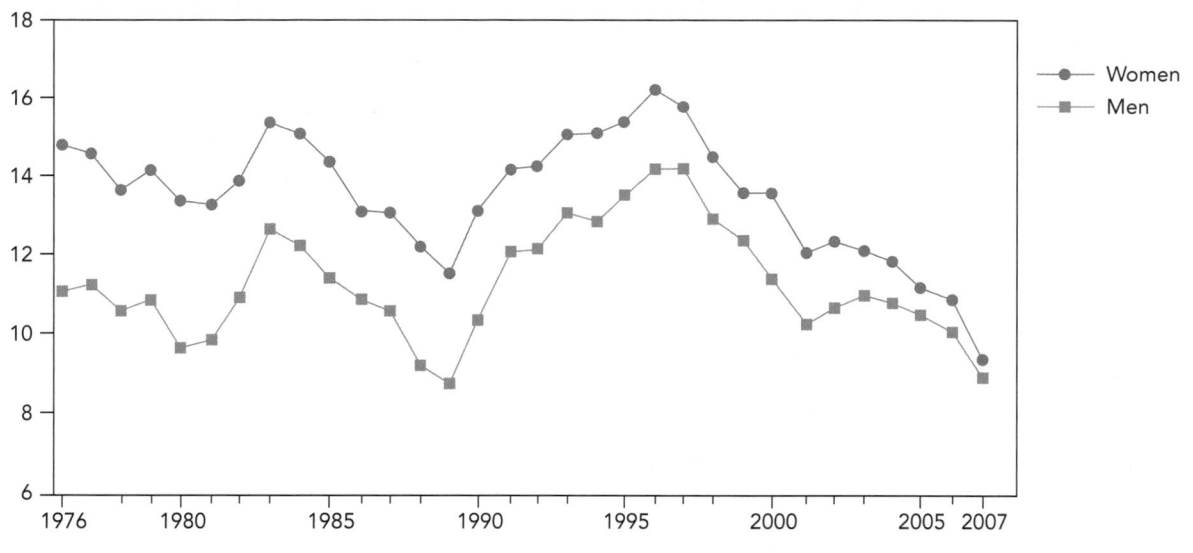

▼ FIGURE 8.3 Low-Income Rate by Gender, 1976–2007 (per cent)

SOURCE: Statistics Canada. 2009. 'Persons in low income, annual'. CANSIM Table 202-0802.

HUMAN DIVERSITY

New Canadians

How many times have we heard this before: New Canadians with professional and medical degrees are driving taxis instead of designing buildings or delivering pizzas instead of delivering babies? The proverbial immigrant taxi driver with a doctorate from a developing country reflects a key contradiction in Canada's immigration trends: Immigrants may be increasingly skilled and highly educated, yet many can't get a break because they are penalized by the very qualifications that gained them entry in the first place (Kunz 2003). And yet despite this untapped potential of underused labour, Canada is experiencing a skills shortage that borders on the scandalous.

Many foreign-trained doctors encounter frustrating roadblocks on the road to accreditation, ranging from costly retraining programs to a restricted number of residencies (from two to seven years of training in hospitals upon graduation), despite a doubling to 200 of residency positions in Ontario for foreign-trained doctors. However, not all foreign-trained medical graduates must jump the hoops to practise in Canada. For example, foreign-trained doctors from the so-called Category 1 countries (New Zealand, Australia, South Africa, England, and the United States) are allowed to bypass the internship requirement and practise medicine immediately after an evaluating exam. By contrast, foreign-trained doctors from elsewhere have to pass equivalence and evaluating exams when they are applying for the limited number of positions in the mandatory internship program (Bains 1999).

Let's be honest: To encourage the highly skilled to come to Canada, then deny them access to good jobs, is an inexcusable waste of human talent. The fact that Canada is poaching talent from those countries who can scarcely afford to lose the brightest and best (whose training costs they have absorbed as well) is unconscionable. For example, for many poor countries, the percentage of skilled nationals residing in the global north is stunning, including 41 per cent from the Caribbean regions (for example, the island of Grenada must train 22 physicians to secure the services of one) and 27 per cent from western Africa, in effect dooming these countries to poverty (Kapur 2005). Soon the word will get out that Canada's welcome mat is not what it seems to be, that Canada is big on seducing immigrants into Canada but then leaving them stranded to fend for themselves.

SOURCE: Excerpt taken from Augie Fleras, 2009, *Unequal Relations*. Reprinted with permission from the author.

Canada 2008e). There has been a dramatic rise in the educational attainment of new immigrants due to changes to Canada's immigrant selection criteria in 1993; however, this has had little impact on poverty outcomes (Collin and Jensen 2009). Of immigrants who arrived in Canada in 2000 and who experienced chronic poverty (low income for four out of the first five years in Canada), 52 per cent were skilled immigrants, and 41 per cent had a university degree (Picot, Hou, and Coulombe 2007).

These statistics suggest that recent immigrants do not reap the same rewards from their educational qualifications and work experience as those who are Canadian-born. Foreign education and experience is less valued, resulting in immigrant populations needing to update their education in Canada and often taking dead-end survival jobs in an effort to gain some type of Canadian experience. Racism, language

difficulties, cultural differences, and poor access to job networks are also used to explain their disproportionate representation among those living with a low income.

People Living with a Disability

Statistics Canada has stated that in 2006, 4.4 million Canadians reported living with a disability. In 2005, disabled Canadians who were of working age (18 to 64 years of age) had an average income 10 per cent lower than those without disabilities, and 17.1 per cent of people with disabilities had earnings below $5000 annually (Collin and Jensen 2009). Human Resources and Skills Development Canada (HRSDC) (2007) reported that people with disabilities are more likely than people without to rely on government supports. In 2005, they found that 59 per cent of people with disabilities relied on a source of income other

(© Enigma/Alamy)

Simple solutions, like the bowling ball guide above, allow individuals with disabilities to take part in social activities that were previously inaccessible.

than wages. The experience of those living with disabilities also varied by gender, with women reporting higher rates of disability as well as lower incomes and employment rates than their male counterparts with disabilities (Collin and Jensen 2009).

Besides physical limitations that may prevent some people with disabilities from carrying out certain types of employment, they are often alienated from the workforce through varying types of discrimination. For example, they are often excluded via institutional discrimination (organizations work or deliver services in a way that prejudices this minority group), environmental discrimination (for example, inaccessible buildings and lack of accessible transportation), and attitudinal discrimination (prevailing attitudes exclude and alienate disabled people). These types of discrimination are forms of ableism, discrimination against people based on preconceived notions about their limitations.

The Working Poor

Although there a number of definitions, the working poor are generally defined as individuals who are not full-time students and who have an income below the low-income threshold (LICO, LIM, or MBM) despite working a minimum of 910 hours in a given year (Collin and Jensen 2009; Fleury and Fortin 2006). Contrary to the impression given by media reports that poor Canadians are unemployed and rely on government transfers, in 2007 working-poor families accounted for almost one in three (31 per cent) of all low-income families (HRSDC 2009). Not only are they employed, they are working on average as many hours as other workers (although fewer have full-time, year-round employment) (Fleury and Fortin 2006). Despite similar work efforts, the working poor earn on average 65 per cent of the wages of other salaried workers (Collin and Jensen 2009). They also tend to hold jobs offering fewer benefits, including limited or no access to health insurance, dental care plans, life and disability insurance, and pension plans (Fleury and Fortin 2006).

Understanding Poverty

Many people believe that we all start out with the same chances of obtaining the 'American Dream'—democracy promises prosperity. This belief system is grounded in **classism**—bias, prejudice, and discrimination on the basis of social class—and often results in **blaming the victim** rather than the system. Coined by William Ryan (1971), blaming the victim is a view that holds individuals entirely responsible for any negative situations that may arise in their lives. Because there is an implicit understanding in Canada that individuals who work hard should be able to prosper, those who do not succeed are seen as at fault and are criticized as lacking motivation or too weak to help themselves.

In a poverty 'classic', anthropologist Oscar Lewis (1966) investigated the lives of the poor and argued that people who live in poverty are a subculture with different value systems from the rest of society. He maintained that those who live in this culture of poverty feel inferior and helpless and have a defeatist attitude, which then limits their ability to improve their situation. In this analysis, the responsibility for changing impoverished circumstances is placed squarely on the individual. Contemporary analysts argue that the classism and blaming-the-victim stance embedded in Lewis's analysis have contributed to

UNDER the WIRE

Technological Inequality

Poverty has wide-ranging consequences. Not the least of these is exclusion from many of the goods and services Canadian society can provide. For example, in 2007 one in five Canadian households was too poor to afford eyeglasses, dental care, or meaningful involvement in community recreational activities (NUPGE 2009). For many poor Canadians, this pattern also extends to access to modern technologies. According to the Canadian Centre for Policy Alternatives' recent report *The Affordability Gap*, only one-third of low-income households have high-speed Internet access, only 40 per cent have cellphones, and only 26 per cent have computers. In contrast, 80 per cent or more of wealthy homes have access to these technologies (Kerstetter 2009). These results confirm earlier research indicating that information and communication technologies are slower to 'penetrate' low-income households (Sciadas 2002). In a society where education and employment are increasingly linked to information technology, the negative potential of this pattern is clear.

'poor-bashing'. Facts about poverty and the poor are ignored, and instead stereotypes are repeated. These stereotypes, in turn, may lead to both verbal and physical assaults on the poor.

This approach to the poor and poverty, of course, completely overlooks the facts that indicate poverty may be the result of institutional arrangements and even legislation. For example, a great deal of Canadian poverty would immediately disappear if legislation on welfare rates and minimum wage were improved. Moreover, this kind of individual-focused thinking about poor people—that they should transcend their social environment—'demands a higher standard of behaviour and sacrifice from people who are poor than from people who are not' (Swanson 2001, 3).

Michael Katz, a well-known poverty researcher in the United States, argues that poverty and inequality is not about specific people and their personal qualities; rather, it is about wealth distribution— some people receive a great deal more than others. He writes, 'descriptions of the demography, behavior or beliefs [of people who are poor] can't explain inequality' (Katz 1989, 7). Jean Swanson (2001) furthers this assertion and urges society to stop blaming individuals for their impoverished circumstances so that 'we can expose the policies, laws, and economic system that force millions of people in Canada and around the world to compete against each other, driving down wages and creating more poverty' (2001, 8).

Katz and Swanson are employing a perspective often referred to as **blaming the system**. This is more consistent with a sociological view, since it recognizes the systemic discrimination that exists within society. People are poor for many reasons that are beyond their control. There are many structural variables that influence poverty levels, such as deindustrialization (the replacement of well-paid manufacturing jobs with lower-paid service-based jobs), rising costs of living, barriers to opportunities (such as increased tuition for education and training programs), limited access to affordable housing, inability to obtain credit, and so on.

TIME to REFLECT

Who or what is to blame for social inequality? Should individuals be blamed for their lack of motivation and effort, or should society take some responsibility for ensuring an equitable standard of living for all citizens?

Policies to Address Poverty

There is a very long tradition of efforts to 'help' the poor. Initially driven by Christian and Jewish religious ethics, early charities in Canada provided food, clothing, and occasionally shelter for those who had fallen on 'hard times'. Then, as now, single mothers (often widows), orphaned children, seniors, and the disabled were routinely marginalized. In the absence of government-funded welfare programs, they were often forced to turn to private and religious agencies to beg for whatever support they could find. Reflective of the brutality of the times, many poor families had little recourse but to give up their hungry children. For example, tens of thousands of poor children were brought over to Canada from shelters in England to work as servants and farm workers. Expected to receive only their room and board in return for their labours, they often experienced brutal living conditions. However, they were at least spared the hunger and malnutrition they had left behind on the streets of their native country.

Throughout the early twentieth century, very little changed for the overwhelming majority of the Canadian population who were not property-owners. Industrial work was harsh, injury and work-related illness were commonplace, and wages were far from reliable. Not surprisingly, increasing numbers of workers and their families rebelled against a system that provided them with so little in the way of personal security. In particular, the Winnipeg General Strike in 1919—a strike that resulted in the deployment of the army to maintain order—suggested to many that Marx's predictions of class conflict were not far off the mark. Then the Great Depression of the 1930s—when unemployment rates soared to 30 per cent—underscored the vulnerability of most workers to impoverishment. Predictably, this constellation of events pushed social activists and politicians to formulate legislation that softened the harshest edges of the capitalist economy. In 1927, the Old Age Pensions Act was introduced, followed by unemployment insurance in 1940, the Family Allowance Act in 1944, and the first hospital insurance plan in 1947. Throughout the postwar period, particularly in the economically buoyant 1960s, a host of legislation was passed—Canada and Quebec Pension Plans (1965), federal education grants (1976), and an improved Unemployment Insurance Act (1971). Many of these initiatives were funded by progressive income tax programs that taxed the well-to-do at a significantly higher rate than less well-off citizens. As a net result, government bodies and agencies were in place to provide at least some assistance to the elderly, ill, and out-of-work.

This levelling of the economic playing field was then generally reversed throughout the 1980s and 1990s. Largely in response to the popularization of neo-liberalism by Ronald Reagan, Margaret Thatcher, and Brian Mulroney, legislators opted to reduce the size of government and place a greater burden directly on individuals and their families. Effectively, there has been a steady dismantling of much of the 'welfare state'. For example, unemployment benefits were increasingly restricted both in terms of eligibility criteria and the length of the benefit period. Similarly, the cash value of welfare benefits across the country was reduced, and workfare provisions (requiring many welfare recipients to accept any paid work or training in exchange for their benefits) were introduced.

As our understanding of the welfare state has changed over time, so too have social policy goals in Canada (Hay 2009). Today, in most circles, the private market is understood as ineffective in ensuring an adequate 'distribution of resources amongst citizens and the state is understood to have an important role in pursuing a number of policies and programs with redistributive and social minimum goals' (Hay 2009, 6). Meeting the challenge of poverty requires action from all sectors of society (families, communities, businesses, and governments). However, in recent years the role of each of these sectors has become blurred or contested (Jenson 2004).

Currently, a number of policies and programs are in place to prevent and reduce poverty.

BENEFITS FOR SENIORS

Canada's retirement income system includes three parts; Old Age Security (OAS), a universal benefit that all people receive once they reach the age of 65; the Canada Pension Plan (CPP); and private pensions and savings. The CPP provides a monthly pension to all people who have worked for wages or were self-employed and thus have paid into the plan. It pays benefits to the surviving spouse and children should a contributor pass away. The amount of the benefit is dependent upon the amount of money the recipient has contributed. The federal government, however, warns Canadians that OAS and CPP are aimed at helping seniors ease into retirement rather than at providing for all their financial needs (HRSDC 2011). Thus, the government instituted Registered Retirement

Savings Plans (RRSPs) and Registered Pension Plans. An RRSP is an investment account within which a person can shelter money from income tax until it is withdrawn, usually in retirement. RRSPs are primarily used by middle- and upper-class individuals, however, since poorer Canadians have little—if any—money to invest (Hay 2009). Registered Pension Plans are employer-sponsored pension funds to which both employer and employees contribute. Since governments and large companies are the only employers in a position to offer such plans, they currently cover less than 50 per cent of Canadian workers.

The Canadian Population Health Initiative (2004) contends that these programs have dramatically reduced poverty among all seniors, and the retirement income system has been labelled a success story (Hay 2009). However, some seniors (especially unattached females) remain disadvantaged, at least in part because much of the retirement income system depends on the degree of a person's attachment to the labour market throughout the life course. Further, the stock market downturn in 2008 triggered great concern about the future economic security of retirees in Canada as RRSPs plunged in value and questions were raised about the long-term viability of various pension funds.

CHILD AND FAMILY BENEFITS

The Canada Child Tax Benefit (CCTB) is a monthly payment given to families with children under the age of 18. The amount given per child is based on income claimed on the previous year's income tax return. The Universal Child Care Benefit (UCCB), introduced in 2006, was promoted as a means of helping Canadians balance their work and family responsibilities. The benefit amounts to $100 per month per child under the age of six to help parents meet their child-care expenses (it is counted as taxable income). While an extra $100 per child undoubtedly helps low-income families provide for their families, it does little to address the intended purpose. The real issue facing low-income families is a lack of affordable child care. The Organisation for Economic Co-operation and Development (OECD) has continually drawn attention to Canada's poor record in this regard compared to that of other industrialized nations.

SOCIAL ASSISTANCE

Social assistance programs are aimed at helping residents of Canada who are in financial need. The program—Canada's social safety net of last resort—provides financial assistance to cover the cost of basic living requirements for an individual or family when all other means of financial support have been exhausted (HRSDC 2011). Assistance is given to families who meet the needs assessment and are unable to adequately provide for themselves and any dependants. All able-bodied recipients are actively encouraged to retrain, seek employment, and accept any reasonable offers of employment as a condition of receiving assistance. Persons with a disability are exempt from this condition. The amount of the benefit is dependent upon many factors and varies across provinces. Table 8.1 shows the average annual benefit for a lone mother with one two-year-old child across the provinces.

While social assistance most certainly has the potential to help prevent families in need from becoming homeless, the low rates do little to redress social inequality. Indeed, the rates are far below the LICO in most provinces. The before-tax LICO in 2007 for a single person ranged from $16,968 for those living in communities with less than 30,000 population to $21,666 for those living in large cities of 500,000 or more (National Council of Welfare 2008). Few provinces provide social assistance anywhere near this poverty threshold even for a family of two, and those that do have a much higher cost of living than the others.

EMPLOYMENT BENEFITS

Canada's Employment Insurance (EI) program offers financial assistance to unemployed Canadians who lose their job involuntarily. There are a number of

TABLE 8.1 ▼	Average Yearly Rate of Social Assistance for a Single Mother with One Two-Year-Old Child
Province	**Basic Social Assistance**
Newfoundland and Labrador	$12,285
Prince Edward Island	$10,927
Nova Scotia	$9,300
New Brunswick	$9,492
Quebec	$8,308
Ontario	$10,883
Manitoba	$9,636
Saskatchewan	$11,093
Alberta	$9,098
British Columbia	$11,347
Yukon	$17,201
Northwest Territories	$20,974
Nunavut	$42,986

SOURCE: Information from National Council of Welfare, 2008, *Welfare Incomes.*

rules regarding the number of hours and weeks a person must work before being qualified to apply, which vary by province. The maximum EI benefit is $435 per week for a maximum of 45 weeks (Hay 2009). The claimant receives 55 per cent of their lost income to a maximum of $41,100. At the maximum benefit level, EI offers a replacement income at about the level of the poverty line (LICO) for a single person in a large city (Hay 2009). Thus, EI is effective in helping individuals to avoid a catastrophe in the event of an unforeseen job loss. However, the benefit is a short-term strategy that helps only those who have been able to secure an attachment to the labour market.

OTHER PROGRAMS AND BENEFITS

Canada also offers a variety of subsidies for housing and child care for low-income families. While these services are often noted as the most helpful to low-income families, in most provinces there are long waiting lists for housing, and the housing is often not well maintained and in neighbourhoods with high crime rates. As discussed above, there are far too few regulated child-care spaces available across Canada, and the requirements for obtaining and maintaining the child-care subsidy are often burdensome.

In addition, almost every community across Canada provides access to food banks, soup kitchens, and shelters to assist the economically marginalized. Indeed, it is commonly remarked that there are more food banks in Canada than there are McDonald's franchises. However, as every food bank manager would be quick to point out, food banks and soup kitchens are a stop-gap measure. They help to ensure that the poor do not starve and are not severely malnourished. Beyond this, they can offer little, and they are hard-pressed to provide adequate food for the poor. Most food banks restrict the number of visits per month and keep a record so that their clients do not double-dip. Further, they generally have little to offer in the way of fresh fruit and vegetables—products that non-poor families recognize as key to good health.

TIME to REFLECT

In the late 1990s, the federal government under Jean Chrétien committed to end child poverty in Canada by 2000. Campaign 2000 has never fulfilled its promise. What is our government's responsibility to end poverty among Canada's 637,000 poor children (Statistics Canada 2009a, 14)?

Conclusion: Social and Economic Inequalities— Future Trends

Various indicators suggest that economic polarization is increasing rather than decreasing (for a global perspective, see Braun 1991). While the incomes (controlling for inflation) of average Canadians tended to either stay the same or decline over the past 25 years, the wealthiest .01 per cent of the Canadian population saw their incomes double, on average rising from $3.6 million to $8.4 million (McQuaig 2009, A19). This trend is also reflected in the following figures: 'In 1995, the average pay of Canada's highest paid 50 CEOs was $2.66 million, 85 times the average worker's pay. By 2008, the average pay of the same group of CEOs had skyrocketed to 234 times the pay of the average worker' (Cartwright 2010, A16; Olive 2009, B4). Even when CEO earnings dropped from 2007 to 2008, they still averaged $7.35 million per year, and the top Canadian CEO was bringing home $36.6 million a year in salary, benefits, bonuses, and stock options (Flavelle 2010, B1). To put it another way, in 2005 the average annual pay among the top 100 CEOs in Canada was the same as that received by 238 Canadians working full-year at an average salary. Moreover, the highest-paid CEO that year made more than 5000 people working full-year at minimum wage did (Mackenzie 2007).

OPEN for DISCUSSION

Measuring Success: Single Mothers and Welfare Reform in Ontario

While dramatic welfare roll reductions have been proclaimed as the measure of success of Canadian welfare reforms in the 1990s and 2000s, little attention has been paid to the *process* of leaving welfare or outcomes after exiting benefits. Cumming, Cooke, and Caragata (2009) use qualitative data to explore how lone mothers, a large component of the welfare caseload, leave social assistance. Using multiple interviews with 36 lone mothers receiving social assistance in 2006, they find that while some left social assistance for employment, other routes to exit are important, including exchanging social assistance for other program supports. Findings point to the importance of interrelationships among programs and cast doubt on the 'independence' achieved by welfare leavers.

During the recession of the early 1990s, social assistance receipt in Canada skyrocketed, with 3.1 million individuals receiving assistance in 1994. The 1996 replacement of the Canada Assistance Plan (CAP) with the Canada Health and Social Transfer (CHST) reduced federal–provincial transfers for social assistance and also freed the provinces from restriction on program design. As a result, all provinces subsequently instituted changes aimed at reducing welfare 'dependency'.

A plethora of research has judged the success of welfare reform on the basis of the observed reduction of welfare rolls. By 2005, social assistance beneficiaries had fallen from 3.1 million in 1994 to 1.7 million individuals, half of whom were children (Finnie and Irvine 2008). While this could indeed be perceived as a 'successful' reform, it does not take into account how these families left social assistance.

In fact, Cumming, Cooke, and Caragata (2008) found that leaving social assistance does not mean leaving dependence on state-provided benefits. Of the 36 lone mothers interviewed, five were attending school and receiving student loans, one was attending school through Employment Insurance, and three had left Ontario Works for the Ontario Disability Support Program. Two other women who had left social assistance told interviewers that the fathers of their children had renewed their commitment to parenting and were assuming some of the responsibilities for child care. While some lone mothers indeed left social assistance, none in this sample were able to secure employment that paid enough for them to live without government transfers or without the help of family. Of the nine women who exited via work or student loans, all were receiving at least one other benefit (subsidized housing and/or subsidized child care).

These data highlight the problems with assuming an easy division between 'dependent' welfare recipients and 'independent' welfare leavers. Moreover, they call into question the 'success' of these reforms and the very definition of success itself. Should success be measured by a mere reduction in rolls, or should it include how a family is actually faring after exiting social assistance? Should the primary goal of government be to get people off assistance and into paid work?

SOURCE: Excerpts from Cumming, Cooke, and Caragata 2008.

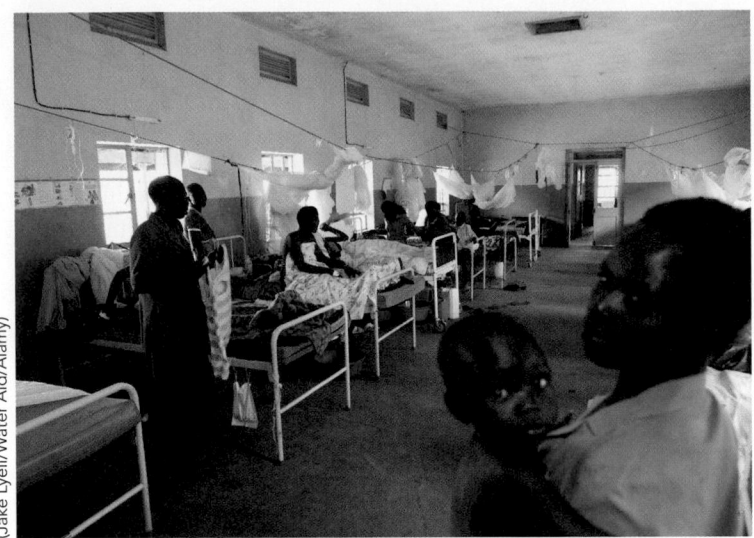

(Jake Lyell/Water Aid/Alamy)

Inequality in health care is an endemic global issue. Widespread poverty and limited access to care, technology, and medication put developing nations at a huge disadvantage to developed nations, which have the infrastructure and systems in place to treat and control health issues.

Class polarization appears to be particularly alarming for the middle and working classes. Deindustrialization has meant the loss of numerous well-paid and secure manufacturing jobs. The manufacturing jobs that remain are primarily the purview of middle-aged males with long seniority rights, and even these jobs are in jeopardy because of plant closures. In their place are a variety of poorly paid, often precarious (part-time, contract, seasonal, temporary) jobs that do not afford job-holders, even in a dual-income family, the prospect of a secure middle-class existence. At the same time, many of the public sector jobs (for example, employment in the federal, provincial, or municipal governments) have also been eroded. Neo-liberalism and a penchant for deficit reduction have meant not only that much of this work has disappeared but also that what work remains has become increasingly part-time, contract, or covered through employment agencies.

At the same time that well-paying, secure, full-time employment has become increasingly scarce, the economic pressures on middle-class and working-class families have grown dramatically. For example, undergraduate university tuition increased by 165 per cent between 1991/2 and 2007/8, while graduate and professional students face even greater increases (CAUT Almanac 2008/9, 38;

GLOBAL ISSUES

Health Inequality

Brenda Marilyn Arndt is currently 52 years old and resides in Ontario. She was diagnosed with chronic myeloid leukemia (CML) in 2006 and has a drug plan available to her through her employer. Brenda was prescribed a drug that cost her insurance company approximately $7000 per month. Brenda is currently in remission and is expected to stay in remission as long as she continues to take her medication.

Bishnu Prasad Paudel is 45 years old and lives in Nepal. He was diagnosed with CML in 2000. Bishnu has no drug plan, and the drug that Brenda takes to fight her CML is not available in Nepal, so Bishnu travelled to India to obtain it. However, there are no drug or government social plans available to help him with the high cost of the drug, which amounts to $5000 per month in India. Bishnu sold most of his property in order to purchase the drug, but after only a few months his money was depleted, and he was no longer able to purchase it. Bishnu, although ill, continues to work and earns approximately US $195 per month. Without the financial means to purchase this drug, Bishnu will die.

A report by the World Health Organization (WHO) (2008) revealed extraordinarily high levels of global health inequality and links poor health to poverty:

Our children have dramatically different life chances depending on where they were born. In Japan or Sweden they can expect to live more than 80 years; in Brazil, 72 years; India, 63 years; and in one of several African countries, fewer than 50 years. And within countries, the differences in life chances are dramatic and are seen worldwide. The poorest of the poor have high levels of illness and premature mortality (2008, 8).

The main message in WHO's report on health inequalities was that the country you are born into is a significant determinant of your life chances, a fact that is tied to poverty rather than to genetics. WHO contends that 'inequities in health, avoidable health inequalities, arise because of the circumstances in which people grow, live, work, and age' (2008, 3).

Statistics Canada 2010f). Predictably, more students are shouldering more debt. In 2005, 57 per cent of the graduating class had student loans (up from 49 per cent in 1995), and average student debt crept up to $18,800. This is not surprising, since the past few summers have recorded some of the highest rates of student unemployment since records were first collected in 1977 (Statistics Canada 2010f). When these students graduate, they enter a paid labour force in which the unemployment rate officially indicates that more than 7 per cent is unemployed and looking for work (Statistics Canada 2010d).

Significantly, in 2007 (before the 2008 recession) the Canadian Centre for Policy Alternatives, in a report entitled *The Rich and the Rest of Us* (Mackenzie 2007), pointed out that almost 80 per cent of Canadian families were working more (longer hours) and earning less of the economic pie than they did in 1977, while the wealthiest families continue to prosper. The report concluded that there are fewer people 'in the middle' and '[w]e ignore these trends at our collective peril' (Mackenzie 2007). Interestingly, after the 2008 recession, the conservative British newspaper *The Guardian* published an article titled 'Is the British middle class an endangered species?' If, as seems likely, the argument applies to Canada, then we may face the same unsettling conclusion: 'the crisis of a generation is just beginning' (Beckett 2010).

SOCIOLOGY in ACTION

Challenging Poverty and Economic Inequality

As discussed throughout this chapter, sociologists have played a leading role in exposing economic inequalities in Canada, in exploring the sources of such inequalities, and in gauging the societal impact of growing gaps between the haves and the have-nots. While it is very important that this sociological research helps to inform all Canadians about the realities of poverty and economic inequality in Canada, it is particularly important that government policy—at the municipal, provincial, and federal levels—is informed by rigorous research information. In this regard, sociologists have played a pivotal role in creating and maintaining a variety of organizations that support research, the dissemination of this research, and lobbying government to incorporate accurate research information into legislation. Through the creation and maintenance of groups such as the Canadian Centre for Policy Alternatives, the Canadian Centre for Social Development, and the Caledon Institute of Social Policy, sociologists and other social scientists have been instrumental in the development of much of the contemporary social policy that addresses poverty and economic inequalities.

The recent (2005) creation of the Centre d'étude sur la pauvreté et l'exclusion (CEPE) (Centre for the Study of Poverty and Social Exclusion) by the government of Quebec is yet another example of the role that sociologists, along with other social scientists and government officials, are playing in the creation of agencies whose mandate is to provide 'reliable and rigorous information' that can then be used to fashion appropriate social policies. CEPE proposes research topics and, in some instances, funds proposals and makes the resultant reports available to the public through its website (www.cepe.gouv. qc.ca). This information—examining both local and international efforts to combat poverty—is then easily available to inform future policy initiatives.

One of the outstanding current efforts to understand and reduce inequalities is the Vibrant Communities program (www.vibrantcommunities.ca). Supported by the Caledon Institute of Social Policy along with a variety of other organizations, Vibrant Communities encourages collaboration between social science researchers, government representatives, businesspeople, and community workers to explore local initiatives to combat poverty. Specific groups involved in the program include, for example, Opportunities Niagara, the Winnipeg Poverty Reduction Council, and the Hamilton Roundtable for Poverty Reduction. The resultant community-level evaluations are now playing a very important role in suggesting specific ways that economically depressed areas can create and apply new strategies to reduce inequalities and improve community strengths. For example, some communities have been effective in mobilizing involvement in arts and crafts to both strengthen community ties and provide an avenue for income production. Significantly, Vibrant Communities identifies 'high use of research to inform the work' as one of the key ingredients in program success.

In short, sociologists are playing an important role both as participants in national as well as grassroots social justice organizations and as researchers whose work is crucial to informed policy initiatives.

questions for critical thought

1. What kinds of social class messages are embedded in your day-to-day experiences at university?

2. Which theoretical paradigm best describes social inequality? Explain why.

3. What is the basis of the feminist critique of stratification theory?

4. Why would private schools and exclusive clubs be important to the Canadian economic elite?

5. Are there policies that could be put in place that could lessen the gap between the haves and the have-nots? What might some of these policies be, and how would they be funded?

6. Do Canadians appear to be heading toward more or less economic and social inequality in the coming decades?

recommended readings

Rosemary Crompton. 1993. *Class and Stratification*. Cambridge: Polity.
Crompton argues that class processes are not the only factors contributing to the maintenance and reproduction of inequality; rather, cultural practices are deeply involved in both its reproduction and its maintenance.

Ralf Dahrendorf. 1959. *Class and Class Conflict in Industrial Society*. Stanford, CA: Stanford University Press.
Dahrendorf believes that in an age in which property and control have become separated in the industrial world, it is not property but the exercise of or exclusion from authoritative control that establishes class position.

Edward G. Grabb. 2007. *Theories of Social Inequality: Classical and Contemporary Perspectives*. 5th edn. Toronto: Harcourt.
Grabb provides an overview and analysis of theories of social inequality, including theorists such as Marx, Weber, Durkheim, Lenski, Poulantzas, Wright, Parking, and Giddens.

Gerhard Lenski. 1966. *Power and Privilege: A Theory of Stratification*. New York: McGraw-Hill.
Lenski contends that classes are aggregations of persons in society who stand in similar positions with respect to some form of power, privilege, or prestige. He shifts from a focus on class structure to processes that generate structure.

Peter S. Li. 1992. 'Race and gender as bases of class fractions and their effects on earnings'. *Canadian Review of Sociology and Anthropology* 29 (4): 488–510.
Li examines the joint effect of race and gender on income and argues that gender fractionalizes class in two important senses: as the occupational structure changes, gender provides the ground for segregating occupations into men's and women's jobs, and women have become more proletarianized than men in the labour force.

J.A. McMullin. 2004. *Understanding Inequality: Intersections of Class, Age, Gender, Ethnicity, and Race in Canada*. Toronto: Oxford University Press.
McMullin outlines the ways in which opportunities are distributed differentially in society on the basis of things such as class, age, gender, ethnicity, and race and argues for a relational understanding of inequality.

John Porter. 1965. *The Vertical Mosaic: An Analysis of Social Class and Power in Canada*. Toronto: University of Toronto Press.
Porter argues that there is a hierarchical structure of material rewards that acts as a barrier to equal opportunity. He argues that in Canada there is an establishment of a vertical mosaic of class and power, with the Anglo-Saxon charter group at the top of the class and power structures, recruiting their offspring and social equals as future elites.

Erik Olin Wright. 1997. *Class Counts: Comparative Studies in Class Analysis*. Cambridge: Cambridge University Press.
Wright outlines his conception of social class and argues that class analysis is based on the notion of exploitation conceived as an antagonistic interdependence of material interests. He contends that there are three basic classes: capitalists (exploiters), workers (exploited), and petty bourgeoisie (neither exploited nor exploiter).

recommended websites

Canada Without Poverty

www.cwp-csp.ca

This organization is a not-for-profit charitable organization dedicated to the elimination of poverty in Canada.

Human Resources and Skills Development Canada (HRSDC)

www.rhdcc-hrsdc.gc.ca/

The site offers information on Aboriginal peoples, children, homelessness, and indicators of well-being as well as many links to information on particular policies.

Make Poverty History

www.makepovertyhistory.ca

This site provides information about poverty across the world and includes information and resources for anti-poverty activism.

National Council of Welfare

http://cnb-ncw.gc.ca

The role of the National Council of Welfare is to advise the minister of Human Resources and Skills Development Canada on matters concerning poverty, low-income Canadians, and related programs and policies.

Ontario Coalition Against Poverty (OCAP)

www.ocap.ca/

OCAP is an anti-poverty organization located in Toronto. It provides advocacy for individuals combating organizations (such as the Ontario Disability Support Program and social assistance and public housing authorities) that make it difficult for poor people to receive what they are entitled to.

Organisation for Economic Co-operation and Development (OECD)

www.oecd.org

The OECD brings together governments from around the world to support sustainable economic growth, boost employment, raise living standards, maintain financial stability, assist other countries' economic development, and contribute to growth in world trade.

2009 Report Card on Child and Family Poverty

www.campaign2000.ca/reportCards/national/2009EnglishC2000NationalReportCard.pdf

This site offers useful data on child and family poverty in Canada.

SocioSite

www.sociosite.net/topics/inequality.php

The section 'Social inequality and class' offers a comprehensive list of sociological resources on social inequality, class, stratification, and poverty.

chapter 9

Gender Relations

Janet Siltanen and Andrea Doucet

In this chapter you will:

▶ Learn four conceptual pillars of sociological thinking about gender relations;

▶ Come to understand the historical development of sociological theory of gender relations;

▶ See how the sociology of gender relations is grounded in everyday life and issues, especially in the sites of school, paid work, and family, as well as in the negotiations of gender identities in an increasingly globalized world;

▶ Make connections between everyday life (especially the everyday lives of young adults in Canada) and sociological thinking about gender relations;

▶ Begin to think sociologically about how gender matters, or not, in your everyday life and how gender intersects with other dimensions of experience such as, for example, race/ethnicity, sexuality, class, age and generation, and locality.

Introduction: Gender Relations and You

When you wake up in the morning and look in the mirror, what do you see? A person? A man or a woman? What else do you see? A black man, a young man? A lesbian, a poor woman? *What* you see and *how* you see it is framed by social norms, institutions, culturally defined roles, and dominant ideologies. In short, it is framed by what we will refer to in this chapter as the 'social construction of gender'.

If you are, for example, a white, middle-class, heterosexual male, and you look in the mirror and simply see yourself as just a 'person', this is probably because you live in a society where these attributes give you taken-for-granted privileges and position you as the social norm. On the other hand, if you are a woman standing in front of this metaphoric mirror, there is a good chance that you will see yourself as a gendered person. If you are a black woman, you may see yourself as a **racialized** and gendered person. If you are a black, working-class woman, you may see yourself as a racialized, classed, and gendered person. In these last three cases, the characteristics reflected in our metaphoric mirror are 'seen' because they stand out as marginal to the norm—not male, not white, not middle-class. We live our lives in particular historical and social contexts, and these contexts have gendered laws, rules, conventions, and expectations that set out opportunities and obstacles in the life paths we follow. Sociology takes as one of its tasks the job of uncovering, and even challenging, the recognized and unrecognized dimensions of structured social experience that place some people in the privileged centre and marginalize others. In the case of gender, men more often than not are in the privileged centre, and women more often than not occupy the margins.

As you will learn in this chapter, gender is not just about how you feel about yourself and your own possibilities. It is also about how society is structured by gender and how gender works in social institutions. That is, the cities you live in, the families who raise you, the television shows you watch, the schools you go to, the computer games you play, and the friends you have all shape your sense of who you are and what you are capable of as a girl, a boy, a man, or a woman. Thinking sociologically about gender means thinking about what it means to interrogate these contexts and expectations and what they mean for you and your peers. As well, it means thinking about whether and how you might want to change the current state of affairs.

We have both been researching, teaching, and working with policy-makers and non-governmental organizations (NGOs) on issues of gender and gender relations for more than two decades. To begin, we'd like to share with you two observations that come from our years in the university classroom. First, it has been a constant struggle to interest men in gender courses. This relates to our point above—those in privileged positions rarely feel the need to interrogate their privilege. As you will read in this chapter, the historical development of the sociology of gender relations began with a struggle to find a place for women's views and female experiences within a very male-defined sociology. More recently, as women's experience and issues of gender inequality have become accepted as important sociological topics, attention is moving toward a closer examination of issues of gender inequality and differences in the lives of men and between men. This is heightening men's interest in the analysis of gender.

The second observation is that many young women seem convinced that the battle for gender equality is won and now passé. In this age of unlimited talk-and-text, iPods and iPads, Twitter and Facebook, young women can be led to feel empowered as they connect with 'friends' from around the globe and post pictures and videos of themselves at any time, day or night. It can thus be difficult to stand at the front of a class and try to convey the historical, cultural, economic, global, and local disparities that women faced and still face. Yet continuing gender inequalities affect us all. They are of great concern to sociologists who study gender relations—and we think that they should be of concern to you as students of sociology. We plan to cover many of these issues in this chapter and to highlight how they connect to your everyday experiences.

As sociologists, we explore gender issues with at least four benchmark ideas in mind. They are:

1. Gender is a vantage point of critique.

2. Gender is a social construction.

3. Gender is realized in social roles and institutions.

4. Gender is a relation of power and inequality.

As you will see from our brief description of each of these benchmark ideas below, they are dynamic tools that have evolved, widened, and deepened over time. Beginning in the 1960s and into the early 1980s, the concept of gender slowly gained a firm grounding as a key feature of sociological theory and research. That is, the argument that gender mattered was accepted in most sociology courses and programs. However, the conceptualization of gender and gender inequality has evolved over time—both as a sociological concept and as a focus of policy development and political action. This change has come about partly from diverse groups of women—and men—putting their concerns onto the sociological agenda; these efforts include critiques and debates from women of colour, women from Third World settings, pro-feminist men, and GLBT (gay, lesbian, bisexual, and transgendered) communities. In addition, changes in sociological thinking have come from post-modern and post-structuralist critiques, which stretch across all scholarly disciplines and have radically transformed ways of thinking and knowing across the academy. Thus, new ways of thinking and speaking about gender have been surfacing alongside of, as well as within, larger intellectual challenges and changes in how individuals, social relations, and societies are understood.

This chapter is also guided by a fundamental pedagogical principle, which is that you will learn more deeply and more meaningfully if you are able to link the material presented to your own life. We hope that in some way you will be able to connect some of what you read in this chapter, which is a sociological narrative on gender relations, with your own political and personal life. In order to do this, we organize the chapter in two parts. The first, 'Thinking sociologically', is linked to our first two chapter objectives. That is, part 1 takes you through four conceptual pillars of sociological thinking about gender relations while also providing a brief overview of some key historical shifts in this thinking. Part 2, 'Living gendered lives', addresses the next three objectives of the chapter. It encourages you to make links between sociological

In the First Person

I remember schooling as a battle over what I was allowed to look like. In Grade 7, I was sent home because I wore a tie. In Grade 10, I was ordered out of math class because I showed up with peace symbols drawn on my legs and purple streaks in my hair. In Grade 12, I was barred from the science lab because I helped to organize a protest against the 'skirts only' dress code for girls in my public high school and showed up for class wearing slacks. Though they were small, I knew they were acts of rebellion. However, I thought the adult reaction was way over the top. What were they so afraid of? My first-year sociology course helped me to answer this question. Societies have deeply entrenched interests in creating and preserving gender differences. Social institutions invest a lot of time, energy, and emotion policing the gender status quo. Flouting gender norms is more than just a matter of fashion choice or personal inclination—it challenges what is understood as normal and pushes at the boundaries of what is thought to be an acceptable way to live. Sociology also helped me to think about why I wanted to resist restrictive gender expectations. I hope that in reading this chapter, you'll find new understandings of how gender is shaping your current experiences and new ways to think about whether or not it will shape your future.

—Janet Siltanen

It's a rainy evening, and I am driving one of my daughters to her soccer game. It seems like a good time to ask a question I've not asked this 16-year-old before: 'Do you think of yourself as a feminist?' She looks puzzled. 'Do you mean that girls can do so many things that boys couldn't even dream of doing?' 'Something like that,' I replied. 'Yes, I am a feminist,' she says as she pulls on her cleats.

Feminism has had an invisible presence in the homes I have lived in. But it quietly runs through generations of women in my family, appearing suddenly at particular junctures. My maternal grandmother (born in 1900) did not know the word, but it comes to mind when I think of how she subversively hid money from my grandfather; she sold her home-made butter and stuffed dollars into a pillow sack for her oldest daughters' education (my grandfather did not believe in education for girls). My mother (who didn't go to university because the butter money ran out) also silently reflected that word as she cleverly stretched my father's paper-mill wages while also insisting that her three daughters be economically self-reliant. Each generation has its own reasons for—and definitions of—this word. What is yours?

—Andrea Doucet

thinking and your own experience of gender by taking you through three social sites where gender relations are very much in play—school, work, and family. We also discuss how young people navigate this daily terrain through the constant negotiations of complex gender identities in a global world.

Thinking Sociologically

GENDER IS A VANTAGE POINT OF CRITIQUE

As indicated by the inclusion of our chapter in this introductory sociology text, the study of gender and gender relations is certainly a central part of sociological study in Canada. Yet not so long ago, sociology was a male-centred discipline, undertaken primarily by men who were studying areas of social life where men were dominant and where women were absent or invisible. This led to two serious consequences. First, women were invisible within the content of sociology; they were the wives, sisters, mothers, teachers, and daughters of the men whose work, concerns, and daily lives populated sociological study and understandings of human life.

The second impact, intensely related to the first, is that women were invisible within the profession of sociology. There were very few female faculty, reinforcing the impression that sociology was a man's business. In the 1960s and 1970s and into the 1980s, the small number of women professors found themselves in male-dominated academic environments where the key tenets of the discipline reflected men's experiences and it was simply assumed that women would fit into this perspective. In her seminal and now classic book, *The Everyday World as Problematic* (1987), one of Canada's best-known sociologists, Dorothy Smith, used telling subtitles that captured well the sense of exclusion, isolation, and invisibility of women's perspectives within the discipline of sociology; these subtitles included, for example, 'A peculiar eclipsing: Women's exclusion from man's culture'; 'Text, talk and power: Women's exclusion'; 'Men's standpoint is represented as universal'; and 'The brutal history of women's silencing'.

Smith, along with other pioneering sociologists of gender relations, began to develop ways of including women's experiences within sociological research and theory. They began to imagine, as Smith put it, 'how a sociology might look if it began from the view of women's traditional place in it and what happens to a sociology that attempts to deal seriously with that' (1974, 7). This early work and the four decades

of sociological research since then have made gender a central concept for sociological investigation and critique. That is, gender became a key vantage point of sociological thinking. A vantage point is defined as 'a place or position affording a good view of something'; we are arguing that gender as a vantage point enables us to 'see' ourselves, the social institutions around us, and our social worlds in ways that attend to both women's and men's experiences.

Today, many sociology departments have an almost 50/50 balance of women and men, and gender is well-placed in sociological theory and research. Nevertheless, sociologists continue to identify refinements and developments in the interrogation of gender. As work on gender in sociology moved through the 1980s and into the 1990s and the new millennium, sociologists turned their attention to widening and deepening the earlier idea of gender as a vantage point of critique so that it *includes a broader array of vantage points*. In Canada, for example, women of colour drew on earlier work conducted by pioneering scholars in the United States and pointed to how the seemingly harmonious Canadian multicultural mosaic was marred by exclusion, discrimination, and, at times, brutality. The work of sociologists Himani Bannerji (York University) and Sherene Razak (University of Toronto) stands out in this regard. Both of these authors employ the evocative metaphor of the *gaze* to capture relations of both oppression and resistance. Bannerji's *Returning the Gaze* (1993) and Razak's *Looking White People in the Eye* (1998) depict the active agency, response, and resistance of racialized women who stare back at and challenge injustices in Canadian society. Work with and by Aboriginal women represents an important project to have the gender struggles of some of the most marginalized individuals in our society recognized and addressed. These efforts have helped to put the many forms of racialization in Canada, and the way they intertwine with gendered hierarchies, on the sociological agenda. They have also pushed sociology toward a more complex and nuanced understanding of gender.

Another interesting development in the push for more varied vantage points in the sociological treatment of gender has come from across the generational divide. Feminist sociologists who struggled in earlier decades for the acknowledgement of gender were also caught up in a politics that reflected the problems and issues of their time. Now the daughters and granddaughters, figuratively speaking, of feminist pioneers in sociology have taken up different issues that are particularly salient to the social and cultural agendas of more recent times. These issues include body image, **transgressive sexualities**, eating disorders, and male sexual responsibility, as well as recent cultural phenomena like Riot Grrrls, the *Bust* and *Bitch* zines, and varied adaptations of Eve Ensler's *Vagina Monologues*. Such concerns have also been highlighted in numerous books by young women focusing on what has been identified as third-wave feminism.[1] At the beginning of the new millennium, Amy Richards and Jennifer Baumgardner, two '30-something' feminists who met while working at *Ms* magazine, wrote a best-selling book that resonated with the experiences of many younger feminists growing up in Canada and the United States who felt alienated from the feminism of the '50-something' generation of their mothers. They noted the divide between older feminists and those their own age 'with tight clothes and streaky hair, who made zines and music and Web sites' (Richards and Baumgardner 2000, 24). This generational divide was also felt within sociology as younger women coming into the discipline began to reflect on and develop their own generation's interpretation and understanding of gender issues.

While scholarly and public commentators have continued to note the recurring generational issues between feminist mothers and their daughters, there has also been a wave of young women reclaiming the word 'feminism' and trying to remind young women that the journey toward gender equality is not yet over. An intriguing part of this conversation over whether 'feminism is dead' (as per the 1990 headline article in the magazine *Newsweek*) is an ongoing conversation about how to wrestle with the contradictory merging of pop culture's emphasis on female empowerment or 'girl power' with its tendency to constantly demean women. This tension for young women is well depicted in American feminist writer Ariel Levy's (2006) provocatively titled book *Female Chauvinist Pigs: Women and the Rise of Raunch Culture*, in which she laments the curious and ironic change as one section of the women's movement seems to have, in just one generation, gone from bra-burning to breast implants within bras as an expression of supposed female empowerment. She writes about her own process of questioning this puzzling shift when she began to ask:

What was going on? . . . Only thirty years (my lifetime) ago, our mothers were 'burning their bras' and picketing *Playboy*, and suddenly we

were getting implants and wearing the bunny logo as supposed symbols of our liberation. How had the culture shifted so drastically in such a short period of time? (Levy 2006, 2).

Levy concludes with a sobering warning: '"Raunchy" and "liberated" are not synonyms. It is worth asking ourselves if this bawdy world of boobs and gams we have resurrected reflects how far we've come, or how far we have left to go' (Levy 2006, 2).

> ## TIME to REFLECT
>
> Do you agree that 'feminism is dead'? Why or why not? If you think that feminism is alive and important, can and should sociologists engage in research that supports its aims? Why or why not? If so, what would you suggest as key topics for research?

As diverse groups of women have, in varied ways across the generations, reflected on what it means to be a feminist and/or to live fulfilling lives with a wide range of opportunities, this prompts the question as to whether and how their boyfriends, brothers, fathers, or sons have been engaged in thinking about gender relations. Men have, in fact, also been part of this conversation. Indeed, increasing attention to gender issues on the part of men is evident in popular writing as well as within sociology, where there has been a growing recognition that men's views on and struggles with masculinity need to be a component of gender studies. One impulse in this direction came from male sociologists themselves who began to theorize and research multiple forms of masculinity.

A male pro-feminist theoretical and activist perspective began to take shape in sociology through the 1980s and 1990s. Male scholars such as Canadian Blye Frank, three American scholars who all share the first name Michael (Messner, Kimmell, and Kauffman), R.W. Connell,[2] and many others have pioneered studies of men and masculinities. In doing so, men have not only added to the rich scholarship of gender studies in sociology but also offered views on where sociological work needs to advance and improve.

One area of contention within feminist scholarship is how, where, and when to include men in discussions that are of particular relevance to women. This sense of ambivalence should not be surprising, given that for centuries most scholarly work, including sociology, had been the study of men's lives and interests; it follows then that men's place, both as researchers and as subjects of study within the study of gender relations, would have to be arrived at through a process of delicate and steady negotiation. Both of us have memories of difficult discussions about whether to accept male colleagues as participants in feminist reading groups and about the appropriateness of male academics teaching courses in women's studies or gender studies. Feminists did and still do come to varied responses and conclusions on such issues. Nevertheless, we would argue that there has been a growing impulse to include men more fully in the sociology of gender relations.

TIME to REFLECT

Can you identify important issues facing young men today that could be explored and understood from the perspective of the sociology of gender relations?

One strong example of ambivalence on the part of women's groups and feminists regarding the role of men is the case of sexual violence. On the one hand, some women's groups and feminists have argued that 'Take Back the Night'[3] events should involve women only; on the other hand, others have argued that men also need to be somewhat involved in these events, since sexual violence is not only a 'women's issue'. At the same time, some men's groups have come up with their own, male-centred approaches to ending sexual violence. Perhaps the best-known is the White Ribbon Campaign, which was started by Toronto sociologist Michael Kauffman. For close to two weeks at the end of every year, men are encouraged to wear a white ribbon as a symbolic pledge never to commit, condone, or remain silent about violence against women. The White Ribbon Campaign is bookended by two days that women have designated as important times to reflect on gender and sexual violence: it starts on 25 November, the International Day for the Eradication of Violence against Women, and it ends on 6 December, Canada's National Day of Remembrance and Action on Violence against Women (and the anniversary of the Montreal Massacre).[4]

Just as new generations of women and sociological scholarship widened and deepened approaches to sexual violence, the work of MA student Elana Finestone typifies how a younger generation of women is seeking to include men in discussions of rape on campus. That is, if sexual violence is understood as a 'women

(Chris Gibson/Alamy)

Sexual assault, especially on university and college campuses, continues to be a widespread problem. This poster puts the onus on the female to be alert and conscious of her safety.

only' problem, this means that men are not called upon to acknowledge or take responsibility for their role. Finestone argues that since men play a strong role in perpetuating sexual violence, they should also be involved in helping to solve the issue of sexual violence on university campuses and elsewhere. The term **hegemonic masculinity** is introduced here to refer to a dominant form of masculinity in society—one espousing that men be strong, assertive, aggressive, self-reliant, and free of traditional feminine characteristics such as a willingness to display emotion and caring.

Recently, Leah McLaren, a columnist for *The Globe and Mail*,[5] commented negatively on the growing number of university courses, conferences, and academic journals devoted to the study of men and masculinity. She argues that whereas women's studies made sense because women's perspectives have historically been invisible or marginalized, men's studies did not. How would you support or challenge McLaren's claim?

GENDER IS A SOCIAL CONSTRUCTION

One of the earliest developments in thinking about gender in sociology was to challenge the notion that gender identities (e.g., masculinity and femininity) could be easily mapped onto biological identities (e.g., male and female). It was argued that ideas about masculinity and femininity, and about appropriate behaviour for women and men, boys and girls, have a social foundation independent of biological necessity. *Gender* was introduced as a term distinct from *sex*: the latter referred to biologically based differences, primarily related to differences in chromosomes and reproductive functions, while *gender* referred to socially produced differences, primarily of character, ambition, and achievement.

British sociologist Ann Oakley's classic book *Sex, Gender and Society* (1972) was one of the first to make the case that distinctions of sex are not as clear-cut or as straightforward as people typically think. She and others argued that the cultural and psychological features of gender are so variable historically

OPEN for DISCUSSION

Sexual Assault on Campus

Elana Finestone

During my six years as a student at Queen's and Carleton, sexual assault posters captured my attention in women's bathroom stalls, the University Centre atrium, and the tunnel system at Carleton. I eyed posters for women's self-defence courses, offering to teach me how to protect myself from sexual assault. Scanning another poster, I learned that I should watch my drink at parties to lessen the risk of date rape.

These posters, like most of the sexual assault posters around campus, exclusively target women. They imply that women are passive in the face of sexual assault and that if they fail to protect themselves, they invite it. This message is not only victim-blaming for sexual assault survivors, but it completely ignores the role men can play in ending sexual assault.

Recently, Carleton University administration tried to target men in their sexual assault posters. Through focus groups with first-year Carleton University men and women, I learned that unfortunately, men did not feel targeted by these posters.

Most of the men *and* women I spoke with felt that men did not feel actively engaged because the posters portrayed all men as exhibiting central traits of hegemonic masculinity: aggressive, emotionless, and slaves to their sex drive. For example, many men and women felt the poster with the catchphrase 'Ask First: Any form of sexual activity without consent is sexual assault' talked down to all men as potential rapists.

While posters depicting certain patterns of hegemonic masculinity were unsuccessful, so too were posters depicting characteristics associated with femininity. For example, since both genders considered expressing feelings to be a feminine trait, posters for events encouraging men to share their feelings were unsuccessful. Students' desire to differentiate masculine from feminine traits suggests they view gender as a dichotomy rather than a continuum.

My discussions with first-year students revealed the importance of sexual assault campaigns on campus that target men in ways that recognize the multiplicities of masculinities and femininities. For example, a poster with the slogan 'Being a friend means stopping him before he does something stupid: . . . Rape is a man's issue too' was most effective for men. It redefined masculinity in positive ways by constructing men as part of the solution to sexual violence; men are good friends who stop each other from hurting women. Similar campaigns can help foster discussion about sexual assault by involving both genders in university-wide sexual assault prevention efforts.

and cross-culturally that it is impossible to map these features onto biological sex difference. While early explorations of the social construction of masculinity and femininity in Canada tended to use terminology such as 'sex role' and 'sex stratification' (e.g., Stephenson 1973), by the end of the 1980s the distinction between sex and gender commonly appeared in first-year sociology textbooks. In short, this principle of gender studies within sociology—that of gender as a social construction—has meant that taking a gender perspective means we regard being masculine or feminine not as a natural phenomenon but as a very *social* achievement requiring intense effort and scrutiny on the part of individuals and societies. Further, attributes and inequalities associated with being male and female came to be seen as socially created consequences of the way society is organized around gendered identities.

The idea of gender as a social construction has not remained static. Its rethinking involved a number of developments, and we would like to identify two of the most significant ones. First, the idea that gender as a social phenomenon can be distinguished from sex as a biological phenomenon underwent an intense interrogation, and scholars began to reconsider what it meant to socially produce gender identities that are fluid and diverse. The decoupling of the sex–gender distinction from a division between the social and the natural also loosened the association of gender and sexuality and ushered in a torrent of research activity on sexuality as a socially constructed phenomenon that may or may not have a gendered form. While much of the sexual revolution of the 1960s and 1970s was very conventional in terms of understandings of women's and men's sexuality, the 1980s saw a stronger development of woman-defined female sexuality and other explorations of sexuality that were not gender-defined. The trend now is to see a more dynamic relationship between identified features of sexed bodies and what these features come to mean in social situations and in personal identities. Thus, with the social construction of gender released from more limited notions of heterosexuality and reproduction, a greater range of behaviours emerged,

HUMAN DIVERSITY

Potatoes and Rice: How Desirability Is Racialized in the Gay Community

Emerich Daroya

Desirability in the gay community is intrinsically linked with race, gender expression, and body types. Whiteness, muscularity, and masculinity intersect in perceptions of desirability, putting gay white, masculine, and muscular men at the top of the gay desirability totem pole, while gay men of colour are considerably lower. Gay Asian men are particularly de-eroticized and inferiorized because they are stereotyped as being feminine and having smaller bodies. In a community that considers them 'unsexy', interracial relationships between white men and Asian men are de-legitimized by the usage of racialized discourses that reflect these stereotypes.

Within gay communities, a white man who is exclusively attracted to Asian men is called a 'rice queen'. A rice queen is stereotypically an older white male who is considered unattractive and goes after younger Asian guys because he is not able to get younger white men. A 'potato queen' is an Asian man who is exclusively attracted to white men, but because the stereotypes attributed to them render them undesirable, they are seen to be 'settling' for older rice queens because they are the closest 'desirable' men they can get. The effeminized Asian is also considered passive, which the rice queen fetishizes

so that Asian men become objects for white men's fantasies of domination (Han 2007, 57).

These discourses are often used to de-legitimize white–Asian relationships in the gay community. This is because when (even younger) Asian–white gay couples are seen, people often see the white man as a rice queen and the Asian man as a potato queen. However, there is no popular discourse to describe a white man who is exclusively attracted to other white men (snow queen?), which tells us that whiteness in the gay community remains unmarked. In my own experience, I have been called a 'potato queen' and my partner a 'rice queen', which essentially shows that race (which is also gendered) is (still) a category of exclusion in the gay community. As Chuang (1999) argues, when a gorgeous white man is seen with an Asian man, people wonder why he would choose to be with an Asian man, since he should be able to attract many desirable white men. It is not yet cool for a white guy to have an Asian boyfriend (Chuang 1999, 38). This indicates that desirability is racialized and that the intra-racial relationship between two white men (after all, they are the 'desirable' ones) in the gay community is seen as the only legitimate relationship, to the exclusion of all others.

which expressed a more varied relationship between sexual orientation and the gendered/sexed body.

This tendency has been increasingly taken up by a new generation of scholars, especially individuals who bring together complex and intriguing intersections between sexuality, embodiment, gender, and ethnicity. Emerich Daroya, an MA student at Carleton University, provides compelling and provocative reflections that typify this move within sociology toward complexity and nuanced discussions of how some of these sociological concepts are lodged in people's everyday lives.

A second issue that has troubled earlier ideas about the need to separate the social experience of gender and the embodied one came when sociologists began to think that perhaps the body *is* important in particular contexts and moments in the life course. Indeed, as early as 1980 Canadian feminist sociologist Margaret Eichler voiced some of the analytical problems that emerge out of a systematic separation of biological sex and social gender. Other feminists from different scholarly disciplines expressed similar concerns that are well captured by American feminist theorist Linda Nicholson, who agreed that the body 'becomes a historically specific variable whose meaning and importance are recognized as potentially different in different historical contexts' (1994, 101).

One example of this focus on the entangling of the social and embodied aspects of gender appears in Andrea's recent work on fathering. In *Do Men Mother?* (2006), her book on stay-at-home fathers and single fathers, she argued that embodiment can matter in particular sites and in particular times when men are caring for children: 'When a father is attending to children—by cuddling, feeding, reading, bathing, or talking to them—gendered embodiment can be largely negligible. But there are also times when embodiment *can* come to matter a great deal, both for the men in these situations as well as for those who are observing them' (Doucet 2006, 712). Here she mentions examples of community spaces where gendered embodiment matters; most notably, they include instances of fathers caring for the children *of others*, single fathers hosting girls' sleepover parties, and community observations of unknown men lingering in sites where children gather (parks, playgrounds, schoolyards).

GENDER IS REALIZED IN SOCIAL ROLES AND INSTITUTIONS

Over the past few decades, as sociologists of gender relations have elaborated on the view that gender is a social construction, there has also been a great deal of conceptual work on the issue of *how* gender relations are socially produced in practice. This terrain of thinking was directed at unpacking how social institutions worked to produce differences and disparities between the two genders and particular types of gender relations.

In attempting to understand how gender relations worked within society, feminist sociologists first turned to available concepts within sociology—such as *social roles* and *social institutions*—to assist them in identifying gender both as an identity and as a property of social structures. Reflecting a strong current in sociology at the time, 'sex role socialization' was identified as a major force. That is, society provides different gender roles or scripts, and boys and girls are socialized into these roles through a process of subtle or explicit sanctions or rewards (Parsons and Bales 1956; Pleck 1976). Social institutions—the family, schools, the media—acting as agents of socialization, were said to reward boys and girls who behaved in ways deemed appropriate to their gender and to punish those who did not. While there would be many critiques of sex role socialization theory in later years, this approach did recognize that individuals become gendered because society is organized in such a way as to make gendered identities meaningful. In other words, gender was identified as a *systemic* feature of society.

It initially seemed like a radical move to suggest that societies had a gender structure as a defining characteristic (which many feminists referred to as **patriarchy**) and that apparently gender-neutral structures (such as the law or the state) were fundamentally shaped by assumptions about available and suitable gender identities and social positions. For the most part, gendered society and structures were seen as *constraints* on individual action and were problematized as external sources of gender oppression. However, as the progress of this gender critique developed within sociology, analysts wanted a more dynamic concept of structure that would have a more direct and reciprocal relationship with the everyday actions of individuals. Thus, social institutions—such as schools, workplaces, the state or government, families, the law—came to be seen not only as sites of the production and reproduction of gender relations but also as more complex sites of negotiation, contestation, resistance, and change. Overall, researchers became increasingly aware that the meaning of gender, and even its salience, can be highly varied depending on context and the particular circumstances being investigated.

(Najiah Feanny/Corbis)

Technology increasingly plays a role in the ways in which students learn and study.

A good contemporary example of the complexity of gender relations within social institutions relates to the social institution of education. As we will discuss in more detail later in the chapter, there have been constantly shifting gender relations within education, with women now outperforming men in overall attainment at almost all levels. Nevertheless, the educational choices of women and men still differ, with men typically pursuing careers that offer greater financial remuneration. Some of these choices relate to the use of technology within the social institution of education. That is, if we look closely at the social institution of education, we can see a complicated terrain of gender differences and similarities. How, for example, are the documented gender differences in the use of communication, information, and gaming technologies affecting the way students relate to academic knowledge production and dissemination (see, for example, France 2007, 119ff; Morrison 2010)? As Erin Murphy's MA research indicates, there can be gender differences in students' comfort level with and use of technology in the university classroom.

TIME to REFLECT

Have you noticed any gender differences in the ways your classmates use technology? Have you noticed whether the increased use of technology in the classroom is having a differential impact on how female and male students experience their education?

Building on some of these insights on gender relations within social institutions, we move now to the fourth conceptual pillar of sociological thinking, which is that gender relations are not only relations of differences but also of inequality.

GENDER RELATIONS ARE RELATIONS OF DIFFERENCES . . . AND INEQUALITY

In the 1970s and 1980s, sociologists of gender argued that a systemic feature of society was that gender relations were marked by persistent inequalities. This point was constantly made by feminist sociologists across these decades as they sought to counter prominent sociological approaches, such as structural functionalism, that insisted on the complementarity of male instrumental and female expressive gender roles. Indeed, feminist sociologists were very active in documenting deep and extensive inequalities associated with gender. They made many claims, two of which we detail below.

First, sociologists of gender argued that 'sex differences' have often been exaggerated and that this often occurs to the detriment of women's opportunities and self-esteem. It was observed that discussions of 'sex differences' in temperament, in attitudes, and in capabilities overstated the extent of difference between men and women as social groups. These overstatements hindered women's abilities and achievements and contributed to men's claims of power and authority in all things and in all places. Countering this, there has been an extensive review and debunking of such research by psychologists (e.g., Maccoby and Jacklin 1974) who helped to fuel arguments made by feminist sociologists against the exaggeration of difference in the analysis of women's and men's character and personalities. As British

feminist Juliet Mitchell famously observed (1980, 234), compared to giraffes, women and men are very much alike! In short, sociologists came to argue that the overstatement of sex difference was becoming itself a form of discrimination and oppression, which largely worked to the advantage of men.

The second claim was that inequality in women's and men's life chances is a consequence of how society is organized and of the particular ways in which gender is created and sustained as a significant feature of social and personal life. In Canada, there were several streams of work developing this idea. One was focused on the more organizational aspects of inequality, with an emphasis on inequalities in education, in employment, in incomes, and in the amount and types of work that women and men do. University of Toronto sociologist Monica Boyd's early work on occupations is a good example of this interest (Boyd 1982; 1985a; 1985b). The second stream of research in Canada was concerned with larger systemic foundations of gender inequality, captured by analyses of capitalism, patriarchy, and what came to be called the sex/gender system. The work

UNDER the WIRE

Gender and Educational Technology

Erin L. Murphy

Do you own a cellphone or a smartphone? What about an MP3 player or an iPod? A laptop? An iPad or a tablet computer? Chances are you own at least one of these mobile technologies and that you bring them everywhere you go—even to class.

Some authors have argued that the current generation of students should be thought of as 'digital natives', because they have grown up surrounded by digital technologies. These authors suggest that this digital saturation has influenced young people's expectations in every aspect of their lives, from how they communicate to how they learn. As a result, technology is becoming increasingly present in university classrooms as students, professors, and administrators find new ways to digitize the learning experience.

To investigate the ways students are using and want to use technology in their classrooms, I asked students in two introductory sociology classes to participate in an online survey. In addition to asking about their use of technology, the survey included basic demographic questions regarding their age and gender. The inclusion of gender on the survey

meant that the final data could then be analyzed to look for similarities and differences in the ways that young men and women responded.

The results of the survey showed many similarities and some interesting differences in how female and male students use the technology they bring to the classroom. For example, the majority of both men and women who bring a laptop to class say that they use their laptops to take notes and follow along with the lecturer's PowerPoint presentation. In addition, women students were more likely to use their laptops to check email and social networking sites while in class. Men, on the other hand, were more likely to use their laptops to google what the professor was lecturing about, play games, or work on assignments.

Perhaps most interestingly, the men were more likely than the women to report that they were very comfortable using technology. This finding is consistent with the stereotype that men possess greater technical skills than women, but we cannot be certain that this result confirms the stereotype or was simply influenced by it.

of Queen's University sociologist Roberta Hamilton captures this research trend very well, both in her early studies of relations of patriarchy and capitalism (1978) and in her well-known revision of Canadian sociologist John Porter's (1965) classic work on the Canadian vertical mosaic. Porter's mosaic was structured along class lines; Hamilton titled her effort and the book she produced as the 'gendering' of the vertical mosaic (Hamilton 1996/2005).

While these efforts are still regarded as foundational ones that cleared paths for later investigations, as with the other pillars of sociological thought on gender relations, scholars of gender increasingly began to attend to diversity, complexity, nuance, and fragmentation within earlier concepts and arguments. Researchers became increasingly aware that the meanings and salience of gender are highly varied depending on context and the particular circumstances being investigated. There was thus a move toward regarding the relevance and substance of gender as questions for, and not assumptions of, research. This point is significant for all investigations of gender, but it has perhaps been most controversial and contested in the analysis of *gender inequality*.

Growing recognition of the significance of race and ethnicity in identifying different experiences of inequality *among* women has ultimately led analysts to realize that they could not be content with a simple, dichotomous presentation of gender. For example, Sedef Arat-Koc (1989) examined the lives of immigrant women working as live-in domestic workers in Canada, a situation in which housework 'becomes the responsibility of *some* [women] with subordinate class, racial and citizenship status, who are employed and supervised by those who are liberated from the direct physical burdens' (1989, 53). She concluded that the domestic service relationship, between female employer and female employee, adds class and race complexities to gender inequalities (see also Stasiulis and Bakan 2005).

Some of this nuance and complexity can also be found in one large debate that has been waged within feminist sociology for the past two decades as to how best to capture the complex sets of inequalities between women. In this vein, the words **interlocking** or **intersectional analysis** came to dominate understandings of women's multiple identities as well as their location in multiple structures of gender, class, race, and sexuality. According to the well-known American sociologist Patricia Hill Collins, intersectionality 'refers to particular forms of intersecting

oppressions, for example, intersections of race and gender, or of sexuality and nation. Intersectional paradigms remind us that oppression cannot be reduced to one fundamental type, and that oppressions work together in producing injustice' (2000, 18). By the 1990s, feminists in Canada and the United States were also arguing that ability/disability as well as age/generation were markers of socially structured identity intersecting with gender inequality.

While the analytical terms themselves seem to be constantly shifting to capture increasing levels of diversity in North American society, the important point to emphasize here is that issues of difference and disadvantage need to be viewed as both relational and structural phenomena. They operate at the level of identity or agency (i.e., 'I feel discriminated against as an Aboriginal woman, not just as a *woman* and not just as an Aboriginal person but as an Aboriginal woman'), and at the same time they are rooted in the ways in which social institutions such as families, workplaces, and governments are set up and function. This point is highlighted in an interesting way in the work of Canadian sociologist Himani Bannerji (2000), who contests the terms 'visible minorities' and 'women of colour' on the grounds that they draw attention to the physical appearance of those subjected to oppression and deflect attention from the institutionalized social practices that oppress them.

The four conceptual pillars of sociological thinking on gender relations we have discussed in this section provide ways of understanding the everyday worlds of women and men. In part 2 of this chapter, we are guided by these ideas as we take you into sites and issues where gender relations are constantly in play, with varied shifts, transitions, and negotiations between women and men.

Living Gendered Lives

GENDER RELATIONS IN SCHOOL, WORK, AND FAMILY

School

Gender differences in the experience of schooling, and in school attainment, have been matters of sociological attention for many decades. Throughout the 1960s and 1970s, the concern was that girls were losing out in education. Sociological research demonstrated that starting as early as elementary school and getting more pronounced as one progressed up the education ladder, classroom experiences, curriculum design, and measurements of student success highly favoured boys. After successful campaigns and educational changes, the tables now seem to have turned. From the early 1980s onwards, the enrolment of women in Canadian universities has exceeded that of men. Statistics Canada has reported that by 2009, women made up close to 60 per cent of undergraduate and MA enrolment. Men still predominate (53 per cent) at the doctoral level, but even here the gender gap appears to be narrowing.[6] Research by Abada and Tenkorang (2009, 201) shows that the gender gap in attending and graduating from university also applies to children of immigrants to Canada, with young immigrant women '60 per cent more likely to have a university education' than their male counterparts. For Aboriginal youths, there is a slight gender gap in favour of women completing a university education, but this is the case in an overall situation of extreme educational disadvantage.[7]

What does the shifting gender imbalance in higher education enrolments mean? Some have noted (e.g., Davies and Guppy 2006) that degree-level qualifications have become important for careers that girls typically pursue more than boys (for example, teaching, nursing, and social work). Thus, enrolment equalization reflects changes in the kinds of programs universities are offering and the inflation in degree qualifications required in certain professions. This dynamic may also be behind the increased proportion of women in master's-level programs. Supporting this 'more apparent than real' idea is the fact that students continue to be gender-segregated in terms of the programs and courses they take as undergraduates. Women have made significant inroads into professions such as law and medicine (now both roughly 50 per cent female) and now dominate some degree programs in which they were previously a minority (for example, pharmacy, education, and veterinary medicine). However, overwhelming imbalances continue to exist in traditionally female programs (nursing, social work, and fine arts), where roughly 66 per cent (or more) of the degrees are granted to women. Some male-dominated programs also seem more resistant to change. Programs like architecture, agriculture, engineering, math, physics, and forestry show an increase in female participation but remain strongly male-dominated. As Davies and Guppy suggest (2006, 114), these are programs (and areas of employment) where a long-standing male-focused culture presents a very chilly climate to women daring enough to cross the gender divide.

TIME to REFLECT

Is the gender imbalance in post-secondary school attendance, programs, and achievement anything to be concerned about? Why or why not?

Work

At university, there are interesting gender differences in students' participation in paid work. Canadian statistics show that female students are more likely to have paid employment than male students but also that male students who are employed work longer hours than female students.[8] The need to combine schooling and employment points to ongoing class differences in access to higher education. These class differences combine with gender (and other dimensions of inequality such as ethnicity, race, and dis/ability) to produce quite complex patterns in how young people experience the school-to-work transition. Wolfgang Lehmann's research (2005) on youth apprenticeships elaborates the way school-to-work transitions happen in contexts structured by gender and class expectations, and he and other researchers show that class and gender dynamics operate strongly in more general life-course transitions (Andres and Adamuti-Trache 2008; Cote and Bynner 2008).

Although women have surpassed men in terms of obtaining education credentials, they are less able to turn this educational advantage into dollars when they hit the labour market. This is partly because the gender segregation in university programs carries female and male students into gender-segregated jobs in the labour market. However, evidence shows that even within the same occupation, young men begin their employment careers with a higher starting salary (MacAlpine 2005).

Karen Foster's discussion of the difference between the perception and the reality of change in women's experience of employment opportunities makes a good case for the continuing need for campaigns for gender equality.

Sociologists have written a great deal about how the experience of employment has changed over the past decades. They have observed a decline in the protection of workers and a corresponding shift to the individualization of risk and responsibility. This has two manifestations that are particularly relevant for the employment risks of young people—and both are gendered.

SOCIOLOGY in ACTION

Paid Work across Generations and Gender Equality

Karen Foster

For my doctoral dissertation, I have been interviewing people of all ages, backgrounds, and occupations about how people relate to work over the course of their lives. For the older women I spoke to, gender mattered. It mattered because in their lifetime, women had flung open the doors of traditionally male occupations and stepped into new, exciting roles as working people. It mattered because they knew their career opportunities had exploded in a single generation. While they still confronted old attitudes about women's roles, they had realized some collective power, as women, to transform the world. They looked at their daughters' working lives and saw a world of boundless opportunity.

Indeed, among the youngest people I interviewed—women and men—there was no discernable gender difference in the way work and life were planned for, imagined, and idealized. Regardless of gender, when these young people transitioned from school to work, they looked for careers that were personally fulfilling and socially useful. The mantra 'work

to live, don't live to work' was ubiquitous. Gender did not appear to be limiting on any level, in the form of outright discrimination and rigid roles or in terms of underlying subjectivities. But from a sociological perspective, a more sobering perspective emerges. My interviews with young parents, only a few years into their careers, highlight the way gender enters their consciousness abruptly when children are born. If they avoid the traditional imbalance (i.e., if Dad stays home), they are subject to judgment by those with traditional values. If they do the reverse, they are impelled to justify their arrangement to those who see it as an affront to women's equality. Inevitably, the mother will face career setbacks even if she only takes a short time off for childbirth; moreover, child-care costs are so high that for many it makes more financial sense to have one parent stay at home. Gender inequality has narrowed, certainly, but the reduction of blatant sex discrimination and gender ideology might be overshadowing persistent structural barriers to balancing work and family.

One type of risk in employment relates to issues of personal safety; that is, young women and men are likely to be in jobs where the sort of risks they must personally shoulder differ. Women are more likely to run the risk of sexual attack and harassment in the workplace. This is in part because they are often employed in situations where they are expected to wear revealing clothing and to present a sexualized image/demeanour as part of their job. Young men, on the other hand, are the most likely of any group of employees to suffer serious work-related injuries—they fall off roofs, have accidents with forklift trucks, and, as in one horrible incident in Ottawa recently, get run over by steamrollers.[9]

The other type of employment risk involves the rise of what is called non-standard or precarious employment. This is employment that has irregular and often inconvenient hours, no benefits, no promotion prospects, and no security. Such jobs are often referred to as **McJobs**. Melinda Mills (2004, 131) observes that as students, young women and men tend to be equally employed in precarious, non-standard jobs. However, once they leave school, gender differences

start to appear, with young women more likely than young men to remain in precarious employment.

Sociological research has established that gender inequality in employment tends to increase over the life course, and this can have consequences for how easy or difficult it is to cope with job loss or demands to keep pace with job changes. Janet's research on inequality in the negotiation of work change shows that intersecting inequalities of gender, race, and class create experiences of work change that are relatively easy for some—they can 'go with the flow' of change—and very difficult for others—they get 'swamped' by change (Siltanen et al. 2009).

Family

While much of university students' time is spent balancing school, paid work, and a social life, some students also juggle family life. Some of you may be living at home. Others may have formed your own households with friends or with a partner and children. Some of you may have worked for years and/or started a family and are now returning to university to upgrade your work credentials or to prepare for a new career. Whatever reason motivates you to read this chapter, you will certainly have had some experience of family life that influences the way you think about gender relations in the family.

Inequalities in the work that occurs within families—care work and domestic labour—were not topics of sociological concern until the 1960s, when a few female sociologists put these topics on the map as very important ones that deserved sociological study. Meg Luxton, one of Canada's leading sociologists of gender, wrote her PhD dissertation on inequalities in domestic labour. Looking back on this experience, she wrote about how the topic was not seen as serious enough for academic study, with one professor in her department telling her that 'he was embarrassed to be part of a department that would even consider such work suitable for a doctorate' (2008, 273). In fact, her thesis produced an award-winning book that is still in print after 30 years and has just been reprinted in a new edition (Luxton 2010).

Much has changed since then. Inequality within the family and the study of **gender divisions of domestic labour** are now burgeoning areas of research within sociology. Indeed, this is a central field within the sociology of gender relations and one to which we have both contributed in varied ways over the past two decades. While this is a large and diverse field of study, we want to underline three features here.

First, it is important to see domestic labour and the care of children not as a small issue of concern only to women. How societies care, or do not care, for dependent others—both old and young—tells us a great deal about the social fabric, social institutions, and political priorities of that nation, province, territory, or city. For more than three decades, feminist social scientists have highlighted the economic, social, political, and personal costs to women of the gender imbalance in what American economist Nancy Folbre (1994) and Pulitzer Prize–winning Anne Crittenden (2001) call the 'costs of caring' for the very young, the very old, the injured, and the sick in all societies. On the other hand, there has also been increasing attention to how *not* caring for others has affected men, including lower lifespans and loneliness or isolation when they live without wives or

(Florian Franke/Alamy)

Gender relations in the family are shifting as more men take on household responsibilities, such as doing laundry.

partners. A position outlining the social and personal costs to men of not caring actually goes back to several feminist classics, perhaps most notably Dorothy Dinnerstein's *The Mermaid and the Minotaur* (1977), in which she outlined the fundamental imbalances that occur in a society when one gender does the metaphoric 'rocking of the cradle' while the other gender 'rules the world'. Some 30 years later, fathers' rights groups and men's groups are considering and promoting emotional connection and responsibilities as important and enriching to men's lives.

A second point relates to something you have probably seen in your own household. If you lived—or live—in a two-parent heterosexual family, who does the unpaid domestic work in the home? If a good portion of it was done by your father—or if it is done by your partner/husband—then you are part of a rising social shift in gender relations in the family whereby men are doing more domestic work. While there are different interpretations of how much housework Canadian men are actually doing, there

is a fairly strong consensus that *more* men are doing *more* housework than they did in previous decades and generations, and much of this increase is in child care–related tasks and activities (Marshall 2006).

Third, there have also been dramatic increases over the past 30 years in the number of men in Canada who are stay-at-home dads; the proportion of single-earner families in which the father is the stay-at-home parent increased from 2 per cent (in 1976) to 10 per cent in 2002 (Statistics Canada 2002a) and 12.5 per cent in 2007 (Statistics Canada 2008e). With women's full-time employment growing over the past few decades, dual-earner households have also increased. In 2007, for example, the proportion of husband/wife families with dependent children (under 16 years of age) who were dual-earner was 70 per cent of all Canadian households, the highest level since 1976. Perhaps the most dramatic social shift in gender relations that straddle family and work is the fact that women are primary breadwinners in nearly one-third of Canadian **two-earner families**

SOCIOLOGY in ACTION

Planning for Parenthood: Gender Differences and Gender Similarities

Elisabeth Wilson

Women and men experience life differently and are expected by society to perform their respective gender roles, resulting in a gendered division of labour. To get a better sense of women's and men's respective life experiences and how they compare to each other, feminist sociologists are applying a 'gender lens' in their research and are collecting and analyzing data on a sex-disaggregate basis. Such gender-sensitive research methodologies reveal existing gender differences between women and men, but sometimes they may also reveal that gender differences are far less pronounced than one would expect.

A case in point is a recent study I conducted with first-year university students between the ages of 18 to 25 with respect to whether or not they would like to have children one day, how many children they would like to have, and how they think they would like to care for them. Given that traditionally, girls are socialized to become mothers and primary caregivers and boys are socialized to become primary breadwinners, I went into my research with the hypothesis that there would be gender differences in how female and male students envision becoming a parent. I expected that women at such a relatively young age would already have more concrete ideas

than men with respect to how many children they would like and at what age.

To my surprise, the data revealed some unexpected similarities between female and male students. In fact, the overwhelming majority of both female and male students had already thought about having children one day (96.4 per cent), and more than 80 per cent also *wanted* to have children one day. There was, however, a gender difference with respect to how old students wanted to be when they became parents for the first time and the number of children they would like. Female students, on average, would like to have their first child at the age of 26½ and would like 2.3 children; male students, on average, would like to have their first child at the age of 28½ and would like 1.8 children.

While the difference in how many children female and male students would like to have may be a point of contention when starting a family, my research revealed that negotiations with respect to who should care for the children should be less controversial. The vast majority of female and male students think that fathers should invest the same amount of time in child care as mothers, and they would opt for an equal child-care model when raising their children.

(Sussman and Bonnell 2006). This trend had been growing over time, but it solidified in 2008 when a widespread economic recession saw male unemployment rise, mainly through the loss of jobs in traditional male fields such as manufacturing. This situation has particular implications for family life and raises intriguing questions about the gender balance of paid and unpaid work.

As you may have gleaned from this section, many interesting issues about gender relations in family life remain for sociologists to explore. As demonstrated in the boxed insert on page 216 by recent MA graduate Elisabeth Wilson, one interesting question is how a new generation of young women and men are envisioning their future family life. Wilson argues that there are some gender differences but also some gender convergence in men's and women's aspirations for parenthood.

In spite of the expressed aspirations of the next generation of parents, the recent work of American sociologist Kathleen Gerson tempers this sense of hopeful optimism with her research on work–family issues for young women and men (aged 18 to 32). Despite a strong consensus among researchers that women still carry much of the responsibility for housework and child care, Gerson notes that there has been a 'revolution' in gender relations in the balance of paid work and care work; nevertheless, she adds that it is an 'unfinished revolution'. In her recent book by the same name, she argues that amid this 'unfinished revolution', a new generation is seeking flexible work options to assist them with the challenges of balancing work and home. She argues that 'young women and men from all family backgrounds are searching for new, more flexible ways to combine love and work' (2010, 99).

While noting this with hopefulness, Gerson also provides a sobering analysis of what she calls the 'fall back' strategy for women and men when they face less than ideal conditions at work or at home. She writes:

> Although young women and men share the ideals of lasting commitment, gender flexibility, and work-family balance, they harbor some different fears about what might happen if these aspirations remain beyond their grasp. Women stress the dangers of depending on someone else for their identity or financial well-being; men focus on the costs of failing at work. These contrasting concerns point toward a gender divide lurking beneath the surface of shared ideals.

They pit 'self-reliant' women, who see personal autonomy as essential for their survival, against 'neotraditional' men, who see work success as a key to self-respect (2010, 105).

TIME to REFLECT

What are your current (or future) aspirations for work–family balance? Do you think there are still gender differences in expectations for career and family? If so, do you think such differences matter?

While, as discussed above, young people are constantly negotiating gender relations at school, at work, and in their families, they are also negotiating a wide range of gender identities in an increasingly globalized world.

NEGOTIATING GENDER IDENTITIES IN A GLOBAL AGE

Negotiating gender identities is a major focus of youth activity—and contexts for this activity vary between and within generations of youths according to historical, social, economic, and political circumstances. A distinguishing aspect of the generational experience of gender negotiation for youths today is that it is being done in a relational context that is *globalized*.

Globalizing processes bring a double-sided dimension to gendered identities. On one side, there is greater exposure to multiple images and practices of masculinity and femininity. On the other, there is a tendency toward a homogenized portrayal of masculinity and femininity, and worse, these portrayals can tend toward conservative, stereotyped, and, at the extreme, exploitative images. As mentioned previously in this chapter, sociologists have used the word 'hegemonic' to identify this sort of dominant form of gender identity. Young people negotiate their own gender identity within the pushes, pulls, and pulses of the tensions between multiplicity, homogeneity, and the hegemonic. Adding further complexity are relationships among gender, class, race, and other intersecting dimensions of identity and social structures, which set parameters and possibilities for experience.

While there is nothing new in young people's using music, fashion, and other cultural resources as key tools in identity construction, the explosion of a globalized youth culture intensifies the negotiation required to make use of such tools in the construction of gendered identities. Sociologists have been critical of notions that youths are simply followers of global fashion. While recognizing youth culture as big business and full of marketing ploys, sociologists nevertheless see youths as capable of discernment when it comes to deciding what to wear, watch, and listen to and how to have fun. What this means is that despite globalizing pressures toward homogenization and hegemony, youths see their fashion, music, and other consumption and activity choices as part of the work they do to construct individualized identities. Nevertheless, the resources youths have to draw on in performing this identity work are highly gendered. In discussing this point, we first want to focus on negotiating multiple genders and then on gender homogenization with particular reference to hegemony.

Through immigration, increased intercultural relationships, and the influence of global media, many Canadian youths find themselves confronted with multiple ways of enacting femininity or masculinity. As Bill Osgerby, a British researcher on youth culture and gender, writes (2004, 181), 'young people's subject positions have come to operate across, and within, multiple cultural sites—their identities are constituted by the intersection of crisscrossing discourses of age, ethnicity, gender, class, sexuality and so on'. If family, friends, and social connections are diverse in cultural traditions, there may be different understandings of appropriate gendered behaviour, and these understandings may very well clash with each other.

Research in Canada has highlighted how young people negotiate gender identities within relational contexts that can include clashes between dominant and subordinated cultures as well as clashes within cultures. For example, Amita Handa (2003) tells of the 'tightrope' that young South Asian girls walk in negotiating the expectations of femininity that vary *within* Asian culture as well as *between* it and the

GLOBAL ISSUES

Body Beautiful: Perfection under Construction

Zainab Amery

Who knew boob jobs and liposuction would be the top two surgical procedures in China and India, according to the 2010 survey of the International Society of Aesthetic Plastic Surgery? With the United States and Brazil taking first and second place as cosmetic surgery nations, China runs a close third, with India not too far behind. There are also reports that customers everywhere are getting younger, with cosmetic surgery becoming one way young women get ready for university.[10]

Guided by image-conscious peers, the media, and celebrities, young people around the globe are surgically altering their appearance. Reality TV and dramas like *Extreme Makeover*, *I Want a Famous Face*, and *Nip/Tuck* have normalized and glorified cosmetic surgery. Supplemented by the constant barrage of image ideals in teen magazines (think *Cosmo Girl* and *Teen Vogue*) and glossy billboards, youths seem to be buying into the attainment of bodily perfection.

But it's not about beauty alone: it's about the rewards that being beautiful can bring. Studies show that attractive people achieve a higher degree of success than their less attractive competitors on the labour market. In Brazil, where the body is worshipped and beauty means access to employment, government-run hospitals offer subsidized procedures for low-income people. In Asian countries, business is booming but is racialized, with many young people opting for double-eyelid surgeries to become 'Caucasian-looking' in order to get the best jobs.

However, it is Lebanon that takes the prize for the first cosmetic loan program. Billboards advertising the First National Bank's Plastic Surgery program read, 'Beauty is no longer a luxury' beside an image of a Caucasian, blonde, blue-eyed, Western-looking woman. Astonishingly, the only stipulations are that the borrower be at least 17 years of age and make a minimum of $600 a month. In a country where beauty is paramount for the purpose of getting (and staying) married, the perfect body can mean the difference between starting a family and remaining a spinster—especially since there is only one eligible male for every five women in the population. The situation is so serious that mothers escort their daughters to surgeons to raise their odds of finding a husband. But the search for perfection is no longer solely a women's quest. Of the 1.5 million surgeries performed yearly in Lebanon, 30 per cent are on young men,[11] with a similar trend occurring globally.

dominant 'white' Canadian culture. She highlights bhangra dances as flashpoints for such clashes. 'Bhangra music and dances stand in opposition to dominant white culture in the struggle for cultural space. . . . They also assert girls' resistance to parental attempts to control their sexuality' (Handa 2003, 116). Murray Forman (2005) describes a similar tension when observing how young Somali men in North American high schools struggle to express their masculinity by customizing elements of hip hop and militarized fashion gear in a context where they are regarded as 'black' by their peers but 'not black' by their own Somali culture.

Stereotypical beauty ideals can be spotted nearly everywhere across the globe, including on this Brazilian newsstand.

(© jeremy sutton-hibbert/Alamy)

In the midst of multiple and fluid gender identities, however, there is also an extraordinary pressure to conform. L. Susan Williams (2002) argues that experimentation with forms of gender identity may be fairly brief. She identifies the early years of high school as the time when a narrowing of possibilities occurs, along with increasing pressure to 'fit in'. While there still may be some room left for experimentation with gender identity as youths move through high school and into jobs or post-secondary education, pressures to conform to the norm exist and increase as youths approach key decision points in their lives.

Becoming oneself also involves confronting the contours and power of hegemonic gender and sexual identities. Although the specific characteristics of hegemonic masculinity can vary according to context, it is usually identified with the traditional masculine qualities of 'being strong, successful, capable, reliable, in control' (Connell 2000, 10). Hegemonic masculinity is distinguished from other expressions of masculinity that are subordinated and/or marginalized, and it is *especially* distinguished from femininity, or what Connell calls **emphasized femininity**. Both hegemonic masculinity and emphasized femininity have taken on different shapes and forms within a globalized context that includes the ubiquitous presence of Internet pornography, the stunning scale of sex tourism, and the increase in human trafficking for the sex trade. All of these are serious examples of where gender, race, class, and historic processes of colonization and contemporary globalization collide to produce extreme, exploitative versions of hegemonic masculinity and emphasized femininity.

TIME to REFLECT

Advertising is full of assumptions and messages about gender. What criteria could you use to determine whether advertising images are depicting hegemonic masculinity or emphasized femininity?

It is also the case that more mundane forms of hegemonic masculinity and emphasized femininity are all around us. One does not have to travel the globe to be affected by globalizing trends in the presentation and enactment of gender identities. Stereotyped presentations of gendered bodies are very familiar on television, in magazines, and on celebrity websites; there are many versions of the thin, tall, perfect-complexioned, blonde young woman and the six-packed, clean-shaven young man with the dazzling white smile. Billions of dollars are spent every day on body products, and this industry caters increasingly to men as well as to women. Also booming is demand for and supply of more permanent techniques for achieving the perfect body—cosmetic surgery. There are four prominent aspects to this worldwide trend. First, although young women are more likely to undergo body-shaping surgery than young men, this gender difference is lessening. Second, both genders tend to have surgery done on the face and on body parts most associated with sexuality and gender identity—lips, hips, breasts, buttocks—and increasingly the genitals themselves. Third, there is a global valorization of white Western characteristics, so body reshaping has both a gendered and racialized dynamic. Finally, the age of cosmetic surgery clients is getting younger and younger.

Surgical modification to enhance and produce gendered standards of beauty is an expensive option, but as Zainab Amery explains, it is becoming a normalized practice in many places around the globe.

On to the Future: Gender Relations and Social Change

In closing, we want to draw your attention to the fact that uncovering strategies and possibilities for change is an important focus of sociological investigations. This brings us to ask the question: what is/are the way(s) forward? How do we identify progress in the ways in which gender is structured in society and negotiated in everyday lives? Is progress the eradication of gender differences or the neutralization of the social, material, and political consequences of gender differences? Ideas about gender equality change with the times, and we need to keep a constant watch on whether our understanding of and platforms for equality are keeping pace with social, economic, and political changes. One thing is for certain: systematic analysis of the presence and consequences of gendered experience is required as an ongoing aspect of tracking whether things are getting better or worse. Many eyes in many places are needed to keep this constant watch. In fact, this has been identified as one of the successes of the second wave of the women's movement—the ubiquity of attention to gender issues (Gibson-Graham 2006).

One example of this ubiquity is the extent to which those responsible for social and public policy development at all levels of government—municipal, provincial, national—are required to systematically examine gender effects of potential and existing policies and programs. Canada has in fact been at the forefront in developing this kind of analysis—called **gender-based analysis**—for use all over the world, and sociologists have been among those closely involved in the development of this form of gender analysis.[12] While this has been a significant development, many eyes in many places are focused on pressing for a more nuanced analysis of gender inequality in government policy and program development—one that attends to the inequalities intersecting with those of gender. Forces pressing for this more conceptually complex and experientially appropriate use of gender are both inside and outside all levels of government in Canada and around the globe.

Another example of ubiquity is the tremendous scope of the activities of social movements and gender activists who formulate demands for change and imagine ways to improve the lives of women and men negatively affected by gender issues. Youths are a key source of this energy for change. Many young people in Canada today are involved in campaigns and other activist work to address gender issues—from Femmetoxic, the campaign to remove toxic chemicals from women's beauty products, to the involvement of male youths in the White Ribbon Campaign.

Equally ubiquitous is the effort made by many people in their homes, workplaces, schools, neighbourhoods, and city streets to address gender inequality in everyday and every-night relationships. The feminist insight that the personal is political continues to reverberate in contemporary gender relations.

The understanding of gender has been closely linked to the development of sociology as a discipline, as well as to the dynamics of feminism and the women's movement. During the 20-plus years that we have been doing academic research, all three of these—sociology, feminism, the women's movement—have changed tremendously. These changes reflect and are a consequence of important shifts in the ways women and men live their lives and understand themselves, their relationships, and their society. In writing this chapter, we wanted to connect with your generational experience by highlighting recent sociological research on how youths of today confront and negotiate gender relations and identities. At the same time, we wanted to demonstrate how sociology might connect with your personal experience of gender relations and identities by including the voices and thoughts of those closer in age to you than ourselves—that is, the voices of several students we have had the pleasure to teach in recent years.

While we have tried to write a sociology of gender relations that speaks to you, at the end of the day it is you who know best how these ideas fit your everyday experience. We encourage you to think critically about how these sociological ideas fit with your own life— how you think of yourself, how you believe others see you, and what is happening in your family, school, workplace, and the wider global community. Where do the ideas fit your experience—where do they not? For sociological understandings and explanations of gender relations to move forward, they must speak to the experiences of your generation—and your input to this development is important. Some of the gendered experiences you will struggle with will be familiar to us—because we've struggled with them too—reproductive responsibility and choice, autonomous sexuality, education, careers, and children. But you may struggle with these issues from different starting points, from within distinct cultural or social contexts, or you may struggle with different issues and with divergent goals in mind. We urge you to bring these struggles into a conversation with sociology—we are sure that both you and the discipline of sociology will benefit from such an engagement.

questions for critical thought

1. Gender differences can be an aspect of our lives but not necessarily matter, in the sense of having consequences for choices and opportunities. Thinking about your own life right now, where would you say gender matters the most? Can you think of areas of your life where gender does not matter at all?

2. 'Women are oppressed.' Discuss.

3. 'Men are oppressed.' Discuss.

4. Are the gender expectations in your family different from those in the families of your friends? If they are, how would you explain this difference?

Is it due to differences in, for example, class, religion, ethnicity? If you haven't noticed a difference, how would you account for the similarity?

5. What evidence is there to support the idea that gender equality has been achieved? What evidence would suggest that this is not the case?

6. Boys are more likely than girls to play computer games that involve physical violence and killing. Do you think this is a gender difference that matters? Should girls be encouraged to play these games as well?

recommended readings

R.W. Connell. 2005. *Masculinities*. 2nd edn. Cambridge, UK: Polity Press.
The latest edition of a classic sociological treatise on masculinities, this book includes a very influential theoretical approach to understanding masculinities, as well as interesting Australian case studies of specific forms of masculinity.

Roberta Hamilton. 2005. *Gendering the Vertical Mosaic: Feminist Perspectives on Canadian Society*. 2nd edn. Toronto: Pearson.
Hamilton provides a great discussion of the relationship between sociology, women's studies, and the women's movement. The 'vertical mosaic' in the title is a reference to one of the most famous books on social inequality in Canada, written by John Porter.

Barbara Marshall. 2000. *Configuring Gender: Explorations in Theory and Politics*. Peterborough, ON: Broadview Press.
This book discusses developments in the sociological analysis of gender in greater detail. It also will introduce you to how gender is understood in newer approaches to sociology influenced by post-modernism and post-structuralism.

Adie Nelson. 2009. *Gender in Canada*. 4th edn. Toronto: Pearson Education.
This book provides very detailed and comprehensive coverage of gender issues with a Canadian focus throughout.

Janet Siltanen and Andrea Doucet. 2007. *Gender Relations in Canada: Intersectionality and Beyond*. Don Mills, ON: Oxford University Press.
This book sets out the value of sociological analysis for understanding gender at different moments of the life course (childhood, adolescence, and adulthood). It explores how the concept of intersectionality can add to understandings of gender both in theory and in research.

Valerie Zawilski and Cynthia Levine-Rasky, eds. 2005. *Inequality in Canada: A Reader on Intersections of Gender, Race and Class*. Don Mills, ON: Oxford University Press.
Zawilski and Levine-Rasky offer an excellent collection of articles by Canadian sociologists examining how gendered experience is shaped by other dimensions of inequality such as race and class.

recommended websites

Canadian Research Institute for the Advancement of Women (CRIAW)
www.criaw-icref.ca
For more than 30 years, CRIAW has been documenting the economic and social situation of women in Canada through groundbreaking research while also making this research accessible for public advocacy and education.

Father Involvement Research Alliance (FIRA)
www.fira.ca
FIRA is a Canadian alliance of individuals, organizations, and institutions dedicated to the development and sharing of knowledge focusing on supporting father involvement.

Feminist Alliance for International Action (FAFIA)
www.fafia-afai.org
FAFIA is a dynamic coalition of more than 75 Canadian women's equality-seeking and related organizations. Its mandate is to further women's equality in Canada through domestic implementation of its international human rights commitments.

Femmes et Villes International
www.femmesetvilles.org
Femmes et Villes International (Women in Cities International) is a global network of city-based initiatives to promote the development of inclusive cities for women and girls. From its headquarters in Montreal, it acts as a knowledge and skills exchange network for those concerned with gender equality issues in cities around the globe.

Sisters in Spirit
www.nwac.ca/programs/sisters-spirit
Sisters in Spirit is a research, education, and policy initiative driven and led by Aboriginal women. Their primary goal is to conduct research on and raise awareness of the alarmingly high rates of violence against Aboriginal women and girls in Canada.

Status of Women Canada
www.swc-cfc.gc.ca/index-eng.html
Status of Women Canada is a federal government organization that promotes the full participation of women in the economic, social, and democratic life of Canada. It places particular emphasis on increasing women's economic security and eliminating violence against women.

The White Ribbon Campaign: Men Working to End Men's Violence against Women
www.whiteribbon.ca
The White Ribbon Campaign (WRC) is the largest effort in the world of men working to end violence against women. It started in 1991 in Canada and is now active in more than 55 countries, where campaigns, led by both men and women, focus on educating men and boys.

Women's Worlds Congress
www.womensworlds.ca
The Women's Worlds Congress has been held every three years since 1981 in different parts of the world. It is a worldwide interdisciplinary gathering that focuses on research and activism pertaining to women's issues. The 2011 congress was held in Ottawa and hosted by Carleton University and the University of Ottawa.

XY
www.xyonline.net
XY is a website focused on men, masculinities, and gender politics. It is a space for the exploration of men's and women's everyday lives, issues of gender and sexuality, and practical discussions of personal and social change.

notes

1. These publications include Amy Richards and Jennifer Baumgardner's *Manifesta: Young Women, Feminism, and the Future* (2000), Lynn Crosbie's *Turbo Chicks: Talking Young Feminists* (2001), and two special issues of *Canadian Woman Studies* on activism in the lives of young feminists (2001) and *Colonize This: Young Women of Colour on Today's Feminism* (2002).

2. R.W. Connell, an Australian sociologist who has made many significant contributions to the study of gender, was male in the 1980s and 1990s. She is now a woman.

3. The first documented Take Back the Night event occurred in Philadelphia in October 1975. Since that time, the slogan 'Take Back the Night' and the marches that have been organized under that name have become internationally renowned as a visible way to take a stand against sexual violence, especially violence against women. In 1981, the Canadian Association of Sexual Assault Centres declared the third Friday of September the evening for Take Back the Night marches nationwide. www.takebackthenight.org/history.html.

4. The Montreal Massacre is the name given to a tragic period of 45 minutes on 6 December 1989 when a 25-year-old gunman named Marc Lepine killed 14 women at Montreal's École Polytechnique. He separated the men from the women, screamed, 'I hate feminists,' and then began shooting female engineering students. The Montreal Massacre brought national and international attention to the tragedy and loss caused by violence against women. The Canadian government has proclaimed 6 December as the National Day of Remembrance and Action on Violence against Women; it is a day of commemorative events across Canada, many of them on university campuses.

5. 'Man, don't I feel like a womyn'. *The Globe and Mail* 11 September 2010, p. L3.

6. www.statcan.gc.ca/daily-quotidien/100714/dq100714a-eng.htm.

7. In 2001, only 2 per cent of Aboriginal youths (aged 20 to 24) had a university degree (Mendelson 2006).

8. For further information, see the Statistics Canada report by Katherine Marshall, 'Employment patterns of post-secondary students' in the September 2010 issue of *Perspectives on Labour and Income* (www.statcan.gc.ca/pub/75-001-x/2010109/article/11341-eng.htm).

9. More information about gender, youth, and workplace safety can be found on the website of the Canadian Centre for Occupational Health and Safety in the 'Young worker zone' (www.ccohs.ca/youngworkers/for_young_workers).

10. 'China, India listed among world's top plastic surgery markets'. www.2point6billion.com/news/2010/08/10/china-india-listed-among-worlds-top-plastic-surgery-markets-6680.html; www.dancewithshadows.com/extra/2010/08/28/china-sees-massive-rise-in-teenadolescent-plastic-surgery; www.guardian.co.uk/world/2010/jul/18/india-students-plastic-surgery.

11. Louisa Ajami. 2008. 'Nose job nation—In Lebanon, surgical bandages are something to flaunt'. *NOW* 2 May. www.nowlebanon.com/NewsArchiveDetails.aspx?ID=40594#ixzz0y6RszOKO.

12. See the Status of Women website (www.swc-cfc.gc.ca/pol/gba-acs) for lots of information about how government departments in Canada analyze policies and programs to see whether they might have adverse gender consequences.

chapter 10

Sexuality

Patrizia Albanese

In this chapter, you will:

▶ Learn what is meant by sexuality and the diversity it encompasses;

▶ Appreciate some of the cross-cultural and historical diversity in attitudes and practices surrounding sexuality;

▶ Read about ways in which sex and sexuality have been studied in the past and across disciplines;

▶ Compare various theoretical approaches to understanding sexuality;

▶ See what contributions sociology has made to the study of sexuality;

▶ Read about changes to our understanding of Canadians as sexual citizens;

▶ Identify contemporary issues and trends in sexual attitudes and practices.

Introduction

Canadians spend a considerable amount of time thinking about sex. A key-word Internet search of 'Sex—Canada' came up with more than 129 million results. The material ranged from Health Canada and CBC reports on changes to same-sex marriage legislation to 'Sex toys mild to wild' at PinkCherry.ca. In almost sharp contrast, sociology as a discipline has done considerably less thinking about sex—until relatively recently.

As a matter of fact, the *Oxford Dictionary of Sociology*, under the heading 'sociology of sex', explains that the study of sexuality 'was not a major concern in sociology until late in the twentieth century' (Scott and Marshall 2005, 595). There is some general agreement that only in the 1960s did sociology begin to develop a stance of its own in regard to studying sexuality. On the other hand, important sociological theorizing some decades earlier built the foundations for this development.

Typically, when sociologists and other social scientists write about 'sex', they are referring to biological facts associated with being born male or female (i.e., anatomical facts, hormonal facts, and so on). According to the World Health Organization (WHO), *sex* refers to the biological characteristics that define humans as female or male, but WHO notes that while these categories are not mutually exclusive, since there are individuals who possess elements of both, we nonetheless continue to differentiate humans as either male or female. In other words, evidence indicates that while we often think of male and female as a binary or dichotomy, meaning that only two categories exist and that someone is born *either* male *or* female, some infants are born with 'ambiguous' genitalia, or genitalia that are 'difficult to understand' (Murray 2009; Morland 2005, 335). Morland, writing on the **intersexed** body, explains that intersexed infants (previously labelled **hermaphrodites**—some 2 to 4 per cent of children; see Gough et al. 2008) are born, as medical historian Alice Domurat Dreger noted, 'with genitals that are pretty confusing to all adults in the room' (Morland 2005, 335). In other words, intersexed bodies have hormonal, anatomical, and genetic configurations that do not fit our traditional discourses of sexual difference. While they are biological exceptions, they nonetheless exist universally. The difference, however, lies in how societies treat these 'exceptions', making sex and sexuality political and social issues rather than mere biological facts or exceptions (see Murray 2009; Maharaj, Dhai, Wiersma, and Moodley 2005).

TIME to REFLECT

How should we 'treat' intersexed babies and bodies? Why do you think this?

Despite scientific evidence of the existence of diversity, we continue to dichotomize sex into male and female as measured by visible genital facts—the presence of a vagina or a penis. And because in our

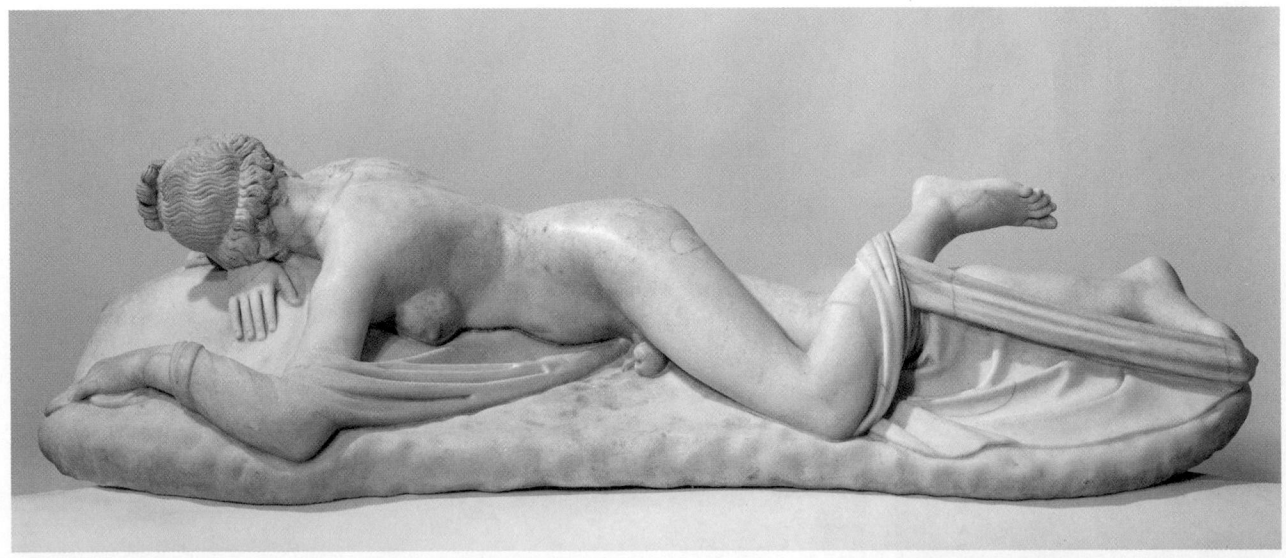

(Réunion des Musées Nationaux/Art Resource, NY)

Hermaphroditos, son of the Greek gods Hermes and Aphrodite. When he met the nymph Salmacis, she was so smitten by his appearance that she begged the gods that they might never be parted. The gods assented, merging the two into a single person, both male and female.

OPEN for DISCUSSION

Vancouver's Heteronormative Nightscape

Boyd (2010), examining Vancouver's entertainment district, argues that the 'mainstream' produces, maintains, and reiterates the moral contours of heterosexuality within the city. Boyd shows that nightclubs, as spaces of hypermasculinity and hyperfemininity, offer a prime example of how governmentality, surveillance, and private enterprise work together in the maintenance and regulation of social/sexual conformity. Boyd's ethnographic study of Vancouver's entertainment district—the Granville Strip—highlights young adults' perceptions of how hegemonic sexuality and nightlife collide.

Boyd explains that through heterosexual hegemony and heteronormativity, heterosexuality has been normalized and is understood as unproblematic and natural rather than revealed as constrained and produced by power relations. Like Adrienne Rich (1980/2003), Boyd argues that heterosexuality is a compulsory fiction and political institution, maintained and enforced through state practices. Like Judith Butler (2006 [1990]) and Foucault (1990), Boyd also theorizes gender as performative in that it re-enacts meaning systems through the reiteration of heterosexual norms, which construct heterosexuality as stable and true. All of this, Boyd (2010) argues, ultimately works to conceal the instability and production of sex and gender, enabling gender inequality to remain unchallenged.

Young people in Boyd's (2010, 183) study commented on the difference in gender dynamics between indie events in the East Side of Vancouver and mainstream events in 'glossy bars' on the Granville Strip. One young woman explained:

> Visually there's more hyper-females in mainstream clubs . . . more Barbie doll style. Not to say that they all look like Barbies. But they often have very long hair, they'll be wearing quite a lot of makeup, they'll be wearing very feminine clothes that show off their physical attributes like their boobs and their bum, their waist and hips and stuff. (Boyd 2010, 182)

Another young woman added:

> You go to other clubs [non-alternative places] and the girls are all dressed up, and they're doing their little body wiggle or whatever. You know there's definitely these very specific gender roles. You know, like the big guys, with the baseball cap who's loud with his beers yelling, being all loud and having possession over his woman who's always wearing tight revealing clothes or hairspray. (Boyd 2010, 182)

Boyd showed that (hyper-)(hetero-)sexualized performance in Granville clubs reinforced hegemonic femininity and masculinity and heterosexual hegemony. In contrast, study participants noted that the indie dance scene was more open to diverse sexualities compared to the Granville scene. One woman explained:

> Well, I think there's more transgendered sort of sexuality happening in the indie scene probably than there is in mainstream scenes. (Boyd 2010, 183)

A male interviewee said that compared to other scenes, the indie scene was fairly diverse in terms of sexuality:

> [Alternative dance spaces are] less homophobic. I'm shocked when I encounter homophobia. But I will encounter it usually at those shitty nightclubs I try to avoid.
> Such as?
> Like Stone Temple, anything on Granville, anything on the downtown. All those places are pretty homophobic. I get it at my bar where I work at The Dodson. I've noticed some of those guys are totally homophobic. (Boyd 2010, 183)

Do you agree with Boyd's analysis? How do your own experiences compare?

---

society people often have little tolerance for ambiguity, we typically reconstruct the genitalia, appearance, and personalities of these infants to fit into one of the two boxes. Once we think of individuals as fitting into one of the two dichotomous sex categories, we also come to see men and women as naturally polar opposites of one another—note the term *opposite sex*—who are assumed to be sexually 'drawn' to one another like magnets. As a result, in dichotomizing sex we have also tended to dichotomize sexuality, sexual identity, and sexual orientation into heterosexual ('normal' sexual attraction to the 'opposite' sex) or **homosexual** (less 'normal' attraction to someone of the same sex) when in fact sexuality is considerably more complex. Thus, many aspects of social life are constructed on the assumption that 'normal' people are heterosexual. This has come to be called **heteronormativity**.

Sexuality has been defined by the World Health Organization as a central aspect of being human, and encompasses sex, gender identities and roles, **sexual orientation**, eroticism, pleasure, intimacy, and reproduction. WHO explains that sexuality is experienced in thoughts, fantasies, desires, beliefs, attitudes, values, behaviours, roles, and relationships. It adds that sexuality is influenced by the interaction of biological, psychological, social, economic, political, cultural, ethical, legal, historical, religious, and spiritual factors.

Jeffrey Weeks (1993, 16) notes that the meanings we give to sexuality are 'socially organized, sustained by a variety of languages, which seek to tell us what sex is, what it ought to be—and what it could be'. These languages of sex are then 'embedded in moral treatises, laws, educational practices, psychological theories, medical definitions, social rituals, pornographic or romantic fictions, popular music and common sense assumptions (most of which disagree) [which] set the horizon of the possible'. Sexuality therefore has to do with who we are and what place we (are allowed to) take within society.

Once people are packaged into boxes—male or female, masculine or feminine, heterosexual or homosexual—**sexism** (the subordination of one sex, usually female, and the perceived superiority of the other) and **homophobia** (an irrational fear and/or hatred of homosexuals and homosexuality) help to reinforce rigid boundaries and keep people in their place. For example, a young person who challenges traditional gender ideology or practices is likely to be teased and taunted ('you fag'). Boys, perhaps more than girls, who cross the gender divide are often harassed back into stereotypically masculine behaviour. Children and youths learn quickly to avoid ridicule by conforming to prescribed gender and sexual norms. As a result, we come to see certain types of behaviour as normal, natural, and inevitable—the core or **essence** of femininity or masculinity, heterosexuality or homosexuality—when in fact we may have been forced to suppress parts of our identities that cross the gender and sexual divide. (We have come to treat as natural, inevitable, or **essential** things that are cultural, learned, and open to change.)

In the First Person

If you asked me in my first year of university what I wanted to do in life, I would probably have said 'teach . . . perhaps.' If you pressed me, you'd find that I had no idea what I wanted to do in life. To keep my options open, I decided to do a double major in history—my first love—and something else.

Psychology required that I take methods and statistics in upper years, so that was out of the question. Sociology seemed straightforward and based on common sense. They study families; I'm part of a family. They study ethnicity; I'm an immigrant. They study social class; I grew up in a working-class family. Sounded easy, and no stats.

Did I say 'no stats'? That was my first mistake. My second was the idea that sociology was based on common sense. I quickly learned that sociology challenges commonsense assumptions about how the world works. That's what got me hooked.

Despite my misinformed and rocky beginning, I find myself teaching sociology at Ryerson (and social research methods, to boot!).

—Patrizia Albanese

Sexuality over the Centuries

SEX AND SPIRITUALITY: EXAMPLES OF CROSS-CULTURAL DIVERSITY AND SOCIAL CHANGE

Sex and spirituality have been closely linked among some cultures and belief systems, and certainly all religions have had something to say about our sexual, procreative 'nature'. Theosophy, for example, is a philosophical system professing to achieve knowledge of the divine through spiritual ecstasy. Dozens of cultures throughout history—including ancient Sumerians, Egyptians, and Greeks and, over the past few centuries, some groups in Borneo, the Ibo of Nigeria, the Ewe-speaking peoples of southeast Ghana, Benin, and Togo, and some groups in pre-conquest South and Meso-America—have had sacred male and/or female prostitutes (Bishop 1996). Many North American Aboriginal societies had transgendered or two-spirited shamans or healers fulfilling religious duties because they were believed to possess spiritual qualities (Miranda 2010; Carocci 2009; Baird 2001).

In ancient China, sex was seen as a form of worship that led to immortality. Using some of the basic principles of Taoism, the *Su-nu Ching* (an ancient text), believed to have been written around the second or third century CE, noted that 'those who know the Tao of loving and harmonize the yin and yang are able to blend the five joys into a heavenly pleasure' (cited in Bishop 1996, 140). The Taoist sex guide advised a man to prolong intercourse as long as possible, to arouse a woman to a state of orgasm, allowing him to absorb more of her *yin* (her natural feminine essence) in order to enhance his masculine *yang* (Bishop 1996).

In India, some time between the third and fifth centuries CE, the *Kama Sutra* was written by a lifelong celibate and sage named Vatsyayana. The *Kama Sutra* has been identified as significant as a text because it gave as much importance to a woman's active sexual involvement and pleasure as it did to a man's. Women were to be seen and treated as sexual equals to men (Doniger 2003). Vatsyayana's classifications and instructive illustrations demonstrated the ritualistic and sacred nature of sexual acts. The *Kama*

Sutra also equated spirituality with spontaneity in sex (Doniger 2007; Morris 1997).

Tantra, a Sanskrit word meaning 'web' or 'weaving', is said to have both a Hindu and a Buddhist variant and involves a process of psychosomatic training in which the tantrika tries to attune his or her body and mind to increasingly higher levels of cosmic energy. The process involves activating the energy of the opposite sex in one's own body through ritualized intercourse. *Tantra*, while largely targeting men, is credited with preserving ancient forms of goddess worship and acknowledging, unlike orthodox thinking at the time, that women could attain transcendental bliss, and many of its practices were designed to break caste barriers and taboos (Bishop 1996).

Judaism, Christianity, and Islam, while seemingly not as open to the exploration of sexuality, have encouraged followers to use sexual acts to enrich their spiritual lives. Indeed, 'The Song of Solomon' in the Old Testament, dating as far back as the tenth century BCE, is one of the earliest writings describing the joy and significance of physical sexuality. The religions that developed in the Middle East particularly

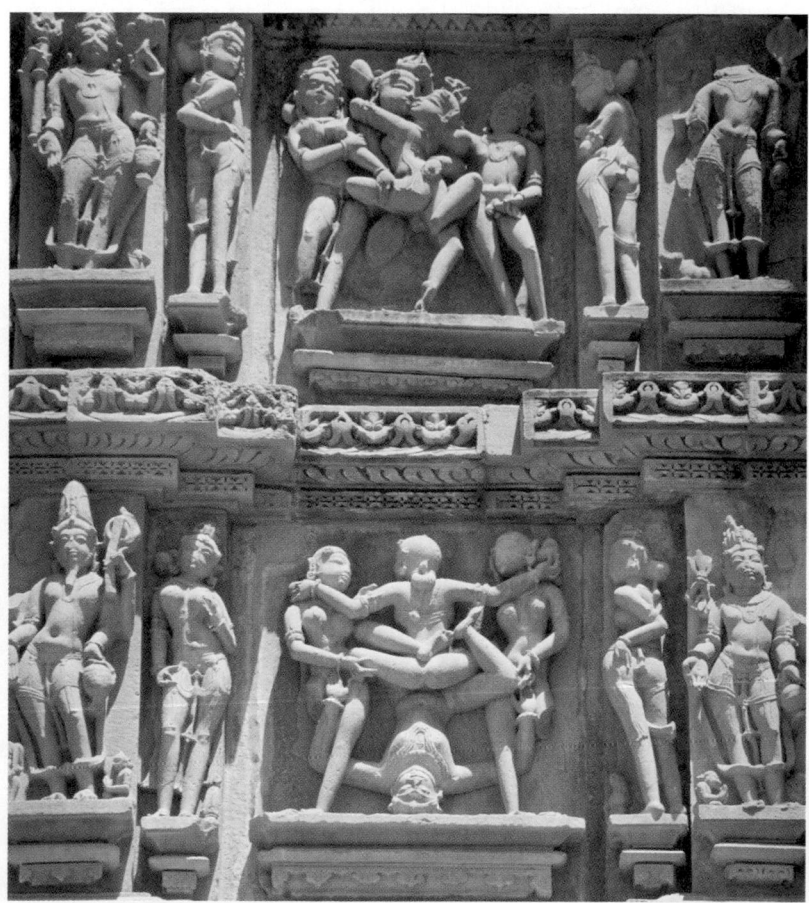

(Borromeo/Art Resource, NY)

Erotic reliefs cover the entire Kandariya Mahadeva Temple in India. The temple was created during the Chandella dynasty, between the years 1025 and 1150.

extolled the value of sex if linked to procreation in marriage. For example, the Talmud notes that sex in marriage and for the purpose of procreation should be enjoyed, and in some branches of Judaism, sex is considered a blessed duty to be carried out the evening before the Sabbath (Bishop 1996). In Islam, in the Qur'an, sex is encouraged, provided that it is preceded by an act of piety, and parts of the Qur'an describe the 'delights' of men who will be waited on in paradise by *houris*—bright-eyed maidens with swelling breasts (Bishop 1996). In Christianity, Jesus is presented in the New Testament as forgiving of sexual sinners as long as they sin no more. Yet the apostle Paul showed a personal distaste for sex—if people must marry, have sex, and procreate, that was fine, but it was better to remain apart from any physical relationship, because it could get in the way of the Christian's total commitment to God. Because of Paul's extensive and influential writings (the epistles of Paul comprise a sizable portion of the New Testament), this stance shaped the sexual moderation embraced by the early Christian church. For the most part, however, Judeo-Christian ideas have done much to forge a sexual morality based on self-denial and on what writer Clifford Bishop (1996, 78) claims is an 'exaggerated or unhealthy contempt for the flesh'.

Over the millennia, religious attitudes toward homosexuality have been quite mixed. Early Buddhist traditions appeared to celebrate Buddha's intimate and loving relationship with his disciple Ananda, as depicted in the *Jataka* tales. Some early Chinese and Japanese Buddhists showed considerable tolerance toward homosexuality among their monks and nuns (Baird 2001). On the other hand, some later variants of Indian Buddhism condemned and expelled homosexual monks and were especially intolerant of transgendered homosexual males, called *pandakas* (Baird 2001). More modern variants of Hinduism also appear hostile to homosexuality, as are some variants of Islam (Sufis are an exception), Judaism, and Christianity. For example, the prophet Muhammad is believed to have said 'no man should look at the private parts of another man and no woman should look at the private parts of another woman.' Similarly, the Old Testament states that 'You shall not lie with a male as with a woman; it is an abomination' (Leviticus 18:22). Christian theologian St Thomas Aquinas identified four categories of vice against nature under the rubric of lust: 'masturbation, bestiality, coitus in unnatural positions and copulation with an undue sex, male with male and female with female' (Baird 2001, 89–96).

The Scientific Study of Sex

BIOMEDICAL/REPRODUCTIVE APPROACH

For most of human history, the scientific processes connected to procreation remained a mystery. It was

(Scala/Art Resource, NY)

Artists like Hieronymus Bosch (c. 1450–1516) were inspired by sex and sexuality. *The Garden of Earthly Delights* is the centre panel of this triptych, on display in the Museo el Prado in Madrid, Spain.

in the late seventeenth and early eighteenth century that the Dutch scientist Anton van Leeuwenhoek observed that sperm 'swam' in human semen. And it was not until 1875 that Oscar Hertwig became the first scientist to observe the fertilization of an egg by sperm—in sea urchins. Much of the early scientific research on sex focused on the biomedical aspects of sex and procreation. Some scientific work shifted focus toward 'sexual deviance', which included any acts that did not have reproduction as a possibility or goal. The development of sexology involved a shift in focus from reproductive processes to the study of sexual practices. For some time, this branch of the field remained medical rather than social in orientation. Beginning in the latter half of the nineteenth century, as physicians sought to strengthen their hold on the medical profession and to extend their professional control over the human body and mind, 'sexual deviance' was seen as a mental illness, to be treated by medical interventions. Indeed, the American Psychological Association only removed homosexuality from its list of psychiatric disorders—in the *Diagnostic and Statistical Manual of Mental Disorders*—in 1973.

This approach to studying sex, which focused on sex for procreation in marriage as normal and all other sexual activity as deviant, was deeply ingrained in social thinking until fairly recently. A 1967 academic textbook titled *Human Sexuality: A Contemporary Marriage Manual* (McCary 1967) included chapters on such topics as 'Fertilization, prenatal development and parturition', 'Techniques in sexual arousal', and 'Positions in sexual intercourse'. The last chapter, 'Sexual aberrations', dealt with what was deemed to be deviant: sexual oralism, sexual analism (among a list of abnormal methods), homosexuality, zoophilia, necrophilia, masturbation (among a list of abnormal choices of sexual partners), and frigidity, promiscuity, and seduction (among a list of abnormal degrees of desire). The message of this scholarly text was clear and not very different from the earliest studies of sex: human sexuality is something that takes place not only between heterosexuals but only within marriage, for procreation, and with prescribed and approved methods and positions.

Throughout the 1800s and into the 1900s, a number of prominent thinkers contributed to a growing body of sex research. Sigmund Freud (1856–1939), for example, the founder of psychoanalysis, produced a comprehensive theory of human development with sex at the centre. According to Freud, the development of a healthy personality depended on the successful navigation through various stages of psychosocial and psychosexual development, each involving the careful management of various aspects of the sexual instinct. Many of these early theories of sexuality used the metaphor of repression, which comes from hydraulics and includes the image of a gushing energy that must be held back and controlled. Sexuality, historically, has been perceived as an innate 'force' that needs to be regulated and successfully manipulated or (re)directed toward acceptable channels.

Scientific and popular examples of this way of thinking about sex and sexuality continue to this day. For example, Jim Popp, head coach and general manager of the Montreal Alouettes of the Canadian Football League, asked his players to refrain from sex during the week leading up to the 2006 Grey Cup championship game (Canadian Press 2006b). Presumably, he wanted their energy and stamina channelled into hard-hitting play that would lead to a Grey Cup victory (they lost the game, by the way!).

Another pioneer in the scientific study of sexuality was Henry Havelock Ellis (1859–1939). His is considered a biological approach to the study of sex; however, unlike others at the time, he tried to demystify sex and challenge many sexual norms of Victorian England, famous for its sexually repressive norms and abundance of clandestine erotic literature (see Kearney 1982). For example, he assured his readers that masturbation did not lead to illness and homosexuality was not a disease. He argued that homosexuality was simply an innate variation from the norm, not a vice or an amoral choice.

SOCIAL SURVEY APPROACH

Like Ellis, the American biologist Alfred Kinsey (1894–1956) broke new ground in the scientific study of sex by challenging some of the accepted norms of his time. He was critical of biologists and psychologists who assumed that heterosexual responses are part of an animal's innate or instinctive equipment and was especially critical of their treatment of non-reproductive sexual activity as perversions of normal instincts (Kinsey et al. 1953).

In 1947, Kinsey founded the Institute for Research in Sex, Gender and Reproduction at Indiana University, now called the Kinsey Institute. Kinsey is famous for surveying approximately 18,000 Americans in the 1940s on their sexual practices. Through the survey, he found, among other things, that there were significant class differences among men in the incidence of masturbation, homosexuality, oral sex, sex with

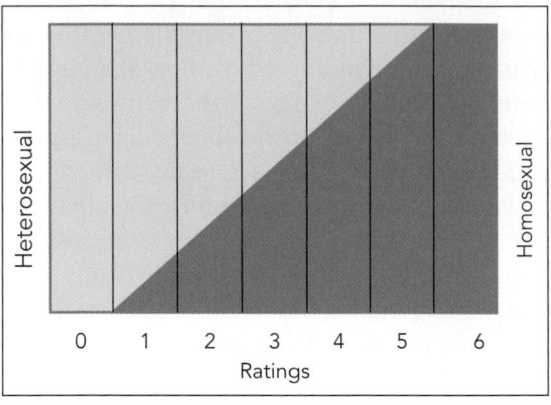

▼ FIGURE 10.1   Kinsey's Heterosexual–Homosexual Rating Scale

Kinsey's 'Heterosexual–Homosexual Rating Scale', a seven-point continuum, was originally published in Kinsey's *Sexual Behavior in the Human Male* (1948, 638). Zero refers to exclusively heterosexual with no homosexual experiences; 1 is predominantly heterosexual and only incidentally homosexual; 2 is predominantly heterosexual but more than incidentally homosexual; 3 is equally heterosexual and homosexual; 4 is predominantly homosexual but more than incidentally heterosexual; 5 is predominantly homosexual and only incidentally heterosexual; and 6 is exclusively homosexual.

SOURCE: Diagram from www.kinseyinstitute.org/about/photo-tour.html. Reprinted by permission of the Kinsey Institute for Research in Sex, Gender and Reproduction, Inc.

prostitutes, and premarital and extramarital sex. For women, their age and gender ideologies were significantly more important than social class in explaining variations in sexual preferences and practices. He is also famous for his 'Heterosexual–Homosexual Rating Scale', a seven-point continuum representing a considerably more complex approach to understanding sexuality and sexual orientation than was typical at the time.

He proposed that males do not represent two discrete populations—heterosexual and homosexual—and that the living world is a continuum in all its aspects (Kinsey et al. 1948). He emphasized a continuity of the gradations (a scale) between exclusively heterosexual and exclusively homosexual life histories. Note that he did not say 'exclusively heterosexual individuals' but rather 'histories'. This was intentional; Kinsey argued that an individual may be assigned a different position on the scale at different periods in life.

LABORATORY APPROACH

Other noted sex researchers include William Masters (1915–2001) and Virginia Johnson (1925–). In 1957, William Masters hired Virginia Johnson as his research assistant in studies of human sexuality. After years of working with Johnson, Masters divorced his wife and married her (Masters and Johnson divorced in 1993). They founded the Reproductive Biology Research Foundation in St Louis, Missouri, later renamed the Masters and Johnson Institute. In their early laboratory research, they recorded data, based on direct observation, on the anatomy and physiology of human sexual response (including the nature of female arousal and orgasm). They began by observing and documenting the stages of sexual arousal and orgasmic responses of 382 women and 312 men. Among other things, they observed and reported on the sexual responsiveness of older and elderly men and women, noting that many older men and women are perfectly capable of excitement and orgasm well

HUMAN DIVERSITY

Categories of Male Sexuality

Fernando Luiz Cardoso (2002) applied the scale in Figure 10.1 to his field study of 41 men living in a remote fishing village in southern Brazil. Cardoso found that many 'straight' men customarily had sex with local 'gay' men, called *paneleiros*. He found at least three different categories of male sexuality: men who have sex only with men, men who have sex only with women, and men who have sex with men and women. Cardoso noted that 'they believe, a "real" man is somebody "who has never been fucked but can fuck whoever is available"' (2002, 57). Masculinity is determined by sexual function and position, not by the sex of the partner.

into their 70s. They observed and measured masturbation and sexual intercourse in laboratory settings and wrote about sex as a healthy and natural activity, enjoyed for pleasure and intimacy. They developed a clinical approach—sex therapy—to the treatment of 'sexual dysfunction', including premature ejaculation, impotence, and female frigidity. Given the voyeuristic nature of this research, it is not surprising that their work was more trendy and 'popular' than of long-lasting importance.

ETHNOGRAPHIC/ANTHROPOLOGICAL APPROACH

Although anthropologists such as Margaret Mead (1901–78) would not be considered 'sexologists', they have nonetheless done extensive research documenting the sexual lives of people across diverse cultures. In doing so, they have contributed a body of research that challenges the view that sex and sexuality are biological (fixed, innate) facts. In *Coming of Age in Samoa* (1928), Mead shocked some of her American readers when she wrote about her observations of young Samoan women who deferred marriage while enjoying casual premarital sex before eventually marrying. She documented the impact of variations in culture (rather than biology) in the construction of sex roles and sexuality in *Sex and Temperament in Three Primitive Societies* (1935). Similarly, Clellan Ford and Frank Beach (1951), in an extensive survey of more than 200 societies, produced anthropological evidence of striking diversity in sexual practices and norms. The amount of variation across and within cultures is one way by which we know that sexual responses are learned and not innate.

Sociology of Sex: Theoretical and Methodological Approaches

Most sociologists today would argue that sexuality involves much more than an understanding of biological aspects of physical attraction. Sociologists, like anthropologists, frequently stress the social and cultural relativity of norms surrounding sexual behaviour and the socio-historical construction of sexual identities and roles. Within sociology, sexuality is typically studied and understood as being intricately connected to cultural, economic, political, legal, moral,

and ethical phenomena. Janice Irvine (2003, 431), for example, notes that from a sociological perspective, 'sexuality is a broad social domain involving multiple fields of power, diverse systems of knowledge, and sets of institutional and political discourses.' While sociology may have been comparatively slow to enter this field of study, it has gone a long way to address, explain, and understand some of these diverse issues and dimensions. But it does so from a number of different theoretical and methodological perspectives.

In tracing the history of sociological theorizing on sexuality, Irvine identified five broad themes in the sociological literature: (1) the denaturalization of sexuality (a shift away from biological explanations); (2) the historicization of sexuality; (3) the analytic shift from the study of 'sexual deviants' (the individuals) to the study of 'sexual deviance' (the rule-making strategies or social systems that define people as deviant or stigmatize them), thus challenging the pathologizing categories of sexuality and blurring the status of insider/outsider; (4) the destabilization of sexual categories and identities, with new emphasis on the fluid and diverse meanings of sex and sexuality; and (5) the theorizing of sexuality (and gender) as performance.

DENATURALIZATION

Structural Functionalist

Early sociologists, especially some structural functionalists, made liberal use of biological models and metaphors but did not wholeheartedly embrace simple biological explanations of social reality. When the American structural functionalist Kingsley Davis wrote about human sexuality, he looked at sexual intercourse as more than a biological exchange or a simple response to natural urges. He instead saw it as a social exchange, often involving 'the employment of sex for non-sexual ends within a competitive-authoritarian system' (Davis 1937, 746). In his classic article 'The sociology of prostitution', Davis asked the question: why is it that a practice so thoroughly disapproved, and widely outlawed in Western civilization, can flourish so universally?

Davis noted that if the family is strong, there tends to be a well-defined system of prostitution. He explained that the family is an institution of status that limits the variety, amount, and nature of a person's satisfaction. Through prostitution, a man is paying for the privilege of demanding what he wants. To a certain extent, then, prostitution serves to keep nuclear families together and 'strong'. When men cannot have their sexual needs met within marriage, prostitution functions to fill that role. Davis added that prostitution serves a number of other functions in economical ways: 'enabling a small number of women to take care of the needs of a large number of men, it is the most convenient sexual outlet for an army, and for the legions of strangers, perverts and physically repulsive in our midst' (1937, 755). He warned, however, that a decline of the family and a decline of prostitution are both associated with a rise of sexual freedom, and explained that 'unrestricted indulgence in sex for the fun of it by both sexes is the greatest enemy, not only of the family, but also of prostitution' (1937, 755).

TIME to REFLECT

What do you think of Davis's argument? Do you have any sense from his work of how a woman's/wife's sexual and marital satisfaction might be constructed?

More recently, using a structural functionalist perspective, Davidson and Hoffman (1986) conducted a survey of 212 married female graduate and undergraduate students at a midwestern commuter university in the United States to see what meaning and function sexual fantasies played in marital satisfaction. They found that, contrary to popular belief and some previous studies, engaging in sexual fantasies did not negatively affect married women's mental health, including level of guilt, sexual adjustment, and overall satisfaction with their current sex life. They found no significant differences between frequency of sexual fantasizing and marital satisfaction. Respondents who reported being satisfied with their current sex life fantasized about their current sex partner, and those who were dissatisfied were much more likely to fantasize about a more affectionate partner. Davidson and Hoffman concluded that their data strongly suggest that sexual fantasies function to help achieve sexual arousal irrespective of satisfaction or dissatisfaction with married women's current sex lives.

Conflict Approaches

We often think of Friedrich Engels in relation to his work with Karl Marx, writing about social class inequality. However, his famous work *The Origin of Family, Private Property and the State* (1990 [1884]) has a great deal to say about sexuality, private property, power, and subordination. Engels noted that in tribal societies with no concept of private property, promiscuous intercourse prevailed so that 'every woman belonged equally to every man and every man to every woman' (1990 [1884], 142). He added that among the Iroquois, a man considered his own and his brother's children his children, and they would all call him father. Paternity was no mere honorary title linked to procreation but rather carried serious mutual obligations, essential for the social constitution of these people (1990 [1884], 141). Engels explained that with the advent of private property, this changed. As the desire for the accumulation of wealth increased, men gained greater status in the family than women. This also created a stimulus for men to overthrow traditional communal forms of inheritance in favour of their own children. To do this, men would have to ensure paternity as a biological rather than social category and did so through the introduction of monogamy, the repression of women's sexual freedom, and the rise of the patriarchal family as a dominant family form (1990 [1884], 164–5).

Engels stated that the final outcome of 3000 years of monogamy was the bourgeois family in which men have exclusive domination over women, including their sexual autonomy. He predicted that monogamy and women's sexual oppression would disappear when the economic cause—private property ownership—disappeared. Engels explained that at that point, 'society takes care of all children equally' and 'the anxiety about the "consequences", which is today the most important social factor—both moral and economic—that hinders a girl from giving herself freely to the man she loves, disappears.' He then proclaimed: 'Will this not be cause enough for a gradual rise of more unrestrained sexual intercourse, and along with it, a laxer public opinion regarding virginal honour and female shame?' (1990 [1884], 183). Engels's views on 'individual love sex' or free love were adopted by a number of others, including left-wing feminists Alexandra Kollontai and Emma Goldman.

More recent conflict theorists have focused on analyzing the processes underlying class differences in sexual behaviour. For example, Higgins and Browne (2008) conducted sexual history interviews with 36 women and men, half middle-class and half working-class or poor. They found that most respondents reported that men have greater sexual appetites than women; however, middle-class respondents were more likely to cite social influences affecting sexual appetites, while working-class/poor respondents ascribed biological origins to sexual desire. They noted that the social construction of sexual controllability among the middle class contributed to perceptions that sex was a containable force. In contrast, poor and working-class women described men's sexual needs as physiologically irrepressible, which then shaped their experiences with and responses to sexual refusal.

Higgins and Browne's (2009, 716) work is not unlike Gonzales and Rolison's (2004, 715), which concluded that sexual behaviour and attitudes reflect patterns of dominance and inequality and these 'structures of sexual inequality are enshrined in taken-for-granted American moral dispositions.' In other words, differences in attitudes and experiences reflect an individual's position in a stratified society in which 'private' choice is conditioned by race, class, and gender dominance.

The Chicago School

In the first four decades of the twentieth century, the Chicago School was famous for its sociological studies of the city. Sociologists such as Robert Park did extensive ethnographic research on urban life. Using the city as a natural laboratory, Park examined, among other things, vice districts, urban environments, and other social structures that produced unique sexual worlds. While Park, like many of his time, believed that sexuality and sex drives were biologically based, he nonetheless noted that they were shaped and constrained by social forces.

Interactionist Sociology and Sexual Scripts

John Gagnon and William Simon, both of whom trained at the University of Chicago and later worked at the Kinsey Institute (1965–8), have been identified as 'fathers' of the sociological study of sex in North America. They openly challenge the biological determinism of most sexologists, arguing that if sex does play an important part in shaping human affairs, it is because societies have created its importance, not because of rigid biological grounding (Simon and Gagnon 2003). For them, sex is neither a dangerous instinct that needs curbing nor a passionate impulse that needs liberating. They further argue that neither sexual activities nor body parts are inherently sexual; rather, they become sexual when social meanings are attributed to them. While sexual activity most often takes place in private settings, they argue that 'the sexual encounter remains a profoundly social act in its enactment and even more so in its antecedents and consequences' (2003, 492). In other words, the

language and actions that make up sexual encounters, and their rules, restrictions, and **taboos**, are socially constructed and part of socially defined **sexual scripts** or road maps for sexual activity (Gagnon and Simon 1986). The script concept implies a complex construction of culturally defined socio-sexual roles. For example, while this may be changing (evidence that these roles are constructed), men are/have been expected to conduct themselves assertively and to make the first move, and women are/have been expected to be passive, compliant, and more responsive as the interaction progresses. Scripts also include internal dialogues about desire and resistance (Weis 1998).

HISTORICAL SOCIOLOGY AND SEX

A number of sociologists have sought to understand sex and love in historical perspective and, in doing so, have argued that sexual desire and intimate relations have not always been understood in the same way (Brickell 2006). Researchers like Jeffrey Weeks (1993) have mapped the historical origins of sexual categories, subcultures, belief systems, and language to understand how we have arrived at the social arrangements that prevail today. Particular emphasis has been placed on how sexual meanings have been negotiated within specific historical moments.

One historical moment of significance, and a focus of study in the West, is the sexual revolution of the 1960s. Leisa Meyer (2006) notes that the history of sexuality was a subfield of the new social history that emerged in the 1960s. This period was characterized by a cultural shift in attitudes, an increased ability to control reproduction (legal changes and the introduction of the birth control pill), challenges to conventional definitions of masculinity and femininity, and the investigation of groups that had been little studied. This took place in a climate of change marked by a new emphasis on questions of power raised by the civil rights, anti-war, feminist, and other movements for social justice. Both Canada and the United States were seemingly becoming more sexually liberal, and the courts reflected some of this through more progressive rulings—for example, in cases that overturned laws against interracial marriages in some American states and those that made contraception legal (Meyer 2006). Sexual freedom became symbolic of other types of freedom, and sexuality became central to many Westerners' understanding of themselves (Meyer 2006). This was especially true among youths. Legal changes that took place in Canada during this period are outlined below.

FROM SEXUAL DEVIANTS TO DEVIANCE TO MARGINS TO MAINSTREAM

From the Chicago School onward, researchers studying sexual minorities argued for the need to study dominant sexual institutions and definitions concurrently. After all, sexuality is governed and regulated by social norms, or shared expectations about what is considered culturally desirable and appropriate and, at the same time, by what is considered culturally deviant. But cultural norms change. Sharon Marcus, for example, notes that library shelves tell interesting stories. Thirty years ago, it took no time to get from feminism to homosexuality in the stacks in libraries, and homosexuality was 'sandwiched between bestiality and incest, on one side, and prostitution, sadism, fetishism, masturbation and emasculation, on the other' (Marcus 2005, 192). However, the work of deviance theorists, some of whom, like Michel Foucault, have been themselves outspoken members of sexual minorities, challenged assumptions about 'deviant' sexual categories and the individuals who inhabited them. Consequently, a large and growing body of literature on sexuality also challenges various systems of oppression. Some writers, for example, examined the role of stigma in the social control of sexuality (Plummer 1975). One of the most (in)famous sociological studies of sexual stigma, Laud Humphreys's *Tearoom Trade* (1970), controversially examined the dehumanizing role of stigma in the lives of men who sought sexual pleasure in public washrooms (see Chapter 7). While some continue to study sexual stigma and our understanding of sexual deviance, many have since come to problematize the very notion of sexual deviance. Plummer (2003), for example, critiques and contests the language of perversion.

DESTABILIZING SEXUAL CATEGORIES: FEMINISTS, QUEER THEORY, AND BEYOND

Many feminists have questioned and challenged, among other things, the social construction of sex and sexuality, the control of women's bodies and reproduction, the objectification of women, sexual double standards, the link between sex and power, and sexual abuse and oppression (Millett 1969; Greer 1984; Weitz 2002). Holly Benkert notes that 'the basis for oppression of women is deeply rooted to our sexuality, the very source of our primary "difference"' (Benkert 2002, 1197). Some, however, have pointed out that since the 1960s, feminists in North America have understood sexuality as both 'an arena for women's liberation' and 'a crucial vector of women's oppression' (Marcus 2005, 193). Some feminists have attempted to deconstruct and then reclaim women's rights to sexual pleasure, autonomy, and knowledge (Bell 1994; Eaves 2002; Miller-Young 2010), while others have challenged the forces that stood against women's autonomy, including pornography, rape, and sexual harassment (Brownmiller 1975; Dworkin 1981; Dworkin and MacKinnon 1988).

TIME to REFLECT

What role do you think pornography plays in our understanding of men's and women's sexuality?

The debates around pornography have been especially divisive (see Miller-Young 2010). Some, like Dworkin and MacKinnon, consider pornography demeaning and degrading to women and representative of male power over women. For others, the freedom to explore diverse representations of sexuality, including pornography or erotica, is seen as liberating to women and challenges restrictions placed on women's sexuality (Miller-Young 2010; Bell 1994; Sprinkle 1991). The African-American social theorist and feminist critic bell hooks explains that many feminists, in fact, stopped talking about sex publicly because it exposed 'our differences' (hooks 1994, 79); however, challenging patriarchal definitions and restrictions on women's sexual autonomy has been a unifying theme within feminism.

Writers such as Judith Butler (2006 [1990]) have argued that categories like 'heterosexual' and 'homosexual' are used to control and constrain individuals and therefore should be challenged on a number of fronts. Some, like Carr, promote the notion of a 'fluid conception of sexual identification' that is subject to 'the flux and flow of life' (1999, 17), allowing for the possibility of individuals to change from one identification to another. Queer theory calls for this type of challenge and change.

The use of the term *queer* within the gay community began as a ploy to reclaim a slur and highlight the multiple ways that sexual practices, sexual fantasy, and sexual identity 'fail to line up consistently' and 'expresses an important insight about the complexity of sexuality' (Marcus 2005, 196). Queer theory derives part of its philosophy from the ideas of Michel Foucault (1990), who saw homosexuality as a strategically situated marginal position from which it may be possible to see new and diverse ways of relating to oneself and others. Queer politics rejects forms of gender and sexual oppression, but it intentionally

SOCIOLOGY in ACTION

Sexual Scripts and Sexual Double Standards in Popular Magazines

A number of content analyses of women's and men's magazines have found that stories, advice columns, and advertising reinforce dominant gender and sexual norms (Carpenter 1998; Reichert and Lambiase 2003; Jackson 2005; McCleneghan 2003). For example, Ménard and Kleinplatz (2008) analyzed the content of the messages regarding how to achieve 'great sex' in popular men's and women's magazines. They found that the magazines overwhelmingly focused on technical, mechanical, and physical factors and variety as the prescribed means to achieve 'great sex'. Most important, they noted that advice on how to achieve 'great sex' tended to be framed in ways that promoted sexual and gender role stereotypes and enforced narrow sexual scripts. For example, men were depicted as sexually wild, aggressive, and animalistic, yet they were defenceless against particular sexual tricks. While women's magazines were full of sex tips, women were not supposed to show that they actually enjoyed sex. Sexual experimentation or 'kink' was to be undertaken strictly for the pleasure of the male partner, reinforcing the idea that women should not be interested in sex for its own sake. Men were believed to prefer 'quickies', and so these were promoted in women's magazines, while men's magazines suggested that women prefer long, drawn-out sex and framed their advice accordingly. Magazines sent out contradictory and conflicting messages, while sexual double standards and stereotypes loomed large.

does so from the margins in order to maintain a critical outsider perspective (Baird 2001). According to Plummer (2003, 520), queer theory is 'poststructuralism (and postmodernism) applied to sexualities and genders'.

TIME to REFLECT

Why do you think the word 'queer' was adopted by some in the gay community? Does this help us to rethink sexuality? How? Why?

Some recent Canadian theorizing on sexuality has been critical of past approaches. Green (2008), for example, argues that the sociology of sexuality continues to gloss over the role of psychodynamic processes and structures in favour of analyses of interactions and institutions. He is critical of scripting theory—grounded in a social learning framework—for not providing a proper conceptual resolution to this problem and, in fact, reproducing it. In response, Green (2008) argues that an effective sociological treatment of desire must incorporate a conception of the 'somatisation' of social relations found in Bourdieu's notion of 'embodiment' and his analysis of habitus. According to Green (2008, 599), for Bourdieu, social structures are not simply external to the individual but, rather, occupy a somatic/embodied relationship to the self—where social structures are 'deposited' as a set of 'embodied' inclinations, dispositions, schemes of actions, and appreciations captured in the concept 'habitus'. From this, Green (2008) developed the concept of 'erotic habitus', which is a socially constituted complex of dispositions and inclinations arising from objective historical conditions (revolving around classifications of race, class, and sex) that mediate the formation and selection of sexual scripts. According to Green (2008), erotic habitus generates sexual fantasies that are subjective but also embedded in social structures, giving sexual fantasies their collective and historical character.

Sexuality and Ethno-racial Diversity

Another critique from the margins comes from those who study the racialization of sexuality and the sexualization of minorities. Benkert (2002, 1205) notes that 'comprehensive conversations about race and sexuality together and separately have been taboo topics universally.' Writers such as bell hooks (1992) and Kamela Kempadoo (1998), for example, have explained that sexuality has been experienced and treated differently when it intersects with race. From this perspective, images of 'the exotic' are critically assessed, and the experiences of gays and lesbians of colour who experience racism within white-dominated gay organizations are also a focus of discussion (Baird 2001). These authors challenge false universalisms and stereotypes that do not reflect the complex reality and lived experiences of minorities (see Miller-Young 2010). Some have also been critical of the cultural imperialism of American media and the Westernization of sexuality (Nelson and Paek 2005; Hesse-Biber 2007). Still others have studied the explicit and implicit sexual dimensions of race, ethnicity, and nationality, arguing that the borderlands at the edges of racial, ethnic, and national boundaries are *ethnosexual frontiers,* patrolled, policed, protected, and at times penetrated by invaders (Nagel 2006; also see Yuval-Davis and Anthias 1989; Yuval-Davis 1997; Gamson and Moon 2004). Emerging from this, a number of feminists and queer theorists have raised radical challenges to mainstream approaches to citizenship (Langdridge 2006).

Sexual Trafficking: An Intersection of Inequalities

> 'Lena' was 26 when she was brought from Thailand to Japan with the promise of a job as an 'entertainer'. But when she arrived her passport was confiscated; her job, she quickly learned, was to have sex with men in seedy hotel rooms. Her good looks made her a popular choice among clients; she earned 1 million yen (US $9,100) in the first month alone. She needed money. She owed 6 million yen ($55,000) to her traffickers and minders in Japan—a debt she was expected to pay back in full under the constant threat of violence. She was allowed to keep just 10,000 yen ($90) a month pocket money and was confined to a cramped one-room flat. (McCurry 2004, 1393)

Lena's story is by no means unique. In 2003, *Lancet*, a UK-based medical journal that features stories on global medical issues, reported that an estimated 800,000 to 900,000 men, women, and children are illegally transported across international borders and forced to work as virtual slaves in the sex trade or in low-paying, dangerous jobs each year (*Lancet* 2003). Another source estimated that 1.2 million women and girls enter the global commercial sex market, within and across borders, every year (Roby 2005). Clearly, the trade in women and children has become a booming industry. And some feminist scholars have developed a political economy of the international sex trade that explicitly analyzes gender, class, ethno-racial, and national inequalities in comparative and global contexts (Limoncelli 2009).

Women and children are often trafficked from Central and South America, west and central Africa, South Asia, the Middle East, eastern Europe, China, and parts of Southeast Asia. Trafficked workers seem to have a few things in common—they are often fleeing poverty, joblessness, and/or the social dislocation characteristic of transitional economies, as seen after the collapse of the Soviet Union (Orlova 2004; Alalehto 2002; Kempadoo and Doezema 1998).

Roby (2005) notes that the factors contributing to the global sex trade include international, national, and local demographic, social, economic, ethnic, and cultural environments, with economic injustice and poverty among the major factors. Berman (2010) explains that the narratives surrounding human trafficking mask its racial and economic aspects by sensationalizing its sexual and criminal aspects, in turn allowing states to pursue political projects under the guise of a benevolent concern for trafficked women and/or the protection of its citizens. Along the same lines, Bertone (2000) suggests that the global sex trade and the trafficking of women are but another manifestation of the international inequality found in the North/South, East/West political–economic divide. She explains that the international subculture of docile and/or exotic women from underdeveloped nations is sustained by the international, patriarchal capitalist market system. While efforts to combat this type of exploitation have been made by supranational bodies like the United Nations (see Roby 2005) and some states (often in the name of national security; see Alalehto 2002), on a smaller scale some trafficked workers and sex workers are organizing to fight forms of exploitation and oppression, both within and outside the sex trade (see Limoncelli 2009; Petzer and Issacs 1998).

TIME to REFLECT

Despite legal changes, in the spring of 2002 the Durham Catholic school board unanimously voted to deny a gay Oshawa high school student, Marc Hall, the right to attend his high school prom with his same-sex partner. Do you support the school board decision? Why? Why not?

Sexual Citizenship and Sexuality in Canada

Former Prime Minister Pierre Trudeau, when he was justice minister in 1967, played a key role in helping to legally redefine sexuality for Canadians. For the first 100 years after Confederation, homosexuality was illegal in Canada and considered to be a mental illness. In December of 1967, Trudeau introduced a controversial omnibus bill in the House of Commons (Bill C-150) that challenged this and some other restrictions placed on sexuality. Trudeau's legislation brought issues like abortion, homosexuality, and the divorce law to the forefront, changing the sexual landscape of Canada. At the time, Trudeau famously stated that 'there's no place for the state in the bedrooms of the nation.' Despite considerable opposition, by 1969 homosexual acts had been decriminalized, and women had more control over sexual reproduction and their bodies. It was not until the year 2000, however, with the passing of Bill C-23, that same-sex couples were granted the same rights and obligations as common-law heterosexual couples. Then in 2005, 'equal marriage' legislation came into force with the passage of Bill C-315, the Civil Marriage Act, which recognized the right of same-sex couples to have access to civil marriage, without discrimination. Some, however, note that Canada has repressed and harassed gays and lesbians in the past as threats to society and to a certain extent still does (Kinsman and Gentile 2010).

Over the past 40 years, alterations have also been made to the laws governing the legal age of consent to have sex in Canada, with major changes in 1988 and 2008. In 1988, the passage of Bill C-15 created the offence of 'sexual interference' and prohibited adults from engaging in virtually any kind of sexual contact with either boys or girls under the age of 14,

regardless of consent. The legislation included the notion of 'sexual exploitation', making it an offence for an adult to have any sexual contact with boys or girls ages 14 to 18 if there was a relationship of trust or authority (e.g., teachers, coaches). The law added that consensual sex with those 12 to 14 'may not be an offence' if the accused was under 16 and less than two years older than the complainant. This was an attempt to reflect some of the trends in early sexual engagement among teenagers. More recently, the Tackling Violent Crime Act, introduced by Stephen Harper's Conservative government on 1 May 2008, raised the age of consent for sexual activity to 16 years.

Even before these changes, Curtis and Hunt (2007), like many others, have noted that official Canadian policy toward adolescent sexuality promotes the provision of information but tends to stress risk, danger, and harm, with no discussion of sexual pleasure. Curtis and Hunt (2007) explain that official strategies either try to bar young people from participation in the 'sexual arts' or lace their discourse with warnings about the risks of sex.

SEXUAL ACTIVITY AMONG YOUTHS

Findings from the National Longitudinal Survey of Children and Youth in Canada revealed that an estimated 12 per cent of boys and 13 per cent of girls have had sexual intercourse by the age of 15

The sex trade and poverty are worldwide phenomena that seem to go hand in hand.

(CP/Paul Chiasson)

Groups like the Coalition for the Rights of Sex Workers—shown here—have been organizing to fight forms of exploitation and oppression.

(Garriguet 2005; Statistics Canada 2005b). The study also found that characteristics associated with early sexual activity differed for boys and girls. Similarly, Gallupe, Boyce, and Fergus (2009), looking at partner influences and social expectations among a sample of Grade 9 and 11 students, found that all variables in their study differed significantly by gender (see Table 10.1). For example, using data drawn from the Canadian Youth, Sexual Health and HIV/AIDS Study, Gallupe, Boyce and Fergus (2009) found that 25.9 per cent of boys reported not using a condom the last time they had sex, compared to nearly 38.6 per cent of girls. Boys were more likely to report having had four or more partners, and girls were more likely than boys to report having had sex when they did not want to. They found that a higher percentage of boys indicated that they would have sex with a partner who did not want to use a condom and boys were less likely to believe that sex without love is not satisfying and more likely to endorse casual sex. Finally, girls were more likely to indicate that they would talk to a partner about using a condom before having sex. Clearly, while some attitudes about sex and sexuality have changed, gender differences persist.

Attitudes toward sexuality seem to have shifted considerably when we look across various generations of Canadians (see Table 10.2).

TIME to REFLECT

Does our society continue to have sexual double standards—that is, do we think differently about sexually active young women compared to sexually active young men? Is virginity still an 'asset' for young women but a liability for young men? Would you react differently to a young woman who bragged about having multiple sexual partners compared to a young man?

While attitudes seem to have liberalized, are young people today actually more sexually active than those in the recent past? Contrary to popular belief, the Sex Information and Education Council of Canada reports that for Grade 9 males, the percentage who reported sexual intercourse actually declined from 31 per cent in 1988 to 23 per cent in 2002 (McKay 2004). Figure 10.2 similarly shows that the numbers have declined for Grade 9 girls. The trend of declining rates of early intercourse is mirrored in US data as well (McKay 2004). There have also been decreases in the percentage of Grade 11 male and female students with multiple sexual partners (see Figure 10.3). Teenagers may well be more cautious, knowing that HIV/AIDS (and other sexually transmitted diseases) disproportionately affect

TABLE 10.1 ▼	Gender Difference in Response to Partner Influence and Social Expectation Questions among Students Who Had Ever Had Sexual Intercourse			
	Male (%)	Female (%)	Chi	df
Behaviour Measures				
Did you use a condom the last time you had sexual intercourse?				
Yes	74.1	61.4	38.2	1
No	25.9	38.6		
Number of intercourse partners				
1	45.3	53.3	31.8	3
2	17.3	18.8		
3	10.1	10.9		
4+	27.2	17.1		
Partner Influence Variables				
Have had sex when did not want to				
Yes	87.5	73.5	67.2	1
No	12.5	26.5		
Been pressured to have sex when did not want to:				
Yes	90.2	72.7	102.5	1
No	9.8	27.3		
Would have sex with partner who didn't want to use condom				
Disagree	35.4	56.6	121.3	2
Neither agree nor disagree	28.4	25.8		
Agree	36.2	17.6		
Social Expectations Variables				
Sex without love is not satisfying				
Disagree	21.9	35.6	99.8	2
Neither agree nor disagree	20.5	28.3		
Agree	57.6	36.1		
It's all right to have casual sex				
Disagree	5.2	23.0	277.3	2
Neither agree nor disagree	12.6	29.5		
Agree	82.2	47.6		
Would talk to partner about condom before sex				
Disagree	62.6	70.8	15.9	2
Neither agree nor disagree	22.1	16.9		
Agree	15.3	12.3		

SOURCE: SIECAN.

young people (McKay 2004). A 20-year study of sexual behaviour among students at Okanagan University College in British Columbia certainly supports this conclusion, reporting a steady increase in safer sexual practices among students (Netting and Burnett 2004).

Sex and the Workplace

A number of scholars of sex and gender have argued that many different workplace and organizational cultures play key roles in creating, maintaining, and undermining sexual identity and inequality at work (Hearn and Parkin 1987; Woods and Lucas 1993; Welsh 1999; Dellinger 2002). Dellinger suggests that instead of simply looking at sexuality as something individuals bring to work, we can examine, and some have examined, how customs and practices

in a workplace constitute a type of **organizational sexuality** or social practice that determines explicit and culturally elaborated rules of behaviour to regulate sexual identities and personal relationships. In other words, different occupational cultures hold different and specific social rules about what constitutes 'appropriate' or acceptable sexuality. Workplace norms about sexuality regulate who we say we are, who we 'date', how we dress, and how we understand and experience sexual harassment in the workplace.

Woods and Lucas (1993) write about the 'corporate closet', which strongly encourages gay men in some professions to keep their sexual identities and relationships hidden. In such work contexts, gays and lesbians intentionally pass as heterosexuals because of the pervasiveness of heteronormative discourses and **heterosexism** in the workplace (Johnson 2002).

TABLE 10.2 ▼ Acceptance of Sexual Relations and Approval of Legal Abortion by Age Cohorts, 1975 and 2000*	1975	2000
Premarital Relations		
Nationally	68%	84%
18–34	90	93
35–54	65	89
55+	42	74
Homosexual Relations		
Nationally	28	73
18–34	42	75
35–54	25	64
55+	12	42
Legal Abortion a Possibility . . .		
'If her own health is seriously endangered'		
Nationally	94	94
18–34	97	98
35–54	92	95
55+	93	90
'If she is married and doesn't want more children'		
Nationally	45	52
18–34	47	51
35–54	43	44
55+	45	48

*Sexual relations: % indicating 'not wrong at all' or 'sometimes wrong' versus 'always wrong' or 'almost always wrong'; legal abortion: % indicating 'yes, it should be possible'.

SOURCE: Bibby 2004. www.vifamily.ca/library/future/section_2.pdf. Reprinted by permission of the Vanier Institute of the Family.

Ironically, heterosexist norms in the workplace may at times make sexual interaction between co-workers of the same sexual orientation somewhat less problematic than sexual interaction between heterosexual co-workers. Some workplaces actively discourage the sexual involvement of heterosexual colleagues even when there is sexual consent between the individuals involved. The issue is, of course, very complex. Williams and colleagues (1999, 76) explain that 'workers themselves often conceive of sexual behaviors at work along a continuum, ranging from pleasurable, to tolerable, to harassing.'

Sandy Welsh (1999), writing on sexual harassment, explains that some organizations actually mandate the sexualization of their workers, and as a result, in some sexually charged work cultures, degrading and/or sexual behaviours become an institutionalized component of work. Thus, for example, a waitress at Hooters is required to wear short shorts and a top that shows cleavage, but this is not considered sexual harassment. Of course, for some women who work at Hooters bar-restaurants, such a requirement may be demeaning and objectifying; for others, however, this might not be the case. That said, research by Lynn (2009) on waitresses found that waitresses' tips varied with age in a negative, quadratic relationship, increased with breast size, increased with having blond hair, and decreased with body size.

Researchers such as Dellinger (2002) have noted that most workplaces, either formally or informally, convey rules of dress and that dress is a well-recognized site of gender construction and sexual identity. Dellinger finds that dress norms and local workplace norms 'influence people's definition of pleasurable, acceptable, and unacceptable sexuality at work' (2002, 23). As a result, workplace norms and organizational culture affect how sexuality is negotiated at work and, in part, determine what counts as sexual harassment.

Welsh (1999) reminds us that heterosexual norms in the workplace often exclude or sexualize women,

▼ FIGURE 10.2 Percentage of Canadian Grade 9 and 11 Students Who Have Had Intercourse, 1988 and 2002

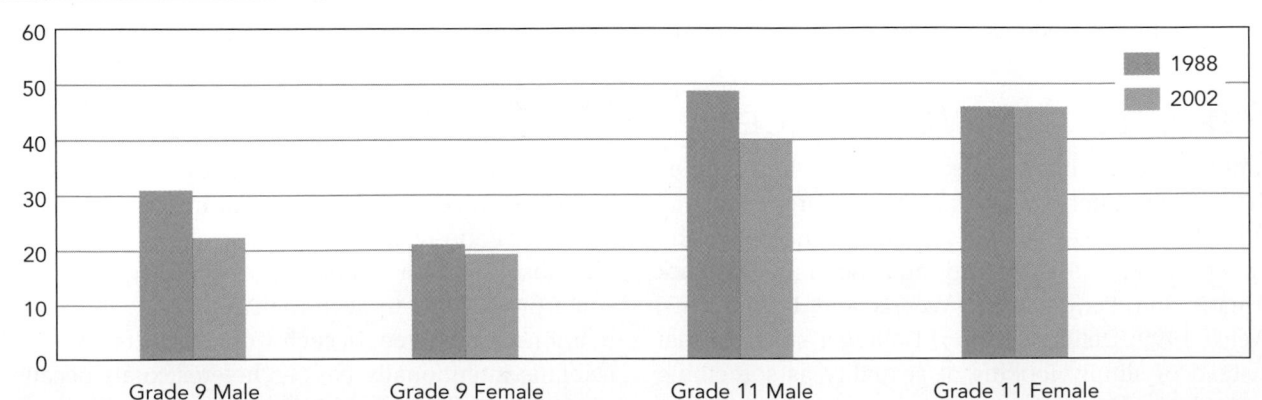

SOURCE: Boyce et al. (cited in McKay 2004, 74, www.sieccan.org/pdf/mckay.pdf).

▼ **FIGURE 10.3**   Number of Sexual Partners among Grade 11 Students Who Have Had Intercourse, 1988 and 2002

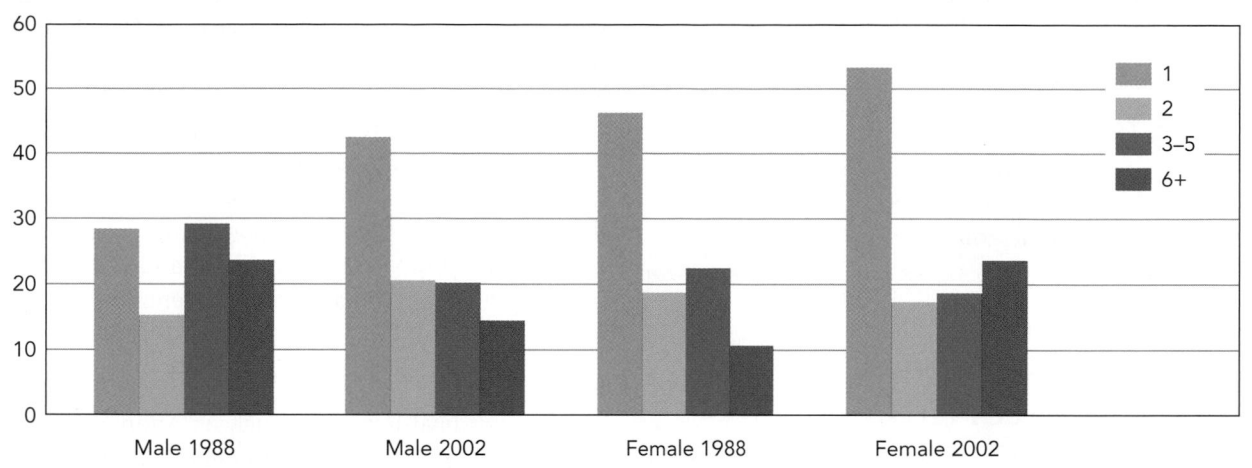

SOURCE: McKay 2004, 74. www.sieccan.org/pdf/mckay.pdf.

GLOBAL ISSUES

HIV/AIDS: A Global Crisis

In 2001, approximately 5 million people were newly infected with HIV, according to UNAIDS and the World Health Organization. UNAIDS (2009) reported that in 2008, 33.4 million (between 31.1 million and 35.8 million) people were living with HIV around the world. It was estimated that of these, 31.3 million were adults and 2.1 million were children under the age of 15. UNAIDS estimated that in 2008, a total of 2.7 million people had been newly infected and that AIDS-related deaths reached 2.0 million that year; children constituted about 280,000 of these deaths (UNAIDS 2009).

Approximately half of the people who become infected do so before they reach 25. Globally, this means that more than one out of 100 adults aged 15 to 49 is HIV-positive. At least 80 per cent of these infections result from heterosexual transmission (Health Canada 2000, in Bulman et al. 2004, 142). In fact, according to the Canadian International Development Agency, 3000 women a day contract HIV, and in Africa there are six HIV-positive women for every five HIV-positive men (CIDA 2000, in Bulman et al. 2004, 141).

On the twentieth anniversary of the release of the first published reports on what later came to be called AIDS, Eleanor Maticka-Tyndale (2001) published a paper highlighting the importance of sociological research to the understanding of AIDS as a global phenomenon. Maticka-Tyndale notes that for sociologists, it is not so much the number of people infected but the distribution of infection that sets the sociological imagination in motion. 'HIV is not an equal-opportunity virus when it comes to gender and age. Nearly 90 percent of new infections are in those in their most productive stage in life, between 15 and 49 years of age' (2001, 17).

She explains that 'psychological perspectives on how attitudes influence behaviours and models of individual learning and behaviour change are the foundations of prevention programming' (2001, 19), yet despite the success of these programs, they fall short when addressing factors that affect large numbers (communities and nations), the relationship between communities, relationships between communities and other organizations, such as corporations, and relationships between nations. That's where sociology comes in. Sociology can help in the understanding 'of vulnerabilities, community systems and structures, social systems, power differentials, cultural contexts, social change, social movements, mobilization' (2001, 19).

Maticka-Tyndale suggests that sociologists from various theoretical backgrounds could contribute to understanding AIDS by drawing on theories of 'collective behaviour and social movements; of social capital's impact on community responses to crisis; of health and well-being in the face of disease and of meaning-making in the face of apparent disaster; of stigma and responses to stigmatization; of power differentials and inequality, their consequences and reversals; of globalization and its impacts; of alternative forms of development; of community and social change' (2001, 19).

silence or closet gay men and lesbians, and work to constrain the behaviour of heterosexual men who are at times labelled 'unmasculine' when they choose not to participate in 'hypermasculine' stereotypical behaviour.

TIME to REFLECT

If an employer required you to dress in a sexually provocative manner, would you consider this a form of sexual harassment? Why? Why not?

Cybersex, Pornography, and the Internet

With the spectacular growth in the availability of sexually explicit material on the Internet (Barak and Fisher 2001), some have claimed that it has revolutionized sex yet again. No doubt, people today have easy (free and for-fee) access to many diverse sites and formats, with varying levels of 'interactivity', featuring very different and at times unusual types of sexual acts and preferences (some legal, others not). Cybersex and Internet porn can be had without leaving the safety and privacy of our homes or offices, all under a perceived cloak of anonymity. Knowing this, some researchers have written about compulsive viewing (Cooper et al. 2000). Others argue that this anonymity expands most people's expression of their sexuality (Barak and Fisher 2001; Fisher and Barak 2001). The question remains, does our access to technology change the nature and our understanding of sexuality?

Sociologists Lisa Byers, Ken Menzies, and William O'Grady (2004) surveyed approximately 500 students at a Canadian university and found, among other things, that male students engaged in significantly more viewing of sexually explicit material on the Internet than female students. They also found that people who viewed more non-Internet pornographic material also viewed substantially more on the Internet, suggesting that our society's existing patterns of sexuality, not Internet technology, will be the dominant determinant of future sexual patterns. Clearly, gendered patterns persist.

Terrie Schauer did a comparative textual analysis of nine pornographic Internet sites, three of which labelled themselves 'for straight women'. She found that the nude solo male in 'soft-core' porn directed toward women generally conforms to the erotic conventions used in gay male porn, which in turn mirrors conventions used to sexualize and objectify women in heterosexual soft-core porn directed toward men. The male models in porn for women also co-opt symbols from traditionally 'manly' or masculine trades—firemen, cowboys, and police (Schauer 2005). With the more 'hard-core' couples photos in porn sites aimed at women, Schauer surmises that many images and scenes were taken directly from heterosexual male sites. In other words, if the Internet is indeed revolutionary, it has not yet been so for heterosexual women, where the visual coding of women's pornography has not yet broken away from traditional forms and developed a unique set of representations or conventions.

Sexual Offences

For some radical feminists, such as Andrea Dworkin and Catherine MacKinnon, pornography on- or offline 'reveals that male pleasure is inextricably tied to victimization, hurting, exploiting' (Dworkin 1981, 69). Others have not gone that far but have pointed out that pornography often contains sexual violence (Barron and Kimmel 2000; Palys 1986) and that this may have a negative effect on men's attitudes toward women. But whether pornography actually causes its consumers to engage in sexual violence is still debated. Yet sexual assault and abuse remain a problem in Canada, especially for younger victims.

In 1983, Bill C-127 introduced a three-tiered structure of sexual assault offences designed to improve legal processing of rape cases (see DuMont 2003). Today, under Canadian criminal law, a broad array of activities qualify as sexual assault, ranging from unwanted touching to sexual violence resulting in serious injury, with penetration not being an essential component (Johnson 2005). Noting that the number of sexual assaults reported to police is likely a considerable undercount of the actual number that occur, a recent Statistics Canada report stated that in 2009, there were almost 21,000 sexual assaults, 98 per cent of which were classified as level 1, the least serious form of sexual assault (Dauvergne and Turner 2010). Since peaking in 1993, the rate of police-reported sexual assault has been steadily declining (by 36 per cent between 1993 and 2002; Kong et al. 2003), including a 4 per cent decrease in 2009 (Dauvergne and Turner 2010).

In contrast, sexual offences against children and youths were alarmingly high. Police reports and victimization surveys reveal that young women and girls are at the highest risk of sexual assault victimization, and rates of sexual offending were highest among male teenagers (Kong et al. 2003). Consequently, a majority of the sexual assaults today are committed

UNDER the WIRE

Internet Porn Harms Men's Self-Esteem?

Internet sexuality and participatory digital media, or DIY pornography, is a popular, multi-billion-dollar industry, with many transnational social networks run by corporate-driven entertainment companies that allow people to buy memberships and upload sexually explicit photos and videos. Jacobs (2010) notes that Friendfinder Inc., for example, was founded in 1996 by a Silicon Valley company that pioneered a number of sex and dating sites. In December 2007, Penthouse bought the site for US $500 million, making it the world's largest corporate network for adult entertainment, with a combined membership of more than 40 million (Jacobs 2010). The network caters to a wide range of cultures and communities based on demographics such as age (seniorfinder .com), religion (BigChurch.com, JewishFriendfinder .com), and ethnicity or nationality (AsiaFriendfinder .com, IndianFriendfinder.com, Amigos.com, GermanFriendfinder.com, FrenchFriendfinder.com, KoreanFriendfinder.com, and Filipino Friendfinder. com) (Jacobs 2010).

Morrison and colleagues note that much of the research on sex on the Internet has taken a harm-based approach, focusing on how it negatively affects male viewers' attitudes toward women. Their study, in contrast, shifts the focus away from the impact of Internet porn on male attitudes toward women to its impact on male attitudes toward themselves. They surveyed close to 200 men enrolled at a Canadian college and found significant negative correlations between exposure to pornographic imagery online and their levels of genital and sexual self-esteem (Morrison et al. 2006). They found that those who reported greater exposure to Internet porn were more likely to have lower self-esteem. To be more precise, findings from Morrison et al.'s study suggest that male participants' level of exposure to sexually explicit material (SEM), especially on the Internet, correlates inversely with genital esteem. Put simply, exposure to pornography on the Internet was inversely correlated with satisfaction with one's penis.

against children and youths (AuCoin 2005). In Canada in 2002, six of every 10 victims of sexual offences reported to police were children and youths under 18 (Statistics Canada 2005a). Girls made up the majority of victims (85 per cent), with the highest rates among those aged 11 to 19 (with the peak at 13 years of age). For male victims, rates were highest for boys aged 3 to 14 (Kong et al. 2003). Boys aged 13 to 14 were at highest risk of committing level 1 sexual offences (level 1: minor or no physical injuries to the victim; level 2: sexual assault with a weapon; level 3: wounding or endangering the life of victim).

Younger victims were more likely to be sexually assaulted by a family member. In fact, only 4 per cent of female victims under the age of six and about 10 per cent of victims aged six to 13 were sexually assaulted by a stranger (Statistics Canada 2005a). Those aged 12 to 17 were most often sexually victimized by peers and acquaintances (Kong et al. 2003). In sum, 86 per cent of the cases of sexual assaults were perpetrated by an individual known to the victim (AuCoin 2005). At the same time, it should be noted that more attention is being paid to the sexual victimization of children and youths by strangers via the Internet. Future research is likely to focus on this relatively new and growing form of sexual exploitation.

Conclusion

This chapter began by noting that sociology has been a relative latecomer to the study of sex and sexuality, but over the course of the chapter we have seen that the discipline has come a long way in (re)defining, theorizing, and researching sex and sexuality.

Most sociologists today, unlike many early sexologists, would argue that sexuality involves much more than an understanding of biological aspects of (heterosexual) physical attraction (in marriage and for procreation) and would, like many anthropologists, stress the social and cultural relativity of norms surrounding sexual behaviour, identities, and roles. As a result, many sociologists would accept a broad definition of sexuality that stresses the importance of understanding sexuality as intricately linked to and influenced by the interaction of biological, psychological, social, economic, political, cultural, ethical, legal, historical, religious, and spiritual factors. Sexuality is therefore understood and studied as a 'broad social domain involving multiple fields of power, diverse systems of knowledge, and sets of institutional and political discourses' (Irvine 2003, 431). Sociological theorizing and research on sexuality have taken a multiplicity of forms, including the study of sexual scripts, sexual stigma, sexual oppression, and the critical destabilization of sexual categories by feminists and queer theorists.

Furthermore, the amount of historical and cross-cultural diversity and change found in sexual attitudes and practices, both within Canada and around the world, should be further evidence of sexuality's socio-cultural dimensions. In other words, nature, to be sure, plays some role in people's experience of sexuality, but nurture or social conditioning is of greater importance than earlier theorists and sexologists suggested. Your own experiences are also evidence of this. For example, if you found yourself reading this chapter and saying 'it's not like that anymore', you too have provided evidence in support of socio-cultural explanations over fixed, biological ones.

Finally, the issues of sex and the media, especially the Internet, and of sexual violence and sexual offences are likely to be areas of debate and research in years to come.

questions for critical thought

1. How has the common, binary (dichotomized) definition of 'sex' affected our understanding of sexuality?

2. According to Plummer (2003, 516), 'There is no essential "sexuality" with a strictly biological base that is cut off from the social.' What does this mean? Can you think of ways it is manifested in your life?

3. There is considerable historical and cross-cultural diversity in regard to sexuality and spirituality. In your opinion, why have sex and sexuality been linked to spirituality? Why do some groups try to keep the two far apart?

4. There are clearly very different approaches to the study of sex and sexuality. How do biological and psychoanalytic explanations compare to sociological approaches? What are the key factors that set them apart?

5. For decades feminists have had deeply divisive debates about the roles of pornographic media in promoting heterosexist institutions and relations. What role do you think pornography plays in our understanding of men's and women's sexuality? Has the Internet changed this?

6. Studies show that adolescent men and women differ significantly in their attitudes and experiences when it comes to sex. Why do gender differences persist? Do you expect these differences will disappear? Why? Why not?

7. Welsh (1999) noted that heterosexual norms in the workplace often exclude or sexualize women, silence or closet gay man and lesbians, and work to constrain the behaviour of heterosexual men. How are these norms manifested? What should be done, if anything, to address these issues?

8. Has the Internet 'revolutionized' sex? What evidence do we have that it has? What evidence is there that it has not?

recommended readings

Mary Louise Adams. 1999. *The Trouble with Normal: Postwar Youth and the Making of Heterosexuality*. Toronto: University of Toronto Press.
Adams writes about the social construction of heterosexuality and the discourse surrounding the notions of 'normal' and 'heterosexuality' as they were imposed on youth in postwar Canada.

Judith Butler. 2006 [1990]. *Gender Troubles: Feminism and the Subversion of Identity*. New York: Routledge Classics.
Butler's groundbreaking work challenges some traditional feminists' assumptions about the 'naturalness' and essentialism of sex and gender. She argues that the masculine and feminine are not biologically fixed categories but rather are culturally determined.

Michel Foucault. 1990. *The History of Sexuality: An Introduction*. New York: Vintage Books.
This is the first volume of Foucault's three-volume study of sexual history, presenting a detailed, critical, and provocative account of the changing attitudes and discourses surrounding sexuality and sexual repression. His purpose is 'to show how deployments of power are directly connected to the body' (p. 151).

Adam Green. 2008. 'Erotic habitus: Toward a sociology of desire'. *Theory and Society* 37 (6): 597–626.
This article provides an overview and critique of past theorizing in the sociology of sexuality. In it, Green develops his concepts: erotic habitus and erotic work, based on Bourdieu's notion of 'embodiment' and habitus.

Gary Kinsman and Patrizia Gentile. 2010. *The Canadian War on Queers*. Vancouver: University of British Columbia Press.
This book draws on official security documents obtained through Access to Information requests and on interviews with gays and lesbians, civil servants, and high-ranking officials to reveal that for nearly 50 years, agents of the Canadian state spied on, interrogated, and harassed gays and lesbians as threats to society.

Mireille Miller-Young. 2010. 'Putting hypersexuality to work: Black women and illicit eroticism in pornography'. *Sexualities* 13 (2): 219–35.
This article focuses on the labour marginalization of black female performers in the pornography industry. It shows that their representations and experiences are shaped by racialized and gendered sexual commerce, stereotypes, structural inequalities, and social biases. Miller-Young shows that black sex workers, while facing multiple axes of discrimination, also employ hypersexuality and illicit eroticism to achieve mobility and erotic autonomy.

Jeffrey Weeks. 1991. *Against Nature: Essays on History, Sexuality and Identity*. London: Rivers Oram Press.
Weeks presents a historical overview of the study and regulation of sexuality in general and homosexuality in particular. He also explores the personal and cultural impact of the AIDS crisis and the politics and values of the post-modern Western world.

recommended websites

Egale Canada
www.egale.ca
Egale Canada is a national organization that advances equality and justice for lesbian, gay, bisexual, and trans-identified people and their families across Canada. The site includes summaries of key court cases, press releases, and information on local, national, and international campaigns and events.

Joint United Nations Program on HIV/AIDS (UNAIDS)
www.unaids.org/en
UNAIDS brings together the efforts and resources of 10 UN organizations to the global AIDS response. The website includes international data on HIV and an extensive range of publications and materials (e.g., research reports, best practices) on a variety of topics related to HIV/AIDS.

Kinsey Institute for Research in Sex, Gender and Reproduction
www.indiana.edu/~kinsey
The Kinsey Institute at Indiana University promotes interdisciplinary research and scholarship in the fields of human sexuality, gender, and reproduction. The website includes an abundance of resources and information on, for example, research and publications, interdisciplinary conferences and seminars, and opportunities for graduate students.

Pivot Legal Society
www.pivotlegal.org/Issues/sextrade.htm
Pivot Legal Society, a non-profit legal advocacy organization located in Vancouver's Downtown Eastside, takes a strategic approach to social change, using the law to address the root causes that undermine the quality of life of those most on the margins. Pivot's main campaigns are directed toward addiction, housing, policing, and sex work; the group helped to co-ordinate sex workers' participation in the parliamentary review of prostitution laws in 2005.

Sex Information and Education Council of Canada (SIECCAN)
www.sieccan.org
SIECCAN is dedicated to informing the public and professionals about diverse aspects of human sexuality. It also publishes the *Canadian Journal of Human Sexuality*.

Sex Workers Education and Advocacy Task Force (SWEAT)
www.sweat.org.za
SWEAT, a not-for-profit organization based in Cape Town, South Africa, works with sex workers around health and human rights issues. The organization provides safer-sex educational outreach as well as legal advice and skills training, and its site contains fact sheets, resources for sex workers, and recent research and publications related to sex work.

Society for the Scientific Study of Sexuality (SSSS)
www.sexscience.org
This international organization dedicated to the advancement of knowledge about sexuality consists of an interdisciplinary group of professionals who believe in the importance of both the production of quality research and the clinical, educational, and social applications of research related to all aspects of sexuality.

World Health Organization (WHO)—Sexual Health
www.who.int/topics/sexual_health/en
The World Health Organization, a UN agency, provides information on a wide range of health topics, including sexual health and sexual violence. Included here are fact sheets, reports, and publications on a wide range of topics, such as adolescent sexual and reproductive health, sexually transmitted diseases, and female genital mutilation.

chapter 11

Ethnic and Race Relations

Nikolaos I. Liodakis

In this chapter, you will:

▶ Learn that the meaning of the terms 'ethnicity' and 'race' are historically specific and are important bases for the formation of social groups;

▶ Discover that ethnic and racial hierarchies exist in society;

▶ Understand how Canada has been shaped by the colonization of Aboriginal peoples, the requirements of 'nation-building', capitalist economic development, and discriminatory immigration policies;

▶ Find out that multiculturalism and interculturalism are ideological frameworks within which government policies and programs attempt to manage ethnic and race relations and provide social cohesion;

▶ Come to appreciate how, despite improvements in immigration policy and government integration efforts, discrimination and racism continue to permeate many aspects of Canadian social, political, and economic life;

▶ Learn about different theoretical approaches that attempt to explain the economic inequalities among and within ethnic and racial groups.

Introduction

The examination of ethnic and race relations is crucial in our understanding of Canadian society. Canada is, demographically, one of the most multicultural countries in the world. With the exception of Aboriginal peoples, everyone else is either an immigrant to this country or the descendant of one. As sociologists, we are interested in analyzing social relations—that is, relations of power (domination and subordination) among individuals and social groups. We cannot understand current economic, political, and social relations or conflicts in Canada without a comprehensive understanding of ethnic and race relations. For example, we cannot understand the current struggles of Aboriginal peoples with various levels of Canadian government over their land and self-determination without examining the legacy of colonization and the long-lasting effects it has had on their cultural, economic, and social lives. We cannot begin to talk about the formation of the Canadian state without reference to historical British–French conflicts. Modern Canada is constituted as a political entity, in large part, as a result of the struggles and the uneasy union between the charter groups—the French and the English—at the expense of Aboriginal peoples. Our existing demographic makeup is a product of our history; it is a reflection of Canada's immigration policies and practices even before Confederation, many of which were blatantly discriminatory and racist at least until the mid-1960s.

Similarly, current problems in the economic, social, and political integration of visible minority groups may be attributed in part to racism. It is not surprising, then, that the field of ethnic and race relations has been central and continues to grow and assume importance within Canadian sociology. Let us begin by briefly examining how sociologists define the concepts of ethnicity and race and how we can approach the study of ethnic and race relations theoretically.

A Brief History of Ethnicity and Race

Sociologists argue that the terms *ethnicity* and *race* have historically specific significations—i.e., they mean different things to different people at different times and in different places. Ethnicity and race

Children play street hockey in Inuvik, Northwest Territories, Canada.

(Lowell Georgia/Corbis)

are not constant or monolithic concepts but rather represent dynamic social relations in flux. Thus, because people's understandings of these two terms vary considerably, they are not readily or succinctly defined. Popular uses of the terms tend to differ from social scientific definitions (Miles and Torres 1996). **Ethnicity** refers to social distinctions and relations among individuals and groups based on their cultural characteristics (language, religion, customs, history, and so on), whereas **race** refers to people's assumed but socially significant physical or genetic characteristics (Satzewich and Liodakis 2010, 11). The term *ethnicity* comes from the Greek word *ethnos* and means a large group of people. The ancient Greek historian Herodotus, in the fifth century BCE, was the first to study **ethnic groups**. In his *Histories*, he described the languages, gods, customs, 'idiosyncrasies', geography, history, politics, other social arrangements, and economies of several groups, including Greeks, Persians, Arabs, Egyptians, Ethiopians, and Libyans (Herodotus 1996, Books 1–9). Herodotus recognized that ethnic groups shared a sense of 'belonging together'.

Émile Durkheim used the concept of collective consciousness as a primary source of identity formation. In *The Division of Labour in Society* (1964 [1893]), he tried to explain what made pre-modern societies so cohesive and emphasized the importance of community or group sentiments over individual ones. Social solidarity is based on sameness and the conformity of individual consciousness to the collective. Similarities among members or sameness within the social group lead members to differentiate between themselves and others (non-members) and to prefer their 'own kind' over others. Durkheim believed that the collective consciousness of people leads them to 'love their country . . . to like one another, seeking one another out in preference of foreigners' (1964 [1893], 60). This 'us' versus 'them' feeling is important in social group formation, reproduction, and maintenance.

Max Weber argued that social group formation is associated with social practices of inclusion/exclusion, important in turn for the production and distribution of scarce valuable resources (goods, services, wages, social status and status symbols, economic and political power, equality, voting rights and citizenship, access to social programs, human rights, self-determination, autonomy, and so on). This practice of inclusion/exclusion constitutes the basis upon which decisions about rewards and sanctions are made. According to Weber (1978 [1908]), common

In the First Person

I am an immigrant to this country. I have always been interested in the study of power relations. After graduating from high school in Greece, I decided to study politics at York University. Canada's international reputation as an ethnically diverse and tolerant society influenced my decision to come here. Fascinated by Canadian society, I decided to pursue graduate work at McMaster University, where I soon developed an interest in the social and economic aspects of inequality and consequently switched to sociology. My research today examines the class and gender dimensions of economic inequalities among and within ethnic groups.

—Nik Liodakis

descent, tribe, culture (which includes language and other symbolic codes), religion, and nationality are important *ethnic markers* and determinants of ethnicity. Ethnicity should be seen as a subjective and presumed identity based on what Weber called a 'folk-feeling', not (necessarily) on any blood ties. Ethnic identity is often linked to people's 'primordial attachment'. Whereas *hard primordialism* holds that people are attached to one another and their communities of origin because of their blood ties, *soft primordialism*, as Weber argues, proposes that people's feelings of affinity, attachment, acceptance, trust, and intimacy toward their 'own kind' are not mediated by blood ties (Allahar 1994).

Weber used the term *race* to denote the common identity of groups based on biological heredity and endogamous conjugal groups. Visible similarities and differences, however minor, serve as potential sources of affection and appreciation or repulsion and contempt. Weber wrote: 'Almost any kind of *similarity or contrast of physical type* and of habits can induce the belief that affinity or disaffinity exists between groups that attract or repel each other' (1978 [1908], 386; emphasis added). Cultural and physical differences, produced and reproduced over time, constitute the foundations upon which a 'consciousness of kind' can be built. Such traits, in turn, 'can serve as a starting point for the familiar tendency to monopolistic closure' (1978 [1908], 386). 'Monopolistic closure' refers to economic, political, and social processes and practices, often institutionalized, whereby members of the in-group ('we'/'Self') have access to the scarce valuable resources mentioned above, while members of the out-group ('they'/'the Other') are excluded. The former monopolize, the latter are left out. Social boundaries, then, have been set and reproduced over time.

Today, sociologists use the term **racialization** to refer to sets of social processes and practices through which social relations among people are structured 'by the signification of human biological characteristics in such a way as to define and construct differentiated social collectivities' (Miles and Brown 2003, 99). Social group labelling creates hierarchical social dichotomies by the attribution of negative intellectual, moral, and behavioural characteristics to subordinate populations and the attribution of positive or not-negative characteristics to the dominant group(s) (those who label or stereotype). Social *positions* of superiority and inferiority are thus created; a social order is built (Li 1999). Ethnicity and race are central factors in power relations; they not only set boundaries but they also designate hierarchical positions of superiority and inferiority among and within social collectivities. The meanings of the categories and the populations they describe or 'contain' are not fixed in time and space.

When I ask my students how many races there are, they usually answer only one: the human race. But when we begin to discuss the legacy of colonialism or issues of inequality among social groups, terms such as 'white', 'black', 'visible minorities', 'Asians', and 'Aboriginals' cannot be avoided. These terms connote race as real. The physical characteristics of humans that have been used to classify social groups have included skin colour, eye colour, hair type, nose shape, lip shape, body hair, and cheekbone structure (Driedger 1996, 234–5). The term *race* as a means of categorizing human populations is linked to European colonization, exploitation, domination, and often extermination of indigenous peoples.

During the advent of capitalism as a new mode of production, a new social dichotomy slowly emerged, one based on definitions of 'Self' and 'Other'. The 'Self' referred to dominant European populations and cultures and was considered superior; the 'Other' referred to non-Europeans, who were seen as inferior and subordinate. Prior to the emergence of capitalism, race was used in a legal sense to describe people with common lineage and as a self-identification label

for the aristocracy (a category that defined the Self), but with the emergence of the bourgeoisie, the term was used to define 'Others', 'others' being 'Negroes', 'Jews', 'Arabs', 'Asiatics', and so on. It became an externally imposed label. The classification of certain groups as races was coupled with negative evaluations of their members' biological and social characteristics, just as Herodotus had characterized the so-called barbarians more than 2000 years earlier.

TIME to REFLECT

Do you believe that races exist? If so, how many races are there? Could all human populations be categorized in terms of inherited physical characteristics? Should they be?

Building One Nation or Two: Canada's Development through Immigration

In the context of global economic competition, Canada needs a growing population to keep labour costs down, increase the tax base, finance social programs, increase international competitiveness, and maintain its comparative advantage in the oil and gas industries and in other resource extraction. It needs immigrants. But what explains the present demographic composition of Canada? What could account for the cultural, ethnic, and racial makeup of the country today? Why are certain ethnic groups in Canada more populous than others? Why have there been dramatic changes in the patterns of immigration to Canada? Some answers to these questions are provided below when we examine briefly the particular role that immigration policies have played in Canadian nation-building.

Until the 1960s, the image of Canada as a nation was based on the notion that the British and French peoples founded this country. These two **charter groups**, by this thinking, built the country; everyone else 'joined in' later. The 'two founding nations' thesis endures even today, but in part it is historically inaccurate. To be sure, the French and the British colonized Canada and sent settlers to this land. But they did this at the expense of the Native peoples who were already here. Not only did they lose their lands through colonization, warfare, and deceitful treaties, but 'efforts to assimilate *them*' into the dominant aspects of the British and French cultures (through, for example, Christianity, private property, and competitive individualism) left Aboriginals with long-lasting cultural trauma and without the communal economies that had sustained them for centuries. In addition, immigration from other countries began in earnest around the time of Confederation. Nation-building required the creation of a national transportation infrastructure (roads, railways, canals), the development of commercial agriculture in western Canada, and capitalist industry in major urban centres. In the minds of government policymakers, these requirements in turn necessitated a large influx of mostly northern and central European and American immigrants (except blacks), since what was left of the once-thriving Aboriginal population, it was believed, either did not have the necessary skills or could not adapt to the British/French 'ways of doing things'. More often than not, Canada's first peoples were seen and treated as 'uncivilized savages'—very much 'Others'. Cultural compatibility was a requirement for immigration to Canada. The offer of free land to European and American settlers (land that was taken away from Aboriginals) resulted in the first wave of immigration to Canada, from 1896

to the beginning of World War I. The federal government passed Immigration Acts (1906, 1910) that set the terms under which immigrants other than the charter groups, called *entrance groups*, were accepted in Canada. In 1913, 400,000 immigrants arrived in Canada. During the years of World War I, the Great Depression, and World War II, immigration almost ceased (see Figure 11.1).

Not everyone has always been welcomed in Canada. The Immigration Act of 1910 prohibited the immigration of people who were considered 'mentally defective', 'idiots, imbeciles, feeble-minded, epileptics, insane, diseased, the physically defective, the dumb, blind, or otherwise handicapped' (McLaren 1990, 56). A 1919 amendment to the Immigration Act decreed that people with 'dubious' political loyalties also be excluded outright or, if they were already in Canada, be subject to deportation (Roberts 1988, 19). Immigrants from China and India were of particular 'concern' to xenophobic immigration authorities, since the former were seen as impossible to assimilate and thus unsuitable for permanent residence (see Sociology in Action box). Until the liberalization of immigration in the 1960s, successive Canadian governments, Liberal and Conservative, exercised exclusionary policies. Some groups were preferred

▼ FIGURE 11.1   Immigration in Historical Perspective, 1860–2008

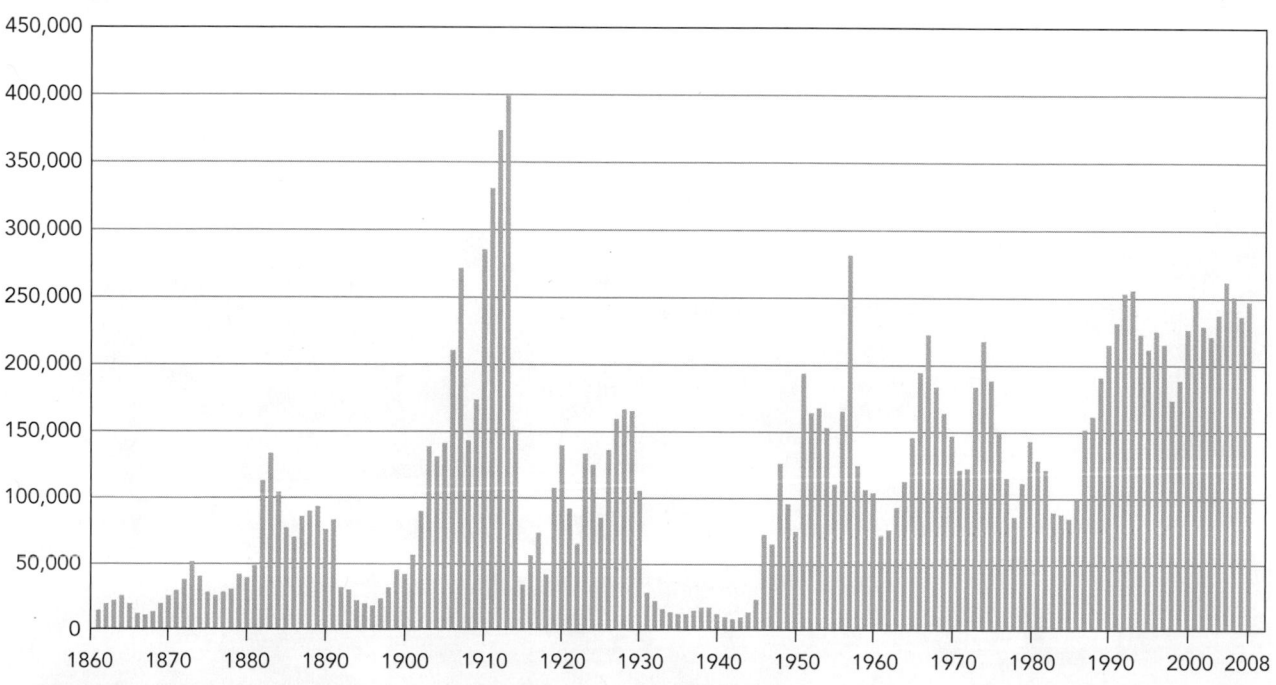

SOURCE: Citizenship and Immigration Canada, *Immigration Overview*, 2008. www.cic.gc.ca/english/resources/statistics/facts2008/permanent.
Reproduced with permission of the Minister of Public Works and Government Services Canada, 2011.

(mostly northern and central Europeans as well as Americans). Others were labelled as 'non-preferred' and were thus systematically excluded (such as the Chinese, black Americans, eastern and southern Europeans, and people from India). Members of the former groups were seen as good workers, law-abiding people, and desirable future citizens, whereas the latter were regarded as culturally, ethnically, or racially 'unsuitable' and would be admitted to Canada only as a last resort.

The end of World War II saw some minor improvements in Canadian immigration policy. It did not end **discrimination**—that process started slowly in the 1960s—but Canada began by repealing the Chinese Immigration Act (1947) and the continuous journey stipulation. Gradually, some non-whites were allowed to immigrate. Small numbers of black women were admitted as domestic workers, typists, and nurses. It was not until 1962 that the government initiated the elimination of racist criteria in the immigrant selection process and not until 1967 that the familiar points system was introduced, which relied on more objective criteria and assigned more weight to the applicant's age, educational credentials, job skills, work experience, and English- and/or French-language abilities, not to a person's country of origin. Family reunification provisions led to a large increase in the numbers of southern Europeans (Italians, Greeks, Portuguese, and, to a smaller extent, Spanish) in the late 1960s and early 1970s and a moderate rise in non-white immigration to Canada.

The New Mosaic: Recent Canadian Immigration Trends

The 'colour' of Canadian immigration has changed since the 1980s. Immigration from the traditional European (white) source countries has diminished substantially since the standard of living in these countries has improved markedly, especially with the emergence of the European Union. There has been a remarkable shift in the geographical regions from which Canada admits immigrants. During the 1950s, for example, the United Kingdom and the rest of Europe accounted for almost 90 per cent of all immigration to Canada, but at the dawn of the twenty-first century, the percentage

Sikh passengers aboard the *Komagata Maru*.

of European immigration had dropped to less than 20 per cent. In 2006, immigration to Canada from Europe and the UK dropped even further, to only 15.1 per cent of total immigration. Immigration from Africa and the Middle East increased to 20.6 per cent and from South and Central America to 9.7 per cent. Immigration from Asia and the Pacific was just over 50 per cent (Statistics Canada 2006c).

Canada admits approximately 260,000 immigrants per year. Table 11.1 lists the top 10 source countries of immigrants to Canada for 2007. Together, these 10 countries accounted for 52.4 per cent of all immigrants. The top source country is China (11.4 per cent of all admitted immigrants that year), and the second highest percentage belongs to India (11 per cent). We have certainly come a long way since the era of the

SOCIOLOGY in ACTION

Is Canada So 'Innocent'? Racist Immigration Policies toward Chinese and Indian Immigrants

I would like you, for a few moments, to assume that your ethnic ancestry is Chinese and/or Indian. Read the following carefully:

In the 1880s, Chinese immigrants were allowed into Canada because of the growing demand for cheap and disposable labour in the building of the trans-continental railway. From 1880 to 1884, approximately 16,000 Chinese immigrants arrived, mainly in British Columbia, to work in railway construction. Employers and contractors saw an opportunity to exploit Chinese labourers and called for increases in Chinese immigration; labour unions, on the other hand, opposed the influx of Chinese workers, since increases in the supply of immigrant labour led to stiffer labour market competition and kept wages low. In 1885, the government imposed the Chinese head tax. It started at $50, then was increased to $100 in 1890 and to $500 per person in 1903. Only Chinese who could afford the head tax could immigrate to Canada. In 1923, Chinese immigration was completely prohibited and remained so until after World War II. There were exceptions, however. Social class and gender were important, since Chinese businessmen who either invested or established businesses in Canada were exempt from the head tax and were allowed to immigrate even after 1923. The immigration of Chinese women was restricted since—it was feared—they would reproduce the 'yellow peril' and lead to the propagation of 'alien' cultures and races, thus undermining the image of Canada as a 'white' settler society. The federal government collected $23 million during the period the head tax was in effect (Satzewich and Liodakis 2010). In today's dollars, this would amount to approximately $1.2 billion.

Here is another case: by the 1910s, around 5000 immigrants from India had arrived in British Columbia to work in the lumber and mining industries. The government could not introduce outright anti-Indian immigration legislation, since India was part of the British Empire, so for the sake of appearances, it opted for the covertly discriminatory policy of the so-called 'continuous journey stipulation'. An order-in-council passed on 9 May 1910 stipulated that only people who had made a non-stop journey from their country of origin to Canada would be allowed in as immigrants (Basran and Bolaria 2003, 99). Not surprisingly, there were no direct sailings between India and Canada, since the government had 'persuaded' Canadian steamship companies to end all such direct travel. Indians who wanted to immigrate to Canada had to travel via Hong Kong or Hawaii, but that did not constitute a continuous journey. Thus, they were not admitted. This policy was tested by an enterprising Sikh, Gurdit Singh, who organized the immigration to Canada of 340 of his fellow Sikhs aboard the Japanese-registered freighter the *Komagata Maru*. The freighter sailed from Hong Kong in April 1914 with stops in Shanghai and Japan but was prohibited from landing at the Port of Vancouver. After a tense standoff lasting two months, involving extensive world press coverage and rapidly deteriorating conditions aboard the vessel anchored in Vancouver harbour, the *Komagata Maru* returned to Asia with its passengers (Buchignani et al. 1985, 54–61).

Now, using the Weberian concept of *verstehen* ('understanding'), and C. Wright Mills's sociological imagination, describe how you do or would feel about Canadian immigration policies and practices of the past. Do the above sociological concepts assist you in understanding and explaining certain aspects of Canadian history and/or current immigration and racism issues? Think of the situation of the Tamil refugees who arrived by ship in Vancouver in August 2010. What connections can you make?

For further information and commentary, follow this link: www.vancouversun.com/life/Public+rage+against+Tamil+refugees+nasty+xenophobic+od our/3426924/story.html.

Chinese head tax and the continuous journey policy. The UK, France, and the US are still found in the top 10 source countries, but their percentage contributions are small (3.4, 2.3, and 4.4 per cent, respectively).

Immigrants are divided into four major immigration classes: skilled workers, business immigrants, the family class, and refugees. Data from 2008 show that contrary to public misconceptions about the people admitted, the largest immigration class is that, of skilled workers. Skilled workers and business immigrants (149,072 people) constituted 60.3 per cent of total immigration in 2008. Skilled workers are independent applicants who are admitted through the use of the points system (see Global Issues box).

The family class in 2008 (65,567 people) ranked second, with 26.5 per cent of total immigration to Canada. These immigrants are admitted if they have close relatives (spouses or parents) in Canada who are willing to sponsor them to come and to support them financially for a period of three to 10 years after they arrive. Refugees (21,860 people) follow, representing 8.8 per cent (Citizen and Immigration Canada 2008). Canada is a signatory to international treaties and is obliged by international law to provide asylum to those who have demonstrably genuine refugee claims.

Once in Canada, most immigrants tend to settle in Ontario, British Columbia, and Quebec. The distribution of immigrants across Canada is decidedly uneven. From Ontario's almost 50 per cent of all immigrants, after Quebec (19.1 per cent) and BC (16.5 per cent), only Alberta (8.8 per cent) and Manitoba (4.6 per cent) attract more than 2 per cent of the total immigration (Citizenship and Immigration Canada 2009). Immigration to other parts of Canada is negligible. This uneven distribution, of course, influences variably the ethnic and racial makeup of certain parts of Canada. There is also a clear urban–rural divide. Immigrants are more attracted to major urban centres because they usually find more economic opportunities there as well as other immigrants from their own part of the world. Today, for example, the majority of Torontonians were born outside Canada, whereas almost all of the residents of Hérouxville, a small farming community in Quebec, are Canadian-born (see Open for Discussion box).

TIME to REFLECT

Why does Canada admit immigrants? If you were to (re)design Canada's immigration policy, what criteria would you use for admitting immigrants? Why are these criteria of most importance?

	Top 10 Source Countries	
TABLE 11.1 ▼	of Immigrants to Canada, 2007	

Source Countries	Number	% of Total Immigrants
People's Republic of China	27,014	11.4
India	26,054	11.0
Philippines	19,064	8.1
United States	10,450	4.4
Pakistan	9,547	4.0
United Kingdom	8,128	3.4
Iran	6,663	2.8
Korea, Republic of	5,864	2.5
France	5,526	2.3
Colombia	4,833	2.0
Top 10 source countries	144,447	52.4
Other countries	117,789	47.6
Total	262,236	100

SOURCE: 'Immigrants by class according to the 10 main countries of birth, Canada, 2005 to 2007'. Based on Citizenship and Immigration Canada, 'Annual report to Parliament on immigration', 2005 to 2007, and Statistics Canada, 'Report on the demographic situation in Canada', catalogue 91-209-XWE, issue 2005 and 2006, release date: 23 July 2008, Table 4.3, www.statcan.gc.ca/bsolc/olc-cel/olc-cel?lang=eng&catno=91-209-X.

Multiculturalism and Its Discontents

Canada may not be as racist as it used to be in the past (Levitt 1994), but it is still racist, capitalist, and patriarchal (Liodakis 2002). For example, evidence suggests that recent immigrants, mostly visible minority working-class women, continue to face discrimination and unequal treatment in the labour market (Galabuzi 2006; Hum and Simpson 2007; Krahn, Lowe, and Hughes 2007). But what makes an individual a member of a visible minority? Who is a 'true' Canadian? Why are Aboriginals not considered visible minorities? Are Aboriginals or the Québécois ethnic groups like all the others? What defines your identity? In the 2006 census, almost one in three respondents identified their ancestry as Canadian. Are Canadians an ethnic group? More important, who decides, who makes the definitional rules, who sets the criteria? These are tough questions, and simplistic approaches usually do not provide satisfactory social scientific answers. Let us examine a few important aspects of the formation and meanings of ethnic and racial identities in the Canadian context of multiculturalism.

GLOBAL ISSUES

The Points System for Skilled Workers: Would You Make It?

Do you think you would qualify as an immigrant to your own country? Test yourself to find out whether you would qualify as a skilled worker for admission to Canada. The table below outlines the various categories of qualification for which points are rewarded. For example, if at present your education consists of a secondary school diploma but no further diplomas, certificates, or degrees, you will receive five points in this category; under language, if you can read, write, speak, and understand English with complete proficiency, you will be granted 16 points, and total proficiency in French will earn you an additional eight points. You will need at least 67 points (out of a possible 100) to be admitted to Canada as a skilled worker. For a complete breakdown of the points system, go to the following website: www.canada-immigration .biz/permanent_skilled.asp.

Selection Criteria	Points Awarded
Education	up to 25
Knowledge of official language(s)	up to 24
Work experience	up to 21
Age: Applicants 21–49 years of age receive maximum points. 2 points are deducted for each year under 21 or over 49, so that someone 16 or younger or 54 or older will receive no points.	up to 10
Arranged employment in Canada	up to 10
Adaptability	up to 10
Spouse's or common-law partner's education	3 to 5
Minimum one year full-time authorized work in Canada	5
Minimum two years post-secondary study in Canada	5
Maximum points awarded	100
Minimum required to pass for skilled worker immigrants	67

SOURCE: www.canada-immigration.biz/permanent_skilled.asp

Dominant cultural values in Canada have changed over time. Prior to the advent of official **multiculturalism** in 1971, which we are so used to today, there had been the long, hard years of anglo- and/or franco-cultural conformity. Canadian society was characterized by ethnocentrism from the era of colonization to the early 1970s. An individual is ethnocentric when he or she evaluates (usually negatively) the culture of others based on criteria derived from his or her own. The charter groups have set the terms for the entrance of all others into the country. There has always been sustained pressure on Aboriginals and minority-group newcomers to adopt the dominant British and French cultural values, customs, and symbols. In short, 'others' had to conform; they had to assimilate to the norms of the 'Self'.

Assimilation, a term always encountered in race and ethnic relations discourses, is usually defined as the processes and social practices by which members of minority groups are incorporated into the dominant culture of a society (Isajiw 1999, 170). Sociologists distinguish between behavioural assimilation ('acquiring' the values of dominant groups) and structural assimilation (the integration of 'others' into the economic, social, and political life of a country). But assimilation has not always been a simple matter of choice for minorities; often it was forceful and violent, as was the case with Aboriginals, for example, through the residential schools and the banning of Native customs and ceremonies such as the thirst dance and the potlatch. Early Canadian government efforts to assimilate Aboriginals and immigrants somewhat resemble American **melting-pot policies**.

Multiculturalism is one of those elusive terms that we use every day, but it means different things to different people. In Canada, we understand the term as having four interrelated meanings (Fleras and Elliott 1996, 325):

It is a demographic reality.

It is part of pluralist ideology.

It is a form of struggle among minority groups for access to economic and political resources.

It is a set of government policies and accompanying programs.

Multiculturalism as policy and ideology gives rise to sets of economic, political, and social practices, which in turn define boundaries and set limits to ethnic and racial group relations in order to either maintain social order or manage social change (Liodakis and Satzewich 2003, 147).

First, when we say that multiculturalism is a fact of Canadian society, we mean that the Canadian population comprises people who come from innumerable ethnic backgrounds. Canadian society has never been ethnically homogeneous, as some would like to believe. Demographically, Canada was a multicultural country long before the implementation of multicultural policy. Second, as an ideology multiculturalism includes normative descriptions about how Canadian society *ought to be*. The basis of multiculturalist ideology is cultural pluralism, which advocates tolerance of cultural diversity and, most important, promotes the idea that such diversity is compatible with national goals, especially those of national unity and socio-economic progress (Fleras and Elliot 1996). The basic principles of multiculturalism rest on the notion of cultural relativism, as opposed to ethnocentrism. Cultural relativism promotes tolerance and diversity in order to achieve the peaceful coexistence of groups in ethnically and

UNDER the WIRE

New Technologies: Keeping in Touch, Accessing Services, and Doing Social Science

In this era of globalization, the advent of new technologies, especially in the field of communication, has helped many immigrants to keep in touch with their culture of origin and loved ones 'back home'. International satellite TV programming, available in Canada, keeps immigrants informed about political, economic, social, and cultural developments in their home countries. It helps them stay connected. Affordable Internet services provide opportunities for easy communication with their loved ones on a daily basis. In addition, more and more government and local services can be found on the Web. For example, information about immigration to Canada (e.g., application forms, regulations, the points system) is readily available at www.cic.gc.ca.

Access to many services, such as English and French language training programs, local employment opportunities, social and health services, child care, education, and transportation, is also available online. Bureaucracy, often a time-consuming and frustrating hurdle for new immigrants, can be now partially avoided if they apply online for social insurance numbers, health insurance, citizenship, and so on. With the use of the Internet, access to information in Canada has come a long way in the past two decades.

Today, doing social science in the field of race and ethnic relations is 'under the wire'. Vast quantities of credible information and research are now available online. Statistics Canada (www.statcan.gc.ca) posts its own research and makes available online large datasets on everything from employment trends, regional unemployment rates, the geographical distribution of new immigrants, and occupations and professions to income disparities, educational attainment, language skills, gender, and marital status. My own research has benefited immensely from such availability. As you will notice, a lot of data contained in this chapter have come from the Statistics Canada website.

For example, rates of ethnic exogamy are increasing. If you want to find information about the current makeup of Canada's mixed ethno-cultural couples, you may visit www.statcan.gc.ca/daily-quotidien/100420/dq100420b-eng.htm.

Tables from the latest census (2006) are available for free to anyone interested at www.statcan.gc.ca/subject-sujet/result-resultat.action?pid=30000&id=-30000&lang=eng&type=CENSUSTBL&pageNum=1&more=0.

If you would like to study Aboriginal people in Canada, you may find useful information at www.statcan.gc.ca/subject-sujet/subtheme-soustheme.action?pid=10000&id=-10000&lang=eng&more=0.

Despite the ills of globalization, new technologies like the Internet offer efficient and useful means for keeping in touch and for being informed, educated, and up-to-date on race and ethnicity.

racially heterogeneous societies. It suggests that we should not judge other cultures by our own norms and criteria. If we recognize the right of all people to self-identify and promote their own culture, then, it is hoped, the same courtesy will be extended to individuals who share different cultural norms and values. Third, multiculturalism is also a process and a terrain of competition among and between minority groups for valuable economic and political resources. As such, it is used by governments as a mechanism for conflict management and resolution. Finally, multiculturalism also refers to all government initiatives and programs aimed at realizing multiculturalism as ideology and transforming it into a concrete form of social intervention.

Multiculturalism was not given as a gift to minority groups; historically, it emerged for several reasons (Fleras and Elliott 1996, 335). First, during the 1960s immense political pressure was exerted on the federal government by the 'other' ethnic groups (e.g., Ukrainians and Germans in western Canada) for recognition of their contributions to Canadian society. Second, it became a political necessity for the federal government to counterbalance western alienation and Quebec nationalism. Third, the Liberal Party of Canada sought to acquire greater electoral support from immigrants in urban centres, and this was a tactical move in that direction.

As policy, it is a relatively recent aspect of Canadian state activity. It was introduced in 1971 by

HUMAN DIVERSITY

The 'Veil Issue'

Many Muslim women in the world, including Canada, choose to cover parts of their face with veils. It is part of their religious tradition, just as southern European Christian women in mourning choose to wear long black dresses and partially cover their heads as well. In the post 9/11 world, xenophobia and Islamophobia are on the rise worldwide. The 'veil issue' has sparked heated debates in the UK, Belgium, and France. Politicians such as former UK Prime Minister Tony Blair and his ex-Minister of Foreign Affairs Jack Straw asked Muslim women to remove their veils in their presence 'in order to improve communication'. A Muslim teacher in the UK was required by law to remove her veil in the classroom, since, it was argued, 'it hindered student learning.' On 12 July 2010, the French lower house of parliament overwhelmingly passed a bill that bans the wearing of the veil in public and stipulates fines of up to 150 euros for women wearing it. In Canada, the veil issue has also attracted media attention.

In September 2007, three federal by-elections took place in Quebec. Marc Mayrand, Canada's chief electoral officer, was under pressure from politicians, the media, and 'concerned citizens' to take a stance against allowing veiled Muslim women to vote unless they first showed their faces. Should women with their faces covered be allowed to vote? How could their identity be verified? He decided, in the spirit of being reasonable, that veiled Muslim women have the same rights as everyone else. After this decision, Mayrand held a press conference in Ottawa to address some of the criticisms he had received. Allegedly, the chief electoral officer had 'flouted the will of Parliament' by his loose interpretation of the Elections Act (*National*

Post 10 Sept. 2007). According to Mayrand, there is nothing in the current electoral law to prevent veiled people from voting. Moreover, the law allows citizens—for religious reasons—to vote with their face covered provided they show two pieces of valid ID and swear an oath. After all, said Mayrand, in the previous federal election, 80,000 people cast votes by mail (*Toronto Star* 10 Sept. 2007). Prime Minister Stephen Harper and Official Opposition Leader Stéphane Dion disagreed with this interpretation of the law and argued that people must show their faces when voting to maintain integrity in the election process. Mayrand countered that if parliamentarians did not like his interpretation, they should have changed the law when they had the chance. On the other hand, John Ivison, a writer for the *National Post*, pointed out that Parliament has not ruled 'on voting by comic book characters but if Batman and Robin turned up in the polling booth, one hopes that Elections Canada staff would force them to reveal their secret identities' (*National Post* 10 Sept. 2007).

How would you feel if you were a Canadian Muslim woman and were not allowed to vote because you wear a veil? Do you think that veiled women want to hide their identities? Canada is considered a tolerant society, has an official policy of multiculturalism, and freedom of religion is protected by the Charter of Rights and Freedoms. So should we allow veiled citizens to cast ballots? Before you grapple with this last question, you should know that both the Canadian Islamic Congress and the Canadian Council of Muslim Women agreed that veiled women should show their faces before voting.

a very charismatic Canadian prime minister, Pierre Elliott Trudeau. Ironically, and contrary to what we may think today, it was not the historical legacy of racism, discrimination, and **prejudice** in Canada that multicultural policy initially aimed to redress. In fact, these issues did not figure at all in the framework for the initial development of multicultural policy. From 1971 to 1980, the policy was essentially folkloric (ethnic food, costume, and dance) and focused on 'celebrating our differences' (Fleras and Elliott 1996). The basic principles that guided federal multiculturalism at this time were the following: support for all of Canada's cultures; assistance to ethnic groups in need to overcome cultural barriers to full participation in Canadian society; promotion of creative encounters and interchange among all Canadian cultural groups in the interest of national unity; and assistance to immigrants in acquiring at least one of Canada's two official languages in order to become full participants in Canadian society (Hawkins 1988, 220).

Rising Quebec nationalism was countered by the repatriation of the Constitution (1982) and the inclusion of the Charter of Rights and Freedoms in the Constitution Act, 1982. Section 27 of the Charter states that its interpretation, i.e., in Canadian courts and legislatures, 'shall be . . . in a manner consistent with the preservation and enhancement of the multicultural heritage of Canadians'. Multiculturalism had become a fundamental and legally contestable characteristic of Canadian society. In 1988, the Progressive Conservative government passed the Multiculturalism Act, which turned a de facto reality into a de jure legal framework, thus elevating multiculturalism to equality with the principle of bilingualism.

Consistent with neo-conservative economic doctrines was the attempt to justify the Multiculturalism Act not only in terms of pluralist ideology but also in terms of potential economic benefits to the country. This involved a shift in emphasis away from a 'culture for culture's sake' perspective toward a more instrumentalist view of the benefits of multicultural policy. Simply put, the Mulroney government of the day strongly believed that multiculturalism could and did mean business—increased business, more economic opportunities, and greater prosperity for all.

Since the 1990s, the development of 'civic multiculturalism' has taken place. Folkloric and institutional multiculturalism have been coupled with notions of social equality and citizenship. The focus of civic multiculturalism is society-building; today, fostering a common sense of identity and belonging is considered essential for the participation and inclusion of all Canadians in national institutions (Fleras and Elliott 1996, 334–5). Governments have moved away from the initial folkloric focus, which has meant a withdrawal from certain programs associated with it (e.g., funding for cultural festivals).

The policy of multiculturalism has been critiqued since its introduction. There has never been agreement about the effectiveness, desirability, or necessity of the policy and its accompanying programs. Simply put, some argue that multiculturalism makes Canada a unique and great country, while others say that multiculturalism is useless, unnecessary, and ineffective. In the post-9/11 context, debates about multiculturalism have acquired renewed political importance.

CRITICISMS OF MULTICULTURALISM

Early criticisms of multiculturalism focused on the policy's inherent inability to deliver the goods or solve the problems it set out to address. Some critics argued that too great an emphasis was placed on depoliticized 'song and dance' activities that were non-threatening to British and/or French economic, political, and cultural hegemony and that the policy mystified social reality by creating the appearance of change without actually changing the fundamental bases of ethnic and racial inequality within Canada (Bolaria and Li 1988; Moodley 1983). Furthermore, the identification of only 'cultural barriers' to the full participation of immigrants in Canadian society precluded the examination of racism and discrimination as barriers (Bolaria and Li 1988). Economic barriers were not recognized, nor were they examined (Stasiulis 1980, 34). The exclusive focus on cultural and linguistic barriers to equality conceals other, perhaps more fundamental social inequalities based on people's property rights, position in the labour market, education, gender, age, and so on. In fact, Canadian society is characterized by ethnic- and gender-based class hierarchies and socio-political struggles, which are not addressed by multiculturalism because such struggles challenge, if not threaten, such hierarchies. Multiculturalism obfuscates these antagonisms and shifts the struggle to the cultural realm.

In the 1990s, critics claimed that the policy of multiculturalism helped to reproduce stereotypes of ethnic groups, undermined Canadian unity, ghettoized minority issues, and took away from the special claims that francophones and Aboriginal peoples have within Canadian society. A policy that had as one of its underlying intentions the improvement of inter-group relations was seen by many as a policy leading to deteriorating inter-group relations and as a threat to the coherence and stability of Canada. It was argued that 'caravans', 'folk fests', and other multicultural festivals do not promote serious cultural exchanges but rather are superficial and have the effect of commodifying cultures and reproducing cultural, ethnic, and racial stereotypes (Bissoondath 1994, 83). According to Neil Bissoondath (1994, 224), we have become a nation of cultural hybrids. He argues that '[w]e are . . . of so many colours, that we are essentially colourless' (1994, 73). Indeed, in this view, no evidence indicates that intercultural exchanges take place or have assisted in the 'harmonization' of racial and ethnic relations in Canada.

Another criticism, developed by sociologist Reginald Bibby (1990), has been that multiculturalism promotes cultural relativism and hence undermines Canadian values and social cohesion. Canadian society has changed as a result of multiculturalism, but not in entirely positive ways. Until the 1950s, Canadians—in policy and practice—had emphasized community and the collectivity, but since then, according to Bibby, the focus has been on the individual. Pluralism, although imperative for coexistence, does not offer a subsequent vision of the country, does not set national goals, and does not have a cause. Canada, in attempting to promote peaceful coexistence, effectively promotes the breakdown of group life. As both Bibby and Bissoondath have viewed it, we have ended up with a value system that contains nothing exclusively Canadian. Multiculturalism does not offer a vision of unity, and it encourages division by ghettoizing people into ethnic groups.

CRITICIZING THE CRITICS: MULTICULTURAL POLICY AS A REFLECTION OF REALITY

These criticisms share an appeal to the 'national' character of Canada, which is never defined. The implication is that the current system is somehow biased in favour of 'non-whites' and 'non-Europeans' and that it should not be. In addition, the critics seem silent or purposefully vague in describing what constitutes Canadian culture, the definition of what and who is or should be Canadian, or what Canadian values are, and there is no definition of what constitutes 'the Canadian nation, culture, or character', who defines it, or whose interests it serves.

Rhoda Howard-Hassmann (1999) has pointed to a basic fault in the critiques of Bibby and Bissoondath: they both assume that Canadian multiculturalism calls for individuals to retain their ancestral identities. But the Canadian policy is 'liberal', not 'illiberal'—that is, it does not impose the idea of maintaining ethnic differences, nor does it force individuals to identify with ancestral cultural groups. In her view:

> Multiculturalism 'normalizes' a wide range of customs and makes the enjoyment of such customs part of what it means to be a Canadian. . . . Liberal multiculturalism acknowledges the social need for difference, for smaller, more close-knit communities separated from the Canadian mainstream. But it does not mandate such difference. (1999, 533)

She also argues that far from promoting disloyalty to Canada and things Canadian, multicultural policy has the seemingly ironic consequences of integrating immigrants to the dominant society, promoting national unity, and encouraging 'a sense of connection with other Canadians' (1999, 534). The rising number of people who identify their ancestry as Canadian in recent censuses tends to support her argument.

ABORIGINAL PEOPLES, QUÉBÉCOIS, AND MULTICULTURALISM

Another criticism is that multiculturalism detracts from the special claims that francophones and Aboriginal peoples have in Canadian society. In Quebec, multiculturalism was seen as an attempt by the federal government to undermine the legitimate Quebec aspirations for 'nationhood'. By severing culture from language, multiculturalism rejected the 'two founding nations' metaphor of Canada's historical development and reduced the status of French Canadians from that of 'founding people' to just another ethnic group (Abu-Laban and Stasiulis 1992, 367). Multiculturalism also became a mechanism to 'buy' allophone votes. Assimilationist language policies in Quebec, directed toward allophones, can be understood in this context.

Successive Quebec governments have pursued a policy of *interculturalism* (prominent in Europe) instead of multiculturalism. According to Kymlicka (1998), interculturalism operates within three important principles: (1) it recognizes French as the language of public life; (2) it respects the liberal-democratic values of political rights and equality of opportunity for all; and (3) it respects pluralism and openness to and tolerance of the differences of others. These principles constitute a 'moral contract' between the province of Quebec and immigrant groups. Interculturalism may sound a lot like the federal policy of multiculturalism, but there are some nuanced differences.

For example, it promotes *linguistic assimilation*. The 'centre of convergence' for different cultural groups in Quebec is the 'collective good' of the French language, which is seen as an indispensable condition for the creation of the *culture publique commune* (common public culture) and the cohesion of Quebec society. The French language needs to be protected and promoted.

Some researchers have argued that interculturalism is the most advanced form of pluralism today (Karmis 2004, 79), since it combines multiculturalism and multinationalism and is more inclusive than either. It does not apply only to ethnic groups or nations but also to 'lifestyle' cultures and world views associated with new social movements, including cultural gay, punk, environmental, feminist, and other non-ethnic-based identities. In principle, no cultural community is excluded from Québécois identity.

TIME to REFLECT

Would you prefer to live in a country without official multiculturalism and/or interculturalism? Would you rather live in the US, France, or Germany? Why?

Canadian Aboriginal peoples and their organizations are similarly critical of and have similar reservations about multiculturalism. Aboriginal leaders argue that multiculturalism reduces them to 'just another minority group' and undermines their aspirations for self-government (Abu-Laban and Stasiulis 1992, 376). They claim that they possess a distinct and unique set of rights—now enshrined in the Constitution—that stem from their being the first occupants of Canada. Since Aboriginal peoples do not consider themselves part of the so-called mainstream Canadian pluralist society but as distinct peoples, multiculturalism is seen as an actual threat to their survival. They prefer to negotiate their futures in a binational framework with federal (and provincial) governments that recognizes their collective rights to special status and distinctiveness (Fleras and Elliott 1996, 343).

MULTICULTURALISM IN A CHANGING WORLD

Many countries now officially celebrate their multicultural makeup, and some have policies designed to promote the peaceful coexistence of diverse groups. However, a number of events over the past few years have provided a context for renewed questions about,

and attacks on, policies of multiculturalism both in Canada and abroad. Certainly, the attacks on the World Trade Center in New York and the Pentagon on 11 September 2001 put many Western governments on alert about the threats that cultural and religious 'Others' may pose to the 'peace and security' of their countries. In the post-9/11 era of Islamophobia and 'big brother' surveillance, two criticisms of multiculturalism are also prevalent: (1) that multiculturalism encourages and tolerates the promotion of cultures and religions that are decidedly intolerant and (2) that multiculturalism is a recipe for homegrown terrorism. Such critiques are often concealed forms of racism. No country has arrived at an ideal management of ethnic and racial diversity. Canada's multicultural approach to diversity issues may not be perfect—indeed, it is rather limited—but many other, far more problematic approaches to diversity exist (the US and France spring readily to mind), and we can take pride that we have avoided them so far. Let us now turn, then, to the unresolved issues of racism.

TIME to REFLECT

Are Aboriginal Canadians and the Québécois just ethnic groups? If not, why not?

Prejudice and Racism

Racism is based on 'othering' (Simmons 1998). According to Stuart Hall, it is

> not a set of false pleas which swim around in the head . . . not a set of mistaken perceptions. . . . [Racist ideas] have their basis in real material conditions of existence. They arise because of the concrete problems of different classes and groups in society. Racism represents the attempt ideologically to construct those conditions, contradictions, and problems in such a way that they can be dealt with and deflected at the same moment (in Li 1999, 325).

Many sociologists have suggested (e.g., Bolaria and Li 1988; Li 1999) that race problems often begin as labour problems. Competition for employment among workers from different ethnic/racialized (and gender) groups keeps wages low and profits for employers high. Workers usually participate in a *split labour market* in which more members of the dominant groups may have more secure, full-time, and high-paying jobs, whereas minorities are found in largely part-time, low-paying, insecure, and menial occupations. Expressions of working-class racism may be attributable to labour market conditions of inequality (see Dunk in Satzewich 1998).

This labour market split develops over long periods of time and is reproduced by prejudice and discrimination. Often, we have preconceived notions about ethnic/racialized groups. Members of some groups are seen through the prism of stereotypes (Driedger 1996). Some are deemed hard-working, law-abiding, smart, moral, and so on. Others are seen as 'lazy', 'smelly', 'dirty', 'stingy', 'criminals', 'promiscuous', 'uncivilized', and the like. Ethnic jokes, which might amuse us uncritically, are based on these stereotypes. Negative stereotypes are often reserved by the majority group for minority groups; positive stereotypes are related to dominant groups, although minority groups use positive self-stereotypes to resist racism. *Discrimination* refers to behaviours and policies that reproduce ethnic and racial social stereotypes as well as economic and political inequalities. We also use the term *prejudice* to refer to the negative views of and attitudes about members of various minority groups. 'Prejudice' comes from the Latin word for 'prejudgment'. Also associated with this term is the *ecological fallacy*—i.e., the assumption that an individual member of a social group (in our case, an ethnic/racialized group) has the social characteristics associated with that group. As such, stereotypes, discrimination, and prejudice maintain and reproduce racism.

We can speak of two types of discrimination against minorities: de jure (i.e., by law) and de facto (in fact). Historically, many groups have experienced varying degrees of racism in their daily lives, but as a group, First Nations people have been singled out for unequal treatment by Canadian governments. Their lands have been taken away (presumably legally, through 'treaties'), they have been forcefully segregated in reserves, and for many, their children were sent to residential schools, depriving generations of their own cultural heritage. During the two world wars of the twentieth century, members of some ethnic groups (Germans, Italians, and Japanese) were singled out for internment by Canadian authorities, and others (e.g., Russians, Ukrainians, Jews) were seen as harbouring communist political beliefs and often were not allowed to immigrate to Canada or, when they were involved in labour strife, were quickly deported. Canada would accept only a relative handful of European Jewish refugees in the 1930s, and as we saw earlier, government policy in the first half of twentieth century excluded Chinese and South Asians.

Today, a more subtle type of discrimination permeates Canadian life. Because it is covert, de facto discrimination is more difficult to resist and combat. Canadian law prohibits overt discriminatory acts in employment, social services, and education, but the reality is that some members of minority groups face issues of *systemic discrimination*—impersonal, covert practices that penalize members of certain groups. Also called *institutional racism*, it is the outcome of the inner workings of institutions (the economy, education systems, the government) that disadvantage particular individuals and groups. For example, in the labour market, a minimum educational requirement of a high school diploma may exclude from unskilled jobs some members of minority groups with low educational attainment. Recent immigrants may be excluded from good jobs when government

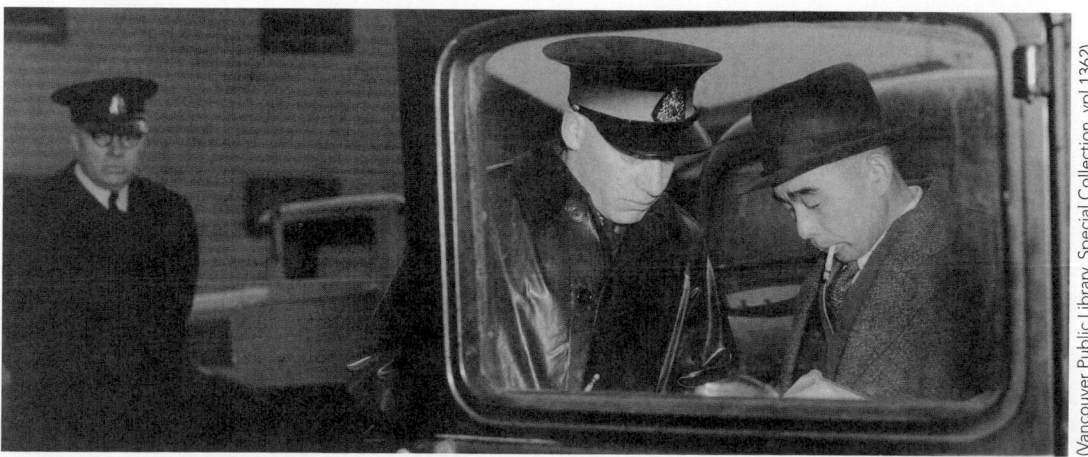

Racism has always been a part of Canadian society. In this historical photograph, taken in 1942, Vancouver police harass a Japanese man and confiscate his vehicle. After Pearl Harbor, Canada expelled or interned 22,000 Japanese immigrants and Japanese Canadians.

(Vancouver Public Library, Special Collection, vpl 1362)

and employers require long years of 'Canadian experience'. Members of some minority groups may be excluded from Canadian police or firefighting forces on the basis of a minimum height requirement (similar regulations have kept many women out of these forces for a long time). Not recognizing educational credentials attained abroad (especially from developing countries) keeps large numbers of visible minority immigrants out of secure, well-paying jobs.

Henry and Tator (2005) argue that today, a peculiar form of racism exists in Canada: *democratic racism*. Democratic racism is not necessarily based on old racist notions of the biological and social superiority of whites over racialized minorities, but rather on contradictions about and conflicts over social values. For example, Canada is supposedly committed to justice, equality, and fairness, but these values coexist with differential treatment and discrimination against minorities. Democratic racism is an ideology and a mechanism for reducing the conflict inherent in maintaining a commitment to both liberal and non-egalitarian

values. It permits and sustains the rationalization, justification, and maintenance of two apparently conflicting sets of values (liberal-democratic versus non-egalitarian with regard to people of colour).

Racism is reflected in the systems of cultural production and representation and in the codes of behaviour of the dominant culture. Henry and Tator argue that they are 'embedded in the values and meanings, policies and practices of powerful institutions' (2005, 90). Society gives voice to racism through words, images, stories, explanations (or silences), categorizations, justifications, and rationalizations, which in turn produce a shared understanding of the world and of the (inferior) status of people of colour in that world (2005, 91). This discourse is used to extend or defend the traditional interests of the dominant culture.

For example, many people, usually members of dominant groups, claim that they do not 'see' colour. This claim of colour-blindness may be true, but it obfuscates the reality of the pervasiveness of the historical 'baggage' of colour in our everyday lives—the

OPEN for DISCUSSION

Reasonable Accommodation, Xenophobia, and Islamophobia

'Reasonable accommodation' is the new mantra used by proponents of Quebec interculturalism. The term implies that government policies and programs will endeavour not just to tolerate but also to accommodate the cultural differences, the 'otherness', of new immigrants in the spirit of pluralism. The debate on the usefulness of this policy is ongoing. The Quebec government has actually institutionalized the debate by holding public hearings. Not all Quebecers agree with reasonable accommodation. In fact, a clear urban/rural cultural split reflects current socio-demographic realities: urban centres like Montreal have sizable immigrant populations and are more accepting of difference; rural areas are largely homogeneous and culturally conservative, and they would like to keep it that way. For example, in January 2007, Hérouxville, a small (population 1338) Quebec farming community of almost exclusively white, francophone, nominally Catholic residents located 180 kilometres north of Montreal, gained notoriety when its town council passed a resolution prescribing a code of conduct for potential immigrants. It set conditions under which new immigrants could be admitted to their town. Specifically, the resolution stated that immigrants who 'cover their face', 'carry weapons to school', 'stone or burn alive women', or 'perform female genital mutilation' were

not welcome in their community. As André Drouin, a town councillor, put it, reasonable accommodation had reached a state of emergency in Quebec. The implication was clear: apparently, interculturalism and reasonable accommodation had gone too far, since, it was presumed, they 'allow everything'. Quebec Premier Jean Charest suggested that Hérouxville's 'measures' might be drastic and exaggerated and not representative of Quebec society.

The reaction of minority communities was swift. A delegation of women from the Canadian Islamic Congress visited Hérouxville and met with the town council and some local residents to discuss the issue in the spirit of cultural understanding. The resolution was clearly directed against Muslims and other people from northern Africa, the Middle East, and Asia. Its intentions were discriminatory. It was, in fact, a concrete example of xenophobia and Islamophobia. After the exchange of niceties and gifts, the town resolution was watered down, but the controversy remained and sparked debates in other parts of the country among politicians, the media, students, professors, and many others.

What do you think are the implications of this issue for the study of race and ethnicity in Canada? Who decides what is 'reasonable' in reasonable accommodation? What are the criteria? What should they be?

policies, programs, and practices that continue to be racist. In addition, the discourse of equal opportunity expresses a value dear to Canadians, but it is often assumed that we do not have to dismantle dominant (white) institutions of power in order to achieve it. Just because we exalt tolerance of others through the language and policy of multiculturalism does not mean that multiculturalism necessarily leads to social harmony. The new buzzword—'reasonable accommodation'—is a hoax (see Open for Discussion box). We continue to use the dominant values, beliefs, and ideas as measuring sticks for evaluating others. In addition, multiculturalism conceals the structural, economic, and political inequalities in Canada. Multiculturalism does not combat racism. The discourse of national identity tends to be racist, since it erases or silences the contributions of ethnic/racialized minorities to the Canadian national identity.

Finally, it should be kept in mind that racism is not found exclusively among members of dominant groups. There also exist intra- and inter-group racisms. For example, some members of the same group may exhibit racism toward other members because of regional, linguistic, religious, or political differences. We often encounter inter–minority group racism. Members of groups who had been marginalized in the past (such as the Irish, Greeks, and Italians) identify today more with their skin colour than with their ethnic background and thus reproduce the racial dichotomy of 'us' versus 'them'. They have changed from being 'micks' and 'wops' and 'macaronis' to being racists. There is a broader theoretical issue here: the tendency to view racism as a binary opposition between racists and those who are racialized.

Culturalism and Political Economy: Explanations for Socio-economic Inequalities

Broadly speaking, two major theoretical frameworks attempt to explain ethnicity and race as social phenomena: culturalism and political economy. The central argument of culturalism can be summarized as follows: ethnic and racial groups share common values, religion, beliefs, sentiments, ideas, languages, historical memories and symbols, leaderships, a common past, and often the same geographical territory. If we want to explain their differential socio-economic achievements, we must look into their culture,

the key to understanding their differences. Culture is considered the *explanans* (that which explains), not the *explanandum* (that which must be explained). Cultural values (often linked to biological traits) affect the psychological composition of their members and produce, it is claimed, 'differences in cognitive perception, mental aptitude, and logical reasoning' (Li 1999, 10). In turn, such differences are thought to affect subsequent educational and economic achievements. Thus, some groups, on average, are doing better than others in school and the labour market. Some cultures foster values conducive to economic achievement (in capitalist conditions); others do not.

The political economy perspective, on the other hand, begins with the tenet that socially constituted individuals belong to inherited social structures that enable but also constrain their social actions. Examples of these structures include those built on social relations of class, gender, race/ethnicity, age, sexual preference, physical ability, mental health/illness, and so on. Societies are characterized by the unequal distribution of property, power, and other resources (both natural and socio-political). Who owns and controls what, when, why, and how are central concerns of political economy (Satzewich 1998, 314). Race and ethnicity are seen as *relational* concepts. Social class, status, race, and ethnicity constitute an 'index of social standing or rank reflected in terms of criteria like wealth, education, style of life, linguistic capacity, residential location, consumptive capacity, or having or lacking respect. Status has to do with one's ranking in a social system *relative to the position of others*, where the ranking involves . . . [positive] self-conception and (de)valuations of others' (Goldberg 1993, 69; emphasis added).

Most immigrant groups have been primarily associated with the lower classes because of the menial (yet highly important) jobs they have done upon arrival. In contrast, members of the charter groups have been associated with the upper classes and with less labour-intensive, more prestigious occupations. Historically, there appeared to be an overlap between lower-class membership and membership in a minority ethnic/racial group. John Porter, in *The Vertical Mosaic* (1965), argued that people's ethnic affiliation was a determinant of their social class membership and prevented the upward mobility of certain groups, partly because they had not *assimilated culturally* to the new conditions of capitalist development in Canada. This is the 'blocked mobility thesis'. We shall examine the economic dimensions of ethnic/racial inequalities below.

The Vertical Mosaic Then and the Colour-Coded Mosaic Today

Over the years, most research on social inequality in Canada has focused on the economic performance of ethnic groups to determine whether Canadian society is hierarchically structured (Agocs and Boyd 1993, 337). Porter argued that immigration and ethnic affiliation were important factors in the process of social class formation in Canada, especially at the bottom and elite layers of the stratification system (Porter 1965, 73). Canadian society, understood as an ethnic mosaic, is hierarchically structured in terms of the differential distributions of wealth and power among its constituent ethnic groups.

Porter argued that the charter groups (British and French) appropriated positions of power and advantage in the social, economic, and political realms and relegated 'entrance status' groups to lower, less preferred positions. 'Less preferred' groups that arrived in Canada after the charter groups were employed in lower-status occupations and were subject to the assimilation processes laid down by the charter groups (Porter 1965, 63–4). Ethnic affiliation implied blocked social mobility. Upward mobility of ethnic groups depended on the culture of the ethnic group in question and the degree to which it conformed to the rules of assimilation set by the charter groups. The improvement in the position of entrance status groups over time could be determined by their 'assimilability' or their behavioural and structural assimilation (1965, 67–73).

Porter found a persistent pattern of ethnic inequality. Canadians of Jewish and British origin were at the top. They were persistently overrepresented in the professional and financial occupations (higher status and income) and under-represented in agricultural and unskilled jobs (lower status and income). The Germans, Scandinavians, and Dutch were closest to the British. Italians, Polish, and Ukrainians were next, with other southern Europeans (Greeks and Portuguese) near the lower end of the spectrum (1965, 90). The French, somewhere between the northern and southern Europeans, were underrepresented in professional and financial occupations and overrepresented in agricultural and unskilled jobs—a result of historical and socio-political factors. Aboriginal people were at the bottom of the hierarchy. Regarding the charter groups, the British were more powerful than the French (1965, 73–103).

In fact, despite the considerable influence exerted on the political system by French Canadians, not only in Quebec but also at the federal level (1965, 417–56), and their access to high-status political positions and the media, the British dominated Canada's economic life and were overrepresented in elite positions (1965, 201–308, 337–416, 520–59).

Even though questions have been raised about the persistence of the **vertical mosaic** for certain European-origin ethnic groups, some suggest that the vertical mosaic persists in a racialized form and that Canada is characterized today by a *colour-coded vertical mosaic* (Galabuzi 2006, 7). In 1984, the Royal Commission on Equality in Employment found substantial income disparities between visible minorities, Aboriginals, and non-visible groups (Royal Commission on Equality in Employment 1984, 84–5). These income disparities were attributed to **systemic discrimination** in the workplace. Visible minorities were sometimes denied access to employment because of unfair recruitment procedures and were more likely to be unemployed. Often, education credentials acquired outside Canada were not recognized in the labour market or by governments. Sometimes, Canadian experience was required unnecessarily (1984, 46–51). For Aboriginal peoples, the situation was even worse: Aboriginal men earned 60 per cent of the earnings of non-Aboriginal men; Aboriginal women made 72 per cent of what non-Aboriginal women earned (1984, 33), and this spoke only of those who had jobs—a high percentage of Aboriginal people then and now, isolated in peripheral locations far from job markets, are unemployed and not seeking employment and therefore are not counted in unemployment statistics.

Lian and Matthews (1998) examined 1991 census data and analyzed ethnic inequalities in earnings, studying the relationship between ethnicity and education and between education and income. They argued that race is now *the* fundamental basis of income inequality in Canada. They found a general trend of convergence of earnings among the European groups, but visible minorities at all educational levels receive lower rewards, substantially below the national average (Lian and Matthews 1998, 471, 475). They concluded that the old ethnic vertical mosaic may be disappearing, but it is being replaced by a strong 'coloured mosaic' (1998, 476). Similar findings were reported by Li (2003), using 1996 census data. Galabuzi (2006) showed that in 2000 the average after-tax income for racialized persons was $20,627, 12.3 per cent less than the average after-tax income of $23,522 for non-racialized persons. Among university degree–holders in 2000, racialized individuals had an after-tax income of $35,617, while their non-racialized counterparts had an after-tax income of $38,919, an 8.5 per cent difference.

Earnings Differentials within Ethnic Groups: The Roles of Class, Gender and Place of Birth

Apart from a few notable exceptions (Li 1988, 1992; Nakhaie 1999, 2000; Liodakis 2002), the class dimension of ethnic earnings inequality in Canada has not been adequately examined. Ethnic groups have internal hierarchies and are themselves stratified. They are not homogeneous; they are differentiated internally by religion, dialect, region of origin, time of arrival to Canada (Porter 1965, 72–3), social class (Li 1988; 1992), gender (Boyd 1992), age, and place of birth (Liodakis 1998; 2002). The vertical mosaic thesis should be questioned, not because we now have more ethnic equality but arguably because inequality

in Canada is still very much based on social class, gender, and place of birth, and ethnicity and race serve as sources of division within the broader class structure (Li 1992). Ethnic inequality cannot be analyzed outside the class context (Li 1988, 141; Nakhaie 1999, 2000; Liodakis 2002; Satzewich and Liodakis 2010).

TIME to REFLECT

Why do some people make more money than others? Do all members of the same ethnic/racialized group have the same income? What do you think causes income inequalities within ethnic and racialized groups?

The terms *ethnicity*, *visible minorities*, *whites*, and *non-whites* are, irreducibly, political categories that construct racial and ethnic groups, often with government approval (Nobles 2000). The term *visible minorities* also homogenizes the 'non-visible' category. Classifications used by Statistics Canada in the census tend to lump together groups from different ethnic backgrounds, often based on racial markers like skin colour. This makes it statistically

(Photos courtesy of TRIEC)

Increasing numbers of immigrants to Canada have impressive credentials and work experience, yet they often are underemployed because their schooling and experience were gained abroad. These posters were part of a public awareness campaign for hireimmigrants.ca, the Toronto Region Immigrant Employment Council (TRIEC).

easier to construct a dichotomy between visible and non-visible groups by adding their 'constituent' parts together, irrespective of their internal ethnic or cultural differences. For example, the category 'Latin American' includes people from diverse cultures and countries such as Brazil, Argentina, Uruguay, Colombia, Venezuela, and so on. 'Arabs' could come from numerous countries across North Africa, the Middle East, the Pacific, or the Balkans. In 1996, Statistics Canada (1996) defined the following groups as part of the 'collectivity' of visible minorities: black, South Asian, Chinese, Korean, Japanese, Southeast Asian, Filipino, Arab/West Asian, Latin American, visible minority not included elsewhere, and multiple visible minority. For the 2006 census, the variable 'visible minority' contained the categories Chinese, South Asian, black, and other visible minority. Such taxonomies create categories so broadly defined that the considerable internal socio-economic heterogeneity within groups is concealed (Boyd 1992, 281; Liodakis 2002; Satzewich and Liodakis 2010).

The colour-coded vertical mosaic thesis does not fully explain the patterns of earnings inequality in Canada. The racialized vertical mosaic thesis overlooks many anomalies that undermine it. In much of the literature on social inequality, southern European groups—the Greeks, the Portuguese, and to a lesser extent the Italians—are not as well educated as the rest of the European groups and do not earn as much; studies have shown that they have lower educational levels and earn less than some visible minority groups (Li 1988, 76, 78, 82, 84, 88, Tables 5.1–5.5). In Boyd's (1992) research, non-visible minority women of Greek, Italian, Portuguese, other European, and Dutch origin made less than the average earnings of all women. Li's data from the 1996 census show that visible minority native-born women actually make more than their non-visible counterparts (Li 2003).

Whereas earlier analyses have tended to emphasize the 'mosaic' dimension of inequality and to examine the earnings inequalities *among* ethnic groups, the 'vertical' dimension is also worth examining to discover the earnings inequalities both among and *within* ethnic groups. We suggest that within each structural locational basis of inequality (ethnicity, gender, or class), the other two coexist. All classes have gender and ethnic segments. Gender groups have class and ethnic segments. All ethnic groups are permeated by class and gender differences.

Table 11.2 shows the social class composition of the several groups and provides contemporary confirmation that ethnic groups are not homogeneous in regard to class composition. If we look at social inequality from the perspective of class composition, it appears that in the case of the proletariat there is no clear-cut visible–non-visible distinction. The Chinese, for example, are less proletarianized than the Portuguese, the Italians, the Greeks, the British, and the French. The Portuguese are more proletarianized than the Chinese, the Caribbeans, and the South Asians. Aboriginal Canadians are more likely to be found in the working class and less likely to

TABLE 11.2 ▼ The Class Composition of Ethnic Groups, 2005 (per cent)

Ethnic Group	Workers	Professionals	Managers, Supervisors	Petite Bourgeoisie	Small Employers
Aboriginal	68.4	17.5	8.1	3.9	2.1
British	51.3	21.2	14.5	7.8	5.3
Caribbean	63.6	19.6	9.3	4.7	2.8
Chinese	46.5	27.7	11.3	7.8	6.8
Filipino	67.3	19.7	7.3	3.6	1.9
French	50.4	24.7	13.4	7.2	4.3
Greek	49.7	20.1	12.7	7.9	9.7
Italian	50.5	21.1	14.7	6.9	6.8
Jewish	28.7	31.3	15.9	11.8	12.3
Portuguese	67.1	11.9	11.6	5.3	4.2
South Asian	61.9	18.6	9.4	5.7	4.4
Canada	52.4	22.3	12.7	7.6	5.0

NOTE: Table includes respondents over 18 years old who worked at least one week in 2005 in Canada, excluding the territories, Atlantic Canada, and inmates. Percentages may not add up to 100 due to rounding error.
SOURCE: Calculated from the Public Use Microdata File on Individuals, 2006 Census.

be found in the other classes, although Aboriginal women are less proletarianized than Aboriginal men (Liodakis 2009). Aboriginal, Caribbean, Filipino, French, South Asian, and Portuguese individuals are under-represented in the ranks of employers, while British, Chinese, Greek, and Jewish individuals are variously overrepresented among the ranks of the petite bourgeoisie. In the professional category, the Chinese are well above the national average, above the Portuguese, the Italians, the Greeks, and the charter groups, second only to Jewish-origin Canadians. In the small-employer category, the Chinese are also overrepresented, but all other visible groups are under-represented, along with the Portuguese, Greeks, and Aboriginals.

As one would expect, there is great diversity among and within ethnic groups in terms of their class composition. In addition, ethnic groups have different gender and nativity compositions. For example, in terms of gender, the Caribbean and Filipino groups have more women than men in the labour market because of immigration patterns. In all visible groups, the percentage of foreign-born exceeds 90 per cent, whereas in the non-visible category, the percentages are much lower (less than 50 per cent). These differences affect the earnings inequalities that exist *within* ethnic groups but are concealed if we only look at them as homogeneous entities (Satzewich and Liodakis 2010). There are considerable differences in the earnings of classes within ethnic groups. Historically, the petite bourgeoisie and the proletarians have mean earnings below the national average, while small employers, semi-autonomous workers, and managers and supervisors have earnings considerably above it (Li 1988, 1992; Liodakis 2002). In addition, women, on average, make less than men in all ethnic groups and in all social classes. In general, internal variations of earnings *within* ethnic/racialized groups are greater than the earnings differentials *among* them (Liodakis 2002; Satzewich and Liodakis 2010).

Moreover, the greater the size of the immigrant component of an ethnic/racialized group, the more likely it is that the group will experience lower earnings. Data from the 2006 Canadian census show that earnings differentials exist among the Canadian-born, immigrants, and recent immigrants. Three patterns have emerged over the years: First, immigrants, as a group, make less than Canadian-born individuals. Second, these differences are greater among those with university education. For example, recent Statistics Canada census data (2006) show that the median 2005 earnings of the Canadian-born with a university degree were $51,656, whereas those of immigrants with a university degree were only $36,451—a difference of $15,205. For those without a university degree, the difference was only $4,801 ($32,499–$27,698). It appears that immigrants with higher education, although they make more than other immigrants without a university education, make a lot less than their Canadian-born counterparts. Third, the 2005 median earnings of recent immigrants (i.e., those who had immigrated to Canada during the five years before the census was taken) were much lower than those of immigrants who had been in Canada longer than five years and the Canadian-born. For example, when compared with the Canadian-born with a university degree, recent immigrants with a university degree earned $27,020 less ($51,656–$24,636). Recent immigrants without a university degree made $13,927 less than their Canadian-born counterparts ($32,499–$18,572) (Satzewich and Liodakis 2010).

TIME to REFLECT

Do you think that there is a link between a person's ethnic/racial background and his or her moral, intellectual, and behavioural characteristics? If so, can you think of specific examples, without resorting to stereotypes?

A closer look at the general trends of earnings of recent immigrants since the 1980s points to a steady decline in their earnings compared to those of the Canadian-born. Table 11.3 shows median earnings differences among male and female recent immigrants with and without a university degree from 1980 to 2005. Whereas in 1980 recent immigrant males with a university degree made 77 cents for every dollar their Canadian-born counterparts made, in 1990 they made 63 cents. In the year 2000, they made 58 cents and in 2005, only 48 cents. Female recent immigrants in 1980 with a university degree made 59 cents for every dollar their Canadian-born counterparts made, 63 cents in 1990, 52 cents in 2000, and only 43 cents in 2005. Male recent immigrants without a university degree in 1980 made 84 cents for every dollar their Canadian-born counterparts made. In 1990, they made 67 cents, in 2000, 65 cents, and in 2005, only 61 cents. Female recent immigrants without a university degree in 1980 made 86 cents for every dollar their Canadian-born counterparts made. In 1990, they made 77 cents, in 2000, 66 cents, and in 2005, only 51 cents.

In short, there has been a steady deterioration of recent-immigrant earnings, irrespective of gender and university education. This is a troubling trend, given that today, most recent immigrants have more educational credentials than those who immigrated to Canada in the 1980s. Although there are individual variations (knowledge of official languages and foreign education play important roles in influencing immigrant earnings), in general, immigrant status has a strong, negative impact on earnings. Recent immigrants experience higher levels of earnings inequality. They are more likely to work part-time than full-time, more likely to face unemployment, and less likely to move up the occupational hierarchy. They earn less than Canadian-born workers, and they also face earnings volatility (instability). Those initial earnings differences tend to persist in later years, especially during times of economic recession, as we experienced in the 1990s (Ostrovsky 2008, 24–5). The most recent recession is likely to have had negative effects on the earnings of all Canadian workers but especially on those of recent immigrants.

Conclusion: The Future of Race and Ethnicity

In this chapter, we have argued that ethnicity and race are social relations. As such, they are about power among individuals and social groups. Notions of ethnicity and race are about domination and subordination; they are rooted in the history of colonialism and associated with the development of capitalism. Historical processes that have made some people 'minorities' have led to and continue to inform and reproduce the formation of the social, political, and economic dichotomies of the 'Self' and the 'Other'. Canada's current socio-demographic makeup is linked to the historical (and ongoing) 'othering' of Aboriginal peoples, the usurpation of their lands, the destruction of their cultures, and government policies of forced assimilation. It is also intertwined with racist immigration policies that, for a long time, excluded visible minorities and other 'non-preferred' groups from immigrating to Canada.

Race and ethnicity are bases of social inequality. They inform and are part of its class and gender dimensions. In Canada, some groups are doing better than others. If we consider ethnic and racial groups as homogeneous entities, there appears to be a binary social hierarchy based on visibility. When we examine the internal class and gender differences among groups, it is apparent that the Canadian-born, males, managers and supervisors, professionals, and small employers do better than the foreign-born, females, workers, and the petite bourgeoisie. Canada now has an official policy of multiculturalism that attempts to integrate minorities into the social fabric. But the policy does little to address the economic inequalities in Canadian society and has not been very successful in combating racism or promoting the institutional integration of minorities.

TABLE 11.3 ▼ Median Earnings of Male and Female Recent Immigrant and Canadian-Born Earners, 1980 to 2005

	Recent Immigrant Earners				Canadian-Born Earners				Recent Immigrant to Canadian-Born Earnings Ratio			
	With a university degree		With no university degree		With a university degree		With no university degree		With a university degree		With no university degree	
	Males	Females	Males	Females	Males	Females	Males	Females	Males	Females	Males	Females
Year	2005 constant dollars								Ratio			
1980	48,541	24,317	36,467	18,548	63,040	41,241	43,641	21,463	0.77	0.59	0.84	0.86
1990	38,351	25,959	27,301	17,931	61,332	41,245	40,757	23,267	0.63	0.63	0.67	0.77
2000	35,816	22,511	25,951	16,794	61,505	43,637	39,902	25,622	0.58	0.52	0.65	0.66

NOTES:
1. The numbers refer to all earners, whether or not they worked on a full-time basis for a full year. Individuals with self-employment income are included, while those living in institutions are excluded.
2. Medians are not available for counts of less than 250. Earnings are in 2005 constant dollars.
3. Recent immigrants in 2005 are those who immigrated between 2000 and 2004; recent immigrants in 2000 are those who immigrated between 1995 and 1999; and recent immigrants in 1995 are those who immigrated between 1990 and 1994.

SOURCE: 'Median earnings, in 2005 constant dollars, of male and female recent immigrant earners and Canadian-born earners aged 25 to 54, with or without a university degree, Canada, 1980 to 2005', adapted from 'Income and earnings, 2006 Census', Catalogue 97-563-XWE2006002, Table 8, www.statcan.gc.ca/bsolc/olc-cel/olc-cel?catno=97-563- XWE2006002&lang=end.

Recent efforts of 'reasonable accommodation' have sparked more debates. This is by no means an exclusively Canadian phenomenon. The wider global context is interesting: in the post-modern, globalized world, the hegemonic economic, political, and cultural powers (e.g., the US, the European Union, Japan) have increasingly pushed for world economic integration through free trade, the free movement of capital across nation-states, the control and surveillance of international labour migration, the weakening of the role of the nation-state, as well as the rise of supranational organizations like the World Bank, the International Monetary Fund, and the World Trade Organization. A trend toward global cultural homogenization is partly attributable to the export of consumer popular culture to developing nations.

In the past two decades, the world has witnessed the destruction of the Soviet Union, the triumph of capitalism, and the dominance of Western culture. And yet the world does not seem to be any more peaceful or egalitarian. Nor have ethnic/racial and cultural identities or racism disappeared. On the contrary, we have witnessed the rise of nationalisms; ethnic cleansing; a new racism, xenophobia, and Islamophobia (especially after 9/11); wars in ex-Yugoslavia, Afghanistan, Iraq, and elsewhere; and a general thrust against the protection of individual and group rights and freedoms in all Western, liberal-capitalist democracies—all in the name of fighting 'terrorism' and 'exporting' what is claimed to be democracy. At the heart of all these matters are race and ethnicity, a major field of study within the social sciences, especially within sociology.

questions for critical thought

1. What criteria would you use to differentiate human populations, and why?

2. What makes you a member (or not) of an ethnic and/or racial group? Should Ontarians be considered an ethnic group? If yes, why? If not, why not? Try to apply the criteria listed in the first part of this chapter to answer these questions.

3. The Canadian policy of multiculturalism is better than the American view of their society as a melting pot. Do you agree or disagree with this statement? Why?

4. With which criticisms of multiculturalism do you agree or disagree, and why?

5. Can the policy of multiculturalism alone provide solutions to the problems of racism and the attendant issues of immigrant and minority group integration into Canadian political, social, and economic institutions? Explain.

6. What else should policy-makers do to address the issues of racism and immigrant integration into Canadian society?

7. What accounts for the earnings differentials among different ethnic/racial groups—cultural or structural differences? Assume, for the sake of argument, that all members of ethnic/racial groups share the same cultural and behavioural characteristics. If culturalist explanations could account for the economic inequalities among ethnic/racial groups, what would explain the marked economic inequalities within ethnic/racial groups?

8. What is the notion of reasonable accommodation? Who decides what is reasonable? With this in mind, try to explain the rise of xenophobia and Islamophobia in the post-9/11 world of control and surveillance in the US, Canada, and other parts of the world. Choose a particular issue (e.g., the veil), and survey the opinions of your friends and family. What do you conclude?

recommended readings

Tanya Basok. 2002. *Tortillas and Tomatoes: Transmigrant Mexican Harvesters in Canada*. Montreal and Kingston: McGill-Queen's University Press.
This book examines the role of Mexican seasonal workers in Canadian agriculture and provides a critique of the ways they have been treated by employers and the Canadian government.

Grace-Edward Galabuzi. 2006. *Canada's Economic Apartheid: The Social Exclusion of Racialized Groups in the New Century*. Toronto: Canadian Scholars' Press.
In this controversial argument that supports the view of Canada as characterized by a new colour-coded vertical mosaic, Galabuzi presents evidence of persistent income inequalities between racialized and non-racialized Canadians.

Frances Henry and Carol Tator. 2005. *The Colour of Democracy: Racism in Canadian Society*. 3rd edn. Toronto: Thomson Nelson.
This thorough and caustic critique of racism in Canadian policies and institutions points to the contradictions of multiculturalism and democratic racism in Canadian society.

Peter Li. 2003. *Destination Canada: Immigration Debates and Issues*. Toronto: Oxford University Press.
This is an excellent and up-to-date review of the major debates about the social and economic consequences of immigration to Canada.

Katharyne Mitchell. 2004. *Crossing the Neoliberal Line: Pacific Rim Migration and the Metropolis*. Philadelphia: Temple University Press.
This is an excellent account of the debates and controversies surrounding Chinese business immigration to British Columbia in the 1980s and 1990s.

Vic Satzewich and Nikolaos Liodakis. 2010. *'Race' and Ethnicity in Canada: A Critical Introduction*. 2nd edn. Toronto: Oxford University Press.
This work summarizes theoretical approaches to the study of race and ethnicity, Canadian immigration policies, Aboriginal–non-Aboriginal relations, economic inequalities among and within ethnic groups, multiculturalism, racism, and transnationalism.

Stephen Steinberg. 1989. *The Ethnic Myth: Race, Ethnicity, and Class in America*. 2nd edn. Boston: Beacon Press.
The author argues that cultural 'traits' often considered 'ethnic' may be more directly related to class, locality, and other social conditions and provides a caustic commentary on the conditions of recent immigrants and a penetrating reappraisal of the black underclass in the United States. You also may want to look at his latest study, *Race Relations: A Critique* (Stanford, CA: Stanford University Press, 2007).

Anthony Synnott and David Howes. 1996. 'Canada's visible minorities: Identity and representation'. In V. Amit-Talai and C. Knowles, eds, *Re-situating Identities: The Politics of Race, Ethnicity and Culture*. Peterborough, ON: Broadview Press.
In this critique of the concept of visible minority, Synnott and Howes question whether it makes sense to lump together so many different groups, with different immigration histories and backgrounds, into a single category of dubious analytical value.

recommended websites

Assembly of First Nations
www.afn.ca
This excellent website of the national organization for status Indians, established in 1982 out of the earlier National Indian Brotherhood, includes press releases, publications, news, policy areas, information on past and future annual assemblies, and links to provincial and territorial organizations. You might also want to check out the fine websites of the other two national Aboriginal organizations in Canada: Inuit Tapiriit Kanatami, at www.itk.ca, and the Métis National Council, at www.metisnation.ca.

Canadian Heritage: Multiculturalism
www.canadianheritage.gc.ca/progs/multi/index _e.cfm
This federal department site includes information on multicultural programs, definitions of multiculturalism and diversity, news releases, publications, and links to numerous Canadian and international organizations.

Canadian Race Relations Foundation (CRRF)
www.crr.ca
The CRRF, established by an act of Parliament in 1991, is the lead government agency that aims to eliminate racism in Canada. Its site outlines programs, includes publications, and has useful links to other sites.

Global Networks
www.globalnetworksjournal.com
Global Networks journal, founded at Oxford University in 2001, provides links to sites on transnational movements of goods and people and on globalization, as well as journal contents.

Greek Community of Toronto
www.greekcommunity.org
Operated by the largest local Greek organization in Canada, this site is representative of ethnic organization at the metropolitan level and provides information on employment opportunities, social happenings, political action of interest to the Greek community, and information on the various departments within the organization, such as cultural affairs, women's issues, education, and social services.

International Organization for Migration
www.iom.ch
This intergovernmental organization, with 120 member countries, provides news of interest and information on policy and research, as well as 'Quick Links' to such topics as international migration law and United Nations resolutions and reports related to migration. The organization is premised on the liberal-capitalist belief that 'humane and orderly migration benefits migrants and society.'

Justicia for Migrant Workers—J4MW
www.justicia4migrantworkers.org
This non-governmental organization, founded in 2001 and based in Toronto with an office in Vancouver, seeks to promote the rights of Mexican and Caribbean migrant workers in Canada. The website, which is bilingual English/Spanish, provides notices of upcoming events, press releases, description of ongoing campaigns, and a 'Wall of Shame' of public statements by politicians and other stakeholders showing 'the face of racism in Canada'.

Québec interculturel
www.quebecinterculturel.gouv.qc.ca/fr/index.html
This Quebec government Ministry of Immigration and Cultural Communities site, in French, provides information on associations in Quebec, employment, the province's ethno-cultural diversity, and much more.

Ukrainian Canadian Congress (UCC)
www.ucc.ca
The Ukrainian Canadian Congress is one of the oldest and largest national ethnic organizations and was among the several older national ethnic groups that profited significantly from the introduction of official multicultural policy in 1971. The UCC site includes background on and history of Ukrainian immigration, information for prospective immigrants from Ukraine, updates on Canada–Ukraine relations, and news of relevance to the Ukrainian-Canadian community.

chapter 12

The Social Aspects of Aging

Lynn McDonald

In this chapter, you will:

▶ Learn what is meant by *aged*, *aging*, and *social gerontology*;

▶ Examine the growth of the aging population in Canada;

▶ Compare various theoretical approaches to understanding social gerontology;

▶ Learn about the health, wealth, and social relations of older Canadians;

▶ Identify some of the contemporary issues in social gerontology, such as health care, retirement, homelessness, and ageism.

Not only are there many more aged people than there were, but they no longer spontaneously integrate with the community: society is compelled to decide upon their status, and the decision can only be taken at government level. Old age has become the object of a policy.

Simone de Beauvoir,
The Coming of Age (1972, 222)

Introduction

Although stated more than 30 years ago, French philosopher Simone de Beauvoir's observation remains relevant today. Now a central issue in most societies, the sociological study of aging focuses on the aging both of individuals and of populations in general. People age physiologically, psychologically, and socially, and the ways in which aging occurs vary from time to time and from society to society. Today, population aging is occurring worldwide at an unprecedented rate. Between 1950 and 2000, the number of people aged 60 and over tripled to 600 million. In 2006, this figure surpassed 700 million, and it is projected to reach 2 billion by the year 2050 (United Nations, Department of Economic and Social Affairs, Population Division 2007). Currently, the median age for the world's population is 28 years, which means that half the global population is below the age of 28 and the other half is above it (2007).

Gerontology, the scientific study of aging and its consequences, is a relatively new field of inquiry (Harris 2007). The term was coined in 1903 by Russian-born zoologist and microbiologist Elie Metchnikoff and represents a combination of the Greek word *geron*, meaning 'old man', and *ology*, meaning 'the study of'. Metchnikoff, usually considered the father of gerontology, stated that the 'scientific study of old age and the means of modifying its pathological character will make life longer and happier' (Metchinkoff 1903).

In the ensuing years, gerontology attracted researchers from around the world and began to build intellectual capital on the shoulders of many scholars in a wide variety of disciplines. In the 1950s, social gerontology emerged as one of the core areas of gerontology, focusing on the study of the social aspects of aging. Firmly rooted in the disciplinary traditions of sociology, psychology, economics, anthropology, and the humanities, it is concerned equally with age and aging. Originally, social gerontology focused solely on older adults (those aged 65 years and older) as a fixed group with similar patterns of behaviour. Today, the concept of **aging** has been added to the focus on the aged and is viewed as a multilevel dynamic process followed over a life course. Individuals are seen to move through normative (e.g., becoming a grandparent) and non-normative life course events (e.g., an epidemic) on their journey to later life, all within a given historical period.

The study of aging is an important issue for all of us because we all age. Today in Canada, nearly everyone, barring a catastrophic death, can expect to live a long life into their 80s, 90s, and even beyond. For example, the number of centenarians in Canada increased to 4635 in 2006, up more than 22 per cent from 2001 (Statistics Canada 2007d) and was estimated at about 6000 in 2009 (Statistics Canada 2009b). From a historical perspective, this is a unique situation, because never before have so many people lived beyond the age of 65. In our lifetime, aging will affect virtually every aspect of society, including family, education, work, business, and the government. Therefore, how Canadian society deals with the challenges posed by a greying population is an important question. Education about aging is crucial for everyone, not just for people who are older or for those who pursue careers in gerontology. People of all ages need to learn about the realities of aging that have long been overshadowed by persistent myths such as 'to be old is to be sick' or by clichés such as 'you can't teach an old dog new tricks'. As the World Health Organization (WHO) states, 'Aging is truly a privilege and a societal achievement. It is also a challenge which will impact all aspects of the 21st century' (www.who.int/ageing/projects/en).

The purpose of this chapter, then, is to examine population aging both in Canada and globally, to review the main theories in gerontology that have their roots in sociology, and to learn about the health, wealth, and social relationships of older adults. We end the chapter with an examination of some of the issues concerning gerontologists today and a glimpse at the future.

TIME to REFLECT

What is your view of the proverb 'Age is a sickness from which everyone must die'?

The Demographic Imperative

Demography is the scientific study of human populations that documents their size, age–sex structure, and how they are distributed geographically. These features of populations and any changes in them are a result of fertility, mortality, and migration rates.

POPULATION AGING

Populations of people can grow older or younger as measured by indicators used to describe the age structure of a population. Two indicators of the age structure of a population are the proportion of persons aged 65 years and over and the proportion of children aged less than 15 years. A population is considered older than another when its proportion of older adults is higher. In contrast, a population is considered younger than another if its proportion of children is higher. According to the 2006 census, the number of Canadians aged 65 and over increased by 11.5 per cent over the previous five years, and the number of children under the age of 15 declined by 2.5 per cent over the same period. Those 65 years and older made up a record 13.7 per cent of the total population of Canada in 2006, representing one in seven Canadians. During

this period, the proportion of the population under the age of 15 fell to 17.7 per cent—the lowest level ever recorded (Statistics Canada 2007d).

The percentage of people over age 65 has been rising steadily since 1966, when it was 7.7 per cent (Figure 12.2). According to the most recent population projections, it is estimated that the proportion of those 65 years and older will represent between 23 and 25 per cent of the population by 2036 and between 24 and 28 per cent of the population by 2061 (Statistics Canada 2010e), while the proportion of children is expected to continue falling. By the year 2015, the number of older adults will exceed the number of children in Canada. In addition, the aging of the population will accelerate over the next three decades, particularly as individuals from the baby-boom years, a large cohort born between 1946 and 1965, begin turning age 65 in 2011.

An informal rule used by demographers is that any population with more than 10 per cent of the population over age 65 is an old population. Canada, therefore, is seen to have an old population and in fact has the oldest population in the Americas. For example, in the United States 12.4 per cent of the population was over age 65, while 6 per cent of the population of Mexico was over 65 years of age as of 2007. Among the G8 countries, however, 20.8 per cent of the population

▼ FIGURE 12.1 Number of Persons Aged 65 years and Over and Number of Children Aged Less Than 15 Years in the Canadian Population, 1956–2017

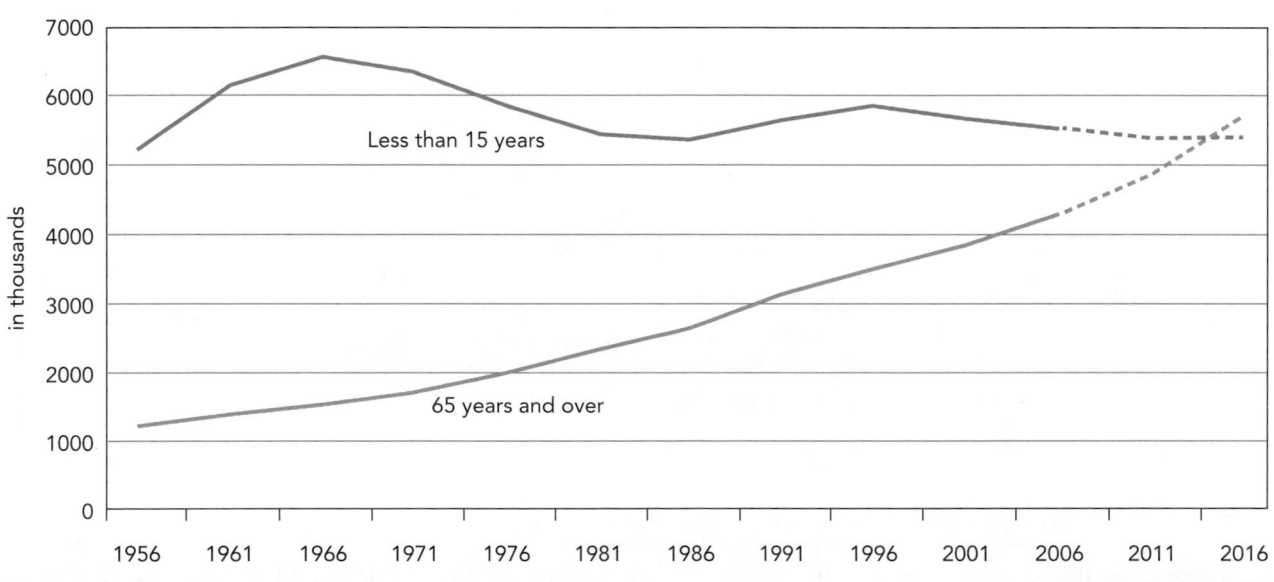

SOURCES: Statistics Canada, Censuses of population, 1956 to 2006; Alain Bélanger, Laurent Martel, and Éric Caron Malenfant, 2005, *Population Projections for Canada, Provinces and Territories 2005–2031*, Statistics Canada Catalogue no. 91-520, scenario 3.

▼ FIGURE 12.2 Proportion of Persons Aged 65 Years and Over in the Canadian Population, 1956 to 2006

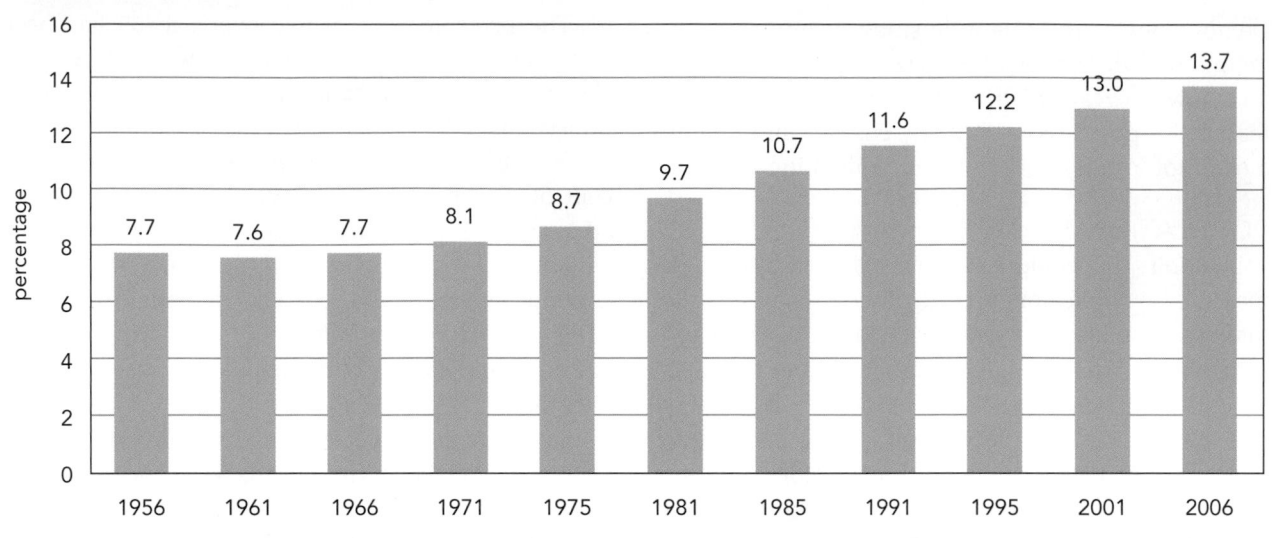

SOURCE: Statistics Canada, Censuses of population, 1956 to 2006.

in Japan is 65 and over, 19.7 in Italy, and 19.3 in Germany (see Figure 12.3). These countries have the highest proportions of older adults in the world.

Not surprisingly, developing countries have the lowest proportions of older adults. The region of Africa, for example, has just 3 per cent of its population over 65 years of age. In short, developing countries are only beginning to experience what developed countries have faced and therefore have more time to prepare for the age wave that will wash over the world.

TIME to REFLECT

Can we afford an aging society?

▼ FIGURE 12.3 Proportion of Persons Aged 65 Years and Over among the G8 Countries, 2006

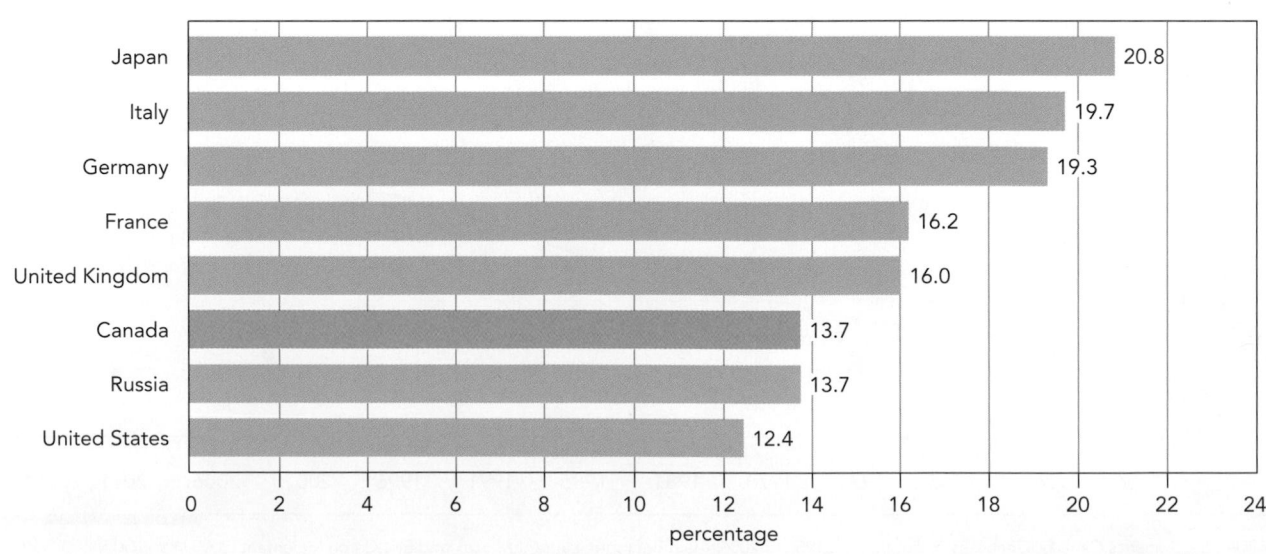

SOURCES: Statistics Canada, Census of population, 2006; US Census Bureau, Population Estimates Program; Istituto nazionale di statistica; Institut national de la statistique et des études économiques; United Kingdom National Statistics; Statistics Bureau of Japan; Federal Statistical Office of Germany; Federal State Statistics of Russia.

Not only is the population of Canada aging, but the population of older adults is itself aging. Among those 65 years of age and older, there is an increasing number and percentage of people who are aged 80 and over. It was estimated in 2009 that there were 1.3 million people over age 80, the largest number in the history of Canada (Statistics Canada 2010e). The proportion of people aged 80 and over in the 65-and-over population was an estimated 27.6 per cent in 2009, up from 26.9 per cent in 2006, 24 per cent in 2001, and 14.6 per cent in 1956 (Statistics Canada 2009b).

Another measure of Canada's aging population is its median age. Canada's median age has been rising steadily since 1966 and reached 39.5 years in 2006. Again, this would mark Canada as an old country, because a population with a median age over 30 is considered old by demographers. Just how old the country is, using this measure, is most apparent when Canada is compared to other countries. As noted earlier, the median age for the world is currently 28 years. The youngest population lives in Uganda, with a median age of 15 years, and the oldest population lives in Japan, with a median age of 43 years (United Nations 2007).

An aging population is commonly believed to be a sign of an imminent crisis that will place a stranglehold on a country's economy, pension system, health and welfare systems, and structures of long-term care for older adults. It could also lead to conflict between generations because of inequities in resources. Terms like the 'demographic time bomb' or the 'agequake' capture the urgency of this so-called crisis. Most recently, the aging population has been termed 'as big a threat as climate change' (Beckford 2008). The labelling of older people as a societal burden is manifest in what gerontologists call 'apocalyptic demography': a belief system that asserts that population aging, with its 'hordes' of older people, will bankrupt society and exact an unfair toll on younger generations (Gee 2000). Apocalyptic demography operates on five assumptions:

1. All older people are pension-rich, 'greedy geezers'.

2. Older adults are responsible for the debt and the imminent collapse of the health-care system.

3. The older population constitutes a social problem that must be solved.

(imagebroker/Alamy)

The HIV/AIDS epidemic and crippling poverty among sub-Saharan nations are major contributors to the low median age in Africa.

TABLE 12.1 ▼ World Population Aging

	Per cent of Persons Ages 65 and Older		
	2007	2025	2050
WORLD	7	10	16
Industrialized Countries	16	21	26
Developing Countries	6	9	15
Europe	16	21	28
North America	12	18	21
Oceania	10	15	19
Latin America & Caribbean	6	10	19
Asia	6	10	18
Africa	3	4	7

SOURCES: C. Haub, 2007 World Population Data Sheet, and United Nations Population Division.

▼ **FIGURE 12.4**   Per Cent of Population Aged 65 Years and Over, 2009

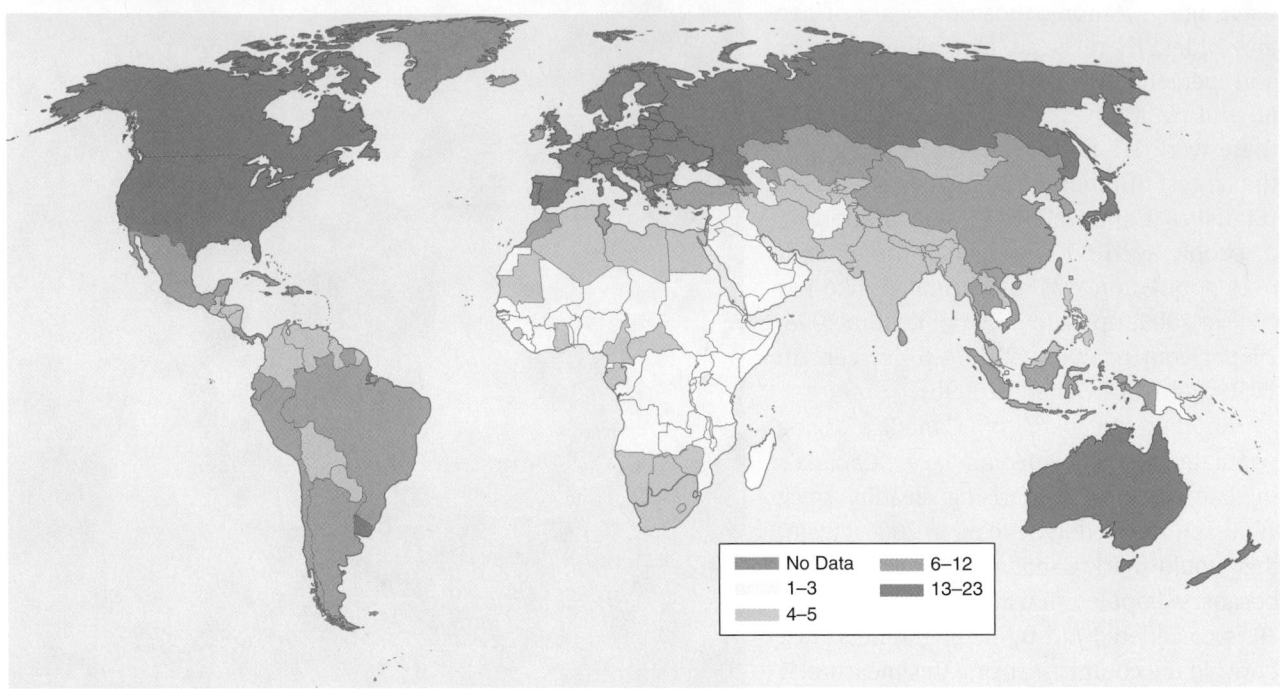

	No Data		6–12
	1–3		13–23
	4–5		

SOURCE: © 2009 Population Reference Bureau and *World Population Data Sheet* for 2009.

GLOBAL ISSUES

Cities Are Growing and Greying

By 2007, more than half of the world's population lived in cities, and by 2030 about three out of every five people will be urban-dwellers. At the same time that cities are growing, the proportion of older people worldwide is rising rapidly. At approximately 600 million today, the number of people aged 60 and older worldwide will double to reach 1.2 billion by 2025. Interestingly, both of these trends are occurring at a much faster rate in developing countries, so it is important to remember that in supportive urban settings, older people are a resource for their families, communities, and economies. To help cities make the most of an increasingly older population, WHO released the Global Age-Friendly Cities Guide on the occasion of the International Day of Older Persons, 1 October 2007. An age-friendly city adapts its structures and services so that they are accessible to, and inclusive of, older people with their varying needs and capacities.

The age-friendly cities project asked about 1500 older people in 33 cities from 22 countries to describe the advantages and barriers they experience in eight areas of city living:

1. outdoor spaces and buildings
2. transportation
3. housing
4. social participation
5. respect and social inclusion
6. civic participation and employment
7. communication and information
8. community support and health services

The issues, concerns, and suggestions voiced by older people were complemented by the views of about 750 people who acted as their caregivers or service providers. From these worldwide consultations, WHO was able to identify the key features of an age-friendly city and developed checklists for each of the eight identified areas.

SOURCE: Adapted from the World Health Organization, www.who.int/ageing/age_friendly_cities_material/en/index.html.

4. Intergenerational conflict will occur as older adults get more than their fair share at the expense of younger generations.

5. Population change and politics will combine when the aging of the population becomes a tool for social policy reform based on cuts to the welfare state. (Gee 2002; Robertson 1997)

This negative view of population aging must be questioned (Eberstadt 2007; Gee 2002; Gee and Gutman 2000; Mullen 2000; National Academy on an Aging Society 1999). There is little evidence that population aging will have dire social and economic consequences for Canada, and any assumptions on that score illustrate how demography can be used to construct a social problem in ways that sometimes serve political agendas, such as cutting health services or pensions.

WHAT CAUSES POPULATION STRUCTURES TO CHANGE?

Demographic change is determined by three factors: fertility or birth rates, mortality or death rates, and net immigration. The main factor accounting for population aging in Canada is declining fertility; as fertility decreases, the percentage of children in the population drops and the percentage of older adults rises. Today, the level of fertility in Canada is 1.5 children per woman, far from the fertility level of 2.1 children per woman that is required for population replacement, a figure last seen in Canada in 1971. Canada coasted for some time as a 'young' country because of the baby boom that occurred from 1946 to 1965, following World War II. For example, in 1959 Canada's fertility level peaked at four children per woman, but it has never reached that level since then.

Declines in mortality and increases in life expectancy have also contributed to the aging of the population. Because of widespread infectious diseases, people used to die at younger ages. In the late nineteenth century, and through the twentieth century, **life expectancy**—the number of years that an average person in a given population can expect to live—increased continuously in developed countries as a result of the conquest of infectious diseases, better nutrition, improved sanitation, and factors related to family structure such as family size. By the end of the twentieth century, the cause of death had shifted toward chronic degenerative diseases, such as heart disease and stroke, and became concentrated at older ages, which fuelled a remarkable decline in mortality

among the most elderly. Over the past 15 years, the mean age at death for both sexes has risen steadily, increasing by 3.5 years for males and by 3.2 years for females. In 2005, the mean age at death of the population in Canada was 74.2. The mean age at death for males was 71.1 and for females 77.4 (Statistics Canada 2008a).

The trends in life expectancy at birth and at age 65 from 1995–7 to 2005–7 differ. In the three-year period 2005–7, life expectancy at birth was 80.7 years—78.3 years for males and 83.0 years for females. At age 65, life expectancy was 19.8 years overall (18.1 years for males and 21.3 years for females). It may seem surprising, but between the periods 1995–7 and 2005–7, life expectancy at birth rose by 2.3 years, from 78.4 to 80.7 years. Males had a greater gain in life expectancy at birth (an increase of 2.9 years) compared to females (1.8 years), meaning that the male–female gap in life expectancy at birth narrowed by 1.1 years from 5.8 years in 1995–7 to 4.7 years in 2005–7 (Statistics Canada 2010c). On an international scale, Canada ranked sixth in terms of life expectancy at birth for males and seventh for females among 15 countries, with Iceland as the top-ranked country and the Czech Republic the lowest (Statistics Canada 2008a).

Whether life expectancy will continue to increase during the twenty-first century is a matter of dispute.

(Tom Craig/Alamy)

The return of young soldiers to North America and the affluence of the postwar years resulted in the baby-boom generation.

Some scientists argue that developed countries are fast approaching a biologically fixed lifespan in the range of 85 to 100 years, while others suggest that the human lifespan could be extended even further as a result of enhancements in molecular medicine and lifestyle improvements, such as healthier diets (National Institute on Aging and Population Reference Bureau 2006).

Because fertility has been below replacement level in Canada for more than three decades, population growth cannot be maintained without sustained immigration. Data confirm that immigration has had a significant effect on the growth and diversity of Canada's population and helps to meet certain labour force requirements. During the 2007–8 period, Canada's overall gain from its global population exchanges came to 257,100—a level that had only been surpassed once before (292,100 in 1988–9). Whether immigration will slow the aging of the population in the future is a matter of statistical projection, a notoriously difficult task because it depends to a considerable extent on the national policies of other countries (Statistics Canada 2005c). It is also important to consider that immigrants typically arrive in Canada when they are about 30 years of age and then continue to age along with the rest of the population.

Thus, while Canada may increasingly look to immigration to help reduce the effects of the aging of its population, immigration can play only a very small part in reversing population aging.

TIME to REFLECT

Immortality. Life extension. The Fountain of Youth. Is it hope, or is it hype?

DIFFERENCES IN POPULATION AGING

The aging population of Canada is very diverse. Some of the main sources of this diversity include gender, indigenous peoples, and ethnicity, all of which we consider next.

Gender

Overall, slightly more than half of the people in Canada are female, and this proportion increases with age. In 2004, 57 per cent of all Canadians aged 65 years and older were women (Statistics Canada 2006e). Statistics Canada also notes that in 2005, women accounted for almost 75 per cent of persons aged 90 or older and 52 per cent of persons aged 65 to 69 years. As we saw earlier, a longer life expectancy among women explains their overrepresentation at older ages (Turcotte and Schellenberg 2007). However, we also saw that the differences in life expectancy between women and men have begun to narrow, and consequently, gender composition at older ages is expected to equalize in the future.

The differential in longevity between women and men has a number of ramifications for women. Because older women live longer than men, they are more likely to be widows and grandparents for a longer period, to live alone, and to face chronic diseases or to have a disability, and they are twice as likely as men to end their lives in an institution. As well, because women live longer, their financial resources have to be spread over a longer period. This is a serious concern, because women typically have lower incomes than men throughout their lives. At older ages, half of women's incomes come from government transfers, such as Old Age Security and the Guaranteed Income Supplement (Statistics Canada, 2006e). There is no doubt that gender is an important factor in the lives of Canadians and crucial to the understanding of aging.

Aboriginals

The Aboriginal population is aging more slowly than the rest of the Canadian population because of a more gradual increase in life expectancy despite declining birth rates (Statistics Canada 2003a). Only 4 per cent of Aboriginal people were 65 years and over in 2001. Of the three Aboriginal groups, the Inuit had the youngest population, with only 3 per cent of Inuit 65 years and over compared to 4 per cent for both the North American Native and Métis populations (Turcotte and Schellenberg 2007).

These aging trends within Aboriginal populations have a number of different causes. According to the 2001 census, more than one in three older Aboriginals are widowed, most live in rural areas or on reserves, and many live in inadequate and crowded housing. As well, the health of Aboriginals is worse than that of the general Canadian population, and their incomes are lower. Of all older groups, Aboriginals have the lowest incomes, well below the **Low Income Cut-off** rate (LICO) set by Statistics Canada. For example, 30.8 per cent of older Aboriginals live below the LICO compared to 17.8 per cent of the older non-Aboriginal population. The aging trends of Aboriginal peoples is of particular concern for many reasons, one of which is that older Aboriginals serve as a very important link to traditional Aboriginal culture through their use of language. The three most common languages spoken by North American Aboriginals are Cree, Inuktitut, and Ojibway; older adults are twice as likely to speak an Aboriginal language as their children or grandchildren (Turcotte and Schellenberg 2007). It is quite possible that as the older generation passes, these languages will become extinct.

Ethnicity and Aging

The aging of ethnic populations that is predicted to occur in major urban centres in Canada is an important issue for several reasons. First, the diversity of old people will change, because there will be a population decline of those of British ancestry and population increases of those of Aboriginal and visible-minority backgrounds among older people, most of whom are predicted to age in a metropolitan environment. Second, the study of older persons both within their own ethnic group and in relation to other ethnic groups is significant, because different societies respond to the challenges of aging in different ways and these differences have serious implications for healthy and successful aging. Third, there will be a whole 'new' older population that will require the creation of health and social services aligned with different ethnic and cultural needs. As a result of these developments, there is some doubt as to the

relevance of past gerontological research into the current and future conditions of older Canadians.

Between 1981 and 2001, the proportion of all immigrants to Canada from western and northern Europe declined from 45.5 per cent to 24.6 per cent, and the proportion of immigrants from Asia increased dramatically from 13.9 per cent to 36.5 per cent. The Employment Equity Act (1995) indicates that 7.1 per cent of the total population 65 years and over were members of a visible-minority group in Canada in 2001 (Chappell, McDonald, and Stones 2007). Of all the visible minorities over age 65, 39 per cent were Chinese, 21 per cent were South Asian, and 13 per cent were black. Older visible minorities live in major cities (Toronto, 43 per cent; Vancouver, 18 per cent; and Montreal, 12 per cent), they have some difficulty speaking one of the official languages of Canada, and they tend to have limited resources. Keeping in mind that the majority of visible minorities are sponsored as family-class immigrants by their relatives, it is important to note that in these cases, a family sponsor must support and house the older adult for up to 10 years. This indicates that while ethnicity can be helpful in weathering aging, it can also contribute to the isolation and dependency of the older adult involved.

(First Light/Alamy)

With an aging Aboriginal population, Native and Inuit communities run the risk of losing an important link to their culture.

POPULATION PYRAMIDS

The age structure and changes to the age structure can be graphically displayed using what demographers call population pyramids.

One of the most interesting aspects of the structure of the Canadian population has to do with the aging of the baby boomers, a cohort who were born between 1946 and 1967. Baby boomers currently make up the largest proportion of the Canadian population because almost one in three were boomers in 2006 (Statistics Canada 2007d). This cohort began to reach age 65 in the year 2011, with the result that the working-age population will slow while the proportion of persons aged 65 and older will increase at a rapid rate. This shift in age structure will of course have implications for the Canadian pension system, health care, and the use of social services. It will also have implications for the size and composition of the labour force and the rate of retirement.

TIME to REFLECT
Do aging baby boomers represent a boom or a bust for Canada?

Social Theories in the Field of Aging

Theories, although not always evident in aging research, are very important. Theories help us to integrate knowledge rather than grappling with bits of seemingly unrelated findings; they offer explanations

HUMAN DIVERSITY

Grandparenting in Malawi

(Antonio Olmos/HelpAge International)

Tubakwerwa at her fruit stall

Before my husband died, I used to work on the farm, growing fruit and vegetables. The produce was just for feeding the family. My husband used to look after cows. When he died, it was up to me to do all the work. I now wake up at 6 a.m. and often don't get to bed until midnight.

Tubakwerwa, 58, lives in a village in Uganda. She cares for seven grandchildren who are all in school. Every day, before heading to work at the market, she wakes the children, makes their lunches, and prepares them for school. Then she walks two kilometres to the market, where she works until eight o'clock at night.

Food prices are a big problem. I try very hard to give the children three meals a day, but it is not always possible. What I can afford is what we eat. Sometimes the children go hungry and complain, but there is nothing I can do.

Without any form of guaranteed pension, older people have no choice but to continue to work into old age. In fact, older people who continue to work past retirement age are often the main breadwinners in their families, and without government policies and programs to benefit them, they are a largely invisible group.

With support from HelpAge International, Tubakwerwa received a loan from Uganda's Reach the Aged Association. She invested it in a market stall, where she sells fruit and vegetables. The money she earns at the stall helps her to feed the seven grandchildren she cares for. But as she gets older, her working life is getting harder and harder.

I think older people should always work, but it does get harder. I have back pains and I am losing my eyesight. What makes it worse is that I am so stressed about all the problems I have to deal with that I get migraines. When I go to the hospital they give me a few tablets, but they don't help much.

I don't get any help from anyone. Yes, I chose this work, but I didn't really have any other options—I have a family to feed.

SOURCE: HelpAge International, www.helpage.org/newsroom/life-stories/work-and-pensions/tubakwerwa-58-uganda/

In the First Person

Living in a family of four generations of women, I was fascinated by my great-grandmother and her boundless curiosity in life. When I went to university, I took an undergraduate course in sociology of the family and thought I would write a paper on aging families, since I had had a very good experience and . . . well, it would be easy. I was astounded to find one book in the University of Manitoba library on aging, the landmark study by British author Peter Townsend, *The Last Refuge* (1962). I was hooked intellectually, and my heart soon followed when I discovered the deplorable poverty and health care older people received back in the 1960s. I promptly went into social work and worked with older adults for the next 10 years, advocating at every turn. One of the leading Canadian sociologists in gerontology, Betty Havens, decided that more researchers were required in gerontology in Canada and convinced a few of us that we had to get PhDs if we really wanted to further the cause. Dragging my feet, I did a PhD in sociology, which turned out to be one of the most interesting journeys of my life. I love the intellectual rigor of sociology and the enormous contributions it has made to gerontology both in theory and research.

—Lynn McDonald

about phenomena, such as why people retire; they help predict what will happen in the future, such as who will enter a nursing home; and they offer practitioners ways of solving problems—for example, discovering the best way to help older adults cope with depression. Theories can be used to deal with individual or societal aging and/or the link between the two.

INDIVIDUAL AGING

Most of the early theories on social aging focus on the adaptation of the individual to aging.

Activity Theory

The first, and still relevant, theory of aging, introduced by Havighurst and Albrecht in 1953, proposes that life satisfaction decreases with age but that this can be remedied by engagement in activities (Havighurst and Albrecht 1953). **Activity theory** suggests that aged individuals who can meet their social and psychological needs by maintaining the level of activity they developed in middle age via similar and new activities will adjust better to old age and be more satisfied with life. Empirical support for activity theory remains mixed. Research supports the positive effects of engaging in informal activities, such as visiting friends, but not the effects of engaging in formal activities, such as joining clubs. Problems with this theory include the idea that a new activity will effectively serve as a replacement for one in the past; social and psychological needs can change over time because of situations such as widowhood; individuals may not be able to pursue any activities because of poverty or disability; and, lastly, it is difficult to conceptualize some activities and hence to measure their influence on aging (Chappell, McDonald, and Stones 2007).

Disengagement Theory

In contrast, **disengagement theory**, proposed in 1961, indicates that aging is accompanied by a mutual withdrawal of individuals and society from each other. This process of withdrawal is beneficial to both parties because the individual wishes to be less involved in societal interaction, and this allows society to experience minimal disruption with the death of the older person (Cumming and Henry 1961). This theory caused quite a stir at the time it was developed because it seems to take a negative view of aging. More important, no evidence was found to back up the theory from either an individual or a societal perspective. It should come as no surprise that defining and measuring the concept of disengagement is a slippery task for researchers, not to mention that there are many types of disengagement, many individuals may re-engage after disengagement, and disengagement levels will vary across groups of individuals. Surprisingly, disengagement theory held sway through the 1980s.

Continuity Theory

Proposed by Atchley in 1989, **continuity theory** suggests that people can best adjust to aging by maintaining a certain continuity with the past as they move into the future as long as the rate of change is in accordance with their own personal choices and societal norms. Though still a popular theory, continuity theory has limitations that are difficult to dispute. If one's middle life was dysfunctional, does this change at an older age? Does dysfunctional behaviour continue? Why is middle age the standard for old age? What happens when structural issues, such as poverty, prevent continuity in engagement?

Social Exchange Theory

Another approach that garnered attention in the field of aging focused on the interaction between individuals (Homans 1961). **Social exchange theory** was initially applied to social gerontology by James Dowd (1980). Dowd argued that people try to maximize rewards and minimize costs in their interactions with others. People attempt to profit from social interactions in which, according to their perception of the profit involved, the rewards outweigh the costs. Older people presumably end up with fewer resources in exchange relationships, because, apparently, they cannot reciprocate and can therefore be forced to comply with the demands of others—an uncomfortable position for most. Their alternative is to withdraw from relationships. The main problem with this theory is the assumption that older persons have fewer resources and also that these resources matter in social interaction. Thus, it has received only moderate, conflicted support. In a study by Kart and Longino (1987) of a sample of retirees, those who gave and received more social support had lower levels of life satisfaction, whereas a more recent study found higher levels of psychological well-being among older people with more exchange resources.

THE STRUCTURAL VIEW OF AGING

Age Stratification Theory

Riley, Johnson, and Foner (1972) expanded the focus of the study of aging from an individual adaptation to aging to consideration not only of all age groups in society but also of aging as an aspect of society at large. The theory is based on the main concept of age strata, which refers to groups of people that are based on age. As people age, they move from one stratum to another, and each stratum has age-related capacities that depend on both biological and social definitions. Each stratum also has social roles based on age, such as the age of graduation from school and the age of retirement, as well as expectations for these age-based roles. Therefore, various age strata have unequal statuses and opportunities and make different contributions to society. Tying together the four structural features of age stratification theory (age strata, age-related capacities, social roles, and age-related expectations) are four processes: cohort flow, individual aging, allocation, and socialization. Cohort flow refers to the fertility, mortality, and migration factors discussed earlier in this chapter as they apply to a cohort as it moves from one age stratum to the next. Individual aging refers to the physiological changes that occur over time. Role allocation by age entails the assigning and reassigning of people of different ages to appropriate roles depending on a variety of social, economic, and cultural factors. In times of recession, for example, young adults may remain in university longer because jobs are harder to secure. Finally, socialization refers to an internalization of the age stratification system, which causes us to adopt our age-based roles as we move through the age strata.

Age stratification theory has been subject to considerable criticism regarding its emphasis on maintaining equilibrium in society and the static nature of age strata. Today, a newer paradigm has been introduced called the **Aging and society paradigm** (Riley, Foner, and Riley 1999), which essentially argues that in the future, age-based distinctions will be removed from society so that we will live in an age-integrated society in which everyone, no matter what their age, can interact freely together (Riley 1997).

Modernization Theory

Aging and modernization theory was originally proposed by Cowgill and Holmes in 1972 and focuses on the status of the aged in society. The most important argument of this theory is that the status of the aged deteriorates with the modernization of society. First, with modernization, health technology improves and increases life expectancy, expanding the aging population of a society. With an older population, intergenerational competition for jobs develops, and the only way to solve this problem is retirement. Retirement, however, marginalizes older adults and reduces their income, thereby decreasing their status within society. Second, the reduced status of the elderly is spurred by economic modernization, which produces specialized jobs that older people cannot fill because of the obsolescence of their skills. These specialized jobs attract younger people with up-to-date skills who can easily migrate to urban centres. Younger people, therefore, tend to secure better jobs, leaving older people with poorer-quality jobs and a strong motivation to retire. Third, the modern migration of the young to cities creates a geographical distance between generations, with nuclear families predominating in cities and the elderly more likely to be left behind in rural areas—a trend that further diminishes their influence. Lastly, older adults have a lower status in modern society because they may have less education than the current high school education level achieved by their children, which means that their previous claim to status based on knowledge and wisdom is no longer valid.

Modernization theory has few followers today because there is little empirical support for the theory. The history of modernization, at least in the West, does not match the timing of the marginalization of the elderly, and the universality of the process is questionable given the variety of cultures in modern society.

Political Economy of Aging

The political economy of aging theory has undergone a number of changes, but at heart is a purely structural theory borrowed from the social sciences. The major proponents of the approach are John Myles (1984) in Canada, Meredith Minkler and Carol Estes (1984) in the United States, and Alan Walker (1981) and Chris Phillipson (1982) in the United Kingdom. The theory contends that the experience of aging is better understood within the context of the economy, social policy, and the social structure of society—namely the intersection of age, gender, class, and ethnicity (Estes, Gerard, Zones, and Swan 1984). Succinctly stated, the political economy theory of aging provides 'an understanding of the character and significance of variations in treatment of the aged' and allows us 'to relate these to polity, economy, and society' (Estes 1986). This perspective suggests that the problems of older adults are socially constructed and not biologically determined. Social structure shapes how older adults are viewed, which in turn influences how they view themselves, and social policy perpetuates wider social and economic inequalities between the aged and the non-aged in society. For example, the state, by institutionalizing retirement, has helped create the structured dependence of older adults through their reliance on public pensions. The usual critique of this perspective of aging is that it leaves little room for human agency because it is overly deterministic of human behaviour.

INTEGRATING THE INDIVIDUAL AND STRUCTURAL LEVELS

The Life Course Perspective

The **life course perspective** emerged as part of a trend toward a contextual understanding of the human developmental processes involved in aging and is sometimes considered the pre-eminent theory in social gerontology (Dannefer and Uhlenberg 1999; Elder, Johnson, and Crosnoe 2003; Settersten, 2003a). The

objective of life course studies was to develop a conceptual framework of social pathways and their relation to socio-historical conditions with an emphasis on human development and aging (Settersten 2003a). The main architect of the life course perspective, Glen Elder, developed five paradigmatic principles that provide a concise, conceptual map of the life course: lifelong development and aging, lives in historical time and place, social timing, linked lives, and human agency (Elder at al. 2003).

The first principle, lifelong development and aging, suggests that individual development does not stop in adulthood but extends from birth to death, has both gains and losses, and occurs along biological, psychological, and social dimensions. All life periods involve unique and significant developmental experiences, and no single experience is more important than another (Settersten 2003a). An addendum to this view is the acknowledgement that the life course is composed of a set of multiple, interdependent trajectories, such as career, school, work, and family (Elder 1998; Settersten 2003a). What happens along one trajectory can have consequences for other trajectories, such as when university decisions reflect work demands (Settersten 2003b). Trajectories are punctuated with events, transitions, and turning points.

The second principle, lives in time and place, suggests that the life course of individuals is shaped both by the historical periods they experience over their lifetime and where they happen to be physically located. If there is rapid change in a society, historical influences are usually expressed as cohort consequences, because social change differentiates the experiences of successive cohorts, as has been the case with different cohorts of baby-boom women. Women born between 1946 and 1955 have a lower labour force participation rate than those born between 1956 and 1965, and this difference has shaped their life course experiences (Galarneau 1994). History can also take the form of a 'period effect' in which the influence of social change is relatively uniform across successive birth cohorts—for example, the effects of the Great Depression.

The third principle outlined by Elder (2001) is the principle of timing. The idea of social timing means that the same transitions may affect people in different ways depending on when they occur during the life course. A recent study shows that transitions into adult statuses that occur earlier than average, such as leaving home, cohabiting, or becoming a parent, have a detrimental effect on the mental health of young people (Harley and Mortimer 2000).

▼ **FIGURE 12.5** Different Age Cohorts among the Age Pyramid of the Canadian Population in 2006

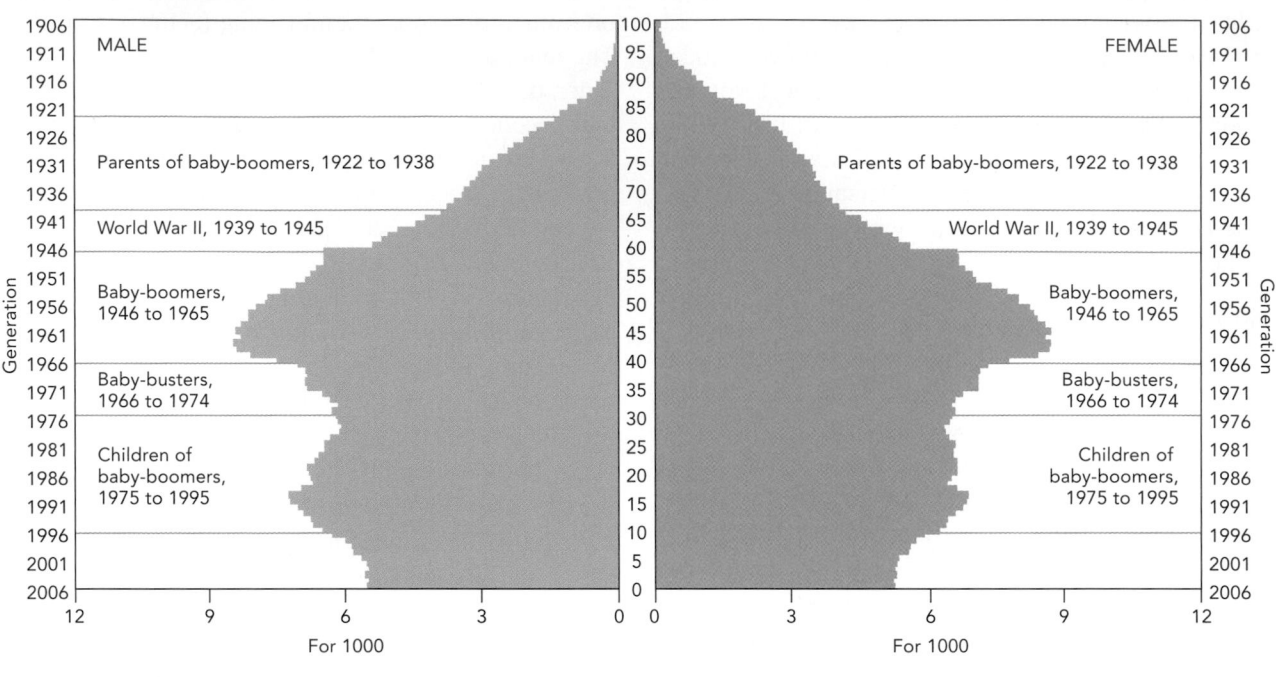

SOURCE: Statistics Canada, Census of population, 2005.

An underlying concept related to time is chronological age, which is considered in several different ways by life course scholars (McDonald 2006). Age is frequently referenced in laws and policies, and there are age-based laws that prescribe when one can vote, drink, marry, be prosecuted for a crime, receive a pension, and so on (Settersten 2006). As well, there are age-based statistical patterns for human development concerning weight, height, or intelligence that are captured under the designation of a 'normal curve'. While not formalized, there are also shared ideas in society about certain transitions, such as when birth should occur, when young adults should finish school or leave home, or when older adults should leave the labour force to retire (Settersten 2006). Generally, age acts as a dimension for organizing families, educational institutions, work organizations, and other spheres of the life course (Settersten 2003a). As a result, age provides the individual with 'mental maps' to help them organize their own lives and the lives of others (Settersten and Hagestad 1996). Age-related expectations are, in turn, influenced by social-structural factors such as gender, cohort, ethnicity, and social class.

The fourth principle, linked lives, states that the individual life course is embedded in relationships with others (Elder 2001). A transition in the life of one individual has repercussions for the lives of others, and this interdependence can provide both challenges to and resources for the individual. Interdependence requires some co-ordination to avoid potential tension, as in the case of the joint retirement of dual-career couples if age-related benefits make mutual retirement difficult when partners are of different ages (Schellenberg, Turcotte, and Ram 2006). When lives are asynchronous, age-related issues can become even more strained, such as one partner wanting to retire in order to care for the other before reaching the age at which that partner can receive a public pension (McDonald, Sussman, and Donahue 2007). However, when hardship strikes an individual, the interdependence of lives allows for economic or emotional support.

The fifth principle developed by Elder (2001) recognizes that the life course is also influenced by the decisions people make and the individual competencies they bring to these decisions. The decision to drop out of university, look for another job, or marry are examples of the principle of human agency. The choices individuals make allow them to construct their own life courses through these decisions within the constraints of history and social conditions. The principle of human agency is based on individual initiative and is a counterpoint to the social patterning and regulation of the life course. It also recognizes that there is a loose connection between social stages and transitions (Elder and O'Rand 1995).

Although it is one of the most significant theories in social gerontology, the life course perspective has a number of limitations. One might ask 'Whose life is this anyway?' The life course perspective is representative of life in Western industrialized nations but not necessarily of life in developing nations (Dannefer 2003). As well, the main focus of the perspective still tends to be on the individual, with a less direct focus on social structure. For example, the principle of linked lives has been studied quite extensively in terms of individual relationships, but the impact of macro-societal factors, such as social policy, have for the most part been ignored (Hagestad and Dannefer 2001).The life course perspective is also problematic because of its lack of regard for the negative side of human agency, an underlying belief that seems to suggest that most self-determined behaviour is positive.

Feminist Theory

There is no particular theory about older women, so the usual feminist theories apply to gerontology, including theories of gender difference, through theories of gender inequality, to theories of gender oppression. Gee and Kimball (1987) in Canada and Browne (1998), Moen (2001), and Calasanti (2004) in the United States have attempted to draw attention to the importance of feminist theory in gerontology, while a handful of feminist gerontologists, such as Bernard (2001), Laws (1995), and Tulle (2000), have expanded our understanding of the relationship between age and gender by drawing on post-modern theory, autobiography, and critical theories of embodiment. Because gender is relational, it is important to remember that feminist gerontology also includes the study of older men.

Common themes found in feminist gerontology are fourfold: gender is a fundamental organizing feature of a society; the experiences and treatment of women are socially produced according to how women and men are viewed in a society; the 'problems' of older women are not a matter of poor individual choices but are structurally manufactured through family, the market, and the state; and the interlocking oppressions of race, ethnicity, and social class combine to contribute to the precarious position of women in a society.

The limitations to this perspective have to do with the overlap with critical and post-modern theory and the omission of gay and lesbian older adults. There is also some question about how one can be a feminist post-modernist when post-modernism does not subscribe to grand narratives (big theories) like feminism.

Critical Theory

Whether critical theory is an integrating theory depends upon the version offered, since critical gerontology represents a number of theories with a broad bandwidth (Katz 2006). Critical gerontology has its roots in the Frankfurt School (neo-Marxism), post-modernism, the humanities, feminism, and, some would argue, political economy. The core of critical theory is based on the importance of evidence. Critical gerontologists offer explanations that are based on evidence about the causes of oppression, such as economic dependence via retirement or ideological beliefs such as those found in anti-aging medicine (Olshansky, Hayflick, and Carnes 2002; Phillipson 1999). They are normative in approach, since they critically evaluate the existing social structure and how aging and the aged are socially constructed within these structures, and they are practical because they provide a better self-understanding for those who might want to improve social conditions, such as the elderly themselves. Critical theorists are self-reflexive, since they have to account for their own conditions of possibility given their own aging and for the potentially transformative effects they have on others. The seeds for a critical gerontology can be found in 'The case for a critical gerontology' by Phillipson and Walker (1987).

Having noted these developments, Minkler draws attention to critical gerontology's post-modern emphasis on meaning, metaphor, textuality, and the imagery of aging and the aged (Minkler and Esters 1999). Today, this trend has been augmented by biographical and narrative perspectives that examine the self, memory, meaning, and wisdom (Katz 2006). Canadian Stephen Katz has used this approach in his understanding of aging, recently seen in his study of 'busy bodies' and how activities are a way to manage everyday life for older people (2000).

Critical theorists run into trouble sometimes because they ignore the need to resolve individual issues; consciousness-raising does little to solve intractable problems like poverty, because the people with the problem have little power. Moreover, a belief in empowerment as the solution can lead to the state's dropping its responsibility to individuals. As an example, cutting back welfare in an attempt to make people independent will not work for people who cannot be independent because of a disability, lack of job skills, or the need to care for a new child.

The Latest Buzz in Theory

Two perspectives, considered themes rather than genuine theories, have become quite popular. **Successful aging**, proposed by Rowe and Kahn (1998), posits that older adults currently have the ability to avoid illness and to minimize losses, including physical losses, and that they should enhance their engagement in life. With little research to support its view, problems with the conceptualization of successful aging, and a very close resemblance to activity theory, the theme of successful aging needs more work before it can become an in-depth gerontological theory. The second theme, **productive aging**, was introduced by Robert Butler in 1982 as a counterpoint concept to ageism, a term he originally coined (Butler and Gleason 1985). Productive aging suggests that to be content, older people must maintain their independence as long as possible by serving in the workforce, volunteering, and acting within the family and community (Taylor and Bengston 2001). Although it is a positive gerontological theory, it is difficult to define productive aging, or even unproductive aging, so there has been little theoretical development using this concept.

TIME to REFLECT

Identify the underlying assumptions about older people that appeal to you from two of the previously mentioned theories. For example, are older people frail and sick or aging well? Are they engaged in society or marginalized? Are they valued or devalued?

The Basics: Health, Income, and Social Connectedness

HEALTHY AGING

Most Canadians would probably prefer the longest life possible, if that longer life was a healthy life, free of illness and disability, and one in which independence in activities of daily living (ADL) was the norm. In 2001, the expectancy of years in good health for people at age 65 was estimated at 12.7 years for men and at 14.4 years for women (Turcotte and Schellenberg 2007). Healthy aging, then, is much more than the absence of illness (what people experience when sick) and disease (what doctors diagnose and treat). Healthy aging usually refers to 'a lifelong process of optimizing opportunities for improving and preserving health and physical, social, and mental wellness, independence, and quality of life, and enhancing successful life-course transitions' (Health Canada 2002a).

Self-perceived health is one of the most reliable indicators of actual health when measured on a Likert-type scale. Although health declines as people age, the majority of older adults still perceive their health to be generally good. The Canadian Community Health Survey 2009 shows that 56 per cent of older adults aged 65 and over rate their health as excellent, very good, or good. Among those aged 65 and over, men are more likely than women to have good health (59 versus 53 per cent, respectively) (Ramage-Morin, Shields, and Martel 2010, 4–5).

Based on four dimensions of health—self-perceived general health, mental health, functional abilities, and independence—Figure 12.6 shows that a considerable proportion of people aged 45 and above are in good health. This figure also shows that from 2000/1 to 2009, the prevalence of good health rose significantly in almost every age group. The four dimensions constituting good health each contributed to the overall increase (Ramage-Morin, Shields, and Martel 2010).

The poor health of older adults can be largely attributed to chronic conditions, given that about 74 per cent of older adults living in the community suffer from at least one chronic disease and 27 per cent have four or more conditions. Of the top five conditions, high blood pressure is the most prevalent, followed by arthritis, back problems, eye problems, and heart disease. Older women are known to endure more chronic health conditions than men. For example, while 55 per cent of women suffer from arthritis or rheumatism, 38 per cent of older men do. It is important to note that chronic diseases affect health in different ways.

TIME to REFLECT

Given the health-care issues facing older people, how might the Canadian health-care system fail to meet their needs?

The majority of older adults adapt to and compensate for their physical limitations to the extent that less than half, or about 41 per cent, of those over age 65

▼ FIGURE 12.6 — Percentage of People in Good Health, by Age Group, Household Population, Aged 45 Years and Over, 2000–2001 and 2009

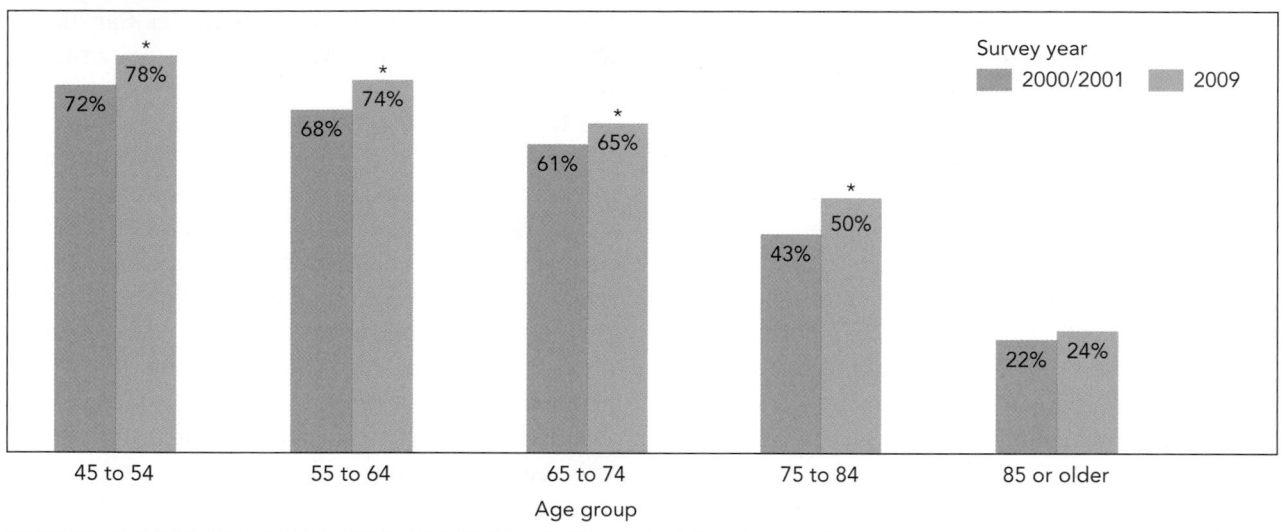

* Significantly different from estimate for 2000/1 (p < 0.05)
NOTE: Depression used instead of self-perceived mental health for 2000/1.
SOURCE: 2009 Canadian Community Health Survey, *Healthy Aging: 2000/2001 Canadian Community Health Survey.*

experience a reduced rate of activity (Statistics Canada 2002). Late-life dementias, of which Alzheimer's disease is the most common, affect only about 8 per cent of adults 65 and over, but this figure rises to 25 per cent of those over the age of 80. Alzheimer's disease is among the conditions that have the most negative impact on the quality of life of older adults.

The majority of older adults rate their mental health as good across all age groups. Using another indicator of mental health produces a similar result. As age increases, psychological stress decreases, and this trend only reverses at the oldest of ages (Turcotte and Schellenberg 2007). Using a well-being scale that measures people's feelings about different parts of their life (e.g., satisfaction with achievement), studies indicate that older adults are more likely than younger adults to have higher scores, indicating greater life satisfaction (Turcotte and Schellenberg 2007). These data challenge the idea that declining life satisfaction is inevitable at older ages, suggesting instead that individuals can maintain a good quality of life as they age.

Despite Canadian seniors' own positive rating of their health, it is estimated that about 20 per cent of those aged 65 and older are living with some type of mental illness (MacCourt 2005), a percentage consistent with the prevalence of mental illness in other age groups. According to Conn (2002), between 10 and 15 per cent of older adults in society suffer from depressive symptoms or clinical depression. These percentages, however, do not reflect the high rates of depression seen in long-term care facilities. Studies suggest that the prevalence rates for all mental disorders among nursing home residents are three or four times higher than those of the general population. This fact is important in that about 7 per cent of older Canadians are in some type of long-term care facility, a percentage that has been consistent since 1981. The possibility that mental illness could develop following institutionalization implies that this health issue may be open to modification with appropriate treatment, especially since about one-third of mental health issues go undetected in people 65 years of age and older (Malach, Conn, and Le Clair 2005).

In Canada, the costs of acute care and physician services are covered by provincial health insurance programs. However, coverage of related health-care services, such as nursing home care, physiotherapy, or podiatry, varies by province. The result is that acute illnesses, rather than degenerative chronic illnesses, are the focus of Canadian health care, creating a mismatch between which illnesses afflict the elderly and what care is offered. Whether this mismatch influences the cost of care is not known, but it is worth remembering that older adults consumed 44 per cent of all health spending in 2006—the highest of all age groups and three times their proportion in the population.

The health of our older population is a matter of concern to all Canadians. Older adults make enormous contributions to Canadian society through paid and unpaid labour. For example, in 1998, 42 per cent of Canadians aged 55 to 64 and 44 per cent of Canadians over the age of 65 spent, on average, 2.2 hours a day as volunteers. These numbers represent an economic value to our society estimated at $60.2 billion each year (Edwards and Mawani 2006). It has also been estimated that it would take about 300,000 full-time employees, at a cost of $6 billion per year, to replace the work of the 2.1 million Canadians who provide unpaid care for older adults with long-term health problems (Keating, Swindle, and Foster 2005). More than 350,000 workers aged 65 and over were in the labour force in 2007, and the trend for people to remain in the labour force to later ages is strengthening (Statistics Canada 2007e; Pignal, Arrowsmith, and Ness 2008).

INCOME

A secure income has been identified as one of the most important determinants of health and affords older adults the ability to own or purchase housing and transportation, food, and uninsured medical care and makes possible the pursuit of leisure activities and social involvement with friends and family. It is therefore remarkable that the relative economic

position of older citizens today is one of the 'biggest success stories' in Canadian social policy. In 1976, a whopping 65 per cent of older Canadians lived below the Low Income Cut-off set by Statistics Canada. In 2006, or 30 years later, that proportion had dropped to 16 per cent.

This success story is attributable to the introduction of the Canadian pension system, which attempts to reduce poverty among older people and help them to maintain an adequate living standard after retirement (Task Force on Retirement Income Policy 1979). The public pension system includes Old Age Security (OAS), started in 1951, which pays a flat rate amount to all eligible Canadians aged 65 or over; the Guaranteed Income Supplement (GIS), introduced in 1966, which provides an income-tested supplement to retirees with few or no private sources of income; and the Spouse's Allowance (SPA), begun in 1976, a controversial program for younger spouses or widows/widowers of GIS recipients. The compulsory Canada and Quebec Pension Plans (C/QPP) pay earnings-related benefits to workers based on contributions made during their working years. This system, implemented in 1966, did not become fully operational until 1976. Thus, the C/QPP is primarily responsible for the rising income of older people. Tax-assisted private pensions

OPEN for DISCUSSION

Study Finds Having Money, Health, Optimism, No Stress, Moderate Drinking, No Smoking Means Longer Life

October 27, 2008—New research was released today with this shocking finding: elderly people who have a positive outlook, lower stress levels, moderate alcohol consumption, abstention from tobacco, moderate to higher income and no chronic health conditions are more likely to thrive in their old age. As my grandchildren say, 'Duh?'

The researchers were from Portland State University, the Kaiser Permanente Center for Health Research, Oregon Health & Science University, and Statistics Canada.

They surveyed 2,432 older Canadians about their quality of life. They then selected the 'few' who maintained excellent health over an entire decade and named them the 'thrivers'.

'Important predictors of "thriving" were the absence of chronic illness, income over $30,000, having never smoked, and drinking alcohol in moderation,' said

Mark S. Kaplan, PhD, lead author and professor of community health at Portland State University.

All Dr Kaplan had to do was give me a call and I could have told him that the old folks who live the longest are generally going to be the happy, healthy, rich guys that never smoked and control their drinking. Was that ever a mystery?

'We also found that people who had a positive outlook and lower stress levels were more likely to thrive in old age.'

And you will probably not be shocked to learn this study was funded by a grant from the National Institute on Aging, part of the US National Institutes of Health. So the US taxpayers financed a study to get the answer to a question that any guy on the street could have answered. And, we did the study in Canada.

Is this the study to nowhere?

SOURCE: Sutherland 2008. The study was published in the October issue of *The Journal of Gerontology: Medical Sciences*.

include employer-sponsored registered pension plans (RPPs) and individually based registered retirement saving plans (RRSPs). This three-tiered structure of the Canadian pension system has remained largely unchanged over time. Figure 12.7 illustrates the improvement in income for senior families since 1976 as compared to other family types.

The median after-tax income of elderly families (family head aged 65-plus) in 2007 was $44,900, compared to $65,500 for non-senior families. The importance of the pension system is captured in the following figures: senior families receive 48 per cent of their income from OAS/GIS/SPA, 43 per cent of their income from C/QPP, and 9 per cent from elsewhere. For unattached seniors who live alone, the figures were 52, 42, and 7 per cent, respectively (Statistics Canada 2008c).

Although the income story for seniors is a victory, there are still some flaws in the system. The incidence of low income was 15.5 per cent for all older adults in 2006, with 14 per cent of unattached older men below the LICOs compared to 16.1 per cent of women. Women tend to carry the burden of poverty as a result of their labour force participation and family histories. Many older women today are poor because they belong to a cohort in which a woman's primary occupation was housewife and mother. While baby-boom women are far more likely to work than other women, they too are at a disadvantage to men in terms of the value of their pensions. First, payments from the private (RPPs) and semi-private aspects of the pension system (C/QPP) reflect the income inequality in a labour market where women earn less than men. Second, women's interrupted work histories as a result of family responsibilities limit their ability to save and accumulate pension benefits. Two subgroups of women stand out as more at risk for low income in old age. Separated and divorced women, 5.1 per cent of the population, have been found to be the poorest among all women over the age of 65 (McDonald and Robb 2004), while Aboriginal women are also one of the poorest groups as well as the most dependent on government pensions within the indigenous population of Canada (McDonald 2006).

Retirement, usually viewed as the transition from paid employment to receiving a pension, is changing in most industrialized countries. A single

▼ **FIGURE 12.7** Median After-Tax Income by Family Types, Canada, 1977 to 2007

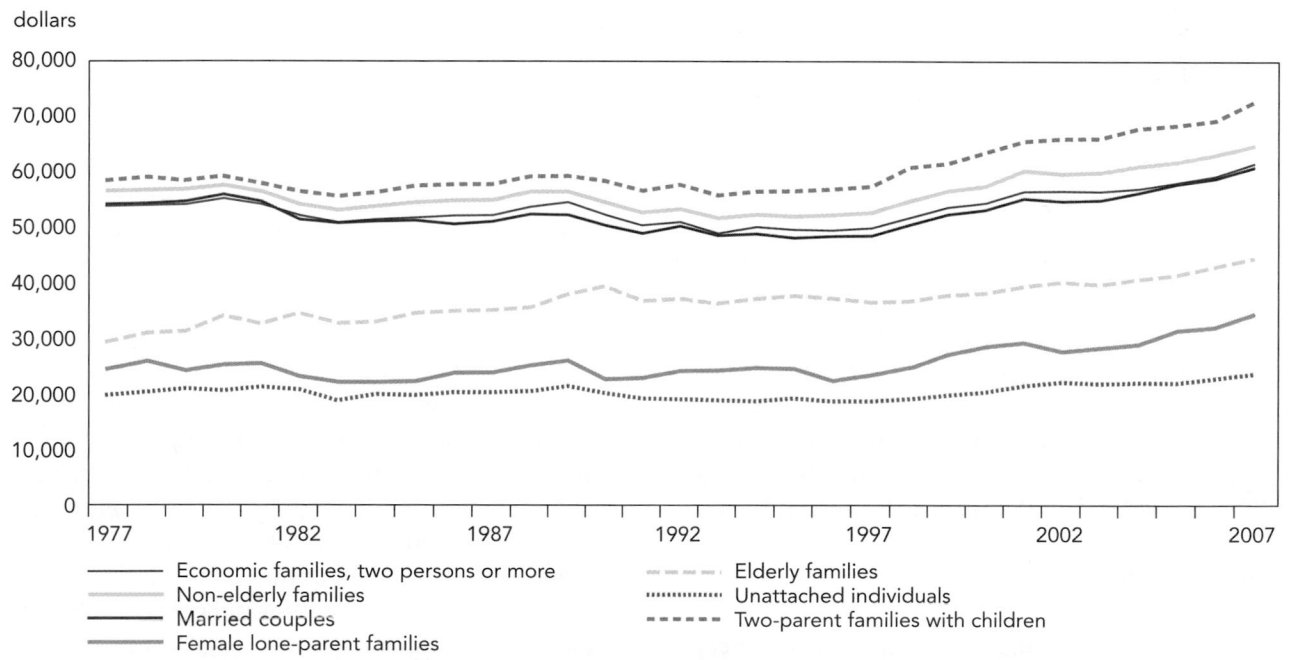

1. All results presented in the analysis are statistically significant at the 5% level.
2. Market income is the sum of earnings from employment, investment income, private retirement income and other income (other employment income and alimony payments).
3. Statistics Canada. Table 380-0017 – Gross domestic product (GDP), expenditure-based, annual (dollars unless otherwise noted), CANSIM (database).

SOURCE: Statistics Canada. 2009a, p. 9.

chronological age no longer dictates the time of the transition. Most Canadians in 2006 retired, on average, at 61.5 years of age, which is before the institutionalized age of 65. As well, retirement occurs before any physical decline. In fact, some people never retire and continue to work all their lives to a very old age. Sometimes older adults are forced into early retirement against their will because of poor health, caregiving responsibilities, job loss, or the retirement of a spouse. Therefore, mandatory retirement makes little difference because most people retire before age 65. Often, retirement is less likely to be subject to pension policy, since some people retire via unemployment or disability programs or because of early retirement packages offered by their firm. Moreover, retirement is not necessarily a one-time event, an abrupt transition from work to non-work. For many people, retirement can be gradual, it can be part-time or on-and-off, and it can involve a number of exits from the labour force (McDonald 2006). Nonetheless, retirement helps to define when old age starts and therefore plays a significant role in changing the nature of the life course. How this reallocation of time and timing occurs also affects the income of older Canadians.

TIME to REFLECT

Do you think that the recent worldwide downturn in the economy has extended or lowered the age of retirement?

SOCIAL CONNECTEDNESS

There is little doubt that social engagement in life on the part of older adults substantially contributes to better physical and mental health and a sense of belonging in the community. Social exclusion, psychological distress, and isolation are known to increase the risk of poor health and, as you would expect, loneliness. What is more, there is increasing evidence that a lack of involvement with others is associated with death (Wilkins 2006). Here we examine the ideas of social networks, social support, and social engagement.

Social Networks

While there are several definitions of 'social networks' in the literature, the term usually refers to the entire set of relationships in which an individual is involved (Pearlin 1985) and which can be described in terms of structure, such as density and size (House and Kahn 1985). Social networks are the source from which older people draw to obtain access to resources, whether they receive or provide care. Many factors affect the size and composition of social networks, such as whether people are married or not, where they live and with whom, and the size of their families. For example, we know that twice as many older women live alone compared to older men and that the baby-boom generation has fewer children than previous generations, which may have implications for caregiving. More than 5 million Canadians are grandparents with, on average, five grandchildren, which affects the size of their social networks. It is important to remember, however, that one can live with someone and have a large social network yet still be isolated, or can live alone yet be actively engaged in a large social network of family, friends, and neighbours.

Overall, the majority of older adults in Canada are not socially isolated. Older adults have smaller social networks than younger adults because they lose relatives and friends with advancing age, but their networks are condensed and exhibit an increased emotional closeness (Schellenberg 2004). Only 10 per cent of older adults say they have no close friends, and only 8 per cent say they do not have a close relative. If anyone could be considered disengaged, it could only be the very small 2 per cent of older adults who report they have no close friends or relatives (Turcotte and Schellenberg 2007). These people are less likely to be in excellent or good health and more likely to report that they are not particularly happy. The problem, however, is somewhat of a chicken-and-egg dilemma. Poor health and unhappiness may hamper their involvement in close relationships, or the lack of close relationships may cause their poor health and low spirits. We don't really know.

Caregiving as Social Support

Social networks often evolve into social support systems or care networks when seniors become ill or frail (Keating, Swindle, and Foster 2005). Therefore, caregiving is an example of social support and is a well-researched area in social gerontology. Caregiving provides support to older adults because their health may be deteriorating and they are generally not as functionally independent as they were when younger. The care provided may be **informal care**, supplied by unpaid family and friends, or **formal care**, provided by paid caregivers who are sometimes family but usually professional health-care workers. In 2002, about 26 per cent of Canadians aged 65 and over stated that they received assistance because they had a chronic health problem or a physical condition that interfered

UNDER the WIRE

Your Grandmother Is Likely Online

According to the 2007 Canadian Internet Use Survey (CIUS), and the 2007 General Social Survey (GSS) that compared the Internet use of baby boomers aged 45 to 64 in 2007 with that of older adults aged 65 and older, seniors were the fastest-growing group of users—probably because they started at a lower baseline. In 2007, most boomers used the Internet, but significantly fewer seniors went online. Since 2000, however, rates of Internet use have been highest among seniors, with usage in 2007 four times higher than in 2000. The study found that 90 per cent of older adults used the Internet for email purposes, compared to 88 per cent of boomers, and surprisingly, 36 per cent of older adults compared to 27 per cent of boomers played games! It is important to remember, however, that there remain significant differences in access to and use of the Internet along socio-economic and demographic lines, with age in particular identified as an important factor (Veenhof and Timusk 2009).

with their activities of daily living. The types of care available consist of indoor or outdoor household work, shopping, transportation, or personal care. The types and amounts of care generally increase with age so that at younger ages people with long-term health limitations require more help with activities like household maintenance and outdoor work, while at older ages help is more focused on personal care, such as bathing and dressing.

Researchers in gerontology agree that even in countries like Canada that have universal health insurance programs, an informal network still provides between 75 to 85 per cent of the care given to older adults (Kane, Evans, and MacFadyen 1990). The reasons why older people receive such overwhelming help from family and friends are complex and connected to many factors: some services offered by family may be better than those offered by paid caregivers; the older person or their family may not be able to afford the necessary services; the older person may live too far from the formal services required; language-appropriate services may not be available; the family may simply want to provide care; or a combination of some or all of these factors may exist. What we know for sure is that 70 per cent of informal care is provided by close family, and family in this case means women. In 2007, 57 per cent of women provided care to an elderly person; six in 10 were looking after an aging parent or parent-in-law, although for the most part, the recipient of care was the woman's mother; one in 10 was caring for a spouse; and more than half were employed while also providing care (Cranswick and Dosman 2008). The provision of care is divided along gender lines, with women and men shouldering different types of tasks. While men are more likely to provide outdoor help, women are more likely to provide indoor help and personal care.

The burden of caring has been studied extensively. This term *burden* generally refers to the physical, social, psychological, and financial challenges close family members face as they provide care for an older relative (Chappell, McDonald, and Stones 2007). Social gerontologists also study the stress generated by caregiving as defined by the caregiver. Some of the negative outcomes related to caregiver burden are depression, anxiety, loneliness, lower social functioning, financial distress, and poorer general health than that of the average person. However, it is important to recognize that caregiving also has a rewarding side, such as the pleasure of seeing a family member doing well and being happy and the caregiver's satisfaction at being helpful and giving back care similar to what they received earlier in life (Chappell and Litkenhaus 1995).

TIME to REFLECT

If you had to assist your family in providing care to an older family member, how do you think this would affect other roles in your life?

ISSUES

At this time, it could be argued that there is one fundamental issue that affects every aspect of aging and that has been poorly studied: namely, ageism. The term was coined by Robert Butler, a psychiatrist who used it in a conflict over high-rise housing for older people in Washington in 1968. **Ageism** has a number

SOCIOLOGY in ACTION

Older Homeless People

Although the number of homeless people in Canada is not known for any age group because of measurement difficulties, older adults are as much in evidence as younger people, according to reports from hospitals, social service agencies, and homeless shelters. The few studies in Canada have shown that older homeless people are unique because their homelessness is chronic over their life course. Once they plunge into homelessness at an earlier age, they have a difficult time extricating themselves, aging in place on the streets, where it becomes more and more difficult for them to cope because of their age. A new phenomenon is older people who become homeless late in life because of poor retirement income, a death in the family, poor health, or an inability to work.

(Lynn McDonald)

Collage created by older, previously homeless people on what it means to live on the streets.

'This collage was created by older, previously homeless people to convey what homelessness means to them. On the reverse side of the collage is a postcard with findings from the study *In from the Streets: The Health and Well Being of Homeless Older Adults*, recommendations for change, and excerpts from in-depth interviews with 53 older, previously homeless people now living in Calgary and Toronto. More than 5000 postcards were sent to decision-makers who had the potential to change the course of homelessness for older adults in Canada.

The Lingering Trauma of Homelessness

Participants conveyed that exits from homelessness were neither discrete nor final. 'Homeless effects' reported by still participants were so profound that several participants still described themselves as 'homeless' even though they were housed. Frequently, housing services were characterized as insensitive to the demands of transitioning from homeless to housed, leaving participants feeling vulnerable. Security, safety, and trust were key aspects that participants felt were lost or damaged and needed to be slowly re-established:

"You lose your dignity, your trust." (Toronto participant)

SOURCE: McDonald, Donahue, Janes, and Cleghorn 2006.

For some, the effects of homelessness were so vivid that they were characterized in terms similar to those used to describe the symptoms of post-traumatic stress disorder:

"I can't go back into that area where I was. Because I went there twice now to see my psychiatrist. And both times I left there—'I can't go back here, I can't.' It was the shelter—that's where the doctor is. So once a month on a Tuesday, I'd go down there and I'd go into a sweat." (Toronto participant)

Frequently, participants invoked the language of recovery to describe the process of healing from the stress induced by the conditions of homelessness:

"Because the transience of the shelter system it takes its toll on your mental, physical, emotional, financial, and spiritual well-being and it's something that takes you so long to recover from that I think unless you go through it you don't understand. When I was finally housed I sat up every night, I didn't sleep for about 4 days. I sat up every night. I was afraid to close my eyes and go to sleep because I had been so accustomed to catnapping in the shelter system." (Toronto participant)'

In From the Streets study

of definitions, but in essence, it refers to 'an alteration in feeling, belief, or behaviour in response to an individual's or group's perceived chronological age' (Levy and Banaji 2002, 50). Levy and Banaji (2002) argue that implicit ageism is the basis of most interactions with older adults and emerges in cultural environments where there is widespread acceptance of negative feelings and beliefs about older people. Stones and Stones (1997) have referred to ageism as a 'quiet epidemic' that could lead to the neglect of older adults as a social category. This prejudice minimizes our expectations of older adults and influences how they are treated, their participation in society, and what steps are taken to enhance their engagement. While supporting evidence is slim, decision-makers in Canada need to consider why we do not have universal homecare, which would greatly ease the situations of older adults and the families who care for them. Again, there are many questions to consider. Why has there not been a rigorous study that considers ageism a causal factor of elder abuse in the community or in institutions? Why are nursing home infractions ever tolerated? Why do older women continue to experience poverty? To answer these questions, one must consider why, at times, society uses ageism to promote specific views of aging in order to relieve itself of any responsibility toward older people and, at other times, individuals use ageism to avoid confronting what they might fear about aging, such as illness and death (Martens, Goldenberg, and Greenberg 2005).

TIME to REFLECT

How can we move away from a focus on the 'elderly' towards a focus on aging in general? Would this combat ageism?

Conclusion

The process of aging is not only about the individual as he or she moves across the life course but also about a societal transition that occurs within a historical, economic, and social/political context that not only shapes individuals and cohorts but, in turn, is shaped by the passage of those cohorts through our society. We have also seen that the behaviours and decisions you make today will have an important effect on your later-life well-being and satisfaction—as the life course perspective would remind us. Given the pace of social and technological change in a global economy, you can also be confident that your own aging will be different from that of your parents and grandparents and that there will always be new social, economic, psychological, and political issues to consider. Based on present research, social gerontologists of the future will likely take a life course view, make these macro/micro links, work in interdisciplinary contexts using mixed research methods, and make their findings accessible to policy decision-makers, practitioners, and the public.

questions for critical thought

1. What factors have contributed to the rapid aging of the Canadian population?

2. Do you think apocalyptic demography is a case of ageism?

3. In light of the aging of the Canadian population, what aspects of aging will you and your family experience?

4. How are the aging experiences of Aboriginals in Canada different from those of mainstream society?

5. What do think would happen if there were no theories about aging?

6. How effective is our current public health-care system in meeting the needs of older adults?

7. Do you think you will retire when you grow older? If you do retire, how much will you depend on the public pension system for your income in retirement?

8. How important is the relationship between social connectedness and well-being?

recommended readings

V.L. Bengston, M. Silverstein, N.M. Putney, and D. Gans, eds. 2009. *Handbook of Theories of Aging.* 2nd edn. New York: Springer.
This volume thoroughly covers aging theory, focusing on the biological, psychological, and social aspects of aging. It addresses how aging theory informs public policy and concludes with a summary of the major themes of aging, offering predictions about the future of theory development.

Robert H. Binstock and Linda K. George. 2006. *Handbook of Aging and the Social Sciences.* 6th edn. Burlington, MA: Academic Press.
This handbook provides a comprehensive summary and evaluation of recent research on the social aspects of aging. Areas addressed include aging and time, aging and social structure, social factors and social institutions, and aging and society.

N.L. Chappell, L. McDonald, and M. Stones. 2007. *Aging in Contemporary Canada.* 2nd edn. Toronto: Pearson Education.
This is a comprehensive introduction to gerontology in Canada, covering attitudes to aging, demography, theory and research, and diversity (including women and ethnic aging); health and well-being (including mental health); social institutions such as families and work, retirement, and pensions; and policy issues related to health care and end-of-life care.

Ingrid Arnet Connidis. 2010. *Family Ties and Aging.* 2nd edn. Thousand Oaks, CA: Sage.
Connidis examines the issues surrounding family ties and aging—e.g., changing family structures and new patterns of work–family balance—and how they are negotiated in the family lives of middle-aged and older adults. The coverage is extensive and reflects contemporary society. Groups and relationships that have been neglected are considered, such as single, divorced, and childless older people, sibling relationships, live-in partnerships, and the family ties of gays and lesbians.

K. Warner Schaie and Ronald P. Abeles, eds. 2008. *Social Structures and Aging Individuals.* New York: Springer.
The final volume of 20 in the Societal Impact on Aging series, this book focuses on challenges for older persons in a rapidly changing society and tries to forecast what may be the next set of issues to lie at the intersection of social structures and the individual aging process.

L. Stone, ed. 2006. *The New Frontiers of Research on Retirement.* Catalogue no. 75-511-XPE. Ottawa: Ministry of Industry.
This volume deals with aspects of retirement that have been outside the main focus of the research literature but that will receive much greater attention as baby boomers retire, such as family dynamics and retirement, dual-career couples, and the retirement of women.

M. Turcotte and G. Schellenberg. 2007. *A Portrait of Seniors in Canada, 2006.* Ottawa: Statistics Canada.
The authors offer a factual overview of aging in Canada, covering demography, health, income, community participation, social support and caregiving, leisure activities, and immigrant seniors.

A.V. Wister. 2005. *Baby Boomer Health Dynamics: How Are We Aging?* Toronto: University of Toronto Press.
This book asks whether baby boomers in Canada are more or less healthy than previous generations. Focusing on four health behaviours that have been proven to be major risk factors for disease (smoking, unhealthy exercise, obesity, and heavy drinking), the book investigates the long-term implications of several lifestyle–health conundrums.

recommended websites

American Sociological Association, Section on Aging and the Life Course

www.pop.psu.edu/asasalc/index.htm

Sociology of Aging and the Life Course provides an analytical framework for understanding the interplay between human lives and changing social structures and is a good source for sociological developments on aging.

Canadian Association on Gerontology (CAG/ACG)

www.cagacg.ca/whoweare/200_e.php

The Canadian Association on Gerontology is a national, multidisciplinary, scientific, and educational association established to provide leadership in matters related to the aging population. The *Canadian Journal on Aging* is a premier publication of this organization.

Commission for Social Development (CSOCD), United Nations

www.un.org/esa/socdev/csd/2009.html

The commission is a functional commission of the Economic and Social Council (ECOSOC) of the United Nations. It consists of 46 members elected by ECOSOC. Most issues cover global aging.

European Sociological Association (ESA) Research Network on Ageing in Europe

www.ageing-in-europe.org

This new network aims at facilitating contacts and collaboration among those with research interests in aging in different countries.

Gerontological Society of America (GSA)

www.geron.org/About%20Us

The GSA is the oldest and largest multidisciplinary organization devoted to research, education, and practice in the field of aging. Its aim is to advance the study of aging and disseminate information among scientists, decision-makers, and the general public. The *Journal of Gerontology* and *The Gerontologist*, the top aging journals, are produced by the GSA.

International Association of Gerontology and Geriatrics (IAGG)

www.iagg.com.br/webforms/index.aspx

The IAGG is an umbrella organization composed of national societies of gerontology and geriatrics. Every four years, the association sponsors the World Congress on Aging, which brings researchers together from around the world. The website has virtual lectures and resources.

International Sociology Association, Research Committee on Sociology of Aging

www.isa-sociology.org/rc11.htm

The purpose of the Research Committee on Sociology of Aging is to encourage research of high quality on aging within and between countries. The committee produces a semi-annual newsletter of research programs among its membership.

Organisation for Economic Co-operation and Development (OECD)

www.oecd.org/home/0,3305,en_2649_201185_
1_1_1_1_1,00.html

The OECD brings together the governments of countries committed to democracy and the market economy from around the world, including Canada. It provides statistics and economic data about aging in its 30 member countries.

World Health Organization Aging and Life Course Programme

www.who.int/ageing/projects/en

WHO is the co-ordinating authority for health within the United Nations and is responsible for providing leadership on global health matters. The aging program deals with issues of aging around the world.

PART IV

Social Institutions

chapter 13

Families and Personal Life

Maureen Baker

In this chapter, you will:

▶ Learn to differentiate popular myths about personal and family life from actual research results.

▶ Gain a clearer understanding of variations in family life.

▶ Understand how sociologists have conceptualized and explained family patterns.

▶ Gain some insight into several concerns about Canadian families.

▶ Identify current demographic and social trends in Canadian families.

▶ Understand how future predictions are made about family life.

Introduction

The media often dwell on the negative side of family life by highlighting violent relationships, custody disputes, and divorce. Yet both public polls and sociological research indicate that most Canadians value their families and that young people expect to have children and live within a stable relationship for most of their lives (Baker 2010). Ideally, family life can contribute to personal development and can provide companionship, love, sexual expression, children, care, a sense of belonging, and shared resources. Yet some people spend years living with people they resent and seem to fight more with family members than with friends or acquaintances.

Governments typically encourage heterosexual marriage and childbearing because they need future citizens, taxpayers, voters, consumers and workers to maintain the nation. Both governments and employers rely on parents to produce children, to socialize and discipline them to become employees and law-abiding citizens, and to provide the necessary recuperation that enables people to return each day to school or work.

Intimate relationships remain important to individuals and the larger society, but family life and gender relations have changed considerably in Canada during the past 30 years (Baker 2010; Siltanen and Doucet 2008). Demographic research indicates that since the 1970s, young people have been delaying marriage while they gain an education and prepare for paid employment. More couples now live together without legal marriage, and couples are producing fewer children, who spend more time with non-family carers while their parents work for pay. Remarriages form a larger percentage of all marriages, since relationships now have a higher probability of ending in separation or divorce, and an increasing proportion of children are raised in stepfamilies. People tend to live longer, but more people are living alone, especially before marriage, after separation, and after widowhood.

This chapter defines families and outlines some of the variations in family structure and practices. The different ways that sociologists have discussed and explained personal life are introduced before we turn our attention to five controversial issues: sharing domestic work, assisted conception, non-family child care, children and divorce, and wife abuse. Some general comments are made about family policies before we turn to predictions about future families.

Family Variations

DEFINING FAMILIES

Many definitions of family have been used in academic and policy research, as well as in the delivery of social programs. Most definitions focus on legal obligations and family structures rather than on feelings of attraction, love, and obligation or the services intimates provide for each other. These definitions always include heterosexual couples and single parents sharing a home with their children, but until recently few definitions encompassed same-sex couples. Most definitions include parents with dependent children but also childless couples or those whose children have left home. Others extend the definition of family to grandparents, aunts, uncles, and cousins who are sharing a dwelling.

Sociologists and anthropologists used to talk about 'the family' as a monolithic **social institution** with one acceptable structure and common behavioural

Families come in different forms and sizes. What unites families is what they do, not how they look.

patterns (Eichler 2005). Academics used to assume that family members were related by blood, marriage, or adoption and that they shared a dwelling, earnings, and other resources; that couples maintained sexually exclusive relationships and reproduced and raised children together; and that family members cherished and protected each other. Nevertheless, academics have always differentiated between **nuclear families**, which consist of parents and their children sharing a dwelling, and **extended families**, which consist of several generations or adult siblings with their spouses and children who share a dwelling and resources. Both kinds of families continue to be a part of Canadian life, although nuclear families are more prevalent.

The most common definition used in policy research is Statistics Canada's **census family**, which includes married couples and cohabiting couples who have lived together for longer than one year, with or without never-married children, as well as single parents living with never-married children. As of 2006, couples can be same-sex or heterosexual, but this definition says nothing about the larger kin group of aunts, uncles, and grandparents or about love, emotion, caring, or providing household services. Yet a common definition must be agreed upon when taking a **census** or initiating policy research.

The Canadian government also uses the concept of **household** in gathering statistics relating to family and personal life. *Household* refers to people sharing a dwelling, whether or not they are related by blood, adoption, or marriage. For example, a boarder might be part of the household but not part of the family. Table 13.1 shows the percentage of Canadians living in various family types in 2006 compared to 1981.

In a culturally diverse society such as Canada, it is inaccurate to talk about 'the family' as though a single type of family exists or ever did exist. In fact, cultural groups tend to organize their families differently, depending on their traditions, religious beliefs, socioeconomic situation, immigrant or indigenous status, and historical experiences, although most Canadians live in nuclear families comprising parents and their children (Vanier Institute of the Family 2004). Extended families, in which several generations (or siblings, their spouses, and children) share a residence and cooperate economically, remain important as a living arrangement and support group, especially among some recent immigrants. Even when family members do not share a residence, relatives may live next door or nearby, visit regularly, telephone daily, assist with child care, provide economic and emotional support, and help find employment and accommodation for one another. When relatives do not share a household but still rely heavily on one another, they are said to be a **modified extended family**.

In the 1950s, American sociologists lamented the isolation of the modern nuclear family, suggesting that extended families used to be more prevalent prior to industrialization (Parsons and Bales 1955). Since then, historians and sociologists have found that nuclear families were always the most prevalent living arrangement in Europe and North America (Goldthorpe 1987; Nett 1981), but extended families were and still are widespread among certain cultural groups, such as some **First Nations** peoples, southern Europeans, and some South and western Asians. They are also more prevalent among those with lower incomes and at certain stages of the family lifecycle, providing low-cost accommodation and practical support for young cash-strapped couples, lone mothers after separation, or frail elderly parents after widowhood.

Many immigrants come from countries where extended families are more prevalent, yet the percentage of 'multi-family households' (a term used by Statistics Canada that approximates an extended family) declined from 6.7 per cent in 1951 to 1.1 per cent in 1986, when immigration rates were high (Ram 1990, 44). The explanation for this decline is that more Canadians began to live alone during that period and that immigrants tend to change their family practices to fit in with the host country. In the 2006 census, only 1.3 per cent of people lived with relatives in family households (Statistics Canada 2007b).

In this chapter, 'families' will be used in the plural to indicate the continued existence of different family structures. Qualifying phrases, such as 'male-breadwinner families', 'lesbian families', and 'stepfamilies', will be used for clarification. Although sociological definitions formerly focused on who

TABLE 13.1 ▼	Percentage of Families in Canada by Type, 1981 and 2006	
Type of Family	**1981**	**2006**
Legally married couples with children	55	39
Legally married couples without children	28	30
Lone-parent families	11	16
Common-law families without children	4	9
Common-law families with children	2	7

SOURCE: Vanier Institute of the Family (www.vifamily.ca, Virtual Library, 2010).

constitutes a family, more researchers now emphasize what makes a family. This approach downplays the couple's sexual preference and the legality of the relationship and focuses instead on patterns of caring and sharing.

MONOGAMY AND POLYGAMY

In all Western countries, it is illegal to marry more than one spouse at a time, but **polygyny**, or having several wives at a time, is practised in some countries in Africa and western Asia, especially those using Islamic law. In sub-Saharan Africa, about half of married women aged 15 to 49 were in polygynous unions throughout the 1990s in Benin, Burkina Faso, and Guinea and over 40 per cent in Mali, Senegal, and Togo (UN 2000a, 28). Wealthy men are more likely than those with fewer resources to take on more than one legal wife (Barker 2003).

Polygynous unions, which lead to a proliferation of stepchildren and step-relatives, tend to be associated with patriarchal authority and wider age gaps between husbands and wives. They are more common among rural and less educated women, as well as among those who do not formally work for pay outside the household (Barker 2003). Multiple wives, who are sometimes sisters, may resent their husband taking a new partner, but they may also welcome her assistance with household work, child care, and

horticulture and may value her companionship in a society where marriage partners are seldom close friends. Furthermore, the husband's second marriage elevates the rank of the first wife, who becomes the supervisor of the younger wife's household work.

Polygamy refers to the practice of having more than one spouse at a time, but polygyny is much more prevalent than **polyandry**, or marriage between one woman and several husbands. When polyandry does occur, the husbands are often brothers, and the practice relates to the desire to keep family land in one parcel (Ihinger-Tallman and Levinson 2003). Most societies prefer polygyny because more children can be born with multiple wives, which could be important if children are the main source of labour for the family or community. Also, the identification of the father is particularly important in patrilineal societies, because children take their father's surname, belong to his kin group, and inherit from him, and married men are responsible for supporting their children. Knowing who the father is would be difficult with multiple husbands, so this is not usually an acceptable form of marriage in patrilineal systems. Since most societies have also been patriarchal, men have ensured that marriage systems suit their interests.

In the First Person

I first became interested in sociology in 1968 after working as a research assistant for a sociologist at the University of Toronto when I was completing my second year of university. That work experience and more courses led to several similar jobs working for sociology professors, who encouraged me to continue with graduate studies. Since completing my doctorate in sociology at the University of Alberta in 1974, I have worked in four different countries (Australia, Canada, the United Kingdom, and New Zealand) as a university teacher/researcher, freelance consultant, parliamentary researcher, and social policy adviser.

—Maureen Baker

TIME to REFLECT

Are polygamous relationships practised anywhere in North America? If so, how do they circumvent the bigamy laws?

ARRANGED VERSUS FREE-CHOICE MARRIAGE

Marriages continue to be arranged in parts of the world in order to enhance family resources, reputation, and alliances and because parents feel more qualified to choose their children's partners. The family of either bride or groom may make initial arrangements, but marriage brokers with extensive contacts are occasionally used to help families find mates for their offspring.

Middle Eastern and South Asian immigrants living in Canada sometimes have their marriages arranged, which may involve returning to the home country to marry a partner selected by family members still living there or being introduced to a suitable partner from the same cultural group living in Canada. Young people expect to have veto power if they object to their family's choice, but in the home

HUMAN DIVERSITY

A Maori Lone-Mother Family on Social Assistance, New Zealand

'. . . from Friday to Sunday it's pretty much mayhem here. I can have anything up to 13 kids. Nieces, nephews, the *mokos* (grandchildren), the neighbours. Last weekend I had their baby, a 15-month-old baby from next door. Because they were having a big party and they were out of babysitters and I said, well just chuck him over the fence and we'll be right and she can sleep here the night so you can . . . pick her up in the morning. So

they did that. My niece had to go to a funeral and she's got a three-week-old baby and she popped her over to me with a little bottle of breast milk as well. So I had those two babies and . . . my son had his friend over for the night because his mother was next door partying, and so while everybody does their thing, I have the kids and I had another little girl 'cause her mother was there too and I don't really know them.'

SOURCE: Excerpt from an interview with a Maori lone mother on social assistance in New Zealand, Maureen Baker, 2002.

country, considerable pressure exists to abide by the judgment of elders (Nanda and Warms 2007).

Family solidarity, financial security, and potential heirs are more important in arranged marriages than sexual attraction or love, but new partners are urged to respect each other, and it is hoped that love will develop after marriage. Often, arranged marriages are more stable than free-choice unions, because both families have a stake in marriage stability. Furthermore, divorce may be legally restricted, especially for women, and may involve mothers relinquishing child custody and struggling to support themselves outside marriage.

In cultures with arranged marriages, dowries have been used to attract a partner for daughters, to cement family alliances, and to help establish new households. Dowries may involve payments of money or gifts that accompany brides into marriage and become part of marriage agreements. Although the types of payment vary considerably, they might include household furnishings, jewels, money, servants, farm animals, or land. If a woman has a large **dowry**, she can find a 'better' husband, which usually means one who is wealthier, healthier, better educated, and from a more respected family. In some cultures, the dowry money becomes the property of the groom's family, and in others, it is used to establish the new household. Dowries have also been used to provide brides with some measure of insurance in case of partner abuse, divorce, or widowhood, but this depends on how much control women have over the money or property (Barker 2003).

In other societies practising arranged marriage (such as eastern Indonesia), the groom's family may be expected to pay the bride's parents a **bride**

price for permission to marry their daughter. If she is beautiful or comes from a wealthy or well-respected family, the price rises. If the groom and his family are short of assets, the bride price can sometimes be paid through the groom's labour. Although dowries and bride prices are associated with arranged marriages, free-choice marriages have retained symbolic remnants of these practices. For example, trousseaus, wedding receptions, and the honeymoon are remnants of dowries, while the engagement ring and wedding band given to the bride by the groom are remnants of a bride price.

PATTERNS OF AUTHORITY AND DESCENT

Most family systems designate a 'head' who makes major decisions and represents the group to the outside world. In both Western and Eastern societies, the oldest male is typically the family head, in a system referred to as **patriarchy**. An authority system in which women are granted more power than men is a **matriarchy**, but these systems are rare. Some black families in the Caribbean have been referred to as matriarchal, or at least **matrifocal** (Smith 1996), since wives and mothers make a considerable contribution to family income and resources as well as to decision-making. Although Canadian families used to be patriarchal, men and women now have equal legal rights, and men are no longer automatically viewed as family heads. However, in some cultural communities, men are still regarded as family heads.

When Canadians marry, they usually consider their primary relationship to be with each other rather than with their parents or siblings. In most cases, however, the newly married pair is expected to maintain contact and to participate in family gatherings and could inherit from either side of the family. This situation is termed a **bilateral descent pattern**. In some other societies, the bride and groom are considered to be members of only one kin group, in a system called patrilineal descent if they belong to the groom's family, matrilineal descent if they belong to the bride's. Patterns of descent may determine where the couple lives, how they address members of each other's family, what surname their children receive, and from whom they inherit.

In Canadian families, bilateral descent is common for kinship and inheritance, but **patrilineal descent** has been retained for surnames in some provinces. The surname taken by a wife and by the couple's children has traditionally been the husband's name—a symbol of his former status as head of the new household. This tradition has been changed in Quebec, where brides are required to retain their family name. In Ontario, brides have a choice between keeping their family name and taking their husband's name. When there is some legal choice, couples may also abide by their cultural traditions.

Explaining Family Patterns and Practices

All social studies are based on underlying philosophical assumptions about what factors are responsible for patterns in social behaviour, what influences social change, and what the focus of social research should be. These assumptions, often called theoretical frameworks, cannot be proven or disproven but guide our research and help to explain our observations (Klein and White 1996; Cheal 1991). In this section, several theoretical frameworks used to study families will be examined, including their premises, strengths, and weaknesses.

THE POLITICAL ECONOMY APPROACH

The basic thesis of the political economy approach is that people's relation to wealth, production, and power influences the way they view the world and live their lives. Family formation, personal life, and well-being are all affected by events in the broader society, such as economic cycles, working conditions,

UNDER the WIRE

Internet Dating

In the past, many people relied on community events and personal introductions to meet potential partners for dating and marriage, although newspaper advertising has been a prevalent practice since the 1960s. Now, more people are using the Internet to search for friends, lasting relationships, and brief sexual encounters.

Relationships formed through cyberspace have been called 'hyperpersonal' because participants can be less concerned about what they look like and how to initiate a conversation with a potential partner. Instead, they can easily filter out undesirable candidates with dissimilar interests or values, carefully plan their email messages, decide how best to portray themselves, and send their message when they are ready. Once they have contacted a potential partner, they may be better able to articulate their needs and express their emotions online than in face-to-face situations. Potentially, dating relationships found through the Internet can more easily be based on common background and interests rather than just on common locality.

One of the differences between initiating a relationship in a face-to-face situation and online may be the depth and breadth of self-disclosure within a short time period. People can quickly reveal to an online partner the many details of their past and present life before they agree to meet, although the accuracy of this information may prove to be a stumbling block for continuing the relationship once they have actually met (McKenna, Green, and Gleason 2002).

Because Internet encounters can be 'disembodied', at least initially, there is more scope for fantasy, deception, and experimentation, making it possible to explore identities and sexualities (Baker 2010, 46). Research on Internet dating has demonstrated that people can be rather strategic and sometimes quite deceptive in the way they present themselves in cyberspace. They can alter details about their gender and sexuality and can pretend to be younger or older, more attractive, wealthier, and more successful in their chosen occupation.

Because of the potential for deception, parents and teachers often worry about the safety of Internet dating, since it can be difficult for anyone to differentiate between sexual predators and people genuinely searching for lasting relationships. To minimize the risk, the initial face-to-face meeting, if there is one, should normally take place in a safe public place.

laws, and government programs. This perspective originated in the nineteenth-century work of German political philosophers Karl Marx (1818–83) and Friedrich Engels (1820–95). In *The Origin of the Family, Private Property and the State* (originally published in 1884), Engels discussed how European family life was transformed as economies changed from hunting-and-gathering to horticultural to pre-industrial and finally to industrial. This approach has been debated and modified over the years. Political economists argue that social life always involves conflict, especially between those who have wealth and power and make social policies and those who do not. Conflicting interests remain the major force behind societal change.

In the nineteenth century, men's workplaces were removed from the home, which gradually eroded patriarchal authority and encouraged families to adapt to employers' needs. Furthermore, many goods and services formerly produced at home for private consumption were eventually manufactured more cheaply in factories. This meant that families eventually became units of shared income and consumption rather than units of production. Once the production of most goods and services took place outside the home, people began to see the family as private and separate from the public world of business and politics. Nevertheless, the two are actually related, because unpaid labour within the family helps keep profits high and wages low in the labour market (Bradbury 2005; Luxton and Corman 2001).

The impact of industrialization and workplace activities on family life became the focal point of the political economy approach, as well as the belief that economic changes transform ways of viewing the world. Political economists would argue, for example, that the surge of married women into the workforce after the 1960s occurred mainly for economic rather than ideological or feminist reasons. The service sector of the economy expanded with changes in domestic and foreign markets, requiring more workers. While married women had always worked as a reserve labour force, the creation of new job opportunities, as well as inflation and the rising living costs, encouraged more wives and mothers to accept paid work. These employment changes then led to new **ideologies** about family and parenting. Political economists focus on the impact of the economy on family life, on relations between the **state** and families, and on social conflict arising from these changes. In doing so, they downplay voluntary behaviour and interpersonal relations.

STRUCTURAL FUNCTIONALISM

The basic assumption of structural functionalism is that behaviour is governed more by social expectations and unspoken rules than by economic changes or personal choices. Individuals cannot behave any way they want but must abide by societal or cultural guidelines learned early in life. Deviant behaviour that violates rules is always carefully controlled.

Within this approach, 'the family' is viewed as the major social institution providing emotional support, love, sexual expression, and children. Parents help to maintain social order through socializing and disciplining their children, while families cooperate economically, share resources, and often protect their members from outsiders. Finally, people acquire money and property through inheritance, suggesting that social **status** is largely established and perpetuated through families.

Talcott Parsons and Robert F. Bales (1955) theorized that the growth of industrialization and the shift to production outside the home encouraged the acceptance of the nuclear family as a living arrangement. These authors assumed that this kind of family has two basic structures: a hierarchy of generations and a differentiation of adults into instrumental and expressive **roles**. Parsons and Bales assumed that the wife accepts the expressive role, maintaining social relations and caring for others, while the husband takes on the instrumental role of earning household money and dealing with the outside world (Thorne 1982).

Structural functionalists have been criticized for their conservative position in that they often assume that there is one acceptable family form rather than many variations. They believe that behaviour is difficult to alter because it is shaped largely by social expectations and family upbringing. Structural functionalists have also implied that a gendered division of labour was maintained throughout history because it was functional for society, when it may actually have benefited heterosexual men more than others (Thorne 1982). In addition, change is seen as disruptive rather than as normal, and individual opposition to social pressure has been viewed as deviant. Consequently, the structural functionalists have not dealt with conflict and change as well as those using a political economy approach have. Nor have they focused on the dynamic nature of interpersonal relations. For these reasons, many researchers who want to examine inequality, conflict, and change find this theoretical perspective less useful than others.

Systems theory accepts many of the assumptions of structural functionalism but focuses on the interdependence of family members and the way that families often close ranks against outsiders, especially in troubled times. This approach has been particularly useful in family therapy (Braithwaite and Baxter 2005).

SOCIAL CONSTRUCTIONIST APPROACH

The **social constructionist** approach refutes the idea that people behave according to social expectations and assumes that we construct our own social reality based on our experiences, insights, and choices (Berger and Luckmann 1966). Life does not just happen to us; we make things happen by exerting our will. This approach, also called **symbolic interactionism**, originated with the work of Americans Charles H. Cooley (1864–1929) and George Herbert Mead (1863–1931), who studied how parents assist children to develop a sense of **self**. Within this perspective, the way people define and interpret reality shapes their actions, and this process is aided by non-verbal as well as verbal cues. Social constructionists also theorized that part of socialization is developing the ability to look at the world through the eyes of others and anticipating a particular role before taking it (called **anticipatory socialization**).

Studies using this approach often occur in a laboratory, using simulations of interaction and decision-making. Researchers observe the interaction between parents and children, among children in a playgroup, and between husbands and wives. Sometimes interaction will be videotaped and participants will be asked to comment on their own behaviour, which is then compared to the researchers' observations. Research is often centred on communication processes during everyday experiences, but it also focuses on perceptions and justifications. Perceptions and 'definitions of the situation' are thought to influence action or behaviour more than objective reality (Holstein and Miller 2006). This perspective could be seen as the precursor of post-modernist theory, discussed later in this chapter.

FEMINIST THEORIES

Feminist theorists have focused on women's experiences, on written and visual representations of women, and on 'gendered' social circumstances. These perspectives developed and proliferated as more researchers concluded that women's experiences and contributions to society had been overlooked, downplayed, or misrepresented in previous social research.

Some feminist researchers have used a **structural approach** to analyze the ways in which inequality is perpetuated through social policies, laws, and employment practices (Baker 2006; O'Connor, Orloff, and Shaver 1999). Others have concentrated on inter-personal relations, examining non-verbal communication, heterosexual practices, and public discourse about women (Krane 2003; Wall 2009). Still other feminist theorists have created a more interpretive feminist analysis that takes women's experiences and ways of thinking and knowing into consideration (Butler 1992; Smith 1999).

Feminists typically argue that **gender** differences are social and cultural, are developed through socialization, and are maintained through institutional structures and practices. Most argue that gender differences in interests, priorities, and achievements grow out of psychological and sexual experiences, which are shaped by different treatment by parents, teachers, and others (Brook 1999). Nancy Chodorow (1989) combined psychoanalysis and feminist theory, showing how unconscious awareness of self and gender, established in earliest infancy, shapes the experiences of males and females as well as the patterns of inequality that permeate our culture. Carol

In addition to working outside the home, many women are still responsible for most of the family's domestic labour and care.

© Neil Fraser/Alamy

Gilligan argued in *In a Different Voice* (1982) that women's moral development differs from men's: while men tend to focus on human rights, justice, and freedom, women's sense of morality is typically based on the principles of human responsibility, caring, and commitment. Feminist scholars continue to argue that whatever is considered 'feminine' in our culture is granted lower status than 'masculine' achievements or characteristics.

Feminists note that housework and child care are unpaid when performed by a wife or mother but paid when done by a non-family member; in both cases, the work retains low occupational status and prestige. Although most adult women now work for pay, they continue to accept responsibility for domestic work in their own homes (Fox 2009; Ranson 2010). The unequal division of labour within families, as well as women's 'double shift' of paid and unpaid work, is seen as interfering with women's employment equity.

Critics of the feminist perspective argue that it glosses over men's experiences, but feminists argue that men's experiences and views are already well represented by traditional social science. The increasing integration of this perspective into mainstream sociology has been promoted by greater acceptance of the ideas that perception and knowledge are relative and that reality is socially constructed.

TIME to REFLECT

What empirical or statistical evidence can you identify that shows that female experiences, activities, and accomplishments tend to be undervalued in Canadian society?

POST-STRUCTURAL APPROACHES

These perspectives argue that knowledge and understanding depend on one's social position, gender, race, and culture. Furthermore, vast differences exist in personal and family life, which is always changing. In Canada and other Western countries, sexuality is increasingly separated from marriage, and marriage is being reconstructed as a contract that can be ended. Childbearing and child-rearing are no longer necessarily linked with legal marriage, and the gendered division of labour is continually negotiated and renegotiated (Beck-Gernsheim 2002). With more people 'writing their own biographies', family and personal life have changed considerably over the twentieth and twenty-first centuries.

Post-structuralists argue that images of gender and family are shaped by everyday language and policy discourse (Giddens 2006). By deconstructing—or analyzing—the origins and intended meanings of beliefs about gender and family, researchers can understand how these images have been socially constructed throughout history. Western welfare systems were premised on the now outdated ideal of the male-breadwinner/female-caregiver family, but labour market changes and new lifestyle possibilities have encouraged both men and women to question this gender order and 'normative heterosexuality'. Some researchers have proposed a reconceptualization of family studies away from the focus on heterosexual affiliation and intimacy between couples toward a focus on caring relationships between parents and children (Fineman 1995).

Post-structuralists tend to focus on images of the body, media representations of gender and family, sexual diversity, families of choice, and the 'performance' of gender (Smart 2007). However, critics of this approach argue that too much emphasis is sometimes placed on personal choices and minority situations rather than on the ways that most people live or the practical constraints on their life choices (Baker 2010).

Recent Issues in Canadian Families

In this section, we consider a number of controversial issues, with specific reference to Canadian families. First, we examine the gendered division of labour at home.

SHARING HOUSEHOLD WORK

Over the past 20 years, gendered patterns of employment have changed dramatically. In 1976, 47 per cent of women 20 to 64 years of age who were either married or living in a common-law relationship participated in the labour force, but by 2009 the percentage had increased to 76 per cent (Lu and Morissette 2010). Especially after parenthood, however, more fathers than mothers work full-time and overtime, and men tend to earn higher wages. Table 13.2 shows that fathers are still more likely than mothers to be working for pay, regardless of the age of their children.

Research typically concludes that most heterosexual couples, especially after parenthood, divide their household labour in such a way that husbands work full-time and perform occasional chores around the

TABLE 13.2 ▼	Canadian Labour Force Participation Rates (%) for Mothers and Fathers with Children under 15 Years of Age, 2006		
Family Type		Mother	Father
Single (never-married) parent with children under 6		61	78
Single (never-married) parent with youngest child age 6–14		80	85
Married parent with children under 6		75	94
Married parent with youngest child age 6–14		82	93
Divorced parents with child under 6		76	90
Divorced parents with youngest child age 6–14		86	90

SOURCE: Vanier Institute of the Family website (www.vifamily.ca), derived from Statistics Canada, 2006 Census.

house, usually in the yard or related to the family car. Most wives are employed for fewer hours per week than their husbands, but they usually take responsibility for routine indoor chores and child care, even when employed full-time. Wives are also expected to be 'kin keepers' (Rosenthal 1985), which includes maintaining contact with relatives and organizing family celebrations. In addition, wives and mothers usually retain responsibility for emotional work, such as soothing frayed nerves, assisting children in building their confidence, and listening to family members' troubles (Ranson 2009).

Despite this prevalent division of labour, wives who are employed full-time tend to perform less housework than those who work part-time or who are outside the labour force, and employed women did less housework in 2005 than in 1986 (Lindsay 2008). Wives employed full-time may lower their housework standards, encourage others to share the work, or hire help. Yet women retain most of the responsibility for indoor housework and child-rearing tasks, including the hiring and supervision of cleaners and care providers (Ranson 2009; Baker 2010). Many employed mothers report feeling exhausted and drained when they shoulder the responsibility for household work while working for pay.

More equitable workloads are apparent among younger, well-educated couples with few or no children. However, some employed wives in dual-earner families retain responsibility for housework, especially older women and those without high school education. I have shown (2010, 123) that even among university professors with doctorates and full-time jobs, wives tend to do a disproportionate amount of housework. Yet wives' bargaining power may increase slightly when they earn an income comparable to their husbands'. These women are better able to persuade their partners to do more housework or to

hire help (Statistics Canada 2006d), and they tend to be less willing to relocate with their husbands' jobs.

Why do employed wives accept responsibility for unpaid work even when they prefer more sharing of housework? Partners' hours of work and income earned seem to be less important than marital power relations in determining the allocation of household tasks (Davies and Carrier 1999). These relations are influenced by gender expectations and by opportunities and experiences in the larger society, but gender also intersects with race, ethnicity, and social class to influence such relations. Dempsey (1999) concluded that most Australian husbands use a variety of tactics to avoid doing housework, such as waiting to be asked by their wives, saying they do not know how to do the task, arguing that it does not really need doing yet, and delaying completion.

Considerable research suggests that even when men have time available, they do not choose to spend it on domestic work (Baker 2010). Husbands will lend a hand if their wives are pressed for time, but housework and child care are viewed as low-status women's work throughout much of the world. Cohabiting couples seem to have a less gendered division of labour than legally married couples (Wu 2000; Baxter, Hewitt, and Haynes 2008), but relative power differences in heterosexual relationships continue to influence patterns of domestic work.

An uneven and gendered division of labour has many implications. More mothers than fathers develop close ties with their children through years of physical and emotional care. In addition, some wives supported by their husbands are able to pursue hobbies and friendships during the day. At the same time, accepting most of the responsibility for household work reduces the likelihood of obtaining employment qualifications, of working full-time or overtime, or of being promoted to higher-paying positions. Furthermore, the consequent lack of income may reduce women's confidence in their ability to earn a living outside marriage. This may translate into reduced decision-making power within marriage and less income in the event of divorce or widowhood. When women accept responsibility for domestic work, they also reinforce traditional role models for their children (Ranson 2009). However, many women feel they have little control over the household division of labour but would like their husbands to accept a larger share of household work.

TIME to REFLECT

Why do wives, and especially mothers, accept responsibility for most of the indoor housework and child care? Do husbands and fathers compensate through their outdoor work and financial contribution to the household?

MEDICALLY ASSISTED CONCEPTION

Cohabitation, same-sex partnerships, divorce, and remarriage have complicated marriage and family relationships in recent decades, but reproductive and genetic technologies may be in the process of fundamentally reshaping families (Eichler 1997). This reshaping includes separating biological and social parenthood, changing generational lines, and enabling sex selection. A wide range of procedures have now become routine, such as donor insemination and in vitro fertilization. Freezing sperm, eggs, and embryos makes conception possible after their donors' death, post-menopausal women can bear children, and potential parents can contract with surrogates to bear children for them (Baker 2005).

Reproductive technologies tend to commercialize human reproduction: we can now buy eggs, sperm, embryos, and reproductive services—all of which are produced and sold for profit (Eichler 1996). These technologies tend to raise the potential for eugenic thinking and enable us to evaluate embryos on their genetic makeup. However, little research has been done on the impact of these technologies on family life, such as how parents involved in artificial insemination reveal their children's background to them and how children deal with this knowledge.

Fertility continues to influence social acceptance and gender identity, and conception problems often contribute to feelings of guilt, anger, and depression and to marital disputes (Doyal 1995). Low fertility may be caused by many factors, including

exposure to sexually transmitted diseases, long-term use of certain contraceptives, environmental pollutants, hormonal imbalances, and lifestyle factors such as tobacco smoking, a large consumption of caffeine or other drugs, and prolonged stress (Bryant 1990). The probability of conception also declines with a woman's age. Some couples spend years trying to conceive, while others place their names on adoption waiting lists. The number of infants available for adoption, however, has dramatically decreased, with more effective birth control, greater access to abortion, and social benefits enabling single mothers to raise their own children. Consequently, more couples with fertility problems are turning to medically assisted conception.

Infertility is usually defined as the inability to conceive a viable pregnancy after one year of unprotected sexual intercourse, but this definition encourages some fertile couples to seek medical attention prematurely. Access to assisted conception is often limited to those considered most acceptable as parents: young heterosexual and childless couples in stable relationships. Private clinics charging fees may be less selective, and many women over 40 years of age approach fertility clinics for assistance. Most treatments last for several months and involve the use of drugs that produce side effects such as depression, mood swings, weight gain, and multiple births. Some treatments continue for years (Baker 2005).

Fertility treatments are expensive, although those who end up with a healthy baby may find these costs acceptable. The chances of complications following in vitro fertilization (IVF) are higher than with natural conception, and the success rate is not always as high as couples anticipate. In Australia, the viable pregnancy rates were 14.9 per cent after one cycle of IVF, 15.9 per cent for insemination with sperm, and 18.1 per cent for egg transfer. If these products are frozen or thawed, the pregnancy rates fall (Ford et al. 2003, 100). Australian research also shows that adverse infant outcomes, such as pre-term delivery, low birth weight, stillbirth, and neonatal death, are higher among assisted conception births compared to all births (Ford et al. 2003). Medically assisted births and their complications tend to consume greater public health resources and also place financial and time constraints on new parents.

Eichler (1997) argues that reproductive and genetic technologies represent a quantum leap in complexity by blurring the role designations of mother, father, and child. For the child in a surrogacy relationship, who is the mother: the woman who gave birth or the woman who was part of the commissioning couple? What does it now mean to be a father? Does a man become a father if he impregnates a woman but has no social contact with the child? Does he become a father when he contracts with another woman to use his sperm to make a baby, which he then adopts with his legal wife? Although sociologists have always been interested in the impact of absent fathers on family life, they are now referring to sperm donors as the new absent fathers (Jamieson 1998, 50). Researchers are also interested in the increasing number of lesbian couples using self-insemination to create families without men.

Sociologists have been ambivalent about assisted conception. Although it offers hope and opportunities for potential parents who might otherwise be excluded, some technologies are experimental and intrusive. They also medicalize the natural act of childbearing, reinforce the pressure on all women to reproduce, and offer costly services that are not available to the poor. Feminist scholars have also been concerned that patriarchal societies will use sex selection to reinforce the cultural preference for males and that working-class women will be exploited

through surrogacy arrangements. Sociologists seem to be most supportive of reproductive technologies when they discuss self-insemination within lesbian relationships (Nelson 2001), perhaps because they assume that unequal power relationships and coercion are minimized.

AFFORDABLE AND REGULATED CHILD CARE

The dramatic increase in employed mothers has led to a higher demand for non-family child care as well as public concerns about the quality of care. The demand still outstrips the supply, the assurance of quality continues to be a problem, and child-care costs are still unaffordable for many parents. Since the 1960s, Canadian governments have subsidized child-care spaces for low-income and one-parent families. However, there were never enough spaces for eligible families (Clevedon and Krashinsky 2001; Beach et al. 2008).

Canadian governments offer two forms of child-care support: a federal income tax deduction of up to $7000 per child for employed parents using non-family care (with no family maximum), which is

OPEN for DISCUSSION

Medically Assisted Conception

'We tried IVF [in vitro fertilization] and we got pregnant the first time. So it was very successful and obviously we are very pleased. I would say the IVF process was very cruel, even though we succeeded. I take my hat off to people who have it two or three times, because it was extremely tough. It is very impersonal. You can't fault the treatment or the staff, but all the injection, different phases—it's like a roller coaster. You think you're ahead. Then you have a setback, bad news. One of the most stressful times for me was when I was in the room and they were harvesting the eggs from [my wife's] ovaries the first time—one egg. It felt terrible because we were hoping to get 12. We got five off the second so that was great but you still feel pretty disappointed. Then they fertilize and you only get two embryos and we were pretty depressed. Then a day later, we were up to four, so we were elated. Their goal was to try for five or six to choose. We had four but one was a bit dodgy so we had three good embryos— you get that news a day later and you're down a little bit. Then during the IVF we had always envisaged that they would insert two embryos, which is extremely

common—most people have two put in. We only had one because [my wife's] uterus is a bit dodgy and they didn't want twins with a uterus shaped like that. They didn't want a prem baby [premature] so all of a sudden you think that your chances are halved, which wasn't quite true. Then we had three great embryos and they chose the best one and it took, which is very pleasing. The other two are cryogenically frozen just like Austin Powers, waiting for their day in the sun.'

Interviewer: 'Amazing isn't it when you think of the technology?'

'Yes it is and we wouldn't have got pregnant otherwise. . . . One of the senior doctors said that this was just the start and there were plenty more ups and downs. We are in the process now where there is lots of worry. [My wife] is worried about what she should and shouldn't eat. She worries that the baby will be born with some fault because she didn't take enough care. I'm very much in the reassuring mode—I'm sure it will be fine. I'm sure that once it is born then more worries start. It is an intriguing game, becoming a parent, I'd say.'

SOURCE: Interviews with couples experiencing fertility treatments in Auckland, New Zealand, Maureen Baker, 2002.

most useful for middle-income families paying higher taxes. In addition, the provincial governments subsidize child-care spaces for low-income families and lone parents (Baker 2006). Unlike other Canadian provinces, Quebec offers heavily subsidized care for all parents who need it, regardless of their household income or work status, at a cost to parents of only $7 per day (Albanese 2006). Not surprisingly, the employment rates of mothers with children under six are much higher in Quebec than in the rest of Canada (Statistics Canada 2006e).

Across Canada, many not-for-profit child-care centres have long waiting lists, and many do not accept children under two unless they are toilet-trained. Even if space is available, parents want to ensure that the centre employs an adequate number of staff to keep the infants clean, fed, and stimulated and worry about the spread of infectious diseases. Finding a qualified babysitter to come to the child's home or who will welcome an extra child in her home is also difficult, although licensed family homes are available in most jurisdictions (Beach et al. 2008).

Sitter care is unregulated by any level of government, yet it remains the most prevalent type of child care for employed parents. Grandparents (usually grandmothers) sometimes provide child care while the parents are working, and care by grandmothers can save money, provide culturally sensitive care, and create a solid bond between generations. Yet it could also lead to disagreements about child-rearing

(Popperfoto/Alamy)

Idealized or romanticized images of family life—so common in the media—give many people the sense that they or their family are failing in some important respect; otherwise, wouldn't they be happier?

techniques between the parents and the grandparent, who is likely to have retained more traditional cultural values. Child-care concerns have encouraged some mothers to remain at home to care for their own children, although most can no longer afford this option.

Centre-based care generally operates during office hours, but some parents need child care in the evening and weekends. In two-parent families, employed parents may be able to share child-rearing responsibilities if they work on different shifts, but it is difficult to maintain their own relationship or to engage in family activities. Parents whose children have special needs also experience problems. Before institutions and hospitals were built for these children in the 1960s, mothers were their main caregivers. With deinstitutionalization policies since the 1980s, greater expectations have been placed on maternal care. Yet because more mothers are now employed, fewer are able to care for children with disabilities without some remuneration and community assistance.

The quality of care remains a concern, both in licensed centres and by babysitters in private homes. In some jurisdictions (such as Alberta), employees of child-care centres and babysitters are not required to have special training. These jobs pay the minimum wage or less and experience difficulty attracting and retaining trained workers. A number of advocacy groups have formed around child-care concerns, asking governments to tighten regulations; to improve training, fringe benefits, and pay for child-care workers; and to allocate more public money to child care for employed parents. However, funding and regulation remain contentious, and child care has become a regular election issue in Canada.

DIVORCE AND RE-PARTNERING: THE IMPACT ON CHILDREN

Since divorce rates began to increase in the 1970s, considerable attention has been devoted to the outcomes for children. Canadian researchers have estimated that about 38 per cent of marriages will end in divorce before their thirtieth anniversary, but only half of these divorces involve children (Vanier Institute of the Family 2004, 33). Nevertheless, children from one-parent families have been found to experience a higher risk of negative outcomes than children from two-parent families, including lower educational attainment, behavioural problems, delinquency, leaving home earlier, premarital pregnancy for girls, and higher divorce rates when they marry.

Despite negative media attention given to one-parent families, most children from these families do not experience problems but only a higher risk. Furthermore, when studies control for changes in household income after parental separation, the incidence of problems declines, although it does not disappear (Kiernan 1997; Lipman, Offord, and Dooley 1996, 88). Does the high risk of negative outcomes result from the parental conflict during marriage and separation, the absence of a father in the household, or some other factor such as poverty? As Figure 13.1 shows, mothers typically retain custody of the children.

In Canada, as well as in other English-speaking countries, many children in one-parent families live in poverty, especially if their mother is not employed, as the table in the Global Issues box indicates. However, one-parent families more often experience economic disadvantage both before and after separation, since people from lower socio-economic groups

▼ FIGURE 13.1 Dependent Children by Party to Whom Custody Was Granted, 1995 and 2002

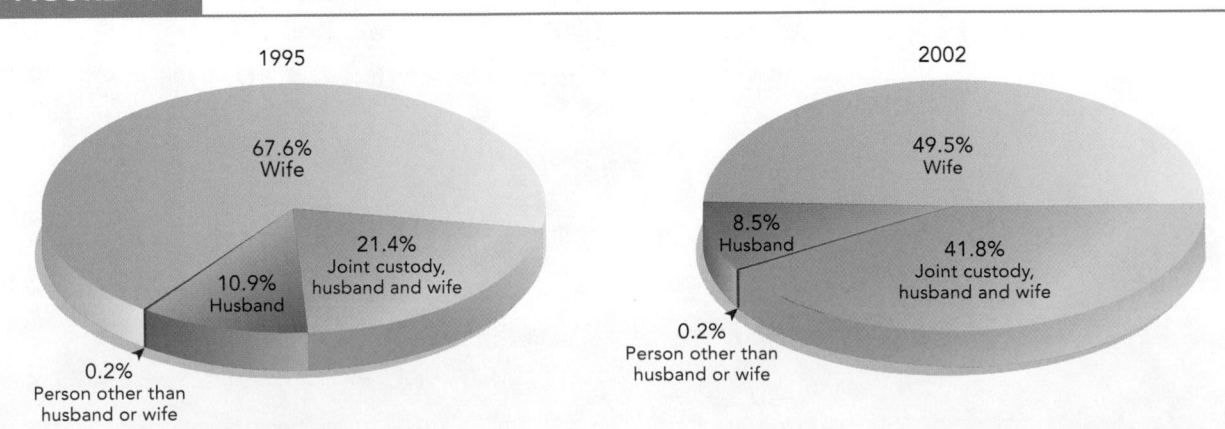

SOURCE: Adapted from Vanier Institute of the Family, 2004, *Profiling Canada's Families III*, p. 38 (Ottawa: Vanier Institute of the Family, 2004), 38. Reprinted by permission of the publisher.

tend to have higher rates of bereavement, separation, and divorce (Pryor and Rodgers 2001). When children are raised in poverty, they are more likely to suffer disadvantages that continue into adulthood.

Distinguishing between the impact of poverty and of parental separation is difficult for researchers. Lower socio-economic status after parental separation seems to be a mediating factor for some negative outcomes but not others; it accounts for a decline in educational attainment but not for rates of delinquency, psychosomatic illnesses, cigarette-smoking, or heavy drinking in adulthood (Hope, Power, and Rodgers 1998). Even children raised in low-income families with two parents are more likely than other children to experience negative behavioural outcomes. These include delayed school readiness, lower educational attainment, a greater number of serious childhood illnesses, higher childhood accident rates, premature death, high rates of depression, high rates of smoking and alcohol abuse as young adults, and more trouble with school authorities and the law, to name only a few (NLSCY 1996; Canadian Institute of Child Health 2002). Consequently, social researchers must consider poverty an important variable in all

discussions of the outcomes for children after separation and divorce.

Many studies also indicate that children who live with their mothers after divorce tend to experience diminished contact with their fathers and to suffer distress from this loss (Smyth 2004; Qu and Weston 2008). As children grow older, the time they spend with the non-resident parent decreases, and about a third lose contact (Amato 2004). However, father–child contact is not the deciding factor in children's adjustment after parental separation, since frequent contact with their father may be detrimental if there is a high degree of conflict surrounding it. If conflict is absent or contained, children both want and benefit from frequent contact with both parents (Amato 2004). In general, a close relationship with both parents is associated with a positive adjustment in children after divorce (Pryor and Rodgers 2001). Whether or not the father continues to pay child support may also influence the children's adjustment and the family's socio-economic status.

Parental separation clearly adds stress to children's lives through changes in relationships, living situations, and parental resources, but few

GLOBAL ISSUES

Comparative Child Poverty Rates

Living in poverty is influenced by household living arrangements and parental employment status but also varies considerably by country of residence. Children living in Canada and the United States have much higher poverty rates than those in the United Kingdom, Finland, and Sweden, especially when they reside with one parent who is not in the workforce. The table below suggests if parents can find and keep well-paid jobs and/or governments provide generous social benefits and services (such as income support and child care), children's poverty rates can be reduced, even when they live with only one parent.

POVERTY RATES of Households with Children in Various Countries, by Working Status of Parents

Country	1 parent, no worker	1 parent, 1 worker	2 parents, no workers	2 parents, 1 worker	2 parents, 2 workers	Poverty rate, all children
United States	92	36	82	27	6	18
Canada	89	32	81	22	4	13
Australia	68	6	51	8	1	10
New Zealand	48	30	47	21	3	13
United Kingdom	39	7	36	9	1	9
Netherlands	62	27	65	12	2	9
Finland	46	6	23	9	1	4
Sweden	18	6	36	14	1	4
OECD Average	54	21	48	16	4	11

SOURCE: Organisation for Economic Co-operation and Development (OECD). 2009. *Society at a Glance*. Based on Table EQ3.2, p. 93. Paris: OECD.

studies conclude that problems are severe or prolonged (Baker 2010). Instead, most research finds that the first two years after separation require adjustments by both parents and children. However, adult children of divorced parents are more likely than those from intact marriages to believe in the fragility of marriage, to have poor relationships with their parents, and to end their own marriages with divorce (Cartwright and McDowell 2008; Cunningham and Thornton 2006; Hughes 2005). This may result from poor parental role models, from a history of family conflict, or from the acceptance of divorce as a potential solution to an unhappy marriage if it has already happened to one's parents.

Never-married mothers who become pregnant before completing their education are particularly vulnerable to low income and child disciplinary problems. These mothers often re-partner within a few years of the child's birth, but the socio-economic disadvantages of bearing a child at a young age may linger (Edin and Kefalas 2005). Their children are most likely to spend their childhood in one or more stepfamilies, which are often conflictual. This helps to explain the higher rates of behavioural problems in the children of never-married mothers (Marcil-Gratton 1998).

Research suggests that stepfamilies are difficult to maintain, requiring considerable negotiation. Children living in these families are at the same risk of behavioural problems and distress as those growing up in one-parent households. Neither a higher income nor two adults in the home ensures good outcomes for these children (Pryor and Rodgers 2001). One explanation is that parental conflict and separation have a lasting effect on children. Another is that stepparents do not relate to their stepchildren with the same warmth and concern as they do with their biological children, because they do not see them as their own children and have not lived with them during their formative years.

Although researchers usually study separation and divorce as negative life events, parents often experience relief and contentment after the initial adjustment of leaving an unhappy marriage. This is reflected in their general outlook and in their interactions with their children. Consequently, most researchers agree that children living in stable one-parent households are better off than children living in conflict-ridden two-parent households (Cavanagh 2008). Furthermore, children of employed lone mothers tend to accept more egalitarian gender roles because they see their mothers managing tasks previously defined as masculine. This suggests that separation and divorce could have positive as well as negative outcomes for parents and children.

WIFE ABUSE

Since the 1980s, when domestic violence appeared to be on the rise, sociologists have studied this phenomenon. The very term 'family violence' implies that the behaviour is randomly distributed within families, when men actually are the perpetrators in the vast majority of cases that come to the attention of police and social workers (Dobash et al. 1992). However, many government reports and academic studies continue to use gender-neutral language.

The Canadian Urban Victimization Study found that in cases of 'spousal violence', physical abuse is not an isolated event. Some abused women are assaulted on numerous occasions by their male partners and have sought help many times from friends, neighbours, social workers, and the police. Furthermore, separated women are more likely to be assaulted than divorced or married women (DeKeseredy 2009). Women are also more vulnerable if they see their partner as the household head, if they are financially dependent on him, or if they live in a housing development with other single mothers. Violence becomes normalized when it is prevalent among friends or within the community or when it is continually viewed as a form of entertainment in films and sports events in the larger society.

Murray Strauss and Richard Gelles (1990) found that marital violence actually decreased in the United States throughout the 1980s even though reporting to authorities increased. They argued that reports to police were influenced by the women's movement, by police campaigns to prosecute perpetrators, and by increasing options for women wishing to leave violent marriages. Yet they also made the controversial claim that women are as likely as men to abuse their partners, although they acknowledged that this behaviour is less likely to be reported to the authorities, is less consequential in terms of physical harm, and is often motivated by self-defence. Walter DeKeseredy (2009) criticized the 'conflict tactics' scale used by Strauss and Gelles, which has counted the incidences of violence without examining the social context. He also argued that women's 'violence' against men is often in self-defence.

The abuse of women may reflect men's rising concern that they are losing authority in their households, especially on the part of men who are experiencing unemployment or other personal problems. This kind of violence is also aggravated by alcohol

SOCIOLOGY in ACTION

The Emotional Scars of Family Violence

'I was raped by my uncle when I was 12 and my husband has beat me for years. For my whole life, when I have gone to the doctor, to my priest, or to a friend to have my wounds patched up, or for a shoulder to cry on, they dwell on my bruises . . . that's for sure. . . . I don't look anything like I did 15 years ago, but it's not my body that I really wish could get fixed. The abuse in my life has taken away my trust in people and in life. It's taken away the laughter in my life. I still laugh, but not without bitterness behind my laughter. It's taken away my faith in God, my faith in goodness winning out in the end, and maybe worse of all, it's taken away my trust in myself. I don't trust myself to be able to take care of the kids, to take care of myself, to do anything to make a difference in my own life or anyone else's. That's the hurt I would like to fix. I can live with my physical scars. It's these emotional scars that drive me near suicide sometimes.'

SOURCE: DeKeseredy and Macleod 1997, 5.

---

and substance abuse but represents much more than an interpersonal problem. The fact that most victims of reported violence are women, and that separated women are often the targets, indicates important social patterns relating to gender and power.

In the past, the police failed to respond seriously to calls about wife abuse because they thought women did not want charges laid or would later withdraw them (DeKeseredy 2009). Policies have now been implemented in most jurisdictions for police to charge men who abuse their female partners. Yet many women remain with abusive partners because of a shortage of affordable housing, an inability to support their children, and a lack of knowledge about where to turn for assistance. Abuse often continues because some women feel that they deserve it, especially those with low-self esteem who were abused as children. In addition, many women fear reprisal from spouses or former spouses who have threatened to kill them if they go to the police or tell anyone about an incident. The enormous publicity given to women killed by their partners indicates that fear of reprisal is often justified.

Women's groups, social service agencies, police, and researchers have developed new ways of dealing with violence against women in intimate relationships. Many programs are crisis-oriented and focus on women, helping them develop a protection plan that could involve laying charges against the perpetrator, finding transitional housing, engaging a lawyer, gaining a protection order, and, if necessary, acquiring welfare to cover living costs. Through individual counselling or group therapy, abused wives and their abusers can be helped to restructure their thinking about violence and to view it as always unacceptable.

The male abuser is now more often charged with an offence. He is also given opportunities for counselling, including accepting responsibility rather than blaming his partner, learning to manage his anger, developing better communication skills, and learning non-violent behaviour from positive male role models. Action against intimate violence has also included sensitization workshops for professionals to increase their knowledge of program options and of the implications of this form of violence for women, families, abusive men, and the wider society. In addition, more support services have been provided for families in high-risk circumstances.

TIME to REFLECT

What sources of support and help are available for someone in an abusive relationship? Which of these might people turn to first, and which might be seen as a last resort? Why?

Although governments have voiced concern about violence against women and children, money is a major impediment to establishing new programs. Transition houses are usually funded by private donations, staffed by volunteers, and operated with short-term resources. Follow-up therapy and counselling may be necessary for the entire family, but these services also cost money to develop and maintain. Despite the serious nature of violence against women, new program funding for the rising number of reported victims and their abusers is difficult to find (Se'ver 2002).

Exposure to violence during childhood has been linked to dating violence as well as spousal victimization and perpetration (Gover, Kaukinen and Fox 2008). Three broad explanations arise from these kinds of findings. The intergenerational theory suggests that solving conflicts through physical or verbal violence is learned from early experiences of witnessing parental conflict or violence against one's mother or being a victim of child abuse or neglect. The solution within this perspective focuses on improving anger management, self-esteem, couple communication, and parenting skills. A second theory sees marital violence as a misguided way of resolving conflicts that is used by husbands who feel that their authority within the family is being threatened. The solution to the problem within this systems framework is to offer therapy sessions to improve couple communication skills, to manage their anger, and to become more assertive about their feelings without resorting to violence.

In contrast, feminist theories reinforce the fact that marital violence is usually violence by men against their female partners. This reflects women's lack of interpersonal power in families, the way that patriarchal states permit husbands to control their wives, and the social acceptability of violence toward those considered most vulnerable (Sev'er 2002). The three theories, however, are not entirely incompatible. Not everyone who has witnessed abuse or who feels threatened by lack of power in their workplace or home becomes abusive. Furthermore, everyone lives in a society that condones certain kinds of violence,

yet only a small minority abuse others. Individually, none of the theories can explain the perpetuation of violence in intimate relationships.

It is clear that domestic violence cuts across national, cultural, class, and age boundaries and that violence against women is not just confined to marriage or cohabitation. Changing public attitudes toward physical and sexual abuse requires that more people report such activity, but authorities also need to take women's reports seriously and provide the necessary protection and social services.

Reforming Canadian Family Policies

When Canada was established as a nation in 1867, jurisdiction over policy was divided between the federal and provincial governments. Over the years, new policies and programs were established to meet the needs of the changing society. In the nineteenth century, the federal government developed ways to count residents and to require them to register marriages, births, adoptions, divorces, and deaths. Most provincial governments also enabled married women to control their property, equalized the guardianship rights of mothers and fathers over their children, and established basic social services in the 1920s. As Table 13.3 indicates, income security programs for families were developed by both levels of

TABLE 13.3 ▼ The Establishment of Social Benefits in Canada

Family Allowance	This universal allowance was created in 1945 and paid to mothers for each child.
Old Age Pension	Originally established in 1926 as a pension for low-income seniors; in 1951 it was converted to a universal pension.
Mothers'/Widows' Pensions	These pensions were developed around 1920, but the date varies by province.
(Un)Employment Insurance	UI was established as a federal social insurance program in 1941; maternity benefits were added in 1971 and parental benefits in 1990 and increased in 2001.
Hospital/Medical Insurance	Hospital insurance was established in 1958; universal medical insurance (medicare) was established in 1966.
Canada Pension Plan	This broad social security program, effectively a retirement program, began in 1966, financed by contributions from employees and employers and from government; CPP also pays survivor benefits and disability benefits to contributors.
Spouses' Allowance	Established in 1975, this is an income-tested pension for spouses (mainly women) aged 60–4 of old-age pensioners.
Child Tax Benefit	The former Family Allowance and tax deductions/credits for children were rolled into this targeted tax benefit for lower- and middle-income families in 1993.
Parliamentary Resolution to End 'Child Poverty'	An all-party agreement was passed in 1989 to end child poverty by the year 2000.
Canada Child Tax Benefit	The Child Tax Benefit and Working Income Supplement were rolled together to form this benefit in 1998, which includes the child disability benefit.

SOURCES: Baker 2006; Baker and Tippin 1999; McGilly 1998.

government, mainly from the 1940s to the 1970s, but they have been modified considerably since then. Provincial governments also tightened abuse and neglect laws as well as the enforcement of child support and eligibility for 'welfare' during the 1980s and 1990s (Baker 2006; Ursel 1992).

The extent of intervention in family life has always been questioned, but governments need to regulate certain aspects, especially to protect vulnerable family members and to assist those in serious financial difficulty. Families also require health and social services to ensure healthy and safe pregnancy, childbirth, and childhood, and these services need government regulation and financial support. Regulation of life events by the state is designed to prevent incestuous and bigamous marriages, adoptions by inappropriate parents, and hasty divorces, and to ensure that spouses and parents understand and fulfill their support obligations. Governments also need to gather basic statistics about populations in order to plan future social services and facilities. They must be able to predict the size and structure of the future labour force and the numbers of future voters, taxpayers, and consumers. Some of these statistics also prove useful to the business sector in their marketing and growth plans.

Since the 1980s, labour markets have been restructured, full-time jobs have become harder to find, and more households now need two or more earners to pay the bills, but growing marriage instability means that more households contain only one parent. These socio-economic trends have raised the cost of social programs for governments, which has led to greater concern about the high level of taxes needed to maintain the welfare state at existing levels and to questioning the effectiveness of anti-poverty strategies. In addition, many politicians and researchers continue to worry about the state's ability to sustain social programs with an aging population, growing structural unemployment, and high rates of separation. While several provincial governments have cut back on income support since the 1990s, the federal government has tightened eligibility for unemployment benefits but enhanced child benefits. However, Canadian parents are increasingly expected to rely on their own resources for family well-being rather than depending on state assistance.

Throughout Canadian history, the desirability of government involvement in family life has been debated, although these debates usually focus on the cost of income support. However, governments have also tightened laws on the enforcement of child

support and on spousal and child abuse, but these laws have been difficult to enforce. The monitoring of families assumed to be 'at risk' suggests that the state regulates family life as a form of social control as well as merely for assistance, information-gathering, and future planning.

TIME to REFLECT

You are a senior policy adviser to the social services minister in your province, and she has asked you to develop a new and comprehensive family policy to be included in the party platform for the upcoming election. What issues will you emphasize, and which ones will you steer away from? Why?

Future Families

Sociologists often use current family trends as an indicator of future patterns, but predicting the future is always complex. Nonetheless, from what we know about patterns in family formation, we might assume that cohabitation will become more prevalent in the future (Baker 2010). As the distinction between cohabitation and legal marriage becomes socially and legally blurred, the average age of legal marriage will increase slightly, and marriage rates will continue to fall. Living together will become more socially acceptable as a preliminary step to the public commitment of legal marriage. Particularly people with divorced parents and older divorced adults will be reluctant to enter legal marriage without some previous experience of living with their partner. Furthermore, those who are ideologically opposed to traditional family obligations and gender roles may continue to see cohabitation as a preferable alternative to legal marriage. However, research suggests that the differences in gender roles between cohabiting and married couples will become minimal over the years (Baxter 2000).

With more dual-earning households, opportunities to move to jobs in new locations may be limited by the partners' employment. Employees increasingly find it harder to encourage their partners to give up their jobs, because households cannot survive on one income. More couples may be forced to live apart for short periods in order to further their education or obtain work. Commuter marriages could become more prevalent as professional and managerial positions become harder to find, as the labour market becomes more globalized, and as women become more career-oriented.

Delayed marriage and child-bearing are definitely on the rise in Canada. This suggests that more individuals and couples will live for longer periods in non-family households in a lifestyle that focuses on work, career development, leisure pursuits, and travel. Some will become accustomed to this lifestyle and will choose a partnership without children. If substantial numbers of Canadians make this choice, policy-makers will become concerned about the future of the nation's population. An increase in child-free marriages could also lead to higher separation and divorce rates, since it is easier to part without children. Higher rates of cohabitation also suggest greater marriage instability in the future, because couples who lived together before legal marriage have a higher probability of divorce than those who have not lived together. In addition, children of divorced parents have a higher probability of divorce, suggesting greater impermanence in future relationships.

Most demographers predict that declining birth rates will persist, because raising children is increasingly costly and because combining work and family life is difficult when both parents are employed. Many working parents experience problems finding affordable and high-quality child-care services. Although government reports have said that children are our greatest future resource, little has been done in many Canadian provinces to help parents combine paid work and child-rearing.

Young people tend to remain at home with their parents for longer periods now than in the 1970s. The greater need for higher education, the longer time required to find a secure job, and the higher cost of housing prolong the period of active parenting. At the same time, more women are employed full-time in mid-life, which means that they have neither the time nor the energy to supervise their young adults adequately.

As **life expectancy** increases and fertility declines, more people will live beyond the age of 75, and more

In the transition from large to small families, many of us have lost the opportunity to experience relationships with cousins, aunts, and uncles.

(David Young-Wolff/Photo Edit Inc.)

frail elderly people will require care. More parents, especially women, may have to provide attention, emotional care, and domestic services for both children and aging parents. Furthermore, middle-aged people will have fewer siblings to help them care for frail parents. In the future, they will also need to remain employed longer to counteract job insecurity throughout their lives and lower employment pensions. To deal with higher public pension costs as well as employees' need for more income and activity in later life, an increasing number of governments have prohibited mandatory retirement and extended the age of eligibility for receiving the public pension.

As more people remarry in later life, attitudes toward aging, marriage, and leisure may gradually change. Family relationships will become more complicated, with stepchildren and ex-partners and with older men re-partnering and reproducing with younger women. A global economy and high rates of migration could also mean that more elders remain in their home communities while their children migrate to find work, taking the grandchildren far away. This may lower the frequency of family activities, but it could also strengthen friendship ties for both generations. Yet keeping in touch will become even easier with email, long-distance telephone, and text messaging, as well as new, as-yet-unforeseen communication technologies.

Conclusion

Intimate relationships remain central to most people, yet families and personal life have changed substantially over recent decades. Cohabitation, separation, and divorce are now more prevalent than they were a generation ago, while legal marriage and fertility rates are declining. Cultural variations in personal and family life are becoming more noticeable with new immigration sources. In addition, more people are creating or modifying their own intimate arrangements in response to changes in the larger society, but governments continue to clarify the rights and responsibilities of family members within these new arrangements.

Social scientists have used different theoretical frameworks to study changes in personal life and family practices, emphasizing different aspects and issues. The five theoretical frameworks presented in this chapter suggest that family theorists differ in their focus. The chapter also discussed five of the many controversial issues in personal and family life. The first is the sharing of domestic work; despite dramatic increases in women's paid work, female partners still do the major portion of housework and child care. The second issue relates to the apparent rise in infertility and to concerns about the inherent contradictions and high costs of medically assisted conception. The third issue relates to the cost and quality of child care for employed parents, which makes it difficult, especially for mothers, to combine paid work and child-rearing. The fourth involves the contradictory evidence of the impact of separation, divorce, and re-partnering on children, suggesting that separation is not always a negative solution but that remarriage is not always the best outcome for children. And the fifth issue relates to wife abuse and why it continues despite public efforts to reduce its prevalence.

This chapter has shown that more people now cohabit, separate, re-partner, and live with more than one partner over their lifetime. Nevertheless, not all of our personal living arrangements represent our own choices. Most Canadians still hope to develop loving and stable intimate relationships and to watch their children grow into adults. However, few people anticipate the ways that work requirements, money problems, social policy changes, prevalent ideas, and the actions of partners and others will shape their personal lives.

questions for critical thought

1. Why are more young people cohabiting before marriage? What does this indicate about the authority of the church and state?

2. Why do different government departments, such as social welfare and immigration, tend to define 'family' in different ways?

3. Would you expect societies that practise arranged marriages to have more stable and happier marriages than societies that allow free-choice marriage? Why or why not?

4. How would you explain the rise in maternal employment with reference to (a) post-structural feminist perspectives, (b) structural functionalism, and (c) political economy theory?

5. Should parents stay together for the sake of their children, or is parental conflict more detrimental to children?

6. Should same-sex couples and women over 40 be permitted to use the services of fertility clinics at the public's expense? Give reasons for your viewpoint.

7. Should parents be required to pay the entire cost of child-care services when they are employed?

8. Is there any reason to believe that legal marriage rates will rise again in the near future?

recommended readings

Patrizia Albanese. 2009. *Children in Canada Today*. Toronto: Oxford University Press.
This book involves an analysis of the sociology of childhood and youth culture, discussing agents of socialization and Canadian social policies to improve children's lives.

Maureen Baker. 2010. *Choices and Constraints in Family Life*. 2nd edn. Toronto: Oxford University Press.
This book argues that our choices about partners and living arrangements are shaped by economic circumstances, cultural expectations, popular culture, and events in the larger society, including labour market changes and political discourse.

David Cheal, ed. 2010. *Canadian Families Today: New Perspectives*. 2nd edn. Toronto: Oxford University Press.
This edited book contains chapters by 18 experts on various aspects of Canadian families.

Walter DeKeseredy. 2009. 'Patterns of family violence'. In Maureen Baker, ed., *Families: Changing Trends in Canada*, 6th edn, 179–205. Toronto: McGraw-Hill Ryerson.
This chapter discusses recent research and theorizing about various forms of family violence, including wife abuse, child abuse, sibling violence, and elder abuse.

Andrea Doucet. 2006. *Do Men Mother? Fathering, Care and Domestic Responsibility*. Toronto: University of Toronto Press.
This book discusses different parenting styles among men and women, based on qualitative interviews with primary-caregiver fathers and some of their partners.

Bonnie Fox. 2010. *When Couples Become Parents: The Creation of Gender in the Transition to Parenthood*. Toronto: University of Toronto Press.
This book examines how couples negotiate their division of household labour and care from late pregnancy through the early stages of parenting, based on qualitative interviews with 40 heterosexual couples.

Gillian Ranson. 2009. 'Paid and unpaid work: How do families divide their labour?' In Maureen Baker, ed., *Families: Changing Trends in Canada*, 6th edn, 108–29. Toronto: McGraw-Hill Ryerson.
This chapter provides an overview of theories and research on the division of labour in Canadian families.

Vanier Institute of the Family. 2004. *Profiling Canada's Families III*. Ottawa: Vanier Institute of the Family.
This book contains numerous tables and charts about family trends and patterns, accompanied by a discussion of their relevance.

recommended websites

Australian Institute for Family Studies
www.aifs.gov.au
This website contains details about new research and recent publications from an Australian government research centre on family issues.

Campaign 2000
www.campaign2000.ca
Campaign 2000, created in 1989 to monitor 'child poverty' in Canada, publishes an annual report card.

Centre for Families, Work and Well-Being
www.worklifecanada.ca
The website of the Centre for Families, Work and Well-Being at the University of Guelph contains information about research projects.

Child and Family Canada
www.cfc-efc.ca
This website offers public education from numerous non-profit organizations.

Childcare Resource and Research Unit
www.childcarecanada.org
The website of the Childcare Resource and Research Unit in Toronto includes Canadian and cross-national research and other material on child-care issues.

National Association for the Education of Young Children (NAEYC)
www.naeyc.org
This US-based association publishes the journal *Young Children*, which includes reviews of research and practical information.

Statistics Canada
www.statcan.gc.ca
Statistics Canada provides a wide range of census documents and statistics relating to families and households.

Vanier Institute of the Family
www.vifamily.ca
The Vanier Institute of the Family in Ottawa provides educational material, news items, and research on Canadian families.

chapter 14

Education

Terry Wotherspoon

In this chapter, you will:

▶ Understand how and why formal education has become a central social institution in Canada and in other nations.

▶ Identify the main dimensions and challenges associated with the growth of formal education systems.

▶ Gain a critical understanding of various forms of lifelong learning beyond formal education.

▶ Understand the major theoretical perspectives and theories that sociologists employ to explain educational institutions, practices, and outcomes.

▶ Understand the relationships between education and social inequality.

▶ Explore how education and educational outcomes are shaped by relationships between educational institutions and participants and how the social contexts within these institutions operate.

▶ Critically evaluate contemporary debates and controversies over major educational issues.

Introduction

The chief economist for the World Bank recently high-lighted the importance of education as 'critical to participation and productivity in economic life'. 'A healthy, literate labor force,' he said, 'will both increase the amount of growth realized from establishing a sound investment climate and strongly re-inforce the poverty reduction benefit from that growth' (Stern 2002, 21). Probably few people would take issue with these comments. What do they mean, though, to people in different situations? Consider the following cases of some students in a Canadian high school:

- One is pressured by parents who feel the school's poor academic and disciplinary standards are impeding the student's ability to gain entry into a prestigious university.

- A recent immigrant speaks little English or French, while another has traumatic memories of war and violent conflict, which make it difficult to concentrate.

- A continuing student has difficulties with reading comprehension but remains in school in order to play on a competitive sports team.

- Another, bored with school, is tempted by high-paying jobs in the resource sector.

- Two new arrivals find the school much larger and more regimented than elementary schools in nearby **First Nations** and rural communities.

- Another has switched schools but finds regular school attendance difficult because of repeated conflicts with students and teachers, along with substance abuse and domestic problems.

These situations reveal much about education's significance in the context of what is often called the *knowledge* or *learning society*. Education has been thrust into a central role as individuals, organizations, and nations struggle to keep pace with demands for new knowledge and credentials regarded as essential for jobs, career advancement, and economic and social development. We expect educational institutions to educate and prepare learners by instilling a wide range of technical, social, and personal competencies so that they can cope with changing, often uncertain futures. We also look to schools to respond to the needs of diverse students and communities.

Sociologists are interested in several issues associated with educational processes and outcomes

A century ago, schools were often segregated by class and age. Here, young ladies are learning to sketch a live model—a skill that will give them cultural capital when they pass into adult social life.

(Courtesy of the Bishop Strachan School Museum & Archives)

and in the environments within which education operates. This chapter examines several key questions that sociology addresses in its concern to understand education:

- Why is formal education so important in contemporary societies, and how did it get to be that way?
- What are the main dimensions of education and education systems in Canada and in other nations?
- How do sociologists explain the growth of education systems and the outcomes associated with education for different groups?
- What are the main educational experiences and outcomes for different social groups?
- What are the main challenges facing education systems in Canada and other nations?

The Changing Face of Education

Education is generally understood as the formal learning that takes place in institutions such as schools, colleges, universities, and other sites that provide specific courses, learning activities, or credentials in an organized way. Informal learning also occurs as people undertake specific activities to learn about distinct phenomena or processes. Both formal and informal education are part of the broader process that sociologists typically call **socialization**, which refers to all direct and indirect learning related to humans' ability to understand and negotiate the rules and expectations of the social world.

Nearly all Canadians engage in formal education for extended periods of time. Just over a century ago, by contrast, many communities lacked schools or qualified teachers, many children did not go to school, and those who did attend typically left by their early teen years (Guppy and Davies 1998). Ian Davey observes that 'in good times more parents sent more of their children to school and sent them more regularly. Yet, lower attendance during bad times resulted largely from a magnification of those factors which caused irregular attendance throughout the nineteenth century—transience and poverty' (1978, 230).

Today's students and teachers typically exhibit a far greater array of personal, stylistic, and cultural variation than was apparent a century ago, and they have access to many more learning and community resources. Despite these changes, the casual visitor to classrooms in either time period is not likely to mistake

schools for other settings. Education is a unique social institution, but it also reveals characteristics that are integral to the society in which it operates.

TIME to REFLECT
What is the difference between education and socialization? How has the introduction of formal schooling influenced the relationship between these two phenomena?

DIMENSIONS OF EDUCATIONAL GROWTH
Table 14.1 provides an overview of the increasing educational attainment of Canadians since the mid-twentieth century. The proportion of the population who had less than a Grade 9 education diminished rapidly, especially in the 1960s and 1970s. Half of the population now hold post-secondary credentials, in stark contrast to the situation at the start of the period. Growing emphasis on the importance of formal education and credentials has been matched by three interrelated factors: the overall expansion of educational opportunities and requirements, increasing levels of educational attainment among people born in Canada, and recent emphasis on the selection of highly educated immigrants.

In the late nineteenth and early twentieth centuries, relatively few occupations required educational credentials. Early advocates of public **schooling** under-took a mission to convince the public, and especially members of influential groups, of the merits of the educational system (for examples, see Lawr and Gidney 1973). They promoted schooling as an efficient enterprise that would serve the public or general interest, unlike narrow, more selective sites such as families, churches, and businesses.

By the mid-twentieth century, more and larger schools were required to accommodate growing educational demands. As credentials became more important for many jobs, people were more likely to extend their schooling into and beyond the high school years. The **baby boom** that occurred after World War II resulted in unprecedented sizes of cohorts of children who were entering and moving through the school system. The figures in Table 14.2 demonstrate that while total enrolment in Canadian public elementary and secondary schools in 1950 was slightly more than twice what it had been in 1900, enrolment doubled again over the next decade and a half. The average number of pupils per school in Canada was 66 in 1925–6, but by the early 1970s

	Less than Grade 9	Grades 9–13	Some Post-secondary	Post-secondary Certificate or Diploma	University Degree
1951	51.9	46.1	–	–	1.9
1961	44.1	53.0	–	–	2.9
1971	32.3	45.9	11.2	5.8	4.8
1981	20.1	44.3	16.1	11.5	8.0
1986	17.7	42.5	19.3	12.3	9.6
1991	14.4	43.8	8.8	21.9	11.4
1996	12.3	39.4	8.9	25.9	13.3
2001	9.7	36.9	9.2	28.3	16.0
2006	8.1	35.0	8.1	29.9	18.9
2009	6.9	34.0	8.3	30.6	20.2

TABLE 14.1 ▼ Educational Attainment in Canada by Percentage of Population Aged 15 and Over, Selected Years, 1951–2009.

NOTE: Figures are rounded.
SOURCE: Compiled from Statistics Canada, census data (1951–86) and CANSIM labour force survey, annual averages (1991–2009).

it had reached the current level of about 350 (Manzer 1994, 131; CMEC 2008, 8).

The data in Table 14.2 demonstrate how formal education has expanded throughout the life course. Just over a century ago, it was not uncommon for children to begin their schooling at seven or eight years of age or even older. Today, most Canadian children attend compulsory kindergarten, sometimes beginning preschool or early childhood education programs as early as at two years of age, continuing their studies well past high school. The number of graduate students alone is now nearly double total university enrolments of the early 1950s. There has also been significant growth in other post-secondary options since the introduction and expansion of the community college system in the 1960s and 1970s, offering students several pathways toward certification in specialized trades or vocations or university degrees.

TIME to REFLECT

To what extent has educational enrolment increased in Canada over the past century? What factors account for this growth?

EDUCATION IN THE LEARNING SOCIETY

The organization and nature of schooling across Canada vary. People typically encounter a diverse array of education and work settings during the course of their lives. Elementary and secondary education is under the jurisdiction of the provinces in accordance with the Canadian Constitution, while other forms of education, including adult and post-secondary education and vocational training, are controlled, operated, or funded by a variety of governments (federal, provincial, and First Nations) and by private sources.

The growing popularity of terms such as *information society*, *learning society*, and **lifelong learning**

In the First Person

I became interested in sociology almost by default. Like many students, I had little concept of the discipline until a friend suggested a sociology course as an interesting elective at university. Sociology proved to offer several useful skills, understandings, and connections to significant issues, but after completing my first degree, I did not return to it until my role as a schoolteacher led me to question many aspects of educational processes and the relationship between schooling and various dimensions of students' lives and life chances. When I returned to university to complete graduate work, I began to take sociology more seriously as I came to apply the discipline to my experiences and emerging interests. Sociology is continually changing, even when focused around a core set of disciplinary focuses and attributes, and I find that I can continually make new linkages as I attempt to connect academic life with communities and policies related to education and other aspects of social life.

—Terry Wotherspoon

signifies the central place that education holds within the context of what is commonly designated as the **new economy** or *knowledge-based economy*. The new economy has gained prominence as a result of rapidly changing information technologies and scientific advancements that have affected not only business and the workplace but virtually every major sphere of social life. Learning is central to these emerging relationships for a variety of purposes: training qualified personnel; researching for continuing innovation; developing, testing, and marketing new products and services; processing the vast quantities of new information being created; and providing people with the capacity to employ new technologies at work and at home (Wolfe and Gertler 2001). In this climate, what counts is not so much the knowledge that we acquire as our capacities to learn, innovate, and apply knowledge to emergent situations. People are expected not simply to learn more but to develop different ways of learning and transferring knowledge.

Nearly 6.5 million Canadians (about one-fifth of the entire population) identified in Table 14.2 are engaged in full-time schooling, and well over 300,000 more are involved in part-time studies. More than half of Canadians are involved in various forms of adult education, such as in-person, correspondence, or private courses; workshops; apprenticeships; or arts, crafts, or recreation programs. Levels of *informal learning* are even higher, as people seek knowledge outside formal schooling for work, personal, or community circumstances, such as learning a new language on one's own or with other people, learning computer skills or software programs, or gaining competencies that can be used for volunteer work or in family situations. Nearly all Canadians are involved in learning activities, although the more formal education they have, the more engaged they are likely to be in adult education and informal learning (Livingstone 2004, 36–7; Desjardins et al. 2005, 89).

TABLE 14.2 ▼ Full-Time Enrolment in Canada by Level of Study, Selected Years, 1870–2007 (000s)

	Pre-elementary	Elementary and Secondary	Non-university Post-secondary	University Undergraduate	University Graduate
1870	–	768	–	2	–
1880	–	852	–	3	–
1890	–	943	–	5	<1
1900	–	1,055	–	7	<1
1910	–	1,318	–	13	<1
1920	–	1,834	–	23	<1
1930	–	2,099	–	32	1
1940	–	2,075	–	35	2
1950	–	2,391	–	64	5
1955	103	3,118	33	69	3
1960	146	3,997	49	107	7
1965	268	4,918	69	187	17
1970	402	5,661	166	276	33
1975	399	5,376	221	331	40
1980	398	4,709	261	338	45
1985	422	4,506	322	412	55
1990	468	4,669	325	468	64
1995	542	4,895	392	481	75
2000	522	5,035	425	505	81
2005*		5,212	458	648	111
2007**		5,119	461 (06)	653	122

*Pre-elementary figures for 2005 and 2007 are included in elementary and secondary total.
**Latest data for non-university post-secondary are for 2006.

SOURCES: Compiled from various editions of Dominion Bureau of Statistics/Statistics Canada census data, 'Education at a Glance', *Education Quarterly Review*, CANSIM data Tables 477-0013 and 477-0015; Brockington 2010, 21).

OPEN for DISCUSSION

Debating Alternatives to Public Schools

Although strong public support exists for public schooling in Canada, many individuals and groups have advocated the need for alternative forms of schooling within public school systems or through various types of privately operated schools.

Some critics view education as a marketplace in which parents should 'shop' for the kinds of education best suited to their children's needs. Private schools, voucher programs, home schooling, and other measures are advanced as mechanisms for schools to become more responsive and accountable to the varied interests of educational consumers (Maguire 2006).

Other reforms are taking place within public education systems. Most provinces now offer a variety of specialized school alternatives as well as programs to link schools with other public and community agencies. Schools under First Nations control have introduced many innovations or modifications to public models of education, while other groups are seeking more specialized schools in response to particular concerns or interests.

Sociologist George Sefa Dei suggests that black-focused schools can address high dropout and failure rates and related social problems such as violence, unemployment, and drug abuse among inner-city black youth. Observing that present educational practices contribute to youth alienation through disconnection from students' real concerns, Dei (2006, 28) sees these African-centred schools as guided by 'a new vision' that stresses

> the development of a culture of youth affirmation that fosters a sense of pride by helping to build strong personal, social and cultural identities. For example, such a culture can be achieved when teachers introduce alternative forms of school discipline to replace suspensions, expulsions or the summoning of law enforcement.

Still other researchers indicate that increasing diversity in Canada's public school populations can be beneficial for all students, creating opportunities for critical reflection and broadened understanding of what it means to live and function in a truly global society.

Canadians are not unique in their growing pursuit of education and training. Emphasis on credentials and lifelong learning is a phenomenon associated with **globalization** and competitiveness across national settings. Throughout the twentieth century, the degree to which a population was educated came to be recognized as a significant indicator of modernization and development status. The more education one has, the higher the chances of having a job, a better income, good health status, and many other factors positively associated with a high standard of living. Conversely, rates of poverty, unemployment, crime, serious illness and injury, and other less desirable characteristics rise when formal education is limited.

Many nations are accelerating the pace of educational advancement as they undertake economic and human resource development strategies aimed at the production of new knowledge and a workforce better trained for the knowledge-based economy. Governments, businesses, and agencies concerned with economic development stress the need to expand education well beyond compulsory levels in order to foster both economic growth and non-economic benefits, such as improved health, the ability to use skills for non-monetary purposes, and the intrinsic desire to learn. Organizations like the Canadian Council on Learning (2010, 10) stress that 'learning and training are more critical now than ever as shifting workforce demographics, rapid advancements in technology and increased global competitive pressures are transforming our society.'

Despite this emphasis on learning, many people encounter significant barriers to educational access, achievement, and success, as noted by the Canadian Council on Learning (2010, 8): 'Canada has much to celebrate with regard to its formal education sector, [but] we cannot afford to remain complacent. Lack of progress in lifelong learning threatens to undermine the development of our greatest asset—the potential of our people.'

Educational disparities are most evident on a global scale. Figure 14.1 illustrates that those living in the Americas, Europe, and parts of Asia and the Pacific regions can expect on average to complete more than 12 years of formal schooling. In other parts of the world, average levels of schooling, particularly among female populations, are below high school completion. Educational attainment levels by girls and women in North America and Europe are more than twice those in sub-Saharan Africa.

TIME to REFLECT

What is meant by the concept of 'lifelong learning'? What is the relationship between lifelong learning and globalization?

Alternative Accounts of Educational Growth and Development

There is no uniform way of understanding education. Various educational **ideologies** pose contrasting views about what kinds of education a society should have, who should control and pay for education, and what should be taught in schools. Sociological theories of education, by contrast, are more concerned with describing and explaining education systems, educational processes, and educational change. Two early influential theoretical perspectives (the structural functionalist approach and the symbolic interactionist–interpretive theories) are contrasted below,

followed by examination of two critical challenges (conflict and feminist theories) and more recent integrative orientations to the analysis of education.

STRUCTURAL FUNCTIONALISM

Structural functionalism is concerned primarily with understanding how different parts of the entire social system are interconnected in order to keep the system going. It addresses questions of how societies perpetuate themselves, how individuals come to be integrated within social frameworks, and how social change can occur without upsetting the social order. Structural functionalism examines education in terms of its contributions to social order and stability. Education gains importance in modern societies as an institution that provides participants with the core understandings, capabilities, and selection criteria necessary to enable them to fit into prescribed social and economic roles. As society becomes more complex and specialized, educational institutions take on many of the functions previously managed by families, communities, and religious organizations to ensure that successive generations are able to make a seamless **transition** from early childhood and family life into schooling and eventually the labour force.

Émile Durkheim (1956 [1922], 123) described education as 'the means by which society perpetually re-creates the conditions of its very existence'. Talcott Parsons (1959) later identified schools' two central

▼ FIGURE 14.1 Average Years of Expected Schooling by Region, 2007

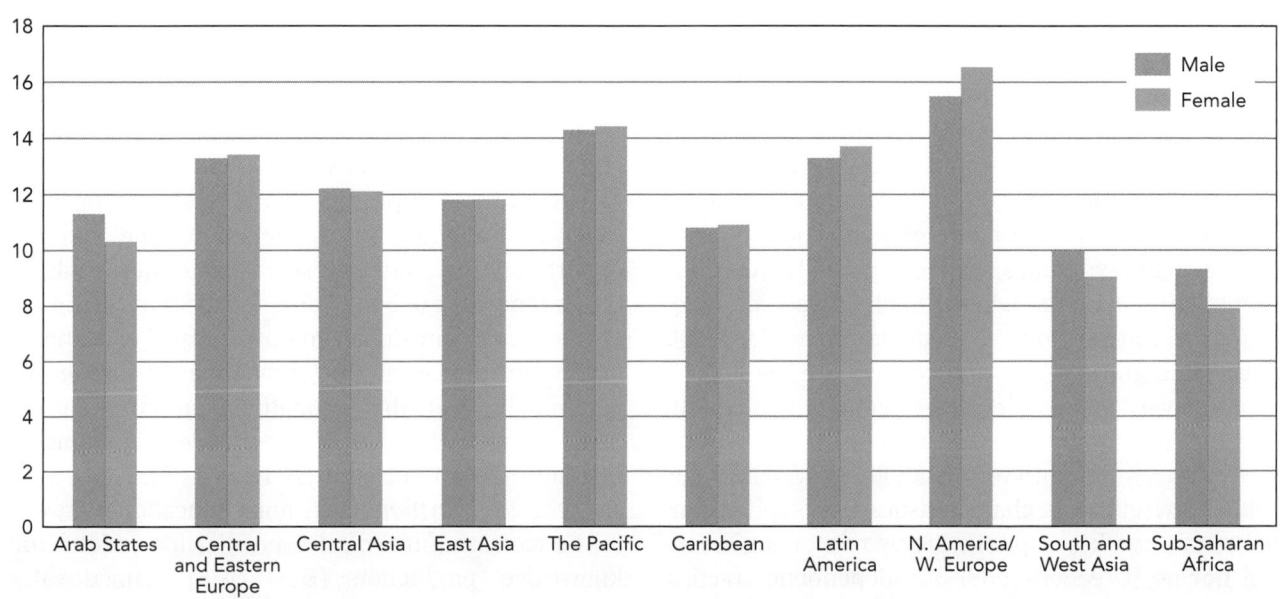

SOURCE: Based on data in UNESCO 2010, 329.

GLOBAL ISSUES

Toward Global Progress on Education

Nations on a global scale have identified the establishment of access to basic education and improvements in educational standards as core priorities. In 2000, an agenda entitled Education for All was adopted by 164 governments co-ordinated through the United Nations Education, Science and Cultural Organization (UNESCO), seeking to achieve targets in six priority areas by 2015: (1) comprehensive early childhood care and education; (2) free, compulsory, high-quality primary education attainable by all children; (3) learning and life-skills programs for all young people; (4) adult literacy, especially for women, and basic and continuing education for all adults; (5) gender equality in basic and continuing education; and (6) ensuring that all people achieve established learning outcomes, especially in literacy, numeracy, and essential life skills.

Subsequently, UNESCO (2010, 5) has observed progress toward the achievement of many of these goals but acknowledges that 'the world is unequivocally off track . . . and the battle to achieve universal primary education by 2015 is being lost.' These discrepancies are produced by poverty and deep socioeconomic inequalities, health issues, war, gender discrimination, and other related factors. For example:

- Malnutrition affects approximately 175 million young children each year, constituting a health and education emergency.
- 72 million children were out of school in 2007. Business-as-usual would leave 56 million children out of school in 2015.

- About 54 per cent of the children out of school are girls. In sub-Saharan Africa, almost 12 million girls may never enrol. In Yemen, nearly 80 per cent of girls out of school are unlikely ever to enrol, compared with 36 per cent of boys.
- Literacy remains among the most neglected of all education goals, with about 759 million adults lacking literacy skills today. Two-thirds are women.
- Millions of children are leaving school without having acquired basic skills. In some countries in sub-Saharan Africa, young adults with five years of education had a 40 per cent probability of being illiterate. In the Dominican Republic, Ecuador, and Guatemala, fewer than half of Grade 3 students had more than very basic reading skills.
- Some 1.9 million new teacher posts will be required to meet universal primary education by 2015.
- Inequalities often combine to exacerbate the risk of being left behind. In Turkey, 43 per cent of Kurdish-speaking girls from the poorest households have fewer than two years of education, while the national average is 6 per cent; in Nigeria, 97 per cent of poor Hausa-speaking girls have fewer than two years of education.
- Failure to address inequalities, stigmatization, and discrimination linked to wealth, gender, ethnicity, language, location, and disability is holding back progress towards Education for All (UNESCO 2010, 1–2).

functions within contemporary societies as selection (allocating individuals with appropriate skills and talents into necessary jobs and social positions) and socialization (providing people with aptitudes and knowledge required for adult roles and specific jobs). In the primary grades, schools partially resemble home environments where highly personal, emotional ties prevail, but as students proceed through successive grade levels, the schools are marked by progressively greater degrees of competition, merit, and instrumentality intended to prepare the individual for integration into work and other social settings.

Schools cultivate characteristics essential for contemporary work and public life by reinforcing essential **norms** (Dreeben 1968) of independence (acting according to expectations without supervision), achievement (actions to meet accepted standards of excellence), universalism (impartial treatment of others based on general categories), and specificity (a focus on selected individual characteristics as opposed to the person as a whole). Teachers are expected not only to convey knowledge and the opportunity to practise these norms to their students but to model these behaviours for them as well.

Structural functionalism rationalizes educational expansion by connecting schooling to the growing complexity of the occupational structure and by highlighting its increasing importance to citizenship in industrialized societies. A related form of analysis, *technical functionalism*, links educational growth to the increasing technical sophistication of jobs and knowledge production (Bell 1973). Functionalist analysis typically assumes a broad social consensus about what should be taught in schools and how educational institutions should be organized. It tends not to question either the legitimacy of educational credentials to determine entry into specified labour market positions or the fairness of the way the education system operates.

Functionalist analysis tends to portray deviation from these ideals as abnormalities or temporary problems that warrant minor reforms rather than as challenges to the education system as a whole. Describing liberal democratic visions of what schools should be like rather than explaining how schooling came about, functionalism presents education as a meritocratic ideal, a means of enabling people to gain opportunities for social or economic success regardless of their social backgrounds. Societies require a careful fit among capability, talent, effort, training, and jobs as social tasks become more complex and specialized. Such claims have led to subsequent research into the definition and measurement of educational inequality, calling into question the degree to which educational realities match the needs of industrial democratic societies.

Human capital theory, an approach with some affinity to structural functionalism, emphasizes education's role as a critical tool for developing human capacities to create and apply new knowledge. The human being is regarded as an input, along with material and economic resources, that contributes to economic productivity and development. **Human capital** can be enhanced when adequate investment is made in the form of proper training, education, and social support; this approach has been used to justify the massive investment by governments that contributed to the significant growth in post-secondary enrolment observed in Table 14.2. Human capital theories have gained renewed currency as attention turns to the importance of advanced training and educational credentials in knowledge-based societies (Heckman and Krueger 2004).

Extensive evidence demonstrates that levels of employment, income, and other benefits improve with educational attainment. However, structural functionalism and related theories, such as human capital theory, are unable to account for the presence of persistent inequalities in educational opportunities, outcomes, and benefits. They often ignore how initial advantage is likely to contribute to ongoing educational and economic success. The theoretical emphasis on consensus limits consideration of differences in educational values, content, and practices; of how some things get incorporated into schooling while others do not; and of how these differences affect people from different social backgrounds. Alternative theoretical approaches to education attempt to address some of these issues.

TIME to REFLECT

What are the main functions of education identified by structural functionalist approaches to the understanding of schooling?

SYMBOLIC INTERACTIONISM AND MICROSOCIOLOGY

In contrast to structural functionalism's focus on education systems and institutional arrangements, *microsociology* or *interpretive theories* are concerned more with interpersonal dynamics and how people make sense of their **social interactions**. **Symbolic interactionism**, one of the most influential branches of microsociology, focuses on how meanings and **symbols** are integral to social activity. Symbolic interactionism draws attention directly to the lives and understandings of social participants.

Interpretive analysis of education examines such questions as how schooling contributes to the development of personality and **identity**, how some forms of knowledge and not others enter into the curriculum, and how students and teachers shape learning processes in and outside of the classroom. It focuses on the meanings and possibilities that social actors bring to social settings. Willard Waller depicts schools as 'the meeting-point of a large number of intertangled social relationships. These social relationships are the paths pursued by social interaction, the channels in which social influences run. The crisscrossing and interactions of these groups make the school what it is' (1965 [1932], 12). Peter Woods (1979) explores schooling as a series of **negotiations** among teachers, students, and parents, expressed in such phenomena as how pupils select the subjects they take, the role of humour and laughter in the classroom and staff room, and teacher reports on student progress. Howard Becker (1952) shows how teachers' backgrounds influence their construction of images of the ideal pupil, which in turn affect how they treat and assess students.

For symbolic interactionists, societies and institutions are fluid rather than fixed entities. Institutional patterns are the result of recurrent daily activity and of people's capacities to shape, interpret, reproduce, and modify social arrangements through their social relations. *Ethnomethodology*, a variant of interpretive sociology, examines in detail the methods or approaches that people draw on to construct a sense of reality and continuity in everyday life. Understood this way, classrooms tend to resemble one another not so much because of a given model of schooling but more likely because people act in accordance with images about what is expected of them.

Symbolic interactionism and ethnomethodology offer interesting insights, but their focus on the details of ongoing social activity can restrict their ability to account for social structures and historical processes. Classroom dynamics or how one interprets the curriculum cannot be understood fully without reference to educational policy, **power** structures, social change, and persistent social inequalities that strongly influence educational processes and outcomes.

Some researchers have combined interpretive sociology, with its insights into practical social activity, with other approaches that pay greater attention to the social contexts in which social action takes place. They are concerned with breaking down barriers between **micro-** and **macrosociology**. Educational knowledge and practices are socially constructed, but they are also shaped by wider relations of power and control as they become part of the taken-for-granted assumptions that guide the actions and understandings of teachers and other educational participants, including notions about 'what it is to be educated' (Bernstein 1977; Blackledge and Hunt 1985, 290). In particular, the early contributions of British sociologist Basil Bernstein (1977) to a systematic understanding of micro and macro levels of analysis have influenced writers working within diverse theoretical traditions (Sadovnik 1995).

TIME to REFLECT

How do symbolic interactionist theories differ from functionalist and conflict approaches to the study of education?

CONFLICT THEORY

Conflict theory highlights inequalities and power relations among social actors, groups, or forces. This approach emphasizes how institutional structures and social inequalities are maintained or changed through conflict and struggle.

Samuel Bowles and Herbert Gintis, like structural functionalists, emphasize schools' role as mechanisms that select and prepare people for different positions in labour markets and institutional life.

However, drawing from Marx, they see the labour market as conditioned more by capitalist interests than by general consensus about social values and needs. Claims that all students have a fair chance to succeed represent a democratic ideology that cannot be fulfilled. Bowles and Gintis (1976, 49) posit education, historically, as 'a device for allocating individuals to economic positions, where inequality among the positions themselves is inherent in the hierarchical division of labor, differences in the degree of monopoly power of various sectors of the economy, and the power of different occupational groups to limit the supply or increase the monetary returns to their services'.

Conflict theorists emphasize the persistent barriers to opportunity and advancement created by deeply rooted relations of domination and subordination. This critical sociological orientation denies the functionalist and human capital theory accounts of educational expansion as being a result of the rising technical requirements of jobs. Different **social groups** are understood to employ education and educational ideologies as tools for pursuing their own interests. Employers rely on formal educational credentials—regardless of the skills demanded by the job—to screen applicants and assess a person's general attributes. Professions control access to education and certification as a way of preserving the status and benefits attached to their occupations. New knowledge and technological advancements in areas such as medicine, nursing, teaching, engineering, and information processing may appear to produce a demand for increasingly more advanced, specialized training. But more often, credential inflation occurs as occupations preserve special privileges by simultaneously claiming the need for superior qualifications and restricting entry into these kinds of jobs (Collins 1979).

Technological developments are not necessarily accompanied by increasing skill requirements for many jobs. Machines and information technology often replace human input for routine technical operations or influence the content of 'new jobs' in which people are required to do little but read gauges, respond to signals, or key in information. Under these conditions, schools may function more as warehouses for delaying people's entrance into the labour force and for dissipating their dissatisfaction with the economy's failure to provide a sufficient number of satisfying jobs than as places where effective learning and occupational training take place. Harry Braverman suggests that 'there is no longer any place for the young in this society other than school. Serving to fill a vacuum, schools have themselves become that vacuum, increasingly emptied of content and reduced to little more than their own form' (1974, 440). **Capitalism**, in this view, has contributed less to skills upgrading through technological advancement than to processes that erode working skills, degrade workers, and marginalize youth.

Other conflict theorists highlight the biases and inequalities that are produced directly or indirectly through the curriculum and classroom practices. Students and parents have different understandings, resources, and time that affect the extent to which they can participate in and benefit from educational opportunities. Government cutbacks and changes to school-funding formulas exacerbate many of these inequalities, posing special difficulties for communities and families unable to provide additional resources to subsidize educational costs.

In post-secondary education, decreased government funding has led to rising tuition fees, which, accompanied by higher costs for textbooks and technology support, higher living expenses, and other factors, make it increasingly difficult for students without sufficient resources or unable or unwilling to take on significant student loans to attend colleges and universities. Conflict analysis also points to concern about the growing reliance of educational institutions on corporate donations and sponsorships to make up for shortfalls in government funding. More generally, processes of commercialization and **marketization**, in which educational institutions and practices become increasingly organized on the basis of consumer choice, competition, and profitability, reveal the continuing expansion of capitalist relations within a global context (Raduntz 2005).

Conflict theories of education, in short, stress that expectations for schooling to fulfill its promise to offer equal opportunity and social benefits to all are unrealistic or unattainable within current forms of social organization. Barriers that exist at several levels—access to schooling, what is taught and how it is taught, ability to influence educational policy and decision-making, and differential capacity to convert education into labour market and social advantage—deny many individuals or groups the chance to benefit from meaningful forms and levels of education. Conflict theories offer varying assessments of what must be done to ensure that education can be more democratic and equitable. Some analysts stress that educational institutions and organizations themselves must be transformed, while others suggest that any

kind of school reform will be limited without more fundamental social and economic changes to ensure that people will be able to use, and be recognized for using, their education and training more effectively.

FEMINIST THEORIES

Feminist analyses of schooling share some of the observations of other conflict theories, though with an explicit emphasis on the existence of and strategies to address social inequalities based on **gender**. Feminist theory stresses that social equity and justice are not possible as long as males and females have unequal power and status through **patriarchy** or gendered systems of domination. In the eighteenth century, Mary Wollstonecraft (1986 [1792]) saw access to education as a fundamental right for women; denied such a right historically, women were degraded as 'frivolous' or a 'backward sex'. Later waves of feminism have continued to look to education as a central institution through which to promote women's rights, opportunities, and interests.

Various forms of feminism pose different questions for educational research and propose different explanations and strategies for change (Gaskell 1993; Weiner 1994). In general, though, feminist analysis shows that influential mainstream studies of schooling have most often concentrated on the lives of boys and men, with little recognition that girls and women have different experiences and little chance to voice their concerns. Much research in the 1970s and 1980s focused on how such things as classroom activities, language use, images and examples in textbooks and curriculum material (including the absence of women and girls in many instances), treatment of students by teachers, and patterns of subject choice reflected gender-based stereotypes and perpetuated traditional divisions among males and females (Kenway and Modra 1992).

Feminist analysis seeks to do more than simply demonstrate how these social processes contribute to inequalities; rather, it aims to change the conditions that bring these practices about. This focus has shifted as some aspects of the agenda on women's rights and issues have advanced successfully, while specific barriers continue to restrict progress on other fronts. For instance, school boards have policies, enforced through human rights legislation, to restrict sexist curricula and to prohibit gender-based **discrimination** in educational programs and institutions. The educational participation rates of and attainment by females have come to exceed those for males. Yet gender parity has not been achieved in several important respects. Female students remain

SOCIOLOGY in ACTION

Dimensions of Educational Participation

In Canada and in most other nations, educational participation rates and attainment levels are increasing, regardless of social background. Sociologists have debated the extent to which these trends reflect the ability of education to fulfill its promise to provide social and economic opportunity, especially in relation to demands associated with labour markets that require more skilled and highly qualified workers.

Exploration of these issues offers a useful opportunity to apply a *sociological imagination*, described by C. Wright Mills (1959) as the ability to link one's personal biography or background and circumstances with historical sensitivity to wider social structures and processes.

Begin by examining your own educational and career pathways. How much formal education have you attained so far? What level would you like to attain? In what kinds of programs or institutions has your education taken place? What other kinds of education (such as informal learning through self-directed or group study, special interest courses, adult education, or on-the-job training) have you engaged in? Have there been any gaps or interruptions in your studies? What experiences (positive or negative) have affected your interest in and ability to gain the level of education you desire or have completed? What is the relationship between your educational background, including any specific skills, knowledge, or credentials you have, and your work and employment experiences? (Examine both the starting qualifications for the job and the actual tasks involved in the job.) What future employment do you desire, and how is it related to your educational plans and qualifications?

Second, consider your educational experiences in relation to your social background and context. How do your own educational and work experiences compare with those of members of your family and other social groups you have associated with (such as your grandparents, parents, childhood peers, and community members) or those you consider yourself part of now? Relate your educational experiences and aspirations to other important characteristics or aspects of your social background (including your gender, race, ethnicity, family income, regional and national origin, place of residence such as urban or rural, age, and other factors you consider important).

Third, engage in broader comparisons between your own education and working experiences and those of others. Examine data from various studies cited in this chapter and records maintained on the Statistics Canada website. What is the relationship between your experiences, those of other people in your family and home community, and wider trends evident from these data?

Finally, explain the major patterns and conclusions derived from your inquiries. What do these findings reveal about the nature of education and its social and economic importance?

---

highly under-represented in important fields such as information technologies, engineering, and some natural sciences, while gender-based barriers exist in other areas of schooling. Moreover, educational achievements do not always contribute equitably to successful social and economic outcomes.

Analysis of the feminization of teaching, as female teachers came to outnumber male teachers by the end of the nineteenth century, illustrates the relationship between education and changing gender relations. Teachers often lack the professional recognition that might otherwise accompany the demands and training their work involves. Teachers—and women teachers in particular—have been heavily regulated by governments and by school boards. During the early part of the twentieth century, guidelines often specified such things as what teachers could wear, with whom they could associate, and how they should act in public (Wotherspoon 1995). Until the 1950s, legislation in many provinces required that women resign their teaching positions upon marriage. Although today's teachers have much greater personal and professional autonomy than those of the past, teachers' lives and work remain subject to various forms of scrutiny, guidelines, and informal practices that carry gender-based assumptions or significance. Female teachers predominate in the primary grades, while men tend to be overrepresented in the upper grades and in post-secondary teaching positions, especially in the most senior teaching and educational administrative positions.

Feminist analysis has increasingly come to address interrelationships among gender, sexuality, and other social factors and personal characteristics.

Gender-based identities, experiences, and opportunities are affected by race, region, social class, and competing expectations and demands that people face at home, in the workplace, and in other social spheres (Arnot 2011; Dillabough, McLeod, and Mills 2011). Students and teachers from different backgrounds encounter diverse experiences, concerns, and options even within similar educational settings, which in turn affect subsequent educational and personal options.

TIME to REFLECT

What are the main bases of inequality emphasized in conflict theories and feminist theories of education, respectively?

EMERGING ANALYSIS AND RESEARCH IN THE SOCIOLOGY OF EDUCATION

Sociologists commonly employ insights from several models or orientations, acknowledging theory as a tool to help understand and explain phenomena and guide social action.

Critical pedagogy is one approach that draws from different theoretical positions, including conflict theory, feminist theory, and post-modernist challenges, both to explore how domination and power enter into schooling and personal life and to seek to change those aspects that undermine our freedom and humanity (Giroux 1997; Darder, Baltodano, and Torres 2003; McLaren and Kincheloe 2007). Anti-racism education shares similar orientations, further stressing the ways in which domination builds on notions of racial difference to create fundamental inequalities among groups that are defined on the basis of biological differences or cultural variations (Dei 1996).

Pierre Bourdieu (1997a; Bourdieu and Passeron 1979) has explored how **social structures** (the primary focus of structural functionalism and conflict theory) become interrelated with the meanings and actions relevant to social actors (the main concern of symbolic interactionism or interpretative sociology). Bourdieu, as a critical theorist, emphasizes that education contributes to the transmission of power and privilege from one generation to another as it employs assumptions and procedures that are to the advantage of some

Much of what is learned in schools occurs in groups as we learn with and from other people, including our classmates. Eventually, group work gives way to individualistic competition.

(O. Bierwagen/Ivy Images)

groups and the disadvantage of others. Educational access, processes, and outcomes are shaped through struggles by different groups to retain or gain advantages relative to one another. However, the mere fact that people hold varying degrees of economic, social, and cultural resources does not guarantee that these resources will be converted automatically into educational advantage. Competition for educational access and credentials increases as different groups look to education to provide a gateway into important occupational and decision-making positions.

Canadian research, influenced by Bourdieu's analysis and other integrative approaches such as life course theory, demonstrates the complex interactions among personal and social structural characteristics that affect the pathways taken by children and youth through education and from schooling into work and other life transitions (Anisef et al. 2000; A. Taylor 2005; Lehmann 2007). Researchers are also especially interested in the complex ways in which education intersects with both local and global dimensions of cultural, economic, and social forces (Apple, Au, and Gandin 2009; Spring 2008). In order to understand schooling fully, it is necessary to take into account several interrelated dimensions:

- how education systems are organized and what happens inside schools

- how school experiences are made sense of and acted on by various educational participants

- the relationships between internal educational processes and external factors, including governments and agencies that set and administer educational policy, employers who demand particular kinds of education and training and recognize particular types of credentials, political frameworks composed of competing values and ideologies about what education should be and about how resources should be allocated for education in relation to other priorities, and broader structures of social and economic opportunity and inequality

- the relations among transformations occurring on a global scale with more specific economic, political, and cultural structures that alternatively provide opportunities for and systematically exclude democratic participation by specific social groups

TIME to REFLECT

What advantages do integrative approaches to the analysis of education offer in comparison to other theories presented earlier in this chapter?

(Carlos Osorio/Torstar Syndication Services)

Modern universities are like factories in their efforts to mass-produce a product—educated graduates. Some believe that in this environment, people mainly learn how to submit to regimentation.

Educational Participants

Educational institutions reveal considerable complexity in their organization and composition. Some colleges and universities exceed the size of small cities. Consequently, sociologists are interested in questions related to the changing nature of who attends and works in these institutions (with respect to gender, racial, ethnic, religious, socio-economic, and other factors), what positions they occupy, and what barriers and opportunities they encounter.

Several factors contribute to increasing diversity in education. The educational participation of girls and women has increased significantly since World War II, especially at the post-secondary level. Immigration has contributed to changing educational profiles, with increasing racial, linguistic, and cultural diversity and variations in educational backgrounds and expectations among groups. Combined processes of rural-to-urban migration, policy changes, and population growth have increased the concentrations of Aboriginal students in elementary and secondary schools, especially in western and northern Canada. Economic changes have exacerbated many inequalities, including the perpetuation or magnification of gaps between high- and low-income families. Poverty and economic marginalization affect up to one-quarter of Canada's children. Classrooms today integrate students who historically have been excluded, such as teen parents or those with physical or learning disabilities. Numerous additional factors, such as religious orientation, the health of regional economies, and distance to essential educational and support services, affect educational participation and outcomes.

Sociologists are interested in issues of diversity beyond simply how the curriculum and formally structured activities affect students' learning and chances for success. Educational organization, rules, expectations, and practices also contain a **hidden curriculum**: the unwritten purposes or goals of school life. School life has a daily rhythm, through repeated variations between structured learning situations and informal interactions, thereby channelling students into selected directions and contributing to taken-for-granted understandings about order, discipline, power relations, and other aspects of social life that favour students from some backgrounds to the detriment of others (Lynch 1989). These processes, though often unintended and sometimes resisted by teachers, are difficult to change, because they are part of highly complex interactions between schooling and other features of social organization. Benjamin

Levin (2007, 75–6) points to overwhelming comparative evidence 'that socioeconomic status remains the most powerful single influence on students' educational and other life outcomes' while cautioning that '[i]n public policy and politics, though, evidence matters only if it affects beliefs, and this does not happen so quickly.' Schooling can also have limited connection with—and produce negative consequences for—the students and communities it is intended to serve (Dei et al. 2000; Stonechild 2006).

Two mechanisms—referred to as *silencing* and the *banking model*—illustrate how common educational practices can have indirect and unequal consequences for students, their identities, and their educational experiences and outcomes. *Silencing* refers to practices that prevent educational participants from raising concerns that are important to them (such as when teachers do not give students the opportunity to talk about current events or matters of student interest), as well as to indirect processes that make students question their own cultural background or that discourage parents from talking to teachers because of their discomfort with the authority represented by the school. The *banking model* of **pedagogy** (Freire 1970) refers to educational practice in which material is pre-packaged and transmitted in a one-way direction, from the educator to the student. This practice limits the forms of knowledge that are presented as valid, leaving students from alternative backgrounds with a sense that their experiences, questions, and capacities are invalid or irrelevant.

TIME to REFLECT

What does an understanding of the hidden curriculum tell us about the nature and purposes of schooling?

Educational Policy, Politics, and Ideologies

Educational policy is established and administered in quite different ways in other countries. Many nations, such as Sweden and Japan, have highly centralized systems of education. Canada and the United States, by contrast, do not have uniform or centralized education systems, contributing to what Paul Axelrod (1997, 126) describes as an 'educational patchwork'. In nearly all nations, however, competing demands for more co-ordinated educational planning, national standards, and consistency coexist with competing reforms driven by greater responsiveness and **accountability** to local concerns (Manzer 1994; Hoffer 2008).

Provincial and territorial governments in Canada have the constitutional authority to create legislation and guidelines that outline virtually all aspects of the education system, including how it is organized, the length of the school year, curriculum and graduation requirements, teacher qualifications and certification, and educational funding. The specific details related to setting and carrying out educational policies and operating schools are normally delegated to elected local school boards or similar regional bodies. In recent years, jurisdictions across Canada have undertaken a variety of initiatives that have modified the ways education is organized, administered, and delivered (CMEC 2008).

Public education at elementary, secondary, and post-secondary levels has experienced significant financial changes since the early 1990s. Education spending exceeds $80 billion, or about 6 per cent of Canada's Gross Domestic Product (GDP, or total expenditures), although it reached levels of 8 per cent in the early 1990s (Statistics Canada 2004, 54; Statistics Canada 2009c, 81). Figure 14.2 reveals that in Canada, as in most other highly developed nations, education continues to be funded primarily by governments but the share of educational spending from private sources, including tuition fees, is growing.

These trends suggest the risk of increased educational inequalities. Along with tuition fees, other educational costs, and rising living expenses, many students encounter substantial financial burdens. About six out of 10 students who graduated from Canadian undergraduate programs in 2009 had some student debt, owing an average of $26,680 (Berger 2009, 185). Elementary and secondary schools also face difficult choices as they weigh the costs and benefits of seeking higher taxes to finance schools, increasing school fees, fundraising, relying on corporate sponsors to cover educational expenses, or cutting school programs and services. The Canadian Teachers' Federation (2010) observes that public schools commonly raise funds for such activities as school trips, library books, and athletic programs (reported by 79 per cent, 49 per cent, and 44 per cent, respectively, of schools surveyed across Canada) but also for academic programs (24 per cent), school supplies (18 per cent), and textbooks (10 per cent). Teachers frequently pay for school materials and activities themselves, while parents or communities 'are having to make up for programs that aren't paid for—so then it depends on where you live and who you are. . . . There is a growing concern about equity. There is

▼ FIGURE 14.2 Public and Private Spending on Education: Relative Share of Public and Private Expenditure on Education in Canada, Selected Years

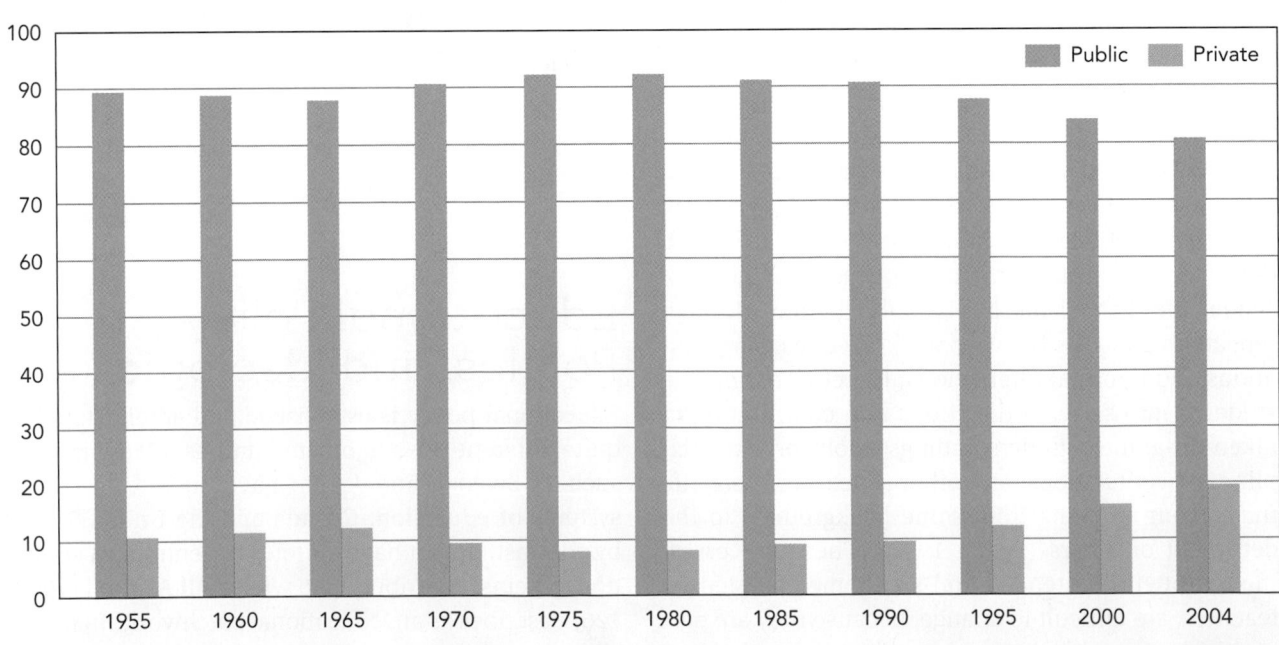

SOURCES: Based on data from Statistics Canada, CANSIM Table 478-0001 and Catalogue no. 81-582X, Tables B2.2 and B2.4.

a growing gap between the "have" and "have-not" schools' (L. Brown 2002, A1, A26).

Educational funding decisions are accompanied by growing concern over the extent to which education systems are able to prepare learners for contemporary economic and social conditions. There are competing viewpoints (often expressed through concerns about educational quality and excellence) about what role governments should play within this changing environment.

Neo-liberal critics promote the application of business or market-based principles to education and other social services. High-quality education is defined in terms of the excellence of educational 'products', measured by such things as standardized test scores, parental choice, and public accountability. Parents and learners become 'consumers' who should have the opportunity to approach education like decision-making about other purchases, with the added importance that humans, not some material object, require the options to make personal choices about educational futures.

Other observers draw attention to the dangers inherent in treating schooling like a market or reducing it to narrowly defined kinds of outcomes. Some suggest that inequality is likely to increase without a true commitment to community participation and high-quality education dedicated to the full range of activities and competencies that schools seek to foster (Osborne 1999). Many initiatives that claim to increase choice and flexibility lead to the concentration of resources, control, and opportunities among a relatively small circle of agencies or participants (Gidney 1999; Kachur and Harrison 1999; Sears 2003).

Education is the focus of intense debate in part because of its social and economic importance. It is both a central institution in the lives of children and youth and a strategic focus for policy related to emerging economic realities.

Education, Work, and Families

Changes in the nature and composition of learners' families and the varied demands from workplaces for particular kinds of qualified labour force participants have made it even more crucial to understand how education systems interact with other institutions.

The nature of childhood and adolescence is changing profoundly as students and their families experience various life challenges. Few people experience 'traditional' linear pathways from home

▼ FIGURE 14.3 Public and Private Spending on Education: Relative Share of Public and Private Expenditure on Education in Canada and Selected Nations, 2007

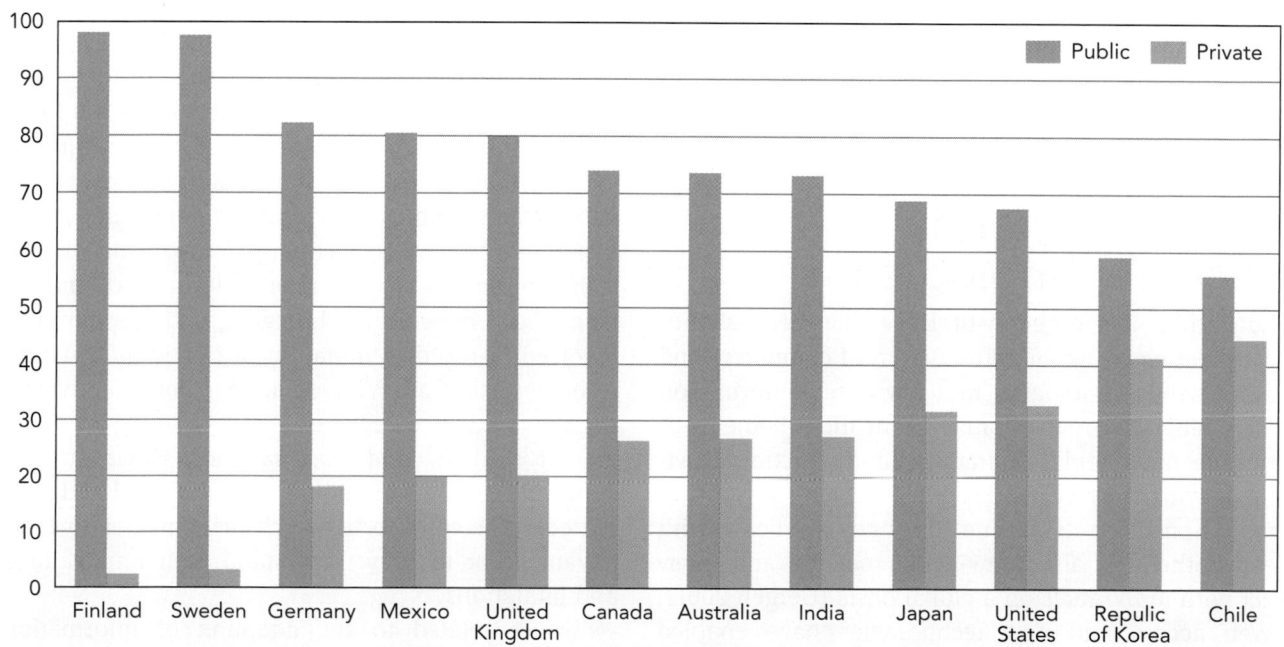

SOURCES: Based on data from UNESCO 2009, 220–3.

to school to work. Periods of work and study often overlap, while family, work, and community responsibilities create multiple demands on both children's and parents' time. Tensions sometimes spill over from one site of social life to another, expressed in public concern over phenomena such as bullying, violence, gang warfare, and 'risk' among children and youths (although it is also important not to overemphasize the dangers while ignoring the positive experiences and contributions often associated with childhood and youth). Taking their cue from the market model of education, many parents view their children's education as an investment. Uncertainty about job futures heightens expectations on learners (seeking high performance to be competitive) and on teachers and educational administrators (in order to deliver a high-quality product that will yield the best results in the marketplace).

Parental education, along with emphasis on early reading and literacy skills, factors heavily as an influence on children's subsequent educational attainment and success (Statistics Canada 2006b). Parents and community members from diverse backgrounds frequently have differing expectations about the way education should be organized and delivered. Some immigrants, for instance, may feel that the Canadian education system is too unstructured and undemanding in comparison to the systems they were familiar with prior to arriving in Canada, while others take the opposite view (Campey 2002). Aboriginal people look to schools to reconcile the need to prepare youth for a meaningful place in global society with the need to make strong connections with indigenous people, their cultural heritage, and their contemporary circumstances (Stonechild 2006).

Education and New Technologies

Education, like other institutions, has been significantly affected by the introduction of computers and other new technologies. In a few cases, information technology has revolutionized education. Some institutions have replaced traditional instructional settings with fully wired teaching/learning centres in which participants can not only communicate with each other but also draw upon material and interact with individuals on a global basis (Gergen 2001). Web access and new technologies have enabled schools and learners in remote regions to gain access to varied learning resources and opportunities in both

accredited and special education programs. Schools and universities are just beginning to explore fully the opportunities that new technologies are making available to them (even though the origins of the World Wide Web lie, in part, in the development of a tool that could be used to produce and share new knowledge among university-based researchers).

New technologies and their use in and impact on education give rise to several important questions. Levin and Riffel raise the still relevant consideration for schools that 'it may be that technology is not living up to its promise because it has been seen as an answer to rather than a reason to ask questions about the purposes of schools and the nature of teaching and learning' (1997, 114). Two issues are especially critical in this respect.

First, significant gaps remain between those who have access to computers and reliable electronic connections—and the skills and know-how to use and take advantage of new technologies—and those who do not. This 'digital divide' is most commonly posed in global terms, distinguishing richer, more technologically developed nations from developing nations. However, even within Canada and other nations that have high rates of computer ownership and extensive Internet services, regular access to computers and mobile technology, along with the ability to use them productively, depend on such factors as a steady job, income and education levels, gender, area of residence and work, social class, and racial characteristics (Statistics Canada 2010a; United Nations 2009, 16–20).

Second, it is important to examine how and why new technologies are being adopted as tools and expectations in education. Clearly, information technology offers many advantages to users, contributes to important educational innovation, and may provide greater employment and economic opportunities outside of school. The rapid expansion of new technologies and applications—from text messaging, social networking, blogging, and communities involved in the dissemination of public information resources such as Wikipedia, to gaming and electronic surveillance devices—is transforming everyday life for students and their families. However, students and teachers are not always equipped and supported sufficiently to use such technologies to their advantage or to fully understand their implications and limitations.

Issues related to the adoption of information technologies in education reflect more enduring concerns about the relationship of what happens

UNDER the WIRE

The One Laptop per Child Project

A global initiative called the One Laptop per Child project seeks to equip students around the world with tools that enable them to become part of the information society. The project aims at providing each child in developing and underdeveloped nations with an inexpensive laptop, giving them access to resources and information that are essential for contemporary education. Widely hailed as a progressive measure, the project nonetheless carries many hidden costs and dangers that illustrate the risk of adopting a one-dimensional orientation to educational problems. Shrestha (2007) cautions that the project appears relatively modest in cost ($100 per computer), but

> for those who really need it . . . there's lots of hidden cost like software cost, maintenance cost,

distribution cost as well. Software should be developed in local languages and the content of the internet and other educational resources should be localized in order to make it accessible to these poor children. Poor countries are not in the condition to buy it on their own. They have to bring in money either as loan or grant through some INGOs (international non-governmental organizations), international banks like World Bank, International Bank . . . [which will] increase the national debt of these countries.

As information technologies become more pervasive, educators and researchers are being called upon to pose similar questions about the true costs and benefits of new applications within specific educational contexts.

in the classroom with structures and processes outside of schooling. Educational practices are strongly influenced by social, technological, and economic developments and innovations, although they also reveal their own peculiarities and rhythms. Demands for education to prepare people for the changing workplace sit side by side with parallel demands for producing better citizens and persons with multiple competencies to function in a global society.

TIME to REFLECT

What are the main ways in which the introduction of new technologies has influenced formal education?

Educational Opportunities and Inequalities

Questions about the relationship of education to social inequality and opportunity structures have long been central to the sociological study of education. This is due in large part to public expectations about education's contributions to social and economic advancement. As educational participation and attainment rates increase across populations, many traditional forms of inequality diminish or disappear altogether.

Data presented in previous sections of this chapter, for example, demonstrate that women's educational attainment levels now match or exceed those of men in Canada and in many other nations. Nonetheless, significant inequalities persist with respect to many dimensions of educational experience and outcomes.

These trends can be illustrated by examining changes in the composition of the young adult population with post-secondary credentials. In 1981, the proportion of Canadians in the 25- to 34-year-old age cohort with a degree, certificate, or diploma from a university or college was about 32 per cent for both males and females. By 2006, the corresponding figures had risen to 63 per cent for women and 49 per cent for men (Statistics Canada 1984; Statistics Canada 2008b). In 2007, 61 per cent of all persons who received university degrees were women, although at the highest end, 55 per cent of doctoral degrees were awarded to men (Statistics Canada 2009c; see also Table 14.3).

The shift in the gender balance of educational attainment has drawn attention to the complex interactions that occur among various socio-economic factors in relation to education. Findings from numerous surveys that girls have begun to outperform boys on a number of indicators, especially in areas like reading, have generated controversy over suggestions that gender inequality has reversed to the point that the education system is now 'failing' boys (Bussière and Knighton 2004, 38). However, major comparative studies from Canada and several other nations also demonstrate the complex nature of gender inequalities in education. For instance, in most provinces and in many dimensions of mathematics and science performance, relatively few pronounced gender differences appear, while in some instances boys outperform girls. Moreover, these surveys highlight how similarities and differences based on gender cannot be understood without reference to a broad array of other family, school, and individual characteristics, notably family socio-economic background and immigration status (Bussière, Knighton, and Pennock 2007, 37–42).

Gender-related differences are often obscured through simple comparisons between boys' and girls' test results (Alloway 2007). Women outnumber men in post-secondary enrolment and graduation, but there are strong gender differences in fields of study and types of training programs (see Table 14.3). Programs in areas such as business, management, and commerce, some arts and social sciences, protection

(Bonnie Jacobs/iStockphoto)

New learning technology rapidly passes into the school curriculum, and children often know more about the technology than their teachers or their parents.

and correction services, and languages are relatively popular among both men and women. Women are much more heavily concentrated in a few fields, such as education, nursing, and social work or social services. Men tend to be more widely dispersed over more fields but outnumber women considerably in areas such as engineering and electrical technologies, computer science, and primary industries.

Differences in fields of study reflect a combination of personal choices and circumstances, institutional characteristics (such as cues or levels of comfort and discomfort that direct students into some areas and away from others or the compatibility between particular programs and responsibilities for the care of dependent children), and broader socio-economic factors (Arnot 2011; Wotherspoon 2000). Employment options and life pathways are generally associated with the kinds of education and credentials that people attain. Nonetheless, women's rising levels of education do not always translate fully into gains in labour market positions, incomes, and other equitable outcomes (Vosko 2010).

Parallel to gender comparisons, educational differences between racial and ethnic groups appear to have disappeared or diminished significantly in recent decades (Davies and Guppy 2006, 116–20). Immigration policies have simultaneously emphasized the recruitment of immigrants with high educational credentials and made Canada less dependent on immigrants from western Europe and the United States. These policies have contributed to a growing proportion of highly educated or professionally qualified visible-minority immigrants who place a high value on their children's educational advancement. Racial diversity has been accompanied by increasing sensitivity to the impact of racial discrimination and other mechanisms that have historically excluded or discouraged racial minority students from advancing through the Canadian education system.

As with gender inequalities, analysis of racial and ethnic inequalities in education reveals a complex series of interactions that do not lead to any straightforward conclusions. The short answer to the question of whether some groups are advantaged or disadvantaged in relation to racial and ethnic criteria is 'it depends'. Davies and Guppy (2006), in common with many other commentators who have reviewed census data and education indicators over time, observe that Canadians in most categories (based on gender, race, region, age, class, and other factors) have benefited from the expansion of education systems. However, outcomes for specific groups, including Aboriginal people, some immigrant and visible-minority populations, those from working-class backgrounds, and many persons with disabilities, continue to be less favourable relative to other groups. Social class has a strong impact on post-secondary attendance and educational attainment. These general trends are compounded by considerable variation in educational success and attainment within groups.

Research on education for Aboriginal people is instructive in this regard. Many First Nations expressed their desire in the nineteenth-century treaty-making process to have access to formal education in order to keep pace with contemporary social and economic demands. However, subsequent developments, including the damaging legacy of residential schooling, lack of acceptance or discriminatory treatment in provincial schools, and other social, cultural, and economic factors, have left Aboriginal people's overall education levels (especially those of registered Indians who live on reserve) well below national levels (Schissel and Wotherspoon 2003). Despite continuing increases in the levels of educational attainment by Aboriginal people, by 2006, Aboriginal people aged 25 to 44 years were 1.8 times less likely than non-Aboriginal people to have a post-secondary degree or diploma and 3.5 times less likely to have graduated from university but nearly three times as likely not to have completed high school (based on data in Statistics Canada 2008d).

Sociologists and other researchers have identified numerous factors, such as cultural differences, lack of individual motivation and family or community support, and social and educational discrimination, to explain these educational inequalities. They

TABLE 14.3 ▼ University Degrees Awarded by Field of Study and Gender, Canada, 2007

Field of Study	Number of Graduates	Rank order for Female Graduates	Female Graduates in Field as % of All Female Graduates	Females as % of All Graduates in Field	Rank order for Male Graduates	Male Graduates in Field as % of All Male Graduates	Males as % of All Graduates in Field
Social and behavioural sciences and law	50,529	1	23.3	67.6	2	17.3	32.4
Business, management, and public administration	48,705	2	17.6	52.9	1	24.2	47.1
Education	27,420	3	14.3	76.4	6	6.8	23.6
Humanities	27,222	5	12.0	64.7	4	10.1	35.3
Health, parks, recreation, and fitness	26,226	4	13.9	78.0	8	6.1	22.0
Architecture, engineering, and related technologies	19,434	8	3.2	23.8	3	15.6	76.2
Physical and life sciences and technologies	18,726	6	7.6	59.2	5	8.1	40.8
Visual and performing arts and communications technologies	8,727	7	4.0	67.3	9	3.0	32.7
Mathematics, computer and information sciences	8,547	9	1.7	29.9	7	6.3	70.1
Agriculture, natural resources, and conservation	3,864	10	1.5	57.9	10	1.7	42.4
Other	2,151	11	1.0	64.9	11	0.8	35.3
All Fields	241,551		100.1	60.7		100.0	39.3

SOURCE: Compiled with data from Statistics Canada 2009c.

typically occur through a complex chain of inter-related cause-and-effect mechanisms. Increasing attention has been paid to the importance of early childhood development and to the family and social environments in which children are raised for the development of literacy and language skills, thinking processes, and other capacities that are central to educational success. These conditions, in turn, depend on the socio-economic circumstances of parents, the availability of support networks in the home and community, labour market opportunities for parents and students coming out of the education system, the extent to which people in particular communities or regions have access to high-quality educational programs and services, and numerous other factors. There are strong associations between social class or socio-economic background and educational attainment. Parents' education levels and household income are strong predictors, both independently and in combination with one another, of the likelihood that a person will continue into post-secondary education (Finnie, McMullen, and Mueller 2010).

Educational institutions are implicated in these broader processes in several ways. Schooling makes a difference in many ways, such as how well institutions are equipped to deal with students from diverse cultural and social backgrounds; the kinds of relationships that prevail between and among teachers, parents, and students; curricular objectives and materials; standards for assessing and evaluating students; and the general social climate within educational institutions. Social class and cultural differences are evident, for instance, in the grouping and streaming of students into specific educational programs that contribute, in turn, to diverse educational pathways.

HUMAN DIVERSITY

Education for Canada's Aboriginal People

The educational experiences of Aboriginal people in Canada are instructive for an understanding of how education can both advance and restrict social and economic opportunities. Historical practices and inequities have contributed to a legacy of widespread failure, marginalization, and mistrust, but considerable optimism also accompanies many new initiatives.

Many Aboriginal people in the late nineteenth century looked to schooling as a way of ensuring integration into contemporary societies. Tragically, while some education-related treaty promises were fulfilled, the residential school system and continuing problems with other forms of educational delivery had devastating consequences that many Aboriginal communities and their members are still struggling to cope with. The report of the Royal Commission on Aboriginal Peoples (1996) endorsed the long-standing principle of First Nations control over education along with other measures to ensure that all educational institutions would provide more receptive schooling for Aboriginal people.

Mixed results have ensued so far, as one of the co-chairs of the royal commission has observed:

Considering the primary importance of children in aboriginal cultures, it is not surprising that education was one of the first sectors where aboriginal nations and communities are now administered locally, and where possible they incorporate aboriginal languages and cultural content in the curriculum. . . . More young people are staying in school to complete a high-school diploma, though a gap still exists between graduation rates of aboriginal and non-aboriginal people. . . . Aboriginal youth are especially vulnerable. They are less likely than mature adults to have attained academic and vocational credentials and they are hardest hit by unemployment. (Erasmus 2002, F6–7)

Accomplishing educational improvement is difficult in the context of considerable diversity among Aboriginal populations and their educational options, aspirations, and circumstances. A 'report card' issued by the Assembly of First Nations (2006, 16–17) on the tenth anniversary of the royal commission report assigned failing grades to all but three of 11 specific recommendations in the section on education. However, several Aboriginal communities and educators suggest that it is also important to reframe success in Aboriginal learning in terms that are not restricted and not oriented to the measurement of learning deficits. Such a model would revitalize holistic notions of education as a lifelong process in which 'Aboriginal learning is a fully integrated and potentially all-encompassing process that permeates all aspects of the learner's life and their community' (Canadian Council on Learning 2009, 11).

There is general agreement, in the context of global economic developments that place a premium on knowledge and learning, that education is important for all people. The same consensus does not exist, however, with regard to how education should be arranged to fulfill its promise on an equitable basis.

Conclusion

This chapter has examined several dimensions of education and its relevance for sociological inquiry. It has highlighted the phenomenal growth of formal systems of education since the nineteenth century and the accompanying increases in general levels of education throughout the population. It has linked that growth to a strong degree of public faith in the ability of education both to contribute to individual development and to address social needs for knowledge, innovation, and credentials. Educational growth, processes, and outcomes have been understood from diverse theoretical perspectives: structural functionalism (focusing on education's contributions to dominant social and economic requirements); symbolic interactionism and microsociology (highlighting understandings and interactions of various participants within educational processes); conflict theories (emphasizing education's contributions to social inequality and power relations); feminist theories (drawing attention to gender-based educational differences); and integrative approaches (linking insights from diverse perspectives). The chapter has also addressed the changing significance of formal schooling to the experiences and social and economic opportunities of different social groups, particularly with respect to gender, race and ethnicity, and social class. All groups have benefited from educational expansion, though in varying degrees. Adequate sociological analysis of education requires an ability to integrate an understanding of what happens in and as a result of formal education with the social context in which education is situated.

questions for critical thought

1. Why is education in most nations organized formally through schools and related institutional structures rather than through some other arrangement, such as families or community-based agencies? To what extent should education be a private as opposed to a public responsibility?

2. Explain how and why employers and other agencies have come to rely on formal educational credentials or qualifications as legitimate mechanisms for determining applicants' eligibility for positions in their organizations.

3. Compare and contrast schooling (formal education) with other major social institutions, including businesses, families, prisons, and religious organizations. Describe and explain the major similarities and differences.

4. What is the impact of emerging emphases on lifelong learning and the new economy on education systems? What kinds of alternatives to formal schooling are being developed in response to increasing demands for lifelong learning? Explain and critically discuss the changes (or lack of change) you have identified.

5. Why have education levels increased across populations in Canada and in most other nations? Which theoretical perspectives offer the most adequate explanation of these trends?

6. To what extent and in what ways have educational institutions been influenced by new information technologies? Discuss the relative strengths and limitations of these changes in terms of schooling's ability to meet the needs of learners in contemporary societies.

7. To what extent has education in Canada fulfilled its promise to provide greater opportunities for social and economic advancement to all social groups? Explain your response with reference to at least three different theoretical frameworks.

8. Whose interests are best served by formal education? Discuss with reference to how particular social groups (such as students, teachers, administrators, parents, policy-makers, corporations, or interest groups) have influenced educational decision-making and processes.

recommended readings

Jeanne H. Ballantine and Joan Z. Spade, eds. 2008. *Schools and Society: A Sociological Approach to Education.* 3rd edn. Los Angeles: Pine Forge Press.
This collection integrates a comprehensive range of classical and contemporary contributions to the sociological analysis of education. Major theoretical frameworks are represented, as well as substantive areas such as student cultures, school organization, stratification, educational reform, and global issues.

Shailaja Fennell and Madeline Arnot, eds. 2009. *Gender Education and Equality in a Global Context.* London: Routledge.
Taking global commitments to achieve equity in basic education by the year 2015 as a starting point, several authors explore theoretical and policy dimensions associated with gender and education. Feminist orientations are integrated with understandings of the ways in which gender inequality is intertwined with poverty, culture, economic systems, and fundamental social conditions, illustrated with research from several developing nations.

Neil Guppy and Scott Davies. 2006. *The Schooled Society: An Introduction to the Sociology of Education.* Toronto: Oxford University Press.
The authors integrate their discussion of core concepts and theories in the sociological analysis of education with material drawn from sociological research and case studies. Both historical and contemporary issues and developments are covered.

Hugh Lauder, Phillip Brown, Jo-Anne Dillabough, and A.H. Halsey, eds. 2006. *Education, Globalization, and Social Change.* Oxford: Oxford University Press.
This is one of the most comprehensive collections of analyses of education from various perspectives in sociology and other disciplines, containing influential chapters from different national settings that examine the impact of political and economic changes on education, cultural diversity, new conceptions of knowledge and curricula, the reshaping of teaching, and the dynamics of inequality and exclusion in relation to formal education.

Cynthia Levine-Rasky, ed. 2009. *Canadian Perspectives on the Sociology of Education.* Toronto: Oxford University Press.
This volume assembles both theoretical and empirical contributions to the analyses of relationships between school and society, with an explicitly Canadian focus. Chapters include analyses of such issues as educational reform, race, gender and identity in education, socio-economic inequality, and rural schooling.

D.W. Livingstone. 2004. *The Education–Jobs Gap: Underemployment or Economic Democracy.* Toronto: Garamond.
Livingstone systematically analyzes the relationships between education and work in the current economic context. Both education and the extent to which it is related to actual employment situations are explored through several dimensions, integrating statistical data with people's accounts of their own education and work experiences.

Terry Wotherspoon. 2009. *The Sociology of Education in Canada: Critical Perspectives.* 3rd edn. Toronto: Oxford University Press.
Various dimensions of Canadian education are explored from a critical orientation that emphasizes inequalities based on class, race, gender, region, and other factors. The book addresses contemporary aspects of Canadian education in the context of various theoretical perspectives and historical factors.

Jon Young, Benjamin Levin, and Dawn Wallin. 2007. *Understanding Canadian Schools: An Introduction to Educational Administration.* 4th edn. Toronto: Thomson Nelson.
Although focused on educational administration and organization, this work offers a useful overview of the main dimensions of Canadian education systems, including the roles, rights, and relationships among parents, teachers, students, administrators, and other educational participants. The authors integrate various data sources with their own insights from experiences as educators, teachers, and administrators in public school and post-secondary systems.

recommended websites

American Sociology Association, Section on Sociology of Education

www2.asanet.org/soe

This site, though directed primarily to professionals and researchers, contains a summary description of the sociology of education and emerging issues and many useful links to other databases and relevant sites.

Canadian Council on Learning

www.ccl-cca.ca

The Canadian Council on Learning, funded by the federal government, was established to promote lifelong learning in Canada, in part by promoting and co-ordinating evidence-based research about various aspects of education, training, and learning. The website includes useful reports and information related to such key themes as Aboriginal learning, post-secondary education, early childhood and adult learning, and public attitudes toward learning.

Canadian Education Association

http://cea-ace.ca/home.cfm

The Canadian Education Association offers information, resources, and services for anyone involved in or concerned about education issues, including parents, students, teachers, governments, researchers, and many others. Its website includes news updates, publications, event listings, and other information, the aim being to promote dialogue and research-based evidence in educational decision-making.

Canadian Teachers' Federation

www.ctf-fce.ca

This website provides an educator's perspective on important educational matters, ranging from factual information on education systems and significant educational developments to position papers and analyses of pressing educational issues.

Council of Ministers of Education Canada

www.cmec.ca

The Council of Ministers of Education provides access to major reports and studies conducted through that organization as well as links to each of the provincial and territorial ministries of education and other important Canadian and international education bodies.

Educational Resources Information Center (ERIC)

www.eric.ed.gov

The ERIC database is a comprehensive collection of information (mostly abstracts of journal articles and reports) on various aspects of and fields related to education, including the sociology of education.

Intute

www.intute.ac.uk

Intute, which includes information previously available through the Social Science Information Gateway and other science consortia, is based in the United Kingdom. It offers a substantial and useful set of links to significant reports, databases, journals, publishers, government bodies, and other organizations pertinent to sociology, education, and other social scientific fields in numerous national settings.

Organisation for Economic Co-operation and Development (OECD)

www.oecd.org

The OECD website provides useful and up-to-date information for major international comparisons and developments. It includes report summaries, statistics, and links to major documents on education and related thematic areas that highlight significant trends and issues for 30 member countries and several dozen other nations.

Statistics Canada

www.statcan.gc.ca

Statistics Canada provides a comprehensive body of data and information on education on its website and through its links with other sites.

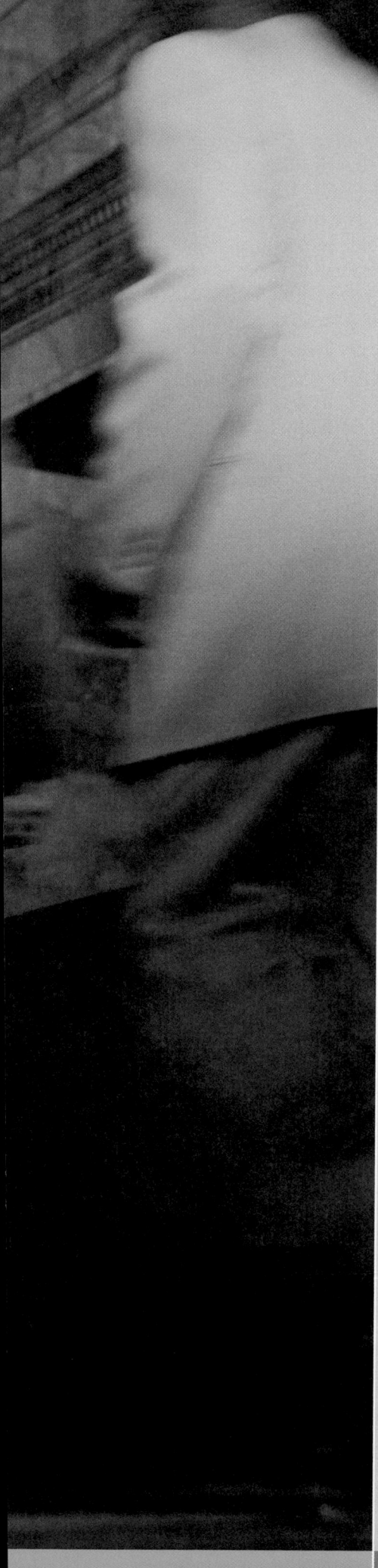

chapter 15

Work and the Economy

Pamela Sugiman

In this chapter, you will:

▶ Be introduced to some of the main concepts that are used in the sociological analysis of work.

▶ Examine the different ways in which work has been socially organized by employers.

▶ Learn about some of the recent trends in employment.

▶ Recognize the impact of flexibility strategies on workers who are located differently in a gendered and racialized capitalist society.

▶ Come to understand the different types of paid and unpaid work that people carry out in this society.

▶ Highlight ways in which workers experience work and sometimes resist.

Introduction

Let us begin this chapter on work and the economy by making three key points. These assertions are the premises on which this chapter is based. First, most of us will spend the better part of our lives working because work is central to our economic well-being. Second, work is a social product and, as such, it is negotiable. Third, people seek meaning in the work that they perform; there is a close relationship between work, life, and **identity**. Let us look more closely at each of these points.

1. *Work is central to our existence.* What would it be like if you never held a job? Would it be possible? Unless you are incredibly wealthy, unable to work as a result of disability or poor health, or willing (or forced) to live on social assistance or handouts on the street, it is unlikely that you could live without work. If you are like most people, you have no choice but to work in order to secure for yourself the basic necessities (food, clothing, a hospitable living environment). Most of us will spend a large chunk of our lives working; the majority will work for someone else, on another's terms. This holds true whether you bus tables, drive a truck, stock grocery shelves, trade on Bay Street, or teach in a school. The very wealthy rely heavily on investment income for their economic well-being, and the extremely poor depend on social welfare (transfer payments). But the majority of people in the middle- and highest-income groups in Canada count on wages and salaries for their existence (Jackson and Robinson 2000, 11). In 2008, close to 8 in 10 persons in Canada aged 15 to 64 participated in the paid labour force (Jackson 2009, 3). It is crucial to think about work, for it has strong implications for how we will live our lives.

2. *Work is a social product.* The second point emerges from the observation that many Canadians view the work they perform as a given. Work is something that we either have or do not have (Gorz 1999), that we go to every day or night, that we leave at the end of a career. But just as we need to face up to the fact that work is what we will do for a good part of our lives, it is important to understand that there is nothing inevitable about the way work is presented and organized. Work is a social product. The way it is structured, the nature of jobs, the rewards of work are all products of **social relationships** between different groups of people. Over time and across cultures, work has taken varied forms. We need to critically examine its current organization with the knowledge that this may be transformed.

3. *People seek meaning in their work.* Although most of us work in order to survive and live comfortably, we also work for more than economic survival or comfort of living. The quality of work matters to workers, young and old. According to a national work ethic study, when Canadian workers were asked what they would do if they won a million dollars, only 17 per cent said that they would quit their jobs and never work again; 41 per cent of respondents claimed that they would keep their job, 17 per cent would embark on a different career, and 24 per cent would start their own business (Lowe 2000, 52). Upon suddenly winning the lottery jackpot, no doubt many of us, over time, would quit our current jobs. But would we abandon work entirely? When faced with job loss and retirement, and even after voluntary early retirement, some Canadians express a desire to return to work. They report a loss of purpose and dignity and feelings of marginality in society. We have a strong attachment to our work.

In order to understand work fully, it is necessary to think about the wider economy in which it is situated. We may define the *economy* as a social institution in which people carry out the production, distribution, and consumption of goods and services. It is critical that we understand how economic systems function, for they have a direct bearing on how we live. The economy and our location in it shape, for instance, the quality of health care, housing, diet and nutrition, consumer spending, and overall lifestyle. The economic system is, furthermore, linked to a nation's political system, to people's conceptions of democracy and citizenship, and to general measures of success and failure.

TIME to REFLECT

In work that you have done for pay, what, besides the money, did you gain from the work? Did you learn anything of value? What did you learn about yourself? How do you think work affects a person's self-esteem and feelings of belonging in society?

World Economic Systems

Economic systems are not abstract and untouchable entities. They are structured and contested, shaped and reshaped, by the people who inhabit them. They further reflect relations of **power** and inequality. In Canada, we currently live in a society that is based on a system of **capitalism**, one in which there are blatant as well as subtle manifestations of inequality. We observe extremes of wealth and poverty every day. On the highway, a shiny new Porsche whirs by a 1996 Chevy Impala. A businessman rushing to pick up a $5000 suit from a Holt Renfrew store walks quickly past a homeless person squatting on the corner. A Filipina nanny on a temporary work permit spends her days taking someone's children to Montessori school, piano lessons, and ballet class. On her way home, she buys their groceries. At night, she returns, tired, to her small room beside the furnace in the basement of the family's well-appointed home. We live in a society in which economic inequalities are complexly wound up with inequalities based on **gender**, **race**, and **ethnicity**.

The power of capitalism is so pervasive that we tend to take for granted many of its central premises. Few of us notice, much less question, the kinds of inequalities that characterize a capitalist society. Concerned about how we can individually climb up the capitalist hierarchy, we seldom stop to question the system itself. But by looking more closely, with a sociological lens, we can see how our present society is the result of historical relationships based on conflict and struggle.

CAPITALISM

Unlike earlier economic systems, capitalism is based on private ownership of the **means of production**, an exchange relationship between owners and workers, an economy driven by the pursuit of profit, and competitive market relations.

In order to understand capitalism, let us turn to the ideas of the classical social theorist Karl Marx. Marx (1967 [1867]) wrote about the profound changes he observed in nineteenth-century England. He witnessed a gradual but dramatic transition from a feudal agricultural society to an industrialized, capitalist economy. Under capitalism, the capitalist class (or bourgeoisie) owns the means of production, while the majority of people, the working class (or proletariat), does not. *Means of production* is a concept that refers to wealth-generating property, such as land, factories, machines, and the capital needed to produce and distribute goods and services for exchange in a market. In a capitalist society, capitalists and workers are engaged in a relationship of unequal exchange. Since workers do not own the means of production, they have no choice but to sell their labour to a capitalist employer in exchange for a wage. Working people are forced into

In the First Person

Historical circumstances denied my parents the opportunity of a university education, and they spent their adult lives toiling in hard jobs, for inadequate pay. Thus they extolled the virtues of a post-secondary education, one that would help me to land a fulfilling, well-paid, and respectably middle-class job. But after high school, I decided I wanted independence, a paycheque, and nice clothes. So, after a week at university, I dropped out and entered the job market. My romantic illusions were soon dispelled. I spent a year in a string of low-end service jobs with demanding bosses, rude customers, imposed overtime, and low pay. By the next September, I was back at school—a scared kid from the working class. But a few good professors propped open the doors of learning. Their lectures inspired, bridged personal experience with sociological insight, and allowed me to view my parents' and my own work experiences with a more critical eye. By second year, I had begun my journey into the sociology of work.

—Pam Sugiman

this relationship because in this type of economy, it is almost impossible to survive without money. One can try to feed a family with the produce of a home vegetable garden, wear home-made clothes, and live without electricity, but at some point it is necessary to purchase market goods and services. For example, you will need to buy fabric, sewing needles, seeds, and a plot of land.

The capitalist class organizes production (work) with the specific goal of maximizing profits for personal wealth. For this reason, it structures work in the most efficient way imaginable, pays workers the lowest possible wages, and extracts the greatest amount of labour from the worker within a working day. Lastly, capitalism is based on a freely competitive market system and therefore a laissez-faire ('hands-off') government. Under capitalism, the market forces of supply and demand are supposed to determine the production and distribution of goods and services, with no government interference.

TIME to REFLECT

Would you characterize your family of origin as 'bourgeoisie' or 'proletariat'? In what ways, if any, do you believe that your family background has affected your choice of occupation and your attitudes toward work and workplace rights?

CAPITALISM AND INDUSTRIALIZATION

While capitalism is a broad economic system, industrialization refers to a more specific process that has consequences for the nature and organization of work as well as for the division of labour. In Canada, as in England, industrialization resulted in a transformation of capitalist production. The rise of industrial capitalism in the late nineteenth and early twentieth centuries constituted one of the most fundamental changes in our society. Industrialization involved the introduction of new forms of energy (steam, electricity) and of transportation (railroads), urbanization, and the implementation of new machine technology, all of which contributed to the rise of the factory system of production and the manufacture and

mass production of goods. These changes greatly facilitated and heightened capitalist production. As well, and in profound ways, they shaped the ways in which people worked and organized their lives.

The proliferation of factories led to the movement of work from homes and small artisanal workshops to larger, more impersonal sites, to the concentration of larger groups of workers under one roof, and to the introduction of *time discipline* (by the clock), in addition to a more specialized division of labour.

During the period of industrial capitalism, economic inequalities became increasingly visible, and conflict between classes grew. While successful capitalists made huge amounts of money, working-class men toiled in factories or mines for a pittance, women combined long hours of domestic drudgery with sporadic income-generating activities, and children were sent off to factories or domestic work. Many people lived in poverty and misery.

FAMILY CAPITALISM

In the mid- to late nineteenth centuries, industrial capitalism was in its early stages. Throughout this period, a small number of individuals and families

(Anthony Bannister/Gallo Images/Corbis)

Family production was key to production in the earliest societies, and though work was differentiated by age and sex, it was shared by all in a familiar household setting.

owned and controlled most of the country's wealth—major companies and financial institutions. Because wealth accrued from business enterprises was passed on within families from generation to generation (for example, the Fords and Rockefellers in the United States and the Eatons and Seagrams in Canada), this era is aptly termed that of *family capitalism.*

CORPORATE CAPITALISM

The subsequent phase of economic development, occurring in the late nineteenth to mid-twentieth centuries, is called *corporate* (or *monopoly*) *capitalism.* This phase witnessed the movement of ownership from individuals and families to modern corporations (and their shareholders). A *corporation* is defined as a legal entity distinct from the people who own and control it. As an entity, the corporation itself may enter into contracts and own property. This separation of enterprise from individuals has served to protect owners and chief executives from personal liability and from any debts incurred by the corporation.

Under corporate capitalism, furthermore, there has been a growing concentration of economic power (that is, power in the hands of a few large corporations). One way in which capitalists have increased their economic power is through mergers. By merging, large corporations have been able to create situations of monopoly and oligopoly. We have a monopoly when one corporation has exclusive control over the market. Obviously, this situation is undesirable for

TABLE 15.1 ▼ The 25 Largest Employers in Canada, 2007

Rank	Company (Year End)	Number of Employees	Location of Head Office	Revenue per Employee ($)	Profit per Employee ($)
1	Onex Corp. (De06)	167,000	Toronto, ON	$120,341	$6,000
2	George Weston (De06)	155,400	Toronto, ON	$207,857	$779
3	Magna International (De06)[a]	83,000	Aurora, ON	$292,253	$6,361
4	Royal Bank of Canada (Oc06)	70,000	Toronto, ON	$514,929	$67,543
5	Metro Inc. (Se06)	65,000	Montreal, QC	$168,398	$3,892
6	Alcan Inc. (De06)[a]	64,700	Montreal, QC	$365,394	$27,604
7	Canadian Tire (De06)	56,559	Toronto, ON	$146,203	$6,270
8	Bombardier Inc. (Ja07)[a]	56,000	Montreal, QC	$267,375	$4,786
9	BCE Inc. (De06)	54,434	Montreal, QC	$326,469	$36,870
10	Bank of Nova Scotia (Oc06)	53,251	Toronto, ON	$422,189	$67,210
11	Toronto-Dominion Bank (Oc06)	51,147	Toronto, ON	$436,037	$89,996
12	Garda World Security (Ja07)	50,000	Montreal, QC	$13,660	$421
13	Jean Coutu Group (My06)[a]	47,115	Longueuil, QC	$236,714	$2203
14	Shoppers Drug Mart (De06)	44,040	Toronto, ON	$176,804	$9,593
15	Sears Canada (De06)	41,107	Toronto, ON	$144,752	$3,712
16	CIBC (Oc06)	40,559	Toronto, ON	$497,202	$65,238
17	Alimentation Couche-Tard (Ap06)[a]	39,500	Laval, QC	$257,438	$4,967
18	Empire Company (My06)	37,000	Stellarton, NS	$356,565	$8,022
19	Quebecor Inc. (De06)	36,588	Montreal, QC	$268,911	$–2,566
20	Bank of Montreal (Oc06)	34,942	Toronto, ON	$519,518	$76,212
21	Nortel Networks (De06)[a]	33,760	Brampton, ON	$343,661	$829
22	Extendicare REIT (De06)	33,700	Markham, ON	$51,794	$–1,060
23	Thomson Corp. (De06)[a]	32,375	Toronto, ON	$205,869	$34,595
24	ACE Aviation Holdings (De06)[b]	32,256	Saint-Laurent, QC	$331,318	$12,649
25	Telus Corp. (De06)	30,000	Vancouver, BC	$289,777	$37,417

[a]Company reports in US dollars.
[b]Figures have been annualized in previous three through five years.

Figures for fiscal periods other than 12 months are annualized for rankings and calculating returns. Foreign currencies are converted into Canadian dollars at the end of the relevant period for balance sheet items and at the average exchange rate for the relevant period for earnings items.

SOURCE: Excerpted and adapted from 'The top 1000', *Globe and Mail Report on Business Magazine* (2007); available at www.reportonbusiness.com/v5/content/tp1000-2007/index.php?view=top_50_employers. Reprinted with permission from *The Globe and Mail.*

GLOBAL ISSUES

ETAG's 2006 Transparency Report Card: Revealing Clothing

The Ethical Trading Action Group's (ETAG's) 2006 Transparency Report Card shows that companies are opening up on labour standards, but tackling labour rights abuses requires more worker involvement.

Revealing Clothing, ETAG's second Transparency Report Card, assesses and compares public reporting on labour standards compliance by 30 top apparel retailers and brands selling clothes in the Canadian market, including Levi Strauss, Nike, adidas, H&M, Mountain Equipment Co-op, Roots, La Senza, Reitmans, and 22 others.

ETAG's 2006 Transparency Report Card assesses companies on the basis of their policies and programs to achieve and maintain compliance with recognized international labour standards in factories around the world where their products are made and the steps they are taking to thoroughly, effectively, and transparently communicate these efforts to the public.

The report card shows that over the past year, some major brands and retailers have improved reporting on labour standards compliance. Canadian companies Mountain Equipment Co-op, Mark's Work Wearhouse, and the Hudson's Bay Company all improved their scores in this year's rating. But despite improvements in reporting, there is still a long way to go to improve actual labour conditions in apparel factories worldwide.

Leading companies have publicly expressed a willingness to discuss root causes of persistent labour rights abuses in their supply chains. More companies are reporting training programs for factory management and other efforts to change persistent bad practices. But ETAG's study shows that companies are less willing to discuss how their own business model of ever-lower prices and highly mobile production might be causing these problems.

ETAG's 2006 Transparency Report Card

Company name	Score	Governance	Code	Engagement	Management	Auditing & Reporting
Reebok	77					
Mountain Equipment Co-op	74					
Adidas	73					
Gap	71					
Levi Strauss	69					
Nike	68					
H&M	64					
Eddie Bauer	63					
Liz Clairborne	62					
Zara (Inditex)	49					
HBC	48					
American Eagle Outfitters	40					
Wal-Mart	40					
Mark's Work Wearhouse	39					
Winners (TJX)	36					
Roots	27					
La Senza	23					
Lululemon	18					
Sears	8					
Northern Group	6					
Reitmans	6					
Polo Ralph Lauren	5					
Boutique Jacob	0					
Forzani	0					
Grafton-Fraser	0					
Harry Rosen	0					
International Clothiers	0					
Le Chateau	0					
Tristan & Amercia	0					
YM Inc.	0					

SOURCE: http://en.maquilasolidarity.org/node/230. Reprinted with permission of Maquila Solidarity Network.

consumers, since it restricts their market 'choices'. The Canadian government has, as a result, implemented various controls to curb the monopolization of an industry.

An *oligopoly* exists when several companies control an industry. The insurance, newspaper, and entertainment industries all are characterized by oligopolistic control. Increased revenues by way of mergers and acquisitions is obviously desirable to corporate owners but may occur at the expense of industrial development, employment, and workers. In 2006, American corporate profits reached a record high of $785 billion. With the deflation of the credit bubble globally in 2008, US corporate profits plummeted to $98.6 billion. However, this economic crisis did not signify the dismantling of corporate capitalism. By 2009, the collective profits of US Fortune 500 companies had risen again to an outrageous $391 billion, more than three times that of the previous year. Why did the corporate sector bounce back with such vigour? Economists tell us that the return was not the result of an increase in sales and new marketing strategies. In fact, during this period revenues fell by 8.7 per cent. Rather, high rates of profit were the result of cost-cutting measures, largely at the expense of workers. In 2009, Fortune 500 firms eliminated 821,000 jobs, close to 3.2 per cent of their payroll, representing the largest job loss in their history (www.cbsnews.com/8301-503983-162-20002576-503983.html). And while the world's most powerful firms provide jobs for million of workers, the corporate elite includes such employers as Wal-Mart, McDonald's, Home Depot, and Target, all of which are retailers known for offering low wages and few opportunities for training and advancement. Workers have not benefited from this growing wealth (http://money.cnn.com/magazines/fortune/fortune500/2010/full_list/index.htm; see also Table 15.1).

The Global Economy

Today, economic activity knows no national borders. Most large companies operate in a global context, setting up businesses in Canada, the United States, and various parts of Asia, Africa, and India. These companies may be called *transnational* or *multinational*. The head offices of transnational corporations are located in one country (often the United States), while production facilities are based in others. We see the products of the global economy everywhere we turn. Look at the clothes you wear, the car you drive, the food you eat. Where are they from? Products of the new global economy typically move through many nations.

Clearly, the goal of transnational corporations is profit. Capitalists are rapidly moving beyond national boundaries in an effort to secure the cheapest available labour, lowest-cost infrastructure (power, water supply, roads, telephone lines), and production unencumbered by health and safety regulations, minimum-wage and hours-of-work laws, maternity provisions, and the like.

Critics have pointed to the negative cultural, social, and economic consequences of **globalization**. Some argue, for example, that globalization has resulted in a homogenization of **culture**. Media giants Time Warner and Disney, for instance, distribute many of the same cultural products (television shows, films, videos, books) to audiences around the globe. Among many other holdings, Time Warner owns well over 1000 movie screens outside of the United States and the second-largest book publishing business in the world, cable and satellite operations, film and TV production/distribution companies, and more than 40 magazines (including *Time, People, Real Simple,* and *Fortune*), in addition to music recording labels, retail parks, and sports teams (http://ketupa.net/time1.htm). The company furthermore boasts that CNN, its popular news network, aggregates approximately 2 billion audience impressions worldwide every day (www.timewarner.com/corp/businesses/detail/turn_broadcasting/index.html). Admittedly, corporate capitalists of the early twentieth century wielded great power, but the power of transnational firms in the current era is immense.

Global capitalism has, furthermore, had an uneven impact on different groups of people around the world. Media exposés of children sewing Nike soccer balls in Pakistani sweatshops for the equivalent of six cents an hour have brought worldwide attention to sweatshop abuses in the garment and sportswear industries. More hidden, says the Maquiladora Solidarity Network, are the adolescent and teenage girls, often single mothers, who sew clothes in the maquiladora factories of Central America and Mexico for major North American retailers such as Wal-Mart and The Gap and the United Kingdom's Marks and Spencer (www.maquilasolidarity.org/resources/child/issuesheet.htm).

Garment manufacturers in Central America's free-trade zones, Mexico's maquiladora factories, and Asia's export-processing zones say that they prefer to hire young girls and women because 'they have nimble fingers. Workers suspect that children and young people are hired because they are less likely to complain about illegal and unjust conditions. And more important, they are less likely to organize unions' (www.maquilasolidarity.org/resources/child/issuesheet.htm). We are seeing the intensification of divisions of labour, globally, along the lines of class, sex, and race.

These developments have direct consequences for the organization of work and for the collective power of working people in Canada as well. Many Canadians now work under a constant threat of company relocation to lower-cost areas. And this has resulted in a weakening of the political power of workers and their unions. In light of this threat, many people in Canada have agreed to concessions (that is, giving up past gains) such as pay cuts, loss of vacation pay, and unpaid overtime. In the long term, the lingering threat of job loss affects the standard of living in the country as a whole.

(Keith Dannemiller/Corbis)

Around the world, modern industrial technology remains central to producing wealth—despite the increased importance of knowledge and information.

The Capitalist Economy: Where People Work

Most of us contribute to the economy in one way or another. Just as the economy undergoes change throughout history, so does our relationship to work. With the expansion of some economic sectors and the contraction of others, our opportunities for certain kinds of jobs also change. We may identify four major economic sectors in which people in this country find employment: primary and resource industries, manufacturing, the service sector, and social reproduction (see Table 15.1).

PRIMARY RESOURCE INDUSTRY

Decades ago, most Canadians worked in the primary (or resource) industry. It is likely that your grandparents or great-grandparents performed primary-sector work. Though not always for pay, Aboriginal peoples have had an important history in the resource industry (Knight 1996). Work in the primary sector involves the extraction of natural resources from our environment. Primary-industry jobs may be found, for instance, in agricultural production (farming, skilled and unskilled agricultural labour), ranching, mining, forestry, hunting, and fishing.

Throughout the eighteenth and nineteenth centuries, the primary sector represented the largest growth area in Canada. However, in the twentieth century it began to experience a dramatic decline. Many forces have contributed to its contraction, notably the demise of small family farms and small independent fishing businesses, along with a corresponding rise in corporate farming (or 'agribusiness') and large fishing enterprises. These developments have resulted in dwindling opportunities for many people. Moreover, because of the geographic concentration of primary-sector jobs, this decline has devastated some towns (for example, Elliot Lake, Ontario) and entire regions (for example, Atlantic Canada).

MANUFACTURING

Into the twentieth century, growing numbers of Canadians began to work in the *manufacturing* (or *secondary*) *sector*. Manufacturing work involves the processing of raw materials into usable goods and services. If you make your living by assembling vans, knitting socks, packing tuna, or piecing together the parts of Barbie dolls, you are employed in manufacturing.

On the whole, the manufacturing sector in Canada has experienced a slower decline than primary industry. The decline in manufacturing began in the early 1950s. In 1951, manufacturing represented 26.5 per cent of employment in Canada, but by 1995, the employment share of manufacturing had been cut nearly in half, to 15.2 per cent (Jackson and Robinson 2000, 11). By 2002, the manufacturing sector had entered a phase of crisis and restructuring. Between 2002 and 2007, more than 30,000 direct manufacturing jobs were lost, in large part because of plant closures and layoffs resulting from lower production, a corporate drive to intensify productivity, and greater outsourcing (Jackson 2009, 262). Economist Andrew Jackson (2009, 262) notes that the impact of such job loss has been greatest in the unionized manufacturing sector. These losses are significant. On the whole, manufacturing jobs are more likely to be full-time, to offer pensions and benefits, and to be unionized.

THE SERVICE SECTOR

Over the past several decades, a massive number of new jobs have been created in the rapidly expanding *service (or tertiary) sector*. Study after study demonstrates

that employees who lost jobs in manufacturing have been absorbed by the service industry. Indeed, many of you are no doubt currently employed in part-time or temporary service jobs. If so, you are not unlike many Canadians.

The rise of the service industry has been linked to the development of a post-industrial, information-based economy and to the growth of a strong consumer culture. All of this has resulted in a growing need for people to work in information processing and management, marketing, advertising, and servicing. In the course of a day, you will encounter dozens of service-sector employees. Airline reservation agents, taxi drivers, teachers and professors, daycare staff, bank employees, computer technicians, crossing guards, librarians, garbage collectors, and Starbucks baristas—all of these people are service workers.

As you can see from these examples, the service sector embraces a wide range of jobs. So dissimilar are these jobs that some people speak of a polarization of work. In other words, there are some 'good', high-skilled, well-paid jobs at one end of the

TABLE 15.2 ▼ Employment by Industry and Sex, 2009 (numbers in thousands)

	2009		
	Number employed		
	Both Sexes	**Men**	**Women**
		thousands	
All industries	16,848.9	8,772.7	8,076.2
Goods-producing sector	3,736.4	2,907.3	829.2
Agriculture	320.5	226.1	94.3
Forestry, fishing, mining, oil and gas	316.2	257.6	58.5
Utilities	147.8	110.0	37.8
Construction	1,161.4	1,032.0	129.3
Manufacturing	1,790.6	1,281.4	509.2
Services-producing sector	13,112.5	5,865.4	7,247.0
Trade	2,639.8	1,346.0	1,293.9
Transportation and warehousing	820.3	622.2	198.1
Finance, insurance, real estate, and leasing	1,099.0	468.3	630.8
Professional, scientific, and technical services	1,201.6	675.4	526.1
Business, building, and other support services[1]	656.5	368.7	287.8
Educational services	1,192.7	395.3	797.4
Health care and social assistance	1,955.0	336.2	1,618.8
Information, culture, and recreation	776.7	401.9	374.8
Accommodation and food services	1,055.9	429.0	626.9
Other services	788.3	357.2	431.0
Public administration	926.6	465.1	461.5

[1]Formerly Management of companies, administrative and other support services.
SOURCE: Statistics Canada, Table 282-0008, Labour force survey estimates (LFS), by North American Industry Classification System (NAICS), sex and age group, annual (persons unless otherwise noted).

spectrum and many 'bad', poorly paid, dead-end jobs at the other. Jobs in retail trade and food services are at the low end of the hierarchy, while those in finance and business, health, education, and public administration tend to be at the high end.

The experience of service work is also qualitatively different from that of manufacturing. Much service employment involves not only the physical performance of a job but also an emotional component. In the face of an intensely competitive market, how does a company vie for customers? Service. And service rests on a big smile and (artificially) personalized interactions. In *The Managed Heart* (1983), Arlie Hochschild explored the emotional work of flight attendants. According to Hochschild, emotional labour, typically performed by women, is potentially damaging to workers precisely because it involves regulating one's emotional state, sometimes suppressing feelings and often inventing them.

Also problematic is the frequently tense relationship between workers and their bosses. Low-end service work is characterized by low-trust relationships. With the expectation that their workforce will have only weak loyalties to the company and its goals, managers attempt to control employees largely through close direction and surveillance (Pupo and Noack 2010; Tannock 2001). It is now common practice for employers to use electronic equipment to monitor telephone conversations between employees and clients and to install video security cameras to keep an eye on retail clerks. Another form of surveillance, more common in the United States than in Canada, is drug testing (through urinalysis) of prospective employees. Such testing is standard, for example, at Wal-Mart stores (Ehrenreich 2001; Featherstone 2004). The most common complaint among workers in low-end service jobs is a high level of stress (Lowe 2007; Tannock 2001).

Social Reproduction

All the work we have discussed so far is conducted in what social scientists call the *sphere of production*. Production typically occurs in the public world of factory, office, school, and store. Moreover, it involves monetary exchange. The study of work in this country has largely been biased toward production.

However, in Canada as well as elsewhere, many people spend hours and hours each day doing work that is not officially recorded as part of the economy. This type of labour may be called **social reproduction**. Social reproduction involves a range of activities for which there is no direct economic exchange. Often, though not always, this work is performed within family households. Typically, it is done by women. We do not view as economic activity the hours women (and, less typically, men) spend buying groceries, planning and cooking meals, folding laundry, chauffeuring children, buying clothes, vacuuming, cleaning the toilet bowl, managing the household budget, caring for aging relatives, and supervising homework. The instrumental value of such activities has long been hidden; rather than being viewed as work, they are deemed a labour of love (Luxton 1980).

But what would happen if women and other family members no longer performed this labour? How would it get done? Equally important, who would pay for it? If capitalist employers or the state had to ensure that workforces got fed, clothed, nurtured, and counselled, what would the cost be? These kinds of questions perplex economists and social statisticians. Says economist Marilyn Waring, breastfeeding, for example, is 'a major reproductive activity carried out only by women, and this thoroughly confuses statisticians' and economists' production models. The reproduction of human life also seems conceptually beyond their rules of imputation. But bodies most certainly have market prices' (1996, 86). In the United States, the cost of reproducing another life through artificial insemination ranges from $1800 for artificial insemination to $10,000 to more than $30,000 for a surrogate mother to carry the child (en. wilkipedia.org/wilki/Surrogatemother). According to the Vanier Institute of the Family, family members in Canada spend 20 billion hours annually performing housework. Unpaid labour is valued at no less than $197 billion, or the equivalent of 10 million full-time jobs (cited in Nelson 2010, 253).

The system of capitalism benefits tremendously from the performance of unpaid labour. Yet not only are the unpaid services of housewives and other family members excluded from traditional economic measures, but for many years sociologists did not even consider them 'work'. This is paradoxical insofar as such work is essential to basic human survival and to the quality of our lives.

THE INFORMAL ECONOMY

Also hidden from official growth figures—as well as from the public conscience—is a wide range of economic activities that are not officially reported to the government. These activities make up the **informal** (or underground) **economy**. Such activities include, for example, babysitting, cleaning homes, sewing clothes, peddling watches, playing music on the streets, gambling, and dealing drugs. As you make your way through the downtown areas of most major cities in Canada, the United States, and almost anywhere in the developing world, you will see people of all ages trying to eke out a living in the informal sector.

Of course, we do not know the precise size of the underground economy. We have only estimates of its share of officially recognized economies and see much variation across the globe. According to the International Labour Organization (ILO), in developing countries as a whole, the informal economy has been estimated to involve one-half to three-quarters of the non-agricultural labour force. In the

developing world, informal employment is generally a larger source of employment for women than for men (www.ilo.org/public/english/employment/gems/download/women.pdf).

Informal economies have flourished for a long time in most nations, but this sector has been growing in importance, largely because of economic hardship related to restructuring, globalization, and their effects of dislocation and forced migration. Increasingly, people are turning to 'hidden work' in order to survive in the midst of contracting opportunities in the formal economy. It has become a safety net of sorts for the poorest groups in society. Without doubt, workers in this sector have had to be enterprising. Some are highly motivated and possess

OPEN for DISCUSSION

Hard Work Never Killed Anyone

Hard work never killed anyone! This is an old and familiar phrase. Teenagers are likely to hear it from their parents when they balk at having to cut the grass or shovel snow. But hard work does injure and kill. In 2004, 340,000 Canadian workers were injured on the job severely enough to lose time from work. In the same year, workers' compensation boards across the country accepted 928 fatality claims. Many would regard these numbers as highly conservative estimates. Tens of thousands of workers who have accidents while at work do not report them, while official statistics of workplace fatalities fail to take into account the daily wear and tear of jobs that can take years off one's life—not to mention the problem of occupational disease.

Canadian workers are injured and killed on the job in at least three ways. First, some jobs such as mining, logging, fishing, and farming are dangerous, and accident and fatality rates are unacceptably high. Second, accident rates are related to how fast and how long a person works at a job. The more hours a person works in a day, the greater the likelihood of an accident. Why? Fatigue. If you get tired shovelling snow, you can stop. If you are paid according to the number of laptops or telephone calls you make, you are likely to push yourself beyond safe limits. And if you are manipulating a fast-paced machine with sharp cutting tools, even a brief lapse in attention can result in serious injury. Third, years of working hard can lead to various occupational diseases that are both debilitating and fatal.

The change from an industrial to a 'post-industrial' or 'information' society has altered patterns of

workplace accidents. More than 30 years ago, the dominant form of injury compensated for by provincial compensation boards involved crushed or severed limbs. Now, one-half of all accident/injury claims are related to strains and sprains, especially of the lower back and upper limbs. Musculoskeletal injuries are on the rise because we are being asked to work harder and faster in jobs that are poorly designed and highly repetitive. Under such conditions, our bodies break down. Many experience chronic neck, shoulder, arm, and back pain.

Numerous studies also link long-term exposure to toxic substances and chemicals such as asbestos, lead, benzene, and arsenic to cancer and other deadly diseases. So, too, waiters in bars and casino workers have to deal with unruly customers and sexual harassment, which can lead to high levels of stress and subsequent health problems such as heart disease. Yet these serious workplace health and safety problems rarely find their way into official accident and compensation statistics.

Workers and unions in Canada have long protested these alarming conditions. In the 1970s, such protests resulted in the passage of occupational health and safety laws that gave workers the right to know about the substances they were working with, the right to participate with management in identifying unsafe and unhealthy working conditions, and the right to refuse work they believed to be unsafe.

—Robert Storey, Labour Studies and Sociology, McMaster University

valuable skills; others lack formally recognized credentials. Unfortunately, most people who rely on the informal economy for a living face precarious, unstable 'careers' in unregulated environments.

TIME to REFLECT

Have you or anyone you know ever worked in the informal or 'hidden' economy? Was this out of choice or necessity? What examples of work in the informal economy can you observe in daily life?

The Social Organization of Work Today

REVOLUTIONARY NEW TECHNOLOGY

Today, popular writers and scholars alike are talking about the emergence of a new world of work, one that is rooted in a 'knowledge society'—a world that offers opportunity, an increase in leisure time, an experience of work that is far more positive than in the past. Are these assertions founded? Do people now have better jobs than their parents and grandparents? Has work been transformed?

Admittedly, most people agree that the new technology may eliminate routine, repetitive tasks, thereby freeing people to perform more challenging work. Think, for example, about preparing a research paper without a computer, printer, or access to the Internet. Moreover, the technology has had a positive impact on job creation. Yet some sociologists argue that at the same time, the technology has created new forms of inequality and exacerbated old ones. While it has resulted in new, more challenging jobs for some people, many others have lost their jobs (or skills) as a result of technological change in the workplace. In the **service economy**, for instance, employers have relied extensively on computers and microelectronics to streamline work processes. In banking, many of the decisions (such as approving a bank loan) that used to be made at the discretion of people are now computer-governed. And the introduction of automated bank machines has made redundant the work of thousands of tellers. As well, in various industries, computers have taken over the supervisory function of employee surveillance. With state-of-the-art computer equipment, and without the direct intervention of a supervisor, firms can now effectively enforce productivity quotas and monitor workers, especially those who perform highly routine tasks (Fox and Sugiman 1999; Lewchuk and Robertson 2006).

Another problem is that the technology is rapidly changing. Competence with the technology thus necessitates continually learning new skills and making ongoing investments in training. Often, workers themselves assume the costs of such training. In the past, says Graham Lowe, employment was based on an implicit understanding of loyalty in exchange for job security; today, this idea has been replaced with a system based on 'individual initiative and merit' (Lowe 2000, 61). Moreover, opportunities for extra job-related training are unequal. Lowe notes that only 3 in 10 Canadian workers annually receive training related to their present or future employment (2000, 65). Currently, employers are far more likely to invest in the training of managers and professionals who already possess relatively high levels of formal education. Thus, rather than equalizing opportunities and outcomes, workplace training systems exacerbate existing inequalities of income and opportunity (Jackson 2009, 59).

FLEXIBLE WORK

Alongside information technology, some writers have extolled the benefits of related trends in management methods. Over the past decade in business circles, one would hear buzzwords such as 'workplace restructuring', 'downsizing', and 'lean production'. All of these concepts are part of a managerial approach called flexibility. One popular practice, termed **numerical flexibility**, involves shrinking or eliminating the core workforce (in continuous jobs and full-time positions) and replacing them with workers in non-standard (or contingent) employment. **Non-standard (or precarious) work** is a term used to describe various employment arrangements such as part-time work, temporary (seasonal and other part-year) work, contracting out or outsourcing (work that was previously done in-house), and self-employment. Non-standard work is, in short, based on an employment relationship that is far more tenuous than those of the past (Jackson 2009; Vosko 2000, 2003, 2006).

In the current economy, non-standard work arrangements now characterize most spheres of employment. We need look no further than the university or college, for example, to see the employment of people in non-standard jobs. In these institutions of higher learning, you may discover that many of your courses are taught by part-time or sessional instructors, some of whom hold PhDs, others of whom are graduate students. These individuals are paid by the university to teach on a course-by-course or session-by-session basis. Sessional or part-time instructors typically do not work on a full-time basis, and they seldom receive assurances of stable employment.

The proportion of Canadian workers in the most precarious forms of work has remained high over the past two decades. Indeed, non-standard labour represents the fastest-growing type of employment in this country. Not unlike the 'reserve army of labour'

described by Karl Marx, non-standard employees provide owners and managers with a ready supply of labour to 'hire and fire' as the market demands. Employers invest minimally in these workers and offer them only a limited commitment. In order to remain competitive in the global market, it is argued, corporations must reduce labour costs through downsizing—that is, laying off permanent, full-time workers and replacing them with part-time, temporary,

TABLE 15.3 ▼ Full-Time and Part-Time Employment by Sex and Age Group, 2002–2006 (000s)

	2002	2003	2004	2005	2006
Both sexes					
Total	15,310.4	15,672.3	15,947.0	16,169.7	16,484.3
15–24 years	2,399.1	2,449.4	2,461.0	2,472.5	2,535.8
25–44 years	7,575.6	7,571.5	7,594.0	7,597.5	7,610.7
45 years and over	5,335.7	5,651.4	5,892.0	6,099.7	6,337.8
Full-time	12,439.3	12,705.3	12,998.1	13,206.2	13,509.7
15–24 years	1,323.1	1,344.3	1,361.4	1370.2	1,419.8
25–44 years	6,627.0	6,624.7	6,671.2	6,684.7	6,730.9
45 years and over	4,489.1	4,736.3	4,965.5	5,151.3	5,359.0
Part-time	2,871.1	2,967.0	2,948.9	2,963.5	2,974.7
15–24 years	1,076.0	1,105.1	1,099.6	1,102.3	1,116.0
25–44 years	948.5	946.8	922.8	912.8	879.9
45 years and over	846.6	915.0	926.5	948.4	978.8
Men					
Full-time	7,287.9	7,423.0	7,559.3	7,664.0	7,781.0
15–24 years	763.9	774.9	781.2	782.5	809.2
25–44 years	3,831.1	3,832.2	3,834.1	3,832.6	3,845.6
45 years and over	2,692.9	2,815.9	2,944.1	3,048.9	3,126.2
Part-time	896.5	925.0	921.3	930.7	946.1
15–24 years	460.4	468.3	467.1	456.5	467.7
25–44 years	197.4	196.9	189.8	199.5	189.7
45 years and over	238.8	259.8	264.4	274.7	288.7
Women					
Full-time	5,151.4	5,282.3	5,438.8	5,542.3	5,728.7
15–24 years	559.2	569.4	580.2	587.8	610.5
25–44 years	2,796.0	2,792.5	2,837.2	2,852.1	2,885.3
45 years and over	1,796.2	1,920.4	2,021.4	2,102.4	2,232.8
Part-time	1,974.6	2,041.9	2,027.6	2,032.8	2,028.5
15–24 years	615.6	636.8	632.4	645.8	648.4
25–44 years	751.2	749.9	733.0	713.3	690.1
45 years and over	607.8	655.2	662.1	673.7	690.0

SOURCE: Statistics Canada, www40.statcan.ca/l01/cst01/labor12.htm.

and contract labour (Cranford, Vosko, and Zukewich 2006; Vosko 2000, 2006).

Furthermore, non-standard workers as a whole receive relatively low wages and few benefits. Consequently, many people who rely on this type of work must resort to holding multiple jobs in an effort to make ends meet. People carve out a living by stringing together a host of low-paying, part-time, and temporary jobs. Often this involves moonlighting or doing shift work, situations that no doubt put added strain on families.

In light of these trends, the concept of a career is a remnant of the past for most Canadians. The gold watch for 50 years of continuous service to the same company is not attainable in the new workplace scenario. Says Richard Sennett, 'flexibility today brings back this arcane sense of the job, as people do lumps of labor, pieces of work, over the course of a lifetime' (1998, 9). Not surprisingly, living in this era of economic uncertainty, with the attendant worry about layoffs and job loss, is a major source of stress for people in Canada (Jackson 2009, 74–7; Lewchuk et al. 2006; World Health Organization 1999).

TIME to REFLECT

Have you ever worked or do you now work at what could be termed a 'McJob'? Did it or does it fulfill your needs? How does it do this? In what ways, if any, did or does it fall short? Would you like to work at a 'McJob' for the rest of your working life? Why or why not?

The Changing Face of Labour: Diversity among Workers

Just as places of work have changed dramatically over time, so too has the workforce itself. Workplaces today are becoming increasingly diverse. Only a minority of families rely on a single paycheque. Aboriginal Canadians make up a growing proportion of the paid labour force in certain geographic areas. People of colour, some of whom are immigrants to

(Amit Bhargava/Corbis)

When you next order a pizza, complain about the non-arrival of a package, or report a lost credit card, you may find yourself talking to someone in India—thanks to the globalization of work and the export of Canadian jobs.

HUMAN DIVERSITY

Offshore Migrant Farm Workers: A New Form of Slavery?

Who, if anyone, are the new slaves in Canada? While they are not really slaves in the classical sense of the term, the 16,000 workers who come to Canada every year from the Caribbean and Mexico to work in Canadian agriculture are a form of unfree labour. *Slaves* may be too strong a term to describe what they are, but their condition of unfreedom does bear a strong resemblance to slavery.

What makes these workers different from other Canadians and from immigrants who come here to build better lives for themselves? Migrant workers from the Caribbean and Mexico come to Canada under labour contracts. These contracts specify how long they can remain in the country and the conditions under which they must work. Workers are allowed to stay in Canada for between three and eight months every year. When their contracts expire, or if they breach one of the terms of their contract, they are expected to leave the country. Workers pay for a portion of their transportation and must pay their employers back to help them cover the costs of accommodation. In some cases, workers bunk five or six to a room and live in hot, overcrowded conditions. However, the main reason that they are considered to be unfree labourers stems from their inability to quit or change jobs in Canada without the permission of their employer and a representative of the federal government. If they do quit their jobs with a Canadian employer without permission, they are subject to deportation from the country.

Why does this condition of unfreedom matter? After all, some people think that migrant workers are lucky to be here, compared to where they come from. They invariably make more money here than they would back home, so they should be grateful for the opportunity to come here to work, even if only temporarily. Yet even though no one is forcing them to sign a labour contract and come to Canada to work, it does matter that they are a form of unfree labour. Their lack of choice when it comes to whom they work for and their inability to vote with their feet and find better-paying jobs in other sectors of the Canadian economy mean that farm employers have a tremendous degree of power over migrant workers. In many cases, workers are fearful of saying 'no' when they are asked to do jobs that are dangerous and might harm their health. And employers who have a captive labour force do not have market incentives to improve wages or working conditions.

—Vic Satzewich, McMaster University

this country, many Canadian-born, currently have a stronger-than-ever presence, particularly in big cities such as Vancouver, Toronto, and Montreal. As well, the workforce has become more highly educated and younger.

GENDERED WORK

The participation of women in the paid labour force has increased steadily over the past four decades. In the mid-1970s, the labour force participation rate of women aged 15 to 64 was slightly over one-half. By 2006, it was close to three-quarters (or 73.5 per cent) (Jackson 2009, 100). Most striking has been a rise in the employment rates of married women and mothers of children under the age of six. The two-breadwinner (also called *dual-earner*) family is now the norm.

Today, many young women and men entering the labour force are unaware of the blatant sexual inequalities of the past. Whether or not they identify themselves as feminists, women today are building their careers on a feminist foundation. If not for the challenges posed by women's rights activists, university lecture halls would be filled exclusively by men, women would not be permitted entry into the professions or management, and paid employment would simply not be an option after marriage.

But just as women's historical breakthroughs are instructive, so too are the persisting inequalities. In spite of a dramatic increase in female labour-force participation, women and men are by no means equal in the labour market. The **social institution** of work is still very much a gendered one. Some women have made inroads in non-traditional fields of manual labour, the professions, and management and administration, but the majority remain concentrated in female-dominated occupations such as retail salesperson, secretary, cashier, registered nurse, elementary school teacher, babysitter, and receptionist, while men are more commonly truck drivers, janitors, farmers, motor vehicle mechanics, and construction trade helpers, for example (Statistics Canada 2008g).

To the extent that occupational segregation by sex has lessened somewhat over time, it is more because of the entry of men into female-dominated occupations than the reverse.

As well, women (in addition to youths of both sexes) are more likely than men to be employed on a part-time and temporary basis. For years now, women have made up approximately 70 per cent of the part-time workforce in Canada. And while the majority of the self-employed are men, the 1990s witnessed a rapid growth in women's self-employment. In comparing the sexes, we also see that self-employed men are more likely than self-employed women to hire others and that businesses operated by men are more likely to be in the goods sector, whereas female-run businesses are likely to be in the less lucrative service sector (Nelson 2010, 246–7).

These trends—labour market segregation by sex and the overrepresentation of women in precarious employment—have contributed to gender-based differences in earnings.

Currently, among all categories of earners (full-time and part-time), we find that a woman earns 64.5 cents for every dollar earned by a man (Statistics Canada, CANSIM Table 202-0102). When we compare women employed on a full-time basis with their male counterparts, the gap narrows—although it does not disappear. In fact, the gap persists even when we control for education. While Canadian women posted better academic achievements than men at all levels, women with a post-secondary education still earn on average 63 per cent of the salary of men who possess similar educational qualifications (www.theglobeandmail.com/news/national/women-at-work-still-behind-on-thebottom-line/article1699176).

The category *woman*, as we know, is not a homogeneous one. Immigrant women, women of colour (or racialized women), and Aboriginal women bear the brunt of income and occupational polarization by sex. In consequence, their average annual earnings are disproportionately low. Statistics Canada reports that of those employed full-time, full-year in 2000, visible-minority women, for example, had average annual earnings of $32,100, roughly 10 per cent less than their 'non-visible' equivalents (www.statcan.ca/english/freepub/89503-IE/0010589-XIE.pdf).

Faced with multiple forms of **discrimination**, working-class women of colour and some female immigrants have come to occupy job ghettos. Indeed, many of the jobs that are typically performed by working-class people of colour have a 'hidden' quality: the work they do is not noticed; the workers are rendered invisible. All too often, we regard private domestic workers and nannies, hotel and office cleaners, taxi drivers, health-care aides, and dishwashers—all of whom perform indispensable labour—as simply part of the backdrop (Arat-Koc 1990; Das Gupta 1996; Sherman 2007). Not only are they physically out of sight (in basements, in kitchens, working at night when everyone else has gone), they are out of mind.

In documenting sex-based inequalities in employment, social scientists have produced reams of statistics. But there are many other ways in which we may speak of the gendering of work, some not easily quantifiable. Joan Acker (1990) writes about the process by which jobs and organizations come to be gendered, regardless of the sex of job-holders. The bureaucratic rules and procedures, hierarchies, and informal organizational culture may rest on a set of gender-biased assumptions, for example. In *Secretaries Talk* (1988), Rosemary Pringle highlights the ways in which gendered family relationships are reproduced in workplace relations between bosses (fathers) and secretaries (wives, mistresses, daughters): 'Male bosses go into their secretaries' offices unannounced, assume the right to pronounce on their clothes and appearance, have them doing housework and personal chores, expect overtime at short notice and assume the right to ring them at home' (Pringle 1988, 51).

Today, many young women plan to both have a professional career and raise a family, but they are not quite sure how they will combine the two. Feminist researchers have demonstrated how the very concept of 'career' is gendered, built on a masculine model. Career success depends on the assumption of a wife at home—a helper who will pick up the children from school, arrange dinner parties, and generally free the 'breadwinner' to work late at nights or on weekends and for out-of-town business travel.

Furthermore, feminist analysis has called attention to the complex link between paid and unpaid labour, employment and family (Corman and Luxton 2007; Eichler et al. 2010; Fox 2009). With two breadwinners, both of whom spend increasing hours in their paid jobs, families are under enormous pressure. While the demands of paid work have risen over time, so too have pressures on family life. Government restructuring and cutbacks in resources have affected public daycare, after-school programs, special needs programs, and care of the elderly and the disabled. Who picks up the slack? The family. One consequence has been an intensification of (unpaid) family work, more stress, and growing tensions within families as people try to cope.

RACE AND RACIALIZED WORK

Although we now have an abundance of research on the gendering of work, sociologists in Canada have paid far less attention to the relationship between race, citizenship, and employment. Barriers faced by people of colour, Aboriginal Canadians, and some immigrant groups are most often demonstrated in unemployment and earnings disparities. Aboriginal people comprise only a tiny percentage of the working-age population, yet this group is growing rapidly and over the next 20 years will constitute a sizable share of new entrants to the labour force in certain parts of Canada (Jackson 2009, 146). Although one should be wary of making broad generalizations about the diverse group of people who identify as Aboriginals, one clear observation is that on the whole, they are disadvantaged in the labour market. Disadvantage is linked to systemic discrimination generally in addition to low levels of education and patterns of geographic residence (Jackson 2009, 146–7).

The **unemployment rate** for Aboriginals is disturbingly high in comparison to that for the Canadian population as a whole (16.1 per cent for Aboriginal men and 13.5 per cent for Aboriginal women). And for Aboriginal youth (aged 20 to 24) specifically, the rate of unemployment is 20.8 per cent (Jackson 2009, 147). In addition, more than half of Aboriginals are in part-time employment—jobs that offer little security—and they are concentrated in marginalized sectors of the economy where they face low pay, seasonal jobs, and high levels of discrimination in hiring (canadianlabour.ca/index.php/Aboriginal Workers/464). The economic prospects for those who live on reserves are even more bleak. Close to half of the on-reserve Aboriginal population live in poverty (Jackson and Robinson 2000, 71).

The category 'people of colour' is likewise quite diverse, containing notable differences according to class, education, and citizenship status. In Canada, about one in 10 workers is defined as being of colour or as a racialized person (the official census term is 'visible minority' and excludes Aboriginal Canadians), and currently, one in five Canadians is foreign-born (Jackson 2009, 135). These gaps remain even when we control for age and education. According to the 2006 census, 'visible minority' workers aged 25 to 44 with a university degree earned 74.6 per cent of the median for the group as a whole, while 'non-visible'-minority earners made 105 per cent of the median.

Even second-generation visible-minority earners in this age and educational category earned less than their non-visible-minority counterparts ($14,675 compared to $46,172) (Jackson 2009, 145–6). Typically, recent immigrants are younger than the labour force as a whole, but they also have more schooling. One problem is that foreign credentials are not always respected in Canada, thus contributing to a high concentration of immigrants of colour in low-wage jobs (Jackson and Robinson 2000, 69–70).

Though telling, statistics reveal only one dimension of the research on disadvantaged groups. It is equally important to recognize that because of racial and cultural differences, people experience the work world in distinct ways. In their classic study *Who Gets the Work*, Frances Henry and Effie Ginzberg (1985) found a striking incidence of discrimination directed at job seekers. For example, when whites and blacks with similar qualifications applied for entry-level positions that had been advertised in a newspaper, jobs were offered three times more often to white than to black applicants. Similarly, of the job seekers who made inquiries by telephone, those who had accents (especially South Asian and Caribbean) were often quickly screened out by employers.

Furthermore, the role of the Canadian state historically in promoting or facilitating racialized work has been documented extensively (Schecter 1998). Agnes Calliste (1993) notes that between 1950 and 1962, Canadian immigration authorities admitted limited numbers of Caribbean nurses but under rules different from those for white immigrant nurses. Black nurses were expected to have nursing qualifications superior to those demanded of white nurses. Several scholars (Arat-Koc 1990; Daenzer 1993; Macklin 1992; Stasiulis and Bakan 2005) have also discussed the role of the Canadian state in addressing the need for cheap child-care workers by importing women from the developing world (the Caribbean and the Philippines in particular) to perform domestic labour without granting them full citizenship rights.

YOUTH

In Canada today, youths (persons 15–24 years old) constitute a much smaller share of the population than in past years. Nevertheless, the youth labour market is expanding at a significant rate. Young people today must confront harsh economic conditions, with the youth unemployment rate roughly 50 per cent higher than that of the population as a whole. The research presents us with a woeful picture. Study after study suggests that young people are in important ways no

TABLE 15.4 ▼ Median Annual Income of All Visible-Minority Persons, 2005

	All	Men	Women	As % of median income of all workers		
				All	Men	Women
All	$25,615	$32,224	$20,460	100.0	125.8	79.9
Not a visible minority	$26,863	$33,898	$21,164	104.9	132.3	82.6
Visible minority	$19,115	$22,670	$16,638	74.6	88.5	65.0

SOURCE: Statistics Canada, Labour Force Survey, 2007.

different from the majority of Canadian workers. They want high-quality work—work that is interesting and challenging and that provides a sense of accomplishment (Lowe 2000). And youths have been increasing their human capital to acquire such jobs. Notably, young people are acquiring more education. (While a university degree does not guarantee a job, young people are still better off if they have the formal credentials.) But while Canadian youths are better schooled on the whole, they are also working less and in jobs for which they feel that they are overqualified. Young people are most likely to be employed in low-paying service-sector jobs such as fast-food restaurants, clothing stores, and grocery stores. For most students, contingent work is all that is available.

Some writers argue that the youth labour market makes a perfect accompaniment to the new goals of managerial flexibility. Employers invest in the belief that young people will have a limited commitment to the goals of the firm and that they expect to stay in jobs temporarily as a stop-gap measure discontinuous with their adult careers and identities (Tannock 2001). Stuart Tannock explains that youth themselves partially accept the popular **ideology** that positions them 'as a separate class of workers who deserve less than adult workers do. Good jobs are predominantly the privilege of adulthood. Young workers must be content at first to spend their time in a tier of lower-quality service and retail employment. Dreams of meaningful work must be deferred' (2001, 109). Many young people compare themselves not to other workers across the spectrum but exclusively to other youth workers (Sennett 1998). Consequently, youths are more pliable and passive. Also, because their jobs are viewed as transient, youths are not as likely to become unionized. All of these features render them an extremely exploitable source of labour.

But as Tannock points out, youths are not stop-gap workers simply because they are young: they are also stop-gap workers because of the poor conditions under which they have to labour—conditions that have been created by employers in the service sector.

Despite the popular view that young people are not especially concerned about their conditions of work, there is now much evidence that points to the contrary: 'Teenagers and young adults working in these industries, who expect to have long lives ahead of them, worry that their jobs, which are supposed to be meaningless, stop-gap places of employment, will have lasting and detrimental effects on their bodies and future life activities' (Tannock 2001, 54).

Workers' Coping and Resistance: The Struggle for Dignity and Rights

FINDING MEANING IN WORK

Regardless of the many differences among Canadian workers today, one point remains clear: most Canadians want work that is personally fulfilling (Lowe 2000). People have a powerful desire to maintain dignity at work (Hodson 2001). Some of us are fortunate enough to hold jobs that offer challenge, jobs in which we can exercise autonomy and from which we can reap fruitful economic rewards. But even the 'good jobs' are not always meaningful, and there are many jobs that are rarely rewarding. How do people cope with their work?

Sociologists have found that no matter how meaningless the job, people seek meaning in their work. Sometimes this is done through the culture of the workplace. People who have boring, routine jobs, for example, may make a game out of their work, varying repetitions, altering pace and intensity, imagining the lives of customers. As well, the social component of work (peer relations) is frequently a source of pleasure. In some workplaces, employees regularly exchange gossip, flirt, engage in sexualized play, share personal problems, debate politics, ridicule management. Relationships with co-workers often make the job itself more bearable if not meaningful. In cases where the organization of work permits such exchanges, the lines between employment and leisure can become blurred.

Job satisfaction studies suggest that work is not all that bad. Most people report that they are generally satisfied with their jobs (Lowe 2000). On close examination, though, discontent broods near the surface. At the same time that they report satisfaction, a majority of workers say that their jobs are somewhat or highly stressful, that they are not sufficiently involved, recognized, and rewarded, and that their talents are underutilized (Lowe 2000). In addition, there are high rates of absenteeism, oppositional attitudes, slacking off, pilfering, and even destruction of company property. Some workers simply quit their jobs. But in the face of a competitive job market, family responsibilities, consumer debt, and, for some, limited marketable skills, this is not always a viable option. Furthermore, it is telling that even though they claim to like their jobs, many people add that they do not want their own children to end up doing the same kind of work (Sennett 1998).

United Steelworkers rally in a show of support for protecting employee rights.

(Dick Loek/GetStock.com)

Faced with unfair, unsafe, and sometimes unchallenging work, workers will be discontented. They will find ways to make changes, to resist. The question is, how? Individual acts of coping and resistance may give workers the feeling of agency and control, but insofar as they are individual acts, they rarely result

UNDER the WIRE

Privacy in the Workplace

We spend so much of our lives at work: what about our right to privacy as employees? . . . Just as employers are demanding 'more work for less pay', they are also demanding or simply taking more and more of our personal information. We are being screened, tested, monitored, and reported like never before. Put simply, organizations have voracious appetites for employee (and lots of other) information, together with the computer communications, surveillance, and other technological means to feed them. Practically all large organizations need to go on personal information diets. . . .

Who knows us better than our employers? Their files and data bases include our hiring, pay and benefits information, banking, insurance, family matters, pictures, personal identifiers and contacts, attendance, sick leave, claim and medical information, performance and other career-related records, grievances and other complaint files, 'challenges' we're having (at work or otherwise), investigations, discipline . . . the list is a long one. It's just about all there.

In larger organizations, employee information is collected, 'shared', used in decision-making, and retained by great numbers of supervisory, personnel, finance, security, and other officials. Increasingly, their operators . . . are computerized. By definition, computers enable far more access to employee information, by far more people, far faster than ever before. The privacy risks of massive breaches are far greater, too, and although violations are usually inadvertent, the damage is sometimes irreparable. . . .

An estimated 75 per cent of companies in the United States electronically monitor their employees. Video surveillance cameras are multiplying like rabbits, and everybody seems to need to wear ID cards these days, to enter or move around unthreatened workplaces. Maybe your employer has gone hi-tech with biometric tracking or radio frequency identification (RFID). One would think we all had 'top secret' jobs and that none of us could be trusted, regardless of long service.

SOURCE: Richard Sharp. 2006. 'Privacy in the workplace'. Canadian Centre for Policy Alternatives, *The Monitor* July. www.policyalternatives.ca/publications/monitor/july-privacy-workplace.

in a fundamental or widespread change in conditions of work. In order to effect large-scale change, people must resort to collective measures.

PROFESSIONS AND NEGOTIATING PROFESSIONAL CONTROL

Securing professional control is an option for middle-class people who possess formally recognized credentials and can claim expertise in an area. When we think of a *professional*, who comes to mind? Physicians, educators, psychiatrists, dentists, lawyers, engineers, accountants. Some sociologists (proponents of trait theory) have attempted to define professionals with reference to a checklist of characteristics (Freidson 1970). This checklist includes, for example, possession of a body of esoteric or abstract knowledge, reliance on a specialized technical language or vocabulary, and membership in associations that control entry and membership in the occupation through licensing, accreditation, and regulation.

Critics, however, argue that trait theory does not fully explain how and why some occupations come to be defined as professional while others do not. Rather than list a series of traits that define a profession, Terence Johnson (1972) highlights the resources available to different occupational groups. These resources have enabled physicians, psychologists, and lawyers to define themselves as distinct from other groups such as managers, clerical workers, and massage therapists. In focusing on the process of professionalization, critical theorists have noted that at the heart of the struggle to professionalize are relations of power and control. Feminist scholars have recently offered a more nuanced analysis of the ways in which **patriarchy** (a system of male dominance), too, structures the process of securing professional authority and control (Adams 2000; Witz 1992).

LABOUR UNIONS AND LABOUR'S AGENDA

But the struggle to professionalize is not one in which many Canadians will be engaged—it is largely an exclusive one. Greater numbers of people in Canada, and globally, turn to another form of collective action to secure their rights and dignity in the workplace: they look to unionization. Just as campaigns to secure professional control have had a middle-class base, the struggle to unionize in this country has traditionally been one of white men in blue-collar jobs. In the latter part of the twentieth century and into the present time, however, increasing numbers of women, people of colour, white-collar workers, and middle-class employees have joined the ranks of the labour movement.

When most of us think of unions, strikes come to mind. Some of us may view trade unionists as just a bunch of greedy, overpaid workers demanding higher wages and in the process disrupting our lives, transportation, communication—even our garbage collection. We may owe this perception to dominant media representations of unions, their members, and their leaders.

The labour movement in this country goes far beyond this narrow and unfair characterization. The basic premise of the organized labour movement is to take collective action through the process of bargaining a contract. This *collective agreement* is the outcome of days, weeks, or even months of negotiations between two parties: worker representatives and company representatives. The contract is a legally binding document, an agreement that has been signed by both the employer and the union. Only if the two parties cannot reach an agreement is there potential for strike action. The actual incidence of strikes in Canada is, in fact, low. Recently, throughout the entire country, there have only been 300 to 400 work stoppages per year, and they involved 100,000 to 400,000 workers. While one-third of all employees are members of unions, annual time lost due to strikes has typically been far less than one-tenth of one per cent of total working time (Jackson 2009, 202). The strike is usually a measure of last resort. The vast majority of contracts that come up for renewal are settled without resorting to strike action. Workers in the nineteenth century first struggled to secure union representation in an effort to protect themselves against excessively long work days, extremely hazardous work environments, low pay, and blatant favouritism on the job. Today, labour–management conflict arises over a host of issues. Not only are wages an item of dispute, but companies and union representatives also negotiate benefits packages, job security, the implementation of technological change, outsourcing, concessions, and anti-harassment policies. Because of the struggles of union members, Canadian workers in offices, stores, and factories now have the right to refuse unsafe work, the right to participate in company-sponsored pension plans, and, in some cases, access to on-site daycare centres.

The gains of unionized workers, moreover, spill over into the wider society. Both unionized and non-unionized workers now have employment standards, (un)employment insurance, a standard work day of eight hours, a five-day work week, overtime premiums, vacation pay, health benefits, and sick-leave provisions. Unions have been pivotal in lobbying governments to introduce worker-friendly provincial and federal legislation.

Union Membership

Today, unions represent roughly one in three workers (31.5 per cent in 2007) and 18.7 per cent of private-sector workers (Jackson 2009, 225). Sex-based differences in union density (membership) have now disappeared. In fact, 2006 marked the first year in Canadian history in which women outnumbered men as union members (Jackson 2009, 227). In part, growth in female membership reflects the high rate of unionization in the (female-dominated) public service (for example, in Crown corporations, education, and health care). It is also, in part, a result of recent union organizing drives in private services (Jackson and Robinson 2000). As well, over the past several decades, the loss of female union members has been less dramatic as compared to that of men. The unionization rate for men has been dropping since the 1960s. This trend is largely attributable to a shrinking proportion of jobs in traditionally male-dominated and heavily unionized sectors, such as primary/resource, manufacturing, and construction (Jackson and Robinson 2000).

The Union Advantage

There is absolutely no doubt that unionization benefits workers (see Table 15.5). Collective bargaining has secured for employees advantages in wages, benefits, job security, and extended health plans. This has been called the *union advantage*. The union wage premium in particular is greatest for (traditionally disadvantaged) workers who would otherwise be low-paid.

SOCIOLOGY in ACTION

Wal-Mart assailed in Supreme Court of Canada

Shutting down the Charter Rights of 200 Canadian Workers

Ottawa (22 Jan. 09)—Wal-Mart violated the rights of workers when it closed a store in Jonquière, Que., after the employees were certified to form a union, the Supreme Court of Canada was told Wednesday.

The country's top court has agreed to hear an appeal by the United Food and Commercial Workers (UFCW Canada) which says the giant retailer's actions violated the right to freedom of association, which workers are guaranteed in Canada by the Charter of Rights and Freedoms. Wal-Mart claims the closure was for economic reasons.

The case is being watched by Wal-Mart employees and business groups worldwide for its potential repercussions. To date, no legitimate union has been able to gain a lasting foothold in any of Wal-Mart's global operations. Once arguments are heard in this case, the court is expected to take several months to bring down a decision.

The case involves a store Wal-Mart opened in Jonquière, Quebec, in 2001 and closed four years later—after employees had unionized and the Quebec ministry of labour had ordered binding arbitration to impose a first contract.

'No employee at Jonquière Wal-Mart should have lost their job because they were exercising their right [to union activity],' lawyer Claude Leblanc argued Wednesday on behalf of the employees. He rejected Wal-Mart's explanation of the closure.

Anti-union Animus?

'We have to look deeper into it and see whether that was a real reason or was it just an excuse; was it just used to cover up the real reason, anti-union animus?' Leblanc said.

In October, Wal-Mart closed an auto service centre at another store in Quebec—in Gatineau—after a collective agreement was imposed by the Quebec labour board. Again, the company claimed the closure was for economic reasons, not bias against unions.

Workers have also been certified at three other Canadian Wal-Mart stores. In none has the company negotiated a collective agreement with employees. The stores include the full Wal-Mart store in Gatineau, where the auto shop was located, a store in Ste-Hyacinthe, Que., and one in Weyburn, Sask.

'This isn't whether Wal-Mart has the right to close a store,' says UFCW Canada president Wayne Hanley.

'It's about Wal-Mart shutting down the Charter rights of 200 Canadian workers. As Canadians, what's more important to us? The business rights of some multinational corporation or the human rights of Canadian workers and their families here at home?'

SOURCE: National Union of Public and General Employees (NUPGE). www.nupge.ca/node/810.

TABLE 15.5 ▼ The Union Average Hourly Wage Advantage, 2007

	Union in $	Non-Union in $	Union Advantage in $	Union Advantage as % of non-union
All	23.58	18.98	4.60	24.2
Men	24.38	21.20	3.18	15.0
Women	22.79	16.71	6.08	36.4
Ages 15–24	14.45	11.63	2.82	24.2
By occupation:				
Professionals in business, finance	30.03	28.94	1.09	3.8
Secretary, administration	22.00	19.05	2.95	15.5
Clerical, supervisors	20.12	15.92	4.20	26.4
Natural sciences	29.61	28.32	1.29	4.6
Heath, nursing	29.96	29.22	0.74	2.5
Assist. health occupation	21.05	18.94	2.11	11.1
Social sciences	26.13	21.64	4.50	20.8
Teacher, professor	30.45	23.44	7.01	29.9
Art, culture, recreation	24.42	19.33	5.09	26.3
Mainly low-wage private services:				
Retail, sales, cashier	13.02	11.40	1.62	14.2
Chefs, cooks	15.22	11.42	3.80	33.2
Protective services	23.68	16.81	6.87	40.9
Child care	17.88	11.56	6.32	54.7
Sales, service, travel	15.18	11.26	3.92	34.8
Blue-collar:				
Construction trades	24.76	18.50	6.26	33.8
Other trades	25.76	19.60	6.16	31.4
Transport equipment	21.62	17.00	4.62	27.2
Trades helpers	21.01	14.41	6.60	45.8
Primary industry	22.29	16.69	5.60	33.5
Machine operators	20.75	17.10	3.65	21.3
Process, manufacturing	17.38	13.07	4.31	32.9

SOURCE: Statistics Canada. 2007. *Labour Force Survey*. Taken from: Andrew Jackson. 2009. *Work and Labour in Canada: Critical Issues*. 2nd edn, p. 207. Toronto: Canadian Scholars' Press.

Unionization tends to compress wage and benefit differentials and thereby promote an equalization of wages and working conditions among unionized workforces. In 2007, for example, average hourly earnings of unionized workers in Canada were $23.48, while for non-unionized workers the average hourly rate was $18.98, thereby representing a union advantage of 24.2 per cent. The difference in wages between unionized and non-unionized women was even greater. In 2007, the union advantage for women was $6.08 as compared to $3.18 for men (Jackson 2009, 208).

TIME to REFLECT

Imagine that you are employed in a non-unionized workplace, and a union organizer seeks to enlist union members and certify your workplace. Would you support this action? Would you if it meant you might be fired from your job?

Conclusion: Work in the Future, Our Future as Workers

Workers and unions, of course, have limited powers. While newspaper headlines promote the 'big' collective bargaining gains of the most strongly organized unions, most unionized workers across the country are still struggling to attain basic rights that others managed to secure years, if not decades, ago. Every day in small workplaces, employees (unionized and non-unionized) negotiate their rights. More often now than in the past, these are women, people of colour, the disabled—not members of the dominant groups in this country.

These struggles have been difficult and continue to be so, particularly in the context of the current assault on unions. Powerful corporations such as Wal-Mart and McDonald's effectively curb workers' rights to organize by simply closing down stores, mounting strong union decertification campaigns, or stalling when it comes time to bargain a first contract. The power of workers and their movements is being even more severely circumscribed by the aggressiveness of global capitalists, many of whom are openly supported by networks of governments in both developing and developed nations. Whether you work part-time at The Gap, labour a 60-hour week in a steel factory, freelance as a consultant, or find sporadic office employment through a temporary help agency, you face a challenge.

Regardless of theoretical perspective or political agenda, scholars today are debating the nature of the challenge of the transformation of work. Young people entering the labour market for the first time and middle-aged people confronting reconfigured jobs and refashioned workplaces both are part of this transformation. Workers, young and old, must work in order to survive, to nurture families, to participate in life. Given this reality, it is crucial to know the debate, engage in it, and perhaps transform the world of work according to your own vision.

questions for critical thought

1. Think about where you are located in the economy. If you are not currently employed, where do you plan to find work? How does this depart from your parents' and grandparents' work histories? What factors have shaped (or constrained) your work-related aspirations?

2. Some employers believe that if you pay a worker more money (and offer better benefits), then you can compensate her or him for the boredom of work, loss of control, and lack of autonomy on the job. What do you think about this belief?

3. Think about the work that you perform in the course of an average day. What proportion of this is paid and what unpaid? Do you believe that we should define unpaid domestic activities as 'work' that is of economic worth? If you were asked to calculate the economic worth of unpaid domestic labour, how would you begin? What factors would you take into account?

4. Most people in Canada today take the new computer technology for granted, but as a sociologist, you must take a closer, critical look. Consider some of the ways in which computer technology has reshaped employment opportunities and the nature of work.

5. A prevailing view is that the youth today are merely 'stop-gap' workers. They are young and resilient. As they mature, they will move on to better, more secure and fulfilling employment. Thus, their conditions of work are not problematic. Young people themselves are not concerned about the nature of the jobs that they perform. Why should sociologists bother writing about youths at work?

6. Do you believe that who you are (that is, female or male, Aboriginal, of Asian or European or African descent, young or middle-aged, working-class, educated) is an important indicator of the type of work you will perform? If so, in what ways? If not, explain.

7. Even though women's workforce participation rate is now almost the same as men's, there are many persisting gender-based inequalities in employment. Identify some of these inequalities. What are some of the formal and informal barriers to equality between women and men in the labour market today? How would you confront them?

8. Labour unions have long faced challenges in capitalist societies. Some people would argue that today, union leaders and members face new challenges, perhaps more formidable than those of the past. Identify and discuss some of the new challenges that confront the labour movement in this country.

recommended readings

Randy Hodson. 2001. *Dignity at Work*. Cambridge: Cambridge University Press.
Based on an examination of 109 organizational ethnographies, Hodson sensitively highlights the ways in which workers search for dignity and self-worth on the job.

Meg Luxton and June Corman. 2001. *Getting by in Hard Times: Gendered Labour at Home and on the Job*. Toronto: University of Toronto Press.
In this study, based on a series of interviews with women and men in families having one member employed at Stelco's manufacturing plant in Hamilton, Ontario, the authors demonstrate how working families are coping in the face of the economic restructuring that began in the 1980s.

Norene J. Pupo and Mark P. Thomas, eds. 2010. *Interrogating the Economy: Restructuring Work in the 21st Century*. Toronto: University of Toronto Press.
Pupo and Thomas offer a timely collection of articles based on Canadian research on the many features that define work today in the context of economic and political change.

Richard Sennett. 1998. *The Corrosion of Character: The Personal Consequences of Work in the New Capitalism*. New York: Norton.
This book provides a meaningful, eloquent critique by one of America's finest sociologists of the consequences of the new flexible workplace on individual lives and moral identity.

Rachel Sherman. 2007. *Class Acts: Service and Inequality in Luxury Hotels*. Berkeley: University of California Press.
This is an engaging ethnographic study of the invisible and semi-visible workers in two luxury hotels that makes a sobering comment on class relations and the normalization of inequality in the service industry in the United States.

Stuart Tannock. 2001. *Youth at Work: The Unionized Fast-Food and Grocery Workplace*. Philadelphia: Temple University Press.
In an excellent, engaging study of youth at work, Tannock gives voice to young people themselves, their experiences and concerns, while offering a rigorous critique of the low-end service economy today.

recommended websites

Canadian Auto Workers (CAW)

www.caw.ca

The CAW is the largest private-sector union in Canada. For most of its history, the union represented only autoworkers. Since the 1980s, however, the CAW has broadened its membership to include fishers, fast-food workers, Starbucks employees, health-care workers, and thousands of other workers in many different sectors.

Canadian Centre for Occupational Health and Safety (CCOHS)

www.ccohs.ca

The CCOHS, based in Hamilton, Ontario, promotes a safe and healthy working environment by providing information and advice about occupational health and safety issues.

Canadian Centre for Policy Alternatives (CCPA)

www.policyalternatives.ca

The CCPA offers an alternative to the message that we have no choice about the policies that affect our lives, undertaking and promoting research on issues of social and economic justice.

Centre for Labour Management Relations (CLMR)

www.ryerson.ca/clmr

The CLMR is a new initiative of Ryerson University and the Ted Rogers School of Management. It offers support to faculty researchers and students doing research in the study of work, labour, and union-management relations. Its website is a valuable resource for anyone interested in the latest developments in the study of paid employment.

International Labour Organization (ILO)

www.ilo.org

The ILO was founded in 1919 and is now an agency of the United Nations. Its mandate is to promote and realize standards, fundamental principles, and rights at work.

Labour/Le Travail

www.mun.ca/cclh/llt

Labour/Le Travail is the leading academic journal for labour studies in Canada. In operation since 1976, the journal publishes historical and contemporary articles on all aspects of work in Canada.

LabourStart

www.labourstart.org

LabourStart is a web-based news organization that provides up-to-the-minute information on a wide variety of labour-related issues and developments around the globe. Visitors to this site can find anything from job advertisements in Australia, to the latest information on strikes in the United Kingdom, to reviews of recently published books.

No Sweat

www.nosweat.org.uk

No Sweat is a UK-based activist organization that fights sweatshops around the world. It stands for a living wage, safe working conditions, and independent trade unions. No Sweat is an open, broad-based campaign that aligns with anti-capitalist protest movements and the international workers' movement.

chapter 16

Health Issues

Juanne Clarke

In this chapter, you will:

▶ See how health, illness, and disease are distinct in the sociology of health, illness, and medicine.

▶ Learn that health, illness, disease, and death are integrally related to social inequality throughout the world through a number of intermeshing levels.

▶ Examine medicare as a system that embodies five principles: portability, universality, comprehensive coverage, public administration, and accessibility.

▶ Consider how privatization is increasing in the Canadian medical system.

▶ Learn that medicalization is a powerful cultural force.

▶ Discover that there are significant problems in the medical profession.

Introduction

Canada is a part of a cross-national health research initiative sponsored by the World Health Organization. Every four years since 1990, a survey called Health Behaviour in School Aged Children (HBSC) has been carried out. It includes answers to questions about the health-related behaviours and attitudes of more than 7000 Canadian students from five grades (6, 7, 8, 9, and 10). Some of you may have been in the survey at one point or another. Taking a **social determinants** approach, the survey explains health in the context of social issues such as socio-economic status, gender, and health or risk-taking behaviours—for instance, cigarette-smoking, unprotected sex, and alcohol consumption. You might not be surprised to learn that while most students said that their emotional health was good, between 20 and 30 per cent said they had some form of emotional problem. Girls were more likely to report depression and headaches increasingly as they aged to Grade 10. Boys and girls reported similar levels of irritability and backache. Emotional suffering seemed to be at its highest level in Grade 7, which usually occurs at a time of transition to middle school. Young people who got along better with their parents and had higher socio-economic status withstood this major transition better than those who did not. As this brief illustration demonstrates, health is linked to social, economic, and school policies (among other social variables).

This chapter is an investigation of health and medical issues from the perspective of sociology. What do you think are some significant health issues facing you today as a student, as a young man or woman, as a Canadian? Do you think immediately of HIV/AIDS, alcoholism, cancer, or heart disease? Or do you think of the medical care system and topics in the news such as long waiting lists for emergency service, the apparent lack of available physicians, or homecare services? This chapter will introduce you to some sociological perspectives on topics such as these that are related to the sociology of health, illness, and medicine.

Health is linked inextricably to the social order. Its very definition, its multitudinous causes, and its consequences all are social. What is considered healthy in one culture or society may not necessarily be considered healthy in another. Rates, as well as understandings, of sickness and death vary across time and place. Social classes differ in their definitions of good health. What a woman considers health may be different from a man's definition of health. Moreover, class and gender differences lead to varying levels of health and sickness and different rates of death.

The first part of this chapter will examine several health issues: the changing health of Canadians over the nineteenth and twentieth centuries; HIV/AIDS; environmental health problems in different parts of Canada, including the Walkerton water crisis; social inequality; **social capital** and health; the sense of coherence; obesity and eating disorders; and health among Aboriginal Canadians.

The sociology of health is not only about health, illness, and disease but also about systems of diagnosis, prognostication, and treatment. Conventional modern medicine—sometimes called **allopathic medicine** because it treats by means of opposites, such as cutting out or killing germs, bacteria, or other disease processes through medications, surgery, or radiation—is taken for granted in much of the Western world. However, naturopathic (treatment through 'natural' remedies and procedures, such as herbs or massage), chiropractic (treatment through spinal adjustment), and homeopathic medicine (treatment with similars) all are examples of CAM, or *complementary and alternative medicines*, and are of increasing importance in the Western world. The second part of this

In the First Person

My initial intention at university was to major in psychology, but I was more intrigued by my first-year course in sociology. Having grown up in one of the suburbs of Toronto, I was naive about the diversity of values and the impact of social structures on people's lives. In particular, I was fascinated and 'empowered' by the sociological perspective on gender differences and was influenced by early papers on pain and culture by Irving Zola and by Ivan Illich's *Limits to Medicine* (1976). Because a number of people in my family suffered with cancer, I have focused much of my research and writing on the social aspects of this disease, including a personal account of childhood cancer (*Finding Strength*, 1999) written with and about my daughter Lauren. At the present time, my research is largely related to the media portrayal of illness and medicine.

—Juanne Clarke

chapter will investigate some of the most important trends and social policy issues in the area of medical sociology, including **medicalization**, the future of the health-care system, and **privatization**.

Theoretical Perspectives

Four theoretical paradigms may be considered the most significant approaches to understanding health and medicine sociologically: structural functionalism, conflict theory, interpretive theory, and feminism/anti-racism. We will discuss texts from these perspectives.

STRUCTURAL FUNCTIONALISM

From the structural functionalist perspective, health is necessary for the smooth running of the social system. In a stable society, all institutional forces work together to create and maintain good health for the population. Your university or college assumes your good health as it organizes its courses and exams—you probably have to get a letter from a doctor for exemption from writing a test or an exam. The smooth functioning of societies depends on the good health of their members. For instance, societies are organized to support a population up to an average **life expectancy** and at a given level of health and ability. This normative standard of health and age at death are reinforced by political, economic, cultural, and educational policy.

Assertions about the interrelationships among institutions all fit within the structural functionalist theoretical perspective. A classic statement of this perspective is found in the work of Talcott Parsons (1951), in particular in his concept of the **sick role**. The sick role is to be thought of as a special position in society. It exists to prevent sickness from disrupting the 'ongoingness' of social life. The sick role also provides a way of institutionalizing what might otherwise become a form of deviant behaviour. It does this by articulating certain rights for those who claim sickness in a society as long as they fulfill certain duties.

There are two rights and two duties for those who want to claim sickness and engage in the sick role. The rights include exemption from normal social roles and freedom from blame or responsibility for the sickness. The duties are to want to get well and to seek and cooperate with technically competent help. However, that these theoretically derived ideas do not always have empirical support is evident in a number of ways. For example, it is well known that the right to be exempt from the performance of social roles depends in part on the nature of the sickness. A hangover, for instance, may not be considered a good enough reason to claim the sick role when proffering an excuse for an exam exemption. There is also a great deal of evidence that many people with AIDS were seen as culpable in the early days of the disease in North America—in fact, it was called the 'gay plague' by some (Altman 1986). And with respect to duties, not everyone is expected to want to get well. Indeed, those with a chronic disease such as diabetes are expected to accept their condition and learn to live with it. Parsons assumed the dominance of allopathic medicine in his statement that a sick person was to get technically competent help. Today, however, many people believe that the best help may not always come from allopathic medicine even though it is the state-supported type of medical care. Indeed, a substantial minority—between one-fifth and one-quarter of Canadians—use alternative health-care providers (Statistics Canada 2006a, 161).

> ## TIME to REFLECT
> Why do you think a growing number of people choose to go to alternative health-care providers? Use sociological reasoning.

CONFLICT THEORY

From the perspective of conflict theory, health and ill-health result from inequitable and oppressive economic conditions. The primary focus of analysis is the distribution of health and illness across the social structure. Questions driving this perspective include: Are the poor more likely to get sick? Is the **mortality rate** (the frequency of death per a specified number of people over a particular period of time) among the poor higher than among the rich? Does racism affect the **morbidity** (sickness) **rate** (the frequency of sickness per a specified number of people over a particular period of time)? In this perspective, health is seen as a good that is inequitably located in society.

A classic statement of this position is found in the work of Friedrich Engels, who often wrote with Karl Marx. In his book *The Condition of the Working Class in England* (1994 [1845]), Engels demonstrates the negative health consequences of early capitalism. He describes how the development of capitalism advanced mechanism in agriculture and forced farm workers off the land and into the cities to survive. Capitalists in the cities sought profit regardless of the cost to the well-being of the workers. Owners maintained low costs through poor wages and long hours of backbreaking labour in filthy and noisy working conditions. Even children worked in these unhealthy circumstances.

As a consequence, poor labourers and their families lived exceedingly rough lives in shelters that offered little or no privacy, cleanliness, or quiet. They had very little money for food, and the quality of the foodstuffs available in the cities was poor. The slum-like living conditions were perfect breeding grounds for all sorts of diseases, and because of the high density of living quarters, the lack of facilities for toileting and washing, and the frequent lack of clean drinking water, the morbidity and mortality rates in the slums were very high. Infectious diseases such as tuberculosis (TB), typhoid, scrofula, and influenza spread quickly and with dire results through these close quarters and malnourished populations.

Epidemics were almost common in nineteenth-century industrial cities, where overcrowding, overflowing cesspits, garbage piled all around, and unsafe water were the norm. It was only after new discoveries in bacteriology, when it became clear that many of the worst diseases were spread by bacteria and viruses in the water, air, and food, that governments enacted public health measures. These new prevention policies included sewage disposal, garbage removal, cleaned and filtered drinking water, and hygienic handling of food. The death rates began to abate (Crompton 2000).

Conflict theory has also been given a feminist emphasis, as in the work of Hilary Graham. In *Women, Health and the Family* (1984), Graham documents how inequality affects the various types of home health-care work done by women in order to protect the good health of their families. In particular, she articulates four different components of women's home health-care work: (1) maintaining a clean, comfortable home with an adequate, safe, and balanced diet as well as supportive social and emotional familial relations; (2) nursing family members when they feel ill or are debilitated; (3) teaching family members about health and hygiene; and (4) liaising with outsiders regarding the health-care needs of family members, such as taking children or a partner to the doctor, clinic, hospital, or dentist.

TIME to REFLECT

There are new infectious diseases occurring around the world today. The H1N1 flu is an example. Discuss contemporary methods of containing such pandemics.

INTERPRETIVE THEORIES

Interpretation and meaning are the hallmarks of sociology within **interpretive theories**. What is the meaning, for example, of anorexia and bulimia? Are they medical conditions? Are they the result of a moral choice? Or could they be considered 'socio-somatic' conditions—that is, caused by society (Currie 1988)? Various authors have attributed them to women's 'hunger strike' against their contradictory positions, against culturally prescribed images, and against lack of opportunities in contemporary society. They have been conceptualized as a means 'through which women, both unconsciously and consciously, protest the social conditions of womanhood' (Currie 1988, 208).

Stigma is often attached to people with HIV/AIDS, lung cancer, depression, inflammatory bowel disease, or fibromyalgia. Some people think these diseases have connotations of morality or immorality. Good health is even associated with being a good person. The following research addresses some of the paradoxes of stigma in respect to Asperger's syndrome, often considered a mild form of autism. The study is based on a qualitative analysis of the blogs of people who self-identify as having Asperger's (AS) and parents or caregivers of those thought to have the disorder (Clarke and Van Amerom 2008). The findings indicate that these two groups held not only different but even oppositional views regarding AS. People who self-identified as having Asperger's rejected the popular but denigrating understanding of AS. They called themselves Aspies. They called others NTs or neurotypicals. They said they were proud of who they were and of the way they thought. They said that they felt the major problems they faced were not due to the 'disorder' or the 'limitations' they suffered because of AS but resulted from the stigma of AS and the way that others perceived and acted toward them. Parents and caregivers, on the other hand, expressed worry about their children's problems in schooling and in their social lives. They tended to accept the dominant and pathologizing view of AS, while the bloggers who self-identified as having AS expressed pride and mutual solidarity. The paper demonstrates the value of an 'up close and personal' investigation of the world views of people, particularly those who are vulnerable to stigmatization or marginalization.

FEMINISM/ANTI-RACISM

Feminist and anti-racist health sociology recognizes the centrality of gender and racialization to social life. Feminist/anti-racist health sociology investigates whether, how, and why people who are racialized and men and women have different health and illness profiles, as well as different causes and average ages of death. It also includes consideration of such things as **ethnicity**, sexual preference, gender identity, and ability/disability as fundamental characteristics of social actors. These axes of inequality, therefore, are central issues to be included in designing research, uncovering social injustice, and planning and making social change.

Women's health has been a central issue and in many ways a major impetus for the recent women's movement. The book *Our Bodies, Our Selves*, when first published in 1971, became a major rallying document for women. Translated into many languages, and more recently revised and updated, this book offers a radical critique of medical practice and medicalization and provides women's views of their own health, sickness, and bodies.

Another example of work within the feminist paradigm is Anne Kasper and Susan Ferguson's *Breast Cancer: Society Shapes an Epidemic* (2000), which suggests that among the reasons for the growing **incidence** (number of new cases in a year) and **prevalence** (number of cases within a given population) of the disease is that it is largely a women's disease and is therefore not given the serious and systematic research attention that it would have received had it been primarily a male disease. Contributions to this book by scholars and practitioners from a wide variety of fields examine the social and political contexts of breast cancer as a social problem, arguing that gender, politics, social class, race, and ethnicity have affected the type of research that is done, the types of treatments that have become dominant, the rates of growth in the morbidity and mortality of breast cancer, and even the ways in which the disease is experienced by women. Indeed, they suggest that one of the reasons for the continuance of the epidemic is that it is not only primarily a women's disease but that it is located in their breasts.

Kasper and Ferguson's collection provides a thought-provoking look at one of the major causes of worry, sickness, and death among women in Canada and the United States. Despite the fact that both heart disease and lung cancer are more frequent causes of death for Canadian women, women in Canada fear breast cancer more and even think of their breasts

GLOBAL ISSUES

Warfare and Human Health

Human health depends on the complex interaction of manifold social determinants that operate across a number of levels, as portrayed in Figure 16.1. An important part of this picture is international and intra-national conflict. The twentieth century was the most violent century in history. Almost three times the number of people who died in conflict during the previous four centuries died in the twentieth century. This is partly due to the huge numbers massacred in World Wars I and II and partly due to the overall frequency of conflict. The Rwandan genocide in 1994, for instance, resulted in the deaths of approximately 1 million people. The civil war in the Democratic Republic of the Congo killed 7 per cent of the population of that country, and the several-decades-old conflict in Sudan has resulted in 2 million deaths and the displacement of 6 million people. Over time, conflict has increasingly occurred in the poorest countries of the world. By the 1990–2003 time period, low-income countries accounted for more than half of the world's conflicts.

The consequences of conflict for health are many. Wars inevitably result in death, disability, and rape. They also result in the destruction of the infrastructure necessary for everyday living for the masses of people affected. This destruction occurs in food production, storage, and distribution systems. It limits or obviates access to potable water, sewage systems, and electricity, not only for homes but for hospitals. Fundamentals such as roads, houses, schools, and health-care facilities are also affected and frequently destroyed or damaged. Other negative consequences include chronic and acute psychological trauma and distress. The majority of the countries that have experienced war have child death rates that have either stagnated or worsened after the conflicts.

The World Bank has suggested that a civil war reduces the growth of a nation's economy by about 2.2 per cent per year and costs an average of $54 billion for a low-income country. The longer a conflict lasts, the greater the toll on all fronts, from the human to the economic. Violence also sets in motion uncertainty about the future that inevitably has long-term health consequences for people within the countries engaged in the conflict as well as in interacting countries around the world.

as essentially flawed and vulnerable to disease (Robertson 2001). This is undoubtedly related to the enormous mass media attention the disease has received in the past 20 years or so. During this time, first in the United States and then in Canada, powerful lobby groups of women activists founded highly successful breast cancer advocacy coalitions, lobbied governments and corporations, and received substantial increases in the funding levels for research into the disease and its treatment. There is, of course, a painful irony in the fact that the increased attention and financial investment have been coupled with a proliferation of stories in the mass media that have served to increase anxiety and fear of risk of disease among women. Today, more white women are diagnosed, but black women are still more likely to die (breast cancer rates by race and ethnicity: www.cdc.gov/cancer/breast/statistics/race.htm).   King, in *Pink Ribbons, Inc.*, also demonstrates how the focus on consumer activism, especially among white women in the breast cancer movement, 'shaped as it is by an ideology of individualism and an imperative for uncomplicated, snappy marketing slogans, has allowed for the emergence of a preoccupation with early detection to the virtual exclusion of other approaches to fighting the epidemic (e.g., prevention) and a failure to address the barriers, financial or otherwise, to treatment' (King 2006, 117–18). She argues that the corporations involved in breast cancer awareness and fundraising have benefited from their involvement. It is possible that this corporate advantage has been to the detriment of women's health.

The Sociology of Health, Illness, Disease, and Sickness

At the broadest level, sociologists compare within and between societies around the world and over time with respect to the rates of, causes of, and treatments for health and sickness and to rates and causes of death. Here, factors such as wars, famine, drought, epidemics, natural disasters, air and water quality, quantity and quality of foodstuffs, transportation safety, level and type of economic development, technology, available birth control, immunization, pharmaceuticals, medicalization, culture, and political economy all are considered relevant.

At the next level, sociologists examine morbidity and mortality within societies and cultures and compare people of different social class, educational

levels, genders, religiosity, rural/urban locations, occupations, ethnicities, family statuses, and so on. A further level of investigation concerns the way socio-psychological factors, such as level of stress and sense of coherence, are implicated in illness, disease, and sickness.

The next level is an examination of the relationships between various 'lifestyle' behaviours, such as smoking, seat-belt use, alcohol consumption, diet, risk-taking behaviours, sexual activity and protection, drug use, and health. Finally, the existential considerations, including the meaning and the experience of morbidity and mortality to individuals, are studied. Figure 16.1 shows these links, beginning from the person.

In this chapter, we will look at specific and limited topics within each of these levels of analysis,

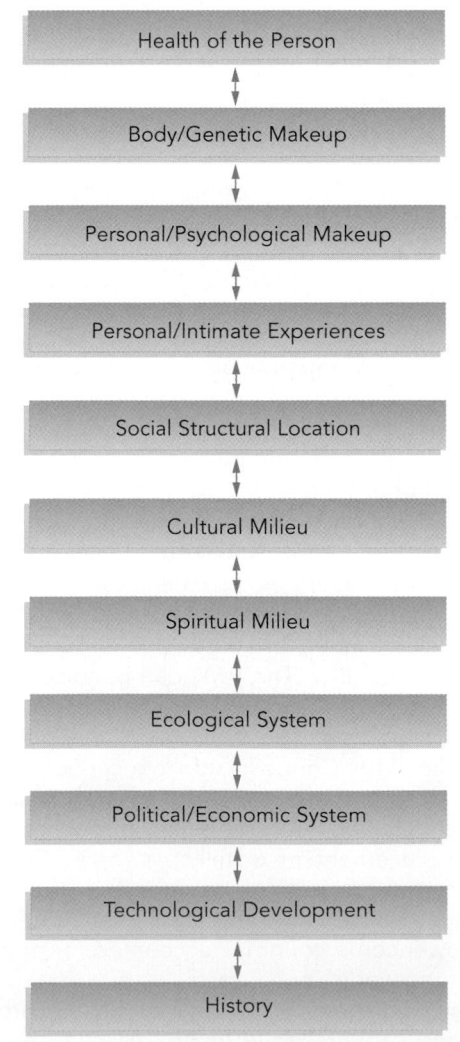

▼ FIGURE 16.1   Components of Health

Health of the Person

Body/Genetic Makeup

Personal/Psychological Makeup

Personal/Intimate Experiences

Social Structural Location

Cultural Milieu

Spiritual Milieu

Ecological System

Political/Economic System

Technological Development

History

beginning with the changing health of Canadians over the nineteenth and twentieth centuries, HIV/AIDS in Canada and around the world, and environmental issues (including the water crisis in Walkerton, Ontario) as examples of first-level concerns. At the second level, we will look at social inequity and social capital, along with Aboriginal health issues. At the third level, the focus will be on sense of coherence. Fourth, we will investigate eating-disordered behaviour and attitudes, including obesity. The fifth and final level will be illustrated by a discussion of popular conceptions of illness as well as an illustration through poetry of the experiences of one mother when her child had cancer.

The contamination of the water supply with *E. coli* in Walkerton, Ontario, resulted in 2500 people falling ill and seven deaths.

(Rick Eglinton/GetStock.com)

You will not be surprised to learn that this separation into levels is artificial and done only for reasons of analytical clarity. In fact, each of the levels influences all the other levels in reciprocal ways.

COMPARATIVE ANALYSES

The Changing Health of Canadians

People generally are living longer and healthier lives today than they did in the past. The increase in health and the decrease in mortality rates over the past 150 years have been substantial. In the nineteenth century, infectious and communicable diseases such as cholera, typhoid, diphtheria, and scarlet fever were responsible for enormous suffering and death for early Canadians. Wound infections and septicemia were frequent results of dangerous and unhygienic working, living, and medical conditions. Puerperal fever killed many women during and after childbirth.

In 1831, the average life expectancy for Canadian men and women was 39.0 years—38.3 for women and 39.8 for men (Clarke 2008, 50). Today, life expectancies are about double this for Canadian men and women. Today's women can expect to live to 83, men to 78.3 (www.statcan.gc.ca/daily-quotidien/100223/dq100223a-eng.htm). What has happened to cause this dramatic shift? You might think first of medical interventions such as antibiotics or immunization. However, the most important causes of the increase in life expectancy are related to public health measures that prevented the spread of disease. These included improved nutrition, better hygiene through sanitation and water purification practices, and advances in birth control. Interventions such as these brought the

average life expectancy to 59 in the 1920s and, largely because of dramatic declines in infant mortality, to 78 in 1990–2 (Crompton 2000, 12).

In the 1920s, the most common causes of death became heart and kidney disease, followed by influenza, bronchitis and pneumonia, and the diseases of early infancy. Widespread use of newly discovered vaccines and antibiotics (vaccines against diphtheria, tetanus, typhoid, and cholera were developed in the late nineteenth and early twentieth century, and antibiotics were introduced in the 1940s and 1950s) made a significant difference in the twentieth century (Crompton 2000, 12). While heart disease remains the most common cause of death, it has declined dramatically over time, probably as a result of lifestyle changes, such as declines in smoking and dietary fat and improvements in exercise, and better medical treatments. Lower infant death rates today have resulted

TABLE 16.1 ▼ Health Indicators, Canada, 2007		
	Males	**Females**
Life expectancy at birth (years)*	78.3	83.0
Infant mortality rate (deaths per 1000 live births)*	5.7	4.8
Babies with low birth weight (%)*	5.4	6.3
Increased body mass index from 1994/1995 to 2006/2007**	31.6%	29.4%
Daily smokers (%)*	15.5	12.9

SOURCES:
*Statistics Canada, 'Canadian vital statistics, birth and death databases', and Appendix 1 (CANSIM Table 102-0512).
**Statistics Canada, www40.statcan.ca/101/cst01/hlth68-eng.htm.

primarily from better nutrition and improved hygiene in pregnancy, secondarily from medical and technological advances. For example, prematurity, a frequent cause of infant death in the past, is now more often prevented through educational programs, prenatal care, and effective management in hospitals.

The incidence of diseases such as measles, scarlet fever, and whooping cough was cut virtually to zero until some people abandoned the vaccines in the 1990s. For a short time, the incidence of these diseases increased, but by the late 1990s it had declined again when public health authorities became aware of the issue and more diligent about universal vaccinations in Canada. However, the safety and acceptability of vaccines is an ongoing issue.

One other important feature of the declines in mortality or gains in life expectancy is the gap between men and women and how this gap has changed over time. From 1920 to 1922, women lived an average of two years longer than men; from 1990 to 1992, they lived an average of six years longer than men. Now women live almost five years longer. Part of the explanation for women's greater benefit from the changes of the twentieth century relates to the decline in maternal mortality over this period. Another part of the explanation has been the greater tendency for men to engage in risk-taking behaviours such as cigarette-smoking and drunk driving.

TIME to REFLECT

Consider the recent introduction of the HPV vaccine for preteen girls. What issues did this raise?

These rates do not take age into account. Potential years of life lost, or PYLL, is a statistical representation of death that does take age into account: the younger the average age of death for a given disease, the greater the number of years of life lost. PYLL allows us to see the years of life lost by disease type, taking 70 years as the cut-off point for age. Here, the importance of suicide and accidents, occurring as they often do among younger people, increases (www.statcan.ca/english/freepub/82-221-XIEI/2004002/hlthstatus/deaths4.htm).

HIV/AIDS

HIV/AIDS is a disease of pandemic proportions around the world today (see Figure 16.2). Many people in North America have come to associate the

disease with certain categories of other people. North Americans tend to think of HIV/AIDS as the male 'gay disease' (Crossley 2002). In an international context, however, HIV/AIDS is almost as common among heterosexual women as among men: in 1997, 59 per cent of those diagnosed internationally were male, and 41 per cent were female; by 2005, 52 per cent were male and 48 per cent female (UNAIDS 2001b, 1). In sub-Saharan Africa, however, more women than men are HIV-positive (this means that they carry the precursor virus but have not yet developed AIDS). In some of the sub-Saharan African countries where the disease has taken its highest toll, teenage girls are five to six times as likely to be infected as boys. Whether because they themselves are diagnosed with the disease or because they are much more likely to act as caregivers to others who are sick, women are substantially more affected by the AIDS epidemic than men in Africa. Already, too, in Africa, more than

TABLE 16.2 ▼ Leading Causes of Death, Canada, 1921–1925 and 2005

Cause of Death[a]	Rate per 100,000
1921–5	
All causes	1030.0
Cardiovascular and renal disease	221.9
Influenza, bronchitis, and pneumonia	141.1
Diseases of early infancy	111.0
Tuberculosis	85.1
Cancer	75.9
Gastritis, duodenitis, enteritis, and colitis	72.2
Accidents	51.5
Communicable diseases	47.1
2005[b]	
All causes	654.4
Cancer (malignant neoplasms)	170.3
Heart disease	121.5
Cerebrovascular disease	32.5
Chronic lower respiratory diseases	25.1
Accidents	25.6
Influenza and pneumonia	13.2
Intentional self-harm (suicide)	10.9
Alzheimer's disease	12.7

[a]Disease categories are not identical over time.
[b]Rates are age-standardized.

SOURCE: Adapted from Statistics Canada, *Canadian Social Trends*, Catalogue no. 11-008, winter 2000, p. 13, and Statistics Canada, CANSIM, Table 102-0552 and Catalogue no. 840209X.

1 million children are living with HIV, and 12.1 million children have been orphaned by the disease. The number of orphans was expected to double by 2010. Children who are orphaned because of AIDS bear the additional burdens of stigma and decreased access to health and education. In turn, they are more susceptible to becoming infected by HIV/AIDS (UNAIDS 2001a, 2). At present, there are estimated to be about 38.6 million people living with HIV, and there were approximately 4.1 million new cases of HIV diagnosed in 2005 (www.unaid.org).

What is this disease that has caused so much havoc over the past 30 years? AIDS (acquired immune deficiency syndrome) is a disease of the immune system that most scientists believe comes from HIV (human immunodeficiency virus). It is spread from person to person through bodily fluids such as semen, blood, vaginal fluid, and breast milk. Sexual intercourse, needle-sharing, and mother-to-fetus transmission are currently the most common means of transmission. In the end, though, it is not HIV/AIDS per se that causes death but rather such diseases as cancer, pneumonia, and tuberculosis that the vulnerable immune system is too weak to resist.

Even though the incidence and prevalence of HIV/AIDS is much lower here, Canada is not exempt from the epidemic. By 2007, this country had experienced a cumulative total of 20,746 AIDS cases. Until the mid-1990s, women made up only 7 per cent of the total AIDS cases in Canada. By 2007, they accounted for about 20 per cent of the new cases each year. Overall, however, the increase in disease incidence reached its peak in 1993, and diagnoses have been gradually declining since then (www.avert.org/Canada-aids.htm).

The most common means of transmission in Canada has been among men who have sex with men (MSM)—75.6 per cent of cases have been attributed to MSM (this category also includes other high-risk behaviour) and an additional 9.9 per cent to men who inject drugs. Among women, 59.6 per cent of cases are attributed to heterosexual contact, 21.9 per cent to injection drug use, and 9.6 per cent to receipt of blood and blood products (UNAIDS 2002, 6).

HIV/AIDS is overrepresented among Aboriginal people and black Canadians. Aboriginal and black people constituted 3.3 per cent and 2.2 per cent of the population, respectively, but 5.4 per cent and 15.0 per cent of the AIDS cases with known ethnicities in

▼ **FIGURE 16.2**   Estimated Adult and Child Deaths Due to AIDS, 2008

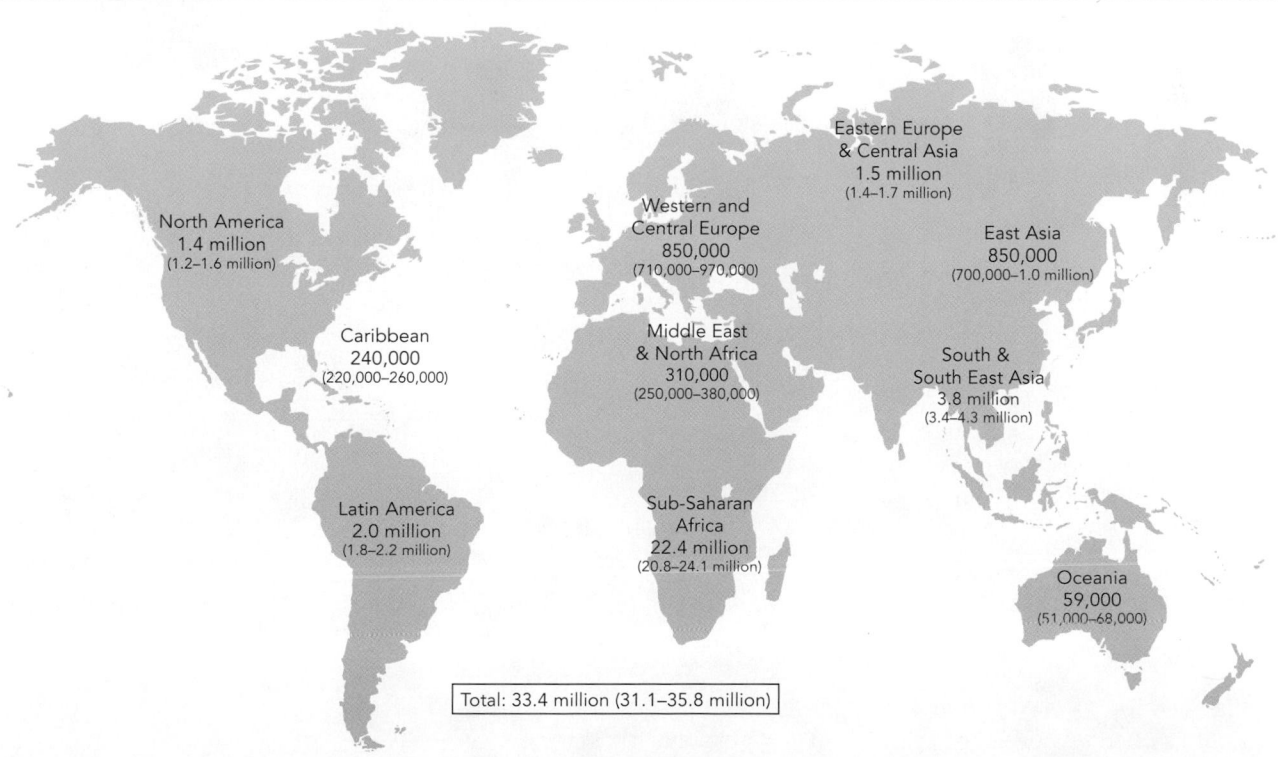

SOURCE: From www.unaids.org/en/KnowledgeCentre/HIVData/EpiUpdate/EpiUpdArchive/2009/default.asp. Reprinted with permission from UNAIDS.

2007 (www.avert.org/Canada-aids.htm). Evidence suggests that the major routes for transmission differ in these two groups. Injecting drugs appear to be the most common cause for Aboriginals, while a large majority of black Canadians appear to have acquired the disease through heterosexual contact (ibid.).

ENVIRONMENTAL ISSUES ACROSS THE COUNTRY

At the broadest level, our health is dependent on the natural environment and the ways in which we as members of societies maintain the health of the natural environment. Water, air, and soil are among the basic and determining factors. Do you know how safe your drinking water is? Do you know how it is purified or how much is available and for how long? I must admit that I don't. Is there enough water for your generation, for your children and their children, and so on? Good and healthy land and soil are prerequisites to health because they are the basis for the production of foodstuffs. However, there are threats to the environment all across Canada.

Alberta Oil Sands

In Alberta, the extraction of oil from the oil sands is an ongoing challenge to the health of the people and the environment. While the amount of oil is immense (there are said to be 175 billion barrels of proven oil, second only to the 260 billion barrels in Saudi Arabia), the processes needed to access this oil essentially require the strip mining and consequent removal of the sand to extract the infused oil. This affects soil organisms and water flow in the Athabasca River and water levels in the river and, downstream, in Lake Athabasca (huge amounts of water are required to extract the oil, but the water cannot be returned to the river because of its toxicity). In addition, mining of the oil sands displaces animals, increases erosion, and decreases carbon sequestration. Moreover, the energy required to access the oil is already

(Christopher Morris/Corbis)

A heroin addict in Vancouver's Downtown Eastside, a neighbourhood notorious for its drug addicts and dealers. The area is also home to North America's first safe injection site, set up to counter the spread of HIV and AIDS among the neighbourhood's drug users.

causing public health problems, air pollution, and global warming. Some argue that the damage will be repaired naturally. Others say that 30 years after mining the oil sands, there is still no reforestation.

The Walkerton Water Crisis

Water is fundamental to our good health. Walkerton, a small town nestled among rolling hills beside the Saugeen River in southwestern Ontario, was, at least from the outside, a perfect picture-postcard town—until May 2000. That was when the drinking water became polluted with a virulent strain of *E. coli* bacteria that was ingested repeatedly by the townspeople, leading to seven deaths and to illness in approximately 2300 to 2500 others. The 2002 report of the Walkerton Inquiry, written by Justice Dennis R. O'Connor, detailed the causes of these tragic deaths and sicknesses and suggested that individual behaviours, cultural values, and social structural arrangements were all responsible for the suffering of the people of Walkerton.

Let us examine these causes. First, what were the causes at the level of individuals? Two brothers, Stan and Frank Koebel, were particularly implicated in the tragedy. According to the O'Connor report, Stan Koebel neglected several essential aspects of his job with the Public Utilities Commission (PUC) as the general manager responsible for water chlorination and safety. In addition, he repeatedly lied to officers in the health unit and, even after many people had taken sick, reported that the water was 'okay'. Some of his actions were the result of his lack of understanding of the health consequences of his work. Some were due to the fact that the norms in the PUC culture had been lax for at least 20 years, before Koebel was hired.

As well, there were structural causes for this tragedy. These included the fact that Stan and Frank Koebel were certified on the basis of their experience ('grandfathered') and were not required to take any courses or pass any examinations for continued certification. Another structural deficiency related to the fact that the Ministry of the Environment failed in its responsibility to regulate or to enforce regulations pertaining to the construction and operation of municipal water systems. Budget reductions at the provincial level were also implicated because they led to the privatization of lab testing and to failure to regulate the reporting responsibilities (to the Ministry of the Environment and the local medical officer of health) of private labs whenever unsafe water was detected and to ensure that proactive water quality interventions were made.

Sydney Tar Ponds

On the other side of the country, the 'tar ponds' at Sydney, Nova Scotia, contain dangerously contaminated soil and sediment as a result of decades of steel and coke production. The process of heating coal to produce coke for the manufacture of steel produces toxic chemicals such as benzene, kerosene, and naphthalene. These chemicals have accumulated in the harbour in Sydney. 'More than 80 years of this type of coke-oven operation left the ground water and surface water in the area seriously contaminated with arsenic, lead and other toxins.' There were about 700,000 tonnes of chemical waste and raw sewage there (www.cbc.tarponds). In 2005, after years of public lobbying, a $400-million cleanup was announced.

Clearly, in the cases of both the oil sands and the tar ponds, economic gains have been privileged over the possibilities of long-term health deficits. This could be seen as an example of hope (i.e., that the damage done will be insignificant or that scientists and others will find a solution to the resulting problems before they cause too much damage) or of greed (i.e., a focus on profit and economic well-being regardless of threats to people or environments along the way).

INTRA-SOCIETAL ANALYSES

The Social Determinants of Health

The degree of economic inequality has been increasing in Canada, especially recently, in the last decade of the twentieth century and into the first decade of the twenty-first century. A report for the Organisation for Economic Co-operation and Development (OECD) pointed out that Canada was, in fact, one of the two (out of 30) wealthy nations characterized by the largest growth in income inequality in the 1990s and 2000s (Mikkonen and Raphael 2010). Consider the following statistics. In the years 1984 to 2005, 30 per cent of Canadians had no net worth and over this time became more indebted. By comparison, the net worth of the top 10 per cent increased over this period to $1.2 million (an increase of $659,000 in constant dollars) (ibid.). Many people are aware of the widening gap between the rich and the poor, the 'haves' and the 'have-nots' in Canada. What might be less familiar to you is that there is a direct link between income inequality and health. A classic illustration of this relationship can be found in the Whitehall studies, which followed the health of more than 10,000 British civil servants for nearly 20 years and found that both the experience of well-being and a decline in mortality rates were associated with upward mobility in the occupational hierarchy of the British civil service (Marmot et al. 1978, 1991). Positive health benefits were found *in each increase in rank*. Remember, too, that this was a study of the civil service. Thus, all of the jobs under scrutiny were white-collar, office jobs with 'adequate' incomes. It is interesting that this finding held true even among people who engaged in health-threatening behaviours such as smoking. Thus, for instance, 'researchers found that top people who smoked were much less likely to die of smoking-related causes' than those nearer the bottom who did not smoke (National Council of Welfare 2001–2, 5).

Poverty exacerbates health problems from birth onward. In 2007, the child poverty rate in Canada was 15 per cent (Report Card on Child and Family Poverty in Canada 2009).

Forty-three per cent of children of visible-minority parents and 52 per cent of Aboriginal children live in poverty (www.cich.ca). Poor women are more likely to bear low-birth-weight babies. Low birth weight is associated with myriad negative health, disability, learning, and behavioural effects. Children born in the poorest neighbourhoods in Canada (the lowest 20 per cent) live shorter lives, by two to 5.5 years. They also tend to spend more of these shorter lives with some degree of disability. Children at the lower end of the social hierarchy have a greater variety of health and developmental deficits than those higher up on the socio-economic status ladder. It is also important to notice that these results, like the Whitehall study results, are situated in the context of a nationally funded medical care system. It is also interesting to note that in a global context, the rate of childhood

HUMAN DIVERSITY

Health and Aboriginal Canadians

One of the most troubling health issues facing Canadians today is the relatively poor health and shorter lives of Aboriginal Canadians. To contextualize this from a social determinants perspective, note that the average income of Aboriginal men and women, at $21,958 and $16,529, respectively, was 58 per cent of the average income of non-Aboriginal men and 72 per cent that of non-Aboriginal women in 2001 (Mikkonen and Raphael 2010). In the Northwest Territories, where more than half of the population is Aboriginal, the life expectancy is more than five years less for men and women, at 75 and 70, respectively (Helwig 2000, 681). The infant mortality rate, one of the most sensitive indicators of the overall health of a people, is 1.5 times higher among First Nations people at eight deaths per thousand as compared to 5.5 for all Canadians (Adelson 2005). While this rate reflects a great improvement over the past 30 years, there is still reason for concern. Moreover, the major causes of death among First Nations people include diseases of development and impoverishment, such as injury, poisoning, and respiratory disease, along with the causes more common among other Canadians such as heart disease and cancer. Suicide and self-inflicted injury are the chief causes of death among those aged 10 to 44. This is followed by motor vehicle accidents, drowning, and homicide. Violence, including physical abuse, sexual abuse, and rape, is also a significant problem. Aboriginal people are three to five times more likely to suffer from diabetes (non-insulin-dependent diabetes mellitus).

Tuberculosis is eight to 10 times more frequent. HIV/AIDS is also increasing dramatically among this population. The rate of infection grew from 1.0 per cent in 1990 to 15.0 per cent in 2007 (www.avert.org/canada-aids.htm).

First Nations people also do very poorly with respect to other social determinants of health, including employment rates, food security, education levels, and crowded housing (Mikkonen and Raphael 2010). This results from the fact that most Aboriginal Canadians live in the rural north of Canada, where there are fewer employment opportunities, as well as from 'the complex interplay of job market discrimination, lack of education, cultural genocide, and loss of land and sovereignty that affects employment status, and, ultimately, the degree of poverty faced by those who are caught in a circle of disadvantage' (ibid.). Aboriginal Canadians also are more likely to live in inadequate and crowded housing. For instance:

1. Homes are twice as likely to need repair.

2. Houses are 90 times more likely (9.4 per cent as compared with 0.1 per cent) to lack piped water.

3. Homes are five times more likely to lack bathroom facilities (3.2 per cent as compared with 0.6 per cent)

4. Households, on average, include four rather than three people, which is the Canadian average today.

The continuation of these inequities in the lives of Aboriginal Canadian reflects their historical position in Canadian governmental policy.

poverty in Canada is high (www.cich.ca)—*higher* than in other developed nations such as Sweden, where the incidence of childhood poverty is 3 per cent; the Netherlands, where it is 6 per cent; France and Germany, 7 per cent; and the United Kingdom, 10 per cent (www.cich.ca). Men and women in neighbourhoods that differ by income have different life expectancies: men and women in the poorest neighbourhoods can expect to live 74.7 and 80.9 years, respectively. Men and women in the richest neighbourhoods can expect to live 79 and 82.8 years, respectively (Wilkins 2007).

Even though there are significant links between income inequality and both ill-health and death, Canadian health policy continues to involve substantial investments in the health-care system rather than in community-level interventions such as a guaranteed annual wage, job creation, a national daycare program, or proactive prenatal care for low-income mothers.

The ways in which economic inequality affects health outcomes are complex and contested. Certainly, material needs are part of the explanation. For example, differential ability to pay for ample healthy foodstuffs and for clean, quiet, and temperature-appropriate living quarters is a part of the explanation. What the Whitehall studies suggest, however, is that there is likely something beyond material differences contributing to the explanation. One finding of the Whitehall studies was that while all levels of the civil service had elevated stress levels while at work, blood pressure levels of the senior administrators dropped when they went home; in contrast, the stress levels remained high for those lower in the hierarchy (National Council of Welfare 2001–2). Socio-psychological issues related to perceptions of well-being appear to be implicated in inequities in health outcomes. In other words, people also feel stress and appear to have consequent health difficulties as a result of comparing their socio-economic positions (negatively) to those of others (Marmot et al. 1978; 1991).

Social Capital

It is clear from all types of research done today and in the past, and in this and in other societies, that social status and health are related. Much of this analysis compares individuals who differ in health and social status. However, when the level of analysis moves from the individual to the society as a whole, the link between status and health remains; societies with greater degrees of inequality have poorer overall health outcomes regardless of their overall wealth.

This interesting paradox needs clarification. Explanations for the individual-level correlation have suggested that people with higher incomes, higher occupational prestige scores, and higher educational levels are more able to prevent ill health through eating and drinking wisely, avoiding serious threats to health such as cigarette-smoking and excessive alcohol consumption, and engaging in prescribed early detection such as mammograms and PSA tests (for prostate cancer). When those in these higher levels are sick, they can get medical attention immediately and take advantage of the most sophisticated and effective new treatments. They are also, as the Whitehall studies intimated, able to maintain a sense of well-being through various socio-psychological processes.

Why would the degree of inequality in a society be important in predicting health and illness outcomes? A recently developed theoretical explanation is that it is the degree of social cohesion, social capital, or trust that is the link between inequity and health (Mustard 1999). A society characterized by inequity is one in which 'there is a pronounced status order' (Veenstra 2001, 74). As people compare themselves to one another, it is possible—indeed, likely—that those lower in the status hierarchy 'will feel this shortcoming quite strongly, given the width of the gap, and consequently will suffer poorer health' (2001, 75). This may result from 'damaging emotions such as anxiety and arousal, feelings of inferiority and low self-esteem, shame and embarrassment, and recognition of the need to compete to acquire resources that cannot be gained by any other means' (2001, 75).

A number of researchers have suggested that societies with high degrees of inequality are also low in *social cohesion* (or *social capital*), and it is social cohesion that mediates between social status and illness. Social cohesion is thought to be evident in societies to the extent that people are involved in public life and volunteer to work together for the good of the whole. A society with little social cohesion might, for instance, be dominated by market values and characterized by transactions in the interest of profit. Current social policies in Canada that favour market dominance over state intervention exacerbate the degree of inequity in society.

SOCIO-PSYCHOLOGICAL FACTORS: THE SENSE OF COHERENCE

The *sense of coherence* is a socio-psychological concept articulated first by Aaron Antonovsky (1979). Rather than asking what makes people sick, Antonovsky wondered what kept people healthy. Having thought about this and reviewed available research, he defined sense of coherence as an orientation to the world and to one's place in it that leads a person to a long-lasting and dynamic feeling of confidence that 'things will work out' because (1) life is basically comprehensible, understandable, and predictable; (2) there are sufficient resources for the individual to be able to cope with whatever circumstances arise; and (3) life makes sense or has meaning. These three components of the sense of coherence enable individuals to manage life experience in a positive manner and to establish a basis for resisting disease and handling suffering. There is a continuing link between the sense of coherence and health outcomes (Kivimaki et al. 2000; Suominen et al. 2001).

TIME to REFLECT

What encourages or discourages your involvement in extracurricular activities (which might enhance your social cohesion while at university)?

LIFESTYLE BEHAVIOURS: OBESITY AND EATING DISORDERS

A number of recent articles in the *Canadian Medical Association Journal*, and indeed in the mass media, have reported on seemingly opposite health concerns: obesity, on the one hand, and eating disorders such as anorexia and bulimia, on the other. It seems that both are increasing among children and adults and that both herald other serious medical problems. Why are so many young people facing such problematic and contradictory issues related to food, body image, and control of eating?

Obesity is now pandemic (Katzmarzyk 2002), affecting hundreds of millions of people around the world, especially in rich countries, where between 10 and 20 per cent of the population are obese. *Obesity* is defined as an excess of fatty or adipose tissue; it 'results from un-balanced energy budgets. An overweight person consumes food energy in excess of expenditure and stores the surplus in body fat' (Obesity Canada 2001). Excess body fat is associated with higher rates of premature morbidity and death from diseases such as coronary heart disease, stroke, type 2 diabetes mellitus, gallbladder disease, and some cancers. The rate of growth from 1981 to 1996 is estimated at 92 per cent in boys and 57 per cent in girls (Obesity Canada 2001). Our sedentary lifestyle,

typified by television viewing and computer games, along with the ubiquitous 'fast' food, has played a role in this growing health concern. Children who watch four or more hours of television a day have higher body mass indices and thicker skin folds than those who watch fewer than two hours per day. In addition, caloric intake is positively associated with television viewing (Obesity Canada 2001). By 2004, 26 per cent of children and adolescents ages 2 to 17 were overweight or obese compared with 15 per cent in 1978–9 (Statistics Canada 2006a, 165). These percentages amount to almost half a million young people between 12 and 17 in Canada who are overweight or obese (Statistics Canada, CANSIM Table 105-0501 and Catalogue no. 82-221-X).

While some children are gaining too much weight, others are losing or trying to lose too much. For example, in a survey of 1739 adolescent females, 23 per cent were dieting to lose weight (Jones et al. 2001, 549). Binge eating was reported by 15 per cent, self-induced vomiting by 8.2 per cent, and the use of diet pills by 2.4 per cent. Disordered attitudes towards eating were found in more than 27 per cent of the young women surveyed. Consistent with other studies, disordered eating behaviours and attitudes seemed to increase gradually during adolescence and were more common among girls with higher body mass indices (BMIs), an international standard for measuring overweight and obesity (Jones et al. 2001, 549–50).

TIME to REFLECT

What other social factors are related to disordered eating behaviours and attitudes among adolescents?

THE EXISTENTIAL LEVEL

How do you experience illness? What sorts of illnesses have you had? Have you always gone to the doctor when you felt ill? One compilation of popular notions of illness includes illness as choice, illness as despair, illness as secondary gain, illness as a message of the body, illness as communication, illness as metaphor, illness as statistical infrequency, and illness as sexual politics (Clarke 2008).

Illness as choice refers to the notion that we choose when to become sick, what type of illness we will have, and so on. In other words, illness episodes are viewed as a reflection of the deep tie between the mind and the body. That illness is a sort of despair is a related notion. Primarily, however, the idea is that illness results from emotional misery. The notion of secondary gain emphasizes the idea that people sometimes benefit from illness—for instance, an ill student might not be able to write an exam for which he or she also happens to be unprepared. Closely related to this notion is the philosophy of illness that suggests that physical symptoms are a means through which the body communicates a message to the consciousness. And related to this, in turn, is the idea that the symptoms are meant to reflect a particular message, a particular set of unmet needs. For example, a cold, with its running nose and eyes, may be said to represent a frustrated desire to cry.

Susan Sontag (1978) has described some of the metaphors attached to diseases such as tuberculosis, AIDS, and cancer. One illustration of disease metaphor is the idea of a disease as an enemy and the subsequent need for a war against it. Illness as statistical infrequency, in contrast, is simply a numerical definition that names as 'illness' a bodily functioning or symptom that is infrequent in the population.

Finally, the idea that illness reflects gender politics is related to the patriarchy of the medical profession and its consequent tendency to see women's bodies as basically flawed and women's behaviours as more likely to be pathological (for example, meriting psychiatric diagnosis) than those of men (see Clarke 2008).

The existential level of analysis also now includes a new way of sharing research findings through art and poetry. The following prose poem was written to express the feelings of a mother whose child had cancer and had symptoms (which immediately after this writing almost killed him) from the side effects of treatment that were not taken seriously. It is based on interview data (Clarke and Fletcher 2004). (Note: The significance of chicken pox is that it is the one disease that parents are told they must avoid or get immediate treatment for during the time that their child is sick with cancer and undergoing chemotherapy.)

(Karen Kasmauski/Corbis)

Thanks to those sweet and salty foods we love to eat, our society has produced the largest-ever crop of overweight and obese people.

A Mother's Lament

> When we were sent home on a weekend
> Benjamin was having abdominal pains
> Severely
> We called and it was the other oncologist
> And it wasn't ours.
> She said,
> He is just constipated
> And
> to keep him at home
> I ended up with a counter full of medications
> and . . .
> Finally
> Rich rushed him into emergency in the
> middle of the night.
> The child
> was
> like a woman
> in labour
> Without an epidural.
> His eyes were dilated and you could see the
> contractions coming on . . .
>
> We brought him in
> And we found chicken pox on his abdomen
> Rich brought him in
> At 4 o'clock
> in the morning . . . and
> Emergency sent him home with a
> prescription . . .
>
> We were sent home.

The experience of illness has been portrayed in theatre (in plays such as M. Edson's *Wit*, 1999), movies (such as *Lorenzo's Oil*, 1992), poetry, and other art forms, such as the quilts made to honour those with breast cancer or HIV/AIDS.

Sociology of Medicine

The sociology of medicine examines the location, definition, diagnosis, and treatments of disease. It includes an examination of the various health-care institutions such as hospitals, clinics, co-operatives, and homecare, along with medically related industries and the training, work, and statuses of medical and nursing professionals and other health-care providers today and in historical context.

Because the history of the twentieth century has been characterized by the increasing dominance of allopathic medicine and its spreading relevance to more and more of life (Zola 1972), the term **medicalization** (the tendency for more and more of life to be defined as relevant to medicine) has provided an important conceptual framework for critical analysis. In this part of the chapter, we will discuss the medical care system in Canada today.

THE CANADIAN MEDICAL CARE SYSTEM

Our present medical care system was first implemented in 1972 after a Royal Commission on Health Care (Hall 1964–5) under Justice Emmett Hall recommended that the federal government work with the provincial governments to establish a program of universal health care. While hospitalization and some medical testing had been covered before that, the new program was designed to cover physicians' fees and other services not already covered under the Hospital Insurance and Diagnostic Services Act (1958).

Four basic principles guided the program. The first was universality. This meant that the plan was to be available to all residents of Canada on equal terms, regardless of prior health record, age, income, non-membership in a group (such as a union or workplace), or other considerations. The second was portability. This meant that individual benefits would travel with the individual across the country, from province to province. The third was comprehensive coverage: the plan was to cover all necessary medical services, including dentistry that required hospitalization. The fourth was administration. This referred to the fact that the plan was to run on a non-profit basis.

The Canada Health Act of 1984 added a fifth principle, accessibility. The costs of the plan were to be shared by the federal and provincial governments in such a way that the richer provinces paid relatively more than the poorer provinces; thus, the plan would also serve to redistribute wealth across Canada. Doctors, with few exceptions (found mostly in community health clinics), were not salaried by the government. Instead, they were and continue to be private practitioners paid by the government on a fee-for-service basis.

PRIVATIZATION

Despite the presence of the universally available and federally supported national medical care system, there is considerable evidence of privatization within the system. At present, approximately 70 per cent of the Canadian system is public and 30 per cent private (Mikkonen and Raphael 2010, 38). The private aspects of the Canadian system are dominated by multinational corporations involved in providing a variety of health-related goods and services, including additional medical insurance, information technology services, food and laundry for hospitals, long-term and other institutional care, drugs, medical devices, and homecare. The most important impetus for growth in the medical system is in the private sector, particularly in drugs and new (and very expensive) technologies such as MRI, CAT scan, and mammography machines and other increasingly popular diagnostic technologies, such as the PSA test for prostate cancer.

Table 16.3 portrays the increase in personal expenditures on medical care from 2000 to 2004. Notice especially the relative rate of growth in expenditures for the often private component of care, pharmaceuticals (both prescribed and over-the-counter), as well as other (unspecified) expenses.

There is considerable debate today about whether or not Canada can continue to afford a publicly funded and universally available medical care system. The mass media are full of stories of overcrowded emergency rooms and long waiting lists. These sorts of concerns often seem to lead to the argument that the problem is the publicly funded system. However, evidence from a wide variety of sources does not necessarily support this point of view (Canadian Health Services Research Foundation 2002; 2005). For example, Calgary recently moved to a degree of privatization: cataract surgery services are now bought from private companies. This has resulted not only in more costly cataract surgery but also in longer waiting times than in the nearby cities of Lethbridge and Edmonton. According to the Canadian Health Services Research Foundation, not only do parallel private health care systems not cut waiting lists, but they seem to lengthen lists for those in the public system (2005).

US studies on the effect of governments' buying medical services from private companies demonstrate other problems with privatization. Johns Hopkins University researchers compared more than 3000 patient records and found that the for-profit centres had higher death rates, were less likely to refer patients for transplants, and were less likely to treat children with the dialysis method most likely to be of benefit to them (Canadian Health Services Research Foundation 2005). Most US-based research suggests that for-profit (private) care costs more, pays lower salaries to staff, and incurs higher administrative costs but does not provide higher-quality care or greater access. Not-for-profits tend to provide higher rates of immunization, mammography, and other preventive services (ibid.). On average, people lose two years of life when they are treated in for-profit hospitals (Devereaux et al. 2002, 1402).

MEDICALIZATION

Medicalization is the tendency that more and more of life is defined as relevant to medicine. Irving Zola (1972) is one of the social theorists who has been critical of this process. He defined medicalization as including the following four components:

1. an expansion of what in life and in a person is relevant to medicine

2. the maintenance of absolute control over certain technical procedures by the allopathic medical profession

3. the maintenance of almost absolute access to certain areas by the medical profession

4. the spread of medicine's relevance to an increasingly large portion of living

The first area of medicalization is the expansion of medicine from a narrow focus on the biomechanics of the human body to a broader concern by medicine with the 'whole' person. The second area refers to the fact that there are things that only doctors are

TABLE 16.3 ▼ Health Expenditures, 2000–2004 and 2008 ($ millions)

	2000	2001	2002	2003	2004	2008
Hospitals	30,554.5	32,199.0	34,375.1	36,808.7	36,896.8	50,947.8
Other institutions	9,331.3	10,104.7	10,776.5	11,547.6	12,456.1	18,276.3
Physicians	12,977.0	13,978.0	15,050.7	16,012.6	16,785.2	25,634.1
Other professionals	11,586.6	12,576.7	13,116.8	13,891.0	14,635.6	20,013.0
Drugs (prescribed and non-prescribed)	15,085.8	16,660.8	18,408.7	20,002.9	21,758.4	34,614.8
Other expenditures	18,368.3	20,791.8	22,313.9	24,741.0	25,743.7	
Total health expenditures*	97,903.4	106,302.0	114,041.6	123,003.7	130,275.2	183,120.9

*Totals may not be precise due to rounding.
SOURCE: Adapted from *Canada Year Book 2009*, 214.

allowed to do to the human body, such as surgery. The third refers to the fact that doctors, through medicine, have been able to transform into medical problems areas of life such as pregnancy and aging that were formerly viewed as normal, neither as pathological nor as medically relevant processes. The fourth pertains to the way in which medicine increasingly has jurisdiction in areas formerly considered of relevance to the criminal justice or religious systems, such as criminality and alcohol addiction.

Medicalization has been shown as evident in the tendency for more and more of life to be defined by the medical profession. Ivan Illich (1976) attributes the growth of medicalization to bureaucratization. Vicente Navarro (1975) claims that medicalization, or medical dominance, is more related to class and class conflict, in particular the upper-class background and position of physicians. He also relates medicalization to the work of physicians who operate as entrepreneurs in the definition of health and illness categories and their relevant treatments.

DISEASE MONGERING

Furthering the argument about the role of capitalism in the growth of medical dominance is the instrumental role that pharmaceutical corporations play in 'disease mongering'. Through a series of suggestive anecdotes, Ray Moynihan, Iona Heath, and David Henry (2002) argue for critically examining the ways in which the pharmaceutical industry plays a significant role in defining as diseases conditions for which they have developed an effective drug. Ostensibly involved in public education about new diseases and treatments, and often working alongside doctors and consumer groups, the pharmaceutical industry has promoted as problematic conditions that may well be better seen as part of life. For example, the medicalization of baldness by Merck occurred after the development of their anti-baldness drug, Propecia, in Australia. Around the time of the patenting of the drug, a major Australian newspaper reported on a new study that indicated that about one-third of men experienced hair loss (Hickman 1998). Further, the article emphasized, hair

OPEN for DISCUSSION

The Power of Medicalization

The power of medicalization can be illustrated by the case of Tyrell Dueck. He was 13 when, in early October 1999, he was diagnosed with osteogenic sarcoma, or bone cancer. Treatment upon diagnosis usually begins immediately, with chemotherapy to shrink the tumour and stop the spread of the disease. Surgery is used next to remove the tumour or sometimes the whole limb if the disease is found to have spread. Chemotherapy may then continue. With immediate treatment and localized disease, the prognosis can be excellent for full recovery.

At the point of diagnosis, Tyrell's father, Tim Dueck, said that the family did not want Tyrell to undergo chemotherapy and surgery but wanted him to try alternative treatments. By 11 December, now a couple of months after the diagnosis, the hospital had gone to court in Saskatchewan and received a court order giving the Saskatchewan minister of social services guardianship over Tyrell. This gave the minister the right to consent to treatment on Tyrell's behalf. After two rounds of chemotherapy, doctors decided Tyrell's leg would have to be amputated and that this surgery would give him a 65 per cent chance of survival. The doctors were also clear that they believed that he would die without the amputation of his leg.

Nevertheless, Tyrell decided that he did not want to have his leg removed. Nor did he want more chemotherapy. This created a new dilemma. The court order of guardianship had taken the power of consent for treatment away from Tyrell's parents. It had not taken it from Tyrell himself. Another hearing began on 13 March. A psychologist and a psychiatrist gave contradictory evidence regarding whether Tyrell was legally mature and thus capable of making the decision against treatment on his own. The media were highly involved by this time, and debate raged about the issue. It became clear that the Duecks believed that a combination of prayer and complementary health care in Mexico would heal Tyrell.

By 18 March, the court had decided that Tyrell was a 'mature minor' and therefore was required to have the prescribed medical treatment. But by this time, further medical investigations had revealed that the time lapse had decreased Tyrell's chances of survival to 10 to 15 per cent. At this point, the minister of social services withdrew the order of treatment, and the Duecks were free to pursue alternative treatment in Mexico. They did so. Tyrell received the treatment. Nonetheless, Tyrell died a few months later (Rogan 1999, 43–52).

Who do you think should have the power to decide in a situation like this: the state, the doctors, the parents, or the young person?

loss sometimes led to panic and other emotional difficulties and had a negative impact on job prospects and well-being. At the same time, the paper featured news of the establishment of an International Hair Study Institute. What the newspaper failed to report was that both the study and the institute were funded by Merck and that the 'expert' quotations were from the public relations firm hired by Merck.

This example is just one illustration of some strategies used by pharmaceutical industries, who have been advised in Britain's *Pharmaceutical Marketing* magazine to 'establish a need and create a desire' (Cook 2001, cited in Moynihan, Heath, and Henry 2002). More and systematic research on the extent of such practices is needed, but even these few examples raise questions about what may be invisible and unregulated attempts to 'change public perceptions about health and illness to widen markets for new drugs' (Moynihan, Heath, and Henry 2002, 891).

TIME to REFLECT

PlosMedicine (www.plosmedicine.org), a peer-reviewed online journal, has a series of critiques of 'pharmaceuticalization', including the development of drugs for male erectile dysfunction, female sexual dysfunction, and ADHD, published in 2006. Take a look at this online journal, and read at least one of the articles on pharmaceuticals. Can you offer a counter-argument to the ones put forward in *PlosMedicine*?

MORALE AND BULLYING AMONG DOCTORS

Despite the power of medicine and the relative power and wealth of doctors, medical work is often difficult. Recent research in the United Kingdom has identified workplace bullying as a major source of stress at work for health-care professionals (Quine 2002). One study among doctors who worked for the National Health Service found that one in three reported that they had been bullied in the year previous to the study (Quine 1999). Another study found that bullying, racial harassment, and discrimination were everyday occurrences in the work lives of Asian and black doctors in the United Kingdom (Coker 2001). In the United States, too, a few studies have identified mistreatment and bullying experienced by medical students, interns, and residents (Kassebaum and Cutler 1998). Doctors are not immune to discrimination on the job. While studies of bullying have not been done in Canada, there is reason to think that the patterns might be generalizable to this country. Such research is particularly important as increasing numbers of Canadian-educated doctors are both female and from visible-minority backgrounds.

Bullying is one of the causes of increasingly poor morale among doctors around the world (Edwards et al. 2002). While declining morale is partly the result of doctors' increasing workload, accompanied by a relative decrease in remuneration, according to extensive research it is also, and perhaps more importantly, related to the changing social compact between doctors and the societies in which they practise. Doctors who were previously sole practitioners and operated as

UNDER the WIRE

Health Care Online

The Internet has changed health care in the modern world. Nettleton and colleagues argue that this is such a profound alteration that it deserves a new name, which they suggest should be called 'e-scaped' medicine (2004). More people go online for medical advice on a daily basis than visit doctors. According to Statistics Canada's Canada Internet Use Survey, 80 per cent of Canadians used the Internet for personal reasons such as email, social networking, and banking in 2009. Almost three-quarters of adult women (74.9 per cent) and about two-thirds of adult men (66 per cent) used the Internet for information about health (www.statcan.gc.ca/daily-quotidien/100510/ dq100510a-eng.htm). This increasing access to and reliance on the medium has led to increased knowledge among patients and future patients in their relationships with doctors. Not all of the information available on the Web is credible, however, and evaluating medical sites is complicated. There are often millions of potential Internet sites for any one diagnosis. Nettleton and her colleagues (2005) have found that people have become 'medicalized' in their use of the Internet and use 'rhetorics of reliability' that suggest there is an increasing level of similarity between the dominant biomedical ideas of what constitutes good information and lay use patterns.

SOURCES: Nettleton et al. 2004; 2005.

and puts it into the hands of epidemiologists and other scientists who determine best-practice principles that doctors are expected to follow. Moreover, best-practice findings are no longer published only in arcane medical journals but often are easily accessible to insurance company personnel and to individual patients through numerous, disease-specific Internet sites.

THE SOCIO-ECONOMIC BACKGROUND OF MEDICAL STUDENTS IN CANADA

There are substantial differences between the backgrounds of medical students and those of the rest of the Canadian population. Doctors are not drawn in a representative way from across the socio-economic and socio-demographic variation of the whole population of citizens.

A half-century ago, medical students were more likely to have fathers who were doctors or who had higher incomes and more graduate degrees than average Canadians, and this remains true today. In 1965, 11.8 per cent of medical students had fathers who were doctors, while today 15.6 per cent do (Dhalla et al. 2002, 1032; Collier 2010). Rural students have also been under-represented; whereas 30.4 per cent of Canadian high school students lived in rural areas in 1965–6, only 8.4 per cent of the medical students did. Rural students are slightly better represented today (Collier 2010). In 1965–6, the fathers of 7.5 per cent of the population had attended university, while 38.0 per

independent entrepreneurs within a private—even, many suggest, 'sacred'—doctor–patient relationship are now under intense surveillance by governments and corporations (such as insurance companies). *Evidence-based medicine* may also contribute to decreasing morale. This new approach to medical practice removes the clinical judgment from the doctor

These elderly people maintain their energy and improve their health by participating in group exercise.

cent of the fathers of medical students had done so. These days, about 39.0 per cent of the fathers of medical students had earned graduate or doctoral degrees compared with only 6.6 per cent of the total population of the same age (Collier 2010). With respect to gender, however, significant changes have occurred. In 1965–6, 11.4 per cent of medical students were female. Now about half are (Collier 2010). However, medical students continue to be less likely than the general population to be Aboriginal or black Canadian. The Canadian Federation of Medical Students has called for change and for an expansion in the degree to which medical students represent a broad selection of the Canadian population (Collier 2010).

TIME to REFLECT

Do you think it is important for the backgrounds of physicians to reflect the cultural, ethnic, and other diverse features of Canada? Explain.

Conclusion

In many ways, health issues are fundamentally social issues. The rates, definitions, and meanings of illness, sickness, disease, and death have varied and continue to vary around the world and over time. Within Canadian society, these differences, particularly in rates, reflect culture and social structure and mirror inequality and marginalization. Medical care is dominated by allopathic medicine today. However, a sizable minority of Canadians are now choosing complementary and alternative care. At the same time, there is increasing evidence of medicalization by the pharmaceutical industry through its entrepreneurial disease-defining work. While there is substantially more privatization in the Canadian medical care system today, it has tended not to reduce costs or to provide better medical service but rather the reverse. Finally, it appears that doctors in the system are experiencing low levels of job satisfaction and morale today.

SOCIOLOGY in ACTION

Population Aging

You have probably heard the concern voiced that the health-care system will be stressed during the next several decades in Canada as a result of the aging of the baby-boom generation. In fact, people over 65 are no more likely to be ill than adults at other ages. They are, however, more likely to use the health-care system when they are ill. Recent research from the Canadian Health Services Research Foundation demonstrates that increasing utilization by seniors is not a result of their increasing numbers in the population but rather their relatively higher utilization rates. There is evidence that healthy, not sick, seniors are responsible for the increase in costs. The rate of doctor visits among the well in Manitoba between the 1970s and 1983, for example, increased by 57.5 per cent for specialists and 32 per cent for general practitioners. Unhealthy seniors' rates increased by only 10 per cent or less. It appears that the elderly routinely receive more care than they formerly did. The cost of increased health care as a result of the simple aging of the population is estimated at only 1 per cent of total health-care costs. The significant impact of population aging, then, appears to be the result of some degree of over-treatment of the elderly. Although such interventions as flu shots, hip replacement, and cataract surgery may be necessary, others may not (www.chsrf.ca/mythbusters).

Examine the following figure. What is the obvious explanation for what is portrayed? In light of the above paragraph, what do you think are the more complex and interesting explanations?

Increase in Medical Use by Seniors in Good and Bad Health, Manitoba, 1971–83

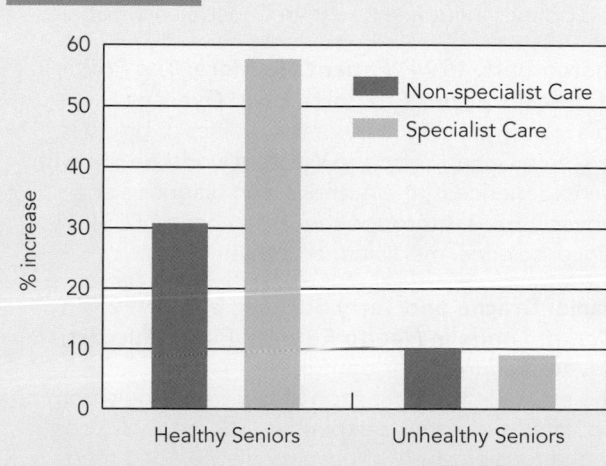

SOURCE: Canadian Health Services Research Foundation, *Mythbusters*.

questions for critical thought

1. What are the three most important health/social policies that you would recommend the government establish to minimize the rate at which Canadians die from car accidents today?

2. Why are more and more people using complementary and alternative health care?

3. Do you think that medicine holds too much power in Canada today? Explain.

4. Assess the opportunities for social cohesion in your college or university. Are there things to do available in class or extra-curricularly that give you chances to get to know and trust people from different programs and years at your university? If not, what changes would you advocate?

5. What challenges to your health are evident in your school? In your answer, include threats from the physical plant, the organization of learning and testing, and the social life available to students.

6. Discuss the water or air quality of the town where you are going to school. How difficult or easy has it been for you to get information about the air or water quality? Assess the quality of the information to which you have gained access.

7. Examine three magazines that you commonly read—for their health-related messages, perhaps—focusing on a particular subject such as depression. Include both articles and advertisements (for anti-depressants) in your analysis. Consider the portrayal of issues such as gender, ethnicity, and social class in your discussion.

8. One of the most dangerous places to work in Canada today is a hospital. Explain.

recommended readings

Pat Armstrong and Hugh Armstrong. 1996. *Wasting Away: The Undermining of Canadian Health Care.* **Toronto: Oxford University Press**
This book is based on interviews, observation, and documentary evidence regarding the effects of the 'cuts' to and privatization of health care in Canada.

Pat Armstrong, Hugh Armstrong, and David Coburn, eds. 2001. *Unhealthy Times: Political Economy Perspectives on Health and Care in Canada.* **Toronto: Oxford University Press.**
This is a fascinating book on the ways that economics and politics influence health in Canada and globally.

Sharon Batt. 1994. *Patient No More: The Politics of Breast Cancer.* **Charlottetown: Gynergy.**
This award-winning story of the politics of breast cancer in Canada and the Western world begins with the experience of the author's own diagnosis and moves from this through the larger world of breast cancer science, medicine, and charities.

Daniel Drache and Terry Sullivan, eds. 1999. *Market Limits in Health Reform: Public Success, Private Failure.* **London: Routledge.**
The essays in this collection are written largely from a political economy perspective. The book focuses on the tensions between a nationally funded medical care system and a growing move toward a market-driven economy.

Colleen Fuller. 1998. *Caring for Profit: How Corporations Are Taking Over Canada's Health Care System.* **Vancouver: New Star.**
Fuller is a health-care activist and researcher who wrote this book while she was a research associate at the Canadian Centre for Policy Alternatives. Fuller charts medicare's history up to the current period of privatization. She documents examples of privatization across the country to demonstrate that medicare is no longer a universal system.

Samantha King. 2008. *Pink Ribbons, Inc.: Breast Cancer and the Politics of Philanthropy.* **Minneapolis: University of Minnesota Press.**
This book examines the corporate face of fundraising for breast cancer with a series of interrelated papers on various aspects of funding medical research. King traces a development of the history of breast cancer from an individually and privately experienced disease to an enormous marketing cause. She offers critical analysis of this process.

recommended websites

Health Canada
www.hc-sc.gc.ca

Health Canada, a government department, provides health-related information on topics such as healthy living, health care, diseases and conditions, health protection, and media stories. You can also find here the latest statistics regarding health, illness, death, PYLL, medical care system characteristics, and current issues.

The Hunger Site
www.thehungersite.com

This is a site for those who are concerned about world hunger and want to keep informed on the situation and take action.

National Institutes of Health
www.nih.gov

The National Institutes of Health in the US provides access to research, health news, and various health-related resources.

National Network on Environments and Women's Health
www.yorku.ca/nnewh/english/nnewhind.html

This is one of the five federally funded Centres of Excellence for Women's Health. It focuses on women's health and workplaces, including paid and unpaid work, unemployment and labour force restructuring and adjustments, health systems (both conventional and unconventional forms of health care), formal and informal practices, women's understandings of health and health risks, and policy.

National Women's Health Information Center
www.4woman.gov

The Office on Women's Health and the Department of Health and Human Services of the US federal government provides an excellent site for the latest findings regarding health research, particularly as it pertains to women (there is also an internal link to men's health issues included on the home page). Topics include screening and immunization information, advice on how to quit smoking, action strategies regarding breastfeeding, information about violence against women, body image, women with disabilities, and health information for 'minorities'. This site would also be of use as a document for content analysis regarding current definitions of significant health issues and their treatments.

Public Health Agency of Canada
www.phac-aspc.gc.ca

The Public Health Agency of Canada offers access to research and working papers on the social determinants of health, health promotion, and population health perspectives.

Statistics Canada
www.statcan.gc.ca

Statistics Canada publishes myriad studies on Canadian society, including statistics relevant to morbidity, mortality, disease incidence, birth rates, and so on.

World Health Organization
www.who.int

The World Health Organization provides information, fact sheets, various reports, and news about health issues around the world. It also covers essential information regarding worldwide epidemics and news about outbreaks of various illnesses in a worldwide context, including epidemic and pandemic alerts and responses.

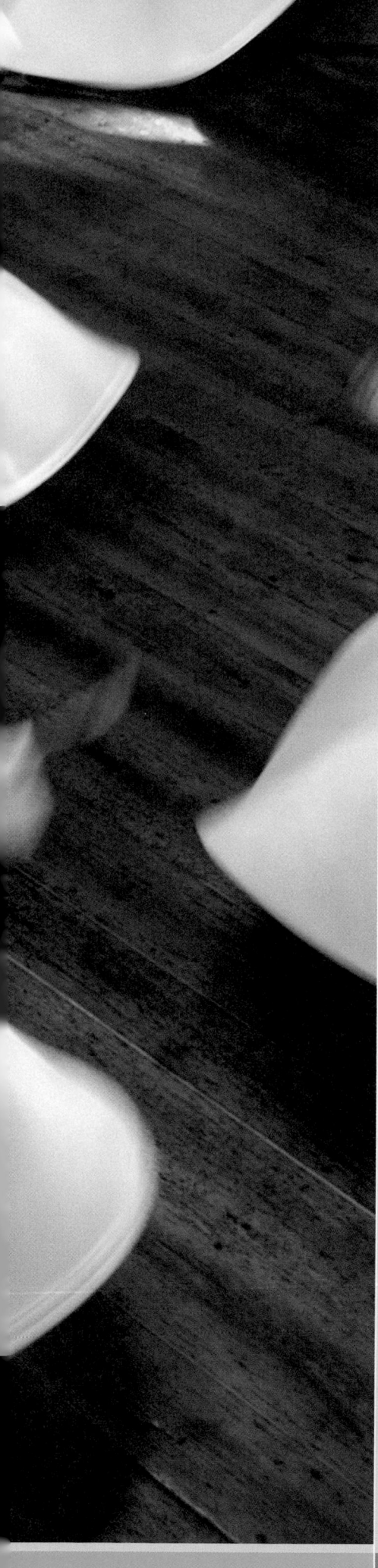

chapter 17

Religion in Canada

Lori G. Beaman

In this chapter, you will:

▶ Explore definitions of religion and spirituality.

▶ Learn about the changing religious demography of Canada and its potential impact.

▶ Consider new religious movements and minority religious groups.

▶ Examine the concept of secularization.

▶ Explore the relationship between law and religion.

▶ Think about the gendered dimensions of religious participation.

Introduction

Religion is an important point of identity for many people in Canada. Almost daily, we hear or read about an aspect of someone's religious beliefs that bumps up against a regulation, law, or policy. Even as this chapter was being written, a reference on the criminality of polygamy, which is practised by some Fundamentalist Latter-day Saints, was before the Supreme Court of British Columbia and has generated considerable discussion and debate, reflected in newspaper and magazine articles, website postings, and letters to the editor (see, for example, Bramham 2010; Burkholder 2009; MacQueen 2009; Marples 2010; Schoof 2010). The case of Naima Ahmed, a 29-year-old Muslim immigrant from Egypt to Quebec, made the headlines and stirred the quiet embers of a fire over reasonable accommodation that had only just calmed in Quebec. Naima was ordered to remove her **niqab** when attending her French class. Her complaint to the Quebec Human Rights Commission became public, and heated debate ensued. For Naima, the choice to wear the niqab was an important part of her religious identity. For those opposed to the niqab in the classroom, it is an important reminder of a past that was dominated by what is perceived by some as religious oppression (Peritz 2010; Montpetit 2010; El Akkad 2010). In Alberta, a small group of Hutterites believe that refusing to have their photographs taken is so core to their religious beliefs and identities that they were willing to fight all the way to the Supreme Court of Canada to be legally exempt from that government requirement (*Alberta v. Hutterian Brethren of Wilson Colony*; for more information, see CBC 2009b). No matter how one may feel personally about religion, it matters to a good many people. For social scientists, religion constitutes an important area of study in our pursuit to better understand social life.

Sociologists of religion do not ask questions about the veracity of particular sets of religious beliefs. In other words, we do not care whether gods exist or whether Raelians have really had contact with extraterrestrial beings. Rather, our concern is with how human beings act out their religious beliefs and practices, as well as how religious beliefs and social institutions intersect. How are certain sets of beliefs legitimized? What is constructed as being a 'religion'? What are the power relations embedded in these processes? In other words, who gets to decide whether a religion is really a religion?

In the First Person

As a sociologist, I have spent my entire career studying religion, using a variety of methodological and theoretical approaches. Through focus groups, face-to-face interviews, and textual analysis, studying the diversity of religious life in Canada offers insight into the normalization of some religious beliefs and practices and the 'making strange' of others. I have written on religion and diversity in numerous books and articles, including *Defining Harm: Religious Freedom and the Limits of Law* (2008); 'Is religious freedom impossible in Canada?' in *Law, Culture and the Humanities* 6 (3) (2010); 'Who decides? Harm, polygamy and limits on freedom', in *Nova Religio* 10 (1) (2006); 'Defining religion: The promise and the perils of legal interpretation', in *Law and Religious Pluralism in Canada* (2008); and 'A cross-national comparison of approaches to religious diversity: Canada, France and the United States', in *Religion and Diversity in Canada* (2008). I am currently the principal investigator of the Religion and Diversity research project.

—Lori Beaman

MARXIST INFLUENCE

Until relatively recently, sociology—as a discipline—did not take the study of religion particularly seriously. There are a number of reasons for this. First, a good number of scholars accepted the popular wisdom that we live in a secular society. However, religion remains an important part of the Canadian social fabric and is likely to continue to do so.

Another reason for the lack of attention to religion within sociology can be explained by the strong Marxist tradition, particularly in Canadian sociology. Marx worried about the power of religion and in fact stated that 'Religion is the sigh of the oppressed creature, the heart of a heartless world' (Marx, as cited in Raines 2002, 167). While he recognized the ability of religion to offer solace in times of trouble, Marx worried that the happiness offered by religion was illusory and that it distracted people from seeking real happiness (which in Marx's view inevitably involved the transformation of economic arrangements and the end of capitalism). Marx concluded that 'the abolition of religion as the illusory happiness of the people is the demand for their real happiness' (Marx, as cited in Raines 2002, 167). Some sociologists understood this statement as a licence to minimize the importance of religion in society, to exclude religion as a variable from research, and to ignore its importance in theoretical work. This hardly reflects the spirit in which Marx wrote (one would think something deemed to be so powerful would need to be studied carefully) and has resulted in a paucity of research about religion and social life.

Moreover, on a substantive note, there have been some significant social movements that have been grounded in a combination of religion and Marxism, most specifically liberation theology, which began in South and Central America in the 1960s and was based on the premise that part of the mission of Christianity is to bring 'justice to the poor and oppressed, particularly through political activism' (Smith 1991, 12). Its goal was to effect socio-economic change, and indeed, it became so sufficiently threatening that it was condemned by the Vatican. Several Latin American bishops were deposed for their continued fight for social and economic justice based on Marxist principles.

Religion in Profile

What does religion in Canada look like from a demographic perspective? Mainstream Christianity has dominated Canada's historic landscape and, to some extent, continues to do so. Mainstream Christianity includes Roman Catholicism and Protestant groups such as the United Church of Canada and the Anglican Church of Canada. Numbers of affiliates have been approximately equal between Roman Catholics and Protestants. Groups outside of those two broad categories make up a relatively small proportion of the religious picture in Canada, but as we will see below, that picture is rapidly changing. For the most part, Canadians have remained affiliated with the religion of their parents and grandparents, even if they do not actually attend church. Social scientists have realized that while church attendance is a measure of religious commitment or participation, it is only one measure and thus offers a fairly limited understanding of religion in Canada.

Statistically speaking, Canada is still dominated by Christianity. The 2001 Statistics Canada General Survey shows that 80 per cent of Canadians identify as 'Christian'. We are fortunate in Canada that the government had collected data on religious affiliation since the late 1800s, thus allowing us to formulate a longitudinal understanding of religious participation in Canada. These data also show the historical presence of Sikhs, Muslims, Buddhists, and Hindus. Moreover, we know that the Canadian Jewish community has roots that date back to the 1700s. Thus, while we hear much about the increasing presence of religious groups who are not Christian, most of these religions have been present since the birth of Canada as a nation.

Reginald Bibby has spent a great deal of time investigating the religious participation of Canadians, focusing primarily on Christian-centred behaviours such as church attendance, belief in biblical teachings, and experiences of god's presence. Bibby has found much evidence of both identification and lack of participation among Canadians in relation to Christian churches. He argues that Christian churches have dominated and will continue to dominate the religious

SOCIOLOGY in ACTION

Capturing the Complexity of Religious Participation

Despite the fact that participation in organized religion has indeed declined in Canada, the vast majority of Canadians still identify with a faith tradition, however sporadic their participation in the formal rituals of that tradition might be. Moreover, Canadians are finding new and interesting ways of expressing their religious interests that do not necessarily manifest as participation in traditional religious groups. The increasing participation in yoga, for example, might arguably be a 'religious practice'. Think too about the creation of sacred space through labyrinths in the past five or so years. For example, on 14 September 2005 the Toronto Public Labyrinth opened. While it is adjacent to the Church of the Holy Trinity, it is fully accessible to the public, and its creation was supported by the City of Toronto. Many people walk this labyrinth, some of whom are connected to faith communities, some not. Many such forms of participation in sacred rituals don't fit into traditional measures of religious behaviour and thus remain undetected in research that measures religious participation.

TABLE 17.1 ▼	Top 10 Religious Denominations, Canada, 2001	
	Number	**%**
Roman Catholic	12,793,125	43.2
No religion	4,796,325	16.2
United Church	2,839,125	9.6
Anglican	2,035,495	6.9
Christian, not included elsewhere[1]	780,450	2.6
Baptist	729,475	2.5
Lutheran	606,590	2.0
Muslim	579,640	2.0
Protestant, not included elsewhere[2]	549,205	1.9
Presbyterian	409,830	1.4

[1]Includes persons who report 'Christian' as well as those who report 'Apostolic', 'Born-again Christian', and 'Evangelical'.
[2]Includes persons who report only 'Protestant'.

SOURCE: Statistics Canada. 2003. '2001 Census: Analysis series. Religions in Canada', p. 20. Catalogue no. 96F0030XIE2001015. Ottawa: Statistics Canada. www12.statcan.ca/english/census01/products/analytic/companion/rel/pdf/96F0030XIE2001015.pdf.

scene in Canada (Bibby 1993; 2002; 2006). Bibby is concerned that while there is still a strong identification with Christian churches among Canadians, there is much less evidence of actual participation.

The statistical data collected by Statistics Canada and researchers like Bibby provide invaluable resources for important information about religious beliefs and behaviours in Canada. However, we need more information about how Canadians 'do' religion and spirituality. In other words, what do religion and spirituality look like in their daily lives? Do some Canadians have shrines at home? What role do they play in everyday life? What is the nature of religious or spiritual practice at home? Unfortunately, statistical data can give us little information about these sorts of questions. Especially lacking are data that gives us insight into minority religious communities.

Since the terrorist attacks on various locations in the United States, or '9/11', and subsequent attacks in the UK, Spain, and Bali, discussions about the link between religion and violence have become more

▼ FIGURE 17.1   Religious Affiliation of Population (by Census Division): No Religion

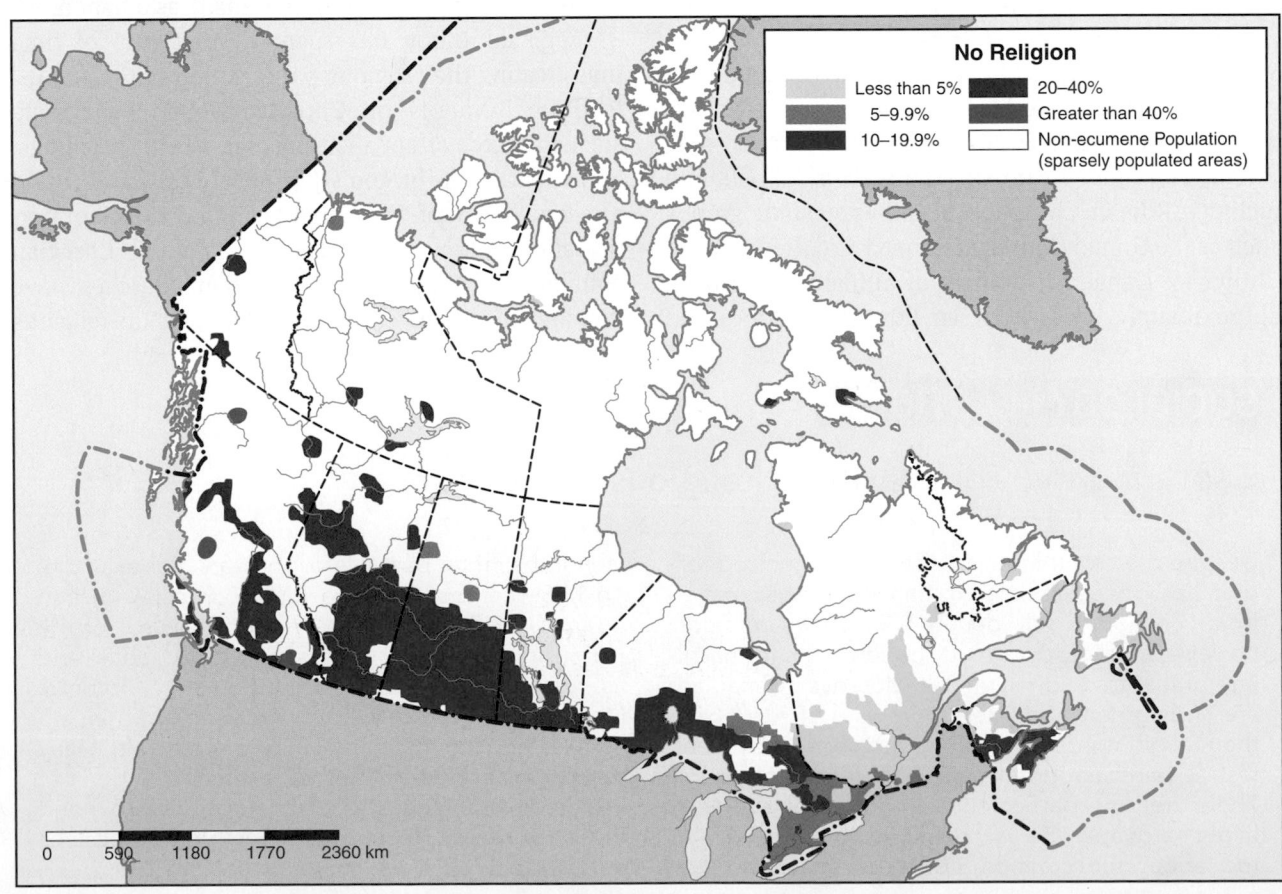

SOURCE: Natural Resources Canada. *The Atlas of Canada.* http://atlas.nrcan.gc.ca/site/english/maps/peopleandsociety/religion/religion01.

frequent. Although this connection might seem recent, it has been present throughout history. We need only to examine Canada's history of violence against and violation of its First Nations peoples to uncover a horrifying picture of the intertwining of religion, political goals, and power relations to understand the intersection of religion and violence. If we look beyond Canada, the Christian crusades (the first was in 1095) provide another example in which the political and economic desires of kings and princes combined with religious ideology to justify a so-called holy war on Muslims, Jews, Orthodox Christians, and many other groups who fell outside of the then-mainstream definitions of Christian religion. More subtle forms of violence can be identified in the anti-condom messages of some Christian missions in an AIDS-ridden Africa.

Does religion cause violence? The answer to this question is not easy. Certainly religion can provide an ideological justification for violent acts or approaches that do violence to people or their culture in more subtle ways.

When we look at the power of religious ideology, we can begin to understand why it is important to have a better sense of the role of religion in the lives of Canadians. Religion can provide a source of comfort, direction, and community for people. It can be both prescriptive, in that it offers people direction on important choices, and explanatory, in that it provides a source of explanation for everyday events. It is often an important influence on how people think about issues like same-sex marriage, abortion, and gender roles. It can influence how and why new Canadians feel welcome and a part of Canadian society or feel excluded and marginalized.

TIME to REFLECT

Think of some examples of the ways that religion is 'prescriptive'. What are some ways in which it is 'explanatory'?

▼ **FIGURE 17.2**   Religious Affiliation of Population (by Census Division): Protestant

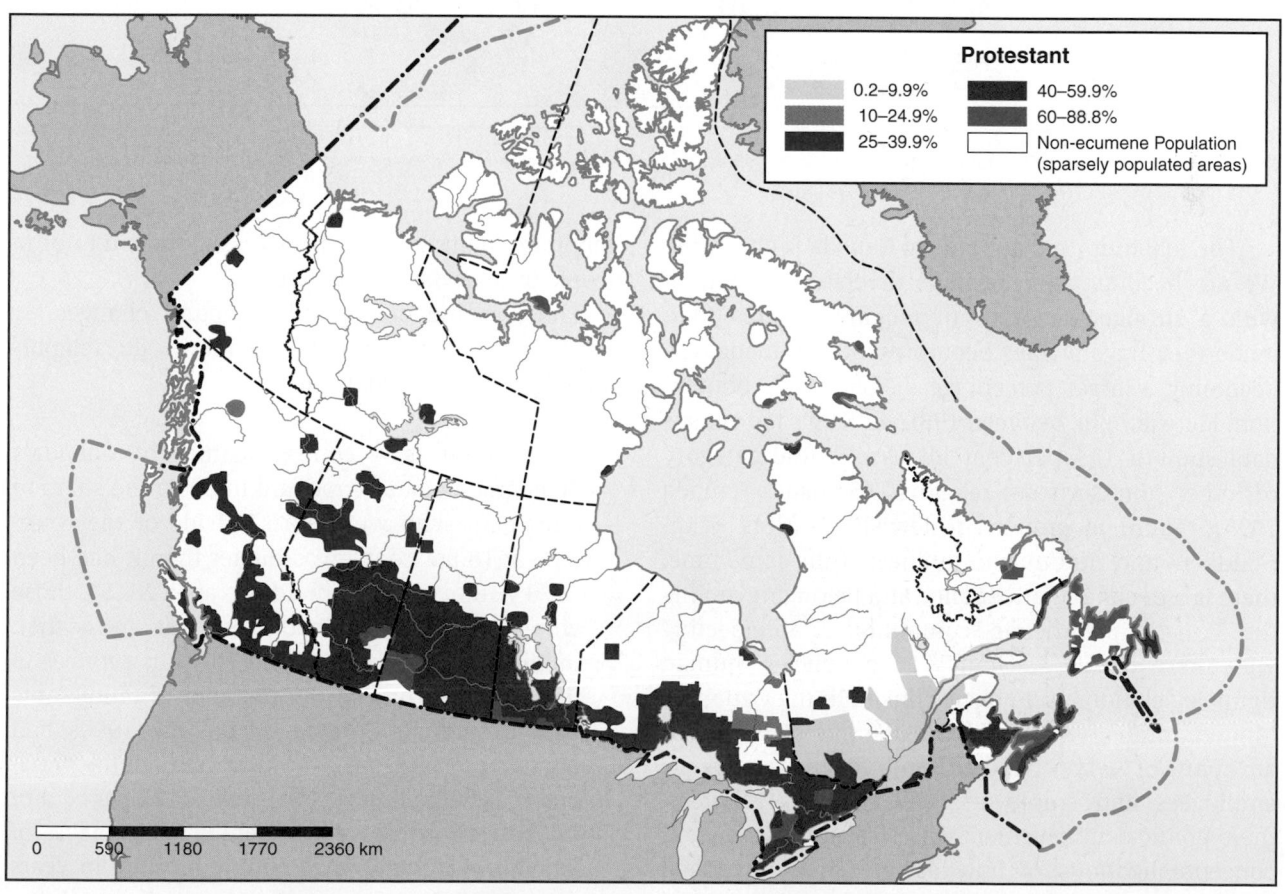

SOURCE: Natural Resources Canada. *The Atlas of Canada.* http://atlas.nrcan.gc.ca/site/english/maps/peopleandsociety/religion/religion01.

▼ FIGURE 17.3 Religious Affiliation of Population (by Census Division): Roman Catholic

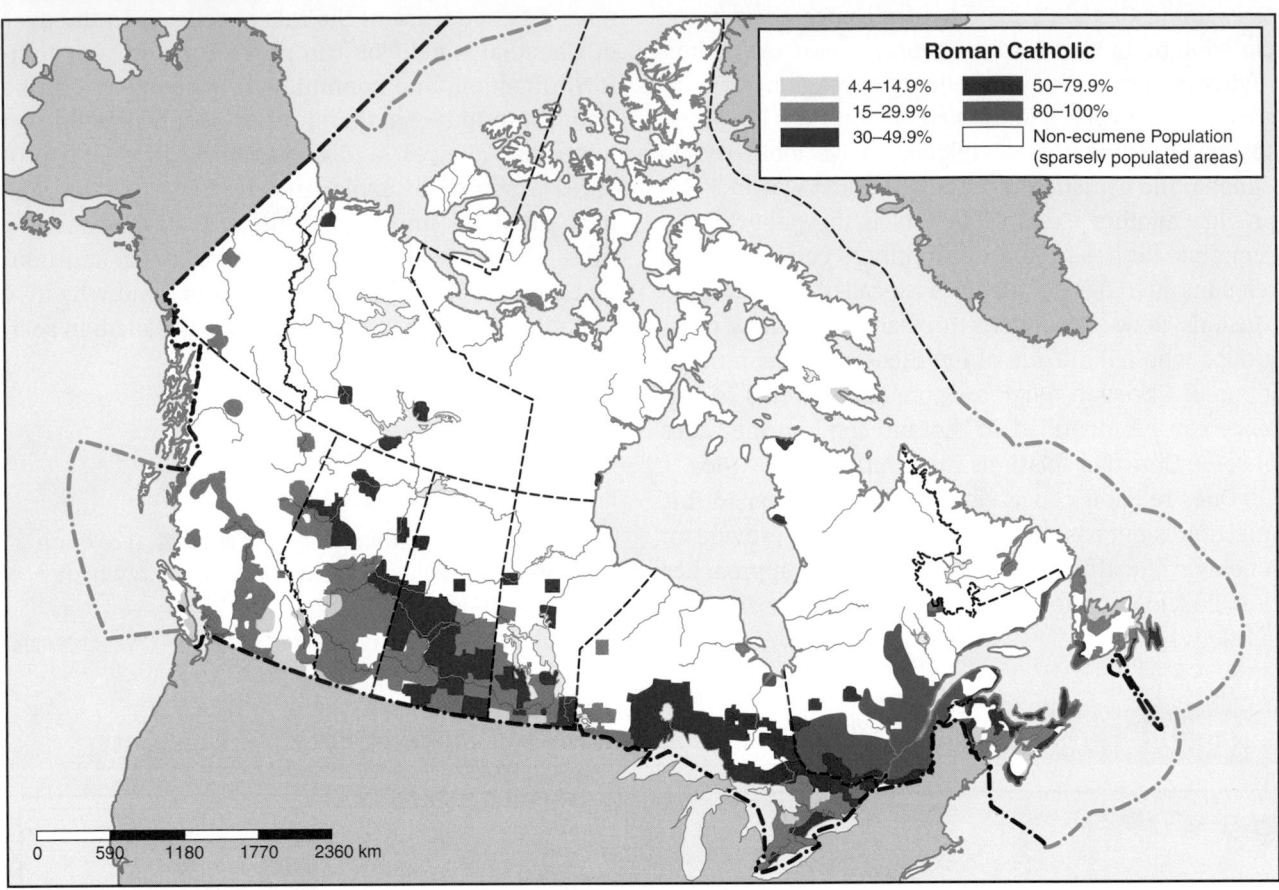

SOURCE: Natural Resources Canada. *The Atlas of Canada.* http://atlas.nrcan.gc.ca/site/english/maps/peopleandsociety/religion/religion01.

The religious demographic in Canada is changing. We are becoming increasingly diverse as a nation, with a stronger presence of religious groups that, while they have always been present in Canada, are becoming a larger percentage of the overall population. For example, between 1981 and 1991, the census data show a 144 per cent increase in the category of 'other non-Christian religions' (Statistics Canada 1993). There is growth in Jewish, Muslim, Sikh, Buddhist, and Hindu communities; at the same time, there is a decline in attendance and belonging among the Canadian Christian communities. Intersecting with these trends are immigration policies, human rights legislation, and policies linked to the Canadian Multiculturalism Act. Religious identity is an important part of what people bring to their roles as employees and employers, their financial choices, their political involvement and decisions, and their conceptualizations of how society should respond to social issues. While religion is, of course, not the only factor in people's decisions, it is nonetheless important to understand the ways in which it informs people in their day-to-day lives.

Thus, as the demographic picture changes, so too does the picture of how Canadians 'do' religion. Statistics Canada (2003b) notes:

Immigration is a central feature of Canada's demographic landscape and in 2001, the share of Canadians who were born outside of the country—at 18 per cent—was higher than it had been in 70 years. Among individuals aged 25–54, those who were born in Canada were less likely than immigrants to have attended religious services in the past year. Indeed, 39 per cent of immigrants who arrived in Canada during the 1990s had attended services on a monthly basis in the previous year while this was the case for 22 per cent of Canadian-born persons. Overall, while places of worship were central to community life in years past, most Canadians today do not have a long-standing attachment to a place of worship. Only

GLOBAL ISSUES

Securitization and the Link between Religion and Violence

The link between religion and violence has become a central preoccupation of many Western governments as they try to develop a deeper understanding of the ways in which religious ideology can motivate violence. Unfortunately for some groups, the actions of marginal minorities come to represent, in the eyes of many, the face of religion. Some people, like Richard Dawkins, argue that religion is responsible for violence in the world and for the wars we currently witness globally. But the quest to understand the link between religion and violence must go deeper. For example, riots in Paris suburbs in October/November 2005 were superficially linked with Muslim youth (for more information, see CBC 2007). But some sociological analyses pointed out that the historical disadvantaging of immigrants coupled with racism resulted in a denial of employment opportunities, which in turn led to a lack of social integration. Is the cause of violence then religion or poverty? Recently, governments have begun to focus on the causes of so-called home-grown terrorism. What, they ask, causes young men, in particular, who are native-born to take up causes that inspire them to commit acts of violence, often in the name of religion or a particular religious group? The answer to this question remains open to debate and to illumination through sociological research.

37 per cent of all Canadians attended religious services or meetings at least once in the previous year and had attended the same place of worship for more than five years.

The meaning of worship service participation rates by new Canadians has yet to be closely studied. Places of worship may be playing an important role in the lives of new Canadians, but the parameters of

As Canada assimilates more people from around the world, many still retain their religious traditions. Here, Sikh men march in the Vaisakhi Festival in Vancouver.

(© Gunter Marx/Alamy)

that impact are little known. So, for example, do religious communities facilitate integration, or are new Canadians somewhat isolated from other parts of Canadian society? Do faith communities play a role in political decisions? What are the generational implications of this involvement?

Peter Beyer, head of the Religion and Immigrant Youth Research Team at the University of Ottawa, situates a sociological understanding of religion in Canada in the context of the global flows of which it is a part. The team has spent the past four years examining the intersection of religion, youth, and Canadian culture. Participants self-identify as having Islam, Hinduism, or Buddhism as part of their religious background, have at least one immigrant parent, and were born in Canada or arrived here as an immigrant when they were under the age of 11. More than 200 youth have participated in the study thus far (the second phase is now underway under the title Religion and Immigrant Young Adults). Results from this research show that the Hindu, Muslim, and Buddhist youth who participated in the study locate their religious or spiritual quests in a complex web of family and cultural reference points. For these youth, spiritual definition is their own prerogative, and while they may rely on their parents and extended families to some extent for spiritual or religious information, they take responsibility for their own spiritual and religious journeys. This group of immigrant youth from Toronto, Montreal, and Ottawa were generally highly integrated into Canadian society, feeling connected to Canada as their country. Certainly, they had experienced incidents of discrimination, but this was generally explained as a product of individual ignorance rather than a reflection of Canadian society (Beyer 2006a; 2006b). For these youth, multiculturalism in Canada is a positive ideal that situates Canada as a progressive nation. Here is how one Muslim participant expressed this idea:

> Give it a couple of . . . generations for people to get . . . out of the shell of their own culture, to mix with the world. Because I believe what we have in Canada is an opportunity that a lot of the world doesn't have, I mean, don't get me wrong, there's a lot of blood on the hands of everybody who lives in this country. But we have an opportunity for people to start fresh. We have people from all different backgrounds, all over the world. We are a representation to the world. . . . There are certain points into staying and understanding your own culture and appreciating your own

culture. But to be able to evolve and to move on with the times . . . we can show the world here how to live amongst people from all different backgrounds (MM26). (Beyer 2008)

TIME to REFLECT

In your opinion, why are many Canadians less tied to religious institutions than they were several decades ago? What are the potential impacts on micro and macro levels of society?

Definitions of Religion

What do we mean when we use the word *religion*? Does it include spirituality? And what do we mean by *spirituality*? Sociologists face an ongoing challenge when they attempt to define religion. Meredith McGuire (2005) very simply categorizes definitions of religion into functional definitions and substantive definitions. In short, functional definitions focus on what religion does for the social group and for the individual. The dominant theme running through most functional definitions is social cohesion—in other words, how religion offers a sense of connectedness to others and to a larger picture. Substantive definitions, on the other hand, examine what religion is and what does not count as religion. Substantive definitions attempt to define religion by examining its core elements, most typically a belief in a higher being, a set of prescribed beliefs and rituals, and so on.

TIME to REFLECT

Is hockey a religion?

Émile Durkheim (1964 [1912]) certainly deserves some of the credit for the shape of functional definitions of religion, but his influence can be seen in substantive understandings as well. Durkheim's preoccupation was with social cohesion, and thus he viewed religion through this lens. For Durkheim, religion contributed to social cohesion in that it was fundamentally a reflection of the society in which it existed and it was, at its core, a social or group phenomenon. He argued that society divided the world into the sacred and the profane and that the former was the focus of religion. Durkheim's work had some powerful effects on how sociologists of religion define and think about religion. The binary between the sacred and the profane has limited conceptual resonance for some cultures, especially, for example, many Aboriginal groups. Moreover, the emphasis on the social aspects of religion as they are highlighted by Durkheim has resulted in a denigration of sacred practices that are not communally oriented. Wiccans, for example, are often sole practitioners, unconnected to a 'faith community' in the traditional Christian sense of the word or in the sense that Durkheim thought was necessary for religious expression.

A contemporary application of Durkheim's functional ideas is reflected in the research of Robert Bellah and his colleagues (1985) in the United States. They spent considerable time exploring the role of religion in social cohesion, or **civil religion**. This elusive and amorphous concept emerged in the American context and was most vocally defended as a 'real' phenomenon by Bellah and his colleagues, who argued that it transcended any specific religious tradition and formed an ethical framework that existed apart from any one religion. However, the strong Christian presence in the United States might belie that claim. Perhaps the most important thing to remember about civil religion, its proponents argue, is that it forms an overarching framework that supports a cohesive society. It is equally important to remember that underlying this notion is the idea that society is based on and functions because of shared values and perspectives, something that is highly contested by many scholars. Much of Bellah's work has focused on understanding how society has departed from those common ideals and the consequences of that, which they identify as being largely negative.

In part, narratives of a cohesive society lost, such as that told by Bellah, return us to debates about the definitions of religion and spirituality. Many people neatly divide the two into religion as 'organized religion' and spirituality as somehow representing something less institutional and more private. This division is arbitrary at best, and hidden behind the categorization are particular power sedimentations that create a hierarchy in which 'religion' is privileged as what counts in terms of spiritual belief and practice. The implications of this are profound. Think, for example, about Wiccans, who are often sole practitioners or part of a very loosely organized group with broadly defined rituals and practices. Further, Canada's First Nations peoples are largely excluded from these conceptualizations of religion. It is important that sociologists think carefully about the work such categories do in the preservation of particular hierarchies of what 'counts' as religion.

Substantive definitions focusing on content are equally vulnerable and are often characterized by a reliance on Christianity to form the basis of the determining criteria. Some important challenges to conventional thinking about definitions of religion have emerged both within sociology and from other disciplines, such as anthropology. For example, Talal Asad (1993) and Naomi Goldenberg (2006), among others, each call into question the work that definitions of religion do. They argue that the separating out of religion as something distinct from everyday life is a decidedly Christian approach to thinking about spirituality and religion. As Asad states, 'It is preeminently the Christian church that has occupied itself with identifying, cultivating and testing beliefs as a verbalizable inner condition of true religion' (1993, 48). These scholars are making important inroads into the ways in which we conceptualize religion, which in turn may eventually affect how religion is measured.

Other scholars are making related arguments. For example, Linda Woodhead (2007) is especially critical of the dominance of functionalist conceptualizations of religion, especially Durkheimian models that privilege religion over magic despite the fact that both are related to the sacred or transcendent. We see among sociologists a tendency to denigrate or minimize religious or spiritual behaviours that don't fit into organized religion patterns like Christianity. Woodhead also argues that 'Religions look remarkably like what Christians think of as religion' (2007, 2).

What might an alternative approach be? How would we think about people's involvement with the sacred if we were able to step outside of Christian thinking about religion? Meredith McGuire (2005) has proposed a methodological strategy for attempting to move outside of the confines of Christianity. She asks us to attempt to bracket assumptions that we hold about religion and to travel with sociological eyes or a sociological imagination to the 'past as another country'. She uses this strategy in order to present the possibility of thinking in a time when there were 'no tidy boundaries between the sacred and profane' (2005, 3). In her more recent work, McGuire (2008) emphasizes the importance of **lived religion**, arguing that it is critical that sociologists pay attention to the ways in which people integrate and practise religion in their day-to-day lives. We will explore the idea of lived religion in greater depth later in this chapter.

Canadian scholar William Closson James (2006) offers a beginning point for the exercise in boundary deconstruction. He argues that 'as religion in Canada in the twentieth century becomes more highly personal and individual, we should expect it to continue to be characterized more by an eclectic spirituality cobbled together from various sources rather than a monolithic and unitary superordinating system of beliefs' (2006, 288). If we only consider religion that looks a particular way (in the view of the scholars mentioned above—that is, religion that looks like Christianity), we will miss a great deal of the richness of Canadians' spiritual lives.

New Religious Movements

There are some very practical implications of decisions about what constitutes a religion. Religious groups in many countries receive privileges simply because they are religious. In Canada, for example, there are certain tax exemptions for religions as charitable organizations. Thus, the determination of a group's status as a religion is not merely an academic discussion. In many countries, religions must register with a central state authority in order to be recognized for some benefits. Moreover, those who are not on official state lists are often persecuted through the harassment of group members, the denial of benefits, and the use of state apparatus such as the criminal justice system to keep groups under close scrutiny. Especially vulnerable are **new religious movements** (NRMs).

The classic study on new religious movements is *The Making of a Moonie*, by Eileen Barker (1984).

Barker was intrigued by the increasing frenzy around new religious movements, particularly the talk of 'cults', 'brainwashing', and 'deprogramming' in the late 1970s and early 1980s. As a social scientist, she decided that she would investigate the workings of the Reverend Sun Myung Moon's Unification Church, also known as the Moonies. As it turned out, the mostly young adults who were joining the Unification Church were not brainwashed, deprived of sleep, or malnourished, as the hysterical discourse around their 'conversion' had suggested. Rather, they were simply middle-class young adults who were seeking a spiritual or religious experience and a sense of community. Barker's research (1984; 2005; 2007) had profound implications and triggered a debate that continues to this day.

TIME to REFLECT

Find an example of the use of 'cult' language in news media coverage of a religious group. How does the article portray the group in question?

Many scholars have taken up the challenge of studying new religious movements, but debates around these very interesting groups rage on, complete with lawsuits from so-called deprogrammers who disagreed with social scientific findings that new religious movements are not dangerous. Occasional dramatic events

UNDER the WIRE

Religion on the Internet

From their early presence on Internet newsgroups to the development of interactive websites, religious people and organizations have taken advantage of and adapted Internet technology. The religious use of the Internet has come to the attention of scholars like Christopher Helland, who ask how religion manifests itself online. Helland, and others who study religion on the Web, note that religion on the Internet exists along a spectrum between religion online, which he defines as websites dedicated to sharing information about largely offline religious organizations, and online religion, which he describes as religious organizations that exist only on the Internet. While it is easy to determine what religion online might look like (take, for example, the Vatican webpage, which offers a great deal of information on the

Roman Catholic church but is not very interactive), it is difficult to gauge just what online religion might be or look like. For religion to be considered an online religion, does all interaction have to take place on the computer? Do rituals have to be virtual rituals, or can there be some offline components if the ritual is led by someone over the Internet? How many members have to be part of the community before it counts as an online religion? How active does the group have to be? Can there be solitary religious practice online? Does it still count as online religion if members tend to drop in and out? The difficulty of defining online religion reflects the debates about the very definition of religion itself. The dynamic nature of religion is demonstrated by its adaptation of new technology.

—Morgan Hunter

such as the mass suicide by 39 members of Heaven's Gate in California in 1997 served to fan the flames of the debates. For the most part, events like this are relatively unusual; however, they often shape public perceptions of new religious movements.

Susan Palmer is a Canadian researcher who is internationally known for her research on new religious movements. She has studied a number of religious groups, including the Raelians, the Quebec-based UFO cult that claimed several years ago that they had successfully cloned a human being. After 15 years of fieldwork with the Raelians, which involved attending their meetings, countless interviews with members and leaders, and examining video and written materials, Palmer wrote a book about Raelian culture. Her findings described a new religious movement and challenged the stereotypes associated with NRMs. For example, she found that 'Raelians with children make no effort to transmit the message to them, true to the Raelian ethic of individual choice' (Palmer 2004, 139). Children cannot be baptized until at least age 15; even when they ask to be baptized as Raelian, they must pass a test 'to prove that their choice was not due to parental influence or pressure' (Palmer 2004, 139). As Palmer discovered, the Raelians have sometimes contradictory beliefs and, as with any social organization, there are power struggles and tensions. In her response to a journalist who wanted to know whether she had observed coercive or manipulative behaviour among the Raelians, Palmer stated: 'Well, sure, but no more so than in my women's Bulgarian choir or my PTA meetings. In any human organization you'll find people who try to control other people. Often they have to, just to get the job done' (2004, 6).

Unfortunately, new religious movements still suffer from a great deal of stigma. The language of cults and brainwashing used in the news media and in day-to-day conversation undergoes little critical examination of why it is that we are sometimes quick to marginalize such groups. After all, what is the difference between brainwashing and socialization? This is where questions of agency come into play. What we mean by agency is the capacity and ability of a human being to freely make decisions. This sounds simple enough, but ultimately none of us make decisions 'freely' or without constraints. Whether it is the influence of parents, friends, economic

constraints, or possibilities, our decisions are shaped by our social world and by social structure. Often, marginal religious groups are conceptualized as exerting 'undue influence' on their members simply because the decisions those members make may be different from the choices we might make.

Theories of Religion and Society

Do we live in a secular society? We frequently hear this question as an affirmative statement with little explanation about what it means. To say a society is secular is to say that it is without religion in its public sphere. **Secularization** is the process by which religion increasingly loses its influence. Whether and how society is secular has occupied a great deal of time and energy among sociologists of religion. The narrative begins like this: Once upon a time, society was very, very religious—everyone participated in religious activity, and religion formed a sacred canopy of meaning over life for the vast majority of people. State and church were one and the same, with no separation between them and no perceived need for a separation. Then along came the Enlightenment, and gradually, science replaced religion (Berger 1967).

To complicate the story, secularization theory developed some very sophisticated versions. In the midst of it all were contested notions of how religion should be defined. They are important, because in order to determine whether religion is on the decline, we must first know what religion is. So if people stop attending church but take up yoga and engage in

Raelian founder and leader Claude Vorilhon.

rituals such as meditative walks in labyrinths, can we say that we live in a more secular world? If we measure secularization as the decline in people's participation in the rituals and practices of organized religion, such as church attendance, marriage, and baptism, then yes, Canada has definitely secularized. But what happens if the population is increasingly made up of people for whom church attendance is not and has never been a measure of religious participation? How then do we think about secularization? So while one measure of secularization can be the level of individual participation in religious activities, we can see that this presents some interesting measurement challenges.

TIME to REFLECT
Do we live in a secular society?

Another measure of secularization exists at the level of institutions. As religion loses its influence, it has less and less presence in social institutions such as law, education, health care, and so on. And in this process, religion loses its influence as an important social voice. The overt involvement of religion in social institutions, to be sure, is different from what it was in other periods in Canadian history. But the religious voice cannot be discounted entirely. Think, for example, about the religious lobby against the same-sex marriage legislation. The *Reference Re Same-Sex Marriage* case was decided in 2004; the legislation passed, and the Civil Marriage Act was approved on 20 July 2005. Groups opposed to the legislation include the Catholic Civil Rights League, the Convention of Atlantic Baptist Churches, the Christian Heritage Party of Canada, and the Roman Catholic Church of Canada. The Canadian Conference of Catholic Bishops, the Ontario Conference of Catholic Bishops, and the Seventh-day Adventist Church acted as religious interveners in opposition to the legislation. Moreover, in some provinces, access to abortion is severely limited because of the insistence of religious lobby groups. In court, witnesses still swear to tell the truth on the bible. Public institutions close on Christian holidays such as Christmas and Easter. In some measure, religious beliefs are so embedded in our social institutions and form part of their histories that it is almost impossible for them to become completely secular or without religious influence.

One of the most important pieces of social scientific research on the idea of secularization in recent years is that of José Casanova (1994), whose work employs a multilevel conceptualization of secularization. Casanova conducted a comparative study of religion using five case studies from two religious traditions (Protestantism and Catholicism) in four countries (Spain, Portugal, Brazil, and the United States). He identified a trend of 'deprivatization' of religion, arguing that beginning in the 1980s, religions began to reassert their intention to have a say over contemporary life. Casanova argues that secularization theory is actually made up of three interwoven strands of argument: (1) secularization as religious decline, (2 secularization as differentiation, and (3) secularization as privatization. While Casanova says that the idea that religion is differentiated (there is a secular and a sacred sphere) is a possible proposition, it does not follow that religion must be marginalized and privatized. Of course, if we think about world events, Casanova's argument seems plausible. Religion is intertwined with many of the major world events we might think about and in sometimes very public ways.

In Canada, we now have what some scholars describe as **believing without belonging**, which means that while many Canadians still cite an affiliation with organized religion at census time, many of them do not have much, or any, contact with the churches to which they say they belong. Given current measures of religious life, it is difficult to determine the parameters of belief. To what extent individuals engage in religious and spiritual activities that are not included in common measures cannot be known. Home-based religious practices, for example, remain largely invisible. Some scholars argue that such 'private' religious behaviours do not really count when thinking about secularization or when measuring religious behaviour. This is a puzzling argument, because such thinking would exclude many religious groups who engage in religious practice almost exclusively in the realm of the so-called private. Scholars like Robert Orsi (2003), who argues that even prayer is public, have challenged this public–private dichotomy. The essence of Orsi's argument is that we cannot create a meaningful dichotomy between the public and the private. In other words, you take your 'private' self with you in the realm of the 'public'.

Also discounted by some scholars are 'seekers', or people who may combine a variety of spiritual practices to create a pastiche of spiritual meaning. Thus, someone may engage in yoga, go to the Valentine's Day labyrinth walk at a neighbourhood church, and do a cleansing ritual of her living space to rid it of bad energy. Canada's First Nations present a complex blending of Christianity and Native spirituality that is difficult to characterize. This blending of traditions remains outside of the focus of study of much of the research on religion in Canada.

Understanding religion in complex ways can give us a rich picture of how people integrate religion and spiritual practices into their daily lives. One of the reasons secularization theory seemed to have so much credibility is the problem of definition. If religion is conceptualized in narrow ways—church attendance, institutional involvement, and so-called other public measures—then without a doubt it has shown a decline of such proportions that it might be reasonable to conclude that it will eventually disappear. But there are alternative practices that are not measured and that form an important part of spiritual identity. Moreover, at an institutional level, much of Canada's Christian heritage remains embedded in day-to-day practices, as we will discover in the next section.

The Quiet Revolution

The province of Quebec deserves special mention in our consideration of secularization. Its unique cultural position has numerous facets, not least of which is the story of religion in that province. If ever there was a classic story of secularization, Quebec seems to tell it. It had what we might consider an established church; historically, the Roman Catholic church played an enormous role in the lives of Quebec citizens at a personal level as well as institutionally (Simpson 2000, 276). Schools, hospitals, and much of public life was intertwined with the church. Public officials were Roman Catholic, as were most members of Quebec society.

In the late 1960s, it seemed that quite suddenly, the church pews were empty. The **Quiet Revolution** had happened. How this seemingly sudden shift came about remains a bit of a mystery, but the perception of the church as anti-modern, oppressive, and representative of an establishment with which the people of Quebec no longer wished to identify combined to create an impetus to abandon what had been a core part of identity in Quebec. David Seljak has argued that the Roman Catholic church did not give up its place in Quebec society; instead, it recreated its public role (2000, 135). Gregory Baum argues that the Quiet Revolution 'initiated a gradual process of secularization' (2000, 151) in Quebec. What this means in practice is still unclear. The influence of the Roman Catholic church in Quebec institutions has not been completely eliminated, and the relationship between

the church and the citizens of Quebec remains a complex one in the process of negotiation.

Recently, a very public debate about the role of religion in society has taken place in Quebec. That debate, in some measure prompted by the Multani case (discussed in detail in the next section), has been framed around the notion of 'reasonable accommodation' and has focused attention primarily on the religious practices of immigrants, particularly Muslims (although some practices of Orthodox Jews have also been the subject of discussion). The government of Quebec formed the Bouchard–Taylor Commission to examine the nature and extent of the problem that seemed to be emerging around accommodation. The question of how much 'accommodation' should be made for religious minorities has resulted in heated debate. For example, Solange Lefebvre has noted that 'when it comes to granting religious accommodations to members of minority religions, Quebeckers are reacting more strongly and publicly than people in other parts of North America' (Lefebvre 2008, 179). To some extent, the impact of the debate and the report prepared by the commission, *Building the Future: A Time for Reconciliation*, released 22 May 2008, is seen as an issue largely confined to Quebec. But this is not so, because we see similar debates arising all across Canada. The issues discussed in the report remain largely unresolved and are likely to remain so as the religious demographics of Canada shift to include a greater percentage of people who practise minority religions.

One of the important issues raised by the Quebec debates is the question of how disputes about religious freedom should be resolved. This is a complicated issue, and the answer to the dilemma tends to go in two directions which some people see as being incompatible. On the one hand, it is argued that people should just work out solutions to disputes among themselves. In the Bouchard–Taylor report, this was called 'concerted adjustment'. In other words, imagine that you are an employee who needs to have Thursday off because that is your holy day. A concerted adjustment approach means that you would just ask your employer and hope that everything could be worked out so that you could observe your holy day. But what if your employer thinks that your religion is silly and stupid or that you aren't really serious about your religion? Or that because Thursday is the busiest day of the week at work, you are just trying to avoid hard work? Maybe you try to work something out with your fellow employees—perhaps you pay someone extra to work for you on that day. Maybe there is no one you can pay extra to work for you. And so on. Then the concerted adjustment approach may not work so well. You are stuck. You can't just work it out. You need this work, but your religion is very important to you. This brings us to the other model, which is also discussed, although less favourably, in the Bouchard–Taylor Commission report. The legal model suggests that such disputes can be resolved by relying on legal rights and obligations, such as those found in the Charter of Rights and Freedoms or provincial human rights legislation, which protect religious freedom. The legal model, in theory, recognizes that religion can be an important part of one's identity and, as such, it should be legally protected.

There are problems with both of the models described above. The concerted adjustment model assumes that people behave reasonably and rationally. It fails to account for structural disadvantage and power differentials between people. Employees, especially in low-wage work, are often very vulnerable and are not in a position to negotiate with their bosses. Students are often vulnerable in educational settings. The model does not account for racism, poverty, and other factors that may make particular groups and individuals especially vulnerable and not in a good position to ask for 'accommodation'. Sociology offers us tools to explore the idea of structural inequality that is caused by factors beyond our control. The legal model too has its own set of problems. Legal solutions turn situations into adversarial matters which can generate or exacerbate hard feelings and misunderstandings. They are often expensive and are not accessible to those who may need them most. They can take months and sometimes years to resolve, and in the meantime, the parties are left without protection or resolution to the issue at hand. In truth, there is no easy mechanism for solving disputes about difference that emerge from religious commitment, and a mixed model, rather than one solution or the other, is the best approach. Laws can act as guidelines to help people understand what might be expected of them, both in terms of how they might be accommodated and how they should accommodate.

The very language of accommodation leads us to another important issue raised by the public discussion around the Bouchard–Taylor report. 'Accommodation' is often used interchangeably with the idea of 'tolerance'. Certainly, liberal principles of diversity management have relied on the notion of tolerance, and it has been a pragmatic strategy used by interfaith and other groups to set the basic ground rules for interacting with those who are different. But as Janet R. Jakobsen and Ann Pellegrini ask, 'what does it feel like to be on the receiving end of this tolerance? Does it really feel any different from contempt or exclusion?' (2004, 14). If you think about it, there is a rather large difference between someone tolerating you and someone thinking that you are equal, and therefore worthy of respect, and your ideas worthy of protection. Understanding religious difference as something that is to be tolerated, dealt with, or managed is therefore problematic. Some people argue that if we eliminate these concepts, then we will be left with nothing, and that even these basic positions are better than nothing. We might think, although perhaps optimistically, that it is time to develop an approach to difference that is rooted in a deep understanding of equality rather than a teeth-gritting tolerance of those who are not like us. In summary, the debates that have been most intensely played out in Quebec raise issues that are crucial to all of us as we figure out what it means to live in an increasingly diverse multicultural nation.

Religion and Law

One important social institution that mediates the ways in which religious beliefs can be expressed through practice is law. For example, if you are a Sikh and you wish to wear your **kirpan** (ceremonial dagger) to school, you may find yourself, as Gurbaj Multani did, arguing before the courts for your right to do so (*Multani v. Commission scolaire Marguerite-Bourgeoys* 2006). Law sets important boundaries on religious practices. Law provides a forum to which people can come to affirm their right to engage in certain religious practices. It is especially important for minority religious groups, whose practices are more likely to be called into question than those of majority religious groups.

Sections 2(a) and 15 of the Charter are the core sections dealing with the protection of religious beliefs in Canada:

2. Everyone has the following fundamental freedoms:
   a) freedom of conscience and religion . . .;

15. (1) Every individual is equal before and under the law and has the right to the equal protection and equal benefit of the law without discrimination and, in particular, without discrimination based on race, national or ethnic origin, colour, religion, sex, age or mental or physical disability.

In addition to these sections, two others have an impact on the shape of religion in the public sphere:

27. This Charter shall be interpreted in a manner consistent with the preservation and enhancement of the multicultural heritage of Canadians . . . .

29. Nothing in this Charter abrogates or derogates from any rights or privileges guaranteed by or under the Constitution of Canada in respect of denominational, separate or dissentient schools. (93)

The broad considerations mandated in s. 27 mean that the intertwining of ethnic and religious interests must receive some consideration in policy matters.

Contrary to popular belief, mostly imported from the United States and France, both of which establish the separation of church and state in their founding constitutional documents, there is no separation of church and state in Canada. Separation of church and state means that the church has no authority over the state or political decisions. You can see, particularly in the United States, that this is more an ideal than a reality. Keep in mind, though, that while Canada does not have a strict separation of church and state, it also does not have an established church or a church that has authority over the state. So where does religion fit in Canada from a socio-legal perspective?

Since the Charter was enacted in 1982, the Supreme Court of Canada has attempted to find a workable definition of religion that can be applied in its considerations of religious freedom. To this end, the Court has attempted to use a comprehensive definition that employs both functional and substantive elements. For example, in the *Syndicat Northcrest v. Amselem* (2004) decision, the Court states:

In order to define religious freedom, we must first ask ourselves what we mean by 'religion'. While it is perhaps not possible to define religion precisely, some outer definition is useful since only beliefs, convictions and practices rooted in religion, as opposed to those that are secular, socially based or conscientiously held, are protected by the guarantee of freedom of religion. Defined broadly, religion typically involves a particular and comprehensive system of faith and worship. Religion also tends to involve the belief in a divine, superhuman or controlling power. In essence, religion is about freely and deeply held personal convictions or beliefs connected to an individual's spiritual faith and integrally linked to one's self-definition and spiritual fulfillment, the practices of which allow individuals to foster a connection with the divine or with the subject or object of that spiritual faith.

Trying to define religion is no easy task, as we have already seen. It becomes especially challenging as courts try to distil a very complex concept into a workable definition that acts as a gatekeeper for claims based on religious identity. Thus, courts are faced with the unenviable task of trying to capture a dynamic idea in a definitional box. Orsi's work on lived religion offers important insight into the vast scale of the task of definitions and why it might pose especially difficult in law: 'The study of lived religion is not about practice rather than ideas, but about ideas, gestures, imaginings, all as media of engagement with the world. Lived religion cannot be separated from other practices of everyday life, from the ways that humans do other necessary and important things, or from other cultural structures and discourses (legal, political, medical, and so on)' (2003, 1972). In the definition quoted above, we see the Court using a substantive understanding of religion in its statement that religion is 'a particular and comprehensive system of faith and worship'. This raises the question of why a religion must be a comprehensive system of faith and worship. We might also argue whether that is even a useful way to conceptualize religion, since the religious behaviour of the vast majority of Canadians does not actually seem to fit with this notion of 'comprehensiveness'. Belief in a divine, superhuman, or controlling power poses similar problems in that some religions do not have what we might call a central authority figure. How is it possible to determine whether a conviction is 'deeply' held? These are the challenges posed by attempts to solidify religious behaviour into manageable definitions for law.

Although the Charter of Rights and Freedoms guarantees religious freedom and equality, the guarantees and rights in the Charter are limited by section 1, which provides a balance of sorts between individual rights and interests and those of society more generally. Section 1 states that the rights and freedoms included in the Charter are subject only to 'such reasonable limits prescribed by law as can be demonstrably justified in a free and democratic society'. This limitation means a court can find that rights have been violated but the offending legislation or policy can remain because it is a reasonable limit on religious freedom by the standards of section 1. Take, for example, a Jehovah's Witness parent who wishes to refuse blood transfusions for her child's cancer treatment. A court may force the child to receive transfusions by assuming temporary custody and overriding the wishes of the parent (and the child). While a court may find that such treatment constitutes a violation of the religious freedom provisions of the Charter, it may also find that such a violation is justifiable under section 1 as representing a societal interest (see Beaman 2008).

To give this discussion a bit more context, let's consider one case in more detail. When we discussed the definition of religion and the matching of definitions and actual religious beliefs and practices as they come before the courts, we questioned whether it is possible to define religion in an inclusive way. The Multani case (2006) provides an example of the subtleties of this process. In that case, 'G' (the son) and 'B' (the father) were fully observant Sikhs, or, as the Court described them, orthodox Sikhs. An arrangement was made to accommodate G's wearing of a kirpan (an article of faith that resembles a dagger) to school; this agreement specified that it be kept under his clothing, sheathed, and sewn shut. The school commission refused to ratify the agreement;

however, the superior court set aside that decision. The court of appeal upheld the school commission, and the Supreme Court of Canada allowed the appeal. We might see the carrying of a kirpan as an issue of safety, but given the extent of the provisions the Multanis made to keep the kirpan relatively inaccessible, this is not a viable argument and was not one that the Supreme Court of Canada accepted. We must ask, therefore, why the kirpan became an issue to the point that the family was forced to go to the Supreme Court of Canada to be able to exercise their religious beliefs. In part, we can identify the very narrow conceptualization we have of religion and the way it is practised. Fortunately, the Supreme Court was not so limited in its approach.

The naming and construction of the kirpan as a weapon is a discursive practice that relies on a socially constructed set of categories. Manjit Singh comments:

> Rather than go into a detailed explanatory meaning of the kirpan, I would like to talk about another weapon with origins in medieval Europe that has been adopted in Canada and Quebec as a symbol of public authority. I am talking about the mace that lies on a table in front of the speaker

(Independent Picture Service/Alamy)

The Hutterites, although a small religious community, have internal differences related to dress. The Dariusleut will fasten their shirts only with buttons, whereas the Schiedeleut, who consider buttons too flashy, opt for hooks, eyes, and snaps.

of the House of Commons in Ottawa as well as in the National Assembly in Quebec City. According to Webster's dictionary, a mace is 'akin to a staff or club used especially in the Middle Ages for breaking armour' and 'an ornamental staff borne as a symbol of authority before a public official or legislative body'. It is clear from the above wording that a mace is a weapon. In the context of the two legislative chambers, however, it is a symbol of state authority. No one has ever questioned that some day, some member of one of these chambers, in a fit of rage, could use this weapon to attack a fellow member. The point of this discussion is that through mutual consent and historical tradition, this lethal weapon has come to represent the authority of the state. (*Multani v. Commission scolaire Marguerite-Bourgeoys* 2006, at para 39)

Singh's insightful commentary points to the constructed nature of religious symbols. James Beckford (2003) employs a moderate social constructionism, which acknowledges the culturally and socially situated position of religion as a concept and as a practice. Beckford notes the definition of religion as shifting throughout time and the link between those shifts and power relations: 'what counts as "really religious" or "truly Christian" are authorized, challenged and replaced over time' (2003, 17).

Religious groups complicate issues around definition because they may disagree among themselves about who counts as a 'real' member of their group. Adherents may challenge others who claim to be members of their group but who don't participate in particular rituals or adhere to specific beliefs fundamental to their religious world view. Thus, we have some Anglicans, for example, who support same-sex marriage while others are opposed to it. Some Muslims support the use of sharia law, others do not, and so on. These tensions can be confusing to those who are outside of the religious group and might be trying to understand the group's religious identity. Some members of a religious group may claim to speak for all members. However, to perceive that all members of a religious group believe and practise in the same way is misled at best. Think about the variations of the faith group with which you are most familiar. You will quickly see that religious identity is a complex factor in the consideration of religious belief and practice and its protection.

Religion and Gender

Why is it important to talk about religion and gender? Because there are some decidedly gendered aspects to religious participation. Women tend to make up the bulk of religious congregations, and some argue that it is women's decisions to stop attending that have been

HUMAN DIVERSITY

The Hutterites

For a small group of Hutterites in Alberta, having one's picture on one's driver's licence is a violation of religious teachings because it contradicts the second commandment, which prohibits making a graven image of oneself. For this group, posing for the picture on the driver's licence violates this teaching and is contrary to their religious beliefs. When the Alberta government revoked its exemption for the Hutterian Brethren in 2003 and insisted that they would need to have their pictures taken in order to obtain a valid driver's licence, the group took the matter to court, eventually going all the way to the Supreme Court of Canada, arguing that the new law violated their religious freedom. Unfortunately, the Supreme Court disagreed and found that the requirement for a photograph was a reasonable requirement that was necessary to protect the integrity of the Alberta system,

which was also directed at preventing identity theft. Not all of the judges agreed, however, with the majority decision. Justice Abella put the matter succinctly:

Unlike the severity of its impact on the Hutterites, the benefits to the province of requiring them to be photographed are, at best, marginal. Over 700,000 Albertans do not have a driver's licence and are therefore not in the province's facial recognition database. There is no evidence that in the context of several hundred thousand unphotographed Albertans, the photos of approximately 250 Hutterites will have any discernable impact on the province's ability to reduce identity theft. (*Alberta v. Hutterian Brethren of Wilson Colony* 2009, at para 115)

the catalyst for the sharp decline in religious participation, at least among Christians. Gender roles are a definite flashpoint in contemporary conversations about the ways in which religion and society intersect.

A number of scholars, most notably Mary Daly (1985) and Naomi Goldenberg (2006), have argued that religion has been a key institutional site of women's oppression. Women have been excluded from positions of power within church structure; they are instead relegated to 'domestic' roles within the church. Some of these same debates are emerging in relation to Muslim women. Interpretations about wearing the **hijab** vary; some define it as a cultural rather than a religious symbol, while others see it as a symbol of women's agency or choice, and still others perceive it as a sign that a woman is oppressed by her religion. Perhaps most important, but sometimes overlooked in this discussion, are the voices of Muslim women themselves. Research reveals the complexity of this issue but, most significantly, clearly demonstrates that women's interpretations cannot be excluded from the interpretation of the meaning of a particular practice.

Homa Hoodfar's research (2003) helps us to better understand the complex ways in which women interpret their own choice to wear a hijab. For many of the young women Hoodfar interviewed, the choice to wear the hijab created a newfound sense of freedom from strict parents. All of a sudden, they were free to engage in activities that their parents had previously forbidden: 'parents seem to be relieved and assured that you are not going to do stupid things, and your community knows that you are acting like a Muslim woman, you are much freer' (2003, 214). Moreover, for some women it was a strategy to generate respect not only among fellow Muslims but also in the broader society: 'I am telling them to see me otherwise. Do not think of my body, but of me as a person, a colleague, and so on' (2003, 221). Many of the women interviewed by Hoodfar were extremely strategic in their choices to wear the hijab, weighing the advantages and disadvantages and often concluding that the advantages outweighed the negatives. For some, it also opened opportunities to discuss their Muslim beliefs and to dispel prejudices and misconceptions.

Ultimately, there are a number of strategic choices involved in women's engagement with religion. Some, like Daly (1985), who is a former Roman Catholic nun, argue that religion is so patriarchal that there is no way that women can freely exercise agency within its confines and that women must therefore abandon traditional religion and create their own spheres for spiritual fulfillment. In part, this has been, if not the goal, the effect of some Wiccan groups. Wendy

Muslim women who choose to wear the hijab find that the head covering allows them greater freedom and opportunity.

(Megapress/Alamy)

Griffin's (2000) research on pagan groups has documented the approach of Dianic Wiccan groups, which is largely separationist, and her research on the Circle of the Redwood Moon found that they had largely abandoned traditional approaches to religion and spirituality. Naming themselves Dianics after the Roman goddess Diana, they base their holy days largely on seasonal cycles. They conceptualize the divine as female rather than male, and they situate their spirituality in feminist analysis that includes political activism. They are largely a women-only group and in this way have separated themselves both from the patriarchy of larger society and from traditional religious practices and beliefs.

Some strategies are less radical and call simply for a reshaping or reframing of religious teachings. For example, while evangelical Christians describe gender roles in rather particular ways (wives, for example, are taught to submit to their husbands), the exact ways in which these roles are interpreted are perhaps not as literal as one might think. Listen to Jane, for example, an evangelical Christian woman who, while she says she is a 'submissive' wife, tells the interviewer that all decisions in her marriage are made jointly:

> I have had some discussions with women who have a real difficult time with that—wives submit to your husbands. Now, I don't have difficulty with that at all, because in the next breath it says 'Husbands love your wives as Christ loved the church.' In my mind, we've got the easy end of the job, they've got the hard one. I mean, they've got to love like Christ. (Beaman 1999, 30–1)

Jane's approach is representative of that of many women who are part of more conservative religious traditions. Women within these groups grow impatient with those who would characterize them as having no choice or agency. While non-members tend to characterize them negatively, they often see themselves as benefiting from the demands of their religious traditions, which create rules for both men and women.

TIME to REFLECT
Does religion cause violence?

One of the most heated areas of discussion in relation to such teachings is violence against women. The relationship between teachings such as submission or male headship and violence against women is rendered even more complex by the valorization of family unity within faith communities (Alkhateeb and Abugideiri 2007; Nason-Clark and Kroeger 2004). This often means that women from conservative faith communities—whether Christian, Muslim, or Jewish—are often especially hesitant to take action if they are abused by their husbands, for fear of disrupting the family unit. Thus, such women are 'more likely stay in abusive relationships, more likely to return to abusive relationships after counseling, and tend to be more optimistic that abuse will stop if the abuser has some form of counseling' (Nason-Clark and Fisher-Townsend 2007; Nason-Clark 2004).

OPEN for DISCUSSION
Polygamy: Do Women Really Choose?

After years of keeping a low profile in BC's lush Creston Valley, the community of Bountiful opened its doors to the public April 21—media and protesters alike—to set some records straight.

A group from the community, calling itself the Women of Bountiful, hosted a press conference at a community centre 10 km away in Creston. Their aim was to show Canada that they are fully aware of their lifestyle choice; they enjoy sharing husbands even though they admit polygamy is illegal in Canada, and they will use Canada's Charter of Rights and Freedoms to argue that plural marriage is covered by their freedom of religion.

'We the women of our community will be silent no more,' said Zelpha Chatwin to the 300 people in attendance. 'I love the fact that my girls and I only have to cook and clean once a week. [Polygamists are] a team of players who care for each other.'

The women also said plural marriages come with various benefits, such as pooling resources and talent and higher household incomes, reported the *National Post*.

—Meghan Wood

Violence against women within religious communities is made even more complex by the fact that such communities are reluctant to admit the existence of violence within the family, perpetuating what Nason-Clark and Kroeger (2004) have called a 'holy hush' of denial. To further muddy the waters, secular agencies for women are often reluctant to include religious resources in their strategies for helping abused women, thus sometimes excluding a resource that is important to abused women from faith communities.

Gender roles within new religious movements are difficult to generalize because new religious movements are immensely varied in their teachings. Further, with the exception of a few pioneering scholars like Susan Palmer (2004), there is little research on new religious movements in Canada. Palmer's research on Raelians, for example, found shifting gender roles during the life course of the group. Before 1998, for example, gender roles were egalitarian; however, following the founding of 'The Order of Rael's Angels', a special women's caucus within the movement that emphasizes women's 'feminine charisma' and highlights free love, Raelian views shifted to place an emphasis on the so-called unique qualities of women, a move, observes Palmer, that has served to polarize the sexes within the movement (2004, 139).

The intersection of religion and gender often triggers interesting debates. The opposition of some religious groups to same-sex marriage was arguably a reaction to what they perceived to be shifting gender roles. Some groups argued, for example, that marriage was solely the terrain of opposite-sex couples and that it was inherently designed (by god) that way. Heather Shipley (2008, 5) notes that 'Defining marriage as an historically religious institution, while inaccurate, is a common argument promoted by religious interest groups who seek to preserve the heterosexual institution of marriage.' For example, in their factum regarding Bill C-23, An Act to Modernize the Statutes of Canada in Relation to Benefits and Obligations, in 2000 the Evangelical Fellowship of Canada argued that 'God instituted marriage for the express purposes of companionship, partnership in the task of procreation, for fulfilling a stewardly responsibility for the Earth, but fundamentally to mirror the intimate relationship which God desires to have with his people' (2000, 3). At the core of their arguments was a sense that there is a divinely mandated purpose to marriage. The Supreme Court of Canada rejected this argument, and it is now legally possible for same-sex couples to marry in Canada. However, the legislation has allowed for 'the freedom of officials of religious groups who choose not to perform marriages that are not in accordance with their religious beliefs' (Canada 2005, 3), preserving a space in which religious voices override basic human rights.

Conclusion

This chapter has considered the contemporary picture of religion in Canada and explored some of the issues that are of key concern to sociologists of religion. We have seen that the religious demographic of Canada is in an interesting period of flux that may have an important impact on Canadian society. Despite their small numbers, religious minorities in Canada play an important role in defining diversity in Canadian society. It is exciting to be a sociologist of religion in Canada in this time of rapid change.

We have already discussed to a great extent the limits of existing social scientific research on religion in Canada. Quantitative measures need to be more comprehensive to reflect Canada's shifting demographic. We need to understand how religious beliefs and practices fit into the day-to-day lives of Canadians and, most especially, the religious and spiritual practices of those who have arrived in Canada more recently. We need definitional and measurement standards that are not based on Christian understandings of religion. In other words, our research needs to extend beyond church attendance and bible belief. Our measures of spiritual practices should take an inclusive turn.

Research on religion in Canada is slowly becoming a priority. Sociology has an important role to play in the study of religion in Canada. No matter which theoretical tradition one uses, it brings tools that are invaluable to research. Ethnographic accounts of religious communities will provide insight that is as important as more detailed survey work. Research into important intersections—like youth and religious practice and belief—is central to predicting the role that religion will play in Canada's future. Key, though, to understanding the role of religion in Canada's future are interdisciplinary approaches that seek to draw on the expertise and insights of various traditions of scholarly thought.

questions for critical thought

1. Critically examine your own religious history. Where do you and your family fit in terms of believing and belonging? Are you part of a religious minority or the religious majority?

2. Do we live in a secular Canada? Are there some areas of life that are more or less removed from the influence of religion?

3. How is religion best defined? Explain your decision.

4. Should religion have a say in public policy issues?

5. Why do you think the language of 'cults' and 'brainwashing' persists?

6. The United States and France have a separation of church and state. What does this mean? Can you think of other examples of states that have such a separation? Examples of states that have a merging of church and state, either now or historically?

recommended readings

James Beckford. 2003. *Social Theory and Religion*. Cambridge: Cambridge University Press.
James Beckford presents an important examination of religion by developing clear links between social theory and the social scientific study of religion. Relying on moderate social constructionism, his theory focuses on the ways in which religion is a complex and social phenomenon.

Peter Beyer. 2006. *Religions in Global Society*. London: Routledge.
In this book, Peter Beyer analyzes religion as a dimension of the historical processes of globalization as a means of understanding religion in a contemporary global society. Beyer uses examples ranging from Islam and Hinduism to African traditional religions, resulting in an overview of how religion has developed in a globalized society.

Meredith B. McGuire. 2008. *Lived Religion: Faith and Practice in Everyday Life*. Oxford: Oxford University Press.
McGuire explores how the concept of lived religion can be used to understand the actual religious experiences and practices of people in everyday life.

William A. Stahl. 2010. 'One-dimensional rage: The social epistemology of the New Atheism and fundamentalism'. In Amarnath Amarasingam, ed., *Religion and the New Atheism: A Critical Appraisal*. Leiden: Brill.
Stahl examines New Atheism and fundamentalism, exploring the similarities within the two movements to argue that they have the same underlying purpose, which is to recreate authority in late modern society.

Joel Thiessen and Lorne L. Dawson. 2008. 'Is there a "Renaissance" of religion in Canada? A critical look at Bibby and beyond'. *Studies in Religion/ Science Religieuses* 37 (3/4): 389–415.
Thiessen and Dawson challenge Bibby's notion that there is an increase in Canadian involvement in organized religion. The authors look to find other ways to understand the expressions of religiosity Bibby finds among the Canadian public.

Michael Wilkinson. 2009. *Canadian Pentecostalism: Transition and Transformation*. Montreal and Kingston: McGill-Queen's University Press.
Wilkinson's collection offers a look at Pentecostalism across Canada, including essays from a range of disciplinary perspectives, and covers both contemporary and historical aspects of this global phenomenon.

recommended websites

CBC Archives on Religion and Spirituality
http://archives.cbc.ca/society/religion_spirituality
The CBC archive contains clips from radio and television coverage of religious issues in Canada over the past 60 years.

Centre for Studies in Religion and Society
http://csrs.uvic.ca
The Centre for Studies in Religion and Society, hosted by the University of Victoria, studies the intersection of religion and public life from an interdisciplinary perspective. The website offers information about religious research in Canada.

RAVe
www.theraveproject.com/index.php
Religion and Violence e-Learning offers support and provides resources for women, clergy, and other service providers on the issue of violence against women. It is based on sociological research conducted by Nancy Nason-Clark and her Religion and Violence research team.

Religion and Diversity Project
www.religionanddiversity.ca
This University of Ottawa–based project brings together 36 researchers from Canada and other nations to research the contours of religious diversity and possible responses to the opportunities and challenges it presents.

Statistics Canada: Religion, guide to the latest information
www.statcan.gc.ca/search-recherche/bb/info/3000017-eng.htm#il
The Statistics Canada guide to the latest information on religion offers links to tables with data on religion in Canada and links to the most recent Statistics Canada publications on religion.

United Church of Canada, Wonder Café
www.wondercafe.ca
This website attracted a great deal of media attention when it was first launched because of its innovative approach to the discussion of religion and spirituality, including one ad used to promote the webpage that featured a can of whipped cream with the caption 'How much fun can sex be before it's a sin?' as a way of introducing questions of religion and sexual behaviour.

chapter 18

Politics and Political Movements

Howard Ramos and Karen Stanbridge

In this chapter, you will:

▶ Learn that sociologists understand politics as encompassing the negotiation and contestation of *power*, the capacity to realize one's will despite the resistance of others.

▶ Understand that power is relational and socially determined.

▶ Discover that the state plays a focal role in the sociological analyses of politics because it shapes the ways in which people and groups exercise and negotiate power in crucial ways.

▶ Learn that political sociologists have analyzed politics in terms of control and contests over material, cultural, social, and institutional resources.

▶ Appreciate that it is most productive to understand politics as the interaction of these processes in the contestation of power.

▶ Understand that transnationalism has begun to challenge the role of states in contemporary society but has not eliminated their legitimacy in governing politics.

Introduction

Politics has had a bad rap. For centuries, the great (and not-so-great) have complained about the incompetence and treachery of politicians, the ineffectiveness of the democratic process, and the inadequacies of government bureaucracies. And things don't seem to have changed much. Hear the word 'politics', and what often comes to mind is a bunch of people in suits, probably corrupt, arguing about matters that are so tedious, complex, or seemingly removed from our own lives that it's easier to just ignore them altogether.

But sociologists interested in politics and political movements believe that we ignore politics to our disadvantage. This is because politics is really about *power* and its *contestation*. It is not only about what happens on Parliament Hill—it is also a part of our daily lives. Sociologists understand politics as endemic to our social existence. It is not only about elections or the 'red tape' and bureaucracy of government but is also about how people negotiate their lives with family, friends, social groups, and institutions of all sorts. These too reflect power relations and challenges to them.

Sociologists use a much broader concept of politics than most of us are used to: one that sees politics playing out in realms of society that we normally consider outside of or unaffected by politics proper (Scott and Marshall 2009). But the reality is that power is exercised and contested in all these areas, formal and informal, and their outcomes interrelate. For these reasons, the sociological conception of politics is complex and multi-faceted. In this chapter, we will provide you with an overview of many of the key perspectives in political sociology.

We will begin by elaborating on the core concept of power and then consider the state, or what is usually referred to as 'the government', and its central, but not exclusive, place in sociological analyses. From there, we will move on to the different ways that political sociologists have conceived of power, looking at who holds it and on what bases it is exercised. At the same time, you will learn about sociological approaches to the *challenge* of existing power arrangements—that is, when and why people come together in political movements, the means by which they undertake their activities, and the barriers or limitations that they face in contesting the status quo.

Power

Max Weber famously defined **power** as the capacity to realize one's will despite the resistance of others. It is the ability to do what you want, when you want, even in the face of opposition. Most people don't go around thinking about and interpreting the world as consisting of power relations. It might even sound somewhat hostile and confrontational to see the world in that way. More often, we go about our lives interacting with family, friends, classmates, or co-workers without considering how those relations and actions are shaped by the distribution of power among us. But the capacity to pursue and realize your ambitions or live in the kind of world you choose has a lot to do with the power you hold and the power of others to prevent or at least place limitations on your doing so.

Most North Americans see themselves as 'free' individuals who are at liberty to make decisions and do what they choose. They generally think of others in this way too. There are even some good reasons to think this way. People who live in Canada, for instance, enjoy rights and freedoms to pursue their self-interests that people in other countries might only dream about. But not everyone, even in Canada, has the same capacity to do what he or she wishes. Wealthy people, for example, tend to be more able to pursue their interests and achieve their aims than poor people; children and youth are much more restricted in their decision-making than adults; and

In the First Person

Howard Ramos is an associate professor at Dalhousie University. He engages issues of social justice and has published on political sociology, Canadian Aboriginal mobilization, transnational human rights, and identity. He continues to research these issues and works on new projects looking at the interacting influences of news media, political opportunities, and social movements, Canadian immigration, and the roles of time and space in social science.

Karen Stanbridge is an associate professor of sociology at Memorial University of Newfoundland. She is the author of *Toleration and State Institutions* (Lexington Books, 2003) and several articles on the intersection of state structures with policy and politicized groups. Her recent research concerns children, childhood, and the state, in particular, the ways in which childhood as a social category interrelates with constructions of the state and nationalism.

immigrants to Canada often have a harder time realizing their ambitions than most people who are born in Canada. Why is this so?

Often, the disparities we observe in the ability of different groups to achieve their aims are understood as stemming from differences in individual effort. In a country like Canada, where the Constitution, under section 15, says that everyone should have equal opportunity, there is a temptation to blame people in a disadvantageous position for their own misfortunes.

Many people believe that one's condition is determined by one's own decisions, choices, and ambition despite circumstances. But sociologists have found that there are other factors that influence people's capacities to pursue and achieve their goals, such as material wealth, cultural and social recognition, or access to institutions. These factors are defined by social structures that shape power relations and ultimately reward some people who possess or control them more than those who do not.

To illustrate, let's look at one example, which is perhaps the most obvious thing that affects people's capacity to do what they want in contemporary industrial societies: *money*.

TIME to REFLECT

What section of the Canadian Constitution protects equality, and which groups are mentioned in it?

People who have more money not only can buy more things but usually hold more power than those who have less money. Wealthy people often have more power to influence who owns and controls material resources, such as businesses or factories and cultural and social resources, such as investing in art and leisure activities or belonging to elite clubs, and because of this, they often have preferred access to institutions of power, such as schools or governments, that set standards for others in a society to follow. At first glance, many people think that merely having more money makes one powerful. However, the situation is not so straightforward.

Money is just paper and bits of metal and has little intrinsic value itself. It is important, however, because people agree that it is so and accept it as payment for work and goods or services delivered. The power of money, then, is socially negotiated and only has value because people agree that it does. Because of this, it doesn't matter what money *is* but

▼ FIGURE 18.1 Section 15 of the Constitution Act, 1982

Equality before and under law and equal protection and benefit of law

15. (1) Every individual is equal before and under the law and has the right to the equal protection and equal benefit of the law without discrimination and, in particular, without discrimination based on race, national or ethnic origin, colour, religion, sex, age or mental or physical disability.

Affirmative action programs

(2) Subsection (1) does not preclude any law, program or activity that has as its object the amelioration of conditions of disadvantaged individuals or groups including those that are disadvantaged because of race, national or ethnic origin, colour, religion, sex, age or mental or physical disability.(84)

Endnote 84:
(84)   Subsection 32(2) provides that section 15 shall not have effect until three years after section 32 comes into force. Section 32 came into force on April 17, 1982; therefore, section 15 had effect on April 17, 1985.

SOURCE: http://laws.justice.gc.ca/en/charter/1.html#anchorbo-ga:l_ l-gb:s_15.

rather that people agree that it is valuable. Believe it or not, in human history, different staples have been used to represent value instead of paper and metals. Rice, cocoa, and salt have all served as forms of currency at one time or another (Allen 2009). And codfish was used as currency by generations of fishers in Newfoundland, who exchanged their catch for goods supplied to them by the fish merchants through what was known as the 'credit system' or 'truck system' (Higgins 2007).

It is in this sense that sociologists understand power as relational—a phenomenon that only manifests in human relationships—rather than intrinsic to its various representations. It is not the paper or metal or food that matters but that other people recognize its importance and the legitimacy of those who try to exchange it.

Because power is relational, it is not fixed. It is a dynamic phenomenon that is under constant negotiation or challenge. Power is continually contested by people who oppose or disagree with its bases and defended by those who benefit from the status quo.

OPEN for DISCUSSION

Ice Pirates

The 1984 sci-fi comedy *Ice Pirates* depicts the hijinks of two space pirates, Jason and Roscoe, in a future where almost all of the sources of fresh water in the galaxy have been destroyed. So scarce is water that it has become the most valuable substance in the universe. Those who have it are all-powerful, and those who don't go to great lengths to get it, setting up the power contests that dominate politics in this movie. Water, in other words, is the new money.

Ice Pirates was not a particularly good or popular film, but it played on the idea that power is socially determined. It accrues to people and groups who possess or control things that others agree are valuable. And if you think the idea that water could become as important a source of power as money in the future is far-fetched, think again: some experts believe that the supply of potable water in the world is shrinking and that in the future, fresh water will be a much more valuable resource than it is today. Maude Barlow is national chairperson of the Council of Canadians and a strong proponent of making access to water a human right. As freshwater sources are polluted through human mismanagement or literally dry up as climate change warms the Earth, Barlow wants world governments to ensure that the water remaining will not become a commodity accessible only to the highest bidders. She warns that the decisions of government officials around the world to privatize water delivery to reduce the public cost of supplying water to their citizens means that freshwater delivery is coming increasingly under the control of private corporations like Vivendi, Suez, and Thames Water. Since private corporations are in the business to make profits, there is a potential for the executives of these companies to increase the cost of water to consumers in order to enhance their bottom lines and their power. Increase it by how much? Hard to say, but Barlow cites water privatization experiences in a number of places in the world where the cost of water to consumers has gone up by as much as 200 per cent, an increase that has, naturally, affected the poorest disproportionately. Money still rules as an important source of power in the world, but might we see a time when potable water trumps material wealth as a power resource? It seems there are some corporations out there that may be gambling on a real Ice Pirates world. Maybe it's time to dust off that video for some pirating tips from Jason and Roscoe!

For this reason, sociologists tend to treat power as a series of social or **political processes**. Power contests can result in new forms of power and modification of old power relations. Yet challenging and changing the bases of existing power relations is often difficult. It took a long time for women in industrial societies, for example, to successfully confront and alter many of the power advantages traditionally held by men and acquire greater freedom and status. Even so, women still face many power disadvantages, despite more than 100 years of gains in terms of voting, recognition of human rights, and emergence of pay equity legislation. To offer just one among many examples, women still earn less, on average, than men.

TIME to REFLECT

What reflects political power? Is it fixed and seen, or is it embedded within material, cultural, social, and institutional processes?

Political sociologists would argue that these inequalities persist in part because men, like most people and groups who hold and exercise power in a society, are generally not ready to give it up or even share it and so have mustered opposition to those seeking to eliminate old bases of power or generate new ones. Existing power-holders usually have material, cultural, social, and institutional resources at their disposal that their challengers do not. These resources are often the mechanisms that shape the political process that sociologists try to understand. You will be introduced to some of them throughout this chapter.

The State

Before examining different political processes in greater detail, however, let's first elaborate very briefly on what is meant by the **state** and why it is of central importance to political sociologists. Perhaps the greatest concentration of power in contemporary Western societies lies in the domain of the state. It

FIGURE 18.2   Average Weekly Earnings by Gender, 1997–2009

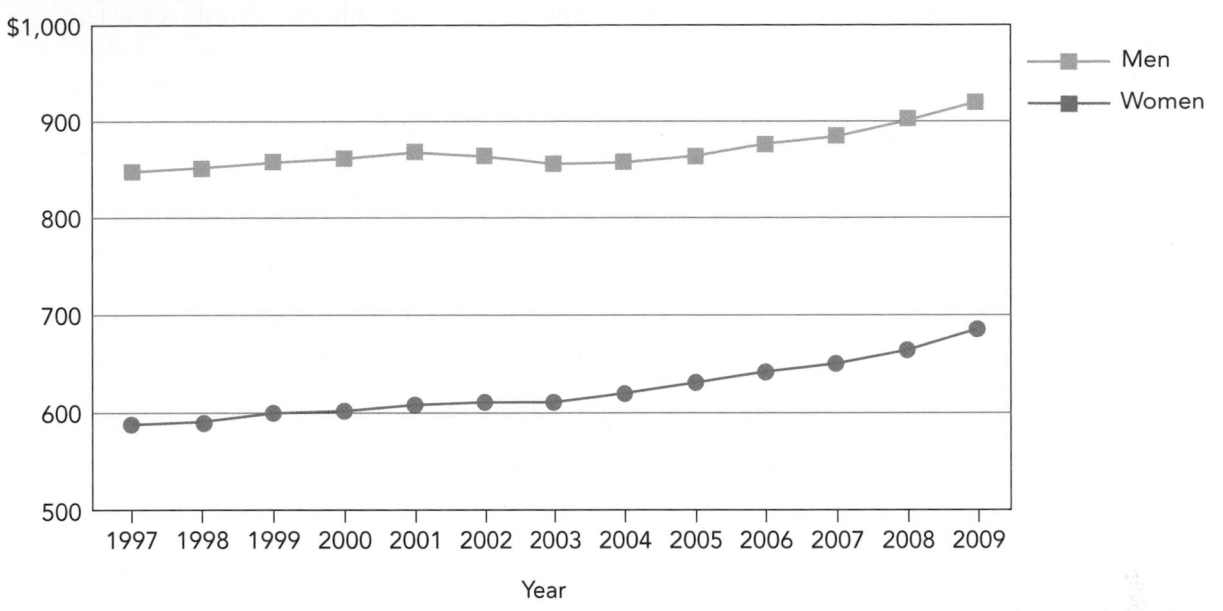

SOURCE: Human Resources and Skills Development Canada. 'Indicators of well-being in Canada'. www4.hrsdc.gc.ca/.3ndic.1t.4r@-eng.jsp?iid=18#M_2.

is the only institution whose officials have the legal right to tax people, to assault them through the police and military, to permit or force people to commit murder when conscripted into military service, to legally detain people in prison, and, in places where capital punishment exists, even kill them. It is also the only institution that can set policies and laws governing personal behaviour, restrict the ways individuals conduct business and their personal and love relationships, authorize credentials, license professions' activities, and so on. Basically, states set the rules by which other social processes are ordered and, in turn, how they are challenged. For these reasons, it is clear that states are the containers of much power and influence many political processes. You might ask: why do states hold so much power?

The shortest answer to the question is because the majority of people governed by a given state accept that it should. In exchange for **citizenship** that grants rights and privileges, people acquiesce to following and supporting the institutions of a state. Just like any other source of power, state power is largely contingent on whether or not people recognize it as legitimate. It is also founded in its ability to legitimately use violence against those who oppose its leaders. A state's power is likewise based on whether or not other states, social groups, and organizations recognize its legitimacy and do not challenge its claim to hold power

over its citizens. For instance, Canada exists because it gained sovereignty from Britain in 1867, was recognized by that country and its neighbouring state, the US, and is internationally recognized by groups like the International Committee of the Red Cross or organizations like the United Nations. If states do not obtain the recognition of their citizens and that of other power-holders, their claims to power are weak, and their ability to shape political processes is negligible.

A fuller answer to the question of why states hold so much power requires that we examine Western history. States emerged as one of the many transitions related to the **Industrial Revolution** and the emergence of modernity. As societies became larger and more urbanized and work became more specialized and mechanized, older forms of social order became unworkable. New urban middle classes that formed in association with these changes challenged the traditional authority of aristocracies, and the nature of negotiations around power began to change. Land ownership, for instance, which was once the epitome of material power, was replaced by capital and money as symbolic forms of material power. Divine and heredity rights, which formed the fundamental bases for power in feudal societies, were questioned as new religions were encountered and families began to marry outside of ethnic and religious lines. The dominance of churches and other religious institutions was also challenged by science and competing belief systems and their organizations. As the traditional foundations of power, knowledge, trust, and obligation weakened, new social configurations developed. They included new systems of governance like the modern state, through which leaders managed the social conflicts that accompanied negotiations around the new foundations of power: capital over land, individual over hereditary rights, science and rationality over religion, and so forth.

A key development that accompanied the emergence of the modern state was **bureaucracy**. It is an organizational form that is ordered by criteria independent of the personal qualities of people holding positions of power. Whereas the authority of organizations of the old order flowed from the personal power of the people who occupied positions in them, bureaucracy enabled power to be exercised through a staff abiding by impersonal practices and procedures less susceptible to individual whims and preferences. This granted modern state leaders a very powerful tool through which they could manage conflicts and set rules around their negotiation without having to worry that their decisions would be challenged or their position usurped by an unruly or disloyal official.

TIME to REFLECT

The state is always a central concern in the research of political sociologists, but it is seldom their only concern. Why?

So the state emerged as a key institution that accompanied the massive changes wrought by the shifts in the primary bases of power that came with industrialization. Through its various branches, state leaders today can shape all manner of social processes. In Canada, for example, sitting governments affect how material processes are negotiated in the country. When the state legalized same-sex marriage in 2005, it became possible for individuals whose same-sex partner died to collect survivor benefits, thus easing the material loss of wealth that accompanies the death of a spouse. It thus changed how material power is distributed. The same example also shows that state leaders influence how cultural and social power is engaged. The legislation redefined who is and is not considered legitimately married. Through the Constitution, the Canadian state also recognizes the cultural prominence of English and

French, not to mention defining who is and who is not Aboriginal. State leaders not only influence cultural and social power but can also direct whether other social institutions are permitted to exist and the scope of their power. In 2005, for example, the premier of Ontario, Dalton McGuinty, and his government ruled that Islamic sharia law could not be used to settle family disputes in that province. That decision eliminated people's access to that institution as a means of resolving conflicts. These examples show why sociologists pay so much attention to states. They are the institutional arenas in which political processes—that is, power relationships—are configured.

Let us now examine three broad groupings of these processes—material, cultural and social, and institutional—and their challenges.

MATERIAL PROCESSES

When it comes to power, money matters. It may or may not buy happiness, but people and groups who have more money certainly have the capacity to exercise more power in the world than those who do not. This is because, as we noted earlier, most others accept, wholeheartedly or begrudgingly, that people who possess more of those scraps of paper and pieces of metal should be permitted to purchase more things, are those who are most rewarded and have the most status, and because of this should have more say in how other relationships are negotiated.

Political sociologists who are concerned with why and how money operates as a source of power are part of a group of scholars who view the possession and control of *material resources* as central to politics. Material resources include money and other kinds of financial resources but also comprise things like property, technology, natural resources, and different means of communication, transportation, organization, and networking. Recall that money is just a symbol of other forms of privilege and represents shared value and worth; possession and control of other material resources have been historically, and still are, quite important. Materialists generally maintain that the more material resources a person, group, institution, or state has, the greater their capacity to realize their will over others and exercise power; they will also be more successful in their challenges to existing power arrangements.

Certainly the most famous scholar to have made the argument that power is determined by possession of material resources is Karl Marx. He said that those who owned and controlled the **means of production** have exercised the most power in human societies. Those who possessed the land to cultivate food, for example, held much power in agrarian societies and the feudal era; they controlled peasants' ability to provide for themselves and their families and dictated much of their daily lives. In the industrial era, which Marx witnessed emerging firsthand, those who owned machinery and factories, the **bourgeoisie**, wielded much power over the workers that kept them running, the **proletariat**. Again, owners governed, and continue to govern, workers' daily routines, setting their wages and determining the conditions under which they work.

For Marx, a person's power is contingent on her class position or her relationship to the means of production. If she owns the means of production, she holds power and can exercise it by influencing who has access to those means, for example, and how and for what purposes they are used. If she does not own the means of production, she is effectively powerless, compelled to submit to the owning class to gain access to the means she needs to survive.

But for Marx, the power of the owning class, the dominant minority, went far beyond the manor estate or the factory. Other institutions that comprise a society, like the state, religion, family, and the education and legal systems—what Marx called the *superstructure*—also reflected and sustained the power of the dominant class. Even the ideas that people held about how the world 'works' sustained the dominance of the owning class. The system Marx described was one in which the majority subordinate class was deceived by the dominant class into thinking that power inequalities of social institutions and the ideologies that sustained them were natural, commonsensical, or at best unchangeable. The subordinate class thus operated under a *false consciousness*, because these phenomena masked the true reasons why elites held power, which was because they controlled or owned the means of production. It was for this reason, Marx argued, that despite being in dialectic, or oppositional, relationship with the proletariat, the bourgeoisie was largely able to maintain its advantage over the working class.

Later writers critiqued Marx for reducing all politics to class politics. They elaborated upon his ideas by looking at how the control of material resources was related to other political processes. One of his most influential critics was Max Weber, another key figure in the study of power and politics. Like Marx, Weber recognized that most societies are **stratified**—that is, hierarchically ordered in terms of power—and that access to material resources is an important determinant of that order. Yet unlike Marx, he believed that material power flowed from non-material processes in addition to the means of production. In this regard, he highlighted status, or a person's social prestige, and party, the organizational and institutional resources that one commanded. We will examine each of these in subsequent sections, but first, let's briefly examine Weber's understanding of class.

Weber held a much broader conception of class and its outcomes than Marx. He said that if we really wanted to understand how material resources shaped the exercise of power, we should look to a person's or group's *market situation*—literally, the circumstances they face in the market economy—rather than their property holdings. This is because people's capacity to participate in the economy, to produce goods or services in exchange for income, ultimately shaped their life chances. These are the opportunities they have to acquire the things and lead the kinds of lives they want. For Weber, it was more useful to think of a class as a group of people who shared a similar relationship to the market and thus similar life chances, rather than just the possession of material resources alone. This makes some intuitive sense: the struggling small business operator has more in common, power- and life-chances-wise, with a warehouse clerk than with the owner of a successful chain of hotels, even though her relationship to the means of production is, strictly speaking, the same as the hotelier's.

Weber thus believed that material power resided not only in the ownership of the means of production but also in the possession of other sorts of material or material-related assets. The idea that such assets combine and overlap with other sorts of resources to affect how power is exercised in a society was expanded upon by C. Wright Mills. In his book *The Power Elite* (1956), he argued that the post–Second World War US was ruled by a higher circle of corporate, military, and political leaders. They held power because they controlled the massive and interlocking hierarchies of the economy, army, and state that made up the means of power.

TIME to REFLECT

In what ways is Weber's conception of class closer to what most of us understand to be constitutive of class position than Marx's definition?

John Porter (1965), a Canadian sociologist in this tradition, witnessed a hierarchical intersection between class, education, political standing, and ethnicity (see Helmes-Hayes 2002, 2009; Helmes-Hayes and Curtis 1998; Satzewich and Liodakis 2010). Porter described Canadian society as a **vertical mosaic** whose own 'power elite' was comprised of political, economic, and cultural leaders from the **charter groups**. These were individuals whose heritage was linked to Canada's English and French 'founders'. Porter argued that these groups used their material and other advantages to sustain their power and hold subsequent migrants to Canada in an *entrance status*. That is, non-charter groups earned lower incomes for their labour, held less prestigious positions, had fewer educational opportunities, and had less access to the state bureaucracy and political system. As a result, there had been little change in the class and ethnic composition of the elites who controlled various power structures between 1931 and 1961. Porter's findings were damning, because they showed little mobility in Canadian society.

Many of Porter's insights were challenged by scholars who came after him. Vic Satzewich and Nikolaos Liodakis (2010) offer an excellent overview of his work and later critiques, some of which we will highlight here. Gordon Darroch (1979), for instance, found that Porter focused too much on the hierarchical ordering of ethnic occupations and not enough on changes over time. When Darroch examined changes in the proportions of ethnic groups in given occupations, he concluded that there was a decreasing association over time. This seriously challenged Porter's blocked mobility conclusion. Others, like Lorne Tepperman (1975), argued that the privileged position of charter groups had been challenged by other European ethnic groups. And Michael Ornstein (1981) found that class, gender, and other labour market variables were more important determinants of earning inequalities than ethnicity. Notwithstanding such critiques, the vertical mosaic hypothesis was

SOCIOLOGY in ACTION

The Electric Car

By 2012, the first fully electric car, the Nissan Leaf, will be available to the global mass market. But did you know that a fully electric car was developed and was on the verge of being made available to the general public in the 1990s?

General Motors introduced a prototype of the EV1 in 1990 and began investigating its commercial viability in California soon after. Despite being wildly popular among the limited number of people the company leased the car to for testing and positive reviews from respected trade magazines, General Motors discontinued production in 1999 and pulled all the existing EV1s off the road in 2003, saying demand for the cars was too low and the costs of its production and maintenance too high. GM executives went so far as to have most of the existing EV1s crushed or shredded in an Arizona desert. Why would they do such a thing?

There is considerable controversy around the decision, but the explanation that some people offer for GM's actions suggest the 'power elite' that C. Wright Mills said controlled the US in the 1950s was alive and well some 40 years later. According to the critically acclaimed 2006 documentary *Who Killed the Electric Car?*, GM executives' decision to scrap production of the EV1 had less to do with the real demand for and cost of the car and more to do with the threat the electric car posed for the bottom line of powerful interests in the oil and gas and hydrogen industries.

The documentary suggests that leaders in these industries, along with executives from GM and other major car companies and officials in George W. Bush's government, purposely undermined the viability of the EV1 themselves, lobbying hard against the zero-emission legislation in California that motivated car companies to develop electric prototypes in the first place, understating consumer interest in the vehicle, overstating production costs, and promoting the hydrogen fuel cell as a better alternative to gas than electricity.

Some experts have dismissed the documentary's claims as mere conspiracy theory, with *Time* magazine confirming GM's decision to scrap the EV1 by declaring it one of 'the 50 worst cars of all time'. And with the major car companies, including GM, now embracing electric and hybrid technology enthusiastically, the suspicions that surrounded the EV1 have faded somewhat. But what *is* known about the episode raises some interesting questions from the perspective of Mills's claims around how the US is run by an 'uneasy alliance' (1956, 231) between political, economic, and military elites.

Why would the Bush government join forces with auto and oil executives to lobby hard for repeal of California emission-standards legislation? Is it significant that a key advisor to Bush on this issue, Andrew Card, was a former GM vice-president and president of the American Automobile Manufacturers Association? Was it purely a business decision to kill the electric car or an example of the power elite in action?

influential in setting the tone of contemporary Canadian political sociology.

Porter's influence has been especially prominent in what has become known as the political economy perspective in Canadian scholarship. This tradition focuses on the politics around the acquisition and negotiation of material power within the economic, geographic, and social conditions characterizing the Canadian experience. The importance of Porter's thesis can be seen in its revisiting over the past 20 years (Helmes-Hayes and Curtis 1998). Many of its tenets have been shown to still hold true, but on racialized rather than ethic grounds (cf. Agócs and Boyd 1993; Helmes-Hayes and Curtis 1998; Gosine 2000; Galabuzi 2006; Nakhaie 2007). In other words, visible minorities, rather than all non-charter ethnics, tend to retain their entrance status in a colour-coded vertical mosaic.

The Winnipeg General Strike, 1919, with strikers about to tear down the sign on the Board of Trade Building.

(Archives of Manitoba, Winnipeg Strike 7, N12298)

As Porter's and others' work has shown, material resources are elements of what constitute power, advantage, and inequality. But they can also be sought as means of *challenging* power and bringing about changes to existing power relations. Karl Marx certainly understood this when he predicted that the global proletariat would eventually experience conditions so dire that they would develop a class consciousness and overthrow the bourgeoisie-dominated capitalist system that so aggrieved them. Although the global revolution that Marx believed was imminent has yet to occur, class's serving as the basis for political movements has been a prominent part of Western social history. The nineteenth and twentieth centuries are littered with examples of workers, male and female, protesting the conditions under which they were compelled to work.

The Winnipeg General Strike is perhaps the most significant instance of organized labour protest seen in Canadian history. On 15 May 1919, approximately 24,000 workers walked off the job, to be joined by about another 6000 soon after. In total, roughly 94 unions participated, representing factory workers but also a wide range of public sector workers like the police and firefighters, among others. The strike lasted six weeks, and strikers demanded the right to collective bargaining, a decent salary, and an eight-hour workday (CBC News 2010). The city was literally brought to a standstill, and the federal government feared that the strike would lead to widespread mobilization across the country. In March of the same year, labour leaders met in Calgary to discuss the creation of 'One Big Union' that would represent workers across the country. All this happened barely two years after (in 1917) the Russian Empire was overthrown in a communist revolution.

But these collective efforts on the part of Canadian labour would not go unchallenged. Local business people and government fired the police and replaced them with 1800 'special' officers, at the hands of whom strikers faced harsh repression and violence. To counter agitation perceived as inspired by the Russian Revolution, the federal government issued a number of anti-immigration policies and threatened a number of eastern European ethnic workers with deportation (see Camfield 2008; Avery 1975). As Clément (2011) notes, after the strike ended, hundreds of trade unionists and communists were deported. The years following the strike witnessed only moderate rates of unionism, especially as the early 1930s saw unprecedented economic depression. It wasn't until after World War II that Canadians began to unionize in large numbers (Bowden 1989). But the Winnipeg strike left a legacy that no doubt contributed to working-class defiance in Canada's west and to the place of labour in Canadian politics more generally. In 1932, the Co-operative Commonwealth Federation (CCF) was founded in Alberta, a political party dedicated to voicing the concerns of labour and other groups committed to socialist ideals. The CCF was the precursor of Canada's current party of the left, the New Democratic Party (NDP), whose members continue to support the same broad principles.

Marx and like-minded scholars understood that the distribution of material resources has given rise to class-based political movements that have affected the progress of history. But political sociologists have also explored how material resources shape political movements in other ways too. *Resource mobilization theory* (RMT), for instance, is a perspective on social movements and collective action that highlights how crucial material assets are for launching successful challenges to power. This means money, obviously, but also includes things like places to meet to organize dissent, Internet space to launch counter-information campaigns and generate list-servs or Facebook sites to get people to mobilize, and 'human capital', person-hours committed to organizing resistance. RMT changed thinking around social movements, because for years prior to its development, political contention had been understood as emerging out of emotion and **contagion**, whereby people fall victim to irrational psychological urges when in group situations or deprivation and grievance. But in the 1970s, RMT theorists such as John McCarthy and Mayer Zald (1977) noted that crowds or feelings of anger and injustice, while important, were not sufficient to motivate people to start or join in collective action to challenge existing power structures. They said that in order to sustain mobilization and generate change, groups also had to draw upon and rely on various resources. Because of this, proponents of RMT studied, among other things, the organizations and networks needed to order and sustain contention, the leaders needed to rally people and pool resources, and the rational and competitive nature of contention.

> ## TIME to REFLECT
>
> Why is access to material resources so important to the success of a social movement? Can a movement with few resources still accomplish anything? Like what?

We have come some way since Marx and his elaboration of the material bases of power and its contestation. But while scholars after him have expanded upon and extended his ideas about how material resources configure power relations, his core claim—that control of material resources fundamentally determines the distribution of power in human societies—remains a key observation around which debates in political sociology are organized. Let us now turn to some of the non-material processes of power that sociologists examine.

Cultural and Social Processes

Clearly, material resources play a significant role in determining who does and does not exercise power in a society and which groups succeed or fail in challenging and changing existing distributions of power. But there are times when people and groups who are not particularly well off materially exercise and challenge power too. Social scientists have found that men, light-skinned people, non-Aboriginals, and able-bodied individuals, to name just some examples, are able to realize their wills, on average, more readily than their female, racialized, Aboriginal, and disabled counterparts, regardless of their material situations. Yet there are occasions when materially challenged groups, such as immigrants, victims of crime, the elderly, children, and homecare workers, defy or confront existing power-holders and achieve their aims. So the power of these groups must derive from sources other than material assets.

Earlier, we mentioned that Weber understood power as also stemming from sources other than material. One of these 'other' sources was *status*, the social prestige held by a person. Social prestige is often connected to material advantage but not always, said Weber; it can serve as a basis for power independent of money, wealth, or other material resources. And this makes sense; otherwise, how do we explain the power of influential people who were famously poor, like Mother Theresa? We also saw how scholars like Mills and Porter looked at overlapping axes of power: how material resources interact with other, non-material factors like ethnicity to shape politics. These scholars and others who followed them recognized that cultural and social processes can likewise affect power relations in ways not captured by purely materialist perspectives.

One of the most notable sociologists to ponder how culture affects politics is Antonio Gramsci. He was an Italian Marxist of the early twentieth century who wondered why the working-class masses did not respond to Marxist calls for a class-based revolution. Gramsci reasoned that it was because elites 'manufactured' the consent of the masses by communicating ideals supportive of the status quo. He argued that this was accomplished through political and cultural mechanisms like the state, schools, religious institutions, and media. He believed that to mobilize masses to contest power, **counter-hegemonic** positions had to be generated. These would lay claim to alternative cultural and social ideals that would reconfigure and challenge the established **hegemonic** order. Unfortunately, many of the intellectuals and activists who Gramsci thought were best positioned to challenge the hegemony of elites had little in common with the masses that were to be mobilized. They had trouble relating to ordinary people's everyday lives. To overcome this barrier, he advocated for 'organic intellectuals' to communicate counter-hegemonic ideals through popular culture in ways that would resonate with average people. Max Horkheimer and Theodor Adorno (2006) wrote about similar ideas when they introduced the notion of the culture industry and the role cultural production played in the rise of fascism and later American consumer culture. However, they were less optimistic about the potential to challenge the cultural power of elites.

Like other forms of power, culture is not equally distributed or valued. Herbert J. Gans (1974, 19), for example, observed something that still holds true today: that objects, ideas, symbols, and activities associated with 'high culture', such as hand-crafted furniture, opera, and classical literature and philosophy are more highly valued in society than things affiliated with 'pop culture', such as fast-food restaurants and clothes from Wal-Mart. Because it is more highly valued, people who have access to, control over, and familiarity with high culture tend to have higher status than those knowledgeable about pop culture. This status can in turn be employed to enhance a person's power.

TIME to REFLECT

In what ways are you exercising power when you choose to buy a Pepsi over a Coke? The Gap over Old Navy chinos? A Mac rather than a PC? In what ways is the concept of 'choice' in these cases illusory?

HUMAN DIVERSITY

Food as a Source of Power

Can food be a source of power? We all need sustenance to do anything, of course, but can the *type* of food we consume allow us to exercise power over other people or challenge existing power arrangements?

A number of political sociologists have begun to explore the political implications of food choice, with some interesting results. Recent years have seen an increase in consumer interest in Canada in organic and 'fair trade' products, as well as in locally grown and processed produce. A few years ago, CBC News (2009a) reported that organic food sales had increased by 20 per cent each year over the past decade. Canadian companies such as Just Us! Coffee Roasters Co-op in Nova Scotia have seen their profits in fair trade coffee grow over the years, and more restaurateurs are embracing the principles of the 'slow food' movement and the '100-mile diet' in response to demands from dining 'locavores'. Some Canadians are becoming small-scale farmers themselves, growing food and raising livestock in their backyards, even in urban areas.

How can we understand this trend in food consumption and production? Businesses that are part of the movement and their consumers often cite ethical, health, and environmental reasons for adopting these practices. But political sociologists have pondered some of the less obvious reasons why people engage with these trends. These scholars reveal that these 'conscientious' businesses and consumers can end up either confirming or confronting the status quo, depending on how they take up the new food trends.

Emily Kennedy and Sara O'Shaughnessy (2010) found that many pursuing these trends understand themselves as activists of sorts by providing 'alternative' examples to others in their communities. However, alternative food choices can also serve as a way to consolidate existing power, as Josée Johnston and Shyon Baumann (2009) contend. They observe that most of the people who embrace recent trends in organic/local/fair trade foods are white and affluent. Although they might deny that they are 'food snobs', these 'foodies' eschew food experiences they deem inauthentic, especially meals from chain restaurants, which the less affluent are more likely to patronize.

The result, say the authors, is that food choice becomes a means by which the privileged distinguish themselves from others, another status marker that signifies their belonging to an economic or cultural elite and differentiates them from less knowledgeable consumers.

Whether one's food choices represent a challenge to existing power arrangements or further evidence of one's privilege veils what is perhaps the most obvious inequity embedded in this discussion: that having any sort of 'relationship' with one's food is in itself a marker of advantage—which is something to think about as we consume our next burger or fair trade organic latte.

The sociologist who perhaps has gone the furthest in articulating how this works is Pierre Bourdieu. He coined what have become key terms in the field: **cultural capital** and **social capital**. Bourdieu's elaboration of these terms helped scholars to articulate how the different components of culture and other social practices operate in much the same ways as material resources. They act as sources of power and the bases of politics.

Simplifying somewhat, Bourdieu said that cultural capital is anything that reflects and facilitates cultural exchange between people. It includes how people see and understand the social world and the interactions that occur within it, as well as cultural symbols that help them to do so, such as language, clothing, customs, and so forth. Social capital is a resource that fosters social relationships and the privileges and obligations that one can draw from them; it is derived from membership in groups. The number and nature of groups to which you belong thus affect the social capital you possess, which in turn can affect the power you are able to exercise over your life and others.

Later scholars, including James Coleman and Robert Putnam, built upon Bourdieu's concepts. They looked at how social capital and social networks influence political participation. Putnam (2000) theorized a connection between membership in voluntary organizations like social clubs and bowling leagues and enhanced social capital and political participation. Informal associations, Putnam argued, build social capital by fostering social interaction that produces reciprocity and trust and confirms social norms. These are qualities that he said are essential in well-functioning communities and democracies. Putnam claimed that the US was experiencing declining rates of participation in these clubs, which he interpreted as an indication of the deterioration of civil society. He lamented that because of this, fewer people were acquiring and cultivating the social capital necessary to sustain an effective political community. Especially troubling was the decline in people's access to the *bridging social capital* that voluntary associations fostered. This is a resource found in the ability to tap into extended social networks and broad, inclusive associations. In contrast, Putnam feared the growth of *bonding social capital*, which is more insular and characteristic of homogeneous groups and fosters societal fragmentation rather than unity.

In his more recent work, Putnam (2007) has continued to warn of the threat to social cohesion posed by the dearth of bridging social capital, this time

among ethnic communities. He cautions against the over-insularization associated with bonding social capital in these communities, saying it could be detrimental to the unity of North American society. These writings reflect the observations of some of his earlier critics, such as Berman (1997), who identify how vertical social ties that generate strong but exclusionary bonds within a group contribute to radicalization, violence, and xenophobia in a society. They also echo concerns expressed many years ago by Raymond Breton (1964), who wrote about the benefits and problems associated with the **institutional completeness** of ethnic minority communities in Canada. But not all scholars see eye-to-eye with Putnam's analyses. A number of sociologists and some policy-makers doubt that (bonding) social capital fosters division and animosity among dominant and minority groups and instead cite an *opposite* trend. They have looked to social capital as a tool for immigrant and ethnic minority communities to overcome inequalities (see Portes 1998; Aizlewood and Pendakur 2005; Couton and Gaudet 2008; Breton 2003; James 2003). Canadian sociologists Baer, Curtis, and Grabb (2001) go even further and call into question Putnam's entire thesis. Their 15-country comparison of membership in voluntary associations and political participation concludes that Putnam's claim that a decline in the former precipitates a drop in the latter does not hold across many countries, including Canada, and that his argument might be exaggerated.

Although cultural and social processes can be used to understand how dominant groups sustain their power over subordinate masses, those same resources can be used to challenge existing power arrangements through political movements. Interestingly, some of the earliest and most influential work on how culture and social factors shape political movements was done by a sociologist who completed his doctoral work in the 1940s in Saskatchewan, Seymour Martin Lipset. Like many political sociologists, Lipset was interested in comparative analysis (Schwartz 2007). One of his focuses was on the similarities and differences between Canada and the United States, especially with respect to unionism, values, and institutions. Lipset (1991; 1964; 1963) observed high rates of union membership in Canada as compared to the US and said that the disparity was linked to value differences between the two populations (1990). Canadians tended to be more conservative and supportive of tradition and collective politics than Americans, owing to Canada's history of commitment to loyalist traditions. The US was born of rebellion and revolution, a history that fostered a culture of individualism that discouraged collective responses such as unionism.

A number of prominent Canadian sociologists have challenged Lipset's theory (Bowden 1990, 1989; Grabb and Curtis 2002; Grabb 1994; Baer et al. 1990, 1987). Many have found that Canadian *attitudes* are not significantly different from those of Americans but rather that Canadian *institutions* have been more supportive of unions than those in the United States. Lipset himself countered these challenges by asking *why* such institutions were more supportive in the first place, suggesting again that Canadian values have encouraged us to elect politicians who institute collective and union-friendly policies (Lipset 1990).

Others, such as Yates (2008, 88–9), have questioned whether Canadian values are as homogeneous as Lipset implied, noting that class politics in Canada have largely been overshadowed by regionalism, language, and nationhood.

If Lipset's thesis was one of the first to ponder the effects of culture and social factors on collective action, it was certainly not the last. A new generation of scholars emerged in the 1970s and 1980s seeking to understand contemporary movements arising around identity or lifestyle concerns. Although the grievances of second-wave feminists, LGBT activists, environmentalists, and so forth sometimes revolved around material matters, many of these movements seemed to be more concerned with acquiring wider

UNDER the WIRE

Knowledge Age?

In 1994, the author and futurist Alvin Toffler and his colleagues wrote *A Magna Carta for the Knowledge Age*. In it, they detailed the implications of what they called the 'Third Wave Economy' for humankind, a world that privileges and values knowledge over the physical resources that held sway in the past.

This new 'Knowledge Age' came about, they said, because communication technology, most significantly the Internet, liberated knowledge from its physical environment. In 'cyberspace', ideas and innovations were now created and intermingled in a way not possible in the old economies. Toffler and his colleagues predicted this knowledge 'free-for-all' would only continue as communication technology became cheaper and more people acquired access to and opportunities to generate knowledge, the new source of power. The result, they said, was 'the potential for vastly increased human freedom', a democratization of power that would shake up the old authority structures that were so dependent on the 'brute force of things'. Such a society would foster new forms of institutions, including the state. They encouraged the creation of governments that would minimize regulation of cyberspace and maximize citizens' access to the Internet to enhance the liberating and levelling potential of the wired Knowledge Economy. Such were the early, heady days of the 'information superhighway'. What happened? Has knowledge really usurped material as the primary power source in the world? Has access to cyberspace resulted in more democratic politics?

According to Darin Barney (2000), it has not. He argues that access to the new technologies is limited to the world's most (materially) privileged

populations, while control over their production and distribution has, with the blessing of most governments, remained in the hands of huge private corporations like Microsoft and Apple. Barney says that the new technologies, by making it easier for people to communicate and gain access to information, might have the potential to enhance participative democracy. But the reality is that they actually sustain, even strengthen, existing power arrangements, allowing the most materially powerful people and private interests to determine who possesses and uses technology and in what sorts of ways.

Manuel Castells (2007), however, has a different take on the issue. He says that the new network technologies allow for interaction among people in an essentially boundless communications space that counters traditional vertical ways of thinking about and structuring power. Although traditional (vertical, material) power-holders continue to try to hold on to and reassert their authority, it is just a matter of time before these new forms of communication will manifest in institutional change.

Who's right? Well, the unrestrained optimism of the early 1990s has been replaced by some sobering reflection on the continued exercise of power by the materially advantaged, and the caveat that any 'brave new world' that may emerge out of the new networked society may not be any better than the last. As we witness and participate in the future's unfolding, however, we can be sure that it will not be technology that will ultimately determine the outcome, but rather the manner in which people, the powerful and the not-so-powerful, take up and use that technology in their pursuits.

recognition of their unique identities and establishing the legitimacy of their alternative cultures or lifestyles than 'old' (labour unions and working-class) movements of the past. Since the intent of these *new social movements* (NSMs) seemed to be to create fresh cultural and social imaginations rather than to transform the dominant economic and political institutions, the political sociologists seeking to explain them did not see how traditional Marxist social science could be of much help.

Instead, some depicted emerging forms of mobilization as manifestations of a burgeoning post-industrial revolution, which Ronald Inglehart (1990a) termed a *culture shift*. These scholars said the NSMs were an outgrowth of the post–World War II baby-boom generation forming a large and unprecedented new middle class and fighting for the advancement of non-traditional goals. Later researchers carried on with the theme of 'newness' in these movements, theorizing that they were associated with the historical peculiarities of late modernity (see Touraine 2003; Habermas 1989; or Giddens 1991), the dawn of the information or discursive age (see Melucci 1996a; 1996b; 1989), and the rise of networked society (see Castells 2004).

Not all scholars agreed with these interpretations, however, and some were quite critical of the assumed newness of NSMs (cf. Calhoun 1993). Out of these debates emerged scholars like Alberto Melucci (1989; 1996a), who instead chose to explore contemporary movements in terms of their *collective identity*. These political sociologists said that what inspired, mobilized, and bound the members of these movements were not their common material conditions but their perception that they shared some relation or status, real or imagined, that connected them cognitively, morally, and emotionally with one another. This collective identity was separate from and distinguishable from members' personal identities but nevertheless fostered feelings of belonging and solidarity among members of the movement (Polletta and Jasper 2001). Since Melucci, many others have elaborated on the concept of collective identity and employed it to better understand 'why' people mobilize. Few see it as trumping previous material and resource-based explanations of challenges to power and instead see it as complementing material perspectives to offer a fuller understanding of mobilization (Tilly 2008).

Whereas collective identity approaches consider how cultural and social resources are used by political movements to challenge prevailing norms and the institutions that sustain them, these processes are also relevant for those challenging the very legitimacy of a state—a far more radical challenge to power. Post-colonial and nationalist movements represent two forms of protest by which people try to acquire greater cultural and social recognition, often in opposition to an existing state. Many sociologists have tried to examine why both have persisted over the past 100 years despite ever-increasing economic integration, communication, and technological development. Some, such as Canadian political philosopher Charles Taylor (1992), believe that much of their staying power derives from what he calls the *politics of recognition*—that is, negotiations around the value or status of subdominant cultures by those who hold power.

One of the most prominent early thinkers to write about post-colonial mobilization was Frantz Fanon (1967; see also 1968). As a black citizen of France in the 1950s, he experienced and documented the double standards of French society. He found that it was hierarchically ordered by class, ethnicity, language, and race. The most important cultural and social markers were largely monopolized by dominant power-holders, in this case almost exclusively a white aristocratic elite. For people from the former French colonies to succeed in France, they had to endure a devaluing of their own cultural and social identities and embrace those of the dominant colonizers. The psychological turmoil such circumstances caused led Fanon to plead for racialized groups to recognize and value *their own* identities as the first step to their emancipation and recognition by dominant power-holders. This line of thinking influenced later activists and anti-colonial thinkers such as Malcolm X in the United States and Steve Biko in South Africa. Similar observations were made by scholars looking at other colonial relationships, such as Edward Said, whose attention was focused on the Middle East, Asia, and the Muslim world. Fanon and Said showed that repression of identity and culture can have damaging influences on entire nations of people and that awareness and celebration of self-identity can be used as a basis for challenging state power.

In Canada, this post-colonial scholarship has resonated most strongly with québécois nationalists and Aboriginal rights activists, many of whom have embraced its endorsement of self-realization as a means to emancipation. Certainly Pierre Vallières, a leader of the Front de libération du Québec (FLQ), was aware of the defiance that could be generated among the repressed through self-identity when he titled his book about the Québécois *Nègres blancs d'Amérique*

Face-to-face: Soldier Patrick Cloutier and Brad Larocque, an Aboriginal from Saskatchewan, come face-to-face in a tense standoff at Kanesatake in Oka, Quebec, on 1 September 1990.

(Shaney Komulainen [1990], Canadian Press)

(*White Niggers of America*). It is also present in Aboriginal mobilization, as seen in Taiaike Alfred's call for a Wasase movement based on indigenous values and traditions (Alfred 2005).

But political sociologists have shown that culture and social factors affect post-colonial movements in other ways too. Staying with Canada's Québécois and Aboriginal movements for the moment, Susan Olzak (1983) pointed to the intersection of cultural changes and command over resources among Quebec's colonized francophone majority as contributing to its mobilization. She observed that as the Catholic church came to play an increasingly smaller role in provincial affairs, and as Quebec experienced the Quiet Revolution, the province witnessed unprecedented educational attainment, middle-class wealth, and urbanization among francophones. Olzak posits that this shift in cultural and material conditions among the francophone population fostered ethnic competition with anglo groups and contributed to mobilization in Quebec. Nagel and Olzak (1982) came to similar conclusions, arguing that these cultural changes, in concert with greater control of material resources, contributed to the emergence of Québécois

nationalism. Rima Wilkes (2004), likewise, partially examined whether the ethnic competition model can be extended to Aboriginal populations in Canada. Such studies suggest that colonial and post-colonial grievances associated with self-identity may be necessary but are not sufficient means to mobilization.

TIME to REFLECT

What are some of the triggers that account for post-colonial and nationalist mobilization?

In challenging a purely materialist understanding of power relations, Max Weber and other political sociologists after him came up with concepts and identified means that have proven useful for analyzing how cultural and social processes are taken up and how they play into the negotiations that characterize politics in societies. While none of these thinkers would abandon the idea that material resources affect how power is exercised and challenged, they recognize that it alone cannot account for all forms of politics.

Institutional Processes

So far, we have seen how political sociologists have depicted power as shaped by differential ownership of material resources or cultural and social recognition. But there are other sociologists who view these approaches to politics as overly reductionist. That is, they say these accounts incorrectly 'reduce' power relations to a function of the assets or qualities that people and groups possess, like money or social and cultural capital, and don't consider how institutions configure the circumstances in which people operate. Institutions set the bounds of social interactions, the 'rules' and guidelines around how social relations unfold, and they delimit the choices and actions that are available to people to employ under different conditions. In this way, institutions may not determine political outcomes, but they will shape them to a significant degree.

What are **institutions**, exactly? Sociologists understand institutions as patterns of behaviour that order people's lives in relatively predictable ways. They can be informal and flexible, like a group of friends meeting for happy-hour drinks on a regular basis, or more formal, such as schools, business associations, or the state. Institutions shape social interactions because they exhibit *inertia* and *path dependence*. The first means that they are relatively stable and require concerted effort to modify. Institutions 'push back' at attempts to change them by discouraging people from acting, or even thinking of acting, in other ways. The second means that once institutions are established, they affect decision-making down the road, encouraging people to proceed down certain 'paths' of action instead of others.

Because of these tendencies, institutions can both help and hinder those engaged in the contestation of power. They help by providing a stable and predictable means through which people can exercise power. They hinder by channelling and delimiting the choices for action available to those who engage with and sustain them. Both can be a boon to power-holders, since it is often difficult, or at least bothersome, for others to change the procedures structuring the distribution of power or even imagine alternatives to the status quo in the first place. But institutions can help challengers too, because once they learn how things operate, challengers can formulate ways of changing the rules, getting around them, or mobilizing them on their own behalf.

The institution that political sociologists are most concerned with is the state. Certainly, many political sociologists analyze how institutions shape formal politics, such as elections, political attitudes, and political parties (see Grabb 1994; Grabb and Curtis 2002), but sociologists also recognize that state institutions influence the politics that happens outside this 'official' sphere too. This is because state institutions shape other social processes, such as who is and who is not considered a citizen. Your capacity to bring your interests to the fore and exercise power can be limited greatly if you are not recognized by state institutions. It is for this reason that so many academics and policy-makers became alarmed in the summer of 2010 when the Canadian federal government moved to eliminate the mandatory completion of the long-form census. It was distributed to one in five Canadians and contained many more questions than the shorter form all other Canadians receive. The importance of such information may not be immediately evident, but think about how important it is for the state and other decision-makers to accurately know, for example, the ethnic composition of Canada. Without a tool to reliably measure how ethnicity is distributed in the country, some of the most marginalized groups in Canadian society could be under-counted or, even worse, missed entirely, with the result that their presence and concerns would disappear as bases for legislation. As Brubaker et al. (2004) note, failure to count certain people can literally mean they are forgotten or treated as if they don't exist. Who gets counted and how it is done are

(Lonely Planet Images/Alamy)

In 2006, the federal government of Canada recognized that 'Québécois form a nation within a united Canada.' What does this mean for Quebec nationalism?

dictated by decisions taken through state institutions, which in turn determine who can and cannot make legitimate claims to grievance.

Sociologists have taken up and applied these observations around institutions, especially state institutions, in many ways. These ideas have had the most impact, however, among proponents of the 'new institutionalism' and social movement theorists who examine how political opportunities affect mobilization efforts. New institutionalists hail from a number of different disciplines, including anthropology, economics, history, and political science. Although the perspective is inherently interdisciplinary, most of the sociologists who subscribe to it are concerned with identifying and tracking the ways that the institutional practices of particular states have affected political outcomes over the longer term. Theda Skocpol is perhaps the best known institutionalist in sociology. In a comparative-historical analysis of the revolutions in China, France, and Russia, she traced the impetus for revolution in each case to the incapacity of existing state structures to meet or resolve crises arising in the international realm, from war, or economic troubles (Skocpol 1979). Whichever powerful groups ended up exploiting these situations and fostering revolution depended on the relationships that had been conditioned by the state institutions in crisis. Although her subject matter may seem obscure, Skocpol's study was significant because it reminded sociologists of the important role that state institutions play in configuring class and other antagonisms in different countries.

Michael Mann has elaborated on the capacity of state institutions to shape power relations. In his recent work on genocide, for example, Mann (1999; 2004) challenges popular explanations of genocide as terrible manifestations of rabid 'tribal' racisms and points to democracy as spawning these atrocities. Democratic state institutions do not always foster peace by permitting everyone access to the state so they can resolve group conflicts in an orderly manner, says Mann. In places inhabited by people with many ethnic backgrounds, they can result in the elite power-holding group defining who has legitimate citizenry in ethnic terms. As non-persons, members of subordinate groups and minorities can then be liable to persecution or, at worst, repression or mass murder.

An institutional perspective can inform examinations of less violent events too, of course. In Canadian sociology, Daniel Béland (2005; 2006; 2008; Béland and Hacker 2004) has explored how political 'institutional legacies' in the US, Canada, Belgium, and other countries have shaped a range of social phenomena, including nationalism, fiscal policy, health and pension programs, and welfare regimes. Dominique Clément (2008) also offers an overview of how federal policy and funding interacted with and influenced the rise of what he calls Canada's 'rights revolution'—that is, the expansion of civil liberty and human rights organizations during the course of the twentieth century. And Jane Jenson and Denis Saint-Martin (2003) draw attention to the important role that state institutions have played in the post–World War II era in helping determine who gets to enjoy rights of citizenship and who does not. In other work, Jenson and Papillon (2000) show the complexities of how citizenship is recognized by looking at Quebec nationalists' attempts to acquire their own state and, James Bay Cree attempts to resist them by appealing to the international community to gain national recognition *within* the Canadian state. Such work highlights how states function in the contestation of ethnic and nationalist demands for power and how their fate is being challenged by it.

One last and related area that has been influenced by the new institutionalism is organizational theory. These scholars elaborate on the inertia and path dependence exhibited by *all* organizations but highlight the influence of states. Paul Dimaggio and Walter Powell's (1983) article is a classic in this field. The article takes up Weber's observations on the modern bureaucracy as an efficient organizational form that would only become more pervasive in the future because of capitalism's demand for rationalization. Dimaggio and Powell note that while Weber was right that bureaucratization has proliferated, it is not because of its efficiency. Rather, it has become the primary way that organizations are structured because there are forces present in modern societies that press them to eventually all look the same. Dimaggio and Powell called this institutional isomorphic change, and it happens in large part because of the omnipresence of bureaucratic state institutions. It is much easier for organizational leaders to communicate with and abide by the directives of the state if their organizations align with the ordering of the state agencies with which they invariably deal. The result is an extraordinary homogeneity of organizational forms across the social spectrum, which invariably conform to the institutional arrangements configuring the state.

Clearly, institutions and the contexts they create influence politics. States adopt constitutions and other means to manage and contest formal politics, including how political parties are established and the processes by which power is exchanged. However, institutions and context also affect challenges to power. Early attempts to theorize the effects of the political climate on political movements focused on political opportunities. These are the changing conditions of the existing political order that provide people with the inspiration and space to take action on an issue of importance to them. Peter Eisinger (1973) coined this term when he observed that American race riots of the 1960s often took place in cities that generated conditions of grievance but did not repress dissenters. Charles Tilly (1978) expanded on these ideas by also considering the choices that challengers make when they face changing political conditions. In both instances, these scholars found that contention emerges when political institutions are 'closed' or unresponsive enough to create dissent but 'open' or accessible enough not to repress it. As a result, contenders have the ability to challenge existing power structures but are also bound by them.

TIME to REFLECT

What are some of the things that prevent modern state leaders from using state institutions entirely in their own interests? Do they work?

Later scholars writing in this tradition focused less on specific opportunities and instead more broadly on the **political processes**, or the ebbs and flows of different sources of power. *Political process theory* (PPT) pays attention to the interaction of political opportunities of mobilizing structures, such as organization, resources, and access to elites, and the conceptual frames of issues and grievances used to challenge and sustain power. PPT includes a complex array of political phenomena that operate together to affect social movement activities and outcomes (McAdam, McCarthy, and Zald 1996). Doug McAdam (1982) showed how the US civil rights movement was shaped by political context, which included the abolition of segregated schooling, and the movement's access to mobilizing resources, such as supportive political elites like John F. Kennedy and Lyndon Johnson, in addition to the obvious grievances suffered by its members. Suzanne Staggenborg (1994), who supervised a number of contemporary Canadian political sociologists while at McGill in the 1990s and 2000s, uses a similar approach to depict

the progress of the pro-choice movement in the US as the outcome of a dynamic interplay of political opportunities, including the 'cycle of protest' underway in the 1960s, the legalization of abortion in 1973, organizational factors like the professionalization of leadership and formalization of organization after the *Roe v. Wade* decision, and actions taken against a hostile counter-movement, the 'pro-life' movement of the late 1970s and 1980s.

The success of PPT is evident in its prominent, if not default, standing among social movement scholars, a prominence that not everyone agrees is deserved (Goodwin and Jasper 1999). But its chief proponents (McAdam, Tarrow, and Tilly 2001; Tarrow and Tilly 2006) recommend that PPT be taken up not in opposition to the more specific approaches to contention that focus on material, cultural and social, or political processes alone, but as a tool for understanding how different social processes interact in the maintenance and challenge of power.

Transnationalism

So far, we have seen how political sociologists depict politics as encompassing contests around material resources, cultural and social factors, and institutions within states or as an attempt to create states. However, with the spread of global capitalism, many have anticipated that we are entering a new era in which old notions of time and social space are collapsing (King 1995; Appadurai 1990) and states are losing their status as the dominant institutions in the world (Jenson and Saint-Martin 2003). In fact, some have surmised that as the world becomes increasingly integrated and political processes become globalized, we are witnessing a massive restructuring of societies rivalling that which accompanied the Industrial Revolution, the emergence of modernity, and the rise of the state. It is a restructuring that these scholars say demands a new form of politics and, perhaps, a new kind of political sociology.

Malcolm Waters (1995) describes **globalization** as a social process in which the traditional constraints posed by geography, economic activity, culture, and social configurations have receded and have been replaced by processes that extend beyond state boundaries. He and other globalization scholars say that old ways of understanding politics and political movements are of little help today, considering that economies have become increasingly integrated, cultures homogenized, networks broadened, and state institutions challenged. Such developments potentially undermine or at least introduce new and problematic twists to traditional perspectives on political processes.

Take material resources, for example. Globalization scholars note that economies have become increasingly integrated, in part because new technologies have made it easier for people to engage in economic exchanges across national boundaries. This has had repercussions for both elites and labourers within individual countries. Increasing foreign direct investment and trade across states, for example, has generated new international flows of wealth that have altered how national elites respond to and take up these resources. It has also fostered the development of new, sometimes more exploitive labour practices, such as use of **export processing zones**, which circumvent the ability of workers to unionize and challenge power-holders.

If some sociologists say that the circumstances surrounding the distribution and exercise of material power have been altered by globalization, others say that claims to power based on culture have also been undermined. They argue that globalization has been accompanied by cultural homogenization. As Hollywood films gain prominence in international movie theatres (Barber 1995), for instance,

and fast-food culture spreads to all corners of the globe (Watson 1997), local cultures are swamped by a hegemonic consumer culture. This Westernized global consumer culture, say some, has even become accepted as 'local culture' to such an extent that people often have a false consciousness about who exactly produced what they are consuming (Robertson 1995). The result is an overall reduction in the range of unique cultural resources to which people seeking power have access.

Even institutions are seen to be internationalizing by these scholars. The past 30 years have witnessed a proliferation of international agreements that challenge the sovereignty of existing states. Take, for example, the emergence of the European Union (EU), which integrated most of the economies and governments of western Europe, or the implementation of the **North American Free Trade Agreement** (**NAFTA**) in Canada, the US, and Mexico. The prominence of these and other international institutions, such as the United Nations, has led some to declare that a 'world society' is arising in which the practices and procedures structuring people's lives and actions are becoming increasingly homogeneous (cf. Meyer, Boli, Thomas, and Ramirez 1997).

In an increasingly globalized world, institutions of global governance such as the UN have become increasingly important.

(ITAR-TASS Photo Agency/A army)

Scholars who study political movements have also begun to look at global processes. In their case, they are concerned with whether and how globalization has shaped political contention. The so-called 'Battle in Seattle' in 1999 saw thousands of individuals and groups from many regions of the world, nurturing a wide range of grievances, come together in opposition to the economic policies being promoted by officials of the World Trade Organization. The protesters believed the organization's policies were at the foundation of a myriad of ecological, economic, cultural, and social ills they wished to correct. Similar large mobilization efforts occurred throughout the first decade of the twenty-first century, as seen in a similar protest against the third summit of the Americas in Quebec City in 2001. Social movement scholar Donatella della Porta and her colleagues (2006) ponder whether or not traditional social

movement perspectives can account for these seemingly novel challenges to globalization. They maintain that contention against global capitalism consists of organizational structures, identities, strategies, and interactions with power-holders that are distinct from old ways of protest, which in turn foster 'democracy from below'.

It is the opinion of these scholars, then, that existing states are facing challenges to their traditional status as the primary institutions that govern political processes and that conventional ways of explaining politics, in material, cultural and social, and institutional terms, are just not up to the task of deciphering new global realities. But as convincing as some of the claims about globalization appear to be, they are far from indisputable. For example, at the very core of the argument of emerging globalization is that it is a *new* process, distinctly different from the

GLOBAL ISSUES

Transnationalism

More than a century ago, Karl Marx and Friedrich Engels (1985) forecast in broad form many of the developments that we now commonly refer to as 'globalization': the spread of capitalism to all corners of the globe, the subjugation of local economies to the demands of capitalist enterprise, the rapid revolutionizing of the techniques and forms of production to enhance profit, the economic interdependence of nations, even the homogenization of cultures. They predicted that a global proletariat created out of the relentless advance of capitalism would eventually come to realize their shared class position, rise up en masse to overthrow the ruling bourgeoisie, and establish the conditions necessary for communism. Developments in transportation and communication technologies would facilitate contact between far-flung workers and foster the unity and common cause required to incite the revolution.

The revolution has yet to come to pass, but is there any sign that globalization has fostered activism among the world's working class? Well, yes and no. As Sidney Tarrow (2005) shows, the number of social movements that transcend local or national boundaries has increased over time, but not all of them are class-based. They involve not just labourers but people from a broad array of social groupings, and they coalesce around a wide range of domestic and international concerns. Furthermore, conceiving, organizing, executing, and maintaining a global movement

to challenge existing power-holders is a complicated business—one that takes more than just a cellphone and access to the Internet. As easy and cheap as communication has become for at least a portion of the world's population, it is still very hard to recruit people to social movements and keep them involved in their efforts. Indeed, Tarrow finds that on balance, virtual contact through the Internet or some other electronic means cannot beat personal contact as a means to inspire and motivate. Accordingly, the most successful transnational movements are led by what he calls 'rooted cosmopolitans': people who are well-informed about global issues and technologically savvy but 'who grow up in and remain closely linked to domestic networks and opportunities' (2005, xiii). So while globalization and communication technology have generated new grievances and created new political opportunities and avenues through which people can challenge power, they have not entirely replaced good old-fashioned face-to-face interaction as the medium through which people are roused to take action.

Tarrow's discovery that the physical proximity of activists is still crucial to the capacity of a social movement to challenge power successfully is something that Marx and Engels (1985, 90) may have recognized in the *Communist Manifesto*, albeit unintentionally. The technology they said would literally help bring the world's proletariat together and incite revolution? The train!

processes of industrialization, modernization, and the emergence of the state that preceded it.

However, throughout recorded human history, there are many examples of international, if not global, exchanges of material, cultural, social, and political resources, all of which have shaped politics in different ways. The spice trade involved regions and empires of the ancient world interacting to facilitate the commercial exchange of spices and other commodities. The modern Olympics has for generations brought the world together to promote global cooperation through sport, and international agreements among states have been taking place at least since 1648 and the Treaty of Westphalia, when, for the first time, warring countries in Europe agreed to respect the 'advantage of the other' as a means of securing their own interests. At the same time, many doubt that globalization has occurred to the degree that some scholars claim.

Peter Urmetzer (2005), for instance, looking at the Canadian case, effectively questions whether or not states have lost control of their economies and the extent to which foreign direct investment is actually globalized. Most trade pacts, and for that matter international agreements, are bilateral, between just two states, rather than multilateral or global. With respect to communication, it is indisputable that we live in an era of unprecedented technological innovation, and for those with material wealth, increased travel and communication are indeed an option. Yet much of the world does not have access to those resources, and in a post-9/11 era, the securitization and monitoring of communication and travel has been unprecedentedly limiting, even for the most privileged populations. When it comes to cultural and social exchange, here too we find that much occurs among wealthy states and the wealthy people within them. Not everyone participates. If culture is becoming so cheerfully homogeneous, then why, as Benjamin Barber (1995) noted, do we at the same time see rising radicalism and nationalist groups seeking independence from, rather than integration into, the international community? Lastly, although there has been a rise of international institutions, few have the force to challenge the will of powerful states. Take, for instance, the US's ignoring UN resolutions when they do not fit that state's agenda, as it did when US leaders decided to launch a war against Iraq in 2003, despite the UN Security Council's decision that such an invasion would be illegal under international law. For all these reasons, it is likely more accurate to acknowledge the rise of transnational processes or 'transnationalism'

between some states and thus pause before calling recent changes truly global.

TIME to REFLECT

Why do some sociologists prefer to characterize the increasing interdependence of the world as transnationalism rather than globalization?

The key difference between globalization and transnationalism is the scale of the processes. The former claims that the entire world is involved, whereas the latter makes a lesser claim, suggesting only that more than one state is involved in an interaction with at least one other state. Scholars of transnationalism want us to remember that, as Sidney Tarrow (2005, 2–3) rightly cautions, it is still states that uphold international norms and implement laws, not international organizations. Moreover, as Clifford Bob (2005) and Margaret Keck and Kathryn Sikkink (1998) note, there are indeed 'activists beyond borders'; however, they still work largely within and against states as the main arenas of power. The ways in which state leaders exercise power, the issues with which they are concerned, and the institutions through which they work have indeed changed over time. But states are by no means withering away.

Conclusion

In this chapter, we have highlighted a number of key perspectives to help you grasp how sociologists understand politics. In doing so, we noted that, unlike many other social sciences and vernacular understandings of politics and political movements, sociologists see these processes as being inherently linked to the negotiation of power. Sociologists, moreover, understand power as a relational concept that involves the overt and hidden interactions of individuals and groups, organizations and institutions. We also highlighted that the state plays a prominent role in political sociology because it remains a key force in structuring political processes and shaping political resources. Throughout the chapter, we illustrated this by looking at material, cultural, social, and institutional resources to understand how they are used to maintain power as well as launch challenges to it through political movements. We hope that we have sparked your curiosity about political sociology by helping you to see politics differently: to see it through a sociological lens.

questions for critical thought

1. We don't usually think of our capacities to pursue our interests in terms of the power we hold relative to others. In what ways is your capacity to 'realize your will' or do what you want enabled by your access to material, cultural and social, and/or institutional resources?

2. Are you 'into politics'? Why or why not? Is it possible for someone to 'not be into politics', as so many people claim these days? What accounts for their apparent disinterest?

3. Does the Canadian state influence how you go about your daily life? How does it affect your power, your rights, and your ability to challenge authority?

4. What can account for Quebec nationalism in the 1970s, 1980s, and 1990s? Why have Aboriginal peoples been active in resisting the colonizing efforts of the Canadian government?

5. Does the food you consume, the clothes you wear, the entertainment you engage in, or the friends you have reflect your power? Why or why not?

6. If you are unhappy with the decisions that powerful people have taken, what factors do you need to consider to mobilize resistance and challenge these decisions? What are some of the barriers you face?

7. Why do many sociologists believe the world is being influenced by the process of globalization? How has it affected material, cultural and social, and institutional resources of power?

8. Do people recognize your identity as unique? If so, does that recognition offer you privilege, or does it present obstacles that you have to manoeuvre around? What resources do you draw upon to maintain your status or challenge its undervaluing?

recommended readings

Doug Baer, ed. 2002. *Political Sociology: Canadian Perspectives*. Oxford: Oxford University Press.
This is an introductory text that overviews Canadian scholarship in political sociology and includes many of the area's key luminaries.

Rick Helmes-Hayes and James Curtis, eds. 1998. *The Vertical Mosaic Revisited*. Toronto: University of Toronto Press.
This book examines the importance of John Porter's vertical mosaic 30 years after its original publication. It includes key Canadian political sociologists commenting on the vertical mosaic thesis and reformulations for the contemporary era.

Richard Lachman. 2010. *States and Power*. Cambridge: Polity Press.
This text offers an overview of how states emerged and how they have come to dominate contemporary political processes. It introduces students to key perspectives on theories of the state.

Kate Nash and Alan Scott, eds. 2001. *Blackwell Companion to Political Sociology*. Malden, MA: Blackwell.
An encyclopaedia of North American political sociology, this book is a one-stop resource that covers the main debates and innovations in this area of sociology.

Suzanne Staggenborg. 2011. *Social Movements*. 2nd edn. Oxford: Oxford University Press.
This text examines a broad range of social movement perspectives and uses Canadian cases to illustrate them. It is a concise reader that is accessible yet sophisticated in its presentation of material.

Karen Stanbridge and Howard Ramos. 2012. *Seeing Politics Differently: An Introduction to Political Sociology*. Oxford: Oxford University Press.
Stanbridge and Ramos provide an introduction to political sociology by challenging the common assumption that people can be 'not that into' politics. They look at much Canadian scholarship and examples to illustrate their points.

Sidney Tarrow. 2005. *The New Transnational Activism*. Cambridge: Cambridge University Press.
A sober examination of activism in the new millennium, this book examines claims of a global justice movement and the challenges of contemporary contentious politics.

recommended websites

Alternet
www.alternet.org
Alternet is an award-winning Web magazine dedicated to issues of social justice and the promotion and amplification of alternative voices.

Canadian Centre for Policy Alternatives
www.policyalternatives.ca
This is the official website of the Canadian Centre for Policy Alternatives, an independent non-profit organization committed to presenting counter-hegemonic information.

Charter of Rights and Freedoms
www.laws.justice.gc.ca/eng/charter/
The Department of Justice hosts this website of the Canadian Charter of Rights and Freedoms, which is the freestanding first 34 sections of the Constitution Act, 1982.

Independent Media
www.independentmedia.ca
This website offers a directory of non-corporate journalism. It lists many news sources that are not widely disseminated or are missed by mainstream audiences.

Nationalism Project
www.nationalismproject.org
This Web-based clearinghouse offers access to the ideas of prominent scholars of nations and nationalism. It also provides subject bibliographies, announces conferences and workshops, and provides links related to research on these topics.

The Onion
www.theonion.com
The Onion is an alternative spoof newspaper and website. It provides humorous political commentary on current events and news and is both informative and entertaining.

Rabble.ca
http://rabble.ca
Rabble.ca is a prominent alternative Web source of information on Canadian news and politics. It regularly presents articles and essays by politicians, academics, and activists.

Radical Information Project
www.bsos.umd.edu/gvpt/davenport/home.htm
This is the website of Christian Davenport's Radical Information Project. He is a leading scholar of state repression, conflict, and human rights. His website offers numerous resources and links to social science data on politics and contention.

chapter 19

Social Movements

John Veugelers and Randle Hart

In this chapter, you will:

▶ See how social movements are studied sociologically.

▶ Review the theoretical approaches to the study of social movements.

▶ Read about key debates within the study of social movements.

▶ See how empirical research is used to test and criticize social movement theories.

▶ Learn how social movements are embedded in national and international (global) politics.

▶ Understand why social movements have historical importance in the process of social change.

Introduction

Late one night during the summer of 2006, a rural road in southern Ontario was the scene of a clash between dozens of protesters. Separated by police officers who stood their ground, Natives and non-Natives hurled taunts and then rocks and golf balls at each other. For the non-Natives, frustration in the face of a standoff between the Native protesters and a local real estate developer was high. That February, members of the Six Nations had occupied land slated for a new housing development. Soon they expanded their protest by mounting a roadblock near the town of Caledonia. Residents and businesses now faced long and inconvenient detours if they wanted to drive into or out of the area. For Native protesters, however, creating trouble for commuters and businesses was meant to right a historical injustice: the loss of land along the Grand River despite an agreement with the British Crown in 1784. By mobilizing not only the members of their community but also supporters and Native peoples in other parts of Canada, the Six Nations protesters at Caledonia were resorting to politics by other means (CBC News 2006).

Writing in the nineteenth century about political associations, Alexis de Tocqueville noted that 'citizens who are individually powerless do not very clearly anticipate the strength that they may acquire by uniting together' (1945 [1835], v. 2, 124). When powerless people lend time, energy, and material resources to a cause, their combined efforts create a **social movement** with a life and force of its own. Examples of present-day social movements include Native, women's, peace, gay and lesbian, anti-nuclear, labour, environmental, ethnic, and regionalist movements.

A social movement is more than the sum of the individuals forming it. As a consequence, social movements may take a direction contrary to their members' wishes. Surprisingly, some social movements even betray their supporters by adapting to a social system once seen as corrupt or unresponsive. Others disappear as a result of internal bickering or government repression even though their members' grievances persist. A psychological analysis of a social movement's members will reveal little about the movement's origins and development, its effect on the social order, or the causes of its successes and failures. A social movement is a distinctive social reality for which sociology offers appropriate tools of analysis.

This chapter opens by looking at the characteristics of social movements. It next considers theoretical approaches to social movements, and then examines examples of social movements in Canada. The chapter ends with a discussion about social movements in today's global context.

What Is a Social Movement?

A social movement depends on the actions of non-elite members of society, those people who have relatively little or no control over major economic, symbolic, political, or military resources—in short, over anything scarce that, if controlled, gives one power over others.

People form a social movement when they voluntarily work together to influence the distribution of social goods. A social good is anything that a particular society values. Familiar examples include money, honour, peace, security, citizenship, leisure time, political power, and divine grace. There are probably no universal social goods, because no two societies have exactly the same set of values. Furthermore, social goods vary historically. They emerge and disappear as values change or traditions lose relevance (Walzer 1983). Social goods are scarce—that, in part, is why they are valuable—and some individuals or groups get more of them than others do.

Social movements try to achieve change through the voluntary cooperation of the relatively powerless. These people may contribute financial or other material resources, recruit new members, or spread a counter-ideology. They may also participate in strikes, sit-ins, boycotts, demonstrations, protest marches, violent action, or civil disobedience. Social movements aim to change attitudes, everyday practices, public opinion, or the policies and procedures of business and government.

Social movements are easier to understand when compared and contrasted with other phenomena studied by sociologists (Diani 1992). A *social trend*, for example, is simply a changing pattern of social behaviour, whereas a social movement is a cooperative effort to achieve social change from below. The rising labour market participation of women is a social trend; a group of volunteers who fight for gender equality is a social movement. A *pressure group* is an organization that aims to influence large institutions, particularly the **state**. A social movement is one kind of pressure group. However, other pressure groups—known as *interest groups*—represent the concerns of specific sets of people. Prominent interest groups include the Canadian Labour Congress, the Canadian Medical Association, Canadian Manufacturers and Exporters, and the Consumers' Association of Canada. Interest groups restrict their membership and rely heavily on a professional staff rather than volunteers. Moreover, lobbying politicians and receiving recognition from government can give them semi-official or even official status. Like social movements, interest groups may use public opinion to put pressure on political or economic elites. But membership in social movements is more open, and their ideologies typically appeal to people from different walks of life.

Since social movements depend on voluntary participation, they are *voluntary associations*. However, not all voluntary associations seek deeper changes in the distribution of social goods. Some provide social or health services; others organize leisure activities or unite the followers of a spiritual doctrine. Examples of voluntary associations are groups that help the homeless, run food banks, offer language classes for immigrants, or mobilize residents for annual clean-ups in their neighbourhood. Voluntary associations that only help people to accept or enjoy the existing social system are not social movements.

While social movements try to change the distribution of social goods, political parties try to win and keep political power. In principle, a social movement becomes a political party when it fields candidates in elections. The Green parties in Canada, Germany, France, and Italy, for example, have grown from environmental movements in these countries. Not all groups with non-elite, voluntary members who aim to reallocate social goods are necessarily social movements. A counter-movement may have all of the characteristics of a social movement but with one important difference: a counter-movement arises in response to a social movement. Three conditions must be met for counter-movements to appear:

1. A social movement must be seen as successful (or as gaining success).

2. A social movement's goals must be seen as a threat to another group.

3. Allies must be available to support mobilization of the counter-movement.

While some counter-movements wish to defend the status quo against a perceived threat by social movements, others (such as the anti-abortion movement) emerge when a state or government agency has ambiguous policies or is internally divided on a particular social issue (Meyer and Staggenborg 1996).

TIME to REFLECT

Do you consider yourself an active participant of any social movement? If not, do you wish you were? What social movement would you become involved in, and what part do you think you could play?

Theoretical Approaches

Different beliefs about society separate the main approaches to social movements. The **resource mobilization approach** assumes that social order is based on competition and conflict and that interests are the fundamental cause of action. On the other hand, the **political process approach** assumes that social order rests on an unsteady resolution of conflict and that culture is the major determinant of action.

These approaches developed in a critical response to earlier approaches, which treated social movements

The iconic image of the Argentine-born Marxist revolutionary Ernesto 'Che' Guevara has been a symbol of struggle for revolutionary movements around the globe, as in this International Labour Day rally, 1 May 2007, in Karachi, Pakistan.

as ailments. This charge arose during the 1960s, a time when social movements supported by mainstream members of society were flourishing in Western democracies. Many sociologists welcomed the new movements against war, racism, sexism, pollution, bureaucracy, and the flaws in the educational system

as positive signs of healthy protest against injustice and alienation.

Earlier approaches that focused on the breakdown of society forgot that value consensus and social stability result partly from relations of domination. As Barrington Moore, Jr, observes, 'To maintain

SOCIOLOGY in ACTION

The Last Days of a Desert Despot?

As the circle of political and military support around him dwindles to an increasingly small and desperate rump and protesters take control of a growing swath of Libya, dictator Moammar Gadhafi has become increasingly erratic and unpredictable, alternating between brutal attacks and attempts at appeasement and cooperation.

On a day when hundreds of thousands of Arabs took to the streets in the capitals of half a dozen countries across the Middle East and North Africa, Libya appeared to be the most volatile and deadly on Friday and its strongman leader the most likely to fall or flee. Instead, Col. Gadhafi remained in his redoubt in Tripoli while the conflict consuming the country escalated, prompting further defections by top regime officials and the imposition of international sanctions organized by the United States.

Residents of key cities to the immediate east and west of Tripoli, Misurata and Zawiyah, interviewed by telephone or after crossing the border into Tunisia, described scenes of violent street combat, with protesters gaining the upper hand despite live-ammunition gunfire from Col. Ghadafi's mercenaries. The opposition, they reported, is firmly in control of the country's eastern third and most of the south.

Most seriously for Col. Gadhafi, personnel at Tripoli's Mitiga air force base—a former US installation seized by Col. Gadhafi in his 1969 coup and considered a key military asset—have defected and joined the opposition. Tripoli residents said that sections of the capital are now held by protesters. After hiding in their houses in the wake of Monday's violent reprisals, people emerged from the city's mosques after Friday prayers to take part in a new round of protests after jeering the regime-written sermons all imams were ordered to deliver.

Col. Gadhafi and his son Saif responded with a series of desperate and sometimes strange statements. First, Col. Gadhafi made a show of encouraging pro-government demonstrators in Tripoli's central Green Square with a televised speech from a rooftop overlooking part of the square, urging people

to 'sing, dance, and prepare yourselves' because 'life without green flags has no value'—the green flag being the symbol of his regime. He then urged the small crowd of loyalists around him and on TV to take up arms and fight the protesters. 'Get ready to fight for Libya!' he yelled. 'Get ready to fight for dignity! Get ready to fight for petroleum! We will fight them and we will beat them. . . . The people are armed and when necessary, we will open the arsenals to arm all the Libyan people and the Libyan tribes.'

That scorched-earth language was contradicted several hours later by Saif Gadhafi, who addressed a hand-picked group of reporters flown to Tripoli (and forbidden from leaving their hotel to report on the protests or shootings), saying that he would be negotiating toward a peace settlement with the protest movement—and appearing to acknowledge that the military is no longer following the family's orders.

Those words reached a world that appeared to have finally lost patience with the Gadhafis. The United States announced the imposition of unilateral and multilateral sanctions against Libya, including an immediate end to the military cooperation between the two countries that had flowered after Col. Gadhafi renounced terrorism and opened his country's economy in the early 2000s. French President Nicolas Sarkozy—who had been late in supporting the uprisings in Tunisia and Egypt—became the first leader to call openly for Col. Gadhafi's departure Friday. As he spoke, the United Nations Security Council was preparing to pass a resolution that would impose further sanctions on Libya, a travel ban on its ruling family and a freeze on any assets—believed to be in the billions—held in foreign banks. Among the most outspoken voices calling for sanctions were those who until a week ago were Col. Gadhafi's loyal ambassadors and ministers. His former deputy UN ambassador, who quit along with a dozen other ambassadors to side with the protesters, delivered an impassioned speech calling for the Libyan leader's resignation on Friday.

SOURCE: Adapted from Doug Saunders, 'The last days of a desert despot', *The Globe and Mail*, 25 Feb. 2011.

and transmit a value system, human beings are punched, bullied, sent to jail, thrown into concentration camps, cajoled, bribed, made into heroes, encouraged to read newspapers, stood up against the wall and shot, and sometimes even taught sociology' (1966, 486). Contrary to the assumptions of past approaches, social conflict may be a normal feature of social life. If this is so, then the breakdown of value consensus and stability may not explain the formation of social movements. Social movements do not result from outbursts of uncontainable emotion. Instead, experience suggests that participation in social movements may involve the same kind of calm and rational decision-making found in other areas of life. This interpretation underlies the resource mobilization approach.

THE RESOURCE MOBILIZATION APPROACH

The resource mobilization approach challenges the image of social movements as unusual, impermanent, or disorderly. Instead, it assumes that social movements are quite similar to other organizations. Some sociologists go as far as treating social movement organizers as calculating 'moral entrepreneurs' with a 'product' to sell (Jenkins 1983).

Proponents of the resource mobilization approach argue that dissatisfaction is built into society. There will always be people with grievances, because social goods are unequally distributed. But grievances alone do not make a social movement. What social movements do is lift grievances out of the shadows, giving them ideological form and propelling them into public life.

The resource mobilization approach puts **power** at the centre of analysis. As the German sociologist Max Weber (1864–1920) put it, power refers to a person's or group's chance of fulfilling their goals even when others would have it otherwise (1978 [1908], 926). The source of power is control over resources. Control creates leverage, the ability to get others to do what one wants. What represents a resource in any given situation varies, but three sources of power stand out: *Economic power* is based on control over the means of material production: land, energy, capital, technology, labour, factories, raw materials, and so forth. *Political power* is based on control over the legitimate means of violence: the police and the armed forces. *Ideological power* is based on control over the means of producing and disseminating **symbols**: schools, churches, newspapers, publishing houses, television and radio, film and advertising companies, and the like. Social movements must compete against other

In the First Person

I was first introduced to the study of social movements by William K. Carroll, a sociology professor at the University of Victoria, and I was lucky enough to be admitted into his senior-level sociology seminar, though I had been studying Canadian literature and had very little experience in sociology. I was soon hooked on the study of activism. I remained at the University of Victoria to study for my master's degree, and I continued to be inspired during my doctoral studies at the University of Toronto. I now teach and conduct research on social movements and Peruvian guest-labour shepherds in the high desert of southwest Utah.

—Randle J. Hart

social institutions for the scarce resources necessary to start and operate an organization.

The free-rider problem is a central concern for resource mobilization theory, in particular the question of how and why selective incentives attract volunteers (see Open for Discussion box).

Criticisms of the Resource Mobilization Approach

Critics have stressed the limited applicability of the free-rider problem. The assumption that social movements attract support only by providing selective incentives may misconstrue people's reasons for joining. Instead, people may join a movement simply because it seems headed for success. Or they may join because they identify with other members of the social movement and believe the group will benefit if its members work together (Barry 1978 [1970]). Finally, norms of fairness may override concerns about efficiency. Pressure to conform may lead people to join social movements, irrespective of selective incentives (Elster 1989). People are ruled by more than self-interest. Social movements are groups, moreover, so they cannot be explained by individualistic decisions alone.

Further, the resource mobilization approach runs the danger of missing some important differences between social movements and other organizations. Few sociologists still believe that social movements

Social movements often mobilize against the state. Here, activists protest the threatened closure of a safe injection site.

are irrational, exotic, or unusual. However, many are beginning to realize that social movements differ from other types of organizations. They have different resources, career cycles, and relationships with government authorities as well as with other social movements. They also exhibit distinctive modes of acting, organizing, and communicating (Tarrow 1988).

The image of human action conveyed by the resource mobilization approach may exaggerate the extent to which social movements reflect careful planning and successful strategy. People often behave with vague or conflicting goals in mind. Like any other social group, a successful social movement probably does things the participants never intended in the first place. Further, the goals of social movements often emerge and change as situations evolve.

CULTURE IN SOCIAL MOVEMENTS

Cultural approaches to the study of social movements assume that neither the goals of social movements

OPEN for DISCUSSION

The Free-Rider Problem

The Fresh Air Coalition is fighting for a reduction in the toxic emissions of Steel City. If the group meets its goal, all of Steel City's citizens will breathe cleaner air: it would be impossible to give the cleaner air only to people who had joined the Fresh Air Coalition, while everyone else—the people who didn't attend the protest rallies, write letters to newspapers and politicians, or contribute to the Fresh Air Coalition's fundraising plant sale—got the same dirty air as before.

Assuming the citizens of Steel City are self-interested, it doesn't make sense for them to join the Fresh Air Coalition. They are free riders: they will benefit even if they don't help, because collective goods such as clean air cannot be divided.

What can the Fresh Air Coalition do? According to free-rider theory, groups can foster cooperation through selective incentives. The Fresh Air Coalition can make cooperation worthwhile by providing rewards. So in addition to fighting for clean air, the leaders of the Fresh Air Coalition will be sure to organize social events—picnics, parties, camping weekends—that attract and keep members by satisfying their immediate self-interest.

nor the ways they calculate the best means of achieving them are self-evident. Since norms and values are created in and by social movements, proponents of the cultural approach believe that the formation of goals needs to be explained (Nedelmann 1991). The resource mobilization approach also takes for granted the sense of community that creates collective identity and a willingness to work together. How people define themselves depends very much on whom they identify with—on what community, with its unique norms, values, and ways of feeling. Social movements also redefine identities by changing or reinforcing people's sense of community and by providing them with opportunities to work together with a shared sense of purpose.

The New Social Movements

European *new social movement* (NSM) theorists (e.g., Melucci 1989; Touraine 1981) propose that structural changes in Western societies have fundamentally altered people's identities and cultures. This gives rise, they argue, to social movements that are distinct from older, class-based movements. NSMs are interested in the politics of cultural recognition; they are concerned less with the redistribution of wealth and status than with securing rights to expressive freedoms, symbolic practices, and/or styles of life. In this sense, the appearance of NSMs may be explained by a value shift (Inglehart 1990b).

The NSM perspective focuses largely on the relationship between culture and collective identity. It proposes that social movements are cultural laboratories where people try out new forms of **social interaction** (Melucci 1989). In many ways, the resource mobilization approach defines the success of social movements in terms of change in economic or political institutions. The NSM approach defines success differently. To be sure, it does not deny the desirability of change in dominant institutions. However, for NSM activists, more important struggles take place in civil society, the areas of social interaction that stand largely outside of the state and the market. In fact, theorists claim that NSMs have come about since the 1960s because state and economic practices have increasingly encroached on people's everyday lives. Slogans such as 'the personal is political' are meant to express how everyday life is pervaded by government and corporate activities as well as by dominant cultural ideas that create inequality.

According to this approach, civil society offers greater chances for freedom, equality, and *participatory democracy*, a system of decision-making in which all members of a group exercise control over group decisions. Indeed, NSMs are, in part, characterized by institutional arrangements whereby their members try to organize according to the ideals of equal participation. This is what social movements are good at, and striving for other kinds of success risks perverting these ideals (Cohen 1985).

Framing Theory

At the same time that European NSM scholars were criticizing resource mobilization theory, so were some North American researchers, albeit in a different way. Rather than assuming that collective action was entirely rational, these scholars were interested in how collective understandings were created, communicated, and used to further a movement's goals. Because they sought to understand the role of cultural meaning in collective action, they looked to the symbolic interactionist tradition for inspiration. The result was **framing theory**, a cultural approach that explains the ways movements create and spread their understandings of the world and how these meanings help to form a sense of collective identity and common purpose.

Drawing on Erving Goffman's ideas, these social movement theorists define collective action frames as 'action-oriented sets of beliefs and meanings that inspire and legitimate the activities and campaigns of a social movement organization' (Benford and Snow 2000, 614). Collective action frames are the communal understandings of a social movement, and these understandings are used to identify and promote grievances.

According to framing theory, a social movement must succeed at three core framing tasks in order to mobilize support. First, an organization must articulate *diagnostic* frames that define social problems (or injustices) and their guilty agents. Second, *prognostic* frames must propose solutions to these social problems. Prognostic frames give meaning to specific strategies and are used to persuade potential recruits and members that these actions are the best way to solve or address particular social problems. Third, since agreement with diagnostic and prognostic frames does not necessarily translate into participation, a social movement organization must provide compelling *motivational* frames that convince people to join.

The process whereby individuals come to adopt the ideology and methods of a particular movement organization is called *frame alignment* (Snow et al. 1986). The alignment of interpretations is a necessary condition for maintaining participation. This is because members can identify with an organization once their cultural understandings are more or less the same as everyone else's in the movement. As Gamson suggests, 'any movement that seeks to sustain commitment over a period of time must make the construction of collective identity one of its most central tasks' (Gamson 1991, 27).

Frame theorists recognize that collective action frames are not simply imposed on members by leaders but are often changed and agreed on through social interaction and discussion. Although disputes over how to frame something inevitably arise, a minimum level of agreement must be maintained. There would be little reason to participate in a social movement that could not agree on how to collectively define a social problem or issue.

Criticisms of the Cultural Approach

Like resource mobilization theory, the cultural approach tends to be voluntaristic in its emphasis on people's potential for actively challenging and changing society. It focuses on altering the shared understandings that maintain patterns of domination. Consequently, it too often ignores the structures of economic and political opportunity that shape the destinies of social movements.

New social movement theory may explain movement emergence, but it tells us very little about how movements themselves operate: the 'how' of social movement theory. It would seem that in trying to address the deficiencies of resource mobilization theory, NSM theory has created its own blind spot.

Finally, framing theory is quite narrow in its focus and does not adequately explain how cultural processes external to a movement may influence internal cultural understandings. Framing theory tends to assume that a social movement's framing activities are fully bounded within the movement (Hart 2008). A better starting point is to recognize that cultural framings flow easily into (and out of) a movement. Cultural interpretations abound, and analysts must examine all sources of meaning to fully understand the relationships between culture and collective action.

THE POLITICAL PROCESS APPROACH

The political process approach assumes that the polity can be characterized by its opportunities and constraints. Opportunities involve almost anything that provides reasons and resources for people to mobilize—as long as the political climate is not so oppressive that people cannot mobilize without fear

or great difficulty. Political opportunities may include economic crises, laws ensuring the right to assemble, a history of previous collective action, even accidents that show the need for social change. Constraints include anything within the polity that may act as a barrier to the mobilization and survival of a social movement. Political constraints include a repressive police state, inexperience with collective action, even a lack of communication among social movement participants. Opportunities and constraints go hand in hand: no polity is completely open or completely closed (Tarrow 1998).

Fluctuations in the opportunities and constraints that influence the incidence of collective action create a cycle of contention. A rise in the cycle means that social movements have created or met new opportunities and have made room for the rise of other movements. For instance, the rise and decline of collective action by Canadian Aboriginal bands from 1981 to 2000 can be seen in Figure 19.1 (Wilkes 2001). Protest events among Native groups in Canada rose dramatically between 1989 and 1991, peaking in the 'Indian summer' of 1990. This increase can be attributed to the 78-day armed uprising at Kanesatake (Oka, Quebec) over municipal plans to convert a Mohawk burial ground into a golf course. In support of the Kahnawake, Akwesasne, and Kanesatake bands, bands across Canada increased their protest activities.

Like the resource mobilization approach, the political process model focuses on institutions. Specifically, this approach looks at mobilizing structures, which include levels of informal and formal organization (McCarthy 1996). An example of informal organization is a friendship network. When the cycle of contention is at its lowest point—when there are relatively few (or no) active social movement organizations—the network of friendships among demobilized movement participants keeps the spirit of collective action alive. These latent networks explain why social movements arise when political opportunities appear and when constraints are eased (Melucci 1989). Although informal communication alone cannot give rise to a social movement, it can become an important resource for mobilization.

The analysis of formal organization looks at the inner dynamics of social movements. These dynamics generally include leadership structures, flows of communication, the entry and exit of members, and the means of identifying, obtaining, and utilizing resources. The study of social movement organizations also includes inter-organization dynamics, such as movement coalitions. A coalition results when

FIGURE 19.1 ▼ Number of Protests by First Nations, 1981–2000, Canada

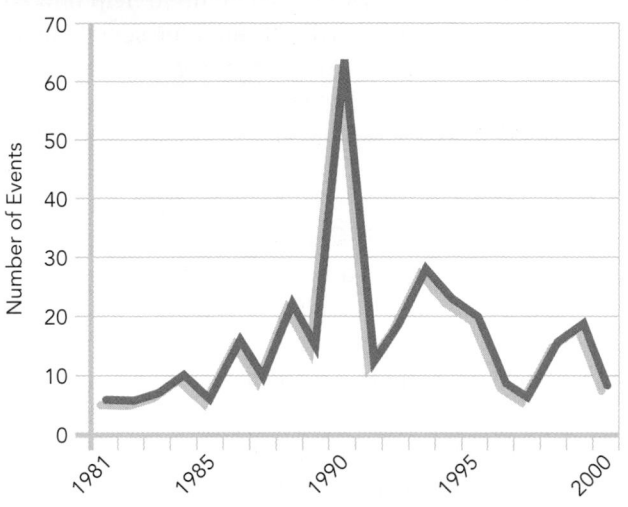

SOURCE: Rima Wilkes. 2001. 'Competition or colonialism? An analysis of two theories of ethnic collective action' (University of Toronto, PhD thesis). Reprinted by permission of the author.

two or more social movement organizations share resources, such as information, in the course of pursuing a common good. Coalitions can be temporary or enduring, and they can bridge different types of movements. Environmental, feminist, gay and lesbian, labour, peace, and anti-poverty organizations in British Columbia, for example, have formed coalitions based on shared understandings of social injustice (Carroll and Ratner 1996).

Incorporating the insights of framing theory, the political process approach assumes that social movements develop their own cultural understandings of the world. These understandings form the basis for identifying and acting on social grievances and provide movement participants with the resources needed to create activist identities. In this view, collective action frames can be used to identify appropriate forms of protest. For example, Table 19.1 shows the frequency of types of protest among Native bands in Canada from 1981 to 2000. Clearly, road blockades were the most common protest strategy during this period. This may be due to framing processes; the popularity of roadblocks as a tactic may arise from the cultural significance of this form of protest. As more Native groups block roadways to express their grievances, this form of protest becomes more strongly associated with their social movement. Other Native bands then become more likely to adopt the same tactic.

Social movements may also use collective action frames as strategic resources. To mobilize general support for their cause, movements promote their own ideologies in the wider culture. If a social movement's framing of injustice and its solution are accepted in society, then it has created its own political opportunities. If a movement is unsuccessful, however, it risks adding to its own difficulties.

Criticism of the Political Process Approach

One strength of the political process approach is its simultaneous focus on structural, institutional, and cultural conditions. Ideally, such a perspective should provide a robust account of social movement processes. As William Gamson and David Meyer observe, however, the political process model has been 'used to explain so much, it may ultimately explain nothing at all' (1996, 275). While somewhat overstating their case, these critics are concerned that if variable after variable is included in the model, the explanation loses simplicity and thus its explanatory appeal.

TABLE 19.1 ▼ Frequency of Types of Protest among Native Bands in Canada, 1981–2000

	Number	Percentage
Road blockade	114	36.08
March/demonstration	86	27.22
Train and boat blockade	19	6.01
Boycott	18	5.70
Occupation of land	17	5.38
Illegal fishing/logging	17	5.38
Occupation of building	11	3.45
Hunger strike	5	1.58
Toll booth	5	1.58
Non-strategic violence	4	1.26
Withdrawal from school	3	0.95
English signs changed	2	0.63
Illegal gambling	2	0.63
Invitation of foreign ambassadors	2	0.63
Eviction of police and non-Natives	2	0.63
Dam diversion	2	0.63
Destruction of property	2	0.63
Other	5	1.58
Total	316	100.00

SOURCE: Rima Wilkes. 2001. 'Competition or colonialism? An analysis of two theories of ethnic collective action' (University of Toronto, PhD thesis). Reprinted by permission of the author.

Doug McAdam (1996) provides two solutions to this problem. First, he suggests restricting the concepts of opportunity and constraint to include only four variables:

1. the openness of the state
2. the stability of alliances among elite members of society
3. support within the elite for a particular movement
4. the level of state repression.

Second, McAdam argues that opportunities and constraints are different for each type of social movement. For example, the elements of the political structure that give rise to revolutionary movements will likely be different from those that give rise to identity-based movements. Revolutionary movements are likely to identify most of their political opportunities within the state system, whereas identity-based movements are likely to find most of their opportunities within the cultural practices of civil society (McAdam 1996).

Political process theory is also too focused on structural dynamics and does not consider how individual agency influences social movement organizations. For a leader to take advantage of political opportunities or to manoeuvre around constraints, for instance, she or he must have tactical skills, and these skills can develop in relation to that leader's unique biography (Hart 2010). If we only examine social structures, movement scholars are at risk of forgetting that individual activists like Dr Martin Luther King, Jr, have had a massive impact on history not solely because of social structural conditions but also because of personal charisma, organizational skills, and even good luck.

TIME to REFLECT

Think about each of the social movement theories. Which one seems the most comprehensive? Can you think of ways to improve or combine these theories?

The Analysis of Social Movements

At one time, sociologists argued that successful movements promote their supporters' interests. Nowadays such explanations are rejected, for they fail to recognize that interests are themselves cultural constructs.

Moreover, a movement's supporters often have only a schematic or confused understanding of its ideology. But the compelling question remains: why do some movements succeed while others fail? To help find an answer to this question, we will examine social movements in Canada: the women's movement, agrarian social movements in the Prairie provinces and New Brunswick, and the separatist movement in Quebec.

UNITY AND DIVERSITY IN THE CANADIAN WOMEN'S MOVEMENT

Social movements need both diversity and unity. In the history of the Canadian women's movement, diversity of membership and experience has helped the movement adapt to a range of situations. Diversity has also encouraged recognition of the many faces of gender inequality. By maintaining a stock of alternative views and ideas, ideological diversity readies the movement for social change. Unity, in turn, gives the movement strength. A one-woman strike, boycott, or sit-in scarcely represents a threat to dominant institutions. Women who are individually powerless gain leverage by acting together. Unified, they can disrupt patriarchal institutions and pressure authorities into finding new solutions.

Though diversity and unity are both beneficial, they pull social movements in opposite directions. Diversity tends to impede unity and may lead to factionalism. Unity tends to suppress diversity and may stifle flexibility and innovation. As in any complex social arrangement, there can be no either/or choice for social movements: survival and efficacy dictate a balance between diversity and unity. The story of the first and second waves of the Canadian women's movement illustrates this dilemma.

The first wave of **feminism** in Canada began in the late nineteenth century and effectively ended in 1918, when women gained the right to vote in federal elections. During this period, women formed organizations for the protection and education of young single women, such as the Anglican Girls Friendly Society and the Young Women's Christian Association (YMCA). Women's groups also protested against child labour and poor working conditions and pressed for health and welfare reforms.

Feminists of the first wave differed in their religious, class, and ethnic backgrounds. While many were Protestant, others were not. Anglo-Saxon women from the middle and upper classes predominated, especially among the leadership, while language divided anglophone and francophone feminists. Moreover, women's organizations had diverse goals. But the battle for women's voting rights unified the movement. One of the earliest women's groups in Canada, the Toronto Women's Literary Club (soon renamed the Women's Suffrage Association), was founded in 1876. By 1916, women had won the right to vote in provincial elections in Alberta, Saskatchewan, and Manitoba. Other provinces soon followed, and Canadian women finally received the federal franchise in 1918.

As with many other social movements, success led to decline. The fight for voting rights had given the women's movement a common goal. When this goal was attained, the movement lost unity and momentum. Certainly, women did not stop pushing for change after winning the right to vote. Some worked within the labour movement; others continued to fight for social reform or female political representation. Yet after 1918, the Canadian women's movement became fragmented, and four decades would pass before it regained strength (Wilson 1991).

The second wave of the movement rose out of the peace, student, and civil rights movements of

HUMAN DIVERSITY

Activism after 11 September 2001

This story starts with planes flying into buildings on the other coast, in another country. It was just over a year ago, and in the heady months that followed, our elected representatives in Ottawa rushed to approve anti-terrorism legislation intended to protect Canada's national security. In recent months, two Vancouver Island activists found themselves on the wrong side of the Royal Canadian Mounted Police unit created to enforce that legislation.

Early in the morning on September 21, the RCMP's anti-terrorism force, the Integrated National Security Enforcement Team (INSET), followed an anonymous tip and raided John Rampanen's family's empty home in a Port Alberni suburb. They were looking for unauthorized guns.

The INSET team didn't discover anything at the home, says Rampanen, but later, when they found him and his family at his parents' house, they made 'veiled threats towards the safety of our children'. He plans to lodge a formal complaint about the threats. While Rampanen sees the need for the police to investigate such serious allegations as possession of illegal weapons, he says, 'It baffles me that they would be so aggressive in their approach.'

As a member of the West Coast Warriors, an Aboriginal activist group, Rampanen has figured prominently in high-profile confrontations at Cheam on the Fraser River and at Burnt Church in New Brunswick. He also delivers drug and alcohol education programs to First Nations communities. The warriors don't shy from confrontation, he says, but they aren't terrorists and it doesn't make sense for them to be under the scrutiny of INSET. 'It's been over three weeks now, and I still don't understand.'

David Barbarash, former spokesperson for the Animal Liberation Front [ALF], had two computers, computer disks, videos, photos, files, papers, and other documents seized from his home and office in Courtenay on July 30, by INSET officers. He says he understands only too well why INSET might take an interest in activists. 'It used to be people could take these kinds of actions and they'd be labelled protesters', he says. 'At some point in time there was a move to criminalize dissent. What's happening now, post–September 11, is it's shifted again. Now we're not even criminals, we're terrorists.'

The July raid on Barbarash's property stemmed from an ALF action in Maine three years earlier. A group there broke into hunting clubs, spray-painted messages on walls, broke windows, and stole stuffed animal heads, which they later 'returned to their natural environment to rest in peace.' Damage was estimated at $8700.

Barbarash acknowledges that the ALF actions were criminal but stresses the group doesn't physically harm anyone and that what he calls 'economic sabotage' has a long, dignified history that can be traced back to the Boston Tea Party and further.

As with other animal rights actions, Barbarash received information from sources he says he does not know, and has no way to contact, and communicated their message to the media. 'I'm not committing any crimes, I'm simply voicing my support for these kinds of activities', he says. 'I don't see why our resources should be spent in this way, as if this is some kind of terrorist activity. I think it's outrageous.'

'Lets face it', says Barbarash. 'This is Canada. There isn't a lot of terrorism.' So instead of infiltrating al-Qaeda sleeper cells, he says, INSET officers are being used to investigate more mundane criminal matters.

'When they're going after people like me who just speak to the media, it's pretty pathetic,' he says. 'They've got to justify their expense account somehow.'

SOURCE: Abridged from Andrew MacLeod, 'Anti-terrorism police harass Island activists', *Monday Magazine*, 24–30 Oct. 2002, pp. 28, 43.

the 1960s. In some cases, organizations advanced the women's cause by branching out. For example, a Toronto organization called the Voice of Women (VOW) was founded in 1960 as a peace group but gradually adopted other women's issues, and by 1964 it was promoting the legalization of birth control.

The social movements of the 1960s spurred women in other ways. Women in the student movement came to realize that many male activists were sexist. This drove home the extent of gender inequality and the need to organize apart from men. Through the New Left movement, women discovered that socialism helped make sense of gender inequality. More generally, the cultural upheaval of the 1960s encouraged women to question their position in private and public life.

As a distinct women's movement emerged in the late 1960s and early 1970s, so did internal diversity. Some members were revolutionary Marxists while others were socialists, liberals, or radical feminists. At times, those who favoured grassroots activism criticized those who worked through high-profile official committees such as the Canadian Advisory Council on the Status of Women. The specific concerns of lesbian, non-white, immigrant, or Native women were often ignored or marginalized by mainstream women's groups. Finally, issues of language and separatism split women's organizations in Quebec from those in the rest of Canada.

Still, the movement found bases for unity. In 1970, a cross-Canada caravan for the repeal of the abortion law attracted much publicity. The caravan collected thousands of petition signatures, showing women what could be achieved through collective action. Other coalitions formed around the issues of daycare, violence against women, labour, and poverty. Women's groups also worked together on International Women's Day celebrations.

To better represent their interests, in 1972 Canadian women formed the National Action Committee on the Status of Women (NAC). NAC grew and by the late 1980s had become an umbrella organization for more than 575 women's groups. At the same time, however, debate over the use of assisted reproductive technologies was growing. These technologies include cloning, surrogacy, assisted insemination, in vitro fertilization, embryo research, and prenatal diagnosis techniques. During a first round of consultations about these technologies with the federal government (1989–1993), NAC adopted a position that dissatisfied many of its members. Leaders of the women's organization argued that reproductive technologies were

being developed not to meet the needs of ordinary women but to further the interests of the scientific community and the biotechnology industry. These technologies, claimed NAC, 'represent the values and priorities of an economically stratified, male-dominated, technocratic science' (NAC 1990, quoted in Montpetit, Scala, and Fortier 2004, 145). Many within NAC disagreed with this position, which was seen as too simple and out of touch with concerns at the grassroots. Those offended included lesbians and infertile women who wished to bear children.

Between 1993 and 1997, therefore, NAC adopted a more open approach to the question. Discussions within the women's organization allowed ample room for the expression of diverse views. Rejecting its earlier stance against science as a result of this more open process, NAC now argued that assisted reproductive technologies are acceptable when they reduce inequalities between women. This stance resulted from a compromise between different perspectives within NAC. But policy-makers in Ottawa were puzzled, because

The women's movement has organized around many different issues, including the objectification of women. These women protested against the Miss America Pageant in Atlantic City, NJ, in 1969.

the translation of this stance into actual public policy was not obvious. Losing influence as a result, thereafter NAC was pushed to the margins relative to other pressure groups involved in making Canadian policy on assisted reproductive technologies (such as the Canadian Bar Association and the Canadian Medical Association; see Montpetit, Scala, and Fortier 2004; Scala, Montpetit, and Fortier 2005).

During both its first and second waves, then, the Canadian women's movement has organized around many issues. The diversity of its concerns and perspectives not only reflects the many faces of gender inequality but also promotes a diffusion of the movement's ideas and its survival in the face of changing social conditions. Internal arguments may exhaust activists, however. Although factions permit the coexistence of different constituencies, they draw attention and energy away from common interests that unify. When the time for action comes, a movement may lose effectiveness if its factions do not set aside their differences. As with all social movements, the success of the women's movement depends on balancing the trade-offs between diversity and unity (Briskin 1992).

TIME to REFLECT

In what ways, if any, do you believe the women's movement has altered your life and your attitudes about gender? Has the environmental movement had a greater or lesser effect on you than the women's movement?

THE ROOTS OF AGRARIAN PROTEST IN CANADA

Regional differences between farming economies have affected agrarian social movements in Canada. The ideology and popularity of these movements and their links with other social groups all depend on the type of farming found in each region. Brym (1980) examined regional differences in agrarian protest by comparing farming economies in Alberta, Saskatchewan, and New Brunswick during the Depression years.

During the 1930s, agrarian protest grew rapidly in the Prairies but not in New Brunswick. Much can be explained by the degree to which farmers' livelihoods were affected by the market. Farmers in the West concentrated on producing beef or wheat, both for the rest of Canada and for export. Hence, western ranchers and wheat farmers faced similar economic pressures.

Eastern Canada set the tariffs on manufactured goods, the rates for railroad freight and bank credit, even the prices of beef and wheat. United by common economic interests, western farmers responded by creating marketing, consumer, and other voluntary associations that stressed cooperation.

In New Brunswick, by contrast, farmers practised mixed agriculture. Their primary productive goal was meeting their economic needs without selling what they produced or buying what they needed—strictly speaking, they were peasants rather than farmers. Since changes in market prices hardly affected them, they had little reason to defend themselves by forming cooperatives. Historical and geographical factors also mattered. While the dominance of shipping and timber interests hampered the commercialization of agriculture in New Brunswick, the province's poor soil and rugged terrain confined farming to river valleys and the coastline. Finally, New Brunswick farmers were more isolated than those in the West and had much smaller debts. Farmers in New Brunswick, therefore, were much less likely to form associations. In 1939, for instance, membership in farmers' cooperatives per 1000 rural residents over 14 years of age was 32 in New Brunswick, compared with 326 in Alberta and 789 in Saskatchewan (Brym 1980, 346).

Thus, the greater radicalism of Prairie farmers stemmed from high solidarity and a loss of control over their means of production. The two western provinces diverged in their approach to agrarian protest, however. Alberta's Social Credit party was right-wing, while Saskatchewan's CCF (Co-operative Commonwealth Federation, the predecessor of today's New Democratic Party) was left-wing. What accounts for this divergence?

In Alberta, a leftist agrarian party known as the United Farmers of Alberta excluded small-town merchants and others seen as exploiters of farmers. During the difficult Depression years of the 1930s, however, cooperation between farmers and merchants increased when they saw that their economic fortunes were connected—if farmers did badly, so would local businesses, and vice versa. With the support of right-wing merchants, teachers, professionals, and preachers, the new Social Credit party spread from Calgary to the small towns of southwestern Alberta. Eventually, Social Credit reached farmers and won their support too, but the party never lost the right-wing ideology of its urban roots.

In Saskatchewan, by contrast, the CCF maintained strong ties between farmers and the left-wing urban working class. Of the CCF leadership, 53 per cent were farmers and 17 per cent workers, while of the Social Credit party 24 per cent were farmers and none were workers (Brym 1980, 350). Thus, the differing class backgrounds of the farmers' allies help to explain differences in the ideologies of agrarian movements in Saskatchewan and Alberta.

Economic factors affect the formation of social movements as well as affecting which ideological direction they take. Agricultural producers such as prairie farmers were more likely to protest during a downturn in the capitalist economy, because their livelihood, unlike that of producers in New Brunswick, depended on the market. Furthermore, the organization of a protest movement is hampered when potential supporters lack pre-existing social ties or work in isolation from other potential supporters. Finally, the alliances of a social movement affect both its ideology and its chances of success.

The history of agrarian protest in Alberta is linked to a later development in Canadian politics: the rise of the Canadian Alliance. This party began as the Reform Party in 1987 and won a staggering 52 seats in the 1993 national election. In many ways, the Reform Party was a protest party. It provided critical opposition to the Progressive Conservatives while appealing to western Canadians' sense of regional pride. Although many factors shaped the rise of the Canadian Alliance, the early organizers of the Reform

Party were able to draw on a right-wing, populist ideology already created and maintained by Alberta's Social Credit party. The ideological conditions that made possible the rise of a new national conservative party thus were set many years earlier. Indeed, the Reform Party and the Social Credit party shared similar bases of support: Alberta's farmers were much more likely than any other group in that province to vote for the Reform Party (Harrison and Krahn 1995).

TIME to REFLECT

Are conservative movements social movements? Or do you think they are best characterized as counter-movements? Why? Are terrorist organizations social movement organizations? Why or why not?

THE SEPARATIST MOVEMENT IN QUEBEC

For five days during 1918, thousands of Quebecers mobilized against Ottawa's decision to impose conscription (mandatory military service) on young men. After Great Britain declared war on Germany and its allies in 1914, Canada had recruited and trained hundreds of thousands of soldiers and sailors. English Canadians saw themselves as British, and

UNDER the WIRE

G20 Protests in Toronto

In June 2010, world officials met in Toronto to discuss the global economy and to negotiate financial plans. Ever since the 1999 protests in Seattle, security has been strong at these economic summits, and Toronto police and private security firms had planned for months how to deal with both peaceful and potentially riotous protests. Canadian and worldwide activists came to Toronto for a week of organized activity to help publicize their issues and grievances: worldwide poverty and growing inequality, the expansion of corporate power, colonization and indigenous rights, women's undervalued global labour, environmental degradation, food security, financial deregulation, and so on. How did so many people get involved in protesting this (and other) financial summits? Why has protest activity now become expected by activists and police forces alike?

The answer is simple: the Internet and the ease of worldwide communication. Now that protest is fully

'wired', activists can maintain a strong sense of collective identity and collective efficacy by staying in contact with one another, by reliving their triumphs and/or sorrows on YouTube, by recruiting and staying in touch through Facebook, by organizing and participating in online activist forums, or simply by adding their email address to a listserv, electronic newsletter, blog, or Twitter.

After more than 900 arrests in Toronto, activists have used online communication technologies to help raise money for legal fees and to publicize what some take to be police brutality or government repression. Toronto police are utilizing modern technologies too as they comb through footage of rioters and use advanced face-recognition software to identify those culpable for damages to property and for endangering public safety.

voluntary enlistment in the Canadian Expeditionary Force was high. By 1917, however, tens of thousands of Canadians had been wounded or killed in action. For Britain and its allies, the situation was becoming precarious. Warfare in the trenches had reached a stalemate, Canadian units were falling below fighting strength, and recruitment was stalling at home.

From the start of the war, a smaller proportion of French Canadians had volunteered for service. This made them a target for resentment among English Canadians: it seemed that French Canada was not shouldering its fair share of the country's burden. But French Canadians were less willing to fight a war on behalf of Great Britain and its empire. They harboured their own resentments too, most recently over legislation that Ontario had introduced in 1912 to limit the use of French in its schools. To put it simply, French Canadians believed 'the enemy was closer to home, that it was the English-speaking Canadians and not the Germans who were the real threat to French Canada's survival' (Auger 2008, 515).

What came to be known as the Easter Riots began on 28 March 1918, when police in Quebec City arrested a young man for avoiding conscription. As he was escorted to the police station, the suspect was followed by a group of angry sympathizers. Soon the police released him, but by then a crowd of some 2000 people had gathered. Instead of dispersing they stormed the police station and gave several officers a beating. The next day, about 8000 people attacked the offices of two pro-conscription newspapers before setting fire to the office of the registrar for conscription.

Authorities in Quebec City and Ottawa responded quickly, partly because they feared a repeat of recent revolution and civil war in Mexico, South Africa, Ireland, and Russia. Bolstered by hundreds of English-speaking troops from Ontario and western Canada, over the next few days the army patrolled the streets of Quebec City, guarding key government buildings such as the Legislative Assembly, the Dominion Arsenal, and the Dominion Rifle Factory. This did not prevent a series of violent clashes that left up to 10 civilians dead and dozens of soldiers and civilians wounded (Auger 2008).

Other episodes of collective action for the French-Canadian cause followed, but until the 1960s they were sporadic. What was then known as French-Canadian 'nationalism' was conservative. Its goal was to preserve the identity of the French by insulating them from outside influences: not just the English language but the world of politics, business, and the mass media. In schools and parishes, the

Catholic church taught that defending the French-Canadian 'race' required not contestation and collective action but conformity with the mores of a society said to be stable, agrarian, and homogeneous (Bélanger and Lemieux 1969). During the 1930s, other organizations promoting 'nationalism' in towns like Drummondville, Quebec, were the trade unions and local associations of property-owners and retail merchants. These organizations urged the populace to put their 'own' people first by favouring French Canadians in hiring and buying from people like themselves (under the slogan *Achetez chez nous*) rather than outsiders like the town's Jewish merchants (Hughes 2009 [1943]).

Social change accelerated after the asbestos strike of 1949, a bitter labour dispute overlaid with ethnic tensions because the miners who walked off the job for four months were mostly francophone, while their managers tended to be English-speaking and the owners of mining companies were American. Through advances in communication—the spread of radio, telephone, and television—during the 1950s, the world outside of Quebec became ever harder to ignore. This encouraged comparison, self-scrutiny, and awareness that more social change was inevitable. Families relying on agriculture were disappearing, and most of the population was now urban. During the Great Depression, citizens had sought help from the Canadian state, whose reach expanded greatly during the two world wars. French Canadians could not escape modernity. Why not therefore assert control over its direction (Balthazar 1992)?

The church was losing its hold over ways of thinking. Formerly a conservative and religious project, promoting the interests of the French minority thus became a secular project consistent with liberalism, the welfare state, and even socialism. As always, of course, matters were complex. Some clergy in the church, for example, did defend the francophone miners during the asbestos strike. In turn, backward-looking myths of historical continuity gave spice to forward-looking claims that Quebec was a distinct society (Bock-Côté 2009). Nonetheless, the magnitude of social change contributed directly to the Quiet Revolution of the late 1950s and 1960s, a period of administrative expansion and heightened intervention in health, education, social welfare, and economic development by the government of Quebec (McRoberts and Postgate 1980; Meadwell 1993).

In a different way, though, the scope of political hopes narrowed. Until the 1950s, the French inhabitants of the province tended to call themselves French

Canadians in recognition of their membership in a larger cultural and linguistic group also living in other parts of Canada (especially New Brunswick, Ontario, and Manitoba, which had sizable French-speaking minorities). Thus, the primary reference of the word *québécois* was formerly to those who lived in Quebec City. But the 1950s brought a growing belief in the need to connect the defence of French-Canadian culture and interests to the territory of Quebec. No other place in North America seemed to offer a realistic possibility of living in French while building a society that was prosperous, well governed, and capable of providing its citizens with equitable rights and opportunities. Quebec thus displaced the French-Canadian people as the marker of a common destiny (Balthazar 1992).

The 1960s and 1970s witnessed a sharp rise in collective action on behalf of the French in Quebec. In 1962, demonstrators in Montreal protested against the absence of any French-speakers on the board of directors of Canadian National Railways. Queen Elizabeth's 1964 visit to Quebec City spurred anti-royalist, anti-British protests. When Prime Minister Trudeau paid a visit to Montreal for the annual Saint-Jean-Baptiste Day parade in 1968, hundreds of hostile and sometimes violent protesters voiced their opposition to the Liberal party and Canadian federalism. That same year, the school board for Saint-Léonard (a Montreal neighbourhood with a significant Italian minority) decided to limit the language of instruction in local schools to French. Immigration to Montreal was increasing, and the school board wanted to halt a trend—the adoption of English by newcomers—that threatened the importance of French. In what came to be known as the Saint-Léonard Crisis, unruly demonstrations continued as backers of Quebec's collective right to protect the French language confronted those who believed parents had a right to choose the language of instruction for their children (Hewitt 1994).

The October Crisis of 1970 is etched in our collective memory. Members of the Front de libération du Québec (FLQ) kidnapped a British diplomat in Montreal, James Cross, and then the provincial minister of labour, Pierre Laporte, who was murdered by his captors. This was the peak of contentious politics by nationalists in Quebec. Since its founding in 1963, the FLQ and allied groups had carried out bombings that targeted property and monuments associated with the federal government, big business, and British colonialism. The extremism of the FLQ alienated the majority of Quebecers, however, and indeed can be taken as a sign of its political weakness (Breton 1972; Hewitt 1994). The events of 1970 were the culmination of a wave of contention that subsided quickly thereafter.

It might be tempting to conclude that all of the collective action described above promoted nationalism (or separatism). However, consider a few important distinctions: *patriotism*—a sentiment involving love of one's country; *ethnic mobilization*—collective action on behalf of an ethnic group; *nationalism*—a political project founded on the premise that a people has an inalienable right to its own state and associated territory. Although we find many examples of patriotism and ethnic mobilization in Quebec history, we cannot conclude that participants were always nationalists. Like Americans in the civil rights movement who marched to achieve civil rights for blacks, Quebecers demanding more French Canadians on the board of directors of Canadian National Railways wanted integration, not separation.

It might also be tempting to conclude that nothing has changed since the 1960s: 'When will Quebecers stop asking for more?' is a question often heard in English Canada. But consider the *political expression* of Québécois nationalism. True, the Parti Québécois remains influential in provincial politics, and the Bloc Québécois altered the balance of power in Ottawa when it won half of the vote in Quebec and 54 seats in Parliament in the 1993 federal election. But these are *parties*. Social movements are recognizable by their use of contentious politics: they operate *outside* the channels of established political institutions. Looking back on the period since the October Crisis of 1970, it is difficult to discern a separatist *movement* in Quebec.

This raises some provocative questions. If political parties are now its main promoters, has separatism become tamed? As organizations that work within the political system, today the Parti Québécois and the Bloc Québécois push less for the independence of Quebec than for the decentralization of power in Canada. Around the globe, moreover, citizens are turning to social movements as they search for grassroots alternatives to parties. What political options does this leave for those in Quebec who are nationalists yet have no separatist movement to join and believe that the Parti Québécois and the Bloc Québécois are no different from other parties—remote, bureaucratic, and unable to fulfil their promises to voters?

TIME to REFLECT

Given what you have learned about the cyclical nature of social movement activity, what do you think the future holds for the separatist movement in Quebec? Can you justify your ideas with any particular social movement theory?

IS THE FUTURE OF SOCIAL MOVEMENTS GLOBAL?

Capital and commodities, information and ideas, people and their cultures are criss-crossing the globe, and these interactions are changing the world's societies. New opportunities and constraints are appearing that force social movements to adapt their strategies, resources, and ideologies. Many social movements recognize that globalization is changing the political terrain. Charles Tilly's research (1978) demonstrates that modern social movements arose at the same time as the nation-state in Europe. Will globalization now give rise to new, global forms of protest?

Some environmental organizations, such as Greenpeace International, Amnesty International, and the Sea Shepherd Society, as well as a variety of anti-globalization and social justice movements, claim to operate in a global polity. These organizations take the globe as their site of struggle while simultaneously operating in specific locations. In other words, organizations such as these claim to 'think globally but act locally'. Their strategy is clear: concerted efforts in locations throughout the world will alter the negative social and environmental effects of globalization.

German sociologist Ulrich Beck (1996) claims that globalization creates opportunities for new forms of collective action that operate outside the politics of the nation-state in the politics of a 'world risk society'.

GLOBAL ISSUES

Japan Cuts Short Whale Hunt over Clash with Activists

Obstruction by a hard-line anti-whaling group has forced Japan to cut short its Antarctic whale hunt, the fisheries minister said on Friday, the first time the fleet is heading home early due to clashes with activists.

Repeated attempts by the Sea Shepherd Conservation Society to block the hunt have caused irritation in Japan, one of only three countries—along with Norway and Iceland—that now hunt whales. The government describes the hunt as an important cultural tradition.

'It has become difficult to secure the fleet's safety,' Fisheries Minister Michihiko Kano told a news conference. 'We have no choice but to cut short our research.'

Sea Shepherd described the decision as a victory for whales and said the movement was proving increasingly effective at disrupting the Japanese fleet's operations.

'We didn't do much differently, but got there early and intercepted them and disrupted them early before they could really begin,' Jeff Hansen, Sea Shepherd's Australian director, said by telephone. 'Each year we're costing them more and more money. It's a heavily subsidized hunt and they're spending and losing millions so we're very much hurting them in the pocket. We'll stay down there in the Southern Ocean, and if they return next season, we'll be there to escort them northward.'

The Japanese fleet, made up of 180 people on four vessels, is heading back from its annual hunt about a month earlier than scheduled after catching 170 minke whales, around a fifth of its target, said Shigeki Takaya, a fisheries ministry official. He said it was unclear how Japan would proceed in the future with the hunt.

Japan had suspended the hunt last week after Sea Shepherd started to harass the fleet's mother ship. Clashes between Japanese whalers and Sea Shepherd activists have escalated in recent years. The group had introduced a new high-speed ship after another vessel sank following a collision last year with a Japanese whaling ship. An activist was given a two-year suspended jail term by a Japanese court in July for boarding a whaling ship.

Japan introduced scientific whaling to skirt the commercial whaling ban under a 1986 moratorium, arguing it had a right to monitor the whales' impact on its fishing industry.

Last year, Australia filed a complaint against Japan at the world court in The Hague to stop Southern Ocean scientific whaling. The decision is expected to come in 2013 or later.

The last time Japan shortened its Antarctic whaling expedition was in 2007, when a fire killed a crew member aboard the flagship and crippled the vessel.

SOURCE: Adapted from Yoko Kubota. 2011. 'Japan cuts short whale hunt over clash with activists'. *Vancouver Sun*, 18 February.

Beck suggests that ordinary people in all societies have been socialized to understand that the modern world is full of human-created hazards. Widely publicized dangers, such as the radioactive cloud that drifted from a nuclear reactor in Chernobyl (in Ukraine, at that time part of the Soviet Union) to the rest of Europe in 1986, have forced people to acknowledge that many political issues transcend borders.

Greenpeace International and the Sea Shepherd Society are good examples of global, or *transnational*, social movement organizations that appear to have adapted to this world risk society.

These organizations have developed their own political opportunities by creating unique forms of global diplomacy and activist campaigns (Beck 1996). They often operate outside the boundaries of the nation-state, such as on the high seas, where individual nations have no clear legal jurisdiction (Magnusson 1990). Conscious of the influence of the international media, these organizations rally support by organizing global boycotts that challenge governments and corporations to change their environmental policies

and practices. Through these media events, they attempt to stir up moral indignation while recognizing that different cultures have various understandings and experiences of global environmental dangers (Eyerman and Jamison 1989).

Not all sociologists agree that globalization has created a fundamentally new political reality. Leslie Sklair (1994) argues that global politics are very much like national politics, simply on a larger scale. For Sklair, organizations such as Greenpeace International mirror the organizational structures of transnational corporations. He suggests that the global environmental movement consists of transnational environmental organizations whose professional members make up a global environmental elite. This elite plays an ideological game with the transnational corporate and governmental elite: each side attempts to have its version of the environmental reality accepted as the truth. For Sklair, this is politics as usual.

Sociologists also question whether the rise of supranational organizations, such as the European Union (EU), will bring about new forms of collective

(Courtesy of Western Wilderness Committee/Jeremy Williams)

Many environmental groups believe that concerted efforts in locations throughout the world will alter the negative social and environmental effects of globalization. Increasingly, ordinary people are becoming involved in the environmental movement and its many issues.

action that link activists across national boundaries. Although the EU does constitute a new political terrain, Doug Imig and Sidney Tarrow (2001) have found that collective action in Europe remains strongly rooted within the nation-state. While Europeans have many grievances against the EU, most protest against it is domestic rather than transnational. This may simply indicate that activists have yet to develop new transnational strategies and linkages. Nevertheless, domestic politics remain a viable political arena for voicing concerns about the EU (Imig and Tarrow 2001).

Today, the world is more intricately connected than in the past. A variety of new social issues have arisen as a result, and there are now social movements that attack globalization. Each has to identify guilty institutions and actors, however, and states and corporations remain the best choice because they are largely responsible for the policies and practices that promote globalization.

Generally, three characteristics are needed for a social movement to be truly global. First, a social movement must frame its grievances as global grievances. Many environmental organizations do this. By framing environmental risks as global risks, the environmental movement hopes to demonstrate that environmental degradation affects everyone. Second, to be global, a social movement needs to have a worldwide membership and organizational structure. On a global scale, membership and frame alignment are probably supported by communication technologies such as email and the World Wide Web. Alternatively, a global movement can arise through a long-term coalition or network of movement organizations. For example, indigenous peoples across North and South America, Australia, and New Zealand have united against the ongoing effects of colonialism and to ensure that the rights of indigenous populations are recognized. Third, collective identity has to be a globalized identity. Global activists throughout the world would have to see each other as serving the same, common goal. Each would also have to identify cognitively and emotionally with that goal, as well as with other global activists and movement organizations.

TIME to REFLECT

Can you think of any causes that you would be prepared to go to jail or die for? Why are some issues more important to you than other issues?

Conclusion

Figure 19.2 depicts the changes in protest repertoires in western Europe and North America. Early forms of collective action were poorly organized and relatively sporadic. Often their grievances were tied to local affairs, and usually their targets were local elites. With the rise of nation-states, however, new kinds of social movements appeared. These movements were highly organized and often identified social issues that stemmed from structural conditions such as economic inequality and narrow political representation. They also routinized protest activities: different social movements learned to apply similar methods of protest, such as the mass demonstration. The rise of NSMs in the second half of the twentieth century marks another change. These movements are more concerned with gaining cultural recognition than with the redistribution of social goods. So even though NSMs tend to use traditional forms of protest, they are more concerned with the politics of everyday life (the politics of recognition) than with the traditional politics of governance (the politics of redistribution).

The success of collective action is always linked to the social and political climate. Social and political changes can create opportunities for social movements, or they can impose constraints. According to a

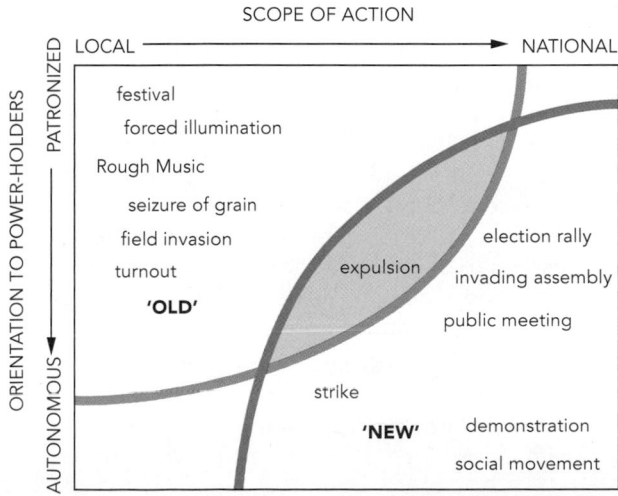

'Old' and 'New' Repertoires in Western Europe and North America

▼ **FIGURE 19.2**

SOURCE: Charles Tilly. 1983. 'Speaking your mind without elections, surveys, or social movements'. *Public Opinion Quarterly* 47 (4): 461–78. Reprinted by permission of Oxford University Press.

theory developed by Herbert Kitschelt (1993), present conditions in Canada have created opportunities that may lead to an increase in social movement activity. Support for social movements usually rises when political parties and interest groups fail to channel citizens' demands. Social movements can then mobilize support, attract resources, and forge alliances among protest groups. However, according to Kitschelt, this surge in social movement activity peaks as resources dwindle, as political parties begin to take up citizens' concerns, and as people's interest in collective mobilization wanes. Social movement activity then falls, only to rise again the next time organizers capitalize on frustration with parties and interest groups. In other words, the short-term pattern of movement activity is cyclical.

The long-term trend, by contrast, is toward an increase in the number of social movements. In the wealthy capitalist democracies, social movement activity has grown steadily since the 1960s. Established parties and politicians have proven increasingly incapable of providing satisfactory solutions to such issues as nuclear power, toxic waste disposal, resource management, and equal rights. Many Canadians now share a distrust of established politicians, political parties, and interest groups. Clearly, the extent of citizen discontent should not be exaggerated: recent federal elections have shown that established parties still attract much support. Nevertheless, many burning public questions—around gender, citizen participation, and environmental, ethnic, and Native rights issues—often elude both parties and interest groups. The current climate in Canada, therefore, favours an expansion of social movement activity. Whether organizers actually will exploit this situation remains to be seen. The outcome will depend on social movement leaders and on the political establishment's ability to co-opt them.

questions for critical thought

1. With regard to the worldwide economic problems that began in 2008, does economic prosperity encourage or hinder the formation of protest movements? Would you expect more protest or less as a result? Which countries do you think will be more likely to experience social movement action? Which would be less likely? Can you draw on social movement theory to justify your answers to these questions?

2. Many people today believe that social movements offer better prospects for democratic participation than political parties or interest groups. However, Roberto Michels's 'iron law of oligarchy' says that organization discourages democratic participation because resources, expertise, and status tend to flow to leaders. If you were organizing a social movement, what kind of safeguards would you put in place to prevent social movement leaders from dominating a movement organization? Are there lessons to be learned from the women's movement about de-centred decision-making or participatory organizational forms?

3. Find an ideological statement from a social movement, such as a flyer or leaflet, a brochure or members' newsletter, a website, or an interview with a movement representative. What are the movement's ideals, goals, and plan of action? What social goods does the statement value, disparage, or neglect? How does the statement use emotional appeals to make its message more persuasive? Is the movement offering selective incentives to attract new members? What kinds of people are most and least likely to be persuaded by this statement?

4. Why do some social issues provoke social movement campaigning while other issues are hardly addressed by social movement activities? Do you think that these 'orphan' issues have something in common that does not resonate with Canadian (or global) society? How might an activist frame an issue to increase the chances that people will pay attention?

5. The political process approach determines how political opportunities and constraints influence social movement activities. What are the most important political opportunities for a social movement organization to succeed? What are the greatest barriers to success? As an aid, look over the case studies in this chapter, and try to identify the most important political opportunities and constraints.

6. Egypt has recently experienced massive social protests, which resulted in Hosni Mubarak's resignation as president, a position he had held since the 1981 assassination of Anwar Sadat. During the Egyptian protests, other, nearby countries started to experience an increase in activism too, especially Libya. Do you think that this is the beginning of a cycle of protest in the Middle East? Or can we understand these forms of activism through another theoretical lens?

7. A counter-movement is generally understood to be a response to a social movement. Find an example of a counter-movement, and compare it with its associated social movement. What do you think are the main differences? Are there similarities that you think are of sociological importance? What social movement theory do you think is the most appropriate for understanding the relationship between these two forms of collective action?

8. Social movements disseminate ideas and hope to influence the general public. What is the role of the media in this process? How might new media technologies change social movement strategies, processes, and internal organization?

recommended readings

Ronald R. Aminzade et al., eds. 2001. *Silence and Voice in the Study of Contentious Politics.* **New York: Cambridge University Press.**
This text is an excellent collection of empirical research testing contentious politics theory.

Christian Davenport, Hank Johnston, and Carol Mueller, eds. 2005. *Repression and Mobilization.* **Minneapolis: University of Minnesota Press.**
A collection of essays on state responses to collective action, this is a must-read for students of social movements.

Donatella della Porta, Massimiliano Andretta, Lorenzo Mosca, and Herbert Reiter. 2006. *Globalization from Below: Transnational Activists and Protest Networks.* **Minneapolis: University of Minnesota Press.**
Challenging the idea that global social movements are merely coalitions of local movements, the authors argue that the global movement against neo-liberalism is a form of collective action that represents important changes in tactics, collective identities, and patterns of organization.

Jeff Goodwin, James M. Jasper, and Francesca Polletta, eds. 2001. *Passionate Politics: Emotions and Social Movements.* **Chicago: University of Chicago Press.**
This is a volume of much-needed research on the role of emotions in social activism.

Jo Reger, Daniel J. Myers, and Rachel L. Einwohner, eds. 2008. *Identity Work in Social Movements.* **Minneapolis: University of Minnesota Press.**
The editors have compiled a reinvigorated and empirically based volume of research on collective identity.

David Snow, Sarah A. Soule, and Hanspeter Kriesi, eds. 2004. *The Blackwell Companion to Social Movements.* **Oxford: Blackwell Publishing.**
This is a comprehensive examination of the state of social movement research and what remains to be studied and theorized. Chapters are written by well-known movement scholars.

Suzanne Staggenborg. 2008. *Social Movements.* **Toronto: Oxford University Press.**
Staggenborg provides a thorough introduction to major social movement theories as well as a comprehensive account of the history of important Canadian social movements.

Sidney Tarrow. 1998. *Power in Movements: Social Movements and Contentious Politics,* **2nd edn. New York: Cambridge University Press.**
Tarrow presents an up-to-date survey of social movement studies written by a scholar versed in theory and empirical work on both sides of the Atlantic.

recommended websites

American Sociological Association, Section on Collective Behavior and Social Movements

www.asanet.org/sectioncbsm

This a good starting place for more information on the sociological study of social movements. Read *Critical Mass*, the section's newsletter, to be informed of new publications, conferences, and the latest research.

Assembly of First Nations

www.afn.ca

This very comprehensive site contains detailed information about social issues pertaining to Canada's First Nations.

Canadian Lesbian and Gay Archives (CLGA)

www.clga.ca/archives

This site provides information that relates to lesbian, gay, bisexual, and transgender movements. Its focus is mostly Canadian, but the archive also provides plenty of information from around the world.

Canadian Race Relations Foundation

www.crr.ca

The Canadian Race Relations Foundation's primary goal is to end race- and ethnic-based discrimination in Canada. This website provides information about current issues and research.

Centre for Social Justice

www.socialjustice.org

This organization was established in 1997 and is based in Toronto. Its goals are to foster national and international social change through research and advocacy.

Global Solidarity Dialogue

www.antenna.nl/~waterman/dialogue.html

This is a good starting place for information on global social movements. The site provides research, news, and discussion on social movements throughout the world, as well as information on globalization.

Greenpeace Canada

www.greenpeace.ca

This site provides information about Greenpeace's past and current campaigns. Peruse the site, and try to establish how this organization frames environmental issues.

National Action Committee on the Status of Women (NAC)

www.nac-cca.ca

NAC is the largest women's movement organization in Canada. This site provides the history of the organization and presents current issues and discussion about equity issues in Canada.

chapter 20

Challenges of Globalization

Pierre Beaudet

In this chapter, you will:

▶ Understand how globalization came about in history;

▶ Learn how it has changed recently under the influence of the United States;

▶ Discuss how it has affected Canadian society;

▶ Understand the conflicts and tensions in the globalization process generated by cultural, political, and economic conflicts;

▶ Examine how globalization is changing under various pressure points;

▶ See the influence of rising powers (like China) while understanding the decline of traditional powers (like the US);

▶ Understand the demands and proposals of various social movements concerned with the impact of globalization.

Introduction

Our planet Earth, Pachamama ('mother world', in the language of the indigenous peoples of South America), now seems very small. Our societies and economies are more and more integrated, worldwide, through trade, capital flows, communication and transportation, migration, environmental connections, and cultural and political exchanges, in a manner that affects our daily lives from the very biological dimension to the ways we understand our reality and try to change it.

The world, our world, is changing fast. For example,

- The work Canadians perform is part and parcel of a 'globalized economy'. A slight variation in the Tokyo or London stock exchanges and our economy goes up or down. Top-notch engineers, autoworkers, or even part-time cooks lose or gain instantaneously because of unpredictable changes in the financial markets. The impact of these changes on the future of our jobs and prosperity might be tremendous.

- Our society in Canada is changing. More people than ever before are richer, while at the same time more people are poorer. Our traditional social 'safety net' is watered down along lines determined by big private (corporate) and public forces, with direct impacts on our health and education.

- The future of our environment depends on what will happen thousands of kilometres away in the Arctic. We know that the food we eat is produced in California or Kenya and transported across the globe. We see oil and gas platforms threatening fragile oceans. Many of us wonder whether that level of degradation is sustainable.

- Key decisions affecting our lives are made by anonymous and invisible technocrats. Sometimes we see how contested our world is becoming, especially when noisy movements take to the streets as in Toronto in June 2010. Local initiatives to claim policy changes become global. It is not clear how our world is governed and by whom.

- Along with so many people throughout the world, north and south, we watch the World Cup games and know the names of Brazilian soccer stars. We watch the same Hollywood 'blockbusters'. This cultural integration has reached an unprecedented level.

Although our globalized world sometimes seems threatening, there is an abundance of awareness and solidarity and stronger flows of ideas and proposals. Creative 'eco-friendly' engineers are inventing new ways of producing and saving energy. Indigenous farmers in South America, with young Canadians, are inventing 'fair trade' to produce and live better. Because of available worldwide information, many people come forward to help populations in crisis, as in Haiti or Pakistan. Isn't humankind always creative, proactive, thinking, imagining, and struggling to find solutions?

In this chapter, we will examine many 'globalizations', many processes involving millions of peoples, many dimensions (positive and negative), many interpretations, and many questions that are important for all of us.

TIME to REFLECT

What do you already know about globalization? How is it changing your life?

Globalizations and Globalization

Very early in the history of humanity, Africans spread out to colonize Europe and Asia. Asians crossed the Bering Sea to populate the Americas. Later, cities and empires developed in Egypt, Mesopotamia (ancient Iraq), India and China, and around the Mediterranean. Slowly and surely, the world became globalized through the 'Silk Road' that connected ancient China with central Asia and even Europe for hundreds of years.

RISE AND FALL OF EUROPEAN CAPITALISM

'Globalization' took another turn with the rise of European powers from the sixteenth century onwards and the emergence of capitalism, a new system that was from the beginning extraordinarily expansive. The 'bourgeoisie', the modern class of entrepreneurs who came to rule the world, was, as Karl Marx (1985 [1848]) explained, 'revolutionary' and 'international':

> The bourgeoisie has through its exploitation of the world market given a cosmopolitan character to production and consumption in every country. . . . All old-established . . . are dislodged by new industries . . . that no longer work up indigenous raw material, but raw material drawn from the remotest zones; industries whose products are consumed, not only at home, but in every quarter of the globe. In place of the old wants, satisfied by the production of the country, we find new wants, requiring for their satisfaction the products of distant lands and climes. In place of the old local and

national seclusion and self-sufficiency, we have intercourse in every direction, universal inter-dependence of nations. The intellectual creations of individual nations become common property. National one-sidedness and narrow-mindedness become more and more impossible.

The expansion of capitalism got a boost from the European conquest of the Americas. Africans were enslaved to man huge plantations of profitable products like sugar and cotton, which were shipped back to Europe along with wood and minerals from Canada. European capitalism became very powerful and was able to divide up the world into 'spheres of influence'. It went on to colonize Africa and Asia, controlling resource flows and taking huge profits from a system of 'unequal exchange' between the powerful and the powerless. Large firms in Britain and France, in particular, became 'multinational corporations' (before the phrase came into use), using new transportation and communication techniques like the steamship and the telegraph. In parallel, the 'modern' state came about with its large bureaucracy and its ability to regulate social tensions and conflicts (Weber 1978 [1908]). To social scientists and policy-makers, it seemed that modern capitalism was going to perform well, even though important reforms were necessary to improve the state and enlarge the social basis of the system. Émile Durkheim, one of the founders of sociology, argued that rigorous social sciences, applying methods pioneered in the natural sciences, should be used to modernize society (Poggi 2000).

TIME to REFLECT

What distinguishes modern capitalism from previous economic and political systems?

AS THE WORLD CHANGED

In the early part of the twentieth century, modern capitalism was put to the test. Many people rebelled because they were angry about the growing gaps between social classes. They became involved in various social movements and political projects that led to uprisings (in Russia) or massive struggles to gain social and economic rights.

Later, the 'dream' of the 'rational' state using science to improve life was blown away when Europe became engulfed in a series of bloody wars, leading to the death of more than 60 million people. After the Second World War, the US and the Soviet Union emerged as two 'superpowers'. At the same time, there was a general rebellion by the peoples of Asia, Africa, and Latin America, which brought an end to most of the remaining colonial outposts.

Globalization then took another spin as the United States was able to impose its new industrial and cultural models. The US saw itself as a 'global' power and was convinced that the key to its continuing domination was to internationalize the transformation worldwide through direct foreign investments from US multi-nationals and access to world markets. Globalization in that period was clearly **Americanization**. US products such as automobiles came to dominate the market, while at the same time the images of Hollywood penetrated every household. The whole world was invited to imitate the 'American way of life'. Canadian scholar Marshall McLuhan argued that in the modern economy, expanding mass communication was creating the foundation of a technology-based 'global village' (McLuhan 1964).

To appease tensions between the dominant United States and the Soviet Union following the Second World War, negotiations brought about the formation of the United Nations (UN) in 1948. Its fundamental objective was to restore peace and to resolve conflicts without war. Later, the UN undertook to help the poorest nations and even allowed these nations to have a say in world politics.

This period also saw the emergence of newly independent countries, the Third World, as it was known, that tried to lessen their dependence on the dominant powers. By the 1970s, the globalized world was composed of three parts: the capitalist West (North America, western Europe, Japan), the socialist East (the Soviet Union and central European nations), and the underdeveloped (sometimes called 'developing') Third World (Asia, Africa, South and Central America).

TIME to REFLECT

Where does the expression 'Third World' come from?

OUR MODERN GLOBALIZED WORLD

In the 1980s, many Third-World countries fell into crisis fuelled by growing debt. The US, in the meantime, was determined to 'win' the Cold War by exhausting the Soviet Union through military and economic competition. Indeed, the US succeeded when the USSR imploded in 1989. Some social scientists declared then that capitalism would henceforth dominate the world, a world that would enter a period of peace and prosperity (Fukuyama 1992).

American power became more assertive. Internally, there was a major policy change toward what was later defined as **neo-liberalism**, which entailed many things. Economic policies were changed for the benefit of the richer components of society, securing corporate power through such means as deregulation, privatization, and corporate tax reduction. Since then, explains David Harvey, the modern state's primary mission was to optimize conditions for capital accumulation, an abrupt change from the period following World War II, when assuring the well-being of citizens was also part of the mission (Harvey 2006). More specifically, according to Harvey, the neo-liberal state works to suppress the right to the commons, to commodify labour power and forms of production, and to appropriate the assets of largely colonized peoples through expropriation.

The Contested Impact of Neo-liberal Globalization

Canada adopted neo-liberal priorities in the late 1980s. This meant allowing more freedom (less control from the state) for private capital, reducing taxes for wealthy individuals and corporations, reducing budgetary transfers for health and education, and narrowing eligibility for important social programs like employment insurance (EI). In the past, more than 80 per cent of unemployed workers were able to collect EI benefits. After changes imposed by successive Conservative and Liberal governments in the late 1980s and early 1990s, only 4 in 10 unemployed men and 3 in 10 unemployed women are eligible for EI benefits, and they receive less income over a shorter time period. Says Bruce Campbell, 'not since the early 1940s, when unemployment insurance was introduced, have Canadian workers been so thinly protected against job loss' (Campbell 2009). Another example of important policy changes in the social domain was the elimination of Canada's family allowance program in 1992, which considerably reduced financial transfers to middle-class families (Olsen 2002).

TABLE 20.1 ▼	Canadian Imports from and Exports to the United States (in $billions), 2003–2008		
	2003	2008	% Change
Exports	329.0	370.0	+ 22.6
Imports	240.7	280.7	+ 16.8

Canada also became more economically integrated with the United States when the North American Free Trade Agreement (NAFTA) came into force in 1994. In 2008, for example, more than 75 per cent of Canadian exports went to the US, compared with 60 per cent in 1960.

Overall, rising trade with the US has made the Canadian economy richer but at the same time more vulnerable and more polarized. Tied to such an extent to one single economy, Canada can benefit or suffer depending on how the US performs. At the same time, following the US 'model' has resulted in widening gaps between social groups and classes (see Chapter 8 on class and inequality).

The same process has had significant impacts elsewhere in the world, such as South America. Writers, including Canadian author Naomi Klein (2002), believe that neo-liberal policies in that part of the world led to a 'lost decade' of impoverishment, social conflict, and environmental neglect. Others, however, argue that these policies led to economic benefits in countries, such as China, that attracted investments (Sachs 2005).

Mexican President Carlos Salinas de Gortari, US President George Bush, and Canadian Prime Minister Brian Mulroney sign the North American Free Trade Agreement in 1992.

(© Bettmann/Corbis)

TIME to REFLECT

Has the impact of neo-liberal policies (privatization, deregulation, tax reduction for the rich) on the whole been mostly positive or mostly negative?

Breaking Barriers: Culture and Technology

These tremendous economic and institutional changes were accompanied by transformations of cultural flows, supported by the expansion of information and communication technologies (ICTs). The expansion began with the development of satellite television in the early 1990s and was facilitated by the expansion of trade, which led to an exchange not only of goods but of lifestyles, fashion, food, and arts and crafts. Because of their economic strength, Western powers were able to dominate these flows through what was described by scholars as the 'McDonaldization' of the world through adoption of Western norms, procedures, and structures (Ritzer 2010). Indeed, the symbol of this process has been the proliferation of McDonald's outlets everywhere in the world, including the Third World, where more often than not, consumers flock not just to buy food but to feel 'modern' and 'Westernized'.

GLOBAL ISSUES

Canada and NAFTA

In 1992, the governments of Canada, the United States, and Mexico signed the **North American Free Trade Agreement**. Officially, NAFTA is intended to facilitate the flow of products and capital across the borders between the three countries by lowering or even eliminating trade barriers such as tariffs. However, NAFTA is more than a tool to improve trade.

The Canadian government has celebrated NAFTA as a major breakthrough for the Canadian economy. Indeed, exports to the US have soared, generating billions of dollars of profits. Influential corporate lobbies like the Canadian Council of Chief Executives are pushing for an extension of NAFTA to other areas of 'mutual concern' such as security, social and environmental policies, and immigration. This 'NAFTA+' would entail setting up a new scheme, the North American Security and Prosperity Partnership (SPP), whereby most of the important policies of the three countries would be harmonized. Currently, this project is officially off the agenda, but it has not been renounced by the Canadian or US government.

Since the signing of NAFTA, corporations have forced Canada to backtrack on measures to restrict the use of certain products that are considered health hazards. A well-known case is that of Ethyl Corporation. When the Canadian Parliament acted in 1998 to ban the import of an Ethyl product—the gasoline additive MMT, which Canada considers a dangerous toxin—Ethyl responded by filing a lawsuit against the Canadian government under NAFTA. In the end, Canada agreed to rescind the MMT ban, pay Ethyl in excess of $19 million, and issue a statement that MMT was neither an environmental nor a health risk.

Opposition parties, labour unions, and citizen groups argue that NAFTA has benefited the few. The 40 largest Canadian corporations have in fact increased their revenue (from $137 billion to $310 billion between 1987 and 2006) but in the same period reduced their workforce by more than 19.6 per cent, according to Bruce Campbell (2007). Other experts point to the growing presence of foreign (read: US) firms in the Canadian economy. Of the 40 largest companies operating in Canada, 20 are foreign-owned. In addition, other policy changes have 'aligned' Canada with the US (corporate taxation levels, for example). The employment insurance program has shrunk to levels similar to what prevails in the US. For Canadian investigative reporter William Marsden, Canada through NAFTA has become an 'energy slave' of the US and, moreover, has been forced to absorb the heavy health and environmental costs involved in projects like the Alberta oil sands (Marsden 2007).

While the economics of US–Canada integration has been widely commented upon, opposition groups fear that the impact could be negative when it comes to some of the basic freedoms and liberties in Canada. They pinpoint, for example, the establishment of 'no-fly' lists whereby the United States can prevent certain Canadians from flying over US territory without any legal process involved: 'Increasingly those people with views the Canadian and US governments do not agree with are being stopped at the Canada–US border, or banned entirely' (Council of Canadians 2010).

Later in the 1990s, the expansion of cellphone and Internet use resulted in another wave of globalized culture. Instant and cheap communication allowed people to absorb information from around the world at any time. Although the Internet was the creation of powerful Western countries and private corporations, it had contradictory effects. Other peoples, other nations, minorities, and social groups began to use it to communicate 'their' message, leading to a sort of polyphony of images and ideas. Arjun Appadurai, in a much quoted study, described this multilateral process as fluid, mobile, partly or wholly independent of nation-states (even the most powerful), producing many things, including 'hybrids' between different world views (Appadurai 1996).

American fast-food franchises are a common sight in Chinese urban centres.

TIME to REFLECT

Is the Internet enlarging space for meaningful and free communication? Or is it mainly a new tool for large companies to market new ideas and goods?

The Great Debate

The previous sections have outlined how our world has changed, especially over the past 30 years. The basic foundations of our societies, worldwide, have changed. The end results are not clear, but we can see that the world is more unequal and more unstable than ever before. However, there are many differing views on the latest globalization process and its impact.

For some, despite problems and shortcomings, the effects of globalization in general, and economic integration specifically, remain highly positive. Thomas Friedman, a columnist with the *New York Times*, says that our world is now 'flat, with fewer barriers, more prosperity and more peace' (Friedman 2005). Poor countries can sell more goods to rich nations, thereby becoming richer. In parallel, expanding financial flows across borders gives states access to much-needed capital. Friedman concludes that globalization is in any case 'inexorable' and that no sane person would want to go backward. Martin Wolf, a well-known economic commentator from Britain, argues that markets, and not incompetent governments, are in a better position to create and distribute wealth (Wolf 2004).

For others, globalization is just another word for imperialism, even if 'modern' imperialism is not the same as the earlier form (Chomsky 2003). In a sarcastic tone, some call this domination 'Wal-Mart-ization' or 'Coca-Colonization' because of the large influence US-based multinational corporations have on the world. In this view, the same pattern remains: resources from the Third World are plundered, not necessarily by colonial conquest, as in the past, but by dominant policies of free trade and liberalization. It is implemented through the might of

UNDER the WIRE

Information and Communication Technologies

The explosion of the Internet and other means of instant communication have changed our societies to an extent comparable to what happened in Europe with the expansion of printing more than 500 years ago. Today, ICTs are used mainly by economic actors to transfer data and control the cycle of production and distribution of goods and services throughout the world 24 hours per day, seven days a week. A modern automobile can be simultaneously designed, produced, assembled, verified, and marketed in dozens of countries, with factory outlets in China or Mexico while technological and financial command centres are located in Japan or Germany. But ICTs are developing in all sectors of society. Instant communication between Philippine nurses and their communities back home are part of the new reality in which millions of people are on the move throughout the world. Educational institutions are rethinking their approaches in a world where information is readily accessible and where millions of people anywhere in the world can 'talk' or listen to experts and academics.

Sociologist Manuel Castells explains that it is not so much information technologies per se that have changed our world, but the fact that they are being used in a 'new' economy or a 'new capitalism, based on international competition, knowledge generation, and information processing and the creation of large networks of multinational firms and institutions, each of which has the capacity to work as a unit in real time, or chosen time, on a planetary scale' (Castells 1996).

In parallel, social movements communicate widely and design strategies to debate with the World Bank or the **G8**, linking scholars, activists, and organizers from all over the world. Arturo Escobar (Escobar and Mignolo 2010) believes that the use of cyberspace contrasts with the traditional and unilateral model of communication: 'it is based on interactivity . . . a relational model in which negotiated views of reality may be built, where all receivers are potential emitters. . . . Cyberspace can be seen as a de-centralized archipelago of relatively autonomous zones in which communities create their own media and process their own information.'

Many other issues of ownership, environmental impact, and regulation concerning ICTs abound. For some writers (Appadurai 1996), the communication 'revolution' allows greater freedom and creativity. Others stress that on the contrary, it leads to greater homogenization and greater power for those who can control the flows (Ritzer 2010). Despite the fact that the Internet is an 'open space' where everyone can, at least in principle, intervene, barriers are abundant, either because of unequal internal access in poorer countries or communities or because of 'firewalls' established by autocratic regimes.

rich nations, which can impose their terms on poorer nations. Indeed, 50 per cent of the large multinational corporations that dominate the economy are based in the United States.

This leads scholars like Peter Urmetzer to deny that the current architecture of the world is very different from what it was in the past. He is one of the people defined as 'globalization sceptics', who believe that our world has not really changed. They argue that internationalization of the economy has not modified in substance the way our societies work. These globalization sceptics point out, for example, that most of the economic transactions (investments, financial flows, trade) occur within rich nations and between rich nations, along with a very small and selective number of Third-World nations. Globalization is a myth, says Peter Urmetzer, designed to make people accept policies that they would otherwise have fought. Now, however, people are opposing the concept of globalization because they think their jobs will migrate to the Third World (Urmetzer 2005).

TABLE 20.2 ▼ Top 10 Firms in the World	
Wal-Mart	US
Exxon Mobil	US
Royal Dutch Shell	UK
BP	UK
Toyota	Japan
Chevron	US
ING	Netherlands
Total	France
GM	US
ConocoPhillips	US

SOURCE: *Fortune* 21 July 2008.

Says Saint Mary's University professor of sociology Henry Veltmeyer, 'In the 1990s we saw a decided shift in the dominant form of imperialism. . . . By most accounts it is a form of imperialism that is not afraid to be named and hide itself behind globalization or development; disposed to use whatever means at its disposal: the coercive apparatus and power of military force' (Veltmeyer 2008).

Yet other scholars, such as Steger, believe that our world is not the same and has indeed become radically different. One of their concerns is what they call the suppression or compression of time and space. Because of modern information and communication technologies, policies and implementation can be diffused through space and time. Our societies are changed by the tremendous intensification of social, political, economic, and cultural interconnections and interdependencies on a global scale (Steger 2002).

A major transnational corporation can operate worldwide 24 hours per day, seven days a week, with production sites in Mexico or Indonesia and command centres in London or Toronto. This, of course, affects the political economy of our world, but also our governance system and our culture (Ritzer 2010). States do not have the same power that they did in the past, because the economy is now more and more international. In many ways, the modern state (in the Weberian sense), which used to have a near-monopoly of power over governance in its political as well as in its security role within a given territory, has seen that situation eroded.

Non-state actors such as transnational corporations sometimes have greater capacity to influence decision-making. For citizens, these changes are also blurring the distinction between what is local or national and what is international or global. Ulrich Beck, a German sociologist, talks about 'deterritorialization', in which traditional boundaries are eroding and becoming permeable to flows of information and capital (Beck 2008). Everyday decisions on such things as food, education, and security are taken in a context in which risks and potentialities are defined beyond the boundary of the national state. Global flows, argues Ritzer, are so intensive, extensive, and rapid that it is necessary to reconceptualize the current era as the 'global era'.

In their extensive work on globalization, British scholars David Held and Anthony McGrew (2000) argue that globalization is a process that transforms social relations. It stretches social activities across frontiers. It interconnects trade, investment, finance, migration, and culture, and it accelerates the pace at which these interactions are conducted. Through these processes, globalization blurs the usual boundaries between 'local' and 'global' matters.

ONE OR MANY GLOBALIZATIONS?

Beyond these controversies, there is a growing trend of thinking that affirms the multiplicity and the plurality of globalization. Antonio Negri and Michael Hardt, for example, argue in their best-seller *Empire* (Negri and Hardt 2000) that there is not just one track of globalization but many. According to these scholars, following the intuitions of Marx, modern capitalism levels off most national barriers that block or limit the expansion of capital. Old-style imperialism linked with nation-states has become outmoded, replaced by a new style that manages social contradictions in an interconnected and internationalized way. They believe that although globalization was designed mainly to serve the needs of capitalism, it is actually 'opening the door' for society to move beyond capitalism. In their view, the massive expansion of social movements and non-governmental organizations (NGOs) everywhere in the world illustrates the rise of a 'globalization from below' that constitutes one of several paths currently 'offered' to modern societies.

TIME to REFLECT

Read the definition of globalization by David Held and Anthony McGrew, below (taken from *The Oxford Companion to the Politics of the World*). Does it reflect the complexity of the matter?

Globalization can be conceived as 'a process (or set of processes) which embodies a transformation in the spatial organization of social relations and transactions, expressed in transcontinental or inter-regional flows and networks of activity, interaction, and power' (Held and McGrew 2001). It is characterized by four types of change. First, it involves a stretching of social, political, and economic activities across frontiers, regions, and continents. Second, it is marked by the intensification, or the growing magnitude, of interconnectedness and flows of trade, investment, finance, migration, culture, and so on. Third, it can be linked to a speeding up of global interactions and processes as the development of worldwide systems of transport and communication increases the velocity of the diffusion of ideas, goods, information, capital, and people. And fourth, the growing extensity, intensity, and velocity of global interactions can be associated with their deepening *impact*, such that the effects of distant events can be highly significant elsewhere and specific local developments can come to have considerable global consequences. In this sense, the boundaries between domestic matters and global affairs become increasingly fluid. Globalization, in short, can be thought of as the widening, intensifying, speeding up, and growing impact of worldwide interconnectedness.

One Global World: Prospects and Constraints

Currently, our society is at best perplexing: Where are we going? Will the 'crisis of crises' that we see in the economy, culture, politics, and security continue provoking more fears? In the Mandarin language, *crisis* is understood differently from the way it is in English and other Western languages because it is composed of two characters (or two ideas), one expressing destruction (which is how we understand 'crisis' generally) and another expressing opportunity, implying that humanity sometimes goes to the brink before it can reset itself and implement needed changes that otherwise would not come about.

ONE DAY IN SEPTEMBER 2001

On 11 September 2001, the two tallest office buildings in the largest city in the United States, which happens to be the financial capital of the world, along with the Pentagon (Department of Defense) in Washington, were hit by a terrorist attack on an unprecedented scale. More than 3000 innocent civilians were killed. The next day, President George W. Bush announced the beginning of an endless war against terrorism and all those who supported it, one way or another. A relatively obscure movement, Al-Qaeda, led by a relatively obscure leader, Osama bin Laden, was accused of masterminding the disaster with the support of a relatively obscure government in Afghanistan. Shortly afterwards, the United States invaded Afghanistan and easily displaced the government; however, they were unable to catch bin Laden. Meanwhile, the Bush administration planned the invasion of Iraq, claiming that it was a threat to world security and another safe haven for terrorists. The real goal, some believed,

Non-governmental organizations funded by donors in the West provide essential services in developing countries. This well at a school in Nepal was funded by Save the Children, an international NGO.

was to reengineer the Middle East and transform it into a gigantic free-trade zone (Klein 2008).

By the end of the decade, US military personnel numbered more than 2.5 million, including 360,000 deployed overseas at 737 foreign military bases in more than 150 countries (Johnson 2007). In 2010, the US military budget exceeded $720 billion.

Beyond that stark reality, the resumption of major conflicts worldwide has shattered the idea that the world would become more integrated and peaceful through the globalization process. Since 2001, military conflicts have expanded from Asia to Africa through a sort of geographic 'arc' concentrated in central Asia and the Middle East and stretching out to eastern, central, and western Africa. There are zones of tensions elsewhere, including southern Europe and South America.

Canada, as part of the Western nations' alliance NATO, is a participant in these wars, particularly in Afghanistan, where thousands of Canadian soldiers have been deployed. According to research conducted by David Macdonald and Stephen Staples, Canadian taxpayers will have paid more than $18 billion for the Afghan war, which represents a very significant percentage of the federal budget in a period when the Canadian government is arguing for fiscal restraint and spending cutbacks (Macdonald and Staples 2008).

'CLASH OF CIVILIZATIONS'?

The descent of huge regions into apparently endless wars has seized the imagination and created a lot of fear. It contradicts earlier views, which dominated in the early 1990s, to the effect that the world would enter an era of peace with the end of the Cold War. The belief was that opening borders to trade and investment would eventually eliminate ideological and political differences, provided that the US could operate as the 'gendarme' of the world.

But some observers were sceptical, arguing that wars and other conflicts would continue in new forms. A proponent of that thesis, Samuel Huntington, a prominent professor and advisor to various US administrations,

OPEN for DISCUSSION

Winner and 'Losers' of Globalization

Globalization, according to its proponents, can help poor nations of the world to emerge from misery and poverty. China is often cited as an example to prove this point. At the same time, it has become very clear that globalization has produced a lot of 'losers'—i.e., nations and states that fall into decline, stagnation, and even conflict. The graph below illustrates these conflicting trends.

It is also significant that the changes of the past decade do not affect *only* the poorest countries. Gaps between the poor and the rich have been aggravated within countries, including rich countries like the United States, where average wages have diminished, as shown in the graph below.

Income per Head as Percentage of North

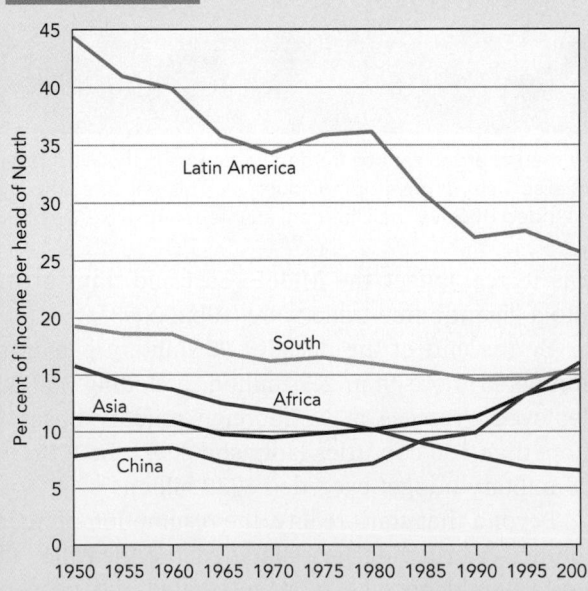

SOURCE: Bob Sutcliffe. 2004. 'World inequality and globalization'. *Oxford Review of Economic Policy* 20 (1).

Income Change for Top and Bottom Earners in the United States in 2006 Dollars

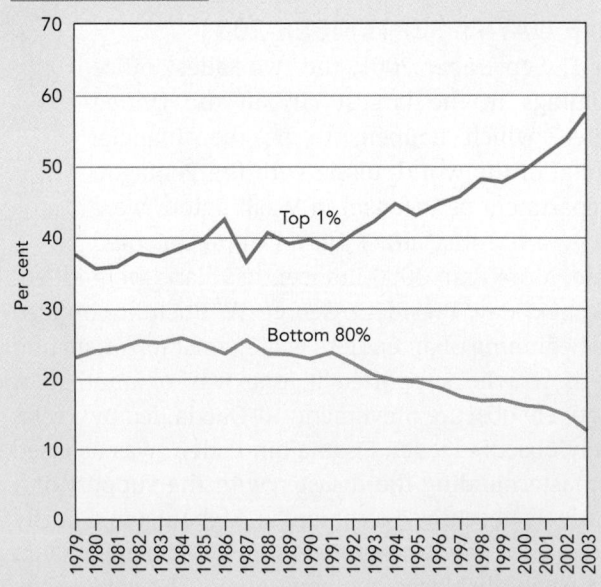

SOURCE: Arthur B. Kennickell. 2006. *Currents and Undercurrents: Changes in the Distribution of Wealth, 1989–2004.* Federal Reserve Board, 30 Jan., Table 1.

wrote in his book *The Clash of Civilizations and the Remaking of the World Order* (2002) that the globalizing world was not flat but, on the contrary, riddled with intractable cultural and political barriers. What he called 'Western civilization' (the US and Canada, western Europe, Japan, Australia) would always be in conflict with other 'civilizations', which were associated, in his view, with China, Russia, the Arab world, Africa, and Latin America. He argued that these areas would never accept the basic foundations of the 'Western world' such as capitalism, free markets, and individual freedoms. In that sense, he stated, current and coming conflicts would proliferate, not diminish. Thus, the 'West' should always be prepared to defend itself.

In the First Person

Back in the 1960s, I was, like many students, hoping to 'change the world'! But one day I listened to a Jesuit missionary who had just come back from Brazil. He described a life of misery and famine affecting dispossessed communities in the rural hinterlands and in the *favelas* (shantytowns) surrounding big cities. But what struck me was that he also spoke of strikes and peasant leagues, as well as of the songs and dances of ex-slaves expressing their determination and hope. Later, in the 1970s, I was lucky enough to pursue this exploration by working 'in the field' with various Canadian NGOs in Africa, South America, and Asia. In the meantime, I obtained my doctorate in sociology, analyzing the struggle against apartheid in South Africa, where I worked with African trade unions. After coming back to Canada, I was able to participate in a broad people's movement that contributed to this anti-apartheid struggle against racial discrimination and exploitation. More recently, I learned from scholars and activists in many countries, such as Brazil, Bolivia, Niger, Palestine, and India, the value of rigorous research related to intervention and popular education in an endless process in which social identities are built in the search for justice and truth. Today, I try to encourage students to work and think that, yes, definitely, 'another world is possible.' There is a 'creative utopia' out there, which sometimes can be observed in gatherings such as the World Social Forum, a network of more than 500,000 social movements and NGOs.

—Pierre Beaudet

This view became popular in the United States at the turn of the 1990s. It was associated with a new group of activists and scholars of the right known as neo-conservatives, who argued, along the lines of Huntington, that the US should prepare for new and more complex wars. In an influential study entitled *The New American Century*,[1] they outlined a comprehensive agenda that slowly trickled down into US policies. It focused on re-arming the US so as to confront the threatening 'civilizations' that inevitably would come knocking on the door or disturb the US model of capitalism. The implications of **neo-conservatism** were important. Its proponents claimed that economic integration was in fact *not* the glorious road to peace and prosperity in a single 'one world'. They contended that despite and beyond globalized economics, distinct politics and culture would not only differentiate peoples on this planet but that such differences would themselves be 'resolved' by conflicts (if not wars) rather than by commerce and diplomacy.

This view has been adopted by the Canadian government over the last little while, particularly by the Conservative Party, which has been in power since 2006. Murray Dobin, a Toronto-based author, argues that this represents a significant change, pointing out that successive increases in the military budget (up 27 per cent between 2001 and 2008) have 'militarized the minds' and weakened the traditional distinction between Canada and the US, with Canada perceived as more 'caring', more 'sharing', and more 'peaceful' than its southern neighbour (Dobin 2008).

TIME to REFLECT

Do you agree with Huntington on the 'clash of civilizations'? Do you believe that the 'Western world' must inevitably protect itself from other 'civilizations'?

Failing Processes

After 9/11, this world view resonated loudly in the US and other Western countries, including Canada. The events in New York and Washington seemed to validate Huntington's perspective: the 'one small world' was exploding. On 12 September, George W. Bush put forward the concept that the United States had the right to intervene anywhere in the world, with or without the consent of other nations. 'We are in a crusade,' said Bush, a statement that many people took to mean that the US wanted to impose its views and power in the Middle East, just as the European crusaders had 1000 years before. The narrative had strong religious overtones. President Bush himself implied that US actions were inspired by God. Neo-conservative policies were strongly supported by right-wing evangelical groups, who thought that Western Christianity should take on the world and resist non-Christian nations, including Muslim countries. This association between neo-conservatism, militarism, and religion is also evident in Canada (Patrick 2009).

Bush's declaration of unilateralism was accompanied by another statement to the effect that US-led wars could be 'pre-emptive'—i.e., that the US could attack a country or a region before actually declaring war, a dramatic departure from previously accepted rules established by the United Nations in 1948, whereby a country could not unilaterally decide to attack another without the authorization of the Security Council.

These new policies had a tremendous impact on the world and on our society. We had already seen a tremendous increase in military capability in the United States and in Canada. But less obvious is a series of changes in other aspects of our society. For example, the US has introduced new legislation and mechanisms to monitor and control citizens and other nations. Individuals considered a 'potential threat' can be detained without trial. Many immigrants and refugees have been placed under surveillance, especially those from certain parts of the world. Torture, seen as necessary to 'fight terrorism', was legalized in the US. Borders that seemed to have been weakened by economic globalization were strengthened by various means, such as the system of trenches and walls separating Mexico from the United States (Dauvergne 2008). In a similar vein, Canada, considered in the past especially generous to migrants and refugees, has also tightened controls, legalized the detention of 'terrorist suspects', and made it more difficult for people claiming refugee status (Canadian Council for Refugees 2005).

With these policies still unfolding today, many questions are being raised about the future of our world and the concept of globalization as it was understood back in the 1990s. The issues are more dramatic now that policy-makers are ambivalent and divided on the way to go. Many people think that the US administration's response to 9/11 was unfounded and ill-conceived and was in fact based on a 'hidden agenda' to control rich oil-producing areas and re-establish a contested US hegemony (Klein 2008).

Other experts believe that the invasion of Iraq and Afghanistan 'went too far' and that the eagerness to wage war is becoming dangerously uncontrollable, even in terms of economics. According to Joseph Stiglitz (Stiglitz and Bilmes 2008), the cost of the intervention in Iraq has surpassed $3 trillion, seriously weakening the US economy. The war was particularly costly, said the former World Bank official, because it was partially 'privatized', with huge (and very lucrative) contracts going to private military corporations that charge high rates, sometimes with the connivance of US officials.

Since the election of Barack Obama as US president in 2008, most of these unilateral policies have continued, even though the tone has changed in Washington. No durable solution appears in sight for major flashpoints such as Iraq, Afghanistan, or Israel and Palestine. Many areas of confrontation persist, sometimes involving other major military powers such as Russia and China. The United Nations has become largely disempowered. The United States, which along with the four other permanent members of the Security Council (China, Russia, France, and the UK) has veto power over it, does not want the UN to 'interfere' in areas where it believes its 'strategic interests' are at stake. Thus, we live in a world without a common platform or accepted institutions for resolving conflicts and with might and force as the main policy tools of big powers.

TIME to REFLECT

Do you think that new policies to restrict and control migrants and refugees in the United States and Canada were necessary to assure the security of the people?

Widening Gaps through the Crisis

As the world seems to be entering a period of intensified conflicts, another major problem has occurred with growing economic instability in some of the most powerful countries on Earth. The serious financial crisis that erupted in the United States in 2008 has affected many other countries. Large financial institutions based on Wall Street were caught with bad, even illegitimate investments, which put the entire financial structure of the country (and of the world) at risk. Technically, the crash originated with many financial institutions taking on bad loans known as 'sub-primes', which had enabled people to obtain easy mortgages.

However, many experts saw the sub-prime crisis coming (Robert Brenner 2003), because it had been preceded by other speculative 'bubbles' that affected many economic sectors and countries. In reality, the financial sector had become overblown, too powerful, a drift that some termed 'financialization'. Finance became the centre of the economy, with banks encouraged to move their money around and create a variety of new financial products.

The other side of the picture was the relative weakening of the 'real' economy related to the production and marketing of goods and services. Many of these industries relocated their operations to countries where labour costs were lower. Others simply went out of business. Among firms that survived, there was a race to the bottom to cut costs in order to preserve profits. The impact of these changes was significant for countries like Canada, where many working families faced declining incomes. People in the higher echelons of the economy did very well, but the majority were at best stagnating and at worst losing ground.

According to the Canadian Centre for Policy Alternatives, Canada, as one of the 30 wealthiest nations, has shown the greatest increase in income inequality (Couturier and Schepper 2010). Since 1980, the richest 20 per cent of Canadians have enjoyed a median earnings increase of 16.4 per cent, but the poorest 20 per cent have experienced a 20.6 per cent drop in earnings. The total average compensation for Canada's 100 highest-paid CEOs was $7.3 million in 2008, compared with an average $42,300 for all Canadians. Some 400,000 families have to get by on an income of about $23,000. And while the average income of the richest 10 per cent was 31 times greater than the average income of the poorest 10 per cent in 1976, that figure has almost tripled, to 82 today (Yalnizyan 2007). Even average families whose income has remained about the same as it was 30 years ago are worse off in that breadwinners are working longer hours.

Since 1995, tax revenue in Canada has dropped from 36 per cent of GDP to 33 per cent. This represents a decline of almost $50 billion a year in public revenue, mostly for the benefit of top income-earners. At the same time, many social programs have been cut, such as subsidies for low-cost housing. According to some estimates, more than 150,000 Canadians are homeless (Echenberg and Jensen 2008). Almost a million people depend monthly on food banks, the majority of them children (Food Banks Canada 2009).

The recent crisis was the most severe since the 1929 stock market crash, when the world economy came to a halt. However, governments were quick to intervene and salvage the financial sector with an infusion of billions of dollars. According to most Canadian and international experts, the economy will probably recover fully, but it will take some time until it returns to where it was before the descent (United Nations Conference on Trade and Development 2009). And even though financial institutions were rescued through public funding, the 'real' economy (production, services, jobs) is likely to stagnate for a longer period because it is not as profitable. Unless structural problems are resolved, the crisis might return. And these problems are huge.

TIME to REFLECT

What is your opinion of the growing gap between income-earners in Canada? Is it 'inevitable' under current economic pressures?

Globalization into the Twenty-First Century

BEYOND THE CRISIS

As explained earlier, the world is going through turbulent times. Strangely enough to the untrained eye, the epicentre of the crisis is the United States, the most powerful country on Earth. But it is not strictly speaking an American crisis. First, the influence of the US is so overwhelming that to varying degrees, every country is necessarily affected. Second, our world is interconnected through the economy, culture, politics, and security. Therefore, no nation or society can claim that they are not concerned when 43 million Americans (15 per cent of the total population of the US) have to use food stamps to sustain themselves.[2]

In Canada, many pressure points have been exacerbated by the crisis. First, our economy is now more than ever 'export-led' (exports of Canadian goods and services play a larger role in the overall economy than transactions in our own internal market). That reality is further complicated by the fact that more than 80 per cent of Canada's total exports go to one single market: the US. Therefore, when the United States goes through a critical phase, Canada is inevitably affected in terms of employment levels and budgetary deficits that could lead to cuts to social and educational programs funded

HUMAN DIVERSITY

Mexican Farm Workers in Canada

Since 1964, Canada has had a Seasonal Agricultural Workers Program in place, allowing employers to bring in workers from foreign countries to work here temporarily. Fewer than 10,000 in 2003, these farm workers, mostly from Mexico, now number more than 30,000. The Canadian government ensures that these temporary workers are protected by Canadian laws. Migrant farm workers are paid the minimum wage, which is of course much more than they would earn in their own country. They also have a legal status, which is more than the millions of undocumented (illegal) Mexican farm workers have in the US (North South Institute 2006).

Hard work

There is a flip side, however. Mexican farm workers are on fixed contracts with a predetermined duration. They are not allowed to apply for permanent residency (unless they return home and start the procedure there). Often they cannot speak English. They live in trailers in isolated rural areas. Working hours on the farms are very long—sometimes up to 100 hours a week without overtime or holiday pay. Farm work can be dangerous because of the widespread use of chemicals, limited safety equipment, and insufficient protection or training. Clearly, most of the farm-owners in Canada try to treat their workers well. Nonetheless, the system is based on the 'good will' of the employer. The worker can be fired without any

process, and that means being sent home immediately. Various Canadian rights organizations and labour unions have asked the federal government to grant these migrants rights equal to those of the rest of the working population. One such right, of course, is the right to join a labour union and negotiate better working conditions and wages. But employers challenged this move in court, arguing that migrant farm workers were unlike other workers and therefore should not be allowed to join labour unions. In 2010, however, the Supreme Court of Canada ruled that these workers did indeed have the right to join a union. Following that decision, many Mexican farm workers joined the United Food and Commercial Workers of America (United Food and Commercial Workers Union 2007).

Documentary: *Los Mexicanos*

In the summer of 2006, Patricia Pérez, a pro-union militant, launched a major drive to organize the workers on several farms south of Montreal. She informed them of their rights, protected them from abuse, and struggled to bring them together within a union that would extend to them the same rights enjoyed by Canadian agricultural workers. The documentary film *Los Mexicanos* (directed by Charles Latour, 2007) is a portrait of a fight against the injustices of globalization—not in the Third World but here on the Canadian farms that provide our families with our daily food.

by the federal government. Supporters of NAFTA and more comprehensive relations with the United States, however, argue that the only viable future for Canada is to continue along the same road of 'deep integration'. More specifically, the C.D. Howe Institute, along with the Canadian Council of Chief Executives (made up of Canada's largest 50 companies), is proposing a full alignment of Canadian policies with US policies on energy and the environment (Dymond and Hart 2003).

The question is then raised: will these changes affect our values and visions? Canadians tend to see themselves as different from their southern neighbours—more 'liberal' and less conflictual, for example. They are less suspicious of the state, which they see as an instrument to 'redress the balance' and help disadvantaged groups, and less inclined to

rely only on individual initiative and private charity (Hood 2008). In the United States over the past while and even in Britain, the idea has been promoted and up to a point entrenched that the 'game' is 'everyone needs to run for themselves'. This radical individualism was related to a famous pronouncement by former British Prime Minister Margaret Thatcher, who said in the early 1980s that 'there is no such thing as society' (Piven 2007).

TIME to REFLECT

Do you think that Canadian values are undermined by neo-liberalism and the pressures stemming from the recent financial crisis?

RETURN OF THE STATE

In the face of such a situation, many people—citizens, stakeholders, policy-setters—are asking: what can be done?

One line of thinking holds that there has to be more control over and discipline of globalized economics, particularly in the realm of finances. The economy may have become global, but in many ways systems that were set up in an earlier stage to control flows have become outdated. For example, the role of the nation-state was redefined by neo-liberalism, and governments were expected not to intervene in the private sector (or 'market'), which was supposed to operate under the guidance of an 'invisible hand', as Adam Smith (1976 [1776]) put it long ago. Neo-liberalism broke with traditional contemporary theories and practices of the 'modern' and 'rational' state, as described by Max Weber (Dusza 1989). In Weber's view, the state was the 'centre of gravity' of society. Because of the recent crisis and a generalized feeling that current governance systems are leading to further chaos, many believe that the state should again become the centre of gravity and impose limits and constraints on all actors and units operating in society and the economy (Stiglitz 2009).

Indeed, many governments have had to intervene to help their declining economies. The US and Canadian governments, for example, granted huge loans to major private-sector businesses such as the automobile industry. At the same time, there is a growing call to 're-regulate' the financial sector. But imposing regulations is not simple. First, financial institutions have become very powerful. Second, no international body has the authority or the means to impose regulations on a worldwide basis, which makes it risky for any national government to restrict its financial sector, since companies can merely move their assets elsewhere instantaneously.

REFORMING CAPITALISM?

One possible solution, hotly contested, is a return to some sort of regulatory capitalism, as proposed by Joseph Stiglitz, a former official of the World Bank and now a well-known critic of current policies. Stiglitz proposes that the **World Bank** and the **International Monetary Fund** (IMF) should control money flows—for example, eliminate financial 'safe havens' (states where individuals and corporations can put their money, not pay tax, and keep their assets hidden). He also advocates elimination of the Third-World debt and fundamentally 'humanizing' globalization

by increasing social programs dedicated to helping poor communities. These reformist proposals also include the reconstruction of the 'social safety net' that existed under Keynesian policies after 1945.

Another version of 'reforming capitalism' emphasizes the need to focus on the environment. The drive to diminish greenhouse gases is seen by many experts as an imperative. Some argue that it would not only be good for the environment but would also serve to kick-start the economy (Sachs 2008). So far, however, Stiglitz and other experts calling for reforms have had little success in influencing the policy process.

TIME to REFLECT

What do you think of Joseph Stiglitz's proposals to 'restore our sense of balance between the market and the state, between individualism and the community, between humankind and nature, between means and ends?'

CHINA RISES

While these debates are taking place mainly in the Global North and to some extent in the international institutions that are struggling to remain relevant (such as the World Trade Organization, the IMF, and the UN), something else is happening in the world: it is now widely recognized that China is rising to become a 'big' player. Chinese society is sophisticated, with an educated population and strong institutions. Internal contradictions abound, however, because the opening of China to the world and profound social changes linked to urbanization and education are leading to more and more demands for political and cultural liberalization, while the traditional system established by the Communist Party in 1949 resists and hopes to remain in control. But in the meantime, a double-digit GDP growth rate has made China one of the largest economies as it becomes the 'workshop of the world' for most consumer goods and, increasingly, for more technological products as well. China is now a major international investor ($150 billion in 2009 alone) and has a huge trade surplus and currency reserve. At the same time, the Chinese population has benefited from this surge out of poverty. Many Chinese people, especially in urban and coastal areas, have grown richer—more or less the opposite of what has happened in the United States. Some scholars argue that China is in the process of becoming not only a very strong economy but a sort of hybrid 'model' between state-led socialist policies and market-driven capitalist development (Arrighi 2007).

In addition, China has suffered less than other leading economies from the slump that followed the financial crash of 2008. First, its economy is still managed from the top by a strong state. Private investors (Chinese and foreign) are welcome to exploit the vast resources of cheap and competent Chinese labour, but under conditions largely determined by the state. Second, despite a great deal of pressure from the United States, China has refused to liberalize its financial sector, which is still 100 per cent controlled by the state, and thus has avoided some of the problems with financial 'bubbles' and speculative 'hot' monies that have plagued many other countries. Third, the ruling group controlling the state knows that capitalist growth needs to 'trickle down' through the population. Income has risen accordingly and new infrastructure has been constructed, even in the vast hinterland, still very underdeveloped compared to the coastal areas. Still, the stability of rising China is not guaranteed, since social polarization continues and other issues are emerging, such as growing environmental problems (Hart-Landsberg 2010).

THE EMERGENCE OF THE 'NEW' SOUTH

In the meantime, China is becoming more active internationally. As a major trading partner of the United States and the European Union, it still wants to maintain good relations. But at the same time, it is diversifying its interactions with the rest of the world. The changing world scene facilitates that process as other Third-World countries emerge, expanding their economic capacities and internal markets. At the core are the 'BRIC', which includes a select number of **emerging countries** (Brazil, Russia, India and China) and other surging nations where conditions are seen as favourable for more growth. According to economic forecasters, the BRIC will account for 40 per cent of global economic growth between 2009 and 2020.[3]

Other rising countries include Turkey, South Korea (already considered a 'developing' country), as well as poorer nations that have increased their economic scope in the past decade, such as Vietnam and Indonesia. Apart from a few exceptions, this 'emerging' South remains centred in Asia, with China and India far ahead.

In South America, the process is different because of the rise of democratic governments over the past two decades. In Brazil, for example, a left-centre government has managed to reduce poverty through financial transfers to the poorest part of the nation (40 million people), which not only reconciled social groups (at least partially) but also helped local industries and markets. This pattern is expanding to countries like Bolivia, Ecuador, Venezuela, and Argentina. Out of it came various projects of regional (South American) integration with the aim of creating a unified platform. What is foreseen is some sort of Latin American union (modelled on the European Union). In the meantime, several processes are underway to integrate South American countries, such as UNASUR (Union of Nations of South America), MERCOSUR (Common Market of South America), and ALBA (Bolivarian Alliance for the Peoples of Our America), led by Venezuela, Cuba, and Bolivia, with the goal of promoting regional economic integration based on a vision of social welfare, bartering, and mutual economic aid rather than strictly trade liberalization. It is notable that all of these projects exclude the United States and Canada, whose own integration project (Free Trade Area of the Americas), initiated in 1994, was rejected by South American nations a decade later.

DECLINE OF THE EMPIRE?

What does this 'emergence' mean for the rest of the world?

With their capacities falling, the leading Western nations, beginning with the United States, are often described by historians and social scientists as in 'decline'. In the United States, the economics of that decline are indeed staggering: a national debt at

A technician checks the cockpit functions of a new airplane at an Embraer factory in Sao Paulo. Diverse resources and investment in technologies, such as aircraft development, have propelled Brazil into becoming an emerging economy.

(© Sue Cunningham Photographic/Alamy)

$11 trillion (of which 25 per cent was owned by foreigners in 2010), a federal annual deficit forecast at $1 trillion for many years to come, and a declining share of world economic output (from 50 per cent in 1950 to 15 per cent predicted by 2020). In a way, explains economic historian Robert Brenner, the US is both a 'winner' and a 'victim' in the globalization process. It has maintained its lead in many sectors by profiting from the liberalization of trade and investments. But it has been out-competed by emerging powers (including China and India) and its traditional 'partners' like Japan and the European Union.[4] 'Excessive' supply and cheaper products coming out of newly industrialized countries such as China and Japan or even Germany have reduced profits to the extent that US firms slow down production, cut wages, and in the process create a vicious circle that leads to crisis.

To these macro-economic indicators one can add the much–commented upon decline in US education and science, which threatens the long-term competitiveness of the national economy, according to President Obama.[5] Not only had enrolment in post-secondary education diminished, but the number of high school graduates in 2009 was the lowest since the 1960s. This situation, according to the president, is 'untenable for our economy, unsustainable for our democracy and unacceptable for our children'.[6]

In the meantime, many people who used to define themselves as 'middle-class', with an 'average' income based on a relatively stable job, are experiencing a dramatic descent in their social position. Many blue-collar workers face long-term unemployment, which can mean losing basic assets like their homes. White-collar workers are also threatened by massive restructuring in commercial and service sectors whereby many high-tech jobs in engineering and ICTs are being relocated to cheaper-wage areas. Still, the US has enormous assets and advantages and remains uncontested in several strategic areas.

There are, however, competing sociological explanations for the new configuration shaping up before our eyes. Many, like Egyptian political economist Samir Amin (2006), say that the US decline is irreversible—that it will inevitably lead to a situation that, paradoxically, will resemble the way the world was 500 years ago, before the ascent of Europe, when China and India were richer and more developed. There are many reasons to believe that, indeed, China and to a lesser extent India have the 'ingredients' to reach the top of the hierarchy. For example, China manages to feed 22 per cent of the world's population with less than 7 per cent of the arable land. It has technological

and productive capacities steered by a strong state that still retains a great deal of legitimacy (contrary to a common belief in the Western world). There are many other emerging poles in the Asia-Pacific region, with vast natural resources and easy access to Europe and Africa.

Indeed, the United States has moved from a position of being able to impose its will on the world, as it did between 1945 and 1970, to a situation of relative impotence, exacerbated in recent times by the wars in Iraq and Afghanistan and the financial crisis of 2008 and its aftermath. According to sociologist Immanuel Wallerstein, the causes of this decline have to be understood in a historical perspective. Rising competition, the decolonization of the Third World, and the rise of a new generation of protest movements all challenge US supremacy (Wallerstein 2009).

But the thesis of the 'inevitable' decline of the United States is contested. Besides a continuing economic gap, China and other BRIC nations are still very much behind the US in terms of military power. The US stands alone in this area, with qualitative and quantitative superiority in weapons of mass destruction (including nuclear weapons), advanced aircraft, and missiles, and the advantage provided by having military personnel stationed throughout the world. This wide gap in respective military capacities certainly works against the thesis of an 'inevitable' American decline.

Whatever the future brings, such trends will have a huge impact on Canada. Because of our close relationship with the United States, our future economic prosperity is intimately tied to the strength of the US economy. In addition, some of the same threatening pressures facing the US (like the relocation of industry and services) are also at play here. For example, the aircraft industry in Montreal might face stiff competition from Brazilian or Chinese firms with access to the skills and lower-cost labour for producing advanced avionics. Sociologist Trevor W. Harrison (Harrison and Friesen 2010) believes that the restructuring of the Canadian economy under the influence of NAFTA is taking Canada back to where it started, as a provider of natural (mineral and agricultural) resources. The effect on Canadians would be significant. Resource-based economic activities tend to employ fewer people than the manufacturing industry, for example. The service sector, while providing high-paying jobs in certain domains (higher education, science, and technology) also entails part-time, insecure, low-paying jobs (Krahn, Lowe, and Hughes 2007).

Anti or Alter Globalization?

In 1994, an indigenous rebellion in southern Mexico sent a message that did not have a great deal of influence at that time, when neo-liberal globalization seemed triumphant. Since then, however, many voices have been raised demanding a change to policies affecting the economy, governance, and rights. A few years later in Seattle, the movement became visible when 50,000 demonstrators tried to block a meeting of the World Trade Organization (WTO) (Klein 2002). That led to more demonstrations elsewhere in the world as people came together to protest against the way trade negotiations were being conducted with a blind eye to social and environmental issues. Protesters also objected to the secrecy involved in the negotiations and the absence of public influence and scrutiny. The same sentiments brought 60,000 people to Quebec City in April 2001, where heads of states were discussing how to extend **free trade** and financial liberalization throughout the hemisphere in a manner similar to the NAFTA treaty between Canada, the US, and Mexico (Beaudet 2009).

After 2002, the 'anti-globalization' movement shifted to South America, where significant movements invaded public places, as in Argentina, where demonstrators forced the resignation of successive governments. In 2002, Brazil elected a left-wing president, Luiz Inácio (Lula) da Silva, a former trade union leader who promised to reorient priorities in favour of the working classes. Later, left-centre governments were elected in other South American countries, with diverse but nevertheless convergent programs of social reforms, democratization, and partial de-linking from northern-led institutions and processes such as the **World Trade Organization** and NAFTA.

These diverse movements subsequently engaged in various dialogues and created new networks such as the **World Social Forum** (WSF), which currently links some 500,000 NGOs and social movements around the world (Sen and Waterman 2007). Although the WSF has been very influenced by and close to South American social and political processes, it subsequently migrated to North America, Europe, Asia, and Africa, where hundreds of international, regional, and national forums have been held. However, the WSF is only a space for debating and exchanging views and has neither the mandate nor the capacity to 'unify' some thousands of initiatives into a program. Moreover, it attempts to maintain a balance among the views of various social movements and avoid becoming overtly partisan, even though the majority of the movements involved are closely linked to left or centre-left political platforms (Worth and Buckley 2009).

However, it is important to note that the WSF and its participating movements and NGOs do not simply represent a repetition of the 'old' left. First and foremost, most of them have refused to be 'instrumentalized' or led by leftist political groups. They believe in the power of horizontal and egalitarian relationships. In reality, explain Dufour and Conway, 'the social forum as a political form represents a break with the disciplines of modern political organization, be it based on representative democracy, democratic centralism or participatory democracy' (Dufour and Conway 2010). The novelty of this organizational space rests on 'inclusiveness and global reach, in the inexistence of leaders and hierarchical organization, in its emphasis on cyberspace networks, and its flexibility and eagerness to engage in experimentation' (Santos 2005).

The development of the WSF has, however, improved the capacity of many social movements to develop new strategies and proposals. One of the notable changes has been the declaration that 'another world is possible'—that the people of the world should not feel resigned to accepting neo-liberal policies. Regarding the environment in particular, movements have become radicalized, contesting capitalism as 'anti-nature' and demanding another development paradigm, as explained by François Houtart (2010), a critical thinker who works closely with the WSF:

> [Under] capitalism, it is only possible to make a profit and accumulate capital on the basis of the exchange of merchandise; the result is the mercantilisation of everything. The systems of production, distribution and transportation are built on the foundation of how to increase exchange value. What we need is a new definition of economy, a different philosophy of economic activity: from production of added value for private interest to activity that produces the basis for life—physical, cultural and spiritual—for all human beings in the world. That concept, however, is in contradiction with the basic definition of the capitalist system. The market can no longer be merely a forum for making a profit for the few, but must rather be a place of mediation between supply and demand. Production of goods and services will be completely different when use value is privileged over exchange value.

In the wake of the WSF, networks like Via Campesina,[7] comprising more than 150 farmer and peasant movements, are fighting against big 'agro-business' multinationals and the spread of genetically modified seeds. They also want ecological family-based productive units to be restored: units that could be better sustained by land reform and extensive agricultural extension schemes. Here, the principle is 'food sovereignty', which is based on local ownership and control over the production of food and the protection of the natural environment. Movements in Canada such as the National Farmers Union (NFU) contend that food should be considered a basic human right, not a commodity.[8]

'ALTER GLOBALIZATION' IN CANADA

The experience of the NFU is similar to that of many social and civil movements that have participated in the **alter globalization** movement (initially known as an **anti globalization** movement) in recent years. Women's organizations in particular have played an important role in raising policy issues in the context of globalization. Quebec's largest women's group, the Fédération des femmes du Québec (FFQ), organized the World March of Women (WNW) in 2000 in which 6000 women's groups, as well as representatives from labour unions and political parties from 163 countries, marched to protest poverty and violence against women.

Perhaps the strongest expression of that initial process came from trade unions as well as environmental and other civil groups that opposed NAFTA in the late 1980s and early 1990s. A coalition was created, later known as Common Frontiers, comprising the Canadian Labour Congress, the Canadian Council for International Co-operation, the ecumenical network KAIROS, the Sierra Club, and several other organizations.[9]

Later, a larger coalition was formed, bringing together civil organizations from Brazil, Mexico, Chile, and the United States. In 1999, the coalition created the Hemispheric Social Alliance, which mobilized several thousand people to oppose a project proposed by the United States and Canada, the Free Trade Area of the Americas (FTAA), which would expand NAFTA to take in the entire hemisphere. The process peaked in Quebec City in April 2001, when more than 50,000 people demonstrated against FTAA. The series of events was organized mainly by Quebec groups working under their own umbrella, the Quebec Network in Regional Integration (Réseau québécois sur l'intégration continentale, or RQIC), with the support of Common Frontiers and the Hemispheric Social Alliance. Apart from organizing street demonstrations, the 'People's Summit of the Americas' encouraged the participation of thousands of people, both activists and ordinary citizens, through popular education and communication efforts (Dufour and Conway 2010). While a small number of demonstrators confronted the police in clashes that were widely seen on TV, the majority of participants worked to express their concerns and to design—'imagine', one might say—a different 'model'.

This led to the formulation of an 'alternative' integration program for the Americas, which went further than the usual anti globalization movement (McNally 2002). The 'alter globalization' platform advocated that labour rights and environmental protection should be at the centre of any economic integration process, that state sovereignty should take precedence over the interests of private firms, and that there should be significant transfers of resources to the poorest (southern) countries of the hemisphere so as to narrow the gaps created during decades of northern domination (Gagnon 2002).

'Globalization from below' has been gaining strength since the historic confrontation in Quebec City (Barlow and Clarke 2001). Many Canadian groups and networks joined the WSF process, attending meetings in different parts of the world and bringing home 'alter' perspectives on social, economic, cultural, and political issues. Local social forums were organized in Vancouver, Edmonton, Toronto, Ottawa, Quebec City, and other locations, using the same 'methodology' of open debates and self-organized activities. The WSF states in its charter of principles that social forums are

> open meeting place[s] for reflective thinking, democratic debate of ideas, formulation of proposals, free exchange of experiences and interlinking for effective action, by groups and movements of civil society that are opposed to neoliberalism and to domination of the world by capital and any form of imperialism, and are committed to building a planetary society directed towards fruitful relationships among Humankind and between it and the Earth.[10]

The WSF Canadian process culminated in Quebec in province-wide forums (2007 and 2009) that attracted thousands of people from a wide range of perspectives and experiences.

SOCIOLOGY in ACTION

Fair Trade

Initiated by North American organizations in the 1940s, the idea to improve the lot of the poor by changing the rules of commerce took off in the 1970s and 1980s through NGO networks in rich countries but also through people's movements in Central America and southern Africa. In practical terms, **fair trade** means buying directly from the producers, usually small farmers and cooperatives, and selling the product in Western markets at a 'fair' (higher) price so that local producers get their 'fair' share of the business and do not remain dependent on or exploited by big multinationals involved in importing and exporting huge volumes of goods.

To make a rough calculation, producers in fair trade schemes receive about 26 cents from consumers instead of 11 cents. It is estimated that 7.5 million producers and their families benefit from fair trade–funded infrastructure, technical assistance, and community development projects. In 2008, fair trade–certified sales amounted to approximately $4.08 billion worldwide, a 22 per cent year-to-year increase. Many Canadians participate, using their skills to upgrade the capacities of fair trade schemes and working with local farmers' organizations in Canada and elsewhere in the world.[11] Students with social sciences or administration backgrounds organize fair trade networks through information programs and popular education. As a result, fair trade has grown tremendously in Canada over the past few years, as the following chart indicates.

According to fair trade campaigner Laure Waridel, the movement is only one strategy among many to fight for a better world, but 'one which we can practice in our day-to-day lives, by buying according to our values (Waridel 2008).

Canadian Sales of (Labelled) Fair Trade–Certified Coffee

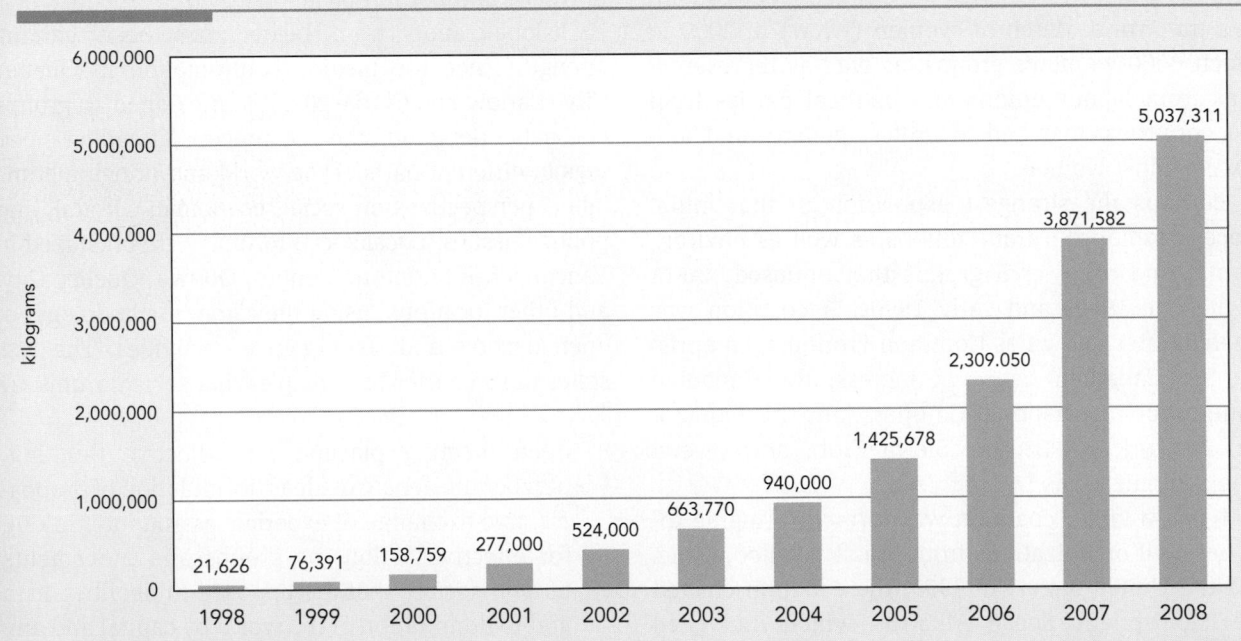

SOURCE: TransFair. 2010. *Facts and Figures about Fair Trade.*

Multitudes and 'Glocal'

Social movements have expanded their reach to mobilizing for environmental and other related issues. They have joined thousands of local, regional, and international initiatives worldwide, taking many forms, autonomous one from another but linked through organized networks or simply through the Internet. This development has triggered new debates and research in sociology and in the social sciences more generally. Is the 'alter globalization' movement really a 'movement'? Does it have a defined identity? What does it represent in comparison to other, 'traditional' vehicles for social transformation?

Antonio Negri and Michael Hardt (2000; 2004) express an optimistic but well articulated view on the matter in a series of books that have been widely commented upon. The 'empire', they say, is the new name for modern capitalism, or globalized capitalism, which is not only spreading its control to all nations but also penetrating the 'fabric of life' through modern science and technology. On the other side stands the multitude, an assemblage of multinational social groups that are fighting exploitation and oppression and, moreover, are able to express their autonomy and alternative utopias independently of states and nation-states. According to Negri, because the multitude works in the production and manipulation of concepts, processes, and symbols that are at the heart of modern capitalism, it has the ability to 'turn the system around' and make it ungovernable. Thus, the multitude can produce a new 'commonwealth', an alternative mode of human development based on the common good, cooperation, and solidarity.

However, this optimistic view is contested. Activists and well-known scholars like Samir Amin argue that globalized capitalism is not likely to 'go away' that easily (Amin 2010). They insist that neo-liberal globalization will not be 'altered' by grassroots pressures but by political upheaval and confrontation led by major states (like China) or coalitions of states.

The world that emerged from the Second Word War and had undergone periods of stability and turmoil since then seems more and more fragile today. The combination of many crises (financial, economic, social, ecological, and political) is not necessarily unique in history, and of course other world systems or vast imperial powers have changed or been defeated in the past. But the complexity, the scope—one could say the 'globality'—of today's crisis makes it a tremendous challenge for social science scholars and citizens alike.

questions for critical thought

1. Is globalization really changing the world, or is it just another name for imperialism or domination by the rich countries over the poor?

2. Why are the gaps between poor and rich nations still widening? Why are such gaps widening in rich countries like Canada as well?

3. Can current neo-liberal policies change? What could replace them?

4. Is war inevitable, considering present circumstances and pressure points?

5. Can NGOs and civil society make a real difference?

6. What is at stake for Canada as a nation?

recommended readings

Jennifer Clapp and Marc J. Cohen, eds. 2009. *The Global Food Crisis: Governance, Challenges and Opportunities.* **Waterloo, ON: Wilfrid Laurier University Press.**
With more than one billion people trapped in absolute poverty if not starvation, the globalized world in which we live seems incapable of satisfying basic human needs. Nowhere is the crisis more acute than in the vast rural areas that have been transformed by global capitalism. Nowhere are the environmental pressures as strong as they are in these areas. This book, with contributions from more than 20 authors, examines this complex and critical crisis.

Thomas Friedman. 2005. *The World Is Flat: A Brief History of the Twenty-First Century.* **New York: Farrar, Straus and Giroux.**
The author describes globalization as a basically positive trend that is reducing differences and conflicts and leading to 'one small world'.

Lui Hebron and John F. Stack. 2011. *Globalization.* **Toronto: Longman.**
This informative book presents the pros and cons of several views and analyzes the globalization process, focusing on contested issues such as markets, growth, democracy, sovereignty, and culture.

Naomi Klein. 2002. *No Logo: No Space, No Choice, No Jobs.* **New York: Picador.**
In this book, Naomi Klein takes a critical view of globalization. She argues that it allows more control by large multinationals, which they extend over the production and the marketing of goods, but also promotes their symbols and images, which become icons or standards for the whole world.

George Ritzer. 2010. *Globalization: A Basic Text.* **Oxford: Wiley-Blackwell.**
This book provides a panoramic view of globalization in all its dimensions (social, political, cultural, environmental). It also analyzes the main 'actors' of globalization, including governments, the United Nations, multinational firms, and civil society.

Jai Sen and Peter Waterman, eds. 2007. *World Social Forum: Challenging Empires.* **Montreal: Black Rose Books.**
This collection of essays by more than 40 authors presents the alternative views on globalization arising from social movements, NGOs, and women's organizations from all over the world. Is there an alternative to 'neo-liberal' globalization?

Manfred Steger. 2009. *Globalization: A Very Short Introduction.* **New York: Oxford University Press.**
Why has globalization become the buzzword of our time? This book provides a short but accurate introduction to the causes and effects of globalization in its historical, economic, political, cultural, and ecological dimensions.

Joseph Stiglitz. 2010. *Freefall: America, Free Markets and the Sinking of the World Economy.* **New York: Norton.**
Stiglitz, a former vice-president of the World Bank, describes the policies that led to the 2008 financial crash. He believes, however, that the situation can be reversed if, and only if, radical reforms are imposed to re-establish the primacy of public interests.

recommended websites

Canadian Centre for Policy Alternatives (CCPA)
www.policyalternatives.ca
CCPA is a progressive think-tank based in Ottawa that studies the impact of globalization on the Canadian economy and society.

Focus on Global South
www.focusweb.org
Focus on Global South is one of the foremost progressive think-tanks in the Third World, covering globalization issues from a critical angle.

The Globalization Website
www.sociology.emory.edu/globalization
This is a solid scholarly website that explains concepts and theories of globalization.

Social Watch
www.socialwatch.org/en/portada.htm.
This website is managed by a vast network of NGOs focusing on social development.

The World Bank
www.worldbank.org
The World Bank is the largest development institution in the world, with a website containing a huge collection of documents on poverty and other social issues.

notes

1. www.newamericancentury.org/aboutpnac.htm.
2. Michael Balad. 2010. 'Food stamp use rises in the US'. *The Globe and Mail* 2 June.
3. 'Global financial power to shift to BRIC: Economic forecast'. *Thaindian News* 22 December 2008.
4. Robert Brenner. 2009. 'The economy in a world of trouble'. *Hankyoreh* (journal published in Seoul) 22 January.
5. Scott Wilson. 2010. 'Obama warns of falling education standards'. *Washington Post* 13 March.
6. 'Obama sees US education in decline'. Knoxnews Sentinel webpage: www.knoxnews.com/news/2009/mar/18/editorial-obama-sees-us-education-decline/ 2009.
7. http://viacampesina.org/en.
8. The National Farmers Union defines itself as a 'non-partisan organization working for the development of economic and social policies that will maintain the family farm as the basic food-producing unit in Canada [protecting] individual farmers to assert their interests in an agricultural industry increasingly dominated by multi-billion-dollar corporations'. More details at www.nfu.ca.
9. Common Frontiers webpage: www.commonfrontiers.ca/index.html.
10. www.forumsocialmundial.org.br/main.php?id_menu=4&cd_language=2.
11. For further information: http://transfair.ca/en.

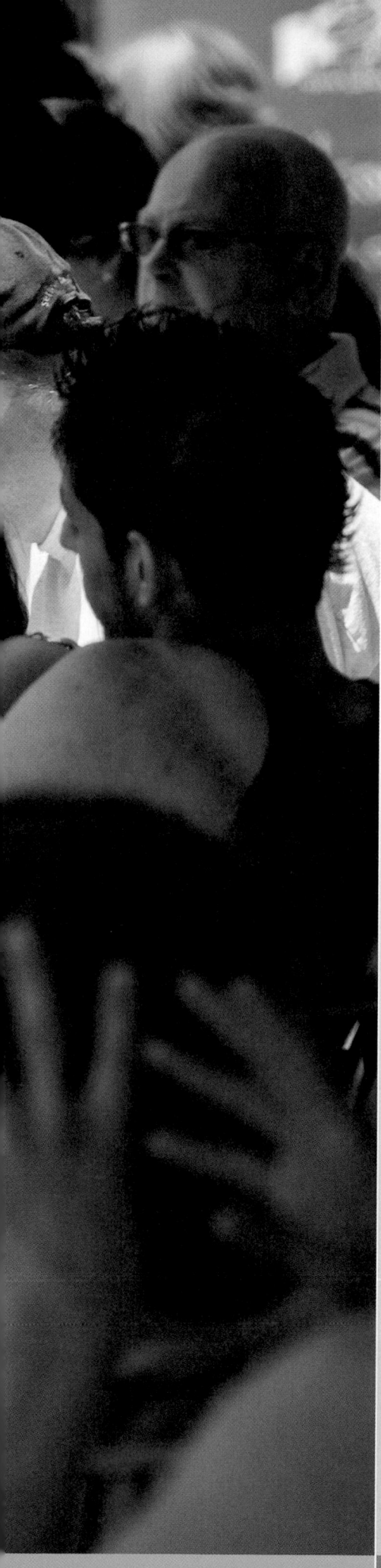

chapter 21

Population and Society

Frank Trovato

In this chapter, you will:

▶ Learn the definition of demography and the components equation for population change;

▶ Explore various demographic phenomena, the aggregate expressions of individual behaviour, conditioned by culture and social structure;

▶ See how the 'population explosion' is a relatively recent phenomenon in human history;

▶ Study the demographic transition theory, which summarizes the long-term historical trends in birth and death rates and population growth in three stages;

▶ Compare the demographic transitions of industrialized and developing countries;

▶ Explore the implications of the Malthusian theory of population growth and available resources;

▶ Consider the Marxist perspective on population;

▶ Examine the growth of Canada's population and associated trends in fertility, mortality, and migration over the past century and into the future.

Introduction

At just over 6.987 billion, the population of the world in 2011 is growing at a rate of 1.2 per cent per annum. This seemingly small rate, if applied to the total world population, would translate into an annual increase of nearly 84 million persons. Should this rate of increase remain constant, the population of the world would double in only 58 years. Some scholars view figures such as these with alarm because of the potentially devastating impact continued growth might have on the long-term sustainability of the planet's environment and resources (Meadows, Randers, and Meadows 2005). More optimistic pronouncements in the literature suggest that population growth may actually serve as a stimulus to economic growth and, indirectly, to human well-being (Lomborg 2001; Simon 1995). As we shall see, a number of competing perspectives exist on the question of population and its relationship to resources and human welfare.

To demographers—scientists who study the census and vital statistics of societies—the term *population* encompasses a number of interrelated dimensions. Samuel Preston, Patrick Heuveline, and Michel Guillot (2001) define *population* as a collection of persons alive at a specified point in time that meets certain criteria, of which national and geographic boundaries are two of the most obvious. As a collectivity, a population persists through time even though its members are continuously changing as a result of *attrition* (losses through out-migration and death) and *accession* (gains through births and immigration). A population also entails the aggregate of persons who have ever been alive in a designated national or geographic area and possibly those yet to be born there. Populations can be projected into the future by using mathematical procedures guided by sound assumptions concerning anticipated changes in fertility, mortality, and migration.

Demographers study the growth, distribution, and development of populations with respect to their geographic concentration and compositional characteristics, of which age, sex, and marital status are especially relevant. The natural processes of fertility and mortality, plus *net migration* (the net exchange between numbers of incoming and outgoing migrants), determine change in population size for a country or specified territory over some defined interval. This is illustrated with the **demographic components equation**:

$$P_{t1} - P_{t0} = (B_{t0,t1} - D_{t0,t1}) + (IN_{t0,t1} - OUT_{t0,t1})$$

Letting P_{t0} and P_{t1} represent the population at the beginning and the end of some specified interval, the change over this period $(P_{t1} - P_{t0})$ is a function of the difference between births and deaths $(B_{t0,t1} - D_{t0,t1})$ plus the net exchange between the numbers of persons moving into the area $(IN_{t0,t1})$ and the number leaving $(OUT_{t0,t1})$ during the interval t_0 to t_1. The component $(B_{t0,t1} - D_{t0,t1})$ represents natural increase because it measures the difference between the two natural processes of births and deaths, and $(IN_{t0,t1} - OUT_{t0,t1})$ measures net migration.

Populations also change as a result of the reclassification of people across distinct statuses (e.g., a change in marital status from single to married or from married to divorced). The distribution of the population in accordance with characteristics such as age and sex is also subject to change as a function of changes in fertility, mortality, and migration.

> **TIME to REFLECT**
>
> Populations most commonly experienced growth during the twentieth century. However, in recent decades, some populations have been experiencing very low rates or even negative natural increase. Can you think of some actual examples of this? Why do you think such populations have very low or negative rates of natural increase? What are the societal factors behind this pattern of population change, and what are the long-term societal consequences of continued low or negative natural growth?

Demographic Change and Social Change

Aggregate demographic phenomena are the collective expression of individual behaviour conditioned by cultural **norms** and **social structure** (Davis and van den Oever 1982). Often, novel behaviours develop into widespread demographic phenomena. One such behaviour concerns the growing preference among recent generations of youth to form cohabiting unions rather than marriages. In Canada during the early part of the 1970s, only about 17 per cent of couples entering their first union were in common-law relationships. By the early 1980s, the corresponding figure had risen to 40.5 per cent. During the latter part of the 1980s, more than half of all first-union couples were common-law (Wu 2000, 51). Coinciding with the generalized acceptance of cohabitation, there has been a precipitous decline in the total marriage rate of single persons, and the median age at marriage has been going up.

Other examples may be used to illustrate the interconnectedness of demographic factors with societal change. One could, for instance, discuss the evolution of Canadian society into a multicultural and multiracial nation as a function of shifts in the regional origins of immigrants from the once-dominant European sending areas to the current predominance of Asia, Africa, and South America as sending regions.

We could also examine how the majority of today's population is alive because of the mortality improvements that took place at the turn of the twentieth century in the industrialized world. As a result of many developments in public health, medicine, and standards of living in the early twentieth century, newborns at that time enjoyed unprecedented gains in survival probabilities over earlier generations in history. Under preceding historical conditions, large proportions of infants and children succumbed to the ravages of infectious and parasitic diseases. Improved socio-economic conditions and public health programs helped babies and young children to overcome the dangerous stages of infancy and early childhood. More infants and children would live to adulthood to eventually have their own children, who would later bear progeny of their own under even more favourable health conditions. Kevin White and Samuel Preston (1996) show that approximately half of the population alive today in the United States owe their existence to twentieth-century mortality improvements. In other words, half of the current population would never have been born had it not been for the significant progress in health and disease prevention that took place around the turn of the twentieth century.

David Herlihy (1997) gives another interesting account of how demography and social-structural change are related. He describes how mortality conditions in European history helped provoke lasting social-structural and cultural transformations in Western society. Among other things, he attributes indirect responsibility for the development of the national university system in Europe and for the rise of nationalism to the Black Death (bubonic plague) that struck Europe in 1348 and several times into the fifteenth century. The population of Europe may have been reduced by two-thirds between 1320 and 1420 as a result of this highly infectious disease. Under the spectre of early death, many wealthy people bequeathed their fortunes to institutions such as the church and the educational system. This helped

In the First Person

I became interested in population studies during my first year of university. This subfield of sociology has held my interest primarily because of its interdisciplinary nature and the many intellectual challenges it offers.

—Frank Trovato

the creation of new national universities throughout Europe. Since many of the teachers in the newly created universities were at first unfamiliar with Latin (the language of higher learning at that time), they used the local languages in their teaching. This, according to Herlihy, stimulated the development of strong nationalistic sentiments in the population and the eventual rise of nationalism in Europe.

World Population

Ansley Coale (1974) divides population history into two broad segments of time: the first, from the beginning of humanity to around 1750 CE, was a very long era of slow population growth; the second, relatively brief in broad historical terms, is one of explosive gains in human numbers. According to Coale (1974, 17), the estimated average growth rate between 8000 BCE and 1 CE was only 0.036 per cent per year. Between 1 CE and 1750, it rose to 0.056, and to 0.44 per cent from 1750 to 1800. In modern times, the trajectory of population growth has followed an exponential pattern. Since the early nineteenth century, each successive billion of world population has arrived in considerably less time than the previous one. It took humanity until about 1750 CE to reach a population size of approximately 800 million. The first billion of population occurred in 1804, the second 126 years later, in 1930. Thirty years beyond that point, the world witnessed its third billion. The 4 billion mark was reached in 1974, only 14 years later, and 13 years passed before the Earth welcomed its five-billionth person in 1987. In 2011, the globe's population became 7 billion (see Figure 21.1).

World population growth rates peaked at just over 2.0 per cent during the 1960s and early 1970s. In recent decades, the rate of natural increase has been declining to its present level of 1.2 per cent (Population Reference Bureau 2010). This trend is expected to proceed into the foreseeable future such that by the year 2050, the growth rate of the world might be as low as 0.34 per cent per year (United Nations, Department of Economic and Social Affairs, Population Division 2009a, 13). This remarkable reduction will come about as a result of anticipated declines in fertility and mortality over the next half-century. The latest medium variant projection of the United Nations (a projection that assumes change in fertility and mortality thought to be most likely, given past trends) assumes that the total fertility rate at the world level will decline from 2.56 children per woman in 2005–10 to about 2.02 children per woman in 2045–50. The change in **life expectancy at birth** during this same period is expected to be from 67.6 years in 2005–10 to 75.5 years in 2045–50. This 'central' scenario projection suggests a population in 2050 of just over 9 billion (United Nations, Department of Economic and Social Affairs, Population Division, 2009a, 1–9).

During this century, most of the projected population growth will take place in the **developing countries**, especially in the poorest nations, where rates of natural increase remain high. Although fertility has been declining in many developing countries, natural increase is high because of the faster pace of the mortality declines. And although some of the regions in Africa (e.g., southern Africa) have been hit hard by the HIV/AIDS epidemic, Africa's overall share of the world's population will increase rapidly and is projected to account for almost 22 per cent of the world's population in 2050, according to the UN's medium variant projection. In sharp contrast, the average annual rate of natural increase for the **developed** regions, currently at 0.2 per cent per year, is expected to continue falling and turn negative between 2020 and 2025. This anticipated pattern of negative growth can be attributed to the contribution of the European region, where natural growth is already negative (United Nations, Department of Economic and Social Affairs, Population Division, 2009a, 13).

Some of the expected population growth for the world over the next half-century is unavoidable. Even if current fertility rates worldwide were to decline suddenly to the replacement level of 2.1 children per woman, substantial population growth would

(© amana images inc./Alamy)

Most industrialized nations have been experiencing a baby dearth in recent years. In Italy, the pope has urged the population to have more children in order to solve the 'birth rate problem'. In Japan, the government has tried to implement policies to encourage young couples to have larger families.

nonetheless occur. This unavoidable growth is due to the powerful effects of **population momentum**: growth that is built into the current age structure of the population (Bongaarts and Bulatao 2000; Lutz 1994). Because of high fertility in the past, the proportion of the world's population in the reproductive ages (roughly ages 15 to 49) is quite large. And notwithstanding declining birth rates, the large parental cohorts will be bringing many babies into the world. Even the latest 'low variant' projection by the United Nations, which assumes substantial declines in total fertility, shows a population of about 7.96 billion in 2050 (considerably larger than the 6.987 billion in 2011).

▼ FIGURE 21.1   The World Population Explosion

50,000+ BCE	Human life
3,500–1000 BCE	First cities, 3500 BCE, in the fertile crescent of Persia; 3000 BCE, Thebes and Memphis, Egypt; 2500 BCE, Indus River settlements, India; 2500 BCE, Yellow River settlements, China; 1000 BCE, Meso-America (Peru)
500 BCE–500 CE	City-states of Carthage, Athens, Sparta, Rome
750–1400	Dark Ages, feudalism (decline of most European cities)
1400–1700	Renaissance, nation-states, capitalism, guilds, revival of trade
1700–1900	Industrialization in parts of the world
1900–1950	Electricity, telephones, radio, automobiles, metropolitan cities, nuclear energy
1950–1975	Megalopolis, television
1975–present	Widespread air travel, electronic data processing, mega-cities, global cities, globalization, CNN, and the 'global village'

Not to scale.

World Population Growth through Broad Historical Periods and Projections

Period	Estimated population at end of period	Estimated average annual growth rate (%) at end of period	Years to add 1 billion
1 million BCE – 8000 BCE	8 million[1]	0.010[1]	
8000 BCE – 1 CE	300 million[1]	0.036[1]	
1 CE – 1750 CE	800 million[1]	0.056[1]	
1804	1 billion[1]	0.440[1]	all of humanity
1930	2 billion[1]	0.540[1]	130
1950	2.5 billion[1]	0.800[1]	—
1960	3 billion[1]	1.700 – 2.000[1]	30
1974	4 billion[2]	2.000 – 1.800[3]	14
1987	5 billion[2]	1.800 – 1.600[3]	13
1999	6 billion[2]	1.600 – 1.400[3]	12
2011	7 billion[2]	1.010[3]	12
2024 (projected)	8 billion[2]	0.850[3]	13
2045 (projected)	9 billion[4]	0.480[3]	21

[1] Population and range of growth rates between specified periods (adapted from Coale 1974 and Population Reference Bureau 2010, 3).
[2] Population Reference Bureau 2010, 3. The projected populations for 2011 and 2024 are based on 'medium range' assumptions.
[3] United Nations, Department of Economic and Social Affairs, Population Division 2009a, 3.
[4] Interpolated figure based on United Nations 'medium range' projection (United Nations, Department of Economic and Social Affairs, Population Division 2009b, x).
SOURCE: Adapted from Glenn Trewartha. 1969. *A Geography of Population: World Patterns*, p. 29. New York: Wiley.

Over the course of this century, all populations in the world will become older as a result of decades of fertility declines worldwide. Countries with more rapid and sustained fertility reductions will experience greater degrees of demographic aging. According to the United Nations, in 2009 the oldest countries were Japan, Germany, and Italy, with median ages exceeding 43 years. In 2050, the median variant projection for Japan sees this country reaching a median age of just under 56 years. Joining Japan, among the oldest countries will be the Republic of Korea (53.7), Singapore (53.5), Hong Kong Special Administrative Region, China (52.7), Bosnia-Herzegovina (52.2), Cuba (51.9), Germany (51.7), Netherlands Antilles (51.1), and Poland (51.0). These countries/areas will see their potential support ratio—the number of persons of working age (15 to 64) per older person—drop substantially. It is anticipated that in 2050 the world will see just over 16 per cent of its population being over the age of 65. Seniors will account for 26 per cent of the population in the more developed regions, while in the less developed regions this proportion will reach almost 15 per cent (United Nations, Department of Economic and Social Affairs, Population Division 2009a, 27; 2009b, 48–59).

(Reuters)

Most of the population growth in the coming years will occur in the poorest countries, where rates of natural increase remain high.

TIME to REFLECT

What does it mean to say that some of the projected population growth of the world into the next half-century or so is 'unavoidable'?

Age Compositions of Developed and Developing Countries

The different demographic experiences of the developed and developing countries are reflected in their respective **age compositions**. Fertility plays a greater role than mortality in determining the age composition of a population (Coale 1964; Coale and Hoover 1958; Frejka 1973; Keyfitz 1968). The higher a population's fertility level, the higher the percentage of the population in the younger ages. Low-fertility populations have proportionately fewer people in the younger ages and relatively more in the older ages (see Figure 21.2).

The age distribution of developed countries manifests a slow-growing population, with relatively few people below age 15. Sustained low-fertility rates in these countries are responsible for this. On the other

TABLE 21.1 ▼ Estimated and Projected Population of the World, Major Development Groups and Major Areas, 1950, 2009, and 2050, according to Different Projection Variants

Major Area	Estimated Population (millions)		Population in 2050 (millions) by Type of Variant			
	1950	2009	Low	Medium	High	Constant
World	2,529 (100%)	6,829 (100%)	7,959 (100%)	9,150 (100%)	10,461 (100%)	11,030 (100%)
More developed regions	812 (32.1%)	1,233 (18.1%)	1,126 (14.1%)	1,275 (13.9%)	1,439 (13.8%)	1,256 (11.4%)
Less developed regions	1,717 (67.9%)	5,595 (81.9%)	6,833 (85.9%)	7,875 (86.1%)	9,022 (86.2%)	9,774 (88.8%)
Least developed countries	200	835	1,463	1,672	1,898	2,475
Other less developed countries	1,517	4,761	5,369	6,202	7,123	7,299
Africa	227 (9.0%)	1010 (14.8%)	1,748 (22.0%)	1,998 (21.8%)	2,267 (21.7%)	2,999 (27.2%)
Asia	1,403 (55.5%)	4,121 (60.3%)	4,533 (57.0%)	5,231 (57.2%)	6,003 (57.4%)	6,010 (54.5%)
Europe	547 (21.6%)	732 (10.7%)	609 (7.7%)	691 (7.6%)	782 (7.5%)	657 (6.0%)
Latin America/Caribbean	167 (6.6%)	582 (8.5%)	626 (7.9%)	729 (8.0%)	845 (8.1%)	839 (7.6%)
Northern America	172 (6.8%)	348 (5.1%)	397 (5.0%)	448 (4.9%)	505 (4.8%)	468 (4.2%)
Oceania	13 (0.5%)	35 (0.5%)	45 (0.6%)	51 (0.6%)	58 (0.6%)	58 (0.5%)

NOTE: The sum of Least Developed and other Less Developed countries adds up to the total population for the less developed countries. The sum of the six regions adds up to the overall world total population. The United Nations classifies countries as follows: More Developed—all of Europe and North America plus Australia, Japan, and New Zealand; Less Developed—all other regions and countries; Least Developed—countries with especially low incomes, high economic vulnerability, and poor human development indicators (UN Statistics Division, http://unstats.un.org/unsd/methods/m49/m49regin.htm#developed).

SOURCE: United Nations, Department of Economic and Social Affairs, Population Division 2009a, 1.

▼ FIGURE 21.2   Comparison of Age Structures: Population by Age, Sex, and Development

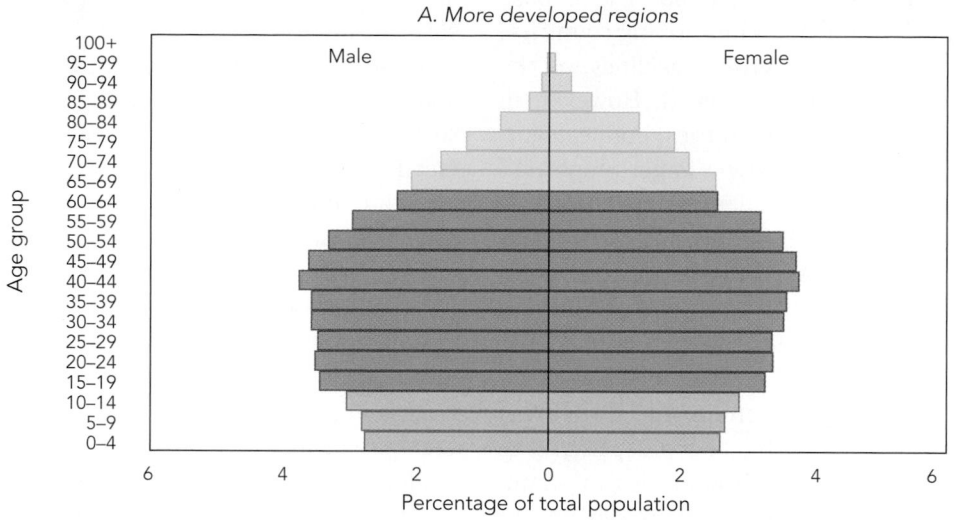

A. More developed regions

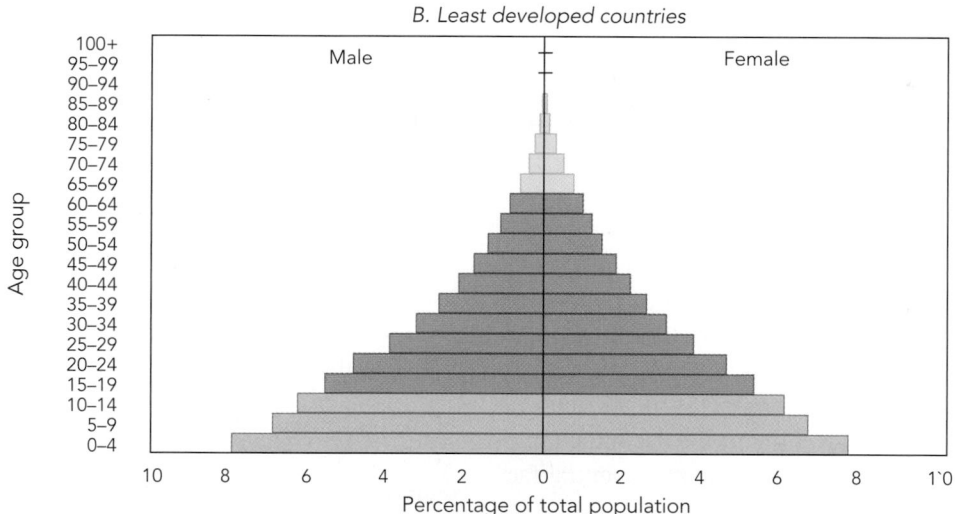

B. Least developed countries

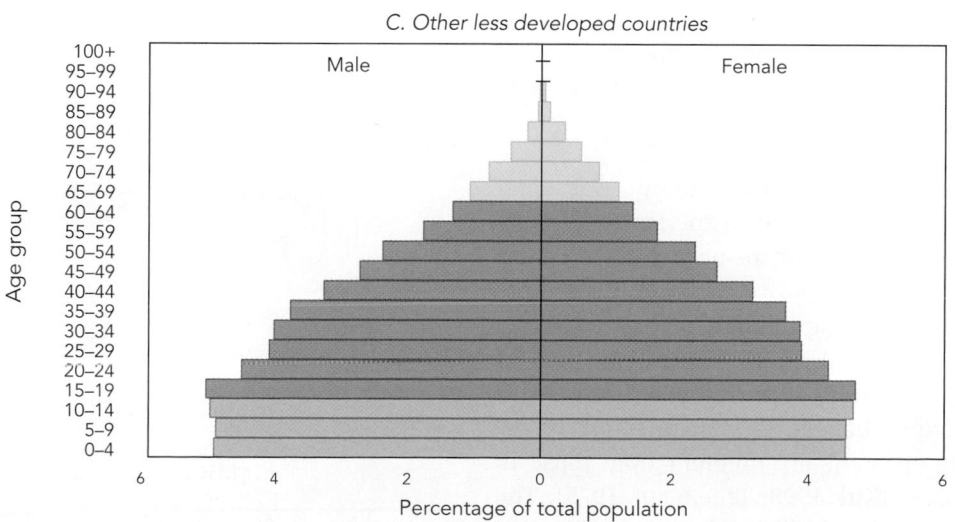

C. Other less developed countries

SOURCE: United Nations, Department of Economic and Social Affairs, Population Bureau 2006a, 27.

hand, the age distribution of the least developed countries characterizes a rapidly growing population, approximating the shape of a pyramid: a wide base and a narrow top. For the other developing countries, there is evidence of recent fertility declines, as the bottom of the pyramid has narrowed. However, in comparison to the developed countries, both categories of developing countries show a larger proportion of their population as being under the age of 15.

Owing to decades of high fertility and rapid population growth, the countries in the developing world must cope with huge waves of young people seeking productive work (Cleland 1996; Bloom, Canning, and Sevilla 2003). Assuming that sufficient work is made available, a youthful labour force could help boost productivity, economic growth, and prosperity. The key is for governments of such countries to stimulate sufficient economic growth and therefore jobs for the burgeoning youthful labour force. Failure to do so may provoke extreme social unrest and political instability (Clarke 1996; Cleland 1996; Homer-Dixon, Boutwell, and Rathjens 1993; McCarthy 2001).

Theories of Population Change

Two influential themes can be identified in the literature regarding the interrelationship of population and resources. The first proposes that curbing population growth is essential for maintaining a healthy balance between human numbers, resources, and the sustainability of the environment; the second characterizes population as a minor or inconsequential factor in such matters. Thomas Malthus and Karl Marx (with Friedrich Engels) are the principal thinkers representing these opposing views. Before examining their ideas, let us review another important theory of population dynamics: the demographic transition.

THE DEMOGRAPHIC TRANSITION

The **demographic transition** theory was first developed on the basis of the experience of western European countries with respect to their historical pattern of change in birth and death rates in the context of socio-economic modernization. In general terms, the theory can also describe the situation of the developing countries, although the structural conditions underlying changes in vital rates are recognized as being substantially different from those in the European case (Kirk 1998; Teitelbaum 1975). The demographic transition of Western societies entailed three successive stages: (1) a pre-transitional period

of high birth and death rates with very low population growth; (2) a transitional phase of high fertility, declining death rates, and explosive growth; (3) a final stage of low mortality and fertility and low natural increase. (The second stage may be divided into early and late Stage 2.) By the early 1940s, most European societies had completed their demographic transitions (see Figure 21.3).

Crude birth and death rates in the ancient world probably fluctuated between 35 and 45 per 1000 population (Coale 1974, 18). With gradual improvements in agriculture and better standards of living, the death rate declined, though fertility remained high. During the second stage, the excess of births over deaths was responsible for the modern rise of population—the 'population explosion' (McKeown 1976). With gradual modernization and socio-economic development during the middle and later years of the nineteenth century, birth rates in Europe began to fall, first in France and then in other countries. In the early 1930s, Western nations had attained their lowest birth rates up to that point in their histories; the death rate had fallen considerably, and a new demographic equilibrium had been reached. In pre-transition times, the low growth rates were the result of humans' lack of control over nature; the end of the demographic transition came from incremental controls over nature—agricultural development, industrialization, **urbanization**, economic growth, modern science, and medicine.

Modifications to Demographic Transition Theory
Coale (1969; 1973) undertook an extensive investigation to re-examine the causes of the European fertility transition. Theorists had proposed that in

▼ **FIGURE 21.3** The Classical Demographic Transition Model and Corresponding Conceptual Types of Society

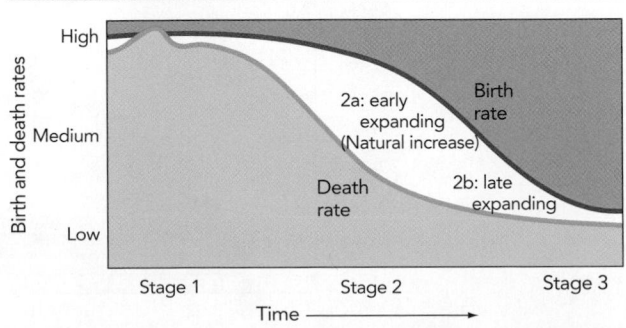

SOURCE: Adapted from Glenn Trewartha. 1969. *A Geography of Population: World Patterns*, 45, 47. New York: Wiley.

OPEN for DISCUSSION

Sex Ratio Imbalances and the Marriage Market

A population's age distribution mirrors its experience with respect to fertility, mortality, and migration. The **sex ratio** measures the balance of males and females in a population (males/females = 100). A sex ratio of 100 indicates an equal number of males and females, whereas a ratio below 100 denotes that the number of women exceeds the number of men. Values in excess of 100 mean there are more men than there are women.

In most national populations, the overall sex ratio is usually just below 100. This is because even though the sex ratio at birth favours males (i.e., for every 100 baby girls, there are usually 105 baby boys), in relation to females, males have higher mortality rates (Perls and Fretts 1998; Kramer 2000; Waldron 1976). Generally, at the national level, the role of migration is minor in affecting the overall sex ratio of the population (at the community level, however, sex differences in migration can affect sex ratios; see, for example, Messner and Sampson 1991).

In some situations, the marriage market (that is, the balance of eligible young men to eligible young women) is affected as a result of either a shortage or a surplus of males or females in the prime marriageable ages (Guttentag and Secord 1983). As a case in point, it has been reported that in China there is a serious deficit of young women for the number of eligible young men to marry—the number of young men is much greater than that of potential brides (Tuljapurkar, Li, and Feldman 1995).

This problem stems from the Chinese government's implementation of the one-child policy, initiated in the late 1970s. The Chinese have a traditional preference for sons. Given the government's one-child directive, over the past several decades many couples, it is suspected, have been resorting to sex-selective abortion to ensure having a sole male child (Jha et al. 2006; Klasen 2003; Tuljapurkar, Li, and Feldman 1995; Banister and Coale 1994). As explained in the following article, over the course of two or more decades, this situation has helped to produce in China a highly distorted marriage market: too few young women for marriage to eligible young men.

With Women So Scarce, What Can Men Do?

Years of Female Infanticide Help Shatter the Taboo on Incestuous Marriages

Liu Dehai and Hai Hongmei's matrimonial bed is laid with a thick quilt embroidered with the characters for 'double happiness', words meant to augur good fortune for just-married Chinese couples. This should be a room of joy and hope, but Liu's mother doesn't want anyone investigating too closely. 'Please,' she says, 'do not speak of this room.' The marriage of Liu and Hai is a subject of shame, for they are not just husband and wife; they are also first cousins.

Their marriage, and others of cousins and even siblings, is the latest consequence of China's profound shortage of females. For two decades, the government has tried to control population by limiting most rural families to one child, two if the first is a girl. Because boys are prized in rural areas—they can work the land and give more support to their families—this has led many couples to abort female fetuses, kill newborn daughters, or neglect them to death. The result: China, according to the World Health Organization, is short 50 million females. The first wave of children born under the policy is reaching marriageable age, and there are far too few brides to go around. The most desperate bachelors have taken to marrying relatives. In a few places, the practice has become so common, the communities are referred to as 'incest villages'.

Liu Dehai never imagined he would marry his shy first cousin Hai. Though intramarriage was common in imperial days, it is taboo in modern China. But at age 20, with his friends already paired off, Liu found himself the odd man out. His parents, farmers in the village of Nanliang in Shaanxi province, could not raise the $2000 required to attract a woman to Nanliang to marry their son. With so many men to choose from, women are loath to settle in hardscrabble villages like Nanliang. Desperate, Liu's mother contacted her sister and requested a favour: Could she ask Hai to be Liu's bride? Young women like Hai are not apt to defy their parents. And so Liu and Hai were wed. While a recent US study concluded that the odds of first cousins producing children with birth defects may have been overstated, the risk is still almost double that for unrelated couples. Denizens of the incest villages see ample evidence of this. Near the city of Yan'an, a brother and sister squat in the mud-brick slums, signing a secret language to each other: both Cao Shuai and Cao Jing were born deaf, to parents who are first cousins. Early this year in Yan'an county, a severely retarded newborn girl was found abandoned beside a road. Her parents, it turned out, were brother and sister.

The female shortage in China is only worsening. In 2000, 900,000 fewer female births were recorded than should have been, based on male births. In 1990, the shortfall was 500,000. Some of that owes to parents giving up daughters for adoption without registering their birth. But population experts at the Chinese Academy of Social Sciences in Beijing estimate that up to one-third of the girls are missing because of gender-based abortions. Rural Chinese women also tend to breast-feed girls for shorter periods, providing less hope for survival. Chinese demographers estimate that in some rural areas, 80 per cent of children ages 5 to 10 are boys.

In Shaanxi's Qiaogou village, children play under a dusty apple tree. The noise is the raucous glee of boys being boys. There is only one girl among them. Asked what he thinks his future will hold, Xiaochun, 7, replies, 'I'll get married and be a good farmer, of course.' Where will he get a wife? 'I think in other villages far away, there are many more girls,' Xiaochun says. 'I will get my wife from there.' Across China, millions of boys are hoping the same thing, but only a few will ever meet the woman of their dreams.

SOURCE: Hannah Beech/Nanliang, 'With Women So Scarce, What Can Men Do?' *Time*, 1 July 2002, 8. ©TIME Inc. Reprinted by permission.

pre-transitional societies, conscious use of family limitation was absent, that economic development and urbanization preceded the onset of fertility declines, and that a drop in mortality always occurred prior to any long-term drop in the birth rate (Davis 1945; Notestein 1945; Thompson 1929, 1944). But some of the empirical evidence uncovered by Coale failed to support some of these propositions. One important discovery was that economic development is not always a precondition for a society to experience the onset of sustained declines in fertility (although economic development would help speed up the transition). Coale concluded that sustained fertility declines in a society would take place when three preconditions were met: (1) fertility decisions by couples must be within the calculus of conscious choice—that is, cultural and religious norms do not forbid couples from practising family planning, nor do they promote large families; (2) reduced fertility must be viewed by couples as economically advantageous; and (3) effective methods of fertility control must be known and available to couples (Coale 1969, 1973; Coale and Watkins 1986).

Having long completed their mortality and fertility transitions, the industrialized countries have gained widespread economic success; their populations enjoy a great deal of social and economic security and well-being. Couples in these societies see little need to have large families. In many developing countries, however, entrenched cultural norms and traditions favour high fertility; parents tend to view children as a source of security in an insecure environment (Cain 1983; Caldwell 1976). Nevertheless, over recent decades, much progress has been made in raising contraceptive prevalence levels. Organized family planning programs have played a major role in this trend (Caldwell, Phillips, and Barkat-e-Khuda 2002; Robey, Rutstein, and Morris 1993). New evidence suggests a growing number of developing nations are now approaching the end of their demographic

transitions and that others in the poorer regions of the world have recently begun their fertility transitions (Bulatao 1998; Bulatao and Casterline 2001).

Unfortunately, there is also evidence that after some improvements, some of the poorest countries have stalled in this transition, particularly in sub-Saharan Africa (Ezeh, Mberu, and Emina 2009; Bongaarts 2008). Two factors in particular may have played key roles in this. First, unlike much of the rest of the world, economic growth in sub-Saharan Africa declined rather than increasing during the 1990s. Life expectancy fell in this region owing to the devastating effects of the HIV/AIDS epidemic while the rest of the world enjoyed longevity improvements. The second factor, according to John Bongaarts (2008), is the lower priority assigned to family planning in sub-Saharan African countries. Ezeh and colleagues (2009) concur with Bongaarts's assessment on the role played by the decline in family planning services in the fertility stall and point out that there have been increases in unwanted fertility and a decline in access to family planning services, especially among adolescents. These stalls in fertility decline will affect the population of sub-Saharan Africa significantly. It is anticipated that its population will likely more than double, from 769 million in 2005 to 1.76 billion in 2050, even after the effect of the AIDS pandemic has been taken into account in the population projections (Bongaarts 2008).

TIME to REFLECT

Do you think it is possible for a country to experience significant economic growth without having completed the demographic transition? Is the completion of demographic transition a prerequisite for economic and social development in a country?

Demographic Transitions of Industrialized and Developing Countries

Figure 21.4 displays in schematic form the demographic transitions of the West and of the contemporary developing countries, the latter subdivided into 'transitional' and 'delayed transition' societies. Examples of transitional populations are India, Turkey, China, Indonesia, Taiwan, Thailand, Mexico, and countries in Latin America and the Caribbean. Delayed transition societies are found in sub-Saharan Africa and southwest Asia (for example, Afghanistan, Pakistan, and Bangladesh). In these cases, mortality reductions have been fairly rapid. In the European historical context, health improvements occurred

HUMAN DIVERSITY

Youth in the International Labour Market

As the world's population surges past the six-billion mark . . . 700 million young people will enter the labour force in developing countries—more than the entire workforce of the developed world in 1990—the United Nations says.

And as the largest-ever group of young people enters its child-bearing and working years, the number of people over the age of 65 continues to swell as health and longevity improve, according to the UN's annual *State of the World Population Report*. . . .

If jobs can be found or created for the global bulge of one billion people between the ages of 15 and 24—the result of past high fertility—there is a chance to increase human capital so that the dependent young and elderly age groups can be better cared for, the report said.

But without investment in jobs for the young, better education for children, especially girls, and better health care for both young and old, social unrest and instability are inevitable.

'The rapid growth of young and old "new generations" is challenging societies' ability to provide education and health care for the young, and social, medical and financial support for the elderly,' the report says. . . .

Some developing nations, particularly in Southern Asia and Northern Africa, could reap an economic windfall in the next couple of decades as the bulge in 15- to 24-year-olds swells the workforce in comparison to dependent age groups, the report says.

To avoid squandering this one-time 'demographic bonus', these countries will have to ensure their young people can find jobs and don't start families too soon.

Stan Bernstein, chief author of the UN report and a research adviser at the UN Population Fund, said developing countries need both private and public, domestic and international investment in their basic social services to ensure they don't miss this 'window of opportunity'.

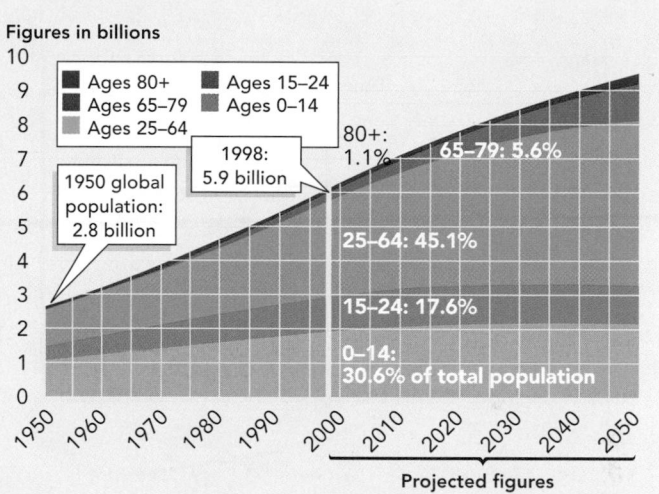

Global Population Growth by Age Group

Figures in billions

Legend:
- Ages 80+
- Ages 65–79
- Ages 25–64
- Ages 15–24
- Ages 0–14

1950 global population: 2.8 billion

1998: 5.9 billion

80+: 1.1%
65–79: 5.6%
25–64: 45.1%
15–24: 17.6%
0–14: 30.6% of total population

Projected figures

SOURCE: United Nations. *State of the World Population Report.*

It is in the interest of wealthier nations and private companies to make this investment, Mr Bernstein said in a phone interview from New York yesterday.

'These countries, if given the opportunity to accelerate their development now, are going to be significant economic, trade, and social partners in the future. It's a win-win situation.'

However, with most developed nations steadily reducing the amount of foreign aid they provide, and cutting their own social-service budgets, other sources of investment are needed. . . .

At the other end of the age spectrum, a rapidly growing population over the age of 60—578 million this year—is seeing more years of healthy life, is able to work longer, and is moving toward greater independence from grown children. . . .

SOURCE: From Jane Gadd, 'Record numbers of youth will seek work: UN', *The Globe and Mail*, 2 Sept. 1998. Reprinted with permission from *The Globe and Mail*.

Schematic Representation of Demographic Transition: Western, Delayed, and Transitional Models

▼ **FIGURE 21.4**

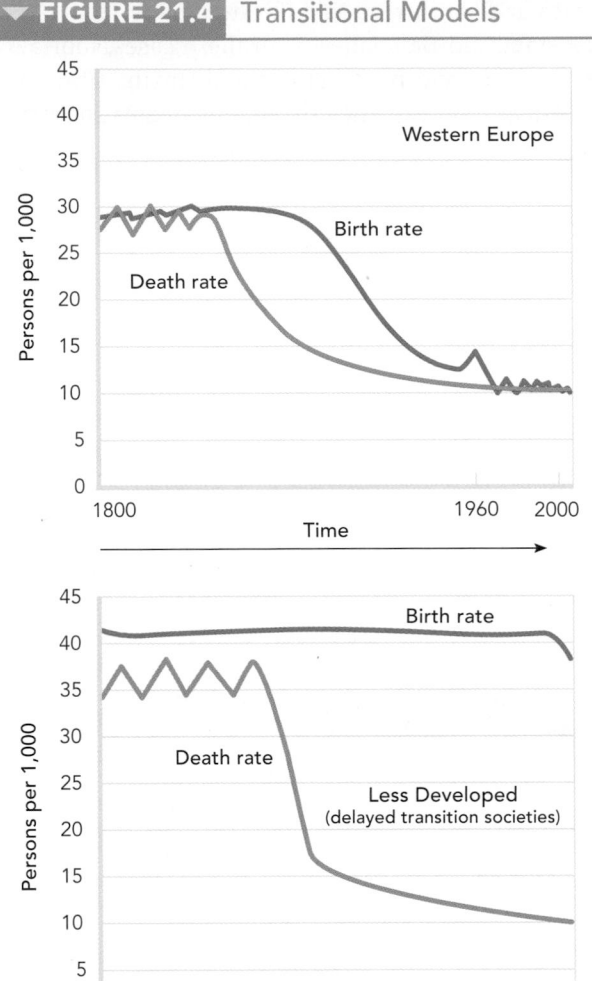

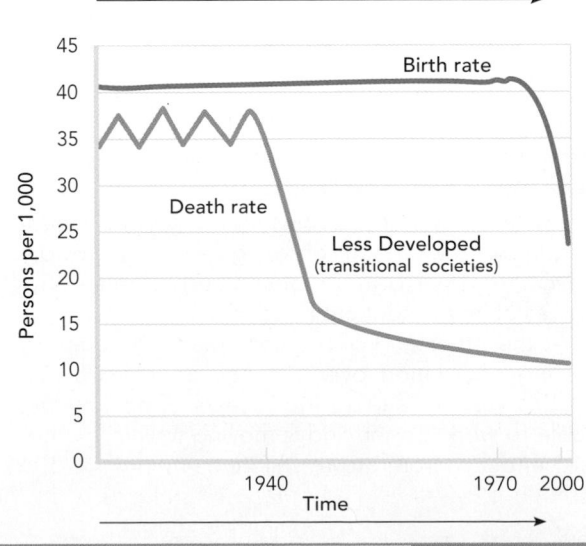

SOURCE: Adapted from Glenn Trewartha. 1969. *A Geography of Population: World Patterns*, 45, 46. New York: Wiley.

more gradually in response to incremental socio-economic advancements and economic modernization. In the developing countries, reduced mortality initially has been achieved to a large extent through public health programs, including family planning and anti-malarial programs, with the help of the industrialized countries (Preston 1986b).

MALTHUSIAN THEORY

Thomas Malthus (1766–1834) was an ordained Anglican minister and professor of political economy in England. His most famous work, *An Essay on the Principle of Population*, was published in 1798. This important treatise has had a lasting influence on subsequent theorizing about population matters (Overbeek 1974).

Citing directly from the work of Malthus, Alan Macfarlane identifies three fundamental themes in the *Essay*:

The first is that human beings are very strongly motivated by a desire for sexual intercourse:

> The passion between the sexes has appeared in every age to be so nearly the same, that it may always be considered, in algebraic language, as a given quantity.
>
> All else being equal, men and women will mate as soon as possible after puberty. If such mating is only permitted within marriage, . . .
>
> Such is the disposition to marry, particularly in very young people, that, if the difficulties of providing for a family were entirely removed, very few would remain single at twenty-two . . .

The second fact is the high fertility of humans. If this high fertility is combined with a reasonable rate of mortality, such early and frequent mating will lead to rapid population growth. . . . The third fact is that economic resources, and in particular food production, cannot keep pace with this population growth within a basically agrarian economy largely dependent on human labour. This is due to the law of diminishing marginal returns. While there may be periods when rates of growth in agriculture rise to three or four per cent per annum, which is equivalent to a doubling of food in a generation, such periods cannot be sustained for more than a few decades. (Macfarlane 1997, 12–13)

Malthus assumed an inherent tendency in humans to increase in numbers beyond the means of subsistence available to them. Mankind, he argued, lives at the brink of subsistence. Population, if left unchecked, tends to double once every generation. Population grows geometrically (1, 2, 4, 8, 16, 32, . . .), but the food supply grows arithmetically (1, 2, 4, 6, 8, 10, . . .). Thus, in the long term population would outstrip food and other essential resources.

Malthus proposed that population could be kept in equilibrium through the activation of two different mechanisms. The first is what he called *positive checks* (i.e., 'vice and misery'). These are conditions that raise the death rate and thus serve to reduce population—famine, pestilence, war, and disease. The second, and more desirable, alternative would be widespread exercise of *preventive checks* (i.e., 'moral restraint') to curb population growth—the imposition of the human will to deliberately curtail reproduction, through celibacy, postponed marriage, and sexual abstinence. By postponing marriage until people were in an adequate position to maintain a family, individuals and society would gain economically. By working longer before marriage, people would save more of their incomes, and this would help reduce poverty and raise the overall level of well-being in society. Malthus considered abortion and contraception immoral.

CRITICISM OF MALTHUSIAN THEORY

Malthus failed to fully appreciate the resilience of humanity when faced with difficult problems. Throughout history, humanity has shown a remarkable ability to solve many of its predicaments. Writers point out that progress in science, technology, and socio-economic well-being has evolved in tandem with explosive population growth and that the agricultural and industrial revolutions took place in response to problems and demands arising from a rapidly growing population (Boserup 1965, 1981; Fogel and Costa 1997; Simon 1995, 1996).

Malthus suggested that population and food supply must be in balance. But exactly what constitutes an 'optimum' population size is difficult, if not impossible, to specify. Perceptions of the 'optimal', in this sense, are highly dependent on resources available to a society, as well as on consumption patterns and the degree of economic activity and production. Consumption and production are closely tied to a society's cultural standards for material comfort and demand for consumer products. As described by Paul Ehrlich and J.P. Holdren (1971), the potential impact of population on the environment is multiplicative. That is, the effect of the population depends partly on its size and growth, but also on a society's level of affluence and technological complexity. In a slow-growing population with strong material expectations, the potential for environmental and resource depletion may be greater than in a fast-growing population with lower levels of material aspirations and technological sophistication. The more developed the society, the higher the expected standard for 'basic' necessities. As material aspirations rise, so do levels of consumption and expenditures by the public. Increased consumer demand for material goods spurs economic activity; greater levels of economic activity heighten the risk of environmental damage because of increased pollution and resource depletion (Ehrlich and Ehrlich 1970; 1990).

Industrial societies may be facing an inversion of the Malthusian scenario, since in these societies, overpopulation is no longer the threat Malthus put forward (Woolmington 1985). The seeds of systemic instability in these societies may lie more in their reliance on endless economic growth and consumption. Rather than population pressing on resources, as was proposed by Malthus, the economy now presses the population to consume. In the long term, the spiral of consumption and production may not be sustainable (Woolmington 1985).

Finally, Malthus's insistence on the unacceptability of birth control is inconsistent with his advocacy of curbing rapid population growth. The reality is that people throughout the centuries have always resorted to some means of birth control at one time or another (Ross 1998, 3).

THE MARXIST PERSPECTIVE ON POPULATION

Marxist scholars have refuted the Malthusian principles. They contend that socio-economic inequality is a root cause of human problems and suffering and therefore population is a secondary issue to these pernicious problems; large families arise from poverty. With his collaborator Friedrich Engels, Marx advocated the radical restructuring of society to ensure an equitable distribution of wealth and resources (Marx 1976 [1867]; 1887).

However, in matters of population control, Marxist scepticism toward the role of population in human problems is now largely ignored (Petersen 1989). For example, China—a communist state—has outwardly rejected the Marxist doctrine of population. Its officials have recognized that slowing population growth through concerted family planning policies is, in the long term, essential to societal well-being (Bulatao 1998; Caldwell, Phillips, and Barkat-e-Khuda 2002; Haberland and Measham 2002). With few exceptions, developing countries embrace population policies that are consistent with neo-Malthusian principles and view family planning and reproductive health programs as being of critical importance.

CONTEMPORARY PERSPECTIVES ON POPULATION

Neo-Malthusian scholars—the contemporary followers of Malthus—believe that the world's population has been growing too fast and that the planet is already close to reaching critical ecological limits. Unlike Malthus, however, neo-Malthusians view contraception and family planning as a key element in population control (Ehrlich and Ehrlich 1990). The neo-Malthusian perspective inherently implies that the world would be a better and safer place if it contained fewer people. An expanding population in conjunction with excessive consumerism and economic production will, in the long term, lead to the depletion of essential resources and to ecological breakdown.

Like Marx and Engels, neo-Marxist scholars place less emphasis on the centrality of population as a source of human predicaments. Focusing on population as the root cause of human suffering obscures the reality that the world is divided into wealthy and relatively poor regions and that this divide is widening rather than narrowing. Neo-Marxists alert us to the extreme consumerism of wealthy regions and their overwhelming economic and political influence over less advantaged nations. They also argue that the **globalization** of capital—seen by many other observers as the key to emulating the 'success story' of the West—often exacerbates, rather than diminishes, socio-economic disparities within and across societies (Gregory and Piché 1983; Wimberley 1990). Andre Gunder Frank (1991) has coined the phrase 'the underdevelopment of development' to refer to the overwhelming influence and control the world's major economic powers hold over developing countries.

(Mariana Bazo/Reuters)

Is the poverty shown in this photograph a result of overpopulation or of unequal access to resources and opportunities? This is a shantytown just outside of the commercial district in Buenos Aires, Argentina.

UNDER the WIRE

The Geography of Poverty

A country's natural endowments, its location, and its climate can help to explain part of the long-standing 'puzzle' first introduced by Adam Smith in his *Inquiry into the Nature and Causes of the Wealth of Nations* (1976 [1776]): Why are some nations wealthy and others not? David Landes (1999) has devoted considerable attention to this question. One of his propositions for the success of Europe in the world economy relates to its luck in geography, which gave this part of the world important natural attributes to catapult it to economic prominence: a temperate climate, warm winds from the Gulf Stream, gentle rains, water in all seasons, and low rates of evaporation, giving Europeans large, dense forests, good crops, and big livestock. Added to these endowments are Europe's many navigable rivers, harbours, and plenty of mineral resources.

Sachs, Mellinger, and Gallup (2001) expand on this theme. Nations in tropical climates and desert zones generally face higher rates of infectious diseases and lower agricultural productivity than do nations in temperate zones. The very poorest regions in the world are those saddled with both handicaps: distance from sea trade and a tropical or desert ecology. Tropical areas are susceptible to persistent endemic diseases such as malaria, whose pathogen is transmitted to humans by mosquitoes, which thrive in tropical environments. Malaria, a debilitating illness for millions in such areas and a major killer, significantly reduces the productivity of a nation, affecting a large part of the working population. Malaria is seldom seen in temperate climates. Thus, economic disparities between nations can be partly attributed to geography and its indirect relationships to factors detrimental to economic growth.

Coastal countries with temperate climates such as Germany have lower transportation costs and higher farm productivity than landlocked tropical countries such as Uganda. Among the high-income economies of the world, only Hong Kong, Singapore, and part of Taiwan are in the tropical climate zone. Sachs and associates also note that almost all the temperate-zone countries have either high-income economies (as in the cases of North America, western Europe, South Korea, and Japan) or middle-income economies burdened by socialist policies in the past (as in the cases of eastern Europe, the former Soviet Union, and China). In addition, there is a strong temperate–tropical divide within countries that straddle both types of climates. Most of Brazil, for example, lies within the tropical zone, but the richest part of the nation—the southernmost states—is in the temperate zone.

Finally, geography and climate affect food production. Tropical environments are plagued by diverse infestations of pests and parasites that can devastate crops and livestock. And environmental variables correlated with geographic location determine a country's ability to exploit its resources. Nations advantaged in geographic location, climate, and resources are able to develop institutions that bolster social well-being and productivity—free markets, equitable tax laws, protection and promotion of private property rights, and universal education and health care.

Other writers concerned with the complex interactions of population, environment, and resources are neither neo-Marxists nor neo-Malthusians (Ahlburg 1998; Cincotta and Engelman 1997; Clarke 1996; Evans 1998; Furedi 1997). Some are quite positive in their outlook. Julian Simon (1995; 1996), for instance, has written that population growth historically has been, on balance, beneficial to humankind—people are the 'ultimate resource', he contends.

Others take a more neutral (or revisionist) position. The US National Research Council's report on population, environment, and resources (1986) exemplifies such a perspective. It proclaimed that in some cases, population may have no discernable relationship to some of the problems often attributed to it. The report also concluded that population's relationship to depletion of exhaustible resources is statistically weak and often exaggerated. Income growth and excessive consumption are more important factors in this sense: a world with rapid population growth but slow increases in income might experience slower resource depletion than one with a stationary population but rapid increase in income. It was also found that reduced rates of population growth would increase the rate of return to labour and help bring down income inequality in a country.

The National Research Council also suggested that while rapid population growth is directly related to the growth of large cities in the **Third World**, its role in urban problems is most likely secondary.

Ineffective or misguided government policies may play a more important role in the development of urban problems (National Research Council 1986).

Canada's Population: An Overview

COMPONENTS OF GROWTH

At the time of its first national census in 1851, Canada had only 2.4 million residents. By 1931, this country recorded its first 10 million inhabitants. And by 1967, the year of its centennial as a country, the population had grown to more than 20 million. In 2010, the population of Canada stood at just over 34 million (Statistics Canada website, www.statcan.gc.ca/start-debut-eng.html).

Table 21.2 looks at the components of population growth from 1851–61 to 2001–6. Historically, natural increase has been the principal driving force behind population growth. However, since the middle of the 1980s, the contribution of net migration has been rising in prominence. Net international migration now accounts for about two-thirds of the growth in Canada's population. This highlights the importance of immigration for the long-term maintenance of Canada's population in a context of continued levels of sub-replacement fertility.

MORTALITY

Since the early 1920s, Canada has witnessed a steady decline in its crude death rate (the number of deaths in a given year divided by the mid-year population, usually expressed per 1000), from 11.6 per 1000 population in 1921 to 10.1 in 1941 and 7.7 in 1961, the latter being very close to the 2010 rate of 7 per 1000.

The crude death rate is not the best indicator of mortality, since it fails to take into account the confounding effects of age composition. Measures such as life expectancy at birth and the infant mortality rate provide more reliable accounts of mortality conditions because they mirror very closely a society's level of socio-economic development and standard of living. In 2010, Canadians enjoyed a life expectancy at birth of 81 years (men 78, women 83). By comparison, the corresponding life expectancy for Mexicans was 76 years (men 74, women 79). Closely tied to this survival gap is the wide discrepancy in infant mortality (the number of deaths of infants in a given year divided by the number of live births in that year) between these countries—5.1 in Canada versus 17.0 in Mexico (Population Reference Bureau 2010).

In general, women live longer than men. The literature emphasizes the interaction of biology and environment (broadly speaking) as the underlying basis of this differential (Kramer 2000; Perls and Fretts 1998; Vallin 1983; Waldron 1976; El-Badry 1969; Madigan 1957). The Population Reference Bureau (2010) reports that life expectancies for men and women in the more developed regions of the world are 74 and 81, respectively. For the less developed countries (excluding China), the corresponding life expectancies are much lower—66 and 69 years, respectively. Since the 1990s, there have been setbacks in life expectancy gains in sub-Saharan Africa (comprising countries in eastern, central, western, and southern Africa but not North Africa) because of the HIV/AIDS epidemic as well as periodic famines and socio-political unrest (Lamptey et al. 2002; UNAIDS 2006; Bongaarts 2008).

Epidemiological Transition

In Canada and in other advanced societies, the majority of deaths on an annual basis are accounted for by a few leading 'killers'—cardiovascular disease, cancer, accidents, and violence. Infectious and parasitic diseases such as typhus, cholera, smallpox, and tuberculosis are rare causes of premature mortality. In the past, these ailments predominated as leading killers over the chronic, degenerative, and 'man-made' causes of death (Omran 1971). In 1921, out of 67,722 deaths in Canada, malignant neoplasms and diseases of the circulatory system accounted for 20.9 per cent of all deaths. Of the 210,733 deaths recorded in 1991, these same diseases represented 66.5 per cent of this total (McVey and Kalbach 1995, 206).

The historical shift in Canada's cause-of-death distribution is part of a general phenomenon that societies are expected to experience as they modernize. Abdel Omran's epidemiological transition theory (1971) states that industrialized societies have gone through three epidemiological stages: (1) During stage one, in prehistoric times, life was ruled by Malthusian 'positive checks'—famine, misery, pestilence. Life was brutish and short. Infectious and parasitic diseases were the leading killers, along with violence

TABLE 21.2 ▼ Canada's Population and Growth Components, 1851–2006[1]

Period	Census Population at End of Period (000s)[1]	Total Population Growth (000s)[1]	Average Annual Growth Rate (%)	Births (000s)	Deaths (000s)	Immigration (000s)	Emigration (000s)	Natural Increase (NI) (000s)	Net Migration (NM) (000s)	Ratio: NI/Total Growth x 100	Ratio: NM/Total Growth x 100
1851–61	3230	793	2.5	1281	670	352	170	611	182	77.0	23.0
1861–71	3689	459	1.2	1370	760	260	410	610	–150	132.6	–32.6
1871–81	4325	636	1.5	1480	790	350	404	690	–54	108.5	–8.5
1881–91	4833	508	1.1	1524	870	680	826	654	–146	128.7	–28.7
1891–1901	5371	538	1.0	1548	880	250	380	668	–130	124.2	–24.2
1901–11	7207	1836	2.5	1925	900	1550	740	1025	810	55.9	44.1
1911–21	8788	1581	1.8	2340	1070	1400	1089	1270	311	80.3	19.7
1921–31	10,377	1589	1.5	2415	1055	1200	970	1360	230	85.5	14.5
1931–41	11,507	1130	1.0	2294	1072	149	241	1222	–92	108.1	–8.1
1941–51[2]	13,648	2141	1.6	3186	1214	548	379	1972	169	92.1	7.9
1951–6	16,081	2433	3.0	2106	633	783	185	1473	598	71.1	28.9
1956–61	18,238	2157	2.4	2362	687	760	278	1675	482	77.7	22.3
1961–6	20,015	1777	1.8	2249	731	539	280	1518	259	85.4	14.6
1966–71[3]	21,568	1553	1.4	1856	766	890	427	1090	463	70.2	29.8
1971–6	23,450	1882	1.6	1755	824	1053	358	931	695	57.3	42.7
1976–81	24,820	1371	1.1	1820	843	771	278	977	493	66.5	33.5
1981–6	26,101	1280	1.0	1872	885	678	278	987	399	71.2	28.8
1986–91	28,031	1930	1.4	1933	946	1164	213	987	986	50.0	50.0
1991–6	29,611	1580	1.1	1936	1024	1118	338	912	908	50.1	49.9
1996–2001	31,021	1410	0.9	1705	1089	1217	376	609	781	43.8	56.2
2001–2006	32,649	1628	1.0	1681	1113	1403	323	548	1080	33.7	66.3

[1]Total population growth is the difference in census population counts between the end and the beginning of each period.
[2]Beginning in 1951, Newfoundland is included.
[3]Beginning in 1971, the population estimates are based on census counts adjusted for net census undercount, and the reference date is 1 July instead of census day. The 1 July 1971 population adjusted for net census undercount is 21,962,000. Immigration figures include immigrants, returning emigrants, and net non-permanent residents. Population growth calculated using the components will produce a different figure from that reported in the table. Prior to 1971, the emigration figures are 'residual' estimates and include the errors in the other three components of growth—births, deaths, and immigration—as well as errors in the census counts. Beginning in 1971, an independent estimate of emigration is produced. From 1991 on, emigration includes emigrants and net temporary emigrants.

SOURCES: Statistics Canada, 'Population and growth components (1851–2001 Censuses)', www.statcan.ca/l01/cst01/demo03.htm; Statistics Canada 2007a.

and accidents. (2) With the advent of agriculture and, later, improved systems of food production and distribution, as well as general advances in the standard of living, humanity developed the ability to resist many infectious diseases, and people lived on average longer than in the preceding stage. This period of epidemiological history in the Western world, 'the stage of receding pandemics', began around 1750 CE and ended around the turn of the twentieth century. (3) The third stage, that of 'man-made and degenerative diseases', began in the 1930s; its main feature was a rising dominance of the chronic and degenerative ailments, such as cancer and heart disease, as the leading killers. Infectious and parasitic diseases receded in relative importance. During this stage, life expectancy rose to unprecedented levels. For instance, by the early 1940s a male newborn in England and Wales could expect to live nearly 60 years, while for a newborn baby girl, life expectancy had already surpassed 60 years. In Canada, life expectancy at birth for men in 1941 had reached almost 62 years and for women, nearly 65.5 years (Preston, Keyfitz, and Schoen 1972).

At present, the industrialized societies of western Europe, North America, Japan, Australia, and New Zealand are situated in the *fourth stage of epidemiological transition* (Olshansky and Ault 1986). Its essential features include (1) life expectancy at birth in excess of 70 years, (2) continuation of cancer and heart disease as the leading causes of death, (3) increased survival by people with these major ailments as a result of effective medical therapies and interventions, (4) unprecedented survival improvements among seniors, and (5) the increased concentration of the majority of deaths on an annual basis in the older ages, since relatively few deaths take place in infancy, early childhood, and young adulthood (Kannisto et al. 1994). Some analysts have posed the question as to whether these societies are approaching a maximum attainable average lifespan (Fries 1980; Olshansky, Carnes, and Cassel 1990; Coale 1996), while others express a more open-ended and optimistic view on this question (Vaupel 2010; Christensen et al. 2009; Manton 2008; Oeppen and Vaupel 2002).

TIME to REFLECT

The Western world has passed through four stages of epidemiological transition. Do you think there will be a fifth stage? If so, what would be the key features of this fifth stage?

Fertility

The most basic measure of fertility is the crude birth rate (CBR): the number of births in a given year divided by the mid-year population times 1000. Currently, the Canadian crude birth rate is 11 per 1000 (Population Reference Bureau 2010). Between the early years of the 1920s and the present, some of the highest crude birth rates were recorded during the peak years of the **baby boom**, the period between the end of World War II and 1966. Between 1947 and 1966, a total of 8,571,376 babies were born in Canada, constituting the largest generation in the history of this country (Foot and Stoffman 1998, 24–5; Romaniuc 1984, 121–2). From 1966 onward, Canada experienced a **baby bust** followed by an ongoing period of sub-replacement fertility beginning in the early 1970s (Bélanger 2006; Grindstaff 1975, 1994; Romaniuc 1984, 1994).

There have been other low-fertility periods in Canadian history. In 1933, births in Canada had fallen to a low point, with only 229,791 registered that year. One year after the end of World War II (that is, in 1947), the number of newborns rose to 372,589, a 62 per cent jump from 1933. A steady increase in births occurred during the 1950s, peaking at 479,275 in 1959, the height of the baby boom. As the 1960s unfolded, Canadians embarked on a long-term trend toward having smaller families. By 1966, the eve of Canada's centennial, the decline was well underway. Only 387,710 babies were born that year, which represented nearly a 20 per cent decline from 1959. The annual numbers of births kept dropping until 1990, when, for the first time since 1966, more than 400,000 babies were born. But this was a temporary surge, as 1993 saw a return to fewer babies being born (388,394), and the number dropped even further in 2000 to 327,882, the lowest since 1946. Between 1 July 2008 and 30 June 2009, there were 337,703 births registered in Canada (Statistics Canada website, 2010: www40.statcan. gc.ca/l01/cst01/demo04a-eng.htm).

The *total fertility rate* (TFR) measures the number of children a woman would bear throughout her reproductive lifetime if she experienced the prevailing age-specific birth rates in a given period. In 1959, at the peak of the baby-boom period, the TFR had climbed to 3.94. But by 1966, it had fallen to 2.81 children per woman. The year 1972 marks an important turning point in Canadian fertility: the TFR fell for the first time to below 2.1, the number of children needed to ensure long-term replacement of the generations. Canada's total fertility rate in 2000 was 1.49, the lowest in the country's history. By 2007, it had

GLOBAL ISSUES

Global Summary of the AIDS Epidemic

The number of people living with HIV worldwide continued to grow in 2008, reaching an estimated 33.4 million. The total number of people living with the virus in 2008 was more than 20 per cent higher than the number in 2000, and the prevalence was roughly three times higher than in 1990.

The persisting rise in the population of people living with HIV reflects the combined effects of continued high rates of new HIV infections and the beneficial impact of antiretroviral therapy. As of December 2008, approximately 4 million people in low- and middle-income countries were receiving antiretroviral therapy—a tenfold increase over five years. In 2008, an estimated 2.7 million new HIV infections occurred. It is estimated that 2 million deaths due to AIDS-related illnesses occurred worldwide in 2008.

The latest epidemiological data indicate that globally, the spread of HIV appears to have peaked in 1996, when 3.5 million new HIV infections occurred. In 2008, the estimated number of new HIV infections was approximately 30 per cent lower than at the epidemic's peak 12 years earlier.

Global HIV estimates, 1990–2008

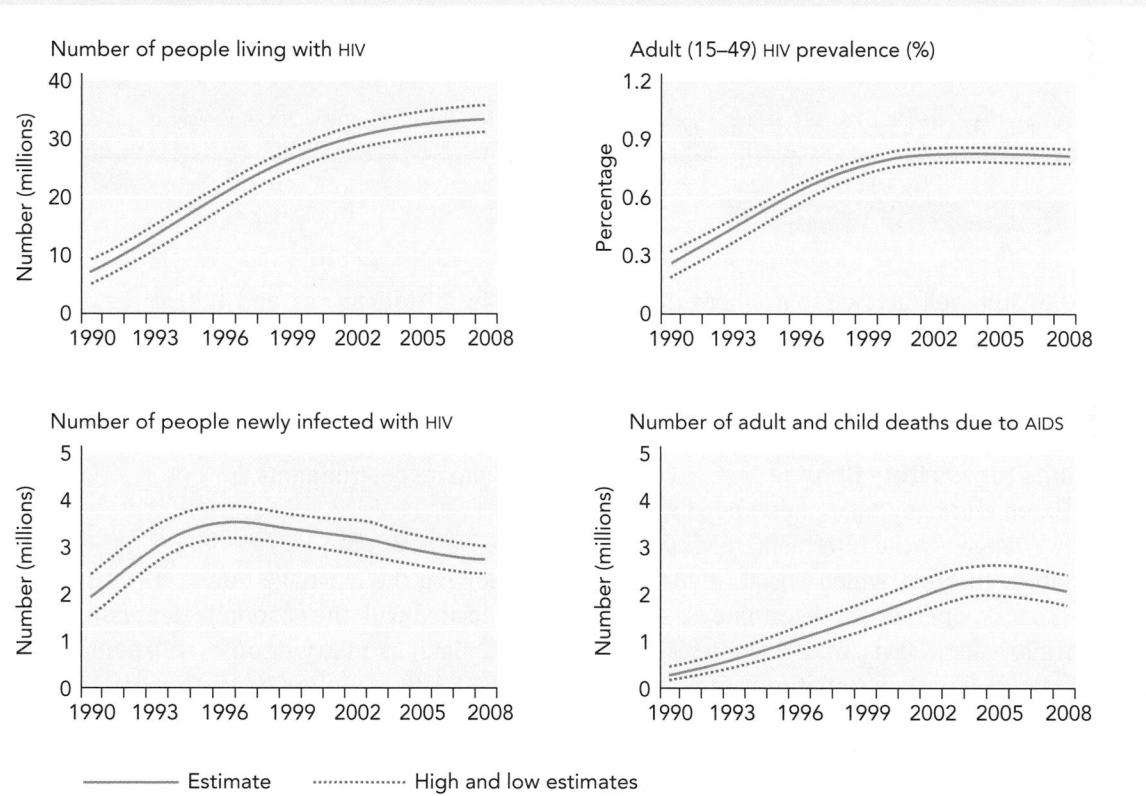

SOURCES: UNAIDS; World Health Organization.

Consistent with the long interval between HIV seroconversion and symptomatic disease, annual HIV-related mortality appears to have peaked in 2004 when 2.2 million deaths occurred. The estimated number of AIDS-related deaths in 2008 was roughly 10 per cent lower than in 2004. An estimated 430,000 new HIV infections occurred among children under the age of 15 in 2008. Most of these new infections are believed to stem from transmission in utero, during delivery, or post-partum as a result of breastfeeding. The number of children newly infected with HIV in 2008 was roughly 18 per cent lower than in 2001.

The epidemic appears to have stabilized in most regions, although prevalence continues to increase in eastern Europe and central Asia and in other parts of Asia because of a high rate of new HIV infections. Sub-Saharan Africa remains the most heavily affected region, accounting for 71 per cent of all new HIV infections in 2008. The resurgence of the epidemic among men who have sex with men in high-income countries is increasingly well-documented. Differences are apparent in all regions, with some national epidemics continuing to expand even as the overall regional HIV incidence stabilizes.

Adults and Children Estimated to Be Living with HIV, 2008

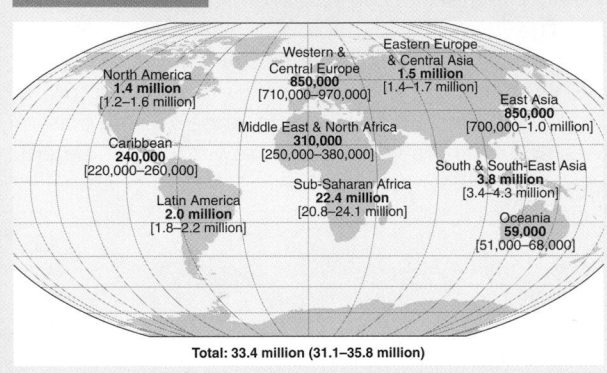

Total: 33.4 million (31.1–35.8 million)

Estimated Adult and Child Deaths Due to AIDS, 2008

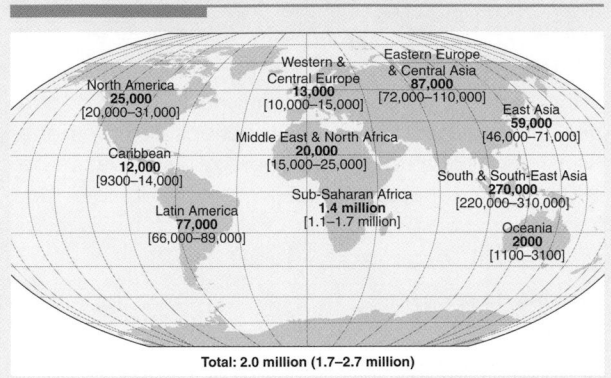

Total: 2.0 million (1.7–2.7 million)

SOURCES: Adapted from UNAIDS. 2009. 'AIDS epidemic update 2009' (UNAIDS/09.36E/JC1700E), pp. 7–11, http://data.unaids.org/pub/Report/2009/JC1700_Epi_Update_2009_en.pdf and 'Epidemiology core slides', www.unaids.org/en/KnowledgeCentre/HIVData/EpiUpdate/EpiUpdArchive/2009/default.asp.

risen to 1.66, still well below replacement (Statistics Canada 2010b). This trend of below-replacement fertility in Canada is also typical of most other industrialized nations.

Explanations for Fertility Change

The baby-boom and baby-bust phenomena were precipitated by changes in the intermediate variables—a set of variables through which social, cultural, and biological factors operate to determine a society's overall fertility rate (Davis and Blake 1956). John Bongaarts (1978) has shown that four of these intermediate variables (also known as *proximate determinants*) account for most of the fertility variation across societies: the extent of non-marriage (the higher the number of people who remain unmarried, the lower the overall fertility), the level of contraceptive use (the greater the use of contraception, the lower the fertility), the degree to which abortion is practised (high rates of abortion translate into reduced overall fertility), and the level of postpartum amenorrhea (the longer women breastfeed their babies, the later the return of ovulation and hence the lower the societal fertility rate). Changes in the extent of marriage and contraceptive use have played leading roles in the historical rise and fall of the Canadian birth rate (Grindstaff 1975, 1994, 1995; Romaniuc 1984). Sociological, cultural, and economic factors have also played an important role in the observed changes in Canadian fertility, though indirectly, through the proximate determinants.

The Baby Boom

Following the marriage downturn of the early 1930s associated with the economic depression, the number of Canadians marrying rose substantially during the 1940s. Although the Canadian marriage rate has followed an uneven historical trend, between 1940 and the late 1950s it was well above the rates recorded for the 1920s and the early 1930s.

According to Richard Easterlin (1969; 1980), the economic recovery and prosperity of the postwar period was a major factor explaining the rise of marriage and early procreation among young couples during the baby-boom period. Young men could find abundant work, and their prospects in the labour market looked exceptionally promising. As well, the major institutions—the church, the economy, and the state—provided the individual with a strong sense of security and stability. Governments were in the midst of creating safety-net systems to enhance the welfare of their populations (universal health care, education, employment insurance, and so on). The church exerted a strong moral influence on the people, reinforcing pro-natalist values. Together, these conditions promoted early marriage and childbearing (Ariès 1980; Lesthaeghe and Surkyn 1988; Simon 1980). As described by Easterlin (1969; 1980), demographic conditions in the postwar period were also favourable to high fertility. The relatively small cohorts of young male workers, born during the low-fertility period of the 1930s, would enter the labour market in the late 1940s and 1950s, a period of increased demand for workers in a rapidly expanding economy.

Changes in **gender roles** also contributed to the baby-boom phenomenon. In her study of the 'feminine mystique', Betty Friedan (1963) outlined the tendency of women in the 1950s to be preoccupied with marriage to a successful husband, motherhood, and a new home in the suburbs. Their self-concept was tied to this 'mystique', presented and reinforced by society. Thus, early marriage and early parenthood were common. The male role was predominantly that of breadwinner. This traditional system of gender roles began to collapse in the 1960s and was eventually supplanted by a more egalitarian system.

At the outset of the 1960s, women began their 'flight' from domesticity, seeking to redefine themselves as full participants in the economic and educational spheres of society. In 1951, only slightly more than 20 per cent of Canadian women of working age held paid jobs. By 1960, the proportion of women in the paid labour force had risen to almost 29 per cent, and it increased to more than 38 per cent in the following decade. With the advent of the 1980s and 1990s, the proportion of women working was fast approaching 60 per cent (Mills and Trovato 2001, 108; Romaniuc 1994, 221). The rise of female employment has been most pronounced among those in the prime childbearing ages of 20 to 34. Improvements in birth-control methods helped women gain greater control of their reproductive and productive lives (Davis 1984; Davis and van den Oever 1982; Murphy 1993).

The Baby Bust

Economists have enunciated theories to help explain the current low-fertility environment of post-industrial societies. A central postulate of economic theories of the family concerns the rising costs of parenting (Becker 1960; Willis 1987). Easterlin's cyclical theory (1969; 1980) incorporates economic and sociological factors. As described by Easterlin, the baby boom and the baby bust represent a natural sequence in a self-regulatory process whereby periods of low fertility give rise to periods of high fertility and vice versa. The driving forces of this cyclical pattern include economic and sociological forces—how well the economy performs in meeting the material aspirations of young adults, the size of one's birth cohort, and the strength of material preferences among young people of parental age.

Consistent with this thesis, the baby bust has resulted from the growing gap between the material aspirations of the baby-boom cohorts and declining socio-economic opportunities for this generation to satisfy their material goals. The numerically large size of the baby-boom cohorts represents a further source of insecurity, because large cohorts do not fare as well as small cohorts in finding permanent work and in advancing in the workplace (too much competition). In general, such conditions have led to delayed marriage and parenthood (Ram and Rahim 1993).

Following theoretical premises developed by Gary Becker (1960), William Butz and Michael Ward (1979) have suggested that low fertility rates in advanced societies in recent decades are linked to the rising value of time for women. Unlike in the past, most women are now gainfully employed in the labour market. This means that the economic value of their time is now greater. Having children therefore means having

(David Tanaka)

The period between the end of World War II and 1966 was one of high birth rates and came to be known as the baby boom. Since then, like many other industrialized countries, Canada has experienced a baby bust.

to forgo not only potential income but also career opportunities. Looking at the United States as a case study, Butz and Ward determined that a rise in the average incomes of males helps to promote increased childbearing among couples because of greater afford-ability. The same trend for women was found to be inversely associated with fertility, overriding the posi-tive income effect of men. This finding is explained in terms of women's rising opportunity costs, such as loss of income and career opportunities in the workplace associated with having children. Besides material considerations, there are also psychological costs—the perceived complications and stresses asso-ciated with having to care for children, often while maintaining a job or career, and the restricted freedom that goes along with being a parent.

Sociologically based explanations of fertility decline are grounded on the assumption that social change is largely a function of diffusion processes whereby new ideas and values gradually spread through-out the society and supplant old ones. Such theor-ies recognize the important interactions of structural and economic forces with ideational factors. The traditional sources of authority in matters of family and procreation—religion, community, extended family—have weakened considerably in recent dec-ades. Contemporary values, attitudes, and lifestyles seem incompatible with early marriage and raising large families as the pro-natalist forces of the past have given way to the small-family ideal of the present. During the baby-boom period, people gener-ally felt optimistic about their economic future, and they tended to marry early and to have more babies. Today, young people are generally sceptical of estab-lished institutions, including traditional marriage; therefore, cohabitation is seen as a less restrictive form of conjugal union. Diverse family forms are increasingly tolerated and accepted; it is no longer unusual for young couples to consciously avoid hav-ing children or for individuals to forgo matrimony altogether (Ariès 1980; Lesthaeghe 2010; Lesthaeghe and Surkyn 1988; van de Kaa 1987).

Demographic and Societal Implications of Sub-replacement Fertility

For a society, one of the most profound long-term outcomes of continued low fertility is demographic aging. The median age of the Canadian population—the point where exactly one-half of the population is older and the other half is younger—is now just under 40 years (Statistics Canada 2009b). In 1901, the median age was only 22.7 years (*National Post*

2002, A9). The Canadian situation is not unique, with other industrialized nations experiencing the same trend. For the world, the median age in 2010 was 28.9 (United Nations, Department of Economic and Social Affairs, Population Division 2009a, 27).

Figure 21.5 shows a typology of age pyramids, juxtaposed with Canada's actual and projected popu-lation age structure to 2036. This typology outlines in schematic form the long-term progression of expected change in the age pyramids of societies as they pass through their demographic transitions. The Canadian population is projected to attain the stationary form around 2036.

Aging societies are bound to face unprecedented pressure on their working-age populations as the proportion in the working ages shrinks while that of old-age dependants keeps growing. Such societies will have to rely increasingly on immigration to help supplement their labour-force deficits (Aysan and Beaujot 2009; Magnus 2009; Lee 2007; Bongaarts 2004; Denton, Feaver, and Spencer 2002).

Canada should more than adequately meet the challenges of an increasingly aging population. The main societal challenge will be to maintain and pro-mote for all age groups and future generations uni-versal principles of equity, freedom, and access to socio-economic opportunities. Demographic aging may turn out to be a problem to the extent that soci-ety fails to operate on these basic principles (Gee and Gutman 2000; Day 1992).

MIGRATION

The history of Canada is closely tied to immigration (C. Brown 2002). Almost one-fifth of the population are first-generation immigrants.

Immigration to Canada has fluctuated signifi-cantly over the years. In 1852, this country welcomed 29,307 people; in 1913, there were 400,000 new-comers—an annual figure not yet surpassed. Since the early 1970s, immigration levels have been in the range of 80,000 to 250,000 per year. From the time of Confederation to 1967, more than 8 million people were admitted to Canada (McVey and Kalbach 1995, 83). Until recently, most immigrants were from Britain, other northwestern European areas, and the United States. After World War II, immigration started to diversify, and Canada welcomed many people from all regions of Europe, including Italians, Germans, Hungarians, Dutch, and Portuguese. Since the early part of the 1970s, the major immigration sources to Canada have been Asia, South America, the Caribbean, Central America, and eastern Europe.

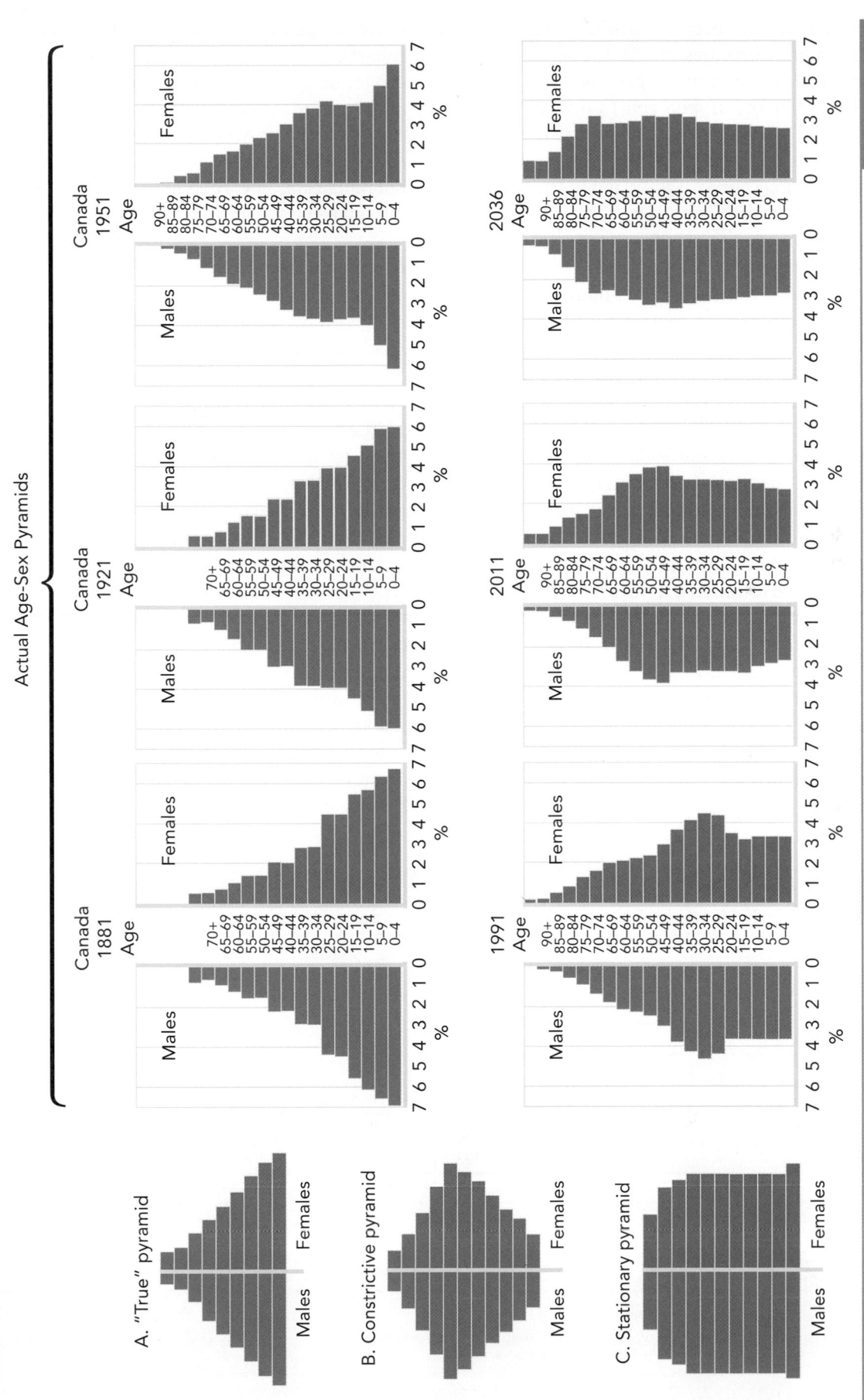

▼ **FIGURE 21.5**   Typological Age-Sex Pyramids and Actual and Projected Age-Sex Pyramids of Canada, 1881, 1921, 1951, 1991, 2011, and 2036

Actual Age-Sex Pyramids

A. "True" pyramid

B. Constrictive pyramid

C. Stationary pyramid

SOURCE: Adapted from Bertrand Desjardins, 1993, *Population Aging and the Elderly: Demographic Analysis*, p. 89 (Chicago: Dorsey). 2nd edn, p. 18 (Ottawa: Statistics Canada); C.W. Kammayer and Helen L. Ginn, 1986, *An Introduction to Population*,

Immigration has been a significant factor in the growth of Canadian cities (Stone 1967). The majority of immigrants settle in the largest metropolitan areas of Ontario (about 50 per cent), British Columbia (15 per cent), and Quebec (15 per cent). This order of preference has remained fairly stable over time (Bélanger 2006; Dumas 1990); it reflects variations in economic and social pulls across Canada's regions, with the more economically advantaged locations receiving the majority of immigrants. In 2006, more than one-third of the Canadian population was concentrated in the **census metropolitan areas** (CMAs) of Toronto (5.1 million), Montreal (3.6 million), and Vancouver (2.1 million) (Statistics Canada 2007c). Along with Ottawa-Gatineau and Calgary, these CMAs welcomed 77 per cent of those who immigrated to Canada between 2001 and 2006 (Fong 2010, 135).

Although the volume of interprovincial migration has been declining since the 1970s, the number of people in Canada who change province annually remains substantial. Almost 18 million interprovincial moves were recorded between 1951 and 2005 (see Table 21.3). Internal migration is predominantly from the economically disadvantaged to the economically prosperous regions—i.e., a core/periphery phenomenon whereby underdeveloped regions (the periphery) export people, and economically developed areas (the core) import labour (Massey et al. 1993). In the Canadian case, the Atlantic region has been a net loser in terms of migration, and throughout different periods, Ontario, British Columbia, and Alberta have shared dominance as net gainers. As economic conditions across these regions fluctuate, boom periods translate into large net migration gains, and economic downturns provoke high rates of out-migration. Over the interval 2001–5, because of its booming economy, Alberta had the largest net migratory gain (i.e., 117,067 people).

Factors of Migration

One of the most predictive variables for migration is age—a variable that is strongly correlated with lifecycle stage and family circumstances (Mincer 1978). The intensity of lifecycle events is greatest in the young adult years, between the ages of 18 and 34 (Lee 1966; Shaw 1975; Ritchey 1976). In the late teens up to the early thirties, this stage of the lifecycle involves a number of important life events for most people, such as graduating from high school, attending university or college, post-secondary graduation, entering the labour force, and marriage. These types of events often dictate having to relocate (for example, for a job in another city). The likelihood of migration diminishes considerably after the late thirties and early forties, since during this part of life people are established with work and family. Beyond this point in the lifecycle, researchers have observed a tendency for migration propensities to increase again in old age. This is most often associated with seniors moving to retirement communities and with

TABLE 21.3 ▼ Interprovincial Migratory Balance in Canada, 1951–1960 to 2001–2005

	1951–60	1961–70	1971–80	1981–90	1991–2000	2001–5
Newfoundland	−9,816	−34,557	−20,840	−30,626	−52,404	−15,352
Prince Edward Island	−7,938	−5,732	2,927	378	1,608	61
Nova Scotia	−28,851	−43,521	4,165	3,331	−8,827	−7,333
New Brunswick	−25,360	−45,277	6,441	−5,915	−10,363	−6,930
Quebec	−72,877	−142,594	−234,163	−122,143	−122,267	−20,529
Ontario	148,036	236,081	−96,391	184,649	15,128	−12,109
Manitoba	−40,587	−64,161	−68,977	−37,968	−46,121	−23,371
Saskatchewan	−87,938	−123,492	−50,603	−67,475	−51,548	−36,385
Alberta	32,858	30,022	244,991	−61,203	140,223	117,067
British Columbia	93,075	192,713	216,486	144,345	142,156	6,546
Yukon and Northwest Territories[a]	−600	519	−4,036	−7,373	−7,585	−1,665
Total movements	2,962,004	3,660,061	3,849,741	3,168,426	2,954,321	1,379,745

[a]Includes data for Nunavut beginning in 1992.
SOURCES: Bélanger 2002, 58; Dumas 1990, 106; Statistics Canada, 2008, 'Report on the demographic situation in Canada: 2005 and 2006', Catalogue no. 91-209-X, www.statcan.gc.ca/pub/91-209-x/2004000/part1/t/t5-1-eng.htm.

those of advanced age resettling in their original communities and rejoining their adult children (Warnes 1992; Silverstein 1995).

TIME to REFLECT

Should the government of Canada increase its immigration targets from their current levels? Can you think of reasons why immigration targets should not be increased?

Conclusion

Population is the foundation of society and its subsystems. The social world cannot be understood in isolation from demography, nor can demography be properly understood devoid of a proper understanding of society. Given the centrality of population to the study of society, sociologists pay particular attention to the dynamics of population change. Fertility, mortality, and migration are the basic demographic variables. In combination, they determine whether a population experiences growth, stability, or decline.

Aggregate change in fertility, mortality, and migration results from change in individual behaviour in accordance with social, cultural, and economic conditions.

The history of the human population can be subdivided into two broad stages: a very long period of very slow growth from the beginning of humankind to about 1750 CE, followed by a relatively recent phase of explosive growth. As a legacy of the population explosion that took place in the modern era and is currently ongoing in many developing countries, the world faces challenges heretofore unforeseen in the history of humanity. The Earth's population (just over 6.8 billion in 2010) is expected to surpass 9 billion by 2050. Most of the projected growth will be in the developing countries, particularly in Africa. Some developing countries are close to completing their demographic transitions; others have recently begun fertility declines after decades of mortality reductions.

It seems inevitable over the course of this new century that the population of the world will become older, developing countries will grow much more than the developed countries, and there will be intense pressure on highly industrialized countries to accommodate an even larger share of immigrants.

questions for critical thought

1. Describe how populations change. What are the demographic components of change?

2. What is atypical about the current stage of the demographic history of the world? Describe the relationship of demographic transition to the history and projected future of the human population. Are there any certainties about the future population of the world?

3. Assess the perspectives of Malthus and Marx on the matter of population and its relationship to contemporary issues concerning resources and the environment. Discuss how contemporary perspectives on population matters relate to the classical Malthusian and Marxist theories of population.

4. What are some of the most important sociological determinants of fertility change over the past half-century in Canada?

5. What is the relative importance of fertility and mortality in determining the age distribution of a population?

6. Explain how the epidemiological profile of Canadian society today differs from that of the early part of the century.

7. Why do immigrants to Canada prefer to settle in the largest metropolitan areas of the country?

8. Regional inequalities in socio-economic development explain differential migratory flows within Canada. Evaluate this proposition.

9. What are some of the social and economic implications of increased immigration for the future of Canadian society and other receiving countries in the Western world?

recommended readings

Roderic Beajot and Don Kerr. 2004. *Population Change in Canada.* **2nd edn. Don Mills, ON: Oxford University Press.**
This important book focuses specifically on the demography of Canada.

Stephen Castles and Mark J. Miller. 2009. *The Age of Migration: International Population Movements in the Modern World.* **4th edn, revised. New York: Guilford Press.**
Castles and Miller have produced one of the most authoritative analyses of international migration trends and their far-reaching socio-economic ramifications.

John I. Clarke. 1997. *The Future of Population.* **London: Phoenix.**
This book is an excellent brief introduction to population.

David K. Foot with Daniel Stoffman. 1998. *Boom, Bust and Echo 2000: Profiting from the Demographic Shift in the New Millennium.* **Toronto: Macfarlane Walter and Ross.**
The revised edition of this best-selling book examines many potential practical applications of demography to business.

Wayne McVey, Jr, and Warren E. Kalbach. 1995. *Canadian Population.* **Toronto: Nelson.**
This is an updated version of a classic Canadian text on population first published in the early 1970s.

George Magnus. 2009. *The Age of Aging: How Demographics Are Changing the Global Economy and Our World.* **Singapore: John Wiley and Sons, Asia.**
This is a current and highly readable study of global demographic aging and its social and economic consequences.

Malcolm Potts and Thomas Hayden. 2008. *Sex and War: How Biology Explains Warfare and Terrorism and Offers a Path to a Safer World.* **Dallas, TX: Benbella Books.**
The authors argue that by understanding that war is part of humanity's biological nature, we can learn to make conflict less likely in the world, since peace and cooperation are also part of our biological roots.

Anatole Romaniuc. 1984. *Fertility in Canada: From Baby-Boom to Baby-Bust.* **Ottawa: Statistics Canada.**
Although this study's primary focus is Canadian fertility, the theories and explanations reviewed by the author have broader application to the other industrialized countries.

W.W. Rostow. 1998. *The Great Population Spike and After: Reflections on the 21st Century.* New York: Oxford University Press.
The author takes a careful look at the populations of industrial and developing countries, how they have been changing, and how they will change over the course of the new millennium.

Jeffrey D. Sachs. 2008. *Common Wealth: Economics for a Crowded Planet.* New York: Penguin.
Sachs's book is a state-of-the-world report on the interrelationships of poverty, climate change, and the environment.

Alan B. Simons. 2010. *Immigration and Canada: Global and Transnational Perspectives.* Toronto: Canadian Scholars' Press.
This is an important book that focuses on Canada as well as on the international view concerning immigration and its social, economic, and demographic implications.

Frank Trovato. 2009. *Canada's Population in a Global Context: An Introduction to Social Demography.* Toronto: Oxford University Press.
This textbook introduces the field of social demography—its concepts, measures, and theories—with a focus on both Canada and the international demographic scene in both historical and contemporary perspectives.

recommended websites

Health Canada: Population Health
www.hc-sc.gc.ca/hppb/phdd/docs/social
Mortality and other population health information on Canadians can be found at the Health Canada website.

POPLINE Digital Services
www.jhuccp.org/popline
POPLINE is an important online bibliographic database on population research and related topics, based at Johns Hopkins University.

Population Reference Bureau
www.prb.org
The Population Reference Bureau issues a series of excellent publications on a regular basis, including *Population Bulletin* and the *World Population Data Sheet*.

Statistics Canada
www.statcan.gc.ca
Statistics Canada is the authoritative source for Canadian census and other population data.

United Nations Population Division
www.unpopulation.org
The annual *United Nations Demographic Yearbook* contains a wealth of demographic information by country. Another important UN publication is the *Human Development Report*.

United States Bureau of the Census
www.census.gov
This is one of the best sources of demographic information for the United States and other countries. Check out their population clock.

Vienna Institute of Demography
www.oeaw.ac.at/vid/index.html
The Vienna Institute of Demography publishes studies on a variety of demographic topics. Among their publications are *The European Fertility Data Sheet* and the *Vienna Yearbook of Demographic Research*.

World Health Organization
www.who.int/en/index.html
The *World Health Organization Statistics Annual* contains deaths by cause, age, and sex for a large number of countries.

Worldwatch Institute
www.worldwatch.org/
The Worldwatch Institute researches global issues of climate change, resources, population growth, poverty, and their interrelationships. See their annual *State of the World* publications.

chapter 22

Cities and Urban Sociology

Louis Guay and Pierre Hamel

In this chapter, you will:

▶ Be introduced to cities (the things out there) and to urban sociology (the words and ideas you will need to understand the things out there);

▶ Aim at an understanding of what cities are, the problems they encounter, the challenges they face, their diverse social groups and activities, forms, infrastructures, and constructions;

▶ Come to understand how sociology and other urban disciplines approach complex systems like cities;

▶ Learn what the industrial city was and how its problems were tackled;

▶ Discuss the new reality of metropolitan areas;

▶ Learn how the environmental question is also an urban question and whether cities are on the path to sustainability;

▶ Understand that city development is a political process and that there are opposing views of how cities should be managed and planned.

Introduction: Understanding Cities

Cities may be among the most complex of human inventions. Almost from its very beginning, sociology has investigated cities as a prime instance of the modern industrial world in the making: their workings, their intrinsic characteristics, their multiple relationships with the countryside and other cities, their governance structures, and, not least, their production of different forms of internal inequality.

What are the main analytical approaches of the sociology of cities? The first is what the French Durkheimians called morphology (materiality, in today's terminology) (Halbwachs 1938). A city is a material phenomenon, with an occupied geographical space and a distribution of people and activities on the ground, but also with erected and constructed structures such as buildings, monuments, infrastructure, and equipment. The morphological approach was attentive to demographic structures as well and considered the economy and technology as morphological variables. Sociology has not devoted too much effort toward understanding urban materiality as such, except in more recent times with the advent of the social construction of technology perspective (Aibar and Bijker 1997; Coutard and Guy 2007).

The city as an economic phenomenon has a long tradition of research. Max Weber (1958 [1921]) considered the Western city an autonomous entity capable of ruling itself, especially economically. It is more recently, however, that the city as an economic phenomenon has become a leading approach in urban sociology. The strong influence of Marxism on urban scholars clearly must be acknowledged (Castells 1972; Dear and Scott 1981; Harvey 1973, 1989; Scott and Soja 1996; Soja 2000).

The leading school of urban sociology has without doubt been the Chicago School (Park and Burgess 1967 [1916]). For about 30 years, the school reigned supreme in urban sociology. It focused on many issues, problems, and features of cities and urban life. Park, one of the school's leaders, wrote an entire program of urban investigation that was gradually and over many decades followed not only by sociologists but by many researchers in urban studies. The school is famous for its interest in urban culture as a whole, contrasting it with rural culture, and in the relationship between ecological variables, such as size, density, and heterogeneity, and the urban way of life (Wirth 1938). Whether there is one overall urban culture or many sub-urban cultures has long been debated, and the Chicago School itself has contributed to the study of urban social areas (Davies and Murdie 1991; Robson 1975; Timms 1971).

TIME to REFLECT

Traditionally, sociology has been tied to modernity. But it is linked as well to cities and their main issues. Why were the pioneers of sociology from the outset interested in cities and urban problems?

Finally, another major approach to cities and their dynamics is to consider their political dimension, which is manifold. It entails governance of urban problems, administrative structures, and the distribution of power in a city. In this category one should include urban planning and urban social movements. The recent theory of **urban regimes** is an interesting evolution of the long tradition of political analysis that Park hoped to launch but was not pursued much by other sociologists (Stone 2005; 2006).

Urban Sociology: Past and Present

THE EMERGENCE OF URBAN SOCIOLOGY

The history of urban sociology is tied to sociology itself in a number of ways. At the outset, urban sociology adopted sociology's general categories of thinking. This is because it is mainly in reference to modern cities that social transformations occurring in industrial and modern societies can be examined. It is not surprising, then, that on both sides of the Atlantic, the first sociologists—we refer here to Max Weber and Georg Simmel as well as to Robert E. Park—were fascinated by the city. If sociology was created above all to understand social changes in the making, one can say that under modernity, these changes were taking place above all in the fabric and culture of the city.

Robert E. Park coined the expression 'urban sociology' in 1925. Dedicated from its inception to the study of social problems in cities, urban sociology took more than two decades to be recognized in sociological discourse (Topalov 2008). But since the 1950s, urban sociology has expanded rapidly both in the United States and elsewhere in the world. It has developed as an intellectually coherent research field and as an area of specialization.

The place of the city in the view of the first European sociologists was influenced by the Western city; the notion that served to designate this place was eventually transformed to mean 'large urban place' (Johnston, Gregory, and Smith 1994, 67). Despite its vague nature, this latter notion captures the characteristics of the classic city—namely, a relationship to the central core and the presence of common spaces.

However, in the discourse of early sociologists, the city was first and foremost a reflection of social transformations. They were interested in the city in order to understand and explain the changes at work in modern urban and industrial society. While the European tradition of sociology, on the one hand, and the North American tradition, on the other hand, raised specific distinct social concerns (Bash 1995), these

In the First Person

Choosing sociology at the end of 1960 was both going against the grain and being in the mainstream. Sociology was not taught at the college level and was a fairly unknown discipline for most students entering university, although some dared to try it. There were great changes in the social and political environment in North America and Europe, which were the two geographical areas a young student living in Montreal was chiefly exposed to in the 1960s. If one was open to these changes and wanted to better understand them, sociology was a natural choice.

What did I learn from sociology? I didn't choose sociology for some political involvement, as many other young sociologists did. If sociologists are often on the side of the underdogs, as some have said, they are on the side of understanding society in all its diversity. I still believe that sociology is a great avenue toward understanding the great variety of social forms, conditions, and changes. This deep belief may come from my minor in anthropology, in which I discovered how diversified human beings and human societies are.

I was particularly attracted to urban sociology. Reading Weber's *The City* and the works of the Chicago School of sociology is, I still believe, a great intellectual experience. But I also had a strong interest in quantitative analysis, which I got into in my PhD research in the tradition of Chicago's urban ecology. I left urban ecology for the sociology of urban planning, because after my studies at the London School of Economics, I believed that urban ecology and urban sociology needed to focus more on the role of planning in shaping urban social relations. British urban researchers—well-known figures such as Peter Hall and Ray Pahl, for example—were influential in the evolution of my urban research interests.

Throughout my doctoral studies, I may have been of two minds, for I read a great deal in the sociology of the sciences, discovering a specialty that I had not studied previously. When we studied Merton, we did not cover his sociology of science. Thus, I hesitated between urban sociology and sociology of the sciences, which I had discovered in the UK, and have since read about and used its perspective in my own research in environmental sociology. I was, and still am, more interested in the relationships between science and society, or science in society, than in the kind of sociology of science that many current sociologists of science do, which may be described as society in science.

—Louis Guay

I was trained in sociology through small doses. It was as though I came to sociology without noticing it. In fact, at the outset I was not certain that I had made a good choice when I decided to undertake university studies in philosophy and literature. But this choice partly explains why I was fascinated by the place of the city in literature and especially in James Joyce's *Ulysses*. Because I wanted to better understand the production of the modern city and its distinctive social life, I chose to do graduate studies in the field of urban planning. Subsequently, I taught planning theory and social aspects of urbanism in a school of planning for 25 years before moving to the Université de Montréal's Department of Sociology. I have always had a strong concern for social justice. This is probably what engaged me in the contested field of sociology and concomitantly in urban sociology and in the study of collective action and social movements.

—Pierre Hamel

two traditions nonetheless converge with respect to the institutionalization of urban sociology.

TIME to REFLECT

Urban sociology as conceived by Robert E. Park took a long time to be recognized as a specialized field. And after it was well integrated in academia, it was challenged by urban studies. Does that mean that sociology is no longer useful in better understanding city life in the making?

Following Saskia Sassen, even though the 'study of cities was at the heart of sociology' (Sassen 2005, 352) in the first half of the twentieth century, it has since been replaced by other concerns. The early European sociologists were also troubled by the possible negative impacts of mass urbanization on social life:

> Mass urbanization was neither progressive nor liberative, but signified a degeneration of social existence. From [the] work [of early European sociologists] stems the perennial theme of the loss, the eclipse of community, and the arrival of the mass society in which political life, culture and personality are in decay. (Mellor 2007, 172)

These sociologists were concerned above all with the moral degeneracy that could come with urban development. This in any case is the reading that Nisbet (1966) made of the intent of the discipline's founding fathers. As a specialized discipline, sociology has nonetheless been able to construct theoretical and methodological instruments with enhanced heuristic range and explanatory capacity for use on empirical phenomena, in comparison to the reflections on humanist thought that preceded it.

Recent urban sociologists have partially turned to these tools once again, reconnecting with a certain critical tradition. Nowadays, even though the changes at play are different, the critical concepts of sociology can continue to provide a better understanding not only of what is happening in the cities but, beyond that, of the forms of social structuring (Sassen 2005). In that respect, the city and, more broadly, the urban phenomenon remain an ideal location for examining the social transformations at play in the process of modernity.

Cities are once again at the top of the public agenda and, to a certain extent, the scientific agenda. This is because the socio-spatial configuration of the city has changed a great deal. The spread-out form that characterizes so many current agglomerations has led to the creation of city-regions encompassing various administrative units over a fairly large territory. In terms of management and/or planning, this development is raising questions that were ignored in the past regarding the co-ordination of public and private interventions at the metropolitan or city-region scale and also with respect to political responsibilities and issues of democratization. Although this development was foreseen by Patrick Geddes (1949) with his concept of conurbation in the mid-twentieth century, governing city-regions is more complex and has to include new concerns about environmental problems, the integration of immigrants, and social justice in terms of access to public services, as well as the capacity to distribute the tax burden for common services over a large territory in an equitable way.

URBAN STUDIES AND URBAN SOCIOLOGY

Urban sociology existed before it was formally recognized as such by academia. However, it took a long time for urban sociology's supremacy to be contested by a new field of study called urban studies. Urban studies goes back to the 1950s, when American cities were facing a series of problems resulting from developments that followed the Second World War. Urbanization and the problems it engendered were at the centre of the political agenda of the American federal government. By getting involved in financing infrastructure, urban renewal, and social housing, the federal government contributed to stimulating research in urban affairs and urban studies. This also encouraged the recognition—indeed, institutionalization—of urban sociology. Similar developments occurred in Canada, Europe, and a host of developing countries.

Urban sociology reigned supreme over urban studies until the 1970s. The criticism it received at the beginning of the 1970s because of its theoretical deficiencies (Castells 1972) did not diminish support for urban sociology. Nonetheless, the rise in influence of urban studies in the 1960s as a result of the creation of a number of specialized academic journals, as well as the formation of networks of international researchers in the field, ultimately challenged the supremacy of urban sociology.

Yet we must remember that urban sociology remained an influential partner in the progress of urban studies. However, some think that urban sociology has seen its leadership wane during the past 20 years (Perry and Harding 2002). But did urban sociology's purported decline in leadership in the field of urban studies indicate a drop in its theoretical, social, and political importance? If it is true that the most important influences in the field of urban studies, as some claim (Savage 2005), no longer come from urban sociology, is this a source of concern?

From the start of the 1970s until the middle of the 1980s, the 'new urban sociology' that attacked the spatial determinism of the first urban sociologists in the Chicago sociological tradition allowed for a renewal of the debate on the city and its issues in striking fashion. Manuel Castells (1972) opened hostilities and provoked a genuine academic debate on the underlying nature of cities and on what 'urbanism as a way of life' means. Aware of some contradictions in capitalism and defining the city around issues of collective consumption, Castells formulated the hypothesis of a possible common ground between the interests of the working class and those of the middle class with respect to problems of housing, transportation, mobility, and access to the urban centrality and to community services. This led him to pay particular attention to urban social movements opposing political elites and raising questions about urban policies.

Castells's critique, inspired by both the structural Marxism of Poulantzas (1968) and the sociology of action developed by Alain Touraine (1965), contributed to the reopening of the theoretical debate on the city and urban sociology. It also fostered a series of empirical studies on social relationships to space, urban policies, and urban movements.

Castells was one of the first to contribute to the emergence of the 'new urban sociology', but he was not the only one. A number of other researchers also furthered its development. Their contributions include the work of Henri Lefebvre as well as the research, often empirical, of Jean Lojkine, David Harvey, Ray Pahl, and Peter Saunders. 'New urban sociology' attracted the attention of researchers examining the transformation of contemporary cities and the economic, political, and cultural processes at work. Nonetheless, it did not manage to remove all the ambiguities surrounding the status of urban sociology and its place within the urban studies field. As Pickvance (1994, 127) has said, 'Urban geography, urban politics, and urban sociology overlap considerably—hence the idea that urban studies is more appropriate than any disciplinary label.' Among researchers in urban studies, there seems to be consensus that in recent years, urban sociology is no longer dominant in research on the city (Perry and Harding 2002).

In joining the field of urban studies, sociology can benefit from other disciplines' contributions. Incorporating theoretical and methodological tools from elsewhere does not imply a questioning of sociology's relevance and its particular contribution. We might even think that it fosters sociology's dynamism rather than representing a threat (May and Perry 2005).

The progress of urban sociology is still ongoing. Despite the arrival of urban studies at the forefront of research on the city, sociology and especially urban sociology continue to provide research tools and elements of knowledge vital to the understanding of modern cities' transformations. In other words, as has been the case in the past, the content and the future of urban sociology are linked to the changes that are taking place in cities.

In the aftermath of the Second World War, North Americans experienced a period of prosperity over more than 20 years that would change 'city lives and city forms', to borrow the title of the book edited by Jon Caulfield and Linda Peake (1996). The transition from a military economy to an economy structured around consumer goods proceeded smoothly. Households had easier access to credit and were able to improve their living and housing conditions. With the help of federal programs in both the US and Canada, a growing number of people chose to become homeowners, and many decided to move out of cities and into suburbs. In addition, the administrative and financial assistance of higher levels of government in the building of highways connecting the new suburbs to the centre also played an important role. This was one of the main factors explaining the large movement of middle-class households to the periphery. The consequences for cities were dramatic in terms of their ability to finance infrastructure and services, accelerating decay in low-income neighbourhoods.

From then on, cities had to face new problems and new issues. The households and businesses that chose to remain in the city had to pay higher taxes for services of lower quality. As the economy became more and more service-oriented, the urban fabric became a factor of production. The quality of urban services and facilities is playing a greater role in attracting investments and people in the new economy. On the east coast of the United States in particular, as well as within old industrial areas in Canada, city centres that were once booming and vibrant tended to undergo a process of decline that made things difficult when it came to improving the built environment (Guay and Hamel 2010). Furthermore, the transition to a knowledge economy proved to be problematic for many of these cities (Hess 2009). In this regard, the problem was no longer adapting to growth and urbanization and regulating their negative impacts, but rather the reverse: stimulating job creation and finding incentives to keep households within the city limits.

The transformations at play were considerable. Central cities tend to lose the advantages they used to have in terms of access to centrality as multiple centres of commerce and services emerge at the periphery. Even though centrality is still connected to specific spatial forms, it is no longer necessarily the historic core that assumes this function. In fact, as city limits are blurred, the many roles the city centre plays are being performed by several nodes dispersed in the territory of the city-region or the **metropolis**. This is the main reason why proponents of the Los Angeles (LA) School of **urbanism** contend that the 'urban question is radically changing' (Dear and Dahmann 2008, 275).

The LA School of urbanism's understanding of cities is multidimensional. Its advocates combine a cultural analysis with economic, geographic, sociological, and political accounts of the city.

From another perspective, one can ask to what extent economic and cultural processes at play in the spatial organization of a city like Los Angeles were not already simmering under the driving forces that produced industrial cities. Is it accurate to equate Los Angeles with a new spatial form corresponding to the social relations of a post-modern era, while Chicago remains stuck in an earlier modernity (Sassen 2008)?

If this discussion overlaps with the debate between modernity and post-modernity, it cannot be reduced to it. This is mainly because with cities and urban forms, we are facing a very dynamic reality, an object in continual transformation. The clash between modernity and post-modernity, on the other hand, is not precisely bound in time, even if the two categories can sometimes be associated with specific periods.

What was the industrial city? What were its problems, challenges, and preferred solutions and policies? The next section looks at the evolution of the industrial city, roughly from 1820 to 1920.

The Rise and Demise of the Industrial City

With the coming and unfolding of the Industrial Revolution, cities grew considerably. An urban industrial model took shape, and the continuing, almost endless, process of first industrial and then world

SOCIOLOGY in ACTION

The Chicago School

The Chicago School of sociology initiated a strong empirical tradition in urban research. Chicago was seen as a social laboratory, spurred by a growing and dynamic urban environment. All industrial cities had developed some kind of research based on themselves. But the Chicago school was more systematic.

It fostered an empirical tradition in which observation, ethnography, quantitative data analysis, and mapping were the preferred tools of the urban sociologist's trade. Sound theoretical propositions had to be grounded in solid empirical observations and data-gathering. Urban studies researchers have all benefited from the pioneer work of the Chicagoans. Walking in a city, one is struck by its great variety. This urban diversity was admired and defended by Jane Jacobs (1916–2006). Jacobs was an American thinker who lived in Toronto and was a key intellectual figure in urban planning and studies, criticizing the modern movement in planning for its lack of proper consideration for urban diversity.

What can be learned from this? Urban sociology and studies are not abstract: they relate to a concrete world of differences and change. Cities are constantly changing: planning decisions and policies change the physical aspects of cities; new migrants concentrate in a few neighbourhoods, which gradually change; large infrastructure deeply changes the urban landscape, and many people are awestruck by new and daring works of architecture. But at the same time, one can be concerned about certain kinds of change: urban motorways have been opposed in many cities; demolition of the architectural heritage has been fiercely contested; poor living conditions in degraded housing are not something urban residents are proud of. No students of cities can afford not to go and see for themselves the different social and physical areas making up a complex city. Direct observations can be combined with official statistics on where people live and who they are. These statistics can be mapped to show wide and deep differentiations in large urban areas. Finally, one should participate in local hearings where planning decisions are prepared and discussed publicly.

urbanization has barely stopped (Bairoch 1995). However, sheer numbers are poor instruments for understanding urban life. Urbanization is more than demographic and spatial change. Rather, it is a social change process in full bloom or, in the words of Marcel Mauss in his essay on the gift published in 1923 and 1924, the urbanization process is a total social phenomenon (Mauss 1978, 274).

URBANIZATION AS A TOTAL SOCIAL PHENOMENON

The double idea of functional differentiation and type of social relationship, which marked classical sociology, was taken up by the Chicago School of sociology and in particular by Louis Wirth (1938) in a celebrated article titled *Urbanism as a Way of Life*. Wirth's essay marked a significant point in the school's history. It defined ecological variables and their consequences on human behaviour. These variables are: size, density, and heterogeneity.

People living in cities chiefly orient their action to other people according to functional needs. Exchanges and social transactions are particularistic and specific and aim at immediate reciprocity governed by rational decisions. Ties and boundaries that are built on family or group solidarity are not as important. For the modern machine to work, people have to accept that weak ties rule their social interactions.

In Wirth's essay, these are irreducible modern facts that no one has deliberately planned, but they remain the most important characteristics of a modern urban society. Although functional differentiation and individualism may serve modern-world ambitions, they are gained at a cost and at a loss. Durkheim, for example, was afraid that anomie and decline in community attachment might provoke great ills, such as deviance, criminality, violence, and other social problems.

The contrast between rural, pre-modern society and the urban, modern one is built on an ideal-type methodology. It involves opposing two abstract cases that can never be met in reality. No city is purely functional; many villages and the countryside are generally differentiated by occupations, by classes, and by a small, or shallow, division of labour. Pre-industrial cities had their own cultural and ethnic areas: 'Added to the strong ecological differentiation in terms of social classes, occupational and ethnic distinctions are solemnly proclaimed in the land use patterns' (Sjoberg 1960, 323). All large cities, perhaps all cities of a certain size, have some sort of economic and social differentiation. Cities of the first Industrial Revolution were socially fairly homogeneous. Only the large industrial cities, like Montreal or Chicago or London, had a rich and diversified population. Moreover, large commercial cities, in the past as in the present, have perhaps hosted a greater cultural diversity than industrial cities as such. But these cities were few. Finally, one should not overlook the fact that imperial, royal, and capital cities of the past were always ports of entry for many people who gave them their creativity (Mumford 1961).

This diversity is wrongly seen as an ecological factor; however, diversity breeds diversity, may be self-perpetuating, and is more a social process than an ecological one. There may not be one urban culture but many, as some sociologists have shown (Fischer 1975, 1984, 1995; Gans 1968; Borer 2006; Sharp 2005). Within a broad urban culture, if it exists at all, there may be many urban subcultures, each one coexisting with the others without mutual penetration, as Robert Park observed long ago (Park and Burgess 1967 [1916]). Recent immigration to Canada demonstrates this very clearly (Charbonneau and Germain 2002; Ley 1999).

THE INDUSTRIAL CITY AND ITS PROBLEMS

Urban problems during this period focused on three urban conditions: public health and sanitary conditions; congestion and urban mobility; housing and urban poverty. Each problem had its own evolving context and set of causes; each had its own social actors and movements that pleaded for them to be solved; each gave rise to a series of innovations, technological, institutional, and administrative. Three main collective actors, jointly or independently, focused on urban transformations. They were: (1) the urban **social movements**, which focused on the industrial question, so to speak; they were concerned with working conditions and wages but were not blind to urban conditions, as Castells (1983) has shown; (2) a variegated professional movement composed of new professions, or new specializations within existing professions, in medicine, education, social services, engineering, and the fledgling urban planning profession; (3) a political and administrative movement that was instrumental in long-term urban transformation.

Why have sanitary conditions been so important in industrial cities? Why have they at times overshadowed other public concerns and political issues? Why have they mobilized so many people and resources? Today, the answers seem obvious. Sanitary conditions—namely, urban population health, waste disposal, air and water quality, and wastewater treatment, to name just the most important—could be deadly or healthy. People can die or live depending on sanitary conditions. The industrial city was plagued with diseases and experimented with new public interventions and methods of disease prevention. For instance, some diseases like cholera, which struck urban populations late in the nineteenth century, were attributed to 'miasmas': foul odours emanating from decaying organic matter. Then medical officials and chemists, by carefully observing the open water supply in different areas of industrial and large cities, demonstrated that the cholera microbe (a bacterium) was borne in unclean water and that odours had nothing to do with the disease, even though they were unpleasant (Hamlin 1990). This bacterial revolution (Melosi 2000; Porter 1997) came about gradually and not without opposition from some leading authorities. What helped public medicine to combat such resistance was that, as Weber and Simmel have shown, a rationalistic mentality had been widely adopted, at least among educated people and certainly within professional groups.

The organization of public health raised challenging problems of collective co-ordination. Churches in Canada, but elsewhere as well, had been prime actors in public health. Private charities played a part as well. Progressively, the responsibility passed to central states, but in the process of this takeover, municipalities, for obvious reasons, came to play a leading role. Canadian cities started a small administrative revolution on the sanitation and public health front (Turmel and Guay 2008). To do so, they had to change their own way of doing things: administrative reform was called for (Dagenais 2000; Germain 1984). There was recognition that a movement had begun and that it was, as Porter has written, the end of a very long era of 'people-killing': 'By 1900 civilization had lost its biological population check: infectious disease' (Porter 1997, 427). Combined with other urban improvements, the urban health revolution was the product of many factors, many social actors, and large and small actions.

In the industrial city, housing was mainly a private and a market choice (Sutcliffe 1981). State intervention was considered unwarranted except in cases where the very poor and the sick had to be housed and cared for (Hall 1980). But a fast-growing city is prone to speculative behaviours. Crowding was endemic in industrial cities. Pre-scientific surveys could, at first more or less rigorously, help to evaluate the extent of the problem, but with the passing of time these surveys improved their methodology. When Booth began his more than decade-long investigation of housing and working-class conditions in London, his survey methods were innovative and clearly superior to journalistic observations such as the ones Ames made in Montreal in *The City below*

A civilian group unofficially opens the 'W.R. Allen Park' in protest against the proposed Spadina Expressway in Toronto. Public opposition to the expressway resulted in its cancellation in 1971.

the Hill or the ones Jacob Riis made in New York in what became a best-seller, *How the Other Half Dies* (Methot 2003; Scott 1969, 6–10). Surveys of urban conditions were also part and parcel of the emerging social sciences, notably sociology, and of administrative modernization (Abrams 1968; Geddes 1904).

What were the public responses to these conditions? Altogether, there were three responses. One was to do nothing, or very little, and let private actors take the lead. The second was regulatory: enact regulations and by-laws to limit crowding in a dwelling or a room and to prohibit some behaviours with dire environmental consequences, such as disposing of waste in the backyard or in the streets. Finally, the third response was to build public housing for the poor and the sick. But there was a large divide between European cities and North American cities in this respect. European cities were a little quicker to use the ways and means of public action, whereas North American cities generally relied on a market solution (Scott 1969). As Peter Hall (1988, ch. 2) has observed, the problem ('an international problem of the giant city itself', p. 44) was clearly seen for the first time; people knew it existed, but mindsets were slow to change, in part because cities had insufficient means for collecting the resources needed for a housing policy (Sutcliffe 1981).

TIME to REFLECT

The problems of industrial cities were staggering. Health issues, for example, came rapidly to the forefront. Why was it so important to take action on these matters?

Public amenities in the industrial city were not at first a great concern. However, Robert Owen and the philanthropic industrial owners were aware of the problem and concerned about public amenities and facilities. Should industrial towns offer their chiefly working-class populations collective goods in the form of parks, green spaces, libraries, theatres, concert halls, sports facilities, and the like? Or should they focus on private wealth creation and profits distribution? Should public amenities come out of economic growth and personal wealth, or should governments be involved? Again, we can discern a contrasting pattern between Europe and North America. North American cities, although somewhat less so in Canada, relied heavily on private initiatives for the funding and selection of projects. In Europe,

governments were more active, and a great number of public amenities were initiated and funded by public money. One area on which public officials on both continents agreed was that green spaces and urban parks should be publicly provided (Benevolo 1993, 200–1; Scott 1969, 10–15). Many, if not virtually all, industrial cities designed in the second half of the nineteenth century had at least one large park and a few smaller ones. The reason for designing parks was to meet a combination of health and recreational needs. Many urban parks originated out of the public health movement. Urban residents needed open green spaces for a healthy life. But recreation was also important, and parks proved a valuable addition to city life when the bicycle was invented and adopted for sports purposes by men and for leisure, social, and even economic reasons by women (Bijker 1997, ch. 2). Without large urban parks where sports and leisure activities could be practised, the popularity of bicycles might have grown much more slowly and their use might have been more limited.

THE RISE OF THE TECHNOLOGICAL CITY

Of the great innovations of the late nineteenth century, a fair number were invented, tested, and adopted in cities (Smil 2005). Because ideas and information travel faster in concentrated and denser human environments, people are exposed to them more rapidly (Bairoch 1995, 420–4). Since urban-dwellers are on the whole richer than rural people, they tend to adopt and buy the new technologies in the first place. With the proliferation of communication technologies, the difference between rural and urban diffusion of new technologies is winding down.

However, the industrial city was the setting for a large diffusion of what might be called modern conveniences, such as electricity for heat and light, water supply, gas for cooking and heating, and the telephone, which together substantially increased the well-being of its inhabitants (Ball 1988; Fougères 2004; Gagnon 2006; Tarr and Dupuy 1988). The urban way of life is also a comfortable way of life. City planning had to accommodate the penetration of these new technologies, in many cases making room for them in crowded areas. Planning had to take into account the new technologies of personal and collective 'comfort', so to speak, and invent means to accommodate them (Poitras 2000). Planners and local authorities also understood that technologies could deface and harm the urban landscape: creative destruction is not only true in industrial and commercial businesses but also in urban spaces (Gilliland 2002).

THE DEMISE OF THE INDUSTRIAL CITY?

In the midst of the industrial city, in large part as a result of all these responses, a new city has gradually emerged. Metropolitan civilization progressively took root as people came to appreciate large urban areas for the benefits they offered. Large cities evolved into even larger urban regions containing millions of inhabitants. New cities arose, and long corridors of joint urbanization grew in size and in importance to form what Gottman called, even in 1961, a megalopolis. The modern city has lived and thrived, but many changes have led to its demise and its replacement by the post-modern city, whose governance has proved difficult and contentious.

Urban Governance and Metropolitan Issues

IN SEARCH OF A 'NEW POLITICAL CULTURE'

In North America and more specifically in Canada, the main changes—defined in economic, social, and cultural terms—that occurred during the second half of the twentieth century and the first decade of the twenty-first century had a huge impact on urban politics. Political life is moving away from a static, hierarchical, and elitist representation toward a more open process in which civil society is becoming an active partner of local governments.

Such a process did not happen overnight. It began in the 1950s and was certainly effective in a more explicit way by the 1960s. The savings that households had made during the war boosted economic growth as consumer goods became available again in abundance. In addition, the demographic increase brought about by the baby boom also triggered economic growth (Linteau 1994). These developments, in conjunction with new industrial investments and a global increase in economic activities, occurred during the first decade after the war and included the revival of suburbanization and new waves of immigration from Europe. To some extent, they all contributed to the emergence of new cultural values in the social landscape.

According to Ulrich Beck (1992), the socio-economic changes after the war disrupted the model of the industrial society, which now relied on more individualized relationships and adopted redistributive and social measures to counter the inequalities of an ill-adapted welfare state. It was in that context that a 'new political culture' emerged. From this perspective, the politico-administrative system is no longer the centre of politics.

Defined in reference to decentralization, the new culture implies the creation of sub-political fields—in

OPEN for DISCUSSION

Governance: Public and Private

In contemporary metropolises, social relationships are increasingly individualized. The consequences are diverse: dispersion of economic activities over the territory; increased mobility of individuals within urban agglomerations; differentiated patterns of interaction with space; and priority to market over state regulation. Contemporary metropolises as shaped by urban sprawl are more difficult than ever to plan in a sustainable way. In that respect, can urban governance provide a useful perspective and an adequate tool to overcome the conflicts of interest that are necessarily emerging among different urban actors? What can urban governance achieve in terms of social and political regulation?

In general, the literature on governance is highly critical about what cooperation between public and private actors can achieve. In other words, governance does not provide a panacea for legitimacy and efficiency in the public realm. Nonetheless, it remains true that governance introduces a valuable criticism regarding the previous hierarchical and substantive planning approach that characterized urban policies of the 1960s and 1970s. It has the advantage of raising the question of the difficulties public authorities have to overcome in order to counter the fragmentation of interests and find a common basis for compromises and cooperation.

Governance remains a multiple-meaning notion. It does not guarantee that all the actors involved have the same resources and power to influence the decision process. Governance mechanisms do not always limit powerful actors' influence on decision-making. This is why one has to be careful with the normative dimension of the notion. It is one thing to look at the way private and public actors cooperate and bring together knowledge and resources. It is another to suggest that the type of cooperation involved in governance will succeed in overcoming fragmented and divergent interests.

which specialized professionals who are connected to pressure groups in a more or less intensive way are playing an increasing role in defining and implementing state programs—but it also implies an openness to citizens' participation in diverse processes of democratization. Decision-making by a centralized authority can no longer impose a unique rationality without the participation of concerned actors. This is why public management must call for the active participation of actors from civil society. Thus, the reality of the 'new political culture' proved to be central to the definition of urban governance as implemented in OECD countries from the 1980s onwards.

Because of the socio-economic changes brought about by advanced modernity—economic globalization, the rise of social fragmentation and pluralism, and the increased role of information and other new technologies in the burgeoning service economy—the term 'urban governance' was coined to define a different approach to co-ordinating public action (Le Galès 1993) by bringing in private actors. This means two things. First, in the turmoil of a globalizing society, the state can no longer act as the only one in charge of establishing societal priorities. It has neither the resources nor the knowledge to assume such a responsibility. Second, partnerships with economic, social, and community actors need to be arranged. In other words, in order to respond to the people's expectations, different categories of actors have to be involved in public policy-making. In this respect, governance, rather than government, means that many actors need to be mobilized in order to achieve the public mission.

Thus, governance proved to be an adaptation of public management and decision-making to contextual changes in regard to three main aspects. The first aspect is related to what was mentioned in the previous section concerning the emergence of a 'new political culture'. The second, a prerequisite to the first aspect, comes from the failure of traditional political institutions to produce a convincing legitimacy for the state. The third aspect implies redesigning a new regime of public action, chiefly defined in pragmatic terms and turned toward partnership with private actors.

As Kevin Cox (1997, 101) has said, governance requires the creation of coalitions and trust relationships capable of strengthening cooperative structures between actors evolving at the same time in the public and the private sectors. But from a more demanding analytical point of view, governance coincides with a frame analysis of power. As Bob Jessop claims, it can be defined as a 'conduct of conduct' (1997, 58). For Jessop, governance is more than a

public discourse about changes in perceptions and mindsets. It can be adapted to different theoretical perspectives and can involve social practices in their empirical manifestations:

> Because state power is inevitably realized through its projection in wider society and its coordination with other forms of power, one must look beyond formal government institutions to a wide range of mechanisms and practices. Likewise, governance is relevant to the day-to-day practices in and through which the various structural forms of regulation are instantiated and reproduced (Jessop 1997, 59).

These practices can be accommodated through diverse forms of cooperation between actors, such as public hearings, consultation mechanisms, public–private partnerships, and so on. What makes governance a success or creates the required conditions for success, whatever the object it is applied to, is primarily related to the availability and mobilization of resources and actors beyond those that are members of the government in a formal way. Joe Painter has summarized this point in the following way: 'Governing a city, particularly in the United States, where the institutions of elected urban government are relatively weak, relies on the ability to form governing coalitions that bring together the formal agencies of government with interest groups from the wider society' (Painter 1997, 128).

Even though a consensus prevails regarding the fact that governance is opposed to the traditional vision of politics, which sees the management of large organizations through the mechanic vision of a guiding centralized and hierarchical structure, it is clear that governance can be tackled from different perspectives. Politicians and scholars who think in terms of 'good governance', as if there were a preferable or optimal model of public action, share a normative vision of what governance is. Governance is, in a non-normative way, a research program indicating that past models of regulation can no longer respond to the current state of public affairs. How is it possible to define collective action when no one has the power to impose their vision and priorities on others and when authority lacks the resources to act in accordance with its intentions (Lefèvre 2009)?

This conundrum is particularly salient in the case of metropolitan regions and metropolitan governments. Two opposing logics prevail in this case: one is tied to a competitive attitude in the context of globalization, whereas the other has to respond to social exclusion (Le Galès 1993). But the urban form and the future of cities are also at stake.

THE METROPOLITAN SCALE

According to Michel Bassand (2001), the analysis of metropolises can be conducted within two perspectives. The first one entails focusing on internal processes of social and spatial structuring, whereas the second considers, first and foremost, the development of a system of metropolises within the context of globalization. It is easy to understand that even though these perspectives are quite distinct, they remain inseparable and interrelated. Now, what are the processes and factors involved in the production of metropolitan regions that Bassand (2001) equates with metropolitanization?

In his analysis of contemporary metropolises, François Ascher (2003) highlights one key element explaining metropolitan areas—namely, mobility. The mobility of individuals and households has increased a great deal in contemporary metropolises because of easier access to efficient means of transportation and communication. The consequence is that urban agglomerations tend to expand by constantly pushing their limits farther. This strong trend would not have been possible if there were not an increased individualization of social relationships (Bourdin 2005). The new spatial configuration of metropolises has to accommodate all the household members, whose daily journeys can be quite different and dispersed from one another.

In *Bourgeois Utopias* (1987), Robert Fishman emphasizes that if one can describe the nineteenth century as the era of great cities, the term 'era of great suburbs' is more appropriate for the period after 1945. In North America (with perhaps the exception of Montreal) in the aftermath of the Second World War, as central cities were stagnating or declining in terms of their population and economic activities, development was occurring mainly—if not exclusively—in the periphery. Detroit is a striking example in that respect; whereas poverty was concentrated in the centre, growth and wealth could be found in the suburbs.

North American cities (including Canadian cities, although with 'shades and degree of differences' [Mercer 1991, 63]) have thus experienced a huge trend reversal. The impact on city forms has been no less impressive. We have witnessed the emergence of a new style of city living in which access to the city centre is no longer the main concern for citizens. In this context, the old notion of centrality defined in connection to the city centre has become obsolete. Within contemporary metropolises and their rapid expansion on the fringe, centrality is dispersed and constantly renewing itself. New developments offer opportunities to workers and consumers that were not previously available. Urban diversity no longer presupposes urban density. In this regard, we can think of the 'technoburb' and the 'techno-city', as described by Fishman (1987).

The 'techno-city' refers to the entire metropolitan region as transformed by the advent of 'technoburbs'. As such, the 'technoburb' is characterized by urban diversity without urban concentration. It is a peripheral zone that can be as large as a county but is viable from a socio-economic standpoint. Its development takes place along highways, producing growth corridors with commercial centres, industrial parks, office space for retail, health care, or education, and diverse types of housing. More often, the 'technoburbs' focus on relationships with one another rather than with central cities.

Fishman's understanding of recent urban changes stresses the pre-eminence of technological factors to explain the new morphology of metropolises. However, in emphasizing such factors, this analysis overlooks the main social factors at play, which were clearly taken into account by Simmel (1950 [1903]) in his analysis of modern metropolises and that remain in effect nowadays—mainly, the processes of specialization and fragmentation that were active in the transformation of modern societies. The technical division of work required by the industrial society, the specialization of activities, and the advent of new professional activities essential to the functioning of large cities all contributed to the birth of new professional categories and professional types. In other words, large cities are not only the product of technological changes; a double process of individualization and levelling was clearly at work at the same time. Consequently, the divisions, polarization, and inequalities among social groups that characterized cities of the past did not disappear with modern metropolises but rather were reframed. In addition, with the new urban forms assumed by contemporary metropolises, we are seeing a radicalization of processes typical of the modern metropolis. This is certainly one of the reasons why the problem of metropolitan governance has come to the fore.

There is no predefined model of planning and management capable of guaranteeing cooperation among social, economic, and political actors at the metropolitan scale. So far, the experiments on both sides of the Atlantic over the past 20 or 25 years with regard to metropolitan governance are not conclusive (Lefèvre 2009). For the multiple political units of a metropolitan region, the temptation remains strong to rely primarily on their own capacity in order to face the new economic challenges. And social problems are also increasingly present and complex. Social segregation and exclusion did not vanish with urban sprawl.

Urban and social problems are increasingly defined at a metropolitan scale. This is clearly recognized by social actors, especially those who are active within civil society (Fontan, Hamel, and Morin 2006). At the same time, this perception does not entail a willingness to act on this scale. Various hypotheses can explain such a discrepancy. First, from an institutional standpoint, it turns out that in most cases, metropolitan organizations were only recently created. Thus, there is a lack of tradition and confidence—a lack of legitimacy, we should say—which are required to convince social actors to call upon these organizations. Second, the resources that social actors are looking for are not primarily located at that scale. Third, political leadership is lacking. Usually, metropolitan governments or the special agencies in charge of metropolitan governance do not have enough support from lower and upper levels of government to take significant initiatives to solve the social and economic problems of cities. Their capacity to take initiatives depends strongly on others. In such circumstances, why should actors take the risk of wasting time and energy?

Faced with such limitations, social actors choose to direct their requests elsewhere. Nevertheless, they are right in identifying the metropolitan scale as an increasingly pertinent scale at which to solve urban problems. With suburban expansion and other forms of urban development shaping new metropolitan regions, public decisions should necessarily be taken at that scale. Social inequalities, economic development strategies, environmental issues, and other related concerns necessarily involve the metropolitan scene.

Cities, the Environment, and Sustainable Development

The industrial city and the suburbanization process that developed out of its expansion have produced a series of new urban—or rather metropolitan—problems. Suburbanization is the flip side of urbanization, when cities grow wider and larger. There is no way of avoiding it if urbanization does not stop (Brugman 2009). Moreover, changing lifestyles and prosperity

have to be taken into account. Numbers and land uses do not tell the whole story; social change bears on the urbanization process, its pace and spread. When people become richer, they tend to consume more dwelling space, and the individual house, including its surrounding area, remains a preferred option for many households and families (Fortin and Bédard 2003). On the other hand, an aging population may choose smaller living spaces and a more central position in an urban area. These two trends may cancel each other out, but other factors may come into the equation. The patterns are not always clear, and one cannot conclude that the larger Canadian urban areas are contracting instead of sprawling, at least for the time being (Bourne and Simmons 2003).

Sprawl may be a loaded term, masking a value judgment. Not only is the term difficult to define, but in the planning professions it has become by and large something that should be avoided. Older urban regions are denser, whereas newer ones tend to be less dense (Newman and Kenworthy 1999). Between old and new urban regions, many social, economic, technological, and cultural changes have occurred.

This section will explore changing urban conditions with respect to two new concerns: the environment and sustainable development. The two ideas are not interchangeable. **Sustainable development** is a much broader idea than environmental protection, but it is meaningless if the environment is not included and, indeed, not a central aspect of sustainable development (Guay 2005; Mancebo 2006). Some authors prefer the expression *sustainability* to *sustainable development* to emphasize that all development is predicated on a sustainable environment (Clark 2007). A sustainable environment, or a sustainable ecosystem, means that its resources, its integrity, and its 'health' are maintained in the long run (Walter-Toews, Kay, and Lister 2008). For instance, the Canadian Federal Sustainable Development Act defines sustainability as follows: 'the capacity of a thing, action, activity, or process to be maintained indefinitely' (Canada 2008, 2).

Now, is a city, whether large or small, an ecosystem? There are two conceptions of cities as ecosystems (Pickett et al. 2001). Cities comprise some specific and particular ecosystems: a pond, a lake, a park, and a wetland enclosed in an urban area are all within-city ecosystems. Is the city as a whole an ecosystem? The ecosystem conception is of limited value, for a city is much more than energy flows and materials cycled; it is a human construct. A city is better described as a socio-technical system than a

GLOBAL ISSUES

Globalization and Urbanity

Urbanization is a long-term trend. Countries and continents, up to now, differed greatly in the rate of urbanization. A longer and wider trend was taking place over many decades in the twentieth century: world urbanization. In the first decade of this century, the world has become urban—that is, more than half of the world population lives in urban areas. This trend is quite phenomenal and unique in history. The United Nations forecast in 2003 that by 2030, 60 per cent of the world population would be living in cities. Africa and Asia would be the least urbanized at 53.5 per cent and 54.5 per cent, respectively. In the developed world, 80 per cent of the population would be living in cities.

With globalization, a complex process of speeding up, deepening, and broadening human exchanges and relations, some cities have come to play a crucial and leading role. Global cities like New York, London, and Tokyo perform a wide range of activities central to the globalization process. Other cities have found a role or a niche in international urban differentiation. Not all possess the same combination of resources in making a mark in the globalization process. New York and London dominate in financial markets, but Paris and Los Angeles are recognized centres of cultural creation and diffusion.

Toronto, Montreal, and Vancouver may occupy an intermediate level, but they have their own attractions. Vancouver is seen (or projects itself) as a leader in smart growth and sustainability. Bilingual cities like Montreal offer a different way of life. Toronto, on the other hand, may be one of the most culturally diverse cities in the world. International migration is both an effect and a cause of globalization.

socio-ecological system. The crucial question is, however, not whether cities are ecosystems but whether they contribute to general sustainability or not and whether they are themselves sustainable.

In North America, large urban areas, dominated by sprawl, low density, and the car and its infrastructures, have come under scrutiny in the context of sustainable development (Boone and Modarres 2006). Although one cannot escape the fact that urban areas are bound to continue to grow and that megacities seem to be consolidating their spatial landmark and ecological footprint, one might ask, with Jesse Ausubel (2001), whether more compact and denser urban areas, with taller buildings, are the way to the future for a sustainable urban civilization. Less space devoted to human habitat means more space devoted to natural habitats. In order to reach such a stage, if ever it can be reached, huge transformations in urban governance and planning, not to mention urban lifestyles and cultures, are required. Apart from undoing existing cities and returning to very small and local communities, the focus of sustainability will have to be on large and complex human settlements called cities. But how can cities contribute to sustainability and sustainable development? A few ideas and programs have been put forward to tackle the question. Two new intellectual and professional programs in particular are worth discussing: the 'smart growth' idea and 'new urbanism' in planning.

(Jim Zuckerman/Alamy)

Millions of travellers flock to Paris to experience art and culture housed in institutions like the Louvre.

TIME to REFLECT

It is now widely accepted that planning and city design should take environmental concerns into account as their main requirement. In that respect, what are the advantages but also the limits of thinking in terms of sustainable development?

SMART GROWTH AS ACCEPTABLE GROWTH?

It may seem ironic to describe growth as smart, for most people believe that growth is far from smart, at least with respect to the environment. A degraded environment may be the price that must be paid for economic growth. The environmental movement may have alarmed the 'growth coalitions' (Logan and Molotch 1987) that ruled without limits for many decades and are now reacting to a no-growth discourse (Meadows et al. 1972).

At the urban scale, growth has enriched many social groups and provided affordable housing and public goods for a large part of the urban population. But growth still seems to be needed, for not all needs and aspirations have been satisfied (World Commission on Environment and Development 1987). If economic growth is still an imperative, is urban growth necessary? It depends on what is meant by urban growth. If one means dynamic cities that overcome the de-industrialization period, open themselves to new global opportunities, invest in new technologies, and attract members of the creative class, building creative cities (Florida 2003; Scott 2006), growth is not only acceptable but also desirable, at least for many urban citizens. On the other hand, if it means doing things the old-fashioned way with respect to planning and urban policies, consensus may break down. Is smart growth a smart response?

What are the key elements of smart growth (Filion 2003; Ouellet 2006; Smart Growth Canada Network)? In a nutshell, smart growth is contained growth; it is a struggle against sprawl in urban 'development'. But to be against something is not the best rallying cry. Smart growth must be *for* something. Two schools of thought may clash here. The first, an heir of the limits to growth perspective, will plead, like good Malthusians, for severe checks on growth because there may be no choice. Smart growth is seen not so much as a choice but as an inescapable constraint. The need to adapt is often a need to limit. This way of thinking is not dominant and has so far not found strong professional or political support. The second school of thought emphasizes the positive side of turning away from urban sprawl and adopting smart growth. Urban sprawl consumes resources and investments. A denser and tighter city will lead to economic gains. Moreover, if compact cities are well served by public transport, air pollution, noise, and greenhouse-gas emissions will be reduced. Also, smart growth may make people less dependent on their cars and more on their own biological means of transport, as well as on smaller technologies like the bicycle. This could lead to improved health. Finally, a more compact city may engender a greater sense of community, the loss of which has plagued many urban neighbourhoods, as predicted by Wirth.

This is the plan. But what is the reality? Large Canadian cities have adopted a smart-growth approach more or less unknowingly and often tepidly. Toronto was an early leader (Filion 2000; 2003); Vancouver was not far behind, perhaps even ahead (Alexander 2000; Alexander and Tomalty 2002; Curran and Tomalty 2003). Montreal has planned new urban developments broadly based on smart-growth principles. Calgary, Quebec City, and Ottawa, which may have been the true pioneer, are initiating policies and actions along a smart-growth path (Ouellet 2006; Guay and Hamel 2010). However, can all the new urban planning policies be considered sustainable development?

To be considered sustainable, urban policies and projects must meet three broad criteria: to respect and protect ecological integrity; to act with economic efficiency; and to foster and promote social equity (Guay 2005). It would be presumptuous to conclude that current policies and urban projects all fall under the banner of sustainable urban development. For instance, the federal government and the provinces recently embarked on large infrastructure investments. The programs and their guidelines are less animated by sustainable development than by other factors such as providing jobs, repairing crumbling infrastructures, and ensuring public safety, not to mention economic stimulus in response to the recession that began in 2008 (Guay and Émond 2010). Also, compared to European cities, Canadian cities are trailing in adopting local agenda 21s (Agenda 21 is a document that attempts to apply the ideas and principles of sustainable development; local agendas are the locally based applications of Agenda 21) and in adopting some homegrown Aarlborg Charter for municipalities (Federation of Canadian Municipalities 2005). (The Aarlborg Charter, signed in 2004, is an engagement by European local authorities in sustainable development.) Although there are signs that things—mindsets and actions—may be changing, there is still a long way to go to achieving sustainable urban policies and practices (Grant 2006). Smart growth might be more amenable to a Canadian policy network—federal as well as provincial—that values growth in the context of globalization and believes that large Canadian cities should be 'active actors' in the globalization process (Wolfe 2003).

NEW URBANISM AND THE SUSTAINABILITY CHALLENGE

If smart growth remains strongly economic, **new urbanism** is strongly 'urbanistic', with its emphasis on urban form. This planning innovation emerged in the 1980s when a diverse group of professionals, planners, architects, and urban analysts, sensing a change in the attitudes of urban residents and seeing problems in the 'natural' evolution of large urban areas, became concerned that urban spatial trends might not be either sustainable or desirable. Not only do growing cities consume space and resources, but they also invest heavily in infrastructure to sustain their growing pace. All this is costly, and many trends in space occupation are wasteful. When public finances are tight, ideas for conserving land and minimizing investment in infrastructure might emerge. The idea of a compact city, or compact neighbourhoods more appropriate for the new urbanism, took root.

Post–Second World War cities were planned very differently. In Europe, governments were involved, in part because of reconstruction requirements (Hall 1982). But in North America, cities—their new developments in particular—were not so much planned as left to private initiatives. Planning was often more regulatory than decisional, although co-ordination requires planning decisions (Hodge and Robinson 2001, ch. 8, 9). The ruling philosophy was that housing and commercial, industrial, and office buildings are best left to the marketplace, while regulation ensures that local nuisances are avoided or minimized. Segregation of urban functions was paramount, despite planning ambitions on the part of some professionals and officials for greater urban diversity and 'mixcity' (Wolfe 1984).

One of the most striking characteristics of the past 60 or so years of urban development is the consolidation of suburbs following the basic layout of individual housing, roads for the private car, commercial centres, and some public spaces in the form of recreation facilities and green spaces. This pattern, planned by nobody in particular, came to represent a way of life, the aspiring 'urbanity' of a growing middle class (Fishman 1987; Jackson 1985). Some planners helped to define some of its features, but many new suburbs were left to the interaction between developers and the consuming families and households. With the emerging environmental movement and new

Effective and efficient public transportation service is a hallmark of smart urban growth.

attitudes toward nature and the environment, including the built environment, the relationship to space and urban resources was gradually redefined. In this context, new ideas were formulated. Can cities consume less space, offer other forms in which urban life can be expressed? Can they be planned differently, and is a rethinking of planning models, either private or public, necessary?

TIME to REFLECT

After the Second World War, industrial and urban changes had huge impacts on cultural and political life. As society was becoming more and more individualized, the old regulation models were revealed as more and more ineffective. Can we say that governance can become the new regulative model of the twenty-first century?

The new urbanism school of **urban planning** thought so. The movement is a professional movement, as was the previous modern movement in planning and architecture that dominated a large part of the twentieth century. It had some resonance in official circles and local governments. Built around a small group of architects and planners whose professional practices were chiefly private, it gained some recognition in public planning (Ouellet 2006).

Like the smart-growth movement, its origins are diverse. Concerned with new issues like protection of the environment, the economic and environmental costs of urban development, and a more or less acknowledged longing for a sense of place and **community**, the new urbanism movement was built around a set of principles that usher in, at least according to its proponents, a new relationship to urban space and, as is often the case with urban planning and architecture, new relationships among people. The basic ideas of new urbanism are the following:

- Walking should play a greater role in urban mobility. The car has relegated walking to short distances and as a secondary mode of mobility. The modern movement has been so obsessed with traffic congestion and the efficiency of goods and people transportation that it has downgraded walking in urban space. Walking has many virtues. When cities and neighbourhoods are planned to allow walking to play a greater role, health and environmental benefits will result. Moreover, when people walk, they may experience social interactions. A greater sense of community could develop.

- New urbanism promotes greater density and deeper diversity. The modern city has segregated urban functions in almost exclusive zones of industry, commerce, and residence, with the result that car mobility became necessary.

- A town or a city should have an identifiable centre and some identifiable peripheries. Although this principle is promulgated by the modern movement, to its extreme, so to speak, as expressed in the Athens Charter (Le Corbusier 1972 [1941]), new urbanism is a reaction to the overall suburbanization of the trends that took root in the 1970s onward, when suburbs urbanized while at the same time city centres lost their attraction and began degrading. Although a large part of urban planning and decisions in the 1980s and 1990s represented a strong public response to this trend, city centres lost some of their *raison d'être* and did not entirely regain their historic role (Guay and Hamel 2010). Citywide spatial organizations are now highly diversified. No models seem to adequately describe a large city's evolution, as the Chicago School of urban sociology hoped to achieve. No concentric model, no sector model can capture the multiple faces of recent urban evolution.

- A challenge to the compact city, denser and more diversified, is transportation and mobility. New urbanism represents a plea for public transport. In the planning of new urban developments, the idea of transit-oriented communities is promoted (Laliberté 2002). But the model can also be applied to older communities that lacked proper public transport systems in the past.

- New urbanism also pursues broader objectives such as environmental sustainability and the improvement of urban life through planned communities.

New urbanism has been criticized (Garde 2004; Ouellet 2006). Some see it as a marketing strategy. Since new urbanism communities are often privately developed, its leading thinkers have a vested interest in emphasizing its merits while muting its limitations. Moreover, new urbanism communities are chiefly intended for the rich or the middle classes, rarely for the poor or degraded central city neighbourhoods, despite some experiments in central-city developments. Finally, new urbanism may be a concept with limited applications outside American and Canadian urban regions.

In relation to smart growth, new urbanism appears local and less ambitious in scale and scope. Smart growth is a long-term strategy for urban development and planning that makes sense only at the regional level (Filion 2003). New urbanism, by contrast, focuses on local interventions and mostly on new urban fringe developments. It is also more spatially oriented and less economic. However, both smart growth and new urbanism seem to agree with sustainability principles, and new urbanism, with its emphasis on liveable communities, permeates the smart-growth project (Curran and Tomalty 2003, 11). But does it really?

IN SEARCH OF THE SUSTAINABLE CITY

The World Commission on Environment and Development devoted a entire chapter to urban challenges (1987, ch. 9). Although a great deal of its analysis and recommendations addressed cities in the developing world, some ideas were highly relevant to the existing developed-world cities. For instance, it highlighted the environmental problems of developed cities that, because they are richer, consume a lot of energy and resources and valuable agricultural areas, as a result of sprawl. But one relevant criticism of this section is that the quality of urban infrastructure remains poor and in urgent need of improvement and restoration to this day and that new infrastructure has to be rethought. Finally, the focus on cars in planning policies leads to pollution, noise, and congestion. Agenda 21 contains a long chapter on human settlements, suggesting that large cities in developed countries have some work to do to get on the sustainable development path.

As mentioned earlier, sustainable development is composed of three broad principles: ecological integrity, economic efficiency, and social justice, all of which are constantly reaffirmed in the official literature, such

as Canadian and provincial legislation on sustainable development (Canada 2008; Québec 2006).

Are there examples of sustainable cities? Perhaps not, but some new urban developments are moving in that direction (McGranathan and Satterthwaite 2003). Successes in ecological planning have been achieved in some European cities, such as Freiburg, Germany, a well-researched and much talked about case, and in new developments in Sweden and the Netherlands. Some are rehabilitated industrial zones (Dumesnil and Ouellet 2002). In 1994, European municipalities adopted a charter on local sustainability called the Aarlborg Charter (European Conference on Sustainable Cities and Towns 1994). They particularly want to do their part—and their best—to fight climate change. They can achieve this in many ways, but a change in transport policy is a key element. Moreover, some cities, such as Stockholm, have embarked on the protection, restoration, and improvement of urban biodiversity by enlarging urban parks and connecting them, thus creating (though it might seem contradictory at first glance) an 'urban biosphere' (Dogsé 2004; Platt 2004).

Much remains to be accomplished on the sustainable development front. Once a city is designed in a particular way, it is difficult to undo what has been done. There is, as Hommels (2005) has aptly put it, obduracy—in other words, physical resistance in the urban form, at times on top of social resistance. The goal of achieving sustainable cities is similar to the goal of achieving the sanitary industrial city, but one and a half centuries later. However, the urban challenge today is compounded by rapid urbanization worldwide, in which economic, social, and environmental problems are tightly linked (Myllyla and Kuvaja 2005).

HUMAN DIVERSITY

Integration in Large Cities

Large cities have always been open to human diversity. This diversity is expressed in many different ways. Because of their differing religious beliefs and practices, people may relate very differently to the material world, to others, and to how god or gods manifest themselves in nature. But language and culture define people as strongly as religion. The modern and post-modern cities received a large influx of people coming from different areas of the world. The modern city conception relies on people integrating into the social and economic structure through education and occupation, whereas the post-modern conception seems to tolerate different communities within larger communities. Can different communities live side by side for long without mingling cultures, habits, and differences? It depends. Some communities prefer to live at least somewhat apart from the main group. This can be a choice or the result of discrimination. Other migrant communities may integrate and assimilate fully. Large diversified cities exhibit all of these cultural paths.

A unique vision of the urban sustainable future may not be warranted, even though the three principles of sustainable development must be upheld.

Conclusion

Cities are difficult to grasp in their diverse manifestations. Beyond their materiality or their distinctiveness in relation to specific historical evolution and geographical conditions, they experience different processes of restructuring in the face of globalization. Urban sprawl, redefinition of centrality, suburban expansion, and the emergence of decentred metropolises are inexorably transforming city forms all over the world in a similar fashion.

It is true that urban problems are defined more than ever at a metropolitan scale. In 2001, 46 per cent of the Canadian population lived in the six largest agglomerations (Bourne and Simmons 2003). This does not mean that other scales are not important. But the increased importance of the metropolitan scale is changing our understanding of urban issues and of social issues more broadly.

Let us recall the two distinct visions of urbanism and social relations to space defined by Sharon Zukin (2009): the 'urban village' model and the 'corporate city' model, an opposition that Jeanne Wolfe (2003) saw clearly at work in Canadian urban policies. While the 'urban village' model is mainly oriented toward use values and living space, the **corporate city** responds first and foremost to expectations for profit and revenue. The corporate city is directly produced through the market and orientated toward the rise in value of economic investments, while the urban village is produced through social interaction, conviviality, and a concern for the common good. Is

(imagebroker/Alamy)

Freiburg, Germany, a sustainable city, features a large pedestrian zone, a vast public transportation network made up of trams, buses, and trains, and a population that embraces recycling and cycling.

the debate about new urban planning options, such as new urbanism and smart growth, a revival of the urban village city? Will the sustainable city move away from the corporate model? There are indications that sustainable development may lead to new

UNDER the WIRE

Changing Social Communications

In contemporary metropolises, as was already the case to a certain extent within modern metropolises of the early nineteenth century, social relationships are increasingly individualized. The consequences are diverse: dispersion of economic activities over the territory; increased mobility of individuals within and among urban agglomerations; differentiated patterns of interaction with space; and priority of market over state regulation. The centrality of the classical city and its core as described by the researchers of the Chicago School of sociology in their ecological model must consequently be rethought. The dispersed and fragmented form of contemporary metropolises has been greatly facilitated in more recent years by the development of media and communication technologies. With people able to interact with one another instantaneously, new social patterns like telework have gradually gained ground, changing social relationships slowly but profoundly in urban space.

social relations, but no one can predict what urban sustainability may result in.

TIME to REFLECT

Cities of tomorrow will probably keep a profound relationship with today's urban agglomerations. What city model do we want to live in? Who is going to decide?

Even if actual urban life is constantly crossing the borders between these two models, it seems that the 'urban village' model is losing ground to the 'corporate city' model. Citizens are increasingly aware of the negative impacts of urban sprawl on the environment. But is this awareness sufficient to convince them to bring to the fore the issue of the urban form and the urban model that sustainable development implies? These questions clearly underline the need to place common urban issues back on the public agenda.

questions for critical thought

1. City forms of today are diverse and complex. They no longer look like the classical city as conceived by the Greeks and Romans and reproduced in medieval times. Therefore, the genetic code of the city, so to speak, remained unchanged for a very long time. The reference to centrality remained an inescapable element in defining the city. How would you define the centrality of contemporary cities?

2. Beyond centrality, can you name and define two useful notions that can help us to understand the contemporary city?

3. To what extent is it possible to define current Canadian cities as areas of cooperation between public and private actors?

4. The industrial city grew rapidly and faced many problems. Can you define the main problems of the industrial city?

5. In solving the industrial city's problems, who were the main actors? Can you explain why they took part in that process?

6. According to the World Commission on Environment and Development, sustainable development is a very large social project. Why do cities have to become sustainable? What are the main schools of thought in planning sustainable cities? Do you think that they all encompass what sustainable development means?

recommended readings

Peter Hall. 2002. *Cities of Tomorrow: An Intellectual History of Urban Planning and Design in the Twentieth Century.* 3rd edn. Oxford: Blackwell.
This book focuses on problems the industrial city met and the different solutions and experiments planners, seers, and utopians have invented and applied. It concentrates on ideas and visions of the good, or at least the liveable, city. Some ideas were far-fetched, such as the non-urban Broadacre City of Frank Lloyd Wright, while others, like the planning ideas the modern movement in architecture and planning put forth, were taken up by professional planners and policy-makers.

Harry H. Hiller, ed. 2010. *Urban Canada.* Toronto: Oxford University Press.
This collection of articles, written by Canadian urban sociologists, represents a variety of approaches to

and themes on the Canadian city. Important aspects of Canadian cities, such as urbanization trends, rural–urban differences, immigration, urban inequalities, gender issues, social problems and housing, planning options, and Aboriginal issues, are introduced and explored at some length.

Gerald Hodge and Ira Robinson. 2001. *Planning Canadian Regions.* Vancouver: University of British Columbia Press.
This book is a broad and comprehensive view of spatial planning in Canada. The authors combine historical and empirical analysis of problems and conditions, ideas and ideals, actions and policies in the planning process. Examples abound and are drawn from all the provinces. The concept of region, urban as well as non-urban, is an organizing principle.

Lewis Mumford. 1961. *The City in History: Its Origins, Its Transformations, Its Prospects.* New York: Harcourt, Brace and World.

Mumford was a polymath, though not a professional historian or social scientist. His histories of the city and of technology are certainly what best define Mumford's intellectual stature. He was wrong on many important points, such as the demise of the metropolis, but his work is still worth reading for his historical and cultural insights as well as for his lively style of writing. He was an intellectual pioneer of ecological planning and bioregionalism. Judgmental and critical, he was and remains a public intellectual.

Robert E. Park, Edgar W. Burgess, and Roderick D. McKenzie. 1925. *The City.* Chicago: University of Chicago Press, reprinted 1967.

All serious study of the modern industrial city should start with this seminal book, written by the three musketeers of American urban sociology who, with Louis Wirth, define the socio-ecological approach to the city. Although the aim is to characterize the industrial city, it is actually based on the city of Chicago, seen as a social laboratory by the Chicago School's members. The book shows one of the first spatial models of a city's evolution. This model was later criticized for lack of generality. A French translation of the school's major papers, together with some European contributions, was published under the title *L'école de Chicago: naissance de l'écologie urbaine* (reprint, Paris: Champs Flammarion, 2004).

Saskia Sassen. 2006. *Cities in a World Economy.* 3rd edn. Thousand Oaks, CA: Pine Forge Press.

The author initiated an ambitious research program on large cities and globalization. Her first book on global cities compared such urban regions as London, New York, and Tokyo as central actors in the globalization process. Her research has encompassed a larger set of cities along the ranking ladder of global and international cities, each playing some functional or regional role. The book can be criticized for focusing too much on finance and not enough on other types of global activities, such as knowledge and cultural production and the diffusion of environmental norms.

A.J. Scott and E. Soja, eds. 1996. *The City: Los Angeles and Urban Theory at the Eve of the Twentieth Century.* Berkeley: University of California Press.

This book is a deferred response and challenge to the program of urban sociology developed by Park, Burgess, and McKenzie. The 'Los Angles School', as it is now known, was influenced by the writings of Henri Lefebvre and by Marxism. The school is less interested in the cultural and ecological aspects of urban life, and more in the economic dimensions and, not surprisingly, in global processes. The authors contrast the modern city of the Chicago School with the 'postmodern' city, of which Los Angles, their own social laboratory, may be the best example.

recommended websites

Atlas Natural Resources Canada

www.atlas.nrcan.gc.ca

This site provides maps of resources and environments in Canada. Together with the Statistics Canada website, it should be consulted regularly for some idea of how the country (people and their environments) is changing.

Cities Alliance

www.citiesalliance.org

Cities Alliance is an organization that connects cities in their common challenges, such as housing for all, climate change mitigation and adaptation, clean water provision, education, and urban planning.

Globalization and World Cities Research Network

www.lboro.ac.uk

This research network studies the formation and evolution of large urban areas and international cities. The network is known for its ranking of global and international cities, using complex methodologies and indicators.

Statistics Canada

www.statcan.gc.ca

This federal agency provides data on almost every kind of activity in Canada. It is responsible for the Canadian census. The website provides useful data on metropolitan areas and mid-sized cities.

UN Habitat

www.unhabitat.org

This site focuses on all forms of human settlement and on settlement problems and principles worldwide. The urbanization of the world is a leading theme, but rural conditions are not ignored.

United Nations Environment Programme (UNEP)

www.unep.org

The environmental organization of the United Nations provides data on global environmental conditions. The approach is global, but UNEP has a special responsibility toward developing countries. Cities have an important role to play in environmental protection.

chapter 23

Mass Media and Communication

David Young

In this chapter, you will:

▶ Learn about different theories sociologists have used to analyze the mass media;

▶ Identify various forms of media ownership and understand concerns about deepening ownership concentration;

▶ Study the role of the state and globalization in relation to the mass media;

▶ Grasp how media content represents the working class, women, and ethno-racial minorities;

▶ Discover how media content is interpreted by audience members;

▶ Examine how the Internet reflects long-standing issues in the sociology of mass media.

Introduction

You are probably exposed to various mass media every day. You might browse through a newspaper over breakfast, listen to rock songs on your iPod as you head to class that morning, access the Internet to check your favourite websites or do research for a paper that afternoon, and relax in the evening by going out to see a movie or by staying at home to watch a television show.

In light of such circumstances, a number of questions may come to mind when you think about the mass media. What is the role of the mass media in society? Who has power over what we read, see, or hear in the mass media? What kinds of messages exist in media content, and how do people react to them? What are the social implications of the Internet, which is associated with the 'new media'? Sociology provides the tools to grapple with such questions. Throughout this chapter, you will be introduced to answers developed by media sociologists.

Sociological Theories of the Media

The classical theorists who have informed sociological research—Karl Marx, Émile Durkheim, and Max Weber—did not write specifically about the mass media. Nevertheless, at least some of these men presented ideas that sociologists have applied to the study of mass communication. Four sociological perspectives on communication and media can be identified: symbolic interactionism, structural functionalism, conflict theory, and feminism.

SYMBOLIC INTERACTIONISM

Symbolic interactionism focuses on the microsociological issue of interaction among individuals through the use of symbols. Symbols can be verbal (such as the words used in spoken language) or non-verbal (including forms of body language such as a smile, frown, or gesture). All of these symbols carry meaning, which the members of a society come to understand and share through the process of socialization. Without this shared understanding of symbols, interaction among people would not be possible.

Since the 1970s, symbolic interactionism has been used in research on communication and media. Faules and Alexander (1978) identified how symbolic interactionist theory could be employed to study different forms of communication. They indicated that symbolic interactionism can be applied to *interpersonal*

communication (face-to-face interaction between two individuals that involves the reciprocal exchange of verbal and non-verbal symbols); *group communication* (face-to-face interaction among several individuals in a context, such as a seminar discussion, that still permits the mutual exchange of verbal and non-verbal cues); *public communication* (sending messages to large groups in a face-to-face setting, as in a lecture, where the participants have a much more limited opportunity for the exchange of verbal and non-verbal symbols); and *mass communication* (which exists through the use of mass media such as newspapers, motion pictures, radio, or television).

Faules and Alexander (1978, 11) noted that mass communication is different from the other forms of communication since it does not involve direct, face-to-face interaction and 'there is no opportunity for immediate mutual exchange of verbal and nonverbal cues between the initiator of a message and the recipients of that message.' However, since much of our interpretation of the world is based on our previous experience with direct and indirect forms of communication, they concluded that 'the media are a prime source of indirect experience and for that reason have impact on the construction of social reality' (1978, 23).

Symbolic interactionism is not one of the principal theories employed by media sociologists. The reason for this is its microsociological focus. An approach that addresses interaction through symbols can be useful in certain areas of sociology, such as the analysis of socialization, and it can also be useful to scholars who study various forms of face-to-face communication. However, in terms of research on mass communication, symbolic interactionist theory has more limited applicability. It does not allow sociologists to address macrosociological concerns like the role of media institutions in society.

STRUCTURAL FUNCTIONALISM

Structural functionalism is inspired by the ideas of Émile Durkheim. As a macrosociological framework, structural functionalism focuses on social order. The order and stability of society are facilitated by consensus among its members regarding norms and values. Stability is also facilitated by the interconnected parts of society.

Structural functionalism was used in studies of the mass media during the 1950s. Unlike symbolic interactionism, this theoretical approach enabled sociologists to examine the role of media institutions. Wright (1959) argued that media institutions contribute to the maintenance and survival of society by performing four functions: surveillance of the environment, correlation of the parts of society, transmission of the social heritage, and entertainment.

Surveillance of the environment involves the collection and distribution of information about events that occur inside and outside a particular society. This surveillance is provided by news media organizations, and it is functional for society in several ways. The flow of information through the news helps institutions or individuals to organize their activities, and the population can be warned of imminent danger (such as a hurricane).

Correlation of the parts of society refers to interpretation of information about the environment and prescriptions for behaviour in response to events. Editorials in the news or commentaries through other media, such as televised speeches by political leaders, are functional for society because they help people to make sense of what is happening. They are also functional because they aim to integrate society by building consensus. A specific example is the patriotic tone of the American media after the terrorist attacks on 9/11.

Transmission of the social heritage involves communicating information, norms, and values from one generation to another or from the members of a group to new members. Various mass media are functional for society, since they contribute to the socialization process and help to ensure that the culture of a society or group will continue across time.

Entertainment involves forms of communication that are mainly intended to provide amusement or diversion. Entertainment is functional for society because it provides relaxation and enables the release of emotional tensions that may generate conflict and threaten social order.

Although structural functionalism offered some important insights, media sociologists had abandoned this theoretical approach by the 1970s. It was seen as having several problems, including a conservative orientation. The conservative nature of structural functionalism is evident in the value it places on the order and stability of society. Structural functionalist theory stresses the role of media institutions (and other social institutions) in maintaining society as it is.

CONFLICT THEORY

Conflict theory is rooted in the work of Karl Marx. Like structural functionalism, it provides a macrosociological framework. However, conflict theory differs from structural functionalism in two important

ways. First, rather than focusing on social order, it emphasizes how social change occurs through conflict between unequal groups such as the capitalist class and the working class. Second, this theoretical approach makes it possible for sociologists to question rather than defend the role of media institutions. For instance, conflict theory enables sociologists to see how media institutions are tied to Marx's concerns about power and inequality in capitalist society.

Many sociologists began turning to conflict theory during the 1960s. This had much to do with the inability of structural functionalism to address the social conflict that had taken shape between social institutions and social movements. In the 1960s, the labour movement was joined by other movements that had emerged to struggle against social inequality (such as the women's movement and the civil rights movement). Concerns about power, inequality, and conflict have informed much media research since that decade. Examples of early research include studies of how media institutions are connected to power in American society (Schiller 1969) and of news coverage of conflict between social institutions and the student movement during the 1960s (Gitlin 1980).

FEMINISM

Feminism can be used to understand microsociological issues associated with the lives of women and their experience with inequality, but it also has a strong focus on addressing macrosociological conditions that account for the oppression of women. Especially with regard to the macro level of analysis,

In the First Person

When I started my first year of studies at Queen's University in Kingston, I had already decided to major in psychology. I also took a course in sociology, even though I had far less idea of what this subject was about. To my surprise, while I found that psychology was nowhere near as interesting as I thought it would be, I was fascinated by sociology—particularly by that guy named Marx. By my second year, my major had switched to sociology. I sampled various areas of sociology over the next three years and was intrigued enough to pursue graduate studies at Queen's. It was while working on my master's degree that I finally found what I wanted to make my life's work: the sociology of mass media.

—David Young

patriarchy is a key concept in feminism. Patriarchy refers to a society or form of social organization based on male domination.

Recent feminist theory has emphasized three themes connected to analysis of media and communication. Wackwitz and Rakow (2004) identify these themes as difference, voice, and representation. *Difference* raises several issues, including the notion of differences between women and men. Feminists have shown that patriarchy justifies inequality between women and men by asserting supposedly biological differences—men being 'naturally' more aggressive than women, for example—even though these differences are actually products of culture and socialization. The theme of difference also refers to differences among women with regard to class, race/ethnicity, sexual preference, or other aspects of identity. Feminists have increasingly examined these elements of difference in order to overcome the deficiencies of earlier feminist theories, which minimized or ignored diversity among women. The second theme identified by Wackwitz and Rakow focuses on exclusion. *Voice* concerns the degree to which women are denied an opportunity to speak in various forms of communication (including interpersonal or group communication) or given a voice only to have their ideas ignored. The third theme considers the portrayal of women in the media. *Representation* draws attention to the way women are depicted in media content and the way this negatively affects them. Wackwitz and Rakow (2004, 9) see these three themes as overlapping because 'systems of difference and exclusion are linked with the process of representation.'

CRITICAL PERSPECTIVES ON THE MEDIA

Conflict theory and feminism have contributed to critical perspectives on the media. Critical perspectives challenge the type of society we have while analyzing the media in relation to power, inequality, conflict, and change.

Critical perspectives on the media are often divided into two categories: political economy and cultural studies. Political economy focuses on ownership and control of the media. It examines private corporations and the state in relation to the media as well as the opposition of subordinate groups to the role of powerful media organizations. In contrast, work within cultural studies addresses the ideological aspects of the media. This approach analyzes the ideology embedded in media content, the interpretation of media content by audience members, and efforts to change media representations or

disseminate alternative media messages. Adapting the themes that Mosco (1989) associated with political economy and cultural studies, we will now turn to these two critical perspectives.

> ## TIME to REFLECT
> Consider how a functionalist or a feminist would analyze your favourite television show.

Political Economy of the Media

While placing a strong emphasis on historical analysis, researchers who specialize in the political economy of media devote particular attention to several issues. The main issues are, forms of media ownership; the state and media policy; globalization and the media; and conflict over ownership, policy, and globalization.

FORMS OF MEDIA OWNERSHIP

We can distinguish between public and private media ownership as well as various types of ownership that are private.

Public ownership—ownership of media by the government—has a long history in Canada. Examples of public media ownership in this country include the National Film Board (NFB), the Canadian Broadcasting Corporation (CBC), and the educational television broadcasters operated by some provincial governments: TV Ontario, Radio-Québec, and the Knowledge Network (in British Columbia). The goal of these organizations is to provide a public service by utilizing the media to satisfy social objectives. Such objectives include providing media that are freely available to citizens, using the media for educational purposes, and ensuring a Canadian voice in the media. Media organizations under public ownership are often supported by government funding, but additional funding may come from advertising or memberships. Public ownership of the media in Canada frequently takes the form of a Crown corporation: a business owned by a government but operating at arm's length (independently) from government (Lorimer and Gasher 2001).

Private ownership refers to ownership of the media by commercial firms, and it too has a long tradition in Canada. Most of the mass media in Canada are under private ownership. The prevalence of private ownership is illustrated by Table 23.1, which presents a list of the leading media organizations in Canada. Only one of these organizations, the CBC, is a public corporation. The goal of private media organizations is 'survival and growth in a marketplace driven by profit' (Lorimer and Gasher 2001, 223).

Critical researchers argue that the pursuit of profit through private ownership has significant implications for media content. The interests of private media companies mean that media content 'is regarded by their management not as a public service, but as a business cost to be met as inexpensively as possible' (Hackett, Pinet, and Ruggles 1996, 260). For example, the private television network CTV can purchase the rights to broadcast American shows for approximately one-tenth the cost of producing a Canadian series. As Taras (2001, 189) has indicated, this explains why private television broadcasters in Canada 'put as little as possible into Canadian content and squeeze the most out of imported Hollywood productions'. While noting that private media have done little to reflect Canadian culture in their programming, critical scholars have also been concerned about several specific types of private ownership.

Independent ownership is the most basic and least problematic of these types. It exists when the owners of a media company confine themselves to that one company and are not involved in the ownership of other firms. Their media company usually operates on a small scale. It is often closely associated with a local community and aims to serve that community. This form of ownership means that the newspaper, radio station, or television station in a small town or city might be owned by an entrepreneur who lives in the area. Independent ownership was once quite common in the Canadian media, but it has diminished as large companies have bought small media firms (Lorimer and Gasher 2001).

Horizontal integration is also known as chain ownership. It exists when one company owns a number of media organizations in different locations that are doing the same type of business. One company may own several newspapers, for example. Critical scholars contend that this form of ownership has negative implications. For instance, if a company owns a chain of newspapers, it could cut costs by using syndicated news stories across the chain and reducing the number of journalists and local stories at each of the newspapers (Hackett and Gruneau 2000). Therefore, when an independently owned newspaper becomes part of a chain, a number of jobs disappear along with some of the newspaper's local flavour.

TABLE 23.1 ▼ The Leading Media Organizations in Canada, 2009	
Organization	**Revenues (2009)**
Rogers Communications Inc.	$11,731,000,000
Quebecor Inc.	$ 3,781,000,000
Shaw Communications Inc.	$ 3,390,913,000
Canwest Global Communications Corp.	$ 2,867,459,000
Lions Gate Entertainment Corp.	$ 1,651,137,000
Torstar Corp.	$ 1,451,259,000
Cogeco Inc.	$ 1,252,794,000
Astral Media Inc.	$ 905,725,000
Corus Entertainment Inc.	$ 788,718,000
Canadian Broadcasting Corp.	$ 612,152,000

SOURCE: Adapted from *Financial Post*, FP500 Database, www.financialpost.com/news/FP500/list.html.

Vertical integration exists when one firm owns media enterprises that link processes such as production, distribution, and exhibition or retail. For instance, Quebecor Media owns the French-language broadcaster TVA as well as Videotron, a cable company in Quebec that carries TVA on its systems. Quebecor Media also owns Select (a distributor of CDs) and Archambault (a music retail chain in Quebec). Vertical integration enables a firm to have a guaranteed outlet for its products. However, critical researchers suggest that this form of ownership can result in content from other sources being shut out (Croteau and Hoynes 2000).

Cross-ownership exists when one company owns organizations that are associated with different types of media. CTVglobemedia is a Canadian example. The holdings of this company include the television network CTV, print media (including *The Globe and Mail*), and specialty television channels (such as CTV News Channel and the Comedy Network). Cross-ownership has certain advantages for a company, including the opportunity to share resources or personnel among its media outlets, but critical researchers argue that it can limit the diversity of journalistic opinions or media messages that are presented (Hackett, Pinet, and Ruggles 1996). When one company owns the number of media outlets that CTVglobemedia has, the same news stories and television shows will appear across the range of its holdings.

Finally, it is necessary to consider **conglomerate ownership**. A conglomerate is a company containing many firms engaged in a variety of (usually) unrelated business activities. This form of ownership may combine different linkages (horizontal and vertical integration and even cross-ownership). There are also different types of conglomerates. A media conglomerate does most of its business in the media. A non-media conglomerate has its foundation in other types of business but it might also own one or more media organizations (Lorimer and Gasher 2001). Critical researchers have been concerned about the content of the news media held by both types of conglomerates. In this regard, Hackett and Gruneau (2000, 60–1) identify 'two worrying implications'. First, news media owned by a conglomerate may be required to carry promotional material for other parts of the company. Second, and even more significant, news stories could be suppressed if they contain negative and damaging information about other aspects of the corporate empire. The suppression of news stories might occasionally stem from orders at the top of a conglomerate, but it is more likely to emerge from self-censorship by journalists or directions from editors when journalists present their stories. These news workers realize that corporate executives do not want such stories broadcast or published and act in a way that protects their jobs (Wasko 2001).

TIME to REFLECT

How much time do you spend with public media such as the CBC? In your view, how important is it to have public ownership of the media?

THE STATE AND MEDIA POLICY

The concept of the **state** is sometimes confused with that of the government. However, for sociologists, *the state* is a much broader term. Cuneo (1990) defines the state in Canada as encompassing various institutions: the federal, provincial, and local levels of government; the administration (including the civil service and regulatory agencies); parliamentary assemblies; the armed forces and police; intelligence agencies such as the Canadian Security Intelligence Service (CSIS); the legal, judicial, and court systems; prisons, reform institutions, and asylums; Crown corporations; and the institutions associated with public education, public health care, and public media that are under different levels of government.

While the state includes public media, other parts of the state have implications for both public and private media. As prepared by governments and passed by parliamentary assemblies, various acts associated with the legal system set out certain requirements for media organizations. For example, the Broadcasting Act indicates what is expected of organizations that provide public and private radio or television in Canada. The legislation makes it clear that these organizations must present Canadian programming. The latest version of the legislation, the 1991 Broadcasting Act, sets out a broadcasting policy that includes the following clause: 'each broadcasting undertaking shall make maximum use, and in no case less than predominant use, of Canadian creative and other resources in the creation and presentation of programming' (Canada 1991, 3.1.f). Regulatory agencies are also components of the state that have consequences for the media. Historically, Canada has had two independent broadcasting regulators. The Board of Broadcast Governors (BBG) was established in 1958 and replaced in 1968 by what is now referred to as the Canadian Radio-television and Telecommunications Commission (CRTC).

These regulatory agencies were created to help ensure that media organizations comply with media legislation by setting specific rules for the organizations to follow. For example, in relation to the Broadcasting Act, the requirement that radio and television undertakings must utilize Canadian resources in programming has been reflected in Canadian content regulations. These regulations were established by the BBG in 1960, and they initially required that at least 45 per cent of television programming be Canadian. The CRTC maintained the regulations for television and established similar regulations for radio in 1970. At first, a minimum of 30 per cent of the music on popular music stations had to be Canadian. Over the years, the CRTC has set various percentages of required Canadian content for different types of radio and television programming, but Canadian content regulations remain a key aspect of the agency's policy.

Analysis of media policy is often based on a key point in Marxist theories of the state. As Gold, Lowe, and Wright (1975, 31) noted, 'Marxist treatments of the state begin with the fundamental observation that the state in capitalist society broadly serves the interests of the capitalist class.' The state in Canada has served the interests of private media companies in a number of ways. For example, the federal government responded to the cable industry's desire for vertical integration by placing a clause in the 1991 Broadcasting Act that identified cable companies as distributors and programmers. While cable companies distributed television channels on their systems, the clause paved the way for these companies to own and program television channels (Raboy 1995). This is now the case, as illustrated by the ownership of the OMNI stations and Rogers Sportsnet by Rogers Communications. The CRTC's regulatory process has also done much to assist private media companies. According to Mosco (1989, 57), 'this formal regulatory process generally serves the interests of communications companies and large corporate users of communications systems.' For instance, since the CRTC has taken 'a permissive attitude to industry mergers' (Mosco 1989, 212), the agency has given regulatory approval to the deepening ownership concentration that worries critical scholars and others. A recent example is the CRTC's decision in 2007 to allow the acquisition of CHUM by CTVglobemedia despite the concerns of unions and citizens (Smith 2007a; 2007b).

Some critical scholars suggest that the role of the state has been decreasing as a result of neo-liberalism. **Neo-liberalism** is an economic doctrine that is favoured by private companies and has been adopted by many governments around the world since the late 1970s. The doctrine of neo-liberalism supports free trade between countries, cuts in social spending, and measures such as deregulation and privatization. Deregulation means that regulatory agencies reduce or eliminate rules they had previously imposed on organizations. For instance, under its 1998 Commercial Radio Policy, the CRTC reduced restrictions on how many radio stations private companies could own in a single market (Canadian Radio-television and Telecommunications Commission 1998). Privatization, which means that organizations under public ownership are transferred to private ownership, has also been apparent in the Canadian media; during the 1990s, the government of Alberta sold the Access Network, its educational television broadcaster, to CHUM (which rebranded the service as Access: The Education Station).

GLOBALIZATION AND THE MEDIA

In recent decades, the issue of globalization has been the focus of much analysis by sociologists. **Globalization** involves the flow of goods, services, media, information, and labour between countries around the world. Researchers often examine different but interrelated aspects of globalization. *Economic globalization* concerns worldwide production and financial transactions, while *cultural globalization* refers to 'the transmission or diffusion across national borders of various forms of media and the arts' (Crane 2002, 1).

Technological factors are associated with the deepening impact of globalization. Computers and telecommunication technologies make possible the instantaneous transfer of data, which has contributed to the formation of a transnational financial system and facilitated worldwide production by the multinational corporations that are central to economic globalization. Linkages between satellite and cable technologies have played an important role in cultural globalization by enabling the news media to make us almost immediately aware of important events as they are occurring even half a world away. However, despite the discourse of *technological determinism* (the notion that technologies themselves cause changes in society), information and communication technologies provide only partial explanations for globalization (Nash 2000; Winseck and Pike 2009).

Globalization is also being driven by a complex mixture of economic and political factors. Economic factors include deepening ownership concentration within countries and across national borders as well as the international impact of neo-liberalism and free trade between countries. These economic factors are tied to political factors. The latter include the role of the state in assisting private capital, partly through negotiation of the North American Free Trade Agreement (NAFTA) and similar international treaties. Political factors also include the emergence of the World Trade Organization (WTO), an international institution that enforces trade rules for member countries and has thereby reduced the control of governments over their own economies (Karim 2002). These various developments have generated a number of issues that concern critical media sociologists.

Cultural Globalization

In relation to cultural globalization, critical researchers are concerned about the historically deepening and worldwide impact of media industries.

We can illustrate this through reference to the American motion picture industry. The dominance of the American film industry was established soon after the earliest Hollywood productions, at the beginning of the twentieth century. By 1939, Hollywood was already supplying 65 per cent of the films shown in theatres worldwide. This export flow expanded dramatically after World War II, and the United States was providing more than 80 per cent of the world's films in the 1990s (Miller et al. 2001). Major importers of Hollywood films include Canada, Japan, the Netherlands, and the United Kingdom. American films generate at least half of the total box-office receipts in all their major markets and sometimes even more than two-thirds of total receipts (Scott 2004). In Canada, Hollywood films accounted for 88.5 per cent of box-office receipts during 2008 (Canadian Film and Television Production Association 2009).

Several factors help to explain the dominance of the American film industry. For instance, the ownership structure of the industry has played a crucial role. The vertical integration of production, distribution, and exhibition during the early history of the Hollywood studios ensured that the films these studios made were seen in the United States and in many international markets. In Canada and elsewhere, this made it more difficult for domestic films to secure theatrical exhibition (Miller et al. 2001; Pendakur 1990). Hollywood production companies usually do not own movie theatres any more, but they still have substantial distribution operations around the world. Furthermore, as the Hollywood studios have increasingly come under conglomerate ownership since the 1980s, massive and often non-American multinational firms such as the Sony Corporation (which owns Columbia Pictures) have developed strategies to pursue global audiences (Miller et al. 2001).

It is also important to note that the state in various countries has contributed to the global dominance of Hollywood. The film industry in the United States has prospered internationally in part because it has 'a willing servant in the state' (Miller et al. 24). Under pressure from the Motion Picture Association of America (MPAA), which represents the major film studios, American federal bureaucracies have pushed governments in other countries to satisfy Hollywood's interests (Scott 2004). In Canada, this has historically resulted in several successful efforts to discourage the federal government from establishing measures to protect the Canadian film industry, including quotas that would have placed limits on the importation of American films (Magder 1993).

GLOBAL ISSUES

The Global Music Industry

Worldwide, the music industry is dominated by four groups. The 'Big Four' are the Universal Music Group, Sony BMG Music Entertainment, the Warner Music Group, and the EMI Group. These groups are made up of record companies, record labels (various brand names on recordings), and music publishers. Music organizations not held by the Big Four are considered to be independent.

Some of the Big Four are part of massive conglomerates. The Universal Music Group is owned by the French conglomerate Vivendi. Since purchasing Universal in 2000, Vivendi has sold most of its interests in Universal Entertainment (which became NBC Universal) while retaining the Universal Music Group. Sony BMG Music Entertainment was created in 2004 through a merger of the music divisions of two companies. Sony Music is part of the Sony Corporation, a Japanese conglomerate, while BMG is a component of the German conglomerate Bertelsmann. The Warner Music Group was once part of the conglomerate AOL/Time Warner, but the music division of the company was purchased in 2004 by a number of investors. Like the Warner Music Group, the EMI Group is not held by a conglomerate (Bishop 2005).

The Big Four control most of the music market in the United States and around the world. The Universal Music Group was in the leading position during 2009, with 30.7 per cent of the American market. It was followed by Sony BMG Music Entertainment with 28.0 per cent, the Warner Music Group with 20.3 per cent, and the EMI Group with 8.8 per cent. The independent sector held the remaining 12.2 per cent of the market (Christman 2010).

The global domination of the Big Four is problematic, because the concentration of *ownership* in the music industry is also the concentration of *power* (Bishop 2005). A few companies get to make decisions about the recorded music heard by people around the world. Many of these decisions are based on generating massive sales from a small number of artists who have enough mainstream appeal to become global superstars. Consequently, the power of the Big Four has implications for the degree of diversity and innovation in music.

Independent record companies are more willing than the Big Four to take risks on less mainstream artists (Bishop 2005). However, these small companies face difficulties stemming from the presence of their multinational competitors. The difficulties are familiar to the Independent Music Publishers and Labels Association (IMPALA), which represents the independent music sector in Europe. Martin Mills, the chair of IMPALA, has argued that 'four big companies can impose their will on retail and media in a way that 15 did not. In battling each other for space and attention—and leveraging their strength—they intentionally or unintentionally reduce opportunities for smaller players' (Mills 2006, 8). He concluded that 'a concentrated market carries dangers to musical diversity, to smaller companies and to music fans' (2006, 8).

Economic Globalization

With regard to economic globalization, critical researchers are concerned about the emergence of an *international division of labour*. Sociologists and other scholars have analyzed how multinational corporations, including those with holdings in the media and information industries, have spread their production operations around the world. The standard view of this process suggests that it involves shifting jobs from developed, rich countries (such as the United States) to developing, poor countries (like India). Although the trend is certainly in the direction of moving jobs to the developing world, Mosco (2005, 52) has pointed to 'an increasingly complex international division of labour involving far more than simply the transfer of service jobs from high- to low-wage nations'.

This is illustrated by the fact that several developed countries, especially Canada and Ireland, have been the recipients of much outsourcing and offshoring of jobs in the media and information industries. *Outsourcing* occurs when a company shifts a portion of its production to another entity, typically independent local companies in a foreign country. *Offshoring* exists when a company has one of its own foreign affiliates handle the production. Although developing countries offer cheaper labour and other advantages, multinational corporations maintain some outsourcing or offshoring in developed countries because they need certain jobs to be filled by workers with higher levels of skill or education (Mosco 2005). Because of the value of the dollar and wage rates in these developed countries, multinationals can still enjoy considerable savings on production costs compared to keeping production in the United States. These factors explain the existence of so-called 'runaway' productions. Hollywood studios have moved a number of film and television productions from Los Angeles to Canadian cities—everything from the *X-Men* movie trilogy to *The X-Files* television series—in order to cut costs while utilizing the expertise of Canadian companies and production crews (Elmer and Gasher 2005).

We can illustrate some aspects of the international division of labour through the case of the Walt Disney Company. Wasko has shown that Disney arranges to have some of its film, television, and animation production done in countries such as Canada and Australia. However, when manual labour is involved, as in the production of toys or clothing that features Disney characters, Disney has most of the work done in less developed countries to take advantage of cheap labour. The products are designed by Disney in the United States, but the actual manufacturing is licensed to independent subcontractors in developing countries. Many of the toys and clothing sold in the Disney Store at your local shopping mall are made 'in Third World countries where workers are paid poverty-level wages and often work in inhumane conditions' (Wasko 2001, 69).

CONFLICT OVER OWNERSHIP, POLICY, AND GLOBALIZATION

Capitalist interests in ownership, policy, and globalization have generated conflict between private companies and various subordinate groups. Although the state serves the interests of the capitalist class, Marxist theory suggests that the state makes some concessions to the working class and its allies. Consequently, despite the power of corporate capital and its influence on the state, subordinate groups have occasionally won victories through their resistance.

We can see this in the historical conflict over public and private ownership of broadcasting in Canada. Broadcasting in this country began during the 1920s with private radio stations, but a federal government commission soon recommended that all broadcasting in Canada be publicly owned and operated by a Crown corporation. Consequently, the

(Ivy Images)

Mickey doesn't have a monopoly yet, but like other media conglomerates, the Walt Disney Company seeks to gain greater market share for its various enterprises.

period from 1930 to 1936 was marked by 'a struggle between the popular forces in Canada fighting for public service broadcasting and those seeking private profit' (Smythe 1981, 165). The 'popular forces' were organized around the Canadian Radio League (CRL). The CRL was joined by trade unions, farm groups, women's organizations, churches, and educational leaders in calling for public broadcasting. The advocates of using broadcasting for private profit were led by the Canadian Association of Broadcasters (CAB), an organization that represented the existing private radio stations. Supporters of the CAB included many companies that wanted to use private radio for advertising (Raboy 1990; Smythe 1981). In the end, the CRL and its allies were successful in their struggle for public broadcasting; the federal government established the Canadian Radio Broadcasting Commission (CRBC) as Canada's first public broadcaster in 1932, and the CRBC paved the way for the emergence of the CBC as a Crown corporation in 1936. However, private broadcasting was allowed to continue and grow in Canada.

The Canadian content regulations for television and radio referred to earlier are important aspects of regulatory policy, and they have long provided the basis for conflict. Private broadcasters do not want the regulations; in the pursuit of profit, they are more interested in maximizing audiences and advertising revenues by offering popular American television shows and music. A different position has been taken by Canadian nationalists as well as unions representing actors, musicians, and other workers in Canadian media industries. They contend that Canadian content regulations are necessary to provide jobs for Canadian artists and ensure that Canadian culture is presented over the Canadian airwaves. In the late 1950s, the BBG's proposal for a minimum percentage of Canadian content on television was opposed by the CAB as well as by advertisers (Peers 1979). However, the proposal was supported by the Canadian Broadcasting League (CBL)—an organization that had been inspired by the CRL—and the Canadian Labour Congress, which represented workers (Raboy 1990). Similarly, while the CAB was against the CRTC's notion of having Canadian content regulations for radio, those in favour of such regulations included labour unions. There is still ongoing conflict that takes the form of debate about whether or not Canadian content regulations are needed—a debate that is especially fierce in relation to radio (Young 2008).

In recent years, much conflict associated with the media has been connected to globalization. This is most obvious in the protests against economic and cultural globalization that various social movements have held at WTO meetings and at the meetings of other international bodies that focus on economic growth, but conflict over globalization also takes other forms. For instance, in developed countries like the United States, trade unions have resisted outsourcing in the media and information industries. While these unions have resisted the loss of jobs to the Third World, they have also challenged the shifting of work to other developed countries (Mosco 2005). For example, there has been considerable opposition from labour in the United States to the movement of film and television production to Canada (Magder and Burston 2001).

(Reprinted with permission. Torstar Syndication Services)

Opponents of Canadian content regulations, especially private broadcasters, claim that CRTC regulations are an attack on the freedom of broadcasters.

TIME to REFLECT

Imagine that you are one of the protesters against economic and cultural globalization. Which particular issues would you like to focus on in your opposition to powerful media corporations, and why would these issues be important to you?

Cultural Studies of the Media

As noted earlier, while political economy focuses on ownership and control of the media, cultural studies address the ideological aspects of the media. Three key issues in cultural studies need to be discussed.

They are representation in mainstream media, interpreting and resisting mainstream media, and opposition through alternative media.

REPRESENTATION IN MAINSTREAM MEDIA

The mainstream media include the newspapers, magazines, radio stations, or television channels that most people are exposed to every day. These means of communication are owned by private companies or by the government. The mainstream media present *texts* (such as newspaper and magazine articles or television shows) that convey certain messages about society and groups in society. Critical media sociologists and other scholars argue that these messages reflect the **dominant ideology**. In other words, the messages express the viewpoints of the capitalist class and other powerful groups. Capitalist, patriarchal, or racist ideologies are some specific forms of the dominant ideology that have been embedded in media texts.

We can investigate aspects of the dominant ideology by considering the representation of the working class, women, and racial or ethnic minorities in the mainstream media. These groups have little power and receive poor representation in media content. This takes the form of *under-representation* (since members of less powerful groups are usually not seen in the media as frequently as they actually exist in society) and *misrepresentation* (because members of these groups often are portrayed in ways that are stereotypical and negative).

OPEN for DISCUSSION

Do We Need CanCon Regulations for Radio?

Canadian content regulations for radio stations in Canada were established in 1970. The regulations are administered by the Canadian Radio-television and Telecommunications Commission (CRTC). Currently, CanCon regulations (as they are often known) stipulate that at least 35 per cent of the music on radio stations featuring popular music must be Canadian. The amount of Canadian content required by the CRTC has varied over time and in accordance with the type of music played by a station. To qualify as Canadian, a piece of music must meet at least two of the following four criteria: the music is composed entirely by a Canadian; the lyrics are written entirely by a Canadian; the music or lyrics are performed principally by a Canadian; or the musical selection consists of a performance that was either recorded wholly in Canada or performed wholly in Canada and broadcast live in Canada.

There has been considerable debate about whether or not we need Canadian content regulations. Opponents of CanCon include private radio broadcasters and some Canadian musicians. Supporters of the regulations include other Canadian musicians, labour unions that represent Canadian musicians, Canadian nationalists, and the owners of Canadian independent record companies. The debate between the opponents and supporters of CanCon has focused on several issues, two of which are outlined below.

The first issue concerns whether or not radio stations should be forced to play Canadian music. Opponents of CanCon, especially owners of private radio stations, contend that CanCon regulations are an attack on the freedom of broadcasters. They maintain that radio broadcasters should have the right to play whatever music they want and whatever music their listeners want to hear. Supporters of CanCon argue that the owners of private radio stations are making money off the Canadian airwaves, which are public property and subject to legal requirements under the Broadcasting Act. In exchange for being given the opportunity to use the public airwaves for profit, they point out, it is reasonable for private radio broadcasters to meet some public service obligations by playing music that is Canadian.

The second issue concerns whether or not Canadian artists need CanCon regulations. Opponents of CanCon argue that if Canadian artists have talent, they will make it on their own without assistance from the CRTC. In their view, radio stations will be happy to play music by Canadian artists as long as their music is good. Supporters of CanCon maintain that Canadian artists will not get on the radio unless they are already well known or unless they are signed to or distributed by one of the big companies that dominate the music industry. They contend that developing artists (those who are unsigned or signed to small, independently owned Canadian record companies) will not receive airplay unless there are regulations that force radio stations to give them a chance to be heard.

Now that you have learned about some of the points made on each side of the debate, which side do you agree with?

SOURCE: Adapted from Young 2008.

Representation of the Working Class

Some research has been done on the problematic representation of the working class, and this work is strikingly illustrated in a series of studies that were conducted on American domestic situation comedies.

After researching all of the domestic situation comedies that appeared on American television between 1946 and 1990, Butsch (1992) identified the class position of the family in each show through the occupation of the lead male character. The family was considered to be working class if the lead character was a blue-collar worker, clerical worker, retail worker, or service worker. The criteria for indicating that the family was middle class included the existence of a professional or manager. Some comedies fell outside these categories because they featured a lead character who was self-employed or independently wealthy, and it was not possible to identify the occupation of the lead character in certain shows. Once he had categorized the comedies, Butsch examined the data.

He demonstrated that there were significant differences in the extent to which social classes were represented in domestic situation comedies. Butsch could not specify the occupation of the household head in 11.6 per cent of the comedies, but he found that 5.7 per cent featured a family that was independently wealthy and 1.1 per cent portrayed a family led by a farmer. However, his key findings were that 70.5 per cent of domestic situation comedies were about a middle-class family and only 11.0 per cent centred on a working-class family. In earlier research, Butsch and Glennon (1983) noted that similar findings were out of line with the existence of both classes in American society; based on census data, 28.7 per cent of actual household heads in the United States were middle-class while 65.0 per cent were working-class. Thus, based on research conducted up to 1990, the middle class was overrepresented in domestic situation comedies and the working class was under-represented. Unfortunately, no studies have been done to update these findings.

In his research on domestic situation comedies, Butsch found that working-class men generally represented negatively. Emphasis was placed on their 'ineptitude, immaturity, stupidity, lack of good sense, or emotional outburst' (Butsch 1992, 391). The women in working-class comedies (and sometimes even the children) were portrayed as being much more intelligent and level-headed. In these comedies, the humorous situation typically involved the husband/father. The situation was often one of his own making, and he was usually helped out of it by his wife (Butsch 1992). As Butsch (1995) pointed out, this scenario describes various working-class domestic situation comedies and their lead male characters in decades from the 1950s through to the 1980s. Consider *The Honeymooners* (Ralph Kramden), *The Flintstones* (Fred Flintstone), *All in the Family* (Archie Bunker), and *The Simpsons* (Homer Simpson).

In contrast, Butsch found that middle-class men often received positive representation. They were portrayed as being 'intelligent, rational, mature, and responsible' (Butsch 1992, 391). The women in middle-class families were shown as also having these characteristics in their roles as wives and mothers. In middle-class comedies, the humorous situation typically involved one of the children. The parents guided the child through the situation, and they often provided a moral lesson in the process. Examples of these comedies from the 1950s to the 1980s include *Father Knows Best*, *My Three Sons*, *The Brady Bunch*, and *The Cosby Show* (Butsch 1992).

In research that extended analysis of domestic situation comedies to the decade of the 1990s, Scharrer (2001, 33) confirmed Butsch's findings. She noted that 'the lower the social class of the sitcom father the more foolish the portrayal will be.' Such findings are significant. Butsch (1995, 404) argued that the stark differences between the representation of working-class and middle-class men ideologically justify inequality in our class-divided society: 'Blue-collar workers are portrayed as requiring supervision, and managers and professionals as intelligent and mature enough to provide it.'

Representation of Women

A great deal of research has been done on the problematic representation of women. We can illustrate some aspects of this work through a few American and Canadian studies.

Dole (2000; 2001) examined the representation of women as law enforcers in American motion pictures that were released during the 1980s and 1990s. Dole (2000, 11) argued that the women in these films had 'types of power culturally coded as masculine'. The women had power because they occupied the position of law enforcer and because they carried a gun (two characteristics socially defined as 'masculine' within our culture). Dole saw the genre of women cop films as emerging in two phases. The earlier films (1987–91), such as *Blue Steel* and *Impulse*, often imitated the physicality and violence of male action films by showing the women using their guns. Because many of these films were commercially unsuccessful, the later films (1991–5) took a softer approach. These films, including *The Silence of the Lambs* and *Copycat*, were more inclined 'to privilege intellectual over physical power' (2000, 12). Rather than using their guns, the female law enforcers in the later films relied on their sleuthing skills.

Several other techniques were employed to play down 'the threatening image of Woman with a Gun' (Dole 2000, 16). These techniques included *domestication* (portraying the female cops as single mothers or at least as women who have 'maternal instincts'), *infantilization* (representing the women as being dependent, vulnerable, helpless, or in need of rescue), and *sexualization* (emphasizing the bodies of the women through the provocative way they are dressed). Finally, the films that focused on intellectual power utilized what Dole called *splitting strategies*. Splitting strategies distribute among multiple characters the power that would otherwise be concentrated in one character. Through the use of splitting strategies, the power of the female law enforcer is reduced. This can be illustrated by *The Silence of the Lambs*. In

that film, intellectual power was split between Clarice Starling (Jodie Foster) and Hannibal Lecter (Anthony Hopkins). Although Starling was intelligent, she needed male assistance in the form of Lecter (Dole 2000, 16). The ideological message was that a woman is incapable of solving the case and catching the killer on her own. Although female cop films gave women more representation than most movies (because they occupied central roles rather than peripheral roles), the stereotypical and patriarchal misrepresentation of women was still quite evident in these films.

It is also important to consider how the news media are connected to the representation of women or issues associated with women, and some Canadian research in this area has focused on news coverage of what is referred to as the Montreal Massacre. In December 1989, Marc Lépine went to the University of Montreal with a semi-automatic rifle. He walked into a classroom, ordered the men to leave, and accused the women of being 'a bunch of feminists' before shooting six of them to death. Lépine then entered other classrooms and murdered eight more women. He also injured nine women and four men. At the end

(Reuters/Ho Old)

FBI agent Clarice Starling, played here by Julianne Moore, wields her firearm in the movie *Hannibal* (2001). Dole, citing films including *Hannibal*'s prequel, *The Silence of the Lambs*, argues that Hollywood has preferred to show women cops using their wits and being dependent on males; this scene of a 'woman with a gun' is an exception to the rule.

of his shooting rampage, he killed himself. In a suicide note found on his body, Lépine cited 'political reasons' for the murders: he blamed 'the feminists, who have always ruined my life' (cited in Eglin and Hester 1999, 256). There was much coverage of the Montreal Massacre in the news media, and some studies have been done of this coverage.

In one study, Hayford compared newspaper coverage of the Montreal Massacre at the time of the murders to coverage of a similar incident in Chicago during 1966, when eight women were killed by a man named Richard Speck. She found that the killings in Chicago were often interpreted by journalists in individualistic terms as the act of a madman. However, by the time the murders in Montreal occurred 23 years later, the women's movement had experienced some success in raising public awareness about the prevalence of wife-beating, rape, and other acts of violence against women in a patriarchal society. Since this had an impact on at least some journalists, there was media debate about individual versus societal explanations for the murders. Hayford (1992, 209) indicated that 'the question of whether Lépine was no more than a demented individual or a reflection of broader social patterns of male violence against women, a question never raised about Speck, became a central issue in coverage of the Montreal killings.'

In another study, Rosenberg examined news coverage of the tenth anniversary of the Montreal Massacre. One aspect of her study addressed the 'emblematic' characteristics the coverage had taken on by this point. She noted that in order to contest individualistic and psychological explanations for Lépine's actions, many feminists had encouraged interpretation of the Montreal Massacre as an emblem or symbol of violence against women. However, by this time, some feminists had come to see the 'emblematic' interpretation of the Montreal Massacre as problematic because of the growing emphasis within feminism on the issue of difference. These feminists argued that taking the murders of 14 middle-class white women as emblematic obscured diversity among women in terms of their experiences with violence through other class positions and racial or ethnic backgrounds. Nevertheless, Rosenberg (2003, 15) found that the 'critiques of emblemization by feminists' were 'largely absent in tenth anniversary coverage'.

Representation of Racial and Ethnic Minorities

We also need to consider the under-representation and misrepresentation of racial and ethnic minorities in the mainstream media.

Many Canadian studies have documented the under-representation of racial and ethnic minorities. Researchers have found that although Canadian society features growing cultural diversity, this diversity is not usually reflected in media content. For example, my own research shows that ethno-racial minorities have rarely been seen at the annual and nationally televised Juno Awards ceremony for the Canadian music industry. Francophones, blacks, and Aboriginal peoples were almost never among the musical artists who appeared on the ceremony or won Junos during the 1970s and 1980s. There has been some improvement since then, but the Juno Awards ceremony still does not adequately reflect the cultural diversity of artists in Canada's music scene (Young 2006). Other scholars have arrived at similar findings. The general absence of racial and ethnic minorities (relative to their existence in the actual population) is apparent in advertisements, magazines, news, television series, and other forms of media content (Fleras and Kunz 2001; Mahtani 2001). The under-representation of these minorities is significant, because it means that their contributions to Canadian society are trivialized and their roles as Canadian citizens are devalued (Mahtani 2001).

Canadian studies have also shown that racial and ethnic minorities have experienced misrepresentation. To the extent that they are seen, ethno-racial minorities are frequently portrayed in stereotypical and negative ways. In the news media, this often takes the form of identifying them as social problems; racial and ethnic minorities are depicted as 'having problems or creating problems in need of political attention or costly solutions' (Fleras and Kunz 2001, 145). Members of these minorities are presented as social problems in a variety of ways. They are seen to be participating in illegal activities, clashing with police, cheating on welfare, creating difficulties for immigration authorities, and having other undesirable effects (2001, 145). Specific groups—including Aboriginal peoples, Asians, blacks, and Muslims—are often singled out. For instance, Karim (2008) demonstrated that editorials and columns opposing multiculturalism have appeared in Canadian English-language newspapers after news reports about the arrests of Muslim men on terrorism-related charges.

EXPLAINING THE REPRESENTATION

How can we explain the under-representation or misrepresentation of the working class, women, and racial or ethnic minorities? First, we must reject the notion that there is a plot or 'conspiracy' by powerful groups against less powerful groups. The circumstances are far more complex than that, and we must return to the concept of the dominant ideology to understand why. According to Hall, the dominant ideology is woven into media texts through **encoding**. Messages are constructed within the economic and technical frameworks of media institutions through a complicated production process that involves (among other things) organizational relations or practices and 'meanings and ideas' drawn from the production structure (the media institutions) and 'the wider socio-cultural and political structure' (Hall 1980, 129). Therefore, in order to grasp the representation of subordinate groups, we need to consider some of these factors in more depth.

Economic factors associated with media institutions help to partially account for the representation of subordinate groups in the entertainment media. For instance, Butsch (1995) noted that the under-representation of the working class in domestic situation comedies has much to do with the need for producers and broadcasters to develop programs that will attract advertisers by providing a good atmosphere for products. There is a tendency, then, to create shows that feature middle-class characters and occupational groups that can afford to buy the products appearing in the ads. Furthermore, to the limited extent that working-class domestic situation comedies have been made, their persistently negative representation of working-class men exists in part because producers and broadcasters avoid financial risk by relying on formulas that have proven to be successful. Thus, the popularity of *The Honeymooners* in the 1950s spawned *The Flintstones* in the 1960s.

Factors associated with production and ideology also help to explain the problematic representation of subordinate groups in the entertainment media. Members of less powerful groups often do not occupy important positions associated with media production. For example, women have little control over production in the film industry; women comprised only 16 per cent of all directors, producers, executive producers, writers, cinematographers, and editors who worked on the top-grossing American films for the year 2009 (Horowitz 2010). Such exclusion from the process of media production can have a substantial impact on media content. Butsch (1995) makes this clear when he notes that the under-representation of the working class in domestic situation comedies, along with the negative representation of working-class men, can partially be explained by the middle-class background of most producers and writers. Middle-class people develop shows based on what is familiar to them, and when they occasionally focus on working-class characters, they rely on the negative stereotypes of the working class that circulate in our culture as part of the dominant system of meanings and ideas.

Such factors also help to account for the representation of subordinate groups in the news media. The need to attract advertising revenues by capturing large numbers of readers or viewers has contributed to the under-representation of the working class and ethno-racial minorities in news stories. For instance, since affluent middle-class audiences are desired by advertisers, the content of the news is designed to attract these audiences by reflecting their interests or concerns (Hackett and Uzelman 2003). The middle-class male, white backgrounds of many news personnel also have at least some consequences for how less powerful groups are covered (Mahtani 2001). Like that of producers and writers in the entertainment media, the work of journalists is partially shaped by their socio-demographic backgrounds and the dominant meanings and ideas they are exposed to. As Mahtani (2001, 115) has indicated, 'journalists are largely bound by the dominant cultures within which they operate, including embedded societal prejudices, stereotypes, and populist frames of thinking.'

INTERPRETING AND RESISTING MAINSTREAM MEDIA

In cultural studies, research has gone beyond studying representation in mainstream media to considering interpretation of this representation and resistance to representation. We have seen that the dominant ideology is embedded in media texts through the process of encoding, but it is also important to consider the **decoding** of media content by audience members. As part of his encoding/decoding model, Hall argued that the dominant ideology is inscribed as the *dominant* or *preferred meaning* within media content.

Since Hall recognized that audience members may not always adopt this meaning when they interpret media messages, he identified three possible ways of decoding media texts. A *dominant-hegemonic* reading involves taking the preferred meaning, while an *oppositional* reading involves resisting a message by interpreting it through an alternative ideological framework. A *negotiated* reading contains a mixture of the dominant-hegemonic and oppositional readings. According to Hall (1980, 137), it reflects 'the dominant definition of events' while refusing to accept every aspect of the definition. Consequently, 'this negotiated version of the dominant ideology is shot through with contradictions' (1980, 137). For example, a worker may accept the argument of government officials (as presented by the news media) that the 'national interest' requires citizens to make economic sacrifices, while opposing the related argument that such sacrifices must be made through legislation imposing wage freezes (1980, 137).

Research has been done on the decoding of media content by relatively powerless groups. In order to examine the approaches to decoding that Hall identified, Morley (1980) investigated how groups interpreted the British current affairs television series *Nationwide*. Morley's seminal study demonstrated that decoding is affected by one's class position. For instance, in relation to a *Nationwide* program about the effects of budget policy on families, Morley found that no middle-class groups adopted an oppositional reading, while working-class groups produced more oppositional and negotiated readings. Although the research on *Nationwide* is often remembered for its analysis of how class position affects interpretation of media texts, Morley (2006) has stressed that his aim was to examine how decoding is also influenced by other types of social position (such as gender, race/ethnicity, and age). These elements of Morley's study have come through more clearly in a

HUMAN DIVERSITY

Differences in Media Consumption

For many decades, media sociologists have theorized and investigated media consumption patterns among groups within societies and between people in different societies.

It had long been thought that individuals who like 'high culture' (such as classical music or opera) look down on and avoid 'popular culture' (such as rock music or country music). However, Peterson (1992) developed the 'highbrow omnivorousness' hypothesis: the empirical expectation that highbrows (those who like high culture) consume a wider variety of culture and have less limited tastes than non-highbrows. Using survey data based on a representative sample of Americans, Peterson and Kern (1996) found empirical evidence to support the hypothesis as it applied to the consumption of music. They discovered that highbrows (people who identified classical music or opera as their best-liked form of music) tended to be white, older, more educated, and have a higher annual family income than other individuals in the sample while also being inclined to like a more varied selection of music than those individuals who identified other types of music as their best-liked form.

Lizardo and Skiles (2009) tested the 'highbrow omnivorousness' hypothesis in relation to both music and television consumption in European countries. They came up with a number of findings based on representative survey data from 2001 that identified the music and television consumption choices of individuals in various European countries. Consistent with the findings of Peterson and Kern, the researchers uncovered strong support for the 'highbrow omnivorousness' hypothesis in the case of

music, because highbrows in most of the European countries under study consumed a wider variety of music than non-highbrows. It was a different story in the case of television, though. There was support for the hypothesis in Greece and Portugal, where highbrows did not limit their television viewing to elite genres such as documentaries or news and instead consumed a more varied selection of television programming than non-highbrows. However, two other findings did not support the hypothesis. The television viewing choices of highbrows were indistinguishable from those of non-highbrows in several countries (Belgium, Germany, Italy, Spain, Ireland, the Netherlands, the United Kingdom, Finland, and Sweden). Furthermore, in some countries (Denmark, France, Luxembourg, and Austria), highbrows were more inclined to limit themselves to elite fare and largely avoided the popular genres (drama series, talk shows, movies, and so on) that were consumed by non-highbrows.

Lizardo and Skiles provided a partial explanation for these findings. They developed the hypothesis that 'highbrow omnivorousness' tends to exist in countries that have less commercialized (i.e., less profit-oriented) television systems rather than in countries that have more commercialized systems; their reasoning was that through less commercialized systems, highbrows may find more programming that appeals to them across the range of television genres. Comparing their data on the degree of 'highbrow omnivorousness' with data on the extent to which television systems in European countries are profit-oriented, Lizardo and Skiles (2009, 13) found results that were 'broadly consistent' with their hypothesis.

SOCIOLOGY in ACTION

Watching Homeless Men Watch *Die Hard*

Fiske and Dawson (1996, 297) investigated 'the process in which audiences selectively produce meanings and pleasures from texts' by examining how homeless men interpreted television.

The researchers approached authorities who ran a homeless shelter in an American city and received permission to conduct their study. Fiske and Dawson then spent time at the shelter until the homeless men felt comfortable with them. Eventually, they started collecting data on how the men used a television set and video-cassette recorder in the shelter's lounge.

Fiske and Dawson gathered data unobtrusively. They just watched television with the homeless men. The researchers carefully observed reactions of the men to what was being watched and made notes afterwards.

During the study, the homeless men borrowed the film *Die Hard* from a library. Fiske and Dawson analyzed reactions of the men to this movie. The movie's main character is John McClane (Bruce Willis), an off-duty detective who happens to be in the office tower of the Nakatomi Corporation at the time thieves take company executives hostage. The thieves want the millions of dollars in the company's vaults. The plot of the film involves McClane's efforts to kill the thieves as the police try to contend with the situation from outside the building.

Fiske and Dawson made interesting observations about reactions of the homeless men to *Die Hard*. They found the men paid attention to 'violence that was directed against the social order' (1996, 301). For instance, the homeless men cheered when the thieves killed the head of the corporation after he refused to give them a computer code to access the company's vaults. They also cheered after the thieves fired a rocket at an armoured police vehicle, destroying the vehicle and killing the policemen in it. In contrast to the way people are intended to interpret the film, the homeless men were enthusiastic about attacks on corporate capital and the police.

What explains the reactions of homeless men to *Die Hard*? Their reactions must be understood in terms of their position in society and the reasons for their position. Although the dominant view of the homeless assumes they are to blame for their situation, Fiske and Dawson stress that homelessness stems from 'the contemporary conditions of US capitalism' rather than individual failings (1996, 301). In the years prior to their study, neo-liberal policies had 'minimized the role of the state in social life and maximized that of capital and the market' (1996, 302). Millions of jobs had disappeared, and government assistance to the poor was reduced. This generated increased homelessness. Fiske and Dawson also note that riot police had cleared the homeless from parks in American cities and confronted activist homeless groups.

The experiences of the homeless with capitalism and the police make it possible to understand their reactions to violence in *Die Hard*. Fiske and Dawson conclude that 'certain representations of violence enable subordinated people to articulate symbolically their sense of opposition and hostility to the particular forms of domination that oppress them' (1996, 304).

---

statistical re-analysis of his data conducted by Kim. For example, in relation to age, Kim (2004, 88, 91) found that 'the *younger* working-class viewers are, the more probability they have of producing *dominant* readings. . . . The more youthful viewers' consent to the preferred meanings can be explained by their relatively low-level of political consciousness, which is commonly found among youths in general.'

TIME to REFLECT

Can you think of some ways in which your class position, gender, race/ethnicity, and age might affect your decoding of media messages?

OPPOSITION THROUGH ALTERNATIVE MEDIA

While relatively powerless groups have engaged in resistance to problematic representation and messages in mainstream media, some of these groups have turned to opposition through alternative media. **Alternative media** are forms of communication used by subordinate groups and social movements to present their own messages, which often involve challenging existing conditions in society.

Many types of alternative media have been employed by groups and movements committed to social change, but community broadcasting has historically played a particularly important role in Canada. During the early 1970s, community broadcasting started operating through small radio stations or 'public access' television channels on privately owned cable systems. This enabled social activists and members of marginalized groups to express themselves. However, as private cable companies started to assert their ultimate control over community channels in the late 1970s, groups committed to social change largely gave up using community television to achieve their goals (Goldberg 1990). Although community television lost its focus on the disadvantaged and social change, community radio still plays an important role in these respects. For instance, Vancouver Co-operative Radio is a non-commercial station that is owned and run by its members. According to its website (www.coopradio.org), the station is 'a voice for the voiceless that strives to provide a space for under-represented and marginalized communities'.

Other types of alternative media have also been used by oppositional movements in Canada. Carroll and Ratner (1999; 2001) considered the use of alternative media when investigating the political strategies of different social movement organizations. They noted that The Centre, an organization for gay men and lesbians in Vancouver, was heavily involved in alternative media. The Centre had established the monthly newspaper *Angles* and *The Coming out Show* on Vancouver Co-operative Radio. Another organization that Carroll and Ratner examined was End Legislated Poverty (ELP). Committed to mobilizing the poor and fighting various state policies that perpetuate poverty, ELP had engaged in popular education through alternative media directed at the poor and the general public. ELP's alternative media included *Fighting Poverty Kits* and a newspaper, *The Long Haul*. Carroll and Ratner found that activists in these movement organizations could talk strategically about how they were using alternative media and other means to struggle for social change. In an interview conducted by Coburn (2010), Carroll suggested that they would have had similar findings if these studies were done now.

There is a continuing need for alternative media. As Hackett, Pinet, and Ruggles (1996, 271) have indicated, 'establishment of alternative media is essential to building popular democratic movements, without which the hope of progressive social transformation is in vain.'

TIME to REFLECT

Have you ever been exposed to alternative media? What types of media might be most effective in helping movements get their messages to you?

The Internet: Extending Political Economy and Cultural Studies

The critical perspectives of political economy and cultural studies have long been utilized by sociologists and other researchers to study so-called 'old media' like newspapers, motion pictures, and television. However, these perspectives can also help us to understand the 'new media' of digital communication such as the Internet (an interconnected computer network) and the World Wide Web (interconnected documents accessible through the Internet).

Through research on the Internet, media sociologists and other scholars have given renewed emphasis to an issue in political economy that was not addressed in our earlier discussion: namely, social inequalities in access to information and communication technologies (ICTs). This issue had been important in political economy up to the late 1980s. Critical scholars investigated how private ownership and control of ICTs helped to 'deepen social class divisions nationally and internationally, as people now divide into those who can afford the technology, services and content—the *information rich*—and those who cannot—the *information poor*' (Mosco 1989, 80). Research addressed social class differences in access to such technologies as cable, video-cassette recorders, direct-to-home satellite receivers, and the initial personal computers.

With the further development of personal computers and the growth of the Internet, scholars in political economy have revived the issue of inequalities in access through research on the **digital divide**. The concept of the digital divide refers to inequalities in access to computers and/or the Internet. Most sociologists view the digital divide in terms of divisions in access by socioeconomic status and social class (Cuneo 2002). The data presented in Table 23.2 support the view of sociologists that there is an economic and class dimension to the digital divide. The table indicates access to the Internet in relation to the personal income of individuals in Canada for the years 2005, 2007, and 2009. The highest quartile (the top 25 per cent of individuals in terms of income) had consistently more access to the Internet than the lowest quartile. While 83.2 per cent of the highest-income group had access in 2005, only 58.7 per cent of the lowest-income group did. However, the gap between these groups has diminished. In 2009, 92.1 per cent of the highest-income group and 76.2 per cent of the lowest-income group had access to the Internet (Statistics Canada 2010a).

TABLE 23.2 ▼ Personal Income of Individuals Using the Internet

Personal Income Quartile	Percentage of Individuals		
	2005	2007	2009
Lowest quartile	58.7	68.8	76.2
Second quartile	56.9	60.7	69.9
Third quartile	71.3	75.5	83.1
Fourth quartile	83.2	87.9	92.1

SOURCE: Adapted from Statistics Canada 2010a.

The Internet can also be understood through other long-standing issues in political economy. Analysis of the Internet must consider the state as well as globalization and private ownership. The state was crucial to the origins of the Internet. The Internet had its beginnings in the late 1950s when the US Department of Defense established the Advanced Research Projects Agency (ARPA) to ensure that the Soviet Union would not develop military superiority over the United States in relation to computers and communication. ARPA developed a means to interconnect computers in such a way that an attack on one server and one part of the network would not knock out other servers or the rest of the network (Cuneo 2002). While the Internet began with the state, its growth has had implications for globalization. The global impact of the Internet is closely tied to private ownership. Like the 'old media', the 'new media' have become associated with deepening ownership concentration. For instance, in 2000 America Online (AOL) took over the global media conglomerate Time Warner to form AOL/Time Warner. This allowed AOL to combine its role in Internet services with Time Warner's involvement in film, television, music, and magazine and book publishing. AOL's objective was to attract advertisers with multimedia deals and promote its own services across a range of media (Taras 2001).

Key issues associated with cultural studies provide a basis for thinking about the Internet as well. In many ways, the Internet represents an extension of mainstream media and the content provided by powerful groups. Mainstream media content and the dominant ideology within this content have spread onto the Internet. Recorded music, motion pictures, and television shows are available for download. The advertising that is crucial for supporting many of the mainstream media is appearing with increasing frequency on websites. Newspapers and television news channels have websites where they can reproduce news stories. Fortunately, as the Under the Wire box makes clear, the Internet can also be seen as a new form of alternative media, which subordinate groups use to challenge the dominant ideology through their own content and messages.

TIME to REFLECT

Think about the divide between the 'information rich' and the 'information poor' as well as the digital divide in relation to your own access to media technologies. For example, when you were growing up, did your family have access to certain media technologies that the families of your friends did not?

(Reuters/Corbis/Mike Hutchings)

School pupils in Alexandra Township, Johannesburg, South Africa, learn computer skills. Poverty and lack of resources often hamper the development of technical skills in the developing world.

UNDER the WIRE

The Internet as Alternative Media

Activists associated with the global justice movement have made considerable use of the Internet to challenge neo-liberalism and other aspects of the agenda that multinational corporations have adopted in relation to globalization. For instance, the Independent Media Centre established a website just before the Ministerial Conference of the World Trade Organization in Seattle during 1999. The aim was to provide coverage and analysis that would counter news of the conference offered by the corporate-dominated mainstream media (Downey and Fenton 2003). Through the website (known as Indymedia), journalists broke stories about the brutality of the police in relation to the demonstrators. The website had received more than 1 million hits from individual users by the end of the conference. The success of the website spawned other Indymedia websites around the world (Pickard 2006). As Carroll has indicated, it is useful to see the Internet as 'a tremendously subversive communications medium' (cited in Coburn 2010, 85).

Conclusion

Several questions were posed at the beginning of this chapter. We are now in a position to review these questions and provide sociological answers to each of them. In the process, we will be reminded of some of the issues that have been addressed.

What is the role of the mass media in society? Structural functionalists suggest that the mass media perform key functions for society by contributing to its order and stability. In contrast, conflict theorists argue that the mass media help to sustain power relations under capitalism and are a basis for conflict between unequal groups. Feminists also focus on power relations while indicating that the mass media are an aspect of patriarchal society.

Who has power over what we read, see, or hear in the mass media? The critical perspective of political economy holds that much of this power is held by private companies through their ownership and control of media organizations. The concentration of ownership by private companies raises concerns about the diversity and suppression of media content and its use for promoting specific agendas. Private ownership also raises concerns about the amount of American media content in Canada and other countries, especially in view of globalization. The state in Canada has countered some of these tendencies through media regulations and public ownership of the media. However, since the state generally serves the interests of the capitalist class, these types of measures were implemented only because groups with little power pushed for them.

What kinds of messages exist in media content, and how do people react to them? According to the critical perspective of cultural studies, the content of mainstream media reflects the dominant ideology in society. Media texts are encoded with capitalist, patriarchal, and racist ideology through a complex mixture of economic and production factors. However, these texts are decoded by audience members in different ways. Dominant, negotiated, and oppositional readings are possible. Oppositional readings involve resistance to media content. Members of subordinate groups have even gone beyond resisting the content of mainstream media to express their own messages through alternative media.

What are the social implications of the Internet, which is associated with the 'new media'? Inequalities in access to media technologies have long existed, and the digital divide indicates that these inequalities are continuing through the Internet. Furthermore, as Taras (2001, 113) has noted, 'the Internet seems more and more to be an instrument that reflects and reinforces the power of the powerful.' This is evident, for instance, in the way it has become tied to deepening ownership concentration as well as to the distribution of mainstream media content and the dominant ideology embedded in this content. However, the Internet clearly presents new opportunities for radical movements to challenge social inequality and generate social change.

questions for critical thought

1. Research the background and current status of a media organization identified in Table 23.1 by visiting its website. What types of media have been associated with the organization? Can you connect the organization to one or more forms of ownership and the concerns raised about these ownership forms?

2. Compare the evening television schedules of the CBC, CTV, and Global by looking at their websites or your local television listings. How much Canadian and American content do you see being scheduled? Do we need more Canadian content on Canadian television? Why or why not?

3. Find a newspaper article that reports on a regulatory decision the CRTC has made about radio or television. Does the CRTC's decision favour powerful groups or less powerful groups? What do sociological ideas about the state suggest about the reasons for the CRTC's decision?

4. Review Butsch's findings on the representation of the working class and the middle class in domestic situation comedies. Can you apply his analysis to more recent shows?

5. Review Dole's analysis of how female law enforcers are represented in motion pictures. Can you think of more recent films or even television shows in which female law enforcers are represented in similar ways?

6. Examine the content in some form of alternative media (such as Vancouver Co-operative Radio or an Indymedia website). Identify specific ways in which the content is different from that of the mainstream media.

7. Examine the content of websites that you often visit. In what ways do they involve an extension of mainstream media and the content provided by powerful groups?

recommended readings

Augie Fleras. 2003. *Mass Media Communication in Canada*. Scarborough, ON: Thomson Nelson.
Fleras presents sociological perspectives on various media issues. His book provides a good next step for students who would like to know more about the sociology of mass media.

Augie Fleras and Jean Lock Kunz. 2001. *Media and Minorities: Representing Diversity in a Multicultural Canada*. Toronto: Thompson Educational Publishing.
Fleras and Kunz offer a sociological approach to the media representation of minorities. Their book is especially appropriate for sociology students who want to learn more about these issues in media or race/ethnicity courses.

Robert A. Hackett and William K. Carroll. 2006. *Remaking Media: The Struggle to Democratize Public Communication*. New York: Routledge.
Hackett and Carroll use interviews with media activists and case studies of social movement organizations in the United States, Britain, and Canada to identify key issues involving the use of media by social movements and the struggles of movements to challenge the mainstream media.

Naomi Klein. 2000. *No Logo: Taking Aim at the Brand Bullies*. New York: Picador.
In this very readable book, Klein examines issues associated with brands and logos (such as the Nike swoosh). She discusses the expansion of branding by companies, the role of corporations in relation to globalization, and forms of resistance to these developments.

David Taras. 2001. *Power and Betrayal in the Canadian Media*. updated edn. Peterborough, ON: Broadview.
Taras examines what he sees as a crisis facing the Canadian media system. His fascinating analysis considers the problems confronting public broadcasting, the detrimental impact of private broadcasting, and the implications of developments such as ownership concentration.

Serra Tinic. 2005. *On Location: Canada's Television Industry in a Global Market*. Toronto: University of Toronto Press.
Tinic examines television production in Vancouver with reference to tensions between Hollywood's needs and the need to reflect Canadian culture. Since her analysis draws on political economy and cultural studies, the book is useful for students who want to learn more about these approaches.

recommended websites

Alliance of Canadian Cinema, Television, and Radio Artists (ACTRA)

www.actra.ca/actra/control/main

ACTRA is a labour union that represents performers working in the English-language media. Among other things, ACTRA's website includes collective agreements the union has with various media organizations.

Canadian Broadcasting Corporation (CBC)

www.cbc.ca

The CBC'S website will enable you to find out more about Canada's national public broadcaster. The website supplies annual reports, corporate documents, and other information.

Canadian Radio-television and Telecommunications Commission (CRTC)

www.crtc.gc.ca

The CRTC's site will help you to learn more about this federal regulatory agency and its various policies.

Department of Canadian Heritage

www.canadianheritage.gc.ca

The Department of Canadian Heritage is the federal government department responsible for broadcasting, film, and other aspects of culture. Its website includes relevant policies and legislation as well as various reports.

Independent Media Center

www.indymedia.org/en/index.shtml

You can find out more about Indymedia through this website, which also provides links to Indymedia websites in various areas of the world.

Vancouver Co-operative Radio

www.coopradio.org

The website for Vancouver Co-operative Radio will enable you to learn more about community radio as a form of alternative media. You can also listen to the station through the website.

chapter 24

The Environment

G. Keith Warriner

In this chapter, you will:

▶ Review the origins of environmental sociology as a new field of study for understanding environmental problems.

▶ Examine global population growth and its relationship to poverty and development.

▶ Differentiate theories of environmental sociology and their basic assumptions.

▶ Critically assess such terms as 'sustainable development', 'scarcity', and 'carrying capacity'.

▶ Understand the theory of 'risk society'.

▶ Learn about the social constructionist perspective as it is applied in environmental sociology.

▶ Appreciate the distribution of environmental benefits and impacts.

▶ Differentiate between the various sides of the environmental movement.

Introduction

Saving the environment is often in the forefront of public concern, but few people associate sociology with the study of environmental problems. Typically, people assume that overcoming these problems requires an understanding of the natural world and knowledge of biology, chemistry, physics, bioengineering, and geography, areas in which sociologists tend to have little expertise. In fact, students may be surprised to learn that a field of sociology associated with the study of environmental issues even exists.

No one denies that science and technology play vital roles in the fight to protect the environment. However, it takes only a little thought to appreciate the very significant connection between social conditions and environmental quality. Consider the ongoing debate over climate change. Global warming may have very significant social implications for the future, eradicating some low-lying Pacific island nations, altering agriculture and food consumption patterns in the world, and triggering catastrophic regional weather disturbances with resulting economic disaster and destruction of human life. The natural sciences of physics, chemistry, biology, and geography are all closely associated with attempts to understand global warming, but almost all agree that human society is among the root causes of global climate change, as well as being significantly affected by it. Industrialization, population growth, and even the eating habits of much of the world are all to a degree responsible for the production of greenhouse gases that lead to global warming. Greenhouse gases, therefore, exist as one of the largely human-induced environmental problems that threaten the **carrying capacity** of the Earth: its ability to provide the resources to sustain all of humankind.

That human beings contribute to environmental problems through our social systems comes as no surprise. But the issue of global warming is sociologically far more profound than simply that humans contribute to it. The debate on the nature and extent of climate change, and its solution, has emerged as among the most significant social controversies of this century, with scientists, business leaders, policy-makers, environmentalists, the media, and members of the public at large all vying to have their version of the evidence accepted, thus providing an example of how even so-called objectively determined environmental problems are social constructions.

In addition, how societies define global warming affects responses. In Canada, some groups argue that achieving emissions reduction targets will be prohibitively expensive and harmful to Canadian economic competitiveness. Finally, the global warming debate is responsible, to a degree, for rekindling long-standing regional tensions in Canada, such as alarm in Alberta over criticisms of oil sands development creating additional greenhouse gases. Hence, sociological implications surround every point in the global climate change question, from debates over whether global warming even exists to who is responsible, what can be done, and how the costs of solutions will be shared. The sociological complexity of the climate change issue means that science and technology alone are not likely to solve the environmental problem of global warming.

The example of climate change is typical of the societal debate around most environmental problems, and over the past several decades, the field of environmental science has become interdisciplinary, spanning the natural and social sciences. Researchers, policy-makers, and environmentalists have all come to accept that the complexity of the environment defies the ability of any single field of science to solve environmental problems. Environmental sociology makes its contribution by seeking to understand those aspects of societies, organizations, and people that contribute to environmental degradation, as well as by assessing the prospects for social change necessary for improving environmental quality.

The Basics of Environmental Sociology

Environmental sociology as a distinct subfield of the larger discipline is generally acknowledged as having originated in the mid-1960s in connection with the rise of the environmental movement. Sociologists of the day were fascinated by this new social movement that emerged in the United States during the late 1960s and rapidly spread worldwide.

Environmentalism among US college students of the 1960s was as much a result of the social climate of American society as of any startling understanding of or revelation about environmental concerns (Hays 1987). The long wave of economic prosperity following World War II helped to deflect the attention of youth from material concerns to social conditions, particularly to the uneven distribution of political and economic power. This new awareness, together with obvious environmental problems such as smog and the influence of Rachel Carson's *Silent Spring* (1962), which detailed the effects of pesticides on the environment and human health, made environmental protection a cause. Two earlier causes, the civil rights and anti-war movements, provided background, helping the environmental movement's leadership to gain experience in activism and providing the organizational framework needed to launch a successful new movement.

As the environmental movement escalated, leading to such events as the first Earth Day in 1970, many sociologists became interested in studying it. The field of social movements has long been important in sociology, and the environmental movement allowed sociologists to bring established theories and approaches to an exciting new social phenomenon. Along with the environmental movement, the development of environmental sociology was aided by the theoretical tradition of **human ecology**. Developed by Robert Park and Ernest Burgess of the Chicago School of sociology of the 1920s and 1930s, human ecology was the prevailing sociological tradition prior to being supplanted by the structural functionalism of Talcott Parsons and Robert Merton in the 1940s (Michelson 1976). For environmental sociology, the legacy of human ecology provided a theoretical perspective capable of being reactivated at the moment when the environmental movement crystallized a broadly based sociological interest in the environment.

As is usually the case for sociologists in other subfields of the discipline, environmental sociologists tend to be diverse with respect to their research interests, theoretical approaches, and methodological practices. Despite this, environmental sociologists share a number of assumptions. Environmental sociology studies the interrelationships between society and the environment. The focus is on the relationship between human social organization and the physical environment. This departs somewhat from the broader discipline of sociology, wherein the term *environment* generally refers to a socio-cultural or symbolic system—a particular social context, its organizational framework, and the relationships among and meanings shared by the individuals involved. Sociologists have been reluctant to extend the term to include the physical environment, preferring to leave this usage to natural scientists. Rather, within the discipline of sociology generally, environmental sociology is unique in seeking to consider 'environment' as *both* a physical entity and a socio-cultural (symbolic) phenomenon. This position is not without controversy, contributing to sociology's 'essential dualism', according to Alan Irwin (2001, 3), meaning that environmental sociologists are critical of their colleagues from other sociological fields for treating the physical environment as little more than a backdrop to social activity, while in turn they are themselves criticized for encroaching on areas of natural science understanding for which they have little training.

A complication arises in that environmental issues are likely to be the focus of conflict between competing social groups. Resources such as air and water tend to be considered common property, freely available to everyone and having economic, recreational, and aesthetic uses. But competition for such resources arises, making solutions to environmental problems all the more difficult, since the usual response of the authorities has been to compromise environmental integrity in order to satisfy competing social demands.

Chief among the social processes and structures often questioned by environmental sociology are the benefits of economic growth. In the past, sociologists, together with other scientists, tended to be enamoured of social progress, which was regarded as the means for achieving uniform prosperity and thereby eliminating class differences. Today's environmental sociologists are of two minds. On the one hand, economic expansion has adverse consequences for the environment: pollution, waste, and the destruction of non-renewable resources. On the other hand, the expansion of economic markets, besides helping to provide economic and social well-being, also provides the prosperity societies need to deal with environmental problems. The conundrum of economic growth versus environmental protection has been recognized by Schnaiberg (1975) as the 'socio-environmental dialectic'. Schnaiberg points out that economic policies that are 'regressive' in their effects—that is, that lead to reductions in economic benefits because of stagnant or negative growth—are more likely to result in environmental policies that are less sensitive to the environment. When times are hard, politicians see stimulating the economy as the first priority, with environmental protection being put on hold. Hence, there are grounds for believing that a healthy, growing economy is needed for environmental preservation.

Support for this viewpoint was offered by the report of the 1987 World Commission on the Environment and Development, *Our Common Future*, which, by and large, was responsible for popularizing the principle of **sustainable development**. According to this principle, only through significant improvements in the economic conditions of developing nations can global ecological disaster be averted. Sustainable development approaches have emerged as indispensable to economic planning at many levels, but at the cost of falling short of providing a set of agreed-on 'best practices', operating principles, and goals. Professing support for the environment while doing little about it is one way many projects gain political and public approval. On these grounds, while still acknowledging the importance of a healthy economy for environmental matters, most environmental sociologists remain sceptical about the wisdom of constant economic expansion, especially in the guise of sustainable development, and advocate both redirection and some curtailment in growth.

TIME to REFLECT

How important is societal organization in terms of contributing to environmental problems? Are such problems inevitable, or can human societies be organized so as not to create environmental problems or perhaps even to enhance environmental conditions? If so, how? Are such solutions practical for saving the planet or mainly utopian, while global environmental problems will continue to develop and get worse?

The Environment and Ecological Scarcity

Issues of ecological scarcity have been of considerable interest to environmental sociologists. Scarcity has to do with problems associated with the overuse

OPEN for DISCUSSION

Environmental Problems as Collective Behaviour

Environmental sociologists agree that environmental issues are social problems, even though these problems often affect the biophysical world while other social problems affect humans almost exclusively. Environmental problems, like other social problems, exist and persist largely because of the way societies and the global social order are organized. Consider the fact that modern industrial and post-industrial societies contain within their central logic the enhancement of such values as individualism, universalism, and achievement. These values are part of the socialization of citizens and have resulted in social differentiation and material abundance. At the same time, the positive values attached to economic growth and structural differentiation fuel an expansionist society. The expansionist tendency, in turn, causes myriad environmental problems. This does not have to be—the environmental problems associated with the 'treadmill of production', as it has been labelled by sociologist Allan Schnaiberg (1980, 227), could be avoided, but to do so would involve a fundamental change in ideology.

Too often there has been a tendency to see environmental problems as inevitable consequences of the process of modernization. From this perspective, many are satisfied with attempting to keep environmental problems in check, without believing that they can be eliminated altogether. Moreover,

societies tend to adapt to scarcities imposed by environmental destruction rather than seeking fundamental social change. Environmental sociologists want to challenge these attitudes. They argue that the environment has fallen victim to a collective social definition that accepts the inevitability of environmental damage and fails to question the social system that allows it—a system that in itself is neither right, best, nor inevitable. Society can choose to deal with environmental problems, as with other social problems, through deliberate social change. Achieving environmental integrity may mean sacrifices, but it is a choice people can make.

A number of years ago, sociologist Herbert Blumer analyzed this common error in reasoning with respect to social problems in general (Blumer 1971), in part arguing that since social problems are mainly matters of shared interpretation, the focus of sociology should be more on investigating how this collective opinion comes to be rather than expecting sociology to either uncover new social problems or to stipulate how they will be solved. If we assume that all social problems, including environmental ones, are really the products of shared collective behaviour around how they are defined, legitimized, and addressed, it can be the basis for new insight together with a very radical reinterpretation of how we view the world.

of natural resources, leading to their exhaustion, or with their waste or destruction by contamination or misuse. The immense reliance of societies on natural resources and the extent to which this reliance influences social arrangements as well as prospects for social change often go unrecognized. Sociological interest in resource scarcity therefore addresses questions of world population growth, the limits of global carrying capacity, and the relationship between development and scarcity.

POPULATION GROWTH

There are currently more people living on the Earth than there have been throughout all of human history. It took more than 1 million years for the population to reach 1 billion, around the mid-nineteenth century. In the century and a half since then, the figure has grown to 6.9 billion (in 2010). Some 79 million people are added each year, at a rate of around 150 people per minute. Although the rate of world population increase is slowing, the absolute amount of growth continues to be substantial. By 2050, the world's population will likely have increased by more than a third of its size today, to 9.1 billion people (United Nations, Department of Economic and Social Affairs, Population Division 2009a). These startling statistics explain why world population has been likened to a time bomb threatening to destroy the planet. Population pressure is regarded as one of the most serious environmental threats, contributing to resource exhaustion, destroying species and habitat, causing pollution, and taxing the capacity of agricultural systems. It is a major factor in such diverse ecological disasters as famines, global warming, deforestation, the garbage crisis, and the spread of disease. While we have not yet arrived at the theoretical limits for food production on the planet, they may be reached by the year 2100, with a projected population of 11.2 billion (World Commission on Environment and Development 1987, 98–9).

The most serious environmental problems mainly affect the more than 5.5 billion people in the developing countries of Africa, Asia, Latin America, and the Caribbean—more than 80 per cent of the global population (see Table 24.1). The 174 countries classified by the United Nations as the 'less developed

Located close to the wealthy suburb of Sandton, Alexandra Township, on the outskirts of Johannesburg, is one of the poorest urban areas in South Africa and suffers from numerous environmental shortcomings and health problems, resulting from a lack of basic infrastructure—clean water, sewerage, and garbage disposal.

(Hervé Col art/Sygma/Corbis)

regions' (LDRs) are projected to account for 98 per cent of global population natural increase—the difference between numbers of births and numbers of deaths—occurring between 2009 and 2050. By the end of the first quarter of the twenty-second century, the world's more developed countries (MDCs) will be experiencing negative natural increase, and all of the global population increase will come from the less developed world. Consequently, by 2050 the populations of the world's least developed regions will have grown dramatically while accounting for more than 86 per cent of total global population, whereas population growth in the more developed countries will be only slight, the result of international migration rather than domestic births (United Nations, Department of Economic and Social Affairs, Population Division 2009a, 1). It is not surprising, then, that environmental scientists and population experts generally agree that the way to avoid reaching the limits of the global carrying capacity is to reduce the birth rates of developing nations.

How can the differences in birth rates between developed and developing nations be accounted for? The **demographic transition** theory is widely used to explain the dynamic relationship between fertility and mortality, which is based on economic and social progress in societies (Notestein 1967). During the first stage, pre-transition, a society experiences high rates of both fertility and mortality. Medicine, science, and agriculture are not sufficiently developed to keep deaths from disease, injury, starvation, or childbirth in check, while lack of contraception means

fertility remains high. Births are offset by deaths, so natural increase is slow and the population remains stable or grows slowly. At the second stage, transition, mortality rates decline because of scientific and technological advances, while fertility rates remain high. Births outstrip deaths, so the population grows rapidly. In the final stage of the demographic transition, post-transition, the death rate remains low but the birth rate also decreases because of contraception and societal changes. Births are again offset by deaths, so natural increase is slow, and the population remains stable or grows slowly.

After the postwar baby boom (1947 to 1966), industrialized nations moved rapidly into the third stage of the demographic transition and remain there today. The social bases for this transition are quite clear. The modernization process in these countries was accompanied by improvements in sanitation, nutrition, water quality, medicine, housing, and social programs, all of which help to increase life expectancy substantially. The reasons for the corresponding decline in fertility are more difficult to pin down. They probably centre on changes in institutional structures and cultural and personal values. Examples of the former are the entry of women into the workforce, which tends to be associated with later marriage and childbearing, and the rise of the nuclear family, which made the services of extended family members for child care less available. Examples of changes in values include the preference of many couples for emphasizing the quality of their children's upbringing instead of having large families,

TABLE 24.1 ▼ Population by Level of Development and World Regions, 1950, 2009, and 2050

	Population (millions)			Percentage Distribution		
	1950	2009	2050	1950	2009	2050
World	2529	6829	9150	100.0	100.0	100.0
Level of Development						
More developed regions	812	1233	1275	32.1	18.1	13.9
Less developed regions	1717	5596	7875	67.9	81.9	86.1
Least developed countries	200	835	1672	7.9	12.2	18.3
Other less developed countries	1517	4761	6202	60.0	69.7	67.8
World Regions						
Africa	227	1010	1998	9.0	14.8	21.8
Asia	1403	4121	5231	55.5	60.3	57.2
Europe	547	732	691	21.6	10.7	7.6
Latin America and the Caribbean	167	582	729	6.6	8.5	8.0
North America	172	348	448	6.8	5.1	4.9
Oceania	13	35	51	0.5	0.5	0.6

SOURCE: United Nations, Department of Economic and Social Affairs, Population Division 2009a.

and the tendency for children to remain dependent on their parents into early adulthood.

Developing countries have benefited to some extent from the same technological and social improvements. However, mortality rates remain high in comparison with those in the developed world. Moreover, these societies are largely agrarian and based on the extended family. Large numbers of children are needed to guarantee the economic survival of the family by working and performing household tasks, as well as to support the parents in old age. In some cases, an entire family may be carried by an exceptional child who is given access to higher education and a secure, well-paying job. Although birth control is widely available and in many cases, strongly encouraged by the state, a variety of compelling cultural, religious, and lifestyle reasons militate against contraception. As a result of all these factors, fertility levels have remained relatively high, and most developing nations remain at the second stage of the demographic transition, with fertility outpacing mortality.

At present, only some 20 to 30 per cent of the world population has attained the stable third stage. A significant question for social scientists concerns the prospects for the remainder of the world reaching post-transition. It is generally felt that for this change to occur, the requisite social and cultural conditions of industrialization must first appear, including a complex division of labour, more labour specialization and increased employment opportunity, and a more efficient system of agricultural production capable of freeing families from the necessity of providing for just their own basic needs. But long-standing relations of inequality between developed and developing countries, stemming from colonialism and from political and economic imperialism, have distorted and weakened the ability of poorer nations to move toward full-fledged industrialization. Without the accompanying fundamental improvements, the chance of reducing birth rates is small.

Some gains have been made through programs of public education about birth control, the dispensing of free contraceptives, and coercion by the state through laws enforcing birth control policies, most notably in China. However, as noted, contraception has had limited success. Poverty is commonly cited as the major reason these countries remain in the transitional stage. Thus, the most direct means of solving the world population problem is to eradicate poverty, as difficult as this may be. Environmental planners have increasingly come to accept this argument. Both the 1987 report of the World Commission on Environment and Development, chaired by Norwegian Prime Minister Gro Harlem Brundtland, and the 1992 and 2002 United Nations conferences on environment and development (the 'Earth Summits') have based their recommendations on the same proposition.

Only improvements to the economies of developing nations that help them move to the final stage of the demographic transition will end the poverty underlying high birth rates. But many factors stand in the way. Industrialized nations have traditionally benefited from access to cheap resources and labour in Third-World countries and so have contributed greatly to their economic problems. Furthermore, wealthier nations laud their overseas investments in developing countries as helpful, but studies show that there is no guarantee that the benefits will be shared equally in the face of historically rigid class divisions and traditional concentrations of wealth among a tiny elite. Industrialized nations may also view economic improvements in less developed countries as a potential threat. For all these reasons, the task of altering socio-economic arrangements in these countries—giving families a greater measure of financial security and more control over their lives—is daunting. But without a more equitable distribution of economic, social, and technological benefits, people in developing nations will have little incentive for having fewer children, and the world will have little hope of achieving environmental integrity by defusing the population bomb.

THE LIMITS TO GROWTH

The concept of limits to growth—the extent of the planet's ability to sustain its population—is an important basis for much environmentalism and for the scientific arguments underlying conservationism. The metaphor of 'Spaceship Earth' readily evokes the image of a shimmering orb floating in the vast void of space, reminding us of the fragility of the planet and of our dependence upon it.

That the Earth's natural resources are finite is accepted by almost everyone. Nonetheless, debate has raged for decades about how to ascertain the ecological limits of the planet and pinpoint their implications for environmental well-being. The controversy is highly significant, since it influences much of present-day resource use and planning. Some observers believe that the Earth's capacity to sustain itself in the face of population expansion and resource exploitation is rapidly nearing the limit and that ecological collapse is a possibility unless growth is curbed very soon (Ehrlich 1981; Catton 1980). Others regard this perspective as needlessly alarmist, arguing against

(US National Aeronautics and Space Administration)

The Earth from space.

any immediate, or even long-term, crisis and claiming that resources are abundant and people have the inventiveness to adapt to shortages (Simon 1981). The continuum of environmental concern defined by these two poles is the basis of calls for either continued economic expansion or economic restriction.

The controversy over the issue of limits to growth was initiated by the publication in 1972 of a study titled *The Limits to Growth* (Meadows et al. 1972). The research team attempted to model the interplay among five factors affected by economic growth on the planet: population, agricultural production, natural resources, industrial production, and pollution. One assumption of the model was that the components of the system grow geometrically (2, 4, 8, 16, 32, . . .). Another was that the five variables affect one another reciprocally through feedback loops of cause and effect. As you can see in Figure 24.1, the model is highly complex.

The model was used to generate the baseline projection of outcomes on the five variables until the year 2100, given the assumption that no significant changes in human values or in global population and economic functioning would occur over the next 50 years. The results were startling. According to the model, sometime before 2100, 'overshoot'—the team's term for ecological disaster—was imminent. By the middle of the twenty-first century, the world's population would overtake food production and resources. The resultant predicted collapse of the industrial system would, in all likelihood, be followed by famine, poverty, war, and significant population loss.

When assumptions about values, population, and the economy were changed, more optimistic models resulted. Still, even when the team assumed inexhaustible natural resources, a 75 per cent reduction in pollution, perfect birth control, and more output from food production, global ecological collapse was still predicted, although it would be delayed until after the twenty-second century. The most optimistic model was premised on significant changes in population and industrial growth occurring before the end of the twentieth century. In this model, equilibrium or sustainability was achieved, and ecological disaster was avoided.

The release of the study unleashed intense debate between its supporters and critics. The latter pointed to the naiveté of certain of the model's assumptions and the inadequacies of many of its measures. Nevertheless, *The Limits to Growth* generated an international furore over environmental issues. In so doing, it had a notable impact as a warning against unchecked growth and was responsible for creating much of today's widespread opposition to economic expansion.

The growth-versus-no-growth debate remains central to discussion on the environment. But the debate has become anything but clear-cut. Some scientists who favour continued expansion have close ties to the environmental movement. Their argument is that growth is essential for ameliorating the conditions of the poor in the Third World, that it will lead to reduced population growth and ultimately to the preservation of soil, water, and other resources. An extension of this argument is that even in industrialized nations, limits on growth will have the most adverse effects on the working class and the poor. Yet another view is that economic growth is necessary to provide the profits to invest in technologies for reducing waste and controlling pollution. Thus, where economic growth and environmental quality were once considered irreconcilable, this is no longer the case.

TIME to REFLECT

Does economic development create environmental problems or solve them? Think of an environmental problem that has occurred largely as a result of economic growth, as well as one that has been mitigated or solved by it. What conditions should apply to sustainable development in order for this to serve as an approach for alleviating environmental degradation?

▶ **FIGURE 24.1**   The World Model from *The Limits to Growth*

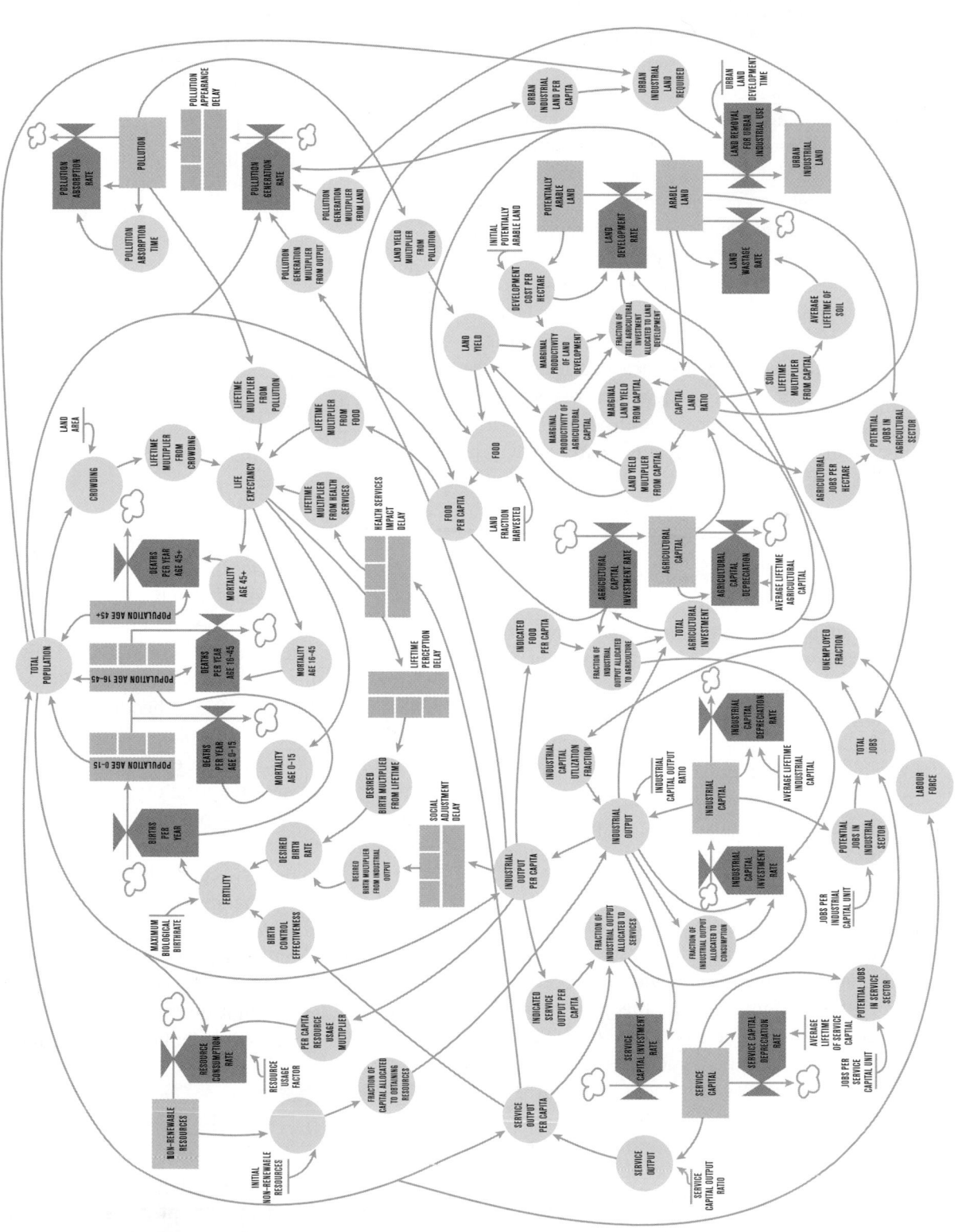

SOURCE: Donella Meadows, Dennis L. Meadows, Jorgen Randers, and William Behrens III. 1972. *The Limits to Growth*, p. 101–3. New York: Universe, 1972. By permission of Donella Meadows.

SUSTAINABLE DEVELOPMENT

The concept of sustainable development grew out of the perspective that economic development and environmental conservation are compatible goals. First appearing during the 1972 United Nations Conference on the Human Environment in Stockholm, the principle gained widespread support over the three decades that followed while being the focus of various international conferences and reviews, including the 1987 Brundtland Commission and the 1992 and 2002 Earth Summits. Sustainable development calls for the reconciliation of several apparently competing ends: environmental integrity; the protection of ecosystems and biodiversity and the meeting of human needs; and positive economic growth and equitable distribution of the benefits of the environment and resources among social classes and across nations. While the idea of the existence of ecological limits is clearly ingrained in sustainable development and while there is an insistence on strict resource husbandry, the principle is unabashedly pro-development. In the words of the World Commission on Environment and Development (1987, 45),

> Growth has not set limits in terms of population or resource use beyond which lies ecological disaster. Different limits hold for the use of energy, materials, water and land. Many of these will manifest themselves in the form of rising costs and diminishing returns, rather than in the form of any sudden loss of a resource base. The accumulation of knowledge and development of technology can enhance the carrying capacity of the resource base. But ultimate limits there are, and sustainability must ensure equitable access to the constrained resource and reorient technological efforts to relieve pressure.

After initial enthusiasm, certain environmentalists have come to regard sustainable development with scepticism. Some see it as no more than a legitimization of development under the guise of environmental stewardship. Philosopher Wolfgang Sachs, for one, states, 'Capital, bureaucracy and science . . . the venerable trinity of Western Modernization declare themselves indispensable to the new crisis and promise to prevent the worst through better engineering, integrated planning and sophisticated models' (1991, 257). Others claim that the principles of ecology, together with the scientific community, are being co-opted to support the further destruction of nature on the grounds of scientific rationality. While such criticisms are undoubtedly sometimes valid, it is nevertheless the case that the concept of sustainable development has become widely adopted as a planning goal and should be considered a benchmark to thinking on human–environment interactions.

The Environment and Social Theory

Environmental sociology has its theoretical bases in several sociological traditions. Among them is the field of human ecology, a sociological perspective with important ties to the work of Émile Durkheim. The division between the order and conflict schools so prevalent elsewhere in sociology is also a characteristic feature of environmental sociology. Contemporary theoretical approaches include the concept of the risk society, developed by Ulrich Beck. Finally, social constructionism, a perspective found elsewhere in sociology, has developed to become a prominent approach within environmental sociology. In this section, we review these theoretical positions.

HUMAN ECOLOGY

The science of ecology is central to the study of environmental issues; human ecology is the application of the same approach to sociological analyses. Human ecology emerged under the direction of Robert Park and Ernest Burgess at the University of Chicago during the 1920s (Park and Burgess 1921; Park and McKenzie 1925; Theodorson 1961, 1982). Much as the science of ecology studies plant and animal communities, human ecology sought to explain human spatial and temporal organization by concentrating on the dynamic processes of competition and succession that influence human social organization. Park and Burgess concentrated on studying how Chicago's rapidly changing society physically accommodated increases in population and changes in the industrial and cultural organization of the city. Their approach was to focus on symbiosis—the dynamic interdependencies that bind people together in communities and lead to particular living arrangements.

TIME to REFLECT

What is your position on the future of the planet in terms of the likelihood of its reaching its ecological limits? Can current patterns of population growth and development be sustained, or will they lead eventually to ecological collapse? If collapse is to be avoided, what steps by societies are required?

In developing the concept of human ecology, Park was greatly influenced by Durkheim's *The Division of Labor in Society* (1964 [1893]). Durkheim addressed the development of social complexity from human population growth and density. As populations grow, the threat to available resources is crucial from a sociological viewpoint because it leads to competition and conflict. Problems of resource scarcity can therefore affect societal organization.

Durkheim was appealing to early human ecologists because of their interest in sustenance activities: the routine functions necessary to ensure the survival of a population from generation to generation (Hawley 1950). Humans have great capacity for adapting

GLOBAL ISSUES

Energetics

Energetics is the study of the flow of solar energy through the biosphere and the various processes by which it is transformed into other forms of energy capable of performing work for humans. As early as 1955, sociologist Fred Cottrell noted that modern societies developed without a full appreciation of their dependence on physical energy resources. Cottrell recognized that the forms of energy used by a given society influence its organization and ideological characteristics. Each transformation from lower to higher energy forms throughout history (for example, wood to coal to hydroelectricity) has been accompanied by fundamental changes in the social, economic, political, and psychological makeup of the society.

In short, most people fail to realize that the survival of any society depends on its characteristic energy flows. Disruptions in these flows foreshadow social dislocation and change. 'Energy crises' are no new phenomenon. In the eighteenth century, for instance, the depletion of wood in England provided the stimulus for the development of coal and steam as energy resources, which in turn led to the Industrial Revolution with its massive social disruptions.

The 1973 oil embargo of the Organization of the Petroleum Exporting Countries (OPEC) was responsible for rekindling sociological interest in energetics. The resulting body of microsociological research focused on energy consumption and programs as sociologists responded to the energy crisis by attempting to demonstrate the relevance of their work in solving or helping to avoid future energy shortages. Other studies examined such matters as the beliefs and attitudes of energy consumers about conservation and pricing programs, the effectiveness of information campaigns on changing attitudes, the social-demographic correlates of reduced energy use, and the relationship between attitudes and conservation behaviours. Still other researchers focused on general issues of energy policy for society, particularly on alternative forms of energy.

In this connection, a distinction is often made between 'hard' and 'soft' energy paths. *Hard energy paths* involve the generation and distribution of energy through large-scale, centralized production systems relying mainly on non-renewable energy forms such as oil, gas, coal, and nuclear energy. Such systems prevail in the industrialized world today. *Soft energy paths* comprise systems relying mainly on renewable energy sources: solar power, wind, tidal power, hydro-electricity, and energy from biomass; these systems also involve conservation and recycling. Because of the nature of these fuels, these systems are more likely to be small and decentralized; examples include solar panels on a building, a community hydroelectric dam, and windmills. The essence of a soft energy path is to use the resources available locally to produce energy to be consumed locally.

(One Word Photography)

to resource scarcity, an ability labelled competitive cooperation. Adaptive responses include reductions in per capita consumption, increases in production through technology or more intensive resource exploitation, changes to distribution networks, and decreases in competition because of emigration from the community or an increased division of labour (Micklin 1973; Schnore 1958). Through such adaptive mechanisms, a state of equilibrium is reached. Park and Burgess (1921) postulated that competition and cooperation are the key forms of human exchange by which organized populations seek to maintain equilibrium within a dynamic environment.

Park and Burgess's theory has undergone significant revision to correct what are commonly regarded as major shortcomings: an overemphasis on the spatial arrangements of populations at the expense of understanding societal–environmental relations and the neglect of culture and values (Dunlap and Catton 1979a, 1979b; Hawley 1981). Still, while highly influenced by the conceptual approaches and terminology of ecology, human ecology continued, even after this reformulation, to apply the concept of environment in mainly socio-cultural or symbolic terms (Dunlap and Catton 1983; Michelson 1976).

HUMAN EXEMPTIONALISM AND THE NEW ENVIRONMENTAL PARADIGM

Paradigms are metatheoretical (that is, broad and comprehensive) frameworks of understanding based on a set of shared assumptions by practitioners in a given field. Because paradigmatic assumptions are widely shared, they tend not to be questioned. At the same time, they influence the direction and scope of the field by defining the nature both of the questions asked and of those that are resisted (Kuhn 1970).

Sociology has long been characterized by paradigmatic divisions that have led to hot debates. Environmental sociology was at the centre of one such paradigmatic clash in the early 1980s that has never been fully resolved. In several influential articles, William Catton and Riley Dunlap (Catton and Dunlap 1978, 1980; Dunlap and Catton 1979a, 1983) forcefully advanced the thesis that 'the numerous competing theoretical perspectives in contemporary sociology—e.g., functionalism, symbolic interactionism, ethnomethodology, conflict theory, Marxism, and so forth—are prone to exaggerate their differences from each other' (1978, 42). That is, Catton and Dunlap argued that while purporting to be paradigms in their own right, all these approaches really were variants of a larger paradigm. The basis of their similarity was

their 'shared anthropocentrism'. The authors argued that this assumption—that humans are separate from and superior to other things in nature—is the product of 500 years of Western culture in which societies have behaved as though nature existed primarily for human use (Dunlap and Catton 1983; White 1967). Catton and Dunlap referred to this world view as the **human exemptionalism paradigm (HEP)**, arguing that it comprises several assumptions that sociologists, regardless of their orientation, implicitly accept. This failure, even in the face of the contradictions suggested by contemporary environmental events, implied that sociologists were not equipped to deal meaningfully with ecological problems.

For Catton and Dunlap, the assumptions and approaches of the emerging environmental sociology constituted a paradigm shift, or a challenge to orthodox sociology. Catton and Dunlap referred to the new approach as the **new ecological paradigm (NEP)**. Its assumptions are compared to those of the human exemptionalism paradigm in Table 24.2. The essential difference between them can be summed up as anthropocentrism (HEP) versus ecocentrism (NEP). According to Catton and Dunlap, the HEP–NEP distinction should be recognized as the principal paradigmatic cleavage in sociology; moreover, the HEP should be considered obsolete.

NEP VERSUS CLASSICAL SOCIOLOGICAL THEORY

The debate over HEP and NEP raised further questions over the relevance of classical sociological theory for understanding environmental problems. One concern is whether the HEP–NEP distinction should be regarded as anything but a further manifestation of the long-running theoretical debate in sociology over order versus conflict. The two approaches offer competing views on both the social origins of environmental problems and their solutions.

The structural functionalist school of sociology, stressing the rational functioning of society, has been summarized by Buttel (1976). The image of society is that of a social system with needs. Individual actors and institutions within the system have competing needs, and the state acts as an impartial arbitrator to smooth out differences and relieve stress and misalignment. From this perspective, environmental problems are associated with the process of modernization or progress. The positive functions associated with the needs of economic growth, abundance, and social stratification sometimes get out of hand, leading to environmental harm. Social reform or adjustment is

TABLE 24.2 ▼ A Comparison of the Human Exemptionalism Paradigm and the New Ecological Paradigm

Assumptions	Human Exemptionalism Paradigm (HEP)	New Ecological Paradigm (NEP)
About the nature of human beings	Humans have cultural heritage in addition to (and distinct from) their genetic inheritance and so are quite unlike all other animal species.	While humans have exceptional characteristics (culture, technology, etc.), they remain one among many species that are interdependently involved in the global ecosystem.
About social causation	Social and cultural factors (including technology) are the major determinants of human affairs.	Human affairs are influenced not only by social and cultural factors but also by intricate linkages of cause, effect, and feedback in the web of nature; purposive human actions therefore have many unintended consequences.
About the context of human society	Social and cultural environments are the crucial context for human affairs, and the biophysical environment is largely irrelevant.	Humans live in and are dependent upon a finite biophysical environment which imposes potent physical and biological restraints on human affairs.
About constraints on human society	Culture is cumulative; therefore, technological and social progress can continue indefinitely, making all social problems ultimately solvable.	Although human inventiveness and the powers derived from it may seem for a while to extend carrying capacity limits, ecological laws cannot be repealed.

SOURCE: William Catton, Jr, and Riley Dunlap. 1980. 'A new ecological paradigm for post-exuberant society'. *American Behavioral Scientist* 24: 34. Copyright 1980 Sage Publications. Reprinted by permission of Sage Publications Inc.

then called for. The public's values must be modified so that society may remain within its survival base and adapt to environmental exigencies. Protective environmental legislation is also enacted. The goal is to create an environmental ethos based on rights and the rational use of resources. Appropriate environmental use is maintained through state laws, social norms, and collective action. Any adverse distributional effects on certain social groups as a result of environmental laws are regarded as the unfortunate, but necessary, trade-off for protection of the resource base.

By contrast, the conflict approach takes the view that environmental problems are irrationalities within the capitalist system leading to societal contradictions (Schnaiberg 1980). The key proposition is that class struggle is the permanent condition of society because the state has favoured and promoted the interests of the upper classes. Growth is mandatory, and the environment is the victim. Arrangements for maintaining growth and profits—such as planned obsolescence, disposable products, private transportation, and the military—promote waste and excessive resource exploitation and lead to environmental destruction. In short, Marxists and neo-Marxists see environmental destruction as inherent in capitalism. Conflict theorists tend to be hostile to reform solutions, arguing that they do not treat the root causes of environmental destruction and mislead the public into believing something is being done. Finally, the costs of such reforms are carried mainly by the working class through the loss of resource jobs and through higher prices and taxes for environmental protection.

THE RISK SOCIETY

Ulrich Beck's conception of the **risk society** was initially outlined in *Risk Society: Towards a New Modernity* (1992). Originally published in German in 1986, this text has had considerable impact in sociology while also drawing wide public debate and media attention.

The central thesis of the risk society concerns the evolution toward a new modernity, referred to by Beck as 'late modernity' (1992, 10), reflecting social change on a global scale. The impetus for this transformation is risk, hence the term 'risk society'. According to this theory, the world has evolved beyond the industrial state, with its successor, late modernity, being essentially the outcome of the success of the period preceding it. For affluent Western societies, the success of industrialization has meant the end of scarcity. Wealth, science, and technology combine to provide for the needs of those in prosperous countries. But at the same time, a multitude of problems, or risks, face individuals on a daily basis, which can be traced directly to industrialization. These involve all sorts of uncertainties—to do with changing workplace and gender roles, the nature of the family, social class relations, crime, environmental dangers, and more—all confronting the individual and for which there are no obvious solutions. This creates uncertainty, doubt, and confusion.

How should people respond? In industrial society, the primary concern involved the distribution of 'goods', with class action as the resulting collective response to inequities. Within late modernity, however, the concern is with the distribution of the 'bads'—the new risks and dangers confronting individuals. Social class relations are now much less relevant. Rather, the defining source of individual well-being and the struggle to exist is with respect to one's risk position. With this comes a decline in the importance of collective, class-based structures as the means for social support and their replacement by an increasing emphasis on the individualization of the actor, who now is forced to choose from a range of ambiguous options.

'Reflexive modernity' is the term used to describe the response to this new reality. Problems faced by individuals are no longer clearly externally imposed (for example, resource scarcities), nor are solutions founded on some kind of 'natural' order (for example, gender roles). Instead, we must reflect on our options while struggling to make the best choice. Uncertainty, alienation, and loneliness may result. Where alliances exist, they are less likely to be defined on socioeconomic grounds, and they are more likely to be ideologically determined and 'strategic', as in the example of the European Green Party, in which the anti-establishment forces of the environmental, feminist, and anti-nuclear movements combine to form a pragmatic alliance in opposition to the traditional parties. Thus, while modernity has freed Western society from material want, we are confronted with new challenges and fewer guideposts to understanding.

Reflexive modernity must also be considered in light of the new and ambiguous role of science. Science and technology are irrevocably tied to the success of industrialization. Hence, science is in part responsible for the growth of hazards and risks, while at the same time it is the body called on to provide knowledge claims needed to overcome or avoid risk (Dietz, Frey, and Rosa 2002). However, in the period of late modernity, the notion of the existence of simple truth and certainty seems naive. One only has to reflect briefly on the scientific debates and controversies that have raged over such issues as global warming, genetically modified foods, ozone depletion, hazardous waste, and the risks of nuclear energy to realize that there is no single scientific position on these issues or even an agreed-to set of facts. Rather, scientific knowledge is often revealed to be a body of contested claims, with the supremacy of any position largely linked to the skills of its advocates and their resources for advancing it. This absence of a clear and unequivocal knowledge system means that science, rather than solving the problems of the risk society, only adds to them by increasing uncertainty.

The premise of the risk society has important implications with respect to the study of environmental problems by environmental sociologists. On the one hand, the theory represents a serious challenge to what has emerged as the most widely supported position on how to deal with the environmental crisis: that of sustainable development. The logic of sustainable development sees the need for fine-tuning the existing system. Industrialization and development are not in themselves considered inimical to environmental preservation. Indeed, the position typically is that economic development needs to be expanded, especially in the case of the developing world, while still adhering to sustainability principles. Thus, the underlying assumption on which sustainable development is based is that the existing system can cope. The risk society represents an entirely different perspective, that of a society imbued with uncertainty. On what grounds, Beck asks, is society likely to right itself and overcome environmental risks? The continued expansion of industry and development in the name of prosperity only increases environmental risk rather than solving it, and science, the henchman of this advancement, increases the odds of risk further while proving incapable of providing solutions or reducing uncertainty.

TIME to REFLECT

To what extent do you feel Beck's description of the 'risk society' accurately reflects reality? Do you think it represents your own life experience? What are the implications for ecological solutions? Does the idea of the 'new environmental paradigm' contradict the risk society or complement it?

SOCIAL CONSTRUCTIONISM AND ENVIRONMENTAL SOCIOLOGY

Social constructionism is a perspective often applied in other areas of sociology and has become established within environmental sociology as well. This has stirred new debates. Social constructionism argues that social reality is more a matter of perception than of objective determination (Best 1989, 1993; Blumer 1971; Spector and Kitsuse 1977). In other words, reality is what we think it is. While most people assume social problems are recognized and dealt with because their existence is obvious, social constructionists argue that such recognition only occurs following a process of negotiation by which the 'reality' of the problem becomes recognized. What is real is contested among parties with competing claims struggling to frame their version of the situation in order that the broader public will come to accept it. In these negotiations, the media play an important role, allowing the means for claims to be reported and providing important interpretation and emphasis, which may assist one or another of the competing parties to be successful in defining the problem and the approaches for dealing with it.

Social constructionism has been widely applied in sociology in areas such as crime and deviance, homelessness, gender inequality, sexual orientation, illness and health care, and race and ethnicity. The perspective is appealing to some environmental sociologists as well. John Hannigan (1995, 55), for example, argues that the 'successful construction' of environmental problems requires that six conditions be met:

1. scientific 'authority for and validation of claim' by parties

2. the existence of 'popularizers' who can bridge environmentalism and science

3. media attention in which the problem is framed as novel and important

4. the dramatization of the problem 'in symbolic and visual terms'

5. economic 'incentives for taking positive action'

6. the emergence of an 'institutional sponsor who can ensure both legitimacy and continuity'.

In other words, an environmental problem is only the result of the success of the claims-making of those who advocate its existence. The implication here concerns the sociological process that underlies the 'discovery' of the environmental problem. Social dynamics replace objective existence of scientific risk as the object of scholarly interest, since the social process is the basis for what we believe the problem to be.

Does this make sense when environmental problems are associated with such catastrophic threats that the very life of the planet is considered at risk? The social constructionist approach has provided abundant insight with respect to a variety of environmental debates, including global warming, the population 'explosion', environmental racism, globalization in general, and a host of local and regional industrial contamination and development debates. Examine any recent environmental controversy in your community, and you likely will be able to apply a social constructionist perspective. Issues such as new industrial development, chemical contamination, or the siting of a landfill invariably pit residents, politicians, factory managers, government officials, and scientific experts against one another in a struggle to get their version of 'the truth' accepted. The media provide the means for broadcasting the competing views while arbitrating what will be presented. Science has an authoritative voice in these debates, but all sides struggle to mount compelling scientific and/or moral/emotional arguments, and it is unlikely that science alone will determine the outcome. The result may be perceived by many as objective truth, but it is clear that such so-called reality is largely a social product.

Concerns over social constructionist analyses as applied to environmental problems have been expressed by a number of environmental sociologists (Benton 1994; Dunlap and Catton 1994; Murphy 1994; Dickens 1996). It is typically argued that there are objective, independent, and physical qualities to environmental problems that cannot be accounted for simply on the grounds of being 'social constructions'. For example, chemical contamination of groundwater from industry and radiation from nuclear energy constitute absolute and deadly threats to individuals and should not be treated simply as perceived concerns. Murphy accuses social constructionists within environmental sociology of having lost touch with nature and 'gone overboard' (1994, 970), and Benton laments the 'over-socialized' view of environmental risks by social constructionists (1994, 44). Further, these 'realist' critics believe that environmental sociology can assist in overcoming environmental problems and express alarm over whether the constructionist perspective is deflecting scholarly interest away from such work by focusing only on the moral and political issues surrounding the way a problem becomes defined.

TIME to REFLECT

What are the implications of the social constructionist approach to defining environmental problems for the identification of legitimate problems? Is society attempting to 'solve' problems that objectively do not exist while failing to recognize other, valid problems? Can you think of examples in each case?

Environmental social constructionists respond by noting that their position is not strict constructionism, as sometimes found in other areas of sociology such as post-modernism (Burningham and Cooper,

1999). Rather, for environmental sociology, the constructionist approach has been a more mild relativism, or contextual constructionism, which attempts to draw attention to the social processes involved in the development of societal recognition and response to environmental problems without insisting that environmental problems do not objectively exist. For example, while previous sociological studies reviewing the political discourse around the late acid rain debate reveal considerable framing with respect to both supporting and denying the reality of acid rain, there was still always recognition of the existence of dying lakes and forests and a desire to help end this (Zehr 1994).

The Environment and Social Movements

Fascinated by the environmental movement from its inception, environmental sociologists continue to be deeply interested in it. It has proved to be among the most successful and enduring social movements of all time. Few other movements can match it in terms of sustained activity, size of following, and ability to affect the lives of so many people. It has even changed our language, with such terms as NIMBY ('not in my back yard') and 'environmentally friendly product' entering the vernacular. The first Earth Day, staged on 22 April 1970, was impressive, drawing some 20 million people (Dunlap and Gale 1972), and Earth Day has since grown to become an international annual event—Earth Week, celebrated in 180 countries. Today, few people admit to not supporting environmentalism; in fact, most people claim to be environmentalists (Dunlap 1992). The environmental lobby, institutionalized as a significant player in government decision-making, is further evidence of the movement's impressive success.

The environmental movement has changed significantly over the years, often appearing to share little with its student–activist beginnings. The movement seems less angry today but at the same time far more meticulous and deliberate in its approaches, often more at home in the corridors of power than on the protest line. The discussion that follows offers a look at the several strands of the contemporary environmental movement.

MAINSTREAM ENVIRONMENTALISM

Contemporary environmentalism traces its roots to the **progressive conservation** movement of the late-nineteenth-century United States (Fox 1985; O'Riordan 1971). Led by such reformers as Gifford Pinchot and John Muir, progressive conservation was a reaction against the unchecked destruction of nature during this period of freewheeling capitalism. A characteristic of this early environmental movement was the distinction between the so-called *preservationists*, who advocated setting aside and protecting wilderness so that its natural, aesthetic, recreational, and scientific values could remain undisturbed, and *consumptive wildlife users*, who promoted conservation for more utilitarian ends such as recreation but also logging, mining, and grazing. In terms of modern environmentalism, a legacy of progressive conservation has been the legitimation of government involvement in the environment along with the responsibility for accommodating both sides of this ongoing debate. Hence, progressive conservation set the scene for the current relationship between business and government. The main beneficiaries of this policy are the large corporations, which, while gaining controlled access to resources, have paid little in resource rents. Some observers regard the sustainable development movement as a new expression of the principle of consumptive wildlife use.

Early preservationists quickly learned that they had to cooperate with the consumptive wildlife users and the Washington administration or they would have little hope of making progress toward environmental protection. By now, environmentalists have become highly skilled at working as partners with government and developers in reaching compromise on environmental decisions. The inevitable result is trade-offs on preferred environmental solutions. Rik Scarce (1990, 15) reports that most environmental organizations admit to having no specific approach or plan for the environment other than saving what they can. Such 'muddling through' has resulted in some checks on development but also in serious environmental losses. Rarely have the mainstream environmental groups been in a position to claim complete victory in their efforts to stop a development or save an ecosystem.

Contemporary mainstream environmentalism is increasingly in the form of inside lobbying, politicking, and consultation and relies mainly on its well-organized bureaucracies for success (Mertig, Dunlap, and Morrison 2002). The leaders tend to be highly educated environmental professionals, often with backgrounds in public administration or environmental law. Fundraising and research are essential to successful competition with large corporations over the fate of resources. The individual member is far more likely to write a cheque or the occasional letter to an elected representative than to take part in a sit-in or blockade.

Many mainstream environmentalists argue that it has only been through these increasingly well-organized, well-funded, professional organizations that environmental review and assessment have become a permanent part of economic planning. Critics such as William Devall (1992) have suggested, however, that these same organizations are too accommodating to development interests, their leaders too close to their opposite numbers in business and government and too secure in their professional status. Still others are critical of mainstream environmentalism in general, arguing that it has long suffered from elitism. Various writers have pointed out the middle- or upper-class origins and high educational levels of environmental leaders and members of mainstream environmental organizations (Humphrey and Buttel 1982; Morrison and Dunlap 1986). However, supporters of the environmental movement—if not those actually involved in it—tend to be drawn widely from across the social-class spectrum (Mertig and Dunlap 2001). A related criticism levelled at mainstream environmentalism is that the programs or policies advocated may lead to reductions in resource-based jobs or even in wholesale plant closures because of the high costs of environmental regulation or the protection of a given wilderness area (Schnaiberg 1975). Such economic events are likely to have the most adverse effects on the working class and the poor.

THE NEW ECOLOGIES

Mainstream environmentalism is one wing of the larger environmental movement, which includes various alternatives. The new ecologies are a range of approaches within environmentalism with a number of common features. First, they are critical of mainstream environmentalism for its failure to address ecological problems by taking into account the systems of dominance in social relations that help to create those problems. Inequality among nations and regions serves to enhance competition for scarce resources and thereby increases environmental harm. The new ecologies argue that central to solving environmental problems is the promotion of social equity and self-determination, which will allow peoples and nations to meet their human needs while maintaining ecological integrity (Gardner and Roseland 1989).

Another distinguishing feature of the new ecologies is their devolved character. Hierarchical relations of authority between the membership and leaders or between the branches of each organization are rejected as being inconsistent with the prevailing thesis of human equality with nature rather than domination over it. This essentially ecocentric (and preservationist) stance is yet another characteristic of these groups, which tend to be sharply critical of any anthropocentric tendency to 'manage' the environment—an approach mainstream environmentalists seem all too willing to accept.

Finally, the new ecologies tend to outline specific principles for environmental reform consistent with their broad vision of the human–nature relationship rather than simply muddling through. They are also far less willing than the mainstream to accommodate solutions in the interest of political and economic expediencies. Indeed, some radical arms of the new ecologies movement advocate the use of illegal, even violent, actions in order to win environmental disputes. While these radicals are in the minority, mainstream environmentalists admit to having been helped by them in reaching compromises more favourable to the environment—they appear reasonable in comparison to the unbending demands and extremism of the radicals (Scarce 1990).

Thus far, we have enumerated the similarities among the new ecologies. Now we look at three of these movements in order to highlight their differences.

Eco-feminism

Eco-feminism represents the partnership of ecology and feminism. It is founded mainly on mutual opposition to hierarchy and domination. Feminists argue that the subordination of women by men has been achieved through the ability of men to employ conceptual frameworks that place women at a disadvantage. According to Val Plumwood (1992), these include hierarchical frameworks that justify inequality; dualism, which justifies exclusion and separation; and rationality, which justifies logic and control. By advancing these three conceptual preferences, men have succeeded in legitimizing their domination over both women and nature.

The logic of domination holds that by virtue of the distinctiveness of men from nature and of men from women, together with the supposed greater rationality of men, the domination of men over both women and nature is 'reasonable'. In other words, eco-feminists argue that exactly the same male-controlled value system is used to justify both patriarchal human relations and the exploitation of nature.

Feminism and environmentalism connect, then, at the point of recognizing the similarities in the ways

UNDER the WIRE

Environmental Cyberactivism

That the environmental movement is a successful social movement is a mark of its adaptability as much as of its resilience. The complexity of the movement, evolving from early preservationists and consumptive wildlife users to today's many-layered and multifaceted environmentalism, reflects a long-standing talent for transforming and redefining itself as circumstances require. Such change keeps it current— able not only to respond to new challenges but also to utilize innovative strategies for spreading its message. Over the years, environmentalism has achieved many successes by drawing on a wide array of means for achieving its ends, from old-fashioned lobbying and politicking, to sit-ins, consumer boycotts, petitions, and lawsuits, to peaceful public protest and even, on occasion, violent direct action.

Recently, environmentalism has begun to draw strength from the new media technologies and social networking, such as Facebook and Twitter. While not exclusive to environmental causes, the environmental movement has embraced these new technologies as the means for inciting, organizing, and directing protest. This strategy, termed cyberactivism, permits environmental organizers to use the capability of these vastly popular technologies to reach far greater numbers of potential adherents as well as new political communities, while transcending both geographical and political boundaries.

Take Greenpeace's campaign to force food giant Nestlé to eliminate palm oils from its candy products like Kit Kat and Coffee Crisp. The production of palm oil contributes to the destruction of the Indonesia rainforest, home of the orangutan, which is being pushed toward extinction. Utilizing social networks, Greenpeace spread the message of potential habitat destruction and succeeded in galvanizing thousands of individuals to send online protest messages to Nestlé, with the company eventually announcing it would no longer buy palm oil products that contribute to the destruction of the rainforest. In Canada, Greenpeace has also used cyberactivism to pressure the British Columbia government into preserving the Great Bear rainforest. Greenpeace reaches out to hundreds of thousands through social networking sites and cellphones, and an individual's protest can be quickly directed to the politician or company in question with just a few mouse clicks on the Greenpeace website. As an environmental organization, Greenpeace has long augmented its success through various high-profile media campaigns and stunts along with other forms of protest, earning the organization and its causes considerable notoriety. Although not intended to replace its impressive arsenal of protest tools, cyberactivism is a significant new and complementary strategy—one that may prove revolutionary.

(Torfino Photography—W.C. Barnes)

A grove of cedars in Hesquiat Lake Creek, Clayoquot Sound, on Vancouver Island. Although Clayoquot Sound's rainforest is a UNESCO Biosphere Reserve, commercial activities such as logging continue.

men treat women and nature. If one form of domination—of men over women—is wrong, then all forms of domination are wrong, including that of humans over nature. To be a feminist, therefore, compels one to be an environmentalist. Moreover, eco-feminists argue, inasmuch as environmentalists recognize and reject the domination of men over nature, they must also reject the domination of men over women. Therefore, all environmentalists must be feminists (Warren 1990).

Social Ecology

Social ecology is a body of philosophical thought appealing to many in the contemporary environmental movement who are seeking to understand the interplay between humans and nature. Founder Murray Bookchin articulated this philosophy during nearly five decades until his death in 2006. Social ecology advances a holistic world view of the human–nature partnership, one based on community. Bookchin identifies the dualism and domination informing current human–nature relations as products of human ideology and culture through which society has come to be defined as distinct from and superior to nature.

While he acknowledges that culture and technology do distinguish society from nature, Bookchin rejects the idea that they are separate. Rather, society springs from nature, reworking it into the human experience. Society always has a naturalistic dimension, and social ecology is largely involved with attempting to describe how both the connectedness and the divergences between society and nature occur. Appropriate technology, reconstruction of damaged ecosystems, and human creativity will combine with equity and social justice to produce an ecological society in which human culture and nature are mutually supportive and evolve together. Social ecology envisions a society in harmony with nature, combining human-scale sustainable settlement, ecological balance, community self-reliance, and participatory democracy. Social ecology is by no means the sole advocate of many of these objectives, which are shared by various environmental schools of thought, including the writings of David Suzuki, European Greens, and several branches of green political thought. What social ecology, along with these other schools, shows is how deeply philosophical various segments of the

The forests, lakes, and rivers north of Kenora, Ontario, have sustained the people of Grassy Narrows First Nations for thousands of years. Clear-cut logging threatens their traditional way of life, as did mercury pollution from a paper mill 40 to 50 years ago. In 2007, Amnesty International (Canada) called for the provincial government to respect a moratorium on logging declared by the people of Grassy Narrows.

environmental movement are while seeking to establish an intellectual foundation for social change and environmental improvement.

Deep Ecology

Deep ecology is among the more intriguing of the new ecologies, as well as the most controversial. The name was coined in 1973 by Norwegian philosopher Arne Naess. Defining contemporary environmentalism as 'shallow' ecology, Naess (1973) argued that its advocacy of social reforms to curb problems of pollution and resource depletion identifies it as concerned mainly with protecting the health and affluence of the developed countries. By contrast, deep ecology is concerned with the root causes of environmental crisis and inspired by the understanding derived through personal experiences as humans in nature. The most distinctive aspect of deep ecology is its biocentric emphasis. Deep ecologists claim to value all forms of life equally but at the same time raise non-human life forms beyond the human. Therefore, while deep ecology shares with the other new ecologies the rejection of anthropocentrism, it goes beyond the humanistic, ecocentric ecology of human–nature coexistence. Deep ecologists desire humans to have the least possible effect on the planet and to respect ecological integrity above all else (Tokar 1988).

Deep ecology also places heavy emphasis on self-realization, the extension of the environmentally conscious individual's self beyond his or her personal needs to include the environment as a whole. An important practical consequence of self-realization is

SOCIOLOGY in ACTION

Becoming an Environmental Activist

Supporters for mainstream environmental organizations are typically people with a high socio-economic background and a high level of education. An exception to this pattern is described in an article by environmental sociologist Hal Aronson titled 'Becoming an environmental activist: The process of transformation from everyday life into making history in the hazardous waste movement'. Aronson documents the process by which ordinary citizens can become career environmental activists. Such people initially have few political aspirations and trust government institutions and elected representatives to solve any problems or risks they and their families face. Reality comes as a shock when their own lives are disrupted by environmental disaster and they discover that the state and its officials have no intention of rescuing them. The experience can be life-altering. Tasked with fighting not only polluters but government indifference as well, citizens who formerly felt little political consciousness become environmental activists.

The case of Brenda Halloran, later elected mayor of the City of Waterloo, Ontario, provides a vivid illustration of Aronson's thesis. In 1969, a home on Ralgreen Crescent in Kitchener exploded as a result of methane build-up. Following this 'accident', the city investigated. While acknowledging in its report that 'a great deal of potential hazard was evident' (*The Record*, 18 October 1978), the city did nothing further to ensure the safety of residents. The neighbourhood of semi-detached homes and townhouses contained residents of mostly modest means without much sway with city politicians. In fact, the subdivision had been built over a former municipal waste disposal site, something the current residents had never been told. By the mid-1990s, problems stemming from the old dump began to be evident, with cracking and slumping of foundations, foul-smelling black sludge in basements, and decomposing garbage buried under lawns. Some residents were experiencing chronic illnesses, with at least one reported case of leukemia.

Despite acknowledging the presence of contaminants lying beneath it, the City of Kitchener refused to assist the residents. A city-commissioned study reported the levels of toxicity to be safe. It was left to the residents to discover that the recommendation was based on standards applying to an industrial site, not a residential neighbourhood. In the face of such government stonewalling, denials, and misinformation, the fight for reparations was left to the residents themselves. Ms Halloran led the campaign on behalf of the Ralgreen Homeowners Association in an extended struggle that captured national media coverage. As a result of such notoriety, as well as a $65-million civil action suit launched by residents in 2000, the city eventually backed down and agreed to purchase the 27 affected homes and to rebuild the neighbourhood once all the contaminated soil had been removed. It was a dramatic victory for the rights of the average citizen and a defining time in the life of Brenda Halloran, who went on to a career as a community and environmental leader, culminating with her election as mayor of Waterloo in 2006.

the obligation to strive actively to prevent environmental destruction. The emphasis on direct action has particularly inspired the best-known of the deep ecology groups, Earth First!, which advocates the use of whatever means are necessary to save wilderness areas. Earth First! has garnered much attention—and criticism—for the use of ecological sabotage ('ecotage')—that is, illegal force intended to block actions perceived as harming the environment (Taylor 1991). 'Monkey wrenching'—disruption by such covert and unlawful means as removing survey stakes, destroying machinery, or spiking trees—is controversial even within Earth First! These tactics stand in sharp contrast with the more widely accepted civil disobedience strategies of other radical environmentalists. Civil disobedience involves public protest for a cause, and while the marches or blockades may result in the protesters being charged with civil crimes, there is a strong commitment to non-violence.

GRASSROOTS ENVIRONMENTALISM

While the roots of environmentalism date back to the preservationist movement of the nineteenth century, it was the publication of Rachel Carson's *Silent Spring* in 1962 that led to human health risks' assuming significance along with conservation and preservation as environmental goals. The current era of environmentalism has increasingly focused on the dangers associated with industrial pollution and the occurrence of pollution sources in residential communities. Recent years have seen the emergence of new grassroots forms of environmentalism with this focus as their mandate.

The Toxic Waste Movement

The **toxic waste movement** is a branch of environmentalism unlike either the mainstream environmental movement or the new ecologies. On the one hand, the well-funded and organized mainstream environmental organizations, such as the Sierra Club, rely on professional leadership and a skilled staff, along with well-placed connections within the power structure and a large public base of followers for financial support. On the other hand, the new ecologies, while far less resourced, are inspired by the ideology and shared values of their members.

The University of Waterloo's Midnight Sun solar car team unveils their entry at the 2007 World Solar Challenge in Australia.

The toxic waste movement reflects few of the tendencies of either of these more general arms of the environmental movement. The movement is, in a sense, all the disputes and protests by myriad groups opposed to perceived environmental threats present in their own communities and neighbourhoods. Diffuse in its focus, the toxic waste movement is associated with all manner of protest against everything from proposed developments, such as a new landfill, factory, or highway, to pollution caused by an existing industry. What unites the toxic waste movement is a common focus on perceived health threats to the community. The movement is intrinsically grassroots in its composition and approach, constituted typically of groups of formerly uninvolved citizens now struggling to stop a development or clean up pollution while facing the efficient and well-funded opposition of industry and/or government.

The toxic waste movement may be the fastest-growing branch of environmentalism (Szasz 1994). It is also in many ways far less distinguishable than the other types of environmentalism. For one thing, there is little in the way of national organizing bodies or even communication among the various local groups. This extreme decentralization means that local protesters have few resources, outside of their own means, on which to draw in developing their plans of opposition. Mainstream environmental organizations typically employ professional social movement organizers in order to guide their agendas, but local toxic waste protesters rarely have the backgrounds or resources required to mount a well-managed and effective campaign. Valuable skills may be learned as the protest develops, but since these campaigns are often short-lived, such knowledge may not be passed on.

What is characteristic of the toxic waste movement is the high proportion of its members who are women and homemakers, minorities, and from lower socio-economic backgrounds (Brown and Masterson-Allen 1994). Toxic waste activists also tend to be older, politically conservative, and trusting of existing institutions, laws, and regulations. The movement's high composition of women and homemakers is in keeping with its principal focus on preserving human health, especially that of children, in the face of an immediate threat from a nearby development or pollution source.

Grassroots toxic waste protest is often dismissed as NIMBYism by those who disagree with its ends, and it is true that it often does appear that self-interest is an underlying motivation on which such protests are based. Nevertheless, if the threat is real, why should self-interest depreciate the legitimacy of the group's goal? It is also the case that a general increase within society in concerns over health risk from pollution and development is helping to move the toxic waste movement toward a more formally defined foundation of support and new allegiances. This is seen, for example, in connection with the general movement to supplant NIMBY with NIABY—'not in anyone's back yard'—indicative of the reduced emphasis on self-interest, as well as with LULU ('locally unwanted land uses'), reflective of the greater sensitivity to the broader public interest currently sought in the development of many municipal land-use plans (Freudenburg and Pastor 1992).

Environmental Justice

Environmental justice (**EJ**) is a grassroots environmental movement with an agenda going beyond the traditional concerns of conservation and preservation common to most environmentalism. The EJ movement has ties to the toxic waste movement but is also altering the focus of environmentalism generally to include broader concerns with regard to the societal inequities that result from industrial facility siting and industrial development.

While the environmental movement has long been concerned with the risks to human health from industrial pollution, only recently has that awareness developed over the distributional risks associated with these effects. Various studies have documented the inequitable distribution of environmental hazard (Bryant and Mohai 1992; Bullard 1990; Hofrichter 1993), showing that, for the most part, people of low income and racial minorities are disproportionately affected by poor environmental quality resulting from exposure to industrial pollution and contaminated water and lands.

Various explanations revolve around the processes that result in the inequitable distribution of environmental burdens. One position reflects economic or market dynamics, suggesting that 'sound' business decisions and the need to reduce costs may be grounds for locating potentially polluting industrial facilities (Kriesel, Centner, and Keeler 1996; Oakes 1996). It points to economic efficiency within the marketplace as the central criterion that guides what results in the unfair distribution of environmental risks to the poor. Industry's desire to minimize costs specifically associated with land or property values is seen as a major contributing factor to the disproportionate exposure to environmental pollutants. The suggestion is that this unequal risk occurs because cost-efficient industrial areas with low property values also are likely to

HUMAN DIVERSITY

Environmental Justice

In a Canadian study, Alice Nabalamba (2001) focuses on southern Ontario, in particular, Toronto, Hamilton, and the Niagara region. Using 1996 data from the Canadian census and municipal records, Nabalamba investigates the link between visible-minority status, socio-economic status, and the location of pollution sources. Five types of polluted sites were included: contaminated sites, industrial discharges, hazardous waste treatment and storage facilities, polychlorinated biphenyl (PCB) storage and treatment facilities, and other waste treatment, disposal, and storage facilities.

Nabalamba found that 'people of lower socio-economic means were more likely than the general population to live near a pollution source and industrial land use' (2001, 141). Visible-minority status was also related to increased exposure to certain types of pollution sources, but the relationship is clouded by the lower socio-economic status of many visible-minority groups. The location of pollution sources was related to decreased real estate values, decreased home ownership (versus renting), and the age of housing. The relationship between pollution siting and older, poorer neighbourhoods reflects a lack of political and organizational clout to defend against these types of uses. Therefore, Nabalamba predicts, future siting of these kinds of facilities will continue to affect these types of neighbourhoods to a greater extent than they will wealthier, newer, and 'whiter' areas.

be near areas with low residential property values or affordable housing and, therefore, a concentration of low-income populations.

Another rationale given for why the poor face greater pollution risk is the 'path of least resistance' argument (Higgins 1994; Hofrichter 1993). It suggests that low-income and minority communities end up with a disproportionate share of disposal and polluting industrial facilities and poor environmental quality in general because they have less political clout than more affluent communities.

Finally, a more contentious explanation cites 'environmental racism' among private industry and government decision-makers as being behind the disparities found in the uneven distribution of polluting industrial facilities (Bryant and Mohai 1992; Bullard 1990). This position draws largely on interpretations of evidence from the United States that show race to be a major factor in who is likely to be exposed to pollution risk. Hence, it is concluded that when race stands out as being significantly associated with the location of new disposal and polluting industrial facilities, it is racism that is influencing the decision-making process.

Conclusion

It is evident that the environment and social change are profoundly intertwined. The relationship cuts both ways: either societies will change to achieve environmental integrity or they will be changed by environmental contamination and resource depletion. Social change on behalf of the environment, therefore, is one of the most pressing global issues.

If sociology can make one substantial contribution to the understanding of ecological crisis, it is the recognition that environmental problems are social products. This understanding goes beyond descriptions of how individuals or firms contribute to environmental degradation, and the solutions suggested involve more than promoting more environmentally responsible behaviours or technologies. While such approaches may help deal with an immediate situation, they ultimately do more harm than good by deflecting attention from the real roots of environmental problems and the discovery of long-term solutions.

TIME to REFLECT

According to many public opinion surveys, most people claim to be environmentalists. What kind of environmentalist are you? In the context of the broad umbrella of the environmental movement encompassing everything from national environmental lobbying groups to local neighbourhood associations, what is your place? Do you see this changing? How?

In short, social systems must change so that global disaster may be averted. One promising development in this direction has been the globalization of environmental discourse. The 1972 UN conference on the environment held in Stockholm initiated the process, and since then there has been a steady proliferation of both high-level conferences, including the Earth Summits in 1992 and 2002, and regular international congresses on specific problems, such as the 2009 Copenhagen Conference on Climate Change attended by 170 countries, as well the creation of various international treaties and protocols.

Almost as important as the actual steps taken to check environmental damage has been the extension of such talks to include global social relations and their connection to environmental threats. It is now widely accepted, for example, that the eradication of poverty in the less developed world is among the keys to solving environmental problems. Yet this realization has served to kindle both old and new tensions between industrialized and developing nations. The former continue to promote such traditional mechanisms for economic improvement as foreign investment, new trading relations, and foreign aid. In some cases, it has been suggested that foreign assistance be tied to population control or environmental improvements. Third World countries tend to be deeply suspicious of such tactics, referring to them as environmental colonialism and arguing that they are little more than a new version of the historical patterns of domination that have been responsible for most underdevelopment.

Developing countries are also sensitive to any threats to their sovereignty perceived to result from intrusive foreign aid or investment and are distrustful of the World Bank, the International Monetary Fund, the Food and Agriculture Organization, and other global bodies traditionally involved with Third-World programs. Finally, many developing countries have crushing foreign debts and are compelled to earn hard currency through the export of raw resources or agricultural products in demand in industrialized nations. The harvesting of the resources and the farming practices used to grow the crops often cause considerable environmental damage while doing little to improve the long-term economic prospects of poorer countries.

TIME to REFLECT

Some argue that the root of global environmental problems is poverty—that overcoming poverty, especially in the developing world, will go a long way toward solving environmental problems as well. Consider the link between poverty and environmental degradation. How important is solving a social problem like being poor with respect to also protecting the natural environment?

On another level, the principle of sustainable development is becoming ingrained in the policy and planning frameworks of developed and less developed countries. Many herald this tendency as a breakthrough in attitudes toward the environment. Others are more sceptical, suggesting that the concept of sustainable development may be hijacked by development interests and used to legitimize unnecessary economic expansion. The potential for deflecting such a hijacking, if one is planned, could rest with the success of the new ecologies and grassroots environmentalism. The intensely participatory focus of these organizations, together with their high levels of commitment to local control of ecosystems, could go far toward curtailing economically driven environmental exploitation. As noted earlier, the influence of such movements in environmental planning is increasing, and if Ulrich Beck's risk society thesis is correct, then such involvement is destined to grow. The less conventional approaches of the new ecologies to environmental planning are also considered as representing the new forms of thinking necessary for achieving true sustainable development (Gardner and Roseland 1989). By reflecting alternatives to current forms of social organization and changing environmental values and aspirations, contemporary environmentalism may be in the vanguard of social change on behalf of the environment and of social improvement generally.

questions for critical thought

1. Despite the recognition that significant environmental problems persist, there still exists a spirit of optimism in many quarters: a feeling that humankind is making progress toward environmental quality. Should we be optimistic or pessimistic about the view that societies are succeeding in overcoming environmental problems?

2. Some individuals argue that environmental problems are inherent in the process of modernization, which involves urbanization, economic progress, and population growth. Do you feel that the inevitability of environmental damage is a valid assumption?

3. The concept of sustainable development is key to much economic planning, but its critics often argue that it is being used mainly as a rationale for allowing more economic growth at the expense of the environment. Can economic growth and environmental quality coexist, in your opinion?

4. When, if ever, is illegal protest such as the ecotage practised by Earth First! justifiable for protecting the environment?

5. Eco-feminists see similarities between environmentalists' fight to preserve the environment and women's struggle to achieve equality. Do you agree that there is a connection between these two social movements? Should all feminists be environmentalists, and vice versa?

6. Population growth is frequently likened to an ecological time bomb and one of the most virulent threats to the future of the planet. Are such claims valid? What are the various sides to this controversy? Discuss ways in which the global population problem could be solved.

7. Social constructionists see environmental problems as something formed by perception—basing their view on the extent to which these problems are recognized and validated by the greater part of society, measured in terms of the responses provided by such groups. How helpful is this perspective in alleviating environmental problems? Consider environmental controversies both at the local and international levels. How does the constructionist viewpoint contribute to their understanding?

8. How valid is the claim that being poor or a member of an ethnic minority increases one's degree of environmental threat in Canada? Aside from these populations, who else might face environmental risk? Are these risks reasonable, or should steps be taken to ensure that when dangers to health exist from the environment, they should be shared equally across the population? Is inequitable distribution of environmental hazard inevitable?

9. Ulrich Beck's theory of the risk society is a challenge to conventional thinking with regard both to the organization of society in the twenty-first century and to how environmental problems are viewed. How could Beck's theory alter the course of environmentalism and the manner in which environmental problems are addressed?

recommended readings

Ronald Bailey. 1993. *Eco-scam: The False Prophets of Ecological Apocalypse*. New York: St Martin's Press.
Not everyone agrees with environmentalism, and some are adamantly critical of its claims, particularly with respect to carrying capacity, or the Earth's ability to sustain its resource base. This book is a response to environmentalism's assumptions, claiming environmental alarmism and extremism are actually serving to undermine the future of the planet.

Jared Diamond. 2005. *Collapse: How Societies Choose to Fail or Succeed*. New York: Viking.
Diamond's writing is thoughtful, comprehensive, and scientific, and he draws on history, natural science, palaeontology, and archaeology to address the issue of why societies fail because of the collapse of the ecological system on which they depend. Diamond introduces the idea of 'progress traps'—innovations that give societies a competitive advantage but ultimately undermine their existence and result in their demise. The volume contains a number of excellent, detailed reviews of societal collapses due to ecocide, with examples of some amazingly enduring civilizations.

Lois Gibbs. 1982. *Love Canal: My Story*. Albany: State University of New York Press.
Love Canal is the most famous incident of chemical hazardous waste contamination in the United States and the struggle of neighbourhood residents to have it recognized. For years, Love Canal, an abandoned hydroelectric canal built in Niagara Falls, New York, in 1890, was used first as a municipal landfill by the City of Niagara Falls and later as a chemical dump by the Hooker Chemical Company. Houses and a school were built on the site after it was closed. In the mid-1970s, contamination and health problems began plaguing the neighbourhood. Lois Gibbs emerged as the leader of the struggle to have the contamination recognized and compensation provided to its victims.

Robert Hunter. 2004. *The Greenpeace to Amchitka: An Environmental Odyssey*. Vancouver: Arsenal Press.
This is the story of the 1971 trip by 12 environmental protesters sailing from Vancouver aboard the vessel *Greenpeace* in an attempt to stop US nuclear weapons testing on the island of Amchitka in Alaska. The protest failed when their ship could not reach the site prior to the time the test took place, but the legacy of that trip was the founding of the international environmental group Greenpeace by Hunter and others taking part in the protest. This highly personal account was published shortly before Hunter's death in 2005. Hunter also wrote an earlier history of Greenpeace, *Warriors of the Rainbow: A Chronicle of the Greenpeace Movement* (New York: Holt, Rinehart and Winston, 1979).

Anthony Sampson. 1975. *The Seven Sisters: The Great Oil Companies and the World They Shaped*. New York: Viking Press.
This book is a fascinating history of the great international oil companies founded in the nineteenth and early twentieth centuries and which still dominate the international oil business today. Highly compelling, it is valuable reading for anyone interested in understanding current geo-political relations and tensions between Middle East and Western societies and the role of Big Oil in formulating them.

Rik Scarce. 1990, *Eco-warriors: Understanding the Radical Environmental Movement*. Chicago: Noble Press.
Scarce reviews the histories, actions, and philosophies of the various groups constituting the radical arm of contemporary environmentalism, including Greenpeace, Earth First!, the Sea Shepherds, and Animal Liberation. Comparisons are made with nonradical grassroots environmentalism and the mainstream environmental lobby.

Stephen Schneider. 2009. *Science as a Contact Sport: Inside the Battle to Save Earth's Climate*. Washington: National Geographic Society.
Schneider offers an insider's examination of the international debate over the science of climate change, undoubtedly the biggest environmental controversy of the past decade or more. Although written by a biologist, the review provides strong support for the social constructionist perspective in terms of how environmental problems get defined. Dr Schneider was a co-recipient of the 2007 Nobel Peace Prize for his work with the Intergovernmental Panel on Climate Change (IPCC).

Paul Wapner. 2010. *Living through the End of Nature: The Future of American Environmentalism*. Cambridge, MA: MIT Press.
This is a provocative book by a respected environmentalist who argues that since the environment has become so modified and tamed, protecting nature from human intervention is no longer possible. He rejects both the traditional premises of humankind's responsibility toward nature, based on either preservation or mastery, and proposes a third option termed co-evolutionary. The book is a good source of critical assessment of the various contemporary branches of environmentalism while offering a direction forward largely reflective of the principles underlying the new ecologies.

recommended websites

Center for Health, Environment and Justice
www.chej.org
The Center for Health, Environment and Justice website is an online extension of the public campaign to promote environmental justice through community actions and public awareness of toxins. The site provides information and practical advice on community mobilization against toxic products, processes, and wastes.

David Suzuki Foundation
www.davidsuzuki.org
Canada's most famous environmentalist, David Suzuki, began a foundation in 1990 which focuses on climate change, biodiversity, and forest and fishery issues. The website offers information on public action and advocacy, a regular column by Dr Suzuki, and media releases and community events information.

Earth Day
www.earthday.net/footprint/index.html
A number of websites allow online calculation of one's ecological footprint: the amount of productive land needed to sustain each individual's own lifestyle. This one from the Earth Day organization calculates your ecological footprint based on information you provide on your consumption, housing, and travel patterns.

Environmental Defence Canada
www.environmentaldefence.ca
Environmental Defence Canada is a charitable organization with broad-based membership of university, private, and corporate sponsor. The organization advocates community and individual actions on environmental problems, and the website contains links to their newsletter, action alerts, and media releases.

Foundation for Deep Ecology
www.deepecology.org
The Foundation for Deep Ecology supports 'education, advocacy, and legal action on behalf of wild Nature and in opposition to the technologies and developments . . . destroying the natural world'. The website describes news, publications, programs, and grants in connection with the deep ecology movement.

Greenpeace
www.greenpeace.ca
Greenpeace's website provides information on the group's various campaigns as well as public information guides and press releases on environmental problems.

National Geographic: EarthPulse
www.earthpulse.nationalgeographic.com
This site provides abundant up-to-date information on the current state of the world with respect to environmental problems and vital signs, along with many striking photographs and maps. As well, there is an online quiz you may take to test your knowledge about the global trends affecting us.

100 Mile Diet
www.100milediet.org
The food eaten daily by typical North Americans travels an average of 2400 kilometres from the farm to the place it is consumed. In 2005, two Vancouver residents, Alisa Smith and James MacKinnon, attempted to see whether they could eat and drink for a year only products from within a radius of 100 miles of their home. The result has been a minor social movement comprising '100 milers', all attempting to replicate this feat in connection to products found in proximity to their own communities. This website recounts their stories while also providing advice and information for those wishing to accept the challenge.

Pembina Institute
www.pembina.org
The Pembina Institute is an independent, not-for-profit environmental policy research and education organization located in Alberta. The website states: 'the Institute's major policy research and education programs are in the areas of sustainable energy, climate change, environmental governance, ecological fiscal reform, sustainability indicators, and the environmental impacts of the energy industry.'

United Nations Environmental Programme (UNEP)
www.unep.org
The UNEP website offers students and other researchers access to maps, UN publications, and many databases from various sources, including the World Bank, UNICEF, UNESCO, and UNPOP. Databases can be downloaded in many different formats for student use and investigation.

glossary

Accountability The expectation that public education, like other state-provided services, has clearly defined objectives that members of the public can identify and use to assess how well and how cost-effectively they are being met.

Activity theory This theory, proposed by Havighurst and Albrecht in 1953, suggests that people in later life who meet their social and psychological needs through maintaining the level of activity they developed in middle age will adjust better to older ages and be more satisfied with life.

Aesthetics A system of rules for the appreciation of the beautiful or for the judgment of and reflection on the value of art and other matters of taste.

Age composition The distribution of the population with respect to age (and usually also sex); typically displayed graphically as a population pyramid.

Ageism An alteration in feeling, belief, or behaviour in response to an individual's or group's perceived chronological age.

Agency The human capacity to interpret, evaluate and choose, and then to act accordingly.

Agents of socialization Organizations and institutions through which culture (norms, values) is transmitted.

Aging The multilevel and dynamic process of getting older in the context of the life course.

Aging and society paradigm This newer paradigm (introduced by Riley, Foner, and Riley in 1999) argues that age-based distinctions will be removed in society so that we will live in an age-integrated society in which everyone, no matter what their age, interacts freely together.

Allopathic medicine Conventional medicine that treats by opposing something, whether viruses, bacteria, cells, organs, or other pathology.

Alter globalization Alternative views, policies, and proposals in support of 'another' globalization promoting human rights and environmental concerns.

Alternative media Types of communication that have been used by subordinate groups and social movements to present their own messages, which often involve challenging existing conditions in society. Examples include community radio and the Internet.

Americanization The United States, because of its huge economic superiority, is sometimes perceived as a global power imposing or promoting itself as a 'model' that other nations should follow and become, in a way, 'Americanized'.

Anomie The condition of modern society in which there are too few moral rules and regulations to guide people's conduct.

Anti globalization Associated with many social movements around the world that protest against policies that in their mind are undermining people's rights and the environment to promote the agenda of free trade and economic liberalization.

Anticipatory socialization Explicit or implicit learning in preparation for a future role; in Merton's definition, the acquisition of values and orientations found in statuses and groups in which one is not yet engaged but that one is likely to enter; socialization directed toward the preparation of future roles.

Baby boom The dramatic rise in the birth rate in Canada following World War II, lasting until well into the 1960s.

Baby bust The continuing decline in fertility following the end of the baby boom in the industrialized world.

Believing without belonging The idea that today many people may hold religious beliefs without actually belonging to or participating in any religious institution.

Bilateral descent pattern A system under which a married couple is considered part of both the female's and the male's kin groups.

Binary The use of either/or concepts (e.g., good/evil, body/mind) in social theory.

Blaming the system Refers to analyses that emphasize the structural and institutional sources of inequality. Unequal access to education would, for example, contribute to poverty patterns.

Blaming the victim Refers to the tendency to hold individuals entirely responsible for any negative situation that may arise in their lives. As applied to poverty, the poor are considered to bepoor because of their lack of ambition.

Bourgeoisie Marx used this term to refer to the capitalist class—that is, those individuals who own the means of production (factory owners), the merchant (economically dominant) or ruling class.

Bride price Money, property, or labour provided by the groom or his family to a bride's family for permission to marry her.

Bureaucracy The most developed, most efficient formal organization, with formal properties that include written rules, protected careers, and a clear chain of reporting relationships. It is ordered by criteria independent of the personal qualities of the people who hold positions of power and is a system that is rationalized and associated with states and the post–Industrial Revolution period.

Capitalism An economic system characterized by a relationship of unequal economic exchange between capitalists (employers) and workers. Because they do not own the means of production, workers must sell their labour to employers in exchange for a wage or salary. Capitalism is a market-based system driven by the pursuit of profit for personal gain.

Carrying capacity The ability of the Earth to provide the resources to sustain all of humankind.

Census A complete count of the population at one point in time, usually taken by a country every five or 10 years. The census is distinguished from the vital statistics system, a continuous registration system of births, deaths, marriages, and divorces.

Census family The Statistics Canada definition of the family used in the census, which usually includes married or long-term cohabiting couples, with or without never-married children, as well as single parents living with never-married children.

Census metropolitan areas (CMAS) Statistics Canada's term for large urban agglomerations of 100,000 or more people, sometimes consisting of more than one political jurisdiction or municipality, interconnected in relatively close proximity by systems of roadways. Today in Canada there are 33 CMAS.

Charter groups Canadians of British and French origin are known as charter groups because they have a special status entrenched in the Canadian Constitution and have effectively determined the dominant cultural characteristics of Canada. Each of these groups has special rights and privileges, especially in terms of the language of the legislature, of the courts, and of education.

Citizenship Rights and privileges people are granted by a recognized state in exchange for their support and loyalty to it.

Civil religion The idea that society is based on and functions because of shared values and perspectives that serve as the foundation of a cohesive society.

Claims-making The social constructionist process by which groups assert grievances about the troublesome character of people or their behaviour.

Class Inequality among groups of people based on the distribution of material resources and social capital.

Class and status Class, also termed socio-economic class, refers to one's position within a society's economic hierarchy. Typical designations include upper, middle, and lower class. Finer discriminations, such as upper middle class, also appear in the literature. In contrast, status refers to one's social position in terms of privilege and esteem. While often based in economic considerations, status suggests a broader lifestyle dimension. Status may be achieved (becoming a CEO) or ascribed (born an 'untouchable').

Class consciousness Put simply, this term refers to the sense of membership in a social class. For Marx, the working class would eventually (as a result of their concentration in factories and oppressive working conditions) recognize their common interests and act in concert to overthrow capitalism.

Classism Refers to the tendency to discriminate based on social class position.

Community As a broad sociological notion, community refers to a group of people living together and sharing common values, a common territory, and a daily life. Often, communities are self-contained, with community members working and living within the same limited geographic area. Community has been opposed to society in a radical way, community being the locales for mechanistic solidarity (as defined by Durkheim) while society involves organic solidarity. But this duality can be considered too simplistic, since communities are always being created or recreated within societies.

Concept An abstract idea that cannot be tested directly. Concepts can refer to anything, but in social research, they usually refer to characteristics of individuals, groups, or artifacts or to social processes. Some common sociological concepts include religiosity (strength of religious conviction), social class, and alienation.

Conflict theory A theoretical paradigm linked to the work of Marx and Weber that emphasizes conflict and change as the regular and permanent features of society because society is made up of various groups that wield varying amounts of power. Conflict theorists often stress the importance of status, economic inequality, and political power.

Conglomerate ownership A form of ownership in which one company has many firms that engage in a variety of often unrelated business activities.

It may combine *horizontal integration*, *vertical integration*, and even *cross-ownership*.

Conspicuous consumption Popularized by Veblen, this term refers to the many ways in which the well-to-do (the leisure class) display their social status by ostentatious display of their possessions.

Contagion The rapid spread of something like a fashion, mob mentality, or riot. It is associated with classic collective behaviour accounts of social movement action.

Continuity theory This theory was proposed by Atchley in 1989. It suggests that people can best adjust to aging by maintaining a certain continuity with the past as they move into the future, as long as the rate of change is in accordance with their own personal choices and societal norms.

Control theory A category of explanation that maintains that people engage in deviant behaviour when the various controls that might be expected to prohibit them from doing so are weak or absent.

Corporate city The corporate city refers to the precedence given to exchange values over use values in planning and/or regulating the location of activities within any given urban area, with the intention of increasing capital accumulation and profits. In such a model, as Sharon Zukin mentions, the opposition to the urban village, where cooperation between citizens and sociality prevails, reveals the essence of the corporate city.

Corporate crime Committed on behalf of a corporation, corporate crime victimizes consumers, competing businesses, or governments. It can lead to major social, financial, or physical harm, although often no criminal law has been violated.

Counter-hegemony see *Hegemony*.

Cross-ownership A form of ownership in which one company owns organizations associated with different types of media. For instance, a company might own a newspaper and a television network.

Cult A popular name for new religious movements. The word 'cult' tends to have a negative connotation, and thus scholars of religion tend to use 'new religious movement'.

Cultural and social capital Value that is embedded in cultural and social resources that can be exchanged and drawn upon. They facilitate exchanges between people, understandings of the world, social networks, trust, and obligations (see also *Social capital*).

Cultural capital A term coined by Pierre Bourdieu for cultural and linguistic competence, such as prestigious knowledge, tastes, preferences, and educational expertise and credentials, that individuals possess and that influences the likelihood of their educational and occupational success.

Cultural support theory An explanation of deviance that emphasizes an understanding of how deviant values lead to deviant behaviour.

Culture At its broadest, the sum total of the human-produced environment (the objects, artifacts, ideas, beliefs, and values that make up the symbolic and learned aspects of human society) as separate from the natural environment; more often refers to norms, values, beliefs, ideas, and meanings; an assumption that different societies are distinguished by their shared beliefs and customary behaviours; the products and services delivered by a number of industries: theatre, music, film, publishing, and so on.

Decoding The process of interpreting or 'reading' media content. It may involve a dominant-hegemonic reading, an oppositional reading, or a negotiated reading. See also *Encoding*; *Encoding and decoding*.

Deep ecology Term coined by Norwegian philosopher Arne Naess referring to a philosophical approach to environmentalism that calls for fundamental social change, in contrast to the more reformist orientation of mainstream environmentalism, referred to by Naess as 'shallow ecology'. Deep ecology has been criticized for its biocentric emphasis in claiming nature as separate from and superior to human society and is used by some environmental organizations to justify any means, even illegal acts, for addressing environmental problems.

Demographic components equation A method of estimating population size by adding births, subtracting deaths, and adding net migration occurring in an interval of time, then adding the result to the population at the beginning of the interval; also known as a balancing equation.

Demographic transition The process by which a country moves from high birth and death rates to low birth and death rates. The shift in fertility rates is often referred to as the fertility transition, while the complementary change in death rates is referred to as the mortality transition. The epidemiological transition theory is a complementary theory to demographic transition theory.

Developed (countries) The most industrialized countries of the world. According to the United Nations, these are the countries in Europe and in North America, as well as Australia, New Zealand, and Japan.

Developing countries All the countries not in the developed world. A subdivision of developing countries is the least developed countries, defined by the United Nations as countries with average annual incomes of less than $9000 (US). See also *Third World*.

Deviance People, behaviours, and conditions subject to social control.

Digital divide Inequalities in access to computers and/or the Internet. Inequalities associated with social class, gender, national origin, and other characteristics are seen as the basis for the digital divide.

Discourse A way of talking about and conceptualizing an issue, presented through ideas, concepts, and vocabulary that recur in texts.

Discrimination An action whereby a person is treated differently (usually unfairly) because of his or her membership in a particular group or category.

Disengagement theory This theory, proposed by Cumming and Henry in 1961, suggested that aging is accompanied by a mutual withdrawal of individuals and society from each other. This normative process of withdrawal was considered to be beneficial to both, since the individual wishes to be less involved in interaction, and society will experience minimal disruption with the death of an older person.

Dominant ideology The ideas and viewpoints held by the capitalist class or other powerful groups in society. Specific forms of the dominant ideology include capitalist, patriarchal, and racist ideology.

Dowry Money or property provided by a bride's family upon her marriage to help obtain a suitable husband and to be used by her new household (or sometimes to support her in case of divorce or widowhood).

Eco-feminism The branch of environmentalism that likens human domination over nature to male domination over women. Therefore, to be a feminist one must also be an environmentalist, and to be an environmentalist one must also be a feminist.

Economic elite This term refers to men and women who hold economic power in a society. Contemporary researchers often operationalize this concept in terms of reported financial assets (wealth) and/or leadership positions on the boards of key (largest 100) corporations.

Education The process by which human beings learn and develop capacities through understanding their social and natural environments, which takes place in both formal and informal settings.

Emerging countries (or BRIC) Countries like China, India, or Brazil are sometimes defined as 'emerging' because their economic and political strength is growing in relation to that of the traditional powers like the United States, Japan, or member states of the European Union. They are sometimes known as BRIC (for Brazil, Russia, India, and China).

Emphasized femininity A form of femininity matched to (and defined by) hegemonic masculinity.

Encoding The process of embedding ideology in media content. Encoding emerges through the complex interplay of economic and technical conditions associated with a media institution, the organizational relations and practices of the institution, and the ideology existing within the institution and the wider society. See also *Decoding*; *Encoding and decoding*.

Encoding and decoding The embedding and subsequent interpretation of cues, meanings, and codes in cultural productions.

Enlightenment An era in the 1700s when theorists believed that human reason could be the instrument of perfecting social life; emotions had to be controlled, and the role of religion, custom, and authority was criticized.

Environmental justice The branch of environmentalism that focuses on the inequitable distribution of environmental risks that affect the poor and racial minorities.

Essence; essentialism; essential nature The idea that a 'true' or core reality lies behind appearances, which makes something what it is and which, once identified, can establish its 'truth'. In the study of sexual and gender identities, for example, many challenge the idea that there is an essence of 'femaleness' (something all women share/are) or 'maleness' that sets females and males apart from each other.

Ethnic group People sharing a common ethnic identity who are potentially capable of organizing and acting on their ethnic interests.

Ethnicity Sets of social distinctions by which groups differentiate themselves from one another on the basis of presumed biological ties. Members of such groups have a sense of themselves as a common 'people' separate and distinct from others.

Export processing zone (EPZ) Special area within a state where its tariff restrictions, laws, and labour practices are not fully enforced and goods are produced for export to other countries. Such zones usually house foreign companies that exploit local workers. The maquiladoras along the northern Mexican border with the US are an example of an EPZ.

Extended family Several generations and/or married siblings and their children sharing a residence and cooperating economically.

Fair trade Set of ideas, proposals, and programs to establish direct relations between farmers in the Third World and consumers in the rich countries so that more income is redirected toward the farmers.

Feminism A theoretical paradigm, as well as a social movement, that focuses on causes and consequences of inequality between men and women, especially patriarchy and sexism.

Feminization of poverty This terms refers to the fact that globally, most women (as girls, adults, and seniors) are at greater risk of impoverishment than their male counterparts.

Financialization Under current economic policies, the financial sector has been growing tremendously at the expense of other sectors of the economy. The financial sector dominates industry and services and absorbs a very large quantity of capital, which is then used to speculate and expand the 'casino' economy.

First Nations 'Indians' in Canadian law; together with Métis and Inuit, they constitute Canada's Aboriginal peoples.

Formal care Caregiving provided by paid caregivers who are sometimes family but mainly professional health-care workers.

Formal organization An organization with clearly specified goals and a high degree of task differentiation among members.

Framing theory Since any issue can be viewed from a variety of perspectives and be construed as having implications for multiple values or considerations, framing refers to the process by which people develop a particular conceptualization of an issue or change their thinking about an issue. Goffman defines frames as definitions of a situation that is built up in accordance with social principles of organization that govern them and our subjective involvement in them. Frame analysis is concerned with the organization of experience.

Free trade Policy of trade liberalization promoted by richer states and international institutions like the World Bank and the International Monetary Fund to open economies to external inputs.

Function The role played by a part of society or a social institution, usually in terms of the importance of the role in maintaining social order and stability.

G8 Annual consultative process involving the most powerful states (the US, the UK, France, Germany, Italy, Canada, Japan, and Russia).

G20 Newly designed consultative process involving members of the G8 and some of the 'emerging' countries such as China, Brazil, and India.

Gender Socially recognized distinctions of masculinity and femininity.

Gender-based analysis A research methodology used in policy deliberations in Canada and internationally to help assess the impact of policies and programs on gender inequality.

Gender division of domestic labour The identification of specific household tasks as appropriate to men (for example, painting, cutting the grass, putting out the garbage) or to women (for example, cooking, cleaning, child care).

Gender mainstreaming An initiative to induce governments at all levels to commit to considering gender in all decisions and actions. The aim is to combat the marginalization of gender in political decision-making.

Gender role A set of behaviour patterns, attitudes, and personality characteristics stereotypically perceived as masculine or feminine within a culture.

Gender role socialization Begins at birth and refers to a set of cultural norms and behaviours considered appropriate for males and females.

Gendered hierarchy An organizational or institutional structure in which gender is mapped on to vertical dimensions of inequality, typically with men assuming the higher positions.

Generalized Other Mead's final developmental stage of the self whereby we learn cultural rules and values.

Gerontology The scientific study of aging.

Globalization A social process in which the constraints of geographic, economic, cultural, and social arrangements have receded and have been replaced by processes that extend beyond state boundaries. The flow of goods, services, media, information, and labour between countries around the world; different but interrelated aspects include economic globalization and cultural globalization; worldwide control and co-ordination by large private-sector interests not constrained by local or national boundaries.

Hawthorne effect Named for the famous studies of workers at Western Electric Company in Hawthorne, Illinois, that revealed that subjects' behaviour changes when they are observed, because they believe they are somehow special. The Hawthorne effect thus refers to the behaviour-modifying effect of surveillance, particularly in research.

Hegemonic masculinity A dominant form of masculinity. What is identified as hegemonic masculinity may vary depending on the social context. However, hegemonic masculinity is typically the valorization of physical strength, economic power, and heterosexuality and the domination of women and subordinate men.

Hegemony The dominance of ideology and culture by an elite group to the point that few alternatives exist or can be imagined. Counter-hegemony is the ability to launch oppositional views that challenge existing power-holders.

Hermaphrodite (hermaphroditism) A term no longer used in the social sciences (but still used in the natural sciences in sexual differentiation) to refer to individuals born with 'ambiguous' genitalia.

Heteronormativity The assumption that heterosexuality is a universal norm, therefore making homosexuality invisible or 'abnormal'.

Heterosexism A set of overt and covert social practices in both the public and private spheres that privileges heterosexuality over other sexual orientations.

Hidden curriculum The understandings that students develop as a result of the institutional requirements and day-to-day realities they encounter in their schooling; typically refers to norms such as competition, individualism, and obedience, as well as to a sense of one's place in school and social hierarchies.

Hijab A headscarf worn by some Muslim women that covers the hair but leaves the face visible.

Homophobia A term coined by George Weinberg in 1972 to refer to the psychological fear of homosexuality; tends to neglect the wider structural sources of the homosexual taboo essence.

Homosexual Someone who has sex with and/or is attracted to a person of the same sex.

Horizontal integration A form of ownership in which one company owns a number of media organizations in different locations that are doing the same type of business; also known as 'chain ownership' (e.g., a company might own several radio stations across Canada).

Household People who share a dwelling, whether or not they are related by blood, adoption, or marriage.

Human capital The notion that education, skills development, and other learning processes are investments that enhance our capacities. Human capital theory builds on this notion.

Human ecology The science of ecology, as applied to sociological analyses.

Human exemptionalism paradigm (HEP) The term used by Catton and Dunlap in arguing that the competing theoretical perspectives in sociology, including functionalism, conflict theory, and symbolic interactionism, all share a world view based on anthropocentrism.

Hypotheses Testable statements composed of at least two variables and how they are related.

Identity How we see ourselves and how others see us. How we view ourselves is a product of our history and of our interpretation of others' reactions to us. How others view us is termed 'placement' and is other people's reactions to our projections of ourselves, which is termed 'announcement'.

Identity politics Activity aimed at addressing how marginalized groups have been defined.

Ideology A system of beliefs, ideas, and norms, reflecting the interests and experiences of a group, class, or subculture, that legitimizes or justifies the existing unequal distribution of power and privilege; ways of seeing and of understanding the world and its actors. Ideologies function by making the social appear natural or functional rather than constructed for partisan interests and advantage.

Impression management The term used by Erving Goffman to describe how individuals try to shape the impression others have of them.

Incidence In epidemiology, the frequency of occurrence or onset of new cases of a disorder as a proportion of a population in a specific time period, usually expressed as the number of new cases per 100,000 per year.

Industrial Revolution A period of rapid social and political transition from feudal to modern forms of governance beginning in the eighteenth century in western Europe.

Informal care Caregiving provided by unpaid family and friends.

Informal economy A wide range of legal and illegal economic activities that are not officially reported to the government.

Informal organization An organization with loosely specified goals and little task differentiation between members.

Institutional completeness The ability of an ethnic community to establish all the institutions needed to satisfy the community's needs so that a member does not have to interact with institutions outside of it. The term was coined by Raymond Breton.

Institutions Patterns of behaviour that order people's lives in relatively predictable ways. Institutions are comprised of norms and social practices that have calcified to the extent that they create a predictable pattern or map of behaviours that people will usually follow.

Interlocking/intersectional analysis A way of understanding inequality that takes into account multiple, connecting dimensions. Which dimensions are significant and how their interconnections shape inequality are matters for research. Used in the analysis of gender to draw attention to the importance of looking at gender inequalities as they are connected with and specified by race, class, (dis)ability, and so on.

International Monetary Fund (IMF) One of the institutions created in 1944 in Bretton Woods, originally designed to lend money to western European countries for their postwar reconstruction. In the past while, the IMF has been involved mainly in 'redressing' Third-World states in economic decline by imposing stiff austerity programs.

Interpretive theory An approach that pays close attention to the cultural meanings held by actors, derived from socialization in the group, which is seen as the key to understanding human behaviour and patterns of action.

Intersectionalities Recent analysts, particularly those working from a feminist perspective, have called attention to the ways in which social inequalities are interwoven in a complex fashion. Gender inequalities, for example, are influenced by social class, disability, sexual orientation, race, ethnicity, age, and immigrant status. Class analysis increasingly pays attention to these intersections.

Intersexed (bodies) Infants born with genetic, hormonal, and anatomical configurations that do not coincide with normative anatomical sexual difference (male/female).

Intersubjectivity We adjust our behaviour according to the back-and-forth interpretation of what we think is on others' minds; for example, what I think *you* think of me.

Kirpan Ceremonial dagger worn by some Sikhs as one of the five practices of being an observant Sikh.

Juggernaut A powerful, unstoppable object that crushes any obstacle in its path. Originally a Hindu idol atop an enormous cart that easily ran out of control.

Labyrinth As distinct from a maze, a labyrinth is a path that serves as a walking meditation. It has only one winding path, which leads to a centre.

Life course perspective This conceptual framework emphasizes social pathways and their relation to socio-historical conditions, with an emphasis on human development and aging. The framework emerged as part of a trend toward a contextual understanding of human developmental processes and is sometimes considered the pre-eminent theory in social gerontology.

Life expectancy The number of years that an average person in a given population can expect to live.

Life expectancy at birth The average number of years left to live for a newborn in a given period. Life expectancy is distinct from life span, which is the oldest age humans can attain.

Lifelong learning The ongoing expectations for people to acquire new knowledge and capacities through learning that occurs in various levels and kinds of formal education as well as in other learning contexts; associated with increasing emphasis on the new economy and the continuing transitions that individuals undergo throughout their lives.

Lifelong socialization Denotes how socialization continues throughout the lifecycle. As we age, changing transitions and roles create new socialization experiences.

Lived religion An understanding of religion that focuses on the actual everyday spiritual practices of people rather than on religious authorities or religious texts.

Low Income Cut-off (LICO) A relative measurement of income used by Statistics Canada. The LICO considers that if a family spends 20 per cent or more of its income on food, shelter, and clothing than the average family, it is in precarious financial circumstances.

McJobs A term used to refer to jobs that do not provide quality employment. Such jobs are characterized by irregular and often unsocial hours, low pay, no benefits, and no job security.

Macrosociology The study of social institutions and large social groups; the study of the processes that depict societies as a whole and of the social-structural aspects of a given society.

Marketization The progressive exposure of the public sector to market forces, such as, for example, increased privatization of schooling or health care.

Mass media The technologies, practices, and institutions through which information and entertainment are produced and disseminated on a mass scale.

Master status A status characteristic that overrides other status characteristics in terms of how others see an individual. When a person is assigned a label of 'deviant' (for example, 'murderer', 'drug addict', 'cheater'), that label is usually read by others as signifying the most essential aspects of the individual's character.

Matriarchy A type of social organization in which mothers head families and descent may be reckoned through them.

Matrifocal A family-life system focused around the women, who earn most of the money and hold the family together (often in the absence of a male breadwinner).

Means of production A term is used by Marxists to refer to wealth-generating property such as land, factories, and machinery; the ways goods are produced for sale on the market, including all the workers, machinery, and capital such production needs.

Medicalization The tendency for more and more of life to be defined as relevant to medical diagnosis and treatment.

Melting-pot policies Policies that are most notable for their failure to recognize difference among communities and ethnic/racial groups, derived from American liberal individualistic ideology that assumes that immigrants should discard all of the traditions and distinctions they brought to the United States with them, such as their ethnic language or national identity, and become nothing but 'Americans'.

Meritocracy This form of social stratification relies on differences in effort and ability rather than ascribed statuses such as gender, age, or race.

Metropolis For the early sociologists like Simmel and Tönnies, as well as for Wirth, the notion of metropolis refers to the great city of their time. In contrast to villages, the metropolis could become an 'inhuman' environment because it could destroy traditional social life. But the metropolis can also offer new opportunities for modern individuals. They are exposed to different ways of life and can express their creativity more freely.

Microsociology The analysis of small groups and of the face-to-face interactions that occur within these groups in the everyday.

Modified extended family Several generations who live near each other and maintain close social and economic contact.

Morbidity rate The sickness rate per a specified number of people over a specified period of time.

Mortality rate The death rate per a specified number of people over a specified period of time.

Multiculturalism In Canada, a government policy to promote tolerance among cultural groups and to assist ethnic groups in preserving the values and traditions that are important to them; multiculturalism became official policy in 1971 following the report of the Royal Commission on Bilingualism and Biculturalism.

Negotiation A discussion intended to produce an agreement.

Neo-conservatism Political ideology based in the United States (and now extending to other nations such as Canada) promoting the return to 'traditional' values and emphasizing the need to oppose any nations or states that could out-compete one's own.

Neo-liberalism Political philosophy that flourished in the 1980s onwards and that promotes privatization, deregulation, and trade liberalization, as well as fiscal reforms to reduce social expenses and lower taxation of the wealthiest. This doctrine supports free trade between countries, cuts in social spending, and measures such as deregulation or privatization.

New ecological paradigm (NEP) The term used by Catton and Dunlap in arguing that environmental sociology constitutes a paradigm shift within general sociology based on the understanding that human societies cannot be separate and distinct from nature.

New economy A term used to highlight the shift in emphasis from industrial production within specific industries, firms, and nations to economic activities driven by information and high-level technologies, global competition, international networks, and knowledge-based advancement.

New religious movements Religious groups whose development is recent and who often have a less established position in society than more mainstream religious groups.

New urbanism An urban design and architectural movement that has flourished during the past 30 years in the US, with an emphasis on ecological concerns, denser cities and neighbourhoods, and accessible public spaces for pedestrians. New urbanism promotes diversity and tries to connect with local history when rebuilding old neighbourhoods.

Niqab A garment worn by Muslim women that covers most of the face but leaves the eyes exposed.

Non-standard (or precarious) work Jobs that are characterized by an increasingly tenuous or precarious relationship between employer and employee, including part-time employment, temporary employment, contract work, multiple job-holding, and self-employment; also termed 'contingent work' and 'casual work'.

Norms The rules and expectations of appropriate behaviour under various social circumstances. Norms create social consequences that have the effect of regulating appearance and behaviour.

North American Free Trade Agreement (NAFTA) Signed in 1992 by the US, Canada, and Mexico, NAFTA is the 'model' of trade liberalization that the most powerful in the world were hoping to extend to the rest of the world. Under NAFTA, capital flows and investments as well as trade have been liberalized. Under chapter 11 of the accord, private entities (companies) are entitled to sue governments if their interests are jeopardized by legislation on issues like labour rights or the environment.

Nuclear family A husband, wife, and their children, sharing a common residence and cooperating economically.

Numerical flexibility Part of a new general managerial approach that rests on flexibility in employment; involves shrinking or eliminating the core workforce (in continuous, full-time positions) and replacing them with workers in non-standard employment.

Objective Something completely unaffected by the characteristics of the person or instrument observing it. 'Objective' observations were used in the past to establish the truth of scientific theories until it became clear that completely objective observations are impossible.

Operationalization The translation of abstract theories and concepts into observable hypotheses and variables. Once abstract ideas are operationalized, we can test them in a study.

Organization A set of people connected by regular relationships that conform to shared norms and values.

Organizational sexuality Social practices that determine explicit and culturally elaborated rules of behaviour to regulate sexual identities and personal relationships in the workplace.

Organized religion Institutionalized and public expressions of religion, as opposed to spirituality.

Patriarchy A society or family system in which men have more authority than women.

Patrilineal descent The tracing of relationships and inheritance through the male line.

Pedagogy Processes associated with the organization and practice of teaching; more generally, various kinds of interactions (and how they are understood and organized) in teaching/learning situations.

Political process(es) The dynamic contestation of power among dominant elites and subordinate challengers. It is also associated with a branch of social movement theory.

Political process approach An approach that assumes that political constraints and opportunities influence the rise and fall of social movements as well as their institutional organization.

Polyandry The practice of being legally married to more than one man at a time.

Polygamy The practice of being legally married to more than one spouse at a time.

Polygyny The practice of being legally married to more than one woman at a time.

Population momentum The tendency for population to keep growing, even when the fertility rate drops to just the replacement level of 2.1 children per woman, as a consequence of a high proportion of persons in the child-bearing ages.

Positivism The assumption that human society can be studied objectively, as in natural science, and that social laws can be discovered to understand how society works.

Power In the classic formulation, power refers to the ability to exercise one's will, even in the face of opposition from others. In Marxist sociology, a social relationship that has a material base. Those who own the means of production have the power to exploit workers through the appropriation of their labour efforts. In Weberian sociology, power is more broadly defined and can reflect an individual's or group's capacity to exert their will over others. Contemporary analysts point out that power may also involve a wide variety of indirect and subtle manifestations, including the ability to mobilize bias or define a situation in one's own interests.

Prejudice An attitude by which individuals are prejudged on the basis of stereotyped characteristics assumed to be common to all members of the individual's group.

Prevalence In epidemiology, the total number of existing cases of a disorder as a proportion of a population (usually per 100,000 people) at a specific time.

Primary socialization Socialization that takes place during the formative period of life.

Privatization The movement away from a completely universally available and state-funded medical system to one that includes profit-making components.

Productive aging This theory was first introduced by Robert Butler in 1982 as a counterpoint concept to ageism, a term he originally coined. It is about older persons serving in the workforce, as volunteers, in the family, in the community, and so on, but first and foremost maintaining their independence and autonomy as long as possible.

Progressive conservation The movement originating in the nineteenth century that sought to check environmental destruction caused by unbridled economic growth and that resulted in the founding of such modern environmental organizations as the Audubon Society and the Sierra Club.

Proletariat Marx popularized this term to refer to those individuals who provided the labour power to capitalism. Lacking property, the proletariat was forced to survive by selling its labour to the bourgeoisie, who in turn exploited workers' efforts in the pursuit of profit.

Quiet Revolution Decline in church attendance and the power of the Roman Catholic church in Quebec, an example of secularization.

Race A group that is defined on the basis of perceived physical differences such as skin colour.

Racialization Sets of social processes and practices whereby social relations among people are structured according to visible physical difference among them to the advantage of those in the visible majority and the disadvantage of those in visible minorities.

Racialized A term that recognizes the social construction of race, it refers to a process whereby race comes to have social significance as well as political and material consequences.

Reciprocal socialization The process of bi-directional socialization whereby we learn from others (e.g., parents, teachers) just as others learn from us.

Relativism The idea that there is no single, unchangeable truth about anything; all things are either true or false only relative to particular standards. Many sociologists who do not view sociology as a science take their stance persuaded by relativism.

Reliability The consistency of a measure, indicator, or study. Note that reliability is different from validity and does not refer to the accuracy of a measure or study.

Resocialization The learning of new roles, norms, and values that are different from those of the past.

Resource mobilization approach An approach that assumes social movements are quite similar to other organizations.

Risk society A theory of the new modernity that argues that perception of risk is modernity's defining feature, creating uncertainty and compelling individuals to seek new strategic allegiances.

Role The specific behaviours, privileges, duties, and obligations expected of one who occupies a specific status.

Role conflict The conflict one experiences when the expectations attached to one role interfere with those attached to another role one is playing.

Role expectations The responsibilities and characteristics of an assumed role that tell someone how to play a role, like that of schoolteacher, and how others should act toward someone playing that role.

Role-making The process by which individuals creatively adapt and modify the roles they play to fit the situations they encounter.

Role strain The strain one experiences as a result of the competing demands built into a single role.

Role-taking The process by which we put ourselves in the position of others and align our actions with theirs.

Sample The group of people or objects drawn from the whole population that will be studied. In quantitative research, a great deal of time and effort is devoted to the selection of truly random samples, while in qualitative research, samples are often selected on the basis of the theoretical importance of the people or objects.

Schooling Processes that take place within formal educational institutions.

Secondary socialization Socialization that takes place in wider society, such as in the educational system, in peer groups, and from the mass media/technological spheres.

Secularization The process by which religion increasingly loses its influence.

Self In Mead's theory, an emergent entity with a capacity to be both a subject and an object and to assign meaning to itself, as reflected upon in one's own mind. In Goffman's dramaturgical theory, the self is a more shifting 'dramatic effect'—a staged product of the scenes one performs in.

Service economy The economic sector in which most Canadians are currently employed. In comparison to primary industry (the extraction of natural resources) and manufacturing (processing raw materials into usable goods and services), the service economy is based on the provision of services rather than on a tangible product, ranging widely from advertising and retailing, to entertaining, to generating and distributing information. Also called the 'tertiary sector'.

Sex ratio; primary sex ratio; secondary sex ratio The number of males in relation to the number of females in a population. The primary sex ratio is the sex ratio at birth, typically in the range of about 105 baby boys per 100 baby girls. The secondary sex ratio is the sex ratio beyond infancy.

Sexism Unfair discrimination on the basis of sex. It ranges from the obvious to the (nearly) hidden.

Sexual orientation An individual's sexual preference(s), which could include partners of the opposite sex, the same sex, both sexes, or neither.

Sexual scripting An approach that argues that socio-cultural processes are fundamental in determining what is perceived as sexual and how individuals should behave sexually.

Sexuality The ways in which we experience and express ourselves as sexual beings.

Sick role A conception of illness as comprising a set of rights and responsibilities, with four main aspects: (1) exemption from social responsibilities, which must be authorized by a proper authority such as a doctor; (2) exemption from blame for the illness; (3) a responsibility to get better; and (4) an expectation that the sick person will seek and follow the advice of outside medical help.

Significant others Individuals with whom a person interacts and who are important in that person's life, such as a parent or teacher.

Situated transaction A process of social interaction that lasts as long as the individuals find themselves in each other's company. As applied to the study of deviance, the concept of the situated transaction helps us to understand how deviant acts are social and not just individual products.

Social capital A concept widely thought to have been developed by American sociologist James Coleman in 1988 but discussed by Pierre Bourdieu in a similar way in the early 1980s; reflects the power that is derived from ties to social networks.

Social constructionism/constructionist The sociological theory that argues that social problems and issues are less objective conditions than they are collective social definitions based on how they are framed and interpreted.

Social control The various and myriad ways in which members of social groups express their disapproval of people and behaviours. They include name-calling, ridicule, ostracism, incarceration, and even killing.

Social determinants of health The factors that contribute to maintaining and improving health and well-being. They include individual lifestyle factors such as diet and smoking, social and community networks, and general socio-economic, cultural, and environmental conditions. According to the World Health Organization, they are the conditions in which people are born, grow, live, work, and age that are shaped by the distribution of money, power, resources, and policy choices. These conditions are considered responsible for health inequities within and between individuals and countries.

Social ecology The deeply philosophical branch of environmentalism founded by Murray Bookchin, which rejects duality between nature and humankind while stressing their connectiveness and potential for harmony and mutual sustainability.

Social exchange theory This theory, applied to social gerontology by James Dowd in 1980, stated that people try to maximize rewards and minimize costs in their interactions with others. Older people presumably end up with fewer resources in exchange relationships. Supposedly, they cannot reciprocate and can be forced to comply with the dictates of others, an uncomfortable position for most, or they can withdraw from relationships.

Social group A number of individuals, defined by formal or informal criteria of membership, who share a feeling of unity or are bound together in stable patterns of interaction; two or more individuals who have a specific common identity and who interact in a reciprocal social relationship. Primary groups are small and involve direct personal contact, whereas in a secondary group, a member may not interact with every other member.

Social institution A stable, well-acknowledged pattern of social relationships that endures over time, including the family, the economy, education, politics, religion, the mass media, medicine, and science and technology. Social institutions are the result of an enduring set of ideas about how to accomplish various goals generally recognized as important in a society.

Social interaction The process by which people act and react in relationships with others.

Social movement The social form taken by collective actors engaged in struggles against domination relations; the co-ordinated, voluntary action of non-elites (people with no control over major resources) for the manifest purpose of changing the distribution of social goods. The outcomes of these struggles are often difficult to grasp. To what extent social actors contribute to social change or toward participating in system regulation remains an open question.

Social network The set of direct and indirect connections among a group of people. Direct connections include links of kinship, friendship, and acquaintance. Information, social support, and other valuable resources flow through incompletely connected, or weakly tied, networks.

Social relationships Interactions of people in a society. Because people share culture and a sense of collective existence, these interactions will, to some extent, be recurrent and predictable.

Social reproduction A range of unpaid activities that help to reproduce workforces daily and over generations; typically, though not exclusively, performed by women in the family household.

Social stratification This term refers to the structured patterns of inequality that often appear in societal arrangements. From a macrosociological perspective, it is possible to discern the hierarchical strata of social classes that characterize most contemporary societies.

Social structure Patterns of behaviour or social relationships developed and accepted through time in a given group, organization, or society.

Socialization A lifelong interactive learning process through which individuals acquire a self-identity and the social skills needed to become members of society. See also *Primary socialization.*

Spontaneous organization An organization that arises quickly to meet a single goal and disbands when the goal is achieved.

State An institution associated with governing over a specific territory as well as establishing and enforcing rules within that territory. The state in a number of countries (including Canada) is involved in providing various public services. It is a formal bureaucracy that largely shapes material, cultural and social, and institutional political processes.

Status A socially defined position that a person holds in a given social group or organization to which are attached certain rights, duties, and obligations; a relational term, since each status exists only through its relation to one or more other statuses filled by other people.

Status degradation ceremony The rituals by which formal transition is made from non-deviant to deviant status. Examples include the criminal trial and the psychiatric hearing.

Status groups Organized groups comprising people who have similar social status situations. These groups organize to maintain or expand their social privileges by excluding outsiders from their ranks and by trying to gain status recognition from other groups.

Strain theory Robert Merton's theory that deviance results when people experience a gap between their aspirations and their opportunities.

Stratified Something is considered stratified when it is hierarchically ordered; in relation to power, stratification occurs when some groups control more material, cultural and social, and institutional resources than others and block others from challenging their control or gaining access to them.

Streaming Also known as tracking. A practice in elementary and secondary school systems aimed at homogenizing classrooms by placing similar students in the same classroom according to criteria that may include performance on standardized aptitude tests, perceived personal qualities and aspirations, or even social class and ethnic origin.

Structural analysis or approach An approach within organizational theory in the Weberian tradition; focuses on the structural characteristics of organizations and their effect on the people within them; in the context of urban studies, the analysis of the functions cities perform, the size and shape of their governments, and who has what bearing on decisions and outcomes involving cities.

Structural functionalism A theoretical paradigm that emphasizes the way each part of a society functions to fulfill the needs of society as a whole.

Subculture A subset of cultural traits of the larger society that also includes distinctive values, beliefs, norms, style of dress, and behaviour.

Subjective The opposite of objective; refers to the observer's mind, to perceptions, intentions, interpretations, and so on that affect how we act in the world.

Successful aging This theory, proposed by Rowe and Kahn in 1998, posits that older adults have significant abilities to prevent illness and minimize losses, including physical losses, and that they should enhance their engagement in life.

Surplus That part of the value of goods produced by working people that is taken by the dominant class and used to maintain social inequalities; a measure of the exploitation of the working class.

Sustainable development Refers mainly to the capacity of creating wealth without destroying the environment and preserving the environment for future uses. In the words of the World Commission on Environment and Development report (*Our Common Future,* 1987), it means to meet 'the needs of the present without compromising the ability of future generations to meet their own needs'. The principle is that economic growth and environmental conservation are compatible goals.

Symbolic interactionism An intellectual tradition in sociology akin to interpretive theory, founded in the early twentieth-century work of Charles Horton Cooley and George Herbert Mead, although the term itself was not coined until years later by Herbert Blumer. Symbolic interactionism emphasizes the importance of understanding the meanings of social action and uses ethnographic methods to discover these meanings for individuals in an effort to explain human conduct.

Symbols The heart of cultural systems, for with them we construct thought, ideas, and other ways of representing reality to others and to ourselves; gestures, artifacts, or uses of language that represent something else.

Systemic discrimination Discrimination that is built into the fabric of Canadian life, as in the case of institutional self-segregation.

Systems theory An approach within organizational theory that sees organizations as open systems and that views organizations and their goals as shaped by the interests of their participants and their environments.

Taboo Behaviour that is prohibited, such as incest and sexual relations with specific categories of kin; from the Tongan word meaning 'sacred' or 'inviolable'.

Third World Poor countries; an element of a classification in which the First World was made up of western Europe, North America, Australia, and New Zealand, the Second World of the various communist countries (the Soviet Union, the numerous Soviet satellites, and China), and the Third World of poorer countries in Asia, Africa, and Latin America. See also *Developing countries.*

Total institution A group or organization that has complete control over an individual and that usually engages in a process of resocialization.

Totalitarian An all-powerful form of government that exerts extreme control over the private lives of its citizens and is often associated with fascist policies of racism and aggressive nationalism.

Toxic waste movement The fast-growing grassroots branch of environmentalism that has largely emerged from opposition to local pollution problems or other environmental risks by neighbourhood residents who typically have not formerly been involved in environmental protest.

Transgressive sexualities Sexual identities and performances that challenge the norm.

Transitions The pathways that people follow from family life, into and out of education, and into various jobs or other social situations throughout their life course.

Two-earner family A family in which both adults are employed while raising their children. Now the most common form of family financing in Canada and other post-industrial countries.

Unemployment rate People are considered to be unemployed only if they do not have a job and are actively looking for a job. The unemployment rate is the number of people who meet those two conditions divided by the labour force (which includes both the employed and unemployed), expressed as a percentage. Those who do not have a job and are not looking for one are considered not in the labour force.

Urban planning This broad notion designates the processes leading to the production of urban spaces that professionals like planners follow. Urban planning is always historically and culturally characterized. Urban planning decisions can be more or less open to public participation.

Urban regime Refers to the governing coalitions that emerge in American cities in order to strongly influence local power-holders and orient urban public policies. Depending on their interests, these coalitions choose diverse courses of action to cope with urban change. In the literature, different types of regimes have been identified: status quo, pro-growth, caretaker, and progressive.

Urbanism Denotes the specific form taken by urban design within a historical moment. But it also includes the processes involved in the production of that form. Urbanism always includes a certain urbanity, defined as the culture of living within a city.

Urbanization The nature, extent, and distribution of cities in the larger society or nation.

Validity The accuracy of a measure, indicator, or study; many different dimensions to validity can be established through formal tests, logic, or depth of understanding.

Variable The operational or observable equivalent of concepts. Many concepts require more than one variable for proper operationalization. The key characteristic of variables is that there must be a range of different values that can be observed.

Vertical integration A form of ownership in which one company owns firms or divisions that are part of the overall process linking production, distribution, and exhibition (e.g., a company that owns a movie studio, a movie distributor, and movie theatres).

Vertical mosaic A view of Canadian society as constituting a materially, educationally, socially, politically, and ethnically divided stratification system, with the charter groups at the top, Native people at the bottom, and other-ethnic immigrant groups fitting in depending on their entrance status; from John Porter's *The Vertical Mosaic* (1965).

World Bank Along with the IMF, an important organization promoting neo-liberal policies through conditional loaning to developing countries.

World Social Forum Annual consultative process managed by more than 500,000 NGOs worldwide working to develop alternative development proposals focusing on rights.

World Trade Organization (WTO) Created in 1995 to facilitate trade liberalization worldwide, this organization has been stalemated in recent times by a growing conflict between the G8 countries and the 'emerging' countries over a wide range of economic issues.

references

Aapola, Sinikka, Marnina Gonick, and Anita Harris. 2005. *Young Femininity: Girlhood, Power, and Social Change*. London: Palgrave McMillan.

Abbott, P., and R. Sapsford. 1987. *Women and Social Class*. London: Tavistock.

Abrams, P. 1968. *The Origins of British Sociology, 1834–1914*. Chicago: University of Chicago Press.

Abu-Laban, Baha, and Daiva Stasiulis. 1992. 'Ethnic pluralism under siege: Popular and partisan opposition to multiculturalism'. *Canadian Public Policy* 27 (4): 365–86.

Acker, Joan. 1990. 'Hierarchies, jobs, and bodies: A theory of gendered organizations'. *Gender and Society* 4: 139–58.

———. 1991. 'Hierarchy, jobs, bodies: A theory of gendered organizations'. In Judith Lorber and Susan Farrell, eds, *The Social Construction of Gender*, 162–79. Newbury Park, CA: Sage.

Adamic, L., and Eytan Adar. 2005. 'How to search a social network'. *Social Networks* 27: 187–203.

Adams, Tracey L. 2000. *A Dentist and a Gentleman: Gender and the Rise of Dentistry in Ontario*. Toronto: University of Toronto Press.

Adelson, N. 2005. 'The embodiment of inequity: Health disparities in Aboriginal Canada'. *Canadian Journal of Public Health* 96: S45.

Adler, Patricia A., and Peter Adler. 1995. 'Dynamics of inclusion and exclusion in preadolescent cliques'. *Social Psychology Quarterly* 58 (3): 145–62.

———. 2006. 'The deviance society'. *Deviant Behavior* 27: 129–48.

Agnew, Robert. 1985. 'A revised strain theory of delinquency'. *Social Forces* 64 (1): 151–67.

———. 2006. 'General strain theory: Current status and directions'. In F.T. Cullen, J.P. Wright, and K.R. Blevins, eds, *Taking Stock: The Status of Criminological Theory*. New Brunswick, NJ: Transaction.

Agocs, Carol, and Monica Boyd. 1993. 'The Canadian ethnic mosaic recast for the 90s'. In James Curtis, Edward Grabb, and Neil Guppy, eds, *Social Inequality in Canada: Patterns, Problems, Policies*, 2nd edn, 330–52. Scarborough, ON: Prentice-Hall Canada.

Ahlburg, Dennis A. 1998. 'Julian Simon and the population growth debate'. *Population and Development Review* 24: 317–27.

Aibar, E., and W.E. Bijker. 1997. 'Constructing a city: The Cerdà plan for the extension of Barcelona'. *Science, Technology, and Human Values* 22: 3–30.

Aizlewood, Amanda, and Ravi Pendakur. 2005. 'Ethnicity and social capital in Canada'. *Canadian Ethnic Studies* 37 (2): 77–102.

Akers, R.L., and G.F. Jensen. 2003. *Social Learning Theory and the Explanation of Crime: A Guide for the New Century*, v. 11, *Advances in Criminological Theory*. New Brunswick, NJ: Transaction.

Alalehto, Tage. 2002. 'Eastern prostitution from Russia to Sweden and Finland'. *Journal of Scandinavian Studies in Criminology and Crime Prevention* 3 (1): 96–111.

Albanese, Patrizia. 2006. 'Small town, big benefits: The ripple effect of $7/day child care'. *Canadian Review of Sociology and Anthropology* 43 (2): 125–40.

———. 2009. *Children in Canada Today*. Toronto: Oxford University Press.

Albas, Dan, and Cheryl Albas. 1993. 'Avoiding the label of cheater during exams'. In Lorne Tepperman and James Curtis, eds, *Sociology of Everyday Life: A Reader*, 2nd edn, 217–23. Toronto: McGraw-Hill.

———. 2003. 'Aces and bombers: The post-exam impression management strategies of students'. In Ramón S. Guerra and Robert Lee Maril, eds, *A Social World: Classic and Contemporary Sociological Readings*, 3rd edn, 27–36. Boston: Pearson Custom Publishing.

Alberta v. Hutterian Brethren of Wilson Colony. 2009. SCC 37, [2009] 2 S.C.R. 567.

Alden, H.L., and K.F. Parker. 2005. 'Gender role ideology, homophobia and hate crime: Linking attitudes to macro-level anti-gay and lesbian hate crimes'. *Deviant Behavior* 26 (4): 321–43.

Alexander, D. 2000. 'The best so far: Vancouver's remarkable approach to the Southeast False Creek redevelopment is a big step towards sustainable development planning for urban sites'. *Alternatives* 26: 10–14.

———, and R. Tomalty. 2002. 'Smart growth and sustainable development: Challenges, solutions, and policy decisions'. *Local Environment* 7: 397–409.

Alfred, Taiaiake. 2005. *Wasase: Indigenous Pathways of Action and Freedom*. Peterborough, ON: Broadview.

Alkhateeb, Maha B., and Salma Elkadi Abugideiri. 2007. *Change from Within: Diverse Perspectives on Domestic Violence in Muslim Communities*. Great Falls, VA: Peaceful Families Project.

Allahar, Anton. 1994. 'More than an oxymoron: The social construction of primordial attachment'. *Canadian Ethnic Studies* 16 (3): 15–63.

Allen, Larry. 2009. *The Encyclopedia of Money*. Santa Barbara, CA: ABC-CLIO.

Alloway, Nola. 2007. 'Swimming against the tide: Boys, literacies, and schooling: An Australian story'. *Canadian Journal of Education* 3 (2): 582–605.

Altman, Dennis. 1986. *AIDS in the Mind of America*. New York: Doubleday-Anchor.

Alzheimer Society of Canada. 2010. *Rising Tide: The Impact of Dementia on Canadian Society*. Toronto: Alzheimer Society of Canada.

Amato, Paul. 2004. 'Parenting through family transitions'. *Social Policy Journal of New Zealand* 23 (Dec.): 31–44.

Ambert, Anne-Marie. 1997. *Parents, Children, and Adolescents*. Binghamton, NY: Hawthorne Press.

Amin, Samir. 2006. *Beyond U.S. Hegemony: Assessing the Prospects for a Multipolar World*. Trans. Patrick Camille. London: Zed Books.

———. 2010. *The World We Wish to See: Revolutionary Objectives in the Twenty-First Century*. New York: Monthly Review Press

Anderson, Benedict. 1991. *Imagined Communities*. London: Verso.

Anisef, Paul, Paul Axelrod, Etta Baichman-Anisef, Carl James, and Anton Turritin. 2000. *Opportunity and Uncertainty: Life Course Experiences of the Class of '73*. Toronto: University of Toronto Press.

Anningson, Brett. 2008. 'The slow food movement picks up'. *Canadian Dimension* 42 (4): 22–4.

Antonovsky, Aaron. 1979. *Health, Stress and Coping*. San Francisco: Jossey-Bass.

Appadurai, Arjun. 1961. Modernity at Large: Cultural Dimensions of Globalization. Minneapolis: University of Minnesota Press.

———. 1990. *Disjuncture and Difference in the Global Cultural Economy*. Durham, NC: Duke University Press.

———. 1996. *Modernity at Large*. Minneapolis: University of Minnesota Press.

Apple, Michael W., Wayne Au, and Luis Amando Gandin, eds. 2009. *The Routledge International Handbook of Critical Education*. New York: Routledge.

Appleton, Lynn M. 1995. 'Rethinking medicalization: Alcoholism and anomalies'. In Joel Best, ed., *Images of Issues: Typifying Contemporary Social Problems*, 2nd edn, 59–80. New York: Aldine de Gruyter.

Arai, A. Bruce. 2000. 'Changing motivations for home-schooling in Canada'. *Canadian Journal of Education* 25 (3): 204–17.

Arat-Koc, Sedef. 1990. 'Importing housewives: Non-citizen domestic workers and the crisis of the domestic sphere in Canada'. In Meg Luxton, Harriet

Rosenberg, and Sedef Arat-Koc, eds, *Through the Kitchen Window: The Politics of Home and Family*, 2nd edn, 81–103. Toronto: Garamond.

Archibald, W. Peter. 1976. 'Face-to-face: The alienating effects of class, status and power divisions'. *American Sociological Review* 41: 819–37.

Ariès, Philippe. 1962. *Centuries of Childhood: A Social History of Family Life*. New York: Vintage Books.

———. 1980. 'Two successive motivations for declining birth rates in the West'. *Population and Development Review* 6: 645–50.

Armstrong, Pat, and Hugh Armstrong. 1994. *The Double Ghetto: Canadian Women and Their Segregated Work*. Toronto: McClelland and Stewart.

Arnot, Madeleine. 2011. *Gender and Education*. New York: Routledge.

Aronson, Hal. 1993. 'Becoming an environmental activist: The process of transformation from everyday life into making history in the hazardous waste movement'. *Journal of Political and Military Sociology* 1: 63–80.

Arrighi, Giovanni. 2007. *Adam Smith in Beijing: Lineages of the Twenty-First Century*. London: Verso.

Arrigo, B.A. 1999. 'Can students benefit from an intensive engagement with postmodern criminology?' In J.R. Fuller and E.W. Hickey, eds, *Controversial Issues in Criminology*, 149–56. Boston: Allyn and Bacon.

———. 2004. 'Theorizing non-linear communities: On social deviance and housing the homeless'. *Deviant Behavior* 25: 193–213.

———, and T.J. Bernard. 1997. 'Postmodern criminology in relation to radical and conflict criminology'. *Critical Criminology* 8 (2): 39–60.

Asad, Talal. 1993. *Genealogies of Religion: Discipline and Reasons of Power in Christianity and Islam*. Baltimore: Johns Hopkins University Press.

Asbridge, M., R.E. Mann, and R. Flam-Zalcman. 2004. 'The criminalization of impaired driving in Canada: Assessing the deterrent impact of Canada's first per se law'. *Journal of Studies in Alcohol* 65 (4): 450–9.

Ascher, F. 2003. 'Métropolisation'. In J. Lévy and M. Lussault, eds, *Dictionnaire de la géographie et de l'espace des sociétés*, 613–15. Paris: Belin.

Assembly of First Nations. 2006. *Royal Commission on Aboriginal People at 10 Years: A Report Card*. Ottawa: Assembly of First Nations.

Atchley, R.C. 1989. 'A continuity theory of normal aging'. *The Gerontologist* 29 (2): 183–90.

Atkinson, Michael. 2006. 'Masks of masculinity: (Sur)passing narratives and cosmetic surgery'. In Dennis Waskul and Phillip Vannini, eds, *Body/Embodiment: Symbolic Interaction and the Sociology of the Body*, 246–62. Hampshire, UK: Ashgate.

———. 2008. 'Exploring male femininity in the "crisis": Men and cosmetic surgery'. *Body and Society* 14 (1): 67–87.

AuCoin, Kathy. 2005. 'Children and youth as victims of violent crime'. *Juristat* 25 (1). Catalogue no. 85–022–XIE. Ottawa: Statistics Canada.

Auger, Martin F. 2008. 'On the brink of civil war: The Canadian government and the suppression of the 1918 Quebec Easter riots'. *Canadian Historical Review* 89 (4): 503–40.

Ausubel, J. 2001. 'The great reversal: Nature's chance to restore land and sea'. *Technology in Society* 22: 289–301.

Avery, Donald. 1975. 'Continental European immigrant workers in Canada, 1896–1919: From "stalwart peasants" to radical proletariat'. *Canadian Review of Sociology* 12 (1): 53–64.

Axelrod, Paul. 1997. *The Promise of Schooling: Education in Canada, 1800–1914*. Toronto: University of Toronto Press.

Aysan, Mehmet F., and Roderic Beaujot. 2009. 'Welfare regimes for aging populations: No single path to reform'. *Population and Development Review* 35 (4): 701–20.

Babbie, Earl R. 1988. *The Sociological Spirit: Critical Essays in a Critical Science*. Belmont, CA: Wadsworth.

Baer, Douglas, James Curtis, and Edward Grabb. 2001. 'Has voluntary association activity declined? Cross-national analyses for fifteen countries'. *Canadian Review of Sociology* 38 (3): 249–74.

Baer, Douglas, Edward Grabb, and William A. Johnston. 1987. 'Class, crisis, and political ideology in Canada: Recent trends'. *Canadian Review of Sociology* 24 (1): 1–22.

———. 1990. 'The values of Canadians and Americans: A critical analysis and reassessment'. *Social Forces* 68 (3): 693–713.

Baird, Vanessa. 2001. *The No-Nonsense Guide to Sexual Diversity*. Toronto: Between the Lines.

Bairoch, P. 1995. *De Jericho à Mexico: villes et économies dans l'histoire*. 2nd edn. Paris: Arcades-Gallimard.

Baker, Maureen. 2005. 'Medically assisted conception: Revolutionizing family or perpetuating a nuclear and gendered model?' *Journal of Comparative Family Studies* 36 (4): 521–43.

———. 2006. *Restructuring Family Policies: Divergences and Convergences*. Toronto: University of Toronto Press.

———. 2009. *Families: Changing Trends in Canada*. 6th edn. Toronto: McGraw-Hill Ryerson.

———. 2010. *Choices and Constraints in Family Life*. 2nd edn. Toronto: Oxford University Press.

———, and David Tippin. 1999. *Poverty, Social Assistance and the Employability of Mothers: Restructuring Welfare States*. Toronto: University of Toronto Press.

Ball, N.R. 1988. *Bâtir un pays: histoire des travaux publics au Canada*. Montréal: Boréal.

Balthazar, Louis. 1992. 'L'évolution du nationalisme québécois'. In Gérard Daigle and Guy Rocher, eds, *Le Québec en jeu: comprendre les grands défis*, 647–67. Montreal: Presses de l'Université de Montréal.

Banister, Judith, and Ansley J. Coale. 1994. 'Five decades of missing females in China'. *Demography* 31: 459–79.

Barak, Azy, and William Fisher. 2001. 'Toward an Internet-driven, theoretically-based, innovative approach to sex education'. *Journal of Sex Research* 38 (4): 324–32.

Barber, Benjamin. 1995. *Jihad vs. McWorld*. New York: Ballentine Books.

Barker, Eileen. 1984. *The Making of a Moonie: Choice or Brainwashing?* Oxford: Blackwell.

———. 2005. 'New religions and cults in Europe'. In L. Jones, ed., *The Encyclopedia of Religion*. New York: Free Press.

———. 2007. 'How do modern European societies deal with new religious movements?' In P. Meusburger, M. Welker, and E. Wunders, eds, *Knowledge and Space: Clashes of Knowledge*, 154–71. Heidelberg: Springer.

Barker, John. 2003. 'Dowry'. In James J. Ponzetti, ed., *International Encyclopedia of Marriage and Family*, 2nd edn, 495–6. New York: Thomson Gale.

Barlow, Maude. 2007. *Blue Covenant: The Global Water Crisis and the Coming Battle for the Right to Water*. Toronto: McClelland and Stewart.

———, and Tony Clarke. 2001. *Global Showdown: How Activists Are Fighting Global Corporate Rule*. Toronto: Stoddard.

Barney, Darin. 2000. *Prometheus Wired: Hope for Democracy in the Age of Network Technology*. Vancouver: University of British Columbia Press.

Baron, S.W. 2003. 'Self-control, social consequences and criminal behavior: Street youth and the general theory of crime'. *Journal of Research in Crime and Delinquency* 40 (4): 403–25.

Barron, Martin, and Michael Kimmel. 2000. 'Sexual violence in three pornographic media: Towards a sociological explanation'. *Journal of Sex Research* 37 (2): 161–8.

Barry, Brian. 1978 [1970]. *Sociologists, Economists and Democracy*. Chicago: University of Chicago Press.

Bash, H.H. 1995. *Social Problems & Social Movements: An Exploration into the Sociological Construction of Alternative Realities*. Amherst, NY: Humanity Press.

Basran, Gurcharn, and B. Singh Bolaria. 2003. *The Sikhs in Canada: Migration, Race, Class and Gender*. New Delhi: Oxford University Press.

Bassand, M. 2001. 'Métropoles et métropolisation'. In M. Bassand, V. Kaufmann, and D. Joye, eds, *Enjeux de la sociologie urbaine*, 3–16. Lausanne: Presses polytechniques et universitaires romandes.

Baudrillard, Jean. 1998 [1970]. *The Consumer Society: Myths and Structures*. Thousand Oaks, CA: Sage.

Baum, Gregory. 2000. 'Catholicism and secularization in Quebec'. In D. Lyon and M. Van Die, eds., *Rethinking Church, State and Modernity: Canada between Europe and America*, 149–66. Toronto: University of Toronto Press.

Baxter, Janine. 2000. 'The joys and justice of housework'. *Sociology* 34 (4): 609–31.

———, Belinda Hewitt, and Michele Haynes. 2008. 'Life course transitions and housework: Marriage, parenthood and time spent on housework'. *Journal of Marriage and Family* 70 (2): 259–72.

Beach, Jane, Martha Friendly, Carolyn Ferns, Nina Prabhu, and Barry Forer. 2008. *Early Childhood Education and Care in Canada*. Toronto: Childcare Resource and Research Unit.

Beaman, Lori G. 1999. *Shared Beliefs, Different Lives: Women's Identities in Evangelical Context*. St Louis: Chalice Press.

———. 2008. *Defining Harm: Religious Freedom and the Limits of Law*. Vancouver: University of British Columbia Press.

Beaudet, Pierre. 2009. 'Asterix on the Saint Lawrence'. In Jai Sen and Peter Waterman, eds, *World Social Forum, Challenging Empires*, 332–42. Montreal: Black Rose Books.

Beck, Ulrich. 1992. *Risk Society: Towards a New Modernity*. Trans. Mark Ritter. London: Sage.

———. 1996. 'World risk society as cosmopolitan society? Ecological questions in a framework of manufactured uncertainties'. *Theory, Culture, and Society* 13 (4): 1–32.

———. 2008. *World at Risk*. Cambridge: Polity Press.

Beck-Gernsheim, Elisabeth. 2002. *Reinventing the Family: In Search of New Lifestyles*. Cambridge: Polity.

Becker, Gary. 1960. 'An economic analysis of fertility'. In *Demographic and Economic Change in Developed Countries: A Conference of the Universities–National Bureau Committee for Economic Research*, 209–40. Princeton, NJ: Princeton University Press.

Becker, Howard. 1952. 'Social class variations in the teacher–student relationship'. *Journal of Educational Sociology* 25: 451–65.

———. 1963. *Outsiders: Studies in the Sociology of Deviance*. New York: Free Press.

———. 1982. *Art Worlds*. Berkeley: University of California Press.

Beckett, Andy. 2010. 'Is the British middle class an endangered species? *The Guardian*. 14 July. www.guardian. co.uk.

Beckford, James A. 2003. *Social Theory and Religion*. Cambridge: Cambridge University Press.

Beckford, M. 2008. 'Britain's ageing population "as big a threat as climate change"'. www.telegraph.co.uk/news/2048794/ Britain's-ageing-population-'as-big-a-threat-as-climate-change'.html.

Béland, Daniel. 2005. 'Ideas and social policy: An institutionalist perspective'. *Social Policy and Administration* 39 (1): 1–18.

———. 2006. 'The politics of social learning: Finance, institutions, and pension reform in the United States and Canada'. *Governance* 19 (4): 559–83.

———. 2008. *Nationalism and Social Policy: The Politics of Territorial Solidarity*. Oxford: Oxford University Press.

———, and Jacob S. Hacker. 2004. 'Ideas, private institutions, and American welfare state "exceptionalism": The case of health and old-age insurance, 1915-1965'. *International Journal of Social Welfare* 13: 42–54.

Bélanger, Alain. 2006. *Report on the Demographic Situation in Canada 2003 and 2004*. Catalogue no. 91–209–XIE. Ottawa: Statistics Canada.

Bélanger, André-J., and Vincent Lemieux. 1969. 'Le nationalisme et les partis politiques'. *Revue d'histoire de l'Amérique française* 22 (4): 539–63.

Bell, Daniel. 1973. *The Coming of Post-industrial Society*. New York: Basic Books.

———, and Irwin Kristol. 1981. *The Crisis in Economic Theory*. New York: Basic Books.

Bell, K., A. Salmon, M. Bowers, J. Bell, and L. McCullough. 2010. 'Smoking, stigma and tobacco "denormalization": Further reflections on the use of stigma as a public health tool'. *Social Science and Medicine* 70 (6): 795–9.

Bell, Shannon. 1994. *Reading, Writing and Rewriting the Prostitute Body*. Bloomington: Indiana University Press.

Bellah, Robert, Richard Madsen, William M. Sullivan, Ann Swidler and Steven M. Tipton. 1985. *Habits of the Heart: Individualism and Commitment in American Life*. Berkeley: University of California Press.

Bemiller. Michelle. 2005. 'Men who cheer'. *Sociological Focus* 38 (3): 205–22.

Benevolo, L. 1993. *The European City*. Oxford: Blackwell.

Benford, Robert D., and David A. Snow. 2000. 'Framing processes and social movements: An overview and assessment'. *Annual Review of Sociology* 26: 611–39.

Bengston, V.L. 1996. 'Continuities and discontinuities in intergenerational relationships over time'. In V.L. Bengston, ed., *Adulthood and Again: Research on Continuities and Discontinuities*, 271–303. New York: Springer.

Benkert, Holly. 2002. 'Liberating insights from a cross-cultural sexuality study about women'. *American Behavioral Scientist* 45 (8): 1197–1207.

Bensman, Joseph. 1987. 'Mediterranean and total bureaucracies: Some additions to the Weberian theory of bureaucracy'. *International Journal of Politics, Culture and Society* 1 (1): 62–78.

Benson, Michael L. 1985. 'Denying the guilty mind: Accounting for involvement in a white-collar crime'. *Criminology* 23 (4): 583–607.

Benton, Ted. 1994. 'Biology and social theory in the environmental debate'. In Michael Redclift and Ted Benton, eds, *Social Theory and the Global Environment*, 28–50. London: Routledge.

Berger, Joseph. 2009. 'Student debt in Canada'. In Joseph Berger, Anne Motte, and Andrew Parkin, eds, *The Price of Knowledge: Access and Student Finance in Canada*, 4th edn, 182–205. Montreal: Canada Millennium Scholarship Foundation.

Berger, Peter L. 1963. *Invitation to Sociology*. New York: Anchor Books.

———. 1967. *The Sacred Canopy: Elements of a Sociological Theory of Religion*. Garden City, NY: Doubleday.

———, and Thomas Luckmann. 1966. *The Social Construction of Reality: Treatise in the Sociology of Knowledge*. Garden City, NY: Anchor.

Berlan, Elise, Heather Corliss, Alison E. Field, Elizabeth Goodman, and S. Bryn Austin. 2010. 'Sexual orientation and bullying among adolescents in the Growing Up Today Study'. *Journal of Adolescent Health* 46 (4): 366–71.

Berlin, Isaiah. 1963. *Karl Marx: His Life and Environment*. New York: Oxford University Press.

Berman, Jacqueline. 2010. 'Biopolitical management, economic calculation and "trafficked women"'. *International Migration* 48 (4): 84–113.

Berman, Sheri. 1997. 'Civil society and the collapse of the Weimar Republic'. *World Politics* 49: 401–29.

Bernard, Andrew B., and J. Bradford Jensen. 2000. 'Understanding increasing and decreasing wage inequality'. In Robert C. Feenstra, ed., *The Impact of International Trade on Wages*, 227–61. Chicago: University of Chicago Press.

Bernard, M. 2001. 'Women ageing: Old lives, new challenges'. *Education and Ageing* 16 (3): 333–52.

Bernard, T.J., J.B. Snipes, and A.L. Gerould. 2009. *Vold's Theoretical Criminology*. 6th edn. Oxford: Oxford University Press.

Bernburg, J.G., M.D. Krohn, and C.J. Rivera. 2006. 'Official labeling, criminal embeddedness, and subsequent delinquency: A longitudinal test of labeling theory'. *Journal of Research in Crime and Delinquency* 43 (1): 67–88.

Bernstein, Basil. 1977. 'Class and pedagogies: Visible and invisible'. In Jerome Karabel and A.H. Halsey, eds, *Power and Ideology in Education*, 511–34. New York: Oxford University Press.

Bertone, Andrea Marie. 2000. 'Sexual trafficking in women: International political economy and the politics of sex'. *Gender Issues* 18 (1): 4–22.

Best, Joel. 1989. 'Extending the constructionist perspective: A conclusion—and an introduction'. In Joel Best, ed., *Images of Issues: Typifying Contemporary Social Problems*, 243–53. New York: Aldine de Gruyter.

———. 1993. 'But seriously folks: The limitations of the strict constructionist interpretation of social problems'. In James A. Holstein and Gale Miller, eds, *Reconsidering Social Constructionism: Debates in Social Problems Theory*, 129–47. New York: Aldine de Gruyter.

———. 2004. *More Damned Lies and Statistics*. Berkeley: University of California Press.

———. 2008. *Social Problems*. New York: Norton.

Beyer, Peter. 2006a. *Religion in Global Society*. London: Routledge.

———. 2006b. 'Religion among immigrant youth in Canada: Research project'. Ottawa: Department of Classics and Religious Studies, University of Ottawa. www.research.uottawa.ca/excellence-discoveries-details_36.html.

———. 2008. 'From far and wide: Canadian religious and cultural diversity in global/local context'. In L.G. Beaman and P. Beyer, eds, *Religion and Diversity in Canada*, 9–40. Leiden: Brill Academic Press.

Bhagwati, Jagdish. 2004. *In Defense of Globalization*. New York: Oxford University Press.

Bibby, Reginald. 1993. *Unknown Gods: The Ongoing Story of Religion in Canada*. Toronto: Stoddart.

———. 2002. *Restless Gods: The Renaissance of Religion in Canada*. Toronto: Stoddart.

———. 2004. 'Section 2: Dating, sexuality and cohabitation'. In *The Future of Families Project: A Survey of Canadian Hopes and Dreams*, 11–24. Ottawa: Vanier Institute of the Family.

———. 2006. *The Boomer Factor: What Canada's Most Famous Generation Is Leaving Behind*. Toronto: Bastion Books.

———, with Sarah Russell and Ron Rolheiser. 2009. *The Emerging Millennials: How Canada's Newest Generation Is Responding to Change and Choice*. University of Lethbridge: Project Canada Books.

Bies, Robert J., and Thomas M. Tripp. 1996. 'Beyond distrust: "Getting even" and the need for revenge'. In Roderick M. Kramer and Tom R. Tyler, eds, *Trust in Organizations: Frontiers of Theory and Research*, 246–60. Thousand Oaks, CA: Sage.

Bijker, W.E. 1997. *Of Bicycles, Bakelites, and Bulbs: Toward a Theory of Sociotechnical Change*. Cambridge, MA: MIT Press.

Bishop, Clifford. 1996. *Sex and Spirit*. Alexandria, VA: Time-Life Books.

Bishop, Jack. 2005. 'Building international empires of sound: Concentrations of power and property in the "global" music market'. *Popular Music and Society* 28 (4): 443–71.

Bissoondath, Neil. 1994. *Selling Illusions: The Cult of Multiculturalism*. Toronto: Penguin.

Blackledge, David, and Barry Hunt. 1985. *Sociological Interpretations of Education*. London: Routledge.

Blanke, D.D., and V. Silva. 2004. *Tobacco Control Legislation: An Introductory Guide*. Geneva: World Health Organization.

Bloom, David E., David Canning, and Jaypee Sevilla. 2003. *The Demographic Dividend: A New Perspective on the Economic Consequences of Population Change*. Santa Monica, CA: Rand Corporation.

Blumer, Herbert. 1969. *Symbolic Interactionism: Perspective and Method*. Englewood Cliffs, NJ: Prentice-Hall; Berkeley: University of California Press.

———. 1971. 'Social problems as collective behavior'. *Social Problems* 8 (3): 298–396.

Bob, Clifford. 2005. *The Marketing of Rebellion: Insurgents, Media, and International Activism*. New York: Cambridge University Press.

Bock-Côté, Mathieu. 2009. 'L'identité occidentale du Québec ou l'émergence d'une "cultural war" à la québécoise'. *Recherches sociographiques* 50 (3): 537–70.

Bolaria, B. Singh, and Peter Li. 1988. *Racial Oppression in Canada*. 2nd edn. Toronto: Garamond.

Bongaarts, John. 1978. 'A framework for analyzing the proximate determinants of fertility'. *Population and Development Review* 4 (1): 105–32.

———. 2004. 'Population aging and the rising cost of public pensions'. *Population and Development Review* 30 (1): 1–24.

———. 2008. 'Fertility transition in developing countries: Progress or stagnation?' *Studies in Family Planning* 39 (2): 105–10.

———, and Rodolfo Bulatao, eds. 2000. *Beyond Six Billion: Forecasting the World's Population*. Washington: National Academy Press.

Bookchin, Murray. 1989. *Remaking Society*. Montreal: Black Rose.

Boone, C.G., and A. Modarres. 2006. *City and Environment*. Philadelphia: Temple University Press.

Borer, M.I. 2006. 'The location of culture: The urban culturalist perspective'. *City and Community* 5: 173–97.

Boserup, Ester. 1965. *The Conditions of Agricultural Growth: The Economics of Agrarian Change under Population Pressure*. Chicago: Aldine.

———. 1981. *Population and Technological Change: A Study of Long-Term Trends*. Chicago: University of Chicago Press.

Bostock, L. 2002. '"God, she's gonna report me": The ethics of child protection in poverty research'. *Children and Society* 16 (4): 273–83.

Bourdieu, Pierre. 1984 [1979]. *Distinction: A Social Critique of the Judgement of Taste*. Trans. Richard Nice. London: Routledge.

———. 1997. 'The forms of capital'. Trans. Richard Nice. In A.H. Halsey, Hugh Lauder, Phillip Brown, and Amy Stuart Wells, eds, *Education: Culture, Economy, and Society*, 46–58. Oxford: Oxford University Press.

———, and Jean-Claude Passeron. 1979. *The Inheritors: French Students and Their Relations to Culture*. Trans. Richard Nice. Chicago: University of Chicago Press.

Bourdin, A. 2005. *La métropole des individus*. Paris: La Tour d'Aigues, Éditions de l'Aube.

Bourne, L.S., and J.W. Simmons. 2003. 'New fault lines? Recent trends in the Canadian urban system and their implications for planning and public policy'. *Canadian Journal of Urban Research/Revue canadienne de recherche urbaine* 12: 22–47.

Bowden, Gary. 1989. 'Labour unions in the public mind: The Canadian case'. *Canadian Review of Sociology* 26 (5): 123–42.

———. 1990. 'From sociology to theology: A reply to Lipset'. *Canadian Review of Sociology* 27 (4): 536–39.

Bowles, Samuel, and Herbert Gintis. 1976. *Schooling in Capitalist America: Education Reform and the Contradictions of Economic Life.* New York: Basic Books.

Boyd, Jade. 2010. 'Producing Vancouver's (hetero)normative nightscape'. *Gender, Place and Culture: A Journal of Feminist Geography* 17 (2): 169–89.

Boyd, Monica. 1992. 'Gender, visible minority, and immigrant earnings inequality: Reassessing an employment equity premise'. In Vic Satzewich, ed., *Deconstructing a Nation: Immigration, Multiculturalism and Racism in '90s Canada*, 279–321. Halifax: Fernwood.

Bradbury, Bettina. 2005. 'Social, economic, and cultural origins of contemporary families'. In M. Baker, ed., *Families: Changing Trends in Canada*, 5th edn, 71–98. Toronto: McGraw-Hill Ryerson.

Braithewaite, John. 1979. *Inequality, Crime and Public Policy.* London: Routledge and Kegan Paul.

Braithwaite, Dawn O., and Leslie A. Baxter. 2005. *Engaging Theories in Family Communication: Multiple Perspectives.* Thousand Oaks, CA: Sage.

Bramham, Daphne. 2010. 'Polygamy is harmful to society, scholar finds'. *Vancouver Sun.* www.vancouversun. com/news/Polygamy+harmful+ society+scholar+finds/3290757/story. html#ixzz0u9ayfizY.

Brannigan, Augustine, and William Zwerman. 2001. 'The real "Hawthorne Effect"'. *Society* 38 (2): 55–60.

Braun, Denny. 1991. *The Rich Get Richer: The Rise of Income Inequality in the United States and the World.* Chicago: Nelson-Hall.

Braverman, Harry. 1974. *Labor and Monopoly Capital: The Degradation of Work in the Twentieth Century.* New York: Monthly Review Press.

Brenner, Robert. 2003. *The Boom and the Bubble: The US in the World Economy.* London: Verso.

Breton, Raymond. 1964. 'Institutional completeness of ethnic communities and the personal relations of immigrants'. *American Journal of Sociology* 18 (5): 193–205.

———. 1972. 'The socio-political dynamics of the October events'. *Canadian Review of Sociology and Anthropology* 9 (1): 33–56.

———. 2003. 'Social capital and the civic participation of immigrants and members of ethno-cultural groups' (Policy Research Initiative. Presented at the Opportunities and Challenge of Diversity: A Role for Social Capital? Conference, Montreal, November).

Brickell, Chris. 2006. 'A symbolic interactionist history of sexuality'. *Rethinking History* 10 (3): 415–32.

Briskin, Linda. 1992. 'Socialist feminism: From the standpoint of practice'. In M. Patricia Connelly and Pat Armstrong, eds, *Feminism in Action: Studies in Political Economy*, 267–93. Toronto: Canadian Scholars' Press.

Brockington, Riley. 2010. *Summary Public School Indicators for Canada, the Provinces and Territories, 2001/2002 to 2007/2008.* Research Paper 81-595-M, no. 083. Ottawa: Minister of Industry, Culture, Tourism and the Centre for Education Statistics.

Bromley, David G., and Anson D. Shupe, Jr. 1981. *Strange Gods: The Great American Cult Scare.* Boston: Beacon.

Brook, Barbara. 1999. *Feminist Perspectives on the Body.* London: Longman.

Brown, Craig, ed. 2002. *The Illustrated History of Canada.* Toronto: Key Porter.

Brown, Louise. 2002. 'Two-tier grade schooling feared'. *Toronto Star* 31 May: A1, A26.

———. 2010. 'Focus on struggling minority students, report recommends'. 14 July. www.parentcentral.ca.

Brown, Phil, and Susan Masterson-Allen. 1994. 'The toxic waste movement: A new type of activism'. *Society and Natural Resources* 7 (3): 269–87.

Browne, C. 1998. *Women, Feminism, and Aging.* New York: Springer.

Brownlee, Jamie. 2005. *Ruling Canada: Corporate Cohesion and Democracy.* Halifax: Fernwood.

Brownmiller, Susan. 1975. *Against Our Will: Men, Women and Rape.* New York: Simon and Schuster.

Brubaker, Rogers. 1984. *The Limits of Rationality: An Essay on the Social and Moral Thought of Max Weber.* London: George Allen and Unwin.

———, et al. 2004. 'Ethnicity as cognition'. *Theory and Society* 33 (1): 31–64.

Brugmann, J. 2009. *Welcome to the Urban Revolution: How Cities Are Changing the World.* New York: Bloomsbury Press.

Bryant, Bunyon, and Paul Mohai, eds. 1992. *Race and the Incidence of Environmental Hazard: A Time for Discourse.* Boulder, CO: Westview.

Bryant, Heather. 1990. *The Infertility Dilemma: Reproductive Technologies and Prevention.* Ottawa: Canadian Advisory Council on the Status of Women.

Brym, Robert J. 1980. 'Regional social structure and agrarian radicalism in Canada: Alberta, Saskatchewan and New Brunswick'. In Alexander Himelfarb and C. James Richardson, eds, *People, Power and Process: A Reader*, 344–53. Toronto: McGraw-Hill Ryerson.

Bryman, Alan, Jane J. Teevan, and Edward Bell. 2009. *Social Research Methods.* 2nd Canadian edn. Toronto: Oxford University Press.

Brzozowski, J.A., A. Taylor-Butts, and S. Johnson. 2006. 'Victimization and offending among the Aboriginal population in Canada'. *Juristat* 26 (3).

Buchignani, Norman, Doreen Indra, and Ram Srivastiva. 1985. *Continuous Journey: A Social History of South Asians in Canada.* Toronto: McClelland and Stewart.

Bulatao, Rodolfo. 1998. *The Value of Family Planning Programs in Developing Countries.* Santa Monica, CA: Rand.

———, and John Casterline, eds. 2001. 'Global fertility transition'. *Population and Development Review* 27 (Suppl.).

Bullard, Robert. 1990. *Dumping in Dixie: Race, Class and Environmental Quality.* Boulder, CO: Westview.

Bullock, Cathy Ferrand, and Jason Culbert. 2002. 'Coverage of domestic violence fatalities by newspapers in Washington State'. *Journal of Interpersonal Violence* 17: 475–99.

Bulman, Donna, Diana Coben, and Nguyen Van Anh. 2004. 'Educating women about HIV/AIDS: Some international comparisons'. *Compare* 34 (2): 141–59.

Burkholder, Carolynne. 2009. 'B.C. to ask court if polygamy violates the Charter'. *National Post.* www. nationalpost.com/news/canada/story. html?id=2133113#ixzz0UiIVKCMg.

Burningham, Kate, and Geoff Cooper. 1999. 'Being constructive: Social constructionism and the environment'. *Sociology* 33: 297–316.

Burt, Ronald S., and Marc Knez. 1996. 'Trust and third-party gossip'. In Roderick M. Kramer and Tom R. Tyler, eds, *Trust in Organizations: Frontiers of Theory and Research*, 68–89. Thousand Oaks, CA: Sage.

Bussière, Patrick, and Tamara Knighton. 2004. *Measuring Up: Canadian Results of the OECD PISA Study: The Performance of Canada's Youth in Mathematics, Reading, Science and Problem Solving, 2003. First Findings for Canadians Aged 15.* Ottawa: Minister of Industry.

Bussière, Patrick, Tamara Knighton, and Dianna Pennock. 2007. *Measuring Up: Canadian Results of the OECD PISA Study: The Performance of Canada's Youth* in Science, Reading and Mathematics 2006. First Results for Canadians Aged 15. Catalogue no. 81-590-XPE, no. 3. Ottawa: Minister of Industry.

Butler, Judith. 2006 [1990]. *Gender Troubles: Feminism and the Subversion of Identity*. New York: Routledge Classics.

———. 1992. 'Contingent foundations: Feminism and the question of "post-modernism"'. In Judith Butler and Joan W. Scott, eds, *Feminists Theorize the Political*, 3–21. New York: Routledge.

Butler, R.N., and H.P. Gleason, eds. 1985. *Productive Aging: Enhancing Vitality in Later Life*. New York: Springer.

Butsch, Richard. 1992. 'Class and gender in four decades of television situation comedy: Plus ça change . . .'. *Critical Studies in Mass Communication* 9: 387–99.

———. 1995. 'Ralph, Fred, Archie and Homer: Why television keeps recreating the white male working-class buffoon'. In Gail Dines and Jean M. Humez, eds, *Gender, Race and Class in Media: A Text-Reader*, 403–12. Thousand Oaks, CA: Sage.

———, and Lynda M. Glennon. 1983. 'Social class: Frequency trends in domestic situation comedy, 1946–1978'. *Journal of Broadcasting* 27 (1): 77–81.

Buttel, Frederick. 1976. 'Social science and the environment: Competing theories'. *Social Science Quarterly* 57: 307–23.

Butters, Jennifer, and Patricia Erickson. 1999. 'Addictions as deviant behaviour: Normalizing the pleasures of intoxication'. In Lori G. Beaman, ed., *New Perspectives on Deviance: The Construction of Deviance in Everyday Life*, 67–84. Toronto: Prentice-Hall Allyn and Bacon.

Butz, William P., and Michael P. Ward. 1979. 'The emergence of counter-cyclical US fertility'. *American Economic Review* 69: 318–28.

Byers, Lisa, Ken Menzies, and William O'Grady. 2004. 'The impact of computer variables on the viewing and sending of sexually explicit material on the Internet: Testing Cooper's "Triple-A Engine"'. *Canadian Journal of Human Sexuality* 13 (3/4): 157–69.

Cain, Mead. 1983. 'Fertility as an adjustment to risk'. *Population and Development Review* 9: 688–702.

Calasanti, T. 2004. 'Feminist gerontology and old men'. *The Journals of Gerontology* 59 (6): S305–S314.

Calavita, K., and H.N. Pontell. 1991. '"Other people's money" revisited: Collective embezzlement in the savings and loan insurance industries'. *Social Problems* 38 (1): 94–112.

Caldwell, John C. 1976. 'Toward a restatement of demographic transition theory'. *Population and Development Review* 2: 321–66.

———, James F. Phillips, and Barkat-e-Khuda, eds. 2002. 'Family planning programs in the twenty-first century'. *Studies in Family Planning* 33 (1) (special issue).

Calhoun, Craig. 1993. '"New social movements" of the early nineteenth century'. *Social Science History* 17 (3): 385–427.

Callero, Peter L. 2008. 'The globalization of self: Role and identity transformation from above and below'. *Sociology Compass* 2 (6): 1972–88.

Calliste, Agnes. 1993. 'Sleeping car porters in Canada: An ethnically submerged split labour market'. In Graham S. Lowe and Harvey Krahn, eds, *Work in Canada: Readings in the Sociology of Work and Industry*, 139–53. Scarborough, ON: Nelson.

Camfield, D. 2008. 'The working class movement in Canada: An overview'. In M. Smith, ed., *Group Politics and Social Movements in Canada*, 61–84. Peterborough, ON: Broadview.

Campbell, Bruce. 2007. *20 Years Later: Has Free Trade Delivered on Its Promises?* Canadian Centre for Policy Alternatives, December.

———. 2009. *Understanding the Canadian Dimension of the World Crisis* (Presentation to the Canadian Council for International Co-operation).

Campbell, Donald T., and Julian C. Stanley. 1970. *Experimental and Quasi-experimental Designs for Research*. Chicago: Rand McNally.

Campey, John. 2002. 'Immigrant children in our classrooms: Beyond ESL'. *Education Canada* 42 (3): 44–7.

Canada. 1991. *Broadcasting Act*. Statutes of Canada 1991, c. 11.

———. 2005. *Civil Marriage Act*. Statutes of Canada 2005, c. 33.

———. 2008. *Federal Sustainable Development Act*. Chapter 33.

Canadian Council for Refugees. 2005. *Closing the Front Door on Refugees: Report on the First Year of the Safe Third Country Agreement*. December. Ottawa: Canadian Council for Refugees.

Canadian Council on Learning. 2010. *State of Learning in Canada 2009–2010: A Year in Review*. Ottawa: Canadian Council on Learning.

Canadian Film and Television Production Association (CFTPA). 2009. *09 Profile: An Economic Report on the Canadian Film and Television Production Industry*. Ottawa: CFTPA.

Canadian Health Services Research Foundation. 2002. *Myth: For-Profit Ownership of Facilities Would Lead to Better Health Care*. Ottawa: Canadian Health Services Research Foundation.

———. 2005. *Myth: A Parallel Private System Would Reduce Waiting Times in the Public System*. Ottawa: Canadian Health Services Research Foundation. www. chsrf.ca/publicationsandresources/ Mythbusters/ArticleView/05-03-01/5bda3483-f97b-4616-bfe7-d55d0d66b9a0.aspx.

Canadian Institute of Child Health. 2002. *The Health of Canada's Children*. 3rd edn. Ottawa: Canadian Institute of Child Health.

Canadian Population Health Initiative. 2004. *Improving the Health of Canadians 2004*. Ottawa: Canadian Institute for Health Information.

Canadian Press. 2006. 'Alouettes coach Popp hoping his players refrain from sex during Grey Cup Week'. 15 Nov. www.cbc.ca/cp/football/051115/ f111527A.html.

Canadian Radio-television and Telecommunications Commission (CRTC). 1998. *Commercial Radio Policy*. Broadcasting Public Notice 1998-41, 30 Apr. Ottawa: CRTC.

Canadian Teachers' Federation. 2010. 'Fundraising'. *Commercialism in Canadian Schools: Who's Calling the Shots?* CTF Fact Sheet. www. ctf-fce.ca/documents/Resources/ en/commercialism_in_school/en/ CISCKitFundraising%28R6%29.pdf.

Cardoso, Fernando Luiz. 2002. '"Fishermen": Masculinity and sexuality in a Brazilian fishing community'. *Sexuality and Culture* 6 (4): 45–72.

Carley, Kathleen. 1989. 'The value of cognitive foundations for dynamic social theory'. *Journal of Mathematical Sociology* 14 (2/3): 171–208.

———. 1991. 'A theory of group stability'. *American Sociological Review* 56: 331–54.

Carnes, B.A., S.J. Olshansky, and D. Grahn. 2003. 'Biological evidence for limits to the duration of life'. *Biogerontology* 4 (1): 31–45.

Carocci, Max. 2009. 'Visualizing gender variability in Plains Indian pictographic art'. *American Indian Culture and Research Journal* 33 (1): 1–22.

Carpenter, Laura. 1998. 'From girls into women: Scripts for sexuality and romance in *Seventeen* magazine, 1974–1994'. *Journal of Sex Research* 35 (2): 158–68.

Carr, Lynn. 1999. 'Cognitive scripting and sexual identification: Essentialism, anarchism, and constructionism'. *Symbolic Interactionism* 22 (1): 1–24.

Carre, Dominique, and Sylvie Craipeau. 1996. 'Entre délocalisation et mobilité: analyse des stratégies entrepreneuriales de télétravail'. *Technologies de l'information et société* 8: 333–54.

Carroll, William K. 2004. *Corporate Power in a Globalizing World: A Study of Elite Organization*. Don Mills: Oxford University Press.

———. 2007. 'From Canadian corporate elite to transnational capitalist class: Transitions in the organization of corporate power'. *Canadian Review of Sociology and Anthropology* 44 (3): 265–88.

———, and R.S. Ratner. 1994. 'Between Leninism and radical pluralism: Gramscian reflections on counter-hegemony and the new social movements'. *Critical Sociology* 36 (5): 3–26.

———. 1996. 'Master framing and cross-movement networking in contemporary social movements'. *Sociological Quarterly* 37: 601–25.

———. 1999. 'Media strategies and political projects: A comparative study of social movements'. *Canadian Journal of Sociology* 24 (1): 1–34.

———. 2001. 'Sustaining oppositional cultures in "post-socialist" times: A comparative study of three social movement organizations'. *Sociology* 35 (3): 605–29.

———. 2006. 'Hegemony, counter-hegemony, anti-hegemony'. *Socialist Studies* 2 (2): 9–43.

Carson, Rachel. 1962. *Silent Spring*. Boston: Houghton Mifflin.

Cartwright, Claire, and Heather McDowell. 2008. 'Young women's life stories and accounts of parental divorce'. *Journal of Divorce and Remarriage* 49 (1/2): 56–77.

Cartwright, John. 2010. 'Executives are overpaid, not workers. *Toronto Star* 6 August: A16.

Casanova, José. 1994. *Public Religion in the Modern World*. Chicago: University of Chicago Press.

Castells, Manuel. 1972. *La question urbaine*. Paris: Maspéro.

———. 1983. *The City and the Grassroots: A Cross-cultural Theory of Urban Social Movements*. London: Arnold.

———. 2004. *The Rise of the Network Society*. Chichester, UK: John Wiley and Sons.

———. 2007. 'Communication, power, and counter-power in the network society'. *International Journal of Communication* 1: 238–66.

Catton, William, Jr. 1980. *Overshoot: The Ecological Basis of Revolutionary Change*. Urbana: University of Illinois Press.

———, and Riley Dunlap. 1978. 'Environmental sociology: A new paradigm'. *American Sociologist* 13: 41–9.

———. 1980. 'A new ecological paradigm for post-exuberant sociology'. *American Behavioral Scientist* 24: 15–47.

Caulfield, J., and L. Peake, eds., 1996. *City Lives and City Forms: Critical Research and Canadian Urbanism*. Toronto: University of Toronto Press.

Cavanagh, Kate, R. Emerson Dobash, Russell P. Dobash, and Ruth Lewis. 2001. '"Remedial work": Men's strategic responses to their violence against intimate female partners'. *Sociology* 35 (3): 695–714.

Cavanagh, Shannon E. 2008. 'Family structure history and adolescent adjustment'. *Journal of Family Issues* 29 (7): 944–80.

CBC. 2010. "Remembering the Winnipeg General Strike." *CBC Digital Archives*. http://archives.cbc.ca/on_this_day/05/15.

CBC News. 2006. 'Caledonia land claim: Timeline'. www.cbc.ca/news/background/caledonia-landclaim.

———. 2007. 'Understanding the violence'. www.cbc.ca/news/background/paris_riots/timeline.html.

———. 2009a. 'Going organic: Growing demand, tougher regulations'. www.cbc.ca/consumer/story/2008/05/07/f-food-organic.html.

———. 2009b. 'Hutterites need driver's licence photos: Top court'. www.cbc.ca/canada/story/2009/07/24/hutterite-supreme-court024.html.

Chalmers, A.F. 1999. *What Is This Thing Called Science? An Assessment of the Nature and Status of Science and Its Methods*. 3rd edn. Indianapolis: Hackett.

Chappell, Allison T., and Lonn Lanza-Kaduce. 2010. 'Police academy socialization: Understanding the lessons learned in a paramilitary-bureaucratic organization'. *Journal of Contemporary Ethnography* 39 (2): 187–214.

Chappell, N.L., and R. Litkenhaus. 1995. *Informal Caregivers to Adults in British Columbia*. Victoria, BC: University of Victoria, Centre on Aging, and the Caregivers Association of British Columbia.

Chappell, N.L., McDonald, and M. Stones. 2007. *Aging in Contemporary Canada*. 2nd edn. Toronto: Person Education.

Charbonneau, J., and A. Germain. 2002. 'Les banlieues de l'immigration'. *Recherches sociographiques* 43: 311–28.

Charon, Joel M. 1979. *Symbolic Interactionism: An Introduction, an Interpretation, an Integration*. Englewood Cliffs, NJ: Prentice-Hall.

Chasteen, Amy L. 2001. 'Constructing rape: Feminism, change, and women's everyday understandings of sexual assault'. *Sociological Spectrum* 21: 101–39.

Cheal, David. 1991. *Family and the State of Theory*. Toronto: University of Toronto Press.

———, ed. 2010. *Canadian Families Today: New Perspectives*. 2nd edition. Toronto: Oxford University Press.

Chilton, R. 2004. 'Regional variations in lethal and non lethal assaults'. *Homicide Studies* 8 (1): 40–56.

Chodorow, Nancy. 1989. *Feminism and Psychoanalytic Theory*. New Haven, CT: Yale University Press.

Chomsky, Noam. 2003. *Hegemony or Survival: America's Quest for Global Dominance*. New York: Metropolitan Press.

Christensen, Kaare, Gabriele Doblhammer, Roland Rau, and James W. Vaupel. 2009. 'Ageing populations: The challenges ahead'. *Lancet* 374: 1196–1208.

Christman, Ed. 2010. '2009 sales rap: Transactions up as digital growth slows'. *Billboard.biz* 6 Jan. www.billboard.biz/bbbiz/content_display/industry/e3ib067cb2aa5cb826b34641dba4e3f0c59.

Chung, Donghum, Brahm Daniel DeBuys, and Chang S. Nam. 2007. 'Influence of avator creation on attitude, empathy, presence, and para-social interaction'. Lecture Notes in Computer Science, *Proceedings of the 12th International Conference on Human-Computer Interaction: Interaction Design and Usability, Beijing, China*. Berlin: Springer-Verlag.

Church, W.T., II, T. Wharton, and J.K. Taylor. 2009. 'An examination of differential association and social theory: Family systems and delinquency'. *Youth Violence and Juvenile Justice* 7 (1): 3–15.

Cincotta, Richard P., and Roberta Engelman. 1997. *Economics and Rapid Change: The Influence of Population Growth*. Washington: Population Action International.

Citizenship and Immigration Canada. *Facts and Figures, 2008*. Ottawa: Citizenship and Immigration Canada.

———. *Facts and Figures, 2009*. Ottawa: Citizenship and Immigration Canada.

Clark, W.C. 2007. 'Sustainability science: A room of its own'. *PNAS* 104: 1737–8.

Clarke, John I. 1996. 'The impact of population change on environment: An overview'. In Bernardo Colombo, Paul Demeny, and Max F. Perutz, eds, *Resources and Population: Natural, Institutional, and Demographic Dimensions of Development*, 244–68. Oxford: Clarendon.

Clarke, Juanne N. 2008. *Health, Illness and Medicine in Canada*. Toronto: Oxford University Press.

———, and P. Fletcher. 2004. 'Parents as advocates: Stories of surplus suffering when a child is diagnosed and treated for cancer'. *Health and Social Work* 39 (3/4): 107–27.

——, and G. Van Amerom. 2008. 'The differences between parents and people with Asperger's'. *Social Work in Health Care* 46 (3): 85–106.

Cleland, John. 1996. 'Population growth in the 21st century: Cause for crisis or celebration?' *Tropical Medicine and International Health* 1 (1): 15–26.

Clément, Dominique. 2008. *Canada's Rights Revolution: Social Movements and Social Change, 1937–82*. Vancouver: University of British Columbia Press.

——. 2011. 'Winnipeg General Strike'. In *Canada's Human Rights History*. www.historyofrights.com/events/1919.html.

Clement, Wallace. 1975. *The Canadian Corporate Elite*. Toronto: McClelland and Stewart.

Clevedon, Gordon, and Michael Krashinsky. 2001. *Our Children's Future: Child Care Policy in Canada*. Toronto: University of Toronto Press.

Cloud, John. 2010. 'Why your DNA isn't your destiny'. Time.com (in partnership with CNN). www.time.com/time/printout/Wed.

Cloward, Richard A., and Lloyd E. Ohlin. 1960. *Delinquency and Opportunity: A Theory of Delinquent Gangs*. New York: Free Press.

CMEC (Council of Ministers of Education Canada). 2008. *The Development of Education: Reports for Canada*. Ottawa: Council of Ministers of Education Canada.

Coale, Ansley J. 1964. 'How a population ages or grows younger'. In Ronald Freedman, ed., *Population: The Vital Revolution*, 47–58. Chicago: Aldine.

——. 1969. 'The decline of fertility in Europe from the French Revolution to World War II'. In S.J. Berhman, Leslie Corsa, and Ronald Freedman, eds, *Fertility and Family Planning: A World View*, 3–24. Ann Arbor: University of Michigan Press.

——. 1973. 'The demographic transition reconsidered'. In International Union for the Study of Population, *Proceedings of the International Population Conference*, v. 1, 53–72. Liège, Belgium.

——. 1974. 'The history of the human population'. *Scientific American* (special issue): 15–25.

——. 1996. 'Age patterns and time sequence of mortality in national populations with the highest expectation of life at birth'. *Population and Development Review* 22 (1): 127–36.

——, and Edgar M. Hoover. 1958. *Population Growth and Economic Development in Low-Income Countries*. Princeton, NJ: Princeton University Press.

——, and Susan Cotts Watkins, eds. 1986. *The Decline of Fertility in Europe: The Revised Proceedings of a Conference on the Princeton European Fertility Project*. Princeton, NJ: Princeton University Press.

Coburn, Elaine. 2010. '"Pulling the monster down": Interview with William K. Carroll'. *Socialist Studies* 6 (1): 65–92.

Cockburn, Cynthia. 1990. 'Men's power in organizations'. In Jeff Hearn and David Morgan, eds, *Men, Masculinity and Social Theory*, 72–89. London: Unwin Hyman.

Cohen, Albert K. 1966. *Deviance and Control*. Englewood Cliffs, NJ: Prentice-Hall.

Cohen, Benjamin. 1998. *The Geography of Money*. Ithaca, NY: Cornell University Press.

Cohen, Jean L. 1985. 'Strategy or identity: New theoretical paradigms and contemporary social movements'. *Social Research* 53: 663–716.

Coleman, James S. 1988. 'Social capital in the creation of human capital'. *American Journal of Sociology* 94: S95–120.

Collier, R. 2010. 'Medical school admission targets urged for rural and low-income Canadians'. *Canadian Medical Association Journal* 182 (8): E327–8.

Collin, C., and H. Jensen. 2009. *A Statistical Profile of Poverty in Canada*. Library of Parliament/Parliamentary Information and Research Services. PRB 09-17E.

Collins, Randall. 1979. *The Credential Society: An Historical Sociology of Education and Stratification*. New York: Academic Press.

——. 2008. *Violence: A Micro-sociological Theory*. Princeton, NJ: Princeton University Press.

Comack, E. 2008. *Out There/In Here*. Halifax: Fernwood.

Comte, Auguste. 1974. 'Positivist philosophy'. In Stanislav Andreski, ed., *The Essential Comte*. London: Croom Helm.

Conn, D.K. 2002. 'An overview of common mental disorders among seniors'. *Writings in Gerontology* 18: 19–32.

Connell, R.W. 1996. 'Teaching the boys: New research on masculinity and gender strategies for schools'. *Teachers College Record* 98: 206–35.

Conrad, Peter, and Joseph Schneider. 1980. *Deviance and Medicalization: From Badness to Sickness*. St Louis: Mosby.

Constant, David, Lee Sproull, and Sara Kiesler. 1996. 'The kindness of strangers: The usefulness of electronic weak ties for technical advice'. *Organization Science* 7 (2): 119–35.

Cook, J. 2001. 'Practical guide to medical education'. *Pharmaceutical Marketing* 6: 14–22.

Cooley, Charles Horton. 1902. *Human Nature and Social Order*. New York: Scribner.

——. 1962 [1909]. *Social Organization: A Study of the Larger Mind*. Glencoe, IL: Free Press.

Cooper, A., D. Delmonico, and R. Burg. 2000. 'Cybersex users, abusers, and compulsives: New findings and implications'. *Sexual Addiction and Compulsivity* 7: 5–30.

Corman, June, and Meg Luxton. 2007. *Getting By in Hard Times: Gendered Labour at Home and on the Job*. Toronto: University of Toronto Press.

Cottrell, Fred. 1955. *Energy and Society*. New York: McGraw-Hill.

Council of Canadians. 2010. *Abandon the SPP and Renegotiate NAFTA*. August. Council of Canadians.

Coutard, O., and S. Guy. 2007. 'STS and the city: Politics and practices of hope'. *Science, Technology, and Values* 32: 713–34.

Couton, Philippe, and Stéphanie Gaudet. 2008. 'Rethinking social participation: The case of immigrants in Canada'. *Journal of International Migration and Integration* 9 (1): 21–44.

Couturier, Eve-Line, and Bernard Schepper. 2010. *Who Is Getting Richer? Who Is Getting Poorer?* Ottawa: Canadian Centre for Policy Alternatives.

Cowgill, D.O., and L.D. Holmes. 1972. *Aging and Modernization*. New York: Appleton-Century Crofts.

Cox, K.R. 1997. 'Governance, urban regime analysis, and the politics of local economic development'. In M. Lauria, ed., *Reconstructing Urban Regime Theory: Regulating Urban Politics in a Global Economy*, 99–121. Thousand Oaks, CA: Sage.

Crain, William C. 1985. 'Kohlberg's stages of moral development'. In *W.C. Crain's Theories of Development*, 118–36. Englewood Cliffs, NJ: Prentice-Hall.

Crane, Diana. 2002. 'Culture and globalization: Theoretical models and emerging trends'. In Diana Crane, Nobuko Kawashima, and Kenichi Kawasaki, eds, *Global Culture: Media, Arts, Policy, and Globalization*, 1–25. New York: Routledge.

Cranford, Cynthia J., Leah F. Vosko, and Nancy Zukewich. 2006. 'The gender of precarious employment in Canada'. In Vivian Shalla, ed., *Working in a Global Era: Canadian Perspectives*, 99–119. Toronto: Canadian Scholars' Press.

Cranswick, K., and D. Dosman. 2008. 'Eldercare: What we know today'. *Canadian Social Trends* 86: 48–56.

Crockett, Lisa, Mike Losoff, and Anne C. Petersen. 1984. 'Perceptions of the peer group and friendship in early adolescence'. *Journal of Early Adolescence* 4 (2): 155–81.

Crompton, Susan. 2000. 'Health'. *Canadian Social Trends* 59: 12–17.

Cromwell, Paul, and Quint Thurman. 2003. 'The devil made me do it: Use of neutralizations by shoplifters'. *Deviant Behavior* 24 (6): 535–50.

Crossley, Michelle L. 2002. 'The perils of health promotion and the "barebacking" backlash'. *Health* 6 (1): 47–68.

Croteau, David, and William Hoynes. 2000. *Media/Society: Industries, Images, and Audiences*. 2nd edn. Thousand Oaks, CA: Pine Forge.

Crozier, Michel. 1964. *The Bureaucratic Phenomenon*. Chicago: University of Chicago Press.

Cullen, F.T., J.P. Wright, and K.R. Blevins, eds. 2007. *Taking Stock: The Status of Criminological Theory*. Edison, NJ: Transaction.

Cumming, E., and W. Henry. 1961. *Growing Old: The Process of Disengagement*. New York: Basic Books.

Cumming, Sara, Martin Cooke, and Lea Caragata. 2008. *Women's Resources and Exits from Social Assistance*. The 2008 Population, Work and Family Research Collaborations Conference, 8–10 December 2008, Gatineau, QC.

Cuneo, Carl. 1990. *Pay Equity: The Labour-Feminist Challenge*. Toronto: Oxford University Press.

———. 2002. 'Globalized and localized digital divides along the information highway: A fragile synthesis across bridges, ramps, cloverleaves, and ladders'. 33rd Annual Sorokin Lecture, University of Saskatchewan, 31 Jan.

Cunningham, Mick, and Arland Thornton. 2006. 'The influence of parents' marital quality on adult children's attitudes toward marriage and its alternatives: Main and moderating effects'. *Demography* 43 (4): 659–72.

Curra, John. 2011. *The Relativity of Deviance*. 2nd edn. Thousand Oaks, CA: Sage.

Curran, D., and R. Tomalty. 2003. 'The wide range of support for smart growth in Canada promises more livable towns and cities'. *Alternatives* 29: 10–18.

Currie, Dawn. 1988. 'Starvation amidst abundance: Female adolescents and anorexia'. In B. Singh Bolaria and Harley D. Dickinson, eds, *Sociology of Health Care in Canada*, 198–215. Toronto: Harcourt Brace Jovanovich.

Curtis, James. 1989. 'On Lipset's measure of voluntary association differences between Canada and the United States'. *Canadian Journal of Sociology* 14 (3): 383–9.

———, and Alan Hunt. 2007. 'The fellatio "epidemic": Age relations and access to the erotic arts'. *Sexualities* 10 (1): 5–28.

Curtiss, Susan. 1977. *Genie: A Psycholinguistic Study of a Modern Day Wild Child*. New York: Academic Press.

Curwood, S. 2009. 'What is poverty?' *Journal of Hunger and Poverty* 1 (December): 9–17.

Daenzer, Patricia. 1993. *Regulating Class Privilege: Immigrant Servants in Canada, 1940s–1990*. Toronto: Canadian Scholars' Press.

Dagenais, M. 2000. *Des pouvoirs et des hommes*. Montréal: Institut d'administration publique du Canada.

Daly, Kerry. 2004. *The Changing Culture of Parenting*. Ottawa: Vanier Institute of the Family.

Daly, Mary. 1985. *Beyond God the Father: Towards a Philosophy of Women's Liberation*. Boston: Beacon Books.

Dannefer, D. 2003. 'Whose life course is it, anyway? "Linked lives" in global perspective'. In R.A. Settersten, Jr, ed., *Invitation to the Life Course: Toward New Understandings of Later Life*. New York: Baywood.

———, and P. Uhlenberg. 1999. 'Paths of the life course: A typology'. In V.L. Bengston and K.W. Schaie, eds, *Handbook on Theories of Aging*, 306–26. New York: Springer.

Darder, Antonia, Marta Baltodano, and Rodolfo D. Torres, eds. 2003. *The Critical Pedagogy Reader*. New York: Routledge Falmer.

Darroch, Gordon. 1979. 'Another look at ethnicity, stratification, and social mobility in Canada'. *Canadian Journal of Sociology* 4 (1): 1–14.

Das Gupta, Tania. 1996. *Racism and Paid Work*. Toronto: Garamond.

Dauvergne, Catherine. 2008. *Making People Illegal: What Globalization Means for Migration and Law*. Cambridge: Cambridge University Press.

Dauvergne, Mia, and John Turner. 2010. 'Police-reported crime statistics in Canada, 2009'. *Juristat* 30 (2): 1–37. Ottawa: Statistics Canada.

Davenport, Christian, Hank Johnston, and Carol Mueller, eds. 2005. *Repression and Mobilization*. Minneapolis: University of Minnesota Press.

Davey, Ian E. 1978. 'The rhythm of work and the rhythm of school'. In Neil McDonald and Alf Chaiton, eds, *Egerton Ryerson and His Times*, 221–53. Toronto: Macmillan.

Davidson, Kenneth, and Linda Hoffman. 1986. 'Sexual fantasies and sexual satisfaction: An empirical analysis of erotic thought'. *Journal of Sex Research* 22 (2): 184–205.

Davies, Lorraine, and Patricia Jane Carrier. 1999. 'The importance of power relations for the division of household labour'. *Canadian Journal of Sociology* 24 (1): 35–51.

Davies, Scott, and Neil Guppy. 2006. *The Schooled Society: An Introduction to the Sociology of Education*. Don Mills, ON: Oxford University Press.

Davies, W.K.D., and R.A. Murdie. 1991. 'Measuring the social ecology of cities'. In L.S. Bourne and D.F. Ley, eds, *The Changing Social Geography of Canadian Cities*, 52–75. Montreal and Kingston: McGill-Queen's University Press.

Davis, Charles R. 1996. 'The administrative rational model and public organization theory'. *Administration and Society* 28: 39–60.

Davis, Kingsley. 1937. 'The sociology of prostitution'. *American Sociological Review* 2 (5): 744–55.

———. 1945. 'The world demographic transition'. *Annals of the American Academy of Political and Social Sciences* 237: 1–11.

———. 1984. 'Wives and work: The sex role revolution and its consequences'. *Population and Development Review* 8: 495–511.

———, and Judith Blake. 1956. 'Social structure and fertility: An analytic framework'. *Economic Development and Cultural Change* 4 (4): 211–35.

———, and Wilbert E. Moore. 1945. 'Some principles of stratification'. *American Sociological Review* 10: 242–9.

———, and Pietronella van den Oever. 1982. 'Demographic foundations of new sex roles'. *Population and Development Review* 8: 495–512.

Davis, Phillip W. 2003. 'The changing meaning of spanking'. In Donileen Loseke and Joel Best, eds, *Social Problems: Constructionist Readings*, 6–12. New York: Aldine de Gruyter.

Day, Lincoln H. 1992. *The Future of Low-Birthrate Populations*. London: Routledge.

Dear, M., and N. Dahmann. 2008. 'Urban politics and the Los Angeles School of Urbanism'. *Urban Affairs Review* 44: 266–79.

Dear, M., and A.J. Scott. 1981. *Urbanization and Urban Planning in Capitalist Society*. London and New York: Methuen.

de Beauvoir, Simone. 1972. *The Coming of Age*. New York: G.P. Putnam.

deGroot-Maggetti. 2002. *A Measure of Poverty in Canada: A Guide to the Debate about Poverty Lines*. Citizens for Public Justice.

Dei, George J. Sefa. 1996. *Anti-racism Education: Theory and Practice*. Halifax: Fernwood.

———. 2006. 'Black-focused schools: A call for re-visioning'. *Education Canada* 46 (3): 27–31.

———, Irma Marcia James, Leeno Luke Karumanchery, Sonia James-Wilson, and Jasmin Zine. 2000. *Removing the Margins: The Challenges and Possibilities of Inclusive Schooling*. Toronto: Canadian Scholars' Press.

DeKeseredy, Walter. 2009. 'Patterns of family violence'. In M. Baker, ed., *Families: Changing Trends in Canada*, 6th edn, 179–205. Toronto: McGraw-Hill Ryerson.

———, and Linda MacLeod. 1997. *Woman Abuse: A Sociological Story*. Toronto: Harcourt Brace.

de la Torre, Isabel. 1997. 'La formacion y las organizaciones: Los acuerdos nacionales de formacion continua'. *Revista espanola de investigaciones sociologicas* 77–8 (Jan.–June): 15–33.

della Porta, Donatella, Massimiliano Andretta, Lorenzo Mosca, and Herbert Reiter. 2006. *Globalization from Below: Transnational Activists and Protest Networks*. Minneapolis: University of Minnesota Press.

Dellinger, Kirsten. 2002. 'Wearing gender and sexuality "on your sleeve": Dress norms and the importance of occupational and organizational culture at work'. *Gender Issues* 20 (1): 3–25.

Dempsey, Ken. 1999. *Resistance and Change: Trying to Get Husbands to Do More Housework*. Paper presented at the Australian Sociologists Association annual meetings, Monash University.

Denton, Frank T., Christine H. Feaver, and Byron G. Spencer. 2002. 'Alternative pasts, possible futures: A "what if" study of the effects of fertility on the Canadian population and labour force'. *Canadian Public Policy* 28 (3): 443–59.

Derrida, J. 1976. *Of Grammatology*. Baltimore, MD: Johns Hopkins University Press.

Dery, David. 1998. '"Papereality" and learning in bureaucratic organizations'. *Administration and Society* 29: 677–89.

Desjardins, Richard, Scott Murray, Yvan Clermont, and Patrick Werquin. 2005. *Learning a Living: First Results of the Adult Literacy and Life Skills Survey*. Ottawa and Paris: Statistics Canada and OECD.

Deutschmann, Linda. 2007. *Deviance and Social Control*. 4th edn. Toronto: Thomson Nelson.

Devall, William B. 1992. 'Deep ecology and radical environmentalism'. In Riley Dunlap and Angela G. Mertig, eds, *American Environmentalism*, 51–62. Philadelphia: Taylor and Francis.

Devereaux, P.J., et al. 2002. 'A systematic review and meta-analysis of studies comparing mortality rates of private for-profit and private not-for-profit hospitals'. *Canadian Medical Association Journal* 166: 1399–406.

Dewing, Michael. 2010. 'Social media 2: Who uses them?' Library of Parliament Background Paper, Publication no. 2010-05-E, 3 February. Ottawa: Library of Parliament.

De Young, Mary. 1988. 'The indignant page: Techniques of neutralization in the publications of pedophile organizations'. *Child Abuse and Neglect* 12 (4): 583–91.

Dhalla, Irfan A., et al. 2002. 'Characteristics of first-year students in Canadian medical schools'. *Canadian Medical Association Journal* 166: 1029–35.

Diani, Mario. 1992. 'The concept of social movement'. *Sociological Review* 40: 1–25.

Dickens, Peter. 1996. *Reconstructing Nature: Alienation, Emancipation and the Division of Labour*. London: Routledge.

Dietz, Thomas, R. Scott Frey, and Eugene A. Rosa. 2002. 'Risk, technology and society'. In Riley Dunlap and William Michelson, eds, *Handbook of Environmental Sociology*, 329–69. Westport, CT: Greenwood.

Dill, Karen E., and Kathryn P. Thill. 2007. 'Video game characters and the socialization of gender roles: Young people's perceptions mirror sexist media depictions'. *Sex Roles* 57: 851–64.

Dillabough, Jo-Anne, Julie McLeod, and Martin Mills, eds. 2011. *Troubling Gender in Education*. New York: Routledge.

Dimaggio, Paul, and Walter Powell. 1983. 'The iron cage revisited: Institutional isomorphism and collective rationality in organizational fields'. *American Sociological Review* 48 (2): 147–60.

Di Martino, Vittorio. 1996. 'Télétravail: à la recherche des règles d'or'. *Technologies de l'information et société* 8: 355–71.

Dobash, R. Emerson, Russell P. Dobash, Margo Wilson, and Martin Daly. 1992. 'The myth of sexual symmetry in marital violence'. *Social Problems* 39: 71–91.

Dobin, Murray. 2008. 'Afghanistan transforms Canada: To play junior partner to empire, we've militarized our identity'. *The Tyee* 11 August.

Dogsé, P. 2004. 'Toward urban biosphere reserves'. *Annals of the New York Academy of Science* 1023: 10–48.

Dole, Carol M. 2000. 'Woman with a gun: Cinematic law enforcers on the gender frontier'. In Murray Pomerance and John Sakeris, eds, *Bang Bang, Shoot Shoot! Essays on Guns and Popular Culture*, 2nd edn, 11–21. Needham Heights, MA: Pearson Education.

———. 2001. 'The gun and the badge: Hollywood and the female lawman'. In Martha McCaughey and Neal King, eds, *Reel Knockouts: Violent Women in the Movies*. Austin, TX: University of Texas Press.

Domhoff, G. William. 2006. *Who Rules America? Power, Politics and Social Change*. 5th edn. Boston: McGraw-Hill.

Doniger, Wendy. 2003. 'The Kamasutra: It isn't all about sex'. *Kenyon Review* 25 (1): 18–37.

———. 2007. 'Reading the "Kamasutra": The strange and the familiar'. *Daedalus* 136 (2): 66–78.

Dowd, J.J. 1980. 'Exchange rates and old people'. *Journal of Gerontology* 35 (4): 596–602.

Downes, D., and P. Rock. 2003. *Understanding Deviance*. 4th edn. Toronto: Oxford University Press.

Downey, John, and Natalie Fenton. 2003. 'New media, counter publicity and the public sphere'. *New Media and Society* 5 (2): 185–202.

Doyal, Lesley. 1995. *What Makes Women Sick: Gender and the Political Economy of Health*. New Brunswick, NJ: Rutgers University Press.

Dreeben, Robert. 1968. *On What Is Learned in School*. Reading, MA: Addison-Wesley.

Driedger, Leo. 1996. *Multi-ethnic Canada: Identities and Inequalities*. Toronto: Oxford University Press.

Driessen, Geert, and Frederik Smit. 2007. 'Effects of immigrant parents' participation in society on their children's school performance'. *Acta Sociologica* 50 (1): 39–56.

Duffy, Ann D. 1986. 'Reconceptualizing power for women'. *Canadian Review of Anthropology and Sociology* 23 (1): 21–46.

Dufour, Pascale, and Janet Conway. 2010. 'Emerging visions of another world? Tensions and collaboration at the Quebec Social Forum'. *Journal of World-Systems Research* 16 (1): 29–47.

Dumas, Jean. 1990. *Report on the Demographic Situation in Canada 1990*. Ottawa: Statistics Canada, Demography Division.

DuMont, Janice. 2003. 'Charging and sentencing in sexual assault cases: An exploratory examination'. *Canadian Journal of Women and the Law* 15 (2): 305–30.

Dunlap, Riley. 1992. 'Trends in public opinion toward environmental issues: 1965–1990'. In Riley Dunlap and Angela G. Mertig, eds, *American Environmentalism*, 89–116. Philadelphia: Taylor and Francis.

———, and W. Richard Catton. 1994. 'Struggling with human

exemptionalism: The rise, decline and revitalization of environmental sociology'. *American Sociologist* 25 (spring): 5–30.

———, and William Catton, Jr. 1979a. 'Environmental sociology'. *Annual Review of Sociology* 5: 243–73.

———. 1979b. 'Environmental sociology: A framework for analysis'. In Timothy O'Riordan and R.C. d'Arge, eds, *Progress in Resource Management and Environmental Planning*, v. 1, 57–85. Chichester, UK: Wiley.

———. 1983. 'What environmental sociologists have in common (whether concerned with "built" or "natural" environments)'. *Sociological Inquiry* 53 (2/3): 113–15.

Dunlap, Riley, and Richard P. Gale. 1972. 'Politics and ecology: A political profile of student eco-activists'. *Youth and Society* 3: 379–97.

Dunleavy, Patrick, and Brendan O'Leary. 1987. *Theories of the State*. London: Palgrave.

Durkheim, Émile. 1887. 'Nécrologie d'Hommay'. durkheim.uchicago.edu/Texts/1887d.html.

———. 1964 [1893]. *The Division of Labor in Society*. Trans. George Simpson. New York: Free Press.

———. 1951 [1897]. *Suicide: A Study in Sociology*. Trans. John A. Spaulding and George Simpson. New York: Free Press.

———. 1964 [1912]. *The Elementary Forms of the Religious Life*. London: Allen and Unwin.

———. 1956 [1922]. *Education and Society*. Trans. Sherwood W. Fox. Glencoe, IL: Free Press.

Durkin, K.F. 2009. 'There must be some kind of misunderstanding, there must be some kind of mistake: The deviance disavowel strategies of men arrested in Internet sex stings (2008 presidential address)'. *Sociological Spectrum* 29 (6): 661–76.

Dusza, Karl. 1989. 'Max Weber's conception of the state'. *International Journal of Politics, Culture and Society*. 3 (1): 71–105.

Dworkin, Andrea. 1981. *Pornography: Men Possessing Women*. New York: Perigee.

———, and Catherine Mackinnon. 1988. *Pornography and Civil Rights: A New Day for Women's Equality, Organizing against Pornography*. Minneapolis, MN: Organizing against Pornography.

Dworkin, Ronald W. 2001. 'The medicalization of unhappiness'. *The Public Interest* (summer): 85–99.

Dymond, Bill, and Michael Hart. 2003. 'Canada and the global challenge: Finding a place to stand'. Border Papers 180. Toronto: C.D. Howe Institute.

Easterlin, Richard A. 1969. 'Towards a socio-economic theory of fertility: A survey of recent research on economic factors in American fertility'. In S.J. Berhman, Leslie Corsa, and Ronald Freedman, eds, *Fertility and Family Planning: A World View*, 127–56. Ann Arbor: University of Michigan Press.

———. 1980. *Birth and Fortune: The Impact of Numbers on Personal Welfare*. New York: Basic Books.

Eaves, Elizabeth. 2002. *Bare: On Women, Dancing, Sex and Power*. New York: Knopf.

Ebaugh, Helen. 1988. *Becoming an Ex: The Process of Role Exit*. Chicago: University of Chicago Press.

Eberstadt, N. 2007. 'China's one-child mistake'. *Wall Street Journal* 17 September: A17.

Echenberg, Havi, and Hilary Jensen. 2008. *Defining and Enumerating Homelessness in Canada*. PRB 08-30E. Ottawa: Library of Parliament, Social Affairs Division.

Edin, Kathryn, and Maria J. Kefalas. 2005. *Promises I Can Keep: Why Poor Women Put Motherhood before Marriage*. Berkeley: University of California Press.

Edwards, Bob, Michael W. Foley, and Mario Diani. 2001. *Beyond Tocqueville: Civil Society and the Social Capital Debate in Comparative Perspective*. Hanover, NH: University Press of New England.

Edwards, Nigel, Mary Jane Kornacki, and Jack Silversin. 2002. 'Unhappy doctors: What are the causes and what can be done?' *British Medical Journal* 324: 835–8.

Edwards, Peggy, and Aysha Mawani. 2006. 'Healthy aging in Canada: A new vision, a vital investment'. A discussion brief prepared for the Healthy Aging and Wellness Working Group of the Federal/Provincial/Territorial (F/P/T) Committee of Officials (Seniors). www.phac-aspc.gc.ca/seniors-aines/publications/public/healthy-sante/vision/vision-bref/index-eng.php.

Eglin, Peter, and Stephen Hester. 1999. '"You're all a bunch of feminists": Categorization and the politics of terror in the Montreal Massacre'. *Human Studies* 22: 253–72.

Ehrenreich, Barbara. 2001. *Nickel and Dimed: On (Not) Getting By in America*. New York: Henry Holt.

Ehrlich, Paul R. 1981. 'Environmental disruption: Implications for the social sciences'. *Social Science Quarterly* 62 (1): 7–22.

———, and Anne H. Ehrlich. 1970. *Population, Resources, Environment: Issues in Human Ecology*. San Francisco: Freeman.

———. 1990. *The Population Explosion*. London: Hutchinson.

Ehrlich, Paul, and J.P. Holdren. 1971. 'The impact of population growth'. *Science* 171: 1212–17.

Eichler, Margrit. 1996. 'The impact of new reproductive and genetic technologies on families'. In Maureen Baker, ed., *Families: Changing Trends in Canada*, 3rd edn, 104–18. Toronto: McGraw-Hill Ryerson.

———. 1997. *Family Shifts: Families, Policies, and Gender Equality*. Toronto: Oxford University Press.

———. 2005. 'Biases in family literature'. In Maureen Baker, ed., *Families: Changing Trends in Canada*, 5th edn, 121–42. Toronto: McGraw-Hill Ryerson.

———, Patrizia Albanese, Susan Ferguson, Nicky Hyndman, Lichun Willa Liu, and Ann Matthews. 2010. *More Than It Seems: Household Work and Lifelong Learning*. Toronto: Women's Press.

Eisinger, Peter K. 1973. 'The conditions of protest behaviour in American cities'. *American Political Science Review* 67: 11–28.

El Akkad, Omar. 2010 'Woman shocked by portrayal as hard-line Islamist'. *The Globe and Mail*. www.theglobeandmail.com/news/national/woman-shocked-by-portrayal-as-hard-line-islamist/article1490612.

El-Badry, M.A. 1969. 'Higher female than male mortality in some countries of South Asia: A digest'. *Journal of the American Statistical Association* 64: 1234–44.

Elder, G.H., Jr. 1998. 'The life course as developmental theory'. *Child Development* 69 (1): 1–12.

———. 2001. 'Families, social change and individual lives'. *Marriage and Family Review* 31 (1): 187–203.

———, M.K. Johnson, and R. Crosnoe. 2003. 'The emergence and development of life course theory'. In J.T. Mortimer and M.J. Shanahan, eds, *Handbook of the Life Course*, 3–19. New York: Plenum.

———, and A.M. O'Rand. 1995. 'Adult lives in a changing society'. In K.S. Cook, G.A. Fine, and J.S. House, eds, *Sociological Perspectives on Social Psychology*, 452–75. Boston: Allyn and Bacon.

Elmer, Greg, and Mike Gasher, eds. 2005. *Contracting out Hollywood: Runaway Productions and Foreign Location Shooting*. Lanham, MD: Rowman and Littlefield.

Elster, Jon. 1989. *The Cement of Society: A Study of Social Order*. Cambridge: Cambridge University Press.

Engels, Friedrich. 1994 [1845]. *The Condition of the Working Class in*

England. Trans. W.O. Henderson and W.H. Chaloner. Stanford, CA: Stanford University Press.

———. 1990 [1884]. *The Origin of the Family, Private Property and the State*, v. 26 of *Karl Marx and Frederick Engels Collected Works*. New York: International Publishers.

Ennett, Susan T., and Karl E. Bauman. 1996. 'Adolescent social networks: School, demographic, and longitudinal considerations'. *Journal of Adolescent Research* 11: 194–215.

Erasmus, Georges. 2002. 'Why can't we talk'. From the 2002 Lafontaine-Baldwin Lecture. *The Globe and Mail* 9 March: F6–7.

Eribon, Didier. 1991 [1989]. *Michel Foucault*. Trans Betsy Wing. Cambridge, MA: Harvard University Press.

Erikson, Erik. 1963. *Childhood and Society*. New York: Norton.

———. 1982. *The Life Cycle Completed: A Review*. New York: Norton.

Erikson, Kai T. 1966. *Wayward Puritans: A Study in the Sociology of Deviance*. New York: Wiley.

Ermann, M. David, and Richard J. Lundman. 1996. 'Corporate and governmental deviance: Origins, patterns, and reactions'. In M. David Ermann and Richard J. Lundman, eds, *Corporate and Governmental Deviance: Problems of Organizational Behavior in Contemporary Society*, 3–44. New York: Oxford University Press.

Escobar, Arturo. 2004. 'Other worlds are already possible: Self organization, complexity and post-capitalist culture'. In Jai Sen and Peter Waterman, eds, *World Social Forum: Challenging Empires*, 393. Montreal: Black Rose Books.

———, and Walter Mignolo, eds. 2010. *Globalization and the Decolonial Option*. London: Routledge.

Estes, C.L. 1979. *The Aging Enterprise*. San Francisco: Jossey Bass.

———. 1986. 'The politics of ageing in America'. *Ageing and Society* 6 (2): 121–34.

———, L. Gerard, J.S. Zones, and J. Swan. 1984. *Political Economy, Health, and Aging*. Boston: Little, Brown.

European Conference on Sustainable Cities and Towns. 1994. *Charter of European Cities and Towns: Towards Sustainability*. http://ec.europa.eu/environment/urban/pdf/aalborg_charter.pdf.

Evangelical Fellowship of Canada. 2000. 'Submission to the Standing Committee on Justice and Human Rights on Bill C-23, An Act to Modernise the Statutes of Canada in Relation to Benefits and Obligations'. 2 March. http://files.

efc-canada.net/si/Sexual%20 Orientation/C23BriefonSameSex%20 BenefitsHofC.pdf.

Evans, L.T. 1998. *Feeding the Ten Billion: Plants and Population Growth*. Cambridge: Cambridge University Press.

Eyerman, Ron, and Andrew Jamison. 1989. 'Environmental knowledge as an organizational weapon: The case of Greenpeace'. *Social Science Information* 28 (1): 99–119.

Ezeh, Alex C., Blessing U. Mberu, and Jacques O. Emina. 2009. 'Stall in fertility decline in eastern African countries: Regional analysis of patterns, determinants and implications'. *Philosophical Transactions of the Royal Society* B 364: 2991–3007.

Fanon, Frantz. 1967. *Black Skin, White Masks*. New York: Grove Press.

———. 1968. *The Wretched of the Earth*. New York: Grove Press.

Faules, Don F., and Dennis C. Alexander. 1978. *Communication and Social Behaviour: A Symbolic Interaction Perspective*. Reading, MA.: Addison-Wesley.

Featherstone, Liza. 2004. *Selling Women Short: The Landmark Battle for Workers' Rights at Wal-Mart*. New York: Basic Books.

Federation of Canadian Municipalities. 2005. 'Sustainable Communities Knowledge Network: Centre for Sustainable Community Development. www.fcm.ca.

Ferguson, Kathy E. 1984. *The Feminist Case against Bureaucracy*. Philadelphia: Temple University Press.

Fielding, K.S., M.A. Hogg, and N. Annandale. 2006. 'Reactions to positive deviance: Social identity and attribution dimensions'. *Group Processes and Intergroup Relations* 9 (2): 199–218.

Filion, P. 2000. 'Balancing concentration and dispersion? Public Policy and urban structure in Toronto'. *Environment and Planning* C 18: 163–89.

———. 2003. 'Towards smart growth? The difficult implementation of alternatives to urban dispersion'. *Canadian Journal of Urban Research/Revue canadienne de recherche urbaine* 12: 48–70.

Financial Post. 2010. 'FP500: Canada's biggest companies by 2009 revenue'. www.financialpost.com/news/FP500/list.html.

Fine, Ben. 2001. *Social Capital versus Social Theory: Political Economy and Social Science at the Turn of the Century*. London: Routledge.

Fineman, Martha A. 1995. *The Neutered Mother, the Sexual Family, and Other Twentieth Century Tragedies*. New York: Routledge.

Finnie, Ross, and Ian Irvine. 2008. 'The welfare enigma: Explaining the dramatic decline in Canadians' use of social assistance, 1993–2005'. Toronto: C.D. Howe Institute.

Finnie, Ross, Kathryn McMullen, and Richard Mueller. 2010. 'New perspectives on access to postsecondary education'. *Education Matters: New Insights on Education, Learning and Training in Canada* 7 (1: 29 April). www.statcan.gc.ca/bsolc/olc-cel/olc-cel?catno=81-004-X&chropg=1&lang=eng.

Firestone, Shulamith. 1970. *The Dialectic of Sex: The Case for Feminist Revolution*. New York: Bantam Books.

Firey, Walter. 1977. *Man, Mind and Land: A Theory of Resource Use*. Glencoe, IL: Free Press.

Fisher, C. 1975. 'Towards a subcultural theory of urbanism'. *American Journal of Sociology* 80: 1319–41.

———. 1984. *The Urban Experience*. 2nd edn. Orlando: Harcourt, Brace, Jovanovich.

———. 1995. 'The subcultural theory of urbanism: A twentieth-year assessment'. *American Journal of Sociology* 101: 543–77.

Fisher, William, and Azy Barak. 2001. 'Internet pornography: A social psychological perspective on Internet sexuality'. *Journal of Sex Research* 38 (4): 312–23.

Fishman, R. 1987. *Bourgeois Utopias: The Rise and Fall of Suburbia*. New York: Basic Books.

Fiske, John, and Robert Dawson. 1996. 'Audiencing violence: Watching homeless men watch *Die Hard*'. In James Hay, Lawrence Grossberg, and Ellen Wartella, eds, *The Audience and Its Landscape*, 297–316. Boulder, CO: Westview.

Flavelle, Dana. 2010. 'Top executives still raking it in'. *Toronto Star* 5 January: B1.

Fleras, Augie, and Jean Leonard Elliott. 1996. *Unequal Relations: An Introduction to Race, Ethnic and Aboriginal Dynamics in Canada*. Toronto: Prentice-Hall.

———. 2009. U*nequal Relations: An Introduction to Race, Ethnic and Aboriginal Dynamics in Canada*. 6th edn. Scarborough, ON: Prentice-Hall.

Fleury, D., and M. Fortin. 2006. *When Working Is Not Enough to Escape Poverty: An Analysis of Canada's Working Poor*. Ottawa: Human Resources and Skills Development Canada.

Florida, R. 2003. 'Cities and the creative class'. *Cities and Communities* 2: 3–19.

Fogel, Robert W., and Dora L. Costa. 1997. 'A theory of technophysio evolution, with some implications for forecasting population, health care costs, and pension costs'. *Demography* 34: 49–66.

Fong, Eric. 2010. 'Immigration and race in the city'. In Harry H. Hiller, ed., *Urban Canada*, 2nd edn, 132–54. Toronto: Oxford University Press.

———, and Emi Ooka. 2002. 'The social consequences of participating in ethnic economy'. *International Migration Review* 36: 125–46.

Fontan, J.-M., P. Hamel, and R. Morin. 2006. 'Le développement local dans un contexte métropolitain: la démocratie en quête d'un nouveau modèle?' *Politique et sociétés* 25: 99–127.

Food Banks Canada. 2009. *Hunger Count 2008: A Comprehensive Report on Hunger and Food Bank Use in Canada*. Toronto: Food Banks Canada.

Foot, David, with Daniel Stoffman. 1998. *Boom, Bust and Echo 2000: Profiting from the Demographic Shift in the New Millennium*. Toronto: Macfarlane Walter and Ross.

Force, William Ryan. 2009. 'Consumption styles and the fluid complexity of punk authenticity'. *Symbolic Interaction* 32 (4): 289–309.

Ford, Clellan, and Frank Beach. 1951. *Patterns of Sexual Behavior*. New York: Harper and Row.

Ford, J., N. Nassar, E. Sullivan, G. Chambers, and P. Lancaster. 2003. *Reproductive Health Indicators, Australia, 2002*. Sydney: Australian Institute of Health and Welfare.

Ford, J.A. 2009. 'Nonmedical prescription drug use among adolescents: The influence of bonds to family and school'. *Youth and Society* 40: 336–52.

Fortin, A., and M. Bédard. 2003. 'Citadins et banlieusards: représentations, pratiques et identités'. *Canadian Journal of Urban Research/Revue canadienne de recherche urbaine* 12: 58–76.

Foschi, Martha, Larissa Lai, and Kirsten Sigerson. 1994. 'Gender and double standards in the assessment of job applicants'. *Social Psychology Quarterly* 57: 326–39.

Foucault, Michel. 1990. *The History of Sexuality: An Introduction*. New York: Vintage Books.

Fougères, D. 2004. *L'approvisionnement en eau à Montréal: du privé au public, 1796–1865*. Quebec City: Septentrion.

Fox, Bonnie. 2001. 'The formative years: How parenthood creates gender'. *Canadian Review of Sociology and Anthropology* 38: 373–90.

———. 2009. *When Couples Become Parents: The Creation of Gender in the Transition to Parenthood*. Toronto: University of Toronto Press.

———, and Pamela Sugiman. 1999. 'Flexible work, flexible workers: The restructuring of clerical work in a large telecommunications company'. *Studies in Political Economy* 60: 59–84.

Fox, James Alan, and Jack Levin. 2001. *The Will to Kill: Making Sense of Senseless Murder*. Boston: Allyn and Bacon.

Fox, Stephen. 1985. *The American Conservation Movement: John Muir and His Legacy*. Madison: University of Wisconsin Press.

Francis, Diane. 1986. *Controlling Interest: Who Owns Canada*. Toronto: Macmillan.

———. 2008. *Who Owns Canada Now: Old Money, New Money and the Future of Canadian Business*. Toronto: HarperCollins.

Frank, Andre Gunder. 1967. *Capitalism and Underdevelopment in Latin America: Historical Studies of Chile and Brazil*. New York: Modern Reader Paperbacks.

———. 1991. 'The underdevelopment of development'. *Scandinavian Journal of Development Alternatives* 10 (3): 5–72.

Frank, Arthur W. 2002. *At the Will of the Body: Reflections on Illness*. New York: Houghton Mifflin.

Freidson, Eliot. 1970. *The Profession of Medicine: A Study in the Sociology of Applied Knowledge*. New York: Harper and Row.

Freire, Paulo. 1970. *Pedagogy of the Oppressed*. Trans. Myra Bergman Ramos. New York: Herder and Herder.

Frejka, Tomas. 1973. *The Future of Population Growth: Alternative Paths to Equilibrium*. New York: Wiley.

Freud, Sigmund. 1938. *The Basic Writings of Sigmund Freud*. Trans., ed. Abraham Brill. New York: Modern Library Press.

Freudenburg, William, and Susan Pastor. 1992. 'NIMBYs and LULUs: Stalking the syndromes'. *Journal of Social Issues* 48 (4): 39–61.

Friedan, Betty. 1963. *The Feminine Mystique*. New York: W.W. Norton.

Friedman, Thomas. 2005. *The World Is Flat: A Brief History of the Twenty-First Century*. New York: Farrar, Straus and Giroux.

Fries, James F. 1980. 'Aging, natural death, and the compression of morbidity'. *New England Journal of Medicine* 303: 130–5.

Fritsche, I. 2005. "Predicting deviant behavior by neutralization: Myths and findings'. *Deviant Behavior* 9 (2): 199–218.

Frytak, J.R., C.R. Harley, and M.D. Finch. 2006. 'Socioeconomic status and health over the life course: Capital as a unifying concept'. In J.T. Mortimer and M.J. Shanahan, eds, *Handbook of the Life Course*, 623–46. New York: Springer.

Fukuyama, Francis. 1992. *The End of History and the Last Man*. New York: Free Press.

Furedi, Frank. 1997. *Population and Development: A Critical Introduction*. New York: St Martin's.

Gagnon, J., and W. Simon. 1986. 'Sexual scripts: Performance and change'. *Archives of Sexual Behavior* 15: 98–104.

Gagnon, Mona Josée. 2002. 'The labor movement and civil society: Reflections on the People's Summit of Quebec City'. *Just Labour* 1: 58–67.

Gagnon, R. 2006. *Questions d'égouts: santé publique, infrastructures et urbanisation à Montréal au XIXe siècle*. Montreal: Boréal.

Galabuzi, Grace-Edward. 2006. *Canada's Economic Apartheid: The Social Exclusion of Racialized Groups in the New Century*. Toronto: Canadian Scholars' Press.

Galarneau, D. 1994. 'Baby boom women'. *Perspectives on Labour and Income* 6 (4): 23–9.

Gallupe, Owen, William Boyce, and Stevenson Fergus. 2009. 'Non-use of condoms at last intercourse among Canadian youth: Influence of sexual partners and social expectations'. *Canadian Journal of Human Sexuality* 18 (1/2): 27–34.

Gamson, Joshua. 2001. 'Normal sins: Sex scandal narratives as institutional morality tales'. *Social Problems* 48 (2): 185–205.

———, and Dawne Moon. 2004. 'The sociology of sexualities'. *Annual Review of Sociology* 30: 47–64.

Gamson, William. 1991. 'Commitment and agency in social movements'. *Sociological Forum* 6: 27–50.

———, and David S. Meyer. 1996. 'Framing political opportunity'. In Doug McAdam, John McCarthy, and Mayer Zald, eds, *Comparative Perspectives on Social Movements: Political Opportunities, Mobilizing Structures, and Cultural Framing*, 275–90. New York: Cambridge University Press.

Gannon, M., K. Mihorean, K. Beatie, A. Taylor-Butts, and R. Kong. 2005. *Criminal Justice Indicators, 2005*. Ottawa: Canadian Centre for Justice Statistics, Statistics Canada.

Gans, Herbert J. 1968. *People and Plans*. New York: Basic Books.

———. 1974. *Popular Culture and High Culture: An Evaluation of Taste*. New York: Basic Books.

Garde, A.M. 2004. 'New urbanism as sustainable growth? A supply side story and its implications for public policy'. *Journal of Planning Education and Research* 24: 154–70.

Gardner, Julia, and Mark Roseland. 1989. 'Thinking globally: The role of social equity in sustainable development'. *Alternatives* 16 (3): 26–35.

Garfinkel, Harold. 1956. 'Conditions of successful status degradation ceremonies'. *American Journal of Sociology* 61: 420–4.

Garriguet, Didier. 2005. 'Early sexual intercourse'. *Health Reports* 16 (3): 9–19. Catalogue no. 82-003-XPE. Ottawa: Statistics Canada.

Gaskell, Jane. 1993. 'Feminism and its impact on educational scholarship in Canada'. In Leonard L. Stewin and Stewart J.H. McCann, eds, *Contemporary Educational Issues: The Canadian Mosaic*, 2nd edn, 145–60. Toronto: Copp Clark Pitman.

Geddes, P. 1904. *Civics: As Applied Sociology*. Dodo Press (reprint).

———. 1949 [1915]. *Cities in Evolution*. London: Williams and Norgate.

Gee, E.M. 2000. 'Population and politics: Voodoo demography, population aging, and Canadian social policy'. In E.M. Gee and G.M. Gutman, eds, *The Overselling of Population Aging: Apocalyptic Demography, Intergenerational Challenges, and Social Policy*, 5–25. Toronto: Oxford University Press.

———. 2002. 'Misconceptions and misapprehensions about population ageing'. *International Journal of Epidemiology* 31 (4): 750–53.

———, and G.M. Gutman, eds. 2000. *The Overselling of Population Aging: Apocalyptic Demography, Intergenerational Challenges, and Social Policy*. Toronto: Oxford University Press.

———, and M.M. Kimball. 1987. *Women and Aging*. Toronto: Butterworths.

Geiger, B. 2006. 'Crime, prostitution, drugs, and malingered insanity: Female offenders' resistant strategies to abuse and domination'. *International Journal of Offender Therapy and Comparative Criminology* 50 (5): 582–94.

Gellner, Ernest. 1983. *Nations and Nationalism*. Oxford: Blackwell.

George, Susan. 2010. 'Converging crises: Reality, fear and hope'. *Globalizations* 7 (1/2): 17–22.

Gergen. Kenneth. 1991. *The Saturated Self: Dilemmas of Identity in Contemporary Life*. New York: Basic Books.

———. 2001. 'From mind to relationship: The emerging challenge'. *Education Canada* 41 (1): 8–11.

Germain, A. 1984. *Les mouvements de réforme urbaine à Montréal au tournant du siècle*. Montreal: Université de Montréal.

Gerth, H.H., and C.W. Mills. 1946. 'Introduction'. In *Max Weber: Essays in Sociology*. New York: Oxford University Press.

Giddens, Anthony. 1991. 'Modernity and self-identity: Self and society in the late modern age'. Stanford, CA: Stanford University Press.

———. 2006. *Sociology*. 5th edn. Cambridge: Polity Press.

Gidney, R.D. 1999. *From Hope to Harris: The Reshaping of Ontario's Schools*. Toronto: University of Toronto Press.

Gilbert, Neil. 1997. 'Advocacy research and social policy'. *Crime and Justice* 22: 101–48.

Gilligan, Carol. 1982. *In a Different Voice: Psychoanalytic Theory and Women's Development*. Cambridge, MA: Harvard University Press.

Gilliland, J. 2002. 'The creative destruction of Montreal: Street widenings and urban (re)development in the nineteenth century'. *Urban History Review/ Revue d'histoire urbaine* 31 (1): 37–51.

Gimlin, D. 2008. 'NAAFA: Attempting to neutralize the stigma of the hugely obese body'. In E. Goode and D.A. Vail, eds, *Extreme Deviance*, 72–80. Los Angeles and London: Pine Forge Press.

Giroux, Henri. 1997. *Pedagogy and the Politics of Hope: Theory, Culture, and Schooling: A Critical Reader*. Boulder, CO: Westview.

Gitlin, Todd. 1980. *The Whole World Is Watching: Mass Media in the Making and Unmaking of the New Left*. Berkeley: University of California Press.

Glassner, Barry. 1999. *The Culture of Fear: Why Americans Are Afraid of the Wrong Things*. New York: Basic Books.

Goffman, Erving. 1959. *The Presentation of Self in Everyday Life*. Garden City, NJ: Doubleday-Anchor.

———. 1961. *Asylums: Essays on the Social Situation of Mental Patients and Other Inmates*. New York: Doubleday.

———. 1963. *Stigma: Notes on the Management of Spoiled Identity*. Englewood Cliffs, NJ: Prentice-Hall.

Gold, David A., Clarence Y.H. Lo, and Erik Olin Wright. 1975. 'Recent developments in Marxist theories of the capitalist state'. *Monthly Review* 27: 29–43.

Goldberg, David Theo. 1993. *Racist Culture: Philosophy and the Politics of Meaning*. Oxford: Blackwell.

Goldberg, Kim. 1990. *The Barefoot Channel: Community Television as a Tool for Social Change*. Vancouver: New Star Books.

Goldenberg, Naomi. 2006. 'What's God got to do with it? A call for problematizing basic terms in the feminist analysis of religion'. Paper presented at the biannual meeting of the Britain and Ireland School of Feminist Theology, Edinburgh, July.

Goldenberg, Sheldon. 1992. *Thinking Methodologically*. New York: HarperCollins.

Goldthorpe, J.E. 1987. *Family Life in Western Societies: A Historical Sociology of Family Relationships in Britain and North America*. Cambridge: Cambridge University Press.

Goode, E., and D.A. Vail. 2008. *Extreme Deviance*. Thousand Oaks, CA: Pine Forge Press.

Goodwin, J., and J.M. Jasper. 1999. 'Caught in a winding, snarling vine: The structural bias of political process theory'. *Sociological Forum* 14 (1): 27–54.

Gordon, Robert M., and Jacquelyn Nelson. 2000. 'Crime, ethnicity, and immigration'. In Robert A. Silverman, James J. Teevan, and Vincent F. Sacco, eds, *Crime in Canadian Society*, 6th edn. Toronto: Harcourt Brace.

Gorski, Philip S. 1995. 'The Protestant ethic and the spirit of bureaucracy'. *American Sociological Review* 60: 783–6.

Gorz, André. 1999. *Reclaiming Work: Beyond the Wage-Based Society*. Trans. Chris Turner. Cambridge: Polity Press.

Gosine, Kevin. 2000. 'Revisiting the notion of a "recast" vertical mosaic in Canada: Does a post secondary education make a difference?' *Canadian Ethnic Studies* 32 (3): 89–104.

Gottfredson, M. 2006. 'The empirical status of social control in criminology'. In F.T. Cullen, J.P. Wright, and K.R. Blevins, eds, *Taking Stock: The Status of Criminological Theory*, 77–100. New Brunswick, NJ: Transaction.

———, and Travis Hirschi. 1990. *A General Theory of Crime*. Stanford, CA: Stanford University Press.

Gottman, J. 1961. *Megalopolis. The Urbanization of the Northeast Seaboard of the United States*. New York: Twentieth Century Fund.

Gough, Brendan, Nicky Weyman, Julie Alderson, Gary Butler, and Mandy Stoner. 2008. '"They did not have a word": The parental quest to locate a "true sex" for their intersex children'. *Psychology and Health* 23 (4): 493–507.

Gover, Angela R., Catherine Kaukinen, and Kathleen A. Fox. 2008. 'The relationship between violence in the family of origin and dating violence among college students'. *Journal of Interpersonal Violence* 23 (12): 1667–93.

Grabb, Edward. 1994. 'Democratic values in Canada and the United States: Some observations and evidence from the past and present'. In J. Dermer, ed., *The Canadian Profile: People, Institutions, and Infrastructure*, 113–39. Toronto: Captus Press.

———. 2007. *Theories of Social Inequality*. 5th edn. Toronto: Thomson Nelson.

———, and James Curtis. 2002. 'Comparing central political values in the Canadian and American democracies'. In D. Baer, ed., *Political Sociology: Canadian Perspectives*, 37–54. Don Mills, ON: Oxford University Press.

Graham, Hilary. 1984. *Women, Health and the Family*. Brighton, UK: Wheatsheaf.

Gramsci, Antonio. 1992. *Prison Notebooks*, v. 1. Trans. Joseph A. Buttigieg and Antonio Callari. New York: Columbia University Press.

Granovetter, Mark S. 1974. *Getting a Job: A Study of Contacts and Careers*. Cambridge, MA: Harvard University Press.

Grant, J. 2006. 'The ironies of new urbanism'. *Canadian Journal of Urban Research/Revue canadienne de recherche urbaine* 15: 158–74.

Grant, Nina, Mark Hamer, and Andrew Steptoe. 2009. 'Social isolation and stress-related cardiovascular, lipid, and cortisol responses'. *Annals of Behavioural Medicine* 37: 29–37.

Gray, Gary, and Neil Guppy. 2008. *Successful Surveys: Research Methods and Practice*. 4th edn. Toronto: Nelson Thomson.

Green, Adam. 2008. 'Erotic habitus: Toward a sociology of desire'. *Theory and Society* 37 (6): 597–626.

Greenglass, Esther R. 1982. *A World of Difference: Gender Roles in Perspective*. Toronto: John Wiley and Sons.

Greer, Germaine. 1984. *Sex and Destiny: The Politics of Human Fertility*. London: Martin Secker and Warburg.

Gregory, J.W., and V. Piché. 1983. 'Inequality and mortality: Demographic hypotheses regarding advanced and peripheral capitalism'. *International Journal of Health Services* 13: 89–106.

Griffin, Wendy. 2000. 'The embodied goddess: Feminist witchcraft and female divinity'. In S.C. Monahan, W.A. Mirola, and M.O. Emerson, eds, *Sociology of Religion: A Reader*. New York: Prentice-Hall/Penguin Putman.

Grindstaff, Carl F. 1975. 'The baby bust: Changes in fertility patterns in Canada'. *Canadian Studies in Population* 2: 15–22.

———. 1994. 'The baby bust revisited: Canada's continuing pattern of low fertility'. In Frank Trovato and Carl F. Grindstaff, eds, *Perspectives on Canada's Population: An Introduction to Concepts and Issues*, 168–72. Toronto: Oxford University Press.

———. 1995. 'Canada's continued trend of low fertility'. *Canadian Social Trends* (winter): 12–16.

Griswold, Wendy. 1987. 'A methodological framework for the sociology of culture'. *Sociological Methodology* 17: 1–35.

Guay, L. 2005. 'Introduction: le concept de développement durable en contexte historique et cognitif'. In L. Guay et al., *Les enjeux et les défis du développement durable: connaître, décider, agir*, 1–31. Quebec City: Les Presses de l'Université Laval.

———, and N. Émond. 2010. 'Infrastructures urbaines et développement durable'. In Institut du nouveau monde, *L'état du Québec 2010*, 109–15. Montreal: Boréal.

———, and P. Hamel. 2010. 'Urban change and policy response in Quebec'. In H.H. Hiller, ed., *Urban Canada*, 2nd edn, 297–322. Toronto: Oxford University Press.

Guillebaud, Jean-Claude. 2002. 'Definition of man: What is left of the Nuremburg Code?' *Diogenes* 195: 7–12.

Guillemard, A.M. 2000. *Aging and the Welfare State Crisis*. Cranbury, NJ: Delaware University Press.

Guppy, Neil, and Scott Davies. 1998. *Education in Canada: Recent Trends and Future Challenges*. Ottawa: Statistics Canada.

Gusfield, Joseph R. 1963. *Symbolic Crusade: Status Politics and the American Temperance Movement*. Urbana: University of Illinois Press.

———. 1981. *The Culture of Public Problems: Drinking-Driving and the Symbolic Order*. Chicago: University of Chicago Press.

———. 1989. 'Constructing the ownership of social problems: Fun and profit in the welfare state'. *Social Problems* 36: 431–41.

Gutman, D. 1976. 'A cross-cultural view of adult life in the extended family'. In K. Riegel and J. Meacham, eds, *Developing Individual in a Changing World*, 107. New Jersey: Aldine de Gruyter.

Guttentag, Marcia, and Paul F. Secord. 1983. *Too Many Women? The Sex Ratio Question*. Beverly Hills, CA: Sage.

Haas, Jack, and William Shaffir. 1987. *Becoming Doctors: The Adoption of a Cloak of Competence*. Greenwich, CT: JAI Press.

Haberland, Nicole, and Diana Measham, eds. 2002. *Responding to Cairo: Case Studies of Changing Practice in Reproductive Health and Family Planning*. New York: Population Council.

Habermas, Jürgen. 1989. *The Structural Transformation of the Public Sphere*. Cambridge: Polity Press.

Hackett, Robert A., and Richard Gruneau. 2000. *The Missing News: Filters and Blind Spots in Canada's Press*. Aurora, ON: Garamond.

Hackett, Robert A., Richard Pinet, and Myles Ruggles. 1996. 'News for whom: Hegemony and monopoly versus democracy in Canadian media'. In Helen Holmes and David Taras, eds, *Seeing Ourselves: Media Power and Policy in Canada*, 2nd edn, 257–72. Toronto: Harcourt Brace Canada.

Hackett, Robert A., and Scott Uzelman. 2003. 'Tracing corporate influences on press content: A summary of recent NewsWatch Canada research'. *Journalism Studies* 4 (3): 331–46.

Hagestad, G.O., and D. Dannefer. 2001. 'Concepts and theories of aging'. In R. Binstock and L. George, eds, *Handbook of Aging and the Social Sciences*, 5th edn, 3–21. New York: Academic Press.

Halbwachs, M. 1938. *Morphologie sociale*. Paris: Armand Colin.

Hall, Emmett. 1964–5. *Report of the Royal Commission on Health Services*. Ottawa: Queen's Printer.

Hall, John A. 1993. 'Nationalisms: Classified and explained'. *Daedelus* 122: 1–28.

———, ed. 1998. *The State of the Nation*. Cambridge: Cambridge University Press.

Hall, M., D. Lasby, S. Ayer, and W.D. Gibbons. 2009. *Caring Canadians, Involved Canadians: Highlights from the 2007 Canada Survey of Giving, Volunteering and Participating*. Ottawa: Statistics Canada.

Hall, Peter. 1982. *Urban and Regional Planning*. Harmondsworth, UK: Penguin.

———. 1988. *Cities of Tomorrow: An Intellectual History of Urban Planning and Design in the Twentieth Century*. Oxford: Blackwell.

Hall, Stuart. 1980. 'Encoding/decoding'. In Stuart Hall, Dorothy Hobson, Andrew Lowe, and Paul Willis, eds, *Culture, Media, Language: Working Papers in Cultural Studies, 1972–79*, 128–38. London: Hutchinson.

Hamilton, Roberta. 1978. *The Liberation of Women*. London: Allen and Unwin.

Hamlin, C. 1990. *A Science of Impurity: Water Analysis in Nineteenth Century Britain*. Berkeley and Los Angeles: University of California Press.

Hannigan, John. 1995. *Environmental Sociology: A Social Constructionist Perspective*. London: Routledge.

Harary, Frank. 1966. 'Merton revisited: A new classification for deviant behaviour'. *American Sociological Review* 31 (5): 693–7.

Harding, Sandra. 1986. *The Science Question in Feminism*. Ithaca, NY: Cornell University Press.

Harley, C., and J.T. Mortimer. 2000. *Social Status and Mental Health in Young Adulthood: The Mediating Role of the Transition to Adulthood*. Paper presented at the biennial meeting of the Society for Research on Adolescence, Chicago, March 30–April 2.

Harris, D.K. 2007. *The Sociology of Aging*. Lanham, MD: Rowman and Littlefield.

Harrison, Trevor W., and John W. Friesen. 2010. *Canadian Society in the Twenty-First Century*. Toronto: Women's Press.

Harrison, Trevor W., and Harvey Krahn. 1995. 'Populism and the rise of the Reform Party in Alberta'. *Canadian Review of Sociology and Anthropology* 32: 127–50.

Hart, Randle. 2008. 'Practicing Birchism in unsettled times: The assumption and limits of idiocultural coherence in framing theory'. *Social Movement Studies* 7 (2): 121–47.

———. 2010. 'There comes a time: Biography and the founding of a movement organization'. *Qualitative Sociology* 33 (1): 55–77.

Hart-Landsberg, Martin. 2010. 'The US economy and China: Capitalism, class and crisis'. *Monthly Review* 61 (9): 14–31.

Harvey, D. 1973. *Social Justice and the City*. Baltimore: Johns Hopkins University Press.

———. 1989. *The Urban Experience*. Baltimore: Johns Hopkins University Press.

Harvey, David. 2006. *A Brief History of Neoliberalism*. Oxford: Oxford University Press.

Hathaway, A.D. 2004. 'Cannabis users' informal rules for managing stigma and risk'. *Deviant Behavior* 25 (6): 559–77.

Havighurst, R.J., and R. Albrecht. 1953. *Older People*. New York: Longmans, Green.

Hawley, Amos A. 1950. *Human Ecology: A Theory of Community Structure*. New York: Ronald Press.

———. 1981. *Urban Society*, 2nd edn. New York: Wiley.

Hay, D.I. 2009. *Poverty Reduction Policies and Programs*. Social Development Report Series. Ottawa: Canadian Council on Social Development.

Hayford, Alison. 1992. 'From Chicago 1966 to Montreal 1989: Notes on new(s) paradigms of women as victims'. In Marc Grenier, ed., *Critical Studies of Canadian Mass Media*, 201–12. Markham, ON: Butterworths.

Hays, Samuel. 1987. *Beauty, Health and Permanence: Environmental Politics in the United States, 1955–1985*. New York: Cambridge University Press.

Health Canada. 2002a. *Dare to Age Well: Workshop on Healthy Aging*. Part 1: Aging and health practices. Ottawa: Health Canada, Division of Aging and Seniors.

———. 2002b. *A Report on Mental Illness in Canada*. Ottawa: Health Canada.

———. 2004. 'Drugs and health products: General questions'. www.hc-sc.gc.ca/dhp-mps/prodnatur/faq/question_general-eng.php.

———. 2009. 'Canadian alcohol and drug use monitoring survey'. www.hc-sc.gc.ca/hc-ps/drugs-drogues/cadums-esccad-eng.php.

Hearn, Jeff, and Wendy Parkin. 1987. *'Sex' and 'Work': The Power and Paradox of Organizational Sexuality*. New York: St Martin's.

Hechter, Michael. 2000. *Containing Nationalism*. Oxford: Oxford University Press.

Heckman, James J., and Alan B. Krueger. 2004. *Inequality in America: What Role for Human Capital Policies?* Cambridge, MA: MIT Press.

Heimer, Robert. 2002. *Social Problems: An Introduction to Critical Constructionism*. New York: Oxford University Press.

Held, David, and Anthony McGrew, eds. 2000. *The Global Transformations Reader*. Oxford: Polity Press.

———, D. Goldblatt, and J. Perraton. 1999. *Global Transformations*. Oxford: Polity Press.

Helmes-Hayes, Rick. 2002. 'John Porter: Canada's most famous sociologist (and his links to American sociology)'. *American Sociologist* 33 (1): 79–104.

———. 2009. *Measuring the Mosaic: An Intellectual Biography of John Porter*. Toronto: University of Toronto Press.

———, and James Curtis, eds. 1998. *The Vertical Mosaic Revisited: Social Inequality and Social Justice in Canada, 1965–1995*. Toronto: University of Toronto Press.

HelpAge International. 2006. *Irene, 80, Malawi*. www.helpage.org/News/Casestudies/HIVAIDS/@24224.

Heltsley, Martha, and Thomas C. Calhoun. 2003. 'The good mother: Neutralization techniques used by pageant mothers'. *Deviant Behavior* 24 (2): 81–100.

Helwig, David. 2000. 'NWT residents are accident prone, live shorter lives'. *Canadian Medical Association Journal* 162: 681–2.

Henry, Frances, and Effie Ginzberg. 1985. *Who Gets the Work: A Test of Racial Discrimination in Employment*. Toronto: Urban Alliance on Race Relations and Social Planning Directorate.

Henry, Frances, and Carol Tator. 2005. *The Colour of Democracy: Racism in Canadian Society*. 3rd edn. Toronto: Thomson Nelson.

Herlihy, David. 1997. *The Black Death and the Transformation of the West*. Cambridge, MA: Harvard University Press.

Herodotus. 1996. *Histories*. Hertfordshire, UK: Wordsworth.

Hess, D.J. 2009. *Localist Movements in a Global Economy*. Cambridge, MA: MIT Press.

Hesse-Biber, Sharlene Nagy. 2007. *The Cult of Thinness*. New York: Oxford University Press.

Hewitt, Christopher. 1994. 'The dog that didn't bark: The political consequences of separatist violence in Quebec, 1963–70'. *Conflict Quarterly* 14 (1): 9–29.

Hewitt, John P. 2000. *Self and Society: A Symbolic Interactionist Social Psychology*. 8th edn. Boston: Allyn and Bacon.

Hey, Shereen. 1997. *The Company She Keeps: An Ethnography of Girls' Friendships*. Buckingham, UK: Open University Press.

Hickman, B. 1988. 'Men wise up to bald truth'. *Australian* 21 (May): 4.

Hier, Sean P. 2002. 'Raves, risks and the ecstasy panic: A case study in the subversive nature of moral regulation'. *Canadian Journal of Sociology* 27: 33–52.

Higgins, Jenny. 2007. 'The truck system'. Newfoundland and Labrador Heritage website, www.heritage.nf.ca/society/truck_system.html.

Higgins, Jenny A., and Irene Browne. 2008. 'Sexual needs, control, and refusal: How "doing" class and gender influences sexual risk taking'. *Journal of Sex Research* 45 (3): 233–45.

Higgins, Robert R. 1994. 'Race, pollution and the mastery of nature'. *Environmental Ethics* 16: 251–64.

Hilberg, Raul. 1996. 'The Nazi Holocaust: Using bureaucracies, overcoming psychological barriers to genocide'. In M. David Ermann and Richard J. Lundman, eds, *Corporate and Governmental Deviance: Problems of Organizational Behavior in Contemporary Society*, 158–79. New York: Oxford University Press.

Hilgartner, Stephen, and Charles Bosk. 1988. 'The rise and fall of social problems: A public arenas model'. *American Journal of Sociology* 94: 53–78.

Hirschi, Travis. 1969. *Causes of Delinquency*. Berkeley: University of California Press.

Hochschild, Arlie. 1983. *The Managed Heart: Commercialization of Human Feeling*. Berkeley: University of California Press.

———. 1997. *The Time Bind: When Work Becomes Home and Home Becomes Work*. New York: Metropolitan Books.

Hodge, G., and I.M. Robinson. 2001. *Planning Canadian Regions*. Vancouver: UBC Press.

Hodson, Randy. 2001. *Dignity at Work*. Cambridge: Cambridge University Press.

Hoffer, Thomas B. 2008. 'Accountability in education'. In Maureen T. Hallinan, ed., *Handbook of the Sociology of Education*, 529–44. New York: Springer.

Hofrichter, Richard, ed. 1993. *Toxic Struggles: The Theory and Practice of Environmental Justice*. Philadelphia: New Society.

Hogeveen, B.R. 2005. '"If we are tough on crime, if we punish crime, then people get the message": Constructing and governing the punishable young offender in Canada during the late 1990s'. *Punishment and Society* 7 (1): 73–89.

Holmes, Malcolm D., and Judith A. Antell. 2001. 'The social construction of American Indian drinking: Perceptions of American Indian and white officials'. *Sociological Quarterly* 42: 151–73.

Holstein, James A., and Jaber F. Gubrium. 2000. *The Self We Live By: Narrative Identity in a Postmodern World*. New York: Oxford University Press.

Holstein, James A., and Gale Miller. 2006. *Reconsidering Social Constructionism: Debates in Social Problems Theory*. New York: Aldine de Gruyter.

Homans, George. 1951. 'The Western Electric researchers'. In Schyler Dean Hoslett, ed., *Human Factors in Management*, 210–41. New York: Harper.

———. 1961. *Social Behavior: Its Elementary Forms*. New York: Harcourt, Brace, and World.

Homer-Dixon, Thomas F., Jeffrey H. Boutwell, and George W. Rathjens. 1993. 'Environmental change and violent conflict'. *Scientific American* February: 38–45.

Hommels, A. 2005. *Unbuilding Cities: Obduracy in Urban Socio-technical Change*. Cambridge, MA: MIT Press.

Hood, Duncan, 2008, 'How Canada stole the American Dream'. *Maclean's* 7 July: 51–4.

Hoodfar, Homa. 2000. 'More than clothing: Veiling as an adaptive strategy'. In S.S. Alvi, H. Hoodfar, and S. McDonough, eds., *The Muslim Veil in North America: Issues and Debates*. Toronto: Women's Press.

hooks, bell. 1992. *Black Looks: Race and Representation*. Toronto: Between the Lines.

———. 1994. *Outlaw Culture*. New York: Routledge.

Hope, Steven, Chris Power, and Bryan Rodgers. 1998. 'The relationship between parental separation in childhood and problem drinking in adulthood'. *Addiction* 93 (4): 505–14.

Horkheimer, Max, and Theodor Adorno. 2006. 'The culture industry: Enlightenment as mass deception'. In Meenakshi Gigi Durham and Douglas M. Kellner, eds, *Media and Cultural Studies: Key Works*, rev. edn. Malden, MA: Blackwell.

Horowitz, Lisa. 2010. 'Despite Bigelow's success, women directors in decline'. *The Wrap* 16 February. www.thewrap.com/movies/article/study-despite-bigelows-success-women-directors-decline-14207.

Horton, A.D. 2009. 'Sin cinema: Images of prostitution in popular film' (unpublished paper, Queen's University, Kingston, ON).

Hou, Feng. 2007. 'Changes in the initial destinations and redistribution of Canada's major immigrant groups: Reexamining the role of group affinity'. *International Migration* 41 (3): 680–705.

House, J.S., and R.L. Kahn. 1985. 'Measures and concepts of social support'. In S. Cohen and L. Syme, eds, *Social Support and Health*, 79–108. New York: Academic Press.

Houtart, François. 2010. 'The multiple crisis and beyond'. *Globalizations* 7 (1): 9–15.

Howard-Hassmann, Rhoda. 1999. '"Canadian" as an ethnic category: Implications for multiculturalism and national unity'. *Canadian Public Policy* 25, 4: 523–37.

Hughes, Everett. 2009 [1943]. *French Canada in Transition*. Toronto: Oxford University Press.

———. 1945. 'Dilemmas and contradictions of status'. *American Journal of Sociology* 50: 353–9.

Hughes, Karen. 2005. 'The adult children of divorce: Pure relationships and family values?' *Journal of Sociology* 41 (1): 69–86.

Hum, Derek, and Wayne Simpson. 2007. 'Revisiting equity and labour: Immigration, gender, minority status, and income differentials in Canada'. In S. Hier and B. Singh Bolaria, eds, *Race and Racism in 21st Century Canada*. Peterborough, ON: Broadview.

Human Resources and Skills Development Canada (HRSDC). 2007. 'Advancing the inclusion of people with disabilities—Chapter 3: Income support, benefits and service delivery'. www.hrsdc.gc.ca/eng/disability_issues/reports/fdr/20007/page05.shtml.

———. 2009. *Low Income in Canada 2000–2007 Using the Market Basket Measure*. Final Report, August.

———. 'Canada Pension Plan and Old Age Security rates'. www.hrsdc.qc.ca/eng/corporate/facts/seniors.shtml.

Hume, Stephen. 2010. 'Celebrity scandal? Nothing new about that'. *Vancouver Sun* 9 July: A13.

Humphrey, Craig R., and Frederick R. Buttel. 1982. *Environment, Energy and Society*. Belmont, CA: Wadsworth.

Humphreys, Laud. 1970. *Tearoom Trade: Impersonal Sex in Public Places*. Chicago: Aldine.

Huntington, Samuel P. 2002 [1997]. *The Clash of Civilizations and the Remaking of the World Order*. London: Simon and Schuster.

Ihinger-Tallman, Marilyn, and David Levinson. 2003. Definition of marriage, rev. by J.M. White. In J. Ponzetti, Jr, ed., *International Encyclopedia of Marriage and Family*, 2nd edn. New York: Macmillan Reference and Thomson Gale.

Illich, Ivan. 1976. *Limits to Medicine*. Toronto: McClelland and Stewart.

Imershein, Allen W., and Carroll L. Estes. 1996. 'From health services to medical markets: The commodity transformation of medical production and the non-profit sector'. *International Journal of Health Services* 26: 221–38.

Imig, Doug, and Sidney Tarrow. 2001. 'Mapping the Europeanization of contention: Evidence from a quantitative data analysis'. In Doug Imig and Sidney Tarrow, eds, *Contentious Europeans: Protest and Politics in an Emerging Polity*, 27–49. New York: Rowman and Littlefield.

Inglehart, Ronald. 1990a. *Culture Shift in Advanced Industrial Society*. Princeton, NJ: Princeton University Press.

———. 1990b. 'Values, ideology, and cognitive mobilization in new social movements'. In R.J. Dalton and M. Kuechler, eds, *Challenging the Political Order*, 23–42. New York: Oxford University Press.

Ingram, R.J., and S. Hinduja. 2008. 'Neutralizing music piracy: An empirical examination'. *Deviant Behavior* 29 (4): 334–66.

Inkeles, Alex. 1960. 'Industrial man: The relation of status to experience, perception and value'. *American Journal of Sociology* 66 (1): 1–31.

———. 1975. 'Becoming modern: Individual change in six developing countries'. *Ethos* 3 (2): 342–3.

Ipsos Reid. 2010. 'Weekly Internet usage overtakes television watching'. www.ipsos-na.com/news-polls/pressrelease.

Irvine, Janice. 2003. '"The sociologist as voyeur": Social theory and sexual research, 1910–1978'. *Qualitative Sociology* 26 (4): 429–56.

Irwin, Alan. 2001. *Sociology and the Environment*. Cambridge: Polity Press.

Irwin, K. 2003. 'Saints and sinners: Elite tattoo collectors and tattooists as positive and negative deviants'. *Sociological Spectrum* 23: 27–57.

Isajiw, Wsevolod. 1999. *Understanding Diversity: Ethnicity and Race in the Canadian Context*. Toronto: Thompson Educational Publishing.

Jackson, Andrew. 2009. *Work and Labour in Canada: Critical Issues*. 2nd edn. Toronto: Canadian Scholars' Press.

———, and David Robinson. 2000. *Falling Behind: The State of Working Canada, 2000*. Ottawa: Canadian Centre for Policy Alternatives.

Jackson, K.T. 1985. *Crabgrass Frontier: The Suburbanization of the United States*. New York: Oxford University Press.

Jackson, Sue. 2005. '"I'm 15 and desperate for sex": "Doing" and "undoing" desire in letters to a teenage magazine'. *Feminism and Psychology* 15 (3): 295–313.

Jacobs, Katrien. 2010. 'Lizzy Kinsey and the Adult Friendfinders: An ethnographic study of Internet sex and pornographic self-display in Hong Kong'. *Culture, Health and Sexuality* 12 (6): 691–703.

Jakobsen, Janet R., and Ann Pellegrini. 2004. *Love the Sin: Sexual Regulation and the Limits of Religious Tolerance*. New York: Beacon Press.

James, Carl E. 2003. 'Schooling, basketball and US scholarship aspirations of Canadian student athletes'. *Race, Ethnicity and Education* 6 (2): 123–44.

James, Daniel Lee, and Elizabeth A. Craft. 2002. 'Protecting one's self from a stigmatized disease . . . once one has it'. *Deviant Behavior* 23: 267–99.

James, William Closson. 2006. 'Dimorphs and cobbles: Ways of being religious in Canada'. In L.G. Beaman, ed., *Religion and Canadian Society: Traditions, Transitions and Innovations*, 119–31. Toronto: Canadian Scholars' Press.

Jamieson, Lynn. 1998. *Intimacy: Personal Relationships in Modern Societies*. Cambridge: Polity Press.

Janis, Irving Lester. 1982. *Groupthink: Psychological Studies of Policy Decisions and Fiascoes*. 2nd edn. Boston: Houghton Mifflin.

Jenkins, J. Craig. 1983. 'Resource mobilization theory and the study of social movements'. *Annual Review of Sociology* 9: 527–53.

Jenkins, Philip. 1994. *Using Murder: The Social Construction of Serial Homicide*. New York: Aldine de Gruyter.

Jenson, Jane. 2004. *Canada's New Social Risks: Directions for a New Social Architecture*. CPRN Social Architecture Papers, Research Report F-43. Ottawa: Canadian Policy Research Networks.

———, and Martin Papillon. 2000. 'Challenging the citizenship regime: The James Bay Crees and transnational action'. *Politics and Society* 28 (2): 245–64.

———, and Denis Saint-Martin. 2003. 'New routes to social cohesion? Citizenship and the social investment state'. *Canadian Journal of Sociology* 28 (1): 77–99.

Jessop, B. 1997. 'A neo-Gramscian approach to the regulation of urban regimes: Accumulation strategies, hegemonic projects and governance'. In M. Lauria, ed., *Reconstructing Urban Regime Theory: Regulating Urban Politics in a Global Economy*, 51–73. Thousand Oaks, CA: Sage.

Jha, Prabhat, Rajesh Kumar, Priya Vasa, N. Dhingra, D. Thiruchelvam, and R. Maoineddin. 2006. 'Low male-to-female sex ratio of children born in India: National survey of 1.1 million households'. *Lancet* DOI: 10.1016/So14-6736(06). Published online, 9 January 2006, www.lancet.com.

Joanisse, Leanne. 2005. '"This is who I really am": Obese women's conceptions of self following weight loss surgery'. In Dorothy Pawluch, William Shaffir, and Charlene Miall, eds, *Doing Ethnography: Studying Everyday Life*, 248–59. Toronto: Canadian Scholars' Press.

Johnson, Carol. 2002. 'Heteronormative citizenship and the politics of passing'. *Sexualities* 5 (3): 317–36.

Johnson, Chalmers. 2007. *Nemesis: The Last Days of the American Republic*. New York: Metropolitan Books.

Johnson, Holly. 1996. *Dangerous Domains: Violence against Women in Canada*. Toronto: Nelson.

———. 2005. 'Assessing the prevalence of violence against women in Canada'. *Statistical Journal of the United Nations ECE* 22: 225–38.

Johnson, Terence. 1972. *Professions and Power*. London: Macmillan.

Johnston, Josée, and Shyon Bauman. 2009. *Foodies: Democracy and Distinction in the Gourmet Foodscape*. New York: Routledge.

Johnston, R.J., D. Gregory, and D.M. Smith, eds. 1994. *The Dictionary of Human Geography*. 3rd edn. Oxford: Blackwell.

Jones, Jennifer M., Susan Bennett, Marion P. Olmsted, Margaret L. Lawson, and Gary Rodin. 2001. 'Disordered eating attitudes and behaviours in teenaged girls: A school-based study'. *Canadian Medical Association Journal* 165: 547–52.

Joyce, James. 1968 [1922]. *Ulysses*. Harmondsworth, UK: Penguin.

Kachur, Jerrold L. 1999. 'Quasi-marketing education: The entrepreneurial state and charter schooling in Alberta'. In Dave Broad and Wayne Antony, eds, *Citizens or Consumers? Social Policy in a Market Society*, 129–50. Halifax: Fernwood.

———, and Trevor W. Harrison. 1999. 'Introduction: Public education, globalization, and democracy: Whither Alberta?' In Trevor W. Harrison and Jerrold L. Kachur, eds, *Contested Classrooms: Education, Globalization, and Democracy in Alberta*, xiii–xxxv. Edmonton: University of Alberta Press and Parkland Institute.

Kane, R.L., J.G. Evans, and D. MacFadyen. 1990. *Improving the Health of Older People: A World View*. Oxford: Oxford University Press.

Kannisto, Vaino, Jens Lauritsen, A.R. Thatcher, and J.W. Vaupel. 1994. 'Reflections in mortality at advanced ages: Several decades of evidence from advanced countries'. *Population and Development Review* 20: 793–810.

Kanter, Rosabeth Moss. 1977. *Men and Women of the Corporation*. New York: Basic Books.

Kanungo, Shivraj. 1998. 'An empirical study of organizational culture and network-based computer use'. *Computers in Human Behavior* 14 (1): 79–91.

Karakayali, Nedim. 2005. 'Duality and diversity in the lives of immigrant children: Rethinking the "problem of the second generation" in light of immigrant autobiographies'. *Canadian Review of Sociology and Anthropology* 42 (2): 325–43.

Karim, Karim H. 2002. 'Globalization, communication, and diaspora'. In Paul Attallah and Leslie Regan Shade, eds, *Mediascapes: New Patterns in Canadian Communication*, 272–94. Scarborough, ON: Thomson Nelson.

———. 2008. 'Press, public sphere, and pluralism: Multiculturalism debates in Canadian English-language newspapers'. *Canadian Ethnic Studies* 40 (1): 57–78.

Karmis, Demetrios. 2004. 'Pluralism and national identity(ies) in contemporary Quebec: Conceptual clarifications, typology, and discourse analysis'. In Alain-G. Gagnon, ed., *Quebec: State and Society*, 69–96. Peterborough, ON: Broadview.

Kart, C.S., and C.F. Longino. 1987. 'The support systems of older people: A test of the exchange paradigm'. *Journal of Aging Studies* 1: 239–51.

Käsler, Dirk. 1988. *Max Weber: An Introduction to His Life and Work*. Chicago: University of Chicago Press.

Kasper, Anne S., and Susan J. Ferguson, eds. 2000. *Breast Cancer: Society Shapes an Epidemic*. New York: St Martin's.

Kassebaum, Donald G., and Ellen R. Cutler. 1998. 'On the culture of student abuse in medical school'. *Academic Medicine* 73: 1149–58.

Katz, M. 1989. *The Undeserving Poor*. New York: Pantheon.

Katz, S. 2000. 'Busy bodies: Activity, aging and the management of everyday life'. *Journal of Aging Studies* 14 (2): 135–52.

———. 2006. 'From chronology to functionality: Critical reflections on the gerontology of the body'. In J. Baars, D. Dannefer, and C. Phillipson, eds, *Aging, Globalization and Inequality: The New Critical Gerontology*. New York: Baywood.

Katzmarzyk, Peter T. 2002. 'The Canadian obesity epidemic: 1995–1998'. *Canadian Medical Association Journal* 166: 1039–40.

Kearney, Patrick. 1982. *A History of Erotic Literature*. Hong Kong: Parragin Books.

Keating, N. 2008. *Rural Ageing: A Good Place to Grow Old?* London: Policy Press.

———, J. Swindle, and D. Foster. 2005. 'The role of social capital in aging well'. In *Social Capital in Action: Thematic Policy Studies*, 24–48. Ottawa: Policy Research Initiative.

Keck, Margaret, and Katherine Sikkink. 1998. *Activists beyond Borders: Advocacy Networks in International Politics*. New York: Cornell University Press.

Kelley, Maryellen R., and Susan Helper. 1997. *Interorganizational Learning and the Environment: The Influences of Regional Agglomeration and Local Institutional Linkages on the Adoption of New Technologies*. Paper presented at the annual meeting of the American Sociological Association.

Kempadoo, Kamala. 1998. 'Introduction: Globalizing sex workers' rights'. In Kamala Kempadoo and Jo Doezema, eds, *Global Sex Workers: Rights, Resistance, and Redefinition*, 1–28. New York: Routledge.

———, and Jo Doezema, eds. 1998. *Global Sex Workers: Rights, Resistance, and Redefinition*. New York: Routledge.

Kennedy, Emily H., and Sara O'Shaughnessy. 2010. *Relational Activism: Gender and Environment Reconsidered*. Presentation to the Canadian Sociological Association, 31 May.

Kenway, Jane, and Helen Modra. 1992. 'Feminist pedagogy and emancipatory possibilities'. In Carmen Luke and Jennifer Gore, eds, *Feminisms and Critical Pedagogy*, 138–66. London: Routledge.

Kerley, Kent R., X. Xu, and B. Sirisunyaluck. 2008. 'Self-control, intimate partner abuse, and intimate partner victimization: Testing the general theory of crime in Thailand'. *Deviant Behavior* 29 (6): 503–32.

Kerr, Clark, Frederick Harbison, John Dunlop, and Charles Myers. 1960. *Industrialism and Industrial Man: The Problem of Labor and Management in Economic Growth*. Cambridge, MA: Harvard University Press

Kerstetter, Steve. 2009. *The Affordability Gap: Spending Differences between Canada's Rich and Poor*. Ottawa: Canadian Centre for Policy Alternatives.

Kessler, R.C., P. Berglund, O. Demler, R. Jin, K.R. Merikangas, and E.E. Walters. 2005. 'Lifetime prevalence and age-of-onset distributions of DSM-IV disorders in the National Comorbidity Survey replication'. *Archive of General Psychiatry* 62: 593–602.

Keyfitz, Nathan. 1968. *Introduction to the Mathematics of Population*. Reading, MA: Addison-Wesley.

Kiernan, Kathleen. 1997. *The Legacy of Parental Divorce: Social, Economic, and Demographic Experiences in Adulthood*. London: Centre for Analysis of Social Exclusion.

Killingsworth, B. 2007. '"Drinking stories" from a playgroup: Alcohol in the lives of middle-class mothers in Australia'. *Ethnography* 7 (3): 357–84.

Kim, Sujeong. 2004. 'Rereading David Morley's The "Nationwide" Audience'. *Cultural Studies* 18 (1): 84–108.

King, Anthony. 1995. 'Globalization, modernity and the spatialization of social theory: An introduction'. In M. Featherstone, S. Lash, R. Robertson, eds, *Global Modernities*. Thousand Oaks, CA: Sage.

King, Samantha. 2006. *Pink Ribbons, Inc.* Minneapolis: University of Minnesota Press.

———. 2010. 'Pink diplomacy: On the uses and abuses of breast cancer awareness'. *Health Communication* 25: 286–9.

Kinsey, Alfred, Wardell Pomeroy, and Clyde Martin. 1948. *Sexual Behavior in the Human Male*. Philadelphia: W.B. Saunders.

———, and Paul Gebhard. 1953. *Sexual Behavior in the Human Female*. Philadelphia: W.B. Saunders.

Kinsman, Gary, and Patrizia Gentile. 2010. *The Canadian War on Queers*. Vancouver: UBC Press.

Kirk, Dudley. 1998. 'Demographic transition theory'. *Population Studies* 50: 361–87.

Kiser, Edgar, and Joachim Schneider. 1995. 'Rational choice versus cultural explanations of the efficiency of the Prussian tax system'. *American Sociological Review* 60: 787–91.

Kitschelt, Herbert. 1993. 'Social movements, political parties, and democratic theory'. *Annals of the American Academy of Political and Social Science* 528 (July): 13–29.

Kivimaki, M., T. Feldt, J. Vahtera, and J.E. Nurmi. 2000. 'Sense of coherence: Evidence from 2 cross-lagged longitudinal samples'. *Social Science and Medicine* 50 (4): 583–97.

Klasen, Stephan. 2003. 'Sex selection'. In Paul Demeny and Geoffrey McNicoll, eds, *Encyclopedia of Population*, 879–81. New York: Macmillan Reference USA, Thomson/Gale.

Klein, David M., and James M. White. 1996. *Family Theories: An Introduction*. Thousand Oaks, CA: Sage.

Klein, Naomi. 2002. *No Logos: No Space, No Choice, No Jobs*. New York: Picador.

———. 2008. *The Shock Doctrine: The Rise of Disaster Capitalism*. Toronto: Vintage Canada.

Knight, Rolf. 1996. *Indians at Work: An Informal History of Native Labour in British Columbia, 1858–1930*. Vancouver: New Star.

Kohlberg, Lawrence. 1969. 'Stage and sequence: The cognitive-development approach to socialization'. In David A. Goslin, ed., *Handbook of Socialization: Theory and Research*, 347–80. Chicago: Rand McNally.

———. 1975. 'Moral education for a society in moral transition'. *Educational Leadership* 33: 46–54.

Kohn, Melvin. 1969. *Class and Conformity*. Homewood, IL: Dorsey.

Kong, Rebecca, Holly Johnson, Sara Beattie, and Andrea Cardillo. 2003. 'Sexual offences in Canada'. *Juristat* 23 (6). Catalogue no. 85–002–XIE. Ottawa: Statistics Canada.

Krahn, Harvey J., Graham S. Lowe, and Karen D. Hughes. 2007. *Work, Industry, and Canadian Society*. 5th edn. Toronto: Thomson Nelson.

Kramer, Sebastian. 2000. 'The fragile male'. *British Medical Journal* 321 (23–30 December): 1609–12.

Krane, Julia. 2003. *What's Mother Got to Do with It? Protecting Children from Sexual Abuse*. Toronto: University of Toronto Press.

Krieger, Joel. 2001. *Oxford Companion to Politics of the World*. 2nd edn. Oxford: Oxford University Press.

Kriesel, Warren, Terrence J. Centner, and Andrew Keeler. 1996. 'Neighborhood exposure to toxic releases: Are there racial inequities?' *Growth and Change* 27: 479–99.

Kuhn, Manford H., and Thomas S. McPartland. 1954. 'An empirical investigation of self-attitudes'. *American Sociological Review* 19: 68–76.

Kuhn, Thomas S. 1970. *The Structure of Scientific Revolutions*. 2nd edn. Chicago: University of Chicago Press.

Kymlicka, Will. 1998. 'The theory and practice of Canadian multiculturalism'. *Canadian Federation of the Social Sciences and Humanities* 23 (Nov.): 1–10. www.fedcan.ca/english/fromold/breakfast-kymlicka1198.cfm.

Laliberté, P. 2002. 'Un développement urbain pour réduire concrètement la dépendance à l'automobile'. *VertigO* 3 (2), online.

Lamptey, Peter, Merywen Wigley, D. Carr, and Y. Collymore. 2002. 'Facing the HIV/AIDS epidemic'. *Population Bulletin* 57 (3).

Lancet. 2003. 'Slavery today'. 361: 2093.

Landes, David S. 1999. *The Wealth and Poverty of Nations: Why Some Are So Wealthy and Others Poor*. New York: W.W. Norton.

Langdridge, Darren. 2006. 'Voices from the margins: Sadomasochism and sexual citizenship'. *Citzenship Studies* 10 (4): 373–89.

Langlois, S., and P. Morrison. 2002. 'Suicide deaths and suicide attempts'. *Health Reports* 13 (2): 9–22.

Latour, Bruno, and Steve Woolgar. 1987. *Laboratory Life: The Construction of Scientific Fact*. 2nd edn. Princeton, NJ: Princeton University Press.

Laufer, William S., and Freda Adler. 1994. *The Legacy of Anomie Theory: Advances in Criminological Theory*. New Brunswick, NJ: Transaction.

Lawr, Douglas, and Robert Gidney, eds. 1973. *Educating Canadians: A Documentary History of Public Education*. Toronto: Van Nostrand Reinhold.

Laws, G. 1995. 'Understanding ageism: Lessons from feminism and postmodernism'. *The Gerontologist* 35 (1): 112–18.

Leavitt, L., and N. Fox, eds. 1993. *The Psychological Effects of War and Violence on Children*. Mahwah, NJ: Lawrence Erlbaum Associates.

Le Corbusier. 1972 [1941]. *La Charte d'Athènes*. Paris: Seuil Points.

Lee, Everet. 1966. 'A theory of migration'. *Demography* 3: 47–57.

Lee, Ronald D. 2007. *Global Population Aging and its Economic Consequences*. Henry Wendt Lecture Series. Washington: AEI Press, Publisher for the American Enterprise Institute.

Lefèvre, C. 2009. *Gouverner les métropoles*. Paris: LGDJ.

Lefevre, Solange. 2008. 'Between law and public opinion: The case of Québec'. In Lori G. Beaman and Peter Beyer, eds, *Religion and Diversity in Canada*. Leiden: Brill.

Le Galès, P. 1993. *Politique urbaine et développement local: une comparaison franco-britannique*. Paris: L'Harmattan.

———. 1995. 'Du gouvernement des villes à la gouvernance urbaine'. *Revue française de science politique* 45 (1): 57–95.

Lehmann, Wolfgang. 2007. '"I just didn't feel like I fit in": The role of habitus in university drop-out decisions'. *Canadian Journal of Higher Education* 37 (2): 89–110.

Lemert, Edwin. 1951. *Social Pathology: A Systematic Approach to the Theory of Sociopathic Behavior*. New York: McGraw-Hill.

Lesthaeghe, Ron. 2010. 'The unfolding story of the second demographic transition'. *Population and Development Review* 36 (2): 211–51.

———, and Johan Surkyn. 1988. 'Cultural dynamics and economic theories of fertility change'. *Population and Development Review* 14: 1–45.

Levin, Benjamin. 2007. 'Schools, poverty, and the achievement gap'. *Phi Delta Kappan* 89 (1): 75–6.

———, and J. Anthony Riffel. 1997. *Schools and the Changing World: Struggling toward the Future*. London: Falmer.

Levitas, Ruth. 1998. *The Inclusive Society? Social Exclusion and New Labour*. Basingstoke, UK: Macmillan.

Levitt, Cyril. 1994. 'Is Canada a racist country?' In Sally F. Zerker, ed., *Change and Impact: Essays in Canadian Social Sciences*, 304–16. Jerusalem: Magnes Press, Hebrew University.

Levy, B., and M.R. Banaji. 2002. 'Implicit ageism'. In T. Nelson, ed., *Ageism: Stereotyping and Prejudice against Older Persons*, 49–75. Cambridge, MA: MIT Press.

Lewchuk, Wayne, and David Robertson. 2006. 'Listening to workers: The reorganization of work in the Canadian motor vehicle industry'. In Vivian Shalla, ed., *Working in a Global Era: Canadian Perspectives*, 53–73. Toronto: Canadian Scholars' Press.

Lewicki, Roy J., and Barbara Benedict Bunker. 1996. 'Developing and maintaining trust in work relationships'. In Roderick M. Kramer and Tom R. Tyler, eds, *Trust in Organizations: Frontiers of Theory and Research*, 114–39. Thousand Oaks, CA: Sage.

Ley, D. 1999. 'Myths and meanings of immigration and the metropolis'. *Canadian Geographer/Le géographe canadien* 43: 2–19.

Li, Peter. 1988. *Ethnic Inequality in a Class Society*. Toronto: Thompson Educational Publishing.

———. 1992. 'Race and gender as bases of class fractions and the effects on earnings'. *Canadian Review of Sociology and Anthropology* 29 (4): 488–510.

———, ed. 1999. *Race and Ethnic Relations in Canada*. 2nd edn. Toronto: Oxford University Press.

———. 2003. *Destination Canada: Immigration Debates and Issues*. Toronto: Oxford University Press.

Lian, Jason Z., and Ralph Matthews. 1998. 'Does the vertical mosaic still exist? Ethnicity and income in Canada, 1991'. *Canadian Review of Sociology and Anthropology* 35 (4): 461–81.

Liebow, Elliot. 1993. *Tell Them Who I Am: The Lives of Homeless Women*. New York: Free Press.

Limoncelli, Stephanie. 2009. 'The trouble with trafficking: Conceptualizing women's sexual labor and economic human rights'. *Women's Studies International Forum* 32 (4): 261–9.

Lindsay, Colin. 2008a. 'Canadians attend weekly religious services less than 20 years ago'. Catalogue no. 89-630-X. *The General Social Survey*, Matter of Fact no. 3. Ottawa: Statistics Canada.

———. 2008b. 'Are women spending more time on unpaid domestic work than men in Canada?' Catalogue 89-630-X. *The General Social Survey*, Matter of Fact. Ottawa: Statistics Canada.

Lindsey, L., and S. Beach. 2003. *Essentials of Sociology*. Upper Saddle River, NJ: Prentice-Hall.

Linteau, P.A. 1994. *Histoire du Canada*. Paris: PUF.

Liodakis, Nikolaos. 1998 'The activities of Hellenic-Canadian secular organizations in the context of Canadian multiculturalism'. *Études helléniques/Hellenic Studies* 6 (1): 37–58.

———. 2002. 'The vertical mosaic within: Class, gender and nativity within ethnicity' (PhD dissertation, McMaster University, Hamilton, ON).

———. 2009. 'The social class and gender differences within Aboriginal groups in Canada: 1995–2000'. In Dan Beavon and Daniel Jetté, eds, 'Journeys of a generation: Broadening the Aboriginal well-being policy research agenda'. *Canadian Issues Journal* (winter): 93–7. Montreal: Association for Canadian Studies.

———, and Victor Satzewich. 2003. 'From solution to problem: Multiculturalism and "race relations" as new social problems'. In Wayne Antony and Les Samuelson, eds, *Power and Resistance: Critical Thinking about Canadian*

Social Issues, 3rd edn, 145–68. Halifax: Fernwood.

Lipman, Ellen L., David R. Offord, and Martin D. Dooley. 1996. 'What do we know about children from single-parent families? Questions and answers from the National Longitudinal Survey on Children'. In *Growing Up in Canada*, 83–91. Ottawa: Human Resources Development Canada.

Lipset, Seymour Martin. 1963. 'The value patterns of democracy: A case study in comparative analysis'. *American Sociological Review* 28 (4): 515–31.

———. 1964. 'Canada and the United States: A comparative view'. *Canadian Review of Sociology* 1 (4): 173–85.

———. 1990. *Continental Divide: The Values and Institutions of the United States and Canada*. New York: Routledge.

———. 1991. 'Canada and the United States: The great divide'. *Current History* 90 (560): 432–7.

Livingstone, D.W. 2004. *The Education–Jobs Gap: Underemployment or Economic Democracy*. Aurora, ON: Garamond.

Lizardo, Omar, and Sara Skiles. 2009. 'Highbrow omnivorousness on the small screen? Cultural industry systems and patterns of cultural choice in Europe'. *Poetics* 37: 1–23.

Logan, John, and Harvey Molotch. 1987. *Urban Fortunes: The Political Economy of Place*. Berkeley: University of California Press.

Lomborg, Bjorn. 2001. *The Skeptical Environmentalist*. Cambridge: Cambridge University Press.

Lopez, S.H., R. Hodson, and V.J. Roscigno. 2009. 'Power, status, and abuse at work: General and sexual harassment compared'. *Sociological Quarterly* 50 (1): 3–27.

Lorimer, Rowland, and Mike Gasher. 2001. *Mass Communication in Canada*. 4th edn. Toronto: Oxford University Press.

Loseke, Donileen R. 2003. *Thinking about Social Problems: An Introduction to Constructionist Perspectives*. New York: Aldine de Gruyter.

———. 2003. 'Constructing conditions, people, morality and emotion: Expanding the agenda of constructionism'. In James A. Holstein and Gale Miller, eds, *Challenges and Choices: Constructionist Perspectives on Social Problems*, 120–9. New York: Aldine de Gruyter.

———. 2009. 'Examining emotion as discourse: Emotion codes and presidential speeches justifying war'. *Sociological Quarterly* 50 (3): 497–524.

———, and Spencer E. Cahill. 1984. 'The social construction of deviance: Experts on battered women'. *Social Problems* 31 (3): 296–310.

Lowe, Graham S. 2000. *The Quality of Work: A People-Centred Agenda*. Toronto: Oxford University Press.

———. 2007. *21st Century Job Quality: Achieving What Canadians Want*. Ottawa: Canadian Policy Research Networks. www.cprn.org.

Lu, Yuqian, and René Morisette. 2010. 'Women's participation and economic downturns'. Catalogue 75-001-X. *Perspectives* (May): 18–22. Ottawa: Statistics Canada.

Luckenbill, David F. 1977. 'Criminal homicide as a situational transaction'. *Social Problems* 25: 176–86.

Lukes, Steven. 1972. *Émile Durkheim: His Life and Work*. New York: Harper and Row.

———. 1974. *Power: A Radical View*. London: Macmillan.

Lutz, Wolfgang, ed. 1994. *The Future Population of the World: What Can We Assume Today?* London: Earthscan.

Luxton, Meg. 1980. *More Than a Labour of Love*. Toronto: Women's Press.

———, and June Corman. 2001. *Getting By in Hard Times: Gendered Labour at Home and on the Job*. Toronto: University of Toronto Press.

Lynch, Kathleen. 1989. *The Hidden Curriculum: Reproduction in Education, A Reappraisal*. London: Falmer.

Lynch, Michael. 1985. 'Discipline and the material form of images: An analysis of scientific visibility'. *Social Studies of Science* 15 (1): 37–66.

Lynn, Michael. 2009. 'Determinants and consequences of female attractiveness and sexiness: Realistic tests with restaurant waitresses'. *Archives of Sexual Behavior* 38: (5) 737–45.

McAdam, Doug. 1982. *Political Process and the Development of Black Insurgency*. Chicago: University of Chicago Press.

———. 1983. 'Tactical innovation and the pace of insurgency'. *American Sociological Review* 48 (6): 735–54.

———. 1996. 'Conceptual origins, current problems, future directions'. In Doug McAdam, John McCarthy, and Mayer Zald, eds, *Comparative Perspectives on Social Movements*, 23–40. New York: Cambridge University Press.

McAdam, Doug, John McCarthy, and Mayer Zald, eds. 1996. *Comparative Perspectives on Social Movements*. New York: Cambridge University Press.

McAdam, Doug, Sidney Tarrow, and Charles Tilly. 2001. *Dynamics of Contention*. Cambridge: Cambridge University Press.

McCarthy, John D. 1996. 'Constraints and opportunities in adopting, adapting, and inventing'. In Doug McAdam, John McCarthy, and Mayer Zald, eds, *Comparative Perspectives on Social Movements*, 141–51. New York: Cambridge University Press.

———, and Mayer N. Zald. 1977. 'Resource mobilization and social movements: A partial theory'. *American Journal of Sociology* 82 (6): 1212–41.

McCarthy, Kevin R. 2001. *World Population Shifts: Boom or Doom?* Santa Monica, CA: Rand.

McCary, James Leslie. 1967. *Human Sexuality: A Contemporary Marriage Manual*. Toronto: D. Van Nostrand.

McClelland, David. 1961. *The Achieving Society*. Princeton, NJ: Van Nostrand.

McCleneghan, Sean. 2003. 'Selling sex to college females: Their attitudes about *Cosmopolitan* and *Glamour* magazines'. *Social Science Journal* 40: 317–25.

McClung, Nellie. 1971. *In Times Like These*. Toronto: University of Toronto Press.

Maccoby, Eleanor, and Carole Jacklin. 1974. *The Psychology of Sex Differences*. Stanford, CA: Stanford University Press.

MacCourt, P. 2005. *Brief to the Senate Standing Committee on Social Affairs, Science and Technology*. Vancouver.

McCurry, Justin. 2004. 'Smuggling for sex'. *Lancet* 364: 1393–4.

Macdonald, David, and Steven Staples. 2008. *The Cost of the War and the End of Peacekeeping: The Impact of Extending the Afghanistan Mission*. Ottawa: Rideau Institute.

MacDonald, John A., and L.D. MacDonald. 1964. 'Chain migration, ethnic neighborhood formation and social networks'. *Milbank Memorial Fund Quarterly* 42: 82–7.

McDonald, L. 2006. 'Gendered retirement: The welfare of women and the "new" retirement'. In L. Stone, ed., *New Frontiers of Research on Retirement*, 137–64. Ottawa: Statistics Canada.

McDonald, L., P. Donahue, J. Janes, and L. Cleghorn. 2006. *In from the Streets: The Health and Well Being of Homeless Older Adults*. Toronto: University of Toronto Press.

McDonald, L., and A.L. Robb. 2004. 'The economic legacy of divorce and separation for women in old age'. *Canadian Journal on Aging* 23 (suppl. 1): S83–S97

McDonald, L., T. Sussman, and P. Donahue. 2007. *When Bad Things Happen to Good People: The Economic Consequences of Retiring to Caregive*. Hamilton, ON: SEDAP, McMaster University.

Macek, S. 2006. *Urban Nightmares*. Minneapolis: University of Minnesota Press.

Macfarlane, Alan. 1997. *The Savage Wars of Peace: England, Japan and the Malthusian Trap*. Oxford: Blackwell.

McGilly, Frank. 1998. *An Introduction to Canada's Public Social Services: Understanding Income and Health Programs*. 2nd edn. Toronto: Oxford University Press..

McGranathan, G., and D. Satterthwaite. 2003. 'Urban centers: An assessment of sustainability'. *Annual Review of Environmental Resources* 28: 243–74.

McGuire, Meredith. 2005. *Rethinking Sociology's Sacred/Profane Dichotomy: Historically Contested Boundaries in Western Christianity*. Paper presented at SISR/ISSR, Zagreb.

———. 2008. *Lived Religion: Faith and Practice in Everyday Life*. Oxford: Oxford University Press.

McKay, Alexander. 2004. 'Adolescent sexual and reproductive health in Canada: A report card in 2004'. *Canadian Journal of Human Sexuality* 13 (2): 67–81.

McKenna, K., A. Green, and M. Gleason. 2002. 'Relationship formation on the Internet: What's the big attraction?' *Journal of Social Issues* 58 (1): 9–31.

Mackenzie, Hugh. 2007. *Timing Is Everything: Comparing the Earnings of Canada's Highest-Paid CEOs and the Rest of Us*. Toronto: Canadian Centre for Policy Alternatives.

McKeown, Thomas. 1976. *The Modern Rise of Population*. London: Edward Arnold.

Mackie, Marlene. 1990. 'Socialization'. In Robert Hagedorn, ed., *Sociology*, 4th edn, 61–96. Toronto: Holt, Rinehart and Winston.

Macklin, Audrey. 1992. 'Foreign domestic worker: Surrogate housewife or mail order bride?' *McGill Law Journal* 37: 681–760.

McLaren, Angus. 1990. *Our Own Master Race: Eugenics in Canada, 1885–1945*. Toronto: McClelland and Stewart.

McLaren, Peter, and Joe L. Kincheloe, eds. 2007. *Critical Pedagogy: Where Are We Now?* New York: Peter Lang.

McLuhan, Marshall. 1964. *Understanding Media: The Extensions of Man*. New York: McGraw Hill.

McNabola, A., and L.W. Gill. 2009. 'The control of environmental tobacco smoke: A policy review'. *International Journal of Environmental Research and Public Health* 6 (2): 741–58.

McNally, David. 2002. *Another World Is Possible: Globalization and Anticapitalism*. Winnipeg: Arbeiter Ring Publishing.

McQuaig, Linda. 2009. 'Ever upward trend for bankers' pay'. *Toronto Star* 20 October: A19.

———, and Neil Brooks. 2010. *The Trouble with Billionaires*. Toronto: Viking Canada.

MacQueen, Ken. 2009. 'Making their bed: Some 16 groups take sides on polygamy in landmark case'. *Maclean's*, www2.macleans.ca/2010/03/17/making-their-bed/#more-115096.

McRoberts, Kenneth, and Dale Postgate. 1980. *Quebec: Social Change and Political Crisis*. Toronto: McClelland and Stewart.

McVey, Wayne, Jr, and Warren E. Kalbach. 1995. *Canadian Population*. Toronto: Nelson.

Madigan, Francis C. 1957. 'Are sex mortality differentials biologically caused?' *Milbank Memorial Fund Quarterly* 35 (2): 202–23.

Magder, Ted. 1993. *Canada's Hollywood: The Canadian State and Feature Films*. Toronto: University of Toronto Press.

———, and Jonathan Burston. 2001. 'Whose Hollywood? Changing forms and relations inside the North American entertainment economy'. In Vincent Mosco and Dan Schiller, eds, *Continental Order? Integrating North America for Cybercapitalism*, 207–34. Lanham, MD: Rowman and Littlefield.

Magnus, George. 2009. *The Age of Aging: How Demographics Are Changing the Global Economy and Our World*. Singapore: John Wiley and Sons (Asia).

Magnusson, Warren. 1990. 'Critical social movements: Decentring the state'. In Alain G. Gagnon and James Bickerton, eds, *Canadian Politics: An Introduction*, 525–41. Peterborough, ON: Broadview.

Maguire, Patrick. 2006. *Choice in Urban School Districts: The Edmonton Experience*. Kelowna, BC: Society for the Advancement of Excellence in Education.

Maharaj, N.R., A. Dhai, R. Wiersma, and J. Moodley. 2005. 'Intersex conditions in children and adolescents: Surgical, ethical, and legal considerations'. *Journal of Pediatric and Adolescent Gynecology* 18 (6): 399–402.

Mahtani, Minelle. 2001. 'Representing minorities: Canadian media and minority identities'. *Canadian Ethnic Studies* 33 (3): 99–133.

Maier, R., W. De Graaf, and P. Frericks. 2007. 'Pension reforms in Europe and life course politics'. *Social Policy and Administration* 41 (5): 487–504.

Malach, F., D.K. Conn, and K. Le Clair. 2005. *Written Submission to the Standing Senate Committee on Social Affairs, Science and Technology*.

Malthus, Thomas R. 1970 [1798]. *An Essay on the Principle of Population*. Harmondsworth, UK: Penguin.

Mancebo, F. 2006. *Le développement durable*. Paris: Armand Colin.

Mankoff, Milton. 1971. 'Societal reaction and career deviance: A critical analysis'. *Sociological Quarterly* 12: 204–18.

Mann, Michael. 1999. *The Dark Side of Democracy*. Cambridge: Cambridge University Press.

———. 2004. *Fascists*. Cambridge: Cambridge University Press.

Manton, Kenneth G. 2008. 'Recent declines in chronic disability in the elderly U.S. population: Risk factors and future dynamics'. *Annual Review of Public Health* 29: 91–113.

Manzer, Ronald. 1994. *Public Schools and Political Ideas: Canadian Educational Policy in Historical Perspective*. Toronto: University of Toronto Press.

Marcil-Gratton, Nicole. 1998. *Growing up with Mom and Dad? The Intricate Family Life Courses of Canadian Children*. Ottawa: Statistics Canada.

Marcus, Sheron. 2005. 'Queer theory for everyone: A review essay'. *Signs* 31 (1): 191–218.

Markin, K.M. 2005. 'Still crazy after all these years: The enduring defamatory power of mental disorder'. *Law and Psychology Review* 29: 155–85.

Marmot, Michael G., Geoffrey Rose, Martin Shipley, and P.J. Hamilton. 1978. 'Employment grade and coronary heart disease in British civil servants'. *Journal of Epidemiological Community Health* 32: 244–9.

Marmot, Michael G., George Davey Smith, Stephen Stansfeld, Chandra Patel, Fiona North, Jenny Head, Ian White, Eric Brunner, and Amanda Feeney. 1991. 'Health inequalities among British civil servants: The Whitehall II Study'. *Lancet* 337: 1387–93.

Marples, James A. 2010. 'If regulated, plural marriage might not be harmful'. *Vancouver Sun*, www.vancouversun.com/life/regulated+plural+marriage+might+harmful/3331184/story.html#ixzz0zAU0M5m3.

Marsden, William. 2007. *Stupid to the Last Drop: How Alberta Is Bringing Environmental Armageddon to Canada*. Toronto: Alfred A. Knopf.

Martens, A., J.L. Goldenberg, and J. Greenberg. 2005. 'A terror management perspective on ageism'. *Journal of Social Issues* 61 (2): 223–39.

Martin, Daniel D. 2010. 'Identity management of the dead: Contests in the construction of murdered children'. *Symbolic Interaction* 33 (1): 18–40.

Marx, Karl. 1959 [1844]. 'Excerpt from *Towards the Critique of Hegel's Philosophy of Right*'. In Lewis Feuer, ed., *Marx and Engels: Basic Writings on Politics and Philosophy*, 262–6. Garden City, NY: Anchor Books.

———. 1976 [1867]. *Capital*, v. 1. Harmondsworth, UK: Penguin.

———. 1887. *Capital: A Critical Analysis of Capitalist Production*, ed. Friedrich Engels. London: Lowry and Co.

———, and Friedrich Engels. 1985 [1848]. *The Communist Manifesto*. New York: Penguin Books.

Maslovski, Mikhail. 1996. 'Max Weber's concept of patrimonialism and the soviet system'. *Sociological Review* 44: 294–308.

Mason-Schrock, Douglas. 1996. 'Transsexuals' narrative construction of the "true self"'. *Social Psychology Quarterly* 59 (3): 176–92.

Massey, Douglas S., Joaquin Arango, Graeme Hugo, Ali Kouauci, Adela Pellegrino, and J. Edward Taylor. 1993. 'Theories of international migration: A review and appraisal'. *Population and Development Review* 19: 431–65.

Maticka-Tyndale, Eleanor. 2001. 'Twenty years in the AIDS pandemic: A place for sociology'. *Current Sociology* 49 (6): 13–21.

Matza, D., and Gresham Sykes. 1957. 'Techniques of neutralization: A theory of delinquency'. *American Sociological Review* 5: 1–12.

Maume, David J. 2006. 'Gender differences in taking vacation time'. *Work and Occupations* 33 (2): 161.

Mauss, M. 1978. *Sociologie et anthropologie*. Paris: Presses universitaires de France.

Maxim, Paul S., and Paul C. Whitehead. 1998. *Explaining Crime*. 4th edn. Newton, MA: Butterworth-Heinemann.

May, T., and B. Perry. 2005. 'The future of urban sociology'. *Sociology* 39 (2): 343–70.

Mead, George Herbert. 1934. *Mind, Self, and Society from the Standpoint of a Social Behaviorist*. Chicago: University of Chicago Press.

Mead, Margaret. 1928. *Coming of Age in Samoa*. New York: William Morrow and Co.

———. 1935. *Sex and Temperament in Three Primitive Societies*. New York: Dell.

Meadows, Donella H., Dennis L. Meadows, and Jorgen Randers. 1992. *Beyond the Limits: Confronting Global Collapse, Envisioning a Sustainable Future*. Post Mills, VT: Chelsea Green.

———, and William H. Behrens III. 1972. *The Limits to Growth*. New York: Universe.

Meadows, Donella H., Jorgen Randers, and Dennis Meadows. 2005. *The Limits to Growth—The 30-Year Update*. London: Earthscan.

Meadwell, Hudson. 1993. 'The politics of nationalism in Quebec'. *World Politics* 45 (2): 203–41.

Mellor, J.R. 2007. *Urban Sociology in an Urbanized Society*. London: Routledge.

Melosi, M.V. 2000. *The Sanitary City: Urban Infrastructure in America from Colonial Times to the Present*. Baltimore: Johns Hopkins University Press.

Melucci, Alberto. 1989. *Nomads of the Present: Social Movements and Individual Needs in Contemporary Society*. Philadelphia: Temple University Press.

——— . 1996a. *Challenging Codes: Collective Action in the Information Age*. Cambridge: Cambridge University Press.

———. 1996b. *The Playing Self: Person and Meaning in the Planetary Society*. Cambridge: Cambridge University Press.

Ménard, A. Dana, and Peggy Kleinplatz. 2008. 'Twenty-one moves guaranteed to make his thighs go up in flames: Depictions of "great sex" in popular magazines'. *Sexuality and Culture* 12 (1): 1–20.

Mercer, J. 1991, 'The Canadian city in continental context: Global and continental perspectives on Canadian urban development'. In T. Bunting and P. Filion, eds, *Canadian Cities in Transition*, 45–68. Toronto: Oxford University Press.

Mertig, Angela, and Riley Dunlap. 2001. 'Environmentalism, new social movements and the new class: A cross-national investigation'. *Rural Sociology* 66 (1): 113–36.

———, and Denton Morrison. 2002. 'The environmental movement in the United States'. In Riley Dunlap and William Michelson, eds, *Handbook of Environmental Sociology*, 448–81. Westport, CT: Greenwood.

Merton, Robert K. 1938. 'Social structure and anomie'. *American Sociological Review* 3: 672–82.

———. 1957. *Social Theory and Social Structure*. Glencoe, IL/New York: Free Press.

Mertus, Julie. 2009. *The United Nations and Human Rights: A Guide for a New Era*. 2nd edn. New York: Routledge

Messner, Steven F., and Robert J. Sampson. 1991. 'The sex ratio, family disruption, and rates of violent crime: The paradox of demographic structure'. *Social Forces* 69 (3): 693–713.

Metchnikoff, E. 1903. *The Nature of Man*. New York: Putnam.

Methot, M. 2003. 'Herbert Brown James: Political reformer and enforcer'. *Urban History Review/Revue d'histoire urbaine* 31 (2), online.

Meyer, David, and Suzanne Staggenborg. 1996. 'Movements, countermovements and the structure of political opportunity'. *American Journal of Sociology* 101: 1628–60.

Meyer, John W., John Boli, George M. Thomas, and Francisco Ramirez. 1997. 'World society and the nation state'. *American Journal of Sociology* 103 (1): 144–81.

Meyer, Leisa. 2006. 'Sexual revolutions'. *OAH Magazine of History* 20 (2): 5–6.

Meyerson, Debra, Karl E. Weick, and Roderick M. Kramer. 1996. 'Swift trust and temporary groups'. In Roderick M. Kramer and Tom R. Tyler, eds, *Trust in Organizations: Frontiers of Theory and Research*, 166–95. Thousand Oaks, CA: Sage.

Michelson, William. 1976. *Man and His Urban Environment: A Sociological Approach*. Reading, MA: Addison-Wesley.

Micklin, Michael, ed. 1973. *Population, Environment and Social Organization: Current Issues in Human Ecology*. Hinsdale, IL: Dryden.

Mihorean, K. 2005. 'Trends in self-reported spousal violence'. In K. AuCoin, ed., *Family Violence in Canada: A Statistical Profile*. Ottawa: Canadian Centre for Justice Statistics, Statistics Canada.

Mikkonen, J., and D. Raphael. 2010. 'Social determinants of health: The Canadian facts'. Toronto: York University, Health Policy and Management. www.thecanadianfacts.org.

Miles, Robert, and Malcolm Brown. 2003. *Racism*. 2nd edn. London: Routledge.

Miles, Robert, and Rudy Torres. 1996. 'Does "race" matter? Transatlantic perspectives on racism after "race relations"'. In V. Amit-Talai and C. Knowles, eds, *Re-situating Identities: The Politics of Race, Ethnicity and Culture*, 24–46. Peterborough, ON: Broadview.

Miller, Gale, and James A. Holstein. 1993. *Constructionist Controversies: Issues in Social Problems Theory*. New York: Aldine de Gruyter.

Miller, J., and C.W. Mullins. 2007. 'The status of feminist theories in criminology'. In F.T. Cullen, J.P. Wright, and K. Blevins, eds, *Taking Stock: The Status of Criminological Theory*, v. 15, *Advances in Criminological Theory*. Piscataway, NJ: Transaction.

Miller, Toby, Nitin Govil, John McMurria, and Richard Maxwell. 2001. *Global Hollywood*. London: British Film Institute.

Miller-Young, Mireille. 2010. 'Putting hypersexuality to work: Black women and illicit eroticism in pornography'. *Sexualities* 13 (2): 219–35.

Millett, Kate. 1969. *Sexual Politics*. New York: Doubleday; Avon.

Mills, C. Wright. 1940. 'Situated actions and vocabularies of motive'. *American Sociological Review* 5 (6): 904–13.

———. 1956. *The Power Elite*. New York: Oxford University Press.

———. 1959. *The Sociological Imagination*. New York: Oxford University Press.

Mills, Martin. 2006. 'Sony-BMG annulment good for business'. *Billboard* 2 (September): 8.

Mills, Melinda, and Frank Trovato. 2001. 'The effect of pregnancy in cohabiting unions on marriage in Canada, the Netherlands, and Latvia'. *Statistical Journal of the United Nations Economic Commission for Europe* 18: 103–18.

Mincer, Jacob. 1978. 'Family migration decisions'. *Journal of Political Economy* 86: 749–73.

Minkler, M., and C.L. Estes. 1984. *Readings in the Political Economy of Aging*. New York: Baywood.

———, eds. 1999. *Critical Gerontology: Perspectives from a Political and Moral Economy*. New York: Baywood.

Miranda, Deborah A. 2010. 'Extermination of the Joyas'. *GLQ: A Journal of Lesbian and Gay Studies* 16 (1/2): 253–84.

Mitchell, Andrew, and Richard Shillington. 2002. 'Poverty, inequality and social inclusion'. In Working Paper Series, vii–18. Toronto: Laidlaw Foundation.

Mitchell, Barbara A. 2006. *The Boomerang Age: Transitions to Adulthood in Families*. New Brunswick, NJ: Aldine Transaction.

———. 2009. *Family Matters: An Introduction to Family Sociology in Canada*. Toronto: Canadian Scholars' Press.

Moen, P. 2001. 'The gendered life course'. In R. Binstock and L. George, eds, *Handbook of Aging and the Social Sciences*, 5th edn, 179–96. New York: Academic Press.

———. 2003. 'Midcourse: Navigating retirement and a new life stage'. In J.T. Mortimer and M.J. Shanahan, eds, *Handbook of the Life Course*, 269–91. New York: Kluwer Academic/Plenum Publishers.

Montpetit, Eric, Francesca Scala, and Isabelle Fortier. 2004. 'The paradox of deliberative democracy: The National Action Committee on the Status of Women and Canada's policy on reproductive technology'. *Policy Sciences* 37: 137–57.

Montpetit, Jonathan. 2010. 'Quebec woman barred from course for second time over refusal to remove niqab'. *Winnipeg Free Press*, www.winnipegfreepress.com/canada/breakingnews/quebec-government-kicks-niqab-wearing-woman-out-of-class-for-second-time-87125232.html.

Moodley, Kogila. 1983. 'Canadian multiculturalism as ideology'. *Ethnic and Racial Studies* 6 (3): 320–31.

Moore, Barrington, Jr. 1966. *Social Origins of Dictatorship and Democracy: Lord and Peasant in the Making of the Modern World*. Boston: Beacon.

Moore, Mike. 2009. *Saving Capitalism*. Singapore: John Wiley and Sons.

Morgan, Stephen, and Aage B. Sorensen. 1999. 'Parental networks, social closure and mathematics learning: A test of Coleman's social capital explanation of school effects'. *American Sociological Review* 64: 661–81.

Morland, Iain. 2005. '"The glans opens like a book": Writing and reading the intersexed body'. *Continuum: Journal of Media and Cultural Studies* 19 (3): 335–48.

Morley, David. 1980. *The 'Nationwide' Audience*. London: British Film Institute.

———. 2006. 'Unanswered questions in audience research'. *Communication Review* 9: 101–21.

Morris, Desmond. 1997. *The Human Sexes: A Natural History of Man and Woman*. New York: St Martin's.

Morris, R.G., and G.E. Higgins. 2009. 'Neutralizing potential and self-reported digital piracy: A multitheoretical exploration among college undergraduates'. *Criminal Justice Review* 34 (2): 173–95.

Morrison, Denton, and Riley E. Dunlap. 1986. 'Environmentalism and elitism: A conceptual and empirical analysis'. *Environmental Management* 10: 581–9.

Morrison, Todd, Shannon Ellis, Melanie Morrison, Anomi Bearben, and Rebecca Harriman. 2006. 'Exposure to sexually explicit material and variations in body esteem, genital attitudes and sexual esteem among a sample of Canadian men'. *Journal of Men's Studies* 14 (2): 209–22.

Mosco, Vincent. 1989. *The Pay-per Society: Computers and Communication in the Information Age*. Toronto: Garamond.

———. 2005. 'Here today, outsourced tomorrow: Knowledge workers in the global economy'. *Javnost—The Public* 12 (2): 39–55.

Moynihan, Ray, Iona Heath, and David Henry. 2002. 'Selling sickness: The pharmaceutical industry and disease mongering'. *British Medical Journal* 324: 886–91.

Mullen, F. 2000. 'Grandparents and welfare reform'. In C.B. Cox, ed., *To Grandmother's House We Go and Stay: Perspectives on Custodial Grandparents*, 113–31. New York: Springer.

Multani v. Commission scolaire Marguerite-Bourgeoys. 2006. 1 S.C.R. 256, 2006 SCC 6.

Mumford, L. 1961. *The City in History*. Harmondsworth, UK : Penguin.

Murphy, Elizabeth. 2000. 'Risk, responsibility and rhetoric in infant feeding'. *Journal of Contemporary Ethnography* 29: 291–325.

———. 2007. 'Images of Childhood in Mothers' Accounts of Contemporary Childrearing'. *Childhood*. 14 (1): 105–27. search1.scholarsportal.info/ids70/view_record.php?id = 2&recnum = 3&SID = dd5cc57671b4400ab13563fc5427839d.

Murphy, Michael. 1993. 'The contraceptive pill and women's employment as factors in fertility change in Britain, 1963–1980: A challenge to the conventional view'. *Population Studies* 7: 221–44.

Murphy, Raymond. 1994. 'The sociological construction of science without nature'. *Sociology* 28: 957–74.

Murray, Samantha. 2009. 'Within or beyond the binary/boundary? Intersex infants and parental decisions'. *Australian Feminist Studies* 24 (60): 265–74.

Mustard, Fraser. 1999. 'Health care and social cohesion'. In Daniel Drache and Terry Sullivan, eds, *Market Limits in Health Reform: Public Success, Private Failure*, 329–50. London: Routledge.

Myer, John, John Boli, George Thomas, and Francisco Ramirez. 1997. 'World society and the nation state'. *American Journal of Sociology* 103 (1): 144–81.

Myles, J. 1984. *Old Age in the Welfare State: The Political Economy of Public Pensions*. Boston: Little Brown.

———. 2005. 'What justice requires: A normative foundation for U.S. pension reform'. In R. Hudson, ed., *The New Politics of Old Age Policy*, 42–64. Baltimore: Johns Hopkins University Press.

———. 2006. 'From pension policy to retirement policy: Towards a new social agenda?' In L. Stone, ed., *New Frontiers of Research on Retirement*, 65–82. Ottawa: Statistics Canada.

Myllyla, S., and K. Kuvaja. 2005. 'Societal premises for sustainable development on large southern cities'. *Global Environmental Change* 15: 224–37.

Nabalamba, Alice. 2001. 'Locating risk: A multivariate analysis of the spatial and socio-demographic characteristics of pollution' (PhD. dissertation, University of Waterloo).

Nack, Adina. 2008. *Damaged Goods? Women Living with Sexually Transmitted Diseases*. Philadelphia: Temple University Press.

Naess, Arne. 1973. 'The shallow and the deep, long range ecology movement'. *Inquiry* 16: 95–100.

Nagel, Joane. 2006. 'Ethnicity, sexuality and globalization'. *Theory, Culture and Society* 23 (2/3): 545–7.

———, and Susan Olzak. 1982. 'Ethnic mobilization in new and old states: An extension of the competition model'. *Social Problems* 30 (2): 127–43.

Nakhaie, M. Reza, ed. 1999. *Debates on Social Inequality: Class, Gender and Ethnicity in Canada*. Toronto: Harcourt Canada.

———. 2000. 'Ownership and management position of Canadian ethnic groups in 1973 and 1989'. In Madeline A. Kalbach and Warren Kalbach, eds, *Perspectives on Ethnicity in Canada*. Toronto: Harcourt Canada.

———. 2002. 'Class, breadwinner ideology and housework among Canadian husbands'. *Review of Radical Political Economics* 34 (2): 137–57.

———. 2004. 'Who controls Canadian universities? Ethnoracial origins of Canadian university administrators and faculty's perception of mistreatment'. *Canadian Ethnic Studies* 36 (1): 92–110.

———. 2007. 'Universalism, ascription, and academic rank: Canadian professors 1987–2000'. *Canadian Review of Sociology* 44 (3): 361–86.

———, Robert A. Silverman, and Teresa C. LaGrange. 2000. 'An examination of gender, ethnicity, class and delinquency'. *Canadian Journal of Sociology* 25: 35–59.

Nanda, Serena, and Richard Warms. 2007. *Cultural Anthropology*. 9th edn. Belmont, CA: Wadsworth.

Nash, Kate. 2000. *Contemporary Political Sociology: Globalization, Politics, and Power*. Oxford: Blackwell.

Nason-Clark, Nancy. 2004. 'When terror strikes at home: The interface between religion and domestic violence'. *Journal for the Scientific Study of Religion* 43 (3): 303–10.

———, and Barbara Fisher-Townsend. 2007. 'Women, gender and feminism in the sociology of religion: Theory, research and social action'. In T. Balsi, ed., *American Sociology of Religion Histories*, 203–21. Leiden: Brill.

———, and Catherine Clark Kroeger. 2004. *Refuge from Abuse: Hope and Healing for Abused Christian Women*. Downers Grove, IL: InterVarsity Press.

National Academy on an Aging Society. 1999. *Demography Is Not Destiny*. Washington: National Academy on an Aging Society.

National Advisory Council on Aging. 2005. *Aging in Poverty in Canada*. Ottawa: Minister of Public Works and Government Services.

National Council of Welfare. 2001–2. *The Cost of Poverty*. Ottawa: National Council of Welfare.

———. 2008. *Welfare Incomes*. Ottawa: Minister of Public Works and Government Services.

National Institute on Aging and Population Reference Bureau. 2006. *The Future of Human Life Expectancy: Have We Reached the Ceiling or Is the Sky the Limit?* Washington: National Institute on Aging and Population Reference Bureau.

National Post. 2002. 'Census: How you fit into the national picture'. 17 July: A9.

National Research Council, Committee on Population and Working Group on Population Growth and Economic Development. 1986. *Population Growth and Economic Development: Policy Questions*. Washington: National Academy Press.

Natural Resources Canada. 2006. 'Atlas of Canada: Religious affiliation 2001'. Ottawa: Natural Resources Canada. http://atlas.nrcan.gc.ca/auth/english/maps/peopleandsociety/religion/religion01.

Navarro, Véase Vicente. 1975. 'The industrialization of fetishism or the fetishism of industrialization: A critique of Ivan Illich'. *Social Science and Medicine* 9: 351–63.

Nedelmann, Birgitta. 1991. 'Review of *Ideology and the New Social Movements*, by Alan Scott'. *Contemporary Sociology* 20: 374–5.

Negri, Antonio, and Michael Hardt. 2000. *Empire*. Cambridge, MA: Harvard University Press.

———. 2004. *Multitude: War and Democracy in the Age of Empire*. New York: Penguin Press.

Nelson, Addie. 2010. *Gender in Canada*. 4th edn. Toronto: Pearson.

Nelson, Fiona. 2001. 'Lesbian families'. In B.J. Fox, ed., *Family Patterns, Gender Relations*, 441–57. Toronto: Oxford University Press.

Nelson, Michelle, and Hye-Jin Paek. 2005. 'Cross-cultural differences in sexual advertising content in a transnational women's magazine'. *Sex Roles* 53 (5/6): 371–83.

Nett, Emily. 1981. 'Canadian families in social-historical perspective'. *Canadian Journal of Sociology* 6 (3): 239–60.

Netting, S. Nancy, and Matthew Burnett. 2004. 'Twenty years of student sexual behaviour: Subcultural adaptations to a changing health environment'. *Adolescence* 39 (153): 19–38.

Nettleton, S., R. Burrows, and L. O'Malley. 2005. 'The mundane realities of the everyday lay use of the Internet for health and their consequences for media convergence'. *Sociology of Health and Illness* 27 (7): 972–92.

Nettleton, S., L. O'Malley, and I. Watt. 2004. 'The emergence of e-scaped medicine'. *Sociology* 38 (4): 661–79.

Newman, David M. 2006. *Sociology: Exploring the Architecture of Everyday Life*. Thousand Oaks, CA: Pine Forge Press.

Newman, P., and J. Kenworthy. 1999. *Sustainability and Cities: Overcoming the Automobile Dependence*. Washington: Island Press.

Newton, Michael. 2002. *Savage Girls and Wild Beasts: A History of Feral Children*. New York: St Martin's.

Nisbet, R.A. 1966. *The Sociological Tradition*. New York: Basic Books.

NLSCY (National Longitudinal Survey of Children and Youth). 1996. *Growing Up in Canada*. Ottawa: Human Resources Development Canada and Statistics Canada.

Nobles, Melissa. 2000. *Shades of Citizenship: Race and the Census in Modern Politics*. Stanford, CA: Stanford University Press.

Noel, Alain. 2009. 'Aboriginal peoples and poverty in Canada: Can provincial government make a difference?' Paper prepared for the annual meeting of the International Sociological Association Research Committee, Montreal, 20 August. www.cccg.umontreal.ca/RC19/PDF/Noel-A_Rc192009.pdf.

North South Institute. 2006. *Migrant workers in Canada*. Ottawa: North South Institute.

Notestein, Frank. 1945. 'Population: The long view'. In Theodore W. Schultz, ed., *Food for the World*, 36–57. Chicago: University of Chicago Press.

———. 1967. 'The population crisis: Reasons for hope'. *Foreign Affairs* 46 (1): 156–80.

NUPGE (National Union of Public and General Employees). 2009. 'What poverty means for Canada's poorest households'. www.nupge.ca.

Nylund, D. 2004. 'When in Rome: Heterosexism, homophobia and sports talk radio'. *Journal of Sport and Social Issues* 28 (2): 136–68.

Oakes, J.M. 1996. 'A longitudinal analysis of environmental equity in communities with hazardous waste facilities'. *Social Science Research* 25: 125–48.

Obesity Canada. 2001. 'What is obesity?' www.obesitycanada.com.

Occhionero, Marisa Ferrari. 1996. 'Rethinking public space and power'. *International Review of Sociology* 6: 453–64.

O'Connor, Julia S., Ann Shola Orloff, and Sheila Shaver. 1999. *States, Markets, Families: Gender Liberalism and Social Policy in Australia, Canada, Great Britain and the United States*. Cambridge: Cambridge University Press.

Oeppen, Jim, and James W. Vaupel. 2002. 'Broken limits to life expectancy'. *Science* 296 (10 May): 1029–31.

Olsen, Gregg. 2002. *The Politics of the Welfare State: Canada, Sweden and the United States*. Oxford: Oxford University Press.

Olshansky, S. Jay, and Brian A. Ault. 1986. 'The fourth stage of the epidemiological transition: The age of delayed degenerative diseases'. *Milbank Memorial Fund Quarterly* 46: 355–91.

Olshansky, S. Jay, Bruce A. Carnes, and Christine Cassel. 1990. 'In search of Methuselah: Estimating the upper limits to human longevity'. *Science* 250: 634–40.

Olshansky, S. Jay, L. Hayflick, and Bruce A. Carnes. 2002. 'No truth to the fountain of youth'. *Scientific American* 286 (6): 92–5.

Olzak, Susan. 1983. 'Contemporary ethnic mobilization'. *Annual Review of Sociology* 9: 355–74.

Omran, Abdal R. 1971. 'The epidemiologic transition'. *Milbank Memorial Fund Quarterly* 49: 509–38.

O'Rand, A.M., and J.C. Henretta. 1999. *Age and Inequality: Diverse Pathways through Later Life*. Boulder, CO: Westview.

O'Riordan, T. 1971. 'The third American environmental conservation movement: New implications for public policy'. *Journal of American Studies* 5: 155–71.

Orlova, Alexandra. 2004. 'Insiders and outcasts: From social dislocation to human trafficking—The Russian case'. *Problems of Post-communism* 51 (6): 14–22.

Ornstein, Michael. 1981. 'The occupational mobility of men in Ontario'. *Canadian Review of Sociology and Anthropology* 18 (2): 181–215.

Orsi, Robert. 2003. 'Is the study of lived religion irrelevant to the world we live in?' *Journal for the Scientific Study of Religion* 42 (3): 169–74.

Osborne, Ken. 1999. *Education: A Guide to the Canadian School Debate: Or, Who Wants What and Why?* Toronto: Penguin.

Ostrovsky, Yuri. 2008. 'Earnings inequality and earnings instability of immigrants in Canada'. Analytical Studies, Research Paper Series no. 309. Catalogue no. 11F0019M. Ottawa: Statistics Canada.

Ouellet, M. 2006. 'Le *Smart Growth* et le nouvel urbanisme: synthèse de la littérature récente et regard sur la situation canadienne'. *Cahiers de géographie du Québec* 50: 175–93.

Overbeek, Johannes. 1974. *History of Population Theories*. Rotterdam: Rotterdam University Press.

Painter, J. 1997. 'Regulation, regime and practice in urban politics'. In M. Lauria, ed., *Reconstructing Urban Regime Theory*, 122–43. Thousand Oaks, CA: Sage.

Palameta, B. 2004. 'Low income amongst immigrant and visible minorities'. Catalogue no. 75-001-X1E. *Perspectives* (April). Ottawa: Statistics Canada.

Palmer, Susan. 2004. *Aliens Adored: Rael's UDO Religion*. New Brunswick, NJ: Rutgers University Press.

Paloni, Alberto, and Maurizio Zanardi. 2006. *The IMF, World Bank and Policy Reform*. New York: Routledge.

Palys, T.S. 1986. 'Testing the common wisdom: The social content of video pornography'. *Canadian Psychology* 27: 22–35.

Park, Kristin. 2002. 'Stigma management among the voluntarily childless'. *Sociological Perspectives* 45 (1): 21–45.

Park, R.E., and E.W. Burgess. 1967 [1916]. *The City*. Chicago: University of Chicago Press.

———. 1921. *Introduction to the Science of Sociology*. Chicago: University of Chicago Press.

———, and Roderick D. McKenzie, eds. 1925. *The City*. Chicago: University of Chicago Press.

Parsons, Talcott. 1937. *The Structure of Social Action*. New York: McGraw-Hill.

———. 1951. *The Social System*. Glencoe, IL: Free Press.

———. 1959. 'The school class as a social system: Some of its functions in American society'. *Harvard Educational Review* 29: 297–318.

———, and Robert F. Bales. 1955. *Family Socialization and Interaction Process*. New York: Free Press.

Patrick, Margie. 2009. 'Political neoconservatism: A conundrum for Canadian evangelicals'. *Studies in Religion/Sciences religieuses* 38: 481.

Paul, Annie Murphy. 2010. 'How to be brilliant'. *The New York Times* 21 March.

Pearce, Frank, and Laureen Snider, eds. 1995. *Corporate Crime: Contemporary Debates*. Toronto: University of Toronto Press.

Pearlin, L.I. 1985. 'Social structure and processes of social support'. In S. Cohen and L. Syme, eds, *Social Support and Health*, 43–60. New York: Academic Press.

Peers, Frank W. 1979. *The Public Eye: Television and the Politics of Canadian Broadcasting, 1952–1968*. Toronto: University of Toronto Press.

Pendakur, Manjunath. 1990. *Canadian Dreams and American Control: The Political Economy of the Canadian Film Industry*. Detroit: Wayne State University Press.

Peritz, Ingrid. 2010. 'Quebec Muslim woman ordered to unveil or leave French Class'. *The Globe and Mail*, www. theglobeandmail.com/news/national/quebec-muslim-woman-ordered-to-unveil-or-leave-french-course/article1530874.

Perls, Thomas T., and Ruth C. Fretts. 1998. 'Why women live longer than men'. *Scientific American Presents* 9 (2): 100–3.

Perreault, S. 2009. 'The incarceration of Aboriginal people in adult correctional services'. *Juristat* 29 (3). Ottawa: Statistics Canada.

Perry, B., and A. Harding. 2002. 'The future of urban sociology: Report of joint sessions of the British and American sociological associations'. *International Journal of Urban and Regional Research* 26: 844–53.

Petersen, William. 1989. 'Marxism and the population question: Theory and practice'. *Population and Development Review* 14 (suppl.): 77–101.

Peterson, Richard A. 1992. 'Understanding audience segmentation: From elite and popular to omnivore and univore'. *Poetics* 21: 243–58.

———. 1994. 'Culture studies through the production perspective: Progress and prospects'. In Diana Crane, ed., *The Sociology of Culture: Emerging Theoretical Perspectives*, 163–89. Oxford: Blackwell.

———, and Roger M. Kern. 1996. 'Changing highbrow taste: From snob to omnivore'. *American Sociological Review* 61 (5): 900–7.

Petzer, Shane, and Gordon Issacs. 1998. 'SWEAT: The development and implementation of a sex-worker advocacy and intervention program in post-apartheid South Africa'. In Kamala Kempadoo and Jo Doezema, eds, *Global Sex Workers: Rights, Resistance, and Redefinition*, 192–6. New York: Routledge.

Pfohl, Stephen J. 1977. 'The discovery of child abuse'. *Social Problems* 24: 310–23.

Pfuhl, Erdwin H., and Stuart Henry. 1993. *The Deviance Process*. 3rd edn. New York: Aldine de Gruyter.

Phillipson, C. 1982. *Capitalism and the Construction of Old Age*. London: MacMillan.

———. 1999. 'The social construction of retirement: Perspectives from critical theory and political economy'. In M. Minkler and C.L. Estes, eds, *Critical Gerontology: Perspectives from Political and Moral Economy*, 315–18. New York: Baywood.

———, and A. Walker. 1987. 'The case for a critical gerontology'. In S. DeGregorio, ed., *Social Gerontology: New Directions*, 1–15. London: Croom Helm.

Piaget, Jean. 1932. *The Moral Judgement of the Child*. London: Routledge and Kegan Paul.

——. 1950. *The Construction of Reality in the Child*. London: Routledge and Kegan Paul.

Pickard, Victor W. 2006. 'Assessing the radical democracy of indymedia: Discursive, technical, and institutional constructions'. *Critical Studies in Media Communication* 23 (1): 19–38.

Pickering, Carmen. 2008. 'Challenges in the classroom and teacher stress'. *Health and Learning* 6: 22–7. www.ctf-fce.ca/publications/health_learning/Issue6_Article8_EN.pdf.

Pickett, S.T.A., et al. 2001. 'Urban ecological systems: Linking terrestrial ecological, physical, and socioeconomic components of metropolitan areas'. *Annual Review of Ecological Systems* 32: 127–57.

Pickvance, C. 1994. 'Extended review: Sociology and spatial development'. *Work, Employment and Sociology* 8 (1): 127–30.

Picot, Garnett, F. Hou, and S. Coulombe. 2007. *Chronic Low Income and Low Income Dynamics among Recent Immigrants*. Analytical Studies Branch, Research Paper Series, no. 294. Catalogue no. 11F0019M1E. Ottawa: Statistics Canada.

Pignal, Jean, Stephen Arrowsmith, and Andrea Ness. 2008. *First Results from the Survey of Older Workers*. Research Paper. Catalogue no. 89-646-X ISBN: 978-1-100-16772-5. Ottawa: Statistics Canada, Special Surveys Division.

Piliavin, Erving, and S. Briar. 1964. 'Police encounters with juveniles'. *American Journal of Sociology* 70: 206–14.

Pines, Christopher L. 1993. *Ideology and False Consciousness: Marx and His Historical Progenitors*. Albany: State University of New York Press.

Piven, Frances Fox. 2007. 'The neoliberal challenge'. *Contexts* 6 (3): 13–15.

Platt, R.H. 2004. 'Regreening the metropolis: Pathway to more ecological cities'. *Annals of the New York Academy of Science* 1023: 49–61.

Plummer, Ken. 1975. *Sexual Stigma: An Interactionist Account*. London: Routledge and Kegan Paul.

——. 2003. 'Queers, bodies and postmodern sexualities: A note on revisiting the "sexual" in symbolic interactionism'. *Qualitative Sociology* 26 (4): 515–30.

Plumwood, Val. 1992. 'Feminism and ecofeminism: Beyond the dualistic assumptions of women, men and nature'. *Ecologist* 22 (1): 8–13.

Poggi, Gianfranco. 2000. *Durkheim*. Oxford: Oxford University Press.

Poitras, C. 2000. *La cité au bout du fil*. Montreal: Les presses de l'Université de Montréal.

Polletta, Francesca, and James Jasper. 2001. 'Collective identity and social movements'. *Annual Review of Sociology* 27: 283–305.

Population Reference Bureau (PRB). 2007. *World Population Data Sheet for 2007*. Washington: PRB.

——. 2009a. *Population Age 65 + (%)*. www.prb.org/Datafinder/Topic/Map.aspx?variable = 119.

——. 2009b. *World Population Data Sheet for 2009*. Washington: PRB.

——. 2010. *World Population Data Sheet for 2010*. Washington: PRB.

Porter, John. 1965. *The Vertical Mosaic: An Analysis of Social Class and Power in Canada*. Toronto: University of Toronto Press.

Porter, R. 1997. *The Greatest Benefit to Mankind. A Medical History of Humanity*. New York: Norton.

Portes, Alejandro. 1998. 'Social capital: Its origins and applications in modern sociology'. *Annual Review of Sociology* 24: 1–24.

Poulantzas, N. 1968. *Pouvoir politique et classes sociales*. Paris: François Maspéro.

Pratt, T.C., and T.W. Godsey. 2003. 'Social support, inequality, and homicide: A cross-national test of an integrated theoretical model'. *Criminology* 44 (3): 611–44.

Preston, Samuel H. 1986. 'Mortality and development revisited'. *United Nations Population Bulletin* 18: 34–40.

——, Patrick Heuveline, and Michel Guillot. 2001. *Demography*. Malden, MA: Blackwell.

——, Nathan Keyfitz, and Robert Schoen. 1972. *Causes of Death: Life Tables for National Populations*. New York: Seminar Press.

Pringle, Rosemary. 1988. *Secretaries Talk: Sexuality, Power and Work*. London: Verso.

Proudfoot, Shannon. 2010. 'Living life by other people's rules'. *Vancouver Sun* 3 June.

Pryor, Jan, and Bryan Rodgers. 2001. *Children in Changing Families: Life after Parental Separation*. Oxford: Blackwell.

Pupo, Norene, and Andrea Noack. 2010. 'Dialling for service: Transforming the public-sector workplace in Canada'. In Norene J. Pupo and Mark P. Thomas, eds, *Interrogating the New Economy: Restructuring Work in the 21st Century*, 111–28. Toronto: University of Toronto Press.

Putnam, Robert D. 2000. *Bowling Alone: The Collapse and Revival of American Community*. New York: Simon and Schuster.

——. 2007. '*E pluribus unum*: Diversity and community in the twenty-first century'. 2006 Johan Skytte Prize Lecture. *Scandinavian Political Studies* 30 (2): 137–74.

Qu, Lixia, and Ruth Weston. 2008. 'Snapshot of family relationships'. *Family Matters* (May).

Quebec. Assemblée nationale. 2006. *Loi sur le développement durable*. Quebec City: Éditeur officiel.

Quine, Lyn. 1999. 'Workplace bullying in NHS community trust: Staff questionnaire survey'. *British Medical Journal* 318: 228–32.

——. 2002. 'Workplace bullying in junior doctors: Questionnaire survey'. *British Medical Journal* 324: 878–9.

Raboy, Marc. 1990. *Missed Opportunities: The Story of Canada's Broadcasting Policy*. Montreal: McGill-Queen's University Press.

——. 1995. 'The role of public consultation in shaping the Canadian broadcasting system'. *Canadian Journal of Political Science* 28 (3): 455–77.

Raduntz, Helen. 2005. 'The marketization of education within the global capitalist economy'. In Michael W. Apple, Jane Kenway, and Michael Singh, eds, *Globalizing Education: Policies, Pedagogies, and Politics*, 231–45. New York: Peter Lang.

Raines, John, ed. 2002. *Marx on Religion*. Philadelphia: Temple University Press.

Ram, Bali. 1990. *New Trends in the Family: Demographic Facts and Figures*. Ottawa: Minister of Supply and Services Canada.

——, and Abdur Rahim. 1993. 'Enduring effects of women's early employment experiences on child-spacing: The Canadian evidence'. *Population Studies* 47: 307–18.

Ramage-Morin, P.L., M. Shields, and L. Martel. 2010. 'Health-promoting factors and good health among Canadians in mid- to late life'. *Health Reports* 21 (3): 1–9.

Ranson, Gillian. 2009. 'Paid and unpaid work: How do families divide their labour?' In Maureen Baker, ed., *Families: Changing Trends in Canada*, 6th edn, 108–29. Toronto: McGraw-Hill Ryerson.

——. 2010. *Against the Grain: Couples, Gender, and the Reframing of Parenting*. Toronto: University of Toronto Press.

Reichert, Tom, and Jacqueline Lambiase. 2003. 'How to get "kissably close": Examining how advertisers appeal to consumers' sexual needs and desires'. *Sexuality and Culture* 7 (3): 120–36.

Reinarman, Craig. 1996. 'The social construction of an alcohol problem'. In Gary W. Potter and Victor E. Kappeler, eds, *Constructing Crime: Perspectives on Making News and Social Problems*,

193–220. Prospect Heights, IL: Waveland.

Report Card on Child and Family Poverty in Canada. 2009. http://intraspec.ca/2009EnglishC2000NationalReportCard.pdf.

Rich, Adrienne Cecile. 2003. 'Compulsory heterosexuality and lesbian existence (1980)'. *Project Muse—Journal of Women's History* 15 (3): 11–48.

Riley, M.W. 1997. 'Rational choice and the sociology of age: Heuristic models'. *American Sociologist* 28 (2): 54–60.

——, A. Foner, and J.W. Riley, Jr. 1999. 'The aging and society paradigm'. In V.L. Bengston and K.W. Schaie, eds, *Handbook of Theories of Aging*, 327–43. New York: Springer.

——, M.E. Johnson, and A. Foner. 1972. *Aging and Society*, v. 3, *A Sociology of Age Stratification*. New York: Russell Sage.

——, and J.W. Riley. 1994. 'Structural lag: Past and future'. In M.W. Riley, R. L. Kahn, and A. Foner, eds, *Age and Structural Lag*, 115–36. New York: Wiley.

Ritchey, Neil P. 1976. 'Explanations of migration'. *Annual Review of Sociology* 2: 363–404.

Ritzer, G. 2000. *Sociological Theory*. 5th edn. New York: McGraw-Hill.

Ritzer, George. 2000. *The McDonaldization of Society*. 3rd edn. Thousand Oaks, CA: Pine Forge.

——. 2010. *Globalization: A Basic Text*. Oxford: Wiley-Blackwell.

Roberts, Barbara. 1988. *Whence They Came: Deportation from Canada, 1900–1935*. Ottawa: University of Ottawa Press.

Robertson, A. 1997. 'Beyond apocalyptic demography: Towards a moral economy of interdependence'. *Ageing and Society* 17 (4): 425–46.

Robertson, Ann. 2001. 'Biotechnology, political rationality and discourses on health risk'. *Health* 5: 293–310.

Robertson, Roland. 1995. 'Glocalization: Time-space and heterogeneity-homogeneity'. In M. Featherstone, S. Lash, and R. Robertson, eds, *Global Modernities*. Thousand Oaks, CA: Sage.

Robey, Bryant, Shea O. Rutstein, and Leo Morris. 1993. 'The fertility decline in developing countries'. *Scientific American* 269 (December): 60–8.

Robson, B. 1975. *Urban Social Areas*. Oxford: Oxford University Press.

Roby, Jini. 2005. 'Women and children in the global sex trade: Toward more effective policy'. *International Journal of Social Work* 48 (2): 136–47.

Rogan, Mary. 1999. 'Acts of faith'. *Saturday Night* 114 (5): 42–51.

Romaniuc, Anatole. 1984. *Current Demographic Analysis: Fertility in Canada: From Baby-Boom to Baby-Bust*. Ottawa: Statistics Canada.

——. 1994. 'Fertility in Canada: Retrospective and prospective'. In Frank Trovato and Carl F. Grindstaff, eds, *Perspectives on Canada's Population: An Introduction to Concepts and Issues*, 213–30. Toronto: Oxford University Press.

Rosenberg, Sharon. 2003. 'Neither forgotten nor fully remembered: Tracing an ambivalent public memory on the 10th anniversary of the Montreal Massacre'. *Feminist Theory* 4 (1): 5–27.

Rosenthal, Carolyn J. 1985. 'Kinkeeping in the familial division of labour'. *Journal of Marriage and the Family* 47: 965–74.

Ross, David, and Richard Shillington. 1994. *The Canadian Fact Book on Poverty*, 3–4. Ottawa: Canadian Council on Social Development.

Ross, Eric B. 1998. *The Malthus Factor: Poverty, Politics and Population in Capitalist Development*. London: Zed Books.

Rothkopf, David. 2008. *Superclass: The Global Power Elite and the World They Are Making*. Toronto: Penguin Canada.

Rowe, J.W., and R.L. Kahn. 1998. *Successful Aging*. New York: Pantheon/Random House.

Royal Commission on Aboriginal Peoples (RCAP). 1996. *Report of the Royal Commission on Aboriginal Peoples*, v. 3, *Gathering Strength*. Ottawa: RCAP.

Royal Commission on Equality in Employment. 1984. *Report*. Ottawa: Supply and Services Canada.

Rudner, L.M. 1999. 'Scholastic achievement and demographic characteristics of home school students in 1998'. *Education Policy Analysis Archives* 7 (8). http://epaa.asu.edu/epaa/v7n8.

Ryan, W. 1971. *Blaming the Victim*. New York: Pantheon.

Rymer, Russ. 1993. *Genie: Escape from a Silent Childhood*. London: Michael Joseph.

Sacco, Vincent F. 1992. 'An introduction to the study of deviance and control'. In Vincent F. Sacco, ed., *Deviance: Conformity and Control in Canadian Society*, 1–48. Scarborough, ON: Prentice-Hall.

——. 2005. *When Crime Waves*. Thousand Oaks, CA: Sage.

——, and K. Ismaili. 2001. 'Social problem claims and the undefended border: The case of Canada and the United States'. In Joel Best, ed., *How Claims Spread: Cross-national Diffusion of of Social Problems*. New York: Aldine de Gruyter.

——, and L.W. Kennedy. 2011. *The Criminal Event: An Introduction to Criminology*. 5th edn. Scarborough, ON: Thomson.

Sachs, Jeffrey. 2005. *The End of Poverty: Economic Possibilities of Our Time*. New York: Penguin.

——. 2008. *Common Wealth: Economics for a Crowded Planet*. New York: Allen Lane.

——, Andrew D. Mellinger, and John L. Gallup. 2001. 'The geography of poverty and wealth'. *Scientific American* (March): 70–5.

Sachs, Wolfgang. 1991. 'Environment and development: The story of a dangerous liaison'. *Ecologist* 21: 252–7.

Sadovnik, Alan R., ed. 1995. *Knowledge and Pedagogy: The Sociology of Basil Bernstein*. Norwood, NJ: Ablex.

Said, Edward W. 1978. *Orientalism*. London: Routledge and Kegan Paul.

Sandstrom, Kent L. 1990. 'Confronting deadly disease: The drama of identity construction among gay men with AIDS'. *Journal of Contemporary Ethnography* 19 (3): 271–94.

——, Dan Martin, and Gary Alan Fine. 2006. *Symbols, Selves, and Social Reality: A Symbolic Interactionist Approach to Sociology and Social Psychology*. 2nd edn. New York: Oxford University Press.

Santos, Bonaventura de Sousa. 2005. 'The future of the World Social Forum'. *Development* 48 (2): 15–22.

Sarlo, Christopher. 1996. *Poverty in Canada*. 2nd edn. Vancouver: The Fraser Institute.

Sassen, Saskia. 2005. 'Cities as strategic sites'. *Sociology* 39 (2): 352–7.

——. 2008. 'Re-assembling the urban'. *Urban Geography* 29 (2): 113–26.

Sasson, Theodore. 1995. *Crime Talk: How Citizens Construct a Social Problem*. Hawthorne, NY: Aldine de Gruyter.

Satzewich, Vic, ed. 1998. *Racism and Social Inequality in Canada*. Toronto: Thompson Educational Publishing.

——, and Nikolaos Liodakis. 2010. *'Race' and Ethnicity in Canada: A Critical Introduction*. 2nd edn. Toronto: Oxford University Press.

Savage, M. 2005. 'Urban sociology in the third generation'. *Sociology* 39 (2): 357–61.

Scala, Francesca, Eric Montpetit, and Isabelle Fortier. 2005. 'The NAC's organizational practices and the politics of assisted reproductive technologies in Canada'. *Canadian Journal of Political Science* 38 (3): 581–604.

Scarce, Rik. 1990. *Eco-warriors: Understanding the Radical Environmental Movement*. Chicago: Noble.

Scharrer, Erica. 2001. 'From wise to foolish: The portrayal of the sitcom father, 1950s–1990s'. *Journal of Broadcasting and Electronic Media* 45 (1): 23–40.

Schauer, Terrie. 2005. 'Women's porno: The heterosexual female gaze in porn sites "for women"'. *Sexuality and Culture* 9 (2): 42–64.

Schecter, Tanya. 1998. *Race, Class, Women and the State: The Case of Domestic Labour*. Montreal: Black Rose.

Schellenberg, G. 2004. *2003 General Social Survey on Social Engagement, Cycle 17: An Overview of Findings*. Ottawa: Statistics Canada.

———, M. Turcotte, and B. Ram. 2006. 'The changing characteristics of older couples and joint retirement in Canada'. In L. Stone, ed., *New Frontiers of Research on Retirement*, 199–218. Ottawa: Statistics Canada.

Schiller, Herbert I. 1969. *Mass Communication and American Empire*. Boston: Beacon.

Schissel, Bernard, and Terry Wotherspoon. 2003. *The Legacy of School for Aboriginal People: Education, Oppression, and Emancipation*. Toronto: Oxford University Press.

Schnaiberg, Allan. 1975. 'Social synthesis of the societal-environmental dialectic: The role of distributional impacts'. *Social Science Quarterly* 56: 5–20.

———. 1980. *The Environment: From Surplus to Scarcity*. New York: Oxford University Press.

Schneider, Christopher J. 2009. 'The musical ringtone as an impression management device: A research note'. *Studies in Symbolic Interaction* 33: 35–46.

Schnore, Leo F. 1958. 'Social morphology and human ecology'. *American Journal of Sociology* 63: 620–34.

Schoof, M. 2010. 'Religions based on inequality have no place here'. *Vancouver Sun*, www.vancouversun.com/news/Religions + based + inequality + have + place + here/3302829/story.html#ixzz0zAUMXcJE.

Schwartz, Mildred. 2007. 'Remembering Seymour Martin Lipset'. *Canadian Journal of Sociology Online* (March-April), www.cjsonline.ca/soceye/lipset.html.

Sciadas, George. 2002. *The Digital Divide in Canada*. Ottawa: Statistics Canada.

Scott, A.J. 2006. 'Creative cities: Conceptual issues and policy questions'. *Journal of Urban Affairs* 28: 1–17.

———, and E. Soja, eds. 1996. *The City: Los Angeles and Urban Theory at the Eve of the Twentieth Century*. Berkeley: University of California Press.

Scott, Allen J. 2004. 'Hollywood and the world: The geography of motion-picture distribution and marketing'. *Review of International Political Economy* 11 (1): 33–61.

Scott, John, and Gordon Marshall. 2005. *A Dictionary of Sociology*. Oxford: Oxford University Press.

———. 2009. 'Political sociology'. In John Scott and Gordon Marshall, *A Dictionary of Sociology*. Oxford: Oxford University Press.

Scott, M. 1969. *American City Planning since 1890*. Berkeley and Los Angeles: University of California Press.

Scott, Marvin B., and Stanford M. Lyman. 1968. 'Accounts'. *American Sociological Review* 33: 46–64.

Scott, Robert A. 1969. *The Making of Blind Men: A Study of Adult Socialization*. New York: Russell Sage Foundation.

Scully, Diana, and Joseph Marolla. 1984. 'Convicted rapists' vocabulary of motive: Excuses and justifications'. *Social Problems* 31 (5): 530–44.

Sears, Alan. 2003. *Retooling the Mind Factory: Education in a Lean State*. Aurora, ON: Garamond.

Seljak, David. 2000. 'Resisting the "no man's land" of private religion: The Catholic church and the public politics in Quebec'. In D. Lyon and M. Van Die, eds., *Rethinking Church, State and Modernity: Canada between Europe and America*, 131–48. Toronto: University of Toronto Press.

Sen, A. 2000. *Social Exclusion: Concept, Application, and Scrutiny*. Social Development Papers no. 1. Manila: Asian Development Bank.

Sen, Jai, and Peter Waterman, eds. 2007. *World Social Forum: Challenging Empires*. Montreal: Black Rose Books.

Sennett, Richard. 1998. *The Corrosion of Character: The Personal Consequences of Work in the New Capitalism*. New York: Norton.

Settersten, R.A., Jr. 2003a. 'Age structuring and the rhythm of the life course'. In J.T. Mortimer and M.J. Shanahan, eds, *Handbook of the Life Course*, 81–98. New York: Kluwer Academic/Plenum Publishers.

———, ed. 2003b. *Invitation to the Life Course: Toward New Understandings of Later Life*. New York: Baywood.

———. 2006. 'Aging and the life course'. In R. Binstock and L. George, eds, *Handbook of Aging and the Social Sciences*, 6th edn, 3–20. San Diego: Academic Press.

———, and G.O. Hagestad. 1996. 'What's the latest? Cultural age deadlines for family transitions'. *The Gerontologist* 36 (2): 178–88.

Sev'er, Aysan. 2002. *Fleeing the House of Horrors: Women Who Have Left Abusive Partners*. Toronto: University of Toronto Press.

Shaffir, William. 1991. 'Conversion experiences: Newcomers to and defectors from Orthodox Judaism (*hozrim betshuvah* and *hozrim beshe'elah*)'. In Z. Sobel and B. Beit-Hallahmi, eds, *Tradition, Innovation, Conflict: Jewishness and Judaism in Contemporary Israel*, 173–202. Albany: State University of New York Press.

———, and Steven Kleinknecht. 2005. 'Death at the polls: Experiencing and coping with political defeat'. *Journal of Contemporary Ethnography* 34 (6): 707–38.

Sharp, E. 2005. 'Cities and subcultures: Exploring validity and predicting connections'. *Urban Affairs Review* 41: 132–56.

Shaw, R. Paul. 1975. *Migration Theory and Fact: A Review and Bibliography of Current Literature*. Philadelphia: Regional Science Research Institute.

Sherman, Rachel. 2007. *Class Acts: Service and Inequality in Luxury Hotels*. Berkeley: University of California Press.

Shibutani, Tamotsu. 1961. *Society and Personality: An Interactionist Approach to Social Psychology*. Englewood Cliffs, NJ: Prentice-Hall.

Shipley, Heather. 2008. 'Accommodating sexuality? Religion, sexual orientation and law in Canada'. Concordia University Graduate Conference, Accommodating Religion? Community, Discourse, Definitions, 7 February.

Shon, P.C.H., and B.A. Arrigo. 2006. 'Reality-based television and police–citizen encounters: The intertextual construction and situated meaning of mental illness as punishment'. *Punishment and Society* 8 (1): 59–85.

Shrestha, Alok. 2007. '$100 computer! Is it worth what it seems to be?' *Panorama* 18 November. TakingIT Global website, www.tigweb.org/express/panorama/article.html?start = 5268&ContentID = 17063.

Siegel, M. 2005. *False Alarm: The Truth about the Epidemic of Fear*. New York: Wiley.

Siltanen, Janet, and Andrea Doucet. 2008. *Gender Relations in Canada: Intersectionality and Beyond*. Toronto: Oxford University Press.

Silverstein, Merril. 1995. 'Stability and change in temporal distance between the elderly and their children'. *Demography* 31 (1): 29–46.

Simmel, Georg. 1950 [1903]. 'The metropolis and mental life'. In K.H. Wolff, ed., *The Sociology of Georg Simmel*. Glencoe, IL: Free Press.

———. 1957. 'Fashion'. *American Journal of Sociology* 62: 541–58.

Simmons, Alan. 1998. 'Racism and immigration policy'. In Vic Satzewich, *Racism and Social Inequality in Canada*. Toronto: Thompson Educational Publishing.

Simon, D. 2007. *Elite Deviance*, 9th edn. Boston: Pearson.

Simon, Julian. 1981. 'Environmental disruption or environmental improvement?' *Social Science Quarterly* 62: 30–43.

———, ed. 1995. *The State of Humanity*. Cambridge, MA: Blackwell.

———. 1996. *The Ultimate Resource 2*. Princeton, NJ: Princeton University Press.

Simon, William, and John Gagnon. 2003. 'Sexual scripts: Origins, influences and changes'. *Qualitative Sociology* 26 (4): 491–7.

Simons, John. 1980. 'Reproductive behaviour as religious sacrifice'. In Charlotte Hohn and Rainer Mackensen, eds, *Determinants of Fertility Trends: Theories Re-examined*, 131–46. Liège, Belgium: Ordina.

Simpson, John H. 2000. 'The politics of the body in Canada and the United States'. In D. Lyon and M. Van Die, eds, *Rethinking Church, State and Modernity: Canada between Europe and America*, 263–82. Toronto: University of Toronto Press.

Singer, Dorothy, and Jerome Singer. 2001. *Handbook of Children and the Media*. Thousand Oaks, CA: Sage.

Sjoberg, Gideon. 1960. *The Preindustrial City: Past and Present*. New York: Free Press.

Skinner, Burrhus F. 1953. *Science and Human Behaviour*. Oxford: Macmillan.

Sklair, Leslie. 1994. 'Global sociology and global environmental change'. In Michael Redclift and Ted Benton, eds, *Social Theory and the Global Environment*, 205–27. London: Routledge.

Skocpol, Theda. 1979. *States and Social Revolutions: A Comparative Analysis of France, Russia, and China*. Cambridge: Cambridge University Press.

———. 1995. *Protecting Soldiers and Mothers*. Cambridge, MA: Harvard University Press.

———. 2004. *Diminished Democracy: From Membership to Management in American Civic Life*. Norman: University of Oklahoma Press.

Smart, Carol. 2007. *Personal Life*. Cambridge: Polity Press.

Smart Growth Canada Network. www. smartgrowth.ca.

Smil, V. 2005. *Creating the Twentieth Century: Technical Innovations of 1867–1914 and Their Lasting Impact*. Oxford: Oxford University Press.

Smith, Adam. 1976 [1776]. *An Inquiry into the Nature and Causes of the Wealth of Nations*. Ed. W.B. Todd. Oxford: Oxford University Press.

Smith, Charlie. 2007a. 'Gorilla media'. straight.com (*Georgia Straight*) 5 April. www.straight.com/article-82806/gorilla-media.

———. 2007b. 'Union submission in CTV takeover bid notes that the CRTC has not published any limits on media concentration'. straight.com (*Georgia Straight*) 5 April. www.straight.com/article-82925/union-submission-ctv-take-over-bid-notes-crtc-has-not-published-any-limits-media-concentration.

Smith, Christian. 1991. *The Emergence of Liberation Theology: Radical Religion and Social Movement Theory*. Chicago: University of Chicago Press.

Smith, Dorothy. 1987. *The Everyday World as Problematic: A Feminist Sociology*. Boston: Northeastern University Press.

———. 1990. *The Conceptual Practices of Power: A Feminist Sociology of Knowledge*. Toronto: University of Toronto Press.

———. 1993. *Earnings of Men and Women*. Ottawa: Statistics Canada.

———. 1999. *Writing the Social: Critique, Theory, and Investigations*. Toronto: University of Toronto Press.

Smith, Philip. 2001. *Cultural Theory: An Introduction*. Oxford: Blackwell.

Smith, Raymond T. 1996. *The Matrifocal Family: Power, Pluralism and Politics*. New York: Routledge.

Smyth, Bruce, ed. 2004. *Parent–Child Contact and Post-Separation Parenting Arrangements*. Research Report no. 9. Melbourne: Australian Institute of Family Studies.

Smythe, Dallas W. 1981. *Dependency Road: Communications, Capitalism, Consciousness, and Canada*. Norwood, NJ: Ablex.

Snow, David A., E. Burke Rochford, Jr, Steven K. Worden, and Robert D. Benford. 1986. 'Frame alignment processes, micromobilization, and movement participation'. *American Sociological Review* 51: 464–81.

Snow, David A., Sarah A. Soule, and Hanspeter Kriesi, eds. 2004. *The Blackwell Companion to Social Movements*. Oxford: Blackwell.

Soja, E. 2000. *Postmetropolis: Critical Studies of Cities and Regions*. Oxford: Blackwell.

Sontag, Susan. 1978. *Illness as Metaphor*. New York: Farrar, Straus and Giroux.

———. 2003. *Regarding the Pain of Others*. New York: Picado.

Spector, Malcolm, and John I. Kitsuse. 1977. *Constructing Social Problems*. Menlo Park, CA: Cummings.

Spencer, Herbert. 1960. 'The sins of legislators'. In Donald MacRea, ed., *Herbert Spencer: The Man versus the State*. Harmondsworth, UK: Penguin Books.

Spittler, Gerd. 1980. 'Abstract knowledge as a basis of power: The history of the evolution of bureaucratic power in the Prussian peasant state'. *Kolner Zeitschrift fur Soziologie und Sozialpsychologie* 32: 574–604.

Spitzer, Steven. 1975. 'Toward a marxian theory of deviance'. *Social Problems* 22: 638–51.

Spreitzer, G.M., and S. Sonenshein. 2004. 'Toward the construct definition of positive deviance'. *American Behavioral Scientist* 47 (6): 828–47.

Spring, Joel. 2009. *Globalization of Education: An Introduction*. New York: Routledge.

Sprinkle, Annie. 1991. *Post Porn Modernism*. Amsterdam: Torch Books.

Staggenborg, Suzanne. 1994. *The Pro-choice Movement: Organization and Activism in the Abortion Conflict*. Oxford: Oxford University Press.

Stankiewicz, Julie M., and Francine Rosselli. 2008. 'Women as sex objects and victims in print advertisements'. *Sex Roles* 58: 579–89.

Stark, Sasha. 2009. 'The gambling research community: A preliminary collaboration network analysis' (unpublished report, Ontario Problem Gambling Research Centre, Guelph, ON).

Stasiulis, Daiva. 1980. 'The political structuring of ethnic community action'. *Canadian Ethnic Studies* 12 (3): 19–44.

———, and Abigail Bakan. 2005. *Negotiating Citizenship: Migrant Women and the Global System*. London: Palgrave-Macmillan.

Statistics Canada. 1984. *School Attendance and Level of Schooling. 1981 Census of Canada. Population*. Catalogue no. 92-914. Ottawa: Minister of Supply and Services Canada.

———. 1993. '1991 Census of Canada highlights: Religion'. *The Daily* 1 June. Ottawa: Minister of Supply and Services Canada

———. 1996. Public Use Microdata File on Individuals User Documentation. Ottawa: Minister of Supply and Services Canada.

———. 2001. *Population Projections of Visible Minority Groups, Canada, Provinces and Regions, 2001 to 2017*. Catalogue no. 91-541-XIE. Ottawa: Minister of Industry.

———. 2002a. *Labour Force Survey*. Unpublished data. Ottawa: Statistics Canada.

———. 2002b. *A Profile of Disability in Canada, 2001*. Ottawa: Statistics Canada.

———. 2003a. *Aboriginal Peoples of Canada: A Demographic Profile, 2001 Census*. Ottawa: Statistics Canada.

———. 2003b. *Canada's Ethnocultural Portrait: The Changing Mosaic*. Ottawa: Statistics Canada.

———. 2004. 'Education at a glance'. *Education Quarterly Review* 9 (4): 53–8. Ottawa: Statistics Canada.

———. 2005a. 'Children and youth as victims of violent crime'. *The Daily* 20 April. Ottawa: Statistics Canada.

———. 2005b. 'Early sexual intercourse, condom use and sexually transmitted diseases'. *The Daily* 3 May. Ottawa: Statistics Canada.

———. 2005c. *Population Projections for Canada, Provinces and Territories, 2005–2031*. Ottawa: Statistics Canada.

———. 2006a. *Canada Year Book*. Ottawa: Statistics Canada.

———. 2006b. *Education in Canada: School Attendance and Levels of Schooling*. Catalogue no. 97F0017XCB2001001. Ottawa: Statistics Canada.

———. 2006c. *Facts and Figures*. Ottawa: Statistics Canada.

———. 2006d. 'General social survey: Paid and unpaid work'. *The Daily* 19 July. Ottawa: Statistics Canada.

———. 2006e. 'Study: Changing patterns of women in the Canadian labour force'. *The Daily* 15 June. Ottawa: Statistics Canada.

———. 2006f. *Women in Canada: A Gender-Based Statistical Report*. Ottawa: Statistics Canada.

———. 2007a. 'Annual demographic estimates: Canada, provinces and territories, 2007, revised'. Catalogue no. 91-215-X. Ottawa: Minister of Industry. http://ivt.crepuq.qc.ca/demographie/documentation2007/91-215-XIE2007000.pdf.

———. 2007b. 'Family portrait: Continuity and change in Canadian families and households in 2006: National portrait: Individuals'. Ottawa: Statistics Canada. www12.statcan.ca/english/census06/analysis/famhouse/ind3.cfm.

———. 2007c. *Portrait of the Canadian Population in 2006, by Age and Sex, 2006 Census*. Ottawa: Statistics Canada.

———. 2007d. 'Study: Participation of older workers'. Ottawa: Statistics Canada. www.statcan.gc.ca/daily-quotidien/070824/dq070824a-eng.htm.

———. 2008a. *Deaths 2005*. Ottawa: Statistics Canada.

———. 2008b. 'Highest certificate, diploma or degree, age groups and sex for the population 15 years and over of Canada, provinces, territories, census divisions and census subdivisions'.

Catalogue no. 97-560-XCB2006008. Ottawa: Statistics Canada, 2006 Census of Population. www12.statcan.gc.ca/census-recensement/2006/dp-pd/tbt/index-eng.cfm.

———. 2008c. *Income in Canada, 2006*. Ottawa: Statistics Canada.

———. 2008d. 'Labour force activity, Aboriginal identity, highest certificate, diploma or degree, area of residence, age groups and sex for the population 15 years and over of Canada, provinces and territories'. Catalogue no. 97-560-XCB2006031. Ottawa: Statistics Canada, 2006 Census of Population. www12.statcan.gc.ca/census-recensement/2006/dp-pd/tbt/Lp-eng.

———. 2008e. *Labour Force Survey*. Unpublished data. Ottawa: Statistics Canada.

———. 2008f. *Low Income Cut-offs for 2007 and Low Income Measures for 2006*. Ottawa: Minister of Industry.

———. 2008g. 'Women in Canada: Paid work'. Ottawa: Statistics Canada. www.statcan.gc.ca/daily-quotidien/1012091/dq101209a_eng.htm.

———. 2009a. *Income in Canada, 2007*. Catalogue 75-202-XIE. Ottawa: Statistics Canada.

———. 2009b. 'Population estimates by sex and age group as of July 1, 2009, Canada'. Ottawa: Statistics Canada. www.statcan.gc.ca/daily-quotidien/091127/t091127b2-eng.htm.

———. 2009c. 'University degrees, diplomas and certificates awarded, 2007'. *The Daily* 13 July. Catalogue no. 11-001-XIE. Ottawa: Statistics Canada.

———. 2010a. 'Characteristics of individuals using the Internet'. *The Daily* 10 May. Ottawa: Statistics Canada. www40.statcan.gc.ca/l01/cst01/comm35a-eng.htm.

———. 2010b. 'Crude birth rate, age-specific and total fertility rates (live births), Canada, provinces and territories, annual'. CANSIM Table 102-4505. Ottawa: Statistics Canada.

———. 2010c. 'Deaths 2007'. Ottawa: Statistics Canada. www.statcan.gc.ca/daily-quotidien/100223/dq100223a-eng.htm.

———. 2010d. 'Labour force survey'. *The Daily*. Ottawa: Statistics Canada. www.statcan.gc.ca/dai-quo/index-eng.htm.

———. 2010e. *Population Projections for Canada, Provinces and Territories: 2009 to 2036*. Ottawa: Statistics Canada.

———. 2010f. 'University tuition fees'. *The Daily*. Ottawa: Statistics Canada. www.statcan.gc.ca/dai-quo/index-eng.htm.

——— and Council of Ministers of Education Canada. 2009. *Education

Indicators in Canada: An International Perspective*. Catalogue no. 81-604-X2009001E. Toronto: Canadian Education Statistics Council.

Steger, Manfred B. 2002. *Globalism: The New Market Ideology*. Lanham, MD: Rowman and Littlefield.

Stern, Nicholas. 2002. 'Keynote address: A strategy for development'. In Boris Pleskovic and Nicholas Stern, eds, *Annual World Bank Conference on Development Economics 2001/2002*, 11–35. Washington and New York: World Bank and Oxford University Press.

Stiglitz, Joseph. 2009. 'Report of the Commission of Experts of the President of the United Nations General Assembly on Reforms of the International Monetary and Financial System'. www.un.org/ga/president/63/interactive/financialcrisis/PreliminaryReport210509.pdf.

———. 2010. *Freefall: America, Free Markets and the Sinking of the World Economy*. New York: Norton.

———, and Linda J. Bilmes. 2008. *The Three Trillion Dollar War*. New York: W.W. Norton.

Stone, C.N. 2005. 'Looking back to look forward: Reflections on urban regime analysis'. *Urban Affairs Review* 40: 309–41.

———. 2006. 'Power, reform, and urban regime analysis'. *City and Community* 5: 23–38.

Stone, Gregory P. 1962. 'Appearance and the self'. In Arnold Rose, ed., *Human Behavior and Social Processes: An Interactionist Approach*, 86–116. Boston: Houghton-Mifflin.

Stone, Leroy O. 1967. *Urban Development in Canada*. Ottawa: Dominion Bureau of Statistics.

Stonechild, Blair. 2006. *The New Buffalo: The Struggle for Aboriginal Post-secondary Education in Canada*. Winnipeg: University of Manitoba Press.

Stones, M.J., and L. Stones. 1997. 'Ageism: The quiet epidemic'. *Canadian Journal of Public Health* 88 (5): 293–6.

Strasburger, Victor C., Barbara J. Wilson, and Amy B. Jordan. 2009. *Children, Adolescents and the Media*. 2nd edn. Thousand Oaks, CA: Sage.

Strauss, Anselm. 1959. *Mirrors and Masks: The Search for Identity*. Chicago: Free Press of Glencoe.

Strauss, Murray A., and Richard J. Gelles. 1990. *Physical Violence in American Families: Risk Factors and Adaptations to Violence in 8,145 Families*. New Brunswick, NJ: Transaction.

Stubera, J., S. Galea, and B.G. Link. 2008. 'Smoking and the emergence of a stigmatized social status'. *Social Science and Medicine* 67 (3): 420–30.

Sumner, C. 1994. *The Sociology of Deviance: An Obituary*. New York: Continuum.

Suominen, S., H. Helenius, H. Blomberg, A. Uutela, and M. Koskenvuo. 2001. 'Sense of coherence as a predictor of subjective state of health: Results of 4 years of follow-up of adults'. *Journal of Psychosomatic Research* 50 (2): 77–86.

Sutcliffe, A. 1981. *Towards the Planned City: Germany, Britain, the United States, and France, 1790–1914*. Oxford: Blackwell.

Sutherland, Edwin. 1940, 1949. *White Collar Crime*. New York: Holt, Rinehart and Winston

———. 1947. *Principles of Criminology*. 4th edn. Chicago: Lippincott.

Sutherland, T. 2008. 'Study finds having money, health, optimism, no stress, moderate drinking, no smoking means longer life'. http://seniorjournal.com/NEWS/Aging/20081027-StudyFindsHavingMoney.htm.

Swain, Jon. 2004. 'The right stuff: Fashioning an identity through clothing in a junior school'. In M. Webber and K. Bezanson, eds, *Rethinking Society in the 21st Century: Critical Readings in Sociology*, 81–92. Toronto: Canadian Scholars' Press.

Swanson, J. 2001. *Poor Bashing: The Politics of Exclusion*. Toronto: Between the Lines.

Sykes, Gresham M., and David Matza. 1957. 'Techniques of neutralization: A theory of deliquency'. *American Sociological Review* 22 (6): 664–70.

Syndicat Northcrest v. Amselem, [2004] 2 S.C.R. 551, 2004 SCC 47.

Szasz, Andrew. 1994. *Ecopopulism: Toxic Waste and the Movement for Environmental Justice*. Minneapolis: University of Minnesota Press.

Tannenbaum, Frank. 1938. *Crime and the Community*. Boston: Ginn.

Tannock, Stuart. 2001. *Youth at Work: The Unionized Fast-Food and Grocery Workplace*. Philadelphia: Temple University Press.

Taras, David. 2001. *Power and Betrayal in the Canadian Media*. Updated edn. Peterborough, ON: Broadview.

Tarr, J., and G. Dupuy, eds. 1988. *Technology and the Rise of the Networked City in Europe and America*. Philadelphia: Temple University Press.

Tarrow, Sidney. 1988. 'National politics and collective action: Recent theory and research in western Europe and the United States'. *Annual Review of Sociology* 14: 421–40.

———. 1998. *Power in Movements: Social Movements and Contentious Politics*.

2nd edn. New York: Cambridge University Press.

———. 2005. *The New Transnational Activism*. New York: Cambridge University Press.

———, and Charles Tilly. 2006. *Contentious Politics*. New York: Cambridge University Press.

Task Force on Retirement Income Policy. 1979. *The Retirement Income System in Canada: Problems and Alternative Policies for Reform*. Ottawa: Supply and Services Canada.

Taylor, Allison. 2005. 'Finding the future that fits'. *Gender and Education* 17 (2): 165–87.

Taylor, B., and V.L. Bengston. 2001. 'Sociological perspectives on productive aging'. In N. Morrow-Howell, J. Hinterlong, and M. Sherraden, eds, *Productive Aging: Concepts and Challenges*, 120–44. Baltimore: Johns Hopkins University Press.

Taylor, Bron. 1991. 'The religion and politics of Earth First!'. *Ecologist* 21: 258–66.

Taylor, Charles. 1992. *Multiculturalism and the Politics of Recognition*. Princeton, NJ: Princeton University Press.

Taylor, Leslie Ciarula. 2009. 'Gravy train shows no sign of slowing'. *Toronto Star* 20 November: B3.

Teitelbaum, Michael S. 1975. 'Relevance of demographic transition theory for developing countries'. *Science* 2 (May): 420–5.

Ten Bos, René. 1997. 'Essai: Business ethics and Bauman ethics'. *Organization Studies* 18: 997–1014.

Tepperman, Lorne. 1975. *Social Mobility in Canada*. Toronto: McGraw-Hill.

Theodorson, George A. 1961. *Studies in Human Ecology*. New York: Harper and Row.

———. 1982. *Urban Patterns: Studies in Human Ecology*. University Park: University of Pennsylvania Press.

Thiessen, Victor. 2009. 'The pursuit of postsecondary education: A comparison of First Nations, African, Asian, and European Canadian youth'. *Canadian Review of Sociology* 46 (1): 5–40.

Thomas, D.S., and W.I. Thomas. 1928. *The Child in America: Behavior Problems and Programs*. New York: Knopf.

Thompson, John B. 1997. 'Scandal and social theory'. In James Lull and Stephen Hinerman, eds, *Media Scandals*, 34–64. New York: Columbia University Press.

Thompson, Warren S. 1929. 'Population'. *American Journal of Sociology* 34: 959–75.

———. 1944. *Plenty of People*. Lancaster, PA: Jacques Cattel.

Thompson, William E. 1991. 'Handling the stigma of handling the dead: Morticians and funeral directors'. *Deviant Behavior* 12 (4): 403–29.

Thorne, Barry. 1982. 'Feminist rethinking of the family: An overview'. In Barry Thorne with Marilyn Yalom, eds, *Rethinking the Family: Some Feminist Questions*, 1–24. New York: Longman.

Tilleczek, K.C., and D.W. Hine. 2006. 'The meaning of smoking as health and social risk and adolescence'. *Journal of Adolescence* 29 (2): 273–87.

Tilly, Charles. 1978. *From Mobilization to Revolution*. Reading, MA: Addison-Wesley.

———. 2008. *Contentious Performances*. New York. Cambridge University Press.

Timms, D. 1971. *The Urban Mosaic*. Cambridge: Cambridge University Press.

Tirone, Susan, and Alison Pendlar. 2005. 'Leisure, place, and diversity: The experience of ethnic minority youth'. *Canadian Ethnic Studies* 37 (2): 32–48.

Titchkosky, Tanta. 2001. 'Disability: A rose by any other name? "People-first" language in Canadian society'. *Canadian Review of Sociology and Anthropology* 38 (2): 125–40.

Tittle, C.R., W.J. Villemez, and D.A. Smith. 1978. 'The myth of social class and criminality: An empirical assessment of the empirical evidence'. *American Sociological Review* 43: 643–56.

Tocqueville, Alexis de. 1945 [1835, 1840]. *Democracy in America*. New York: Vintage.

Toffler, Alvin, Esther Dyson, George Gilder, and George Keyworth. 1994. 'Cyberspace and the American Dream: A Magna Carta for the Knowledge Age'. www.alamut.com/subj/ideologies/manifestos/magnaCarta.html.

Tokar, Brian. 1988. 'Exploring the new ecologies'. *Alternatives* 15 (4): 31–43.

Tönnies, Ferdinand. 1957 [1887]. *Community and Society (Gemeinschaft und Gesellschaft)*. New York: Harper and Row.

Topalov, C. 2008. 'Sociologie d'un étiquetage scientifique: Urban sociology, Chicago, 1925'. *L'année sociologique* 58 (1): 203–34.

Toronto Star. 2009. 'Time for CEOs to pay piper'. 26 July: A13.

———. 2010. 'Millionaires' riches return close to pre-crisis levels'. 23 June: B4.

Touraine, Alain. 1981. *The Voice and the Eye: An Analysis of Social Movements*. Cambridge: Cambridge University Press.

———. 2003. 'Sociology without societies'. *Current Sociology* 51 (2): 123–31.

Townsend, P. 1962. *The Last Refuge: A Survey of Residential Institutions and Homes for the Aged in England*

and Wales. London: Routledge and Kegan Paul.

Trautner, Mary Nell, and Jessica L. Collett. 2010. 'Students who strip: The benefits of alternate identities for managing stigma'. *Symbolic Interaction* 33 (2): 257–79.

Travisano, R. 1970. 'Alteration and conversion as qualitatively different transformations'. In G.P. Stone and H. Farberman, eds, *Social Psychology through Symbolic Interaction*, 594–606. Waltham, MA: Ginn-Blaisdell.

Troyer, Ronald, and Gerald Markle. 1983. *Cigarettes: The Battle over Smoking*. New Brunswick, NJ: Rutgers University Press.

Tuggle, Justin L., and Malcolm D. Holmes. 1997. 'Blowing smoke: Status politics and the Shasta County smoking ban'. *Deviant Behavior* 18: 77–93.

Tuljapurkar, Shripad, Nan Li, and Marcus W. Feldman. 1995. 'High sex ratios in China's future'. *Science* 10 February: 874–6.

Tulle, E. 2000. 'Old bodies'. In M. Tyler, ed., *Body, Culture and Society*, 64–83. Buckingham, UK: Open University Press.

Turcotte, M., and G. Schellenberg. 2007. *A Portrait of Seniors in Canada, 2006*. Ottawa: Statistics Canada.

Turk, Austin T. 1976. 'Law as a weapon in social conflict'. *Social Problems* 23: 276–92.

Turkle, Sherry. 1995. *Life on the Screen: Identity in the Age of the Internet*. New York: Simon and Schuster.

Turmel, A., and L. Guay. 2008. 'Une sociologie historique des problèmes urbains: la montée de l'état aménagiste'. In D. Fyson and Y. Rousseau, eds, *L'État au Québec: Perspectives d'analyse et experiences historiques*, 33–8. Quebec City: Ciéq.

Twining, Hillary, Arnold Arluke, and Gary Patronek. 2000. 'Managing the stigma of outlaw breeds: A case study of pit bull owners'. *Society and Animals* 8 (1): 1–28.

Tyrell, Hartmann. 1981. 'Is Weber's type of bureaucracy an objective, true type? Remarks on a thesis by Renate Mayntz'. *Zeitschrift fur Soziologie* 10 (1): 38–49.

Tyyskä, Vappu. 2009. *Youth and Society: The Long and Winding Road*. 2nd edn. Toronto: Canadian Scholars' Press.

UNAIDS. 2001a. *Children and Young People in a World of AIDS*. Geneva: Joint United Nations Programme on HIV/AIDS.

———. 2001b. *Gender and HIV Fact Sheet*. Geneva: Joint United Nations Programme on HIV/AIDS.

———. 2002. *Epidemiological Fact Sheets on HIV/AIDS and Sexually Transmitted Infections: Canada*. Geneva: Joint United Nations Programme on HIV/AIDS.

———. 2006. *2006 Report of the Global AIDS Epidemic*. Geneva: Joint United Nations Programme on HIV/AIDS. http://data.unaids.org/pub/GlobalReport/2006/2006_GR_CH02_en.pdf.

———. 2009. 'AIDS epidemic update, 2009'. Geneva: Joint United Nations Programme on HIV/AIDS and World Health Organization. http://data.unaids.org/pub/Report/2009/JC1700_Epi_Update_2009_en.pdf.

UNESCO. 2009. *Global Education Digest 2009: Comparing Education Statistics across the World*. Montreal: UNESCO Institute for Statistics.

———. 2010. *Reaching the Marginalized: EFA Global Monitoring Report 2010*. Paris: UNESCO; Oxford: Oxford University Press.

United Food and Commercial Workers Union (UFWUC). 2007. *The Status of Migrant Farm Workers in Canada*. Ottawa: UFWUC.

United Nations. 2000. *The World's Women: Trends and Statistics*. New York: United Nations.

———. 2007. *World Population Ageing*. New York: Department of Economic and Social Affairs, Population Division, United Nations.

———. 2009a. *Information Technology Report 2009: Trends and Outlook in Turbulent Times*. New York and Geneva: United Nations Conference on Trade and Development.

———. 2009b. *World Population Ageing, 2009*. New York: Department of Economic and Social Affairs, Population Division, United Nations.

United Nations Conference on Trade and Development (UNCTAD). 2009. *Trade and Development Report*. New York: UNCTAD.

United Nations Department of Economic and Social Affairs, Population Division. 2009a. *World Population Prospects: The 2008 Revision, Highlights*. SA/P/WP.210. New York: United Nations.

———. 2009b. *World Population Prospects: The 2008 Revision*. v. 1: *Comprehensive Tables*. ST/ESA/SER.A/287. New York: United Nations.

Urmetzer, Peter. 2005. *Globalization Unplugged: Sovereignty and the Canadian State in the Twenty-first Century*. Toronto: University of Toronto Press.

Ursel, Jane. 1992. *Private Lives, Public Policy: 100 Years of State Intervention in the Family*. Toronto: Women's Press.

Vallières, Pierre. 1967. *Nègres blancs d'Amérique*. Montreal: Éditions Parti pris.

Vallin, Jacques. 1983. 'Sex patterns of mortality: A comparative study of model life tables and actual situations with special reference to the case of Algeria and France'. In Alan D. Lopez and Lado T. Ruzicka, eds, *Sex Differences in Mortality*, 443–76. Canberra: Australian National University.

Van de Kaa, Dirk J. 1987. 'Europe's second demographic transition'. *Population Bulletin* 42 (1).

Vanier Institute of the Family. 2004. *Profiling Canadian Families III*. Ottawa: Vanier Institute of the Family.

Vaupel, James W. 2010. 'Biodemography of human ageing'. *Nature* 464 (25 March): 536–42.

———, J.R. Carey, K. Christensen, T.E. Johnson, A.I. Yashin, et al. 1998. 'Biodemographic trajectories of longevity'. *Science* 280 (5365): 855–60.

Veblen, T. 1899. *The Theory of the Leisure Class*. New York: Penguin.

Veenhof, B., and Peter Timusk. 2009. 'Online activities of Canadian boomers and seniors'. *Canadian Social Trends*, 25–32. Catalogue no. 11-008-X. Ottawa: Statistics Canada.

Veenstra, Gerry. 2001. 'Social capital and health'. *Canadian Journal of Policy Research* 2: 1672–81.

Veltmeyer, Henry. 2008. 'Development and globalization imperialism'. In Mark Charlton and Paul Rowe, *Crosscurrents: International Development*, 14. Toronto: Nelson.

Venkatesh, S.A. 2008. *Gang Leader for a Day: A Rogue Sociologist Takes to the Streets*. New York: Penguin Books.

Vosko, Leah. 2000. *Temporary Work: The Gendered Rise of a Precarious Employment Relationship*. Toronto: University of Toronto Press.

———. 2003. 'Gender differentiation and the standard/non-standard employment distinction in Canada, 1945 to the present'. In Danielle Juteau, ed., *Patterns and Processes of Social Differentiation: The Construction of Gender, Age, 'Race/Ethnicity' and Locality*. Toronto: University of Toronto Press.

———. 2006. *Precarious Employment: Understanding Labour Market Insecurity in Canada*. Montreal: McGill-Queen's University Press.

———. 2010. *Managing the Margins: Gender, Citizenship, and the International Regulation of Precarious Employment*. Oxford: Oxford University Press.

Wackwitz, Laura A., and Lana F. Rakow. 2004. 'Feminist communication theory: An introduction'. In Lana F. Rakow and Laura A. Wackwitz, eds, *Feminist Communication Theory: Selections in Context*, 1–10. Thousand Oaks, CA: Sage.

Waldron, Ingrid. 1976. 'Why do women live longer than men?' *Social Science and Medicine* 10: 349–62.

Walker, A. 1981. 'Towards a political economy of old age'. *Ageing and Society* 1: 73–94.

Wall, Glenda. 2009. 'Childhood and childrearing'. In Maureen Baker, ed., *Families: Changing Trends in Canada*, 6th edn, 91–107. Toronto: McGraw-Hill Ryerson.

Waller, Willard. 1965 [1932]. *The Sociology of Teaching*. New York: Wiley.

Wallerstein, Immanuel. 1974. *The Modern World-System*, v. 1, *Capitalist Agriculture and the Origins of the European World-Economy in the Sixteenth Century*. New York/London: Academic Press.

———. 2009. *The Essential Wallerstein*. New York: W.W. Norton.

Walter-Toews, D., J. Kay, and N.M. Lister. 2008. *The Ecosystem Approach: Complexity, Uncertainty, and Managing for Sustainability*. New York: Columbia University Press.

Walzer, Michael. 1983. *Spheres of Justice: A Defense of Pluralism and Equality*. New York: Basic Books.

Waridel, Laure. 2008. 'Coffee with pleasure: Just java and world trade'. In Mark Charlton and Paul Rowe, *Crosscurrents: International development*. Toronto: Nelson.

Waring, Marilyn. 1996. *Three Masquerades: Essays on Equality, Work and Human Rights*. Toronto: University of Toronto Press.

Warnes, Anthony M. 1992. 'Age-related variation and temporal change in elderly migration'. In Andrei Rogers, ed., with W.H. Frey, A. Speare, Jr, P. Rees, and A.M. Warnes, *Elderly Migration and Population Redistribution: A Comparative Study*, 35–55. London: Belhaven.

Warren, Carol A.B., and Tracy Xavia Karner. 2009. *Discovering Qualitative Methods*. 2nd edn. New York: Oxford University Press.

Warren, Karen. 1990. 'The power and promise of ecological feminism'. *Environmental Ethics* 12 (2): 125–46.

Wasko, Janet. 2001. 'The magical-market world of Disney'. *Monthly Review* 52 (11): 56–71.

Waskul, Dennis, and Phillip Vannini, eds. 2006. *Body/Embodiment: Symbolic Interaction and the Sociology of the Body*. Hampshire, UK: Ashgate.

Waters, Malcolm. 1995. *Globalization: Key Ideas*. London: Routledge.

Watson, James L., ed. 1997. *Golden Arches East: McDonald's in East Asia*. Stanford, CA: Stanford University Press.

Weber, Max. 1958 [1904]. *The Protestant Ethic and the Spirit of Capitalism*. Trans. Talcott Parsons. New York: Scribner.

———. 1978 [1908]. *Economy and Society*. Trans. Ephraim Fischoff, eds Guenther Roth and Claus Wittich. Berkeley: University of California Press.

———. 1946 [1915]. 'Religious rejections of the world and their directions'. In H.H. Gerth and C. Wright Mills, *From Max Weber: Essays in Sociology*, 323–59. New York: Oxford University Press.

———. 1964 [1920]. *The Theory of Social and Economic Organization*. New York: Free Press.

———. 1958 [1921]. *The City*. New York: Free Press.

———. 1958 [1922]. *Essays in Sociology*. Trans. H.H. Gerth and C. Wright Mills. New York: Oxford University Press.

Weeks, Jeffrey. 1993. *Sexuality*. London: Routledge.

Weinberg, Martin S., Colin J. Williams, and Douglas W. Pryor. 1994. *Dual Attraction: Understanding Bisexuality*. New York: Oxford University Press.

Weis, David. 1998. 'Conclusion: The state of sexual theory'. *Journal of Sex Research* 35 (1): 100–14.

Weitz, Rose, ed. 2002. *The Politics of Women's Bodies: Sexuality, Appearance and Behaviour*. 2nd edn. Oxford: Oxford University Press.

Welsh, Sandy. 1999. 'Gender and sexual harassment'. *Annual Review of Sociology* 25 (1): 169–90.

Westhues, Kenneth. 1982. *First Sociology*. New York: McGraw-Hill.

White, David Manning. 1950. 'The "Gatekeeper": A case study in the selection of news'. *Journalism Quarterly* 27: 383–90.

White, Kevin M., and Samuel H. Preston. 1996. 'How many Americans are alive because of twentieth-century improvements in mortality?' *Population and Development Review* 22: 415–30.

White, Lynn, Jr. 1967. 'The historical roots of our ecological crisis'. *Science* 155: 1203–7.

Whyte, William Foote. 1943. *Street Corner Society: The Social Structure of an Italian Slum*. Chicago: University of Chicago Press.

Wilkes, Rima. 2001. 'Competition or colonialism? An analysis of two theories of ethnic collective action' (PhD dissertation, University of Toronto).

———. 2004. 'First Nation politics: Deprivation, resources and participation in collective action'. *Sociological Inquiry* 74 (4): 570–89.

Wilkins, K. 2006. 'Predictors of death in seniors: How healthy are Canadians?' *Health Reports* 16: 57–66.

Wilkins, R. 2007. *Mortality by Neighbourhood Income in Urban Canada from 1971 to 2001*. HAMG Seminar, 16 January 2007. Ottawa. Statistics Canada.

Williams, Christine, Patti Giuffre, and Kirsten Dellinger. 1999. 'Sexuality in the workplace: Organizational control, sexual harassment and the pursuit of pleasure'. *Annual Review of Sociology* 25 (1): 73–93.

Williams, F.P., and M. McShane. 2009. *Criminological Theory*. 5th edn. Englewood Cliffs, NJ: Prentice-Hall.

Willis, J.R. 1987. 'What have we learned from the economics of the family?' *American Economic Review* 77 (2): 68–81.

Wilson, Sue. 2008. 'Socialization'. In L. Tepperman, J. Curtis, and P. Albanese, eds, *Sociology: A Canadian Perspective*, 2nd edn, 98–127. Toronto: Oxford University Press.

Wilson, Susannah J. 1991. *Women, Families, and Work*. 3rd edn. Toronto: McGraw-Hill Ryerson.

Wimberley, Dale W. 1990. 'Investment dependence and alternative explanations of Third World mortality'. *American Sociological Review* 55: 75–91.

Winseck, Dwayne, and Robert M. Pike. 2009. 'The global media and the empire of liberal internationalism, circa 1910–30'. *Media History* 15 (1): 31–54.

Wirth, Louis. 1938. 'Urbanism as a way of life'. *American Journal of Sociology* 44: 1–24.

Witz, Anne. 1992. *Professions and Patriarchy*. London: Routledge.

Wolf, Martin. 2004. *Why Globalization Works*. New Haven, CT: Yale University Press.

Wolfe, David A., and Meric S. Gertler. 2001. *The New Economy: An Overview*. Discussion paper produced for the Social Sciences and Humanities Research Council of Canada.

Wolfe, J. 1984. 'Our common past: An interpretation of Canadian planning history/Retour sur le passé: un survol historique de l'urbanisme canadien'. *Plan Canada* (July): 12–34.

———. 2003. 'A national urban policy for Canada? Prospects and challenges'. *Canadian Journal of Urban Research/ Revue canadienne de recherche urbaine* 12: 1–21.

Wolfgang, Marvin, and Franco Ferracuti. 1967. *The Subculture of Violence: Towards an Integrated Theory in Criminology*. Beverly Hills, CA: Sage.

Wollstonecraft, Mary. 1986 [1792]. *Vindication of the Rights of Women*. Middlesex, UK: Penguin.

Wong, Janelle, and Vivian Tseng. 2008. 'Political socialization in immigrant families: Challenging top-down parental socialization models'. *Journal of Ethnic and Migration Studies* 34 (1): 151–68.

Woo, J.K.H., S.H. Ghorayeb, C.K. Lee, H. Sangha, and S. Richter. 2004. 'Effect of patient socio-economic status perceptions of first and second year medical students'. *Canadian Medical Association Journal* 170 (13): 1915–19.

Woodhead, Linda. 2007. *Religion as Normative, Spirituality as Fuzzy: Questioning Some Deep Assumptions in the Sociology of Religion*. Paper presented at SISR/ISSR, Zagreb.

Woods, James, with Jay Lucas. 1993. *The Corporate Closet: The Professional Lives of Gay Men in America*. New York: Free Press.

Woods, Peter. 1979. *The Divided School*. London: Routledge and Kegan Paul.

Woolmington, Eric. 1985. 'Small may be inevitable'. *Australian Geographical Studies* 23: 195–207.

World Commission on Environment and Development. 1987. *Our Common Future*. New York: Oxford University Press.

World Health Organization (WHO). 2008. 'Closing the gap in a generation: Health equity through action on the social determinants of health'. New York: WHO. www.who.int/en.

Worth, Owen, and Karen Buckley. 2009. 'The World Social Forum: Postmodern prince or court jester?' *Third World Quarterly* 30 (4), 649–61.

Wortley, Scott. 1999. 'A northern taboo: Research on race, crime and criminal justice in Canada'. *Canadian Journal of Criminology* 41: 261–74.

Wotherspoon, Terry. 1995. 'The incorporation of public school teachers into the industrial order: British Columbia in the first half of the twentieth century'. *Studies in Political Economy* 46: 119–51.

———. 2000. 'Transforming Canada's education system: The impact on educational inequalities, opportunities, and benefits'. In B. Singh Bolaria, ed., *Social Issues and Contradictions in Canadian Society*, 250–72. Toronto: Harcourt Brace.

Wright, Charles R. 1959. *Mass Communication: A Sociological Perspective*. New York: Random House.

Wrong, Dennis. 1961. 'The oversocialized concept of man in modern sociology'. *American Sociological Review* 26: 183–93.

Wu, Zheng. 2000. *Cohabitation: A New Form of Family Living*. Toronto: Oxford University Press.

Yalnizyan, Armine. 2007. *The Rich and the Rest of Us: The Changing Face of Canada's Growing Gap*. Ottawa: Canadian Centre for Policy Alternatives.

Yates, Charlotte. 2008. 'Organized labour in Canadian politics: Hugging the middle or pushing the margins?' In Miriam Smith, ed., *Group Politics and Social Movements in Canada*. Toronto: University of Toronto Press.

Yeates, N. 2001. *Globalization and Social Policy*. London: Sage.

Young, David. 2006. 'Ethno-racial minorities and the Juno Awards'. *Canadian Journal of Sociology* 31 (2): 183–210.

———. 2008. 'Why Canadian content regulations are needed to support Canadian music'. In Josh Greenberg and Charlene D. Elliott, eds, *Communication in Question: Competing Perspectives on Controversial Issues in Communication Studies*, 216–21. Toronto: Thomson Nelson.

Yuval-Davis, Nira. 1997. *Gender and Nation*. London: Sage.

———, and Floya Anthias, eds. 1989. *Women-Nation-State*. New York: St Martin's.

Zehr, Stephen. 1994. 'The centrality of scientists and the transition of interests in the U.S. acid rain controversy'. *Canadian Review of Sociology and Anthropology* 31 (3): 325–53.

Zola, Irving Kenneth. 1972. 'Medicine as an institution of social control'. *Sociological Review* 20: 487–504.

Zucker, Lynne G., Michael R. Darby, Marilynn B. Brewer, and Yusheng Peng. 1996. 'Collaboration structure and information dilemmas in biotechnology'. In Roderick M. Kramer and Tom R. Tyler, eds, *Trust in Organizations: Frontiers of Theory and Research*, 90–113. Thousand Oaks, CA: Sage.

Zukin, S. 2009. 'Changing landscapes of power: Opulence and the surge for authenticity'. *International Journal of Urban and Regional Research* 33: 543–53.

Zurcher, Louis A. 1977. *The Mutable Self: A Self-Concept for Social Change*. Beverly Hills, CA: Sage.

index

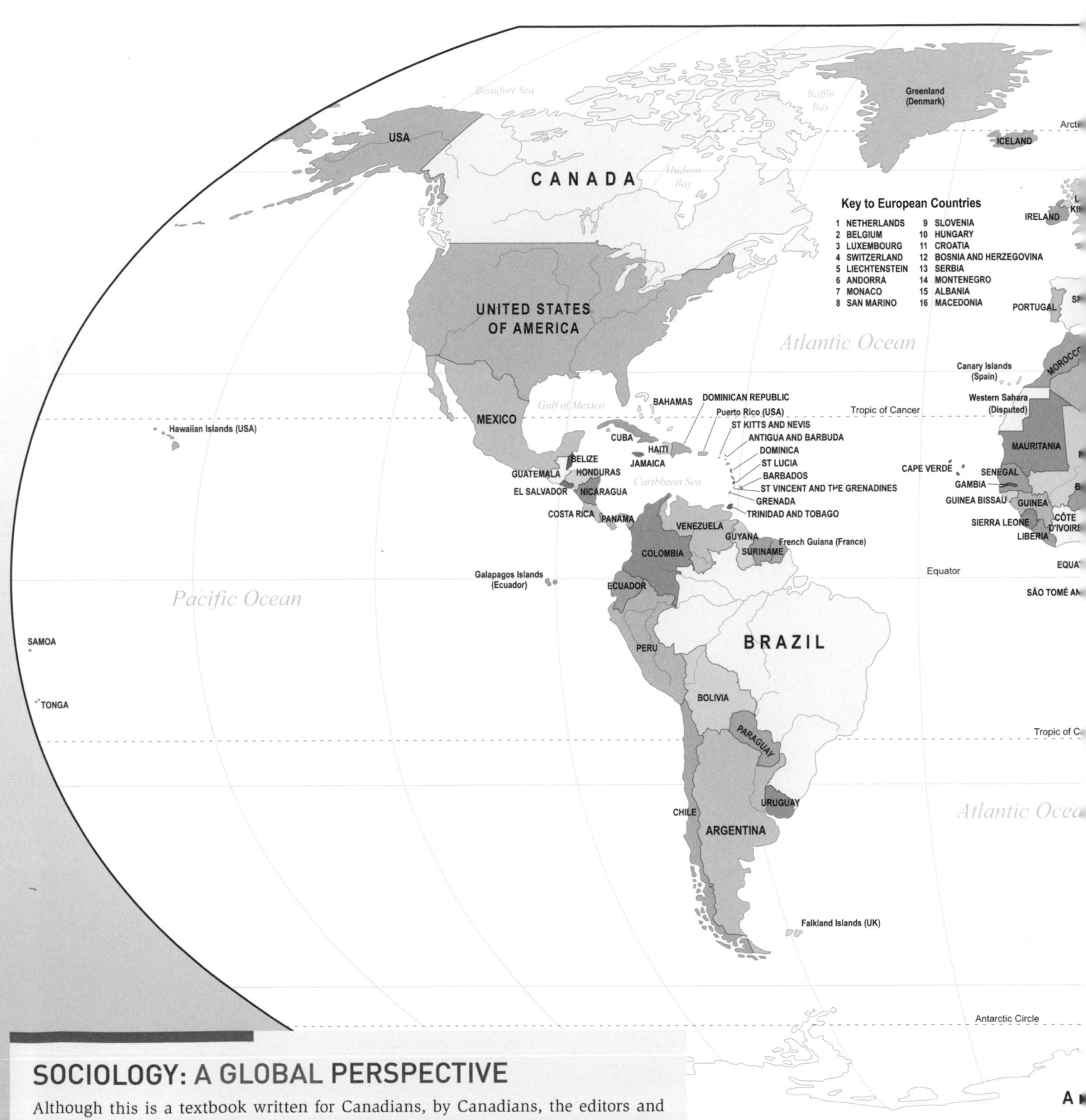

Key to European Countries

1	NETHERLANDS	9	SLOVENIA
2	BELGIUM	10	HUNGARY
3	LUXEMBOURG	11	CROATIA
4	SWITZERLAND	12	BOSNIA AND HERZEGOVINA
5	LIECHTENSTEIN	13	SERBIA
6	ANDORRA	14	MONTENEGRO
7	MONACO	15	ALBANIA
8	SAN MARINO	16	MACEDONIA

SOCIOLOGY: A GLOBAL PERSPECTIVE

Although this is a textbook written for Canadians, by Canadians, the editors and authors never lose sight of the fact that sociology is very much a global discipline. Along with Canadian data, examples, and illustrations, a wealth of information about how humans live and interact around the world is presented in every chapter. This map will help you situate these references in their global context.

Cartography by Douglas Fast